ENCYCLOPEDIA OF
Physical Science
AND Technology

THIRD EDITION

Ste-To

Volume 16

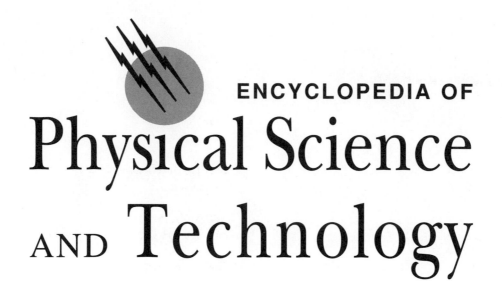

ENCYCLOPEDIA OF
Physical Science
AND Technology

THIRD EDITION

Ste-To

Volume 16

Editor-in-Chief

Robert A. Meyers, Ramtech, Inc.

ACADEMIC PRESS
A Harcourt Science and Technology Company

San Diego   San Francisco   Boston   New York   London   Sydney   Tokyo

Academic Press
A Harcourt Science and Technology Company
525 B Street, Suite 1900, San Diego, California 92101-4495, USA
http://www.academicpress.com

Academic Press
Harcourt Place, 32 Jamestown Road, London NW1 7BY, UK
http://www.academicpress.com

Library of Congress Catalog Card Number: 2001090661

International Standard Book Number:

0-12-227410-5 (Set)	0-12-227420-2 (Volume 10)
0-12-227411-3 (Volume 1)	0-12-227421-0 (Volume 11)
0-12-227412-1 (Volume 2)	0-12-227422-9 (Volume 12)
0-12-227413-X (Volume 3)	0-12-227423-7 (Volume 13)
0-12-227414-8 (Volume 4)	0-12-227424-5 (Volume 14)
0-12-227415-6 (Volume 5)	0-12-227425-3 (Volume 15)
0-12-227416-4 (Volume 6)	0-12-227426-1 (Volume 16)
0-12-227417-2 (Volume 7)	0-12-227427-X (Volume 17)
0-12-227418-0 (Volume 8)	0-12-227429-6 (Index)
0-12-227419-9 (Volume 9)	

PRINTED IN THE UNITED STATES OF AMERICA
01 02 03 04 05 06 MM 9 8 7 6 5 4 3 2 1

SUBJECT AREA EDITORS

Ruzena Bajcsy
University of Pennsylvania
Robotics

Allen J. Bard
University of Texas
Analytical Chemistry

Wai-Kai Chen
University of Illinois
Circuit Theory and Circuits

John T. Christian
Consultant, Waban, Massachusetts
Mining Engineering
Snow, Ice, and Permafrost
Soil Mechanics

Robert Coleman
U.S. Geological Survey
Earth Science
Geochemistry
Geomorphology
Geology and Mineralogy

Elias J. Corey
Harvard University
Organic Chemistry

Paul J. Crutzen
Max Planck Institute for Chemistry
Atmospheric Science

Gordon Day
*National Institute of Standards
and Technology*
Lasers and Masers
Optical Materials

Jack Dongarra
University of Tennessee
Computer Hardware

Gerard M. Faeth
University of Michigan
Aeronautics

William H. Glaze
University of North Carolina
Environmental Chemistry, Physics,
and Technology

Tomifumi Godai
National Space Development Agency of Japan
Space Technology

Barrie A. Gregory
University of Brighton
Instrumentation and Special Applications
for Electronics and Electrical Engineering

Gordon Hammes
Duke University
Biochemistry

Frederick Hawthorne
University of California, Los Angeles
Inorganic Chemistry

Leonard V. Interrante
Rensselaer Polytechnic Institute
Materials Science and Engineering

Bruce H. Krogh
Carnegie Mellon University
Control Technology

Joseph P. S. Kung
University of North Texas
Mathematics

William H. K. Lee
U.S. Geological Survey
Geodesy
Solid Earth Geophysics
Terrestrial Magnetism

Yuan T. Lee
Academica Sinica
Physical Chemistry
Radio, Radiation, and Photochemistry

Robert W. Lenz
University of Massachusetts
Polymer Chemistry

Irvin E. Liener
University of Minnesota
Agricultural and Food Chemistry

Larry B. Milstein
University of California, San Diego
Communications

Donald R. Paul
University of Texas
Applied Chemistry and Chemical
Engineering

S. S. Penner
University of California, San Diego
Energy Resources
Power Systems

George S. Philander
Princeton University
Dynamic Oceanography
Physical and Chemical Oceanography

Peter H. Rose
Orion Equipment, Inc
Materials Science for Electrical and
Electronic Engineering

Jerome S. Schultz
University of Pittsburgh
Biotechnology

Melvin Schwartz
Columbia University
Physics

Phillip A. Sharp
Massachusetts Institute of Technology
Molecular Biology

Martin Shepperd
Bournemouth University
Computer Software

Steven N. Shore
Indiana University
Astronomy

Leslie Smith
University of British Columbia
Hydrology and Limnology

Stein Sture
University of Colorado
Mechanical, Industrial, and Civil
Engineering

Robert Trappl
University of Vienna
Systems Theory and Cybernetics

Nicholas Tsoulfanidis
University of Missouri
Nuclear Technology

M. Van Rossum
Catholic University of Leuven
Components, Electronic Devices, and
Materials

R. H. Woodward Waesche
Science Applications International Corporation
Propulsion and Propellants

Contents

Foreword

The editors of the *Encyclopedia of Physical Science and Technology* had a daunting task: to make an accurate statement of the status of knowledge across the entire field of physical science and related technologies.

No such effort can do more than describe a rapidly changing subject at a particular moment in time, but that does not make the effort any less worthwhile. Change is inherent in science; science, in fact, seeks change. Because of its association with change, science is overwhelmingly the driving force behind the development of the modern world.

The common point of view is that the findings of basic science move in a linear way through applied research and technology development to production. In this model, all the movement is from science to product. Technology depends on science and not the other way around. Science itself is autonomous, undisturbed by technology or any other social forces, and only through technology does science affect society.

This superficial view is seriously in error. A more accurate view is that many complex connections exist among science, engineering, technology, economics, the form of our government and the nature of our politics, and literature, and ethics.

Although advances in science clearly make possible advances in technology, very often the movement is in the other direction: Advances in technology make possible advances in science. The dependence of radio astronomy and high-energy physics on progress in detector technology is a good example. More subtly, technology may stimulate science by posing new questions and problems for study.

The influence of the steam engine on the development of thermodynamics is the classic example. A more recent one would be the stimulus that the problem of noise in communications channels gave to the study of information theory.

As technology has developed, it has increasingly become the object of study itself, so that now much of science is focused on what we have made ourselves, rather than only on the natural world. Thus, the very existence of the computer and computer programming made possible the development of computer science and artificial intelligence as scientific disciplines.

The whole process of innovation involves science, technology, invention, economics, and social structures in complex ways. It is not simply a matter of moving ideas out of basic research laboratories, through development, and onto factory floors. Innovation not only requires a large amount of technical invention, provided by scientists and engineers, but also a range of nontechnical or "social" invention provided by, among others, economists, psychologists, marketing people, and financial experts. Each adds value to the process, and each depends on the others for ideas.

Beyond the processes of innovation and economic growth, science has a range of direct effects on our society.

Science affects government and politics. The U.S. Constitution was a product of eighteenth century rationalism and owes much to concepts that derived from the science of that time. To a remarkable extent, the Founding Fathers were familiar with science: Franklin, Jefferson, Madison, and Adams understood science, believed passionately in empirical inquiry as the source of truth, and felt that government should draw on scientific concepts for inspiration. The concept of "checks and balances" was borrowed from Newtonian physics, and it was widely believed that, like the orderly physical universe that science was discovering, social relations were subject to a series of natural laws as well.

Science also pervades modern government and politics. A large part of the Federal government is concerned either with stimulating research or development, as are the National Aeronautics and Space Administration (NASA) and the National Science Foundation, or with seeking to regulate technology in some way. The reason that science and technology have spawned so much government activity is that they create new problems as they solve old ones. This is true in the simple sense of the "side effects" of new technologies that must be managed and thus give rise to such agencies as the Environmental Protection Agency (EPA). More importantly, however, the availability of new technologies makes possible choices that did not exist before, and many of these choices can only be made through the political system.

Biotechnology is a good example. The Federal government has supported the basic science underlying biotechnology for many years. That science is now making possible choices that were once unimagined, and in the process a large number of brand new political problems are being created. For example, what safeguards are necessary before genetically engineered organisms are tested in the field? Should the Food and Drug Administration restrict the development of a hormone that will stimulate cows to produce more milk if the effect will be to put a large number of dairy farmers out of business? How much risk should be taken to develop medicines that may cure diseases that are now untreatable?

These questions all have major technical content, but at bottom they involve values that can only be resolved through the political process.

Science affects ideas. Science is an important source of our most basic ideas about reality, about the way the world is put together and our place in it. Such "world views" are critically important, for we structure all our institutions to conform with them.

In the medieval world view, the heavens were unchanging, existing forever as they were on the day of creation. Then Tycho Brahe observed the "new star"—the nova of 1572—and the inescapable fact of its existence forced a reconstruction of reality. Kepler, Galileo, and Newton followed and destroyed the earth- and human-centered universe of medieval Christianity.

Darwin established the continuity of human and animal, thus undermining both our view of our innate superiority and a good bit of religious authority. The germ theory of disease—made possible by the technology of the microscope destroyed the notion that disease was sent by God as a just retribution for unrepentant sinners.

Science affects ethics. Because science has a large part in creating our reality, it also has a significant effect on ethics. Once the germ theory of disease was accepted, it could no longer be ethical, because it no longer made sense, to berate the sick for their sins. Gulliver's voyage to the land of the Houynhyms made the point:

By inventing a society in which individuals' illnesses were acts of free will while their crimes were a result of outside forces, he made it ethical to punish the sick but not the criminal.

Knowledge—most of it created by science—creates obligations to act that did not exist before. An engineer, for instance, who designs a piece of equipment in a way that is dangerous, when knowledge to safely design it exists, has violated both an ethical and a legal precept. It is no defense that the engineer did not personally possess the knowledge; the simple existence of the knowledge creates the ethical requirement.

In another sense, science has a positive effect on ethics by setting an example that may be followed outside science. Science must set truth as the cardinal value, for otherwise it cannot progress. Thus, while individual scientists may lapse, science as an institution must continually reaffirm the value of truth. To that extent science serves as a moral example for other areas of human endeavor.

Science affects art and literature. Art, poetry, literature, and religion stand on one side and science on the other side of C. P. Snow's famous gulf between the "two cultures." The gulf is largely artificial, however; the two sides have more in common than we often realize. Both science and the humanities depend on imagination and the use of metaphor. Despite widespread belief to the contrary, science does not proceed by a rational process of building theories from undisputed facts. Scientific and technological advances depend on imagination, on some intuitive, creative vision of how reality might be constructed. As Peter Medawar puts it: [Medawar, P. (1969). Encounter 32(l), 15-23]: All advance of scientific understanding, at every level, begins with a speculative adventure, an imaginative preconception of what might be true—a preconception that always, and necessarily, goes a little way (sometimes a long way) beyond anything that we have logical or factual authority to believe in.

The difference between literature and science is that in science imagination is controlled, restricted, and tested by reason. Within the strictures of the discipline, the artist or poet may give free rein to imagination. Although we may critically compare a novel to life, in general, literature or art may be judged without reference to empirical truth. Scientists, however, must subject their imaginative construction to empirical test. It is not established as truth until they have persuaded their peers that this testing process has been adequate, and the truth they create is always tentative and subject to renewed challenge.

The genius of science is that it takes imagination and reason, which the Romantics and the modern counterculture both hold to be antithetical, and combines them in a synergistic way. Both science and art, the opposing sides of the "two cultures," depend fundamentally on the creative use of imagination. Thus, it is not surprising that many mathematicians and physicists are also accomplished musicians, or that music majors have often been creative computer programmers.

Science, technology, and culture in the future. We can only speculate about how science and technology will affect society in the future. The technologies made possible by an understanding of mechanics, thermodynamics, energy, and electricity have given us the transportation revolutions of this century and made large amounts of energy available for accomplishing almost any sort of physical labor. These technologies are now mature and will continue to evolve only slowly. In their place, however, we have the information revolution and soon will have the biotechnology revolution. It is beyond us to say where these may lead, but the implications will probably be as dramatic as the changes of the past century.

Computers affected first the things we already do, by making easy what was once difficult. In science and engineering, computers are now well beyond that. We can now solve problems that were only recently impossible. Modeling, simulation, and computation are rapidly becoming a way to create knowledge that is as revolutionary as experimentation was 300 years ago.

Artificial intelligence is only just beginning; its goal is to duplicate the process of thinking well enough so that the distinction between humans and machines is diminished. If this can be accomplished, the consequences may be as profound as those of Darwin's theory of evolution, and the working out of the social implications could be as difficult.

With astonishing speed, modern biology is giving us the ability to genuinely understand, and then to change, biological organisms. The implications for medicine and agriculture will be great and the results should be overwhelmingly beneficial.

The implications for our view of ourselves will also be great, but no one can foresee them. Knowledge of how to change existing forms of life almost at will confers a fundamentally new power on human beings. We will have to stretch our wisdom to be able to deal intelligently with that power.

One thing we can say with confidence: Alone among all sectors of society and culture, science and technology progress in a systematic way. Other sectors change, but only science and technology progress in such a way that today's science and technology can be said to be unambiguously superior to that of an earlier age. Because science progresses in such a dramatic and clear way, it is the dominant force in modern society.

Erich Bloch
National Science Foundation
Washington, D.C.

Preface

We are most gratified to find that the first and second editions of the *Encyclopedia of Physical Science and Technology* (1987 and 1992) are now being used in some 3,000 libraries located in centers of learning and research and development organizations world-wide. These include universities, institutes, technology based industries, public libraries, and government agencies. Thus, we feel that our original goal of providing in-depth university and professional level coverage of every facet of physical sciences and technology was, indeed, worthwhile.

The editor-in-chief (EiC) and the Executive Board determined in 1998 that there was now a need for a Third Edition. It was apparent that there had been a blossoming of scientific and engineering progress in almost every field and although the World Wide Web is a mighty river of information and data, there was still a great need for our articles, which comprehensively explain, integrate, and provide scientific and mathematical background and perspective. It was also determined that it would be desirable to add a level of perspective to our Encyclopedia team, by bringing in a group of eminent Section Editors to evaluate the existing articles and select new ones reflecting fields that have recently come into prominence.

The Third Edition Executive Board members, Stephen Hawking (astronomy, astrophysics, and mathematics), Daniel Goldin (space sciences), Elias Corey (chemistry), Paul Crutzen (atmospheric science), Yuan Lee (chemistry), George Olah (chemistry), Melvin Schwartz (physics), Edward Teller (nuclear technology), Frederick Seitz (environment), Benoit Mandelbrot (mathematics), Allen Bard (chemistry) and Klaus von Klitzing (physics)

concurred with the idea of expanding our coverage into molecular biology, biochemistry, and biotechnology in recognition of the fact that these fields are based on physical sciences. Military technology such as weapons and defense systems was eliminated in concert with present trends moving toward emphasis on peaceful uses of science and technology. Aaron Klug (molecular biology and biotechnology) and Phillip Sharp (molecular and cell biology) then joined the board to oversee their fields as well as the overall Encyclopedia. The Advisory Board was completed with the addition of John Bollinger (engineering), Michael Buckland (library sciences), Jean Carpentier (aerospace sciences), Ludwig Faddeev (physics), Herbert Friedman (space sciences), R. A. Mashelkar (chemical engineering), Karl Pister (engineering) and Gordon Slemon (engineering).

A 40 page topical outline of physical sciences and technology was prepared by the EiC and then reviewed by the board and modified according to their comments. This formed the basis for assuring complete coverage of the physical sciences and for dividing the science and engineering disciplines into 50 sections for selection of section editors. Six of the advisory board members decided to serve also as section editors (Allen Bard for analytical chemistry, Elias Corey for organic chemistry, Paul Crutzen for atmospheric sciences, Yuan Lee for physical chemistry, Phillip Sharp for molecular biology, and Melvin Schwartz for physics). Thirty-two additional section editors were then nominated by the EiC and the board for the remaining sections. A listing of the section editors together with their section descriptions is presented on p. v.

The section editors then provided lists of nominated articles and authors, as well as peer reviewers, to the EiC based on the section scopes given in the topical outline. These lists were edited to eliminate overlap. The Board was asked to help adjudicate the lists as necessary. Then, a complete listing of topics and nominated authors was assembled. This effort resulted in the deletion of about 200 of the Second Edition articles, the addition of nearly 300 completely new articles, and updating or rewrite of approximately 480 retained article topics, for a total of over 780 articles, which comprise the Third Edition. Examples of the new articles, which cover science or technology areas arising to prominence after the second edition, are: molecular electronics; nanostructured materials; image-guided surgery; fiber–optic chemical sensors; metabolic engineering; self-organizing systems; tissue engineering; humanoid robots; gravitational wave physics; pharmacokinetics; thermoeconomics, and superstring theory.

Over 1000 authors prepared the manuscripts at an average length of 17-18 pages. The manuscripts were peer reviewed, indexed, and published. The result is the eighteen volume work, of over 14,000 pages, comprising the Third Edition.

The subject distribution is: 17% chemistry; 5% molecular biology and biotechnology; 11% physics; 10% earth sciences; 3% environment and atmospheric sciences; 12% computers and telecommunications; 8% electronics, optics, and lasers; 7% mathematics; 8% astronomy, astrophysics, and space technology; 6% energy and power; 6% materials; 7% engineering, aerospace, and transportation. The relative distribution between basic and applied subjects is: 60% basic sciences, 7% mathematics, and 33% engineering and technology. It should be pointed out that a subject such as energy and power with just a 5% share of the topic distribution is about 850 pages in total, which corresponds to a book-length treatment.

We are saddened by the passing of six of the Board members who participated in previous editions of this Encyclopedia. This edition is therefore dedicated to the memory of S. Chandrasekhar, Linus Pauling, Vladimir Prelog, Abdus Salam, Glenn Seaborg, and Gian-Carlo Rota with gratitude for their contributions to the scientific community and to this endeavor.

Finally, I wish to thank the following Academic Press personnel for their outstanding support of this project: Robert Matsumura, managing editor, Carolan Gladden and Amy Covington, author relations; Frank Cynar, sponsoring editor; Nick Panissidi, manuscript processing; Paul Gottehrer and Michael Early, production; and Chris Morris, Major Reference Works director.

Robert A. Meyers, Editor-in-Chief
Ramtech, Inc.
Tarzana, California, USA

FROM THE PREFACE
TO THE FIRST EDITION

In the summer of 1983, a group of world-renowned scientists were queried regarding the need for an encyclopedia of the physical sciences, engineering, and mathematics written for use by the scientific and engineering community. The projected readership would be endowed with a basic scientific education but would require access to authoritative information not in the reader's specific discipline. The initial advisory group, consisting of Subrahmanyan Chandrasekhar, Linus Pauling, Vladimir Prelog, Abdus Salam, Glenn Seaborg, Kai Siegbahn, and Edward Teller, encouraged this notion and offered to serve as our senior executive advisory board.

A survey of the available literature showed that there were general encyclopedias, which covered either all facets of knowledge or all of science including the biological sciences, but there were no encyclopedias specifically in the physical sciences, written to the level of the scientific community and thus able to provide the detailed information and mathematical treatment needed by the intended readership. Existing compendia generally limited their mathematical treatment to algebraic relationships rather than the in-depth treatment that can often be provided only by calculus. In addition, they tended either to fragment a given scientific discipline into narrow specifics or to present such broadly drawn articles as to be of little use to practicing scientists.

In consultation with the senior executive advisory board, Academic Press decided to publish an encyclopedia that contained articles of sufficient length to adequately cover a scientific or engineering discipline and that provided accuracy and a special degree of accessibility for its intended audience.

This audience consists of undergraduates, graduate students, research personnel, and academic staff in colleges and universities, practicing scientists and engineers in industry and research institutes, and media, legal, and management personnel concerned with science and engineering employed by government and private institutions. Certain advanced high school students with at least a year of chemistry or physics and calculus may also benefit from the encyclopedia.

Robert A. Meyers
TRW, Inc.

Guide to the Encyclopedia

Readers of the *Encyclopedia of Physical Science and Technology (EPST)* will find within these pages a comprehensive study of the physical sciences, presented as a single unified work. The encyclopedia consists of eighteen volumes, including a separate Index volume, and includes 790 separate full-length articles by leading international authors. This is the third edition of the encyclopedia published over a span of 14 years, all under the editorship of Robert Meyers.

Each article in the encyclopedia provides a comprehensive overview of the selected topic to inform a broad spectrum of readers, from research professionals to students to the interested general public. In order that you, the reader, will derive the greatest possible benefit from the *EPST*, we have provided this Guide. It explains how the encyclopedia was developed, how it is organized, and how the information within it can be located.

LOCATING A TOPIC

The *Encyclopedia of Physical Science and Technology* is organized in a single alphabetical sequence by title. Articles whose titles begin with the letter A are in Volume 1, articles with titles from B through Ci are in Volume 2, and so on through the end of the alphabet in Volume 17.

A reader seeking information from the encyclopedia has three possible methods of locating a topic. For each of these, the proper point of entry to the encyclopedia is the Index volume. The first method is to consult the alphabetical Table of Contents to locate the topic as an article title; the Index volume has a complete A-Z listing of all article titles with the appropriate volume and page number.

Article titles generally begin with the key term describing the topic, and have inverted word order if necessary to begin the title with this term. For example, "Earth Sciences, History of" is the article title rather than "History of Earth Sciences." This is done so that the reader can more easily locate a desired topic by its key term, and also so that related articles can be grouped together. For example, 12 different articles dealing with lasers appear together in the La- section of the encyclopedia.

The second method of locating a topic is to consult the Contents by Subject Area section, which follows the Table of Contents. This list also presents all the articles in the encyclopedia, in this case according to subject area rather than A-Z by title. A reader seeking information on nuclear technology, for example, will find here a list of more than 20 articles in this subject area.

The third method is to consult the detailed Subject Index that is the essence of the Index volume. This is the best starting point for a reader who wishes to refer to a relatively specific topic, as opposed to a more general topic that will be the focus of an entire article. For example, the Subject Index indicates that the topic of "biogas" is discussed in the article Biomass Utilization.

CONSULTING AN ARTICLE

The First Edition of the *Encyclopedia of Physical Science and Technology* broke new ground in scholarly reference publishing through its use of a special format for articles.

The purpose of this innovative format was to make each article useful to various readers with different levels of knowledge about the subject. This approach has been widely accepted by readers, reviewers, and librarians, so much so that it has not only been retained for subsequent editions of *EPST* but has also been adopted in many other Academic Press encyclopedias, such as the *Encyclopedia of Human Biology*. This format is as follows:

- Title and Author
- Outline
- Glossary
- Defining Statement
- Main Body of the Article
- Cross References
- Bibliography

Although it is certainly possible for a reader to refer only to the main body of the article for information, each of the other specialized sections provides useful material, especially for a reader who is not entirely familiar with the topic at hand.

USING THE OUTLINE

Entries in the encyclopedia begin with a topical outline that indicates the general content of the article. This outline serves two functions. First, it provides a preview of the article, so that the reader can get a sense of what is contained there without having to leaf through all the pages. Second, it serves to highlight important subtopics that are discussed within the article. For example, the article "Asteroid Impacts and Extinctions" includes subtopics such as "Cratering," "Environmental Catastophes," and "Extinctions and Speciation."

The outline is intended as an overview and thus it lists only the major headings of the article. In addition, extensive second-level and third-level headings will be found within the article.

USING THE GLOSSARY

The Glossary section contains terms that are important to an understanding of the article and that may be unfamiliar to the reader. Each term is defined in the context of the article in which it is used. The encyclopedia includes approximately 5,000 glossary entries. For example, the article "Image-Guided Surgery" has the following glossary entry:

Focused ultrasound surgery (FUS) Surgery that involves the use of extremely high frequency sound targeted to highly specific sites of a few millimeters or less.

USING THE DEFINING STATEMENT

The text of most articles in the encyclopedia begins with a single introductory paragraph that defines the topic under discussion and summarizes the content of the article. For example, the article "Evaporites" begins with the following statement:

EVAPORITES are rocks composed of chemically precipitated minerals derived from naturally occurring brines concentrated to saturation either by evaporation or by freeze-drying. They form in areas where evaporation exceeds precipitation, especially in a semiarid subtropical belt and in a subpolar belt. Evaporite minerals can form crusts in soils and occur as bedded deposits in lakes or in marine embayments with restricted water circulation. Each of these environments contains a specific suite of minerals.

USING THE CROSS REFERENCES

Though each article in the *Encyclopedia of Physical Science and Technology* is complete and self-contained, the topic list has been constructed so that each entry is supported by one or more other entries that provide additional information. These related entries are identified by cross references appearing at the conclusion of the article text. They indicate articles that can be consulted for further information on the same issue, or for pertinent information on a related issue. The encyclopedia includes a total of about 4,500 cross references to other articles. For example, the article "Aircraft Aerodynamic Boundary Layers" contains the following list of references:

Aircraft Performance and Design • Aircraft Speed and Altitude • Airplanes, Light • Computational Aerodynamics • Flight (Aerodynamics) • Flow Visualization • Fluid Dynamics

USING THE BIBLIOGRAPHY

The Bibliography section appears as the last element in an article. Entries in this section include not only relevant print sources but also Websites as well.

The bibliography entries in this encyclopedia are for the benefit of the reader and are not intended to represent a complete list of all the materials consulted by the author in preparing the article. Rather, the sources listed are the author's recommendations of the most appropriate materials for further research on the given topic. For example, the article "Chaos" lists as references (among others) the works *Chaos in Atomic Physics, Chaos in Dynamical Systems,* and *Universality in Chaos.*

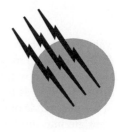

Steam Tables

Allan H. Harvey

National Institute of Standards and Technology

GLOSSARY

Critical point For a pure substance, the upper limit of the vapor–liquid saturation curve where the equilibrium vapor and liquid phases become identical and the compressibility of the fluid becomes infinite. For water, the critical point occurs at a temperature of approximately 374°C and a pressure of approximately 22 MPa.

Formulation A mathematical equation or set of equations from which a desired quantity or set of quantities (such as the thermodynamic properties of water) can be calculated.

Saturation A condition where two phases of a substance (in most common usage, the vapor and the liquid) are in thermodynamic equilibrium. Some thermodynamic variables, such as the temperature and pressure, have identical values in the two phases at saturation; other properties, such as density and enthalpy, have different values in each phase.

Skeleton tables Accepted values of properties presented at specific values (usually round numbers) of temperature and pressure. These are based on analysis and interpolation of data at nearby conditions, and an esti-

mate of the uncertainty of each value at each point is usually included.

STEAM TABLES is the traditional name for tabulations of the thermodynamic properties of water in both its vapor (steam) and liquid states. This information is of particular importance for the steam power-generation industry, but many other industrial processes make use of water in some form and therefore need reliable values of its properties. In addition, water is widely used in research, and some of these uses require highly accurate representations of its properties. Modern "steam tables" are for the most part no longer printed tables, but are mathematical formulations implemented in computer programs.

I. INTRODUCTION

The need for standardized representation of properties of water and steam is most apparent in the power industry, where electricity is generated by passing large amounts of steam through turbines. The thermodynamic properties of water and steam are vital to the design of equipment and to the evaluation of its performance. Small differences in the properties used will produce small changes in calculated quantities such as thermal efficiencies; however, because of the magnitude of the steam flows, these differences can translate into millions of dollars. It is therefore essential that all parties in the steam power industry, particularly bidders and purchasers in contracts, use a uniform set of properties in order to prevent any party from having an unfair advantage.

For other industries, such as petroleum refining and chemical manufacturing, standardization of water properties is less important, but there is still a need for reliable thermodynamic values for process design, optimization, and operation. In some cases, the complexity of the formulation is an issue, because the properties must be evaluated within iterative calculations. Thus, there is an incentive to keep the formulations as simple as possible without sacrificing significant accuracy or consistency.

Scientific and engineering research also requires highly accurate values for properties of water. This is not only for work where water is used directly, but also because of the widespread use of water as a calibration fluid.

In this article, we review the historical development of steam tables and then describe the current international standards as maintained by the International Association for the Properties of Water and Steam (IAPWS). While the primary focus of steam tables (and therefore of this article) is thermodynamic properties (density, enthalpy, entropy, etc.), other properties (such as viscosity, thermal conductivity, and dielectric constant) are also of some importance and will be mentioned briefly.

II. EARLY HISTORY OF STEAM POWER AND STEAM TABLES

About 2000 years ago, a figure showing a workable steam reaction turbine was included in a book by Hero of Alexandria. He showed other ways in which steam, or other hot gases, could be used to do mechanical work, and used the boiler, valve, and piston (basic components of a steam engine) at various places in his book. In spite of this promising start, the use of steam for power never advanced significantly in antiquity; it remained for later generations to develop its potential.

Although early engines of Papin, Savery, and Newcomen paved the way, it was James Watt in the late 1700s who made steam power an industrial success. Watt greatly improved the design of steam engines and took advantage of the improved metal-working techniques then available. Early in his career, Watt measured the temperature and pressure of saturated steam and constructed a curve through the points to permit interpolation. In a sense this curve was the first steam table; however, the science of Watt's day was inadequate to make much use of such data.

In the 1840s, V. Regnault (with some assistance from a young William Thomson, who later became Lord Kelvin) produced a set of careful measurements of the properties of steam. These data and others from Regnault's laboratory provided a foundation for the development and application of the new science of thermodynamics by Thomson, Clausius, and Rankine. By the late 19th century, steam tables based on Regnault's data began to appear, and in 1900 Callendar devised a thermodynamically consistent set of equations for treating steam data. Further steam tables soon appeared based on Callendar's equations; Mollier published the first steam tables in modern form.

The proliferation of steam tables soon became a problem, because different tables used different data. The differences between tables were particularly serious at high temperatures and pressures, where industrial interest was concentrated in the quest for increased thermodynamic efficiency. In addition to uncertainties in design, the different tables made it difficult to compare designs from different manufacturers. It became clear that international standardization was necessary in order to put all parties in the industry on a fair, consistent, and physically sound basis.

III. INTERNATIONAL STANDARDIZATION

The first conference directed at producing international agreement on steam tables was held in London in 1929. After a second (Berlin, 1930) and a third (New York, 1934) conference, agreement was reached on a set of "skeleton tables." These tables consisted of a rectangular grid of temperatures and pressures, with values of specific volume and enthalpy determined at each point by interpolation of the surrounding data. Values were similarly determined along the vapor–liquid saturation curve. An estimated uncertainty was assigned to each value at each point. These tables and their supporting data were the basis for the widely used steam tables book of J. H. Keenan and F. G. Keyes, published in 1936, which became the *de facto* standard for engineering calculations for many years.

After World War II, international standardization efforts resumed. Improvements in power-generation technology required reliable properties at pressures and temperatures beyond the range covered by Keenan and Keyes, and new data (notably from laboratories in the Soviet Union) became available. At the Sixth International Conference on the Properties of Steam (New York, 1963), a new set of skeleton tables was approved, covering an expanded range of pressures and temperatures.

By 1963, it was recognized that the growing use of computers for engineering calculations made it desirable to represent the properties of water and steam by equations in addition to skeleton tables. An International Formulation Committee (IFC) was appointed to develop a consistent set of equations that reproduced the 1963 skeleton table values within their tolerances. The main product of this effort was "The 1967 IFC Formulation for Industrial Use" (known as IFC-67). This formulation, and the printed steam tables based on it, replaced the Keenan and Keyes tables as the standard for industrial calculations for the next 30 years. Because of its association with the book of the same name, it is sometimes called the 1967 "ASME Steam Tables" formulation, although the American Society of Mechanical Engineers was only one of several participants in the international effort, and steam tables based on IFC-67 were published in several countries in addition to the United States.

At about the same time, the need was recognized for a permanent organization to manage the international conferences and the maintenance and improvement of property standards. In 1968, the International Association for the Properties of Steam (IAPS) was established; in 1989 the name was changed to the International Association for the Properties of Water and Steam (IAPWS). IAPWS, which now has 11 member countries, continues to work to improve international standards for water and steam properties for use in science and industry. It meets annually, and sponsors an International Conference on the Properties of Water and Steam every five years. Some of the activities of IAPWS are discussed in the following sections.

IV. STANDARDS FOR GENERAL AND SCIENTIFIC USE

There are actually two different audiences for steam tables. The steam power industry, which historically provided the impetus for standardized steam tables, needs a formulation that can be calculated relatively quickly by computer for use in iterative design calculations. It also needs a standard that is fixed for many years, because switching from one property formulation to another involves much adjustment of software that uses steam properties and impacts areas such as contracting and testing of equipment where large sums of money are at stake. Once the accuracy attains a level sufficient for most engineering purposes, the industry is willing to forego additional accuracy for the sake of speed and stability.

However, there are other users in industry for whom speed and stability are not significant issues. In addition, researchers need to have the most accurate properties available for their work. IAPWS therefore has two separate tracks of standards. Formulations "for industrial use," such as IFC-67, are designed for computational speed and are intended to remain the standard for use in the power in-

dustry for decades. In contrast, formulations "for general and scientific use" are intended to be kept at the state of the art, giving the best possible representation of the best experimental data in conjunction with theoretical constraints regardless of how complicated they must be or how frequently they are updated.

The first thermodynamic property formulation for general and scientific use was adopted in 1968, but was never widely used. In 1984, IAPS replaced it with a formulation developed by L. Haar, J. S. Gallagher, and G. S. Kell. The Haar–Gallagher–Kell (HGK) formulation, codified in the *NBS/NRC Steam Tables*, saw widespread use as a standard for scientific work and for some engineering applications. It used a single thermodynamic function (the Helmholtz energy as a function of temperature and density) to cover a wide range of temperatures and pressures. This guarantees thermodynamic consistency and prevents the discontinuities inherent in formulations that use different equations for different pressure/temperature regions. While the HGK formulation was not the first to take this approach, it was the first such approach to be adopted as an international standard.

As better data became available and small flaws were found in the HGK formulation, IAPWS in the early 1990s began an organized effort to produce a replacement. A task group, led by W. Wagner, evaluated the available experimental data and worked on producing an improved formulation; others tested the formulation exhaustively. The final result was the "IAPWS Formulation 1995 for the Thermodynamic Properties of Ordinary Water Substance for General and Scientific Use," which we shall refer to as IAPWS-95.

The structure of IAPWS-95 is a single equation for the Helmholtz energy as a function of temperature and density. All thermodynamic properties can be obtained in a consistent manner from differentiation and manipulation of that equation. The equation was also forced to meet certain theoretical constraints such as a correct approach to the ideal-gas limit at low densities, and it closely approximates the correct behavior as water's critical point is approached.

IAPWS-95 is now the state of the art and the international standard for representing water's thermodynamic properties at temperatures from its freezing point to 1000°C and at pressures up to 1000 MPa. It also extrapolates in a physically meaningful manner outside this range, including the supercooled liquid water region. The uncertainties in the properties produced by IAPWS-95 are comparable to those of the best available experimental data; this is quite accurate in some cases (for example, relative uncertainty of 10^{-6} for liquid densities at atmospheric pressure and near-ambient temperatures) and less so where the data are less certain (for example, relative uncertainty of 2×10^{-3} for most vapor heat capacities). Values of properties for saturation states and the property change upon vaporization, as generated by IAPWS-95, are shown in a typical steam tables format in Table I.

Formulations for general and scientific use have also been adopted for other properties of water. The most industrially important of these are probably the viscosity and the thermal conductivity, although some other properties such as the static dielectric constant and the refractive index are important in research. There are also IAPWS formulations for some properties of heavy water (deuterium oxide, D_2O).

Table II shows all the properties for which IAPWS has thus far adopted official formulations (known as "releases"). Copies of specific IAPWS releases may be obtained at no charge from www.iapws.org or by requesting them from the Executive Secretary of IAPWS. Currently, the Executive Secretary is:

Dr. R. B. Dooley
Electric Power Research Institute
3412 Hillview Avenue
Palo Alto, California 94304

V. STANDARDS FOR INDUSTRIAL USE

As mentioned earlier, the steam power industry requires a standard formulation that is both stable (in the sense of not changing for tens of years) and computationally fast. For 30 years, the IFC-67 formulation mentioned in Section III fulfilled that need. Through the years, however, some deficiencies in IFC-67 became apparent. Probably the worst problems related to inconsistencies at the boundaries between the regions of pressure–temperature space in which different equations defined the formulation. These inconsistencies can cause problems in iterative calculations near the boundaries. Also, with improvements in optimization methods and computer technology, it was believed that the computational speed of IFC-67 could be surpassed. In addition, there was a desire (driven by the increasing use of combustion turbines) to add standard properties for steam at temperatures higher than the upper limit of 800°C for IFC-67.

Therefore, in parallel to (and slightly behind) the development of the IAPWS-95 standard for general and scientific use, IAPWS undertook an effort to develop a new formulation for industrial use that would replace IFC-67. This effort, led by a development task group chaired by W. Wagner and a testing task group chaired by K. Miyagawa, resulted in the adoption of a new standard in 1997 called "IAPWS Industrial Formulation 1997 for the Thermodynamic Properties of Water and Steam" (abbreviated IAPWS-IF97).

The structure of IAPWS-IF97 is shown in Fig. 1. It consists of five regions defined in terms of pressure and temperature. The heavy solid line (Region 4) is the vapor–liquid saturation curve, represented by a single equation giving the saturation pressure as a function of temperature (and vice versa). The compressed liquid (Region 1) and the superheated vapor (Region 2) are represented by equations giving the Gibbs energy as a function of pressure and temperature (the most convenient independent variables for typical power-industry calculations). Other thermodynamic functions are obtained by appropriate differentiation of the Gibbs energy function. A Gibbs energy equation is also used in Region 5, which covers the high-temperature range needed for combustion turbines. In Region 3, which includes the area around the critical point, a Helmholtz energy function is used with density and temperature as independent variables (because pressure and temperature do not work well as independent variables near the critical point). Careful efforts were made to ensure that the values of the thermodynamic properties at either side of the region boundaries matched within tight tolerances.

Figure 1 also indicates the so-called "backward" equations in Regions 1 and 2, which allow the temperature to be obtained directly from pressure and enthalpy, or pressure and entropy, without iteration. The backward equations were made to reproduce the results from iterative solution of the "forward" equations (which have p and T as independent variables) within close tolerances. The backward equations in IAPWS-IF97 provide a great increase in speed for calculations (common in the power industry) where pressure is known in combination with either entropy or enthalpy. The backward equations necessarily introduce some inconsistency compared to exact solution of the forward equations, but this is negligible for most purposes. If greater consistency is desired at the expense of speed, the backward equations can be used as initial guesses for iterative solution of the forward equations to the required precision.

The accuracy of IAPWS-IF97 is for the most part only slightly less than that of the IAPWS-95 formulation for general and scientific use. In fact, rather than being fitted to experimental data, IAPWS-IF97 was fitted to the IAPWS-95 formulation and therefore agrees with it closely.

For industrial users, switching from IFC-67 to IAPWS-IF97 can be a major effort. Especially in the design and testing of large power-generation equipment, the relatively small changes in properties can produce numbers sufficiently different to have a large economic impact. Other aspects of the design and testing process, including software, which have been "tuned" to give the right results for IFC-67 properties, must therefore be readjusted to be

TABLE I Thermodynamic Properties of Water in Saturated Liquid and Vapor States as Calculated from the IAPWS-95 Formulation

t (°C)	Pressure MPa	Volume, cm³/g			Enthalpy, kJ/kg			Entropy, kJ/(kg·K)			t (°C)
		v_L	Δv	v_V	h_L	Δh	h_V	s_L	Δs	s_V	
0.01	0.000 612	1.0002	205 990	205 991	0.00	2500.9	2500.9	0.0000	9.1555	9.1555	0.01
5	0.000 873	1.0001	147 010	147 011	21.02	2489.0	2510.1	0.0763	8.9486	9.0248	5
10	0.001 228	1.0003	106 302	106 303	42.02	2477.2	2519.2	0.1511	8.7487	8.8998	10
15	0.001 706	1.0009	77 874	77 875	62.98	2465.4	2528.3	0.2245	8.5558	8.7803	15
20	0.002 339	1.0018	57 756	57 757	83.91	2453.5	2537.4	0.2965	8.3695	8.6660	20
25	0.003 170	1.0030	43 336	43 337	104.83	2441.7	2546.5	0.3672	8.1894	8.5566	25
30	0.004 247	1.0044	32 877	32 878	125.73	2429.8	2555.5	0.4368	8.0152	8.4520	30
35	0.005 629	1.0060	25 204	25 205	146.63	2417.9	2564.5	0.5051	7.8466	8.3517	35
40	0.007 385	1.0079	19 514	19 515	167.53	2406.0	2573.5	0.5724	7.6831	8.2555	40
45	0.009 595	1.0099	15 251	15 252	188.43	2394.0	2582.4	0.6386	7.5247	8.1633	45
50	0.012 352	1.0121	12 026	12 027	209.3	2381.9	2591.3	0.7038	7.3710	8.0748	50
60	0.019 946	1.0171	7666.2	7667.2	251.2	2357.7	2608.8	0.8313	7.0769	7.9081	60
70	0.031 201	1.0228	5038.5	5039.5	293.1	2333.0	2626.1	0.9551	6.7989	7.7540	70
80	0.047 414	1.0291	3404.1	3405.2	335.0	2308.0	2643.0	1.0756	6.5355	7.6111	80
90	0.070 182	1.0360	2358.0	2359.1	377.0	2282.5	2659.5	1.1929	6.2853	7.4781	90
100	0.101 42	1.0435	1670.7	1671.8	419.2	2256.4	2675.6	1.3072	6.0469	7.3541	100
110	0.143 38	1.0516	1208.2	1209.3	461.4	2229.6	2691.1	1.4188	5.8193	7.2381	110
120	0.198 67	1.0603	890.15	891.21	503.8	2202.1	2705.9	1.5279	5.6012	7.1291	120
130	0.270 28	1.0697	666.93	668.00	546.4	2173.7	2720.1	1.6346	5.3918	7.0264	130
140	0.361 54	1.0798	507.37	508.45	589.2	2144.3	2733.4	1.7392	5.1901	6.9293	140
150	0.476 16	1.0905	391.36	392.45	632.2	2113.7	2745.9	1.8418	4.9953	6.8371	150
160	0.618 23	1.1020	305.68	306.78	675.5	2082.0	2757.4	1.9426	4.8066	6.7491	160
170	0.792 19	1.1143	241.48	242.59	719.1	2048.8	2767.9	2.0417	4.6233	6.6650	170
180	1.0028	1.1274	192.71	193.84	763.1	2014.2	2777.2	2.1392	4.4448	6.5840	180
190	1.2552	1.1415	155.22	156.36	807.4	1977.9	2785.3	2.2355	4.2704	6.5059	190
200	1.5549	1.1565	126.05	127.21	852.3	1939.7	2792.0	2.3305	4.0996	6.4302	200
210	1.9077	1.1727	103.12	104.29	897.6	1899.6	2797.3	2.4245	3.9318	6.3563	210
220	2.3196	1.1902	84.902	86.092	943.6	1857.4	2800.9	2.5177	3.7663	6.2840	220
230	2.7971	1.2090	70.294	71.503	990.2	1812.7	2802.9	2.6101	3.6027	6.2128	230
240	3.3469	1.2295	58.476	59.705	1037.6	1765.4	2803.0	2.7020	3.4403	6.1423	240
250	3.9762	1.2517	48.831	50.083	1085.8	1715.2	2800.9	2.7935	3.2785	6.0721	250
260	4.6923	1.2761	40.897	42.173	1135.0	1661.6	2796.6	2.8849	3.1167	6.0016	260
270	5.5030	1.3030	34.318	35.621	1185.3	1604.4	2789.7	2.9765	2.9539	5.9304	270
280	6.4166	1.3328	28.820	30.153	1236.9	1543.0	2779.9	3.0685	2.7894	5.8579	280
290	7.4418	1.3663	24.189	25.555	1290.0	1476.7	2766.7	3.1612	2.6222	5.7834	290
300	8.5879	1.4042	20.256	21.660	1345.0	1404.6	2749.6	3.2552	2.4507	5.7059	300
310	9.8651	1.4479	16.887	18.335	1402.2	1325.7	2727.9	3.3510	2.2734	5.6244	310
320	11.284	1.4990	13.972	15.471	1462.2	1238.4	2700.6	3.4494	2.0878	5.5372	320
330	12.858	1.5606	11.418	12.979	1525.9	1140.2	2666.0	3.5518	1.8903	5.4422	330
340	14.601	1.6376	9.143	10.781	1594.5	1027.3	2621.8	3.6601	1.6755	5.3356	340
350	16.529	1.7400	7.062	8.802	1670.9	892.7	2563.6	3.7784	1.4326	5.2110	350
360	18.666	1.8954	5.054	6.949	1761.7	719.8	2481.5	3.9167	1.1369	5.0536	360
370	21.044	2.215	2.739	4.954	1890.7	443.8	2334.5	4.1112	0.6901	4.8012	370
t_c[a]	22.064	3.106	0	3.106	2084.3	0	2084.3	4.4070	0	4.4070	t_c

[a] $t_c = 373.946$°C.

TABLE II IAPWS Releases for Calculating Properties of Water and Heavy Water[a]

Property	Date of latest version
Thermal conductivity[b]	1998
Viscosity[b]	1997
Refractive Index	1997
Static dielectric constant	1997
Thermodynamic properties (industrial use)	1997
Thermodynamic properties (general and scientific use)	1996
Surface tension	1994
Surface tension (D_2O)	1994
Melting and sublimation pressures	1993
Critical point properties	1992
Thermodynamic properties (D_2O)	1984
Viscosity and thermal conductivity (D_2O)	1984
Ion product	1980

[a] Copies of IAPWS Releases may be obtained by writing to the IAPWS Executive Secretary: Dr. R. B. Dooley, Electric Power Research Institute, 3412 Hillview Ave., Palo Alto, CA 94304.

[b] These releases contain formulations both for industrial use and for general and scientific use.

consistent with IAPWS-IF97. IAPWS, upon adopting the formulation, recommended a waiting period (which expired at the beginning of 1999) in which IAPWS-IF97 should not be used for contractual specifications, in order to allow users time to adjust. Similar adjustments and therefore a similar waiting period will likely be required whenever a successor to IAPWS-IF97 is adopted; however, the intention is for IAPWS-IF97 to remain the standard in the power industry for at least 20 years.

Property formulations for industrial use have also been generated for the viscosity and the thermal conductivity, as mentioned in Table II.

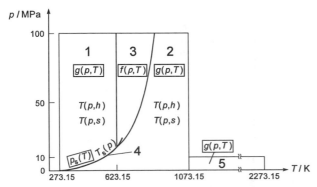

FIGURE 1 Regions of the IAPWS-IF97 standard for thermodynamic properties of water and steam for industrial use.

A final question to be addressed is when it is appropriate to use the industrial formulation (IAPWS-IF97), as opposed to the formulation for general and scientific use (IAPWS-95). In general, since IAPWS-95 is the state of the art, it (or any formulation for general and scientific use that might replace it in the future) should be used in all cases except those where IAPWS-IF97 is specifically required or preferable. IAPWS-IF97 is preferred in two cases:

- In the steam power industry, the industrial formulation (now IAPWS-IF97) is the industry standard for contracting and testing purposes. It therefore makes sense to use IAPWS-IF97 for calculations in all facets of the power industry.
- In any application where computing time is at a premium (and where the calculation of water properties consumes most of that time), IAPWS-IF97 may be preferred because of its much faster computing speed. An example would be finite-element calculations of steam flow in a turbine. In some cases, even IAPWS-IF97 might be too slow; an alternative in such cases (especially if the calculations are confined to a narrow range of conditions) is to generate a table of properties in advance and then use a table interpolation algorithm in the computations.

VI. FUTURE DIRECTIONS

Probably the most notable current direction for steam tables is the migration from printed tables to computer software. Most engineering design and research now uses computer-generated properties. Printed tables and charts formerly had to be detailed enough to permit interpolation with an accuracy suitable for design; now they are mostly relegated to the role of an auxiliary to be used for quick estimates when the computer is not handy, and therefore can be much less detailed. With the increased use of computers, it has become increasingly important to present water property standards in user-friendly software, either for standalone calculations or as something to be plugged into a spreadsheet or other application. There is also demand for implementations of steam property formulations in computer languages other than the traditional FORTRAN and for access to properties via the World Wide Web. Figure 2 shows a window from some modern "steam tables" software.

As new data are obtained and new theoretical understanding is gained, work will continue to improve the existing formulations for the properties of water and steam. Much of this work is organized by IAPWS. Current areas of focus include improvement of the formulations

	Temperature [K]	Pressure [MPa]	Density (L) [kg/m³]	Density (V) [kg/m³]	Enthalpy (L) [kJ/kg]	Enthalpy (V) [kJ/kg]
1	300.0	0.003537	996.5	0.02559	112.6	2550
2	320.0	0.01055	989.4	0.07166	196.2	2586
3	340.0	0.02719	979.5	0.1744	279.9	2621
4	360.0	0.06219	967.4	0.3786	363.8	2654
5	380.0	0.1289	953.3	0.7483	448.1	2686
6	400.0	0.2458	937.5	1.369	533.0	2716
7	420.0	0.4373	919.9	2.352	618.6	2742
8	440.0	0.7337	900.6	3.833	705.3	2765
9	460.0	1.171	879.6	5.983	793.4	2783
10	480.0	1.790	856.5	9.014	883.3	2796
11	500.0	2.639	831.3	13.20	975.4	2802

FIGURE 2 Sample screen from a modern "steam tables" database implementing the IAPWS-95 formulation (From A. H. Harvey, A. P. Peskin, and S. A. Klein, "NIST/ASME Steam Properties," NIST Standard Reference Database 10, Standard Reference Data Office, NIST, Gaithersburg, MD 20899; information also available at srdata@nist.gov or http://www.nist.gov/srd/nist10.htm).

for viscosity and thermal conductivity, and representation of water's thermodynamic behavior in accordance with theoretical constraints in the vicinity of its critical point.

IAPWS is also increasing its activities in application areas where water properties play a role. In the physical chemistry of aqueous solutions, efforts are devoted not only to those properties of pure water (such as the dielectric constant and ionization constant) that are important for solution chemistry, but also to the description of key properties of aqueous mixtures. Areas of interest include the partitioning of solutes between liquid water and steam and the properties of water/ammonia mixtures. In power-plant chemistry, IAPWS seeks to help industrial users apply fundamental knowledge and identifies key areas requiring further research. Documents called IAPWS Certified Research Needs (ICRN's) describe these needs; these documents are intended to serve as evidence to those who set research priorities that work in an area would be of significant use to industry. More information on current ICRN's may be obtained from the IAPWS Executive Secretary (see Section IV) or on the IAPWS Website (see following).

New data, new scientific capabilities, and new industrial needs will continue to shape the international steam properties community (and specifically IAPWS) as it seeks to maintain its core mission of developing steam tables and other water property standards while expanding into related areas to meet the needs of the power industry and of others who require accurate knowledge of the properties

of water and aqueous mixtures. Up-to-date information on the activities of IAPWS may be found on its Website at www.iapws.org.

SEE ALSO THE FOLLOWING ARTICLES

CRITICAL DATA IN PHYSICS AND CHEMISTRY • THERMO-DYNAMICS • THERMOMETRY • WATER CONDITIONING, INDUSTRIAL

BIBLIOGRAPHY

Haar, L., Gallagher, J. S., and Kell, G. S. (1984). "NBS/NRC Steam Tables," Hemisphere Publishing Corporation, New York.

Harvey, A. H., and Parry, W. T. (1999). Keep Your "Steam Tables" Up to Date. *Chemical Eng. Progr.* **95**(11), 45.

Parry, W. T., Bellows, J. C., Gallagher, J. S., and Harvey, A. H. (2000). "ASME International Steam Tables for Industrial Use," ASME Press, New York.

Tremaine, P. R., Hill, P. G., Irish, D. E., and Balakrishnan, P. V. (eds.) (2000). "Steam, Water, and Hydrothermal Systems: Physics and Chemistry Meeting the Needs of Industry, Proceedings of the 13th International Conference on the Properties of Water and Steam," NRC Research Press, Ottawa.

Wagner, W., and Kruse, A. (1998). "Properties of Water and Steam," Springer-Verlag, Berlin.

White, Jr., H. J., Sengers, J. V., Neumann, D. B., and Bellows, J. C. (eds.) (1995). "Physical Chemistry of Aqueous Systems: Meeting the Needs of Industry, Proceedings of the 12th International Conference on the Properties of Water and Steam," Begell House, New York.

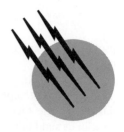

Stefan Problems

A. D. Solomon

Negev Academic College of Engineering

GLOSSARY

Ablation Melting process in which liquid formed is driven away by flow conditions at the surface, a key example being an ablating surface of a rocket during its passage through air.

Interface Surface separating the solid and liquid phases of a material undergoing melting or freezing.

Interface condition Condition, usually representing energy conservation, holding at the interface between liquid and solid.

Latent heat of melting Difference of specific internal energies of solid and liquid at the melt temperature.

Marangoni convection Convection of a fluid driven by the surface tension effects of any interfaces in the fluid. Its effect is dominant in microgravity when gravity-driven natural convection becomes ignorably small.

Melt temperature Temperature for equilibrium between liquid and solid states.

Moving boundary problem Mathematical problem of finding a function and a domain in which the function satisfies a differential equation and the domain is partly described by interface conditions holding at its bounding surface.

Mushy region Region consisting of slush (liquid mixed with fine solid particles, uniformly at the melt temperature).

Supercooled liquid Liquid at a temperature below its melt temperature.

STEFAN PROBLEMS are a class of mathematical problems arising from macroscopic models of melting and freezing. They are the mathematical models for simple phase change processes. In this article, we will describe their form by means of a specific example, their history, recent developments, and anticipated future work.

I. A CLASSICAL STEFAN PROBLEM

A. Physical Setting

The Stefan problem arises from a mathematical model of melting and freezing of materials. In their solid phase, the materials are assumed to have an ordered crystalline structure maintained by intermolecular bonds. Liquid, on the other hand, is marked by the molecules essentially maintaining their close contact, but no longer having the crystalline structure of the solid. The energy associated with this structure is the latent heat of the material (in, for example, joules per gram), and it must be added to the solid in order to melt the solid and turn it into liquid.

Melting and freezing occur at a temperature, the melt temperature, characteristic of the material. At atmospheric pressure, water, for example, melts and freezes at 0°C.

Consider a large block of material in its solid phase and at its melt temperature everywhere. Assume that all but one of the faces of the block are insulated. At this face let us suddenly impose a temperature above the melting point. Instantaneously, we will see liquid form on the heated face. If the block is contained in a thin-walled container, for example, then the liquid will remain in the container and it will constitute an ever-growing layer between the heated container wall and the solid beyond. If the block is not in any container, then the liquid will flow away from the solid, whose melting will be driven by the imposed hot temperature at its surface. This latter process is known as ablation. In the former case the solid and liquid phases meet at an interface. The interface will be parallel to the heated face, and as time progresses, the melted region will continue to grow at the expense of the frozen region. Eventually, all of the solid will have melted and the container will be filled only with liquid.

In our experiment, the temperature of the solid remains at the melt temperature for all time: indeed, it could not decrease, since temperature declines only when heat is extracted; similarly, it could not increase without the material having melted. Thus, ice above 0°C becomes water. On the other hand, the temperature in the liquid will always rise and lie between the melt temperature and the imposed face temperature.

For our experiment of melting a block of material, the Stefan problem can be stated as follows:

- At all times after the moment when the face temperature was imposed, find the temperature at all points in the block together with the solid and liquid regions.

In the melting process the interfacial surface separating liquid and solid regions is always moving. Problems in which the boundary of the region of study is moving with time and is an unknown of the problem are called "moving boundary problems." The Stefan problem is such a moving boundary problem.

B. Mathematical Formulation

The mathematical formulation of the Stefan problem in its classical form consists in relationships between the variables of the melting process. These are the temperature distribution at all times, the location of the boundary between solid and liquid regions, the thermophysical properties of the material (solid and liquid conductivities, specific heats, etc.), and the imposed conditions (for example, the face temperature) at the exterior fixed boundary

of the block. These relationships take the form of partial differential equations in the solid and liquid regions and additional conditions at the moving boundary expressing local energy conservation and temperature behavior at the interface of solid and liquid material. The resulting problem is extremely difficult, since while the temperature is varying, so is the region in which it is to be found. Thus, for our case of melting a block of material, the temperature remains constant in the solid region, while in the liquid, it must be found by solving an equation within a time-varying region whose determination is part of the problem.

To illustrate the mathematical formulation, consider the process of melting a slab of material, initially solid, occupying the interval $0 < x < \delta$. Suppose that the initial temperature of the material is the melt temperature T_m, and at an initial time $t = 0$, the temperature at the face $x = 0$ of the material is set for all time as the constant value $T_L > T_m$. As a result a melting process takes place resulting in a boundary, located at a point $x = X(t) > 0$, separating liquid to its left from solid at the melt temperature to its right. With passing time, the moving boundary $X(t)$ moves to the right until all the material is melted. Within the liquid region, the temperature $T(x, t)$ at a point x and time t obeys the heat equation

$$T_t(x, t) = \alpha T_{xx}(x, t),$$

with the subscripts denoting the corresponding partial derivatives. At the front $x = X(t)$ the temperature is identically equal to the melt temperature:

$$T(X(t), t) \equiv T_m, \qquad t > 0,$$

while for $x = 0$,

$$T(0, t) \equiv T_L.$$

Conservation of energy at the moving boundary states that the rate at which energy arrives at the front from the left (by conduction) is equal to the rate at which heat is absorbed by the material as its heat of melting,

$$\rho L X'(t) = -k T_x(X(t), t), \qquad t > 0.$$

This condition is referred to as the Stefan condition. Here α is the thermal diffusivity of the liquid, L is the latent heat of melting, k is the liquid thermal conductivity, and ρ is the liquid density.

II. HISTORY OF THE PROBLEM

The first serious discussion of a melting/freezing problem was by G. Lame and B. Clapeyron in 1831 and focused on the scenario of our example. No solution to the problem was found, but they did state that the front depth from the wall was proportional to the square root of the elapsed time measured from the imposition of the wall temperature.

In 1889, J. Stefan examined the process of freezing of the ground. We note that this problem is still one of intensive study, with importance in such areas as storage of heat, proper design of roadways, and the performance of heat exchangers for heat pumps. Stefan studied the problem posed by Lame and Clapeyron as well as the question of what would happen if solid and liquid blocks of the same material, the first at a temperature below the melt temperature, the second above, were brought rapidly into contact: would the solid melt, or would the liquid freeze? Each was solved in the sense that formulas for the temperature distribution and for the location of the moving boundary between phases were found for all times. We note that credit for the solution of the first problem is given to F. Neumann, who is said to have done this already in the 1860s but did not publish his result.

There are two major difficulties that must be overcome in order to find the location of solid and liquid regions and the temperature distribution as time evolves. The first is the varying nature of the regions: the second is the nonlinearity of the problem. Because of this, the linear techniques of applied mathematics of the late 19th and early 20th centuries, such as Fourier and Laplace transforms, were essentially inapplicable to the problem studied by Stefan. For this reason, no significant advances in the study or solution to problems of this type were made over a 40-year period following Stefan's work.

From the time of Stefan's work until the 1930s two kinds of activities were taking place that would bring the Stefan problem into sharper focus and render it more tractable. The first was a growing understanding of so-called generalized functions, first treated rigorously by D. Hilbert in the early 1900s. Unlike the strictly continuous and differentiable functions to which mathematical analysis had been limited before, these functions could be undefined everywhere but still, in some integrated sense, would have mathematical and physical meaning. The most famous of such functions is, of course, the delta function $\delta(x - x_0)$ representing a source of finite strength located at the single point x_0. The second development relevant to the Stefan problem was a growing tendency toward placing problems in an algorithmic setting, with a view toward computing solutions that would become a reality with the advent of computing machines.

Beginning in the 1930s, modest advances in the understanding and theoretical development of the Stefan problem began to be made. Existence, uniqueness, and continuous dependence on input parameters were proved for a number of standard problems in one space dimension. This work reached its fruition in the 1970s when for the case of one space dimension it was proved that the Stefan problem was mathematically well posed. Similarly, beginning with the space program in the 1950s, significant work was done on analytical and computational methods for melting and freezing processes, aimed at such applications as the use of waxes as thermal buffers for electronic equipment and the treatment of ablation problems. This work, performed in the engineering community, produced approximation techniques such as the Goodman and Megerlin methods that could provide accurate estimates of front location and temperature for one space dimension.

The numerical solution of Stefan problems began to be of great interest to researchers in the oil industry in the mid-1950s. Initial efforts produced specific schemes relevant to methods for driving oil to desired areas by pumping other fluids into the ground. In the 1960s, a number of researchers developed so-called weak methods based on the use of the enthalpy as the key variable (in place of temperature) for the numerical solution of the Stefan problem in any number of space dimensions for arbitrary boundary conditions. These methods continue to form the general-purpose numerical methodology for phase change problems, while such approaches as finite element methods are commonly used for specific processes of interest.

The theoretical treatment of Stefan problems in two and three dimensions remains essentially undone. Unlike one space dimension, two or more dimensions make possible almost unlimited degrees of discontinuity. Thus, for example, a sculptured ice unicorn melting in a pool of heated water can be reduced to any number of small chunks of ice at a later time.

III. RECENT WORK

In recent years, fueled by physical problems of interest and an exponentially expanding computing capability, the community of interest in the Stefan problem has expanded to include such far-flung areas as geophysics, food processing, medicine, metallurgy, and economics. This in turn has added challenging complexities to the form of the original problem explored by Stefan. We now describe some of them.

A. Mushy Zones

Suppose that a block of ice and a pool of water are both at the uniform temperature of $0°C$ and that the ice is ground into arbitrarily small chunks and thrown into the water. Since the chunks can be made as small as we wish, the resulting mixture, viewed on a normal length scale, is a slush that is neither solid nor liquid. In particular, its average enthalpy lies between that of solid and that of liquid at $0°C$. We call such material mushy.

Mushy zones arise in a variety of situations. Certain materials solidify in such a way that the interface between solid and liquid is long and dendritic, with many treelike or columnar branches of arbitrarily small size. In this case, there is effectively a mushy zone between solid and liquid. Similarly, radioactive material is heated by its own decay heat. If a lump of solid radioactive metal is placed in an insulated container, the metal will be heated by its own internally generated heat until it reaches its normal melt temperature. Further heating will not affect its state (in principle) until a quantity of heat equal to the latent heat has been generated. Until this happens, the material is entirely mushy. In the same way, materials heated by electrical resistance, such as by welding or *in situ* vitrification, pass through a mushy stage when their enthalpy lies between that of solid and that of liquid at the melt temperature.

The theoretical treatment of mushy zones is to a large extent open. Each material solidifies in a manner characteristic of its crystal structure. Some salts, for example, exhibit very broad mushy zones under ordinary freezing while others do not; some have crystals that are long and columnar with liquid trapped in between; others have filmy dendritic fronts, again with liquid trapped in between. The use of generalized functions has enabled us to obtain general existence results even in the presence of possible mushy zones, but more precise results may well require a better understanding of the materials involved and the material-dependent aspects of the mushy zone phenomenon.

B. Supercooled Solidification

If we slowly cool a liquid in a very smooth container under very quiet conditions, we may observe that the temperature of the liquid goes below the normal melt temperature for a period of time; it then rises to the melt temperature and the liquid freezes rapidly. A liquid whose temperature is below its normal melting point is said to be supercooled. On thermodynamic grounds, a material in the liquid state but at a temperature below the normal melt temperature is unstable, and given any kind of physical or thermal perturbation, will quickly freeze. Water can be supercooled to temperatures as low as $-40°C$, while other materials, such as hydrated salts, can be cooled to temperatures much further below their normal melt temperature.

Numerical methods have been developed for simulating the solidification of supercooled material based on the weak solution approach.

C. Natural Convection

The density of a liquid depends on its temperature, generally decreasing with increasing temperature. Thus, any temperature gradient in the vertical direction with temperature increasing with increasing depth usually will induce a flow, or natural convection, much like that seen when heating a thick soup. Thus, the liquid formed in the course of melting a solid from the vertical side or from below will ordinarily begin to flow through natural convection, greatly enhancing heat transfer and altering the form of the relations for heat transfer in the liquid phase. The development of effective numerical methods for simulating this effect in three-dimensional space is an active area of research. In the casting of special metals and alloy systems (see Section III.D), natural convection may be a highly undesired effect. For this reason, there is a long-standing effort to explore casting and metal processing in space, where the effects of gravity are not present. However, in the microgravity of space, convection driven by surface tension, a form of natural convection known as Marangoni convection, occurs at the solid/liquid interface.

D. Alloy Solidification

In recent years, high technology in both civilian and military spheres has pushed us to develop alloys of ever-increasing precise constitution. For example, our needs for special sensors and photovoltaics have resulted in a need for very precisely structured alloy systems in their solid phase.

An alloy is a material formed by the bonding of its constituent elements and differing from both. Thus, a binary alloy of copper and nickel is defined by the relative amounts of copper and nickel present, and for each such composition it is a different material in terms of its thermophysical properties. Moreover, the liquid and solid phases of an alloy have an additional interesting property related to their coexistence. Let us briefly describe this by considering a pure material such as water.

A flat piece of ice and an adjacent body of water can coexist, that is, the ice will remain ice and the water will remain water if the temperature at the interface between them is $0°C$. We say that this is a state of thermodynamic equilibrium of the two materials. In contrast, on the same thermodynamic grounds, a solid piece of an alloy with one composition is not generally in thermodynamic equilibrium with a liquid alloy of the same composition. Rather, for each composition of solid, there will be a necessarily different composition needed by a liquid to be in equilibrium with the solid. The implication of this fact for the Stefan problem is clear. When a liquid alloy solidifies, the liquid alloy adjacent to the moving interface must be of a different composition from the solid on the other side of the interface. If not, the system is unstable, as in the case of a supercooled liquid. However, in order that the composition of adjacent solid and liquid be different there must be diffusion of the component materials, and the diffusion process must be as rapid as that of the heat transfer. Since this is generally not possible, solidification processes for alloy systems occur in the presence of a phenomenon known as constitutional supercooling, where in place of a sharp interfacial surface between solid and liquid we find a generally thick mushy zone of intermixed solid and liquid. Knowing how to adequately model this phenomenon thermodynamically, analytically, and numerically requires significant additional effort, both theoretical and numerical.

E. Ultrarapid Solidification and Melting Processes

In recent years, the tools of thermal processes have evolved from flames at relatively low temperatures and propagation speeds to lasers, which are capable of inputting large amounts of energy in time spans that can be measured in picoseconds. On mathematical grounds, this creates difficulties for the heat equation as a model of heat transfer. The heat equation has the mathematical property of transmitting thermal signals at infinite speed, an impossibility because, of course, they cannot move faster than the speed of light. This is no problem for thermal processes that have been used up to present times, for the predicted amount of heat moving at unrealistically high speeds is ignorably small. This does become a potential problem for such thermal processes as laser pulsing, and will present an ever-larger problem as mechanisms for heat transfer develop further. With this in mind, alternative mechanisms for modeling heat transfer have been developed (e.g., hyperbolic heat transfer), our understanding of which are correct is as yet poor. In addition, because rapid processes are often applied in melting situations, the correct formulation of a Stefan problem in terms of a more realistic heat transfer mechanism is of particular importance.

F. Voids

Density changes accompanying phase change processes can increase or reduce the volume of the material undergoing these changes. In the former case, one sees containers broken (such as water pipes in the winter). In the latter case, one observes containers that have buckled or the formation of voids. Voids are vapor bubbles of material formed as a result of volume reduction accompanying a phase change. For most materials, the density change upon freezing or melting is of the order of 10%. For some materials, such as lithium fluoride, they may be as great as 30%, with their formation inhibiting heat transfer and producing other problems as well.

G. Control

Since melting and freezing processes are of vital importance, their control is of great interest. A typical control situation involves "backtracking" from moving boundary location to boundary conditions (an inverse problem). Research in this area focuses largely on numerical procedures and inverse problems, and is largely open.

IV. FUTURE ACTIVITIES

In the next decades, the Stefan problem and the modeling of melting and freezing processes will be of ever-increasing interest. Future developments can be anticipated in the following areas.

A. Thermodynamics of Phase Change Processes

We will make significant progress in understanding the underlying mechanisms of freezing and melting for pure materials and alloy systems. We will also learn to better understand the thermodynamics of the interface. We note that the interface is itself an important factor in freezing and melting processes. The fact that energy is needed to maintain an interface results in a depression of the equilibrium temperature between a locally spherical crystal and adjacent liquid, a result due to Gibbs-Thomson. These efforts will encompass ultrarapid processes as well.

B. Control of Phase Change Processes

The automation of a broad variety of freezing processes with real-time control of thermal conditions at the boundary will take place in the coming decades under both gravity and microgravity conditions. The mathematics, software, and hardware needed for this control will be developed. One aspect of this capability will be in the area of latent heat thermal energy storage, that is, the storage of heat derived from intermittent sources, such as the sun, for use at a later time or in a different place. Other areas include food preservation, manufacturing of photovoltaics, and metal casting.

C. Large-Scale Simulation

The rapid development of distributed and massive computing capabilities will make it possible to perform simulations of large-scale, long-term phase change processes in three dimensions, including effects of natural convection in the melt, rapidly changing boundary conditions, and local surface effects.

D. Mathematical Analysis of the Stefan Problem

The dimensionality barrier of the analysis of the Stefan problem will be breached with well-posed results obtained for higher dimensional problems arising from pure heat transfer. In addition, significant advances will be made in coupled problems, where the effects of convection, composition, void formation, etc., are felt, and inverse problems for boundary conditions and thermophysical and diffusion parameters.

E. Testing of Simulation Software and Its Implications

The development of increasingly massive and sophisticated software products leads directly to the question of how to test this software. Thus, for example, a software package simulating loss of coolant accidents in nuclear reactors will need to include modules whose purpose is to simulate the melting of metals that come into contact with other molten metals or radioactive metals whose melting is induced by their own internally generated heat. Such modules are small parts of large codes, and the importance of the overall software package will demand their unquestioned correctness. The formulation of tests for these modules based on data input and output analysis is a major challenge because it encompasses all the material-related phenomena noted above (e.g., how to model ultrarapid melting) together with numerical analysis pitfalls (e.g., the consistency or Du-Fort Frankel-like schemes) as well as theoretical considerations (e.g., the proper formulation of inverse problems in which inputs are to be found on the basis of observed melt/freeze processes). We anticipate that the question of software testing will lead necessarily to the need for a more fundamental understanding of all aspects of the Stefan problem.

SEE ALSO THE FOLLOWING ARTICLES

CHEMICAL THERMODYNAMICS • PHASE TRANSFORMATIONS, CRYSTALLOGRAPHIC ASPECTS

BIBLIOGRAPHY

Carslaw, H. S., and Jaeger, J. C. (1959). "Conduction of Heat in Solids," 2nd ed., Oxford University Press, London.

Cheng, K., and Seki, N. (eds.). (1991)."Freezing and Melting Heat Transfer in Engineering: Selected Topics on Ice-Water Systems and Welding and Casting Processes," Hemisphere Publishers, Washington, DC.

Goldman, N. L. (1997). "Inverse Stefan Problems," Kluwer, Dordrecht, The Netherlands.

Hill, J. M. (1987). "One-Dimensional Stefan Problems: An Introduction," Longman, London.

Meirmanov, A., and Crowley, A. (1992). "The Stefan Problem," De Gruyter, Berlin.

Rubinstein, L. I. (1971). "The Stefan Problem," American Mathematical Society, Providence, RI.

Sarler, B., Brebbia, C., and Power, H. (eds.). (1999). "Moving Boundaries V: Computational Modelling of Free and Moving Boundary Problems," WIT Publishers, Wessex, England.

Solomon, A. D., Alexiades, V., and Wilson, D. G. (1992). "Mathematical Modeling of Melting and Freezing Processes," Hemisphere, New York.

Van Keer, R., and Brebbia, C. (eds.). (1997). "Moving Boundaries IV: Computational Modelling of Free and Moving Boundary Problems," Computational Mechanics, Wessex, England.

Wilson, D. G., Solomon, A. D., and Boggs, P. (eds.). (1978). "Moving Boundary Problems," Academic Press, New York.

Wrobel, L. C., and Brebbia, C. A. (eds.). (1993). "Computational Methods for Free and Moving Boundary Problems in Heat and Fluid Flow," Elsevier, Amsterdam.

Wrobel, L. C., Sarler, B., and Brebbia, C. (eds.). (1995). "Computational Modelling of Free and Moving Boundary Problems III," Computational Mechanics, Wessex, England.

Zerroukat, M., and Chatwin, C. (1994). "Computational Moving Boundary Problems," Research Studies Press, London.

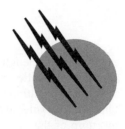

Stellar Spectroscopy

Jason P. Aufdenberg

Harvard–Smithsonian Center for Astrophysics

I. Formation of a Spectrum and Spectral Lines
II. Spectral Classification
III. Additional Spectroscopic Information
IV. Modern Developments

GLOSSARY

Binary system Two stars that are bound together by their mutual gravitational forces. They each orbit about the center of mass of the system. The study of such systems provides direct information about stellar masses.

Chromosphere Upper part of the atmosphere of a solar-type star where the temperature starts to rise with increasing height.

Corona Tenuous outer part of the atmosphere of a solar-type star where the temperature may exceed a million degrees.

Curve of growth Relationship between the equivalent width and the corresponding total abundance of an element.

Doppler imaging Use of the Doppler effect to determine the distribution of features on the surface of a star.

Effective temperature Representative temperature of the stellar atmosphere. A perfect radiator at this temperature would emit as much radiation per unit area as does the star. Abbreviated T_{eff}.

Equivalent width Integrated absorption of a spectral line produced by the gas in the stellar atmosphere.

Microturbulence Small-scale random motions in the stellar atmosphere that are invoked to obtain consistent abundance determinations from both weak and strong lines.

Radial velocity Component of motion that is toward or away from the earth.

Spectrum synthesis Wavelength-by-wavelength computation of the stellar spectrum, including all known sources of absorption, for comparison with the observed spectrum. Abundances can be derived by using this method when the spectrum is too complicated to be studied in any other way.

Stellar atmosphere Thin layer of gas at the stellar surface from which light can escape.

Surface gravity Acceleration of gravity at the surface of a star. This quantity sets the pressure in the stellar atmosphere through the requirement of pressure equilibrium.

THE LIGHT from a star contains many different colors or wavelengths of light, but these wavelengths are mixed together so thoroughly that the overall impression is generally a color close to white; the amount or brightness of the light creates a greater impression than does the underlying mix of colors. It is possible, however, to break the white light into its constituent colors, which then can be displayed in order of increasing or decreasing wavelength; the wavelength provides the quantitative means of arranging this order. Simple devices such as glass prisms or diffraction gratings can accomplish this decomposition. The resulting ordered display of light by wavelength is the spectrum of the light. The measurement and analysis of the stellar spectrum, wavelength by wavelength, are the activities of stellar spectroscopy. Temperatures, luminosities, and chemical signatures are just a few examples of the wealth of information available from the analysis of stellar spectra. It has been said that: "A picture may be worth a thousand words, but a spectrum is worth a thousand pictures." First, let us examine in some detail the formation of a spectrum.

I. FORMATION OF A SPECTRUM AND SPECTRAL LINES

In the 1820s, Joseph Fraunhofer used high-resolution observations of sunlight for testing and improving the quality of his prisms. He dispersed the light to a degree that enabled him to see for the first time that the sun's spectrum is not a continuous, smooth function of wavelength. Instead, he found its brightness greatly reduced at certain, precise, narrow places in the spectrum, referred to as "lines."

What is the origin of these spectral lines and how does a spectrum form? An understanding of the formation of a spectrum requires a description of the physics of radiative transfer. For this problem the fundamental macroscopic observable quantity is the monochromatic intensity, I_λ,

$$dE_\lambda = I_\lambda \, d\omega \, d\sigma \, d\lambda \, dt \tag{1}$$

defined as the energy E at a wavelength λ per second passing through area $d\sigma$ into a solid angle $d\omega$. Therefore, an intensity is specified at a point, in a particular direction, and at a particular wavelength. As radiant energy is transported through a medium, its intensity can be altered by absorption, re-emission, and scattering in both wavelength and direction through interactions with matter. The microscopic inputs to radiative transfer theory center on the description of the matter using electrons, ions, atoms, and molecules and the quantum mechanical processes required to understand the interaction of these particles with light.

To begin, it is instructive to consider the transfer of radiation along a path s:

$$\frac{dI_\lambda}{ds} = -\kappa_\lambda I_\lambda + \epsilon_\lambda. \tag{2}$$

This equation describes the absorption and emission of radiation at the wavelength λ over a small path length ds. Here, $-\kappa_\lambda I_\lambda$ specifies the amount of radiation removed

from the beam by absorption. The absorption process is directly proportional to the intensity where κ_λ is the absorption coefficient, which in linear theory is independent of the intensity. The emission coefficient ϵ_λ specifies the amount of radiation added to the beam along the path length.

Classically, a basic problem in spectrum formation was to understand the connection between light and heat by understanding the emission by solids. It is illustrative to consider the radiation inside a cavity of uniform temperature T. Kirchhoff's radiation laws, discovered about 30 years after Fraunhofer's solar studies, state that (1) the intensity is everywhere independent of location and direction within the cavity; (2) the intensity must be equal to the ratio of the emission and absorption coefficients, namely $I_\lambda = \epsilon_\lambda / \kappa_\lambda$, as $dI_\lambda / ds = 0$; and (3) the ratio $\epsilon_\lambda / \kappa_\lambda$ must equal the intensity of blackbody radiation at each wavelength. Following its successful derivation from quantum theory in 1900, blackbody radiation is described by the Planck function,

$$B_\lambda(T) = \frac{2hc^2}{\lambda^5} \frac{1}{e^{hc/\lambda kT} - 1} \tag{3}$$

where h is Planck's constant, c is the speed of light, and k is Boltzmann's constant. The law (3) is commonly called *Kirchhoff's law* or the *Kirchoff–Planck law* and can be expressed as:

$$\epsilon_\lambda = \kappa_\lambda B_\lambda \tag{4}$$

One way to read Kirchoff's law is "a good emitter is a good absorber, independent of its composition or geometry." This condition must be met in order to satisfy conservation of energy. Furthermore, it must be strictly true in detail at every wavelength so that temperature gradients do not appear without work being done.

In the laboratory, a good emitting substance is neutral sodium, Na I; this was one of the first substances used by Bunsen and Kirchhoff in their spectrochemical studies. When light emitted from heated sodium vapor is collimated and passed through a prism, two bright yellow lines are observed at 5890 Å and 5895 Å. These lines are due to the resonance transitions of Na I between two closely separated energy upper levels and the ground level. However, when the heated sodium vapor is placed between a bright incandescent light source and the prism, strong absorption lines at the same wavelengths are seen superimposed against a bright continuum. Comparing these two cases with Kirchhoff's law shows that because Na I shows strong emission near 5890 Å and 5895 Å in the first case, that Na I must show strong absorption at these same wavelengths in the second case.

But, why does an absorption spectrum result when a bright light source is placed behind the vapor and an emission spectrum result otherwise? We turn again to the transfer equation. First, dividing by κ_λ,

$$\frac{dI_\lambda}{\kappa_\lambda \, ds} = -I_\lambda + \frac{\epsilon_\lambda}{\kappa_\lambda} \qquad (5)$$

then

$$\frac{dI_\lambda}{d\tau} = -I_\lambda + S_\lambda \qquad (6)$$

Here, $\kappa_\lambda \, ds$ is defined as the optical depth $d\tau_\lambda$, which measures the mean free path of a photon before it interacts with matter. The ratio of the emission and absorption coefficients is called the source function, $S_\lambda = \epsilon_\lambda / \kappa_\lambda$. So, in the case of a uniform temperature cavity $I_\lambda = S_\lambda = B_\lambda$.

The solution of the transfer equation for an isothermal, homogeneous slab of gas is

$$I_\lambda(\tau_\lambda) = I_\lambda(0) \, e^{-\tau_\lambda} + S_\lambda(1 - e^{-\tau_\lambda}). \qquad (7)$$

This provides the intensity as a function of depth and is a crude but useful approximation for light passing into a planetary atmosphere. The term $I_\lambda(0)e^{-\tau_\lambda}$ is the intensity removed from the beam and $S_\lambda(1 - e^{-\tau_\lambda})$ is the intensity added to the beam. The intensity at zero optical depth, $I_\lambda(0)$, specifies the light source entering the slab and $I_\lambda(\tau_\lambda)$ is intensity at some optical depth of interest—for example, on the other side of the slab.

Returning to the first case with no light source, $I_\lambda(0) = 0$, and a thin slab so $\tau_\lambda \ll 1$, the intensity is

$$I_\lambda(\tau_\lambda) = S_\lambda \tau_\lambda \qquad (8)$$

as $e^{-\tau}$ becomes $1 - \tau$ for very small τ. Since the medium is hot, it emits its own light that might otherwise be dominated by an external brighter source. This relation shows that the intensity is greatest at wavelengths where the optical depth is highest—in other words, where κ_λ is large. This is just Kirchoff's law from Eq. (4). If the optical depth is not small then $e^{-\tau}$ goes to zero and the intensity is simply the source function.

In the second case a light source is required, so $I_\lambda(0) > 0$; along with a thin slab, again $\tau_\lambda \ll 1$. The solution of the transfer equation now becomes:

$$I_\lambda(\tau) = I_\lambda(0)(1 - \tau_\lambda) + S_\lambda(\tau_\lambda) \qquad (9)$$

At wavelengths where the S_λ is small compared with the light source the intensity is reduced after passing through the slab, $I_\lambda(\tau_\lambda) < I_\lambda(0)$, and an absorption spectrum is produced. Alternatively, at wavelengths where $S_\lambda > I_\lambda(0)$ then $I_\lambda(\tau_\lambda) > I_\lambda(0)$ and an emission spectrum results. These basic results can be used to understand the formation of the solar spectrum. As a first approximation, it is not bad to assume that at a particular depth τ the solar atmosphere can be characterized by a single temperature. This assumption is known as local thermodynamic equilibrium (LTE) and, just like the isothermal

cavity, the source function equals the Planck function, $I(\tau) = S(\tau) = B(\tau)$. Equation (9) can be written, dropping the λ subscripts for the moment,

$$I(\tau) = I(0) + \tau(S(\tau) - I(0)) \qquad (10)$$

which in LTE can be written:

$$I(\tau) = B_{\text{deep layer}} + \tau(B_{\text{outer layer}} - B_{\text{deep layer}}) \qquad (11)$$

In contrast to the continuous spectrum, which in the earliest treatments of sunlight was assumed to come from a hot, dense interior, the absorption spectrum was treated as a cool, thin, isothermal gaseous atmosphere called a "reversing layer" with the same chemical composition as the rest of the star. In detail, however, the temperature of the solar atmosphere increases with depth below the visual surface, the *photosphere*, and also increases with height into the chromosphere and corona. This temperature rise is likely due to nonthermal processes connected with magnetic fields that structure the tenuous outer atmosphere. Consider the visible photosphere. At optical wavelengths, deeper layers are brighter than the outer layers,

$$B_\lambda(T)_{\text{deep layer}} > B_\lambda(T)_{\text{outer layer}} \qquad (12)$$

Therefore, according to Eq. (11) the intensity is reduced at wavelengths of appreciable optical depth—for example, in spectral lines. Thus, the visible solar spectrum is an absorption spectrum, hence the term *reversing*. In contrast, at ultraviolet wavelengths, the high temperature of chromosphere and corona satisfy:

$$B_\lambda(T)_{\text{deep layer}} < B_\lambda(T)_{\text{outer layer}} \qquad (13)$$

Then Eq. (11) shows that as the intensity increases with height, an emission spectrum is produced.

It is important to point out that at any wavelength an observer sees to an optical depth of about unity, which means the chance is better than even that a photon will be absorbed or scattered before exiting from that depth. As pointed out above, when the optical depth becomes large the intensity reduces simply to the source function. The intensity the observer sees at any wavelength is essentially the source function at an optical depth of unity (see Fig. 1). In the cores of strong photospheric absorption lines, for example, one sees to shallower physical depths, thus to fainter, cooler layers than one sees outside of the line core. For this reason, the line cores are dark. Similarly, when we observe the sun near the limb, an optical depth of unity is reached at a shallower physical depth, due to the longer geometric path length, compared with the physical depth seen at the center of the solar disk. Thus, the limb of the sun appears less bright than the center.

The shape of the overall stellar spectrum and the character and strength of its spectral lines are most sensitive to the temperature and pressure structure of atmosphere.

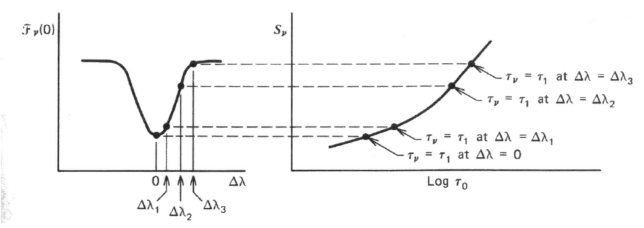

FIGURE 1 Mapping between the line profile (left) and the source function (right). It is the decline of the source function outward through the photosphere that produces absorption lines. [From Gray, D. F. (1992). "Observation and Analysis of Stellar Photospheres," Cambridge University Press, Cambridge, U.K., p. 271. Reprinted with the permission of Cambridge University Press.]

How and what we know about stellar temperatures and pressures are the subjects of the next two sections.

A. Stellar Temperatures

To first order, in between the spectral lines, the radiative energy distribution from a stellar atmosphere follows the Planck function (3) of a characteristic temperature known as the effective temperature, T_{eff}. Formally, the effective temperature does not correspond to a particular temperature in the stellar atmosphere, but to the total radiative energy or the bolometric flux, $\mathcal{F}$,

$$\mathcal{F} = \int_0^\infty F_\lambda \, d\lambda = \int_0^\infty B(T_{\text{eff}})_\lambda \, d\lambda = \sigma T_{\text{eff}}^4 \quad (14)$$

where F_λ is monochromatic flux at each wavelength λ, and σ is the Stefan–Boltzmann constant. The stellar continuum from a geometrically thin photosphere emanates from layers near where $\tau = 1$ in the continuum; consequently, these layers have temperatures very near T_{eff}.

A qualitative estimate of T_{eff} can be made by observing the color of a star. The measurement of the ratio of blue to red light will capture this brightness distribution, the measurement being made either through filters transmitting broad bands of the blue and red spectral regions or with instruments designed to sample the light monochromatically from widely separated wavelengths. As observed in the laboratory and described by the Planck function, incandescent objects radiate most brightly at the short wavelengths, blue or violet colors, whereas cool stars radiate most strongly at longer wavelengths, red colors. The wavelength at the peak intensity of a blackbody spectrum must satisfy:

$$\left. \frac{dB_\lambda(T)}{d\lambda} \right|_{\lambda_{\max}} = 0 \quad (15)$$

which yields the relation:

$$\lambda_{\max} T = 0.294 \, \text{cm K} \quad (16)$$

known as Wien's displacement law. Low temperatures yield large values for $\lambda_{\max}$ and vice versa, consistent with observations.

To determine T_{eff} quantitatively the spectral energy distribution of the star must be known relative to some reference standard such as a calibrated oven. Near the end of the 19th century, ultra-sensitive alcohol thermometers were being employed to study the sun's heat distribution between 3000 Å and 3 μm. For other stars, visual comparisons were made between starlight and a lamp calibrated on an oven. Assuming that the shape the spectrum obtained in this way was a blackbody, a value of T_{eff} could be determined. Stellar continua are blackbodies only to first order, so these first measurements were subject to considerable systematic error. Modern spectrophotometric observations of stars are calibrated relative to the bright star Vega, whose spectral energy distribution has been derived relative to precisely calibrated freezing-point copper blackbodies. If both the absolute flux distribution, corrected for extinction by the interstellar medium and the earth's atmosphere, and the angular diameter of a star can be measured precisely, a fundamental effective temperature can computed from:

$$T_{\text{eff}} = \left[\frac{4\mathcal{F}}{\sigma \theta^2} \right]^{\frac{1}{4}} \quad (17)$$

where θ is the angular diameter. Currently, it is possible to measure θ accurately for only bright stars, but the situation will improve soon with optical interferometers and space-based astrometric missions set to make precise measurements of fainter stars. In the absence of accurate angular diameter measurements, synthetic stellar continua from model atmospheres are employed to estimate T_{eff}.

The line spectrum is also used to estimate T_{eff}. This is possible because the state of a gas, the excitation and ionization of the constituent atoms and molecules, is governed by radiation, collisions, and the principle of *detailed balance*. Detailed balance in a thermal plasma occurs when any reaction is balanced by an inverse reaction—for example, the balance between ionization (by radiation and collisions) and recombination of an ion and electron. When collisions dominate, this principle can be used to derive the *Boltzmann distribution*:

$$\left(\frac{n_2}{n_1}\right) = \frac{g_2}{g_1} e^{-E_{21}/kT} \tag{18}$$

for the relative population, n_2/n_1, of two energy levels in a atom separated by an energy E_{21}. The ratio statistical weights, g_2/g_1, depends on the quantum mechanical properties of the two levels. In a similar way, the *Saha equation*,

$$\log \frac{N^+}{N^0} n_e = \frac{2u^+}{u} \lambda_e^{-3} e^{-\chi_{ion}/kT} \tag{19}$$

where

$$\lambda_e = \frac{h}{(2\pi m_e kT)^{1/2}} \tag{20}$$

can be derived which gives ratio of the total number of ions N^+ to the total number of atoms N^0. Here, n_e is the electron density, m_e is the electron mass, χ_{ion} is the ionization potential, and u and u^+ are the partition functions which depend of the energy level structure of the atom and ion.

From these equations it is clear that the excitation and ionization of atoms and molecules in a stellar atmosphere will be quite sensitive to temperature. In cooler stars, from the sun, near 5000 K, down to the star/planet boundary, near 1000 K, absorption lines from molecules and neutral atoms are observed. The hydrocarbon radical, CH, is observed in the solar spectrum, and the increasing presence of molecular spectral features is observed toward cooler temperatures where molecules such as TiO dominate the spectrum below 3500 K. Dissociation energies of molecules are generally less than 10 eV and typically a few eV. This energy corresponds to a wavelength of about 1240 Å. Any atmosphere with a significant radiation field below this wavelength will photodissociate molecules. From Wien's law (15), a star with $T_{eff} = 23,000$ K has a spectrum which peaks near this wavelength. From here down to the temperature of the sun, stars have sufficient

flux shortward of the flux peak and relatively high collisional temperatures to prohibit the significant formation of molecules. For the same reason, neutral metals with ionization energies of a few eV, such as Fe I and Cr I, are only prominent in the cooler stars.

In hotter stars, with strong radiation fields and high collision energies, metals are ionized to an ever increasing degree as T_{eff} increases. In stars just hotter than the sun, lines from Ca II, Fe II, and Cr II are present. Above 7500 K, lines from Mg II, Si II, and Ti II are visible with lines from neutral atoms showing weakly. Above $T_{eff} = 10,000$ K, the optical hydrogen lines begin to fade as hydrogen is ionized and lines from O II, Si III, and He I are present. He I has an atomic structure which places the lower levels of optical transitions 19 eV above the ground state. Its ionization energy is 25 eV. For this reason, He lines are an important temperature diagnostic in hot stars. Above $T_{eff} = 30,000$ K, lines from He II are seen, along with lines from N III, Si IV, and C IV. Even hotter stars show progressively higher ionization stages. Figure 2 shows the blue spectral region of stars spanning the entire range of normal surface temperatures.

To determine a quantitative value for T_{eff} from spectral lines alone, from a spectral energy distribution, or both, a synthetic stellar spectrum from a model atmosphere is required. Standard model atmospheres are generally parameterized by an effective temperature, T_{eff}, and a gravity, g. The fundamental constraints on such a model are that it must conserve energy and be in pressure balance. The energy constraint means that the radiation emitted to space must be exactly equal to the energy coming up from the deep interior of the star. The luminosity,

$$L = 4\pi R^2 \sigma T_{eff}^4 \tag{21}$$

therefore, must be independent of depth. For a geometrically thin atmosphere, where the radius is essentially constant throughout the atmosphere, T_{eff}, is a well-defined quantity and the radiative flux:

$$\mathcal{F} = \frac{L}{4\pi R^2} = \sigma T_{eff}^4 \tag{22}$$

is a constant. A constant flux implies *thermal equilibrium*, which means the temperature structure is not changing with time. This is *not* the same as *thermodynamic equilibrium*, which would exist only if every layer of the atmosphere had the same temperature. Nuclear energy production in the stellar interior prevents this from happening. If thermal equilibrium exists and the energy is exclusively transported by radiation, then the medium is said to be in *radiative equilibrium*. In the tenuous outer layers of stars, the radiation is by far the dominant energy transport mechanism so radiative equilibrium is a good approximation. To solve the direct radiative transfer problem, the

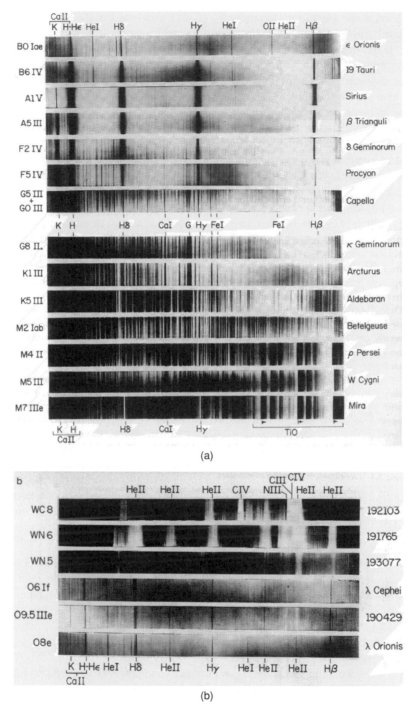

FIGURE 2 The spectral sequence. (a) Representative stellar spectra in the blue and violet spectral region, obtained at the University of Michigan by W. C. Rufus and R. H. Curtiss. The Balmer series of hydrogen is conspicuous in the upper stars. The H and K lines at 3968 Å and 3934 Å are produced by ionized calcium, and the G band arises from the CH molecule. Other identifications are neutral He at 4026 and 4472 Å, ionized oxygen at 4649 Å, neutral calcium at 4227 Å, and neutral iron at 4384, 4405, and 4668 Å. In the cool M stars the violet edges of molecular TiO bands lie at 4585, 4752, and 4954 Å. Classification is traditionally done in this blue–violet part of the spectrum (where early photographic plates were most sensitive to light), but the yellow–red can also be employed, and recently the far (satellite) ultraviolet and the IR have become useful, as well. (b) Spectra for the hot O stars, plus examples of carbon- and nitrogen-rich (WC and WN) Wolf–Rayet stars, which are often included in class O. Their emission lines show that they are surrounded by outflowing matter. [From Kaler, J. (1989). "Stars and Their Spectra," Cambridge University Press, Cambridge, U.K., pp. 70–71. Reprinted with the permission of Cambridge University Press.]

source function as a function of depth must be known. The condition of radiative equilibrium is used to reach this goal. For the sun, it is possible to derive not only T_{eff}, but also the temperature as a function of depth in the atmosphere. Limb darkening, as discussed above, results because we observe shallower physical depths, and cooler layers, as we look closer to the limb. The intensity of the sun has been measured as a function of angle and this relationship can be inverted to yield $T(\tau)$. In 1906, K. Schwarzschild showed that the temperature distribution derived in this way was consistent with radiative equilibrium.

Figure 3 compares the observed energy distribution of the bright star Spica (T_{eff} =23,000 K) with the synthetic spectrum from a model stellar atmosphere. Spica is one of the few hot stars for which T_{eff} can be found directly from observations; the empirical and the model T_{eff} agree to within the uncertainty of each method. Modeling of this kind shows that stars span a wide range of effective temperatures nearly two orders in magnitude, from brown dwarfs near 1000 K to white dwarfs hotter than 100,000 K.

B. Surface Gas Pressure

In an atmosphere model, the pressure structure of the atmosphere is computed from the condition of hydrostatic equilibrium,

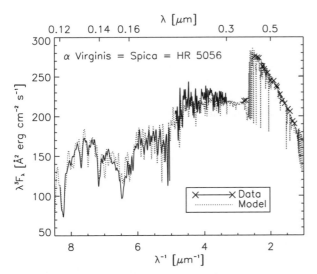

FIGURE 3 The observed energy distribution of Spica (α Virginis), represented by the thick line and crossmarks, is shown between 1250 Å in the far-ultraviolet to 8000 Å in the near-infrared. A model spectrum (dotted line) is shown for comparison. Data at wavelengths less than 0.3 μm were obtained by the International Ultraviolet Explorer, while data at longer wavelengths were obtained by ground-based telescopes. Differences between the observed and model energy distributions reflect poor assumptions or inaccurate input data for the model and uncertainties in the measurement of the absolute flux.

$$\frac{dP}{dr} = -g\rho(r) \qquad (23)$$

The pressure gradient, dP/dr, where P is the sum of the gas pressure and the radiation pressure, is balanced at every depth r by the gravitational force gradient, the product of the gravity g and the density ρ. In this way, the gravity controls the extension of the atmosphere. Lower gravities result in more extended, lower pressure atmospheres for the same T_{eff}. As a result of stellar evolution, stars with the same T_{eff} can have very different gravities, depending on their stage of evolution. The extended atmospheres of supergiant stars may have gravities as much as 10^8 times smaller than the compact white dwarfs. The gravity can be determined directly for members of an eclipsing binary systems where the mass and radius are measured directly.

The pressure structure of the atmosphere affects a star's spectral energy distribution. For stars with T_{eff} above 7000 K, the principal deviation of their spectral energy distribution from the Planck function visible to ground-based telescopes is the discontinuity near 3650 Å. A major opacity source for photons immediately below this wavelength is the ionization of neutral hydrogen in the first excited state ($n = 2$). The strength of the Balmer discontinuity depends on the total amount of H I and the fraction of H I in the first excited state. In LTE, these are described by the Saha (19) and Boltzmann (18) equations, respectively. The Saha equation shows that for a fixed temperature the number of neutral atoms N^0 is directly proportional to the electron density, n_e. Therefore, in low-gravity, low-pressure atmospheres where n_e will be smaller, the ratio of (the ion fraction) N^+/N^0 will be larger. More hydrogen ions (free protons) mean less H I opacity and consequently more flux below 3650 Å. Effectively, with less hydrogen opacity, we are able to see to deeper, hotter layers of the atmosphere which are brighter. In this way, the value of $\log(g)$ is correlated with the strength of the Balmer discontinuity. The temperature also regulates the degree of ionization and the fraction of H I in the $n = 2$ level. The Balmer discontinuity is greatest for high-gravity stars near 10,000 K. Low-resolution spectrophotometry between 3100 and 4800 Å, with careful corrections made for the ultraviolet extinction caused by the earth's atmosphere, can thus be used to estimate both temperature and gravity from a comparison with model predictions.

Analogous to the diagnostic property of the Balmer discontinuity, one pressure diagnostic in cooler stars is the comparison of absorption lines from a neutral atom and an ion of the same element. Consider the comparison of Fe I and Fe II lines in the spectrum of the sun and in the spectrum of a star with a very similar T_{eff} but much lower gravity—for example, the supergiant β Draconis. From the similar temperature of these stars, we expect the level excitation as described by the Boltzmann equation to be similar. However, as described by the Saha equation, the ionization ratio Fe I/Fe II will be different because of the different pressures in the two stars' atmospheres. Therefore, in the solar spectrum the line ratio Fe I/Fe II should be larger because of the high densities and the greater possibility of ion–electron recombination.

The pressure structure of the atmosphere also affects the shapes of lines in the spectrum. Spectral lines are fundamentally broadened by the quantum mechanical natural width and by the integrated effect of thermal motions by individual atoms along a line-of-sight. Pressure broadening refers to the collisional interaction between the atoms absorbing light and other particles, mainly protons and electrons, in hot stars with $T_{eff} > 9000$ K. The frequency of collisions depends on the particle density and hence the pressure. The *Stark effect* describes the effect on an atom placed in an electric field. The charged particles in a stellar atmosphere provide a randomly fluctuating electic field that is approximately uniform localy.

It is here that knowledge gained from laboratory studies becomes of great importance. The hydrogen lines, which are so useful in the hotter stars, have been studied extensively in laboratory plasmas as well as with the theoretical methods of quantum physics. As a result, the response of the lines to various environments is well known. With this knowledge it is possible to compute the appearance of the hydrogen lines from a model of the stellar atmosphere having a certain effective temperature and a variation of the pressure with depth.

In the laboratory, the energy levels of H I are shifted in direct proportion to the strength of the electric field (linear Stark effect). The energy levels of other ions, such as He I, respond quadratically to the strength of the electric fields (quadratic Stark effect). In cooler stars, too few charged particles are present for the Stark effect to be important. Here, collisions of molecules and atoms with neutral hydrogen (van der Waals broadening) are important. Since the wavelengths of spectral lines arise from the difference in energy between upper and lower levels, ions with perturbed energy levels absorb light at wavelengths shifted from the zero-field case. The resulting spectral line, which forms along a line-of-sight into the atmosphere, is then shifted and broadened as the result of ions that are pertubed to varying degrees.

The strong H I Balmer lines probe a large range of pressures in the atmosphere. Because a spectral line absorbs most strongly in the line core, one sees to shallow depths and therefore lower pressures in the line core. Outside the line core, the line wings have much lower opacity and one sees to greater depths and higher pressures. As a result, the wings of the Balmer lines form in layers where the

pressure broadening is the greatest, while the core forms where pressure broadeing is much less significant. The resultant line profile is a narrow, deep core with broad wings extending up to ± 100 Å from the core. Figure 4 shows that the width of the Balmer lines varies with stellar luminosity. The low-pressure, extended atmospheres of luminous stars have narrower lines in their spectra relative to less luminous stars. High-resolution spectra can be compared with synthetic spectra from a model atmosphere in order to determine the pressure structure of the atmosphere and therefore an estimate of the gravity, g. Figure 5 shows the observed and computed $H\gamma$ profiles for B-type giant β Canis Majoris.

II. SPECTRAL CLASSIFICATION

As discussed above, the sensitivity of a star's color and line spectrum to its effective temperature results in conspicuous changes in the appearance and complexity of stellar spectra over a range of temperatures. At a given temperature, the atmospheric pressure also affects the appearance of a stellar spectrum. Spectral classification is the ordering or ranking of stellar spectra in relation to the strength and width of particular spectral lines. Particular classes are assigned within this ranking relative to standard stars. Early systems of classification ranked the stellar spectra according to their color and the general complexity of the spectra.

Fraunhofer not only analyzed the solar spectrum, but was also the first to observe the spectra of other stars. The first spectrum of a star other than the sun, that of the bright star Sirius, was viewed by Fraunhofer in 1814. He recognized that stellar spectra differ in varying degrees from the solar spectrum. Starting in the 1860s, improvements in prismatic spectroscopes led to the classification of thousands of bright stars, first visually and than with the aid of photography. Angelo Secchi introduced a classification system for stellar spectra which was to lay the groundwork for modern stellar spectral classification. Most of the approximately 400 stellar spectra that Secchi classified fell into three groups: I, stars with strong H I lines, such as Sirius and Vega; II, stars with numerous metallic lines and weak H I lines, such as the sun, Capella, and Arcturus; and III, stars with prominent bands of molecular lines, such as Betelgeuse and Antares, where the lines get darker toward the blue end of the spectrum. Soon after Secchi's work, Henry Draper photographed the spectrum of Vega, the first photographic recording of a stellar spectrum. Inspired by Bunsen and Kirchhoff's spectrochemical analysis and their explanation for the solar line spectrum, William Huggins took up stellar spectroscopy in the early 1860s. By comparing these stellar

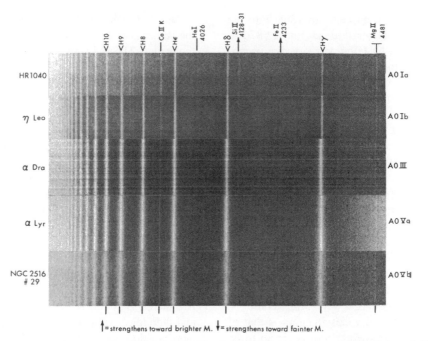

FIGURE 4 The blue–violet spectral region for stars with effective temperatures near 9500 K, but of different luminosities. The spectra are arranged with the most luminous stars at the top of the figure. Note the dramatic change in the width and strength of the hydrogen lines as the luminosity and atmospheric pressure vary. [From Morgan, W. W., Abt, H. A., and Tapscott, J. W. (1978). "Revised MK Spectral Atlas for Star Earlier than the Sun," Yerkes Observatory, The University of Chicago, and Kitt Peak National Observatory. With permission.]

spectra to reference spectra, he discovered that the same chemical elements were common to the earth, sun, and other stars. These studies led to the production of a photographic atlas of representative stellar spectra produced in collaboration with Lady Margaret Huggins. The atlas was made possible by Lady Huggins' expertise in photography

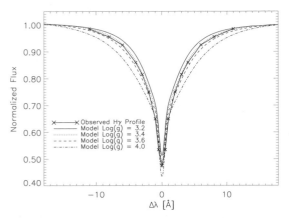

FIGURE 5 A comparison of an observed Hγ line profile of star β Canis Majoris, represented by crossmarks, and computed profiles for a range of surface gravities. Larger surface gravities, with correspondingly higher gas pressures, produce broader wings in the line profile. A best match between the computed and observed profiles provides an estimate of this star's surface gravity.

and improvements in dry photographic plates which soon led to the large-scale collection of stellar spectra.

In 1885 the enormous task of classifying the then largest collection of stellar spectra began at Harvard College Observatory under the direction of Edward Pickering. Under this program, the first general photographic classification of stellar spectra was undertaken by Williamina Fleming. Of the 10,000 stars classified, the large majority fell into a few typical classes. However, Fleming discovered many peculiar spectra of hot stars and novae with bright lines. Over the next 40 years, supported by the estate of Henry Draper, the Observatory published catalogues of stellar spectral types culminating in the Henry Draper (HD) catalogue and later its extension. Larger telescopes at Harvard and in the southern hemisphere facilitated the acquisition of an additional 400,000 stellar spectra, which were chiefly classified by Annie Jump Cannon. Antonia Maury's great contribution to the work was the introduction of a second dimension to spectral classification, that of luminosity, through her careful examination of spectral line widths.

While the basic spectral sequence OBAFGKM, still in use today (see Fig. 1), was adopted by Cannon for the publication of the HD catalogue, the ordering of the first three spectral classes was complicated in earlier work by the presence of then unidentified lines from He I and

He II. The discovery of helium came in 1868 with the observation by Norman Lockyer of the D$_3$ line at 5876 Å, as distinct from the nearby Na I doublet, in a solar prominence spectra. However, the isolation of terrestrial helium did not come until 1895. Following this Lockyer was able to show that He I was responsible for other chromospheric lines in the solar spectrum and absorption lines in B-type stars. Pickering, upon examining the spectrum of O-type star ζ Puppis, discovered the He II absorption line series in 1897. However, this series was not identified with helium at high temperature until its emission spectrum was identified in the laboratory 15 years later. Along similar lines, in the 1870s Lockyer's laboratory experiments with the spectrum of CaCl showed the calcium contained two separate line spectra, one at low temperature (Ca I) and one at high temperature (Ca II) which matched the Franhaufer H and K lines. The idea of ionization was there, but the discovery of the electron and the development of atomic theory to the Saha equation were still to come.

The modern system of stellar classification is that of Morgan–Keenan (MK). The MK system is set up to be independent from theoretical considerations regarding the physics of stellar atmospheres. Each spectral type is assigned in comparison to adopted standard stars. In this way, a star of, say, spectral type A2 I, an A-type supergiant, keeps its classification even as the our understanding of the physical conditions in its atmosphere develops or changes. The spectra are classified using broadened, low-dispersion (typically 125 Å mm^{-1}) photographic spectra spanning a spectral region from the Ca II K line (3934 Å) to the Hβ line (4861 Å). What follows are some of the standard spectral features examined to classify stellar spectra.

The O-, B-, and A-type stars are primarily classified by the appearance of their hydrogen and helium lines. The O-type stars and their spectral subtypes O9–O3 are classified on the basis of the He II λ4541/He I λ4471 line ratio which increases with increasing temperature until the He I λ4471 line disappears at O3. The He II λ4686 line can be used as a luminosity indicator, the line starting out in absorption for the dwarfs and switching into emission for the giants and supergiants. Numerous resonance lines in the ultraviolet show the same effect. The spectral types B9–B0 are characterized by the appearance of He I at B9 which strengthens toward B0 and the appearance of He II at O9. Weak lines of O II, N II, Si II, Si IV, and Mg II define the subclasses B0.5–B9.5. The hydrogen lines and line ratios of Si III, Si IV, and O II to He I all weaken with increasing luminosity. For the A-type stars, characterized by the strongest H I Balmer lines, the Ca II K line is a useful indicator of temperature, while blends of Ti II and Fe II lines at λ4172 and λ4178 strengthen with higher luminosities. Narrower and weaker hydrogen lines characterize the supergiants relative to lower luminosity classes.

The F-, G-, K-, and M-type stars are characterized by the transition from ionized metals to neutrals and molecules. The H I Balmer lines fade as the $n = 2$ level becomes difficult to excite at lower temperatures. At F0, Ca I λ4227 appears and at F5 the "G" band, λ4314 from the CH molecule, appears and strengthens toward later types. The ratio of Mn I λ4030 to Si II λ4128, which increases toward later F subclasses, also demonstrates the transition to lower ionization species. Line ratios of singly ionized metals and neutrals are used as luminosity indicators. From F5 to G5 the ratio of Fe I λ4045 and λ4226 to Hγ and Hδ increases as well. The G band reaches its maximum strength at G5. The luminosity of G- and K-type stars is deduced from the presence and strength of an emission component in the Ca II K line, which is formed in chromospheres of these stars. Giants and supergiants show stronger emission components. The strength of the CN band λ4215 also increases with luminosity. At K0, the Ca II H and K lines attain their greatest strength and fade towards later types as Ca I λ4227 continues to increase along with the Na I D lines. Bands of TiO first appear at K5 and strengthen towards M0. The TiO bands continue to strengthen toward later types, and more fragile molecular species such as VO, MgH, and H$_2$ appear. The spectral appearance of luminous M-type and similar very cool stars is quite sensitive the chemical composition of the atmosphere. In particular, when the C/O ratio is large the formation of metallic oxides such as TiO will be shut off as the oxygen is locked up in CO. The appearance of strong C$_2$ bands signifies the carbon stars. Bands of ZrO also show up in metal-rich cool stars designated type S.

Temperatures, luminosities, and chemical signatures are just a few examples of the wealth of information available from the analysis of stellar spectra. So, let us examine in some detail the information that stars reveal through their spectra.

III. ADDITIONAL SPECTROSCOPIC INFORMATION

A. Radial Velocities

In addition to information on the physical conditions in the stellar atmosphere, the spectrum contains information about the motions of the star and the motion of the atmosphere relative to the center of the star. The motion of a light-emitting source produces a shift in the spectrum with wavelength as measured by an observer. This is known as the Doppler effect. The Doppler effect is widely employed in studying the stellar spectra because the spectral lines, particularly stellar absorption lines, provide a precise template of accurately known reference wavelengths. If the measurement of a stellar spectrum shows that all the spectral lines are displaced to either longer or shorter wavelengths, we can conclude that the star and the earth are in relative motion away from or toward each other, respectively. In practice the observed spectrum of a source is cross-correlated against an interposed template spectrum. The wavelength displacement is related to the speed of radial velocity according to:

$$\lambda = \lambda_0[(1 + v/c)/(1 - v/c)]^{1/2} \qquad (24)$$

where v is the relative radial velocity, c is the speed of light, λ_0 is the wavelength emitted by the source, and λ is the observed wavelength. When $v \ll c$, this relation reduces to:

$$\Delta\lambda/\lambda_0 = v/c \qquad (25)$$

where $\Delta\lambda$ is the wavelength displacement $\lambda - \lambda_0$.

The smallest detectable radial velocity depends on the smallest measurable wavelength displacement. In turn, this depends on the elimination of all systematic instrumental complications and scrupulous attention to experimental detail. Extremely careful measurements with a precision as small as 1 m s^{-1} have been reported for stellar sources. Without exercising extraordinary effort, many studies have routinely measured velocities with an accuracy of 1 km s^{-1}. The spectral resolution of an instrument over a range of wavelengths is specified by $R = \lambda/\Delta\lambda$ where $\Delta\lambda$ is resolution at a given wavelength λ.

A major application of the Doppler effect is the determination of stellar masses. Most stars in space seem to occur in bound systems of two or more stars instead of as single stars, such as the sun appears to be. The stars in such systems orbit about their center of mass and the system as a whole has a motion through space. The existence of a bound system can be detected by repeatedly measuring the radial velocity of the stars. At one time we observe a component of one star's motion approaching us while the component of the other star's motion is receding from us. At another time the two stars might both be moving perpendicular to our line-of-sight in opposite directions; consequently, they would have no radial velocity apart from the center of mass. The periodic variation of the radial velocities is the proof that the two stars from a bound system. Figure 6 displays the spectrum of a binary system at two different phases.

As the two stars orbit their center of mass, they are held in the system by the gravity of their companion; therefore, one star becomes a measure of the strength of the gravity, and hence the mass, of the other star. In particular, the semi-major axis and the period of the orbit give the sum of the masses of the two stars by Kepler's third law:

$$(M_1 + M_2)P^2 = a^3 \qquad (26)$$

where the sum of the masses is measured in solar masses, the period, P is measured in years, and the semi-major axis

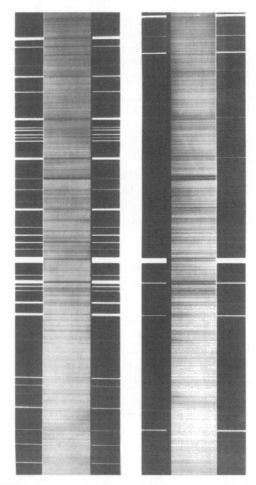

FIGURE 6 Spectra of the double star Mizar at two different phases. The left spectrum shows the system when the two stars are moving perpendicular to our line of sight, so the spectral lines are coincident. The right spectrum shows the system when the star is approaching us, so its lines are shifted to shorter wavelengths, while its companion is moving away from us, producing a shift toward longer wavelengths. [From Shu, F. (1982). "Physical Universe," University Science Books, SSD Gate Fire Road, Sausalito, CA, p. 181. With permission.]

a is measured in astronomical units (AU). The individual masses are found from observing the orbital speeds of the two stars. The more massive star is closer to the center of mass of the system. The two stars must complete one orbit around the center of mass in the same length of time, so the more massive star moves more slowly because it has a shorter distance to travel. The ratio of masses is the inverse ratio of the orbital speeds,

$$M_1/M_2 = v_2/v_1 \qquad (27)$$

From the sum of the masses and the ratio of the masses it is possible to find the mass of each star.

To actually apply the method outlined in the preceding discussion, one additional piece of information is essential: We must know the angle between our line-of-sight and the orbital axis of the system. This information is needed to convert the observed velocities, which are projections, to the true orbital velocities. Most binary systems contain stars that appear too close together in the sky for their orbit to be resolved and therefore the inclination cannot be determined as with visual binaries. Binaries are detected purely by the radial velocity variations in their spectra and are known as spectroscopic binary systems. Outside of the special case of eclipsing binary systems ($i = 90°$), the inclination is generally not known and only $M_1 \sin^3 i$ and $M_2 \sin^3 i$ can be recovered. Furthermore, in many spectroscopic binary systems we observe absorption lines for only one of the components. In this case, the mass function,

$$f(M_1, M_2) = M_2^3 \sin^3 i /(M_1 + M_2)^2$$
$$= 1.0385 \times 10^{-7} (1 - e^2)^{3/2} K_1^3 \quad (28)$$

can be computed and this provides a lower limit on the mass of the unseen component, M_2. K_1 (in km s^{-1}) is half the total range of the radial velocity variation of the brighter star, P (in years) is the orbital period, and e is the eccentricity of the orbit.

The measurement of radial velocities plays a central role in many interesting research projects in stellar spectroscopy. One of these is the detection of planets outside the solar system. It is extremely difficult to directly detect planets, which are only visible by reflected light. The associated star emits the vast majority of light. However, since 1995, spectroscopic methods have led to a steady discovery of planets orbiting more than 40 other stars. A majority of these extra solar planets, which range in mass from sightly less than Saturn to about 15 times more massive than Jupiter, orbit their parent stars within a few percent of 1 AU. Such a massive planet orbiting so close to its parent star was not predicted before the discoveries, but it is exactly these types of planetary systems that spectroscopic search methods are sensitive to. From Eq. (27) the velocity of the parent star about the center of mass is greatest for massive planets having large orbital speeds and therefore close orbits. Take the first system detected, 51 Pegasi, which lies 45 light-years from the sun. Here a solar-type star and a planet of approximately 0.45 Jupiter masses orbit about their center of mass with a period of only 4.6 days. The parent star has a regular radial velocity variation with an amplitude of 55 m s^{-1} and the planet lies only 0.05 AU from the star.

A very similar application is the detection of massive but compact objects that might be black holes. The theory of stellar evolution predicts that neutron stars can have masses up to three times the sun's mass; stars above this limit can only be black holes. The issue, then, is to de-

termine the masses of the compact companions by finding the effect of the compact star on the other star of the system by measuring the periodic variation of the other star's radial velocity. The analysis is complicated for various reasons, but a conservative assessment is that at least 10 stars appear to have compact companions that are too massive to be neutron stars. This finding is of great importance for astronomy because of the information it provides about stellar structure and the development of double star systems. However, it is also of great importance for our understanding of the properties of matter and gravity under extreme conditions that cannot be reproduced in the physics laboratory.

In addition to the stellar components of binary systems, there is sometimes a gas component in the form of a stream or disk. Many types of close binary systems undergo mass exchange and/or mass loss. The presence of hot gas in the system can be detected by the emission. Radial velocity measurements from emission lines from the stream or disk yield information about the speed of mass transfer and the sizes of accretion disks. The combined effect of emission from the intervening gas plus the absorption spectrum from the stars can produce either an emission above the surrounding wavelengths or a partial filling in of the absorption line, depending on the amount of gas between the stars. This combination changes as the stars revolve about the center of mass, presenting different perspectives to our line-of-sight.

Radial velocity measurements also show that stars experience complex atmospheric motions. Some stars exhibit periodic changes in both radial velocity and brightness, swelling and shrinking periodically, with their surfaces moving at speeds up to 50 km s^{-1}. One type of pulsating star, the Cepheid variable, has played a particularly large role in astronomy through its use in the distance scale. When high-resolution (< 10 km s^{-1}) spectra of Cepheids are obtained throughout a pulsation cycle, the radial velocity curve constructed from these observations can be used to solve for the mean radius of the star. This is the Baade–Wesselink method. The change in the radius of the star between two phases is determined by integrating the radial velocity curve from the first phase to the second, with a correction factor to account for spectral line formation integrated through a spherical atmosphere. The ratio of the radii at the two phases is related to ratios of T_{eff} and L, via Eq. (21), which are determined observationally from the color and brightness differences at the two phases. In this way the mean radius can be solved for and from this the mean luminosity and distance.

Radial velocities also provide dynamical analysis of stellar atmospheres beyond simple radial pulsations. Studies of stellar spectra have shown that stars also exhibit nonradial pulsation in which a wave moves across the surface

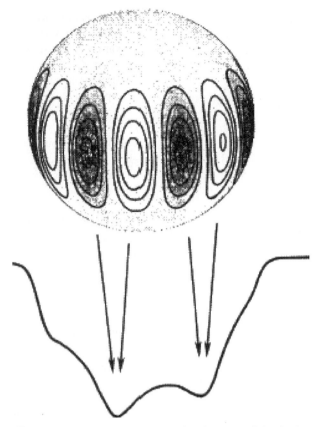

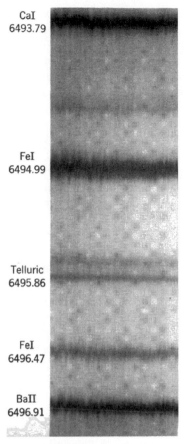

CaI
6493.79

FeI
6494.99

Telluric
6495.86

FeI
6496.47

BaII
6496.91

FIGURE 7 Depiction of a star experiencing nonradial pulsation and a schematic line profile showing the effect of the nonradial motions on the line shape. The shaded regions on the stellar surface represent motions away from the observer relative to the center of the star, and the contours are drawn at 5-km s^{-1} intervals. [From Vogt, S. S., and Penrod, G. D., (1983). *Astrophys. J.* **275**, 661. With permission.]

FIGURE 8 Solar spectra show the individual Doppler shifts of convective cells according to how the cells are imaged on the entrance slit of the spectrograph. Notice that the telluric line, which forms in the earth's atmosphere, has no wiggles. [From Gray, D. F. (1992). "Observation and Analysis of Stellar Photospheres," Cambridge University Press, Cambridge, U.K., p. 406. Reprinted with the permission of Cambridge University Press.]

of the star. This motion is detected as a Doppler shift because the wave motion is primarily a vertical displacement. The observation of nonradial pulsations requires high spectral resolution and low noise data. Figure 7 shows a representation of the composite profile of a single spectral line whose shape varies in a systematic way across the stellar surface. The sun was the first star observed to have wave motions in its atmosphere. As Fig. 8 shows, recording the solar spectrum with high spectral and spatial resolution reveals that absorption lines are wiggly; different places on the solar surface show small wavelength shifts because of vertical motions. These motions are produced by rising and sinking elements in the deep solar atmosphere, called granules, which are interpreted as the top of a region of convective motion at the base of the atmosphere. The best-studied solar oscillation has a period of about 5 minutes, produced by localized wave motion extending over about 10,000 km of the solar surface. Various oscillations are currently topics of intense study, a

field known as helioseismology, because they provide a means of determining the condition in regions of the sun that had been completely unobservable. Helioseismology has revealed much about the interior of the sun including the density of solar material as a function of depth, the depth of the solar convection zone, and the sun's rotation as a function of depth. These phenomena are revealed by radial velocity variations of about 1 m s^{-1}.

Because of the localized nature of solar oscillations, the detection of an analogous phenomenon in the integrated light from a star is unlikely. Other stars, however, appear to have wave motions covering more of their surfaces, presenting a coherent motion strong enough to be seen in the spectrum. With the ability to achieve radial velocity measurements precise to a few m s^{-1}, the field of asteroseismology continues to open up. Spectroscopic observations of rapidly oscillating, peculiar A-type stars with radial velocity amplitudes of 50 to 400 m s^{-1} and periods of several minutes are now possible. Most of what we know about these types of stars comes from studies of their light variations. Recently, however, it has become possible to add rich and precise spectroscopic information to our knowledge of these objects. In single G- and K-type giant stars oscillations are expected to produce radial velocity variations as high as 60 m s^{-1}. In the last decade ordinary K giant stars have been discovered to be a new class of low-amplitude radial velocity variables. Researchers are just beginning to interpret these precise radial velocity measurements and are faced with disentangling the effects of pulsation, rotation, convective cell motion, and the possible presence of a faint stellar or planetary companions.

Another important diagnostic provided by radial velocity analysis is the speed of a stellar wind from a luminous star. In the outer atmospheres of luminous stars, hydrostatic equilibrium, Eq. (23), breaks down and radiative forces dominate. The force provided by the absorption of the stellar radiation field by spectral lines is sufficient to force matter out of the gravitational well of the star. Radiative forces become increasingly important for hot stars, since their luminosity scales as T_{eff}^4 and the spectrum peaks at far ultraviolet wavelengths where the bulk of line opacity resides. The classic spectroscopic signature of a stellar outflow is the P-Cygni profile. Following Kirchoff's law this profile consists of a line-of-sight absorption component formed by the gas in the outflowing wind against the hotter layers of the photosphere below and an emission component formed by the tenuous gas in the wind projected against the cold interstellar medium. The absorption component is broadened by the velocity gradient in the wind from the photosphere outward. The overall absorption component is blue-shifted by the Doppler effect. Strong resonance lines, particularly in the ultraviolet, show deep absorption components with a steep edge on the blue side. From these lines the so-called terminal velocity of the wind is measured. Figure 9 shows some of these strong lines for one O-type supergiant star. The velocity as a function of depth is probed by velocity progressions seen in series of absorption lines. In the Balmer series of hydrogen, the weakest lines form in the inner wind where the velocity is low, while the progressively stronger lines form farther out in the wind where the velocity is greater. Furthermore, the analysis of stellar wind velocity fields can be greatly enhanced in some binary systems. The light of one star can be used as a background source to probe the wind of its companion from angles not available to the observer in the case of a single star. In this respect it is helpful to have a source that is compact and hot, such as a white dwarf or neutron star.

Finally, an important spectral line shift that mimics a radial velocity is the gravitational redshift predicted by

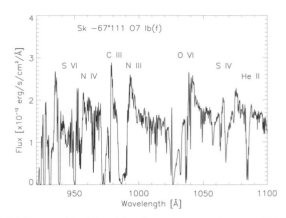

FIGURE 9 A Far Ultraviolet Spectroscopic Explorer (FUSE) spectrum of the O-type supergiant Sk −67°111 in the Large Magellanic Cloud. Important stellar transitions are labeled. The strongest spectral features (e.g., C III and N III) show both absorption and emission components characteristic of a strong stellar wind. The spectrum also shows many interstellar lines of H I and H$_2$. [Data courtesy of A. W. Fullerton.]

Einstein's general theory of relativity. The conservation of energy dictates that the energy of a photon must decrease just as that of a particle does when it climbs out of a gravitational field. The loss of energy results in a redshift of the photon. A collection of photons, a stellar spectrum, emerging from the gravitational well of a star are in this way redshifted. The first attempts to measure the gravitational redshift from a star focused on the white dwarf companion of the bright star Sirius, Sirius B. White dwarfs are earth-sized stars which weigh roughly as much as our sun but have surface gravities over 10,000 times greater than the sun and therefore have correspondingly larger gravitational redshifts. Since Sirius B's orbital motion (and therefore its radial velocity) is known, any redshift measured for Sirius B in excess of its orbital motion can to be attributed to the gravitational redshift. For Sirius B this redshift is measured to be 89 km s^{-1}, in excellent agreement with theory. If the radius of a star is known, its mass can be determined directly from this kind of measurement.

In order to exploit this method of measuring masses of isolated stars, one must determine the radial velocity of a star without the use of spectroscopy. This has been done by the High-Precision Parallax Collecting Satellite (HIPPARCOS) for a select number of nearby stars. By measuring the changing distance and proper motion (motion in the plane of the sky) of these stars, HIPPARCOS has pinned down their absolute space velocities to better than 100 m s^{-1}. Previously, such absolute stellar velocities were only known for components in a few well-determined binary systems, like Sirius. When the radial velocity is known independently to high accuracy, spectroscopic observations have the potential to reveal spectral line shifts previously hidden, such as the gravitational redshift. Other shifts include blueshifted lines from gas caught up in the convective motions of a star's atmosphere, line shifts due to atmospheric oscillations, shifts do to previously unknown companions, and radial velocity changes correlated with the stellar activity cycle similar to the solar cycle. The combination of high-precision stellar spectroscopy and measurements from astrometric space missions following in the footsteps of HIPPARCOS has the great potential to improve our understanding of fundamental stellar parameters and stellar physics.

B. Rotation

In addition to their radial velocities, orbital motion, pulsation, or complex atmospheric oscillation, all stars exhibit another motion: They rotate. The solar rotation can be detected easily by tracking the positions of sunspots, but the spectrum also shows this rotation. The equator at the east limb of the sun is approaching us at 2 km s^{-1}, which corresponds to a rotation period of 27 days. As a result, the spectral lines from the eastern solar hemisphere are systematically blueshifted and the lines from the western hemisphere are all redshifted. In fact, the first strong confirmation of the Doppler effect for light came in the 1880s, when the of solar rotation rates as a function of latitude from the sunspot and spectroscopic methods were shown to be in agreement.

The same systematic velocity shifts are of course present in other stars, but, with few exceptions, we see only the light integrated over the whole disk and we observe spectral line profiles that are broadened compared with the nonrotating case. It is necessary to use a model to derive the stellar rotation speed. The model of stellar rotation must take several factors into account. First, the maximum speed occurs at the equator because the matter must travel the greatest distance during the rotation period; moving away from the equator the rotational speed drops, reaching zero at the poles. Second, we only observe part of the rotational speed of any point on the stellar disk, the component that is moving toward or away from us that can produce a Doppler shift; the material at the limbs, whose motion is most closely aligned with our line-of-sight, produces the largest shifts. This reduces to the fact that any chord on the apparent stellar disk that is parallel to the central meridian is a line of constant radial velocity. The wavelength shift is

$$\Delta\lambda = \frac{v_{eq}\lambda_0}{c}(1-\mu^2)^{1/2}\sin i \qquad (29)$$

for a given equatorial rotation velocity, v_{eq} and inclination angle i, and $\mu = \cos\theta$, where θ is the angle of the chord from the center meridian: $0°$ at the center of the disk and $90°$ at the limb.

A third factor is the fraction of the star's light that has a particular projected velocity. Matter at the equatorial limbs has the largest radial velocity, but this represents a tiny fraction of the stellar disk. On the other hand, matter along the rotation axis represents the largest portion of the disk, but it has no velocity shift because the material is moving perpendicular to our line of sight. This is accounted for by weighting the contribution from each chord by its length, approximately 2μ. The residual flux of the line, the ratio of the flux at λ to the flux of the adjacent continuum,

$$r_\lambda = F_\lambda/F_{cont} \qquad (30)$$

becomes

$$r_\lambda = 1 - (1-r_0)\left(1 - \frac{\Delta\lambda^2 c^2}{\lambda_0^2 v_{eq}^2 \sin^2 i}\right)$$

$$\Delta\lambda \leq \frac{v_{eq}\lambda_0}{c}\sin i \qquad (31)$$

where r_0 is the residual flux depth of the line in the absence of rotation. A rapidly rotating star will have shallow line cores which appear to be washed out relative to the slowly rotating case. To incorporate limb darkening into this model for rotational line broadening, a rotation broadening function,

$$A(y) = \frac{3}{3+2\beta}\left[\frac{2}{\pi}(1-y^2)^{1/2} + \frac{\beta}{2}(1-y^2)\right] \qquad (32)$$

where β is the limb darkening coefficient, is convolved with the intrinsic residual profile, $r'(x)$:

$$1 - r(x) = \int_{-1}^{+1}\left[1 - r'(x-y)\right]A(y)dy \qquad (33)$$

where x is measured in units of the equatorial velocity shift, $\Delta\lambda c/\lambda_0 v_{eq}$.

As indicated by the appearance of $\sin i$ above, the rotation axis of the star is not necessarily perpendicular to our line of sight. So in general we only see some unknown projection of the equatorial rotation speed. The projected rotation speed of the star is found by computing the broadened profile for various values of the equatorial rotation speed and interpolating to find the value giving the closest match to the observations. Figure 10 illustrates this process.

The discussion of rotation to this point, as well as the description of the method by which rotation is calculated, has assumed that a star can be characterized by a single equatorial rotation speed. As noted above, the sun's surface rotates differentially, with the rotation period increasing from about 27 days at the equator to more than 35 days near the poles. In addition, the sun's rotation appears to vary with time and to exhibit a north–south asymmetry. These facts all show that the solar rotation is a much more complex phenomenon than might be apparent at first. Studies of the differential rotation in other stars

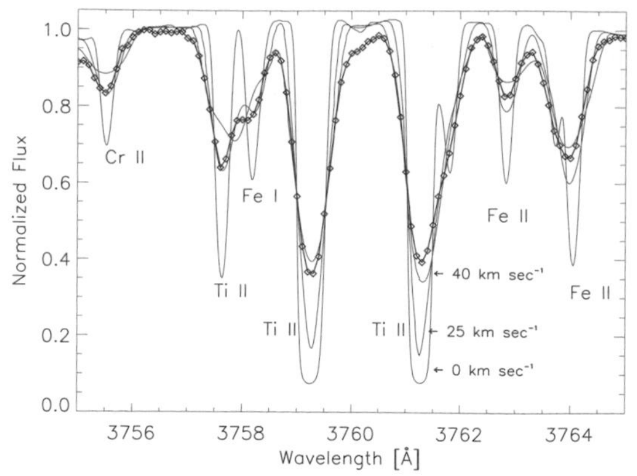

FIGURE 10 A portion of an observed spectrum of the bright A-type star Deneb showing absorption lines of Fe I and Fe II, Ti II, and Cr II. Also shown is the computed spectrum for three different assumed values of the star's projected equatorial rotation velocity. Note the blended Ti II and Fe I lines near 3758 Å. Comparision of the blend with the computed spectra suggests that the projected rotational velocity of Deneb is close to 25 km s^{-1}. [Deneb spectrum kindly provided by A. Kaufer.]

are just beginning and are closely linked to the study of stellar surface structure, which is described next.

C. Surface Structure

An important advantage of studying the sun is the ability to observe its surface point-by-point in addition to using the integrated light from the whole disk. Because of this we know of sunspots, with their concentrated magnetic fields, as well as many other solar phenomena. It is still not possible to directly image the disk of any other star with sufficient resolution to see detailed surface structure, although many stars have been resolved and their angular diameters have been measured by interferometers. Stellar spectroscopy, however, already provides a means of deriving information about the surfaces of some stars. The method, Doppler imaging, has been applied to cool stars that have dark spots analogous to sunspots.

A starspot is an area of decreased atmospheric brightness, probably associated with a localized magnetic field. If a large spot is located on the surface of a star, the star's light will be diminished by an amount that depends on the temperature decrease in the spot and on its area relative to the entire disk. If the star were not rotating, the diminished brightness would always make the same contribution to every measurement of the stellar spectrum. Due to rotation, however, the decreased brightness will be concentrated at the shifted wavelength appropriate for the projected rotation speed of the spot. As the spot is carried across the stellar disk, it will have different projected speeds and, therefore, different wavelength shifts. By making repeated observations of the line profiles, we

can follow the progress of the spot across the stellar surface. Of course, if several spots are present simultaneously, the interpretation becomes more complicated, but some progress can be made, especially if the spots last for more than one stellar rotation period. These observations, however, have demanding requirements: high spectral resolution to isolate different wavelengths shifts cleanly, low noise data so that variations in brightness at a given wavelength in a line profile are true signals, and continuous nightly coverage for several days to several weeks. In the interest of low noise data, long exposures are required, but these cannot be too long or else serious smearing of the surface features occurs in the reconstructed images.

The construction starspot maps is possible for several active, rapidly rotating main sequence stars and several dozen active evolved stars with temperatures near to and slightly cooler than the sun. One important example, a K0 main-sequence star, AB Doradus, with X-ray and ultraviolet variability, has starspots which have been sufficiently resolved (2° resolution in latitude) such that, for the first time, the differential rotation of a star other than the sun has been reliably measured. The lower latitude spots are seen to pull ahead of the higher latitude spots with time. Figure 11 shows one reconstruction of the surface structure of AB Doradus. Starspot maps of some active evolved stars reveal spots whose size is a large fraction of the stellar disk.

D. Turbulence

As might be expected, the gases in stellar atmospheres are not quiescent. In addition to the ordered, large-scale motions of rotation and pulsation, there is disordered motion on a variety of smaller scales that is referred to as turbulence. This is not used in the precise hydrodynamical sense, but just to suggest the existence of quasi-random motions.

Motions such as those observed in the sun are at the large end of the scale. They are called macroturbulence because the moving units are large compared to the average distance traveled by light before it is reabsorbed by the gas of the stellar atmosphere. These motions, of course, are unresolved in the light of a stars; their effect can only be seen in the changes to the line profile. Because we see the combined light of many moving elements, the total effect is similar to that of rotation. For a particular star it is extremely difficult to decide whether a small amount of rotation or a large amount of turbulence is present, although statistical studies provide evidence for macroturbulence. These studies show that among some groups of stars there are no cases of negligibly small rotation. Because some stars should be viewed from directly above the rotation pole, the line width from the stars showing the least amount of broadening is attributed to turbulence.

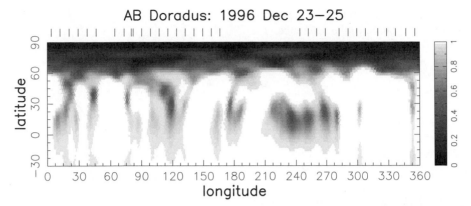

FIGURE 11 This map shows the surface brightness as a function of latitude and longitude on the active star AB Doradus. A polar spot is stretched out along the top of the map in this projection. The rotation axis has an inclination of 30°, which means the pole is tilted 60° toward the observer. This image was constructed by modeling spectral line variations compiled from many observations of the star during its approximately 12-hour rotation period. Comparison of such maps taken over the course of a week reveals that the lower latitude features pull ahead of the higher latitude features. This is explained by differential rotation of the stellar surface as a function of latitude like that observed on the sun. [From Donati, J. F. et al. (1999). *Mon. Notices R. Astron. Soc.* **302**, 437. With permission.]

There is also evidence for small-scale turbulence that cannot be spatially resolved even on the sun (note that the smallest structure visible on the sun is about 500 km). However, the next generation of solar telescopes, with adaptive optics technology, promises to greatly improve this situation. The limiting case of this motion, called microturbulence, is pictured as moving elements that are small compared to the average distance between absorptions of light. The main effect of microturbulence is to broaden the cores of strong lines so that they become more effective absorbers. As discussed in the next section, by using microturbulence it is possible to match both weak and strong absorption lines with the same elemental abundance. Even the analysis of a single point on the solar disk requires a small-scale turbulence of 0.5 to 1.0 km s^{-1} to achieve consistency. Typical values for other stars are about 2 km s^{-1}, and sometimes values as large as 10 to 20 km s^{-1} are needed. These large values are suspect because they exceed the sonic velocity in the stellar atmosphere. Motions this fast would be energetic enough to alter the populations of the atomic levels. The small values, however, are reasonable given the likely presence of mechanical energy from various sources in the atmosphere. More realistic models of the velocity fields present in stellar atmospheres are under development. Future work with high-precision radial velocities should help to constrain such models.

E. Abundances

The determination of stellar abundances is a primary application of spectroscopy. The analysis goes beyond the mere identification of absorption lines in the spectrum to a quantitative measurement of the amounts of various elements. The first breakthroughs in stellar abundance analysis came in the 1920s, starting with Meghnad Saha and his theory of the ionization and excitation of gases (see Eq. (19)). He realized that different temperature and pressure conditions in a stellar atmosphere excited spectral lines from different chemical species. Next, the work of R. Fowler and E. Milne provided the connection between the number of absorbing ions which form a spectral line and the strength of that spectral line, at a particular temperature and pressure. Cecilia Payne applied this understanding to the analysis of stellar spectra and worked out for the first time the relative abundances of the chemical elements in stellar photospheres. The major result of this work was that the derived chemical abundances for the brightest stars were very similar, despite the significant differences in the appearance of their spectra. The techniques applied by Payne for her stellar abundance analysis were refined and applied by Henry Norris Russell and collaborators to a detailed analysis of the solar spectrum. From this work it was realized the composition of the solar photosphere is dominated by hydrogen. The concepts of equivalent width and the curve of growth were developed by Marcel Minnaert and colleagues and form the basis for much stellar abundance work performed today. The field continues to advance for several reasons: The quality of the data is constantly improving, our knowledge of atomic structure is becoming increasingly accurate, and the growth of computer power permits more detailed studies.

The basic method of determining abundances is built on the measurements of the strength of an absorption line. The integrated line absorption, known as the equivalent width, is found first by estimating the spectrum's shape in the absence of the absorption line and then by measuring the sum of the absorption relative to the reference. The equivalent width, W_λ, is related to the residual flux of a spectral line (30) by:

$$W_\lambda = \int (1 - r_\lambda)\, d\lambda \qquad (34)$$

where W_λ (measured in Å) of a completely dark line has the same total absorption as the observed line. The strongest lines can have equivalent widths of more than 10 Å. At the other extreme, the weakest detectable line depends on the resolution and the amount of noise in the data, but equivalent widths weaker than 0.01 Å have been measured with high-quality observations.

A model of the stellar atmosphere is needed to interpret the equivalent width. This requires a knowledge of the atom's intrinsic ability to absorb at this particular wavelength as well as the density of atoms in the appropriate absorbing state. Computing the line strength for different assumed values of the total abundance of the element gives a relation known as the curve of growth. The shape of the curve exhibits three different segments: linear, flat, and square-root. The equivalent widths of weak lines increase directly as the abundance is increased; lines on the linear portion of the curve of growth give the most reliable results because they are free of many complications that plague stronger lines. Lines of medium strength are saturated, and the equivalent widths of these lines increase very slowly as the abundance is increased; here, the curve of growth is nearly flat. On this portion of the curve line, profiles have flat cores from which the sides rise almost vertically to the continuum level. Increasing the abundance has little influence on the equivalent width because the line only broadens its saturated core. As a result, these lines give uncertain results; in addition, they are subject to the presence of microturbulence, small-scale motions that desaturate the profile and shift the flat portion of the curve to larger abundances. The equivalent widths of strong lines are proportional to the square root of the abundance, but the line profiles also have extensive wings due to pressure broadening; as such, a good estimate of the surface gravity is needed to correctly interpret these lines. Figure 12 shows both a typical curve of growth and the kinds of lines that are found on its three main segments.

In the simple case of a homogeneous slab of uniform temperature the equivalent width of a weak line transition from level i to j is given by:

$$W_\lambda = \frac{\pi e^2}{mc^2}\lambda^2 N_i f_{ij} \qquad (35)$$

where N_i is the number of ions in level i of element Z, and f_{ij} is the atomic oscillator strength of the transition. The Boltzmann (18) and Saha (19) equations relate N_i to the

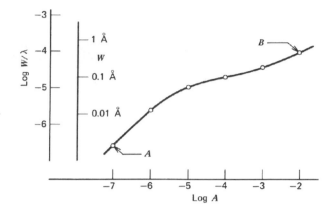

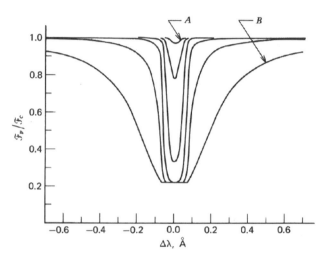

FIGURE 12 (Top) The curve of growth; the equivalent width (and equivalent width divided by the wavelength) of a line (y-axis) as a function of the chemical abundance of the absorbing species (x-axis). "A" labels a point on the "linear" portion of the curve. "B" labels a point on the "square-root" portion of the curve. The transition region between "A" and "B" is referred to as the "flat" portion of the curve. (Bottom) The shapes of spectral lines are different portions of the curve. Weak lines form at "A" and strong lines with pressure-broadened wings form at "B." Line profiles in the transition region are also shown. [From Gray, D. F. (1992). "Observation and Analysis of Stellar Photospheres," Cambridge University Press, Cambridge, U.K., p. 292. Reprinted with the permission of Cambridge University Press.]

total abundance of the element Z at a given temperature and pressure. The first segment of the curve of growth is established by the linear relationship between W_λ and N_i.

The equivalent width of a given line, therefore, is dependent on the abundance, the oscillator strength, and the thermal excitation. For a given line, the f value and excitation potential are fixed and our model temperature and pressure structure (constant for the homogeneous slab) sets the ionization and excitation state of the gas. From here the value for the abundance is adjusted until the observed equivalent width is reproduced. Abundances derived from lines (1) with different excitation potentials, (2) with different strengths, and (3) of several different ionization stages of the same element should yield the same abundance in a chemically homogeneous atmosphere. If they do not, the model parameters (T_{eff}, $\log g$, abundance, turbulent line width) are adjusted and the analysis proceeds in an iterative fashion until the abundances derived from different lines produce consistent results.

The use of equivalent widths discards all the shape information present in the line profile: Only the integrated strength is used. This is appropriate for observations made with moderate wavelength resolution because the equivalent width is not altered by the resolution, expect for possible systematic effects related to estimation of the continuum reference level, whereas the line shape is dominated by instrumental effects. Measurements made with high-resolution, however, minimize the instrumental contamination, permitting a direct comparison of the computed and observed line shapes. Both the equivalent width and the line profile approaches assume that the absorption line is isolated in the stellar spectrum. This is generally not true, especially for cool stars and at blue and ultraviolet wavelengths where severe crowding and overlapping of lines occur. Important absorption lines are frequently badly blended together, and there is no clean spectrum of the measurement of an equivalent width or a line profile. In such cases, it is necessary to compute the predicted spectrum including all the overlapping and blended absorption lines, though this pushes our knowledge of atomic physics to the limit. If the synthetic spectrum does not match the observed spectrum, the abundances of the elements are adjusted and the calculation is repeated until agreement is achieved. Figure 13 exhibits a comparison of the observed and synthesized infrared spectra for a K giant, showing the close agreement that is possible.

The combined results from various sources and using various techniques give the standard solar abundances shown in Table I. Even a brief inspection of Table I shows that the standard composition is different from our experience of terrestrial compositions. In every thousand atoms in a typical star there are about 910 atoms of H, 89 He atoms, and 1 atom drawn from the other elements. More subtle differences are also present in Table I. For example, cosmically the abundance of Fe is comparable to Si, whereas on the earth's surface Si is much more abundant. This is attributed to the differentiation of the early earth, with the heavy elements such as Fe sinking to the core. Another result is the sawtooth variation in cosmic abundances superimposed on the general trend. For example, Cr, Fe, Ni, and Zn have greater abundances than their neighboring elements of V, Mn, Co, Cu, and Ga. This is a

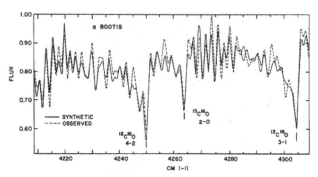

FIGURE 13 A high-resolution IR spectrum of the star α Bootis (dashed line) and a computed spectrum that best matches the observations (solid line). This portion of the spectrum displays the many lines associated with the CO molecule, with the isotopes ^{12}CO and ^{13}CO giving distinctly different wavelengths of absorption. The fit to the observations requires a ^{12}CO:^{13}CO ratio of about 10, indicating that the ^{13}CO abundance in the atmosphere has been enhanced by mixing of material from the interior. [From Ridgway, S. T. (1974). *Astrophys J.* **190,** 591. With permission.]

signature of the particular nuclear reaction pathways that form these elements in stellar interiors.

Generally, due to the minuteness of isotopic wavelength shifts, only elemental abundances can be obtained from the analysis of stellar spectra. An important exception to this is the detection of isotope shifts in the absorption lines of molecules. The spectrum of the molecule CO can yield isotopic information. Referring again to Fig. 13, we see that the absorptions by ^{12}CO and ^{13}CO are cleanly separated, permitting a determination of the abundance of each isotope. This is important because isotopic abundances show the physical conditions under which the element was fused by nuclear reactions. The relative amount of the isotope ^{13}C is attributed to mixing between the stellar atmosphere and interior. A sure sign of mixing is the spectroscopic detection of an element with no stable isotopes. Such an isotope, ^{99}Tc, has a half-life of only 2×10^5 years, which is much shorter than stellar lifetimes.

IV. MODERN DEVELOPMENTS

Stellar spectroscopy is a mature field of astronomy, but it continues to develop in several areas: the expansion into new spectral regions available for analysis, the development of new instruments and detectors, and the use of more realistic models to interpret the data.

A. Ultraviolet and High Energy Stellar Spectroscopy

Throughout most of it history, astronomy has been restricted to light visible to the eye, which corresponds

TABLE I Solar System Abundances[a]

Element	Photosphere	Meteorites	Element	Photosphere	Meteorites
1 H	12.00	—	42 Mo	1.92 ± 0.05	1.97 ± 0.02
2 He	10.93 ± 0.004	—	44 Ru	1.84 ± 0.07	1.83 ± 0.04
3 Li	1.10 ± 0.10	3.31 ± 0.04	45 Rh	1.12 ± 0.12	1.10 ± 0.04
4 Be	1.40 ± 0.09	1.42 ± 0.04	46 Pd	1.69 ± 0.04	1.70 ± 0.04
5 B	2.55 ± 0.30	2.79 ± 0.05	47 Ag	0.94 ± 0.25	1.24 ± 0.04
6 C	8.52 ± 0.06	—	48 Cd	1.77 ± 0.11	1.76 ± 0.04
7 N	7.92 ± 0.06	—	49 In	1.66 ± 0.15	0.82 ± 0.04
8 O	8.83 ± 0.06	—	50 Sn	2.0 ± 0.3	2.14 ± 0.04
9 F	4.56 ± 0.3	4.48 ± 0.06	51 Sb	1.0 ± 0.3	1.03 ± 0.07
10 Ne	8.08 ± 0.06	—	52 Te	—	2.24 ± 0.04
11 Na	6.33 ± 0.03	6.32 ± 0.02	53 I	—	1.51 ± 0.08
12 Mg	7.58 ± 0.05	7.58 ± 0.01	54 Xe	—	2.17 ± 0.08
13 Al	6.47 ± 0.07	6.49 ± 0.01	55 Cs	—	1.13 ± 0.02
14 Si	.755 ± 0.05	7.56 ± 0.01	56 Ba	2.13 ± 0.05	2.22 ± 0.02
15 P	5.45 ± 0.04	5.56 ± 0.06	57 La	1.17 ± 0.07	1.22 ± 0.02
16 S	7.33 ± 0.11	7.20 ± 0.06	58 Ce	1.58 ± 0.09	1.63 ± 0.02
17 Cl	5.5 ± 0.3	5.28 ± 0.06	59 Pr	0.71 ± 0.08	0.80 ± 0.22
18 Ar	6.40 ± 0.06	—	60 Nd	1.50 ± 0.06	1.49 ± 0.02
19 K	5.12 ± 0.13	5.13 ± 0.02	62 Sm	1.01 ± 0.06	0.98 ± 0.02
20 Ca	6.36 ± 0.02	6.35 ± 0.01	63 Eu	0.51 ± 0.08	0.55 ± 0.06
21 Sc	3.17 ± 0.10	3.10 ± 0.01	64 Gd	1.12 ± 0.04	1.09 ± 0.02
22 Ti	5.02 ± 0.06	4.94 ± 0.02	65 Tb	—	0.35 ± 0.02
23 V	4.00 ± 0.02	4.02 ± 0.02	66 Dy	1.14 ± 0.08	1.17 ± 0.02
24 Cr	5.67 ± 0.03	5.69 ± 0.01	67 Ho	0.26 ± 0.16	0.51 ± 0.02
25 Mn	5.39 ± 0.03	5.53 ± 0.01	68 Er	0.93 ± 0.06	0.97 ± 0.02
26 Fe	7.50 ± 0.05	7.50 ± 0.01	69 Tm	—	0.15 ± 0.02
27 Co	4.92 ± 0.04	4.92 ± 0.01	70 Yb	1.08 ± 0.15	0.96 ± 0.02
28 Ni	6.25 ± 0.04	6.25 ± 0.01	71 Lu	0.06 ± 0.10	0.13 ± 0.02
29 Cu	4.21 ± 0.04	4.29 ± 0.04	72 Hf	0.88 ± 0.08	0.75 ± 0.02
30 Zn	4.21 ± 0.04	4.29 ± 0.04	74 W	1.11 ± 0.15	0.69 ± 0.03
31 Ga	2.88 ± 0.10	3.13 ± 0.02	75 Re	—	0.28 ± 0.03
32 Ge	3.41 ± 0.14	3.63 ± 0.04	76 Os	1.45 ± 0.10	1.23 ± 0.02
33 As	—	2.37 ± 0.02	77 Ir	1.35 ± 0.10	1.37 ± 0.02
34 Se	—	3.41 ± 0.03	78 Pt	1.8 ± 0.3	1.69 ± 0.04
35 Br	—	2.63 ± 0.04	79 Au	1.01 ± 0.15	0.85 ± 0.04
36 Kr	—	3.31 ± 0.08	80 Hg	—	1.13 ± 0.08
37 Rb	2.60 ± 0.15	2.41 ± 0.02	81 Tl	0.9 ± 0.2	0.3 ± 0.04
38 Sr	2.97 ± .07	2.92 ± 0.02	82 Pb	1.95 ± 0.08	2.06 ± 0.04
39 Y	2.24 ± 0.03	2.23 ± 0.02	83 Bi	—	0.71 ± 0.04
40 Zr	2.60 ± 0.02	2.61 ± 0.02	90 Th	—	0.09 ± 0.02
41 Nb	1.42 ± 0.06	1.40 ± 0.02	92 U	<−0.47	−0.50 ± 0.04

[a]These are logarithmic (base 10) values normalized to hydrogen, which is 12.00.

[From Gervesse, N., and Sauval, A. J. 1998, *Space Sci. Rev.*, **85**, 161. Some of these data are also available in Grevesse, N. et al. (1996). In "Cosmic Abundances" (S. S. Holt, and G. Sonneborn, eds.), p. 117., ASP Conf. Ser. 99, San Francisco, CA]

to wavelengths from about 3900 Å to nearly 7000 Å. The development of photographic detection altered this somewhat; it was possible to detect near ultraviolet (UV) radiation down to 3000 Å and also infrared radiation up to 10,000 Å. The primary range of photography, however, is from 5000 Å down to the limit of atmospheric transparency at 3000 Å. The photographic sensitivity in this spectral region is so low that 99% or more of the

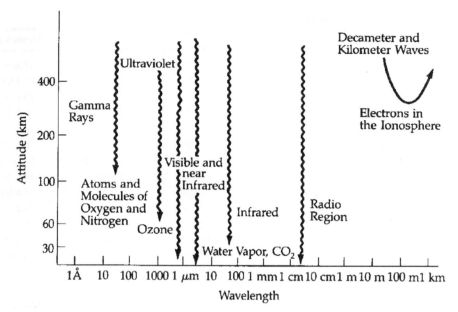

FIGURE 14 Transmission of the earth's atmosphere to the complete electromagnetic spectrum. Radiation only reaches the earth's surface in the visible, radio, and narrow bands in the IR. Additional parts of the IR can be studied by observing at high altitudes and in extremely dry locations. The expansion of astronomy into spectral bands that are completely blocked by the atmosphere has led to the rapid advances of recent years. [From M. Zeilik and E. v. P. Smith (1987). "Introductory Astronomy and Astrophysics," Saunders College Publishing, Philadelphia, PA, p.170. With permission.]

light is not used; the efficiency at wavelengths longer than 5000 Å is even lower. Therefore, most astronomy was done in the blue–violet spectral region to make the most of the available light.

This situation has changed dramatically. The placement of telescopes above the earth's atmosphere is of principal importance. Our atmosphere absorbs much of the radiation that strikes it from space. Below 3000 Å all radiation— UV, extreme ultraviolet (EUV), X-ray, and gamma ray—is prevented from penetrating the atmosphere due to opacity from O_3, N_2, and O_2. This protects life on the earth from harmful radiation but blocks a large portion of the spectrum. Figure 14 displays the total spectrum of light, showing the limited spectral regions where the earth's atmosphere is transparent. Satellite-borne telescopes circumvent this obstruction. Observing in the UV down to 1200 Å alone more than doubles the amount of spectral coverage compared to the traditional blue–violet band. This greatly increases the amount of astrophysical information available to us.

The expansion of spectral coverage to the UV was strongly motivated by solid expectations. As described approximately by Wien's law (16), the peak of a star's spectral energy distribution shifts towards shorter wavelengths as T_{eff} increases. For all stars with $T_{eff} > 10,000$ K, a significant fraction of their flux is emitted in the UV. The direct measurement of T_{eff} via Eq. (17) requires measuring the bolometric flux, so UV observations are crucial espe-

cially for O-, B-, and A-type stars. Notable contributions were first made in this area by the Orbiting Astronomical Observatory-2 (OAO-2) and the European Astronomical Satellite TD-1.

Another clear motivation for obtaining stellar spectra in the UV is rooted in atomic structure. Most elements produce their strongest absorption lines, the resonance lines in the UV. If an element is rare cosmically, the resonance lines may be the only ones we can hope to observe; without them we would have no knowledge of the abundance of that element. Other elements have weak lines in the visible spectral region, but we often lack enough atomic information to analyze these line accurately. However, we do know enough to analyze the stronger UV lines.

With these motivations, a high priority of space astronomy over the last three decades has been the construction of astronomical satellites capable of measuring the UV spectral region. Even before it was possible to launch satellites carrying astronomical instruments, space astronomy went forward by using sounding rockets to capture a few minutes of data above the atmosphere before dropping back to the earth. The sun was an early target, but so were hot, bright stars. Through these early observations, the detection of P-Cygni profiles in the UV resonance lines of O and B star spectra were the first indicatation that mass loss was commonplace in hot, luminous stars.

The sun has been the target of a long series of orbiting observatories, most recently by the Solar and Heliospheric

Observatory (SOHO), in part because it occupies a central place in the study of stellar properties. Not only is the sun very bright, allowing more detailed spectral study, and large, enabling astronomers to study surface structures, but also its light does not suffer absorption by interstellar gas. The interstellar medium is filled by low-density gas, on average 1 atom cm^{-3}. Although better than any terrestrial vacuum, this sparse gas can have a tremendous effect on starlight because of the enormous distances between stars. This fact is particularly critical at EUV wavelengths, where the photoionization of H I can absorb light at wavelengths below 912 Å. As a result, interstellar space becomes opaque at wavelengths immediately below this limit, which limits investigations in the EUV to nearby stars, typically hot white dwarfs and chromospherically active cool dwarfs within 150 light-years of the sun. Below 100 Å the absorption of light by interstellar gas becomes less severe and this enables the investigation of X-rays from more distant stars by satellite observatories. Generally, however, only the sun is close enough to explore these spectral regions thoroughly.

In 1978, following several successful UV satellite missions, the International Ultraviolet Observatory (IUE) satellite was launched. IUE became the workhorse of UV astronomy through the 1980s and much of the 1990s. This satellite, a collaborative effort between NASA, the European Space Agency, and the U.K. Science Research Council, carried a 45-cm telescope equipped with instrumentation to observe the spectral region 1150 to 3200 Å at both high and low resolution. This was the first satellite with this dual capacity, making it applicable to a much greater range of astronomical topics. It was also the first astronomical satellite to be placed in a geosynchronous orbit. The earlier satellites in low orbits had periods of about 90 minutes. Because of this, their exposures had to be limited to avoid contamination by earthshine. The IUE, on the other hand, had an orbital period of 24 hours, placing it always above one part of the earth. This allowed it to continue one exposure for many hours, reaching objects fainter than had ever been studied before in the UV. With IUE it became routine to measure the energy distributions for both bright and faint stars and to observe the detailed absorption line spectra for the brighter objects.

Early last decade saw the launch of the Hubble Space Telescope (HST). The size of its 2.4-m-diameter primary mirror gives HST a tremendous advance over all previous satellite telescopes. The Goddard High-Resolution Spectrograph (GHRS), which operated at UV wavelengths, was the first instrument aboard HST to make significant advances in UV stellar spectroscopy. With the ability to obtain very high signal-to-noise observations and a spectral resolution over four times greater than that of IUE, GHRS

was used to study the detailed chemical compositions of hot stars and the chromospheres of cool stars. GHRS has since been replaced by the Space Telescope Imaging Spectrograph (STIS). STIS has two-dimensional detectors that provide simultaneous wavelength coverage (1150 to 10,300 Å) that is at least a factor of 10 improvement over GHRS. A high-resolution mode ($R \simeq 100,000$) also exists in the UV with STIS.

Our understanding of stars has much improved as a result of UV stellar observations. One extremely important result, mentioned earlier, was the discovery that hot, luminous stars are losing large amounts of mass. Continued monitoring of hot stars has revealed that their stellar winds are not steady, but structured in space and variable in time (see Fig. 15). Intensive time series observations with IUE have demonstrated correlations between wind variability and these stars' projected rotation velocity. These observations have encouraged many studies seeking to understand the physical origin of these correlations. These observations also call into question the reliability of stellar mass-loss rate determinations and further complicate the accurate inclusion of stellar mass loss in stellar evolution calculations.

Diffusion in stellar atmospheres, a process by which radiation pressure can selectively transport particular atoms and ions between different layers of atmosphere, has also been studied. Heavier atoms and ions, which in the absence of a mixing mechanism such as convection would otherwise settle to deep layers within the photosphere, may be kept aloft by radiation pressure. One such star where diffusion is thought to be very important is χ Lupi. A chemically peculiar star with very sharp spectral lines due to extremely slow rotation, this star has been investigated in great detail by GHRS. These analyses have quadrupled the number of elements for which accurate abundances in the photosphere of this star are known. χ Lupi is recognized as a standard astrophysical source for atomic physics, and the very precise wavelength calibration of GHRS spectra is superior to most available laboratory atomic wavelength data. These astronomical spectra have encouraged the improvement in laboratory data for many elements. This is another example of the link between laboratory data and astronomical observations resulting in the birth of astrophysics.

Observations at shorter wavelengths by the Extreme Ultraviolet Explorer (EUVE) have shown that the radiative levitation of heavy elements is important for the hottest white dwarfs. The stratification of hydrogen and helium in atmospheres of these stars has also been studied. EUVE spectra of the hottest white dwarfs, whose spectra reach peak intensity at EUV wavelengths, have also been particularly important for accurate determinations of effective temperature. There is one line-of-sight through the local

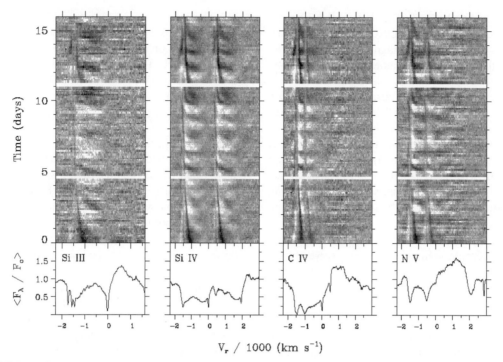

FIGURE 15 Comparison of the dynamic spectra for the Si III, Si IV, C IV, and N V resonance lines observed during the IUE MEGA Campaign for B-type supergiant HD 64760. The mean line profiles are shown below; the variation of the line profile with time is shown above. These spectra demonstrate the cyclic variability of stellar winds from hot luminous stars. [From Fullerton, A. W. et al. (1997). *Astron. Astrophys.* **327**, 699. With permission of the European Southern Observatory.]

interstellar medium where the average density of atoms is only 0.001 cm^{-3}. In this direction EUVE was able to observe two luminous B-type stars and obtain direct measurements of their stellar spectra below 912 Å. Surprisingly, these stars turned out to be much brighter at these wavelengths than expected. Model atmospheres had under-predicted the stellar flux by at least a factor of five. Recent atmosphere models, which include spherical geometry, appear to resolve this problem.

Spectroscopic observations of hot stars also continue with the Far Ultraviolet Spectrscopic Explorer (FUSE) which has wavelength coverage from the Lyman limit at 912 Å to about 1250 Å. Strong lines from species such as C III and O VI in the FUSE region combined with other UV lines provide diagnostics for the ionization state of hot star winds. Earlier IUE observations revealed the presence of anomalously ionized elements, uncharacteristic of a star's T_{eff}, in the atmospheres of some hot stars. The anomaly is observed for ions of nitrogen and oxygen as demonstrated by the spectrum shown in Fig. 9. This O-type star has an effective temperature of about 30,000 K, but the spectrum shows a strong line of O VI, inconsistent with this temperature. It is generally agreed that the presence of X-ray radiation is required to produce these anomalous ions. The X-rays could be produced within an unsteady stellar wind, where portions of the wind collide,

generate shock waves, and heat portions of the wind to high temperatures. Alternatively, the X-rays may originate from deeper layers closer to the photosphere. With the recent launch of the Chandra X-Ray Observatory and the X-Ray Multi-Mirror (XMM) satellite, high-resolution X-ray spectra of hot stars will soon be available and these new observations should help further constrain models for X-ray production in hot stars.

Ultraviolet spectroscopy has also made important contributions to the study of chromospheres and coronae. Above the sun's atmosphere the density of the gas decreases with height, but the temperature begins to increase after reaching a minimum value of about 4300 K. The temperature rises gradually to a value of about 7000 K, and then rapidly heats up to more than 1 million K. These regions are both visible during total solar eclipses. It is possible to detect some of these emission lines by careful observations made from the earth's surface, but they are not prominent. UV observations made from space, however, show the chromospheric emission much more clearly for two reasons. First, the background brightness of the cool solar disk diminishes rapidly at shorter wavelengths. At wavelengths below 2000 Å, the spectrum becomes dominated by the chromosphere and corona. Second, the strong lines of the ions present at the higher temperature of the chromosphere and the corona are concentrated in the UV. The Ultraviolet Choronagraph Spectrometer aboard SOHO measures far-UV spectral lines which provide diagnostics of the physical state of the corona: its temperature, density, and velocity.

In the choromsphere, a rich array of UV emission lines are present including H I, O I, Mg II, Si II, and Fe II. The transition region between the chromosphere and the corona is characterized by temperatures in the range of 50,000 to 250,000 K. Because ions that are present at these temperatures do not emit at optical wavelengths, the presence of transition regions in stars was not shown until UV observations were possible. These observations found emission lines of ions of He, C, N, O, and Si. The presence of a transition region implies the existence of a corona, too, and coronal emission lines have been observed by the EUVE. Temperature-sensitive lines from Fe X to Fe XXIII in the range of 100 to 350 Å probe temperatures up to 10 million K. At even shorter wavelengths, *Chandra* has obtained high-resolution spectra from 2 to 175 Å of the active G-type giant binary Capella (see Fig. 16). Lines from different elements and different ionization stages are used as diagnostics for the temperatures and densities of Capella's coronae which are characterized by temperatures between 2 and 20 million K.

All science data, including stellar spectrocopy, from most space-based UV, optical, and near-IR observatories are available on the Internet at the multi-mission archive of the Space Telescope Science Institute (http://archive.stsci.edu/mast.html).

B. Infrared Wavelengths and Beyond

The UV spectral region has clearly been a major area of development in the past two decades. There has also been a continuing effort to push stellar spectroscopy to longer, infrared (IR), wavelengths. While there are some partially transparent bands, atmospheric absorption, primarily due to water vapor, has severely hampered the exploration of this region. Ground-based observations must be made from the tops of mountains located in arid regions to get about as much of the atmospheric blocking as possible. Observations are also made from aircraft, high-altitude balloons and now from space with successful missions such as the Infrared Astronomical Satellite (IRAS) and the Infrared Space Observatory (ISO). The coming decade promises significant advances with the launch of the Space Infrared Telescope Facility (SIRTF) and the Stratospheric Observatory for Infrared Astronomy (SOFIA), a 747 aircraft carrying a 2.5-m telescope. The Next Generation

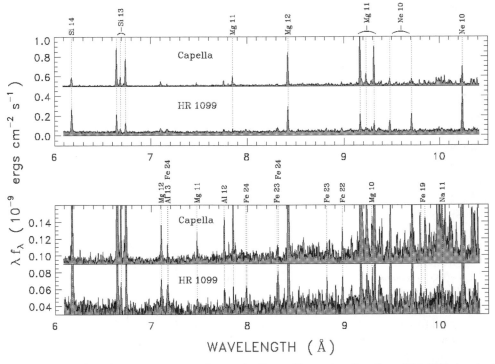

FIGURE 16 Chandra X-ray Observatory spectra from active G-type binary stars Capella and HR 1099 showing X-ray emission lines. The emission arises from the coronae of these stars which contain plasma at temperatures of several million degrees. The plasma temperature is estimated from the ratios of emission line fluxes, such as those of Mg XI near 9.2 Å. The ionization state of the coronal gas is examined through ratios of different ionization stages, such as Mg XI and Mg XII. [From T. Ayres *et al.* (2001) *Astrophys. J.* **549**, 554. With permission of the American Astronomical Society.]

Space Telescope (NGST), launched later in the decade, will also focus on observations at IR wavelengths.

Limited atmospheric transparency is not the only obstacle to observing in the IR. The detectors pick up signals from any source of heat, including the telescope and it housing. Because of this, the detector, and sometimes the whole telescope, is actively cooled to temperatures approaching 0 K. The coolant is an expendable material that limits the life of a satellite mission. Another complication of IR spectroscopy is the difficulty in achieving high spectral resolution. Instruments that work well at visible and UV wavelengths are not appropriate for the IR. New techniques have been required to work at the longer wavelengths.

Despite the difficulties, there is strong motivation to do IR stellar spectroscopy. The natural targets for IR observations are the cool stars that radiate most strongly at long wavelengths. Measurements of the IR energy distribution of a cool star give a firm grip on the effective temperature. This is particularly true because the H^- ion, the major source of atmospheric opacity in cool stars, displays it greatest variation in the near-IR, reaching a minimum at a wavelength near 1.6 μm. The amount of H^- in a stellar atmosphere changes rapidly with effective temperature, so the IR observations have a sensitive temperature indicator. Because only the shape of the energy distribution is needed for this type of determination, low-spectral-resolution measurements are adequate.

Low-resolution IR measurements have also detected excess amounts of radiation from some stars. This is a signature of cool circumstellar material or a very extended stellar atmosphere in connection with stellar mass loss. In the case of mass loss from hot stars, the IR energy distribution complements UV spectral line data, and it is often more straightforward to derive mass-loss rates from the IR data because the detailed velocity distribution of the outward flowing gas is not required. Recently, submillimeter observations beyond 850 μm and radio observations at centimeter wavelengths have helped further constrain mass-loss rates in this way.

Recent projects such as the very sensitive 2-Micron All Sky Survey (2MASS) have discovered very cool stars and brown dwarfs. In turn, the largest telescopes have engaged in a spectroscopic study of these objects. Isolated, very cool stars, once thought to be unrelated to normal stars, have been shown to be prototypes of two newly proposed spectral classes, L and T. The L dwarfs are now recognized to include low-mass stars and warm brown dwarfs just cooler than M dwarf stars. Over 120 L dwarfs have now been identified. L dwarfs lack the molecular bands of TiO and VO common in M stars and instead show strong bands of metal hydrides such as FeH and CrH and strong lines of neutral Na, K, Cs, Rb, and sometimes Li as shown in

Fig. 17. In the near-IR they show strong bands of H_2O along with bands of FeH and CO and lines of neutral Na and K. Brown dwarf spectra with methane band signatures and weak or absent bands of FeH and CO have been found to be just cooler the L dwarfs and are referred to as T dwarfs. T dwarfs, in addition to methane absorption, are characterized by strong H_2O bands throughout the far-red and near-IR regions.

The high-resolution observation of molecular absorption lines also provides a wealth of information. For example, the fundamental bands of CO are located in the IR at 5 μm. As described earlier, in most stars oxygen is more abundant than carbon. Because CO is the dominate molecule of these two elements, the amount of CO determines the availability of carbon to form other molecules. Therefore, it is necessary to understand the CO formation correctly before it is possible to interpret properly other molecules such as CN, C_2, and CH. IR spectroscopy provides this crucial piece of information. In addition to CO, the IR is also rich with bands of molecules formed from the other common elements: C_2, CN, H_2O, SiO, OH, NH, and MgH. Because these are the most important elements, the IR spectrum provides an unprecedented opportunity to derive accurate abundances, both for the total element and for the isotopes of each element.

IR stellar spectroscopy is capable of measuring radial velocities with high precision. This is because of the many individual absorption lines associated with a particular vibration-rotation molecular band. The wavelengths of the individual lines are known with great accuracy, and the large number of lines reduce the accidental error of measurement. Furthermore, because different molecules can form in different strata of the stellar atmosphere, it is possible to study the relative motions of these levels.

Stellar spectroscopy at still longer wavelengths moves from the IR into the millimeter and radio regions. The transition between IR and radio occurs in a region where the earth's atmosphere is opaque, but at wavelengths greater than 1 mm the atmosphere is increasingly transparent and beyond 1 cm is totally transparent out to the long wavelength cutoff of the ionosphere. Therefore, radio spectroscopy enjoys a significant advantage over IR and UV observation in being able to make unobstructed observations from the earth's surface. The spectral features present at radio wavelengths tend to be again molecular lines, and this selects cool stars as the most likely targets of study. Therefore, radio spectroscopy is best suited to studying the cooler circumstellar gas surrounding stars as the result of mass loss, mass exchange among double stars, or remnants of formation. In recent years several submillimeter telescopes have come online and soon arrays of such telescopes will operate together. Access to the submillimeter

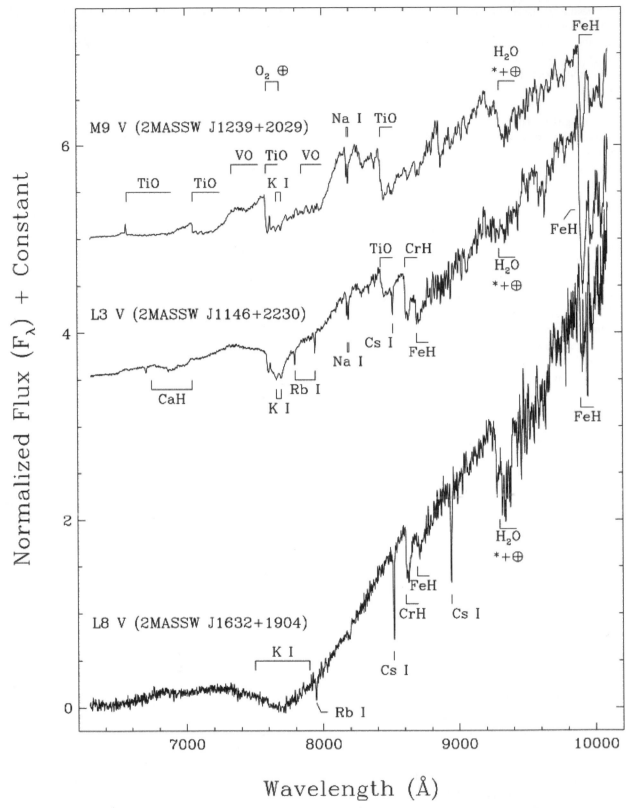

FIGURE 17 Spectra of late-M, early- to mid-L, and late-L dwarfs. Prominent features are marked. Note the absence of oxide absorption in the L dwarfs along with the dominance of alkali lines and hydride bands. [From Kirkpatrick, J. D. et al. (1999). *Astrophys. J.* **519,** 802. With permission of the American Astronomical Society.]

band allows study of a greater variety of molecular species and transitions in circumstellar envelopes.

C. Instrumentation

The expansion of spectroscopy into new spectral regions has been paralleled by the development of new instruments and detectors. The spectrographs used to make high-resolution measurements of line profiles and equivalent widths have traditionally been large, so large that they are housed in stationary rooms away from the telescope, to which the light must be brought by a chain of mirrors. This design has restricted these powerful instruments to just the largest telescopes because of the need to focus a large amount of light down the long mirror train and because of the high cost involved. Manufacturing methods can now produce high-quality components for a new form of high-resolution spectrograph using an echelle in place of the ordinary grating. The echelle differs from a conventional grating by having coarse grooves, typically 50 mm^{-1}, which are tilted at a steep angle. The advantage of the echelle is that it can produce high resolution in a compact instrument. This advantage has been realized for a long time, but production techniques now can turn this realization into a reality. Also, because the echelle spectrograph is smaller, it can be used with telescopes of modest size, greatly increasing the number of locations at which high-resolution spectroscopic observations can be made. An echelle spectrograph was used in the IUE satellite to fit a high-resolution instrument into the available space.

During the 1990s there was a steady increase in the degree of precision and accuracy achieved in the measurement of stellar radial velocities. This was made possible largely through the employment of fiberoptic cables, now ubiquitous in high-speed computer networks, in the acquisition of the stellar spectrum. With transmission losses as low at 20%, fibers can deliver light from the focal plane of the telescope to a spectrograph located in a room far apart from the telescope. The use of the fiber allows the spectrograph to be mechanically isolated from the telescope environment and thermally controlled, thereby drastically improving the repeatability of radial velocity measurements. The measurement of minute shifts in the stellar spectrum can be mimicked by slight motions of the spectrograph elements and changes in the refractive index of air due to temperature fluctuations. The degree to which these systemic instrumental shifts can be isolated and controlled improves stellar radial velocity measurements.

Radial velocity precision achieved by the best fiber-fed spectrographs is about 6 to 10 m s^{-1}, a 100-fold increase from conventional echelle spectrographs. Although the precision of radial velocity measurements has improved

greatly, the accuracy, meaning the radial velocity relative to a standard reference frame, is still problematic. The accuracy of radial velocity measurements is generally uncertain to hundreds of meters per second. This is in part due to the fact that many radial velocity standard stars, once thought to be constant in the era of 1-km s^{-1} precision, are now known to be radial velocity variables and are not useful as standards. Revised lists of radial velocity standard stars have recently been compiled where no velocity variations above the 100-m s^{-1} level are seen.

The Fourier transform spectrometer (FTS) is another recent addition to the array of spectroscopic instruments. It is a Michelson interferometer with the means to vary the path difference between two arms of the instrument. It has an impressive list of attractive features, including variable resolution from high to low, excellent efficiency, high signal-to-noise properties, the absence of scattered light, and high photometric accuracy. Its only drawback is that it uses a single detecting element. When the random fluctuations in the arriving light dominate the noise, the FTS is slower than instruments that use multiple detecting elements, such as a conventional photograph, although this may be more than offset by its good features. This is the situation in the visible spectral band, where the FTS has been used extensively for solar observations, but not for stars because it slowness becomes a disadvantage for the fainter sources. In the IR, however, the dominant source of noise is usually in the detector. The use of a single detector becomes an advantage, and the FTS is the instrument of choice for spectroscopic studies of the sun, stars, planets, and other astronomical objects. It is now possible to make spectroscopic measurements of bright stars (Arcturus, Procyon) that are comparable to the quality that could only be obtained for the sun a few years ago, and both solar and stellar spectra approach the quality of laboratory measurements. This is an area of considerable potential.

The development of new detectors has paralleled the development of new instruments. As mentioned earlier, the photograph has been the workhorse for astronomical detection for more than a century. Even now it is unrivaled if a detector of large area and information capacity is required; emulsions covering an area of 20×25 cm are routinely available, and larger sizes can be manufactured. There are, however, several serious limitations to photographic detection. In addition to the low efficiency of the photographic process, there is the need to calibrate the sensitivity of the emulsion to different light intensities. This calibration, which is crucial to the measurement of line profiles and equivalent widths, depends on the wavelength of the light being studied as well as the way the emulsion is developed, and many errors have been traced to this step in the acquisition of research data.

For these reasons, astronomers have always sought new detectors with improved properties. A significant advance has been the development of solid-state silicon photodiode detectors. These devices, through the photoeletric effect, have the property of producing an electrical current when exposed to light; the greater the exposure, either from a brighter source or a longer exposure time, the greater the signal. Moreover, there is a direct one-to-one relationship between the exposure and the signal. The signal can then be converted to a digital representation for immediate computer manipulation. In addition to the considerable convenience of digital output, these devices enjoy several other advantages. First, they have efficiencies reaching 90% in the red part of the spectrum. This is an enormous gain compared to the small fraction of a percent for photography at the same wavelengths. Another advantage is a well-defined, stable geometry. Each photodiode, or pixel, has a fixed width of about 0.02 mm. A third advantage is the ability to measure a large range of intensities in one exposure. Not only is it possible to measure deep spectral lines in this way, but the noise can also be suppressed to a tiny fraction of the signal; it is possible to make observation where the noise is less than 0.5% of the signal being studied. Such observations show spectral features, such as the signature of pulsations shown in Fig. 6, which had been lost in the noise previously. The main disadvantages of these detectors are their need to be cooled to low temperatures (-100 to $-150°C$) to suppress thermal contamination during exposures of several hours' duration and the need to calibrate accurately the pixel-to-pixel variation in sensitivity.

The silicon photodiodes currently used for astronomical observations come in different forms. Devices consisting of a single row of pixels have been used successfully for spectroscopy for several years. Two-dimensional arrays of diodes are now being used for direct imaging and or spectroscopic detection with both conventional and echelle spectrographs. Over the last decade the size of the silicon wafers that constitute these detectors has continued to increase. Furthermore, these larger wafers are commonly placed together to form a mosaic, typically 4096 pixels on a side, providing 16 times the detector area that was available just ten years ago. These larger detectors make it possible to record the full echelle spectrum of a star from 4000 to 11000 Å in a single exposure. Because of commercial and industrial interest in compact two-dimensional detectors, there has been extensive of development of these devices. These charge-coupled devices (CCDs) can now be found in home video recorders, digital cameras, and other popular electronics.

The advantage of solid-state detectors are now being extended to IR observations. Pure silicon becomes transparent at wavelengths greater than 1 μm, but if the silicon is combined with other elements or if substances other than silicon are used, successful array detectors can be constructed. The use of array detectors for IR spectroscopy promises to be an area of rapid development because of the greatly reduced observing time that can be achieved by recording many spectral elements simultaneously.

D. Modeling

A third area in which rapid advances have been occurring is the realism of the models used to represent the stellar atmosphere and the formation of the spectrum.

Until recently, a number of simplifying assumptions were routinely made in the construction of stellar atmospheres and in the analysis of stellar spectra. Among these were the assumptions that the stellar atmosphere is a plane parallel slab that varies only with depth; that the atmosphere is static because it is in strict pressure balance; that only slowly varying, continuous sources of absorption need be included; and that the state of the gas at any location in the stellar atmosphere is in thermodynamic equilibrium at the local temperature. These simplifications were considered essential, either because it was not known how to solve the problem presented by the actual situation, or because the computing time needed to solve the true problem was beyond the bounds of acceptability.

A combination of greatly increased computer power and new numerical methods has made it possible to remove or reduce these assumptions. For example, it is now possible to represent a model of the stellar atmosphere by a spherically symmetric geometry (that is, variation in two spatial dimensions) instead of a parallel slab. This improvement is not important for the photosphere of a star such as the sun because its atmosphere is so thin that the spherical shape of the sun is not sensed by the radiation flowing up out of the atmosphere. For both giants and supergiants, however, the effect can be significant, both for the structure of the atmosphere and for the formation of the spectrum. Additionally, techniques have been developed to accurately solve for the transfer of radiation through the nonstatic layers of an expanding model atmosphere. These techniques have made it possible to model the spectra of luminous stars with stellar winds as well as the spectra from very rapidly expanding atmospheres, such as novae and supernovae. It is now possible to determine the elemental abundances from the spectra of novae and supernovae using methods originally limited to spectra of stars with thin, static atmospheres.

A second area of major improvement is the representation of stellar opacity. New statistical and averaging methods have made it almost as easy to include sources of opacity that vary rapidly with wavelength, such as many individual spectral lines, as it is to use slowly changing sources of absorption and scattering. The use of the more complete and accurate opacity changes the manner in which the temperature varies with depth in the atmosphere which can then change the appearance and interpretation of the spectrum.

A third assumption that can now be removed concerns the validity of thermodynamic equilibrium at the local value of the temperature at each position in the atmosphere. This assumption is extremely powerful because it implicitly increases the importance of collisions between the atoms and molecules while reducing the significance of radiative interactions. As a consequence, a single quantity, the local temperature, governs everything. The assumption also increases the roles of absorption and emission at the expense of scattering. To avoid these biases, it is necessary to treat the interaction between radiation and the plasma at the microscopic level, and methods have now been developed that can do this rapidly and accurately. This treatment is known as non-local thermodynamic equilibrium, or non-LTE. Non-LTE calculations involve the construction model ions, where the population of each energy level of each ion is regulated by both collisional and radiative processes. For complex model ions, such as Fe II, this means keeping track of over 600 energy levels and the 13,000 strongest spectral lines so they can be treated in non-LTE. Modern atmosphere codes include model ions for all the important atomic species which total to more than 10,000 energy levels and 1 million spectral lines. The new frontier to non-LTE computations is the inclusion of detailed molecular models to remove the assumption of thermodynamic equilibrium in molecular opacity computations. The molecular species to be treated first in non-LTE will be the most important opacity sources in cool stars such as CO, TiO, and H_2O.

The use of realistic opacity from spectral lines and the microscopic treatment of the interaction of radiation with the gas make tremendous demands on our knowledge of atomic and molecular physics. It is necessary to know the intrinsic ability of atoms, ions, and molecules to change between discrete and continuum energy levels by both radiative and collisional processes. Fortunately, the methods are being developed to meet this challenge. It is now possible to calculate many of the quantities of interest, and accurate experiments are being done to test the validity of the theoretical results. For example, the past decade has seen an approximately 50-fold increase in the amount of atomic and molecular data available to use as opacities, so that there are now more than 50 million spectral lines with established parameters. These data are having a marked impact on the ability of the models to match the properties of real stars. Modern stellar atmosphere codes use a technique called opacity sampling to include about 50 million atomic lines and about 500 million molecular

lines as sources of opacity. More complete and accurate spectral line lists of atomic and molecular lines are being compiled and will continue to improve the accuracy of line opacity in model stellar atmospheres, particularly for very cool stars where molecular opacity in so important.

The extension of the standard spectral classes below class M to the L- and T-type dwarfs leads to new challenges for modelers of stellar atmospheres. For these coolest stars, all opacity sources including solid condensates, such as dust grains, must be included in the atmosphere structure and spectrum calculations. This involves keeping track of over 650 species including atoms, ions, molecules, and grains and their formation and destruction, as a function of depth in the atmosphere.

With the development of new detectors, more efficient spectrometers, the use of the entire stellar spectrum from the X-ray to the radio, and the ability to construct more realistic models, the old field of stellar spectroscopy continues to face a bright future.

SEE ALSO THE FOLLOWING ARTICLES

GALACTIC STRUCTURE AND EVOLUTION • INFRARED ASTRONOMY • SOLAR PHYSICS • SOLAR SYSTEM, MAGNETIC AND ELECTRIC FIELDS • STELLAR STRUCTURE AND EVOLUTION • TELESCOPES, OPTICAL • ULTRAVIOLET SPACE ASTRONOMY

BIBLIOGRAPHY

Allard, F. et al. (1997). "Model atmospheres of very low mass stars and brown dwarfs," *Annu. Rev. Astron. Astrophys.* **35**, 137.

Chaffee, F. H., Jr. and Schroeder, D. J. (1976). "Astronomical application of Echelle spectroscopy," *Annu. Rev. Astron. Astrophys.* **14**, 23.

Garrison, R. F., ed. (1984). "The MK Process and Stellar Classification," David Dunlap Observatory, Toronto.

Gray, D. F. (1992). "The Observation and Analysis of Stellar Photospheres," Cambridge University Press, Cambridge, U.K.

Gustafsson. B. (1989). "Chemical analyses of cool stars," *Annu. Rev. Astron. Astrophys.* **27**, 701.

Hearnshaw, J. B. (1990). "The Analysis of Starlight: One Hundred and Fifty Years of Astronomical Spectroscopy," Cambridge University Press, Cambridge, U.K.

Hearnshaw, J. B., and Scarfe, C. D. eds. (1999). "Precise stellar radial velocities." IAU Colloquium 170, A.S.P. Conf. Ser. Vol. 185, San Francisco, CA.

Jordan, S. D., ed. (1981). "The Sun as a Star," National Aeronautics and Space Administration, Washington, D.C.

Kaler, J. B. (1989). "Stars and Their Spectra: An Introduction to the Spectral Sequence," Cambridge University Press, Cambridge, U.K.

Kondo, Y., Boggess, A., and Maran, S. P. (1989). "Astrophysical contributions of the International Ultraviolet Explorer," *Annu. Rev. Astron. Astrophys.* **27**, 397.

Kudritzki, R.-P., and Puls, J. (2000). "Winds from hot stars," *Annu. Rev. Astron. Astrophys.* **38**, 613.

Linsky, J. L. (1980). "Stellar chromospheres," *Annu. Rev. Astron. Astrophys.* **18,** 439.

Morgan, W. W., and Keenan, P. C. (1973). "Spectral classification," *Annu. Rev. Astron. Astrophys.* **11,** 29.

Ridgway, S. T., and Brault, J. W. (1984). "Astronomical Fourier transform spectroscopy revisited," *Annu. Rev. Astron. Astrophys.* **22,** 291.

Rieke, G. H. (1994). "Detection of Light: From the Ultraviolet to the Submillimeter," Cambridge University Press, Cambridge, U.K.

Stellar Structure and Evolution

Peter Bodenheimer

Lick Observatory, University of California,
Santa Cruz

GLOSSARY

Brown dwarf Object in the substellar mass range (0.01–0.075 solar masses) which, during its entire evolution, never attains interior temperatures high enough to account for 100% of its luminosity by nuclear conversion of hydrogen (^{1}H) to helium.

Degenerate gas Gas in which the elementary particles of a given type fill most of their available momentum quantum states as determined by the Pauli exclusion principle. Electron degeneracy occurs in the cores of highly evolved stars and in white dwarfs; neutron degeneracy occurs in neutron stars.

Effective temperature Surface temperature of a star calculated from its luminosity and radius under the assumption that it radiates as a black body.

Galactic cluster Gravitationally bound group of a few hundred or a few thousand stars found in the disk of the galaxy. In a given cluster all stars are assumed

to have been formed at about the same time, but a wide range of ages is represented among the various clusters.

Globular cluster Compact, gravitationally bound group of 10^5–10^6 stars, generally found in the halo of a galaxy and formed early in the history of the galaxy.

Helioseismology Study of the internal structure of the sun through observation and analysis of the small oscillations at its surface.

Hertzsprung-Russell diagram A plot whose ordinate indicates the luminosity of a star and whose abscissa indicates the effective temperature of the star. A given star at a given time is represented by a point in this diagram. The evolution of a star is represented by a curve (or *track*) in this diagram.

Horizontal branch Sequence of stars on the Hertzsprung-Russell diagram of a globular cluster, above the main sequence. The stars all have approximately the same luminosity and are in the evolutionary phase where helium is burning in the core.

Encyclopedia of Physical Science and Technology, Third Edition, Volume 16
Copyright © 2002 by Academic Press. All rights of reproduction in any form reserved.

Luminosity Total rate of radiation of electromagnetic energy from a star, in all wavelengths and in all directions. Unit: energy per unit time.

Main sequence Sequence, or band, of stars in the Hertzsprung-Russell diagram, on which a large fraction of stars fall, running diagonally from high luminosity and high effective temperature to low luminosity and low effective temperature and associated with the evolutionary phase in which the stars burn hydrogen to helium in their cores.

Neutrino Subatomic neutral particle produced in stars chiefly in beta-decay reactions, but also by other processes, which travels at the speed of light and interacts only very weakly with matter.

Neutron star Highly compressed remnant of the evolution of a star of high mass. Its main constituent is free neutrons, the degenerate pressure of which supports the star against gravitational collapse. The mean density is comparable to that of nuclear matter, 10^{14} g cm^{-3}.

Nova Sudden, temporary, but recurrent brightening of a star by factors ranging from 10 to 10^6, occurring in a binary system where, in most cases, a white dwarf is accreting mass from a main-sequence companion. The outbursts are caused either by instability in the accretion disk surrounding the white dwarf (*dwarf nova*) or by nuclear reactions in the material recently accreted onto its surface (*classical nova*).

Nucleosynthesis Production of the elements through nuclear reactions in stars or in the early universe.

Protostar Object in transition between interstellar and stellar densities, during which it undergoes hydrodynamic collapse and is observable primarily in the infrared part of the spectrum.

Red giant Post-main-sequence star whose radius is much larger than its main-sequence value and whose effective temperature generally falls in the range 3000–5000 K.

Supernova Sudden increase in luminosity of a star, by a factor of up to 10^{10}, followed by a slower decline over a time of months to years. The event is caused by explosion of the star and dispersal of much of its matter.

White dwarf Compact star (mean density, 10^6 g cm^{-3}) representing the final stage of evolution of a star of low to moderate mass. It is supported against its gravity by the pressure of degenerate electrons.

Zero-age main sequence Line in the Hertzsprung-Russell diagram corresponding to the points where stars of different masses first arrive on the main sequence and representing the first moment, for a given star, when 100% of its luminosity is provided by the fusion of protons to helium.

STELLAR STRUCTURE is the study of the internal properties of a star, such as its temperature, density, and rate of energy production, and their variation from center to surface. The structure can be determined through a combination of theoretical calculations, based on known physical laws, and observational data. Stellar evolution refers to the change in these physical properties with time, again as determined from both theory and observation. The three main phases of stellar evolution are (1) the pre-main-sequence phase, during which gravitational contraction provides most of the star's energy; (2) the main-sequence phase, in which nuclear fusion of hydrogen to helium in the central region provides the energy; and (3) the post-main-sequence phase, in which hydrogen burning away from the center, as well as the burning of helium, carbon, or heavier elements, may provide the energy. The evolutionary properties are strongly dependent on the initial mass of the star and, to some extent, on its initial chemical composition. The end point of the evolution of a star can be a white dwarf, a neutron star, or a black hole. In accordance with this picture, a star can be defined as a gaseous object that, at some point in its evolution, obtains 100% of its energy from the fusion of protons (^{1}H) to helium nuclei. The boundary in mass between stars and substellar objects (also known as brown dwarfs) falls at about 0.075 solar masses ($M_\odot$), below which the 100% energy condition is never fulfilled.

I. INTRODUCTION

The study of the structure and evolution of the stars, which constitute the major fraction of the directly observable mass in the universe, is of critical importance for the understanding of the production of the chemical elements heavier than helium, of the evolution of the solar system and other planetary systems, of the structure and energetics of the interstellar medium, and of the evolution of galaxies as a whole. The structure of a star is determined by the interaction of a number of basic physical processes, including nuclear fusion; the theory of energy transport by radiation, convection, and conduction; atomic physics involving especially the interaction of radiation with matter; and the equation of state and thermodynamics of a gas. These principles, combined with basic equilibrium relations and assumed mass and chemical composition, allow the construction of mathematical models of stars that give, as a function of distance from the center, the temperature, density, pressure, and rate of change of chemical composition by nuclear reactions. A star is not static, however; it must evolve in time, driven by the loss of energy from its surface, primarily in the form of radiation. This energy is provided by two fundamental sources—nuclear energy and gravitational energy—and in the process of providing this energy the star undergoes major changes in its structure. To follow this evolution mathematically requires the solution of a complicated set of equations, a solution that requires the use of high-speed computers to obtain sufficient detail. The goal of the calculations is to obtain a complete evolutionary history of a star, as a function of its initial mass and chemical composition, from its birth in an interstellar cloud to its final state as a compact remnant or possibly as an object completely disrupted by a supernova explosion.

The heart of the study of stellar structure and evolution is, however, the comparison of models and evolutionary tracks with the observations. There are numerous ways in which such comparisons can be made, for example, by use of the Hertzsprung-Russell (H–R) diagrams of star clusters, the mass–luminosity relation on the main sequence, the abundances of the elements at different phases of evolution, and the mass–radius relation for white dwarfs. There are many exotic stars that the theory is not yet able to fully explain, such as pulsars, novae, X-ray binaries, some kinds of supernovae, or stars showing rapid mass loss. These systems provide a challenge for the future theorist. However, the general outline of the phases of stellar evolution has by now fallen into place through a complex interplay between theoretical studies, observations of stars, and laboratory experiments, particularly those required to determine nuclear reaction rates. The period of development of ideas concerning the structure of stars extends more than 100 years into the past. Among the noteworthy historical developments were the clarification by Sir Arthur Eddington (1926) of the physics of radiation transfer; the development by S. Chandrasekhar (1931) of the theory of white dwarf stars and the derivation of their limiting mass; and the work of H. Bethe (1939) and others, which established the precise mechanisms by which the fusion of hydrogen to helium provides most of the energy of the stars. However, much of the detailed development of the subject has occurred since 1955, spurred by the availability of high-speed computers and by the extension of the observational database from the optical region of the spectrum into the radio, infrared, ultraviolet, X-ray, and gamma-ray regions. Numerous scientists have collaborated to advance our knowledge of the physics of stars in all phases of their evolution.

II. OBSERVATIONAL INFORMATION

The critical pieces of observational data include the luminosity, effective temperature, mass, radius, and chemical composition of a star. A further fundamental piece of information that is required to obtain much of this data is the distance to the star, a quantity that in general is diffi-cult to measure because even the nearest star is 2.6×10^5 astronomical units (AU) away, where the AU, the mean distance of the earth from the sun, is 1.5×10^{13} cm. The AU, measured accurately by use of radar reflection off the surface of Venus, is used as the baseline for trigonometric determinations of stellar distances. The apparent shift (*parallax*) in the position of a star, against the background defined by more distant stars or galaxies, when viewed from different points in the earth's orbit, allows the distance to be determined. The parsec (pc) is defined as the distance of a star with an apparent positional shift of 1 sec of arc on a baseline of 1 AU and has the value of 3.08×10^{18} cm. The nearest stars (the Alpha Centauri triple system) are about 1.3 pc away. The most accurate available database of parallaxes was obtained from the Hipparcos satellite, which measured 120,000 stars with an accuracy down to 0.001 arcsec. Thus, the distance out to which accurate distance measurements are available is about 150 pc, still small compared with the distance to the center of our galaxy ($\approx$8000 pc). Thus, for most stars indirect determinations of distances must be used, based, for example, on period–luminosity relations for variable stars or on properties of the spectrum and apparent luminosity of a star compared to those of a star with a very similar spectrum and known distance.

A. Luminosity

The standard unit of luminosity is that of the sun, which is obtained by a direct measurement of the amount of energy $S_\odot$, received per unit area per unit time, over all wavelengths, outside the earth's atmosphere, at the mean distance of the earth from the sun. This quantity, known as the solar constant, is then converted into the solar luminosity by using the formula

$$L_\odot = 4\pi d_\odot^2 S_\odot = 3.86 \times 10^{33} \text{ erg/sec},$$

where $d_\odot = 1$ AU. For other stars, in principle, the stellar flux S (energy per unit area per unit time) received at the earth is corrected for absorption in the earth's atmosphere and in interstellar space and is extended to include all wavelengths of radiation. If the star's distance d is known, its luminosity follows from $L = 4\pi d^2 S$.

B. Effective Temperature

A number of different methods are used to determine the effective (surface) temperature, most of which are based on the assumption that the stars radiate into space with a spectral energy distribution that approximates a black body. For a few stars, such as the sun, whose radius R can be measured directly, the value of T_{eff} is obtained from L and R by use of the black-body relation $L = 4\pi R^2 \sigma T_{\text{eff}}^4$, where σ is the Stefan-Boltzmann constant. Otherwise the temperature may be estimated by four different methods.

1. The detailed spectral energy distribution is measured, and the temperature of the black-body distribution that best fits it is found.

2. The wavelength λ_{max} of maximum intensity in the spectrum is measured, and the temperature is found from Wien's displacement law for a black body: $\lambda_{max} T = 2.89 \times 10^7$, if λ_{max} is expressed in angstroms.

3. The "color" of the star is obtained by measurement of the stellar flux in two different wavelength bands. For example, the color "B $-$ V" is obtained by comparing the star's flux in a wavelength band centered at 4400 Å and about 1000 Å broad (blue) with that in a band of similar width centered at 5500 Å (visual). In principle, the transmission properties of the B and V filters could be used in connection with black-body curves to determine the temperature. In practice, stars are not perfect black bodies, and the temperature is determined by comparison with a set of standard stars.

4. From the strength in the stellar spectrum of absorption lines of various chemical elements, the spectral type and thereby the temperature can be determined by comparison with a set of standard stars. This method does not depend on the black-body assumption, and the temperature is, in principles, determined from the degree of excitation and ionization of the atoms.

C. Radius

The radius can be measured directly for only a few stars. In the case of the sun, the angular size can be measured and the distance is known. In the case of Sirius and a few other stars, the angular diameter can be measured by use of interferometry or by high-resolution direct imaging with the Hubble Space Telescope. For certain eclipsing binary systems, if the orbital parameters are known and the light variation with time can be accurately measured, the radii of both components can be determined. In all other cases the radius must be estimated from measurements of L and T_{eff} and the formula $L = 4\pi R^2 \sigma T_{eff}^4$.

D. Mass

A stellar mass can be measured directly only if the star is a member of a binary system, in which case Kepler's third law can be applied. If M_1 and M_2 are the stellar masses, in units of the solar mass $M_\odot$, P is the orbital period of the system in years, and a is the semimajor axis of the relative orbit in astronomical units, then the law states that $(M_1 + M_2)P^2 = a^3$. In the case of the sun, the orbital periods and distances of the planets can be used for an accurate determination of solar masses. If both components of a binary are visible (visual binary) and if the angular separation as well as P can be determined, then the sum of the masses follows from Kepler's law as long as the distance is known. The individual masses can be found if the relative distances of the stars from the center of mass can be measured. If the binary system has such a close separation that the components cannot be visually resolved, it may be possible to resolve them spectroscopically. If the Doppler shifts of spectral lines as a function of time can be measured for both components, then the period can be determined as well as the mass ratio and minimum masses of the components. If in addition the system shows an eclipse, then the individual masses can be determined. There are relatively few systems with the required orbital characteristics to allow reasonably accurate mass determinations.

E. Abundances

The relative abundances of the elements in the solar system (at approximately the time of its formation) have been compiled by E. Anders and N. Grevesse, based in most cases on laboratory measurements of the oldest meteoritic material and in some cases from the strengths of absorption lines in the solar atmosphere. Abundances for selected elements are given in Table I. In stellar evolution theory, the fractional abundance of H by mass is known as X, that of He is known as Y, and that of all other elements are known as Z. In stars, the abundances are determined

TABLE I Solar System Abundances

Element	Log abundance by number of atoms ($H = 10^{12}$)	Fractional abundance by mass
1 H	12.00	0.704
2 He	11.00	0.279
3 Li	3.31	9.89×10^{-9}
6 C	8.55	2.97×10^{-3}
7 N	7.97	9.12×10^{-4}
8 O	8.87	8.28×10^{-3}
10 Ne	8.07	1.66×10^{-3}
11 Na	6.33	3.43×10^{-5}
12 Mg	7.58	6.45×10^{-4}
13 Al	6.47	5.56×10^{-5}
14 Si	7.55	6.96×10^{-4}
16 S	7.21	3.63×10^{-4}
20 Ca	6.36	6.41×10^{-5}
24 Cr	5.67	1.70×10^{-5}
26 Fe	7.51	1.26×10^{-3}
28 Ni	6.25	7.29×10^{-5}

from the strengths of absorption lines in the spectrum combined with other parameters in the stellar atmosphere and compared with solar values. In practically all measured systems, the values of X and Y are deduced to be very similar to the solar values. However, Z can vary, and in the oldest stars and globular clusters, it can fall below the solar value by a factor of more than 100.

F. Hertzsprung-Russell Diagram

The Hertzsprung-Russell (H–R) diagram is a plot of luminosity versus surface temperature for a set of stars. Although the data can be plotted in various forms, the sam-

ple H–R diagram shown here (Fig. 1) gives data converted from observed quantities to L and T_{eff}. Most of the stars lie along the main sequence, which represents the locus of stars during the phase of hydrogen burning in their cores, with increasing T_{eff} corresponding to increasing mass. The stars well below the main sequence are in the white dwarf phase, having exhausted their nuclear fuel. The stars in the upper-right part of the diagram are red giants (e.g., Aldebaran); these are stars that have exhausted their central hydrogen and are now burning hydrogen in a shell region around the exhausted core. Some of them are also burning helium in a core or in a shell. The number of stars in a given region of the H–R diagram is roughly proportional

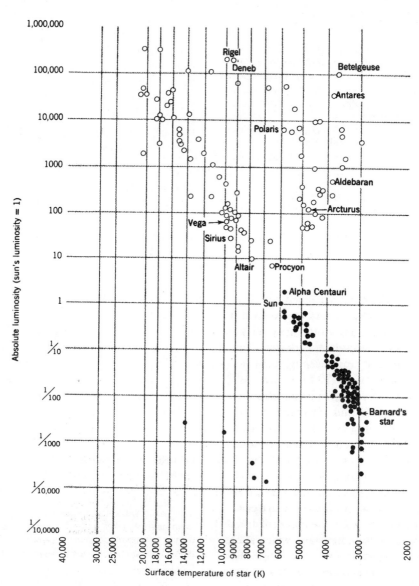

FIGURE 1 H–R diagram for the 100 brightest stars (open circles) and the 90 nearest stars (filled circles). [Reprinted with permission from Jastrow, R., and Thompson, M. H. (1984). "Astronomy: Fundamentals and Frontiers," 4th ed., Wiley, New York. © 1984, Robert Jastrow.]

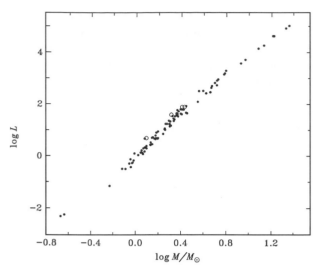

FIGURE 2 Observed masses of main-sequence stars in units of the solar mass as determined from binary orbits plotted against their luminosities in units of the solar luminosity (filled circles). The open circles are subgiants or giants. The observational errors in mass are less than 2%. [Reprinted with permission from Andersen, J. (2000). In "Unsolved Problems in Stellar Evolution" (M. Livio, ed.), Cambridge Univ. Press, Cambridge, UK. Reprinted with permission of Cambridge, University Press.]

to the evolutionary time spent in that region; thus, the main-sequence or core hydrogen-burning phase is that in which stars spend most of their lifetime.

G. Mass–Luminosity Relation

For stars known to be on the main sequence, the observational information on M and L can be combined to produce a reasonably smooth mass–luminosity relation (Fig. 2). The slope of this relation is not constant, however. Empirically, the luminosity increases approximately as M^3 for stars in the region of 10–20 $M_\odot$ and as $M^{4.5}$ for stars between 1–2 $M_\odot$. A few additional masses are available for stars less massive than the sun, but the accuracy is not as good as for those shown in Fig. 2. Theoretical models, with composition close to that of the sun at the zero-age main sequence, agree well with the observed points. There are two reasons for the scatter in the observed points: first, some of the observed stars are not strictly at zero age but have evolved slightly, with a corresponding increase in L; and second, there are small differences in metal abundance among the stars, which affect the luminosities. Because stars evolve and change their luminosity considerably while retaining the same mass, a mass–luminosity relation cannot be specified for most regions of the H–R diagram, only near the zero-age main sequence.

H. Stellar Ages

Although the age of the solar system (and therefore presumably the sun) can be determined accurately from radioactive dating of the oldest moon rocks and meteorites, the ages of other stars cannot be determined directly. Several indirect methods exist, which depend for the most part upon the theory of stellar evolution, and generally, there is some uncertainty in the derived ages.

1. The H–R diagrams of galactic or globular clusters can be compared with evolutionary calculations. The stars in a cluster are assumed to be all of the same age; to have the same composition; and, except for the very nearest clusters, to all lie at the same distance from the earth. The observed cluster diagram is compared with a theoretical line of constant age, known as an *isochrone*, obtained by calculating the evolution in the H–R diagram of a set of stars with different masses but the same composition and by connecting points on their evolutionary tracks that correspond to the same elapsed times since formation. This procedure is illustrated in Figs. 3 and 4, which show how the age is determined for two different clusters. Fig. 3 shows the Hyades, a nearby galactic cluster with a distance of 46.3 pc. The age is determined from the positions in the

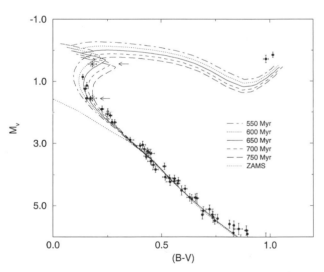

FIGURE 3 H–R diagram for the Hyades cluster. The luminosity is given in terms of the absolute visual magnitude M_V, and the observed color (B − V) is an indicator of surface temperature (increasing to the left). The filled circles with error bars correspond to the observations of the individual single stars, whose distances have been determined from Hipparcos parallaxes. The symbols with arrows correspond to the two components of a spectroscopic binary. The dotted curve corresponds to the theoretical zero-age main sequence for the composition of this cluster. Other curves are theoretical isochrones, that is, predictors of how the H–R diagram should look at the indicated ages. [Reprinted with permission from Perryman, M. A. C., *et al.* (1998). *Astron. Astrophys.* **331**, 81. © European Southern Observatory.]

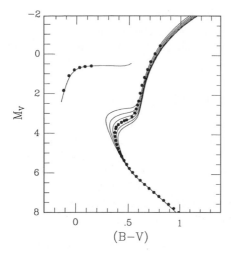

FIGURE 4 H–R diagram for the stars in the globular cluster M92. The axes have the same meaning as in Fig. 3. The filled circles are averaged observed stellar positions. The solid curves are theoretical isochrones for the ages (top to bottom) of 8, 10, 12, 14, 16, and 18 Gyr. The evolved stars in the upper-left part of the diagram are on the "horizontal branch," where they are burning He in their cores. [Adapted with permission from Pont, F., Mayor, M., Turon, C., and VandenBerg, D. A. (1998). *Astron. Astrophys.* **329**, 87. © European Southern Observatory.]

H–R diagram of the stars near the "turnoff" point, usually identified as the point of highest effective temperature on the main sequence. The time spent by a star on the main sequence decreases with increasing mass; thus, stars with masses above that corresponding to the turnoff mass have already left the main sequence and have evolved to the red giant region. In this case the age is determined to be 625 Myr, with an error of ± 50 Myr. Fig. 4 shows the globular cluster M92, one of the oldest objects in the galaxy. Its distance is about 9000 pc, and the abundance of metals, such as iron, is less than 1/100 that of the sun. The distance to the cluster is determined by fitting the location of its main sequence in the H–R diagram to the locations of nearby stars of the same (reduced) metallicity whose distances are known from Hipparcos parallaxes. The age derived from isochrone fitting is 14 ± 1.2 Gyr. Note that the luminosity of the turnoff is much fainter than that for the Hyades cluster. In the cluster fitting method, the age of the stars is really being determined from the nuclear burning time scale.

2. Main-sequence stars like the sun have chromospheric activity, analogous to solar activity, such as flares, prominences, and chromospheric emission lines, which declines with age. An often-used indicator of chromospheric activity is the strength of two emission features in the Ca lines between 3900 and 4000 Å. The age-activity relation is calibrated by use of objects whose ages have been determined by other methods, such as the sun and the Hyades cluster.

3. Certain special types of stars are known to be young, that is, with ages of 1 to a few million years. One example is the T Tauri stars, which are identified by their high lithium abundances, the presence of emission lines of hydrogen, their irregular variability, their association with dark clouds (star-forming regions) in the galaxy, and their location in the H–R diagram above and to the right of the main sequence. Typical masses are 0.5–2 $M_\odot$, and typical values of T_{eff} are around 4000 K. A second example is the massive stars near the upper end of the main sequence, with $L/L_\odot \approx 10^5$. They are also known to be young because their nuclear burning time scale is only a few Myr.

4. The stars in the galaxy have been roughly divided, according to age, into two populations. The Population I stars are associated with the galactic disk, have relatively small space motions with respect to the sun's, and have metal (such as iron) abundances similar to the sun's. Although precise ages cannot be determined, this population is younger as a group than the Population II stars, which are distributed in the galactic halo and in globular clusters and have high space velocities and low metal abundances. These objects were formed early in the history of the galaxy (see Fig. 4), and their low metal abundance supports the point of view that the elements heavier than helium have been synthesized in the interiors of the stars. Subsequently, some of the synthesized material was ejected from the stars in supernova explosions and in winds from evolved stars, so that later generations of stars form from interstellar material that has been gradually enriched in the heavy elements.

III. PHYSICS OF STELLAR INTERIORS

A. Time Scales

The dynamical time scale, the contraction time scale, the cooling time scale, and the nuclear time scale are important at different stages of stellar evolution. If the star is not in hydrostatic equilibrium, it will evolve on the dynamical time scale given by $t_{\text{ff}} = [3\pi/(32G\rho)]^{1/2}$; this is, in fact, the free-fall time from an initial density ρ. Examples of unstable stars that evolve on this time scale are protostars or the evolved cores of massive stars, which undergo photodissociation of the iron nuclei and consequently are forced into gravitational collapse. The resulting supernova outburst, of course, also takes place on a dynamical time scale. Another example is the light variation in Cepheid variables, which is caused by radial oscillations about an equilibrium state; the oscillation period is of the same order as t_{ff}. Characteristic time scales are 10^5 years for the protostar collapse, 0.1 sec for the collapse of the iron core, and 10 days for the pulsation period of a Cepheid.

Stars in hydrostatic equilibrium but without a substantial nuclear energy source undergo a slow contraction with the release of gravitational energy. The associated time scale, known as the Kelvin-Helmholtz time scale, is given by the gravitational energy divided by the luminosity: $t_{KH} \approx GM^2/(RL)$, where R is the final radius and L is the average luminosity. Stars contracting to the main sequence evolve on this time scale as do, for example, stars at the end of the main-sequence phase when they run out of hydrogen fuel in the core. For the present sun, the value of t_{KH} is about 3×10^7 years, which represents the time required for it to contract to its present size without any contributions from nuclear reactions. Because t_{KH}, as well as the other time scales discussed later, depend on L, they are controlled by the time required for energy to be transported from the interior of the star to the surface.

Stars that derive all their energy from nuclear burning evolve on a nuclear time scale, which can be estimated from the total available nuclear energy divided by the luminosity. For the case of hydrogen burning, four protons, each with a mass of 1.008 atomic mass units (amu), combine to form a helium nucleus which has 4.0027 amu. The difference in mass of 0.0073 amu per proton is released as energy. The corresponding time scale is $t_{nuc} = 0.007Mc^2/L$, where M is the amount of mass of H that is burned and c is the velocity of light. For the sun on the main sequence, $X \approx 0.71$, $L \approx L_\odot$, and about 14% of the hydrogen is actually burned, so the estimate gives $t_{nuc} \approx 1. \times 10^{10}$ years. In subsequent evolutionary phases the sun continues to burn hydrogen but at a higher rate. For a star of 30 $M_\odot$, $L \approx 2 \times 10^5 L_\odot$, about half of the hydrogen is burned, and $t_{nuc} \approx 5 \times 10^6$ years. For helium burning, where three helium nuclei combine to produce carbon, the energy release per atomic mass unit is a factor 10 smaller than for hydrogen burning, and correspondingly, the time spent by a star in the core helium burning phase is a factor of 5–10 shorter than its main sequence lifetime, depending on the luminosity and the amount of hydrogen burning going on in a shell at the same time.

A final time scale associated with stellar evolution is the cooling time scale. For white dwarf stars that can no longer contract and also have no more nuclear fuel available, the radiated energy must be supplied by cooling of the hot interior. The electrons by this time are highly degenerate (see later discussion), so the available thermal energy is only that of the ions. The time scale can be estimated from

$$t_{cool} \approx E_{thermal}/L = 1.5 R_g T M/(\mu_A L),$$

where T is the mean internal temperature, R_g is the gas constant, and μ_A is the mean atomic weight (in amu) of the ions. For example, the cooling time of a white dwarf of 0.7 $M_\odot$ composed of carbon from the beginning of the white dwarf phase to a luminosity of $L = 10^{-3} L_\odot$ is 1.15×10^9 years. The cooling time scale also applies to substellar objects once they have contracted to their limiting radius.

B. Equation of State of a Gas

In the pre-main-sequence and main-sequence stages of the evolution of most stars, the ideal gas equation holds. The pressure of the gas is given by $P = NkT$, where k is the Boltzmann constant and N is the number of free particles per unit volume. If X, Y, and Z are the mass fractions of H, He, and heavy elements, respectively, and the gas is fully ionized, then

$$N = (2X + 0.75Y + 0.5Z)\rho/m_H,$$

where m_H is the mass of the hydrogen atom. It has been assumed that the number of particles contributed per nucleus is 2 for H, 3 for He, and $A/2$ (an approximation) for the heavy elements, where A is the atomic weight. In the outer layers of the star, N must be adjusted to take into account partial ionization. The internal energy for an ideal gas is $1.5kT$ per particle or $1.5kTN/\rho = 1.5P/\rho$ per unit mass. The equation of state can also be written $P = R_g \rho T/\mu$, where μ, the mean atomic weight per free particle, is given by $\mu^{-1} = 2X + 0.75Y + 0.5Z$ for a fully ionized gas.

Under conditions of high temperature and low density, the pressure of the radiation must be added. According to quantum theory, a photon has energy $h\nu$, where ν is the frequency and h is Planck's constant, and momentum $h\nu/c$. The radiation pressure is the net rate of transfer of momentum per unit area, normal to an arbitrarily oriented surface. Under conditions in stellar interiors where the radiation is nearly isotropic and the system is near thermodynamic equilibrium, it can be shown that the radiation pressure is given by $P_R = \frac{1}{3}aT^4$, where a, the radiation density constant, equals 7.56×10^{-15} erg cm^{-3} degree^{-4}.

At high densities an additional physical effect must be considered, the phenomenon of degeneracy. The effect is a consequence of the Pauli exclusion principle, which does not permit more than one particle to occupy one quantum state at the same time; it applies to elementary particles with half-integral spin, such as electrons or neutrons. In the case of electrons bound in an atom, the principle governs the distribution of electrons in various energy states. If the lowest energy states are filled, any electrons added subsequently must go into states of higher energy; only a discrete set of states is available. The same principle applies to free electrons. In a momentum interval $dp_x\, dp_y\, dp_z$ and in a volume element $dx\, dy\, dz$, the number of states is $(\frac{2}{h^3})\, dp_x\, dp_y\, dp_z\, dx\, dy\, dz$, where the factor of 2 arises from the two possible directions of the electron spin. Degeneracy occurs when most of these states, up to some limiting momentum p, are occupied by particles. Complete degeneracy is defined as the situation in which all states are occupied up to the limiting momentum p_0 and all higher states are empty. Under this assumption, if the particles are electrons, their pressure can be derived to be $P_e = 1.0036 \times 10^{13} (\rho/\mu_e)^{5/3}$ dyne cm^{-2}, where μ_e, the mean atomic weight per free electron, is $2/(1 + X)$ for a fully ionized gas. This formula is valid if p_o is low enough so that the electron velocities are not relativistic. In the limiting case that all electrons are moving with the velocity of light, the corresponding expression is $P_e = 1.2435 \times 10^{15} (\rho/\mu_e)^{4/3}$ dyne cm^{-2}. In either case most of the electrons are forced into such high momentum states that their pressure, defined again as the rate of transport of momentum per unit area, is much higher than the ideal gas pressure. The electrons in stellar interiors become degenerate when the density reaches 10^3–10^6 g cm^{-3}, depending on T. The protons or neutrons become degenerate only at much higher densities, about 10^{14} g cm^{-3}; these densities are characteristic of neutron stars, and it is the degenerate pressure of the neutrons that supports the star against gravity. The regions in density and temperature where ideal gas, radiation pressure, and electron degeneracy dominate in the equation of state are shown in Fig. 5.

C. Energy Sources

From the previous discussion of time scales, it is clear that only two fundamental energy sources are available

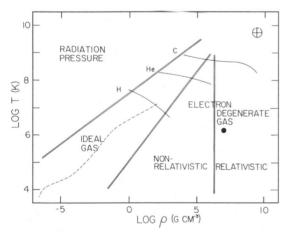

FIGURE 5 Density–temperature diagram. The double lines separate regions in which different physical effects dominate in the equation of state. Solid lines, denoted by H, He, and C, respectively, indicate the central conditions in stars that are undergoing burning of hydrogen, helium, and carbon in their cores. The dashed line represents the structure of the present sun, while the solid dot gives typical central conditions in a white dwarf of 0.8 $M_\odot$. The circled cross gives central conditions in a star of 25 $M_\odot$ just before the collapse of the iron core.

to a star. Gravitational energy can be released either on a dynamical or a Kelvin-Helmholtz time scale. Nuclear energy in moderate-mass stars is produced by conversion of hydrogen to helium or by conversion of helium to carbon and oxygen. Only in the most massive stars do nuclear reactions proceed further to the synthesis of the heavier elements up to iron and nickel. During cooling phases stars draw on their thermal energy, which, however, has been produced in the past as a consequence of nuclear reactions or gravitational contraction. As the star evolves, the energy is taken up by radiation, heating of the star, expansion, neutrino emission, mass loss, ionization of atoms, and dissociation of nuclei. Here, we discuss in somewhat more detail the nuclear energy source.

The nuclear processes involve reactions between charged particles. At the temperatures characteristic of nuclear burning regions in stars, the gas may safely be assumed to be fully ionized. The Coulomb repulsive force between particles of like charge presents a barrier for reactions between the ions. Particles must come within 10^{-13} cm of each other before the strong (but short-range) nuclear attractive force overcomes the Coulomb force. The energy that is required for particles to pass through the Coulomb barrier is $E_{\text{coulomb}} = Z_1 Z_2 e^2 / r$, where r is the required separation of 10^{-13} cm, e is the electronic charge, and Z_1 and Z_2 are, respectively, the charges on the two particles involved, in units of the electronic charge. Evidently, favorable conditions for reactions will occur most readily for particles of small charge, but even for two protons $E_{\text{coulomb}} = 1000$ keV. The only source for the energy is the thermal energy of the particles, which, however, in the temperature range around 10^7 K is only $\frac{3}{2}kT \approx 1$ keV. It would seem that nuclear reactions under these conditions would not be possible.

However, three factors contribute to a small but nonzero probability of reaction. (1) The particles have a Maxwell velocity distribution, so that a small fraction of them have thermal energies much higher than the average. (2) According to the laws of quantum mechanics, there is a small probability that a particle with much less energy than required can actually "tunnel" through the Coulomb barrier. (3) A star has a very large number of particles, so that even if an individual particle has a very small probability of reacting, enough reactions do occur to supply the required energy. It turns out that in main-sequence stars particles with energies of about 15–30 keV satisfy the two requirements of existing in sufficient numbers and having enough energy to have a reasonable change of passing through the Coulomb barrier.

At temperatures of 1 to 3×10^7 K, which characterize the interiors of most main-sequence stars, only reactions between particles of low atomic number Z need be considered, usually those involving protons. However, even if the Coulomb barrier is overcome, there is still only a small probability that a nuclear reaction will occur. This nuclear probability varies widely from one reaction to another, and it can be determined theoretically for only the simplest systems; in general, these probabilities must be measured for each reaction in laboratory experiments at low energy. An intense effort was made during the period 1960–1980 to make measurements of reasonable accuracy of all the nuclear reactions that are important in stars. Nevertheless, in some cases measurements cannot be made at the low energies appropriate for main-sequence stars, and the probabilities have to be extrapolated from measurements at higher energy. The group headed by W. A. Fowler at the California Institute of Technology led in the task of carrying out these difficult measurements.

Two reaction sequences have been identified that result in the conversion of four protons into one helium nucleus, the proton-proton (pp) chains and the CNO cycle. The reactions of the main branch of the pp chains (PP I) are as follows:

$$^1\text{H} + {}^1\text{H} \rightarrow {}^2\text{D} + e^+ + \nu \tag{1}$$

$$^2\text{D} + {}^1\text{H} \rightarrow {}^3\text{He} + \gamma \tag{2}$$

$$^3\text{He} + {}^3\text{He} \rightarrow {}^4\text{He} + {}^1\text{H} + {}^1\text{H} \tag{3}$$

The symbol e^+ denotes a positron; ν is the neutrino; and ^2D is the deuterium nucleus, composed of one proton and one neutron. The positron immediately reacts with an electron, with the annihilation of both and the production of energy. The total energy production of the sequence is 26.7 MeV or 4.27×10^{-5} erg, corresponding to the mass difference between four protons and one ^4He nucleus, times c^2. Each sequence produces two neutrinos, which immediately escape from the star, as they interact only very weakly with matter, carrying with them a certain fraction of the energy. The remainder of the energy is deposited locally in the star, typically in the form of gamma rays.

Reaction (1) turns out to be very improbable, as the conversion of a proton into a neutron requires a beta decay. The reaction rate is so slow that it cannot be measured experimentally and must be calculated theoretically. At the center of the sun, a proton has a mean lifetime of almost 10^{10} years before it reacts with another proton; this reaction controls the rate of energy production, as subsequent reactions are more rapid. The neutrino in Reaction (1) carries away an average energy of 0.265 MeV. Once ^2D is formed, it immediately (within 1 sec) captures another proton to form ^3He. This reaction can be measured in the laboratory down to energies close to those where reactions can occur in stars. Any ^2D initially present in the star will be burned by this reaction at temperatures around 10^6 K. When Reactions (1) and (2) have occurred twice,

the two resulting nuclei of ^3He combine to form ^4He and two protons. There is perhaps a 10% uncertainty in this reaction rate and in most others in the pp chains.

The second branch of the pp chains (PP II) occurs when the ^3He nucleus produced in Reaction (2) reacts with ^4He. The sequence of reactions is then

$$^3\text{He} + {}^4\text{He} \rightarrow {}^7\text{Be} + \gamma \tag{4}$$

$$^7\text{Be} + e^- \rightarrow {}^7\text{Li} + \nu \tag{5}$$

$$^7\text{Li} + {}^1\text{H} \rightarrow {}^4\text{He} + {}^4\text{He} \tag{6}$$

Under conditions of the solar interior, Reaction (4) proceeds at approximately one-sixth of the rate of Reaction (3), and as the temperature increases it becomes relatively more important. Reaction (5) involves an electron capture on the ^7Be and the production of a neutrino with energy 0.86 MeV (90% of the time) or 0.38 MeV (10% of the time). The ^7Li nucleus immediately captures a proton and produces two nuclei of ^4He. This reaction also results in the destruction of any ^7Li that may be present at the time of the star's formation, since it becomes effective at temperatures of 2.8×10^6 K or above. Note that the second branch of the pp chains has an average neutrino loss of about 1.1 MeV, but that otherwise the net result is the same: the conversion of four protons to a ^4He nucleus.

The third branch of the pp chains (PP III) occurs if a proton, rather than an electron, is captured by ^7Be:

$$^7\text{Be} + {}^1\text{H} \rightarrow {}^8\text{B} + \gamma \tag{7}$$

$$^8\text{B} \rightarrow e^+ + \nu + {}^8\text{Be} \rightarrow {}^4\text{He} + {}^4\text{He} \tag{8}$$

Less than 0.1% of pp-chain completions in the sun occur through Reactions (7) and (8); however, they are very important because the neutrino produced has an average energy of 6.7 MeV, high enough to be detectable in all current solar neutrino experiments.

The total rate of energy generation by the pp chains involves a complicated calculation of the rates of all the reactions given above, which are functions of the temperature, density, and concentration of the species involved. If T_6 is the temperature in units of 10^6 K, then under the assumption of equilibrium, that is, the abundances of the intermediate species ^2D, ^3He, ^7Be, and ^7Li reach steady state, the energy generation rate for the pp chains can be written

$$\epsilon_{\text{pp}} = 2.38 \times 10^6 \psi g_{11} \rho X^2 T_6^{-2/3}$$
$$\times \exp(-33.80 T_6^{-1/3}) \text{erg g}^{-1} \text{sec}^{-1}, \tag{9}$$

where ρ is the density, X is the fraction of hydrogen by mass, and

$$g_{11} = 1 + 0.0123 \, T_6^{1/3} + 0.0109 \, T_6^{2/3} + 0.0009 \, T_6.$$

The function ψ is a correction factor, between 1 and 2, that accounts for (1) the increase by a factor of 2 in the energy production rate, with respect to that for PP I, that occurs when the reactions go through PP II or PP III, and (2) the different neutrino energies in the three chains. For the center of the present sun, $\psi \approx 1.5$. For temperatures lower than and densities higher than those of the present solar center, the effect of electron shielding as well as departure from equilibrium may have to be taken into account.

Protons can also interact with the CNO nuclei, but because the Coulomb barrier is higher these reactions become more important than the pp chains only at temperatures above 2×10^7 K. The reactions are the following:

$$^{12}\text{C} + {}^1\text{H} \rightarrow {}^{13}\text{N} + \gamma \qquad (10)$$

$$^{13}\text{N} \rightarrow {}^{13}\text{C} + e^+ + \nu \qquad (11)$$

$$^{13}\text{C} + {}^1\text{H} \rightarrow {}^{14}\text{N} + \gamma \qquad (12)$$

$$^{14}\text{N} + {}^1\text{H} \rightarrow {}^{15}\text{O} + \gamma \qquad (13)$$

$$^{15}\text{O} \rightarrow {}^{15}\text{N} + e^+ + \nu \qquad (14)$$

$$^{15}\text{N} + {}^1\text{H} \rightarrow {}^{12}\text{C} + {}^4\text{He} \qquad (15)$$

The net effect is the conversion of four protons into a helium nucleus, two positrons (which annihilate), and two neutrinos (which carry away energies of 0.71 and 1.0 MeV, respectively). The CNO nuclei simply act as catalysts; however, their relative abundances change as a result of the operation of the cycle. Note from Table I that ^{12}C is considerably more abundant than ^{14}N in the solar system. However, the rate of Reaction (10), which uses up ^{12}C, under stellar conditions is about 100 times faster than that of Reaction (13), which uses up ^{14}N. All other reactions are much faster than these two. The reaction chain tends to reach equilibrium, in which each CNO nuleus is produced as fast as it is destroyed. The equilibrium is obtained only when most of the ^{12}C and other participating nuclei are converted to ^{14}N. This change in relative abundances could be observed if the layers in which the reactions occur could later be mixed to the surface of the star.

A secondary branch of the CNO cycle affects the abundance of ^{16}O. Starting at ^{15}N, the reactions are

$$^{15}\text{N} + {}^1\text{H} \rightarrow {}^{16}\text{O} + \gamma \qquad (16)$$

$$^{16}\text{O} + {}^1\text{H} \rightarrow {}^{17}\text{F} + \gamma \qquad (17)$$

$$^{17}\text{F} \rightarrow {}^{17}\text{O} + e^+ + \nu \qquad (18)$$

$$^{17}\text{O} + {}^1\text{H} \rightarrow {}^{14}\text{N} + {}^4\text{He} \qquad (19)$$

This branch accounts for only 0.1% of the completions, but the ^{16}O is generated less rapidly than it is converted into ^{14}N by Reactions (17)–(19), so that when the branch comes into equilibrium the abundance ratio O/N will be reduced. The overall effect of the CNO cycle, apart from its energy generation, is the conversion of 98% of the CNO isotopes into ^{14}N. Once the cycle reaches equilibrium, the energy generation can be calculated by the rate of the slowest reaction in the main chain (Reaction 13) multiplied by the energy released over the whole cycle (24.97 MeV, after neutrino losses):

$$\epsilon_{\text{CNO}} = 8.67 \times 10^{27} g_{14,1} \rho X X_{\text{CNO}} T_6^{-2/3}$$

$$\times \exp\left(-152.28 T_6^{-1/3}\right) \text{erg g}^{-1} \text{sec}^{-1}, \qquad (20)$$

where X_{CNO} is the total mass fraction of all CNO isotopes and

$$g_{14,1} = 1 + 0.0027\, T_6^{1/3} - 0.00778\, T_6^{2/3} - 0.000149\, T_6.$$

At $T_6 \approx 25$, the energy generation goes approximately as T^{17}.

The synthesis of elements heavier than helium has proved in the past to be a considerable problem, primarily because there is no stable isotope of atomic mass 5 or 8. Thus, if two ^4He nuclei react to produce ^8Be, the nucleus will immediately decay back to He. If a proton reacts with ^4He, the result is ^5Li, which also is unstable. Helium burning must, in fact, proceed through a three-particle reaction, which can be represented as follows:

$$^4\text{He} + {}^4\text{He} \leftrightarrow {}^8\text{Be} \qquad (21)$$

$$^8\text{Be} + {}^4\text{He} \leftrightarrow {}^{12}\text{C} + \gamma + \gamma \qquad (22)$$

Although the ^8Be is unstable, its decay is not instantaneous. Under suitable conditions, e.g., $T = 10^8$ K and $\rho > 10^5$ g cm^{-3}, it was shown by E. Salpeter that a sufficient equilibrium abundance exists so that a third ^4He nucleus can react with it [Reaction (22)]. It was also predicted by F. Hoyle that in order for this triple-alpha process to proceed at a significant rate, Reaction (22) must be resonant; that is, there must be an excited nuclear state accessible in the range of stellar energies whose presence greatly enhances the probability that the reaction will occur. This resonance was later found in experiments performed at the Kellogg Radiation Laboratory at the California Institute of Technology. The amount of energy produced per reaction is 7.275 MeV, or 0.606 MeV per atomic mass unit, a factor of 10 less than in hydrogen burning (per atomic mass unit). An expression for the rate of energy production is

$$\epsilon_{3\alpha} = 5.09 \times 10^{11} \rho^2 Y^3 T_8^{-3}$$

$$\times \exp\left(-44.027 T_8^{-1}\right) \text{erg g}^{-1} \text{sec}^{-1}, \qquad (23)$$

where $T_8 = T/10^8$ K, Y is the mass fraction of ^4He, and electron screening has not been included. This reaction has a very steep temperature sensitivity, with $\epsilon \propto T^{40}$ at $T_8 = 1$. A further helium burning reaction, which occurs at slightly higher temperatures than the triple-alpha process, is

$$^{12}\text{C} + {}^4\text{He} \rightarrow {}^{16}\text{O} + \gamma, \tag{24}$$

with an energy production of 7.162 MeV per reaction and an uncertain reaction rate. Reactions (21), (22), and (24) probably produce most of the carbon and oxygen in the universe.

D. Energy Transport

The transport of energy outward from the interior of a star to its surface depends in general on the existence of a temperature gradient. Heat will be carried by various processes from hotter regions to cooler regions; the processes that need to be considered include (1) radiation transport, (2) convective transport, and (3) conductive transport. Neutrino transport must be considered only in exceptional circumstances when the density is above 10^{10} g cm^{-3}; at lower densities, neutrinos produced in stars simply escape directly without interacting with matter. In each case a relation must be found between the energy flux F_r, defined as the energy flow per unit area per unit time at a distance r from the center of the star, and the temperature gradient dT/dr.

Energy transport by radiation depends on the emission of photons in hot regions of the star and absorption of them in slightly cooler regions. The radiation field may be characterized by the *specific intensity* I_ν, which is defined so that $I_\nu \cos\theta \, d\nu \, d\omega \, dt \, dA$ is the energy carried by a beam of photons across an element of area dA in time dt in frequency interval ν to $\nu + d\nu$ into an element of solid angle $d\omega$, in a direction inclined by $\cos\theta$ to the normal to dA. The equation of transfer shows how the intensity of a beam is changed as it interacts with matter. The *mass emission coefficient* j_ν is defined so that $j_\nu \, \rho \, dV \, d\nu \, d\omega \, dt$ is the energy emitted by the volume element dV into $d\omega$ in time dt in the frequency range $d\nu$. The corresponding *mass absorption coefficient* κ_ν is defined so that the energy absorbed in the same intervals is $\kappa_\nu \, \rho \, I_\nu \, dV \, d\nu \, d\omega \, dt$. If we consider both the absorption and the emission of the radiation passing through a cylinder with length ds and cross section dA, the equation of transfer becomes

$$\frac{dI_\nu}{ds} = -\kappa_\nu \rho I_\nu + j_\nu \rho. \tag{25}$$

Suppose that the direction ds is inclined by an angle θ to the radial direction in the star so that the projected distance element $dr = ds \cos\theta$. Also, define the optical depth τ_ν by

$$d\tau_\nu = -\kappa_\nu \rho dr \tag{26}$$

and the equation of transfer becomes

$$\cos\theta \frac{dI_\nu}{d\tau_\nu} = I_\nu - \frac{j_\nu}{\kappa_\nu}. \tag{27}$$

In the stellar interior the mean free path of a photon before it is absorbed is only 1 cm or less. Thus, a typical photon is absorbed at practically the same temperature as it is emitted. These conditions are so close to strict thermodynamic equilibrium, where the specific intensity is given by the Planck function $B_\nu(T)$, that the ratio j_ν/κ_ν can be shown to be the same as it is in strict thermodynamic equilibrium, namely, $j_\nu/\kappa_\nu = B_\nu(T)$. The equation of transfer in stellar interiors can thus be expressed as

$$\cos\theta \frac{dI_\nu}{d\tau_\nu} = I_\nu - B_\nu(T). \tag{28}$$

This equation can be integrated over all frequencies and solved for the flux in terms of the temperature gradient for conditions appropriate to the stellar interior, that is, where the temperature change is negligible over the mean free path of a photon, to give

$$F_r = -(4ac)(3\kappa\rho)^{-1}T^3(dT/dr), \tag{29}$$

where c is the velocity of light and a is the radiation density constant. Here, κ is the absorption coefficient averaged over frequency according to the so-called *Rosseland mean*:

$$\frac{1}{\kappa} = \frac{\int_0^\infty \frac{dB_\nu(T)/dT \, d\nu}{\kappa_{\nu,a}[1 - \exp(-h\nu/kT)] + \kappa_{\nu,s}}}{\int_0^\infty dB_\nu(T)/dT \, d\nu}. \tag{30}$$

Here, $\kappa_{\nu,a}$ refers to processes of true absorption, which have to be corrected for induced emission, and $\kappa_{\nu,s}$ refers to scattering processes. Equation (29) is known as the diffusion approximation for radiation transfer, an appropriate nomenclature because a photon is absorbed almost immediately after it is emitted, so that an enormous number of absorptions, re-emissions, and scatterings must occur before the energy of a photon as transmitted to the surface. The time required for energy to diffuse in this manner from the center of the sun to its surface is about 10^5 years. The quality of the radiation changes during this process. As the photons diffuse to lower temperatures, their energy distribution corresponds closely to the Planck distribution at the local T, because matter and radiation are well coupled. Thus, the gamma rays produced by nuclear reactions at the center are gradually transformed into optical photons by the time the energy reaches the surface.

As Eq. (29) indicates, the energy transport in a radiative star is controlled by the opacity κ. Numerous atomic processes contribute to this quantity, and in general, the structure of a star can be calculated only with the aid of detailed tables of the opacity, calculated as a function of ρ, T, and the chemical composition. Starting at the highest temperatures characteristic of the stellar interior and proceeding to lower temperatures, the main processes are (1) electron scattering, also known as Thomson scattering, in which a proton undergoes a change in direction but no change in frequency during an encounter with a free electron; (2) free-free absorption, in which a photon is absorbed by a free electron in the vicinity of a nucleus, with the result that the photon is lost and the electron increases its kinetic energy; (3) bound-free absorption on metals, also known as photoionization, in which the photon is absorbed by an atom of a heavy element (e.g., iron) and one of the bound electrons is removed; (4) bound-bound absorption of a heavy element, in which the photon induces an upward transition of an electron from a lower quantum state to a higher quantum state in the atom; (5) bound-free absorption on H and He, which generally occurs near stellar surfaces where these elements are being ionized; (6) bound-free and free-free absorption by the negative hydrogen ion H−, which forms in stellar atmospheres in layers where H is just beginning to be ionized (example, the surface of the sun); (7) bound-bound absorptions by molecules, which can occur only in the atmospheres of the cooler stars ($T_{eff} < 4000$ K, although even the sun shows a few molecular features in its spectrum); and (8) absorption by dust grains, which can occur in the early stages of protostellar evolution and possibly in the atmospheres of brown dwarfs, at temperatures below the evaporation temperature of grains (1400–1800 K).

The Thomson scattering from free electrons (in units of cm^2 g^{-1}) is given by $\kappa = 0.2(1 + X)$. For stars this process is important generally in regions above the uppermost double line in Fig. 5, corresponding to the interiors of massive stars. The free-free absorptions and the bound-free absorptions on heavy elements have the approximate dependence $\kappa \propto \rho T^{-3.5}$. As one moves outward in a star from higher to lower T, the number of bound electrons per atom increases and, correspondingly, κ increases. A maximum in κ occurs in the range 10^4 K $< T < 10^5$ K depending on density; this range corresponds to the zones where H and He, the most abundant elements, undergo ionization. Below this range in T, the above dependence is no longer valid, and κ drops rapidly with decreasing T, reaching a mimimum of about 10^{-2} at $T = 2000$ K. At lower T, where grains exist, the opacities are higher, on the order of 1 cm^2 g^{-1}.

Under certain conditions, the temperature gradient given by Eq. (29) can be unstable, leading to the onset of convection. Suppose that a small element of material in a star is displaced upward from its equilibrium position. It may reasonably be assumed that the element remains in pressure equilibrium with its surroundings. Thus, if the material has the equation of state of an ideal gas, an upwardly displaced element with density less than that of the surroundings will have a temperature greater than that of the surroundings. If, as the element rises, its density decreases more rapidly than that of the surroundings, it will continue to feel a buoyancy force and will continue to rise.

In this case the layer is unstable to convection, and heat is transported outward by the moving elements themselves. If, however, the density of the upward-moving element decreases less rapidly than that of the surroundings, the densities soon will be equalized, there will be no buoyancy force, and the layer will be stable.

The condition for occurrence of convection in a star can be expressed in another form. If the temperature gradient in a layer, calculated under the assumption of radiation transfer, is steeper than the adiabatic temperature gradient, then convection will occur:

$$|dT/dr| = 3\kappa \rho F_r (4acT^3)^{-1} > |dT/dr|_{ad}. \qquad (31)$$

A layer with high opacity, or opacity rapidly increasing inward, is therefore likely to be unstable; this situation can occur near the surface layers of cool stars. A layer with a high rate of nuclear energy generation in a small volume is also likely to be unstable because of high F_r; this situation is likely to occur in the cores of stars more massive than the sun, where the temperature-sensitive CNO cycle operates.

Convection involves complicated, turbulent motions with continuous formation and dissolution of elements of all sizes. The existence of convection zones is important in the calculation of stellar structure because the overturning of material in almost all phases of evolution is much more rapid than the evolution time, so that the entire zone can be assumed to be mixed to a uniform chemical composition at all times. It is also important for heat transport, which is accomplished mainly by the largest elements. No satisfactory analytic theory of convection exists, although three-dimensional numerical hydrodynamical simulations, for example, of the outer layers of the sun, have been able to attain spatial resolution high enough so that they can reproduce many observed features. For long-term stellar evolution calculations, a very simplified "mixing length" theory is employed. It is assumed that a convective element forms, travels one mixing length vertically, and then dissolves and releases its excess energy to the surroundings. The mixing length is approximated by αH, where α is a parameter of order unity and H is the distance over which the pressure drops by a factor e. By use of this theory the average velocity of an element, the excess thermal energy, and the convective flux can be estimated. In most situations in a stellar interior the estimates show that it is adequate to assume that $dT/dr = (dT/dr)_{ad}$. Although an excess in dT/dr over the adiabatic value is required for convection to exist, convection is efficient enough to carry the required energy flux even if this excess is negligibly small. There is a small uncertainty in stellar models because of possible overshooting, and resulting mixing, of convective elements beyond the boundary of a formal convection zone; otherwise, the uncertainty in the use of the mixing length theory is limited to the surface layers of stars, where in fact the actual $|dT/dr|$ in a star can be significantly greater than $|dT/dr|_{\text{ad}}$. The parameter α can be calibrated by comparison of calculated mixing lengths with the size of the granular elements in the solar photosphere that represent solar convective elements. Also, the radius of a theoretical model of the sun, which is sensitive to α, can be compared with and adjusted to the observed radius. In this manner it has been determined that the parameter α should be in the range 1–2. Calibration can be also made for stars in other phases of evolution by the use of numerical hydrodynamical models; however, it is often assumed that stars evolve with the same α that has been obtained for the sun.

Conduction of heat by the ions and electrons is, in general, inefficient in stellar interiors because the density is high enough that the mean free path of the particles is small compared with the photon mean free path. In the interiors of white dwarfs and the evolved cores of red giants, however, the electrons are highly degenerate, their mean free paths are very long, and conduction becomes a more efficient mechanism than radiative transfer. The flux is related to the temperature gradient by $F_r = -K \, dT/dr$, where K, the conductivity, is calculated by a complicated theory involving the velocity, collision cross section, mean free path, and energy carried by the electrons. Numerical tables are constructed for use in stellar structure calculations.

E. Basic Equilibrium Conditions

During most phases of evolution, a star is in hydrostatic equilibrium, which means that the force of gravity on a mass element is exactly balanced by the difference in pressure on its upper and lower surface. In spherical symmetry, an excellent approximation for most stars, this condition is written

$$\frac{\partial P}{\partial M_r} = -\frac{GM_r}{4\pi r^4}, \qquad (32)$$

where M_r is the mass within radius r, ρ is the density, and P is the pressure. A second equilibrium condition is the mass conservation equation in a spherical shell, which simply states that

$$\frac{\partial r}{\partial M_r} = \frac{1}{4\pi r^2 \rho}. \qquad (33)$$

The third condition is that of thermal equilibrium, which refers to the situation where the star is producing enough energy by nuclear processes to exactly balance the loss of energy by radiation at the surface. Then,

$$L = \int_0^M \epsilon \, dM_r, \qquad (34)$$

where M is the total mass and ϵ is the nuclear energy generation rate per unit mass after subtraction of neutrino losses. This condition holds on the main sequence, but in many stages of stellar evolution it does not, because the star may be expanding, contracting, heating, or cooling. The more general equation of conservation of energy, effectively the first law of thermodynamics, must then be used:

$$\frac{\partial L_r}{\partial M_r} = \epsilon - \frac{\partial E}{\partial t} - P\frac{\partial V}{\partial t}, \qquad (35)$$

where $V = 1/\rho$; E is the internal energy per unit mass; and $L_r = 4\pi r^2 F_r$, the total amount of energy per unit time crossing a spherical surface at radius r. If $\epsilon = 0$, the star contracts and obtains its energy from the third term on the right-hand side.

To obtain a detailed solution for the structure and evolution of a star, one must solve Eqs. (32), (33), and (35) along with a fourth differential equation which describes the energy transport. Equation (29) for radiation transport can be re-written, with M_r as the independent variable, in a form which can be used for all three types of transport:

$$\frac{\partial T}{\partial M_r} = -\frac{GM_r T}{4\pi r^4 P}\nabla, \qquad (36)$$

where, if the energy transport is by radiation,

$$\nabla = \nabla_{\text{rad}} = \frac{3}{16\pi Gac}\frac{\kappa L_r P}{M_r T^4} = \left(\frac{\partial \log T}{\partial \log P}\right)_{\text{rad}}. \qquad (37)$$

If transport is by conduction, an equivalent "conductive opacity" κ_{cond} can be obtained from the conductivity K. However, each point in the star must be tested to see if the condition for convection is satisfied, and if so, ∇ in Eq. (36) is replaced by the adiabatic gradient $\nabla_{\text{ad}} = (\partial \log T/\partial \log P)_{\text{ad}}$ in the interior or by a "true" ∇ obtained from mixing-length theory in the surface layers. The four differential equations are supplemented by expressions for the equation of state (see Section III.B), nuclear energy generation (see Section III.C), and opacity κ (see Section III.D) as functions of ρ, T, and chemical composition.

Four boundary conditions must be specified. At the center the conditions $r = 0$ and $L_r = 0$ at $M_r = 0$ are applied. At the surface ($M_r = M$) the photospheric boundary conditions can be used for approximate calculations:

$$L = 4\pi R^2 \sigma T_{\text{eff}}^4 \quad \text{and} \quad \kappa P = \frac{2}{3}g, \qquad (38)$$

where g is the surface gravity. For detailed comparison with observations, however, model atmospheres are used as surface boundary conditions and are joined to the interior calculation a small distance below the photosphere.

To start an evolutionary calculation, one specifies an initial model by giving its total mass M and the distribution of chemical composition with mass fraction. A typical starting point is the pre-main-sequence phase where the composition is uniform. A sequence of models is calculated, separated by small intervals of time. If nuclear burning is important, the change of composition caused by reactions (e.g., conversion of H to He) is calculated at each layer. The system of equations is solved numerically, generally by a method developed by L. G. Henyey, in which first-order corrections to a trial solution are determined simultaneously for all variables at all grid points; the process is repeated until the corrections become small.

A final important equilibrium condition is the virial theorem, which, for a non-rotating, non-magnetic spherical star in hydrostatic equilibrium, can be written

$$2T_i + W = 0. \tag{39}$$

Here, W is the gravitational energy, $-qGM^2/R$, where R is the total radius and q is a constant of order unity that depends on the mass distribution, and T_i is the total internal energy $\int_0^M E \, dM_r$. This expression can be used to estimate stellar internal temperatures for the case of an ideal gas. Then $T_i = 1.5(R_g/\mu)TM$, where T is the mean temperature, and $T \approx GM\mu/(3RR_g) \approx 5 \times 10^6$ K for the case of the sun. This expression also shows that a contracting, ideal-gas star without nuclear sources heats up, with T increasing approximately as $1/R$.

IV. STELLAR EVOLUTION BEFORE THE MAIN SEQUENCE

Early stellar evolution can be divided into three stages : star formation, protostar collapse, and slow (Kelvin-Helmholtz) contraction. The characteristics of these three stages are summarized in Table II. As the table indicates, there is a vast difference in physical conditions between the time of star formation in an interstellar cloud and the time when a star arrives on the main sequence and begins to burn hydrogen. An increase in mean density by 22 orders of magnitude and in mean temperature by 6 orders of magnitude occurs. In the star formation and pro-

tostar collapse stages, it is not sufficient to assume that the object is spherical, and a considerable variety of physical processes must be considered. The problem involves solution of the equations of hydrodynamics, including rotation, magnetic fields, turbulence, molecular chemistry, gravitational collapse, and radiative transport of energy. Some two- and three- dimensional hydrodynamic simulations have been performed for the star formation phase and for the protostar collapse phase, with the aim of investigating the following, as yet unsolved, problems: (1) What initiates collapse? (2) What is the rate and efficiency of the conversion of interstellar matter into stars? (3) What determines the distribution of stars according to mass? (4) What is the mechanism for binary formation, and what determines whether a star will become a member of a double or multiple star system or a single star?

A. Star Formation

It is clear that star formation is limited to regions of unusual physical conditions compared with those of the interstellar medium on the average. The only long-range attractive force to form condensed objects is the gravitational force. However, a number of effects oppose gravity and prevent contraction, including gas pressure, turbulence, rotation, and the magnetic field. The chemical composition of the gas, the degree of ionization or dissociation, the presence or absence of grains (condensed particles of ice, silicates, carbon, or iron with characteristic size 5×10^{-5} cm), and the heating and cooling mechanisms in the interstellar medium all have an important influence on star formation. The regions where star formation is observed to occur is in the molecular clouds, where they can form in a relatively isolated mode, as in the Taurus region or, more commonly, in clusters (Fig. 6).

We consider first only the effects of thermal gas pressure and gravity in an idealized spherical cloud with uniform density ρ, uniform temperature T, and mass M. The self-gravitational energy is $W = -0.6GM^2/R$, where R is the radius. The internal energy is $T_i = 1.5R_g TM/\mu$, where μ is the mean molecular weight per particle. Collapse will be

TABLE II Major Stages of Early Stellar Evolution[a]

Phase	Size (cm)	Observations	Density (g cm^{-3})	Internal temperature (K)	Time (year)
Star formation	10^{20}–10^{17}	Radio	10^{-22}–10^{-19}	10	10^7
Protostellar collapse	10^{17}–10^{12}	Infrared	10^{-19}–10^{-3}	10–10^6	10^6
Slow contraction	10^{12}–10^{11}	Optical	10^{-3}–1.0	10^6–10^7	4×10^{7b}

[a]Reproduced with permission from Bodenheimer, P. (1983). "Protostar collapse." *Lect. Appl. Math.* **20,** 141. ©American Mathematical Society.

[b]For 1 solar mass.

FIGURE 6 The Trifid Nebula in Sagittarius (M20, NGC 6514), a region of recent star formation (Shane 120-in. reflector). (Lick Observatory photograph.)

possible roughly when $|W| > T_i$; that is, when the cloud is gravitationally bound, a criterion that has been verified by numerical calculations. Thus, for collapse to occur the radius must be less than $R_J = 0.4 G M \mu/(R_g T)$, known as the Jeans length. By eliminating the radius in favor of the density, we obtain the Jeans mass, which is the minimum mass a cloud with density ρ and temperature T must have for collapse to occur:

$$M_J = \left(\frac{2.5 R_g T}{\mu G}\right)^{3/2} \left(\frac{4}{3}\pi\rho\right)^{-1/2}$$

$$= 8.5 \times 10^{22} \left(\frac{T}{\mu}\right)^{3/2} \rho^{-1/2} \text{g}. \quad (40)$$

This condition turns out to be quite restrictive. A typical interstellar cloud of neutral H has $T = 50$ K, $\rho \approx 1.7 \times 10^{-23}$ g cm^{-3}, and $\mu \approx 1$, and the corresponding $M_J = 3600\, M_\odot$. The actual masses are far below this value, so the clouds are not gravitationally bound. On the other hand, typical conditions in an observed molecular cloud are $T = 10$ K, $\rho \approx 1.7 \times 10^{-21}$ g cm^{-3}, and $\mu \approx 2$, with $M_J \approx 10\, M_\odot$. Fragmentation into stellar mass objects seems quite possible.

However, this analysis must be modified to take into account rotation and magnetic fields. The angular momentum problem can be stated in the following way. The angular momentum per unit mass of a typical molecular cloud with mass $10^4\, M_\odot$ and a radius of a few parsecs is 10^{24} cm^2 sec^{-1}. The rotational velocities of young stars that have

just started their pre-main-sequence contraction indicate that their $J/M \approx 10^{17}$ cm^2 sec^{-1}. Evidently, some fragments of dark cloud material must lose 7 orders of magnitude in J/M before they reach the stellar state. The angular momentum reduction may occur at several different evolutionary stages and by different physical processes. During the star formation stage, it has been proposed that if the matter is closely coupled to the magnetic field, the twisting of the field lines can generate Alfvén waves that would transfer angular momentum from inside the cloud to the external medium. It has been estimated that significant reduction of angular momentum could take place in 10^6–10^7 years. Once the density increases to about 10^{-19} g cm^{-3}, the influence of the magnetic force on the bulk of the matter becomes relatively small, because the density of charged particles becomes very low. However, observations of rotational motion at this density indicate that J/M has already been reduced to $\approx 10^{21}$ cm^2 sec^{-1}. These dense regions, known as *molecular cloud cores*, have sizes of about 0.1 pc and temperatures of about 10 K; they are observed in the radio region of the spectrum, often through transitions of the ammonia molecule. They are the regions where the onset of protostellar collapse is likely to occur.

However, the magnetic force also opposes collapse, and a magnetic Jeans mass can be estimated by equating the magnetic energy of a cloud with its gravitational energy:

$$M_{J,m} \approx 10^3\, M_\odot \left(\frac{B}{30\,\mu G}\right) \left(\frac{R}{2\,\text{pc}}\right)^2, \quad (41)$$

where B is the magnetic field. In typical molecular cloud regions, with $\rho \approx 10^{-21}$ g cm^{-3}, $R \approx 2$ pc, and observed $B \approx 30\,\mu G$, magnetic effects are seen to be much more significant than thermal effects in preventing collapse. However, in the cores, the observed field, although not accurately determined, is still $\approx 30\,\mu G$ and the radius is only 0.1 pc, giving a value of $M_{J,m} \approx 2.5\, M_\odot$, while $M_J \approx 1\, M_\odot$. The cores actually have masses of a few solar masses, so, again, they are likely regions to collapse. The value of B does not increase in the cores as compared with the average cloud material because during the compression the predominantly neutral matter is able to drift relative to the field lines.

There are actually two scenarios which could explain how collapse is initiated. The first is based on the discussion above: molecular cloud material gradually diffuses with respect to magnetic field lines until a situation is reached, in a molecular cloud core, where the core mass is about the same as both the thermal and the magnetic Jeans masses. Rotation is not important at this stage because of magnetic braking, and collapse can commence. Of course, if angular momentum is conserved, rotational effects will soon become very important as collapse proceeds. The second picture, known as *induced* star formation, relies on a more impulsive event, such as the collision between two clouds or the sweeping of a supernova shock across a cloud, to initiate collapse. The resulting compression can, under the right conditions, lead to more rapid diffusion of the magnetic field, more rapid decay of turbulence, and enhanced cooling of the cloud, thereby reducing the magnetic, turbulent, and thermal energies, respectively, and increasing the likelihood of collapse.

B. Protostar Collapse

Once collapse starts, the subsequent evolution can be divided into an isothermal phase, an adiabatic phase, and an accretion phase. A typical initial condition is a cloud core with radius 2×10^{17} cm, $T = 10$ K, $M = 2\,M_\odot$, and mean density $\approx 1 \times 10^{-19}$ g cm^{-3} (cloud cores are actually observed to be somewhat centrally condensed). During the isothermal phase, the cloud is optically thin to its own infrared radiation. It tends to heat by gravitational compression, but the cooling radiation generated by gas-grain collisions rapidly escapes the cloud, and the temperature remains at about 10 K. The pressure cannot increase rapidly enough to bring the cloud to hydrostatic equilibrium, and instability to collapse continues. Numerical simulations show that even if the cloud initially has uniform density, a density gradient is soon set up the propagation of a rarefaction wave inward from the outer boundary; soon the denser central regions are collapsing much faster than the outer regions, and the protostar becomes highly condensed toward the center with a density distribution close to $\rho \propto r^{-2}$. During this process the thermal Jeans mass (Eq. 40) rapidly decreases with the increase in density. Thus, in principle, fragmentation of the cloud into smaller masses can occur. Calculations that include rotation show that the cloud tends to flatten into a disklike structure, and once it has become sufficiently flat, the central region of the cloud can break up into orbiting subcondensations with orbital specific angular momentums of 10^{20}–10^{21} cm^{-2} sec^{-1}, comparable to those of the wider binary systems (Fig. 7). That fragmentation during the protostar collapse is a reasonable mechanism for binary formation is supported by the observational evidence that many, if not most, very young stars are already in binary systems at the early phases of the pre-main-sequence contraction, with binary frequency similar to that observed for stars on the main sequence. A few binary protostars have also been detected. In the fragmentation process, each of the fragments still retains some angular momentum of spin and is unstable to collapse on its own, but most of the spin angular momentum of the original cloud goes into the orbital motion of the binary; the process thus provides another mechanism for solving the angular momentum problem.

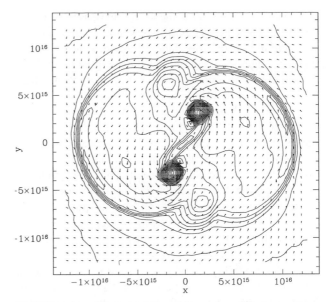

FIGURE 7 Numerical simulation of binary formation in a collapsing, rotating cloud. Contours of equal density are shown in the equatorial (x, y) plane of the inner part of the cloud. The maximum density in the fragments is 10^{-12} g cm^{-3}, and the minimum density at the outer edge is 10^{-18} g cm^{-3}. Velocity vectors have length proportional to speed, with a maximum value of 2.8×10^5 cm sec^{-1}. The forming binary has a separation of about 600 AU. The length scales are given in centimeters.

Once the density in the center of the collapsing cloud reaches about 10^{-13} g cm^{-3}, that region is no longer transparent to the emitted radiation, and much of the released gravitational energy is trapped and heats the cloud, marking the beginning of the adiabatic phase. By this time the Jeans mass (Eq. 40) has decreased to about 0.01 $M_\odot$. The heating results in an increase in the Jeans mass above this minimum value. The collapse time is still close to the free-fall time, and, as the density increases further, this time becomes short compared with the time required for radiation of diffuse out of the central regions. The collapse then becomes nearly adiabatic, and the pressure increases relative to gravity quite rapidly, so that a small region near the center approaches hydrostatic equilibrium. As this region continues to compress, the temperature rises to 2000 K. At that point the molecular hydrogen dissociates and thereby absorbs a considerable fraction of the released gravitational energy. As a result, the center of the cloud becomes unstable to further gravitational collapse, which continues until most of the molecules have dissociated at a density of 10^{-2} g cm^{-3} and a temperature of 3×10^4 K. The collapse is again decelerated and a core forms in hydrostatic equilibrium. Initially, this core contains only $\approx 1\%$ of the protostellar mass; it serves as the nucleus of the forming star.

Even if the cloud had previously fragmented, the fragments will collapse through the adiabatic phase. Once the final core has formed, the accretion phase begins. At first, infalling material has relatively little angular momentum, and it joins the core, falling through a shock front at its outer edge. However, material arriving later will have higher and higher angular momentum, so after some time it will not be able to join the core but will start forming a flattened disk surrounding the core, which can grow to a size of several hundred astronomical units. Thus, the protostar, at a typical stage of its development, consists of a slowly rotating central star, a surrounding disk, and an opaque envelope of collapsing material which accretes primarily onto the disk. Once the disk becomes comparable in mass to the central core, it can become gravitationally unstable. In some situations the instability could possibly result in the formation of subcondensations, with masses in the brown dwarf range, in orbit around the core. In other cases, the instability would produce spiral density waves, which would result in transport of angular momentum outward and mass inward, with the result that disk material collects onto the core and allows it to grow in mass. Once the disk mass decreases to less than 10% of the mass of the core, it becomes gravitationally stable. However, observations of young objects show that their disks, which are in the mass range of 1–10% that of the star, still continue to accrete onto the star. The mechanism by which they do so is not fully understood, and hydrodynamic as well as hydromagnetic instabilities are under investigation. Associated with the accretion of mass onto the star is the generation of jets and outflows, generally perpendicular to the plane of the disk, observed in most protostars, and originating through a complicated interaction between rotation, accretion, and magnetic effects near the interface between star and disk.

As material accretes onto the star, either directly or through a disk, it liberates energy at a rate given approximately by $L = G M_c \dot{M} / R_c$, where M_c and R_c are the core mass and radius, respectively, and $\dot{M}$ is the mass accretion rate. This energy is transmitted radiatively through the infalling envelope, where it is absorbed and re-radiated in the infrared. The typical observed protostar has a spectral energy distribution peaking at 60–100 μm, although the peak wavelength and the intensity of the radiation will depend on the viewing angle with respect to the rotation axis. Protostars viewed nearly pole-on will have bluer and more intense radiation than those viewed nearly in the plane of the disk. The time for the completion of protostellar evolution is essentially the free-fall time of the outer layers, which lies in the range 10^5–10^6 years, depending on the initial density. As the accretion rate slows down in the later phases and the infalling envelope becomes less and less opaque, the observable surface of the protostar declines in luminosity and increases in temperature. Once the infalling envelope becomes transparent, the observer can see

through to the core, which by now can be identified in the H–R diagram as a star with a photospheric spectrum. The locus in the diagram connecting the points where stars of various masses first make their appearance is known as the *birth line*.

C. Pre-main-Sequence Contraction

Once internal temperatures are high enough (above 10^5 K) that hydrogen is substantially ionized, the star is able to reach an equilibrium state with the pressure of an ideal gas supporting it against gravity. Rotation, which was very important during the protostar stage, has now become negligible. The star radiates from the surface, and this energy is supplied by gravitational contraction, which can be regarded as a passage through a series of equilibrium states. The virial theorem shows that half of the released gravitational energy goes into radiation and the other half into heating of the interior, as long as the equation of state remains ideal. The calculation of the evolution can now be accomplished by solution of the standard structure equations (32), (33), (35), and (36).

The solutions are shown in the H–R diagram in Fig. 8. These tracks start at the birth line for each mass. For earlier times, a hydrostatic stellar-like core may exist, but it is hidden from view by the opaque infalling protostellar envelope. Note that the sun arrives on the H–R diagram

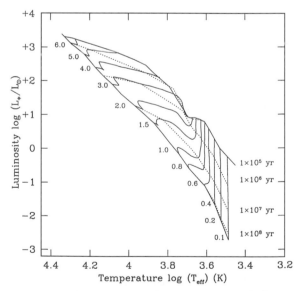

FIGURE 8 Evolutionary tracks for various masses during the pre-main-sequence contraction. Each track is labeled by the corresponding stellar mass (in solar masses, $M_\odot$). For each track, the evolution starts at the birth line (upper solid line) and ends at the zero-age main sequence (lower solid line). Loci of constant age are indicated by dotted lines. [Reproduced by permission from Palla, F., and Stahler, S. (1999). *Astrophys. J.* **525,** 772. © The American Astronomical Society.]

with about five times its present luminosity, and the higher mass stars actually do not appear until they have already reached the main sequence. The results of the calculations show in general that the stars first pass through a convective phase (vertical portions of the tracks) and later a radiative phase (relatively horizontal portions of the tracks). The relative importance of the two phases depends on the stellar mass. During the convective phase, energy transport in the interior is quite efficient, and the rate of energy loss is controlled by the thin radiative layer right at the stellar surface. The opacity is a very strongly increasing function of T in that layer, and that fact combined with the photospheric boundary conditions (38) can be shown to result in a nearly constant T_{eff} during contraction. As the surface area decreases, L drops and T_{eff} stays between 2000 and 4000 K, with lower masses having lower T_{eff}. As the star contracts, the interior temperatures increase and in most of the star the opacity decreases as a function of T. The star gradually becomes stable against convection, starting at the center, because the radiative gradient [Eq. (37)] drops along with the opacity and soon falls below ∇_{ad}. When the radiative region includes about 75% of the mass, the rate of energy release is no longer controlled by the surface layer, but rather by the opacity of the entire radiative region. At this time, the tracks make a sharp bend to the left, and the luminosity increases gradually as the average interior opacity decreases.

Contraction times to the main sequence, starting at the birth line, for various masses are given in Table III. A summary of the evolution of the sun during the pre-main-sequence, main-sequence, and early post-main-sequence phases is given in Fig. 9. The higher mass stars have relatively high internal temperature and therefore relatively low internal opacities and are able to radiate rapidly. The contraction times are short, and because the luminosity is relatively constant during the radiative phase, during which they spend most of their time, the contraction time is well approximated by the

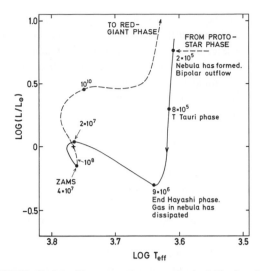

FIGURE 9 Sketch of the pre-main-sequence (solid line) and post-main-sequence (dashed line) evolution of a star of 1 $M_\odot$. Evolutionary times (years) are given for the filled circles. The present sun is indicated by a cross. The term "nebula" refers to the circumstellar disk. [Adapted with permission from Bodenheimer, P. (1989). In "The Formation and Evolution of Planetary Systems" (H. Weaver and L. Danly, eds.), p. 243, Cambridge Univ. Press, Cambridge, UK. Reprinted with permission of Cambridge University Press.]

Kelvin–Helmholtz time $t_{\text{KH}} \approx GM^2/(RL)$. A star of 1 $M_\odot$ spends 10^7 years, on the vertical track, known as the Hayashi track after the Japanese astrophysicist who discovered it. He also pointed out that the region to the right of the Hayashi track is "forbidden" for a star of a given mass in hydrostatic equilibrium, at any phase of evolution. For the next 2×10^7 years, the star is primarily radiative, but it maintains a thin outer convective envelope all the way to the main sequence. The final 10^7 year of the contraction phase represents the transition to the main sequence, during which the nuclear reactions begin to be important at the center, the contraction slows down, and, as the energy source becomes more concentrated toward

TABLE III Evolutionary Times (Year)

Mass ($M_\odot$)	Pre-main-sequence contraction	Main-sequence lifetime	To onset of core He burning	Core He lifetime
30	—	4.8×10^6	1.0×10^5	5.4×10^5
15	—	1.0×10^7	3.0×10^5	1.6×10^6
9	—	2.1×10^7	9.1×10^5	3.8×10^6
5	2.3×10^5	6.5×10^7	4.8×10^6	1.6×10^7
3	2.0×10^6	2.2×10^8	2.9×10^7	6.6×10^7
2	8.5×10^6	8.4×10^8	3.1×10^8	—
1	3.2×10^7	9.2×10^9	2.5×10^9	1.1×10^8
0.5	1.0×10^8	1.5×10^{11}	—	—
0.3	1.8×10^8	7.1×10^{11}	—	—
0.1	3.7×10^8	5.9×10^{12}	—	—

the center, the luminosity declines slightly. For the lower mass stars, the evolution is entirely along the Hayashi track. Stars of 0.3 $M_\odot$ or less remain fully convective all the way to the main sequence. Because the luminosity varies continuously during the contraction, t_{KH} is determined by integration along the track. It takes a star of 0.1 $M_\odot$ about 4×10^8 years to reach the main sequence.

A number of comparisons can be made between the contraction tracks and the observations.

1. As predicted by Hayashi, no stars in equilibrium are observed to exist in the forbidden region of the H–R diagram.

2. The lithium abundances of stars that have just reached the main sequence can be compared with calculations of the depletion of lithium in the surface layers during the contraction. Lithium is easily destroyed by reactions with protons at temperatures above about $T = 2.8 \times 10^6$ K. The lower mass stars have outer convection zones extending down to this T during the contraction; the higher mass stars do not. Because the material in the convection zone is completely mixed, the low-mass stars would be expected on the average to have less surface (observable) Li when they arrive at the main sequence, in agreement with observations.

3. The youngest stars, known as the T Tauri stars, have high lithium abundances; are associated with dark clouds where star formation is taking place; display irregular variability in light, as well as mass loss that is much more rapid than that of most main-sequence stars; and have excess infrared emission over that expected from a normal photosphere, indicating the presence of a surrounding disk, and excess ultraviolet radiation, indicating accretion onto the star. All these characteristics are consistent with youth; their location in the H–R diagram falls along Hayashi tracks for stars in the mass range 0.5–2.0 $M_\odot$. The disk characteristics vanish after ages of a few million to 10^7 years, putting a constraint on the time available for the formation of gaseous giant planets.

4. Dynamical masses obtained from pre-main-sequence eclipsing binaries are an important tool for calibration of the pre-main-sequence theoretical tracks.

5. The H–R diagrams for young stellar clusters can be compared with lines of constant age (Fig. 8). The results show that there is a considerable scatter of observed points about a single such line, indicating that star formation in a given region probably occurs continuously over a period of about 10^7 years.

V. THE MAIN SEQUENCE

A. General Properties

The zero-age main sequence (ZAMS) is defined as the time when nuclear reactions first provide the entire luminosity radiated by a star. The chemical composition is assumed to be spatially homogeneous at this time, although actually in the centers of higher mass stars, some ^{12}C is converted to ^{14}N by Reactions (10)–(12) before that time. Later, the nuclear reactions result in the conversion of hydrogen to helium, with the change occurring most rapidly at the center. The characteristics of a range of stellar masses on the ZAMS are given in Table IV; the assumed composition is $X = 0.71$, $Y = 0.27$, $Z = 0.02$ (subscripts c refer to central values). The following general points can be made.

1. The equation of state in the interior is close to that of an ideal gas. There are deviations only at the high-mass end, where radiation pressure becomes important, and at the low-mass end, where electron degeneracy begins to be significant.

2. The radius increases roughly linearly with mass at the low-mass end and roughly as the square root of the mass above 1 $M_\odot$.

3. The theoretical mass–luminosity relation agrees well with observations (Fig. 2); the location of the theoretical ZAMS in the H–R diagram is also in good agreement with the observations.

TABLE IV **Zero-Age Main Sequence**

Mass ($M_\odot$)	log $L/L_\odot$	Log T_{eff} (K)	$R/R_\odot$	T_c (10^6 K)	ρ_c (g cm^{-3})	M_{core}/M	M_{env}/M	Energy source
30	5.15	4.64	6.6	36	3.0	0.60	0	CNO
15	4.32	4.51	4.7	34	6.2	0.39	0	CNO
9	3.65	4.41	3.5	31	10.5	0.30	0	CNO
5	2.80	4.29	2.3	27	17.5	0.23	0	CNO
3	2.00	4.14	1.7	24	40.4	0.18	0	CNO
2	1.30	4.01	1.4	21	68	0.12	0	CNO
1	−0.13	3.76	0.9	14	90	0	0.02	PP
0.5	−1.44	3.56	0.44	9	74	0	0.4	PP
0.3	−2.0	3.53	0.3	7.5	125	0	1.0	PP
0.1	−3.3	3.43	0.1	5	690	0	1.0	PP

4. The central temperature increases with mass, as expected from the virial theorem. The higher mass stars burn hydrogen on the CNO cycle, and because of the steep T dependence of these reactions and the corresponding strong degree of concentration of the energy sources to the center, they have convective cores. The fractional mass of these cores increases with total mass. Comparisons with cluster H–R diagrams suggests that a small amount of convective overshooting occurs at the edge of these cores. The lower mass stars run on the pp chains, and because of the more gradual T dependence, they do not have convective cores. However, because they have cooler surface layers and therefore steeply increasing opacity going inward from the surface, they have convection zones in their envelopes, which become deeper as the mass decreases. At 0.3 $M_\odot$, this convection zone extends all the way to the center. The lowest mass stars do not have high enough T_c to allow ^{3}He to react with itself [Reaction (3)]; therefore (at least on the ZAMS), the pp chain proceeds only through Reactions (1) and (2).

5. There is no substantial evidence for stars above 100 $M_\odot$. The upper limit for stellar masses arises either from the fact that radiation pressure from the core can prevent further accretion of envelope material during the protostellar phase when the core reaches some critical mass or from the fact that main-sequence stars above a given mass are pulsationally unstable which, in combination with strong radiation pressure, tends to result in rapid mass loss from their surfaces.

6. The lower end of the main sequence occurs at about 0.075 $M_\odot$. Below that mass nuclear reactions cannot provide sufficient energy. Physically, the limit arises from the fact that as low-mass objects contract, they approach the regime of electron degeneracy before T_c becomes high enough to start nuclear burning. As the electrons are forced into higher and higher energy states, much of the gravitational energy supply of the star is primarily used to provide the required energy to the electrons. The ions, whose temperatures determine nuclear reaction rates, are still and ideal gas, but they reach a maximum temperature and then begin to cool. Their thermal energy is required, along with the gravitational energy, to supply the radiated luminosity. Substellar objects below the limit simply contract to a limiting radius and then cool. Extensive observational searches for such objects, known as "brown dwarfs," have resulted in the discovery of many strong candidates.

B. Evolution on the Main Sequence

During the burning of H in their cores, stars move very slowly away from the ZAMS. The main-sequence lifetimes, up to exhaustion of H at the center, are given in Table III. High-mass stars go through this phase in a few million years, while stars below about 0.8 $M_\odot$ (with solar composition) have not had time since the formation of the galaxy to evolve away from the main sequence. The rate at which energy is lost at the surface, which determines the evolutionary time scale, is controlled by the radiative opacity through much of the interior. The stellar structure adjusts itself to maintain thermal equilibrium, which means that the energy production rate is exactly matched by the rate at which energy can be carried away by radiation. If for some reason the nuclear processes were producing energy too rapidly, some of this energy would be deposited as work in the inner layers, these regions would expand and cool, and the strongly temperature-dependent energy production rate would return to the equilibrium value.

The structure of the star on the main sequence changes during main-sequence evolution as a consequence of the change in composition. In the case of upper main-sequence stars the H is depleted uniformly over the entire mass of the convective core, while in the lower mass stars, which are radiative in the core, the H is depleted most rapidly in the center. The conversion of H to He results in a slight loss of pressure because of a decease in the number of free particles per unit volume; as a result, the inner regions contract very slowly to maintain hydrostatic equilibrium. The conversion also reduces the opacity somewhat, which tends to cause a slow increase in L. The outer layers see no appreciable change in opacity; a small amount of the energy received from the core is deposited there in the form of work, and these layers gradually expand. For example, in the case of the sun, L has increased from 0.7 to 1.0 $L_\odot$ since age zero, R has increased from 0.9 to 1.0 $R_\odot$, T_c has increased from 14×10^6 to 15.7×10^6 K, and ρ_c has increased from 90 to 152 g cm^{-3}. In the case of high-mass, stars when X goes to zero at the center, it does so over the entire mass of the convective core, and the star is suddenly left without fuel. A rapid overall contraction takes place, until the layers of unburned H outside the core reach temperatures high enough to burn. For solar-mass stars, the main-sequence evolution does not end so suddenly, because the H is depleted only one layer at a time, and the transition to the red giant phase takes place relatively slowly.

An important calibration for the entire theory of stellar evolution is the match between theory and observation of the sun. The procedure is to choose a composition and to evolve the model sun from age zero to the present solar age of 4.56×10^9 years . The value of Z is well constrained by observations of photospheric abundances; thus, if the solar L does not match the model L, the abundance of He in the model can be used as a parameter to be adjusted to provide agreement. Once L of the model is satisfactory, the radius can be adjusted to match the observed R by varying the mixing-length parameter α that is used to calculate the structure of the outer part of the convection zone. A further check is provided by the results of helioseismology, a method of probing the interior structure of the sun by making precise measurements of the frequencies of its oscillations at the surface. The best-known periods are of the order of 5 min. The technique is able to measure the sound speed as a function of depth far into the sun, as well as the location of the bottom of the convection zone. Models that start with the composition consistent with that of Table I ($X = 0.70$, $Y = 0.28$, $Z = 0.02$) and $\alpha = 1.8$, and which include the slow settling of He downward during the evolution, provide excellent agreement with the results of helioseismology and the observed Z, L, and R. The present sun (see Table V) has $Y = 0.24$ in the outer layers; the convection zone includes the outer 29% of the radius with $T = 2.2 \times 10^6$ K at its lower edge.

C. The Solar Neutrino Problem

The results of terrestrial solar neutrino detection experiments can be summarized as follows:

1. The chlorine experiment, $^{37}\text{Cl} + \nu \rightarrow {}^{37}\text{Ar} + e^-$ with detection threshold 0.81 MeV, detects primarily neutrinos produced by ^{8}B decay [Reaction (8)]. A long series of experiments going back to 1970 shows a detection rate only one-third that predicted by the standard solar model.

2. The neutrino-electron scattering experiment carried out by the Superkamiokande project has a high detection threshhold (about 6 MeV) and can detect only neutrinos from Reaction (8). Results show about half the expected detection rats, but the experiment can determine from which direction the neutrinos arrive, and it is clear that they come from the sun.

3. Two independent gallium experiments, the SAGE project and the GALLEX project, detect neutrinos by $^{71}\text{Ga} + \nu \rightarrow {}^{71}\text{Ge} + e^-$ with detection threshhold 0.234 MeV. The experiments can detect neutrinos from all three branches of the pp chain; in particular, about half of the neutrinos produced by Reaction (1), which are by far the most numerous of the solar neutrinos, are detectable by this experiment. The actual detection rate in these experiments, which have been calibrated through the use of a terrestrial neutrino source of known production rate, is slightly more than half of the expected rate. The discrepancies in all case are far greater than the experimental or theoretical uncertainties.

The following possibilities have been considered as solutions to the solar neutrino problem.

1. The solar model is inappropriate. The production rate of the neutrinos produced by ^{8}B is very sensitive to the temperature near the center of the sun; those produced

TABLE V Standard Solar Model

$M(r)/M_\odot$	$r/R_\odot$	T (10^6K)	ρ (g cm^{-3})	P (erg cm^{-3})	$L(r)/L_\odot$	X
0.000	0.00	15.7	152	2.33×10^{17}	0.0	0.34
0.009	0.045	15.0	132	2.05×10^{17}	0.073	0.40
0.041	0.078	13.9	104	1.63×10^{17}	0.282	0.48
0.117	0.120	12.3	73.3	1.10×10^{17}	0.602	0.58
0.217	0.159	10.8	51.5	7.17×10^{16}	0.822	0.65
0.308	0.191	9.69	38.3	4.89×10^{16}	0.920	0.68
0.412	0.226	8.60	26.9	3.08×10^{16}	0.971	0.69
0.500	0.255	7.79	19.5	2.03×10^{16}	0.990	0.70
0.607	0.299	6.82	12.2	1.12×10^{16}	0.998	0.71
0.702	0.345	5.96	7.30	5.85×10^{15}	1.000	0.71
0.804	0.411	4.97	3.48	2.32×10^{15}	1.000	0.71
0.902	0.518	3.81	1.13	5.82×10^{14}	1.000	0.72
0.927	0.562	3.42	0.726	3.35×10^{14}	1.00	0.72
0.962	0.656	2.69	0.304	1.11×10^{14}	1.000	0.72
0.976[a]	0.714	2.18	0.185	5.53×10^{13}	1.000	0.74
0.990	0.804	1.33	0.088	1.59×10^{13}	1.000	0.74
0.998	0.900	0.60	0.026	2.10×10^{12}	1.000	0.74
0.9997	0.949	0.28	0.008	3.06×10^{11}	1.000	0.74
1.0000	1.000	0.0058	2.8×10^{-7}	1.20×10^5	1.000	0.74

[a]Base of convection zone.

by the p + p reaction are also sensitive, but less so. A decrease in this temperature in the models would bring the predictions more into line with the experiments. One of the main parameters in the solar model is the abundance of He. A decrease in its assumed value (and replacement by H) results in an increase in the opacity, so that a compensating decrease in Z is required so that the solar luminosity will be matched at the solar age. The value of T_c is reduced because in the presence of the increased value of X, Eq. (9) shows that the energy requirements of the sun can be met with a lower temperature and density. However, the value of Z is well enough known observationally that a change in Y sufficient to bring about agreement would result in a required value of Z that is outside the observational uncertainties. Furthermore, the value of Y is well constrained through helioseismology as a function of radius almost all the way to the center of the sun. The agreement between the observationally determined sound speed as a function of radius and the standard solar model is within 1%. In general, a change in the solar model sufficient to bring about consistency with the solar neutrino experiments will result in inconsistency with at least one other well-observed property of the sun.

2. The nuclear reaction rates, when extrapolated to stellar energies, are incorrect. The experimental rate of Reaction (7) is particularly difficult to determine, and it is directly proportional to the production rate of neutrinos from ^{8}B. However, the quoted uncertainty in all the rates, typically 10%, is not large enough to solve the problem.

3. Through so-called "neutrino oscillations," the electron neutrinos produced in the sun are converted to muon or tau neutrinos before they reach the earth; the detection experiments now running are insensitive to these other types. This possibility is the subject of intensive research by particle physicists. There is some direct experimental evidence that muon neutrinos undergo oscillations, but none so far for electron neutrinos.

4. The sun undergoes periodic episodes of mixing of the chemical composition in the interior. Some hydrogen outside the burning region would be mixed downward, providing extra fuel and increasing the energy generation rate given by Eq. (9). Some energy is deposited in the form of work in the burning layers, they expand and cool to re-establish the balance between L and the nuclear energy generation rate. The cooler sun then produces fewer detectable neutrinos. It is also possible that the sun is continuously mixed by a slow diffusive process; in either case, the mechanism for mixing is not understood, and there is no evidence for it in the helioseismological results. The solution to the solar neutrino experiment, may come about through different experimental techniques. For example, the SNO experiment, which uses heavy water (D_2O) rather than ordinary water as a detecting agent, in principle will be able to determine whether solar neutrinos have undergone oscillations.

VI. STELLAR EVOLUTION BEYOND THE MAIN SEQUENCE

Following the exhaustion of H at the center of a star, major structural adjustments occur; the physical processes that result in the transition to a red giant were first made clear by the calculations of F. Hoyle and M. Schwarzschild. The central regions, and in some cases the entire star, contract until the hydrogen-rich layers outside the exhausted core can be heated up to burning temperatures. Hydrogen burning becomes established in a shell source, which becomes thinner in terms of mass and hotter as the star evolves. The central region, inside the shell, continues to contract, while the outer layers expand considerably as the star evolves to the red giant region. The value of T_{eff} decreases to 3500–4500 K, and a convection zone develops in the outer layers and becomes deeper as the star expands. When T in the core reaches 10^8 K, helium burning begins, resulting in the production of ^{12}C and ^{16}O. Evolutionary time scales up to and during helium burning are given in Table III. Whether further nuclear burning stages occur that result in the production of still heavier elements depends on the mass of the star. For masses less than 9 $M_\odot$, interior temperatures never become high enough to burn the C or O. The star ejects its outer envelope, goes through a planetary nebula phase, and ends its evolution as a white dwarf. The more massive stars burn C and O in their cores, and nucleosynthesis proceeds to the production of an iron-nickel core. Collapse of the core follows, resulting in a supernova explosion, the ejection of a large fraction of the mass, and the production of a neutron star remnant for masses up to about 25 $M_\odot$ and possibly a black hole for higher masses. The details of the evolution are strongly dependent on mass. The following sections describe the post-main-sequence evolution of stars of Population I with masses of 1, 5, and 25 $M_\odot$ and of a Population II star of about 0.8 $M_\odot$, which represents a typical observable star in a globular cluster. Evolutionary tracks for the Population I stars are shown in Fig. 10.

A. Further Evolution of 1 $M_\odot$ Stars

The transition phase to the red giant region, during which the star evolves to the right in the H–R diagram, takes place on a time scale that is short compared with the main-sequence lifetime, but long enough so that stars undergoing the transition should be observable in old clusters. As the convective envelope deepens, the evolutionary track changes direction in the H–R diagram, and the luminosity increases considerably while T_{eff} decreases only slowly.

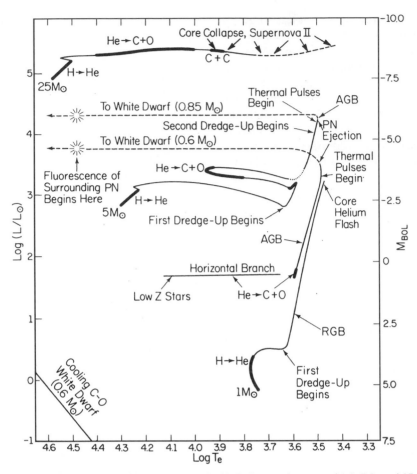

FIGURE 10 Post-main-sequence evolutionary tracks in the H–R diagram for stars of 1.0, 5.0, and 25 $M_\odot$ under the assumption that the metal abundance Z is comparable to that in the present sun. The heavy portions of each curve indicate where major nuclear burning stages occur in the core. The label RGB refers to the red giant branch, AGB refers to the asymptotic giant branch, and PN refers to the planetary nebula phase. Helium burning occurs on the horizontal branch only if Z is much less than the solar value. [Reproduced with permission from Iben, I., Jr. (1991). *Astrophys. J. Suppl.* **76**, 55. © The American Astronomical Society.]

The inner edge of the convection zone moves inward to a point just outside the hydrogen-burning shell; inside the shell is the dense, burned-out core consisting mainly of He and increasing in mass with time. Temperatures in the shell increase to the point where the CNO cycle takes over as the principal energy source. The outer convective envelope becomes deep enough so that it reaches layers in which C has been converted to N by Reactions (10)–(12), which proceed at a temperature slightly less than that required for the full cycle to go into operation. The modification of the C to N ratio at the surface of the star, which is expected because of convective mixing, has been verified in some cases by observations of abundances in red giant stars. The oxygen surface abundance is not affected. This process is known as the *first dredge-up*. As the sun evolves to high luminosity on the red giant branch, it undergoes a mass loss episode, which, depending on the uncertain mass loss rate, could result in the loss of the outer 25% of the mass.

Helium burning in the core finally begins when its mass is about 0.45 $M_\odot$. At this time, $L \approx 2 \times 10^3 L_\odot$, $T_c = 10^8$ K, and $\rho_c = 8 \times 10^5$ g cm^{-3}. At the center, the electron gas is quite degenerate; thus, the total gas pressure depends strongly on density but not on temperature. The helium-burning reaction rate is very sensitive to T; when the reaction starts, the local region heats somewhat and the reaction rate increases, resulting in further heating. Under normal circumstances, the increased T would result in increased pressure, causing the region to expand to the point where the rate of energy generation matched the rate at which the energy could be carried away. However, under degenerate conditions the pressure does not respond, only very slight expansion occurs, and the region simply heats, resulting in a runaway growth of the energy generation. This thermal instability is known as the *helium flash*, during which the luminosity can increase to $10^{11} L_\odot$ for a brief period at the flash site. This enormous luminosity is absorbed primarily in the gradual expansion of the core, whose density decreases by a factor 40, and the luminosity at the surface does not increase. On the contrary, the expansion of the inner regions results in a decrease in the temperature at the hydrogen shell source, a reduction in the energy generation, a reduction of the surface luminosity, and a contraction of the outer layers to compensate. The temperature in the core rises to the point where it is no longer degenerate, cooling can occur, and the nuclear reaction rate once more becomes regulated. Relatively little He is actually burned during the flash, and the star settles down near the red giant branch, but with a reduced $L = 40 L_\odot$ (see Fig. 10). Energy production comes from both core helium burning and shell hydrogen burning. Although a convection zone develops in the region of the helium flash, it does not link up with the outer convection zone, and so the products of helium burning are not mixed outward to the surface of the star at this time.

Helium burns in the core, converting it to C and O. When the helium is exhausted in the core, that region begins to contract until helium burning is established in a shell region. The hydrogen-burning shell is still active; thus, the star now has a double shell source surrounding the core, and the convective envelope still extends inward to a point just outside the hydrogen-burning shell. The star resumes its upward climb in luminosity along the asymptotic giant branch (AGB; see Fig. 10). The shell sources become thinner, hotter, and closer together. The core becomes degenerate, and the increase in its temperature stops because the energy released by contraction must go into lifting the electrons into higher and higher energy states and because of neutrino losses. The structure of the star is divided into three regions: (1) the degenerate C/O core, which grows in mass to about $0.6 M_\odot$, in ρ_c to over 10^6 g cm^{-3}, and maintains a radius of about 10^9 cm; (2) the hydrogen- and helium-burning shells, which are separated by a very thin layer of He and which contain only a tiny fraction of the total mass; and (3) the extended convective envelope, which still has essentially its original abundances of H and He and a mean density of only about 10^{-7} g cm^{-3}, expands to a maximum size of about $200 R_\odot$ (see Fig. 11). The helium-burning shell is subject to a thermal instability during which the energy generation increases locally to peak values of $10^6 L_\odot$. These events, known as helium shell flashes, repeat with a period of about 10^5 years and result in surface luminosity changes of factors of 5–10.

As the star increases in luminosity, up to a maximum of about $5000 L_\odot$, it again develops an increasingly strong stellar wind, as a result of which it undergoes further mass

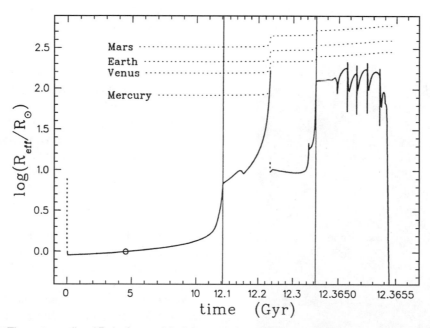

FIGURE 11 The outer radius (R_{eff}) of a model of the evolution of the sun, as a function of time, starting at the zero-age main sequence (solid line). The dashed line refers to the pre-main-sequence phase. The present sun is indicated by an open circle. Dotted lines indicate the orbital distances of the inner planets. These orbits move outward at times when the sun undergoes rapid mass loss. At the end of the simulation, the sun's mass has been reduced to $0.54 M_\odot$. The oscillations in the right-hand part of the diagram are caused by helium shell flashes. [Adapted with permission from Sackmann, I.-J., Boothroyd, A. I., and Kraemer, K. E. (1993). *Astrophys. J.* **418,** 457. © The American Astronomical Society.]

loss. Although the mechanism behind this final mass ejection is not completely understood, it is probable that the star first becomes pulsationally unstable. In fact, many of the stars in this part of the H–R diagram are variable in light with periods of about a year. As the luminosity increases, the amplitude of the oscillations increases and eventually grows to the point where a small amount of mass at the outer edge is brought to escape velocity. The formation of grains in the outer, very cool layers of these pulsating stars may contribute to the mass loss; the radiation pressure on them can result in a stellar wind. In any case, practically the entire hydrogen-rich envelope is ejected, leaving behind the core, the shell sources, and a thin layer of hydrogen-rich material on top. The star now enters the planetary nebula phase.

The system now consists of a compact central star and an expanding diffuse envelope, known as the planetary nebula. The star evolves to the left in the H–R diagram at a nearly constant luminosity of about 5000 $L_\odot$, with T_{eff} increasing from the red giant value of about 4000 to over 10^5 K. Once T_{eff} exceeds 30,000 K, the ultraviolet radiation results in fluorescence in the nebula, causing it to glow in optical light (Fig. 12). The evolution time from the red giant region to maximum T_{eff} is about 10^5 years, during which time the radius decreases from its maximum value to about 0.1 $R_\odot$ (Fig. 11). The nuclear energy source from one or both shells is still active. The hydrogen-rich outer envelope decreases in mass as the fuel is burned, and eventually, when the envelope mass decreases to $\sim 10^{-4}$ $M_\odot$,

FIGURE 12 The giant planetary nebula NGC 7293 (Shane 120-in. reflector). (Lick Observatory photograph.)

it is no longer hot enough to burn. The outer layers contract until the radius approaches a limiting value, at which point only a very slight amount of additional gravitational contraction is possible because practically the entire star is highly electron degenerate, and the high electron pressure supports it against gravity. Essentially, the only energy source left is from the cooling of the hot interior. From this point the star evolves downward and to the right in the H–R diagram and soon enters the white dwarf region. Thus, the final state of the evolution of 1 $M_\odot$ is a white dwarf of about 0.6 $M_\odot$.

B. Evolution of 5 $M_\odot$ Stars

The evolution of higher mass stars differs from that of the lower mass stars in several respects. On the main sequence the star of 5 $M_\odot$ burns hydrogen on the CNO cycle and has a convective core that includes about 23% of the mass at first, but which decreases in mass with time. The H is uniformly depleted in the core. With the exhaustion of H in the core, the star undergoes a brief overall contraction that leads to heating and ignition of the shell source; the convection zone has now disappeared. As the shell narrows, the central regions interior to it contract, while the outer regions expand. The star rapidly crosses the region between the main sequence and the red giant branch. Because of this rapid crossing, very few stars would be expected to be observed in this region, and for that reason it is known as the Hertzsprung gap. As T_{eff} decreases, a convection zone develops at the surface and advances inward; when it includes about half the mass, the star turns upward in the H–R diagram. When log $L/L_\odot = 3.1$ and the hydrogen-exhausted core includes about 0.75 $M_\odot$, helium burning begins at the center. Because this region is not yet degenerate, the helium flash does not occur and the burning is initiated smoothly. A central convection zone develops because of the strong temperature dependence of the triple-alpha reaction. The structure of the star now consists of (1) the convective helium-burning region, (2) a helium-rich region that is not hot enough to burn He, (3) a narrow hydrogen-burning shell source, and (4) an extended outer envelope with essentially unmodified H and He abundances. However, the outer convection zone has extended downward into layers where some C has been converted into N through previous partial operation of the CNO cycle. The relative depletion of C and enhancement of N could now be observable at the stellar surface. This first dredge-up is analogous to that which occurs in 1 $M_\odot$ (see Fig. 13 at a time of 6×10^7 years).

As helium burning progresses, the central regions expand, resulting in a decrease in temperature at the hydrogen-burning shell and a slight drop in luminosity. The expansion of the outer layers is reversed, and the

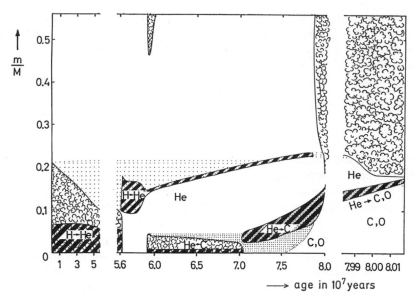

FIGURE 13 The evolution of the internal structure of a star of 5 $M_\odot$ with composition similar to that of the sun. The abscissa gives the age, starting from the zero-age main sequence. A vertical line through the diagram corresponds to a model at a given time. The figure shows only the inner half of the mass of the star. Cloudy regions indicate convection zones. Hatched regions are those with significant nuclear energy production. Regions which are non-uniform in composition are dotted. Breaks in the diagram correspond to changes in the time scale. [Adapted with permission from Kippenhahn, R., and Weigert, A. (1990). "Stellar Structure and Evolution," Springer-Verlag, Berlin. © Springer-Verlag.]

star evolves to higher T_{eff}. The energy production is divided between the core and the shell, with the former becoming relatively more important as the star evolves. The total lifetime during helium burning is about 1.4×10^7 years, about 20% of the main-sequence lifetime. As Fig. 10 shows, most of this time is spent in a region where $T_{eff} \approx 6000 - 8000$ K and $\log L/L_\odot \approx 3$, where the Cepheid variables are observed; they are interpreted as stars of 5–9 $M_\odot$ in the phase of core helium burning. The evolutionary calculations show that for solar metallicity only stars in this mass range make the excursion to the left in the H–R diagram during He burning. Lower mass stars stay close to the giant branch during this phase.

As He becomes depleted in the core, the region contracts and heats in order to maintain about the same level of energy production. The contraction results in a slight increase in T at the hydrogen-burning shell. The energy production increases there, and a small amount of this excess energy is deposited in the outer envelope in the form of mechanical work, causing expansion. The evolution changes direction in the H–R diagram and heads back toward the red giant region. The He is used up at the center, a helium-burning shell source is established, and the star resumes its interrupted climb up the giant branch. The surface convection zone develops again and reaches deep within the star to layers that have previously been enriched in He. In this *second dredge-up* the ratio of He

to H is increased at the surface, and both C and O are decreased with respect to N. As the star rises to higher L, its structure includes the following regions: (1) a degenerate core consisting of a mixture of ^{12}C and ^{16}O, (2) a narrow helium-burning shell source, (3) a thin layer composed mostly of He, and (4) the extended outer convective envelope, somewhat enriched in He and with modified CNO ratios relative to the original composition. The H-burning shell goes out, to become re-established later on during He shell flashes. The structure of the interior of the star as a function of time up to about this point is illustrated in Fig. 13.

The star increases in L up to about 2×10^4 $L_\odot$, and the mass of the C/O core increases to 0.85 $M_\odot$. A series of thermal flashes, analogous to those found for 1 $M_\odot$, occurs in the He-burning shell; during the flash cycle, the main energy production alternates between an H-burning shell and an He-burning shell. The flashes occur in nondegenerate layers, unlike the He core flash in the lower mass stars. After several flashes, the *third dredge-up* occurs. Some of the C produced in the He-burning shell can be mixed to the surface of the star, eventually producing a C star. Ejection of the envelope now takes place by processes similar to those described for 1 $M_\odot$. The total mass lost during the evolution of the star is estimated to be 4.15 $M_\odot$; the remaining C/O white dwarf has a mass of 0.85 $M_\odot$. The same result will occur for stars up to 9 $M_\odot$, with the maximum core mass being about 1.1 $M_\odot$. The second dredge-up occurs only for stars in the mass range from about 4.5 to 9 $M_\odot$; the third dredge-up occurs only in stars from about 2 to 9 $M_\odot$.

C. Evolution of 25 $M_\odot$ Stars

The high-mass stars evolve similarly to the stars of 5 $M_\odot$ during main-sequence and immediate post-main-sequence evolution, except that the fraction of the mass contained in the convective core is larger, and the very highest mass stars may lose mass even on the main sequence. The 25-$M_\odot$ star evolves toward the red giant region, but helium burning at the center occurs when T_{eff} is still in the range 10^4 to 2×10^4 K, and carbon burning starts soon afterward. The star expands to become a red giant with $L/L_\odot = 2 \times 10^5$ and $T_{eff} = 4500$ K. A sequence of several new nuclear burning phases now occurs in the core, rapidly enough that little concurrent change occurs in the surface characteristics. Carbon burns at about 9×10^8 K, neon burns at 1.75×10^9 K, oxygen burns at 2.3×10^9 K, and silicon burns at 4×10^9 K. A central core of about 1.5 $M_\odot$, composed of iron and nickel, builds up, surrounded by layers that are silicon rich, oxygen rich, and helium rich, respectively. Outside these layers is the envelope, still with its original composition. The layers are separated by active shell sources. The temperature at the center reaches 7×10^9 K and the density is 3×10^9 g cm^{-3} (see Fig. 5). However, the sequence of nuclear reactions that has built the elements up to the iron peak group in the core can proceed no further. These elements have been produced with a net release of energy at every step, a total of 8×10^{18} ergs g^{-1} of hydrogen converted to iron. However, to build up to still heavier elements, a net input of energy is required. Furthermore, the Coulomb barrier for the production of these elements by reactions involving charged particles becomes very high. Instead, an entirely different process occurs in the core. The temperature, and along with it the average photon energy, becomes so high that the photons can react with the iron, breaking it up into helium nuclei and neutrons. This process requires a net input of energy, which must ultimately come from the thermal energy of the gas. The pressure therefore does not rise fast enough to compensate for the increasing force of gravity and the core begins a catastrophic gravitational collapse. On a time scale of less than 1 sec, the central density rises to 10^{14} g cm^{-3} and the temperature rises to 3×10^{10} K. As the density increases, the degenerate free electrons are captured by the nuclei, reducing the electron pressure and further contributing to collapse. At the same time, there is very rapid energy loss from neutrinos. The point is reached where most of the matter is in the form of free neutrons, and when the density becomes high enough, their degenerate pressure increases rapidly enough to stop the collapse. At that point a good fraction of the original iron core has collapsed to a size of 10^6 cm and has formed a neutron star, nearly in hydrostatic equilibrium, with a shock front on its outer edge through which material from the outer parts of the star is falling and becoming decelerated. The core collapse is thought to be the precursor to the event known as a supernova of type II.

The question of what happens after core collapse is one of the most interesting in astrophysics. Can at least part of the gravitational energy released during the collapse be transferred to the envelope and result in its expansion and ejection in the form of a supernova? Present indications are that it is possible and that the shock will propagate outward into the envelope. A large fraction of the gravitational energy is released in the form of neutrinos, produced during the neutronization of the core. Most of these neutrinos simply escape, but the deposition of a small fraction of their energy and momentum in the layers just outside the neutron star is crucial to the ejection process. Assuming that the shock does propagate outward, it passes through the various shells and results in further nuclear processing, including production of a wide variety of elements up to and including the iron peak. It also accelerates most of the material outside the original iron core outward to escape velocities. When the shock reaches the surface of the red giant star, the outermost material is accelerated to 10,000 km sec^{-1}, and the deeper layers reach comparable but somewhat smaller velocities. Luminosity, velocity, and T_{eff} as a function of time in numerical simulations of such an outburst agree well with observations. The enormous luminosity arises, in the earlier stages, from the rapid release of the thermal energy of the envelope. At later times, most of the observed radiation is generated by the radioactive decay of the ^{56}Ni that is produced mainly by explosive silicon burning in the supernova shock. Supernova observations are best fit with total explosion energies of about 10^{51} erg. The Crab Nebula (Fig. 14) is consistent with this energy and an expansion velocity of 10,000 km sec^{-1}. Another good test of the theory of stellar evolution is the calculation of the relative abundances of the elements between oxygen and iron in the ejected supernova envelope. Integration of the yields of these elements over the range of stellar masses that produce supernovae gives values that are consistent with solar abundance ratios.

D. Evolution of Low-Mass Stars with Low-Metal Abundance

The oldest stars in the galaxy are characterized by metal abundances that are 0.1–0.01 that of the sun and by a

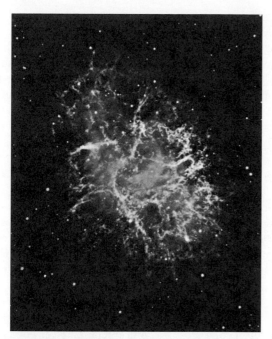

FIGURE 14 Crab Nebula in Taurus (M1, NGC 1952) in red light, the remnant of the supernova explosion in AD 1054 (Shane 120-in. reflector). (Lick Observatory photograph.)

distribution in the galaxy that is roughly spheroidal rather than disklike. These stars can be found in globular clusters or in the general field; in particular, the H–R diagrams of globular clusters (Fig. 15) give important information on the age of the galaxy and the helium abundance at the time the first stars were formed. There are observational difficulties in determining the properties of these stars, since most of them are very distant. For example, it has not been possible to determine observationally the mass–metallicity–luminosity relation for their main sequence. However, observations of the locations in the H–R diagram of a few nearby low-metal stars show that they fall below the main sequence, defined by stars of solar metal abundance. This information, combined with detailed analysis of the H–R diagrams of globular clusters, indicates that the helium abundances in the old stars are slightly less than that of the sun: $Y = 0.24 \pm 0.02$.

The age estimates of globular clusters, based on detailed comparisons of observed H–R diagrams with theoretical evolutionary tracks (see Fig. 4), range from 10 to 15 $\times$ 10^9 years. Recent work, based on improved observations, Hipparcos parallaxes for nearby low-metal stars which are used to calibrate the distances to the globular clusters, and improved theoretical tracks, favors ages in the range 13 to 14 $\times 10^9$ years for the clusters of lowest metallicity. The high-mass stars in these clusters have long ago evolved to their final states, mostly white dwarfs. The mass of the stars that are now evolving off the main sequence and be-

coming relatively luminous red giants is about 0.8 $M_\odot$. On the main sequence, however, these stars have approximately the solar luminosity because of their lower metal content, and hence lower opacity, when compared with stars of normal metal abundance. The evolution during the main-sequence phase and the first ascent of the giant branch is very similar to that of 1-$M_\odot$ stars described previously. Hydrogen burning occurs on the pp chains; there is no convection in the core; and when hydrogen is exhausted in the center, the energy source shifts to a shell and the star gradually makes the transition to the red giant region. The core contracts and becomes degenerate, and the envelope expands and becomes convective. The first dredge-up occurs, and a comparison can be made between theoretical and observed element abundances at this stage. Indications are that there is a disagreement in the sense that observed abundances change at earlier phases of evolution than the theory suggests. The implication is that some additional mechanism besides convection is causing mixing into the deep interior. A helium flash in the core occurs when its mass is 0.45 $M_\odot$ and the total luminosity is above $10^3 L_\odot$.

When the helium flash is completed and the core is no longer degenerate, the star settles down to a stable state with He burning in the core and H burning in a shell. However, the location in the H–R diagram during this phase differs from that of the low-mass stars with solar metals. The star evolves rapidly to a position on the horizontal branch (HB), with $T_{\rm eff}$ considerably higher than that on the giant branch and with L at about 40 $L_\odot$, considerably less than that at the He flash (see Figs. 10 and 15). Calculations show that as the assumed value of Z is decreased for a given mass, the position on the model on the HB shifts to the left. This behavior is in agreement with observations of globular clusters, which show that the leftward extent of the HB is well correlated with observed Z, in the sense that smaller Z corresponds to an HB extending to higher $T_{\rm eff}$. In theoretical models, as Z is increased to 10^{-2} the HB disappears altogether, and the He-burning phase occurs very close to the red giant branch. This calculation is also in agreement with the observed fact that old clusters with solar Z do not have an HB. The HB itself is not an evolutionary track. The spread of stars along the HB in a cluster of given Z represents a spread in stellar mass, ranging in a typical case from about 0.6 $M_\odot$ at the left-hand edge to about 0.8 $M_\odot$ at the right. The mean HB mass is less than the mass of the stars that are just leaving the main sequence in a given cluster. The implication is that stars lose mass on the red giant branch during the period of their evolution just prior to the He flash and that the total mass lost varies from star to star. The mass loss probably occurs by a mechanism similar to that which drives the solar wind, but much stronger.

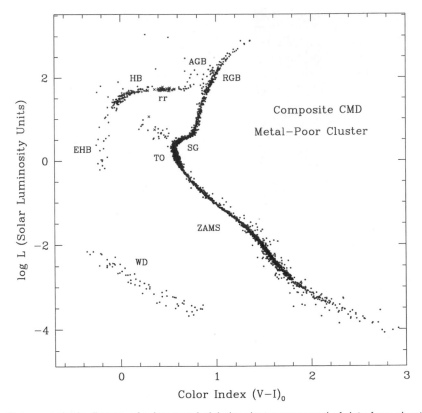

FIGURE 15 Color-magnitude diagram of a low-metal globular cluster composed of data from about five different actual clusters. The various features are ZAMS, zero-age main sequence; WD, white dwarfs; TO, turnoff; SG, subgiant branch; RGB, red giant branch; HB, horizontal branch; rr, RR Lyrae variables; EHB, extreme horizontal branch; and AGB, asymptotic giant branch. The location of the various features can be used to compare theory with observations. [Figure courtesy of William E. Harris.]

The phase of core helium burning on the HB lasts about 10^8 years, after which the star develops a double shell structure and evolves back toward the red giant branch, which during the following phase of evolution is referred to as the asymptotic giant branch. The evolution from this point on is very similar to that of the star of $1 \, M_\odot$ with $Z = 0.02$; ascent of the asymptotic giant branch, development of a carbon/oxygen core that becomes increasingly degenerate, development of a pulsational instability in and a wind from the convective envelope, ejection of the envelope, evolution through the planetary nebula phase, and, finally, the transition to the white dwarf phase.

VII. FINAL STATES OF STARS

A. Brown Dwarfs

Although objects below $0.075 \, M_\odot$ are not, strictly speaking, stars, their general observational parameters are now becoming available, and the physics that must be included in theoretical models is very similar to that for the lowest

mass stars. For example, the equation of state must include non-ideal Coulomb interactions as well as partial electron degeneracy, and the surface layer calculation must include the effects of molecules. Substellar objects never attain internal temperatures high enough that nuclear burning can supply their entire radiated luminosity. They can, however, burn deuterium by Reaction (2), and this energy source can be significant for a short period of time. Below about $0.06 \, M_\odot$, internal temperatures never become high enough to burn Li by Reaction (6); thus, most brown dwarfs should show Li lines in their spectra. However, above that mass, the Li does burn near the center, and because the objects are fully convective, the depletion of Li becomes evident at the surface. This *lithium test* has been successfully used observationally to identify brown dwarfs.

The brown dwarfs contract to the point where the electrons become degenerate. For $0.07 \, M_\odot$, the value of T_{eff} during contraction is about 2900 K; for $0.02 \, M_\odot$, it is about 2500 K. Following that time the contraction becomes very slow and the objects approach an asymptotic value of the radius, depending on mass and composition. At $0.07 \, M_\odot$, the time to contract to maximum central temperature, which corresponds to significant electron degeneracy, is about 3×10^8 years; for lower masses, it is less. Beyond this point the object cools, with decreasing T_{eff} and L. A typical brown dwarf after an evolution time of 3×10^9 years will have $T_{eff} = 1000$–1500 K and $L/L_\odot = 10^{-5}$ to 10^{-6} and is therefore very difficult to observe. Nevertheless, a number of them have been identified, for example, the companion to Gl 229, which falls exactly into this range and is thought to have a mass of about 0.03–$0.05 \ M_\odot$. The brown dwarfs are distinguished from white dwarfs first by their very cool surface temperatures and second by the fact that their internal composition is practically unchanged from the time of formation; that is, it is about 71% hydrogen by mass. As a result, the radii are larger than those of white dwarfs, with values near the end of contraction of about $0.1 \ R_\odot$ over the mass range 0.01–$0.08 \ M_\odot$. The dividing line between brown dwarfs and planets is arbitrary. One possibility is the mass below which deuterium burning is no longer possible, about $0.012 \ M_\odot$. Or they could be distinguished by their formation mechanism: planets form by accretion in protostellar disks, while brown dwarfs form in the same way as stars, by fragmentation in collapsing interstellar clouds.

B. White Dwarfs

Observational parameters of the most commonly observed white dwarfs are $T_{eff} = 10,000$–$20,000$ K and $\log L/L_\odot = -2$ to -3. The objects, therefore, lie well below the main sequence, and the deduced radii fall in the range $10^{-2} \ R_\odot$, with mean densities above 10^6 g cm^{-3}. At these high densities, most of the mass of the object falls in the region of complete electron degeneracy; only a thin surface shell is non-degenerate. Only a few white dwarfs occur in binary systems with sufficiently accurate observations so that fundamental mass determinations can be made. Examples are Sirius B, Procyon B, and 40 Eridani B, with masses of 1.05, 0.63, and 0.43 $M_\odot$, respectively.

The structure of a white dwarf can be determined in a straightforward way under the assumptions of hydrostatic equilibrium and uniform composition, with the pressure supplied entirely by the degenerate electrons. The ion pressure and the effect of the thin surface layer are small corrections. Because the pressure is simply a function of density—for example, $P \propto \rho^{5/3}$ in the nonrelativistic limit—it can be eliminated from Eq. (32). The structure can then be calculated from the solution of Eqs. (32) and (33). The results show that for a given composition there is a uniquely defined relation between mass, radius, and central density. In other words, if the central density is considered to be a parameter, for each value of it there corresponds a unique value of the mass and of the radius. This theoretical mass–radius relation, appropriate for a

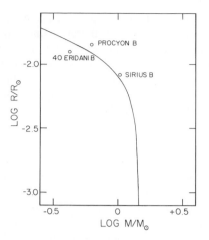

FIGURE 16 Theoretical mass–radius relation derived by S. Chandrasekhar for fully degenerate white dwarfs (solid line) compared with a few observed points (open circles). The masses of Sirius B and 40 Eridani B are known to within 5%; all radii and the mass of Procyon B are less accurately known.

typical C/O white dwarf or for a helium white dwarf, is shown in Fig. 16, where it is compared with the observations of a few white dwarfs. If a more accurate equation of state is included, the line is shifted slightly, depending on composition. Note that the radius decreases with increasing mass and vanishes altogether at $1.46 \ M_\odot$.

This upper limit to the mass of a white dwarf was first derived by S. Chandrasekhar and corresponds to infinite central density in the formal solution of the equations. In fact, however, the physical process that limits the mass of a white dwarf is capture of the highly degenerate electrons by the nuclei, predominantly carbon and oxygen. Suppose that a white dwarf increases in mass upward toward the limit. The central density goes up, the degree of electron degeneracy goes up, and a critical density is reached, about 10^{10} g cm^{-3} for carbon, where electron capture starts. As free electrons are removed from the gas, the pressure is reduced and the star can no longer maintain hydrostatic equilibrium. Collapse starts, more electron capture takes place, and (assuming no nuclear burning takes place) neutron-rich nuclei are formed. Collapse stops when the neutron degeneracy results in a very high pressure, in other words, when a neutron star has been formed. The white dwarf mass limit is reduced by about 10%, depending upon composition, from the value of $1.46 \ M_\odot$ given earlier.

The theoretical mass–radius relation agrees, within the uncertainties, with the observations. The location of the theoretical white dwarfs in the H–R diagram also agrees with observations. From the L and T_{eff} of observed white dwarfs, one can obtain radii and from them the corresponding masses, using the M–R relation. The masses of white dwarfs, determined in this manner, range from 0.42 to 0.7 $M_\odot$. These objects have presumably evolved from stars of somewhat higher original mass, as indicated in Fig. 10. Their composition is therefore that of the evolved cores of these stars, mainly carbon and oxygen with only a thin layer of hydrogen or helium on the outside. The internal temperatures are not high enough to burn nuclear fuel and contraction is no longer possible, so the only energy source available is the thermal energy of the ions in the interior. The star's L, T_{eff}, and internal temperature all decrease as the star evolves at a constant radius. The energy transport in the interior is by conduction by the free electrons, which are moving very rapidly and have long collisional mean free paths. The conduction is efficient with the result that the interior is nearly isothermal; the energy loss rate is controlled by the radiative opacity in the thin outer nondegenerate layers. These loss rates are sufficiently low that the cooling times are generally several billion years.

C. Neutron Stars

These objects, whose existence as condensed remnants left behind by supernova explosions was predicted by F. Zwicky and W. Baade in 1934, were first discovered by Bell and Hewish in 1967 from their pulsed radio radiation. The typical mass is 1.4 $M_\odot$, corresponding to the mass of the iron core that collapses for stars in the 10- to 20-$M_\odot$ range, the typical radius is 10 km, and the mean density is above 10^{14} g cm^{-3}, close to the density of the atomic nucleus. At this high density the main constituent is free neutrons, with a small concentration of protons, electrons, and other elementary particles. The surface temperature is very difficult to determine directly from observations. Neutron stars form as collapsing cores of massive stars and reach temperatures of 10^9–10^{10} K during their formation. However, they then cool very rapidly by neutrino emission; within one month T_{eff} is less than 10^8 K, and within 10^5 years it is less than 10^6 K. Their luminosity soon becomes so low ($\sim 10^{-5}$ $L_\odot$) that they would not be detected unless they were very nearby. The fact that neutron stars are observable arises from the remarkable properties that (1) they are very strongly magnetic, (2) they rotate rapidly, and (3) they emit strong radio radiation in a narrow cone about their magnetic axes. The rotation axis and the magnetic axis are not coincident, and so if the star is oriented so that the radio beam sweeps through the observer's instrument, he or she will observe radio pulses separated by equal time intervals. The observed pulsars are thus interpreted as rapidly spinning neutron stars with rotation periods in the range 4.3 to 1.6 msec, with an average of 0.79 sec. The pulsar in the Crab nebula (Fig. 14) has a period of 1/30 sec. The origin of the beamed radiation, which is also observed in the optical and X-ray regions of the spectrum, is not known. The electromagnetic energy is emitted at the expense of the rotational energy, and the rotation periods are gradually becoming longer.

D. Black Holes

Neutron stars also have a limiting mass, but it is not known as well as that for white dwarfs because the physics of the interior, in particular the equation of state, is not fully understood. A reasonable estimate is 2–3 $M_\odot$. If the collapsing core of a massive star exceeds this limit, the collapse will not be stopped at neutron star densities because the pressure gradient can never be steep enough to balance gravity. The collapse continues indefinitely, to infinite density, and a black hole is formed. Stars with original masses in the range 10–20 $M_\odot$ form cores of around 1.4 $M_\odot$; they will become supernovae and leave a neutron star as a compact remnant. Stars above 25–30 $M_\odot$ have larger cores at the time of collapse, and the amount of mass that collapses probably exceeds the neutron star mass limit; the result is a black hole.

E. Supernovae

Only very small fraction of stars ever become supernovae. Two main types are recognized. The supernovae associated with the collapse of the iron core of massive stars are called type II. They tend to be observed in the spiral arms of galaxies, where recent star formation has taken place. The luminosity at maximum light is about 10^{10} $L_\odot$, and the decline in luminosity with time takes various forms with a typical decline time of a few months. The observed temperature drops from about 15,000 to 5000 K during the first 30–60 days, and the expansion velocity decreases from 10,000 to 4000 km sec^{-1} over the same time period. The abundances of the elements are similar to those in the sun, with hydrogen clearly present. The type I supernovae, on the other hand, tend to be observed in regions associated with an older stellar population, for example, in elliptical galaxies. The luminosity at maximum is about the same as that for type II, and the curves of the decline in light with time are all remarkably similar, with a rapid early decline for 30 days and then a more gradual exponential decay in which the luminosity decreases by a factor 10 in typically 150 days. Temperatures are about the same as in type II, the expansion velocities are somewhat higher, and the element abundances are quite different: there is little evidence for hydrogen, or helium and the spectrum is dominated by lines of heavy elements. The origin of the type I supernova is controversial, but one promising model, which applies to the so-called type Ia, is that it represents the end point of the evolution of a close binary system. Just before the explosion the system consists of a white dwarf near the limiting mass and a companion of relatively low mass, which lasts for several Gyr on the main sequence. When the companion evolves, it expands and at some point transfers mass to the white dwarf, which therefore is accreting material that is hydrogen rich. The hydrogen burns near the surface, producing a layer of helium that increases in mass. By the time the helium is hot enough to burn, it does so under degenerate conditions. The temperature increases rapidly to the point where the carbon ignites explosively. Much of the material is processed to heavier elements such as iron, nickel, cobalt, and silicon. It still is not entirely certain whether the whole star blows up or whether a neutron star or white dwarf remnant is left. The light emitted is produced primarily from the radioactive decay of ^{56}Ni. In any case, the type Ia supernova is essentially a nuclear explosion, while the type II supernova is driven ultimately by the rapid release of gravitational energy during core collapse.

VIII. SUMMARY: IMPORTANT UNRESOLVED PROBLEMS

The study of stellar structure and evolution has resulted in major achievements in the form of quantitative and qualitative agreement with observed features, in the pre-main-sequence, main-sequence, and post-main-sequence phases. A number of aspects of the physics need to be improved, and improvement in many cases means extending the theory from the relatively simple one-dimensional calculations into two- and three-dimensional hydrodynamic simulations. Considerable uncertainty is caused by a lack of a detailed theory of convection that can be applied in stellar interiors. On the main sequence this uncertainty could result in relatively minor changes in the surface layers of stars, mainly in the mass range 0.8–1.1 $M_\odot$. Three-dimensional hydrodynamic simulations of convection near the surface of sunlike stars provide very good agreement with the observed properties of solar granulation and can provide a calibration for the value of the mixing length to be used in long-term calculations over billions of years. However, the question of overshooting at the edges of interior convection zones still needs to be solved. In pre-main-sequence and post-main-sequence stars, which have deep extended convection zones, a better theory could result in some major changes. There are several examples of situations where observations suggest mixing of the products of nucleosynthesis to the surface of stars, but the theoretical models of convection zones do not provide sufficiently deep mixing. Other possible mixing mechanisms, including overshoot, need to be considered. Improvement in laboratory measurements of nuclear reaction rates, particularly for the pp chains, could reduce the theoretical error bar in the rate of solar neutrino pro-

duction. The continued discrepancy in the solar neutrino detection rates, along with the excellent agreement between the solar model and the helioseismological results, suggests that neutrino oscillations may actually be the solution to the puzzle. New experiments under way, such as SNO, should provide further clues on this question.

The assumption of mass conservation is built into much of the theory discussed in this article. However, it is known from observation that various types of stars are losing mass, including the T Tauri stars in the early pre-main-sequence contraction phase, the O stars on the upper main sequence, and many red giant stars. The physics of the ejection process is not well explained, especially in the case of rapid mass loss leading to the planetary nebula phase. Also, in the early phases of evolution, the mass loss, often in the form of bipolar ouflows and jets, is associated with simultaneous accretion of mass onto the star through a disk. The physics of the mass and angular momentum transfer at this stage needs to be clarified.

Binary star evolution has not been treated here, but numerous fascinating issues remain for further study. If the two components interact, the problem again involves three-dimensional hydrodynamics. The first difficulty is the formation process. Most stars are found in double or multiple systems, but their origin, presumably a result of fragmentation of collapsing, rotating interstellar clouds, is not well understood, particularly for the close binaries. A further problem is the explanation of short-period systems consisting of a white dwarf and a nearby low-mass main-sequence companion. These systems are associated with nova outbursts and possibly supernovae of type I. The problem is that the orbital angular momentum of such a binary is much smaller than that of the original main-sequence binary from which the system must have evolved. A goal of current research is to uncover the process by which the loss of angular momentum occurs. A likely candidate is the "common envelope" phase. An expanding red giant envelope interacts with its main-sequence companion, so that the latter spirals into the giant, losing angular momentum and energy because of frictional drag and gravitational torques. The energy lost from the orbital motion is deposited in the envelope, resulting in its ejection, leaving the main-sequence star in a short-period orbit about the white dwarf core of the giant. Clarification of this process requires detailed three-dimensional hydrodynamic simulations.

The effects of rotation and magnetic fields have also not been discussed in this article. However, their interaction in the context of stellar evolution is certain to produce some very interesting modifications to existing theory. During the phase of star formation these processes are central, and consideration of them is crucial for the clarification of the processes of binary formation and generation of jets.

During main-sequence evolution, rotational and magnetic energies, although present, are almost certainly small compared with gravitational energy, and their overall effect on the structure is therefore insignificant. However, the circulation currents induced by rotation could be important with regard to mixing and the exchange of matter between the deep interior and the surface layers. During post-main-sequence phases, the cores of stars contract to very high densities, and if angular momentum is conserved in them from the main-sequence phase, they would be expected to rotate very rapidly by the time the core becomes degenerate. However, white dwarfs are rotating slowly, suggesting that at some stage almost all of the angular momentum is transferred, by an as yet unexplained process, out of the cores of evolving stars. The same problem applies to the rotation of neutron stars.

A number of other problems present themselves. The complicated physical processes involved in the generation of supernovae explosions and the formation of neutron stars require further study, in some cases involving three-dimensional hydrodynamics. Here, rotational and magnetic effects again become important, as well as neutrino transfer, development of jets, and the possible generation of gamma-ray bursts. The observational and theoretical study of stellar oscillations can provide substantial information on the structure and evolution of stars. For example, the interior of the sun has been probed through the analysis of its short-period oscillations, and the techniques can be extended to other stars. For many types of stars (including the sun!), the physical mechanism producing the oscillations is not understood. The determination of the ages of globular clusters, presumably containing the oldest stars in the galaxy, through the use of stellar evolutionary tracks, is a particularly interesting problem as it helps to constrain cosmological models. It is clear that the study of stellar evolution involves a wide range of interacting physical processes and that the findings are of importance in numerous other areas of astronomy and physics.

SEE ALSO THE FOLLOWING ARTICLES

BINARY STARS • DARK MATTER IN THE UNIVERSE • GALACTIC STRUCTURE AND EVOLUTION • NEUTRINO ASTRONOMY • NEUTRON STARS • STAR CLUSTERS • STARS, MASSIVE • STARS, VARIABLE • STELLAR SPECTROSCOPY • SUPERNOVAE

BIBLIOGRAPHY

Bahcall, J. N. (1989). "Neutrino Astrophysics," Cambridge Univ. Press, Cambridge, UK.

Bowers, R., and Deeming, T. (1984). "Astrophysics I. Stars," Jones & Bartlett, Boston.

Clayton, D. D. (1983). "Principles of Stellar Evolution and Nucleosynthesis," Univ. of Chicago Press, Chicago.

Goldberg, H. S., and Scandron, M. (1981). "Physics of Stellar Evolution and Cosmology," Gordon & Breach, New York.

Kippenhahn, R. (1983). "100 Billion Suns," Basic Books, New York.

Kippenhahn, R., and Weigert, A. (1990). "Stellar Structure and Evolution," Springer-Verlag, Berlin.

Phillips, A. C. (1994). "Physics of Stars," Wiley, Chichester.

Shklovskii, I. S. (1978). "The Stars: Their Birth, Life, and Death," Freeman, San Francisco.

Tayler, R. J. (1994). "The Stars: Their Structure and Evolution," Cambridge Univ. Press, Cambridge, UK.

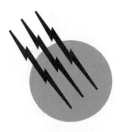

Stereochemistry

Ernest L. Eliel

University of North Carolina

GLOSSARY

Angle strain Excess potential energy of a molecule caused by deformation of an angle from the normal, e.g., of a C–C–C angle from the tetrahedral. Also called "Baeyer strain."

Anti or antiperiplanar Conformation of a segment X–C–C–Y in which the torsion angle (which see) is 180° or near 180°. X and Y are said to be anti or antiperiplanar (to each other).

Cahn–Ingold–Prelog (C-I-P) Descriptors Descriptors *R* and *S* (and others) used to describe the spatial arrangement (configuration) of ligands at a chiral center or other chiral element (chiral axis, chiral plane).

Chiral center An atom to which a C-I-P descriptor can be attached, usually a group IV (tetrahedral) or group V (pyramidal) atom. Reflection of the molecule must reverse the descriptor.

Chirality Handedness; the property of molecules or macroscopic objects (such as crystals) of not being superposable with their mirror images. Substances may be called chiral if the constituent molecules are chiral, even if the substance is racemic.

Configuration The spatial arrangement of atoms by which stereoisomers are distinct, but disregarding facile rotation about single bonds.

Conformation The (differing) spatial array of atoms in molecules of given constitution and configuration produced by facile rotation about single bonds.

Conformers Stable conformations (located at energy minima). Isomers which differ by virtue of facile rotation about single bonds, usually readily interconvertible.

Constitution (of a molecule) The nature of its constituent atoms and their connectivity.

Diastereomers (diastereoisomers) Stereoisomers that are not mirror images of each other.

Eclipsed conformation Conformation of a nonlinear array of four atoms X–C–C–Y in which the torsion angle (which see) is zero or near zero (also "synperiplanar"). X and Y are said to be eclipsed.

Enantiomer(ic) excess In a partially or completely

resolved substance, the excess of one enantiomer over the other as a percentage of the total.

Enantiomers Stereoisomers that are mirror images of each other.

Enantiomorphs Macroscopic objects (e.g., crystals) whose mirror images are not superposable with the original entity.

Factorization The analysis, where possible, of chirality in terms of individual chiral elements, such as chiral centers, chiral axes, and chiral planes.

Gauche Conformation of a segment X–C–C–Y in which the torsion angle (which see) is ±60° or near ±60° (also "synclinal"). X and Y are said to be gauche (to each other).

Kinetic resolution Separation of enantiomers by virtue of their unequal rates of reaction with a nonracemic chiral reagent.

Optical activity The property of chiral assemblies of molecules or chiral crystals of rotating the plane of polarized light.

Optical purity The ratio of the observed specific rotation of a substance to the specific rotation of the enantiomerically and chemically pure substance.

Optical rotation Rotation of the plane of polarized light produced by the presence of chiral molecules or chiral crystals in the light path.

Racemic mixture (racemate) A composite of equal amounts of opposite enantiomers.

Racemization Conversion of individual enantiomers into a racemic mixture.

Resolution Separation of enantiomers from a racemic mixture.

Staggered conformation Conformation of a nonlinear array of four atoms (A–B–C–D) in which the torsion angle (which see) is near 60°. Ligands A and D are said to be *gauche* or *synclinal* to each other.

Stereochemistry Chemistry in three dimensions; topographical aspects of chemistry.

Stereogenic center An atom where exchange of two of its ligands changes the configuration. Such an atom is usually, but not necessarily, a chiral center.

Stereoisomers Isomers of the same constitution but differing in spatial arrangement of their constituting atoms.

Torsion angle In a nonlinear array of four atoms A–B–C–D, the angle between the plane containing A, B, and C and the plane containing B, C, and D.

Torsional strain Excess potential energy of a molecule caused by incomplete staggering of pertinent bonds (e.g., deviation of the torsion angle in H–C–C–H from the normal 60°). Also called "Pitzer strain."

van der Waals or compression strain Excess potential energy of a molecule caused by approach of two or more of its constituting atoms within the repulsive regime of the van der Waals potential. Also called "nonbonded (repulsive) interaction."

I. INTRODUCTION

This article deals with the stereochemistry of organic compounds, although many of the general principles also apply to organometallic and inorganic compounds.

The term "stereochemistry" is derived from the Greek "stereos" meaning solid—it refers to chemistry in three dimensions. Since nearly all organic molecules are three dimensional (with the exception of some olefins and aromatics to be discussed later), stereochemistry cannot be considered a branch of chemistry. Rather it is an aspect of all chemistry, or, to put it differently, a point of view which has become increasingly important and, as will be shown, essential for the understanding of chemical structure and function.

II. HISTORY. OPTICAL ACTIVITY

Historically, stereochemical thinking developed from the observations of J. B. Biot in 1812–1815 that quartz crystals as well as solutions of certain organic substances rotate the plane of polarized light; this phenomenon is called "optical rotation." When the beam of light coming toward the observer rotates the plane to the right (clockwise) we call the rotation positive (+); when the rotation is to the left or counterclockwise, it is negative (−). The angle of rotation α is proportional to the concentration c (conventionally expressed in grams per milliliter, i.e., density, for solids and pure liquids and gases) and the thickness of the layer—usually the length of the cell in which the liquid or solution is contained (conventionally expressed in decimeters): $\alpha = [\alpha] \cdot l \cdot c$ ("Biot's law"). The proportionality constant $[\alpha]$ is called "specific rotation": $[\alpha] = \alpha / l \, (\text{dm}) \cdot c \, (\text{g/cm}^3)$. For solutions, where the concentration c' is conventionally expressed in g/100 ml, $[\alpha] = 100\alpha / l \, (\text{dm}) \cdot c' \, (\text{g/100 ml})$. The constant $[\alpha]$ (conventionally reported without units) is used for the characterization of "optically active" substances (i.e., which display optical rotation). It varies with temperature and with the wavelength of the light used in the observation, which need to be indicated (as subscripts and superscripts, respectively): thus, $[\alpha]_{\lambda}^{t}$, where t is the temperature in °C and λ is the wavelength in nanometers (nm). Unfortunately, since most substances are prone to solvation and intermolecular association in solution or in the liquid state, and since such association varies with concentration and solvent, these items must also be recorded. A typical specific rotation might thus read $[\alpha]_D^{20} + 57.3 \pm 0.2$ (95% EtOH, $c = 2.3$), denoting measurement at 20°C at the sodium D line (589 nm) in 95% ethanol at a concentration of 2.3 g/100 ml. (Monochromatic light of this wavelength is easily generated in a sodium vapor lamp and has thus classically been used for polarimetric measurements.)

In 1822 Sir John Herschel found that mirror-image crystals of quartz (discovered by R. J. Haüy in 1801 and called "enantiomorphs") rotate the plane of polarization in opposite directions. This provided the first correlation of optical rotation with enantiomorphism, in this case of crystals. In 1848 Louis Pasteur achieved separation of the enantiomorphous crystals he detected in the sodium–ammonium salt of the (optically inactive) paratartaric acid (today called racemic tartaric acid, see below) and thereby obtained two different substances, one of which rotated polarized light to the right, the other to the left, even in aqueous solution. Pasteur concluded that the enantiomorphous crystals were made up of molecules that themselves differed as object and (reflection) image on the molecular scale. Such mirror-image molecules are called "enantiomers" and the separation which Pasteur accomplished is called "resolution." Since the difference between enantiomers resembles the difference between a right and a left hand (which are also mirror images of each other, but otherwise essentially identical in their dimensions), we call such molecules "chiral" (Greek "cheir" = hand) and we call the property of certain molecules to display enantiomerism "chirality" (terms coined by Lord Kelvin in 1893). While every molecule has a mirror image, chirality exists only if the image is nonsuperposable with the original molecule, just as a right shoe is not superposable with a left one. (In contrast, socks worn on right and left feet are superposable; they are "achiral.")

III. STEREOISOMERISM

The understanding of molecular structure was not well enough advanced in 1848 for Pasteur to explain enantiomerism (chirality) in terms of atomic arrangement. The basis for that was laid only a decade or so later when, in separate publications, A. S. Couper, F. A. Kekulé, J. Loschmidt, and A. Crum Brown illuminated the structure of molecules in terms of the connectivity between their constituent atoms. Then, in 1874, J. H. van't Hoff in the Netherlands and J. A. Le Bel in France simultaneously proposed the structural basis for chirality: When four different atoms or groups (jointly called "ligands") are attached *tetrahedrally* to a given carbon atom, two mirror-image arrangements are possible (Fig. 1) corresponding to the two enantiomers. The enantiomers are said

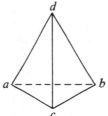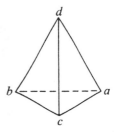

FIGURE 1 Tetrahedral molecule Cabcd (the carbon atom at the center of the tetrahedron is conventionally not shown). [Reprinted with permission from Eliel, E. L., and Wilen, S. H. (1994). "Stereochemistry of Organic Compounds," Wiley, New York.]

to differ in "configuration." Thus they explained not only the stereoisomerism of such simple molecules as CHF-BrI or $CH_3CHOHCO_2H$, but also that of more complex molecules such as tartaric acid (Fig. 2). The two tartaric acids, A and B in Fig. 2 (separated by Pasteur as sodium–ammonium salts), are enantiomers. Pasteur's starting salt came from a 1:1 mixture of the two, called "racemic tartaric acid." A "racemate" is a mixture of equal quantities of corresponding enantiomers. Since the numerical values for the optical rotations of the two enantiomers (negative, or −, for one; positive, or +, for the other) are equal and opposite, the racemate displays no optical activity.

In later work Pasteur discovered a fourth species of tartaric acid [if we count the enantiomers A and B and the racemate (A + B) as three distinct species]. He was not able to explain the nature of this optically inactive isomer, but the Le Bel–van't Hoff theory led to assignment of the structure C (Fig. 2) to this stereoisomer. It has the same connectivity (constitution) as A and B but differs in configuration. The reason for its lack of optical activity is that it is superposable with its mirror image D; i.e., it is achiral. It is a stereoisomer of A and B, but not an enantiomer. Such stereoisomers that are not mirror images of each other are called "diastereomers"; thus C is a diastereomer of A and of B (and vice versa), whereas A is an enantiomer of B (and vice versa). Compound C is called "meso-tartaric acid," the prefix "meso" indicating that it is the achiral member in a set of diastereomers that also contains chiral members.

IV. STEREOISOMERISM OF ALKENES AND CYCLANES

So far it would appear that stereoisomerism is dependent on three-dimensional structure, but van't Hoff recognized that stereoisomers can also exist in two dimensions, as in *cis*- and *trans*-disubstituted ethylenes (Fig. 3). The substituents may be on the same side (*cis*) or on opposite sides (*trans*) at the two ends of the (planar) double bond.

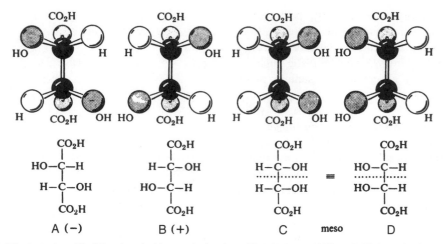

A (−)   B (+)   C   meso   D

FIGURE 2 The tartaric acids. [Reprinted with permission from Eliel, E. L., and Wilen, S. H. (1994). "Stereochemistry of Organic Compounds," Wiley, New York.]

The classical *cis/trans* nomenclature for alkenes works well for 1,2-disubstitued ethenes (A, B), but with tri- (C) and tetrasubstituted species it is not unequivocal. Thus in the propenoic acid (C), Br is *trans* to CO_2H but *cis* to Cl. For an unequivocal description, substituents at the same terminus are ordered by the Cahn–Ingold–Prelog system (see below; for the present purpose it suffices to recognize that substituents of higher atomic number have priority over those of lower atomic number). Descriptors in this currently used system, are *Z* (for the German *zusammen*) if the higher priority ligands are on the same side of the double-bond system and *E* (for *entgegen*) when they are on opposite sides. Thus C in Fig. 3 is (*Z*)-2-chloro-3-bromopropenoic acid.

Cis–trans isomerism is also found in cyclanes, which, for the purpose of counting stereoisomers, may be considered planar (but see below). Figure 4 shows the *cis–trans* isomerism of 1,2- (A–C) and 1,3- (D, E) dichlorocyclobutanes. The situation in the 1,2 isomers is similar to that in the tartaric acids: there are two enantiomers (A, B; *trans*) plus an (achiral) meso isomer (C; *cis*), which is a diastereomer of A and B. The situation in the 1,3 isomers (D, *cis*; E, *trans*) is different. While D and E are diastereomers, neither of them is chiral (each one is superposable with its own mirror image). Carbons 1 and 3 are not chiral centers since there is no chirality in the molecule, and yet, changing their relative position (*cis* or *trans*) gives rise to (dia)stereoisomers. Carbons 1 and 3 are therefore called "stereogenic." All chiral centers are stereogenic, but, as seen in this case, not all stereogenic centers are chiral centers. The *E/Z* system is *not* used for cycloalkanes.

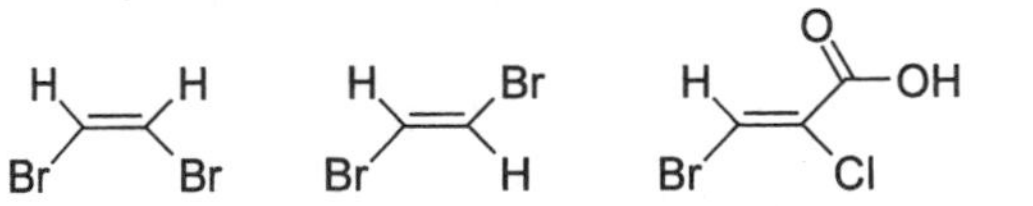

V. PROPERTIES OF STEREOISOMERS

Enantiomers, though not superposable, resemble each other very closely (as do right and left hands). The distances (both bonded and nonbonded) between corresponding constituent atoms are the same, and thus

A   B   C

Z (cis)   E (trans)   Z

FIGURE 3 The *E, Z* nomenclature for *cis* and *trans* substituted alkenes.

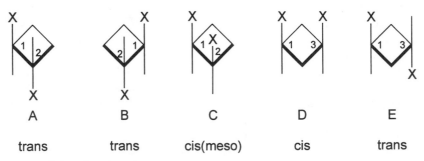

A   B   C   D   E

trans   trans   cis(meso)   cis   trans

FIGURE 4 Stereoisomers of 1,2- and 1,3-disubstituted cyclobutanes.

enantiomeric species have been called "isometric." In the absence of a second source of chirality their physical and chemical properties are identical. This is true of melting and boiling points, density, refractive index, a whole host of spectral properties (UV, IR, NMR, mass spectrum, etc.), thermodynamic properties, chromatographic behavior, and reactivity with achiral reagents (thus, enantiomeric esters are hydrolyzed in the presence of achiral acids or bases at the same rate.) A macroscopic analogy is that of right and left feet (chiral) which fit equally into either one of a pair of socks (achiral).

However, when the chemical or physical agents are themselves chiral, this equivalence breaks down, just as left and right feet do not fit equally well into a right shoe. Examples are esterification of racemic α-methyl benzyl alcohol, $C_6H_5CHOHCH_3$, with the 2,4-dichlorophenyl ether of (+)-lactic acid, $Cl_2C_6H_3OCH(CH_3)CO_2H$ [the (−) enantiomer reacts faster], irradiation of racemic α-azidopropanoic acid dimethylamide, $CH_3CH(N_3)CON(CH_3)_2$, with left-circularly polarized light (chiral; see below), which leads to faster photochemical destruction of the (+) enantiomer and chromatography on chiral stationary phases.

The situation is otherwise with diastereomers, which are not isometric and therefore do not have identical physical or chemical properties even in the absence of external chirality. Thus, meso-tartaric acid (C in Fig. 2) melts at 140°C, whereas the diastereomeric (+) and (−) acids A and B both melt at 170°C (as explained above, A and B have the same melting point even though their optical rotations are opposite).

VI. SIGNIFICANCE

Optical rotation is useful to characterize enantiomers of chiral substances and has played an important historical role in the development of stereochemistry. However, the real significance of stereoisomerism lies in the concept of fit or misfit, epitomized by the fit of a right glove with a right hand and the corresponding misfit of a left glove. In 1858, Pasteur discovered that, when a solution of racemic tartaric acid is inoculated with the mold *Penicillium glaucum*, the (+) acid is metabolized (and thereby destroyed) by the microorganism and the (−) acid remains behind. This selectivity is due to the organism's enzymes being able to interact with the naturally occurring (+) enantiomer, but not with the other (−). A further example is the hydrolysis of the acetyl derivative of a racemic α-amino acid, $RCH(NHCOCH_3)CO_2H$, promoted by the enzyme hog kidney acylase. Only the acyl derivative of the (naturally occurring) L-amino acid (see below for notation) is hydrolyzed to $L\text{-}RCH(NH_2)CO_2H$, which can then be readily separated from the unhydrolyzed

$D\text{-}RCH(NHCOCH_3)CO_2H$. The latter can separately be hydrolyzed (with aqueous acid) to $D\text{-}RCH(NH_2)CO_2H$; both pure D and L acids can thus be prepared from the racemate. This process is called "kinetic resolution" ("kinetic" because it depends on reaction rate; one enantiomer reacts relatively rapidly, the other much more slowly or not at all). The hand-and-glove analogy applies here: One enantiomer fits into the active site of the acylase enzyme and its hydrolysis is thus promoted; the other enantiomer does not fit, just as a right glove (substrate) fits on a right hand (enzyme analog), whereas a left glove does not fit. Considering the matter more generally, the right-hand/right-glove combination is diastereomeric to the right-hand/left-glove combination; diastereomers have different free energies and so, presumably, have the transition states leading to their formation. The effectiveness of the kinetic separation (resolution) depends on the difference in activation energies (and the associated difference in reaction rate): The greater the difference, the better the resolution. Enzymes tend to excel here because their relatively complex topography usually leads to a large difference in activation energy between enantiomeric substrates (fit vs. misfit in the transition state) and hence a high degree of discrimination.

A similar discrimination is seen in pharmaceuticals, whose action depends on their interaction with enzymes (if they are enzyme inhibitors) or other receptors. By way of example, only the (−) enantiomer of α-methyldopa (α-methyl-3,4-dihydroxyphenylalanine) is an antihypertensive since it is enzymatically converted in the body to the pharmacologically active α-methylnorepinephrine. The (+) enantiomer is inactive. We call the active enantiomer "eutomer" and the inactive one "distomer." Since most drugs have some side effects and therefore should be used at the minimal effective dose, the pharmaceutical industry is much interested in specifically preparing the eutomer, free of the ineffectual and potentially detrimental distomer (which constitutes 50% of the racemate). Other areas where enantiomers differ in function are in the agriculture and flavor industries. Thus in the case of the agrochemical paclobutrazol, the (+) enantiomer is a fungicide, whereas the (−) enantiomer is a plant growth regulator. Concerning flavor, the (−) enantiomer of carvone (2-methyl-5-isopropylidenecyclohexanone) has the odor of spearmint, whereas its (+) enantiomer has the odor of caraway.

VII. SEPARATION (RESOLUTION) AND RACEMIZATION OF ENANTIOMERS

Since enantiomers are so similar to each other, it is not surprising that their separation requires special methodology, which can be summarized as follows:

84

1. Separation by crystallization (Pasteur's first method, 1848)
2. Separation by formation and separation of diastereomers (Pasteur's second method, 1853): (a) separation by crystallization, (b) separation by chromatography
3. Asymmetric transformation (a) of diastereomers, (b) of enantiomers
4. Kinetic resolution: (a) chemical, (b) enzymatic (Pasteur, 1858)
5. Separation by chromatography on chiral stationary phases
6. Enantioselective synthesis
7. Synthesis from enantiomerically pure precursors (sometimes called enantiospecific synthesis)
8. Miscellaneous methods

Pasteur's method of manually separating enantiomorphous crystals of enantiomers is obviously not practical; it is also not general. Racemic mixtures can lead to three different kinds of crystals: conglomerates, racemic compounds, and racemic solid solutions. Compounds (where the unit cell, the smallest unit of a crystal, contains equal numbers of enantiomeric molecules) are unsuitable for Pasteurian resolution, as are racemic solid solutions. Only when the enantiomers crystallize in discrete crystals (i.e., as a macroscopic mixture called a "conglomerate") can the two types of crystals be separated even in principle. But only ca. 10% of racemic mixtures (ca. 20% in the case of salts) crystallize as conglomerates. When they do, separation can be achieved by a modification of Pasteur's technique (called the "method of entrainment") involving alternate seeding with one enantiomer, separating the additional crystalline material formed, replenishing the solution with racemate, then seeding with the opposite enantiomer, thereby inducing crystallization of the second enantiomer. Large quantities of enantiomers, for example, of glutamic acid, have been separated in this manner commercially.

Pasteur's second method, formation of diastereomers by reaction of a racemate with an optically active auxiliary or "adjuvant" ("resolving agent"), is much more common. Common resolving agents are naturally occurring chiral acids [such as (−)-malic acid, $HO_2CCHOHCH_2CO_2H$] for chiral bases, and naturally occurring chiral bases, such as (−)-quinine, for chiral acids. The salts formed are often crystalline and can be separated by fractional crystallization. After separation, the resolved acid or base is liberated by treatment of the salt with mineral acid or base, respectively. Alternatively, covalent diastereomers may be formed [e.g., by esterification of a racemic acid with (−)-menthol (2-isopropyl-5-methylcyclohexanol)] and separated by some type of chromatography on an

achiral stationary phase. After separation the chiral auxiliary is removed (e.g., by hydrolysis in the case of an ester).

To understand the third method, "asymmetric transformation," we must first take up "racemization," i.e., the conversion of one of the enantiomers into a racemate. This apparently counterproductive process is actually useful: In resolution the undesired enantiomer is produced as an equimolar by-product. Racemization of that enantiomer allows one to start the resolution process over. Actually, racemization involves converting one enantiomer into the opposite one, but since enantiomers have the same free energy, the equilibrium constant between them is unity, i.e., the product of equilibration is the racemate. However, racemization, to be feasible, requires a chemical pathway. For example, a chiral ketone, $RR'CHCOR''$ may be racemized by base via the resonance-stabilized achiral enolate anion $RR'C^-COR'' \Leftrightarrow RR'C=CR''O^-$.

If such equilibration occurs concomitant with resolution, it is sometimes possible to convert the racemate entirely into one of the enantiomers. This process, called "crystallization-induced asymmetric transformation," may be observed during crystallization of diastereomers when the stereoisomers to be resolved can simultaneously be equilibrated. An example is phenylglycine, $C_6H_5CH(NH_2)CO_2H$, required as the (−) isomer in manufacture of the antibiotic ampicillin. Equilibration of the enantiomers is effected by adding benzaldehyde to the racemic material (resulting in reversible formation of a Schiff base which is readily racemized) and precipitation of the desired (−) acid as its salt with (+)-tartaric acid. The (+) isomer is concomitantly reconverted to the racemate; in the end, nearly the entire amino acid crystallizes as the (+)-tartrate of the (−) acid.

Kinetic resolution has already been discussed. Purely chemical approaches (exemplified by the resolution of chiral allylic alcohols, e.g., $C_6H_{11}CHOHCH=CHCH_3$, with Sharpless' reagent, which contains isopropyl tartrate as the chiral constituent) are currently rare; enzymatic methods (e.g., hydrolysis of esters of chiral alcohols catalyzed by lipase enzymes) are more common since enzymes are frequently highly selective for one enantiomer over the other and the effectiveness of kinetic resolution depends on the degree of selectivity.

While ordinary chromatography does not separate enantiomers (though it can lead to separation of diastereomers), enantiomers can be separated by chromatography employing a chiral stationary phase enriched in a single enantiomer. In that case, the interactions between the two enantiomers of the analyte and the chiral stationary phase are diastereomeric in nature and therefore often differ in strength, the stronger interaction leading to longer retention time.

Asymmetric (or enantioselective) synthesis is a targeted method for obtaining individual enantiomers from achiral precursors. Ordinarily, the introduction of a chiral center (or other chiral element) in the course of synthesis leads to equal formation of the two enantiomers, i.e., to a racemate. Selectivity can, however, be achieved by using a chiral (enantiomeric) reagent or catalyst, in which case the transition states leading to the two enantiomeric products are diastereomeric, or by attaching a chiral auxiliary (see above) to the starting material so that the products are diastereomers rather than enantiomers, in which case their free energies and the activation energies for their formation also differ and one isomer is formed in preference over the other. At the end of the reaction, the chiral auxiliary is chemically removed. By way of example: Reduction of pyruvic acid, CH_3COCO_2H, to lactic acid with an achiral reagent such as sodium borohydride gives racemic lactic acid, $CH_3CHOHCO_2H$, since approach of hydride from either face of the keto-carbonyl group is equally likely. But when reduction (by a reducing coenzyme) is carried out in the presence of the chiral enzyme lactic acid dehydrogenase, only (+)-lactic acid is formed.

A distinct process, sometimes called "enantiospecific synthesis," is one in which an enantiomerically pure starting material (natural or man-made) is converted by standard reactions into an enantiomerically pure product.

Among other methods are diffusion through chiral membranes and partition methods involving chiral solvents.

VIII. STRUCTURE, CONFIGURATION, NOTATION

By "structure" of a molecule we understand the totality of the nature and array of its atoms. This comprises the identity and connectivity of these atoms ("constitution") and their arrangement in space ("configuration," "conformation"; see below). Structure may be determined by X-ray, electron, or neutron diffraction of a crystal of the substance in question.

Constitution can usually be inferred from elemental analysis and chemical degradation or by various spectroscopic methods, such as nuclear magnetic resonance. Constitutional isomers, such as butane and 2-methylpropane, have the same elemental composition but differ in connectivity. In contrast, the two enantiomers of lactic acid, $CH_3CHOHCO_2H$, identical in constitution, differ in the spatial disposition of the ligands at C(2): They are said to differ in "configuration" and may be called configurational isomers. While configuration is a property of the molecule as a whole, it is convenient to "factorize" it into elements of chirality, notably the "center of chirality" [e.g., at C(2)

FIGURE 5 Ordering substituents (A > B > C > D) according to the Cahn–Ingold–Prelog chirality rule.

in lactic acid]. This "factorization" allows us to specify configuration by assigning a configurational "descriptor" to each chiral center or other chiral element.

The configurational descriptors now universally used are "R" (for *rectus*, right, in Latin) and "S" (for *sinister*, left) (Cahn *et al.*, 1966); they are sometimes called "CIP descriptors" after their proponents. To assign a descriptor to a chiral center, assume that its four ligands, a, b, c, and d, can be ordered by some convention: a > b > c > d. The molecule is then viewed from the side away from the lowest ranked ligand (d) such that the other three ligands (a, b, c) lie in a plane (Fig. 5). If a–b–c then describe a clockwise array, the descriptor is R; if the array is counterclockwise, the descriptor is S. To establish priorities in a real molecule one orders the ligands by atomic number, thus in CHFClBr, Br > Cl > F > H. For lactic acid the priority is O > C > H, but no immediate decision is reached for CH_3 versus CO_2H. Here one goes out to the next atom away from the chiral center: O for CO_2H and H for CH_3 and since O > H, CO_2H has priority over CH_3. Where still no decision is reached, one goes out one tier more, thus $–CH_2CH_2OH$ has priority over $–CH_2CH_2CH_2OH$. Once a decision is reached, the process stops. In the outward path, one always gives preference to the atom of higher priority; thus $–NHCl$ has priority over $–N(CH_3)_2$: Cl > C overrides C > H. All ligands on the atom reached must be probed, thus $–CO_2H$ > $–CHO$ (two O's over one O). When the lack of a decision is caused by a doubly bonded ligand (e.g., CH=O vs. CH_2OH as in glyceraldehyde, $HOCH_2CHOHCH=O$) the double bond is replaced by two single bonds with the ligands "complemented" at either end; thus $–CH=O$ is considered as O–CH–(OC) and thus has priority over CH_2OH. An absent ligand (as the lone pair in N: or :O:) is considered to have atomic number zero. When chirality is due merely to the presence of an isotope, as in C_6H_5CHDOH, the ligand of higher atomic weight is given priority: O > C > D > H. For more complicated cases, the reader is referred to standard texts.

There are two exceptions to the current use of CIP descriptors, α-amino acids and sugars, where an older nomenclature is often used. Before considering this point, a discussion of projection formulas is required. Since molecules are three dimensional but paper is planar,

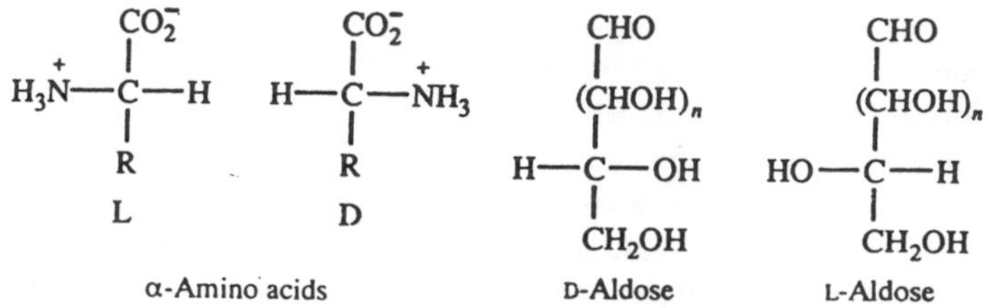

FIGURE 6 The D and L nomenclature for α-amino acids and aldohexoses. [Reprinted with permission from Eliel, E. L., and Wilen, S. H. (1994). "Stereochemistry of Organic Compounds," Wiley, New York.]

several conventions have been developed to represent molecules (or rather their three-dimensional models) in two dimensions. In one of them, the molecule is written with as many atoms as possible in the plane of the paper and with additional atoms being connected with a heavy line (—) when the attached ligand is in front of the plane and with a dashed line (- - -) when it is behind. In another, the so-called "Fischer projection" (after its originator Emil Fischer) the tetrahedron is oriented so that two of the groups (top and bottom) are pointed away from the observer, with the other two (sideways groups) pointing toward the observer, and the molecule is projected in this fashion. Figure 6 shows Fischer projections of α-amino acids and of monosaccharides (aldoses) with the further proviso that the most oxidized group is to be placed at the top and the NH_2 or OH and H ligands on the side. In α-amino acids the symbol D is used when the NH_2 group is on the right, L if it is on the left; it turns out that all naturally occurring α-amino acids are L. In the case of monosaccharides, the chiral center furthest away from the aldehyde or ketone function determines the descriptor and the symbol used is D when the OH is on the right in the Fischer projection formula, L if it is on the left, independent of the configuration of any of the other chiral centers.

The Fischer projection formulas shown above, while useful for the assignment of the descriptors, do not correspond to the actual shape of most molecules. As explained below under the topic of conformation, most molecules exist in "staggered" (Fig. 7) rather than the "eclipsed" conformations implied in Fischer projections. A more realistic representation of, say, (R,R)-tartaric acid is shown in Fig. 7, which, in addition to the unrealistic three-dimen-

sional formula corresponding to the Fischer projection, displays the staggered conformation in a three-dimensional, so-called "saw-horse" formula and its projection (seen from one end of the molecule) in a so-called Newman projection. [This staggered representation is generated from the eclipsed one by rotation about the C(2)–C(3) bond.]

IX. DETERMINATION OF CONFIGURATION

Since configuration is an integral part of structure, the determination of the architecture of any molecule, naturally occurring or synthetic, is not complete until its configuration is known. For example, the fit of a drug with its receptor or of an inhibitor with an enzyme cannot be understood (or modeled, or rationally improved) absent information on its configuration.

Configuration may be relative or absolute. To say that a right hand fits a right glove is to make a statement of the *relative* configuration of the two. Even a small child may be able to make this correlation. But to recognize that a picture of a glove is that of a right glove in an *absolute* sense is more difficult. The same applies to the determination of the absolute configuration (or sense of chirality) of a molecule.

There are many ways of determining relative configuration (Eliel and Wilen, 1994). The most straightforward one, where accessible, is an X-ray crystal structure; since an X-ray (or neutron or electron) diffraction picture leads to the positions of the constituting atoms in space, their relative orientation (i.e., configuration) can be determined. Relative configuration thus correlates one chiral center

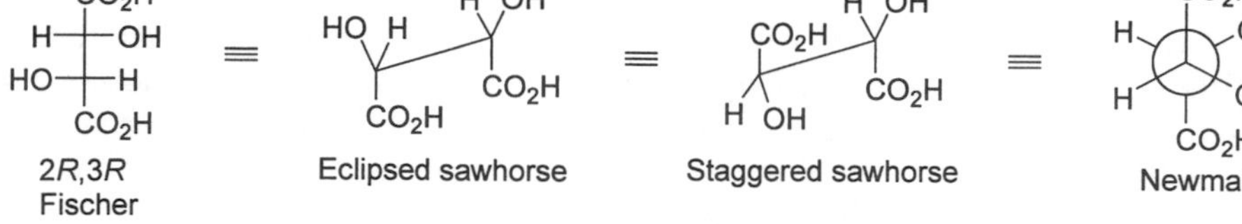

FIGURE 7 Fischer, sawhorse, and Newman representations of $(2R, 3R)$-tartaric acid.

with another. The two chiral centers need not be in the same molecule; as will be shown later, configurational correlations between two similar molecules can sometimes be based on comparison of their optical rotations, optical rotatory dispersion, or circular dichroism spectra (see below). It is also possible to tie two chiral centers, one of known absolute configuration, the other unknown, together chemically by either ionic or covalent bonds and determine their relative configuration by X-ray diffraction or other means. Since the absolute configuration of one of the chiral centers is known, that of the other can then be deduced. If the compound containing that center can then be separated by an appropriate chemical reaction, its optical rotation can be measured and thus the necessary correlation between optical rotation ($+$ or $-$) and configuration (S or R) is established. [There is no general relation whatever between $+$ and $-$ (experimental quantities) and R and S (descriptors).]

Determination of the absolute configuration (R or S) of an isolated species is more difficult. Enantiomers are indistinguishable in their physical behavior in scalar measurements (i.e., in measurements not involving absolute orientation in space); ordinary X-ray diffraction is of this type. However J. M. Bijvoet, in the Netherlands, found in 1951 that this impediment can be circumvented by employing X-rays of a wavelength close to the absorption edge of one of the constituent atoms in the molecule to be examined. This specific absorption (usually by a relatively heavy atom, such as sulfur or bromine introduced in the species to be examined by chemical transformation if necessary) leads to a phase shift of the wavefront diffracted by this particular atom. This phase shift causes a pair of spots in the normally centrosymmetric diffraction pattern (so-called "Bijvoet pairs") to become unequal in intensity; from the relative intensity of these spots the absolute configuration of the compound under investigation can be inferred. Thanks to the availability of powerful computers, it has also become increasingly feasible to derive absolute configuration from optical rotation or circular dichroism spectra (see below) by theoretical computation. In some cases, absolute configuration can also be established by examining the crystal habit (macroscopic dimensions) of a crystal in the presence of certain impurities (Addadi et al., 1986).

Once the absolute configurations of a few chiral molecules are known, those of others can be established by correlation.

X. CHIRALITY IN ABSENCE OF CHIRAL CENTERS

Although Le Bel's and van't Hoff's understanding of chirality rested on the concept of tetrahedral carbon or, more generally, of what is now called a chiral center, chirality is not dependent on the existence of chiral atoms. Any molecule that is not superposable with its mirror image is chiral. An example, shown in Fig. 8E, is twistane. A secondary criterion for chirality is the absence of a plane of symmetry or a point of inversion. However, chiral molecules may contain simple (proper) axes of symmetry; twistane, in fact, has three mutually perpendicular twofold symmetry axes. One class of chiral molecules already foreseen by van't Hoff (though obtained as individual enantiomers only much later) are appropriately substituted allenes, as shown in Fig. 8A. The orbitals are so disposed that the two double bonds are perpendicular to each other, and so two mutually different substituents at the two termini will give rise to chirality (in contrast to the cis–trans isomerism of alkenes, Fig. 3).

Related chiral molecules are appropriately substituted spiranes (Fig. 8B) and alkylidenecycloalkanes (Fig. 8C). These molecules are said to possess chiral axes (along the

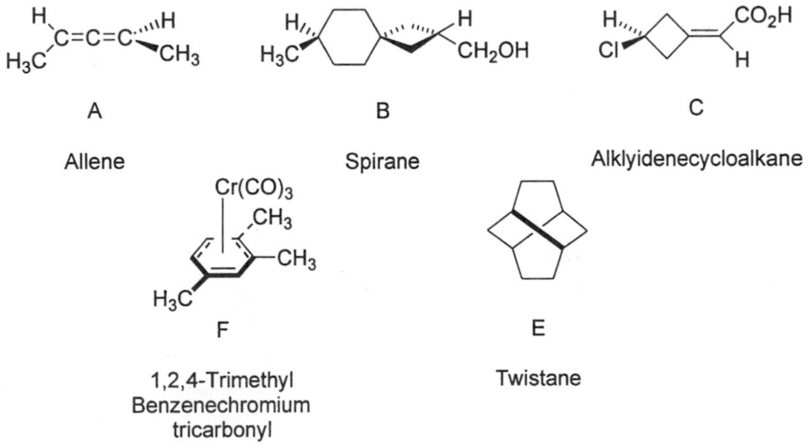

FIGURE 8 Chiral compounds lacking chiral centers.

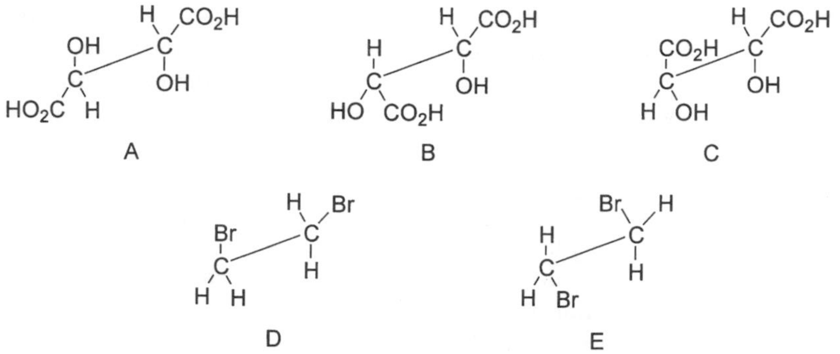

FIGURE 9 Conformers of meso-tartaric acid and of 1,2-dibromoethane.

line of the double bonds or bisecting the rings). Also in this category are the biphenyls, to be discussed later. Yet another common type of chirality due to "chiral planes" is seen in the benzenechromium complex in Fig. 8F. (The normal symmetry plane of the benzene rings is abolished by the out-of-plane coordinated chromium atom.)

XI. CONFORMATION

van't Hoff's counting of stereoisomers was based on the assumption that rotation about single bonds was "free," otherwise there should be at least three isomers of meso-tartaric acid shown in Fig. 9A–C two of which (Fig. 9B, C) would be chiral and enantiomeric. (This is on the assumption that the substituents are staggered— see below—otherwise many more isomers could exist.) In 1932, However, S.-I. Mizushima discovered by vibrational spectroscopy that there are, in fact, two isomers of 1,2-dibromoethane (Fig. 9D, E; vibrational spectroscopy cannot distinguish between enantiomers). It was established later that the bromine substituents are "staggered" rather than "eclipsed," meaning that the torsion angle Br–C–C–Br is 60° (Fig. 9D) or 180° (Fig. 9E) rather than 0°. The structures in Fig. 9D, E are said to differ in "conformation" (rotation about single bonds).

Why are there differing isomers for 1,2-dibromoethane even though they cannot be isolated? The answer to this question was provided by K. S. Pitzer in 1936 when he discovered that ethane itself (Fig. 10) exists in staggered conformation and that the three possible staggered conformations are separated by energy barriers of 12.1 kJ/mole (corresponding to the eclipsed conformation as the energy maximum). Such barriers are high enough to allow detection of the individual conformers by vibrational spectroscopy but far too low to allow chemical separation. What one sees in chemical behavior (and also in physical measurements involving "slow" time scales, such as measurement of dipole moments or of electron diffraction patterns), is an average of the contributing stable conformations (called "conformers") produced by their rapid interconversion.

The staggered conformations of butane (in Newman projections) are shown in Fig. 11 and resemble those in 1,2-dibromoethane (Fig. 9). There are three; in two of them (the enantiomeric *gauche* conformers in Fig. 11A, C the terminal methyl groups are close enough together to give rise to van der Waals repulsion. Thus these conformers are less stable (by 4 kJ/mole) than the anti conformer (Fig. 11B), in which the methyl groups are remote from each other. [The relative instability of the gauche (Fig. 9D) relative to the anti confomer (Fig. 9E) is even greater in $BrCH_2CH_2Br$ because of additional dipole repulsion in Fig. 9D and its enantiomer.] The conformations of Fig. 11A, C in which the torsion angle $\omega[C(1)–C(2)–C(3)–C(4)]$ is 60° are called "gauche"

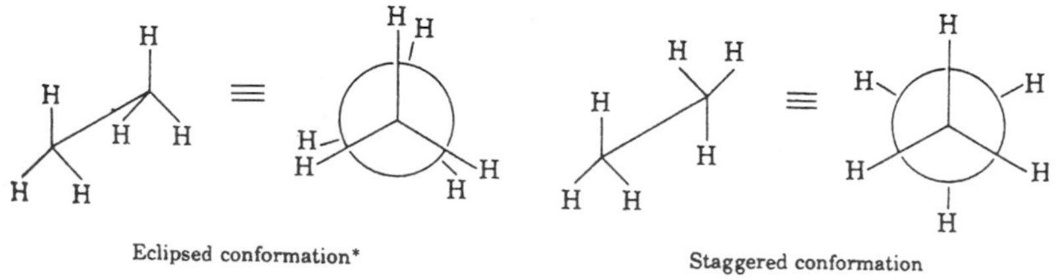

Eclipsed conformation*   Staggered conformation

FIGURE 10 Eclipsed and staggered conformations of ethane. Since the hydrogen atoms on the front carbon obscure those in the rear in the eclipsed conformation, the torsion angle is offset by a few degrees in the Newman formula. [Reprinted with permission from Eliel, E., Allinger, N. L., Morrison, G. A., and Angyal, S. J. (1981). "Conformational Analysis," American Chemical Society, Washington, DC. Copyright 1981 American Chemical Society.]

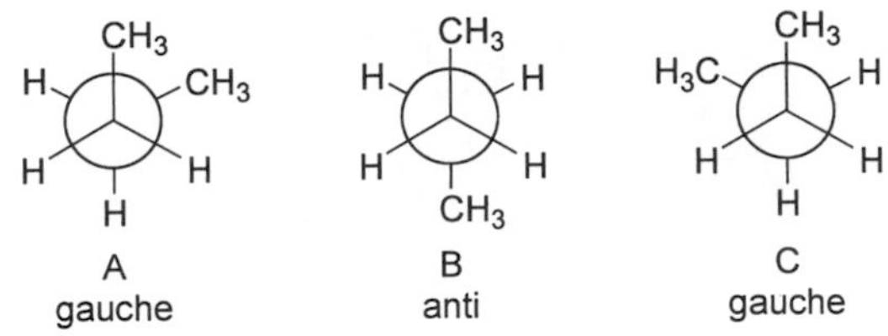

FIGURE 11 Stable conformers of butane.

(French for "skew"). In Fig. 11B the torsion angle is 180°; this conformation is called "anti." Torsion angles often deviate from the ideal values (60° or 180°) for staggered conformations; thus it may be desirable to specify the exact torsion angle ω when known (e.g., from X-ray structure determination). When C(1)–C(2)–C(3)–C(4) describe a right-handed helical turn, ω is positive; for a left-handed turn, it is negative. As an alternative, a system of semiquantitative conformational descriptors more detailed than gauche and anti has been developed by Klyne and Prelog (1960) and is described in the original reference and in Eliel and Wilen (1994).

In straight-chain hydrocarbons larger than butane, rotation about each single bond is possible, giving rise to a large number of conformations; this situation exists especially in linear polymers, where it was studied early by P. Flory. Even when conformations in which the chain coils upon itself in such a way as to generate excessive van der Waals (steric) repulsion are excluded, the number of low-energy conformers will be quite large and a Monte Carlo approach may have to be used to find the family of populated (low-energy) conformers. (Conformers which lie 12 kJ/mole or more above the lowest energy conformer are populated to the extent less than 1% of the total and may be neglected for most purposes.) By way of an example, in linear polyethylene, only a minor fraction of the molecules will be in the most stable zigzag (all-anti) conformation.

These considerations are important in the conformation of proteins (natural polymers). When polypeptides are synthesized—in the laboratory and perhaps also *in vivo*—they are first formed as linear strings (so-called "random coil" conformations, similar to those of a polyethylene). But in polypeptides and proteins additional considerations come into play, notably hydrogen bonding between amino acid residues and hydrophobic forces generated by the reluctance of hydrocarbon side chains to be in contact with the common water solvent. Additional interactions between nonadjacent amino acids may come about because of oxidation of cysteine to cystine residues (2 –SH → –S–S–). These various interactions lead to a folding of the chain into so-called secondary structures, which include a helical (α-helix) conformation (Pauling *et al.*, 1951) stabilized mainly by hydrogen bonding between nonadjacent but close amino acids in the polymer chain, and a doubled-

up conformation called a β or pleated sheet stabilized by hydrogen bonding between rather more distant members of the chain folded onto each other. "Tertiary structure" of proteins comprises the combination of α-helices, pleated sheets, and some random-coil areas, which gives rise to their three-dimensional shape.

The distinction between configuration and conformation is usually based on whether the interconversion of the pertinent stereoisomers is slow or fast. Since a barrier of 84 kJ/mole between two species corresponds to an interconversion rate at 25°C of 1.3×10^{-2} sec^{-1}, i.e., a half-life of 1 min, making isolation of the individual species quite difficult, one might say that the division between configuration and conformation comes at barriers of about 84 kJ/mole. However, such a precise distinction is problematic. At lower temperatures, interconversion rates decrease and isomers that differ only in conformation (cf. Fig. 9) may become isolable. Also, the technique of observation matters. Infrared and Raman spectroscopy are "very fast" and thus the vibrational spectra of the conformational isomers of 1,2-dibromoethane are distinct. Nuclear magnetic resonance (NMR) is intermediate, and there are numerous instances where an averaged NMR spectrum is seen at one temperature but spectra for the individual conformers emerge at lower temperatures.

An interesting example of the fluidity of the delineation between configuration and conformation is seen in the biphenyls (Fig. 12). In biphenyl itself, rotation is fast; thus a 3,3′,5,5′-tetrasubstituted biphenyl (Fig. 12A) cannot be resolved into enantiomers, even though conformations in which the two rings are not coplanar are chiral. However, as soon as sizable substituents are introduced at positions 2, 2′, 6, and 6′ (Fig. 12B, X ≠ Y) the compounds become resolvable; they display axial chirality. When X and Y (in Fig. 12B) are different and other than F or CH$_3$O, the enantiomers are stable. When one of the four substituents is H, however, the compounds are resolvable but usually racemize readily either at room temperature or above by rotation about the Ar–Ar bond. Biphenyls with only two ortho substituents are generally not resolvable unless the substituents are very bulky, as in 1,1′-dinaphthyl (Fig. 12C, Z = H). (The enantiomeric 2,2′-dihydroxybinaphthyls, Fig. 12C, Z = OH, have found manifold uses, e.g., as parts of chiral reagents and chiral catalysts.)

XII. CYCLOALKANES AND THEIR CONFORMATIONS

Before considering conformation in cyclic molecules (which is more complex since rotations about individual

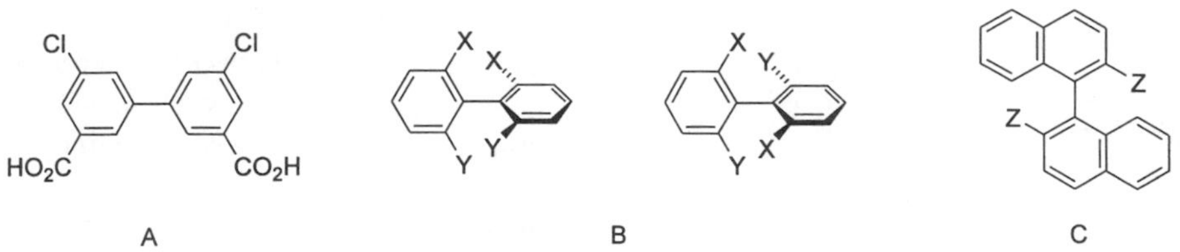

FIGURE 12 Stereoisomerism in biphenyls.

bonds are not independent of each other, unlike in alkane chains) we need to consider the topic of "strain." Already A. von Baeyer in 1885 realized that formation of small rings, such as cyclopropane or cyclobutane, required deformation of the normal tetrahedral or near-tetrahedral bond angle of 109°28′. Thus in cyclopropane the internuclear bond angle is 60°, i.e., it deviates 49°28′ from the normal and this causes angle strain, which in turn destabilizes the cyclopropane molecule. If one takes the contribution of a CH_2 group to the heat of combustion as 658.7 kJ per group [this is the difference in heat of combustion between two large homologous alkanes $CH_3(CH_2)_nCH_3$ and $CH_3(CH_2)_{n+1}CH_3$], the calculated heat of combustion of cyclopropane is $3 \times 658.7 = 1976$ kJ/mole, whereas the experimental value is 2091 kJ/mole; the difference of 115 kJ/mole is a measure of the strain in cyclopropane. Corresponding values are, for cyclobutane, 110 kJ/mole; for cyclopentane, 26.0 kJ/mole; and for cyclohexane, 0.5 kJ/mole. The low value for cyclohexane is at first sight surprising; Baeyer thought that cyclohexane was planar, with bond angles of 120°, and should therefore show strain of $120° - 109°28′$ or 10°32′, though this strain would be due to enlargement rather than diminution of the bond angle. There are two ways of accounting for this discrepancy. First, while strain increases again for the so-called "medium rings" (C_7, 26.2 kJ/mole; C_8, 40.5 kJ/mole; C_9, 52.7 kJ/mole; C_{10}, 51.8 kJ/mole; C_{11}, 47.3 kJ/mole), it diminishes thereafter for the "large rings," e.g., to 8.0 kJ/mole for C_{14}. This is due to the fact that rings other than cyclopropane are actually not planar and there are different sources of strain in these rings (and actually even in cyclopropane). One source is strain due to eclipsing of bonds ("torsional strain"), as explained above for ethane. In planar cyclobutane and cyclopentane, this strain (due to four or five pairs of eclipsed hydrogen atoms, respectively) is large enough to cause these species to be nonplanar, even though this increases angle strain. (A nonplanar polygon has smaller angles than a planar one.) In the larger cycloalkanes, however, where (if they were planar) the angles would be expanded beyond the tetrahedral, puckering actually diminishes not only the torsional or eclipsing strain (see the discussion on ethane above) but also the angle strain. In fact, much of the strain in medium rings is due to nonbonded atoms getting too close to each

other; this causes so-called "nonbonded" or van der Waals strain (steric repulsion).

Cyclohexane, which is virtually strain-free, is a special case. In 1890 (only 5 years after Baeyer proposed his strain hypothesis) H. Sachse realized that C_6H_{12} is not planar, but can be constructed from tetrahedral carbon atoms, either in the shape of a chair (Fig. 13A) or that of a boat (Fig. 13B). Today we know that, because of steric repulsion between the hydrogen atoms at C(1) and C(4) pointing inside plus eclipsing strain at C(2, 3) and C(5, 6), the shape in Fig. 13B is actually deformed to a "twist-boat" form (Fig. 13C) and that the chair (Fig. 13A) is the most stable conformer. But it took some 60 years after Sachse for the physical and chemical consequences of the chair shape of cyclohexane to be recognized, by K. Pitzer, O. Hassel, and D. H. R. Barton. Chemically speaking, axial substituents are more hindered (crowded) than equatorial ones and therefore generally less stable, and react more slowly (e.g., in the esterification of acids and alcohol and the hydrolysis of the corresponding esters). Also, the bimolecular elimination reaction (e.g., of H_2O in cyclohexanols or HX in cyclohexyl halides and toluenesulfonates) proceeds more readily when the substituent (OH or X) is axial than when it is equatorial. Barton saw these consequences (and others) of cyclohexane conformation (Eliel *et al.*, 1965) in the rigid cyclohexane systems of steroids and terpenes. Thus in 3-cholestanol (Fig. 14) the equatorial or β isomer is more stable than the axial one (designated α, meaning that the substituent is on the side opposite to the angular methyl groups, whereas β implies that it is on the same side). Also, the β isomer is more easily esterified than the α, but elimination of water to give a cholestene is more facile for the axial α isomer.

In monocyclic cyclohexanes the situation is more complex since the barrier to interconversion of the ring is only

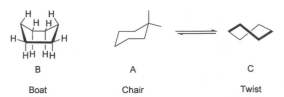

FIGURE 13 Conformations of cyclohexane.

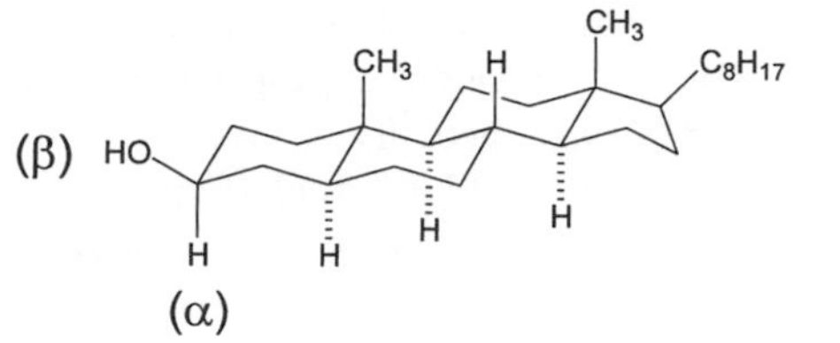

FIGURE 14 Conformation of 3β-cholestanol.

about 42 kJ/mole and interconversion of the two conformers is extremely rapid at room temperature. Thus chlorocyclohexane (Fig. 15, X = Cl) exists in rapid equilibrium between axial and equatorial conformers which differ in free energy by only about 25 kJ/mole, corresponding to 74% of the equatorial and 26% of the axial isomer at 25°C. As expected, in the infrared spectrum there are two C—Cl stretch frequencies, but in the laboratory, chlorocyclohexane, even though a mixture of two conformers, appears as a homogeneous substance with the average properties (such as chemical shifts in NMR) of the two conformers. When one cools the substance to ca. −60°C (the exact temperature required depends on the operational frequency of the NMR instrument), however, two different NMR spectra begin to emerge, and at −150°C the equatorial conformer has actually been crystallized from trideuteriovinyl chloride solution, with concomitant enrichment of the axial isomer in solution.

Equilibria for a large number of monosubstituted cyclohexanes have been determined and tabulated; they were mostly determined by low-temperature ^{13}C NMR spectroscopy (Eliel and Wilen, 1994).

The conformations of piperidine (azacyclohexane) and tetrahydropyran (oxacyclohexane) are qualitatively similar to those of cyclohexane. (Some quantitative differences are seen, for example, in the equatorial preferences of some substituents resulting from dipolar interactions with the ring hetero atom and from the fact that C–N and C–O distances are shorter than C–C in cyclohexane.) These ring systems are important, being found in alkaloids and hexose sugars, respectively.

Because of torsional (eclipsing) strain, cyclobutane and cyclopentane are not planar. Cyclobutane is wing-shaped; cyclopentane oscillates among a number of low-energy

conformations of which the envelope (four carbon atoms in a plane, one out of plane) and the half-chair (three adjacent carbons in a plane, the fourth above that plane and the fifth below) are the most symmetrical. The barrier between these conformations is very low; thus their rapid interconversion, which involves up-and-down motions of successive adjacent carbon atoms, has the appearance of a bulge moving around the rings; this process has therefore been named "pseudorotation." Higher cycloalkanes from C(7) on display families of conformations which are separated by barriers similar to those in cyclohexane, but within a given family there may be several individual members interconverted by pseudorotation. This subject is discussed in detail in Eliel and Wilen (1994).

XIII. CHIROPTICAL PROPERTIES. ENANTIOMERIC PURITY

By "chiroptical properties" are meant optical properties that differ between enantiomers and can be used to characterize them. They comprise optical rotation, optical rotatory dispersion (ORD), and circular dichroism (CD).

Optical rotation has already been discussed. Because of its critical dependence on solvent (including cosolvents, such as ethanol in chloroform), temperature, concentration, and the potential presence of impurities, especially chiral impurities, in the sample, experimental determination of [α] requires considerable care and many of the values given in the literature cannot be trusted. This is unfortunate since it is often desirable to determine the "enantiomeric purity" of a sample to see whether the desired enantiomer is obtained free of the other. (For example, in pharmaceutical chemistry one wants to obtain the pure eutomer free of the distomer; see above.) Since in all but a few cases optical rotation is proportional to the fraction of the major enantiomer in the total substance, one might expect enantiomeric or optical purity to be equal to $100[\alpha_{obs}]/[\alpha_{max}]\%$, where $[\alpha_{max}]$ is the (presumed known) specific rotation of an enantiomerically pure sample. However, this is true only if both values have been accurately determined under exactly the same conditions (solvent, temperature, etc.).

Because of this difficulty, other, more reliable methods of determining enantiomeric purity (now no longer called "optical purity") have been developed. Basically these depend on converting the enantiomers into diastereomers, either by covalent chemical bonding or by complexing in some fashion, with another, usually enantiomerically pure chiral auxiliary. (For some methods the auxiliary need not even be enantiomerically pure.) Once the enantiomers have been converted into diastereomers, their ratio can be determined by NMR or chromatographic methods,

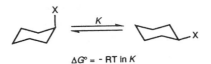

$$\Delta G° = - RT \ln K$$

FIGURE 15 Conformational inversion of substituted cyclohexane. [Reprinted with permission from Eliel, E. L., and Wilen, S. H. (1994). "Stereochemistry of Organic Compounds," Wiley, New York.]

including NMR in a chiral solvent or with a chiral complexing agent, or by chromatography of unmodified enantiomers on a chiral stationary phase. The methods just described provide the mole fractions of each individual enantiomer, which may be expressed as an enantiomer ratio (n_+/n_- or n_-/n_+). However, because optical purity had been used in the past, it is more common to use an expression for "enantiomeric excess" (e.e.): e.e. = $|(n_+ - n_-)|/(n_+ + n_-)$. Since this represents the mole fraction of the major enantiomer diminished by that of the minor one, it is equal to the above-defined optical purity.

Although there is no direct relation between sign of rotation and configuration, it may now become possible to infer configuration from optical rotation data (Kondra, Wipf, and Beratan, 1998). The method is based on van't Hoff's "optical superposition" rule, which says that in a molecule containing several chiral centers, the total molar rotation ($\Phi = 100\alpha/MW$) is the sum of the contributions of each chiral center. As originally proposed the rule had several shortcomings, including (1) that it would require model compounds of known absolute configuration to determine the contribution of a given chiral center and (2) it does not hold when the centers are close to each other and thus influence each other's contributions. This second limitation means that for closely connected chiral centers, it is necessary to determine the contribution of a segment containing all of these centers. The first problem is being attacked by performing *a priori* computations of the molar rotation contribution of individual chiral centers or appropriate groupings thereof (Kondra, Wipf, and Beratan, 1998). This is becoming possible due to the advances in quantum chemical calculations (e.g., by density functional methods) and the increasing power of computers to handle computationally demanding problems.

Other chiroptical techniques to infer configuration are optical rotatory dispersion (ORD) and circular dichroism (CD). ORD relates to the change in optical rotation with the wavelength of the light employed in the measurement. Normally the absolute value of rotation increases as wavelength becomes shorter; observation at shorter wavelengths is thus a convenient way to increase rotation (and thereby the accuracy of measuring it) when α_D is small. However, as the wavelength approaches that of a UV absorption band (e.g., of C=O in a ketone), its absolute value (whether positive or negative) suddenly drops precipitously, passes through zero near the UV absorption maximum, reverses sign rapidly approaching another extremum (of opposite sign to the first), and then gradually declines. This phenomenon of rapid change at the UV maximum is called the "Cotton effect" after the French scientist who discovered it. Similarity in Cotton effects of related compounds, one of known and one of unknown configuration, can sometimes be used to assign the configuration of the unknown. However the use of ORD has largely been superseded by the simpler to interpret CD (Nakanishi *et al.*, 2000). While it might appear that plane-polarized light is achiral, it may actually be considered to be a superposition of right- and left-circularly polarized light in which the sense of polarization changes as a right-handed or left-handed helix along the direction of propagation of the light beam. If the right- and left-circularly polarized beams proceed at the same speed, the result is light polarized in an unchanging plane, but if one of the two generating beams moves faster than the other, the plane of polarization will keep turning as the light propagates, i.e., there will be optical rotation. Since the speed of light depends on the refractive index of the medium it traverses, the polarization is thus caused by unequal refractive indices for right- and left-circularly polarized light. There are devices that can produce right- and left-circularly polarized light beams separate from each other. Using such beams, it is found that not only the refractive indices, but also the absorption coefficients for the two beams differ. The phenomenon resulting from this difference in absorption is called "circular dichroism" and manifests itself in what looks similar to an absorption curve in the UV, except that it is signed. (In fact its maximum or minimum occurs at the wavelength of the UV maximum.) Comparison of CD absorption spectra can be used to infer configuration similarly as was the case for ORD; however, CD spectra are better resolved than ORD spectra since there is less band overlap resulting from multiple UV absorption maxima.

CD is also very useful in throwing light on conformation. Thus the (weak) CD absorption spectrum of a random-coil polypeptide chain is essentially a superposition of the spectra of the individual constituting amino acids. However, when secondary structure comes into play, as in an α helix of β-pleated sheet, a large and characteristic increase in CD absorption is observed, which, in turn, allows one to infer the nature of the secondary structure, if any. CD is used to infer not only protein but also nucleic acid and polysaccharide conformation (Fasman, 1996).

XIV. PROCHIRALITY

The phenomenon of prochirality (Mislow and Raban, 1967) is important both in NMR spectroscopy and in enzyme chemistry. An atomic center (e.g., a tetrahedral carbon atom) in a molecule is considered "prochiral" if replacing one of two identical ligands at the center by a different one not previously attached to that center produces a chiral center. Thus the carbon atom in CH_2ClBr is prochiral since hypothetical replacement of one of the hydrogen atoms by deuterium yields $CHDClBr$, which is chiral. The apparently identical (or "homomorphous," from Greek "homos," same, and "morphe," form) hydrogen atoms in CH_2ClBr are in fact distinct; they are called "heterotopic" (from Greek "heteros," different, and "topos," place). In contrast, the carbon atom in CH_3Cl is not prochiral since replacement of H by, say, D would produce CH_2DCl, which remains achiral.

In the former case (CH_2ClBr) replacement of one or other of the two hydrogen atoms gives rise to enantiomeric products. The hydrogen atoms are therefore called "enantiotopic." In the latter case, replacement gives the same compound and the hydrogens in CH_3Cl are called "homotopic." In a molecule such as $CH_2BrCHOHCO_2H$, replacement of one of the terminal hydrogens by, say, chlorine would give one or other of the diastereomers of $CHBrClCHOHCO_2H$; in this case the terminal hydrogens are said to be diastereotopic. These definitions of homotopic, enantiotopic, and diastereotopic ligands also provide a means for their recognition: Replacement of one of two or more homotopic ligands by a different ligand gives identical products, analogous replacement of enantiotopic ligands gives enantiomeric products, and such replacement of diastereotopic ligands gives diastereomeric products. There is also a symmetry criterion which may be applied to the appropriate molecules above: Homotopic ligands in a molecule are interchanged by operation of both simple symmetry axes and symmetry planes; enantiotopic ligands are interchanged by operation of a symmetry plane but not by operation of a simple symmetry axis, and diastereotopic ligands are interchanged neither by symmetry axes nor by symmetry planes.

Diastereotopic ligands (e.g., protons or C-13 atoms) generally display distinct signals in NMR spectra, but homotopic and enantiotopic ligands have coincident (identical) signals (except possibly in the case of enantiotopic ligands, in a chiral solvent, or in the presence of a chiral complexing agent) since NMR is an achiral technique. Both enantiotopic and diastereotopic ligands may be distinguished by enzymes (which are chiral). Thus in citric acid, $HO_2CCH_2C(OH)(CO_2H)CH_2CO_2H$, all four methylene hydrogen atoms are distinguished by enzymes in the citric acid cycle (to demonstrate this distinction, they must be individually labeled as deuterium atoms). On the other hand, the CH_2 groups are pairwise identical in NMR (e.g., C-13) but the geminal hydrogen atoms in each are diastereotopic and provide an $(AB)_2$ system in the proton NMR spectrum. Further details may be found in Eliel (1982) and Eliel and Wilen (1994).

SEE ALSO THE FOLLOWING ARTICLES

BIOPOLYMERS • ENZYME MECHANISMS • NUCLEAR MAGNETIC RESONANCE • ORGANIC CHEMICAL SYSTEMS, THEORY • ORGANIC CHEMISTRY, SYNTHESIS • PHYSICAL ORGANIC CHEMISTRY • PROTEIN STRUCTURE • PROTEIN SYNTHESIS • RHEOLOGY OF POLYMERIC LIQUIDS

BIBLIOGRAPHY

Addadi, L., Berkovitch-Yellin, Z., Weissbuch, I., Lahav, M., and Leiserowitz, L. (1986). "A link between macroscopic phenomena and molecular chirality: Crystals as probes for the direct assignment of absolute configuration of chiral molecules. *In* "Topics in Stereochemistry," Vol. 16, pp. 1–85, Wiley, New York.

Cahn, R. S., Ingold, C., and Prelog, V. (1966). "Specification of molecular chirality," *Angew. Chem. Int. Ed. Engl.* **5**, 385–415 .

Eliel, E. L. (1982). "Prostereoisomerism (prochirality)."*In* "Topics in Current Chemistry," Vol. 105, pp. 1–76, Springer-Verlag, Heidelberg.

Eliel, E. L., and Wilen, S. H. (1994). "Stereochemistry of Organic Compounds," Wiley, New York.

Eliel, E. L., Allinger, N. L., Angyal, S. J., and Morrison, G. A. (1965). "Conformational Analysis," Wiley, New York [reprinted (1981), American Chemical Society, Washington, DC.]

Fasman, G. D. (ed.). (1996). "Circular Dichroism and the Conformational Analysis of Biomolecules," Plenum Press, New York.

Gawley, R. E., and Aubé, J. (1996). "Principles of Asymmetric Synthesis," Pergamon Press, Oxford.

Hegstrom, R. A., and Kondepudi, D. K. (1990). "The handedness of the universe," *Sci. Am.* **262**(January), 108–115.

Jacques, J., Collet, A., and Wilen, S. H. (1981). "Enantiomers, Racemates and Resolutions," Wiley, New York.

Juaristi, E. (ed.). (1995). "Conformational Behavior of Six-Membered Rings," VCH, New York.

Kagan, H. B., and Fiaud, J. C. (1988). "Kinetic resolution." *In* "Topics in Stereochemistry," Vol. 18, pp. 249–330, Wiley, New York.

Klyne, W., and Prelog, V. (1960). "Description of stereochemical relationships across single bonds," *Experientia* **16**, 521–523.

Kondru, R. K., Wipf, P., and Beratan, D. N. (1998). "Atomic contributions to the optical rotation angle as a quantitative probe of molecular chirality," *Science* **282**, 2247–2250; *id.* (1998). Theory-assisted determination of absolute stereochemistry for complex natural products vid computation of molecular rotation angle, *J. Am. Chem. Soc.* **120**, 2204–2205.

Mislow, K., and Raban, M. (1967). "Stereoisomeric relationships of groups in molecules." *In* "Topics in Stereochemistry," Vol. 1, pp. 1–38, Wiley, New York.

Nakanishi, K., Berova, N., and Woody, R. W. (eds.). (2000). "Circular Dichroism: Principles and applications," Wiley-VCH, New York.

Pauling, L., Corey, R. B., and Branson, H. R. (1951). "The structure of proteins: Two hydrogen bonded helical configurations of the polypeptide chain," *Proc. Natl. Acad. Sci. U.S.A.* **37**, 205–211.

Prelog, V., and Helmchen, G. (1982). "Basic principles of the CIP system and proposals for a revision," *Angew. Chem. Int. Ed. Engl.* **21**, 567–583.

Ramsay, O. B. (1981). "Stereochemistry," Heyden & Son, Philadelphia.

Sih, C. J., and Wu, S.-H. (1989). "Resolution of enantiomers via biocatalysis." *In* "Topics in Stereochemistry," Vol. 19, pp. 63–125, Wiley, New York.

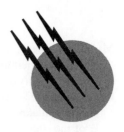

Stochastic Description of Flow in Porous Media

Ghislain de Marsily

Université Pierre et Marie Curie and
Ecole des Mines de Paris

I. Definition of the Average Properties
of a Porous Medium
II. Darcy's Law and Averaging Permeabilities
in Heterogeneous Porous Media
III. Estimation by Kriging of Heterogeneous
Permeability Fields
IV. Stochastic Flow Equations, Analytical
and Numerical Solutions

GLOSSARY

Averaging Estimating physical quantities defined at a macroscopic scale for a continuum from microscopic measurements. Spatial averaging is done by integration over space and ensemble averaging by integration over the space of realizations.

Conditional simulation See simulation.

Covariance Second moment of a stationary random function.

Ergodicity Assumption that the spatial distribution of a realization of a stationary random function has the same probability distribution function as the ensemble of realizations of the same random function.

Geostatistics A theory that applies to variables extending over space, which are spatially correlated. Most geologic quantities (thickness of a layer, concentration of an element in the rock or the water, permeability, porosity, etc) display such spatial correlation. Geostatistics offer the means to estimate the value of the variable in space, based on local measurements (see kriging) or to simulate the value of the variable in space (see simulations and conditional simulation).

Kriging A geostatistical technique that is used to estimate the value of a Random Function (e.g., a parameter such as the permeability) in space, based on a set of local measurements. Kriging produces contour maps of the parameter and can be extended to generate conditional simulations of the same parameters.

Macroscopic scale Level at which the porous medium is considered as an equivalent continuum without distinguishing between the pores and the grains.

Microscopic scale Level at which the pores and the grains of a porous medium are distinguished and their respective roles analyzed.

Random function (RF) Function $Z(x, \xi)$ of both the spatial coordinates x and a state variable ξ. For each value ξ_i of the state variable, the spatial function $Z(x, \xi_i)$ is completely defined in space. Each value of ξ_i represents a different spatial function; for a position x_j, $Z(x_j, \xi)$ is a random variable.

Random variable Variable Z, which can take an infinite number of values. In general, the statistical properties of these values are defined: mean, variance, probability density function.

Realization of an RF Spatial function $Z(x, \xi_i)$ for a given value of the state variable ξ_i.

Simulation of an RF To simulate an RF is to produce a (large) number of realizations of the spatial function $Z(x, \xi_i)$ (i.e., for different ξ), which have the correct statistical properties of Z (same mean, variance, covariance). A **conditional simulation** is a simulation that has not only the correct statistical properties of Z, but also takes the measured values Z_i of Z at a discrete set i of measurement points.

Stationarity Assumption that any statistical property of a random function $Z(x, \xi)$ (e.g., mean, variance, covariance, higher-order moments or probability distribution functions) is stationary in space (i.e., does not vary with a translation).

THE STOCHASTIC DESCRIPTION of porous media is an approach that has been developed in recent years to take into account the very complex nature of the void space in such media and to derive an exact phenomenological description of the macroscopic flow, coherent with the classical empirical approach represented by Darcy's law. On a larger scale, it makes it possible to quantify the uncertainty in the macroscopic flow predictions, given the inherent uncertainty and spatial variability of the properties of porous media in natural systems. The concepts of the change from the microscopic pore scale to the macroscopic one are outlined and followed by a description of the mathematics of developing Darcy's law and the flow equation, and of the estimate of average properties of porous media in one, two, or three dimensions in spatially variable systems. The problem of estimating spatially variable parameters, generally known as Geostatistics, is then summarized for the example of permeabilities; kriging equations are given and simulation methods are outlined. Stochastic partial differential flow equations are presented and analytical or numerical techniques to solve them are briefly discussed.

I. DEFINITION OF THE AVERAGE PROPERTIES OF A POROUS MEDIUM

A porous medium is a very complex assembly of grains, solids, and pores containing fluids. The major conceptual obstacle to the understanding of the porous medium behavior is that of change of scale: at the microscopic pore scale, the physics of the interactions between the fluid and the solid are relatively well understood and documented, but at any larger scale, it is necessary to perform an integration in order to define the "average" properties of the porous medium, regarded as a continuum. This problem of scale change is discussed in the following.

A. Representative Elementary Volume

The classical approach is that of the representative elementary volume (REV) (see, e.g., Bear, 1972, 1979). Consider the case of one very simple property of a porous medium, its porosity ω, defined as the ratio of the pore volume to the total volume of a sample. If a porous medium is regarded as a continuum, we must be able to define its porosity at a point x in space. The REV method consists in saying that we give to the mathematical point x the porosity $\langle \omega \rangle$ of a certain volume of material surrounding this point, the REV, which will be used to define (and possibly measure) the "average" property of the volume in question. Here, $\langle . \rangle$ stands for "volume average." Conceptually, this definition involves integration in space or taking a representative sample of the medium to measure the property. The size of the REV is constrained in two ways.

The first one assumes that the REV is sufficiently large to contain a sufficiently great number of pores to allow us to define a mean global property, while ensuring that the effects of the fluctuations from one pore to another are negligible. One may, for example, take 1 cm^3 or 1 dm^3.

The second one assumes that the REV is small enough for the parameter variations from one domain to the next to be approximated by continuous functions, so that we may use infinitesimal calculus, without introducing any error that may be picked up by the measuring instruments at the macroscopic scale, where meters and hectometers are the usual dimensions.

The size of the REV (e.g., measured by one of its characteristic dimensions l, such as the radius of a sphere or the side of a cube) is determined by a flattening of the curve representing the variation of the porosity with l (Fig. 1). However, nothing allows us to assert that such a flattening always exists, and the size and existence of a REV is thus quite arbitrary. A further limitation of this concept is that it gives no basis for studying the variation of the porosity in space.

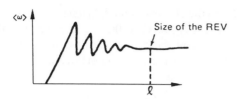

FIGURE 1 Porosity (ω) as a function of the dimension 1 of the sample. [From G. de Marsily (1986). "Quantitative Hydrogeology, Groundwater Hydrology for Engineers," Academic Press, Orlando, FL.]

B. Stochastic Approach

The most recent definition of the average properties of a porous medium is the stochastic approach (e.g., Dagan, 1989; Gelhar, 1993). The microscopic porosity $\omega(x)$ is then considered as a random function (RF). This approach consists in saying that the studied porous medium is a "realization" of a random process.

Let us try to visualize the concept. Suppose that we create, in the laboratory, several sand columns, each one filled with the same type of sand. Each column represents the same porous medium but is somewhat different from the others. Each column is a "realization" of the same porous medium, defined as the ensemble of all possible realizations (infinite in number) of the same process.

A property such as the porosity can then be defined, at a given geometrical point in space, as the average over all possible realizations of its point value (defined as zero in a grain and one in a pore). One speaks of "ensemble averages" instead of "space averages." For the previously mentioned sand columns, it is obvious that the ensemble average (or expected value E) of these point porosities will be identical to the space average defined by taking the column itself as the REV. Furthermore, this ensemble average will be the same for any point of the column. We will define later the conditions necessary for this to be true.

In more general terms, a property Z will be called a random function (RF) $Z(x, \xi)$ if it varies both with the spatial coordinate system x and with the "state variable" ξ in the ensemble of realizations. Then, $Z(x, \xi_1)$ is a realization of Z; $Z(x_0, \xi)$ is a random variable (i.e., the ensemble of the realizations of the RF Z at location x_0); and $Z(x_0, \xi_1)$ is the single value of Z at x_0 for realization ξ_1. To simplify the notations, the variable ξ is generally omitted.

A more realistic example of realizations of random porous media is given by sand dunes in a desert: each of the dunes can be seen as a "realization" of the same process, which here is genetically the accumulation of grains of sand transported by the wind. It would be meaningful to describe the properties of the dunes statistically (ensemble averages), as their number can indeed be very large.

The immense advantage of the stochastic approach over the REV concept is that one can study other statistical properties of the porous medium in the ensemble of realizations than just the expected value. One very often uses the variance of the property, which characterizes the magnitude of the fluctuations with respect to the mean, and the autocovariance (or simply covariance), which characterizes the correlation between the values taken by the property at two neighboring points in space.

However, when a given real porous medium is studied, there will be only one realization of the conceptual random medium. Some assumptions are necessary to make this concept useful. The most common ones are stationarity and ergodicity.

Stationarity assumes that any statistical property of the medium (mean, variance, covariance, or higher-order moments, defined by the "ensemble averaging" concept) is stationary in space (i.e., does not vary with a translation): it will be the same at any point of the medium. Weak stationarity refers to a medium where only the first two moments are stationary: if $Z(x)$ is the studied property, where x is the coordinates in one, two, or three dimensions and the state variable ξ is omitted, the random function (RF) $Z(x)$ satisfies: (1) expected value:

$$E[Z(x)] = m$$

not a function of x, and (2) covariance:

$$E\{[Z(x) - m][Z(x + h) - m]\}$$

not a function of x, but a function of only the lag h, a vector in two or three dimensions. By developing and labeling this covariance $C(h)$,

$$C(h) = E[Z(x) - Z(x + h)] - m^2$$

By definition,

$$C(0) = E[(Z(x) - m)^2] = \sigma_z^2$$

is the variance of Z.

In more rigorous terms, true stationarity means that all the probability distribution functions (PDF) of the random function $Z(x)$ are invariant under translation, whether we consider one point $p[Z(x)]$ or n points $p[Z(x_1) \ldots Z(x_n)]$.

Ergodicity implies that the real unique realization available behaves in space with the same PDF (and with the same moments) as the ensemble of possible realizations. In other words, by observing the variation in space of the property on the only available realization, it is possible to determine the PDF of the random function for all realizations. This is called the "statistical inference" of the PDF of the RF $Z(x)$.

In the vocabulary of stochastic processes, a phenomenon that is "stationary" and "ergodic" is called "homogeneous." We would then use "uniform" to describe a medium in which a property does not vary in space. Geologists traditionally call it "homogeneous." Other, less stringent hypotheses, can also be defined (e.g., stationarity of increments of Z).

II. DARCY'S LAW AND AVERAGING PERMEABILITIES IN HETEROGENEOUS POROUS MEDIA

A. At the Microscopic Level

Darcy's law is an empirical linear relationship between the macroscopic filtration velocity $\langle u \rangle$, and the macroscopic pressure gradient $\langle \text{grad } p \rangle$: $\langle u \rangle = -(k/\mu)\langle \text{grad } p \rangle$. Here, $\langle . \rangle$ stands for a spatial or an ensemble average. k is called the permeability, a parameter specific to each soil or rock, which can be a scalar or a second-order tensor if the medium is anisotropic, and μ is the dynamic viscosity. Physically, Darcy's law is the result of the integration of the Navier–Stokes equations in the very complex geometry of the pore space. Navier–Stokes is the general equation of fluid mechanics for Newtonian fluids. However, since this pore geometry is, in general, unknown, it is impossible to systematically derive Darcy's law and the value of the tensor k from Navier–Stokes, except when a very simple geometry of the pore space is assumed (e.g., cylindrical tubes or fissures of constant aperture (see, e.g., Marsily, 1986). The dimension of k is length (squared).

The general linear form of Darcy's law, and some properties of the permeability tensor k, can, however, be rigorously established. This is done for an incompressible fluid in steady-state flow while assuming that the microscopic velocity u is small enough to neglect the inertial term in the Navier–Stokes equations (this is quite acceptable in practice since the flow velocity in porous media is, in general, very small). Furthermore, these equations are written without the body forces F, only for the sake of simplicity. Navier–Stokes then reduces to

$$\mu \nabla^2 \mathbf{u} = \text{grad } p \quad \text{and} \quad \text{div } \mathbf{u} = 0.$$

Here, ∇^2 is the Laplace differential operator $\Sigma_i \partial^2 / \partial x_i^2$, p is the microscopic fluid pressure, and $\mathbf{u}$ the microscopic fluid velocity vector. Let the porous medium be considered as a stationary and ergodic random ensemble, and let $\omega(x)$ be the microscopic porosity ($\omega(x) = 1$ in the pores and 0 in the grains). The solution of the microscopic flow problem in the entire domain can be described as finding a stationary random velocity $\mathbf{u}(x)$ that satisfies

$$\mu \nabla^2 \mathbf{u} = \omega(x) \text{ grad } p; \quad \text{div } \mathbf{u} = 0$$

$$\mathbf{u}(x) = \omega(x)\,\mathbf{u}(x); \quad E(\mathbf{u}) = \langle \mathbf{u} \rangle.$$

To establish the existence and uniqueness of this solution, Matheron (1967) proposed the use of a variational principle to represent the energy dissipation by the viscous forces. The power per unit volume of this energy at the microscopic level is

$$W = -\mathbf{u} \cdot \text{grad } p.$$

It is then possible to show that the random velocity $\mathbf{u}$, which minimizes the mathematical expectation $E(W) = -E(\mathbf{u} \cdot \text{grad } p)$ while satisfying div $\mathbf{u} = 0$ and $\mathbf{u}(x) = \mathbf{u}(x)\omega(x)$, is also the solution of the Navier–Stokes equation. If $\mathbf{u}$ is then extended to the grains ($\mathbf{u} = 0$ in the grains), the relationship $E(W) = -E(\mathbf{u} \cdot \text{grad } p) = -\mu E(\mathbf{u} \cdot \nabla^2 \mathbf{u})$ can be extended over the whole space. Furthermore, for a stationary grad p and a stationary $\mathbf{u}$ with div $\mathbf{u} = 0$, we have

$$E(\mathbf{u} \cdot \text{grad } p) = E(\mathbf{u}) \cdot E(\text{grad } p).$$

Then

$$E(W) = -E(\mathbf{U} \cdot \text{grad } p) = \langle \mathbf{u} \rangle \cdot \langle \text{grad } p \rangle,$$

which means that the averaging conserves the energy: the average of the microscopic energy dissipation is equal to the energy dissipation at the macroscopic level.

It can then be shown that the macroscopic Darcy law $\langle \mathbf{u} \rangle = -(k/\mu)\langle \text{grad } p \rangle$ derives from the linearity of the Navier–Stokes equation $\mu \nabla^2 \mathbf{u} = \text{grad } p$ and from this conservation of energy. Furthermore, it can be shown that the permeability tensor k is symmetric and positive definite.

B. At the Macroscopic Level

At the macroscopic level, the permeability $k(x)$ can also be regarded as a RF, for studying the behavior of heterogeneous porous media where the macroscopic parameter $k(x)$ varies in space. It is then of interest to again average the permeability in order to obtain the equivalent homogeneous permeability at a larger scale. Again, using the variational principle, it is possible to give an upper and a lower bound to this average permeability $\langle k \rangle$ at the large scale:

$$[E(k^{-1})]^{-1} < \langle k \rangle < E(k).$$

Expressed in words, the average $\langle k \rangle$ always lies between the harmonic and the arithmetic mean of the local permeability value. The harmonic mean is the obvious average for one-dimensional flow.

Furthermore, it is possible to show (Matheron (1967) for porous media; Landau and Lifschitz (1960) in electrodynamics) that, in two dimensions, and for macroscopic parallel flow conditions, the average permeability $\langle k \rangle$ is exactly the geometric mean if the probability distribution function of k is log-normal:

$$\text{Ln}\langle k \rangle = E(\text{Ln}\,k).$$

The geometric mean, which is an arithmetic averaging in the log space, always lies between the harmonic and the arithmetic means. This result has been extended to three dimensions, using a perturbation approach, initially as a conjecture, e.g., by King (1987), Dagan (1993), Indelman and Abramovich (1994), de Wit (1995), and now demonstrated by Noetinger (2000). The general expression for the average value of a log-normal permeability distribution is thus, for an isotropic medium in one, two, or three dimensions :

$$\begin{aligned} 1/\langle k \rangle &= \langle 1/k \rangle &&\text{in 1-D, the harmonic mean} \\ \text{Ln}\langle k \rangle &= \langle \text{Ln}\,k \rangle &&\text{in 2-D, the geometric mean} \\ \langle k \rangle &= \langle k^{1/3} \rangle^3 &&\text{in 3-D, a power average with} \\ & &&\quad \text{exponent } 1/3 \end{aligned}$$

Most field studies show that the experimental PDF of the permeability of rocks is indeed log-normal, therefore these expressions are commonly used. For radial flow systems, or for transient conditions, average permeabilities have not yet been established theoretically but the previous results have been shown to be applicable by numerical experiments with flow models. See also Renard and Marsily (1997), Meier *et al.* (1998), and Dagan (2001), for the problem of up-scaling permeabilities.

III. ESTIMATION BY KRIGING OF HETEROGENEOUS PERMEABILITY FIELDS

Kriging, which is a part of the theory of Geostatistics, is a very powerful tool for estimating random functions (RF), which are defined by a set of local measurements and by their statistical properties (mean, variance, covariance), see, e.g., Matheron (1973), Journel and Huijbregts (1978), Marsily (1986), Deutsch and Journel (1992), Chilés and Delfiner (1999). We will restrict this presentation to the case of a second-order stationary random function Y with a constant mean, a constant variance and a stationary covariance. We consider the hydraulic conductivity $K = k\rho g/\mu$, where k is the macroscopic, spatially variable permeability, g is the acceleration due to gravity, and μ the dynamic viscosity or, in two dimensions, the transmissivity $T = Ke$ of an aquifer, with e the thickness of the aquifer. As already stated, K or T are generally log-normally distributed, so that we will first transform them into their logarithm Y, which will be the RF on which we will work. Kriging is indeed a better estimator if the variable to be estimated is approximately normally distributed, although this is not an absolute requirement. Let us assume that a set of N measurements of Y have been made at N locations x_i in space. We call them Y_i. Here, x_i represents the one, two or three coordinates in space of point i.

The problem that we want to solve is to *estimate* the value of Y at locations x_j, which have not been measured, based on the knowledge of the N measurements Y_i of Y and of the statistical properties of Y. These estimations are noted $Y^*(x_j)$, and will always be an approximation of reality. Kriging is a method for estimating Y^*; it is both an *unbiased* and an *optimal* estimator. "Unbiased" means that the error of estimation will be, on average, zero (no systematic underestimation or over-estimation). "Optimal" means that the variance of the estimation error will be minimum. Lastly, Kriging is a linear estimator. It is written:

$$Y^*(x_j) = \Sigma_i \lambda_i Y_i,$$

where the summation over i is, generally, extended to all the N measurements Y_i, but in some cases only to a subset of measurements surrounding the point x_j to be estimated (e.g., if the number N is very large). The N λ_i's are the "kriging weights," they are the unknowns of the kriging problem, which we will now calculate. There are as many kriging equations as there are locations x_j where Y^* will be estimated and for each such location, the N weights λ_i will be different. In general, kriging is performed on a regular square grid, and the set of estimated values on that grid is used, e.g., as input in a flow model or for automatic contouring of the field Y.

The set of weights λ_i are determined by successively applying the two conditions defined above: (1) Unbiased estimation: one writes that $E[Y^*(x_j)] = Y'$, where Y' is the true (unknown) mean of the RF Y. By developing the kriging equation:

$$E[Y^*(x_j)] = Y' = E[\Sigma_i \lambda_i Y_i] = \Sigma_i \lambda_i E[Y_i] = \Sigma_i \lambda_i Y'$$

or

$$\Sigma_i \lambda_i = 1.$$

(2) Optimal estimation: one writes that the variance of the estimation error is minimum, i.e., $\text{Var}[Y^*(x_j) - Y(x_j)]$ minimum, which, since $E[Y^*(x_j) - Y(x_j)] = 0$, is written: $E[(Y^*(x_j) - Y(x_j))^2]$ minimum.

By substituting the kriging equation and after some rearranging, one gets

$$E[(Y^*(x_j) - Y(x_j))^2] = \Sigma_i \Sigma_k \lambda_i \lambda_k E\{[Y_i - Y(x_j)]$$
$$\times [Y_k - Y(x_j)]\}.$$

By definition, the covariance of Y is defined as $E\{[Y(a) - Y'][Y(b) - Y']\} = C(a-b)$, where a and b are the coordinates of any points in space, and $(a-b)$ the distance (or vector) between these two points. We can rewrite $E[(Y^*(x_j) - Y(x_j))^2]$ with the objective of developing it in terms of the covariance function C:

$$E\big[\big(Y^*(x_j) - Y(x_j)\big)^2\big] = \Sigma_i \Sigma_k \lambda_i \lambda_k E\{([Y_i - Y']$$
$$- [Y' - Y(x_j)])([Y_k - Y'] - [Y(x_j) - Y'])\},$$

which, after some developments and taking into account that $\Sigma_i \lambda_i = 1$, gives

$$E\big[\big(Y^*(x_j) - Y(x_j)\big)^2\big] = \Sigma_i \Sigma_k \lambda_i \lambda_k [C(x_i - x_k)$$
$$- 2C(x_i - x_j) + C(0)]$$
$$= \Sigma_i \Sigma_k \lambda_i \lambda_k C(x_i - x_k)$$
$$- 2\Sigma_i \lambda_i C(x_i - x_j) + C(0).$$

This is where the major "trick" of kriging comes in: the expression which gives the variance of the estimation error is no longer a function of the unknown "real" value $Y(x_j)$ but only of the covariance function C of the RF Y. The rest of the story is straightforward. The variance of the estimation error subject to the constraint that $\Sigma_i \lambda_i = 1$ is minimized with the Lagrange multiplier method: one minimizes $E[(Y^*(x_j) - Y(x_j))^2] + 2\mu[\Sigma_i \lambda_i - 1]$. Here, μ is the Lagrange multiplier and the factor 2 is just used to simplify the result. The minimization of this quadratic form in λ is obtained by setting to zero all its partial derivatives with respect to the N λ's and μ, which provides a linear system with $N + 1$ equations and $N + 1$ unknowns:

$$\Sigma_k \lambda_k C(x_i - x_k) + \mu = C(x_i - x_j) \quad \text{for } i = 1, \ldots, N$$
$$\Sigma_i \lambda_i = 1.$$

Once this linear system is solved, e.g., by Gauss elimination, the λs can be used to estimate $Y^*(x_j)$ with the first kriging equation and furthermore, it is possible to estimate the variance of the estimation error by some simple substitutions of the expression given previously:

$$\text{Var}\big[Y^*(x_j) - Y(x_j)\big] = C(0) - \Sigma_i \lambda_i C(x_i - x_j) - \mu.$$

In the next section, we will describe how a variable can be simulated rather than just estimated. We conclude this section by some indications on how to determine the statistical properties of a variable Z, which has been measured in the field at N different locations. This step is called the "statistical inference." The type of PDF can be estimated by building the histogram of the observed values and trying to fit it with a usual PDF (e.g., normal). On a normal distribution paper, this histogram should plot as a straight line; one can see how much the data deviate form this line. The mean can be estimated by averaging the data, in some cases with weighting coefficients, to eliminate clustering effects. A kriging equation can also be written to estimate the mean. Once the mean is known, the calculation of the experimental variance is straightforward. The covariance is more elaborate. The N measurement points are grouped into pairs, there are $N(N - 1)/2$ pairs that can be built

with N points. Let a_i and b_i be the two points of pair i. These pairs are then grouped into classes of distances, i.e., $h_i = a_i - b_i$ falls into a class of distance h_k. For each distance class h_k, the value of the experimental covariance C^* is estimated as

$$C^*(h_k) = (1/N_k)\{\Sigma_i[Z(a_i) \cdot Z(b_i)] - Z'^2\},$$

where N_k is the number of pairs a_i and b_i in the class of distance h_k, and Z' is the mean of Z. The experimental covariance C^* is plotted versus the distances h_k, and a general function $C(h)$ is fitted on the experimental covariance C^*, e.g., an exponential function, etc. Any function cannot be used to represent a covariance; it has to be a positive definite function (see, e.g., Matheron, 1973; Chilès and Delfiner, 1999).

A more common approach to kriging and statistical inference is to assume that Z is not necessarily stationary, but that the first-order increments of Z are stationary; this is called the intrinsic hypothesis. One then defines a new function, the variogram of Z, which is more general than the covariance, and is written:

$$\gamma(h) = \tfrac{1}{2}E\{[Z(x + h) - Z(x)]^2\}.$$

It can be shown that, when both functions exist, the variogram γ is related to the covariance C by

$$\gamma(h) = C(0) - C(h).$$

Kriging equations can be written by using the variogram rather than the covariance and the inference of γ is easier than that of C as the mean Z' does not need to be known (see, e.g., Marsily, 1986; Chilés and Delfiner, 1999).

IV. STOCHASTIC FLOW EQUATIONS, ANALYTICAL AND NUMERICAL SOLUTIONS

The diffusivity equation in hydrogeology is the result of combining (i) the macroscopic mass-balance equation, (ii) the macroscopic Darcy law, and (iii) the state equations for both the fluid and the porous medium, involving the compressibility of the fluid and that of the medium. Expressed in terms of the hydraulic head H rather than of the pressure p, its is similar to the heat equation:

$$\text{div}(K \operatorname{grad} H) = S_s \partial H/\partial t,$$

where H is the hydraulic head, equal to $p/\rho g + z$ (p is the fluid pressure, ρ is the fluid mass per unit volume, z is the elevation above the datum, e.g., the mean sea level); K is the macroscopic hydraulic conductivity, equal to $k\rho g/\mu$; k is the macroscopic permeability, g is acceleration due to gravity, and μ the dynamic viscosity; S_s is the specific storage coefficient equal to $\rho \omega g(\beta + \alpha/\omega)$, with ω

the macroscopic porosity, β the compressibility of water, and α the compressibility of the porous medium; t is time.

This equation can be solved analytically, when the parameters K and S_s are constant over space (or when equivalent average parameters are used), and when the boundary and initial conditions are sufficiently simple. In more complex or heterogeneous cases, numerical solutions (finite differences, finite elements, boundary element, etc.) are commonly used (see e.g., Anderson and Woessner, 1992; Kinzelbach, 1986; Marsily, 1986). However, when the parameters of this equation are considered as random functions, the method of solution is quite different from the usual approach. Two methods of solving this problem are described below.

A. Analytic Solutions of the Stochastic Diffusivity Equation

The stochastic definition of the properties of porous media has led to the concept in which the diffusivity equation is considered as a stochastic partial differential equation (PDE). The parameters (K and S_s) of the equation are considered as stationary random functions, and therefore the solution H of the diffusivity equation is also an RF. Solving the stochastic PDE means determining the statistical properties of the solution H, given the statistical properties of the parameters K and S_s.

The advantage of this approach is that it makes it possible to take into account, in the field, the spatial variability and the uncertainty in the values of the parameters of the diffusivity equation. Local measurements of these parameters in the field (e.g., on samples or through local *in situ* hydraulic tests) show that they are, in general, highly variable in natural media. Rather than attempting to estimate them deterministically at all locations in the domain of interest, it is more relevant to try to estimate their variability statistically and to relate this variability to the uncertainty of the solution of the stochastic PDE.

We shall give a simple example of one method for solving analytically a stochastic PDE, namely the method of perturbation. This method is only valid if the variance of the stochastic process of interest is small. We shall solve a one-dimensional steady-state flow equation, without any source term, written as

$$d/dx[K(x)\,dH/dx] = 0.$$

The hydraulic conductivity $K(x)$ is assumed to be a stationary RF, the first two moments of which are known

$$E(K) = K'$$

$$E\{[K(x+s) - K'][K(x) - K'] = C_K(s).$$

The variance of K, $C_k(0)$, is assumed small. Since K is stationary, K' is not a function of x, and $C_K(s)$, the covariance of K is only a function of the lag vector s, not of x. The stochastic PDE will be solved if we can determine the same first two moments of the solution H, its mean and covariance. To use the method of perturbation, one assumes that both K and H can be decomposed into a mean plus a perturbation:

$$K = K' + k \qquad \text{with} \qquad E(K) = K', \ E(k) = 0$$
$$H = H' + h \qquad \text{with} \qquad E(H) = H', \ E(h) = 0.$$

K and H are then developed to the first order (i.e., by adding to K' and H' a small fraction β of their fluctuation, which thus requires the perturbation to be small, i.e., the variance of K is small):

$$K = K + \beta k \qquad \text{and} \qquad H = H + \beta h.$$

We introduce this definition into the original PDE, develop in β and disregard terms in β^2:

$$K'\,d^2H'/dx^2 + \beta[K'\,d^2h/dx^2 + dk/dx \cdot dH'/dx$$
$$+ k\,d^2H'/dx^2] = 0.$$

If this is to hold for any small β, each one of the two terms must be equal to zero:

(i) $$K'd^2H'/dx^2 = 0$$

i.e.,

$$dH'/dx = 2q/K' \qquad \text{and} \qquad H' = 2qx/K' + a,$$

which gives us the required first moment of H, which, in this case, is not stationary. The two constants q and a are determined by the boundary conditions.

(ii) $$K'\,d^2h/dx^2 + dk/dx \cdot dH'/dx + k\,d^2H'/dx^2 = 0$$

substituting the solution H' in this equation gives

$$d^2h/dx^2 = (q/K'^2)\,dk/dx$$

or

$$dh/dx = (q/K'^2)k + b$$

where b is a constant. We take the expected value of this expression:

$$E[dh/dx] = d/dx[E(h)] = (q/K'^2)E(k) + E(b).$$

As $E(h) = E(k) = 0$, we can see that $E(b) = b = 0$.

Then the relation

$$dh/dx = (q/K'^2)k$$

gives directly:

$$C_{dh/dx}(s) = (q^2/K'^4)C_k(s) = (q^2/K'^4)\,C_K(s).$$

However, for a stationary RF with a differentiable covariance, one can write

$$C_{dh/dx}(s) = -d^2/ds^2[C_h(s)].$$

Thus, if we can assume that

$$d/ds[C_h(s)]|_{s \to -\infty} = 0 \quad \text{and} \quad C_h(s)|_{s \to -\infty} = 0.$$

Then with two integrations, we find

$$C_h(s) = -(q^2/K'^4) \int_{-\infty}^{s} \int_{-\infty}^{y} C_k(u) \, du \, dy.$$

If we take, for instance, a modified exponential covariance for K:

$$C_k(s) = \sigma_K^2 \, \text{Exp}(-|s|/l)(1 - |s|/l),$$

where l is a characteristic length and σ_K^2 is the variance of K, then we find, by integration,

$$C_h(s) = l^2(q^2/K'^4)\sigma_K^2 \, \text{Exp}(-|s|/l) \, (1 + |s|/l).$$

We can, for instance, determine the variance of h:

$$\sigma_h^2 = C_h(0) = l^2 \, (q^2/K'^4) \, \sigma_K^2.$$

We have thus found the first two moments of the solution H of the stochastic PDE. Other methods for solving the stochastic PDE are the spectral method (see, e.g., Gelhar, 1993) and the Monte Carlo simulation method (see following, and in, e.g., Delhomme, 1979; Freeze, 1975; Lavenue et al., 1995; Ramarao *et al.*, 1995; Rubin and Gomez-Hernandez, 1990; Smith and Freeze, 1979).

B. Monte Carlo Simulation Method for Solving the Stochastic Flow Equations

This is probably the most powerful method, where fewer assumptions are required. However, it is a numerical method that may require much central processing unit (CPU) time and a careful examination of the results. The principle of the method is very simple. Let $Z(x, \xi)$ be a stochastic process, x being the coordinates in space and ξ the state variable, for instance, the permeability. One first builds a numerical model of the problem, in each mesh of which a value of the permeability Z is required to solve the diffusivity equation. One then generates "simulations" $Z(x, \xi_i)$ of Z in the probabilistic sense, i.e., values of $Z(x_j, \xi_i)$ for each mesh x_j of the numerical model. A large number N of realizations ξ_i, $i = 1, \ldots, N$, of Z are generated. To do so, one must know the probability distribution function of Z, its mean and its variance and covariance (if Z is spatially correlated). Note that the knowledge of the probability distribution function of Z was not necessary in the previous analytical method, only the mean, variance and covariance. Then for each of these realizations ξ_i, the parameter represented by $Z(x_j, \xi_i)$ in each mesh of the model is completely determined and known (e.g., the permeability). Thus, the diffusivity equation representing the flow in the porous medium can be solved numerically for each realization, providing the value of the dependent variable [e.g., $H(x_j, \xi_i)$ for the same realization ξ_i and

for all the x_j meshes]. It is then possible to statistically analyze the ensemble of calculated solutions $H(x, \xi_i)$ for $i = 1, \ldots, N$: expected value of H at any location x_j, variance, histogram, and distribution function. It is no longer necessary to assume that H is stationary; these statistics can be calculated at each point. The covariance can also be determined if H is found to be stationary.

There are, however, some difficulties associated with the simulation method. First, a large number N of realizations is necessary in order to get meaningful statistics: from 50 to several hundreds or thousands. Second, as N is necessarily finite, one can always calculate an experimental variance or covariance, even for a phenomenon where they do not exist. It is preferable to check that when N increases, these statistics become, in fact, constant. Third, the solution can be a function of the mesh size: because the numerical solution requires that an average of $Z(x, \xi_i)$ be estimated over a mesh, this estimate becomes less variable as the mesh size increases, simply because of the spatial integration over each mesh. Thus, the variability of the solution $H(x, \xi)$ will also be affected. Furthermore, one must realize that if C_Z is the correlation structure of Z in space, then the correlation structure of the average of Z over a mesh will be the integral of C_Z over the mesh.

To generate realizations of $Z(x, \xi_i)$, several methods are available. Let us first assume that Z is stationary with a given PDF, but that Z is not spatially correlated. Most computer software provides a random number generator, which usually gives real numbers with a uniform distribution between 0 and 1, with mean 0.5 and variance 1. Standard routines can be used to transform these numbers into random numbers with a given PDF (e.g., normal), with mean 0 and variance 1. Let us call such random numbers z_j with, e.g., a normal PDF, mean 0 and variance 1. Let us assume that the parameter Z that we want to simulate is the permeability, which has a log-normal distribution, i.e., $Y = \text{Ln} \, Z$ is normal with mean Y' and variance σ_Y^2. The uncorrelated values of Y and thus of Z at each mesh x_j ($j = 1, \ldots M$, with M the number of meshes of the model) for a realization ξ_i are simply generated by

$$\text{For } j = 1, \ldots M: \quad Y(x_j) = Y' + z_j \sigma_Y$$

$$\text{and thus} \quad Z(x_j) = \text{Exp}[Y(x_j)].$$

For a new realization ξ_k, a new set of M random numbers z_j are used. Note that in the case of permeability, the value of Z is simply the exponential of Y, with no additional term to account for the variance of Y, as is sometimes done for such a transformation (see, e.g., Marsily, 1986).

Let us now assume that Y is spatially correlated. The simplest method for generating an RF Y with a spatial correlation is called the Nearest Neighbor (e.g., Smith and

Freeze, 1979). First an uncorrelated RF Y is generated, as outlined above. Then, in each mesh of the model, the correlated RF X is calculated as the weighted mean of the values of Y at mesh x_j and at surrounding meshes of x_j: for instance, on a regular square grid, the four meshes above, below, to the right, and to the left. Such averaging will keep the mean of Y, change the variance and generate a spatial correlation, i.e., a covariance for X. This covariance can be determined experimentally; it is a function of the number of neighbors used and of the weights assigned to each neighbor. By trial and error, one can very approximately fit a given covariance function and variance to X. Z is then again generated as $Z = \mathrm{Exp}[X]$. But there are more elaborate methods that can generate directly correlated RFs with a prescribed covariance, e.g., the turning bands method (Matheron, 1973; Mantoglu and Wilson, 1982), the spectral method (e.g., Gelhar, 1993), the Cholesky decomposition of the covariance matrix (e.g., Neuman, 1984) or the sequential simulation (e.g., Deutsch and Journel, 1992).

Conditional simulations of $Z(x, \xi_i)$ instead of simple simulations can also be generated. This is a great improvement on the Monte Carlo method for practical problems. Indeed, the stochastic process Z is then said to be conditioned by the measurements $Z(x_j)$ in space if all the realizations $Z(x, \xi_i)$ have the measured values $Z(x_j)$ at each point x_j, where a measurement has been made. In nonconditional simulations, as discussed earlier, there is no way of insuring that the random value $Z(x)_j$, generated by the algorithm in mesh x_j, has the desired observed value. The method used to generate these conditional simulations is based on kriging, as shown in Section III, and works as follows.

Let us take the example of permeabilities Z, where $Y = \mathrm{Ln}[Z]$ is a normal RF with a given mean and covariance. First, given the measurements Y_i, a kriged field Y^* is calculated in each mesh x_j of the model, using ordinary kriging:

$$Y^*(x_j) = \Sigma_i \lambda_i Y_i.$$

Then, an unconditional simulation $U(x_j)$ of a new RF U is generated in each mesh x_j of the model, with the same mean and covariance as Y, using one of the standard methods presented above. At each location x_i, where a measurement of Y is available, a fictitious "measurement" is taken of U; let us call these measurements U_i. It is then possible to produce a kriged field U^* of U, based on the "measurements" U_i only:

$$U^*(x_j) = \Sigma_i \lambda_i U_i.$$

The conditional simulation of Y is then given by

$$Y(x_j) = Y^*(x_j) + \left[U(x_j) - U^*(x_j) \right].$$

This expression holds because the deviation $U-U^*$ (or $Y-Y^*$) around the kriged estimate U^* (or Y^*) is independent of U^* (or Y^*). An alternative to this "double kriging" method is the sequential simulation algorithm (e.g., Deutsch and Journel, 1992).

Nonconditional simulations are suitable for studying the theoretical variability of a process: the statistics of Z are assumed to be known, but no measured values are available. On the contrary, conditional simulations take the measured values into account and the considered variability and uncertainty are only those stemming from the residual uncertainty in the estimation of Z between measurement points.

The stochastic approach outlined here for the single-phase flow problem, can be extended to multiphase flow or to multi-component single-phase flow (i.e., petroleum reservoir engineering, hydrodynamic dispersion and pollution problems (see, e.g., Dagan, 1989, Gelhar, 1993 Lavenue *et al.*, 1995).

SEE ALSO THE FOLLOWING ARTICLES

HYDROGEOLOGY • MESOPOROUS MATERIALS, SYNTHE-SIS AND PROPERTIES • SOIL MECHANICS • SOIL PHYSICS • STOCHASTIC PROCESSES O STREAMFLOW

BIBLIOGRAPHY

Anderson, M. P., and Woessner, W. W. (1992). "Applied Groundwater Modeling: Simulation of Flow and Advective Transport," Academic Press, San Diego, CA.

Bear, J. (1972). "Dynamics of Fluids in Porous Media," American Elsevier, New York.

Bear, J. (1979). "Hydraulics of Groundwater," McGraw-Hill, New York.

Chilès, J. P., and Delfiner P. (1999). "Geostatistics: Modeling Spatial Uncertainty," Wiley, New York.

Dagan, G. (1989). "Flow and Transport in Porous Formations," Springer-Verlag, New York.

Dagan, G. (1993). "High-order correction of effective permeability of heterogeneous isotropic formations of lognormal conductivity distribution," *Transp. Porous Media* **12**, 279-290.

Dagan, G. (2001). "Effective, equivalent and apparent properties of heterogeneous media." "Proc. 20th Intern.Congress of Theoretical and Applied Mechanics" (H. Aref and J. W. Phillips, eds.), Kluwer Academic Press, Dordrecht, The Netherlands.

Delhomme, J. P. (1979). Spatial variability and uncertainty in groundwater flow parameters: a geostatistical approach. *Water Resour. Res.* **15**(2), 269–280.

Deutsch, C. V., and Journel, A. (1992). "GSLIB, Geostatistical Software Library and Users' Guide," Oxford University Press, New York.

de Wit, A. (1995). "Correlation structure dependence of the effective permeability of heterogeneous porous media," *Phys. Fluids* **7**(11), 2553–2562.

Freeze, R. A. (1975). "A stochastic-conceptual analysis of one-dimensional groundwater flow in non-uniform homogeneous media," *Water Resour. Res.* **11**(5), 725–741.

Gelhar, L. W. (1993). "Stochastic Subsurface Hydrology," Prentice-Hall, Englewood Cliffs, NJ.

Indelman, P., and Abramovich, B. (1994). "A higher order approximation to effective conductivity in media of anisotropic random structure," *Water Resour. Res.* **30**(6), 1857–1864.

Journel, A., and Huijbregts, C. (1978). "Mining Geostatistics," Academic Press, New York.

King, P. (1987). "The use of field theoretic methods for the study of flow in heterogeneous porous media," *J. Phys. A. Math. Gen.* **20**, 3935–3947.

Kinzelbach, W. (1986). "Groundwater Modeling," Elsevier, Amsterdam, The Netherlands.

Landau, L. D., and Lifschitz, E. M. (1960). "Electrodynamics of Continuous Media," Pergamon Press, Oxford, United Kingdom.

Lavenue, A. M., Ramarao, B. S., Marsily, G., de, and Marietta, M. G. (1995). "Pilot point methodology for an automated calibration of an ensemble of conditionally simulated transmissivity fields," *Appl. Water Resour. Res.* **31**(3), 495–516.

Mantoglu, A., and Wilson, J. L. (1982). "The turning bands method for the simulation of random fields using line generation by a spectral method," *Water Resour. Res.* **18**(5), 645–658.

Marsily, G. de (1986). "Quantitative Hydrogeology, Groundwater Hydrology for Engineers," Academic Press, Orlando, FL.

Matheron, G. (1967). "Eléments pour une théorie des milieux poreux," Masson, Paris.

Matheron, G. (1973). "The intrinsic random functions and their applications," *Adv. Appl. Prob.* **5**, 438–468.

Meier, P. M., Carrera, J., and Sanchez-Vila, X. (1998). "An evaluation of Jacob's method for the interpretation of pumping tests in heterogeneous formations," *Water Resour. Res.* **34**(5), 1011–1025.

Neuman, S. P. (1984). "Role of geostatistics in Subsurface Hydrology. In Geostatistics for Nature Resources Characterization, NATO-ASI" (Verly *et al.*, eds.), Part 1, pp. 787–816, Reidel, Dordrecht, The Netherlands.

Noetinger, B. (2000). "Computing the effective permeability of lognormal permeability fields using renormalization methods," *C.R. Acad. Sci. Paris/Earth Plan. Sci.* **331**, 353–357.

Ramarao, B. S., Lavenue, A. M., Marsily, G., de, and Marietta, M. G. (1995). "Pilot point methodology for an automated calibration of an ensemble of conditionally simulated transmissivity fields. 1. Theory and computational experiments," *Water Resour. Res.* **31**(3), 475–493.

Renard, P., and Marsily, G. de (1997). Calculating equivalent permeability: A review. *Adv. Water Resour.* **20**(5–6), 253–278.

Rubin, Y., and Gomez-Hernandez, J. (1990). "A stochastic approach to the problem of upscaling the conductivity in disordered media: Theory and unconditional numerical simulations," *Water Resour. Res.* **26**(3), 691–701.

Smith, L., and Freeze, R. A. (1979). "Stochastic analysis of steady-state groundwater flow in a bounded domain. 1. One-dimensional simulations. 2. Two-dimensional simulations," *Water Resour. Res.* **15**(3), 521–528; **15**(6), 1543–1559.

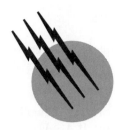

Stochastic Processes

Yûichirô Kakihara
California State University, San Bernardino

GLOSSARY

Probability space A triple consisting of a sample space Ω, a certain collection Σ of subsets of Ω, and a probability P defined on Σ.

Random variable A numerical function defined on the probability space.

Sample space The set of all possible outcomes of an experiment.

Stochastic process A collection of random variables often indexed by time.

IN AN EMPERICAL SENSE, a stochastic process is considered as the description of a random phenomenon evolving in time that is governed by certain laws of probability. Stochastic processes model fluctuations in economic behavior or stock markets, the path of a particle in a liquid, outputs of physical systems, and, in fact, most phenomena exhibiting unpredictable fluctuations. A mathematical abstraction of a stochastic process is any indexed collection of random variables defined on a fixed probability space. Probability theory is a branch of mathematics and the theory of stochastic processes is a part of it. Modern probability theory is based on Kolmogorov's axiomatic treatment using measure theory.

I. BASICS OF STOCHASTIC PROCESSES

The concept of a stochastic process is obtained as a collection of random variables, usually indexed by a time parameter. A random variable is a real- (or complex-) valued function defined on a probability space. Hence the notion of a stochastic process is a generalization of the idea of a random variable.

First we define a probability space according to Kolmogorov's axiomatic formulation. Thus a *probability space* consists of a triple (Ω, Σ, P), where Ω is a *sample space*, Σ is a *σ-algebra* of events, and P is a *probability* on Σ. Each $\omega \in \Omega$ represents an outcome of some experiment and is called a *basic event*. Each $A \in \Sigma$ is a subset of Ω, called an *event*. Σ satisfies the following conditions:

(i) $\Omega \in \Sigma$.
(ii) $A \in \Sigma$ implies $A^c \in \Sigma$, A^c being the complement.
(iii) If $A_1, A_2, \ldots \in \Sigma$, then $\cup_{n=1}^{\infty} A_n \in \Sigma$.

P satisfies the following:

(iv) $0 \le P(A) \le 1$ for any $A \in \Sigma$.
(v) $P(\Omega) = 1$.
(vi) If $A_1, A_2, \ldots \in \Sigma$ are disjoint, then
$P(\cup_{n=1}^{\infty} A_n) = \sum_{n=1}^{\infty} P(A_n)$.

In other words, the σ-algebra Σ is a collection of subsets of the sample space Ω that contains the entire event Ω and is closed under complementation and countable union. We also can say that the probabilty P is a measure taking values in $[0, 1]$ and countably additive.

Consider an experiment of rolling a die. The sample space Ω consists of six numbers 1, 2, 3, 4, 5, and 6. The σ-algebra Σ is then the set of all subsets of Ω, consisting of $2^6 = 64$ subsets. Probability is assigned $1/6$ to each $\omega = \{1\}, \ldots, \{6\}$ if the die is fair. If A is an event that the die faces up with a number less than or equal to 4, then $P(A) = 2/3$. A mapping $X: \Omega \to \mathbb{R} = (-\infty, \infty)$ is a (real) *random variable* if X is measurable, i.e., for any $x \in \mathbb{R}$,

$$\{\omega \in \Omega: X(\omega) < x\} \equiv \{X < x\} \in \Sigma. \qquad (1)$$

If we denote the Borel σ-algebra of $\mathbb{R}$ by $\mathfrak{B} = \mathfrak{B}(\mathbb{R})$, then (1) is equivalent to

$$\{\omega \in \Omega: X(\omega) \in A\} \equiv \{X \in A\} \in \Sigma \qquad (2)$$

for every $A \in \mathfrak{B}$. [Similarly, a complex random variable is defined as in (2) using the Borel σ-algebra $\mathfrak{B}(\mathbb{C})$ of the complex number field $\mathbb{C}$.] In this section, we shall consider real random variables unless otherwise stated. The *expectation* of a random variable X is defined by the integral of X over Ω if $E\{|X|\} < \infty$, where

$$E\{X\} = \int_{\Omega} X(\omega)\, P(d\omega).$$

Then the *distribution* of X is obtained as a probability on $(\mathbb{R}, \mathfrak{B})$:

$$P_X(A) = P(\{X \in A\}), \qquad A \in \mathfrak{B}.$$

In many cases, the distribution of a random variable determines many of its properties. The *distribution function* F_X of a random variable X is obtained as

$$F_X(x) = P_X((-\infty, x)), \qquad x \in \mathbb{R}.$$

Using the distribution function, we can write the expectation of X as the Stieltjes integral

$$E\{X\} = \int_{-\infty}^{\infty} x\, dF_X(x).$$

In a special case, there exists a *density function* p_X such that

$$F_X(x) = \int_{-\infty}^{x} p_X(r)\, dr, \qquad x \in \mathbb{R},$$

so that $p_X(x) = dF_X(x)/dx$. If

$$p(x) = \frac{1}{\sqrt{2\pi}\,\sigma} \exp\left\{-\frac{(x-m)^2}{2\sigma^2}\right\}, \qquad x \in \mathbb{R}, \quad (3)$$

where $\sigma > 0$ and $m \in \mathbb{R}$, then p is called a *Gaussian* (or *normal*) density function and X is called a *Gaussian* (or *normal*) random variable. The probability distribution given by (3) is denoted as $N(m, \sigma^2)$. Random variables $X_1, \ldots, X_n$ are said to be *mutually independent* if their distribution functions satisfy

$$F_{X_1,\ldots,X_n}(x_1, \ldots, x_n) = F_{X_1}(x_1) \cdots F_{X_n}(x_n) \qquad (4)$$

for any $x_1, \ldots, x_n \in \mathbb{R}$, where $F_{X_1,\ldots,X_n}$ is called the *joint distribution function* of $X_1, \ldots, X_n$ defined by

$$F_{X_1,\ldots,X_n}(x_1, \ldots, x_n) = P(\{X_1 < x_1, \ldots, X_n < x_n\}).$$

If the density functions exist, then (4) is equivalent to

$$p_{X_1,\ldots,X_n}(x_1, \ldots, x_n) = p_{X_1}(x_1) \cdots p_{X_n}(x_n),$$

where $p_{X_1,\ldots,X_n}$ is the density function of $F_{X_1,\ldots,X_n}$. The *characteristic function* of a random variable X is defined by

$$\Phi_X(\lambda) = E\{e^{i\lambda X}\}, \qquad \lambda \in \mathbb{R}.$$

A *stochastic process* is a collection of random variables on the same probability space indexed by a subset T of real numbers. In practice, each $t \in T$ represents a time. We usually denote a stochastic process on T by $\{X(t)\}_{t \in T}$ or simply by $\{X(t)\}$, where the underlying probability space (Ω, Σ, P) is fixed. Thus, for each $t \in T$, $X(t)$ is a random variable on (Ω, Σ, P) and its value at $\omega \in \Omega$ is denoted by $X(t, \omega)$. For each $\omega \in \Omega$, $X(\cdot, \omega)$ represents a function on T and is called a *sample path* or *realization* or *trajectory*. If T is a closed interval such as $\mathbb{R}$, $\mathbb{R}^+ = [0, \infty,)$, or $[0, 1]$, then we consider *continuous time* stochastic processes. If T is a discrete set such as $\mathbb{Z} = \{0, \pm 1, \pm 2, \ldots\}$ or $\mathbb{Z}^+ = \{0, 1, 2, \ldots\}$, then $\{X(t)\}$ is called a *discrete time* stochastic process or a *time series*.

Let $\{X(t)\}$ be a stochastic process on $T \subseteq \mathbb{R}$. Then, for $t_1, \ldots, t_n \in T$, the joint distribution function of $X(t_1), \ldots, X(t_n)$ is denoted by $F_{t_1,\ldots,t_n}(x_1, \ldots, x_n)$ and the joint density function by $p_{t_1,\ldots,t_n}(x_1, \ldots, x_n)$, if it exists, so that

$$F_{t_1,\ldots,t_n}(x_1,\ldots,x_n)$$
$$= \int_{-\infty}^{x_1} \cdots \int_{-\infty}^{x_n} p_{t_1,\ldots,t_n}(r_1,\ldots,r_n)\,dr_1 \cdots dr_n.$$

Then, distribution functions should satisfy the following conditions:

$$p_{t_1}(x_1) = \int_{-\infty}^{\infty} p_{t_1,t_2}(x_1, x_2)\,dx_2,$$

$$p_{t_1,\ldots,t_m}(x_1,\ldots,x_m)$$
$$= \int_{-\infty}^{\infty} \cdots \int_{-\infty}^{\infty} p_{t_1,\ldots,t_n}(x_1,\ldots,x_n)\,dx_{m+1} \cdots dx_n \tag{5}$$

for $m < n$, etc. This means that, if we know the higher joint distribution functions, then the lower joint distribution functions are obtained as marginal distribution functions. *Kolmogorov's* (1933) *consistency theorem* states that, if we are given a system of joint density functions $\{p_{t_1,\ldots,t_n}(x_1,\ldots,x_n)\}$ for which (5) is true and it holds under permutations of time, then there exists a real stochastic process $\{X(t)\}$ whose joint density functions are exactly this system.

II. STATISTICS OF STOCHASTIC PROCESSES

Consider a stochastic process $\{X(t)\}$ on $T \subseteq \mathbb{R}$. The *expectation* of a stochastic process $\{X(t)\}$ is a function of $t \in T$ given by

$$m(t) = E\{X(t)\}.$$

For an integer $k \geq 1$, the *kth absolute moment* of $\{X(t)\}$ is defined by a function of t as

$$M_k(t) = E\{|X(t)|^k\}.$$

$m(t)$ is sometimes called an *ensemble average* at time t. Another type of average is obtained. The *time average* of $\{X(t)\}$ on $\mathbb{R}$ is defined by

$$\overline{X}(\omega) = \lim_{s \to \infty} \frac{1}{2s} \int_{-s}^{s} X(t, \omega)\,dt.$$

$\overline{X}(\omega)$ exists for $\omega \in \Omega$ such that $X(\cdot, \omega)$ is bounded and measurable on $\mathbb{R}$ or, in particular, $X(\cdot, \omega)$ is bounded and continuous on $\mathbb{R}$. Moreover, if this is true for all $\omega \in \Omega$, then $\overline{X}(\omega)$ defines a random variable. If $m(t) \equiv m = \overline{X}(\omega)$ P-a.e., then the process $\{X(t)\}$ is said to be *ergodic*. Thus, a stochastic process is ergodic if its ensemble average is equal to its time average. Here, P-a.e. refers to P-almost everywhere.

If $M_2(t) < \infty$ for all $t \in T$, then we say that $\{X(t)\}$ is a *second-order* stochastic process. The set of all complex random variables of finite second moment forms a Hilbert

space $L^2(\Omega, \Sigma, P) = L^2(P)$, where the inner product and the norm are respectively defined by

$$(X, Y)_2 = E\{X\overline{Y}\}, \qquad \|X\|_2 = E\{|X|^2\}^{1/2},$$

for $X, Y \in L^2(P)$, where $\overline{z}$ is the complex conjugate of $z \in \mathbb{C}$. The *mixed moment function* of $\{X(t)\}$ is defined by

$$R(s, t) = (X(s), X(t))_2, \qquad s, t \in T.$$

Similarly, the *covariance function* of $\{X(t)\}$ is defined by

$$K(s, t) = (X(s) - m(s), X(t) - m(t))_2, \qquad s, t \in T.$$

K (and also R) is a *positive-definite kernel* on $\mathbb{R} \times \mathbb{R}$, i.e.,

$$\sum_{j,k=1}^{n} \alpha_j \overline{\alpha}_k K(t_j, t_k) \geq 0$$

for any integer $n \geq 1$, $t_1, \ldots, t_n \in T$, and $\alpha_1, \ldots, \alpha_n \in \mathbb{C}$. Moreover, a Schwarz-type inequality holds:

$$|K(s, t)|^2 \leq K(s, s)K(t, t), \qquad s, t \in T.$$

A stochastic process $\{X(t)\}$ is said to be *centered* if $m(t) \equiv 0$, in which case we have $K(s, t) = R(s, t)$. The *characteristic function* of $\{X(t)\}$ is defined by

$$\Phi_t(\lambda) = E\{e^{i\lambda X(t)}\} = \int_{-\infty}^{\infty} e^{i\lambda x}\,dF_t(x),$$

$$\Phi_{t_1,\ldots,t_n}(\lambda_1,\ldots,\lambda_n) = \int_{-\infty}^{\infty} \cdots \int_{-\infty}^{\infty} e^{i(\lambda_1 x_1 + \cdots + \lambda_n x_n)}$$
$$\times dF_{t_1,\ldots,t_n}(x_1,\ldots,x_n).$$

It is easily seen that for any $\lambda, \lambda_1, \ldots, \lambda_n \in \mathbb{R}$

$$|\Phi_t(\lambda)| \leq \int_{-\infty}^{\infty} dF_t(x) = \Phi_t(0) = 1,$$

$$|\Phi_{t_1,\ldots,t_n}(\lambda_1,\ldots,\lambda_n)| \leq \Phi_{t_1,\ldots,t_n}(0,\ldots,0) = 1.$$

III. SOME EXAMPLES OF STOCHASTIC PROCESSES

We now examine some special stochastic processes which are used to model physical phenomena practically.

A. Gaussian Processes

A real stochastic process $\{X(t)\}$ on $\mathbb{R}$ is said to be *Gaussian* if for every integer $n \geq 1$ its n-dimensional joint distribution is Gaussian, i.e., the density function $p_{t_1,\ldots,t_n}(x_1,\ldots,x_n)$ is of the form

$$p_{t_1,\ldots,t_n}(x_1,\ldots,x_n) = \frac{1}{(2\pi)^{n/2}|V|^{1/2}}$$
$$\times \exp\left\{-\frac{1}{2}(\mathbf{x} - \mathbf{m})V^{-1}(\mathbf{x} - \mathbf{m})^{t}\right\},$$

where $\mathbf{x} = (x_1, \ldots, x_n)$, $\mathbf{m} = (m(t_1), \ldots, m(t_n)) \in \mathbb{R}^n$,

$$V = ((X(t_i) - m(t_i), X(t_j) - m(t_j))_2)_{i,j}$$

is an invertible $n \times n$ positive-definite Hermitian matrix, and $|V|$ is the determinant of V, with the superscript t being the transpose. One important characteristic of Gaussian processes is that if we are given a function $m(t)$ and a positive-definite function $K(s, t)$, then we can find a (complex) Gaussian process for which $m(t)$ is its expectation and $K(s, t)$ is its covariance function. As is seen from the Central Limit Theorem, the Gaussian distribution is universal and it is reasonable to use the Gaussian assumption on a stochastic process unless there is another specific choice.

B. Markov Processes

A real stochastic process $\{X(t)\}$ on $T = \mathbb{R}, \mathbb{R}^+, \mathbb{Z}$, or $\mathbb{Z}^+ = \{0, 1, 2, \ldots\}$ is said to be a *Markov process* if for any $s_1 < \cdots < s_n < t$ it holds that

$$P\big(X(t) \in A | X(s_1) = x_{s_1}, \ldots, X(s_n) = x_{s_n}\big)$$
$$= P\big(X(t) \in A \mid X(s_n) = x_{s_n}\big)$$

P-a.e. for any $A \in \Sigma$, where $P(\cdot|\cdot)$ is the conditional probability. In this case, there exits a function $P(s, x, t, A)$, called the *transition probability,* for $s, x, t \in T$ and $A \in \Sigma$, such that the following hold:

(i) For fixed $s, t \in T$, $P(s, x, t, \cdot)$ is a probability for each $x \in T$, and $P(s, \cdot, t, A)$ is $\mathfrak{B}$-measurable for each $A \in \Sigma$.

(ii) $P(s, x, s, A) = \begin{cases} 1, & x \in A, \\ 0, & x \notin A. \end{cases}$

(iii) $P(X(t) \in A \mid X(s) = x_s) = P(s, x_s, t, A)$, P-a.e.

(iv) (Chapman–Kolmogorovs equation) For fixed $s, u, t \in \mathbb{R}$

$$P(s, x, u, A) = \int_{-\infty}^{\infty} P(s, x, t, dy) P(t, y, u, A),$$

P_{X_s}-a.e.x.

Conversely, if a family of functions $\{P(s, x, t, A)\}$ is given such that (i), (ii), and (iv) hold, and a probability P_0 on $(\mathbb{R}, \mathfrak{B})$ is given, then there exists a Markov process $\{X(t)\}$ for which $P(s, x, t, A)$ is the transition probability and P_0 is the distribution of $X(0)$, i.e., the initial distribution.

There are some special Markov processes. A Markov process $\{X(t)\}$ is said to be *temporally homogeneous* if the transition probability depends on the difference of t and s, i.e., $P(s, x, t, A) = P(t - s, x, A)$. A Markov process $\{X(t)\}$ is said to be *spatially homogeneous* if $P(s, x, t, A) = P(s, t, A - x)$, where $A - x = \{y - x : y \in A\}$. A real stochastic process $\{X(t)\}$ on $\mathbb{R}^+$ is called an *additive process* if $X(0) = 0$ and for any $t_1, \ldots, t_n \in \mathbb{R}^+$ with $t_1 < \cdots < t_n$, $X(t_2) - X(t_1), \ldots, X(t_n) - X(t_{n-1})$ are mutually independent. As can be seen, each spatially homogeneous Markov process is an additive process. A stochastic process $\{X(t)\}$ on $\mathbb{R}^+$ is called a *diffusion process* if it is a Markov process such that the sample path $X(\cdot, \omega)$ is a continuous function on $\mathbb{R}^+$ for almost every $\omega \in \Omega$. If $\{X(t)\}$ is a Markov process and the range of $X(t, \cdot)$, $t \in T$, is a countable set, it is called a *Markov chain*. A *random walk* is a special case of a Markov chain.

C. Brownian Motion

The botanist R. Brown was observing pollens floating on the surface of water and discovered that these particles were moving irregularly all the time. If we consider a particle floating in the air or in the liquid as above, its motion caused by collisions with molecules of the air or the liquid is extremely irregular and regarded as a stochastic process. More precisely, if we denote the position of the particle at time t by $\mathbf{X}(t) = (X_1(t), X_2(t), X_3(t))$, then $\mathbf{X}(t)$ is a three-dimensional random variable such that $\mathbf{X}(t) - \mathbf{X}(s)$ obeys a three-dimensional Gaussian distribution. This kind of motions was considered as a class of stochastic processes and investigated by A. Einstein, N. Wiener, and P. Levy.

Here is a precise definition. A stochastic process $\{\mathbf{X}(t)\}$ on $T \subseteq \mathbb{R}$ is said to be a *Wiener process* or a *Brownian motion* if the following conditions hold:

(i) $\mathbf{X}(t, \omega) \in \mathbb{R}^d$ (d is a positive integer).
(ii) For $t_1 < \cdots < t_n$, $\mathbf{X}(t_2) - \mathbf{X}(t_1), \ldots, \mathbf{X}(t_n) - \mathbf{X}(t_{n-1})$ are mutually independent.
(iii) If $\mathbf{X}(t) = (X_1(t), \ldots, X_d(t))$, then $\{X_1(t)\}, \ldots,$ $\{X_d(t)\}$ are mutually independent stochastic processes such that $X_i(t) - X_i(s) \approx N(0, |t - s|)$ for each i, i.e., $X_i(t) - X_i(s)$ obeys the Gaussian distribution $N(0, |t - s|)$.

It follows from the definition that a Wiener process is a temporally homogeneous additive process. Moreover, a Wiener process is a special case of a diffusion process since $S = \mathbb{R}^d$ is the state space [i.e., the range of the random variables $X(t)$, $t \in \mathbb{R}$] and the transition probabilities are given by

$$P(t, \mathbf{x}, B) = \int_B \frac{1}{(2\pi t)^{d/2}} \exp\left\{-\frac{1}{2t}|\mathbf{x} - \mathbf{y}|^2\right\} d\mathbf{y},$$

where $t > 0$, $\mathbf{x}, \mathbf{y} \in \mathbb{R}^d$, $B \in \mathfrak{B}(\mathbb{R}^d)$, and $|\mathbf{x}| = \sqrt{x_1^2 + \cdots + x_d^2}$, the norm of $\mathbf{x} = (x_1, \ldots, x_d)$.

D. Stationary Processes

Stationarity is considered as an invariance under the time shift. There are two kinds of stationarity, weak and strong. A stochastic process $\{X(t)\}$ is said to be *strongly stationary* or *stationary in the strict sense* if the joint distribution is invariant under the time shift, i.e., for any $t, t_1, \ldots, t_n \in T$ and $E \in \mathfrak{B}(\mathbb{C}^n)$, the Borel σ-algebra of $\mathbb{C}^n$, it holds that

$$P((X(t_1 + t), \ldots, X(t_n + t)) \in E)$$
$$= P((X(t_1), \ldots, X(t_n)) \in E).$$

A second-order stochastic process $\{X(t)\}$ is said to be *weakly stationary* or *stationary in the wide sense* if its average is constant, if its covariance function $K(s, t)$ depends only on the difference $s - t$, and if K is continuous as a two-variable function. Clearly, if the process is of second order and the covariance function is continuous, then strong stationarity implies weak stationarity. The converse is true for Gaussian processes. In general, a strongly stationary process need not have any moment (e.g., a Cauchy stationary process). Weak stationarity will be singled out and discussed in the next section.

E. White Noise

A discrete or continuous time stochastic process $\{X(t)\}$ is called a *white noise* if $E\{X(t)\} \equiv 0$ and $K(s, t) = \delta(s - t)$, the Kronecker delta function, i.e., $= 1$ for $s = t$ and $= 0$ for $s \neq t$. A white noise does not exist in a real world. But as a mathematical model it is used in many fields. It is known that the derivative of a Wiener process as a generalized stochastic process is a white noise (cf. Section VIII).

IV. WEAKLY STATIONARY STOCHASTIC PROCESSES

Weak stationarity for second-order stochastic processes was initiated by Khintchine (1934). Let us begin with the definition again. A stochastic process $\{X(t)\}$ on $T = \mathbb{R}$ or $\mathbb{Z}$ is said to be *weakly stationary* or *wide sense stationary* if its expectation is constant and if the covariance function is continuous and depends only on the difference of the variables. In other words, $\{X(t)\}$ is weakly stationary if the following hold:

(i) $m(t) = E\{X(t)\} \equiv m$ for $t \in T$.
(ii) $K(s, t) = \tilde{K}(s - t)$ for $s, t \in T$.
(iii) $\tilde{K}(\cdot)$ is continuous on T.

If $T = \mathbb{Z}$, the condition (iii) is automatically satisfied. We may assume that $m = 0$ in (i) [since otherwise we can let $Y(t) = X(t) - m$] and consider pro-

cesses on $\mathbb{R}$. Denote by $L_0^2(P)$ the closed subspace of $L^2(P)$ consisting of functions with zero expectations, i.e., $L_0^2(P) = \{X \in L^2(P): E\{X\} = 0\}$. So we consider centered processes $\{X(t)\} \subseteq L_0^2(P)$.

Let us proceed to obtain integral representations of a weakly stationary process and its covariance function, and the Kolmogorov isomorphism theorem, which states that the time and the spectral domains are isomorphic.

Thus, suppose $\{X(t)\}$ is a weakly stationary process on $\mathbb{R}$ with the (one-variable) covariance function $K(t)$. Since $K(t)$ is continuous and positive definite, there exists, by Bochner's theorem, a finite positive measure ν on $(\mathbb{R}, \mathfrak{B})$ such that

$$K(t) = \int_{-\infty}^{\infty} e^{itu} \nu(du), \qquad t \in \mathbb{R}. \tag{6}$$

ν is called the *spectral measure* of the process. The *time domain* $H(X)$ of $\{X(t)\}$ is defined by

$$H(X) = \mathfrak{S}\{X(t): t \in \mathbb{R}\},$$

the closed subspace of $L_0^2(P)$ spanned by the set $\{X(t): t \in \mathbb{R}\}$. To derive an integral representation of the process itself, define an operator $U(t)$ by

$$U(t)X(s) = X(s + t), \qquad s \in \mathbb{R},$$

or more generally by

$$U(t)\left(\sum_{k=1}^{n} \alpha_k X(t_k)\right) = \sum_{k=1}^{n} \alpha_k X(t_k + t),$$

where $\alpha_1, \ldots, \alpha_n \in \mathbb{C}$, $t_1, \ldots, t_n \in \mathbb{R}$. Then because of stationarity we see that $U(t)$ is well defined and norm preserving, and can be extended to a unitary operator on the time domain $H(X)$. Hence, by Stone's theorem, there is a spectral measure (i.e., an orthogonal projection-valued measure) E on $(\mathbb{R}, \mathfrak{B})$ such that

$$U(t) = \int_{-\infty}^{\infty} e^{itu} E(du), \qquad t \in \mathbb{R}.$$

Consequently we have that

$$X(t) = U(t)X(0) = \int_{-\infty}^{\infty} e^{itu} E(du) X(0), \qquad t \in \mathbb{R}.$$

If we let $\xi(A) = E(A)X(0)$ for $A \in \mathfrak{B}$, then the random quantity ξ is an $L_0^2(P)$-valued, bounded, countably additive measure, and is *orthogonally scattered*, i.e., $(\xi(A), \xi(B))_2 = 0$ if $A \cap B = \emptyset$. Therefore, we get an integral representation of the process $\{X(t)\}$ as

$$X(t) = \int_{-\infty}^{\infty} e^{itu} \xi(du), \qquad t \in \mathbb{R}. \tag{7}$$

By (6) and (7) we see that $\nu(A) = (\xi(A), \xi(A))_2$ for $A \in \mathfrak{B}$.

Consider the Hilbert space $L^2(\nu) = L^2(\mathbb{R}, \mathfrak{B}, \nu)$, which is called the *spectral domain* of the process $\{X(t)\}$. Then the *Kolmogorov isomorphism theorem* states that the time domain $H(X)$ and the spectral domain $L^2(\nu)$ are isomorphic as Hilbert spaces. Moreover, the isomorphism $V: L^2(\nu) \to H(X)$ is given by

$$V(f) = \int_{-\infty}^{\infty} f(u)\,\xi(du), \qquad f \in L^2(\nu). \qquad (8)$$

The importance of this theorem is that each random variable in $H(X) \subseteq L_0^2(P)$ is expressed as a vector integral of a usual function and, especially, $X(t)$ is a Fourier transform of a vector measure ξ. Furthermore, the measure ξ can be obtained from the process $X(t)$ by inversion as

$$\xi((u, v)) = \lim_{t \to \infty} \int_{-t}^{t} \frac{e^{-ivs} - e^{-ius}}{-is} X(s)\,ds, \qquad (9)$$

where $u < v$, $\xi(\{u\}) = \xi(\{v\}) = 0$, and the right-hand side is a Bochner integral (cf. Section VI). Thus (9) is called the *inversion formula*.

V. CLASSES OF NONSTATIONARY STOCHASTIC PROCESSES

Stationarity in the strict sense or wide sense is very restrictive in a practical application and somewhat weaker conditions are needed in many cases. For the sake of simplicity we consider only centered second-order stochastic processes and use the space $L_0^2(P)$. The following are classes of nonstationarity.

A. Karhunen Class

Karhunen (1947) introduced a class of stochastic processes given by

$$X(t) = \int_{-\infty}^{\infty} g_t(u)\,\xi(du), \qquad t \in \mathbb{R}, \qquad (10)$$

where ξ is an $L_0^2(P)$-valued orthogonally scattered measure and $\{g_t: t \in \mathbb{R}\} \subseteq L^2(\nu)$ with $\nu(\cdot) = \|\xi(\cdot)\|_2^2$. In this case the covariance function is given by

$$K(s, t) = \int_{-\infty}^{\infty} g_s(u)\overline{g_t(u)}\,\nu(du), \qquad s, t \in \mathbb{R}.$$

When $g_t(u) = e^{itu} (t \in \mathbb{R})$, $\{X(t)\}$ reduces to a weakly stationary process.

B. Harmonizable Class

Harmonizability was first introduced by Loève in the middle 1940s. Later a weaker harmonizability was defined by Rozanov (1959). These two notions were distinguished by calling them weak and strong harmonizabilities by Rao (1982).

A second-order stochastic process $\{X(t)\}$ on $\mathbb{R}$ is said to be *strongly harmonizable* if its covariance function K has a representation

$$K(s, t) = \int_{-\infty}^{\infty} \int_{-\infty}^{\infty} e^{i(su-tv)}\,\beta(du, dv), \qquad s, t \in \mathbb{R}, \qquad (11)$$

for some *positive-definite bimeasure* $\beta: \mathfrak{B} \times \mathfrak{B} \to \mathbb{C}$ of bounded variation, where $\mathfrak{B} \times \mathfrak{B} = \{A \times B: A, B \in \mathfrak{B}\}$ and the total *variation* ($=$ *Vitali variation*) is given by

$$|\beta|(\mathbb{R}, \mathbb{R}) = \sup \sum_{j=1}^{\ell} \sum_{k=1}^{n} |\beta(A_j, B_k)|,$$

the supremum being taken for all finite measurable partitions $\{A_1, \ldots, A_\ell\}$ and $\{B_1, \ldots, B_n\}$ of $\mathbb{R}$. Here, $\beta: \mathfrak{B} \times \mathfrak{B} \to \mathbb{C}$ is a *bimeasure* if $\beta(A, \cdot)$ and $\beta(\cdot, A)$ are $\mathbb{C}$-valued measures on $\mathfrak{B}$ for each $A \in \mathfrak{B}$. When β is of bounded variation, i.e., $|\beta|(\mathbb{R}, \mathbb{R}) < \infty$, β can be extended to an ordinary measure on $\mathfrak{B}(\mathbb{R}^2)$ (the Borel σ-algebra of $\mathbb{R}^2$) and (11) becomes the Lebesgue integral. When β is of unbounded variation, we need to interpret (11) as a bimeasure integral or (a weaker) MT-integral (after Morse and Transue). Since every bimeasure satisfies $\sup\{|\beta(A, B)|: A, B \in \mathfrak{B}\} < \infty$, we can show that (11) is well defined. So, $\{X(t)\}$ is said to be *weakly harmonizable* if its covariance function K is expressed as (11) for some positive-definite bimeasure β. Since the integrand $e^{i(su-tv)} = \varphi_1(u)\varphi_2(v)$, say, in the integral (11) is the product of bounded continuous functions of one variable, we can interpret the MT-integral as follows. For $A, B \in \mathfrak{B}$, φ_1 is $\beta(\cdot, B)$-integrable and φ_2 is $\beta(A, \cdot)$-integrable. Hence, letting

$$\mu_1(A) = \int_{-\infty}^{\infty} \varphi_2(v)\,\beta(A, dv), \qquad A \in \mathfrak{B},$$

$$\mu_2(B) = \int_{-\infty}^{\infty} \varphi_1(u)\,\beta(du, B), \qquad B \in \mathfrak{B},$$

we see that μ_1 and μ_2 are $\mathbb{C}$-valued measures on $\mathfrak{B}$, and φ_j is μ_j-integrable ($j = 1, 2$). Moreover, when they have the same integral, i.e.,

$$\int_{-\infty}^{\infty} \varphi_1(u)\,\mu_1(du) = \int_{-\infty}^{\infty} \varphi_2(v)\,\mu_2(dv),$$

we denote the value by $\int_{-\infty}^{\infty} \int_{-\infty}^{\infty} \varphi_1(u)\varphi_2(v)\,\beta(du, dv)$.

An integral representation of a weakly harmonizable process $\{X(t)\}$ is obtained:

$$X(t) = \int_{-\infty}^{\infty} e^{itu}\,\xi(du), \qquad t \in \mathbb{R}, \qquad (12)$$

for some $L_0^2(P)$-valued bounded measure ξ, not necessarily orthogonally scattered. The measure ξ is called the *representing measure* of $\{X(t)\}$ and the inversion formula (9) is also valid. Furthermore, comparing (11) and (12), we see that

$$\beta(A, B) = (\xi(A), \xi(B))_2, \qquad A, B \in \mathfrak{B}.$$

As is easily seen, a weakly harmonizable process is weakly stationary if and only if the representing measure is orthogonally scattered. If a weakly stationary process is projected onto a closed subspace of its time domain, then the resulting process is not necessarily weakly stationary but always weakly harmonizable. The converse is known as a *stationary dilation*. That is, each weakly harmonizable process can be obtained as an orthogonal projection of some weakly stationary process.

The Kolmogorov isomorphism theorem holds for a weakly harmonizable process $\{X(t)\}$. The spectral domain of $\{X(t)\}$ is defined to be the space $\mathcal{L}_*^2(\beta)$ of all functions $f(u)$ on $\mathbb{R}$ that are *strictly integrable* with respect to the bimeasure β, in a somewhat "restricted sense." Then, the time domain $H(X)$ and the spectral domain $\mathcal{L}_*^2(\beta)$ are isomorphic Hilbert spaces, where the isomorphism is given by

$$V(f) = \int_{-\infty}^{\infty} f(u)\,\xi(du), \qquad f \in \mathcal{L}_*^2(\beta),$$

which is similar to (8). The weakly harmonizable class is a fruitful area of study with many applications because each process in this class is a Fourier transform of a vector measure and is relatively simple to handle.

C. Cramér Class

A further generalization of the Karhunen class was introduced by Cramér (1951). A stochastic process $\{X(t)\}$ is said to be *of strong Cramér class* if its covariance function is written as

$$K(s, t) = \int_{-\infty}^{\infty} \int_{-\infty}^{\infty} g_s(u)\overline{g_t(v)}\,\beta(du, dv), \qquad s, t \in \mathbb{R}, \tag{13}$$

for some positive-definite bimeasure β of bounded variation and some family $\{g_t : t \in \mathbb{R}\}$ of bounded Borel functions on $\mathbb{R}$. When the bimeasure β in (13) is simply bounded, $\{X(t)\}$ is *of weak Cramér class* relative to the family $\{g_t : t \in \mathbb{R}\}$. Integral representations of such processes are available. If $\{X(t)\}$ is of weakly or strongly Cramér class, then there exists an $L_0^2(P)$-valued measure ξ such that (10) is true. As a harmonizable process has a stationary dilation, each process of Cramér class has a *Karhunen dilation*, i.e., it is expressed as an orthogonal projection of some process of Karhunen class.

D. KF Class

The concept of the spectral measure can be generalized. In the late 1950s, Kampe de Fériet, and Frankiel, and independently Parzen and Rozanov, defined the "associated spectrum" for a process $\{X(t)\}$ for which

$$\tilde{K}(h) = \lim_{t \to \infty} \frac{1}{t} \int_0^t K(s, s + h)\,ds, \qquad h \in \mathbb{R},$$

exists and is positive definite and continuous in h, where K is the covariance function of the process. In this case, by Bochner's theorem there exists a finite positive measure ν on $(\mathbb{R}, \mathfrak{B})$, called the *associated spectrum* of $\{X(t)\}$, such that

$$\tilde{K}(h) = \int_{-\infty}^{\infty} e^{ihu}\,\nu(du), \qquad h \in \mathbb{R}.$$

Such a process is said to be *of KF class* or an *asymptotically stationary* process. Unfortunately, not every process of KF class allows an integral representation of the process itself. A strongly harmonizable process is necessarily of KF class, but a weakly harmonizable process is not in general.

E. Periodically Correlated Class

The number of black spots in the surface of the Sun increases periodically every 11 years. To describe such a phenomenon we use a periodically correlated process. Here, we consider only discrete time processes, namely those on $\mathbb{Z}$. A process $\{X(t)\}$ on $\mathbb{Z}$ is said to be *periodically correlated with period $p > 0$* if the covariance function K satisfies

$$K(s, t) = K(s + p, t + p), \qquad s, t \in \mathbb{Z}.$$

If $p = 1$, then the process is weakly stationary. So we assume $p \geq 2$.

Consider the Cartesian product space $[L_0^2(P)]^p = L_0^2(P) \times \cdots \times L_0^2(P)$. For $\mathbf{X} = (X_1, \ldots, X_p)$ and $\mathbf{Y} = (Y_1, \ldots, Y_p) \in [L_0^2(P)]^p$ define

$$(\mathbf{X}, \mathbf{Y})_{2,p} = \sum_{j=1}^{p} (X_j, Y_j)_2,$$

so that $(\cdot, \cdot)_{2,p}$ is an inner product in $[L_0^2(P)]^p$. Also define a new process $\{\mathbf{X}(t)\}$, which is $[L_0^2(P)]^p$-valued, by

$$\mathbf{X}(t) = (X(t), X(t+1), \ldots, X(t+p-1)), \qquad t \in \mathbb{Z}.$$

Then we can verify that an $L_0^2(P)$-valued process $\{X(t)\}$ is periodically correlated with period p if and only if the $[L_0^2(P)]^p$-valued process $\{\mathbf{X}(t)\}$ is weakly stationary. Moreover, note that $L_0^2(P)$ can be regarded as a closed subspace of $[L_0^2(P)]^p$ if we identify $L_0^2(P)$ with $L_0^2(P) \times \{0\} \times \cdots \times \{0\}$. Hence, if $\{X(t)\}$ is periodically correlated with period p and $J: [L_0^2(P)]^p \to L_0^2(P)$ is the orthogonal projection, then $X(t) = J\mathbf{X}(t)$, $t \in \mathbb{Z}$. Therefore, $\{X(t)\}$ is weakly harmonizable and $\{\mathbf{X}(t)\}$ is its stationary dilation. Furthermore, it is known that $\{X(t)\}$ is strongly harmonizable.

VI. THE CALCULUS OF STOCHASTIC PROCESSES

A. Convergence

In order to talk about continuity, differentiability, and integrability of stochastic processes on $\mathbb{R}$, we need to present the notion of convergence of a sequence of random variables. Let $\{X_n\}$ be a sequence of random variables and X a random variable on a probability space (Ω, Σ, P). We say that $\{X_n\}$ converges to X:

1. *Everywhere* if $X_n(\omega) \to X(\omega)$ for every $\omega \in \Omega$
2. *Almost everywhere* (a.e.), *almost surely* (a.s.), or *with probability* 1, if $P(\{\omega \in \Omega: X_n(\omega) \to X(\omega)\}) = 1$
3. *In mean-square* or *in L^2-mean* if $\|X_n - X\|_2 \to 0$
4. *In probability* if for any $\varepsilon > 0$, $P(\{|X_n - X| > \varepsilon\}) \to 0$
5. *In distribution* if $F_n(x) \to F(x)$ for each $x \in \mathbb{R}$ for which F is continuous, where F_n and F are distribution functions of X_n and X, respectively.

When the density functions exist, the last condition is equivalent to $p_n(x) \to p(x)$ for almost all $x \in \mathbb{R}$, where p, p_n are density functions.

The interrelationship among these convergences is as follows:

everywhere convergence

$\Downarrow$

a.e. convergence   mean-square convergence

$\Downarrow$ $\Downarrow$

convergence in probability

$\Downarrow$

convergence in distribution.

B. Continuity

Consider a continuous time second-order stochastic process $\{X(t, \omega)\} = \{X(t)\}$ on $T \subseteq \mathbb{R}$. If we choose an $\omega \in \Omega$, then $X(\cdot, \omega)$ is a function of $t \in T$ (a sample path) and its continuity is discussed. We say that $\{X(t)\}$ is *continuous* a.e. or has a *sample path continuity* a.e. if, for a.e. $\omega \in \Omega$, the scalar function $X(\cdot, \omega)$ is a continuous function on T.

Another type of continuity is in the mean-square sense. $\{X(t)\}$ is said to be *continuous in mean-square* at $t_0 \in T$ if

$$\|X(t) - X(t_0)\|_2 \to 0 \quad \text{as} \quad t \to t_0,$$

which is equivalent to the continuity of the covariance function $K(s, t)$ at (t_0, t_0). Since

$$|E\{X(t)\} - E\{X(t_0)\}| \leq E\{|X(t) - X(t_0)|\}$$
$$\leq \left[E\{|X(t) - X(t_0)|^2\}\right]^{1/2}$$
$$= \|X(t) - X(t_0)\|_2,$$

we see that the expectation of $X(t)$ is continuous at t_0 if $\{X(t)\}$ is continuous in mean-square at t_0.

C. Differentiability

Let $\{X(t, \omega)\} = \{X(t)\}$ be a second-order stochastic process on an interval $T \subseteq \mathbb{R}$. First consider a sample path differentiability. $\{X(t)\}$ has *a.e. differentiable sample path* if, for a.e. $\omega \in \Omega$, the sample path $X(\cdot, \omega)$ is differentiable. Another type of derivative is in the mean-square sense. $\{X(t)\}$ is *differentiable in mean-square* at t_0 if there is a random variable $X'(t_0)$ such that

$$\left\| \frac{X(t_0 + h) - X(t_0)}{h} - X'(t_0) \right\|_2 \to 0 \quad \text{as} \quad h \to 0.$$

If $\{X(t)\}$ on $\mathbb{R}$ is weakly stationary with the covariance function $K(t)$, then it is differentiable in mean-square at t_0 if and only if K has the second derivative at t_0. When $\{X(t)\}$ is nonstationary with the covariance function $K(s, t)$, it is differentiable in mean-square at t_0 if $\partial^2 K / \partial s \partial t$ exists at $s = t = t_0$.

D. Integrability

Consider a second-order stochastic process $\{X(t)\}$ on $[a, b]$ and its integral

$$\int_a^b X(t, \omega) \, dt. \tag{14}$$

If, for all $\omega \in \Omega$, the sample path $X(\cdot, \omega)$ is continuous or bounded and measurable, then the integral (14) is defined as a Riemann or Lebesgue integral and defines a random variable $Y(\omega)$. If this is not the case, however, we can consider (14) in a mean-square sense. Then, the integrand is an $L_0^2(P)$-valued function $X(t)$, the measure is the Lebesgue measure dt, and the integral (14) is a *Bochner integral*. The definition of this integral is as follows: Let

$$X_n(t) = \sum_{k=1}^{k_n} 1_{A_{n,k}}(t) X_{n,k},$$

where $X_{n,k} \in L_0^2(P)$, $A_{n,k} \in \mathcal{B}$, and $1_{A_{n,k}}(t)$ is the indicator function of $A_{n,k}$, i.e., $= 1$ if $t \in A_{n,k}$ and $= 0$ otherwise. Such a function is called a *finitely valued* measurable function. If there exists a sequence $\{X_n(t)\}$ of such functions for which

$$\|X_n(t) - X(t)\|_2 \to 0 \qquad (n \to \infty)$$

for almost all $t \in [a, b]$ in the Lebesgue measure and $\int_a^b \|X(t)\|_2 \, dt < \infty$, then we can define the Bochner integral as

$$\int_a^b X(t) \, dt = \lim_{n \to \infty} \int_a^b X_n(t) \, dt$$

$$\equiv \lim_{n \to \infty} \sum_{k=1}^{k_n} X_{n,k} L(A_{n,k}),$$

$L(A_{n,k})$ being the Lebesgue measure of $A_{n,k}$, where the limit is in mean-square sense.

VII. EXPANSIONS OF STOCHASTIC PROCESSES

Since a stochastic process is regarded as a two-variable function $X(t, \omega)$ on $T \times \Omega$, it might be interesting and useful to express it as a sum of products of single-variable functions of t and ω:

$$X(t, \omega) = \sum_n \phi_n(t) X_n(\omega), \qquad (15)$$

where the right-hand side is convergent in some sense, so that $\phi_n(t)$ is a time function and $X_n(\omega)$ is a random variable for $n \geq 1$. We consider three types of expressions of stochastic processes given by (15).

A. Karhunen–Loève Expansion

Let $\{X(t)\}$ be a second-order stochastic process on a finite closed interval $[a, b]$ with the covariance function $K(s, t)$. Assume that the process $\{X(t)\}$ is mean-square continuous, which is the same as saying that $K(s, t)$ is continuous as a two-variable function on $[a, b]^2$. Then it follows that $K(s, t)$ is measurable on $[a, b]^2$ and

$$\int_a^b \int_a^b |K(s, t)|^2 \, ds \, dt < \infty,$$

i.e., $K \in L^2([a, b]^2)$. Moreover, K defines an integral operator $\mathbf{K}$ on $L^2([a, b])$ in such a way that

$$(\mathbf{K}\varphi)(t) = \int_a^b K(s, t)\varphi(s) \, ds \qquad (16)$$

for $\varphi \in L^2([a, b])$. Now Mercer's theorem applied to $K(s, t)$ gives an expansion of $K(s, t)$:

$$K(s, t) = \sum_{k=1}^{\infty} \lambda_k \varphi_k(s)\overline{\varphi_k(t)}, \qquad s, t \in [a, b], \qquad (17)$$

where the φ_k are eigenfunctions of the operator $\mathbf{K}$ defined by (16) and the λ_k are corresponding eigenvalues:

$$(\mathbf{K}\varphi_k)(t) = \lambda_k \varphi_k(t), \qquad t \in [a, b], \qquad (18)$$

and the convergence in (17) is uniform on $[a, b]^2$ and in $L^2([a, b]^2)$, i.e.,

$$\sup_{s,t \in [a,b]} \left| K(s, t) - \sum_{k=1}^{n} \lambda_k \varphi_k(s)\overline{\varphi_k(t)} \right| \to 0,$$

$$\left\| K - \sum_{k=1}^{n} \lambda_k \varphi_k \otimes \overline{\varphi_k} \right\|_2^2$$

$$\equiv \int_a^b \int_a^b \left| K(s, t) - \sum_{k=1}^{n} \lambda_k \varphi_k(s)\overline{\varphi_k(t)} \right|^2 \, ds \, dt \to 0$$

as $n \to \infty$, where $(\varphi_k \otimes \overline{\varphi_k})(s, t) = \varphi_k(s)\overline{\varphi_k(t)}$. Note that φ_k is continuous on $[a, b]$ and $\lambda_k > 0$ for each $k \geq 1$. For instance, if $K(s, t) = \min\{s, t\}$ on $[0, T]$, the covariance function of a one-dimensional Wiener process, then

$$\min\{s, t\} = \sum_{k=0}^{\infty} \frac{2}{T} \left(\frac{2T}{(2k+1)\pi} \right)^2 \sin\left(\frac{(2k+1)\pi}{2T}s \right)$$

$$\times \sin\left(\frac{(2k+1)\pi}{2T}t \right),$$

the uniform convergence holding on $[0, T]^2$.

Now for each integer $k \geq 1$ define

$$X_k = \frac{1}{\sqrt{\lambda_k}} \int_a^b X(t)\overline{\varphi_k(t)} \, dt,$$

which is a well-defined random variable in $L^2(P)$. Here, the functions φ_k satisfy (18). Moreover, we see that $\{\varphi_k\}_{k=1}^{\infty}$ forms an orthonormal set in $L^2([a, b])$, i.e.,

$$(\varphi_j, \varphi_k)_2 = \int_a^b \varphi_j(t)\overline{\varphi_k(t)} \, dt = \delta_{jk},$$

the Kronecker delta, and that $\{X_k\}_{k=1}^{\infty}$ also forms an orthonormal set in $L^2(P)$. Finally we obtain an expansion of $X(t)$:

$$X(t) = \sum_{k=1}^{\infty} \sqrt{\lambda_k}\varphi_k(t)X_k, \qquad (19)$$

where the convergence is in mean-square. Equation (19) is called the *Karhunen–Loéve expansion*. The characteristic of this expansion is that we are using two orthonormal sets $\{\varphi_k\}$ and $\{X_k\}$, and that $\{X_k\}$ is the set of Fourier coefficients of $X(t)$ with respect to $\{\varphi_k\}$. Moreover, $\{\varphi_k\}$ is obtained as the eigenfunctions of the operator $\mathbf{K}$ associated with the covariance function of the process.

B. Sampling Theorem

Shannon's (1949) sampling theorem was obtained for deterministic functions on $\mathbb{R}$ (signal functions). This was extended to be valid for weakly stationary stochastic processes by Balakrishnan (1957).

First we consider bandlimited and finite-energy signal functions. A signal function $X(t)$ is said to be *of finite energy* if $X \in L^2(\mathbb{R})$ with the Lebesgue measure. Then its Fourier transform $\mathcal{F}X$ is defined in the mean-square sense by

$$(\mathcal{F}X)(u) = \frac{1}{\sqrt{2\pi}} \int_{-\infty}^{\infty} X(t)e^{iut}\,dt = \underset{T \to \infty}{\text{l.i.m.}} \int_{-T}^{T} X(t)e^{iut}\,dt,$$

where l.i.m. means "limit in the mean" and $\mathcal{F}: L^2(\mathbb{R}) \to L^2(\mathbb{R})$ turns out to be a unitary operator. A signal function $X(t)$ is said to be *bandlimited* if there exists some constant $W > 0$ such that

$$(\mathcal{F}X)(u) = 0 \qquad \text{for almost every } u \text{ with } |u| > W.$$

Here, W is called a *bandwidth* and $[-W, W]$ a *frequency interval*.

Let $W > 0$ and define a function S_W on $\mathbb{R}$ by

$$S_W(t) = \begin{cases} \frac{W}{\pi} \frac{\sin Wt}{Wt}, & t \neq 0 \\ \frac{W}{\pi}, & t = 0. \end{cases}$$

Then, S_W is a typical example of a bandlimited function and is called a *sample function*. In fact, one can verify that

$$(\mathcal{F}S_W)(u) = \frac{1}{\sqrt{2\pi}} 1_W(u), \qquad u \in \mathbb{R},$$

where $1_W = 1_{[-W, W]}$, the indicator function of $[-W, W]$. Denote by $\mathbf{BLW}$ the set of all bandlimited signal functions with bandwidth $W > 0$. Then it is easily seen that $\mathbf{BLW}$ is a closed subspace of $L^2(\mathbb{R})$. Now the *sampling theorem* for a function in $\mathbf{BLW}$ is stated as follows: Any $X \in \mathbf{BLW}$ has a sampling expansion in L^2- and L^∞-sense given by

$$X(t) = \sum_{n=-\infty}^{\infty} X\left(\frac{n\pi}{W}\right) \sqrt{\frac{W}{\pi}} \varphi_n(t), \qquad (20)$$

where the φ_n are defined by

$$\varphi_n(t) = \sqrt{\frac{\pi}{W}} S_W\left(t - \frac{n\pi}{W}\right), \qquad t \in \mathbb{R}, \quad n \in \mathbb{Z},$$

$\{\varphi_n\}_{-\infty}^{\infty}$ forms a complete orthonormal system in $\mathbf{BLW}$, and it is called a *system of sampling functions*. We can say that a sampling theorem is a Fourier expansion of an L^2-function with respect to this system of sampling functions.

A sampling theorem holds for some stochastic processes. Let $\{X(t)\}$ be an $L_0^2(\Omega)$-valued weakly harmonizable process with the representing measure ξ, i.e.,

$$X(t) = \int_{-\infty}^{\infty} e^{itu}\,\xi(du), \qquad t \in \mathbb{R}.$$

We say that $\{X(t)\}$ is *bandlimited* if there exists a $W > 0$ such that the support of ξ is contained in $[-W, W]$, i.e., $\xi(A) = 0$ if $A \cap [-W, W] = \emptyset$. If this is the case, the *sampling theorem* holds:

$$X(t, \omega) = \sum_{n=-\infty}^{\infty} X\left(\frac{n\pi}{W}, \omega\right) \frac{\sin(W(t - n\pi/W))}{W(t - n\pi/W)},$$

$$t \in \mathbb{R},$$

where the convergence is in $\|\cdot\|_2$ for each $t \in \mathbb{R}$.

C. Series Representation

The above two expansions have some restrictions. Namely, the KL expansion is valid for stochastic processes on finite closed intervals, and the sampling theorem requires bandlimitedness. To relax these conditions we consider separability of the time domain $H(X)$, which is a fairly general condition. Thus let us consider the series representation of a second-order stochastic process $\{X(t, \omega)\}$:

$$X(t, \omega) = \sum_{n=1}^{\infty} \varphi_n(t) X_n(\omega). \qquad (21)$$

It is easily verified that $\{X(t, \omega)\}$ has a series representation of the form (21) if and only if the time domain $H(X)$ is separable, i.e., there exists a countable orthonormal base $\{Y_k\}_{k=1}^{\infty}$ for $H(X)$. If we choose one such base $\{X_k\}_{k=1}^{\infty}$, then the $\varphi_n(t)$ are the Fourier coefficients, i.e.,

$$\varphi_n(t) = (X(t), X_n)_2, \qquad t \in \mathbb{R}, \quad n \geq 1.$$

Moreover, the covariance function $K(s, t)$ is expressed as

$$K(s, t) = \sum_{n=1}^{\infty} \varphi_n(s)\overline{\varphi_n(t)}, \qquad s, t \in \mathbb{R}.$$

There are some sufficient conditions for a process $\{X(t)\}$ on T to have such a series representation:

1. $\{X(t)\}$ is weakly continuous on T, i.e., $(X(t), Y)_2$ is a continuous function of t for any $Y \in L_0^2(\Omega)$.
2. $\{X(t)\}$ is mean-square continuous on T.
3. The σ-algebra Σ has a countable generator.
4. T is a countable set, e.g., $\mathbb{Z}$ or $\mathbb{Z}^+$.

Since the covariance function $K(s, t)$ is positive definite, we can associate the RKHS (reproducing kernel Hilbert space) $\mathcal{H}_K$ consisting of $\mathbb{C}$-valued functions on T in a way that the following hold:

(i) $K(\cdot, t) \in \mathcal{H}_K$ for every $t \in T$.
(ii) $f(t) = (f(\cdot), K(\cdot, t))_K$ for every $f \in \mathcal{H}_K$ and $t \in T$.

Here $(\cdot, \cdot)_K$ is the inner product in $\mathcal{H}_K$. Then, an interesting fact is that, if $\{X_k\}_{k=1}^\infty$ is an orthonormal base in the time domain $H(X)$ and $\varphi_k(t) = (X(t), X_k)_2$ for $k \geq 1$, then $\{\varphi_k\}_{k=1}^\infty$ is also an orthonormal base in the RKHS $\mathcal{H}_K$. Hence we have that $H(X)$ and $\mathcal{H}_K$ are isomorphic Hilbert spaces.

VIII. EXTENSIONS OF STOCHASTIC PROCESSES

A. Finite-Dimensional Extension

So far we have considered mainly one-dimensional second-order stochastic processes $\{X(t)\}$. If there are two such processes $\{X_1(t)\}$ and $\{X_2(t)\}$, we can describe these processes as a single process by letting

$$\mathbf{X}(t) = (X_1(t), X_2(t)).$$

Hence, $\{\mathbf{X}(t)\}$ is regarded as an $[L^2(P)]^2$-valued process. If $k \geq 2$ is an integer and $\{X_j(t)\}$ $(j = 1, \ldots, k)$ are one-dimensional second-order stochastic processes, then

$$\mathbf{X}(t) = (X_1(t), \ldots, X_k(t))$$

is a k-dimensional stochastic process, or an $[L^2(P)]^k$-valued process. For such a process $\{\mathbf{X}(t)\}$ the average $\mathbf{m}(t)$ is a k-dimensional vector given by

$$\mathbf{m}(t) = E\{\mathbf{X}(t)\} = (m_1(t), \ldots, m_k(t)), \qquad t \in T,$$

where $m_j(t) = E\{X_j(t)\}$, $1 \leq j \leq k$, and the *scalar covariance function* $K(s, t)$ is given by

$$
\begin{aligned}
K(s, t) &= (\mathbf{X}(s) - \mathbf{m}(s), \mathbf{X}(t) - \mathbf{m}(t))_{2,k} \\
&= \sum_{j=1}^k (X_j(s) - m_j(s), X_j(t) - m_j(t))_2 \\
&= \sum_{j=1}^k K_j(s, t),
\end{aligned}
$$

where $(\cdot, \cdot)_{2,k}$ is the inner product in $[L^2(P)]^k$ and $K_j(s, t)$ is the covariance function of $\{X_j(t)\}$. However, the scalar covariance function does not well reflect the correlatedness of $\{X_j(t)\}$ and $\{X_\ell(t)\}$ for $1 \leq j, \ell \leq k$. The *matricial covariance function* $\mathbf{K}(s, t)$ is defined by

$$\mathbf{K}(s, t) = (K_{j\ell}(s, t))_{j, \ell},$$

where $K_{j\ell}(s, t) = (X_j(s) - m_j(s), X_\ell(t) - m_\ell(t))_2$, the *cross-covariance* of $\{X_j(t)\}$ and $\{X_\ell(t)\}$, and it expresses correlatedness among all the processes $\{X_j(t)\}$, $1 \leq j \leq k$. The k-dimensional weakly stationary class and nonstationary classes (given in Section IV and V) are defined using the matricial covariance function $\mathbf{K}(s, t)$, and for many processes in these classes, the integral representations of the processes are obtained in terms of $[L_0^2(P)]^k$-valued measures.

B. Infinite-Dimensional Extension

Finite-dimensional extension is rather straightforward, as seen above. However, infinite-dimensional extension needs considerably more work. One way to do this is to identify $[L_0^2(P)]^k$ with $L_0^2(P; \mathbb{C}^k)$ for each $k \geq 1$, where the latter is the space of all $\mathbb{C}^k$-valued random vectors with finite second moment and zero expectation. Then, when $k = \infty$, we replace $\mathbb{C}^k$ by an infinite-dimensional Hilbert space H. Thus we consider the Hilbert space $L_0^2(P; H)$ of all H-valued random variables X on (Ω, Σ, P) with finite second moment $\|X\|_2^2 = \int_\Omega \|X(\omega)\|_H^2 P(d\omega) < \infty$ and zero expectation, where $\|\cdot\|_H$ is the inner product norm in H. As in the finite-dimensional case, we have two kinds of covariance functions, the scalar and the operator (instead of matricial) ones. Then, we obtain the stationary and nonstationary classes mentioned above.

C. Stochastic Fields

Another type of extension of stochastic processes is to consider a time parameter set T to be other than a subset of $\mathbb{R}$. It is sometimes appropriate to replace $\mathbb{R}$ by $\mathbb{R}^k$ ($k \geq 2$, an integer) and consider stochastic processes $\{X(\mathbf{t})\}$ on $\mathbf{T} \subseteq \mathbb{R}^k$. These processes are called *stochastic fields*. When the stochastic process is depending on time t and two-dimensional location (x_1, x_2), then by letting $\mathbf{t} = (t, x_1, x_2) \in \mathbb{R}^3$, we may describe the process as $\{X(\mathbf{t})\}$. (The more abstract case is obtained by replacing $\mathbb{R}^k$ by a locally compact Abelian group G.) Most of the results obtained so far can be extended to processes on $\mathbb{R}^k$. For instance, if $\{X(\mathbf{t})\}$ is a weakly harmonizable process on $\mathbb{R}^k$ with the covariance function $K(\mathbf{t})$, then there exists an $L_0^2(\Omega)$-valued measure ξ and a positive-definite bimeasure β on $\mathfrak{B}(\mathbb{R}^k) \times \mathfrak{B}(\mathbb{R}^k)$ such that

$$X(\mathbf{t}) = \int_{\mathbb{R}^k} e^{i(\mathbf{t}, \mathbf{u})} \xi(d\mathbf{u}), \qquad \mathbf{t} \in \mathbb{R}^k,$$

$$K(\mathbf{s}, \mathbf{t}) = \int_{\mathbb{R}^{2k}} e^{i((\mathbf{s}, \mathbf{u}) - (\mathbf{t}, \mathbf{v}))} \beta(d\mathbf{u}, d\mathbf{v}), \qquad \mathbf{s}, \mathbf{t} \in \mathbb{R}^k,$$

where $(\mathbf{s}, \mathbf{u}) = \sum_{j=1}^k s_j u_j$, the inner product in $\mathbb{R}^k$.

D. Generalized Stochastic Processes

Generalized stochastic processes were initiated by Itô (1953) and Gel'fand (1955). Let $\{X(t)\}$ be a stochastic process on $\mathbb{R}$ and $\mathfrak{D}$ be the set of all *test functions* on $\mathbb{R}$, i.e., $\phi \in \mathfrak{D}$ if ϕ is infinitely many times differentiable and is zero outside of some finite closed interval. Hence the dual space of $\mathfrak{D}$ is the set of *distributions* in the sense of Schwartz and not probability distributions. If

$$X(\phi) = \int_{-\infty}^{\infty} X(t)\phi(t)\,dt \tag{22}$$

is defined for each $\phi \in \mathfrak{D}$, so that $X(\phi)$ represents a random variable, then $\{X(\phi)\}$ is regarded as a stochastic process on $\mathfrak{D}$, called a *generalized stochastic process* induced by an ordinary stochastic process $\{X(t)\}$.

Here is a formal definition. Let $X(\phi, \omega)$ be a function of $\phi \in \mathfrak{D}$ and $\omega \in \Omega$ such that the following hold:

(i) $X(\cdot, \omega)$ is a distribution for a.e. $\omega \in \Omega$.
(ii) $X(\phi, \cdot)$ is a random variable for every $\phi \in \mathfrak{D}$.

Then $\{X(\phi)\}$ is called a *generalized stochastic process* on $\mathfrak{D}$. It is known that not every such process is induced by an ordinary stochastic process as in (22).

The derivative of a generalized stochastic process $\{X(\phi)\}$ can be obtained using that of distribution:

$$X'(\phi) = \langle \phi, X' \rangle = \langle -\phi', X \rangle = X(-\phi')$$

for $\phi \in \mathfrak{D}$, where $\langle \cdot, \cdot \rangle$ is a duality pair. In general, we have

$$X^{(p)}(\phi) = (-1)^p \langle \phi^{(p)}, X \rangle = (-1)^p X(\phi^{(p)})$$

for $p \geq 2$ and $\phi \in \mathfrak{D}$. If $\{B(t)\}$ is a Brownian motion (or a Wiener process) on $\mathbb{R}$, then this induces a generalized stochastic process $\{B(\phi)\}$ on $\mathfrak{D}$. Its derivative $\{B'(\phi)\}$ is regarded as a white noise Gaussian process.

SEE ALSO THE FOLLOWING ARTICLE

CALCULUS ● EVOLUTIONARY ALGORITHMS AND META-HEURISTICS ● MATHEMATICAL MODELING ● OPERATIONS RESEARCH ● PROBABILITY ● QUEUEING THEORY ● STOCHASTIC DESCRIPTION OF FLOW IN POROUS MEDIA

BIBLIOGRAPHY

Balakrishnan, A. V. (1957). A note on the sampling principle for continuous signals, *IRE Trans. Inform. Theory* **IT-3,** 143–146.

Cramér, H. (1951). A contribution to the theory of stochastic processes, *Proc. Second Berkeley Symp. Math. Statist. Probab.* (J. Neyman, ed.) Univ. of California Press, Berkeley, pp. 329–339.

Gel'fand, I. M. (1955). Generalized random processes, *Dokl. Acad. Nauk SSSR* **100,** 853–856.

Itô, K. (1953). Stationary random distributions, *Mem. Coll. Sci. Univ. Kyoto* **28,** 209–223.

Kakihara, Y. (1997). "Multidimensional Second Order Stochastic Processes," World Scientific, Singapore.

Karatzas, I., and Shreve, S. E. (1991). "Brownian Motion and Stochastic Calculus," Springer, New York.

Karhunen, K. (1947). Über lineare Methoden in der Wahrscheinlichkeitsrechung, *Ann. Acad. Sci. Fenn. Ser. A. I. Math.* **37,** 1–79.

Khintchine, A. Ya. (1934). Korrelationstheorie der stationäre stochastischen Prozesse, *Math. Ann.* **109,** 605–615.

Kolmogorov, A. N. (1933). "Grundbegriffe der Wahrscheinlichkeitsrechung," Springer-Verlag, Berlin.

Loéve, M. (1948). Fonctions aléatoires du second ordre, Appendix to P. Lévy, "Processus Stochastiques et Movement Brownian," Gauthier-Villas, Paris, pp. 299–352.

Rao, M. M. (1982). Harmonizable processes: structure theory, *L'Enseign. Math.* **28,** 295–351.

Rao, M. M. (1984). "Probability Theory with Applications," Academic Press, New York.

Rao, M. M. (1995). "Stochastic Processes: General Theory," Kluwer Academic, New York.

Rao, M. M. (2000). "Stochastic Processes: Inference Theory," Kluwer Academic, New York.

Rozanov, Yu. A. (1959). Spectral analysis of abstract functions, *Theory Probab. Appl.* **4,** 271–287.

Shannon, C. E. (1949). Communication in the presence of noise, *Proc. IRE* **37,** 10–21.

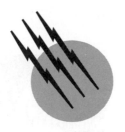

Strategic Materials

John D. Morgan
Industrial College of the Armed Forces

GLOSSARY

Strategic and critical materials Materials required by attackers or defenders in overt or covert hostilities, including declared wars, guerilla wars, insurgencies, sabotage, terrorism, cyberwars, disinformation, and espionage.

WATER and food are daily essentials. Most wars have been fought to gain access to resources. The 41st chapter of Genesis describes how Egypt amassed a huge stockpile of grain 3700 years ago. In recent centuries colonial wars were intended to increase access to needed materials. In the 20th century World War I lasted more than 4 years. Then, counting from Japan's seizure of Manchuria in 1931, World War II lasted 14 years. In both wars expanding land, sea, and air forces required full mobilization of the resources of the major belligerents and use of unprecedented quantities of fuels and materials. Figure 1 illustrates some major applications of materials in military material. Assured supplies of many materials are essential in peace or in war.

I. CURRENT U.S. LAWS AND AGENCIES

The Defense Production Act of 1950, as amended [50 USC, App. 2061 *et seq.* (DPAct)], and The Strategic and Critical Materials Stock Piling Act of 1946, as amended [50 USC, Sect. 98 *et seq.* (SPAct)], establish basic policies and programs. Title I of the DPAct authorizes use of priorities and allocations to direct inadequate supplies to essential uses, and Title III authorizes expansion of supplies. The DPAct designates energy as a strategic and critical material and states that the term materials includes "any raw materials (including minerals, metals, and advanced processed materials), commodities, articles, components (including critical components), products, and items of supply; and any technical information or services ancillary to the use of any such materials, commodities, articles, components, products, or items." The DPAct further states that the term national defense means "programs for military and energy production or construction, military assistance to any foreign nation, stockpiling, space, and any directly related activity" including "emergency preparedness activities conducted pursuant to Title VI of The Robert T. Stafford Disaster Relief and Emergency

FIGURE 1 Strategic and critical materials.

Assistance Act" (42 USCA, No. 5195 *et seq.*). The SPAct states that the term strategic and critical materials means "materials that would be needed to supply the military, industrial, and essential civilian needs of the United States during a national emergency, and are not found or produced in the United States in sufficient quantities to meet such need." The SPAct further states that the term national emergency means "a general declaration of emergency with respect to the national defense made by the President or by the Congress." The SPAct authorizes stockpiling, resource investigations, supply expansions, and research to conserve strategic materials and to develop substitutes and alternates.

By Executive Orders the President has delegated responsibilities under the above acts to agencies with related functions, expertise, and funds. The Departments of Agriculture, Commerce, Energy, and the Interior collect and publish information on resources, production, imports, exports, and uses of materials. These four plus the Department of Defense (DoD), the National Aeronautics and Space Administration, and the National Science Foundation sponsor materials research. The Department of Commerce also supervises export/import controls. The DoD's Defense Logistics Agency (DLA) manages the National Defense Stockpile. The Department of Energy concentrates and stores uranium, deuterium, tritium, and plutonium, develops nuclear weapons, and manages the Strategic Petroleum Reserve (crude oil stored in Gulf of Mexico salt domes). The Department of the Interior's U.S. Geological Survey makes topographic and geologic maps, investigates mineral resources, and collects and publishes information on worldwide minerals, metals, and uses. The Department of the Interior's Bureau of Land Management oversees public lands and manages the government/industry Helium Stockpile in a gastight reservoir near Amarillo, Texas. The Department of Agriculture encourages the production of certain commodities and holds stocks of some. The Treasury Department holds stocks of precious metals for coinage and backing the currency. Working with federal, state, and local governments, the Federal Emergency Management Agency is responsible for emergency planning, preparedness, mitigation, response, and recovery.

II. EXPANDING SUPPLIES

At the end of World War II, Russia, in control of most of the northern part of the Eurasian landmass, became increasingly hostile to the Western democratic nations, initiating the Cold War. In the Far East, China's Red Army extended its control over the mainland. In Southeast Asia returning colonial powers were resisted, impeding a return to prewar

levels of production of raw materials. By 1949 Russia had developed atomic weapons, just 4 years after the United States. Consequently, the United States, Canada, and 10 European nations formed the North Atlantic Treaty Organization (NATO) for mutual security. On June 25, 1950, North Korea attacked U.S.-occupied South Korea. Fears that escalation of the Korean War could lead to World War III increased defense readiness. On September 8, 1950, the DPAct became law. Priorities, allocations, economic controls, and supply expansions soon followed. On October 26, 1950, strengthening earlier arrangements, the United States and Canada agreed that "the production and resources of both countries be used for the best combined results." And in 1950 the United States had begun directly supplying the French in Vietnam, where insurgency had spread. U.S. involvement in Vietnam would increase gradually for 25 years.

By mid-1955,

- the DoD had spent $110 billion (U.S. billion $= 10^9$) for hard and soft goods and construction;
- $8 billion of DPAct financing expanded basic industries, some in foreign nations;
- Rapid Tax Writeoffs under the Internal Revenue Code (26 USC) encouraged private investment of $37 billion to create new or to enlarge 225 basic industries;
- $13 billion of Marshall Plan funding expanded Allied essential industries, including strategic materials production;
- production of strategic minerals was further assisted by exemptions from the excess profits tax, and, still in effect, depletion allowances scaled to reflect relative criticality; and
- SPAct stocks at 270 sites rose to 22 million short tons, valued at $6 billion.

As a result, communist aggression was halted, and, except for the 1973 Arab Oil Embargo, there have been no major materials shortages. Table I includes all materials listed under the SPAct from 1946 to 1999. Most, but not all, were physically stockpiled during the Cold War.

III. CHANGING REQUIREMENTS

Space vehicles, multistage rockets, ballistic, cruise, and guided missiles, supersonic aircraft, remotely piloted vehicles, urban robots, and smart bombs and mines all require strong lightweight metal/ceramic/plastic components and sophisticated electronics with microchips. Observation, aiming, and automatic target recognition rely on infrared and laser devices. Position determination and missile guidance rely on orbiting satellites coupled into

TABLE I   All Materials Listed under the Strategic and Critical Materials Stock Piling Act, 1946–1999

Abrasives: Abrasive bauxite, crude aluminum oxide, corundum, emery, silicon carbide, industrial diamonds and diamond dies

Agar

Aluminum: Alumina, aluminum metal, metal-grade bauxite

Antimony

Base metals: Copper, iron ore and iron scrap, lead, tin, zinc

Beryllium: Beryl, beryllium/copper master alloy, beryllium metal

Bismuth

Cadmium

Chromium: Chemical, metallurgical, and refractory ores; chromium metal; ferrochromium; ferrosilicochromium

Cobalt

Columbium (niobium): Columbite, columbium metal, columbium carbide powder, ferrocolumbium

Forest products: Balsa, cork, mahogany

Germanium

Indium

Iodine (for chemical uses)

Jewel bearings and natural sapphires and rubies

Leather: Calf and kip skins, cattle hides

Loofa sponges

Magnesium

Manganese: Battery, chemical, and metallurgical ores; ferromanganese; ferrosilicomanganese; manganese metal

Medicinals: Emetine, hyoscine, iodine, morphine sulfate, opium, quinidine, quinine

Mercury

Molybdenum: Ferromolybdenum, molybdenum disulfide

Monazite and rare earths

Natural fibers: Abaca (manila), extralong-staple cotton, hemp, henequin, hog bristles, jute and burlap, kapok, raw silk and silk waste and noils, sisal, waterfowl feathers and down, wool

Natural oils: Castor, coconut, palm, rapeseed, sesame, sperm, tung

Nickel

Nonmetallic minerals: Asbestos—amosite, chrysotile, and crocidolite; barite, calcite, celestite, cryolite, English chalk, fluorspar—chemical and metallurgical; graphite—amorphous, flake, and lubricating and packing; kyanite, muscovite and phlogopite mica, quartz crystals, refractory bauxite, steatite talc

Optical glass

Pepper

Petroleum and petroleum products

Platinum-group metals: Iridium, osmium, palladium, platinum, rhodium, ruthenium

Pyrethrum

Radium

Rayon fiber (aerospace grade)

Rubber: Crude rubber, latex, guayule seeds and seedlings

Selenium

Shellac

Silver

Tantalum: Tantalite, tantalum carbide powder, tantalum oxide, tantalum metal powder, tantalum ingots

Thorium nitrate

Titanium: Rutile, titanium metal sponge

Tungsten: Tungsten ores, tungsten carbide powder, ferrotungsten, tungsten metal powder

Uranium

Vanadium: Ferrovanadium, vanadium pentoxide

Vegetable tannins: Chestnut, quebracho, wattle

Zirconium: Baddeleyite, zircon

the Global Positioning System (GPS) and the Geographical Information System (GIS). Underwater observation is assisted by multibeam Side Scanning Sonar. Assuring compatibility of communications, ordnance, and material across just the U.S. Army, Navy, Air Force, Marines, and Coast Guard and the many Reserve Components is a major challenge, while the 1999 Kosovo pacification demonstrated that much needs to be done to achive interoperability among the 19 NATO nations.

U.S. World War II tanks weighed 30 tons; the present MlAl (HA) Abrams Main Battle Tank is up to 70 tons and consumes 6 gal of fuel per mile. Now tanks weighing less than 30 tons are desired to facilitate airlift and mobility over poor terrain. But improved armor penetrators include shaped-charge explosives, projectiles with tungsten carbide or depleted uranium cores, and self-forging tantalum rounds. Lighter bridging equipment is also needed. Mortars are simple effective weapons in the hands of ground troops. The U.S. World War II 81-mm mortar had a separate cast-steel baseplate weighing 48 lb. Now under development are much lighter mortars with composite baseplates and thin steel barrels encased in titanium. A ground force combatant carrying water, food, bandoliers of ammunition, and a dual-caliber automatic rapid-fire weapon with infrared scanning and laser sights, wearing appropriate clothing plus body armor and a helmet with two way TV and multichannel voice radio, and carrying the batteries needed to power the electronics may have a load of up to 100 lb—twice what would be desirable.

The availability of new materials has reduced older requirements for many common materials. U.S. 1950 plastics production was 1 million tons. Now it is about 46 million tons, 40% of the weight of steel but more than three times the volume of steel. Plastics are now widely used in weapons, ammunition, vehicles, ships, boats, construction, containers, and packaging. Synthetic minerals replace natural diamonds, sapphires, rubies, mica, and quartz crystals. Small battery-powered timepieces replace larger, less accurate mechanical ones. Fiber-optic cables and wireless networks speed communications. Synthetics replace cork and balsa in flotation applications, abaca in marine hawsers, hog bristles in brushes, feathers and down in jackets, leather in footwear, and extralong-staple cotton in parachute shrouds. Synthetic rubber replaces natural, although for carrier-based aircraft and huge earthmovers, natural rubber is preferred. Chemicals replace pyrethrins in insecticides and tung oil and shellac in sealants. Research showed that pepper, once considered for stockpiling, was not a preservative in military rations, although pepper is still widely used as a flavoring agent. New uses increase requirements for some materials. Rhenium is now added to chromium/nickel/cobalt jet engine alloys. Titanium is increasingly used in airframes. Once used un-

separated as "mischmetall" (mixed metals), each of the separated 17 rare earth metals has specialized uses. First recovered from natural gas during World War I to serve as a lifting gas in barrage balloons and blimps, helium is now used in shielded-arc welding, deep diving atmospheres, and reactor cooling. Liquefaction of air made possible tonnage oxygen for blast furnaces, nitrogen in portable refrigerators, and argon in welding. Worldwide production of basic commodities and technological advances developing new and improved materials have blunted threats of materials shortages in recent years.

The shift of DoD concern from bulk materials to high-technology materials is illustrated by the DoD's DPAct Title III Program Office's list of year 2000 projects: solid-state power semiconductor switching devices, silicon-on-insulator wafers, silicon carbide substrates, high-purity float zone silicon, flat panel displays, continuously reinforced silicon carbide fiber/titanium metal matrix composites, selectively reinforced silicon carbide whisker/aluminum metal matrix composites, semi-insulating indium phosphide wafers, silicon carbide electronics, and eyewear for protection from laser beams.

Environmental concerns and regulations also influence materials requirements. Formerly widely used poisonous metals and compounds are curtailed or eliminated in traditional uses: lead in gasoline, paints, cable covering, roofing, solders, and containers; cadmium in colors, plating, and plastics; mercury in fungicides and antifouling paints; and cyanide in metallurgy. Curtailment of solid, liquid, and gaseous emissions boosts requirements for corrosion-resistant materials, absorbents, and precipitators. At the same time recycling of metals, paper, and plastics adds to the supplies of basic commodities.

Following the breakup of the USSR and the decline of the Cold War, U.S. stockpile goals for most materials have been sharply reduced or zeroed, reflecting the continued reliance on nuclear deterrence, desire for highly mobile armed forces-in-being with state-of-the-art material, assumed ready access to many foreign sources with only minor shipping losses, and curtailment of production of some civilian goods in an emergency. In the last decade the DLA disposed of stockpiled materials valued in excess of $3 billion by sales in an orderly manner to avoid market disruption and by transfers of materials to other agencies, e.g., titanium to the Army Tank and Automotive Center and precious metals to the Treasury Department. As of September 30, 1999, in fiscal year 1999 the DLA had disposed of materials valued at $446 million, leaving in inventory more than 70 commodities valued at $3375 million, including the 12 underlines in Table I not authorized for disposal, valued at $582 million. Despite stockpile disposals, and even if a material had never been listed in Table I, *all* materials are still covered by the broad DPAct.

IV. INTERNATIONAL CONCERNS

Convertible currencies are highly strategic to all nations, and even well-designed counterfeits are strategic to evildoers. Most materials are of strategic importance to industrialized nations. Some maintain stockpiles or encourage domestic industries to hold larger-than-normal stocks. But most rely on many imports. Even Russia, with a land area greater than that of the United States and Canada combined, and espousing autarky for 70 years, imported many manufactured items and significant quantities of bauxite, alumina, barite, cobalt, fluorspar, molybdenum, tin, and tungsten. Petroleum and gas resources are particularly attractive to investors. Rogue states, insurgents, and guerillas find some materials to be highly strategic—coca and opium for conversion to cocaine, morphine, and heroin and small-volume, high-value drugs easy to smuggle to obtain money or weapons. Even tonnage lots of bulky marijuana are smuggled. Illicit drugs pose serious threats to the stability of developed nations, just as in the 19th century opium imports forced on China led to its collapse. Insurgents and guerillas use primitive artisinal hand mining to recover gold, diamonds, sapphires, and rubies—all high-value, low-volume items. Chemical, biological, and nuclear (CBN) materials are highly strategic to rogue nations, terrorists, and saboteurs. Common chemical agents include ricin, cyanide, mustard gas, phosgene, and explosives. Two common materials—fertilizer-grade ammonium nitrate and fuel oil—when properly mixed, make the powerful explosive ANFO. Common biologic agents include anthrax and smallpox. Any radioactive substances can be used for nefarious purposes. Indeed, just the threat posed by unaccompanied packages possibly containing CBN materials causes urban panics.

Although industrialized nations have been able to obtain supplies of most materials is recent years and at reasonable prices, there can be no cause for complacency as the world enters the 21st century. Billions of people in less developed nations are now able to use battery-powered small TVs that depict highly exaggerated views of life in the developed nations, and these people are demanding higher standards of living for themselves. As the population surges, pressure on the resources of the earth will increase and the demand for fuels and materials will accelerate. And among the most strategic materials will be water and food.

SEE ALSO THE FOLLOWING ARTICLES

ENERGY RESOURCES AND RESERVES • FUELS • GEOSTATISTICS

BIBLIOGRAPHY

Department of Defense (annual). "Strategic and Critical Materials Report to the Congrese," U.S. Government Printing Office, Washington, DC.

Energy Information Administration (annual). "Annual Energy Review," U.S. Government Printing Office, Washington, DC.

Joint Committee on Defense Production, U.S. Congress (1951–1976). "Annual Report of the Activities of the Joint Committee on Defense Production," U.S. Government Printing Office, Washington, DC.

Morgan, J. D. (1949). "The Domestic Mining Industry of the United States in World War II," U.S. Government Printing Office, Washington, DC.

National Defense Industrial Association (monthly). *National Defense*, NDIA, Arlington, VA.

U.S. Geological Survey (annual). "Mineral Commodity Summaries," U.S. Government Printing Office, Washington, DC.

Stratigraphy

Derek Ager
University College of Swansea

GLOSSARY

Biostratigraphy Correlation of sedimentary rocks by the fossils they contain.

Chronostratigraphy Correlation of sedimentary rocks on the basis of their supposed ages, based on paleontological and other evidence.

Diachronism Phenomenon of apparently similar strata being shown to cut across the time planes from place to place.

Event stratigraphy Correlation of sedimentary rocks on the basis of events (such as marine transgressions or volcanic eruptions) recorded in them.

Lithostratigraphy Correlation of rocks simply by their physical properties (i.e., rock type).

Regression Major withdrawal of the sea from a continental area.

Sequence stratigraphy Integration of seismic, borehole, and surface data to recognize major sedimentary units delimited by unconformities.

Transgression Major extension of the sea over a formerly continental area.

Unconformity Break in the record of the rocks representing a considerable interval in time between the deposition or emplacement of adjacent rock bodies.

STRATIGRAPHY is the branch of geological science that is concerned with the correlation and interpretation of rock strata. Similar strata separated by distance can be identified by similar appearance, shared fossils, or other methods. The job of the stratigrapher is to identify the layers of rock in a given area and to corrclate them with other regions.

I. INTRODUCTION

As usual, the ancient Greeks thought of it first, back in the 6th, century BC. Writers such as Empedocles, Pausanius, Theophrastus, and Anaximander recognized the remains of marine creatures in rocks far from the sea. Though they had no concept of geological time, they concluded that at some time in the past, those regions had been covered by salt water.

In the 15th century, the incomparable Leonardo da Vinci (1452–1519) brought these ideas forward again on the basis of his own observations in Italy, notably during the digging of a canal near Milan. The standard views of the Middle Ages, however, explained such things as "sports of nature" or even as phenomena put there by the Almighty to torment doubting humans.

Paradoxically, however, the Renaissance of learning drove science, including geology, back into the library and laboratory and away from the real evidence of the rocks themselves in the field.

There was a long argument between what are generally known as the plutonists (or vulcanists) and the neptunists. Put simply, the former thought that the rocks had a fiery origin, that they had poured out onto the surface of the earth in a molten condition and had then cooled down. The rival school thought rather that the rocks had all been laid down at the bottom of the sea. Feelings got very strong. Thus the great German writer Goethe, who was quite knowledgeable as an amateur geologist, sided with the neptunists. One suspects that this may in part have been for emotional reasons, since he preferred the peaceful processes of Neptune to the fiery violence of Pluto. Of course, both sides were right in some respects.

James Hutton (1726–1797), a Scot, was really the man who ended the argument. Although he was generally recognized as a plutonist, he was also the first man to appreciate fully the vastness of geological time and that to understand the past we must look at what is going on today.

However, the science of stratigraphy really started with a humble man with little education and mercifully, therefore, no preconceived ideas. He was the Englishman William Smith (1769–1839), who was above all a typically English pragmatist. He was a simple man who did not theorize. He was born on the glorious Jurassic rocks of the English Cotswolds, son of the village blacksmith and a very practical man who earned his living as a surveyor, first in the coal mines and then along the courses of the canals that were spreading all over England in their brief heyday between the start of the Industrial Revolution and the coming of the railways. It was along the sections made for the new canals (curiously reminiscent of da Vinci's observations centuries before) that Smith recognized that the rock strata followed each other in a regular succession and that each was characterized by a particular group of fossils. He lived to be hailed as the "Father of English Geology," and as a German writer said of him, he "had at last opened the book of earth's history."

While William Smith was working in England, a much more aristocratic and educated man was working in the Paris Basin. Georges Cuvier (1769–1832) is remembered as the "Father of Vertebrate Palaeontology" and as the leader and founder of the "catastrophist" school of stratigraphers. He was German/Burgundian by birth and his chief links were with Switzerland. Although one of France's greatest sons, he only became French after the revolution. He lived through the revolutionary wars and the long Napolionic Wars that followed. He saw high office under the Emperor but lived on to see the restoration of the French monarchy and finished up a baron, the leading naturalist of his day, and President of the Council of State. It is hardly surprising, therefore, that he saw the earth's history, like his own, as passing through a series of catastrophes and triumphs.

He bitterly resisted ideas of evolution as then proclaimed by Lamarck, though he recognized the extinction of species. He was particularly unkind to Lamarck, but later suffered himself the general and often scathing rejection of his ideas on stratigraphy. He was probably the first person to record in detail the bed-by-bed succession of strata. He noted in the early Tertiary strata of the Paris Basin that there are frequent and sudden changes, from freshwater deposits and faunas to evaporites, to brackish water sediments, and to those that were laid down in the sea. He observed the detail rather than the generality, and the detail showed a very episodic and frequently interrupted record.

Charles Lyell (1797–1875), on the other hand, was a theorizer who did not trouble himself too much with the minutiae of strata, but showed brilliantly how the processes seen to be going on at the present day could be used to explain the stratigraphical record as it is preserved in the rocks. His great principle was that "the present is the key to the past" and his book *The Principles of Geology* was perhaps the most important ever published on the subject. His friend Charles Darwin carried it with him on his voyage round the world on the *HMS Beagle*, and perhaps it is not without significance that Vladimir Ilyich Ulyanov (later called Lenin) had a copy in his childhood study bedroom at Simbirsk on the Volga. It is interesting to put the ideas of Lyell in their context of Victorian thought in the 19th century. The new liberalism was opposing the old tyrannies; the search for natural processes and natural explanations was replacing the old ideas of divine intervention and predetermination. Inevitably the general public took sides in the scientific debates, since they mirrored the contemporary political struggles of the age of revolutions and the overthrowing of the old absolute and conservative monarchies. Every educated man of those days knew something of geology, and William Wordsworth might have written of the rapidly expanding science of stratigraphy (as he did of the French Revolution): "Bliss was it in that morn to be alive, But to be young was very heaven."

The whole history of the earth, with all its exciting changes in lands and seas, volcanoes and mountains, animals and plants, was rapidly being unraveled by the young geologists striding over the hills. Stratigraphy was the exciting new subject that caught the public's imagination as is exemplified by the vast concourse that hailed Sir Roderick Murchison as "King of Siluria" at Birmingham, England, in 1849.

Most stratigraphical work, since those early days, has consisted of piecing together the story all over the world and dividing the stratigraphical record into finer and finer divisions. The subject of stratigraphy itself, however, is now commonly divided into different supposed disciplines.

II. THE DIVISIONS OF STRATIGRAPHY

American stratigraphical thought insists on dividing stratigraphy into an unholy trinity of lithostratigraphy, biostratigraphy, and chronostratigraphy. This is not universally accepted, however. Lithostratigraphy, or rock stratigraphy, is what one does first in the field. It is the recognition of lithological units that can be mapped or recognized in bore holes, with formations and groups as the most commonly used units. Lithostratigraphy has been called prestratigraphy; that is to say it is what one does before one attempts the real business of stratigraphy, which is correlation and interpretation. Biostratigraphy, or fossil stratigraphy, is one of the methods of correlation, usually in terms of biozones, and obviously is most useful in the Phanerozoic part of the record, where fossils are abundant. It is discussed below. Chronostratigraphy, or time stratigraphy, is neither one thing nor the other, since it relates to rocks with a time connotation. It may be said that in effect there are only two elements involved in stratigraphy—the rocks themselves (which accumulated or were emplaced and/or altered) and time or chronology. Chronostratigraphy is the business of recognizing time–rock divisions such as systems and stages, which have reality in their rocks but only a tentative and unprovable reality in their ages. These divisions are, by definition, the rocks laid down during a certain period of time that is not yet defined in actual years (see Table I).

III. STRATIGRAPHICAL METHODOLOGY

The business of stratigraphy is correlation, and there are many methods for doing this. The simplest is correlating rocks of similar appearance. Thus the Upper Cretaceous Chalk—a pure coccolith limestone—looks remarkably similar whether it comes from the Mississippi Embayment of the United States, right across northern Europe from Ireland to Georgia in Russia or Gingin in Western Australia. Similarly, the Triassic red mudstones of the Moenkopi Formation in Arizona look exactly like the equivalent strata resting on much older rocks in the Harz Mountains in Germany.

On the other hand, though the Upper Cretaceous Chalk is so very similar across northern Europe, rocks of the same age in southern Europe are completely different as, for ex-

TABLE I   Divisions of the Stratigraphical Time Scale[a]

Eras	Periods	Approximate age (millions of years)
Cenozoic		
Quaternary	Holocene	
	Pleistocene	1.5
Tertiary	Pliocene	13
	Miocene	25
	Oligocene	36
	Eocene	54
	Paleocene	64
Mesozoic	Cretaceous	135
	Jurassic	180
	Triassic	230
Paleozoic	Permian	280
Upper	Carboniferous	
	Pennsylvanian	315
	Mississippian	345
	Devonian	400
Lower	Silurian	425
	Ordovician	500
	Cambrian	570
Precambrian	Proterozoic	2000
	Archean	4500

[a] There are some differences in usage, for example, with the term Neogene, and the period names Mississippian and Pennsylvanian are not normally used outside North America.

ample, the deeper water "flysch" deposits at the west end of the Pyrenees in northern Spain. That is what is called a change in facies—that is, the total characters of a sedimentary rock—brought about by the different environments in which it was deposited. Thus in eastern Canada and parts of New England, Devonian rocks are in a continental facies, with freshwater fish and land plants, comparable to the classic "Old Red Sandstone" of northern Europe, whereas farther west, for example, in Iowa, equivalent Devonian sediments were laid down as calcareous muds in the sea and contain abundant marine fossils.

So, clearly lithostratigraphical correlation, by similarity of rock type, is not enough of itself. Rocks equivalent in age may be quite different in appearance, while conversely rocks that look very similar (such as shelf limestones or deltaic sandstones) may be of completely different ages.

It is therefore necessary to use methods of correlation that relate to time rather than to frequently repeated patterns of sedimentation or rock types. The most obvious, and the most widely used, is correlation by means of fossils, as first demonstrated by William Smith. This is based on the truism that evolution is unidirectional and irreversible. Fossil assemblages change with time and so provide an age indicator in the rocks like the date on a newspaper. Obviously, some fossils evolved more rapidly than others and are therefore more useful for detailed dating. Some fossils last little more than a million years, whereas others hardly seem to have changed in more than 500 million years. Other factors that make a fossil useful for correlation are abundance, independence of facies, and size. This last consideration is especially important in connection with oil exploration. Obviously, you cannot fit many dinosaurs in a bore hole, but you can find thousands and thousands of minute sea creatures, such as foraminifera, in every centimeter of sediment. Especially useful in recent years has been the use of pollen grains and other microscopic plants and plant remains. Apart from the advantage of their small size and great abundance, they have proved especially valuable because they are spread widely in the air and fall everywhere, whether it be on land or sea. Thus they can be used to correlate marine and nonmarine deposits.

However, it is important to emphasize that biostratigraphy is only one method among many for correlating the rocks, albeit still the most accurate for the vast majority of Phanerozoic sediments. Among the many others that might be mentioned are chromostratigraphy, seismic stratigraphy, climatostratigraphy, magnetostratigraphy, radiometric stratigraphy, and event stratigraphy. The first two are just variants of lithostratigraphy. Chromostratigraphy, which uses just the color of rocks, as distinguishing the dark reds of Permian continental sediments from the paler reds of the Triassic, can generally be ignored, though it is locally useful. Seismic stratigraphy, on the other hand, has seen a great burst of popularity, especially among oil geologists, in recent years. It uses the methods of geophysical seismology to delimit sedimentary units (formations or groups) underground, but it is based on the principle that bedding planes between strata are more likely to be isochronous, that is, time markers, than the limits of the units as a whole. An important aspect of seismic stratigraphy, however, is recognizing the relationships between discrete bodies of sediment when successive stratified bodies cut across each other as the centers of sedimentation moved one way or another. Unconformities, where there have been erosional breaks between successive sedimentary sequences, have been recognized for their importance since the days of Hutton and his "succession of former worlds." "Sequence stratigraphy" has developed comparatively recently from seismic methods combined with surface and borehole data; it has become very important in the oil industry. The basic unit is a sequence which is defined as a succession of strata bounded by unconformities marking transgressive-regressive episodes resulting from changes in sea level and thought to make possible continent or even world-wide correlation. All such relationships may be important in oil exploration, since any useful oil or gas accumulation needs three sedimentary units in the right relationship to each other. One needs a source rock to produce the useful hydrocarbons when they are buried to the right depth and raised to the right temperature. One needs a reservoir rock with the right porosity to hold the oil or gas, and one needs a cap rock or impervious seal to prevent the hydrocarbons from escaping to the surface. Much oil and gas is found in so-called stratigraphical traps, where the relationships provide source rock, reservoir, and cap rock in the right places. However, seismic stratigraphy, though of great practical value, tells us nothing about the ages of the rocks concerned and does not help us very much in correlating them from place to place.

Climatostratigraphy is simply correlation by means of the climatic record in the strata concerned. Its most obvious application is in the sediments of Quaternary times, with repeated glaciations and interglacial episodes and many minor oscillations. But the time intervals are so brief and the record so confused that it cannot be said that the method is perfect. On the other hand, by Quaternary times we are so close to the present day that there is little in the way of evolutionary paleontology to help us. Fossils are used, of course (notably pollen, spores, and minute snails), but chiefly as indicators of the climatic changes and even the clearing of the forests by early humans.

Magnetostratigraphy records the changing polarization of the earth's magnetic field in the later part of the Phanerozoic record, notably in the igneous rocks of the ocean floors, but is limited in its application elsewhere. It provides a record only in collaboration with the next method to be considered.

Radiometric stratigraphy provides actual ages in terms of years by means of the known rate of isotopic decay of elements. However, it provides far too crude a scale to permit the use of such dates in detailed stratigraphy. Obviously, it is useful, or at least interesting, to know the approximate ages in years of our stratigraphical divisions. But the margins of error are considerable and the rocks amenable for such study are limited. It is used particularly in the vast thickness of igneous and metamorphic rocks in the Precambrian shields. In the Phanerozoic, the isotope ratios may occasionally provide the age in years, but the fossils provide the stratigraphical age in terms of correlation with standard sections. Thus the Palisades Sill, up the Hudson River from New York City, was formerly thought to be Triassic in age. This fits with the general picture of tension at this time as the Atlantic began to open and lavas poured out of fissures as on the other side of the Atlantic in Morocco. Radiometric dating, however, now shows that the Palisades Sill is younger and falls within the age bracket assigned to the Jurassic Period on the basis of standard reference sections in Europe.

Another stratigraphical method that has become very popular in recent years is what has been called event stratigraphy. In this it is not the sediments themselves that are correlated or the fossils they contain, but the events that they record. Thus a marine transgression in one place may be recorded by a color change in land-deposited sediments somewhere else, even though the sea and its fossils never reached the second locality.

Some events were so sudden and widespread that they provide very good time markers in the stratigraphical record. A widespread fall of volcanic ash would be an obvious example (though these rarely come singly), and it should be mentioned that volcanicity and igneous rocks generally, together with episodes of metamorphism, are also part of earth history and therefore of stratigraphy in the broad sense.

Some stratigraphers see event stratigraphy as a return to old-fashioned catastrophism and therefore (unjustifiably) as suspect. However, it is being recognized more and more that rare episodic events, such as hurricanes and earthquakes, do play an important part in earth history. Parts of the stratigraphical column in some places consist of no more than a record of occasional happenings, such as storms hitting a low-lying coastline or turbidity currents carrying rushes of sediment into the deep sea.

The most spectacular and fashionable events in earth history in the recent literature have been the sudden collisions with the earth of large meteorites or asteroids. More than a hundred astroblemes about the earth's surface have now been blamed on such impacts, and it must be presumed that many more extraterrestrial objects fell into the sea. Whether they are recorded in the stratigraphical record is another matter. Great publicity was given to the finding of high levels of iridium (a platinum-related element associated with celestial bodies) in a clay seam at the Cretaceous–Tertiary boundary in Italy and Denmark, roughly marking the horizon where many organisms (including the overpublicized dinosaurs) became extinct. More important than the extinction of the vulnerable reptiles was the complete collapse of the marine ecosystem, though it has recently been suggested that the extinction caused the iridium rather than the other way around. In other words, the extinction of most of the coccoliths and many of the other limestone-forming microorganisms in the sea caused a long halt in deposition and therefore the accumulation of meteoric and other extraneous material.

High iridium levels have now been found at other levels characterized by mass extinctions, but it is remarkable how one finds a thing where and when one expects to find it. The author is not convinced that enough care has been taken in sampling other thin clay seams within thick carbonate sequences, where slow deposition is bound to show a concentration of trace elements.

It must be said, however, that there are not enough records of extraterrestrial impacts or of mass extinctions (whether the two are or are not connected) to build a stratigraphy.

IV. DEFINING THE LIMITS

The stratigraphical column arose in a somewhat haphazard manner and would probably have looked very different if it had originated in North America rather than in Europe. There is a particular difficulty in defining the limits between the divisions that have originated in this way. Some boundaries were better than others when it came to ease of recognition, but it soon became assumed that there was some kind of logic about them rather than pure chance. Thus the top of the Devonian might have been better defined at the top of the Frasnian Stage (where there was one of the mass extinctions already mentioned) rather than at the top of the Famennian Stage above it. It was clearly important for all geologists to speak the same stratigraphical language, and there soon developed great controversies about where boundaries should be drawn and the meaning (in terms of rocks and time) of the various stratigraphical names.

Many of the great arguments were only solved (or are still in the process of being solved) by the creation of the Commission on Stratigraphy of the International Union of Geological Sciences and its various subcommissions. Matters of principle were involved in the discussions and even matters of national pride with the proud claiming of type sections or stratotypes to define the different divisions. It seemed to be assumed subconsciously by many geologists that there was an absolute reality about the various boundaries with new faunas, seemingly without ancestors, marine transgressions, and even fanfares of trumpets perhaps to herald in the new period. Such dogmatism seems now to be at last being replaced by a new pragmatism as it becomes more and more generally recognized that the boundaries are humanmade conveniences rather than the record of great events in the history of the earth. To define boundaries in such a pragmatic way there has grown up the principle of what is popularly known as the golden spike. This is to be driven in, hypothetically at least, at a point in a rock sequence, accepted internationally, to define the base of a given division. The top of the previous division is then automatically defined by the base of the one above it and any gaps or any missing strata subsequently found elsewhere automatically go into the lower division. The first such golden spike was hypothetically hammered in, by international agreement, at the base of a thin limestone in a section at Klonk in the Czech Republic. This now defines for all time the base of the Devonian System; it frees us from further tiresome squabbles and enables us to get on with the real business of stratigraphy, which is correlation and interpretation.

SEE ALSO THE FOLLOWING ARTICLES

EARTH SCIENCES, HISTORY OF • GEOLOGIC TIME • METAMORPHIC ROCK SYSTEMS • PALYNOLOGY • ROCK MECHANICS • SEDIMENTARY PETROLOGY • ULTRA HIGH-PRESSURE METAMORPHIC ROCKS

BIBLIOGRAPHY

Ager, D. V. (1981). "The Nature of the Stratigraphical Record," 2nd ed., Wiley, New York.

Boggs, S. (1994). "Principles of Sedimentology and Stratigraphy," 2nd ed., Prentice Hall, New York.

Condie, K., and Sloan, R. (1998). "Origin and Evolution of Earth: Principles of Historical Geology," Prentice Hall, New York.

Moullade, M., and Nairn, A. E. M., eds. (1978). "The Phanerozoic Geology of the World," Elsevier, New York.

Noble, J. P. A., and Cotillon, P. (1992). "Stratigraphy," Springer-Verlag, New York.

Payton, C. E., ed. (1977). Seismic stratigraphy—Applications to hydrocarbon exploration." *Am. Assoc. Petrol. Geol.* Memoir 26.

Seilacher, A. Richen, W., and Einsel, □. (1991). "Cycles and Events in Stratigraphy," 2nd ed., Springer-Verlag, New York.

Weller, J. M. (1960). "Stratigraphic Principles and Practice," Harper, New York.

Wicander, R., and Monroe, J. S. (2000). "Historical Geology (International Version): Evolution of the Earth and Life Through Time," 3rd ed., Brooks/Cole, Belmont, CA.

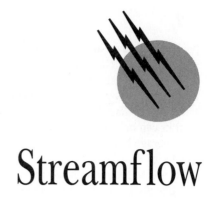

Streamflow

A. I. McKerchar

National Institute of Water and Atmospheric Research Ltd.

GLOSSARY

Current meter A precision instrument with a rotating propeller or set of cups for measuring velocity at a point in flowing water. The speed of rotation is proportional to the mean velocity at the point.
Discharge The total volume of water flowing past a point in a river at a specified time. Discharge is measured as volume per unit time.
River basin The area of land draining to a particular location on a river. Synonymous terms are "catchment" and "watershed."
Runoff The average depth of water yielded from the catchment in a specified time interval, expressed in the same units as precipitation and evaporation. Runoff is calculated by dividing the average discharge issuing from a catchment for a specified time interval by the area of the catchment.
Streamflow The flow of water in a stream or river.

STREAMS of flowing water are a visible and enduring feature of the hydrological cycle. Streams provide ready sources of freshwater that are essential for communities and, in inland areas, are typically the receivers of wastewaters.

In hydrological terms, streamflow is an outflow from a catchment. The inflow is provided by precipitation (rain and snow) received over the area of the catchment. The water provided by precipitation moves through the catchment, some seeping into deep groundwater, some evaporating from the vegetation canopy or from the ground surface, and some being held in the soil from where it meets the transpiration requirements of plants. Streamflow is the residual after the demands for seepage, evaporation, and transpiration have been met.

I. STREAMFLOW MEASUREMENT

Quantitatively, streamflow rate is measured as discharge in units of volume per unit time. The most common units are cubic meters per second (m^3/s, also known as cumecs) and liters per second (L/s) for smaller streams. Time units of seconds are appropriate because streamflow can vary over short time intervals. Other units include megaliters per day (ML/day) and, in the United States, cubic feet per second (ft^3/s or cusecs) and acre-feet per hour.

Runoff expresses discharge as an average depth of water running off over the area of the river basin. It is calculated by dividing discharge by the area of the catchment. Runoff can be compared with the mean annual precipitation averaged over catchment. The precipitation should be greater by an amount depending on the losses to groundwater and to evaporation and transpiration. In areas where rain gauges are sparse, or where precipitation varies, especially in hilly and mountainous terrain, the runoff derived from streamflow measurements with an estimate of mean evaporation and transpiration can be a useful indicator of the mean precipitation over the river basin.

Streamflow is measured using a two-step process. First, the water level, or stage, (defined as height above a datum) is measured at regular time intervals. Then the series of stage values are converted to a series of discharges using a transformation that relates stage level to discharge. The transformation is termed a stage/discharge rating curve. A series of measurements of discharge, typically using current meters, must be undertaken to establish the stage/discharge rating. The shape of this curve is determined by the hydraulic conditions at the measurement site on the stream channel.

The site for measurement of stage, commonly termed a gauging station, must be carefully selected to ensure that the downstream hydraulic conditions will be suitable. The prime attributes sought are stability and sensitivity. Typically, stability is provided by a rock bar across the stream, or a gorge section. Stability of the site ensures that the hydraulic relationship represented by the rating curve is well defined and does not have to be continually recalibrated. This can be particularly difficult to achieve in rivers that convey substantial sediment loads and suffer periodic aggradation and scour of the channel. The second attribute required is sensitivity, such that a measurable change in stage is accompanied by a measurable change in discharge. Other desirable attributes are ease of access, even in extreme flood conditions; freedom from aquatic and terrestrial vegetation; and freedom from variable backwater effects (due to tides or high water levels in tributaries or lakes or reservoirs downstream).

A desirable feature sought for the discharge measurement site is relatively uniform flow across the river channel, and this is normally achieved in a long, straight section of channel. Areas of high velocity, changing current angles, and rapid changes in depth of water should be avoided.

On smaller streams and rivers, weirs or flumes with known stage/discharge rating curves are often used for flow measurement. Weirs and flumes have the advantages that once installed, and once the calibrations have been verified, the data can be more reliable. However, regular checks are still necessary to ensure that flow through the structure is not impeded by sediment deposits or by vegetation growth. Dams and hydroelectric power stations also provide stable hydraulic conditions that enable reliable measurement of flow.

A. Measurement of Stage

Early measurements of water level, typically at daily intervals, were made using a staff gauge fixed in a secure position. With water level recorders, a stilling well, connected to the river or lake by narrow pipes, is commonly used to provide a calm, undisturbed area of water to enable a reliable measurement. A float is connected to a rotating shaft on the recorder by a metal tape or chain. Early recorders used pens drawn across a clock-driven chart. These have been largely replaced by electronic recorders that store regular values for the stage, as determined by the rotation of the float-driven shaft. These modern measurement methods have better resolution than charts and avoid the labor-intensive work necessary to digitize charts.

Typically, the recording interval used is 15 min, but a finer time resolution may be necessary on small, fast responding streams and on lakes and reservoirs where seiching, i.e., oscillations of the water surface caused by wind or barometric pressure gradients, is of interest. An example of a stage record for a river over a 2-month period is shown in Fig. 1A.

Other devices for water level measurement include pressure transducers and gas-purge manometers. Experience with pressure transducers shows that because their calibration is prone to drift over long time periods, they are not a suitable primary sensing mechanism for water level measurement. Nevertheless, they can provide measurements of the rate of change of water level, which is useful for flood warning purposes.

B. Measurement of Discharge

To establish the stage/discharge rating curve, discharge measurements must be carried out for a range of river levels. Measurement of discharge is an exacting task if good quality measurements are to be achieved. Commonly, these measurements are undertaken by dividing a river cross section into 20 or more panels, each of similar area, and measuring the velocity through each of the panels. Total discharge is assessed as the sum of the product of the panel area multiplied by the mean velocity across the panel.

The width of each panel is measured with a tape stretched across the river, and the depth is determined by

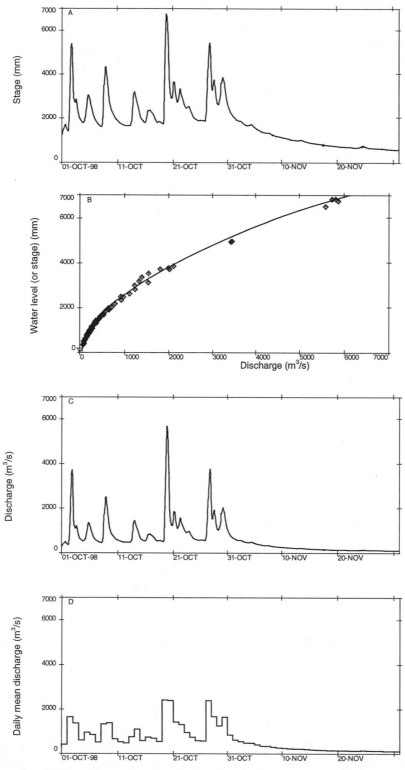

FIGURE 1 Illustration of the measurement of streamflow data for a 2-month interval for the Grey River near Greymouth, New Zealand. (A) Stage data recorded at 15-min intervals. (B) Stage/discharge rating curve used to transform the stage data to discharge, with the calibration measurements (obtained over a period of years using current meters) plotted. (C) Discharge hydrograph resulting from application of the rating curve in B to the stage data in A. (D) Daily mean discharge series calculated from C. Note that the daily averaging eliminates much of the detail of the floods.

sounding with a weight or, in the case of a small stream that can be waded, with a rod. The velocity at points within each panel is measured with a current meter. Typically, in shallow streams, the velocity measured at 0.6 of depth provides a reasonable estimate of the mean velocity in the cross section. In deeper rivers, the mean of the velocity at 0.2 and 0.8 of depth is used as an estimate of the mean velocity for the cross section.

Current meters have propellers or cups with rates of rotation proportional to the water velocity. In small streams that can be waded, the current meter can be mounted on a rod. In larger rivers, current meters are suspended with winches from cableways, bridges, or boats. The meters are precision instruments that require regular maintenance and calibration in special towing tanks.

Other less frequently used methods for discharge measurement include the use of floats (which enable assessment of velocity on the surface), ultrasonic equipment, chemical methods based on determining the extent of dilution of a concentrated solution, and measurement of volume change.

Measurement of the rate of volume change is useful for calibrating very low flow rates, provided the streamflow can be diverted into a container, and for assessment of total streamflow into lakes. In the case of lakes, it can be impractical to measure flows in all the streams entering the lake. If the total inflow for all streams is of interest, this can be assessed from the rate of change of the lake volume and the outflow. The rate of change of volume is estimated as the rate of change of lake level multiplied by the lake surface area. Precise measurement of the lake level is necessary to assess the changes in level and to distinguish these changes from the effects of seiching, which is the oscillation of the lake surface caused by wind blowing over the lake or by barometric pressure gradients.

C. Stage/Discharge Rating Curve

The quality of the stage/discharge rating curve (see example in Fig. 1B) determines the reliability of the streamflow record. Ideally, the calibration measurements should cover a full range of flow conditions from droughts to floods to ensure that extrapolation of the curve beyond the range of supporting measurements is minimized. However, achieving gaugings over a full range of flows is difficult where the gauging site is remote or where the river basin is small so that the river responds rapidly to infrequent storms. For this reason some degree of extrapolation of the rating curve is often necessary. Extrapolation tends to introduce error, and, consequently, larger measurement errors generally accompany estimates of flood flows.

D. Calculation of Streamflow Record

Application of the rating curve (Fig. 1B) to the record of stage (Fig. 1A) yields the record of flow (Fig. 1C). Although the flow data are of prime interest and are used for a wide range of purposes, there are reasons (see below) why the archives of data should retain the original measurements of stage.

II. STREAMFLOW DATA ARCHIVES

Streamflow is a natural phenomenon. It is not controllable in the same way that a laboratory experiment is controllable. When extremes of flood and drought occur, only one opportunity is presented to measure the event, and that is as it happens. Thus, measurements of streamflow and especially measurements of extremes are critically important because they cannot be repeated. Ready access to records of extremes is essential to ensure that societies adequately manage the risk of extremes.

Most countries have recognized the importance of archives of hydrometric data. National hydrometeorological services have been established to gather records of streamflow and to maintain and review archives to ensure that the data collected are correct and are securely archived and that copies are available for use.

With the advent of computers, archiving operations have moved from the publication of yearbooks to reliance on electronic data held in computer archives. Desirable features of the archives and the computer software used to submit and retrieve data include:

1. The inclusion of the original field measurements (stage, rating curves, calibration gaugings) in the archive: This offers the advantages that the details of flow hydrographs can be retrieved, the quality of rating curves and the extent of extrapolations can be assessed, and the rating curves can be adjusted retrospectively as new measurements of extreme discharge come to hand. These advantages are lost when the data are archived in a processed form, for example, as daily mean discharge. Compare the detail of Fig. 1C (based on 15-min data) with daily mean discharges for the same period in Fig. 1D: the flood peak that occurred was 5670 m^3/s, whereas the maximum daily mean is only 2400 m^3/s.

2. The ability to resolve units of time over hundreds of years to minutes or seconds: Details of extreme flood measurements retain their importance and validity over long periods, and accurate timing detail is necessary for detailed hydraulic study of the movement of flood waves.

FIGURE 2 Rees Street in Queenstown on 29 September 1878 (above) and 16 November 1999 (below). In both cases, the water level is nearly at the windowsills of the hotel in the center of the photographs. The maximum lake level in 1999 was 15 cm above the 1878 maximum. The mountain range in the background of the 1878 photograph is obscured by clouds in the 1999 picture. (Photos: Otago Daily Times Dunedin, New Zealand.)

3. Ready access to the archive: In some countries this is being achieved by making streamflow data available through the Internet.

4. Timeliness: Use of data for forecasting and management during floods and droughts requires data up to the present time. The facility for the archive to be regularly updated with the most recent data is highly desirable. This access to data in near real time requires an electronic system for retrieval of data from remote stations. American data provided by the U.S. Geological Survey are available via the Internet (website: http://water.usgs.gov/realtime.html).

Archives are important. This is illustrated by the pair of photographs in Fig. 2 which show flooding of the town of Queenstown in New Zealand in 1878 and 1999. Better use of knowledge that the site is flood prone could have guided the development of the town since 1878 to lessen the impact of the 1999 flood.

III. STREAMFLOW IN THE HYDROLOGICAL CYCLE

A. Simple Water Budget

The entry of water into a catchment and the movement through it to become streamflow is indicated in the schematic diagram in Fig. 3, which represents the terrestrial phase of the hydrological cycle. Precipitation (rain, snow, hail, dew) arriving at the catchment surface can be intercepted by the vegetation canopy and then can evaporate back to the atmosphere or drip though onto the ground.

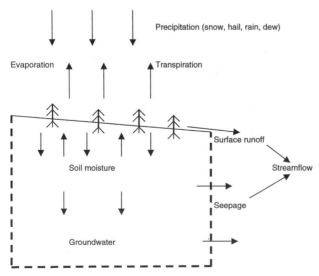

FIGURE 3 Schematic diagram to show the movement of water through a catchment to produce streamflow.

Water arriving at the ground surface can evaporate, infiltrate into the soil, or flow over the ground surface into a stream channel if it cannot infiltrate.

Water seeping into the soil forms a reservoir of soil moisture. This is depleted by losses to groundwater, abstractions by plant roots, and capillary action moving water to the ground surface where it evaporates.

Streamflow is supplied by surface and near-surface runoff and by seepage from groundwater. During high flows and floods, surface and near-surface runoff supplies most of the water appearing as streamflow, whereas in the absence of precipitation, flows are sustained largely by seepage from groundwater.

The size of flood peak flows is determined by

- The size of the catchment
- The intensity of the storm rainfall and its distribution over the catchment
- The vegetation cover
- The hydraulic properties of the soil of the catchment
- The status of the soil moisture storage at the start of the storm
- The geometry and hydraulic properties of the stream channel network of the catchment

Low flows occurring during prolonged dry spells when the soil moisture becomes depleted are dictated mainly by the rate at which seepage occurs from groundwater. This is determined by the nature and size of the groundwater reserves and by the extent to which the groundwater zones are intercepted by the stream channels. These, in turn, are set by the geological conditions. The magnitudes and rates of recession of low streamflows are indicators of the soil moisture and groundwater conditions.

Understanding, quantifying, and modeling the translation of water through a catchment to become streamflow is a challenging task that forms much of the science of hydrology. It is particularly difficult to define and deal with the spatial variation of the catchment processes described above. With the ready availability of powerful computers that provide access to databases which characterize the surface shape of large land areas, modeling methods have advanced dramatically in recent years. Such models are critically important for assessing extreme flood size, developing flood forecasts, predicting the effects of land use changes, and quantifying the components of the hydrological cycle to assess impacts of climate variability and change.

When studying catchments and comparing rainfall and streamflow, it is useful to convert streamflow from the units of discharge measurement (volume per unit time) into runoff (average depth of water per unit time flowing from the catchment area). For example, the Grey River near Greymouth on the west coast of New Zealand has a mean discharge of 364 m^3/s. At the stream gauge the catchment area is 3830 km^2, and the catchment is hilly to mountainous. The average annual runoff rate is calculated as

$$\text{Runoff} = 364 \text{ m}^3/\text{s} \times 86,400 \text{ s/day}$$

$$\times 365 \text{ day/year}/(3830 \times 10^6 \text{ m}^2)$$

$$= 3.00 \text{ m/year}$$

$$= 3000 \text{ mm/year}.$$

From Fig. 3, a simple budget for the inputs and outputs to a catchment can be expressed as

$$P = E + T + \Delta S + \Delta G + \Delta L + R, \qquad (1)$$

where

P = precipitation
E = evaporation
T = transpiration
ΔS = change in soil moisture storage
ΔG = change in groundwater storage
ΔL = loss of deep groundwater out of the catchment
R = runoff

Over a time interval of 1 year, the changes in soil moisture and groundwater storage ($\Delta S + \Delta G$) typically may be negligible compared with the first three terms in Eq. (1). Also, if the loss to deep groundwater (ΔL) is minimal, Eq. (1) simplifies to

$$P = E + T + R, \qquad (2)$$

which can be rewritten as

$$R = P - (E + T). \qquad (3)$$

This emphasizes that runoff is the residual term after the demands of evaporation and transpiration have been satisfied. Evaporation and transpiration are determined by the solar radiation, the wind, the relative humidity, the roughness of the vegetation surface, the vegetation type, the soil type, and the rooting depth. It varies widely. Values range from more than 2000 mm/year in arid subtropical zones to less than 200 mm/year in high-latitude regions. In dry regions where evaporation and transpiration are high, runoff may be only a few millimeters per year and highly variable from year to year.

For much of New Zealand, evaporation and transpiration is thought to be in the range 500–800 mm/year. For the Grey catchment, runoff of 3000 mm/year implies that average precipitation should be about 3500–3800 mm/year.

Only 1 of the 11 long-term rain gauges in the Grey catchment has a mean rainfall approaching these figures. This is because the mountainous high rainfall zones in the catchment are sparsely inhabited and have few rain gauges: most of the gauges are in the drier areas where

people live. This illustrates a widespread problem of adequately sampling rainfall in hilly and mountainous areas. In fact, the streamflow record for the Grey River is a better indicator of the total precipitation received by the catchment than the estimate inferred from the network of rain gauges. For this reason, investing in streamflow measurement rather than rainfall measurement can be a prudent use of limited resources.

IV. STREAMFLOW VARIABILITY

A. Scales of Time Variability

Streamflow varies at different time scales, and all these scales are of interest. Graphs are an excellent way to display the variability of streamflow. As examples, Figs. 4–6 show the variability of a single streamflow record at three different time scales: short, medium, and long. At each scale there are interesting features that have an impact on a range of issues. The streamflow record used is from the Clutha River at Balclutha in New Zealand. The catchment area for this river is 20,582 km^2 and the streamflow record commenced in 1954. Upstream of the recorder there are two hydroelectric dams and one controlled lake that provides limited regulation of the streamflow, storing water in summer for release during winter. Several other small hydroelectric developments have negligible influence on the flows in the main river. There are also minor withdrawals for irrigation in spring and summer.

1. Short-Term Variation

At a time scale of hours, Fig. 4 shows an example of the variation of flows at the stream gauge. The flow has a daily (diurnal) cycle caused by the daily variation of power generation at the hydroelectric stations, the first of which is 90 km upstream of the recorder. Typically, the hydroelectric station has a fairly constant minimum outflow in the early morning from 12 AM to 7 AM. The downstream data recorded at Balclutha data show that the daily minima occurs as a distinct minima at about 9 PM, reflecting the attenuation of the changing flow as it moves down the river to the gauge at Balclutha. The hydrograph is modified in the translation down the river channel, with the higher flows moving more quickly and gradually overtaking the slower moving lower flows and eliminating the 7-h interval of low flow.

In the case of the Clutha River, the diurnal variations in streamflows in the river have adverse consequences for

- Instream biota which tend to favor relatively constant flows

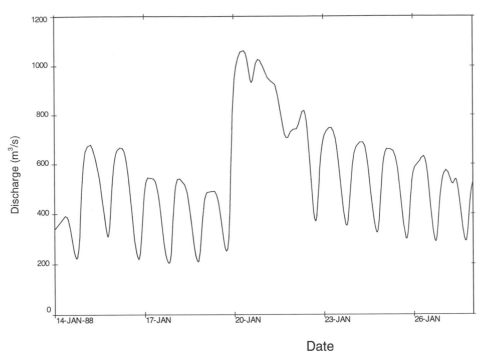

FIGURE 4 Example of short-term streamflow variations caused by the daily variations in generation of a hydroelectric power station (Clutha River at Balclutha, New Zealand).

- The stability of the river bank which suffers greater erosion as a result of constantly fluctuating flows and levels
- The opening and closing of the mouths of the branches of the river as it enters the sea; increased occurrences of mouth closure have impeded drainage of productive low-lying land near the river mouth.

2. Seasonal Variation

Seasonal variation is evident in most streamflow records and occurs because of seasonal variations in the precipitation, evaporation and transpiration, and soil moisture and groundwater levels. Snow and ice accumulation in winter and snowmelt in spring and summer contribute to the seasonal variation in cooler regions. Where rivers are sources of water for irrigation, the abstractions typically have a strong seasonal pattern, which impacts on the flow remaining in the river. Other human impacts are regulating reservoirs that store water at times of high flows for release when natural flows are lower.

All these factors contribute to the seasonal variations. For the Clutha River, seasonal variation is displayed as the distributions of monthly mean flows in Fig. 5. This figure is a boxplot of the monthly mean flows for each month from 1955–1999. For each month it shows the minimum monthly mean streamflow, the value exceeded in 75, 50, and 25% of years, and the maximum monthly

mean streamflow. (The 50 percentile value is also known as the median.) The seasonal variation is modest, but the austral winter values (in June, July, and August) are typically lower than the austral spring and summer values (in September, October, November, December, January, and February). The lower winter flows and the higher spring and summer flows are due primarily to winter snow accumulation and spring and summer snowmelt.

3. Long-Term Variation

Annual mean flows for the Clutha River are displayed in Fig. 6. These means display an interesting pattern of variability. The year 1958 had exceptionally high flows, but otherwise the flows up to and including 1978 tended to be less than those after 1978. The mean for the whole series is 572 m^3/s, and the standard deviation is 112 m^3/s. The ratio of the standard deviation to the mean, known as the coefficient of variation, is 0.195. (This dimensionless quantity is useful because it enables comparison of the variability of the series for different rivers.)

Means for the two periods shown on Fig. 6 differ by 16%. This difference is so large that it is very unlikely to arise through chance variation on the expectation that mean flow in every year is the same. The difference is consistent with fluctuations noted in some long-term rainfall records, and it appears to be related to the frequencies of occurrence of phases of the El Niño Southern

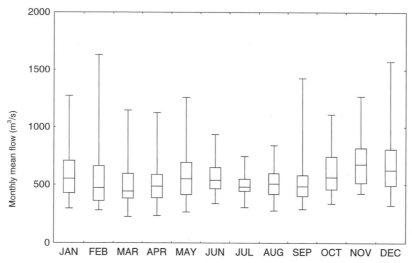

FIGURE 5 Seasonal variation in streamflow illustrated using boxplots of monthly mean flows for the Clutha River at Balclutha, New Zealand, for 1954–1999. For each month, the plots show the minimum monthly streamflow, the value exceeded in 75, 50, and 25% of years, and the maximum monthly streamflow.

Oscillation phenomenon. These appear to have changed in 1978.

Climate change studies examine long time series to detect shifts and trends. The apparent shift in Fig. 6 raises interesting questions about the causes of fluctuations in long-term records.

B. Variability between Rivers

The mean annual discharge varies greatly between river basins. The variation is driven primarily by the difference between total precipitation and the evaporation and transpiration occurring over the basin. Any thorough assessment of streamflow has to look beyond the mean annual discharge to the expected variation in annual discharge and the expected variation in seasonal streamflow.

Seasonal variation is driven by the seasonal variability in the water balance. In tropical areas and subtropical conditions, the patterns are driven largely by seasonality in rainfall, and in monsoon-type climates, most of the stream-

flow occurs over a few months, with low flows occurring for the remainder for the year. In mid- and high-latitude zones, seasonality may or may not be present in precipitation, but it occurs in streamflow because of the strong seasonal patterns affecting transpiration of plants and evaporation. In colder continental climates and in mountainous areas, snow accumulation in winter and snowmelt in spring and summer are primary sources of seasonal variation.

To illustrate the contrasts experienced in a small country, Fig. 7 shows the monthly flow regime for another New Zealand river basin that experiences conditions differing from the Clutha River. This river is the Motu, in the North Island. The catchment area is 1393 km^2, and the influence of snow is negligible. Figure 7 shows that the highest monthly flows tend to occur in the (austral) winter (June, July and August) and that the lowest flows tend to occur in the late (austral) summer and early fall (January, February, March). Higher winter flows occur partly because rainfall is somewhat higher in winter, but

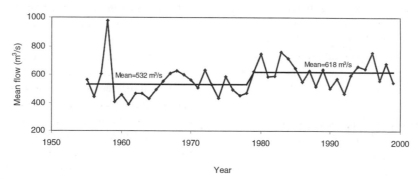

FIGURE 6 Annual mean flows for the Clutha River at Balclutha, New Zealand, for 1954–1999. The mean for the whole record is 572 m^3/s.

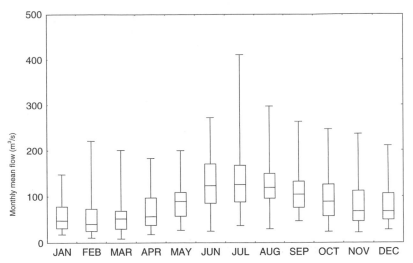

FIGURE 7 Seasonal variation in streamflow illustrated using boxplots of monthly mean flows for the Motu River, New Zealand, for 1957–1999. For each month, the plots show, the minimum monthly streamflow, the value exceeded in 75, 50, and 25% of years, and the maximum monthly streamflow.

mainly because evaporation and transpiration demands are suppressed. The mean annual discharge is 92 m^3/s, the runoff is 2080 mm/year, and the coefficient of variation for annual streamflow is 0.210. Although the coefficient of variation is practically identical to the Clutha River, the seasonal patterns differ significantly. Higher streamflow in winter would be important if, for example, this river basin was to be developed for hydroelectric power generation. High winter flows would tend to match the peak energy demand in winter for heating, whereas, without construction of a storage dam, the relatively lower summer flows reduce the attractiveness of the river as a source of irrigation water. This contrasts with the Clutha River, where the high season flows are out of phase with the seasonal energy loads.

Global runoff studies use long streamflow records from around the world to examine runoff patterns. The coefficient of variation of annual discharge is an indicator of the expected year-to-year variation of streamflow. In general, a spatial averaging applies, with larger river basins having lower coefficients of variation than smaller basins. The relationships differ by continental regions, with characteristically higher values in Australia and Southern Africa. The higher values relate to the general aridity of these regions and indicate that they are relatively more prone to extremes of streamflow.

V. IMPORTANCE OF STREAMFLOW

A. Spiritual and Cultural

Streamflow has always been important for people, and many of the great civilizations of the world, e.g., Egypt,

Mesopotamia, India, and China, have been located by large rivers.

In the Judeo-Christian tradition, streamflow figures prominently in the historical narrative of the Israelite people. For example, the 7 years of plenty followed by 7 years of famine in the Exodus story relate to the annual flow volumes of the Nile River (Genesis 41). (It is now known that a portion of the Nile River flow variability is determined by the state of the El Niño Southern Oscillation phenomenon.) The plagues that were experienced in Egypt as a prelude to the release of the enslaved Israelites also appear to relate to abnormal Nile River flows, which were followed by a set of environmental catastrophes stemming from a toxic algal bloom in the river (Exodus 7).

Other groups, who determine their identity in relation to the country and the landscape, see streamflow differently. For example, the Maori people of New Zealand define their identity in relation to particular mountains and rivers of their traditional tribal area. Rivers, and the streamflow they convey, are seen as having a life force contributing to personal and tribal identity and status. In addition, the river is a provider of food from the fish and bird life and a traditional route for navigation by canoe.

Water resource developments have often upset this relationship between people and rivers. Damming of rivers and impeding access for migrating fish, manipulating river flows and diverting them to other catchments, and discharging effluents into rivers are all at variance with this life view. Some conventional developments can thereby ignore and belittle these deeply held views.

Resolving development with these alternative views will be a significant challenge for river basin management in the 21st century. An example of resolving these issues is the disposal of sewage effluent of an inland city by spray irrigation in a forest, rather than into a river noted for its eel fishery. Other examples include the construction of an artificial wetland to remove contaminants from stormwater drainage from an industrial site before it enters an estuary used for gathering shellfish and the provision of fish passes to enable elvers to migrate upstream over dams.

B. Water Supply and Effluent Disposal

Every community needs a clean water supply and safe effluent disposal. Streamflow is frequently used as a water source. Determining whether a river has the capacity to meet a demand requires analysis of the streamflow records to establish the low-flow characteristics of the stream. Specific quantities of interest are the low flows expected in the stream, their probabilities of occurrence, and the residual flows required to be maintained in the stream after abstractions have occurred. The study of low flows will determine whether a storage is needed to sustain a supply during dry periods.

A reliable record covering a number of years is necessary to provide a comprehensive understanding of the seasonal and year-to-year variability of low flows. Where no such record exists, hydrological models that encapsulate the dynamics of water movement through the catchment, as illustrated in Fig. 3, can be used with long records of rainfall to extend streamflow records. This presumes that suitably representative long rainfall records are available. Constructing, calibrating, and verifying a hydrological model normally requires that at least a short-term measured streamflow record is available.

Disposal of sewage is a complement of water supply. Communities normally dispose of sewage by treating it to various levels. The treated effluent, or wastewater, is then disposed of into a suitable stream, lake, estuary, or the sea. Varying levels of treatment are available, but even highly treated effluent can have elevated levels of nitrates and phosphates. The nutrients can lead to enhanced growth of aquatic plants in the receiving water. The eventual die-off and decay of these plants causes depleted dissolved oxygen levels, leading to the die-off of fish populations.

Analysis of low streamflow quantities and the quality characteristics of the water, such as the nitrate levels, the biochemical oxygen demand, and the dissolved oxygen levels, assist in determining the capacity of a stream to receive treated effluents.

Where water supplies are sparse, or where effluent disposal into receiving water is offensive to a section of the community, on-land disposal by irrigation or by recharge of groundwater aquifers may sometimes be a viable solution.

C. Irrigation

Streamflow is a primary source of water for irrigation. Where streamflows are highly variable, storage reservoirs are often necessary to sustain flows for irrigation during dry periods.

Careful design and operation of irrigation schemes are necessary to manage the delivery of water. Excessive water supplied for irrigation can lead to high water tables and waterlogging, which causes a range of adverse effects. Drainage of surplus irrigation water back into the stream can convey contaminants into the stream below the irrigation area. In the lower Murray River in Australia, for example, elevated levels of salt, due partially to drainage of water from irrigation areas, limits the usefulness of the river water.

Depletion of streamflows due to irrigation abstractions can also affect a river as a biological habitat. Prominent effects include

- Reduction of the areas of water with depths and velocities suitable for fish habitat
- Alterations to algal communities, with changes from thin diatom layers to thick growths of filamentous algae
- Alterations to invertebrate and, ultimately, fish communities
- Disturbance of the movement of migratory fish (e.g., salmon)
- Elevated water temperatures

The impacts of these deleterious effects can be managed when streamflow data are readily available and when a community has agreed on rules about reducing irrigation abstractions when streamflows become low.

D. Energy Generation

1. Hydroelectric Power

Hydroelectric power development is an attractive source of renewable energy. Hydroelectric power generation does not have the emissions that accompany fossil-fueled thermal power stations or the waste-disposal problems that accompany nuclear power stations. The "fuel," the flowing water, is continually replenished, though its supply is subject to the variability described above. In a sense the water is free, but it can have a value imputed to it on the basis of its potential energy, which is given by prevailing energy prices. On the other hand, when areas of land are flooded during the filling of reservoirs, methane, a powerful greenhouse gas, is released by decaying vegetation.

Hydroelectric power stations can have very long lifetimes. Many stations constructed early in the 20th century generate power today and will do so for the foreseeable future.

Conventional hydroelectric developments normally require large dams to provide the head of water for generation. Large dams are expensive to construct, and the economics can be marginal if the transmission line losses are high because the center for electricity demand is a long distance from power station.

Streamflow data are necessary to establish the patterns of generation that will be available from the station. It is highly desirable that the streamflows are well sustained in dry periods. This is illustrated in Fig. 8, which compares the flow variability for two contrasting rivers, the Clutha and the Motu. Figure 8 plots the cumulative flows for the two rivers, as percentages of their respective means, against the probability of flow being greater than a given value. For example, the plot shows that for the Clutha River the flow exceeded 80% of the time is 63% of the mean, whereas for the Motu River the corresponding figure is only 30% of the mean. In other words, the Motu River flows are much more variable than the Clutha River. The lower variability of the Clutha River occurs because of three large natural lakes, one with controlled outflows, in the headwaters that moderate the flow extremes. No such moderation of the Motu River flows occurs, and, not surprisingly, hydroelectric development of the Motu River has not proceeded.

Consequences of the changes in streamflow regime caused by hydroelectric power development can be severe. They can include

- The loss of wild and scenic features when reaches of a river are inundated by a reservoir.
- The loss through inundation of productive land and the forced relocation of communities.
- The restriction of access to the river system for fish such as salmon and eels whose life cycle is spent partially in the sea. Fish passages typically present only partial solutions.

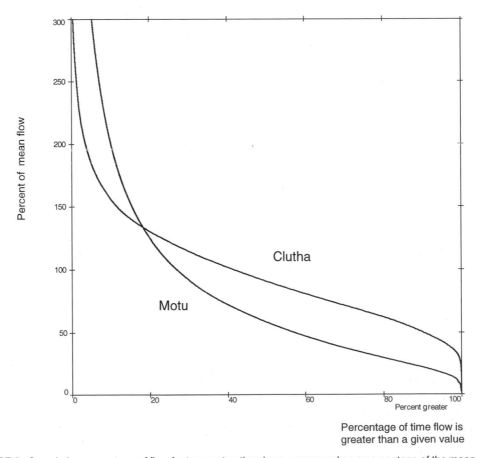

FIGURE 8 Cumulative percentage of flow for two contrasting rivers, expressed as a percentage of the mean, plotted against the probability of flow being greater than a given value.

- The reduction of downstream sediment loads, resulting in increased channel scouring and bank erosion. Furthermore, the reduction of sediment entering the coastal zone can result in shifts in the coastal geomorphology.
- The diminution of floods that have periodically flushed sediment and reshaped and redefined the river channel.

Where the environmental costs are seen as being too high, the desire to restore a river to a more natural flow regime has resulted in the closure and removal of some early hydroelectric schemes where the power output is relatively small.

2. Cooling Water

Rivers are also used as sources of cooling water for thermal power stations, both fossil fueled and nuclear. The capacity of the streamflow to receive waste heat from these plants is determined by the discharge in dry conditions, the prevailing stream temperatures, and the maximum temperature desired for the stream. In cases where low river flows occur at times of high power demands, the capacity of the river to receive waste heat can be a constraint on the operation of the plant.

E. Aquatic Habitat

Many biological communities live in flowing water. The composition and structure of these communities are affected by many aspects of streamflow, including the water quality and the nutrients and sediments that it carries, the temperature, and the scales of variability in the streamflow quantities.

The scales of variation are conveniently considered as large (macro) scale, medium (meso) scale, and small (micro) scale.

1. Large-Scale Flow Variation

Large-scale variations in flows regimes caused by seasonal and annual variability affect populations at a community level and are the ultimate regulators of species composition. Communities in a particular flow regime with regular seasonal variability tend to have become adapted to that regime, whereas erratic patterns of flood flow can cause considerable disturbance to a community.

The number of high-flow events and the durations between events are useful measures in assessing biological habitats. In streams that flood frequently, smaller "weedy" taxa tend to dominate among aquatic plant species, whereas larger more competitive taxa prefer environments that flood rarely, such as the outflows from lakes.

Low flows also affect biological communities. Algal growth commonly increases during low-flow episodes, and algal communities change from thin diatom films to thick growths of filamentous algae. These thick growths tend to smother instream habitats and have adverse impacts on invertebrate and fish communities. Other deleterious effects are a reduction in stream habitat by exposure and desiccation of a normally inundated streambed. Consideration of these impacts is necessary to establish the residual flows required to maintain the health of the stream when use of a stream is being considered for water supply, effluent disposal, irrigation, or hydroelectric power generation.

2. Medium-Scale Flow Variation

The variability encountered along individual reaches of streams, with the length scales in the range of meters to kilometers, exerts influences over the composition of local populations of biota and often dictates the differences observed in species composition along a stream. An individual stream reach can be classified into riffles, runs, and pools, and the differing hydraulic properties (depths and velocities) and the differing substrate composition are important factors determining the biological variability. Algal and invertebrate communities differ between riffles, pools, and runs.

Substrate stability is also important for stream communities. Stability is determined by the composition of the substrate (silt, sand, gravel), the flow velocity, imbrication, and the relative bed roughness. Macrophytes (larger aquatic plants) and bryophytes (mosses) can stabilize substrates in lowland and upland streams, respectively, and thereby play an important role in determining the structure and function of other instream components.

The hydraulic habitat conditions for plants are controlled by flood frequency, the velocities occurring between spates, and substrate stability. Invertebrate communities often respond to similar variables.

3. Small-Scale Flow Variation

Small spatial scales (millimeters to meters) are important for the control that they exert on the generation and dissipation of turbulence. The extent of flow velocity variation and the associated turbulence directly influences individual organisms in streams and is often responsible for the observed inherent variability of populations in streams.

Turbulence is manifest as the variation over short time scales of velocity in a steady flow. The variability of the three spatial components of flow velocity with time can now be measured with acoustic Doppler velocity meters. On a fixed streambed, the velocity must be zero, and in the zone near the bed (the boundary layer), the velocity changes rapidly. Measurements of velocity close to the bed of a stream provide a measure of the near-bed stream turbulence. Understanding the near-bed turbulence is necessary to estimate the fluxes of oxygen and nutrients available to biota within the slow-flowing, near-bed boundary layer. Organisms that grow too large and intrude into faster flowing water can get washed away. Small-scale variability in invertebrate and algal communities often reflects changes in the velocity and turbulence regimes. The growth of organisms on stream boundaries can also affect the turbulence in streams.

VI. HAZARDS OF EXTREME STREAMFLOW

Unusually high streamflows are described as floods. The area of land beside a river inundated during floods is termed the floodplain. Floodplains are naturally fertile and are often the most productive land in a region. When inundation occurs on floodplains that are closely settled, the damage and suffering can be enormous. Most countries have developed a suite of measures to limit flood damage, including regulations and rules that keep people away from flood-prone areas and controls and structures such as levees and control dams that keep the flood water away from people. On large river basins, where severe precipitation can take days to weeks to progress down a river system to cause flooding, national hydrometeorological services can provide forecasts, enabling some reduction in damages.

Rivers draining mountainous areas tend to have steep channels where water flows swiftly. The response times are short, and flooding can occur within hours of the heavy rain. So-called flash-flood forecasting is a challenging area.

VII. CONCLUSION

Fresh, clean water is essential for life. Streamflow is a major source of this water, and streams are a vital resource.

Understanding streamflow, quantification of its magnitude, and variability is essential for sustainable utilization of this precious natural resource.

ACKNOWLEDGMENT

Funding of the streamflow data used in the figures is from the Foundation for Research Science and Technology, contract CO1815 (Information on New Zealand's freshwater). I thank Dr A. Suren and other colleagues of the National Institute of Water and Atmospheric Research, Christchurch, New Zealand, for advice and support.

SEE ALSO THE FOLLOWING ARTICLES

COASTAL GEOLOGY • DAMS, DIKES, AND LEVEES • DRINKING WATER QUALITY AND TREATMENT • FLOW VISUALIZATION • GLACIOLOGY • HYDROGEOLOGY • HYDROLOGIC FORECASTING • SOIL AND GROUNDWATER POLLUTION • STOCHASTIC DESCRIPTION OF FLOW IN POROUS MEDIA • WATER RESOURCES

BIBLIOGRAPHY

Ackers, P., White, W., Perkins, J., and Harrison, A. (1978). "Weirs and Flumes for Flow Measurement," Wiley, Chicester.

Harper, D., and Ferguson, A. (eds). (1995). "The Ecological Basis for River Management," Wiley, Chicester.

Herschy, R. (1985). "Streamflow Measurement," Elsevier, London.

Herschy R. (1998). "Hydrometry: Principles and Practices," 2nd ed., Wiley, Chicester.

Hosking, J., and Wallis, J. (1998). "Regional Frequency Analysis: An Approach Based on L-Moments," Cambridge Univ. Press, Cambridge, UK.

Maidment, D., ed. (1993). "Handbook of Hydrology," McGraw-Hill, New York.

McMahon, T., Finlayson, B., Haines, A., and Srikanthan, R. (1992). "Global Runoff: Continental Comparisons of Annual Flows and Peak Discharges," Catena Verlag, Cremlingen-Destedt.

Sutcliffe, J., and Parks, Y. (1999). "The Hydrology of the Nile," International Association of Hydrological Sciences, Wallingford.

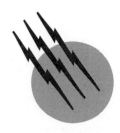

Stress in the Earth's Lithosphere

Mark D. Zoback

Stanford University

Mary Lou Zoback

U.S. Geological Survey

GLOSSARY

Critically stressed crust Stress magnitudes in the brittle crust are equal to the crust's frictional strength.
***In situ* stress** Forces in the lithosphere.
Sources of stress The mechanisms responsible for stress in the lithosphere.
Stress map Map showing the orientation and relative magnitude of horizontal principal stress orientations.

THE STATE OF STRESS in the lithosphere is the result of the forces acting upon and within it. Knowledge of the magnitude and distribution of these forces can be combined with mechanical, thermal, and rheological constraints to examine a broad range of lithospheric deformational processes. For example, such knowledge contributes to a better understanding of the processes that both drive and inhibit lithospheric plate motions, as well as the forces responsible for the occurrence of crustal earthquakes—both along plate boundaries and in intraplate regions.

While our topic is the state of stress in the earth's lithosphere, the comments below come primarily from the perspective of the state of stress in the brittle upper crust. As defined by the depth of shallow earthquakes, the brittle crust extends to a depth of ∼15–20 km at most continental locations around the world. We adopt this perspective because nearly all the data available on lithospheric stress comes from the upper crust of continents. Furthermore, in the sections that follow, we argue that, to first order, the state of stress in the brittle crust results from relatively large-scale lithospheric processes, so that knowledge of crustal stress can be used to constrain the forces involved in these processes.

I. BASIC DEFINITIONS

Stress is a tensor which describes the density of forces acting on all surfaces passing through a point. In terms of continuum mechanics, the stresses acting on a homogeneous, isotropic body at depth are describable as a second rank tensor, with nine components (Fig. 1, left).

Encyclopedia of Physical Science and Technology, Third Edition, Volume 16
Copyright © 2002 by Academic Press. All rights of reproduction in any form reserved.

$$\bar{S} = \begin{vmatrix} S_{11} & S_{12} & S_{13} \\ S_{21} & S_{22} & S_{23} \\ S_{31} & S_{32} & S_{33} \end{vmatrix} \qquad (1)$$

The subscripts of the individual stress components refer to the direction that a given force is acting and the face of the unit cube upon which the stress component acts. Thus, in simplest terms, any given stress component represents a force acting in a specific direction on a unit area of given orientation. As illustrated in the left side of Fig. 1, a stress tensor can be defined in terms of any arbitrary reference frame. Because of equilibrium conditions,

$$S_{12} = S_{21}$$

$$S_{13} = S_{31} \qquad (2)$$

$$S_{23} = S_{32}$$

so that the order of the subscripts is unimportant. In general, to fully describe the state of stress at depth, one must

estimate six stress magnitudes or three stress magnitudes and the three angles that define the orientation of the stress coordinate system with respect to a reference coordinate system (such as geographic coordinates, for example).

We utilize the convention that compressive stress is positive because *in situ* stresses at depths greater than a few tens of meters in the earth are *always* compressive. Tensile stresses do not exist at depth in the earth for two fundamental reasons. First, because the tensile strength of rock is generally quite low, significant tensile stress cannot be supported in the earth. Second, because there is always a fluid phase saturating the pore space in rock at depth (except at depths shallower than the water table), the pore pressure resulting from this fluid phase would cause the rock to hydraulically fracture should the least compressive stress reach a value even as low as the pore pressure.

Once a stress tensor is known, it is possible to evaluate stresses in any coordinate system via tensor transformation. To accomplish this transformation, we need to specify the direction cosines (a_{ij}, as illustrated in Fig. 1, center)

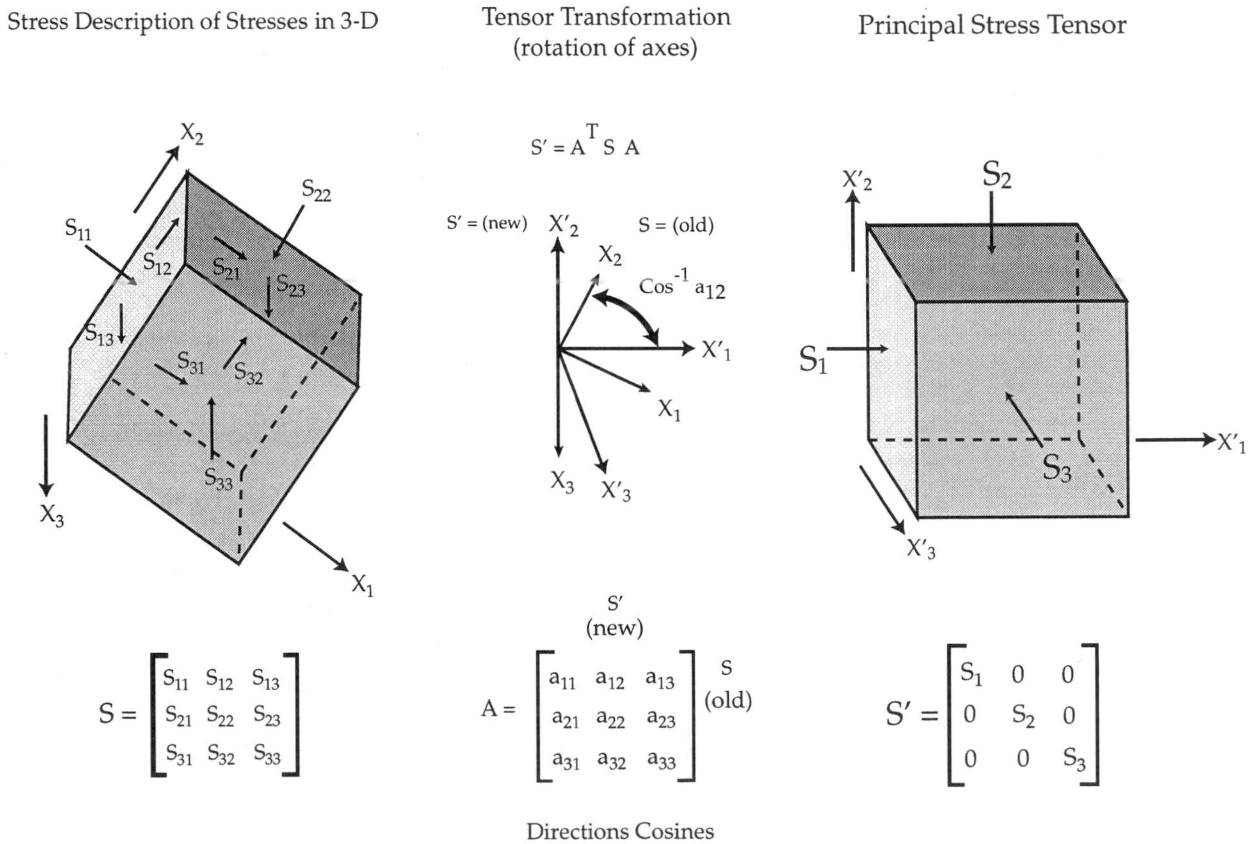

Stress Description of Stresses in 3-D   Tensor Transformation (rotation of axes)   Principal Stress Tensor

$$S = \begin{bmatrix} S_{11} & S_{12} & S_{13} \\ S_{21} & S_{22} & S_{23} \\ S_{31} & S_{32} & S_{33} \end{bmatrix} \qquad A = \begin{bmatrix} a_{11} & a_{12} & a_{13} \\ a_{21} & a_{22} & a_{23} \\ a_{31} & a_{32} & a_{33} \end{bmatrix} \qquad S' = \begin{bmatrix} S_1 & 0 & 0 \\ 0 & S_2 & 0 \\ 0 & 0 & S_3 \end{bmatrix}$$

Directions Cosines

FIGURE 1 Definition of stress tensor in an arbitrary Cartesian coordinate system (left) and the rotation of stress coordinate systems through tensor transformation (middle) and principal stresses (right) as defined in a coordinate system in which shear stresses vanish.

that describe the rotation of the coordinate axes between the old and new coordinate systems. Mathematically, the equation which accomplishes this is

$$\bar{S}' = \bar{A}^T \bar{S} \bar{A}, \qquad (3)$$

where

$$\bar{A} = \begin{vmatrix} a_{11} & a_{12} & a_{13} \\ a_{21} & a_{22} & a_{23} \\ a_{31} & a_{32} & a_{33} \end{vmatrix}.$$

The ability to transform coordinate systems is of interest here because we can choose to generally describe the state of stress in terms of the principal coordinate system. The principal coordinate system is the one in which shear stresses vanish and only three principal stresses, $S_1 \geq S_2 \geq S_3$, fully describe the stress field (as illustrated in right side of Fig. 1). Thus, we have diagonalized the stress tensor such that the principal stresses correspond to the eigenvalues of the stress tensor and the principal stress directions correspond to its eigenvectors.

$$\bar{S}' = \begin{vmatrix} S_1 & 0 & 0 \\ 0 & S_2 & 0 \\ 0 & 0 & S_3 \end{vmatrix} \qquad (4)$$

The reason this concept is so important is that as the earth's surface is in contact with a fluid (either air or water) and cannot support shear tractions, it is a principal stress plane. Thus, one principal stress is generally expected to be normal to the earth's surface with the other two principal stresses acting in an approximately horizontal plane. While it is clear that this must be true very close to the earth's surface, compilation of earthquake focal mechanism data and other stress indicators (described below) suggest that it is also generally true to the depth of the brittle-ductile transition in the upper crust (Zoback and Zoback, 1989; Zoback, 1992). Assuming this is the case, we must define only four parameters to describe the state of stress at depth; one stress orientation (usually taken to be the azimuth of the maximum horizontal compression, S_{Hmax}) and three principal stress magnitudes: S_v, the vertical stress, corresponding the weight of the overburden; S_{Hmax}, the maximum principal horizontal stress; and S_{hmin}, the minimum principal horizontal stress. This obviously helps make stress determination in the crust a tractable problem.

In applying these concepts to the earth's crust, it is helpful to consider the magnitudes of the greatest, intermediate, and maximum principal stress at depth (S_1, S_2, and S_3) in terms of S_v, S_{Hmax}, and S_{hmin} in the manner originally proposed by E. M. Anderson. (1951). This is illustrated in Fig. 2. There are a number of simple but fundamental points about these seemingly straightforward relations.

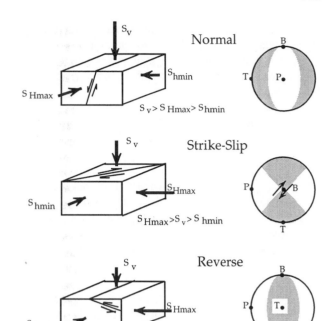

FIGURE 2 E. M. Anderson's classification scheme for relative stress magnitudes in normal, strike-slip, and reverse faulting regions. Corresponding focal plane mechanisms are shown to the right.

First, the two horizontal principal stresses in the earth, S_{Hmax} and S_{hmin}, can be described relative to the vertical principal stress, S_v, whose magnitude corresponds to the overburden. Mathematically, S_v is equivalent to integration of density from the surface to the depth of interest, z. In other words,

$$S_v = \int_0^z \rho(z) g \, dz \approx \bar{\rho} g z, \qquad (5)$$

where $\rho(z)$ is the density as a function of depth, g is the gravitational acceleration, and $\bar{\rho}$ is the mean overburden density. It is, of course, necessary to add atmospheric pressure and the pressure resulting from the weight of water at the earth's surface, as appropriate.

Second, the horizontal principal stresses are almost never equal and may be less than or greater than the vertical stress. In fact, the relative magnitudes of the principal stresses can be simply related to the faulting style currently active in a region. As illustrated in Fig. 2, (and following Anderson, 1951), characterizing a region by normal, strike-slip, or reverse faulting is equivalent to defining the horizontal principal stress magnitudes with respect to the vertical stress. When the vertical stress dominates in extensional deformational regions ($S_1 = S_v$), gravity drives normal faulting. Conversely, when both horizontal stresses exceed the vertical stress ($S_3 = S_v$), compressional deformation (shortening) is accomodated through reverse faulting. Strike-slip faulting represents an intermediate stress state $(S_2 = S_v)$, where the maximum horizontal stress is greater than the vertical stress and the minimum horizontal stress is less than the vertical stress $(S_{Hmax} \geq S_v \geq S_{hmin})$.

The concept of effective stress is used to incorporate the influence of pore pressure at depth where a component of effective stress σ_{ij} is related to the total stress S_{ij} via

$$\sigma_{ij} = (S_{ij} - \delta_{ij}P_p), \qquad (6)$$

where δ_{ij} is the Kronecker delta and P_p is the pore pressure.

Laboratory studies of the frictional strength of faulted rock carried out over the past several decades indicate that the Coulomb criterion describes the frictional strength of faults. That is, fault slippage will occur when

$$\tau = S_o + \mu\sigma_n, \qquad (7)$$

where τ is the shear stress acting on the fault, S_o is the fault cohesion, μ is the coefficient of friction on the fault, and σ_n is the effective normal stress acting on the fault plane. The maximum shear stress is given by $1/2(S_1 - S_3)$.

Using the concept of effective stress at depth, we can extend Anderson's faulting theory to predict stress magnitudes at depth through utilization of simplified two-dimensional Mohr-Coulomb failure theory. Two-dimensional faulting theory assumes that failure is only a function of the difference between the last and greatest principal effective stresses σ_1 and σ_3 as given by Jaeger and Cook (1971)

$$\sigma_1/\sigma_3 = (S_1 - P_p)/(S_3 - P_p) = ((\mu^2 + 1)^{1/2} + \mu)^2. \quad (8)$$

Thus, a third point about crustal stress that can be derived from Andersonian faulting theory is that the magnitudes of the three principal stresses at any depth are limited by the strength of the earth's crust at depth. In the case of normal faulting, stress magnitudes are controlled by σ_v, and σ_{hmin}, which correspond to σ_1 and σ_3, respectively σ_{Hmax}, which corresponds to σ_2 is intermediate in value between σ_v and σ_{hmin}, does not influence faulting. Coulomb failure theory indicates that frictional sliding occurs when the ratio shear stress to effective normal stress on preexisting fault planes is equal to the coefficient of friction. As the coefficient of friction is relatively well defined for most rocks and ranges between ~0.6 and 1.0 (Byerlee, 1978), Eq. (8) demonstrates that frictional sliding will occur when $\sigma_1/\sigma_3 \sim 3.1$–5.8. For the case of hydrostatic pore pressure and commonly observed friction coefficients of 0.6 (e.g., Townend and Zoback, 2000), in extensional areas $S_{hmin} \sim 0.6\ S_v$, in reverse faulting areas $S_{Hmax} \sim 2.3\ S_v$, and in strike-slip faulting areas (when $S_v \sim 1/2(S_{Hmax} + S_{hmin}))S_{Hmax} \sim 2.2\ S_{hmin}$. As discussed below, these simple relationships have been confirmed by *in situ* stress measurements to depths of almost 8 km at a number of sites in intraplate areas.

II. INDICATORS OF CONTEMPORARY STRESS

Information on the state of stress in the lithosphere comes from a variety of sources—earthquake focal plane mechanisms, young geologic data on fault slip and volcanic alignments, *in situ* stress measurements, stress-induced wellbore breakouts, and drilling-induced tensile fractures. A stress measurement quality criterion for different types of stress indicators was developed by Zoback and Zoback (1989, 1991). This quality criterion was subsequently utilized in the International Lithosphere Program's World Stress Map Project, a large collaborative effort of data compilation and analyses by over 40 scientists from 30 different countries (Zoback, 1992). A special issue of the *Journal of Geophysical Research* (v. 97, pp. 11,703–12,014, 1992) summarized the overall results of this project and presented the individual contributions of many of these investigators in various regions of the world. Today, the World Stress Map (WSM) database has more than 9100 entries and is maintained at the Heidelberg Academy of Sciences and Humanities (Mueller *et al.*, 1997; http://www-wsm.physik.uni-karlsruhe.de/).

Zoback and Zoback (1991) discussed the rationale for the WSM quality criterion used in the WSM project in detail. The success of the WSM project demonstrates that with careful attention to data quality, coherent stress patterns over large regions of the earth can be mapped with reliability and interpreted with respect to large-scale lithospheric processes.

A. Earthquake Focal Mechanisms

While earthquake focal plane mechanisms are the most ubiquitous indicator of stress in the lithosphere, the determination of principal stress orientations and relative magnitudes from these mechanisms must be done with appreciable caution. The pattern of seismic radiation from the focus of an earthquake permits construction of earthquake focal mechanisms (right column of Fig. 2). Perhaps the most simple and straightforward information about *in situ* stress that is obtainable from focal mechanisms and *in situ* stress is that the type of earthquake (i.e., normal, strike-slip, or reverse faulting) defines the relative magnitudes of S_{Hmax}, S_{hmin}, and S_v. In addition, the orientation of the fault plane and the auxiliary plane (which bind the compressional and extensional quadrants of the focal plane mechanism) defines the orientation of the P (compressional), B (intermediate), and T (extensional) axes. These axes are sometimes incorrectly assumed to be the same as the orientation of S_1, S_2, and S_3.

For cases in which laboratory-measured coefficients of fault friction of ~0.6–1.0 are applicable to the crust, there is a nontrivial error of ~15–20° if one uses the P, B, and T

axes as approximations of average principal stress orientations, especially if the orientation of the fault plane upon which the earthquake occurred is known. If friction is negligible on the faults in question (but higher in surrounding rocks), there can be a considerable difference between the P, B, and T axes and principal stress directions. An earthquake focal plane mechanism always has the P and T axes at 45° to the fault plane and the B axis in the plane of the fault. With a frictionless fault the seismic radiation pattern is controlled by the orientation of the fault plane and not the *in situ* stress field (McKenzie, 1969). One result of this is that just knowing the orientation of the P axis of earthquakes along weak, plate-bounding, strike-slip faults (like the San Andreas) does not allow one to define principal stress orientations from the focal plane mechanisms of the strike-slip earthquakes occurring on the fault (Zoback *et al.*, 1987). For this reason, it is common practice to omit plate-boundary earthquakes from regional stress compilations (see below).

Principal stress directions can be determined directly from a group of earthquake focal mechanisms (or set of fault striae measurements) through use of inversion techniques that are based on the slip kinematics and the assumption that fault slip will always occur in the direction of maximum resolved shear stress on a fault plane. Such inversions yield four parameters, the orientation of the three principal stress and the relative magnitude of the intermediate principal stress with respect to the maximum and minimum principal stress.

The analysis of seismic waves radiating from an earthquake also can be used to estimate the magnitude of stress released in an earthquake (stress drop), although not absolute stress levels. In general, stress drops of crustal earthquakes are on the order of 1–10 MPa. Equation (8) can be used to show that such stress drops are only a small fraction of the shear stresses that actually causes fault slip if pore pressures are approximately hydrostatic at depth and Coulomb faulting theory (with laboratory-derived coefficients of friction) is applicable to faults *in situ*. This is discussed in more detail below.

B. Geologic Stress Indicators

There are two general types of *relatively young* geologic data that can be used for *in situ* stress determinations: (1) the orientation of igneous dikes or cinder cone alignments, both of which form in a plane normal to the least principal stress, and (2) fault slip data, particularly the inversion of sets of striae (i.e., slickensides) on faults as described above. Of course, the term "relatively young" is quite subjective, but it essentially means that the features in question are characteristic of the tectonic processes currently active in the region of question. In most cases, the WSM database utilizes data which are Quaternary in age,

but in all areas represent the youngest episode of deformation in an area.

C. *In Situ* Stress Measurements

Numerous techniques have been developed for measuring stress at depth. Because we are principally interested here in regional tectonic stresses (and their implications) and because there are a variety of nontectonic processes that affect *in situ* stresses near the earth surface, we do not utilize near-surface stress measurements in the WSM or regional tectonic stress compilations (these measurements are given the lowest quality in the criteria used by WSM, as they are not believed to be reliably indicative of the regional stress). In general, we believe that only *in situ* stress measurements made at depths greater than 100 m are indicative of the tectonic stress field at midcrustal depths. This means that techniques utilized in wells and boreholes, which access the crust at appreciable depth, are especially useful for stress measurements.

When a well or borehole is drilled, the stresses that were previously supported by the exhumed material are transferred to the rock surrounding the hole. The resultant stress concentration is well understood from elastic theory. Because this stress concentration amplifies the stress difference between far-field principal stresses by a factor of four, there are several ways in which the stress concentration around boreholes can be exploited to help measure *in situ* stresses. The hydraulic fracturing technique takes advantage of this stress concentration and, under ideal circumstances, enables stress magnitude and orientation measurements to be made to about 3 km depth.

The most common method of determining stress orientation from observations in wells and boreholes is stress-induced wellbore breakouts. Breakouts are related to a natural compressive failure process that occurs when the maximum hoop stress around the hole is large enough to exceed the strength of the rock. This causes the rock around a portion of the wellbore to fail in compression. For the simple case of a vertical well drilled when S_v is a principal stress, this leads to the occurrence of stress-induced borehole breakouts that form at the azimuth of the minimum horizontal compressive stress. Breakouts are an important source of crustal stress information because they are ubiquitous in oil and gas wells drilled around the world and because they also permit stress orientations to be determined over a great range of depth in an individual well. Detailed studies have shown that these orientations are quite uniform with depth and are independent of lithology and age.

Another form of naturally occurring wellbore failure is drilling-induced tensile fractures. These fractures form in the wall of the borehole at the azimuth of the maximum horizontal compressive stress when the circumferential stress acting around the well locally goes into tension; they are not seen in core from the same depth.

III. DISTRIBUTION OF CRUSTAL STRESSES

Figure 3 shows maximum horizontal stress orientations for North America taken from the WSM database. The legend identifies the different types of stress indicators; because of the density of data, only highest quality data are plotted (A and B quality, shown by lines of different length). The tectonic regime (i.e., normal faulting, strike-slip faulting, or reverse faulting), where known, is given by color, as explained in the legend. The data principally come from wellbore breakouts, earthquake focal mechanisms, *in situ* stress measurements greater than 100 m depth, and young (<2 Ma old) geologic indicators. These data, originally presented and described by Zoback and Zoback (1989, 1991), demonstrate that large regions of the North American continent (most of the region east of the Rocky Mountains) are characterized by relatively uniform horizontal stress orientations. Furthermore, where different types of stress orientation data are available, see, for example, the eastern United States, the correlation between the different types of stress indicators is quite good.

Two straightforward observations about crustal stress can be made by comparison of these different types of stress indicators. First, no major changes in the orientation of the crustal stress field occur between the upper 2–5 km, where essentially all of the wellbore breakout and stress measurement data come from, and 5–20 km, where the majority of crustal earthquakes occur. Second, the criterion used to define reliable stress indicators appear to be approximately correct. In other words, data badly contaminated by nontectonic (near surface) sources of stress appear to have been effectively eliminated from the compilations.

IV. FIRST-ORDER GLOBAL STRESS PATTERNS

Figure 4 shows global maximum horizontal compressive stress orientations based on the 1997 WSM database. As with Fig. 3, only data qualities A and B are shown, and the symbols utilized in Fig. 4 are the same as those in Fig. 3. While global coverage is quite variable, the relative uniformity of stress orientation and the relative magnitudes in different parts of the world are striking and permit mapping of regionally coherent stress fields. Figure 5 presents a generalized version of the World Stress Map

that is quite similar to that presented by Zoback (1992), showing, using large arrows, mean stress directions and stress regime based on averages of clusters of data shown in Fig. 4. Tectonic stress regimes are indicated in Fig. 5 by color and arrow type. Blue inward pointing arrows indicate S_{Hmax} orientations in areas of compressional (strike-slip and thrust) stress regimes. Red outward pointing arrows give S_{hmin} orientations (extensions direction) in areas of normal faulting regimes. Regions dominated by strike-slip tectonics are distinguished with green, thick inward pointing and orthogonal, thin outward pointing arrows. Overall, arrow sizes on Fig. 5 represent a subjective assessment of "quality" related to the degree of uniformity of stress orientation and also to the number of density of data.

A number of first-order patterns can be observed in Figs. 4 and 5.

1. In many regions a uniform stress field exists throughout the upper brittle crust, as indicated by consistent orientations from the different techniques that sample very different rock volumes and depth ranges.
2. Intraplate regions are dominated by compression (thrust and strike-slip stress regimes) in which the maximum principal stress is horizontal.
3. Active extensional tectonism (normal faulting stress regimes) in which the maximum principal stress is vertical generally occurs in topographically high areas in both the continents and the oceans.
4. Regional consistency of both stress orientations and relative magnitudes permits the definition of broad-scale regional stress provinces, many of which coincide with physiographic provinces, particularly in tectonically active regions. These provinces may have lateral dimensions on the order of 10^3–10^4 km, many times the typical lithosphere thickness of 100–300 km. These broad regions of the earth's crust subjected to uniform stress orientation or a uniform pattern of stress orientation (such as the radial pattern of stress orientations in China) are referred to as "first-order" stress provinces (Zoback, 1992).

V. SOURCES OF CRUSTAL STRESS

As alluded to above, stresses in the lithosphere are of both tectonic and nontectonic, or local, origin. The regional uniformity of the stress fields observed in Figs. 3 and 4 argue for tectonic origins of stress at depth for most intraplate regions around the world. For many years, numerous workers suggested that residual stresses from

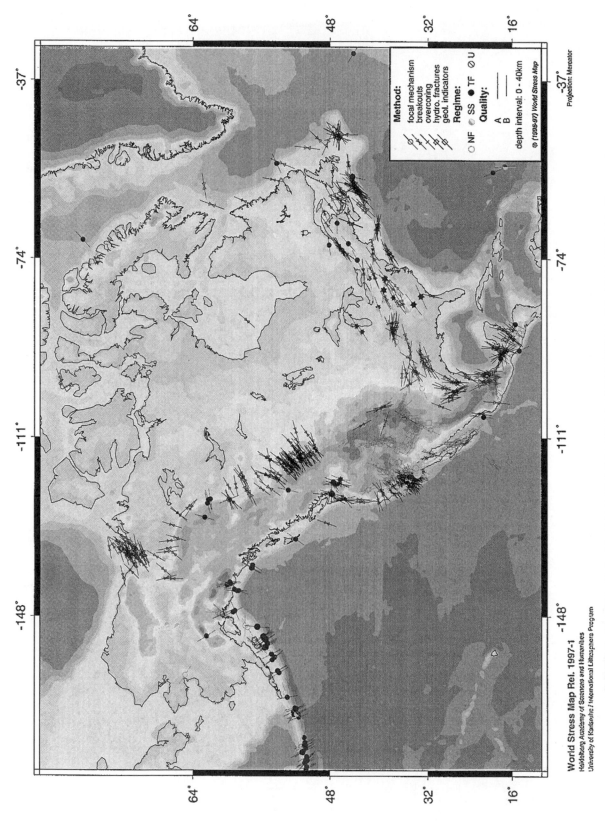

FIGURE 3 Directions of maximum horizontal stress from the WSM database superimposed on global topography and bathymetry. Only A and B quality data are shown. Data points characteristic of normal faulting are shown in red, strike-slip areas are shown in green, reverse faulting areas are shown in blue, and unknown stress indicators are shown in black.

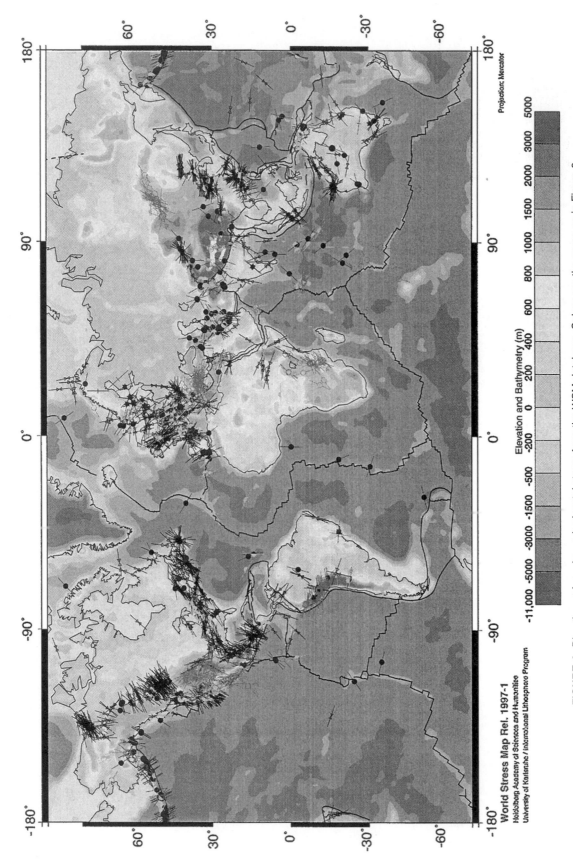

FIGURE 4 Directions of maximum horizontal stress from the WSM database. Colors are the same as in Figure 3. Only A and B quality data are shown.

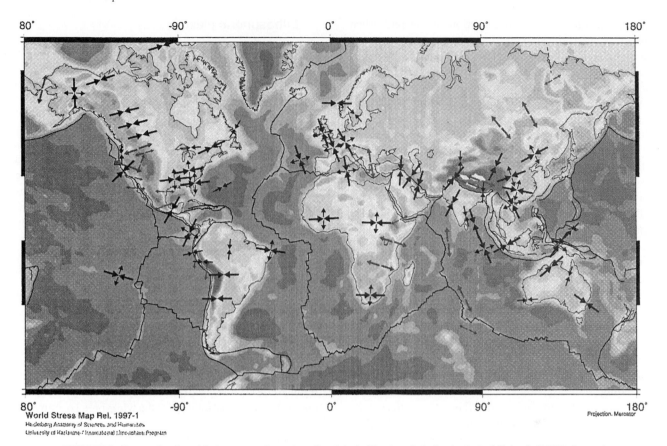

FIGURE 5 Generalized world stress map based on the data in Fig. 4 and similar to that of Zoback (1992). Inward pointed arrows indicate high compression as in reverse faulting regions. Paired inward and outward arrows indicate strike-slip faulting. Outward directed red arrows indicate areas of extension. Note that the plates are generally in compression and that areas of extension are limited to thermally uplifted areas.

past tectonic events may play an important role in defining the tectonic stress field. We have found no evidence for significant residual stresses at depth. If such stresses exist, they seem to be only important in the upper few meters or tens of meters of the crust where tectonic stresses are very small.

Similarly, no evidence has been found that indicates that horizontal principal stresses result simply from the weight of the overlying rock. This oversimplified theory is based on the "bilateral constraint," with the supposition that as a unit cube is stressed due to imposition of a vertical stress it cannot expand horizontally (because a neighboring unit cube would be attempting to expand in the opposite direction). If the bilateral constraint was applicable to the crust, the two horizontal principal stresses at depth would be equal and would always be less than the vertical stress . As demonstrated above, the dominance of compressional intraplate stress fields indicates that one or both horizontal stresses exceed the vertical stress. The broad regions of well-defined S_{hmax} orientations are clear evidence of horizontal stress anisotropy. Thus, the predicted bilateral

stress state is generally not found in the crust. In fact, the assumptions leading to the prediction of such a stress state are unjustified, as the analysis assumes that an elastic crust exists in the absence of gravity (or any other forces) before gravity is instantaneously "switched" on.

In the sections below, the primary sources of tectonic stress are briefly discussed. Although it is possible to theoretically derive the significance of any particular source of stress, because the observed tectonic stress state is the result of superposition of a variety of forces acting on and within the lithosphere, it is difficult to define the relative importance of any one stress source. This can only be resolved by utilizing careful modeling and well-constrained observations.

A. Plate-Driving Stresses

Sources of the broad-scale regions of uniform crustal stress that immediately come to mind are the same forces that drive (and resist) plate motions (e.g., Forsyth and Uyeda, 1975). A ridge push compressional force is associated with the excess elevation of the mid-ocean ridges, whereas the slab pull force results from the negative buoyancy of downgoing slabs. Both of these sources contribute to plate motion and tend to act in the direction of plate motion. If there is flow in the upper asthenosphere, a positive drag force could be exerted on the lithosphere that would tend to drive plate motion, whereas cold, thick lithospheric roots (such as beneath cratons) may be subject to a resistive drag forces that would act to inhibit plate motion. In either case, the drag force would result in stresses being transferred up into the lithosphere from its base. There are also collisional resistance forces resulting either from the frictional resistance of a plate to subduction or from the collision of two continental plates. As oceanic plates subduct into the viscous lower mantle, additional slab resistive forces add to the collision resistance forces acting at shallow depth. Another force resisting plate motion is that due to transform faults, although, as discussed below, the amount of transform resistance may be negligible. Finally, it has been proposed that a suction force may act on the overriding lithosphere in a subduction zone. This force may tend to "suck" the overriding lithosphere toward the trench and result in back-arc spreading.

While it is possible to specify the various stresses associated with plate movement, their relative and absolute importance in plate movement is not understood. While many researchers believe that either the ridge push or slab pull force is most important in causing plate motion, it is not clear that these forces are easily separable or that plate motion can be ascribed to a single dominating force.

B. Topography and Buoyancy Forces

Numerous workers have demonstrated that topography and its compensation at depth can generate sizable stresses capable of influencing the tectonic stress state and style. Density anomalies within or just beneath the lithosphere constitute major sources of stress. The integral of anomalous density times depth characterizes the ability of density anomalies to influence the stress field and to induce deformation. In general, crustal thickening or lithospheric thinning (negative density anomalies) produces extensional stresses, while crustal thinning or lithospheric thickening (positive density anomalies) produces compressional stresses. In more complex cases, the resultant state of stress in a region depends on the density moment integrated over the entire lithosphere. In a collisional orogeny, for example, where both the crust and the mantle lid are thickened, the presence of the cold lithospheric root can overcome the extensional forces related to crustal thickening and maintain compression.

C. Lithospheric Flexure

Loads on or within an elastic lithosphere cause deflection and induce flexural stresses which can be quite large (several hundred megaPascals) and can perturb the regional stress field with wavelengths as much as 1000 km (depending on the lateral extent of the load). Some potential sources of flexural stress influencing the regional stress field include sediment loading, particularly along continental margins; glacial rebound; seamount loading; and the upwarping of oceanic lithosphere oceanward of the trench, the "outer arc bulge." Sediment loads as thick as 10 km represent a potential significant stress on continental lithosphere. Zoback (1992) suggested that a roughly 40° counterclockwise rotation of horizontal stresses on the continental shelf offshore of eastern Canada was due to superposition of a margin-normal extensional stress derived from sediment load flexure.

VI. THE CRITICALLY STRESSED CRUST

Three independent lines of evidence indicate that intraplate continental crust is generally in a state of incipient, but slow, frictional faulting: (1) the widespread occurrence of seismicity induced by either reservoir impoundment or fluid injection (Zoback and Harjes, 1997), (2) earthquakes triggered by small stress changes associated with other earthquakes, and (3) *in situ* stress measurements in deep wells and boreholes. The *in situ* stress measurements further demonstrate that the stress magnitudes derived from Coulomb failure theory utilizing laboratory-derived frictional coefficients of 0.6–1.0 predict stresses that are consistent with measured stress magnitudes (Townend and Zoback, 2000). This is well illustrated in Fig. 6 by the stress data collected in the KTB borehole in Germany to ~8 km depth. Measured stresses are quite high and consistent with the frictional faulting theory [Eq. (8)] with a frictional coefficient of ~0.7. Further demonstration of this "frictional failure" stress state was the fact that a series of earthquakes could be triggered at ~9 km depth in rock surrounding the KTB borehole by extremely low perturbations of the ambient, approximately hydrostatic, pore pressure (Zoback and Harjes, 1997).

That the state of stress in the crust is generally in a state of incipient frictional failure might seem surprising, especially for relatively stable intraplate areas. However, a reason for this can be easily visualized in terms of a simple cartoon as shown in Fig. 7. The lithosphere as a whole (shown simply in Fig. 7 as three distinct layers: the brittle upper crust, the ductile lower crust, and the ductile uppermost mantle) must support plate-driving forces. The figure indicates a power-law creep law typically used to

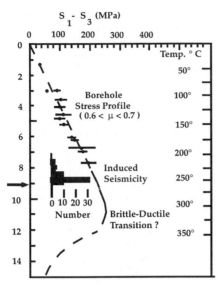

FIGURE 6 Stress measurements in the KTB scientific research well indicate a "strong" crust, in a state of failure equilibrium as predicted by Coulomb theory and laboratory-derived coefficients of friction of 0.6–0.7. [Modified after Zoback, M. D., and Harjes, H. P. (1997). *J. Geophys. Res.* **102**, 18,477–18,491.]

characterize the ductile deformation of the lower crust and upper mantle. Because the applied force to the lithosphere will result in steady-state creep in the lower crust and upper mantle, as long as the "three-layer" lithosphere is coupled, stress will build up in the upper brittle layer due to the creep deformation in the layers below. Stress in the upper crust builds over time, eventually to the point of failure. The fact that intraplate earthquakes are relatively infrequent simply means that the ductile strain rate is low in the lower crust and upper mantle. Zoback and Townend (2001) discuss the fact that at the relatively low strain rates characterizing intraplate regions, sufficient plate-driving force is available to maintain a "strong" brittle crust in a state of frictional failure equilibrium.

While appreciable evidence suggests that the Coulomb criterion and laboratory-derived coefficients of friction are applicable to plate interiors, major plate boundary faults such as the San Andreas fault (and other plate-bounding faults) appear to slip at very low levels of shear stress. Appreciable heat flow data collected in the vicinity of the San Andreas show no evidence of frictionally generated heat (Lachenbruch and Sass, 1980). This appears to limit average shear stresses acting on the fault to depths of ~15 km to about 20 MPa, approximately a factor of 5 below the stress levels predicted by the Coulomb criterion assuming that hydrostatic pore pressure and laboratory-derived friction coefficients are applicable to the fault at depth. Zoback *et al.* (1987) and Townend and Zoback (2001) present evidence that the direction of maximum horizontal stress in the crust adjacent to the San Andreas is at an extremely high angle to fault plane. Like the heat flow data, the stress orientation data imply low resolved shear stresses on the fault at depth. Thus, it appears that the frictional strength of plate boundary faults is distinctly lower than intraplate faults, which thus enables them to accommodate hundreds of kilometers of relative fault offset and correspondingly high strains.

VII. SUMMARY

Large portions of intraplate regions (lateral dimensions of 10^3–10^4 km) are characterized by relatively uniform stress fields, suggesting that the state of stress in the brittle upper crust is dominated by large-scale tectonic processes. In general, principal stresses in the crust are in vertical and horizontal planes. In intraplate areas, *in situ* stress measurements and inferences based on topography and flexure all suggest that shear stresses in the upper lithosphere is fairly large and seem to be controlled by the frictional strength of the faulted crust. Unlike the interiors of plates, plate boundary faults (like the San Andreas

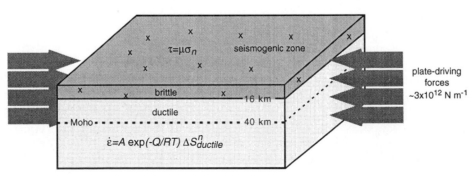

FIGURE 7 Schematic cartoon illustrating how the forces acting on the lithosphere keep the brittle crust in frictional equilibrium through creep in the lower crust and upper mantle. [Modified after Zoback, M. D., and Townend, J. (in press). *Tectonophysics.*]

fault and subduction zones) appear to be quite weak and slip at low levels of shear stress. At sufficient depth below the brittle upper lithosphere, stresses are likely controlled by the ductile flow properties of the constituent rocks and minerals. Plate-driving forces are sufficient in magnitude to maintain the intraplate lithosphere in a state of frictional failure. This frictional failure is manifest as slow, steady-state creep deformation in the lower crust and upper mantle and brittle deformation in the upper crust.

ACKNOWLEDGMENT

We thank Peter Bird for his comments on an early draft of this article.

SEE ALSO THE FOLLOWING ARTICLES

CONTINENTAL CRUST • EARTHQUAKE MECHANICS AND PLATE TECTONICS • EARTHQUAKE ENGINEERING • EARTHQUAKE PREDICTION • EARTH'S CORE • EARTH'S MANTLE • GEOLOGY, EARTHQUAKE • OCEANIC CRUST • PLATE TECTONICS • SEISMOLOGY, ENGINEERING • SEISMOLOGY, OBSERVATIONAL • SEISMOLOGY, THEORETICAL

BIBLIOGRAPHY

Anderson, E. M. (1951). "The Dynamics of Faulting and Dyke Formation with Applications to Britain," Oliver & Boyd, Edinburgh.

Byerlee, J. D. (1978). "Fusion of rock," *Pure and Applied Geophysics*, **116,** 615–626.

Forsyth, D., and Uyeda, S. (1975). "On the relative importance of the driving forces of plate motion," *Geophys. J. R. Astr. Soc.* **43,** 163–200.

Jaeger, J. C., and Cook, N. G. W. (1971). "Fundamentals of Rock Mechanics," 515 pp., Chapman and Hall, London.

Jones, C. H., Unruh, J. R., and Sonder, L. J. (1996). "The role of gravitational potential energy in active deformation in the southwestern United States," *Nature* **381,** 37–41.

Lachenbruch, A. H., and Sass, J. H. (1980). "Heat flow and energetics of the San Andreas fault zone," *J. Geophys. Res.* **85,** 6185–6223.

McKenzie, D. P. (1969). "The relationship between fault plane solutions for earthquakes and the principal stresses," *Sels. Soc. Amer. Bull.* **59,** 591–601.

Mueller, B., Wehrle, V., and Fuchs, K. (1997). "The 1997 Release of the World Stress Map," http://www-wsm.physik.uni-karlsruhe.de/.

Townend, J., and Zoback, M. D. (2000). "How faulting keeps the crust strong," *Geology* **28,** 399–402.

Zoback, M. D., and Townend, J. (2001). "Implications of hydrostatic pore pressures and high crustal strength for the deformation of interplate lithosphere," *Tectonphysics* **336,** 19–30.

Zoback, M. D., and Harjes, H. P. (1997). "Injection induced earthquakes and crustal stress at 9 km depth at the KTB deep drilling site, Germany," *J. Geophys. Res.* **102,** 18,477–18,491.

Zoback, M. D., and Zoback, M. L. (1991). Tectonic stress field of North America and relative plate motions. *In* "The Geology of North America: Neotectonics of North America" (D.B.Spemmons *et al*, ed.), pp. 339–366, Geol. Soc. Am., Boulder, CO.

Zoback, M. D., *et al.* (1987). "New evidence on the state of stress of the San Andreas fault system," *Science* **238,** 1105–1111.

Zoback, M. L., *et al.* (1989). "Global patterns of intraplate stresses; a status report on the world stress map project of the International Lithosphere Program," *Nature* **341,** 291– 298.

Zoback, M. L. (1992). "First and second order patterns of tectonic stress: The World Stress Map Project," *J. of Geophys. Res.* **97,** 11,703–11,728.

Zoback, M. L., and Zoback, M. D. (1989). "Tectonic stress field of the conterminous United States," *Geol. Soc. Am. Memoir* **172,** 523–539.

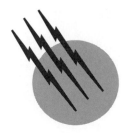

Structural Analysis, Aerospace

David H. Allen
Texas A&M University

I. Historical Introduction
II. The Role of Structural Analysis
III. Mechanics of Aerospace Structures
IV. Analysis of Advanced Beams
V. Advanced Analysis Techniques

GLOSSARY

Advanced beam Long, slender structural member, possibly of nonhomogeneous material makeup, subjected to multiaxial bending, extension, and/or thermal loads.

Aerospace structure Structure whose usefulness diminishes significantly with increasing weight.

Composite Structural component composed of two or more materials bonded together in some fabrication process, thus resulting in a composite material with a high strength-to-weight ratio.

Constitution That set of constraints that distinguish a body from any other body of identical geometric shape; also called material properties.

Continuum mechanics Approach to the analysis of media that ignores the microstructural features such as molecular bonds and seeks to model the bodies as continuous.

Elastic Constitutive term meaning that the stress and strain are uniquely related at a material point.

Finite element method Most powerful structural analysis technique currently in use; obtained by discretizing the actual structure into subcomponents (called finite elements) for purposes of analysis.

Statically determinate Term used when the internal loads can be determined at all points in a body by using Newton's laws of motion and without recourse to the material makeup of the body.

Statically indeterminate Converse of statically determinate.

Structural analysis Step in the structural design process whereby predictions of the temperature, deformation, and internal loads of some structure of predetermined shape are obtained.

Structural design Process whereby a structure is iteratively modified until all design constraints are satisfied.

Variational methods Class of mathematical techniques that can be utilized for performing structural analysis.

AEROSPACE structures include atmospheric flight vehicles, such as shown in Fig. 1, orbital vehicles, such as shown in Fig. 2, and interplanetary probes, such as Explorer I. Analysis of this class of structures consists of the theoretical prediction of deformations, stresses, and temperatures throughout the vehicle, given the external thermomechanical loading conditions and the geometric shape of the structure. The object of this analysis is to

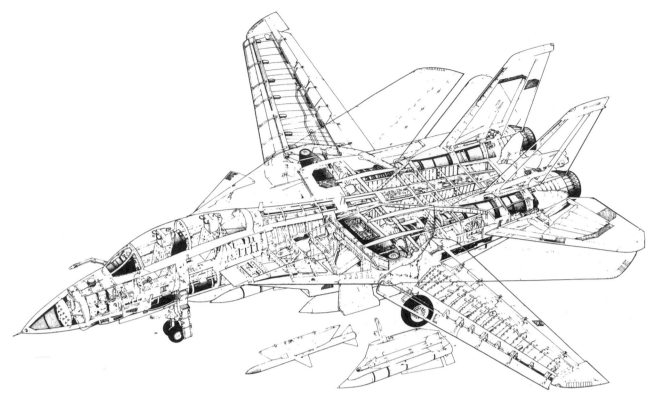

FIGURE 1 Grumman F-14. [Courtesy of the U.S. Navy.]

predict analytically such catastrophic events as excessive deformations, buckling, and fracture, which can lead to structural failure. Because it is extremely important to minimize mass in these structures, relatively small margins of safety are allowed between analytical predictions and the criteria for structural failure. Therefore, it is necessary that the theoretical tools used in the analysis result in accurate predictions. This requirement often necessitates the development of highly complex algorithms requiring numerous well-trained structural analysts and a large capital outlay compared to the actual fabrication cost of the structure.

I. HISTORICAL INTRODUCTION

The first manned flight of a motorized vehicle by the Wright Brothers at Kitty Hawk, North Carolina, on December 17, 1903, is probably the most technologically significant achievement in history. It marked the transition from an age of travel on foot, and by horse and wagon, to Neil Armstrong's footprints on the Moon little more than 65 years later. However, the concepts embodied in the design and analysis of flight structures date back hundreds and in some cases even thousands of years.

The analysis of aerospace structures is rooted in the fundamental axioms of continuum mechanics. The first

exposition of one of these axioms was published in 1676 by Robert Hooke (1635–1703). He stated that the power of any spring is in the same proportion with the tension thereof, thus stating the first constitutive equation for linear elastic bodies.

A short time later, Sir Isaac Newton (1642–1727), probably the greatest scientist of all time, reported his axioms governing conservation of momentum. These were expostulated in his text "Philosophiae Naturalis Principia Mathematica," published in 1687. The elastic field problem was finally reported by Navier (1785–1836) in 1821 and by Cauchy (1789–1857) in 1822. At the same time that Navier and Cauchy were investigating the elastic field problem, Fourier (1768–1830) was constructing the equations modeling the temperature field in a body of general shape. His results on heat conduction were reported in his classical treatise, "La Theorie Analytique de la Chaleur," in 1822. The remainder of the nineteenth century was marked by numerous achievements in elasticity theory by such great scientists as Poisson (1781–1840), Navier (1785–1836), Green (1793–1841), Clapeyron (1794–1864), Saint-Venant (1797–1886), Duhamel (1797–1872), Airy (1801–1892), Helmholtz (1821–1894), Kelvin (1824–1907), Maxwell (1831–1879), Mohr (1835–1918), and Rayleigh (1842–1919).

The twentieth century is marked by numerous milestones in aerospace technology such as the first liquid-fuel

FIGURE 2 Artist's concept of orbiting space station. [Courtesy of the National Aeronautics and Space Administration.]

rocket flight (1926), the first jet engine-powered flight (1939), the first supersonic flight (1947), the first artificial satellite launch (1957), the first manned flight (1961), and the first space shuttle flight (1981). All of these achievements posed many engineering problems and challenges, but for the structural analyst the foremost of these was to achicve minimum weight with maximum safety. Many of the advances have come with the development of advanced materials such as fibrous composites, which have very high strength-to-weight ratios.

Improvements in analytical techniques were basically driven by the advent of the high-speed digital computer, thus making the stiffness method, which has now evolved into the finite element method, more computationally efficient and usable in everyday structural analysis.

II. THE ROLE OF STRUCTURAL ANALYSIS

Structural design is the process whereby the structural engineer begins with little more than a set of loads and design constraints and proceeds iteratively to obtain a structural configuration that satisfies all of these constraints. As shown in Fig. 3, structural analysis is one step in the structural design process. Structural analysts are usually given a well-defined structural geometry. It is their task then to determine whether the current configuration satisfies the

design constraints. These design constraints may include such things as fracture criteria, excessive deformations, size and weight constraints, and even cost constraints. If the analysis results in predicted failure of the structure due to any of the design constraints, then the design geometry must be modified. This process is then carried out iteratively until all of the design constraints are satisfied.

As indicated in Fig. 3, it is possible to check the design experimentally rather than analytically. However, for

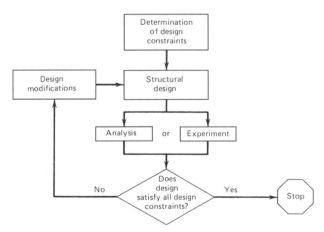

FIGURE 3 The structural design process. [From Allen, D. H., and Haisler, W. E. (1985). "Introduction to Aerospace Structural Analysis," Wiley, New York.]

aerospace structures this is often too costly. Therefore, analysis is often a necessary step in the structural design procedure.

Because one can never completely nullify all error sources, it is customary to design structures to withstand some ultimate loading configuration more stringent than the actually expected allowable load. Thus, a factor of safety SF is defined by

$$SF = \text{ultimate failure load/allowable load}, \qquad (1)$$

where the ultimate failure load is that load that will cause the structure to fail to satisfy any of the design constraints. Another term that is often used is the margin of safety MS, defined by

$$MS = SF - 1. \qquad (2)$$

The value of the safety factor used in a particular analysis will depend on many factors, such as the accuracy of the theory involved, the range of variability of material properties, and the cost of materials. In aerospace structures, where, because of mass production, engineering analysis may be budgeted at hundreds of times the cost of a single aircraft, relatively sophisticated analytic techniques are within cost constraints, and safety factors may be as low as from 1.25 to 1.5 for manned vehicles and from 1.1 to 1.25 for unmanned vehicles.

Although the basic design process described in Fig. 3 applies in the aerospace industry, the procedure is quite involved because of the complexity of the structures produced. There are usually several large groups of engineers, often numbering in the hundreds, who work in several separate groups. These include the design, aerodynamics, aeroelasticity, materials, and weights groups, as well as the structures group. The design process is generally interactive in nature and can be complicated by this interaction. For example, in a typical design process the design group might conceive of a horizontal tail unit for a fighter aircraft. This design would then be forwarded to the aerodynamics group to determine the external lifting and drag loads on the structure. Given these loads, the materials and structures group would then produce an internal structure capable of withstanding the loads. The weights group would then determine the weight of the structure as designed. If, for example, the weight is found to exceed the originally conceived weight designed for, it would then be necessary to modify the design. The decision as to how this modification should be made might involve one or all of the groups. The materials group might propose a lightweight composite material, but if this exceeds the cost constraint, even further design modification might be necessary. A complete design of the external geometry would certainly affect all groups and might lead to a new configuration that exceeds some other design constraint. Therefore, in complex aerospace structures there must be a great deal of interaction between the various engineering groups.

III. MECHANICS OF AEROSPACE STRUCTURES

Aerospace structures are analyzed by using field equations obtained from continuum mechanics and thermodynamics. These equations are algebraic or partial differential equations that apply at all points x_j and for all times t in a continuous body with interior V and surface S, as shown in Fig. 4. (Lowercase subscripts imply three components ranging from 1 to 3, thus representing Euclidean space.) The equations are in general (1) conservation of momentum (kinetics), (2) strain–displacement relations (kinematics), (3) stress–strain–temperature relations (constitution), and (4) conservation of energy (thermodynamics).

The difficulty of the analytic scheme will depend in large part on the level of complexity at which the preceding equations must be applied in order to obtain accurate theoretical models. For example, the shuttle may be considered to be a rigid body for purposes of calculating launch performance characteristics, thus simplifying this portion of the analysis. However, this assumption will give no insight about whether structural components within the shuttle can be expected to fail during launch.

For most aerospace vehicles it is sufficiently accurate to assume that the motions are infinitesimal and the constitution is linear elastic. These assumptions will in most cases produce a linear field problem when items 1 through 4 are cast in a mathematical context. Furthermore, it is usually acceptable to uncouple the conservation of energy from the remaining equations by assuming that the body of interest is rigid for purposes of calculating temperature. In the case where boundary conditions are radiative (as is the case in space), the heat conduction problem resulting from

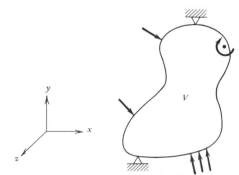

FIGURE 4 General body subjected to complex loading. [From Allen, D. H., and Haisler, W. E. (1985). "Introduction to Aerospace Structural Analysis," Wiley, New York.]

the conservation of energy will be the only nonlinearity in the analysis. Otherwise, for convective boundary conditions (as is usually the case in atmospheric conditions), the entire thermomechanical problem is linear and has a unique solution.

A. Field Variables

The quantities to be derived from the analysis are called field variables, and the analysis will result in a prediction of these quantities at all points x, y, z and for all times t of interest in V. The field variables of interest are stress, strain, displacement, and temperature. Once the analysis is complete these variables may be substituted into any constraint equations (such as failure criteria, mass restrictions, and deformation constraints) to ensure that the proposed structural configuration satisfies all design requirements.

1. Stress Tensor

Consider a point 0 in the interior of a body V. Newton's third law requires that the internal reactions at point 0 must be replaced with external force $\mathbf{F}$ if a cutting plane A is passed through point 0, as shown in Fig. 5. The concept of stress is now introduced. The normal stress on plane A and at point 0 is defined by

$$\sigma_n \equiv \lim_{\Delta A \to 0} (\Delta F_n / \Delta A) = dF_n / dA, \qquad (3)$$

where the symbol $\equiv$ means defined to be.

The shear stress is defined by

$$\sigma_s \equiv \lim_{\Delta A \to 0} (\Delta F_s / \Delta A) = dF_s / dA. \qquad (4)$$

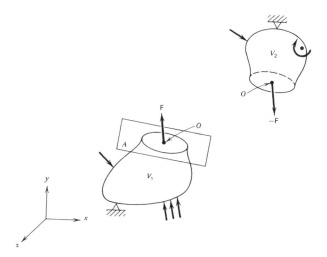

FIGURE 5 Free-body diagram showing internal forces. [From Allen, D. H., and Haisler, W. E. (1985). "Introduction to Aerospace Structural Analysis," Wiley, New York.]

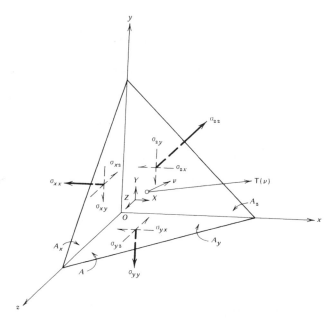

FIGURE 6 Description of the stress tensor at material point 0. [From Allen, D. H., and Haisler, W. E. (1985). "Introduction to Aerospace Structural Analysis," Wiley, New York.]

It can be shown that if three mutually perpendicular cutting planes are passed through point 0, the resulting nine components of stress, σ_{xx}, σ_{yy}, σ_{zz}, σ_{yz}, σ_{zy}, σ_{xz}, σ_{zx}, σ_{xy}, and σ_{yx}, as shown in Fig. 6 are sufficient to determine the components of the stress tensor on any other plane A at point 0. Therefore, there are nine unique components of stress to be determined at each point in the body V. The stress tensor represents a measure of the load intensity, or pressure, at each point in a body.

2. Strain Tensor

Consider again the point 0 in body V, as shown in Fig. 7. For infinitesimal deformations, the strain tensor is defined by

$$\varepsilon_{xx} \equiv \frac{\partial u}{\partial x}, \qquad \varepsilon_{yy} \equiv \frac{\partial v}{\partial y}, \qquad \varepsilon_{zz} \equiv \frac{\partial w}{\partial z},$$

$$\varepsilon_{yz} \equiv \frac{1}{2}\left(\frac{\partial v}{\partial z} + \frac{\partial w}{\partial y}\right), \qquad \varepsilon_{xz} \equiv \frac{1}{2}\left(\frac{\partial u}{\partial z} + \frac{\partial w}{\partial x}\right), \qquad (5)$$

$$\varepsilon_{xy} \equiv \frac{1}{2}\left(\frac{\partial u}{\partial y} + \frac{\partial v}{\partial x}\right),$$

where u, v, and w are components of the displacement vector. Thus, there are six unique components of strain at each point in the body V. The strain tensor represents a measure of the deformation intensity, or deformation per unit length, at each point in a body.

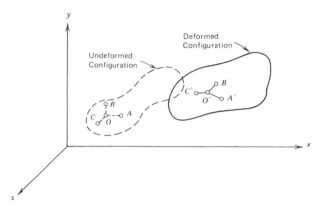

FIGURE 7 Three-dimensional body in undeformed and deformed configurations. [From Allen, D. H., and Haisler, W. E. (1985). "Introduction to Aerospace Structural Analysis," Wiley, New York.]

B. Conservation Laws

For nonablating bodies undergoing infinitesimal deformations, it can be shown that conservation of mass is identically satisfied. Therefore, the nontrivial conservation laws in most aerospace structural applications are conservation of momentum and conservation of energy.

1. Conservation of Momentum

Conservation of linear momentum will lead to the following equations at every point in V:

$$\frac{\partial \sigma_{xx}}{\partial x} + \frac{\partial \sigma_{yx}}{\partial y} + \frac{\partial \sigma_{zx}}{\partial z} + X = \rho \ddot{u},$$

$$\frac{\partial \sigma_{xy}}{\partial x} + \frac{\partial \sigma_{yy}}{\partial y} + \frac{\partial \sigma_{zy}}{\partial z} + Y = \rho \ddot{v}, \qquad (6)$$

$$\frac{\partial \sigma_{xz}}{\partial x} + \frac{\partial \sigma_{yz}}{\partial y} + \frac{\partial \sigma_{zz}}{\partial z} + Z = \rho \ddot{w},$$

where X, Y, and Z are the components of the body force vector per unit volume (such as gravitational loads) and ρ is the mass density. For quasi-static conditions, the inertial term on the right-hand side of (6) may be neglected.

Furthermore, conservation of angular momentum on the surface S will yield

$$T_x = \sigma_{xx} v_x + \sigma_{yx} v_y + \sigma_{zx} v_z,$$

$$T_y = \sigma_{xy} v_x + \sigma_{yy} v_y + \sigma_{zy} v_z, \qquad (7)$$

$$T_z = \sigma_{xz} v_x + \sigma_{yz} v_y + \sigma_{zz} v_z,$$

where T_x, T_y, and T_z are the components of the traction vector (force per unit area) and v_j the components of a unit outer normal at every point on S.

Finally, conservation of angular momentum will result in the following at every point in $V + S$:

$$\sigma_{yz} = \sigma_{zy}, \qquad \sigma_{xz} = \sigma_{zx}, \qquad \sigma_{xy} = \sigma_{yx}. \qquad (8)$$

(This equation assumes that body moments may be neglected.) Thus, there are only six unique components of the stress tensor, which is said to be symmetric.

2. Conservation of Energy

For most realistic applications involving elastic materials (sometimes excluding dynamic circumstances), conservation of energy will lead to the Fourier heat conduction equation:

$$\frac{\partial}{\partial x}\left(k\frac{\partial T}{\partial x}\right) + \frac{\partial}{\partial y}\left(k\frac{\partial T}{\partial y}\right) + \frac{\partial}{\partial z}\left(k\frac{\partial T}{\partial z}\right) = \rho C_v \dot{T} - \dot{r},$$
$$(9)$$

where k is the thermal conductivity, C_v the specific heat at constant volume, and r the internal heat source per unit volume.

C. Constitution

The description of the field problem is completed with the specification of constitutive equations. These equations reflect the material properties of the various components utilized throughout the structure. In general, these are relations among stress, strain, and temperature that apply at each material point in the structure. For example, if a typical structural component is subjected to a prescribed uniaxial deformation history, as shown in Fig. 8, a cross-plot uniaxial stress–strain diagram will indicate that the relationship between stress and strain is highly nonlinear.

Further testing will reveal history dependence, strong temperature dependence, and rate dependence in most structural materials. However, as shown in Fig. 8, there is a linear relation between the stress and the strain for a large portion of the load-carrying capability of the component. This range is called the linear elastic range, and the stress level at which this nonlinearity becomes significant is called the yield point. As a margin of safety against structural failure, it is customary to design most aerospace components such that at every point in the structure the yield point is not exceeded.

Most aerospace structural components are constructed using metals such as aluminum, stainless steel, titanium, and Inconel. For all practical purposes these materials are isotropic in their elastic range; that is, their properties at each material point are independent of the coordinate axis direction.

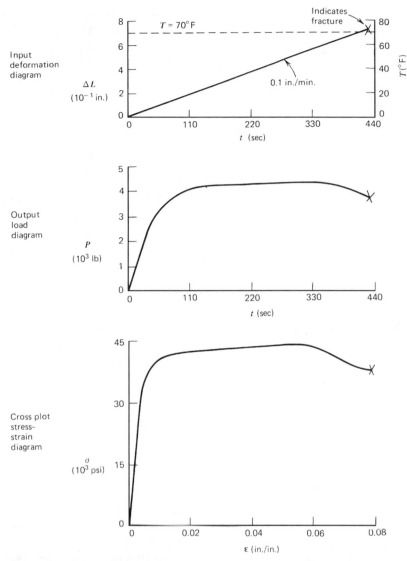

FIGURE 8 Uniaxial bar (Al 6061-T6) with cross-sectional area $A = 0.098$ in.^{2} subjected to monotonic deformation ΔL. (Results obtained on an Instron 1125 testing machine, Texas A&M University Mechanics and Materials Center.) [From Allen, D. H., and Haisler, W. E. (1985). "Introduction to Aerospace Structural Analysis," Wiley, New York.]

In the past two decades one of the most significant advances in aerospace structures has resulted from the development of composite materials. These materials are generally made of two or more materials, with one embedded in the other during some complex forming process. A wide variety of materials is currently under development. The embedding material may be continuous fibers, chopped fibers, or particulates. When fibers are used, they are most often made of graphite, boron, or fiberglass. Where particulates or chopped fibers are used, they may be boron, silicon carbide, or graphite. These materials are embedded in a matrix that is usually a polymer material such as epoxy. Metal matrix composites have been developed by utilizing materials such as aluminum for the matrix. Although composite materials are very costly to fabricate compared to pure metals, their improved strength-to-weight ratios can make them cost effective when weight is a critical factor.

Materials of this type are being utilized in many critical areas on advanced tactical fighter aircraft, the shuttle and its companion launch vehicle, and even bicycle frames. Because of the orientation of the second-phase particles or fibers, these materials are inherently anisotropic.

1. Isotropic Elastic Materials

For isotropic elastic materials it can be shown by using symmetry constraints that

$$
\begin{bmatrix} \sigma_{xx} \\ \sigma_{yy} \\ \sigma_{zz} \\ \sigma_{yz} \\ \sigma_{xz} \\ \sigma_{xy} \end{bmatrix} = \frac{E}{(1+\nu)(1-2\nu)} \begin{bmatrix} 1-\nu & \nu & \nu & 0 & 0 & 0 \\ \nu & 1-\nu & \nu & 0 & 0 & 0 \\ \nu & \nu & 1-\nu & 0 & 0 & 0 \\ 0 & 0 & 0 & \frac{1-2\nu}{2} & 0 & 0 \\ 0 & 0 & 0 & 0 & \frac{1-2\nu}{2} & 0 \\ 0 & 0 & 0 & 0 & 0 & \frac{1-2\nu}{2} \end{bmatrix} \begin{bmatrix} \varepsilon_{xx} \\ \varepsilon_{yy} \\ \varepsilon_{zz} \\ \varepsilon_{yz} \\ \varepsilon_{xz} \\ \varepsilon_{xy} \end{bmatrix} - \frac{E\alpha(T-T_0)}{(1-2\nu)} \begin{bmatrix} 1 \\ 1 \\ 1 \\ 0 \\ 0 \\ 0 \end{bmatrix}, \quad (10)
$$

where E, ν, and α are materials constants obtainable directly from two laboratory tests and T_0 is the reference temperature at which no stress is observed at zero strain.

2. Orthotropic Elastic Materials

Composite materials generally exhibit three mutually perpendicular planes of material symmetry. These materials are therefore called orthotropic. For orthotropic materials it can be shown by using symmetry constraints that

$$
\begin{bmatrix} \sigma_{11} \\ \sigma_{22} \\ \sigma_{33} \\ \sigma_{23} \\ \sigma_{13} \\ \sigma_{12} \end{bmatrix} = \begin{bmatrix} D_{11} & D_{12} & D_{13} & 0 & 0 & 0 \\ D_{12} & D_{22} & D_{23} & 0 & 0 & 0 \\ D_{13} & D_{23} & D_{33} & 0 & 0 & 0 \\ 0 & 0 & 0 & D_{44} & 0 & 0 \\ 0 & 0 & 0 & 0 & D_{55} & 0 \\ 0 & 0 & 0 & 0 & 0 & D_{66} \end{bmatrix}
$$

$$
\times \begin{bmatrix} \varepsilon_{11} \\ \varepsilon_{22} \\ \varepsilon_{33} \\ \varepsilon_{23} \\ \varepsilon_{13} \\ \varepsilon_{12} \end{bmatrix} + \begin{bmatrix} \beta_{11}(T-T_0) \\ \beta_{22}(T-T_0) \\ \beta_{33}(T-T_0) \\ 0 \\ 0 \\ 0 \end{bmatrix}, \quad (11)
$$

where D_{ij} and β_{ij} are material constants that may be obtained from laboratory tests.

D. Field Problem Description

The field problem describing the analysis of aerospace structures can now be defined. There are 16 field variables to be determined at all points in the body of interest: σ_{xx}, σ_{yy}, σ_{zz}, σ_{yz}, σ_{xz}, σ_{xy}, ε_{xx}, ε_{yy}, ε_{zz}, ε_{yz}, ε_{xz}, ε_{xy}, u, v, w, and t. The governing field equations are six strain–displacement equations [Eqs. (5)], three conservation of momentum equations [Eqs. (6)], one conservation of energy equation [Eq. (9)], and six constitutive equations [Eqs. (10) for isotropic materials or Eqs. (11) for orthotropic materials]. Thus, there is a total of 16 equations in 16 unknowns. With the imposition of physically admissable boundary conditions on the surface S, the problem is completely specified. The difficulty in obtaining solutions to problems of this type arises due to the fact that aerospace structures typically have extremely complex geometric shapes (boundaries).

The standard procedure for solving this problem is first to obtain the temperature field $T = T(x, y, z, t)$ by using the thermodynamic equation (9), since it is uncoupled from the remaining mechanical equations. The resulting temperature field may be used as input to solve the remaining 15 equations in 15 unknowns. For isothermal conditions the necessity to solve the heat transfer problem is obviated.

IV. ANALYSIS OF ADVANCED BEAMS

The geometric shape and/or loading conditions in the vast majority of structures are so complex that it is not possible to obtain exact solutions to the thermoelastic field problem. Therefore, a field of approximate techniques has been developed that is called strength of materials or structural mechanics. In these approximate methods it is generally assumed that the structural response is constrained by simplifying kinematic assumptions that are at least approximately correct. The resulting simplifications allow for the construction of approximate solutions for a wide variety of aerospace structures. One such simplified structure is the advanced beam. As shown in Fig. 9, the beam may be subjected to complex load conditions. In addition, the beam may be of heterogeneous material makeup, such as a composite or bonded component. Furthermore, the beam cross section normal to the longitudinal or x coordinate may be variable.

All of these complexities lead to very intricate beams that are nevertheless found in large space structures and in aircraft fuselages and empennages.

A. Equilibrium of Advanced Beams

Consider again Fig. 9. Suppose that a cutting plane is passed through the cross section, as shown, so that the

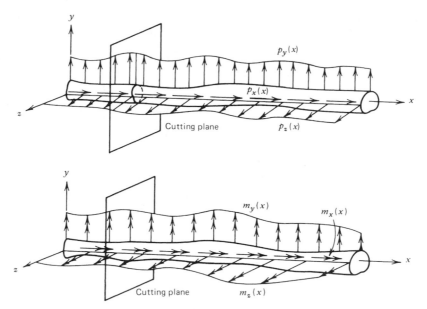

FIGURE 9 Advanced beam with externally applied forces and moments per unit length. [From Allen, D. H., and Haisler, W. E. (1985). "Introduction to Aerospace Structural Analysis," Wiley, New York.]

resulting free-body diagram is as shown in Fig. 10. We now define the following six resultants on any cross section:

$$P = P(x) \equiv \int_z \int_y \sigma_{xx} \, dy \, dz = \int_A \sigma_{xx} \, dA, \quad (12a)$$

$$V_y = V_y(x) \equiv \int_z \int_y \sigma_{xy} \, dy \, dz = \int_A \sigma_{xy} \, dA, \quad (12b)$$

$$V_z = V_z(x) \equiv \int_z \int_y \sigma_{xz} \, dy \, dz = \int_A \sigma_{xz} \, dA, \quad (12c)$$

$$M_x = M_x(x) \equiv \int_z \int_y \sigma_{xz}(y - \sigma_{xy}z) \, dy \, dz, \quad (12d)$$

$$M_y = M_y(x) \equiv \int_z \int_y \sigma_{xx}z \, dy \, dz = \int_A \sigma_{xx}z \, dA, \quad (12e)$$

$$M_z = M_z(x) \equiv -\int_z \int_y \sigma_{xx}y \, dy \, dz = -\int_A \sigma_{xx}y \, dA. \quad (12f)$$

It can be shown for quasi-static conditions ($\ddot{u} \approx 0$, $\ddot{v} \approx 0$, $\ddot{w} \approx 0$) by integrating the differential of equations of motion (6) over the cross section that

$$dP/dx = -p_x(x), \quad (13a)$$

$$dV_y/dx = -p_y(x), \quad (13b)$$

$$dV_z/dx = -p_z(x), \quad (13c)$$

$$dM_x/dx = -m_x(x), \quad (13d)$$

$$dM_y/dx = -m_y(x) + V_z(x), \quad (13e)$$

$$dM_z/dx = -m_z(x) - V_y(x), \quad (13f)$$

where the lowercase quantities on the right-hand side are input loads, as shown in Fig. 9. Thus, with appropriate boundary conditions, Eqs. (13) may be integrated to construct resultant load diagrams, which are statements of internal equilibrium at every point in the beam. By convention, $P = P(x)$ is called the axial load, $V_y = V_y(x)$ and $V_z = V_z(x)$ are called the shear loads, $M_x = M_x(x)$ is called the torque load, and $M_y = M_y(x)$ and $M_z = M_z(x)$ are called bending moments. The response of the beam to each of the resultants P, M_y, and M_z is detailed in the following sections. In many cases the response of advanced beams to the resultants V_y, V_z, and M_x is negligible. For that reason and in the interest of space, the reader is referred to the Bibliography for coverage of these resultants.

B. Bending and Extension of Advanced Beams

The subject of this section is the determination of the axial components of stress σ_{xx} and strain ε_{xx} due to the axial load P and bending moments M_y and M_z.

There are two important simplifying assumptions in the theory of advanced beams: (1) the transverse components of normal stress σ_{yy} and σ_{zz} are assumed to be negligible compared to the axial stress σ_{xx}, and (2) cross sections are assumed to remain planar and normal to the longitudinal axis of deformation, as shown in Fig. 11. The second assumption, called the Euler–Bernoulli assumption, may be represented mathematically by

$$u(x, y, z) = u_0 - \theta_z(x)y + \theta_y(x)z, \quad (14)$$

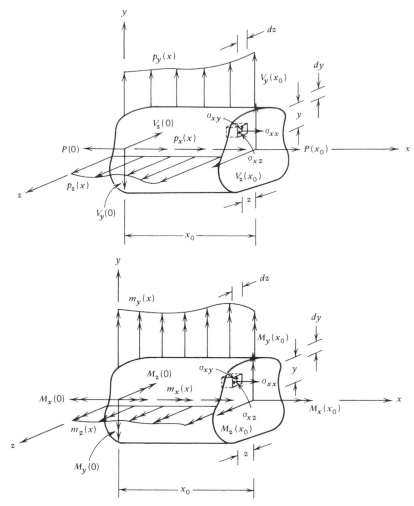

FIGURE 10 Free-body diagram of cut advanced beam. [From Allen, D. H., and Haisler, W. E. (1985). "Introduction to Aerospace Structural Analysis," Wiley, New York.]

where u_0 is the midplane axial deformation and θ_z and θ_y are components of rotation of the cross section, as shown in Fig. 11. Substitution of Eq. (14) into the strain displacement relation (5), and this result into the stress–strain equation (11), and, finally, this result into the resultant load equations (12a), (12e), and (12f) will give, upon rearrangement,

$$
\begin{aligned}
\sigma_{xx} = {} & \frac{E(P + P^T)}{E_1 A^*} \\
& - \frac{E}{E_1} \left[\frac{\left(M_z - M_z^T\right)I_{yy}^* + \left(M_y + M_y^T\right)I_{yz}^*}{\left(I_{yy}^* I_{zz}^* - I_{yz}^{*2}\right)} \right] y \\
& + \frac{1}{E_1} \left[\frac{\left(M_y + M_y^T\right)I_{zz}^* + \left(M_z - M_z^T\right)I_{yz}^*}{\left(I_{yy}^* I_{zz}^* - I_{yz}^{*2}\right)} \right] z,
\end{aligned}
$$

$$(15)$$

where the following quantities are called modulus-weighted section properties:

$$
A^* \equiv \int_A \frac{E}{E_1} \, dA, \tag{16a}
$$

$$
\bar{y}^* \equiv \frac{1}{A^*} \int_A \frac{E}{E_1} y \, dA, \tag{16b}
$$

$$
\bar{z}^* \equiv \frac{1}{A^*} \int_A \frac{E}{E_1} z \, dA, \tag{16c}
$$

$$
I_{yy}^* \equiv \int_A \frac{E}{E_1} z^2 \, dA, \tag{16d}
$$

$$
I_{yz}^* \equiv \int_A \frac{E}{E_1} yz \, dA, \tag{16e}
$$

$$
I_{zz}^* \equiv a \mid b \int_A \frac{E}{E_1} y^2 \, dA. \tag{16f}
$$

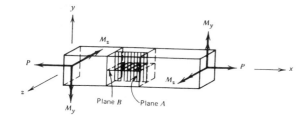

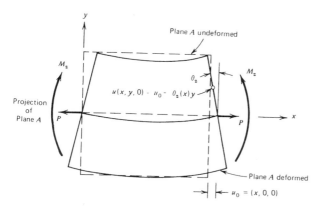

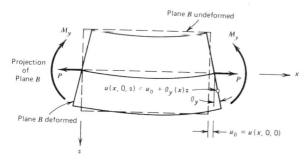

FIGURE 11 Kinematics of an advanced beam. [From Allen, D. H., and Haisler, W. E. (1985). "Introduction to Aerospace Structural Analysis," Wiley, New York.]

Here E_1 is an arbitrary constant and A the area of the cross section normal to the x axis. Furthermore,

$$P^T \equiv \int_A E\alpha(T - T_0)\,dA, \tag{17a}$$

$$M_y^T \equiv \int_A E\alpha(T - T_0)z\,dA, \tag{17b}$$

$$M_z^T \equiv \int_A E\alpha(T - T_0)y\,dA. \tag{17c}$$

Equation (15) will give the axial stress component σ_{xx} at any point in the beam in terms of the resultants P, M_y, and M_z, which may be determined by using Eqs. (13).

It is possible to determine the three components of displacement at all points in an advanced beam. However, due to the Euler–Bernoulli assumption [Eq. (14)], it is of practical importance to determine only the three components of displacement of the centroidal axis; that is, $u_0(x) \equiv u(x, 0, 0)$, $v_0(x) \equiv v(x, 0, 0)$, and $w_0(x) \equiv w(x, 0, 0)$.

Using the Euler–Bernoulli assumption, it can be shown that for small deformations

$$\frac{d\theta_z}{dx} = \frac{d^2 v_0}{dx^2}, \qquad \frac{d\theta_y}{dx} = \frac{-d^2 w_0}{dx^2}. \tag{18}$$

With the preceding equations it can be shown in the process of deriving Eq. (15) that

$$\frac{du_0}{dx} = \frac{P + P^T}{E_1 A^*}, \tag{19a}$$

$$\frac{d^2 v_0}{dx^2} = \frac{\left(M_z - M_z^T\right)I_{yy}^* + \left(M_y + M_y^T\right)I_{yz}^*}{E_1\left(I_{yy}^* I_{zz}^* - I_{yz}^{*2}\right)}, \tag{19b}$$

$$\frac{d^2 w_0}{dx^2} = \frac{-\left(M_y + M_y^T\right)I_{zz}^* - \left(M_z - M_z^T\right)I_{yz}^*}{E_1\left(I_{yy}^* I_{zz}^* - I_{yz}^{*2}\right)}. \tag{19c}$$

Since all of the quantities on the right-hand side of (19) can be determined from the external loads, geometry, material properties, and temperature field, they can be used as input to obtain the displacements u_0, v_0, and w_0.

C. Example

Given: Suppose that the beam shown in Fig. 12 is a composite structure with properties as shown below.

Portion (i)	E_i (psi)	α_i (per °F)
1	30×10^6	5×10^{-6}
2	10×10^6	6.5×10^{-6}
3	10×10^6	6.5×10^{-6}
4	10×10^6	6.5×10^{-6}

Required:

(a) Determine modulus-weighted section properties about the modulus-weighted centroid.
(b) Transform the y' axis to the modulus-weighted centroid and construct resultant load diagrams.
(c) Determine the axial components of the stress and strain at the following points:
   (1) $(x, y, z) = (200, 2, 0)$,
   (2) $(x, y, z) = (200, -2, 0)$,
   (3) $(x, y, z) = (200, 0, 11.12)$, and
   (4) $(x, y, z) = (200, 0, -5.88)$,
   where the coordinate locations are measured with respect to the modulus-weighted axes.
(d) Rework part (c) assuming that, in addition to the mechanical loads, the beam is subjected to a thermal load $T - T_0 = 0.10x$ (°F), where x is in inches.
(e) Obtain the following displacements for the case where the blade is subjected to both thermal and mechanical loads:

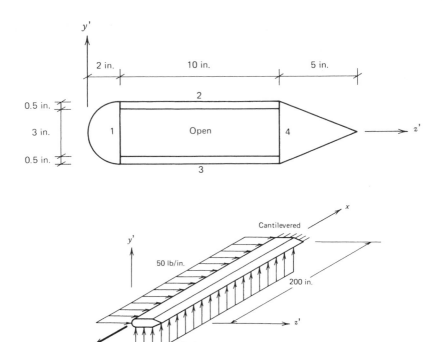

FIGURE 12 Composite helicopter rotor blade subjected to the loading shown. [From, Allen, D. H., and Haisler, W. E. (1985). "Introduction to Aerospace Structural Analysis," Wiley, New York.]

(1) $u_0(x)$, $v_0(x)$, $w_0(x)$, and
(2) $u_0(0)$, $v_0(0)$, $w_0(0)$.

Solution:

(a) By symmetry $\bar{y}' = 0$, $I^*_{yz} = 0$.

Now let $E_1 = 10 \times 10^6$ psi and tabulate Eq. (16c), as shown in Table I. Now tabulate Eqs. (16e) and (16f) to find the modulus-weighted moments of inertia, as shown in Table II.

(b) The applied loads must now be resolved to the modulus-weighted centroid. The statically equivalent loading is shown in Fig. 13. The internal resultants may be obtained from Eq. (13) as follows:

$$P(x) = \overset{10000}{\cancel{P}(0)} - \int_0^x \overset{0}{\cancel{p}(x)}\, dx$$

$$\Rightarrow P(x) = 10,000\, \text{lb},$$

$$V_y(x) = \overset{0}{\cancel{V}_y(0)} - \int_0^x \overset{25}{\cancel{p}_y(x)}\, dx$$

$$\Rightarrow V_y(x) = -25x\, \text{lb},$$

$$V_z(x) = \overset{0}{\cancel{V}_z(0)} - \int_0^x \overset{50}{\cancel{p}_z(x)}\, dx$$

TABLE I

Portion (i)	E_i (10^6 psi)	A_i (in.^{2})	$\dfrac{E_i}{E_1} A_i$ (in.^{2})	$\bar{z}'_i$ (in.)	$\dfrac{E_i}{E_1}\bar{z}'_i A_i$ (in.^{3})
1	30	6.28	18.84	1.15	21.67
2	10	5.00	5.00	7.00	35.00
3	10	5.00	5.00	7.00	35.00
4	10	10.00	10.00	13.67	136.70

$$A^* = \sum \frac{E_i}{E_1} A_i = 38.84 \qquad\qquad \sum \frac{E_i}{E_1}\bar{z}'_i A_i = 228.37$$

$$\bar{z}'^* = \frac{1}{A^*} \sum \frac{E_i}{E_1}\bar{z}'_i A_i = \frac{228.37}{38.84} \Rightarrow$$

$$\bar{z}'^* = 5.88\, \text{in.}$$

TABLE II

Portion (i)	A_i (in.^{2})	E_i (10^6 psi)	$\bar{y}_i'$ (in.)	$\bar{z}_i'$ (in.)	$I_{y_0 y_{0i}}$ (in.^{4})	$I_{z_0 z_{0i}}$ (in.^{4})	$\dfrac{E_i}{E_1}\left(I_{y_0 y_{0i}} + \bar{z}_i'^2 A_i\right)$ (in.^{4})	$\dfrac{E_i}{E_1}\left(I_{z_0 z_{0i}} + \bar{y}_i'^2 A_i\right)$ (in.^{4})
1	6.28	30	0.00	1.15	1.76	6.28	30.21	18.25
2	5.00	10	1.75	7.00	41.67	0.10	286.67	15.42
3	5.00	10	−1.75	7.00	41.67	0.10	286.67	15.42
4	10.00	10	0.00	13.67	13.89	6.67	1882.58	6.67
							$I_{y'y'}^* = 2486.13$	$I_{z'z'}^* = 56.36$

$$I_{yy}^* = I_{y'y'}^* - (\bar{z}'^*)^2 A^* = 2486.13 - 5.88^2 \cdot 38.84 \Rightarrow$$

$$I_{yy}^* = 1143.3 \text{ in.}^4$$

$$I_{zz}^* = I_{z'z'}^* - (\bar{y}'^*)^2 A^* = 56.36 - 0^2 \cdot 38.84 \Rightarrow$$

$$I_{zz}^* = 56.36 \text{ in.}^4$$

$$\Rightarrow V_z(x) = -50x \text{ lb,}$$

$$M_x(x) = M_x^{\,0}(0) - \int_0^x m_x^{\,-28}(x)\,dx$$

$$\Rightarrow M_x(x) = 28x \text{ in. lb,}$$

$$M_y(x) = M_y^{\,-58,800}(0) - \int_0^x \left[m_y^{\,0}(0) - V_z^{\,-50\times}(x)\right]dx$$

$$\Rightarrow M_y(x) = -58,800 - 25x^2 \text{ in. lb,}$$

$$M_z(x) = M_z^{\,0}(0) - \int_0^x \left[m_z^{\,0}(x) + V_y^{\,-25\times}(x)\right]dx$$

$$\Rightarrow M_z(x) = 12.5x^2 \text{ in. lb,}$$

where x is measured in inches. This is shown graphically in Fig. 14.

(c) Utilizing Eqs. (10) and (15) gives

$$\varepsilon_{xx}(x, y, z) = \frac{P}{E_1 A^*} - \frac{M_z y}{E_1 I_{zz}^*} + \frac{M_y z}{E_1 I_{yy}^*} \Rightarrow,$$

$$\varepsilon_{xx}(200, 2, 0) = \frac{10,000}{10^7 \cdot 38.84} - \frac{500,000 \cdot 2}{10^7 \cdot 56.36}$$

$$+ \frac{(-1,058,800) \cdot 0}{10^7 \cdot 1143.3} \Rightarrow,$$

$$\varepsilon_{xx}(200, 2, 0) = -0.00175 \text{ in./in.,}$$

$$\sigma_{xx} = E\varepsilon_{xx} \Rightarrow \sigma_{xx}(200, 2, 0)$$

$$= 10^7 \cdot (-0.00175) \Rightarrow,$$

$$\sigma_{xx}(200, 2, 0) = -17,500 \text{ psi.}$$

Similarly,

$$\varepsilon_{xx}(200, -2, 0) = 0.00180 \text{ in./in.,}$$

$$\sigma_{xx}(200, -2, 0) = 18,000 \text{ psi,}$$

$$\varepsilon_{xx}(200, 0, 11.12) = -0.00100 \text{ in./in.,}$$

$$\sigma_{xx}(200, 0, 11.12) = -10,000 \text{ psi,}$$

$$\varepsilon_{xx}(200, 0, -5.88) = 0.000570 \text{ in./in.,}$$

$$\sigma_{xx}(200, 0, -5.88) = 17,100 \text{ psi.}$$

(d) Next,

$$P^T = \int_A E\alpha\,\Delta T\,dA = \sum_{i=1}^4 E_i \alpha_i A_i\,\Delta T,$$

$$M_y^T = \int_A E\alpha\,\Delta T z\,dA = \sum_{i=1}^4 E_i \alpha_i \bar{z} A_i\,\Delta T,$$

$$M_z^T = \int_A E\alpha\,\Delta T y\,dA = \sum_{i=1}^4 E_i \alpha_i \bar{y}_i A_i\,\Delta T.$$

Table III presents this in tabular form.
Utilizing Eqs. (10) and (15) gives

$$\varepsilon_{xx}(x, y, z) = \frac{(P + P^T)}{E_1 A^*} - \frac{1}{E_1}\left(\frac{M_z - M_z^T}{I_{zz}^*}\right)y$$

$$+ \frac{1}{E_1}\left(\frac{M_y + M_y^T}{I_{yy}^*}\right)z \Rightarrow,$$

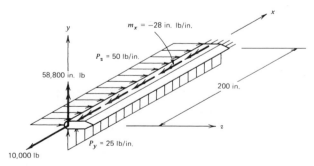

FIGURE 13 External loading configuration for composite beam. [From Allen, D. H., and Haisler, W. E. (1985). "Introduction to Aerospace Structural Analysis," Wiley, New York.]

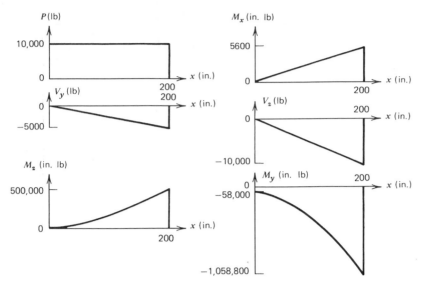

FIGURE 14 Resultant load diagrams for composite beam. [From Allen, D. H., and Haisler, W. E. (1985). "Introduction to Aerospace Structural Analysis," Wiley, New York.]

$$\varepsilon_{xx}(200, 2, 0) = \frac{(10,000 + 224.2 \cdot 200)}{10^7 \cdot 38.84}$$
$$- \frac{(500,000 - 0) \cdot 2}{10^7 \cdot 56.36}$$
$$+ \frac{(-1,058,800 + 133.6 \cdot 200) \cdot 0}{10^7 \cdot 1143.3} \Rightarrow,$$

$$\varepsilon_{xx}(200, 2, 0) = -0.00163 \text{ in./in.},$$
$$\sigma_{xx} = E(\varepsilon_{xx} - \alpha \, \Delta T) \Rightarrow,$$
$$\sigma_{xx}(200, 2, 0) = 10^7 \cdot (-0.00163$$
$$- 6.5 \cdot 10^{-6} \cdot 0.10 \cdot 200) \Rightarrow,$$
$$\sigma_{xx}(200, 2, 0) = -17,600 \text{ psi}.$$

Similarly,

$$\varepsilon_{xx}(200, -2, 0) = 0.00192 \text{ in./in.},$$
$$\sigma_{xx}(200, -2, 0) = 17,900 \text{ psi},$$
$$\varepsilon_{xx}(200, 0, 11.12) = -0.000863 \text{ in./in.},$$
$$\sigma_{xx}(200, 0, 11.12) = -9930 \text{ psi},$$
$$\varepsilon_{xx}(200, 0, -5.88) = 0.000672 \text{ in./in.},$$
$$\sigma_{xx}(200, 0, -5.88) = 5420 \text{ psi}.$$

(e) For this case Eq. (19) reduces to

$$\frac{du_0}{dx} = \frac{P + P^T}{E_1 A^*} = \frac{10,000 + 224.2x}{10^7 \cdot 38.84}$$
$$= 2.575 \cdot 10^{-5} + 5.772 \cdot 10^{-7} x,$$

TABLE III

Portion (i)	A_i (in.²)	E_i (10^6 psi)	α_i ($10^{-6}/°F$)	ΔT (°F)	$E_i \alpha_i A_i \Delta T$ (lb)
1	6.28	30	5.0	0.10x	94.2x
2	5.00	10	6.5	0.10x	32.5x
3	5.00	10	6.5	0.10x	32.5x
4	10.00	10	6.5	0.10x	65.0x
					$P^T = 224.2x$

$\bar{z}_i$ (in.)	$E_i \alpha_i \bar{z} A_i \, \Delta T$ (in. lb)	$\bar{y}_i$ (in.)	$E_i \alpha_i \bar{y}_i A_i \, \Delta T$ (in. lb)
−4.73	−445.6x	0.00	0.0
1.12	36.4x	1.75	56.9x
1.12	36.4x	−1.75	−56.9x
7.79	506.4x	0.00	0.0
	$M_y^T = 133.6x$		$M_z^T = 0.0$

$$\frac{d^2 v_0}{dx^2} = \frac{\left(M_z - M_z^T\right)}{E_1 I_{zz}^*} = \frac{(12.5x^2 - 0)}{10^7 \cdot 56.36}$$

$$= 2.218 \cdot 10^{-8} x^2,$$

$$\frac{d^2 w_0}{dx^2} = \frac{-\left(M_y + M_y^T\right)}{E_1 I_{yy}^*}$$

$$= -\frac{(-81400 - 25x^2 + 133.6x)}{10^7 \cdot 1143.3}$$

$$= 7.120 \cdot 10^{-6} - 1.169 \cdot 10^{-8} x + 2.187 \cdot 10^{-9} x^2.$$

Integrating the preceding equations and utilizing the boundary conditions described in part (a) will result in

$$u_0(x) = 2.886 \cdot 10^{-7} x^2 + 2.575 \cdot 10^{-5} x - 1.669 \cdot 10^{-2},$$

$$v_0(x) = 1.848 \cdot 10^{-9} x^4 - 5.915 \cdot 10^{-2} x + 8.873,$$

$$w_0(x) = 1.822 \cdot 10^{-10} x^4 - 1.948 \cdot 10^{-9} x^3$$
$$+ 3.560 \cdot 10^{-6} x^2 - 7.022 \cdot 10^{-3} x + 0.986.$$

Therefore,

$$u_0(0) = -1.669 \cdot 10^{-2} \text{ in.},$$

$$v_0(0) = 8.873 \text{ in.},$$

$$w_0(0) = 0.986 \text{ in.}$$

The preceding stresses and displacements may be checked against design constraints in order to determine whether the structure design is acceptable.

V. ADVANCED ANALYSIS TECHNIQUES

The analysis techniques described in the previous section can be applied to cantilever beams such as wings and fuselages because they are statically determinate structures. However, for more detailed analysis of aerospace structures, the input resultant loads described in Eq. (10) cannot be determined a priori because the structures are statically indeterminate. For these structures more complex analysis techniques have been developed. The most powerful and commonly used technique for statically indeterminate structures involves the use of variational or energy methods.

Variational methods typically convert all of the pointwise governing differential equations for the problem at hand into a single global equation for the entire body that is amenable to approximation and has the capability to handle complex boundary conditions.

The advent of the high-speed digital computer has brought the capability to perform structural analysis on extremely complicated vehicles, such as that shown in Fig. 1. In almost all cases the variational equations governing structural response have been discretized using the finite element method. This method utilizes assumed solution forms over small subsets of the structure of interest to obtain an accurate approximation of the stresses, strains, and displacements throughout the structure. The choice of the assumed solution form results in a set of coupled algebraic equations that are particularly amenable to solution on a large digital computer.

Because of the power of variational methods and the finite element method, they are reviewed in some detail in the following sections.

A. Variational Methods

Although variational methods may be utilized to solve extremely complex problems, for the purpose of demonstration, a simple example is chosen. Consider a homogeneous beam at a constant temperature and subjected to bending in the x–z plane (see Fig. 10). For this special case M_z is the only nonzero resultant load. Therefore, the displacement equation (19c) simplifies to

$$d^2 w_0 / dx^2 = -M_y / E I_{yy}. \tag{20}$$

Substitution of the preceding into differential equations of equilibrium [(13b) and (13f)] therefore results in (assuming $m_z = 0$)

$$\frac{d^2}{dx^2}\left(E I_{yy}\frac{d^2 w_0}{dx^2}\right) - p_z(x) = 0. \tag{21}$$

The preceding equation can be utilized to obtain the beam transverse displacement field $w_0 = w_0(x)$ in terms of the applied loading $p_z(x)$.

In many cases it is possible to integrate the preceding differential equation directly to obtain an exact solution. However, suppose that this cannot be accomplished. One alternative is to construct a variational principle. To do this, first multiply Eq. (21) against a small but arbitrary variation in the displacement field $\delta w_0 = \delta w_0(x)$ and integrate over the length L of the beam to obtain

$$\int_0^L \left[\frac{d^2}{dx^2}\left(E I_{yy}\frac{d^2 w_0}{dx^2}\right) - p_z(x)\right]\delta w_0(x)\, dx = 0. \tag{22}$$

The preceding is called a variational principle. It has the property that any solution to the pointwise differential equation (21) is also a solution to the global equation (22). In most structural analyses it is customary to obtain a weakened form of (3) by integrating by parts twice to obtain

$$\int_0^L E I_{yy}\frac{d^2 w_0}{dx^2}\,\delta\frac{d^2 w_0}{dx^2}\, dx$$

$$= \int_0^L p_z\,\delta w_0\, dx - V_z(0)\,\delta w_0(0) - M_y(0)\,\delta\frac{dw_0}{dx}(0)$$

$$+ V_z(L)\,\delta w_0(L) + M_y(L)\,\delta\frac{dw_0}{dx}(L), \tag{23}$$

where the equilibrium equations (12) have been used to obtain the boundary conditions. [The resulting equation is called weakened because less smoothness is required of the dependent variable $w_0(x)$.] It can be shown that the left side of Eq. (23) represents the internal work (δW_i) done on the structure, and the right side represents the external work (δW_e) due to the virtual displacement field $\delta w_0(x)$. Thus, Eq. (23) is also an energy equation, often called the principle of virtual work, and often written

$$\delta W = \delta W_i + \delta W_e = 0. \tag{24}$$

Because Eq. (23) is a global statement that explicitly contains the boundary conditions, it is particularly amenable to approximate analysis techniques. One technique, due to Lord Rayleigh (1842–1919), is called the Rayleigh–Ritz method. To see how this method may be used, consider the cantilever beam subjected to an evenly distributed load p_0 shown in Fig. 15.

We take as an initial approximation to $w_0(x)$ the quadratic polynomial

$$w_0(x) = \bar{a}_1 + \bar{a}_2 x + \bar{a}_3 x^2.$$

Because the beam is cantilevered at $x = 0$, the geometric boundary conditions are

$$w_0(0) = dw_0(0)/dx = 0.$$

Satisfaction of these geometric boundary conditions requires that

$$\bar{a}_1 = \bar{a}_2 = 0.$$

Hence, $w_0(x)$ becomes

$$w_0(x) = a_1 x^2,$$

where, for convenience, $\bar{a}_3$ is redefined to be a_1. We must choose a virtual displacement field $\delta w_0(x)$. It has been noted previously that the virtual displacement is actually a variation or differential of the actual displacement. Hence, we write

$$\delta w_0(x) = (\delta a_1)x^2,$$

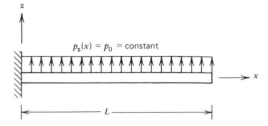

$p_z(x) = p_0$ = constant

FIGURE 15 Cantilever beam subjected to an evenly distributed load p_0. [From Allen, D. H., and Haisler, W. E. (1985). "Introduction to Aerospace Structural Analysis," Wiley, New York.]

which is consistent with the geometry constraints; that is, $\delta w_0(x)$ satisfies the boundary conditions. Since by inspection $V_z(L) = M_y(L) = 0$, the total virtual work becomes

$$\delta W = 0$$

$$= -\int_0^L EI_{yy}\frac{d^2 w_0}{dx^2}\frac{d^2(\delta w_0)}{dx^2}\,dx + \int_0^L p_z \delta w_0\,dx$$

$$= -\int_0^L EI_{yy}(2a_1)(2\delta a_1)\,dx + \int_0^L p_0(\delta a_1)x^2\,dx$$

$$= -4EI_{yy}La_1\,\delta a_1 + \left(p_0 L^3/3\right)\delta a_1$$

$$= \left(-4EI_{yy}La_1 + p_0 L^3/3\right)\delta a_1.$$

For the virtual displacement to be arbitrary, $\delta a_1 \neq 0$, and hence

$$a_1 = p_0 L^2/12EI_{yy}.$$

Therefore,

$$w_0(x) = p_0 L^2/12EI_{yy}.$$

Evaluating $w_0(x)$ at $x = 0.75L$, we find

$$w_0(0.75L) = 0.0469\left(p_0 L^4/EI_{yy}\right).$$

This can be compared to the exact displacement, which can be obtained by double integrating the governing differential equation,

$$w_0(0.75L)_{\text{exact}} = 0.0835\left(p_0 L^4/EI_{yy}\right),$$

which indicates an error of 44% at $x = 0.75L$.

Suppose we take the two-term approximation

$$w_0(x) = a_1 x^2 + a_2 x^3$$

in order to improve our approximate results. The discretized virtual work expression becomes

$$\delta W = 0$$

$$= -\int_0^L EI_{yy}(2a_1 + 6a_2 x)(2\delta a_1 + 6\delta a_2 x)\,dx$$

$$\quad + \int_0^L p_0\left(\delta a_1 x^2 + \delta a_2 x^3\right)dx$$

$$= \left[-EI_{yy}\left(4La_1 + 6L^2 a_2\right) + \frac{p_0 L^3}{3}\right]\delta a_1$$

$$\quad + \left[-EI_{yy}\left(6L^2 a_1 + 12L^3 a_2\right) + \frac{p_0 L^4}{4}\right]\delta a_2.$$

Requiring δa_1 and δa_2 to be arbitrary gives the two equations

$$\begin{bmatrix} 4EI_{yy}L & 6EI_{yy}L^2 \\ 6EI_{yy}L^2 & 12EI_{yy}L^3 \end{bmatrix}\begin{bmatrix} a_1 \\ a_2 \end{bmatrix} = \begin{bmatrix} p_0 L^3/3 \\ p_0 L^4/4 \end{bmatrix}$$

with the solution

$$a_1 = \tfrac{5}{24}\left(p_0 L^2/EI_{yy}\right),$$

$$a_2 = -\tfrac{1}{12}(p_0 L/EI_{yy}).$$

Thus, $w_0(x)$ is given by

$$w_0(x) = (p_0/24EI_{yy})(5L^2x^2 - 2Lx^3).$$

At $x = 0.75L$, the preceding equation gives

$$w_0(0.75L) = 0.08203\left(p_0 L^4/EI_{yy}\right),$$

which is in error by approximately 2%.

For the three-term approximation

$$w_0(x) = a_1 x^2 + a_1 x^3 + a_4 x^4,$$

we obtain

$$w_0(x) = (p_0/24EI)(6L^2x^2 - 4Lx^3 + x^4),$$

which is the exact solution. If additional terms are added to $w_0(x)$, the solution remains the same.

It is instructive to compare displacements and bending moments given by the three approximations with the exact solution. We know that $M_y = -EI_{yy}(d^2w_0/dx^2)$ from strength of materials. Table IV shows the results.

Because an exact solution exists for the preceding problem, it is not necessary to use approximate methods such as the Rayleigh–Ritz. However, the power of this technique is demonstrated by the example. Furthermore, in real aerospace structures, $p_z(x)$ and $I_{yy}(x)$ may be complicated functions, and the structure may be statically indeterminate. For these applications variational methods may be the most powerful tool to obtain approximate results.

In the case of a beam subjected to planar bending, it has been shown that the governing differential equations resulted in the variational principle given by Eq. (23). For more complex problems similar variational principles may be constructed. For example, for an isotropic elastic three-dimensional body, Eqs. (5)–(8) may be combined to give

$$\delta W = \delta W_i + \delta W_e = 0, \tag{25}$$

where

$$\delta W_i = -\int_V (\sigma_{xx}\,\delta\varepsilon_{xx} + \sigma_{yy}\,\delta\varepsilon_{yy} + \sigma_{zz}\,\delta\varepsilon_{zz} + 2\sigma_{yz}\,\delta\varepsilon_{yz}$$
$$+ 2\sigma_{xz}\,\delta\varepsilon_{xz} + 2\sigma_{xy}\,\delta\varepsilon_{xy})\,dV \tag{26}$$

and

$$\delta W_e = \int_S (T_x\,\delta u + T_y\,\delta v + T_z\,\delta w)\,dS$$
$$+ \int_V (X\,\delta u + Y\,\delta v + Z\,\delta w)\,dV \tag{27}$$

and the body forces X, Y, and Z may contain inertial effects for dynamic applications.

It can be seen that the general form of Eqs. (25)–(27) is similar to Eqs. (23) and (24). Furthermore, the Rayleigh–Ritz method may be used in conjunction with Eqs. (25)–(27) to obtain approximate solutions for three-dimensional aerospace structures.

B. The Finite Element Method

For complex structures the Rayleigh–Ritz procedure may break down owing to the fact that sufficient accuracy can be obtained only with an insoluble assumed solution form. An alternative method has therefore evolved in the twentieth century. Although initially called the stiffness method, this method was dubbed the finite element method (FEM) in 1960.

In the FEM, as in the Rayleigh–Ritz procedure, an assumed solution form is used with the variational principle (or some other similar global equation). However, in contrast to the Rayleigh–Ritz method, the FEM utilizes the assumed solution over a subdomain of the structure called a finite element. The approximate equations for this one element are then assembled with the governing equations for all of the other elements to obtain an approximate solution for the entire structure. Because of the way in which the element properties are assumed, the resulting equations are particularly amenable to solution on a large-core mainframe digital computer. Furthermore, the observed

TABLE IV

$w_0(x)$	$w_0(0.75L)\left(\dfrac{EI_{yy}}{p_0L^4}\right)$	$M_y(x)$	$M_y(0)$	$M_y(L)$
One term	0.0469	$\dfrac{1}{6}p_0L^2$	$\dfrac{1}{6}p_0L^2$	$\dfrac{1}{6}p_0L^2$
Two terms	0.0820	$\dfrac{p_0L}{12}(5L-6x)$	$\dfrac{5}{12}p_0L^2$	$\dfrac{-1}{12}p_0L^2$
Three terms	0.0835	$\dfrac{p_0}{2}(L-x)^2$	$\dfrac{1}{2}p_0L^2$	0
Exact	0.0835	$\dfrac{p_0}{2}(L-x)^2$	$\dfrac{1}{2}p_0L^2$	0

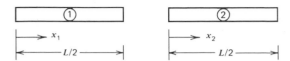

FIGURE 16 Two finite elements used to approximate the beam shown in Fig. 12. [From Allen, D. H., and Haisler, W. E. (1985). "Introduction to Aerospace Structural Analysis," Wiley, New York.]

kinematics of various components can be used to construct highly accurate elements for the various subcomponents of the structure. For example, the highly complex large space structure shown in Fig. 2 might be modeled as an assemblage of beam elements for the trusslike appendages, connected to plate elements representing the solar collectors, and rigid bodies approximating the crew compartments.

To see how the FEM may be employed, consider again the beam example problem shown in Fig. 15. Suppose, however, that in the current section the variational principle is applied twice, each time over one-half the length of the beam ($0 \leq x_1 \leq L/2$ and $L/2 \leq x_2 \leq L$), as shown in Fig. 16.

If a quadratic displacement field is assumed over each subdomain, the result of applying the variational principle [Eq. (23)] over each range and solving for the resulting unknowns will be

$$w_0(x_1) = \frac{7}{48} \frac{p_0 L^2}{E I_{yy}} x_1^2,$$

$$w_0(x_2) = \frac{7}{192} \frac{p_0 L^4}{E I_{yy}} + \frac{7}{48} \frac{p_0 L^3}{E I_{yy}} x_2 + \frac{1}{48} \frac{p_0 L^2}{E I_{yy}} x_2^2.$$

It is instructive to compare the accuracy of the one- and two-element approximations. For example, evaluating the deflection at $x = 0.75L$ (three-quarter span point) yields the following.

$w_0(x)$ approximation	$w_0(0.75L) \dfrac{E I_{yy}}{p_0 L^4}$	$M_y(0)$
One-element, quadratic	0.0469	$0.167 p_0 L^2$
One-element, cubic	0.0820	$0.417 p_0 L^2$
One-element, quartic	0.0835	$0.5 p_0 L^2$
Two-element, quadratic	0.0742	$0.292 p_0 L^2$
Exact	0.0835	$0.5 p_0 L^2$

The preceding results indicate that when the simple quadratic approximation is used, the solution improves significantly when two elements are used instead of one. One would expect that an idealization containing additional (shorter) elements would improve the results and that, in the limit as the number of elements is increased, the solution approaches the exact one. Herein lies the inherent strength of the FEM. Through so-called mesh refinement, either increasing the number of elements or redistribut-

ing them, the FEM can be utilized to obtain virtually any desired degree of accuracy (subject to computational cost constraints) even though an exact solution cannot be obtained. For example, analysis can be performed using brick elements and variational principle [(25)] on the complex fitting shown in Fig. 17.

The FEM is now an indispensable analysis tool in every structural design group in the aerospace industry and the National Aeronautics and Space Administration. Numerous extremely large and flexible computer codes such as NAS-TRAN and ANSYS now are available for analyzing such complex structures as the F-14 shown in Fig. 1. These computer codes are endowed with many user-friendly capabilities such as internal mesh optimization, banded equation solvers, and extremely complex boundary conditions. In addition, they contain large libraries of element types, including plate, shell, and higher-order continuum elements. Often they are capable of solving coupled thermomechanical problems, as well as handling material and geometric nonlinearities.

It must be remembered that the analytic capability of these computer codes is only as useful as the input

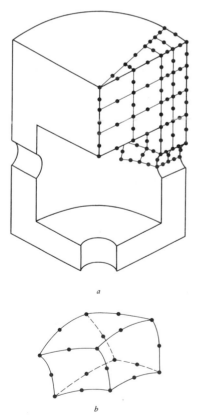

FIGURE 17 Three-dimensional structure modeled with solid brick elements. [From Allen, D. H., and Haisler, W. E. (1985). "Introduction to Aerospace Structural Analysis," Wiley, New York.]

assumptions. Since all analytical tools in general entail a certain number of simplifying assumptions, the results of these analyses must always be compared against independently obtained experimental results. However, with accurate analysis tools it is possible to reduce the margins of safety and thus achieve weight savings, the critical constraint in aerospace structures.

C. Life Prediction Methodologies

Until recently it was believed that the primary structure for flight vehicles could be designed to produce essentially infinite life by limiting the design stresses so that some allowable stress is never exceeded in the structure. However, several commercial airline crashes in the last half of the 1980s have forced the structural analysis community to revise its approach to stress analysis. It was found that these aircraft underwent long-term fatigue-induced damage, which eventually lead to major structural damage. This damage can be exacerbated by adverse environmental conditions such as saltwater surroundings. This is due primarily to the fact that the joining of dissimilar structural parts such as thin skins and stringers causes stress concentrations that can induce microcracks that can grow as a result of long-term fatigue and eventually cause rupture of the part and failure of the entire air- or spacecraft.

As shown in Fig. 18, engineers are now developing a methodology that implements a fatigue crack growth prediction model to aerospace structure finite element computer codes. Although this procedure is still undergoing development at this time, it has reached a fairly mature state for metals. Considerably more research will be necessary before accurate life predictions can be made for composite media.

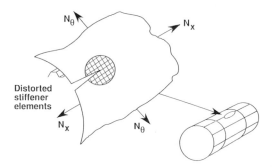

FIGURE 18 Global/local analysis scheme for a fuselage structure with an imbedded crack.

SEE ALSO THE FOLLOWING ARTICLES

AIRCRAFT PERFORMANCE AND DESIGN • COMPOSITE MATERIALS • COMPUTATIONAL AERODYNAMICS • FLIGHT (AERODYNAMICS) • FRACTURE AND FATIGUE • MECHANICS OF STRUCTURES • RELIABILITY THEORY

BIBLIOGRAPHY

Allen, D. H., and Haisler, W. E. (1985). "Introduction to Aerospace Structural Analysis," Wiley, New York.

Bathe, K. J. (1982). "Finite Element Procedures in Engineering Analysis," Prentice–Hall, Englewood Cliffs, NJ.

Beer, F. P., and Johnston, E. R., Jr. (1977). "Vector Mechanics for Engineers Statics and Dynamics," 3rd ed. McGraw–Hill, New York.

Bruhn, E. F. (1965). "Analysis and Design of Flight Vehicle Structures," Tri-State Offset, Cincinnati, OH.

Harris, C. E. (1990). "NASA Airframe Structural Integrity Program," NASA TM 102637.

Jones, R. M. (1975). "Mechanics of Composite Materials," McGraw–Hill, New York.

Lo, D. C., Allen, D. H., and Harris, C. E. (1990). *In* "Inelastic Deformation of Composite Materials" (G. J. Dvorak, ed.). Springer-Verlag, New York.

Malvern, L. E. (1967). "Introduction to the Mechanics of a Continuous Medium," Prentice–Hall, Englewood Cliffs, NJ.

Meriam, J. L. (1978). "Engineering Mechanics Statics and Dynamics," Wiley, New York.

"Metallic Materials and Elements for Flight Vehicle Structures" (1976). Military Standardization Handbook, MIL-HDBK-5C, Sept.

Newman, J. C., Jr. (1981). *In* "Methods and Models for Predicting Fatigue Crack Growth Under Random Loadings," ASTM STP 748, pp. 53–84, American Society for Testing and Materials.

Oden, J. T., and Ripperger, A. E. (1981). "Mechanics of Elastic Structures," 2nd ed., Hemisphere, Washington, DC.

Peery, D. J. (1950). "Aircraft Structures," McGraw–Hill, New York.

Peery, D. J., and Azar, J. J. (1982). "Aircraft Structures," 2nd ed., McGraw–Hill, New York.

Reddy, J. N. (1984). "An Introduction to the Finite Element Method," McGraw–Hill, New York.

Reismann, H., and Pawlik, P. S. (1980). "Elasticity Theory and Applications," Wiley, New York.

Rivello, R. M. (1969). "Theory and Analysis of Flight Structures," McGraw–Hill, New York.

Timoshenko, S. P., and Goodier, J. N. (1970). "Theory of Elasticity," 3rd ed., McGraw–Hill, New York.

Zienkiewicz, O. C. (1977). "The Finite Element Method," McGraw–Hill, New York.

Superacids

George A. Olah
G. K. Surya Prakash
University of Southern California

I. Acid Strength and Acidity Scale
II. Superacid Systems
III. Application of Superacids

GLOSSARY

Brönsted superacids Proton donor acids stronger than 100% sulfuric acid.

Carbenium ions Compounds containing a trivalent, tricoordinate carbon bearing a positive charge. Also called "classical cations."

Carbocations Compounds containing carbon bearing a positive charge which encompass both carbenium and carbonium ions.

Carbonium ions Compounds containing high coordinate carbon bearing a positive charge with multicenter bonding. Also called "nonclassical" cations.

Conjugate Brönsted–Lewis superacids Superacidic proton donor acids comprised of a combination of Brönsted and Lewis acids.

Hammet's acidity constant, H_0 A logarithmic thermodynamic scale used to relate acidity of proton donor acids.

Immobilized superacids Superacids (both Brönsted and Lewis types) bound to inert supports such as graphite, fluorinated graphite, etc.

Lewis superacids Electron acceptor acids stronger than aluminum trichloride.

Solid superacids Solid materials possessing superacid sites. May be of the Brönsted or the Lewis superacid type.

Superacids Acid systems that encompass both Brönsted and Lewis superacids as well as their conjugate combinations.

Superelectrophiles Electrophiles that are further activated by Brönsted or Lewis superacid complexation.

CHEMISTS long considered mineral acids such as sulfuric and nitric acids to be the strongest protic acids to exist. More recently this view has changed considerably with the discovery of extremely strong acid systems that are hundreds of millions, even billions, of times stronger than 100% sulfuric acid. Such acid systems are termed "superacids." The term "superacids" was first suggested by Conant and Hull in 1927 to describe acids such as perchloric acid in glacial acetic acid that were capable of protonating certain weak bases such as aldehydes and ketones.

Superacids encompass both Brönsted (proton donor) and Lewis (electron acceptor) acids as well as their conjugate pairs. The concept of acidity and acid strength can be defined only in relation to a reference base. According to an arbitrary but widely accepted suggestion by Gillespie, all Brönsted (protic) acids stronger than 100% sulfuric acid are classified as superacids. Various methods are available to measure protic superacid strengths (*vide infra*). Lewis acids also cover a wide range of acidities extending beyond the strength of the most frequently used systems such as $AlCl_3$ and BF_3. Olah *et al.* (1985) suggested the use of anhydrous aluminum trichloride, the most widely used Friedel–Crafts catalyst, as the arbitrary unit to define Lewis superacids. Lewis acids stronger than anhydrous aluminum trichloride are considered Lewis superacids. There remain, however, many difficulties in measuring the strength of Lewis acid (*vide infra*).

The high acidity and the extremely low nucleophilicity of the counterions of superacidic systems are especially useful for the preparation of stable, electron-deficient cations, including carbocations. Many of these cations, which were formerly suggested only as fleeting metastable intermediates and were detectable only in the gas phase in mass spectrometric studies, can be conveniently studied in superacid solutions. New chemical transformations and syntheses that are not possible using conventional acids can also be achieved with superacids. These include transformations and syntheses of many industrially important hydrocarbons. The unique ability of superacids to bring about hydrocarbon transformations, even to activate methane (the principal component of natural gas) for electrophilic reactions, has opened up a fascinating new field in chemistry.

I. ACID STRENGTH AND ACIDITY SCALE

The chemical species that plays the key role in Brönsted acids is the hydrogen ion, that is, the proton: H^+. Since the proton is the hydrogen nucleus with no electron in its $1s$ orbital, it is not prone to electronic repulsion. The proton consequently exercises a powerful polarizing effect. Due to its extreme electron affinity, proton cannot be found as a free "naked" species in the condensed state. It is always associated with one or more molecules of acid or the solvent (or any other nucleophile present). The strength of protic acid thus depends on the degree of association of the proton in the condensed state. Free protons can exist only in the gas phase and represent the ultimate acidity. Due to the very small size of a proton (10^5 times smaller then any other cation) and the fact that only $1s$ orbital is used in bonding by hydrogen, proton transfer is a very facile reaction, reaching diffusion-controlled rates, and does not necessitate important reorganization of the electronic valence shells. Understanding the nature of the proton is important when generalizing quantitative relationships in acidity measurements.

A number of methods are available for estimating acidity of protic acids in solution. The best known is the direct measurement of the hydrogen ion activity used in defining pH [Eq. (1)].

$$pH = \log a_{H^+}. \tag{1}$$

This can be achieved by measuring the potential of a hydrogen electrode in equilibrium with a dilute acid solution. In highly concentrated acid solutions, however, the pH concept is no longer applicable, and the acidity must be related very closely to the degree of transformation of a base with its conjugate acid, keeping in mind that this will depend on the base itself and on medium effects. The advantage of this method was shown in the 1930s by Hammett and Deyrup, who investigated the proton donor ability of the H_2O–H_2SO_4 system over the whole concentration range by measuring the extent to which a series of nitroanilines were protonated. This was the first application of the very useful Hammett acidity function [Eq. (2)].

$$H_0 = pK_{BH^+} - \log \frac{BH^+}{B}. \tag{2}$$

The pK_{BH^+} is the dissociation constant of the conjugate acid (BH^+) and BH^+/B is the ionization ratio, which is generally measured by spectroscopic means [ultraviolet, nuclear magnetic resonance (NMR), and dynamic NMR]. Hammett's "H_0" scale is a logarithmic scale on which 100% sulfuric acid has an H_0 value of -12.0.

Various other techniques are also available for acidity measurements of protic acids. These include electrochemical methods, kinetic rate measurements, and heats of protonation of weak bases. Even with all these techniques it is still difficult to measure the acidity of extremely acidic superacids, because of the unavailability of suitable weak reference bases.

In contrast to protic (Brönsted) acids, a common quantitative method to determine the strength of Lewis acids does not exist. Whereas the Brönsted acid–base interaction always involves a common denominator—the proton (H^+) transfer, which allows direct comparison—no such common relationship exists in the Lewis acid–base interaction. The result is that the definition of "strength" has no real meaning with Lewis acids.

The "strength" or "coordinating power" of different Lewis acids can vary widely against different Lewis bases. Despite the apparent difficulties, a number of qualitative relationships have been developed to characterize Lewis acids. Schwarzenbach and Chatt classified Lewis acids into two types: *class a* and *class b*. Class *a* Lewis acids form their most stable complexes with the donors in the first row of the periodic table—N, O, and F. Class *b* acids, on the other hand, complex best with donors in the second or subsequent row—Cl, Br, I, P, S, etc. Guttmann has introduced a series of donor numbers (DN) and acceptor numbers (AN) for various solvents in an attempt to quantify complexing tendencies of Lewis acids. Based on a similar premise, Drago came up with parameter E, which measures the covalent bonding potential of each series of Lewis acids as well as bases.

Pearson has proposed a qualitative scheme in which a Lewis acid and base are characterized by two parameters, one of which is referred to as strength and the other as softness. Thus, the equilibrium constant for a simple Lewis acid–base reaction would be a function of four parameters, two for each partner. Subsequently, Pearson introduced the hard and soft acids and bases (HSAB) principle to rationalize behavior and reactivity in a qualitative way. Hard acids correspond roughly in their behavior to Schwarzenbach and Chatt's class a acids. They are characterized by small acceptor atoms that have outer electrons that are not easily excited and that bear a considerable positive charge. Soft acids, which correspond to class b acids, have acceptor atoms of a lower positive charge and a large size, with easily excited outer electrons. Hard and soft bases are defined accordingly. Pearson's HSAB principle states that hard acids prefer to bind to hard bases and soft acids prefer to bind to soft bases. The principle has proved useful in rationalizing and classifying a large number of chemical reactions involving acid–base interactions in a qualitative manner, but it gives no basis for quantitative treatment.

Many attempts have been made in the literature to rate qualitatively the activity of Lewis acid catalysts in Friedel–Crafts-type reactions. However, such ratings depend largely on the nature of the reaction for which the Lewis acid catalyst is employed.

Thus, the classification of Lewis superacids as those stronger than anhydrous aluminum trichloride is only arbitrary. Just as in the case of Gillespie's classification of Brönsted superacids, it is important to recognize that acids stronger than conventional Lewis acid halides exit, with increasingly unique properties.

Another area of difficulty is measuring the acid strength of solid superacids. Since solid superacid catalysts are used extensively in the chemical industry, particularly in the petroleum field, a reliable method for measuring the acidity of solids would be extremely useful. The main difficulty to start with is that the activity coefficients for solid species are unknown and thus no thermodynamic acidity function can be properly defined. On the other hand, because the solid by definition is heterogeneous, acidic and basic sites can coexist with variable strength. The surface area available for colorimetric determinations may have acidic properties widely different from those of the bulk material; this is especially true for well-structured solids such as zeolites.

The complete description of the acidic properties of a solid requires the determination of the acid strengths as well as the number of acid sites. The methods that have been used to answer these questions are basically the same as those used for the liquid acids. Three methods are generally quoted: (1) rate measurement to relate the catalytic activity to the acidity, (2) the spectrophotometric method to estimate the acidity from the color change of adequate indicators, and (3) titration by a strong enough base for the measurement of the amount of acid. The above experimental techniques vary somewhat, but all the results obtained should be interpreted with caution because of the complexity of the solid acid catalysts. The presence of various sites of different activity on the same solid acid, the change in activity with temperature, and the difficulty of knowing the precise structure of the catalyst are some of the major handicaps in the determination of the strength of solid superacids.

II. SUPERACID SYSTEMS

Following Conant's early work, the field of superacids, which had been dormant till the late 1950s, started to undergo rapid development in the early 1960s, involving the discovery of new systems and an understanding of their nature as well as their chemistry. As mentioned, superacids encompass both Brönsted and Lewis types and their conjugate combinations.

A. Brönsted Superacids

Using Gillespie's arbitrary definition, Brönsted superacids are those with an acidity exceeding that of 100% sulfuric acid (H_0, -12). These include perchloric acid ($HClO_4$), fluorosulfuric acid (FSO_3H), trifluoromethanesulfonic acid (CF_3SO_3H), and higher perfluoroalkanesulfonic acid ($C_nF_{n+2}SO_3H$). Physical properties of some of the most commonly used superacids are listed in Table I.

Studies by Gillespie have shown that truly anhydrous hydrogen fluoride (HF), which is extremely difficult to obtain in the pure form, has a Hammett acidity constant (H_0) of -15.1 rather than the -11.0 found for the usual anhydrous acid. However, traces of water impurity drop the acidity to the generally observed value. Thus for practical purposes, hydrogen fluoride, which always contains some water impurity, is not discussed here, as its acidity of $H_0 = -11.0$ is lower than that of H_2SO_4.

Teflic acid (TeF_5OH) has been suggested to have an acidity comparable to that of fluorosulfuric acid. However, no concrete acidity measurements are available to support such a claim. A number of carbocationic salts bearing carborane anions [$CB_{11}H_6Cl_6^-$, etc.] have been studied. However, their parent Bronsted acids,

TABLE I Physical Properties of Brönsted Superacids

Property	$HClO_4$	$ClSO_3H$	HSO_3F	CF_3SO_3H
Melting point (°C)	−112	−81	−89	−34
Boiling point (°C)	110 (Explosive)	151–152 (Decomposing)	162.7	162
Density (25°C), g/cm^3	1.767^a	1.753	1.726	1.698
Viscosity (25°C), cP	—	3.0^b	1.56	2.87
Dielectric constant	—	60 ± 10	120	
Specific conductance (20°C), $\Omega^{-1} \cdot cm^{-1}$	—	0.2–0.3 × 10^{-3}	1.1 × 10^{-4}	2.0 × 10^{-4}
$-H_0$ (neat)	≈13.0	13.8	15.1	14.1

^{a} At 20°C.
^{b} At 15°C.

which can be considered as potential superacids, are still unknown.

1. Perchloric Acid ($HClO_4$)

Commercially, perchloric acid is manufactured by either reaction of alkali perchlorates with hydrochloric acid or direct electrolytic oxidation of 0.5 N hydrochloric acid. Another commercially attractive method is the direct electrolysis of chlorine gas (Cl_2) dissolved in cold, dilute perchloric acid. Perchloric acid is commercially available in a concentration of 70% (by weight) in water, although 90% perchloric acid also had limited availability (due to its explosive hazard, it is no longer provided at this strength); for 70–72% $HClO_4$, an azeotrope of 28.4% H_2O, 71.6% $HClO_4$, boiling at 203°C is safe for usual applications. It is a strong oxidizing agent, however, and must be handled with care. Anhydrous acid (100% $HClO_4$) is prepared by vacuum distillation of the concentrated acid solution with a dehydrating agent such as $Mg(ClO_4)_2$. It is stable only at low temperatures for a few days, decomposing to give $HClO_4 \cdot H_2O$ (84.6% acid) and ClO_2.

Perchloric acid is extremely hygroscopic and a very powerful oxidizer. Contact of organic materials with anhydrous or concentrated perchloric acid can lead to violent explosions. For this reason, the application of perchloric acid has serious limitations. The acid strength, although not reported, can be estimated to be around $H_0 = -13$ for the anhydrous acid.

Although various cation salts can be prepared with perchlorate gegen ions, the ionic salts tend to be unstable (explosive) due to their equilibria with covalent perchlorates. The main use of perchloric acid is in the preparation of its salts, such as $NH_4^+ ClO_4^-$, a powerful oxidant in rocket fuels and pyrotechnics.

2. Chlorosulfuric Acid ($ClSO_3H$)

Chlorosulfuric acid, the monochloride of sulfuric acid, is a strong acid containing a relatively weak sulfur–chlorine bond. It is prepared by the direct combination of sulfur trioxide and dry hydrogen chloride gas. The reaction is very exothermic and reversible, making it difficult to obtain chlorosulfuric acid free of SO_3 and HCl. On distillation, even in a good vacuum, some dissociation is inevitable. The acid is a powerful sulfating and sulfonating agent as well as a strong dehydrating agent and a specialized chlorinating agent. Because of these properties, chlorosulfuric acid is rarely used for its protonating superacid properties.

Gillespie and co-workers have measured systematically the acid strength of the H_2SO_4–$ClSO_3H$ system using aromatic nitro compounds as indicators. They found an H_0 value of −13.8 for 100% $ClSO_3H$.

3. Fluorosulfuric Acid (HSO_3F)

Fluorosulfuric acid, HSO_3F, is a mobile colorless liquid that fumes in moist air and has a sharp odor. It may be regarded as a mixed anhydride of sulfuric and hydrofluoric acid. It has been known since 1892 and is prepared commercially from SO_3 and HF in a stream of HSO_3F. It is readily purified by distillation, although the last traces of SO_3 are difficult to remove. When water is excluded, it may be handled and stored in glass containers, but for safety reasons the container should always be cooled before opening because gas pressure may have developed from hydrolysis.

$$HSO_3F + H_2O \rightleftharpoons H_2SO_4 + HF$$

Fluorosulfuric acid generally also contains hydrogen fluoride as an impurity, but according to Gillespie the hydrogen fluoride can be removed by repeated distillation under anhydrous conditions. The equilibrium $HSO_3F \rightleftharpoons SO_3 + HF$ always produces traces of SO_3 and HF in stored HSO_3F samples. When kept in glass for a long time, SiF_4 and H_2SiF_6 are also formed (secondary reactions due to HF).

Fluorosulfuric acid is employed as a catalyst and chemical reagent in various chemical processes including alkylation, acylation, polymerization, sulfonation, isomerization, and production of organic fluorosulfates. It is insoluble in carbon disulfide, carbon tetrachloride, chloroform, and tetrachloroethane, but it dissolves most organic compounds that are potential proton acceptors. The acid can be dehydrated to give $S_2O_5F_2$. Electrolysis of fluorosulfuric acid gives $S_2O_6F_2$ or $SO_2F_2 + F_2O$, depending on the conditions employed.

HSO_3F has a wide liquid range (mp $= -89°C$, bp $= +162.7°C$), making it advantageous as a superacid solvent for the protonation of a large variety of weak bases.

4. Trifluoromethanesulfonic Acid (CF₃SO₃H)

Trifluoromethanesulfonic acid (CF_3SO_3H, triflic acid), the first member in the perfluoroalkanesulfonic acid series, has been studied extensively. Besides its preparation by electrochemical fluorination of methanesulfonyl halides, triflic acid may also be prepared from trifluoromethanesulfenyl chloride.

$$CF_3SSCF_3 \xrightarrow{Cl_2} CF_3SCl \xrightarrow[H_2O]{Cl_2} CF_3SO_2Cl$$

$$\xrightarrow{\text{aq. KOH}} CF_3SO_3H$$

CF_3SO_3H is a stable, hygroscopic liquid that fumes in moist air and readily forms the stable monohydrate (hydronium triflate), which is a solid at room temperature (mp, 34°C; bp, 96°C/1 mm Hg). The acidity of the neat acid as measured by UV spectroscopy with a Hammett indicator indeed shows an H_0 value of -14.1. It is miscible with water in all proportions and soluble in many polar organic compounds, such as dimethylformamide, dimethylsulfoxide, and acetonitrile. It is generally a very good solvent for organic compounds that are capable of acting as proton acceptors in the medium. The exceptional leaving-group properties of the triflate anion, $CF_3SO_3^-$, make triflate esters excellent alkylating agents. The acid and its conjugate base do not provide a source of fluoride ion even in the presence of strong nucleophiles. Furthermore, as it lacks the sulfonating properties of oleums an HSO_3F, it has gained a wide range of application as a catalyst in Friedel–Crafts alkylation, polymerization, and organometallic chemistry.

5. Higher Homologous Perfluoroalkanesulfonic Acids

Higher homologous perfluoroalkanesulfonic acids (see Table II) are hygroscopic oily liquids or waxy solids. They are prepared by the distillation of their salts from H_2SO_4, giving stable hydrates that are difficult to dehydrate. The acids show the same polar solvent solubilities as trifluoromethanesulfonic acid but are quite insoluble in benzene,

TABLE II Characteristics of Perfluoroalkanesulfonic Acids

Compound	bp (°C) (760 mm Hg)	Density (25°C)	H_0 (22°C)
CF_3SO_3H	161	1.70	−14.1
$C_2F_5SO_3H$	170	1.75	−14.0
$C_4F_9SO_3H$	198	1.82	−13.2
$C_5F_{11}SO_3H$	212		
$C_6F_{13}SO_3H$	222		−12.3
$C_8F_{17}SO_3H$	249		
(cyclohexyl, CF₃, F, SO₃H)	241		
(cyclohexyl, C₂F₅, F, SO₃H)	257		

heptane, carbon tetrachloride, and perfluorinated liquids. Many of the perfluoroalkanesulfonic acids have been prepared by the electrochemical fluorination reaction of the corresponding alkanesulfonic acids (or conversion of the corresponding perfluoroalkane iodides to their sulfonyl halides). α,ω-Perfluoroalkanedisulfonic acids have been prepared by aqueous alkaline permanganate oxidation of the compounds, $R_fSO_2(CF_2CF_2)_n$–SO_2F. $C_8F_{17}SO_3H$ and higher perfluoroalkanesulfonic acids are surface-active agents and form the basis for a number of commercial fluorochemical surfactants.

B. Lewis Superacids

Lewis superacids are arbitrarily defined as those stronger than anhydrous aluminum trichloride, the most commonly used Friedel–Crafts catalyst. Some of the physical properties of the commonly used Lewis superacids are given in Table III.

1. Antimony Pentafluoride (SbF₅)

Antimony pentafluoride is a colorless, highly viscous liquid at room temperature. Its viscosity is 460 cP at 20°C, which is close to that of glycerol. The pure liquid can be handled and distilled in glass if moisture is excluded.

TABLE III Physical Properties of Some Lewis Superacids

Property	SbF_5	AsF_5	TaF_5	NbF_5	$B(OSO_2CF_3)_3$
mp (°C)	7.0	−79.8	97	72–73	43–45
bp (°C)	142.7	−52.8	229	236	68–83 (0.5 Torr)
Specific gravity at 15°C (g/cc)	3.145	2.33[a]	3.9	2.7	—

[a] At the bp.

The polymeric structure of the liquid SbF$_5$ has been established by ^{19}F NMR spectroscopy and is shown to have the following frameworks: a *cis*-fluorine bridged structure is found in which each antimony atom is surrounded by six fluorine atoms in an octahedral arrangement.

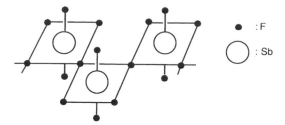

● : F

○ : Sb

Antimony pentafluoride is a powerful oxidizing and a moderate fluorinating agent. It readily forms stable intercalation compounds with graphite, and it spontaneously inflames phosphorus and sodium. It reacts with water to form SbF$_5 \cdot$ 2H$_2$O, an unusually stable solid hydrate (probably a hydronium salt, H$_3$O$^+$SbF$_5$OH) that reacts violently with excess water to form a clear solution. Slow hydrolysis can be achieved in the presence of dilute NaOH and forms Sb(OH)$_6^-$. Sulfur dioxide and nitrogen dioxide form 1:1 adducts, SbF$_5$:SO$_2$ and SbF$_5$:NO$_2$, as do practically all nonbonded electron-pair donor compounds. The exceptional ability of SbF$_5$ to complex and subsequently ionize nonbonded electron-pair donors (such as halides, alcohols, ethers, sulfides, and amines) to carbocations, first recognized by Olah in the early 1960s, has made in one of the most widely used Lewis halides in the study of cationic intermediates and catalytic reactions.

Vapor density measurements suggest a molecular association corresponding to (SbF$_5$)$_3$ at 150°C and (SbF$_5$)$_2$ at 250°C. On cooling, SbF$_5$ gives a nonionic solid composed of trigonal bipyramidal molecules. Antimony pentafluoride is prepared by the direct fluorination of antimony metal or antimony trifluoride (SbF$_3$). It can also be prepared by the reaction of SbCl$_5$ with anhydrous HF, but the exchange of the fifth chloride is difficult, and the product is generally SbF$_4$Cl.

As shown by conductometric, cryoscopic, and related acidity measurements, it appears that antimony pentafluoride is by far one of the strongest Lewis acids known. Antimony pentafluoride is also a strong oxidizing agent, allowing, for example, preparation of arene dications. At the same time, its easy reducibility to antimony trifluoride represents a limitation in many applications, although it can be easily refluorinated.

2. Arsenic Pentafluoride (AsF$_5$)

Arsenic pentafluoride (AsF$_5$) is a colorless gas at room temperature, condensing to a yellow liquid at −53°C. Vapor density measurements indicate some degree of association, but it is a monomeric covalent compound with a high degree of coordinating ability. It is prepared by reacting fluorine with arsenic metal or arsenic trifluoride. As a strong Lewis acid fluoride, it is used in the preparation of ionic complexes and, in conjunction with Brönsted acids, forms conjugate superacids. It also forms, with graphite, stable intercalation compounds that show an electrical conductivity comparable to that of silver. Great care should be exercised in handling any arsenic compound because of its potential high toxicity.

3. Tantalum and Niobium Pentafluoride

The close similarity of the atomic and ionic radii of niobium and tantalum are reflected by the similar properties of tantalum and niobium pentafluorides. They are thermally stable white solids that may be prepared either by the direct fluorination of the corresponding metals or by reacting the metal pentachlorides with HF. Surprisingly, even reacting metals with HF gives the corresponding pentafluorides. They both are strong Lewis acids, complexing a wide variety of donors such ethers, sulfides, amines, and halides. They both coordinate with fluoride ions to form anions of the type (MF$_6$)$^-$. TaF$_5$ is a somewhat stronger acid than NbF$_5$, as shown by acidity measurements in HF. The solubility of TaF$_5$ and NbF$_5$ in HF and HSO$_3$F is much more limited than that of SbF$_5$ or other Lewis acid fluorides, restricting their use to some extent. At the same time, their high redox potentials and more limited volatility make them catalysts of choice in certain hydrocarbon conversions, particularly in combination with solid catalysts.

4. Boron tris(Trifluoromethanesulfonate) [B(OSO$_2$CF$_3$)$_3$]

Boron tris (trifluoromethanesulfonate) was first prepared by Engelbrecht and Tschager in trifluoromethanesulfonic acid solution (*vide infra*) as a conjugate acid system. Olah and co-workers have isolated B(OSO$_2$CF$_3$)$_3$ in pure form by treating boron trihalides (chlorides, bromides) with 3 equiv of triflic acid in Freon 113 or SO$_2$ClF solution.

$$BX_3 + 3CF_3SO_3H \rightarrow B(OSO_2CF_3)_3 + 3HX$$

Boron tris(trifluoromethanesulfonate) is a colorless low-melting compound [mp, 43–45°C; bp, 68–73°C (0.5 Torr)] which decomposes on heating above 100°C at atmospheric pressure. It is extremely hygroscopic and is readily soluble in methylene chloride, 1,1,2-trifluorotrichloroethane (Freon 113), SO$_2$, and SO$_2$ClF. Boron tris(trifluoromethanesulfonate) is a strong nonoxidizing Lewis acid and an efficient Friedel–Crafts catalyst.

Apart from the discussed Lewis acids, other highly acidic systems such as Au(OSO$_2$F)$_3$, Ta(OSO$_2$F)$_5$, Pt(OSO$_2$F)$_4$, and Nb(OSO$_2$F)$_5$ have been reported as their conjugate acids in HSO_3F solution. All the above conjugate superacids were found to be highly conducting and strongly ionizing over the entire conecentration range.

C. Conjugate Brönsted–Lewis Superacids

1. Oleums: Polysulfuric Acids

SO_3-containing sulfuric acid (oleum) has long been considered the strongest mineral acid and one of the earliest superacid systems to be recognized. The concentration of SO_3 in sulfuric acid can be determined by weight or by electrical conductivity measurement. The most accurate H_0 values for oleums so far have been published by Gillespie and co-workers (Table IV).

The increase in acidity on the addition of SO_3 to sulfuric acid is substantial, and an H_0 value of -14.5 is reached with 50 mol% SO_3. The main component up to this SO_3 concentration is pyrosulfuric (or disulfuric) acid $H_2S_2O_7$. On heating or in the presence of water, it decomposes and behaves like a mixture of sulfuric acid and sulfur trioxide. In sulfuric acid, it ionizes as a stronger acid:

$$H_2S_2O_7 + H_2SO_4 \rightleftharpoons H_3SO_4^+ + HS_2O_7^-$$

$$(K = 1.4 \times 10^{-2})$$

At higher SO_3 concentrations, a series of higher polysulfuric acids such as $H_2S_3O_{10}$ and $H_2S_4O_{13}$ is formed and a corresponding increase in acidity occurs. However, as can be seen from Table IV, the acidity increase is very small after reaching 50 mol% of SO_3, and no data are available beyond 75%.

Despite its high acidity, oleum has found little application as a superacid catalyst, mainly because of its strong oxidizing power. Also, its high melting point and viscosity have considerably hampered its use for spectroscopic study of ionic intermediates and in synthesis, except as an oxidizing or sulfonating agent.

2. Tetra(hydrogensulfato)Boric Acid–Sulfuric Acid

$HB(HSO_4)_4$ prepared by treating boric acid $[B(OH)_3]$ with sulfuric acid ionizes in sulfuric acid as shown by acidity measurements.

$$HB(HSO_4)_4 + H_2SO_4 \rightleftharpoons H_3SO_4^+ + B(HSO_4)_4^-$$

The increase in acidity is, however, limited to $H_0 = -13.6$ as a result of insoluble complexes that precipitate when the concentration of the boric acid approaches 30 mol%.

3. Fluorosulfuric Acid–Antimony Pentafluoride (Magic Acid)

Of all superacids, "Magic Acid," a mixture of fluorosulfuric acid and antimony pentafluoride, is probably the most widely used medium for the spectroscopic observation of stable carbocations. The fluorosulfuric acid–antimony pentafluoride system was developed in the early 1960s by Olah for the study of stable carbocations and was studied by Gillespie for electron-deficient inorganic cations. The name Magic Acid originated in Olah's laboratory at Case Western Reserve University in the winter of 1966. The $HSO_3F:SbF_5$ mixture was used extensively by his group to generate stable carbocations. J. Lukas, a German postdoctoral fellow, put a small piece of Christmas candle left over from a lab party into the acid system and found that it dissolved readily. He then ran a ^1H NMR spectrum of the solution. To everybody's amazement, he obtained a sharp spectrum of the t-butyl cation. The long-chain paraffin, of which the candle was made, had obviously undergone extensive cleavage and isomerization to the more stable tertiary ion. It impressed Lukas and others in the laboratory so much that they started to nickname the acid system Magic Acid. The name stuck, and soon others started to use it too. It is now a registered trade name and has found its way into the chemical literature.

The acidity of the Magic Acid system as a function of the SbF_5 content has been measured successively by Gillespie, Sommer, Gold, and their co-workers. The increase in acidity is very sharp at a low SbF_5 concentration ($\approx 10\%$) and continues up to the estimated value of $H_0 = -26.5$ for a 90% SbF_5 content.

The initial ionization of $HSO_3F:SbF_5$ is as follows.

$$2HSO_3F + SbF_5 \rightleftharpoons H_2SO_3F^+ + SbF_5(SO_3F)^-$$

At higher concentrations of SbF_5, complex polyantimony fluorosulfate ions are formed.

$$SbF_5 + SbF_5(SO_3F)^- \rightleftharpoons Sb_2F_{10}(SO_3F)^-$$

Due to these equilibria, the composition of the HSO_3F: SbF_5 system is very complex and depends on the SbF_5 content. Aubke and co-workers have investigated the structures of complex anions in the Magic Acid system by modern ^{19}F NMR studies.

The major reason for the wide application of the Magic Acid system compared with others (besides its very high acidity) is probably the large temperature range in which

TABLE IV H_0 Values for the H_2SO_4–SO_3 System

Mol% SO_3	H_0	Mol% SO_3	H_0	Mol% SO_3	H_0
1.00	-12.24	25.00	-13.58	55.00	-14.50
2.00	-12.42	30.00	-13.76	60.00	-14.74
5.00	-12.73	35.00	-13.94	65.00	-14.84
10.00	-13.03	40.00	-14.11	70.00	-14.92
15.00	-13.23	45.00	-14.28	75.00	-14.90
20.00	-13.41	50.00	-14.44		

it can be used. In the liquid state, NMR spectra have been recorded from temperatures as low as $-160°C$ (acid diluted with SO_2F_2 and SO_2ClF) and up to $+80°C$ (neat acid in a sealed NMR glass tube). Glass is attacked by the acid very slowly when moisture is excluded. The Magic Acid system can also be an oxidizing agent that results in reduction to antimony trifluoride and sulfur dioxide. On occasion this represents a limitation.

4. Fluorosulfuric Acid–Sulfur Trioxide

Freezing-point and conductivity measurements have shown that SO_3 behaves as a nonelectrolyte in HSO_3F. Acidity measurements show a small increase in acidity that is attributed to the formation of polysulfuric acids HS_2O_6F and HS_3O_9F up to $HS_7O_{21}F$. The acidity of these solutions reaches a maximum of -15.5 on the H_0 scale for 4 mol% SO_3 and does not increase any further.

5. Fluorosulfuric Acid–Arsenic Pentafluoride

AsF_5 ionizes in FSO_3H, and the $AsF_5FSO_3^-$ anion has an octahedral structure. The H_0 acidity function increases up to 5 mol% AsF_5, with a value of -16.6.

6. Hydrogen Fluoride–Antimony Pentafluoride (Fluoroantimonic Acid)

The $HF:SbF_5$ (fluoroantimonic acid) system is considered the strongest liquid superacid and also the one that has the widest acidity range. Due to the excellent solvent properties of hydrogen fluoride, $HF:SbF_5$ is used advantageously for a variety of catalytic and synthetic applications. Anhydrous hydrogen fluoride is an excellent solvent for organic compounds with a wide liquid range. The acidity of HF, initially estimated as $H_0 \approx -11$, has now been revised to an H_0 of -15.1 for highly purified anhydrous HF. A dramatic increase in acidity ($H_0 \approx -20.5$) is observed when 1 mol% SbF_5 is added to anhydrous HF. The initial sharp increase in acidity is apparently due to the removal of residual moisture impurity. For more concentrated solutions, only kinetic data are available, mainly from the work of Brouwer and co-workers, who estimated the relative acidity ratio of 1:1 $HF:SbF_5$ and 5:1 $HSO_3F:SbF_5$ to be $5 \times 10^8:1$. This means an H_0 value in excess of -30 on the Hammett scale for the 1:1 composition. The acidity may increase still further for higher SbF_5 concentrations. It has been shown by infrared measurements that an 80% SbF_5 solution has the maximum concentration of H_2F^+. In any case, even for the composition range of 1–50% SbF_5, this is the largest range of acidity known. The same infrared study has also shown that the predominant cationic species (i.e., solvated proton) in 0–40 mol% SbF_5

is the $H_3F_2^+$ ions. The H_2F^+ ion is observed only in highly concentrated solutions (40–100 mol% SbF_5), contrary to the widespread belief that it is the only proton-solvated species in $HF:SbF_5$ solutions.

Ionization in dilute HF solutions of SbF_5 (1–20% SbF_5) is thus

$$SbF_5 + 3HF \rightleftharpoons SbF_6^- + H_3F_2^+$$

The structure of the hexafluoroantimonate and of its higher homologous anions $Sb_2F_{11}^-$ and $Sb_3F_{16}^-$, which are formed when the SbF_5 content is increased, have been determined by ^{19}F NMR studies.

7. HSO₃F:HF:SbF₅

When Magic Acid is prepared from fluorosulfuric acid not carefully purified (which always contains HF), on addition of SbF_5 the ternary superacid system $HSO_3F:HF:SbF_5$ is formed. Because HF is a weaker Brönsted acid, it ionizes fluorosulfuric acid, which, on addition of SbF_5, results in a high-acidity superacid system at low SbF_5 concentrations. ^{19}F NMR studies on the system have indicated the presence of SbF_6^- and $Sb_2F_{11}^-$ anions, although these can result from the disproportionality of $SbF_5(FSO_3)^-$ and $Sb_2F_{10}(FSO_3)^-$ anions.

8. HSO₃F:SbF₅:SO₃

When sulfur trioxide is added to a solution of SbF_5 in HSO_3F, there is a marked increase in conductivity that continues until approximately 3 mol of SO_3 has been added per mol of SbF_5. This increase in conductivity has been attributed to an increase in $H_2SO_3F^+$ concentration arising from the formation of a much stronger acid than Magic Acid. Acidity measurements have confirmed the increase in acidity with $SO_3:SbF_5$ in the HSO_3F system. This has been attributed to the presence of a series of acids of the type $H[SbF_4(SO_3F)_2]$, $H[SbF_3(SO_3F)_3]$, $H[SbF_2(SO_3F)_4]$ of increasing acidity. Of all the fluorosulfuric acid-based superacid systems, sulfur trioxide-containing acid mixtures are, however, difficult to handle and cause extensive oxidative side reactions on contact with organic compounds.

9. HSO₃F–Nb(SO₃F)₅ and HSO₃H–Ta(SO₃F)₅

The *in situ* oxidation of niobium and tantalum metals in HSO_3F by bis(fluorosulfuryl)peroxide, $S_2O_6F_2$, gives the solvated Lewis acids $M(SO_3F)_5$, $M = Nb$ or Ta. These acid systems have been shown to be highly acidic by conductivity studies.

10. $HSO_3F-Au(SO_3F)_3$ and $HSO_3F-Pt(SO_3F)_4$

These superacids based on gold and platinum have been developed. They show a high acidity and good thermal stability. However, the high cost of the metals involved precludes their widespread use.

11. Perfluoroalkanesulfonic Acid-Based Systems

a. $CF_3SO_3H:SbF_5$. $CF_3SO_3H:SbF_5$ ($n = 1$) was introduced by Olah as an effective superacid catalyst for isomerizations and alkylations. The composition and acidity of systems where $n = 1, 2, 4$ have been studied by Commeyras and co-workers. The change in composition of the triflic acid–antimony pentafluoride system depending on the SbF_5 content has been studied. For the 1:1 composition, the main counteranion is $[CF_3SO_3SbF_5]^-$, and for the 1:2 composition $[CF_3SO_3(Sb_2F_{11})]^-$ is predominant. With increasing SbF_5 concentration, the anionic species grow larger and anions containing up to 5 SbF_5 units have been found. In no circumstances could free SbF_5 be detected.

b. $CF_3SO_3H:B(SO_3CF_3)_3$. The acidity of triflic acid can also be substantially increased by the addition of boron triflate $B(OSO_2CF_3)_3$ as indicated by Engelbrecht and Tschager. The increase in acidity is explained by the ionization equilibrium:

$$B(OSO_2CF_3)_3 + 2HSO_3CF_3$$
$$\rightleftharpoons 2HSO_3CF_3^+ + {}^-B(SO_3CF_3)_4$$

The measurements were limited due to the lack of a suitable indicator base, and even 1,3,5-trinitrobenzene the weakest base used, was fully protonated ($H_0 \approx -18.5$) in a 22 mol% solution of boron triflate. The acid system has found many synthetic applications, due mainly to the efforts of Olah and co-workers.

12. Hydrogen Fluoride–Tantalum Pentafluoride

$HF:TaF_5$ is a catalyst for various hydrocarbon conversions of practical importance. In contrast to antimony pentafluoride, tantalum pentafluoride is stable in a reducing environment. The $HF:TaF_5$ superacid system has attracted attention mainly through the studies concerning alkane alkylation and aromatic protonation. Generally, heterogeneous mixtures such as 10:1 and 30:1 $HF:TaF_5$ have been used because of the low solubility of TaF_5 in HF (0.9% at 19°C and 0.6% at 0°C). For this reason, acidity measurements have been limited to very dilute solutions, and an H_0 value of -18.85 has been found for the 0.6% solution. Both electrochemical studies and aromatic protonation studies indicate that the $HF:TaF_5$ system is a weaker superacid than $HF:SbF_5$.

13. Hydrogen Fluoride–Boron Trifluoride (Tetrafluoroboric Acid)

Boron trifluoride ionizes anhydrous HF as follows:

$$BF_3 + 2HF \rightleftharpoons BF_4^- + H_2F^+$$

The stoichiometric compound exists only in an excess of HF or in the presence of suitable proton acceptors. The $HF:BF_3$ (fluoroboric acid)-catalyzed reactions cover many of the Friedel–Crafts type reactions. One of the main advantages of this system is the high stability of HF and BF_3 and their nonoxidizing nature. Both are gases at room temperature and are easily recovered from the reaction mixtures. Acidity measurements of the $HF:BF_3$ system have been limited to electrochemical determinations, and a 7 mol% BF_3 solution was found to have an acidity of $H_0 = -16.6$. This indicates that BF_3 is a much weaker Lewis acid compared with either SbF_5 or TaF_5. Nevertheless, the $HF:BF_3$ system is strong enough to protonate many weak bases and is an efficient and widely used catalyst.

14. Conjugate Friedel–Crafts Acids ($HBr:AlBr_3$, $HCl:AlCl_3$, Etc.)

The most widely used Friedel–Crafts catalyst systems are $HCl:AlCl_3$ and $HBr:AlBr_3$. These systems are indeed superacids by Gillespie's definition. However, experiments directed toward preparation from aluminium halides and hydrogen halides of the composition $HAlX_4$ were unsuccessful in providing evidence that such conjugate acids are formed in the absence of proton acceptor bases.

D. Solid Superacids

The acidic sites of solid acids may be of either the Brönsted (proton donor, often OH group) or the Lewis type (electron acceptor). Both types have been identified by IR studies of solid surfaces absorbed with pyridine. Various solids displaying acidic properties, whose acidities can be enhanced to the superacidity range, are listed in Table V.

1. Immobilized Superacids (Bound to Inert Supports)

Ways have been found to immobilize and/or to bind superacidic catalysts to an otherwise inert solid support. These include graphite intercalated superacids. Graphite possessing a layered structure can form intercalation compounds with Lewis acids such as AsF_5 and SbF_5. These

TABLE V   Solid Acids

1. Natural clay minerals: kaolinite, bentonite, attapulgite, montmorillonite, clarit, Fuller's earth, zeolites, synthetic clays or zeolites

2. Metal oxides and sulfides: ZnO, CdO, Al_2O_3, CeO_2, ThO_2, TiO_2, ZrO_2, SnO_2, PbO, As_2O_3, Bi_2O_3, Sb_2O_5, V_2O_5, Cr_2O_3, MoO_3, WO_3, CdS, ZnS

3. Metal salts: $MgSO_4$, $CaSO_4$, $SrSO_4$, $BaSO_4$, $CuSO_4$, $ZnSO_4$, $CdSO_4$, $Al_2(SO_4)_3$, $FeSO_4$, $Fe_2(SO_4)_3$, $CoSO_4$, $NiSO_4$, $Cr_2(SO_4)_3$, $KHSO_4$, $(NH_4)_2SO_4$, $Zn(NO_3)_2$, $Ca(NO_3)_2$, K_2SO_4, $Bi(NO_3)_3$, $Fe(NO_3)_3$, $CaCO_3$, BPO_4, $AlPO_4$, $CrPO_4$, $FePO_4$, $Cu_3(PO_4)_2$, $Zn_3(PO_4)_2$, $Mg_3(PO_4)_2$, $Ti_3(PO_4)_4$, $Zr_3(PO_4)_4$, $Ni_3(PO_4)_2$, AgCl, CuCl, $CaCl_2$, $AlCl_3$, $TiCl_3$, $SnCl_2$, CaF_2, BaF_2, $AgClO_4$, $Mg(ClO_4)_2$

4. Mixed oxides: SiO_2:Al_2O_3, SiO_2:TiO_2, SiO_2:SnO_2, SiO_2:ZrO_2, SiO_2:BeO, SiO_2:MgO, SiO_2:CaO, SiO_2:SrO, SiO_2:ZnO, SiO_2:Ga_2O_3, SiO_2:Y_2O_3, SiO_2:La_2O_3, SiO_2:MoO_3, SiO_2:WO_3, SiO_2:V_2O_5, SiO_2:ThO_2, Al_2O_3:MgO, Al_2O_3:ZnO, Al_2O_3:CdO, Al_2O_3:B_2O_3, Al_2O_3:ThO_2, Al_2O_3:TiO_2, Al_2O_3:ZrO_2, Al_2O_3:V_2O_5, Al_2O_3:MoO_3, Al_2O_3:WO_3, Al_2O_3:Cr_2O_3, Al_2O_3:Mn_2O_3, Al_2O_3:Fe_2O_3, Al_2O_3:Co_3O_4, Al_2O_3:NiO, TiO_2:CuO, TiO_2:MgO, TiO_2:ZnO, TiO_2:CdO, TiO_2:ZrO_2, TiO_2:SnO_2, TiO_2:Bi_2O_3, TiO_2:Sb_2O_5, TiO_2:V_2O_5, TiO_2:Cr_2O_3, TiO_2:MoO_3, TiO_2:WO_3, TiO_2:Mn_2O_3, TiO_2:Fe_2O_3, TiO_2:Co_3O_4, TiO_2:NiO, ZrO_2:CdO, ZnO:MgO, ZnO:Fe_2O_3, MoO_3:CoO:Al_2O_3, MoO_3:NiO:Al_2O_3, TiO_2:SiO_2:MgO, MoO_3:Al_2O_3:MgO

5. Cation-exchange resins, polymeric perfluorinated resinsulfonic acids

6. Heteropolyacids (Keggin type)

7. Bis(perfluorosulfonyl)imides, bis- and tris(trifluoromethylsulfonyl) methanes

intercalates are not very stable, however, as the Lewis acid tends to leach out. Similar intercalates have been obtained with other Lewis acids such as $AlCl_3$, $AlBr_3$, NbF_5, and TaF_5 and conjugate acid systems such as HF:SbF_5.

Flourine-complexed acids such as SbF_5-fluorinated graphite and SbF_5^--fluorinated alumina have been used for hydrocarbon isomerizations.

III. APPLICATION OF SUPERACIDS

A. Preparation of Stable Trivalent Carbocations

Superacids such as Magic Acid and fluoroantimonic acid have made it possible to prepare stable, long-lived carbocations, which are too reactive to exist as stable species in more basic solvents. Stable superacidic solutions of a large variety of carbocations, including trivalent cations (also called carbenium ions) such as t-butyl cation **1** (trimethylcarbenium ion) and isopropyl cation **2** (dimethylcarbenium ion), have been obtained. Some of the carbocations, as well as related acyl cations and acidic carboxonium ions and other heteroatom stabilized carbocations, that have been prepared in superacidic solutions or even isolated from them as stable salts are shown in Fig. 1.

Spectroscopic techniques such as ^1H and ^{13}C NMR, infrared, ultraviolet, and X-ray photoelectron spectroscopy have been employed to characterize carbocations. In many cases cation salts can be isolated with the superacid gegen ion, and some of them are structurally characterized by X-ray crystallography.

B. Aromatic and Homoaromatic Cations and Carbodications

According to Hückel's $(4n + 2)$ electron rule, if a carbocation has an aromatic character, it is stabilized by resonance.

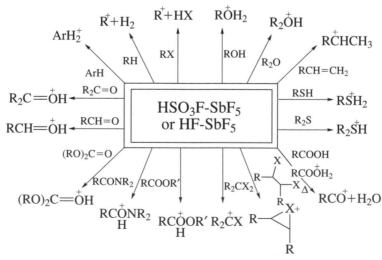

FIGURE 1 Some ways of generating carbocations generated in superacids.

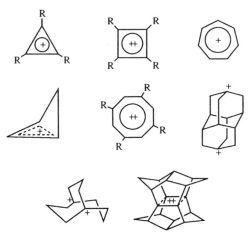

FIGURE 2 Aromatically stabilized cations and dications and some bridgehead dications.

Some aromatically stabilized Hückeloid systems generated in superacid media along with some carbodications are shown in Fig. 2.

C. Static-Bridged or Equilibrating Carbocations

Some carbocations tend to undergo fast degenerated rearrangements through intramolecular hydrogen or alkyl shifts to the related identical (degenerate) structures. The question arises whether these processes involve equilibrations between limiting "classical" ion intermediates (trivalent carbenium ions), whose structures can be adequately described by using only Lewis-type two-electron, two-center bonds separated by low-energy transition states, or whether intermediate "nonclassical" hydrogen- or alkyl-bridged carbonium ions (higher coordinate carbonium ions) are involved, which also require the presence of two-electron bonds between three or more centers for their description. It is difficult to answer this question by NMR spectroscopy because of its slow time scale; however, NMR has been used to delineate structures where degenerate rearrangements lead to averaged shifts and coupling constants.

Solid-state ^{13}C NMR (using cross-polarization magic-angle spinning techniques), isotopic substitution, and faster methods such as infrared, Raman, and, especially, X-ray photoelectron spectroscopy (ESCA) are particularly useful in investigating these systems. Some typical examples are depicted in Fig. 3.

D. Hydrocarbon Transformations

The astonishing acidity of Magic Acid and related superacids allows protonation of exceedingly weak bases. Not only all conceivable π-electron donors (such as olefins, acetylenes, and aromatics) and n-donors (such

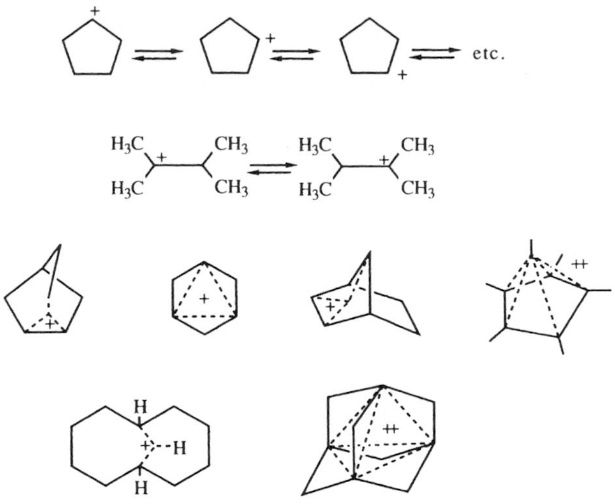

FIGURE 3 Degenerate classical (carbenbium) and nonclassical (carbonium) carbocations.

as ethers, amines, and sulfides) but also weak σ-electron donors such as saturated hydrocarbons including the parent alkane and methane are protonated. The ability of superacids to protonated saturated hydrocarbons (alkanes) rests on the ability of the two-electron, two-center covalent bond to share its bonded electron pair with empty orbitals (p or s) of a strongly electron-deficient reagent such as a protic acid:

$$R\text{–}H + H^+ \rightleftharpoons \left[R\text{--}\underset{\diagdown H}{\overset{\diagup H}{<}} \right]^+$$

Superacids are suitable reagents for chemical transformation, particularly of hydrocarbons.

E. Isomerization

The isomerization of hydrocarbons is of practical importance. Isomeric dialkylbenzenes, such as xylenes, are starting materials for plastics and other products. Generally, the need is for only one of the possible isomers, and thus there is a potential for intraconversion (isomerization). Straight-chain alkanes with five to eight carbon atoms have considerably lower octane numbers than their branched isomers, and hence there is a need for higher-octane branched isomers. Isomerizations are generally carried out under thermodynamically controlled conditions and lead to equilibria. The ionic equilibria in superacid systems generally favor increasing amounts of the higher-octane branched isomers at lower temperatures.

Lewis-acid-catalyzed isomerization of alkanes can be effected with various systems. Superacid-catalyzed reactions can be carried out at much lower temperatures, even at or below room temperature, and thus provide more of the branched isomers. This is of particular importance in preparing lead-free gasoline. Increasing the octane number by this means is preferable to the addition of higher-octane aromatics or olefins, which may pose environmental or health-hazard problems.

Of the many important superacid-catalyzed isomerizations, the isomerization tricyclo [5.2.1.0^{2,6}] decane to adamantane is unique, a reaction discovered by Schleyer.

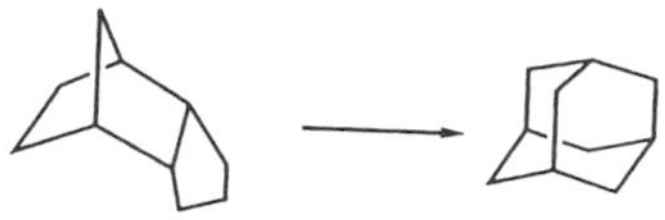

Isomerization of alkylaromatics can also be effectively carried out with superacids.

F. Alkylation

Alkylation of aromatics is carried out industrially on a large scale; an example is the reaction of ethylene with benzene to produce ethylbenzene, which is then dehydrogenated to styrene, the monomer used in producing polystyrene. Traditionally, these alkylations have been carried out in solution with a Friedel–Crafts acid catalyst such as AlCl$_3$. However, these processes are quite energy consuming and form a complex mixture of products requiring large amounts of catalyst, most of which is tied up as complexes and can be difficult or impossible to recover. The use of solid superacidic catalyst permits clean, efficient heterogeneous alkylations with no concomitant complex formation.

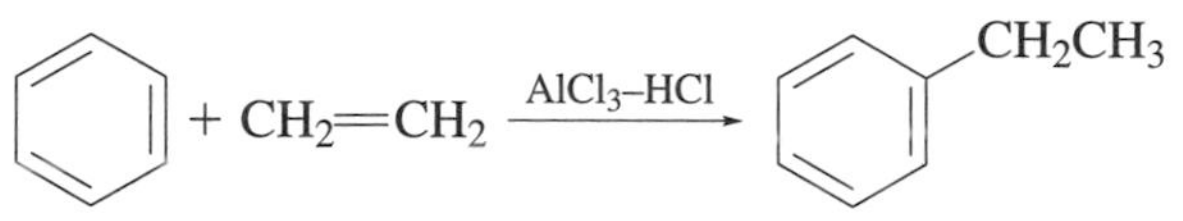

Aliphatic alkylation is widely used to produce high-octane gasolines and other hydrocarbon products. Conventional paraffin (alkane)–olefin (alkene) alkylation is an acid-catalyzed reaction; it involves the addition of a tertiary alkyl cation, generated from an isoalkane (via hydride abstraction) to an olefin. An example of such a reaction is the isobutane–ethylene alkylation, yielding 2,3-dimethylbutane.

The great interest in strong-acid chemistry is further exemplified by the discovery that lower alkanes such as methane and ethane can be polycondensed in Magic Acid at 50°C, yielding mainly C$_4$ to C$_{10}$ hydrocarbons of the gasoline range. The proposed mechanism (Fig. 4) necessitates the intermediacy of protonated alkanes (pentacoordinate carbonium ions), at least as high-lying intermediates or transition states. Hydrogen must be oxidatively removed (by either the excess superacid or added oxidants) to make the condensation of methane thermodynamically feasible.

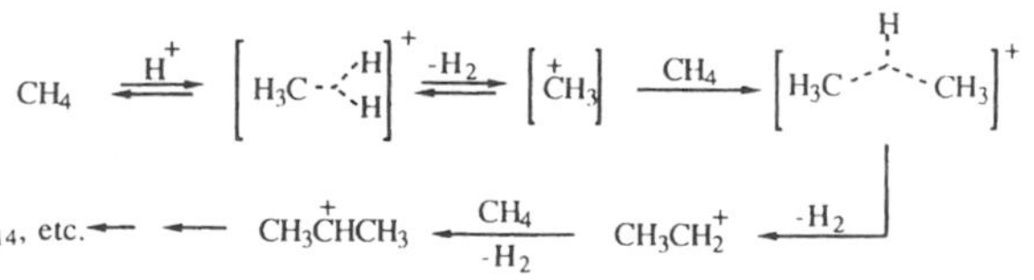

FIGURE 4 Mechanism of oxidative methane oligocondensation.

Because of the high reactivity of primary and secondary ions under these conditions, the alkylation reaction is complicated by hydride transfer and related competing reactions. However, in this mechanism it is implicit that an energetic primary cation will react directly with methane or ethane. This opens the door to new chemistry through activation of these traditionally passive molecules.

A convenient way to prepare an energetic primary cation is to react ethylene with superacid. This has been used with HF–TaF$_5$ catalyst to achieve ethylation of methane in a flow system at 50°C. With a methane–ethylene mixture (85:14), propane is the major product.

G. Polymerization

The key initiation step in cationic polymerization of alkenes is the formation of a carbocationic intermediate, which can then interact with excess monomer to start propagation. The mechanism of the initiation of cationic polymerization and polycondensation has been extensively studied. Trivalent carbenium ions play the key role, not only in acid-catalyzed polymerization of alkenes, but also in polycondensation of arenes (π-bonded monomers), as well as in cationic polymerization of ethers, sulfides, and nitrogen compounds (nonbonded electron-pair donor monomers). Pentacoordinated carbonium ions, on the other hand, play the key role in the electrophilic reactions of σ-bonds (single bonds), including the oligocondensation of alkanes and the cocondensation of alkanes and alkenes.

Alkylation and oligocondensation reactions of alkanes giving higher molecular weight alkanes have been achieved under superacid conditions.

H. Superacids in Organic Syntheses and Superelectrophilic Activation

Since the discovery of stable carbocations, they were known to be readily quenched by various nucleophiles. These reactions, which were first used to confirm the structure of the ions, proved to be very useful in organic synthesis. The selectivity of the reactions is based on the fact that generally only thermodynamically more stable ions are formed under the reaction conditions, resulting in a high selectivity. The new functional group created in a superacid medium will itself undergo protonation and, thus, be protected against any further electrophilic attack. In this way, a number of new selective reactions were achieved in a high yield, as shown by the examples below. Furthermore, electrophiles that contain nonbonded electron pairs, π-bonds, or even σ-bonds can be further activated by protonation or Lewis acid complexation leading to superelectrophiles. Such activated species react with many deactivated aromatic as well as aliphatic compounds.

1. Phenol–Dienone Rearrangement

This isomerization is of substantial importance in natural product syntheses, usually catalyzed by a strong base. The reaction occurs with good yields in polycyclic systems under superacidic conditions, as shown by Gesson and Jacquesay.

2. Reduction

Hydride ion transfer to carbocations is a well-known reaction in hydrocarbon chemistry. This reaction has been used successfully in superacid to reduce α,β-unsaturated ketones with methylcyclopentane as the hydride donor. Superacid-catalyzed reduction of aromatics, as shown by Wristers, requires both a hydride donor and hydrogen.

3. Carbonylation

The reaction between carbocations and carbon monoxide affording oxocarbenium ions (acyl cations) is a key step in the well-known Koch–Haaf reaction for preparing carboxylic acids from alkenes. This reaction has been extensively studied under superacidic conditions. An example is indicate below.

Olah *et al.* have developed direct carbonylation of isoalkanes that lead to ketones in high conversion and high selectivity under $HF:BF_3$ catalysis. The chemistry is unlike the Koch reaction and involves activated formyl cation inserting directly into the C–H σ-bond of isoalkanes, followed by strong acid-catalyzed rearrangement.

4. Oxidation

Novel oxidations of hydrocarbons in superacids with ozone or hydrogen peroxide have been investigated. Protonated ozone (O_3^+H) or hydrogen peroxide ($H_3O_2^+$) attacks the single σ-bond, resulting in oxygen insertion. These can be followed by protolytic transformation, such as the conversion of isobutane into acetone and methyl alcohol.

By similar procedures aromatics are also hydroxylated in high yields at low temperatures.

5. Superelectrophilic Activation

Electrophiles such as NO_2^+, CH_3CO^+, and H_3O^+ can be further activated in strong protic acids to their respective dications: NO_2H^{2+}, CH_3COH^{2+}, and H_4O^{2+}. Such superelectrophiles are responsible for the high electrophilic reactivity in superacids. For example, acetyl cation is a poor acetylating agent for chlorobenzene in trifluoroacetic acid. However, in superacidic trifluoromethanesulfonic acid medium, acetylation takes place with ease.

in CF₃COOH at 60°C, 24 hr, <1%
in CF₃SO₃H at 60°C, 30 min, 81%

Similar activations have been proposed for Lewis acid complexations.

I. Miscellaneous Reactions

Many acid-catalyzed reactions can be advantageously carried out using solid superacids instead of conventional acid systems. The reactions can be carried out in either the gaseous or the liquid phase. Using the example Nafion-H (a perfluoroalkane resin sulfonic acid, developed by DuPont) solid acid, several simple procedures were reported to carry out alkylation, transbromination, nitration, acetalization, hydration, and so on.

J. Superacids in Inorganic Chemistry

1. Halogen Cations

It has often been postulated that the monoatomic ions I^+, Br^+, and Cl^+ are the reactive intermediates in halogenation reactions of aromatics and alkenes. The search for the existence of such species has led to the discovery of I_2^+ and other related halogen cations, which are stable in superacids. The I_2^+ cation may be generated by the oxidation of I_2 with $S_2O_6F_2$ in HSO_3F solution,

$$2I_2 + S_2O_6F_2 \rightarrow 2I_2^+ + 2SO_3F^-$$

and a stable blue solution of this cation can also be obtained by oxidizing iodine with 65% oleum. In a less acidic medium, the I_2^+ cation disproportionates to more stable oxidation states. The electrophilic Br_2^+ cation is obtainable only in the very strong superacid Magic Acid or fluoroantimonic acid, and it disproportionates in HSO_3F. The Cl_2^+ cation, which is much more electrophilic, has not yet been observed in solution. Monoatomic halogen cations seem to be too unstable for direct observation.

2. The Trihydrogen Cation, H_3^+

The H_3 ion, 3, was first discovered by Thompson in 1912 in hydrogen discharge studies. Actually, it was the first observed gaseous ion–molecule reaction product.

Olah and co-workers have shown hydrogen–deuterium exchange of molecular H_2 and D_2, respectively, with 1:1 $HF:SbF_5$ and $HSO_3F:SbF_5$ at room temperature. The facile formation of HD does indicate that protonation or deuteration occurs involving 3 at least as a transition state in the kinetic exchange process. The H_3 ion, 3, is the simplest two-electron, three-center bonded entity.

3

3. Cations of Other Nonmetallic Elements

Elemental sulfur, selenium, and tellurium give colored solutions when dissolved in a number of strongly acidic media. It has been shown that S_{16}^{2+}, S_8^{2+}, S_4^{2+}, Se_8^{2+}, Te_4^{2+}, and Te_6^{2+} are present in such solutions. These cations are formed by the oxidation of elements by $H_2S_2O_7$ or $S_2O_6F_2$; for example,

$$4S + 6H_2S_2O_7 \rightarrow S_4^{2+} + 2HS_3O_{10}^- + 5H_2SO_4 + SO_2$$

Like the halogen cations, the sulfur, selenium, and tellurium cations are highly electrophilic and undergo disproportionation in media with any appreciable basic properties, although, as would be anticipated, the ease of disproportionation increases in series tellurium < selenium < sulfur.

4. Noble Gas Cations

Noble gas cationic salts of xenon and krypton have also been isolated from superacid medium. The examples include XeF^+, $Xe_2F_3^+$, $HCNXeF^+$, $XeOF_3^+$, KrF^+, and $Kr_2F_3^+$.

SEE ALSO THE FOLLOWING ARTICLES

NOBLE-GAS CHEMISTRY • ORGANOMETALLIC CHEMISTRY • PHYSICAL ORGANIC CHEMISTRY

BIBLIOGRAPHY

Gillespie, R. J., and Peel, T. E. (1971). *Adv. Phys. Org. Chem.* **9**, 1.
Jost, R., and Sommer, J. (1988). *Rev. Chem. Int.* **9**, 171.
Olah, G. A. (1993). *Angew. Chem. Int. Ed. Engl.* **32**, 767.
Olah, G. A., Prakash, G. K. S., and Sommer, J. (1985). "Superacids," Wiley, New York.
Tanabe, K. (1970). "Solid Acids and Bases; Their Catalytic Properties," Academic Press, New York.
Vogel, P. (1985). "Carbocation Chemistry," Elsevier, Amsterdam.

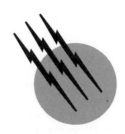

Supercomputers

Aad J. van der Steen

Utrecht University

I. The Use of Supercomputers
II. The Evolution of Supercomputers
III. New Developments
IV. Software for Supercomputers

GLOSSARY

Cache memory Small fast memory located near a CPU. Used to mitigate the difference in speed between a CPU and main memory.

ccNUMA system Cache coherent nonuniform memory access system. A computer system in which not all memory locations can be reached in the same time span.

Chaining Direct transmission of a functional unit's result to another functional unit before storing it a register. In this way the pipelining in both functional units is combined.

Clock cycle Basic unit of time in a computer systems. All processes take place in a whole number of clock cycles. A clock cycle is of the order of a few nanoseconds (10^{-9} sec). Related is the clock frequency, the number of clock cycles per second, usually expressed as MHz.

Computer architecture High-level description of the components and their interconnections in a computer as experienced by the user.

Directive Information given to a compiler to induce a certain behavior. Directives take the form of special comment lines in a program, for instance, to force parallelization of a program fragment.

Distributed-memory system Computer in which the memory is divided over the processing nodes. Each node can address only its own memory and is not aware of the nonlocal memories.

Flop/sec Floating-point operations per second. The speed of a supercomputer is often expressed as megaflop/sec (Mflop/sec) or gigaflop/sec (Gflop/sec).

Functional unit Part of a CPU that performs a definite function such as floating-point addition or calculation of memory addresses.

Interconnection network Network that interconnects memory modules and/or processing nodes in a parallel computer.

Latency Time from the initiation of an operation to the actual start of producing the result.

Parallel process Part of a program that may be executed independently. A process possesses its own program counter and address space.

Pipelining Organization of a functional unit in stages such that it can accept a new set of operands every clock cycle similar to an assembly line in a factory.

RISC processor Reduced instruction set computer processor. A processor with an instruction set that is small compared to that of the earlier complex instruction set processors. Presently, apart from the Intel IA-32-like processors, virtually only RISC processors are built.

Shared-memory system Computer in which all of the

Encyclopedia of Physical Science and Technology, Third Edition, Volume 16

memory is adressable by all processing nodes in the system.

Thread Independent subprocess in a program. In contrast to a process, a thread does not have its own address space: it is subordinate to the process it stems from.

VLIW computer Very large instruction word computer. A computer in which long instruction words cause many functional units to execute their instructions in parallel.

OVER the years many definitions for the notion of "supercomputer" have been given. Some of the most well-known are "the fastest existing computer at any point in time" and "a computer that has a performance level that is typically hundreds of times higher than that of normal commodity computers." Both definitions have their drawbacks. In the first definition the object in question is a moving target because of the fast rate at which new computers are concieved and built. Therefore, with this definition it is hard to know whether a certain computer is still *the* supercomputer or a new, even faster one has just emerged.

The second definition is vague because it presupposes that one can easily determine the performance level of a computer, which is by no means true, and furthermore, the performance factor that should discriminate between supercomputers and commodity computers is also not easily established. Indeed, it is not even straightforward to define what is meant by the term "commodity computer." Should a supercomputer be measured against a PC, used mainly for word processing, or against a workstation used for technical computations?

Still, it is obvious that, whatever definition is used, one expects supercomputers to be significantly faster on any task than the computers to which one is normally exposed. In that sense the second definition is more appropriate. Therefore, we adhere mainly to this rather vague definition, with the addition that supercomputers have a special *architecture* to enable them to be faster than the standard computing equipment we use every day. The architecture, that is, the high-level structure in terms of its processors, its memory modules, and the interconnection network between these elements, largely determines its performance and, as such, whether or not it is a supercomputer. Other defining features of the architecture are the instruction set of the computer and the accessibility of the components in the architecture from the programmer's point of view.

It is good to realize that even for commodity computers the speed is continuously increasing because of the processor speed, which, according to Moore's law, is doubling every 18 months. Therefore, supercomputers need to be at least at the same technology curve with respect to the processor speed and, in addition, employ their architectural advantage to stay ahead of commodity computers. This is also evident from the clock cycle in both commodity computers and supercomputers: in both types of systems the clock cycle is in the range of 1–3 nsec, and it is not likely that future supercomputers will contain processors with significantly faster processors because of the enormous additional costs incurred in the development and fabrication.

As it stands, nowadays the architectural advantage of supercomputers is due almost entirely to *parallelism*, i.e., many processors in a supercomputer are commonly involved in a single computational task. Ideally, the speedup that is achieved increases linearly with the number of processors that are contributing to such a computational task. Because of the time spent in the coordination of the processors, this linear increase in speed is seldomly observed. Nevertheless, parallelism enables us to tackle computational problems that would be simply unthought of without it.

As early as 1972 Flynn made a classification of computer architectures that largely determines if and how parallelism can be employed within these architectures. Commodity computers are of the SISD type. SISD stands here for single-instruction stream, single-data stream computer. Parallel computers are either of the single-instruction stream, multiple-data stream (SIMD) type, in which a single instruction gives rise to executing this instruction on many data items in parallel, or of the multiple-instruction stream, multiple-data stream (MIMD) type, in which various instructions can be executed at the same time, each operating on its own data items. Another very important distinction can be made with respect to the organization of the memory: largely it can be either *shared* or *distributed*. In the former case all data reside in a common memory that is accessible by all processors in the computer. In the latter case one or several processors can normally access only their local part of the memory. In the case that such a processor or group of processors needs data that do not reside in the local part of the memory, it has explcitly to import these data from a nonlocal part of the memory. This has a huge impact on the way programs are written for the respective types of machines. According to Flynn's classification and the way memory is organized, the terms SM-SIMD, DM-SIMD, SM-MIMD, and DM-MIMD computers are often used, where SM stands for shared memory and DM for distributed memory.

In the following we discuss both the use and the workings of supercomputers.

I. THE USE OF SUPERCOMPUTERS

The first electronic computers were designed for military use around 1950 and were used to calculate ballistic orbits.

Soon afterward (1951) the first commercial computers emerged, to be employed for all kinds of scientific and technical computations. This is still the most important realm for today's supercomputers, although the application field has diversified greatly over the years.

An area in which supercomputers have become indispensible is weather forecasting and research in climatology. The quality of weather forecasts has been increasing steadily with that of the numerical models that describe the motions of air, moisture, and driving forces such as sunshine and temperature differences. The price to be paid for these more refined models is the increased amount of computation to be done. In particular, for weather forecasting the timeliness of the results is of obvious importance, so for more intricate weather models, computers have to be faster to meet the time requirements.

Weather models are only one of the manifestations of computer models that deal with phemonema of the flow of gases and fluids, free, in pipes or in ocean beds, or around bodies such as aircraft or cars. This large family of models is the realm of computational fluid dynamics (CFD). In the CFD field vast amounts of supercomputer power are used to investigate climate change, the optimal shape of an aircraft wing (the computer model constituting a "numerical windtunnel"), or the behavior of heated plasma in the Sun's corona.

Also, safety issues for complicated building structures and, again, aircraft and cars are amenable to computer modeling. This area is called structural analysis. The structures under investigation are divided into thousands to millions of subregions, each of which is subjected to forces, temperature changes, etc., that cause the deformation of these regions and stresses in the materials used. Car crash analysis is, in this respect, an obviously important topic that requires supercomputer power to model the deformation of a car's body at impact with other bodies of various sizes, at different speeds, and under different angles. A subject where the CFD and structural analysis fields are combined in a highly complicated way is the Nuclear Stockpile Stewardship, in which it is attempted to replace actual testing of nuclear weapons, sometimes by explosion, by simulation of the testing circumstances with numerical models that integrate the computational components. Especially, the stockpile stewardship has been an enormous incentive for the development of new and faster supercomputers via the Accelerated Strategic Computer Initiative (ASCI) in the United States.

At a more fundamental level, computer models are used to investigate the structure of materials to help understand superconductivity and other phenomena that require knowledge about how electrons behave in semiconducting and conducting materials. Also, the reaction mechanisms in organic and bioorganic molecules and, indeed, their three-dimensional shape depend on the electronic structure in these molecules and, ultimately, their activity in biological systems as building material for living cells or as the key components for medicines. The computer models describing and evaluating all these aspects on an atomic molecular scale are in the field of numerical quantum physics and quantum chemistry. Vast amounts of supercomputing time are spent in this broad field, looking for new medicines, unraveling the structure of viruses, or trying to find higher-temperature superconductors.

In recent years parallel supercomputers have penetrated in areas where computing formerly had no or only a marginal role, for instance, in analysis and prediction of stock exchange rates. Large amounts of computing power are used here to evaluate the many time series and differential equations that model these rates. Furthermore, massive data processing in the form of advanced database processing and data mining relies these days on parallel supercomputers to deliver the timely results required in controlling stock, building customer profiles, and building new knowledge based on hidden patterns in data that were almost-inaccessible before because of their shear size. A subfield that depends critically on this massive data processing is the Human Genome Project, which adds vast amounts of DNA data to the part of the genome that has already been mapped. The amount of data is so large that databases of databases are necessary to handle it. In addition, not all these data are entirely reliable, and they have to be screened and reinserted as the knowledge in this huge project grows. Still, the raw data thus becoming available are just the starting point of all that can be known and done with it. Often quantum chemical and molecular dynamics techniques are used to discover the function and importance of the DNA sequences, and matching with known, almost-identical but subtly different sequences has to be done to assess the influence that, for instance, the exact form has on their activity and function.

Although this is by no means an exhaustive recounting of all the situations that call for the use of supercomputers, it illustrates that supercomputers are wonderfully flexible research instruments. Two general statements about their use can be safely made. First, the number of application fields will expand still more in the near-future, and, second, there is an insatiable need for even higher computing speeds. So we may assume that the notion of supercomputers will be with us for the forseeable future.

II. THE EVOLUTION OF SUPERCOMPUTERS

It is not straightforward to determine how and when the era of supercomputing began. Perhaps it was when the realization of computer architectures that were not necessarily the most cost-effective, but that had the highest possible performance as their primary goal, was explicitly chosen. This was certainly true for the Burroughs ILLIAC IV, a machine that originally was to have 256 processors divided into four 64-processor quadrants. Only one quadrant was ever built and delivered to NASA Ames in 1972. Its 64 processors were executing the same instruction in parallel on data of which each of the processors had a part, and as such, it was of the DM-SIMD machine type. With its 80-nsec clock cycle it was able to reach a speed of 50 Mflop/sec, an impressive performance for that time. A few years later, in 1979, the first Cray-1 vector supercomputer was delivered to Lawrence Livemore National Laboratories. With its conveyor belt-like processing of operands and its 12.5-nsec clock, it was able to produce two results per clock cycle in the right circumstances, at a peak speed of 160 Mflop/sec. This machine was designed by Seymour Cray, formerly of Control Data Corporation, where he was responsible for its predecessor, the STAR-100. With the advent of the Cray machines, the supercomputer era had really begun. A very good account of the early days of supercomputing and the developments that made it possible is given by Hockney and Jesshope (1987). We now turn to the main players in the supercomputing field in the beginning of this period.

A. Vector Processors

In the supercomputing field the early days were dominated by vector processors. Strictly speaking, the results that are produced on a single-processor vector system are not parallel but "almost parallel": after a startup phase each functional unit can deliver a result every cycle. This is brought about by pipelining their operations (see Glossary). Although vector processors are not the only ones that employ pipelining, in a vector processor everything is geared to using it with the highest possible efficiency, such as vector registers and vector instructions, which operate on the data items in these registers. Figure 1 is generic block diagram of a vector processor.

The single-processor vector machine will have only one of the vector processors depicted, and the system may even have its scalar floating-point capability shared with the vector processor. The early vector processors indeed possessed only one VPU, while present-day models can house up to 64 feeding on the same shared memory. It may be noted that the VPU in Fig. 1 does not show a cache. The majority of vector processors do not employ a cache anymore. In many cases the vector unit cannot take advantage of it and the execution speed may even be unfavorably affected because of frequent cache overflow.

Although vector processors have existed that loaded their operands directly from memory and stored the results

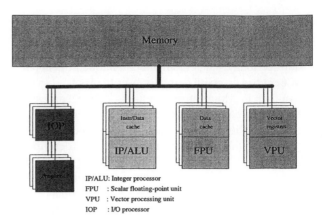

IP/ALU: Integer processor
FPU : Scalar floating-point unit
VPU : Vector processing unit
IOP : I/O processor

FIGURE 1 Block diagram of a vector processor.

again immediately in memory (CDC Cyber 205, ETA-10), all present-day vector processors use vector registers. This usually does not impair the speed of operations, while it provides much more flexibility in gathering of operands and manipulation with intermediate results.

Because of the generic nature of Fig. 1, no details of the interconnection between the VPU and the memory are shown. Still, these details are very important for the effective speed of a vector operation: when the bandwidth between the memory and the VPU is too small, it is not possible to take full advantage of the VPU because it has to wait for operands and/or has to wait before it can store results. When the ratio of arithmetic-to-load/store operations is not high enough to compensate for such situations, severe performance losses may be incurred. The influence of the number of load/store paths for the dyadic vector operation $c = a + b$ (a, b, and c vectors) is depicted in Fig. 2.

Because of the high costs of implementing these data paths between the memory and the VPU, compromises are often sought, and the number of systems that have the full required bandwidth (i.e., two load operations and

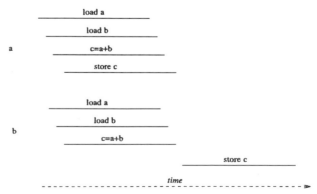

FIGURE 2 Schematic diagram of a vector addition: (a) when two load pipes and one store pipe are available; (b) when two load/store pipes are available.

one store operation at the *same* time) is limited. In fact, the vector systems marketed today no longer have this large bandwidth. Vendors rather rely on additional caches and other tricks to hide the lack of bandwidth.

The VPUs are shown as a single block in Fig. 1, yet there is considerable diversity in the structure of VPUs. Every VPU consists of a number of vector functional units, or "pipes," that fulfill one or several functions in the VPU. Every VPU has pipes that are designated to perform memory access functions, thus assuring the timely delivery of operands to the arithmetic pipes and storing of the results in memory again. Usually there are several arithmetic functional units for integer/logical arithmetic, for floating-point addition, for multiplication, and sometimes for a combination of these, a so-called compound operation. Division is performed by an iterative procedure, table lookup, or a combination of both, using the add and multiply pipe. In addition, there is almost always a mask pipe to enable operation on a selected subset of elements in a vector of operands. Finally, such sets of vector pipes can be replicated within one VPU (2- to 16-fold replication occurs). Ideally, this will increase the performance per VPU by the same factor, provided that the bandwidth to memory is adequate.

The proportion of vector processors in the present-day supercomputer arena is declining rapidly. The reason is the relatively small number of these systems, with their specialized processor architecture, that can be sold. This makes it impossible to amortize the high development and fabrication costs over a large user community. Therefore, nowadays these systems are often replaced by RISC-based parallel machines with a lower effective performance per processor but with more less costly processors.

B. Processor-Array Machines

In processor-array systems all the processors operate in lock-step, i.e., all the processors execute the same instruction at the same time (but on different data items), and no synchronization between processors is required. This greatly simplifies the design of such systems. A *control processor* issues the instructions that are to be executed by the processors in the processor array. All currently available DM-SIMD machines use a front-end processor to which they are connected by a data path to the control processor. Operations that cannot be executed by the processor array or by the control processor are offloaded to the front-end system. For instance, I/O may be through the front-end system, by the processor array machine itself, or both. Figure 3 shows a generic model of a DM-SIMD machine, from which actual models will deviate to some degree. Figure 3 might suggest that all processors in such systems are connected in a two-dimensional (2-D) grid,

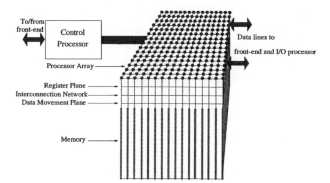

FIGURE 3 A generic block diagram of a distributed-memory SIMD machine.

and indeed, the interconnection topology of this type of machine always includes a 2-D grid. As the opposing ends of each grid line are also always connected, the topology is rather that of a torus. For several machines this is not the only interconnection scheme: they might also be connected in 3-D, diagonal, or more complex structures.

It is possible to exclude processors in the array from executing an instruction under certain logical conditions, but this means that for the time of this instruction these processors are idle (a direct consequence of the SIMD-type operation), which immediately lowers the performance. Another situation that may adversely affect the speed occurs when data required by processor i reside in the memory of processor j—in fact, as this occurs for all processors at the same time, this effectively means that data will have to be permuted across the processors. To access the data in processor j, the data will have to be fetched by this processor and then sent through the routing network to processor i. This may be fairly time-consuming. For both reasons mentioned, DM-SIMD machines are rather specialized in their use when one wants to employ their full parallelism. Generally, they perform excellently on digital signal and image processing and on certain types of Monte Carlo simulations where virtually no data exchange between processors is required and exactly the same type of operations is done on massive data sets of a size that can be made to fit comfortably in these machines.

The control processor as depicted in Fig. 3 may be more or less intelligent. It issues the instruction sequence that will be executed by the processor array. In the worst case (which means a less autonomous control processor), when an instruction is not fit for execution in the processor array (e.g., a simple print instruction), it might be offloaded to the front-end processor, which may be much slower than execution in the control processor. In the case of a more autonomous control processor, this can be avoided, thus saving processing interrupts in both the front-end and the control processor. Most DM-SIMD systems have the capability to handle I/O independently from the front-end processors. This is favorable not only because the communication between the front-end and the back-end systems is avoided. The (specialized) I/O devices for the processor-array system are generally much more efficient in providing the necessary data directly to the memory of the processor array. Especially for very data-intensive applications such as radar and image processing, such I/O systems are very important.

Processor-array machines were first introduced in the 1980s and have not seen much development beyond the first models, except in the overall technology speedup that applies to all computer systems. They fit well in their particular application niche and they will not easily be replaced by radically new architectures.

C. MPP Systems

MPP systems, where MPP stands for massively parallel processors, can be of both the shared-memory type and the distributed-memory type. In both of these types, one is confronted with the problem of how to deliver the data from the memory to the processors. An *interconnection network* is needed that connects the memory or memories, in the case of a distributed memory machine, to the processors. Through the years many types of networks have been devised. Figure 4 shows some types of networks used in present-day MPP systems.

When more CPUs are added, the collective bandwidth to the memory ideally should increase linearly with the number of processors. Unfortunately, full interconnection is quite costly, growing with $\mathcal{O}(n^2)$, while the number of processors increases with $\mathcal{O}(n)$. As shown in Fig. 4, this is exactly the case for a crossbar: it uses n^2 connections, and an Ω-network uses $n \log_2 n$ connections, while with the central bus there is only one connection. This is reflected in the use of each connection path for the different types of interconnections: for a crossbar each data path is direct and does not have to be shared with other elements. In the case of the Ω-network there are $\log_2 n$ switching stages and as many data items may have to compete for any path. For the central data bus all data have to share the same bus, so n data items may have to compete at any time.

The bus connection is the least expensive solution, but it has the obvious drawback that bus contention may occur, thus slowing down the traffic between the begin and end points of the communication. Various intricate strategies have been devised using caches associated with the CPUs to minimize the bus traffic. This leads, however, to a more complicated bus structure, which raises the costs. In practice it has proved to be very hard to design buses that are fast enough, especially with the speed of processors increasing very quickly; this imposes an upper bound on the number of processors thus connected, which appears not to exceed 10–20.

A multistage crossbar is a network with logarithmic complexity, and it has a structure which is situated somewhere between that of a bus and that of a crossbar with respect to potential capacity and costs. The Ω-network as depicted in Fig. 4 is an example. Commercially available machines such as the IBM RS/6000 SP and the SGI Origin2000 use such a network structure. For a large number of processors the $n \log_2 n$ connections quickly become more attractive than the n^2 used in crossbars. Of course, the switches at the intermediate levels should be sufficiently fast to cope with the bandwidth required.

Whichever network is used, the type of processors in principle could be arbitrary for any topology. In practice, however, bus-structured machines do not have vector processors, as the speeds of these would grossly mismatch any bus that could be constructed at a reasonable cost. All available bus-oriented systems use RISC processors. The local caches of the processors can sometimes alleviate the bandwidth problem if the data access can be satisfied by the caches, thus avoiding references to the memory.

DM-MIMD MPP machines are undoubtedly the fastest-growing class in the family of supercomputers, although this type of machine is more difficult to deal with than shared-memory machines and processor-array machines. For shared-memory systems the data distribution is completely transparent to the user. This is quite different for DM-MIMD systems, where the user has to distribute the data over the processors, and also the data exchange between processors has to be performed explicitly. The initial reluctance to use DM-MIMD machines has decreased lately. This is partly due to the now-existing standards for communication software such as MPI (message passing interface) and PVM (parallel virtual machine) and is partly because, at least theoretically, this class of systems is able to outperform all other types of machines.

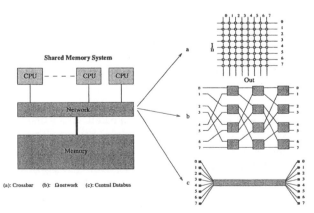

FIGURE 4 Some examples of interconnection structures, used here in a shared-memory MIMD system. The same networks may be applied in DM-MIMD systems.

DM-MIMD systems have several advantages: the bandwidth problem that haunts shared-memory systems is avoided because the bandwidth scales up automatically with the number of processors. Furthermore, the speed of the memory, which is another critical issue with shared-memory systems (to get a peak performance that is comparable to that of DM-MIMD systems, the processors of shared-memory machines should be very fast and the speed of the memory should match it), is less important for DM-MIMD machines because more processors can be configured without the aforementioned bandwidth problems.

Of course, DM-MIMD systems also have their disadvantages: the communication between processors is much slower than in SM-MIMD systems, and so, the synchronization overhead for communicating tasks is generally orders of magnitude higher than in shared-memory machines. Moreover, the access to data that are not in the local memory belonging to a particular processor have to be obtained from nonlocal memory (or memories). This is, again, slow in most systems slow compared to local data access. When the structure of a problem dictates a frequent exchange of data between processors and/or requires many processor synchronisations, it may well be that only a very small fraction of the theoretical peak speed can be obtained. As already mentioned, the data and task decompositions are factors that mostly have to be dealt with explicitly, which may be far from trivial.

Nowadays, processors are mostly off-the-shelf RISC processors. A problem for DM-MIMD MPP systems is that the speed of these processors increases at a fast rate, doubling in speed every 18 months. This is not so easily attained for the interconnection network. So a mismatch of communication vs computation speed may occur, thus turning a computation-bound problem into a communication-bound problem.

D. Clustered Systems

Recently a trend can be observed toward building systems that have a rather small number (up to 16) of RISC processors that are tightly integrated in a cluster, a symmetric multiprocessing (SMP) node. The processors in such a node are virtually always connected by a one-stage crossbar, while the clusters themselves are connected by a less costly network. Such a system may look like that depicted in Fig. 5. Note that in Fig. 5 all CPUs in a cluster are connected to a common part of the memory.

Some vendors have included hardware assistence such that all of the processors can access all of the address space. Therefore, such systems can be considered SM-MIMD machines. On the other hand, because the memory is physically distributed, it cannot be guaranteed that a

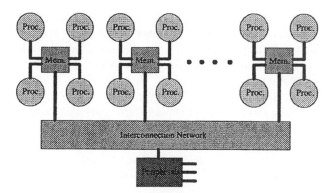

FIGURE 5 Block diagram of a system with a "hybrid" network: clusters of four CPUs are connected by a crossbar. The clusters are connected by a less expensive network; e.g., a butterfly network.

data access operation will always be satisfied within the same time. Therefore such machines are called ccNUMA systems, where ccNUMA stands for cache coherent non-uniform memory access. The term "cache coherent" refers to the fact that for all CPUs any variable that is to be used must have a consistent value. Therefore, it must be assured that the caches that provide these variables are also consistent in this respect by additional hardware and operating system functions.

For all practical purposes we can classify these systems as SM-MIMD machines also because special assisting hardware/software (such as a directory memory) has been incorporated to establish a single system image, although the memory is physically distributed.

III. NEW DEVELOPMENTS

The need for an ever-higher computing speed is insatiable and therefore new ways are continuously sought to increase. As may be clear from the discussion in the former section, the main problem that must be overcome is the growing difference in speed between CPUs and memories. A possible way of hiding the speed difference lies in *multithreading*. A thread can be seen as an independent subtask in a program that requires its own data and produces its own distinct results. Most programs contain many threads that can be executed independently. In a multithreaded machine many threads are executed in parallel, and when the data for a certain thread are not yet available another thread can become active within one or a few clock cycles, thus giving the abandoned thread the opportunity to get its data in place. In this way the slowness or *latency* of memory requests can be hidden as long as enough independent threads are available. The Tera Corporation has actually built such a system, the Tera MTA, which is presently evaluated. Also, the miniaturization of the various parts on chips makes it possible to include a limited amount of multithreading supporting hardware on the chip. So, this latency hiding mechanism will very probably turn up in many more processors in the coming years.

Another way to approach the latency problem is by not shipping the data from memory at all: some intelligence could be built into the memory itself, thus allowing for processing-in-memory (PIM). Research into PIM is quite active, and speed gains of factors of several hundreds and more have been demonstrated. One may expect that this kind of processing will appear in supercomputers within the next 5 years.

A different development that already existed in past supercomputer models is getting renewed attention: the principle of the very long instruction word (VLIW) machine. In these systems a large number of functional units is activated at the same time by many instructions packed into a long instruction word. The late multiflow trace system, which was of this type, used instruction words of 1024 bits that, in optimal circumstances, caused 28 functional units to execute. Note that in this approach the memory latency is not addressed, but one tries to speed up the computation by executing more instructions at the same time. The renewed insterest stems from the fact that in VLIW systems the instructions are scheduled statically, i.e., the CPUs are not involved in deciding which instruction should be executing at what moment. This requires extensive hardware support and the results of this dynamic scheduling are not always satisfactory. Furthermore, when the number of functional units increases beyond five or six, the complexity of dynamic scheduling becomes so high that for most programs no reasonably optimal schedule can be found. Also, because the dynamic scheduling is an intricate process, the clock cycle of the processor may be lengthened to accommodate it. In static scheduling the decision process has already taken place by means of the compiler, which produces the instruction schedule. Consequently, the clock cycle may be lowered and more functional units can be put to work at the same time. The downside of this is that very high demands on the quality of the compilers for VLIW systems must be made to assure that the instruction schedules employ the functional units efficiently. The Intel IA-64 chip family can be seen as a modest form of a VLIW chip, using instruction words of 64 bits, although the concept is renamed EPIC (Explicitly Parallel Instruction Computer) by Intel.

IV. SOFTWARE FOR SUPERCOMPUTERS

To take advantage of the speed that supercomputers can offer, one must be able to put to work as much of an available system as possible for the largest possible part of the computational task at hand. For some system types this is done largely automatically, while for others this is very inefficient or very hard. Over the years much effort has been invested in vectorizing and autoparallelizing compilers that can take off much of the burden that otherwise would befall the user of the machine. Fortunately, in the last few years many standardization efforts have also been finished, the outcomes of which have been widely accepted. This means that programs for one supercomputer can be ported to other ones with minimal effort once a certain progamming model has been chosen. Below we discuss some software developments that complement the supercomputers that are presently marketed.

A. Software for Shared-Memory Systems

Parallelization for shared-memory systems is a relatively easy task, at least compared to that for distributed-memory systems. The reason lies in the fact that in shared-memory systems the user does not have to keep track of where the data items of a program are stored: they all reside in the same shared memory. For such machines often an important part of the work in a program can be parallelized, vectorized, or both in an automatic fashion. Consider, for instance, the simple multiplication of two rows of numbers several thousand elements long. This is an operation that is abundant in the majority of technical/scientific programs. Expressed in the programming language Fortran 90, this operation would look like

```
do i = 1,  10000
   a(i) = b(i)*c(i)
end do
```

and would cause rows b and c, each 10,000 elements long, to be multiplied and the row of 10,000 results to be named a. Most parallel systems have a Fortran 90 compiler that is able to divide the 10,000 multiplications in an even way over all available processors, which would result, e.g., in a 50-processor machine, in a reduction of the computing time of almost a factor of 50 (there is some overhead involved in dividing the work over the processors). Not all compilers have this ability. However, by giving some *directives* to the compiler, one still may induce the compiler to spread the work as desired. These directives are defined by the OpenMP Consortium [4] and they are accepted by all major parallel shared-memory system vendors. The program fragment above would look like

```
!$omp  parallel  do
do i = 1, 10000
   a(i) = b(i)*c(i)
end do
!$omp end parallel do
```

informing the compiler that this fragment should be parallelized. The lines starting with !$omp are OpenMP directive lines that guide the parallelization process. There are many less simple situations where OpenMP directives may be applied, sometimes helping the compiler, because it does not have sufficient knowledge to judge whether a certain part of a program can safely be parallelized or not. Of course this requires an intimate knowledge of the program by the user, to know where to use the appropriate directives. A nice feature of the directives is that they have exactly the same form as the commentary in a normal nonparallel program. This commentary is ignored by compilers that do not have OpenMP features. Therefore, programs with directives can be run on parallel and nonparallel systems without altering the program itself.

Apart from what the programmer can do to parallelize his or her programs, most vendors also offer *libraries* of subprograms for operations that will often occur in various application areas. These subprograms are made to run very efficiently in parallel in the vendor's computers and, because every vendor has about the same collection of subprograms available, does not restrict the user of these programs to one computer. A foremost example is LAPACK, which provides all kinds of linear algebra operations and is available for all shared-memory parallel systems.

B. Software for Distributed-Memory Systems

As remarked in the former section, the paralellization of applications in distributed-memory systems is less simple than in their shared-memory counterparts. The reason is that in distributed-memory systems not all data items will reside in the same memory, and the user must be aware where they are and explicitly move or copy them to other memories if this is required by the computation. As for shared-memory systems, there has been a significant standardization effort to ensure that programs written for one computer also work in systems of other vendors. The most important ones are the communication libraries MPI (Message Passing Interface) and PVM (Parallel Virtual Machine). Both define subprograms that enable copying data from one memory to another, broadcasting data to the memories in a predefined collection of processors, or gathering data from all or a subset of processors to a specified processor. As in the former section we imagine, again, that rows b and c, of length 10,000, should be multiplied and the result should be stored in row a. This time, however, rows b and c are scattered over the memories of 50 processors. For each of the processors the program fragment is very similar to that of the shared-memory version, but each processor contains only 200 of the 10,000 elements of b and c, and therefore the program fragment would read

```
do i = 1, 200
   a(i) = b(i)*c(i)
end do
```

However, when we need to have the total result in one processor, say processor 0, we have to collect all these subresults explicitly in this processor. We can achieve this by using the MPI communication library, and the program fragment should be modified in the following way:

```
do i = 1, 200
   a(i) = b(i)*c(i)
end do
call mpi_gather(a, 200, mpi_real, &
                a, 200, mpi_real, &
                0, mpi_world_comm, ierr)
```

The last line causes all processors in the processor set mpi_world_comm to send their partial results to processor 0, where they will be placed in a row named a in the correct order. One can imagine that in many programs the sending and receiving of data from other processors can quickly become complicated. Furthermore, in contrast with the shared-memory program, we have to alter the program with respect to the nonparallel version to obtain the desired result.

Apart from MPI and PVM, there are programming models that attempt to hide the distributed nature of the machine at hand from the user. One of these is HPF (High Performance Fortran). In this Fortran 90-like language one can specify how the data are spread out over the processors, and all communication that results from this is taken care of by the HPF compiler and run time system. HPF, however, can be applied only in relatively easy and very regular cases. Such applications are characterized as *data parallel*. Therefore, the use of HPF is fairly limited.

As with shared-memory systems, also for distributed-memory systems application software libraries have been, and are being, developed. ScaLAPACK is a distributed-memory version of the LAPACK library mentioned above. Other application libraries may in turn rely on ScaLA-PACK, for instance, in solving partial differential equations as in the PETSc package. Much of this software can be found on the World Wide Web, e.g., via http://www.netlib.org.

SEE ALSO THE FOLLOWING ARTICLES

COMPUTER ARCHITECTURE • COMPUTER NETWORKS • DATABASES • IMAGE PROCESSING • INTELLIGENT CONTROL • MOLECULAR ELECTRONICS • PARALLEL

COMPUTING • QUANTUM CHEMISTRY • SIGNAL PROCESS-
ING, DIGITAL • SOFTWARE ENGINEERING • WEATHER
PREDICTION, NUMERICAL

BIBLIOGRAPHY

Culler, D. E., Singh, J. P., and Gupta, A. (1998). "Parallel Computer
Architecture: A Hardware/Software Approach," Morgan Kaufmann,
San Francisco, CA.

Hockney, R. W., and Jesshope, C. R. (1987). "Parallel Computers II,"
Adam Hilger, Bristol.
Hwang, K., and Xu, Z. (1998). "Scalable Parallel Computing: Technology, Architecture, Programming," WCB/McGraw–Hill, New York.
OpenMP Forum (1997). "Fortran Language Specification, Version 1.0,"
www.openmp.org/, Oct.
van der Steen, A. J. (2000). "Overview of Recent Supercomputers,"
Yearly updated report for the Dutch National Science Foundation,
10th ed., Feb. [The compressed PostScript version can be downloaded
from www.euroben.nl/reports. The web version can be viewed via
www.euroben.nl/.]

Superconducting Cables

E. B. Forsyth
Consultant, Brookhaven, New York

GLOSSARY

Base load Minimum load over a given period of time; the terms base voltage, base current, etc., are used to define the rated values for a system.

Charging current Current that flows into the capacitance of a transmission line when voltage is applied at its terminals.

Circuit breaker Device for interrupting a circuit between separable contacts under normal or abnormal conditions; they are ordinarily required to operate only infrequently, although some classes of breakers are suitable for frequent operation.

Connected load Sum of the continuous ratings of the load-consuming apparatus connected to the system or a part of the system under consideration.

Critical length That length of a cable for which the charging current equals the rated current at the input terminals.

Critical temperature T_c The maximum temperature at which a material remains superconductive under given conditions of magnetic field and current.

Cryostabilization Stabilization obtained if, after a temperature rise caused by either internal or external perturbations, the ohmic dissipative energy caused by the electrical current can be removed by heat transfer to a cryogen more rapidly than it is generated. The system will then not suffer total thermal runaway but instead will ultimately decrease in temperature until the superconductive material again operates at the design point.

Direct current Unidirectional current in which changes in value are either zero or so small that they may be neglected. [Note: As ordinarily used, the term designates a practically nonpulsating current. Power transmitted by dc transmission lines is measured in watts (W) rather than in volt-amperes (VA) as for ac; these are numerically the same for a unity power factor in ac circuits.]

Fault In a wire or cable a partial or total local failure in the insulation or continuity of a conductor.

Ground A conducting connection, whether intentional or accidental, between an electric circuit or equipment and the earth or to some conducting body that serves in place of the earth.

Encyclopedia of Physical Science and Technology, Third Edition, Volume 16

199

High-temperature superconductor (HTS) Ceramic-like compounds which, in practical terms, possess a critical temperature above the boiling point of nitrogen ($\sim$77 K).

Impedance In a two-terminal circuit, the ratio of applied voltage to input current with ac excitation, with due account taken of the phase relationship between these two quantities. Impedance can be expressed as a complex number with real and imaginary parts.

Interconnection tie Feeder interconnecting two electric supply systems. The normal flow of energy in such a feeder may be in either direction.

Interrupting time Interval existing between energizing of the trip coil at the rated voltage and interruption of the circuit.

Load factor Ratio of the average load over a designated period of time to the peak load occurring in that period.

Low-temperature superconductor (LTS) Any superconductor of the classes discovered before 1985 with T_c less than 23 K.

Peak load Maximum power consumed or produced by a unit or group of units in a stated period.

Reactive power The product of the voltage and the imaginary part of the current vector (often expressed as VAR).

Resistance The physical property of a network element that accounts for permanent energy loss in the circuits, such as heat generation. It is the real part of the impedance.

Short circuit Abnormal connection of relatively low resistance, whether made accidentally or intentionally, between two points of different potential in a circuit.

Splice Cable joint between two or more separate lengths of cable, with the conductors in one length and with the protecting sheaths connected so as to extend protection over the joint.

Stability When used with reference to a power system, the attribute of the system or part of the system that enables it to develop restoring forces among the elements thereof equal to, or greater than, the disturbing forces so as to restore a state of equilibrium between the elements.

Substation, electric power An assemblage of equipment for purposes other than generation or utilization through which electric energy in bulk is passed for the purpose of switching or modifying its characteristics. A substation is of such size or complexity that it incorporates one or more buses and a multiplicity of circuit breakers and usually either is the sole receiving point of commonly more than one supply circuit or sectionalizes the transmission circuits passing through it by means of circuit breakers.

Superconductor, type II Material with perfect electric conductivity for direct current (at least for low currents) that does not possess perfect diamagnetism (i.e., flux penetration of the material is possible); most metal alloys and compounds that are superconductors are of type II.

Surge Transient variation in the current and/or potential at a point in the circuit.

Transient stability Condition that exists in a power system if, after an aperiodic disturbance has taken place, the system regains steady-state stability.

SUPERCONDUCTING cables are power transmission cables that can be buried underground and cooled to temperatures near absolute zero, which reduces heat dissipation to a very low level. This enables the conductors to carry about five times more current than conventional cables. The technical characteristics will permit very long cables to operate at high power levels. This technology provides an alternative to overhead transmission in situations where new overhead transmission is unacceptable, such as the penetration of suburban areas around a load center.

I. DEVELOPMENT OF POWER TRANSMISSION TECHNOLOGY

Superconducting power transmission cables are the latest innovation in a technology that is as old as electric power engineering. The construction of central electricity generating stations by Thomas Edison in the United States and by Sebastian Ferranti in England in the 1880s immediately posed the problem of how customers could be connected to the power source. Distribution by means of wires suspended from poles was tried briefly, but the densely populated areas chosen as sites for the early generators soon forced the distribution system underground. Edison's low-voltage direct current (dc) system was a technological deadend, but by 1890, Ferranti had built a 7-mi-long underground cable system from the generating plant at Deptford to central London that operated at the then unprecedented level of 10,000 V alternating current (ac). Ferranti was remarkably prescient in his choice of wrapped brown paper for the cable insulation, a material that is still used in this application today. Paper was chosen for the insulation because it gave good operating performance at a low cost compared to other insulating materials then available, such as rubber and gutta percha. Economic considerations must be weighed carefully in the design of underground power transmission systems, and cost has been a compelling factor in the pattern of development from the turn-of-the century systems to the advanced superconducting systems being tested today.

Three-phase transmission was introduced in 1893, and ultimately this method became standard for all generation and transmission. Centrally located generating stations were soon supplemented by remote generators, particularly hydroelectric projects; these sources spurred the development of overhead transmission lines on metal pylons or wooden poles. By the 1920s, a trend developed to interconnect isolated power-company networks; this development gave further impetus to the growth of long-distance, high-voltage overhead transmission. Lines operating at 69 kV and higher are traditionally considered to be transmission lines, although nowadays by far the largest amounts of electric power transfers occur on lines rated at 230 kV and higher. This level was introduced in the 1930s and by the late 1960s the highest voltage in service had risen to 765 kV.

Direct current transmission is used in special circumstances that are justified when the savings in line cost outweigh the cost of conversion or when an asynchronous tie is required between pools. The largest system in operation in the world is in East Africa. A line about 894 mi long operating at ±533 kV transmits about 2 GW generated by a hydroelectric complex on the Zambezi River to the Republic of South Africa. In the United States, a somewhat smaller system operating at ±400 kV connects the Mead Substation at Hoover Dam to Celilo in Oregon [the Pacific interconnection tie (intertie)]. This line is rated at 1.4 GW and can be used to store energy by pumping water on the Columbia River system as well as by transmitting power from hydrogenerators when conditions are appropriate. A dc transmission corridor 500 mi (800 km) long in Brazil connects in the huge hydroelectric station at Itaipu to São Paulo; the capacity when finished will be 6.3 GW.

Considerable strengthening of transmission interconnection occurred following the infamous northeastern blackout in 1965. The North American Electric Reliability Council (NERC) was formed in 1968 to augment the reliability and adequacy of bulk power supply in the electric utility systems of North America. Essentially, this body encompasses all the electric utility systems in the United States and the system in Canada adjacent to the border except for the provinces of Alberta, Saskatchewan, and Quebec. Regional councils have been set up to coordinate generation and transmission between electric companies and thus maintain the highest possible level of reliability within the region.

Although the creation of the NERC gives an appearance of orderliness to the U.S. system, it must be stressed that the delivery of reliable electric power to over 100 million consumers is an incredibly complex operation and is simply not subject to monolithic control. At any one time,

over 6000 generators are delivering power over hundreds of transmission lines totaling hundreds of thousands of miles in length and under the control of 117 dispatch centers. It is amazing that the system works as well as it does; there are theoretical indications that a unified control could never be made to work. When an emergency arises, three major constraints restrict the corrective options: time, rotating generating reserve, and intertie capacity. Sometimes events cascade into the failure of a complete system, as illustrated by the blackout that occurred in New York City in July 1977. The development of underground transmission systems has proceeded at a slower pace than the development of overhead mileage. The maximum transmission voltage in common use for underground cables is 345 kV in the United States and 400 kV in Europe. The corresponding maximum circuit power is about 800 MVA, considerably less than the maximum power that can be delivered by overhead transmission lines. Two major reasons account for the preponderance of overhead transmission.

1. Underground transmission systems are relatively expensive compared to overhead systems. Exact comparisons are difficult to make because overhead transmission is usually located in open terrain and underground transmission in urban regions; when specific costs are compared (i.e., the cost of moving 1 MVA for 1 mi), underground transmission construction costs are often quoted being as much as 10 times those of overhead construction.
2. The electrical characteristics of overhead transmission circuits are especially suitable for long-distance interconnections. The longest ac underground transmission circuits are no more than 20 mi, compared to many hundreds of miles of overhead lines. The reason for this difference is a property called impedance matching. For long-distance transmission, the impedance of the transmission line should match fairly closely the impedance of the load at the receiving end. This match can be achieved with overhead lines, but a close match is impossible with underground circuits using conventional cables.

These and other factors have shaped the transmission systems operating today. In the United States, the majority of underground transmission circuits is located in the urban areas east of the Mississippi River. The lengths of underground cable circuits have been kept to a minimum; at the outskirts of cities, the transmission circuits revert to overhead lines. Much underground cable consists of flexible copper or aluminum conductors wrapped with wood-pulp paper layers and impregnated with mineral oil, a technique pioneered (1917) by Luigi Emanuelli of the Pirelli

Company of Milan, Italy. In Europe, such cables are usually manufactured with a jacket suitable for direct burial in the ground. In the United States, the cables are usually installed in steel pipes placed underground, and the pipes are filled with oil under pressure after the cables have been drawn in. This technique was developed by C. E. Bennett of the Okonite Cable Company in the early 1930s. The divergence of these practices is due to familiarity with oil pipelines in the United States and the more rapid development of the automobile; trenching for pipe-type cables is much less obtrusive and disruptive to traffic than the long trenches needed for directly buried cables.

By the 1970s shortcomings in the system were becoming apparent. The shortcomings have both technical and social causes. At voltages above 345 kV, the dielectric losses and high charging current associated with paper insulation become increasingly onerous. On the social side, the energy crisis and greater control of environmental pollution have had an impact on the design of transmission systems. For example, older downtown generating plants have been shut down and the power replaced by transmission from outside the immediate area. In the northeastern United States, relatively cheap hydroelectric power from Canada is being purchased to the maximum amount that can be carried on the transmission system. This procedure has led to problems typified by those experienced by the Consolidated Edison Company, which provides electric power to New York City. Due to geographic and political factors, almost all transmission into the Consolidated Edison service area passes through Westchester County to the north. Opposition from Westchester County residents and lack of space on existing right of way have virtually stopped expansion of overhead transmission facilities.

A great deal of development has occurred in the field of underground transmission and some changes have been made in the past two decades to installed equipment. The introduction of rigid conductors insulated with SF_6 gas has seen limited use. In recent years, Kraft paper has been replaced by a laminate of paper and polypropylene tapes known as PPP. It is impregnated with oil in the conventional manner. The incorporation of plastic reduces dielectric losses and improves breakdown strength. The paper is present to permit thorough impregnation by the oil.

The use of extruded cross-linked polyethylene (XLPE) insulation is very common for distribution cables. In Europe and Japan, it is becoming a dominant method of insulating transmission cables. This type of cable is now entering service in the United States at 138 kV. Furthermore, Hydro Quebec has installed a 345-kV XLPE cable. Direct current transmission is frequently used for submarine cables. England and France exchange power at the level of 2 GW via dc cables under the Channel. A factor which will undoubtably affect the future of transmission and distribution facilities is deregulation of the electric utility industry. The supply of electric energy will no longer be under the control of one vertically integrated company from generation to consumer.

II. APPLICATION OF SUPERCONDUCTORS TO CABLE DESIGN

The development of superconducting power cables can be divided into two distinct phases.

1. Application of superconductors with a critical temperature below ~18 K. The cables were envisaged as elements of a transmission network with projected circuit ratings of the order of 1000 MVA or above. Starting in the late 1960s and finishing in the mid-1980s, several prototype systems were constructed and cables tested at voltages up to 138 kV.
2. Application of superconductors with critical temperatures above 77 K. Steady progress since the discovery of these compounds in 1986 has led to improvements in the critical current density, although values are still well below those achievable with some "low-temperature" superconductors (LTS). Fairly large test facilities and electric utility company demonstrations are nearing completion as of 1999.

The early work in this field, using LTS, such as the compound niobium–tin, was aimed at the design of transmission-line components. This was due to several factors.

1. The current rating of comparably sized conductors was about five times greater than that for copper or aluminum.
2. The impedance of the cable corresponded to the surge impedance loading, which promised efficient transmission for many miles without significant VAR generation.
3. The coolant (helium), under supercritical conditions, permitted long runs between cooling stations due to the low pressure drop.

A niche seen for "high-temperature" superconductors (HTS) is at the distribution power level using voltages below 33 kV. This represents a clever adaption to both the advantages and the shortcomings of HTS, as follows.

1. Voltages and fault currents are lower at the distribution than at the transmission level. As a result, there is a better match to the properties of the material. The use of lower voltages reduces the

development time. The current rating for a given conductor diameter is about a factor of two higher for copper than for aluminum.

2. The pressure drop associated with liquid nitrogen-cooled circuits is not so important in the relatively short distances required for distribution. The use of numerous manholes will permit the introduction of intercoolers or circulation pumps into the cooling channel.

3. The presence of existing distribution ducts yields the possibility of retrofitting HTS circuits in place of conventional cables. This procedure greatly reduces the cost of introducing the new technology into the utility industry.

The phenomenon of superconductivity was discovered by the Dutch physicist Heike Kamerlingh Onnes in 1911. He made his historic discovery some 3 years after his pioneering work on the liquefaction of helium at cryogenic temperatures. He was investigating the resistivity of metals in this new low-temperature region because he had a theory that the resistivity of metals should become zero at temperatures somewhat higher than absolute zero. He found that the resistivity did decrease rapidly with temperature but that it always leveled off at a finite value for very low temperatures. Since the magnitude of this "residual" resistance was strongly dependent on the purity of the material under test, he decided to try mercury, which is a liquid at room temperature and can be distilled and redistilled to very high purity. His results showed that the resistivity of mercury was immeasurably low at about 4 K, as his theory had predicted. During the next month, however, more careful experiments revealed that the resistivity did not go smoothly to zero as his theory required, but dropped sharply, as shown in Fig. 1. He had discovered a completely unexpected phenomenon whose detailed understanding concerns physicists to this day.

Kamerlingh Onnes realized the tremendous significance of his discovery, which promised lossless high-field magnets and electric machines. He soon discovered, however, that superconductivity could not be sustained in any of his materials operating in a magnetic field above a few hundred gauss, well below that possible with permanent magnets. Hence there were no practical applications for the phenomenon at that time.

The basic properties of superconducting materials are shown in Fig. 2. Superconductivity can exist only between certain limits of current, magnetic field, and temperature. The operating conditions of the superconductor must lie below the surface shown. The highest temperature at which a materials remains superconducting is termed the critical temperature T_c; the corresponding critical current density is termed J_c. Similarly, the highest magnetic

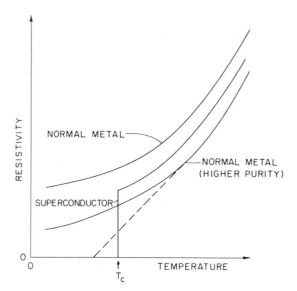

FIGURE 1 Change in resistance near absolute zero.

field in which a material can remain superconducting is termed the critical field H_c. Superconductivity was found to be a common phenomenon (superconductive metals far outnumber nonsuperconductive metals), and niobium, the metal element with the highest critical field, has an H_c of only about 0.5 tesla (T) or 5000 G.

For 50 years after the original work, great strides were made in the theoretical understanding of superconductivity, but virtually no progress was made in the development of superconductive materials. Then, in 1961, workers at Bell Telephone Laboratories announced that they had tested superconducting alloys and compounds of niobium that could support large currents in high magnetic fields. The dramatic improvement of some alloys and compounds compared to the elements is shown in Table I; only NbTi and Nb$_3$Sn are commercially available LTS.

In 1986, Bednorz and Müller in Germany discovered a material which was superconducting at 30 K. Although not of practical interest, the subsequent flurry of development soon produced a compound, yttrium barium copper oxide (YBCO), which remained in the superconducting state at the temperature of boiling nitrogen. Since 1987, a strenuous effort has been made to develop conductors incorporating HTS for practical applications. Although the cryogenic engineering of HTS applications promises a greatly improved Carnot efficiency, the materials tend to suffer from brittleness and, compared to NbTi and Nb$_3$Sn, a relatively low critical current density.

When choosing the operating regime for a superconductor, it is desirable to maximize the margins between the critical values of current and temperature and the working points. If the margins are too small, the superconductor may be forced out of the superconducting envelope during

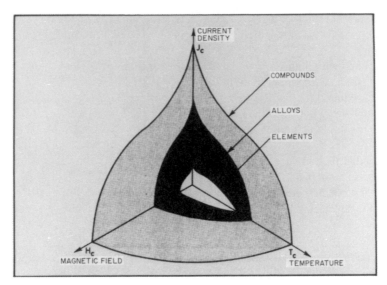

FIGURE 2 The basic requirements to achieve superconductivity are that the temperature, current density, and magnetic field at the conductor must lie below the three-dimensional surface shown above. Elements, alloys, and compounds all lie below certain limiting surfaces. Compounds exhibit the highest critical temperature T_c. The Brookhaven cable was made with Nb_3Sn superconductor, which has a T_c of about 18 K.

some abnormal event such as a short circuit, and at some level of fault, cryostability may be lost. A problem for HTS and niobium–tin is brittleness—the materials suffer from a marked degradation of critical current for strain in the range 0.1 to 0.5%. The superconductor can be laminated to other metals to keep it near the neutral axis during bending. However, the additional materials may exhibit eddy-current losses that must be absorbed by the cooling system.

Despite these engineering problems in the application of superconductors to power cables, the technology holds great promise to provide significant improvements over conventional cables, as follows.

1. By operating at higher currents, the power transmitted is greatly increased at a given voltage. This leads to the possibility of a lower specific cost (i.e., a lower cost per mile) if other cost increases do not cancel out this advantage.
2. A superconducting cable circuit can be designed to feed a load resistance that matches the cable impedance.

This attribute is shared with overhead lines; hence transmission circuits with superconducting cables could be made hundreds of miles long without causing serious difficulties in operating the network.

Before the many technical options open to the designers of superconducting power transmission cables are described, it is useful to derive the fundamental relationships between magnetic and electric fields present in an ac cable. The gross electrical operating characteristics of a transmission system can be predicted with reasonable accuracy from a simple model of the cable, regardless of details concerning the superconductor itself or the type of dielectric insulation. The model consists of two concentric cylinders representing the inner and outer conductor of an isolated phase system with equal but opposite transport currents carried by the two conductors. The annulus between the cylinders is filled with the insulating medium of given permittivity. If the current-carrying surfaces of the

TABLE I Properties of Selected Superconductors

Material	T_c (K)	H_c (T)	Critical current density J_c, order of magnitude (A/cm^2)
Elements			
Pb	7.2	0.1	—
Nb	9.3	0.5	10^6
V	5.3	0.1	
LTS alloys			
NbTi	10.0	12	10^6
NbZr	10.0	9	10^6
LTS compounds			
Nb$_3$Sn	18.3	23	10^7
V$_3$Ga	15–16	22	10^7
Nb$_3$Ge	23	37	$>10^7$
HTS compounds			
YBa$_2$Cu$_3$O$_7$	92	~4 (77 K)	$>10^6$
B$_1$Sr$_2$Ca$_2$O$_{10}$	112	~0.3 (77 K)	$>5 \times 10^4$

inner and outer conductors are at radii r_1 and r_2, respectively, then the important characteristic of the cable can be expressed in terms of the surface electric field E and the surface current density J_s, defined by

$$E = (V/r_1)[\ln(r_2/r_1)]^{-1}, \qquad (1)$$

where V is the line-to-neutral voltage, and

$$J_s = I/2\pi r_1, \qquad (2)$$

where I is the phase current. The inductive and capacitive reactance per meter may be expressed as

$$Z_L = (\omega\mu_0/2\pi)\ln(r_2/r_1)\ \Omega/\text{m}, \qquad (3)$$

$$Z_c = [2\pi\omega\varepsilon/\ln(r_2/r_1)]^{-1}\ \Omega\text{m}, \qquad (4)$$

where ω is the angular frequency of the current, ε the permittivity of the insulation, and μ_0 the permeability of free space. The characteristic or surge impedance of the cable is given by

$$Z_0 = \sqrt{Z_L Z_c}. \qquad (5)$$

Substituting from Eqs. (3) and (4),

$$Z_0 = [\ln(r_1/r_2)/2\pi]\sqrt{\mu_0/\varepsilon} \qquad (6)$$

The per-unit impedance normalized to the base impedance Z_b is

$$\bar{Z}_0 = Z_0/Z_b = (J_s/E)\sqrt{\mu_0/\varepsilon}, \qquad (7)$$

where

$$Z_b = V/I. \qquad (8)$$

Equation (7) is an extremely important guide to the electrical characteristics of the cable. It is, of course, highly desirable that a transmission line operate under matched conditions ($\bar{Z}_0$ should approach unity), i.e.,

$$(J_s/E)\sqrt{\mu_0/\varepsilon} = 1 \qquad (9)$$

or

$$E/J_s = \sqrt{\mu_0/\varepsilon}. \qquad (10)$$

For insulation with $\varepsilon_R = 1.0$, we find

$$E/J_s = 377\ \Omega; \qquad (11)$$

for polymeric insulation ($\varepsilon_R = 2.2$),

$$E/J_s = 254\ \Omega. \qquad (12)$$

The value 377 Ω is familiar to radio engineers as the impedance of free space, and it should come as no surprise to find it to be the ratio of the inner conductor surface electric and magnetic fields for a cable operating under matched conditions with vacuum or gas insulation. It should be stressed that 377 Ω is the ratio of the desirable operating fields and not the impedance of the cable itself,

which is usually much lower. The ratio of E/J_s given in Eq. (10) enables the designer to choose suitable values for these quantities so that other characteristics dependent on the choice can be evaluated. For example, a value of 10 MV/m is typical for the maximum electrical field at the surface of the inner conductor using polyethylene impregnated with supercritical helium at a temperature of $\sim$7 K. From Eq. (12) the value of J_s to match E is 393 A/cm, thus a surface stress of 10 MV/m must be associated with an acceptable dielectric loss and operating lifetime. At the same time, the current-dependent losses and cryostable overload current dependence must be acceptable at an operating field for the conductor of $\sim$400 A/cm at the supply frequency (60 Hz in the United States, 50 Hz in Europe). A cable with a conductor carrying 400 A/cm will deliver about five times more power than a copper conductor of the same size. Therein lies one of the potential advantages of superconducting cables.

III. SUPERCONDUCTING CABLE PROJECTS

Serious development of superconducting power transmission cables started in the 1960s. One of the earliest projects with laboratory experiments was carried out under the direction of P. A. Klaudy at the Institute for Low-Temperature Research (Anstalt für Tieftemperaturforschung) in Graz, Austria. From this work evolved a design for an ac cable with niobium as the conductor. The enclosure was a flexible, thermally insulated jacket rather similar in concept to self-contained cables. The flexible enclosure is based on a proprietary method of making concentric corrugated pipes developed by Kabelmetal Electro GmbH, Hannover, Germany. This type of enclosure was ultimately used by several other groups. There was some doubt in the 1960s that the current-dependent losses of a superconductor could be low enough at 50 or 60 Hz to be cooled by refrigerators of an acceptable size, and thus there was also some development of dc superconducting cables in West Germany and, in the early 1970s, in the United States.

One of the earliest projects in the United States was started by the Linde Division of the Union Carbide Company under the sponsorship of both federal and private research institutions. This design was a cryogenic analogue of gas-insulated cables, with helium performing the dual roles of coolant and dielectric insulation. Another major project in the United States was started at Brookhaven National Laboratory in 1972. A form of niobium–tin was developed for the conductor of a flexible cable using a rigid enclosure, a cryogenic analogue of pipe-type conventional practice. It may be seen from

Table I that a useful increase in operating temperature is possible with niobium–tin compared to niobium; this increase greatly improves the efficiency of the cooling system. Earlier work in the USSR was concentrated on a transmission system to operate at generator bus-bar voltage (typically 15–20 kV) and very high currents. This design was superseded by a design very similar to that developed in Graz but using a niobium–tin conductor.

Since the discovery of HTS, the focus of development has shifted to nitrogen-cooled systems. As shown in Table I, the critical current of HTS is relatively low compared to those of Nb_3Sn and NbTi. Considerable work has gone into developing multiple layer conductors to achieve the highest possible cable current. This approach is complicated by the tendency of the current to flow in the outer layers, thus underutilizing the inner layers and resulting in higher losses. Cables up to 50 m long have been tested (without dielectric insulation) in the United States, Germany, China, Italy, and Japan. In Italy the Pirelli Company tested both LN_2 impregnant dielectric insulation and a warm insulation design intended for use with HTS. A mock-up cable (i.e., Al tapes in place of HTS) with a 115-kV rating was built and successfully tested, including a 700-kV lightning impulse surge.

Before the discovery of HTS two projects reached the stage of testing cables in outdoor facilities on a scale comparable to electric utility company need:

1. A single-phase demonstration at the Arnstein hydroelectric station of an Nb-based cable developed by the Institute for Low-Temperature Research in Austria
2. The operation for 4 years of two Nb_3Sn-based cables installed in a 100-m cryostat at Brookhaven National Laboratory in the United States.

At the time of writing (1999), two sites are under construction to demonstrate the application of HTS to distribution cables:

1. Insertion of a 100-ft (30-m) section of three-phase cables in the 12.4-kV feeder to the Southwire Company plant at Carrollton, Georgia
2. The retrofit of three cables in existing ducts at the Frisbie substation of the Detroit Edison Company

These projects are described more fully in Section V.

A. Cryogenic Engineering for Superconducting Systems

Once the mechanical and electrical design of a cable has been proved, there remains the task of designing a cable enclosure and cooling system that will maintain the cable within the desired temperature range under all operating conditions. These systems must always conform to the severe economic and reliability criteria set by an electric utility power company. The four major areas of cryogenic engineering are (1) design of the refrigerator, (2) thermodynamic and hydrodynamic performance of the system under all transient and steady-state conditions, (3) design of the cable enclosure, and (4) design of the cable terminations, splices, and feeder joints.

There are no insuperable technical problems in cooling long cables with either helium or nitrogen. The two gases, however, possess radically different characteristics which will have profound effects on the specific engineering design of actual transmission or distribution circuits. For example, during the development of a "low-temperature" system, the relatively poor Carnot efficiency of a helium-cooled system operating near 7 K prompted the investigation of both rigid and cryogenic enclosures with an internal nitrogen-cooled shield which intercepted a radial heat leak at about 77 K. The prototype enclosures were also used in cable tests. The laboratory setup is shown in Fig. 3. It was found in the course of overall systems analysis that low-heat in-leak designs with internal shields were not cost-effective, and a change in the method of refrigeration was adopted to reduce total costs. The design is illustrated in Fig. 4. The first design was based on a conceptually simple scheme in which the cooling gas flowed through the cable and inner pipe of the cryostat for a sufficient

FIGURE 3 The testing of both rigid and flexible cryogenic enclosures at Brookhaven. The rigid enclosure on the left was also used for experimental cables carrying current, and the flexible enclosure on the right was used for cable testing with high voltage. [Courtesy of Brookhaven National Laboratory.]

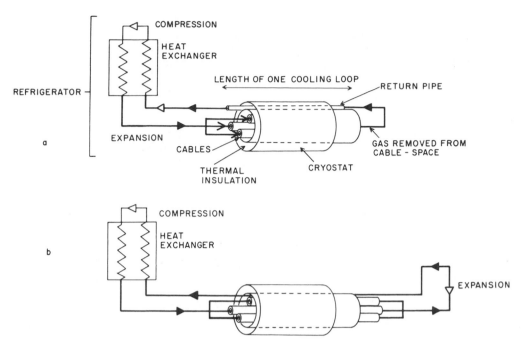

FIGURE 4 (a) Simplified diagram of a cooling loop using a thermally insulated return pipe. (b) A method of reducing cost by forming a counterflow heat exchanger between the cable and the enclosure.

distance to absorb enough heat to cause the maximum allowable temperature rise of 2 K. At this point, the cooling gas was removed from the cable and returned to the refrigerator through insulated pipes contained within the annulus between the inner and outer pipes of the cryostat (see Fig. 4a). These pipes were connected as each section of the cryostat was installed, an expensive operation that also increases the likelihood of leaks at the numerous welds. The system was improved by making the cable and enclosure a heat exchanger that formed a part of the refrigeration system, as shown in Fig. 4b. An expansion engine is required at the end of the cooling loop. Numerous computer-based studies have shown that the revised system is greatly superior to the original design. Increased conductance of the helium flow channels permits greater spacing between refrigerators, Also, an unexpected bonus turned up—smaller compressors are needed for the same heat load. The heat load expressed as compressor horse power required to keep the system at operating temperature can be regarded as the loss or inefficiency of transmission.

The losses in a typical system arise from a variety of sources:

1. heat in-leak into the enclosure;
2. current-dependent losses due to conductor loss, eddy currents, and resistive joints; and
3. voltage-dependent losses (for a warm dielectric insulation these losses are negligibly small).

In a perfect system, if losses at the operating temperature T_2 are W_L, then the input to the refrigerator that is rejecting heat at T_1 is W_R, as given by the Carnot relation, viz.,

$$W_R = W_L[(T_1 - T_2)/T_2]. \qquad (13)$$

For example, at an operating temperature of 7 K with rejection at 300 K, every watt dissipated requires about 42 W of refrigeration input to maintain a constant temperature. In practice, no machine approaches the Carnot efficiency. For practical designs, 1 W of dissipation at 7 K would require about 250 W of refrigeration. For a nitrogen machine operating at 77 K, the corresponding Carnot efficiency is 2.9, but in practice about 10 W of refrigeration would be needed to sustain a 1-W load. This is, of course, a great improvement over a helium-cooled system and explains the intense interest in HTS-based superconducting cables.

The cooling system that evolved at Brookhaven used helium as the working fluid. The coolant was supplied to the load in the supercritical regime as a single-phase fluid at 225 psia (1.55 MPa) and a temperature between 6.5 and 8.5 K. "Supercritical" means that the helium is operating at a temperature above 5.2 K and a pressure of 34 psia (0.23 MPa). The temperature and pressures in the high-pressure and low-pressure spaces of a system similar to that shown in Fig. 4b are given in Fig. 5 for a cooling-loop length of 30 km (i.e., a refrigerator spacing of 60 km). The cryogenic envelope ultimately perfected at Brookhaven contained a thermally insulated space that is evacuated

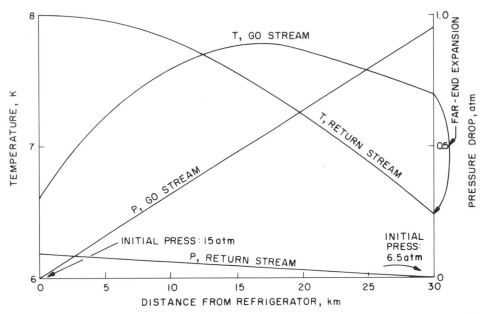

FIGURE 5 Pressures and temperatures in go and return coolant streams in a system similar to that in Fig. 4b.

and sealed at the time of manufacture. It is not practical to envisage external vacuum pumps paralleling the cable route. Both rigid and flexible sections were made in this way. The actual system designs studied at Brookhaven used rigid sections for 90% of the distance and flexible sections for 10%.

The technology for nitrogen-cooled systems is mature and bulk liquid nitrogen is used commercially in many industrial processes. A typical system for cooling a short HTS line, such as those envisaged for demonstration at the distribution voltage, is shown is Fig. 6. A large insulated vessel containing 10,000 gal (37,800 liters) of liquid nitrogen is connected to an evacuated heat exchanger. By lowering the pressure, the boiling point is reduced to about 66 K, causing subcooling of the liquid nitrogen in the cooling loop of the transmission system. The flow rate must be adjusted to provide an acceptable range of temperatures along the HTS cable. The heat input will vary with the electrical load carried by the cable. Calculations of temperature rise for a typical heat flux are shown graphically in Fig. 7. It should be noted that the pressure drop along a loop even as short as 30 m is quite severe, which illustrates the problem of designing nitrogen-cooled systems with lengths of kilometers. The temperature rise in a long length of an HTS cable would be controlled by means of intercoolers at perhaps 500-m intervals. This is a technically feasible solution but likely to prove expensive.

The terminations are a technical challenge comparable to the designs of cables. They perform many functions, including (1) providing a fixed anchor point for the cables, (2) carrying the current conductor and voltage insulation

across the temperature differential between the bus bars at the substation and the superconducting cable, and (3) allowing the introduction and removal of cooling gas from the system.

The termination for an LTS cable design may be divided into three major portions as shown in Fig. 8. The first part provides the anchor point for the cable armor and allows the outer conductor to be attached to a large copper cylinder, which carries the outer conductor current. Inside the cylinder, the cable insulation is formed into a stress cone to grade the horizontal component of the electric field so that the inner conductor is exposed. A perforated joint is made to the inner conductor so that coolant gas can be forced down the inner core of the cable. In the second region, the inner and outer conductors are joined to a device known as the cryogenic bushing. This is a coaxial current-carrying bushing that is cooled by cold helium gas. By correct choice of the heat-transfer rate and gas flow, a temperature gradient is established along the cryogenic bushing between the ambient and operating temperature of the cable. Both parts are cryogenic and need an insulating vacuum space around them. In the Brookhaven design, both of these regions are horizontal, mainly for ease of assembly. The third portion is a large but fairly conventional air-entrance bushing, which is vertical. The insulating medium is SF_6 gas. The joint region at the ambient end of the cryogenic bushing and the SF_6 elbow is cooled by a water–glycol mixture fed from a pump at the top of the elbow. A view of the two air-entrance bushings at the east end of the outdoor test facility is shown in Fig. 9.

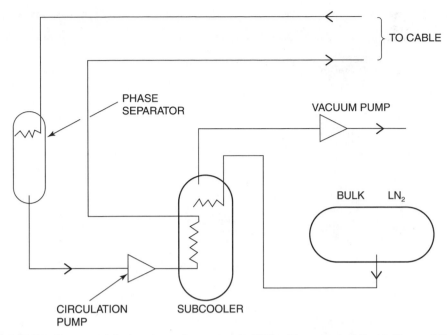

FIGURE 6 Conceptual design of a cooling system for HTS cables using bulk liquid nitrogen (LN_2).

IV. OUTDOOR TEST FACILITIES

An enormous gulf exists between laboratory-scale experiments and apparatus actually intended for service in the field. In the power engineering industry, this is especially true because of the huge capital investment in existing de-

signs and the overriding need for extreme reliability. Thus, the progression of new developments to service is long and tedious. The first step in the path is to build a trial system with sufficiently high electrical ratings and physical size to make a start on the solution of practical problems rather than intrinsic design.

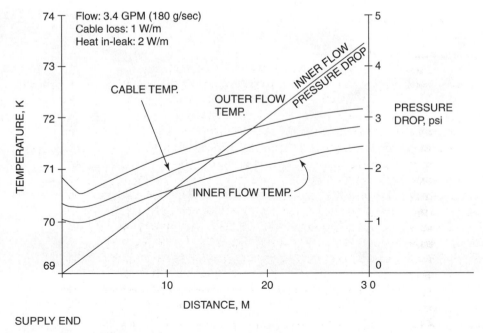

FIGURE 7 Calculated values of temperature profile and pressure drop for a short HTS cable cooled by LN_2.

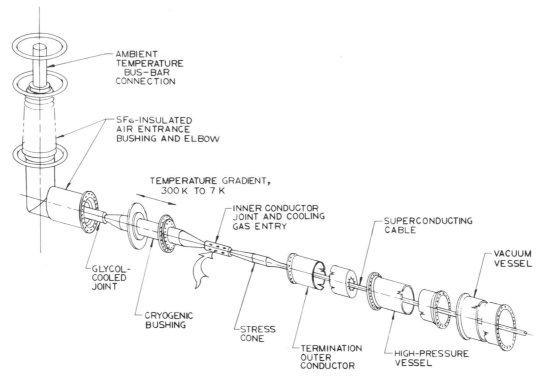

FIGURE 8 Termination for an LTS cable. The cryogenic portion of the termination is horizontal. The cable end is on the right, and the inner conductor is exposed by terminating the dielectric insulation with a hand-built stress cone. Both inner and outer conductors are attached to the cryogenic bushing, which makes the connection across the temperature difference of 300 K between the cable and the ambient-temperature bus bars. The ambient temperature portion is an SF_6-filled elbow that is vertical. A pump at the top of the ceramic bushing circulates a water-glycol mixture to cool the inner conductor of the elbow.

FIGURE 9 The east end of the cable enclosure at the Brookhaven 1000-MVA test facility. The white pipe on the left is the cable enclosure. The device on the right-hand side of the yard is a capacitive divider that permits cooling gas and optical fibers to cross the voltage interface to ground potential. The boxlike structure between the terminations is a power supply used to balance inner and outer conductor currents. [Photograph courtesy of Brookhaven National Laboratory.]

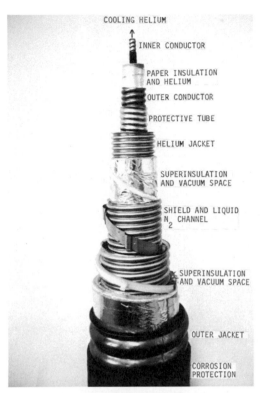

FIGURE 10 Construction details of the fully flexible superconducting developed at the Anstalt für Tieftemperaturforschung and made by Kabelmetal. The assembly contains both the single-phase cable and the thermally insulated cryogenic enclosure. [Photograph courtesy of Anstalt für Tieftemperaturforschung.]

A. Arnstein Test Facility

The Arnstein Test Facility consisted of a 50-m-long, fully flexible conductor and cryogenic envelope made by Kabelmetal. A view of the cable and envelope is shown in Fig. 10. The dimensions and construction details are given in Table II. The way in which the test cable was connected to the three-phase overhead line is shown in Fig. 11. A photograph of the site is shown in Fig. 12. The electrical characteristics of the Arnstein feeder and some details of the test results are given in Table III. R&D on superconducting cables at the Institute for Low-Temperature Research stopped in 1983 on completion of the testing schedule at Arnstein.

B. Brookhaven 1000-MVA Test Facility

The Brookhaven facility provided all the equipment and auxiliary subsystems necessary for testing 138-kV, 1000-MVA cables (three-phase rating) in a single-phase mode. The Brookhaven site was the only site in the world capable of simultaneously exciting cables with rated current and voltage. The test facility comprised an enclosure containing two superconducting cables with terminations at each end. The superconducting load was cooled by means of a 700-W (nominal) supercritical helium refrigerator at one end, with a small auxiliary refrigerator for lead cooling at the other end. Electric equipment provided simultaneous or independent voltage and current excitation at 60 Hz; a separate Marx generator was available for impulse testing.

The 60-Hz excitation equipment consisted of a capacitor insulated from ground, which tuned the cable inner conductor loop to parallel resonance, and a variable inductor, which tuned the cable insulation capacitance into series resonance. In addition to the main supplies, a small resonant supply was used to adjust the balance between inner and outer conductor currents. The facility enabled the three major components of the underground transmission system, namely, the cables, terminations, and cable enclosure, to be tested under realistic conditions. (A view of the test site is shown in Fig. 9.)

After each major component such as the refrigerator or enclosure had been installed, the system was operated to measure the cryogenic performance. The last cryogenic

TABLE II Design Features of the Single-Phase Superconducting Cable Installed at Arnstein

Tube No.	Material	Inner/Outer diameter (mm)	Wall thickness (mm)	Corrugation pitch (mm)	Material surrounding
1	Cu–Nb (Nb outside)	9.8/13.2	0.4	6.5	2 layers carbonized paper, 116 layers insulating paper, 2 layers H paper
2	Cu–Nb (Nb inside)	40.3/46.5	0.5	11.0	Tube No. 3
3	Stainless steel AISI 304	47.6/53.2	0.5	10.5 (left)	Tube No. 4
4	Stainless steel AISI 3044	76.2/65.6	0.6	10.0	Teflon spacer and 5 layers superinsulation
5	Stainless steel AISI 304	98.0/107.2	0.8	11.8	Flanged steel spacer
6	Stainless steel AISI 304	127/143	0.8	16.8	Polyethylene spacer and 50 layers superinsulation
7	Cu–Se F22	195/216	1.3	36.0	Polyment and polyethylene jacket

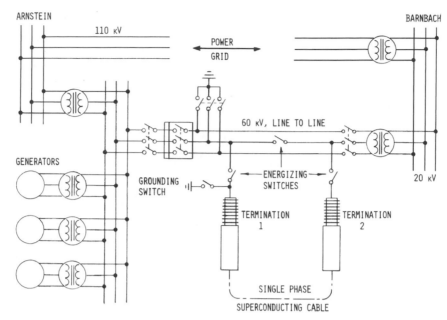

FIGURE 11 Simplified circuit diagram of the test arrangement at Arnstein. By means of the switches, the single-phase superconducting cable could be placed in series with a phase of the 60-kV overhead feeder.

test performed before the cables were energized was carried out in May 1981, and the first electrical excitation occurred in October 1982. The cable design is shown in Fig. 13 and major characteristics are listed in Table IV.

The cables, enclosure section, and terminations were essential designs that would be required in an ultimate utility company application. For example, the enclosure design used in the Philadelphia Electric Company study

FIGURE 12 View of a cable termination at the Arnstein superconducting cable test site. [Photograph courtesy of Anstalt für Tieftemperaturforschung.]

TABLE III Thermal and Electrical Characteristics of the Arnstein Test Cable

Arnstein line rating	
Voltage	60 kV, three-phase; 40 kV, l-n
Current	80 A, 50 Hz
Cable length	50 m, single-phase
Enclosure heat in-leak	0.15 W/m (at 6 K)
Dielectric loss (40 kV)	0.02 W/m (at 6 K)
Cable current rating	1000 A

TABLE IV Characteristics of the Brookhaven Nb$_3$Sn-Based Cable Design

Length installed (each)	~115 m
Outside diameter (over armor)	5.84 cm
Inner conductor diameter	2.95 cm
Operating temperature	7 K average, 8 K maximum
Operating helium density	100 kg/m^3 minimum
Operating pressure	225 psia (1.55 MPa) typical
Rated voltage, 60 Hz	80 kV, l-n; 138 kV, three-phase
Rated current, 60 Hz	4100 A, continuous; 6000 A, 60 min
Operating voltage stress	10 MV/m
Operating surface current density	442 A/cm
Maximum continuous	1000 MVA, three-phase
Surge impedance load	872 MVA, three-phase
Cable conductor loss at 4100 A (7 K)	0.2 W/m
Cable dielectric loss at 80 kV	<0.06 W/m
Cable impedance	24 Ω (calc.), 25 Ω (meas.)

was scaled up from the one at the test site. Three other subsystems were needed for the tests, but they were not directly applicable to the final design; however, a great deal of work during the development phase was needed to build them at the test facility.

The first auxiliary subsystem built was the cooling refrigerator. This machine supplied supercritical helium gas to the cable enclosure at a pressure of about 225 psia (1.55 MPa) and an average temperature of 7 K. The refrigerator used a 350-hp (260-kW) screw compressor. Four very high-speed gas-bearing turbines were used to cool the gas by expansion. The system could absorb about 700 W at the 7-K operating temperature. In addition, a smaller refrigerator supplied helium to cool the current-carrying conductors of the terminations. The characteristics of the refrigerator and enclosure are listed in Table V.

Operation of the cooling plant would have been impossible without the second major subsystem to be built, namely, a computer-based data acquisition, analysis, and control system. About 100 transducers were monitored

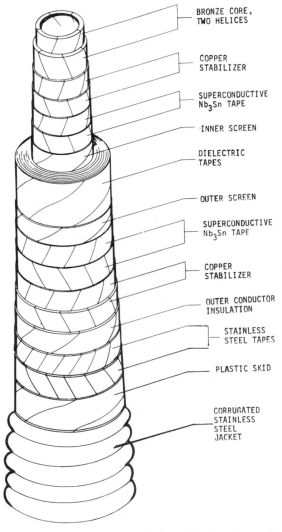

FIGURE 13 General assembly diagram of the Brookhaven flexible superconducting power transmission cable design.

BRONZE CORE, TWO HELICES

COPPER STABILIZER

SUPERCONDUCTIVE Nb$_3$Sn TAPE

INNER SCREEN

DIELECTRIC TAPES

OUTER SCREEN

SUPERCONDUCTIVE Nb$_3$Sn TAPE

COPPER STABILIZER

OUTER CONDUCTOR INSULATION

STAINLESS STEEL TAPES

PLASTIC SKID

CORRUGATED STAINLESS STEEL JACKET

TABLE V Summary of the Cryogenic System at the Brookhaven 1000-MVA Test Facility

Enclosure	20-cm i.d., 40-cm o.d.; five sections; with permanent (nonpumped) insulating vacuum
System length	130 m, termination to termination
Enclosure heat in-leak	50 W for five sections at ~7.5 K (after 150 hr)
Operating pressure	1.55 MPa (225 psia)
Operating temperature	6.5–8.5 K
Helium density	140–110 kg/m^3
Refrigerator	350-hp oil-lubricated screw compressor; three gas-bearing turbine expanders plus a far-end turbine expander
Capacity	770 W without lead flow
Mass flow	70 g/sec (both cables)
Lead flow	0–0.3 g/sec per lead

through the system, many of them in inaccessible places such as the low-temperature cable enclosure or the top of a high-voltage bushing. All of the data gathered were reduced to units of direct interest and displayed on color video monitors that depicted a graphic representation of a section of the system of interest to the operator. The data were placed at the appropriate location on the diagram. The system was designed to permit unattended operation of the test facility. When help was required, the computer telephoned standby operators and delivered a voice-synthesized message. Incoming telephone calls were answered with a voice-synthesized summary of the status of the system. Cables were energized for eight test runs during a 4-year period from 1982 to 1986. These runs are summarized in Table VI.

The same cables were tested during the 4 years. General characteristics of results are shown in Table IV. Impulse tests that simulated very high-voltage transients caused by lightning strikes were included in the test series. The withstand voltage required for conventional cables designed for 138-kV circuits is 650 kV; one cable failed at 488 kV due to a flashover along the cold stress cone of the termination. Subsequent testing indicated that the cable itself was undamaged. The stress cone was rewound but was never retested, as funding by the Department of Energy (DOE) was withdrawn in 1986 and the project was shut down.

Another operating condition that was investigated during this series of tests was the ability of the cables to carry power after a shutdown of the cooling system. Such a condition may arise in practice if distribution power to the refrigerating equipment is interrupted; it would be most undesirable to lose transmission if this contingency arose. During the fourth and fifth runs, the main compressor was stopped to simulate a cooling-system failure. The cables

were carrying 4100 A and continued to do so for more than 2 hr.

Loss measurements for both cables were made during all runs over the range 250 to 6000 A between temperatures of 7.2 and 13.2 K (see Fig. 14). These measurements are not easy to make, and inaccuracies no doubt exist in the data gathered. Typically a loss measurement for these cables corresponds to measuring a power factor angle of 135 μrad. This task is complicated by the magnetic field present near the terminations where the instrumentation wires emerge; typically, fields of 20 to 100 G are present. These fields give rise to induced voltage in the loss-measurement instrumentation that is synchronous in frequency with the signal of interest. Errors also exist in the temperature measurements. The goal for these was an accuracy of about ±200 mK. In addition, the temperature varies along the cables under test and as a function of

TABLE VI Summary of Operating Runs and Life Tests

Number of operating runs	8
Total time cold[a]	2727 hr
60-Hz electrical tests	
Voltage and current at 249 MVA/cable	89 hr
Voltage and current at 330 MVA/cable	169 hr
Total time with voltage or current	441 hr
Emergency rating of 6 kA and 80 kV 1-n (480 MVA/cable)	1 hr
Maximum power test, 6 kA and 110 kV 1-n (660 MVA/cable)	20 min
Overvoltage at 90 kV 1-n	22 hr
Overvoltage at 100 kV 1-n	19 hr
Overvoltage at 110 kV 1-n	13 hr

[a] Includes cooldown at start of each run (≈100 hr for the first seven runs and ≈76 hr for the eighth run).

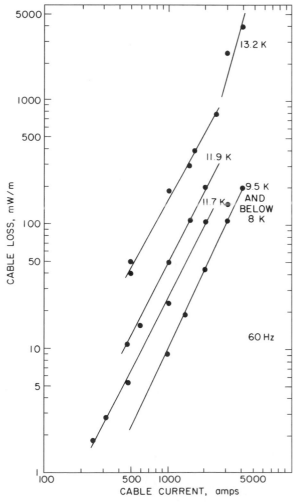

FIGURE 14 Conductor-dependent loss as a function of cable current. The break at the upper end of the curve for 13.2 K corresponds to a quench.

time. At a rated current of 4100 A, the losses are about 200 mW/m. This value is consistent with measurements in the laboratory on 10-m-long cables and is about a factor of two higher than could be obtained with ideal current distribution in the conductor. No doubt future cables will be designed to improve the losses by this factor.

The voltage-dependent characteristics of gas-impregnated tape insulation are critically affected by the electric-field enhancement in the butt gaps between adjacent tapes. In a properly made cable, partial discharge in the butt gaps is the first sign of distress as the voltage is raised. The design of good screens and control of surface irregularities are particular attributes of a "properly made" cable. The onset of partial discharge leads to an increase in dielectric loss and, ultimately, to failure of the cable insulation. Clearly the limiting factor in the allowable stress imposed on the cable insulation is the intrinsic breakdown characteristics of helium gas under the operating conditions of the transmission system. The onset of partial discharge in the cables was measured at a low helium density during the cooldown. When extrapolated to the worst-case operation conditions, there is a margin of 50% between the operating stress and the onset of measurable partial discharge activity. During quenched operation at rated current, the cables carried rated voltage without distress. During the second run, a 110-kV line to ground was applied for 13 hr, 40% over rating. During the emergency-level test at 6000 A, a voltage of 110 kV was sustained for 20 min, corresponding to 660 MVA per phase.

C. Detroit Edison Company Demonstration

The Detroit Edison company is planning to upgrade distribution cables at the Frisbie substation. Older cables will be removed from a duct bank and three ducts made available for an experimental HTS cable installation. Three single-phase cables will be pulled into the ducts with an overall length of 400 ft (~120 m). There will be a splice in the middle. At a voltage of 24 kV (three-phase, line-to-line) and a cable rating of 2400 A, the three-phase circuit will carry 100 MVA. The characteristics are summarized in Table VII. The cables will be cooled by pressurized liquid nitrogen supplied by a refrigerator using helium as the working fluid. The inner cables will carry a central cooling channel with the thermal insulation outside the conductor. The HTS is multifilamentary Bi 2223 in a silver matrix, manufactured by the American Superconductor Company. The electrical insulation will be applied to the thermal jacket and operate at ambient temperature. The cable is being designed and fabricated by Pirelli Cable and Systems of Italy. Cryogenic cooling equipment is being supplied by the Lotepro Division of Linde. Cable conductor characteristics have been measured by personnel from the

TABLE VII Characteristics of the Detroit Edison Test System[a]

Length	400 ft (120 m)
Cable inner conductor diameter	1.4 in. (36 mm)
Overall cable diameter	3.9 in. (100 mm)
Current	2400 A rms
Linear current density	212 A/cm
Voltage (line to line)	24 kV
Insulation stress	13 kV/mm
Superconductor	Multifilamentary Bi 2223 in silver matrix
Dielectric insulation	Paper–polypropylene–paper laminate (PPP) with oil impregnation (at room temperature)
Thermal insulation	Multilayer insulation in vacuum, gettered

[a] At the time of writing, the design is ongoing. Characteristics are approximate.

Los Alamos National Laboratory. A graph of conductor loss vs transport current is shown in Fig. 15. The data are typical for tests of many HTS cables reported in the literature. The consequences of the relatively low critical current density of Bi 2223 compared to Nb_3Sn can be observed by noting the loss per meter in Fig. 15. At 1500 A, the HTS dissipates about 1 W/m at 77 K, and the LTS dissipates about 25 mW/m at 8 K. When the efficiencies of practical refrigerators are considered, losses for either cable at the specified current would require 7 to 10 W of input to the refrigeration compressor. First tests at Frisbie were scheduled for the year 2000. The project is funded privately, by the U.S. DOE and by the Electric Power Research Institute (EPRI).

D. The Southwire Company, Georgia, USA

Using private funding in conjunction with support from the DOE, the Southwire Company has developed an HTS cable using cold dielectric insulation. Prototypes have been tested up to 2000 A (60 Hz) and the construction of a three-phase demonstration site is well advanced. The site is chosen so that the cables can be switched in series with a section of the 12.4-kV overhead feeder which powers the Southwire plant at Carrollton. The cable rating is 1250 A (60 Hz). The cable is a fully shielded, flexible design. The full-current outer shield conductor prevents extra losses caused by the coupling of an unshielded conductor to other cables or metallic ducts. The length of the HTS section will be 100 ft (~30 m). The dielectric insulation is plastic tape impregnated with LN_2 under pressure and the HTS is Bi 2223 multifilamentary tapes manufactured by Intermagnetics General Corporation. Testing of cable prototypes has been carried out at Oak Ridge National

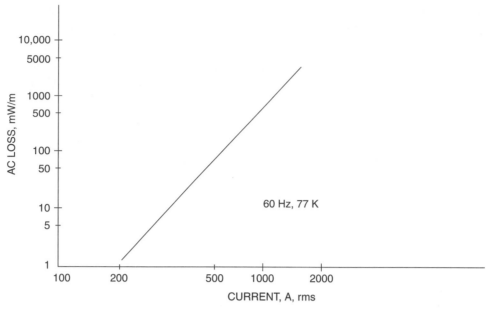

FIGURE 15 AC loss of an HTS conductor made with eight layers of Bi 2223 tapes. The slope of 2.6 suggests that the outer layer(s) is (are) saturated, i.e., the transport current density is J_c (square-law current dependence) and the inner layers exhibit hysteretic loss (cubic current dependence).

Laboratory. An isometric view of the cable design is shown in Fig. 16 and a view of the test site under construction is shown in Fig. 17.

V. DESIGN AND COST OF ELECTRIC UTILITY COMPANY SYSTEMS

All new developments in the field of power engineering must ultimately produce designs that are cost-effective. In addition, the technical performance must be satisfactory under normal and abnormal operating conditions, and the reliability must be acceptable over the projected life of the equipment. These are very stringent requirements that must be considered early in the development stages so that design changes to meet them can be made in time.

Numerous cost studies of low-temperature systems were made during the period 1973 to 1990. These analyses received special emphasis when a study funded by the DOE was carried out by the Philadelphia Electric Company (PEC) to cost several technical approaches then under development for the same complete system 66 mi (106 km) in length. Development funding for the more expensive designs was reduced or discontinued. The flexible cable design using Nb_3Sn superconductor under development at Brookhaven was estimated to have the lowest cost

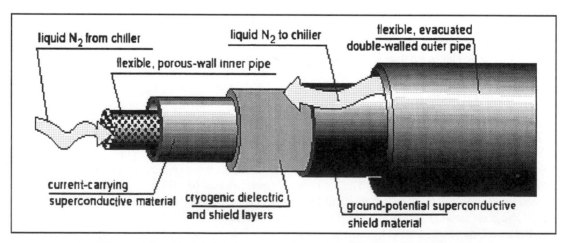

FIGURE 16 Assembly drawing of the HTS cable designed by the Southwire Company.

FIGURE 17 The test site at Carrollton, Georgia, under construction at the Southwire plant. [Photograph courtesy of the Southwire Company.]

when capitalized operating expenses were included. At this distance in time from the PEC study the actual cost estimates are not particularly relevant, but the relative contribution of the major components to the overall cost is still instructive. A breakdown such as that in Table VIII highlights which subcomponents of a system are the most expensive and gives direction for subsequent R&D.

In the mid-1990s the Pirelli Cables and Systems Company in cooperation with Electricite de France (EDF) studied the cost of two designs using HTS conductors. The parameters are shown in Table IX. A cold dielectric insulation of PPP impregnated by LN_2 coolant was considered. Experimental trials of the insulation were carried out to determine the operating stress vs lifetime. The cost of flexible cryostats for each phase was compared with the cost of a rigid cryostat containing three-phase conductors.

Costs were determined for the HTS conductor with two assumed values of critical current, 6×10^4 and 1×10^6 A/cm². The former represents material available today and the latter is a goal for future development. As most cost studies tend to be site-specific, the results are compared to cable insulation (XLPE), a proven circuit element for which installation costs are well known. Multiple circuits of conventional cable would be required to match the system power rating. The conclusion of the study was that the cost of XLPE cables was about the same as an HTS installation using the lower value of critical current.

TABLE VIII Relative Cost of a Low-Temperature Superconducting System Rated at 800 MVA, 40 km in Length

Component	% cost
1. Cable	19.7
2. Cryogenic enclosure	21.0
3. Joints and splices	0.5
4. Terminations	1.2
5. Refrigerator and cryogen	7.6
6. Installation	29.0
7. Losses[a]	21.0

[a] Present worth based on 40-year life: $0.08/kWh, $2000/kW demand charge.

TABLE IX System Parameters for the Pirelli–EDF Study

Power rating (MVA)	3000	
Nominal voltage (kV)	90	225
No. of circuits	6	4
Line current per (n-1) circuits (kA)	3.85	2.57
Power per (n-1) circuits (MVA)	600	1000
Circuit length (km)	10	10
BIL (KV)	459	1050
Short-circuit current (kA)	10.3	31.5
Short-circuit duration (sec)	1.7	0.5

The design with futuristic HTS (J_c of 10^6 A/cm^2) came in at about 60% of the conventional cable installation. It was assumed in both cases that losses were about the same, i.e., the compressor input power of the superconducting design matches the resistive and eddy current losses of the XLPE cables.

VI. SUMMARY

Although enormous progress has been made in this field both in the laboratory and at large outdoor test sites since the late 1960s, superconducting cables have yet to see routine use in electric utility operation. To the Low-Temperature Research Institute in Graz goes credit for the first experimental installation on an electric utility company system. The extensive 4-year testing period at Brookhaven National Laboratory on Long Island demonstrated that LTS cables were astonishingly rugged and could function under a wide range of normal and abnormal conditions. The construction of two test sites to evaluate HTS-based cables at distribution power levels is nearing completion. Their performance will be crucial for the ongoing development of this technology.

It should be noted that the total loss for a superconducting system will be set largely by economic rather than technical factors. For example, the losses of a conventional transmission system could be reduced by a factor of roughly four by halving the rated current. On the other hand, capital would be needed to build another transmission circuit to replace the lost capacity. It is generally more cost-effective to accept the higher losses, but this trade-off depends on the cost of energy and the cost of money. In Section IV, the decision by Brookhaven workers to eliminate the nitrogen-cooled shield of the cryogenic enclosure is an example of this optimization. Thus, a higher heat in-leak was traded for a lower capital cost. Studies at Brookhaven and by Pirelli seem to point to comparable transmission efficiencies for both LTS and HTS cables and are about the same as for conventional facilities.

The test sites now under construction to evaluate HTS cables at the distribution-level voltage point to a relatively inexpensive way to introduce the technology into the utility industry. However, it must be borne in mind that these are highly specialized niche applications. At some stage, the technology will have to be shown to be cost-effective. It seems unlikely that separate cryogenic envelopes for each phase, which severely limit the amount of cable on a reel, will meet this criterion as installation lengths are increased. In addition, the present fabrication cost of HTS is high. It is to be hoped that costs will decrease dramatically as the production volume increases. Finally, the higher the critical current density the better. A minimum value of 10^6 A/cm^2 with $T_c > 112$ K must remain an important goal for researchers.

Author is formerly of Brookhaven National Laboratory, operated by Brookhaven Science Associates, Inc., under contract to the U.S. Department of Energy.

SEE ALSO THE FOLLOWING ARTICLES

CRYOGENIC PROCESS ENGINEERING • CRYOGENICS • DIRECT CURRENT POWER TRANSMISSION, HIGH VOLTAGE • POWER TRANSMISSION, HIGH VOLTAGE • SUPERCONDUCTING DEVICES • SUPERCONDUCTIVITY

BIBLIOGRAPHY

Black, R. M. (1983). "The History of Electric Wires and Cables," Peter Peregrinus, London.

Dahl, P. (1992). Superconductivity: Its Historical Roots and Development from Mercury to the Ceramic Oxides," American Institute of Physics, New York.

Ginsberg, D. M. (ed.) (1996). "Physical Properties of High Temperature Superconductors V," World Scientific, Singapore.

Luhman, T., and Dew-Hughes, D. (1979). "Treatise on Materials Science and Technology," Vol. 14, Academic Press, New York.

Tanaka, T., and Greenwood, A. (1983). "Advanced Power Cable Technology," Vols. I and II, CRC Press, Boca Raton, FL.

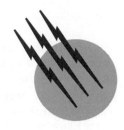

Superconducting Devices

Jukka Pekola
University of Jyväskylä

GLOSSARY

Cooper pair Current carrying condensate of a superconductor formed of pairs of electrons.

Josephson junction Insulating barrier separates two superconductors. Cooper pairs can tunnel through.

NIS Normal metal—superconductor—normal metal tunnel junction, which can be used for sensitive thermometry and for on-chip cooling.

RSFQ Rapid Single Flux Quantum devices for ultrafast superconducting digital electronics.

Single cooper pair devices New set of devices based on transport and storage of single charges. Possible digital, sensor, and quantum computing applications.

SQUID Superconducting Quantum Interference Device, a sensitive detector of magnetic flux and its derivatives.

TES Transition Edge Sensor. Usually a transition of superconductivity is used for sensitive energy and power measuring.

I. INTRODUCTION, ON-CHIP SUPERCONDUCTING DEVICES

In this article we deal with superconducting *on-chip* and mainly *planar* devices. Thus we do not discuss bulky, non-planar components or instruments, such as cables, magnets, motors, etc.

One of the main ingredients in superconducting electronics and devices is the Josephson tunnel junction, and the superconducting quantum interference device (SQUID), with Josephson junctions forming essential parts of it. Thus, much of the history of superconducting electronics and a good fraction of this article also deal with SQUIDs, either as devices as such, or as sensitive detectors integrated in the device. Josephson junctions, forming the key details in the SQUIDs, have been in the focus of even large industrial programs to investigate and promote the possibilities of superconducting electronics.

Encyclopedia of Physical Science and Technology, Third Edition, Volume 16
Copyright © 2002 by Academic Press. All rights of reproduction in any form reserved.

In the present text we discuss, besides Josephson junctions and SQUIDs in their traditional framework, Josephson-junction-based rapid single flux quantum devices (RSFQs), thermal radiation detectors based on superconductivity properties, normal-insulator-superconductor tunneling in thermometry and in superconducting on-chip coolers, and future devices based on single Cooper pair tunneling in very small Josephson junctions.

II. SQUIDS

A. Josephson Junctions

There are principally two different kinds of SQUIDs: superconducting loops interrupted by either one or two Josephson junctions. These SQUIDs are called, mainly for historical reasons, an rf SQUID and a dc SQUID, respectively. Before detailing the SQUID operation, let us discuss the basic properties of a Josephson junction. A Josephson junction consists of two superconductors separated by a thin insulating barrier. The two superconductors both have their macroscopic wavefunctions with phases φ_i, $i = 1, 2$, and the phase difference $\varphi = \varphi_1 - \varphi_2$ determines the current I_s, i.e., the flow rate of Cooper pairs through the barrier as

$$I_s = I_c \sin \varphi, \tag{2.1}$$

where I_c is the critical current of the junction, which is the maximum supercurrent that the junction can sustain without dissipation. Upon increasing the current I through the junction from zero up to $I > I_c$, there is initially no voltage across the junction. But at about $I = I_c$ voltage V appears, and φ starts to evolve with time according to

$$\dot{\varphi} = 2eV/\hbar = 2\pi V/\Phi_0, \tag{2.2}$$

where $\Phi_0 = h/2e \simeq 2.07 \cdot 10^{-15}$ Wb is the so called flux quantum.

To avoid the inevitable, and in usual operation harmful hysteresis in sweeping the current from zero up to above I_c and back to zero again (voltage does not drop back to zero at I_c on return but at current much below I_c), one typically shunts the junction by an external resistance R. In this case, the cross dynamics of the junction can be described according to the so-called "resistively and capacitively shunted junction" (RCSJ) model, as depicted in Fig. 1A. In this model an ideal junction [Eq. (2.1)], is connected in parallel with its self capacitance C and the shunt resistance R.

Including the noise current I_N, which arises from the shunt resistance, we can write the equation of motion of the junction as

$$I_c \sin \varphi + C\dot{V} + V/R = I + I_N, \tag{2.3}$$

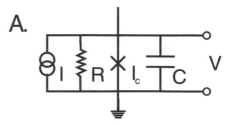

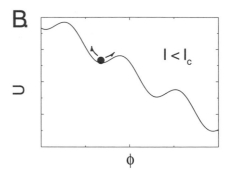

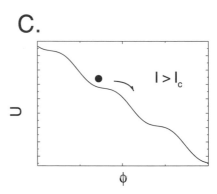

FIGURE 1 The resistively and capacitively shunted Josephson junction (RCSJ) model. The RSJ circuit is seen in (A), the cross depicts the Josephson junction itself. The tilted washboard model for $I < I_c$ in (B) and for $I > I_c$ in (C).

where now I is the total current through the junction-shunt system. Neglecting I_N and making use of Eq. (2.2), we have

$$\frac{\hbar C}{2e}\ddot{\varphi} + \frac{\hbar}{2eR}\dot{\varphi} = -\frac{2e}{\hbar}\frac{\partial U}{\partial \varphi}. \tag{2.4}$$

Here $U = -\frac{\Phi_0}{2\pi}(I\varphi + I_c \cos \varphi)$ acts as a potential for the variable φ. This model [Eq. (2.4)] presents an analogy between the phase difference and a particle in the tilted washboard potential U, where the average slope of the potential is proportional to $-I$.

When $I < I_c$ (Fig. 1B), the "particle" is confined to one of the wells, and oscillates therein at the plasma (angular) frequency $\omega_p = \sqrt{2\pi I_c/\Phi_0 C}[1 - (I/I_c)^2]^{1/4}$. Then the average of $\dot{\varphi}$ and hence the average voltage across the junction are zero. When I exceeds I_c (Fig. 1C), the "particle" starts to roll down the washboard, the average of $\dot{\varphi}$ and the average voltage across the junction are nonzero. This is the dissipative regime of the RCSJ. If the junction is nonhysteretic, the "particle" gets trapped when current drops below the critical value, and V drops to zero again. In order for this to happen we require $\beta_c \equiv (2\pi I_c R/\Phi_0)RC \equiv \omega_J RC \leq 1$. Here ω_J is the Josephson (angular) frequency at the voltage $I_c R$.

One of the consequences of Eqs. (2.1) and (2.2) is that when the Josephson junction is subjected to radiofrequency (rf) irradiation, there are so-called Shapiro steps in the current voltage characteristics (IVC) at voltages $V = nhf/2e$, where n is an integer and f is the frequency. This has made it possible, among other things, to develop a *Josephson volt standard* based on a series connection of more than 10,000 nonhysteretic junctions. The present definition of volt (V) in the SI system of units is based on this Josephson standard.

B. The dc SQUID

In a dc SQUID (Fig. 2) the two junctions interrupting the superconducting loop are shunted resistively as described above, this way eliminating the hysteresis of the current voltage curve. The IVC of the dc SQUID depends on the external flux Φ_{ext} applied to the loop, as shown schematically for two values of Φ_{ext} in Fig. 2B.

The critical current of the dc SQUID, I_m depends periodically on the flux threading it as $I_m = 2I_c |\cos(\pi \Phi/\Phi_0)|$. Here we assume identical junctions with critical current of I_c for both. If the SQUID is biased at a current I_b above $2I_c$, which is the maximum supercurrent through the SQUID, the voltage will therefore oscillate as a function of Φ with period Φ_0 (Fig. 2C). When operated in the steep part of the $V - \Phi_{\text{ext}}$ curve, the SQUID produces an output voltage in response to a small flux $\delta\Phi_{\text{ext}} \ll \Phi_0$. It is thus essentially a flux-to-voltage converter.

C. The rf SQUID

The one junction loop, the rf SQUID, can also be employed as a sensitive detector of flux. Assume a loop with inductance L, and a junction in it with critical current I_c again. The flux threading the loop, Φ, is then, by adding up the total flux due to external flux and the flux due to the circulating current, given by

$$\Phi = \Phi_{\text{ext}} - LI. \qquad (2.5)$$

Here I is the current circulating in the loop. The Faraday induction law, $V = -d\Phi/dt$ for the almost closed SQUID loop, yields a relation between the magnetic flux and the Josephson phase difference

$$\phi = 2\pi\Phi/\Phi_0 \qquad (2.6)$$

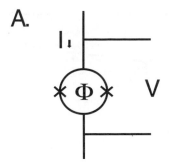

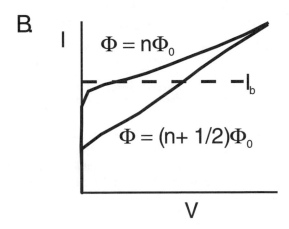

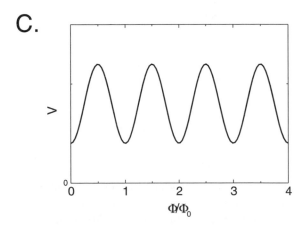

FIGURE 2 The dc SQUID. The SQUID loop is shown in (A); the junctions are shunted resistively as described above, but this has not been shown in the figure. Schematic current (I)-voltage (V) characteristics at two different values of the applied external magnetic flux Φ_{ext} in (B). V at a constant bias current $I > 2I_c$ in (C).

by integration of the ac Josephson relation [Eq. (2.2)]. Using the dc Josephson relation [Eq. (2.1)] and combining it with Eqs. (2.5) and (2.6) we finally obtain

$$\Phi + LI_c \sin\left(2\pi \frac{\Phi}{\Phi_0}\right) = \Phi_{\text{ext}}. \qquad (2.7)$$

Equation (2.7) presents a nonlinear dependence of the flux threading the loop as a function of the external flux. In particular, for parameter values $LI_c \geq \Phi_0/2\pi$ the $\Phi_{\text{ext}} - \Phi$

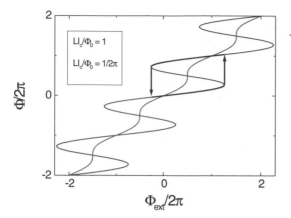

FIGURE 3 The flux Φ through a SQUID loop against the applied magnetic flux Φ_{ext} at two different values of the loop and junction parameters, $L I_c/\Phi_0 = 1/2\pi$ and $L I_c/\Phi_0 = 1$. In the latter case the backbending of the curve induces hysteresis. A typical hysteresis loop on excursion of the external flux is indicated by arrows.

characteristics show re-entrant behavior, and the dynamics are hysteretic on changing Φ_{ext} (see Fig. 3).

The sections with positive slope are stable and the ones with negative slope are unstable. With $L I_c \simeq \Phi_0$ the energy dissipated upon traversing the simplest hysteresis loop indicated in Fig. 3 is about $I_c \Phi_0$.

The radio frequency (rf) operation of the SQUID means connecting an RLC tank circuit with inductive coupling to the SQUID loop (see Fig. 4).

The detection of flux is possible, since the (amplitude of the) current in the inductor of the tank circuit at which the hysteresis loop first gets traversed depends on the bias point, i.e., the quasistatic value of Φ_{ext}. When the flux becomes hysteretic it can be detected as energy loss, i.e., change of rf voltage on the tank circuit.

D. Noise in SQUIDs

It is typical to express the noise in SQUIDs in terms of flux noise spectral density $S_\Phi(f)$, where f is frequency. The most useful quantity indicating the flux noise, $S_\Phi(f)^{1/2}$, is usually expressed in units $\Phi_0/\sqrt{\text{Hz}}$. A convenient way

of characterizing the flux noise is the noise energy per unit bandwidth, $\varepsilon(f) = S_\Phi(f)/2L$, which, in turn, can be expressed in units $\hbar/\text{Hz}$.

Present day dc SQUIDs operating at 4 K typically have flux noise of order $S_\Phi(f)^{1/2} \simeq 10^{-6}\Phi_0/\sqrt{\text{Hz}}$, and $\varepsilon(f) \sim 10^2\hbar$. By cooling the dc SQUID and the shunt resistors down to well below 1 K, one can achieve $\varepsilon(f)$ of less than $10\,\hbar$. One limitation of dc SQUIDs is the $1/f$ noise, which depends on the quality and material of the Josephson junctions. $1/f$ noise is an issue especially in biomagnetic measurements where good resolution at frequencies down to below 0.1 Hz is required.

In contrast to the dc SQUID, noise in the rf SQUID depends also on the pre-amplifier and the tank circuit. Today rf SQUIDs can, when operated at >1 GHz frequencies, achieve flux noise of the order of few times $10^{-6}\Phi_0/\sqrt{\text{Hz}}$. The corresponding noise energy is typically $\varepsilon(f) \sim 10^3\hbar$, which can be improved, however, by using a cooled pre-amplifier.

E. SQUID Instruments

The SQUIDs can be used in a large variety of ultrasensitive measurements where the signal to be detected can be converted into magnetic flux. The basic measurement by a SQUID is to measure magnetic field. A schematic illustration of a dc SQUID as a *magnetometer* is shown in Fig. 5A, where a flux transformer circuit with a pick-up loop is connected to the input coil to make a superconducting flux transformer. The inductances of the pick-up coil

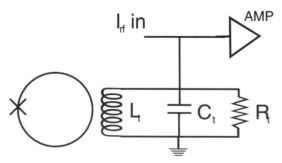

FIGURE 4 An LRC resonant tank circuit coupled via the inductance L_t to an rf SQUID.

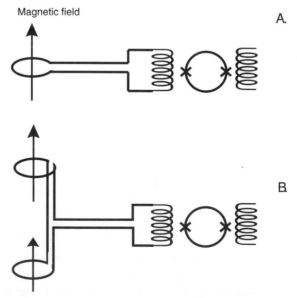

FIGURE 5 SQUID in measuring magnetic fields. Flux transformer as a magnetometer in (A) and as a first-derivative gradiometer in (B).

and the input coil determine the flux into the SQUID loop corresponding to the change in magnetic field at the measurement position (pick-up loop). The third coil is used to set the optimum bias point of the device, and to operate in the flux-locked mode. SQUID-magnetometers are used in geophysics and in various research purposes. The flux transformers have typically involved loops made of Nb. A typical noise level of a superconducting magnetometer is of the order of 10^{-11} mT/$\sqrt{\text{Hz}}$.

The next step is to measure spatially varying magnetic fields. A *gradiometer* measures basically the gradient of magnetic field along the axis of a pick-up coil with two oppositely wound loops (Fig. 5B). With a uniform axial field the gradiometric measurement yields a zero signal if the coils are equal. The input coil and the SQUID operate like a magnetometer. Highly symmetric gradiometer coils can be made by thin film techniques. An important application of gradiometry is biomagnetism, where spatially varying magnetic fields from local sources (e.g., human brain or heart) can be discriminated against virtually uniform fields from distant sources.

A gradiometer set-up of Fig. 5 can be used as a *susceptometer* by inserting the sample whose susceptibility χ is to be determined inside one of the halves of the pick-up coil. The signal of the susceptometer is proportional to χ in the presence of a static uniform magnetic field. SQUID-susceptometers are available commercially.

SQUIDs can be easily configured as very sensitive *ammeters* or *voltmeters* also. For example, in measuring the current through a voltage-biased superconducting transition edge desensor, TES, as will be described in Sections III.C and D, the current is allowed to pass through a series connection of several (in some cases 100) input coils of individual dc SQUIDs; this way the current to be determined is transformed into flux through the SQUID loop.

A variation of susceptometry is the measurement of *displacement*. Position of a magnetic object in the vicinity of one-half of the pick-up coil can be detected with high accuracy due to the change of the "filling factor" of this object with nonzero χ. This way, besides conventional magnetic objects, position of a superconductor with perfect diamagnetism ($\chi = -1$) or of a nonmagnetic object attached to a magnetic one can be measured.

F. RSFQ, Principle and Devices

Binary "0" and "1" are often realized by latching circuits in semiconductor logics. This was how researchers attempted to mimic the realizations of superconducting electronics in the 1970s and 1980s. With superconductors, in particular in Josephson junctions, this unnatural choice of representing the logic states corresponds to the two voltage states of the junction when current through it

is below the critical value I_c. The "0" could be presented by the supercurrent branch, where current flows with zero voltage, whereas "1" corresponds to the dissipative branch with voltage $V \simeq 2\Delta/e$. Because of the complicated dynamics of the junction, especially upon reducing the current, the speed of switching is limited to a few GHz in such devices. Latching circuits also lead to severe cross-talk problems when integrated. Therefore, two extensive programs to develop and promote superconducting electronics, one by IBM in the United States (1969–1983) and the second one, MITI, in Japan (1981–1990) were terminated without commercial continuation, mainly not because of unsuitability of the technology, but because of low operation frequency due to the principally hopeless attempt to operate Josephson junctions in an analogous mode as semiconductor-based digital electronics. Comparable performance of superconductor-based electronics could not justify their use against conventional semiconductor technology: cooling of superconductors to helium temperature (4.2 K) is a handicap anyway.

The development of RSFQs, or alternatively dynamic single flux quantum (SFQ) digital technology presents a new attempt to develop superconductor-based fast digital electronics. The speed of the Josephson junctions can be fully benefited in the SFQ devices. Besides the speed of the active device, one should realize that superconductors also provide the fastest passive transmission lines across the chip essentially without cross talk. This is made possible by the fact that superconductors provide very little AC-attenuation up to the frequency $f = 2\Delta/h$, which is of the order of 700 GHz for Nb, the most common superconductor in planar technology. (Above this frequency attenuation increases due to Cooper pair breaking.) The low cross talk, in turn, is guaranteed by the very thin insulator gap between the conductor and the ground plane allowed by the short penetration depth in a superconductor, the *London depth* which is ~ 0.1 μm for Nb.

Single flux quantum logic is based again on the fundamental property of superconductors to quantize magnetic flux through a loop in multiples of the flux quantum. The most elementary example of an SFQ circuit is provided by the rf SQUID with external current input I_{ext}, see Fig. 6. The external flux $\Phi_{ext} = MI_{ext}$, induced by the inductance M between the current path and the SQUID loop, affects the total flux Φ through the loop according to Eq. (2.5). If we now return to Fig. 3, with $LI_c = 1$, we see that with a flux bias of $\Phi_{ext} \simeq \Phi_0/2$, we have two flux states of equal stability, separated from each other by $\delta\Phi \simeq \Phi_0$.

The "quantization" of the flux difference above is accurate only for very inductive SQUID loops, $LI_c \gg 1$, but it holds approximately also for $LI_c = 1$, in which case only two flux states are simultaneously stable. The RSFQ logics are fundamentally based on this switching between

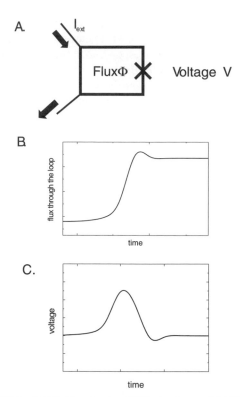

FIGURE 6 RSFQ. The simplest SFQ circuit, an rf SQUID, with an external current input I_{ext} to couple flux into the loop in (A). Schematic presentation of the time evolution of flux through the loop (B) and voltage across the junction (C) when the external current reaches the threshold value corresponding to different flux branches of Fig. 3.

the two neighboring flux states in the SQUID loop. During this switching, induced, for example, by a current pulse taking the Φ_{ext} beyond the stability limit in Fig. 3, a short voltage $V(t)$ pulse, typically a few picoseconds long, is formed across the junction. By Faraday induction law we obtain that the time integral of the pulse is

$$\int V(t)\,dt \simeq \Phi_0 \simeq 2\ \text{mV} \cdot \text{ps}. \qquad (2.8)$$

Digital bits are thus coded in statics by the single quanta of magnetic flux, whereas in dynamics the data are transferred as picosecond SFQ voltage pulses with quantized area. It is expected that RSFQ-based VLSI circuits may operate at clock frequencies well above 100 GHz, which is at least an order of magnitude higher than those of the semiconductor-based devices fabricated with similar design rules.

The high speed of the RSFQ circuits makes them attractive for special applications, despite the psychological barrier posed by the necessity of operating them at helium temperatures. Such recent practical devices include analog-to-digital and digital-to-analog converters in radar technology, digital SQUIDs in supersensitive magnetometry, and digital autocorrelators in radio astronomy. It should be emphasized that no true commercial RSFQ system exists at the moment. Within a decade goals to implement RSFQ techniques include digital signal processing and petaflop-scale computing.

III. THERMAL DETECTORS

A. General

Thermal sensing of radiation in both X-ray and sub-millimeter wave ranges is a technique that is being actively developed presently. Essential elements of a thermal detector are the absorber of radiation and the ultrasensitive temperature sensor. Superconductors and semiconductors are generically well suited as base materials for temperature measuring, because they exhibit strong, sometimes even abrupt, changes in their electrical conductance properties against temperature. In this article we focus on two different types of superconductor-based thermometers: TES and normal metal-insulator-superconductor (NIS) junctions. Recent advances in microlithography, micromachining, and sub-kelvin cooling techniques have made it possible to make integrated detectors with ultrahigh resolution.

B. TES for Microcalorimetry and Microbolometry

A natural choice for a very sensitive thermometer in a narrow temperature interval is that based on a phase transition in a solid (TES). To avoid hysteresis in the transition one should restrict to second order transitions with vanishing latent heat. Normal to superconductor transition of a metal film is a good candidate. The resistance at this transition has been used to measure small temperature excursions over many years. In a textbook the transition to the superconducting state is infinitely sharp at the transition temperature $T = T_c$, but in practical samples it is spread over a small temperature interval due to material inhomogeneities, stress, and dimensional effects. The width of the transition is characterized by the parameter α, defined by

$$\alpha \equiv d \ln R / d \ln T. \qquad (3.1)$$

Here R is the temperature dependent resistance of the film. By this definition α is a dimensionless parameter which gives the inverse of the relative temperature width of the transition. Thin film transitions have typically α's in the range from several tens up to several thousands.

Besides α, or the temperature width of the transition, ΔT, another parameter characterizing the TES is the mean transition temperature T_c, defined usually as the temperature where R has dropped down to one-half of its value in the normal state just above the transition. With this definition we can write $\Delta T \simeq T_c / \alpha$. In addition to high α and practically negligible hysteresis, another advantage of a superconducting TES is that there are a number of ways either to select or tune T_c of the detector. This is necessary because the thermometer should work at a designed temperature interval and the superconducting TES has no sensitivity ($dR/dT \equiv 0$) below the transition and almost no sensitivity (dR/dT positive but very small) in the normal state above the transition.

A traditional way of tuning T_c is to apply an external magnetic field to the metal sample. This is, in many cases inconvenient, e.g., because it involves extra apparatus for producing this field. A recently developed method to set the transition temperature by fabrication is the implantation of magnetic ions to otherwise clean metal films. Magnetic impurities lower the T_c of the virgin metal film. By this method T_c can be tuned accurately, for example, with $^{56}Fe^+$ ion implantation in W films transition temperature can be tuned highly reproducibly in the range of 40–140 mK.

The most widely used method of tuning T_c is to make use of the superconducting proximity effect in metallic thin film bilayers. In the proximity effect a normal metal in good metallic contact lowers the transition temperature of the composite film from that of the bare superconductor component. This method also yields relatively predictable and reproducible T_c's but the method is not free from material problems. The key issue is the quality of the interface between the two metals. An involved theory of the proximity effect has been developed. The results are readily comparable to the measured T_c's without uncertain fit parameters in the case of an absolutely clean and abrupt interface between the metals. In practice, however, the metals tend to interdiffuse forming an alloy layer, or the interface between them cannot be made into a low-resistance metallic contact from the outset: in the worst case it can be a Schottky or a tunnel barrier. For example, Al/Cu bilayer results in a predictable T_c initially right after its fabrication, but these thin films form an alloy in just a few weeks' time in the ambient air. Yet stable material pairs are provided by Ti/Au, Mo/Cu, and Al/Ag with T_c's in the 0.1 K range.

C. Thermal Properties and Operation of TES in a Calorimeter

The thermal properties of a TES detector are largely characterized by two quantities. The first one is the thermal conductance G between the detector at temperature T_s and the heat bath at temperature T_0. In steady state G can be defined as $G \equiv P/(T_s - T_0)$, where P is the (small) constant input power to the detector. The second one is the heat capacity C of the absorbing element, which is directly integrated to the detector itself. The natural thermal time constant τ of the detector is $\tau = C/G$. The parameters of the detector are chosen based on the needs of the calorimetric detection. For example, in single X-ray quantum detection, C has to be set such that the initial rise during the temperature excursion, $\delta T = E/C$, after the hit of the X-ray quantum of energy E on the absorber does not bring the TES into saturation. This condition can be written as $\delta T \ll \Delta T$. Another design consideration of a calorimeter comes from its speed: τ has to be much shorter than the inverse of the count rate of X-ray pulses.

The measurement of a superconducting TES is best done in the constant voltage bias mode. This is because of the positive temperature coefficient (PTC) of the detector at the transition, $dR/dT > 0$. If, instead, the TES with a PTC is set into constant current bias, the sensor can experience thermal runaway and get damaged eventually. In other words, a sudden temperature rise leads to increased voltage and power for a detector with PTC and current bias. With constant voltage bias, increase of temperature leads to decrease in current and consequently the power dissipation diminishes and temperature tends to drop again: the system is thermally stable. Furthermore, in voltage bias the detector becomes faster than what the natural thermal time constant C/G might expect. This is due to the so-called *electrothermal feedback*.

D. TES X-Ray Calorimeter

Transition edge detectors can be used in soft X-ray calorimeters, e.g., in astronomy or elemental microanalysis. A typical TES X-ray detector is shown schematically in Fig. 7.

The TES is grown on a thin silicon nitride membrane window with bulk silicon etched away in this area. This window, often further sliced into a mesh, provides low G between the TES film and the heat bath of the bulk silicon outside the window. The TES film is electrically connected to its bias circuitry via thin film leads made of a higher T_c superconductor, like Nb, to avoid electrical resistance and to eliminate thermal conduction along them. The *quantum efficiency*, i.e., the relative number of detected X-ray quanta is very low if the bilayer TES film also acts as the absorber: the film is thin and often made of not very heavy metals. To increase the quantum efficiency from the level of only a few per cent close to unity, one typically covers the TES film by a relatively thick, 1–3 μm, Bi film. The advantages of using Bi are that it does not strongly influence the proximity effect of the underlying bilayer because Bi is a semimetal, it does not increase the heat capacity too much, and this way the detector will not be too slow.

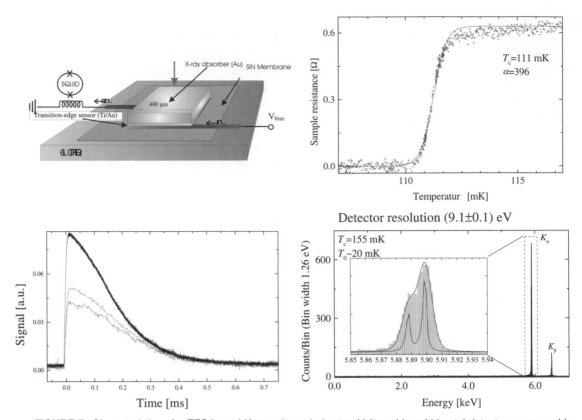

FIGURE 7 Characteristics of a TES-based X-ray microcalorimeter. Voltage-biased bimetal detector, measured by a SQUID ammeter on the top left. A typical superconductivity transition of a Ti/Au film on the top right. Measured SQUID current on X-ray hits on the detector on the bottom left, and the corresponding X-ray spectrum on the bottom right.

E. Antenna Coupled TES Infrared Bolometer

Besides soft X-rays, one can use TES detection in infrared and sub-millimeter wave micro-bolometry, primarily in astronomy. Figure 8 shows an antenna-coupled bolometer. It has been fabricated on silicon substrate. The spiral antenna is used to absorb the THz radiation from a weak source and the resulting ac current heats the TES micro-bridge in the center. It can be seen from the expanded view in Fig. 8 that the submicron wide TES strip is "airborne," i.e., it is etched from the substrate to improve thermal isolation and thus increase the sensitivity. Such devices, and many variations of them, are currently under intense investigation.

IV. NIS TUNNELING DEVICES

A. General

An NIS tunnel junction can be readily made between a number of material combinations. The simplest example with nearly ideal performance is provided by aluminum-aluminum oxide-copper, where aluminum is the super-

conductor at sub-kelvin temperatures and copper is one of the possible normal metals. (Also gold and silver are good examples of a normal metal.) A tunnel barrier of aluminum oxide forms easily on pure aluminum under an oxygen atmosphere at room temperature. Besides normal metal, also heavily doped semiconductors, e.g., Si, can be used in combination with aluminum, in which case the tunnel junction forms as a Schottky barrier even without oxidizing aluminum. Other superconductors can be used as well, for example, Nb with again higher T_c.

B. Transport Through a NIS Junction

The useful properties of NIS junctions stem from the non-constant density of quasiparticle states n_S in a superconductor. There is the gap with width of twice the energy gap, 2Δ, with forbidden single particle states around the Fermi level E_F, at which the Cooper pairs sit. The density of levels diverges at $\pm\Delta$. Explicitly we can write

$$n_S(E) = n(0)|E|/\sqrt{E^2 - \Delta^2}, \qquad |E| \geq \Delta \qquad (4.1)$$

$$= 0, \qquad |E| \leq \Delta. \qquad (4.2)$$

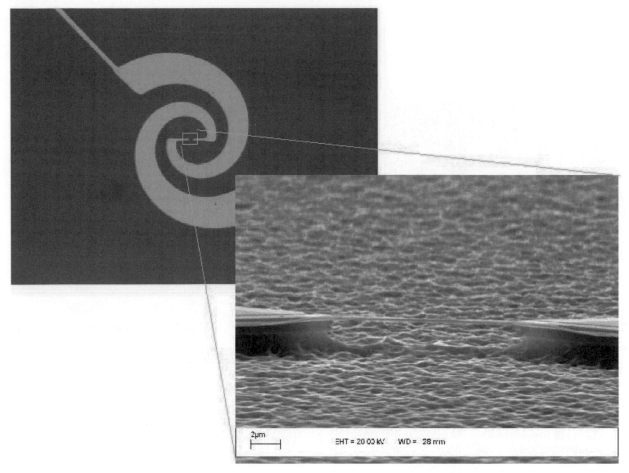

FIGURE 8 An antenna coupled infrared microbolometer. The spiral antenna picks radiation which induces ac current in the superconducting TES nanobridge lifted from the substrate (expanded view). The detector senses very weak IR radiation as temperature rise of the poorly thermalized TES bridge.

Here $n(0)$ is the corresponding (almost) constant density of electron levels in the normal metal state near the Fermi level, which is assigned to be the zero of energy. E is the energy of electrons counted from the Fermi level. The occupation of the levels is given by the Fermi distribution, $f(E) = 1/(1 + \exp(E/k_B T))$, whereby at zero temperature only states below the gap are occupied, and at nonzero temperatures there are thermally occupied states above the energy gap.

Figure 9 is the "semiconductor" energy diagram of an NIS tunnel junction, with occupied states shaded at a nonzero temperature. The bias voltage V lifts the Fermi level of the superconductor with respect to the normal electrode by the amount eV. If we make the typically very well justified approximation of low environmental impedance of the conductors, tunneling represents a horizontal crossing between S and N through the insulator I from an occupied state into an empty one, but not into the forbidden gap region. Therefore, at low voltages, $|eV| < \Delta$,

and at low temperatures, $k_B T \ll \Delta$ or $T \ll T_c$, virtually no current flows through the junction. Above the threshold voltage of $|eV| = \Delta$ the current increases rapidly and soon resumes the ohmic value determined by the normal state tunnel resistance of the junction, R_T. If the temperature is not very low as compared to T_c, current "leaks" at sub-gap voltages due to the smeared Fermi distribution in the normal electrode with tail extending up to energies above the top of the energy gap in the superconductor.

At low temperatures the full current-voltage curve can be calculated approximately by

$$I(V) = (eR_T)^{-1} \int_{\Delta}^{+\infty} [f(E - eV) - f(E + eV)] n_S(E) \, dE.$$

(4.3)

Figure 10A displays a few IVCs of NIS junctions calculated at different temperatures T/T_c. The measured ones closely follow this simple theoretical prediction.

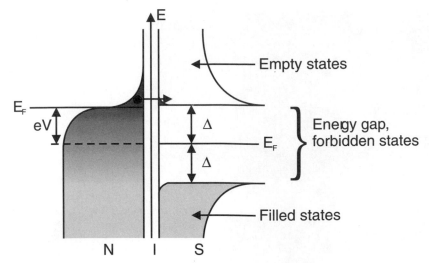

FIGURE 9 The energy diagram of a biased (at voltage *V*) NIS tunnel junction. The shaded area in the normal metal (N) gives the electron occupation according to the Fermi distribution. In the superconductor (S) there is the gap of width 2Δ with forbidden levels.

C. NIS Junction as a Sensitive Thermometer

Based on Fig. 10 it is obvious that an NIS junction can be used as a sensitive thermometer by biasing the junction at a constant current *I* or at a constant voltage. The first method is usually preferable because of the very strong temperature dependence of voltage at constant current within the gap region. In this bias mode the slope in the relatively wide temperature range with linear $V - T$ dependence is

$$dV/dT \simeq (k_{\mathrm{B}}/e) \ln(I/I_0), \qquad (4.4)$$

where the scaling current is $I_0 \equiv (eR_{\mathrm{T}})^{-1} \sqrt{\pi k_{\mathrm{B}} T \Delta^{3/2}/2}$. The NIS thermometer is typically characterized by a large impedance: especially the dynamic resistance in the most sensitive gap region is very high, whereby measurement using a SQUID-based amplifier is not especially advantageous. Therefore measurement by a high input impedance voltage amplifier is usually preferred.

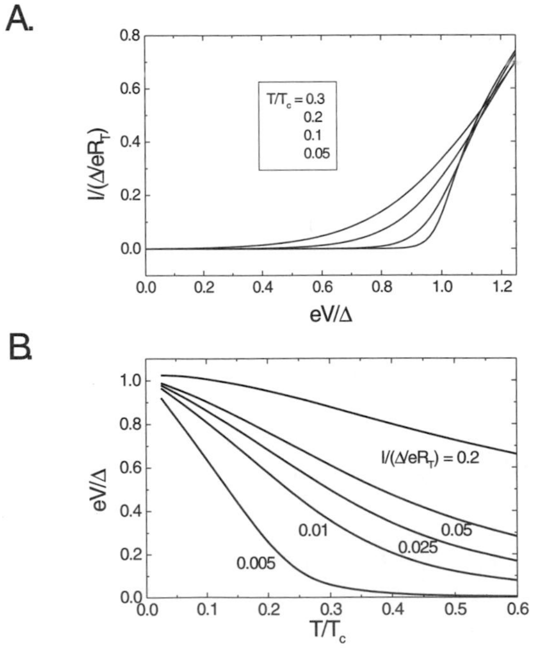

V. NIS MICROCOOLING

A. Thermal Considerations

Besides the fact that current is preferably transported only at voltages exceeding the gap voltage, there is also a nonvanishing net heat current through the junction at nonzero bias voltages. This makes NIS junctions a very attractive candidate for on-chip cooling. When *V* has a value close to Δ/e, it is obvious that only the hot electrons in the tail of the Fermi distribution can be transported across the barrier. This way the average energy and thus the temperature of the normal metal electrode drops. The cooling power P_{NIS} of one NIS junction can be calculated analogously to what we did to determine the IVC [Eq. (4.3)] as

$$P_{\mathrm{NIS}}(V) = \left(e^2 R_{\mathrm{T}}\right)^{-1} \int_{\Delta}^{+\infty} (E - eV)[f(E - eV)$$
$$- f(E + eV)]n_{\mathrm{S}}(E)\, dE. \qquad (5.1)$$

FIGURE 10 Current through an NIS junction. IVC of an NIS junction at different temperatures in (A). Voltage at a few fixed values of current providing a thermometer in (B).

Figure 11 shows P_{NIS} calculated at a few values of temperature T/T_{c}. P_{NIS} reaches its maximum value (optimum

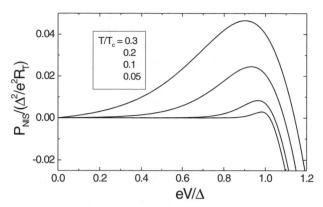

FIGURE 11 Cooling power of an NIS junction against bias voltage calculated at a few values of temperature. Maximum cooling power is reached at V slightly below Δ/e. At large values of V the curve approaches that of Joule heating with resistance R_T.

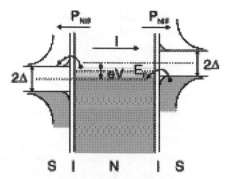

FIGURE 12 The energy diagram of a symmetric SINIS cooler at a bias voltage $V \simeq 2\Delta/e$ across the two junctions. Symmetry of the Fermi distribution leads to equal cooling power in the two junctions: the current is flowing unidirectionally through the SINIS pair (top), but heat is flowing out through both the two junctions.

bias point) just slightly below Δ/e. The value of this maximum cooling power is approximately

$$P_{\mathrm{NIS}} \simeq 0.6 \frac{\Delta^2}{e^2 R_T} \left(\frac{k_B T}{\Delta} \right)^{3/2} \quad (5.2)$$

at temperatures $T \ll T_c$.

If the normal electrode is not connected to a heat bath by a metallic contact, the electrons in it can really cool down, because the electron system decouples efficiently thermally from the underlying lattice due to the very weak electron-phonon (e-p) coupling at low temperatures. Heat current $\dot{Q}$ between the electrons and lattice is given by $\dot{Q} = \Sigma \Omega (T_e^5 - T_p^5)$, where Σ is a material dependent e-p coupling constant, Ω is the volume of the electrode, and T_e and T_p are the electron and lattice temperatures, respectively.

There are at least two ways of isolating the normal metal electrode from its environment in order to allow it to cool down. The N electrode can be made with an NS "Andreev" contact at its one end and the refrigerating NIS junction at the opposite one. A metallic NS contact does not allow heat to be transported electrically through. Yet this configuration can be replaced by a more efficient, stable, and easy-to-fabricate symmetric SINIS cooler. In SINIS the N electrode is symmetrically terminated at both its ends by an NIS junction. Figure 12 shows the energy level diagram of the SINIS cooler with a bias of approximately $2\Delta/e$ connected across.

In a relatively symmetric SINIS pair the voltage drop across each of the two junctions is close to Δ/e, thus allowing optimum cooling in the NIS junction. In the opposite SIN junction the cooling power is equal. The simple explanation of this is that while the hot electrons are skimmed away through the NIS junction, as explained above, they are replaced by cold ones entering into energy levels below E_F in the N electrode through the SIN

junction. Since the Fermi distribution is symmetric around E_F, cooling power of the NIS and SIN junctions are the same, and the resulting total cooling power is doubled with respect to the simple NIS cooler.

One of the most important considerations in designing an NIS cooler is how to get rid of the heat dissipated in the secondary side of the cooler: the superconductor receives heat and as a very poor thermal conductor it tends to warm up whereby the cooling power of the NIS cooler diminishes. The coefficient of performance, COP, of the NIS junction, i.e., the ratio of the heat extracted from the normal electrode as compared to the total heat dissipated at the optimum bias point can be approximated by

$$\mathrm{COP} = \frac{P_{\mathrm{NIS}}}{IV} \simeq \frac{0.6 \frac{\Delta^2}{e^2 R_T} \left(\frac{k_B T}{\Delta} \right)^{3/2}}{(\Delta/e) I_0} \simeq 0.3 T/T_c. \quad (5.3)$$

For an aluminium-based cooler with (thin film) T_c of about 1.3 K, we have $\mathrm{COP} \simeq 0.02$ at $T = 0.1$ K, i.e., approximately 50 times more heat is dissipated in the superconductor than what is removed from the N electrode. This can lead to degraded cooling performance since the cooling power with a "hot" superconductor is reduced from that of Eq. (5.2) to

$$P_{\mathrm{NIS}} \simeq 0.6 \frac{\Delta^2}{e^2 R_T} \left(\frac{k_B T_n}{\Delta} \right)^{3/2} - \frac{\Delta^2}{e^2 R_T} \sqrt{\frac{2\pi k_B T_s}{\Delta}}$$
$$\times \exp \left(-\frac{\Delta}{k_B T_s} \right), \quad (5.4)$$

where we have now indicated by the subscripts n and s if the temperature T is that of the N or S electrode, respectively.

To avoid or at least to reduce heating of the superconductor, i.e., to reduce excess population of hot quasiparticles, so-called *quasiparticle traps* can be used. These are typically normal metal films in contact with the super-conductor near the junction, so that the excess heat in the superconductor can be conducted into this normal metal which has superior heat transport capability. The trap can be made of copper, and it can be either in direct metal-to-metal contact with the superconductor, or it can contact through a thin insulator layer. The most efficient geometrical arrangement may turn out to be a sandwich cooler, where both the NIS junction and the trap are all on top of each other. Such a trap geometry has been illustrated in Fig. 13. At the time of writing this volume, the best achieved performance using quasiparticle traps is a cooling power of about 15 pW per NIS junction (junction area $\sim 20~(\mu m)^2$) at $T_n = 0.1$ K and $T_s = 0.3$ K.

To use the NIS cooler as a conventional refrigerator one needs to provide an intermediate cooled heat bath that is only weakly coupled to the heat bath of the precooling stage. The object to be cooled, e.g., a microcalorimeter or a bolometer, then sits at this intermediate heat bath, but not electrically connected to the N electrode

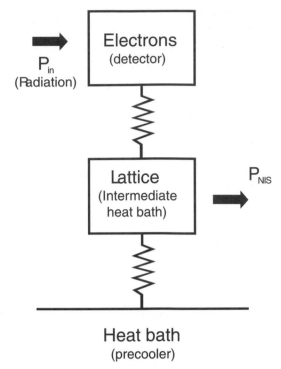

FIGURE 14 Thermal diagram of the indirect cooling method, where SINIS is used to lower the lattice temperature of an intermediate heat bath, formed by, e.g., a silicon nitride membrane with a cooled copper cold finger.

of the cooler. Figure 14 illustrates this *indirect* cooling method.

In some cases it might be tempting to cool down only the electrons directly, for example, if the N electrode could act as a detector as such. This *direct* cooling is, however, not usually a good solution, because the noise properties of the detector to be cooled depend not only on its electron temperature but usually equally as much on the temperature of the lattice. For example, the fundamental limit of NEP, the noise equivalent power, i.e., the minimum resolvable power of a radiation bolometer assumes a form

$$\text{NEP} = \sqrt{5 \Sigma \Omega \left(T_e^6 + T_p^6 \right)}. \qquad (5.5)$$

Thus, by cooling down the electrons only, one reduces the NEP by not more than about 30%.

Figure 15 shows one possible solution of an on-chip SINIS refrigerator which is capable of cooling down a micromachined silicon nitride membrane on which the object to be cooled can be processed also. The cooling junctions are patterned on bulk silicon, and the heat is flowing out from the membrane toward the cooling junctions via a copper cold finger.

A.

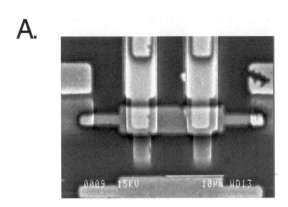

B.

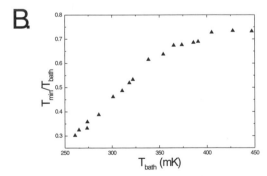

FIGURE 13 A SINIS cooler with quasiparticle traps. In (A) the NIS junctions of the SINIS pair are seen in the central area. The superconductors extend up vertically in the figure, and the quasiparticle copper traps are sandwiched down to the junction area. The horizontal bar is the N electrode cooled, with SINIS thermometers at the two ends. The relative temperature reduction of N as a function of the starting temperature is shown in (B). The cooling power is about 15 nW per junction with N electrode at 100 mK.

| 100μm | EHT = 0.00 kV | WD = 30 mm | Signal A = SE1 | Date :16 Feb 2001 |
| | | | Photo No. = 46 | Time :13:18:27 |

FIGURE 15 An example of an SINIS cooler in an indirect cooling configuration. The cooling junctions are at the top and bottom of the figure. The central area is the thermally semi-isolated silicon nitride membrane, which forms the intermediate heat bath together with the copper cold fingers extending from the cooling junctions to the membrane. Separate sensors, in this case NIS thermometers, can be placed on the cooled membrane in the central area of the membrane.

VI. SUPERCONDUCTING SINGLE CHARGE DEVICES

A. Basics of Single Electron and Single Cooper Pair Tunneling

A family of new generation devices, including single-charge transistors and logic, metrological current and charge standards, a Coulomb blockade absolute thermometer, and eventually a fully solid-state quantum computer, is expected to emerge along the technical development after the feature size in patterning of components has dropped down to deep below the 1-μm range. What governs charge transport then, either electron or Cooper pair current, is the charging energy by single electrons or Cooper pairs. These effects emerge, in particular, when temperature is low such that thermal energy $k_B T$ is much smaller than the charging energy needed to add or remove one elementary charge carrier to or from the circuit. So far, in respect of superconducting devices, particular attention has been paid to aluminium-based, single-charge

devices. These are typically structures where two lithographically patterned thin film conductors are separated from each other by an oxide barrier forming a tunnel junction.

B. Fabrication Issues

The exclusive technique to produce superconducting single-charge devices is electron beam lithography combined with shadow mask evaporation. The aim is to obtain superconducting thin film conductors with 100 nm order line widths connected to each other by a high quality oxide barrier. The standard technique to do this is illustrated in Fig. 16.

An electron-beam-sensitive, double-layer resist has been spun on a flat, preferably semiconducting substrate. The bottom resist layer is more sensitive to the electron shower than the top one, and by this action the developed pattern after electron beam lithography produces an undercut profile as demonstrated in Fig. 16. This way the

Shadow angle evaporation

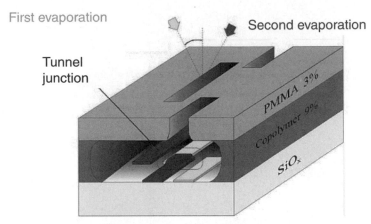

FIGURE 16 Schematic presentation of a resist mask in electron beam lithography to allow multiple angle deposition in fabricating tiny tunnel junction structures for single electron and single Cooper pair devices. Typical feature sizes are of the order of 100 nm.

patterned resist forms a suspended mask which can be used in multi-angle deposition of the superconducting metal. When the neighboring conductors are near enough the undercut in these areas connects the pattern in the lower resist layer. Now, by depositing superconducting metal in the two angles indicated, and by oxidizing the first aluminium layer inbetween the metal depositions, an ultra-small superconductor-insulator-superconductor (SIS) tunnel junction can be formed. The computer-aided pattern generation allows easy design of complicated multijunction circuits as well: new designs are easy to realize in electron beam lithography but the patterning itself is slow for industrial purposes. Figure 17 illustrates an example of a simple double junction structure fabricated by the technique described above.

C. Single Cooper Pair Box, Transistor, and Multijunction Circuits

The basic device concept in single electronics is a single electron box (SEB) and its superconducting counterpart single Cooper pair box (SCB), schematically presented in Fig. 18A. A small superconducting island between the gate capacitor C_g and the "semitransparent" tunnel junction capacitor C can house an integer number, N of Cooper pairs.

Electrostatically, the energy of the system, E_{ch}, corresponding to each N and gate voltage V_g can be written as

$$E_{ch} = (2eN - C_g V_g)^2/[2(C + C_g)]. \qquad (6.1)$$

For each N, Eq. (6.1) presents a parabola against V_g, centered at $C_g V_g = 2eN$ (Fig. 19). Each state $|N\rangle$ gives the minimum energy in the interval

$$2e(N - 1/2) < C_g V_g < 2e(N + 1/2). \qquad (6.2)$$

When temperature is low, $k_B T \ll (2e)^2/[2(C + C_g)] \equiv E_c$, N is determined to within one integer value by the setting of the gate voltage within this interval, and N changes by one upon crossing the limits of this range. This change is allowed by Cooper pair tunneling through the junction.

Each Josephson junction has, in addition to the classical charging energy above, which is important for small junctions, the more familiar Josephson coupling energy,

$$E_J = -E_{J0} \cos \varphi, \qquad (6.3)$$

where φ is the phase difference across the junction and the strength of the Josephson coupling is given by $E_{J0} = \hbar I_c/2e$. The total Hamiltonian of a Cooper pair box, in the absence of quasiparticle (single electron) excitations and in the environment with vanishing impedance, is the sum of E_{ch} and E_J.

The net effect of the two energy contributions is the formation of a *band structure*, where, instead of sharp crossing of the parabolas of E_{ch}, a gap of magnitude E_{J0} opens near the degeneracy points of succeeding states $|N\rangle$. This means, in turn, that upon cranking V_g across one of the degeneracy points, the eigenstate evolves smoothly as a superposition of the two "nearest" charge states from one pure charge state into another in the two-state

AFM image of a SET structure

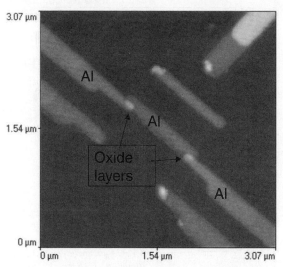

FIGURE 17 A single electron transistor fabricated through a resist stencil mask of the type shown in Fig. 16. The shadows are of aluminum.

approximation. In particular, at the exact degeneracy value of the gate voltage, the two-charge states exist at an equal probability (1/2). Figure 20 demonstrates the influence of E_J on both the eigenstates and the probabilities, i.e., occupations p of different charge states in these eigenstates. The lowest band corresponds to the ground state, the higher ones are naturally excited states, which can be reached, e.g., by so-called Landau-Zener crossing upon fast enough (nonadiabatic) sweep of V_g.

The interplay between the charging energy and the Josephson coupling provides the possibility of using the SCB with a *small Josephson junction as a quantum bit*, possibly in future quantum computation.

More complex circuits with small Josephson junctions can be described in a similar manner. The charging Hamiltonian is technically more complicated and the Josephson energy is the sum of the contributions of the type given by Eq. (6.3), but no conceptual complications arise. Perhaps the best known "device" is the single Cooper pair transistor (SCT) consisting of two series Josephson junctions and of a gate electrode (see Fig. 18B). Josephson current

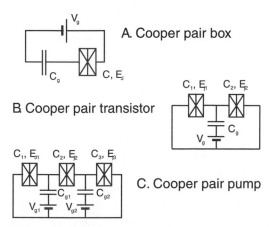

FIGURE 18 (A) A single Cooper pair box connected to a gate bias V_g through gate capacitance C_g. The crossed square depicts a small Josephson junction with capacitance C. (B) A single Cooper pair transistor consisting of two Josephson junctions and a gate. (C) A three-junction linear Josephson junction array with gates for the two islands.

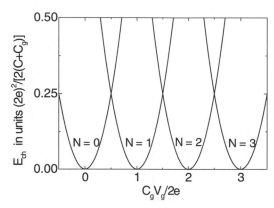

FIGURE 19 Energy diagram including just the charging energy of a Cooper pair box against the gate voltage for states with different number of Cooper pairs on the island.

A.

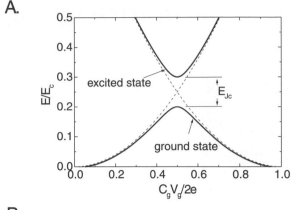

B.

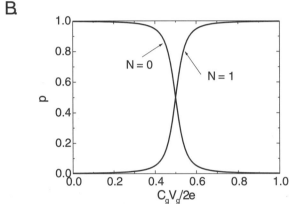

FIGURE 20 Cooper pair box in presence of Josephson coupling. (A) Eigenstates as a function of gate voltage form a band structure with a gap of magnitude $E_{Jc} = E_{jo}$ opening at the degeneracy point. $E_c \equiv (2e)^2/[2(C + C_g)]$. (B) Occupation p of different charge states against gate voltage. On crossing the degeneracy point the state evolves from an almost pure charge state into another. As a controllable two-state system the Cooper pair box forms a candidate of a solid-state quantum bit.

through the SCT can be modulated by the gate voltage periodically. Multijunction arrays of small junctions with gates connected to the intermediate islands (Fig. 18C) can be used for quantized charge transport by Josephson tunneling.

SEE ALSO THE FOLLOWNG ARTICLES

SUPERCONDUCTING CABLES • SUPERCONDUCTIVITY • SUPERCONDUCTIVITY MECHANISMS • SUPERCONDUCTORS, HIGH TEMPERATURE

BIBLIOGRAPHY

Averin, D. V., and Likharev, K. K. (1991). Single electronics: correlated transfer of single electrons and Cooper pairs in systems of small tunnel junctions. *In* "Mesoscopic Phenomena in Solids," B. L. Altshuler, P. A. Lee, and R. A. Webb eds., Elsevier, New York.

Bunyk, P., Likharev, K., and Zinoviev, D. (2001, to appear). "RSFQ Technology: Physics and Device," *Int. J. High Speed Electron. Systems*, special issue, M. Rodwell (ed.)

Clarke, J. (1996). SQUID fundamentals. *In* "SQUID sensors: Fundamentals, Fabrication and Applications," H. Weinstock, ed., Kluwer, The Netherlands.

de Korte, P., and Peacock, T., eds. (1999). "Proceedings of the 8th International Workshop on Low Temperature Detectors, LTD-8," Elsevier, The Netherlands.

Leivo, M. (1999). On-chip Cooling by Quasiparticle Tunnelling below 1 Kelvin, Ph.D. thesis, University of Jyväskylä, Finland, unpublished.

Tinkham, M. (1996). "Introduction to Superconductivity," 2nd ed., McGraw-Hill, New York.

Van Duzer, T., and Turner, C. (1998). "Principles of Superconductive Devices and Circuits," 2nd ed., Prentice Hall, New Jersey.

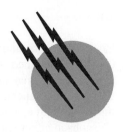

Superconductivity

H. R. Khan
FEM and University of Tennessee at Knoxville

GLOSSARY

Coherence length Correlation distance of the superconducting electrons.

Critical magnetic field Above this value of an externally applied magnetic field, a superconductor becomes non-superconducting (normal).

Energy gap Gap in the low-energy excitations of a superconductor.

Type I superconductor When an external magnetic field is applied on this superconductor, the transition from a superconducting to a normal state is sharp.

Type II superconductor When an external magnetic field is applied, the transition from a superconducting to a normal state occurs after going through a broad "mixed-state" region.

SUPERCONDUCTORS are materials that lose all their electrical resistivity below a certain temperature and become diamagnetic. High values of an externally applied magnetic field are required to destroy the superconductivity. These electrical and magnetic properties of superconducting materials have found applications in lossless electrical transmission and generation of high-magnetic fields. Superconducting magnets are used where normal iron magnets are inadequate. These magnets are used as exciter magnets for homopolar generators or rotors in

Encyclopedia of Physical Science and Technology, Third Edition, Volume 16

large alternators, and much gain in efficiency and power density is obtained. Future fusion reactors will use superconducting magnets for confining deuterium and tritium plasma. Superelectron pairs in a superconductor can tunnel through a nonconducting thin layer. Based on this "Josephson effect," superconducting Josephson junctions are used as sensors, as high-energy electromagnetic radiation detectors, and in high-speed digital signal and data processing.

I. INTRODUCTION

A. Discovery of Liquid Helium Gas and Superconductivity

In 1908 Kammerlingh Onnes succeeded in liquifying helium gas, and this enabled him to measure the electrical resistivity of metals at lower temperatures down, to 4.2 K. The boiling temperature of liquid helium is 4.2 K. He measured the electrical resistivity of gold, platinum, and mercury and found that the electrical resistivity of mercury disappeared almost completely below 4.2 K. As shown in Fig. 1, the electrical resistivity of mercury is almost zero below 4.2 K. This state of a material in which the resistance is zero is called the *superconduct-*

ing state. The current flows without any attenuation in this state, and it has been estimated that the decay time of a current in a superconductor is about 100,000 years. The temperature below which a material loses its resistance is called the *superconducting transition* or *critical temperature T_c*.

B. Effect of a Magnetic Field on Superconductivity and the Meissner Effect

Meissner discovered that a bulk superconducting material behaves like a perfect diamagnet with a zero magnetic induction in its interior. If a paramagnetic material is placed in a magnetic field, then the magnetic lines of force penetrate through the material. But when the same material is made superconducting by cooling to lower temperatures, then all the lines of force are expelled from the interior of this material. This is called the *Meissner effect*. Figure 2 shows a material in the normal and superconducting states in an externally applied magnetic field. When the strength of this externally applied magnetic field is increased slowly, a value is reached where the magnetic lines of force begin to penetrate the material and it becomes nonsuperconducting or normal. This particular value of the magnetic field above which the superconductivity is destroyed is called the *critical magnetic field $H_c(T)$* and is also a function of the temperature of the material. A typical example of the variation of the $H_c(T)$ with temperature T is shown in Fig. 3 for the metal mercury (Hg).

The variation of $H_c(T)$ with temperature can be expressed by the equation

$$H_c(T) = H_c(0)\left[1 - (T/T_c)^2\right],$$

which has a parabolic form. This expression can also be derived using thermodynamics.

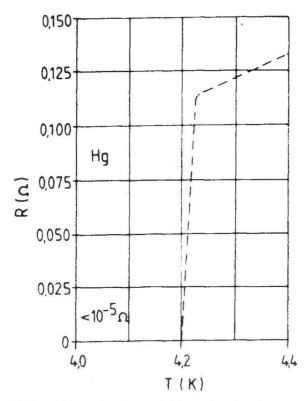

FIGURE 1 Electrical resistance $R\,(\Omega)$ as a function of temperature for mercury metal (Hg).

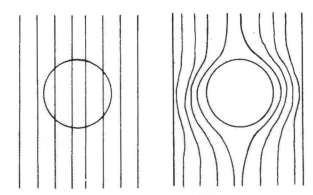

FIGURE 2 Normal and superconducting states of a material in an external magnetic field.

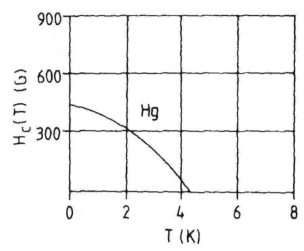

FIGURE 3 Variation of the superconducting transition temperature of mercury in an externally applied magnetic field.

C. Type I and Type II Superconductors

Based on the Meissner effect, superconducting materials are classified as type I and type II. When the magnetic induction $4\pi M$ of a superconducting material in the form of a cylinder with its axis parallel to the applied magnetic field is measured with an increasing magnetic field and if there is a sharp transition to the normal state above a certain value of the magnetic field B_a as shown in Fig. 4, then this type of material is called a *type I superconductor*. This kind of behavior is shown in general by pure metals. On the contrary, the $4\pi M$ versus B_a behavior of a *type II superconductor* is shown in Fig. 5. The magnetic flux

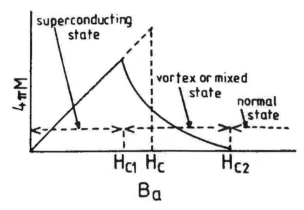

FIGURE 5 Magnetization $4\pi M$ as a function of applied magnetic field B_a of a type II superconductor.

penetrates the material slowly at a field value of H_{c1} and continues up to H_{c2}, where the material is transformed to a normal state. The superconducting state between the field value H_{c1} and the value H_{c2} is called the *vortex* or *mixed state*. The H_{c2} value can be 100 or more times greater than H_c. This type II superconducting behavior is shown in general by alloys and compounds that are called *dirty superconductors*.

Some of the superconducting alloys and compounds of special structures possess very high values of H_{c2}. For example, the H_{c2} value of a compound of composition $Pb_1Mo_{5.1}S_6$ with a Cheveral phase structure is about 51 T. Very high magnetic fields can be generated by the solenoids of the wires made of superconducting materials of high H_{c2}. Commercial superconducting magnets capable of producing magnetic fields of more than 10 T are available and use wires of Nb–Ti and Nb–Sn alloys. The variation of $H_{c2}(T)$ with temperature of some high-H_{c2} alloys is shown in Fig. 6. A type I superconductor can be transformed to a type II superconductor by alloying. A typical example is shown in Fig. 7. Here lead (Pb) is a type I superconductor, and when it is alloyed with indium (In), the alloys show type II behavior and the values of H_{c1} and H_{c2} are a function of the composition. The current flowing through the superconducting wire produces a magnetic field, and when the value of the current is increased slowly, a value is reached where the magnetic field becomes equal to the critical magnetic field. This value of current is called the *critical current*.

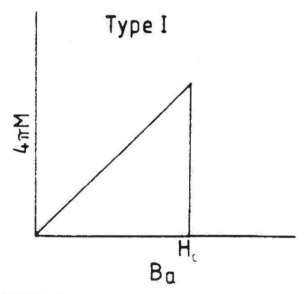

FIGURE 4 Magnetization $4\pi M$ as a function of externally applied magnetic field B_a of a type I superconductor.

II. SUPERCONDUCTING MATERIALS

A. Elements, Compounds, and Alloys

The distribution of superconducting elements in the periodic system is shown in Fig. 8. Some elements do not

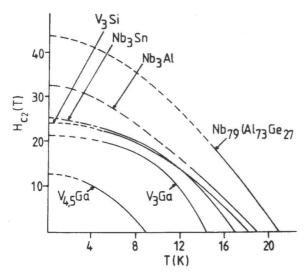

FIGURE 6 Upper critical magnetic fields H_{c2} as a function of temperature for some superconductors of β-W structure. [From Berlincourt, T. G., and Hake, R. H. (1963). *Phys. Rev.* **131**, 140; Fonner, S., McNiff, E. J., Jr., Matthias, B. T., Geballe, T. H., Willens, R. H., and Corenzwit, E. (1970). *Phys. Lett.* **3IA**, 349.]

become superconducting at all and others become so only under pressure. The superconducting transition temperatures along with their crystal structures and melting temperatures are listed in Table I. Niobium (Nb) metal has the highest superconducting transition temperature (9.2 K). The elements that become superconducting under pressure are listed in Table II.

The magnetic elements Mn, Fe, Co, and Ni do not become superconducting down to the lowest temperature available. Also, their presence in small amounts (of the order of parts per million) suppresses the superconducting transition temperature of other superconducting materials. The superconducting elements are classified into two groups. One group is the nontransition elements and consists of the elements Si, Ge, P, As, Sb, Bi, Se, and Tc, which are not superconducting under normal conditions but under pressure become superconducting. The other group consists of the transition elements and have un-

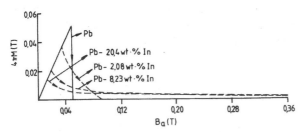

FIGURE 7 Magnetization $4\pi M$ as a function of applied magnetic field B_a and change of type I (Pb) to type II (Pb–In) alloy superconductors.

filled $3d$, $4d$, and $5d$ shells. The crystal structure plays an important role in superconductivity. For example, as shown in Tables I and II, Bi is not superconducting, but different crystal modifications of it obtained by applying pressure exhibit superconductivity at temperatures ranging between 3.9 and 8.5 K.

Multicomponent alloys and compounds of different crystal structures exhibit superconductivity. High-transition temperature superconductivity occurs with cubic structure, and the most favorable is the one with the β-W structure. Compounds and alloys with superconducting transition temperatures above 20 K form this structure. The β-W structure is cubic and is shown in Fig. 9. Each face of the cubic lattice is occupied by two atoms that form orthogonal linear atomic chains. The highest superconducting transition temperature is 23 K, and it is exhibited by a compound of composition Nb_3Ge with the β-W structure. Here the Nb atoms form the linear atomic chains, and the Ge atoms occupy the center and corner sites of the cubic lattice.

The reason the materials of this particular structure show such high superconducting transition temperatures is explored by Labbé and Friedel. Their theoretical calculations based on tight binding approximation suggest that materials of this structure possess an unusually high electron density of states at the Fermi surface; this is also experimentally confirmed. In addition, the d-band of these materials is narrower and taller compared with that of the transition metals. These are the factors that cause the enhancement of the superconducting transition temperature. Some of the high superconducting transition temperature materials are listed in Table III.

There are other kinds of superconducting materials, including low-carrier density superconductors (semimetal or semiconductor), intercalated compounds, amorphous superconductors, and organic superconductors, and they are described separately as follows.

B. Low-Carrier Density Superconductors

A class of materials that have carrier densities in the range of 10^{18} to 10^{21} are called *semimetals* because their carrier densities are between those of metallic conductors and semiconductors. Many of these materials are superconducting. For example, Fig. 10 shows that La_3Se_4, GeTc, SnTc, and $SrTiO_3$ are superconducting, and the superconducting transition temperature increases with increasing carrier density except in the case of $SrTiO_3$. For $SrTiO_3$, the superconducting transition temperature begins to decrease above a carrier density of 10^{20}. This decrease is explained by the occurrence of the magnetic effect. All the above-mentioned materials investigated were in the form of single crystals.

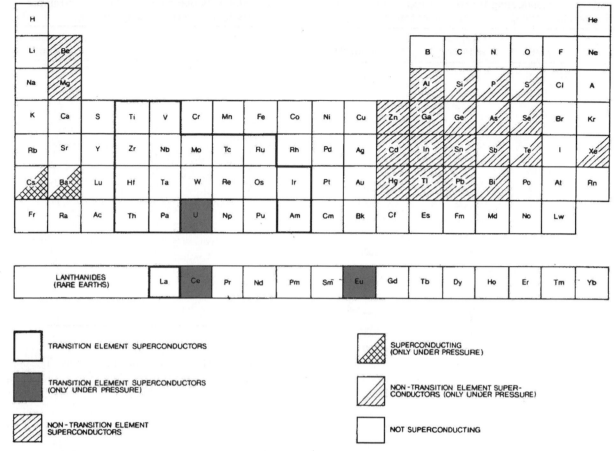

FIGURE 8 Distribution of the superconducting elements in the periodic table. [From Khan, H. R. (1984). *Gold Bull.* **17**(3), 94.]

C. Intercalated Compounds

A typical example of this class of materials is $TaS_2(C_5H_5N)_{1/2}$. This compound is formed when TaS_2 is intercalated with pyridine (C_5H_5N), and metallic layers about 6 Å thick are separated by pyridine layers of the same thickness. This intercalated compound becomes superconducting at 3.5 K. A large number of transition metal chalcogenides exist that crystallize in the layered structures. These types of compounds show anisotropic superconducting properties parallel and perpendicular to the layer surface. The critical magnetic field is about 30 times higher in the direction parallel to the layer surface compared with the perpendicular direction, as shown in Fig. 11.

D. Amorphous Superconductors

Unlike crystalline materials, amorphous or noncrystalline materials consist of atoms that do not form regular arrays and are randomly distributed. These amorphous materi-

als can be obtained in the form of thin films by evaporation deposition on cold substrates. Amorphous materials in bulk can also be obtained by rapidly cooling an alloy melt. The amorphous materials obtained in this way are called *metallic glasses*. Materials of this class also exhibit superconductivity but are completely different from their crystalline counterparts. The superconducting transition temperature T_c and the electron per atom ratio (e/a) of some amorphous nontransition metals and alloys are listed below. One sees that the T_c values in the amorphous state are higher than those in the crystalline state.

Alloy	T_c (K)	e/a
Be	9.95	2.0
$Be_{90}A_{10}$	7.2	2.1
Ga	8.4	3.0
$Pb_{90}Cu_{10}$	6.5	3.7
Bi	5	5

Amorphous films of transition metals and alloys have also been obtained, and the T_c values are lower than those of the

TABLE I Superconducting Transition Temperature T_c, Melting Temperature, and Crystal Structure of Elements

Element	T_c (K)	Crystal structure[a]	Melting temperature (°C)
Al	1.19	f.c.c.	660
Be	0.026	Hex.	1283
Cd	*0.55*	Hex.	321
Ga	1.09	Orth.	29.8
	(6.5, 7.5)		
Hg	*4.15*	Rhom.	−38.9
	(3.95)		
In	3.40	Tetr.	156
Ir	0.14	f.c.c.	2450
La	4.8	Hex.	900
	(5.9)		
Mo	0.92	b.c.c.	2620
Nb	9.2	b.c.c.	2500
Os	0.65	Hex.	2700
Pa	1.3	—	—
Pb	7.2	f.c.c.	327
Re	1.7	Hex.	3180
Ru	*0.5*	Hex.	2500
Sn	3.72	Tetr.	231.9
	(5.3)	Tetr.	
Ta	4.39	b.c.c.	3000
Tc	7.8	Hex.	
Th	1.37	f.c.c.	1695
Ti	0.39	Hex.	1670
Tl	2.39	Hex.	303
U(α)	0.2	Orth.	1132
V	5.3	b.c.c.	1730
W	0.012	b.c.c.	3380
Zn	0.9	Hex.	419
Zr	*0.55*	Hex.	1855

[a] f.c.c., face-centered cubic; hex., hexagonal; orth., orthorhombic; rhom., rhombohedral; tetr., Tetrahedral; b.c.c., body-centered cubic.

same alloys in crystalline form. Metallic glass superconductors are classified into two main groups. One group consists of metal–metal compositions and the other of metal–metalloid compositions. The superconducting transition temperatures of some of the metallic glass superconductors are listed in Table IV. These metallic glass superconductors show some desirable properties. For example, they are ductile and possess a high strength, whereas their crystalline counterparts are brittle. The metallic glass superconductors also possess very high values of the critical magnetic field.

A practical superconductor capable of producing a magnetic field of about 10 T should have a critical current density of about 10^6 A/cm^2. In general, the amorphous super-

TABLE II Superconducting Transition Temperature T_c of Elements under Pressure

Element	T_c (K)	Pressure (kbar)
As	0.5	120
Ba	5.1	140
	(1.8)	55
Bi II	3.9	26
Bi III	7.2	27
Bi V	8.5	78
Ce	1.7	50
Cs	1.5	1000
Ge	5.4	110
Lu	0.1–0.7	130
P	4.6–6.1	100
Sb	3.6	85
Se	6.9	130
Si	6.7	120
Te	4.5	43
Y	1.5–2.7	120–160

conductors have low critical current densities. The current density can be increased by introducing some kind of inhomogeneities into the amorphous matrix. Some binary and ternary pseudoamorphous alloys of vanadium, hafnium, and zirconium metals possess reasonably high superconducting transition temperatures and very high values of critical magnetic fields and critical current densities. At the same time they also have good mechanical properties such as ductility and high tensile strength. These kinds of materials are promising future superconducting materials for generating magnetic fields above 10 T.

E. Organic Superconductors

In 1964 Little proposed that an organic polymer can also become a superconductor. His theory of superconductivity is based on a mechanism entirely different from that of the Bardeen *et al.* (1957) theory of metals and alloys. Little suggested that the electrons on the spine of the polymer chain are attracted to each other by an indirect process

TABLE III Superconducting Transition Temperature T_c of Some β-W Structure Compounds

Compound	T_c (K)
V_3Ga	14.2–14.6
V_3Si	17.1
Nb_3Au	11.0–11.5
Nb_3Sn	18.0
$Nb_3Al_{0.8}Ge_{0.2}$	20.7

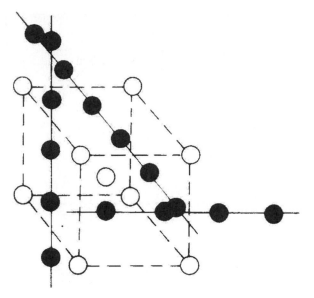

FIGURE 9 The β-W lattice structure.

involving the fixed polar groups on the side branches of the polymer. He predicted a superconducting transition temperature of about 100 K using the molecular polarization mechanism. Experimentally, there are indications of a polymer becoming a superconductor. For example, a polymer of formula $(SN)_x$ shows superconductivity at 0.3 K and another organic compound called tetramethyltetrasulfofluoride (TMTSF) at 1 K under a pressure of 12 kbar.

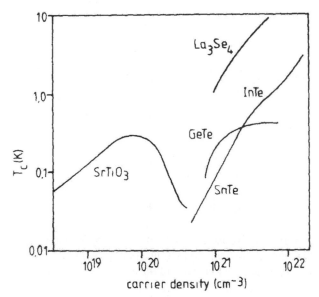

FIGURE 10 Superconducting transition temperature T_c as a function of carrier density. [From Hulm, J. K., Ashkin, M., Deis, D. W., and Jones, C. K. (1970). *Prog. Low Temp. Phys.* **VI**, 205.]

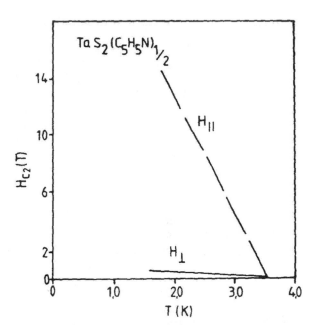

FIGURE 11 Upper critical magnetic field H_{c2} as a function of temperature parallel and perpendicular to the layered surface in $TaS_2(C_5H_5N)_{1/2}$. [From Gamble, F. R., *et al.* (1971). *Science* **174**, 493.]

III. CORRELATION: T_c WITH THE ELECTRONIC STRUCTURE OF A SOLID

Matthias proposed empirically that the superconducting transition temperature T_c and the electron per atom ratio e/a of a solid are related. This Matthias empirical rule suggests that the maximum values of T_c for transition metals occur at e/a values of 5 and 7, as shown in Fig. 12. In the case of solid solutions of transition metals, a slight shift of the first maximum to an e/a value of 4.5 occurs, as shown in Fig. 13. Amorphous materials consisting of transition metals show a different behavior. Amorphous materials based on the transition metals of an unfilled 4d shell show only one maximum at an e/a ratio of 6.4, whereas materials

TABLE IV Superconducting Transition Temperature T_c of Some Metallic Glasses

Metallic glass	T_c (K)
$La_{80}Au_{20}$	3.5
$La_{80}Ga_{20}$	3.8
$Zr_{75}Rh_{25}$	4.55
$Zr_{70}Pd_{30}$	2.4
$Nb_{60}Rh_{40}$	4.8
$(Mo_{0.8}Ru_{0.2})_{80}P_{20}$	7.31
$(Mo_{0.6}Ru_{0.4})_{82}B_{18}$	6.05
$(Mo_{0.8}Ru_{0.2})_{80}P_{10}B_{10}$	8.71

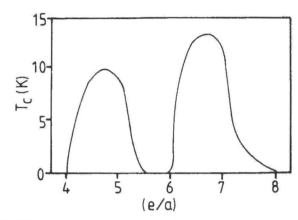

FIGURE 12 Superconducting transition temperature T_c as a function of electron per atom ratio e/a for the transition elements.

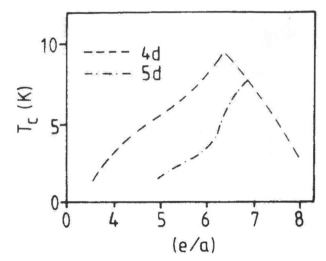

FIGURE 14 Superconducting transition temperature T_c as a function of the electron per atom ratio e/a for the amorphous 4*d* and 5*d* transition metals.

based on transition metals with an unfilled 5*d* shell have a maximum at an e/a of 7, as shown in Fig. 14. At these peak values of the superconducting transition temperatures, the values of the electron density of states are also maximum.

IV. FLUX QUANTIZATION

In 1950 F. London suggested that the magnetic flux trapped by a superconducting ring is quantized and the flux quantum is given by $\Phi_0 = ch12e = 2 \times 10^{-7}$ G/cm², where c is the velocity of light, h is Planck's constant, and e is the electronic charge. The flux trapped in a superconductor is quantized and is equal to $n\Phi_0$. In the case of type I superconductors where the Meissner effect is perfect, the value of n is zero. The flux quantization is observed only in the case of multiply connected geometries such as a superconducting ring. When the external magnetic field is removed, the magnetic flux trapped is equal to $n\Phi_0$. The flux quantization is expected to exist even in singly con-

nected geometries in the case of type II superconductors because a mixed state exists in which the superconducting regions surround the lines of force and form a multiply connected system of filaments.

V. LONDON EQUATION AND COHERENCE LENGTH

The magnetic field H and the supercurrent J_s in a superconductor are related by the equation

$$\nabla \times H = \frac{4\pi}{c} J_s,$$

where c is the velocity of light. The free energy F of a system is given as

$$F = F_s + E_{kin} + E_{mag},$$

where $|F_s|$ is the free energy of the electrons in the superconducting state. The kinetic energy E_{kin} is

$$E_{kin} = \frac{1}{2} \int_{vol} m v^2 n_s d\mathrm{r},$$

where v is the drift velocity of a parabolic band, n_s is the number of superconducting electrons per unit volume, and m is the effective mass of the electrons. The magnetic energy E_{mag} in a magnetic field H is

$$E_{mag} = \int \frac{H^2}{8\pi} d\mathrm{r}.$$

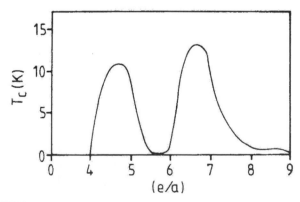

FIGURE 13 Superconducting transition temperature T_c as a function of electron per atom ratio e/a for solid solutions of the transition elements.

The free energy F can be written

$$F = F_s + \frac{1}{8\pi} \int \left[H^2 + \lambda_L^2 |\nabla \times \mathbb{H}|^2 \right] d\tau,$$

where λ_L is a constant,

$$\lambda_L = \left| \frac{mc^2}{4\pi n_s e^2} \right|^{1/2},$$

and is called the London penetration depth. London obtained an equation by minimizing the free energy with respect to the field distribution:

$$\mathbb{H} + \lambda_L^2 \nabla \times \nabla \times \mathbb{H} = 0.$$

When the current flows in the y direction in a superconductor, then the magnetic field in the z direction is

$$H_z = H(0) e^{-z/\lambda_L},$$

which shows that the magnetic field falls off exponentially inside a superconductor. The London penetration depth $\lambda(0)L$ at $T = 0K$ in terms of the Fermi velocity v_F and the electron density of states $N(0)$ is

$$\lambda(0)L = \left[\frac{3C^2}{8\pi N(0) v_F^2 e^2} \right]^{1/2}.$$

VI. COHERENCE LENGTH AND ENERGY GAP

Coherence length is a measure of the correlation distance of the superconducting electrons and is denoted ξ_0. The coherence length in terms of the Fermi velocity v_F, Boltzmann constant K_B, and superconducting transition temperature T_c is

$$\xi_0 = \frac{h v_F}{K_B T_c},$$

where h is Planck's constant.

One of the important features of superconductivity is the existence of a gap in the low-energy excitations, which is denoted ε. In most superconductors, an external energy E must be supplied to create an electron-hole pair close to the Fermi surface.

This energy E is

$$E \geq 2\varepsilon.$$

The coherence length ξ_0 and the energy gap are related by the equation

$$\xi = \frac{h v_F}{\pi \varepsilon}.$$

Bardeen, Cooper, and Schrieffer (BCS), in 1957, related this energy gap to the formation of Cooper pairs. The formation of an energy gap in a superconductor is depicted in Fig. 15 for free electrons. According to the BCS theory,

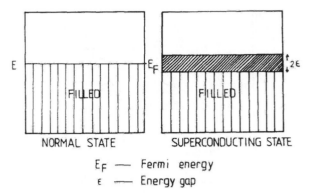

E_F —— Fermi energy
ε —— Energy gap

FIGURE 15 Formation of an energy gap in the superconducting state.

electrons that have energies close to the Fermi energy form Cooper pairs easily. The paired states have a lower energy than the unpaired electrons that form them. The electron density of states $n(E)$-versus-energy E curve of a normal metal as shown in Fig. 16 changes to the curve in Fig. 17 when the normal metal becomes a superconductor. From the BCS theory, a relationship among the energy gap at 0 K $\varepsilon(0)$, the Boltzmann constant K_B, and the superconducting transition temperature T_c is

$$\varepsilon(0) = 1.76 K_B T_c.$$

The variation of the energy gap with temperature is shown in Fig. 18. The BCS theory is discussed later.

The energy gap of a superconductor can be measured experimentally as follows. The absorption coefficients of longitudinal ultrasonic waves in the normal state α_n and superconducting state α_s are related by

$$\frac{\alpha_s}{\alpha_n} = \frac{2}{1 + \exp(\varepsilon/K_B T)}.$$

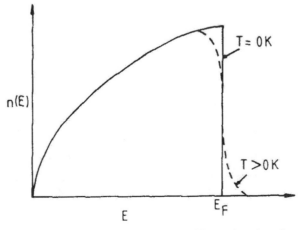

FIGURE 16 Electron density of states $n(E)$ as a function of energy E for a normal metal.

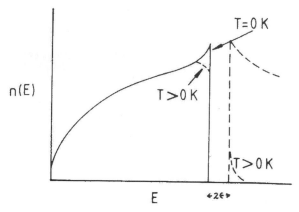

FIGURE 17 Electron density of states $n(E)$ as a function of energy E for an ideal superconductor.

The experimental determination of α_n and α_s at a particular temperature T enables one to determine the value of ε.

Absorption of the electromagnetic waves in the far-infrared ($\lambda \approx 1$ mm) region occurs for photons of energy $hv = 2\varepsilon$. Determination of the frequency of absorption directly gives an energy gap as shown in Fig. 19. The specific heat is proportional to $\exp(-\varepsilon/K_B T)$, and the value of ε can also be obtained from the specific heat measurements.

Tunneling experiments also give the value of ε directly. An experimental arrangement for the determination of ε is shown in Fig. 20. A superconductor with an energy gap ε is depicted as A and is separated from a normal conductor C through a thin insulating layer B. The shaded areas represent the occupied states. The Fermi level is at the center of the energy gap in the case of the superconductor. When a potential difference is applied across the insulat-

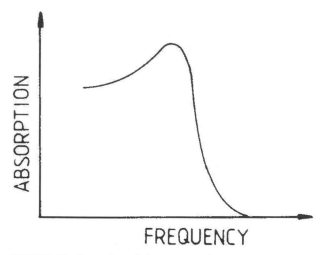

FIGURE 19 Absorption of electromagnetic waves as a function of frequency in a superconductor.

ing layer, electrons tunnel through the barrier B from C to A. This potential difference that causes the onset of the tunneling current is a direct measure of the energy gap ε. When both of the materials across the insulating layer are superconductors, the energy gaps of these two superconductors can be measured simultaneously from the potential difference-versus-current curve. A typical arrangement for these kinds of measurements is shown in Figs. 21 and 22.

VII. THERMODYNAMICS OF SUPERCONDUCTIVITY

From the basic thermodynamic considerations, we derive relations among the critical magnetic field of a

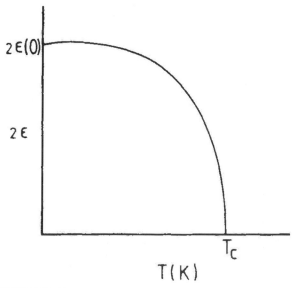

FIGURE 18 Energy gap 2ε as a function of temperature T for a superconductor.

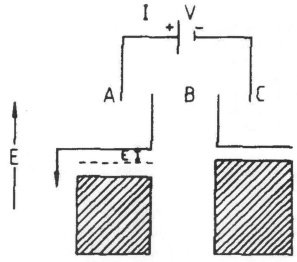

FIGURE 20 Tunneling of electrons through a thin insulating layer, B, between two superconductors, A and C.

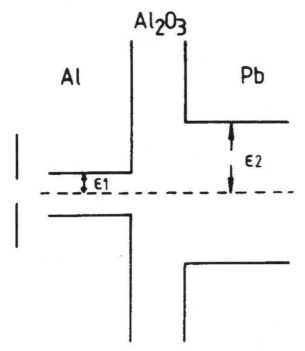

FIGURE 21 Tunneling of electrons through a thin insulating layer of Al_2O_3 between the two superconductors lead (Pb) and aluminum (Al).

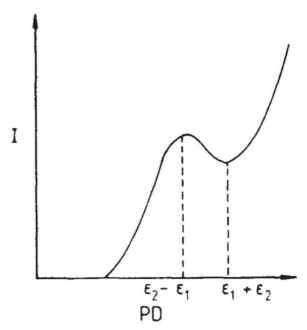

FIGURE 22 Current I-versus-potential difference PD plot of a tunnel junction consisting of two superconductors with energy gaps ε_1 and ε_2.

superconductor and the specific heats in the normal and superconducting states as well as the critical magnetic field and superconducting transition temperature T_c. Let us consider a material in the normal state with a negligible magnetization. Its Gibbs energy function G_n in the normal state is

$$G_n = U - TS + PV,$$

where U is the internal energy, T the temperature, S the entropy, P the pressure, and V the volume. For a superconductor in the presence of an external magnetic field, the magnetization is not negligible and the magnetic induction $\boldsymbol{B}$ is

$$B = H + 4\pi I,$$

where H is the applied magnetic field and I is the intensity of magnetization. In the case of a sharp superconducting transition for a long thin rod parallel to the field $B = 0$,

$$I = -H/4\pi.$$

The Gibbs function G_s in the superconducting state per unit volume is

$$G_s(H) = U - TS + PV - \int_0^H I\, dH$$

$$= G_s + \int_0^H H\, dH/4\pi$$

$$= G_s + H^2/8\pi.$$

Assuming a negligible volume change at the transition,

$$G_n = G_s(H_c),$$

where H_c is the critical magnetic field, and

$$G_n - G_s = H_c^2/8\pi.$$

Because

$$G = U + PV - TS \qquad (1)$$

and

$$dG = dU + P\,dV + V\,dP - T\,dS - S\,dT,$$

using the first law of thermodynamics,

$$dQ = dU + P\,dV = T\,dS,$$

one obtains

$$dG = V\,dP - S\,dT,$$

which gives

$$S = -(\partial G/\partial T)P. \qquad (2)$$

Combining Eqs. (1) and (2),

$$S_n - S_s = -H_c/4\pi \cdot \partial H_c/\partial T. \qquad (3)$$

The difference of the normal-state and superconducting-state entropies is expressed in terms of the critical field H_c and its slope $\partial H_c/\partial T$. The specific heat per unit volume is

$$C = dQ/dT = T\,dS/dT,$$

so that

$$C_n - C_s = T \partial(S_n - S_s)/\partial T,$$

and Eq. (3) reduces to

$$C_s - C_n = T H_c/4\pi \cdot \partial 2H/\partial T^2 + T/4\pi \cdot (\partial H_c/\partial T)^2.$$

At the transition $T = T_c$ and $H_c = 0$,

$$C_s - C_n = T_c/4\pi \cdot (\partial H_c/\partial T)^2. \qquad (4)$$

where C_s and C_n are the specific heats in the superconducting and normal states. The specific heat C_s follows the relationship

$$C_s = BT^3.$$

where B is a constant, whereas C_n is given as

$$C_n = AT^3 + \gamma T.$$

Because

$$dQ = dU + PdV = TdS,$$

$$S = \int dQ/T = \int C \, dT/T.$$

Therefore,

$$S_s = \int BT^2 \, dT = (BT^3/3)$$

and

$$S_n = \int (AT^2 + \gamma) \, dT = (AT^3/3) + \gamma T,$$

and the difference is

$$S_n - S_s = \frac{1}{3}(A - B)T^3 + \gamma T.$$

At the transition temperature T_c (in zero field)

$$S_n = S_s,$$

thus

$$\frac{1}{3}(B - A)T_c^2 = \gamma$$

and

$$S_n - S_s = \gamma\left(T - T^3/T_c^2\right). \qquad (5)$$

From Eq. (4) it follows that

$$-H_c/4\pi \cdot \partial H_c/\partial T = \gamma\left(T - T^3/T_c^2\right)$$

or

$$\partial/\partial T/\left(H_c^2\right) = 8\pi\gamma\left(T^3/T_c^2 - T\right).$$

Since $H_c = H_0$ at $T = 0$ K and integrating,

$$H_c^2 = 8\pi\gamma\left(T^4/4T_c^2 - T^2/2\right) + H_0^2 \qquad (6)$$

when $T = T_c$ and $H_c^2 = 0$, therefore

$$8\pi\gamma\left(T_c^2/2 - T_c^2/4\right) = H_0^2$$

or

$$\gamma = H_0^2/2\pi T_c^2. \qquad (7)$$

Combining Eqs. (6) and (7),

$$H_c = H_0\left(1 - (T/T_c)^2\right).$$

This equation relates the critical magnetic field H_c of a superconductor with the critical temperature T_c and has a parabolic form. This conforms to the experimental observation shown in Fig. 3 for a type I superconductor for which the relationship between H_c and T_c was

$$H_c \cong H_0\left(1 - (T/T_c)^2\right).$$

VIII. MAGNETIC SUPERCONDUCTORS

Ferromagnetism and superconductivity have been considered to be mutually exclusive phenomena. It was assumed that the large internal magnetic field present in a ferromagnetic material would not allow it to become a superconductor. This is true, and so far none of the magnetic elements (for example, chromium, manganese, iron, cobalt, and nickel) have exhibited superconductivity. A search was made to find a material that exhibits superconductivity and ferromagnetism at different temperatures. Among the rare-earth elements, lanthanum is superconducting at 6 K. The other rare-earth elements are either paramagnetic or ferromagnetic, with magnetic moments that are due to $4f$ electrons. Matthias and co-workers dissolved gadolinium metal in lanthanum and measured the superconducting transition temperatures as a function of dissolved gadolinium. Figure 23 shows a plot of the superconducting transition temperature as a function of gadolinium dissolved

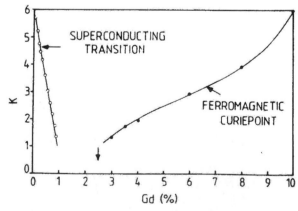

FIGURE 23 Superconducting transition temperature and ferromagnetic Curie point as a function of gadolinium (Gd) in La–Gd alloys. [From Matthias, B. T., and Suhl, H. (1960). *Phys. Rev. Lett.* **4**, 51.]

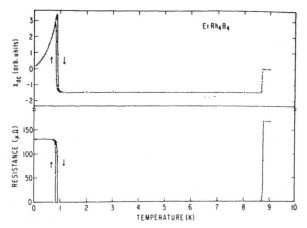

FIGURE 24 Transitions to superconducting and ferromagnetic states in $ErRh_4B_4-$. [From Fertig, W. A., Johnston, D. C., Maple, M. B., and Matthias, B. T. (1977). *Phys. Rev. Lett.* **38**, 987.]

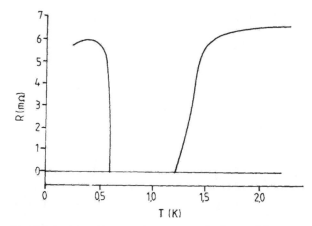

FIGURE 25 Transitions to the superconducting and ferromagnetic states in $HoMo_6S_8-$. [From Ishikawa, M., and Fischer, O. (1977). *Solid State Commun.* **23**, 37.]

in lanthanum. The depression of T_c is a linear function of gadolinium dissolved to approximately 1% gadolinium. More than 2.5% gadolinium in lanthanum makes it a ferromagnetic material.

These data suggest that an exchange interaction over conduction electrons leading to ferromagnetism is easy to bring about in an element that is itself a superconductor. This points to a possibility of a magnetic superconductor in which the phenomena of superconductivity and ferromagnetism overlap. A material of composition $ErRh_4B_4$ has been discovered that becomes superconducting at 8.7 K and shows ferromagnetic ordering at 0.93 K. The measurements of resistance and magnetic susceptibility as a function of temperature for $ErRh_4B_4$ are shown in Fig. 24. The anomalies at the temperatures of 0.93 and 8.7 K in the resistance and magnetic susceptibility curve correspond to the ferromagnetic ordering and superconducting transition temperatures.

Another Chevrel phase compound of composition $HoMo_6S_8$ exhibits superconducting and ferromagnetic transitions at 2.15 and 0.6 K, as shown in Fig. 25. The discovery of the coexistence of ferromagnetism and superconductivity in these ternary rare-earth molybdenum chalcogenides and rare-earth rhodium borides has opened a new field of investigation on the interactions responsible for ferromagnetism and superconductivity.

IX. TUNNELING AND THE JOSEPHSON EFFECT

In 1962 Josephson predicted theoretically that if two superconductors were separated by a thin ($\sim$10-Å) insulating film, then the superconducting electron pairs would tunnel through this junction. The tunneling current would flow without any voltage across the junction between the two superconductors. When the dc current is exceeded, a dc voltage would develop across the junction. This voltage is $2eV1\eta$, where e is the charge on the electron, η Planck's constant, and V the frequency of the photon radiated by the electron pair while tunneling across the junction. The maximum zero-voltage current J across the junction is

$$J = J_0 \sin(\delta_0 + 2e/c\eta) \int A \, ds,$$

where δ_0 is a constant. This shows that the current is a periodic function of the flux passing through the junction at right angles to the current and that the period is equal to the quantum of flux $\eta c/2e$. The Josephson prediction was experimentally proved by Anderson and Rowell, who showed that a zero-voltage current would flow through a thin insulating layer between the two superconductors. The maximum value of this current oscillates with the external magnetic field, as shown in Fig. 26.

X. THEORY OF SUPERCONDUCTIVITY

In 1950 Ginzburg and Landau proposed a model for superconductors in which an order process in a superconductor is described in terms of an order parameter ψ, where ψ represents the fraction of conduction electrons in the superconducting momentum state. This model contained expressions for the momentum and kinetic energy of superelectrons and described the magnetic behavior of superconductors very well, but a basic interaction mechanism was still lacking.

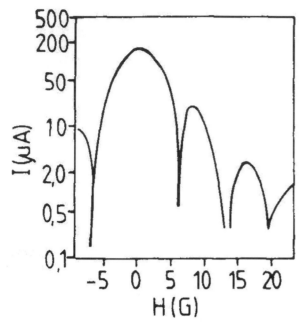

FIGURE 26 Tunneling current as a function of applied magnetic field H on a tunnel junction consisting of two superconductors. [From Rowell, J. M. (1963). *Phys. Rev. Lett.* **11**, 200.]

In 1957 Bardeen *et al.* proposed a theory of superconductivity in which they expressed the superconducting transition temperature in terms of an interaction between the electrons and the lattice vibrations of a solid. The quanta of lattice vibrations in a solid are called *phonons*. According to this theory, when the temperature of a solid is lowered, an interaction between the electrons and the phonons causes an attractive force between the conduction electron pairs called *Cooper pairs*. These Cooper pairs are paired states with equal and opposite momentum at zero supercurrent. When a current is applied to a superconductor, all the electron pairs have the same momentum directed parallel to the electric field. Due to this coherent motion, the pairs do not collide with the lattice and there is no electrical resistance. The expression for the superconducting transition temperature T_c is

$$K_B T_c = 1.14 \eta \omega_c \exp(-1/(N(0)V)). \qquad (8)$$

This equation is valid for $N(0) \ll 1$. Here $N(O)$ is the electron density of states, V the net attractive potential between the electrons, and ω_c the principal phonon frequency. The temperature T_c is extremely sensitive to small changes in V. This theory successfully explains most of the physical property changes associated with the superconducting transition. It is rather difficult to calculate the superconducting transition temperature itself using this theory. It should be mentioned that in all critical phenomena, the critical temperatures are most difficult to calculate. For example, it is not easy to calculate the freezing or boiling point of water.

It has been observed experimentally that the superconducting transition temperature of an element varies with the isotope mass. For example, for the isotopes of mercury, T_c varies between 4.185 and 4.146 K, whereas the average atomic mass M varies between 199.5 and 203.4. In Eq. (8) ω_c is proportional to $1/\sqrt{M}$, where M is the atomic mass and V is independent of M in the BCS equation.

Therefore T_c should be proportional to $1/\sqrt{M}$, and this dependence has been observed in the case of several elements such as tin, mercury, and indium. The term $N(0)V$ occurring in the BCS theory can be further expressed in terms of two parameters: the electron–phonon interaction parameter, λ; and μ^*, which describes the normalized coulomb repulsion of electrons. This modification of the BCS theory was suggested by McMillan for the strong coupling superconductors $\lambda \gg \mu^*$, where the original BCS theory is not valid. The modified expression for T_c is given by the expression

$$T_c = \Theta_D/1.45 \exp\left[-\frac{1.04(1+\lambda)}{\lambda - \mu^*(1+0.62\lambda)}\right], \qquad (9)$$

where Θ_D is the Debye temperature. The electron–phonon interaction parameter λ is

$$\lambda = \eta/M\omega_c^2,$$

where η is a constant for a given structure class. Maximization of T_c in Eq. (9) with respect to ω_c gives

$$T_c(\text{max}) = (\eta/2M)^{1/2} \exp(-3/2). \qquad (10)$$

Substituting suitable parameters into Eq. (10), a maximum T_c value of 35 K is calculated. It must be mentioned that, at present, a maximum T_c value of 23 K exists for Nb_3Ge.

XI. APPLICATIONS OF SUPERCONDUCTIVITY

Since its discovery, superconductivity has found many applications in technology. Because the electrical resistance in a superconductor is almost zero, large and homogeneous fields can be generated simply by winding the coils of the wires made from the high critical transition temperature and critical magnetic field superconducting materials. In the last decade much effort has gone into the development of these superconducting materials. Magnetic fields below and above 10 T can be produced using superconducting wires made from Nb–Ti and Nb–Sn alloys. These superconducting magnets have found a broad range of applications where normal iron magnets are inadequate. These superconducting magnets can be used as exciter magnets for homopolar generators or rotors in large alternators, and a large gain in efficiency and power density is obtained.

Fusion reactors will employ superconducting magnets to confine plasma in which deuterium and tritium will be fused to produce energy. Soon the six D-shaped superconducting coils of dimensions 2.5×3.5 m manufactured in the United States and in European countries will be tested to produce a magnetic field of 8 T, which will be used to confine deuterium–tritium plasma to produce fusion energy. Superconducting magnets have found use in particle beam accelerators for high-energy particle physics research. Another application of superconducting magnets is in nuclear magnetic resonance tomography, which requires a homogeneous magnetic field; superconducting magnets are ideal for this purpose.

Another application of superconductors is as magnetic sensors. As mentioned earlier, a Josephson junction is an extremely nonlinear detector which, when connected to a loop of a superconducting wire, forms a superconducting quantum interference device (SQUID). These SQUIDs are extremely sensitive to small changes in magnetic fields. Based on the Josephson junction, high-frequency electromagnetic radiation detectors for frequencies in the range of microwaves have been developed. Josephson junction technology also finds applications in digital signal and data processing due to the high-speed and low-power dissipation compared to semiconductor technology. The Josephson junction can replace semiconductor technology where high speed, ultrahigh performance, reliability, lower power, and compactness are required. Other applications of superconductors include lossless transport of electrical energy and generation of magnetic fields for levitation and propulsion for high-speed ground transportation.

XII. RECENT DEVELOPMENTS: HIGH-TRANSITION TEMPERATURE SUPERCONDUCTIVITY

Until April 1986, the maximum superconducting transition temperature measured in Nb_3Ge was ~ 23 K. This limited superconducting transition temperature allowed large- and small-scale applications of superconductors only with the use of liquid helium. Decades of experimental and theoretical research work showed that the phenomenon of superconductivity could be explained by the attraction of electrons caused by electron–phonon interaction (BCS theory). It was suggested that, based on this mechanism, a superconducting transition temperature of ~ 35 K could be achieved. These conclusions were based on research on about 24,000 superconducting inorganic phases. In 1986, J. G. Bednorz and K. A. Müller published a paper in *Zeitschrift für Physik* on the possibility of a superconducting transition temperature as high as 30 K in a mixture of lanthanum and barium–copper oxide $(La_{2-x}Ba_x–CuO_x)(x \sim 0.15)$ of tetragonal K_2NiF_4 structure. This discovery broke all previous records and received world attention, and the two authors received the 1987 Nobel Prize.

In a short time, superconducting oxides in the ranges 30–40, 90–100, and above 100 K were discovered. At present, the highest achievable superconducting transition temperature under normal conditions is about 133 K. The superconducting oxides of ~ 90 K superconducting transition temperature are rare-earth barium–copper oxides of orthorhombic structure. The oxygen content in these oxides plays a major role in the superconductivity. When the oxygen content is reduced, the oxides transform to a tetragonal structure and become semiconducting. Superconducting transition temperatures above 100 K are observed in thalium-, bismuth-, strontium-, calcium-, and copper-based oxides.

All these materials are ceramics and brittle, not ductile like metals or alloys, and the electronic properties are highly anisotropic. The critical current density is high in one direction and low in the other, perpendicular, direction. The epitaxial thin films of some of these oxides show critical current densities of 10^6 A/cm^2 at liquid nitrogen temperature. The critical current density of polycrystalline materials in the polycrystalline state is very low and not suitable for technical applications. The coherence length in these ceramic superconductors is quite small and is comparable to the lattice constants. These materials show rather strong electron–electron interactions, for example, as reported by Steiner *et al.* (1988). Therefore there is increasing evidence that the electron pairing in the superconducting state is of a pure electronic nature as suggested by Anderson (1987), and not caused by electron–phonon interaction.

The mechanical properties of these ceramic superconductors as well as their superconducting properties may be improved by the addition of silver metal as reported by Khan *et al.* At present, a worldwide effort is ongoing to improve the mechanical properties and to increase the critical current densities of these materials for large-scale applications. Once the mechanical properties of ceramic superconductors are improved and the critical current density is increased to a practical value, it is expected that these superconducting materials will revolutionize various technologies by working at liquid nitrogen, rather than liquid helium, temperatures.

SEE ALSO THE FOLLOWING ARTICLES

CRYOGENIC PROCESS ENGINEERING • CRYOGENICS • FERROMAGNETISM • RARE EARTH ELEMENTS AND MATERIALS • SUPERCONDUCTING CABLES • SUPERCONDUCTING DEVICES • SUPERCONDUCTIVITY MECHANISMS • SUPERCONDUCTORS, HIGH TEMPERATURE • THERMOELECTRICITY

BIBLIOGRAPHY

Anderson, P. W. (1987). *Science* **235,** 1196.

Bardeen, J., Cooper, L. N., and Schrieffer, J. R. (1957). *Phys. Rev.* **108,** 1175.

Barone, A., and Paterno, G. (1982). "Physics and Applications of the Josephson Effect," Wiley, New York.

Bednorz, J. G., and Müller, K. A. (1986). *Z. Phys.* **B64,** 189.

Buckel, W. (1972). "Supraleitung," Physik Verlag GmbH, Weinheim, Germany.

Khan, H. R. (1984). *Gold Bull.* **17**(3), 94.

Khan, H. R. (1998). *J. Superconduct* **11,** 1.

Khan, H. R., and Loebich, O., (1995). *Physica C.* **254,** 15.

Khan, H. R., and Raub, C. J. (1985). *Annu. Rev. Mater. Sci.* **15,** 21.

Kittel, C. (ed.) (1976). "Introduction to Solid State Physics," 5th ed. Wiley, New York.

Newhouse, V. L. (1964). "Applied Superconductivity," Wiley, New York.

Putlin, S. N., and Antipov, E. V. (1993). *Nature* **362,** 226.

Roberts, B. W. (1976). *J. Phys. Chem. Data* **5**(3), 581–821.

Saint-James, D., Sarma, G., and Thomas, E. J. (1969). "Type II Superconductivity," Pergamon, Oxford.

Steiner, P., *et al.* (1988). *Z. Phys.* **B69,** 449.

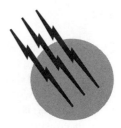

Superconductivity Mechanisms

Jozef Spałek

Jagiellonian University and Purdue University

I. Introduction
II. The Bardeen–Cooper–Schrieffer (BCS) Theory:
A Brief Summary
III. Normal and Magnetic States of Correlated
Electrons
IV. Novel Mechanisms of Electron Pairing
V. Conclusions

GLOSSARY

Almost-localized Fermi liquid A metallic system which, under a relatively small change of an external parameter such as temperature, pressure, or composition, undergoes a transition to the Mott insulating state. In such a metal electrons have a large effective mass. At low temperatures the system may order antiferromagnetically or undergo a transition to the superconducting state. Both nonstoichiometric oxides (such as V_2O_{3-y}) and heavy-fermion systems (e.g., UPt_3) are regarded as almost-localized Fermi liquids.

Bardeen–Cooper–Schrieffer (BCS) theory Theory describing properties of superconductors in terms of the concept of pairing of electrons with opposite spins and momenta. The pairing of electrons is mediated by a dynamic positive-ion lattice deformation, which produces resultant attractive interaction overcoming their mutual coulomb repulsion. At a critical temperature the electron system undergoes a phase transition to a condensed state of pairs which is characterized by a zero dc electrical resistance and a strong diamagnetism

(Meissner–Ochsenfeld effect). The condensed state is destroyed by the application of an applied magnetic field (the critical fields H_c and H_{c2} for superconductors of the first and second kinds, respectively.

Correlated electrons Electrons with their kinetic (or band) energy comparable to or lower than the magnitude U of electron–electron repulsion. This situation is described by the condition $U \gtrsim W$, where W is the width of a starting (bare) energy band. Strictly speaking, we distinguish between the limits of almost–localized Fermi liquids, for which $U \lesssim W$, and the limits of strongly correlated electrons (Tomonaga–Luttinger or spin liquids), for which $U \gg W$. The term "correlated electrons" means that the motion of a single electron is correlated with that of others in the system (for example, its effective mass depends on the two-particle correlation function).

Exchange interaction Part of the coulomb interaction between electrons which depends on the resultant spin state of their partially filled d or f shells. If the spin–singlet configuration is favored in the ground state, then the interaction is called antiferromagnetic. The

Encyclopedia of Physical Science and Technology, Third Edition, Volume 16

exchange interaction provides a mechanism of magnetic ordering in Mott insulators; it may also correlate electrons into singlet or triplet pairs in the metallic state, particularly when the pair-exchange coupling J of an electron pair is comparable to the kinetic energy of each of its constituents, as is the case for strongly correlated electrons. The superexchange (or kinetic exchange) is induced by a strong electron correlation.

Fermi liquid Term describing the state of interacting electrons in a metal. Equilibrium properties of such systems are modeled by a gas of electrons with renormalized characteristics such as the effective mass (they are called *quasiparticles*). The properties at low temperatures are determined mainly by electrons near the Fermi surface. The electron–electron interactions lead to specific contributions to the transport properties of such a system producing, e.g., sound-wave and plasmon excitations.

High-temperature superconductors Oxide materials of the type $La_{2-x}Sr_xCuO_4$ or $YBa_2Cu_3O_{7-x}$, which have a layer structure, with the principal role of electrons confined to the CuO_2 planes. The term "high-temperature superconductors" (HTS) was coined to distinguish these and other oxide superconductors with a critical temperature $T_c \gtrsim 20$ K from "classical" superconductors, which comprise metals and intermetallic compounds such as Nb_3Ti with $T_c \lesssim 23$ K. At present, this class of materials (HTS) is characterized by the quasi-two-dimensonal structure of the normal metallic state above T_c and strong deviations from either the normal Fermi-liquid or the BCS superconducting type of behavior in the corresponding temperature regimes $T > T_c$ and $T < T_c$, respectively.

Hubbard subband Term describing each of the two parts of an energy band in a solid which splits when the electron–electron repulsion energy is comparable to (or larger than) their kinetic (band) energy. The Hubbard splitting of the original band induced by the interaction explains in a natural way the existence of the Mott insulating state in the case of an odd number of electrons per atom (that is, when the atomic shells would normally form an only half-filled band; cf., e.g., CoO).

Mott insulator An insulator containing atoms with partially filled $3d$ or $4f$ shells. These systems order magnetically (usually antiferromagnetically) when the temperature is lowered. Thus, they differ from ordinary (Bloch–Wilson) or band insulators, which are weakly diamagnetic, and are characterized by filled atomic shells, separated from empty states by a gap. In the antiferromagnetic phase of the Mott insulators each electron with its (frozen) spin oriented up is surrounded by electrons with their spin down, and vice versa. The parent stochiometric materials for high-temperature superconductors (e.g., La_2CuO_4 and $YBa_2Cu_3O_6$) are antiferromagnetic Mott insulators with Néel temperatures ($T_N = 250$ and 415 K, respectively).

Real-space pairing Source of attraction or superconducting correlations that is not induced by lattice deformation (phonons). Such pairing may be provided by the density fluctuations within the interacting electron subsystem (e.g., by spin fluctuations or other excitations). By real-space pairing we mean the pairing of electron spins in correlated metals caused by exchange interactions (e.g., kinetic exchange) among electrons in coordinate space. The essence of the real-space pairing, not resolved as yet, is contained in the question, Can a strong short-range part of the coulomb repulsion (of range a_0) lead to an attraction (an effective binding) at intermediate distances ($2 \div 10a_0$), where strong singlet–spin correlations prevail?

Strongly correlated electrons Electrons describing the metallic state of high-temperature superconductors, some heavy-fermion systems (**non-Fermi liquids**), and, particularly, systems of low dimensonality, $d = 1$ and 2. In these systems, the concept of a Fermi liquid is inapplicable, and for $d = 1$, at least, the charge and spin degrees of freedom lead to separate quasiparticle representations—*holons* and *spinons*, respectively. The quantum liquid describing strongly correlated electrons composes a new quantum macrostate.

I. INTRODUCTION

Superconductivity remains among the most spectacular manifestations of a macroscopic quantum state of electrons in a metal or plasma. Experimentally, one observes below a characteristic temperature T_c a transition to a phase with nonmeasurable dc resistance (or with a persistent current), a perfect diamagnetism of bulk samples in a weak magnetic field, and quantum tunneling between superconductors separated by an insulating layer of mesoscopic ($\sim$1-nm) thickness. In the theoretical domain, one studies the quantum–mechanical (nonclassical) mechanisms of pairing of the microscopic particles (fermions) at a macroscopic scale. Here, we summarize briefly our present understanding of the Bardeen–Cooper–Schrieffer (BCS) theory of "classical" superconductors (see Section II) and we review the current theoretical approaches to new superconductors: the heavy-fermion materials and the high-T_c magnetic oxides. The latter subject is discussed in Section IV, after we summarize normal-state properties of correlated electrons in Section III.

A brief characterization of the recent studies of super-conductivity is in order. From the time of the first discovery (1911) of superconductivity in mercury (at temperature $T_c \simeq 4.2$ K) by Kammerlingh Onnes until 1986, studies were limited to low temperatures, $T < 25$ K. During the next 5 years, six classes of new superconducting compounds with critical temperatures $T_c = 30$ K (for $Ba_{1-x}K_xBiO_3$), 40 K (for $La_{2-x}Sr_xCuO_4$), 90 K (for $YBa_2Cu_3O_{7-\delta}$), 110 K (for $Bi_2Sr_2CaCu_2O_8$), and 125 K (for $Tl_2Ca_2Ba_2Cu_3O_{10-y}$), and 135 K ($HgBa_2Ca_2Cu_3O_8$) were discovered and/or thoroughly studied in a number of laboratories. In recent years the idea has also been applied to new systems such as Fermi condensated dilute gases and quark–gluon plasma in high-energy physics. Apart from the discovery of spin–triplet pairing in liquid ^{3}He, evidence for it in Sr_2RuO_4 also opens new possibilities for pairing studies.

The starting point in both classical and new superconducting materials is the electronic structure that determines the metallic properties in the normal phase (that is, that above T_c). In this respect, the classical superconductors are well described by band theory and, in some cases, starting from the concept of the Fermi–liquid concept. In contrast, the new materials are characterized as those whose electrons are close to localization, that is, those close to the metal–insulator transition of the Mott–Hubbard type. The latter transition may be induced by a relatively small change in compound composition (cf. the behavior of $La_{2-x}Sr_xO_4$ or $YBa_2Cu_3O_{7-x}$ as a function of x). It is quite interesting to note that oxides such as $YBa_2Cu_3O_{7-x}$ may be synthesized in either insulating ($x \gtrsim 0.65$) or metallic states. Additionally, antiferromagnetic ordering of the $3d$ electrons is observed close to the insulator–metal transition; the magnetic insulating state transforms into a superconducting state when $0 \leq x \lesssim 0.65$. Therefore, an account of our understanding of the almost-localized metallic state in a normal or magnetic (that is, nonsuperconducting) phase is highly desirable and summarized in Section III. The antiferromagnetic insulating, normal metallic, and superconducting states must all be treated on the same footing for a proper characterization of high-T_c oxides. In this manner, the studies of those systems must incorporate the description of different quantum phase transitions. One can say that the theory of strongly correlated electrons and of the superconductivity in those systems poses one of the most challenging problems for physics of the 21st century.

Details of the electronic structure in high-T_c oxides are also important for two additional reasons. First, as discussed later, in these superconductors the coherence length is quite small, that is, comparable to the lattice constant. Hence, the details of the wave function on the atomic scale become crucial. Second, a whole class of models (discussed in Section IV) relies on the electron pairing induced by short-range electron–electron interactions. These interactions are strong and also present in the normal phase. This is the reason one must develop a coherent theoretical picture of the correlated metallic state that undergoes a transformation either to the Mott insulating or to the superconducting state. Such a theory does yet not exist.

In this chapter, the properties of correlated electrons in normal, insulating, magnetic, and superconducting phases are reviewed and related to the parametrized models, starting from either Hubbard or Anderson-lattice Hamiltonians. These are the models that describe the properties of correlated metallic systems in terms of a few parameters, such as the band width W of starting (uncorrelated, bare) electrons, the magnitude U of short-range (intraatomic) coulomb interactions, etc. Such models provide an overall understanding of both the nature of correlated metallic and insulating ground states and the underlying thermodynamic properties of these systems. However, the guidance of detailed band structure calculations is often needed in choosing appropriate values for the microscopic parameters, as well as to understand the specific features of the compounds.

II. THE BARDEEN–COOPER–SCHRIEFFER (BCS) THEORY: A BRIEF SUMMARY

The BCS theory [1–10] relies on three features of metallic solids: (1) the electron–lattice interaction; (2) the formation of an electron-pair bound state (the so-called Cooper pair state) due to the coupling of the electrons to the lattice; and (3) the instability of the normal metallic state with respect to the formation of a macroscopic condensed state of all pairs ($\mathbf{k}\uparrow, -\mathbf{k}\downarrow$) with antiparallel spins in momentum ($\mathbf{k}$) space. The condensed state exhibits the principal properties of superconductors, such as a perfect diamagnetism, zero dc resistance, etc. We first discuss these three features briefly and then summarize some consequences of the BCS theory. The BCS theory not only deals with one of the possible (phonon-mediated) mechanisms for superconductivity, but also provides proper language for the description of such a condensed state in general terms, independent of the particular pairing mechanism. One should also remark at the beginning that such a condensed state of pairs *cannot* be regarded as a Bose condensed state if the size of the bound-state wave function ξ (*the coherence length*) is much larger than the interparticle distance $a = (V/N)^{\frac{1}{3}}$; this happens for the "classic" superconductors.

A. From Electron–Phonon Coupling to the Effective Attractive Interaction between Electrons: Virtual Exchange of Phonons

The electron–lattice interaction can be described by introducing phonons as quasiparticles representing vibrational modes of the lattice. In this picture, an electron moving in a solid and scattering on the lattice vibration absorbs or emits a phonon with energy $\hbar\omega_q$ and quasi-momentum $\hbar q$. If during such processes the energy of the incoming electron (with energy ϵ_k and momentum $\hbar k$) and the scattered electron (with energy $\epsilon_{k'}$) is conserved, then a real scattering process has taken place. Such events lead to the nonzero resistivity of metals at temperature $T > 0$. For these processes

$$\epsilon_{k'} - \epsilon_k = \pm\hbar\omega_q, \tag{1}$$

where $-$ corresponds to the emission and $+$ to the absorption of the phonon. However, in the quantum-mechanical description of scattering processes, there also exist virtual processes that do not conserve energy. Such events involve the emission and subsequent reabsorption of a phonon in a time interval Δt such that the uncertainty principle $\Delta E \cdot \Delta t \geq \hbar$ is not violated. The uncertainty of particle energies ΔE is related to the magnitude of the electron–phonon interaction. In effect, this leads to the following effective electron–electron interaction energy involving a pair (k, k') of electrons:

$$V_{kk'q} = |W_q|^2 \frac{\hbar\omega_q}{(\epsilon_{k'} - \epsilon_k)^2 - (\hbar\omega_q)^2}, \tag{2}$$

where $(k' - k) = q$, and W_q is the electron–phonon matrix element characterizing the process of single emission or absorption of the phonon by the electron subsystem. In many electron systems, one represents Eq. (2) by an effective electron–electron interaction, which can be written

$$H' = \sum_{kk'q} V_{kk'q} c_{k+q\sigma}^+ c_{k'-q\sigma'}^+ c_{k'\sigma'} c_{k\sigma}. \tag{3}$$

This is a phonon—mediated contribution to the interaction between electrons. More precisely, in this expression $c_{k\sigma}$ symbolizes a destruction or annihilation of an electron in the initial single—particle state $|k\sigma\rangle$, whereas $c_{k'\sigma'}^+$ is the creation of an electron in the state $|k'\sigma'\rangle$ after the scattering process has taken place. The processes represented in Eq. (3) of destruction of the electron pair in the states $|k\sigma\rangle$ and $|k'\sigma'\rangle$ and their subsequent reestablishment in the final states $|k + q\sigma\rangle$ and $|k' - q\sigma'\rangle$ are customarily represented by a diagram of the type in Fig. 1b. It symbolizes the phonon exchange between the two electrons moving through crystal. The virtual processes are composed of two parts: one describing phonon emission and the subsequent reabsorption process and one describing the reverse process.

One should note that if in Eq. (2) $|\epsilon_{k'} - \epsilon_k| < \hbar\omega_q$, then $V_{kk'q} < 0$, that is, the interaction is attractive. This happens, for example, on the Fermi surface, where $\epsilon_k = \epsilon_{k'} = \mu$. The sign of the interaction changes rapidly once we depart from the Fermi surface, since the electronic energies present are much higher than that of phonons. Hence, if only the magnitude of attraction overcomes the magnitude of the coulomb repulsion between the electrons in a given medium, this leads to a net attraction between the electrons. Such a net attractive interaction results in a stable superconducting state, as we shall see next.

B. Instability of the Electron Gas State in the Case of Attractive Interaction between Electrons: Cooper Pairs

Following Fröhlich's discovery [11] that the electron–electron attraction can be mediated by phonons (cf. the previous discussion), the next step was taken by

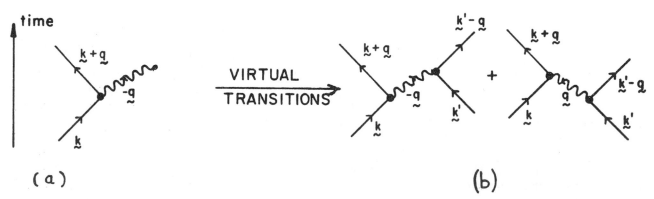

FIGURE 1 (a) Scattering diagram of electrons with wave vectors $k \rightarrow k + q$, accompanied by emission of the phonon of wave vector $-q$. (b) Virtual emission and subsequent reabsorption of the phonon by electrons. The two processes drawn combine into the contribution [Eq. (2)] leading to the effective electron–electron attraction.

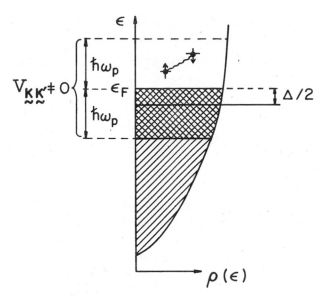

FIGURE 2 Schematic representation of the conduction band filled with electrons up to the Fermi level ϵ_F. The density of states is $\rho(\epsilon)$ (per one spin direction). The two electrons added to the system attract each other with the potential $V_{kk'} = -V$ if placed within the energy interval $\hbar\omega_p$ counting from ϵ_F. The attraction leads to a binding energy Δ below ϵ_F for the pair configuration $(k\uparrow, -k\downarrow)$.

Cooper [12], who asked what happens when two electrons are added to an electron gas at $T = 0$. Because of the Pauli exclusion, they must occupy the states above the Fermi level, as shown in Fig. 2. Cooper showed that if the attractive potential in Eq. (2) is approximated by a negative nonzero constant $(-V)$ in the energy interval $2\hbar\omega_p$ around the Fermi level ϵ_F (cf. Fig. 2), then such a potential introduces a binding between these two electrons with a binding energy

$$\Delta = -2\hbar\omega_p \left[\exp\left(\frac{2}{\rho V}\right) - 1 \right]^{-1} \simeq -2\hbar\omega_p \exp\left(-\frac{2}{\rho V}\right),$$

(4)

relative to the energy $2\epsilon_F$ of those two particles placed at the Fermi energy. In this expression, $\hbar\omega_p$ represents the average phonon energy (related to the Debye temperature θ_D through $\hbar\omega_p = k_B\theta_D$), and ρ is the density of free—particle states at the Fermi energy ϵ_F for the metal under consideration.

A few important features of the bound state represented by Eq. (4) should be mentioned. First, the binding energy Δ is largest for the state of the pair at rest, that is, with the total pair momentum $\mathbf{k}_1 + \mathbf{k}_2 = 0$. Thus, Δ represents the binding energy of the pair $(\mathbf{k}, -\mathbf{k})$. Second, the spin of the pair is compensated, that is, a singlet state is assumed. Finally, the bound state has a lower energy than a pair of

free particles placed at the Fermi level. Hence, the electron gas state is unstable with respect to such pair formation. A system of such pairs may condense into a superfluid state. However, the situation is not so simple since the size of the pair is of the order

$$\xi = \langle r^2 \rangle^{\frac{1}{2}} \approx \frac{2\hbar V_F}{\Delta} \approx \frac{\hbar V_F}{k_B T_C} \sim 10^{+4} \,\text{Å},$$

(5)

where V_F is the Fermi velocity for electrons. The quantity ξ thus exceeds by far the average classical distance between the electrons, which is comparable to the interatomic distance $a \sim \frac{1}{2} 2 \,\text{Å}$. In other words, the wave functions of the different pairs overlap very strongly, forming a condensed and coherent state of pairs in the superconducting phase. The properties of this condensed phase are discussed next. The new length scale ξ appearing in the system when electrons are bound into Cooper pairs is called the coherence length.

C. Properties of the Superconducting State: The Pairing Theory

The BCS theory [1] provides a method of calculating the ground state, thermodynamic, and electromagnetic properties of a superconductor treated as a condensed state of electron pairs with opposite momenta and spins. The starting microscopic Hamiltonian is

$$H = \sum_{k\sigma} \epsilon_k n_{k\sigma} + \sum_{kk'} V_{kk'} c_{k\uparrow}^+ c_{k\downarrow}^+ c_{-k'\downarrow} c_{k'\uparrow}.$$

(6)

The first term describes the single-particle (band) energy, ϵ_k being the energy per particle and $n_{k\sigma} = c_{k\sigma}^+ c_{k\sigma}$ the number of particles in the state $|k\sigma\rangle$. The second term describes the pairing part [Eq. (3)] for the system of pairs that scatters from the state $(\mathbf{k}', -\mathbf{k}')$ into the state $(\mathbf{k}, -\mathbf{k})$. This term describes the dominant contribution of all processes contained in Eq. (3) (cf. Ref. 10).

To obtain eigenenergies of the Hamiltonian [Eq. (6)], one can use either the variational method due to Schrieffer [2], the transformation method developed by Bogoliubov and Valatin [13], or the two-component method due to Nambu [13]. To obtain quasiparticle states in the superconducting phase, one has to combine an electron in the state $|k\uparrow\rangle$ with one in the time-reversed state $|-k\downarrow\rangle$. More precisely, one defines new quasiparticle operators λ_{k0}^+ and λ_{k1}^+, which are expressed by the operators c^+ and c in the following manner:

$$\lambda_{k0}^+ = u_k c_{k\uparrow}^+ - v_k c_{-k\downarrow},$$

and

$$\lambda_{k1}^+ = v_k c_{k\uparrow}^+ + u_k c_{-k\downarrow}.$$

The coefficients of the transformation fulfill the condition $u_{\mathbf{k}}^2 + v_{\mathbf{k}}^2 = 1$. One should note that the transformation *does not* conserve the particle number, so one has to add the term $(-\mu N)$ to the Hamiltonian (b), where $\mu \equiv \epsilon_F$ is the chemical potential in the superconducting state.

The single-particle excitations in the superconducting phase are specified by

$$E_{\mathbf{k}} = \left[(\epsilon_{\mathbf{k}} - \mu)^2 + |\Delta_{\mathbf{k}}|^2\right]^{\frac{1}{2}}, \qquad (7)$$

where μ is the chemical potential of the system and $|\Delta_{\mathbf{k}}|$ is the so-called superconducting gap determined from the self—consistent equation

$$\Delta_{\mathbf{k}} = -\sum_{\mathbf{k}} V_{\mathbf{k}\mathbf{k}'} \frac{\Delta_{\mathbf{k}}}{2E_{\mathbf{k}'}} \tanh\left(\frac{\beta E_{\mathbf{k}'}}{2}\right), \qquad (8)$$

with $\beta \equiv (k_B T)^{-1}$. One should note that if $V_{\mathbf{k}\mathbf{k}'}$ is approximated by a negative constant, then $\Delta_{\mathbf{k}} = \Delta$; Eq. (8) then yields as a solution either $\Delta \equiv 0$ or $\Delta \neq 0$, obeying the equation

$$1 = \frac{V}{N} {\sum_{\mathbf{k}}}' \frac{1}{2E_{\mathbf{k}}} \tanh\left(\frac{\beta E_{\mathbf{k}}}{2}\right), \qquad (9)$$

where now

$$E_{\mathbf{k}} = \left[(\epsilon_{\mathbf{k}} - \mu)^2 + \Delta^2\right]^{\frac{1}{2}}. \qquad (10)$$

The primed summation in Eq. (9) is restricted to the regime of $\mathbf{k}$ states where $V \neq 0$. Equations (9) and (10) constitute the simplest BCS solution for an isotropic ($\mathbf{k}$-independent) gap. One sees that $E_{\mathbf{k}}$ is always nonvanishing and reaches a minimum $E_{\mathbf{k}} = \Delta$ for electrons placed on the Fermi level, where $\epsilon_{\mathbf{k}} = \mu$. Thus, the meaning of the gap becomes obvious: it is the gap for the single—electron excitations from the superconducting (condensed) phase to a free—particle state. The presence of a gap $\Delta > k_B T_c$ in the spectrum of single—particle excitations suppresses the scattering of electrons with acoustic phonons. The thermally excited electrons across the gap do not yield nonzero resistivity because their contribution is short-circuited by the presence of the pair condensate that carries a current with no resistance. The same holds true even for the superconducting systems for which the gap vanishes along some lines or at some points in $\mathbf{k}$ space.

One should emphasize that all thermodynamic properties are associated with thermal excitations; the energies that are specified by Eq. (7) contain $|\Delta_{\mathbf{k}}|$ or Δ as a parameter to be determined self-consistently from Eq. (8) or (9), respectively. Next, we provide a brief summary of the results that may be obtained within the BCS theory.

D. Summary of the Properties: The Homogeneous State

The solution of Eq. (9) provides the following properties.

1. At $T = 0$, Eq. (1) reads
$$1 = (V/2N) {\sum_{\mathbf{k}}}' E_{\mathbf{k}}^{-1}. \qquad (11)$$

The value of $\Delta \equiv \Delta(T = 0)$ for $\rho V \ll 1$ is given by

$$\Delta_o = \frac{\hbar \omega_p}{\sinh(1/\rho V)} \approx 2\hbar \omega_p \exp\left(-\frac{1}{\rho V}\right), \qquad (12)$$

where $\hbar \omega_p \approx k_B \theta_D$[1] and ρ is the density of States at the Fermi energy. One notes a striking similarity between Eq. (12) and Eq. (4), particularly for $\rho V \ll 1$ (this condition represents the so-called weak—coupling limit); The absence of factor 2 in (12) provides an enchancement of the gap in the condensed state due to the presence of other electrons.

2. We can choose the origin of energy at μ. Then Eq. (9) can be transformed into an integral form:

$$1 = V \int_0^{\hbar \omega_p} \frac{\rho(\epsilon)\, d\epsilon}{(\epsilon^2 + \Delta^2)^{\frac{1}{2}}} \tanh\left(\frac{\beta}{2}\sqrt{\epsilon^2 + \Delta^2}\right). \qquad (13)$$

Since $\hbar \omega_p \ll \mu$, we may take $\rho(\epsilon) \approx \rho(\epsilon_F) \equiv \rho$ within the range of integration. This allows for an analytic evaluation of the critical temperature for which $\Delta = 0$:

$$T_c = 1.13 \theta_D \exp\left(-\frac{1}{\rho V}\right). \qquad (14)$$

In all these calculations, it is implicitly assumed that $\rho V \ll 1$. Because of the presence of the exponential factor in Eq. (14), the critical temperature T_c is much lower than the Debye temperature characterizing the average energy of acoustic phonons. This is the principal theoretical reason that T_c is so low in the superconductors discovered in the period 1911–1986. The exponential dependence of T_c on the electronic parameter ρV also explains why the parameters pertaining to the electronic structure, which are of the order of 1 eV or more, respond to phase transitions on an energy scale that is three orders of magnitude smaller ($k_B T_c \sim 1$ meV). Effects with such a nonanalytic dependence of transition temperature on the coupling constant cannot be obtained in any order of perturbation theory starting with the normal state as an initial state. A similar type of effect is obtained in the studies of the Kondo effect (cf. Section IV).

3. Combining Eqs. (14) and (1.12) one obtains the universal ratio

$$\frac{2\Delta_o}{k_B T_c} = 3.53, \qquad (15)$$

---

[1]In actual practice, one assumes that $\hbar \omega_p \approx 0.75 k_B \theta_D$ (cf. Meservey and Schwartz in Ref. 9).

which is frequently used as a test for the applicability of the BCS model. However, this value can also be obtained in the strong-coupling limit [15] for a particular strength of electron–phonon coupling.

4. By regarding energies E_k as representing electron excitations across the gap, one can write the expression for the entropy of a superconductor in the standard form:

$$S = -2k_B \sum_k [f_k \ell n f_k + (1 - f_k) \ell n(1 - f_k)], \quad (16)$$

where $f_k \equiv f(E_k) = [1 + \exp(\beta E_k)]^{-1}$ is the Fermi–Dirac distribution function $[1 + \exp(\beta E_k)]^{-1}$. Hence, the free energy of the superconducting state is

$$F_S = 2 \sum_k E_k f_k - TS. \quad (17)$$

One should note that the thermodynamic properties are determined fully only if the chemical potential $\mu = \mu(T)$ and the temperature dependence of the superconducting gap $\Delta_k = \Delta_k(T)$ are explicitly determined, since only then is the spectrum of single—particle excitations (characterized by the energies $\{E_k\}$) uniquely determined. The quantity $\Delta(T)$ is determined from Eq. (13). The chemical potential is determined from the conservation of the number N_e of particles, that is, from the condition $\sum_k f_k = N_e$. The temperature dependence of the gap in the isotropic case is shown schematically in Fig. 3.

5. By calculating this difference of the free energies $F_S - F_N$ in superconducting (F_S) and normal (F_N) phases and equating the difference with the magnetic free energy $H_c^2 V/8\pi$ (V is the volume of the system), one can obtain an approximate relation of the form

$$\frac{H_c(T)}{H_c(0)} \approx 1 - \left(\frac{T}{T_c}\right)^2. \quad (18)$$

For the applied field $H > H_c$, superconductivity is destroyed because in the thermodynamic critical field H_c

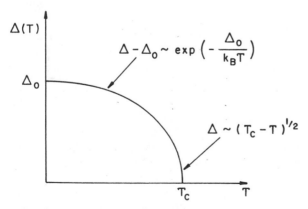

FIGURE 3 Schematic representation of the temperature dependence of the superconducting gap for the isotropic change. T_c is the critical temperature for the transition, and $\Delta_0 \equiv \Delta(T=0)$.

the spin–singlet bound state is destroyed by the thermal fluctuations. The pair binding energy is then effectively overcome by the magnetic energy, so that the pairs break up into single particles. Strictly speaking, this type of behavior characterizes the so-called superconductors of the first kind.

6. By calculating the specific heat from the standard thermodynamic analysis $[C_S = -T(\partial^2 F_S/\partial T^2)_V]$, one obtains at $T = T_c$ a discontinuity of the form

$$\frac{C_S - C_N}{C_N} = 1.43, \quad (19)$$

where C_N is the specific heat at T_c for the material in its normal state. At low temperatures, the specific heat decreases exponentially:

$$C_S \sim \exp\left(-\frac{\Delta_0}{k_B T}\right), \quad (20)$$

for the special case of an isotropic gap. However, if the gap is anisotropic $[\Delta = \Delta_k(T)]$ and has lines of zeros (along which $\Delta_k = 0$), then the low—temperature dependence of C_S does not follow Eq. (20) but rather a power law T^n, with n depending on the details of the gap anisotropy.

The specific heat grows with T because the number of thermally broken pairs increases with rising temperature; eventually, at $T = T_c$ ($k_B T_c \sim \Delta_0$), all bound pairs dissociate thermally, at which point C_s reaches a maximum. If the temperature is raised further (above T_c), the specific heat drops rapidly to its normal-state value since no pairs are left to absorb the energy. This type of behavior is observed in superconductors with an isotropic gap (cf., e.g., Hg and Sn). One should note that this interpretation of the thermal properties is based on the single—particle excitation spectrum [Eq. (10)]; we have disregarded any fluctuation phenomena near T_c, as well as collective excitations of the condensed system. It can be shown that the large coherence length $\xi \sim 10^3/10^4$ Å encountered in classic superconductors [8] is related to the absence of critical behavior near T_c. This is not the case in high-T_c superconductors (discussed in Section IV); hence, the new materials open up the possibility of studies of critical phenomena in superconducting systems.

7. The spin part of the static magnetic susceptibility vanishes as $T \to 0$. This is a direct consequence of the binding of electrons in the condensed state into singlet pairs. Therefore, the Meissner effect (the magnetic flux expulsion from the bulk of the sample) at $T = 0$ is present because the orbital part of the susceptibility is diamagnetic (roughly, it represents an electron-pair analogue of the Landau diamagnetism of single electrons in a normal electron gas). The expulsion of the magnetic flux from the bulk is measured in terms of the so-called London penetration depth $\lambda = \lambda(T)$, which characterizes the decay of

the magnetic induction inside the sample. It decays according to

$$B(z) = H_a \exp(-z/\lambda),$$

where the z direction is perpendicular to the sample surface and the applied magnetic field H_a is parallel to it. The temperature dependence of the penetration depth is given by

$$\frac{\lambda(T)}{\lambda(0)} = \left[\frac{\Delta(T)\tanh(\Delta/2k_BT)}{\Delta_o} \right],$$

$$\approx \left[1 - \left(\frac{T}{T_c} \right)^4 \right]^{-\frac{1}{2}}. \tag{21}$$

This result has been derived under the assumption that the coherence length[2] $\xi_o \simeq \hbar V_F/\Delta_o$ is much larger than λ. One should note that for a bulk sample of dimension $d \gg \lambda$ the induction $B \equiv 0$ almost everywhere. This condition determines the magnetic susceptibility χ of a superconductor regarded as an ideal diamagnet; in cgs units, $\chi \equiv M/H = -1/(4\pi)$.

8. The relative ratio of the two characteristic lengths $\kappa \equiv \lambda/\xi$ determines the type of superconductivity behavior in a magnetic field. From the dependence $\xi \sim \Delta^{-1}$, we infer that as $T \to T_c$, $\xi \sim (T_c - T)^{-\frac{1}{2}}$. The same type of dependence for $\lambda(T)$ can be inferred from Eq. (21) when $T \to T_c$. Within the phenomenological theory of Ginzburg and Landau (which can be derived from the BCS theory as shown by Gorkov [8]), one can show that if $\kappa \lesssim 1/\sqrt{2}$, then the material is a superconductor of the first kind; if $\kappa \gtrsim 1/\sqrt{2}$, then the material is of the second kind. The value of κ is directly related to the penetration depth $\lambda(T)$. The thermodynamic critical magnetic field Eq. (18) has the form

$$H_c(T) = \Phi_0 \frac{\sqrt{2}}{\kappa \lambda^2(T)} \tag{22a}$$

or, equivalently,

$$H_c(T) = \frac{\Phi_o}{2\pi\sqrt{2}\xi(T)\lambda(T)}, \tag{22b}$$

where $\Phi_0 = \hbar c/2e$ is the magnetic—flux quantum. This value of the field terminates superconductivity of the first kind. For superconductors of the second kind, the corresponding field is given by

$$H_{c2} = \kappa\sqrt{2}H_c = \frac{\Phi_0}{2\pi\xi(T)^2}i. \tag{22c}$$

[2]The coherence length in a superconductor can be estimated by using the uncertainty relation $\Delta_p \cdot \xi_o = \hbar$ where Δ_p is a change of the electron momentum (at $\epsilon = \epsilon_F$) due to the attractive interaction, which can be estimated from the corresponding change of the particle kinetic energy $\Delta E = v_F \Delta_p$. Taking $\Delta E \simeq \Delta_0$, we obtain the desired estimate of ξ_o.

For fields $H_{c1} < H_a < H_{c2}$ [with $H_{c1} \equiv H_c(0)\ln\kappa/(\sqrt{2}\kappa)$], the superconducting phase is inhomogeneous, composed of the lattice of vortices, each of the form of a tube containing one flux quantum, penetrating the sample. All of the newly discovered high-T_c superconductors are of the second kind, with very small values of H_{c1} and very large values of H_{c2}. This means that the value of the coherence length ξ is very small in those systems.

9. The sound absorption coefficient α_s in the superconducting phase is related to that in the normal phase α_N by

$$\frac{\alpha_S}{\alpha_N} = \frac{2}{1 + \exp(\Delta/k_BT)}.$$

This is a very simple result; hence, experimental results for (α_S/α_N) are used to determine the temperature dependence of the gap Δ.

A complete discussion of superconducting states within the BCS theory is provided in Refs. 1–10.

E. Strong—Coupling Effects: The Eliashberg Approach

The BCS theory provides a complete though approximate theory of both thermal and dynamic properties of superconductors in the weak-coupling limit $\rho V \ll 1$. The electron–electron interactions deriving from the electron–lattice interaction are treated in the lowest order and the electron–electron correlations are decoupled in the mean field-type approximation. Generalizations of the BCS treatment concentrate on two main problems—(1) inclusion of the repulsive coulomb interaction between the electrons [14] and (2) extension of the BCS theory to the situation with arbitrarily large electron–phonon coupling [15]—by generalizing the treatment of normal metals, with electron–lattice interactions incorporated in a systematic fashion [16]. Both of these factors have been included in the Eliashberg approach to superconductivity [15].

The coulomb repulsive interaction reduces the effective attractive interaction between the electrons, so that, instead of Eq. (14), one obtains in the BCS approximation

$$T_c = 1.14\theta_D \exp\left(-\frac{1}{\lambda - \mu^*}\right), \tag{23}$$

where $\lambda = \rho V$ is the effective electron–phonon coupling and μ^* is the so-called coulomb pseudopotential [14] multiplied by ρ.

The Eliashberg correction to the BCS theory must be evaluated numerically. The numerical solution of the Eliashberg equation representing higher-order corrections to the BCS theory may be represented by [17]

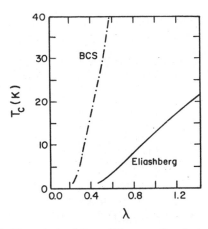

FIGURE 4 Numerical solution of T_c versus the electron–phonon coupling constant λ for the coulomb pseudopotential $\mu^* = 0.1$. The other parameters are taken as for the superconducting element niobium. Note that the Eliashberg theory gives a much slower increase in T_c than does the BCS theory.

$$T_c = \frac{\theta_D}{1.45} \exp\left[-\frac{1.04(1+\lambda)}{\lambda - \mu^*(1+0.62\lambda)}\right]. \quad (24)$$

Figure 4 illustrates the difference in the values of T_c obtained by the BCS vs the Eliashberg theory [18]. We see that the repulsive coulomb interaction and the higher-order electron–phonon effects combine to reduce the superconducting transition temperature drastically. This and other results [19] have led to the conclusion that the value of T_c determined within the phonon–mediated mechanism has an upper limit of the order of 30 K.

One should mention a very important feature of the phonon-mediated electron pairing. Namely, the transition temperature is proportional to the Debye temperature Θ_D. Hence, T_c given by expression (23) depends on the mass M of the atoms composing the lattice. In the simplest situation we expect that $T_c \sim M^{-\frac{1}{2}}$. A dependence of T_c on the mass M was demonstrated experimentally [20] by studying the isotope influence on T_c. These observations provided a crucial argument in favor of the lattice involvement in the formation of superconducting state. If the Coulomb repulsion between electrons is taken into account, then the relation is $T_c \sim M^{-\alpha}$ with [17]

$$\alpha = \frac{1}{2}\left\{1 - \frac{(1+\lambda)(1+0.62\lambda)}{[\lambda - \mu^*(1+0.62\lambda)]^2}\right\}.$$

In the strong coupling limit ($\lambda \geq 1$) the exponent α is largely reduced from its initial value $\frac{1}{2}$. Therefore, if the value of α is small, one may interpret this fact as the evidence for either strong electron–phonon coupling or that a new nonphonon mechanism is needed to explain the superconductivity.

F. Where Do We Go from Here?

The BCS theory is a microscopic theory providing a description of thermodynamic and electrodynamic properties as a function of two parameters: T/T_c and H_a/H_c, where T_c contains the effective attraction strength $|V|$. Such a simple approach is not possible for high-temperature superconductors, as one can see from already existing books and review articles [cf. Refs. 21a–k]. In the next two sections we summarize briefly the principal features of strongly correlated systems. This discussion provides us with new phenomena and some new terms describing them. This overview by no means contains a full discussion of papers published during the last 12 years. Rather, we sketch different paths of approaching the problems encountered in dealing with strongly correlated fermions.

III. NORMAL AND MAGNETIC STATES OF CORRELATED ELECTRONS

A. Narrow—Band Systems

The modern theory of metals derives from the concept of a free electron gas, which obeys the Pauli exclusion principle. The principal influences of the lattice periodic potential on the individual electron states are to renormalize their mass and to change the topology of the Fermi surface. Landau [22] was the first to recognize the applicability of the electron–gas concept to the realistic situation where the repulsive coulomb interaction between particles is not small compared to the kinetic energy of electrons near the Fermi surface. He incorporated the interaction between electrons into a further (many–body) renormalization of the effective mass and investigated the physical properties, such as specific heat, magnetic susceptibility, sound propagation, and thermal and electric conductivities in the terms of quasiparticle contributions.

An important next development was contributed by Mott [23], who pointed out that if the coulomb interaction between the electrons is sufficiently strong (that is, comparable to the band energy of the quasiparticles), then electrons in a solid would have to localize on the atoms, e.g., with one valence electron per atom. This qualitative change of the nature of single–electron states from those for a gas to those for atoms is called the metal–insulator or the Mott transition. An empty (unoccupied) state in the Mott insulator (that is, that without electrons available) will act as a mobile hole. In these circumstances, the transport of charge takes place via the correlated hopping of electrons through such hole states. In the Mott insulator limit, those hole states play a crucial role in establishing the superconductivity of oxides, as discussed in Section IV.

The paramagnetic or magnetically ordered states of electrons comprising the Mott insulator distinguish this class of materials from ordinary band (Bloch–Wilson) insulators or intrinsic semiconductors; the latter are characterized at $T = 0$ by a filled valence band and an empty conduction band, separated by a gap. The electrons in the filled valence band are spin-paired into $|\mathbf{k}\uparrow, \mathbf{k}\downarrow\rangle$ singlets; hence Bloch–Wilson insulators are diamagnetic.

The basic question now arises whether one can treat Mott insulators and metals within a single microscopic description of electron states by generalizing the band theory of electron states so as to describe Mott insulators within the same microscopic model. The first step in this direction was proposed by Hubbard [25], who showed by the use of a relatively simple model that as the interaction strength (characterized by the magnitude U of the intraatomic coulomb repulsion) increases and becomes comparable to the band energy per particle (characterized by the bare bandwidth W), the original band of single-particle states splits into two halves. Thus, the Mott insulator may be modeled by a lattice of hydrogeniclike atoms with one electron per atom, placed in the lowest $1s$ state. The distinction between the normal metallic and the Mott insulating states is shown schematically in Figs. 5a and b, where the metal (a) is depicted as an assembly of electrons represented by the set of plane waves characterized by the wave vector $\mathbf{k}$ and spin quantum number $s = \sigma/2$, where $\sigma \equiv \pm 1$.

The transformation to the Mott localized state may take place only if the number of electrons in the metallic phase is equal to the number of parent atoms, that is, when the starting band of free electron-like states is half-filled. The collection of such unpaired spin moments will lead to the paramagnetic Curie–Weiss behavior at high temperatures. As the temperature is lowered, the system undergoes a magnetic phase transition; in the case of the Mott insulators, the experimentally observed transition is almost always to antiferromagnetism, as shown in Fig. 5b, where each electron with its spin moment up is surrounded by electrons on nearest-neighboring sites with spins in the opposite direction (down). Such a spin configuration reflects a two–sublattice (Néel) antiferromagnetic state. The actual magnetic structure of Cu^{2+} ions in $La_2Cu\,O_4$, taken from Ref. 24, is shown in Fig. 6. The expectation value of the spin is reduced by 40% from the value $s^z = \frac{1}{2}$.

If the number of electrons in the band is smaller than the number of available atomic sites, then electron localization cannot be complete because empty atomic sites are available for hopping electrons. However, for the half-filled band case, as the ratio U/W increases, half of the total number of single-particle states in the starting band is gradually pushed above the Fermi level ϵ_F. An increase in the ratio U/W may be achieved by lengthening the

(a)   Normal metal

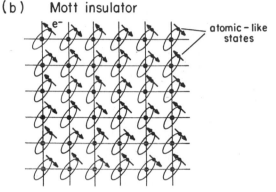

(b)   Mott insulator

FIGURE 5 (a) Schematic representation of a normal metal as a lattice of ions and the plane waves, with wave vector $\mathbf{k}$ representing free electron states. (b) Model of the Mott insulator as a lattice of atoms with electrons localized on them. Note that the ground–state configuration is usually antiferromagnetic (with the spins antiparallel to each other).

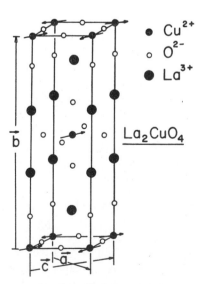

FIGURE 6 The magnetic structure of La_2CuO_4. The neighboring Cu^{2+} ions in the planes have their spins (each representing the $3d^9$ configuration) antiparallel to each other. The antiferromagnetic structure is three–dimensional. (From Endoh *et al.* [24].)

interatomic distance, thus reducing W, which is directly related to the wave–function overlap for the two states located on the nearest-neighboring sites. The splitting of the original band into two Hubbard subbands eliminates the paired (spin–singlet) occupations of the same energy state ϵ. Effectively, this pattern reflects the situation of electrons being separated from each other as far as possible; however, the correspondence between the Hubbard split–band situation, shown schematically in Fig. 7, and the electron disposition in the spin lattice in real space (cf. Fig. 5b) is by no means obvious and requires a more detailed treatment that relates these two descriptions of the Mott insulator. This problem is dealt with in the following section.

1. The Hubbard Model

In discussing narrow band systems, one usually starts from the model Hamiltonian due to Hubbard [25], which appears to be complicated but can really be interpreted in simple terms, namely,

$$H = \sum_{\mathbf{k}\sigma} \epsilon_{\mathbf{k}\sigma} n_{\mathbf{k}\sigma} + U \sum_i n_{i\uparrow} n_{i\downarrow}, \qquad (25)$$

where $\epsilon_{\mathbf{k}}$ is the single-particle (band) energy per electron with the wave vector $\mathbf{k}$, U is the magnitude of intraatomic coulomb repulsion between the two electrons located on the same atomic site i, $n_{\mathbf{k}\sigma}$ is the number of electrons in the single–particle state $|\mathbf{k}\sigma\rangle$, and $n_{i\sigma}$ is the corresponding

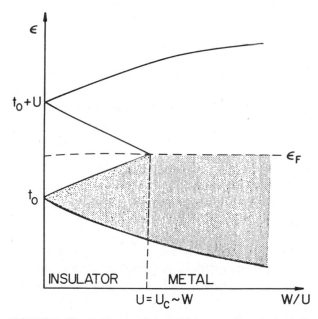

FIGURE 7 The Hubbard splitting of the states in a single, half–filled band for the strength of the intraatomic coulomb interaction $U > U_c$. The state with a filled lower Hubbard subband for $U > U_c$ is identified with that of the Mott insulator. [From Ref. 25.]

quantity for the atomic state $|i\sigma\rangle$. This simple Hamiltonian describes the localization versus delocalization aspect of electron states since the first term provides the gain in energy ($\epsilon_{\mathbf{k}} < 0$) for electrons in the band state $|\mathbf{k}\sigma\rangle$, whereas the second accounts for an energy loss ($U > 0$) connected with the motion of electrons throughout the system that is hindered by encounters with other electrons on the same atomic site. The competitive aspects of the two terms are expressed explicitly if the first term in Eq. (24) is transformed by the so–called Fourier transformation to the site $\{|i\sigma\rangle\}$ representation. Then Eq. (24) may be rewritten

$$H = \sum_{ij\sigma} t_{ij} a_{i\sigma}^+ a_{j\sigma} + U \sum_i n_{i\uparrow} n_{i\downarrow}, \qquad (26)$$

where

$$t_{ij} = \frac{1}{N} \sum_{\mathbf{k}} \epsilon_{\mathbf{k}} \exp[i\mathbf{k} \cdot (\mathbf{R}_j - \mathbf{R}_i)], \qquad (27)$$

is the Fourier transform of the band energy $\epsilon_{\mathbf{k}}$ and $a_{i\sigma}^+$ ($a_i\sigma$) is the creation (annihilation) of electrons in the atomic (Wannier) state centered on the site $\mathbf{R}_j$. The first term in Eq. (25) represents the motion of an electron through the system by a series of hops $j \to i$, which are described in terms of destruction of the particle at site j and its subsequent recreation on the neighboring site i. The width of the corresponding band in this representation is given by

$$W = 2 \sum_{j(i)} |t_{ij}| \approx 2z|t|, \qquad (28)$$

where z is the number of nearest neighbors (n.n.), and t is the value of t_{ij} for the n.n. pair $\langle ij \rangle$. Thus, the Hamiltonian [Eq. (25)] is parameterized through the bandwidth W and the magnitude U. In actual calculations, it is the ratio U/W that determines the localized versus collective behavior of the electrons in the solid.

2. Hubbard Subbands and Hole States

The normal–metal case is represented in Eq. (24) by the limit $W/U \gg 1$; the first (band) term then dominates. On the other hand, the complementary limit $W/U \ll 1$ corresponds to the limit of well-separated atoms, since the excitation energy of creating double occupancy on a given atom (with the energy penalty $\epsilon \sim U$) far exceeds the band energy of individual particles. The transition from the metallic to the atomic type of behavior takes place when $W \sim U$; this is also the crossover point where the single band in Fig. 7 splits in two. The actual dependence of the density of states for interacting particles is shown in Fig. 8 (taken from Ref. 26). These curves were drawn for the Lorentzian shape of the density of states (DOS), that is, for a starting band with a characteristic width Δ:

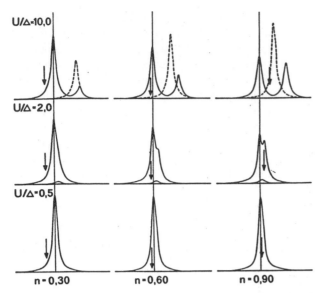

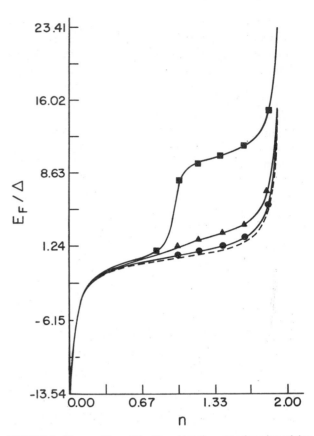

FIGURE 8 The Hubbard splitting of the states for different band fillings, $n = 0.3$, 0.6, and 0.9, and for different U/W ratios, 0.5, 2, and 10, respectively. The x axis is the particle energy value; the y axis is the density of states value. The arrow indicates the position of the Fermi energy, whereas the dashed line represents the inverse lifetime of the quasiparticle state in the pseudogap. [From Ref. 26.]

$$p^{\circ}(\epsilon) = \frac{\Delta}{\pi} \frac{1}{(\epsilon - t_0)^2 + \Delta^2}, \qquad (28a)$$

where t_0 determines the position of the center of the band (usually chosen as $t_0 = 0$). Detailed calculations [28] show that with a growing magnitude of interaction (U/Δ), the DOS [Eq. (28a)] splits into two parts described by the density of states:

$$\rho(\epsilon) = \frac{\Delta}{\pi} \left[\frac{1 - (n/2)}{(\epsilon - t_0)^2 + \Delta^2} + \frac{n/2}{(\epsilon - t_0 - U)^2 + \Delta^2} \right]. \qquad (28b)$$

The first term describes the original DOS [Eq. (28a)], with the weighting factor $(1 - (n/2))$, whereas the second represents the upper subband (on the energy scale), with the weighting factor $(n/2)$ and shifted by an amount U. These two terms and the corresponding two parts of the DOS in Fig. 8 describe the Hubbard subbands. The dashed line in Fig. 8 represents the inverse lifetime of single–electron states placed in the pseudogap, while the arrows point to the position of the Fermi energy in each case. For $n = 1$, the Fermi level falls in a pseudogap, where the lifetime of those quasiparticle states is very short. This is reminiscent of the behavior encountered in an ordinary semiconductor, where the states in the band gap are those with a complex wave vector **k**. The lifetime may qualitatively simulate the atomic disorder-producing spread (Lorentian-shape) form of the bare density states.

FIGURE 9 The position of the Fermi level ϵ_F as a function of the band filling n for different values of interaction (from the bottom to the top curve), $U/\Delta = 0, 0.5, 2$, and 10. For $U/\Delta = 10$, the Fermi level jumps between the subbands when $n \approx 1$. [From Ref. 26.]

To display the of Mott insulator as a two-band system in which the Hubbard subbands assume a role similar to that of the valence and conduction bands in an ordinary semiconductor, we have plotted in Fig. 9 the position of the Fermi level as a function of the numbers of electrons n per atom in the system. As n moves past unity, a jump in ϵ_F occurs for $U/\Delta \gg 1$. This is exactly what happens in the ordinary semiconductor when the electrons are added to the conduction band. This feature shows once more that the states near the upper edge of the lower Hubbard subband (that is, the states near ϵ_F for n close to but less than unity) can be regarded as hole states. We will see that those states are the ones with a high effective mass.

It should be emphasized that the Hubbard subband structure is characteristic of magnetic insulators and cannot be obtained with a standard band theoretical approach to the electron states in solids. The N states in the lower Hubbard subband are almost singly occupied; this is directly related to the picture of unpaired spins in Fig. 5b and is one of the reasons for calling the electron states for such interacting systems correlated electronic states.

The other reason (discussed in detail later) arises because the proper description of electronic states near the localization threshold (the Mott transition) requires that one incorporates two-particle correlations into the quasiparticle states. The Hubbard split-band picture is only the first step in the proper description of the electron states. Those additional correlations will lead to a very heavy mass of quasiparticles near the Mott transition; the heavy mass indicates a strong reduction of the bare bandwidth W as the localization threshold is approached from the metallic side.

3. Localized versus Itinerant Electrons: Metal–Insulator Transitions

The Hubbard split–band picture of unpaired electronic states in a narrow band, shown in Figs. 7 and 8, provides a rationale for the existence of a paramagnetic insulating ground state of the interacting electron system. The corresponding experimentally observed metal–insulator transitions (MITs) at a finite temperature are very spectacular, as demonstrated in Fig. 10, where the resistivity (on a logarithmic scale) is plotted as a function of the inverse temperature for a canonical system $(V_{1-x}Cr_x)_2O_3$ (the data are from Ref. 27). The number of transitions (one, two, or three) depends on the Cr content. Note the presence of an intervening metallic state between the antiferromagnetic insulating (AFI) and the paramagnetic insulating (PM) states, as well as the reentrant metallic behavior at high temperatures for $0.005 \lesssim x \lesssim 0.0178$. To rationalize these data, we discuss the physical implications of a model of interacting narrow-band electrons for $U \sim W$ starting from the Hamiltonian [Eq. (25)]. We summarize here the main features of the detailed discussion presented in Refs. 28–30, which provide the main features of the ground-state and thermodynamic properties.

In the absence of interactions ($U = 0$), the band energy per particle is $\bar{\epsilon} = -(W/2)n(1-n)/2)$, where $0 \leq n \leq 2$ is the degree of band filling; for $n = 1$, this reduces to $\bar{\epsilon} = -W/4$. When the interactions are present, the band narrows; this is because of a restriction on the electron motion caused by their repulsion, as described earlier. One way of handling this restriction is to adjoin to the bare bandwidth a multiplying factor Φ. This leads to a renormalized DOS for quasiparticles, as illustrated in Fig. 11. The factor Φ is a function of the particle–particle correlation function $\eta \equiv \langle n_{i\uparrow}n_{i\downarrow}\rangle$, the expectation value for the double occupancy of a representative lattice site. The quantity η is calculated for $T = 0$ self-consistently by minimizing the total energy E_G (per site), composed of the band energy $E_B = \Phi\epsilon$ and the coulomb repulsion energy $U\eta$, where the parameter is specified by $\Phi = 8\eta(1 - 2\eta)$ [28, 29]. These two energies represent the expectation values of the two terms in Eq. (25) for the case

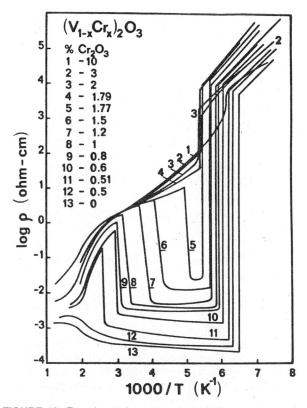

FIGURE 10 Experimental measurements [27] pertaining to the variation of resistivity ρ on a logarithmic scale with inverse temperature $1000/T$ for the $(V_{1-x}Cr_x)_2O_3$ system. The atomic content of Cr_2O_3 in V_2O_3 for each curve is specified.

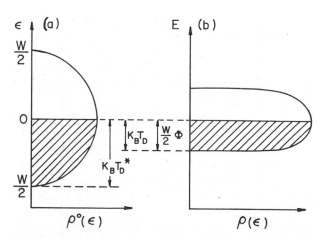

FIGURE 11 Schematic representation of the bare (ρ_0) and quasiparticle (ρ) densities of states. The band narrowing factor Φ for interacting electrons (b) is specified. The degeneracy temperature T_D for the interacting electrons and that corresponding to noninteracting electrons (T_D^*) are also indicated. The situation drawn corresponds to the half–filled case ($n = 1$), for which the Fermi energy can be chosen as $\epsilon_F = 0$.

of a half-filled band. The optimal values of the quantities are given by

$$\eta_0 = \frac{1}{4}(1 - (U/U_c)), \qquad (29a)$$

$$\Phi_0 = 1 - (U/U_c)^2, \qquad (29b)$$

and

$$E_G = \left(1 - (U/U_c)^2\right)\bar{\epsilon}, \qquad (29c)$$

with $U_c = 8|\bar{\epsilon}| = 2W$. Thus, as U increases, η_0 decreases from $\frac{1}{4}$ to 0. At the critical value $U = U_c$, $E_B = 0$ and there are no double occupancies for the same lattice site; this signals the crossover by the system from the itinerant (band) to the localized (atomiclike) state. The point $U = U_c$ corresponds to a true phase transition at $T = 0$; the last statement can be proved by calculating the static magnetic susceptibility, which is [29]

$$\chi = \chi_0 \Big/ \left\{ \Phi_0 \left[1 - U\rho \frac{(1 + I)/2}{(1 + I)^2} \right] \right\}, \qquad (30)$$

where $I \equiv U/U_c$, ρ is the density of bare band states at $\epsilon = \epsilon_F$, and χ_0 is the magnetic susceptibility of band electrons with energy ϵ_k at $U = 0$. As $\Phi \to 0$ (that is, $U \to U_c$), the susceptibility diverges. The localized electrons are represented in this picture by noninteracting magnetic moments for which the susceptibility is given by the Curie law $\chi = C/T \to \infty$ as $T \to 0$. Thus, the MIT is a true phase transition; η_0 may be regarded as an order parameter, and the point $U = U_c$ as a critical point. We concentrate now on a more detailed description of the metallic phase, which permits a generalization of the previous results to the case $T > 0$. First, as has been said, the increase in magnitude of interaction U reduces the band energy according to $E_B = -W\Phi_0/4$. Eventually, E_B becomes comparable to the interaction part $U\eta$; they exactly compensate each other at $U = U_c$. The resultant electronic configuration (localized versus itinerant) is then determined at $T > 0$ by the very low entropy and the exchange interaction contributions. The entropy of the metallic phase in the low–temperature regime may be estimated by using the linear specific heat expression for electrons in a band narrowed by correlations, namely, $C_v \equiv \gamma T = (\gamma_0 | \Phi_0)T$, where $\gamma_0 = 2\pi^2 k_B^2 \rho/3$ is the linear specific heat coefficient (per one atom) for uncorrelated electrons (that is, $U = 0$). Hence, the entropy $S = \gamma T = C_v$. Combining this relation with the resultant energy at $T = 0$, given by Eq. (29c), one can write an explicit expression for the free energy of the metallic phase [30]:

$$\frac{F}{N} = \left(1 - \frac{U}{U_c}\right)^2 \bar{\epsilon} - \frac{1}{2}\frac{\gamma_0}{\Phi_0}T^2. \qquad (31)$$

This is the free energy per one atomic site. On the other hand, if the exchange interaction between the localized

moments is neglected, then each site in the paramagnetic state is randomly occupied by an electron with its spin either up or down. The free energy F_I for such an insulating system of N moments is provided by the entropy term for randomly oriented spins, that is,

$$\frac{F_I}{N} = -k_B T \ell n\, 2 \qquad (32)$$

Now, a system in thermodynamic equilibrium assumes the lowest F state. The condition for the transition from the metallic to the local-moment phase is specified by $F = F_I$. The phase transition determined by this condition can be seen explicitly when we note that the free energy varies with T either parabolically [Eq. (31)] or linearly [Eq. (32)], depending on whether the system is a paramagnetic metallic (PM) or a paramagnetic simulating (PI) phase. As illustrated in Fig. 12, several of those curves

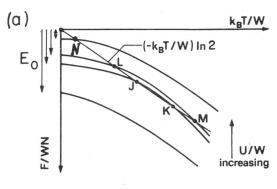

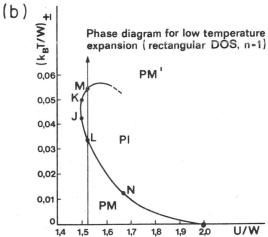

FIGURE 12 (a) Plots of the free energies for the paramagnetic Mott insulator (the straight line starting from the origin) and the correlated metal (the parabolas). The parabolic curves' points of crossing at L and J correspond to a discontinuous metal–insulator transition, while those crossing at K and M correspond to the reverse. (b) Schematic representation of the phase diagram between paramagnetic metallic (PM and PM' and paramagnetic insulating (PI) phases. The points of crossing from a are also shown. The vertical arrow represents a sequence of the transitions shown in Fig. 10 for $0.005 \lesssim x \lesssim 0.018$ and in the paramagnetic phase.

intersect at one or two points depending on the value U/W. These intersection points determine the stability limits of the PM and PI phases. The lowest curve for the PM phase lies below the straight line for the PI state; there is no transition, that is, the metallic Fermi liquid state with the effective mass enhancement $m^*/m_0 \lesssim 2.5$ is stable at all temperatures. As U/W increases, the parabolas fall higher on the free energy (F/WN) scale and the possibilities for transitions open up. The higher two curves illustrate the case in which the intersections with the straight lines occur at J and K and at L and M, respectively; at low and high temperatures the parabola lies below the straight line for F_I/WN, so that the metallic phase is stable in those T regions. At intermediate temperatures, the PI phase is stable. The loci of the intersections move farther apart on the $k_B T/W$ scale as U/W is increased, as shown in Fig. 12b, where the phase boundaries are drawn; this part of the figure represents the temperature of the transitions (the intersection points in Fig. 12a) versus the relative magnitude of interaction U/W. We see that the PM phase is stable at low temperatures; thus, reentrant metallic behavior is encountered at high T. The explicit form of (the curve in Fig. 12b) is obtained from the coexistence condition $F = F_I$, which leads to the following expression for the transition temperatures [29]:

$$\frac{k_B T_+}{W} = \frac{3}{2\pi^2}\left[1 - \left(\frac{U}{U_c}\right)^2\right]\left\{\ell n\, 2\right.$$

$$\left. \pm \left[(\ell n\, 2)^2 + \frac{\pi^2}{3}\frac{1 - (U/U_c)}{1 + (U/U_c)}\right]^{\frac{1}{2}}\right\}. \quad (33)$$

The root T_- represents the low-temperature part, that is, that for $k_B T/W \leq 0.049$. The T_+ part is the one above the point where both curves meet; this takes place at the lower critical value of $U = U_{\ell c}$ such that

$$\frac{U_{\ell c}}{U_c} = 1 - \frac{3\sqrt{2}}{2\pi}\frac{1}{(\rho|\bar{\epsilon}|)^{\frac{1}{2}}} \approx 0.75. \quad (34)$$

Below the value of $U = U_{\ell c}$, the correlated Fermi liquid is stable at all temperatures. Ultimately, for $1.58 \leq U/W \leq 2.0$, only one intersection (at low T) of the curves remains. This means that in this regime of U/W the reentrant metallic behavior is achieved gradually as the temperature increases. The above-described transitions are observed when changing the magnitude of interaction (U/W ratio). In the case of high-temperature superconductors we observe the transition from a Mott insulator to a superconductor as a function of doping (carrier concentration). This case is discussed next.

Note added in August 2000. In recent years, the Mott localization in the limit of infinite dimension has been discussed extensively (e.g., Gebhardt, Ref. 30). A central peak is located between the Hubbard subbands, which carries the main part of the quasiparticle weight. There are two problems with the application of this solution to concrete systems. First, the upper critical dimensonality is not known for Mott systems. Second, the disappearance of the central peak at the localization threshold is being debated.

4. Strongly Correlated Electrons: Kinetic Exchange Interaction and Magnetic Phases in Three-Dimensonal Space

In the limit $W \ll U$, the ground state of the interacting electron system will be metallic only if the number of electrons N_e in the system differs from the number N of atomic sites. Simply, only then can charge transport take place via the hole states in the lower Hubbard subband (for $N_e < N$), that is, when the transport of electrons can be represented via hopping from site to site, avoiding the doubly occupied configurations on the same site. This restriction on the motion of individual electrons is described above in terms of the band narrowing factor Φ, which, in the normal phase, is now of the order [28, 29] $\Phi = (1 - n)/((1 - n)/2)$. This shows that the effective quasiparticle bandwidth $W^* \equiv W\Phi$ is nonzero only if the number of holes $\delta \equiv 1 - n > 0$.

For $W \ll U$, there is one class of dynamic processes that is important in determining the magnetic interactions between strongly correlated itinerant electrons, namely, the virtual hopping processes, with the formation of a doubly occupied site configuration in the intermediate state. Such processes are depicted in Fig. 13, where one electron hops onto the site occupied by an electron with opposite spin and then hops back to the original site. During such processes, the electrons can exchange positions (and the yields to the spin reversal of the pair with respect to the original configuration) or the same electron can hop back and forth. The corresponding effective Hamiltonian, including the virtual-hopping processes in first nontrivial order, has the form

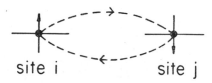

FIGURE 13 Virtual hopping processes between singly and doubly occupied atomic sites that lead to an antiferromagnetic exchange interaction between the neighboring sites. This interaction is responsible for the antiferromagnetism in most of the Mott insulators.

$$H = \sum_{\mathbf{k}\sigma} \Phi_\sigma \epsilon_{\mathbf{k}} n_{\mathbf{k}\sigma} + \sum_{ij} \left(2t_{ij}^2/U\right)\left(\mathbf{S}_i \cdot \mathbf{S}_j - \frac{1}{4}n_i n_j\right),$$

$$(35)$$

where in general the band narrowing factor $\Phi_\sigma = (1-n)/(1-n_\sigma)$, $n_\sigma = \langle n_{i\sigma}\rangle$ is the average number of particles per site with the spin quantum number σ, and $n_i \equiv n_{i\uparrow} + n_{i\downarrow}$ is the operator of the number of particles on given site i. Note that in the paramagnetic state $n_\sigma = n_{-\sigma} = n/2$, and Φ_σ reduces to $\Phi = (1-n)/((1-n)/2)$, the value for the normal state.

One should note that the effective Hamiltonian [Eq. (35)] represents approximately the original Hubbard Hamiltonian for $W \ll U$ (for more precise treatment, see Ref. 31 and Section IV.A). When $n \to 1$, $\phi \to 0$, and Eq. (35) reduces to the Heisenberg Hamiltonian with antiferromagnetic interaction, which is the reason why most Mott insulators order antiferromagnetically. In the limit of a half–filled band, we also find that the effective bandwidth $W^* \equiv W\Phi = 0$, thus proving that the electrons in that case are localized on atoms. The nature of the wave function for these quasi–atomic states has not yet been satisfactorily analyzed, though some evidence given later shows that they should be treated as soliton states.

For $n < 1$, the normalized band (the first term) and the exchange parts in Eq. (35) do not commute with each other. This means that for the narrow-band system of electrons represented by the spin dynamics influences the nature of itinerant quasiparticle states of energies $\Phi\epsilon_{\mathbf{k}}$. What is even more striking is that, as $n \to 1$, the two terms in Eq. (35) may contribute equally to the total energy. The critical concentration of electrons n_c for which these two terms are comparable is

$$n_c \simeq 1 - \frac{1}{2z}\frac{W}{U} \sim 0.02 \div 0.05.$$

In Fig. 14, we have plotted schematically the commonly accepted phase diagram for three dimensional systems describing the possible magnetic phases on the plane $n - (U/W)$. Close to the case of one electron per atom, the antiferromagnetic (AF) phase is stable for any arbitrary strength of interaction. At intermediate filling, the ferromagnetic (F) phase may be stable. On the low-interaction side ($W/U > 1$), the ferromagnetic phase terminates at points where the Stoner criterion is met, that is, when $\rho^0(\epsilon_F)U = 1$, where $\rho^0(\epsilon_F)$ is the value of the bare density of states (per spin) at the Fermi level ϵ_F.

Peculiar features appear in the corner where $n \approx 1$, and $W/U \ll 1$, that is, where the number of holes is small, so that the exchange interaction contribution to the total system energy is either larger than or comparable to the band energy part $\Phi\bar{\epsilon}$. In such a situation, a mixed ferromagnetic–antiferromagnetic phase is possible [32]. When the number of holes is very small, each hole may form a magnetic polaron with a ferromagnetic cloud accompanying it: the hole is self-trapped within the cloud of ferromagnetic polarization it created. We consider those objects next.

B. Magnetic Polarons

1. The Classical Approach

It has been proved by Nagaoka [32] that in the limit $W/U \to 0$ the ground state of the Mott insulator with one hole involves ferromagnetic ordering of spins. This is because in this limit the antiferromagnetic exchange term

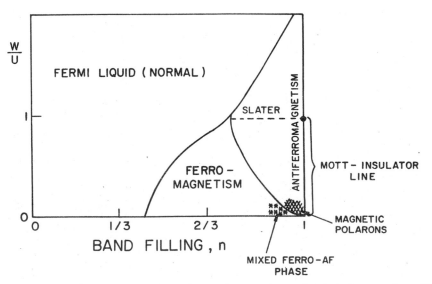

FIGURE 14 Commonly accepted magnetic phase diagram for strongly correlated electrons on the n–(W/U) plane.

in Eq. (35) vanishes and the band energy is lowest when $\Phi_{\sigma=\uparrow} = 1$ and $\Phi_{\sigma=\downarrow} = 1 - n$. We can thus choose an equilibrium state with $n_\uparrow = n$, $n_\downarrow = 0$, that is, a state with all spins pointing up.

Mott, and Hertier and Lederer [23], has pointed out that if W/U is small but finite, a hole may create locally a ferromagnetic polarization of the spins in a sphere of radius R, surrounded by a reservoir of antiferromagnetically ordered spins. The situation is shown schematically in Fig. 15. The energy of such a hole accompanied by a cloud of saturated polarization can be estimated roughly as

$$E(R) = -\frac{W}{2} + \pi^2 |t| \left(\frac{a}{R}\right)^2 + \frac{4\pi}{3} \left(\frac{R}{a}\right)^3 \frac{zt^2}{U}, \quad (36)$$

where a is the lattice constant and t is the hopping integral t_{ij} between the z nearest neighbors. In this expression, the first term is the band energy of a free hole in a completely ferromagnetic medium, the second represents the kinetic energy loss due to the hole confinement, and the third involves the antiferromagnetic exchange energy penalty paid by polarizing the spins ferromagnetically within a volume $4\pi R^3/3$. Minimizing this equation with respect to R, we obtain the optimal number of spins contained in the cloud,

$$N = \frac{4\pi}{3} \left(\frac{\pi U}{W}\right)^{3/5}, \quad (37a)$$

and the polaron energy,

$$E_0 = -\frac{W}{2} \left[1 - \frac{5\pi^2}{3z} \left(\frac{W}{U}\right)^{2/5}\right]. \quad (37b)$$

3d HOLE POLARON

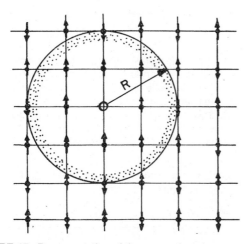

FIGURE 15 Representation of the magnetic polaron state, that is, one hole in the antiferromagnetic Mott insulator. This hole produces ferromagnetic polarization around itself and may become self-trapped.

Equation (36) holds for a three-dimensional system; for a planar system, the factor $(4/3)\pi (R/a)^3$ in the last term should be replaced by the area $\pi (R/a)^2$. One then obtains the corresponding optimal values,

$$N = \pi \left(\frac{2\pi U}{W}\right)^{\frac{1}{2}} \quad (38a)$$

and

$$E_0 = -\frac{W}{2} \left[1 - \frac{2\pi^2}{z} \left(\frac{W}{2\pi U}\right)^{\frac{1}{2}}\right]. \quad (38b)$$

These size estimates will be needed later when discussing the hole states at the threshold for the transition from antiferromagnetism to superconductivity in high-T_c oxide materials. One should note that U/W must be appreciably larger than unity to satisfy the requirement $N \gg 1$. In other words, the condition $R \gg a$ must be met, so that the spin subsystem (and the hole dynamics) may be treated in the continuous–medium approximation, the condition under which Eq. (36) can be derived.

2. The Quantum Approach: Two Dimensions

The motion of a single hole in the Mott insulator is much more subtle than the formation of the polaron discussed above. Namely, if we consider n holes in the lower Hubbard subband, then the probability of electron hopping around is $\approx n(1 - n)$, so effectively, the bandwidth of such itinerant states is $W_{\text{eff}} = zt(1 - n)$. For small n, we have $W \le J$, where J is the magnitude of the kinetic exchange. In the limit of a single hole the dynamics is determined by the magnitude of exchange interactions J, since $W_{\text{eff}} \to 0$. In effect, we have a hole moving slowly in the background of antiferromagnetically ordered spins. This picture seems to be a good representation of the hole motion in highly insulating magnetic oxides such as NiO and CoO. Instead, in high-temperature superconductors individual polaronic states must overlap appreciably for $(n_c \sim 0.95)$ when the magnetic insulator $\to$ metal transition takes place. Therefore, some sort of homogeneous state must be formed in the metallic phase. This is particularly so since high-temperature superconductors evolve from a *charge-transfer insulator*, for which the gap for $2p \to 3d^{n+1}$ ($O^{2-} \to Cu^{1+}$) transitions is smaller than the Hubbard gap $\Delta \simeq U - W$. In effect, the hole states are hybridized $3d$–$2p$ states (in proportions 2:1), not pure $3d$ states due to copper ions. As a result, few alternative pictures of the fermionic liquid of strongly correlated electrons in the normal (metallic) phase arise, starting from the phenomenological pictures of marginal Fermi liquid (Varma *et al.* [33a]) and nearly antiferromagnetic Fermi liquid (NAFL) [33b, 33c] to a mean-field picture of strongly correlated electrons coupled to a gauge field (cf. Lee and Nagaosa [33d]). A separate class of models composes cluster calculations including realistic structures of CuO_2 planes (cf. Ref. 33e). Another class of models forms bosonic models with preformed pairs of bound bipolarons [33f]. The latter class of models requires that the bipolaron radius is $R < (a_0/x)^{\frac{1}{2}}$, where x is the hole concentration, and a_0 is the Cu–Cu interatomic distance. This, in turn, requires a rather strong attractive interaction, which most probably can be furnished only by the combined effect of magnetic and rather strong electron–phonon interactions. Finally, there exists a substantial number of papers on numerical diagonalization for small clusters [33g]. A separate class of models are those involving the stripe structure [33h].

The lack of microscopic theory of normal properties transforms into the arbitrariness in selecting the pairing potential, as we shall see in the next section. The linear resistivity in the full temperature range at optimal doping [34a], the spin-gap existence [34b] in underdoped systems, and the anomalous (non-Drude) form of the optical conductivity all speak in favor of the non-Fermi-liquid (absence of quasiparticles) behavior of correlated electrons [34c] in two spatial dimensions. The role of disorder has not been explained properly either.

3. The Spin Liquid

The difference between an electron liquid of strongly correlated electrons (represented, for example, by the holes in the lowest Hubbard subband) and a Fermi liquid can be shown clearly in the limit of relatively high temperatures $W^* \ll k_B T \ll U$, where the quasiparticle band states with energies ($\Phi \epsilon_{\mathbf{k}}$) are populated equally, independent of their energy. Namely, if N_e electrons are placed into N available states of almost the same energy, then the number of configurations for a phase with excluded double occupancies of each state is [34]

$$2^{N_e} \frac{N!}{N_e!(N - N_e)!}. \tag{39a}$$

The first factor is the number of spin configurations for the singly occupied sites, while the second specifies the configurational entropy—the number of ways to distribute N_e spinless particles among N states. This leads to molar entropy in the form

$$S_L = R[n \, \ell n \, 2 - n \, \ell n \, n - (1 - n) \, \ell n (1 - n)], \tag{39b}$$

where $n = N_e/N$ is the degree of subband filling and R is the gas constant. The above reduces to $S_L = R \, \ell n \, 2$ for $n = 1$, that is, to the entropy of the N spins ($\frac{1}{2}$) on the lattice. In contrast, in a Fermi liquid that obeys the Fermi–Dirac distribution, double occupancies are not excluded,

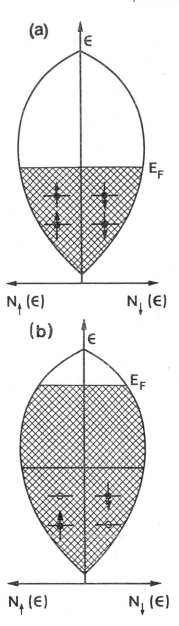

FIGURE 16 Schematic representations of the difference in the **k**–space occupation for ordinary fermions (a) and strongly correlated electrons (b) The spin subbands with $\sigma = \uparrow$ and $\downarrow$ are drawn. Note that the holes drawn in b do not appear; they are shown only to indicate the single occupancy of each single–particle state. The position of the Fermi level is different for the same number of electrons in the two situations. [From Ref. 34.]

as illustrated in Fig. 16a. The corresponding number of configurations is then

$$\left(\frac{N}{N_e/2}\right)^2 = \frac{N!}{(N_e/2)!(N - N_e/2)!}, \tag{40a}$$

with the corresponding molar entropy,

$$S_F = R[2 \, \ell n \, 2 - n \, \ell n \, n - (2 - n) \, \ell n (2 - n)]. \tag{40b}$$

Hence, for $n = 1$, $S_F = 2S_L = 2R \ell n\, 2$. One should emphasize that only the value for S_L reproduces correctly the entropy of N localized paramagnetic spins (the electronic part of the entropy for magnetically disordered states of the Mott insulator). Hence, in accord with intuitive reasoning, the Fermi–Dirac distribution, which allows for double state occupancy, cannot be applied to a strongly correlated electron liquid, which we call a spin liquid. The state of such a liquid reduces to that of the spin system on the lattice if $N = N_e$ (for the Fermi-liquid case, the ground state is then a metal with a half–filled band).

One should now ask how these results may be generalized to handle the regime of low temperatures and of an arbitrary number of holes. One observes that in Fig. 8 the band states for $U \ll W$ are split for any arbitrary degree of band filling [cf. also Eq. (28b)]. Therefore, in enumerating the distribution of particles in the lower Hubbard subband, one must exclude double occupancies of the same energy (ϵ) state. Since the quasiparticle energy is labeled by the wave vector $\mathbf{k}$, one can equivalently exclude the double occupancies of given state $|\mathbf{k}\rangle$. Under this assumption, the statistical distribution is given by [34]

$$\bar{n}_{\mathbf{k}\sigma} = (1 - \bar{n}_{\mathbf{k}-\sigma})\frac{1}{1 + \exp[\beta(E_{\mathbf{k}\sigma} - \mu)]}, \quad (41a)$$

where $\beta = (k_B T)^{-1}$, $\bar{n}_{\mathbf{k}\sigma}$ is the average occupancy of the state $|\mathbf{k}\sigma\rangle$, and μ is the chemical potential that is determined from the conservation of the total number of particles

$$N_e = \sum_{\mathbf{k}\sigma} \bar{n}_{\mathbf{k}\sigma}. \quad (41b)$$

The corresponding molar entropy is now given by

$$S_L = -\frac{R}{N}\sum_{\mathbf{k}}[(1 - \bar{n}_{\mathbf{k}})\, \ell n(1 - \bar{n}_{\mathbf{k}})$$
$$+ \bar{n}_{\mathbf{k}\uparrow} \ell n\, \bar{n}_{\mathbf{k}\uparrow} + \bar{n}_{\mathbf{k}\downarrow}\, \ell n\, \bar{n}_{\mathbf{k}\downarrow}], \quad (41c)$$

with $\bar{n}_{\mathbf{k}} = \bar{n}_{\mathbf{k}\uparrow} + \bar{n}_{\mathbf{k}\downarrow}$.

One should note that the distribution function [Eq. (41c)] differs from the ordinary Fermi–Dirac formula by the factor $(1 - \bar{n}_{\mathbf{k}-\sigma})$, which expresses the conditional probability that there should exist no second particle with the spin quantum number $\mathbf{k}(-\sigma)$ if the state $\mathbf{k}\sigma$ is to be occupied by an electron, as shown in Fig. 16b. If $E_{\mathbf{k}\sigma} \equiv E_{\mathbf{k}}$ (that is, when the particle energy does not depend on its spin direction), Eq. (41a) reduces to

$$\bar{n}_{\mathbf{k}} = \frac{1}{1 + \left(\frac{1}{2}\right)\exp[\beta(E_{\mathbf{k}} - \mu)]}. \quad (41d)$$

This is the same type of formula that applies to the occupation number of simple donors, if the index $\mathbf{k}$ is dropped and ϵ represents the position of the donor level with respect to the bottom edge of the conduction band.

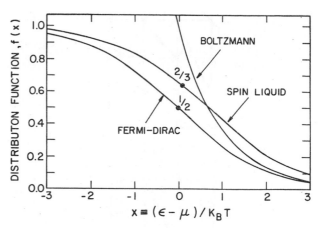

FIGURE 17 Comparison of the Fermi–Dirac and Boltzmann distributions for $\bar{n}_{\mathbf{k}\sigma}$ with that for strongly correlated electrons (the spin—liquid phase); the total occupancy $n_{\mathbf{k}} = n_{\mathbf{k}\uparrow} + n_{\mathbf{k}\downarrow}$ is taken in the latter case.

At $T = 0$, each state is singly occupied. This is the principal feature by which the present formula differs from the Fermi–Dirac distribution at $T = 0$, as illustrated in Fig. 17. The distribution [Eq. (41a)] leads to a doubling of the volume enclosed by the Fermi surface in the spin-liquid state compared to the Fermi-liquid state. At low temperatures, application of the distributions [Eq. (41a) or (41c)] yields Fermi liquid-like properties: a linear T dependence of the specific heat (of large magnitude if $n \to 1$) of the entropy. At high temperatures, the new distribution leads to entropy of the form of Eq. (39b) and local–moment behavior in the form of the Curie–Weiss law for susceptibility. Hence, the properties of the spin liquid governed by the distribution [Eq. (41a) or (41d)] interpolate between those of a metal and those of local moments. Such behavior is observed in many correlated systems, for example, in heavy fermions.

One should note that the entropy expression [Eq. (41c)] can be rewritten for the paramagnetic state in the following form:

$$S_L = -nR\ell n\, 2 - k_B \sum_{\mathbf{k}}[n_{\mathbf{k}}\, \ell n\, n_{\mathbf{k}} + (1 - n_{\mathbf{k}})\, \ell n(1 - n_{\mathbf{k}})].$$
$$(42)$$

The first part represents the entropy of spin moments; the second, the entropy of spinless fermions. An alternative decomposition has been put forward [35] in which the dynamics of correlated electrons is decomposed into that of neutral fermions called spinons and the charged bosons called holons. Within this picture, the onset of superconductivity is considered as a combined effect of Bose condensation of the holons with the simultaneous formation of a coherent paired state by the fermion counterpart [36–38]. This problem is discussed in more detail in Section IV.A.

The above treatment of the spin liquid deals only with its statistical properties in the $U \to \infty$ limit. The problem

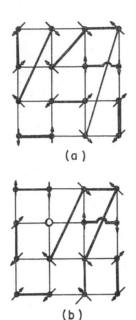

(a)

(b)

FIGURE 18 Schematic representation of singlet–spin pairing forming the RVB state. All paired configurations should be taken to calculate the actual ground state. (a) The RVB state for the Mott insulator; (b) that with one hole. The latter case will contain an unpaired spin, as indicated.

now arises as to what happens when the spin part of the form of the second term in Eq. (35) is explicitly included. The problem of the resultant quantum ground state of holes in a Mott insulator is a matter of intensive debate [36–38]. The state called the resonating valence-bond (RVB) state has been involked [36] specifically to deal with this problem; this state is shown schematically in Fig. 18 for the case without holes (a) and with one hole (b). The connecting lines represent bonds, across which the two electron from spin–singlet pairs. The resonating nature of bonds is connected with the idea that the RVB ground state is a coherent superposition of all such paired configurations. The dynamic nature of this spin dimerization is connected with the terms $(S_i^+ S_j^- + S_i^- S_j^+)$ in the exchange part of the Hamiltonian [Eq. (35)]. There is the possibility that the RVB state [which, for obvious reasons, differs from the ordinary (Néel) antiferromagnet] is a ground state for the planar CuO_2 planes in high-T_c oxides, such as $La_{2-x}Sr_xCuO_4$, where the long–range magnetic order is destroyed for $x \approx 0.02 \div 0.03$. We return to this problem in Section IV when discussing the boundary line between antiferromagnetism and superconductivity for high-T_c oxides.

C. Hybridized Systems

Most of the strongly correlated systems are encountered in oxides and in several classes of organic and inorganic com-

pounds. In oxides the $3d$ orbitals of cations such as Cu^{2+} and Ni^{2+} hybridize with the $2p$ orbitals of oxygen, particularly if the atomic $3d$ states are energetically close to the $2p$ states. The properties of correlated and hybridized states can be properly discused in terms of the Anderson lattice model Hamiltonian, which is of the form

$$H = \epsilon_f \sum_{i\sigma} N_{i\sigma} + \sum_{k\sigma} \epsilon_k n_{k\sigma} + U \sum_i N_{i\uparrow} N_{i\downarrow}$$

$$+ \frac{1}{\sqrt{N}} \sum_{k\sigma} \left(V_k e^{i\mathbf{k}\cdot\mathbf{R}_i} a_{i\sigma}^+ c_{k\sigma} + \text{H.C.} \right). \quad (43)$$

In this Hamiltonian, the first term describes the energy of atomic electrons positioned at ϵ_f, the second represents the energy of band electrons, the third represents the intraatomic coulomb repulsion between two electrons of opposite spins, and the last describes the mixing of atomic with band electrons due to the energetic coincidence (degeneracy) of those two sets of states (H.C. refers to the Hermitian conjugate part of the hybridization part). In heavy fermions, the atomic states are $4f$ states, whereas they are $3d$ states of Cu^{2+} ions in high-T_c systems; the band states are $5d$–$6s$ and $2p$ states, respectively. Note that $N_{i\sigma} = a_{i\sigma}^+ a_{i\sigma}$ and $n_{k\sigma} = c_{k\sigma}^+ c_{k\sigma}$ are the number of particles on given atomic (i) or $\mathbf{k}$ states, respectively. In this Hamiltonian, the following parameters appear: the atomic–level position ϵ_f, the width W of starting band states with energies $\{\epsilon_k\}$, the magnitude U of the coulomb repulsion for two electrons located in the same atomic site, and the degree of hybridization (mixing), V_k, characterized by its magnitude V.

Two completely different situations should be distinguished from the outset: (1) $U > W > |\epsilon_f| \gg |V|$, and (2) $U > W > |V| \gtrsim |\epsilon_f|$. Case 1 applies when the starting (bare) atomic level is placed deeply below the Fermi level and the atomic states admix weakly to the band states. In case 2, the hybridization is large and is responsible for strong mixing of the two starting sets of states. The band structure corresponding to the hybridized band states in the absence of electron–electron interactions (that is, $U = 0$) is depicted in Fig. 19. We observe a small gap in the hybridized band structure; it occurs around the bare atomic level position ϵ_f and separates two hybridized bands. Those two bands, which have the energies

$$E_{k\pm} = \frac{\epsilon_k + \epsilon_f}{2} \pm \left[\left(\frac{\epsilon_k - \epsilon_f}{2} \right) + |V_k|^2 \right],$$

correspond to the bonding and antibonding types of states in molecular systems. The structure of the hybridized bands is demonstrated explicitly in Fig. 20 of the DOS for each band. One sees that strongly peaked structures occur in the regions near the gap. If the Fermi level falls within these peaks, a strong enhancement of the effective mass

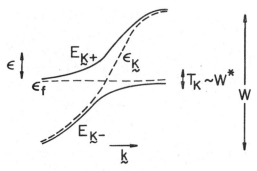

FIGURE 19 Schematic representation of the hybridized bands with energies $E_{k\pm}$, which are formed by mixing the band states (with energy ϵ_k) and atomic states (located at $\epsilon = \epsilon_f$). The original band has width W, much wider than the peaked structure, of width W^*.

should takes place solely because of these peculiarities of the band structure. In some situations only a pseudogap caused by the hybridization is formed, as shown in Fig. 21. This is so if the hybridization matrix element V depends on the wave vector $\mathbf{k}$ and if, along some directions in reciprocal space, $V_k = 0$.

The inclusion of the interaction term in Eq. (43) renders the treatment of the Anderson lattice Hamiltonian much more complicated; up to now this problem has not been solved rigorously. A large variety of approximate treatments has been proposed and reviewed recently [39–42],

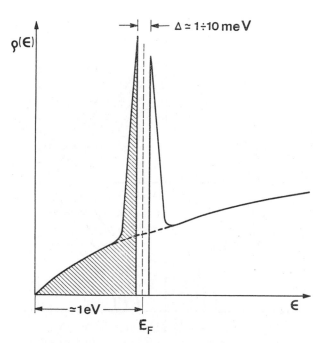

FIGURE 20 Density $\rho(\epsilon)$ of hybridized states versus particle energy ϵ. Note that the hybridization gap Δ_h may be very small compared to the total width of the band states. The position of the Fermi level ϵ_F corresponds to the filled lower band.

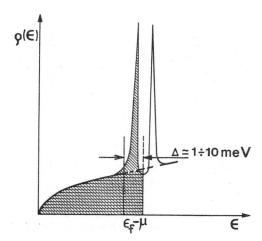

FIGURE 21 Same as Fig. 20 but with the pseudogap among the hybridized bands.

in all of which the principal task was to provide a satisfactory description of heavy-fermion materials [43]. In effect, the limiting case of almost localized strongly correlated electrons was studied, which, among others, provides a quasiparticle electronic structure similar to that shown in Fig. 20, with a very strong enhancement of the DOS near the Fermi surface. This yields to very heavy quasiparticles, which, in some systems, may undergo transitions either to antiferromagnetism or to superconducting states. In this respect, heavy-fermion materials are analogous to high-T_c systems, though with much lower transition temperatures.

D. The Electronic States of Superconducting Oxides

The high-T_c superconducting oxides, such as $La_{2-x}Sr_x CuO_4$ (the so-called 214 compounds) and $YBa_2Cu_3O_{7-\delta}$ (the so-called 123 compounds), have one common structural unit: the quasi–two–dimensional structure that is approximated by CuO_2 planes, one of which is shown schematically in Fig. 22. We discuss mainly the role of these planes since it is widely accepted that the electronic properties of these subsystems are the main factor determining the observed superconductivity, antiferromagnetism, and localization effects in those materials. In stoichiometric La_2CuO_4 or $YBa_2Cu_3O_7$, the formal valence of Cu is 2+, that is, it corresponds to a one-hole ($3d^9$) electron configuration. In a strictly cubic structure, with Cu^{2+} surrounded by O^{2-} ions in an octahedral arrangement, the highest band is doubly degenerate and of e_g symmetry, that is, composed of $d_{x^2-y^2}$ and $d_{3z^2-r^2}$ orbitals. However, in high-T_c materials, the octahedra are largely elongated in the direction perpendicular to the CuO_2 planes, so that the bands are further split; it is commonly assumed that the

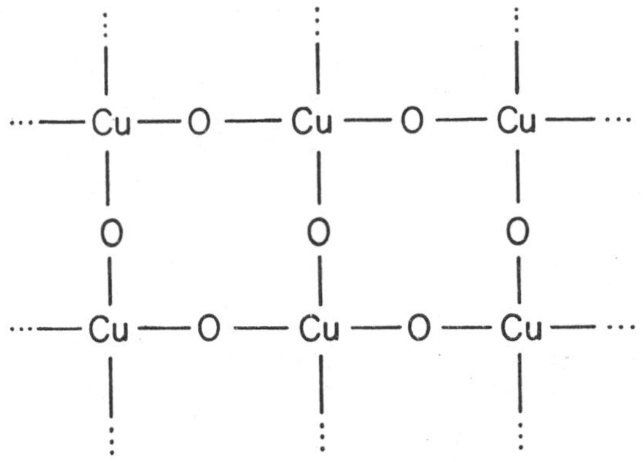

$$\epsilon_{\mathbf{k}} = 2t(\cos k_x a + \cos k_y a), \qquad (44)$$

where t is the so–called hopping or Bloch integral $\langle i|V|j\rangle$ between the nearest neighboring ions i and j, and a is the Cu–Cu distance. For La_2CuO_4 and $YBa_2Cu_3O_{6.5}$, this band is half–filled, with the Fermi surface for bare (noninteracting) electrons determined from the condition $\epsilon_{\mathbf{k}} = \mu = 0$. As shown in Fig. 24, this leads to a square in reciprocal space connecting the points $(\pi/a)(\pm 1, 0)$ with the points $(\pi/a)(0, \pm)$. The oxygen electrons in the $2p$ states are regarded as playing only a passive role of a transmitter of the individual d electrons from one d_{x^2-y} state to its neighbor (note that the O^{2-} valence state has completely filled p shells). If the number of electrons in that band is decreased (for example, by substituting Sr for La in 214 compounds), then the Fermi surface shrinks and gradually transforms into a circle, as shown in Fig. 24 [44]. Within such a model, La_2CuO_4 should be metallic. However, at $T < T_N \simeq 240$ K, this compound orders antiferromagnetically [24], and the ground state is then insulating. The fact that this system remains insulating above the Néel temperature T_N means that the stoichiometric La_2CuO_4 and $YBa_2Cu_3O_{6.5}$ are Mott insulators, not a Slater split–band antiferromagnet; for the latter, the split–band structure for $T < T_N$ should coalesce into one band as $T \to T_N$. The presence of the paramagnetic insulating state for both La_2CuO_4 and $YBa_2CuO_{6+\delta}$ supports the view that those oxides should be regarded as

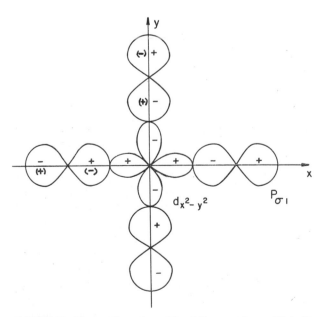

FIGURE 22 Schematic representation of the CuO_2 planes in superconducting oxides in the tetragonal phase. The Cu–Cu distance is ≈ 1.9 Å for La_2CuO_4.

antibonding orbital $d_{x^2-y^2}$ is higher in energy and hence half–filled. These d states hybridize with the oxygen $2p_x$ and $2p_y$ orbitals of σ type, as shown schematically in Fig. 23; both the bonding and the antibonding configurations are shown; the latter corresponds to the signs of the two p orbitals shown in parentheses.

A simple description of the electronic states for the planar CuO_2 system is obtained by introducing a single band representing Cu d electrons in the tight-binding approximation. For the square configuration of the Cu atoms (which reflects the tetragonal structure of La_2CuO_4), such a dispersion of band energies has the form

FIGURE 23 The configuration of the $3d_{x^2-y^2}$ and p_σ orbitals for bonding configurations. The reverse signs for the two p—orbitals (that is, those in parentheses) represent the hybridized configuration for the antibonding state.

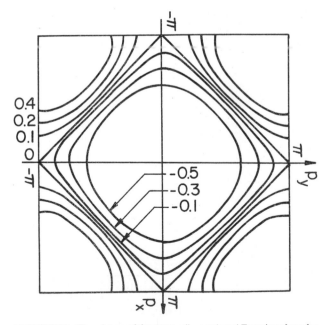

FIGURE 24 The shape of the two—dimensional Fermi surface for band energy of the form of Eq. (44). The values specified represent μ/t as a parameter. The square shape corresponds to $\mu = 0$ or, equivalently, to $n = 1$. [From Ref. 44.]

narrow–band systems characterized by strong electron–electron interactions ($U > U_c$), as originally proposed by Anderson [45]. An antiferromagnetic ground state is then expected since the kinetic exchange interaction between the strongly correlated electrons takes place [31].

A principal problem appears when holes occur in the Mott insulator, that is, when we consider a real situation $La_{2-x}Sr_xCuO_4$ or $YBa_2Cu_3O_{6.5+x}$. We have already seen (at the end of Section III.A) that for small x the kinetic energy of the holes and the exchange energy of electrons may become comparable or the latter may become even larger than the former. In such situations, the motion of the holes will be influenced by the setting-in of almost-instantaneous spin–spin correlations. This means that such metallic states (if formed) cannot be regarded as the Fermi liquid with slowly evolving spin fluctuations; instead, the resonance between various spin configurations must be built into the electron wave function characterizing its itinerant state. The decomposition of the resonating spin configurations into spin pair–singlet configurations constitutes an important characteristic of the RVB theory of the normal state [36, 45]. Some experimental evidence for the quantum spin–liquid state above the Néel temperature has been provided by neutron quasi–elastic scattering [46]; these results were subsequently interpreted [47].

The interpretation of the metallic state in terms of a single, narrow band requires the presence of both $3d^9$ (Cu^{2+}) as well as $3d^8$ (Cu^{3+}) states. Most of the X–ray spectroscopical studies [48] conclude that the satellite peak corresponding to a $3d^8$ configuration is actually absent. Therefore, to explain both the insulating properties of $La_2Sr_xCuO_4$ and the metallic properties of $La_{2-x}Sr_xCuO_4$ for $0.04 \div 0.05$, one introduces hybridized $2p$–$3d$ states for the holes introduced by the doping. The proper model of such states is then the Anderson lattice type of model [Eq. (43)]. Band–structure calculations by Mattheis [49] for $La_{2-x}Sr_xCuO_4$, shown in Fig. 25, justify a reasonable description within a simple two-dimensional tight-binding model with only the Cu $3d_{x^2-y^2}$ and p_σ orbitals on oxygens taken into account. Namely, the structures denoted A and B in Fig. 25 correspond, respectively, to antibonding and bonding hybridized bands, with respective band energies

$$E_{k\pm} = \frac{\epsilon_p + \epsilon_d}{2} \pm \left[\left(\frac{\epsilon_p - \epsilon_d}{2} \right)^2 + 4V^2 \left(\sin^2 \frac{k_x a}{2} + \sin^2 \frac{k_y a}{2} \right) \right]^{\frac{1}{2}}, \quad (45)$$

where ϵ_p and ϵ_d are atomic level positions for the $3d$ and $2p$ states, respectively. Detailed calculations [49] lead to a nonzero bandwidth of the $2p$ band because of the p–p overlap; then $\epsilon_p \rightarrow \epsilon_p + \epsilon_k$, with

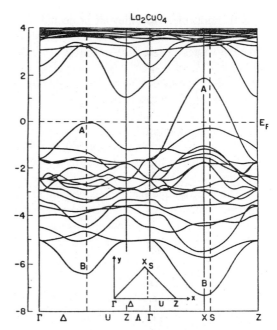

FIGURE 25 Energy bands for La_2Cu_4 calculated within a local–density approximation for the assumed crystal structure are body–centered tetragonal. A portion of the x–y plane in the extended Brillouin zone scheme is shown in the inset. Portions B and A correspond to the bonding and antibonding parts of the hybridized band discussed in the text. [From Ref. 49.]

$$\epsilon_k = 2t_p[\cos(k_x a/\sqrt{2}) + \cos(k_y a/\sqrt{2})].$$

The band structure calculations should be regarded as providing input parameters for the parametrized models which include electron correlations more accurately. On the basis of various estimates [50] of those parameters, one can assume that they fall in the range

$$|\epsilon_p - \epsilon_d| \simeq 3.6 \text{ eV}, \quad |V| \approx 1.3 \div 1.5 \text{ eV}, \quad |t| \simeq 0.5 \text{ eV},$$

$$|t_p| \simeq 0.6 \text{ eV}, \quad \text{and} \quad U \simeq 8\text{–}10 \text{ eV}.$$

From these estimates of the parameters, one sees that $|V| \sim |\epsilon_p - \epsilon_d|$. Hence, one may not be able to use the perturbation expansion in $V/(\epsilon_p - \epsilon_d)$ of the Anderson lattice Hamiltonian. Such a perturbation expansion was used [51] when transforming the hybridized model represented by the p–d Hamiltonian into an effective narrow–band model, which is represented by the effective Hamiltonian for the electrons in the CuO_2 plane,

$$H = t \sum_{i,j=\text{n.n},\sigma} \tilde{a}_{i\sigma}^+ \tilde{a}_{j\sigma} + t' \sum_{i,j=\text{n.n.n},\sigma} \tilde{a}_{i\sigma}^+ \tilde{a}_{j\sigma} + J \sum_{i,j=\text{n.n}} \left(\mathbf{S}_i \cdot \mathbf{S}_j - \frac{1}{4}\tilde{n}_i\tilde{n}_j \right),$$

where t and t' are the hopping integrals between the nearest (n.n.) and the next-nearest neighbors, J is the value of exchange integral for the kinetic superexchange, and the tilded operators mean that they are projected onto subspace of singly occupied lattices sites. The three parameters take the values $t \simeq -0.5$ eV, $t' \simeq 0.05$–0.1 eV, and $J \simeq 0.13$ eV.

On the basis of the facts that the present-day band calculations do not provide paramagnetic insulating states for stochiometric materials, such as La_2CuO_4, and that the antiferromagnetic ground state is difficult to achieve within the local–density approximation [49], we conclude that an approach based on the parametrized models discussed in the preceding two sections should be treated in detail. The microscopic parameters obtained from the band–structure calculations should be treated as input parameters in those models. A review of properties obtained within the parametrized models and relevant to high-T_c systems is given in Section IV.

IV. NOVEL MECHANISMS OF ELECTRON PAIRING

The binding of two fermions into either a bosonic or a more complicate bound state is a prerequisite for the condensation of microscopic particles into a coherent (superfluid) macroscopic state. This condensation may take the form of Bose–Einstein condensation if the interaction energy between the pairs is much lower than the binding energy of a single pair (also, the pairs must be well separated spatially). Such a Bose condensed state of charged particles may exhibit the principal properties of the superconducting state such as the Meissner–Ochsenfeld effect [52]. In the BCS theory (discussed in Section II), pair condensation occurs under a completely different condition, namely, when the states of different pairs overlap strongly so that the motion of one widely separated pair takes place in the mean field of almost all other pairs.

The pairing of particles in the BCS theory is described in momentum (reciprocal) space, where it is assumed that the quasiparticle states with a well-defined Fermi surface are formed first; the pairing involves electrons from the opposite points on the Fermi surface ($\mathbf{k}$, $-\mathbf{k}$) and generates either a simple spin–singlet state (as in the classic superconductors) or a higher angular-momentum state, e.g., $L = S = 1$ (as in superfluid ^{3}He [53]). Because of the small coherence length ($\xi \sim 10$ Å), the new superconductors offer an opportunity for exploring the possibility of pairing in real (coordinate) space. Moreover, since the carrier concentration determined from the Hall-effect measurements [54] for high-T_c oxides is at least one order of magnitude lower than that for ordinary metals, it is tempting to describe the onset of the superconducting state as a Bose condensation of preexisting pairs. In fact, the situation is not that simple. For example, in $La_{2-x}Sr_xCu_4$, with $x = 0.04$,

the average distance between holes in the normal phase is $\approx 5a$, a magnitude comparable to ξ. These circumstances, combined with antiferromagnetism and localization effects, render the new superconducting materials unique in the sense that their description requires a unification of theoretical approaches to phenomena previously regarded as disparate.

The accumulated evidence for rather strong electron–electron interaction in high-T_c oxides [36,48] and in heavy–fermion systems [39–43] makes it unlikely that electron pairing in these materials is caused by extremely strong electron–phonon interaction. Furthermore, the electron–phonon interaction does not allow for a connection (or, strictly speaking, competition) between the observed superconductivity and antiferromagnetism [24]. This is one of the reasons for an intensive search for a purely electronic mechanism of pairing. We now discuss some of the mechanisms that have been proposed. The main emphasis so far has been placed on an exchange–mediated pairing for strongly correlated electrons [45], since for such systems, the pairing, antiferromagnetism, and MITs to the Mott localized phase are derived from a single theoretical scheme. The latter two phases have been discussed in Section III; here, we concentrate on the spin–singlet pairing among strongly correlated electrons. Later we discuss charge transfer- and phonon-mediated pairings. Finally, we classify the types of correlated states and metallic states in solids. This classification provides a concise way of characterizing specific properties of these systems by which the almost-localized systems differ from ordinary metals.

A. Exchange Interactions and the Real-Space Pairing

1. Narrow–Band Systems

In Section III.A we provided an approximate Hamiltonian [Eq. (34)], which includes the antiferromagnetic exchange interactions between the correlated electrons in the limit $U/W \gg 1$. The precise form of this Hamiltonian to second order in W/U is [31, 55]

$$H = \sum_{ij\sigma}' t_{ij} b_{i\sigma}^+ b_{j\sigma} + \sum_{ij}' \frac{2t_{ij}^2}{U}\left(\mathbf{S}_i \cdot \mathbf{S}_j - \frac{1}{4}\nu_i\nu_j\right)$$
$$+ \text{(three–site terms)}, \tag{46}$$

where the primed summation means that $i \neq j$. In this Hamiltonian, doubly occupied site i configurations $|i\uparrow\downarrow\rangle$ are excluded. This exclusion is reflected by the presence of creation ($b_{i\sigma}^+$) and annihilation ($b_{i\sigma}$) operators for electrons in the state $|i\sigma\rangle$, which are defined as

$$b_{i\sigma}^+ \equiv a_{i\sigma}^+(1 - n_{i-\sigma}) \quad \text{and} \quad b_{i\sigma} \equiv a_{i\sigma}(1 - n_{i-\sigma}), \quad (47)$$

so that

$$v_{i\sigma} = b_{i\sigma}^+ b_{i\sigma} \quad \text{and} \quad v_i = \sum_\sigma v_{i\sigma}. \quad (48)$$

The spin operator is defined as

$$\mathbf{S}_i \equiv (S_i^+, S_i^-, S_i^z) \equiv [a_{i\uparrow}^+ a_{i\downarrow}, a_{i\downarrow}^+ a_{i\uparrow}, (n_{i\uparrow} - n_{i\downarrow})/2].$$

Note that the same representation of the operator $\mathbf{S}_i$ can be written in terms of projected operators $b_{i\sigma}^+$ and $b_{i\sigma}$. The factor $(1 - n_{i-\sigma})$ in Eq. (47) imposes explicitly the restriction that the creation or the annihilation of electrons in the state $|i\sigma\rangle$ can take place only if there is no second electron already on the same site. Thus, $v_i = \sum_\sigma n_{i\sigma}(1 - n_{i-\sigma})$ enumerates only the singly occupied sites ($v_i = 0$ or 1). In other words, the N states corresponding to the doubly occupied site configurations have been projected out. Thus, Eq. (46) describes the dynamics of strongly correlated electrons for $N_e \leq N$ of electrons. Also, in performing the summations in Eq. (46), one usually considers only the pairs $\langle ij \rangle$ of nearest neighbors; in this approximation the parameters $J_{ij} = J$ and $t_{ij} = t$ can be chosen as constants.

The first term in Eq. (46) describes the single–particle hopping of electrons from the singly occupied to the empty atomic sites; the second describes the exchange interaction induced by virtual hopping between site i and site j, while the three-site part describes the motion of electron with spin σ from the singly occupied site located at i to the next-nearest neighboring empty site k via the occupied configuration (with electron of opposite spin) located at site j. The various contributions to Eq. (46) are represented graphically in Fig. 26.

If one introduces a new pair of creation and annihilation operators in coordinate space by

$$\tilde{b}_{ij}^+ = \frac{1}{\sqrt{2}}(b_{i\uparrow}^+ b_{j\downarrow}^+ - b_{i\downarrow}^+ b_{j\uparrow}^+) \quad (49a)$$

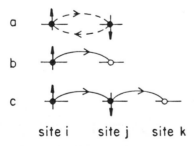

FIGURE 26 Various hopping processes in narrow-band systems in a partial band-filling case: (a) virtual hopping processes leading to a kinetic exchange interaction; (b) single-particle hopping representing the band energy of correlated electrons; (c) contribution to the pair hopping—this process gives the pairing contribution in Eq. (50) with $k \neq i$.

and

$$\tilde{b}_{ij} = \frac{1}{\sqrt{2}}(b_{i\downarrow} b_{j\uparrow} - b_{i\uparrow} b_{j\downarrow}), \quad (49b)$$

then the Hamiltonian [Eq. (46)] with inclusion of the three–site part can be written in the following very suggestive closed form [55]:

$$H = {\sum_{ij\sigma}}' t_{ij} b_{i\sigma}^+ b_{j\sigma} - \sum_{ij}(2t_{ij}t_{jk}/U)\tilde{b}_{ij}^+ \tilde{b}_{kj}. \quad (50)$$

The first term represents, as before, the dynamics of single electrons moving between the empty sites regarded as holes; the second term combines the last two terms in Eq. (46) and expresses the dynamics of the singlet pairs [cf. Eqs. (49a) and (49b)]. The division in Eq. (50) into single–particle and pair parts is in analogy to the BCS Hamiltonian; however, here, the operators are expressed in coordinate space. The term with $i = k$ in the pairing part enumerates the spin–singlet pairs of neighboring spins; the terms with $i \neq k$ represent pair hopping of such singlet pair bonds. Thus, in the language of operators [Eqs. (49)], one adds the bond dynamics to that of single electrons. Moreover, the forms of Eqs. (46) and (50) are completely equivalent; hence, the pairing effect and the antiferromagnetism should be directly linked within this formalism (they are two different expressions of the same part of H).

It is difficult to diagonalize the Hamiltonian [Eq. (50)] to obtain the eigenvalues of the system. Part of the problem arises from the fact that the single–particle operators $b_{i\sigma}$ and $b_{j\sigma}^+$ do not obey the fermion anticommutation relation and that the pair operators $\tilde{b}_{ij}^+$ and $\tilde{b}_{ij}$ do not obey boson commutation relations. Additionally, the two terms in (3.5) do not commute, so that the itinerant characteristics of the electrons and the pair-binding effects combine and produce a paired metallic phase, particularly if the two terms are of comparable magnitude. We have seen in Section III that if the number of holes $\delta \equiv 1 - n < \delta_c \sim 0.02$, then the pairing (or exchange) part dominates and antiferromagnetism sets in. Detailed calculations [32] lead to the boundary line between the antiferromagnetic and the ferromagnetic phase, as shown in Fig. 27. The energy of the completely saturated ferromagnetic phase (CF) indicated does not depend on the value of exchange integral $J_{ij} \equiv 2t_{ij}^2/U$.

We now discuss the superconducting phase for which the pairing part in Eq. (50) plays a crucial role. To make the problem tractable at this point, one replaces the operators [Eqs. (49)] by fermion operators [45, 56], that is,

$$b_{i\sigma}^+ \to a_{i\sigma}^+, \quad b_{i\sigma} \to a_{i\sigma}, \quad (51a)$$

and introduces the replacement

$$\tilde{b}_{ij}^+ \to b_{ij}^+ = \frac{1}{\sqrt{2}}(a_{i\downarrow}^+ a_{j\uparrow}^+ - a_{i\uparrow}^+ a_{j\uparrow}^+) \quad (51b)$$

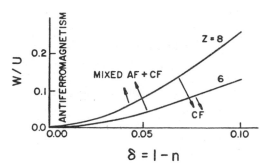

FIGURE 27 Phase boundary between the mixed ferromagnetic (CF)–antiferromagnetic (AF) phase and the pure ferromagnetic phase for simple cubic ($z = 6$) and b.c.c. cubic ($z = 8$) structures. A similar type of phase boundary can be obtained for other structures. [From Ref. 63.]

and

$$\tilde{b}_{ij} \rightarrow b_{ij} = \frac{1}{\sqrt{2}}(a_{i\downarrow}a_{j\uparrow} - a_{i\uparrow}a_{j\downarrow}). \qquad (51c)$$

Simultaneously, one renormalizes the parameters t_{ij} and J_{ij} in such a manner that they contain the restrictions on particle dynamics due to the projection of doubly occupied site configurations in the expression for the ground-state energy. Within the Gutzwiller–Ansatz approximation [29], Eqs. (49) reduce the starting Hamiltonian to the form

$$H = \delta \sum_{ij\sigma} t_{ij}a_{i\sigma}^{+}a_{j\sigma} \sum_{ijk}(2t_{ij}t_{jk}/U)b_{ij}^{+}b_{kj}, \qquad (52)$$

where $\delta = 1 - n$. This Hamiltonian has been solved within the mean-field approximation equivalent to the BCS approximation [56, 57] and with neglect of the pairing terms with $k \neq i$. This leads to the following self-consistent equations for $\Delta_k \neq 0$:

$$\frac{J}{N}\sum_{k}\frac{\gamma_k^2}{E_k}\tanh\left(\frac{\beta E_k}{2}\right) = 1, \qquad (53)$$

with $J = 2t^2/U$, $E_k = [(\epsilon_k - \mu)^2 + |\Delta_k|^2]^{\frac{1}{2}}$, and $\gamma_k = \cos(k_x a) + \cos(k_y a)$ for a planar configuration of the lattice. This equation must be supplemented with the equation for the chemical potential in the superconducting phase of the form

$$\frac{1}{N}\sum_{k}\left(1 - \frac{\epsilon_k}{E_k}\right)\tanh\left(\frac{\beta E_k}{2}\right) = n. \qquad (54)$$

In solving Eq. (53), solutions of the following type have been considered:

1. extended s–wave [56, 58],

$$\Delta_k^{(s)} = \Delta[\cos(k_x a) + \cos(k_y a)]; \qquad (55a)$$

2. d–wave [57, 58],

$$\Delta_k^{(d)} = \Delta[\cos(k_x a) - \cos(k_y a)]; \qquad (55b)$$

3. mixed s and d phases [59],

$$\Delta_k^{(sd)} = s\Delta_k^{(s)} + d\Delta_k^{(d)}. \qquad (55c)$$

The mixed phase was found to be the most stable close to the half-filled band case. For the half-filled band case, the ground-state energies for s– and d–wave states are the same.

The type of solution obtained within the mean-field approximation (cf. Section II for details) is illustrated in Fig. 28, where the temperature dependence of the specific heat is shown for a different number of holes δ and for $|t|/U = 0.1$ and with the inclusion of the nearest-neighbor repulsive coulomb interaction V. A discontinuity of $C(T)$ at $T = T_c$ takes place for each δ. For comparison, the dotted lines represent the specific heat for the normal phase.

There is a major problem with the standard mean-field solutions discussed in Refs. 55–59, namely, it yields a nonzero (in fact, maximal or almost-maximal) value of the superconducting transition temperature T_c for the half-filled band case, which corresponds to the Mott insulating state. This is a spurious result; it appears because by performing the transformation Eqs. (51), the double-site occupancies reappear again for $n < 1$. To remove some of the unphysical features of the mean-field solution, a new formalism has been proposed [58–60] in which auxiliary (slave) bosons are introduced. In this formalism, some

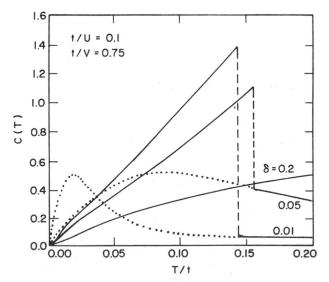

FIGURE 28 Temperature dependence of the specific heat $C(T)$ within the mean-field approach to the exchange-mediated pairing in a narrow band. The dotted line represents $C(T)$ for the normal phase, while the discontinuity occurs at the transition to the superconducting phase. [From Ref. 58.]

of the properties of the projected operators [Eqs. (47) and (49)] are already preserved in the mean field-type approximation involving boson and fermion fields on the same footing. The transition temperature T_c now vanishes, as it should, in the limit $n = 1$. According to Ref. 61, the slave bosons represent holes in the Mott insulators and are regarded as charged, while the fermions are neutral. These entities are called holons and spinons, respectively.

The holon–spinon language is introduced formally by noting that the projected operators [Eq. (47)] are represented as

$$b_{i\sigma}^{+} \equiv b_i f_{i\sigma}^{+} \qquad \text{and} \qquad b_{i\sigma} = b_i^{+} f_{i\sigma}, \qquad (56)$$

where b_i and b_i^{+} are annihilation and creation boson operators located at the atomic site i, while $f_{i\sigma}^{+} \equiv a_{i\sigma}^{+}$ and $f_{i\sigma} \equiv a_{i\sigma}$ are the commonly used fermion operators. Substituting Eq. (56) into Eq. (46), one obtains

$$H = \sum_{ij\sigma}{}' t_{ij} b_i b_j^{+} f_{i\sigma}^{+} f_{j\sigma} - \sum_{ij}{}' \left(2t_{ij}^2 / U\right) b_{ij}^{+} b_{ij}$$

$$+ \sum_i \lambda_i \left(b_i^{+} b_i + \sum_\sigma f_{i\sigma}^{+} f_{i\sigma} - 1 \right)$$

$$- \mu \sum_{i\sigma} n_{i\sigma} + (\text{three–site terms}). \qquad (57)$$

The first two terms represent, respectively, the coupled holon–spinon hopping and the binding of spinons into singlet pairs. The third term expresses the fact that the number of holons and spinons is equal to unity on each site; the Lagrange multiplier λ_i thus explicitly provides formally the removal of double occupancies. The fourth term represents the conservation of the number of electrons. Now, in a further approximation, one decouples fermions from bosons and then solves the two parts self-consistently. The mean-field treatment discussed earlier corresponds to the approximation in which λ_i is taken as the same at each site ($\lambda_i \to \lambda$) and in which one introduces the replacement $\langle b_i b_i^{+} \rangle = \langle b_i \rangle \langle b_i^{+} \rangle = |\langle b \rangle|^2 = 1 - n$. The superconducting solution is described in terms of two correlation functions: $\Delta_B \equiv \langle b_i^{+} b_j^{+} \rangle \approx \langle b^{+} \rangle^2$, characterizing the Bose condensation of holons, and $\Delta_F \equiv \langle f_{i\uparrow} f_{j\downarrow} \rangle$, characterizing the gap in the spectrum of fermion excitations (the site indices i and j denote a pair $\langle ij \rangle$ of nearest neighbors). The nonzero Δ_B occurs only below a temperature T_B, which we call the Bose condensation temperature, whereas the nonzero Δ_F appears only below $T = T_{RVB}$, characterizing the mean-field solution within the RVB theory [56]. The superconducting phase is characterized by nonzero values of both Δ_B and Δ_F simultaneously. This is because in the mean—field approximation $\langle b_{i\uparrow} b_{i\downarrow} \rangle = \Delta_B \Delta_F$. Hence, the lower of the two temperatures (T_B and T_{RVB}) determines the superconducting transition temperature. In Fig. 29, taken from Ref. 63, we have plotted these two tempera-

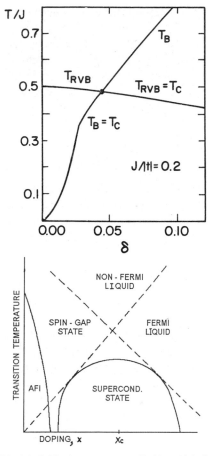

FIGURE 29 (a) Critical temperature T_c (the thick line) versus $\delta = 1 - n$. The temperatures T_{RVB} and T_B are those characterizing the onset of coherency for spinons and the Bose condensation of holons. Note that T_c is determined by the lower of the two temperatures. [From Ref. 62.] (b) Schematic theoretical phase diagram obtained within the gauge theory.

tures as a function $\delta = 1 - n$. One should note that to have $T_B \neq 0$, a small nonzero overlap $t_z = 0.1t$ was taken in the direction perpendicular to the square planar configuration of the atoms. We see that $T_c \to 0$ as $\delta \to 0$, as should be the case.

Note added in August 2000. In the last 10 years the slave-boson approach evolved into the gauge-theory approach to doped Mott insulators [33d]. This approach leads to the phase diagram shown schematically in Fig. 29b. The details of this phase diagram go beyond the scope of this article. The other factors are the detailed role of the van Hove singularity in the two-dimensional idensity of states for a square lattice [67a, b] in increasing even the BCS value of the critical temperature, as well as the determined d-wave symmetry reflecting the strong on-site Coulomb repulsion, which produces a node in the spatial dependence of the gap $\Delta(\mathbf{r})$ [67c, d]. Finally, the role of the interlayer Josephson tunneling has been stressed [67e, f], although there is some discussion [67g] about the magnitude of the condensation energy due to the formation of (a) truly three-dimensonal paired state from a two-dimensonal normal metal. This means that a simple type of Lawrence–Doniach–Ginzburg–Landau approach [67h] and other models based on the interplanar Josephson tunneling [67e, f] may not be sufficient.

2. Hybridized Systems

The electron states near the Fermi surface in high-T_c oxides such as La_2CuO_4 involve hybridization of electrons of atomiclike $3d_{x^2-y^2}$ states of copper with $2p_\sigma$ of oxygen (cf. Fig. 23 and Section III.D). These electron states can be described by the Anderson lattice Hamiltonian of the type of Eq. (43), with a width of the bare p–band $W \approx 4$ ev, the position of the $3d^9$ level at $\epsilon_f \equiv \epsilon_d - \epsilon_p \sim 1$ eV, $U \leq 10$ eV, and hybridization magnitude $|V| \simeq 1.5$ eV [68]. The hybridization is intersite in nature, that is, it involves the $2p$ and $3d$ orbitals located on different sites. Therefore, the effective hybridization energy is $Vz \simeq 6$ eV, where $z = 4$ is the number of nearest-neighboring O atoms in the plane for a given Cu atom. We see that $Vz > \epsilon_f$; hence, the $3d$ and $2p$ states mix strongly, that is, the d electrons can be promoted to $2p$–hole states, and vice versa. Additionally, $2p$ electrons can be promoted to form the $3d^{10}$ configurations of the excited states. If $Vz \gtrsim \epsilon_f$, but $|V|z \ll \epsilon_f + U$, the above two promotion–mixing events are low– and high–energy processes, respectively. The situation is shown schematically in Fig. 30, where the parameter U is assumed to be by far larger than $|\epsilon_f|$, W, or $|V|z$. We consider this limiting situation first [68].

The high–energy processes take place only as virtual events, that is, with electron hopping from the p state to the highly excited $3d$ state and back. Such virtual p–d–p processes are shown schematically in Fig. 31, where site m labels the $2p_\sigma$ state of the oxygen anion O^{2-} centered at $\mathbf{R_m}$ and site i labels $3d_{x^2-y^2}$ due to the Cu^{2+} ion centered at $\mathbf{R_i}$. Then the effective Hamiltonian can be rewritten in the real–space language and for large U reads [68]

$$H = \sum_{\mathbf{k}\sigma} \epsilon_\mathbf{k} n_{\mathbf{k}\sigma} + \epsilon_f \sum_{i\sigma} b_{i\sigma}^+ b_{i\sigma} + \sum_{im\sigma} V_{im}\big(b_{i\sigma}^+ c_{m\sigma}$$
$$+ V_{im}^* c_{m\sigma}^+ b_{i\sigma}\big) - \sum_{imn} \frac{2V_{mi}^* V_{im}}{U + \epsilon_f} B_{im}^+ B_{in}. \quad (58)$$

The first term describes the band energy of itinerant ($2p_\sigma$) electrons, while the third represents the residual mixing pairing since, as in the case of narrow-band electrons, the operators $(b_{i\sigma}^+)$ and $(b_{i\sigma})$ are projected operators [Eq. (47)] for the starting $3d$ states. The last term represents the so-

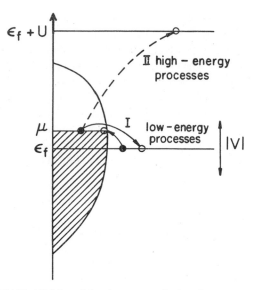

FIGURE 30 Division of the charge-transfer (p–d) processes into low- and high-energy parts. The processes labeled II give rise to Kondo and superexchange interactions when treated perturbationally to second and fourth order, respectively.

called hybrid interorbital pairing with the pairing operators

$$B_{im}^+ = \frac{1}{\sqrt{2}}\big(b_{i\uparrow}^+ c_{m\downarrow}^+ - b_{i\downarrow}^+ c_{i\uparrow}^+\big) \quad (59a)$$

and

$$B_{im} = \frac{1}{\sqrt{2}}(b_{i\uparrow} c_{m\downarrow} - b_{i\downarrow} c_{i\uparrow}). \quad (59b)$$

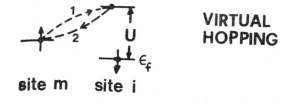

VIRTUAL HOPPING

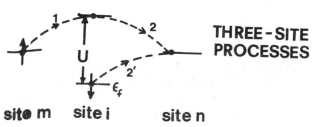

THREE-SITE PROCESSES

FIGURE 31 Schematic representation of the hopping processes induced by high-energy mixing processes. The hoppings labeled 2 and 2′ are alternative processes.

The meaning of the effective Hamiltonian [Eq. (58)] is as follows. The first three terms provide eigenvalues representing the hybridized quasiparticle states with the structure discussed in Section III.C. The last term provides a singlet pairing for those hybridized states. It expresses (for $m = n$) the Kondo interaction between the p and the $3d$ electrons of the form

$$\sum_{im} \frac{2|V_{im}|^2}{U + \epsilon_f} \left(\mathbf{S}_i \cdot \mathbf{s}_m - \frac{1}{4} \nu_i n_m \right).$$

It is antiferromagnetic in nature, with the exchange integral

$$J_{im} \equiv \frac{2|V_{im}|^2}{U + \epsilon_f} \sim 0.5 \text{ eV},$$

hence, the pairing results in a spin–singlet state. It must be underlined that Eq. (58) represents hybridized correlated states in the so-called fluctuating-valence regime in which $U \gg |V_{im}| \gtrsim \epsilon_f$. This is the reason why we cannot completely transform out the hybridization. Also, the occupancy n_f of the atomic level is a noninteger because the strong hybridization induces a redistribution of the particles among starting atomic and band states.

When both U and $|\epsilon_f|$ are much larger than $|V_{im}|$, one can transform out the hybridization completely and obtain, instead of Eq. (58), the following effective Hamiltonian:

$$H = \sum_{\mathbf{k}\sigma} \epsilon_{\mathbf{k}} n_{\mathbf{k}\sigma} + \epsilon_f \sum_{i\sigma} b_{i\sigma}^+ b_{i\sigma} + \sum_{ijm\sigma} \frac{V_{mi}^* V_{mi}}{\epsilon_f}$$

$$\times b_{i\sigma}^+ b_{j\sigma}(1 - n_{m\sigma}) \sum_{im\sigma} \frac{2V_{mi}^* V_{in} U}{\epsilon_f(\epsilon_f + U)} b_{im}^+ b_{in}. \quad (60)$$

We now have a two–band system: the $3d$ electrons acquire a bandwidth $W^* \sim (V^2/\epsilon_f)(1 - \langle n_{m\sigma} \rangle)$. The spin–singlet pairing is again of the interband type. The part with $m = n$ in the last term is equivalent to the Kondo interaction derived a long time ago for magnetic impurities [69]. Here, the lattice version of this Hamiltonian provides both pairing and itinerancy to the bare atomic electrons.

Note that the hybrid pairing introduced in this section expresses both the Kondo interaction (the two–site part) and pair hopping. It is therefore suitable for a discussion of the superconductivity of Kondo lattice effects in heavy-fermion systems. The pairing part supplements the current discussions of the Anderson lattice Hamiltonian in the $U \to \infty$ limit [40–42]. One may state that the Kondo interaction-mediated pairing introduced above represents the strong–coupled version of spin fluctuation-mediated pairing for almost-localized systems introduced previously [70].

An approach using the slave-boson language for hybridized systems has also been formulated [71] and contains a principal feature of the effective Hamiltonian

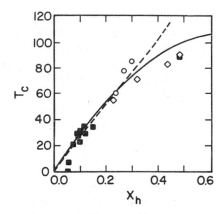

FIGURE 32 Superconducting transition temperature T_c versus hole concentration x_h. Squares, experimental data for $La_{2-x}Sr_xCuO_4$; circles and diamonds, data for $YBa_2Cu_3O_{7-y}$. [From Ref. 71.]

[Eq. (58)]; the solution in the mean-field approximation has been also discussed. Figure 32 illustrates the dependence of the superconducting transition temperature T_c versus the hole concentration x_h; this is compared with experimental data [72]. Dependence of T_c over the full concentration range of holes is shown in Fig. 33. The superconductivity appears for $La_{2-x}Sr_xCuO_4$ only for $0.04 \lesssim x_h < 0.34$. The full phase diagram comprising localization and antiferromagnetism (LM phase) and superconductivity (SC) is provided in Fig. 33.

3. An Overview

Two alternative models and mechanisms of exchange–mediated pairing have been discussed so far: the narrow–band model, with d–d kinetic exchange-mediated pairing, and the hybridized model, with d–p Kondo interaction-mediated pairing. The hybridized model should be regarded as a basis of narrow–band behavior in real oxides and in heavy–fermion systems since the direct d–d (or f–f) overlap of the neighboring atomic wave functions is extremely small. Next, we give a brief overview of the

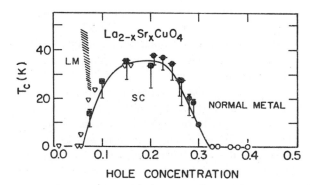

FIGURE 33 Superconducting transition temperature T_c versus hole concentration for $La_{2-x}Sr_xCuO_4$ over the full range. LM, the regime of local moments (insulating phase). [From Ref. 72.]

narrow–band properties of the correlated electrons starting from the hybridized (Anderson lattice) model.

First, we discuss the quasiparticle states in the $U \to \infty$ limit. The simplest approximation is to reintroduce ordinary fermion operators $a_{i\sigma}^+$ and $a_{i\sigma}$ in Eq. (58) and readjust the hybridization accordingly [73]. In effect, one obtains the hybridized bands of the form of Eq. (45), that is,

$$E_{\mathbf{k}\pm} = \frac{\epsilon_f + \epsilon_\mathbf{k}}{2} \pm \left[\left(\frac{\epsilon_f - \epsilon_\mathbf{k}}{2} \right)^2 + 4|\tilde{V}_\mathbf{k}|^2 \right]^{\frac{1}{2}}, \quad (61)$$

where $\tilde{V}_\mathbf{k} \equiv q^{\frac{1}{2}} V_\mathbf{k}$, and $q \equiv (1 - n_f)/(1 - n_f/2)$ for $0 \leq n_f \leq 1$, while $V_\mathbf{k}$ is the space Fourier transform of V_{im}. For the case of the CuO_2 layers [74],

$$|\tilde{V}_\mathbf{k}|^2 = qV^2 \left[\sin^2 \left(\frac{k_x a}{2} \right) + \sin^2 \left(\frac{k_y a}{2} \right) \right]. \quad (62)$$

If the Fermi level falls into the lower hybridization band and $n_f = 1 - \delta$, with $\delta \ll 1$, then it can be shown that the quasiparticles describing the hybridized states are of mainly quasi-atomic character. In other words, the effective Hamiltonian [Eq. (58)] is approximately of the narrow-band form [Eq. (52)]. The pairing takes place between heavy quasiparticles. This limiting situation describes qualitatively the situation in heavy fermions with Kondo interaction mediating the pairing. In contrast, if the Fermi level falls close to the top of the upper hybridization band (as is the case for high-T_c superconductors, since the p band is almost full and the $3d$ level is almost half-filled), then the pairing is due mainly to the band electrons ($2p$ holes in the case of high-T_c oxides). These results are obtained by constructing explicitly the eigenstates corresponding to the eigenvalues [Eq. (61)] and taking the limits corresponding to heavy fermions ($n_f \to 1$) and high-T_c systems ($n = n_d + n_p \approx 3$, which also corresponds to the situation of one hole in the system).

a. Mott–Hubbard insulators, charge–transfer insulators, and mixed–valent systems.

The next problem concerns the Mott localization in systems with hybridized $d-p$ states. The systems such as NiO, CoO, and MnO regarded as classic Mott insulators are, strictly speaking, hybridized $3d$–$2p$ systems. However, these cases are, to a good approximation, ionic systems in the sense that the electronic configuration in, for example, NiO, is $Ni^{2+}O^{2-}$. Then, the valence $2p$ band is completely full and plays only a passive role in effective $d-d$ charge transfer processes [75], since a $2p \to 3d$ transfer is followed by $3d \to 2p$ transfer from the neighboring $3d$ shell of Ni^{2+}. In effect, the antiferromagnetic exchange interaction in Eq. (46) expresses formally the superexchange interaction that has been known for a long time [75, 76]. In this approach, the kinetic exchange interaction between d electrons (induced

by virtual d–d transitions; cf. Section III.A) is expressed as a fourth-order effect in the hybridization V since the virtual d–d transition involves a sequence of d–p and p–d transitions in the fourth order.

The possible macroscopic states of hybridized systems are illustrated in Fig. 34 as a schematic classification of possible states of hybridized systems modeled by the periodic Anderson Hamiltonian [Eq. (43)]. The parameter W/U characterizes the degree of correlation of quasi–atomic electrons that may acquire a nonzero bandwidth due mainly to hybridization; the parameter V/ϵ_f characterizes the degree of mixing of the states involved. If the d (or f) atomic level lies deeply below the top of the valence band ($V/\epsilon_f \ll 1$), then we have either Mott–Hubbard (M–H) or charge–transfer (C–T) insulators; for the former the band gap Δ is due to $d^n \to d^{n+1}$ excitations (that is, $\Delta \sim U - W$), whereas for the latter it is due to $d^n p^2 \to d^{n+1} p^1$ charge–transfer transitions. The atomic $3d$ (or $4f$) electrons are unpaired in both the C–T and the M–H states. If $V/\epsilon_f \gtrsim 1$, and $W/U \ll 1$ then we enter mixed valent (M–V) and (close to the border with M–H) heavy–fermion regimes. On the other hand, if $W/U \gtrsim 1$, then irrespective of the value of V/ϵ_f, we encounter the correlated–metal regime that we call an almost–localized Fermi liquid (AL–FL). Both heavy–fermion and high-T_c systems are close to the line separating M–H and M–V regimes. Such a classification scheme for transition-metal oxides has been proposed in Ref. [77].

The classification shown schematically in Fig. 34 provides only a distinction between insulating and metallic states. A complete magnetic phase diagram for the high-T_c system $La_{2-x}Sr_xCuO_4$ is shown schematically in Fig. 35 (taken from Ref. 79a). Stoichiometric or doped La_2CuO_4,

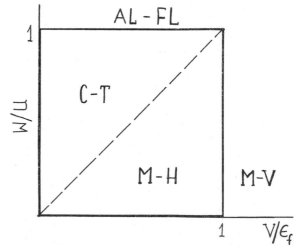

FIGURE 34 Schematic representation of the regimes of stability of the charge-transfer (C–T) and Mott–Hubbard (M–H) insulating states, as well as of the mixed-valent (M–V) and almost-localized Fermi-liquid (AL–FL) metallic states.

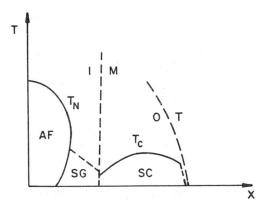

FIGURE 35 Schematic phase diagram on the plane T–x for $La_{2-x}Sr_xCuO_4$. Antiferromagnetic (AF), spin-glass (SG), superconducting (SC), insulating (I), and metallic phases are drawn, as well as the boundary between the orthorhombic (O) and the tetragonal (T) crystallographic phases. [From Ref. 80.]

with $x \lesssim 0.02$, exhibits antiferromagnetism (AF). In the regime of $0.02 \lesssim x \lesssim 0.04$, the inhomogeneous (SG) magnetic insulating phase sets in, while for $x \gtrsim 0.04$, a transition from insulating (I) to metallic (M) takes place and the system is superconducting until a transition from an orthorhombic (O) to a tetragonal (T) crystallographic structure occurs. A similar phase diagram was established for $YBa_2Cu_3O_{6+x}$ [77b]. Those phase diagrams combine all the features we have discussed separately so far. The main features of this phase diagram are explained next.

b. Magnetic interactions hybrid, polarons, and pairing. To address the phase diagram shown in Fig. 35 within the hybridized p–d model, we note first that antiferromagnetism is stable only close to the half-filling of the d–band (cf. Fig. 27 and the discussion in Section III.A). In the case of the hybridized model, one has to calculate explicitly the contributions to the d–p and d–d interactions. Within the perturbation expansion for the Anderson lattice model but with only the high–energy mixing processes (cf. Fig. 30) treated in this manner [68, 70], we obtain the magnetic part of the effective Hamiltonian to fourth order as

$$\mathcal{H}_m \simeq J_{pd} \sum_{im} \left(\mathbf{S}_i \cdot \mathbf{s}_m - \frac{1}{4} n_i n_m \right) + J_{dd} \sum_{\langle ij \rangle} \left(\mathbf{S}_i \cdot \mathbf{S}_j \right.$$
$$\left. - \frac{1}{4} N_i N_j \right) + J_{pp} \sum_{\langle mm' \rangle} \left(\mathbf{s}_m \cdot \mathbf{s}'_m - \frac{1}{4} n_m n'_m \right),$$
$$(63)$$

where the first term represents the p–d Kondo–type interaction, with the exchange integral

$$J_{pd} \approx \frac{2|V|^2}{U + \epsilon_f} \left[1 - \frac{|V|^2}{U + \epsilon_f} (n_d + n_p + 1) \right]. \quad (64)$$

The second term expresses the d–d (kinetic exchange) interaction, with $J_{dd} = |V|^4/(U + \epsilon_f)^3$, and the last term represents the interaction between p holes, with $J_{pp} = |V|_{n_d}^4/(U + \epsilon_f)^3 \approx J_{dd}$. The antiferromagnetic p–d and d–d interactions are not compatible; in the hole language, the p hole polarizes its surroundings ferromagnetically, as shown in Fig. 16 (note that the hole may be located in any O^{1-} ion, so its position with the volume of radius R is not fixed). A simple estimate [79] of the canting angle θ between the neighboring $3d$ spins $\mathbf{S}_i$ and $\mathbf{S}_j$ caused by the hole polarization gives

$$\cos \frac{\theta}{2} \approx \frac{J_{pd} - 2J_{dd}}{2J_{dd}}. \quad (65)$$

Taking $J_{pd} \approx 0.5$ eV and $J_{dd} \simeq 50$ K, we obtain the average canting angle θ through the relation $\cos(\theta/2) \approx 25x_p$. The energy E_c of the system with a single hole canting the surrounding spins is

$$E_c = -\frac{1}{2} \frac{(J_{pd} - 2J_{dd})^2}{J_{dd}} z - J_{dd} z. \quad (66)$$

This energy is lower than the energy ($-J_{dd}z$) of the antiferromagnetic (Néel) state of antialigned d spins due to Cu^{2+} copper ions. Next, we estimate the radius R of the hole polaron with aligned spins, as depicted in Fig. 36. Applying the same type of reasoning as in Section III.A, we obtain the expression for the energy E_p of a single polaron:

$$E_p = \frac{E_0}{(R/a)^2} - \frac{1}{2} \frac{(J_{pd} - 2J_{dd})^2}{J_{dd}} z \cdot \bar{x}_p^2, \quad (67)$$

where now $\bar{x} = (a/R)^2$ is the probability of finding a p hole on a given oxygen atomic site within the radius R. Minimization with respect to R for the two–dimensional case leads to

$$\frac{R}{a} = \left[\frac{(J_{pd} - 2J_{dd})^2 z}{J_{dd} E_0} \right]^{\frac{1}{2}} \approx J_{pd} \sqrt{\frac{z}{J_{dd} e_0}} \approx 4. \quad (68)$$

A MIT takes place when the neighboring polarons overlap, that is, when $R x_{pc}^{-1/2} = 1$; this yields the critical hole concentration $x_c \approx 0.07$. One can also estimate this critical concentration by equating the band energy of holes, which is $-(W/2)x_p(1 - x_p)$, with the magnetic energy gain per hole due to aligning the neighboring d spins $(-J_{pd}^2/2J_d d)zx_p^2$. This leads again to $x_c \approx 0.068$, in rough agreement with the observed value $x_c \simeq 0.04 \div 0.05$. For $x > x_c$, the ground state of the system is metallic, and the pairing described in Sections IV.A.1 and IV.A.2 can take place. Within the exchange–mediated mechanism, all interactions in Eq. (63) are antiferromagnetic. Hence, in general, one has p–d pairing characterized by the operators of Eqs. (59), d–d pairingcharacterized by the operators

2p HOLE POLARON

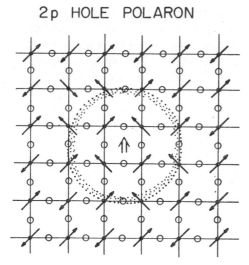

FIGURE 36 Schematic representation of a 2p-hole polaron in a planar CuO_2 structure. Cu^{2+} ions are indicated by arrows, while O^{2-} ions are indicated by open circles. The hole creates a canted spin configuration with resultant ferromagnetic polarization and autolocalizes in it. This is the reason the high-T_c oxides remain insulating when the concentration of the hole does not exceed $x_c \sim 0.04 \div 0.05$.

of Eqs. (49), and p–p pairing [80] characterized by the operators

$$p_{mm'}^+ \equiv \frac{1}{\sqrt{2}}\left(c_{m\uparrow}^+ c_{m'\downarrow}^+ - c_{m\downarrow}^+ c_{m'\uparrow}^+\right) \tag{69a}$$

and

$$p_{mm'} \equiv \frac{1}{\sqrt{2}}\left(c_{m\downarrow} c_{m'\uparrow} - c_{m\uparrow} c_{m'\downarrow}\right). \tag{69b}$$

All three types of pairing may contribute to the superconducting ground state. However, the d–p interaction is much stronger, hence, the d–p hybrid type of pairing is in the limit $U > W > |V| \gtrsim \epsilon_f$, the dominant one. As stated, this type of pairing may appear effectively as a d–d or p–p type of pairing in the hybridized basis, depending on whether the Fermi level lies close to the top of the lower or upper hybridized bands, respectively. For the sake of completeness, we write down the full effective Hamiltonian with all the pairings specified, namely,

$$\mathcal{H} = \sum_{\mathbf{k}\sigma} \epsilon_{\mathbf{k}} n_{\mathbf{k}_\sigma} + \epsilon_f \sum_{i\sigma} b_{i\sigma}^+ b_{i\sigma} + \sum_{im\sigma} \left(V_{im} b_{i\sigma}^+ c_{m\sigma} \right.$$
$$+ V_{im}^* c_{m\sigma}^+ b_{i\sigma} \left.\right) + J_{pd} \sum_{\langle im \rangle} \tilde{B}_{im}^+ \tilde{B}_{in}$$
$$+ J_{dd} \sum_{\langle ijk \rangle} \tilde{b}_{ij}^+ \tilde{b}_{kj} + J_{pp} \sum_{\langle mm' \rangle} p_{mm'}^+ p_{mm'}. \tag{70}$$

In deriving this result, one does not assume that $|V| \ll \epsilon_f$; therefore Eq. (70) is applicable to the situation

with fluctuating valence. Next, by introducing a slave-boson representation [Eq. (56)], we obtain the most general Hamiltonian for treatment of pairing in correlated systems [79]. We should be able to witness a decisive progress in the near-future concerning the relative role of hybrid p–d, d–d, and p–d pairings in high-T_c systems using the slave-boson or gauge-field approaches to Eq. (70). Also, Eq. (70) should serve as a basis for the discussion of antiferromagnetism and superconductivity in heavy–fermion systems; in that situation, the role of itinerant 2p states is played by hybridized 5d–6s conduction bands, while the role of 3d electrons is played by the 4f electrons due to Ce or by 5f electrons in uranium compounds.

c. Coexistence of antiferromagnetism and superconductivity.

In the previous analysis, above we have treated antiferromagnetism (AF) and superconductivity (SC) separately. Detailed calculations [43, 62, 81], within the mean–field theory discussed, point to the possibility of the coexistence of AF and SC phases. It is possible to visualize this coexistence by considering a narrow-band model with the two-dimensional (almost-square) Fermi surface as shown in Fig. 37. Namely, the band energy of electrons located on the Fermi surface has the property $\epsilon_{\mathbf{k}+\mathbf{Q}} = -\epsilon_{\mathbf{k}}$, where $\mathbf{Q} \equiv (\pi/a, \pi/a) = 2\mathbf{k}_F$. This is the so-called nesting condition; any system with this property is unstable with respect to the formation of the spin density-wave (SDW) state with the wave vector $\mathbf{Q}$. One should note (cf. Fig. 37) that $\mathbf{Q}$ connects two single-particle states on the opposite sides of the Fermi surface since $-\mathbf{k}_F + \mathbf{Q} = \mathbf{k}_F$. Furthermore, both SDW and SC states couple electrons with the opposite spins. This is why two sublattice AF and SC states are compatible only for $n \approx 1$, i.e., for the half-filled band. There is no

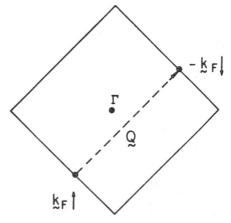

FIGURE 37 Two-dimensional Fermi surface for a half-filled band. The opposite points of the surface are related by the wave vector $\mathbf{Q} = (\pi/a)(1, 1)$.

clear experimental evidence that these two phases coexist in a high-T_c system, though there is some evidence from muon–spin rotation that it is so [82]. A clear detection of such coexistence would demonstrate directly the importance of exchange interactions in a superconducting phase. Namely, within exchange-mediated superconductivity, one can show [55] that close to the half-filled narrow-band case, $T_N/T_c \sim 6 \div 8$ (this is a mean-field-approximation estimate). The analysis of AF–SC coexistence conditions within the Anderson lattice Hamiltonian has not yet been performed satisfactorily, even though those two phases coexist in heavy-fermion compounds such as UPt_3 and URu_2Si_2.

B. Phonons and Bipolarons

After the discoveries of superconductivity in the 40 and 90 K ranges [83], the obvious question was posed whether the phonon–mediated mechanism of pairing, so successful in the past, can explain the superconductivity with such a high value of T_c. It was realized from the outset that one should include specific properties of these compounds, such as the quasi-planar (CuO_2) structure with a logarithmic (Van Hove) singularity in the density of states $\rho(\epsilon)$ at the middle point of the two–dimensional band [83–85], the polar nature of the CuO bonds rendering applicable the tight–binding representation of the electronic states [49, 84], and strong electron lattice coupling [85–87], leading to the local formation of small bipolarons (that is, two–electron pairs) [89] that may undergo Bose condensation when the metallic state is reached [in a more refined version a mixed-fermion model is used (cf. Ranniger, Ref. 90)].

There is no clear evidence for the phonon–mediated mechanism of pairing in classic high-T_c superconductors since the isotope effect in both La systems [90] and Y systems [91] is quite small. However, the recently discovered superconductors $Ba_{1-x}K_xBiO_3$ [92] exhibit a large isotope effect [93] and superconductivity with $20\ K \leq T_c \lesssim 30\ K$ in the concentration range $0.25 \lesssim x \lesssim 0.4$. Also, the proximity of superconductivity and the charge density—wave (CDW) state is observed [94].

The last property, as well as the observed diamagnetism in the insulating phase $x \lesssim 0.25$, is very suggestive [95] that small trapped polarons are formed before the electron subsystem condenses into a superconducting phase. Condensation takes place when the percolation threshold for the insulator–metal transition is reached[3] (at $x \sim 0.2$).

---

[3]The actual percolation threshold for the onset of the metallic phase is $x_c/2 \sim 0.12$ since the bipolarons reside on every alternate Bi lattice site. Also, the holes introduced by K doping must be present in a Bi–O hybridized band for $x > x_c$ to render the bipolarons mobile for $x > x_c$.

Three specific features of $Ba_{1-x}K_xBiO_3$ compounds should be noted. First, the diamagnetic nature of the parent compound $BaBiO_3$ distinguishes the systems from the parent compounds La_2CuO_4 and $YBa_2Cu_3O_7$, which are both antiferromagnetic. Second, the $Ba_{1-x}K_xBiO_3$ systems are copper-free and have a truly three–dimensional cubic structure in the SC phase [92]. Third, their main superconducting properties are in accordance with the prediction of the standard BCS theory [96].

The theory of the $Ba_{1-x}K_xBiO_3$ compound must incorporate three additional obvious facts. First, the pairing process $2Bi^{4+} \rightarrow Bi^{3+} + Bi^{5+}$ is possible when the electron–lattice coupling leads to an attraction overcoming the e–e repulsion in the Bi^{3+} state relative to the Bi^{5+} state [89]. It involves a relaxation of the O^{2-} octahedra, that is, an optical, almost dispersionless, breathing mode. This can provide a local (on-site) attractive interaction between $6s$ electrons of the type $\lambda n_{i\uparrow}n_{n\downarrow}$, which leads to a scalar ($\mathbf{k}$–independent) pairing potential $V_{\mathbf{kk'}} = \lambda$, which, in turn, provides a justification for the observed properties reflecting an isotropic shape of the gap ($\Delta_{\mathbf{k}} \equiv \Delta$), as in the standard BCS theory (cf. Section II).

Second, from the fact that the parent compound $BaBiO_3$ is an insulator, we conclude that either the magnitude V of the coulomb repulsion between the electrons on nearest-neighboring Bi atoms exceeds the width W^* of the bipolaron band [96] or the small bipolarons are self-trapped in the potential created by interaction with nearest-neighboring oxygens. The onset of the metallic phase at concentrations near the percolation threshold $x_c \sim 0.1$ for n.n. interaction means that both effects may be important. In either case, the CDW state will set in, so the entropy of the bipolaron lattice vanishes at $T = 0$ (at least, for $x = 0$). The CDW phase plays the same role here as does AF ordering in La_2CuO_4 and $YBa_2Cu_3O_6$. The properties of $Ba_{1-x}K_xBiO_3$ are instead similar to those of the $Ba_{1-x}Pb_xBiO_3$ compounds discovered over a decade earlier [98].

Third, the fact that the onset of the superconductivity coincides with the transition from the CDW insulator to an SC metal speaks in favor of preexisting electron pairs already present in the insulating phase. However, the bipolaron concentration is large, and hence, the interpretation of the superconducting transition as Bose condensation of bipolarons may be inapplicable even when the coherence length is small. The overall theoretical situation is nonetheless much clearer for $Ba_{1-x}K_xBiO_3$ compounds than for either the $La_{2-x}Sr_xCuO_4$ or the $YBa_2Cu_3O_{7-\delta}$ series since the accumulated (so far) experimental evidence indicates that (optical?) phonon–mediated pairing takes place [99].

The $Ba_{1-x}K_xBiO_3$ compounds seem to be natural candidates for a bipolaronic mechanism of electron pairing

[100]. This is because the diamagnetic (and charge–ordered) parent system $Ba_2Bi^{3+}Bi^{5+}O_3$ can be regarded as an ordered lattice of locally bound two–electron pairs (bipolarons) located on alternate Bi^{3+} sites; these pair states are stabilized by a strong relaxation of the surrounding oxygen anions. Effectively, the two electrons are attracted to each other. The effect of potassium doping is to make these pairs mobile by diminishing the number of bipolarons per Bi site from the value $(1/2)$ [101]. In essence, the lattice distortion is responsible for the bipolaron formation in the same manner as in the case of the copper pairs; the difference is due to the circumstance that the bipolarons are locally bound complexes in a direct space that undergo a Bose condensation from an incoherent state of preexisting and moving pairs. The temperature of such condensation is $T_c \sim x^{\frac{2}{3}}$, where x is the dopant (K) concentration [102]. A key feature of the bipolaron theory of superconductivity is that the Bose-condensed state develops from the CDW insulating state, not from the SDW (antiferromagnetic) state; the latter situation takes place for the cuprates. Further studies are necessary to calculate the physical properties of a bipolaron superconductor and, in particular, the differences from an ordinary (phonon–mediated) superconductor.

C. Charge Excitations

In 1964 Little [103] introduced the idea that virtual electron–hole (exciton) excitations may lead to a pairing with a high value of T_c. This idea has been reformulated recently in the context of high-T_c superconductivity by considering the role of charge transfer ($P \rightarrow d$ and $d \rightarrow p$) fluctuations [104], as well as of intraatomic (Cu $d \rightarrow d$) excitations [105]. The charge–transfer fluctuations involve both Cu^{2+}–O^- and Cu^{3+}–O^{2-} low–energy configurations and Cu^+O^- states. The former two configurations are particularly important if the energy difference $|\epsilon_p - \epsilon_d|$ is comparable to the magnitude $|V|$ of the $2p$–$3d$ hybridization. This is the limit we have considered within the hybridized model in Section IV.A, probably extended to include the $3d$–$2p$ coulomb interaction directly. The method of approach is therefore similar to that in Section IV.A in the limit of strongly correlated electrons. In the limit of weakly interacting electrons (that is, for $U \ll W$), the perturbation expansion in the powers of U provides an effective pairing potential in an explicit form. The processes leading then to the pairing are virtual exciations involving charge and antiferromagnetic spin fluctuations [106]. At the moment, it is difficult to see clearly the difference between exchange-mediated and charge transfer-mediated types of pairing for strongly correlated hybridized systems.

V. CONCLUSIONS

In this article we have concentrated mainly on reviewing the properties of correlated electrons in normal, antiferromagnetic, and superconducting phases, in copper-containing systems in which the last two are phases caused by antiferromagnetic exchange interactions. Two theoretical models have been discussed in detail: the Hubbard model of correlated narrow-band ($3d$) electrons and the Anderson lattice model of correlated and hybridized electrons, involving $2p$ and $3d$ states in the case of high-T_c oxides. The latter model is regarded as more general and applicable to both high-T_c and heavy–fermion systems; in some limiting situations discussed previously, hybridized bands exhibit a narrow–band behavior.

The principal novel feature of the metallic phase involving either $3d$ (in high-T_c oxides) or $4f$ (in heavy–fermion systems) electrons is that for the half–filled band configuration the itinerant electron states transform into a set of localized states constituting the Mott insulator. The difference between the Fermi liquid (FL) and the liquid of correlated electrons [the statistical spin liquid (SL)] is illustrated in Fig. 38, where the high–temperature value of the entropy has been plotted for these two phases as a function of the number n of electrons per atom [the statistical distribution Eq. (41d) was used to calculate the entropy $S(n)$ for the latter phase]. Only the spin–liquid case correctly reproduces the entropy of localized moments when the Mott insulator limit is reached for $n \rightarrow 1$. This limiting value of the entropy per mole, $S = R \ln 2$ for $n \rightarrow 1$, represents one of the necessary conditions to be fulfilled by any theory claiming to describe properly the situation near the Mott insulator limit. Additionally, those systems are characterized by pseudo-particles with a very heavy

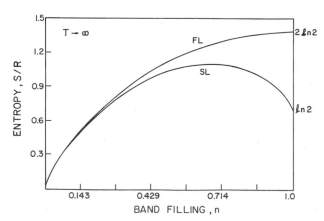

FIGURE 38 The high-temperature limiting value for the entropy (in units of the gas constant R) as a function of n for the Fermi liquid (FL) and the spin liquid (SL). Note the difference in the values of a factor of two in the limit $n \rightarrow 1$.

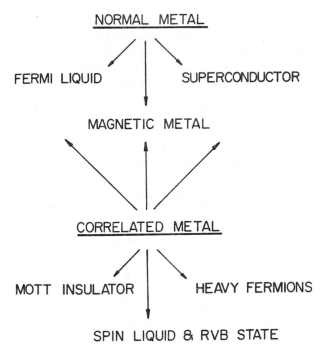

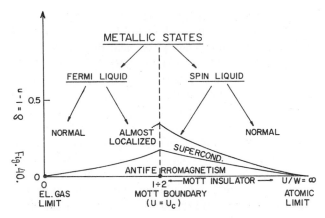

FIGURE 40 Qualitative distinction between the Fermi-liquid and the spin-liquid states. The Mott boundary $U = U_c$ roughly separates the two limiting phases.

FIGURE 39 Schematic representation of the difference between a normal metal and a correlated metal. Only the latter state may lead to Mott localization, as well as to heavy-fermion and spin-liquid metallic phases.

effective mass $m^* \sim \delta^{-1}$ or $W^* \sim \delta$. For $\delta \ll 1$, the band energy becomes comparable to the kinetic exchange characterized by $J = W^2/(Uz)$. Itinerant systems for which $J \gtrsim W^*$ are called quantum spin–liquid systems. The Mott insulator, the spin–liquid, and the heavy–fermion states are the primary phases of correlated electrons different from the normal–metal state. This difference is sketched out in Fig. 39, where the arrows point both to common features for normal and correlated metals and to those specific to the correlated systems.

The correlated systems that interested us here may also be called almost-localized systems. As discussed in Section II, there are two classes of such systems, separated roughly by the Mott–Hubbard boundary $U = U_c \sim W$: those for which the coulomb interaction $U < U_c$ are regarded as Fermi liquids have been treated extensively in Refs. 29 and 30, while those systems for which $U > U_c$ are the spin liquids. This qualitative division is sketched in Fig. 40, where the various thermodynamic phases have been specified for each class (cf. also Fig. 14 for all magnetic phases). The complementary regimes are those with $U/W \ll 1$ and $U/W \gg 1$. Most of the metallic systems can be located between these two limiting situations. It remains to be proven more precisely that the Mott–Hubbard boundary separating, for $n = 1$, Fermi liquid from the Mott insulator extends to the part of the diagram with $n \neq 1$,

where Fermi liquid transforms with increasing interaction into non-Fermi liquid. This is a fundamental problem, related, in the case of strongly interacting systems, to the question of the validity of the Luttinger theorem[4] and to the problem of the existence of local magnetic moments in the itinerant–electron picture, that is, to the problem of the validity of the Bloch theorem for a correlated metal. Also, the question of applicability of the Fermi–liquid concept in the limit $U/W \gg 1$ is connected with that concerning the properly defined existence of fermion quasiparticles,[5] interacting only weakly among themselves. One should emphasize that the discussion of the standard mean-field treatment of superconductivity presented in Section IV reduces the whole problem to the single-particle approach with a self-consistent field $\sim \Delta_{\mathbf{k}}$. It is not yet completely clear what types of collective excitations (antiferromagnetic spin fluctuations? stripes?) are needed to make the theory complete. The introduction of holons as bosons and spinons as fermions [35] seems to be just one possibility; more natural seems to be a treatment of holons as spinless fermions and of spinons as boson operators that reflect magnonlike properties of local moments.

Early studies of high-T_c oxides revealed that some of their characteristics are close to those provided by the BCS theory. Namely, the value of $2\Delta_0/k_bT_c \simeq 4 \div 6$ is

---

[4]The Luttinger theorem states that, as long as the metallic state is stable, the volume encircled by the Fermi surface remains independent of the strength of the electron–electron interaction. This theorem is not valid when the Mott transition takes place, as the Fermi surface then disappears. The volume also doubles when metal is described by a statistical spin liquid discussed in Section II (cf. Figs. 16a and b).

[5]The holons and spinons cannot be regarded as quasiparticles, since the Green function describing them has branching cuts rather than poles.

indicated [107], the temperature dependence of the London penetration depth is close to $[1 - (T/T_c)^4]^{-\frac{1}{2}}$ over a wide temperature range [108], and the electron pairing is in the spin-singlet state [109]. Additionally, the shape of the Fermi surface for $YBa_2Cu_3O_7$, as determined by the positron annihilation technique [110], agrees with the predictions of the band-structure calculations for an even ($n \approx 4$) number of electrons. These results do not necessarily eliminate the principal features obtained from the theory of strong electron correlations. We think that before discarding the theory based on electron correlations, we must show clearly that the stoichiometric La_2CuO_4 or $YBa_2Cu_3O_6$ compounds are not insulating in the paramagnetic phase; actually, they seem to be paramagnetic insulators with well–defined magnetic moments (that is, with the Curie–Weiss law for the magnetic susceptibility obeyed), which supports strongly the view that they are Mott insulators. In this respect, the situation in heavy–fermion systems is rather clear since the recent theoretical results [39–42] based on the theory of strongly correlated and hybridized states provide a reasonable rationalization of most of the properties of their normal state. The mechanism of pairing in superconducting heavy–fermion systems has not yet been determined fully; but in view of the circumstances that some of the superconductors (for example, UPt_3) are antiferromagnetic and exhibit pronounced spin fluctuations in the normal state, the spin–fluctuation mechanism in the version outlined in Section IV.A.2 is a strong candidate [111]. In the coming years one should be able to see a clarification of these problem.

Let us end with a methodological remark concerning the analogy of the studies of magnetism and superconductivity. In 1928, Heisenberg introduced the exchange interaction $J_{ij}\mathbf{S}_i \cdot \mathbf{S}_j$ between the magnetic atoms with spins $\{\mathbf{S}_i\}$. The ferromagnetic state was understood in terms of a molecular field $\mathbf{H}_i \sim \langle \mathbf{S}_i \rangle$ which was related to the direct exchange integral J_{ij}. Later, various other exchange interactions have been introduced, such as superexchange, double exchange, RKKY interaction, the Bloembergen–Rowland interaction, Hund's rule exchange, and kinetic exchange, to explain magnetism in specific systems, such as oxides, rare-earth metals, and transition metals. However, all these new theories provided a description in terms of a single–order parameter—the magnetization $\langle \mathbf{S}_i \rangle$; the particular feature of the electron states in each case (localized states, itinerant states, or a mixture of the two states) is contained only in the way of defining this order parameter or the exchange integral. By analogy, the BCS theory provided a concept of a superconducting order parameter ($\Delta_{\mathbf{k}}$), which is universal for all theories of singlet superconductivity. New mechanisms of pairing should provide a novel interpretation to the coupling constant $V_{\mathbf{kk}'}$ as well

as supplying some details concerning the specific features of the system under consideration: the gap anisotropy, the role of hybridization, etc. It remains to be seen if some qualitative differences arise if superconductivity should occur as a result of Bose condensation of the preexisting pairs. This question is particularly important in the case when the coherence length is small, as in high-T_c systems.

In the coming years, one should see detailed calculations within the exchange mechanism and comparisons with the experiment concerning the complete phase diagram, as well as the thermodynamic and electromagnetic properties of the new superconductors $La_{2-x}Sr_xCuO_4$ and $YBa_2Cu_3O_{7-\delta}$. It would not be surprising if the final answer for these systems came from a detailed analysis of the model outlined in Section IV.A. The systems $Ba_{1-x}K_xBiO_3$ will probably be described satisfactorily within the standard phonon–mediated mechanism. On the other hand, it is too early to say anything definite about Bi and $T\ell$ compounds with $T_c > 100$ K, though the suggested influence of the electronic structure near ϵ_F by the CuO_2 planes seems to indicate a nontrivial role of the exchange interactions also in those systems when coupled with interlayer pair tunneling. One of the missing links between the properties of the last two classes of compounds and those of $La_{2-x}Sr_xCuO_4$ is the conspicuous lack of evidence for antiferromagnetism in the $Bi_2Sr_2CaCu_2O_8$ and the $Tl_2Ca_2Ba_2Cu_3O_{10-y}$ compounds.

Note added in August 2000. This article was originally written almost 12 years ago. During those years a tremendous number of papers has been published, but the questions concerning either the pairing mechanism or the non-Fermi liquid behavior have not been clearly resolved, either for high-T_C or for heavy-fermion superconductors. Nonetheless, a number of experimental results have been obtained in a clear form for high-T_C systems. Let us mention two additional results. First, the role of the hopping between the next neighbors is important for obtaining the open Fermi surface (cf. Fig. 41). This Fermi surface is obtained from the photoemission and encompasses all electrons [112], i.e., not only the hole states in the doped Mott insulator. Thus, the principal question is how to reconcile the strong-correlation nature of the electrons in the CuO_2 plane, as reviewed above with the Luttinger theorem, which seems to be obeyed in optimally doped and overdoped systems, as concluded from the photoemission data. Does this mean that the photoemission experiment samples states physically different from those involved in thermally induced transport properties? The Fermi-liquid features seem clearly to break down in underdoped systems [113], where a pseudogap related to the superconducting gap is also observed [114].

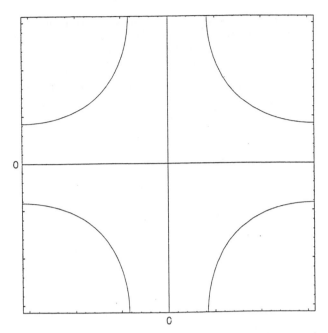

FIGURE 41 Schematical representation of a two-dimensonal Fermi surface for a nonzero amplitude of second-nearest-neighbor hopping.

ACKNOWLEDGMENTS

I would like to thank Leszek Spalek for technical help. This work was supported by KBN (Poland) Project No. 2PO3B O92 18.

SEE ALSO THE FOLLOWING ARTICLES

ELECTRONS IN SOLIDS • SUPERCONDUCTING CABLES • SUPERCONDUCTING DEVICES • SUPERCONDUCTIVITY • SUPERCONDUCTORS, HIGH TEMPERATURE

BIBLIOGRAPHY

1. Bardeen, J., Cooper, L. N., and Schrieffer, J. R. (1957). *Phys. Rev.* **106**, 162. Bardeen, J., Cooper, L. N., and Schrieffer, J. R. (1957). *Phys. Rev.* **108**, 1175.
2. Schrieffer, J. R. (1964). "Theory of Superconductivity," W. A. Benjamin, Reading, PA.
3. De Gennes, P. G. (1966). "Superconductivity of Metals and Alloys," W. A. Benjamin, Reading, PA.
4. Tinkham, M. (1996). "Introduction to Superconductivity," 2nd ed., McGraw–Hill, New York.
5. Rickayzen, G. (1965). "Theory of Superconductivity," Wiley, New York.
6. Blatt, J. M. (1964). "Theory of Superconductivity," Academic Press, New York.
7. Barone, A., and Paterno, G. (1982). "Physics and Applications of the Josephson Effect," John Wiley, New York.
8. Landau, L. D., Lifshitz, E. M. (1950). "Statistical Physics," 2nd ed., Part 2, Chap. 5, Pergamon, Oxford, Abrikosov, A. A., Gorkov, L. P., and Dzyaloshinski, I. E. (1963). "Methods of Quantum Field Theory in Statistical Physics," Chap. 7, Dover, New York.
9. Parks, R. D. (ed.) (1969). "Superconductivity" (2-vols.), Dekker, New York.
10. Kuper, C. G. (1968). "An Introduction to the Theory of Superconductivity," Clarendon Press, Oxford, New York. Rose—Innes, A. C., and Rhoderick, E. H. (1969). "Introduction to Superconductivity," Pergamon Press, Oxford.
11. Fröhlich, H. (1952). *Proc. Roy. Soc. A* **215**, 291.
12. Cooper, L. N. (1956). *Phys. Rev.* **104**, 1189.
13. Bogoliuboy, N. N. (1958). *Nuovo Cimento* **7**, 794. Valatin, J. G. (1958). *Nuovo Cimento* **7**, 843. Nambu, Y. (1960). *Phys. Rev.* **117**, 648.
14. Morel, P., and Anderson, P. W. (1962). *Phys. Rev.* **125**, 1263. See also Scalapino, D. J., in Ref. 9, Chapt. 10.
15. Eliashberg, G. M. (1966). *Zh. Eksp. Teor. Fiz.* **38**, 966 [*Sov. Phys. JETP* **11**, 696 (1960)]. For review see Allen, P. B., and Mitrovic, B. (1982). *In* "Solid State Physics" (H. Ehrenreich, F. Seitz, and D. Turnbull, eds.), pp. 2–92, Academic Press, New York.
16. Migdal, A. B. (1958). *Zh, Eksp. Teor. Fiz.* **34**, 1438 [*Sov. Phys. JETP* **7**, 996(1958)].
17. McMillan, W. L. (1968). *Phys. Rev.* **167**, 331.
18. Khan, F. S., and Allen, P. B. (1980). *Solid State Commun.* **36**, 481.
19. Cohen, M. L., and Anderson, P. W. (1972). *AIP Conf. Proc. No. 4* (D. H. Douglass, ed.), p. 17, (AIP, New York).
20. Maxwell, E. (1950). *Phys. Rev.* **78**, 477. Reynolds, C. A., Serin, B., Wright, W. H., and Nesbitt, L. B. (1950). *Phys. Rev.* **78**, 487.
21. Some of the monographs and reviews which appeared during the last decade are as follows. (a) Ginsberg, D. H. (ed.) (1989–1995). "Physical Properties of High Temperature Superconductors," Vols. 1–5, World Scientific, Singapore. (b) Battlogg, B., *et al.*, (1996). "Proceedings of the 10th Anniversary Workshop on Physics, Materials and Applications, World Scientific, Singapore. (c) Tsuneto, T. (1998). "Superconductivity and Superfluidity," Cambridge University Press, Cambridge. (d) Cyrot, M., and Pavuna, D. (1992). "Introduction to Superconductivity and High-T_C Materials," World Scientific, Singapore. (e) Carbotte, J. P. (1990). *Rev. Mod. Phys.* **62**, p. 1027ff. (f) Waldram, J. R. (1996). "Superconductivity of Metals and Cuprates," IOP, Bristol, Philadelphia. (g) Narlikar, A. (ed.) (1990–). "Studies in High Temperature Superconductors," Vols. 1–16, Nova, Science, New York, Budapest. (h) Anderson, P. W. (1997). "The Theory of Superconductivity in High-T_C Cuprates," Princeton University Press, Princeton, NJ. (i) Bisarsh, A. (ed.) (1999). "Superconductivity: An Annotated Bibiliography with Abstracts," Nova, Science, New York. (j) Poole, C. P. (1999). "Handbook of Superconductivity," Academic Press, San Diego, CA. (k) Plakida, N. M. (1995). "High Temperature Superconductivity," Springer Verlag, Berlin.
22. For review see, e.g., Baym, G., and Pethick, C. (1978). *In* "The Physics of Liquid and Solid Helium" (K. H. Bennemann and J. B. Ketterson, eds.), Chap. 3, Wiley, New York; Pines, D., and Nozieres, P. (1996). "The Theory of Quantum Liquids," W. A. Benjamin, New York.
23. Mott, N. F. (1974). "The Metal–Insulator Transitions," Taylor and Francis, London. Héritier, M., and Lederer, P. (1977). *J. Phys. (Paris)* **38**, L209.
24. Vaknin, D., *et al.* (1987). *Phys. Rev. Lett.* **58**, 2802. Endoh, Y., *et al.* (1987). *Phys. Rev. B* **37**, 7443.
25. Hubbard, J. (1964). *Proc. R. Soc. London A* **281**, 401.
26. Acquarone, M., Ray, D. K., and Spałek, J. (1982). *J. Phys. C* **15**, 959.
27. Kuwamoto, H., Honig, J. M., and Appel, J. (1980). *Phys. Rev. B* **22**, 2626.

28. Spałek, J., Oleś, A. M., and Honig, J. M. (1980). *Phys. Rev. B* **28,** 6802.

29. Brinkman, W. F., and Rice, T. M. (1970). *Phys. Rev. B* **2,** 4302.

30. Spałek, J., Datta, A., and Honig, J. M. (1987). *Phys, Rev. Lett.* **59,** 728. Spałek, J., Kokowski, M., and Honig, J. M. (1989). *Phys. Rev. B* **39,** 4175. For $d = \infty$ solution see, e.g., Gebhard, F. (1997). "The Mott Metal–Insulator Transition," Springer Verlag, Berlin.

31. For the Mott insulators, the kinetic exchange interaction was introduced by in Anderson, P. W. (1959). *Phys. Rev.* **115,** 2. This result was extended to the case of a strongly correlated metal by Chao, K. A., Spałek, J., and Oleś, A. M. (1977). *J. Phys. C* **10,** L271. Cf. also Spałek, J., and Oleś, A. M. (1976). Jagiellonian University preprint SSPJU–6/76, Oct. Zaanen, J., and Sawatzky, G. A. (1990). *J. Solid State Chem.* **88,** 8ff.

32. Vischer, P. B. (1974). *Phys. Rev.* **10,** 943. Spałek, J., Oleś, A. M., and Chao, K. A. (1981). *Phys. Stat. Sol. (b)* **108,** 329. Nagaoka, Y. (1966). *Phys. Rev.* **147,** 392.

33. (a) Varma, C. M., *et al.* (1989). *Phys. Rev. Lett.* **63,** 1996; (1989). *Int. J. Mod. Phys. B* **3,** 2083–2118; Littlewood, P. B., and Varma, C. M. (1991). *J. Appl. Phys.* **69,** 4979. (b) For review see Pines, D. (1998). *In* "The Gap Symmetry and Fluctuations in High-T_C Superconductors" (J. Bok *et al.*, eds.), Plenum Press, New York, NATO ASI Series, Ser. B, Vol. 371, and references therein. (c) Moriya, T., and Ueda, K. (2000). *Adv. Phys.* **49,** 555–606, and references therein. (d) Lee, P. A., and Nagaosa, N. (1992). *Phys. Rev. B* **46,** 5621; for review see Nagaosa, N. (1999). "Quantum Field Theory in Strongly Correlated Electronic Systems," Chap. 5, Springer Verlag, Berlin; Lee, P. A. (1996). *J. Low Temp. Phys.* **105,** 581. (e) van den Brink, J., *et al.* (1995). *Phys. Rev. Lett.* **75,** 4658; (1996). *Phys. Rev. Lett.* **76,** 2826. (f) Bała, J., *et al.* (1995). *Phys. Rev. B* **52,** 4597. (g) Dagotto, E. (1994). *Rev. Mod. Phys.* **66,** 763, and references therein; Dagotto, E., and Rice, T. M. (1996). *Science* **271,** 618; Dagotto, E. (1998). *J. Phys. Chem. Solids* **59,** 1699. (h) Zaanen, J. (1998). *J. Phys. Chem. Solids* **59,** 1769; Castellani, C., *et al.*, *ibid.*, p. 1694; Kivelson, S. A., *ibid.* p. 1705, and references therein.

34. (a) Takagi, H., *et al.* (1992). *Phys. Rev. Lett.* **69,** 2975. (b) Rossat-Mignot, J., *et al.* (1991). *Physica C* **185–189,** 86; Bourges, P., *et al.* (1997). *Phys. Rev. B* **56**(11) 439. (c) Anderson, P. W. (1997). "The Theory of Superconductivity in the High-T_C Cuprates," Princeton University Press, Princeton, NJ; Byczuk, K., Spałek, J., and Wójcik, W. (1998). *Acta Phys. Polonica B* **29,** 3871; Varma, C. M. (1997). *Phys. Rev. B* **55**(14), 554.

35. Spałek, J., and Wójcik, W. (1988). *Phys. Rev. B* **37,** 1532.

36. Kivelson, S. A., Rokhsar, D. S., and Sethna, J. P. (1987). *Phys. Rev. B* **35,** 8865. Anderson, P. W., and Zou, Z. (1988). *Phys. Rev. Lett.* **60,** 132.

37. Anderson, P. W. (1988). Cargese 1988, Lecture Notes. Also (1987). *In* "Proceedings of the International School 'Enrico Fermi'—1987: Frontiers and Borderlines in Many—Particle Physics," North-Holland, Amsterdam.

38. For a recent review, see Fukuyama, H., Hasegawa, Y., and Suzumura, Y. (1988). *Physica C* **153–155.**

39. Anderson, P. W. (1988). *Physica C* **153–155.**

40. Lee, P. A., Rice, T. M., Serene, J. W., Sham, L. J., and Wilkins, J. W. (1986). *Comments Condens. Matter. Phys.* **12,** 99.

41. Rice, T. M. (1987). "*In* Proceedings of the International School of Physics 'Enrico Fermi'—1987: Frontiers and Borderlines in Many—Particle Physics," North-Holland, Amsterdam.

42. Newns, D. M., and Read, N. (1987). *Adv. Phys.* **36,** 799.

43. Fulde, P., Keller, J., and Zwicknagl, G. (1988). *In* "Solid State Physics" (H. Ehrenreich and D. Turnbull, eds.), Vol. 41, Academic Press, San Diego, CA.

44. For a review of experimental properties of heavy fermions, see Stewart, G. R. (1984). *Rev. Mod. Phys.* **56,** 755. Fisk, Z., *et al.* (1986). *Nature (London)* **320,** 124. Steglich, F. (1955). *In* "Theory of Heavy Fermions and Valence Fluctuations" (T. Kasuya and T. Saso, eds.), p. 23ff, Springer-Verlag, Berlin, New York.

45. Wróbel, P., and Jacak, L. (1988). *Mod. Phys. Lett. B* **2,** 511.

46. Anderson, P. W. (1987). *Science* **235,** 1196.

47. Shirane, G., *et al.* (1988). *Phys. Rev. Lett.* **59,** 1613.

48. Chakravarty, S., *et al.* (1988). *Phys. Rev. Lett.* **60,** 1057.

49. Nucker, M., *et al.* (1987). *Z. Phys. B* **67,** 9. Fujimori, A., *et al.* (1987). *Phys. Rev. B* **35,** 8814. Steiner, P., *et al.* (1988). *Z. Phys. B* **69,** 449.

50. Mattheis, L. F. (1987). *Phys. Rev. Lett.* **58,** 1028. Yu, J., Freeman, A. J., and Xu, J.-H. (1987). *Phys. Rev. Lett.* **58,** 1035. Szpunar, B., and Smith, V. H., Jr. (1988). *Phys. Rev. B* **37,** 2338. For review see Hass, K. C. (1989). *In* "Solid State Physics" (H. Ehrenreich and D. Turnbull, eds.), Vol. 42, Academic Press, San Diego, CA.

51. Fulde, P. (1988). *Physica* **153–155,** 1769. Hybertsen, M. S., *et al.* (1994). *Phys. Rev. B* **41,** 11068.

52. Zhang, F. C., and Rice, T. M. (1988). *Phys. Rev. B* **37,** 3759.

53. Schafroth, M. R. (1955). *Phys. Rev.* **100,** 463.

54. For a review see Leggett, A. (1975). *Rev. Mod. Phys.*

55. Shafer, M. W., Penney, T., and Olson, B. L. (1987). *Phys. Rev. B* **36,** 4047.

56. Spałek, J, (1988). *Phys. Rev. B* **37,** 533. Acquarone, M. (1988). *Solid State Commun.* **66,** 937.

57. Baskaran, G., Zou, Z., and Anderson, P. W. (1987). *Solid State Commun.* **63,** 973.

58. Cyrot, M. (1987). *Solid State Commun.* **62,** 821.

59. Ruckenstein, A. E., Hirschfeld, P. J., and Appel, J. (1987). *Phys. Rev. B* **36,** 857.

60. Kotliar, G. (1988). *Phys. Rev. B* **37,** 3664.

61. Isawa, Y., Maekawa, S., and Ebisawa, H. (1987). *Physica* **148B,** 391.

62. Zou, Z., Anderson, P. W. (1988). *Phys. Rev. B* **37,** 627.

63. Inui, M., Doniach, S., Hirschfeld, P. J., and Ruckenstein, A. E. (1988). *Phys. Rev. B* **37,** 2320.

64. Suzumura, Y., Hasegawa, Y., and Fukuyama, H. (1988). *J. Phys. Soc. Jpn.* **57,** 2768.

65. Kotliar, G., and Liu, J. (1988). *Phys. Rev. B* **38,** 5142.

66. Nagaosa, N. (1966). *In* "Proceedings of the 10th Anniversary HTS Workshop on Physics, Materials, and Applications" (B. Batlogg *et al.*, eds.), pp. 505ff, World Scientific, Singapore, and references therein.

67. (a) Labbé, J., and Bok, J. (1987). *Europhys. Lett.* **3,** 1225; Newns, D. M., *et al.* (1992). *Comments Cond. Mat. Phys.* **15,** 273. (b) Bouvier, J., and Bok, J. (1998). *In* "The Gap Symmetry and Fluctuations in High-T_C Superconductors" (J. Bok *et al.*, ed.), pp. 37–54, Plenum Press, New York. (c) Van Harlingen, D. J. (1995). *Rev. Mod. Phys.* **67,** 515. (d) Annett, J. F., Goldenfeld, N., and Legett, A. J. (1996). *In* "Physical Properties in High-Temperature Superconductors," Vol. 5 (D. M. Ginsberg, ed.), pp. 375–461, World Scientific, Singapore. (e) Chakravarty, S., *et al.* (1993). *Science* **261,** 337. (f) Byczuk, K., and Spałek, J. (1996). *Phys. Rev. B* **53,** R518. (g) Leggett, A. J. (1998). *J. Phys. Chem. Solids* **59,** 1729; Tsvetkow, A. A., *et al.* (1998). *Nature* **395,** 360. (h) Lawrence, W. E., and Doniach, S. (1971). *In* "Proc. 12th Int. Conf. Low Temp. Phys." (E. Kanda, ed.), Keigaku, Tokyo; see also Tinkham, M., in Ref. 4.

68. Mila, F. (1988). *Phys. Rev. B* **38,** 11358, and references therein.

69. Spałek, J. (1988). *Phys. Rev. B* **38,** 208. Spałek, J. (1988). *J. Solid State Chem.* **76,** 224.

70. Schrieffer, J. R., and Wolff, P. A. (1966). *Phys. Rev.* **149,** 491.

71. Anderson, P. W. (1984). *Phys. Rev. B* **30,** 1549. Miyake, K., Schmitt–Rink, S., and Varma, C. M. (1986). *Phys. Rev. B* **34,** 6554. Scalapino, D. J., Loh, E., and Hirsch, J. E., (1986). *Phys. Rev. B* **34,** 8190. Norman, M. R. (1987). *Phys. Rev. Lett.* **59,** 232. Norman, M. R. (1988). *Phys. Rev. B* **37,** 4987.

72. Newns, D. M. (1987). *Phys. Rev. B* **36,** 5595. Newns, D. M. (1988). *Phys. Scripta T* **23,** 113.

73. Torrance, J. B., *et al.* (1988). *Phys. Rev. Lett.* **61,** 1127.

74. Rice, T. M., and Ueda, K. (1985). *Phys. Rev. Lett.* **55,** 995. Rice, T. M., and Ueda, K. (1986). *Phys. Rev. B* **34,** 6420.

75. Miyake, K., Matsuura, T., Sano, K., and Nagaoka, Y. (1988). *J. Phys. Soc. Jpn.* **57,** 722.

76. Anderson, P. W. (1959). *Phys. Rev.* **115,** 2. Also (1963). *In* "Solid State Physics" (F. Seitz and D. Turnbull, eds.), Vol. 14, pp. 99–213, Academic Press, New York. Cf. also Zaanen, J., and Sawatzky, G. A. (1990). *J. Solid State Chem.* **88.**

77. For a review see, e.g., Vonsovskii, S. V. (1974). "Magnetism" Wiley, New York.

78. Zaanen, J., Sawatzky, G. A., and Allen, J. W. (1985). *Phys. Rev. Lett.* **55,** 418; (1986). *J. Magn. Magn. Mat.* **54–57,** 607.

79. (a) Aharony, A., *et al.* (1988). *Phys. Rev. Lett.* **60,** 1330. (b) Tranquada, J. M., *et al.* (1988). *Phys. Rev. Lett.* **60,** 156.

80. Spałek, J., and Honig, J. M. (1990). *In* "Studies of High-Temperature Superconductors" (A. Narlikar, ed.), Vol. 4, Nova Science, New York.

81. The p–p pairing was discussed first by Emery, V. J. (1987). *Phys. Rev. Lett.* **58,** 2794, and also by Emery, V. J., and Reiter, G. (1988). *Phys. Rev. B* **38,** 4547.

82. The coexistence of SDW and SC states has also been discussed within BCS theory: Baltensperger, W., and Strassler, S. (1963). *Phys. Kondens. Mater.* **1,** 20; Nass, M. J., *et al.* (1981). *Phys. Rev. Lett.* **46,** 614. Overhauser, A. W., and Daemen, L. (1988). *Phys. Rev. Lett.* **61,** 1885. The corresponding problem for exchange–mediated superconductivity has been outlined by Parmenter, R. H. (1987). *Phys. Rev. Lett.* **59,** 923.

83. Weidinger, A., *et al.* (1989). *Phys. Rev. Lett.* **62,** 102. Brewer, J. H., *et al.* (1988). *Phys. Rev. Lett.* **60,** 1073.

84. Bednorz, J. G., and Muller, K. A. (1986). *Z. Phys. B* **64,** 189. Chu, C. W., *et al.* (1987). *Phys. Rev. Lett.* **58,** 405. Uchida, S., *et al.* (1987). *Jpn. J. Appl. Phys.* **26,** L1. Wu, M. K., *et al.* (1987). *Phys. Rev. Lett.* **58,** 908.

85. abbé, J., and Bok, J. (1987). *Europhys. Lett.* **3,** 1225.

86. Prelovšek, P., Rice, T. M., and Zhang, F. C. (1987). *J. Phys. C* **20,** L229.

87. Jorgensen, J. D., *et al.* (1988). *Phys. Rev. Lett.* **58,** 1024.

88. Weber, W. (1987). *Phys. Rev. Lett.* **58,** 1371. Barisic, S., Batistic, J., and Friedel, J. (1987). *Europhys. Lett.* **3,** 1231.

89. For review of phonon- and bipolaron-mediated superconductivity see, e.g., Oguri, A. (1988). *J. Phys. Soc. Jpn.* **57,** 2133; de Jongh, L. J. (1988). *In* "Proc. 1st Int. Symp. Superconduct.," Nagoya, Springer—Verlag, New York.

90. Alexandrov, A., and Ranninger, J. (1981). *Phys. Rev. B* **24,** 1164. Alexandrov, A., Ranninger, J., and Robaszkiewicz, S. (1986). *Phys. Rev. B* **33,** 4526. For review see Micnas, R., Ranninger, J., and Robaszkiewicz, S. (1990). *Rev. Mod. Phys.* **62,** 113; Ranninger, J. (1998). *J. Phys. Chem. Solids* **59,** 1759, and references therein.

91. Batlogg, B., *et al.* (1987). *Phys. Rev. Lett.* **59,** 912. Faltens, T. A., *et al.* (1987). *Phys. Rev. Lett.* **59,** 915.

92. Leary, K. J., *et al.* (1987). *Phys. Rev. Lett.* **59,** 1236.

93. Mattheis, L. F., Gyorgy, E. M., and Johnson, D. W., Jr. (1988). *Phys. Rev. B* **37,** 3745. Cava, R. J., *et al.* (1988). *Nature* **332,** 814. Hinks, D. G., *et al.* (1988). *Nature* **333,** 6176.

94. Hinks, D. G., *et al.* (1988). *Nature* **335,** 419.

95. Pei, S., *et al.*, preprint.

96. Rice, T. M. (1988). *Nature* **332,** 780.

97. Dąbrowski, B., personal communication.

98. Varma, C. M. (1988). *Phys. Rev. Lett.* **61,** 2713.

99. Sleight, A. W., Gillson, J. J., and Bierstedt, P. E. (1975). *Solid State Commun.* **17,** 27.

100. For critical estimates of isotope shifts of the T_c value, see Allen, P. B. (1988). *Nature* **335,** 396.

101. Chakraverty, B. K. (1979). *J. Phys. Lett.* **40,** L99, Alexandroy, A. S., and Ranninger, J. (1981). *Phys. Rev. B* **23,** 1796. Alexandrov, A. S., Ranninger, J., and Robaszkiewicz, S. (1986). *Phys. Rev. B* **33,** 4526.

102. Rice, T. M. (1988). *Nature* **332,** 780.

103. Prelovšek, P., Rice, T. M., and Zhang, F. C. (1987). *J. Phys. C* **20,** L229.

104. Little, W. A. (1964). *Phys. Rev.* **134A,** 1416. Ginzburg, V. L. (1970). *Sov. Phys. Uspekhi* **13,** 335.

105. Varma, C. M., Schmitt-Rink, S., Abrahams, E. (1987). *In* "Proceedings of the Conference on Novel Mechanisms of Superconductivity" (S. A. Wolff and V. Z. Kresin, eds.), p. 355, Plenum Press, New York; (1987). *Solid State Commun.* **62,** 681.

106. Weber, W. (1988). *Z. Phys. B* **70,** 323.

107. Scalpino, D. J., Loh, E., Jr., and Hirsch, J. E. (1987). *Phys. Rev. B* **35,** 6694. Schrieffer, J. R., Wen, X. G., and Zhang, S. C. (1988). *Phys. Rev. Lett.* **60,** 944. White, S. R., and Scalpino, D. J. (1997), *Phys. Rev. B* **55,** 6504.

108. For a critical review see Little, W. A. (1988). *Science* **242,** 1390.

109. Fiory, A. T., Hebard, A. F., Mankiewich, P. M., and Howard, R. E. (1988). *Phys. Rev. Lett.* **61,** 1419.

110. Niemeyer, J., Dietrich, M. R., and Politis, C. (1987). *Z. Phys. B* **67,** 155.

111. Smedskjaer, L. C., Liu, J. Z., Benedek, R., Legnini, D. G., Lam, D. J., Stahulak, M. D., and Bansil, A. (1988). *Physica C* **156,** 269.

112. Cf. also Mathur, N. D., *et al.* (1998). *Nature* **394,** 39; Fisk, Z., and Pines, D. *ibid.*, p. 22.

113. For review see Shen, Z.-X., and Dessau, D. S. (1995). *Phys. Rep.* **253,** 1ff; Ding, H., *et al.* (1995). *Phys. Rev. Lett.* **74,** 2784; (1996). *Phys. Rev. B* **54,** R9878; Campuzano, J.-C., *et al.* (1998). *In* "The Gap Symmetry and Fluctuations in High-T_C Superconductors" (J. Bok *et al.*, eds.), Plenum Press, New York.

114. Fujimori, A., *et al.* (1998). *J. Phys. Chem. Solids* **59,** 1892, and references therein.

115. Ding, H., *et al.* (1998). *J. Phys. Chem. Solids* **59,** 1888, and refrences therein; Norman, M. R., *et al.*, *ibid.* p. 1902.

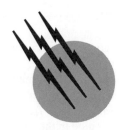

Superconductors, High Temperature

John B. Goodenough
University of Texas

I. Introduction
II. Superconductivity
III. Metallic Oxides
IV. High-T_c Superconductors

GLOSSARY

Bohr magneton μ_B Magnetic moment of an electron.

Brillouin zone Volume in **k** space containing two valence electrons per atomic valence orbital per atom of a primitive unit cell of crystal lattice.

Correlation energy Electrostatic electron-electron interactions not accounted for in Hartree-Fock one-electron band theory of an itinerant electron.

Debye temperature Θ_D Characteristic temperature proportional to maximum vibrational frequency of atoms of a solid ($k\Theta_D = \hbar\omega_{max}$).

Isotope Same element with different nuclear masses.

k-space Momentum (or reciprocal-lattice) space in which electron momenta and energies can be plotted.

Magnetic flux Lines of magnetic-field strength defining field direction; their density defines field strength.

Phonon Quantum of lattice vibrational energy $\hbar\omega$.

Quasi-particle Electron of a one-electron energy band renormalized by electron-electron and/or electron-lattice interactions.

Wave function ψ Quantum-mechanical descriptor of an electron; $|\psi(\mathbf{r})|2$ is the probability of finding an electron at position $\mathbf{r}$.

I. INTRODUCTION

In 1908, the Dutch physicist Heike Kammerlingh Onnes succeeded in liquifying helium. This accomplishment made possible the exploration of the low-temperature properties of matter; and in 1911 he reported a phase transition in metallic mercury from a normal state to a superconductive state below a critical temperature T_c. What Kammerlingh Onnes observed was an abrupt change in the direct-current (dc) resistance of mercury at a $T_c = 4.15$ K; the normal state exhibited an electrical resistance R_n with attendant joule heating $I^2 R_n$ on passing a current I, whereas the superconductive state was a "perfect" conductor with no measurable resistance ($R_s = 0$). Moreover, in the absence of a magnetic field, T_c is independent of the shape or the size of the sample; superconductivity is an intrinsic property of the material.

Since 1911, many materials have been surveyed to determine whether they are superconductors and, if so, the value of their T_c. Although there is still little understanding to guide in the search for, or design of, high-T_c superconductors, extensive investigations of the elements, of alloys, of compounds, and of polymers had, by 1985, resulted in several empirical guidelines.

1. Only metals are superconductors.
2. Superconductivity is associated with a dynamic electron–phonon coupling.

3. The highest values of T_c are associated with partially filled d bands, and T_c varies sensitively with the electron–atom ratio for the partially filled band.
4. T_c is suppressed where the conduction electrons exhibit magnetic order at low temperatures.
5. T_c is suppressed where the electron–phonon coupling becomes static, inducing a phase transformation to a nonmetallic state.

From 1911 to 1986, the critical temperature T_c remained below 25 K, increasing by less than 0.3 K per decade (see Fig. 1). Moreover, the existing theory—applicable to nearly all known superconductors—predicted a ceiling

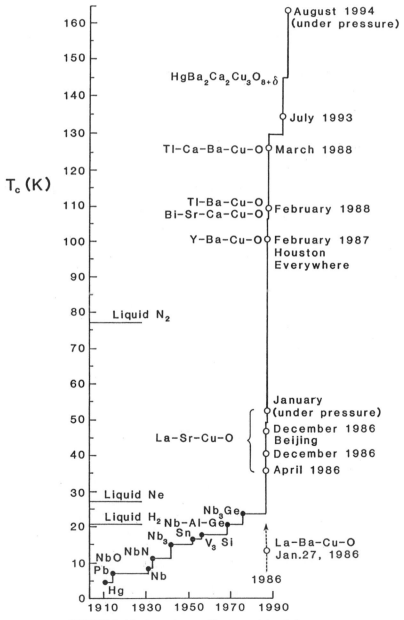

FIGURE 1 Maximum known T_c versus date of discovery.

for T_c in the neighborhood of 30 K. Nevertheless alternate mechanisms for enhancing T_c had been suggested, and a few experimentalists persisted in the hope of finding a material that exhibited such an enhancement.

Bednorz and Müller, of IBM Zürich, were two who persisted, and in 1986 they reported the existence of superconductivity above 30 K in a multiphase oxide containing Ba, La, and Cu. Although their discovery was initially overlooked within their own corporation, it was immediately pursued by Kitazawa and Uchida of the University of Tokyo, who identified the superconductor phase as $La_{2-x}Ba_xCuO_{4-y}$ having a well-known intergrowth structure. Announcement of this identification in December 1986 at conferences in Boston, Massachusetts, and Bangalore, India, triggered an excited effort to reproduce, extend, and "explain" this breakthrough. Within weeks, substitution of Sr for Ba had increased the T_c to 40 K and attempts to substitute Y for La had resulted in a polyphase mixture containing a new superconductor with a $T_c \approx 90$ K. First announced in the *New York Times* by Chu of the University of Houston— but found independently at the same time by workers in Tokyo, Peking, and Bangalore—the latter discovery electrified the entire solid-state community. A T_c higher than 77 K, the boiling point of nitrogen, introduced an entirely new technical dimension, and conventional theory clearly could not be stretched to include this new finding without some radical modification. A race to articulate this theoretical modification, to establish it, and to use it to find new high-T_c superconductors had begun. Simultaneously, the problem of processing these new materials for technological exploitation began to be addressed in more than 1000 laboratories around the world. After 12 years of intensive effort by many groups, there is no consensus yet even on the character of the normal state out of which the superconductive pairs condense, and processing the brittle ceramic materials into flexible wires or tapes that can remain superconducting in high magnetic fields remains a technical challenge. This article can be only a personal commentary on this activity.

II. SUPERCONDUCTIVITY

A. Phenomenology

1. Nomenclature

A superconductor is any material that undergoes a transition from the normal state to the superconductive state below a critical temperature T_c. It is superconducting when it is carrying a resistance-free ($R_s = 0$) current (i.e., a supercurrent) in the superconductive state.

2. The Normal State

Superconductors are metallic in the normal state. Each conduction electron of a metal is said to be itinerant because it belongs equally to all like atoms at energetically equivalent lattice positions in a crystal; each may also belong, to a lesser extent, to other atoms in the crystal. Because their position in real space is poorly defined, itinerant electrons are characterized by their momentum vector $\mathbf{k}$, where the momentum $\mathbf{p}$ transforms to $\hbar\mathbf{k}$ ($\hbar = h/2\pi$, where h is Planck's constant) in the absence of a magnetic field. Where the like-atom interatomic interactions are much stronger than the intraatomic electron–electron interactions, each itinerant electron may be described as a single particle moving in the average electrostatic potential created by the atomic nuclei and all the other electrons; they therefore occupy one-electron states, each having an energy $\varepsilon_\mathbf{k}$ and, in the absence of a magnetic field, a twofold spin degeneracy. Moreover, the one-electron energies for an N-atom array are grouped into energy bands containing $2N/n$ states per atomic orbital, where $n \geq 1$ is an integer that depends on the translational symmetry of the crystal. The density of one-electron states $N(\varepsilon)$ per infinitesimal energy interval $d\varepsilon$ is a fundamental parameter; so also is the effective mass entering the relationship $\varepsilon_\mathbf{k} - E_0 = h^2k^2/2m^*$, where E_0 is a band-edge reference energy. The Pauli exclusion principle allows one electron per state, so at $T = 0$ K the electron states are successively occupied from the bottom of an energy band until all the electrons are accounted for.

What distinguishes a metal from a semiconductor such as silicon is that, in a metal, an occupied band of itinerant-electron states is only partially filled (Fig. 2). In this case there is an abrupt change in the electron population at a surface in $\mathbf{k}$ space; this surface is called the Fermi surface, and the energy of the Fermi surface is called the Fermi energy E_F. At finite temperatures, electrons are thermally

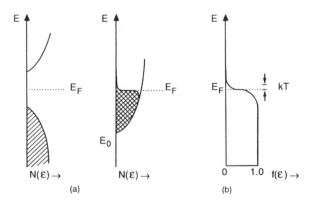

FIGURE 2 Energy versus (a) the density $N(\varepsilon)$ of one-electron states for a semiconductor and a metal and (b) the Fermi–Dirac distribution function $f(E)$.

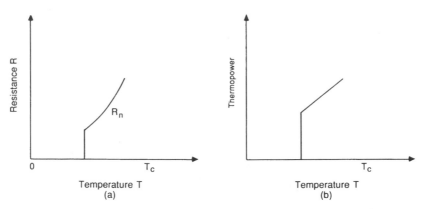

FIGURE 3 Temperature dependence of (a) the resistance and (b) the thermopower of a superconductor.

excited from occupied to unoccupied states in the band: this process gives rise to a Fermi–Dirac statistical distribution that, for $k_B T \ll E_F$ (k_B is Boltzmann's constant), leaves intact the concept of a Fermi surface (Fig. 2b).

As the ratio of interatomic/intraatomic interaction energies decreases toward unity, the band of one-electron energies narrows, increasing $N(E_F)$ for a given E_F; single electrons become transformed into quasiparticles that cannot be described by the average potential of all other electrons because correlations between the electrons on nearest like neighbors reflect the intraatomic electron–electron interactions and/or because electron–phonon interactions "dress" the electrons in local crystallographic distortions. Although these interactions reduce the discontinuity in the electron population at E_F, an identifiable Fermi surface remains as long as the specimen is metallic.

Partially occupied, narrow energy bands of quasiparticle states may lose their Fermi surface at E_F by inducing a diffusionless phase transition at low temperatures that splits the band into bands of occupied and empty states separated by a finite energy gap E_g. Three types of transitions cause such a splitting: (1) an atomic clustering that changes the translational symmetry of the crystallographic structure, (2) a magnetic ordering that changes the translational symmetry and/or the degree of electron localization at atomic positions, and (3) the onset of superconductivity caused by a pairing of one-particle states having energies near E_F into an ordered condensate of two-particle states. Since the first two processes compete with the onset of superconductivity, any realization of a high T_c must involve a mechanism that suppresses the stabilization of atomic clustering and of magnetic ordering of the conducting electrons.

3. The Superconductive State

The critical temperature T_c marks the boundary between two distinguishable thermodynamic states of the material,

each with its own set of properties. The superconductive state is distinguished from the normal state by its electric, magnetic, thermodynamic, and tunneling properties.

a. Electric. The dc resistance R of a superconductor wire drops abruptly at T_c, from $R_n > 0$ in the normal state to $R_s = 0$ in the superconductive state (Fig. 3a).

In the normal state, the potential difference V between the ends of a wire of length l and resistance R_n is, by Ohm's law,

$$V = IR_n \tag{1}$$

if the wire carries a current I. By definition, a constant electric field $E = V/l$ then exists in the wire. In the superconductive state, on the other hand, $R_s = 0$ makes $V = E = 0$. There is no constant electric field in, or potential difference across, a superconducting wire. Consequently all the thermoelectric effects present in the normal state vanish abruptly at T_c. For example, in the normal state an applied temperature gradient ΔT gives rise to an electric field E in the conductor; the thermoelectric power, defined as $E/\Delta T$, vanishes with E below T_c (Fig. 3b).

The resistance R_s of the superconductive state is strictly zero only for direct currents of a constant value. If the current changes with time, as in an alternating-current (ac) application, then R_s is not zero. Nevertheless at temperatures $T \ll T_c$, R_s remains much less than the resistance R_n of the normal state for frequencies $\nu < E_g/h$, where E_g is the energy gap (see Section II.C) at the Fermi energy of the superconductive state. The ratio R_s/R_n increases from a small value to nearly 1 in a finite frequency interval $\Delta \nu$ (Fig. 4). The width $\Delta \nu$ broadens and its midpoint shifts to a lower frequency as T increases to T_c.

b. Magnetic. The magnetization M of a substance is defined as its magnetic moment per unit volume. The magnetic susceptibility per unit volume is defined as

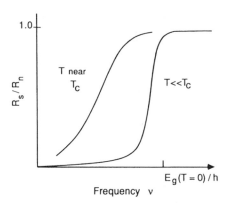

FIGURE 4 Ratio of superconductive-to-normal resistance versus frequency for different values of $T < T_c$.

$$\chi = \mu_0 M / B_a, \tag{2}$$

where B_a/μ_0 is the intensity of an applied magnetic field.

Substances with a negative magnetic susceptibility are called diamagnetic; those with a positive susceptibility are called paramagnetic. Diamagnetism reflects changes in electron motion that oppose the applied magnetic field; paramagnetism reflects an increase in the populations of electron spins (or of localized atomic moments) oriented parallel to the applied magnetic field. The inner, closed-shell atomic cores retain spin-paired electrons; they always give a small, temperature-independent diamagnetic contribution $\chi_{core} < 0$ to the susceptibility. However, in superconductors the dominant contribution is made by the conduction electrons.

In the normal state of a superconductor, an applied magnetic field $\mathbf{B}_a$ defines the orientations of the one-electron spin states and stabilizes the parallel spin states relative to the antiparallel spin states of the conduction band by an energy $2\mu_B B_a/\mu_0$, where μ_B is the magnetic moment imparted by a single-electron spin (the Bohr magneton) (Fig. 5). The resulting change in electron population of

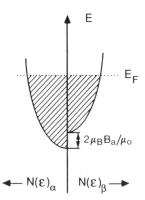

FIGURE 5 Shifting of α-spin and β-spin energies in an applied magnetic field strength B_a/μ_0.

parallel (α) versus antiparallel (β) spin states creates, to lowest order, a paramagnetic magnetization

$$M \approx \mu_B^2 N(E_F) \cdot B_a/\mu_0, \tag{3}$$

where $N(E_F)$ is the density of one-electron states at E_F. First derived by Pauli, this contribution to the total magnetization is called the Pauli spin magnetization. Equation (3) applies to broad energy bands with $E_F \gg k_B T$.

Changes in the motions of the conduction electrons introduce a diamagnetic contribution. Landau has shown that where Eq. (3) applies, this contribution to M is minus one-third the Pauli spin magnetization, so the total conduction-electron contribution to the normal-state susceptibility is paramagnetic and temperature independent:

$$\chi_{cond} = \mu_0 M/B_a = (2/3)\mu_B^2 N(E_F) > 0. \tag{4}$$

It generally dominates the total temperature-independent susceptibility,

$$\chi = \chi_{cond} + \chi_{core}. \tag{5}$$

If the energy bands are narrow, it is necessary to introduce into χ_{cond} a temperature-dependent enhancement factor.

In the superconductive state the situation is quite different. Meissner and Ochsenfeld found that, if a superconductor is cooled in a magnetic field to below T_c, the magnetic flux within the superconductor in the normal state is pushed out of the superconductive state as illustrated in Fig. 6. This phenomenon is called the Meissner effect.

The extent to which the internal magnetic flux is expelled by the Meissner effect depends not only on the temperature and the magnitude of the applied magnetic field $\mathbf{B}_a$, but also on the sample shape and its orientation with respect to $\mathbf{B}_a$. A long, thin cylinder (or wire) oriented with its long axis parallel to $\mathbf{B}_a$ has a negligible demagnetizing field within it, and the internal magnetic field is

$$B = B_a + \mu_0 M, \tag{6}$$

where M is the induced magnetization. Complete expulsion of B would make $B = 0$ inside the superconductor, to give perfect diamagnetism, with

$$M = -B_a/\mu_0 \quad \text{and} \quad \chi = -1. \tag{7}$$

The magnetization curve for such a situation at $T < T_c$ is illustrated in Fig. 7a. It is found to apply quantitatively to pure specimens for applied fields less than a critical field strength $H_c(T)$:

$$B_a/\mu_0 \leq H_c. \tag{8}$$

Two types of superconductors can be distinguished. Type I, originally termed soft superconductors, exhibit an abrupt loss of the Meissner effect at H_c. Type II superconductors exhibit two critical field strengths (Fig. 7b): H_{c1},

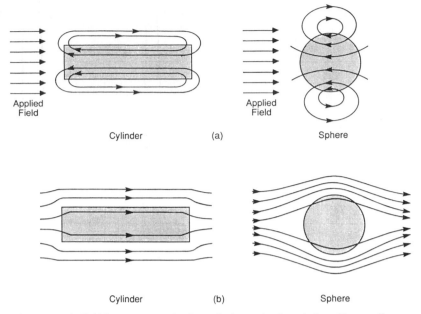

FIGURE 6 (a) The magnetic field in a superconductive cylinder and sphere induced by a uniform applied field B_a. (b) The total induced plus applied field.

beyond which the Meissner effect is less than complete; and H_{c2}, beyond which it disappears completely.

In the absence of a magnetic field, the transition at T_c is always sharp; the material is, essentially, wholly superconductive at temperatures $T < T_c$ and wholly normal at $T > T_c$. At temperatures $T < T_c$, the change from the superconductive to the normal state at the critical field strength $H_c(T)$ is not sharp in type II superconductors, and for most geometries it is not sharp even in type I superconductors. Penetration of magnetic flux occurs between H_{c1} and H_{c2} in Fig. 7b; in this range of applied field, parts of the specimen are in the normal state and parts are in the superconductive state. It is possible to distinguish type I

from type II superconductors by the way in which the normal-state regions penetrate the superconductive state with increasing B_a/μ_0.

The distinction between the two types of superconductors is illustrated in Fig. 8 for the case of a $\mathbf{B}_a$ applied perpendicular to a plane slab of a superconductor. If the superconductor is type I, the normal regions enter as relatively thick, parallel laminae; and if both normal and superconductive states coexist, the superconductor is said to be in an intermediate state. If the superconductor is type II, the normal regions enter as numerous, extremely thin tubular filaments separated by small distances ($\leq 10^{-5}$ mm), and for $H_{c1} < H < H_{c2}$ the superconductor is said to be in a

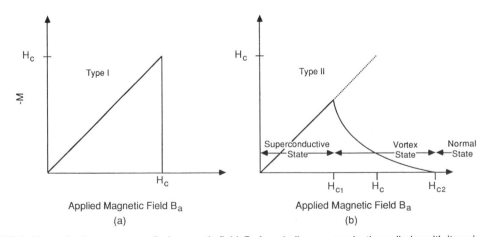

FIGURE 7 Magnetization versus applied magnetic field B_a for a bulk superconductive cylinder with its axis parallel to B_a for (a) type I and (b) type II superconductivity.

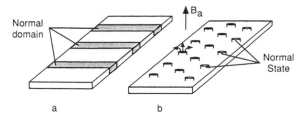

FIGURE 8 (a) Intermediate state of a type I superconductor. (b) Vortex state of a type II superconductor.

vortex state (or in a mixed state). The field H_{c2} in Fig. 7b depends on the mobility of the vortices. Large values of H_{c2} are obtained by introducing crystalline imperfections that pin the vortices; flux pinning introduces hysteresis in Fig. 7b between the curve obtained by increasing and that obtained by decreasing B_a. A hard superconductor is a type II superconductor exhibiting a large magnetic hysteresis due to vortex pinning.

c. Thermodynamic: Order parameter. In the absence of an applied magnetic field, the transition from the superconductive to the normal state is second order: there is no discontinuity at T_c in either entropy (no latent heat) or volume (no thermal hysteresis), but there is a sharp discontinuity ΔC in the heat capacity C (Fig. 9).

A decrease in entropy on going from the normal to the superconductive state shows that the superconductive state is more ordered and can be described by an order parameter that varies smoothly with temperature from unity at 0 K to 0 at $T = T_c$. A natural choice for the order parameter in classical physics is n_s/n_0, the local density of superconductive electrons normalized to its value at 0 K. However, superconductivity is a quantum—not a classical—phenomenon, and a more profound choice is the corresponding quantum physics wave function

$$\psi(r) = |\psi(r)| \exp[i\phi(r)], \qquad (9)$$

where $\phi(r)$ is a phase factor and $n_s \equiv |\psi|^2$.

i. Persistent currents and flux Quantization. The most basic implication of the existence of a phase factor in $\psi(r)$ is the quantization of magnetic flux in a superconducting ring. Consider first the macroscopic ring in Fig. 10. The application perpendicular to the ring of a uniform magnetic field of flux density B_a that varies with time t creates a voltage that induces a current $I(t)$ to circulate in the ring. According to Lenz's law

$$-A_r \frac{dB_a}{dt} = RI(t) + L\frac{dI(t)}{dt}, \qquad (10)$$

where A_r is the area enclosed by the ring, R the resistance of the ring, and L the inductance of the ring. If there is no applied magnetic field ($B_a = 0$), then the solution of Eq. (10) is

$$I(t) = I(0) \exp(-Rt/L), \qquad (11)$$

which shows that any initial current circulating in the ring decays exponentially to zero in the normal state. However, in the superconductive state an $R = R_s = 0$ makes $I(t) = I(0)$, and the initial current $I(0)$ continues to circulate around the ring without any change in its magnitude. Such currents are known as persistent currents; and any current I circulating around the ring produces a magnetic flux threading the ring equal to LI. In the presence of B_a, the total flux Φ threading the ring is $\Phi = A_r B_a + LI$, and Eq. (10) reduces to

$$d\Phi/dt = -IR. \qquad (12)$$

In the superconductive state, $R = 0$ gives

$$\Phi = \text{constant} = A_r B_s, \qquad (13)$$

where B_s is the ring magnetic field.

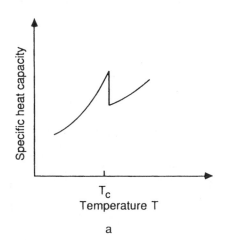

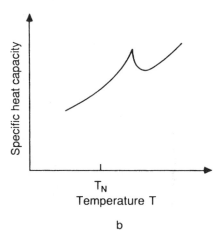

FIGURE 9 Temperature dependence of the specific heat capacity for (a) a BCS superconductor and (b) a typical antiferromagnet.

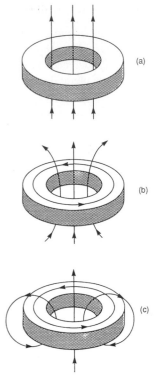

FIGURE 10 Production of persistent current in a superconductive ring: (a) B_a applied normal to a ring at $T > T_c$; (b) ring cooled to $T < T_c$; (c) B_a removed, leaving persistent current.

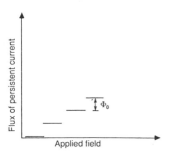

FIGURE 11 Quantum steps in flux of persistent current versus applied field for a type II superconductor.

A persistent current and its associated magnetic field are established in a macroscopic ring by introducing a current $I(t)$ into the ring before it is cooled to below T_c; the external circuit is switched off only after the ring is in its superconductive state. In a type II superconductor below T_c, the penetration of flux in a $B_a/\mu_0 > H_{cl}$ is accomplished by the movement of a normal-state filament into the superconductive state; the normal-state region contains flux, and its movement into the superconductor creates a microscopic persistent current within the surrounding superconductive state that traps the flux within the normal-state core of the vortex. The amount of flux within a microscopic vortex is quantized because of the phase factor ϕ in Eq. (9).

For a superconducting ring, the single-valuedness of ψ requires that $\phi(r)$ return to itself modulo 2π on going once around the circuit; that is, if the orbit of a superconductive electron is quantized to a path length that is an integral number of electron wavelengths, then the electron neither gains nor loses energy. For a superconductive particle in an orbit of radius r, the condition for quantization, and hence the existence of a supercurrent, is

$$\mathbf{p} \cdot 2\pi r = Nh, \qquad (14)$$

where h is Planck's constant, N is an integer, and the canonical momentum in a local magnetic field $\mathbf{B} = \nabla \times \mathbf{A}$ is

$$\mathbf{p} = h(\mathbf{k}_1 + \mathbf{k}_2 \cdots \mathbf{k}_n) + ne\mathbf{A} \qquad (15)$$

for a superconductive particle consisting of n electrons. If the superconducting particle consists of a pair of electrons having opposite momentum vectors $\mathbf{k}$ and $-\mathbf{k}$, then $n = 2$ and $\mathbf{p} = 2e\mathbf{A}$. Moreover, the flux enclosed by a vortex is $\Phi = 2\pi r A$, so that Eq. (14) reduces to

$$\Phi = Nh/2e. \qquad (16)$$

This result is of outstanding importance. It means that if the superconductive state consists of paired electrons, then in a closed superconducting circuit the flux is quantized in units of

$$\Phi_0 = h/2e = 2.07 \times 10^{-15} \text{ Wb}. \qquad (17)$$

The existence of flux quantization and the magnitude of Φ_0 have been confirmed experimentally (Fig. 11); these experiments demonstrate not only the quantum character of the ordering, but also that ordering in the superconductive state consists of the formation of pairs of electrons having opposite momentum vectors $\mathbf{k}$ and $-\mathbf{k}$. Moreover, the perfect diamagnetism associated with this order indicates pairing of $s = \frac{1}{2}$ and $s = -\frac{1}{2}$ spins of a superconductive electron pair. The critical field strength H_c is that required to decouple the spin pairing of a superconductive particle.

A localized atomic moment interacts with the conduction-band electrons via spin–spin "exchange" to produce a local magnetic field; if the local field strength exceeds H_c, superconductivity is suppressed. Ferromagnetic ordering of localized moments generally suppresses superconductivity, but antiferromagnetic ordering may not be incompatible with superconductivity.

ii. Energy gap. The electronic heat capacity in the superconductive state C_{es}, normalized to its value γT_c in the normal state at $T = T_c$, is commonly found to vary exponentially as $-1/T$ at temperatures $T \ll T_c$:

$$C_{es}/\gamma T_c = a \exp[-bT_c/T]$$
$$= a \exp[-\Delta_0/k_B T], \qquad (18)$$

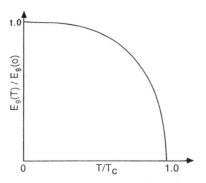

FIGURE 12 Normalized energy gap $E_g(T)/E_g(0)$ versus normalized temperature T/T_c.

which is suggestive of an excitation of electrons across an energy gap $E_g = 2\Delta_0$. An energy gap E_g has been measured independently by spectroscopic techniques; it compares favorably at the lowest T with the calorimetric data (i.e., $b \approx \Delta_0/k_B T_c$). Moreover, $E_g = 2\Delta(T)$ is found to decrease with increasing temperature, from $2\Delta_0$ at 0 K to 0 at $T = T_c$ (Fig. 12) with a Brillouin-function dependence, which makes $\Delta(T)/\Delta_0$ a measure of the order parameter in a mean-field description of the transition. In addition, the presence of an energy gap at the Fermi energy E_F shows that the superconductive electron pairs have been formed by condensing out single-electron states from the vicinity of E_F.

iii. Isotope effect. The T_c for mercury, and most other elemental superconductors, varies smoothly with the average atomic mass M as the isotope mix is varied:

$$M^\alpha T_c = \text{constant.} \qquad (19)$$

This correlation of T_c and M is known as the isotope effect. This early observation shows that for conventional superconductors, electron–phonon interactions play an important role in the binding of superconductive pairs of electrons. In the simplest theory, only the electronic states within an energy $k_B \Theta_D$ of E_F, where Θ_D is the Debye temperature, can be coupled by electron–phonon interactions. This simplest theory limits the magnitude of the energy gap to a specific multiple of $k_B T_c$ and predicts, for an elemental superconductor, an $\alpha = \frac{1}{2}$ in Eq. (19) [see Eqs. (42) and (43)]. Although an $\alpha = \frac{1}{2}$ has been observed for mercury, there is nothing sacred about this value even for the elements; for example, an $\alpha = 0$ for Zr and Ru does not signal the absence of a phonon mechanism in these two superconductors.

An electron–lattice mechanism for binding a superconductive pair leads to an upper limit for T_c of about 30–40 K; a higher T_c requires either another type of superconductive pair, the bipolaron, which is stabilized in the limit of strong electron–lattice coupling, or an electronic enhancement of the electron–lattice mechanism. An electronic en-

hancement would replace Θ_D with $\Theta_e \approx \hbar\omega_e/k_B$, where $\hbar\omega_e \ll E_F$ is the energy of the electronic excitations that enhance the pairing potential energy.

iv. Many-body condensate. Significantly, the usual superconductor transition is much sharper than other second-order transitions. A second-order magnetic-ordering transition, for example, exhibits a substantial temperature range of short-range order above the critical temperature for long-range order. In this case, each atomic moment interacts strongly with only a few near neighbors, so thermodynamic fluctuations not treated by a mean-field theory play an important role. In conventional superconductors, only small vestiges of superconductivity remain above T_c, and any resistivity remaining in the superconductive state is infinitesimally small. This observation indicates that each electron pair in the superconductive state is strongly coupled to all the other pairs in a many-body condensate. To break the binding of a given electron to the condensate costs a minimum energy Δ_0. This many-body aspect of the superconductive condensate makes it difficult to depict in real space the nature of the electron ordering that is occurring. An inability to picture the condensate in real space has hindered formulation of a chemical guide for the search for new high-T_c materials.

v. Temperature dependence of H_c. In the presence of a magnetic field B_a, the transition at T_c becomes first order. In a type I superconductor, the increase in free energy at $B_a/\mu_0 = H_c$ is, from Eqs. (7) and (8),

$$\Delta G = \int_0^{H_c} \mu_0 M dH = \tfrac{1}{2}\mu_0 H_c^2, \qquad (20)$$

and the latent heat at the transition becomes

$$Q = T\,\Delta S = T\frac{d(\Delta G)}{dT} = \mu_0 T H_c \frac{dH_c}{dT}, \qquad (21)$$

which vanishes at $T = T_c$ where $H_c = 0$. It is found empirically that at temperatures below T_c, the entropy difference is described by

$$\Delta S = \gamma T\left[1 - (T/T_c)^2\right]. \qquad (22)$$

Equating Eqs. (21) and (22) and integrating with respect to the boundary conditions $H_c = 0$ at $T = T_c$ and $H_c = H_0$ at $T = 0$ K gives the relation

$$H_c = H_0\left[1 - (T/T_c)^2\right] \qquad (23)$$

for the transition between the superconductive and the normal state in the presence of an applied field (Fig. 13). The extent of the intermediate-state region depends on the shape of the sample and its orientation with respect to $\mathbf{B}_a$.

d. Tunneling. If two metals are separated by an insulator, the insulator acts as a barrier to electron flow from

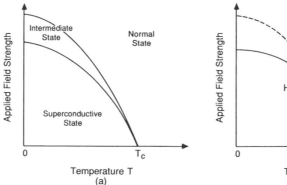
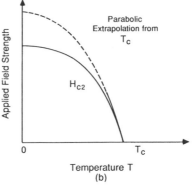

FIGURE 13 Variation with temperature of (a) H_c of type I and (b) H_{c2} of type II superconductors.

one metal to the other. However, the conduction-electron wave functions extend beyond the metal surface, decaying exponentially in magnitude with the distance from the surface. If the insulating layer is thin enough (less than 10 to 20 Å), a significant amplitude extends through the insulating layer into the other metal. If an empty electronic state of equal energy is also available in the other metal, then there is a finite probability that an electron impinging on the barrier will pass through the insulating layer. This phenomenon is called tunneling.

If both metals are superconductors, two types of particles may tunnel: single quasiparticles and paired superconductive particles. Tunneling of single quasiparticles has been used to measure the energy gap in the superconductive state; tunneling of superconductive particles—called Josephson tunneling—exhibits unusual quantum effects that have been exploited in a variety of quantum devices.

In 1962, Josephson proposed that a tunnel junction between two superconductors—each in their superconductive state—should exhibit a zero-voltage supercurrent in the direction x perpendicular to the junction,

$$I_x = I_{0x} \sin \gamma, \tag{24}$$

due to the tunneling of superconductive electron pairs. Both the phase differences $\phi_2 - \phi_1$ of the wave function on either side of the insulating layer and the canonical momentum of Eq. (15) in the presence of a magnetic flux enter into

$$\gamma = (\phi_2 - \phi_1) - \frac{2\pi}{\Phi_0} \int_1^2 A_x \, dx. \tag{25}$$

A maximum dc flows in the absence of any electric or magnetic field. This is the dc Josephson effect.

Josephson further predicted that if a voltage difference V is applied across the junction, the parameter γ becomes time dependent,

$$\gamma(t) = \gamma(0) - (4\pi e V t / h), \tag{26}$$

which means that the current oscillates with a frequency

$$\nu = 2eV/h. \tag{27}$$

This is the ac Josephson effect.

These predictions have been verified experimentally and shown to apply to any sufficiently thin "weak link" in a superconducting circuit. A weak link can be any planar defect at which T_c is sharply reduced from its value in the bulk superconductor. Such weak links appear to limit the supercurrents in the new high-T_c superconductors.

B. Applications

The technical applications of superconductivity have exploited all its basic properties. However, an extensive commercial potential has been made possible only by the discovery of type II superconductors and Josephson tunneling.

1. High Magnetic Field, High Direct Current

The discovery of zero dc resistance, which makes possible macroscopic persistent currents, immediately raised the hope of building a solenoid magnet of superconductive wire capable of generating an intense magnetic field at manageable power levels. Although no energy is expended by a static magnetic field, the energy required to create and sustain an intense magnetic field with a normal conductor is prohibitive.

Attempts to exploit this concept encountered the intrinsic limitation imposed by H_c. A cylindrical wire of radius r_w carrying a current I has, at its surface, a magnetic field strength produced by the current

$$H_{\text{surf}} = I/2\pi r_w. \tag{28}$$

A supercurrent may increase until $H_{\text{surf}} = H_c$; for any current higher than the critical current,

$$I_c = 2\pi r_w H_c. \tag{29}$$

The surface of the wire is transformed to the normal state. In type I superconductors, critical field strengths $H_0 \leq 80$ kA/m (1 kOe) are not sufficient to replace an iron-core magnet. However, a type II hard superconductor is capable of remaining superconductive to high magnetic fields H_{c2}; and the generation of high magnetic fields with type II superconductors is now used in a wide range of applications.

2. Alternating-Current Devices

A low R_s/R_n ratio requires ac operation at $T \ll T_c$ and $\nu \ll E_g/h$. Type I superconductors may retain an $R_s \approx 0$ up to 100 MHz. This property has enabled the realization of very high-frequency linear electron accelerators with magnifications up to 10^{10}; they can operate continuously with only a fraction of the power requirements of conventional accelerators.

Larger energy gaps E_g in type II superconductors permit low-loss ac transmission over superconductive strip lines to even higher frequencies. On the other hand, attempts to use superconductors in ac power devices remain restricted to specialty applications such as space vehicles where, with type II superconductors, high current densities in high fields permit significant reductions in weight and size.

3. Levitation

The Meissner effect is demonstrated in the classroom by levitation of a bar magnet over a superconductive bowl. The experiment begins with a bar magnet resting on the bottom of a shallow bowl of superconductor in its normal state. The bowl is then cooled to below T_c; expulsion of a magnetic field from the bowl creates an "image" magnet that exerts a repulsive force on the real magnet, causing it to rise until this force is balanced by the weight of the magnet.

A most spectacular application of this principle is the "levitated train," which requires high magnetic fields and, therefore, a hard, type II superconductor.

4. Bolometer

A bolometer detects electromagnetic radiation by an absorption of radiation that increases its temperature. The temperature increase ΔT is related to the energy ΔE absorbed per unit mass via the specific heat capacity C_v:

$$\Delta T = \Delta E / C_v. \qquad (30)$$

At low temperatures, a low C_v enhances ΔT for a given ΔE. A type I superconductive detector is designed to operate in the intermediate state where a small ΔT gives rise to a large resistance change ΔR as illustrated in Fig. 14.

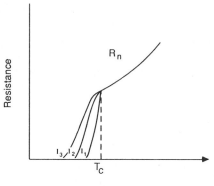

FIGURE 14 Temperature variation of resistance for a wire carrying a current $I_1 < I_2 < I_3$.

The ability to amplify ΔR electronically makes the superconductive bolometer an extremely sensitive radiation detector; it is particularly important in the far-infrared region of the spectrum, where most other types of radiation detectors are inoperative.

5. Josephson Tunneling

Practical application of the dc Josephson effect has been realized in very sensitive galvanometers and magnetometers. The SQUID (superconducting quantum interference device) magnetometer, for example, is used for measuring small magnetic fields, with extensive use in geological surveying. A laboratory SQUID was the first practical demonstration of the high-T_c superconductor oxides.

The ac Josephson effect has been used in precision determinations of the value of h/e.

Applications in the computer field promise higher-density, lower-power components; however, their realization in practice requires an exquisite control of materials processing that is particularly demanding for the new high-T_c superconductor oxides.

C. Theory

1. History

Once the basic phenomena of zero resistance and the Meissner effect had been established, the experimental strategies responsible for our understanding of superconductivity were guided by theory. The theory began with purely phenomenological equations; these equations introduced fundamental length parameters as well as the order parameter, and their application permitted Abrikosov to explain the distinction between type I and type II superconductivity.

The quantum theory of Bardeen, Cooper, and Schrieffer (BCS) introduced numerical values for three universal ratios relating T_c to Δ_0, H_0, and the jump $\Delta C(T_c)$ in the specific heat at T_c. Refinement of the BCS theory by Eliashberg has expressed these ratios in terms of two parameters, and with this refinement the agreement between theory and experiment for the three universal ratios is truly remarkable over a wide range of conventional superconductors.

The phenomenon of high-T_c superconductivity cannot be accounted for with the Eliashberg theory if the theory is restricted to binding of the superconductive electron pairs by a dynamic coupling to phonons. This breakdown of existing theory has split theorists into two camps: one camp would extend the BCS or Eliashberg theory by introducing into the pair-binding potential energy an electronic mechanism together with the phonon mechanism; the other camp would break from the "weak-coupling" theory to construct a "strong-coupling" theory in which electron pairs form as a disordered array of "bipolarons" at temperatures $T > T_c$, ordering into the superconductive state occurring only below T_c.

2. London Equation

In order to account for zero resistance and the Meissner effect in the superconductive state, the London brothers postulated that the local current density in the superconductive state is proportional to the vector potential $\mathbf{A}$, where $\mathbf{B} = \nabla \times \mathbf{A}$:

$$\mathbf{j}_s = \left(1/\mu_0 \lambda_L^2\right)\mathbf{A}. \qquad (31)$$

Applying the Maxwell equation $\nabla \times \mathbf{B} = \mu_0 \mathbf{j}_0$, applicable under static conditions, to Eq. (31) gives, on taking the curl of both sides of Maxwell's equation,

$$\nabla^2 \mathbf{B} = \mathbf{B}/\lambda_L^2 \qquad (32)$$

for a superconductive state. This equation accounts for the Meissner effect because it does not allow a solution uniform in space unless $B = 0$. Moreover, Maxwell's equation shows that $j = 0$ wherever $B = 0$.

On the other hand, Eq. (32) does allow a solution for B that is nonuniform in space. If a field B_a is applied parallel to an external surface, as illustrated in Fig. 15, then Eq. (32) gives the solution

$$B(x) = B(0)\exp(-x/\lambda_L), \qquad (33)$$

where x is the vertical distance into the superconductor from the surface and $B(0)$ is the value of B_a at $x = 0$. Thus λ_L measures the depth of penetration of the magnetic field; it is known as the London penetration depth. The current flowing in the superconductor responsible for expelling B is confined to a thin surface layer.

A measure of the magnitude of λ_L can be obtained from the canonical momentum $\mathbf{p} = m\mathbf{v} + e\mathbf{A}$. If the average superconductive-particle momentum is zero in the

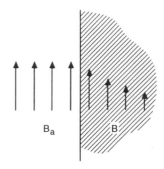

FIGURE 15 Penetration of an applied magnetic field into a semi-infinite superconductor. The penetration depth λ_L is the distance at which B decays to B_a/e.

ground state, then the average velocity is $\langle \mathbf{v}_s \rangle = -e\mathbf{A}/m$. If the number density of electrons participating in the rigid ground state is n_s, then the local superconductive current density becomes

$$\mathbf{j}_s = n_s e \langle \mathbf{v}_s \rangle = -n_s e^2 \mathbf{A}/m, \qquad (34)$$

and comparison of Eq. (34) with Eq. (31) gives

$$\lambda_L = \left(m^*/\mu_0 n_s e^2\right)^{1/2}. \qquad (35)$$

However, careful measurements of $\lambda_L(T)$ near $T = 0$ K indicate that $\lambda_L(0)$ is larger than the prediction of Eq. (35), which suggests a reduced n_s and hence a rigidity of the condensate only over a finite volume defined by a characteristic length ξ_0.

3. Coherence Length

The concept of a characteristic dimension ξ_0 was introduced by Pippard to formulate a nonlocal generalization of the London equation. He estimated this length from the Heisenberg uncertainty principle: only electrons having energies within $\sim k_B T_c$ of E_F can play a major role in a phenomenon that sets in at T_c; these electrons have a momentum range $\Delta p \approx k_B T_c/v_F$, where $v_F = hk_F/m^*$ is the velocity of an electron with Fermi energy E_F. From the uncertainty principle, $\Delta x \geq h/\Delta p \approx hv_F/k_B T_c$ defines a characteristic length

$$\xi_0 = ahv_F/k_B T_c, \qquad (36)$$

where a is a numerical constant of order unity. The length ξ_0 plays a role analogous to the mean free path l in the nonlocal electrodynamics of normal metals; and in the presence of scattering, the characteristic length is called the coherence length, where

$$(1/\xi) = (1/\xi_0) + (1/l). \qquad (37)$$

In fact, Ginzburg and Landau were the first to introduce the idea of a characteristic length. In 1950, they

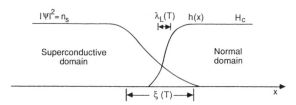

FIGURE 16 Interface between superconductive and normal domains in the intermediate state; $h(x)$ is the local magnetic-field strength.

introduced the order parameter $\psi(r)$ in Eq. (9) defined by $n_s = |\psi|^2$ to obtain an equation for the supercurrent density $\mathbf{j}_s$ in terms of $\psi(r)$. (This theory was later shown to be a limiting form of the microscopic BCS theory first presented in 1957.) With this formalism they were able to treat two features that were beyond the scope of the London formalism: (1) nonlinear effects in fields strong enough to change n_s and (2) the spatial variation of n_s.

A major triumph of the formalism was its description of the intermediate state of a type I superconductor in which superconductive and normal domains coexist in an applied field strength $B_a/\mu_0 \approx H_c$ (Fig. 16). The interface between the two domains is characterized by two lengths: the penetration depth $\lambda_L(T)$, over which the local magnetic field is varying, and the coherence length

$$\xi_{GL}(T) = h/|2m^*\alpha(T)|^{1/2}, \qquad (38)$$

over which $\psi(r)$ can vary without an undue energy increase. In pure superconductors at $T \ll T_c$, the Ginzburg–Landau coherence length approaches the Pippard coherence length [i.e., $\xi_{GL}(T) \approx \xi_0$], but $\xi_{GL}(T)$ diverges as $(T_c - T)^{-1/2}$ near T_c since α vanishes as $(T_c - T)$. Since $\lambda_L(T)$ also diverges as $(T_c - T)^{-1/2}$, the ratio λ_L/ξ_{GL} is nearly independent of temperature. Therefore the Ginzburg–Landau parameter is

$$\kappa = \lambda_L/\xi_{GL}. \qquad (39)$$

In type I superconductors, a $\kappa \ll 1$ results in a positive interface energy, which stabilizes a macroscopic domain pattern.

Abrikosov investigated what would happen if the Ginzburg–Landau parameter is greater than unity. He found that for $\kappa > 1/\sqrt{2}$, the energy of the interface between normal and superconductive domains becomes negative and the superconductor is type II. With a negative interface energy, field strengths $B_a/\mu_0 \geq H_{c1}$ perpendicular to a superconducting slab cause flux to penetrate within cylindrical, normal-state domains; persistent supercurrents surrounding the normal-state regions form vortices. The vortex concentration increases with $B_a/\mu_0 \geq H_{c1}$ until $H_{c2} = \sqrt{2}\kappa H_c$; above H_{c2} the vortices are merged into a single normal-state phase.

4. BCS Theory

In 1956, Cooper showed that so long as there exists an attractive interaction between pairs of electrons, the Fermi sea of electrons is unstable against the formation of at least one bound pair formed from states with $\mathbf{k} > \mathbf{k}_F$. Moreover, he argued that the two-electron wave function for a superconductive pair is a singlet (paired spins), spherical state containing a weighted sum over $\mathbf{k} > \mathbf{k}_F$ of product wave functions with momentum $\mathbf{k}$, $-\mathbf{k}$ for each product and that the maximum contribution comes from states with $\mathbf{k} \approx \mathbf{k}_F$. The two-electron binding energy relative to $2E_F$ was shown to be ($h\omega_c = h\nu_c$)

$$E_{\text{bind}} \approx 2h\omega_c \exp[-2/VN(E_F)] \qquad (40)$$

in the weak-coupling limit $VN(E_F \ll 1$. In this derivation, Cooper made the approximation that the coupling energy V is a constant for all values of $\mathbf{k}$ out to a cutoff energy $h\omega_c$ away from E_F. Since E_{bind} is of the order of k_BT_c, an argument similar to that preceding Eq. (36) suggests that the size of the Cooper-pair state is approximately ξ_0, which is much larger than the interparticle distance. Thus the Cooper pairs are strongly overlapping, which is why they form a rigid condensate.

The interaction between electrons of a pair always contains an electrostatic repulsive energy U_p between the two electrons; a high dielectric constant introduces an electronic screening that reduces U_p, but it is always repulsive. The problem is to identify an attractive mechanism.

In 1950 Frölich suggested that electron–lattice interactions were responsible for the attractive potential, and this idea was confirmed experimentally with the discovery of the isotope effect. The physical idea in the BCS treatment of this suggestion is that the first electron polarizes the crystal by attracting the positive atomic cores; the polarization in turn attracts a second electron provided that it arrives in the polarized region of the crystal before the lattice has had time to relax to its initial state. This time constraint limits the size of a Cooper pair to a characteristic length of order ξ_0. Moreover, the cutoff energy $h\omega_c$ in Eq. (40) is, for this mechanism, the Debye energy $h\omega_D = k_B\Theta_D$, which characterizes the cutoff of the phonon spectrum. If the attractive energy V_C exceeds the electrostatic repulsive energy in magnitude, then the net BCS potential

$$V_{\text{BCS}} = V_C - U_p \qquad (41)$$

is attractive.

The BCS theory involves a calculation of the ground state of the system in the presence of a net attractive potential V_{BCS}. Condensation of Cooper pairs changes the state of the Fermi sea (the collection of one-particle states), and at some point the binding energy for an additional pair has gone to zero. The simplest form of the theory contains one adjustable parameter, V_{BCS}; all other parameters entering the theory are independently measurable. Its principal deductions are the following:

$$T_c \approx 1.13\Theta_D \exp[-1/V_{BCS}N(E_F)], \qquad (42)$$

where $\Theta_D \approx M^{-1/2}$ is the Debye temperature, and

$$2\Delta_0/k_BT_c = 3.53, \qquad (43)$$

$$\Delta C(T_c)/\gamma T_c = 1.43, \qquad (44)$$

$$\gamma T_c^2/H_0^2 = 0.168, \qquad (45)$$

where Δ_0 and H_0 are the gap parameter $E_g = 2\Delta$ and the critical field strength H_c at $T = 0$ K, $\Delta C(T_c)$ is the discontinuity in the specific heat at $T = T_c$, and γ is the Sommerfeld constant of the electron gas in the normal state (i.e., γT_c is the electronic specific heat of the normal state at $T = T_c$). It is found experimentally that α in Eq. (19) is not universally $\frac{1}{2}$ and that Eqs. (43) to (45) do not hold quantitatively in many high- and intermediate-T_c materials, which indicates a need to extend the simplest BCS theory.

5. Beyond BCS

The limitation of the BCS theory is that it is a one-parameter theory in which V_{BCS} is assumed to be constant in an energy region about E_F of width $\pm h\omega_D$; and there is no prescription available for calculating V_{BCS} from microscopic theory. An important extension of BCS theory has been given by Eliashberg.

Whereas BCS theory simply postulates an attractive potential V_{BCS}, Eliashberg theory treats properly the microscopic electron–phonon interactions responsible for the pairing potential in conventional superconductors. The Eliashberg theory contains two parameters. One is the pseudopotential μ^* for the Coulomb electron–electron repulsions; it is adjusted to give the correct value of T_c for a given electron–phonon interaction. The other is the electron–phonon spectral density $\alpha^2 F(\omega)$, where $F(\omega)$ is the number of phonon modes (lattice vibrations) having an energy between $h\omega$ and $h(\omega + d\omega)$; $\alpha^2 F(\omega)$ is a phonon frequency distribution weighted by the strength of the electron–phonon interaction for that mode. It is possible to measure $\alpha^2 F(\omega)$ accurately with superconductive-state–insulator–normal-state tunneling experiments.

The Eliashberg equations determine not only the tunneling characteristic of a tunnel diode, but also all of the thermodynamics of a particular superconductive material provided that $\alpha^2 F(\omega)$ and μ^* are known. In this theory,

the BCS universal ratios [Eqs. (43)–(45)] have become transformed to

$$2\Delta_0/k_BT_c = 3.53[1 + 12.5m^2 ln(1/2m)], \qquad (46)$$

$$\Delta C(T_c)/\gamma T_c = 1.43[1 + 53m^2 ln(1/3m)], \qquad (47)$$

$$\gamma T_c^2/H_0^2 = 0.168[1 - 12.2m^2 ln(1/3m)], \qquad (48)$$

where $m = k_BT_c/h\omega_{ln}$ contains a parameter ω_{ln} that represents a weighted measure of the significant phonon frequencies appearing in $\alpha^2 F(\omega)$. The agreement between theory and experiment for elemental and alloy superconductors, as given by Carbotte, is displayed in Fig. 17 for conventional weak to strong coupling regimes $m \le 0.25$.

Within the Eliashberg theory,

$$2\Delta_0/k_BT_c < 9 \qquad (49)$$

would reach its maximum value only in the unrealistic situation that the entire spectral weight occurs at an optimum vibrational energy $h\omega_E = 0.75$ meV. The critical temperature

$$T_c = C(\mu^*)A_p/k \qquad (50)$$

increases with the strength of the effective electron–phonon interaction

$$A_p = \int_0^\infty \alpha^2 F(\omega)\, d(h\omega), \qquad (51)$$

where $C(\mu^*)$ decreases smoothly with increasing μ^*. T_c increases with A_p until a lattice instability freezes out a static distortion of the structure. Thus the theory of superconductivity itself does not put an upper limit on T_c; however, the conditions for a high T_c appear to be the same as those for the stabilization of competitive mechanisms.

The electronic density of states $N(E_F)$ at the Fermi energy plays an important role, as in the BCS theory, since

$$A_p = N(E_F)\langle g^2 \rangle, \qquad (52)$$

where $\langle g^2 \rangle$ involves a double Fermi-surface average of the square of the electron–atomic core interaction. In the absence of any physical intuition as to how to enhance $\langle g^2 \rangle$, efforts to increase T_c have traditionally concentrated on increasing $N(E_F)$, but this strategy is frustrated by the appearance of spontaneous magnetism or atomic-clustering lattice instabilities as competing processes. Significantly the high-T_c copper oxides have a relatively small $N(E_F)$, which implies that a large $\langle g^2 \rangle$ is enhancing the ratio m into a very strong-coupling regime. This observation requires, in turn, either an electronic enchancement of the electron–phonon interaction or an entirely novel mechanism for the formation of electron pairs.

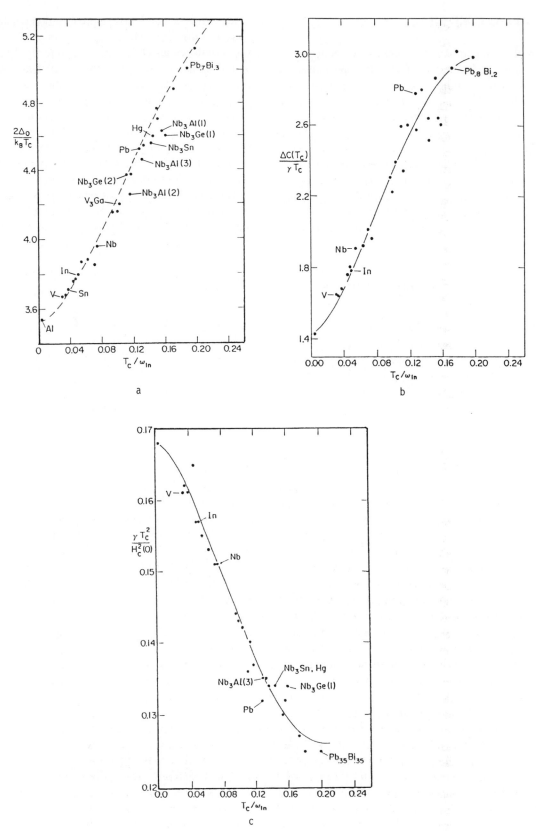

FIGURE 17 The ratios (a) $2\Delta_0/k_B T_c$, (b) $\Delta C(T_c)/\gamma T_c$, and (c) $\gamma T_c^2/H_0^2$ versus $m = kBT_c/\hbar\omega_{ln}$. [After Carbotte, J. P. (1987). *Sci. Prog. Oxf.* **71**, 327.]

III. METALLIC OXIDES

The new high-T_c superconductors are oxides. Since only metals are superconductors, this fact may appear anomalous, as most of the common oxides are either insulators or low-mobility semiconductors. However, many oxides are metallic, and a few of these have extensive commercial application. For example, the metallic cathode of the lead–acid battery is PbO_2; and $Bi_2Ru_2O_7$ has a temperature-independent resistivity in the vicinity of room temperature, which makes it an important resistor material in the electronics industry. Discussion of the high-T_c superconductors rightfully begins, therefore, with a review of the conditions that must be satisfied if an oxide is to be metallic.

A. Electron Energies in a Typical Insulator

Figure 18 shows schematically the construction of an energy diagram for the insulator MgO. The O^{2-} ion is not stable in free space; a negative electron affinity places the $O^{-/2-}$ redox energy above the lowest energy E_{vac} for a free electron in a vacuum. An energy E_1 is required to remove the outer $Mg = 3s^1$ electron from a free Mg^+ ion to a free O^- ion to create free Mg^{2+} and O^{2-} ions. This cost in energy is more than compensated by the electrostatic energy E_M gained by ordering the Mg^{2+} and O^{2-} ions into a crystal structure; the Madelung energy E_M is calculated for a lattice of point charges. The crystalline electric field raises the $Mg^{2+/+}$ level and lowers the $O^{-/2-}$ level; crossing of these two energies ensures stabilization of the crystalline phase with a charge transfer from magnesium to oxygen.

In the real MgO crystal, transfer of an integral electronic charge does not occur; a quantum-mechanical covalent component in the Mg–O bond transfers a fraction of the O^{2-}-ion electronic charge back onto the Mg^{2+} ion. However, the reduction in E_M caused by this lowering of the effective ionic charges is compensated by the quantum-mechanical covalent-mixing repulsion between the two ionic energy levels. Therefore the point-charge model gives a good first approximation to the binding energy of the solid. The covalent component to the bonding introduces an O-$2p$ character into the Mg-$3s$ states and a Mg-$3s$ character into the O-$2p$ states, but without changing the number of electron states at each energy level. Even where this "mixing" is large, it is customary to identify the energy levels by their ionic component only (i.e., as $O^{2-}:2p^6$ and $Mg^{2+}:3s^0$ levels); the wave functions describing the mixed Mg-$3s$ and O-$2p$ states are referred to as crystal-field orbitals so as to distinguish them from the atomic orbitals of a point-charge model.

The crystal-field energies reflect the point-group symmetry of the near-neighbor Mg–O interactions. The final step is to introduce the like-atom interatomic interactions, which broaden the energy levels of the crystal-field orbitals into energy bands of one-electron states. Whereas the crystal-field orbitals are localized to discrete atomic sites, the one-electron states are itinerant with a well-defined momentum $\mathbf{p} = \hbar\mathbf{k}$ in the absence of a magnetic field $\mathbf{B}_a = \nabla \times \mathbf{A}$. Each band contains $2N/n$ one-electron states per atomic orbital (the factor 2 reflects the twofold spin degeneracy of an orbital) for an array of N like atoms containing n atoms per primitive unit cell. Thus the band states reflect the translational space–group symmetry of the crystal.

In MgO there is one magnesium and one oxygen atom per primitive unit cell, so that $O^{2-}:2p^6$ and $Mg^{2+}:3s^0$ levels are each broadened into single bands with a bandwidth much broader than the small spin-orbit splitting of the threefold degeneracy of the oxygen $2p$ crystal-field orbitals. Therefore the highest occupied band of one-electron states is represented as an orbitally threefold-degenerate $O:2p^6$ band, which is full; the lowest unoccupied band is identified as the $Mg^{2+}:3s^0$ band.

In tight-binding theory, the width of a band of one-electron states is

$$W \cong 2zb, \tag{53}$$

where z is the number of like nearest neighbors on energetically equivalent lattice sites and

$$b \equiv (\psi_i, H'\psi_j) \cong \varepsilon_{ij}(\psi_i, \psi_j) \tag{54}$$

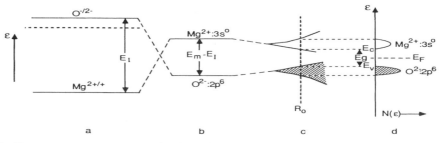

FIGURE 18 Electron energies for MgO: (a) free ions; (b) point-charge model; (c) band model; (d) density of states.

is a measure of the strength of the interatomic interactions between nearest-neighbor like atoms at positions R_i and R_j in the lattice. The perturbation H' of the potential at R_j by the presence of a like atom at R_i factors out as a one-electron energy ε_{ij}, which increases with the overlap integral (ψ_i, ψ_j) for crystal-field wave functions ψ_i and ψ_j at R_i and R_j. Therefore, the overlap integral becomes the guiding qualitative indicator of the strength of the interatomic-interaction parameter b. (Chemists call this parameter a resonance integral; physicists, an electron-energy transfer integral.)

Neglected in the band description of crystalline orbitals is the intraatomic electron–electron electrostatic energy U associated with the creation of polar states (e.g., O^- and O^{3-} in the $O^{2-}:2p^6$ band); the tight-binding theory admixes polar and nonpolar states equally in a first-order perturbation theory. So long as the condition

$$W \gg U \qquad (55)$$

is valid, the assumption of $U \approx 0$ is a useful approximation. Outer s and p electrons participating in near-neighbor chemical bonding satisfy Eq. (55).

The right-hand side in Fig. 18 indicates the density $N(\varepsilon)$ of one-electron states versus the energy E for the equilibrium lattice constant. Since the $O^{2-}:2p^6$ band is filled and the $Mg:3s$ band is empty, the Fermi energy lies near the middle of a large energy gap $E_g = E_c - E_v$ between the two bands, which makes MgO an insulator. The highest occupied band $O^{2-}:2p^6$ is called the valence band; the lowest unoccupied band $Mg^{2+}:3s^0$ is called the conduction band.

Attempts to render MgO conducting by doping with aliovalent impurities, as in semiconductor technology, are frustrated by the energetic inaccessibility of both E_c and E_v; it is energetically favorable for the crystal to incorporate a native defect that charge compensates for the dopant so as to retain E_F near the middle of the energy gap E_g. The cost of introducing a native defect is less in an ionic crystal than in a covalent solid, which is why oxides with large band gaps E_g tend to be good insulators. It follows that the first requirement for metallic conduction in an oxide is the introduction of energetically accessible electron energies.

B. Problems with 5s and 6s Electrons

Heavy group B metals such as Sn and Pb have a relatively large separation of $5s$ from $5p$ or $6s$ from $6p$ states; it is therefore chemically straightforward to stabilize $Sn^{2+}:5s^2$ and $Pb^{2+}:6s^2$ configurations in oxides, which demonstrates that with these cations the outer $5s$ or $6s$ states have become energetically accessible. Therefore, electrons can be introduced into the $5s$ band of SnO_2 and the $6s$ band of PbO_2. In SnO_{2-x} (Fig. 19a) the $5s$

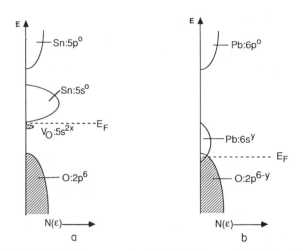

FIGURE 19 Electron energies for (a) SnO_{2-x} and (b) PbO_2. V_O is an oxygen vacancy trapping two electrons from the Sn:5s band.

conduction band is over 3 eV above E_v; but an oxygen deficiency introduces oxygen vacancies that trap out from the conduction band Sn-5s electrons in shallow, two-electron Sn-5s donor states. In PbO_2 (Fig. 19b) the Pb-6s conduction band appears to overlap the O-2p valence band $(E_v > E_c)$, thus eliminating E_g. On the other hand, introduction of additional conduction electrons, as is done by hydrogen insertion into PbO_2 on battery discharge, renders the system unstable with respect to a disproportionation reaction represented by

$$2Pb^{3+} \rightarrow Pb^{2+} + Pb^{4+} \qquad (56)$$

Lattice instabilities associated with trapping out the conduction electrons as pairs at specific Pb^{2+} sites plague efforts to increase the conduction-electron density to any significant concentration. In SnO_{2-x} they are already trapped as pairs at oxygen vacancies.

C. Problems with Valence-Band Holes

Alternatively it is possible to gain access to the $O^{2-}:2p^6$ valence band with strongly electropositive cations such as the alkali-metal ions A^+ and the larger alkaline-earth ions Sr^{2+} and Ba^{2+}. However, in this case holes introduced into the valence band become trapped as pairs in the homopolar O—O bonds of peroxide ions $(O_2)^{2-}$. Only where the covalent component of the M—O bond is strong and there is some overlap of the conduction and valence bands, as in PbO_2, are the valence-band holes not trapped out by O—O dimerization.

D. Transition-Metal Oxides

Transition-metal cations may have d^n or f^n configurations with energies lying within E_g that offer the possibility of obtaining partially filled d or f bands. However, the wave functions of partially filled d or f shells have smaller radial extensions, which reduces the interatomic-interaction parameter b of Eq. (54) and increases the intraatomic-interaction parameter U, so Eq. (55) may no longer be satisfied.

1. Problems with Outer $4f$ Electrons

The $4f$ electrons of a rare earth ion are tightly bound to their atomic nucleus and they are screened from near neighbors by closed $5s^2 5p^6$ shells, so the condition

$$W \ll U \tag{57}$$

prevails in the rare earth oxides. Under these conditions the $4f$ orbitals remain as crystal-field orbitals; they are not transformed into itinerant-electron states. Therefore $4f^n$ configurations remain localized; they impart localized atomic moments to the ions that are essentially identical to the localized atomic moments they impart to the free ions or atoms. In this situation, successive redox potentials are separated by a large energy $U > E_g$, which restricts the accessible $4f^n$ configurations on a given rare earth cation to at most two, and two only if one happens to lie within E_g.

If a $4f^n/4f^{n+1}$ redox couple does lie within E_g, as illustrated in Fig. 20a, then it is possible to obtain mixed-valent compounds in which E_F intersects the redox energy. In this case, metallic conduction could be expected were $b > \hbar\omega_R$, where ω_R is the frequency of a breathing-mode lattice vibration. However, b is so small that the near neighbors have time to relax about a mobile electron, thereby trapping it in a local potential well. These lattice relaxations stabilize the occupied states at the expense of unoccupied states, as does the molecular reorganization in a liquid for a given redox couple. The mobile electrons of the

mixed-valent state thus become "dressed" in a local lattice deformation, which introduces an activation energy into their mobility. These "dressed" electrons are called small polarons; they move diffusively, so $\mathbf{k}$ is no longer a good quantum number. The rare earth mixed-valent systems are not metallic.

On the other hand, where a broad conduction band overlaps a $4f^n$ energy level and E_F intersects the $4f^n$ energy level to give a $4f^{n+1}/4f^n$ mixed valence (in this case denoted an "intermediate valence"), hybridization of the $4f$ wave functions with the conduction-band wave functions may lead to "heavy-fermion" metallic behavior. Superconductivity has been observed in some heavy-fermion compounds (not oxides), but in all of them T_c is low.

2. Outer d Electrons

a. General considerations. The $4f$ electrons of a rare earth ion in an oxide are only weakly perturbed from their free-ion behavior; the outer s and p electrons are so strongly perturbed that they are transformed into itinerant electrons. The perturbations of the d wave functions in a transition-metal oxide are of intermediate strength. In this case also it is convenient to consider first the perturbations imposed by the nearest-neighbor metal–oxygen (M–O) interactions; these give rise to crystal-field orbitals containing the quantum-mechanical covalent mixing between overlapping cation and oxygen orbitals. Whether the crystal-field orbitals remain localized or are transformed into itinerant-electron band states depends on the relative strengths of the crystal-field intraatomic interactions U and the bandwidth W due to interatomic interactions between crystal-field wave functions on neighboring metal atoms M.

Covalent mixing between cation d and oxygen $2s$ and $2p$ orbitals in a transition-metal oxide has two important consequences. First, the fivefold orbital degeneracy of the free ion is at least partially removed by a crystal-field splitting of the energies of the crystal-field orbitals. Second, mixing of oxygen wave functions with the d wave functions spreads the crystal-field orbitals of d wave-function symmetry out over the oxygen atoms, which both reduces the intraatomic energy U and increases the bandwidth W; it also allows M–O–M as well as M–M interactions.

For half-filled crystal-field orbitals, the addition of one more electron to a cation costs an intraatomic energy

$$U = U' + \Delta_{ex}, \tag{58}$$

where U' is the energy required to add an electron to an empty orbital and Δ_{ex} is the additional electrostatic energy required to add it to a half-filled orbital. The term

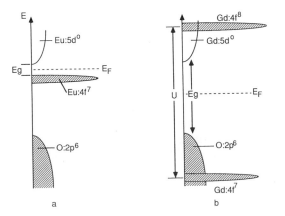

FIGURE 20 Electron energies for (a) EuO and (b) Gd$_2$O$_3$. E_F moves into the $4f^7$ energy level in Eu$_{1-\delta}$O containing a Eu$^{3+/2+}$ mixed valence.

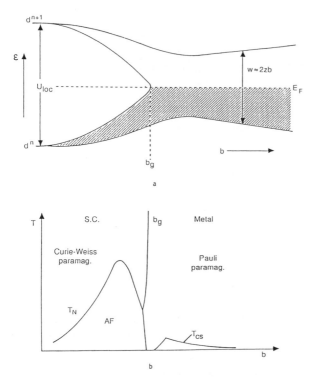

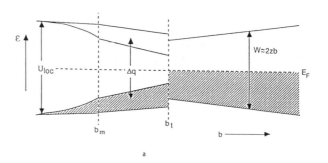

FIGURE 21 (a) Evolution of electron energies versus resonance integral b for interactions between nearest-neighbor like atoms on energetically equivalent sites. (b) Corresponding phase diagram for a single-valent system with a large separated atom U. (Half-filled band illustrated.)

Δ_{ex} enters U wherever an electron is added to a half-filled manifold. Figure 21 shows the evolution of U with increasing interatomic-interaction parameter b, where $W \approx 2zb$ in the tight-binding approximation with $W \gg U$. As b increases, screening of the electrons of a given manifold by electrons on neighboring atoms causes U to decrease rapidly with b in the vicinity where $W \approx U$ causes a transition from semiconducting to metallic behavior. In the domain $W < U$, the crystal-field orbitals remain sufficiently localized to impart an atomic magnetic moment, whereas in the domain $W > U$ the compound not only is metallic, but also has no spontaneous atomic magnetic moment. Clearly a necessary criterion for metallic conductivity in a single-valent transition-metal compound is the condition

$$W > U. \tag{59}$$

If the initial U at small b is relatively small, as may occur where

$$U = U', \tag{60}$$

then the bands may be so narrow at $W \approx U$ that the occupied states become split from the unoccupied states (Fig. 22) by a displacement transition that changes the

translational symmetry of the structure. Such displacement transitions exhibit atomic clustering: like-atom clustering where M—M or O—O interactions are important, M—O clustering in a disproportionation reaction where M—O—M interactions are dominant. Cooperative transitions are to static charge-density wave (CDW) states that compete with the superconductive state. This type of transition is not restricted to a single-valent situation; a longer wavelength CDW may be stabilized where like cations are present on energetically equivalent sites with a mixed valence.

For the case of a mixed valence on energetically equivalent lattice sites, a narrow band may also be split by small-polaron formation as discussed for the mixed-valent rare earth ions. Elimination of small polarons, a necessary criterion for metallic conductivity in mixed-valent systems, requires a bandwidth

$$W > h\omega_R, \tag{61}$$

where ω_R is the frequency of the optical-mode vibration that traps the charge carrier in a local lattice deformation.

These several considerations are best illustrated by some specific examples.

b. MnO, a single-valent compound with $W \ll U$.

Figure 23 illustrates the construction of an energy diagram for the antiferromagnetic insulator MnO. Comparison with Fig. 18 shows that it is similar to the construction

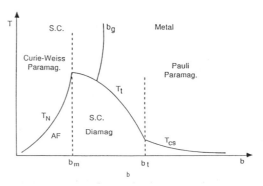

FIGURE 22 Same as Fig. 21 for a small separated atom U.

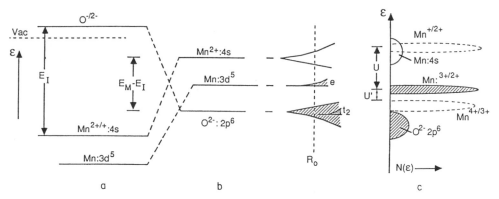

FIGURE 23 Electron energies for MnO: (a) free ions; (b) point charge; (c) crystal.

for MgO except for the appearance of a filled crystal-field d^5 configuration within the large energy gap between the empty Mn:$4s^0$ and the filled O:$2p^6$ bands. Because E_F lies above the top of the O:$2p^6$ band, the formal valence Mn(II) unambiguously assigns five d electrons per Mn.

The Mn atom sits in an octahedral interstice of oxygen atoms in the rock-salt structure of MnO; in this configuration the fivefold d-orbital degeneracy is split in two by a cubic crystalline field. The wave functions for the two crystal-field manifolds have the form

$$\psi_e = N_\sigma[f_e - \lambda_s\phi_s - \lambda_\sigma\phi_\sigma], \tag{62}$$

$$\psi_t = N_\pi[f_t - \lambda_\pi\phi_\pi], \tag{63}$$

where, from Fig. 24, the twofold-degenerate e orbitals of symmetry $x^2 - y^2$ and $[(z^2 - x^2) + (z^2 - y^2)]/\sqrt{2}\sigma$-bond with $2p_\sigma$ and $2s$ orbitals at neighboring oxygen atoms but are orthogonal to the O-$2p_\pi$ orbitals; the threefold-degenerate t_2 orbitals of symmetry xy, yz, zx π-bond with

the O-$2p_\pi$ orbitals but are orthogonal to the O-$2p_\sigma$ and O-$2s$ orbitals. The admixture wave functions ϕ_s, ϕ_σ, and ϕ_π are, respectively, linear combinations of nearest-neighbor O-$2S$, O-$2p_\sigma$, and O-$2p_\pi$ orbitals having the same symmetries as the atomic f_e or f_t orbitals with which they mix. The covalent-mixing parameters are defined as

$$\lambda_\sigma \equiv |b_\sigma^{ca}|/\Delta E \quad \text{and} \quad \lambda_\pi \equiv |b_\pi^{ca}|/\Delta E, \tag{64}$$

where $\Delta E = (E_d - E_p)$ is the energy required to transfer an electron from an O-$2p$ orbital to an empty d orbital at the point-charge Mn(I) = $3d^6$ energy. Because the overlap integrals entering

$$b^{ca} \equiv (\psi_{cat}, H'\psi_{anion}) \approx \varepsilon(\psi_{cat}, \psi_{anion}) \tag{65}$$

are larger for the σ-bonding orbitals, a $\lambda_\sigma > \lambda_\pi$ raises the energy of the antibonding crystal-field e orbitals relative to that of the t_2 orbitals. The crystal-field splitting

$$10Dq \equiv \Delta_c = \Delta_m + \tfrac{1}{2}\left(\lambda_\sigma^2 - \lambda_\pi^2\right)(E_d - E_p)$$
$$+ \tfrac{1}{2}\lambda_s^2(E_d - E_s) \tag{66}$$

between e and t_2 crystal-field energies contains only a relatively small electrostatic term Δ_m.

The energy difference between the Mn(I):$3d^6$ and the Mn(II):$3d^5$ manifolds is

$$U = U_t + \Delta_{ex}, \tag{67}$$

where U_t is the energy required to add an electron to an empty t_2 orbital. In MnO, a $\Delta_{ex} \approx 3$ eV makes both U and ΔE large, and a large ΔE makes λ_σ and λ_π relatively small. Therefore a $W \ll U$ and a $\Delta_c < \Delta_{ex}$ stabilize a localized $t_2^3e^2$ configuration at a Mn(II) ion, and a direct exchange interaction between spins in orthogonal orbitals couples the spins parallel—in accordance with Hund's highest multiplicity rule for free ions—to give a localized Mn(II)-ion magnetic moment of $5\mu_B$.

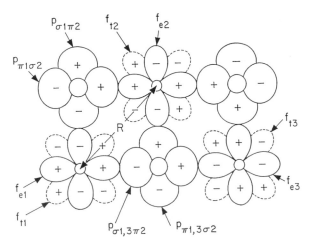

FIGURE 24 Illustration of cation d and anion p orbitals in the (001) plane of the rock-salt structure.

In the presence of a $W \ll U$, the interactions between half-filled orbitals on like atoms are treated in second-order perturbation theory:

$$\Delta\varepsilon \approx |t|^2/U, \qquad (68)$$

where the spin-dependent resonance integral is

$$t = b\sin(\Theta/2) \qquad (69)$$

because the rotation of a spin through an angle Θ transforms as

$$\alpha = \alpha'\cos(\Theta/2) + \beta'\sin(\Theta/2),$$
$$\beta = -\alpha'\sin(\Theta/2) + \beta'\cos(\Theta/2), \qquad (70)$$

and the Pauli exclusion principle allows transfer only of an antiparallel spin to an orbital that is already half-filled. Substitution of Eq. (69) into Eq. (68) gives a spin–spin contribution to the interatomic interaction of the form

$$H_{ex} = -2J_{ij}(\mathbf{s}_i \cdot \mathbf{s}_j),$$
$$J_{ij} = 2b^2/U, \qquad (71)$$

for $s = \frac{1}{2}$. This superexchange interaction has the form of the Heisenberg spin–spin interatomic-exchange interaction. It is responsible for long-range antiferromagnetic order below a Néel temperature T_N.

In MnO, 180° Mn—O—Mn interactions compete with Mn—Mn interactions. The magnetic order and the exchange striction below $T_N = 118$ K demonstrate that Mn—O—Mn interactions are dominant in this compound. Placement of E_F in a large energy gap between the Mn:$4s^0$ band and the Mn:$t_2^3e^2$ level makes MnO an antiferromagnetic insulator.

c. Li[Mn₂]O₄, a mixed-valent compound with W < ħω_R.

Because the Mn(II):$t_2^3e^2$ energy lies above the O:$2p^6$ bands, it is possible to oxidize Mn(II) to Mn(III) by the removal of a single e electron per Mn atom. The intraatomic electrostatic energy separating the Mn(II):$t_2^3e^2$ and Mn(III):$t_2^3e^1$ manifolds does not contain either Δ_{ex} or Δ_c; it is

$$U = U_e, \qquad (72)$$

which is small enough to retain the Mn(III) level above the top of the O:$2p^6$ band. Consequently it is also possible to remove the remaining e electron at a Mn(III) ion to oxidize it to Mn(IV):$t_2^3e^0$. The Mn(IV):$t_2^3e^0$ level, on the other hand, lies well below the top of the O:$2p^6$ band because it is separated from the Mn(III):$t_2^3e^1$ level by the relatively large energy

$$U = U_e + \Delta_c. \qquad (73)$$

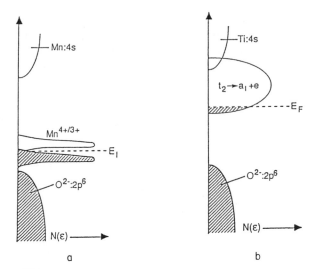

FIGURE 25 Electron energies for two spinels: (a) Li[Mn₂]O₄ and (b) Li[Ti₂]O₄.

Therefore an octahedral-site Mn(V) valence is not stabilized in oxides.

The cubic spinel Li[Mn₂]O₄ illustrates a mixed Mn(III) + Mn(IV) valence configuration on energetically equivalent octahedral sites. In this compound, the Fermi energy intersects the Mn(III):$t_2^3e^1$ energy level (Fig. 25a), and electron transport can occur via the reaction

$$t_2^3e^1 + t_2^3e^0 = t_2^3e^0 + t_2^3e^1. \qquad (74)$$

However, the bandwidth of the $t_2^3e^1$ level is so narrow that the time $\tau_h \approx h/W$ for an electron to hop to a neighboring site is long compared to the period ω_R^{-1} of the optical-mode lattice vibration that traps it as a small polaron. Therefore, the Mn:d^4 level is split by a polaron energy ε_p into occupied Mn(III):$t_2^3e^1$ and empty Mn(IV):$t_2^3e^0$ states in a manner analogous to the splitting of a Mn$^{4+/3+}$ redox couple in a liquid electrolyte. Small-polaron formation introduces an activation energy into the charge-carrier mobility, so Li[Mn₂]O₄ is a magnetic semiconductor; it is not a superconductor.

d. Li[Ti₂]O₄, a mixed-valent compound with W > ħω_R.

In contrast to the manganese oxides, which generally have localized d^n configurations, the titanium oxides generally have itinerant d electrons. For example, the cubic spinel Li[Ti₂]O₄ is a superconductor with $T_c = 13.7$ K. In this compound, the Ti—Ti interactions are strong enough to make $\tau_h < \omega_R^{-1}$ (i.e., $W > \hbar\omega_R$) for the electron-transfer reaction

$$\text{Ti(III):}t_2^1e^0 + \text{Ti(IV):}t_2^0e^0 = \text{Ti(IV):}t_2^0e^0 + \text{Ti(III):}t_2^1e^0, \qquad (75)$$

so band theory becomes applicable (Fig. 25b).

e. Ti_4O_7 and the bipolaron.

The mixed-valent compound Ti_4O_7 is obtained by removing oxygen from TiO_2. The oxygen vacancies order into shear planes so as to leave a close-packed tetragonal oxide-ion array as in TiO_2, but with the Ti atoms arranged to give TiO_2 slabs connected every four Ti atoms along the TiO_2 c axis by a shear plane across which Ti atoms share common octahedral-site faces rather than octahedral-site edges. A strong electrostatic repulsion between cations across a shear plane displaces the shear-plane Ti atoms away from each other. The possibility of a ferroelectric-type displacement of a Ti(IV) ion in an octahedral site stabilizes the shear-plane structure and locates the Ti–$3d$ electrons within the slabs.

At room temperature, both the Ti—Ti and the Ti—O—Ti interactions within a rutile slab satisfy the condition $W > h\omega_R$, and Ti_4O_7 is metallic. However, a first-order semiconductor–metal transition occurs at a $T_t \approx 150$ K due to Ti—Ti dimerization within the slabs; the electrons are trapped out as pairs within specific Ti—Ti homopolar bonds. Such a transition would be typical of cation clustering except that in the mixed-valent compound Ti_4O_7 the homopolar bonds are mobile in the temperature range $130 < T < 150$ K; they become stationary only below 130 K. These mobile homopolar bonds represent strongly coupled, localized electron pairs that, like small polarons, move diffusively in the crystal. Such a mobile electron pair is called a bipolaron. Formation of spin-paired bipolarons causes a sharp drop in the paramagnetic susceptibility. But the compound becomes a semiconductor, not a superconductor, and there is no Meissner effect.

f. TiO, a single-valent compound with $W > U$.

The titanium atom may be stabilized in single-valent oxides as Ti(II) in TiO, Ti(III) in Ti_2O_3, and Ti(IV) in TiO_2. This is possible, even though TiO_2 is an insulator with empty d orbitals some 3 eV above the top of the $O:2p^6$ band, because the energy separating the Ti(III):$t_2^1 e^0$ and Ti(II):$t_2^2 e^0$ levels is a relatively small $U = U_t$. Moreover, the absence of antibonding e electrons allows the Ti—O bond to be short relative to the radial extension of the t_2 crystal-field orbitals, so the overlap of t_2 orbitals on neighboring Ti atoms sharing a common octahedral-site edge is large enough to ensure a $W > U_t$ for Ti—Ti interactions and the overlap of Ti-t_2 and O-$2p_\pi$ orbitals is large enough to make $W \approx U_t$ for 180° Ti(III)—O—Ti(III) interactions.

The corundum structure of Ti_2O_3 contains, on a hexagonal basis, c-axis pairs of cations sharing a common octahedral-site face. Although Ti_2O_3 is a metal above room temperature, it becomes a semiconductor at lower temperatures because the d electrons become trapped in homopolar Ti—Ti bonds within the c-axis pairs.

TiO, on the other hand, is metallic and a superconductor ($T_c = 1.5$ K), even though it contains about 15% cation

and anion vacancies in its rock-salt structure that become ordered at lower temperatures. Here, also, the intraatomic energy $U = U_t$ contains no additional term Δ_c or Δ_{ex}, and some hybridization of titanium $4s$ and $3d$ orbitals increases W.

g. Oxides with only M—O—M interactions.

The perovskite and pyrochlore structures provide systems having 135 to 180° M—O—M interactions and no M—M interactions. A survey of the oxides with these structures shows that the condition $W > U$ may be fulfilled in single-valent oxides and the condition $W > h\omega_R$ may be found in mixed-valent oxides with these structures.

The relevant bandwidth $W \cong 2zb$ arising from M—O—M interactions is proportional to either λ_π^2 or λ_σ^2. A large covalent-mixing parameter λ_π or λ_σ requires, according to Eq. (64), a small $\Delta E = (E_d - E_p)$ and/or a large b^{ca}. The energy ΔE decreases with increasing formal charge on the cation and, for a given charge, on going to the right in any long period, provided that Δ_c or Δ_{ex} is not introduced into U on adding another d electron to compensate for the increased nuclear charge. However, b^{ca} also decreases on going to heavier atoms, but it increases on going from $3d$ to $4d$ to $5d$ orbitals. The overlap integrals are also sensitive to the M—O—M angle and to the character of the countercation A in the AMO_3 perovskites and the $A_2M_2O_7$ pyrochlores.

Perovskite and pyrochlore oxides containing partially filled $5d$ orbitals are generally metallic, whether stoichiometric as in ReO_3 or mixed-valent as in the Na_xWO_3 bronzes; but single-valent metallic conductivity, as in ReO_3, does not guarantee that the compounds are superconductors.

Perovskite and pyrochlore oxides containing partially filled $4d$ orbitals are intermediate in character; the $4d$ electrons are itinerant, but a $W \approx U$ may result in spontaneous magnetism.

The perovskite and pyrochlore oxides containing partially filled $3d$ orbitals commonly contain localized $3d^n$ configurations; however, several of these oxides have a $W \geq U$ and are either metallic or stabilize itinerant-electron antiferromagnetic order. In the perovskites, the homogeneous electronic picture in Fig. 21 does not hold where $W \approx U$; a static CDW and/or spin-density wave (SDW) may be stabilized as indicated in Fig. 22. Recently, dynamic displacive phase segregations have been identified in the metallic systems $Sr_{1-x}Ca_xVO_3$, $La_{1-x}Nd_xCuO_3$, and $Ln_{1-x}Ca_xMnO_3$ and the $LnNiO_3$ family (Ln = lanthanide). These dynamic segregations occur because there is a first-order change in the equilibrium M—O bond length on going from localized $3d$ electrons on the M atoms ($W < U$) to itinerant $3d$ electrons ($W > U$); the equilibrium M—O bond is longer for localized than for itinerant $3d$ electrons. In the LnNiO₃ family, the onset of a static CDW/SDW below a critical temperature T_t has been shown to be an order–disorder transition of localized-electron fluctuations in an itinerant-electron host; it is not due to either Fermi-surface nesting in a new Brillouin zone created by a change in lattice symmetry or to the homogeneous Mott–Hubbard transition illustrated in Fig. 21. In the $Ln_{1-x}Ca_xMnO_3$ system, all the Mn atoms carry a localized t^3 configuration with spin $S = 3/2$, and a single electron per Mn(III) occupies a narrow $(W \approx U)$ σ^* band of e-orbital parentage. In this case, the transition from localized e to itinerant σ^* electrons is approached from the localized-electron side, and ordering of the twofold-degenerate e orbitals promotes e-electron localization and charge ordering. In the absence of charge and orbital ordering, the system becomes ferromagnetic below a Curie temperature T_c, and a dynamic phase segregation in compositions with $W \approx U$ results in a "colossal magnetoresistance" (CMR) at temperatures $T \geq T_c$; a dynamically segregated, ferromagnetic, metallic phase of higher Curie temperature grows in an applied magnetic field at the expense of the host phase until it reaches a percolation threshold. This evidence of vibronic (hybridization of electronic and vibrational state) phenomena in oxides with perovskite-related structures is of great significance for our understanding of the high-temperature superconductivity in copper oxides.

h. Summary.

A review of the known properties of transition-metal oxides reveals the following generalizations that apply to compounds having an E_F above the top of the $O:2p^6$ bands.

1. Formal valences provide a count of the number of crystal-field d electrons per transition-metal ion. Any ambiguity in the distribution of d electrons among different transition-metal ions or between crystallographically inequivalent lattice sites can generally be resolved.

2. Single-valent oxides require a $W > U$ to be metallic, and W is larger for $5d$ than for $4d$ or $3d$ electrons. Itinerant $3d$ and $4d$ electrons are found only where U contains neither a Δ_c nor a Δ_{ex}.

3. Mixed-valent oxides require a $W > \hbar\omega_R$ to be metallic. Metallic mixed-valent oxides are more commonly superconductors.

4. Single-valent oxides having $W \approx U$ are not superconducting; electron correlations that introduce an enhancement of the magnetic susceptibility, even if they do not induce magnetic order at low temperatures, compete with superconductivity. Moreover, where a small U permits a large $N(E_F)$ compatible with $W > U$, static charge-density waves compete with superconduc-

tivity. In fact, CDWs may also compete in mixed-valent oxides.

5. At the crossover from localized to itinerant electronic behavior $(W \approx U$ or $W \approx \hbar\omega_R)$, a first-order change in the equilibrium M—O bond length can give rise to vibronic phenomena in both single-valent and mixed-valent transition-metal arrays.

i. Peculiarity of copper oxides.

Copper oxides are unusual in two respects. First, octahedral-site $Cu(II):t^6e^3$ contains a single e hole in the $3d$ shell, which makes it orbitally degenerate and therefore a strong Jahn–Teller ion; consequently, Cu(II) ions normally occupy octahedral sites that are deformed to tetragonal $(c/a > 1)$ symmetry by Jahn–Teller orbital ordering. However, in the absence of a cooperativity that stabilizes long-range orbital ordering, the electrons may couple locally to E-mode vibrations, forming vibronic states in a dynamic Jahn–Teller coupling. Second the $Cu(II):3d^9$ energy level lies below the top of the $O^{2-}:2p^6$ valence band in an ionic model; the introduction of covalent bonding creates states of e-orbital symmetry at the top of the $O^{2-}:2p^6$ bands that have a large $O\text{-}2p_\sigma$ component (see Fig. 26). Locally this $O\text{-}2p_\sigma$ component increases dramatically on oxidation of Cu(II) to Cu(III). The change in hybridization represents a polarization of the oxygen atoms that decreases the equilibrium Cu—O bond length, but the change in polarization is fast

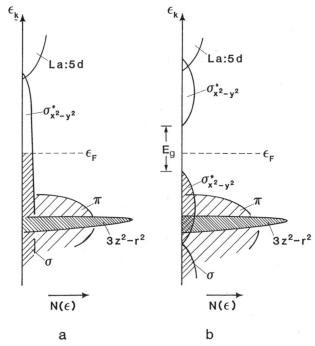

FIGURE 26 Schematic energy density of one-electron states for La_2CuO_4 with bandwidth x^2-y^2. (a) $W_\sigma > U$ and (b) $W_\sigma < U$, $E_g = U - W_\sigma$.

relative to the motion of the oxygen nucleus. Therefore, a dynamic vibronic phenomenon may reflect coupling to the polarization cloud of the oxygen atoms rather than to significant oxygen-atom displacements. Nevertheless, hybridization with a polarization wave on the oxygen-atom array would significantly increase the effective mass m^* of an itinerant electron.

The copper-oxide superconductors all contain CuO_2 sheets in which apical Cu—O bonds perpendicular to a sheet are significantly longer than the in-plane Cu—O bonds. This structural feature signals full occupancy of the $(3z^2-r^2)$ orbitals of an e-orbital pair. The parent compounds of the superconductive systems contain all Cu(II) in the CuO_2 sheets, which leaves the in-plane (x^2-y^2) orbitals half-filled with a

$$U = U_e + \Delta_{ex}. \qquad (76)$$

A $W < U$ results in localized (x^2-y^2) electrons that interact with one another on nearest neighbors by superexchange to give antiferromagnetic order within a semiconductive CuO_2 sheet. On oxidation of the CuO_2 sheets, the system undergoes a crossover from localized to itinerant electronic behavior, and a thermodynamically distinguishable p-type superconductive phase is found at crossover with a hole concentration x per Cu atom of the CuO_2 sheets in the range $0.14 \leq x \leq 0.22$. Superconductivity has also been observed on reduction of the CuO_2 sheets, but n-type superconductivity is more difficult to stabilize and has been studied much less.

IV. HIGH-T_c SUPERCONDUCTORS

A. System BaPb$_{1-x}$Bi$_x$O$_3$

1. Structure

The high-T_c superconductors are oxides having structures related to the cubic perovskite. The ideal cubic perovskite has the composition ABX_3, where A is a large cation, B is a smaller cation, and X is an anion. As illustrated in Fig. 27, the BX_3 array consists of a framework of corner-shared octahedra, and the large A cation occupies the center of each "cage" of the framework. For the A cation to fit easily into the cage, the A—X and B—O bond lengths must satisfy the following relation among the "ionic radii" R_A, R_B, and R_X:

$$t \lesssim (R_A + R_X)/\sqrt{2}(R_B + R_X), \qquad (77)$$

where t is called the Goldschmitt tolerance factor. Since the A—X and B—X bond lengths can, at best, be optimized simultaneously only at a single temperature for a fixed pressure, it is common in AMO_3 perovskites to find

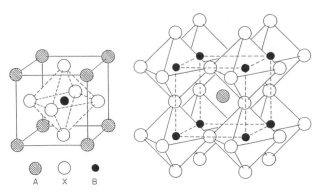

FIGURE 27 Two views of the ABX_3 cubic perovskite structure.

that optimization of the A—O interactions induces a distortion of the cubic MO_3 cage; these distortions—to orthorhombic, rhombohedral, or tetragonal symmetry—are accomplished by a cooperative rotation of the $MO_{6/2}$ octahedra that somewhat reduces the M—O—M bond angles from 180°.

The "cubic" perovskite structure can sustain a wide range of compositional variations. Partial substitution of any of the ions is possible: $A_{1-x}A''_xMO_3$, $AM_{1-x}M'_xO_3$, and $AMO_{3-x}F_x$ are known, for example. Removal of A cations and anions is also possible: the cubic bronzes Na_xWO_3 contain an A-cation deficiency, and the ReO_3 structure consists of just the cubic MO_3 array. Small concentrations of oxygen deficiency may be disordered, as in the superconductor $SrTiO_{3-x}$; but a large electrostatic repulsion between oxygen vacancies tends to introduce short-range order, at least, and commonly a long-range order that defines a new structural type.

2. Superconductive versus CDW State

Pure $BaPbO_3$ is a pseudocubic, metallic perovskite that is distorted to orthorhombic symmetry by a cooperative rotation of the $PbO_{6/2}$ octahedra; all the Pb(IV) ions are in energetically equivalent octahedral sites. Oxygen-deficient $BaPbO_{3-y}$ is an n-type metal and a conventional superconductor.

Pure $BaBiO_3$, on the other hand, is monoclinic, with two distinguishable bismuth octahedra obtained by a cooperative shifting of the oxygen atoms away from one near-neighbor bismuth toward the other so as to make the Bi—O bonds short and long in alternate octahedra. Such a "breathing-mode" oxygen displacement is indicative of a disproportionation reaction

$$2Bi(IV) \rightarrow Bi(III) + Bi(V) \qquad (78)$$

in which the energy gained by stronger covalent mixing in a $Bi(V)O_6$ more than compensates for the electrostatic energy U required to transfer an electron from one Bi(IV)

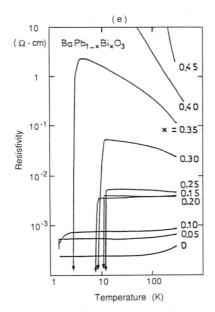

(e)

$BaPb_{1-x}Bi_xO_3$

0.45

0.40

x = 0.35

0.30

0.25
0.15
0.20

0.10
0.05

0

FIGURE 28 Temperature variation of the resistivity for various values of x in the system $BaPb_{1-x}Bi_xO_3$. [After Thanh, T. D., Koma, A., and Tanaka, S. (1980). *Appl. Phys.* **22**, 205.

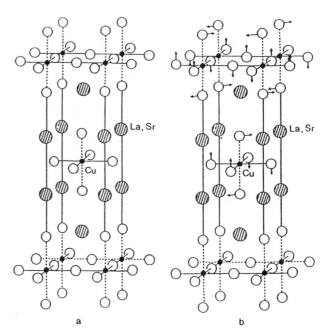

a   b

FIGURE 29 Structure of La_2CuO_4: (a) tetragonal and (b) orthorhombic cooperative $CuO_{6/2}$ rotations.

to the other. Physicists refer to such a spontaneous disproportionation as a "negative U" reaction.

The system $BaPb_{1-x}Bi_xO_3$ is pseudocubic in the compositional range $0.05 \leq x \lesssim 0.3$, and it is an unconventional superconductor with a T_c that increases with x to about 13 K at the limiting composition of the pseudocubic single-phase field (Fig. 28).

Superconductivity was first discovered in the $BaPb_{1-x}Bi_xO_3$ system by Sleight of DuPont. Although this perovskite system reaches a maximum T_c of only about 13 K, it is considered unconventional because such a T_c requires a large $V_{BCS}N(E_F)$ product [Eq. (42)], and a small measured $N(E_F)$ then requires an exceptionally large pairing potential V_{BCS}. Therefore, the system has been examined for clues to the strong coupling mechanism operative in the higher-T_c copper oxides (Fig. 1).

Superconductivity also appears on suppression of the static CDW of $BaBiO_3$ by substitution of more than 12% K^+ for Ba^{2+} in $Ba_{1-x}K_xBiO_3$; a maximum $T_c = 32$ K is found near $x = 0.4$, where the system becomes cubic. For $x > 0.47$, the system behaves as a normal metal without any superconductivity. The effect on T_c of substituting ^{18}O for ^{16}O gave a conventional isotope shift, $\alpha = 0.4$ to 0.5, which indicates that the BCS phonon-mediated pairing mechanism is operative in these systems. On the other hand, Kumal, Hall, and Goodrich have shown that the transition at T_c is fourth-order, not second-order, in the Ehrenfest classification.

B. Copper-Oxide Superconductors

1. Structure

Where the tolerance factor in Eq. (77) approaches unity, epitaxial (001) interfaces between an AX rock-salt layer and an ABX_3 perovskite layer are lattice matched. Nature recognizes this fact by stabilizing intergrowth structures $(AX)(ABX_3)_n$ in which perovskite layers alternate with rock-salt layers along an [001] axis. The La_2CuO_4 structure in Fig. 29, for example, is tetragonal at high temperatures, with LaO rock-salt layers alternating with $LaCuO_3$ perovskite layers on traversing the c-axis. Lattice matching requires a 45° rotation of the [100] axis of a rock-salt layer relative to that of a perovskite layer. As in the perovskite structure itself, the Goldschmitt tolerance factor of Eq. (77) is a measure of the mismatch of the equilibrium A—X and B—X bond lengths. Since the A—X and B—X bonds have different thermal expansions and compressibilities, matching ($t = 1$) of the bond lengths can be perfect only at a specific temperature for a given pressure, and the value of t calculated from tabulated ionic radii corresponds to room temperature at ambient pressure. On cooling, t decreases, and a $t < 1$ is compensated by a cooperative rotation of the $CuO_{6/1.5}$ octahedra about a tetragonal [110] axis (Fig. 29b), to lower the symmetry to orthorhombic. These rotations buckle the Cu–O–Cu bonds from 180° to $(180° - \phi)$, so the CuO_2 planes become CuO_2 sheets.

In the case of La_2CuO_4, the crystallographic c/a ratio is anomalously high because the Cu(II) ions distort their

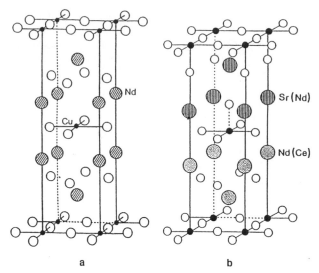

FIGURE 30 (a) T'-tetragonal structure of Nd_2CuO_4. (b) T^*-tetragonal structure of $Nd_{2-x-y}Ce_ySr_xCuO_4$.

octahedra to tetragonal ($c/a > 1$) symmetry with their long apical Cu—O bond along the c-axis. This c-axis ordering of the filled ($3z^2-r^2$) orbitals has an important structural consequence: the apical oxygen atoms are not strongly bound to the Cu atoms and may be removed from the oxygen coordination at a Cu(II). Consequently, the Cu(II) ion may be found in six-, five-, or fourfold oxygen coordination, but the strong square-coplanar bonding within a CuO_2 sheet is always maintained.

Replacement of La with a smaller trivalent rare earth ion illustrates well the weak bonding of the apical oxygen. Substitution of a smaller A cation lowers t, and the structure accommodates the bond-length mismatch by displacing the apical oxygen to tetrahedral interstices of an (001) La bilayer to form a fluorite Ln—O_2—Ln layer (Ln = Pr to Gd) as illustrated in Fig. 30a. This structure is labeled T'-tetragonal to distinguish it from the T-tetragonal phase of high-temperature La_2CuO_4. An important consequence for the chemistry of these phases is that the displacement of the apical oxygen in the T' phase places the CuO_2 planes under tension, whereas the CuO_2 sheets of La_2CuO_4 are under compression as a result of the bond-length mismatch. A tensile stress is relieved by adding antibonding (x^2-y^2) electrons to the CuO_2 planes; a compressive stress is relieved by removing antibonding (x^2-y^2) electrons from the CuO_2 sheets. As a result, the T' phase can only be doped n-type to give n-type superconductivity, whereas La_2CuO_4 can only be doped p-type. In fact, care must be exercised in the preparation of La_2CuO_4, as it may accept interstitial oxygen in the tetrahedral sites of the rock-salt bilayers to give $La_2CuO_{4+\delta}$; this composition phase segregates below room temperature to give filamentary p-type superconductivity in the oxygen-rich phase. In

the p-type system $La_{2-x}Sr_xCuO_4$, the larger Sr^{2+} ion relieves the compressive stress, and oxygen stoichiometry is more easily achieved as x increases.

In the T^*-tetragonal structure of $Nd_{2-x-y}Ce_ySr_xCuO_4$ (Fig. 30b), the larger Sr^{2+} ions order into alternate A-cation bilayers; the Sr^{2+} ions stabilize rock-salt bilayers, whereas the alternate bilayers have the fluorite structure. As a result, the Cu(II) are fivefold coordinated. Whether the Cu(II) are six-, five-, or fourfold coordinated, superconductivity requires preservation of the translational symmetry within a CuO_2 plane or sheet and, therefore, the same nearest-neighbor oxygen coordination for every copper atom within a plane or sheet.

The variable oxygen coordination at a Cu(II) also makes it possible to remove the apical oxygen atoms from a perovskite multilayer if the A-site cations of the perovskite block are stable in eightfold oxygen coordination. In fact, all the copper-oxide superconductors that would contain perovskite multilayers contain eightfold coordination of the A' cations (A' = Ca, Y, or a trivalent lanthanide) to form an $A'_{m-1}(CuO_2)_m$ layer, with integral $m \geq 2$; these superconductive layers alternate with AO—Φ—AO layers (A = La, Sr, Ba) that have a rock-salt AO interface. The intralayer composition Φ has a variable oxygen content and may be quite varied, as illustrated in Figs. 31 to 35.

The nonsuperconductive layer may act as a charge reservoir for the holes in the p-type superconductive layers. This situation is found, for example, in the $YBa_2Cu_3O_{6+x}$ system, where the $0 \leq x < 1$ oxygen in the BaO—CuO_x—BaO layers order into Cu—O—Cu chains

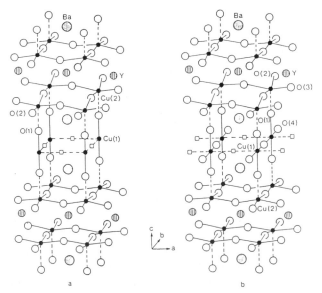

FIGURE 31 Structures of (a) tetragonal $YBa_2Cu_3O_6$ and (b) orthorhombic, ideal $YBa_2Cu_3O_7$.

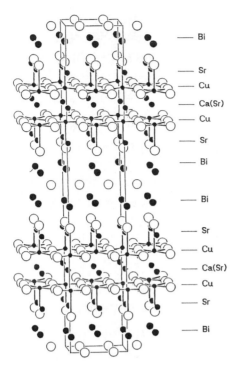

FIGURE 32 Tetragonal subcell of $Bi_2Sr_{3-x}Ca_xCu_2O_{8+y}$ showing the CuO_2 layers. Metal atoms are shaded and only Cu–O bonds are indicated. Oxygen atom positions for the Bi layers are idealized. [After Subramanian, M. A., *et al.* (1988). *Science* **239**, 1015.]

for $x > 0.4$; the fully formed chains are more conductive than the superconductive CuO_2 sheets and they become superconductive with the CuO_2 sheets. Moreover, displacement of the apical oxygen regulates the distribution of holes between the chains and the sheets. On

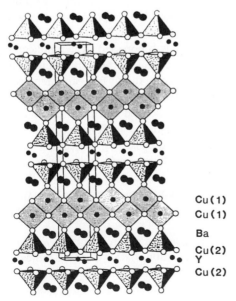

Cu(1)
Cu(1)
Ba
Cu(2)
Y
Cu(2)

FIGURE 33 Structure of $YBa_2Cu_4O_8$ showing double chains in the BaO–Cu_2O_2–BaO layer.

the other hand, the superconductive ($T_c = 16$ K) $RuSr_2$ $GdCu_2O_8 = (CuO_2GdCuO_2)(SrORuO_2SrO)$; is also ferromagnetic, with a Curie temperature of 133 K. As in the ferromagnetic perovskite $SrRuO_3$, the Ru atoms carry a magnetic moment $\mu_{Ru} \approx 1\ \mu_B$. Clearly the nonsuperconductive, ferromagnetic layer in this compound must be electronically isolated from the superconductive layer, but the internal magnetic field lowers T_c.

2. System $La_{2-x}Sr_xCuO_4$

a. Phase identification. The intergrowth structure of La_2CuO_4 (Fig. 29) is the simplest that exhibits p-type superconductivity. Holes may be introduced into the (x^2-y^2) band by creating A-site vacancies or interstitial oxygen; more useful is the substitution of an alkaline-earth ion A^{2+} for La^{3+} in oxygen-stoichiometric $La_{2-x}A_xCuO_4$. The $La_{2-x}A_xCuO_4$ system is of particular interest for two reasons: (1) the number x of holes per formula unit is unambiguously introduced into the (x^2-y^2) band of the CuO_2 sheets; and (2) the solid-solution range $0 \leq x \leq 0.3$ spans the entire range of superconductive compositions, as shown in the phase diagram in Fig. 36.

Crystallographically, there are two distinguishable phases in Fig. 36, a high-temperature tetragonal (HTT) and a low-temperature orthorhombic (LTO) phase, resulting from cooperative rotations of the $CuO_{6/1.5}$ octahedra. In the $La_{2-x}Ba_xCuO_4$ system, cooperative rotations about the [100] and [010] axes in alternate CuO_2 sheets produce a low-temperature tetragonal (LTT) phase below about 60 K in the range $0.12 \leq x \leq 0.15$. In Fig. 36, the LTO–HTT transition temperature T_t is seen to drop with increasing x, crossing T_c near $x = 0.22$.

The transport data distinguish three electronic phases below room temperature: an antiferromagnetic phase in the range $0 \leq x \leq 0.02$, a superconductive phase in the range $0.1 < x < 0.22$, and (3) an n-type metallic phase for $0.26 < x \leq 0.30$. At $x \approx 0.125$, there is a weak suppression of T_c vs x; in the $La_{2-x}Ba_xCuO_4$ system, superconductivity is completely suppressed in the range $0.12 \leq x \leq 0.13$ by the stabilization of a static CDW. Tranquada *et al.* have shown that the LTT phase of $La_{2-x}Ba_xCuO_4$ stabilizes at $x \approx 0.125$, a static CDW having the form of alternating hole-rich and antiferromagnetic stripes running parallel to the tetragonal [100] and [010] axes, respectively, in alternate CuO_2 sheets. Under hydrostatic pressure, superconductivity is restored to the compositions where it was suppressed by the static CDW. From X-ray absorption fine structure (XAFS), Bianconi *et al.* previously found evidence suggesting mobile stripes in superconductive samples, and an open question is whether the stripes are mobile in the superconductive phase or whether a related vibronic coupling characterizes the charge carriers.

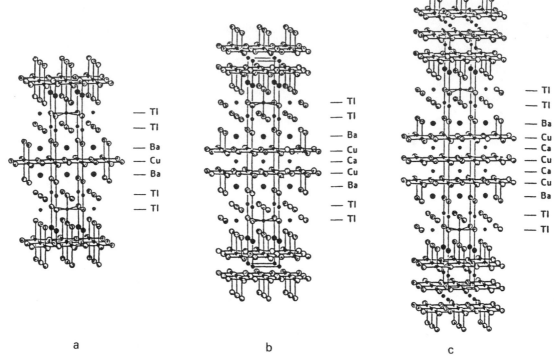

FIGURE 34 Structures of (a) $Tl_2Ba_2CuO_6$, (b) $Tl_2Ba_2CaCu_2O_8$, and (c) $Tl_2Ba_2Ca_2Cu_3O_{10}$.

b. Underdoped. The "underdoped" region $0 < x < 0.1$ supports polaronic conduction, but the thermoelectric power indicates that the nonadiabatic polorans are not small, centered on one Cu atom, but embrace about five Cu centers. A pseudo Jahn–Teller deformation of the in-plane square-coplanar coordination at a Cu(III) would be resisted by an elastic energy, but this energy would be reduced by cooperative deformations over several Cu centers. Calculation has shown that the gain in elastic energy would result in a polaron that embraced five to seven Cu centers. However, some other vibronic mechanism may be responsible for preventing the polaron collapse to a single Cu center. Within the polarons, antiferromagnetic order is suppressed, which indicates that the hole occupies a molecular orbital that includes all the Cu centers of the polaron. In this respect, it represents a mobile metallic phase in the antiferromagnetic matrix. As the volume of this second electronic phase increases, it breaks up the long-range antiferromagnetic order of the parent phase, which causes T_N to decrease precipitously with x from 340 K at $x = 0$. However, localized spins in regions of short-range order persist into the superconducting compositions; they give rise to a maximum in the paramagnetic susceptibility at a T_{max} that decreases with increasing x.

The appearance of a superconductive T_c that increases with x for compositions $0.05 \leq x \leq 0.10$ indicates that the polarons condense at lower temperatures into superconductive filaments. The transition temperatures T_F and T_ρ in Fig. 36 mark anomalies in the temperature dependence of the transport properties; others have noted anomalies in

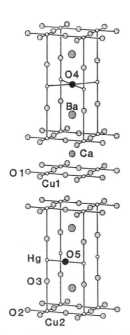

FIGURE 35 Structure of $HgBa_2Ca_2Cu_3O_{8+\delta}$. The O_4 and O_5 are the δ interstitial oxygen.

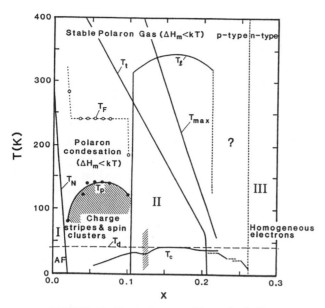

FIGURE 36 Phase diagram of La$_{2-x}$Sr$_x$CuO$_4$.

NMR and thermodynamic measurements at similar temperatures T^* (not shown in Fig. 36) that indicate the opening of a psuedogap, i.e., the lowering of the density of states, at ε_F. These anomalies appear to reflect interactions between polarons and their ordering into hole-rich stripes. Hunt *et al.* have used ^{63}Cu nuclear quadruple resonance (NQR) to reveal the presence of slowly fluctuating, quasitatic charge (i.e., hole-rich) stripes in the range $\frac{1}{16} \leq x \leq \frac{1}{8}$; the stripes become increasingly ordered on lowering the temperature, but the ordering temperature decreases with increasing x, vanishing at $x = \frac{1}{8}$. The polaron ordering apparently occurs within a parent phase that decreases in volume with increasing x; it is replaced by a single superconductive phase.

c. Optimally doped. The superconductive compositions $0.14 \leq x \leq 0.20$ exhibit a nearly temperature-independent thermoelectric power $\alpha(T)$ above T_l; there is no dramatic change in the evolution of $\alpha(T)$ with x above T_l. However, below T_l there is an abrupt change in the character of $\alpha(T)$ between $x = 0.10$ and $x = 0.15$. On cooling below T_l in the range $0.15 \leq x \leq 0.22$, $\alpha(T)$ increases relatively steeply to a maximum value at about 140 K, too high a temperature to be due to phonon drag. Zhou and Goodenough have shown that this unusual features is present in all the single-phase copper-oxide superconductors, and only where there are superconductive CuO$_2$ sheets. This feature reflects the appearance of itinerant quasiparticles of momentum $\hbar\mathbf{k}$ that have an unusual dispersion $\varepsilon_k(\mathbf{k})$ of their one-particle energies. Mihailovic *et al.* have used femtosecond time-domain spectroscopy to demonstrate a change from a polaronic to an itinerant character of the mobile holes on passing from the underdoped

to the optimally doped compositions in the YBa$_2$Cu$_3$O$_{6+x}$ system, and the Fermi surface of the itinerant quasiparticles in optimally doped CuO$_2$ sheets has been mapped with photoemission spectroscopy (PES). Most significant, angle-resolved PES as a function of temperature by Norman *et al.* and Dessau *et al.* has revealed a massive transfer of spectral weight on cooling from the π, π to the π, 0 directions within a CuO$_2$ sheet. These data indicate a progressive stabilization of the itinerant quasiparticles with $\mathbf{k}$ vectors along the Cu—O—Cu bonds of a CuO$_2$ sheet relative to those directed along a tetragonal [110] direction. The $\varepsilon_k(\mathbf{k})$ dispersion becomes extremely flat at the Fermi energy ε_F in the direction of the Cu—O—Cu bonds, indicating that the quasiparticles of the dominant population at ε_F have an unusually heavy mass m^*.

The origin of the heave mass m^* has not been resolved. T_c is expected to reach a maximum at the crossover from Cooper pairing to Bose condensation of bipolarons. Alexandrov and Mott have explored the bipolaron option most thoroughly. Markiewicz has argued extensively for trapping of the Fermi energy in a van Hove singularity. Although there is considerable evidence that the cuprates are close to the Bose–Einstein condensation regime, the PES data show the charge carriers are itinerant in the superconductive phase. Goodenough and Zhou have suggested that the large m^* is due to an unusual electron–lattice (or electron–polarization) interaction that gives rise to vibronic itinerant quasiparticles. The transfer of spectral weight in the angle-resolved PES spectra are consistent with the latter view, as are the data on the pressure dependence of T_c. A vibrational or polarization wave that is hybridized with a traveling-electron wave would be sensitive to changes in the bending angle ϕ of a $(180° - \phi)$ Cu—O—Cu bond. The hydrostatic pressure P decreases ϕ, and the T_c of the LTO phase increases with P, whereas a $dT_c/dP = 0$ is found for the tetragonal ($\phi = 0°$) phase. Moreover, epitaxial La$_{1.85}$Sr$_{0.15}$CuO$_4$ films on SrTiO$_3$ have their CuO$_2$ sheets under tension and the T_c is lowered; those on LaSrAlO$_4$ have their CuO$_2$ sheets under compression and the T_c is raised. The compressive stress built into the films on LaSrAlO$_4$ allows an added hydrostatic pressure achievable in a Cu—Be pressure cell to access at low temperature the tetragonal phase of the optimally doped La$_{1.85}$Sr$_{0.15}$CuO$_4$; T_c increased with P to 47 K, where it became P-independent on going from the orthorhombic to the tetragonal phase.

Below T_c, NMR Knight shift and other measurements have established that superconductive particles consist of two spin-paired electrons as in a conventional superconductor. However, a short coherence length $\xi_o \approx 15$ Å means that the coulomb repulsion between paired electrons is much stronger in the copper oxides. In a conventional superconductor, weak coulomb interactions result in pair wave functions with s-wave symmetry; the superconductive energy gap $2\Delta_0$ is finite over the entire Fermi surface. The pair wave functions in the CuO_2 sheets of the copper oxides have (x^2-y^2) d-wave symmetry; the energy gap $2\Delta_0$ has nodes along the tetragonal [110] and [1$\bar{1}$0] axes. This symmetry reduces the coulomb repulsion between the paired electrons.

d. Overdoped. The overdoped compositions $x > 0.25$ are not superconductors, and a change from p-type to n-type conduction signals a transfer of spectral eight from the lower and upper Hubbard bands of the $x = 0$ parent compound to Fermi-liquid states in the gap $(U-W)$ of the parent. Nevertheless, the metallic resistivity remains high with an anomalous temperature dependence, which indicates that the transition from vibronic to Fermi-liquid states may not be complete by $x = 0.3$.

The decrease in T_c with increasing x in the range $0.22 < x < 0.26$ is not smooth; it is characterized by a series of steps typical of phase segregation. Since T_t crosses T_c near this compositional range, these steps may reflect segregation of orthorhombic and tetragonal phases.

3. System YBa$_2$Cu$_3$O$_{6+x}$

The possibility of practical superconductive devices operating at the boiling point of liquid nitrogen (77 K) captured the imagination of the technical community on the discovery of a superconductive critical temperature of 90 K in YaBa$_2$Cu$_3$O$_{6.95}$. Although other superconductors with a higher T_c, a greater chemical stability, and cleavage planes that simplify fabrication into tapes and wires have since been discovered, the YBa$_2$Cu$_3$O$_{6+x}$, $0 \leq x < 1$, system continues to be of technical importance because, so far at least, films of YaBa$_2$Cu$_3$O$_{6.95}$ have been able to sustain the highest critical currents.

The structure, shown in Fig. 31, contains CuO_2−Y−CuO_2 layers and BaO−CuO_x−BaO layers. The oxygen atoms of the BaO buckled planes are c-axis apical oxygen atoms of the Cu in the CuO_2−Y−CuO_2 layers; these Cu all have fivefold oxygen coordination. The Cu of the BaO−CuO_x−BaO layers bridge the apical oxygen atoms with 180° O−Cu−O bonds oriented parallel to the c-axis. The x interstitial oxygen atoms in the BaO−CuO_x−BaO layers are mobile above 300°C, and their equilibrium concentration depends on the temperature and atomsphere. An air anneal at 400°C is used to obtain the optimally doped YaBa$_2$Cu$_3$O$_{6.95}$ composition. Since the apical oxygen atoms participate in the interstitial oxygen diffusion, it is important to ensure that the thermal history does not leave apical oxygen vacancies, which perturb the periodic potential of the

superconductive CuO_2−Y−CuO_2 layers and suppress T_c. It is interesting that the Y^{3+} ion may be substituted by any trivalent rare earth ion (with the exception of Pr) without influencing T_c significantly. Only in the case of Pr is there an important interaction between the lanthanide $4f$ orbitals and the (x^2-y^2) band of the CuO_2 sheets.

YBa$_2$Cu$_3$O$_6$ is an antiferromagnetic insulator with Cu(II) in the CuO_2−Y−CuO_2 layers and Cu(I) in the BaO−Cu−BaO layers; antiferromagnetic order between Cu(II) ions sets in at a $T_N > 500$ K. The initial interstitial oxygen atoms enter the BaO−Cu−BaO layers randomly and oxidize the neighboring Cu(I) to Cu(II). However, a threefold-coordinated Cu(II) attracts a second interstitial oxygen atom to form square-coplanar coordination, which initiates the formation of a chain segment. The twofold-coordinated Cu remain Cu(I), so the formation of chain segments initiates oxidation of the CuO_2−Y−CuO_2 layers. For $x < 0.3$, the chain segments remain disordered, so the crystallographic symmetry is tetragonal; but T_N drops precipitously with increasing oxidation of the CuO_2−Y−CuO_2 layers in the interval $0.1 < x < 0.25$ (see Fig. 37). For $x \geq 0.4$, the crystallographic symmetry is

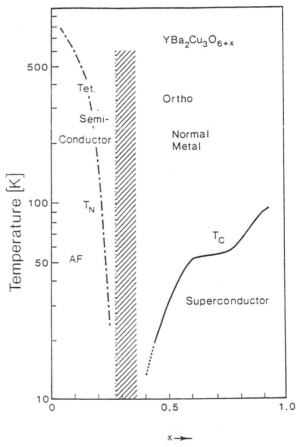

FIGURE 37 Phase diagram for the system YBa$_2$Cu$_3$O$_{6+x}$.

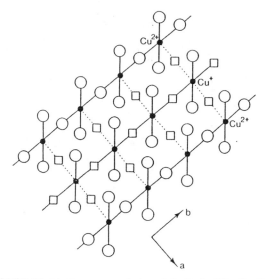

FIGURE 38 Ideal oxygen ordering of chains in $YBa_2Cu_3O_{6.5}$.

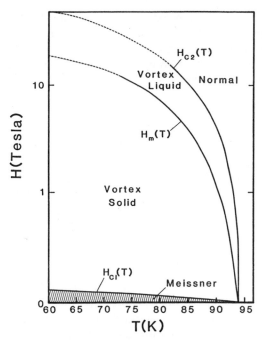

FIGURE 39 Upper and lower critical magnetic fields $H_{c2}(T)$ and $H_{c1}(T)$ and critical field for vortex melting $H_m(T)$ for $YBa_2Cu_3O_{6.95}$.

orthorhombic as a result of an alignment of the chains along the orthorhombic b-axis. The orthorhombic compositions are superconductors with a T_c that increases with x in two steps, a $T_c \approx 60$ K plateau appearing in the interval $0.6 < x < 0.8$. At $x = 0.5$, the chains order, alternating with Cu(I) b-axis rows as illustrated in Fig. 38. If the fully formed chains contained all Cu(II), the number of holes per Cu atom in the CuO_2—Y—CuO_2 layers would be 0.15 at $x = 0.5$, close to the optimal doping. However, the fully formed chains are oxidized beyond Cu(II), which reduces the number of holes per Cu atom in the CuO_2—Y—CuO_2 layers and makes the chains metallic conductors. As first pointed out by Cava, the chains act as charge reservoirs for the superconductive CuO_2—Y—CuO_2 layers. In $YBa_2Cu_3O_{6.95}$, the CuO_2—Y—CuO_2 layers are optimally doped with about 0.18 hole/Cu atom and the chains in the BaO—$CuO_{0.95}$—BaO layers are also superconductive.

Orthorhombic symmetry and superconductive chains are not determinants of the superconductivity of the CuO_2—Y—CuO_2 layers. By doping an equal amount of Ca for Y and La for Ba, the total hole concentration is kept constant. By 40% doping of Ca and La, the La in the BaO—CuO_x—BaO layers breaks up the chains into randomly oriented chain segments, changing the symmetry to tetragonal and suppressing superconductivity in the chain segments. As a result, the mobility of H_2O or CO_2 species in the nonsuperconductive layers is reduced, which suppresses chemical degradation at room temperature on exposure to the atmosphere, but the superconductive transition temperature is reduced only from 90 to 78 K.

Figure 39 shows the phase diagram of applied magnetic field H vs temperature T for clean $YBa_2Cu_3O_{6.95}$ with H parallel to the c-axis. The copper-oxide superconductors are all strongly Type II, and $H_{c1}(T)$ marks the transition between the Meissner phase and the vortex state. The vortex solid consists of an array of stationary vortices. The ability to grow crystals with a good surface quality has allowed imaging of the vortex lattice with scanning tunneling microscopy (STM). The vortex lattice does not show long-range order into the hexagonal close-packed structure; it represents a glassy state rather than a regular lattice. However, locally the flux lines are arranged in an oblique lattice, with approximately equal primitive lattice vectors forming an angle between them of $77 \pm 5°$. Moreover, the shape of the vortex cores is elliptical, not circular, with the long axis along an in-layer orthorhombic axis. These features reflect the anisotropy within an a–b plane that is induced by orientation of the chains of the BaO—CuO_x—BaO layers along the orthorhombic b-axis.

The strength of the pinning of the vortex solid in copperoxide superconductors depends on the coupling between layers. With weak coupling, the vortex lattice of one layer may be displaced relative to that of an adjacent layer, thereby bending the flux trajectory through the vortex cores. The interlayer vortex coupling is relatively strong in $YaBa_2Cu_3O_{6.95}$. Nevertheless, the vortex solid melts at an $H_m(T) < H_{c2}(T)$; the melting transition is weakly first-order. The vortices of the vortex liquid are mobile; moving vortices dissipate energy and introduce a finite resistance. Associated with $H_m(T)$ is an irreversibility line $H_{irr}(T) < H_m(T)$. Both $H_{irr}(T)$ and $H_m(T)$ decrease sharply with loss of oxygen from the BaO—CuO$_x$—BaO layers.

Pinning of the vortex solid is an extrinsic phenomenon, and considerable effort has been given to finding ways to increase the pinning. In films, surface roughness and the deposition of a protective $(Y_{1-x}Ca_x)(Ba_{2-x}La_x)Cu_3O_{7-\delta}$ overlayer increase $H_m(T)$ and therefore the critical current.

The brittleness of the ceramics and the two-dimensional superconductive layers makes fabrication of flexible tapes or wires a formidable challenge. Alignment of the layers from grain to grain is a critical requirement that is most easily achieved by deposition of films on a flexible metallic tape.

4. Layers with Three CuO$_2$ Sheets

The highest values of T_c have been obtained with structures containing layers with three CuO$_2$ sheets. Figure 35 shows the structure of HgBa$_2$Ca$_2$Cu$_3$O$_{8+\delta}$, which has a $T_c = 135$ K at ambient pressure and a $T_c = 164$ K under a quasi-hydrostatic pressure of 30 Gpa. The lattice oxygen atoms of the BaO—HgO$_\delta$—BaO nonsuperconductive layers supply the apical oxygen atoms of the two outer sheets of the CuO$_2$—Ca—CuO$_2$—Ca—CuO$_2$ superconductive layers; the Cu atoms of the inner CuO$_2$ plane have square-coplanar oxygen coordination. The Cu—O—Cu bond angles within the outer CuO$_2$ sheets approach the optimal 180°. The value of T_c at ambient pressure decreases from 135 K to 94 K as the number of O$_4$ interstitial oxygen in the HgO$_\delta$ planes decreases from 0.18 to 0.10 per Hg atom. The O$_4$ oxygen atoms oxidize the superconductive layers, the O$_5$ interstitial oxygen of the HgO$_\delta$ planes do not. This result implies near-optimal doping with 0.36 hole per formula unit in the superconductive layers. These holes would be distributed predominantly in the outer sheets with Cu in fivefold oxygen coordination, which would give a maximum hole concentration of 0.18/Cu atom in these sheets. Optimal doping is thus seen to correspond well with that in other p-type copper-oxide superconductors. Why pressure increases the T_c in HgBa$_2$Ca$_2$Cu$_3$O$_{8+\delta}$ is not known, but it is reasonable to assume that a redistribution of holes within the superconductive layers increases the coupling between the outer CuO$_2$ sheets.

5. Electron Superconductors

The equilibrium Cu—O bond length for Cu(II) ions in square-coplanar oxygen coordination is about 1.93 Å. To dope n-type a CuO$_2$ plane without apical oxygen atoms by reducing Cu(II) to Cu(I), it has been necessary to place the CuO$_2$ planes under tension so as to make the Cu—O bond of the compound larger than 1.93 Å. This feature is

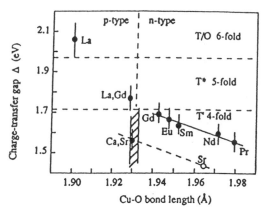

FIGURE 40 Charge-transfer gap vs Cu—O bond length for Ln$_2$CuO$_4$ oxides.

illustrated by the two copper-oxide structures that exhibit n-type superconductivity.

a. T'-Ln$_{2-x}$Ce$_x$CuO$_4$. The parent T' phases Ln$_2$CuO$_4$ have been prepared at atmospheric pressure for Ln = Pr,..., Gd; they have the structure of Fig. 30a, which has isolated CuO$_2$ planes having no apical oxygen. However, care must be taken to order the oxygen in the fluorite layers, as fivefold oxygen coordination at a few Cu atoms would perturb the periodic potential. Sensitivity to perturbations of the periodic potential is another indicator that the charge carriers are itinerant in both the p-type and the n-type superconductors.

The parent compounds contain only Cu(II) and are antiferromagnetic insulators such as the parent La$_2$CuO$_4$ compound with the T/O structure. Figure 40 shows, for room temperature, the magnitude of the energy gap $E_g = U - W$ vs the Cu—O bond length, which remains longer than the equilibrium bond length over the entire series of n-type superconductors. In each T' system Ln$_{2-x}$M$_x$CuO$_4$, M = Ce or Th, n-type superconductivity is found only in a narrow compositional range $0.10 \leq x \leq 0.18$, and at larger x a nonsuperconductive metallic state persists to lowest temperatures. Thus the n-type superconductors, like the p-type superconductors, appear as a distinguishable thermodynamic phase at a crossover from an antiferromagnetic-insulator to a metallic phase (Fig. 41). However, there are also significant differences between the p-type and n-type superconductors. For example, the charge carriers in the underdoped T' systems are conventional small polarons and T_N decreases only slowly with increasing x in a manner typical of a simple dilution with nonmagnetic Cu(I) ions. Moreover, the transition from antiferromagnetic semiconductor to superconductor appears to be a conventional first-order phase change occurring at a critical charge-carrier concentration x_c. With decreasing Cu—O bond length, there is a systematic increase in x_c

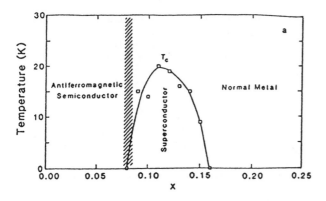

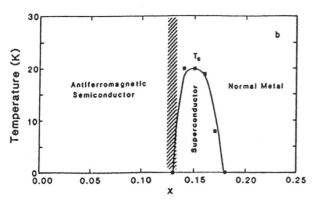

FIGURE 41 Variations of T_c with x for (a) $LaNd_{1-x}Ce_xCuO_4$ and (b) $Nd_{2-x}Ce_xCuO_4$. Shaded areas refer to two-phase regions.

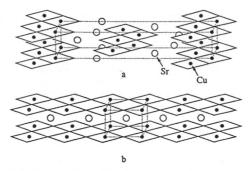

FIGURE 42 Comparison of the (a) atmospheric-pressure and (b) high-pressure forms of $SrCuO_2$.

and a decrease in the Ce solubility limit x_l that results in a narrowing of the superconductive phase field until, in Gd_2CuO_4, it disappears altogether. The appearance of n-type superconductivity is restricted not only to compounds with Cu in fourfold, square-coplanar coordination, but also to those where the Cu—O bonds of the CuO_2 planes are stretched sufficiently beyond 1.93 Å to give an $x_c < x_l$.

b. Infinite layers. The infinite-layer structure in Fig. 42 was first stabilized in the compound $Ca_{0.86}Sr_{0.14}CuO_2$; it has an equilibrium Cu—O bond length of 1.93 Å at room temperature. Synthesis at atmospheric pressure allows little variation in the Ca/Sr ratio, and attempts to dope the compound either p-type or n-type were unsuccessful. On the other hand, $SrCuO_2$ can be prepared under high pressure; it has a Cu—O bond length stretched to 1.965 Å, which satisfies the criterion for n-type doping. Therefore, $Sr_{1-x}Ln_xCuO_2$ (Ln $=$ La, Pr, Nd) were prepared under high pressure; they proved to be n-type superconductors with a $T_c \approx 30$ K.

6. Mechanism: An Open Question

The pairing mechanism in the copper-oxide superconductors remains an open question; a consensus on the char-

acter of the charge carriers in the normal state has yet to be reached. Most theorists have investigated the role of spin–spin exchange interactions without consideration of electron coupling to the lattice or to the oxygen polarization. Since the Cu(III) are diamagnetic, these efforts have been able to justify the separation of holes into charge stripes, but the charge separation can be achieved by other forces. To date, a convincing description of the high-T_c phenomenon has yet to emerge. Nevertheless, experiment has shown that spin fluctuations persist into the superconductive phase, and inelastic neutron scattering has revealed a commensurate (π, π) resonance peak in the spectrum of the antiferromagnetic susceptibility $\chi(q, \omega)$ that has a half-width in momentum space that varies linearly with T_c. These measurements define a characteristic velocity that is lower than a typical electron velocity at the Fermi energy and an order of magnitude smaller than a spin-wave velocity. These spin fluctuations could be associated with either mobile stripes or with slowly moving electron-density fluctuations as in a vibronic state. The data do not reveal what is the driving force for the formation of these density fluctuations.

ACKNOWLEDGMENT

Support of this work by the R. A. Welch Foundation, Houston, Texas, is gratefully acknowledged.

SEE ALSO THE FOLLOWING ARTICLES

BONDING AND STRUCTURE IN SOLIDS • ELECTRONS IN SOLIDS • SUPERCONDUCTING DEVICES • SUPERCONDUCTIVITY

BIBLIOGRAPHY

Alexandrov, A. S., and Mott, N. F. (1996). "Polarons and Bipolarons," World Scientific, Singapore.

Blatt, J. M. (1964). "Theory of Superconductivity," Academic Press, New York.

de Gennes, P. G. (1966). "Superconductivity of Metals and Alloys," Benjamin, New York.

Ginsberg, D. M. (ed.) (1996). "Physical Properties of High-Temperature Superconductors," Vol. 5, World Scientific, Singapore.

Ginzburg, V. L., and Kirzhmits, D. A. (eds.) (1982). "High Temperature Superconductivity" (A. K. Agyei, transl.; J. L. Birman, transl. ed.), Pergamon, Oxford.

Goodenough, J. B. (1972). *Prog. Solid State Chem.* **5,** 145.

Goodenough, J. B., and Longo, J. M. (1970). "Crystallographic and Magnetic Properties of Perovskite and Perovskite Related Compounds, in Landolt–Bornstein Tabellen," New Series Group III/4a, No. 126, Springer-Verlag, Berlin, New York.

Kaldis, E. (ed.) (1994). "Materials and Crystallographic Aspects of HT$_c$—Superconductivity," NATO ASI, Series E: Applied Sciences, Vol. 263, Kluwer Academic, Dordrecht.

Kittel, C. (1976). "Introduction to Solid State Physics," 5th ed., Wiley, New York.

Kulik, I. O., and Yanson, I. K. (1972). "Josephson Effect in Superconductive Tunneling Structures," Halsted, New York.

Kuper, C. G. (1968). "Introduction to the Theory of Superconductivity," Oxford University Press, London, New York.

London, F. (1950). "Superfluids," Vol. I, Wiley, New York.

Lynton, E. A. (1971). "Superconductivity," 3rd ed., Halsted, New York.

Markiewicz, R. S. (1997). *J. Phys. Chem. Solids* **58,** 1179.

McMillan and Rowell (1969). *In* "Superconductivity" (R. D. Parks, ed.), p. 561, Dekker, New York.

Manousakis, E. (1991). *Rev. Mod. Phys.* **63**(1), 1–62.

Mendelssohn, K. (1966). "Quest for Absolute Zero," McGraw–Hill, New York.

Newhouse, V. L. (ed.) (1975). "Applied Superconductivity," Academic Press, New York.

Rickayzen, G. (1965). "Theory of Superconductivity," Wiley (Interscience), New York.

Saint-James, D., Sarma, G., and Thomas, E. J. (1969). "Type II Superconductivity," Pergamon, Oxford.

Scalapino, D. J. (1995). *Phys. Rev.* **250,** 329.

Schrieffer, J. R. (1964). "Theory of Superconductivity," Benjamin, New York.

Solymar, L. (1972). "Superconductive Tunnelling and Applications," Halsted, New York.

Taylor, A. W. B. (1970). "Superconductivity," Wykeham, London, Winchester.

Tinkham, M. (1975). "Introduction to Superconductivity," McGraw–Hill, New York.

Wallace, P. R. (ed.) (1969). "Superconductivity," Gordon and Breach, New York.

Williams, J. E. C. (1970). "Superconductivity and Its Applications," Arrowsmith, Bristol, England.

Ziman, J. M. (1972). "Principles of the Theory of Solids," 2nd ed., Cambridge University Press, London, New York.

Superlattices

Michael J. Kelly
University of Surrey

GLOSSARY

Electron volt (eV) Energy change of one electron whose potential changes by 1 V. It has a value of 1.6×10^{-19} J. Visible light has an energy of 1.5–3 eV per photon.

Epitaxy Growth of a crystal upon a substrate whose own crystal structure is continued in the overlayer.

Heterojunction Abrupt interface within a single crystal between two materials of different composition.

Homojunction Abrupt interface within a single crystal between two regions of different doping levels.

Metal–organic chemical vapor deposition (MOCVD) Cracking of gases over a substrate to form semiconductor overlayers, e.g., trimethyl-gallium and arsine to form gallium arsenide (and methane).

Molecular beam epitaxy (MBE) Deposition of semiconductor thin films under ultrahigh vacuum conditions using beams of constituent species, e.g., gallium and arsenic to form gallium arsenide.

Quantum well Thin layer of a narrower-gap semiconductor sandwiched between thicker layers of a wider-gap semiconductor, presenting a potential well for electrons in the conduction band and holes in the valence band. Quantum size effects are seen in the properties of these electrons and holes if the thin layer is only a few atomic layers thick. Multiple quantum wells refer to repeated thin layers where the intervening wide-gap semiconductor layers are relatively thick.

Superlattice A single crystal formed by the periodic repetition of a set of multilayers whose individual thicknesses range from a few to about 10 atomic layers.

Tunnel barrier A thin layer of a wider-gap semiconductor sandwiched between thicker layers of a narrower-gap semiconductor. Low-energy incident electrons can tunnel through the thin layer rather than be reflected as would happen classically.

THE DEMAND for ever-increasing performance from semiconductor devices has led to the development of technologies to grow semiconductor crystals an atomic layer at a time. Note that one atomic layer is about 0.3 nm thick. Within each layer, the key semiconductor properties of the band gap (difference in energy between the top of the valence bands filled with electrons and the bottom of the empty conduction bands) and the level of doping can be controlled independently. Atomically abrupt interfaces between two semiconductors, called *heterojunctions*, have applications in lasers, transistors, and microwave and optoelectronic devices. In the limit, the periodicity of multilayers can be engineered during growth, resulting in the formation of artificial crystal structures, *superlattices*, with useful properties not found in naturally occurring semiconductors. Research over the last decade has concentrated on achieving ever-greater control over growth and also the formation of patterns within layers with nanometer-scale precision. Instead of thin layers, thin wires and dots have been engineered where quantum size effects are observed in two and all three spatial dimensions (as opposed to the quantum size effects in one spatial dimension in quantum wells).

I. SEMICONDUCTOR HETEROJUNCTIONS AND SUPERLATTICES: NEW PHYSICS AND NEW DEVICES

The electronic and optical properties of crystals are determined by the chemistry of the different atomic constituents, the crystal structure (i.e., the relative positions of the different atoms), and the distance over which a basic unit of the crystal structure is repeated. In semiconductors such as silicon and gallium arsenide these repeat distances are of the order of two atomic spacings, and the naturally occurring forms of these materials at room temperature are in the diamond and wurtzite crystal structures, respectively. It is now possible to grow semiconductor crystals atomic layer by atomic layer, controlling the chemical composition within each layer, and, for example, to interleave two semiconductor compositions with prescribed repeat distances (say repeats of two atomic layers of one and four of the other). The electronic and optical properties of such engineered crystals, called *superlattices*, now depend in part on the repeat distances, and the semiconductor device performance can be tailored. For example, while gallium arsenide lasers operate in the infrared, and aluminium arsenide does not lase at all, repeated multilayers of the two lase in the red. During the 1970s and 1980s, the physics and applications of epitaxially grown semiconductor multilayers came to dominate much of semiconductor science and technology. Even a single interface between semiconductors—a *heterojunction*—can be used to trap

electrons and force them to move in two, rather than three, dimensions. The physics of such constrained motion is quite different from that of the motion in bulk semiconductors, and lower-noise, more temperature-stable devices have emerged.

The intrinsic quantum wavelength (the de Broglie wavelength) associated with an electron in a semiconductor at room temperature is of the order of 20 nm, or 60 atomic spacings. Once the dimensions of active regions of structures or devices become this small, one can anticipate quantum effects being important in determining the electronic and optical properties. Furthermore, the typical distance that an electron moves in a semiconductor device between scattering events is also about 20 nm: one might anticipate ballistic motion, i.e., motion without scattering, and recover vacuum valve-like behavior in small regions of a solid.

At the same time that multilayers are controlled to atomic layer precision, the lithography used to pattern surfaces is also reaching down to the nanometer scale. Clearly electronic devices cannot be miniaturized forever and still retain recognizable macroscopic behavior. Few electrons will be present if we make a device very small, and research has been active for a decade on structures which carry on average at most one current-carrying electron: their resistance at low temperatures can take only the value of $h/2e^2 = 12.91$ kΩ divided by an integer, and this is qualitatively different from anything macroscopic.

II. SEMICONDUCTOR CRYSTAL GROWTH AND TECHNOLOGIES

In the decades since the 1950s, the advances in semiconductor-based technologies (e.g., computation performance per unit area on a chip or per unit cost) has been exponentially rapid. This has been possible in part because of the ability to define the internal interfaces of device structures with ever-growing precision. The earliest p–n–p and n–p–n transistors of the 1950s were made by successive alloying of the dopants into the semiconductor, resulting in interfaces several micrometers (μm) thick and nonplanar (i.e., the interfaces followed the melt fronts). In the following decade, implantation of the dopants as high-energy ions, followed by annealing of the crystal to remove damage, resulted in planar interfaces, but not more than about 1.5 μm below the surface and with an interface thickness of $\sim$0.1 μm. Liquid-phase epitaxy was introduced in the 1970s. Here the growing crystal is pulled from a melt whose concentration of dopants can be changed. The interfaces are planar and of the order of 0.01 μm thick, and a much greater control and range of doping values can be achieved. Finally, in the 1980s, two forms of epitaxial growth came to dominate the methods of fabricating sophisticated semiconductors crystals.

Molecular beam epitaxy (MBE) is basically an ultra-precise form of evaporation, in which atomic or molecular beams impinge on a heated semiconductor substrate in a very high vacuum, at such a rate as to produce mono-layer coverage in about 1 sec. The most common material system with which MBE has been developed is the alloy $Al_xGa_{1-x}As$ (including the binary end points gallium arsenide and aluminium arsenide), which can be grown by controlling the relative fluxes of Al and Ga (in parallel with a flux of As_2 or As_4 molecules) to produce the desired alloy composition. The Al/Ga ratio can be altered in typically 0.1 sec, so producing an interface that is abrupt on the scale of atomic diameters. [In practice, the details of surface coverage of the molecular beams and the limited surface mobility of the added species means that the layers are not continuous over the surface but have regions differing by a monolayer with atomic steps between them. These steps can be frozen into growth, giving an effectively greater interface thickness for device applications. This produces problems for tunnel devices (see Section III.A).] One can also change the flux ratio continuously and produce linear and other composition gradings over prescribed distances. Fluxes of dopant species (Be, Si, etc.) can also be introduced for engineering the doping profiles. A schematic of an MBE chamber is shown in Fig. 1a.

In contrast, metalorganic chemical vapor deposition (MOCVD) is basically a highly controlled form of cracking, whereby $Ga(CH_3)_3$ and AsH_3 gases are carried in a stream of H_2 over a heated substrate to form GaAs and CH_4. The alloy and doping are produced with $Al(CH_3)_3$ and with $Zn(CH_3)_2$ and SiH_4 gases, respectively. The growth rates are typically 10 times faster than for MBE, and without extraordinary effects at gas flow control (which are possible); the interfaces between the layers of different alloy compositions are thicker, at two to six atomic monolayers, than the single monolayer from MBE. A schematic of an MOCVD reaction cell is shown in Fig. 1b.

The abruptness of the doping profiles is more problematic, as while, in principle, one can deliver dopants to a single atomic plane, the detailed thermodynamics and kinetics of growth in some cases mean that the dopant can rise with the growth surface. Remembering that dopant concentrations are typically at the parts per million level, for many applications this effect can be allowed for without degrading the device performance.

These two epitaxial techniques have approached the ultimate limits in tailoring multilayer materials. Early work concentrated on the growth of heterojunctions that were closely lattice matched, and the $Al_xGa_{1-x}As$ alloy system

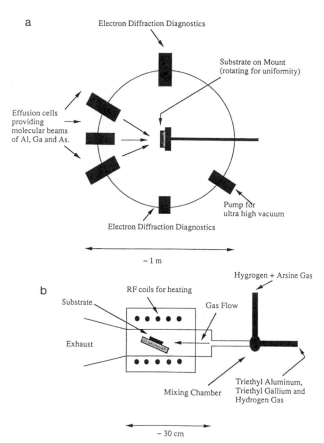

FIGURE 1 Schematics of (a) an MBE growth chamber and (b) an MOCVD reaction cell.

has been particularly convenient, as the lattice constant varies by only <0.2% between $x = 0$ and $x = 1$. Crystal growers have become increasingly bold, so that large strains (even higher than 10%) can be incorporated in very thin layers in semiconductor systems. In addition to the strained layer systems in the group III–V semiconductors, the 1990s saw many research advances in the group IV system Si/Si_xGe_{1-x} and in the group II–VI systems such as CdTe/ZnTe. There have also been efforts at growing metallic multilayers (in this case mainly by sputtering) and metal–semiconductor and semiconductor dielectric multilayers. Some attempts have also been made to combine elements of MBE and MOCVD to get the best from both.

III. PHYSICS AND APPLICATIONS OF SINGLE AND MULTIPLE HETEROJUNCTIONS

At room temperature, intrinsic gallium arsenide has a band gap of 1.42 eV separating filled valence and empty conduction bands. The alloy $Al_xGa_{1-x}As$ (at least for $x < 0.4$) has a direct band gap of $1.42 + 1.247x$ eV. For $x > 0.4$ the

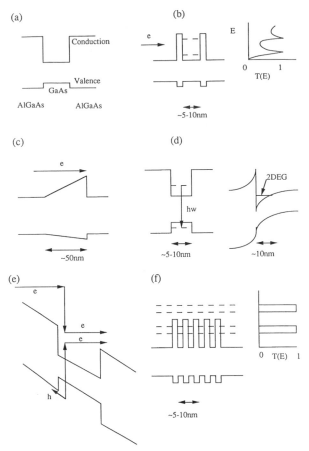

FIGURE 2 The physics from semiconductor multilayers. (a) Typical conduction and valence band profiles for the GaAs/AlGaAs materials system; (b) the double-barrier structure for resonant tunneling; (c) a graded gap layer for hot electron injection; (d) quantum confinement of electrons in a quantum well and at a heterojunction interface; (e) avalanching at a heterojunction in a multilayer structure; (f) energy filtering of electrons in a periodic multilayer (superlattice).

alloy is an indirect-gap semiconductor with more complicated energy bands, which also means that it is not optically active. At the abrupt interface between GaAs and $Al_xGa_{1-x}As$, about 65% of the band-gap discontinuity appears as a potential step in the conduction band. Thus for $x = 0.3$, we have a potential step of ~ 0.24 eV for electrons. This step can be used in several ways (see Fig. 2) to generate and exploit physics not available in bulk uniform semiconductors. Our discussion here is confined to the conduction bands and excess electrons in n-doped material: the equivalent physics applies to holes in valence bands but the energy band structure is much more complicated.

A. Tunneling

In classical mechanics, an incident electron is reflected by a potential barrier if its height is greater than the electron kinetic energy. In quantum mechanics, an electron has a small but finite probability of penetrating into the barrier. If the barrier is sufficiently thin (i.e., of the order of the relevant de Broglie wavelength in the barrier), the electron has a finite probability of emerging from the other side and the electron is said to have tunneled through the barrier. Tunneling has been studied widely for several decades: in many cases it is manifest as a type of leakage current through thin layers. Resonant tunneling is a more striking phenomenon. If two thin barriers are separated by a thin intervening layer (see Fig. 2b), there are discrete energies at which the electron transmission grows to unity: a careful analysis shows that there is a constructive interference between electrons traveling forward and those traveling backward in the central layer. If such a structure is made in the GaAs/AlGaAs materials system, the barrier layers typically have $x \sim 0.4$ and the barrier and central layers on either side are of the order of 3–10 nm thick. The GaAs layers on either side are doped to provide electrical contacts, and under an applied bias, electrons tunnel through the multilayer. When, under bias, electrons in the contacts have the same energy for the high transmission, the current rises (see Fig. 3). It falls for a further increase in bias, because the transmission falls, and so the structure provides a negative differential resistance, the basic prerequisite for oscillator circuits. Indeed these structures have operated at up to 420 GHz in the GaAs/AlAs materials system and up to 720 GHz in the InAs/GaSb system, which represents the fastest frequency achieved by a solid-state device. Variations on this simple structure have been used extensively to clarify the fundamentals of quantum transport of electrons and holes in semiconductors.

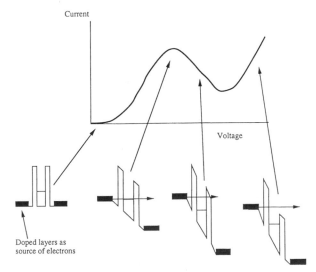

FIGURE 3 The current–voltage characteristics of a resonant tunnel diode.

If one of the two barriers in Fig. 3 is removed, we have an n–i–n structure with a tunnel barrier toward one end. This structure acts as a microwave detector diode (called an ASPAT diode) with a number of attractive properties including a low sensitivity to ambient temperature, a wide dynamic range, and low associated noise. However, this device depends very sensitively on the barrier parameters, and any roughness on interfaces contributes to an unacceptably wide spread in electrical performance.

B. Hot Electron Injection

With reference to Fig. 2c, which typically has the aluminium fraction rising from 0 to 40% over 50 nm, the tunneling probability through this barrier is very low. Under bias, an applied electric field can reduce the "quasi-electric field" represented by the rising conduction band edge (see Fig. 4). Even when the electrons are in the AlGaAs layers, they have the typical kinetic energy associated with thermal electrons, just as in regions away from the barrier. Once they cross the heterojunction they gain typically 0.3 eV as excess kinetic energy. Such electrons are called hot, as this excess energy is typical of that if they were in a solid at 3000 K. The potential energy the electrons receive as they move against the quasi-electric field is converted into forward kinetic energy, implying that the electron motion is strongly forward collimated. The transit time for such electrons over short distances is reduced by having crossed the heterojunction.

Heterostructures such as this have several device application. In heterojunction bipolar transistors the reduced transit time across the base makes for faster transistor action. The heterojunctions also allow heavier doping to be tolerated without unwanted minority carrier effects coming into play, and this also helps increase the speed. In the Gunn diode used as the source of microwave power, electrons are accelerated (heated) in a strong electric field so that peculiarities of the gallium arsenide band structure at $x > 0.3$ eV can be exploited. In a uniform piece of gallium arsenide, electrons with excess energy can lose that excess to the lattice. The structure in Fig. 4 allows the electrons to remain "cold" until they receive all their excess energy as they cross the heterojunction. The layer of graded composition alloy acts as a hot electron launcher and, in this mode of operation, has led to a new generation of Gunn diodes with greater efficiency, lower noise, and reduced sensitivity to operating temperature. The quest for a unipolar hot electron transistor has continued without success since the 1960s, and the hot electron launcher looked to be a promising candidate in the early 1980s. The way in which the hot electrons relax when injected into a heavily doped GaAs base has proved to be an interesting field of physics: in the process of such studies, the relaxation rate was found to be too fast for useful transistor action. Research has continued with other materials, particularly InAlAs/InGaAs, where high-speed transistor action and small-scale integration have been demonstrated: different band-gap offsets allow even higher-energy injection, and so more energy can be lost in the base while still being able to be collected.

C. Quantum Confinement

The two structures shown in Fig. 2d each form the basis of successful commercial devices, namely, the quantum well laser and the heterojunction field-effect transistor, which have superior properties by virtue of the quantum confinement of electrons into quasi-two-dimensional layers. The electron states in a thin layer sandwiched between layers of a wider band gap have a kinetic energy of confinement in the direction perpendicular to the layers of $\sim h^2/2m^*t^2$, where t is the thickness of the layer, m^* is the effective mass of the electrons in the semiconductor, and h is Planck's constant. This result is a simple application of the particle-in-a-box problem in elementary quantum mechanics. For $t = 10$ nm in gallium arsenide, this extra energy is of the order of 50 meV, an appreciable fraction of the potential step at the heterojunction. (There are added complications for thinner wells when this energy equals or exceeds the potential step, but these are understood.) For holes in the valence band, the appropriate hole mass must be used. Both electrons and holes are free to move in the plane of the layer, i.e., in two dimensions. Whereas in bulk semiconductors an electron and hole at the conduction and valence band edges could recombine and emit

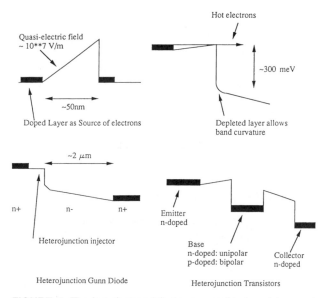

FIGURE 4 The hot electron injector as used in transistors and Gunn diodes.

a photon of light with an energy equal to the band gap of the semiconductor, here we see from the diagram that the extra kinetic energy of confinement for both electrons and holes must be added on. This was the first example of tailoring the optical properties of semiconductors, not by bulk crystal structure and chemistry, but by local structure via the thickness t. The additional energies are sufficient to decrease the wavelength of light emitted from a gallium arsenide layer from 0.86 to as little as 0.66 μm, i.e., from the infrared into the visible red. This principle and indeed other attractive, but technical, features of electrons in two dimensions have led to the current generation of quantum well lasers which have a greater efficiency, lower threshold currents, a reduced dependence on temperature, and shorter wavelengths than the more conventional lasers in gallium arsenide.

The second structure in Fig. 2d results from the growth of a wider-gap n-doped layer of semiconductor on a lower-gap layer of undoped semiconductor. Under thermal or optical activation, the donors give up their electrons as in bulk materials. Here it is possible for them to cross the heterojunction, beyond which they lose their ~0.25–0.3 eV of excess energy before coming into equilibrium. The potential step at the heterojunction prevents them from returning to their donor sites. At the same time the coulomb attraction between the positive donor ion and the electron remains. The electrons are thus trapped at the interface but are free to move parallel to it. The electrons are said to form a quasi-two-dimensional electron gas, which has been the basis of exciting physics arising from the strong reduction in coulomb scattering and the reduction of lattice scattering at low temperatures. Electrons at low temperatures can propagate macroscopic distances (~1 mm) without scattering: regarded as waves, they maintain phase memory over these distances. The low-temperature low-electric field mobility is as much as 1000 times greater than can be achieved in high-quality bulk semiconductors. The quantum Hall effect is seen in this system when a magnetic field is applied perpendicular to the layers: the Hall resistance (the lateral voltage induced by the magnetic field divided by the current) takes on the values of $h/e^2 i = 28.82 \ldots / i$ kΩ, where h is Planck's constant, e the charge on the electron, and i an integer. This relation is held to a precision of 1 part in 10^8, so that it is now a resistance standard. A new and complex many-electron fluid is formed when very high magnetic fields are applied—the so-called fractional Hall regime. A range of quantum interference effects is also being discovered where electron currents are split and recombined all within a distance over which electron phase coherence is maintained. (See Section V.)

The same two-dimensional electron gas is at the heart of the high-electron mobility transistor (HEMT; see Fig. 5),

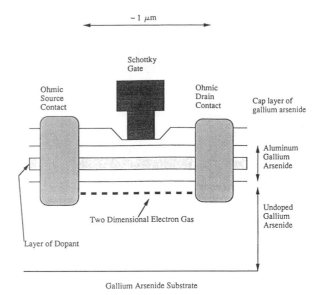

FIGURE 5 The high-electron mobility transistor.

which is a heterojunction version of the GaAs MESFET (metal–semiconductor field effect transistor, which uses a layer of bulk doped GaAs instead of the two-dimensional electron gas in Fig. 5). The process of placing the doping only in the wider-gap semiconductor and set away from the heterojunction is known as modulation doping. The new transistor has two advantages over the MESFET: (1) the parasitic resistances in the low-field regions not under the gate are lower, leading to higher-frequency operation if all other geometric features are the same; and (2) the noise of amplifier operation is much lower, due in part to the restricted final states into which two-dimensional electrons might be scattered. The HEMT is in widespread use as a low-noise amplifier.

In both the quantum well laser and the high-electron mobility transistor, the heterojunction between the two semiconductors is used as a new design feature to improve the device operation in terms of greater efficiency, lower noise, and reduced sensitivity to temperature.

IV. PHYSICS AND APPLICATIONS OF SUPERLATTICES

A. Multiple Quantum Wells

Quantum wells that are separated by more than about 15–20 nm are effectively noninteracting. A structure with typically 50–100 quantum wells over a thickness of 1–3 μm behaves as an unusual solid. For the optical absorption, the effects of each well add together. For transport we have effectively a two-dimensional solid, with easy conduction parallel to the wells but not perpendicular.

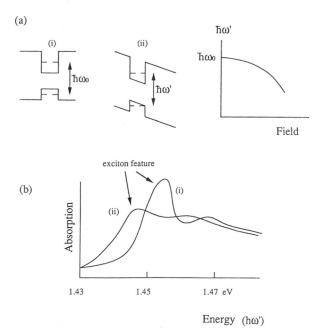

(a)

(b)

FIGURE 6 Optical properties of quantum wells in an electric field. (a) Effect of the field on electron and hole states and the principal absorption features; (b) schematic of actual absorption curves.

In multiple-quantum well lasers, the optical output is a complex trade-off between competing electronic effects associated with getting electrons and holes into the wells and optical effects: about five quantum wells is optimal. Under the application of an electric field normal to the layers, the optical absorption changes markedly in both strength and wavelength. This phenomenon, the quantum confined Stark effect, has important applications in optical modulation, and its origin is shown in Fig. 6. In elementary quantum mechanical terms, the applied electric field tilts the conduction and valance band edges as shown, and so the energy levels move. Since these levels depend on the effective mass (see Section III.C), and the hole and electron masses differ, the separation in energy levels for optical absorption will depend on the applied bias. If one works at a fixed wavelength, the absorption can be increased or decreased by the applied bais.

A second effect is also very important. An exciton is an excitation of a valence electron to the conduction band, leaving a positively charged hole behind. The coulomb attraction between the electron and the hole results in a bound-state energy lower than the conventional semiconductor band gap. Bound electron–hole pairs, excitons, share all the physics of a hydrogen atom scaled by the reduced effective mass of the electron and hole ($1/\mu = 1/m_e + 1/m_h$) and the dielectric constant (ε) of the semiconductor. In gallium arsenide, the exciton

binding energy is only $E_b = \mu e^4 / h^2 \varepsilon^2 \sim 5$ meV, and its diameter is ~ 20 nm (compared with 13.6 eV and 0.1 nm for a hydrogen atom). Excitons are formed in quantum wells, but if the well is less than 20 nm thick, the electron and hole wave functions must fit within the quantum well: the electron and hole are constrained to be closer to each other than is the case in a bulk semiconductor, and the exciton binding energy increases by as much as a factor of four. Whereas thermal energy at room temperature is sufficient to ionize a bulk exciton, quantum well excitonic absorption is still strong. In practical applications it is the exciton absorption that is moved by an applied bias. Another attractive feature of excitons in quantum wells is that the heterojunctions confine the electrons and holes in the direction perpendicular to the layers. Quite modest electric fields applied to a bulk exciton result in field ionization, with the electrons and holes moving apart, but up to 40 times these fields can be applied to a quantum well exciton before such ionization occurs, and electrons tunnel out of the quantum well. Note that if the field is applied in the plane of the quantum well, ionization still occurs easily at low fields.

If a multiple-quantum well structure is built into the intrinsic (undoped) region of a p–i–n diode), an applied bias can be used to modulate the optical absorption. If this diode is placed in a feedback circuit with a series resistance, the excited electrons can in turn lead to a photocurrent which then alters the bias across the diode and so modifies the optical absorption. Under appropriate conditions, optical bistability between states of high and states of low absorption for the same incident light intensity and controlled switching between such states can be achieved. Neither effect occurs in bulk semiconductors. There are near-term applications in optical signal transmission and modulation using multiple quantum wells.

Under very strong electric fields, some electrons are excited into the conduction bands with sufficiently high energy that they can excite another electron across the band gap and still remain in the conduction band. This process is known as impact ionization and avalanching (because of the rapid multiplication over distance of the number of electrons and holes) and is the basis of both powerful solid-state microwave sources and sensitive photodetectors. The spatial localization of the ionization process is not controlled, and these devices have undesirable noise levels as a result. In Fig. 2e we show the impact ionization in a quantum well, where the sudden increase in electron energy at the heterojunction can result in local impact ionization. This localisation would reduce the noise: such effects are seen for photodetectors, although they are not as dramatic as had been hoped—the band-gap energy difference is only a small fraction of the excess energy required for ionization.

B. Superlattices

When the barriers between adjacent quantum wells are sufficiently thin, electrons in bound states in one well can tunnel into adjacent wells. For two identical wells separated by a barrier only two or three atomic layers thick, the interaction is strong enough to induce analogues of the hydrogen molecule, namely, symmetric and antisymmetric combinations of the wave functions from adjacent wells, high polarizability under electric fields, etc. With a periodic repetition of many GaAs wells and 1- to 5-nm AlGaAs alloy barriers, a completely new crystal structure, a *superlattice*, and electronic band structure results. The electron propagation in the plane of the quantum wells is unaffected, but in the direction of the multilayer growth it is very sensitive to the barrier and well parameters. Because the repeat distance of the crystal is several layers, the energy bands are generally narrow and are known as minibands. They range from 1 to 150 meV. Viewed as a resonant tunneling system, the transmission of electrons through a superlattice is either near unity, if the electron is in a miniband, or zero otherwise (see Fig. 2f). Doped superlattices act like metals with partially filled energy bands. The characteristic distances in superlattices are comparable with the electron wavelength so there are electron analogues of antireflection coatings and optical filers can be realized. It was the possibility of new and tailored energy bands, and the prediction in 1970 of a novel form of negative differential resistance in superlattices under applied electric fields, that set off the whole scientific research enterprise described in this article. Only in the last decade has this effect been realized in practice, and it has proven to be rather delicate (i.e., susceptible to destruction by thermal vibrations above low temperatures or by imperfections in the crystals). The effect is very fast, operating at 1–3 tHz, 10 times faster then the resonant tunnel diode.

V. MESOSCOPIC SYSTEMS

The previous topics have concentrated on quasi-two-dimensional physics in multilayers. During the 1990s much effort has concentrated on defining patterns in these layers using electron beam lithography to achieve the same length scales. From quantum wells one progresses to quantum wires and quantum dots, where carriers are constrained in two and all three spatial dimensions, respectively. In such small structures, only a few electrons are present, and they are known as mesoscopic systems (meso = middle, i.e., between macroscopic and microscopic). These systems to date suffer from very large

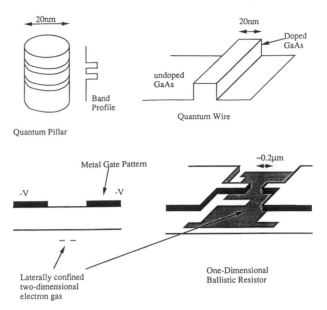

FIGURE 7 A collage of mesoscopic systems.

fluctuations in conductance, optical absorption, etc., as the detailed control over size is not adequate, and when structures get small there are intrinsic statistical variations of small numbers of atoms and electrons. The predicted advantages of quantum dots in laser structures are all lost if the size fluctuations are too great. With nanoampere currents in structures less than 0.1 μm in diameter and $\sim$1 μm long, there can often be less than one transport or optically excited electron present (time averaged) in these structures.

One fertile area of research is based on the so-called split-gate technology—a patterned gate (see Fig. 7) over a two-dimensional electron gas at a high-quality buried heterojunction. A negative bias on the gate depletes the carriers underneath, and where there is a split in the gate a quasi-one-dimensional conducting wire is formed. As electrons can move several microns without scattering, a short split-gate structure can give rise to ballistic motion of electrons in the wire. This is the ballistic resistor mentioned in Section I, taking the resistance value of $h/2e^2$ divided by an integer. This integer is given by the number of quasi-one-dimensional energy bands that are occupied. An increasing gate bias narrows the channel, increases the quantum energy of confinement in the lateral direction, reduces the number of occupied subbands, and increases the resistance in sharp steps. Under a longitudinal bias a new form of negative differential resistance has been predicted and observed. With more elaborate gate patterns it is possible to split an electric current in two and recombine it within 1 μm and see quantum interference phenomena (the so-called Aharonov–Bohm effect) whereby a magnetic field perpendicular to the electron gas induces different phase differences in the two arms of current.

The quantum pillar in Fig. 7, made from resonant tunneling multilayers, has a capacitance of the order of 10^{-18} F, so that charging by even one electron requires a bias on the scale of a volt. This is the co-called coulomb blockade regime. There is now evidence of electrons passing through this structure in a correlated manner—the noise is below the shot noise levels that would apply if the electrons were not correlated.

VI. THE COMING DECADE

In recent years, the rate of discovery of qualitatively new phenomena associated with low-dimensional structures has slowed, and those that are discovered are more delicate (i.e., observed only at very low temperatures and under very high magnetic fields). The emphasis has shifted to extending the range of new materials under investigation. Gallium nitride crystals have now been grown that act as light-emitting diodes and lasers in the blue. This material has a higher dielectric strength (i.e., can withstand larger fields before avalanche breakdown) and the electrons have a higher saturated drift velocity (i.e., maximum speed under bias) than the other group III–V semiconductors. These are the materials parameters that determine transistor performance, and so faster and more power transistors are being developed. Metallic multilayers are giving rise to new magnetic structures and giant magnetoresistance effects that are exploited in magnetic recording. The range of mesoscopic systems under investigation is ever widening. Greater emphasis will be placed on control of these systems, either during fabrication or with clever circuit configurations, so that some applications may be realized. Finally, the advances in polymer-based electronic and optoelectronic devices are exploiting heterojunction concepts.

SEE ALSO THE FOLLOWING ARTICLES

CRYSTAL GROWTH • EPITAXIAL TECHNOLOGY FOR INTEGRATED CIRCUIT MANUFACTURING • EXCITONS, SEMICONDUCTOR • INTEGRATED CIRCUIT MANUFACTURE • MOLECULAR BEAM EPITAXY, SEMICONDUCTORS • SEMICONDUCTOR ALLOYS • THIN-FILM TRANSISTORS

BIBLIOGRAPHY

Kelly, M. J. (1995). "Low Dimensional Semiconductors," Clarendon Press, Oxford.
Luryi, S. A., Xu, J., and Zaslavsky, A. (eds.) (1999). "Future Trends in Microelectronics: The Road Ahead," Wiley, New York.
Sze, S. M. (ed.) (1998). "Modern Semiconductor Device Physics," Wiley, New York.

Supernovae

David Branch

University of Oklahoma

GLOSSARY

Absolute magnitude The apparent magnitude that a star would have if it were at a distance of 10 pc.

Apparent magnitude A logarithmic measure of the brightness of a star as measured through, e.g., a blue or visual filter. A difference of five magnitudes corresponds to a factor of 100 in observed flux; one magnitude corresponds to a factor of 2.512. Increasing magnitude corresponds to decreasing brightness.

Light curve Apparent magnitude, absolute magnitude, or luminosity, plotted against time.

Luminosity The rate at which a star radiates energy into space. The luminosity of a supernova usually refers only to the thermal radiation in the optical, ultraviolet, and infrared bands, and does not include nonthermal γ rays, X-rays, and radio waves.

Neutron star A very compact star that has a radius of the order of 10 km and a density that exceeds that of nuclear matter. The maximum mass is thought to be less than 3 solar masses. A neutron star is one possible final state of the core of a massive star (the other being a black hole).

Parsec The distance at which a star would have an annual trigonometric parallax of 1 sec of arc. One parsec (pc) equals 206,625 astronomical units, or 3.1×10^{18} cm. Distances to the nearest stars are conveniently expressed as pc; distances across our Galaxy, as kiloparsecs (kpc); distances to external galaxies, as megaparsecs (Mpc).

Photosphere The layer in a star from which photons escape to form a continuous spectrum; equivalently, the layer at which the star becomes opaque to an external observer's line of sight.

Stellar wind The steady flow of matter from a (nonexploding) star into surrounding space.

White dwarf A compact star whose internal pressure is provided by degenerate electrons. A typical white dwarf has a radius comparable to that of Earth but a density of the order of 10^7 g/cm^3. The maximum mass, the Chandrasekhar mass, is 1.4 solar masses. A white

Encyclopedia of Physical Science and Technology, Third Edition, Volume 16

dwarf is the final state of a star that forms with less than 8 solar masses.

A SUPERNOVA is a catastrophic explosion of a star. The luminosity of a bright supernova can be 10^{10} times that of the Sun. The explosion throws matter into space at a few percent of the speed of light, with a kinetic energy of 10^{51} erg—the energy equivalent of 10^{28} Mtons of TNT. The ejected matter is enriched in heavy elements, some that were synthesized by nuclear fusion reactions during the slow preexplosion evolution of the star and some that were synthesized during the explosion itself. Thus, supernovae drive the nuclear evolution of the universe.

Some supernovae are the complete thermonuclear disruptions of white dwarf stars; others are initiated by the gravitational collapse of the cores of massive stars. The latter eject only their outer envelopes and leave behind compact stellar remnants—neutron stars and black holes. The ejected matter of a supernova sweeps up and heats interstellar gas to form an extended *supernova remnant*.

Supernovae are valuable indicators of extragalactic distances. They are used to establish the extragalactic distance scale (the Hubble constant) and to probe the history of the cosmic expansion (deceleration and acceleration).

I. BASIC OBSERVATIONS

A. Discovery

At least five of the temporary "new" stars that suddenly appeared in the sky during the last millennium—in 1006, 1054, 1181, 1572, and 1604—are now known to have been supernova explosions in our Galaxy. Some of these were bright enough to be seen by eye even in daylight. Radio, X-ray, and optical emission from the extended remnants of these historical Galactic supernovae is now detected. Hundreds of other extended remnants of supernovae that occurred in the Galaxy within the last 100,000 years also are recognized.

No Galactic supernova has been discovered since 1604, just a few years before the invention of the telescope. The supernova remnant known as Cassiopeia A, the brightest radio source in the sky beyond the solar system, was produced by a supernova that occurred near 1680, but the event was subluminous and either was not noticed or was dismissed as an ordinary variable star. Most of the supernovae that occur in our Galaxy are not detected by observers on Earth, owing to extinction of their light by dust grains that are concentrated within the disk of the Galaxy.

On February 24, 1987, an extraordinary astronomical event took place when a supernova appeared in the Large Magellanic Cloud (LMC), a small irregular satellite galaxy of our Galaxy. At a distance of only 50 kpc (160,000 light years), the LMC is the nearest external galaxy. At its brightest, SN 1987A reached the third apparent visual magnitude and was easily visible to the naked eye (from sufficiently southern latitudes). SN 1987A was the brightest supernova since 1604, and it became by far the most well-observed supernova. Its aftermath will be observed long into the future, perhaps as long as there are astronomers on Earth.

Until another outburst is seen in our Galaxy, or in one of our satellite galaxies, studies of the explosive phases of supernovae must be based on more distant events. The study of extragalactic supernovae began in August 1885, when a star of the sixth apparent visual magnitude appeared near the nucleus of the Andromeda Nebula, now known to be the nearest large external galaxy, at a distance of less than 0.8 Mpc. It was not until 1934 that the titanic scale of such events was generally recognized, and the term *supernova* began to be used. Systematic photographic searches for supernovae began in 1936. The following 35 years were primarily exploratory, as some of the basic characteristics of the various supernova types were established, and statistical information was gathered. A convention for designating each event eventually was adopted: e.g., SN 1987A was the first supernova to be discovered in 1987, and SN 1979C was the third of 1979. By the end of 1989, 687 extragalactic supernovae had been seen. During the late 1990s the discovery rate increased dramatically, owing mainly to searches with modern detectors (CCDs) for very distant supernovae, which can be discovered in batches. SN 1999Z was followed by SN 1999aa to SN 1999az, then SN 1999ba to SN 1999bz, and so on, to SN 1999gu. By the end of 1999, 1447 supernovae had been found (Fig. 1).

B. Types

The classification of supernovae is based mainly on the appearance of their optical spectra (Figs. 2 and 3). Type I supernovae (so named because the first well-observed events of the 1930s were of this kind) show no conspicuous spectral features produced by hydrogen; Type II supernovae do show obvious hydrogen lines. Type I supernovae are subdivided according to other spectroscopic characteristics: those of Type Ia have a very distinctive spectral evolution that includes the presence of a strong red absorption line produced by singly ionized silicon; the spectra of Type Ib supernovae include strong lines of neutral helium; neither ionized silicon nor neutral helium is strong in the spectra of Type Ic. Type II supernovae that have especially *narrow* lines are called Type IIn. Type II events also are divided into Type II-P and Type II-L according to the shapes of their light curves.

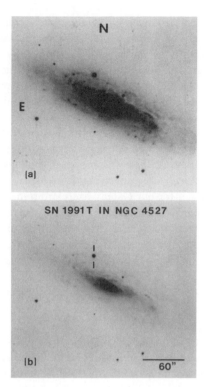

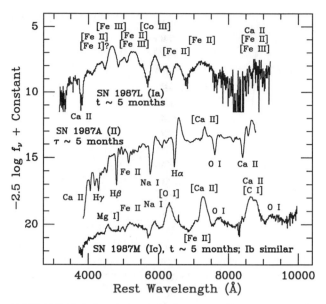

FIGURE 3 Nebular-phase optical spectra of supernovae, obtained 5 months after the time of maximum brightness (5 months after the time of explosion in the case of SN 1987A). [Reproduced by permission from Branch, D., Nomoto, K., and Filippenko, A. V. (1991). *Comments Astrophys.* **XV,** 221.]

FIGURE 1 CCD images of SN 1991T in its parent galaxy NGC 4527, displayed at two contrast levels. [Reproduced by permission from Filippenko, A. V., *et al.* (1991). *Astrophys. J.* **384,** L15.]

C. Light Curves

A typical supernova reaches its maximum brightness about 20 days after explosion. At its brightest, a normal Type Ia supernova (SN Ia) reaches an absolute visual magnitude of -19.5 and has a luminosity exceeding 10^{43} erg/sec, billions of times that of the Sun. SNe Ia (plural) are highly homogeneous with respect to peak absolute magnitude as well as other observable properties. Supernovae of the other types show more observational diversity, and almost all of them are less luminous than SNe Ia. The time-integrated luminosity of a bright supernova exceeds 10^{49} erg, while the typical kinetic energy is 10^{51} erg; thus the radiative efficiency of a supernova is low.

Characteristic light-curve shapes of the various supernova types are compared in Fig. 4. The light curve of a SN Ia consists of an initial rise and fall (the early *peak*) that lasts about 40 days, followed by a slowly fading *tail*. The tail is nearly linear when magnitude is plotted against time, but since magnitude is a logarithmic measure of brightness, the tail actually corresponds to an exponential decay of brightness, The rate of decline of the SN Ia tail corresponds to a half-life of about 50 days.

The light curves of many SNe II interrupt the initial decline from their peaks to enter a "*plateau*" phase of nearly constant brightness for several months; these are designated Type II-P. Others show a nearly *linear* decline (in magnitudes) from peak, with little or no plateau; these are

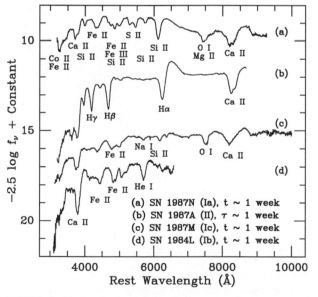

FIGURE 2 Photospheric-phase optical spectra of supernovae of various types, obtained 1 week after the time of maximum brightness (1 week after the time of explosion in the case of SN 1987A). [Reproduced by permission from Branch, D., Nomoto, K., and Filippenko, A. V. (1991). *Comments Astrophys.* **XV,** 221.]

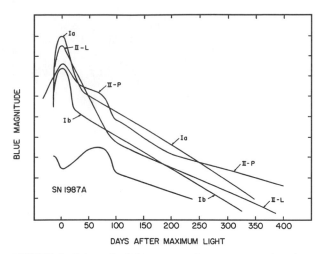

FIGURE 4 Schematic light curves of supernovae of various types. [Reproduced by permission from Wheeler, J. C. (1990). *In* "Supernovae" (J. C. Wheeler, T. Piran, and S. Weinberg, eds.), p. 1, World Scientific, Singapore.]

called Type II-L. Some SNe II, especially the SNe IIn, have slowly fading light curves that fit neither the SN II-P nor the SN II-L categories. Most SNe II eventually enter a linear tail phase, many with a rate of decline that corresponds to a half-life near 77 days.

SN 1987A was subluminous, and its light curve was unusual. It was observationally conspicuous only because it occurred in a very nearby galaxy; most such subluminous events in distant galaxies go undiscovered. This illustrates the importance of a very strong observational selection effect: in our observational sample of supernovae, luminous events are highly overrepresented relative to subluminous ones (Fig. 5).

D. Sites, Rates, and Stellar Populations

Galaxies are classified as spirals and ellipticals. SNe II, SNe Ib, and SNe Ic are found in spirals, usually in the spiral arms. SNe Ia are found in both kinds of galaxies, and those in spirals show little, if any, tendency to concentrate to the arms.

To estimate the relative production rates of the different types of supernovae in the various kinds of galaxies, several observational selection effects must be taken into account. The most important is the bias in favor of the discovery of luminous events. For example, because of their high peak luminosities, SNe Ia are the most numerous type in the observational sample, but they actually occur less frequently than SNe II. Another important effect is that it is difficult to discover supernovae in spiral galaxies whose disks are oriented such that we observe them from the side. When these and additional selection effects are allowed for, it is found that in spirals, SNe II are

most frequent, followed by SNe Ic, SNe Ib, and SNe Ia, which occur at comparable rates. In spirals, the rates of all supernova types are correlated with galaxy color: the bluer the galaxy—and, by inference, the higher the star formation rate within the last billion years—the higher the supernova rate. Elliptical galaxies produce only SNe Ia, at a lower rate than spirals.

Absolute supernova rates are more uncertain than relative rates because the relevant selection effects are still more difficult to evaluate. In large spiral galaxies, the mean interval between supernovae is about 25 years. A similar rate is derived for our Galaxy itself, from the number of historical supernovae that have been seen, but this involves a large correction for events that have been missed owing to our location in the dusty disk of the Milky Way. The fact that the remnants of all of the historical Galactic supernovae are within a few kiloparsecs of the Sun, while the radius of the galaxy is 15 kpc, confirms that only (some of) the nearest Galactic supernovae of the past millennium have been noticed by observers on Earth.

The observation that SNe II, SNe Ib, and SNe Ic appear in the arms of spirals indicates that they are produced by massive stars. Only stars that are formed with a mass that is greater than about 8 solar masses have nuclear lifetimes short enough—less than about 30 million years—that they explode before drifting out of the arms in which they were born.

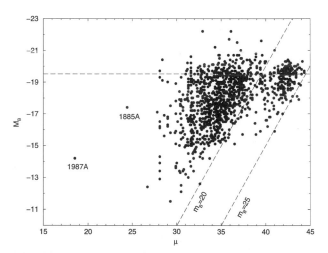

FIGURE 5 Absolute blue magnitudes of supernovae are plotted against the distance modulus of their parent galaxies. The distance modulus is five times the logarithm of the distance, expressed in units of 10 pc. The slanted dashed lines correspond to apparent blue magnitudes of 20 and 25. The horizontal dashed line is at the absolute magnitude of a normal SN Ia, $M_B = -19.5$. For each supernova, the absolute magnitude is derived from the brightest apparent magnitude that was observed, which was not in all cases at the time of maximum brightness. [Courtesy of D. Richardson, University of Oklahoma.]

The observation that SNe Ia occur in elliptical galaxies, but also in spirals in proportion to the recent rate of star formation, appears at first to be paradoxical. Elliptical galaxies stopped forming stars billions of years ago and now contain few stars that are much more massive than the Sun. On the other hand, the correlation between the SN Ia rate and the recent star-formation rate in spirals implies that most SNe Ia in spirals are produced by stars that are fairly short-lived and, therefore, born moderately massive. The resolution of the paradox is that a SN Ia is produced by a white dwarf that accretes matter from a binary companion star until it is provoked to explode. This can account for the low but nonzero SN Ia rate in ellipticals as well as for the correlation of the SN Ia rate with galaxy color in spirals (Section III.C).

II. THE INTERPRETATION OF SPECTRA

A. The Continuous Spectrum

The shape of a star's thermal continuous spectrum is determined primarily by the temperature at the photosphere. The absolute brightness depends also on the radius of the photosphere. In a supernova, the temperature and radius of the photosphere change with time. When a star explodes, its matter is heated and thrown into rapid expansion. The luminosity abruptly increases in response to the high temperature, but most of this energy is radiated as an X-ray and ultraviolet "flash" rather than as optical light. As the supernova expands, it cools. For about 3 weeks the optical light curve rises as the radius of the photosphere increases, and an increasing fraction of the radiation from the cooling photosphere goes into the optical band. Maximum optical brightness occurs when the temperature is near 10,000 K. At this time the radius of the photosphere is 10^{15} cm, or 70 AU, almost twice the radius of Pluto's orbit. The matter density at the photosphere is low, near 10^{-16} g/cm^3.

After maximum light, further cooling causes the light curve to decline. Owing to the expansion and consequent geometrical dilution, the photosphere recedes with respect to the matter, i.e., matter flows through the photosphere, and deeper and deeper layers of the ejecta are gradually exposed. Eventually, at a time that depends mainly on the amount of matter ejected but that ordinarily is a matter of months, the ejected matter becomes optically thin, the continuous spectrum fades away, and the "photospheric phase" comes to an end. The supernova then enters the transparent "nebular" phase.

B. The Spectral Lines

Supernova spectral lines carry vital information on the temperature, velocity, and composition of the ejected

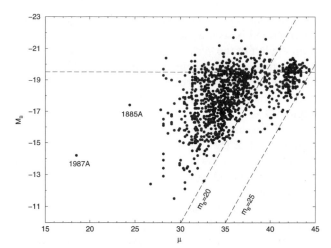

FIGURE 6 Top: A schematic drawing of the photosphere (shaded) and the surrounding atmosphere of a supernova. An observer to the left would see an emission line formed in region E and an absorption line formed in region A; region O would be occulted. Bottom: The shape (in flux versus wavelength) of a characteristic supernova spectral line is shown. [Courtesy of K. Hatano, University of Oklahoma.]

matter. However, because of the fast ejection velocities, typically 10,000 km/sec—3% of the speed of light—the spectral features are highly Doppler-broadened and overlapping, and extracting information from the spectra is difficult.

During the photospheric phase, broad absorption and emission lines form outside the photosphere and are superimposed on the continuous spectrum. A schematic model that is useful for understanding the formation of spectral lines during the photospheric phase is shown at the top in Fig. 6. The central region is the photosphere, and outside the photosphere is the line-forming region. The entire supernova is in differential expansion, with velocity proportional to distance from the center. This simple velocity law is a natural consequence of matter being ejected with a range of velocities and then coasting without further acceleration; after a time, each matter element has attained a distance from the center in proportion to its velocity.

As a photon travels through the expanding atmosphere, it redshifts in wavelength with respect to the matter that it is passing through (as a photon redshifts in the expanding universe). Thus a photon can redshift into resonance with an electronic transition in an atom or ion and be absorbed. In the very useful approximation of pure scattering, the absorbed photon is immediately reemitted, but in a random direction. From the point of view of an external observer, a photon scattered by a moving atom will be Doppler-shifted with respect to the rest wavelength of the transition. A photon scattered into the observer's line of sight from the right of the vertical line in Fig. 6 comes from an atom that has a component of motion away from the observer and is seen to be redshifted; a photon scattered from the left of the vertical line is seen to be blueshifted. Therefore the region labeled E produces a broad, symmetrical emission line superimposed on the continous spectrum and centered on the rest wavelength of the transition. Region O is occulted by the photosphere. In region A, photons originally directed toward the observer can be scattered out of the line of sight. Because it has a component of motion toward the observer, matter in region A produces a broad, asymmetric, blueshifted, absorption component.

The shape of the full line profile is as shown at the bottom in Fig. 6. This kind of profile, having an emission component near the rest wavelength of the transition and a blueshifted absorption component, is characteristic of expanding atmospheres and is referred to as a "P Cygni profile" (after a bright star whose atmosphere is in rapid but nonexplosive expansion). The precise shape of a line profile in a supernova spectrum depends primarily on how the matter density decreases with radius and on the velocity at the photosphere. The higher the velocity, the broader the profile and the more the absorption component is blueshifted.

In supernovae the velocities are so high that line profiles usually overlap. Physically, this corresponds to multiple scattering in the atmosphere: after a photon is scattered by one transition, it continues to redshift and may come into resonance with another transition and be scattered again and again before it escapes from the atmosphere.

During the nebular phase that follows the photospheric phase, there is no continuum to serve as a source of photons to be scattered. However, an ion can be excited by a collision with a free electron and respond by emitting a photon. Thus the spectrum during the nebular phase consists of broad, symmetric, overlapping emission lines.

C. Composition

The spectrum of a SN II near the time of its maximum brightness is almost a smooth featureless continuum. Subsequently, as the temperature falls and the ionized hydrogen begins to recombine, the Balmer lines of hydrogen strengthen. Further cooling causes the spectrum to become much more complicated as many additional lines develop, from ions such as singly ionized calcium, singly ionized iron, and neutral sodium. Detailed analysis of the strengths of the spectral features during the early photospheric phase indicates that the composition of the outer layers of a SN II resembles that of the Sun and ordinary stars: hydrogen and helium are most abundant and only a small fraction of the matter is in the form of heavy elements. The deeper, slower-moving matter, which becomes observable at later times, shows strong lines of elements such as oxygen, calcium, and magnesium, and analysis reveals that the composition is dominated by such elements. In general, the deeper the layer, the heavier the elements of which it consists.

The spectra of SNe Ib lack hydrogen lines but they do develop strong lines of neutral helium. The composition structure is inferred to be much like that of a SN II, except for the absence of the outer hydrogen-rich layers. In SNe Ib, the temperature is not high enough to produce optical lines of helium by ordinary thermal excitation; a nonthermal source is required. This is provided by the radioactive decay of ^{56}Ni that was synthesized in the explosion, and then the decay of its daughter nucleus, ^{56}Co (Section III.A); γ rays from the radioactive decays scatter off thermal electrons to produce a population of fast (nonthermal) electrons, which in turn cause the excitation of helium. The spectra of SNe Ic do not have strong helium lines, either because the outer helium layer is absent (in which case the outer layers consist mainly of a mixture of carbon and oxygen) or because nonthermal excitation is ineffective. Otherwise the composition structure of SNe Ic resembles that of SNe II and Ib.

The spectral evolution of SNe Ia is quite different. Near the time of its maximum brightness the spectrum of a SN Ia is dominated by lines of singly ionized silicon, sulfur, and calcium. Weeks later the spectrum becomes dominated by overlapping lines from various ionization states of iron and cobalt. The inferred composition structure is very different from that of ordinary stellar matter; the outer layers of a SN Ia are a mixture of elements of intermediate mass, from oxygen to calcium, and the deeper layers consist mainly of iron and neighboring elements in the periodic table ("iron-peak" elements).

Because the line blending in supernova spectra is so severe, the process of extracting information from observed spectra usually entails calculating "synthetic spectra" of model supernovae and comparing them with observation. The parameters of the model are varied until the synthetic and observed spectra are in satisfactory agreement; at that point, the physical conditions and composition of the model are accepted as an approximation to the actual conditions and composition of the supernova. An example of a comparison of a synthetic spectrum and an observed one is shown in Fig. 7.

D. Polarization

Information about the shape of a supernova can be obtained by observing the extent to which its light is polarized, but the number of supernovae for which polarization measurements have been made is rather small. If a supernova is spherically symmetric, its light will be observed to be unpolarized, and this appears to be the case for

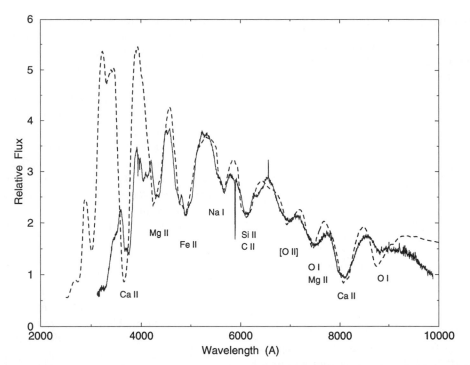

FIGURE 7 A synthetic spectrum (dashed line) is compared to an optical spectrum of the Type Ic SN 1994I obtained 4 days before the time of maximum brightness. The narrow absorption line near 5900 Å was produced by interstellar sodium atoms in the parent galaxy, M51. The synthetic spectrum has excess flux at short wavelengths because not all relevant spectral lines were included in the calculation. [Reproduced by permission from Millard, J., *et al.* (1999). *Astrophys. J.* **527**, 746.]

SNe Ia. The light from core-collapse supernovae, however, is observed to be polarized, which indicates that core-collapse supernovae are asymmetric. The observed degree of polarization, which is typically at the level of 1%, indicates that if the shape is ellipsoidal, then the axial ratio is about 1.5.

III. THERMONUCLEAR SUPERNOVAE

A. Nickel and Cobalt Radioactivity

When a star explodes its matter is heated suddenly, but expansion then begins to cause cooling, approximately adiabatically, with the temperature inversely proportional to the radius. Unless the initial radius of the progenitor star is much larger than that of the Sun, the ejected matter cools too quickly ever to attain the large (10^{15} cm), hot (10^4 K) radiating surface that is required to account for the peak optical brightness of a supernova. Some continuing source of energy must be provided to the expanding gas, to offset partially the adiabatic cooling. For SNe Ia, the continuing energy source is the radioactive decay of about 0.6 solar mass of ^{56}Ni, and then its daughter nucleus, ^{56}Co. The former has a half life of 6 days and decays by electron capture with the emission of a γ ray; the mean energy per decay is 1.7 MeV. Then ^{56}Co decays with a half-life of 77 days to produce stable ^{56}Fe, the most abundant isotope of iron in nature. The ^{56}Co decay is usually by electron capture with γ-ray emission, but sometimes it is by positron emission. The mean energy per decay is 3.6 eV, with 4%, on average, carried by the kinetic energy of the positrons.

Numerical hydrodynamical calculations have shown that the light curve of a SN Ia can be reproduced provided that a total ejected mass of somewhat more than 1 solar mass includes 0.6 solar mass of ^{56}Ni that was freshly synthesized by nuclear reactions during the explosion. During the first month after the explosion, the supernova is optically thick to the γ rays from radioactivity; they are absorbed and thermalized in the ejected matter and provide the continuing source of energy to keep it hot. As the supernova expands, the γ rays have a higher probability of escaping rather than being absorbed. The positrons from ^{56}Co decay can be trapped by even very small magnetic fields, long enough to deposit their kinetic energy into the gas by means of collisions before they mutually annihilate with electrons to produce more γ rays. The fraction of the γ-ray energy that is absorbed, together with the kinetic energy of the positrons, powers the tail of the SN Ia light curve. Because the fraction of the γ rays that is absorbed decreases with time, the optical light curve declines more rapidly than the 77-day half-life of ^{56}Co.

B. The Fate of Low-Mass Stars

The sites of SNe Ia indicate that they are produced by stars that were born with less than 8 solar masses. Such a star fuses hydrogen into helium in its core during its main-sequence phase, at a temperature of about 10^7 K. When the hydrogen in the core is exhausted, the star expands its envelope to become a red giant; during this phase the core contracts until it is able to fuse the helium to a mixture of carbon and oxygen, at 10^8 K. When helium is exhausted, the core contracts again. The carbon–oxygen core would fuse to heavier elements if the temperature approached 10^9 K, but this is not achieved because the density in the core becomes sufficiently high (10^6 g/cm^3) that gravity is balanced by the pressure of degenerate electrons—electrons that strongly resist further compression because their momentum distribution is determined by the Pauli exclusion principle. The star becomes a stable carbon–oxygen white dwarf that has a radius of only a few percent of the radius of the Sun, comparable to the radius of Earth. The maximum mass of a white dwarf is 1.4 solar masses—the Chandrasekhar mass. It appears that most, if not all, stars that are born with less than 8 solar masses manage to lose enough mass during their red giant phases, by means of stellar winds and planetary-nebula ejection, to become white dwarfs. Thus it is difficult to account for SNe Ia as the explosions of single stars.

C. Accreting White Dwarfs

A more promising explanation for the origin of SNe Ia appeals to the more complicated evolution of stars in close binary systems. A pair of stars forms, and the more massive one evolves first to become a white dwarf. At some point in the evolution of the second star, it expands and transfers matter to the surface of the white dwarf, eventually provoking it so explode. The time delay between the formation of the binary system and the explosion is determined by the initial mass of the less massive star. This model can account for the occurence of SNe Ia in elliptical galaxies (the second star has low mass, so binary systems that formed long ago are producing SNe Ia now) as well as for the correlation between the SN Ia rate and the recent rate of star formation in spirals (binaries in which the less massive star's mass is not so low also are producing SNe Ia now). A related possibility is that *both* members of the binary form white dwarfs, which gradually spiral together because of orbital decay caused by the emission of gravitational radiation, and eventually merge to temporarily assemble a super-Chandrasekhar mass. Whether such a configuration would explode as a SN Ia rather than

collapse to a neutron star is not yet known. In this model, the spiral-in phase would add an additional time delay between formation and explosion.

Computer simulations of the response of a carbon–oxygen white dwarf to the accretion of matter lead to a variety of outcomes, depending on the initial mass of the white dwarf and the composition and rate of arrival of the accreted matter. Certain combinations of these parameters do lead to predictions that correspond well to the observed properties of SNe Ia. If the accretion rate of hydrogen-rich matter is of the order of 10^{-7} solar masses per year, fusion reactions in the accreted layers convert hydrogen to helium, and then helium to carbon and oxygen, thus producing an increasingly massive carbon–oxygen white dwarf. As its mass approaches 1.4 solar masses, the white dwarf contracts and heats until carbon begins to fuse near its center. The ignition of a nuclear fuel in a gas that is supported by the pressure of degenerate electrons causes a thermonuclear instability, because the increasing temperature of the nuclei fails to raise the pressure enough to cause a compensating expansion, while it does cause the fuel to burn at an increasing rate. The inner portions of the white dwarf are incinerated to a state of nuclear statistical equilibrium, becoming primarily ^{56}Ni because this is the most tightly bound nucleus that has equal numbers of protons and neutrons, and there is not enough time for weak interactions to change the proton-to-neutron ratio of the carbon–oxygen fuel. A nuclear flame front propagates outward through the white dwarf in about a second, heating and accelerating the matter, and the entire star is disrupted. The outer layers, because of their lower densities, are fused only to elements of intermediate mass, such as magnesium, silicon, sulfur, and calcium (Fig. 8). Thus the thermonuclear white dwarf model can account for the SN Ia composition structure that is inferred from spectroscopy. The uniformity, or near–uniformity, of the amount of mass ejected (1.4 solar masses) is responsible for the remarkable observational homogeneity of SNe Ia.

In this model of a SN Ia, the energy produced by fusion, less the energy needed to unbind the star, results in a final kinetic energy of about 10^{51} erg. However, for reasons given earlier, the explosion becomes optically bright only because of the smaller amount of energy that is released gradually by the radioactive decay of ^{56}Ni and ^{56}Co. In short, energy released promptly by nuclear fusion explodes the star, while energy released slowly by radioactivity makes the explosion shine. Because the final composition of a SN Ia is largely iron and neighboring elements in the periodic table, SNe Ia make an important contribution to the amount of iron-peak elements in the universe.

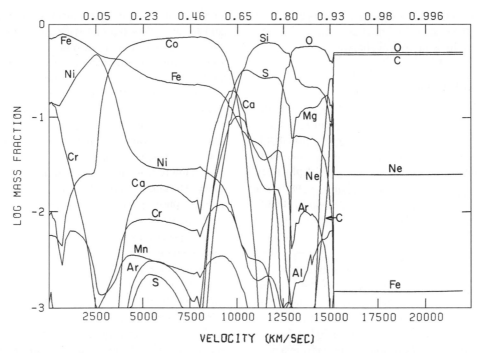

FIGURE 8 The calculated composition of an exploded white dwarf, displayed as element mass fraction plotted against ejection velocity. The fraction of the ejected mass that is interior to each velocity is shown at the top. This figure is for 32 days after the explosion; the composition below 10,000 km/sec changes with time owing to radioactive decays, and most of the cobalt eventually decays to iron. [Reproduced with permission from Branch, D., Doggett, J. B., Nomoto, K., and Thielemann, F. K. (1985). *Astrophys. J.* **294**, 619.]

IV. CORE-COLLAPSE SUPERNOVAE

A. Supergiant Progenitors

SNe II, SNe Ib, and SNe Ic occur in spiral arms and therefore come from massive stars. In its late evolution, a star having an initial mass in the range of 8 to about 30 solar masses becomes a red supergiant, developing a compact, dense, heavy-element core surrounded by an extended hydrogen-rich envelope that may swell to 1000 times the solar radius, approaching 10^{14} cm. If an instability in the core suddenly releases a large amount of energy, a shock wave may form and propagate outward, heating and ejecting the star's envelope. As it expands, the envelope cools nearly adiabatically. Nevertheless, because the initial radius of a red supergiant is so large, the envelope can attain the large, hot radiating surface that is required to produce the peak optical brightness of a supernova, without requiring a delayed input of energy from radioactivity. Hydrodynamical computer simulations of the response of a red supergiant to a sudden release of energy at its center predict light curves and other properties that are in good agreement with observations of SNe II-P, the most common kind of SN II. The light-curve plateau is a phase of diffusive release of thermal energy deposited in the envelope by the shock. To account for the observed peak luminosity, the expansion velocity, and the duration of the plateau phase, the progenitor needs to have a large radius and eject roughly 10 solar masses that carries a kinetic energy of 10^{51} erg. After the plateau phase, the light-curve tail is powered by ^{56}Ni and ^{56}Co decay. The γ rays are efficiently trapped and thermalized by the large ejected mass of a SN II-P, so the decline rate of the optical tail corresponds closely to the ^{56}Co half-life of 77 days. To account for the light curve of a SN II-L, the ejected mass should be only a few solar masses, and as in a SN Ia, γ rays increasingly escape so the tail declines somewhat faster than the 77-day half-life.

The progenitor of a SN Ib is a massive star that loses its hydrogen-rich envelope prior to core collapse, either by means of a stellar wind (if the initial mass is more than about 30 solar masses) or by mass transfer to a binary companion star. A star that lacks a hydrogen envelope has a smaller radius than a red supergiant, so the SN Ib light curve, like that of a SN Ia, is powered primarily by radioactive decay. Because the amount of ^{56}Ni that is ejected by an SN Ib is typically only 0.1 solar mass, a SN Ib is not as bright as a SN Ia. Similar statements apply to the progenitors of SNe Ic. Some of the SN Ic progenitors may be bare carbon–oxygen cores that lose most or all of their helium layers prior to collapse, and others may eject helium but lack the nonthermal excitation that is needed to produce helium lines in optical spectra.

B. Collapse and Explosion

The basic explosion mechanism of SNe II, Ib, and Ic is the same. The sudden energy release at the center of the progenitor star is caused by the inevitable gravitational collapse of its highly evolved core. In massive stars, carbon is nondegenerate when it ignites, and it fuses nonexplosively to a mixture of oxygen, neon, and magnesium. Subsequent nuclear burning of these and heavier elements is unable to disrupt the increasingly tightly bound stellar core, and the core eventually becomes iron (Fig. 9). Iron is the ultimate nuclear "ash," from which no nuclear energy can be extracted. Instead, high-energy photons characteristic of the high core temperature photodisintegrate the iron endothermically, decreasing the energy and pressure in the core and causing it to collapse at a fraction of the speed of light, from the size of a planet to a radius of 50 km. During the collapse electrons are forced into nuclei to form a fluid of neutrons, and as the density exceeds the nuclear density (more than 10^{14} g cm^{-3}) the pressure rises enormously to resist further compression.

The collapse releases a gravitational potential energy of more than 10^{53} erg. The resulting temperatures and densities are so high that neutrinos are created copiously, and because they interact with matter only weakly, they diffuse out of the core in seconds and carry away almost all of the energy. To produce a supernova, however, about 1% of the gravitational energy must somehow go into ejecting the matter outside the core. Just how this occurs is not yet understood. One possibility is a "prompt" hydrodynamical explosion. Owing to its infall velocity, the collapsing core overshoots its equilibrium density, rebounds ("bounces"), transfers energy mechanically to matter that is falling in more slowly from above, and produces an outgoing shock wave. The energy in the shock tends to be consumed, however, in photodisintegrating heavy elements just outside the core, causing the shock to stall before it can eject the outer layers of the star. A related possibility is a "delayed" hydrodynamical explosion, in which the small fraction of neutrinos that are absorbed in the matter just outside the core deposits enough energy to revive the stalled shock. Or it may be that the explosion mechanism is fundamentally different, e.g., a magnetohydrodynamical event in which the increasingly rapid rotation and the amplifying magnetic field of the collapsing core combine to cause the ejection of relativistic jets of matter along the rotation axis, which drive shocks into the envelope and cause ejection. This could be the reason for the significant departures from spherical symmetry that are implied by the observed polarization of the light from core collapse supernovae. It

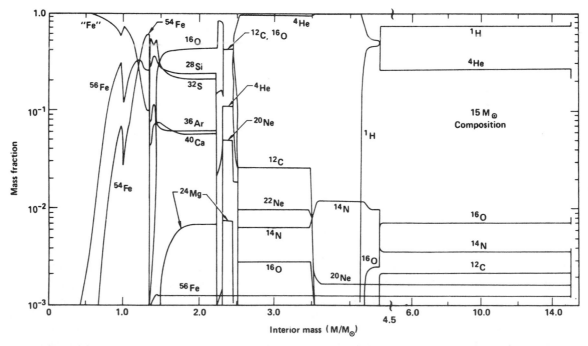

FIGURE 9 The calculated composition of a star of 15 solar masses, just before core collapse. The amount of mass interior to each point is plotted on the horizontal axis; note the change of scale at 4.5 solar masses. [Courtesy of S. E. Woosley, University of California at Santa Cruz, and T. A. Weaver, Lawrence Livermore National Laboratory.]

also could be related to the association of at least some core-collapse supernovae with at least some γ-ray bursts, which became apparent with the spatial and temporal coincidence of the peculiar Type Ic SN 1998bw and the γ-ray burst of April 25, 1998. In SN 1998bw and several other core-collapse events, the kinetic energy carried by the ejected matter appears to have been more than 10 times that of a typical supernova. Such "hyperenergetic" events may be produced by the collapse of the cores of very massive stars (more than 30 solar masses) to form black holes of about 7 solar masses.

Shock waves raise the temperature of the deep layers of the ejected matter high enough to cause fusion reactions. As a result of this "explosive nucleosynthesis" the innermost layers of the ejected matter are fused to ^{56}Ni and other other iron-peak isotopes, and the surrounding layers are fused to elements of intermediate mass. The low-density outer layers of the progenitor star—whether hydrogen, helium, or carbon and oxygen—are not fused. Core-collapse events are the main producers of intermediate-mass elements in nature, and they also make a significant contribution to the production of iron-peak elements.

Whether the collapsed star ends as a neutron star or a black hole depends on whether its final mass exceeds the maximum mass of a neutron star.

C. Circumstellar Interaction

The progenitor of a core-collapse supernova loses mass by means of a stellar wind. For a steady-state wind that has a constant velocity and a constant mass-loss rate, the density of the circumstellar matter is proportional to the mass-loss rate, inversely proportional to the wind velocity, and inversely proportional to the square of the distance from the star. The wind velocity ordinarily is comparable to the escape velocity from the photosphere of the star—of the order of 10 km/sec for a red supergiant and 1000 km/sec for a blue one—so the circumstellar matter of a red supergiant is much more dense than that of a blue one. When a core-collapse supernova occurs, both the ejected matter and the radiated photons interact with the circumstellar matter. The violent collision between the high-velocity ejecta and the circumstellar matter is called a hydrodynamic interaction. The interaction between the supernova photons and the circumstellar matter is called a radiative interaction. These interactions can have observable consequences in various parts of the electromagnetic spectrum.

Hydrodynamic circumstellar interaction has been detected most often by observations at radio wavelengths. The spectrum of the radio emission indicates that it is synchrotron radiation from electrons that are accelerated to relativistic velocities in a very hot (up to 10^9 K) interaction

region. The higher the density of the circumstellar matter, the longer the radio "light curve" takes to rise to its peak, because initially the circumstellar matter outside the interaction region, which has been ionized by the X-ray and ultraviolet flash, absorbs the radio photons by free–free processes. Thermal X-rays from the hot interaction region also have been detected in a smaller number of cases. The fraction of a supernova's total luminosity that is emitted in the radio and X-ray bands ordinarily is very small, but observation of the radio and X-ray emission combined with modeling of the interaction provides valuable information on the mass-loss rate of the progenitor star. Rates in the range of 10^{-6} to 10^{-4} solar masses per year have been inferred. Optical light radiated from the interaction region is responsible for the slow decay rate of the light curves of SNe IIn. Unlike most supernovae, which seldom are followed observationally for more than a year after explosion, some circumstellar-interacting supernovae are observed decades after explosion, at optical, radio, and X-ray wavelengths.

Radiative interactions are manifested most clearly by relatively narrow emission and absorption lines that appear in optical spectra (which cause the supernova to be classified Type IIn) and, especially, in ultraviolet spectra [which can only be obtained from above Earth's atmosphere with instruments such as the *Hubble Space Telescope* (*HST*); Fig. 10]. The widths of the circumstellar lines provide information on the wind velocity of the progenitor, and analysis of the line strengths provides information on the relative abundances of the elements in the wind. The ionization state of the circumstellar matter also provides information on the amount of ionizing radiation that was emitted by the explosion, something that is not observable directly.

Some supernovae emit excess infrared radiation, well beyond the amounts expected from their photospheres, by thermal radiation from cool (1000 K), small (10^{-5} cm) solid particles (dust grains) in the circumstellar medium. The grains are heated by absorbing optical and ultraviolet radiation from the supernova and respond by emitting in the infrared. Infrared observations provide information on the nature and spatial distribution of the dust grains.

D. Supernova 1987A

SN 1987A was discovered during the predawn hours of February 24, 1987, as a star of the fifth apparent magnitude in place of a previously inconspicuous twelfth-magnitude star known as Sanduleak (Sk) $-69\,202$ (star number 202 near $-69°$ declination in a catalog of stars published by N. Sanduleak in 1969). Sk $-69\,202$ was the first supernova progenitor star whose physical properties could be determined from observations that had been made prior to

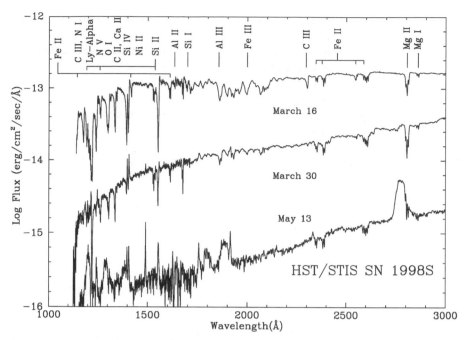

FIGURE 10 Ultraviolet spectra of the Type IIn SN 1998S obtained with the *Hubble Space Telescope*. [Courtesy of P. Challis, Harvard–Smithsonian Center for Astrophysics and Supernova INtensive Study (SINS) team.]

its explosion. It was a supergiant star of spectral type B3, with an effective temperature of 16,000 K and a luminosity 10^5 times that of the Sun. The luminosity indicates that the mass of its core of helium and heavier elements was 6 solar masses, which in turn implies that the initial mass of the star was about 20 solar masses. By the time of the explosion, presupernova mass loss had reduced the mass to about 14 solar masses. From the luminosity and effective temperature, the radius of Sk −69 202 was inferred to be 40 times that of the Sun.

As discussed above, intrinsically luminous SNe II are the explosions of red supergiants, which have very large radii. It is well understood why an explosion of a less extended star such as Sk −69 202 would produce a subluminous supernova (see Section IV.A). A more difficult question is, Why did Sk −69 202 explode as a relatively compact blue supergiant rather than as a very extended red one? Observations of the circumstellar matter of SN 1987A indicate that its progenitor did go through a red supergiant phase, but it ended some 40,000 years prior to the explosion. There are several possible reasons why the progenitor star contracted and heated to become a less extended blue supergiant, including the possibility that the evolutionary history of Sk −69 202 involved a *merger* of two stars in a close binary system.

Within days of the optical discovery of SN 1987A, it was realized that a neutrino burst had been recorded by undergound neutrino detectors in Japan and the United States, a few hours before the optical discovery. Because

neutrinos interact only weakly with matter (each person on Earth was harmlessly perforated by 10^{14} neutrinos from SN 1987A), only a total of 18 neutrinos, with arrival times spread over 12 sec, was detected with certainty. The total neutrino energy emitted by SN 1987A was 2×10^{53} erg, and the mean neutrino energy was 12 MeV, as expected, so the neutrino signal confirmed the basic theory of core collapse. It also provided valuable information for particle physics, including an upper limit of 16 eV to the mass of the electron neutrino. The 12-sec duration of the neutrino burst suggested that the core initially formed a neutron star, because black-hole formation would have terminated the burst earlier. However, the neutron star has not yet been detected directly. It is possible that "fall-back" of matter that was ejected very slowly has caused the neutron star to collapse to a black hole.

The spectrum of SN 1987A has been observed far into the nebular phase, and over a very broad range of wavelengths. Analysis of optical and infrared spectra of SN 1987A has established that, as expected, the composition varied from primarily hydrogen and helium in the outermost layers to mainly heavier elements in the innermost layers. The brightness during the tail of the light curve indicated that the amount of ejected ^{56}Ni was 0.07 solar mass. The optical and infrared spectra, as well as the unexpectedly early leakage of γ rays and X-rays (produced by γ rays that transferred much of their energy to electrons), indicated that a sustantial amount of composition mixing (light elements into the deeper layers and heavier elements, including ^{56}Ni, into the outer layers) took place. The element mixing process still is not well understood. An excess of infrared radiation that began to develop a year after explosion indicated that a small fraction of the heavy elements in the ejected matter had condensed into dust grains.

SN 1987A showed various signs of early circumstellar interaction, at ultraviolet and radio wavelengths, but the interaction was weak because the wind of a blue supergiant is fast and the circumstellar density is correspondingly low. After a few years, however, it became clear that SN 1987A was at the center of a ring of higher density matter, which appears elliptical on the sky but actually is circular, and tilted with respect to our line of sight by 42°. The matter in the ring, which is expanding at 10 km/sec, presumably originated from the slow, dense wind of Sk −69 202 when it was in its red supergiant phase. The reason that it is confined to a ring rather than a spherical shell is not well understood. The later discovery of two larger, fainter rings (Fig. 11), as well as additional complexity in the circumstellar environment, is suggestive of a complicated binary-star history of Sk −69 202 involving interacting stellar winds. The time variation of ultraviolet emission lines emitted by ions in the inner ring, as they recombined following ionization by the supernova radiation, indicated that the radius of the ring is 0.6 light year; when combined with the observed angular size of the ring (0.9 sec of arc), this provides a valuable independent confirmation that the distance to the LMC is 50 kpc. In the late 1990s, the first signs of an imminent collision between the high-velocity supernova ejecta and the slowly expanding inner ring began to appear. This interaction will cause the emission from SN 1987A to increase dramatically, across the electromagnetic spectrum, and provide an opportunity to probe further the distribution of the circumstellar matter of SN 1987A.

V. SUPERNOVA REMNANTS

The matter ejected by a supernova sweeps up interstellar gas. After a time of the order of a century, the mass of the swept-up matter becomes comparable to the mass of the ejected matter, and the latter begins to decelerate, with energy being converted from kinetic into other forms. Thus, a hot region of *interstellar* interaction develops, and an

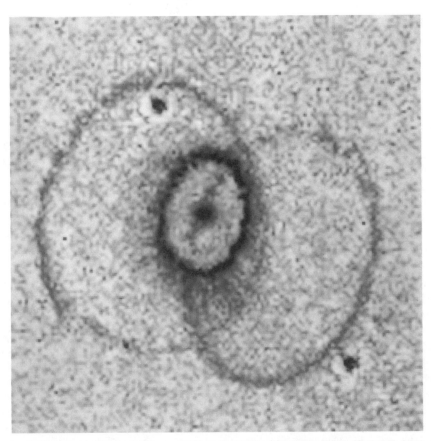

FIGURE 11 A *Hubble Space Telescope* image of SN 1987A (center), its inner ring, and its two fainter outer rings. [Reproduced by permission from Burrows, C. J., *et al.* (1995). *Astrophys. J.* **452,** 680.]

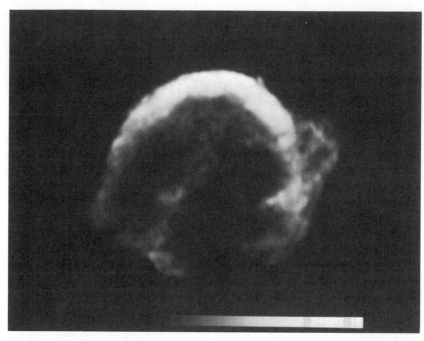

FIGURE 12 A radio image of the shell-type remnant of Kepler's supernova of 1604. The image was made at a wavelength of 20 cm using the Very Large Array radiotelescope at the National Radio Astronomy Observatory. [Reproduced by permission from Matsui, Y., *et al.* (1984). *Astrophys. J.* **287,** 295.]

extended "shell-type" supernova remnant (SNR) forms. The remnants of the supernovae of 1006, 1572 (Tycho Brahe's supernovae), 1604 (Johannes Kepler's), and 1680 (Cas A) are well-observed examples of shell-type SNRs (Fig. 12).

Galactic supernova remnants are among the brightest radio sources on the sky. The emission is synchrotron radiation produced by relativistic electrons that are accelerated in the hot interaction region. Because radio waves are unaffected by interstellar dust grains, radio SNRs are useful for statistical studies of the spatial distribution and rate of supernovae in the Galaxy. Thermal X-rays also are detected from some SNRs, but most SNRs are not detected in this way because of absorption by interstellar matter; the X-ray spectra provide important information on the composition of the ejected matter.

The famous Crab Nebula, the remnant of the supernova of 1054, is the prototype of a less common kind of supernova remnant, called a "filled center" SNR, or a *plerion* (from the Greek word for full). The emission from the Crab includes spectral lines from matter that is expanding at only about 1000 km/sec. Another component of emission, which extends from the radio through the optical to the X-ray region, is nonthermal synchrotron radiation. This indicates the presence of highly relativistic electrons and the need for continued energy input into the central regions of the remnant; the source of this is a pulsar that rotates 30 times per second and generates the relativistic electrons. Searches for signs of high-velocity

matter associated with the Crab Nebula have not yet been successful.

Some SNRs also can be detected in relatively nearby external galaxies, but because of their small angular sizes they cannot be studied in as much detail.

VI. SUPERNOVAE AS DISTANCE INDICATORS FOR COSMOLOGY

A. The Extragalactic Distance Scale

The universe is expanding. The radial velocity of a galaxy relative to us is proportional to the distance of the galaxy from us; thus the cosmic expansion can be represented by the "Hubble law": $v = H_0 d$, where v is the radial velocity (ordinarily expressed as km/sec), d is the distance (as Mpc), and H_0 is the Hubble constant (as km/sec/Mpc) at the present epoch. The reciprocal of the Hubble constant gives the Hubble time—the time since the Big Bang origin of the expansion, assuming deceleration to have been negligible. Radial velocities are easily measured, so determining the value of H_0 reduces to the problem of measuring distances to galaxies. In practice, because galaxies have random motions, it is necessary to determine distances to fairly remote galaxies, at distances of hundreds of megaparsecs. Measuring the distances to supernovae provides an alternative to the classical approach of estimating the distances to galaxies themselves.

There are several ways to estimate distances to supernovae that require some physical understanding of the events but that have the advantage of being independent of all other distance determinations in astronomy. For example, the observed properties of SNe Ia indicate that they eject about 0.6 solar mass of ^{56}Ni, so the peak luminosity of an SN Ia can be calculated from the decay rate of ^{56}Ni and compared with the observed flux to yield the distance. Alternatively, the luminosity of a supernova can be calculated from its temperature (as revealed by its spectrum) and the radius of the photosphere (as the product of the expansion velocity and the time elapsed since explosion). Both methods usually have given longer distances and therefore lower values of H_0, near 60 km/sec/Mpc, than most other methods.

Another important approach takes advantage of the near homogeneity of the peak absolute magnitudes of SNe Ia, to use them as "standard candles" for distance determinations. The top panel in Fig. 13 illustrates the near-homogeneity, and the bottom panel shows how well SNe Ia can be standardized even further by correcting the peak magnitudes with the help of an empirical relation

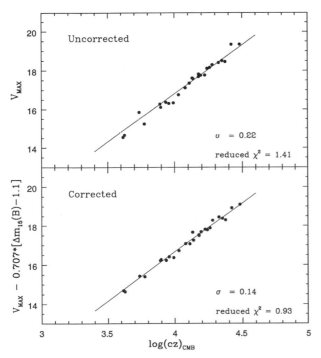

FIGURE 13 A "Hubble diagram" for a sample of well-observed SNe Ia. Top: The apparent visual magnitude of the supernova at maximum brightness is plotted against the logarithm of the parent-galaxy radial velocity (km/sec). If SNe Ia had identical peak absolute magnitudes, the points would fall on the straight line. Bottom: A correction for a relation between the peak magnitudes and the light-curve decay rate has been applied, to reduce the scatter about the straight line. [Reproduced by permission from Hamuy, M., *et al.* (1996). *Astron. J.* **112**, 2398.]

between the peak brightness and the rate of decline of the light curve. The *HST* has been used to find Cepheid variable stars in nearly a dozen galaxies in which SNe Ia were observed in the past. The period–luminosity relation for Cepheids gives the distance to the parent galaxy; the distance and the apparent magnitude of the supernova gives the absolute magnitude of the supernova ($M_V = -19.5$); and that absolute magnitude, when assigned to SNe Ia that appear in remote galaxies, gives their distances. In this way, H_0 has been determined to be about 60 km/sec/Mpc, in good agreement with the supernova physical methods. The corresponding Hubble time is 17 Gyr.

B. Deceleration and Acceleration

The self-gravity of the mass in the universe causes a deceleration of the cosmic expansion. If the matter density and the deceleration exceeded a critical amount, the expansion eventually would cease, and be followed by contraction and collapse. A way to determine the amount of deceleration that actually has occurred is to compare the expansion rate in the recent past, based on observations of the "local" universe, to the expansion rate in the remote past, based on extremely distant objects ("high-redshift" objects, because a large distance means large radial velocities and Doppler shifts toward longer wavelengths). Because SNe Ia are such excellent standard candles, and because they are so luminous that they can be seen at high redshifts, they are suitable for this task. This has been the main motivation for searching for batches of remote SNe Ia, beginning in the mid-1990s. Results based on the first dozens of high-redshift SNe Ia to have been discovered (which actually occurred deep in the past, when the distances between galaxies were only one-half to two-thirds of what they are now) indicate that deceleration is insufficient to bring the expansion to a halt. Evidently, the universe will expand forever. In fact, the SN Ia data indicate that although the cosmic expansion was decelerating in the past, it is *accelerating* now. An acceleration would require the introduction of Einstein's notorious "cosmological constant," or some other even more exotic agent that would oppose the deceleration caused by gravity. The acceleration is inferred from the observation that high-redshift SNe Ia are 20 to 50% dimmer, observationally, than they would be in the absence of acceleration. Observations of many more SNe Ia, and an improved understanding of the physics of SNe Ia, are needed to draw a firm conclusion that acceleration really is occurring.

VII. PROSPECTS

Our understanding of supernovae began to develop only after about 1970. Observations with modern detectors, primarily in the optical region but increasingly in other wavelength bands as well, provided high-quality data for comparison with the results of detailed numerical models. A basic understanding of the nature of supernova spectra and light curves, and detailed computer simulations of the explosions, emerged from the interplay between observation and theory. SN 1987A attracted much attention to the study of supernovae. During the 1990s dramatic progress was made in using SNe Ia to measure the Hubble constant and to probe the cosmic deceleration, and the discovery of an apparent acceleration has caused excitement among cosmologists and particle physicists, and even more interest in supernovae.

Further important observational advances are forthcoming. Ground-based optical astronomy will contribute automated searches with supernova-dedicated telescopes, to increase the discovery rate greatly. New powerful instruments such as the *Next Generation Space Telescope* (the *NGST*, successor to the *HST*), and perhaps even a large orbiting supernova-dedicated observatory, will provide observations of both nearby and very remote events. New sensitive instruments also will detect large numbers of neutrinos, and perhaps even gravitational waves, from supernovae in our Galaxy and the nearest external galaxies. The flow of observational data will stimulate further theoretical modeling. Improved physical data will be fed into new generations of fast computers to produce increasingly realistic models of the explosions. Astronomers will intensify their efforts to understand better which kinds of stars produce supernovae, and how they do it; to evaluate more precisely the role of supernovae in the nuclear evolution of the cosmos; and to exploit supernovae to probe the history and future expansion of the universe with ever-greater precision. These are grand goals—and they can be achieved.

SEE ALSO THE FOLLOWING ARTICLES

COSMOLOGY • GAMMA-RAY ASTRONOMY • GRAVITATIONAL WAVE ASTRONOMY • INFRARED ASTRONOMY • NEUTRINO ASTRONOMY • NEUTRON STARS • STELLAR SPECTROSCOPY • STELLAR STRUCTURE AND EVOLUTION • ULTRAVIOLET SPACE ASTRONOMY

BIBLIOGRAPHY

Arnett, D. (1996). "Supernovae and Nucleosynthesis," Princeton University Press, Princeton NJ.

Barbon, R., Buondi, V., Cappellaro, E., and Turatto, M. (1999). *Astron. Astrophys. Suppl. Ser.* **139,** 531.

Branch, D. (1998). *Annu. Rev. Astron. Astrophys.* **36,** 17.

Filippenko, A. V. (1997). *Annu. Rev. Astron. Astrophys.* **35,** 309.

Mann, A. K. (1997). "Shadow of a Star: the Neutrino Story of Supernova 1987A," W. H. Freeman, San Francisco.

Ruiz–Lapuente, P., Canal, R., and Isern, J. (eds.) (1997)."Thermonuclear Supernovae," Kluwer, Dordrecht.

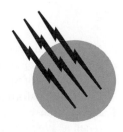

Superstring Theory

John H. Schwarz

California Institute of Technology

I. Supersymmetry
II. String Theory Basics
III. Superstrings
IV. From Superstrings to M-Theory

GLOSSARY

Compactification The process by which extra spatial dimensions form a very small (compact) manifold and become invisible at low energies. To end up with four large dimensions, this manifold should have six dimensions in the case of superstring theory or seven dimensions in the case of M theory.

D-brane A special type of *p*-brane that has the property that a fundamental string can terminate on it. Mathematically, this corresponds to Dirichlet boundary conditions, which is the reason for the use of the letter D.

M-theory A conjectured quantum theory in eleven dimensions, which is approximated at low energies by eleven-dimensional supergravity. It arises as the strong coupling limit of the type IIA and E8 × E8 heterotic string theory. The letter M stands for magic, mystery, or membrane according to taste.

p-brane A dynamical excitation in a string theory that has *p* spatial dimensions. The fundamental string, for example, is a 1-brane. All of the other p-branes have tensions that diverge at weak coupling, and therefore they are nonperturbative.

S duality An equivalence between two string theories (such as type I and *SO*(32) heterotic) which relates one at weak coupling to the other at strong coupling and vice versa.

String theory A relativistic quantum theory in which the fundamental objects are one-dimensional loops called strings. Unlike quantum field theories based on point particles, consistent string theories unify gravity with the other forces.

Supergravity A supersymmetric theory of gravity. In addition to a spacetime metric field that describes spin 2 gravitons, the quanta of gravity, these theories contain one or more spin 3/2 gravitino fields. The gravitino fields are gauge fields for local supersymmetry.

Superstring A supersymmetric string theory. At weak coupling there are five distinct superstring theories, each of which requires ten-dimensional spacetime (nine spatial dimensions and one time dimension). These five theories are related by various dualities, which imply that they are different limits of a single underlying theory.

Supersymmetry A special kind of symmetry that relates bosons (particles with integer intrinsic spin) to fermions (particles with half-integer intrinsic spin). Unlike other symmetries, the associated conserved charges transform as spinors. According to a fundamental theorem, supersymmetry is the unique possibility for a nontrivial extension of the known symmetries of spacetime (translations, rotations, and Lorentz transformations).

T duality An equivalence between two string theories (such as type IIA and type IIB) which relates one with a small circular spatial dimension to the other with a large circular spatial dimension and vice versa.

MANY of the major developments in fundamental physics of the past century arose from identifying and overcoming contradictions between existing ideas. For example, the incompatibility of Maxwell's equations and Galilean invariance led Einstein to propose the special theory of relativity. Similarly, the inconsistency of special relativity with Newtonian gravity led him to develop a new theory of gravity, which he called the general theory of relativity. More recently, the reconciliation of special relativity with quantum mechanics led to the development of quantum field theory. We are now facing another crisis of the same character. Namely, general relativity appears to be incompatible with quantum field theory. Any straightforward attempt to "quantize" general relativity leads to a nonrenormalizable theory. This means that the theory is inconsistent and needs to be modified at short distances or high energies. The way that string theory does this is to give up one of the basic assumptions of quantum field theory, the assumption that elementary particles are mathematical points. Instead, it is a quantum field theory of one-dimensional extended objects called strings. There are very few consistent theories of this type, but superstring theory shows great promise as a unified quantum theory of all fundamental forces including gravity. So far, nobody has constructed a realistic string theory of elementary particles that could serve as a new standard model of particles and forces, since there is much that needs to be better understood first. But that, together with a deeper understanding of cosmology, is the goal. This is very much a work in progress.

Even though string theory is not yet fully formulated, and we cannot yet give a detailed description of how the standard model of elementary particles should emerge at low energies, there are some general features of the theory that can be identified. These are features that seem to be quite generic irrespective of how various details are resolved. The first, and perhaps most important, is that general relativity is necessarily incorporated in the theory. It gets modified at very short distances/high energies but at ordinary distances and energies it is present in exactly the form proposed by Einstein. This is significant, because it is arising within the framework of a consistent quantum theory. Ordinary quantum field theory does not allow gravity

to exist; string theory requires it. The second general fact is that Yang–Mills gauge theories of the sort that comprise the standard model naturally arise in string theory. We do not understand why the specific Yang–Mills gauge theory based on the symmetry group $SU(3) \times SU(2) \times U(1)$ should be preferred, but (anomaly-free) theories of this general type do arise naturally at ordinary energies. The third general feature of string theory solutions is that they possess a special kind of symmetry called supersymmetry. The mathematical consistency of string theory depends crucially on supersymmetry, and it is very hard to find consistent solutions (i.e., quantum vacua) that do not preserve at least a portion of this supersymmetry. This prediction of string theory differs from the other two (general relativity and gauge theories) in that it really is a prediction. It is a generic feature of string theory that has not yet been observed experimentally.

I. SUPERSYMMETRY

Even though supersymmetry is a very important part of the story, the discussion here will be very brief. Like the electroweak symmetry in the standard model, supersymmetry is necessarily a broken symmetry. A variety of arguments, not specific to string theory, suggest that the characteristic energy scale associated to supersymmetry breaking should be related to the electroweak scale, in other words, in the range 100 GeV–1 TeV. (Recall that the rest mass of a proton or neutron corresponds to an energy of approximately 1 GeV. Also, the masses of the $W^{\pm}$ and Z^0 particles, which transmit the weak nuclear forces, correspond to energies of approximately 100 GeV.) Supersymmetry implies that all known elementary particles should have partner particles whose masses are in this general range. If supersymmetry were not broken, these particles would have exactly the same masses as the known particles, and that is definitely excluded. This means that some of these superpartners should be observable at the CERN Large Hadron Collider (LHC), which is scheduled to begin operating in 2005 or 2006. There is even a chance that Fermilab Tevatron experiments could find superparticles before then. (CERN is a lab outside of Geneva, Switzerland and Fermilab is located outside of Chicago, IL.)

In most versions of phenomenological supersymmetry there is a multiplicatively conserved quantum number called R-parity. All known particles have even R-parity, whereas their superpartners have odd R-parity. This implies that the superparticles must be pair-produced in particle collisions. It also implies that the lightest supersymmetry particle (or LSP) should be absolutely stable. It is not known with certainty which superparticle is the LSP, but one popular guess is that it is a "neutralino."

This is an electrically neutral fermion that is a quantum-mechanical mixture of the partners of the photon, Z^0, and neutral Higgs particles. Such an LSP would interact very weakly, more or less like a neutrino. It is of considerable interest, since it has properties that make it an excellent dark matter candidate. There are experimental searches underway in Europe and in the United States for a class of dark matter particles called WIMPS (weakly interacting massive particles). Since the LSP is of an example of a WIMP, these searches could discover the LSP some day. However, the current experiments might not have sufficient detector volume to compensate for the exceedingly small LSP cross sections, so we may have to wait for future upgrades of the detectors.

There are three unrelated arguments that point to the same 100 GeV–1 TeV mass range for superparticles. The one we have just been discussing, a neutralino LSP as an important component of dark matter, requires a mass of about 100 GeV. The precise number depends on the mixture that comprises the LSP, what their density is, and a number of other details. A second argument is based on a theoretical issue called the hierarchy problem. This is the fact that in the standard model quantum corrections tend to renormalize the Higgs mass to an unacceptably high value. The way to prevent this is to extend the standard model to a supersymmetric standard model and to have the supersymmetry be broken at a scale comparable to the Higgs mass, and hence to the electroweak scale. This works because the quantum corrections to the Higgs mass are more mild in the supersymmetric version of the theory. The third argument that gives an estimate of the supersymmetry-breaking scale is based on grand unification. If one accepts the notion that the standard model gauge group is embedded in a larger group such as $SU(5)$ or $SO(10)$, which is broken at a high mass scale, then the three standard model coupling constants should unify at that mass scale. Given the spectrum of particles, one can compute the variation of the couplings as a function of energy using renormalization group equations. One finds that if one only includes the standard model particles this unification fails quite badly. However, if one also includes all the supersymmetry particles required by the minimal supersymmetric extension of the standard model, then the couplings do unify at an energy of about 2×10^{16} GeV. This is a very striking success. For this agreement to take place, it is necessary that the masses of the superparticles are less than a few TeV.

There is other support for this picture, such as the ease with which supersymmetric grand unification explains the masses of the top and bottom quarks and electroweak symmetry breaking. Despite all these indications, we cannot be certain that supersymmetry at the electroweak scale really is correct until it is demonstrated experimentally.

One could suppose that all the successes that we have listed are a giant coincidence, and the correct description of TeV scale physics is based on something entirely different. The only way we can decide for sure is by doing the experiments. I am optimistic that supersymmetry will be found, and that the experimental study of the detailed properties of the superparticles will teach us a great deal.

A. Basic Ideas of String Theory

In conventional quantum field theory the elementary particles are mathematical points, whereas in perturbative string theory the fundamental objects are one-dimensional loops (of zero thickness). Strings have a characteristic length scale, which can be estimated by dimensional analysis. Since string theory is a relativistic quantum theory that includes gravity it must involve the fundamental constants c (the speed of light), $\hbar$ (Planck's constant divided by 2π), and G (Newton's gravitational constant). From these one can form a length, known as the Planck length

$$\ell_p = \left(\frac{\hbar G}{c^3}\right)^{3/2} = 1.6 \times 10^{-33}\,\text{cm}. \tag{1}$$

Similarly, the Planck mass is

$$m_p = \left(\frac{\hbar c}{G}\right)^{1/2} = 1.2 \times 10^{19}\,\text{GeV/}c^2. \tag{2}$$

Experiments at energies far below the Planck energy cannot resolve distances as short as the Planck length. Thus, at such energies, strings can be accurately approximated by point particles. From the viewpoint of string theory, this explains why quantum field theory has been so successful.

As a string evolves in time it sweeps out a two-dimensional surface in spacetime, which is called the world sheet of the string. This is the string counterpart of the world line for a point particle. In quantum field theory, analyzed in perturbation theory, contributions to amplitudes are associated to Feynman diagrams, which depict possible configurations of world lines. In particular, interactions correspond to junctions of world lines. Similarly, perturbative string theory involves string world sheets of various topologies. A particularly significant fact is that these world sheets are generically smooth. The existence of interaction is a consequence of world-sheet topology rather than a local singularity on the world sheet. This difference from point-particle theories has two important implications. First, in string theory the structure of interactions is uniquely determined by the free theory. There are no arbitrary interactions to be chosen. Second, the occurrence of ultraviolet divergences in point-particle quantum field theories can be traced to the fact that interactions are associated to world-line junctions at specific spacetime points. Because the string world sheet is smooth, without any singular behavior at short distances, string theory has no ultraviolet divergences.

B. A Brief History of String Theory

String theory arose in the late 1960s out of an attempt to describe the strong nuclear force, which acts on a class of particles called hadrons. The first string theory that was constructed only contained bosons. The construction of a better string theory that also includes fermions led to the discovery of supersymmetric strings (later called superstrings) in 1971. The subject fell out of favor around 1973 with the development of quantum chromodynamics (QCD), which was quickly recognized to be the correct theory of strong interactions. Also, string theories had various peculiar features, such as extra dimensions and massless particles, which are not appropriate for a hadron theory.

Among the massless string states there is one that corresponds to a particle with two units of spin. In 1974, it was shown by Joël Scherk and the author (Scherk and Schwarz, 1974), and independently by Yoneya (1974), that this particle interacts like a graviton, so that string theory actually contains general relativity. This led us to propose that string theory should be used for unification of all elementary particles and forces rather than as a theory of hadrons and the strong nuclear force. This implied, in particular, that the string length scale should be comparable to the Planck length, rather than the size of hadrons (10^{-13} cm), as had been previously assumed.

In the period now known as the "first superstring revolution," which took place in 1984–1985, there were a number of important developments (described later in this article) that convinced a large segment of the theoretical physics community that this is a worthy area of research. By the time the dust settled in 1985 we had learned that there are five distinct consistent string theories, and that each of them requires spacetime supersymmetry in the ten dimensions (nine spatial dimensions plus time). The theories, which will be described later, are called type I, type IIA, type IIB, $SO(32)$ heterotic, and $E_8 \times E_8$ heterotic. In the "second superstring revolution," which took place around 1995, we learned that the five string theories are actually special solutions of a completely unique underlying theory.

C. Compactification

In the context of the original goal of string theory—to explain hadron physics—extra dimensions are unacceptable. However, in a theory that incorporates general relativity, the geometry of spacetime is determined dynamically. Thus one could imagine that the theory admits consis-

tent quantum solutions in which the six extra spatial dimensions form a compact space, too small to have been observed. The natural first guess is that the size of this space should be comparable to the string scale and the Planck length. Since the equations of the theory must be satisfied, the geometry of this six-dimensional space is not arbitrary. A particularly appealing possibility, which is consistent with the equations, is that it forms a type of space called a Calabi–Yau space (Candelas et al., 1985).

Calabi–Yau compactification, in the context of the $E_8 \times E_8$ heterotic string theory, can give a low-energy effective theory that closely resembles a supersymmetric extension of the standard model. There is actually a lot of freedom, because there are very many different Calabi–Yau spaces, and there are other arbitrary choices that can be made. Still, it is interesting that one can come quite close to realistic physics. It is also interesting that the number of quark and lepton families that one obtains is determined by the topology of the Calabi–Yau space. Thus, for suitable choices, one can arrange to end up with exactly three families. People were very excited by this scenario in 1985. Today, we tend to make a more sober appraisal that emphasizes all the arbitrariness that is involved, and the things that don't work exactly right. Still, it would not be surprising if some aspects of this picture survive as part of the story when we understand the right way to describe the real world.

D. Perturbation Theory

Until 1995 it was only understood how to formulate string theories in terms of perturbation expansions. Perturbation theory is useful in a quantum theory that has a small dimensionless coupling constant, such as quantum electrodynamics, since it allows one to compute physical quantities as power series expansions in the small parameter. In quantum electrodynamics (QED) the small parameter is the fine-structure constant $\alpha \sim 1/137$. Since this is quite small, perturbation theory works very well for QED. For a physical quantity $T(\alpha)$, one computes (using Feynman diagrams)

$$T(\alpha) = T_0 + \alpha T_1 + \alpha^2 T_2 + \cdots. \tag{3}$$

It is the case generically in quantum field theory that expansions of this type are divergent. More specifically, they are asymptotic expansions with zero radius convergence. Nonetheless, they can be numerically useful if the expansion parameter is small. The problem is that there are various nonperturbative contributions (such as instantons) that have the structure

$$T_{NP} \sim e^{-(\text{const.}/\alpha)}. \tag{4}$$

In a theory such as QCD, there are problems for which perturbation theory is useful (due to asymptotic freedom) and other ones where it is not. For problems of the latter type, such as computing the hadron spectrum, nonperturbative methods of computation, such as lattice gauge theory, are required.

In the case of string theory the dimensionless string coupling constant, denoted g_s, is determined dynamically by the expectation value of a scalar field called the dilaton. There is no particular reason that this number should be small. So it is unlikely that a realistic vacuum could be analyzed accurately using perturbation theory. More importantly, these theories have many qualitative properties that are inherently nonperturbative. So one needs nonperturbative methods to understand them.

E. The Second Superstring Revolution

Around 1995 some amazing and unexpected "dualities" were discovered that provided the first glimpses into nonperturbative features of string theory. These dualities were quickly recognized to have three major implications.

The dualities enabled us to relate all five of the superstring theories to one another. This meant that, in a fundamental sense, they are all equivalent to one another. Another way of saying this is that there is a unique underlying theory, and what we had been calling five theories are better viewed as perturbation expansions of this underlying theory about five different points (in the space of consistent quantum vacua). This was a profoundly satisfying realization, since we really didn't want five theories of nature. That there is a completely unique theory, without any dimensionless parameters, is the best outcome for which one could have hoped. To avoid confusion, it should be emphasized that even though the theory is unique, it is entirely possible that there are many consistent quantum vacua. Classically, the corresponding statement is that a unique equation can admit many solutions. It is a particular solution (or quantum vacuum) that ultimately must describe nature. At least, this is how a particle physicist would say it. If we hope to understand the origin and evolution of the universe, in addition to properties of elementary particles, it would be nice if we could also understand cosmological solutions.

A second crucial discovery was that the theory admits a variety of nonperturbative excitations, called p-branes, in addition to the fundamental strings. The letter p labels the number of spatial dimensions of the excitation. Thus, in this language, a point particle is a 0-brane, a string is a 1-brane, and so forth. The reason that p-branes were not discovered in perturbation theory is that they have tension (or energy density) that diverges as $g_s \to 0$. Thus they are absent from the perturbative theory.

The third major discovery was that the underlying theory also has an eleven-dimensional solution, which is

called M-theory. Later, we will explain how the eleventh dimension arises.

One type of duality is called S duality. (The choice of the letter S has no great significance.) Two string theories (let's call them A and B) are related by S duality if one of them evaluated at strong coupling is equivalent to the other one evaluated at weak coupling. Specifically, for any physical quantity f, one has

$$f_A(g_s) = f_B(1/g_s). \qquad (5)$$

Two of the superstring theories—type I and $SO(32)$ heterotic—are related by S duality in this way. The type IIB theory is self-dual. Thus S duality is a symmetry of the IIB theory, and this symmetry is unbroken if $g_s = 1$. Thanks to S duality, the strong coupling behavior of each of these three theories is determined by a weak-coupling analysis. The remaining two theories, type IIA and $E_8 \times E_8$ heterotic, behave very differently at strong coupling. They grow an eleventh dimension.

Another astonishing duality, which goes by the name of T duality, was discovered several years earlier. It can be understood in perturbation theory, which is why it was found first. But, fortunately, it often continues to be valid even at strong coupling. T duality can relate different compactifications of different theories. For example, suppose theory A has a compact dimension that is a circle of radius R_A and theory B has a compact dimension that is a circle of radius R_B. If these two theories are related by T duality this means that they are equivalent provided that

$$R_A R_B = (\ell_s)^2, \qquad (6)$$

where ℓ_s is the fundamental string length scale. This has the amazing implication that when one of the circles becomes small the other one becomes large. Later, we will explain how this is possible. T duality relates the two type II theories and the two heterotic theories. There are more complicated examples of the same phenomenon involving compact spaces that are more complicated than a circle, such as tori, K3, Calabi–Yau spaces, etc.

F. The Origins of Gauge Symmetry

There are a variety of mechanisms than can give rise to Yang–Mills type gauge symmetries in string theory. Here, we will focus on two basic possibilities: Kaluza–Klein symmetries and brane symmetries.

The basic Kaluza–Klein idea goes back to the 1920s, though it has been much generalized since then. The idea is to suppose that the ten- or eleven-dimensional geometry has a product structure $M \times K$, where M is Minkowski spacetime and K is a compact manifold. Then, if K has symmetries, these appear as gauge symmetries of the effective theory defined on M. The Yang–Mills gauge fields arise as components of the gravitational metric field with one direction along K and the other along M. For example, if the space K is an n-dimensional sphere, the symmetry group is $SO(n+1)$, if it is CP^n—which has $2n$ dimensions—it is $SU(n+1)$, and so forth. Elegant as this may be, it seems unlikely that a realistic K has any such symmetries. Calabi–Yau spaces, for example, do not have any.

A rather more promising way of achieving realistic gauge symmetries is via the brane approach. Here the idea is that a certain class of p-branes (called D-branes) have gauge fields that are restricted to their world volume. This means that the gauge fields are not defined throughout the ten- or eleven-dimensional spacetime but only on the $(p+1)$-dimensional hypersurface defined by the D-branes. This picture suggests that the world we observe might be a D-brane embedded in a higher dimensional space. In such a scenario, there can be two kinds of extra dimensions: compact dimensions along the brane and compact dimensions perpendicular to the brane.

The traditional viewpoint, which in my opinion is still the best bet, is that all extra dimensions (of both types) have sizes of order 10^{-30}–10^{-32} cm corresponding to an energy scale of 10^{16}–10^{18} GeV. This makes them inaccessible to direct observation, though their existence would have definite low-energy consequences. However, one can and should ask "what are the experimental limits?" For compact dimensions along the brane, which support gauge fields, the nonobservation of extra dimensions in tests of the standard model implies a bound of about 1 TeV. The LHC should extend this to about 10 TeV. For compact dimensions "perpendicular to the brane," which only support excitations with gravitational strength forces, the best bounds come from Cavendish-type experiments, which test the $1/R^2$ structure of the Newton force law at short distances. No deviations have been observed to a distance of about 1 mm so far. Experiments planned in the near future should extend the limit to about 100 μ. Obviously, observation of any deviation from $1/R^2$ would be a major discovery.

G. Conclusion

This introductory section has sketched some of the remarkable successes that string theory has achieved over the past 30 years. There are many others that did not fit in this brief survey. Despite all this progress, there are some very important and fundamental questions whose answers are unknown. It seems that whenever a breakthrough occurs, a host of new questions arise, and the ultimate goal still seems a long way off. To convince you that there is a long way to go, let us list some of the most important questions.

- What is the theory? Even though a great deal is known about string theory and M-theory, it seems that the optimal formulation of the underlying theory has not yet been found. It might be based on principles that have not yet been formulated.

- We are convinced that supersymmetry is present at high energies and probably at the electroweak scale, too. But we do not know how or why it is broken.

- A very crucial problem concerns the energy density of the vacuum, which is a physical quantity in a gravitational theory. This is characterized by the cosmological constant, which observationally appears to have a small positive value—so that the vacuum energy of the universe is comparable to the energy in matter. In Planck units this is a tiny number ($\Lambda \sim 10^{-120}$). If supersymmetry were unbroken, we could argue that $\Lambda = 0$, but if it is broken at the 1 TeV scale, that would seem to suggest $\Lambda \sim 10^{-60}$, which is very far from the truth. Despite an enormous amount of effort and ingenuity, it is not yet clear how superstring theory will conspire to break supersymmetry at the TeV scale and still give a value for Λ that is much smaller than 10^{-60}. The fact that the desired result is about the square of this might be a useful hint.

- Even though the underlying theory is unique, there seem to be many consistent quantum vacua. We would very much like to formulate a theoretical principle (not based on observation) for choosing among these vacua. It is not known whether the right approach to the answer is cosmological, probabilistic, anthropic, or something else.

II. STRING THEORY BASICS

In this section we will describe the world-sheet dynamics of the original bosonic string theory. As we will see this theory has various unrealistic and unsatisfactory properties. Nonetheless it is a useful preliminary before describing supersymmetric strings, because it allows us to introduce many of the key concepts without simultaneously addressing the added complications associated with fermions and supersymmetry.

We will describe string dynamics from a first-quantized world-sheet sum-over-histories point of view. This approach is closely tied to perturbation theory analysis. It should be contrasted with "second quantized" string field theory, which is based on field operators that create or destroy entire strings. To explain the methodology, let us begin by reviewing the world-line description a massive point particle.

A. World-Line Description of a Massive Point Particle

A point particle sweeps out a trajectory (or world line) in spacetime. This can be described by functions $x^\mu(\tau)$ that describe how the world line, parameterized by τ, is embedded in the spacetime, whose coordinates are denoted x^μ. For simplicity, let us assume that the spacetime is flat Minkowski space with a Lorentz metric

$$\eta_{\mu\nu} = \begin{pmatrix} -1 & 0 & 0 & 0 \\ 0 & 1 & 0 & 0 \\ 0 & 0 & 1 & 0 \\ 0 & 0 & 0 & 1 \end{pmatrix}. \tag{7}$$

Then, the Lorentz invariant line element is given by

$$ds^2 = -\eta_{\mu\nu}\, dx^\mu\, dx^\nu. \tag{8}$$

In units $h = c = 1$, the action for a particle of mass m is given by

$$S = -m \int ds. \tag{9}$$

This could be generalized to a curved spacetime by replacing $\eta_{\mu\nu}$ by a metric $g_{\mu\nu}(x)$, but we will not do so here. In terms of the embedding functions, $x^\mu(\tau)$, the action can be rewritten in the form

$$S = -m \int d\tau \sqrt{-\eta_{\mu\nu}\dot{x}^\mu \dot{x}^\nu}, \tag{10}$$

where dots represent τ derivatives. An important property of this action is invariance under local reparametrizations. This is a kind of gauge invariance, whose meaning is that the form of S is unchanged under an arbitrary reparametrization of the world line $\tau \to \tau(\tilde{\tau})$. Actually, one should require that the function $\tau(\tilde{\tau})$ is smooth and monotonic ($\frac{d\tau}{d\tilde{\tau}} > 0$). The reparametrization invariance is a one-dimensional analog of the four-dimensional general coordinate invariance of general relativity. Mathematicians refer to this kind of symmetry as diffeomorphism invariance.

The reparametrization invariance of S allows us to choose a gauge. A nice choice is the "static gauge"

$$x^0 = \tau. \tag{11}$$

In this gauge (renaming the parameter t) the action becomes

$$S = -m \int \sqrt{1 - v^2}\, dt, \tag{12}$$

where

$$\vec{v} = \frac{d\vec{x}}{dt}. \tag{13}$$

Requiring this action to be stationary under an arbitrary variation of $\vec{x}(t)$ gives the Euler–Lagrange equations

$$\frac{d\vec{p}}{dt} = 0, \tag{14}$$

where

$$\vec{p} = \frac{\delta S}{\delta \vec{v}} = \frac{m\vec{v}}{\sqrt{1 - v^2}}, \tag{15}$$

which is the usual result. So we see that standard relativistic kinematics follows from the action $S = -m \int ds$.

B. World-Volume Actions

We can now generalize the analysis of the massive point particle to a p-brane of tension T_p. The action in this case involves the invariant $(p + 1)$-dimensional volume and is given by

$$S_p = -T_p \int d\mu_{p+1}, \tag{16}$$

where the invariant volume element is

$$d\mu_{p+1} = \sqrt{-\det\left(-\eta_{\mu\nu}\partial_\alpha x^\mu \partial_\beta x^\nu\right)}\, d^{p+1}\sigma. \tag{17}$$

Here the embedding of the p-brane into d-dimensional spacetime is given by functions $x^\mu(\sigma^\alpha)$. The index $\alpha = 0, \ldots, p$ labels the $p + 1$ coordinates σ^α of the p-brane world-volume and the index $\mu = 0, \ldots, d - 1$ labels the d coordinates x^μ of the d-dimensional spacetime. We have defined

$$\partial_\alpha x^\mu = \frac{\partial x^\mu}{\partial \sigma^\alpha}. \tag{18}$$

The determinant operation acts on the $(p + 1) \times (p + 1)$ matrix whose rows and columns are labeled by α and β. The tension T_p is interpreted as the mass per unit volume of the p-brane. For a 0-brane, it is just the mass. The action S_p is reparametrization invariant. In other words, substituting $\sigma^\alpha = \sigma^\alpha(\tilde{\sigma}^\beta)$, it takes the same form when expressed in terms of the coordinates $\tilde{\sigma}^\alpha$.

Let us now specialize to the string, $p = 1$. Evaluating the determinant gives

$$S[x] = -T \int d\sigma\, d\tau \sqrt{\dot{x}^2 x'^2 - (\dot{x} \cdot x')^2}, \tag{19}$$

where we have defined $\sigma^0 = \tau$, $\sigma^1 = \sigma$, and

$$\dot{x}^\mu = \frac{\partial x^\mu}{\partial \tau}, \qquad x'^\mu = \frac{\partial x^\mu}{\partial \sigma}. \tag{20}$$

This action, called the Nambu–Goto action, was first proposed in 1970 (Nambu, 1970 and Goto, 1971). The Nambu–Goto action is equivalent to the action

$$S[x, h] = -\frac{T}{2} \int d^2\sigma \sqrt{-h}\, h^{\alpha\beta} \eta_{\mu\nu} \partial_\alpha x^\mu \partial_\beta x^\nu, \tag{21}$$

where $h_{\alpha\beta}(\sigma, \tau)$ is the world-sheet metric, $h = \det h_{\alpha\beta}$, and $h^{\alpha\beta}$ is the inverse of $h_{\alpha\beta}$. The Euler–Lagrange equations obtained by varying $h^{\alpha\beta}$ are

$$T_{\alpha\beta} = \partial_\alpha x \cdot \partial_\beta x - \frac{1}{2} h_{\alpha\beta} h^{\gamma\delta} \partial_\gamma x \cdot \partial_\delta x = 0. \qquad (22)$$

The equation $T_{\alpha\beta} = 0$ can be used to eliminate the world-sheet metric from the action, and when this is done one recovers the Nambu–Goto action. (To show this take the determinant of both sides of the equation $\partial_\alpha x \cdot \partial_\beta x = \frac{1}{2} h_{\alpha\beta} h^{\gamma\delta} \partial_\gamma x \cdot \partial_\delta x$.)

In addition to reparametrization invariance, the action $S[x, h]$ has another local symmetry, called conformal invariance (or Weyl invariance). Specifically, it is invariant under the replacement

$$h_{\alpha\beta} \to \Lambda(\sigma, \tau) h_{\alpha\beta}$$
$$x^\mu \to x^\mu. \qquad (23)$$

This local symmetry is special to the $p = 1$ case (strings).

The two reparametrization invariance symmetries of $S[x, h]$ allow us to choose a gauge in which the three functions $h_{\alpha\beta}$ (this is a symmetric 2×2 matrix) are expressed in terms of just one function. A convenient choice is the "conformally flat gauge"

$$h_{\alpha\beta} = \eta_{\alpha\beta} e^{\phi(\sigma, \tau)}. \qquad (24)$$

Here $\eta_{\alpha\beta}$ denotes the two-dimensional Minkowski metric of a flat world-sheet. However, because of the factor e^ϕ, $h_{\alpha\beta}$ is only "conformally flat." Classically, substitution of this gauge choice into $S[x, h]$ yields the gauge-fixed action

$$S = \frac{T}{2} \int d^2\sigma \, \eta^{\alpha\beta} \partial_\alpha x \cdot \partial_\beta x. \qquad (25)$$

Quantum mechanically, the story is more subtle. Instead of eliminating h via its classical field equations, one should perform a Feynman path integral, using standard machinery to deal with the local symmetries and gauge fixing. When this is done correctly, one finds that in general ϕ does not decouple from the answer. Only for the special case $d = 26$ does the quantum analysis reproduce the formula we have given based on classical reasoning (Polyakov, 1981). Otherwise, there are correction terms whose presence can be traced to a conformal anomaly (i.e., a quantum-mechanical breakdown of the conformal invariance).

The gauge-fixed action [Eq. (25)] is quadratic in the x's. Mathematically, it is the same as a theory of d free scalar fields in two dimensions. The equations of motion obtained by varying x^μ are simply free two-dimensional wave equations:

$$\ddot{x}^\mu - x''^\mu = 0. \qquad (26)$$

This is not the whole story, however, because we must also take account of the constraints $T_{\alpha\beta} = 0$. Evaluated in the conformally flat gauge, these constraints are

$$T_{01} = T_{10} = \dot{x} \cdot x' = 0$$
$$T_{00} = T_{11} = \frac{1}{2}(\dot{x}^2 + x'^2) = 0. \qquad (27)$$

Adding and subtracting gives

$$(\dot{x} \pm x')^2 = 0. \qquad (28)$$

C. Boundary Conditions

To go further, one needs to choose boundary conditions. There are three important types. For a closed string one should impose periodicity in the spatial parameter σ. Choosing its range to be π (as is conventional)

$$x^\mu(\sigma, \tau) = x^\mu(\sigma + \pi, \tau). \qquad (29)$$

For an open string (which has two ends), each end can be required to satisfy either Neumann or Dirichlet boundary conditions (for each value of μ).

$$\text{Neumann:} \quad \frac{\partial x^\mu}{\partial \sigma} = 0 \quad \text{at} \quad \sigma = 0 \text{ or } \pi \quad (30)$$

$$\text{Dirichlet:} \quad \frac{\partial x^\mu}{\partial \tau} = 0 \quad \text{at} \quad \sigma = 0 \text{ or } \pi. \quad (31)$$

The Dirichlet condition can be integrated, and then it specifies a spacetime location on which the string ends. The only way this makes sense is if the open string ends on a physical object—it ends on a D-brane. (D stands for Dirichlet.) If all the open-string boundary conditions are Neumann, then the ends of the string can be anywhere in the spacetime. The modern interpretation is that this means that there are spacetime-filling D-branes present.

Let us now consider the closed-string case in more detail. The general solution of the two-dimensional wave equation is given by a sum of "right-movers" and "left-movers":

$$x^\mu(\sigma, \tau) = x_R^\mu(\tau - \sigma) + x_L^\mu(\tau + \sigma). \qquad (32)$$

These should be subject to the following additional conditions:

1. $x^\mu(\sigma, \tau)$ is real
2. $x^\mu(\sigma + \pi, \tau) = x^\mu(\sigma, \tau)$
3. $(x_L')^2 = (x_R')^2 = 0$; these are the $T_{\alpha\beta} = 0$ constraints in Eq. (28)

The first two conditions can be solved explicitly in terms of Fourier series:

$$x_R^\mu = \frac{1}{2}x^\mu + \ell_s^2 p^\mu(\tau - \sigma) + \frac{i}{\sqrt{2}}\ell_s \sum_{n \neq 0} \frac{1}{n}\alpha_n^\mu e^{-2in(\tau - \sigma)}$$

$$\tag{33}$$

$$x_L^\mu = \frac{1}{2}x^\mu + \ell_s^2 p^\mu(\tau + \sigma) + \frac{i}{\sqrt{2}}\ell_s \sum_{n \neq 0} \frac{1}{n}\tilde{\alpha}_n^\mu e^{-2in(\tau + \sigma)},$$

where the expansion parameters α_n^μ, $\tilde{\alpha}_n^\mu$ satisfy

$$\alpha_{-n}^\mu = \left(\alpha_n^\mu\right)^\dagger, \qquad \tilde{\alpha}_{-n}^\mu = \left(\tilde{\alpha}_n^\mu\right)^\dagger. \tag{34}$$

The center-of-mass coordinate x^μ and momentum p^μ are also real. The fundamental string length scale ℓ_s is related to the tension T by

$$T = \frac{1}{2\pi\alpha'}, \qquad \alpha' = \ell_s^2. \tag{35}$$

The parameter α' is called the universal Regge slope, since the string modes lie on linear parallel Regge trajectories with this slope.

D. Quantization

The analysis of closed-string left-moving modes, closed-string right-moving modes, and open-string modes are all very similar. Therefore, to avoid repetition, we will focus on the closed-string right-movers. Starting with the gauge-fixed action in Eq. (25), the canonical momentum of the string is

$$p^\mu(\sigma, \tau) = \frac{\delta S}{\delta \dot{x}^\mu} = T\dot{x}^\mu. \tag{36}$$

Canonical quantization (this is just free two-dimensional field theory for scalar fields) gives

$$[p^\mu(\sigma, \tau), x^\nu(\sigma', \tau)] = -i\hbar\eta^{\mu\nu}\delta(\sigma - \sigma'). \tag{37}$$

In terms of the Fourier modes (setting $\hbar = 1$) these become

$$[p^\mu, x^\nu] = -i\eta^{\mu\nu} \tag{38}$$

$$\left[\alpha_m^\mu, \alpha_n^\nu\right] = m\delta_{m+n,0}\,\eta^{\mu\nu},$$
$$\left[\tilde{\alpha}_m^\mu, \tilde{\alpha}_n^\nu\right] = m\delta_{m+n,0}\,\eta^{\mu\nu}, \tag{39}$$

and all other commutators vanish.

Recall that a quantum-mechanical harmonic oscillator can be described in terms of raising and lowering operators, usually called $a^\dagger$ and a, which satisfy

$$[a, a^\dagger] = 1. \tag{40}$$

We see that, aside from a normalization factor, the expansion coefficients α_{-m}^μ and α_m^μ are raising and lowering operators. There is just one problem. Because $\eta^{00} = -1$, the time components are proportional to oscillators with the

wrong sign ($[a, a^\dagger] = -1$). This is potentially very bad, because such oscillators create states of negative norm, which could lead to an inconsistent quantum theory (with negative probabilities, etc.). Fortunately, as we will explain, the $T_{\alpha\beta} = 0$ constraints eliminate the negative-norm states from the physical spectrum.

The classical constraint for the right-moving closed-string modes, $(x_R')^2 = 0$, has Fourier components

$$L_m = \frac{T}{2}\int_0^\pi e^{-2im\sigma}(x_R')^2\,d\sigma = \frac{1}{2}\sum_{n=-\infty}^\infty \alpha_{m-n} \cdot \alpha_n, \tag{41}$$

which are called Virasoro operators. Since α_m^μ does not commute with α_{-m}^μ, L_0 needs to be normal-ordered:

$$L_0 = \frac{1}{2}\alpha_0^2 + \sum_{n=1}^\infty \alpha_{-n} \cdot \alpha_n. \tag{42}$$

Here $\alpha_0^\mu = \ell_s p^\mu/\sqrt{2}$, where p^μ is the momentum.

E. The Free String Spectrum

Recall that the Hilbert space of a harmonic oscillator is spanned by states $|n\rangle$, $n = 0, 1, 2, \ldots$, where the ground state, $|0\rangle$, is annihilated by the lowering operator ($a\,|0\rangle = 0$) and

$$|n\rangle = \frac{(a^\dagger)^n}{\sqrt{n!}}\,|0\rangle. \tag{43}$$

Then, for a normalized ground-state ($\langle 0 | 0 \rangle = 1$), one can use $[a, a^\dagger] = 1$ repeatedly to prove that

$$\langle m | n \rangle = \delta_{m,n} \tag{44}$$

and

$$a^\dagger a\,|n\rangle = n\,|n\rangle. \tag{45}$$

The string spectrum (of right-movers) is given by the product of an infinite number of harmonic-oscillator Fock spaces, one for each α_n^μ, subject to the Virasoro constraints (Virasoro, 1970)

$$(L_0 - q)\,|\phi\rangle = 0$$
$$\tag{46}$$
$$L_n\,|\phi\rangle = 0, \quad n > 0.$$

Here $|\phi\rangle$ denotes a physical state, and q is a constant to be determined. It accounts for the arbitrariness in the normal-ordering prescription used to define L_0. As we will see, the L_0 equation is a generalization of the Klein–Gordon equation. It contains $p^2 = -\partial \cdot \partial$ plus oscillator terms whose eigenvalue will determine the mass of the state.

It is interesting to work out the algebra of the Virasoro operators L_m, which follows from the oscillator algebra. The result, called the Virasoro algebra, is

$$[L_m, L_n] = (m - n)L_{m+n} + \frac{c}{12}(m^3 - m)\delta_{m+n,0}. \quad (47)$$

The second term on the right-hand side is called the "conformal anomaly term" and the constant c is called the "central charge" or "conformal anomaly." Each component of x^μ contributes one unit to the central charge, so that altogether $c = d$.

There is a more sophisticated way to describe the string spectrum (in terms of BRST cohomology), but it is equivalent to the more elementary approach presented here. In the BRST approach, gauge-fixing to the conformal gauge in the quantum theory requires the addition of world-sheet Faddeev-Popov ghosts, which turn out to contribute $c = -26$. Thus the total conformal anomaly of the x^μ and the ghosts cancels for the particular choice $d = 26$, as we asserted earlier. Moreover, it is also necessary to set the parameter $q = 1$, so that mass-shell condition becomes

$$(L_0 - 1)|\phi\rangle = 0. \quad (48)$$

Since the mathematics of the open-string spectrum is the same as that of closed-string right-movers, let us now use the equations we have obtained to study the open-string spectrum. (Here we are assuming that the open-string boundary conditions are all Neumann, corresponding to spacetime-filling D-branes.) The mass-shell condition is

$$M^2 = -p^2 = -\frac{1}{2}\alpha_0^2 = N - 1, \quad (49)$$

where

$$N = \sum_{n=1}^{\infty} \alpha_{-n} \cdot \alpha_n = \sum_{n=1}^{\infty} n a_n^\dagger \cdot a_n. \quad (50)$$

The $a^\dagger$'s and a's are properly normalized raising and lowering operators. Since each $a^\dagger a$ has eigenvalues $0, 1, 2, \ldots$, the possible values of N are also $0, 1, 2, \ldots$. The unique way to realize $N = 0$ is for all the oscillators to be in the ground state, which we denote simply by $|0; p^\mu\rangle$, where p^μ is the momentum of the state. This state has $M^2 = -1$, which is a tachyon (p^μ is spacelike). Such a faster-than-light particle is certainly not possible in a consistent quantum theory, because the vacuum would be unstable. However, in perturbation theory (which is the framework we are implicitly considering) this instability is not visible. Since this string theory is only supposed to be a warm-up exercise before considering tachyon-free superstring theories, let us continue without worrying about the vacuum instability.

The first excited state, with $N = 1$, corresponds to $M^2 = 0$. The only way to achieve $N = 1$ is to excite the first oscillator once:

$$|\phi\rangle = \zeta_\mu \alpha_{-1}^\mu |0; p\rangle. \quad (51)$$

Here ζ_μ denotes the polarization vector of a massless spin-one particle. The Virasoro constraint condition $L_1|\phi\rangle = 0$ implies that ζ_μ must satisfy

$$p^\mu \zeta_\mu = 0. \quad (52)$$

This ensures that the spin is transversely polarized, so there are $d - 2$ independent polarization states. This agrees with what one finds for a massless Maxwell or Yang–Mills field.

At the next mass level, where $N = 2$ and $M^2 = 1$, the most general possibility has the form

$$|\phi\rangle = \left(\zeta_\mu \alpha_{-2}^\mu + \lambda_{\mu\nu} \alpha_{-1}^\mu \alpha_{-1}^\nu\right)|0; p\rangle. \quad (53)$$

However, the constraints $L_1|\phi\rangle = L_2|\phi\rangle = 0$ restrict ζ_μ and $\lambda_{\mu\nu}$. The analysis is interesting, but only the results will be described. If $d > 26$, the physical spectrum contains a negative-norm state, which is not allowed. However, when $d = 26$, this state becomes zero-norm and decouples from the theory. This leaves a pure massive "spin two" (symmetric traceless tensor) particle as the only physical state at this mass level.

Let us now turn to the closed-string spectrum. A closed-string state is described as a tensor product of a left-moving state and a right-moving state, subject to the condition that the N value of the left-moving and the right-moving state is the same. The reason for this "level-matching" condition is that we have $(L_0 - 1)|\phi\rangle = (\tilde{L}_0 - 1)|\phi\rangle = 0$. The sum $(L_0 + \tilde{L}_0 - 2)|\phi\rangle$ is interpreted as the mass-shell condition, while the difference $(L_0 - \tilde{L}_0)|\phi\rangle = (N - \tilde{N})|\phi\rangle = 0$ is the level-matching condition.

Using this rule, the closed-string ground state is just

$$|0\rangle \otimes |0\rangle, \quad (54)$$

which represents a spin 0 tachyon with $M^2 = -2$. (The notation no longer displays the momentum p of the state.) Again, this signals an unstable vacuum, but we will not worry about it here. Much more important, and more significant, is the first excited state

$$|\phi\rangle = \zeta_{\mu\nu}\left(\alpha_{-1}^\mu |0\rangle \otimes \tilde{\alpha}_{-1}^\nu |0\rangle\right), \quad (55)$$

which has $M^2 = 0$. The Virasoro constraints $L_1|\phi\rangle = \tilde{L}_1|\phi\rangle = 0$ imply that $p^\mu \zeta_{\mu\nu} = 0$. Such a polarization tensor encodes three distinct spin states, each of which plays a fundamental role in string theory. The symmetric part of $\zeta_{\mu\nu}$ encodes a spacetime metric field $g_{\mu\nu}$ (massless spin two) and a scalar dilaton field ϕ (massless spin zero). The

$g_{\mu\nu}$ field is the graviton field, and its presence (with the correct gauge invariances) accounts for the fact that the theory contains general relativity, which is a good approximation for energies well below the string scale. Its vacuum value determines the spacetime geometry. Similarly, the value of ϕ determines the string coupling constant ($g_s = \langle e^\phi \rangle$).

$\zeta_{\mu\nu}$ also has an antisymmetric part, which corresponds to a massless antisymmetric tensor gauge field $B_{\mu\nu} = -B_{\nu\mu}$. This field has a gauge transformation of the form

$$\delta B_{\mu\nu} = \partial_\mu \Lambda_\nu - \partial_\nu \Lambda_\mu, \qquad (56)$$

(which can be regarded as a generalization of the gauge transformation rule for the Maxwell field: $\delta A_\mu = \partial_\mu \Lambda$). The gauge-invariant field strength (analogous to $F_{\mu\nu} = \partial_\mu A_\nu - \partial_\nu A_\mu$) is

$$H_{\mu\nu\rho} = \partial_\mu B_{\nu\rho} + \partial_\nu B_{\rho\mu} + \partial_\rho B_{\mu\nu}. \qquad (57)$$

The importance of the $B_{\mu\nu}$ field resides in the fact that the fundamental string is a source for $B_{\mu\nu}$, just as a charged particle is a source for the vector potential A_μ. Mathematically, this is expressed by the coupling

$$q \int B_{\mu\nu} \, dx^\mu \wedge dx^\nu, \qquad (58)$$

which generalizes the coupling of a charged particle to a Maxwell field

$$q \int A_\mu \, dx^\mu. \qquad (59)$$

F. The Number of Physical States

The number of physical states grows rapidly as a function of mass. This can be analyzed quantitatively. For the open string, let us denote the number of physical states with $\alpha' M^2 = n - 1$ by d_n. These numbers are encoded in the generating function

$$G(w) = \sum_{n=0}^{\infty} d_n w^n = \prod_{m=1}^{\infty} (1 - w^m)^{-24}. \qquad (60)$$

The exponent 24 reflects the fact that in 26 dimensions, once the Virasoro conditions are taken into account, the spectrum is exactly what one would get from 24 transversely polarized oscillators. It is easy to deduce from this generating function the asymptotic number of states for large n, as a function of n

$$d_n \sim n^{-27/4} e^{4\pi\sqrt{n}}. \qquad (61)$$

This asymptotic degeneracy implies that the finite-temperature partition function

$$\mathrm{tr}\,(e^{-\beta H}) = \sum_{n=0}^{\infty} d_n e^{-\beta M_n} \qquad (62)$$

diverges for $\beta^{-1} = T > T_H$, where T_H is the Hagedorn temperature

$$T_H = \frac{1}{4\pi\sqrt{\alpha'}} = \frac{1}{4\pi\ell_s}. \qquad (63)$$

T_H might be the maximum possible temperature or else a critical temperature at which there is a phase transition.

G. The Structure of String Perturbation Theory

As we discussed in the first section, perturbation theory calculations are carried out by computing Feynman diagrams. Whereas in ordinary quantum field theory Feynman diagrams are webs of world lines, in the case of string theory they are two-dimensional surfaces representing string world-sheets. For these purposes, it is convenient to require that the world-sheet geometry is Euclidean (i.e., the world-sheet metric $h_{\alpha\beta}$ is positive definite). The diagrams are classified by their topology, which is very well understood in the case of two-dimensional surfaces. The world-sheet topology is characterized by the number of handles (h), the number of boundaries (b), and whether or not they are orientable. The order of the expansion (i.e., the power of the string-coupling constant) is determined by the Euler number of the world sheet M. It is given by $\chi(M) = 2 - 2h - b$. For example, a sphere has $h = b = 0$, and hence $\chi = 2$. A torus has $h = 1$, $b = 0$, and $\chi = 0$, a cylinder has $h = 0$, $b = 2$, and $\chi = 0$, and so forth. Surfaces with $\chi = 0$ admit a flat metric.

A scattering amplitude is given by a path integral of the schematic structure

$$\int Dh_{\alpha\beta}(\sigma) Dx^\mu(\sigma) e^{-S[h,x]} \prod_{i=1}^{n_c} \int_M V_{\alpha_i}(\sigma_i) \, d^2\sigma_i \prod_{j=1}^{n_o}$$

$$\times \int_{\partial M} V_{\beta_j}(\sigma_j) \, d\sigma_j. \qquad (64)$$

The action $S[h, x]$ is given in Eq. (21). V_{α_i} is a vertex operator that describes emission or absorption of a closed-string state of type α_i from the interior of the string world-sheet, and V_{β_j} is a vertex operator that describes emission of absorption of an open-string state of type β_j from the boundary of the string world-sheet. There are lots of technical details that are not explained here. In the end, one finds that the conformally inequivalent world-sheets of a given topology are described by a finite number of parameters, and thus these amplitudes can be recast as finite-dimensional integrals over these "moduli." (The momentum integrals are already done.) The dimension of the resulting integral turns out to be

$$N = 3(2h + b - 2) + 2n_c + n_o. \qquad (65)$$

As an example consider the amplitude describing elastic scattering of two open-string ground states. In this case

$h = 0, b = 1, n_c = 0, n_o = 4$, and therefore $N = 1$. In terms of the usual Mandelstam invariants $s = -(p_1 + p_2)^2$ and $t = -(p_1 - p_4)^2$, the result is

$$A(s, t) = g_s^2 \int_0^1 dx \, x^{-\alpha(s)-1} (1 - x)^{-\alpha(t)-1}, \qquad (66)$$

where the Regge trajectory $\alpha(s)$ is

$$\alpha(s) = 1 + \alpha' s. \qquad (67)$$

This integral is just the Euler beta function

$$A(s, t) = g_s^2 B(-\alpha(s), -\alpha(t)) = g_s^2 \frac{\Gamma(-\alpha(s))\Gamma(-\alpha(t))}{\Gamma(-\alpha(s) - \alpha(t))}. \qquad (68)$$

This is the famous Veneziano amplitude (Veneziano, 1968), which got the whole subject started.

H. Recapitulation

This section described some of the basic facts of the 26-dimensional bosonic string theory. One significant point that has not yet been made clear is that there are actually a number of distinct theories depending on what kinds of strings one includes.

- Oriented closed strings only
- Oriented closed-strings and oriented open-strings; in this case one can incorporate $U(n)$ gauge symmetry
- Unoriented closed strings only
- Unoriented closed-strings and unoriented open-strings; in this case one can incorporate $SO(n)$ or $Sp(n)$ gauge symmetry

As we have mentioned already, all the bosonic string theories are unphysical as they stand, because (in each case) the closed-string spectrum contains a tachyon. A tachyon means that one is doing perturbation theory about an unstable vacuum. This is analogous to the unbroken symmetry extremum of the Higgs potential in the standard model. In that case, we know that there is a stable minimum, where the Higgs fields acquires a vacuum value. Recently, there has been success in demonstrating that open-string tachyons condense at a stable minimum, but the fate of the closed-string tachyon is still an open problem.

III. SUPERSTRINGS

Among the deficiencies of the bosonic string theory is the fact that there are no fermions. As we will see, the addition of fermions leads quite naturally to supersymmetry and hence superstrings. There are two alternative formalisms that are used to study superstrings. The original one, which grew out of the 1971 papers by Ramond and by Neveu and Schwarz (1971) is called the RNS formalism. In this approach, the supersymmetry of the two-dimensional world-sheet theory plays a central role. The second approach, developed by Michael Green and the author in the early 1980s (Green and Schwarz, 1981), emphasizes supersymmetry in the ten-dimensional spacetime. Which one is more useful depends on the problem being studied. Only the RNS approach will be presented here.

In the RNS formalism, the world-sheet theory is based on the d functions $x^\mu(\sigma, \tau)$ that describe the embedding of the world-sheet in the spacetime, just as before. However, in order to supersymmetrize the world-sheet theory, we also introduce d fermionic partner fields $\psi^\mu(\sigma, \tau)$. Note that x^μ transforms as a vector from the spacetime viewpoint, but as d scalar fields from the two-dimensional world-sheet viewpoint. The ψ^μ also transform as a spacetime vector, but as world-sheet spinors. Altogether, x^μ and ψ^μ described d supersymmetry multiplets, one for each value of μ.

The reparametrization invariant world-sheet action described in the preceding section can be generalized to have local supersymmetry on the world-sheet, as well. (The details of how that works are a bit too involved to describe here.) When one chooses a suitable conformal gauge ($h_{\alpha\beta} = e^\phi \eta_{\alpha\beta}$), together with an appropriate fermionic gauge condition, one ends up with a world-sheet theory that has global supersymmetry supplemented by constraints. The constraints form a super-Virasoro algebra. This means that in addition to the Virasoro constraints of the bosonic string theory, there are fermionic constraints, as well.

A. The Gauge-Fixed Theory

The globally supersymmetric world-sheet action that arises in the conformal gauge takes the form

$$S = -\frac{T}{2} \int d^2\sigma \left(\partial_\alpha x^\mu \partial^\alpha x_\mu - i \bar{\psi}^\mu \rho^\alpha \partial_\alpha \psi_\mu \right). \qquad (69)$$

The first term is exactly the same as in Eq. (25) of the bosonic string theory. Recall that it has the structure of d free scalar fields. The second term that has now been added is just d free massless spinor fields, with Dirac-type actions. The notation is that ρ^α are two 2×2 Dirac matrices and $\psi = \binom{\psi_-}{\psi_+}$ is a two-component Majorana spinor. The Majorana condition simply means that ψ_+ and ψ_- are real in a suitable representation of the Dirac algebra. In fact, a convenient choice is one for which

$$\bar{\psi} \rho^\alpha \partial_\alpha \psi = \psi_- \partial_+ \psi_- + \psi_+ \partial_- \psi_+, \qquad (70)$$

where $\partial_\pm$ represent derivatives with respect to $\sigma^\pm = \tau \pm \sigma$. In this basis, the equations of motion are simply

$$\partial_+ \psi_-^\mu = \partial_- \psi_+^\mu = 0. \tag{71}$$

Thus ψ_-^μ describes right-movers and ψ_+^μ describes left-movers.

Concentrating on the right-movers ψ_-^μ, the global supersymmetry transformations, which are a symmetry of the gauge-fixed action, are

$$\delta x^\mu = i\epsilon \psi_-^\mu$$
$$\delta \psi_-^\mu = -2\partial_- x^\mu \epsilon. \tag{72}$$

It is easy to show that this is a symmetry of the action [Eq. (69)]. There is an analogous symmetry for the left-movers. (Accordingly, the world-sheet theory is said to have $(1, 1)$ supersymmetry.) Continuing to focus on the right-movers, the Virasoro constraint is

$$(\partial_- x)^2 + \frac{i}{2}\psi_-^\mu \partial_- \psi_{\mu-} = 0. \tag{73}$$

The first term is what we found in the bosonic string theory, and the second term is an additional fermionic contribution. There is also an associated fermionic constraint

$$\psi_-^\mu \partial_- x_\mu = 0. \tag{74}$$

The Fourier modes of these constraints generate the super-Virasoro algebra. There is a second identical super-Virasoro algebra for the left-movers.

As in the bosonic string theory, the Virasoro algebra has conformal anomaly terms proportional to a central charge c. As in that theory, each component of x^μ contributes $+1$ to the central charge, for a total of d, while (in the BRST quantization approach) the reparametrization symmetry ghosts contribute -26. But now there are additional contributions. Each component of ψ^μ gives $+1/2$, for a total of $d/2$, and the local supersymmetry ghosts contribute $+11$. Adding all of this up, gives a grand total of $c = \frac{3d}{2} - 15$. Thus, we see that the conformal anomaly cancels for the specific choice $d = 10$. This is the preferred critical dimension for superstrings, just as $d = 26$ is the critical dimension for bosonic strings. For other values the theory has a variety of inconsistencies.

B. The R and NS Sectors

Let us now consider boundary conditions for $\psi^\mu(\sigma, \tau)$. (The story for x^μ is exactly as before.) First, let us consider open-string boundary conditions. For the action to be well-defined, it turns out that one must set $\psi_+ = \pm\psi_-$ at the two ends $\sigma = 0, \pi$. An overall sign is a matter of convention, so we can set

$$\psi_+^\mu(0, \tau) = \psi_-^\mu(0, \tau), \tag{75}$$

without loss of generality. But this still leaves two possibilities for the other end, which are called R and NS:

$$\text{R: } \psi_+^\mu(\pi, \tau) = \psi_-^\mu(\pi, \tau)$$
$$\text{NS: } \psi_+^\mu(\pi, \tau) = -\psi_-^\mu(\pi, \tau). \tag{76}$$

Combining these with the equations of motion $\partial_- \psi_+ = \partial_+ \psi_- = 0$, allows us to express the general solutions as Fourier series

$$\text{R: } \psi_-^\mu = \frac{1}{\sqrt{2}} \sum_{n \in \mathbf{Z}} d_n^\mu e^{-in(\tau - \sigma)}$$
$$\psi_+^\mu = \frac{1}{\sqrt{2}} \sum_{n \in \mathbf{Z}} d_n^\mu e^{-in(\tau + \sigma)}$$
$$\text{NS: } \psi_-^\mu = \frac{1}{\sqrt{2}} \sum_{r \in \mathbf{Z} + 1/2} b_r^\mu e^{-ir(\tau - \sigma)}$$
$$\psi_+^\mu = \frac{1}{\sqrt{2}} \sum_{r \in \mathbf{Z} + 1/2} b_r^\mu e^{-ir(\tau + \sigma)}. \tag{77}$$

The Majorana condition implies that $d_{-n}^\mu = d_n^{\mu\dagger}$ and $b_{-r}^\mu = b_r^{\mu\dagger}$. Note that the index n takes integer values, whereas the index r takes half-integer values $(\pm\frac{1}{2}, \pm\frac{3}{2}, \ldots)$. In particular, only the R boundary condition gives a zero mode.

Canonical quantization of the free fermi fields $\psi^\mu(\sigma, \tau)$ is very standard and straightforward. The result can be expressed as anticommutation relations for the coefficients d_m^μ and b_r^μ:

$$\text{R: } \{d_n^\mu, d_n^\nu\} = \eta^{\mu\nu}\delta_{m+n,0} \qquad m, n \in \mathbf{Z}$$
$$\text{NS: } \{d_r^\mu, d_s^\nu\} = \eta^{\mu\nu}\delta_{r+s,0} \qquad r, s \in \mathbf{Z} + \frac{1}{2}. \tag{78}$$

Thus, in addition to the harmonic oscillator operators α_m^μ that appear as coefficients in mode expansions of x^μ, there are fermionic oscillator operators d_m^μ or b_r^μ that appear as coefficients in mode expansions of ψ^μ. The basic structure $\{b, b^\dagger\} = 1$ is very simple. It describes a two-state system with $b \mid 0\rangle = 0$, and $b^\dagger \mid 0\rangle = \mid 1\rangle$. The b's or d's with negative indices can be regarded as raising operators and those with positive indices as lowering operators, just as we did for the α_n^μ.

In the NS sector, the ground state $\mid 0; p\rangle$ satisfies

$$\alpha_m^\mu \mid 0; p\rangle = b_r^\mu \mid 0; p\rangle = 0, \qquad m, r > 0 \tag{79}$$

which is a straightforward generalization of how we defined the ground state in the bosonic string theory. All the excited states obtained by acting with the α and b raising operators are spacetime bosons. We will see later that the ground state, defined as we have done here, is again a tachyon. However, in this theory, as we will also see, there is a way by which this tachyon can (and must) be removed from the physical spectrum.

In the R sector there are zero modes that satisfy the algebra

$$\{d_0^\mu, d_0^\nu\} = \eta^{\mu\nu}. \tag{80}$$

This is the d-dimensional spacetime Dirac algebra. Thus the d_0's should be regarded as Dirac matrices and all states in the R sector should be spinors in order to furnish representation spaces on which these operators can act. The conclusion, therefore, is that whereas all string states in the NS sector are spacetime bosons, all string states in the R sector are spacetime fermions.

In the closed-string case, the physical states are obtained by tensoring right- and left-movers, each of which are mathematically very similar to the open-string spectrum. This means that there are four distinct sectors of closed-string states: NS $\otimes$ NS and R $\otimes$ R describe spacetime bosons, whereas NS $\otimes$ R and R $\otimes$ NS describe spacetime fermions. We will return to explore what this gives later, but first we need to explore the right-movers by themselves in more detail.

The zero mode of the fermionic constraint $\psi^\mu \partial_- x_\mu = 0$ gives a wave equation for (fermionic) strings in the Ramond sector, $F_0|\psi\rangle = 0$, which is called the Dirac–Ramond equation. In terms of the oscillators

$$F_0 = \alpha_0 \cdot d_0 + \sum_{n \neq 0} \alpha_{-n} \cdot d_n. \tag{81}$$

The zero-mode piece of F_0, $\alpha_0 \cdot d_0$, has been isolated, because it is just the usual Dirac operator, $\gamma^\mu \partial_\mu$, up to normalization. (Recall that α_0^μ is proportional to $p_\mu = -i\partial_\mu$, and d_0^μ is proportional to the Dirac matrices γ^μ.) The fermionic ground state $|\psi_0\rangle$, which satisfies

$$\alpha_n^\mu |\psi_0\rangle = d_n^\mu |\psi_0\rangle = 0, \qquad n > 0, \tag{82}$$

satisfies the wave equation

$$\alpha_0 \cdot d_0 |\psi_0\rangle = 0, \tag{83}$$

which is precisely the massless Dirac equation. Hence the fermionic ground state is a massless spinor.

C. The GSO Projection

In the NS (bosonic) sector the mass formula is

$$M^2 = N - \frac{1}{2}, \tag{84}$$

which is to be compared with the formula $M^2 = N - 1$ of the bosonic string theory. This time the number operator N has contributions from the b oscillators as well as the α oscillators. (The reason that the normal-ordering constant is $-1/2$ instead of -1 works as follows. Each transverse α oscillator contributes $-1/24$ and each transverse b oscillator contributes $-1/48$. The result follows since the

bosonic theory has 24 transverse directions and the superstring theory has 8 transverse directions.) Thus the ground state, which has $N = 0$, is now a tachyon with $M^2 = -1/2$.

This is where things stood until the 1976 work of Gliozzi, Scherk, and Olive. They noted that the spectrum admits a consistent truncation (called the GSO projection), which is necessary for the consistency of the interacting theory. In the NS sector, the GSO projection keeps states with an odd number of b-oscillator excitations and removes states with an even number of b-oscillator excitation. Once this rule is implemented the only possible values of N are half integers, and the spectrum of allowed masses are integral

$$M^2 = 0, 1, 2, \ldots. \tag{85}$$

In particular, the bosonic ground state is now massless. The spectrum no longer contains a tachyon. The GSO projection also acts on the R sector, where there is an analogous restriction on the d oscillators. This amounts to imposing a chirality projection on the spinors.

Let us look at the massless spectrum of the GSO-projected theory. The ground-state boson is now a massless vector, represented by the state $\zeta_\mu b_{-1/2}^\mu |0; p\rangle$, which (as before) has $d - 2 = 8$ physical polarizations. The ground-state fermion is a massless Majorana–Weyl fermion which has $\frac{1}{4} \cdot 2^{d/2} = 8$ physical polarizations. Thus there are an equal number of bosons and fermions, as is required for a theory with spacetime supersymmetry. In fact, this is the pair of fields that enter into ten-dimensional super Yang–Mills theory. The claim is that the complete theory now has spacetime supersymmetry.

If there is spacetime supersymmetry, then there should be an equal number of bosons and fermions at every mass level. Let us denote the number of bosonic states with $M^2 = n$ by $d_{NS}(n)$ and the number of fermionic states with $M^2 = n$ by $d_R(n)$. Then we can encode these numbers in generating functions, just as we did for the bosonic string theory

$$f_{NS}(w) = \sum_{n=0}^{\infty} d_{NS}(n)w^n = \frac{1}{2\sqrt{w}} \left(\prod_{m=1}^{\infty} \left(\frac{1 + w^{m-1/2}}{1 - w^m} \right)^8 \right.$$
$$\left. - \prod_{m=1}^{\infty} \left(\frac{1 - w^{m-1/2}}{1 - w^m} \right)^8 \right) \tag{86}$$

$$f_R(w) = \sum_{n=0}^{\infty} d_R(n)w^n = 8 \prod_{m=1}^{\infty} \left(\frac{1 + w^m}{1 - w^m} \right)^8. \tag{87}$$

The 8's in the exponents refer to the number of transverse directions in ten dimensions. The effect of the GSO projection is the subtraction of the second term in f_{NS} and the reduction of the coefficient in f_R from 16 to 8. In 1829, Jacobi discovered the formula

$$f_R(w) = f_{NS}(w). \qquad (88)$$

(He used a different notation, of course.) For him this relation was an obscure curiosity, but we now see that it tells us that the number of bosons and fermions is the same at every mass level, which provides strong evidence for supersymmetry of this string theory in ten dimensions. A complete proof of supersymmetry for the interacting theory was constructed by Green and the author five years after the GSO paper (Green and Schwarz, 1981).

D. Type II Superstrings

We have described the spectrum of bosonic (NS) and fermionic (R) string states. This also gives the spectrum of left- and right-moving closed-string modes, so we can form the closed-string spectrum by forming tensor products as before. In particular, the massless right-moving spectrum consists of a vector and a Majorana–Weyl spinor. Thus the massless closed-string spectrum is given by

$$(\text{vector} + \text{MW spinor}) \otimes (\text{vector} + \text{MW spinor}). \quad (89)$$

There are actually two distinct possibilities, because the two MW spinors can have either opposite chirality or the same chirality.

When the two MW spinors have opposite chirality, the theory is called type IIA superstring theory, and its massless spectrum forms the type IIA supergravity multiplet. This theory is left-right symmetric. In other words, the spectrum is invariant under mirror reflection. This implies that the IIA theory is parity conserving. When the two MW spinors have the same chirality, the resulting type IIB superstring theory is chiral, and hence parity violating. In each case there are two gravitinos, arising from vector ⊗ spinor and spinor ⊗ vector, which are gauge fields for local supersymmetry. Thus, since both type II superstring theories have two gravitinos, they have local $\mathcal{N} = 2$ supersymmetry in the ten-dimensional sense. The supersymmetry charges are Majorana–Weyl spinors, which have 16 real components, so the type II theories have 32 conserved supercharges. This is the same amount of supersymmetry as what is usually called $\mathcal{N} = 8$ in four dimensions, and it is believed to be the most that is possible in a consistent interacting theory.

The type II superstring theories contain only oriented closed strings (in the absence of D-branes). However, there is another superstring theory, called type I, which can be obtained by a projection of the type IIB theory, that only keeps the diagonal sum of the two gravitinos. Thus, this theory only has $\mathcal{N} = 1$ supersymmetry (16 supercharges). It is a theory of unoriented closed strings. However, it can

be supplemented by unoriented open strings. This introduces a Yang–Mills gauge group, which classically can be $SO(n)$ or $Sp(n)$ for any value of n. Quantum consistency singles out $SO(32)$ as the unique possibility. This restriction can be understood in a number of ways. The way that it was first discovered was by considering anomalies.

E. Anomalies

Chiral (parity-violating) gauge theories can be inconsistent due to anomalies. This happens when there is a quantum mechanical breakdown of the gauge symmetry, which is induced by certain one-loop Feynman diagrams. (Sometimes one also considers breaking of global symmetries by anomalies, which does not imply an inconsistency. That is not what we are interested in here.) In the case of four dimensions, the relevant diagrams are triangles, with the chiral fields going around the loop and three gauge fields attached as external lines. In the case of the standard model, the quarks and leptons are chiral and contribute to a variety of possible anomalies. Fortunately, the standard model has just the right particle content so that all of the gauge anomalies cancel. If one omits the quark or lepton contributions, it does not work.

In the case of ten-dimensional chiral gauge theories, the potentially anomalous Feynman diagrams are hexagons, with six external gauge fields. The anomalies can be attributed to the massless fields, and therefore they can be analyzed in the low-energy effective field theory. There are several possible cases in ten dimensions:

- $\mathcal{N} = 1$ supersymmetric Yang–Mills theory. This theory has anomalies for every choice of gauge group.
- Type I supergravity. This theory has gravitational anomalies.
- Type IIA supergravity. This theory is nonchiral, and therefore it is trivially anomaly-free.
- Type IIB supergravity. This theory has three chiral fields each of which contributes to several kinds of gravitational anomalies. However, when their contributions are combined, the anomalies all cancel. (This result was obtained by Alvarez-Gaumé and Witten, 1983.)
- Type I supergravity coupled to super Yang–Mills. This theory has both gauge and gravitational anomalies for every choice of Yang–Mills gauge group except $SO(32)$ and $E_8 \times E_8$. For these two choices, all the anomalies cancel. (This result was obtained by Green and Schwarz, 1984a.)

As we mentioned earlier, at the classical level one can define type I superstring theory for any orthogonal or symplectic gauge group. Now we see that at the quantum level, the only choice that is consistent is $SO(32)$. For any other choice there are fatal anomalies. The term $SO(32)$ is used here somewhat imprecisely. There are several different Lie groups that have the same Lie algebra. It turns out that the particular Lie group that is appropriate is Spin $(32)/\mathbb{Z}_2$. It contains one spinor conjugacy class in addition to the adjoint conjugacy class.

F. Heterotic Strings

The two Lie groups that are singled out—$E_8 \times E_8$ and Spin $(32)/\mathbb{Z}_2$—have several properties in common. Each of them has dimension $= 496$ and rank $= 16$. Moreover, their weight lattices correspond to the only two even self-dual lattices in 16 dimensions. This last fact was the crucial clue that led Gross, Harvey, Martinec, and Rohm (1985) to the discovery of the heterotic string soon after the anomaly cancellation result. One hint is the relation $10 + 16 = 26$. The construction of the heterotic string uses the $d = 26$ bosonic string for the left-movers and the $d = 10$ superstring for the right movers. The 16 extra left-moving dimensions are associated to an even self-dual 16-dimensional lattice. In this way one builds in the $SO(32)$ or $E_8 \times E_8$ gauge symmetry.

Thus, to recapitulate, by 1985 we had five consistent superstring theories, type I [with gauge group $SO(32)$], the two type II theories, and the two heterotic theories. Each is a supersymmetric ten-dimensional theory. The perturbation theory was studied in considerable detail, and while some details may not have been completed, it was clear that each of the five theories has a well-defined, ultraviolet-finite perturbation expansion, satisfying all the usual consistency requirements (unitarity, analyticity, causality, etc.). This was pleasing, though it was somewhat mysterious why there should be five consistent quantum gravity theories. It took another ten years until we understood that these are actually five special quantum vacua of a unique underlying theory.

G. T Duality

T duality, an amazing result obtained in the late 1980s, relates one string theory with a circular compact dimension of radius R to another string theory with a circular dimension of radius $1/R$ (in units $\ell_s = 1$). This is very profound, because it indicates a limitation of our usual motions of classical geometry. Strings see geometry differently from point particles. Let us examine how this is possible.

The key to understanding T duality is to consider the kinds of excitations that a string can have in the presence of a circular dimension. One class of excitations, called Kaluza–Klein excitations, is a very general feature of any quantum theory, whether or not based on strings. The idea is that in order for the wave function e^{ipx} to be single

valued, the momentum along the circle must be a multiple of $1/R$, $p = n/R$, where n is an integer. From the lower dimension viewpoint this is interpreted as a contribution $(n/R)^2$ to the square of the mass.

There is a second type of excitation that is special to closed strings. Namely, a closed string can wind m times around the circular dimension, getting caught up on the topology of the space, contributing an energy given by the string tension times the length of the string

$$E_m = 2\pi R \cdot m \cdot T. \tag{90}$$

Putting $T = \frac{1}{2\pi}$ (for $\ell_s = 1$), this is just $E_m = mR$.

The combined energy-squared of the Kaluza–Klein and winding-mode excitations is

$$E^2 = \left(\frac{n}{R}\right)^2 + (mR)^2 + \cdots, \tag{91}$$

where the dots represent string oscillator contributions. Under T duality

$$m \leftrightarrow n, \qquad R \leftrightarrow 1/R. \tag{92}$$

Together, these interchanges leave the energy invariant. This means that what is interpreted as a Kaluza–Klein excitation in one string theory is interpreted as a winding-mode excitation in the T-dual theory, and the two theories have radii R and $1/R$, respectively. The two principle examples of T-dual pairs are the two type II theories and the two heterotic theories. In the latter case there are additional technicalities that explain how the two gauge groups are related. Basically, when the compactification on a circle to nine dimensions is carried out in each case, it is necessary to include effects that we haven't explained (called Wilson lines) to break the gauge groups to $SO(16) \times SO(16)$, which is a common subgroup of $SO(32)$ and $E_8 \times E_8$.

IV. FROM SUPERSTRINGS TO M-THEORY

Superstring theory is currently undergoing a period of rapid development in which important advances in understanding are being achieved. The focus in this section will be on explaining why there can be an eleven-dimensional vacuum, even though there are only ten dimensions in perturbative superstring theory. The nonperturbative extension of superstring theory that allows for an eleventh dimension has been named *M-theory*. The letter M is intended to be flexible in its interpretation. It could stand for *magic*, *mystery*, or *meta* to reflect our current state of incomplete understanding. Those who think that two-dimensional supermembranes (the M2-brane) are fundamental may regard M as standing for *membrane*. An approach called *Matrix theory* is another possibility. And, of course, some view M-theory as the *mother* of all theories.

In the first superstring revolution we identified five distinct superstring theories, each in ten dimensions. Three of them, the type I theory and the two heterotic theories, have $\mathcal{N} = 1$ supersymmetry in the ten-dimensional sense. Since the minimal ten-dimensional spinor is simultaneously Majorana and Weyl, this corresponds to 16 conserved supercharges. The other two theories, called type IIA and type IIB, have $\mathcal{N} = 2$ supersymmetry (32 supercharges). In the IIA case the two spinors have opposite handedness so that the spectrum is left-right symmetric (nonchiral). In the IIB case the two spinors have the same handedness and the spectrum is chiral.

In each of these five superstring theories it became clear, and was largely proved, that there are consistent perturbation expansions of on-shell scattering amplitudes. In four of the five cases (heterotic and type II) the fundamental strings are oriented and unbreakable. As a result, these theories have particularly simple perturbation expansions. Specifically, there is a unique Feynman diagram at each order of the loop expansion. The Feynman diagrams depict string world-sheets, and therefore they are two-dimensional surfaces. For these four theories the unique L-loop diagram is a closed orientable genus-L Riemann surface, which can be visualized as a sphere with L handles. External (incoming or outgoing) particles are represented by N points (or "punctures") on the Riemann surface. A given diagram represents a well-defined integral of dimension $6L + 2N - 6$. This integral has no ultraviolet divergences, even though the spectrum contains states of arbitrarily high spin (including a massless graviton). From the viewpoint of point-particle contributions, string and supersymmetry properties are responsible for incredible cancellations. Type I superstrings are unoriented and breakable. As a result, the perturbation expansion is more complicated for this theory, and various world-sheet diagrams at a given order have to be combined properly to cancel divergences and anomalies.

As we explained in the previous section, T duality relates two string theories when one spatial dimension forms a circle (denoted S^1). Then the ten-dimensional geometry is $R^9 \times S^1$. T duality identifies this string compactification with one of a second string theory also on $R^9 \times S^1$. If the radii of the circles in the two cases are denoted R_1 and R_2, then

$$R_1 R_2 = \alpha'. \tag{93}$$

Here $\alpha' = \ell_s^2$ is the universal Regge slope parameter, and ℓ_s is the fundamental string length scale (for both string theories). Note that T duality implies that shrinking the circle to zero in one theory corresponds to decompactification of the dual theory.

The type IIA and IIB theories are T dual, so compactifying the nonchiral IIA theory on a circle of radius R and letting $R \to 0$ gives the chiral IIB theory in ten dimensions. This means, in particular, that they should not be regarded as distinct theories. The radius R is actually the vacuum value of a scalar field, which arises as an internal component of the ten-dimensional metric tensor. Thus the type IIA and type IIB theories in ten dimensions are two limiting points in a continuous moduli space of quantum vacua. The two heterotic theories are also T dual, though (as we mentioned earlier) there are additional technical details in this case. T duality applied to the type I theory gives a dual description, which is sometimes called type I' or IA.

A. M-Theory

In the 1970s and 1980s various supersymmetry and supergravity theories were constructed. In particular, supersymmetry representation theory showed that the largest possible spacetime dimension for a supergravity theory (with spins ≤ 2) is eleven. Eleven-dimensional supergravity, which has 32 conserved supercharges, was constructed in 1978 by Cremmer, Julia, and Scherk (1978). It has three kinds of fields—the graviton field (with 44 polarizations), the gravitino field (with 128 polarizations), and a three-index gauge field $C_{\mu\nu\rho}$ (with 84 polarizations). These massless particles are referred to collectively as the *supergraviton*. Eleven dimension supergravity is nonrenormalizable, and thus it cannot be a fundamental theory. However, we now believe that it is a low-energy effective description of M-theory, which is a well-defined quantum theory. This means, in particular, that higher dimension terms in the effective action for the supergravity fields have uniquely determined coefficients within the M-theory setting, even though they are formally infinite (and hence undetermined) within the supergravity context.

Intriguing connections between type IIA string theory and eleven dimension supergravity have been known for a long time, but the precise relationship was only explained in 1995. The field equations of eleven dimension supergravity admit a solution that describes a supermembrane. In other words, this solution has the property that the energy density is concentrated on a two-dimensional surface. A three-dimensional world-volume description of the dynamics of this supermembrane, quite analogous to the two-dimensional world volume actions of superstrings [in the GS formalism (Green and Schwarz, 1984b)], was constructed by Bergshoeff, Sezgin, and Townsend (1987) The authors suggested that a consistent eleven dimension quantum theory might be defined in terms of this membrane, in analogy to string theories in ten dimensions. (Most experts now believe that M-theory cannot be defined as a supermembrane theory.) Another striking result was that a suitable dimensional reduction of this supermembrane gives the (previously known) type IIA superstring world-volume action. For many years these facts remained unexplained curiosities until they were reconsidered by Townsend (1995) and by Witten (1995). The conclusion is that type IIA superstring theory really does have a circular eleventh dimension in addition to the previously known ten spacetime dimensions. This fact was not recognized earlier because the appearance of the eleventh dimension is a nonperturbative phenomenon, not visible in perturbation theory.

To explain the relation between M-theory and type IIA string theory, a good approach is to identify the parameters that characterize each of them and to explain how they are related. Eleven-dimensional supergravity (and hence M-theory, too) has no dimensionless parameters. The only parameter is the eleven-dimensional Newton constant, which raised to a suitable power $(-1/9)$, gives the eleven-dimensional Planck mass m_p. When M-theory is compactified on a circle (so that the spacetime geometry is $R^{10} \times S^1$) another parameter is the radius R of the circle. Now consider the parameters of type IIA superstring theory. They are the string mass scale m_s, introduced earlier, and the dimensionless string coupling constant g_s.

We can identify compactified M-theory with type IIA superstring theory by making the following correspondences:

$$m_s^2 = 2\pi R m_p^3 \qquad (94)$$

$$g_s = 2\pi R m_s. \qquad (95)$$

Using these one can derive $g_s = (2\pi R m_p)^{3/2}$ and $m_s = g_s^{1/3} m_p$. The latter implies that the eleven-dimensional Planck length is shorter than the string length scale at weak coupling by a factor of $(g_s)^{1/3}$.

Conventional string perturbation theory is an expansion in powers of g_s at fixed m_s. Equation (95) shows that this is equivalent to an expansion about $R = 0$. In particular, the strong coupling limit of type IIA superstring theory corresponds to decompactification of the eleventh dimension, so in a sense M-theory is type IIA string theory at infinite coupling.* This explains why the eleventh dimension was not discovered in studies of string perturbation theory.

These relations encode some interesting facts. For one thing, the fundamental IIA string actually *is* an M2-brane of M-theory with one of its dimensions wrapped around the circular spatial dimension. Denoting the string and membrane tensions (energy per unit volume) by T_{F1} and T_{M2}, one deduces that

$$T_{F1} = 2\pi R T_{M2}. \qquad (96)$$

However, $T_{F1} = 2\pi m_s^2$ and $T_{M2} = 2\pi m_p^3$. Combining these relations gives Eq. (94).

B. Type II *p*-branes

Type II superstring theories contain a variety of p-brane solutions that preserve half of the 32 supersymmetries. These are solutions in which the energy is concentrated on a p-dimensional spatial hypersurface. (The world volume has $p + 1$ dimensions.) The corresponding solutions of supergravity theories were constructed by Horowitz and Strominger (1991). A large class of these p-brane excitations are called *D-branes* (or *Dp*-branes when we want to specify the dimension), whose tensions are given by

$$T_{Dp} = 2\pi m_s^{p+1}/g_s. \qquad (97)$$

This dependence on the coupling constant is one of the characteristic features of a D-brane. Another characteristic feature of D-branes is that they carry a charge that couples to a gauge field in the RR sector of the theory (Polchinski, 1995). The particular RR gauge fields that occur imply that p takes even values in the IIA theory and odd values in the IIB theory.

In particular, the D2-brane of the type IIA theory corresponds to the supermembrane of M-theory, but now in a background geometry in which one of the transverse dimensions is a circle. The tensions check, because [using Eqs. (94) and (95)]

$$T_{D2} = 2\pi m_s^3/g_s = 2\pi m_p^3 = T_{M2}. \qquad (98)$$

The mass of the first Kaluza–Klein excitation of the eleven-dimensional supergraviton is $1/R$. Using Eq. (95), we see that this can be identified with the D0-brane. More identifications of this type arise when we consider the magnetic dual of the M-theory supermembrane, which is a five-brane, called the M5-brane.* Its tension is $T_{M5} = 2\pi m_p^6$. Wrapping one of its dimensions around the circle gives the D4-brane, with tension

$$T_{D4} = 2\pi R T_{M5} = 2\pi m_s^5/g_s. \qquad (99)$$

If, on the other hand, the M5-frame is not wrapped around the circle, one obtains the NS5-brane of the IIA theory with tension

$$T_{NS5} = T_{M5} = 2\pi m_s^6/g_s^2. \qquad (100)$$

To summarize, type IIA superstring theory is M-theory compactified on a circle of radius $R = g_s \ell_s$. M-theory is believed to be a well-defined quantum theory in eleven-dimension, which is approximated at low energy by eleven-dimensional supergravity. Its excitations are the

---

*The $E_8 \times E_8$ heterotic string theory is also eleven-dimensional at strong coupling.

*In general, the magnetic dual of a p-brane in d dimensions is a $(d - p - 4)$-brane.

massless supergraviton, the M2-brane, and the M5-brane. These account both for the (perturbative) fundamental string of the IIA theory and for many of its nonperturbative excitations. The identities that we have presented here are exact, because they are protected by supersymmetry.

C. Type IIB Superstring Theory

Type IIB superstring theory, which is the other maximally supersymmetric string theory with 32 conserved supercharges, is also ten-dimensional, but unlike the IIA theory its two supercharges have the same handedness. At low-energy, type IIB superstring theory is approximated by type IIB supergravity, just as eleven-dimensional supergravity approximates M-theory. In each case the supergravity theory is only well-defined as a classical field theory, but still it can teach us a lot. For example, it can be used to construct p-brane solutions and compute their tensions. Even though such solutions are only approximate, supersymmetry considerations ensure that the tensions, which are related to the kinds of conserved charges the p-branes carry, are exact. Since the IIB spectrum contains massless chiral fields, one should check whether there are anomalies that break the gauge invariances—general coordinate invariance, local Lorentz invariance, and local supersymmetry. In fact, the UV finiteness of the string theory Feynman diagrams ensures that all anomalies must cancel, as was verified from a field theory viewpoint by Alvarez-Gaumé and Witten (1983).

Type IIB superstring theory or supergravity contains two scalar fields, the dilation ϕ and an axion χ, which are conveniently combined in a complex field

$$\rho = \chi + ie^{-\phi}. \tag{101}$$

The supergravity approximation has an $SL(2, R)$ symmetry that transforms this field nonlinearly:

$$\rho \to \frac{a\rho + b}{c\rho + d}, \tag{102}$$

where a, b, c, d are real numbers satisfying $ad - bc = 1$. However, in the quantum string theory this symmetry is broken to the discrete subgroup $SL(2, Z)$ (Hull and Townsend, 1995), which means that a, b, c, d are restricted to be integers. Defining the vacuum value of the ρ field to be

$$\langle \rho \rangle = \frac{\theta}{2\pi} + \frac{i}{g_s}, \tag{103}$$

the $SL(2, Z)$ symmetry transformation $\rho \to \rho + 1$ implies that θ is an angular coordinate. Moreover, in the special case $\theta = 0$, the symmetry transformation $\rho \to -1/\rho$ takes $g_s \to 1/g_s$. This symmetry, called *S duality*, implies that coupling constant g_s is equivalent to coupling constant

$1/g_s$, so that, in the case of type II superstring theory, the weak coupling expansion and the strong coupling expansion are identical. (An analogous S-duality transformation relates the type I superstring theory to the $SO(32)$ heterotic string theory.)

Recall that the type IIA and type IIB superstring theories are T dual, meaning that if they are compactified on circles of radii R_A and R_B one obtains equivalent theories for the identification $R_A R_B = \ell_s^2$. Moreover, we saw that the type IIA theory is actually M-theory compactified on a circle. The latter fact encodes nonperturbative information. It turns out to be very useful to combine these two facts and to consider the duality between M-theory compactified on a torus ($R^9 \times T^2$) and type IIB superstring theory compactified on a circle ($R^9 \times S^1$).

A torus can be described as the complex plane modded out by the equivalence relations $z \sim z + w_1$ and $z \sim z + w_2$. Up to conformal equivalence, the periods w_1 and w_2 can be replaced by 1 and τ, with $\mathrm{Im}\ \tau > 0$. In this characterization τ and $\tau' = (a\tau + b)/(c\tau + d)$, where a, b, c, d are integers satisfying $ad - bc = 1$, describe equivalent tori. Thus a torus is characterized by a modular parameter τ and an $SL(2, Z)$ modular group. The natural, and correct, conjecture at this point is that one should identify the modular parameter τ of the M-theory torus with the parameter ρ that characterizes the type IIB vacuum (Schwarz, 1995 and Aspinwall, 1996). Then the duality of M-theory and type IIB superstring theory gives a geometrical explanation of the nonperturbative S-duality symmetry of the IIB theory: the transformation $\rho \to -1/\rho$, which sends $g_s \to 1/g_s$ in the IIB theory, corresponds to interchanging the two cycles of the torus in the M theory description. To complete the story, we should relate the area of the M theory torus (A_M) to the radius of the IIB theory circle (R_B). This is a simple consequence of formulas given above

$$m_p^3 A_M = (2\pi R_B)^{-1}. \tag{104}$$

Thus the limit $R_B \to 0$, at fixed ρ, corresponds to decompactification of the M-theory torus, while preserving its shape. Conversely, the limit $A_M \to 0$ corresponds to decompactification of the IIB theory circle. The duality can be explored further by matching the various p-branes in nine-dimensions that can be obtained from either the M-theory or the IIB theory viewpoints. When this is done, one finds that everything matches nicely and that one deduces various relations among tensions (Schwarz, 1996).

Another interesting fact about the IIB theory is that it contains an infinite family of strings labeled by a pair of integers (p, q) with no common divisor (Schwarz, 1995). The $(1, 0)$ string can be identified as the fundamental IIB string, while the $(0, 1)$ string is the D-string. From this viewpoint, a (p, q) string can be regarded as a bound state of p fundamental strings and q D-strings (Witten, 1996). These strings have a very simple interpretation in the dual M-theory description. They correspond to an M2-brane with one of its cycles wrapped around a (p, q) cycle of the torus. The minimal length of such a cycle is proportional to $|p + q\tau|$, and thus (using $\tau = \rho$) one finds that the tension of a (p, q) string is given by

$$T_{p,q} = 2\pi |p + q\rho| m_s^2. \tag{105}$$

Imagine that you lived in the nine-dimensional world that is described equivalently as M-theory compactified on a torus or as the type IIB superstring theory compactified on a circle. Suppose, moreover, you had very high energy accelerators with which you were going to determine the "true" dimension of spacetime. Would you conclude that ten or eleven is the correct answer? If either A_M or R_B was very large in Planck units there would be a natural choice, of course. But how could you decide otherwise? The answer is that either viewpoint is equally valid. What determines which choice you make is which of the massless fields you regard as "internal" components of the metric tensor and which ones you regards as matter fields. Fields that are metric components in one description correspond to matter fields in the dual one.

D. The D3-Brane and $\mathcal{N} = 4$ Gauge Theory

D-branes have a number of special properties, which make them especially interesting. By definition, they are branes on which strings can end—D stands for *Dirichlet* boundary conditions. The end of a string carries a charge, and the D-brane world-volume theory contains a $U(1)$ gauge field that carries the associated flux. When n Dp-branes are coincident, or parallel and nearly coincident, the associated $(p + 1)$-dimensional world-volume theory is a $U(n)$ gauge theory (Witten, 1996). The n^2 gauge bosons A_μ^{ij} and their supersymmetry partners arise as the ground states of oriented strings running from the ith Dp-brane to the jth Dp-brane. The diagonal elements, belonging to the Cartan subalgebra, are massless. The field A_μ^{ij} with $i \neq j$ has a mass proportional to the separation of the ith and jth branes.

The $U(n)$ gauge theory associated with a stack of n Dp-branes has maximal supersymmetry (16 supercharges). The low-energy effective theory, when the brane separations are small compared to the string scale, is supersymmetric Yang–Mills theory. These theories can be constructed by dimensional reduction of ten-dimensional supersymmetric $U(n)$ gauge theory to $p + 1$ dimensions. A case of particular interest, which we shall now focus on, is $p = 3$. A stack of n D3-branes in type IIB superstring theory has a decoupled $\mathcal{N} = 4$, $d = 4$ $U(n)$ gauge theory associated to it. This gauge theory has a number of special features. For one thing, due to boson–fermion cancellations, there are no UV divergences at any order of perturbation theory. The beta function $\beta(g)$ is identically zero, which implies that the theory is scale invariant. In fact, $\mathcal{N} = 4$, $d = 4$ gauge theories are conformally invariant. The conformal invariance combines with the supersymmetry to give a superconformal symmetry, which contains 32 fermionic generators. Another important property of $\mathcal{N} = 4$, $d = 4$ gauge theories is an electric-magnetic duality, which extends to an $SL(2, Z)$ group of dualities. Now consider the $\mathcal{N} = 4$ $U(n)$ gauge theory associated to a stack of n D3-branes in type IIB superstring theory. There is an obvious identification that turns out to be correct. Namely, the $SL(2, Z)$ duality of the gauge theory is induced from that of the ambient type IIB superstring theory. The D3-branes themselves are invariant under $SL(2, Z)$ transformations.

As we have said, a fundamental $(1, 0)$ string can end on a D3-brane. But by applying a suitable $SL(2, Z)$ transformation, this configuration is transformed to one in which a (p, q) string ends on the D3-brane. The charge on the end of this string describes a dyon with electric charge p and magnetic charge q, with respect to the appropriate gauge field. More generally, for a stack of n D3-branes, any pair can be connected by a (p, q) string. The mass is proportional to the length of the string times its tension, which we saw is proportional to $|p + q\rho|$. In this way one sees that the electrically charged particles, described by fundamental fields, belong to infinite $SL(2, Z)$ multiplets. The other states are nonperturbative excitations of the gauge theory. The field configurations that describe them preserve half of the supersymmetry. As a result their masses are given exactly by the considerations described above. An interesting question, whose answer was unknown until recently, is whether $\mathcal{N} = 4$ gauge theories in four dimensions also admit nonperturbative excitations that preserve $1/4$ of the supersymmetry. The answer turns out to be that they do, but only if $n \geq 3$. This result has a nice dual description in terms of three-string junctions (Bergman, 1998).

E. Conclusion

In this section we have described some of the interesting advances in understanding superstring theory that have taken place in the past few years. The emphasis has been on the nonperturbative appearance of an eleventh dimension in type-IIA superstring theory, as well as its implications when combined with superstring T dualities. In particular, we argued that there should be a consistent quantum vacuum, whose low-energy effective description is given by eleven-dimensional supergravity.

What we have described makes a convincing self-consistent picture, but it does not constitute a complete formulation of M-theory. In the past several years there have been some major advances in that direction, which we will briefly mention here. The first, which goes by the name of *Matrix Theory*, bases a formulation of M-theory in flat eleven-dimensional spacetime in terms of the supersymmetric quantum mechanics of N D0-branes in the large N limit (Banks *et al.*, 1997). Matrix Theory has passed all tests that have been carried out, some of which are very nontrivial. The construction has a nice generalization to describe compactification of M-theory on a torus T^n. However, it does not seem to be useful for $n > 5$, and other compactification manifolds are (at best) awkward to handle. Another shortcoming of this approach is that it treats the eleventh dimension differently from the other ones.

Another proposal relating superstring and M-theory backgrounds to large N limits of certain field theories has been put forward by Maldacena (1997) and made more precise by Gubser, Klebanov, and Polyakov (1998), and by Witten (1998). [For a review of this subject, see (Aharony *et al.*, 2000).] In this approach, there is a conjectured duality (i.e., equivalence) between a conformally invariant field theory (CFT) in d dimensions and type IIB superstring theory or M-theory on an Anti-de-Sitter space (AdS) in $d + 1$ dimensions. The remaining $9 - d$ or $10 - d$ dimensions form a compact space, the simplest cases being spheres. Three examples with unbroken supersymmetry are $AdS_5 \times S^5$, $AdS_4 \times S^7$, and $AdS_7 \times S^4$. This approach is sometimes referred to as *AdS/CFT duality*. This is an extremely active and very promising subject. It has already taught us a great deal about the large N behavior of various gauge theories. As usual, the easiest theories to study are ones with a lot of supersymmetry, but it appears that in this approach supersymmetry breaking is more accessible than in previous ones. For example, it might someday be possible to construct the QCD string in terms of a dual AdS gravity theory, and use it to carry out numerical calculations of the hadron spectrum. Indeed, there have already been some preliminary steps in this direction.

To sum up, I would say that despite all of the successes that have been achieved in advancing our understanding of superstring theory and M-theory, there clearly is still a long way to go. In particular, despite much effort and several imaginative proposals, we still do not have a convincing mechanism for ensuring the vanishing (or extreme smallness) of the cosmological constant for nonsupersymmetric vacua. Superstring theory is a field with very ambitious goals. The remarkable fact is that they still seem to be realistic. However, it may take a few more revolutions before they are attained.

ACKNOWLEDGMENTS

This article is based on lectures presented at the NATO Advanced Study Institute *Techniques and Concepts of High Energy Physics*, which took place in St. Croix, Virgin Islands during June 2000. The author's research is supported in part by the U.S. Dept. of Energy under Grant No. DE-FG03-92-ER40701.

SEE ALSO THE FOLLOWING ARTICLES

Field Theory and the Standard Model • Group Theory, Applied • Perturbation Theory • Quantum Theory • Relativity, General

BIBLIOGRAPHY

Aharony, O., Gubser, S. S., Maldacena, J., Ooguri, H., and Oz, Y. (2000). *Phys. Rep.* **323**, 183.

Aspinwall, P. S. (1996). *Nucl. Phys. Proc. Suppl.* **46**, 30, hep-th/9508154.

Alvarez-Gaumé, L., and Witten, E. (1983). *Nucl. Phys.* **B234**, 269.

Banks, T., Fischler, W., Shenker, S., and Susskind, L. (1997). *Phys. Rev.* **D55**, 5112, hep-th/9610043.

Bergman, O. (1998). *Nucl. Phys.* **B525**, 104, hep-th/9712211.

Bergshoeff, E., Sezgin, E., and Townsend, P. K. (1987). *Phys. Lett.* **B189**, 75.

Candelas, P., Horowitz, G. T., Strominger, A., and Witten, E. (1985). *Nucl. Phys.* **B258**, 46.

Cremmer, E., Julia, B., and Scherk, J. (1978). *Phys. Lett.* **76B**, 409.

Gliozzi, F., Scherk, J., and Olive, D. (1976). *Phys. Lett.* **65B**, 282.

Goto, T. (1971). *Prog. Theor. Phys.* **46**, 1560.

Green, M. B., and Schwarz, J. H. (1984a). *Phys. Lett.* **149B**, 117.

Green, M. B., and Schwarz, J. H. (1984b). *Phys. Lett.* **136B**, 367.

Green, M. B., and Schwarz, J. H. (1981). *Nucl. Phys.* **B181**, 502; *Nucl. Phys.* **B198**, (1982) 252; *Phys. Lett.* **109B**, 444.

Green, M. B., Schwarz, J. H., and Witten, E. (1987). "Superstring Theory," in 2 vols., Cambridge Univ. Press, U.K.

Gross, D. J., Harvey, J. A., Martinec, E., and Rohm, R. (1985). *Phys. Rev. Lett.* **54**, 502.

Gubser, S. S., Klebanov, I. R., and Polyakov, A. M. (1998). *Phys. Lett.* **B428**, 105, hep-th/9802109.

Horowitz, G. T., and Strominger, A. (1991). *Nucl. Phys.* **B360**, 197.

Hull, C., and Townsend, P. (1995). *Nucl. Phys.* **B438**, 109, hep-th/9410167.

Maldacena, J. (1998). *Adv. Theor. Phys.* **2**, 231, hep-th/9711200.

Nambu, Y. (1970). Notes prepared for the Copenhagen High Energy Symposium.

Neveu, A., and Schwarz, J. H. (1971). *Nucl. Phys.* **B31**, 86.

Polchinski, J. (1995). *Phys. Rev. Lett.* **75**, 4724, hep-th/9510017.

Polchinski, J. (1998). "String Theory," in 2 vols., Cambridge Univ. Press, U.K.

Polyakov, A. M. (1981). *Phys. Lett.* **103B**, 207.

Ramond, P. (1971). *Phys. Rev.* **D3**, 2415.

Scherk, J., and Schwarz, J. H. (1974). *Nucl. Phys.* **B81**, 118.

Schwarz, J. H. (1995). *Phys. Lett.* **B360**, 13, Erratum: *Phys. Lett.* **B364**, 252, hep-th/9508143.

Schwarz, J. H. (1996). *Phys. Lett.* **B367**, 97, hep-th/9510086.

Townsend, P. K. (1995). *Phys. Lett.* **B350**, 184, hep-th/9501068.

Virasoro, M. (1970). *Phys. Rev.* **D1**, 2933.

Veneziano, G. (1968). *Nuovo Cim.* **57A**, 190.

Witten, E. (1995). *Nucl. Phys.* **B443**, 85, hep-th/9503124.

Witten, E. (1996). *Nucl. Phys.* **B460**, 335, hep-th/9510135.

Witten, E. (1998). *Adv. Theor. Math. Phys.* **2**, 253, hep-th/9802150.

Yoneya, T. (1974). *Prog. Theor. Phys.* **51**, 1907.

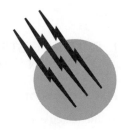

Surface Chemistry

Simon R. Bare
G. A. Somorjai

University of California, Berkeley, and
Lawrence Berkeley Laboratory

I. Surface Structure of Clean Surfaces
II. Surface Structure of Adsorbates
   on Solid Surfaces
III. Thermodynamics of Surfaces
IV. Electrical Properties of Surfaces
V. Surface Dynamics

GLOSSARY

Adsorbate Adsorbed atom or molecule.

Adsorption Process by which molecules are taken up on the surface by chemical or physical action.

Chemisorption Binding of molecules to surfaces by strong chemical forces.

Desorption Process by which molecules are removed from the surface.

Heat of adsorption Binding energy of the adsorbed species.

Physisorption Binding of molecules to surfaces by weak chemical forces.

Sticking probability Ratio of the rate of adsorption to the rate of collision of the gaseous molecule with the surface.

Surface free energy Energy necessary to create a unit area of surface.

Surface reconstruction Equilibration of surface atoms to new positions that changes the bond angles and rotational symmetry of the surface atoms.

Surface relaxation Equilibration of surface atoms to new positions that changes the interlayer distance between the first and second layers of atoms.

Surface state Electronic state localized at the surface.

Surface unit cell Two-dimensional repeating unit that fully describes the surface structure.

Work function Minimum energy required to remove an electron from the surface into the vacuum outside the solid.

SURFACES constitute the boundaries of condensed matter, solids, and liquids. Surface chemistry explores the structure and composition of surfaces and the bonding and reactions of atoms and molecules on them. There are many macroscopic physical phenomena that occur on surfaces or are controlled by the electronic and physical properties of surfaces. These include heterogeneous catalysis, corrosion, crystal growth, evaporation, lubrication, adhesion, and integrated circuitry. Surface chemistry examines the science of these phenomena as well.

Encyclopedia of Physical Science and Technology, Third Edition, Volume 16

I. SURFACE STRUCTURE OF CLEAN SURFACES

A. Introduction

To the naked eye the surface of a single crystal of a metal looks perfectly planar with no imperfections. If this crystal is now examined with an optical microscope features of the surfaces down to the wavelength of visible light (~5000 Å) can be resolved. The surface will look granular with distinct regions of crystallinity separated from each other by boundaries or dislocations. These dislocations indicate areas on the surface where there is a mismatch of the crystalline lattice, and they can take several forms, for example, edge dislocations or screw dislocations. The presence of these dislocations or defects can dominate certain physical properties of the material. Dislocation densities of the order of $10^6–10^8$ cm^{-2} are commonly found on metal single crystals, whereas the number is lower ($10^4–10^6$ cm^{-2}) on semiconductor surfaces due to the different nature of the bonding. These defect densities must be compared with the total concentration of surface atoms (about 10^{15} cm^{-2}). On further magnification, for example, using a scanning electron microscope, features can be resolved down to about 1000 Å, and our view of the surface changes further. The surface will look pitted, with distinct planar areas (terraces) bounded by walls many atomic layers in height. Thus, on the microscopic and submicroscopic scale the surface morphology appears to be heterogeneous, with many different surface sites that differ by the number of neighboring atoms surrounding them. What about the nature of the surface on an atomic scale?

In order to be able to discuss and understand the structure of surfaces it is necessary to understand the techniques that are capable of viewing the surface on an atomic scale. We briefly describe such techniques, illustrating their capabilities with pertinent examples. The techniques more commonly used are field-ion microscopy (FIM), low-energy electron diffraction (LEED), helium atom diffraction, and high-energy ion scattering. In addition, the relatively new technique of scanning tunneling microscopy (STM) is proving to be a very promising tool. Only brief descriptions are given here and the reader is referred to some of the excellent books on the subject given in the bibliography.

B. Techniques Sensitive to Surface Structure

1. Field-Ion Microscopy

Field-ion microscopy is one of the oldest techniques used for surface structure determination, having been invented by Müller in 1936. The basic microscope can be very simple. In an ultrahigh vacuum cell, a potential of about 10,000 V is applied between a hemispherical tip of refractory metal of radius ~10^{-4} cm and a fluorescent screen. The tip is charged positively, and a gas (usually helium) is allowed to impinge on the surface. Under the influence of the very strong electric field helium atoms that are incident on the tip are ionized. The positive ions thus created are repelled radially from the surface and accelerated onto the fluorescent screen, where a greatly magnified image of the crystal tip is displayed. The ionization probability depends strongly on the local field variations induced by the atomic structure of the surface—protruding atoms generate stronger ionization than atoms embedded in close-packed planes, thereby producing individual bright spots on the screen. The small radius of the tip is needed to produce the large fields necessary for ionization, but it also permits the immense magnification of the microscope. The tip surface is directly imaged with magnification of about 10^7. Figure 1 depicts the image of a tungsten field-ion tip. Well-defined atomic planes of the crystal tip can be readily identified, indicating that there is order of the atomic scale, i.e., most of the surface atoms in any crystal face are situated in ordered rows separated by well-defined interatomic distances. The technique is limited to the refractory metals (W, Ta, Ir, and Re) which can withstand the strong electric field at the tip without desorption or evaporation from the surface. However, its great advantage is that individual atoms can be imaged on the screen, which also allows studies of surface diffusion.

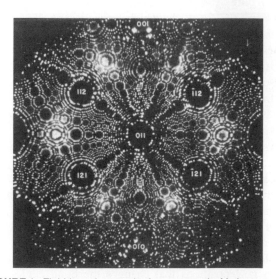

FIGURE 1 Field-ion micrograph of a tungsten tip. Various crystal planes are labeled. (Courtesy of Lawrence Berkeley Laboratory.)

2. Low-Energy Electron Diffraction

Another method which has demonstrated that crystal surfaces are ordered on an atomic scale is low-energy electron diffraction (LEED). This method is the most frequently used technique, such that virtually all modern surface science laboratories now rely on it for surface structural information.

In order to obtain diffraction from surfaces, the incident wave must satisfy the condition $\lambda \leq d$, where λ is the wavelength of the incident beam and d the interatomic distance. In addition, the incident beam should not penetrate much below the surface plane but should backdiffract predominantly from the surface so that the scattered beam reflects the properties of the surface atoms and not those of the bulk. The deBroglie wavelength λ of electrons is given by λ (in Å) $= \sqrt{150}/E$, where E is in electron volts. Thus, in the energy range 10–500 eV, the wavelength varies from 3.9 to 0.64 Å, which is smaller or equal to most interatomic distances, and the escape depth of the backscattered electrons in this energy range is 5–10 Å, thereby providing surface sensitivity. These elastically scattered low-energy electrons yield surface structural information. The technique of LEED is depicted schematically in Fig. 2, while a schematic of the LEED apparatus is shown in Fig. 3. A collimated primary beam of electrons with a diameter of 0.1–1 mm at energies of 15–350 eV is impinged on a surface and the elastically backscattered electrons, after traveling through a field-free region, are spatially analyzed. This is achieved most commonly (see Fig. 3) by passing the scattered electrons through four hemispherical grids. The first grid is at the crystal potential while the second and third are at a retarding voltage to eliminate inelastic electrons, and the fourth is at ground. After passing through these grids the diffracted beams are accelerated onto a hemispherical phosphor screen.

If the crystal surface is well-ordered, a diffraction pattern consisting of bright, well-defined spots will be displayed on the screen. The sharpness and overall intensity of the spots are related to the degree of order of the surface. When the surface is less ordered, the diffraction beams broaden and become less intense, while some diffuse intensity appears between the beams. A typical set of diffraction patterns from a well-ordered surface is shown in Fig. 4. The presence of the sharp diffraction spots clearly indicates that the surface is ordered on an atomic scale. Similar LEED patterns have been obtained from solid single-crystal surfaces of many types including metals, semiconductors, alloys, oxides, and intermetallics.

Due to the importance of LEED in surface chemistry we briefly discuss other aspects of the technique which make it one of the most powerful surface sensitive tools. It is convenient to subdivide the technique into two-dimensional LEED and three-dimensional LEED. In two-dimensional LEED we observe only the symmetry of the diffraction pattern on the fluorescent screen. The bright spots which correspond to the two-dimensional reciprocal lattice belonging to the repetitive crystalline surface structure yield immediate information about the size and orientation of the surface unit cell, i.e., the geometry of the surface layer. This is important information since reconstruction-induced and adsorbate-induced new periodicities are immediately visible. The diffuse background intensity also contains information about the nature of any disorder present on the surface. In three-dimensional LEED the information gained from the two-dimensional pattern is supplemented by the intensities of the diffraction spots which are measured as a function of incident electron energy. By comparing these intensity-versus-voltage curves [$I(V)$ curves] with those simulated numerically with the help of a suitable theory, the precise location of atoms or molecules in the surface with respect to their neighbors is determined. Thus, the bond length and bond angles in the surface layer are calculated. It should be mentioned, however, that the analysis of the LEED beam intensities requires a theory of the diffraction process which is a nontrivial point due to multiple scattering of LEED electrons by the surface, and this is not simple to represent in a theory.

3. Atomic-Beam Diffraction

Another technique that utilizes the principle of diffraction is atomic- or molecular-beam diffraction. The deBroglie wavelength λ associated with helium atoms is given by the following:

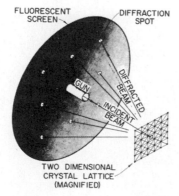

FIGURE 2 Scheme of the low-energy electron diffraction experiment. (Courtesy of Lawrence Berkeley Laboratory.)

$$\lambda(\text{Å}) = \frac{h}{(2ME)^{1/2}} = \frac{0.14}{E(\text{eV})^{1/2}}, \tag{1}$$

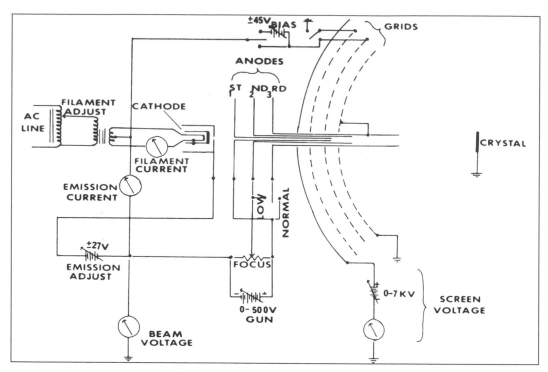

FIGURE 3 Scheme of the low-energy electron diffraction apparatus employing the postacceleration technique.

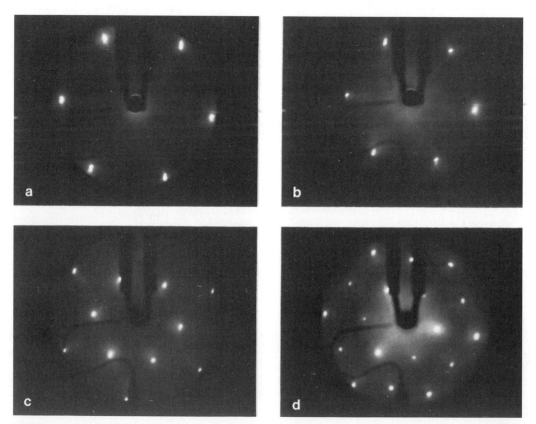

FIGURE 4 LEED pattern from a Pt(111) crystal surface at (a) 51 eV, (b) 63.5 eV, (c) 160 eV, and (d) 181 eV incident electron energy. (Courtesy of Lawrence Berkeley Laboratory.)

where h is Planck's constant and M and E are the mass and energy, respectively, of the helium atom. Atoms with a thermal energy of $\sim$20 meV have $\lambda = 1$ Å and can readily diffract from surfaces. The information obtained from atomic-beam diffraction is similar to that from LEED, but there are differences between the two techniques. In LEED the relatively high-energy (20–200 eV) electrons used penetrate the crystal, multiple scattering events are important, and the LEED electrons are scattered primarily from the ion cores of the crystal lattice. In atom diffraction, there is virtually no penetration of the low-energy (10–200 meV) atomic beam, making it much more surface sensitive than the electron beam. The atomic beam is primarily scattered from the valence electrons of the surface atoms. In fact, their scattering is usually simulated by a "hard wall" around the atoms in the top layer of the surface so that diffraction is from a "corrugated hard wall" with the periodicity of the surface mesh. As in LEED the location of the diffracted beams indicates the surface periodicity. Their intensities are related to the structure of the scattering potential within a unit mesh, in this case to the relative amplitude and positions of corrugations around the surface atoms.

The essential elements of the apparatus necessary to perform atomic-beam diffraction are an atomic beam of gas and a detector. The atomic beam is usually generated from a nozzle source incorporating several skimmers. The energy (wavelength) of the beam is varied by either heating or cooling the nozzle. The detector usually employed is a mass spectrometer, mounted on a rotatable device to enable it to be movable over a large range of scattering angles. The atomic beam is chopped with a variable-frequency chopper before it impinges on the surface. In this way, an alternating intensity of the beam is generated at the mass spectrometer detector, which is readily separated from the noise due to helium atoms in the background.

To illustrate the type of data that is obtained, Fig. 5 shows the He diffraction traces from a Au(110)-(1 × 2) surface at two different wavelengths.

Helium diffraction is especially sensitive to surface order on an atomic scale. On scattering from a well-ordered single crystal surface nearly 15% of the scattered helium atoms appear in the specular helium beam whereas this fraction can drop to 1% when the surface is disordered. Measurements of the fraction of specularly scattered helium can therefore provide information on the degree of atomic disorder in the solid surface.

4. Scanning Tunneling Microscopy

The relatively new technique of scanning tunneling microscopy also clearly demonstrates order on an atomic

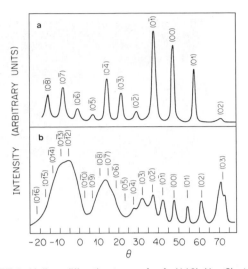

FIGURE 5 Helium diffraction traces for Au(110)-(1 × 2) at a surface temperature of 100 K with incident angle $\Theta_i = 48°$. The wavelength λ_{He} is (a) 1.09 Å and (b) 0.57 Å.

scale on single-crystal surfaces. It images surface topographies in real space with a lateral resolution of $\sim$2 Å and vertical resolution of $\sim$0.05 Å. The technique utilizes the tunnel effect. Due to the wave nature of electrons, they are not strictly confined to the interior bounded by the surface atoms. Therefore, the electron density does not drop to zero at the surface but decays exponentially on the outside with a decay length of a few angstroms. If two metals are approached to within a few angstroms, the overlap of their surrounding electron clouds becomes substantial, and a measurable current can be induced by applying a small voltage between them. This tunnel current is a measure of the wave-function overlap, and depends very strongly on the distance between the two metals. This is the physical basis of the scanning tunneling microscope. Experimentally one of the electrodes is sharpened to a pointed tip which is scanned over the surface to be investigated (the other electrode) at constant tunnel current. The tip thus traces contours of constant wave-function overlap, and in the case of constant decay length, the trace is an almost true image of the surface atomic positions, i.e., the surface topography. An example is shown in Fig. 6.

5. Ion Scattering

Ion scattering from surfaces is usually subdivided into two scattering regions: low-energy ion scattering (LEIS), energies typically $\sim$1 keV, and high-energy scattering (HEIS), energies 0.1–1 MeV. High-energy ion scattering is a probe that tests the local position of surface atoms relative to their bulklike sites. In HEIS the velocity of the ion is such that it is moving fast compared to the thermal motions of the

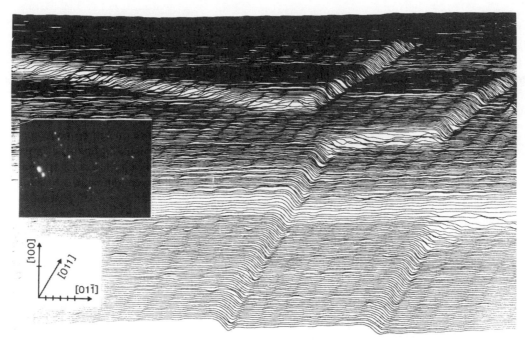

FIGURE 6 Scanning tunneling microscope picture of a clean (1 × 5) reconstructed Au(100) surface with monatomic steps. Divisions on the crystal axes are 5 Å, with approximately 1.5 Å from scan to scan. The inset shows the LEED pattern of the predominant (1 × 5) corrugation. (Courtesy of Lawrence Berkeley Laboratory.)

atoms in the solid, thus the beam senses a frozen lattice. If the target is amorphous each atom would sense a uniform distribution of impact parameters of the ions and diffuse scattering results. However, the scattering spectrum from a single crystal aligned with a major symmetry axis parallel to the beam is drastically modified from that of the amorphous target. The impact parameter distribution is also uniform at the first monolayer, but the first atom shadows the second from the beam, and small angle scattering events determine the impact parameter distribution at the second atom. This results in a unique (nonuniform) flux distribution at the second atom. Figure 7 illustrates the effect of small-angle scattering in a two-atom model. Ions incident at the smallest impact parameter undergo large-angle scattering, those at large impact parameter suffer small deflections which determine the flux distribution of ions near the second atom. The closest approach of the ion to the second atom, R, can be approximated assuming Coulomb scattering as follows:

$$R = 2(Z_1 Z_2 e^2 d/E)^{1/2}, \qquad (2)$$

where Z_1 and Z_2 are the masses of the incident and target atoms, respectively; d is the atomic spacing; and E is the incident ion energy. This gives rise to a shadow cone beneath the surface atom as illustrated in Fig. 7. The flux distribution at the second atom, within the Coulomb ap-

proximation, can be written analytically which leads to an estimate of the two-atom surface peak intensity I:

$$I = 1 + \left(1 + \frac{R^2}{2\rho^2}\right) \exp\left(\frac{-R}{2\rho^2}\right), \qquad (3)$$

where ρ is the two-dimensional root mean square thermal vibrational amplitude. The first term represents the unit contribution from the first atom in the string, the second term represents the variable contribution from the second

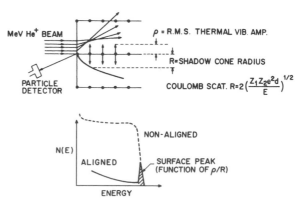

FIGURE 7 Schematic showing the interactions at the surface of an aligned single crystal and the formation of the shadow cone. The energy spectra for the aligned and nonaligned case are also shown.

atom. While the two-atom Coulomb approximation is not adequate enough to compare to experiment, it illustrates that the surface peak intensity is a function of one parameter, ρ/R.

The intensity of the surface peak is thus sensitive to the atomic arrangement on the surface, i.e., the positions of the surface atoms with respect to their bulklike positions. The effect on the surface peak of different surface structures is depicted in Fig. 8. The nature of reconstructed, relaxed, and adsorbate-covered surfaces are discussed in the sections below.

High-energy ion scattering is also a sensitive tool for answering other important questions in surface chemistry, namely what type of atoms are present on the surface and how many are present.

All of the techniques discussed so far indicate that the solid surface is ordered on an atomic scale. Most of the surface atoms occupy equilibrium atomic positions that are located in well-defined rows separated by equal interatomic positions. This atomic order is predominant despite the fact that there are large numbers of atomic positions on the surface where atoms have different numbers of neighbors. A pictorial representation of the topology of a monatomic crystal on an atomic scale is shown in

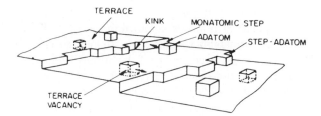

FIGURE 9 Model of a heterogeneous solid surface, depicting different surface sites. These sites are distinguishable by their number of nearest neighbors.

Fig. 9. The surface may have atoms in any of the positions shown in the figure. There are atoms in the surface at kink positions and in ledge positions, and there are adatoms adsorbed on the surface at various sites. Atomic movement from one position to another proceeds by surface diffusion. To the first approximation, the binding energy of the surface atoms is proportional to the number of nearest and next-nearest neighbors. Therefore, for example, atoms at a ledge are bound more strongly than are adatoms. In equilibrium there is a certain concentration of all these surface species, with those species predominating whose binding energies are greatest. Thus, the adatom concentration on clean well-equilibrated surfaces should be very small indeed. However, while these surfaces are ordered on an atomic scale their structure is not always one of simple termination of the bulk unit cell, relaxation or reconstruction being common.

C. Surface Relaxation

Generally, the surface unit cells of clean metal surfaces have been found to be those expected from the projection of the bulk X-ray unit cell onto the surface, referred to as a (1×1) structure (in Miller index notation), and the uppermost layer z spacing (spacing in the direction normal to the surface plane) is equal to the bulk value within about 5%. Such surfaces include the (111) crystal faces of face centered cubic aluminum, platinum, nickel, and rhodium, and the (0001) crystal faces of hexagonal close packed cadmium and beryllium. This information has almost exclusively been determined by a detailed intensity analysis of the diffraction beams in LEED as a function of incident electron energy, and the interatomic positions in the surface layer are calculated to within 0.1 Å. The Al (110) surface shows a 5–15% contraction, the Mo(100) surface a 11–12% contraction, and the W(100) surface a 6% contraction of the top-layer z spacing with respect to the bulk, while retaining the (1×1) surface unit cell. Generally, crystal planes whose atoms are less densely packed [for example, bcc (100) and fcc (110) planes] will be more likely to show relaxation than the more densely

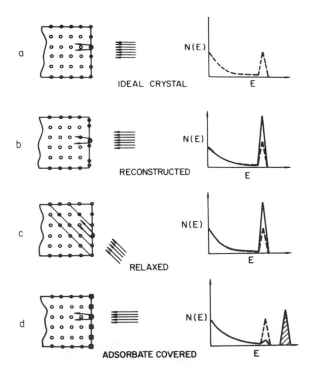

FIGURE 8 Schematic of the dependence of the intensity of the surface peak (SP) on different crystal surface structures. (a) The ideal crystal SP from "bulklike" surface, (b) enhanced SP observed in normal incidence for a reconstructed surface, (c) enhanced SP observed in nonnormal incidence for a relaxed surface, and (d) reduced SP observed in normal incidence for a registered overlayer.

packed planes. In forming a surface of the less densely packed planes it is necessary to remove a larger number of nearest-neighbor atoms. Thus, to minimize the surface free energy a relocation of the surface atoms from their bulk positions is quite likely. There are several explanations as to why surface relaxation is prevalent on the more open surfaces. First, it can be imagined that the electron cloud attempts to smooth its surface, thereby producing electrostatic forces that draw the surface atoms toward the substrate, the effect being stronger the less closely packed the surface. Second, with fewer neighbors the two-body repulsion energy is smaller, allowing greater atomic overlap and therefore more favorable bonding at shorter bond lengths. Third, for surface atoms the bonding electrons are partly redistributed from the broken bonds to the remaining unbroken bonds, thereby the charge content of the latter is increased, reducing the bond length.

D. Surface Reconstruction

Surface atoms in any crystal are in an anisotropic environment which is very different from that around bulk atoms. The crystal symmetry that is experienced by each surface atom is markedly lower than when the atom is in the bulk. This symmetry change and lack of neighbors in the direction perpendicular to the surface allows displacement of surface atoms in ways which are not allowed in the bulk. Surface relaxation is one consequence of this, the other major consequence being surface reconstruction. Here the two-dimensional surface unit cell is different from that given by the termination of the bulk structure on the plane of interest. Surface reconstruction can give rise to a multitude of different structures depending on the electronic structure of a given substance. The phenomenon is more frequent on semiconductor surfaces than on metal surfaces. While the geometry is readily observed in the LEED pattern, the actual elucidation of the real-space reconstructed surface structure is often extremely difficult, and in some cases even after years of study and many proposals of the structure, the true structure is still not known. Such a system is the Si(111) surface. Upon cleaving in UHV at room temperature, the surface exhibits a (2×1) surface structure. On heating to about 700 K the surface structure changes to one with (7×7) periodicity. This (7×7) structure is then the stable structure of the (111) face. While this surface has been studied by a multitude of techniques, including LEED, STM, and He atom diffraction there is still no generally accepted structure of either the (2×1) or (7×7) reconstructions.

One of the best known examples of reconstruction of metallic surfaces is that for the (100) faces of three $5d$ transition metals that are neighbors on the periodic table:

gold, platinum, and iridium. The ideal unreconstructed surfaces have a square net of atoms. Surface reconstruction produces a superlattice that is basically five times larger in one direction than for the ideal surface. For Ir(100) the superlattice is denoted (5×1), for Pt(100) by the matrix notation $\left(\begin{smallmatrix} 5 & 1 \\ -1 & 14 \end{smallmatrix} \right)$, and for Au(100) by the superlattice (5×20). From LEED $I(V)$ analyses, evidence indicates that the nature of the reconstruction is similar on all three metals and consists of a close-packed hexagonal top layer that is positioned in slightly different ways on the square net substrate. These reconstructions are consistent with the knowledge that the (111) face of fcc metals is energetically the most favorable. It is worth noting here that the adsorption of gases such as oxygen, carbon monoxide, or hydrogen, or the presence of impurities can inhibit these surface reconstructions. On the other hand, the presence of such adsorbates can also induce different surface reconstructions.

The nature and cause of these surface phase transformations are not well established at present. The case of structural change from metastable to stable on adsorption or removal of adsorbates indicates the likelihood of electronic transitions that accompany reconstruction. At the surface there are fewer nearest neighbors as compared to atoms in the bulk. The electronic structure that is the most stable in this reduced-symmetry environment may be substantially different from that of the bulk metal. Since the surface atoms are surrounded by atoms only on one side and there is vacuum on the other, they may change their coordination number by slight relocation with simultaneous changes in the electronic structure. It is indeed surprising that more surfaces do not show reconstruction.

E. Stepped and Kinked Surfaces

The close-packed faces of solids (low-Miller-index faces) have the lowest surface free energy, and therefore they are the most stable with respect to rearrangement on disordering up to or near the melting point. However, stepped and/or kinked surfaces (high-Miller-index faces), although of higher surface free energy, are very important. They are known to play important roles during evaporation, condensation, and melting. Steps and kinks are sites where atoms break away as an initial process leading to desorption, or where atoms migrate during condensation to be incorporated into the crystal lattice. Theories of crystal growth, evaporation, and the kinetics of melting have identified the significance of these lower coordination-number sites in controlling the rate processes associated with phase changes. In addition, studies of chemisorption and catalysis using single-crystal surfaces have revealed different binding energies and enhanced chemical activity at steps and kinks on high-Miller-index

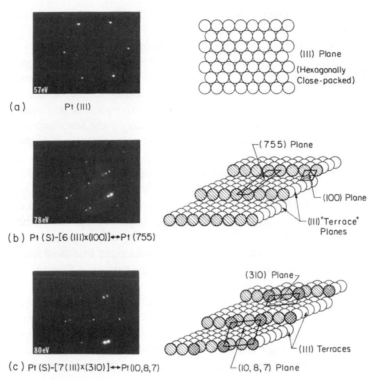

FIGURE 10 LEED patterns (left) and surface structures (right) of (a) flat, (b) stepped, and (c) kinked platinum surfaces. (Courtesy of Lawrence Berkeley Laboratory.)

transition metal surfaces as compared to low-Miller-index surfaces. Adsorption of diatomic or polyatomic molecules frequently leads to dissociation with greater probability of steps and kinks than on flat atomic terraces.

The presence of steps on a single crystal surface is readily discernible by LEED. The LEED patterns differ from those expected from crystals with low-index faces in that the diffraction beams are split into doublets. This splitting (Fig. 10) is a function of ordered steps on the surface. The distance between the split beams is inversely related to the distance between the steps, i.e., the terrace width. From the variation of the intensity maximum of the doublet spots with electron energy the step height can be determined.

Many stepped surfaces exhibit high thermal stability. In particular the one-atom-height step periodic terrace configuration appears to be the stable surface structure of many high-Miller-index surfaces. While most of the stepped surfaces are stable when clean in their one-atom-height step ordered terrace configuration, in the presence of a monolayer of carbon or oxygen many stepped surfaces undergo structural rearrangement. The step height and terrace width may double, or faceting may occur. Faceting occurs when the step orientation becomes as prominent as that of the terrace and new diffraction features become rec-

ognizable in LEED. Upon removing the impurities from the surface, the original one-atom-height step ordered terrace surface structure is usually regenerated.

II. SURFACE STRUCTURE OF ADSORBATES ON SOLID SURFACES

A. Introduction

While the knowledge of the structure of clean solid surfaces is important in its own right for determining various properties of those surfaces, many phenomena are associated with the presence of adsorbates on the surfaces. In fact, in the natural environment of our planet, surfaces are never truly free of adsorbates. On approaching a surface each atom or molecule encounters a net attractive potential. This results in a finite probability that it will be trapped on the surface. This trapping, adsorption, is always an exothermic process. At the low pressure of 1×10^{-6} torr, approximately 1×10^{15} gas molecules collide with each square centimeter of surface per second. Since the surface concentration of atoms is about 10^{15} cm^{-2}, at this pressure the surface may be covered with a monolayer of gas within seconds; this is the major reason why surface studies are performed under ultra-high vacuum conditions ($P < 1 \times 10^{-8}$ torr). The very low pressure is needed to maintain clean surface conditions for a time long enough to perform experimental measurements. At atmospheric pressure the surface will be covered within a fraction of a second. The constant presence of the adsorbate layer influences the chemical, mechanical, and surface electronic properties. Adhesion, lubrication, and resistance to mechanical or chemical attack or photoconductivity are just a few of the many macroscopic surface processes that are controlled by various properties of monolayers.

Two macroscopic experimentally determinable parameters characterize the adsorbed monolayer: the coverage and the heat of adsorption. The coverage Θ is defined as the ratio of the number of adsorbed atoms or molecules to the total number of adsorption sites (usually taken as the number of atoms in the surface plane). The heat of adsorption ΔH_{ads}, is implicitly linked to the strength of the adsorbate–substrate bond. Knowledge of both parameters often reveals the nature of bonding in the adsorbed layer.

Atoms or molecules that impinge on the solid surface from the gas phase will have a residence time τ on the surface. If the impinging molecules achieve thermal equilibrium with the surface atoms $\tau = \tau_0 \exp \Delta H_{ads}/RT$, where τ_0 is related to the average vibrational frequency associated with the adsorbate. The heat of adsorption is always positive and is defined as the binding energy of the adsorbed species. A larger ΔH_{ads} and lower temperature increase the residence time. For a given incident flux, larger ΔH_{ads} and lower temperature yield higher coverages.

It is conventional to divide adsorption into two categories: physisorption and chemisorption. Physisorption (or physical adsorption) systems are characterized by weak interactions ($\Delta H_{ads} < 15$ kcal mol^{-1}, accompanied by short residence times) and require adsorption studies to be performed at low temperature and relatively high pressure (high flux). Adsorbates that are characterized by stronger chemical interactions ($\Delta H_{ads} \geq 15$ kcal mol^{-1}), where near-monolayer adsorption commences even at room temperature and at low pressures ($\leq 10^{-6}$ torr), are called chemisorption systems. While the two names imply two distinct types of adsorption, there is a gradual change from the physisorption to the chemisorption regime.

One of the most fascinating facts about the structure of these physisorbed and chemisorbed overlayers in the submonolayer to few monolayer regime is the preponderance of the formation of long-range ordered structures. Well over 1000 two-dimensional unit cells have been documented in the literature. While only the shape, size, and orientation of the cells are known for most of them, the adsorption site and bond lengths have been determined for about 500 of them.

B. The Ordering of Adsorbed Monolayers

The ordering process in the adlayer is due to an interplay of the bonding with the substrate and the bonding between the adatoms or admolecules. Once a molecule adsorbs it may diffuse on the surface or remain bound at a specific site during most of its residence time. Thermal equilibration among the adsorbate and between the adsorbate and substrate atoms (i.e., adsorption) is assured if ΔH_{ads} and $\Delta E^*_{D(bulk)}$, the activation energy for bulk diffusion, are high enough as compared to kT ($\geq 10kT$). However, ordering primarily depends on the depth of the potential energy barrier that keeps an atom or molecule from hopping to a neighboring site along the surface. The activation energy for surface diffusion, $\Delta E^*_{D(surface)}$, is an experimental parameter that is of the magnitude of this potential energy barrier. ΔE^*_D can be experimentally determined on well-characterized surfaces by field-ion microscopy, for example, and for Ar and W adatoms and O atoms on tungsten surfaces has the value 2, 15, and 10 kcal mol^{-1}, respectively. For small values of $\Delta E^*_{D(surface)}$ ordering is restricted to low temperatures, since as the temperature is increased the adsorbate becomes very mobile. As the value of $\Delta E^*_{D(surface)}$ increases, ordering cannot commence at low temperature since the adsorbate atoms need to have a considerable mean free path along the surface to find their equilibrium position once they land on the surface at a different location. Naturally, if the temperature is too high, the adsorbate will desorb or vaporize.

The binding forces of adsorbates on substrates have components perpendicular and parallel to the surface. The perpendicular component is primarily responsible for the binding energy (ΔH_{ads}), while the parallel component often determines the binding site on the surface. The binding site may also be affected by adsorbate–adsorbate interactions, which produce ordering within an overlayer. These interactions may be subdivided into direct adsorbate–adsorbate interactions (not involving the substrate at all) and substrate mediated interactions. The latter are complicated many-atom interactions, for example dipole–dipole interactions.

The adsorbate–adsorbate interactions may be repulsive; they always are repulsive at sufficiently small adsorbate–adsorbate separations. At larger separations they may be attractive, giving rise to the possibility of island formation. They may also be oscillatory, moving back and forth between attractive and repulsive as a function of adsorbate–adsorbate separation, with a period of several angstroms giving rise, for example, to non-close-packed islands.

Except for the strong repulsion at close separations, the adsorbate–adsorbate interactions are usually weak compared to the adsorbate–substrate interactions, even when we consider only the components of the forces parallel to the surface. In the case of chemisorption, where the adsorbate–substrate interaction dominates, the adsorbates usually choose an adsorption site that is independent of the coverage and of the overlayer arrangement (the positions the other adsorbates choose). This adsorption site is usually that location which provides the largest number of nearest substrate neighbors, which is independent of the position of other adsorbates. Adsorbates with these properties normally do not accept close packing; the substrate controls the overlayer geometry and imposes a unique adsorption site. Close packing of an adsorbate layer is also observed. In this case the overlayer chooses its own lattice (normally a hexagonal close-packed arrangement) with its own lattice constant independent of the substrate lattice and results in the formation of incommensurate lattices. In this case no unique adsorption site exists: each adsorbate is differently situated with respect to the substrate. This situation is especially common in the physisorption of rare gases. Their relatively weak adsorbate–substrate interactions allow the adsorbate–adsorbate interactions to play the dominant role in determining the overlayer geometry.

The chemisorption case is exemplified by oxygen and sulfur on metals; the physisorption case by krypton and xenon on metals and graphite. Intermediate cases exist. Although undissociated CO on metals is not physisorbed but chemisorbed, it sometimes produces incommensurate close-packed hexagonal overlayers.

1. The Effect of Temperature on Ordering

Temperature has a major effect on the ordering of adsorbed monolayers: all of the important ordering parameters (the rates of desorption and surface and bulk diffusion) are exponential functions of the temperature.

The influence of temperature on the ordering of C_3–C_8 saturated hydrocarbons on the Pt(111) crystal face is shown in Fig. 11. At the highest temperatures, adsorption may not take place, since under the exposure conditions the rate of desorption is greater than the rate of condensation of the vapor molecules. As the temperature is decreased, the surface coverage increases and ordering becomes possible. First, one-dimensional lines of molecules form; then at lower temperatures ordered two-dimensional surface structures form. Not surprisingly, the temperatures at which these ordering transitions occur depend on the molecular weights of the hydrocarbons, which also control their vapor pressure, heats of adsorption, and activation energies for surface diffusion. As the temperature is further decreased, multilayer adsorption may occur and epitaxial growth of crystalline thin films of hydrocarbon commences.

Figure 11 clearly demonstrates the controlling effect of temperature on the ordering and the nature of ordering of

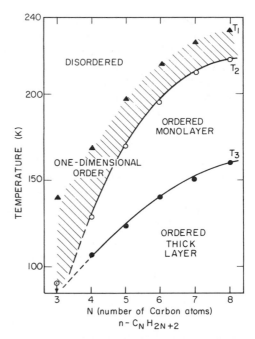

FIGURE 11 Monolayer and multilayer phases of the *n*-paraffins C_3 to C_8 on Pt(111) and the temperatures at which they are observed at 10^{-7} torr.

the adsorbed monolayer. Although changing the pressure at a given temperature may be used to vary the coverage by small amounts and thereby change the surface structures in some cases, the variation of temperature has a much more drastic effect on ordering.

Temperature also markedly influences chemical bonding to surfaces. There are adsorption states that can only be populated if the molecule overcomes a small potential energy barrier. The various bond-breaking processes are similarly activated. The adsorption of most reactive molecules on chemically active solid surfaces takes place without bond breaking at sufficiently low temperatures. As the temperature is increased, bond breaking occurs sequentially until the molecule is atomized. Thus, the chemical nature of the molecular fragments will be different at various temperatures. There is almost always a temperature range for the ordering of intact molecules in chemically reactive adsorbate–substrate systems. It appears that for these systems ordering is restricted to low temperatures below 150 K, and consideration of surface mobility becomes, perhaps, secondary.

2. The Effect of Surface Irregularities on Ordering

A solid surface exhibits a large degree of roughness on a macroscopic scale. Therefore, it is to be expected that if nucleation is an important part of the ordering process, surface roughness is likely to play an important role in preparing ordered surface structures. The transformation temperature or pressure at which one adsorbate surface structure converts into another can also be affected by the presence of uncontrolled surface irregularities (surface defects). Other causes that could influence ordering are the presence of small amounts of surface impurities that block nucleation sites or interfere with the kinetics of ordering, or impurities below the surface that are pulled to the surface during adsorption and ordering.

The effect of surface irregularities on ordering can be investigated in a more controlled way using stepped crystal surfaces. In general, the smaller the ordered terrace between the steps, the stronger the effect of steps on ordering. For instance, the ordering of small molecular adsorbates on a high-Miller-index $Rh(S)-[6(111) \times (100)]$ is largely unaffected by the presence of steps whereas on the $Rh(S)[3(111) \times (111)]$ the ordering is influenced by the higher step density. Steps can also affect the nucleation of ordered domains. It is frequently observed on W and Pt stepped surfaces that when two or three equivalent ordered domains may form in the absence of steps, only one of the ordered domains grows in the presence of steps.

3. The Effect of Coadsorbates on Ordering

It has been found that although certain molecules may not order when present on their own on a surface they can be induced to order by coadsorption of either carbon monoxide or nitric oxide. For example Table I summarizes the ordered structures that have been observed by the coadsorption of alkylidynes, acetylene, aromatics, and alkalis with CO on $Rh(111)$.

At low temperature both Na and ethylidyne form (2×2) overlayers on $Rh(111)$ but with increasing temperature begin to disorder. If CO is coadsorbed then the adsorbates can be reordered into a $c(4 \times 2)$ unit cell.

It is thought that the nature of this type of ordering process is due to adsorbate–adsorbate interactions: A molecule that might not otherwise order due to weak adsorbate–adsorbate interactions is ordered by coadsorb-

ing a molecule such as CO, which has interactions strong enough to induce ordering in the overlayer. Similar phenomena have been observed on $Pt(111)$, and it is thought that this coadsorbate-induced ordering may prove to be a very general phenomenon.

C. Ordered Adsorbate Structures

As mentioned in the introduction to this section, well over 1000 ordered adsorbate structures have been observed with LEED. A full listing and discussion of these structures is outside the scope of this article. Instead, one example of each of the three following categories of adsorption are presented to give an illustrative indication of the types of structures found: (i) an ordered monolayer of atoms, (ii) an ordered organic monolayer, and (iii) an ordered molecular monolayer. First, a few generalities of the ordered adsorbate structures are discussed, based on the large number of LEED observations: the so-called "rules of ordering":

1. The rule of close-packing. Adsorbed atoms or molecules tend to form surface structures characterized by the smallest unit cell permitted by the molecular dimensions and adsorbate–adsorbate and adsorbate–substrate interactions. They prefer close-packing arrangements. Large reciprocal unit meshes are uncommon and the most frequently observed meshes are the same size as the substrate mesh, i.e., (1×1) or are approximately twice as large, e.g., (2×2), $c(2 \times 2)$, (2×1), $(\sqrt{3} \times \sqrt{3})$.
2. The rule of rotational symmetry. Adsorbed atoms or molecules form ordered structures that have the same rotational symmetry as the substrate surface. If the surface unit mesh has a lower symmetry than the substrate, then domains of the various possible mesh orientations are to be expected on different areas of the surface with a resulting increase in symmetry.
3. The rule of similar unit cell vectors. Adsorbed atoms as molecules in monolayer thickness tend to form ordered surface structures characterized by unit cell vectors closely related to the substrate unit cell vectors. The surface structure bears a closer resemblance to the substrate structure than to the structure of the bulk condensate.

These are not hard-and-fast rules but rather are generalizations of a great many systems.

1. Ordered Atomic Monolayers

Some important conclusions can be drawn from the known structures of atomic adsorbates on single-crystal surfaces. First, the adsorbed atoms tend to occupy sites where they are surrounded by the largest number of substrate atoms

TABLE I Ordered Structures Induced by CO on $Rh(111)$

Type of molecule	LEED pattern	System
Alkylidynes	$c(4 \times 2)$	$CCH_3 + CO$
	$(2\sqrt{3} \times 2\sqrt{3})R30°$	$3CCH_2CH_3 + CO$
Acetylene	$c(4 \times 2)$	$C_2H_2 + CO$
	(3×3)	$C_6H_5F + 2CO$
Aromatics	(3×3)	$C_6H_6 + 2CO$
	$c(2\sqrt{3} \times 4)Rect$	$C_6H_6F + CO$
	$c(2\sqrt{3} \times 4)Rect$	$C_6H_6 + CO$
Alkalis	$(\sqrt{3} \times 7)Rect$	$Na + 7CO$
	$c(4 \times 2)$	$Na + CO$

(largest coordination number). This site is usually the one that the bulk atoms would occupy in order to continue the bulk lattice into the overlayer. The tendency toward occupying the site with the largest coordination number during adsorption on metals holds independently of the crystallographic face of a given metal, the metal for a given crystallographic face, and the adsorbate for a given substrate. Second, the adsorbed atom–substrate atom bond lengths are similar to the bond lengths in organometallic compounds that contain the atom pairs under consideration.

The most common adsorption geometries are displayed in Fig. 12. The threefold hollow sites on the fcc(111) and hcp(0001) and bcc(110) are shown both in top and side views. Similarly, the fourfold hollow sites on the fcc(100) and bcc(100) crystal faces are shown. Finally, the center, long-bridge, and short-bridge sites on the fcc(100) crystal face and the location of atoms in an underlayer in the hcp(0001) crystal face are also displayed.

In addition to the situations discussed, there exist some unique atomic adsorbate geometries. For example, small atoms such as nitrogen and hydrogen often prefer to sit below the surface, as in the case of titanium single-crystal surfaces. Also in the presence of strong chemical inter-

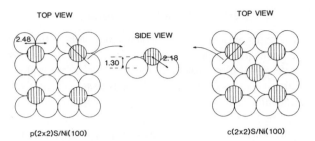

FIGURE 13 Structure of the $p(2 \times 2)$ and $c(2 \times 2)$ sulphur overlayers on Ni(100).

actions there may be a rearrangement of the substrate layer (an adsorbate-induced reconstruction). One example is oxygen on the Fe(100) crystal face.

As an example of ordered atomic adsorption Fig. 13 portrays the two structures of sulfur on Ni(100). At a coverage of one-quarter of a monolayer of S, a $p(2 \times 2)$ overlayer is formed, and at one-half of a monolayer a $c(2 \times 2)$ structure is observed. In both cases the S sits in the fourfold hollow site (highest coordination). A LEED intensity analysis has been performed for both structures, and within experimental error the bond lengths are the same for both structures.

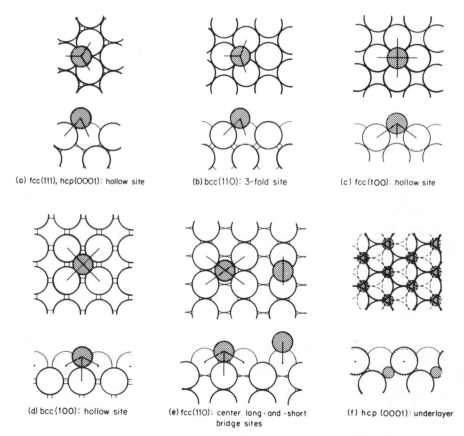

FIGURE 12 Top and side views (in top and bottom sketches of each panel) of adsorption geometries on various metal surfaces. Adsorbates are drawn shaded. Dotted lines represent clean surface atomic positions; arrows show atomic displacements due to adsorption.

2. Ordered Molecular Monolayers

Molecules adsorbed on surfaces may retain their basic molecular identity, bonding as a whole to the substrate. They may dissociate into their constituent atoms, which bond individually to the substrate. Alternatively, molecules may break up into smaller fragments which become largely independent or recombine into other configurations. There are also cases of intermediate character where relatively strong bonding distorts the molecule.

One example of ordered molecular monolayers is CO on Pd(111). The (111) surface of fcc metals is the close-packed plane and shows similar ordering for adsorbed CO for a variety of transition metals. That is the $(\sqrt{3} \times \sqrt{3})$-R30° structure, formed at a coverage of one-third of a monolayer on the (111) faces of Pd, Ni, Pt, Ir, Cu, and Rh. This similarity in ordering is probably due to their surfaces being rather smooth with respect to variations in the CO adsorption energy. Smaller diffusion barriers between different adsorption sites are to be expected, and for large coverages repulsive interactions will be mainly responsible for the arrangement of the adlayer. Figure 14 shows a schematic representation of this $(\sqrt{3} \times \sqrt{3})$-R30° CO structure on Pd(111) and the corresponding observed LEED pattern. A LEED structural analysis has not been performed for this structure, but supporting evidence using infrared spectroscopy indicates that the CO molecules sit in the threefold hollow sites. As the coverage Θ is increased to one-half of a monolayer the LEED pattern transforms to a $c(4 \times 2)$. In this structure the CO molecules all sit in twofold bridge sites. If the adsorption takes place at low temperature (90 K), increasing the CO coverage beyond $\Theta = 0.5$ leads to the appearance of a series of LEED patterns arising from hexagonal superstructures, which by a continuous compression and rotation of the $c(4 \times 2)$ unit cell lead to a (2×2) coincidence pattern at a coverage of $\Theta = 0.75$. These transformations are also shown schematically in Fig. 14.

3. Ordered Organic Monolayers

The adsorption characteristics of organic molecules on solid surfaces are important in several areas of surface science. The nature of the chemical bonds between the substrate and the adsorbate and the ordering and orientation of the adsorbed organic molecules play important roles in adhesion, lubrication, and hydrocarbon catalysis.

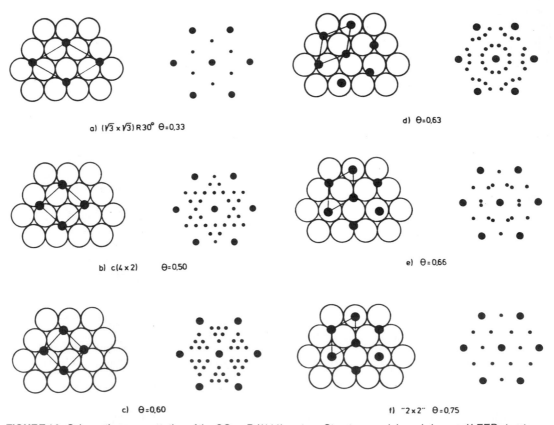

a) $(\sqrt{3} \times \sqrt{3})$ R30° $\Theta = 0.33$

d) $\Theta = 0.63$

b) $c(4 \times 2)$ $\Theta = 0.50$

e) $\Theta = 0.66$

c) $\Theta = 0.60$

f) "2 × 2" $\Theta = 0.75$

FIGURE 14 Schematic representation of the CO on Pd(111) system. Structure models and observed LEED structures for the various CO coverages Θ are shown.

There are many examples in the literature of ordered structures observed by LEED, but only a few of these structures have been calculated from the diffraction beam intensities. However, the ordering characteristics and size and orientation of the unit cells have been determined from the geometry of the LEED patterns. By studying the systematic variation of their shape and bonding characteristics correlations can be made between these properties and their interactions with the metal surfaces.

Examples of ordered organic monolayers are normal paraffins on platinum and silver (111) surfaces. If straight-chain saturated hydrocarbon molecules from propane (C_3H_8) to octane (C_8H_{18}) are deposited from the vapor phase onto Pt or Ag (111) between 100 and 200 K ordered monolayers are produced. As the temperature is decreased a thick crystalline film can condense. The paraffins adsorb with their chain axis parallel to the platinum substrate, and their surface unit cell increases smoothly with increasing chain length as shown in Fig. 15.

Multilayers condensed on top of the ordered monolayers maintain the same orientation and packing found in the monolayers. The monolayer structure determines the growth orientation and the surface structure of the growing organic crystal. This phenomenon is called pseudomorphism, and as a result, the surface structures of the growing organic crystals do not correspond to planes in their reported bulk crystal structures.

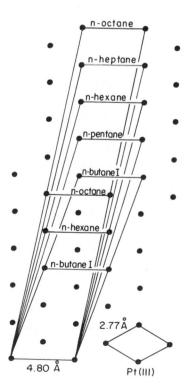

FIGURE 15 Observed surface unit cells for *n*-paraffins on Pt(111).

D. Vibrational Spectroscopy

Vibrational frequencies have been used for many years by chemists to identify bonding arrangements in molecules. Each bond has its own frequency, so the vibrational spectrum yields information on the molecular structure. This same information can now be obtained when molecules are adsorbed on single-crystal surfaces and, when combined with another surface-structure-sensitive technique (e.g., LEED), gives a very powerful combination of surface-structure determination. Vibrational spectroscopy also provides significant information on the identity of the surface species; its geometric orientation; the adsorption site; the adsorption symmetry; the nature of the bonding involved; and, in some cases, bond lengths, bond angles, and bond energies. For example, if CO is adsorbed and we observe the C—O stretching mode the adsorption is molecular, whereas if the individual modes of metal-C and metal-O are observed then dissociation has taken place. In addition, each of these vibrational modes (C, O, and CO) has a different frequency for each bonding site. The intensities also relate to the concentration of each species on the surface.

Electrons scattering off surfaces can lose energy in various ways. One of these ways involves excitation of the vibrational modes of atoms and molecules on the surface. The technique to detect vibrational excitation from surfaces by incident electrons is called high-resolution electron energy loss spectroscopy (EELS). This is the most common type of vibrational spectroscopy used for studying surface–absorbate complexes on single-crystal surfaces.

Experimentally, a highly monoenergetic beam of electrons is directed toward the surface, and the energy spectrum and angular distribution of electrons backscattered from the surface is measured. In a typical experiment the kinetic energy of the incident electron beam is in the range of 1–10 eV. Under these conditions the electrons penetrate only the outermost few layers of the crystal, and the backscattered electrons contain only surface information. The incident electrons, monochromatized typically between 3 and 10 meV ($\sim$25–80 cm^{-1}, 1 meV $= 8.065$ cm^{-1}) and with energy E_i, can lose energy $h\omega$ upon exciting a quantized vibrational mode. These backscattered electrons of energy $E_i - h\omega$ produce the vibrational spectrum. There are several designs of electron monochromator and electron energy analyzers for performing EELS, and one of the most common designs, that of a single-pass 127° cylindrical electrostatic deflector, is shown in Fig. 16. A typical EELS spectrum, that of CO on Rh(111) is shown in Fig. 17.

The sensitivity of EELS in detecting submonolayer quantities of adsorbates on the sample depends on the particular parameters of the spectrometer, the sample, and the

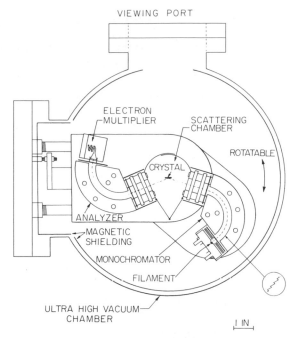

FIGURE 16 Schematic diagram of an EELS spectrometer of the single-pass 127° cylindrical electrostatic deflector type.

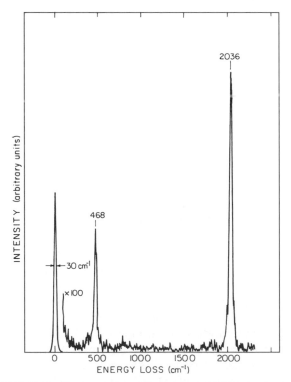

FIGURE 17 Electron energy loss spectrum of CO adsorbed on Rh(111). The loss peak at 468 cm^{-1} is due to the Rh–CO symmetric stretch, and that at 2036 cm^{-1} is due to the C–O symmetric stretch. The spectrum was recorded at a resolution of 30 cm^{-1}.

adsorbate. However, typical sensitivity is quite high due in part to the high inelastic electron cross section. A detection limit of ~10^{-4} monolayers can be achieved for a strong dipole scatterer such as CO. In addition, unlike many other surface spectroscopies, EELS is also capable of detecting adsorbed hydrogen, although at a lower sensitivity (typically 10^{-1}–10^{-2} monolayers). It is a nondestructive technique and can be used to explore the vibrational modes of weakly adsorbed species and those susceptible to beam damage, such as hydrocarbon overlayers.

The spectral range accessible with high-resolution EELS is quite large. Typical experiments examine between 200 and 4000 cm^{-1}, but much larger regions can be analyzed. Vibrational modes as far out as 16,000 cm^{-1} have been examined. Besides fundamentals, energy losses due to overtones, combination bands, and multiple losses are distinguishable.

A distinct advantage of EELS is that electrons can excite the vibrational modes of the surface by three different mechanisms: dipole scattering, impact scattering, and resonance scattering. By analyzing the angular dependence of the inelastically scattered electrons a complete symmetry assignment of the surface–adsorbate complex can be made.

The restrictions on the adsorption system are minimal: ordered or disordered overlayers can be examined, as can either well-structured single crystal samples or optically rough surfaces. Hence, chemisorption on evaporated films can be studied, as can the nature of metal overlayer–semiconductor interactions. In addition, coadsorbed atoms and molecules can be studied without difficulty.

The major disadvantage of EELS, especially compared to optical techniques, is the relatively poor instrumental resolution, which usually varies between 3 and 10 meV (25–80 cm^{-1}). The spectral resolution hinders assignment of vibrations due to individual modes, although peak assignments can be made to within 10 cm^{-1}. The high sensitivity of EELS coupled with the advantages discussed above has encouraged rapid development and use of this technique, despite resolution limitations, such that it has now been used to study hundreds of adsorptions systems.

As an example of the type of surface chemistry that can be followed using EELS, Fig. 18 shows a series of EELS spectra of the adsorption and thermal decomposition of ethylene on Rh(111).

III. THERMODYNAMICS OF SURFACES

A. Introduction

The environment of atoms in a surface is substantially different to that of atoms in the bulk of the solid. Surface

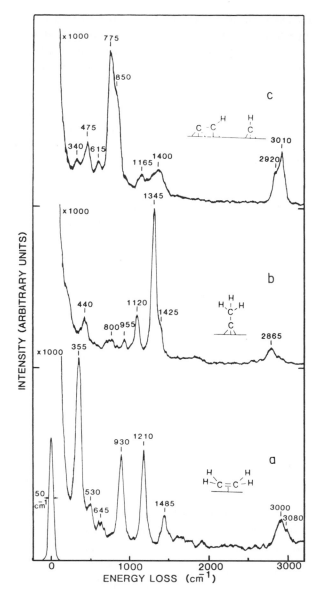

FIGURE 18 EELS spectra of the adsorption and decomposition of ethylene (C_2H_4) on Rh(111) at (a) 77 K, (b) 220 K, and (c) 450 K.

atoms are surrounded by fewer nearest neighbors than bulk atoms, and these neighbors are not distributed evenly around the surface atoms. An atom in the interior experiences no net forces, but these forces become unbalanced at the surface. Consequently the thermodynamic parameters used to describe surfaces are defined separately from those that characterize the bulk phase. The specific surface energy E^S, the energy per surface area, is related to the total energy E by the following equation:

$$E = NE^0 + AE^S, \tag{4}$$

where A is the surface area of a solid composed of N atoms, and E^0 is the energy of the bulk phase per atom.

Therefore, E^S is the excess of total energy that the solid has over E^0, which is the energy that the system would have if the surface were in the same thermodynamic state as the interior.

The other surface thermodynamic functions are defined similarly, for example, the specific surface free energy G^S is given by the following:

$$G^S = H^S - TS^S, \tag{5}$$

where H^S and S^S are the specific surface enthalpy and entropy, respectively.

B. Surface Tension in a One-Component System

Creating a surface involves breaking chemical bonds and removing neighboring atoms, and this requires work. Under conditions of constant temperature and pressure at equilibrium, the surface work δW^S is given by the following:

$$\delta W^S_{T,P} = d(G^S A), \tag{6}$$

where A is the increase in the surface area. If G^S is independent of the surface area, surface work is as follows:

$$\delta W^S_{T,P} = G^S \, dA. \tag{7}$$

In a one-component system the specific surface free energy, G^S, is frequently called the surface tension or surface pressure and is denoted by γ. Here γ may be viewed as a pressure along the surface opposing the creation of new surface. It has dimensions of force per unit length (dynes per centimeter, ergs per square centimeter, or newtons per meter). The surface tension γ for an unstrained phase is also equal to the increase of the total free energy of the system per unit increase of the surface area as follows:

$$\gamma = G^S = \left(\frac{\partial G}{\partial A}\right)_{T,P}. \tag{8}$$

The free energy of formation of a surface is always positive, since work is required in creating a new surface, which increases the total free energy of the system. In order to minimize their free energy solids or liquids assume shapes in equilibrium with the minimum exposed surface area as possible. For example, liquids tend to form a spherical shape and crystal faces which exhibit the closest packing of atoms tend to be the surfaces of lowest free energy of formation and thus the most stable. Surface tension is one of the most important thermodynamic parameters characterizing the condensed phase. Table II lists selected experimentally determined values of surface tensions of liquids and solids that were measured in equilibrium with their vapor.

Comparing the surface tension values of metals and oxides in Table II it can be seen that oxides have in general

TABLE II   Selected Values of Surface Tension of Solids and Liquids

Material	γ (ergs cm^{-2})	T (°C)
He (1)	0.308	−270.5
N$_2$ (1)	9.71	−195
Ethanol (1)	22.75	20
Water	72.75	20
Benzene	28.88	20
n-Octane	21.80	20
Carbon tetrachloride	26.95	20
Bromine	41.5	20
W (s)	2900	1727
Nb (s)	2100	2250
Au (s)	1410	1027
Ag (s)	1140	907
Ag (l)	879	1100
Fe (s)	2150	1400
Fe (l)	1880	1535
Pt (s)	2340	1311
Cu (s)	1670	1047
Ni (s)	1850	1250
Hg (l)	487	16.5
NaCl (s)	227	25
KCl (s)	110	25
CaF$_2$ (s)	450	−195
MgO (s)	1200	25
SiO$_2$ (s)	307	1300
Al$_2$O$_3$ (s)	690	2323
Polytetrafluoroethylene	18.5	20
Polyethylene	31	20
Polystyrene	33	20
Poly(vinyl chloride)	39	20

a low surface tension. Therefore, a reduction in the total free energy of the system can be achieved by oxidation of the surface and a uniform oxide layer covering the surface is expected under conditions near thermodynamic equilibrium. Similarly, deposition and growth of a metal film on a metallic substrate of higher surface tension should yield a uniform layer that is evenly spread to completely cover the substrate surface. Likewise, a very poor spreading of the film is expected on deposition of a metal of high surface tension on a low-surface-tension substrate. This latter condition results in "island growth" and the deposited high-surface-tension metal will grow as whiskers to expose as much of the low-surface-tension substrate during the growth as possible. These, of course, are surface thermodynamic predictions and may be overridden by the presence of impurities at the surface or difficulties of nucleation.

Since atomic bonds must be broken to create surfaces, it is expected that the specific surface free energy will be related to the heat of vaporization, which is related to the energy input necessary to break all the bonds of atoms in the condensed phase. In fact, it has been found experimentally that the molar surface free energy of a liquid metal can be estimated by the following:

$$\gamma_{lm} = 0.15\Delta H_{vap}, \qquad (9)$$

where ΔH_{vap} is the heat of vaporization of the liquid, and the molar surface free energy of a solid metal is given by the following:

$$\gamma_{sm} = 0.16\Delta H_{sub}, \qquad (10)$$

where ΔH_{sub} is the heat of sublimation of the solid.

For other materials, oxides, or organic molecules, such a simple relationship does not work due to the complexity of bonding and the rearrangement or relaxation of surface atoms at the freshly created surfaces.

C. Surface Tension of Multicomponent Systems

In many important surface phenomena, such as heterogeneous catalysis or passivation of the surface by suitable protective coatings, the chemical composition of the topmost layer controls the surface properties and not the composition in the bulk. Thus, investigations of the physical–chemical parameters that control the surface composition are of great importance. One of the major driving forces for the surface segregation of impurities from the bulk and for the change of composition of alloys and other multicomponent systems is the need to minimize the surface free energy of the condensed phase system.

The change of the total free energy of a multicomponent system can be expressed with the inclusion of the surface term as follows:

$$dG = S\,dT + V\,dP + \gamma\,dA + \sum_i \mu_i\,dn_i, \qquad (11)$$

where μ_i is the chemical potential of the ith component and dn_i is the change in the number of moles of the ith component. At constant temperature and pressure, Eq. (11) can be rewritten as follows:

$$dG_{T,P} = \gamma\,dA - \sum_i \mu_i\,dn_i, \qquad (12)$$

where the minus sign indicates the decrease of the bulk concentration of the ith component. Comparing this equation with Eq. (8), the surface tension γ is no longer equal to the specific surface free energy per unit area for a multicomponent system. Using simple arguments in which number of moles of the condensed phase are transferred to the freshly created surface, the Gibbs equation can be derived as follows:

$$d\gamma = -S^S\,dT = \sum \Gamma_i\,d\mu_i, \qquad (13)$$

where Γ_i is the excess number of moles of compound i at the surface. Just like the free energy relations for bulk phases, the Gibbs equation predicts changes in surface tension as a function of experimental variables such as temperature and surface concentration of various components. As a result of the Gibbs equation, the surface composition in equilibrium with the bulk for a multicomponent system can be very different from the bulk composition.

As an example we discuss the surface composition of an ideal binary solution. For such a solution at a constant temperature the Gibbs equation can be expressed as follows:

$$d\gamma_T = -\Gamma_1 \, d\mu_1 - \Gamma_2 \, d\mu_2 \tag{14}$$

and it has been shown that the surface tension of component 1 in an idea dilute solution is given by the following:

$$\gamma = \gamma_1 + (RT/a) \ln \left(X_1^S / X_1^b \right), \tag{15}$$

where γ_1 is the surface tension of the pure component and a the surface area occupied by one mole of component 1. Perfect behavior is assumed, i.e., the surface areas occupied by the molecules in the two different components are the same ($a_1 = a_2 = a$); X_1^S and X_1^b are the atom fractions of component 1 in the surface and in the bulk, respectively. It is also assumed that the surface consists of only the topmost atomic layer. For a two-component system, Eq. (15) can be rewritten in the following form:

$$\frac{X_1^S}{X_2^S} = \frac{X_1^b}{X_2^b} \exp\left[\frac{(\gamma_2 - \gamma_1)a}{RT} \right], \tag{16}$$

where X_1^S, X_2^S, X_1^b, X_2^b have their meaning defined above; γ_1 and γ_2 are the surface tensions of the pure components; and the other symbols have their usual meaning. From Eq. (16), it can be seen that the component that has the smaller surface tension will accumulate on the surface.

Equation (16) also predicts that the surface composition of ideal solutions should be an exponential function of temperature. While the bulk composition of a multicomponent system is little affected by temperature, the surface concentration of the constituents may change markedly.

The surface segregation of one of the constituents becomes more pronounced the larger the difference in surface tensions between the components that make up the solution. Surface segregation is expected to be prevalent for metal solutions, since metals have the highest surface tensions.

In reality, however, metallic alloys are not ideal solutions since they have some finite heat of mixing. In such a case the surface composition can be approximated in the regular solution monolayer approximation

$$\frac{X_2^S}{X_1^S} = \frac{X_2^b}{X_1^b} \exp\left[\frac{(\gamma_1 - \gamma_2)a}{RT} \right] \exp\left\{ \frac{\Omega(l + m)}{RT} \right.$$

$$\left. \times \left[\left(X_1^b \right)^2 - \left(X_2^b \right)^2 \right] + \frac{\Omega l}{RT} \left[\left(X_2^S \right)^2 - \left(X_1^S \right)^2 \right] \right\}, \tag{17}$$

where Ω is the regular solution parameter and is directly related to the heat of mixing ΔH_m by the following:

$$\Omega = \frac{\Delta H_m}{X_1^b \left(l - X_1^b \right)}.$$

Here l is the fraction of nearest neighbors to an atom in the plane and m is the fraction of nearest neighbors below the plane containing the atom. In this approximation the surface composition becomes a fairly strong function of the heat of mixing, its sign, and its magnitude in addition to the surface tension difference and temperature.

Auger electron spectroscopy (AES) and ion scattering spectroscopy (ISS) are two experimental techniques which are most frequently used for quantitative determination of the surface composition. Figure 19 shows the surface atom fraction of gold, determined by AES and ISS, plotted as a function of the bulk atom fraction for the Ag–Au system. The solid line gives the calculated surface composition using the regular solution model and the dashed line indicates the curve that would be obtained in the absence of surface enrichment. The regular solution model appears to overestimate somewhat the surface segregation in this case, although the surface is clearly enriched in silver.

Table III lists several binary alloy systems that have been investigated experimentally by AES or ISS and

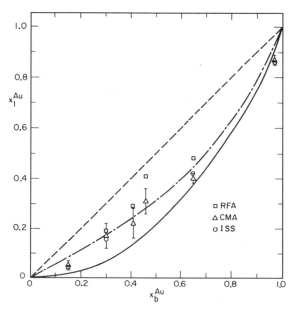

FIGURE 19 Surface phase diagram of Au–Ag alloy.

TABLE III Surface Composition of Alloys: Experimental Results and Predictions of the Regular Solution Model

Alloy system	Segregating constituent	
	Predicted regular solution	Experimental
Ag–Pd	Ag	Ag
Ag–Au	Ag	Ag
Au–Pd	Au	Au
Ni–Pd	Pd	Pd
Fe–Cr	Cr	Cr
Au–Cu	Cu	Au, none, or Cu depending on composition
Cu–Ni	Cu	Cu
Au–Ni	Au	Au
Au–Pt	Au	Au
Pb–In	Pb	Pb
Au–In	In	In
Al–Cu	Al	Al
Pt–Sn	Sn	Sn
Fe–Sn	Sn	Sn
Au–Sn	Sn	Sn

the segregating components that were experimentally observed and also predicted by the regular solution model. The agreement is certainly satisfactory. It appears that for binary metal alloy systems that exhibit regular solution behavior there are reliable methods to predict surface composition.

So far we have discussed the surface composition of multicomponent systems that are in equilibrium with their vapor or in which clean surface–bulk equilibrium is obtained in ultrahigh vacuum. In most circumstances, however, the surface is covered with a monolayer of adsorbates that frequently form strong chemical bonds with the surface atoms. This solid–gas interaction can markedly change the surface composition in some cases. For example, carbon monoxide, when adsorbed on the surface of a Ag–Pd alloy, forms much stronger bonds with Pd. While the clean surface is enriched with Ag, in the presence of CO, Pd is attracted to the surface to form strong carbonyl bonds. When the adsorbed CO is removed, the composition returns to its original Ag-enriched state. Nonvolatile adsorbates, such as carbon or sulfur, may have a similar influence on the surface composition as long as their bonding to the various constituents of the multicomponent system is different.

Adsorbates should therefore be viewed as an additional component of the multicomponent system. A strongly interacting adsorbate converts a binary system to a ternary

system. As a result, the surface composition may markedly change with changing ambient conditions.

The mechanical properties of solids, embrittlement, and crack propagation, among others, depend markedly on the surface composition. These studies indicate that the surface composition and the mechanical properties of structural steels may change drastically when the ambient conditions are changed from reducing to oxidizing environments.

D. Equilibrium Shape of a Crystal or a Liquid Droplet

In equilibrium the crystal will take up a shape that corresponds to a minimum value of the total surface free energy. In order to have the equilibrium shape, the integral $\int \gamma \, dA$ over all surfaces of the crystal must be a minimum. Crystal faces that have high atomic density have the lowest surface free energy and are therefore most stable. The plot of the surface free energy as a function of crystal orientation is called the γ plot.

Solids and liquids will always tend to minimize their surface area in order to decrease the excess surface free energy. For liquids, therefore, the equilibrium surface becomes curved where the radius of curvature will depend on the pressure difference on the two sides of the interface and on the surface tension as follows:

$$(P_{in} - P_{ext}) = 2\gamma/r, \qquad (18)$$

where P_{in} and P_{ext} are the internal and external pressures, respectively, and r is the radius of curvature. In equilibrium a pressure difference can be maintained across a curved surface. The pressure inside the liquid drop or gas bubble is higher than the external pressure, because of the surface tension. The smaller the droplet or larger the surface tension, the larger is the pressure difference that can be maintained. For a flat surface $r = \infty$, and the pressure difference normal to the interface vanishes.

Let us now consider how the vapor pressure of a droplet depends on its radius of curvature r. We obtain the following:

$$\ln(P/P_0) = 2\gamma V_{in}/RTr, \qquad (19)$$

where V_{in} is the internal volume. This is the well-known Kelvin equation for describing the dependence of the vapor pressure of any spherical particle on its size. Small particles have higher vapor pressures than larger ones. Similarly, very small particles of solids have greater solubility than large particles. If we have a distribution of particles of different sizes, we will find that the larger particles will grow at the expense of the smaller ones. Nature's way to avoid the sintering of small particles that would occur according to the Kelvin equation is to produce a system with particles of equal size. This is the world of colloids where particles are of equal size and therefore stabilized and are usually charged either all negative or all positive or to repel each other by long-range electrostatic forces. Milk and our blood are only two examples of systems that contain colloids.

E. Adhesion and the Contact Angle

Let us turn our attention to the interfacial tension, that is, the surface tension that exists at the interface of two condensed phases. Let us place a liquid droplet on a solid surface. The droplet either retains its shape and forms a curved surface or it is spread evenly over the solid. These two conditions indicate the lack of wetting or wetting of the solid by the liquid phase, respectively. The contact angle between the solid and the liquid, to a large extent, permits us to determine the interfacial tension between the solid and the liquid. The contact angle is defined by Fig. 20. If the contact angle is large (Θ approaching $90°$), the liquid does not readily wet the solid surface. If Θ approaches zero, complete wetting of the solid surface takes place. For Θ larger than $90°$ the liquid tends to form spherical droplets on the solid surface that may easily run off, i.e., the liquid does not wet the solid surface at all. Remembering that the surface tension always exerts a pressure tangentially along a surface, the surface free energy balance between the surface forces acting in opposite directions at the point where the three phases solid, liquid, and gas meet is given by the following:

$$\cos \Theta = (\gamma_{sg} - \gamma_{sl})/\gamma_{lg}. \quad (20)$$

Here γ_{lg} is the interfacial tension at the liquid–gas interface and γ_{sg} and γ_{sl} are the interfacial tensions between the solid–gas and the solid–liquid interfaces, respectively. Knowing γ_{lg} and the contact angle in equilibrium at the solid–liquid–gas interface, we can determine the difference $\gamma_{sg} - \gamma_{sl}$ but not their absolute values. Since the wet-

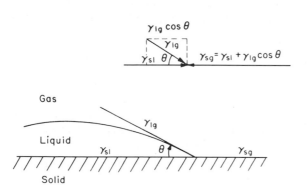

FIGURE 20 Definition of the contact angle between a liquid and solid and the balance of surface forces at the contact point among the three phases (solid, vapor, and liquid).

ting ability of the liquid at the solid interface is so important in practical problems of adhesion or lubrication, there is a great deal of work being carried out to determine the interfacial tensions for different combinations of interfaces.

The usefulness of a lubricant is determined by the extent to which it wets the solid surface and maintains complete coverage of the surface under various conditions of use. The strength of an adhesive is determined by the extent to which it lowers the surface free energy by adsorption on the surface. The work of adhesion is defined as follows:

$$W_A^s = \gamma_{l,0} + \gamma_{s,0} - \gamma_{sl}, \quad (21)$$

where $\gamma_{l,0}$ and $\gamma_{s,0}$ are the surface tensions in vacuum of the liquid and solid, respectively. In general, solids and liquids that have large surface tension form strong adhesive bonds, i.e., have large work of adhesion. The work of adhesion is in the range of 40–150 ergs/cm^2 for solid–liquid pairs of various types. Organic polymers often make excellent adhesives because of the large surface area covered by each organic molecule. The adhesive energy per mole is much larger than that for adhesion between two metal surfaces or between a liquid and a solid metal because of the many chemical bonds that may be formed between the substrate and the adsorbed organic molecule.

F. Nucleation

Another important phenomenon that owes its existence to positive surface free energy is nucleation. In the absence of a condensed phase, it is very difficult to nucleate one from vapor atoms because the small particles that would form have a very high surface area and dispersion and, as a result, a very large surface free energy. The total energy of a small spherical particle has two major components: its positive surface free energy, which is proportional to $\pi r^2 \gamma$, where r is the radius of the particle, and its negative free energy of formation of the particle with volume V. The volumetric energy term is proportional to $-r^3 \ln (P/P_{eq})$, where P is the pressure over the system and P_{eq} is the equilibrium vapor pressure:

$$\Delta G(\text{total}) = -\left(\frac{4\pi r^3}{3V_m}\right) RT \ln \left(\frac{P}{P_{eq}}\right) + 4\pi r^2 \gamma, \quad (22)$$

where V_m is the molar volume of the forming particle. Initially, the atomic aggregate is very small and the surface free energy term is the larger of the two terms. In this circumstance a condensate particle cannot form from the vapor even at relatively high saturation ($P > P_{eq}$). Similarly, a liquid may be cooled below its freezing point without solidification occurring.

Above a critical size of the spherical particles the volumetric term becomes larger and dominates since it

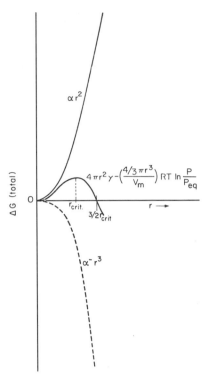

FIGURE 21 Free energy of homogeneous nucleation as a function of particle size.

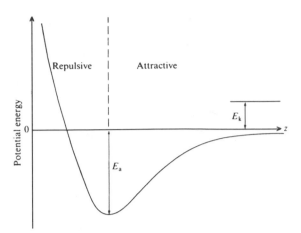

FIGURE 22 One-dimensional potential energy of an adatom in a physisorbed state on a planar surface as a function of its distance z from the surface.

decreases as $\sim r^3$, while the surface free energy term increases only as $\sim r^2$. Therefore, a particle that is larger than this critical size grows spontaneously at $P > P_{eq}$. This is shown in Fig. 21. Because of the difficulty of obtaining this critical size, which involves as many a 30–100 atoms or molecules, homogeneous nucleation is very difficult indeed. To avoid this problem we add to the system particles of larger than critical size that "seed" the condensation or solidification. The use of small particles to precipitate water vapor in clouds to start rain and the use of small crystallites as seeds in crystal growth are two examples of the application of heterogeneous nucleation.

G. Physical and Chemical Adsorption

The concepts of physical adsorption (physisorption) and chemical adsorption (chemisorption) were introduced above. The nature of the two classifications is linked to the heat of adsorption, ΔH_{ads}, which is defined as the binding energy of the adsorbed species. Physical adsorption is caused by secondary attractive forces (van der Waals) such as dipole–dipole interaction and induced dipoles and is similar in character to the condensation of vapor molecules onto a liquid of the same composition.

The interaction can be described by the one-dimensional potential energy diagram shown in Fig. 22.

An incoming molecule with kinetic energy E_k must lose at least this amount of energy in order to stay on the surface. It loses energy by exciting lattice phonons in the substrate, for example, and the molecule comes to equilibrium in a state of oscillation in the potential well of depth equal to the binding energy or adsorption energy $E_a = \Delta H_{ads}$. In order to leave the surface (desorb) the molecule must acquire enough energy to surmount the potential-energy barrier E_a. The desorption energy is equal to the adsorption energy. The binding energies of physisorbed molecules are typically ≤ 15 kcal mol^{-1}.

Chemisorption involves chemical bonding; it is similar to a chemical reaction and involves transfer of electronic charge between adsorbent and adsorbate. The most extreme form of chemisorption occurs when integral numbers of electrons are transferred, forming a pure ionic bond. More usually there is an admixture of the wave functions of the valence electrons of the molecule with the valence electrons of the substrate into a new wave function. The electrons responsible for the bonding can then be thought of as moving in orbitals between substrate and adatoms and a covalent bond has been formed. Two examples of the potential energy diagrams for chemisorption are shown in Fig. 23. Some of the impinging molecules are accommodated by the surface and become weakly bound in a physisorbed state (also called a precursor state) with binding energy E_p. During their stay time in this state, electronic or vibrational processes can occur which allow them to surmount the energy barrier, and electron exchange occurs between the adsorbate and substrate. The molecule, or adatom in the case of dissociative chemisorption, now finds itself in a much deeper well; it is chemisorbed. Figure 23a shows the case in which the energy barrier for chemisorption is less than E_p, so there is no overall activation energy to chemisorption. Figure 23b illustrates

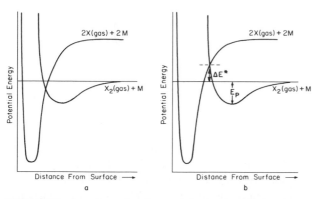

FIGURE 23 One-dimensional potential energy curves for dissociative adsorption through a precursor or physisorbed state: (a) adsorption into the stable state with no activation energy and (b) adsorption into the chemisorption well with activation energy ΔE^*.

the case in which there is an overall activation energy ΔE^* to chemisorption. In the former case the activation energy for desorption E_d is equal to the heat of adsorption, while in the latter case the heat of adsorption is given by the difference between the heat of desorption and the activation energy. The occurrence of an activation energy to chemisorption is by far the exception rather than the rule.

From these considerations it is expected that to a first approximation physisorption will be nonspecific, any gas will adsorb on any solid under suitable circumstances. However, chemisorption will show a high degree of specificity. Not only will there be variations from metal surface to metal surface, as would be expected from the differences in chemistries between the metals, but also different surface planes of the same metal may show considerable differences in reactivity toward a particular gas.

H. Adsorption Isotherms

An adsorption isotherm is the relationship at constant temperature between the partial pressure of the adsorbate and the amount adsorbed at equilibrium. Similarly an adsorption isobar expresses the functional relationship between the amount adsorbed and the temperature at constant pressure, and an adsorption isostere relates the equilibrium pressure of the gaseous adsorbate to the temperature of the system for a constant amount of adsorbed phase. Usually it is easiest from an experimental viewpoint to determine isotherms. The coordinates of pressure at the different temperatures for a fixed amount adsorbed can then be interpolated to construct a set of isosteres, and similarly to obtain an isobaric series.

Adsorption isotherms can be used to determine thermodynamic parameters that characterize the adsorbed layer (heats of adsorption and the entropy and heat capacity changes associated with the adsorption process), and in

the case of adsorption isotherms for physical adsorption, to determine the surface area of the adsorbing solid.

Consider a uniform surface with a number n_0 of equivalent adsorption sites. The ratio of the number of adsorbed atoms or molecules n to n_0 is defined as the coverage, $\Theta = n/n_0$. Atoms or molecules impinge on the surface from the gas phase, where they establish a surface concentration $[n_a]_s$ (molecules per square centimeter). Assuming that only one type of species of concentration $[n_a]_g$ (molecules per cubic centimeter) exists in the gas phase the adsorption process can be written as follows:

$$A_{gas} \underset{k'}{\overset{k}{\rightleftharpoons}} A_{surface}$$

and the net rate of adsorption as

$$F \text{ (molecules cm}^{-2}\text{ sec}^{-1}) = k[n_a]_g - k'[n_a]_s, \quad (23)$$

where k and k' are the rate constants for adsorption and desorption, respectively. Starting with a nearly clean surface far from equilibrium, the rate of desorption may be taken as zero and Eq. (23) becomes the following:

$$F \text{ (molecules cm}^{-2}\text{ sec}^{-1}) = k[n_a]_g, \quad (24)$$

where k, derived from the kinetic theory of gases, equals $\alpha(RT/2\pi M)^{1/2}$ cm sec^{-1}, α is the adsorption coefficient, and M the molecular weight of the impinging molecules. The surface concentration $[n_a]_s$ under these conditions is the product of the incident flux F and the surface residence time τ:

$$[n_a]_s = F\tau. \quad (25)$$

Here τ is the surface residence time, given by:

$$\tau = \tau_0 \exp(\Delta H_{ads}/RT). \quad (26)$$

Replacing $[n_a]_g$ by the pressure using the ideal gas law, Eq. (25) can be rewritten as follows:

$$[n_a]_s = \frac{\alpha P N_A}{(2\pi MRT)^{1/2}} \tau_0 \exp\left(\frac{\Delta H_{ads}}{RT}\right). \quad (27)$$

The simplest adsorption isotherm is obtained from Eq. (27), which can be rewritten as

$$\Theta = k'' P \quad (28)$$

where

$$k'' = \frac{1}{n_0} \frac{\alpha N_A}{(2\pi MRT)^{1/2}} \tau_0 \exp\left(\frac{\Delta H_{ads}}{RT}\right). \quad (29)$$

The coverage is proportional to the first power of the pressure at a given temperature provided that there are an unlimited number of adsorption sites available and ΔH_{ads} does not change with coverage.

The isotherm of Eq. (28) is unlikely to be suitable to describe the overall adsorption process, but the Langmuir isotherm is a simple modification which represents a more real situation. The Langmuir isotherm assumes that adsorption is terminated on completion of one molecular adsorbed gas layer (monolayer) by asserting that any gas molecule that strikes an adsorbed atom must reflect from the surface. All the other assumptions used to derive Eq. (28) are maintained (i.e., homogeneous surface and noninteracting adsorbed species). If $[n_0]$ is the surface concentration of a completely covered surface, the number of surface sites available for adsorption, after adsorbing $[n_a]_s$ molecules is $[n_0] - [n_a]_s$. Of the total flux incident on the surface, a fraction $([n_a]_s/[n_0])F$ will strike molecules already adsorbed and, therefore, be reflected. Thus, a fraction $(1 - [n_a]_s/[n_0])F$ of the total incident flux will be available for adsorption. Equation (25) should then be modified as follows:

$$[n_a]_s = \left(1 - \frac{[n_a]_s}{[n_0]}\right)F\tau, \tag{30}$$

which can be rearranged to give

$$[n_a]_s = \frac{[n_0]F\tau}{[n_0] + F\tau} = \frac{[n_0]kP}{[n_0] + kP} \tag{31}$$

from which

$$\Theta = \frac{k'P}{1 + k'P}, \tag{32}$$

where $k' = k/[n_0]$.

Equation (32) is the Langmuir isotherm. The adsorption of CO on Pd(111) obeys the Langmuir isotherm, and typical isotherms from this system are shown in Fig. 24. It can readily be shown that in the case of dissociative adsorption the Langmuir isotherm becomes

$$P = \frac{1}{k'}\left(\frac{\Theta}{1 - \Theta}\right)^2$$

or

$$\Theta = \frac{(k'P)^{1/2}}{1 + (k'P)^{1/2}}. \tag{33}$$

A clear weakness of the Langmuir model is the assumption that the heat of adsorption is independent of coverage. Several other isotherms have been developed which are all modifications of the Langmuir model. For example, the Temkin isotherm can be derived if a linearly declining heat of adsorption is assumed, i.e., $\Delta H = \Delta H_0(1 - \beta\Theta)$, where ΔH_0 is the initial enthalpy of adsorption. The isotherm is

$$\Theta = \frac{RT}{\beta\Delta H_0}\ln AP, \tag{34}$$

where A is a constant related to the enthalpy of adsorption.

The possibility of multilayer adsorption is envisaged in the Brunauer–Emmett–Teller (BET) isotherm. The assumption is made that the first layer is adsorbed with a heat of adsorption H_1 and the second and subsequent layers are all characterized by heats of adsorption equal to the latent heat of evaporation, H_L. By considering the dynamic equilibrium between each layer and the gas phase the BET isotherm is obtained,

$$\frac{p}{V(p_0 - p)} = \frac{1}{V_m c} + \frac{c - 1}{V_m c}\frac{p}{p_0}. \tag{35}$$

In this equation V is the volume of gas adsorbed, p the pressure of gas, p_0 the saturated vapor pressure of the liquid at the temperature of the experiment, and V_m the

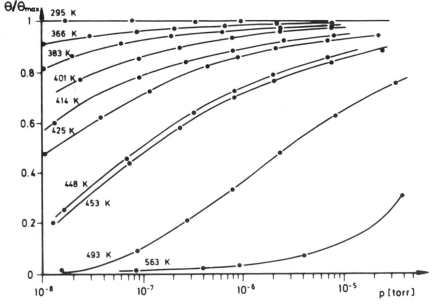

FIGURE 24 Adsorption isotherms for CO on Pt(111) single-crystal surfaces.

volume equivalent to an adsorbed monolayer. The BET constant c is given by the following:

$$c = \exp(H_1 - H_L)/RT. \tag{36}$$

The BET equation owes its importance to its wide use in measuring surface areas, especially of films and powders. The method followed is to record the uptake of an inert gas (Kr) or nitrogen at liquid nitrogen temperature ($-195.8°C$). A plot of $p/V(p_0 - p)$ versus p/p_0, usually for p/p_0 up to about 0.3, yields V_m, the monolayer uptake. This value is expressed as an area by assuming that the area per molecule for nitrogen is 16.2 Å and 25.6 Å for krypton.

In general, the BET isotherm is most useful for describing physisorption for which H_1 and H_L are of the same order of magnitude while the preceding isotherms are more useful for chemisorption. It is worth noting that the BET isotherm reduces to the Langmuir isotherm when $H_1 \gg H_L$.

I. Heat of Adsorption

An important physical–chemical property that characterizes the interaction of solid surfaces with gases is the bond energy of the adsorbed species. The determination of bond energy is usually made indirectly by measuring the heat of adsorption (or heat of desorption) of the gas. The heat of adsorption can be determined readily in equilibrium by measuring several adsorption isotherms. The Clausius–Clapeyron equation

$$\left(\frac{\partial(\ln P)}{\partial T}\right)_\Theta = \frac{\Delta H_{ads}}{RT^2} \tag{37}$$

can be integrated to give

$$\ln\left(\frac{P_1}{P_2}\right)_\Theta = \frac{-\Delta H_{ads}}{R}\left(\frac{1}{T_1} - \frac{1}{T_2}\right).$$

Measuring the adsorption isotherm at two different temperatures, provided that proper equilibrium is established between the adsorbed and gas phase, yields the heat of adsorption.

The heat of adsorption can also be obtained by direct calorimetry. The method most commonly used consists of measuring the temperature rise caused by the addition of a known amount of gas to a film of the metal prepared by evaporation *in vacuo*. This measurement will yield the differential heat of adsorption q_d at the particular value of Θ. The differential heat of adsorption is related to the isosteric heat of adsorption by the following:

$$q = q_d + RT; \tag{38}$$

the difference is only RT, which is within experimental error.

The last, and most common, method of determining the heat of adsorption is a kinetic method called temperature programmed desorption (TPD). The method is as follows. The sample is cleaned in ultrahigh vacuum and a gas is allowed to adsorb on the surface at known pressures while the surface is kept at a fixed temperature. The sample is then heated at a controlled rate, and the pressure changes during the desorption of the molecules are recorded as a function of time and temperature. The pressure–temperature profile is usually referred to as the desorption spectrum. The desorption rate $F(t)$ is commonly expressed as follows:

$$F(t) = vf(\sigma)\exp\left(-\frac{E_{des}}{RT}\right), \tag{39}$$

where v is the preexponential factor and $f(\sigma)$ an adsorbate concentration-dependent function. The various procedures for determining these parameters are well described in the literature.

Assuming that v and E_{des} are independent of the adsorbate concentration σ and t, E_{des} can be obtained for zeroth-, first-, and second-order desorption, respectively, as follows:

$$\frac{E_0}{R} = \frac{v_0}{\sigma a}\exp\left(-\frac{E}{RT_p}\right) \tag{40}$$

$$\frac{E_1}{RT_p^2} = \frac{v_1}{\alpha}\exp\left(-\frac{E_1}{RT_p}\right) \tag{41}$$

$$\frac{E_2}{RT_p^2} = \frac{v_2\sigma}{\alpha}\exp\left(-\frac{E_2}{RT_p}\right), \tag{42}$$

where T_p is the temperature at which a desorption peak is at a maximum and σ is the initial adsorbate concentration. The subscripts 0, 1, and 2 denote the zeroth-, first-, or second-order desorption processes; α is a constant of proportionality for the temperature rise with time, usually $T = T_0 + \alpha t$; that is, the temperature of the sample is raised linearly with time. As seen from the equations, T_p is independent of σ for the first-order desorption process. Alternatively, T_p is increased or is decreased with σ_0 for the zeroth- and second-order process, respectively. Equations (40)–(42) allow us to determine the activation energy and the preexponential factor and also to distinguish between zeroth-, first-, and second-order desorption processes from the measurements of the dependence of the peak temperatures on initial adsorbate concentrations and heating rate α. A typical TPD spectrum is shown in Fig. 25.

The bond energy ΔH_{bond} is readily extracted from the heat of adsorption. In the case of the chemisorption of a diatomic molecule X_2 onto a site on a uniform solid surface M the molecule may adsorb without dissociation

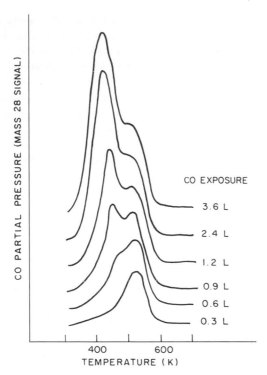

FIGURE 25 Typical thermal desorption spectra of CO from a Pt(553) stepped crystal face as a function of coverage. The two peaks are indicative of CO bonding at step and terrace sites. The higher temperature peak corresponds to CO bound at step sites.

to form MX_2. In this case, the heat of adsorption, ΔH_{ads} is defined as the energy needed to break the M—X_2 bond

$$MX_{2(ads)} \xrightarrow{\Delta H_{ads}} M + X_{2(gas)}$$

If the molecule adsorbs dissociatively, the heat of adsorption is defined as follows:

$$2MX_{(ads)} \xrightarrow{\Delta H_{ads}} 2M + X_{2(gas)}$$

The energy of the surface chemical bond is then given by

$$\Delta H_{bond}(MX_2) = \Delta H_{ads}$$

for associative adsorption or

$$\Delta H_{bond}(MX) = \left(\frac{\Delta H_{ads} + D_{X_2}}{2} \right)$$

for dissociative adsorption, where D_{X_2} is the dissociation energy of the X_2 gas molecule.

The heat of adsorption is not a constant, quantity for a particular adsorbate–substrate system; there are several factors which affect the value of ΔH_{ads}. First, the heat of adsorption can change markedly with the coverage Θ of the adsorbed pahse. An example of this is shown in Fig. 26 for CO on a Pd(111) surface. Decreasing values of ΔH_{ads}

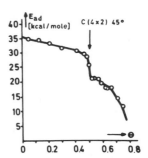

FIGURE 26 Isoteric heat of adsorption for CO on Pd(111) crystal face as a function of coverage.

with increasing adsorbate coverage are commonly observed due to repulsive adsorbate–adsorbate interactions.

Second, the surface is heterogeneous by nature. There are many sites where the adsorbed species have different binding energies. Perhaps the most striking effect is that for adsorption on stepped and kinked platinum and nickel single crystal surfaces where molecules dissociate in the presence of these surface irregularities while they remain intact on the smooth low-Miller-index surfaces. If a polycrystalline surface is utilized for chemisorption studies instead of a structurally well-characterized single-crystal surface the measured ΔH_{ads} will be an average of adsorption at the various binding sites. In fact, even on the same crystal surface molecules may occupy several different adsorption sites with different coordination numbers and rotational symmetries, and each site may exhibit a different binding energy and therefore a different heat of chemisorption. For example, on the (111) face of fcc metals the adsorbates may occupy a three-fold site, a twofold bridge site, or an on-top site. Figure 27 shows the measured heats of adsorption of CO or single-crystal surfaces for many different transition metals while Fig. 28 shows the heats of adsorption of CO on polycrystalline transition metal surfaces. The heats of chemisorption on single-crystal planes indicate the presence of binding sites on a given surface which differ by ~20 kcal mol^{-1}. It is not possible to identify one value of the heat of chemisorption of an adsorbate on a given transition metal unless the binding state is specified or it is certain that only one binding state exists. A polycrystalline surface however exhibits all the adsorption sites of the faces from which it is composed. Since these sites are present simultaneously heats of chemisorption for these surfaces represent an average of the binding energies of the different surface sites. As a result the measured heats of adsorption of Fig. 28 do not show the large structural variations that can be seen in Fig. 27.

The adsorbate may also change bonding as a function of temperature as well as the adsorbate concentration. For example, oxygen may be molecularly adsorbed at low temperatures while it dissociates at higher temperatures.

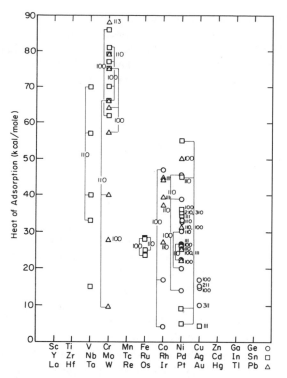

FIGURE 27 Heats of adsorption of CO on single-crystal surfaces of transition metals.

IV. ELECTRICAL PROPERTIES OF SURFACES

A. Introduction

Many of the physical and chemical properties of solid surfaces are directly influenced by the concentration of mobile charge carriers (electrons and diffusing ions). The concentration of these free charge carrier varies widely

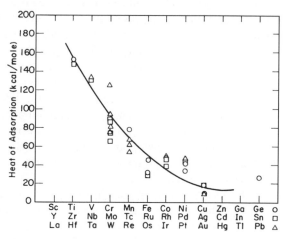

FIGURE 28 Heats of adsorption of CO on polycrystal-line transition-metal surfaces.

for materials of different types. Metals, which are good conductors of electricity with resistivities in the 10^{-4} Ω m range, have large free electron concentrations; almost every atom contributes one electron to the lattice as a whole. For insulators, with a resistivity of 10^9 Ω m, and semiconductors with intermediate values, often less than 1 of every 10^6 atoms may contribute a free electron. The temperature dependence of the carrier concentration and the conductivity may be different for different materials depending on the mechanism of excitation by which the mobile charge carriers are created.

Under incident radiation or bombardment by an electron beam surfaces emit photons, electrons, or both. The emission properties of solid surfaces differ widely, just as their mechanisms or relaxation after excitation by high-energy radiation differ. Many surface-sensitive experimental techniques providing information related to the electronic properties of surfaces are based on these processes, for example, Auger electron spectroscopy (AES), X-ray photoelectron spectroscopy (XPS), and ultraviolet photoelectron spectroscopy (UPS). These are discussed below.

The underlying reason for the differences of the conductivity mechanisms and emission properties on the surfaces of the different materials lies in the differences in their electronic band structure. The band structure model of solids has been successful in explaining many solid-state properties, and we may apply it with confidence in studies of solid surfaces. There are many excellent textbooks on the subject of solid-state physics giving detailed descriptions of the band theory of solids, and a description is not presented here. In the following section a basic understanding of electron bands is assumed.

B. The Energy Level Diagram

For many purposes, in analyzing the electrical properties of metals or semiconductors, we are not concerned with the detailed shape of the electronic bands. We may conveniently represent schematically the electronic bands by straight lines where the potential energy of the electron near the top of the valence band and at the bottom of the conduction band is plotted against distance x through the crystal starting from the surface ($x = 0$). The energy gap represents the minimum potential energy difference between the two bands. In this type of diagram the electron energy increases upward and the energy of the positive hole increases downward, as indicated in Fig. 29. For a homogeneous crystal the bands may be horizontal, as shown in this figure. At the surface the bands may vary in energy with respect to their value in the bulk of the solid since the free carrier concentrations at the surface may be different from those in the bulk of the crystal.

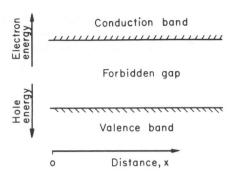

FIGURE 29 Energy-level diagram as a function of distance x from the surface ($x = 0$).

C. Surface Dipole and Surface Space Charge

The anisotropic environment of surface atoms not only gives rise to such processes as surface relaxation and surface reconstruction but also to a redistribution of charge density. For a metal this redistribution can be explained as follows. In the bulk of a metal each electron lowers its energy by "pushing" the other electrons aside to form an "exchange correlation hole." This attractive interaction V_{exch} is lost when the electron leaves the solid, so there is a sharp potential barrier V_s at the surface. In a quantum mechanical description, the electrons are not totally trapped at the surface and there is a finite probability for them to spread out into the vacuum. This is depicted in Fig. 30. This charge redistribution induces a surface dipole V_{dip} that modifies the barrier potential. The work function ϕ (which will be discussed in detail below) is the minimum energy necessary to remove an electron at the Fermi energy E_F from the metal into the vacuum. The magnitude of this induced surface dipole is different at various sites on the heterogeneous metal surface. For example, a step site on a tungsten surface has dipole of 0.37 Debye (D) per edge atom as measured by work function studies. At a tungsten adatom on the surface there is a dipole moment of 1 D.

At semiconductor and insulator surfaces the separation of negative and positive charges leads to the formation of a space-charge region. This space-charge region near the surface is formed by the accumulation or depletion

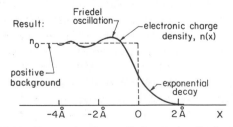

FIGURE 30 Charge density oscillation and redistribution at a metal–vacuum interface.

of charge carriers in the surface with respect to the bulk carrier concentration. Such a space charge may also be induced by the application of an external electric field or by the presence of a charged layer on the surface such as adsorbed ions or electronic surface states which act as a source or sink of electrons. The height of the surface potential barrier V_s and its distance of penetration into the bulk, d, depend on the concentration of mobile charge carriers in the surface region.

It can be shown that

$$d \approx \left(\frac{2 \varepsilon \varepsilon_0 V_s}{e n_e (\text{bulk})} \right)^{1/2}, \tag{43}$$

where ε is the dielectric constant in the solid, ε_0 the permittivity of free space, and $n_e(\text{bulk})$ the bulk carrier concentration. The higher the free carrier concentration in the material, the smaller is the penetration depth of the applied field into the medium. Using a typical value of $\varepsilon = 16$, for electron concentrations of 10^{17} cm^{-3} or larger, the space charge is restricted to distances on the order of one atomic layer or less. This is due to the large free carrier density screening the solid from the penetration of the electrostatic field caused by the charge imbalance. In most metals almost every atom contributes one free valence electron and since the typical atomic density is of the order of 10^{22} cm^{-3} the free carrier concentration in metals is in the range of 10^{20}–10^{22} cm^{-3}. Thus, V_s and d are so small that they can usually be neglected. For semiconductors, or insulators on the other hand, typical free carrier concentrations at room temperature are in the range of 10^{10}–10^{16} cm^{-3}. Therefore, at the surfaces of these materials, there is a space-charge barrier of appreciable height (several electron volts) and penetration depth that could extend over thousands of atomic layers ($\approx 10^4$ Å) into the bulk. This is the reason for the sensitivity of semiconductor devices to ambient changes that affect the space-charge barrier height. There is an induced electric field at the surface under most experimental conditions due to the adsorption of gases or because of the presence of electronic surface states. The electronic and many other physicochemical properties of semiconductor and insulator surfaces depend very strongly on the properties of the space charge. For example, the conduction of free carriers across the solid or along its surface could become space-charge-limited. The rate of charge transfer from the solid to the adsorbed gas, which results in chemisorption or chemical reaction, can become limited by the transfer rate of electrons over the space-charge barrier.

When the energy level diagram was introduced, it was assumed that the electron energy levels remained unchanged right to the surface ($x = 0$). However, the presence of the space charge (and also surface states) leads to a bending of the bands. If the surface region becomes

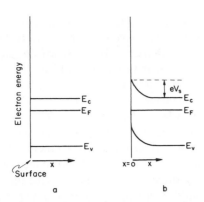

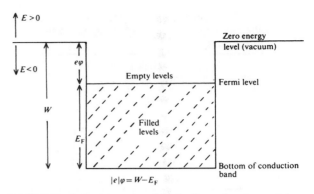

FIGURE 31 Energy-level diagram (a) in the absence of any space charge and (b) with a surface space charge due to depletion of electrons in the surface region.

FIGURE 32 Potential energy diagram illustrating the work function. E_F is the Fermi energy, ϕ is the work function, and W is the potential well bonding the conduction band electrons into the solid.

depleted of electrons it would require more energy to transfer an electron to the conduction band from, for example, the reference state E_F, due to the space charge potential barrier. This is depicted schematically in Fig. 31. Conversely, it is now easier to transfer a hole to the surface since the difference between E_F and E_V becomes smaller.

It is very likely that there is accumulation or depletion of charges at semiconductor or insulator surfaces under all ambient conditions. For surfaces under atmospheric conditions, adsorbed gases or liquid layers at the interface provide trapping of charges or become the source of free carriers. For clean surfaces in ultrahigh vacuum, there are electronic surface states that act as traps or sources of electrons and produce a space-charge layer of appreciable height. Thus, the mobile carriers from the surface layer are swept into the interior or are trapped at the surface as the space-charge layer consists dominantly of static charges, the one most frequently encountered in experimental situations.

We have so far considered the space-charge layer properties only in the insulating solid, assuming that the surface layer that acts as a donor or the electron trap is of monolayer thickness. However, considering the properties of solid–liquid interfaces or semiconductor–insulator contacts, it should be recognized that the space-charge layer may extend to effective Debye lengths on both sides of the interface. This is a most important consideration when we investigate the surface properties of colloid systems or of semiconductor–electrolyte interfaces.

D. Work Function and Contact Potential

The work function of a solid is a fundamental physical property of the solid which is related to its electronic structure. It is defined as the potential that an electron at the Fermi level must overcome to reach the level of zero kinetic energy in the vacuum. In semiconductors and insulators it can be regarded as the difference in energy between an electron at rest in the vacuum just outside the solid (i.e., at the level of zero kinetic energy) and the most loosely bound electrons in the solid. Thus, the work function is evidently an important parameter in situations where electrons are removed from the solid.

A schematic energy level diagram assuming the free-electron model of a metal showing the work function is depicted in Fig. 32.

From the figure it can be seen that the value of ϕ depends on W, the depth of the potential well bonding the conduction electrons into the solid. Here W is a bulk property determined by the attraction for its electrons of the lattice of positive ions as a whole; it has an energy of the order of a few electron volts.

The origin of the work function itself can be considered as being due to the image potential of the escaping electron. Electrostatic theory shows that a charge $-e$ outside a conductor is attracted by an image charge $+e$ placed at the position of the optical image of $-e$ in the conducting plane. If $-e$ is a distance x from the plane the image force is $e^2/16\pi\varepsilon_0 x^2$. This force is experienced by the electron escaping into the vacuum and is negligible beyond 10^{-6}–10^{-5} cm away from the surface.

The image potential is a specific surface contribution to W, and a second surface contribution is the existence of a surface double layer or dipole layer. Surface atoms are in an unbalanced environment, they have other atoms on one side of them but not on the other; thus, the electron distribution around them will be unsymmetrical with respect to the positive ion cores. This leads to the formation of a double layer. Two important effects emanate from this; the work function is sensitive to both the crystallographic plane exposed and to the presence of adsorbates.

The orientation of the exposed crystal face affects ϕ because the strength of the electric double layer depends on the density of positive ion cores which in turn will vary

TABLE IV Work Functions Measured from Different Crystal Faces of Tungsten and Molybdenum

Crystal face	Work function (eV)	
	Tungsten	Molybdenum
(110)	4.68	5.00
(112)	4.69	4.55
(111)	4.39	4.10
(001)	4.56	4.40
(116)	4.39	—

from one face to another. The work function of various crystal planes of tungsten and molybdenum are listed in Table IV. It can be seen that there is more than 0.3 eV difference in work function values. This variation of work function from one crystal face to another can clearly be demonstrated using a field emission microscope (FEM). This microscope is identical in construction to the FIM described earlier. However, instead of having helium or another imaging gas in the vacuum, no gas is admitted. The potential on the sample tip is reversed so that electrons are accelerated out of it by a very high local electric field ($\sim 4 \times 10^7$ V cm^{-1}). The current emitted from the tip surface where the work function is ϕ is approximately proportional to $\exp(-A\phi^{3/2})$ and is a very fast function of ϕ. The brightness observed on the fluorescent screen is a function of the value of ϕ at that place on the tip, and the FEM image will consist of darker and brighter areas, the brightness depending on the work function of each crystal face exposed. An image is shown in Fig. 33 which is the FEM image of

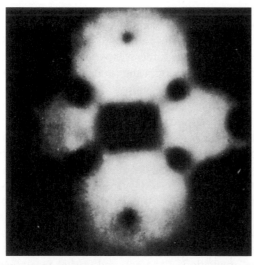

FIGURE 33 Field emission pattern of a tungsten tip. The (011) plane is in the center. (Courtesy of Lawrence Berkeley Laboratory.)

a tungsten tip. The changes in ϕ produced by adsorbed atoms or molecules can be followed in the FEM.

The work function of a solid is also sensitive to the presence of adsorbates. In fact, in virtually all cases of adsorption the work function of the substrate either increases or decreases; the change being due to a modification of the surface dipole layer. The formation of a chemisorption bond is associated with a partial electron transfer between substrate and adsorbate and the work function will change. Two extreme cases are (i) the adsorbate may only be polarized by the attractive interaction with the surface giving rise to the build up of a dipole layer, as in the physisorption of rare gases on metal surfaces; and (ii) the adsorbate may be ionized by the substrate, as in the case of alkali metal adsorption on transition metal surfaces. If the adsorbate is polarized with the negative pole toward the vacuum the consequent electric fields will cause an increase in work function. Conversely, if the positive pole is toward the vacuum then the work function of the substrate will decrease.

The work function is a rather complicated (and not fully understood) function of the surface composition and geometry. Nevertheless, general systematic observations of $\Delta\phi$ are quite helpful. For example, the sign of $\Delta\phi$ for atomic adsorption is mostly that implied by the magnitude of the ionization potential, electron affinity, or dipole moment of the adsorbates as one would expect. The most common usage of work function changes in surface chemistry is in the monitoring of the various stages of adsorption as a function of coverage. Often the work function change will go through a maximum or minimum at particular coverages corresponding to the completion of an ordered atomic arrangement.

Experimentally, the most accurate way of measuring changes in work function is by the Kelvin method, which uses a vibrating probe as a variable capacitor. A contact potential difference is set up between two conductors connected externally and the sample and a reference electrode form a parallel plate condenser. The distance between the two is periodically varied, thus generating an alternating current in the connecting wire. If a voltage source is placed in the connecting circuit just balancing out the contact potential difference, no current will flow. Once this situation has been achieved for the clean surface, a change in work function due to adsorption is simply the additional voltage which needs to be applied to compensate the change and keep the current zero. Accuracies of $\Delta\phi$ to within ± 1 meV are obtainable.

Intimately linked to the concept of work function is the process of thermionic emission. Thermionic emission is, as the name suggests, the phenomenon whereby electrons are ejected from a metal when it is heated in vacuum. The electrons that require the least amount of thermal energy to overcome their binding energy in the solid and escape are those in the high-energy tail of their equilibrium distribution in the metal. Thermionic emission is the most frequently used method to produce electron beams, for instance, in oscilloscope tubes and electron microscopes. Refractory metals (e.g., W) have traditionally been used as filaments in electron guns, mainly due to the fact that they can be heated to high temperatures and thus produce a relatively intense thermionic current. Since the work function of W is relatively high W filaments are often coated with a metal for lower work function, for example, Th ($\phi = 2.7$ eV) to enable them to be operated at lower temperature for the same current thereby extending their lifetime.

E. Surface States

In a bulk solid the infinite array of ion cores in crystallographic sites leads to a potential that varies in a three-dimensionally periodic manner. The solutions to Schrödinger's equation for such a potential lead to allowed energy bands, which are occupied by the electrons in the solid, and to particular values of the wave vector k of the electron where no traveling-wave solutions exist. The absence of eigenstates for these values of k leads to the band gaps in the electronic structure of the solid. The solid, however, is not infinite but is bounded by surfaces. In turn, surface atoms have fewer nearest neighbors and are in an asymmetric environment. The introduction of such a discontinuity at the surface perturbs the periodic potential and gives rise to solutions of the wave equation that would not have existed for the infinite crystal. These are derived by using appropriate boundary conditions to terminate the crystal and are called surface-state wave functions. These special solutions are waves which can travel parallel to the surface but not into the solid. They are localized at the surface and can have energies within the band gap of the bulk band structure. These states can trap electrons or release them into the conduction band. The concentration of electronic surface states in clean surfaces can be equal to the concentration of surface atoms ($\sim 10^{15}$ cm^{-2}). Impurities or adsorbed gases can reduce the surface state density.

One important consequence of the presence of electronic surface states is that the electron bands are modified at the surface even in the absence of a space charge or electron acceptor or donor species (such as adsorbed gases). The shape of the conduction band at the surface of an intrinsic semiconductor in the presence of electron-donor and electron-acceptor surface states is shown in the energy level diagrams in Figs. 34a and 34b.

Surface states can be associated not only with the termination of a three-dimensional potential at a perfect clean bulk exposed plane but also with changes in the potential

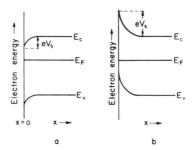

FIGURE 34 Energy-level diagrams for an intrinsic semiconductor in the presence of (a) electron–donor or (b) electron–acceptor surface states.

due to relaxation, reconstruction, structural imperfections, or adsorbed impurities. If the charge associated with any of these surface states is different from the bulk charge distribution then band bending will occur.

Surface states can be observed, for example, using ultraviolet photoelectron spectroscopy, which is discussed below.

F. Electron Emission from Surfaces

The most important methods of analyzing the surface electronic and chemical composition involve energy analysis of electrons emitted from a surface during its bombardment with electrons, ultraviolet photons, or X-ray photons. For example, part of the experimental verification of the band theory of metals comes from the measured intensity and energy distribution of electrons emitted under excitation by photons. It should be remembered that we have already mentioned two ways in which electrons can be emitted from surfaces; (i) by applying a very high electric field ($\sim 10^7$ V cm^{-1}) which pulls electrons from the surface, as used in FEM, and (ii) by heating the solid as in thermionic emissions.

Before discussing the two major electron emission techniques from surfaces, photoelectron spectroscopy and Auger electron spectroscopy (AES), it is pertinent to briefly discuss the surface sensitivity of the interaction of electrons with solids. Figure 35 shows the mean free path of electrons in metallic solids as a function of the electron energy. This curve is often called the "universal curve," and shows a broad minimum in the energy range between 10 and 500 eV with the corresponding mean free path on the order of 4–20 Å. Electron emission from solids with energy in this range must originate from the top few atomic layers. Therefore, all experimental techniques involving the incidence and/or convergence from surfaces of electrons having energy between 10 and 500 eV are surface sensitive. For incident electrons of higher energy (1–5 kV) the surface sensitivity can be enhanced by having the electron beam impinging on the surface at grazing

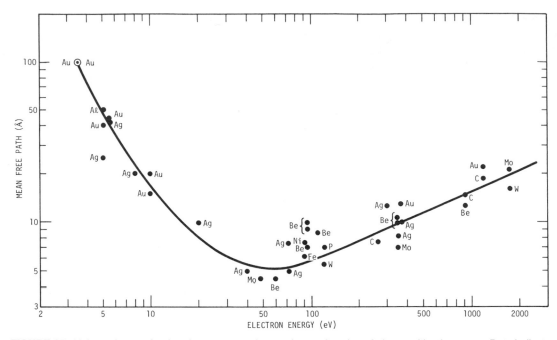

FIGURE 35 Universal curve for the electron mean free path as a function of electron kinetic energy. Dots indicate individual measurements.

incidence. Photons have a much larger penetration depth into the solid due to the much smaller scattering cross section. However, electrons created by excitation below a few atomic layers from the surface cannot escape due to inelastic scattering within the solid. If a monoenergetic beam of electrons of energy E_p strikes a metal surface then a typical plot of the number of scattered electrons $N(E)$ as a function of their kinetic energy E is shown in Fig. 36. The curve is dominated by a strong peak at low energies due to secondary electrons created as a result of inelastic collisions between the incident electrons and electrons bound to the solid. Other features in the spectrum include (i) the elastic peak at E_p that is utilized in LEED, (ii) inelastic peaks at loss energies of 10–500 meV which provide information about the vibrational structure of the surface–adsorbate complex utilized in EELS, (iii) inelastic peaks at greater loss energies (plasmon losses) which provide information about the electronic structure of surface atoms, and (iv) small peaks on the large secondary electron peak due to Auger electrons which provide information on the chemical composition of the surface.

Auger Electron Spectroscopy

Auger electron spectroscopy is the most common technique for determining the composition of solid and liquid surfaces. Its sensitivity is about 1% of a monolayer, and it is a relatively simple technique to perform experimentally. Auger electron emission occurs in the following manner. When an energetic beam of electrons or X-rays (1000–

5000 eV) strikes the atoms of a material, electrons that have binding energies less than the incident beam energy may be ejected from the inner atomic level. By this process a singly ionized, excited atom is created. The electron

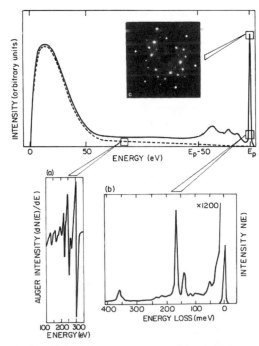

FIGURE 36 Experimental number of scattered electrons N(E) of energy E versus electron energy E curve. (Courtesy of Lawrence Berkeley Laboratory.)

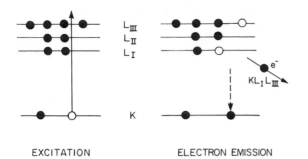

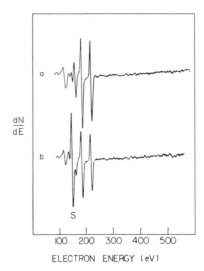

FIGURE 37 Scheme of the Auger electron emission process.

FIGURE 38 Typical Auger spectra from (a) a clean Mo(100) single-crystal and (b) a Mo(100) surface contaminated with sulfur.

vacancy formed is filled by deexcitation of electrons from other electron energy states. The energy released in the resulting electronic transition can, by electrostatic interaction be transferred to still another electron in the same atom or in a different atom. If this electron has a binding energy that is less than the energy transferred to it from the deexcitation of the previous process that involves the filling of the deep-lying electron vacancy, it will be ejected into vacuum, leaving behind a doubly ionized atom. The electron that is ejected as a result of the deexcitation process is called an Auger electron, and its energy is primarily a function of the energy-level separations in the atom. These processes are schematically displayed in Fig. 37. To a first approximation the energy of the Auger electron depicted in Fig. 37 is given by

$$E_{\text{Auger}} = E_{\text{K}} - E_{\text{L}_{\text{I}}} - E_{\text{L}_{\text{III}}} \qquad (44)$$

and is independent of the energy of the incident beam. This is an important difference between AES and photoelectron spectroscopy and means that it is not necessary to monochromatize the electron beam which adds to the experimental convenience.

There are two major experimental designs for AES. One is using the retarding grid analyzer which uses the same electron optics as LEED, thus LEED and AES can be performed using the same apparatus. The second is the cylindrical mirror analyzer (CMA) which has an inherently better signal-to-noise ratio. Scanning Auger microprobes are now in widespread use in the microelectronics industry for spatial chemical analysis of surfaces. With the exception of hydrogen and helium, all other elements are detectable by Auger electron spectroscopy.

The Auger spectrum is usually presented as the second derivative of intensity, $d^2 I/dV^2$, as a function of electron energy (eV). This way the Auger peaks are readily separated from the background, due to other electron loss processes that occur simultaneously. A typical Auger spectrum of molybdenum is shown in Fig. 38.

By suitable analysis of the experimental data, as well as by the use of suitable reference surfaces, the Auger electron spectroscopy study can provide quantitative chemical analysis in addition to elemental compositional analysis of the surface. It is possible to separate the surface composition from the composition of layers below the surface by appropriate analysis of the Auger spectral intensities. In this way the surface composition as well as the composition in the near-surface region can be obtained.

When chemical analysis is desired in the near-surface region, AES may be combined with ion sputtering to obtain a depth-profile analysis of the composition. Using high-energy ions, the surface is sputtered away layer by layer while, simultaneously, AES analysis detects the composition in depth. Sputtering rates of 100 Å/min are usually possible and the depth resolution of the composition is about 10 Å, which is mainly determined by the statistical nature of the sputtering process.

A different aspect of AES concerns shifts in the observed peak energies that are due to chemical shifts of atomic core levels (in a way analogous to X-ray photoelectron spectroscopy). For example, studies of different oxidation states of oxygen at metal surfaces have shown chemical shifts that grow with the formation of higher oxidation states.

G. Photoelectron Spectroscopy

Photoelectron spectroscopy is a technique whereby electrons directly ejected from the surface region of a solid by incident photons are energy analyzed and the spectrum is then related to the electron energy levels of the system. The field is usually arbitrarily divided into two classes: ultraviolet photoelectron spectroscopy (UPS) and X-ray photoelectron spectroscopy (XPS). The names derive from the energies of the photons used in the particular spectroscopy. Ultraviolet photoelectron spectroscopy studies the properties of valence electrons that are in the outermost shell of the atom and utilizes photons in the vacuum ultraviolet region of the electromagnetic spectrum [He I (21.22 eV), He II (40.8 eV), and Ne I (16.85 eV) resonance lamps are the most commonly used photon sources]. X-ray photoelectron spectroscopy investigates the properties in the inside shells of atoms and uses photons in the X-ray region [Mg K_α (1253.6 eV) and Al K_α (1486.6 eV) being the most common]. With the advent of synchrotron radiation, a polarized, tunable light source covering the entire useful energy range, the division is now somewhat redundant.

In both types of spectroscopy, if the incident photon has enough energy $h\nu$ it is able to ionize an electronic shell and an electron which was bound to the solid with energy E_B is ejected into vacuum with kinetic energy E_k. By conservation of energy:

$$E_k = h\nu - E_B. \qquad (45)$$

If the incident radiation is monochromatic and of known energy, and if E_k can be measured using a high-resolution energy analyzer (such as either a concentric hemispherical or cylindrical mirror analyzer), then the binding energy E_B can be deduced.

Equation (45) gives a highly simplified relationship between the kinetic energy, E_k, of the emitted photoelectrons and their binding energy; E_k is modified by the work function of the energy analyzer and by several atomic parameters that are associated with the electron emission process. The ejection of one electron leaves behind an excited molecular ion. The electrons in the outermost and in other orbitals experience a change in the effective nuclear charge due to an alteration of screening by other electrons. This gives rise to satellite peaks near the main photoelectron peaks. Several other effects, including spin-orbit splitting, Jahn–Teller effect, and resonant absorption of the incident photon by the atom, influence the detected photoelectron spectra.

One of the most important applications of XPS is the determination of the oxidation state of elements at the surface. The electronic binding energies for inner-shell electrons shift as a result of changes in the chemical environment. An example of these shifts can be seen in nitrogen, indicating the photoelectron energy for various chemical environments (Fig. 39). These energy shifts are closely related to charge transfer in the outer electronic level. The charge redistribution of valence electrons induces changes in the binding energy of the core electrons, so that information on the valence state of the element is readily obtainable. A loss of negative charge (oxidation)

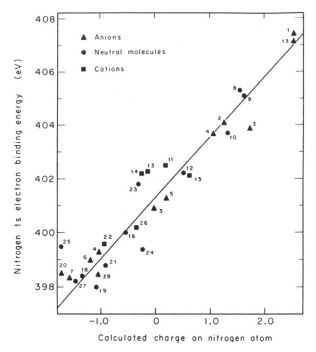

FIGURE 39 Is electronic binding energy shifts in nitrogen, indicating the different photoelectron energies observed in various chemical environments.

is in general accompanied by an increase in the binding energy E_B of the core electrons.

Relative surface coverages can also be obtained with XPS by monitoring the intensities of the core level peaks. Absolute coverages can be obtained from the core level intensities, but it is usual to calibrate against another technique.

As mentioned above UPS probes the valence electrons of the solid. It is these electrons which form a chemisorption bond and a knowledge of electronic density of states at a surface is of vital importance in attempts to understand the formation of chemical bonds between solid surfaces and adsorbed atoms or molecules; UPS can provide even more information about the system if the emitted electrons are both energy and spatially analyzed. This is known as angle-resolved UPS (ARUPS). Using ARUPS the band structures of clean and adsorbate-covered surfaces have been determined, mapping out the dispersion of electronic states; ARUPS also reveals directional effects due to the spatial distribution of electronic orbitals of atoms and molecules at the surface. By changing the angle of incidence and the angle of detection, the electronic orbitals from which the photoelectrons are ejected can be identified. In addition, ARUPS provides detailed information about the surface chemical bond including the direction of the bonding orbitals and the orientation of the molecular orbitals of adsorbed species on the surface.

V. SURFACE DYNAMICS

A. Atomic Vibrations

Until now it has been convenient to discuss both the properties and methods in terms of rigid lattices of atoms or molecules. In reality, the atoms are in motion and this motion should be included in a complete treatment of any properties it may affect.

In X-ray diffraction experiments it is well known that the intensity of the scattered rays decreases as the temperature is increased. Simultaneously, the intensity of the diffuse background of the diffraction pattern increases. The simplest explanation for this observation is that the atoms are not rigid, but are vibrating about their equilibrium positions, and as a result, the exact Bragg condition is not met. Scattered waves from the rigid lattice that were adding up in phase now have a phase difference fluctuating with time due to the atomic motion. The effect of this motion on the intensity of the elastically diffracted beams is described in most good solid-state physics texts. Briefly, if I_0 is the intensity elastically diffracted by a rigid lattice then the intensity I due to scattering by the vibrating lattice in the direction determined by Bragg scattering due to a reciprocal-lattice vector $\bar{\mathbf{g}}$ is given by the following:

$$I = I_0 \exp(-\alpha \langle u^2 \rangle |g|), \qquad (46)$$

assuming that the atoms are in simple harmonic motion. $\langle u^2 \rangle$ is the mean-square amplitude of vibration in the direction $\bar{\mathbf{g}}$ and α is a constant related to the number of dimensions in which the atoms are allowed to vibrate. In one dimension $\alpha = 1$; in three dimensions $\alpha = \frac{1}{3}$. The exponential factor in Eq. (46) is called the Debye–Waller factor and is often denoted as $\exp(-2M)$.

The same kind of effect is observed in LEED only because LEED intensities arise from the just few atomic layer of a crystal the value of $\langle u^2 \rangle$ is that for the surface atoms. Because of the absence of nearest neighbors on the vacuum side we expect that $\langle u^2 \rangle$ at the surface will be greater than in the bulk.

By using the Debye model of the solid it is possible to relate the observed intensity of the elastically scattered electrons in LEED to measurable quantities. We obtain the following:

$$I_{00}(T) = I_{00}(0) \exp\left\{ \left| -\frac{12h^2}{mk} \left(\frac{\cos \phi}{\lambda} \right)^2 \frac{T}{\Theta_D^2} \right| \right\}, \quad (47)$$

where $I_{00}(T)$ is the temperature-dependent intensity of the (0, 0) beam resulting from a beam of electrons of wavelength λ incident on the surface at an angle Θ relative to the surface normal. $I_{00}(0)$ is the specularly reflected intensity from a rigid lattice, h is Planck's constant, m is the atomic mass, k is Boltzmann's constant, T is the temperature, and Θ_D is the Debye temperature. (The Debye temperature is associated with the energy ω_{max} of the highest frequency phonon mode possible in the Debye model of vibrations in the solid, $h\omega_{max} = k\Theta_D$.)

Equation (47) implies that a plot of the logarithm of the intensity at a given energy (wavelength) as a function of temperature is a straight line, the slope of which yields Θ_D, a measure of the surface vibrational amplitude perpendicular to the surface.

In reality, the electron-beam penetration varies as a function of energy, so that Eq. (47) provides, at any given energy, an effective Debye temperature, which is some average of the surface and bulk layers. In empirical fashion, however, we may arrive at a surface Debye temperature from the low-energy limit of this effective Debye temperature.

Adsorbates should have a marked influence on surface-atom vibrations, since they change the bonding environment with respect to that on the clean surface. The adsorption of oxygen on tungsten increases the surface Debye temperature with respect to the bulk value due to the stronger W—O bond as compared to the W—W bond. Studies of surface-atom vibrations in the presence of adsorbates provide information on the nature of the surface bond.

B. Surface Diffusion

As discussed above, at any finite temperature the atoms at the surface of a crystal are vibrating at some frequency ν_0. Thus, ν_0 times every second each atom strikes the potential-energy barrier separating it from its nearest neighbors (Fig. 40). The thermal energy causing the atoms to oscillate with increasing amplitude as the temperature is increased is not sufficient to dislodge most of them from their equilibrium positions. The thermal energy ($3RT \approx 1.8$ kcal mol^{-1} at 300 K) tied up in lattice vibrations is only a small fraction of the total energy necessary

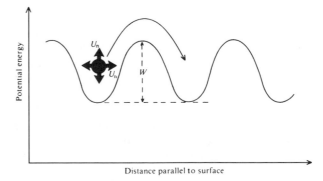

FIGURE 40 One-dimensional potential energy diagram parallel to the surface plane.

to break an atom from its neighbors and to move along the surface. This bond breaking energy is of the order 15–50 kcal mol^{-1} for many metal surfaces. As the temperature of the surface is increased, more and more surface atoms may acquire enough activation energy to break bonds with their neighbors and move along the surface. Such surface diffusion plays an important role in many surface phenomena involving atomic transport, e.g., crystal growth, vaporization, and adsorption. The migration of atoms or molecules along the surface is one of the most important steps in surface reactions and has proved to be the rate-limiting step for many reactions that have been studied at low pressures.

A surface contains many defects on an atomic scale. Atoms in different surface sites have different binding energies. Surface diffusion can be considered as a multistep process whereby atoms break away from their lattice position (e.g., a kink site at a ledge) and migrate along the surface until they find a new equilibrium site.

The frequency f with which an atom will escape from a site will depend upon the height, ΔE_{D}^*, of the potential energy barrier it has to climb in order to escape as follows:

$$f = z\nu_0 \exp\left(-\frac{\Delta E_{\mathrm{D}}^*}{k_{\mathrm{B}}T}\right),$$ (48)

where z is the number of equivalent neighboring sites. For a (111) face of an fcc metal, $z = 6$, the vibration frequency is of the order of 10^{12} sec^{-1}. Assuming that ΔE_{D} is 20 kcal mol^{-1}, at 300 K the atom makes one jump in every 50 sec, while at 1000 K one in 10^{-8} sec. Thus, the rate of surface diffusion varies rapidly with temperature. This is the case for a single jump to a neighboring equilibrium surface site. What is of great importance is the long-distance motion of a surface atom. The result is derived from considering a mathematical treatment of an atom executing a random walk for a time t over a mean-square distance $\langle X^2 \rangle$.

For a sixfold symmetrical surface, we obtain the following:

$$\langle X^2 \rangle = ftd^2/3.$$ (49)

The value of fd^2 is a property of the material that characterizes its atomic transport. Its value provides information about the mechanism of atomic transport, and it is customary to define the diffusion coefficient D as follows:

$$D = fd^2/2b,$$ (50)

where b is the number of coordinate directions in which diffusion jumps may occur with equal probability.

Equation (50) can therefore be rewritten as

$$D = D_0 \exp\left(-\Delta E_{\mathrm{D}}^*/k_{\mathrm{B}}T\right),$$ (51)

where $D_0 = (\nu_0 d^2/6)$, $(\nu_0 d^2/4)$, for sixfold or fourfold symmetry, respectively; D is usually given in units of square centimeters per second. If D is determined experimentally as a function of temperature, then a plot of $\ln D$ versus $1/T$ will yield us the activation energy of the diffusion process, provided that the diffusion occurs by a single mechanism.

The rms distance $\langle X^2 \rangle^{1/2}$ can be expressed in terms of the diffusion coefficient by substitution of Eq. (50) into Eq. (49) to give for $b = 6$ as follows:

$$\langle X^2 \rangle^{1/2} = (4Dt)^{1/2}.$$ (52)

From measurements of the mean travel distance of diffusing atoms the diffusion coefficient can be evaluated. Conversely, knowledge of the diffusion coefficient allows us to estimate the rms distance or the time necessary to carry out the diffusion. For example, the diffusion coefficients of silver ions on the surface of silver bromide can be estimated to be 10^{-19} and 10^{-13} cm^2/sec at 300 and 100 K, respectively. Assuming that a rms distance of 10^{-4} cm is required for silver particle aggregation (printout) to commence, of what duration are the light-exposure times required? Using Eq. (52) we have $t = 5$ sec and $t = 5 \times 10^4$ sec at 300 and 100 K, respectively. The exponential temperature dependence of D is, of course, the reason that silver bromide photography cannot be carried out at low temperatures (much below 300 K) but is easily utilized at about room temperature. We can also see that at slightly elevated temperature ($\sim$450 K) the thermal diffusion of silver particles should be rapid enough ($D \approx 3 \times 10^{-7}$ cm^2 sec^{-1}) so that their aggregation will take place rapidly even in the dark ($t \approx 10^{-2}$ sec) in the absence of any photoreaction.

Surface diffusion has so far been discussed in terms of a single surface atom. However, on a real surface many atoms diffuse simultaneously and in most diffusion experiments the measured diffusion distance after a given diffusion time is an average of the diffusion lengths of a large, statistical number of surface atoms. A thermodynamic treatment in terms of macroscopic parameters can be followed to yield the following:

$$D = D_0 \exp(-Q/RT),$$ (53)

where Q is the total activation energy for the overall diffusion process and only one diffusion mechanism is involved.

Experimentally, the diffusion coefficient D is obtained by using a relationship between the diffusion rate and coverage gradient, namely Fick's second law of diffusion in one dimension:

$$\partial c/\partial t = D(\partial^2 c/\partial x^2),$$ (54)

where c is the concentration of adatoms, t the time, and x the distance along the surface. In most surface diffusion studies the surface concentration of diffusing atoms, c, is measured as a function of distance x along the surface, and Eq. (54) is solved by the use of boundary conditions that approximate the experimental geometry. These experiments are by no means trivial, and many novel experimental techniques have been applied to study surface diffusion on single crystals.

A technique which has been used to measure surface diffusion rates is scanning Auger electron spectroscopy, which can follow adsorbate diffusion. A particular Auger transition of the adsorbate under investigation is used as a monitor of relative concentration versus distance scanned across the surface. Profiles are recorded after heating periods to observe the change in concentration profile as a function of time and temperature.

While this technique monitors mass transport, and values of D and Q are averaged values, field ion microscopy can be used to follow the diffusion of individual atoms across a surface. To study diffusion, the metal is vapor deposited onto the tip. The tip is then heated to remove evaporated adatoms until only one or two remain on the surface plane of interest. The diffusion is then examined by photographically recording the position of the adatom at low temperatures, removing the applied field, and heating to the desired temperature for a given time. The tip is then cooled, the field reapplied, and the field ion image examined to see if the atom has moved to a neighboring site. This process is then repeated many times to obtain useful values of diffusion rates, and by examining the diffusion over a temperature range, the activation barrier to surface diffusion can be determined. Figure 41 shows a series of field ion images of a Rh atom on the W(112)

plane at 327 K. The field ion images are taken at 1-min intervals and the Rh atom can clearly be seen to have diffused across the surface. Unfortunately because of the high field strengths employed, adsorbates such as O or N tend to be stripped from the surface as ions, so their microscopic diffusion cannot be studied by this method. An interesting result from FIM studies of metal adatoms on metals is the recognition of clusters as important contributions to material transport. It has been found that rhenium dimers diffuse more rapidly than single Rh atoms on the W(112) plane, and Rh trimers diffuse at roughly the same rate as dimers. This is not a general trend, however, as iridium dimers move much more slowly than singles.

While the single adatom diffusion technique gives us detailed microscopic information, the mass transport techniques are of use as they help to give understanding of the technologically important processes such as sintering and creep.

C. Surface Reactions

Heterogeneous catalysis, corrosion, photosynthesis, and adhesion are examples of chemical processes that are partially or fully controlled by reactions at surfaces. For the case of gas–solid reactions the surface reactions can be divided into two major categories: (i) stoichiometric surface reactions where the solid surface participates directly in the reaction by compound formation and (ii) catalytic surface reactions where the reaction occurs at the solid surface but the surface does not undergo any net chemical change. In both cases gaseous molecules impinge on the surface, adsorb, react, and form various intermediates of varying stability, and then the products desorb into the gas phase if they are volatile.

All surface reactions involve a sequence of elementary steps that begins with the collision of the incident atoms or molecules with the surface. As the gas species approaches the surface it experiences an attractive potential whose range depends upon the electronic and atomic structures of the gas and surface atoms. A certain fraction of the incident gas molecules is trapped in this attractive potential well with a sticking probability given by the following:

$$S(\Theta, T) = S_0(1 - \Theta)\exp(-E_a/RT), \qquad (55)$$

where S_0 is the initial (zero coverage) sticking coefficient, Θ the surface coverage ($0 < \Theta < 1$), and E_a the activation energy for adsorption. If this force attraction is due to a van der Waals interaction, the trapping is due to physical adsorption. If the attraction is much stronger, having the character of chemical bonding then we have chemisorption and the process is known as sticking. The boundary between the two types of bonding is usually

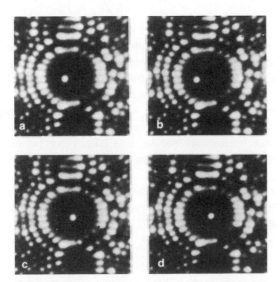

FIGURE 41 Diffusion of rhenium atoms on W(211) at 327 K. Field ion images are taken after 60 sec diffusion intervals. (Courtesy of Lawrence Berkeley Laboratory.)

set at a binding energy of 15 kcal mol^{-1}. Sticking by chemisorption is often preceded by trapping into a physisorbed state, in which case the physisorbed state is known as a precursor state for chemisorption. The presence of a precursor state is indicated by a sticking coefficient that remains almost constant as surface coverage increases until a saturation coverage is reached, when it rapidly falls to zero. This behavior arises because molecules in the relatively mobile precursor state diffuse to parts of the surface which are not covered by chemisorbed molecules. In direct chemisorption the sticking coefficient varies strongly with coverage and with ordering of the chemisorbed layer.

The adsorbed species may also desorb from the surface if its energy overcomes the attractive surface forces. When a surface reaction occurs a certain proportion of the adsorbed species either decomposes (unimolecular reaction) or reacts with a second adsorbed species (bimolecular reaction) before the product desorbs.

During the initial interaction of the gas molecule with the surface as the incoming molecule falls into a potential well the kinetic energy normal to the surface increases. Unless this energy is transferred to some other degree of freedom the molecule will simply bounce off; there will be no trapping or sticking. In the case of physisorption energy transfer via phonons is usually most important while for chemisorption electronic excitation via electron–hole pairs is thought to be important.

The exchange of translational energy T with the phonons V_s is called $T - V_s$ energy exchange. The gas molecule may also exchange internal energy, rotation R or vibration V with the vibrating surface atoms. In this case there are also $R - V_s$ and $V - V_s$ energy transfer processes.

In order to understand the dynamics of gas–surface interaction, it is necessary to determine how much energy is exchanged between the gas and surface atoms through the various energy-transfer channels. In addition the kinetic parameters (rate constants, activation energies, and preexponential factors) for each elementary surface step of adsorption, diffusion, and desorption are required in order to obtain a complete description of the gas–surface energy transfer process.

Most surface reactions take place at high pressures (1–100 atm) either because of the chemical environment of our planet or to establish optimum reaction rates in chemical processing. Under these conditions, surfaces are usually covered by at least one monolayer of adsorbed species. Since activation energies for adsorption and surface diffusion are generally small (a few kT), equilibrium among the different surface species, reactants, reaction intermediates, and products, is readily established. In the simplest (but not general and important) case of localized, associative adsorption into a single state, the surface coverage by adsorbed species is given in terms of the gas pressure P by the Langmuir isotherms:

$$\Theta = KP/(1 + KP), \tag{56}$$

where K is an equilibrium constant. Catalyzed surface reactions usually take place between two or more coadsorbed species which compete for adsorption sites on the surface. When j gases adsorb competitively and associatively, the surface coverage by species i is given by the following:

$$\Theta = K_i P_i \bigg/ \left(1 + \sum_j K_j P_j\right). \tag{57}$$

Many catalyzed surface reactions can be treated as a two-step process with an adsorption equilibrium followed by one rate-determining step (diffusion, surface reaction, or desorption). The surface reaction kinetics are usually discussed in terms of two limiting mechanisms, the Langmuir–Hinshelwood (LH) and Eley–Rideal (ER) mechanisms. In the LH mechanism, reaction takes place directly between species which are chemically bonded (chemisorbed) on the surface. For a bimolecular LH surface reaction. $A_{ads} + B_{ads} \rightarrow$ products, with competitive chemisorption of the reactants, the rate of reaction is given by the following expression:

$$\text{Rate} = k_R \Theta_A \Theta_B$$
$$= k_R K_A K_B P_A P_B/(1 + K_A P_A + K_B P_B)^2. \tag{58}$$

The reaction rate is proportional to the surface coverages Θ_A and Θ_B and to the reaction rate constant k_R. For noncompetitive adsorption, the rate expression becomes the following:

$$\text{Rate} = k_R \Theta_A \Theta_B = \frac{k_R K_A K_B P_A P_B}{(1 + K_A P_A)(1 + K_B P_B)}. \tag{59}$$

General rate expressions of the form given in equations and have been experimentally verified for many types of LH reactions. Similar but more complicated rate expressions are easily derived assuming different (non-Langmuir) isotherms, higher-order reaction steps, or dissociative chemisorption of the reactants. In the ER mechanism, surface reaction takes place between a chemisorbed species and a nonchemisorbed species, e.g., $A_{ads} + B_g \rightarrow$ products. The nonchemisorbed species may be physisorbed or weakly held in a molecular precursor state. In this case, the rate expression for the surface reaction becomes

$$\text{Rate} = k_R \Theta_A P_B = k_R K_A P_A P_B/(1 + K_A P_A). \tag{60}$$

Presently no proven examples exist in which surface reaction occurs by the ER mechanism.

Surface reaction kinetics determined experimentally are often expressed in the form of a power rate law as follows:

$$\text{Rate} = k_R \prod_i P_i^{\alpha_i}, \tag{61}$$

where k_R is the apparent rate constant and α_i is the experimental order of the reaction (positive, negative, integer, or fraction) with respect to the reactants and products. The apparent rate constant in Eq. (61) is not that of an elementary reaction step (it contains adsorption equilibrium constants), but it can usually be represented by an Arrhenius equation as follows:

$$k_R = A \exp(-E_R/RT), \tag{62}$$

where A is an apparent preexponential factor and E_R is the apparent activation energy for the surface reaction. The magnitude of A and E_A can provide important information about the rate-determining step of a surface reaction, and very frequently k_R and A display a compensation effect. A related quantity is the reaction probability, $\lambda_i = (2\pi m K T)^{1/2} \nu_R / P_i = \text{rate/flux}$; that is, the probability that an incident reactant molecule will undergo reaction.

The simplified isotherms and rate expressions developed in this section are extremely useful despite the implicit assumption that a single state exists for the adsorbed species. Real surfaces are heterogeneous on an atomic scale with a variety of distinguishable adsorption sites. Gas molecules adsorbed at each type of site may display a wide distribution of excited rotational, vibrational, and electronic states. Experimentally, we can measure meaningful rate and adsorption equilibrium constants provided that adsorption and desorption are fast compared with surface reactions so that an adsorption equilibrium exists. In this circumstance the kinetic parameters are an ensemble average over all surface sites and states of the system.

1. Molecular-Beam Scattering

The most powerful experimental technique for investigating the dynamics of the gas–solid interaction is molecular-beam surface scattering (MBS). The experimental arrangement is similar to that already described for helium atom diffraction. Instead of using an atomic beam of a light molecular weight gas and observing diffraction effects, a well-collimated beam of molecules strikes the oriented, preferably single-crystal, surface, and the species that are scattered at a specific solid angle are detected by a mass spectrometer. The angular distribution of the scattered molecules can be obtained by rotation of the mass spectrometer about the sample. The velocity distribution of the molecules after scattering is deduced by chopping the scattered molecules and thereby measuring their time of flight to the detector. The surface residence times of the

molecules, together with their angular and velocity distributions provide detailed information about the $T-V_s$ energy transfer processes that occur during the gas–surface interaction. A complete dynamical description for this interaction ($T-V_s$ plus $R-V_s$ and $V-V_s$) can be determined if the distribution of internal energy states for the product molecules is determined simultaneously with their velocity distributions. This type of detection is known as state selective detection.

The angular distribution of scattered molecules is usually displayed by plotting the intensity of detected molecules per unit solid angle versus the angle of scattering Θ_r that is measured with respect to the surface normal. Angular distributions in the two limiting cases of gas–surface interaction, cosine and specular scattering, are shown in Fig. 42. The scattered intensity for the cosine distribution decreases as $\cos\Theta$ with respect to the surface normal. Cosine scattering is expected when the adsorbed species have long residence times or are strongly coupled to the vibrational states of the surface atoms. It is a necessary criterion for complete thermal accommodation, a situation in which the molecules desorb with a kinetic temperature or velocity distribution that is the same as the temperature of the solid surface. Specular scattering occurs when the scattered intensity is sharply peaked at the angle of incidence (specular angle). In this case the interaction is elastic or quasielastic and little or no energy transfer takes place between the incident gas molecules and the surface. Sharply peaked angular distributions for surface reaction products ($I(\Theta) \sim \cos^m\Theta$, $m > 1$) indicate that a repulsive barrier exists in the exit channel. Measurements of velocity distributions provide more

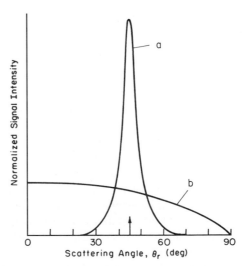

FIGURE 42 Rectilinear plot displaying the (a) specular scattering and (b) cosine angular distribution of scattered beams. The arrow indicates the angle of incidence.

direct information on inelastic scattering than angular distributions alone. Although considerable information can be gained from such studies it has been impossible to get state specific information and indeed it is often unclear whether internal states present more efficient energy transfer channels than phonons or vice versa. The difficulties in studying internal energy transfer in molecular collisions with surfaces can be resolved by the application of state-specific detection techniques. Laser-induced fluorescence, multiphonon ionization, IR excitation with bolometric detection, and IR emission techniques have all been used to obtain state-resolved measurements of the internal energy distributions of molecules scattering from surfaces. It has also been possible to separate experimentally direct inelastic and trapping-desorption scattering. In the direct inelastic scattering of diatomics, coupling to rotational energy has been found to be very important and to exhibit several interesting phenomena: rotational rainbows and the production of rotationally aligned molecules in scattering.

2. Molecular-Beam Reactive Scattering

While molecular-beam scattering has made great advances in our understanding of the energy exchange processes during the gas–surface collision, molecular-beam techniques have also made important contributions to the understanding of the mechanisms of chemical reactions occurring at surfaces in the form of molecular-beam reactive scattering (MBRS). The use of time-of-flight techniques permits measurement of product velocity distributions and the detailed time resolution of fast transient reactions. Also of great value is the use of state-specific detection methods to determine product vibrational and rotational states. Although MBRS can only be utilized at low pressures ($\leq 10^{-4}$ torr) its pressure range permits wide variations of surface coverages. The reaction probabilities on a single scattering can be determined together with the surface residence times of adsorbates. The surface kinetic information is obtained by measurements of the intensity and the phase shift of the product molecules with respect to the reactant flux. Residence times in the range 10^{-6}–1 sec can be monitored with relative ease, and activation energy is determined from the temperature dependences of the intensities and the phase shifts. The phase shift of the product molecules is usually measured at different chopping frequencies of the incident beam. At a given chopping frequency, only those product molecules are detected that are formed in the surface process and desorbed in less time than the chopping period.

As an example of an investigation of the dynamics of a catalyzed surface reaction studied by MBRS we will consider the isotope exchange reaction, H_2–D_2.

Exchange of hydrogen and deuterium to form HD is one of the simplest reactions that can be catalyzed on clean metal surfaces at temperatures as low as 100 K. The same reaction is immeasurably slow in the gas phase due to the very high dissociation energies of the reacting molecules (103 kcal mol^{-1}). the H_2–D_2 exchange reaction has been studied over flat (111) and stepped (332) single crystal surfaces of platinum. The Pt(332) surface contains high concentrations of periodic surface irregularities (steps) that are one atom in height. Reaction probabilities averaged over the cosine HD angular distributions were 0.07 on the (111) surface and 0.35 on the (332) surface under identical experimental conditions ($T_s = 1100$ K, $T_g = 300$ K). The reaction probability on the stepped surface varied markedly with the angle of incidence of the mixed H_2–D_2 molecular beam. This is shown in Fig. 43. The reaction probability was highest when the beam was incident on the open edge of the step and lowest when the bottom of the step was shadowed

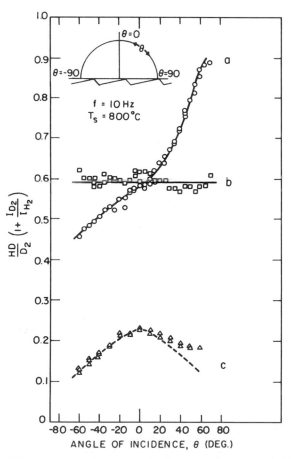

FIGURE 43 HD production as a function of angle of incidence Θ of the molecular beam, normalized to the incident D_2 intensity. (a) Pt(332) surface with the step edges perpendicular to the incident beam ($\phi = 90°$); (b) Pt(332) where the projection of the beam on the surface is parallel to the step edges ($\phi = 0°$); and (c) Pt(111).

(curve a). When the H_2–D_2 beam was incident parallel to the steps, the rate of HD production was independent of the angle of incident at all angles of crystal rotation (curve b). These results indicate that the atomic step sites are about seven times more active than the (111) terrace sites for the dissociative chemisorption of hydrogen and deuterium molecules. Detailed analysis of the scattering data revealed a barrier height of 4–8 kJ mol^{-1} for dissociative H_2 chemisorption on the (111) surface. On the other hand, this barrier did not exist ($E_a = 0$) on the stepped surface. This difference in activation energy alone accounts for the different reaction probabilities of the step and terrace sites. While the dissociation probability of hydrogen molecules was higher on the stepped surface than on Pt(111), the kinetics and mechanism of HD recombination appear to be identical over both surfaces once dissociation takes place. On both surfaces, HD formation follows a parallel LH mechanism with one of the reaction branches operative over the entire temperature range of 300–1075 K. This branch has an activation energy and pseudo-first-order preexponential factor of $E_a = 54$ kJ mol^{-1} and $A_1 = 8 \times 10^4$ sec^{-1} for the stepped surface and $E_a = 65$ kJ mol^{-1} and $A_1 = 3 \times 10^5$ sec^{-1} for the Pt(111) surface. A second branch is observed for temperatures above 575 K, but the kinetic parameters for this pathway could not be accurately determined.

D. Stoichiometric Surface Reactions

Stoichiometric surface reactions are those in which the surface participates directly in the reaction by compound formation. Oxidation and corrosion are the two most important classes of such reactions.

Surface oxidation of metals encompasses a series of at least three reaction steps that include (1) dissociative chemisorption of oxygen on the metal surface, (2) rearrangement of the surface atoms with dissolution of oxygen into the near surface region, and (3) nucleation of oxide islands which grow laterally and eventually condense to produce continuous oxide films. The oxide islands appear to precipitate suddenly once a critical oxygen concentration is reached in the near surface region. Nucleation takes place most readily at surface irregularities such as atomic steps, dislocations, and stacking faults. At room temperature, noble metals such as Rh, Ir, Pd, and Pt display little tendency for oxygen incorporation or surface rearrangement. Initial heats of oxygen chemisorption on these metals are much greater than the heats of formation of the corresponding bulk oxides. Other metals such as Cr, Nb, Ta, Mo, W, Re, Ru, Co, and Ni, dissolve surface oxygen by a place exchange mechanism in which oxygen atoms interchange positions with underlying metal atoms. These metals display heats of adsorption for oxygen that are comparable to the heats of formation of the stable metal oxides. Metals such as Ti, Zr, Mn, Al, Cu, and Fe dissolve oxygen more readily and form stable oxide films even at room temperature. At low oxygen pressures these films often assume a crystalline structure, whereas at higher pressures ($>10^{-3}$ atm) the films tend to be amorphous. At higher temperatures (400–1000 K), oxide formation occurs readily on the surfaces of nearly all metals.

Growth of surface oxide films takes place only if cations, anions, and electrons can diffuse through the oxide layer. The growth kinetics of very thin films ($\sim$10–50 Å) often follow the Mott or Cabrera–Mott mechanisms in which electrons tunnel through the film and associate with oxygen atoms to produce oxide ions at the surface. A large local electric field (10^6–10^7 V/cm) results at the surface which facilitates cation diffusion from the metal–oxide interface to an interstitial site of the oxide. The film thickness Z at time t is given by

$$Z = \alpha_1 \ln(\alpha_2 t + 1) \tag{63}$$

or inverse logarithmic

$$1/Z = \alpha_3 - \alpha_4 \ln t \tag{64}$$

law of growth depending on whether electron tunneling or cation diffusion is rate limiting. The constants α_1–α_4 are determined by the material, its structure, and the reaction conditions. The electron field strength and rate of growth decrease exponentially as the film thickens, resulting in an effective limiting thickness for the surface oxide layer.

In addition to surface oxides, a vast array of surface compounds can be produced from the reactions of halogens, chalcogenides, and carbon-containing molecules with metal surfaces. Chemisorption of chlorine near 300 K on Cu, Ti, W, Mo, Ta, Ni, Pd, and Au, for example, results in the formation of stable surface compounds which often evaporate as molecular chlorides upon heating at elevated temperatures. Chemisorption of chlorine at 300 K on Ag(100) and Ag(111) produces chemisorbed chlorine overlayers which react irreversibly at about 425 K to produce AgCl with an activation energy of 56 kJ mol^{-1}. Upon heating, AgCl desorbs at about 830 K with a desorption activation energy of 192 kJ mol^{-1}.

MBRS has been used to investigate the dynamics of several surface corrosion reactions at low reactant pressures. Systems studied include the oxidation Si, Ge, Mo, and graphite, and the halogenation of Si, Ge, Ta, and Ni. With the exception of silicon and germanium oxidation, where dissociative chemisorption of oxygen is apparently rate limiting, the kinetics of these surface reactions generally appear to be controlled by surface or bulk diffusion of the reacting species.

E. Catalytic Surface Reactions

A major goal of basic surface chemistry is in trying to understand heterogeneous catalysis on an atomic scale. Virtually all chemical technologies and many technologies in other fields use catalysis as an essential part of the process. The most important catalytic processes are summarized in Table V. These processes are listed together with the pertinent chemical reactions, widely used catalysts, and typical reaction conditions. There are several definitions of a catalyst, one general definition being that it is a substance that accelerates a chemical reaction without visibly undergoing chemical change. Indeed, a major role of a catalyst is in accelerating the rate of approach to chemical equilibrium. However, a catalyst cannot change the ultimate equilibrium determined by thermodynamics.

Another major function of a catalyst is to provide reaction selectivity. Under the conditions in which the reaction is to be carried out, there may be many reaction channels, each thermodynamically feasible, that lead to the formation of different products. The selective catalyst will accelerate the rate of only one of these reactions so that only the desired product molecules form with near-theoretical or 100% efficiency. One example is the dehydrocyclization of n-heptane to toluene:

$$CH_3-(CH_2)_5-CH_3 \longrightarrow C_6H_5CH_3 + 4H_2$$

This is a highly desirable reaction that converts aliphatic molecules to aromatic compounds. The larger concentration aromatic component in gasoline, for example, greatly improves its octane number. However, n-heptane may participate in several competing simpler reactions. These include hydrogenolysis, which involves C—C bond scission to form smaller molecular weight fragments (methane, ethane, and propane); partial dehydrogenation, which produces various olefins; and isomerization, which yields branched chains. All of these reactions are thermodynamically feasible, and since they appear to be less complex than dehydrocyclization, they compete effectively. A properly prepared platinum catalyst surface catalyzes the selective conversion of n-heptane to toluene without permitting the formation of other products. The catalyst selectivity is equally important for the reactions of small molecules (such as the hydrogenation of CO to produce a desired hydrocarbon) or very large molecules of biological importance, where enzyme catalysts provide the desired selectivity.

Catalysis is a kinetic phenomenon; we would like to carry out the same reaction at an optimum rate over and over again using the same catalyst. In most cases such a steady-state operation is desirable and aimed for. In the sequence of elementary reactions that include adsorption, surface migration, chemical rearrangements, and reactions in the adsorbed state, and desorption of the products, the rate of each step must be of steady state. The rate of the overall catalytic reaction per unit area catalyst surface can be expressed as (moles of product/catalyst area × time). Another expression for catalytic rate is the turnover number or turnover frequency. This is the (number of molecules of product/number of catalyst sites × time). For most heterogeneous catalyzed small molecule reactions the turnover number varies between 10^{-2} and 10^2 sec^{-1}. The calculation of the turnover number is limited by the difficulty of determining the true number of active sites.

The reaction probability reveals the overall efficiency of a catalyst. It is defined as follows:

reaction probability

$$= \frac{\text{rate of formation of product molecules}}{\text{rate of incidence of reactant molecules}}.$$

The determination of the rates of the net catalytic reactions and how the rates change with temperature and pressure is of great practical importance. Although there are many excellent catalysts that permit the achievement of chemical equilibria (for example, Pt for oxidation of CO and hydrocarbons to CO_2 and H_2O), most catalyzed reactions are still controlled by the kinetics of one of the surface processes. From the knowledge of the activation energy and the pressure dependencies of the overall reaction, the catalytic process can be modeled and the optimum reaction conditions can be calculated. Such kinetic analysis, based on the macroscopic rate parameters, is vital for developing chemical technologies based on catalytic reactions.

The rates of reactions are extremely sensitive to small changes of chemical bonding of the surface species that participate in the surface reaction. Since the energy necessary to form or break the surface bonds appears in the exponent of the Arrhenius expression for the rate constant for the overall reaction, it can increase or decrease the rate exponentially. For example, a change of 3 kcal in the activation energy alters the reaction rate by over an order of magnitude at 500 K. Small variations of chemical bonding at different surface irregularities, steps, and kinks, as compared to atomic terraces, can give rise to a very strong structure sensitivity of the reaction rates and the product distribution. Rate measurements exponentially magnify the energetic alterations that occur on the surface and could provide a very sensitive probe of structural and electronic changes at the surface and changes of surface bonding on the molecular scale.

One of the most important considerations in catalysis is the need to provide a large contact area between the

TABLE V   Chemical Processes Based on Heterogeneous Catalysis

Processes	Typical reactions	Catalyst	Reaction conditions
Ammonia synthesis	$N_2 + 3H_2 \rightarrow 2NH_3$	Triply promoted iron (Fe—K_2O—Al_2O_3—CaO);	720–800 K 40–100 atm
Dehydrogenation		Fe_2O_3—Cr_2O_3—K_2O mixed metal oxides	800–900 K 10–50 atm
Epoxidation	$C_2H_4 + \frac{1}{2}O_2 \rightarrow C_2H_4O$	AgCl—K_2O/Al_2O_3	520–600 K
Fischer–Tropsch synthesis of hydrocarbons	$CO + H_2 \rightarrow$ alkanes olefins aromatics	Fe_3O_4—K_2O/Al_2O_3 supported Co, Ru, Ni, Rh	500–700 K 10–50 atm
Fischer–Tropsch synthesis of oxygenates	$CO + H_2 \rightarrow$ aldehydes acids alcohols	$Rh_2O_3 \cdot H_2O$—K_2O $LaRhO_4$ supported Pd, Pt	500–700 K 10–50 atm
Hydrotreating (desulfurization and denitrification)	$R\text{—}S\text{—}R + H_2 \rightarrow 2RH + H_2S$ $R\text{=}N\text{—}R + \frac{3}{2}H_2 \rightarrow 2RHH + NH_3$	Co—Mo, Ni—Mo, Ni—Co—Mo/Al_2O_3 Ni—W/Al_2O_3, MoS_2, WS_2	570–770 K 30–200 atm
Isomerization			
Olefins		Solid acids, zeolites Group VIII metals	270–470 K 1–5 atm
Xylenes		ZSM-5-zeolites	480–580 K 2–5 atm
Alkanes		Zeolites, Pt/Al_2O_3	570–770 K 5–50 atm
Methanol synthesis	$CO + 2H_2 \rightarrow CH_3OH$	$ZnCrO_3$ ZnO—Cu_2O—Cr_2O_3 ZnO—Cu_2O—Al_2O_3	570–670 K 100–600 atm
Methanol to gasoline	$CH_3OH \rightarrow$ aromatics olefins, H_2O	ZSM-5-zeolites	480–540 K 2–15 atm
NO_x Reduction	$NO + \frac{5}{2}H_2 \rightarrow NH_3 + H_2O$ $2NO + 2H_2 \rightarrow N_2 + 2H_2O$ $2CO + 2NO \rightarrow 2CO_2 + N_2$	Ru, Rh, Pd, Pt/SiO_2 Ru, Rh, metal oxides	370–520 K 1–10 atm 450–650 K 1–10 atm
Oxidation	$\left.\begin{array}{l}\text{Olefins} +\\ \text{Alkanes}\end{array}\right\} + O_2 \rightarrow CO_2 + H_2O$ $2NH_3 + \frac{5}{2}O_2 \rightarrow NO + 3H_2O$ $CO + \frac{1}{2}O_2 \rightarrow CO_2$	Group VIII metals	370–670 K
Partial oxidations			
Alcohols	$CH_3OH + \frac{1}{2}O_2 \rightarrow H_2CO + H_2O$	Ag, $Fe_2(MoO_4)_3$	550–570 K
o-Xylene		V_2O_5	1–10 atm
Olefins	$C_2H_4 + \frac{1}{2}O_2 \rightarrow CH_3CHO$ 	V_2O_5 $SnO_2 \cdot MoO_3$ $Bi_2O_3 \cdot MoO_3$	
Reforming			
Dehydrogenation		Pt, Pt–Re, Pt–Ge	700–800 K

continues

TABLE V (*Continued*)

Processes	Typical reactions	Catalyst	Reaction conditions
Dehydrocyclization	$\triangle\!\!\triangle\!\!\triangle \longrightarrow \bigcirc\!\!\!\bigcirc + 3H_2$	Pt—Au, Pt—Re—Cu	5–50 atm
Dehydroisomerization	$\pentagon \longrightarrow \bigcirc\!\!\!\bigcirc + 3H_2$	Ir—Au/Al$_2$O$_3$	
Isomerization	$\triangle\!\!\triangle\!\!\triangle \longrightarrow \bowtie$		
Hydrogenolysis	$\triangle\!\!\triangle\!\!\triangle + H_2 \longrightarrow 2C_3H_8$		
Hydrogenation	$\hexagon + H_2 \longrightarrow \hexagon$		
Selective			
Hydrogenation	$H_2 + \bigwedge\!\!\!\bigwedge \longrightarrow \bigwedge\!\!\bigwedge$	NiS	420–500 K
Olefins			1–10 atm
Alkynes	$R\!-\!C\!\equiv\!CH + H_2 \rightarrow RHC\!=\!CH_2$	Pt/Al$_2$O$_3$	220–250 K
			1–10 atm
Steam reforming	$CH_4 + H_2O \rightarrow CO + 3H_2$	Ni—K$_2$O/Al$_2$O$_3$	850–1100 K
			30–100 atm
Water gas	$CO + H_2O \rightarrow CO_2 + H_2$	Fe$_2$O$_3$·Cr$_2$O$_3$ZnO—Cu$_2$O	650–800 J
			20–50 atm

reactants and the surface. The total rate (moles of product per time) is proportional to the surface area. As a consequence, a lot of effort is expended to prepare large surfaces area catalysts and to measure the surface area accurately. One example of high-surface-area catalysts is the group of catalysts known as zeolites, which are aluminosilicates used for the cracking of hydrocarbons. They have crystal structures full of pores 8–20 Å in size. The structure of one of the many zeolites used for catalysis, faujasite, is shown in Fig. 44. Since the catalytic reactions occur inside the pores, an enormous inner surface area, of several hundred

square meters per gram of catalyst, is available in these catalyst systems. Transition-metal catalysts are generally employed in a small, 10- to 100-Å-diameter particle form dispersed on large-surface-area supports. The support can be a specially prepared alumina or silica framework (or a zeolite) that can be produced with surface areas in the 10^2-m^2/g range. These supported metal catalysts are often available with near-unity dispersion (dispersion is defined as the number of surface atoms per total number of atoms in the particle) of the metal particles and are usually very stable in this configuration during the catalytic reaction. The metal is frequently deposited from solution as a salt and then reduced under controlled conditions. Alloy catalysts and other multicomponent catalyst systems can also be prepared in such a way that small alloy clusters are formed on the large-surface-area oxide supports.

Most catalytic reactions take place via the formation of intermediate compounds between the reactants or products and the surface. The surface atoms of the catalyst form strong chemical bonds with the incident molecules, and it is this strong chemical surface–adsorbate interaction which provides the driving force for breaking high-binding-energy chemical bonds (C—C, C—H, H—H, N—N, and C=O bonds), which are often an important part of the catalytic reaction.

A good catalyst will also permit rapid bond breaking between the adsorbed intermediates and the surface and

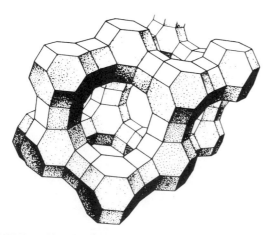

FIGURE 44 Line drawing of the structure of the zeolite faujasite.

the speedy release or desorption of the products. If the surface bonds are too strong, the reaction intermediates block the adsorption of new reactant molecules, and the reaction stops. For too-weak adsorbate-catalyst bonds, the necessary bond-scission processes may be absent. Hence, the catalytic reaction will not occur. A good catalyst is thought to be able to form chemical bonds of intermediate strength. These bonds should be strong enough to induce bond scission in the reactant molecules. However, the bond should not be too strong to ensure only short residence times for the surface intermediates and rapid desorption of the product molecules, so that the reaction can proceed with a large turnover number.

Of course, activity is only one of many parameters that are important in catalysis. The selectivity of the catalyst, its thermal and chemical stability, and dispersion, are among the other factors that govern our choices. While macroscopic chemical-bonding arguments can explain catalytic activity in some cases, atomic-scale scrutiny of the surface intermediates, catalyst structure, and composition, and an understanding of the elementary rate processes are necessary to develop the optimum selective catalyst for any chemical reaction.

One of the important directions of research in catalysis is the identification of the reaction intermediates. The surface residence times of many of these species are longer than 10^{-5} sec under most catalytic reaction conditions (as inferred from the turnover frequency). They may be detected by suitable spectroscopic techniques either during the steady-state reaction or when isolated by interrupting the catalytic process.

The concept of active sites is an important one in catalysis. A surface generally possesses active sites in numbers that are smaller than the total number of surface atoms. The presence of unique atomic sites of low coordination and different valency that are very active in chemical reactions has been clearly demonstrated by atomic-scale studies of metal and oxide surfaces. A catalytic reaction is defined to be structure sensitive if the rate changes markedly as the particle size of the catalyst is changed. Conversely, the reaction is structure insensitive on a given catalyst if its rate is not influenced appreciably by changing the dispersion of the particles under the usual experimental conditions. In Table VI we list several reactions that belong to these two classes. Clearly, variations of particle size give rise to changes of atomic surface structure. The relative concentrations of atoms in steps, kinks, and terraces are altered. Nevertheless, no clear correlation has been made to date between variations of macroscopic particle size and the atomic surface structure.

Most surface reactions and the formation of surface intermediates involve charge transfer, either an electron transfer or a proton transfer. These processes are often

TABLE VI Structure-Sensitive and Structure-Insensitive Catalytic Reactions

Structure sensitive	Structure insensitive
Hydrogenolysis	Ring opening
Ethane: Ni	Cyclopropane: Pt
Methylcyclopentane: Pt	
Hydrogenation	Hydrogenation
Benzene: Ni	Benzene: Pt
Isomerization	Dehydrogenation
Isobutane: Pt	Cyclohexane: Pt
Hexane: Pt	
Cyclization	
Hexane: Pt	
Heptane: Pt	

viewed as modified acid–base reactions. It is common to refer to an oxide catalyst as acidic or basic according to its ability to donate or accept electrons or protons.

The electron transfer capability of a catalyst is expressed according to the Lewis definition. A Lewis acid is a surface site capable of receiving a pair of electrons from the adsorbate. A Lewis base is a site having a free pair of electrons that can be transferred to the adsorbate. The proton-transfer capability of a catalyst is expressed according to the Brønsted definition. A Brønsted acid is a surface site capable of losing a proton to the adsorbate while a Brønsted base is a site that can accept a proton from the adsorbed species.

Perhaps the most widely used catalysts, the zeolites, best represent the group of oxides that exhibit acid–base catalysis. Zeolites are alumina silicates, some of which are among the more common minerals in nature. Modern synthesis techniques permit the preparation of families of zeolite compounds with different Si/Al ratios. Since the Al^{3+} ions lack one positive charge in the tetrahedrally coordinated silica, Si^{4+}, framework, they are sites of proton or alkali–metal affinity. Variation of the Si/Al ratio gives rise to a series of substances of controlled but different acidity. By using various organic molecules during the preparation of these compunds that build into the structure, subsequent decomposition leaves an open pore structure, where the pore size is controlled by the skeletal structure of the organic deposit. Very high internal surface area catalysts (10^2 m^2/g) can be obtained this way with controlled pore sizes of 8–20 Å and controlled acidity [(Si/Al) ratio]. These catalysts are utilized in the cracking and isomerization of hydrocarbons that occur in a shape selective manner as a result of the uniform pore structure and are the largest volume catalysts in petroleum refining. They are also the first of the high-technology catalysts in which the chemical activity is tailored by atomic-scale study and control of the internal surface structure and composition.

A catalyst used in industry is very rarely a pure element or compound. Most catalysts contain a complex mixture of chemical additives or modifiers that are essential ingredients for high activity and selectivity. Promoters are beneficial additives that increase activity, selectivity, or useful catalyst lifetime (stability). Structural promoters inhibit sintering of the active catalyst phase or present compound formation between the active component and the support. The most frequently used chemical promoters are electron donors such as the alkali metals or electron acceptors such as oxygen and chlorine. For example, in the petroleum industry, chlorine and oxygen are often added to commercial platinum catalysts used for reforming reactions by which aliphatic straight-chain hydrocarbons are converted to aromatic molecules (dehydrocyclization) and branched isomers (isomerization).

These additives accomplish several tasks during the reaction. By changing the chemical bonding of some of the surface intermediates, the steady-state concentration of these intermediates may be altered, and thus a somewhat higher concentration of the catalytically active species is obtained. In this way the rate of the reaction is increased and the selectivity may be improved.

Often multicomponent catalyst systems are utilized to carry out reactions consisting of two or more active metal components or both oxide and metal constituents. For example, a Pt–Rh catalyst facilitates the removal of pollutants from car exhausts. Platinum is very effective for oxidizing unburned hydrocarbons and CO to H_2O and CO_2, and rhodium is very efficient in reducing NO to N_2, even in the same oxidizing environment. Dual functional or multifunctional catalysts are frequently used to carry out complex chemical reactions. In this circumstance the various catalyst components should not be thought of as additives, since they are independently responsible for different catalytic activity. Often there are synergistic effects, however, whereby the various components beneficially influence each other's catalytic activity to provide a combined additive and multifunctional catalytic effects.

It should be clear from this discussion that the working, active, and selective catalyst is a complex, multicomponent chemical system. This system is finely tuned and buffered to carry out desirable chemical reactions with high turnover frequency and to block the reaction paths for other thermodynamically equally feasible but unwanted reactions. Thus, an iron catalyst or a platinum catalyst is composed not only of iron or platinum but of several other constituents as well to ensure the necessary surface structure and oxidation state of surface atoms for optimum catalytic behavior. Additives are often used to block sites,

prevent side reactions, and alter the reaction paths in a variety of ways.

While industrial catalytic systems are complex and are not readily suited to basic science studies to understand how they work on an atomic scale, one approach to their understanding is the synthetic approach. In this approach we begin with a very simple system then synthesize complexity from this. The catalyst particle is viewed as composed of single crystal surfaces, as shown in Fig. 45. Each surface has different reactivity and the product distribution reflects the chemistry of the different surface sites. We may start with the simplest single crystal surface [for example, the (111) crystal face of platinum] and examine its reactivity. It is expected that much of the chemistry of the dispersed catalyst system would be absent on such a homogeneous crystal surface. Then high-Miller-index crystal faces are prepared to expose surface irregularities, steps, and kinks of known structure and concentration, and their catalytic behavior is tested and compared with the activity of the dispersed supported catalyst under identical experimental conditions. If there are still differences, the surface composition is changed systematically or other variables are introduced until the chemistries of the model system and the working catalyst become identical. This approach is described by the following sequence:

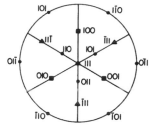

Standard Cubic (III) Projection

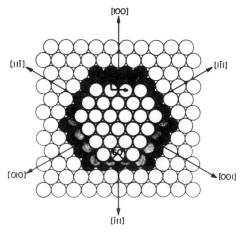

FIGURE 45 Catalyst particle viewed as a crystallite, composed of well-defined atomic planes. (Courtesy of Lawrence Berkeley Laboratory.)

structure of crystal surfaces and adsorbed gases

↓ ↑

surface reactions on crystals at low pressures

($\leq 10^{-4}$ torr)

↓ ↑

surface reactions on crystals at high pressures

($10^{+3}-10^{+5}$ torr)

↓ ↑

reactions on dispersed catalysts

Investigations in the first step define the surface structure and composition on the atomic scale and the chemical bonding of adsorbates. Studies in the second step, which are carried out at low pressures, reveal many of the elementary surface reaction steps and the dynamics of surface reactions. Studies in the third and fourth steps establish the similarities and differences between the model system and the dispersed catalyst under practical reaction conditions.

The advantage of using small-area catalyst samples is that their surface structure and composition can be prepared with uniformity and can be characterized by the many available surface diagnostic techniques.

In this approach to catalytic reaction studies the surface composition and structure are determined in the same chamber where the reactions are performed, without exposing the crystal surface to the ambient atmosphere. This necessitates the combined use of an ultrahigh vacuum enclosure, where the surface characterization is carried out, and a high-pressure isolation cell, where the catalytic studies are performed. Such an apparatus is shown in Fig. 46. The small-surface-area (approximately 1-cm^2) catalyst is placed in the middle of the chamber, which can be evacuated to 10^{-9} torr. The surface is characterized by LEED and AES and by other surface diagnostic techniques. The

lower part of the high-pressure isolation cell is then lifted to enclose the sample in a 30-cm^3 volume. The isolation chamber can be pressurized to 100 atm if desired and is connected to a gas chromatograph that detects the product distribution as a function of time and surface temperature. The sample may be heated resistively both at high pressure or in ultrahigh vacuum. After the reaction study the isolation chamber is evacuated, opened, and the catalytic surface is again analyzed by the various surface-diagnostic techniques. Ion bombardment cleaning of the surface or means to introduce controlled amounts of surface additives by vaporization are also available. The reaction at high pressures may be studied in the batch or the flow mode.

Typical catalytic reactions that have been investigated, in some detail, using this approach include hydrocarbon conversion on platinum and modified platinum surfaces (isomerization, hydrogenolysis, hydrogenation, dehydrogenation and cyclization), dehydrosulfurization on molybdenum, ammonia synthesis on iron, and carbon monoxide hydrogenation on iron.

F. Photochemical Surface Reactions

Photochemical surface reactions form their own class due to the fact that a thermodynamically uphill reaction ($\Delta G > 0$) may be carried out with the aid of an external source of energy, light. In fact, one of the most important chemical reactions of our planet, photosynthesis, requires the input of 720 kcal/mol of energy to convert carbon dioxide and water to one mole of sugar:

$$6CO_2 \rightarrow 6H_2O \xrightarrow[\text{chlorophyll}]{\text{light}} 6H_{12}O_6 + 6O_2 \qquad (65)$$

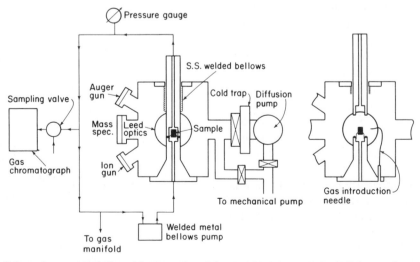

FIGURE 46 Schematic representation of the experimental apparatus to carry out catalytic reaction-rate studies on single-crystal surfaces of low surface area at low and high pressures in the range 10^{-7} to 10^4 torr.

It is useful to consider light as one of the reactants in photosynthesis. By adding the light energy to Eq. (65), the reaction becomes athermic or even exothermic if excess light energy is utilized,

$$hv + H_2O + CO_2 = -CH_2- + \tfrac{3}{2}O_2$$

We may consider photon-assisted or photochemical reactions of many types that lead to the formation of lower molecular weight hydrocarbons and of other products. One of the simplest of these important new classes of reactions leads to the dissociation of water:

$$hv + H_2O = H_2 + \tfrac{1}{2}O_2 \qquad (66)$$

Another leads to the formation of methane:

$$hv + CO_2 + 2H_2O = CH_4 + \tfrac{3}{2}O_2 \qquad (67)$$

or to the fixation of nitrogen:

$$hv + 3H_2O + N_2 = 2NH_3 + \tfrac{3}{2}O_2 \qquad (68)$$

Light as a reactant may be employed in two ways. The adsorbed molecules can be excited directly by photons of suitable energy to a higher vibrational or electronic states. The excited species then may undergo chemical rearrangements or interactions that are different from those in the ground vibrational or electronic states. Alternatively, the solid can be excited by light in the near-surface region. Photons of band-gap or greater energy may excite electron–hole pairs at the surface. As long as these charge carriers have a relatively long lifetime (i.e., they are trapped at the surface, so that their recombination is not an efficient process), there is a high probability of their capture by the adsorbed reactants. These, in turn, can undergo reduction or oxidation processes using the photogenerated electrons and holes, respectively. The photographic process is one example of this type of surface photochemical reaction. However, we would like the photogenerated electrons and holes to be captured by the adsorbed molecules in order to carry out photochemical surface reactions of the adsorbates instead of the photodecomposition of the solid at the surface. The cross sections for adsorption of band-gap or higher-than-band-gap energy photons are so large that the photogeneration of electron–hole pairs is a most efficient process. At present, this cannot be readily matched by the efficiency of direct photoexcitation of vibrational or electronic energy states of the adsorbed molecules.

Many solid surfaces efficiently convert light to long-lived electron–hole pairs that can induce the chemical changes leading to the reactions in Eqs. (66)–(68). In fact, inorganic photoreaction is one of the exciting new fields of surface science and heterogeneous catalysis.

It is important to distinguish between thermodynamically uphill photochemical reactions and thermodynam-ically allowed photon-assisted reactions. The latter reactions are thermodynamically feasible without any external energy input, but light is used to obtain certain product selectively. Excitation of selected vibrations, rotations, or electronic states of the incident or adsorbed molecules by light permits us to change the reaction path or increase the reaction rate. For example, the hydrogenation of acetylene or the oxidation of ammonia can be photon-assisted, leading to different reaction rates than in the absence of light.

As an example of a photocatalyzed surface reaction we discuss the photoelectrochemical dissociation of water. It was shown in 1972 that upon illumination of reduced titanium oxide (TiO_2), which served as the anode in basic electrolyte solution, oxygen evolution was detectable at the anode, and hydrogen evolved at a metal (platinum) cathode. This reaction requires an energy of 1.23 V/electron (a two-electron process per dissociated water molecule). In the presence of light of energy equal to or greater than the band-gap energy of titanium oxide (3.1 eV), an external voltage as low as 0.2 V was sufficient to dissociate water. The process stopped as soon as the light was turned off, and started again upon reillumination. Shortly after, several other systems showed the ability to carry out photon-assisted dissociation of water. When p-type gallium phosphide, GaP, was used as a cathode instead of platinum upon illumination of the TiO_2 anode, O_2 and H_2 could be generated at the semiconductor anode and cathode, respectively, without the need of applying any external potential. When strontium titanate, $SrTiO_3$, was substituted for TiO_2 as the anode, H_2O photodissociation was found to take place without external potential even when a platinum cathode was employed.

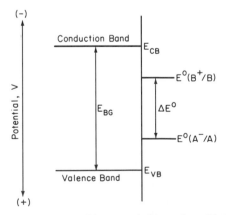

FIGURE 47 Energy conditions needed to reduce B^+ to B and oxidize A^- to A at a semiconductor surface. Electrons that are excited by photons into the conduction band E_{CB} must be able to reduce B^+, and electron vacancies (holes) in the valence band E_{VB} must be able to oxidize A^-.

Figure 47 shows a schematic energy diagram to indicate the conditions necessary to carry out photoelectrochemical reactions efficiently. If the band-gap energy is greater than the free energies for the reduction and oxidation reactions, the photoelectron that is excited into the conduction band by light could reduce B^+ to B by electron transfer from the surface to the molecule. The photogenerated electron vacancies (holes) could also oxidize the A^- anions to A by capturing the electron. For the photodissociation of water, the conduction band must be above the H^+/H_2 potential and the valence band below the O_2/OH^- potential to be able to carry out the photoreaction without an external potential. The band gap must be greater than 1.23 V and the flat-band potential of the conduction and valence bands energetically well placed with respect to the (H^+/H_2) and O_2/OH^- couples. The flat-band potentials can be obtained by capacitance measurements as a function of external potential.

There is, of course, considerable band bending of the conduction and valence bands of any semiconductor at the surface. This is due to the presence of localized electronic surface states and to charge transfer between the adsorbates and semiconductor. Potential-energy diagrams that show the band positions schematically at an n-type or p-type semiconductor liquid interface are shown in Fig. 48. The band bending provides an efficient means of separating electron–hole pairs, since the potential gradient as shown for the n-type semiconductor drives the electrons away from the semiconductor surface while it attracts the holes in the valence band toward the semiconductor electrolyte interface. As a result, the oxidation reaction takes place at the oxide anode while the reduction reaction takes place at the cathode to which the photoelectron migrates along the external circuit. The magnitude of the band bending at the surface depends primarily on the carrier concentration in the semiconductor and on the electron-donating or -accepting abilities of the adsorbates at the surface. Semiconductors that are not likely to carry out the photodissociation of water, according to the location of their

flat-band potential, may become photochemically active as a result of strong band bending at the surface.

Often the oxidation or reduction photoreactions lead to the decomposition of the semiconductor electrode material. Instead of the photoreactions of adsorbate ions or molecules, a solid-state photoreaction occurs. This is particularly noticeable at the surfaces of illuminated CdS, Si, and GaP. Much of the research is therefore directed toward stabilizing these photoelectrode materials by suitable adsorbates that could prevent the occurrence of photodecomposition by providing an alternative chemical route for the photoreduction or photooxidation.

ACKNOWLEDGMENT

This work was supported by the Assistant Secretary for Energy Research, Office of Basic Energy Sciences, Materials Sciences Division of the U.S. Department of Energy under Contract No. DE-AC03-76SF00098.

SEE ALSO THE FOLLOWING ARTICLES

ADHESION AND ADHESIVES • ADSORPTION • AUGER ELECTRON SPECTROSCOPY • BONDING AND STRUCTURE IN SOLIDS • CATALYSIS, INDUSTRIAL • CATALYST CHARACTERIZATION • CHEMICAL THERMODYNAMICS • CRYSTALLOGRAPHY • PHOTOCHEMISTRY, MOLECULAR • PHOTOELECTRON SPECTROSCOPY • SOLID-STATE ELECTROCHEMISTRY • TRIBOLOGY

BIBLIOGRAPHY

Adamson, A. W. (1982). "Physical Chemistry of Surfaces," 4th ed., Wiley, New York.
Anderson, J. R., and Boudart, M. (1981). "Catalysis Science and Technology," Vols. 1–7, Springer-Verlag, Berlin/New York.
Ertl, G., and Gomer, R., eds. (1983). "Springer Series in Surface Sciences," Vols. 1–4, Springer-Verlag, Berlin/New York.
Ertl, G., and Kuppers, J. (1979). "Low Energy Electrons and Surface Chemistry," Verlag Chemie, Weinheim.
Feuerbacher, B., Fitton, B., and Willis, R. F. (1979). "Photoemission and the Electronic Properties of Surfaces," Wiley, New York.
King, D. A., and Woodruff, W. P., eds. (1983). "The Chemical Physics of Solid Surfaces and Heterogeneous Catalysis," Vols. 1–4, Elsevier, New York.
Morrison, S. R. (1977). "The Chemical Physics of Surfaces," Plenum, New York.
Roberts, M. W., and McKee, C. S. (1978). "Chemistry of the Metal-Gas Interface," Oxford Univ. Press, London.
Somorjai, G. A. (1981). "Chemistry in Two Dimensions: Surfaces," Cornell Univ. Press, Ithaca, NY.
Tompkins, F. C. (1978). "Chemisorption of Gases on Metals," Academic Press, NY.
Vanselow, R., and Howe, R., eds. (1979). "Chemistry and Physics of Solid Surfaces," Vols. 1–6, Springer-Verlag, Berlin/New York.

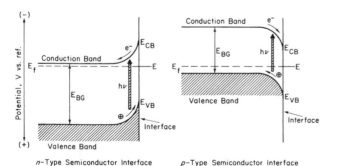

FIGURE 48 Band bending at the n-type and p-type semiconductor interfaces.

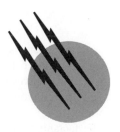

Surfactants, Industrial Applications

Tharwat F. Tadros
Imperial Chemical Industries

GLOSSARY

Cloud point Temperature at which a surfactant solution of a given concentration becomes turbid (cloudy).

Critical micelle concentration (cmc) Concentration at which the physical properties of a surfactant solution show an abrupt change.

Emulsion Dispersion of a liquid in a liquid.

Krafft temperature Temperature at which the surfactant shows a sudden increase in solubility.

Micelles Association units of surfactant molecules.

Microemulsion Thermodynamically isotropic system of water, oil, and surfactant.

Solubilization Incorporation of an insoluble substance in a surfactant solution.

Suspension Dispersion of a solid in a liquid.

SURFACE-ACTIVE agents (usually referred to as surfactants) are amphipathic molecules consisting of a nonpolar hydrophobic portion, usually a straight or branched hydrocarbon or fluorocarbon chain containing 8–18 carbon atoms, which is attached to a polar or ionic portion (hydrophilic). The hydrophilic portion can, therefore, be nonionic, ionic, or zwitterionic, accompanied by counterions in the last two cases. The hydrocarbon chain interacts weakly with the water molecules in an aqueous environment, whereas the polar or ionic head group interacts strongly with water molecules via dipole or ion–dipole interactions. It is this strong interaction with the water molecules which renders the surfactant soluble in water. However, the cooperative action of dispersion and hydrogen bonding between the water molecules tends to "squeeze" the hydrocarbon chain out of the water and hence these chains are referred to as hydrophobic. The balance between hydrophobic and hydrophilic parts of the molecule gives these systems their special properties, e.g., accumulation at various interfaces and association in solution.

Surfactants find application in almost every chemical industry, of which the following are worth mentioning: detergents, paints, dyestuffs, personal care and cosmetics, pharmaceuticals, agrochemicals, ceramics, fibres, plastics, and paper coating. Moreover, surfactants play a major role in the oil industry, for example, in enhanced and tertiary oil recovery. They are also occasionally used for environmental protection, e.g., in oil slick dispersants. Therefore, a fundamental understanding of the physical chemistry of surface active agents, their unusual properties, and their phase behavior is essential for most industrial chemists. In addition, understanding the basic phenomena involved in the application of surfactants such as in the preparation of emulsions and suspensions and their subsequent stabilization, in microemulsions, and in wetting, spreading, and adhesion, is of vital importance in arriving at the right composition of many industrial formulations.

I. INTRODUCTION

In this overview, I start with the general classification of surfactants and their unusual properties. This is followed by some examples to illustrate the application of surfactants in some chemical industries.

II. GENERAL CLASSIFICATION OF SURFACTANTS

A simple classification of surfactants based on the nature of the hydrophilic group is commonly used. Four main classes may be distinguished, namely, anionic, cationic, zwitterionic, and nonionic. A useful technical reference is McCutchen. Another useful text, by van Oss et al., gives a list of the physicochemical properties of selected anionic, cationic, and nonionic surfactants. The handbook by Porter is also a useful book for classification of surfactants. Another important class of surfactants, which has attracted considerable attention in recent years, is the polymeric type. A brief description of the various classes is given below.

A. Anionic Surfactants

This is the most widely used class of surfactants in industrial application, due to their relatively low cost of manu-

facture and their wide application in detergents. For optimum detergency, the hydrophobic chain is a linear alkyl chain with a length in the region of 12–16 C atoms. Linear chains are preferred since they are more effective and more degradable than branched ones. The most commonly used hydrophilic groups are carboxylates, sulfates, sulfonates and phosphates. A general formula may be ascribed to anionic surfactants as follows.

Carboxylates: $C_nH_{2n+1}COO^-X$
Sulfates: $C_nH_{2n+1}OSO_3^-X$
Sulfonates: $C_nH_{2n+1}SO_3^-X$
Phosphates: $C_nH_{2n+1}OPO(OH)O^-X$

n is usually in the range 8–18 C atoms, and the counterion X is usually Na^+.

Several other anionic surfactants are commercially available, such as sulfosuccinates, isethionates, and taurates, and these are sometimes used for special applications. The carboxylates and sulfates are sometimes modified by the incorporation of a few moles of ethylene oxide (referred to as ether carboxylates and ether sulfates, respectively).

B. Cationic Surfactants

The most common cationic surfactants are the quaternary ammonium compounds with the general formula $R'R''R'''R''''NX^-$, where X is usually chloride ion and R represents alkyl groups. A common class of cationics is the alkyl trimethyl ammonium chloride, where R contains 8–18 C atoms, e.g., dodecyl trimethyl ammonium chloride, $C_{12}H_{25}(CH_3)_3NCl$. Another widely used cationic surfactant class is that containing two long-chain alkyl groups, having a chain length of 8–18 C atoms. These dialkyl surfactants are less soluble in water and they produce multilamellar structures (vesicles or liposomes). They are commonly used as fabric softeners.

A widely used cationic surfactant is benzalkonium chloride, where one of the methyl groups is replaced by a benzyl group, $C_6H_5-CH_2$. This surfactant is commonly used as a bactericide. Cationic surfactants can also be modified by the incorporation of ethylene oxide chains in the molecules.

C. Amphoteric Surfactants

These are surfactants containing both cationic and anionic groups. The most common amphoterics are the N-alkyl betaines, which are derivatives of trimethylglycine, $(CH_3)_3NCH_2COOH$ (which was described as betaine). An example of a betaine surfactant is laurylamidopropyldimethylbetaine, $C_{12}H_{25}CON(CH_3)_2CH_2COOH$.

The main characteristics of amphoteric surfactants is their dependence on the pH of the solution in which they are dissolved. In acid solutions, the molecule acquires a positive charge and it behaves like a cationic, whereas in alkaline solutions they become negatively charged and behave like an anionic. A specific pH can be defined at which both ionic groups show equal ionization (the isoelectric point of the molecule).

D. Nonionic Surfactants

The most common nonionic surfactants are those based on ethylene oxide, referred to as ethoxylated surfactants. Several classes can be distinguished: alcohol ethoxylates, alkyl phenol ethoxylates, fatty acid ethoxylates, sorbitan ester ethoxylates, fatty amine ethoxylates, and ethylene oxide–propylene oxide copolymers (sometimes referred to as polymer surfactants). Another important class of nonionics are the multihydroxy products such as glycol esters, glycerol (and polyglycerol) esters, glucosides (and polyglucosides), and sucrose esters. Amine oxides and sulfinyl surfactants represent nonionic with a small head group.

The critical micelle concentration (cmc) of nonionics (see below) is about two orders of magnitude lower than the corresponding anionics with the same alkyl chain length. Molecules with an average alkyl chain length of 12 C atoms and containing more than 5 ethylene oxide (EO) units are usually soluble in water at room temperature. However, as the temperature of the solution is gradually increased, the solution becomes cloudy (as a result of dehydration of the PEO chain) and the temperature at which this occurs is referred to as the cloud point (CP) of the surfactant solution. At any given alkyl chain length the CP increases with increase in the number of EO units in the molecule. The CP is also affected by the presence of electrolytes. Generally, the CP decreases with an increase in electrolyte concentration. The reduction in cloud point depends also on the nature of the electrolyte added.

E. Fluorocarbon and Silicone Surfactants

These surfactants can lower the surface tension of water, γ, to values below 20 mN m^{-1} (most surfactants lower γ to values in the region of 30 mN m^{-1}) and hence they are sometimes described as superwetters. They are very useful for enhancing the wetting and spreading of liquids on solid substrates.

F. Polymeric Surfactants

Polymeric surfactants are specially designed to produce excellent dispersing agent (for suspensions) and emulsi-

fiers (for emulsions). Several molecules have been introduced such as the block copolymers of polyethylene oxide (PEO) and polypropylene oxide (PPO). These are A–B–A block copolymers with the structure PEO–PPO–PEO and molecules are commercially available with various proportions of PEO and PPO. Other types of A–B block copolymers are those based on polystyrene (PS) and PEO. Graft copolymers with one B chain and several A chains also exist such as a graft of poly(methylmethacrylate) with a number of PEO side chains. These polymeric surfactants have been applied for the preparation of highly concentrated suspensions and emulsions to enhance their long term stability.

G. The Hydrophilic–Lipophil Balance (HLB)

A useful index for choosing surfactants for various applications is the hydrophilic–lipophilic balance (HLB), which is based on the relative percentage of hydrophilic-to-lipophilic groups in the surfactant molecule(s). Surfactants with a low HLB number normally form W/O emulsions, whereas those with a high HLB number form a O/W emulsion. A summary of the HLB range required for various purposes is given in Table I.

Griffin developed a simple equation for calculation of the HLB number of certain numbers of nonionic surfactants such as fatty acid esters and alcohol ethoxylates. For the polyhydroxy fatty the HLB number is given by the equation

$$HLB = 20\left(1 - \frac{S}{A}\right), \tag{1}$$

where S is the saponification number of the ester and A is the acid number of the acid. Thus, a glycerol monostearate, with $S = 161$ and $A = 198$, will have an HLB number of 3.8, i.e., it is suitable for a W/O emulsifier.

For the simpler alcohol ethoxylates, the HLB number can be calculated from the weight percentages of oxyethylene E and polyhydric alcohol P, i.e.,

$$HLB = \frac{(E + P)}{5}. \tag{2}$$

TABLE I Summary of HLB Ranges for Various Applications

HLB range	Application
3–6	W/O emulsifier
7–9	Wetting agent
8–18	O/W emulsifier
13–16	Detergent
15–18	Solubilizer

III. PHYSICAL PROPERTIES OF SURFACTANT SOLUTIONS

The physical properties of surface active agents differ from those of smaller or nonamphipathic molecules in one major aspect, namely, the abrupt changes in their properties above a critical concentration. This is illustrated in Fig. 1, in which a number of physical properties (surface tension, osmotic pressure, turbidity, solubilization, magnetic resonance, conductivity, and self-diffusion) are plotted as a function of concentration. All these properties (interfacial and bulk) show an abrupt change at a particular concentration, which is consistent with the fact that above this concentration, surface active ions or molecules in solution associate to form larger units. These association units are called micelles and the concentration at which this association phenomenon occurs is known as the critical micelle concentration (cmc).

Each surfactant molecule has a characteristic cmc value at a given temperature and electrolyte concentration. A compilation of cmc values was given in 1971 by Mukerjee and Mysels. As an illustration, the cmc values of a number of surfactants are given in Table II, to show some of the general trends. Within any class of surface active agents, the cmc decreases with an increase in chain length of the hydrophobic chain. With nonionic surfactants, increasing the length of the hydrophilic (PEO) chain causes an increase in the cmc. In general, nonionic surfactants have lower cmc values than their corresponding ionic surfactants with the same chain length. Incorporation of a phenyl group in the alkyl chain increases its hydrophobicity to a much smaller extent than increasing its chain length with the same number of C atoms.

The presence of micelles can account for many of the unusual properties of solutions of surfactants. For example, it can account for the near-constant surface tension

TABLE II The cmc Values of Some Surfactants

Surface-active agent	cmc (mol dm^{-3})
(A) Anionic	
Sodium octyl-1-sulfate	1.30×10^{-1}
Sodium decyl-1-sulfate	3.32×10^{-2}
Sodium dodecyl-1-sulfate	8.39×10^{-3}
Sodium tetradecyl-1-sulfate	2.05×10^{-4}
(B) Cationic	
Octyl trimethyl ammonium bromide	1.30×10^{-1}
Decyl trimethyl ammonium bromide	6.46×10^{-2}
Dodecyl trimthyl ammonium bromide	1.56×10^{-2}
Hexadecyl trimethyl ammonium bromide	9.20×10^{-4}
(C) Nonionic	
Octyl hexaoxyethylene glycol monoether, C_8E_6	9.80×10^{-3}
Decyl hexaoxyethylene glycol monoether, $C_{10}E_6$	9.00×10^{-4}
Decyl nonaoxyethylene glycol monoether, $C_{10}E_9$	1.30×10^{-3}
Dodecyl hexaoxyethlene glycol monoether, $C_{12}E_6$	8.70×10^{-5}
Octylphenyl hexaoxyethylene glycol monoether, $C_8\phi E_6$	2.05×10^{-4}

value above the cmc. It can also account for the reduction in molar conductance above the cmc, the rapid increase in turbidity above the cmc, etc.

The size and shape of micelles have been a subject of several debates. It is now generally accepted that three main shapes of micelles are present, depending on the surfactant structure and the environment in which they are dissolved, e.g., electrolyte concentration and type, pH, and presence of nonelectrolytes. The most common shape of micelles is a sphere with the following properties: (i) an association unit with a radius approximately equal to the length of the hydrocarbon chain (for ionic micelles); (ii) an aggregation number of 50–100 surfactant monomers; (iii) bound counterions for ionic surfactants; (iv) a narrow range of concentrations at which micellization occurs; and (v) a liquid interior of the micelle core.

Two other shapes of micelles may be considered, namely, the rod-shaped micelle suggested by Debye and Anacker and the lamellar micelle suggested by McBain. The rod-shaped micelle was suggested to account for the light-scattering results of cetyl trimethyl ammonium bromide in KBr solutions, whereas the lamellar micelle was considered to account for the X-ray results in soap solutions. A schematic picture of the three type of micelles is shown in Fig. 2.

One of the characteristic features of solutions of surfactants is their solubility–temperature relationship, which is illustrated in Fig. 3 for an anionic surfactant, namely, sodium decyl sulfonate. It can be seen that the solubility of the surfactant increases gradually with an increase in temperature, but above 22°C there is a rapid increase in

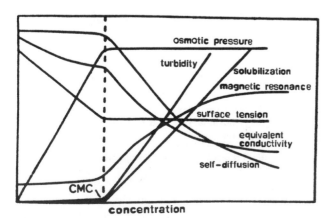

FIGURE 1 Changes in the concentration dependence of a wide range of physicochemical changes around the cmc. [From Lindman, B. (1984). *Surfactants.* Academic Press, London.]

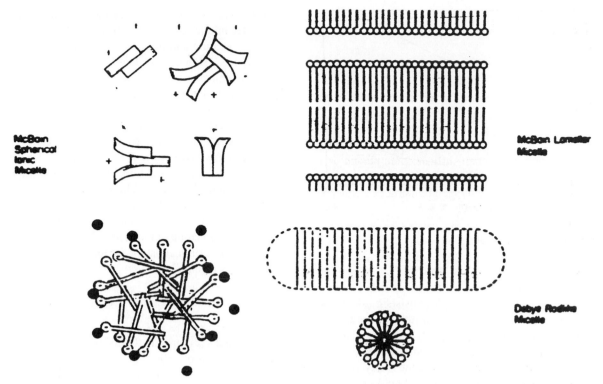

FIGURE 2 Various shapes of micelles following McBain (II). [Adapted from Hartley (1936) and Debye and Anaker (1951).]

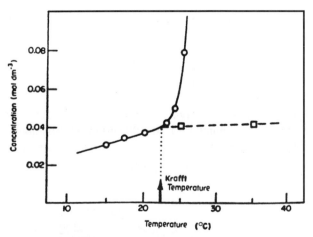

FIGURE 3 Solubility (O) and cmc (□) versus temperature for the sodium decyl sulfonate–water system.

solubility with further increases in temperature. The cmc of the surfactant also increases slowly with increases in temperature. The point at which the solubility curve intersects with the cmc curve (i.e., solubility = cmc) is referred to as the Krafft temperature of the surfactant. At the Krafft temperature, there is an equilibrium among solid hydrated surfactant, micelles, and monomer. The Krafft temperature of an ionic surfactant increases with an increase in the alkyl-chain length. For that reason, surfactants with

an alkyl chain longer than 12–14 C atoms are generally not very useful for application, since a concentrated solution can be prepared only at temperatures significantly higher than room temperature. One useful way to reduce the Krafft temperature is to use an alkyl chain with a wide chain length distribution. Indeed most commercial surfactants have this wide range since they are produced from natural fats and oils.

The solubility–temperature relationship for nonionic surfactants is different from that of ionic surfactants. This is illustrated in Fig. 4 which, shows the phase diagram for the binary system, dodecyl haxaoxyethylene glycol monoether ($C_{12}E_6$)–water. This phase diagram shows the various phases that are formed when the surfactant concentration and temperature are changed. The cloud-point curve at the top of the phase diagram separates the 2L (two liquid phases appear above the cloud point, one rich in surfactant and the other rich in water) from the I isotropic solution. Below the cloud-point curve, the phase diagram shows some characteristic regions at high surfactant concentrations, namely, the M and N region. The M-phase is the region of hexagonal or middle phase, which consists of cylindrical units that are hexagonally close-packed. In this region, the viscosity of the surfactant solution is extremely high and the system appears like a transparent gel. However, when viewed under the polarizing microscope,

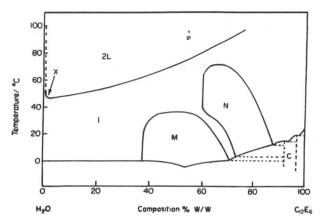

FIGURE 4 Phase diagram for the dodecyl hexaoxyethylene glycol monoether–water system.

it shows some texture ("fan-like structure"), which is due to the anisotropy of the units formed. The N-phase is the lamellar or neat phase, which consists of sheets of molecules in a bimolecular packing with head groups exposed to the water layers between them. This phase is less viscous than the middle phase and it shows different textures under the polarizing microscope ("oily streaks" and "maltese crosses"). Several other liquid crystalline phases may be identified with other nonionic surfactants, such as the cubic viscous isotropic phase (which shows no texture under the polarizing microscope). Figure 5 shows schematic pictures of the three phases described above.

A. Thermodynamics of Micellization

Micellization is a dynamic phenomenon in which n monomeric surfactant molecules S associate to form a micelle S_n,

$$nS \rightleftharpoons S_n. \tag{3}$$

Hartley envisaged a dynamic equilibrium whereby surface-active agent molecules are constantly leaving the

micelles while other molecules enter the micelle. Experimental techniques using fast kinetic methods such as stop flow, temperature and pressure jumps, and ultrasonic relaxation have shown that there are two relaxation processes for micellar equilibrium. The first relaxation time τ_1 is of the order of 10^{-7} sec (10^{-8}–10^{-3} sec) and represents the lifetime of a surface active molecule in the micelle, i.e., it represents the association and dissociation rate for a single molecule entering and leaving the micelle. The second relaxation time τ_2 corresponds to a relatively slow process, namely, the micellization–dissolution process represented by Eq. (3). The value of τ_2 is of the order of milliseconds (10^{-3}–1 sec).

The equilibrium aspect of micelle formation can be considered by application of the second law of thermodynamics. The equilibrium constant for the process represented by Eq. (3) is given by

$$K = \frac{[S_n]}{S^n} = \frac{C_m}{C_s^n}, \tag{4}$$

where C_s and C_m represent the concentration of monomer and micelle respectively.

The standard free energy of micellization, ΔG^0, is then given by

$$-\Delta G_m^0 = RT \ln K = RT \ln C_m - nRT \ln C_s, \tag{5}$$

and the free energy per monomer, $\Delta G^0 (= \Delta G_m^0/n)$, is given by

$$-\Delta G^0 = \left(\frac{RT}{n}\right) \ln C_m - RT \ln C_s. \tag{6}$$

For many micellar systems, n is a large number (>50), and therefore, the first term on the right-hand side of Eq. (6) may be neglected:

$$\Delta G^0 = RT \ln C_s = RT \ln [\text{cmc}]. \tag{7}$$

Middle Phase   Viscous Isotropic Phase   Neat Phase

FIGURE 5 Schematic representation of the structures found in concentrated surfactant solutions.

ΔG^0 is always negative and this shows that micelle formation is a spontaneous process. For example, for $C_{12}E_6$, the cmc is 8.70×10^{-5} mol dm^{-3} and $\Delta G^0 = -33.1$ KJ mol^{-1} (expressing the cmc as the mole fraction).

The enthalpy of micellization ΔH^0 can be measured either from the variation of cmc with temperature or directly by microcalorimetry. From ΔG^0 and ΔH^0, one can obtain the entropy of micellization ΔS^0,

$$\Delta G^0 = \Delta H^0 - T \Delta S^0. \tag{8}$$

Measurement of ΔH^0 and ΔS^0 showed that the former is small and positive and the second is large and positive. This implies that micelle formation is entropy driven and is described in terms of the hydrophobic effect (14). Then hydrophobic chains of the surfactant monomers tend to reduce their contact with water, whereby the latter form "icebergs" by hydrogen bonding. This results in reduction of the entropy of the whole system. However, when the monomers associate to from micelles, these "icebergs" tend to melt (hydrogen bonds are broken), and this results in an increase in the entropy of the whole system.

IV. ADSORPTION OF SURFACTANTS AT VARIOUS INTERFACES

The adsorption of surfactants at the liquid/air interface, which results in surface tension reduction, is important for many applications in industry such as wetting, spraying, impaction, and adhesion of droplets. Adsorption at the liquid/liquid interface is important in emulsification and subsequent stabilization of the emulsion. Adsorption at the solid/liquid interface is important in wetting phenomena, preparation of solid/liquid dispersions, and stabilization of suspensions. Below a brief description of the various adsorption phenomena is given.

A. Adsorption at Air/Liquid and Liquid/Liquid Interfaces

Gibbs derived a thermodynamic relationship between the surface or interfacial tension γ and the amount of surfactant adsorbed per unit area at the A/L or L/L interface, Γ (referred to as the surface excess),

$$\frac{d\gamma}{d \ln C} = -\Gamma RT, \tag{9}$$

where C is the surfactant concentration (mol dm^{-3}).

Equation (9) allows one to obtain the surface excess from the variation of surface or interfacial tension with surfactant concentration. Γ can be obtained from the slope of the linear portion of the $\gamma - \log C$ curve as illustrated in Fig. 6 for A/L and L/L interfaces.

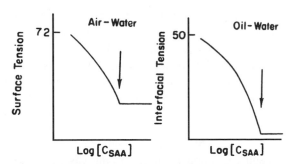

FIGURE 6 Variation of surface and interfacial tension with $\log [C_{SAA}]$ at the air–water and oil–water interface.

It can be seen that for the A/W interface γ decreases from the value for water ($\sim$72 mN m^{-1}), reaching about 25–30 mN m^{-1} near the cmc. For the O/W interface γ decreases from $\sim$50 mN m^{-1} (for a pure hydrocarbon–water interface) to $\sim$1–5 mN m^{-1}. Clearly the rate of reduction of γ with $\log C$ below the cmc and the limiting γ reached at and above the cmc depend on the nature of surfactant and the interface.

From Γ, the area per surfactant ion or molecule can be calculated:

$$\text{area/molecule} (A) = \frac{1}{\Gamma N_{av}} (m^2) = \frac{10^{18}}{\Gamma N_{av}} (nm^2). \tag{10}$$

The area/molecule A gives information on the surfactant orientation at the interface. For example, for an anionic surfactant such as sodium dodecyl sulfate, A is determined by the area occupied by the alkyl chain and head group, if these molecules lie "flat" at the interface. For a vertical orientation, A is determined by the area of the head group ($-O-SO_3^-$), which, at a low electrolyte concentration, is in the region of 0.4 nm^2. This area is larger than the geometrical area occupied by a sulfate group, as a result of the lateral repulsion between the head groups. On the addition of electrolyte, this lateral repulsion is reduced and A reaches a smaller value ($\sim$0.2 nm^2). For nonioinc surfactants, A is determined by the area occupied by the polyethylene oxide chain and A increases with an increase in the number of EO units (values in the region of 1–2 nm^2 are common with ethoxylated surfactants).

An important point can be made from the $\gamma - \log C$ curve. At a concentration just below the cmc, the curve is linear, indicating that saturation adsorption is reached just below the cmc. Above the cmc, the slope of the $\gamma - \log C$ curve is nearly zero, indicating a near-constant activity of the surfactant ions or molecules just above the cmc.

B. Adsorption of Surfactant at the Solid/Liquid Interface

The adsorption of surfactants at the S/L interface involves a number of complex interactions, such as hydrophobic, polar, and hydrogen bonding. This depends on the nature of the substrate as well as that of the surfactant ions or molecules. Generally speaking, solid substrates may be subdivided into hydrophobic (nonpolar) and hydrophilic (polar) surfaces. The surfactants can be ionic or nonioic, and they interact with the surface in a specific manner. The adsorption of ionic surfactants on hydrophobic surfaces (such as C black, polystyrene, and polyethylene) is determined by hydrophobic bonding between the alkyl chain and the nonpolar surface. In this case, the charged or polar head groups play a relatively smaller role, except in their lateral repulsion, which reduces adsorption. For this reason, the addition of electrolyte to ionic surfactants generally results in an increase in adsorption. The same applies for nonionic surfactants, which also show an increase in adsorption with increasing temperature.

The adsorption of surfactants on solid substrates may be described by the Frumkin–Fowler–Guggenheim equation,

$$\frac{\theta}{(1 - \theta)} \exp(A\theta) = \frac{C}{55.5} \exp\left(\frac{-\Delta G^0_{ads}}{kT}\right), \quad (11)$$

where θ is the fractional surface coverage, which is given by Γ/N_s (where Γ is the number of moles adsorbed per unit area and N_s is the total number of adsorption sites as moles per unit area for monolayer saturation adsorption), C is the bulk solution concentration as moles ($C/55.5$ gives the mole fraction of surfactant), A is a constant that is introduced to account for lateral interaction between the surfactant ions or molecules, and ΔG^0_{ads} is the standard free energy of adsorption, which may be considered to consist of two contributions, an electrical term ΔG^0_{elec} and a specific adsorption term ΔG^0_{spec}. The latter may consist of various contributions arising from chain–chain interaction, ΔG^0_{cc}, chain–surface interaction, ΔG^0_{cs}, and head group–surface interaction, ΔG^0_{hs}.

In many cases, the adsorption of surfactants on hydrophobic surfaces may follow a Langmuir-type isotherm,

$$\Gamma_2 = \frac{\Delta C}{mA} = \frac{abC_2}{1 + bC_2}, \quad (12)$$

where ΔC is the number of moles of surfactant adsorbed by m grams of adsorbent with surface area A (m^2 g^{-1}), C_2 is the equilibrium concentration, a is the saturation adsorption, and b is a constant related to the free energy of adsorption ($b \infty - \Delta G^0_{ads}$). The saturation adsorption a can be used to obtain the area per molecule A, as discussed above ($A = 1/aN_{av}$ m^2 or $10^{18}/a N_{av}$ nm^2).

The adsorption of ionic or polar surfactants on charged or polar surfaces involves coulombic (ion–surface charge interaction), ion–dipole, and/or dipole–dipole interaction. For example, a negatively charged silica surface (at a pH above the isoelectric point of the surface, i.e., pH >2–3) will adsorb a cationic surfactant by interaction between the negatively charged silanol groups and the positively charged surfactant ion. The adsorption will continue till all negative charges on silica are neutralized and the surface will have a net zero charge (the surface becomes hydrophobic). When the surfactant concentration is further increased, another surfactant layer may build up by hydrophobic interaction between the alkyl chain of the surfactant ions, and the surface now acquires a positive charge and it become hydrophilic. However, the adsorption of ionic surfactants on hydrophilic surfaces may acquire additional features, whereby the surfactant ions may associate on the surface, forming "hemimicelles." An example of this behavior is the adsorption of sodium dodecyl sulfonate (an anionic surfactant) on a positively charged alumina surface (at a pH below its isoelectric point, i.e., pH 7). Initially, the adsorption occurs by a simple ion-exchange mechanism whereby the surfactant anions exchange with the chloride counterions. In this region, the adsorption shows a slow increase with an increase in surfactant concentration. However, above a certain surfactant concentration (that is, just above that for complete ion exchange), the adsorption increases very rapidly with further increases in surfactant concentration. This is the region of hemimicelle formation, whereby several surfactant ions associate to form aggregation units on the surface.

The adsorption of nonionic surfactants on polar and nonpolar surfaces also exhibits various features, depending on the nature of the surfactant and the substrate. Three types of isotherms may be distinguished, as illustrated in Fig. 7. These isotherms can be accounted for by the different surfactant orientations and their association at the solid/liquid interface as illustrated in Fig. 8. Again, bilayers, hemimicelles, and micelles can be identified on various substrates.

V. SURFACTANTS AS EMULSIFIERS

Emulsions are a class of disperse systems consisting of two immiscible liquids, one constituting the droplets (the disperse phase) and the second the dispersion medium. The most common class of emulsions is those whereby the droplets constitute the oil phase and the medium is an aqueous solution (referred to as O/W emulsions) or where the droplets constitute the disperse phase, with the oil being the continuous phase (W/O emulsions). To disperse a liquid into another immiscible liquid requires a third component, referred to as the emulsifier, which in most cases is a surfactant. Several types of emulsifiers may be used to prepare the system, ranging from anionic, cationic, zwitterionic, and nonioinic surfactants to more specialized emulsifiers of the polymeric type, referred to as polymeric

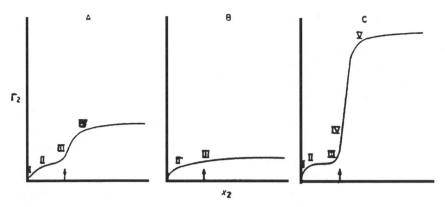

FIGURE 7 Adsorption isotherms corresponding to the three adsorption sequences shown in Fig. 8 (the cmc is indicated by the arrow).

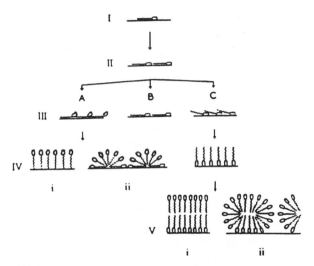

FIGURE 8 Model for the adsorption of nonionic surfactants showing the orientation of the molecules at the surface.

surfactants (see above). As discussed before, W/O emulsions require a low-HLB number surfactant, whereas for O/W emulsions a high-HLB number surfactant (8–18) is required.

The emulsifier plays a number of roles in the formation of the emulsion and subsequent stabilization. First, it reduces the O/W interfacial tension, thus promoting the formation of smaller droplets. More important is the result of the interfacial tension gradient $d\gamma/dz$, which stabilizes the liquid film between the droplets, thus preventing film collapse during emulsification. Another important role for the emulsifier is to reduce coalescence during emulsification as the result of the Gibbs–Marangoni effect. As a result of the incomplete adsorption of the surfactant molecules, an interfacial tension gradient $d\gamma/dA$ is present, and this results in a Gibbs elasticity, ε_f,

$$\varepsilon_f = \frac{2\gamma(d\ln\Gamma)}{1 + \left(\frac{1}{2}\right)h(dC/d\Gamma)}, \qquad (13)$$

where h is the film thickness. As shown in Eq. (13), ε_f will be highest in the thinnest part of the film. As a result, the surfactant will move in the direction of highest γ and this motion will drag liquid along with it. The latter effect is referred to as the Marangoni effect, which reduces further thinning of the film and hence will reduce coalescence during emulsification.

Another role of the surfactant is to initiate interfacial instability, e.g., by creating turbulence and Raykleigh and Kelvin–Helmholtz instabilities. Turbulence eddies tend to disrupt the interface since they create local pressures. Interfacial instabilities may also occur for cylindrical threads of disperse phase during emulsification. Such cylinders undergo deformation and become unstable under certain conditions. The presence of surfactants will accelerate these instabilities as a result of the interfacial tension gradient.

VI. SURFACTANTS AS DISPERSANTS

Surfactants are used as dispersants for solids in liquid dispersions (suspensions). The latter are prepared by two main procedures, namely, condensation methods (that are based on building up particles from molecular units) and dispersion methods, whereby larger "lumps" of the insoluble solid are subdivided by mechanical or other methods (referred to as comminution). The condensation methods involve two main processes, nucleation and growth. Nucleation is a spontaneous process of the appearance of a new phase from a metastable (supersaturated) solution of the material in question. The initial stages of nucleation result in the formation of small nuclei where the surface-to-volume ratio is very high and hence the role of specific surface energy is very important. With the progressive increase in the size of nuclei, the ratio becomes lower and eventually larger crystals appear, with a corresponding reduction in the role played by the specific surface energy.

The addition of surfactants, which can either adsorb on the surface of a nucleus or act as a center for inducing nucleation, can be used to control the process of nucleation and the stability of the resulting nuclei. This is due to their effect on the specific surface energy, on the one hand, and their ability to incorporate the material in the micelles, on the other.

Surfactants play a major role in the preparation of suspensions of polymer particles by heterogeneous nucleation. In emulsion polymerization, the monomer is emulsified in a nonsolvent (usually water) using a surfactant, whereas the initiator is dissolved in the continuous phase. The role of surfactants in this process is obvious since nucleation may occur in the swollen surfactant micelle. Indeed, the number of particles formed and their size depend on the nature of surfactant and its concentration (which determines the number of micelles formed).

Dispersion polymerization differs from emulsion polymerization in that the reaction mixture, consisting of monomer, initiator, and solvent (aqueous or nonaqueous), is usually homogeneous. As polymerization proceeds, polymer separates out and the reaction continues in a heterogeneous manner. A polymeric surfactant of the block or graft type (referred to as "protective colloid") is added to stabilize the particles once formed.

The role of surfactants in the preparation of suspensions by dispersion of a powder in a liquid and subsequent wet milling (comminution) can be understood by considering the steps involved in this process. Three steps may be distinguished: wetting of the powder with the liquid, breaking of aggregates, and agglomerates, and comminution. Surfactants play a crucial role in every step. For wetting the powder with the liquid, it is necessary to lower its surface tension and also reduce the solid/liquid interfacial tension by surfactant adsorption. The latter results in reduction of the contact angle of the liquid on the solid substrate.

The work of dispersion, W_d, involved in wetting a unit area of the solid substrate is given by the difference between the interfacial tension of the solid/liquid interface, γ_{SL}, and that of the solid/vapor interface, γ_{SV},

$$W_d = \gamma_{SL} - \gamma_{SV}. \tag{14}$$

Using Young's equation,

$$\gamma_{SV} - \gamma_{SL} = \gamma_{LV} \cos\theta, \tag{15}$$

one obtains

$$W_d = -\gamma_{LV} \cos\theta. \tag{16}$$

Thus, the work of dispersion depends on γ_{LV} and θ, both of which are reduced by the addition of surfactant.

Breaking of aggregates (clusters joined at their particle faces) and agglomerates (clusters joined at the corners of the particles) is also aided by the addition of surfactants.

Surfactants also aid the comminution of the particles by bead milling, whereby adsorption of the surfactant at the solid/liquid interface and in "cracks" facilitates their disruption into smaller units.

VII. ROLE OF SURFACTANTS IN STABILIZATION OF EMULSIONS AND SUSPENSIONS

Surfactants are used for stabilization of emulsions and suspensions against flocculation, Ostwald ripening, and coalescence. Flocculation of emulsions and suspensions may occur as a result of van der Waals attraction, unless a repulsive energy is created to prevent the close approach of droplets or particles. The van der Waals attraction G_A between two spherical droplets or particles with radius R and surface-to-surface separation h is given by the Hamaker equation,

$$G_A = -\frac{AR}{12h}, \tag{17}$$

where A is the effective Hamaker constant, which is given by the difference of the sum of all dispersion forces of the particles, A_{11}, and the medium, A_{22},

$$A = \left(A_{11}^{1/2} - A_{22}^{1/2}\right)^2. \tag{18}$$

Equation (17) shows that G_A increases with a decrease in h, and at small distances it can reach very large values (several hundred kT units). To overcome this everlasting attractive force and hence prevent flocculation of the emulsion or suspension, one needs to create a repulsive energy that "shields" the van der Waals energy. Two main types of repulsion may be distinguished. The first is the result of the presence of double layers, as, for example, when using ionic surfactants. The latter become adsorbed on the droplet or particle surface, and this results in the formation of a surface charge (which is characterized by a surface potential ψ_o). This surface charge is neutralized by counterions (which have a sign opposite that of the surface charge) which extend a large distance from the surface (which depends on the electrolyte concentration and valency). Around the particle surface, there will be an unequal distribution of counterions and co-ions (which have the same charge sign as the surface). The surface charge plus the counter- and co-ions form the electrical double layer, which may be characterized by a thickness $(1/\kappa)$ that increases with a decrease in electrolyte concentration and valency.

When two droplets or particles approach a distance h that is smaller than twice the double-layer thickness, repulsion occurs due to double-layer overlap (the double layers on the two particles cannot develop completely).

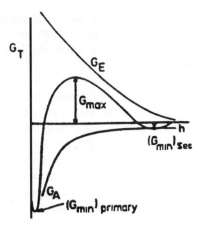

FIGURE 9 Form of the interaction energy–distance curve according to the DLVO theory.

The electrostatic energy of repulsion is given by the expression

$$G_E = 2\pi R\varepsilon_r\varepsilon_0\Psi_0^2 \ln[1 + \exp(-\kappa h)], \quad (19)$$

where ε_r is the relative permittivity and ε_0 is the permittivity of free space.

It is clear from Eq. (19) that G_E increases with an increase in ψ_0 (or zeta potential) and a decrease in κ (i.e., a decrease in electrolyte concentration and valency). Combination of G_E and G_A forms the basis of the stability of lyophobic colloids proposed by Deryaguin and Landau and Verwey and Overbeek, referred to as the DLVO theory. The energy–distance curve based on the DLVO theory is represented schematically in Fig. 9. It shows two minima, at long and short distances, $(G_{min})_{sec}$ and $(G_{min})_{primary}$ respectively, and an energy maximum G_{max} at intermediate distances. If G_{max} is high (>25 kT) the energy barrier prevents close approach of the droplets or particles, and hence irreversible flocculation into the primary minimum is prevented. This high-energy barrier is maintained at low electrolyte concentrations ($<10^{-3}$ mol dm^{-3}) and high surface (or zeta) potentials.

The second repulsive energy (referred to as steric repulsion) is produced by the presence of adsorbed surfactant layers of nonionic surfactants, such as alcohol ethoxylates or A–B, A–B–A block, or BA$_n$ graft copolymers, where B is the "anchor" chain and A is the stabilizing chain [mostly based on polyethylene oxide (PEO) for aqueous systems]. When two droplets or particles with adsorbed PEO chains of thickness δ approach a separation distance h such that $h < 2\delta$, repulsion occurs as a result of two main effects. The first arises as a result of the unfavorable mixing of the PEO chains, when these are in good solvent conditions. This is referred to as G_{mix} and is given by the following expression:

$$\frac{G_{mix}}{KT} = \frac{4\pi\phi_2^2}{3V_1}\left(\frac{1}{2} - \chi\right)\left(\delta - \frac{h}{2}\right)^2\left(3R + 2\delta + \frac{h}{2}\right), \quad (20)$$

where ϕ_2 is the volume fraction of the chains in the adsorbed layer, V_1 is the molar volume of the solvent, and χ is the chain–solvent (Flory–Huggins) interaction parameter. It is clear from Eq. (20) that when $\chi < 0.5$ (i.e., the chains are in good solvent conditions), G_{mix} is positive and the interaction is repulsive. In contrast, when $\chi > 0.5$ (i.e., the chains are in poor solvent conditions), G_{mix} is negative and the interaction is attractive. The condition $\chi = 0.5$, referred to as the θ-point, represents the onset of flocculation.

The second contribution to the steric interaction arises from the loss of configurational entropy of the chains on significant overlap. This effect is referred to as entropic, volume restriction, or elastic interaction, G_{el}. The latter increases very sharply with a decrease in h when the latter is less than δ. A schematic representation of the variation of G_{mix}, G_{el}, G_A, and G_T ($=G_{mix} + G_{el} + G_A$) is given in Fig. 10. The total energy–distance curve shows only one minimum, at $h \sim 2\delta$, the depth of which depends on δ, R, and A. At a given R and A, G_{min} decreases with an increase in δ. With small particles and thick adsorbed layers ($\delta > 5$ nm), G_{min} becomes very small ($<kT$) and the dispersion approaches thermodynamic stability. This shows the importance of steric stabilization in controlling the flocculation of emulsions and suspensions.

Surfactants are also used for reduction of Ostwald ripening. The latter arises from the difference in solubility between small and large particles. The smaller particles will have a higher solubility compared with larger ones. This is the result of the higher radius of curvature of smaller particles (note that the solubility of any droplet or particle S is inversely proportional to its radius R; $S = 2\gamma/R$). With time, smaller droplets or particles dissolve and their molecules diffuse and become deposited on larger droplets

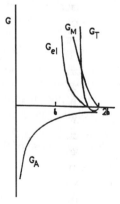

FIGURE 10 Variation of G_{mix}, G_{el}, G_A, and G_T with h for a sterically stabilized dispersion.

or particles. Surfactants reduce Ostwald ripening by two main mechanisms. First, by reduction of interfacial tension, the rate of Ostwald ripening is reduced. Second, as a result of interfacial tension gradients, the Gibbs elasticity causes a significant reduction of Ostwald ripening. The most effective surfactants are those with strong adsorption at the interface.

Surfactants also reduce the coalescence of emulsion droplets. The latter process occurs as a result of thinning and disruption of the liquid film between the droplets on their close approach. The latter causes surface fluctuations, which may increase in amplitude and the film may collapse at the thinnest part. This process is prevented by the presence of surfactants at the O/W interface, which reduce the fluctuations as a result of the Gibbs elasticity and/or interfacial viscosity. In addition, the strong repulsion between the surfactant layers (which could be electrostatic and/or steric) prevents close approach of the droplets, and this reduces any film fluctuations. In addition, surfactants may form multilayers at the O/W interface (lamellar liquid crystalline structures), and this prevents coalescence of the droplets.

VIII. ROLE OF SURFACTANTS IN SOLUBILIZATION AND MICROEMULSIONS

A. Solubilization

Solubilization is the formation of a thermodynamically stable, isotropic solution of a substance (the solubilizate), normally insoluble or slightly soluble in water, by the addition of a surfactant (the solublizer). The micelles of the surfactant cause solubilization of the substrate, producing an isotropic solution of the chemical. The solubilizate can be incorporated in the surfactant micelle in different ways, depending on the nature of the substrate and the surfactant micelles. For hydrophobic substrates, the molecules become incorporated in the hydrocarbon core of the micelle. With more polar substrates, the molecules may become incorporated in the hydrophilic PEO chains of the micelle or they may be simply adsorbed at the micelle surface.

Solubilization is applied in many industrial processes for the administration of insoluble chemicals, e.g., in dying, in drug administration, and in agrochemical applications. The process of solubilization is also important in detergency, whereby fats and oils are removed by incorporation into the hydrocarbon core of the micelles.

B. Microemulsions

Microemulsions are isotropic systems consisting of oil, water, and surfactants(s) which are stable in the thermo-

dynamic sense, consisting of nearly isodisperse, small droplets (usually in the range of 5–50 nm) in another liquid. The small size of the droplets results in the transparent or translucent appearance of microemeulsion.

Several theories have been proposed to account for the thermodynamic stability of microemulsions. The most recent theories showed that the driving force for microemulsion formation is the ultralow interfacial tension (in the region of 10^{-4}–10^{-2} mN m^{-1}). This means that the energy required for formation of the interface (the large number of small droplets) $\Delta A\gamma$ is compensated by the entropy of dispersion $-T\Delta S$, which means that the free energy of formation of microemulsions ΔG is zero or negative.

The ultralow interfacial tension can be produced by using a combination of two surfactants, one predominantly water soluble (such as sodium dodecyl sulfate) and the other predominantly oil soluble (such as a medium-chain alcohol, e.g., pentanol or hexanol). In some cases, one surfactant may be sufficient to produce the microemulsion, e.g., Aerosol OT (dioctyl sulfosuccinate), which can produce a W/O microemulsions. Nonionic surfactants, such as alcohol ethoxylates, can also produce O/W microemulsions, within a narrow temperature range. As the temperature of the system increases, the interfacial tension decreases, reaching a very low value near the phase inversion temperature. At such temperatures, an O/W microemulsion may be produced.

Microemulsions have attracted considerable attention for application in industry. In the early days of their discovery, microemulsions were used in the leather industry, cutting oils, dry cleaning, flavorings, agrochemicals, and pharmaceuticals. However, the main potential application of microemulsions will be in tertiary oil recovery and as reaction media for enzymes and production of nanoparticles (e.g., for application in the electronic industry). Another application of microemulsions is in the field of solar energy production, e.g., production of hydrogen by decomposition of water using UV light.

IX. SURFACTANTS IN FOAMS

The role of surfactants in stabilization/destabilization of foam (air/liquid dispersions) is similar to that for emulsions. This is due to the fact that foam stability/instability is determined by the surface forces operative in liquid films between air bubbles. In many industrial applications, it is essential to stabilize foams against collapse, e.g., with many food products, foam in beer, fire-fighting foam, and polyurethane foams that are used for furniture and insulation. In other applications, it is essential to have an effective way of breaking the foam, e.g., in distillation columns, crude oils, and effluent streams. In the case where foam stability is desirable, it is essential to choose surfactants that enhance the Gibbs–Marangoni effect and produce a viscoelastic film that provides a mechanical barrier preventing foam collapse. This explains the application of protein film for fire-fighting foams. A particularly important process in foam formation and stabilization is Ostwald ripening, which results from gas diffusion from smaller air bubbles to larger ones. This is the result of the higher Laplace pressure of smaller air bubbles. As mentioned in the section on stabilization of emulsions, this process is opposed by the reduction of the interfacial tension and creation of an interfacial tension gradient (Gibbs elasticity).

In the case where foam instability is desirable, it is essential to choose surfactants that weaken the Gibbs–Marangoni effect. A more surface-active material such as a poly(alkyl) siloxane is added to destabilize the foam. The siloxane surfactant adsorbs preferentially at the air/liquid interface, thus displacing the original surfactant that stabilizes the foam. In many cases, the siloxane surfactant is produced as an emulsion which also contains hydrophobic silica particles. This combination produces a synergetic effect for foam breaking.

X. SURFACTANTS IN WETTING PHENOMENA

Wetting is important in many industrial systems, e.g., mineral flotation, detergency, crop protection, dispersion of powders in liquids, and coatings. When a drop of a liquid is placed on a solid surface, the liquid either spreads, forming a thin uniform film (complete wetting with zero contact angle θ), or remains as a discrete droplet with a measurable contact angle (partial wetting). The value of the contact angle is used as a measure of wetting: when $\theta = 0°$, complete wetting occurs; when $\theta = 180°$, the surface is described as nonwettable. When $\theta < 90°$, the surface is described as as being partially wetted, whereas when $\theta > 90°$ the surface is described as as being poorly wetted by the liquid. Thus, to enhance the wetting of an aqueous solution on a hydrophobic substrate, one adds a surfactant, which lowers the surface tension of water and adsorbs on the hydrophobic substrate in a specific manner, i.e., with the hydrophobic alkyl chain being attached to the substrate, leaving the polar head group in the aqueous medium. In contrast, to reduce the wetting of an aqueous solution on a hydrophilic surface (e.g., in waterproofing), one adds a surfactant with the opposite orientation, i.e., the polar head group being attached to the surface, leaving the hydrophobic alkyl chain pointing to the aqueous medium. An example of the latter process is

waterproofing of fabrics, whereby a cationic surfactant is sometimes used. The positive head group of the surfactant is attached to the negative charges on the fabric, leaving the hydrophobic alkyl chains pointing to the solution. The same process applies for fabric softeners, which usually consist of dialkyl quaternary ammonium surfactants.

XI. APPLICATION OF SURFACTANTS IN COSMETICS AND PERSONAL CARE PRODUCTS

Cosmetic and personal care products are designed to deliver a functional benefit and to enhance the psychological well-being of consumers by increasing their aesthetic appeal. Many cosmetic and personal care formulations are designed to clean hair, skin, etc., and impart a pleasant odor, make the skin feel smooth, provide moisturizing agents, provide protection against sunburn, etc. Most cosmetic and personal care products consist of complex systems of emulsions, creams, lotions, suspoemulsions (mixtures of emulsions and suspensions), multiple emulsions, etc. All these complex systems consist of several components of oil, water, surfactants, coloring agents, fragrants, preservatives, vitamins, etc. The role of surfactants in these complex formulations is crucial in designing the system, in achieving long-term physical stability and the required "skin-feel" on application. Conventional surfactants of the anionic, cationic, amphoteric, and nonionic types are used in cosmetics and personal care applications. These surfactants may not cause any adverse toxic effects. Besides the synthetic surfactants used in the preparation of systems such as emulsions, creams, lotions, and suspensions, several other naturally occurring materials have been introduced and there is a trend in recent years to use such natural products in the belief that they are safer for application. Several synthetic surfactants that are applied in cosmetics and personal care products may be listed, such as carboxylates, ether sulfates, sulfates, sulfonates, quaternary amines, betaines, and sarcosinates. The ethoxylated surfactants are probably the most widely used surfactants in cosmetics. Being uncharged, these molecules have a low skin sensitization potential. This is due to their low binding to proteins. Unfortunately, these nonionic surfactants are not the most friendly materials to produce (the ethoxylation process is rather dangerous), and one has to ensure a very low level of free ethylene oxide, which may form dioxane (that is carcinogenic) on storage. Another problem with ethoxylated surfactants is their degradation by oxidation or photooxidation processes. These problems are reduced by using sucrose esters obtained by esterification of the sugar hydroxyl group with fatty acids such as lauric and stearic acid. In this case, the problem of contamination is reduced and the surfactants are still mild to the skin since they do not interact with proteins.

Another class of surfactants that are used in cosmetics and personal care products is the phosphoric acid esters. These molecules are similar to the phospholipids that are the building blocks of the stratum corneum (the top layer of the skin, which is the main barrier for water loss). Glycerine esters, in particular, triglycerides, are also frequently used. Macromolecular surfactants of the A–B–A block type [where A is PEO and B is polypropylene oxide (PPO)] are also frequently used in cosmetics. Another important naturally occurring class of polymeric surfactants is the proteins, which can be used effectively as emulsifiers.

In recent years, there has been a great trend toward using volatile silicone oils in many cosmetic formulations. Due to their low surface energy, silicone oils help spread the various active ingredients over the surface of the skin, hair, etc. While many silicone oils can be emulsified using conventional hydrocarbon surfactants, several silicone-type surfactants have been introduced for their effective emulsification and long-term stability. These silicone surfactants consist of a methyl silocone backbone with pendent groups of PEO and PPO. These polymeric surfactants act as steric stabilizers.

XII. APPLICATION OF SURFACTANTS IN PHARMACEUTICALS

Surfactants play an important role in pharmaceutical formulations. A large number of drugs are surface active, e.g., chloropromazine, diphenyl methane derivatives, and tricyclic antidepressants. The solution properties of these surface-active drugs play an important role in their biological efficacy. Surface-active drugs tend to bind hydrophobically to proteins and other biological macromolecules. They tend to associate with other amphipathic molecules such as other drugs, bile salts, and receptors. Many surface-active drugs produce intralysosomal accumulation of phospholipids which are observable as multilamellar objects within the cell. The interaction between surfactant drug molecules and phospholipid renders the phospholipid resistant to degradation by lysosomal enzymes, resulting in their accumulation in the cell.

Many local anesthetics have significant surface activity and it is tempting to correlate such surface activity with their action. Other important factors such as partitioning of the drug into the nerve membrane may also play an important role. Accumulation of drug molecules in certain cites may allow them to reach concentrations whereby micelles are produced. Such aggregate units may cause significant biological effects.

Several naturally occurring amphipathic molecules (in the body) exist, such as bile salts, phospholipids, and cholesterol, which play an important role in various biological processes. Their interactions with other solutes, such as drug molecules, and with membranes are also very important. The most important surface-active species in the body are the phospholpids, e.g., phosphatidylcholine (lecithin). These lipids (which may be produced from egg yolk) are used as emulsifiers for many intravenous formulations, such as fat emulsions and anesthetics. Lipids can also be used to produce liposomes and vesicles which can be applied for drug delivery. When dispersed into water, they produce lamellar structures, which then produce multilamellar spherical units (liposomes). On sonication of these multilamellar structures, single spherical bilayers or vesicles (10–40 nm) are produced. Both lipid-soluble and water-soluble drugs can be entrapped in the liposomes. Liposoluble drugs are solubilized in the hydrocarbon interiors of the lipid bilayers, whereas water-soluble drugs are intercalated in the aqueous layers.

One of the most important application of surfactants in pharmacy is to solubilize insoluble drugs. Several factors may be listed that influence solubilization such as the surfactant and solubilizate structure, temperature, and added electrolyte or nonelectrolyte. Solubilization in surfactant solutions above the cmc offers an approach to the formulation of poorly insoluble drugs. Unfortunately, this approach has some limitations, namely, the finite capacity of the micelles for the drug, the possible short- or long-term adverse effects of the surfactant on the body, and the concomitant solubilization of other ingredients such as preservatives and flavoring and coloring agents in the formulation. Nevertheless, there is certainly a need for solubilizing agents for increasing the bioavailability of poorly soluble drugs. The use of cosolvents and surfactants to solve the problem of poor solubility has the advantage that the drug entity can be used without chemical modification and toxicological data on the drug may not be repeated.

Surfactants are also used for general formulation of drugs, e.g., as emulsifying agents, dispersants for suspensions, and wetting agents for tablets. Surfactant molecules incorporated in the formulation can affect drug availability in several ways. The surfactant may influence the disintegration and dissolution of solid dosage forms or control the rate of precipitation of drugs administered in solution form, by increasing the membrane permeability and affecting membrane integrity. Release of poorly soluble drugs from tablets and capsules for oral use may be increased by the presence of surfactants, which may decrease the aggregation of drug particles and, therefore, increase the area of the particles available for dissolution. The lowering of surface tension may also be a factor in aiding the penetration of water into the drug mass. Above the cmc, the increase in solubilization can result in more rapid rates of drug dissolution.

XIII. APPLICATION OF SURFACTANTS IN AGROCHEMICALS

Besides the use of surfactants for formulation of all agrochemical formulations (suspensions, emulsions, microemulsion, microcapsules, water-dispersible grains, granules, etc.), these molecules play a major role in optimization of biological efficacy. This can be understood if one considers the steps during application of the crop spray, which involve a number of interfaces. The first interface during application is that between the spray solution and the atmosphere (air), which governs the droplet spectrum, rate of evaporation, drift, etc. In this respect, the rate of adsorption of the surfactant molecules at the air/liquid interface is of vital importance. In a spraying process a fresh liquid surface is continuously being formed. The surface tension of this liquid (referred to as the dynamic surface tension) depends on the relative ratio between the time taken to form an interface and the rate of adsorption of the surfactant from the bulk solution to the air/liquid interface, which depends on the rate of diffusion of the surfactant molecule. The rate of diffusion is directly proportional to the diffusion coefficient, D, of the molecule (which is inversely proportional to its radius) and the surfactant concentration. Thus, for effective lowering of the dynamic surface tension during a spraying process, one needs surfactants with a high D and sufficiently high concentrations. However, the actual situation is not simple since one has an equilibrium between surfactant micelles and monomers. The latter diffuse to the interface and become adsorbed, and hence the equilibrium between micelles and monomers is disturbed. Surfactant micelles then break to supply monomers in the bulk. Thus, the dynamic surface tension also depends on the lifetime of the micelle.

Surfactants also have a large influence on spray impaction and adhesion, which is very important for maximizing capture of the drops by the target. For adhesion to take place, the difference in surface energy of the droplet in flight, E_o ($=4\pi R^2 \gamma$), and that at the surface, E_s (which depends on the contact angle, θ, of the drop on the substrate), should balance the kinetic energy of the drop ($\frac{1}{2}mv^2$, where m is the mass of the drop and v its velocity). For adhesion to occur, $E_o - E_s > \frac{1}{2}mv^2$. Surfactants clearly enhance adhesion by lowering γ and θ.

Surfactants also play a major role in reducing droplet sliding and increasing spray retention. When a drop impinges on an inclined surface (such as a leaf surface), it starts to slide as a result of gravity. During this process,

the droplet produces an advancing contact angle θ_A and a receding contact angle θ_R. The latter is lower than the former, and the difference between the two angles ($\theta_A - \theta_R$) is referred to as contact angle hysteresis. As a result of this sliding process, an area of the surface becomes dewetted (at the back) and an equal area becomes wetted at the front. When the difference between the work of dewetting and that of wetting (which is determined by the contact angle hysteresis) balances the gravity force, sliding stops and the droplet stays retained on the surface. Thus, surfactants which affect the surface tension of the liquid and give this contact angle hysteresis reduce drop sliding and enhance spray retention.

Another role of surfactants in crop sprays is to enhance the wetting and spreading of the droplets on the target surface. This process governs the final distribution of the agrochemical over the area to be protected. The optimum degree of coverage in any spray application depends on the mode of action of the agrochemical and the nature of the pest to be controlled. On evaporation of the drops, deposits are produced whose nature depends on the nature of the surfactant and interaction with the agrochemical molecules or particles. These deposits may contain liquid crystalline phases when the surfactant concentration reaches high values. In many cases, long-lasting deposits are required to ensure supply of the agrochemical, e.g., with systemic fungicides. These deposits may enhance the tenacity of the agrochemical on the leaf surface and hence they enhance rain-fastness.

Finally, surfactants may have a direct effect on the biological efficacy by enhancing the penetration of agrochemical molecules through various barriers, such as plant cuticle and various other membranes. This enhanced penetration may be caused by solubilization of the active ingredient by the surfactant micelles. The latter may enhance flux of the chemical through the plant by increasing the concentration gradient at the interface.

XIV. APPLICATION OF SURFACTANTS IN THE FOOD INDUSTRY

The use of surfactants in the food industry has been known for centuries. Naturally occurring surfactants such as lecithin from egg yolk or soybean and various proteins from milk are used for the preparation of many food products, such as mayonnaise, salad creams, dressing, and desserts. Polar lipids such as monoglycerides have been introduced as emulsifiers for food products. More recently, synthetic surfactants such as sorbitan esters (Spans) and their ethoxylates (Tweens), sucrose esters, have been used in food emulsions. It should be mentioned that the structures of many food emulsions is complex, and in many cases several phases may exist. Such structures may exist under nonequilibrium conditions and the state of the system may depend to a large extent on the process used for preparing the system, its prehistory, and the conditions to which it is subjected.

Food grade surfactants are, in general, not soluble in water, but they can form association structures in aqueous medium that are liquid crystalline in nature. These liquid crystalline structures are produced by heating the solid emulsifier (which is dispersed in water) to a temperature above its Krafft temperature. On cooling such a system, a "gel" phase is produced which becomes incorporated with the emulsion droplets. These gel phases produce the right consistency for many food emulsions.

Proteins, which are also surface active, can be used to prepare food emulsions. The protein molecules adsorb at the O/W interface and they may remain in their native state (forming a "rigid" layer of unfolded molecules) or undergo unfolding, forming loops, tails, and trains. These protein molecules stabilize the emulsion droplets, either by a steric stabilization mechanism or by producing a mechanical barrier at the O/W interface.

SEE ALSO THE FOLLOWING ARTICLES

CHEMICAL THERMODYNAMICS • MESOPOROUS MATERIALS, SYNTHESIS AND PROPERTIES • MICELLES • SILICONE (SILOXANE) SURFACTANTS

BIBLIOGRAPHY

Tadros, Th. F. (1984). "Surfactants," Academic Press, London.

McCutchen (published annually). "Detergents and Emulsifiers," Allied, NJ.

van Os, N. M., Haak, J. R., and Rupert, L. A. (1993). "Physico-Chemical Properties of Selected Anionic, Cationic and Nonionic Surfactants," Elsevier, Amsterdam.

Porter, M. R. (1991). "Handbook of Surfactants," Chapman and Hall, London.

Tadros, Th. F. (1999). *In* "Principles of Polymer Science and Technology in Cosmetics and Personal Care" (E. D. Goddard and J. V. Gruber, eds.), Chap. 3, Marcel Dekker, New York.

Griffin, W. C. (1954). *J. Cosmet. Chem.* **5**, 249.

Lindman, B. (1984). *In* "Surfactants" (Th. F. Tadros, ed.), Academic Press, London.

Mukerjee, P., and Mysels, K. J. (1971). "Critical Micelle Concentrations of Aqueous Surfactant Systems," National Bureau of Standards, Washington, DC.

Hartley, G. S. (1936). "Aqueous Solutions of Paraffin Chain Salts" (Hermann and Cie, Paris).

Debye, P., and Anaker, E. W. (1951). *J. Phys. Colloid Chem.* **55**, 644.

McBain, J. W. (1950). "Colloid Science," Heath, Boston.

Clunies, J. S., Goodman, J. F., and Symons, P. C. (1969). *Trans Faraday Soc.* **65**, 287.

Rosevaar, F. B. (1968). *J. Soc. Cosmet. Chem.* **19**, 581.

Anaisson, E. A. G., and Wall, S. N. (1974). *J. Phys. Chem.* **78**, 1024; (1975) **79**, 857.

Tanford, C. (1980). "The Hydrophobic Effect," 2nd ed., Wiley, New York.

Gibbs, J. W. (1928). "Collected Works," Vol. 1, Longman, New York.

Hough, D. B., and Randall, H. M. (1983). *In* "Adsorption from Solution at the Solid/Liquid Interface" (G. D. Parfitt and C. H. Rochester, eds.), p. 247, Academic Press, London.

Clunie, J. S., and Ingram, B. T. (1983). *In* "Adsorption from Solution at the Solid/Liquid Interface," (G. D. Parfitt and C. H. Rochester, eds.), p. 105, Academic Press, London.

Walstra, P. (1980). *In* "Encyclopedia of Emulsion Technology" (P. Becher, ed.), Chap. 2, Marcel Dekker, Naw York.

Davies, J. T. (1972). "Turbulence Phenomenon," Chaps. 8–10. Academic Press, New York.

Chandrosekhav, S. (1961). "Hydrodynamics and Hydrodynamic Instability," Chaps. 10–12, Cleeverdon, Oxford.

Gibbs, J. W. (1906). "Scientific Papers," Vol. 1, Longman Green, London.

Volmer, M. (1939). "Kinetic der Phase Bildung," Steinkopf, Dreseden.

Blakely, D. (1975). "Emulsion Polymerization," Applied Science, London.

Barrett, K. E. J. (1975). "Dispersion Polymerization in Organic Media," John Wiley and Sons, London.

Hamaker, H. C. (1937). *Physica (Utrecht)* **4**, 1058

Deryaguin, B. V., and Landau, L. (1939). *Acta Phy. Chem. USSR* **10**, 33.

Verwey, E. J., and Overbeek, J. Th. G. (1948). "Theory of Stability of Lyophobic Colloids," Elsevier, Amsterdam.

Napper, D. H. (1983). "Polymeric Stabilization of Colloidal Dispersions," Academic Press, London.

Danielsson, I., and Lindman, B. (1981). *Colloids Surf.* **3**, 391.

Overbeek, J. Th. G. (1978). *Faraday Disc. Chem. Soc.* **65**, 7.

Overbeek, J. Th. G., de Bruyn, P. L., and Verhoecks, F. (1984). *In* "Suractants" (Th. F. Tadros, ed.), p. 111, Academic Press, London.

Breuer, M. M. (1985). *In* "Encyclopedia of Emulsion Technology" (P. Becher, ed.), Vol. 2, Chap. 7, Marcel Dekker, New York.

Attwood, D., and Florence, A. T. (1983). "Surfactant Systems, Their Chemistry, Pharmacy and Biology," Chapman and Hall, New York.

Tadros, Th. F. (1987). *Aspects Appl. Biol.* **14**, 1.

Tadros, Th. F. (1994). "Surfactants in Agrochemicals," Marcel Dekker, New York.

Krog, N. J., and Riisom, T. H. (1985). *In* "Encyclopedia of Emulsion Technology" (P. Becher, ed.), Vol. 2, p. 321, Marcel Dekker, New York.

Surveying

Jack B. Evett
The University of North Carolina at Charlotte

I. Introduction
II. Basic Measurements in Surveying
III. Basic Operations of Surveying
IV. Types of Surveying
V. Additional Topics in Surveying

GLOSSARY

Azimuth Indicator of the direction of a line by giving the clockwise angle the line makes with a north line.

Bearing Indicator of the direction of a line by giving the acute angle between the line and a north–south line and the quadrant in which the line lies.

Error Deviation of a measured value from the true value as a result of conditions basically beyond the control of the person making the measurement.

Geodetic surveying Surveying in which the approximately spheroidal shape of the earth is taken into account.

Local attractions Electromagnetic phenomena that can cause a compass needle to deviate from north.

Magnetic declination Difference between the direction of a compass needle when lined up in the earth's magnetic field and a true north line.

Mistake Deviation of a measured value from the true value as a result of conditions basically within the control of the person making the measurement.

Photogrammetry Means of surveying by photographing land from an airplane and preparing maps from the photographs obtained.

Plane surveying Surveying in which the approximately

spheroidal shape of the earth is not taken into account and the earth is assumed to be flat over the area being considered.

Transit Instrument used in surveying that can perform virtually all of the operations normally done in surveying.

Traversing Surveying of any number of consecutive lines by determining the length and direction of each line.

SURVEYING can be broadly defined as the science and art of determining relative locations of points above, on, or below the earth's surface. It involves, basically, the measurement of distances and determination of directions. Surveyors make certain types of measurement by various methods, use such measurements to perform appropriate computations, and often display results in the form of maps, plats, diagrams, and so forth.

I. INTRODUCTION

A. History of Surveying

Surveying originated many centuries ago (most likely around 1400 B.C. in Egypt) when land was divided into

Encyclopedia of Physical Science and Technology, Third Edition, Volume 16

tracts for delineation of ownership and taxation purposes. Such surveying was, of course, crude by today's standards. Some improvement in surveying techniques occurred during the heyday of the Roman Empire in association with the large-scale construction of that period.

It was not, however, until much later—the eighteenth century—that significant advances in surveying techniques began to occur. The need for more accurate nautical maps for navigation may have provided the initial impetus, but the necessity for better delineation of property boundaries as the value of property increased, as well as the construction of canals and railroads and, later, highways, also contributed to the development of new surveying techniques.

The twentieth century has seen rapid and major improvements in surveying techniques, largely as a result of military requirements during the two World Wars and two "conflicts" (Korea and Vietnam) and of the space program. Modern surveying techniques include the use of electronic measuring devices, lasers, aerial photogrammetry, satellites, and computers, to name a few.

B. Plane Surveying versus Geodetic Surveying

It is important to differentiate between plane surveying and geodetic surveying. In *geodetic surveying*, the approximately spheroidal shape of the earth is taken into consideration, and horizontal distances and angles are projected onto the surface of a spheroid. In actuality, all surveying should be considered geodetic surveying. If, however, a given surveying job covers only a very small portion of the earth's total surface (as is often the case), that surface may be assumed to be flat without appreciable error. Surveying done under this assumption is called *plane surveying*, and horizontal distances and angles are projected onto a horizontal plane.

Inasmuch as plane surveying is much simpler than geodetic surveying, local surveying is generally done under the concept of plane surveying (i.e., a "flat earth"). Geodetic surveying is required for surveys over larger areas (such as a state or country).

C. Units of Measurement

In the United States, the unit for measuring distance has long been the *foot* (ft), with distances given in feet and decimal fractions thereof. In other countries, where the International System of Units (SI) is used, the unit for measuring distance is the *meter* (m). For long distances, units of *miles* (mi) or *kilometers* (km) may be used, where 1 mi = 5280 ft and 1 km = 1000 m.

If distances are measured in feet, a bounded area can be computed in *square feet* (sq ft). Land in the United States is often bought and sold, however, on the basis of area expressed as *acres*, where 1 acre = 43,560 sq ft. For distances measured in meters, a bounded area can be computed in *square meters* (sq m). Large areas may be expressed as *hectares* (ha), where 1 ha = 10,000 sq m.

If distances are measured in feet, an associated volume could be computed in *cubic feet* (cu ft). In common surveying usage, however, volumes are often expressed as *cubic yards* (cu yd). When SI units are used, volume can be computed in *cubic meters* (cu m).

Angular measurement is commonly determined in units of *degrees*, *minutes*, and *seconds*, where 1 degree is $\frac{1}{360}$ of the total angle (i.e., one complete rotation) about a point, 1 minute is $\frac{1}{60}$ of a degree, and 1 second is $\frac{1}{60}$ of a minute. Degrees, minutes, and seconds are commonly denoted by the symbols °, ′, and ″, respectively. Hence, an angle of 22 degrees 19 minutes 45 seconds can be written 22°19′45″.

D. Errors and Mistakes

Surveying measurement and computation must be accurate. Land is often bought and sold for thousands of dollars per acre. It is the surveyor's job to determine the area of land, and any inaccuracies can be costly to the buyer or seller. Similarly, landowners may be compensated for earth being removed from their property for use in nearby highway construction on the basis of so much money per cubic yard of earth removed. Again, it is the surveyor's job to determine the volume of earth removed, and inaccuracies can be costly.

Inaccuracies occur in surveying measurements as a result of errors and mistakes. An *error* is a deviation of a measured value from the true value as a result of conditions generally beyond the control of the person making the measurement. In other words, the person did the best job possible in the circumstances, but the measured value is not the true value. Errors can result from instrumental causes (e.g., a defect in the scale used to evaluate angles), personal causes (e.g., one's inability to maintain the required steady tension on a tape when determining a horizontal distance), and natural causes (e.g., inaccuracy in observing a line of sight through a transit or level telescope as a result of heat waves). Errors tend to be either systematic or random. In the case of systematic errors, the total error in a measurement can be evaluated and the measured value corrected to give the presumed true value. Random errors, on the other hand, tend to be compensative.

A *mistake* is a deviation of a measured value from the true value as a result of conditions basically within the control of the person making the measurement. In other words, the person "goofed." Examples of mistakes are misreading a tape, level rod, or vernier scale (e.g., reading

4.97 instead of the correct value of 3.97 on a level rod) and transposing digits when recording measured values (e.g., recording a measured distance of 324.7 ft as 342.7 ft).

Surveyors should always strive to minimize errors and avoid mistakes by concentrating on the job at hand, exercising care, using good judgement, and checking and rechecking whenever possible.

E. Surveying as a Profession

Although some individuals have, in the past, become surveyors through the "school of hard knocks," becoming a professional surveyor today generally requires some higher education as well as practical experience. The necessary higher education is often obtained as part of a baccalaureate engineering program, notably in civil engineering. An alternative is completion of a 2-year associate degree program emphasizing surveying.

To perform surveying and be in responsible charge, a person is required by the states to be a registered (licensed) land surveyor. Registration (licensing) is carried out by the states. Although specific requirements vary somewhat from one state to another, the requirements generally involve satisfactory education, experience (under the supervision of a registered land surveyor), and good character.

Some surveyors are self-employed; others work for surveying companies, engineering consulting companies, construction companies, and the like. Also, surveyors are employed by various government agencies, such as the Coast and Geodetic Survey and highway departments. Surveying is a good career choice for individuals who are seeking a challenge, are willing to work hard, and enjoy being out-of-doors.

II. BASIC MEASUREMENTS IN SURVEYING

A. Horizontal Measurement

Perhaps the most basic operation in surveying is the measurement of *horizontal distances*. In addition to simply measuring horizontal distances, horizontal angles can be evaluated by measuring three horizontal distances and employing the law of cosines. Hence, a considerable amount of surveying can be done with only a horizontal distance measurement device.

As related in Section I, in plane surveying horizontal distances are projected onto a horizontal plane. A piece of land, no matter what its relief may be, is analyzed and evaluated as if it were projected onto a horizontal plane. Accordingly, whenever the linear distance between any two points on the earth's surface is referred to, it is understood

that it is the horizontal distance between the points, regardless of the elevations of the end points of the line, unless specified otherwise. Therefore, in measuring the distance between two points (such as the length of a side of a lot), either the measurement is made along a horizontal line between the points or the measurement is made along an inclined line between the points, with the horizontal distance computed and recorded. If the first method is used, the position of one or both end points may have to be projected vertically upward to the horizontal line of measurement. This can be achieved in some cases by means of a plumb bob.

Horizontal distances can be determined in a variety of ways. Some common means are pacing, stadia, taping, and electronic measurement.

Pacing is a crude means of finding horizontal distances, but it may be useful if rough values of distances are acceptable. The first step in pacing is to calibrate one's pace (step). The typical adult's pace ranges from about $2\frac{1}{2}$ to 3 ft. Once the pace is determined, a distance can be estimated simply by walking along the line, counting the number of paces (steps), and multiplying that number by the length of one's pace. An experienced pacer should be able to determine the value of a distance of $\sim$100 ft within 2 or 3 ft either way.

Stadia is a means of determining horizontal distances that is more accurate than pacing, although it is not accurate enough for some purposes. Stadia is carried out by viewing a leveling rod through a transit's telescope. Two crosshairs in the telescope appear to intersect the rod at two different points, and the distance between these two points on the rod is directly proportional to the distance from the transit to the rod. Hence, by noting where the two crosshairs intersect the rod, determining the distance between these points, and making appropriate computations, one can evaluate a horizontal distance. Stadia is generally accurate to the nearest foot.

Taping is a common means of determining horizontal distances. The general procedure, which is well known to laypeople, is to stretch a calibrated tape along the line to be measured and read the value of the distance off the tape. Although this is a simple procedure that anyone can do, several points must be considered in order to attain the accuracy generally required in surveying.

To begin with, as discussed previously, in measuring the distance between two points, either the measurement is made along a horizontal line between the points or the measurement is made along an inclined line between the points, with the horizontal distance computed and recorded. If a measurement is being made along a straight, horizontal surface, it is no problem to keep the tape horizontal, since it can be stretched tightly along the horizontal surface. For other situations, one can achieve horizontal alignment of the tape, at least approximately, by using a plumb bob at one or both ends of the tape and stretching the tape tightly.

Another point to consider is that the tape must be stretched in a straight line between the end points of the line. Although this would appear to be a simple matter, it can be difficult in some cases, such as in measuring distances through heavy underbrush. One must also exercise care when measuring distances longer than a single tape length. When this occurs, temporary points must be marked as successive end points of the tape are encountered, and each such marking presents an opportunity to incur error.

If extremely accurate measurement must be done by taping, it may be necessary to compute corrections for various phenomena and apply them to a measured length to obtain the correct length. For example, a given tape may be slightly too short or too long as a result of imperfect manufacture, wear, and so on. Obviously, a distance measured with a tape that is too short or too long will give an erroneous value. In addition, a measured distance is subject to error caused by expansion and contraction of the tape as a result of change in temperature. Error in measurement also results when the tape is pulled either too tightly or not tightly enough. Finally, error in measurement occurs if the tape sags when it is supported only at its end points.

In each of these cases, the actual amount of error in a measured value is ordinarily quite small and is often neglected. However, if extremely accurate measurement is required, formulas are available for computing corrections for each case, and the total correction can be applied to the measured distance to determine the correct value.

More recently developed devices for measuring distances are known as *electronic distance-measuring* (EDM) *devices*. They find lengths based on phase changes that occur as electromagnetic energy (a "signal") is sent from one point to a second point and back to the first point. With the wavelength and length of time of travel known, the distance between the points can be computed. In practice, an EDM device, an example of which is shown in Fig. 1, is placed over one point and a reflecting target is placed over the other point. The signal is sent, the time recorded, and the distance computed. (It should be understood that this description of the operation of an EDM device is oversimplified.)

In the first EDM devices, the "signal" was a light beam. Subsequently, high-frequency microwaves and lasers were used. More recent advances have made the devices smaller, lighter, easier to use, and capable of giving a direct readout of the distance.

EDM devices can be used to measure very long distances while giving extraordinarily accurate results—of

FIGURE 1 Electronic distance-measuring device.

the order of 50 mi ± 2.6 ft (80 km ± 0.80 m). They are exceedingly useful in measuring long distances and others that would be difficult to measure otherwise (such as the distance across a large lake).

B. Vertical Measurement

Another basic operation performed in surveying is the measurement of *vertical distances*. This type of information is often needed in engineering applications to find differences in elevations of various ground points as a means of computing slopes. Determinations of elevations and slopes are important in many contexts, including, to name a few, topographic mapping, design of highways, drainage systems, and sewers.

In some cases, vertical measurements can be made in the same manner as horizontal ones. For example, the height of a building might be found by stretching a tape from the roof to the ground. Often, however, this kind of direct measurement of vertical distance is not feasible because the end points are not located directly above and below one another. This is true, for example, in determining the difference in elevation between two ground points at different locations on the earth's surface. Differences in elevation can be evaluated by a procedure known as *differential leveling*.

Differential leveling is carried out using two devices: a level and a level rod. The basic element of a level is a telescope that can be adjusted so that it is always aligned in a level plane with its line of sight along a level line. It is mounted on a tripod when in use. A level rod is simply a rod that is calibrated for easy viewing and reading through the level's telescope. Examples of levels and a level rod are given in Figs. 2 and 3, respectively.

To begin the leveling procedure, a level rod is held vertically over a point of known elevation (a bench mark). With a level set up and properly leveled, a sighting is made on the rod and a reading made. This reading is known as a backsight; and when it is added to the elevation of the bench mark, the sum gives the height of the instrument

FIGURE 2 Top: Level; Bottom: automatic level.

(i.e., the telescope). The level rod is then moved to a point the elevation of which is desired, a sighting is made on the rod, and a reading made. This reading is known as a foresight; subtracting it from the height of the instrument gives the ground elevation where the rod was held.

The preceding describes the standard procedure for determining the elevation of a second point, given the elevation of a first point. This procedure can be repeated (taking a backsight on the point the elevation of which was just determined) as many times as necessary to obtain elevations of additional points. A check for accuracy is commonly made by continuing levels so as to finish on the starting point. The elevation of the starting point as computed from running the levels should, of course, be the same as its beginning (known) elevation.

Other methods of determining elevations include stadia and the use of EDM devices (see Section II.A) as well as barometers.

C. Directional Measurement

Sections II.A and II.B discussed the basic measurement of distances—including both horizontal and vertical ones. It is often not sufficient, however, to know only how far away something is or how long a line is; one must also know "in what direction it lies." Hence, another basic measurement made in surveying is that of direction. *Direction* is documented by measuring horizontal angles, vertical angles, bearings, and azimuths.

The instrument used in surveying for measuring angles, bearings, and azimuths is known as a *transit*, an example of which is shown in Fig. 4. A transit is a versatile device and exceedingly important to surveyors. It is probably the one piece of surveying equipment most commonly identified with surveyors by laypeople. Transits can be used to perform virtually all basic operations done in surveying, including measuring horizontal distances (by stadia), vertical distances (by stadia and running levels), angles in both horizontal and vertical planes, and directions (bearings and azimuths) as well as aligning points on a straight line.

Like levels, transits have a telescope that can be adjusted so it is aligned in a level plane; hence, transits can be used to run levels. Transits are far more than leveling devices, however. A transit's telescope can be rotated upward and downward to enable surveyors to evaluate angles in a vertical plane as determined from a circular scale. Transits

FIGURE 3 Level rod.

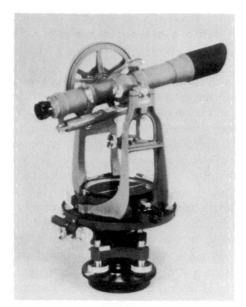

FIGURE 4 Transit.

have another circular scale that makes it possible to measure angles in a horizontal plane. Also, transits have a compass needle and associated scale, which allow evaluation of magnetic bearings and azimuths.

The transit shown in Fig. 4 is an *American transit*. A *European transit* is shown in Fig. 5; it is commonly known as a *theodolite*. In addition to the obvious differences in appearance between the two instruments (see Figs. 4 and 5), theodolites have horizontal and vertical circular scales made of glass, with graduation lines and numerals etched on the glass surface, as opposed to the metal scale

FIGURE 5 Theodolite.

on American transits. As a result, more accurate angular determinations can be made with a theodolite.

An example of a *horizontal angle* in a surveying context is the angle formed by two adjacent property lines that meet at a property corner. Such an angle can be measured by setting up a transit over the corner, sighting first along one property line and then along the other, and reading the value of the angle off the appropriate circular scale.

Another means of indicating the direction of a line is by use of *bearings*. The bearing of a line is indicated by giving the acute angle between the line and a north–south line (meridian) and the quadrant in which the line lies. For example, a line making an angle of 44°30′ with a north–south meridian and being in the northeast quadrant would have a bearing of N 44°30′ E. A line making the same angle with a north–south meridian but being in the northwest quadrant would have a bearing of N 44°30′ W.

Bearings are evaluated by reading a compass needle that rests on a pivot in a recessed well in the horizontal plate of a transit. To determine the bearing of a line, the transit is set up over one end of the line and a sighting is taken along the line. The compass needle is then allowed to swing back and forth until it comes to rest. At this point, the needle is pointing toward north and the telescope is pointing in the direction of the line. The value of the bearing is read where the end of the needle intersects the associated circular scale.

Bearing determinations are subject to error as a result of what is known as *local attraction*. This refers to various electromagnetic phenomena that can cause compass needles to deviate from north. Examples are electric power lines, local ore deposits, and magnetic effects of fences, automobiles, and wristwatches.

Another problem encountered with bearings is known as *magnetic declination*, which results because compass needles do not generally point to the true (geographic) north pole. Instead, they align themselves in the earth's magnetic field, which varies over a period of time. The difference between the direction of a compass needle when aligned in the earth's magnetic field and a true north line is the magnetic declination. It varies in amount from one location to another, and as noted above, it varies over a period of time for a specific location. If the amount of magnetic declination is known, a true bearing can be converted to a magnetic bearing, and vice versa. The direction of true north (and therefore the declination) can be determined by sighting on the North Star (Polaris).

A final means of indicating the direction of a line is by use of *azimuths*. The azimuth of a line is indicated by giving the clockwise angle the line makes with a north line. Hence, the azimuth of a line with a bearing of N 44°30′ E is simply 44°30′; the azimuth of a line with a bearing of N 44°30′ W is 315°30′. Inasmuch as azimuths are tied to compass readings, most of the comments concerning bearings (local attraction, magnetic declination) are applicable to azimuths.

III. BASIC OPERATIONS OF SURVEYING

A. Traversing

Traversing refers to the surveying of a number of consecutive lines by determining the length and direction of each line. There are basically two kinds of traverse in surveying: open and closed. An *open traverse* is one that does not return to its starting point; a *closed traverse* does return to its starting point. Examples of open traverses are surveys to lay off highways, pipelines, and power transmission lines, none of which generally returns to its starting point. An example of a closed traverse is a boundary survey of a piece of property.

Field surveying involved in traversing is essentially the measurement of lengths and directions, in accord with procedures related in Section II. Points where adjacent lines connect are marked by "hubs"; often, a hub is a wooden stake with a tack on it for open traverses or a driven metal pipe or rod in the case of closed traverses.

For closed traverses, there is an excellent scheme for checking the accuracy of field measurements made. The first step is to compute latitudes and departures for all traverse boundary lines. The *latitude* of a line is the component of the line in the north direction; it is computed by multiplying the length of the line by the cosine of its bearing angle. The *departure* of a line is the component in the east direction; it is the product of the length of the line and the sine of the bearing angle. The check on accuracy of field measurements is made by finding the sums of both latitudes and departures, for they should, in theory, both add to zero. In actuality they may very well not sum to exactly zero because surveyors, being human, cannot make every measurement perfectly. The question then becomes: How far from zero may the sums be with the survey results still considered to be within acceptable accuracy? This is answered by computing the value of a parameter called the *precision* and seeing if it falls within reasonable limits.

If a latitude and departure analysis indicates unsatisfactory results, field measurements should be checked. If results are deemed to be acceptable, the survey can be "balanced" by adjusting latitudes and departures to force them to sum to zero. From adjusted latitudes and departures, revised values of lengths and bearings of all lines can be computed.

B. Triangulation

Triangulation provides a convenient means of establishing precise locations of a number of points (i.e., providing horizontal control) over a large area. From these points, future surveys can be begun. For example, triangulation might be carried out over an area of considerable length through which a new highway is to be built, or it could be used to provide horizontal control over an entire county or state.

As suggested by its name, triangulation involves setting up a system of overlapping triangles covering an area under consideration. By measuring a number of angles and sides of these triangles, one can compute precise locations of successive points by trigonometric means. Upon occasion, more angles than sides are measured because it is sometimes easier and more expedient to measure angles than distances over large areas. (This was particularly true before the advent of EDM devices.)

To attain the high accuracy desired in triangulation, it is, of course, necessary that measurements of angles and distances be made with extreme care. As one means of securing the required accuracy, measurements are sometimes made at night to avoid certain atmospheric conditions that occur during sunlight hours and cause problems in sighting long distances through a telescope.

C. Mapping

A great deal of surveying is done for the purpose of preparing maps for certain areas or plats of property. Topographic maps show, in addition to natural and artificial features of the land, relief (variation of elevation of the earth's surface). They are widely used in engineering work.

Anyone needing a map or plat of an area should be aware of the fact that many maps are readily available at nominal cost. The U.S. Geological Survey has large-scale topographic maps covering virtually all of the United States. State highway departments are good sources of state, county, and other maps. County offices contain thousands of plats, which are a matter of public record. If, however, a map or plat of a specific area is not available for a given purpose or need, a surveyor may be employed to collect necessary field data and prepare the required map or plat.

The first step in preparing a map is to establish some kind of horizontal control. This can be done by traversing (Section III.A) or triangulation (Section III.B). The results can then be plotted on the map.

Another step is to locate and plot any significant "details," that is, important points or features on the earth's surface. Whereas maps of single lots might show houses, driveways, power poles, and individual trees, maps of larger areas would probably not feasibly show such detail.

As indicated previously, topographic maps indicate ground relief. This is most commonly done using contour lines. A *contour line* is an imaginary line along the earth's surface, all points on which are at the same elevation. This imaginary line is drawn as a solid line on a topographic map. A series of contour lines provides invaluable information for engineers, planners, and others in many kinds of analysis and design problems. For example, if one traces along a contour line on a map, all points covered are at the same elevation. Hence, it is known, in this case, that water would not flow along the corresponding line on the ground. On the other hand, a line traced on a map from a contour line labeled 97 ft to one labeled 96 ft represents a vertical drop of 1 ft on the ground. By determining the horizontal distance between the two contour lines, which can be scaled off the map, the slope of the ground can be computed. Furthermore, it is clear that closely spaced contour lines indicate steep slopes, whereas widely spaced ones indicate flat terrain.

To draw contour lines on a map, it is necessary first to determine ground elevations of a number of points over the area to be mapped. This can be accomplished by differential leveling or stadia. Then these points must be plotted accurately and the ground elevations written beside respective points on the map. The next steps are to select a suitable contour interval and then draw the actual contour lines. Contour lines are drawn freehand among the plotted points, with ground elevations written beside them. For example, if two adjacent points have elevations of 97.5 and 93.8 ft and contour lines are being drawn at 1-ft intervals, it is clear that four contour lines (the 94-, 95-, 96-, and 97-ft ones) must pass between these points. With practice, one can become proficient at drawing contour lines among labeled, plotted ground points. (Contour lines can also be drawn from aerial photographs.)

Once boundaries, details, contour lines, and so on have been plotted, a map is completed by including a title, the date, a legend, a scale, the name of the person (or agency) preparing the map, and any other information that may be of interest.

Nowadays, many maps are prepared by computers. However, it remains the surveyor's responsibility to provide accurate input data to a computer and to verify the accuracy of maps thus prepared.

D. Area Determination

As related in Section I.D, an important task for surveyors is to determine the area of land, which must be done with considerable accuracy. There are several methods for evaluating area; these are summarized here.

If a parcel of land is relatively simply shaped, such as triangular, rectangular, or trapezoidal, or can be partitioned into several such shapes, and if sufficient data are known, the parcel's area can be computed simply by applying appropriate formulas.

If latitudes and departures (Section III.A) of a tract have been determined, its area can be computed by the *double meridian distance* (DMD) method. The meridian distance

of a line is the distance from the line's midpoint to the north–south meridian. Obviously, the DMD is twice the meridian distance. The DMD method for computing area involves finding the DMD of each line, multiplying each line's DMD by its latitude, summing the products, and taking half the sum. The result is the tract's area in square feet (or square meters) if the lengths of the lines are in feet (or meters). In practice, the DMD method is carried out by a systematic procedure with the data in tabular form.

If coordinates of a tract have been determined, its area can be computed by the *coordinate method*. This is achieved by multiplying the x coordinate of each point by the difference between adjacent y coordinates, summing the products, and taking half the sum. The y coordinates must be taken in the same order around the traverse when obtaining the difference between adjacent y coordinates. As with the DMD method, the result is the tract's area in square feet (or square meters) if the coordinates are in feet (or meters). The coordinate method is also carried out in a systematic procedure with the data arranged in a certain format.

Both the DMD and the coordinate methods require straight-line boundaries. Sometimes, one or more of a tract's boundaries are curved lines. For example, a meandering creek or river may form one boundary of a tract. If such a tract is drawn to scale on a map, its area can be determined using a device called a *planimeter*.

An example of a mechanical planimeter is shown in Fig. 6 (top). It is a machine that has two arms, one of

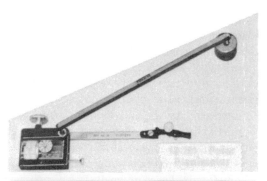

FIGURE 6 Top: Mechanical planimeter. Bottom: Electronic planimeter.

which has an anchor point at one end, while the other has a pointer. It has a wheel on its underside that is rotated as the operator runs the pointer around the boundary of a tract. The area is found by reading a value from a scale on the planimeter. In practice, one generally traces the boundary with the planimeter several times and takes an average value of planimeter readings as the tract's area. (Sometimes planimeter readings must be scaled to obtain the correct area.) More modern planimeters, like the electronic one shown in Fig. 6 (bottom), give direct readouts of areas.

E. Volume Determination

Another important task for surveyors is computation of earth volumes. Many earth volume determinations are associated with highway construction. Two examples, which are covered here, are volumes of cuts and fills in a roadway between adjacent cross sections and volumes of earth taken from borrow pits.

The volume of cut or fill in a roadway segment between adjacent cross sections can be computed by multiplying the average of the crosssectional areas at the ends of the segment by the segment's length. This computation of volume assumes a linear variation in cross-sectional area from one end of the segment to the other. A more accurate determination can be made by adding the two end areas to the cross-sectional area at the middle of the segment multiplied by four, dividing the sum by six, and multiplying the quotient by the segment's length.

Sometimes, earth to be used in a highway fill can be obtained from nearby land. Of course, the landowner must be compensated for earth removed, and such compensation is usually based on the volume of earth removed. The area from which earth is removed is known as a *borrow pit*.

The volume of earth removed from a borrow pit can be determined by laying off a grid and running levels before and after earth is removed to find the initial and final elevations of each grid point. The depth of cut at each grid point can be computed by subtracting its final elevation from its initial elevation. The approximate volume of earth removed within each of the grid's squares or rectangles can be computed by multiplying the average of the depths of cut at each corner of the square or rectangle by the area of the square or rectangle. The sum of volumes from all squares or rectangles gives, of course, the total volume of earth removed from the borrow pit.

IV. TYPES OF SURVEYING

A. Topographic Surveying

Topographic surveying is used to gather data for preparation of topographic maps (see Section III.C). Topographic surveying is conducted by field or aerial methods. Aerial methods (Section V.C) are often used when large areas are to be mapped; field methods (Section III.C), for smaller areas. Whenever aerial methods are used, some fieldwork is usually done to establish control and check for accuracy.

B. Construction Surveying

Probably a majority of all surveying done is associated with construction. Surveying of some type is required in several phases of most kinds of construction. To begin with, a topographic survey and map showing the location of whatever is to be constructed are required before construction begins. During the construction process, surveyors lay out the structure and perform other tasks as needed. After construction is completed, a final map may be prepared to document any changes that may have been made in original design plans.

Several examples of specific types of construction surveys are the staking-out of buildings, highways, and pipelines. Staking out a building involves precisely locating each corner and setting its floor elevation. These tasks are effected and documented by placing batter boards. Staking out a highway entails setting sufficient stakes along the center line and "slope stakes" to each side to indicate how much cut or fill must be made by a contractor in shaping the highway. Staking out a pipeline involves locating the pipeline, such as by placing "line stakes," and setting "grade stakes," which the contractor can use to find the proper pipeline grade.

C. Land Surveying

Land surveying is employed primarily to delineate boundary corners and lines of tracts of land. Specifically, it involves finding old property corners or establishing new ones and then marking the boundary lines that connect the corners. After surveying a tract of land, a surveyor will often prepare a map (plat) showing the tract. He or she may prepare a legal description of the tract if it is to be sold.

In theory, land surveying generally requires only the knowledge and ability to measure and lay off distances, angles, and bearings. In practice, however, good land surveyors must have experience, patience, good judgment, and some knowledge of legal principles and local customs. It is essential that a novice work under an experienced land surveyor for a period of time to become a true professional.

D. Mining Surveying

Mining surveying is a specialized type of surveying done primarily beneath the earth's surface, usually in mines. It is performed to determine locations of underground features relative to aboveground features, to measure quantities of material removed, to establish line and grade for construction or mining operations, and so on.

Like other types of surveying, mining surveying involves measurements of distances—both vertical and horizontal—and directions. Because of the underground nature of mining surveying, however, some unusual features are encountered and must be dealt with. For example, underground traverses sometimes follow very steeply inclined courses, and traverse segments are often straight for only short distances. Traverse hubs are commonly placed in a mine's roof (e.g., a hook driven or screwed into a timber) to prevent their being disturbed by the mining process. Sometimes a line of levels must be carried down a long, vertical shaft; this can be achieved by lowering a tape vertically. Cramped conditions may hamper instrument setups, and problems of darkness and dampness can cause further difficulties. Mining surveying is probably performed under the most dangerous conditions ordinarily encountered in any type of surveying.

E. Route Surveying

Route surveying refers to the surveying necessary for locating and constructing highways, railroads, canals, pipelines, and transmission lines. Initially, topographic surveys must be done and maps prepared for preliminary route location and project design. Land surveying is required in connection with land and right-of-way acquisition. Before construction begins, the final route must be staked out, grade stakes set, and so on. During construction a surveyor must be available to replace line and grade stakes that are disturbed by construction or provide additional stakes that may be needed. Final survey effort is necessary on completion of construction to document work done.

Route surveying also involves making earth volume computations (Section III.E) and design and field layout of vertical and horizontal curves. Vertical curves are generally designed as parabolas; horizontal ones, as spirals and/or parts of a circle.

Actual surveys to stake out the center line of a route (e.g., a highway or railroad) constitute open traverses. Positions along an open traverse are referred to by their distance from the starting point, with each hundred feet (or meters) being designated as a "station" and the distance beyond the last full station as a "plus distance." Hence, the location of a point along a route at a distance of 2265.4 ft from the starting point would be identified as station $22 + 65.4$.

F. Hydrographic Surveying

The prefix *hydro* refers, in general, to water; hence, *hydrographic surveying* is done on, in, or near water to secure information about the physical features of the water

area. A major purpose of hydrographic surveying is to obtain data necessary for preparing nautical charts, which show water depths, navigation channels, structures (such as piers), breakwaters, and so on and which are used by mariners. Hydrographic surveys are also useful in determining information needed for design and construction of structures adjacent to or under water. Such information is also needed when dealing with silting and dredging of lakes and channel bottoms. Hydrographic surveying is also employed in work with drilling platforms, underwater cables, and so on during offshore oil and gas exploration. The latter case is referred to as *marine surveying*.

The main measurement performed in hydrographic surveying is water depth. It can be determined in several ways. If the depth is not too great, a calibrated stick or pole can be lowered in the water until it strikes the bottom. Such a stick or pole is known as a *sounding pole*. For greater depths, a chord that is calibrated and has a weight attached can be lowered until it strikes the bottom. It is called a *leadline*. Water depths can also be determined with an echo sounding device known as a *fathometer* and a laser device called *LIDAR*.

If water depths are determined in a lake one day and then again on another day when the water surface has risen several feet, the difference in surface elevation must certainly be considered in mapping the lake's bottom. Hence, some type of vertical control is necessary. It can be accomplished by finding the surface elevation with respect to some reference datum whenever water depths are being determined and taking this into account when analyzing the raw data.

Some kind of horizontal control is also necessary when water depths are being taken. In mapping a lake, it is essential that the location (in a horizontal plane) of each point for which a water depth is determined be known to plot the map or nautical chart. Horizontal control can be accomplished by traversing or triangulation, but the actual procedure is complicated by the fact that depth determinations are generally made from nonstationary boats, precise locations of which may be difficult to determine over larger water bodies. A navigational device known as a *sextant* is used to measure horizontal angles from a boat to various objects on land, and from these values various locations of the boat (i.e., of the depth soundings) can be determined.

Often associated with surveying and mapping of water areas is the topographic surveying of the shoreline needed to show it on the map or nautical chart. This can be done by transit–stadia ground surveying or by aerial photogrammetry.

G. Municipal Surveying

Municipal (or *city*) *surveying* is an overall term used to describe both surveying done in a municipal setting and that performed by a municipality in fulfilling its service functions. Examples of the former are surveys for delineating property lines and for locating buildings and retaining walls that are to be constructed. Instances of the latter are surveys for streets, curbs, bridges, sewers, and storm drainage systems.

Of primary importance in municipal surveying is a horizontal and vertical control system. This calls for the establishment of numerous monuments, precise locations and elevations of which are known, throughout the municipal area. They provide numerous potential starting points from which future municipal surveys can begin with the assurance that all surveying done will be tied in to a common reference.

Also of importance in the realm of municipal surveying are the gathering of data for and preparation and dissemination of maps of various kinds. These include base maps, topographic maps, property maps, and subdivision maps, as well as maps showing locations of utility lines and underground features (sewers, water pipes, gas mains, subways, etc.), to name a few.

V. ADDITIONAL TOPICS IN SURVEYING

A. State Plane Coordinate Systems

The precise location of any point on the earth's surface can be specified by giving its latitude, longitude, and elevation above mean sea level. Thousands of points across the United States have had their latitudes and longitudes determined accurately and marked with ground monuments by the National Geodetic Survey (formerly the Coast and Geodetic Survey). Latitude and longitude are spherical coordinates, however, and therefore indicative of geodetic positions, and referencing of local surveys to them can be troublesome. Realizing this difficulty, the Coast and Geodetic Survey established in 1935 the *State Plane Coordinate Systems* (SPCS), which relate the horizontal locations of points in terms of plane coordinates. With these available, surveyors can tie in local surveys to the SPCS with the relatively simple performance of and computations involving plane surveying (i.e., assumption of a "flat earth") and thereby document forever the exact location of a local survey.

In the SPCS, each state has a separately established coordinate system, but all such state systems are tied together. The coordinate system for each state was established by projecting, mathematically, its surface onto the surface of a cone or cylinder (or, in some cases, both), which surface can then be developed into a plane. The projection onto a cone is known as the *Lambert conformal conic projection*. Distortions of this projection occur in the north–south direction; hence, it is used for states with relatively short north–south dimensions (such as Tennessee

and North Carolina). The projection onto a cylinder is known as the *transverse Mercator projection*; it is used for states with relatively long north–south dimensions (e.g., Illinois and Mississippi). Some states utilize more than one projection of either kind, and some use both kinds of projection. For example, New York uses a Lambert projection for Long Island and three Mercator projections for the rest of the state. Florida uses a Lambert projection for the northwestern ("panhandle") part of the state and two Mercator projections for the lower ("peninsula") part of the state.

Some advantages of the SPCS, in addition to that of documenting forever exact locations of local surveys, are tying together separate surveys to a common system and providing checking capabilities to prevent excessive accumulation of errors of measurement in large-scale surveys.

B. Astronomical Observations

Astronomical observations can be used in surveying to establish directions by sighting on celestial bodies. Probably the most commonly performed observations are ones made to determine the direction of true north or of a line with reference to true north.

The true azimuth of a line can be determined by astronomical observation by setting up a transit over one end of the line, sighting on a particular celestial body, and reading the vertical and horizontal angles to the line in question. If the time of observation and location (latitude and longitude) of the observation point are known, the exact position of the celestial body can be determined from an almanac of celestial body positions (known as an *ephemeris*). From this information, the true azimuth of the line to the celestial body and then the true azimuth of the line in question can be calculated.

The celestial body most commonly used for determining the direction of true north is the North Star (Polaris). Fortuitously, it is located almost directly above the earth's geographic north pole; hence, a sighting on Polaris gives (approximately) the true north direction. In actuality, inasmuch as Polaris is not exactly directly above the geographic north pole, it (Polaris) appears to the viewer looking through a fixed telescope to move slightly back and forth (left to right to left, etc.) over a period of time. True north is, in effect, the average of the extreme left and right sightings of Polaris.

C. Photogrammetry

Various measuring techniques for gathering data to be used in preparing topographic maps have already been discussed. Although they are widely used and provide adequate results, they have at least two limitations, particularly when applied to surveys over large areas. One is the time (and cost) involved in obtaining required data;

the other is limitation on the actual amount of data that can be obtained. Both of these can be overcome by using *photogrammetry*.

In simple terms, photogrammetry consists of photographing the property in question from an airplane, viewing the aerial photographs, and preparing a topographic map therefrom. It should be clear that substantial savings in time are realized in gathering data by means of photogrammetry compared to more conventional means; and although the costs of operating an aircraft may be high, they are more than offset by savings in time and salaries paid. The fact that the camera "sees everything" ensures that a vast amount of data will be obtained and virtually no detail will be overlooked in the field.

In practice, a number of photographs are made as the airplane flies a prescribed course so that they will overlap one another. Maps are prepared by specialists who view the photographs through a stereoscope. Specialized plotting machines may also be used. As in other cases, adequate horizontal and vertical control systems must be employed to ensure sufficient accuracy.

Photogrammetry is used extensively nowadays by the U.S. Geological Survey in compiling its quadrangle maps. Significant improvements in cameras, film, plotting instruments, and other devices have made it possible to produce maps that meet high accuracy standards.

It is noteworthy that certain situations preclude the use of photogrammetry. For example, for small areas, it is generally not cost-effective. Also, in areas covered by sand (deserts or beaches) or snow, lack of change in color and texture of the ground makes interpretation of the ground surface difficult. Another problem is encountered in an area of extraordinarily dense growth (e.g., a rain forest), which, of course, makes it difficult or impossible to see the ground surface.

D. Recent Developments

Many recent developments in surveying, some of which have been mentioned, provide measurement with greater accuracy than could have even been anticipated a few years ago. Many developments occurred as a result of the space program, which required, on the one hand, very accurate determination of relative locations of space tracking stations around the world and, on the other hand, detailed mapping of the moon as well as planets and other stars. Whereas photogrammetry has been used for many years to "survey from above," nowadays this can be accomplished with earth-orbiting satellites.

Various modern devices utilizing lasers, infrared light, and so on provide extremely accurate determinations of elevations, distances, and directions. The computer revolution has affected surveying also. Microcomputers can automatically record data, make appropriate computations, and prepare maps.

Undoubtedly, efforts will continue into the future to perform surveying with even greater accuracy, in less time, and at a lower cost. Throughout it all, however, surveyors must be knowledgeable in the basic fundamentals of surveying and must constantly exercise good professional judgment in interpreting and utilizing results from the sophisticated instrumentation available.

SEE ALSO THE FOLLOWING ARTICLES

GEODESY • MINING ENGINEERING • NUMBER THEORY, ELEMENTARY • RAILWAY ENGINEERING

BIBLIOGRAPHY

Allan, A. L. (1997). "Practical Surveying and Computations," Butterworth–Heinemann, London.

Anderson, J., and Mikhail, E. (1997). "Surveying: Theory and Practice," 7th ed., McGraw–Hill Higher Education, New York.

Bannister, R. (1998). "Surveying," 6th ed., Addison–Wesley, Reading, MA.

Herubin, C. (1998). "Principles of Surveying," 4th ed., Prentice Hall, Englewood Cliffs, NJ.

Kavanagh, B. F. (2001). "Surveying with Construction Applications," 3rd ed., Prentice Hall, Englewood Cliffs, NJ.

McCormac, J. C. (1999). "Surveying," 4th ed., John Wiley & Sons, New York.

Moffit, F., and Bossler, J. (1998). "Surveying," 10th ed., Addison–Wesley, Reading, MA.

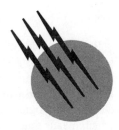

Synthetic Aperture Radar

Jakob J. van Zyl
Yunjin Kim

Jet Propulsion Laboratory
California Institute of Technology

GLOSSARY

Antenna A device that is designed to radiate or receive electromagnetic waves.

Doppler frequency The change of frequency of a received signal due to the relative velocity of a transmitting antenna with respect to an illuminated object.

Polarization The property of a radiated electromagnetic wave describing the time-varying direction and amplitude of the electric field vector.

Radar An acronym for R̲adio D̲etection a̲nd R̲anging. A radar system measures the distance to an object by transmitting an electromagnetic signal and receiving an echo reflected from the object.

Radar interferometer A radar system that determines the angle of arrival of a radar signal by phase comparison of the signals received at separate antennas.

Synthetic aperture radar A radar system that synthesizes the effect of a long antenna through the motion of a small antenna relative to the target.

THE TERM "RADAR" is an acronym for Radio Detection and Ranging. A radar measures the distance, or *range*, to an object by transmitting an electromagnetic signal and receiving an echo reflected from the object. Since electromagnetic waves propagate at the speed of light, one only has to measure the time it takes the radar signal to propagate to the object and back to calculate the range to the object. The total distance traveled by the signal is twice the distance between the radar and the object, since the signal travels from the radar to the object and then back from the object to the radar after reflection. Therefore, once we have measured the propagation time (t), we can easily calculate the range (ρ) as

$$\rho = \frac{1}{2}ct, \qquad (1)$$

where c is the speed of light. The factor $1/2$ accounts for the fact that the radar signal actually traveled twice the distance; first from the radar to the object and then from the object to the radar.

Encyclopedia of Physical Science and Technology, Third Edition, Volume 16

Radars provide their own signals to detect the presence of objects. Therefore, radars are known as *active* remote sensing instruments. Because the radar provides its own signal, it can operate during day or night. In addition, radar signals typically penetrate clouds and rain, which means that radar images can be acquired not only during day or night, but also under (almost) all weather conditions. For this reason, radars are often referred to as *all-weather* instruments. Imaging remote sensing radars such as synthetic aperture radars produce high-resolution (from submeter to a few tens of meters) images of surfaces. The geophysical information can be derived from these high-resolution images by using proper post-processing techniques.

This article focuses on a specific class of implementation of radar known as *synthetic aperture radar*, or SAR, with particular emphasis on spaceborne SAR. As mentioned above, SAR is a way to achieve high-resolution *images* using radio waves. Here, we shall first describe the basics of radar imaging and follow that with a description of the synthetic aperture principle. We will then look at some advanced SAR implementations such as SAR polarimetry and SAR interferometry. We will also briefly discuss some examples of civilian spaceborne SAR missions.

I. BASIC PRINCIPLES OF RADAR IMAGING

Imaging radars generate surface images that are, at first glance, very similar to the more familiar images produced by instruments that operate in the visible or infrared parts of the electromagnetic spectrum. However, the principle behind the image generation is fundamentally different in the two cases. Visible and infrared sensors use a lens or mirror system to project the radiation from the scene on a "two-dimensional array of detectors," which could be an electronic array or, in earlier remote sensing instruments, a film using chemical processes. The two dimensionality can also be achieved by using scanning systems or by moving a single line array of detectors. This imaging approach, like that we are all familiar with when taking photographs with a camera, conserves the relative angular relationships between objects in the scene and their images in the focal plane as shown in Fig. 1. Because of this conservation of angular relationships, the resolution of the images depends on how far away the camera is from the scene it is imaging. The closer the camera is, the higher the resolution and the smaller the details that can be recognized in the images. As the camera moves farther away from the scene, the resolution degrades, and only larger objects can be discerned in the image.

Imaging radars use a quite different mechanism to generate images, with the result that the image characteristics

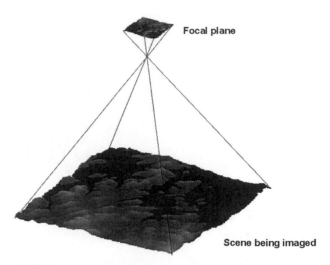

FIGURE 1 Passive imaging systems in the visible and infrared part of the electromagnetic spectrum conserve the angular relationships between objects in the scene and their images in the focal plane of the instrument.

are also quite different from that of visible and infrared images. There are two different ways radars can be used to produce images. These two types of radars are broadly classified as *real aperture* and *synthetic aperture* radars. We shall discuss the differences between these two types in more detail later in this article. In order to separate objects in radar images in the cross-track direction and the along-track direction, two different methods must be implemented. The *cross-track* direction, also known as the *range* direction in radar imaging, is the direction perpendicular to the direction in which the imaging platform is moving. In this direction, radar echoes are separated using the *time delay* between the echoes that are backscattered from the different surface elements. This is true for both real aperture and synthetic aperture radar imagers. The *along-track* direction, also known as the *azimuth* direction, is the direction parallel to the movement of the imaging platform. The angular size (in the case of the real aperture radar) or the Doppler history (in the case of the SAR) is used to separate surface pixels in the along-track dimension in the radar images. As we will see later, only the azimuth imaging mechanism of real aperture radars is similar to that of regular cameras. Using the time delay and Doppler history results, SAR images have resolutions that are independent of how far away the radar is from the scene it is imaging. This fundamental advantage enables high-resolution, spaceborne SAR without requiring an extremely large antenna.

Another difference between images acquired by cameras operating in the visible and near-infrared part of the electromagnetic spectrum and radar images is the way in which they are acquired. Cameras typically look straight

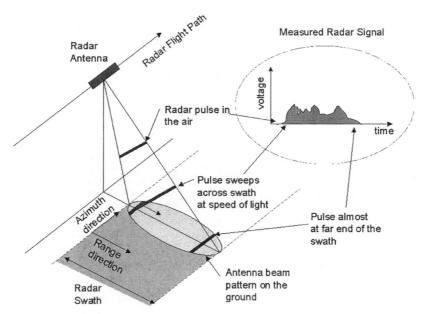

FIGURE 2 Imaging geometry for a side-looking radar system.

down or at least have no fundamental limitation that prevents them from taking pictures looking straight down from the spacecraft or aircraft.This is not so for imaging radars. To avoid so-called *ambiguities*, which we will discuss in more detail later, the imaging radar sensor has to use an antenna which illuminates the surface to one side of the flight track. Usually, the antenna has a fan beam which illuminates a highly elongated elliptical-shaped area on the surface, as shown in Fig. 2. The illuminated area across track defines the image *swath*.

Within the illumination beam, the radar sensor transmits a very short effective pulse of electromagnetic energy. Echoes from surface points farther away along the cross-track coordinate will be received at a proportionally later time (see Fig. 2). Thus, by dividing the receive time in increments of equal time bins, the surface can be subdivided into a series of *range bins*. The width in the along-track direction of each range bin is equal to the antenna footprint along the track x_a. As the platform moves, the sets of range bins are covered sequentially, thus allowing strip mapping of the surface line by line. This is comparable to strip mapping with a so-called pushbroom imaging system using a line array in the visible and infrared part of the electromagnetic spectrum. The brightness associated with each image pixel in the radar image is proportional to the echo power contained within the corresponding time bin. As we will see later, the real difference between real aperture radars and SARs lies in the way in which the azimuth resolution is achieved.

This is also a good time to point out that there are two different meanings for the term *range* in radar imaging.

The first is the so-called *slant range* and refers to the range along the radar line-of-sight, as shown in Fig. 2 in the way that pulses propagate. Slant ranges are measured directly from the radar. The second use of the term range is for the *ground range*, which refers to the range along a smooth surface (the ground) to the scatterer. The ground range is measured from the so-called *nadir track*, which represents the line described by the position directly underneath the radar imaging platform. One has to be careful to take topography into account when resampling radar images from slant range to ground range.

Before looking at radar resolutions, let us define a few more terms commonly encountered in radar imaging. The *look angle* is defined as the angle between the vertical direction and the radar beam at the radar platform, while the *incidence angle* is defined as the angle between the vertical direction and the radar wave propagation vector at the surface. When surface curvature effects are neglected, the look angle is equal to the incidence angle at the surface when the surface is flat. In the case of spaceborne systems, surface curvature must be taken into account, which leads to an incidence angle that is always larger than the look angle for flat surfaces. It is quite common in the literature to find authors using the terms look angle and incidence angle interchangeably; this is only correct for low-flying aircraft and only when there is no topography present in the scene. As we will see next, if topography is present, i.e., if the surface is not flat, the local incidence angle may vary in the radar image from pixel to pixel.

Consider the simple case of a single hill illuminated by a radar system as shown in Fig. 3. Also shown is the

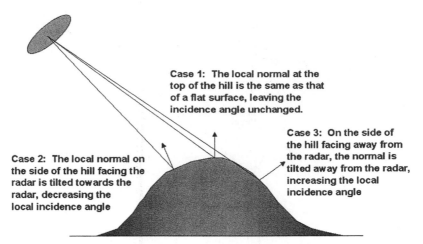

Case 1: The local normal at the top of the hill is the same as that of a flat surface, leaving the incidence angle unchanged.

Case 3: On the side of the hill facing away from the radar, the normal is tilted away from the radar, increasing the local incidence angle

Case 2: The local normal on the side of the hill facing the radar is tilted towards the radar, decreasing the local incidence angle

FIGURE 3 Topographic variations in the image will cause the local incidence angle to be different from that expected for a flat surface with no relief.

local normal to the surface for several positions on the hill. Relative to a flat surface, it is clear that for points on the hill facing the radar, the local normal tilts more toward the radar. Therefore, the local incidence angle will be smaller than for a point at the same ground range, but on a flat surface.

A term commonly encountered in military literature is *depression angle*. This is the angle between the radar beam and the horizontal at the radar platform. The depression angle is therefore related to the look angle in that one is equal to 90° minus the other. A small look angle is equivalent to a large depression angle and vice versa. Similarly, one often finds the term *grazing angle* describing the angle between the horizontal at the surface and the incident wave in the military literature. The grazing angle is therefore related to the incidence angle in the same way that the depression angle is related to the look angle. In this article, we shall use look angle and incidence angle to describe the imaging geometry.

A. Radar Resolution

The *resolution* of an image is defined as the separation between the two closest features that can still be resolved in the final image. First, consider two point targets that are separated in the slant range direction by x_r. Because the radar waves propagate at the speed of light, the corresponding echoes will be separated by a time difference Δt equal to

$$\Delta t = 2x_r/c, \qquad (2)$$

where c is the speed of light and the factor 2 is included to account for the signal round trip propagation as described before. Radar waves are usually not transmitted continuously; instead, radar usually transmits short bursts of

energy known as radar *pulses*. The two features can be discriminated if the leading edge of the pulse returned from the second object is received later than the trailing edge of the pulse received from the first feature, as shown in Fig. 4. Therefore, the smallest separable time difference in the radar receiver is equal to the effective time length τ of the pulse. Thus, the slant range resolution of a radar is

$$2x_r/c = \tau \Rightarrow x_r = \frac{c\tau}{2}. \qquad (3)$$

Now let us consider the case of two objects separated by a distance x_g on the ground. The corresponding echoes will be separated by a time difference Δt equal to

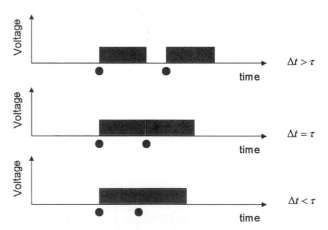

FIGURE 4 If the radar echoes from two point targets are separated in time by more than or equal to the length of the radar pulse, it is possible to recognize the echoes as those from two different scatterers, as shown in the top two panels. If the time difference between the echoes is less than the radar pulse length, it is not possible to recognize two distinct scatters, as in the case of the bottom panel.

$$\Delta t = 2x_r \sin\theta/c. \tag{4}$$

The angle θ in Eq. (4) is the local incidence angle. (This should actually be called the incident angle, or angle of incidence. Since incidence angle is used almost universally in the literature, we shall continue to use that term to avoid confusion.) As in the case of the slant range discussed above, the two features can be discriminated if the leading edge of the pulse returned from the second object is received later than the trailing edge of the pulse received from the first feature. Therefore, the ground range resolution of the radar is given by

$$2x_r \sin\theta/c = \tau \Rightarrow x_r = \frac{c\tau}{2\sin\theta}. \tag{5}$$

In other words, the range resolution is equal to half the footprint of the radar pulse on the surface.

Sometimes the effective pulse length is described in terms of the system bandwidth B. As we will show in the next section, to a good approximation,

$$\tau = 1/B. \tag{6}$$

The $\sin\theta$ term in the denominator of Eq. (5) means that the ground range resolution of an imaging radar will be a function of the incidence angle.

A pulsed radar determines the range by measuring the round trip time by transmitting a pulse signal. In designing the signal pattern for a radar sensor, there is usually a strong requirement to have as much energy as possible in each pulse in order to enhance the signal-to-noise ratio. This can be done by increasing the transmitted peak power or by using a longer pulse. However, particularly in the case of spaceborne sensors, the peak power is usually strongly limited by the available power sources. On the other hand, an increased pulse length leads to a poorer range resolution [see Eq. (5)]. This dilemma is usually resolved by using *modulated* pulses which have the property of a wide bandwidth even when the pulse is very long. After so-called pulse compression, a short effective pulse length is generated, increasing the resolution. One such modulation scheme is the linear frequency modulation or *chirp*.

In a chirp, the signal frequency within the pulse is linearly changed as a function of time. If the frequency is linearly changed from f_0 to $f_0 + \Delta f$, the effective bandwidth would be equal to

$$B = |(f_0 + \Delta f) - f_0| = |\Delta f|, \tag{7}$$

which is independent of the pulse length. Thus, a pulse with long duration (i.e., high energy) and wide bandwidth (i.e., high range resolution) can be constructed. The instantaneous frequency for such a signal is given by

$$f(t) = f_0 + \frac{B}{\tau'}t \quad \text{for} \quad -\tau'/2 \le t \le \tau'/2, \tag{8}$$

and the corresponding signal amplitude is

$$A(t) \sim \Re\left\{\exp\left[-i2\pi \int f(t)\,dt\right]\right\}$$
$$= \cos\left[2\pi\left(f_0 t + \frac{B}{2\tau'}t^2\right)\right], \tag{9}$$

where $\Re(x)$ means the real part of x. Note that the instantaneous frequency is the derivative of the instantaneous phase. A pulse signal such as that shown in Eq. (9) has a physical pulse length τ' and a bandwidth B. The product $\tau' B$ is known as the *time bandwidth product* of the radar system. In typical radar systems, time bandwidth products of several hundred are used.

At first glance it may seem that using a pulse of the form in Eq. (9) cannot be used to separate targets that are closer than the projected physical length of the pulse as shown in the previous section. It is indeed true that the echoes from two neighboring targets, which are separated in the range direction by much less than the physical length of the signal pulse, will overlap in time. If the modulated pulse, and therefore the echoes, have a constant frequency, it will not be possible to resolve the two targets. However, if the frequency is modulated as described in Eq. (8), the echoes from the two targets will have different frequencies at any instant of time and therefore can be separated by frequency filtering.

In actual radar systems, a matched filter is used to *compress* the returns from the different targets. Suppose we transmit a signal of the form described in Eq. (9). The signal received from a single point scatterer at a range ρ is a scaled replica of the transmitted signal delayed by a time $t = 2\rho/c$. The output of the matched filter for a such a point scatterer is mathematically described as the convolution of the returned signal with a replica of the transmitted signal. Being careful about the limits of the integration, one finds that for large time bandwidth products,

$$V_0(t) = \tau' E_r \exp(iwt)\exp(-i4\pi\rho/\lambda)$$
$$\times \frac{\sin(\pi B(t - 2\rho/c))}{\pi B(t - 2\rho/c)}. \tag{10}$$

This compressed pulse has a half power width of $1/B$, and its peak position occurs at time $2\rho/c$. Therefore, the achievable range resolution using a modulated pulse of the kind given by Eq. (9) is a function of the chirp bandwidth and not the physical pulse length. In typical spaceborne and airborne SAR systems, physical pulse lengths of several tens of microseconds are used, while bandwidths of several tens of megahertz are no longer uncommon for spaceborne systems, and several hundreds of megahertz are common in airborne systems.

So far we have seen the first major difference between radar imaging and that used in the visible and infrared part of the spectrum. The cross-track resolution in the radar case is independent of the distance between the scene and the radar instrument and is a function of the system bandwidth. Before looking at the imaging mechanisms in the along-track direction, we will examine the general expression for the amount of reflected power that the radar receiver would measure. This is described through the so-called radar equation, which we will examine in the next section.

B. Radar Equation

One of the key factors that determine the quality of the radar imagery is the corresponding *signal-to-noise* ratio, commonly called SNR. This is the equivalent of the brightness of a scene being photographed with a camera versus the sensitivity of the film or detector. Here, we consider the effect of thermal noise on the sensitivity of radar imaging systems. The derivation of the radar equation is graphically shown in Fig. 5.

The total received power is equal to the power intercepted by the receiving antenna and is given by

$$P_r = \frac{P_t G_t}{4\pi\rho^2} s\sigma_0 \frac{\lambda^2 G_r}{(4\pi\rho)^2}, \qquad (11)$$

where G_t is the gain of the transmitting antenna, G_r is the gain of the receiving antenna, and λ is the wavelength of the radar signal. σ_0 is the surface *backscattering cross section* which represents the efficiency of the surface in re-emitting back toward the sensor some of the energy incident on it. It is similar to the surface albedo at visible wavelengths, except that it is designed to scatter the en-

ergy over a sphere rather than a hemisphere as in the case of the albedo. Here, the area responsible for reflecting the scattered power is denoted by s. In Eq. (13) we explicitly show that the transmit and receive antennas may have different gains. This is important for the more advanced SAR techniques, like polarimetry, where antennas with different polarizations may be used during transmission and reception.

In addition to the target echo, the received signal also contains noise which results from the fact that all objects at temperatures higher than absolute zero emit radiation across the whole electromagnetic spectrum. The noise component that is within the spectral bandwidth B of the sensor is passed through with the signal. The receiver electronics also generates noise that contaminates the signal. The thermal noise power is given by

$$P_N = kTB, \qquad (12)$$

where k is Boltzmann's constant ($k = 1.6 \times 10^{-23}$ W/K/Hz) and T is the total equivalent noise temperature in Kelvin. The resulting SNR is then

$$\text{SNR} = P_r/P_N. \qquad (13)$$

One common way of characterizing an imaging radar sensor is to determine the surface backscatter cross section σ_N which gives an SNR $= 1$. This is called the *noise equivalent backscatter cross section*. It defines the weakest surface return that can be detected and, therefore, identifies the range of surface units that can be imaged.

Typical spaceborne SAR frequencies are shown in Table I. Notice that each frequency band has the frequency range allocated for remote sensing.

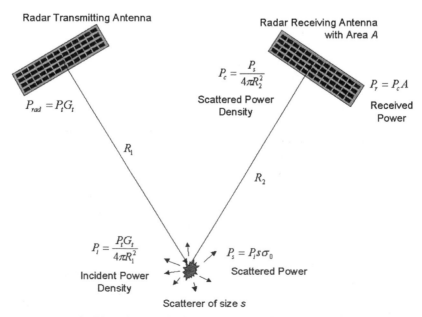

FIGURE 5 Schematic of the derivation of the radar equation.

TABLE I Frequency Allocation at Typical SAR Frequency Bands

Frequency band	Allocated frequency range for remote sensing (MHz)
L-band	1215–1300
S-band	3100–3300
C-band	5250–5460
X-band	8550–8650
	9500–9800

II. REAL AND SYNTHETIC APERTURE RADAR

A. Real Aperture Radar

The real aperture imaging radar sensor also uses an antenna which illuminates the surface to one side of the flight track. As mentioned before, the antenna usually has a fan beam which illuminates a highly elongated elliptical-shaped area on the surface, as shown in Fig. 2. As shown in Fig. 2, the illuminated area across the track defines the image swath. For an antenna of width W operating at a wavelength λ, the beam angular width in the range plane is given by

$$\theta_r \approx \lambda/W, \tag{14}$$

and the resulting surface footprint or swath S is given by

$$S \approx \frac{h\theta_r}{\cos^2\theta} = \frac{\lambda h}{W\cos^2\theta}, \tag{15}$$

where h is the sensor height above the surface, θ is the angle from the center of the illumination beam to the vertical (the *look angle* at the center of the swath), and θ_r is assumed to be very small. Note that Eq. (15) ignores the curvature of the earth. For spaceborne radars, this effect should not be ignored, especially if the antenna beamwidth is large. In that case, one needs to use the law of cosines to solve for the swath width.

A real aperture radar relies on the resolution afforded by the antenna beam in the along-track direction for imaging. This means that the resolution of a real aperture radar in the along-track direction is driven by the size of the antenna as well as the range to the scene. Assuming an antenna length of L, the antenna beamwidth in the along-track direction is

$$\theta_a \approx \frac{\lambda}{L}. \tag{16}$$

At a distance ρ from the antenna, this means that the antenna beamwidth illuminates an area with the along-track dimension equal to

$$x_a \approx \rho\theta_a \approx \frac{\lambda\rho}{L} \approx \frac{\lambda h}{L\cos\theta}. \tag{17}$$

To illustrate, for $h = 800$ km, $\lambda = 23$ cm, $L = 12$ m, and $\theta = 20°$, then $x_a = 16$ km. Even if λ is as short as 2 cm and h is as low as 200 km, x_a will still be equal to about 360 m, which is considered to be a relatively poor resolution, even for remote sensing. This has led to very limited use of the real aperture technique for surface imaging, especially from space. A real aperture radar uses the same imaging mechanism as a passive optical system for the along-track direction. However, because of the small value of λ (about 1 μm), resolutions of a few meters can be achieved from orbital altitudes with an aperture only a few tens of centimeters in size. From aircraft altitudes, however, reasonable azimuth resolutions can be achieved if higher frequencies (typically X-band or higher) are used. For this reason, real aperture radars are not commonly used in spaceborne remote sensing, except in the case of scatterometers and altimeters that do not need high-resolution data.

In terms of the radar equation, the area responsible for reflecting the power back to the radar is given by the physical size of the antenna illumination in the along-track direction and by the projection of the pulse on the ground in the cross-track direction. This is shown in Fig. 2 for the pulses in the radar swath. The along-track dimension of the antenna pattern is given by Eq. (17). If the pulse has a length τ in time, and the signal is incident on the ground at an angle θ_i, the projected length of the pulse on the ground is

$$l_g = \frac{c\tau}{2\sin\theta_i}. \tag{18}$$

Therefore, the radar equation in the case of a real aperture radar becomes

$$P_r = \frac{P_t G_t G_r \lambda^2}{(4\pi)^3 \rho^4} \frac{\lambda\rho}{L} \frac{c\tau}{2\sin\theta_i} \sigma_0. \tag{19}$$

This shows that for the real aperture radar, the received power decreases as the range to the third power. In terms of the physical antenna sizes, we can rewrite this expression as

$$P_r = \frac{P_t W^2 L c\tau \sigma_0}{8\pi\lambda\rho^3\sin\theta_i}. \tag{20}$$

This is the radar equation for a so-called distributed target for the real aperture radar case. From Eq. (20) it is clear that the received power increases as the square of the width of the antenna. However, increasing the antenna width also decreases the swath width. The received power only increases linearly with an increase in antenna length. Increasing the antenna length also increases the along-track resolution of the real aperture radar. For this reason, real aperture radars usually operate with antennas that are the longest that could be practically accommodated.

In summary, a real aperture radar uses the same imaging mechanism as passive imaging systems to achieve along-track resolution. The practically achievable resolutions are usually poorer than what is generally required for remote sensing applications. Real aperture radars are therefore not commonly used for remote sensing applications.

B. Synthetic Aperture Radar (SAR)

SAR refers to a particular implementation of an imaging radar system that utilizes the movement of the radar platform and specialized signal processing to generate high-resolution images. Prior to the discovery of SAR, principle imaging radars operated using the real aperture principle and were known as side-looking aperture radars (SLAR).

Carl Wiley of the Goodyear Aircraft Corporation is generally credited as the first person to describe the use of Doppler frequency analysis of signals from a moving coherent radar to improve along-track resolution. He noted that two targets at different along-track positions will be at different angles relative to the aircraft velocity vector, resulting in different Doppler frequencies. (The Doppler effect is the well-known phenomenon that causes a change in the pitch of a car horn as it travels past a stationary observer.) Using this effect, targets can be separated in the along-track direction on the basis of their different Doppler frequencies. This techniques was originally known as Doppler beam sharpening, but later became known as SAR.

The main difference between real and synthetic aperture radars is therefore in the way in which the azimuth resolution is achieved. The range resolution and radar equation derived previously for a real aperture radar are still valid here. The along-track imaging mechanism and the resulting along-track resolution are, however, quite different for the real and synthetic aperture radar case.

As the radar moves along the flight path, it transmits pulses of energy and records the reflected signals, as shown in Fig. 2. When the radar data are processed, the position of the radar platform is taken into account when adding the signals to integrate the energy for the along-track direction. Consider the geometry shown in Fig. 6. As the radar moves along the flight path, the distance between the radar and the scatterer changes, with the minimum distance occurring when the scatterer is directly broadside of the radar platform. The *phase* of the radar signal is given by $-4\pi\rho/\lambda$. The changing distance between the radar and the scatterer means that after range compression, the phase of the signal will be different for the different positions along the flight path.

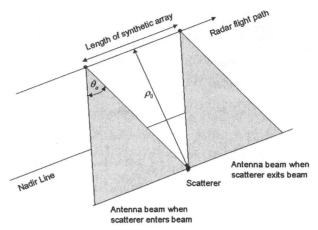

FIGURE 6 The SAR integrates the signal from the scatter for as long as the scatterer remains in the antenna beam.

The range between the radar and the scatterer as a function of position along the flight path is given by

$$\rho(s) = \sqrt{\rho_0^2 + v^2 s^2}, \tag{21}$$

where ρ_0 is the range at closest approach to the scatterer, v is the velocity of the radar platform, and s is the time along the flight path (so-called slow time) with zero time at the time of closest approach. To a good approximation for remote sensing radars, we can assume that $vs \ll \rho_0$. (This may not be true for very high resolution radars, but the basic principle remains the same.) In this case, we can approximate the range as a function of slow time as

$$\rho(s) \approx \rho_0 + \frac{v^2}{2\rho_0} s^2. \tag{22}$$

The phase of the signal after range compression as shown in Eq. (10) then becomes

$$\phi(s) = -\frac{4\pi\rho(s)}{\lambda} \approx -\frac{4\pi\rho_0}{\lambda} - \frac{2\pi v^2}{\rho_0\lambda} s^2. \tag{23}$$

The instantaneous frequency of this signal is

$$f(s) = \frac{1}{2\pi}\frac{\partial\phi(s)}{\partial s} = -\frac{2v^2}{\rho_0\lambda} s. \tag{24}$$

This is the expression of a linear frequency chirp. To find the bandwidth of this signal, we have to find the maximum time that we can use in the signal integration. This maximum "integration time" is given by the amount of time that the scatterer will be in the antenna beam. For an antenna with a physical length L, the horizontal beamwidth is $\theta_a = \lambda/L$, so that the scatterer at the range of closest approach ρ_0 is illuminated for a time

$$s_{tot} = \frac{\lambda\rho_0}{Lv}. \tag{25}$$

Half of this time occurs when the radar is approaching the range of closest approach, and half of it is spent traveling away from the range of closest approach. Therefore, the bandwidth of the signal shown in Eq. (24), which is the Doppler bandwidth of the SAR signal, is

$$B_D = \frac{2v}{L}. \tag{26}$$

If this signal is filtered using a matched filter as described earlier under signal modulation, the resulting compressed signal will have a width in time of $1/B_D$. Since the radar platform moves at a speed of v, this leads to an along-track resolution of

$$x_a = \frac{v}{B_D} = \frac{L}{2}. \tag{27}$$

This result shows that the azimuth (or along-track) surface resolution for an SAR is equal to half the size of the physical antenna and is *independent of the distance between the sensor and the surface*. At first glance this result seems most unusual. It shows that a smaller antenna gives better resolution. This can be explained in the following way. The smaller the physical antenna is, the larger its footprint. This allows a longer observation time for each point on the surface, i.e., a longer array can be synthesized. This longer synthetic array allows a larger Doppler bandwidth and hence a finer surface resolution. Similarly, if the range between the sensor and surface increases, the physical footprint increases, leading to a longer observation time and larger Doppler bandwidth which counterbalances the increase in the range.

As mentioned earlier, the imaging radar transmits a series of pulsed electromagnetic waves. Thus, the Doppler history from a scatterer is not measured continuously, but is sampled on a repetitive basis. In order to get an accurate record of the Doppler history, the Nyquist sampling criterion requires that sampling occurs at least at twice the highest frequency in the Doppler bandwidth. Thus, the pulse repetition frequency, usually called PRF, must be larger than

$$\mathrm{PRF} \geq 2B_D = \frac{2v}{L}. \tag{28}$$

Note that we used half the Doppler bandwidth as the highest frequency in the Doppler bandwidth in Eq. (28). The reason for this is that the Doppler frequency varies linearly from $-B_D/2$ to $+B_D/2$. Therefore, even though the total bandwidth of the signal is B_D, the highest frequency in the bandwidth is only $B_D/2$.

Equation (28) means that at least one sample (i.e., one pulse) should be taken every time the sensor moves by half an antenna length. As an example, for a spaceborne imaging system moving at a speed of 7 km/sec and using an antenna 10 m in length, the corresponding minimum PRF is 1.4 kHz. As we will see in the next section, the requirement to cover a certain swath size provides an upper bound on the PRF. Interpreted in a different way, the requirement to adequately sample the signal bandwidth limits the size of the swath that could be imaged.

III. RADAR IMAGE ARTIFACTS AND NOISE

Radar images could contain a number of anomalies which result from the way imaging radars generate the image. Some of these are similar to what is encountered in optical systems, such as blurring due to defocusing or scene motion, and some such as range and azimuth ambiguities are unique to radar systems. This section addresses the anomalies which are most commonly encountered in radar images.

A. Range and Azimuth Ambiguities

As mentioned earlier (see Fig. 2), a radar images a surface by recording the echoes line by line with successive pulses. The leading edge of each echo corresponds to the near edge of the image scene, and the tail end of the echo corresponds to the far edge of the scene. The length of the echo (i.e., swath width of the scene covered) is determined by the antenna beamwidth or the size of the data window used in the recording of the signal. The exact timing of the echo reception depends on the range between the sensor and the surface being imaged. If the timing of the pulses or the extent of the echo are such that the leading edge of one echo overlaps with the tail end of the previous one, then the far edge of the scene is folded over the near edge of the scene. This is called *range ambiguity*. The temporal extent of the echo is equal to

$$T_e \approx 2\frac{\rho}{c}\theta_r \tan\theta = 2\frac{h\lambda}{cW}\frac{\sin\theta}{\cos^2\theta}. \tag{29}$$

To avoid overlapping echoes, this time extent should be shorter than the time separating two pulses (i.e., 1/PRF). Thus, we must have

$$\mathrm{PRF} < \frac{cW}{2h\lambda}\frac{\cos^2\theta}{\sin\theta}. \tag{30}$$

In addition, the sensor parameters, specifically the PRF, should be selected such that the echo is completely within an interpulse period, i.e., no echoes should be received during the time that a pulse is being transmitted. The above equation gives an upper limit for the PRF as mentioned before. The SAR designer has to trade off system parameters to maximize the swath, while at the same time transmitting a high enough PRF to adequately sample the signal Doppler spectrum.

Another kind of ambiguity present in SAR imagery also results from the fact that the target's return in the azimuth direction is sampled at the PRF. This means that the azimuth spectrum of the target return repeats itself in the frequency domain at multiples of the PRF. In general, the azimuth spectrum is not a band limited signal; instead, the spectrum is weighted by the antenna pattern in the azimuth direction. This means that parts of the azimuth spectrum may be aliased, and high-frequency data will actually appear in the low-frequency part of the spectrum. In actual images, these *azimuth ambiguities* appear as ghost images of a target repeated at some distance in the azimuth direction, as shown in Fig. 7. To reduce the azimuth ambiguities, the PRF of an SAR has to exceed the lower limit given by Eq. (28).

In order to reduce both range and azimuth ambiguities, the PRF must therefore satisfy both the conditions expressed by Eqs. (28) and (30). Therefore, we must insist that

$$\frac{cW}{2h\lambda}\frac{\cos^2\theta}{\sin\theta} > \frac{2v}{L},$$ (31)

from which we derive a lower limit for the antenna size as

$$LW > \frac{4vh\lambda}{c}\frac{\sin\theta}{\cos^2\theta}.$$ (32)

FIGURE 7 Azimuth ambiguities result when the radar pulse repetition frequency is too low to sample the azimuth spectrum of the data adequately. In this case, the edges of the azimuth spectrum fold over themselves, creating ghost images.

A word of caution about the use of Eq. (32). This expression is derived assuming the SAR processor uses the full Doppler bandwidth in the processing, and also a swath that covers the full antenna beam width is imaged. This may not always be the case. For many reasons, SAR images are sometimes processed to only a fraction of the achievable resolution, or the swath width may be limited artificially by recording only that section of the returned echo that falls in a so-called *data window*. When either of these conditions are used, it is not appropriate to limit the antenna size as given by Eq. (32). In fact, in the case where both the swath is artificially limited and the resolution is increased, antennas significantly smaller that that given by Eq. (32) may be used with perfectly good results.

Another type of artifact in radar images results when a very bright surface target is surrounded by a dark area. As the image is being formed, some spillover from the bright target, called side lobes, although weak, could exceed the background and become visible. It should be pointed out that this type of artifact is not unique to radar systems. They are common in optical systems, where they are known as the side lobes of the point spread function. The difference is that in optical systems, the side lobe characteristics are determined by the characteristics of the imaging optics, i.e., the hardware, whereas in the case of an SAR, the side lobe characteristics are determined mainly by the characteristics of the processing filters. In the radar case, the side lobes may be therefore reduced by suitable weighting of the signal spectra during matched filter compression. The equivalent procedure in optical systems is through apodization of the telescope aperture.

The vast majority of these artifacts and ambiguities can be avoided with proper selection of the sensor and processor parameters. However, the interpreter should be aware of their occurrence because in some situations they might be difficult, if not impossible, to suppress.

B. Geometric Effects and Projections

The time delay/Doppler history basis of SAR image generation leads to an image projection different than in the case of optical sensors. Even though at first look radar images seem very similar to optical images, close examination quickly shows that geometric shapes and patterns are projected in a different fashion by the two sensors. This difference is particularly acute in rugged terrain. If the topography is known, a radar image can be reprojected into a format identical to an optical image, thus allowing image pixel registration. In extremely rugged terrain, however, the nature of the radar image projection leads to distortions which sometimes cannot be corrected.

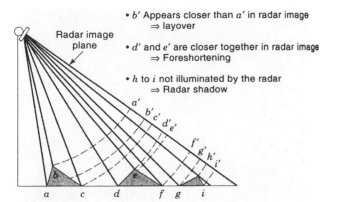

- *b'* Appears closer than *a'* in radar image
  ⇒ layover

- *d'* and *e'* are closer together in radar image
  ⇒ Foreshortening

- *h* to *i* not illuminated by the radar
  ⇒ Radar shadow

FIGURE 8 Radar images are cylindrical projections of the scene onto the image plane, leading to characteristic distortions.

In the radar image, two neighboring pixels in the range dimension correspond to two areas in the scene with slightly different range to the sensor. This has the effect of projecting the scene in a cylindrical geometry on the image plane, which leads to distortions as shown in Fig. 8. Areas that slope toward the sensor look shorter in the image, while areas that slope away from the sensor look longer in the image than horizontal areas. This effect is called *foreshortening*. In the extreme case where the slope is larger than the incidence angle, *layover* occurs. In this case, a hill would look as if it is projected over the region in front

of it. Layover cannot be corrected and can only be avoided by having an incidence angle at the surface larger than any expected surface slopes. When the slope facing away from the radar is steep enough, such that the radar waves do not illuminate it, *shadowing* occurs and the area on that slope is not imaged. Note that in the radar images, shadowing is always away from the sensor flight line and is not dependent on the time of data acquisition or the sun angle in the sky. As in the case of optical images, shadowing can be beneficial for highlighting surface morphologic patterns. Figure 9 contains some examples of foreshortening and shadowing.

C. Signal Fading and Speckle

A close examination of an SAR image shows that the brightness variation is not smooth, but has a granular texture which is called *speckle*. Even for an imaged scene which has a constant backscatter property, the image will have statistical variations of the brightness on a pixel-by-pixel basis, but with a constant mean over many pixels. This effect is identical to when a scene is observed optically under laser illumination. It is a result of the coherent nature (or very narrow spectral width) of the illuminating signal.

Rigorous mathematical analysis shows that the noise-like radar signal has well-defined statistical properties.

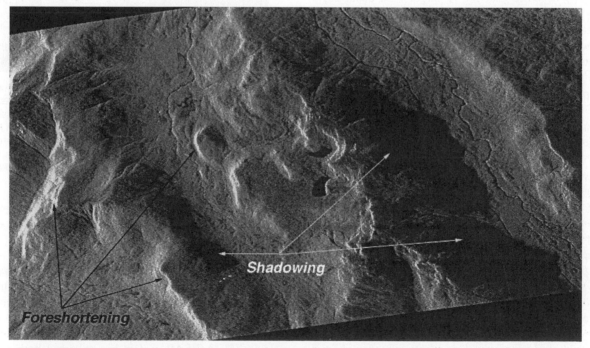

FIGURE 9 This NASA/JPL AIRSAR image shows examples of foreshortening and shadowing. Note that since the radar provides its own illumination, radar shadowing is a function of the radar look direction relative to the terrain and does not depend on the sun angle. This image was illuminated from the left.

The measured signal amplitude has a Rayleigh distribution, and the signal power has an exponential distribution. In order to narrow the width of these distributions (i.e., reduce the brightness fluctuations), successive signals or neighboring pixels can be averaged incoherently (i.e., their power values are added). This would lead to a more accurate radiometric measurement (and a more pleasing image) at the expense of degradation in the image resolution.

Another approach to reduce speckle is to combine images acquired at neighboring frequencies. In this case the exact interference patterns lead to independent signals but with the same statistical properties. Incoherent averaging would then result in a smoothing effect. In fact, this is the reason why a scene illuminated with white light does not show speckled image behavior.

In most imaging SARs, the smoothing is done by averaging the brightness of neighboring pixels in azimuth, or range, or both. The number of pixels averaged is called the number of looks N. It can be shown that the signal standard deviation S_N is related to the mean signal power $\bar{P}$ by

$$S_N = \frac{1}{\sqrt{N}} \bar{P}. \qquad (33)$$

The larger the number of looks N, the better the quality of the image from the radiometric point of view. However, this degrades the spatial resolution of the image. It should be noted that for N larger than about 25, a large increase in N leads to only a small decrease in the signal fluctuation. This small improvement in the radiometric resolution should be traded off against the large increase in the spatial resolution. For example, if one were to average 10 resolution cells in a 4-look image, the speckle noise will be reduced to about 0.5 dB. At the same time, however, the image resolution will be reduced by an order of magnitude. Whether this loss in resolution is worth the reduction in speckle noise depends on both the aim of the investigation, and the kind of scene imaged.

Figure 10 shows the effect of multilook averaging. An image acquired by the NASA/JPL AIRSAR system is shown displayed at 1, 4, 16, and 32 looks, respectively. Figure 10 clearly illustrates the smoothing effect, as well as the decrease in resolution resulting from the multilook process. In one early survey of geologists, the results showed that even though the optimum number of looks depended on the scene type and resolution, the majority of the responses preferred 2-look images. However, this survey dealt with images that had rather poor resolution to begin with, and one may well find that with today's higher resolution systems, analysts may be asking for a larger number of looks.

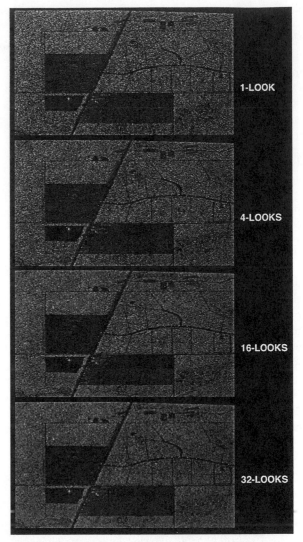

FIGURE 10 The effects of speckle can be reduced by incoherently averaging pixels in a radar image, a process known as multilooking. Shown in this image is the same image processed at 1 look, 4 looks, 16 looks, and 32 looks. Note the reduction in the granular texture as the number of looks increase, while at the same time the resolution of the image decreases. Some features, such as those in the largest dark patch, may be completely masked by the speckle noise.

IV. SAR POLARIMETRY AND INTERFEROMETRY

The field of SAR has advanced dramatically over the past two decades with the introduction of SAR polarimetry and SAR interferometry. Many airborne and spaceborne SAR instruments were developed to demonstrate various applications of both techniques. As an example, the NASA/JPL AIRSAR system has collected both polarimetric and interferometric SAR data as an airborne research tool from the 1980s. Two Shuttle Imaging Radar-C flights in 1994 were the first spaceborne polarimetric SAR at L- and C-band. The Shuttle Radar Topographic Mission (SRTM) launched in 2000 was the first spaceborne implementation of a single pass SAR interferometer to produce a three-dimensional map of 80% of the earth's land surface. One of the most exciting applications of SAR interferometry is differential interferometry that measures minute surface changes (millimeter to centimeter scale) by subtracting two interferometric pairs separated in time.

A. SAR Polarimetry

Since an electromagnetic wave is polarized, electromagnetic wave propagation is a vector phenomenon. Both electric and magnetic fields associated with an electromagnetic wave are perpendicular to the propagation direction. Hence, we can decompose the wave polarization into a vector with two components perpendicular to each other. Mathematically, the electric field ($\vec{E}(t)$) of an electromagnetic wave can be written as

$$\vec{E}(t) = \hat{h} E_h(t) + \hat{v} E_v(t), \tag{34}$$

where $\hat{h}$ represents a unit vector in the horizontal direction that is parallel to the illuminated ground and $\hat{v}$ is a unit vector in the vertical direction that is perpendicular to both $\hat{h}$ and the propagation direction. If this electromagnetic wave is scattered by an object, the polarization of the scattered wave depends on the electrical and geometrical properties of the scattering object. One can consider the scatterer as a mathematical operator which takes one two-dimensional complex vector (incident wave) and changes it into another two-dimensional vector (scattered wave). Therefore, a scatterer can be characterized by a complex 2×2 scattering matrix.

The typical hardware implementation of a radar polarimeter involves transmitting a wave of one polarization and receiving echoes in both polarizations simultaneously. This is followed by transmitting a wave with a second polarization and again receiving echoes with both polarizations simultaneously. In this way, all four elements of the scattering matrix are measured. In order to use the polarimetric data derived from the scattering matrix for science applications, it is necessary to calibrate the data. Polarimetric calibration usually involves four steps: cross-talk removal, phase calibration, channel imbalance compensation, and absolute radiometric calibration.

The availability of calibrated polarimetric SAR data allowed researchers to perform quantitative analysis of the data. Significant progress has been made in the classification of different types of terrain using polarimetric SAR data. One of the most active areas of research in polarimetric SAR involves estimating geophysical parameters such as forest biomass, surface roughness, and soil moisture. Polarimetric SAR data may be optimal for flood monitoring and mapping, especially in the presence of vegetation canopies. Since the water surface is flat and a radar signal can penetrate into vegetation canopies, the inundated area can be identified using the copolarization ratio that depends on the strength of tree-water double bounce scattering.

B. SAR Interferometry

A conventional SAR measures the along-track and cross-track location of an object by projecting topographic relief information into a two-dimensional image plane. SAR interferometry is capable of measuring the third dimension to generate three-dimensional images. The basic principles of SAR interferometry can be explained using the geometry shown in Fig. 11. The slant range difference ($\delta\rho$) of two interferometric antennas can be calculated as

$$\delta\rho \approx -B_L \sin(\theta - \alpha), \tag{35}$$

where B_L is the interferometric baseline length and α is the baseline tilt angle shown in Fig. 11. An interferometric phase difference can be derived from $\delta\rho$ as

$$\delta\phi \approx \frac{a2\pi}{\lambda} B_L \sin(\theta - \alpha), \tag{36}$$

where $a = 1$ for the case where signals are transmitted by one antenna and received through both antennas simultaneously and $a = 2$ for the case where the signal is alternatively transmitted and received through one of the two antennas only. It can be noticed from Eq. (36) that all parameters are known except for θ. Therefore, the angle θ can be derived from the measured differential phase. Then, the elevation of the point being imaged is given by

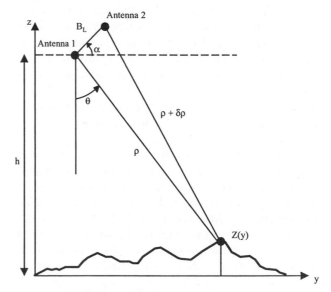

FIGURE 11 SAR interferometry geometry.

$$z(y) = h - \rho \cos\theta, \tag{37}$$

where h denotes the height of the imaging antenna. Since the measured phase varies between 0 and 2π, it must be unwrapped to retrieve the original phase by adding and subtracting multiples of 2π. Phase unwrapping algorithm development remains one of the most active areas of research in SAR interferometry, and many algorithms are under development.

Using differential SAR interferometry, minute surface deformations can be measured with unprecedented accuracy. This differential interferometry is implemented by subtracting two interferometric pairs separated in time. This technique was first demonstrated using SEASAT data to measure millimeter-scale ground motion in agricultural fields. Since then, this technique has been applied to measure co-seismic displacement and volcanic deflation. Differential SAR interferometry has been used for polar ice sheet research by providing information on ice deformation at great spatial details. SAR interferometry can also be used for mapping ocean surface movement by imaging the same area twice within ocean decorrelation time. In this case, the interferometer is implemented in such a way that one antenna images the scene a short time before the second antenna with the exact same viewing geometry. This technique is known as along-track SAR interferometry.

V. SPACEBORNE SAR EXAMPLES

The SEASAT SAR is the first civilian spaceborne SAR system designed to acquire high-resolution images of the earth surface. The SEASAT SAR was launched by NASA in 1978, and it operated successfully for 100 days and acquired approximately 50 h of radar data. The spare hardware of the SEASAT SAR system was modified and flown on the space shuttle in November 1981. This flight is known as the SIR-A (Shuttle Imaging Radar-A) experiment. After several modifications to the SIR-A system, the SIR-B system was flown in 1984, and it was capable of varying the look angle from 20° to 60°, while the SIR-A system operated at a fixed 50° look angle. The characteristics of SEASAT, SIR-A, and SIR-B SAR systems are summarized in Table II.

The SIR-C/X-SAR experiment is a joint U.S./German/Italian project. It is the first spaceborne multifrequency and multipolarization SAR system. The instrument, flown aboard the space shuttle Endeavour, collected valuable multiparameter SAR data in April and October 1994. Two flights provided the opportunity for assessment of change with time due to seasons and other factors. During the second flight, the orbit was trimmed for the last 3 days to demonstrate SAR interferometry to generate large-scale topographic maps. The SIR-C instrument used

TABLE II SEASAT, SIR-A, and SIR-B Radar System Characteristics

SAR parameter	SEASAT	SIR-A	SIR-B
Frequency band	L-band	L-band	L-band
Bandwidth (MHz)	19	6	12
Altitude (km)	790	245	225
Polarization	Horizontal	Horizontal	Horizontal
Pulse length (μsec)	33.9	33.9	33.9
Peak power (W)	1000	1000	1000
Antenna width (m)	2.16	2.16	2.16
Antenna length (m)	10.7	9.36	10.7
Swath width (km)	100	38	30
Look angle (degrees)	20.5	50	20–60
Launch date	1978	1981	1984

phased array antennas to provide the sensitivity to detect weak cross-polarization signals, while maintaining an acceptable transmit power level by using many transmit/receive modules close to the antenna radiating elements. SIR-C also demonstrated, for the first time, the concept of ScanSAR from space. The ScanSAR mode is implemented by moving the antenna beam sequentially in the across-track direction. A wider swath is accomplished at the expense of azimuth resolution. The SIR-C/X-SAR characteristics are shown in Table III.

The SRTM was launched in February 2000 to produce a topographic map of 80% of the earth's land surface in a single 10-day space shuttle flight. The mission was highly successful in collecting all the planned data. The addition of C-band and X-band receiver antennas, extended from the space shuttle bay on a 60-m mast and operating in concert with the existing SIR-C/X-SAR antennas, forms a single pass interferometric radar system. It is estimated that the height accuracy of the C-band SRTM data is

TABLE III SIR-C/X-SAR Radar System Characteristics[a]

SAR parameter	SIR-C (L-band)	SIR-C (C-band)	X-SAR
Frequency band	L-band	C-band	X-band
Bandwidth (MHz)	10, 20, and 40	10, 20, and 40	10 and 20
Altitude (km)	225	225	225
Polarization	Fully polarimetric	Fully polarimetric	Vertical
Pulse length (μsec)	8.5, 17, and 34	8.5, 17, and 34	40
Peak power (W)	4000	1200	3300
Antenna width (m)	2.9	0.74	0.4
Antenna length (m)	12	12	12
Swath width (km)	15–80	15–80	15–80
Look angle (degrees)	20–60	20–60	20–60
Launch date	1994	1994	1994

[a] The fully polarimetric mode measures a complex 2×2 scattering matrix.

less than 16 m absolute and 11 m relative, where these values are the 90% linear error after processing both ascending and descending passes. The elevation posting of the SRTM data is 30 m.

The international use of spaceborne imaging SAR for long-term earth observation dramatically increased in the 1990s. While NASA missions were space shuttle based, four SAR imaging satellites were launched by Europe, Japan, and Canada. The European Space Agency (ESA) launched an SAR instrument aboard the European Remote-Sensing Satellite (ERS) in August 1991 and another in April 1995. Pairs of ERS images, including tandem ERS-1 & 2, have been used to generate interferometric products of surface topography and deformation. The Japanese National Space Development Agency (NASDA) launched an SAR instrument aboard the Japanese Earth Resources Satellite-1 (JERS-1) in February 1992. JERS SAR has provided a global land cover map, and its repeat pass interferometric data have been used to measure surface deformations. The Canadian Space Agency (CSA) launched an SAR instrument aboard RADARSAT in September 1995. The RADARSAT antenna is capable of electronic beam steering that enables very large area mapping using the ScanSAR mode. The characteristics of these four SAR systems are shown in Table IV.

Unlike earth-observing satellites, planetary radar operation is severely limited by available mass, power, and data downlink rate. The concept of an imaging radar mission to Venus was first considered in the late 1960s to penetrate the dense cloud cover of Venus. In 1989, Magellan was launched from the space shuttle Atlantis to image the Venus surface. The radar operates at S-band with low transmit peak power. The Magellan mission imaged 98% of the surface of Venus, far exceeding its requirement of

TABLE V Magellan Radar System Characteristics

SAR parameter	Magellan
Frequency band	S-band
Bandwidth (MHz)	2.26
Altitude (km)	275–2100 (elliptical orbit)
Polarization	Horizontal
Pulse length (μsec)	26.5
Peak power (W)	350
Antenna diameter (m)	3.7
Swath width (km)	20–25
Look angle (degrees)	13–47
Launch date	1989

70% coverage. The characteristics of the magellan radar system are shown in Table V.

VI. CONCLUSIONS AND OUTLOOK

SAR is now a mature imaging technique that has found widespread application in studying the Earth and other planets. Several countries are planning to launch next-generation SAR satellites in the next decade or so. Among them are the ESA with the ASAR system on the Envisat satellite, the CSA with Radarsat-2, and the NASDA with the PALSAR on the ALOS spacecraft. In parallel, many scientists are starting to use SAR images more routinely in their analysis of the earth as an integrated system, particularly in the fields of plate tectonics and glaciology, where the use of differential interferometry has revolutionized our understanding of the processes involved. As science matures into a complete analysis utilizing data from complimentary sensors that cover a wide range of the electromagnetic spectrum, it is expected that radar will play an increasing role in remote sensing of the Earth and other planets.

SEE ALSO THE FOLLOWING ARTICLES

COMMUNICATION SATELLITE SYSTEMS • RADAR • REMOTE SENSING FROM SATELLITES • SATELLITE COMMUNICATIONS • SIGNAL PROCESSING, ANALOG

TABLE IV ERS-1, ERS-2, JERS-1, and RADARSAT Radar System Characteristics

SAR parameter	ERS-1	ERS-2	JERS-1	RADARSAT
Frequency band	C-band	C-band	L-band	C-band
Bandwidth (MHz)	15.5	15.5	15	11.6, 17.3, and 30
Altitude (km)	780	780	568	800
Polarization	Vertical	Vertical	Horizontal	Horizontal
Pulse length (μsec)	37	37	35	42
Peak power (W)	4800	4800	1500	5000
Antenna width (m)	1	1	2.2	1.5
Antenna length (m)	10	10	11.9	15
Swath width (km)	100	100	75	10–500
Look angle (degrees)	23	23	39	20–50
Launch date	1991	1995	1992	1995

BIBLIOGRAPHY

Curlander, J. C., and McDonough, R. N. (1991). "Synthetic Aperture Radar System and Signal Processing," Wiley, New York.

Elachi, C. (1988). "Spaceborne Radar Remote Sensing: Applications and Techniques," IEEE Press, New York.

Henderson, F. M., and Lewis, A. J. (1998). "Principles & Applications of Imaging Radar," Wiley, New York.

Ulaby, F. T., and Elachi, C. (1990). "Radar Polarimetry for Geoscience Applications," Artech House, Norwood, MA.

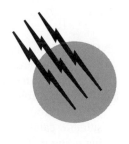

Synthetic Fuels

Ronald F. Probstein
R. Edwin Hicks
Massachusetts Institute of Technology

I. Coal, Oil Shale, and Tar Sand Conversion
II. Thermal Conversion Processes
III. Technologies
IV. Biomass Conversion
V. Outlook

GLOSSARY

Biomass Any material directly or indirectly derived from plant life that is renewable in time periods of less than about 100 years.

Coal Solid fossil hydrocarbon typically composed of from 65 to 75 mass% carbon and about 5 mass% hydrogen, with the remainder oxygen, ash, and smaller quantities of sulfur and nitrogen.

Coprocessing Processing of coal and oil simultaneously with the objective of liquefying the coal and upgrading the oil.

Direct hydrogenation Exposure of a carbonaceous raw material to hydrogen at a high pressure.

Direct liquefaction Hydrogenation of a carbonaceous material, usually coal, to form a liquid fuel by direct hydrogen addition in the presence of a catalyst or by transfer of hydrogen from a solvent.

Gasification Conversion of a carbonaceous material into a gas, with the principal method to react steam with coal in the presence of air or oxygen in a vessel called a gasifier.

Hydrotreating Catalytic addition of hydrogen to liquid fuels to remove oxygen, nitrogen, and sulfur and to make lighter fuels by increasing the hydrogen-to-carbon ratio.

Indirect hydrogenation Reaction of a carbonaceous raw material with steam, with the hydrogen generated within the system.

Indirect liquefaction Combination of a synthesis gas composed of carbon monoxide and hydrogen over a suitable catalyst to form liquid products such as gasoline and methanol.

Oil shale Sedimentary rock containing kerogen, a high molecular mass hydrocarbon, that is insoluble in common solvents and is not a member of the petroleum family.

Pyrolysis Reduction of the carbon content in a raw hydrocarbon by distilling volatile components to yield solid carbon, as well as gases and liquids with a higher hydrogen fraction than the original material.

Reactor Vessel used for gasification, liquefaction, and pyrolysis, with the three main types the moving packed bed, the entrained flow, and the fluidized bed reactor.

Retorting Pyrolysis of oil shale to produce oil in a vessel called a retort.

SNG Substitute natural gas that consists primarily of methane manufactured mainly by the catalytic synthesis of carbon monoxide and hydrogen.

Synthesis Combination of a gas whose major active components are carbon monoxide and hydrogen over a suitable catalyst to form a large number of products including methane, methanol, gasoline, and alcohols.

Synthetic fuels Gaseous or liquid fuels manufactured by hydrogenating a naturally occurring carbonaceous raw material or by removing carbon from the material.

Tar sands Mixture of sand grains, water, and a high-viscosity hydrocarbon called bitumen, which is a member of the petroleum family.

SYNTHETIC FUELS may be gaseous, liquid, or solid and are obtained by converting a carbonaceous material to another form. The most abundant naturally occurring materials for producing synthetic fuels are coal, oil shale, tar sands, and biomass. The conversion of these materials is undertaken to provide synthetic gas or oil to replace depleted or unavailable natural resources and also to remove sulfur or nitrogen, which, when burned, gives rise to undesirable air pollutants. The manufacture of synthetic fuels can be regarded as a process of hydrogenation since common fuels have a higher hydrogen content than the raw materials. All synthetic fuel processes require an energy input to accomplish the conversion. Most of the thermal conversion processes are applicable to all carbonaceous materials. The biochemical processes of fermentation and biological decomposition are specific to biomass.

I. COAL, OIL SHALE, AND TAR SAND CONVERSION

A. Synthetic Fuel Manufacture and Properties

To manufacture synthetic fuels, hydrogenation of the naturally occurring raw materials, or carbon removal, is usually required since common fuels such as gasoline and natural gas have a higher hydrogen content than the raw materials. The source of the hydrogen that is added is water. A typical bituminous coal has a carbon-to-hydrogen mass ratio of about 15, while methane, which is the principal constituent of natural gas, has a carbon-to-hydrogen mass ratio of 3. In between, the corresponding ratio for crude oil is about 9, and that for gasoline 6.

The organic material in both tar sands and high-grade oil shale has a carbon-to-hydrogen mass ratio of about 8, which is close to that of crude oil. However, the mineral content of rich tar sands in the form of sand or sandstone is about 85 mass%, and that of high-grade oil shale, in the form of sedimentary rock, is about the same. Therefore, very large volumes of solids must be handled to recover

relatively small quantities of organic matter from oil shale and tar sands. On the other hand, the mineral content of coal in the United States averages about 10% by mass.

In any conversion to produce a fuel of a lower carbon-to-hydrogen ratio, the hydrogenation of the raw fossil fuel may be direct, indirect, or by pyrolysis, either alone or in combination. Direct hydrogenation involves exposing the raw material to hydrogen at a high pressure. Indirect hydrogenation involves reacting the raw material with steam, with the hydrogen generated within the system. In pyrolysis the carbon content is reduced by heating the raw hydrocarbon until it thermally decomposes, distilling off the volatile components to yield solid carbon, together with gases and liquids having higher fractions of hydrogen than the original material.

Fuels that will burn cleanly require that sulfur and nitrogen compounds be removed from the gaseous, liquid, and solid products. As a result of the hydrogenation process, sulfur and nitrogen, which are always present to some degree in the original raw fossil fuel, are reduced to hydrogen sulfide and ammonia, respectively. Hydrogen sulfide and ammonia are present in the gas made from coal or released during the pyrolysis of oil shale and tar sands and, also, are present in the gas generated in the hydrotreatment of pyrolysis oils and synthetic crude oils.

Synthetic fuels include liquid fuels such as fuel oil, diesel oil, gasoline, and methanol, clean solid fuels, and low-calorific value, medium-calorific value, and high-calorific value gas. The gas is referred to here as low-CV, medium-CV, and high-CV gas, respectively. In British units, which are still used interchangeably, the corresponding reference is to low-Btu, medium-Btu, and high-Btu gas. Low-CV gas, often called producer or power gas, has a calorific value of about 3.5 to 10 million joules per cubic meter (MJ/m^3) or, in British units, 90 to 270 British thermal units per standard cubic foot (Btu/scf). This gas is an ideal turbine fuel. Medium-CV gas is loosely defined as having a calorific value of about 10 to 20 MJ/m^3 (270–540 Btu/scf). This gas is also termed power gas and, sometimes, industrial gas, as well as synthesis gas. It may be used as a fuel gas, as a source of hydrogen for direct liquefaction, or for the synthesis of methanol and other liquid fuels. Medium-CV gas may also be used for the production of high-CV gas, which has a calorific value in the range of about 35–38 MJ/m^3 (940–1020 Btu/scf) and is normally composed of more than 90% methane. This gas is a substitute for natural gas and suitable for economic pipeline transport. For these reasons it is referred to as substitute natural gas (SNG) or pipeline gas.

B. History

Synthetic fuel manufacture, although often thought of as a modern technology, is not new, nor has it been limited in the past to small-scale development. What is different today is the increased fundamental chemical and physical understanding of the complex conversion processes that is built into the technologies, the application of modern engineering, and systems designed to ensure environmentally sound operation. What is not different is the history of synthetic fuel manufacture, whose on-again, off-again commercialization since the beginning of the nineteenth century has been buffeted by the real or perceived supply of natural resources of oil and gas. In the late 1970s, following the Arab oil embargo of 1973, worldwide commercial synthetic fuel manufacture appeared to be on the verge of reality. By the 1990s, however, there seemed scant likelihood for this to take place in the twentieth century, with most, though not all, work in the field reduced to a small research and development level. Historical evidence, however, indicates that any prediction of full-scale commercialization is at best risky and more likely unreliable.

As early as 1792, Murdoch, a Scottish engineer, distilled coal in an iron retort and lit his home with the coal gas produced. By the early part of the nineteenth century, gas manufactured by the distillation of coal was introduced for street lighting, first in London in 1812, following which its use for this purpose spread rapidly throughout the major cities of the world. This coal gas contained about 50% hydrogen and from 20 to 30% methane, with the remainder principally carbon monoxide. Its calorific value was about 19 MJ/m^3 (500 Btu/scf), and this value served as the benchmark for the "town gas" industry. In the latter part of the nineteenth century, gasification technologies, employing the reaction of air and steam with coal, were developed and the use of "synthetic" gas for domestic and industrial application became widespread. Commercial "gas producers" yielded a gas with a low calorific value of about 5–6.5 MJ/m^3 (130–160 Btu/scf) and were used on-site to produce gas for industrial heating. In the early part of the twentieth century the availability of natural gas with a calorific value of 37 MJ/m^3 (1000 Btu/scf) began to displace the manufactured gas industry, which, subsequent to the end of World War II, virtually disappeared worldwide.

Following the Arab oil embargo of 1973, construction of a number of commercial-scale coal conversion plants was undertaken in the United States to produce SNG on scales up to 7 million m^3/day (250 million scf/day). The largest project to be completed was the Great Plains coal gasification plant in North Dakota, which has a design capacity of 3.9 million m^3/day (138 million scf/day) of SNG. The plant started up in 1984 and was operated in turn by the U.S. Department of Energy and the Dakota Gasification Company. Production rates have increased beyond the design rate and, by 1991, had reached 4.5 million m^3/day (160 million scf/day) of SNG. Several smaller coal gasi-

fication processes remained in operation. The Cool Water plant in California, which shut down at the beginning of 1989, manufactured 2 million m^3/day (72 million scf/day) of 9 MJ/m^3 (250 Btu/scf) gas. The gas was used to produce about 100 MW (net) of electric power in combustion and steam turbine generators. The plant was reopened in 1993 using a mixed feed of coal and sewage sludge.

The history of coal liquefaction is considerably more recent than that of coal gasification. Direct liquefaction, in which the coal is exposed to hydrogen at a high pressure, can be traced to the work of Bergius in Germany from 1912 to 1926. Commercial-size hydrogenation units for the production of motor fuels began in Germany in 1926, and by 1939 the output was estimated to be 4 million liters of gasoline per day. Liquid fuel production and, in particular, oil production are most frequently quoted in barrels per day, where 1 barrel (bbl) equals 42 U.S. gal or about 160 liters. The German production of gasoline was therefore about 250,000 bbl/day. During World War II this direct liquefaction production from some 12 plants expanded to about 1 million bbl/day of gasoline. Activity in direct coal liquefaction paralleled that in gasification. Most of the work in the 1970s and early 1980s centered about the development of second-generation plants to run at lower pressures of from 10 to 20 million pascals (MPa). The original German Bergius-type units were run at from 25 to 70 MPa. It may be noted that 1 million Pa is about 10 atm or 147 lb/in.^{2}. The larger of the pilot plants in the United States and Germany operated at about 200 to 250 ton-per-day coal feeds, equal to about 550 to 700 bbl/day of synthetic crude oil output. Throughout this chapter, ton (t) refers to the unit in use with SI units.

The principal method of indirect liquefaction is to react carbon monoxide and hydrogen produced by coal gasification in the presence of a catalyst to form hydrocarbon vapors, which are then condensed to liquid fuels. This procedure for synthesizing hydrocarbons is based on the work of Fischer and Tropsch in Germany in the 1920s. Just prior to and during World War II, Germany produced oil and gasoline by this process at a maximum rate of only about 15,000 bbl/day because of the small output of the individual reactors compared to that obtainable at the time with direct liquefaction. Development of the Fischer–Tropsch process has been pursued in South Africa from 1955 to the time of writing and continues to be worked on. The Sasol plants in that country employ the largest coal gasification banks in the world and produce over 100 million m^3/day (3500 million scf/day) of medium-CV gas. The plants produce about 100,000 bbl/day of motor fuels, employing individual reactors with capacities about 100 times greater than those of the original commercial units in Germany.

Oil shale production has also had a long history, with the earliest shale oil industry started in France in 1838, where oil shale, which is a sedimentary rock containing an insoluble hydrocarbon, was crushed and distilled to make lamp fuel. Its operation was intermittent until the late 1950s, when it was terminated. In 1862, production of oil from shale was begun in Scotland, where it ran for about a hundred years. It reached its peak in 1913, with the production of about 6 thousand bbl/day of shale oil. Many countries have had shale oil retorting (distilling) facilities including the United States, which has particularly large reserves of oil shale. But as a result of the volatile economics of production, oil shale development has been turned on and off in the United States for more than a century. In 1991 there was only one major commercial retorting facility, that of the Unocal Corp. at Parachute Creek, Colorado. Constructed in the 1980s to produce 10,000 bbl/day of shale oil, by 1991 the plant was producing shale oil at a rate of 6000 to 7000 bbl/day when running, which was about two-thirds of the time.

Tar sands, also called oil sands, which are a mixture of sand grains, water, and a high-viscosity crude hydrocarbon called bitumen, are found in every continent. The most sizable reserves are found in Canada and in Venezuela. Between 1930 and 1960 commercial enterprises were formed and reformed with regularity to exploit the large Athabasca deposits in Alberta, Canada. In 1965 commercial production was begun in integrated surface plants that extracted the bitumen from the tar sands with hot water and upgraded it by distillation and hydrogen addition (hydrotreating). The upgrading procedure is much the same as that used in the refining of natural crude oil, and by this means a high-quality synthetic crude is produced. In 1990 Canadian commercial production from its two largest surface plants amounted to over 210,000 bbl/day of synthetic crude.

II. THERMAL CONVERSION PROCESSES

A. Pyrolysis

Pyrolysis refers to the decomposition of organic matter by heat in the absence of air. A common synonym for pyrolysis is devolatilization. Thermal decomposition and destructive distillation are frequently used to mean the same.

When coal, oil shale, or tar sands are pyrolyzed, hydrogen-rich volatile matter is distilled and a carbon-rich solid residue is left behind. The carbon and mineral matter remaining behind is the residual char. Pyrolysis is one method to produce liquid fuels from coal, and it is the principal method used to convert oil shale and tar sands to liquid fuels. Moreover, as gasification and liquefaction are carried out at elevated temperatures, pyrolysis may be considered the first stage in any conversion process.

The use of pyrolysis for the production of liquid products is illustrated in the block diagram in Fig. 1. The py-

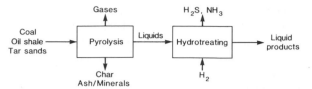

FIGURE 1 Pyrolysis. [Reprinted with permission from Probstein, R. F., and Hicks, R. E. (1990). "Synthetic Fuels," pH Press, Cambridge, MA.]

rolysis vapors, consisting of condensable tar, oil, and water vapor, and noncondensable gases, consisting mainly of hydrogen (H_2), methane (CH_4), and oxides of carbon (CO, CO_2), are produced by heating of the raw material. The char, ash, and minerals left behind are rejected. The hydrocarbon vapors are treated with hydrogen to improve the liquid fuel quality and to remove the sulfur and nitrogen which came from the original raw material. The sulfur and nitrogen are removed as hydrogen sulfide (H_2S) and ammonia (NH_3) gases which form as a result of the hydrogenation.

The composition of the raw material is important in determining the yield of volatile matter, while the pyrolysis temperature affects both the amount and the composition of the volatile yields. When coal, oil shale, and tar sand bitumen are heated slowly, rapid evolution of volatile products begins at about 350 to 400°C, peaks sharply at about 450°C, and drops off very rapidly above 500°C. This is termed the stage of "active" thermal decomposition. There are three principal stages of pyrolysis. In the first stage, above 100°C and below, say, 300°C, the evolution of volatile matter is not large and what is released is principally gas composed mainly of carbon dioxide (CO_2), carbon monoxide (CO), and water (H_2O). In the active or second stage of decomposition, about three-quarters of all the volatile matter ultimately released is evolved, with methane the principal noncondensable gas. The third stage is the one most appropriately defined for coal, in which there is a secondary degasification associated with the transformation of the char, accompanied by the further release of noncondensable gases, mainly hydrogen.

The total volatile matter yield, and hence the yield of tar plus light oils, is proportional to the hydrogen-to-carbon ratio in the raw material. On the other hand, the chemically formed water vapor that distills off during pyrolysis in an inert atmosphere is proportional to the oxygen-to-carbon ratio. The yields and product distributions also depend on the rate of pyrolysis.

B. Gasification

Gasification is the conversion of a solid or a liquid into a gas. In a broad sense it includes evaporation by heating, although the term is reserved for processes involving

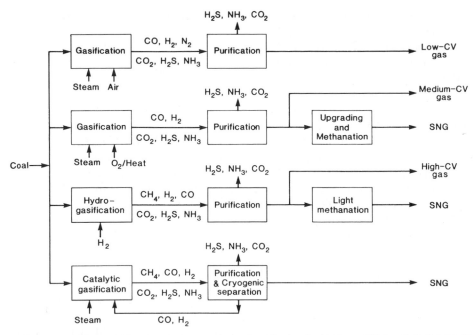

FIGURE 2 Gasification of coal. [Reprinted with permission from Probstein, R. F., and Hicks, R. E. (1990). "Synthetic Fuels," pH Press, Cambridge, MA.]

chemical change. The primary raw material for gasification is normally considered to be coal, although the use of oil shale for gasification has been discussed. Pyrolysis of coal is one method of producing synthetic gas and was the method pioneered in the early nineteenth century. Today the principal methods considered or in use for the gasification of coal to produce synthetic gases are shown in Fig. 2.

The most widely used technologies for the manufacture of gas employ indirect hydrogenation by reacting steam with coal in the presence of either air or oxygen. When air is used, the product gas will be diluted with nitrogen (N_2) and its calorific value will be low in comparison with that of the gas manufactured using oxygen (O_2). The dilution of the product gas with nitrogen can be avoided by supplying the heat needed for the gasification from a hot material that has been heated with air in a separate furnace or in the gasifier itself before gasification. In all of the cases, the gas must be cleaned prior to using it as a fuel. This purification step involves the removal of the hydrogen sulfide, ammonia, and carbon dioxide, which are products of the gasification. As with pyrolysis, the hydrogen sulfide and ammonia are formed from the hydrogenation of the sulfur and the nitrogen that were originally in the coal.

Medium-CV gas, consisting mainly of carbon monoxide and hydrogen, can be further upgraded by altering the carbon monoxide-to-hydrogen ratio catalytically and then, in another catalytic step, converting the resulting "synthesis" gas mixture to methane. A high-CV gas can

be produced by direct hydrogenation, termed hydrogasification, in which hydrogen is contacted with the coal. A procedure that allows the direct production of methane is catalytic gasification. In this method the catalyst accelerates the steam gasification of coal at relatively low temperatures and also catalyzes the upgrading and methanation reactions at the same low temperature in the same unit.

A simplified representation of steam–oxygen or steam–air gasification of coal is shown in Fig. 3. The gasifier represented is termed a moving bed gasifier, in that crushed coal enters the top of the gasifier and moves downward at the same time that it is being reacted, eventually being

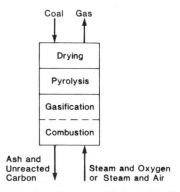

FIGURE 3 Schematic of a moving bed gasifier. [Reprinted with permission from Probstein, R. F., and Hicks, R. E. (1990). "Synthetic Fuels," pH Press, Cambridge, MA.]

removed from the bottom as ash and any unreacted coal. The coal that enters the gasifier at the top is first dried by rising hot gases. Further heating results in devolatilization and pyrolysis. The next stage is the gasification zone, where the temperatures are typically above about 800°C and below about 1500°C. The temperatures are controlled by the temperatures in the combustion zone, which are a function primarily of the relative level of oxygen put into the gasifier.

The chemical reactions that take place in the gasifier are most easily presented by representing the coal by pure carbon (C). In the combustion zone the carbon is burned with the oxygen to produce carbon monoxide and carbon dioxide with the release of heat. The gasification chemistry is more complex but may be represented by a few principal reactions, as oxygen may be assumed not to be present beyond the combustion zone. In the gasification zone the carbon reacts with the steam put into the gasifier to produce carbon monoxide and hydrogen, using the heat released by the combustion; that is, it is an endothermic reaction. This endothermic reaction is known as hydrolysis, although it is more frequently referred to as the carbon–steam or gasification reaction. It may be written in chemical notation as

$$C + H_2O \text{ (steam)} \rightarrow CO + H_2 \qquad (1)$$

Atoms are conserved in a chemical reaction so that the number of carbon atoms, oxygen atoms, and hydrogen atoms must be the same on each side of the equation. The formula states that one atom of carbon reacts with one molecule of water to form one molecule of carbon monoxide and one molecule of hydrogen gas. The relative amounts of the substances participating in a reaction are given by the coefficients in the reaction formula, termed stoichiometric coefficients. In this case all the stoichiometric coefficients are one.

The carbon will also react with the hydrogen produced, to form methane. This reaction is termed hydrogenolysis or, more often, the carbon–hydrogen or hydrogenation reaction. It releases heat; that is, it is exothermic and may be written

$$C + 2H_2 \rightarrow CH_4 \qquad (2)$$

In the oxygen-depleted gasification zone, the coal may also "burn" in the carbon dioxide and form carbon monoxide following the endothermic Boudouard reaction

$$C + CO_2 \rightarrow 2CO \qquad (3)$$

Other reactions take place but are not discussed here. It is noted only that even at equilibrium the relative amounts of different gases produced will depend on the temperature and pressure in the gasifier and on the amount of steam and oxygen relative to the amount of carbon put into the system.

C. Synthesis

The raw gas produced on gasification of coal has a low to medium calorific value, depending on whether air or oxygen is used as the oxidant in directly heated gasifiers. The product from indirectly heated gasifiers, in which an inert material is typically used to transfer the heat from an external source, is generally a medium-CV gas. One of the major reasons for producing synthetic fuels is to replenish dwindling natural supplies of traditional fuels such as natural gas and gasoline. A second reason is to eliminate pollutants to provide a clean-burning fuel. Removal of ammonia, hydrogen sulfide, and inert gases is an obvious requirement and has been noted. The principal gaseous products from gasifiers are carbon monoxide and hydrogen, which, although useful as a fuel, are not direct replacements for natural gas. These products can, however, be reacted with steam to produce substitute natural gas (SNG) as indicated by the methanation blocks in Fig. 2. These same products can also be reacted to produce gasoline, methanol, and other liquid fuels. Production of liquid fuels from coal after first completely breaking down the coal structure in a gasification step is known as "indirect liquefaction" and is shown schematically in Fig. 4.

A gas in which the major active components are carbon monoxide and hydrogen is called a synthesis gas, as these two compounds can be made to combine, or synthesize, to form a large number of products. The products formed from the synthesis gas depend both on the hydrogen-to-carbon monoxide ratio in the gas and on the catalyst and reactor conditions. Hydrogen-to-carbon monoxide mole ratios range from 3, for methane production with the rejection of water, to 1 to 0.5, for gasoline production with the rejection of carbon dioxide. The required hydrogen-to-carbon monoxide ratio can sometimes be achieved directly in the gasifier, although a H_2/CO ratio as high as 3 is normally not produced in commercial systems. In fact, many gasifiers produce a gas having a H_2/CO ratio of less than 1. In these cases an adjustment to the H_2/CO ratio is normally required, and is done by adding steam to the synthesis gas

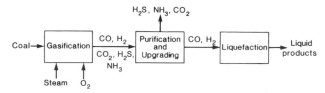

FIGURE 4 Indirect liquefaction of coal. [Reprinted with permission from Probstein, R. F., and Hicks, R. E. (1990). "Synthetic Fuels," pH Press, Cambridge, MA.]

and reacting it with carbon monoxide to form hydrogen and carbon dioxide:

$$CO + H_2O \text{ (steam)} \rightarrow CO_2 + H_2 \qquad (4)$$

This is called the water–gas shift reaction or, frequently, just the shift reaction. The shift reaction is moderately exothermic; that is, it releases heat. Optimum operating temperatures are low, usually below about 225°C. The reaction will not proceed appreciably unless catalyzed, traditionally by reaction over an iron/chromium catalyst.

The need to "shift" the gas introduces an additional process step, so increasing overall the process complexity. In cases where the required synthesis gas composition can be achieved directly in the gasifier, this may be preferred in the interest of reducing the complexity.

Of interest in synthetic fuel manufacture is the production of SNG, which is principally methane. Methane (CH_4) does not contain oxygen, and the oxygen in the carbon monoxide may be rejected as either water or carbon dioxide. Typically water is rejected following the reaction

$$CO + 3H_2 \rightarrow CH_4 + H_2O \text{ (steam)} \qquad (5)$$

In this reaction a H_2/CO ratio of 3 is required in the synthesis gas, and one-third of the hydrogen content is wasted in rejected steam. This reaction is carried out over a zinc/chromium catalyst and is highly exothermic.

Perhaps the simplest synthesis reaction is the combination of one molecule of carbon monoxide with two molecules of hydrogen to form methanol (CH_3OH)

$$CO + 2H_2 \rightarrow CH_3OH \text{ (liquid)} \qquad (6)$$

The catalyst used for this reaction is a copper-containing one, with reaction temperatures of about 260°C and pressures down to about 5 MPa. As with methane manufacture, the reaction is an exothermic one.

Finally, we note the commercially important Fischer–Tropsch synthesis reaction for gasoline manufacture, mentioned in Section I.B. The reaction formula may be written

$$CO + 2H_2 \rightarrow CH_2 \text{ (liquid)} + H_2O \text{ (liquid)} \qquad (7)$$

Here the chemical formula is written CH_2, which is one-eighth of a typical gasoline molecule (C_8H_{16}). The reaction is catalyzed by a number of metal-based catalysts including iron, cobalt, and nickel. The reactors in which the synthesis takes place operate within a temperature range of 225 to 365°C and at pressures from 0.5 to 4 MPa. It should also be noted that the Fischer–Tropsch reactions produce a wide spectrum of oxygenated compounds such as alcohols.

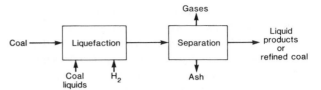

FIGURE 5 Direct liquefaction of coal. [Reprinted with permission from Probstein, R. F., and Hicks, R. E. (1990). "Synthetic Fuels," pH Press, Cambridge, MA.]

D. Direct Liquefaction

The two principal routes for the direct hydrogenation of coal to form a liquid involve the addition of hydrogen to the coal either directly from the gas phase or from a donor solvent. When the hydrogen is added directly from the gas phase, it is mixed together with a slurry of pulverized coal and recycled coal-derived liquid in the presence of suitable catalysts. This is called hydroliquefaction or catalytic liquefaction and is essentially the Bergius technology mentioned in Section I.B. In the donor solvent procedure a coal-derived liquid, which may or may not be separately hydrogenated, transfers the hydrogen to the coal without external catalyst addition. These procedures are illustrated schematically in the block diagram in Fig. 5.

The direct liquefaction of coal may be simplistically modeled by the chemical reaction

$$C + 0.8H_2 \rightarrow CH_{1.6} \qquad (8)$$

Direct liquefaction processes under development are typically carried out at temperatures from about 450 to 475°C and at high pressures from 10 to 20 MPa and up to 30 MPa. Despite the slow rate at which liquefaction proceeds, the process itself is thermally rather efficient, since it is only slightly exothermic. However, hydrogen must be supplied and its manufacture accounts for an important fraction of the process energy consumption and cost of producing the liquid fuel. The hydrogen itself may be produced, for example, by the gasification of coal, char, and residual oil.

III. TECHNOLOGIES

A. Gas from Coal

The three principal reactor types employed in coal gasifier design are the moving packed bed, the entrained flow, and the fluidized bed reactor. In the discussion of gasification principles the moving packed bed (Fig. 3) was used to illustrate steam–oxygen or steam–air gasification of coal.

The reactor type strongly influences the temperature distribution and, in this way, the gas and residue products. The reaction temperature typically varies from about 800 to 1500°C, and up to a maximum of about 1900°C in entrained flow oxygen reactors. Each type of gasifier covers a specific temperature range. At high temperatures a synthesis gas is produced and at low temperatures methane formation is favored. Gasifiers in which the temperature is low enough that the residual ash does not melt are sometimes referred to as "dry ash gasifiers." High-temperature gasifiers in which molten ash (slag) is formed are called "slagging gasifiers." The slagging temperature is dependent on the ash composition but, for most coals, lies roughly in the range 1200 to 1800°C.

Moving bed coal gasifiers (see Fig. 3) operate with countercurrent flow and use either steam and oxygen or steam and air, and the residue may be either slag or dry ash plus any unconverted carbon. Coal particles in the size range of 3–50 mm are fed into the top of the gasifier. The coal passes downward, with an average linear bed velocities of the order of 0.5 m/hr in atmospheric steam/air gasifiers and 5 m/hr in high-pressure steam/oxygen gasifiers.

Representative of the moving bed gasifer are the Lurgi dry ash and slagging gasifiers. The Lurgi dry ash gasifier was the first high-pressure gasifier and was introduced in commercial operation in Germany in 1936. Nominal operating pressures of present commercial units are about 3 MPa, although they have been run at 5 MPa, with projected operating pressures up to 10 MPa. Temperatures in the combustion zone range from about 1000 to 1400°C, and those in the gasification zone from about 650 to 800°C. Typical coal throughputs are 800 t/day. The gasifiers are about 4 to 5 m in diameter and about three times as high, excluding the coal feed and ash lock hoppers that are attached to the top and bottom, respectively, and that more than double the height. A schematic drawing of the Lurgi dry ash gasifier is shown in Fig. 6.

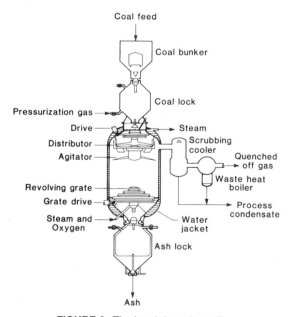

FIGURE 6 The Lurgi dry ash gasifier.

Not shown in Fig. 6 are the gas cleaning and purification units for the product gases leaving the gasifier. As discussed previously, in the manufacture of synthetic fuels any hydrogen sulfide, ammonia, and carbon dioxide present in the product (or byproduct) gases from a reactor usually must be removed. Ammonia is very soluble in water and is generally removed by washing the gases with water. Most of the hydrogen sulfide and carbon dioxide must be removed by other means. The procedure generally used is to remove the hydrogen sulfide and carbon dioxide, which are called acid gases, by absorption into an appropriate liquid solvent. The gases are subsequently desorbed from the liquid by heating and/or pressure reduction. The hydrogen sulfide is then converted to elemental sulfur, in part by burning it in a procedure known as the Claus process.

Entrained flow gasifiers all use coal (or char) pulverized to a size of the order of 75 μm. Oxygen or air together with steam generally is used to entrain the coal, which is injected through nozzles into the gasifier burner. Hot product gas may also be employed to entrain the coal and at the same time gasify it. In the Texaco gasifier, which is the gasifier in use at the Cool Water plant mentioned in Section I.B, the solids are carried in a water slurry, pumped up to gasification pressure (4 MPa), transported to the top of the gasifier, and injected through a burner into the reactor together with oxygen. The most important feature of entrained flow gasifiers is that they operate at the highest temperatures under conditions where the coal slags. In the Texaco gasifier, for example, the gasification temperature is about 1400°C

Fluidized bed gasifiers are fed with pulverized or crushed coal that is lifted in the gasifier by feed and product gases. In single-stage, directly heated gasifiers a steam/oxygen or steam/air mixture is injected near the bottom of the reactor, either cocurrently or countercurrently to the flow of coal (or char). The rising gases react with the coal and, at the same time, maintain it in a fluidized state. Fluidization refers to the case in which the force of the gas on the particles lifts them so that they are in balance against their own weight. The particle "bed" is then expanded typically to twice its settled height, and is in a locally stable arrangement which resembles a boiling liquid. As the coal is gasified, the larger-size mineral particles, which are about twice as dense as the carbonaceous material, fall down through the fluidized bed together with the larger char particles. The advantage of this procedure is that it provides for good mixing and uniform temperatures in the reactor. Although fluidized bed gasifiers are thought of as a relatively recent development, work on the Winkler fluidized bed gasifier began in Germany in 1921, and the first commercial unit went into operation in 1926.

One gasification procedure that is markedly different, although the chemistry is not, is that of underground, or *in situ*, gasification. In this method, the gasification is carried out directly in the unmined coal deposit, which, by appropriate preparation, is turned into a fixed packed bed. The reactants are brought down to the coal bed and the gases formed are brought up to the surface through holes drilled into the deposit.

B. Liquids and Clean Solids from Coal

The three principal routes by which liquid fuels can be produced from coal have been noted to be pyrolysis, direct liquefaction, and indirect liquefaction. A clean fuel that is a solid at room temperature can also be produced by direct liquefaction processes.

In pyrolysis processes the main limitation is that the principal product is char, so that the effectiveness of any technology rests on the ability to utilize the char, for example, to produce gas or electricity. A wide number of technologies were under large-scale development through the early 1980s. One that attained commercial status was the Lurgi–Ruhrgas process, which feeds finely ground coal and hot product char to a chamber containing a variable-speed mixer with two parallel screws rotating in the same direction. Temperature equalization and devolatilization are very rapid due to the uniform mixing and high rates of heat transfer. The pyrolysis liquids and gases are removed overhead from the end of the chamber. Some of the new product char is burned in a transfer line and recycled to the reactor to provide the heat for the pyrolysis.

One process that was developed but not commercialized was the TOSCOAL process, in which crushed coal is fed to a horizontal rotating kiln. There it is heated by hot ceramic balls to between 425 and 540°C. The hydrocarbons, water vapor, and gases are drawn off, and the char is separated from the ceramic balls in a revolving drum with holes in it. The ceramic balls are reheated in a separate furnace by burning some of the product gas.

A process pioneered by the National Coal Board in England that has not reached the fully developed stage but that has considerable potential is supercritical gas extraction. In this process the coal is pyrolized at a relatively low temperature, around 400°C, in the presence of a compressed "supercritical gas," that is, a gas whose temperature is above the critical temperature at which it can be liquefied. Suitable gases are, for example, a number of petroleum fractions. Under these conditions at high pressures, around 10 MPa, the gas density is like that of a liquid, and the gas acts like a strong solvent that causes the liquids to volatilize and be taken up by the vapor. By transferring the gas to a vessel at atmospheric pressure, the density of the solvent gas is reduced and the extracted

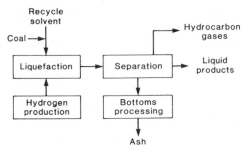

FIGURE 7 Generalized direct liquefaction process train. [Reprinted with permission from Probstein, R. F., and Hicks, R. E. (1990). "Synthetic Fuels," pH Press, Cambridge, MA.]

tar precipitates out. The product is a low-melting glassy solid that is essentially free of mineral matter and solvent, and contains less nitrogen and sulfur than the coal.

Processwise two principal methods of direct liquefaction have been distinguished, in which the hydrogen may be added directly from the gas phase or a coal-derived liquid transfers the hydrogen to the coal. Despite the seeming difference, the major elements of both processes are similar as illustrated in the block diagram in Fig. 7. Coal is slurried with recycled oil or a coal-derived solvent, mixed with hydrogen, and liquefied at high pressures—in the case of hydroliquefaction, in the presence of an externally added catalyst. The resulting mixture is separated into gas and liquid products and a heavy "bottoms" slurry containing mineral matter and unconverted coal. Generally a large fraction of the carbon in the coal ends up in the bottoms, and most processes gasify this slurry to produce fuel gas and hydrogen.

Representative of the hydroliquefaction procedures in which hydrogen is added to the coal in the presence of a catalyst in the H-Caol process developed by Hydrocarbon Research, Inc. This procedure went through a pilot plant development capable of processing 530 t/day of dry coal to about 1350 bbl/day of low-sulfur fuel oil or 190 t/day of coal to a synthetic crude before operation was terminated. The difference in feed rates results from the fact that, to produce a synthetic crude, more hydrogen must be added, resulting in an increase in the residence time in the reactor and hence a decrease in the coal feed rate. In the process, coal crushed to less than 0.2 mm and dried is slurried with recycle oil at a ratio typically between 2 and 3 to 1, and then pumped to a pressure of around 21 to 24 MPa. Compressed hydrogen produced by gasification is added to the slurry and the mixture is preheated to 340 to 370°C. The mixture is passed upward into a reactor vessel operated at temperatures of about 450°C. The reactor contains an active, bubbly bed of catalyst, which is kept in a fluidized state by internally recycling slurry. The reactor is called an "ebullated bed" reactor because there is no locally stable fluidized arrangement in it but, instead, a fluidized bed with an active and ebullient character.

Illustrative of a plant in which a coal-derived liquid or "donor" solvent transfers hydrogen to the coal is the Advanced Coal Liquefaction Research and Development Facility at Wilsonville, Alabama. The nominal coal feed of the pilot plant is 5.4 t/day, and in 1986 the facility was operational. Other plants in the United States have been run on a considerably larger scale but have been shut down. The plant was originally constructed to study the Solvent Refined Coal process for manufacturing a clean solid fuel in one stage. It then evolved to a facility to study two-stage liquefaction processes for making liquid fuels. The principal product from the single-stage process is an ash-free, low-sulfur, pitch-like extract that is a solid at room temperature. The product was formerly called "solvent refined coal," or SRC, and is now called "thermal resid," or TR.

In the process, coal dried and pulverized to less than 3 mm is mixed with recycle solvent at a mass ratio of about 1.5 solvent-to-coal. The slurry is pumped together with hydrogen at from 10 to 14 MPa and preheated to 400 to 450°C. It then enters the thermal liquefaction unit, which is a vertical tube in which the three phases flow cocurrently upward. The residence time in the unit is typically about 30 min, and under these conditions most of the carbonaceous material dissolves. The ash and undissolved coal are separated from the product liquid by a procedure developed by the Kerr–McGee Corp. termed "critical solvent deashing." The principal is similar to that of supercritical gas extraction, discussed above, in that it employs the increased dissolving power of a solvent near its critical temperature and pressure. The solvent is mixed with the slurry and dissolves the product liquid. The solids settle out and the heavy product is subsequently recovered by decreasing the solvent density by heating. In the two-stage operation the product is upgraded by catalytic hydrogenation to light liquid hydrocarbons. The reactor employed for this is the ebullated bed H-Oil reactor developed by Hydrocarbon Research, Inc., which is similar to the H-Coal reactor.

A modification of the solvent extraction process that was investigated extensively in the 1980s is called *coprocessing*, in which the coal is processed together with a crude oil. The objective is to upgrade the oil and to simultaneously liquefy the coal. The fraction of coal in the feed may be less than 10%, in which case the major objective is to upgrade the oil using the coal as a catalyst, or more than 60%, in which case the process more closely resembles a solvent extraction process such as described above for the Wilsonville plant but without recycle of the solvent.

Coprocessing reactors are designed to operate at temperatures of 400 to 500°C and at pressures from 8 to 30 MPa; a catalyst may be added to increase yields. Feed

oils may be residues from petroleum refining processes or even from other synthetic fuel processes. Under the action of high pressure and temperature, the large oil molecules are ruptured to light products, and the sulfur atoms may be removed as hydrogen sulfide. The liquefaction of the coal molecules occurs by a process of extraction into the oil or, at higher temperatures, may involve thermal rupturing of the bonds. If hydrogen is added to the reactors, the latter route is termed hydrothermal processing.

Several processes have been investigated at the pilot and process development scale, and although some plans were made for commercial demonstration, there were no major developments anticipated in 1991.

The last major category for the manufacture of liquid fuels is the indirect liquefaction procedures. The most extensive production of synthetic liquid fuels today is that being carried out by Fischer–Tropsch reactions at the South African Sasol complexes, with a combined output of over 100,000 bbl/day of motor fuels and other liquid products. The two largest plants, each with an output of about 50,000 bbl/day, employ 36 Lurgi dry ash, oxygen-blown gasifiers (see Fig. 6) apiece for the synthesis gas production. The gas is scrubbed with water for removal of particulate matter, tar, and ammonia, following which hydrogen sulfide and carbon dioxide are removed by absorption in cold methanol. The latter process is proprietary to Lurgi and is termed the Rectisol process.

The principal reactors used are fluidized bed reactors, called Synthol reactors, in which the feed gas entrains an iron catalyst powder in a circulating flow. The suspension enters the bottom of the fluidized bed reaction section, where the Fischer–Tropsch and the gas shift reactions proceed at a temperature of from 315 to 330°C. These reactions are highly exothermic, as described previously, and the large quantity of heat released must be removed. The products in gaseous form together with the catalyst are taken off from the top of the reactor. By decreasing the gas velocity in another section, the catalyst settles out and is returned for reuse. The product gases are then condensed to the liquid products.

Of the indirect liquefaction procedures, methanol synthesis is the most straightforward and well developed [Eq. (6)]. Most methanol plants use natural gas (methane) as the feedstock and obtain the synthesis gas by the steam "reforming" of methane in a reaction that is the reverse of the methanation reaction in Eq. (5). However, the synthesis gas can also be obtained by coal gasification, and this has been and is practiced. In one modern "low-pressure" procedure developed by Imperial Chemical Industries (ICI), the synthesis gas is compressed to a pressure of from 5 to 10 MPa and, after heating, fed to the top of a fixed bed reactor containing a copper/zinc catalyst. The reactor temperature is maintained at 250 to 270°C by injecting part of the relatively cool feed gas into the reactor at various levels. The methanol vapors leaving the bottom of the reactor are condensed to a liquid.

An indirect liquefaction procedure of relatively recent origin is the Mobil M process for the conversion of methanol to gasoline following the reaction

$$CH_3OH \rightarrow CH_2 + H_2O \tag{9}$$

The key to the process was the development by Mobil of a size-selective zeolite catalyst, whose geometry and pore dimensions have been tailored so that it selectively produces hydrocarbon molecules within a desired size range. This is a highly exothermic reaction and the major problem in any plant design is the reactor system to effect the necessary heat removal. A plant completed in 1985 in New Zealand uses about 4 million m³/day of natural gas as the feedstock to produce the methanol by the ICI procedure described above. In 1990 the plant produced about 16,000 bbl/day of gasoline, which is somewhat above its design output.

C. Liquids from Oil Shale and Tar Sands

Oil shale is a sedimentary rock containing the hydrocarbon "kerogen," a high molecular mass organic material that is insoluble in all common organic solvents and is not a member of the petroleum family. Oil shale deposits occur throughout the world and may, in fact, represent the most abundant form of hydrocarbon on earth. The United States has by far the largest identified shale resource suitable for commercial exploitation in the Green River Formation in Colorado, Utah, and Wyoming.

Oil shale is characterized by its grade, that is, its oil yield, expressed as liters per ton (liters/t) in British units or as U.S. gallons per ton (gal/ton), as determined by a standard "Fischer" assay in which a given amount of crushed shale is pyrolyzed in a special vessel in the absence of air at 500°C. By definition, oil shale yields a minimum of 42 liters/t (10 gal/ton) of oil and may be found up to 420 liters/t (100 gal/ton). Lower-grade shale yields are below 100 liters/t. Commercially important western United States shales have an amount of organic matter (as mass% of the shale) of from 13.5 to 21%. In comparison, the organic matter in coal typically ranges from 75 to more than 90%, by mass. Consequently a significantly larger amount of oil shale must be processed compared to coal to obtain an equivalent hydrocarbon throughput. The inorganic content of oil shales is a mix of carbonates, silicates, and clays.

The principal method for producing oil from shale is by pyrolysis carried out in a vessel called a "retort," with the process called "retorting" when applied to the commercial-scale recovery of shale oil. As with coal gasi-

fication, oil shale may be mined and retorted on the surface, or it may be retorted *in situ* and the released oil collected and pumped to the surface. Commercial-scale retorts are generally either moving packed beds or solids mixers. An *in situ* retort is in effect a moving bed reactor, but with the retorting zone moving through the stationary shale.

Oil shale retorts, like coal gasifiers, are classified according to whether they are directly or indirectly heated. In directly heated processes, heat is supplied by burning a fuel, which may be recycled retort off gas, with air (or oxygen) within the bed of shale. Some portion of either the coke residue or the unretorted organic matter may be burned as well. Not infrequently, most or even all the heat is provided by combustion of the kerogen. In indirectly heated processes a separate furnace is used to raise the temperature of a heat transfer medium, such as gas or some solid material such as ceramic, which is then injected into the retort to provide the heat. Whether the shale is heated by a gas or a solid defines two subclasses of the indirectly heated retort. The fuel that fires the furnace may be retort off gas or crude shale oil. The three heating methods for oil shale retorting are shown in Fig. 8.

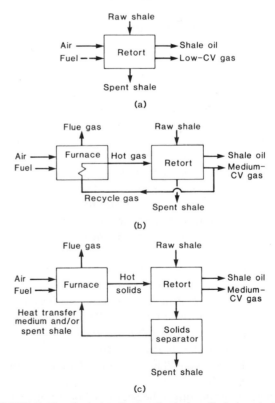

FIGURE 8 The three heating methods for oil shale retorting. (a) Directly heated retort; (b) indirectly heated retort, gas-to-solid heat exchange; (c) indirectly heated retort, solid-to-solid heat exchange. [Reprinted with permission from Probstein, R. F., and Hicks, R. E. (1990). "Synthetic Fuels," pH Press, Cambridge, MA.]

All surface processing operations involve mining, crushing, and then retorting. The liquid product of retorting is too high in nitrogen and sulfur to be used directly as a synthetic crude for refining and requires upgrading, for example, by treating with hydrogen, as discussed in connection with liquefaction, and/or by removal of carbon in a thermal distillation process termed coking. The spent shale remaining after retorting amounts to 80 to 85%, by mass, of the mined shale, so solid waste disposal is a major activity.

Most effort in the commercialization of oil shale processes has centered upon retort development. Over the years numerous technologies have been demonstrated for the surface retorting of oil shale, many of which have been discarded and then resurrected with modification, paralleling the on-again, off-again character of oil shale commercialization itself. Only a few will be mentioned here.

Two indirectly heated oil shale retorting technologies employing solid-to-solid heat transfer have been described in connection with coal pyrolysis. They are the TOSCOAL process, called the TOSCO process when used with oil shale, and the Lurgi–Ruhrgas process. The former process was fully developed before operations were terminated, and the latter has been commercialized in connection with coal devolatilization and hydrocarbon pyrolysis.

A retort that can be operated in either the direct or the indirect mode, which uses gas-to-solid heat transfer, is one developed by the Union Oil Co., now Unocal Corp. In this retort shale is charged into the lower and smaller end of a truncated cone and is pushed upward by a piston referred to as a "rock pump." In the indirect mode recycle gas that has been heated in a furnace flows in from the top countercurrent to the upward-moving shale. Combustion does not occur within the retort. As the shale moves upward it contacts the hot gas and is pyrolized. The shale oil flows down through the upward moving cooler fresh shale and is withdrawn from the bottom of the truncated cone together with the retort gas. In the directly heated mode, which has been demonstrated but discontinued, air without recycle gas is used, and nearly all of the energy of the residual carbon is recovered by combustion within the retort. The retort operating in the indirectly heated mode is the one being used in Unocal's 10,000 bbl/day facility at Parachute Creek, Colorado. Although toward the end of 1986 the plant was off-line for technical reasons, it was operating about one-half to two-thirds of the time between 1988 and 1991.

In situ retorting offers the possibility of eliminating the problems associated with the disposal of large quantities of spent shale that occur with surface retorting. *True in situ* (TIS) retorting involves fracturing the shale in place, igniting the shale at the top of the formation, and feeding in air to sustain the combustion for pyrolysis. The combustion zone moves downward, ahead of which is the retorting zone, and below that the vapor condensation zone. The gases and condensed oil and water are then pumped up from the bottom. Oil shale is not porous and generally does not lie in permeable formations, so adequate flow paths are difficult to create. To overcome this difficulty, an alternative approach known as *modified in situ* (MIS) has been developed. In this procedure a portion of the shale is mined out and the remaining shale is "rubblized" by exploding it into the mined void volume. The resulting oil shale rubble constitutes the retort.

Tar sands are normally a mixture of sand grains, water, and a high-viscosity crude hydrocarbon called bitumen. Unlike kerogen, bitumen is a member of the petroleum family and dissolves in organic solvents. At room temperatures the bitumen is semisolid and cannot be pumped, but at temperatures of about 150°C it will become a thick fluid. In the Alberta deposits of Canada, the bitumen is present in a porous sand matrix in a range up to about 18 mass%, although the sum of bitumen and water generally totals about 17%.

Two options for the recovery of oil from tar sands are of importance: mining of the tar sands, followed by aboveground bitumen extraction and upgrading; and *in situ* extraction, in which the bitumen is released underground by thermal and/or chemical means and then brought to the surface for processing or upgrading. Because the processes of *in situ* recovery are similar to those employed in the enhanced recovery of crude oil, they are not discussed.

Two surface extraction, full-scale commercial facilities are presently in operation to produce synthetic crude oil from the Alberta deposits. One, the Suncor, Ltd., facility was built in the late 1960s and, in 1991, was producing synthetic crude at a rate of about 58,000 bbl/day. The second and larger one, built in the mid to late 1970s, is the facility of Syncrude Canada, Ltd. In 1991 it was producing synthetic crude at a rate of about 156,000 bbl/day.

Both of the Canadian plants use the technique of hot water extraction to remove the bitumen from the tar sand. In this procedure the tar sand, steam, sodium hydroxide, and hot water are mixed and tumbled at a temperature of around 90°C. Layers of sand pull apart from the bitumen in this process. Additional hot water is added and the bitumen–sand mixture is separated into two fractions by gravity separation in cells in which the bitumen rises to the top and is skimmed off, while the sand settles to the bottom. The upgrading of the bitumen to a synthetic crude is then accomplished by oil refinery procedures including coking, in which carbon is removed by thermal distillation and hydrotreating.

IV. BIOMASS CONVERSION

A. Biomass as a Fuel Source

Biomass is any material that is directly or indirectly derived from plant life and that is renewable in time periods of less than about 100 years. More conventional energy resources such as oil and coal are also derived from plant life but are not considered renewable. Typical biomass resources are energy crops, farm and agricultural wastes, and municipal wastes. Animal wastes are also biomass materials in that they are derived, either directly or via the food chain, from plants that have been consumed as food.

As with conventional fuels, the energy in biomass is the chemical energy associated with the carbon and hydrogen atoms contained in oxidizable organic molecules. The source of the carbon and hydrogen is carbon dioxide and water. Both of these starting materials are in fact products of combustion, and not sources of energy in the conventional sense. The conversion by plants of carbon dioxide and water to a combustible organic form occurs by the process of photosynthesis. Two essential ingredients for the conversion process are solar energy and chlorophyll. The chlorophyll, present in the cells of green plants, absorbs solar energy and makes it available for the photosynthesis, which may be represented by the overall chemical reaction

$$n\text{CO}_2 + n\text{H}_2\text{O} \xrightarrow[\text{chlorophyll}]{\text{sunlight}} (\text{CH}_2\text{O})_n + n\text{O}_2 \quad (10)$$

$(\text{CH}_2\text{O})_n$ is used here to represent the class of organic compounds called carbohydrates or "hydrates of carbon," several of which are made in the course of the reaction. Carbohydrates include both sugars and cellulose, which is the main constituent of the cell wall of land plants and the most abundant naturally occurring organic substance.

About one-quarter of the carbohydrate formed by photosynthesis is later oxidized in the reverse process of respiration to provide the energy for plant growth. The excess carbohydrate is stored. The plant typically contains between 0.1 and 3% of the original incident solar energy, which is a measure of the maximum energy recoverable from the plant if converted into a synthetic fuel. Some of this energy may, however, be degraded in the formation of intermediate products, and there will be additional losses in converting the biomass material into a conventional form.

One of the reasons for the great interest in biomass as a fuel source is that it does not affect atmospheric carbon dioxide concentrations. This is because carbon dioxide, which is formed by respiration, biological degradation, or combustion, is eventually reconverted to oxidizable organic molecules by photosynthesis. Therefore, no net change in atmospheric carbon dioxide levels takes place provided an equivalent quantity of vegetation is replanted. More important, perhaps, is that this energy source is renewable. In addition, biomass fuels are clean-burning, in that sulfur and nitrogen concentrations are low, and because the hydrogen-to-carbon ratio is generally high. However, it is not expected that biomass will make a major contribution to overall energy requirements in the near-future. The principal limitations of extensive biomass development are its high land and water requirements and the competition with food production.

B. Conversion Processes

The potential biomass conversion processes are shown in Fig. 9. They include biochemical conversion by fermentation and anaerobic digestion and the thermal processes of combustion, pyrolysis, and gasification. Fermentation produces mainly liquids, in particular, ethanol; pyrolysis results in both liquid and gaseous products; and gasification and anaerobic digestion produce gaseous fuels. Most biomass materials can be gasified, and the resulting gas may be used for synthesis of liquid fuels or substitute natural gas. Direct combustion of biomass is always an option and, in some instances, may be the only viable approach.

In principle, biomass resources can be converted using any of the biochemical or thermal conversion processes.

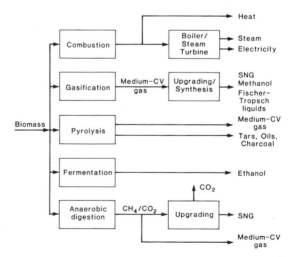

FIGURE 9 Biomass conversion processes. [Reprinted with permission from Probstein, R. F., and Hicks, R. E. (1990). "Synthetic Fuels," pH Press, Cambridge, MA.]

However, some processes can be expected to be more effective than others in recovering energy from specific resources. Wood is perhaps the most versatile resource, with the greatest potential. It is suitable for use on a large scale by combustion, or for air or oxygen gasification, and for pyrolysis. Municipal solid wastes, which are suitable for combustion and gasification on a large scale, also are considered to have potential. However, despite the many advantages of biomass, it is likely that only a small fraction of the world's energy needs could come from this source by the end of the twentieth century.

V. OUTLOOK

Most of the processes discussed either have been or are being used to supply synthetic fuels on a commercial basis. There is, therefore, little question as to the feasibility of these processes. In most cases, however, these ventures have proved and continue to prove economically unattractive in the face of abundant supplies of cheap natural gas and oil. When supplies dwindle and prices escalate, as is likely to happen eventually, specific processes can be expected to become marginally attractive. In the United States, probably the most competitive of the synthetic fuels are shale oil and low-CV and medium-CV gas. The more complex routes to liquid transportation fuels from coal can be expected to be more costly. In all cases a reduction in costs will occur as experience is gained from initial plants. Coal and, eventually, oil shale reserves will, however, also become depleted. Because biomass can probably make only a limited contribution to the total energy demand, other sources of energy will have to be harnessed. The development of synthetic fuels will probably be necessary to obtain the time needed for the evolution of such alternative energy sources.

SEE ALSO THE FOLLOWING ARTICLES

BIOENERGETICS • BIOMASS, BIOENGINEERING OF • BIOMASS UTILIZATION, LIMITS OF • BIOREACTORS • CATALYSIS, INDUSTRIAL • COAL STRUCTURE AND REACTIVITY • COMBUSTION • ENERGY EFFICIENCY COMPARISONS AMONG COUNTRIES • ENERGY FLOWS IN ECOLOGY AND IN THE ECONOMY • ENERGY RESOURCES AND RESERVES • RENEWABLE ENERGY FROM BIOMASS • WASTE-TO-ENERGY SYSTEMS

BIBLIOGRAPHY

Beghi, G. E. (ed.) (1985). "Synthetic Fuels," D. Reidel, Hingham, MA.

Elliott, M. A. (ed.) (1981). "Chemistry of Coal Utilization: Second Supplementary Volume," Wiley, New York.

Gaur, S., and Reed, T. (1998). "Thermal Data for Natural and Synthetic Fuels," Marcel Dekker, New York.

Klass, D. L. (1998). "Biomass for Renewable Energy, Fuels, and Chemicals," Academic Press, New York.

Meyers, R. A. (ed.) (1984). "Handbook of Synfuels Technology," McGraw–Hill, New York.

National Academy of Sciences (1980). "Energy in Transition 1985–2010," Final Report, Committee on Nuclear and Alternative Energy Systems, National Research Council, 1979, W. H. Freeman, San Francisco.

Perry, R. H., and Green, D. W. (eds.) (1984). Fuels. In "Perry's Chemical Engineers' Handbook," 6th ed., pp. 9-3–9-36. McGraw–Hill, New York.

Probstein, R. F., and Gold, H. (1978). "Water in Synthetic Fuel Production," MIT Press, Cambridge, MA.

Probstein, R. F., and Hicks, R. E. (1990). "Synthetic Fuels," pH Press, MIT Branch P.O., Box 195, Cambridge, MA 02139. (First published 1982 by McGraw–Hill, New York.)

Romey, I., Paul, P. F. M., and Imarisio, G. (eds.) (1987). "Synthetic Fuels From Coal. Status of the Technology," Graham and Trotman, Norwell, MA.

Speight, J. G. (ed.) (1990). "Fuel Science and Technology Handbook," Marcell Dekker, New York.

Supp, E. (1990). "How to Produce Methanol from Coal," Springer-Verlag, New York.

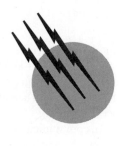

System Theory

F. R. Pichler

Johannes Kepler University

John L. Casti

International Institute for Applied Systems Analysis

GLOSSARY

Canonical (minimal) models Internal descriptions containing no unobservable or unreachable states.

External description Mathematical description of a system's behavior involving only the physically accessible system inputs and outputs.

Inputs Physical actions (controls, decisions, stimuli) that can be applied by the system controller (decision maker, manager) to affect the behavior of the system states.

Internal description Mathematical description of a system's behavior utilizing observable inputs and outputs, as well as physically inaccessible state variables.

Observable state Initial state that can be uniquely identified from knowledge of a system's input and output sequences.

Optimality criteria Scalar functions of controls that are minimized (or maximized) by an optimal control sequence.

Outputs Physically accessible responses (behaviors, observations) emitted by the system in response to applied inputs.

Reachable state Element of the set of states that can be attained from the initial system state by application of an admissible sequence of system inputs.

Realizations Internal system descriptions generated from external descriptions.

THE ORIGIN OF SYSTEMS THEORY lies in the kind of complex problems experienced by engineers (specifically in the field of communications and control) and scientists (in biology and ecology) in the mid-20th century. Later it became obvious that the usual mathematical modeling concepts based on analysis (differential equations) and linear algebra (linear equations and matrices) were no longer appropriate. New mathematical methods for dealing with actual problems in formal modeling tasks were required.

In communications engineering Karl Küpfmüller, well-known engineer and professor, suggested modeling transmission lines and related components not on the level of electrical networks (and related differential equation systems) but on the higher level of functional frequency descriptions (by the transfer function) and looking for ("top-down") means of computational determination of the physical realization on the level below.

He called this approach to modeling "systems theoretical." His book, "Die Systemtheorie der Elektrischen Nachrichtenübertragung," of 1949 can be considered the "birth" of systems theory for information technology. From that it should be clear that systems theory should not be considered as a "theory of systems" but rather as a collection of useful concepts, methods, and tools for the support of top-down multilevel modeling in engineering and science. By this definition it is obvious that formal concepts and methods will play an important role in a systems theory. It is likewise quite clear that, to deal with complex design and simulation problems in engineering and science, computer assistance for the effective application of systems-theoretical methods is a necessity. The field of computer-aided systems theory (CAST), which should provide the proper software tools for the application of systems theory, therefore deserves the vital interest of engineers and scientists working on complex design and simulation projects.

I. MODEL COMPONENTS

The implications of the existing knowledge in fields such as biology, psychology, business, economics, and political science, not to mention "hard" scientific disciplines such as physics and chemistry, specifically in the field of engineering, are so complex that it is no longer possible for the human mind to digest them without extensive abstraction, that is, without mathematics. Here, by *mathematics* we do not mean data analysis, numerical formulas, graphical methods, and other pedestrian (although often useful) tools frequently termed mathematics but, rather, the use of conceptual ideas from set theory, algebra, topology, and analysis to construct and analyze abstract versions of real-world situations with the goal of understanding the essential relations among their constituent parts. Such constructions and analyses are the province of the mathematical modeler (system theorist)—the keeper of the abstract processes.

When confronted by a particular process whose behavior is of interest, the modeler's first task is to separate

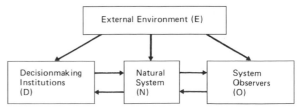

FIGURE 1 High-level system decomposition.

the various aspects of the process into major subsystems, which can then be modeled at a more detailed level. A convenient high-level decomposition of a general situation is depicted in Fig. 1. Part of the art of system modeling lies in the establishment of useful boundaries between the major components indicated in Fig. 1. In actuality, these boundaries are totally artificial, and what constitutes a useful separation is, in general, highly context dependent, as we shall see. Nevertheless, the divisions indicated in Fig. 1 do provide a helpful guideline on which to focus the remainder of the modeling effort.

According to current system-theoretic thinking, the natural system (N) is usually considered to be that part of a given process that is not *directly* accessible to external influence or observation. From a certain point of view, one might say that only the *physically* observable causes and effects reside in the decision-making (D) and observing (O) components, with the natural system playing the role of a mediator. An alternate interpretation is to regard the natural system as a "black box," the inner workings of which we attempt to explore by applying inputs from the decision-making or external (E) component (or both) and measuring outputs in the observing component.

We can loosely delineate the model components N, D, O, and E as follows:

1. *Natural system*: A collection of variables and relationships perceived by D and O as having "internal" dynamics and couplings to E through measuring the apparatus, control mechanisms, and "forcing functions." This is an open-system definition, and there is no pretense that the boundaries separating N from D, O, and E have been chosen in a knowledgeable or even an intelligent fashion.

2. *Decision-making institution*: A system that processes information, develops models, and exerts controlling actions on N, chosen with respect to the models and to objectives that may be at least partially established by E.

3. *System observer*: A component that monitors both N and E and provides information to D about the behavior of the system N.

4. *External environment*: A collection of relationships that affect and are affected by both N and D yet are not generally perceived as "part of the problem" by D. In a very real sense, E can be viewed as "everything else that goes on in the world."

The foregoing definitions are far from entirely satisfactory, but the only crucial point is that natural systems are arbitrary objects of analysis whose formalization in N–D–O–E terms hinges critically on the questions the model is required to answer.

Probably the best way to gain a feel for the decomposition of a given problem into the components just outlined is to consider a few representative examples.

A. Fishery Management

A simple model of interspecific competition between two species is provided by the logistic equations

$$dx/dt = rx(1 - (x/K)) + \alpha xy - h_1(t),$$

$$dy/dt = sy(1 - (y/L)) + \beta xy - h_2(t),$$

where x and y are the two fish populations, r and s are growth rates, K and L are maximum population levels the environment can support, α and β are measures of the extent to which each species interferes with the other's use of the external resource (food supply), and $h_1(\cdot)$ and $h_2(\cdot)$ are harvesting functions.

A plausible separation of this model into the macro-components outlined earlier is to consider the variables x and y, together with their growth rates r and s and interference parameters α and β, as the natural system N. The decision-making institution D clearly is composed of the harvesting functions h_1 and h_2, whereas the observer O may be thought of as the variables x and y, since it is reasonable to suppose that the fish populations can be measured directly. Finally, the environmental carrying capacities K and L comprise the external environment E.

This decomposition of the model variables illustrates the important point that the system components N, D, O, and E are not necessarily disjoint; here we see that the variables x and y belong naturally to both components N and O.

B. National Income Dynamics

Consider a vastly simplified picture of the dynamics of national income in which the total national income in year k is denoted y_k and is the sum of the consumption expenditures w_k and the investment expenditures u_k, that is,

$$y_k = w_k + u_k.$$

We assume that consumption expenditures depend on the national income of the previous year as

$$w_k = by_{k-1},$$

where b is a constant measuring the marginal propensity to consume. Clearly,

$$y_k = by_{k-1} + u_k = u_k + bu_{k-1} + b^2 y_{k-2}$$
$$= u_k + bu_{k-1} + \cdots + b^k y_0.$$

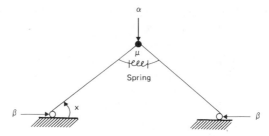

FIGURE 2 Model of an electrical network.

This relation defines an elementary input–output model of national income dynamics, in the sense that the output (total income) is determined as a (linear) function of past inputs (investment expenditures).

In terms of our earlier system components, it would be natural to regard y as the system observer O, u as the decision-making body D, and b as comprising the external environment E. Of interest here is that the natural system N is only implicitly determined as some "mechanism" that generates the measured output y from the inputs u. Thus, although it is impossible to escape the feeling that N must be part of the problem, it is not explicitly represented by the variables defining the input–output relation but, rather, must be mathematically *inferred* from them. We return to this "realization" problem later.

C. Electrical Networks

Consider a given physical network made up of resistor, capacitor, and inductor elements interconnected as shown in Fig. 2. A voltage source S generates u as input to the network; a load L is connected to the output and receives a current i_5, which is driven by a voltage y.

By the well-known Kirchhoff laws and the elementary electrical laws for network elements (Ohm's law, Faraday's law for the inductor, the law for loading a capacitor), we are able to express the relations between the voltages on the capacitors and the coil current of the network by the following set of differential equations.

$$\frac{du_1(t)}{dt} = \frac{1}{C_1} u_1(t),$$

$$\frac{du_2(t)}{dt} = -\frac{1}{C_2(R_1 + R_2)} u_2(t) - \frac{R_1}{C_2(R_1 + R_2)} i_2(t)$$
$$+ \frac{1}{C_2(R_1 + R_2)} u(t),$$

$$\frac{di_2(t)}{dt} = -\frac{1}{L} u_1(t) - \frac{R_1 R_2}{L(R_1 + R_2)} i_2(t)$$
$$+ \frac{R_1}{L(R_1 + R_2)} u_2(t) + \frac{R_2}{L(R_1 + R_2)} u(t).$$

Here u denotes the input voltage of the network; u_1 and u_2 are the voltages on the capacitors C_1 and C_2, respectively; and i_2 is the current in the inductor.

For simplicity let us assume that the load of the network does not consume any energy from the network (then we have $i_5 = 0$ for the output current). Then the only observed output variable is y, which is identical to the voltage u_2 on the capacitor C_2; we get

$$y(t) = u_2(t)$$

as the output equation.

Consequently the set of differential equations above gives a complete description for the input–output behavior of the network.

In systems theory it is usual to write differential equations and the output equation in the following vector form:

$$x' = Fx + Gu,$$
$$y = Hx,$$

where x is given by the vector

$$x := \begin{bmatrix} x_1 \\ x_2 \\ x_3 \end{bmatrix},$$

with

$$x_1 := u_1, \qquad x_2 := u_2, \qquad x_3 := i_2,$$

and

$$F = \begin{bmatrix} \frac{1}{C_1} & 0 & 0 \\ 0 & -\frac{1}{C_2(R_1+R_2)} & -\frac{R_1}{C_2(R_1+R_2)} \\ -\frac{1}{L} & -\frac{R_1 R_2}{L(R_1+R_2)} & \frac{R_1}{L(R_1+R_2)} \end{bmatrix},$$

$$G = \begin{bmatrix} 0 \\ \frac{1}{C_2(R_1+R_2)} \\ \frac{R_2}{L(R_1+R_2)} \end{bmatrix}, \qquad H = [0 \quad 1 \quad 0].$$

The electrical network and the associated environment as described by the input voltage u and the output voltage y now have a mathematical form (F, G, H) which is suitable for the application of linear systems theory in the classical sense. In engineering modeling it is known that many problems associated with mechanical, hydraulic, or pneumatic systems can be described by networks similar to electrical networks. As a consequence, they result in the same mathematical form (F, G, H) and can be treated by (classical) linear system theory.

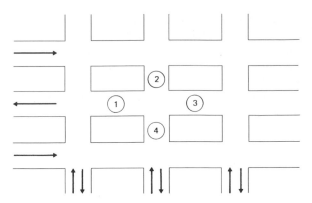

FIGURE 3 An urban traffic network.

D. Transportation Networks

In Fig. 3 we display a section of a typical urban street network. The arrows indicate the allowable directions of traffic flow within the network. For purposes of analyzing traffic flow through such a network, it is convenient to represent it abstractly by the directed graph (diagraph) in Fig. 4. Here nodes 5 and 6 have been added to account for trip initiation or termination. An arc connects nodes i and j if it is possible to pass from node i to node j in a trip of at most one block.

If costs c_{ij} are assigned to each arc (i, j), representing the travel time, say, between node i and node j, a number of questions related to the assignment of network traffic, the regulation of traffic signals, bottleneck intersections, and so forth can be approached through the digraph model in Fig. 4.

In N, D, O, E terms, the foregoing transportation network might be decomposed as follows:

N is the set of streets, together with the allowable directions of traffic flow and the time of traverse along each link.

D is the assignment of automobile trips to the streets of the network by, say, regulation of the traffic signals.

O is the measured traffic flow along each arc of the network or, equivalently, the measurement of traffic passing through each intersection.

E is the traffic flowing into and out of the system through nodes 5 and 6.

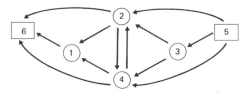

FIGURE 4 Graphic representation of the urban traffic network in Fig. 3.

Examination of the preceding examples led to the observation that each system model exhibits the characteristic features of inputs (decisions), outputs (measurements), and states. The explicit separation of variables into these categories forms the cornerstone of modern mathematical modeling and distinguishes the current view of modeling from an earlier, semiarchaic approach pioneered in operations research and mathematical programming, in which all variables are treated equally, with no explicit acknowledgment of their individual roles in the problem. We continually emphasize the importance of this point throughout this chapter.

II. TYPES OF MATHEMATICAL DESCRIPTIONS

A. General Formal Model Components and Transformations

We distinguish five general types of formal models: the black-box type, the generator type, the dynamics type, the algorithm type, and the network type. A black box describes the model by its interaction with the environment. In the most simple case a black-box model consists of a set of variables which describe the attributes of the model that are essential for interaction with the environment. These variables may depend on the kind of space or time set in which the model is embedded (space variables, time variables). More specific examples of black-box models are given in cases where the variables have a well-defined value set (analogue variables, digital variables) or possess a direction of their interaction with the environment (input variables, output variables).

A generator reflects local laws which are valid for the state variables of a model. Specific generators are given, by systems of differential equations, which reflect fundamental physical laws of nature. For models with discrete time variables difference equations and finite-state machines are well-known models of the generator type. Other generator-type models are given by discrete-event systems or by marked petri nets.

Dynamics (or dynamical systems) describe the global change of states. Since a state reflects the most important properties of a model, global changes of states are very often an essential part of a desired solution. This type of model therefore deserves great interest. In many cases a dynamical system is reached from a generator-type model by "integration."

Algorithms consist basically of operations of computation (the processing part of an algorithm) and of an organizational scheme (the control part) which determines at which step a certain operation has to be applied. Algorithms must possess a well-defined set of initial input data

to specify the starting position and a well-defined stop rule to finish computation. Algorithms obviously have a wide field of application in science and engineering. Besides computing the value for a function, they are equally important for providing a protocol for the cooperation of different component models of an overall system. In this sense they determine the control structure for a network model (discussed next) which uses a discrete-event type of coupling.

Networks are model types which consist of components (determined by the model specifications introduced above) together with a set of coupling rules which interface the components. Networks therefore define a model type which possesses a topology. A network type of model is also called "structured." Different specializations of networks are possible. An important example is recursive networks, where each component is again allowed to be of the network type. Recursive networks give cause to hierarchical multilevel models. The counterpart of recursive networks is, in some respects, flat networks, where each component is considered to be atomic (not decomposable).

The different types of models available constitute a conceptual framework which can be of great help in model building. The general types of models which we have introduced, together with the known specializations of them, allow us, as part of the problem-definition phase, to construct a model (in the first step). Usually this model will not yet be suited for application in the sense that the computation of the wanted solution is directly derivable from it. Generally it will be necessary to take more steps and "process" the model to other representations.

This idea leads to the concept of model transformation. If $\mathbf{M}$ denotes a model, then a model transformation (which fits to $\mathbf{M}$) is an operation T which maps $\mathbf{M}$ to another model, $\mathbf{M}' = T(\mathbf{M})$. $\mathbf{M}'$ is assumed here to have the same ability for problem solving as $\mathbf{M}$. However, $\mathbf{M}'$ should already be somewhat better suited for computation of the solution than $\mathbf{M}$. In general it will be necessary to apply a sequence of model transformations $T_1, T_2, \ldots, T_n$ (starting from the initially constructed model $\mathbf{M}$) to reach, after n steps with $\mathbf{M}' = T_n(T_{n-1} \ldots (T_1(\mathbf{M})) \ldots)$, a model from which an algorithm to compute the desired solution can be directly derived. Finding the right type of model transformations T is an "art" rather than a science. Furthermore, even when a large set of such transformations is known to the modeler, it requires expert knowledge to select the proper one to form the right sequence.

B. Input–Output Descriptions

Closely related to the black-box type of description is that in which we describe the system inputs and outputs by elements in certain spaces, Ω and Γ, say, and define a map

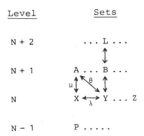

FIGURE 5 Hierarchical levels of sets and relations.

$f: \Omega \rightarrow \Gamma$, which associates inputs with the corresponding outputs. Such a description differs from the black-box type only in that additional algebraic or topological structure (or both) is usually imposed on the sets Ω and Γ, depending on the application. Most commonly, Ω and Γ are assumed to be finite-dimensional vector spaces of some sort, with f a *linear* map. Such is the case, for example, in the so-called input–output models in the Leontief theory of global economic processes (Fig. 5).

Unfortunately, both the sets–relations and the input–output types of mathematical descriptions, though of considerable value in analyzing certain structural and connective features of large systems, are somewhat deficient in dealing with dynamic considerations. Furthermore, because these system descriptions are basically phenomenological, as expressed through the map f, such models are inherently limited in their predictive powers; that is, they offer no real *explanation* of the means by which inputs are transformed into observable outputs. Thus, the need arises for a more detailed description accounting for the "inner workings" of the system under study.

C. Potential Functions

Occupying an intermediate position between purely phenomenological descriptions and detailed internal descriptions of system behavior are potential (or energy) function descriptions, which have as their basis the teleological principle that a system's dynamic is such that the system "moves" to a minimum of a suitably defined energy function.

Such models, of course, have a long tradition in classical mechanics, arising from the well-known variational principles of Fermat, d'Alembert, Hamilton, Lagrange, and others. Considerable ingenuity, imagination, and wishful thinking have been expended in recent years in an attempt to develop corresponding variational principles for more general processes occurring in biology, ecology, and the social sciences. The basic problem, of course, is to find some invariants of motion for such processes. Various thermodynamic arguments interspersed with concepts from information theory have also been employed in this regard. Perhaps surprisingly, there has been some limited

success in such modeling efforts, with interesting results having been reported in population dynamics, cell differentiation, and chemical reactions.

Mathematically, a potential function description of a process assumes that there exists a function $V(x_1, x_2, \ldots, x_n)$, where the x_i are microscopic system variables, such that the equilibrium states x^* of the process are given by the set $M = \{x: \partial V/\partial x_i = 0\}$. Dynamically, this means that the transient motion of the system variables is described by the set of differential equations

$$dx_i/dt = -\partial V/\partial x_i, \qquad i = 1, 2, \ldots, n,$$

or, more compactly,

$$dx/dt = -\nabla V.$$

Thus, we see that the existence of a potential function V induces a dynamic on the system. The converse question, namely, given a dynamic

$$\dot{x} = f(x),$$

whether a potential function V exists such that

$$f(x) = -\nabla V(x),$$

is of some importance also, especially in view of the dependence of Thom's theory of catastrophes on such "gradient" systems. The answer to the above question is provided by the following simple test. The system $\dot{x} = f(x)$ is a gradient system if and only if the Jacobian matrix $J(x) = \partial f/\partial x$ is symmetric, that is, $\partial f_i/\partial x_j = \partial f_j/\partial x_i$ for all x in the region of interest.

D. Internal Descriptions

In the following we discuss some specific descriptions of formal models which can be based on generators, dynamics, or algorithms as formal model types.

Passing from a *local* dynamic induced from a *global* system potential function, we next consider a system description based entirely on local interactions, a so-called state variable model. In continuous time, such a model takes the form of a system of differential equations

$$\dot{x} = g(x, u, t), \qquad x(0) = x_0,$$

where $x(t) \in R^n$ is the system state, whereas $u(t) \in R^m$, is the vector of system inputs. Usually, there is also an output $y(t) \in R^p$, generated by the states (and possibly the inputs), given as

$$y(t) = h(x, u, t).$$

The critical point here is the way in which a mathematical artifice, the state vector x, has been introduced as a vehicle to mediate between the inputs and the outputs.

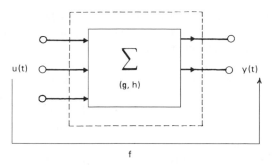

FIGURE 6 Black-box model of the system.

The importance of this observation cannot be overemphasized since it is precisely the introduction of the notion of state that enables an internal description to provide a predictive model of system behavior. Schematically, we have the "black-box" situation in Fig. 6, with the system input–output model being described by the map f. The internal system description is provided by the two maps (g, h), which together give an "explanation" of f in the sense that the input–output behavior of Σ generated by (g, h) agrees with that of f. The question of how to determine (g, h), given f, is discussed in the next section.

The attraction of internal system models is quite clear: if the functions (g, h) can be obtained on the basis of physical laws and observed data, the internal description allows us to predict the future behavior of the process as a function of the inputs (decisions) applied. In short, the functions (g, h) describe the internal "wiring diagram" of the system Σ and uniquely determine its future outputs, given the current state and future inputs.

For a number of technical reasons, the problem of determining an internal model, given an input–output map f, is complicated when the problem state space is infinite (even if finite dimensional), except in certain cases where special structure (e.g., linearity) is present. In addition, there are a number of practical situations in which it is natural to consider a *finite* state-space model, as, for example, when modeling the workings of a digital computer. Such considerations lead to finite-state descriptions of dynamic processes. The usual ingredients of a finite-state description are U, a finite set of admissible inputs; Y, a finite set of admissible outputs; Q, the finite set of states; $\lambda: Q \times U - Q$, the next-state function; and $\delta: Q \times U \to Y$, the output function.

Since the finite sets U, Y, and Q have no interesting topological structure, the analysis of a system described in the above terms is a purely *algebraic* matter, relying heavily on the theory of finite lattices, which are given by certain dynamic-preserving equivalence relations on the state space. As computational considerations ultimately force us to reduce all descriptions of systems to the above terms, it is of considerable importance to understand as much as possible about the underlying structure and behavior of such finite-state descriptions. The famous Hartmanis–Stearns decomposition method provides a starting point for analyzing the inherent structure of the finite-state models, because it ensures the existence of certain coordinatizations of the state set Q, which are advantageous for computation of the action of system inputs on the states.

E. Structural Descriptions

In mathematics the existence of structure allows the development of a theory which can serve as a basis for solving problems. Similarly in systems theory a structural description enables the "problem solver" (a designer or analyst dealing with real-life problems in engineering and science) to develop effective methods for the different tasks of model transformation (model reduction, model decomposition, development of effective algorithms in dealing with model problems).

In engineering the concept of the "architecture" of a real system is strongly related to the general idea of structure. In the simplest case the architecture of a system consists of components which are coupled together to form a network. In systems theory, similarly, of such networks, complex models are "structured" into formal models which are networks. The components are allowed to have the different descriptions which we discussed above. The coupling of such formal models has to be defined by compositions in the mathematical sense. Examples are given here by the concept of cascade composition or parallel composition of finite-state machines or linear differential systems or by composition of input/output descriptions by "joining" input and output variables to form a network.

Besides such "bottom-up" methods for defining the structure of a formal model, in systems theory it is also very important to develop "top-down" means for structuring formal models by decomposition methods. The basic concept for decomposition of formal models is given by morphisms which map a formal model to a "congruent" part that reflects certain properties. A decomposition is reached in the case of a collection of parts derived by morphisms which cover all the properties of the original model. The repetition of such morphism operations and related decomposition operations results in a refinement hierarchy (a multistratum hierarchy in the sense of Mesarovic–Takahara). In the case that an order relation between the components of a decomposition is defined, we get a decision hierarchy (a multilayer hierarchy in the sense of Mearovic–Takahara). The effective computation of structural descriptions of models of this kind is considered to be of central interest in systems theory.

F. Operations Research Problems

As already noted, the typical type of mathematical description arising in game theory, mathematical programming, decision analysis, and other fields, which we collectively term operations research, generally consists of an exhaustive listing of all variables relevant to the system at hand, an account of various identities, inequalities, and constraints existing between the variables, and the presentation of a cost function involving some subset of the variables. A fundamental conceptual deficiency in such a description is the lack of discrimination among inputs, outputs, and states. Although by *a posteriori* analysis one can usually separate the original variables into the appropriate classes, the fact that such a separation has not been carried out before the analysis of the process prejudices, in our view, the entire methodological approach to the problem.

As an illustration of such a modeling approach, consider an elementary linear programming problem. A certain factory manufactures two products, gadgets and gizmaccis. In each case, first the product is processed on a cutting machine, then a hole is drilled into it on a drill press. The times required for these operations, the total time available per week, and the profit per gadget or gizmacci are shown in the following tabulation:

Machine	Gadgets	Gizmaccis	Time available
Cutter	3	5	15
Drill press	5	2	10
Profit per unit	5	3	—

How can the manufacturer maximize the profit?

Introducing the variables x_1 and x_2 (x_1 = the number of gadgets to produce and x_2 = the number of gizmaccis to produce), we have the relations

$$3x_1 + 5x_2 \leq 15,$$

$$5x_1 + 2x_2 \leq 10,$$

where $x_1 \geq 0$, $x_2 \geq 0$, with the profit being

$$5x_1 + 3x_2.$$

Elementary graphic means or a routine application of standard algorithms yields the optimal solution $x_1^* = 20/19$, $x_2^* = 45/19$. (Here, for simplicity, we assume that gadgets and gizmaccis are continuously divisible.)

Although the preceding model certainly deals with the problem as stated, the distinction of what constitutes the decisions, the states, and the outputs is clearly blurred, at best. One might ask, "So what?" The reply is that by neglecting the basic distinction among variables, it is very difficult to incorporate dynamic considerations *naturally*

into the model, and, what is worse, without the concept of a system state, it is next to impossible to consider feedback decision-making or stochastic effects.

As an illustration of how the preceding problem could have been formulated in more system-theoretical terms, let us introduce the variables u, the amount of time available on the cutting machine, and v, the amount of time available on the drill press, and the function $f_n(u, v)$, the profit obtained when u units of cutter time and v units of drill press time are available, n types of items are to be produced, and an optimum decision rule is employed. Then it is easy to see that

$$f_2(u, v) = \max_{\substack{0 \leq 5x_2 \leq u \\ 0 \leq 2x_2 \leq v}} [3x_2 + f_1(u - 5x_2, v - 2x_2)],$$

$$f_1(u, v) = \max_{\substack{0 \leq 3x_1 \leq u \\ 0 \leq 5x_1 \leq v}} [5x_1].$$

Computation of the functions f_1 and f_2 for all values of (u, v) in the range $0 \leq u \leq 15$, $0 \leq v \leq 10$ enables us to solve a *family* of problems for all cutting and drilling times in the indicated ranges. Furthermore, the concept of a system dynamic is introduced through the idea of manufacturing gadgets, followed by gizmaccis. Thus, the solution proceeds one item at a time; there is no attempt to compute all production levels in one fell swoop. This dynamic approach is a direct consequence of introducing the state variables u and v along with the decision variables x_1 and x_2.

The disadvantage of the dynamic programming (DP) formulation just given is that the computational algorithms are not nearly as efficient as for the previous linear programming (LP) setup, which can employ the simplex method. However, if the costs and constraints are nonlinear or if stochastic effects enter, the DP formulation comes into its own. The point here is that it is a fundamental mistake to swear religious adherence to any one orthodoxy: flexibility in modeling must be maintained if the best results are to be hoped for.

III. LOCAL CONSIDERATIONS

Once a particular mathematical description has been chosen for a given process, a number of important system-theoretical questions involving both local and global phenomena present themselves. The manner in which these questions appear in the model depends, of course, on the type of description employed; however, the abstract phenomena are relatively invariant under a change of description, so we attempt to discuss the main issues in as context-free a manner as possible in the next two sections.

One set of system phenomena that any model must cope with is issues that are best termed local system properties.

Here, by *local* we mean that the phenomena either manifest in some restricted region of the system model or, in some meaningful sense, can be analyzed by considering only interactions between system components in the immediate neighborhood of a restricted piece of the entire system. On the other hand, global aspects require consideration of the entire system for their analysis; no smaller subsystem will suffice. We consider global properties later. For now, let us examine some of the local considerations more closely.

A. Stochastic Effects

Typically, the influence of the external part of a system, which we do not fully understand or cannot account for in the system descriptions sketched above, is often assumed to be a random perturbation whose effect is locally felt on the system.

For illustration, assume that we have modeled a process by an internal description as

$$(\Sigma)\, x(t + 1) = g(x(t), u(t), t),$$

$$y(t) = h(x(t), t).$$

To account for the fact that the functions g and h may not be known exactly, this model can be replaced by

$$(\Sigma)\, z(t + 1) = g(z(t), u(t), t) + r(t),$$

where $r(\cdot)$ is a stochastic process with appropriate statistical properties. Here the local effect of the noise r is felt by the system in state $z(t)$; that is, r acts as a perturbation in a local neighborhood of the state $z(t)$. Such a disturbance is in the E part of the system Σ.

Another manner in which stochastic effects locally influence Σ is through the D component. Here we assume exact knowledge of the dynamics g and the observation function h, but the theoretically desirable control law $u(t)$ cannot be applied because of computational inaccuracies or otherwise. Similarly, it may not be possible to measure the system output with complete precision owing to noise corruption in the measuring apparatus. These are, again, local effects in the sense that they affect only a neighborhood of a point in the control or output space.

With a more elementary level of system description, the stochastic features assume a somewhat different form. For instance, if the description is sets and relations, there may be uncertainty as to whether a particular pair $(x_i, y_j) \in \lambda$, or if an input–output model is used, uncertainties in the map f may arise. What is important here is the fact not that stochastic influences appear, but that their influence is exerted at a point of the system, not throughout the entire system simultaneously. This is the essence of what we mean by a local effect.

B. Constraints

Restrictions on the system inputs or states (or both) come in two varieties: local, in which the immediate decision or state is constrained to lie in some admissible region; and global, in which some overall function of the control or state must remain bounded within given limits. We illustrate both types.

Consider an internal system model

$$\dot{x} = x + u, \qquad x(0) = x_0$$

and assume that we wish to transfer this system to the origin. Further, assume that the magnitude of controlling action is limited by

$$|u(t)| \le M.$$

A limitation of this sort may arise as a result of considerations such as maximum stress factors, finite resource availability, or maximum tolerable unemployment rates. In any case, the constraint *locally* restricts the amount of control action that can be exerted to modify the system's dynamic behavior. Here *local* is interpreted in the temporal sense, since the magnitude limit M must be obeyed at each time instant t.

Now consider the same problem with the local constraint replaced by the condition

$$\int_0^\infty u^2(t)\, dt \le K.$$

Here we have an example of a global constraint. There is now no restriction on the instantaneous value of the control u, only the condition that the total control energy expended in transferring the system to the origin remain bounded by K. Thus, we have now traded a constraint on the local *value* of a control function for a condition on the entire function u itself.

If the system description is of the sets and relations type, the constraints are almost automatically built into the schema through the I/O relation. This would be regarded as a global constraint because it restricts those elements of the basic sets X and Y that can be related. In other descriptions, as in operations research, the constraints enter in both a local and a global form, as indicated in the LP example in the preceding section, where we had the local nonnegativity constraints $x_1 \ge 0$, $x_2 \ge 0$, with the global resource constraints on the available cutting machine and drill press time.

C. Time Lags

A fundamental principle of large systems is that *control takes time*. Thus, the theoretical assumption implicitly built into most internal models, that the system output measurement and determination and application of the control action take place coterminously, must be regarded as only a convenient mathematical idealization in real problems. Fortunately, such an approximation works well in many cases, especially in classical physics and engineering. However, in decentralized processes with many components and decision makers, the "simultaneity" hypothesis can no longer be accepted, and explicit account must be taken of the timelag effect. This is particularly true in models arising from social science situations.

As an illustration of the manner in which time lags can affect a control law, consider the internal model

$$\dot{x} = -2x(t) + u(t),$$

$$x(0) = 1, \qquad 0 \le t \le 2,$$

where it is desired to choose the input $u(t)$ so that the terminal state $x(2)$ is as small as possible, subject to the constraint

$$|u(t)| \le 1, \qquad 0 \le t \le 2.$$

It is an elementary exercise to ascertain that the optimal choice is

$$u^*(t) = -1, \qquad 0 \le t \le 2.$$

Now consider the same problem with the sole change that the system dynamics have a unit time lag in the state; that is, the dynamics are

$$\dot{x} = -2x(t - 1) + u(t),$$

with the initial condition now being

$$x(t) = 1, \qquad -1 \le t \le 0.$$

It is a somewhat less elementary, although straightforward, exercise to determine the optimal control for this problem as

$$u(t) = \begin{cases} 1, & 0 \le t \le \frac{1}{2} \\ -1, & \frac{1}{2} \le t \le 2 \end{cases}.$$

Thus, we find that the introduction of a time lag has resulted in a *qualitative* change in the structure of the optimal decision by introducing a switching point from maximum control to minimum control at $t = \frac{1}{2}$. Furthermore, we note that this change in control strategy is a local effect in that it is applied to the system when it is in its state $x(\frac{1}{2})$ (which happens to be $x = \frac{1}{2}$ in this case).

It is a common feature of processes involving time lags that the presence of a delay may cause the appearance of self-exciting oscillations, then exertion of too much control, followed by complete instability of the system. Such undesirable phenomena can be avoided only by the application of inputs timed in such a manner as to counteract the influence of the aftereffects on the state due to the delay. This is an important aspect of controlling large, complex

systems and one to which considerable theoretical work is currently being directed.

IV. HOLISTIC ASPECTS OF SYSTEM STRUCTURE

In contrast to the local issues involving system structure and behavior in a restricted region of its definition, we must also consider aspects of the system that cannot be confined to any particular part of the structure but that are properties of the entire system. Whereas the preceding section dealt with topics associated with a "reductionist" view of the system, this section looks at the system from a "holist's" vantage point, taking into account properties possessed by no single component or subsystem of the total system. We have in mind such system properties as conservation–dissipation laws, hierarchical structure, singularities, and process time scales. Let us examine each of these global features in more detail.

A. Conservation–Dissipation Laws

A good part of mathematical physics is anchored by the laws of conservation of mass, energy, charge, baryon number, and so on. These are all restrictions imposed on the global behavior of physical processes. On the other hand, equally basic principles involving dissipation effects also pervade classical physics. Here we refer to the increase in entropy in closed systems as dictated by the second law of thermodynamics and the transformation of mechanical energy to heat via various functional effects. Again, such dissipative principles are constraints to which the global dynamic behavior of a process must adhere. The conservation–dissipation laws impose no restriction on the local behavior of a process; they simply say that the total motion must be such that certain functions of that motion are invariant, or nondecreasing.

The search for extensions of the conservation laws of classical physics for more general systems has been the topic of much study in the systems literature. Generally speaking, the farther the given system is from a classical physical process, the more fanciful the proposed conservation principles seem. Nonetheless, interesting results have been obtained in some areas. For instance, for the well-known Lotka–Volterra predator–prey dynamics

$$\dot{x} = (a - by)x,$$

$$\dot{y} = (cx - d)y,$$

it can be shown that the following function is constant along solution curves:

$$H(x, y) = cx + by - d \log x - a \log y.$$

The constancy of H imposes certain global structural features on the dynamics of the system, for example, no limit cycles, each trajectory is a closed orbit, and so on.

In a quite different direction, R. Ashby developed his law of requisite variety in an attempt to introduce thermodynamic considerations into system theory. The basic idea is to define the variety of a finite set A to be $\log_2(\text{card } A)$, where card A is the number of elements in A. Then if Ω and E are, respectively, the set of inputs (decisions) and external disturbances for a given system, the law of requisite variety states that only variety in Ω can force down variety due to E. In other words, if the variety in the control is $\log r$ and that of E is $\log c$, the variety in the output is *at least* $\log r - \log c$. An account of the derivation of this basic rule, together with its connections to entropy and information theory, can be found in the works of R. Ashby.

The preceding examples show that conservation principles can yield important information on the structure and behavior of a given system if we are either clever or lucky enough to find them. Regrettably, as yet there appears to be no uniform procedure for generating such laws for general classes of processes.

B. Hierarchical Structure

Almost all large systems—biological, engineering, business, economic, political, or sociological—share the property of hierarchical organization. Decision makers exist on all levels, communicating instructions and receiving information from subordinate levels. From a modeling stand point, we are interested in such questions as how the hierarchical structure influences the flow of information throughout the system, what effect the hierarchical organization has on the system's ability to react to external disturbances, and the sensitivity of the system output to changes in the connective structure of the hierarchy.

The hierarchical organization of a given system is clearly a global feature that cannot be analyzed by local tools. In mathematical system studies, it appears that ideas taken from algebra and geometry will prove most effective in studying questions related to system hierarchy. As indicated earlier, techniques from algebraic topology can be employed in a sets–relations description to study hierarchical organization quantitatively. It is tempting to conjecture that, once the global tools have mapped out the over all hierarchical structure and connective pattern for a system, local tools from analysis can be brought to bear on considerations such as system stability. However, since we do not wish to begin dreaming in print, we leave this as only a speculative possibility and move on to other global system properties.

C. Singularities

For systems modeled by differential (or difference) equations, perhaps the most noticeable feature is the set of points in state, parameter, or control space at which qualitative changes in system behavior occur. For instance, consider the internal model

$$\dot{x} = g(x, a), \qquad x(0) = c,$$

where a is a vector of parameters. The steady-state equilibrium of the system will be the state $x(\infty) = x^*(c, a)$, where we explicitly indicate the dependence of x^* on the parameters a and the initial state c. Furthermore, $x^*(c, a)$ will be a solution of the equation

$$g(x, a) = 0.$$

For *fixed* a and c, any equilibrium x^* can be regarded as a singularity of the process. This is the view taken in classical stability theory. On the other hand, we may also consider the map

$$x^*: (c, a) \rightarrow R^n,$$

$$(c, a) \rightarrow x^*(c, a),$$

where a and c are regarded as variables. In this setting, those values of c and a at which the map x^* is discontinuous or multivalued are also considered to be singularities of the system, although of a very different type.

In either of the above cases, it can be shown that the singularities of the system determine to a large degree its entire dynamic behavior, and it is only a small exaggeration to think of the transient motion as being forced on the system by the structure of its singularities. In addition, one should note that no local coordinate changes can remove singularities of the above type: they are global invariants of the process. Thus, it is of the utmost importance to understand the number and nature of all system singularities if we wish to control any system in an effective manner.

D. Time Constants

An almost-universal feature of large systems is that the variables seem to separate into a fast-time/slow-time dichotomy. This qualitative distinction between variables is so pervasive that, for most engineering problems, the "fast" variables are usually considered state variables, whereas the "slow" variables are generally treated as parameters. Here, again, we see a system property that cannot be localized to a particular region of variable definition.

In mathematical modeling, it is of some importance to isolate the slow variables since, depending on the application, it may be possible to "factor" them out of the problem, at least in a computational sense. For example, if there are $n + m$ variables that describe the evolution of the system and m of them are slow variables that can be regarded as parameters, we have the option of considering a single problem with a state space of dimension $n + m$, or m problems of dimension n. In many cases, the first version may be computationally intractable, or at least impractical. Analysts familiar with DP procedures for control processes will recognize the fast–slow separation of variables as one way in which we may hope to lift the "curse of dimensionality." The catastrophe theory applications of Thom, Zeeman, and others are also a good illustration of time-constant exploitation wherein the slow variables are regarded as inputs (or decisions), whereas the fast variables are the observed outputs. All intermediate-speed variables (the states) are suppressed in what is essentially an input–output theory.

Having now had a look at some of the main local and global aspects of large-scale systems, we turn to a more extensive discussion of the basic system-theoretical questions the mathematical model must address.

V. BASIC QUESTIONS AND PERSPECTIVES OF SYSTEM THEORY

Models are constructed because there are aspects of the system we do not understand and wish to explore. What type of questions can models of the foregoing type address? and To what extent do system-theoretical tools enable us to speak with confidence about the connection between the system model and the system? These issues lie at the heart of the "systems approach," and it is possible to provide only a partial glimpse of the overall situation in such a brief article. So in this section we sketch a broad array of questions that can be approached using mathematical models. This overview should provide the needed perspective for the reader to pursue the technical literature with some confidence.

The basic questions of theoretical and applied system science, when broadly interpreted, are surprisingly few in number, and, in one way or another, all center about the interaction of the system with its environment. For purposes of exposition, it is convenient to group the questions into the following main categories.

1. *Reachability and controllability*: The identification of system behaviors that can be achieved by the application of admissible inputs
2. *Observability and detectability*: The determination of system behaviors that can be identified by measured physical outputs
3. *Realization and identification*: The generation of the class of models that could "explain" a given set of input–output data

4. *Optimality*: The determination of the efficiency with which a system can perform a specified function, subject to physical and theoretical operating constraints
5. *Stability and sensitivity*: The calculation of the way in which errors and disturbances affect the equilibrium behavior of a system

We now examine each of these categories in the context of specific types of mathematical descriptions. As we proceed, it will become evident that the type of mathematical description employed will strongly flavor the precise technical form of the question, but the invariant essence of the problem will be sufficiently clear as to leave no doubt about the category in which the question belongs.

A. Reachability and Controllability

Abstractly, the question of reachability can be formulated in the following terms. Given a set Ω of admissible inputs and a set X of system states, the transition map ϕ of the system associates a particular state with each element $\omega \in \Omega$, assuming that the system starts in some agreed-upon initial state x_0 (usually taken to be the origin if X is a vector space). Thus, the map $\phi: \Omega \to X$ determines the effect of the input ω on the system, transferring x_0 to the state $\phi(x_0; \omega)$. The problem of reachability is to characterize the range of ϕ. In the event that the map ϕ is *onto*, that is, the range of ϕ is all of X, we say that the system is completely reachable; that is, any state in X can be "reached" by the application of some admissible input from Ω. The corresponding problem of controllability is similar: Given that the system is in a state $x \in X$, does there exist an input from Ω that transfers the system to x_0? If so, x is called a controllable state. If all $x \in X$ are controllable, the system is said to be completely controllable. (*Note*: The preceding definitions are incomplete in the sense that the initial and terminal time should also be taken into account. We omit this aspect for two reasons. First, it is relevant only for nonautonomous systems described in internal form, and, second, it requires an extended notation that is needlessly elaborate for our current needs. The technical treatments of mathematical system theory cited in the Bibliography will supply the reader with all relevant definitions and details.)

Certainly, the best-structured concretization of the above abstract setup is for a linear system given in internal form. Here the system dynamics are

$$\dot{x} = Fx + Gu, \qquad x(0) = x_0 = 0,$$

and it is a well-known result that the set of reachable states $\mathcal{R}$ is precisely the set of elements in R^n spanned by the vectors $G, FG, F^2G, \ldots, F^{n-1}G$, that is,

$$\mathcal{R} = \mathrm{span}\{G, FG, F^2G, \ldots, F^{n-1}G\}.$$

Furthermore, a consequence of linearity and continuous time is that, if a state $x \in \mathfrak{R}$, then x can be reached in an arbitrarily short time. As an illustration of this result, consider the system in R^4 given by

$$F = \begin{bmatrix} 3 & 1 & 0 & 0 \\ -4 & -1 & 0 & 0 \\ 6 & -1 & 2 & 1 \\ -14 & -5 & -1 & 0 \end{bmatrix}, \qquad G = \begin{bmatrix} 1 \\ 0 \\ 0 \\ 0 \end{bmatrix}.$$

Here the set

$$\mathfrak{R} = \text{span} \left\{ \begin{bmatrix} 1 \\ 0 \\ 0 \\ 0 \end{bmatrix} \begin{bmatrix} 3 \\ -4 \\ 6 \\ -14 \end{bmatrix} \begin{bmatrix} 5 \\ -8 \\ 20 \\ -28 \end{bmatrix} \begin{bmatrix} 7 \\ -12 \\ 50 \\ -50 \end{bmatrix} \right\} = R^4$$

since the vectors G, FG, F^2G, and F^3G are linearly independent. Hence, this system is completely reachable.

For more general processes given in internal form, the situation cannot usually be resolved by linear algebraic techniques alone. For instance, the reachable set of the nonlinear system

$$\dot{x} = f(x, u), \qquad x(0) = x_0,$$

with f analytic in x and u, can be characterized using techniques from differential geometry and Lie algebras of vector fields.

Systems described by finite-state machines have a reachability theory paralleling that for internal descriptions; it is usually termed *strongly connected* in the automata literature.

In the event that we have a system description of a more general type, for instance, sets and relations, we may no longer have as much structure in the sets Ω and X, and consequently, it may be somewhat more complicated to characterize reachability. Say, for example, that in a sets–relations description we take the state space X to consist of a vector Q whose components (positive integers) characterize the number of connected components that exist at dimension level q in the simplicial complex associated with the relation λ. (The notion of q connectivity is elaborated in detail in the works cited in the Bibliography.) The set Ω may consist of various modifications that one could make to the incidence matrix Λ, for example, addition or deletion of vertices, modifications of entries from 0 to 1, or vice versa. Then the reachability problem would be to ask if a prescribed structure vector Q could be obtained by admissible changes in the relation λ. This is a technical problem far different from that sketched earlier. Nonetheless, the abstract structure of the question is the same.

B. Observability and Detectability

The question of reachability revolves about what can be accomplished using admissible *inputs*. Problems of observability focus on what can be done with system *outputs*. More precisely, each state $x \in X$ of a system generates a certain output via the system output map

$$\eta: X \to \Gamma,$$

where Γ is the output set. Questions of observability deal with the issue of whether two (or more) distinct states x and x' give rise to the same output. In set-theoretical terms, we are concerned with whether the map η is 1–1. In most practical problems, it is impossible physically to monitor the entire system state. We must settle for measurements of accessible variables or aggregates such as sums of various state components. Observability properties of the system then determine whether it is theoretically possible to reconstruct the entire state from output measurements.

As one might suspect, for the constant linear system

$$\dot{x} = Fx, \qquad x(0) = x_0, \qquad y = Hx,$$

the observability question can be settled by purely algebraic means. It is an easy exercise to verify that a given state x^* is *unobservable* if and only if $\eta(x^*) = 0$. Thus, the unobservable states are precisely those elements forming the kernel of the matrix

$$[H', F'H', (F^2)'H', \ldots, (F^{n-1})'H'].$$

In other words, x_0 is unobservable if and only if it is mapped to zero by the above matrix. Otherwise, measurement of the output $y(t)$ over an arbitrarily short interval will suffice to determine x_0 uniquely.

The foregoing result strongly suggests a dual relationship between the concept of reachability and that of observability on making the transformations $F \to F'$, $G \to H'$, $p \to m$ (recall that H is $p \times n$ and G is $n \times m$). A precise duality theory (in the vector space sense) can be developed by following up this observation, and it can be seen that a system is completely reachable if and only if its dual is completely observable. Heuristically, this result is equivalent to interchanging the system inputs and outputs and reversing the flow of time.

As usual, the observability question for more general processes is not so well understood, and its very discussion would require more mathematics than we have room for here. In more general contexts, such as potential functions and sets–relations, even a precise statement of the problem remains to be formulated, although the general notion of the output map being 1–1 provides a starting point.

C. Realizations and Identification

The construction of an internal description from input–output data is the very essence of mathematical modeling.

In technical terms, this is the "realization" (electrical engineering terminology) of the data. A special subcase of the general problem is when the model structure is given and only the values of parameters within the model have to be determined by the data. This is the parameter identification problem. In either case, the objective is to provide a model that, in some sense, "explains" the observed data.

The form of the realization depends, of course, on the type of model one is attempting to obtain. Generally, we are given an external description, that is, a map

$$f: \Omega \to \Gamma,$$

where Ω and Γ are the system input and output spaces, respectively, and the task is to construct an internal model whose input–output behavior reproduces that of the map f. If f is linear, the problem is remarkably easy: there are an infinite number of nonequivalent internal models that will have external behavior identical to that of f. However, all ambiguity is removed (modulo a coordinate change in the state space X) if we further demand that the realization be both completely reachable and completely observable. Such a realization is called canonical and is equivalent to demanding that the dimension of the state space X be as small as possible.

EXAMPLE. Suppose that a single-input–single-output linear system is presented with the input

$$u(t) = \begin{cases} 1, & t = 0 \\ 0, & t > 0 \end{cases},$$

and the observed output is the sequence of natural numbers, that is,

$$y(t) = t, \qquad t = 1, 2, \ldots.$$

The problem is to realize an internal model

$$x(t + 1) = Fx(t) + Gu(t),$$

$$y(t) = Hx(t)$$

whose input–output behavior generates the natural numbers starting with a unit input. Application of standard algorithms soon yields the canonical model

$$F = \begin{bmatrix} 0 & 1 \\ -1 & 2 \end{bmatrix}, \qquad G = \begin{bmatrix} 1 \\ 2 \end{bmatrix}, \qquad H = [1 \quad 0],$$

which can easily be seen to be reachable and observable.

The realization problem becomes much more complicated once we pass out of the realm of linear theory. For some classes of nonlinear processes, some procedures exist that mimic the linear case as long as sufficient structure is present in the input–output map (e.g., multilinear or polynomial). However, almost everything remains to be done to make these methods practically operational.

If the model we are trying to generate is not an internal "differential-equation type," the realization problem enters the realm of system-theory research. For instance, in a sets–relation context, the realization problem would be that of generating the system incidence matrix Λ, given the two finite sets X and Y together with an input–output relation between them. Here it is not even entirely clear what constitutes the input–output map f, but a plausible beginning would be to take the structure vector Q mentioned above. Difficult mathematical questions then arise as to whether Q contains sufficient information to determine Λ (up to a permutation matrix).

In another direction, we might wish to determine a potential function such that observed system equilibria (the measured data) agree with the stationary states of the potential function. Depending on the setting, this is equivalent to solving the so-called inverse problem of the calculus of variations. Much work has been done in this area, but the problem is by no means completely settled.

D. Optimality

The imposition of some measure of system performance on a process changes our view of the choice of system inputs dramatically. Now, instead of choosing an input to transfer the system to some specified state, we select the input to minimize a measure of system cost. (Of course, the reachability problem for a *fixed* terminal state could be viewed as a special case of the optimality problem by introducing a distance measure from the desired state as the cost function; however, it is generally more illuminating to regard the reachability issue separately, as we have done.)

In general terms, the optimality problem is as follows. We are given a cost measure $J: \Omega \to R$, which associates a real number (the process cost) with each admissible input. The problem is to determine those inputs (controls) that yield the minimal cost. The existence and uniqueness of optimal controls for various classes of maps J and various types of internal and external dynamics have been studied for many years, and a considerable body of knowledge, termed optimal control theory, has arisen as a modern outgrowth of the classical calculus of variations, within which the results and techniques are codified. Again, the most extensive results are available for those processes described in internal form, as we now illustrate.

Consider the problem of minimizing

$$J = \int_0^T q(x, u, t)\, dt$$

over all piecewise-continuous functions $u(t)$ on $[0, T]$. Assume that the system dynamics are

$$\dot{x} = f(x, u, t), \qquad x(0) = c.$$

It has been shown that a necessary condition that any candidate optimal control must satisfy is that it yield the pointwise minimum of the system Hamiltonian

$$H(x, u, t) = q(x, u, t) + \lambda(t) f(x, u, t),$$

where $\lambda(t)$ is an arbitrary piecewise-continuous multiplier function to be determined. This is a scalar version of the famous Pontryagin maximum principle. Under convexity conditions on q, this principle can be shown to be a sufficient condition for optimality as well. With minimal analysis, it can be shown that the solution of this problem reduces to solving the nonlinear two-point boundary-value problem

$$\dot{x} = \partial H / \partial \lambda, \qquad x(0) = c,$$
$$-\dot{\lambda} = \partial H / \partial x, \qquad \lambda(T) = 0,$$

with the minimizing control $u^*(t)$ belonging to the set

$$\mathcal{U} = \{u: \partial H / \partial u = 0\}.$$

Thus, the Pontryagin principle is an updated extension of Hamilton's equations from classical mechanics. We note in passing that the same problem can also be approached via DP or even nonlinear programming methods.

In the event that the system is described by a potential function, the dynamics themselves are governed by a variational principle and we can express them as

$$\dot{x} = -\nabla_x V(x, u),$$

where $V(x, u)$ is the appropriate potential. Here we generally regard the inputs u as parameters, and the optimal control problem might be posed as the nonlinear programming problem of finding the best set of parameter values, with the above system dynamics as a constraint. However, if the inputs are functions, the Pontryagin approach sketched above could also be employed.

The more general setting of sets and relations or a graph description introduces the problem of a suitable definition of a criterion, together with the serious technical difficulties of determining the type of inputs that will optimize the chosen performance measure. The difficulty is one of a lack of continuity, a typical obstacle in combinatorial problems. Since there is no notion of "nearness" on which one can construct a variational theory, it is necessary to employ various algebraic means to attempt to isolate the best system input. Unfortunately, these methods are still in their infancy, and nothing approaching a comprehensive set of results is yet available.

E. Stability and Sensitivity

One of the most fundamental of all system-theoretical questions is that of determining the effect of changes in the model on the system structure and observed behavior. Such stability problems take on myriad forms, depending on the type of system disturbance, the observed output, the structural feature under consideration, the type of mathematical description chosen, and so forth. Here we indicate only a few of the common types of stability problems.

Consider a system whose dynamics are described by the potential function $V(x, u)$, where the components of the vector u are a set of system parameters; that is, the system evolves according to the gradient dynamics

$$\dot{x} = -\nabla_x V(x, u).$$

The equilibrium states M of such a system correspond to the critical points of the potential V and the particular location $x^* \in M$ of the equilibria depends on the vector u, that is, $x^* = x^*(u)$, where $M = \{x: \nabla_x V = 0\}$. An important stability problem is to determine those values of u such that the map $u \to x^*(u)$ is discontinuous. Such values of u are called "catastrophe" points and are the focus of the recent catastrophe theory of Thom and Zeeman.

The catastrophe theory setup is a special case of another type of stability concept, structural stability, in which one studies how changes in the system dynamics themselves influence the geometric character of the system trajectories. For example, consider the damped oscillator described by the equation

$$\ddot{u} + a\dot{u} + u = 0,$$
$$u(0) = c, \qquad \dot{u}(0) = 0, \qquad a \geq 0.$$

If $a > 0$, the phase-plane portrait of the trajectories is as in Fig. 7a. For the undamped case $a = 0$, we have the situation depicted in Fig. 7b. The equilibrium at the origin is of an entirely different topological character in the above two cases: In the first case (Fig. 7a) the origin is a focus, whereas in the second (Fig. 7b) it is a center. Thus, the *undamped* harmonic oscillator is not structurally stable with respect to perturbations in the damping coefficient a, since any departure from $a = 0$ changes the character of the system trajectory. On the other hand, the *damped* oscillator is structurally stable with respect to changes in a, since for any $a > 0$, there is a nearby value of a such that the system trajectory is still a focus. Higher-dimensional generalizations of this idea form the essence of multiparameter bifurcation theory, of which catastrophe theory is an important special case.

The most classical stability questions involve a system given in internal form

$$\dot{x} = f(x), \qquad x(0) = c,$$

where it is assumed that $f(0) = 0$; that is, the origin is an equilibrium point. If the initial state $c \neq 0$, it is of interest

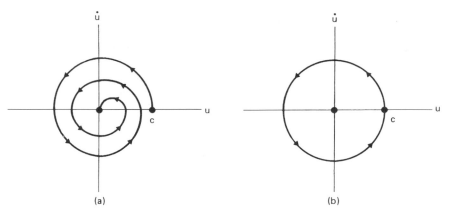

FIGURE 7 Damped and undamped harmonic oscillators.

to know if the system state $x(t) \to 0$ as $t \to \infty$ and, if so, at what rate the state approaches the origin. These are stability problems in the sense of Lyapunov, and many effective techniques exist for answering these questions and many more.

From an applied systems analysis viewpoint, perhaps the most interesting aspect of classical stability is the determination of the domain of attraction of the origin, that is, the determination of those initial states c that will eventually go to the origin. If the system dynamic f contains parameters, that is, $f = f(x, u)$, the variation of the boundary of the domain of attraction with changes in u brings us back to the catastrophe theory setting under appropriate hypotheses on the analytic structure of f.

If the basic system model is not of the internal type but is, say, a graph or simplicial complex, the stability problems are of a somewhat different sort. For instance, consider the energy demand model characterized by the graph in Fig. 8. Here a plus sign on a directed arc from node i to node j means that an increase in the value of variable i tends to increase the value of variable j, all other factors being held

constant, whereas a minus sign means that an increase in i tends to reduce the level of j. This is an example of a signed digraph. A stability question of interest in connection with such a situation is whether a unit pulse introduced into the system at a given node (e.g., population) results in the value of any variable ultimately becoming unbounded. If not, we say that the system is value stable. A related concept, called pulse stability, looks at basically the same question but with respect to the sequence of *changes* in values at a vertex from one time period to another. Both of these stability concepts can be attacked by algebraic means, utilizing the connection between the properties of a planar digraph and the properties of associated matrices.

In the more general case of a system described by a simplicial complex, the stability problems center on changes in the connection pattern induced by perturbations of vertex values or changes in the defining relation λ (or both). To illustrate, consider a pair of sets $X = \{x_1, x_2, x_3, x_4\}$, $Y = \{y_1, y_2, y_3, y_4\}$, with defining relation $\lambda \subset Y \times X$ being characterized by the incidence matrix

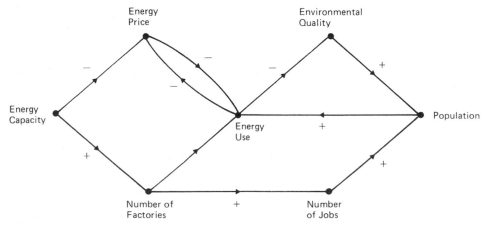

FIGURE 8 Graph model of energy demand.

FIGURE 9 Simplicial complex of the relation A.

$$
\Lambda = \begin{array}{c|cccc}
 & x_1 & x_2 & x_3 & x_4 \\
\hline
y_1 & 1 & 1 & 0 & 0 \\
y_2 & 0 & 0 & 0 & 1 \\
y_3 & 0 & 0 & 0 & 0 \\
y_4 & 0 & 0 & 0 & 1
\end{array} \cdot
$$

Decomposing this complex into its dimensional components, we find that there are three distinct components at the 0 level and one component at the 1 level. This is easily seen from the geometric representation of the complex shown in Fig. 9. The first structure vector of this complex is then $Q = (1\ \ 3)$, indicating a low level of connection at the zero-dimensional level.

A stability problem that may arise in connection with a problem of this sort is whether the components of Q remain unchanged if we vary some elements in Λ. When stated in this form, it is also clear that the stability and reachability problems are related, because we may wish to arrange the modifications in Λ to achieve the structure vector $Q = (1\ \ 1)$, which would indicate a more tightly connected system. Regrettably, a systematic methodology for answering this type of question remains to be developed.

VI. MODEL CHOICE

In the preceding sections we have presented a number of alternative descriptions for modeling applied system processes and discussed a variety of basic questions the models address. However, in the final analysis the modeler must choose one type of description, which then constrains the type of question with which the modeler can effectively deal. As a guide to the selection of a system description, we present Table I, in which the strengths and weaknesses of the model classes that have been presented here are summarized. The reader should consider the table to be only a rough guide, since in any individual case, peculiarities of the problem may require a modeling approach departing from the general guidelines in the table.

VII. COMPUTER-AIDED SYSTEMS THEORY (CAST)

Today the computer is the main tool used in realizing engineering problem-solving environments. Computer-aided design (CAD) tools exist for many engineering problem-solving tasks. They support an expert in certain specific modeling activities such as model building (e.g., drawing a specific diagram of electronic circuitry which uses predefined building blocks) and simulation as part of model application (e.g., computation of the I/O behavior of a specific electronic circuit). CAD tools are usually tailored to a specific domain of application (for example "digital electronic circuits" or "mechanical machinery

TABLE I Relative Merits and Shortcomings of Model Types

Model type	Strengths	Weaknesses
External (input–output)	Deals only with observed data; does not require introduction of state variables, thereby reducing the computational burden	Provides no explanatory mechanism or prediction procedure; reachability and observability questions difficult to formulate
Internal (state variable)	Explicitly postulates a mechanism whereby inputs are transformed into outputs; highly developed mathematical theory for analyzing the most basic system questions; not difficult to naturally incorporate global system constraints such as conservation laws, nonlocal effects, and connectivity structures	Requires detailed knowledge of dynamics and the way system inputs and outputs are processed; computational burden very high unless special structure (e.g., linearity) present; difficult to model nondynamic situations, e.g., art, music, game playing
Potential functions	Easy to synthesize local dynamics from global variational principle	Difficult to justify in cases where no apparent variational principle exists; difficult to formulate meaningful reachability and observability problems
General	Can be employed in very general settings; easy to analyze overall connection pattern between system components; readily accommodates hierarchical decomposition of system structure; computational aspects relatively straightforward	Difficult to incorporate dynamic effects in a natural way; provides little predictive power
Operations research (mathematical programming)	Can manage very large problems if sufficient structure (e.g., linearity) is present; computational procedures well advanced; requires relatively modest mathematical background to understand and employ	Fails to distinguish among inputs, outputs, and states; makes no distinction between open-loop and feedback control; no natural way to include stochastic and adaptive aspects; notions of reachability and observability nonexistent

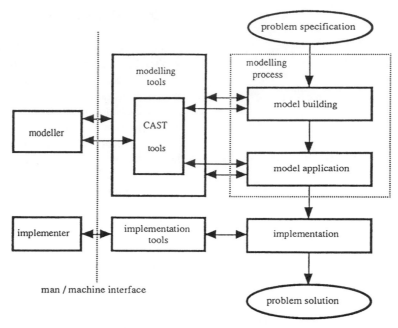

FIGURE 10 CAST tools in modeling.

components") and automate classical engineering approaches such as engineering drawing and investigation of a design by models of reduced size and reduced functionality (e.g., a wooden model of a bridge). CAD tools are supplemented by computer-aided manufacturing (CAM) tools, which help to automate implementation.

The field of CAD tools is still expanding very rapidly. Besides expanding into new technological areas, for example, microsystems technology, there is a tendency to develop CAD tools for early steps in modeling such as problems definition (building a model of the first kind) and model application (which is based on theoretical results) available for specific model types.

In current developments of CAD tools in engineering, systems-theory instruments are needed to support modeling. CAST has as its goal the supplementation of current and future CAD tools by systems-theory software (CAST tools), which can be applied in model building and in model application (Fig. 10). CAST tools allow the design engineer to apply different theoretically based model transformations as part of the problem-solving process. They allow for application of the theoretical knowledge needed to reach optimal results in modeling complex systems. They are the proper means of realizing a systems theory-instrumented modeling philosophy.

SEE ALSO THE FOLLOWING ARTICLES

CONTROLS, LARGE-SCALE SYSTEMS • CYBERNETICS AND SECOND ORDER CYBERNETICS • DIFFERENTIAL EQUATIONS, ORDINARY • DIFFERENTIAL EQUATIONS, PARTIAL • FUZZY SETS, FUZZY LOGIC, AND FUZZY SYSTEMS •

MATHEMATICAL MODELING • OPERATIONS RESEARCH • SET THEORY • SIMULATION AND MODELING • STOCHASTIC PROCESSES

BIBLIOGRAPHY

Ashby, R. (1956). "Introduction to Cybernetics," Chapman & Hall, London.

Atkin, R. H. (1973). "Mathematical Structure in Human Affairs," Heinemann, London.

Bellman, R., and Dreyfus, S. (1962). "Applied Dynamic Programming," Princeton University Press, Princeton, NJ.

Brockett, R. (1970). "Finite-Dimensional Linear Systems," Wiley, New York.

Casti, J. (1985). "Nonlinear System Theory," Academic Press, Orlando, FL.

Casti, J. (1987). "Linear Dynamical Systems," Academic Press, Orlando, FL.

Danzig, G. (1963). "Linear Programming and Extensions," Princeton University Press, Princeton, NJ.

Klir, G. (1985). "The Architecture of Systems Problem Solving," Plenum, New York.

Luenberger, D. G. (1979). "Introduction to Dynamic Systems," John Wiley & Sons, New York.

Mesarovic, M. D., and Takahara, Y. (1989), "Abstract Systems Theory," Springer-Verlag, Berlin, Heidelberg, New York.

Measarovic, M. D., Macko, D., and Takahara, Y. (1970). "Theory of Hierarchical, Multilevel Systems," Academic Press, New York.

Pichler, F., and Schwärtzel, H. (eds.) (1992). "CAST—Methods in Modelling," Springer-Verlag, Berlin, Heidelberg, New York.

Poston, T., and Stewart, I. (1978). "Catastrophe Theory and Its Applications," Pitman, London.

Vemuri, V. (1978). "Modeling of Complex Systems. An Introduction," Academic Press, New York.

Weinberg, G. (1975). "An Introduction to General Systems Thinking," Wiley, New York.

Tectonophysics

Donald L. Turcotte
Cornell University

GLOSSARY

Elastic rebound Relative motion between plates causes elastic deformation of the plates adjacent to a fault; when slip occurs on the fault, the plates rebound.

Fractals Statistical distribution in which the number of objects has a power-law dependence on their size.

Lithosphere Cool rigid outer shell of the earth that is capable of transmitting elastic stresses.

Mantle convection The solid interior of the earth flows like a fluid in response to gravitational buoyancy forces.

Plate tectonics The lithosphere of the earth is broken into a series of plates that are in relative motion with respect to each other.

Plume Quasi-cylindrical flows in the mantle responsible for hotspot volcanism.

Stick-slip Behavior of faults that causes earthquakes.

Subduction zone Region adjacent to an ocean trench where the oceanic lithosphere bends and sinks into the interior of the earth.

TECTONOPHYSICS is the branch of geophysics that deals with the deformation of the solid earth. The solid earth is composed of the mantle and the crust. The crust is a thin surface layer (6–70 km thick) made up of rocks derived from the mantle by partial melting. The earth is a heat engine. Heat is produced within the earth due to the decay of radioactive elements. The loss of this heat to the surface drives solid-state thermal convection in the earth's mantle. The surface plates of plate tectonics are part of this convection system. New surface plates are created at the global midocean ridge system (accretional plate margins). The plates move away from the ocean ridges at velocities of a few centimeters per year in a process known as seafloor spreading. The plates behave rigidly because the rocks that make up the plates are cold and strong. The plates are also known as the lithosphere and have a typical thickness of 100 km. Since new plates are continuously being created, old plates must be destroyed. This occurs at ocean trenches (subduction zones) where the plates bend and sink into the earth's interior. The relative

Encyclopedia of Physical Science and Technology, Third Edition, Volume 16
Copyright © 2002 by Academic Press. All rights of reproduction in any form reserved.

motion between plates at plate boundaries results in volcanism, earthquakes, and mountain building.

The earth's surface is made up of ocean basins and continents. The ocean basins participate in the plate tectonic cycle but the continents do not. The continental crust is thicker and less dense than the oceanic crust. Plates with continental crust are gravitationally stable and cannot be subducted. However, continents ride along with the relative motions of the plates resulting in continental drift. At times, these motions result in continental collisions, a major source of mountain building. Mountain ranges are extremely complex, with deformation occurring on a wide range of scales involving both brittle and fluid-like deformation. However, the statistical aspects of this deformation appear to obey simple fractal relationships. Continental tectonics can certainly exhibit deterministic, chaotic behavior and may involve examples of self-organized criticality.

I. MANTLE CONVECTION

A fluid layer that is heated from below or within and cooled from above is likely to convect. The near-surface fluid is cooler and more dense than the fluid at depth; the surface fluid will tend to sink and the hotter, less dense fluid at depth will rise. A simple example of a fluid layer heated from below is illustrated in Fig. 1; the temperature T_0 of the upper boundary is lower than the temperature T_1 of the lower boundary. Cooling from above creates a cold thermal boundary layer adjacent to the upper boundary that is gravitationally unstable and forms a cold descending plume. Similarly, a hot thermal boundary layer is created adjacent to the lower boundary that is also gravitationally unstable and forms a hot ascending plume. The gravitational body forces in the plumes drive a cellular convective flow with a wavelength λ. Thermal gradients are restricted to the boundary layers and plumes; the isothermal cores have a temperature T_c halfway between T_0 and T_1.

When the earth formed by accretion some 4.5 billion years ago (4.5 Ga) it was hotter than it is today. The earth is cooling at a rate of about 100 K/Ga. This may not seem like a very large temperature change, but this secular cooling results in a substantial heat loss to the surface. In addition, large amounts of heat are generated within the earth by the decay of the radioactive isotopes of uranium, thorium, and potassium. It is estimated that about one-half of the present heat loss from the interior of the earth is due to secular cooling, and one-half due to the decay of radioactive elements. Thus the solid mantle of the earth is heated from within and from below (due to heat loss from the core) and would be expected to convect if it were a fluid.

The acceptance that the solid interior of the earth behaves as a fluid due to solid-state creep was slow in coming. Although a minority of earth scientists had long advocated continental drift, general acceptance of mantle convection and large-scale surface displacements came only in the early 1970s. The geometrical similarity between the east coasts of North, Central, and South America and the west coasts of Europe and Africa was striking. Arthur Holmes, one of the leading British geologists during the first half of this century, advocated thermal convection in the mantle as the driving force for continental drift in 1931, but he was ridiculed by the leading geophysicists of the day. The primary objection was that solid rocks could not possibly have a fluid behavior. Yet conclusive evidence that mantle rocks behaved as a fluid was available in the last half of the nineteenth century. Gravity surveys in India had shown that the Himalayas had roots. That the light crustal rocks that created the highest topography on the planet floated like blocks of wood in water. How could this occur if the mantle rocks did not have a fluid behavior?

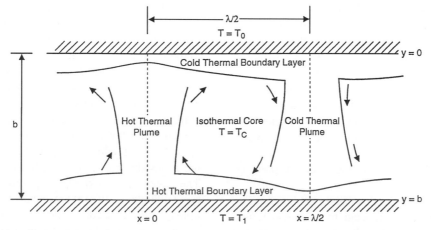

FIGURE 1 Thermal boundary-layer structure of two-dimensional thermal convection in a fluid layer heated from below.

Further evidence that the solid rocks of the mantle behaved as a fluid came from studies of postglacial rebound. In Scandinavia, the thick glacial ice cover depressed the area during the last ice age. The result was that, after the ice cover melted, the area rebounded and shorelines were elevated. In 1937 N. A. Haskell used this rate of elevation to quantify the fluid behavior of the solid mantle. Although his results were not generally recognized for another 30 years, today his values are still accepted as being basically correct. But the question remained, Why should a solid exhibit a fluid-like behavior? In the 1950s, laboratory studies showed that solids near their melting temperature behave as fluids. The flow of crystalline ice in glaciers is one example. It was recognized that the diffusion of vacancies (vacant lattice sites) and the movement of dislocations (crystal irregularities) in stress fields could lead to the very slow displacements associated with the flow of glaciers and the mantle. Today our concepts of solid-state creep, rebound of depressed areas, and mantle convection are all completely consistent with each other.

Mantle convection carries heat upward through the interior of the earth. The required velocities are a few centimeters per year: an apparently low velocity but, on geological time scales, capable of drifting continents. Continental drift is a natural consequence of mantle convection.

The earth behaves like a heat engine. Thermal convection converts heat into flows. These flows are responsible for plate tectonics and, either directly or indirectly, volcanism, earthquakes, and mountain building.

II. RHEOLOGY

Rheology is the science of deformation. Rocks can exhibit a wide range of rheologies including elastic, fracture, plastic, and viscous. Tectonic consequences include faults and folds as well as mantle convection.

At atmospheric pressure and room temperature most rocks are brittle; that is, they behave nearly elastically until they fail by fracture. Cracks or fractures in rock along which there has been little or no relative displacement are known as joints. They occur on a wide range of scales in all types of rocks. Joints are commonly found in sets defining parallel or intersecting patterns of failure related to local stress orientations. The breakdown of surface rocks by erosion and weathering is often controlled by systems of joints along which the rocks are particularly weak and susceptible to disintegration and removal. These processes in turn enhance the visibility of the jointing. Faults are fractures across which there has been a relative displacement.

Although fracture is important in shallow crustal rocks at low temperatures and pressures, there are many circumstances in which rocks behave as a ductile material.

In determining the transition from brittle to ductile behavior, pressure, temperature, and strain rate are important. If the confining pressure of rock is of the order of the brittle strength of the rock, a transition from brittle to ductile behavior will occur. This transition typically occurs at a depth of about 10 km. To model the ductile behavior of crustal and mantle rocks, it is often appropriate to use an idealized elastic-perfectly plastic rheology. An elastic-perfectly plastic material exhibits a linear elastic behavior until a yield stress is reached. The material can then be deformed plastically an unlimited amount at this stress.

At temperatures that are a significant fraction of the melt temperature the atoms and dislocations in a crystalline solid become sufficiently mobile to result in creep when the solid is subjected to deviatoric stresses. At very low stresses diffusion processes dominate, and the crystalline solid behaves as a Newtonian fluid with a viscosity that depends exponentially on the pressure and the inverse absolute temperature. At higher stresses the motion of dislocations becomes the dominant creep process, resulting in a non-Newtonian or nonlinear fluid behavior that also has an exponential pressure and inverse absolute temperature dependence. Mantle convection and continental drift are attributed to these thermally activated creep processes as discussed above.

Rocks can behave elastically on short time scales but as a fluid on long time scales. Such behavior can be modeled with a rheological law that combines linear elastic and viscous rheologies. A material that behaves both elastically and viscously is known as a viscoelastic medium.

Folding is evidence that crustal rocks also exhibit ductile behavior under stress. Pressure solution creep is a mechanism that can account for the ductility of crustal rocks at relatively low temperatures and pressures. This process involves the dissolving of minerals in regions of high pressure and their precipitation in regions of low pressure. As a result, creep of the rock occurs. Folding can also result from the plastic deformation of rock.

III. PLATE TECTONICS

Plate tectonics is a model in which the outer shell of the earth is broken into a number of thin rigid plates that move with respect to one another. The relative velocities of the plates are of the order of a few centimeters per year. The basic hypothesis of plate tectonics was given by Jason Morgan in 1968. The concept of rigid plates with deformations primarily concentrated near plate boundaries provides a comprehensive understanding of the global distribution of earthquakes, volcanism, and mountain building.

The distribution of the major surface plates is given in Fig. 2; the ridge axes, subduction zones, and transform

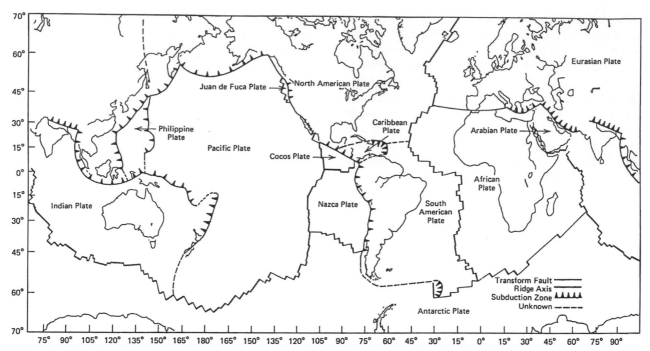

FIGURE 2 Distribution of the major surface plates. The ridge axes, subduction zones, and transform faults that make up the plate boundaries are shown.

faults that make up the plate boundaries are shown. The outer portion of the earth, termed the lithosphere, is made up of relatively cool, stiff rocks and has an average thickness of about 100 km. The lithosphere is divided into a small number of mobile plates that are continuously being created and consumed at their edges. At ocean ridges, adjacent plates move apart in a process known as seafloor spreading. As the adjacent plates diverge, hot mantle rock ascends to fill the gap. The hot, solid mantle rock behaves like a fluid because of solid-state creep processes. As the hot mantle rock cools, it becomes rigid and accretes to the plates, creating new plate area. For this reason ocean ridges are also known as accretionary plate boundaries.

Because the surface area of the earth is essentially constant, there must be a complementary process of plate consumption. This occurs at ocean trenches. The surface plates bend and descend into the interior of the earth in a process known as subduction. At an ocean trench the two adjacent plates converge, and one descends beneath the other. For this reason ocean trenches are also known as convergent plate boundaries. A cross-sectional view of the creation and consumption of a typical plate is illustrated in Fig. 3.

Plate tectonics is directly associated with mantle convection. The pattern of thermal convection illustrated in Fig. 1 can provide a direct understanding of why the earth has plate tectonics. The thermal boundary layer at the surface of the earth is the lithosphere. This is the cold thermal

boundary layer associated with the loss of heat to the surface of the earth. Because the viscosity of mantle rock is exponentially temperature dependent, the cold lithosphere is essentially rigid and behaves as a series of nearly rigid plates. Ascending convection is associated with ocean ridges. New seafloor is created at ocean ridges and the seafloor spreads away from the ridge axis at a velocity u as illustrated in Fig. 3.

As the ocean lithosphere moves away from the ocean ridge where it was created, it cools and becomes gravitationally unstable with respect to the rock beneath. The

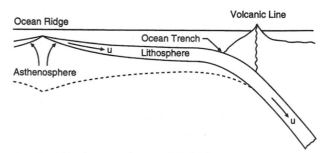

FIGURE 3 Accretion of a lithospheric plate at an ocean ridge (accretional plate margin) and its subduction at an ocean trench (subduction zone). The asthenosphere, which lies beneath the lithosphere, and the volcanic line above the subducting lithosphere are also shown. The plate migrates away from the ridge crest at the seafloor spreading velocity μ.

oceanic lithosphere bends and sinks into the interior of the earth at an ocean trench as illustrated in Fig. 3.

A. The Lithosphere

An essential feature of plate tectonics is that only the outer shell of the earth, the lithosphere, remains rigid during long intervals of geologic time. Because of their low temperatures, rocks in the lithosphere resist deformation on time scales of up to billion years. In contrast, the rock beneath the lithosphere is sufficiently hot that solid-state creep occurs. The lithosphere is composed of both mantle and crustal rocks. The oceanic lithosphere has an average thickness of 100 km, with the uppermost 6 to 7 km being the oceanic crust. The oceanic lithosphere participates in the plate tectonic cycle. The continental lithosphere has a typical thickness of about 200 km. Typically, the upper 30 km of the continental lithosphere is continental crust. Because of the buoyancy of the continental crust, the continental lithosphere does not subduct, although it does participate in plate motions.

The elastic rigidity of the lithosphere also allows it to flex when subjected to a load. An example is the load applied by a volcanic island. The load of the Hawaiian Islands causes the lithosphere to bend downward around the load, resulting in a moat, a region of deeper water around the islands. The elastic bending of the lithosphere under vertical loads can also explain the structure of ocean trenches and some sedimentary basins. However, the entire lithosphere is not effective in transmitting elastic stresses. Only about the upper half of it is sufficiently rigid that elastic stresses are not relaxed on time scales of a billion years. This fraction of the lithosphere is referred to as the elastic lithosphere. Solid-state creep processes relax stresses in the lower, hotter part of the lithosphere. This relaxation can be understood in terms of a viscoelastic rheology. This lower part of the lithosphere, however, remains a coherent part of the plates.

The strength of the lithosphere allows the plates to transmit elastic stresses over geologic time intervals. The plates act as stress guides. Stresses that are applied at the boundaries of a plate can be transmitted through the interior of the plate. The ability of the plates to transmit stress over large distances is a key factor in driving tectonic plates. These stresses are also responsible for some intraplate earthquakes and small amounts of intraplate deformation.

B. Accretional Plate Margins (Ocean Ridges)

Lithospheric plates are created at ocean ridges. The two plates on either side of an ocean ridge move away from each other at nearly steady velocities of a few centimeters per year. As the two plates diverge, hot mantle rock flows upward to fill the gap. The upwelling mantle rock cools by conductive heat loss to the surface. The cooling rock accretes to the base of the spreading plates, becoming part of them; the structure of an accreting plate margin is illustrated in Fig. 4.

As the plates move away from the ocean ridge, they continue to cool and thicken. Seafloor depth as a function of age is shown in Fig. 5. As the lithosphere cools, it contracts thermally and becomes denser; as a result, its upper surface—the ocean floor—sinks relative to the ocean surface. The topographic elevation of the ocean ridge is due to the lower-density, thinner, and hotter lithosphere near the axis of accretion at the ridge crest. A simple heat loss model (half-space cooling model) predicts that the subsidence is proportional to the square root of age. This is a good approximation for young seafloor as shown in Fig. 5 but overestimates the subsidence for seafloor older than about 100 Ma. This deviation can be attributed to the heating of the base of the oceanic lithosphere by mantle plumes. This heating is approximated by assuming a plate model with a specified lithosphere thickness. The data appear to favor a maximum lithosphere (plate) thickness of 125 km as shown in Fig. 5. The elevation of the ridge also exerts a gravitational body force that drives the lithosphere away from the accretional boundary; it is one of the important forces driving the plates and is known as gravitational sliding or ridge push.

Ocean ridges generate a large fraction of the Earth's volcanism. Because almost all the ridge system is below sea level, only a small part of this volcanism can be readily observed. Ridge volcanism can be seen in Iceland, where

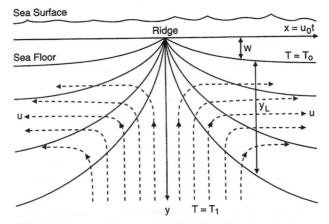

FIGURE 4 Structure of an accretional plate margin (x is the horizontal coordinate and y the vertical coordinate). The rigid lithosphere, thickness y_L, spreads away from the ridge axis at velocity u_0. The solid contours are isotherms; the seafloor has a temperature T_0 and the mantle beneath the lithosphere has a temperature T_1. Mantle material flows along the dashed lines to fill the gap created by the spreading lithospheres. The depth of the subsiding seafloor relative to the ridge axis is w.

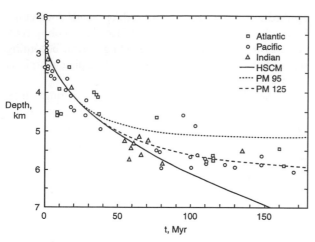

FIGURE 5 Seafloor depth as a function of age in the Atlantic, Pacific, and Indian oceans. Comparisons are made with the half-space cooling model (HSCM) and with plate models with plate (lithosphere) thicknesses of 95 km (PM 95) and 125 km (PM 125).

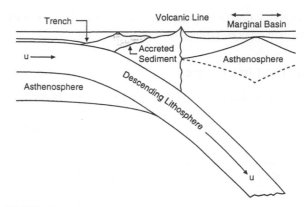

FIGURE 6 Illustration of the subduction of the oceanic lithosphere at an ocean trench. The line of volcanic edifices associated with most subduction zones is shown. A substantial fraction of the sediments that coat the basaltic oceanic crust is scraped off during subduction to form an accretionary prism of sediments. In some cases, back-arc spreading forms a marginal basin behind the subduction zone.

the oceanic crust is sufficiently thick that the ridge crest rises above sea level. The volcanism at ocean ridges is caused by pressure-release melting. The diverging plates induce an upwelling in the mantle. The temperature of the ascending rock decreases slowly with decreasing pressure. The solidus temperature for melting decreases with decreasing pressure at a much faster rate. When the temperature of the ascending mantle rock equals the solidus temperature, melting begins. The ascending mantle rock contains a low-melting-point basaltic component; this component melts first to form the oceanic crust. The region where partial melting is occurring is known as the asthenosphere.

C. Subduction

As the oceanic lithosphere moves away from an ocean ridge, it cools, thickens, and becomes more dense because of thermal contraction. Even though the basaltic rocks of the oceanic crust are lighter than the underlying mantle rocks, the colder mantle rocks in the lithosphere become sufficiently dense to make old oceanic lithosphere heavy enough to be gravitationally unstable with respect to the hot mantle rocks beneath the lithosphere. As a result of this gravitational instability the oceanic lithosphere founders and sinks into the interior of the earth, creating the ocean trenches. This process is known as subduction and is illustrated schematically in Fig. 6.

The excess density of the rocks of the descending lithosphere results in a downward buoyancy force. Because the lithosphere behaves elastically, it can transmit stresses, i.e., it can act as a stress guide. A portion of the negative buoyancy force acting on the descending plate is transmit-

ted to the surface plate, which is pulled toward the ocean trench. This is slab pull, one of the important forces driving plate tectonics.

Ocean trenches are the sites of most of the largest earthquakes. Earthquakes occur on the dipping fault plane that separates the descending lithosphere from the overlying lithosphere. Earthquakes at ocean trenches can occur to depths of 660 km. This seismogenic region, known as the Wadati–Benioff zone, delineates the approximate structure of the descending plate.

Volcanism is also associated with subduction. A line of regularly spaced volcanoes closely parallels the trend of almost all the ocean trenches. These volcanoes may result in an island arc or they may occur within continental crust. The volcanoes generally lie above where the descending plate is 125 km deep, as illustrated in Fig. 6. It is far from obvious why volcanism is associated with subduction. The descending lithosphere is cold compared with the surrounding mantle, and thus it acts as a heat sink rather than as a heat source. The downward flow of the descending slab is expected to entrain flow in the overlying mantle wedge. However, this flow will be primarily downward; thus, magma cannot be produced by pressure-release melting. One possible source of heat is frictional heating on the fault plane between the descending lithosphere and the overlying mantle.

When a subduction zone is adjacent to a continent, as in the case of South America, subduction zone volcanism can form great mountain belts, for example, the Andes. In some subduction zones tensional stresses can result in back-arc, seafloor spreading and the formation of a marginal basin as illustrated in Fig. 6. An example is the Sea of Japan.

IV. HOTSPOTS AND PLUMES

Not all volcanism and tectonism is restricted to the plate margins. Hotspots are anomalous areas of surface volcanism that cannot be directly associated with plate tectonic processes. The term hotspot is used rather loosely. It is applied to any long-lived volcanic center that is not part of the global network of midocean ridges and island arcs. The prototype example is Hawaii. Anomalous regions of thick crust on ocean ridges are also considered to be hotspots. Several hotspot lists have been published, and the number of volcanic centers included on these lists ranges from about 20 to more than 100. Figure 7 shows the locations of 38 prominent hotspots. In many cases hotspots have well-defined tracks associated with volcanic ridges or lines of volcanic edifices; these are also shown in Fig. 7.

In 1971, Jason Morgan attributed hotspot volcanism to mantle plumes. Mantle plumes are quasi-cylindrical concentrated upflows of hot mantle material; they represent the ascending plumes from a basal thermal boundary layer as illustrated in Fig. 1. Pressure-release melting in the hot ascending plume rock produces the basaltic volcanism that is forming the Hawaiian Island chain. The hypothesis of fixed mantle plumes beneath overriding plates explains the systematic age progression of the Hawaiian-Emperor island-seamount chain, the hotspot track extending from Hawaii to the Aleutian Islands.

Most hotspots are also associated with topographic swells. Hotspot swells are regional topographic highs with widths of about 1000 km and up to 3 km of anomalous elevation. The swell associated with the Hawaiian hotspot is roughly parabolic in planform and it extends upstream of the active hotspot some 500 km. The excess elevation associated with the swell decays slowly down the track of the hotspot. Hotspot swells are attributed to the buoyancy of the hot, low-density plume rock impinging on the base of the lithosphere.

Numerical and laboratory studies on the initial ascent of a low-viscosity buoyant plume through a high-viscosity fluid have shown that the plume consists of a large leading diapir or plume head followed by a thin conduit connecting the diapir with the source region. It has been proposed that massive flood basalt eruptions are the result of pressure-release melting in the plume head as it impinges on the lithosphere from below. According to this model, flood basalt eruptions mark the initiation of hotspot tracks as illustrated in Fig. 7. Specifically the Deccan, Tertiary North Atlantic, Parana, and Karoo flood basalts represent the onset of the currently active hotspots at Reunion, Iceland, Tristan de Cunha, and Prince Edward. In each case nearly 2×10^6 km^3 of magma erupted within a few million years of hotspot initiation as the plume head reached the base of the lithosphere.

V. CONTINENTS

The basic facets of plate tectonics do not require continents. But without continents little or no land would rise above sea level and life as we know it would not exist.

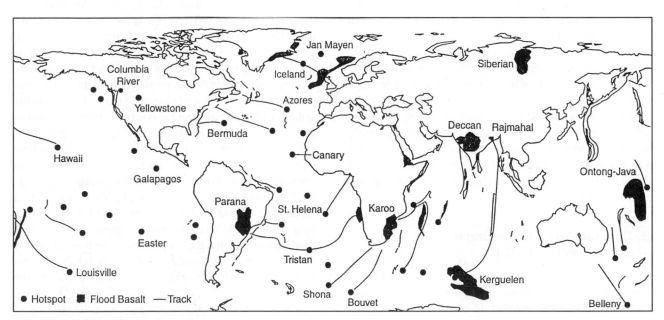

FIGURE 7 Locations of 38 prominent hotspots are shown. In some cases, the associated hotspot tracks and flood basalt provinces are also shown.

The continental crust is much thicker than the oceanic crust ($\approx$40 vs $\approx$6 km) and contains primarily silicic rocks that are less dense than the basaltic rocks of the oceanic crust. The result is that the continental lithosphere is gravitationally stable and resists subduction. Because of the plate tectonic cycle, the seafloor has an average age of only about 100 Ma and the oldest seafloor has an age of about 200 Ma. The mean age of the continents is greater than 2 Ga and some parts have ages greater than 3 Ga.

A. Continental Drift

The continents are rafted about on the plates, resulting in continental drift. The earliest arguments for continental drift were based largely on the fit of the continents. Ever since the first reliable maps were available, the remarkable fit between the east coast of South America and the west coast of Africa has been noted. The fit was pointed out as early as 1620 by Francis Bacon. North America, Greenland, and Europe also fit as illustrated in Fig. 8. Detailed arguments supporting continental drift were given by the well-known German meteorologist Alfred Wegener in 1915. Wegener's book included his highly original picture of the breakup and subsequent drift of the continents and his recognition of the supercontinent Pangea. Later it was argued that there had formerly been a northern continent, Laurasia, and a southern continent, Gondwanaland, separated by the Tethys ocean. Wegener assembled a formidable array of facts and conjectures to support his case, including the match between mountain belts in South America and Africa; similar rock types, rock ages, and fossil species are found on the two sides of the Atlantic Ocean. Tropical climates had existed in polar regions at the same times that arctic climates had existed in equatorial regions. Also, the evolution and dispersion of plant and animal species were best explained in terms of ancient land bridges, suggesting direct connections between now widely separated continents.

Although the qualitative arguments favoring continental drift appear convincing today, they were summarily rejected by the vast majority of earth scientists during the first half of the 20th century. Only with the acceptance of mantle convection and plate tectonics did continental drift receive general acceptance.

B. Delamination and the Origin of the Continental Crust

There is no evidence that the continental lithosphere is subducted. This is attributed to the buoyancy of the continental crust, which results in the continental lithosphere being gravitationally stable. However, the mantle portion of the continental lithosphere is sufficiently cold and dense to be gravitationally unstable. Thus it is possible for the lower part of the continental lithosphere, including the lower continental crust, to delaminate and sink into the lower mantle. It is widely accepted that delamination of the continental lithosphere is presently occurring beneath the Himalayas and the Alps. Delamination also plays an important role in the origin of the continental crust.

The rocks of the continental crust cannot be formed directly from magmas that rise from the mantle. A more complex, three-step hypothesis is required. (1) Basaltic volcanism from the mantle associated with subduction zone volcanics, continental rifts, and hotspots is responsible for the formation of the continental crust. (2) Intracrustal melting and high-temperature metamorphism are responsible for the differentiation of the crust so that the upper crust becomes more silicic and the lower crust becomes more basic. Basaltic magmas from the mantle intruded into a basaltic continental crust in the presence of water can produce the granitic rocks associated with the continental crust. (3) Delamination of substantial quantities of the continental lithosphere, including the mantle and lower crust, returns a substantial fraction of the more basic lower crust to the mantle. The residuum, composed primarily of the upper crust, thus becomes more silicic.

C. The Wilson Cycle

J. Tuzo Wilson proposed in 1966 that continental drift is cyclic. In particular, he proposed that oceans open and close; this is now known as the Wilson cycle and was based on the opening and closing of the Atlantic Ocean. The Wilson cycle, in its simplest form, is illustrated in Fig. 9. The first step in the Wilson cycle is the breakup of a continent. This occurs on continental rift zones. Present examples are the East African Rift system and the Rio Grande graben. These may or may not break apart to form future oceans. Aulacogens (triple junctions with three rifts connected at about 120°) are believed to play a key role in the initiation of rifting and the breakup of continents. Aulacogens are the surface expressions of the impingement of mantle plumes on the base of the continental lithosphere and are associated with lithospheric swells. An example of a lithospheric swell on a continent is the Ethiopian swell on the East African Rift. An example of a triple junction is at the southern end of the Red Sea, the Gulf of Aden, and the East African Rift. When a continent opens, two of the rifts separate and become part of an ocean. The third rift aborts and is known as a "failed" arm. Examples of failed arms associated with the opening of the Atlantic Ocean are the St. Lawrence River Valley Rift and the Niger Rift in Africa.

The second step in the Wilson cycle is the opening of the ocean as illustrated in Fig. 9b. The rift valley splits apart and oceanic crust is formed at an accretional plate

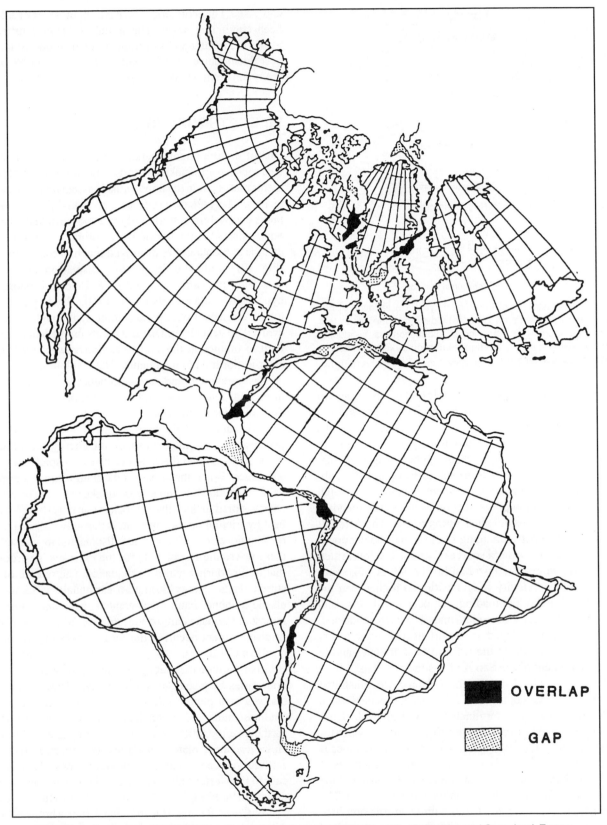

FIGURE 8 The remarkable "fit" between the continental margins of North and South America and Greenland, Europe, and Africa is illustrated. This fit was one of the primary early arguments for continental drift.

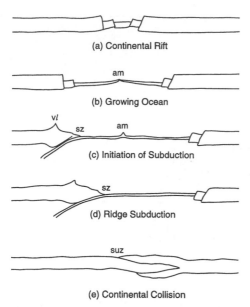

FIGURE 9 Illustration of the Wilson cycle. (a) Initiation of new ocean at a continental rift zone. (b) Opening of the ocean (am, accretional margin). (c) Initiation of subduction (sz, subduction zone; vl, volcanic line). (d) Ridge subduction. (e) Continental collision (suz, suture zone).

boundary. The Red Sea is an example of the initial stages of the opening of an ocean, while the Atlantic Ocean is an example of a mature stage. The margins of an opening ocean are known as passive continental margins, in contrast to active continental margins, where subduction is occurring.

The third step in the Wilson cycle is the initiation of subduction (Fig. 9c). A passive continental margin is a favored site for the initiation of subduction because it is already a zone of weakness established during rifting. The differential subsidence between aging seafloor and the continental lithosphere provides a source of stress. The fourth step in the Wilson cycle, illustrated in Fig. 9d, is ridge subduction. If the velocity of subduction is higher than the velocity of seafloor spreading, the ocean will close and eventually the accretional plate margin will be subducted. Ridge subduction played an important role in the recent geological evolution of the western United States and in the development of the San Andreas fault system.

The fifth and final stage in the Wilson cycle, illustrated in Fig. 9e, is the continental collision that occurs when the ocean closes. This terminates the Wilson cycle. Continental collision is one of the primary mechanisms for the creation of mountains in the continents; the other is subduction. The Himalayas and the Alps are examples of mountain belts caused by continental collisions, and the Andes is a mountain belt associated with subduction. The boundary between the two plates within the collision zone is known as a suture zone. The Himalayas are the result of the continental collision between the Indian subcontinent

and Asia. This collision occurred about 45 Ma and has been continuing since. The initial collision resulted in a major global reorganization of plate motions that is best documented by the bend in the Hawaiian-Emperor seamount chain shown in Fig. 7.

VI. EARTHQUAKES

One of the important phenomena associated with active tectonics is earthquakes. A large fraction of the displacements that occur in the upper crust is associated with earthquakes. The understanding of the stick-slip behavior of faults evolved from studies of the 1906 earthquake on the San Andreas fault in northern California. This earthquake and the subsequent fire destroyed much of San Francisco (Fig. 10). Studies of the geodetic displacements associated with this earthquake led H. F. Reid to propose the hypothesis of elastic rebound in 1910. This hypothesis is totally consistent with plate tectonics, although the latter evolved 60 years later. Displacements on the San Andreas fault accommodate the relative motion between the Pacific and the North American plates.

Elastic rebound and stick-slip behavior are illustrated in Fig. 11. Faults lock, and a displacement occurs when the stress across the fault builds up to a sufficient level to cause rupture of the fault. This is known as stick-slip behavior. When a fault sticks, elastic energy accumulates in the rocks around the fault because of displacements at a distance. When the stress on the fault reaches a critical value, the fault slips and an earthquake occurs. The elastic energy stored in the adjacent rock is partially dissipated as heat by friction on the fault and is partially radiated away in seismic waves. The surface displacements caused by these waves are responsible for the extensive destruction that occurs during major earthquakes. The release of the stored elastic energy during an earthquake is known as elastic rebound. Fault displacements associated with the largest earthquakes are of the order of 30 m. About 4 m of displacement occurred during the 1906 earthquake in northern California.

Great earthquakes are generally associated with the boundaries between the surface plates. They occur regularly where plates slide past each other (e.g., the San Andreas fault) and in subduction zones (e.g., the 1960 earthquake in Chile and the 1964 earthquake in Alaska). However, some plate boundaries are rather diffuse and earthquakes can occur over broad regions. This is the case in the western United States, where deformation and mountain building occur from the Rocky Mountains to the Pacific Coast. It is also true in China, where a broad zone of deformation extends through the entire country, resulting in many disastrous earthquakes including the T'angshan

FIGURE 10 Destruction in San Francisco caused by the magnitude 8.3 earthquake, April 18, 1906, and the subsequent fire. It is estimated that there was 3000 deaths and about 28,000 buildings were destroyed.

earthquake in 1976, which killed some 500,000 people. Broad zones of deformation are required by the evolving motion of the plates; in some cases new boundaries evolve to accommodate the required relative velocities. Earthquakes can also occur in the interior of apparently rigid plates. An example is the series of three large earthquakes that occurred near New Madrid, Missouri, in 1811–1812. These intraplate earthquakes are attributed to the large stresses that are transmitted through the interior of plates; they generally do not result in the development of significant mountain belts.

VII. FRACTALS, CHAOS, AND SELF-ORGANIZED CRITICALITY

The scale invariance of geological phenomena is one of the concepts taught to a student of geology. It is pointed out that an object that defines the scale, i.e., a coin, a rock hammer, or a person, must be included whenever a photograph of a geological feature is taken. Without the scale, it is often impossible to determine whether the photograph covers 10 cm or 10 km. For example, self-similar folds occur over this range of scales. Another example is be an aerial photograph of a rocky coastline. Without an object with a characteristic dimension, such as a tree or house, the elevation of the photograph cannot be determined. It was in this context that Benoit Mandelbrot introduced the concept of fractals in 1967. The definition of a fractal distribution is

$$N_i = C/r_i^D, \tag{1}$$

where N_i is the number of objects with a linear size r_i, C is a constant, and D is the fractal dimension. Mandelbrot showed that the perimeter P_i of a rocky coastline (e.g., Maine) or a contour on a topographic map satisfies Eq. (1)

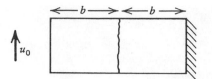

(a) After a major earthquake the fault sticks

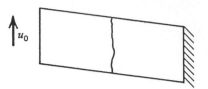

(b) Just prior to the next major earthquake

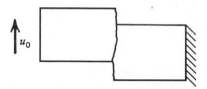

(c) After this major earthquake the fault locks
and the cycle repeats

FIGURE 11 Illustration of the stick-slip and elastic-rebound behavior of faults. (a) After an earthquake the fault sticks and the relative plate velocity u_0 causes an elastic deformation of the plates. (b) As elastic distortion occurs the stress builds up in the plates until the fault slips. (c) Slip on the fault results in elastic rebound, the fault sticks, and the process repeats.

if r_i is the length of the step used in measuring the perimeter and N_i the number of steps:

$$P_i = N_i r_i = C/r_i^{D-1}. \qquad (2)$$

The shorter the step, the longer the perimeter; D is usually about 1.25. Because of scale invariance, the length of the coastline increases as the length of the measuring rod decreases according to a power law; the power determines the fractal dimension of the coastline. It is not possible to obtain a specific value for the length of a coastline, owing to all the small indentations, down to a scale of millimeters or less.

Many geological phenomena are scale invariant. Examples include the frequency-size distributions of rock fragments, faults, earthquakes, volcanic eruptions, and oil fields. The empirical applicability of power-law statistics to geological phenomena was recognized long before the concept of fractals was conceived. A striking example is the Gutenberg–Richter relation for the frequency-magnitude statistics of earthquakes. The proportionality factor in the relationship between the logarithm of the number of earthquakes and earthquake magnitude is known as the b-value. It has been recognized for nearly

50 years that, almost universally, $b = 0.9$. It is now accepted that the Gutenberg–Richter relationship is equivalent to a fractal relationship between the number of earthquakes and the characteristic size of the rupture; the value of the fractal dimension D is simply twice the b-value; typically $D = 1.8$ for distributed seismicity.

An example for earthquakes in southern California is given in Fig. 12. The fact that the distribution of earthquakes is a fractal is evidence that the distribution of faults on which the earthquakes are occurring is also a fractal. Crustal deformation is occurring on all scales in a scale-invariant manner. Although the deformation is complex and chaotic, the deformation satisfies scale-invariant fractal statistics.

Fractal concepts can also be applied to continuous distributions; an example is topography. Mandelbrot has used fractal concepts to generate synthetic landscapes that look remarkably similar to actual landscapes. The fractal dimension is a measure of the roughness of the features. The earth's topography is a composite of many competing influences. Topography is created by tectonic processes including faulting, folding, and flexure. It is modified and destroyed by erosion and sedimentation. There is considerable empirical evidence that erosion is scale invariant and fractal; a river network is a classic example of a fractal tree. Topography often appears to be complex and chaotic, yet there is order in the complexity. A standard approach to the analysis of a continuous function such as topography along a linear track is to determine the coefficients A_n in a Fourier series as a function of the wavelength λ_n. If the amplitudes A_n have a power-law dependence on the wavelength λ_n, a fractal distribution may result. For topography and bathymetry it is found that, to a good approximation, the Fourier amplitudes are proportional to

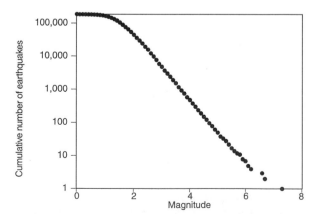

FIGURE 12 Number of earthquakes that occurred in southern California with a magnitude greater than a specified value from 1985 to 1999. Between a magnitude of 1 and a magnitude of 6, the data correlate with $b = 0.98$ and $D = 1.96$.

the wavelengths. This is also true for a Brownian walk, which can be generated as follows. Take a step forward and flip a coin; if tails occurs, take a step to the right, and if heads occurs, take a step to the left; repeat the process. The divergence of the walk or signal increases in proportion to the square root of the number of steps. A spectral analysis of the random walk shows that the Fourier coefficients A_n are proportional to the wavelengths λ_n.

Although fractal distributions would be useful simply as a means of quantifying scale-invariant distributions, their applicability to geological problems has a more fundamental basis. Ed Lorenz, in 1963, derived a set of nonlinear differential equations that approximate thermal convection in a fluid. This set of equations was the first to be shown to exhibit chaotic behavior. Infinitesimal variations in initial conditions led to first-order differences in the solutions obtained. This is the definition of chaos. The equations are completely deterministic; however, because of the exponential sensitivity to initial conditions, the evolution of a chaotic solution is not predictable. The evolution of the solution must be treated statistically and the applicable statistics are often fractal. Mantle convection is one example of a chaotic process in nature.

Slider-block models have long been recognized as a simple analogue for the behavior of a fault. The block is dragged along a surface with a spring and the friction between the surface and the block results in the stick-slip behavior that is characteristic of faults. It has been shown that a pair of slider blocks exhibits chaotic behavior. The two slider blocks are attached to each other by a spring, and each is attached to a constant-velocity driver plate by another spring. As long as there is any asymmetry in the problem, for example, unequal block masses, chaotic behavior can result. This is evidence that the deformation of the crust associated with displacements on faults is chaotic and, thus, is a statistical process. This is entirely consistent with the observation that earthquakes obey fractal statistics.

The concept of self-organized criticality was introduced by Per Bak and colleagues in 1988 in terms of a cellular automaton model for avalanches on a sand pile. A natural system is said to be in a state of self-organized criticality if, when perturbed from this state, it evolves naturally back to the state of marginal stability. The input to the system is slow and steady and the output is in avalanches which satisfy fractal frequency-size statistics. Earthquakes are an example of such a system. The slow tectonic motion of the plates is the input and the earthquakes are the avalanches which satisfy fractal statistics as shown in Fig. 12.

As discussed above, a pair of interacting slider blocks can exhibit chaotic behavior. Large numbers of driven slider blocks are an example of self-organized criticality. A two-dimensional array of slider blocks is considered. Each block is attached to its four neighbors and to a constant-velocity driver plate by springs. Slip events occur chaotically and the frequency–size statistics of the events are generally fractal. By increasing the number of blocks considered, the low-order chaotic system is transformed into a high-order system that exhibits self-organized criticality.

VIII. CONCLUSIONS

Mantle convection and plate tectonics provide a general framework for understanding tectonophysics. Transport of heat from the interior of the earth drives solid-state convection. Plate tectonics is a direct consequence of this convection. The relative velocity between plates causes crustal deformation at the boundaries between plates. In some cases this deformation is diffuse and is spread over a broad area. Volcanism occurs at most plate boundaries and is also responsible for crustal deformation.

Although we now have a general understanding of tectonophysics, we are still not able to predict earthquakes. Deformation on a local scale is extremely complex. In fact, it is quite likely that local deformation is so complex and chaotic that it is fundamentally impossible to make predictions of earthquakes. Only risk assessments will be possible. There is increasing evidence that scale-invariant, fractal statistics are applicable to a variety of tectonophysics problems. One possible application is the direct association of large earthquakes with small earthquakes; a risk of a great earthquake is present only where small earthquakes are occurring and the level of local seismicity can be used to assess the seismic hazard.

SEE ALSO THE FOLLOWING ARTICLES

CHAOS • CONTINENTAL CRUST • EARTHQUAKE MECHANISMS AND PLATE TECTONICS • EARTH'S MANTLE (GEOPHYSICS) • GEOLOGY, EARTHQUAKE • FRACTALS • PLATE TECTONICS • VOLCANOLOGY

BIBLIOGRAPHY

Fowler, C. M. R. (1990). "The Solid Earth," Cambridge University Press, Cambridge.
Press, F., and Siever, R. (1997). "Understanding Earth," W. H. Freeman, San Francisco.
Turcotte, D. L. (1997). "Fractals and Chaos in Geology and Geophysics," Cambridge University Press, Cambridge.
Turcotte, D. L., and Schubert, G. (1982). "Geodynamics," Wiley, New York.

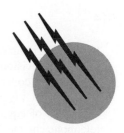

Telecommunications

John N. Daigle
University of Mississippi

Katherine H. Daigle
KHD Systems

GLOSSARY

Baseband The collection of frequencies over which a signal exists in its natural setting; for example, the baseband of a voice signal is the range of frequencies below 20 KHz or 20,000 cycles/sec.

Carrier system A transmission facility that provides multiple equivalent circuits between two geographically distant points. There are two basic classes of carrier systems, short haul and long haul, designed to connect end points at less than or more than about fifty miles, respectively.

Circuit switching A method of communication in which a physical circuit is established between two end systems before communication begins to take place. In circuit switching, specific resources are set aside for the exclusive use of the end systems for the duration of the connection.

Communication circuit The set of equipment and transmission facilities that are used to provide information transmission between two end systems.

Computer communication network A collection of applications hosted on different end systems and interconnected by an infrastructure that provides intercommunications. The infrastructure consists of computer-based end systems, packet switches, and transmission facilities connecting end systems to switches and switches to each other. The Internet is a computer communication network.

End system The equipment used directly by an end user

to facilitate communications; examples include a telephone, a television, or a personal computer.

Equipment A collection of electrical components and circuitry put together in a single physical package. Examples of equipments are telephones, televisions, telephone channel banks, test sets, and radios. This descriptive word is commonly used in any systems engineering discipline in preference to the word *component* to avoid confusion. For example, systems engineers would describe a radio as equipment, but transistors, power cords, transformers, and speakers in a radio as components.

Message switching A service technology used to transfer messages between end systems on a message-by-message basis. Immediate delivery is not guaranteed, and there is no notion of a direct connection between end systems. Message switching is an electronic analog of the postal system.

Packet switching A method of communication in which messages are exchanged between end systems via the exchange of a sequence of fragments of the messages called packets. Specific resources required for transmission of packets are requested as needed during the call. Availability of the resources requested is not guaranteed.

Router This is another name for a packet or message switch. Sometimes a router is called an intermediate system.

SS7 network The packet-switched network that provides communications and control of the call-carrying, circuit-switched telephone network.

Switch A system consisting of collections of input ports, output ports, switching circuitry, and control software that provides the capability to transfer information arriving at an input port to an arbitrary output port or subset of the output ports. A circuit switch literally connects an incoming trunk to an outgoing trunk electronically for the duration of a call. A packet switch transfers information packet-by-packet to possibly different outputs based on addressing information contained in the packet.

Tariff A set of rules governing the pricing structure for a collection of services.

TCP/IP The combination of the Transmission Control Protocol and the Internet Protocol; TCP provides the rules and mechanisms for exchange of information between end systems, and the IP protocol determines how packets are routed through a communication network. TCP/IP is the protocol of choice in the Internet.

Telephone network The world-wide network that provides circuit switched connections between arbitrary end systems around the world, primarily over circuits

suitable for telephone conversations and low speed data.

T1 carrier system A short-haul digital transmission system consisting of a collection of channel banks as terminal equipment and digital transmission facilities between the channel banks. Each channel bank provides input and output circuitry to support 24 voice calls, and the digital lines between the channel banks operate at 1.544 megabits/sec. A T1 system transmits between the channel banks over copper, twisted-pair cable, features baseband transmission, and has repeaters spaced at about one-mile intervals.

Trunk A circuit between two circuit switches that provides the capacity to carry the equivalent of one telephone call in one direction.

World-wide web The collection of web sites and web browsers interconnected over the world-wide Internet.

TELECOMMUNICATIONS literally means *communications at a distance*, the usual connotation being *communications at a distance by electrical means*. The array of public and private and telecommunication systems and services spanning the globe today is the result of amazing technological advances that have taken place over the last two centuries. This article focuses on those aspects of telecommunications concerned with providing the public telecommunications services that a growing proportion of the world's population uses every day.

I. PRIMARY NETWORKS AND TECHNOLOGIES

The purpose of a telecommunication system is to exchange information among users of the system. This information exchange can take place in a variety of ways, for example, multiparty voice communications, television, electronic mail, and electronic message exchange.

The world-wide telecommunications infrastructure in the year 2001 consists largely of two interrelated major infrastructures: the telephone network, which is a *circuit-switched* network, and the Internet, which is a *packet-switched computer communications* network. Today, control of the telephone network occurs mostly over computer communication networks, and transmission facilities that interconnect the major components of the Internet are derived from the same facilities that interconnect telephone switches. The state of the art today is the result of the tremendous growth in scientific knowledge and significant advances over the last two centuries of a wide variety of technologies. Indeed, it is not easy to identify areas of science and technology that do not have some impact upon the state of the art in telecommunications. Any brief discussion of telecommunications must, then, of necessity, omit significant portions of the story.

The telecommunications infrastructure of today provides two fundamentally different types of service: message-oriented and real-time communication-oriented. Examples of message-oriented services include file transfer and inquiry-response. Such a service is naturally implemented over a computer communications network using *packet switching*. Examples of real-time communications are voice telecommunications and television, which are more naturally implemented over the telephone network using *circuit switching* (television is, of course, broadcast locally, but national programming reaches local stations over permanent circuit-switched connections). However, most services that naturally fit into one of these paradigms are provided over the other to some limited extent today.

Since at least as early as 1975, both consumers and providers of telecommunications services have been dreaming of a day when telecommunications services can be obtained in a unified way. As the twenty-first century dawns, both packet- and circuit-switched networks enjoy ubiquitous deployment, many believe this dream of integrating services may become a reality within the next few years. Among those who believe there will be exactly one telecommunications infrastructure, there are credible views based on each technology.

Telecommunications systems, be they packet- or circuit-switched, have a number of elements in common. Those elements are introduced in the following section.

II. BASIC POINT-TO-POINT COMMUNICATION SYSTEMS

Telecommunications was born out of the need that occurs in ordinary life to communicate messages over distances at a reasonable cost. This was an economic need in the most basic sense; survival depended upon the reliable transfer of the message. For example, in the case of a battlefield situation, the reliability of having messages delivered by runners crossing battle lines would be relatively low and the cost would be high; multiple runners would be required and many of them would lose their lives.

Once ways of encoding messages had been invented and a basic understanding of electrical circuits was developed, the way was paved for electrical communications. Electrical wires could be strung between two distant points. Through the process of opening and closing switches, or making and breaking contacts, hisses and pops could be sent and then interpreted as different signals. Using such a means, any desired message could be sent without actually having to carry the physical message. Furthermore, delivery could be fast and more or less private, and receipt could be verified.

To use this astounding technological advance in conveying messages, all one needed was a wire strung between the two distant points, a simple mechanism having the appropriate switches, and a trained person at each end. Samuel Morse invented such a mechanism in 1837. He later invented a scheme for encoding messages, and then, under contract with the government of the United States of America, built the first telegraph system in 1843.

The distances over which signals can be transmitted over electrical circuits without excess quality degradation are limited. More distant places could be reached by means of relaying. A mechanism for automatically relaying messages was introduced, also with Morse's involvement, soon after the introduction of the first telegraph system, thus allowing a single pair of operators to handle the entire message transfer. Acknowledgements could then be sent from the most distant receiving end to the sending end, thus verifying message delivery. Obviously, much could be gained from automating the process of repeating the messages at intermediate points.

This point-to-point system between the end points can be thought of as a basic communications channel. Among its many properties, a channel has a transmitter-receiver pair at each end and a capacity, which defines the volume of information that can be conveyed over the channel.

Once the system is in place to take care of necessary communications, any excess capacity could be used for messages having lower priority; that is, messages for which people would place a lower value of delivery. Spreading the cost among more messages would then lower the per-message cost, and this would increase the amount demanded. To further increase the amount demanded, the per-message cost would have to be lowered. The first step along the way was to raise the number of messages that could be sent simultaneously over the same wire from one to two. Next the capacity could be raised to four, and so forth. The practice of using the same transmission system for multiple simultaneous communications is called *multiplexing*, and the system that implements multiplexing is called a *multiplexing system*.

A multiplexing system supports a number of *communication channels*, each of which carries the equivalent of one conversation. A multiplexing system consists of two terminal equipments, one for each end of the connection, and an electrical transmission circuit that interconnects the terminal equipments. Each terminal equipment has a *drop* side, which faces individual *channels*, and a *line* side, which faces the transmission circuit between the two terminal equipments. In addition, each terminal equipment has a multiplexor, which combines signals of individual channels for transmission over the line side, and a de-multiplexer, which separates the signal received over the line side for transmission over the individual channels of the drop side. The input (output) circuitry associated with each channel at each multiplexer is called an *input (output) port*.

The initial multiplexing systems were based on *frequency division multiplexing*. In a frequency division multiplexing system, there is a collection of single frequency signals called *carriers*. The signal for each call to be carried, for example, the signal from a telegraph transmitter, enters the multiplexor via one of the ports on the drop-side of the multiplexor. This signal, which is called the baseband signal, modulates one of the carriers. The multiplexor then combines the collection of modulated carriers into a single electrical signal that is sent over the transmission line connecting the multiplexing equipment. At the receiving end, the demultiplexer separates the combined line signal, through the use of demodulating and filtering circuitry, into the signals that belong to each of the individual channels and then routes the signal to the correct output port.

Because frequency division multiplexing systems are based on carrier signals, they are called *carrier systems*. A carrier system is analogous to the radio broadcast system, where each radio station uses a different carrier frequency and the listener tunes to the desired station. The electrical signal representing the voice messages and music broadcast over each radio channel is the baseband signal. And, when we tune into "870 on the radio dial," we are tuning to the station whose carrier frequency is 870 KHz. The airways provide the communication link between the radio stations and the radios.

Thus, a carrier system consists of terminating equipment that is used to combine and separate signals and an electrical circuit of some kind between its end points. Individual signals enter the multiplexing systems via input ports and exit from the opposite end via output ports. Multiple parallel communication channels can be achieved over the same electrical circuit through the use of carrier systems.

Generally speaking, if a carrier system is designed to interconnect switches over a great distance, then that carrier system is called a *long-haul carrier system*, otherwise, it is called a *short-haul carrier system*. Both short- and long-haul carrier systems can have a variety of capacities in terms of the number of calls that they can handle simultaneously. In fact, the capacities of the various carrier systems range from a few circuits to many thousands of circuits.

With increased multiplexing capability, a natural improvement would be in the area of message entry and delivery. Once a machine could fill the available channels, one would want to increase the number of channels by using better circuitry at the end points or better wires and so forth. This process could continue indefinitely with advances in wire manufacturing technology as well as transmitter and receiver circuitry, including the circuitry in repeaters at intermediate points. This explains, to a large extent, the tremendous investment in basic and applied research in so many different areas of technology by the telecommunications industry.

With increased message carrying capability, the per-message cost would drop and quantity demanded would increase. This, too, would go on in a continuous cycle. In addition, it is not hard to see that significant savings in cost might be achieved by replacing wires by radio transmission and receiving equipment. For example, to send signals across a large body of water, it may be less costly to use radios than to develop and deploy underwater cables. Guglielmo Marconi developed such a radio system in the mid-1890s.

Once a carrier system is in place, it is natural to separate the location of the end-user equipment from that of the carrier system's terminating equipment. The user's equipment is then connected to the carrier system through an access line, which is often referred to as a *local loop*.

So far, then, we have a point-to-point system for transmitting messages. This system consists of the end-user equipment, a local loop at each end, and a carrier system. The carrier system, in turn, consists of its terminating equipment, repeaters needed between the terminating equipment, and electrical circuits interconnecting the repeaters and the terminating equipment.

In the case of telegraphy, the objective of the system is to convey messages from some source to some destination, much as the objective of the postal service is to deliver letters. The primary difference is that the objective is accomplished electronically, and therefore, more quickly. The *service model* is that the system takes in messages at one end and delivers them to the other end; the service itself is *message delivery*.

III. SWITCHED CONNECTIONS AND NETWORKS

Once point-to-point connections are understood, it is not difficult to see why one would want to build a network so that point-to-point connections are not required between every pair of end points. For example, to cover connections among five major cities in the United States, say, New York, St. Louis, Dallas, Seattle, and Los Angeles, there are many options. One option would be to build one carrier system between each possible pair of the five cities and deliver each message using exactly one carrier system. A second option would be to build one carrier system each between New York and St. Louis, St. Louis and Dallas, Dallas and Seattle, Dallas and Los Angeles. In the second option, messages from New York bound for Los Angeles would go first to St. Louis, then to Dallas, then to Los Angeles; that is, the messages would be delivered by a network of carrier systems.

In the previous case, each message coming off of the carrier system would be examined to determine its appropriate next hop, and the messages could be routed hop-by-hop to their destinations. The service rendered would still be message delivery, but it would be done by a message relaying, or *message switching*, system. This system is the electronic equivalent of a postal delivery system.

A second option would be to provide a means of cross-connecting channels at the intermediate points to create a semipermanent connection between each possible source and destination pair. For example, consider the connection from New York to Los Angeles mentioned above. In St. Louis and Dallas, each input and output port of the carrier system would be terminate at a cross-connect. If there are twelve channels per carrier system, then a cross-connect would have 24 *appearances* for each carrier system, 12 for input and 12 for output. Suppose the New York to Los Angeles message stream is to use channel 1 from New York to St. Louis, channel 4 from St. Louis to Dallas, and channel 7 from Dallas to Los Angeles. Then, on the cross-connect in St. Louis, a patch cord would be placed between the channel 1 output appearance of the New York to St. Louis system and the channel 4 input of the St. Louis to Dallas system. Similarly, on the cross-connect in Dallas, a patch cord would be placed between the channel 4 output appearance of the St. Louis to Dallas system and the channel 7 input of the Dallas to Los Angeles system.

This second option provides essentially the same effect as having a point-to-point carrier system between each city pair, but its implementation better utilizes available capacity because there is broad flexibility in assigning circuits between each city pair. This option provides the same service as before (message delivery), but it provides the service over semipermanent *circuit-switched* connections, where the connection is accomplished manually. The system could be rearranged on request using the same system that transfers regular messages; such requests would be a form of *signaling*.

A third option would be to replace the manual cross-connect by some kind of electromechanical switch and allow the circuits to be built on-the-fly by the introduction of *electrical signaling*. That is, a party in New York wishing to send a message to Los Angeles would have the ability to signal his desire to the system, the system would automatically build the connection from New York to Los Angeles, the party would send its message, and then the system would tear the circuit down. The service

is still message delivery, but it is delivered over a temporary circuit-switched connection; the connection lasts only as long as it takes to transmit the message.

A fourth option would be to replace the cross-connects by computers and to terminate each of the channels of the carrier system at the computer. The destination address of each message would be included with the message, and the computer would determine the next hop to which the message would be sent. Thus, as in the manual system, the message would be conveyed hop-by-hop until it reached its destination. As in the first option, this is a message switching system, but now the routing is handled by a computer called a *message switch*.

The introduction of switches into a communication system results in fundamental changes. First, it results in a natural separation of access lines from carrier systems, with the switch being at the interface; the purpose of the access line then is to connect to the switch, and the purpose of the carrier system is to interconnect switches. Second, signaling is required within the system to enable it to respond dynamically to the desires of the end user. A third, less obvious, change is that the introduction of switches enables tremendous flexibility in allocating resources to end users; that is, users can be allocated resources dynamically for short periods of time on demand. This change opens the door for a pay-by-use service policy, a policy that is absent in point-to-point, permanently connected systems.

A switched telecommunication network then has at least the following components: the end user's terminal equipment, switches, local loops between the customer premises and the switches, carrier systems between the switches, and a control system to process the signaling information.

IV. EARLY EVOLUTION OF THE TELEPHONE NETWORK

Voice telecommunications began with the invention of the telephone in 1876. First, we learned to talk between two end points at short distances; end users rented their telephones, but provided their own wires. Next, telephone companies began providing wires, and manual cross-connects were added to allow a calling party to connect to any other party within a local area through the assistance of an operator. Finally, the scattered local switches were connected together in a network, thus allowing subscribers to connect to other subscribers at distant locations.

There is a fundamental difference between a person-to-person voice service and a message-switching service. Unlike message switching, a voice conversation is carried out in real time; the parties at each end of the connection are present and participating simultaneously. This means that the system must be able to set up a communication path between the end points quickly and also guarantee quality of service for that connection for the duration of the call.

The problem posed by these factors is that a separate circuit is required for each pair of individuals involved in a conversation. Between a person's residence and a switch, the problem is solved by the installation of a local loop. Between two switches, the problem is solved through carrier systems, as described in the previous section.

As the capacity of carrier systems was being improved, there was also a persistent advance in switching technology, including signaling capability. As the signaling capabilities were improved, end customers gained the ability first to dial their own connections within neighborhoods, then within metropolitan areas, then within countries, and finally to anywhere in the world.

The initial carrier systems were based on frequency division multiplexing as described in Section III. In the long-haul part of the network, the frequency division multiplexing system was organized in a hierarchy called the *L-multiplex hierarchy*. At the lowest level, twelve voice circuits were multiplexed together to form a group. Then five groups were multiplexed together to form a *supergroup*. Ten supergroups were combined to form a *mastergroup*. The earliest long-haul transmission systems used coaxial cable, and they carried one mastergroup per cable. Similarly, the microwave radio systems carried one mastergroup per channel. Later, these capacities were raised considerably.

A channel capable of carrying one telephone conversation between two circuit switches is called a *trunk*. The number of trunks between two switches is then equivalent to the maximum number of calls that can be handled simultaneously between the switches. The actual number of trunks installed between switches can be in the hundreds of thousands and is dependent upon the desired rate of calls to be handled by the trunks, the holding time of the calls, and the target quality of service. Quality in this context is measured in terms of the proportion of failed call setup attempts, called *blocking probability*. A tight balance between the number of trunks, placement of switches, and the sizes of the switches is required to achieve high quality at minimal cost. The art of designing a network of switches and trunks is called *traffic engineering*.

The actual placement of multiplexers and switches is determined by cost considerations. For example, suppose that instead of a residence, the telephone service is terminated at a company location that has 1000 employees, all of whom need to use telephone on a regular basis. It would then seem reasonable to develop a class of small switches that can be located at the company premises; such a small switch is called a *branch exchange*, and it may be privately owned, in which case it is called a *private branch exchange*. If this location is at some distance from a telephone switching office and there is a demand for a large number of simultaneous calls, it would also seem reasonable to use a special carrier system, a *loop-multiplex carrier system*, rather than a collection of local loops to interconnect the telephone company's switch and private company's branch exchange.

So far, then, we have the following system components: switches located in strategic places, carrier systems between the switches, local loops between switches and residences, and telephones at the residences. If the switches are far apart, the carrier systems are long-haul, and if the switches are close together, the carrier systems are short-haul.

This model of a network of switches, carrier systems and trunks, and local loops generalizes to almost any conceivable combination in the real world. If one can draw a diagram of any shape, size, or form using these components, then someone has probably built it to meet some need at some time.

The system just described in terms of voice communications provides a service called *circuit switching*. A message switching system could be built over a circuit-switching system; one would simply place message switches wherever one chose and then connect the message switches together using circuits. Therefore, circuit switching can be viewed as a *transmission technology* while voice and message switching can be considered *service technologies*.

Morse's system of dots and dashes for encoding characters and words is related to the system of ones and zeros, called *bits*, that are now familiar in computing systems. Messages consist of strings of characters, characters consist of strings of ones and zeros, and the representation of the message by the string of ones and zeros is referred to as *digital* representation. There would be a strong motivation to invent a machine that could convert digital representations to electrical signals that resemble voice signals so that messages could be transmitted over ordinary voice circuits; such a machine is called a modulator. Once the leap is made to convert data-like signals to voice-like signals, the necessity for a demodulator to perform the reverse process is obvious; otherwise one cannot recover the original signal. The combined machine is called a *modem*. And, once the process of converting between voice-like and data-like signals is well understood, the desirability of converting actual voice signals to digital representation for transmission naturally arises.

A continuous signal, such as voice, in its natural setting is called an *analog signal*, and the process of converting an analog signal to a digital representation is called *analog-to-digital*, or A/D, *conversion*. The combined system that performs A/D conversion, transmission of digital signals between the end points, and D/A conversion is called a *digital transmission system.*

The question of relative efficiency of analog versus digital transmission of all kinds of signals then arises in a natural way. The first step in settling this question was ushered in with the advent of the digital transmission system in the late 1950s. This first commercially successful digital carrier system was a short-haul carrier system having a capacity of 24 voice circuits per channel, and it was called the *T1 Carrier System.*

Unlike the earlier analog transmission systems, which use *frequency division multiplexing* based on carrier modulation, the T1 Carrier System uses *time-division multiplexing* and transmits bits at baseband. In time-division multiplexing systems, time is organized as a sequence of frames, and frames are divided into slots. Each circuit is allocated one time slot in each frame, and the sequence of slots in successive frames gives the illusion of a full-time circuit, provided that the slots have sufficient capacity.

Analogous to the L-multiplexing hierarchy there is a digital hierarchy, which developed between 1960 and 1980. The lowest level is DS-1, and it carries 24 voice calls operating at 1.5 Mbps. The next level is DS-2, and it carries 96 voice calls and operates at 6.2 Mbps. The next level is DS-3, and it carries 572 voice calls and operates at 45 Mbps. Because T1 carrier systems have the same bit rate as a DS-1 system, lines having the DS-1 rate are sometimes called T1 lines, and similarly, DS-3 lines are sometimes called T3 lines.

By the 1960s, computer concepts were integrated into virtually all parts of circuit-switched networks, and the computer and communications businesses were on a collision course. Switching and carrier systems had been improved to the point where they had enormous capacities, cost had come down significantly and were continuing to fall, and demand for telecommunications services, including data, was increasing.

V. EARLY DATA COMMUNICATION NETWORKS

During the late 1950s, the line between data processing and and telecommunications became increasingly blurred. Telephone companies had significant computer skills resulting both from the fact that telephone switches had long since been controlled by (permanently) stored programs and the fact that expertise had become necessary to manage large businesses. Telephone companies had also realized that once voice is in digital format, one can store it to provide voice mail and also process it in many other ways.

This brought up the issue of whether or not the telephone company could process data commercially, and the resolution of this issue was part of the now well-known consent decree of 1956. In that decree, the Bell System was restricted from entering computer and information services markets.

There are a number of motivations for improving technologies, all of them driven by a desire to lower the cost of providing current services and to increase the possibility of being able to offer new services. In the early 1960s, motivated by the desire on the part of a large number of individuals to access a relatively small number of serious computing resources, the idea of time-sharing was conceived. Time-sharing became commonplace by the late 1960s.

Considerable interest in alternate forms of data transmission technology was evident during the 1960s. L. Kleinrock's pioneering work on *message-switching* systems in the early 1960s is generally acknowledged to be a key catalyst in this movement. One advantage of message-switching systems over circuit-switched connections is *statistical time-division multiplexing.* At each message switch along a route, messages can come in from a number of incoming lines or terminals simultaneously, and it is not always possible that these messages could go out at the same time. The solution is to form a waiting line for those messages that are to be transmitted over the same line, and then transmit them in sequence as time becomes available.

Packet switching, a name coined by D. W. Davies in 1965, is a natural extension of message switching, although message switching is really a service technology and packet switching is a transmission technology. In packet switching, messages are divided into smaller units called *packets* for transmission. Time savings are achieved in multihop connections by transmitting packets in parallel over the various hops of the network.

As an illustration of the gains that can be achieved from packet switching, consider a three-hop connection. If a long message is transmitted as a single unit, then the transmission time will be three times the time required to transmit over one hop. If the message is broken into ten small units, then the total transmission time will be the time required to transmit the entire message over one hop plus the time required to transmit one-tenth of the message over the second and third hops. The reason for the savings is that earlier packets can be transmitted over the second and third hops while later packets are being transmitted over the first hop. Thus the total time required is reduced from three message transmission times to approximately 1.2 transmission times. The relative savings increase when messages are divided into even smaller units up to a certain extent or if the route has more hops.

During the 1960s, engineers and scientists were beginning to gain tremendous insights into the savings in transmission costs and increases in system capacity that could be achieved by digitally processing signals. For example, by including special purpose circuitry at the interface between modems and telephone lines called *equalizers*, more bits per second could be pushed through voice telephone lines. Also, by removing redundancy in data files before transmission, the data transferred per bit transmitted could be maximized.

Significant advances were made in the understanding of data transmission, information theory, coding theory, and signal processing. At the end of the 1960s, many of the advancements in computers and communications were brought together when the world's first packet-switching network, the ARPAnet, came into existence.

As the per-unit cost of electronics decreased in the early 1970s, the proliferation of computer equipment increased. Spurred by the demand for increased services, mainframe divisions and minicomputer divisions within and across company boundaries invented their own techniques for implementing those services. Consulting and maintenance expertise were spread thin throughout the computer industry, and this led major players to try to standardize techniques, at least within their own organizations. The situation was so bad that the very survival of some companies was at stake.

As equipment became more and more distributed within company locations, the demand for sharing peripherals and exchanging information among the computers increased. This led to the development of proprietary means of exchanging data among computers over small areas. The resulting on-campus networks were the earliest examples of *local area networks* (LANs). Numerous LAN protocols were invented, with Ethernet and its derivatives being the most commonly recognized and popular survivor.

The mid-1970s was a period of rapid change in telecommunications. First of all, computer communications was emerging as a field; the idea of development of standard protocols to handle just about any communication need was being formalized. Second, the use of long distance had increased to the point where managing the long-haul system on a heavy calling day, such as Mothers' Day, was no longer possible; the only workable solution was to disallow any alternate routing within the long-distance network during such periods. Third, miniaturization of electronic circuits was progressing, and the first microcomputer appeared.

To simplify software development and facilitate system deployment, communication protocols are organized in *layers*, and the collection of layers is called the *protocol stack*. At the top of the stack is the software that interfaces to the application, and at the bottom is the software that deals with the transmission of bits over a physical facility. A computer program that implements a layer of a protocol stack in a system is called an *entity*. In general, from a service point of view, end systems implement all layers of a protocol stack and intermediate systems implement only the bottom three layers. Layer 3 is called the *packet layer* because it provides a packet transfer service between the layer 4 entities implemented in different end systems. Thus, the packet-switching part of a communications network consists of the interconnection of the layer 3 entities in all the systems of the network.

With miniaturization and integrated manufacturing of electronic circuitry, the mid-1970s saw the advent of the microprocessor. This innovation led to inexpensive processing and resulted in technical revolution in many areas. Among the most revolutionary event was the advent of the personal computer and the accompanying explosion in the number of computers owned by individuals. By the late 1970s, personal computers were already in their second generation, and there were numerous players in the market, led by Apple Computer. User groups, focusing on various personal computers, had begun to emerge and they communicated with each other, mainly by exchanging e-mail over message-switching systems and by posting messages to bulletin boards. Many of these bulletin boards were implemented on personal computers attached to networks via modems.

At the same time, companies were developing products such as data processing platforms, front-end processors, packet switches, retail store controllers, electronic cash registers, and anything else that could conceivably be built around microprocessors. This raised the demand for increased processing power, which was met by an evolution from the 8-bit to the 16-bit processor by the end of the 1970s. The demand for low-cost memory shot through the roof, and this led to enormous efforts to advance memory manufacturing, which at that time seemed to have dead ended at about 16 kilobytes per chip. In addition, special-purpose microprocessors were beginning to appear in the marketplace, and developers began to implement the ideas that had been published in the 1960s.

During the mid-1970s, LANs began to realize success in the marketplace. At this time, their protocols were proprietary, and, with the success of LANs came the demand for standardized products in the LAN area. This demand led to the formation of the IEEE 802 Committee, which developed a set of three standards for LANs by the mid-1980s, one each for *token bus*, *token ring*, and *carrier sense multiple access* (CSMA). While all of these standards were widely deployed, the CSMA version, of which Ethernet is an example, dominated the market by the end of the 1980s.

In the late 1970s and through the 1980s, there were two popular views on how packet-switching networks should be implemented: datagram and virtual circuit. In a datagram service, each individual packet makes its own way across the network, with routing decisions from source to destination being made on each individual packet. Each packet, then, must contain complete addressing information, as in a postal letter, but the switches do not need any state information about the service being rendered. Datagram service was the service of choice for the ARPAnet.

By contrast, a virtual circuit connection uses a call connection phase in which the path that the packets will be taking throughout the duration of the call is set up. Thus, information overhead for each packet can be minimized, but the packet switches must maintain state information on the calls. As mentioned earlier, communication protocols are organized in layers. The three lower layers of the protocol stack that implement this idea are collectively referred to as *X.25*. The third, or highest, layer of X.25 is the packet layer, as mentioned above in the context of general protocols. Layer 2 is called the *frame layer* because layer 2 provides a frame transfer service between the layer 3 entities of two systems. Virtual circuit service was the service of choice of the International Standards Organization (ISO), whose approach was called *Open Systems Interconnection* (OSI).

The 1980s saw continued effort in the networking, personal computer, and workstation areas. A fundamental breakthrough in deployment came in the beginning of 1983 when TCP/IP was deployed throughout the ARPAnet. Although it was not clear at the time, this change clearly delineated the networking element from the end-system element, and ultimately resulted in the IP-network, or Internet. Nodes of the network could now be thought of as coming in two flavors: end systems (hosts) and intermediate systems (routers). That is, the network more or less became a network of routers with the end systems being attachments to the networks or routers.

This separation of functions was extremely important because it partitioned the networking software development effort into manageable portions. End systems ran the TCP protocol, and applications could be built above TCP in the end systems. Both routers and hosts ran IP, and protocols could be developed below IP to provide transmission between routers or between hosts and routers.

At the same time, the *Domain Name System*, which allowed translation between descriptive names for hosts and their Internet addresses, was introduced.

Layers below the IP layer gradually evolved during the 1980s. While X.25 was a very successful protocol, it was somewhat too heavy in that it required packet acknowledgements over each hop of the connection. It was realized that intermediate acknowledgements could be eliminated with very little loss in performance and, instead of involving a third layer, frames could simply be relayed at the second layer. The protocol that implements this idea, *frame relay*, was developed in the late 1980s and deployed in the 1990s. It is used to transfer frames containing packets of any type between packet switches. The net effect is that packet transmission facilities of appropriate capacities between packet switches of networks, especially intranets, can be built economically using frame relay services.

Rapid deployment of computer communication networking products continued throughout the 1980s. Much of this was attributable to two factors: free availability of the Unix operating system on a wide variety of platforms, and free availability of Unix-based networking software. That is, the Unix operating system had become the operating system of choice among software developers in university and research environments, and the advancements in communications protocol development became immediately available to those using the Unix operating system.

Much of the computer communications networking activity of the 1980s focused on building routers to facilitate deployment of networks. Information and people on different networks could be reached if one could figure out the right combination for getting from one network to another. Thick books that gave the relationships among networks became available, and knowledgeable people could move information from a host on one network to a host on a different network; internetworking was beginning to arrive. While computer communications was advancing, the telephone network was not standing still. This topic will be addressed in the next section.

There were a huge number of interconnected networks by the end of the 1980s, mostly deployed in high-tech companies and in academia, and with a significant proportion of total bandwidth in the United States funded by the federal government. Yet, at this time, this technology was largely unavailable to the average person.

From this point, it became natural to move to the point where all networks were considered to be domains of the same network and the domain name system could become ubiquitous. This brought us to the early 1990s, when the potential of networking as a commercial enterprise first began to be realized.

VI. COMMON CHANNEL SIGNALING IN THE TELEPHONE NETWORK

During the early 1970s, it was discovered that the overload problem in the long-distance network on heavy calling days was an artifact of the signaling methodology.

Specifically, signaling took place over the same circuits that would eventually carry the call. Constructing a circuit was done sequentially; attempts to find an open circuit would use virtually any available outgoing path. Two unfortunate results were that signaling consumed a disproportionate amount of the trunk resources and the call itself consumed many times the minimum resources required. For example, a call from Miami to Atlanta might consume any number of links, including some to and from the West coast, before finally reaching its destination. Signaling messages themselves also consumed trunking resources so that the effective holding time of the call was increased, and, in addition, signaling might involve numerous trunks that ultimately would not be part of the call, thus adding additional load to the network.

With alternate routing turned off, the situation would be drastically different. For example, a semipermanent route might be set from Miami to Jacksonville and then to Atlanta. If either Miami to Jacksonville or Jacksonville to Atlanta had no available trunks, a call from Miami to Atlanta would be blocked and the caller would be given a busy signal.

A solution to this problem was conceived in the early 1970s and its implementation began in the mid-1970s with the introduction of *common channel interoffice signaling*, or CCIS. Telecommunications engineers had realized that signaling and network management functions could be separated from the call-carrying functions, and separate networks could be designed for each. The call-carrying network would still be circuit switched, but it was decided to implement common channel signaling over a packet-switched network, which was called the *common channel signaling network*. This network was implemented over a protocol stack *signaling system* 7, or SS7.

Thus, from the late-1970s onward, the telephone system became two integrated networks: the SS7 network, which is packet-switched, and the call-carrying network, which is circuit-switched. In order to set up a call, packets are sent around the network, and when it is ascertained that the called party can accept a call and a path of trunks connecting the calling party to the called party is available, the call is connected. No call-carrying capacity is wasted in trying to set up calls for which there were no resources.

By the mid-1980s, new ways of using the SS7 network had emerged. One such use is the introduction of new services, such as 1-800 and call-forwarding, into the telephone system. This was accomplished through enhancements to the control processes of the electronic switches attached to the SS7 network and the introduction of *service control points* attached to that network. The new control software in the switches enabled the switches to communicate with the service control points when necessary to realize the new service. This concept is called the *intelligent network*, and originally, each service control point was dedicated to a specific service. Later the role of service control points was expanded to handle collections of services, and the new concept was called the *advanced intelligent network*.

VII. RECENT DEVELOPMENTS IN TELECOMMUNICATIONS

The U.S. Department of Justice filed an antitrust suit against AT&T in 1974. This suit was settled in 1982 leaving a fragmented U.S. telecommunications business in place in 1984. In the meantime, many major problems standing in the way of the deployment of fiber-optic communications systems had been solved starting in the mid-1970s, and significant advancements had occurred in fiber-optic technology by 1984. These advancements made it possible for more companies to deploy fiber-optic systems at a reasonable cost, and, hence, from a technological point of view, opened the way for more competition in the long-distance market.

Other innovations of the 1980s include the widespread availability of community antenna television (CATV), or cable television, an increase in the use of communication satellites, and the advent of cellular telephony. Cable television, begun as a niche business in the late 1940s to provide television services in remote or mountainous regions, had evolved a new role, where it facilitated the delivery of a wide choice of television programming to its subscribers. Deployment of CATV continued throughout the 1980s and 1990s, and, in addition, satellite-based television was widely deployed. In the latter half of the 1990s, satellite-based digital TV with 18-inch receiving antennas was deployed.

As the cost of electronics and electronic systems plummeted during the 1970s, demand for mobile communications, especially telephone, increased. Because additional frequency spectrum to support the increased demand was not available, techniques for using the existing spectrum more efficiently were aggressively sought. With the improvement in networking technology and electronic switching, the concept of cellular mobile telephony soon developed.

In cellular telephony, a metropolitan area is partitioned into a grid of smaller areas called *cells*. At the center of each cell is a *cell site*, which consists of antennas, transmitters and receivers, and miscellaneous control systems. Each cell is allocated a subset of the available frequency channels, with multiple cell sites within the same metropolitan area having the same sets of frequencies. The system is designed so that transmissions from cell sites having the same collection of frequencies do not interfere with each other; cell sites having the same frequencies are farther apart than those having different sets of frequencies, and, in addition, transmitter power levels are carefully controlled.

The system just described uses analog transmission and frequency division multiplexing. Other systems using digital transmission were developed during the 1990s. One of these digital systems uses a combination of frequency and time-division multiplexing and the other uses a combination of frequency division multiplexing and *spread spectrum multiple access*. These new technological innovations have raised both the quality and the voice call-carrying capacity of cellular telephony. In addition, these innovations have lead to serious research in the area of wireless data access.

An advantage of wireless cellular access over traditionally wired local access is that it can be deployed to an area quickly. Specifically, once a telephone switch is set up in a local area, a cellular telephony system can be deployed, and telephone service can be provided in the local area without installing any local loops. After a long-distance infrastructure is deployed, worldwide telephone service can be realized almost overnight. Because of these properties, cellular telephony is finding a market in countries where very little traditional telephone infrastructure exists today.

The late 1980s and early 1990s also witnessed continuing advances in fiber-optic technology. With the rapid development of optical fiber-based systems, a third multiplexing hierarchy, the *synchronous digital hierarchy* or SDH, was developed. The lowest level of the SDH is OC-1, which has a capacity of about 50 megabits/second and carries the equivalent of 672 voice calls. Popular levels of SDH include OC-3 at about 150 megabits/second, OC-12 at about 600 megabits/second, OC-48 at about 2300 megabits/second, and OC-192 at about 9600 megabits/second. At least one commercial packet switch that can handle OC-192 lines exists today, and the potential to develop packet switches that can handle line rates of OC-768, the equivalent of 500,000 voice calls, is a frequent topic of discussion in research circles. Today, one can buy off-the-shelf equipment to demultiplex an OC-192 stream into 16 OC-12 streams.

During the 1990s, a number of technological refinements of and extensions to basic internetworking capabilities were introduced. Among these were 100 megabit/second Ethernet, wireless LANs, Ethernet hubs and switches, and gigabit/second Ethernet.

But the big news of the 1990s was the widespread deployment and commercialization of the Internet. One driving force was the advent of the *World Wide Web*, or just Web. The Web consists of a collection of server locations called web sites, which are accessed over TCP/IP connections using a web browser, the two most popular of which are Netscape's Navigator and Microsoft's Internet Explorer. *Hypertext mark-up language* (HTML) is used to format web sites, and the *hypertext transfer protocol* (HTTP) is used for interaction between clients and servers, or equivalently, between web sites and web browsers.

Because a web browser's interface is graphical and intuitive, anyone can learn to use the Web quickly. This has led to the creation of many new companies, the development of many web sites, and the expansion of networking services into many millions of homes worldwide. Thus, in the 1990s, packet-based networking services spread from the academic and research environments into the world's businesses and homes, becoming a common household utility on par with the telephone and television.

With the widespread availability of packet-based services, new developments began to occur to provide higher speed data access from the home. Part of this expansion took place with *Integrated Services Digital Network* (ISDN) service, which provides integrated voice and digital services at approximately 144 kilobits/second. This technology developed in the 1980s but did not see broad acceptance in the market until higher speed data access was needed.

The use of SS7 in controlling and managing circuit-switched networks makes it necessary for telephone companies to maintain leading edge competence in both circuit- and packet-switching technology. One consequence of this expertise was the proposal of *asynchronous transfer mode* (ATM), as a solution for integrating multimedia communications using a packet-based technology.

Initially, it was thought that the ATM protocol suite would span the entire range of services from application to application. That is, ATM would be deployed in end systems, in intermediate systems, and everywhere in between. Hundreds of companies worked to develop ATM standards at all layers, and many good ideas came out of those efforts. However, probably because of the broad deployment of TCP/IP based applications and continuing advancements in Ethernet technologies in the local area, ATM has been more successful as a transmission technology than as an end-to-end service technology. Indeed, much of the world's IP datagram transmission takes place over frame relay systems implemented over ATM.

Additional high-speed access from the home was offered through a cable-modem service built over cable television facilities. Still more digital capacity to the home was offered by *Asymmetric Digital Subscriber Service* (ADSL), which uses ordinary twisted pair telephone cable, and which became widely available at the end of the 1990s. As the year 2001 gets under way, there is a continuing effort to increase the rate at which data can enter the home, especially using technologies that can be implemented over either the existing telephone wires or cable television facilities.

As the technologies already discussed were introduced, improved, or supplanted, the firms founded to develop and market these technologies also changed. Before discussing future trends, we present a brief overview of the evolution of the telegraph and telephone industries in the United States from 1837 to mid-February, 2001.

VIII. EVOLUTION OF THE UNITED STATES TELECOMMUNICATIONS INDUSTRY: 1837–2000

The electric telegraph, a system of communication employing electrical apparatus to transmit and receive information using a code of electrical pulses, was invented in 1837 by Samuel Finely Bresse Morse, an American artist and inventor. In Great Britain in the same year, the British physicist Sir Charles Wheatstone, in collaboration with the British engineer Sir William F. Cooke, also invented an electric telegraph. However, Morse was the first to construct an experimental telegraph line and on May 24, 1844, Morse sent the first message between Washington, D.C., and Baltimore, MD. Morse filed for a patent on his invention in 1838. The patent was granted in 1848.

In 1851, the New York and Mississippi Valley Printing Telegraph Company was formed to build a telegraph line from Buffalo, NY, to St. Louis, Mo. In 1856, the company was reorganized and renamed the Western Union Telegraph Company. By the end of 1861, Western Union had built the first transcontinental telegraph line and there were 2250 telegraph offices in operation nationwide.

In 1854 Cyrus West Field, an American merchant and financier, obtained a charter giving him the exclusive right for the next 50 years to use the coast of Newfoundland for a transoceanic cable terminal. In collaboration with American manufacturer Peter Cooper and others, he formed two companies to lay a transatlantic cable: The New York, Newfoundland and London Telegraph Company and The Atlantic Telegraph Company, The British and United States governments helped fund Field's efforts. In July 1866, the first transatlantic cable was complete.

The telegraph was a very successful technology and Western Union's business grew rapidly throughout the early 1900s. By 1943, after acquiring Postal Telegraph and some 500 other competitors, it was the largest company in its field. However, telegraph rates rose over time while the rates of its major competitor, the telephone, declined. During the 1960s, the company moved into a number of other

related services, such as time-sharing computer systems, teleprinters, and satellite communications. Western Union Telegraph Co. was reorganized in 1988. The reorganized firm, Western Union Corp., focused on money transfers and other financial services. In 1991, parts of the company were sold and the company's name was changed to New Valley Corp. New Valley Corp. entered bankruptcy proceedings in 1993 and sold the last of its major holdings, Western Union Financial Services Inc., to First Financial Management Corp.

Alexander Graham Bell, an inventor and teacher of the deaf, was born in Scotland in 1847. Bell immigrated to Canada in 1870 and then to the United States in 1871. Bell was working on a multiple telegraph when he developed the basic ideas for the telephone in 1874. On March 10, 1876, he successfully transmitted the first telephone message. Bell patented the telephone in 1876 and, facing financial difficulty, he offered to sell his patent to Western Union Telegraph Company for $100,000. Western Union declined. In 1877, Bell and three associates, Thomas Watson, Thomas Sanders, and Gardiner Hubbard formed the Bell Telephone Company as a voluntary association. In 1878, the company was incorporated.

By 1877 there were four basic Bell patents. These initial patents protected telephone instruments, not wires or switches, and originally the company rented instruments to individuals who were responsible for providing their own connecting wire. This proved to be an inefficient method for connecting a large number of individuals so subscriber service was initiated. Subscribers paid to be connected to one another through equipment provided entirely by the various telephone companies. Calls were placed through manual exchanges. When a subscriber wanted service they picked up their handset and a small light on a switchboard alerted an operator that the subscriber wanted service. The operator then inserted a cable into the jack on the switchboard that corresponded to the caller. The caller told the operator the called party's name. The operator used another cord adjacent to the first to plug into the called party's jack and then operated an electrical switch that connected ringing current to the called party's telephone. When the called party answered, the operator disconnected leaving the caller and the called party connected. The first commercial exchange, which came into service in New Haven, CT in 1878, connected 21 subscribers. Although these initial exchanges provided only intracity service, telephone service expanded rapidly. There was one exchange with 21 customers in 1878 and there were 400 exchanges with 140 subscribers in 1885. Telephones per capita increased as well. In 1880 there were 1.1 telephones per thousand and by 1894 there were 4.1 telephones per thousand.

In 1878, Western Union Telegraph and the Bell corporation both operated telephone and telegraph services. The Bell Company filed a patent infringement suit against Western Union and won the suit in 1897. As a result of the settlement, Western Union sold its 55-city telephone system to Bell Telephone and agreed to leave the telephone business while Bell Telephone agreed to leave the telegraph business and pay Western union 20% of their telephone rentals for the life of the Bell patents.

In 1882, Bell Telephone acquired Western Electric Company, the largest manufacturer of electrical equipment in the United States, from Western Union. Bell Telephone's long-distance service was expanding and in 1885, Bell Telephone created a subsidiary, the American Telephone and Telegraph Company (AT&T), to finance, build, and operate the long-distance system. In 1899, the company was reorganized, and AT&T became the parent company for the entire Bell system.

Bell's fundamental equipment patents expired in 1893 and 1894. This led to the entry of competitors, first in areas not served by Bell Telephone, and then in major cities where they were in direct competition with Bell. In 1907, a group of investors led by American financier J. P. Morgan took control of the company and Morgan appointed Theodore Vail president. Vail, in an effort to restore AT&T's market share, began an aggressive program of patent control, buyouts of independent companies, and refusal to allow competitors to connect to AT&T's long-distance network. In 1909, AT&T purchased Western Union. This acquisition, combined with Vail's earlier actions, caused AT&T's competitors to charge that it had become a trust, and the Department of Justice (DOJ) threatened antitrust action. In response to this threat, AT&T entered into an agreement with the DOJ known as the Kingsbury Commitment. AT&T agreed to sell Western Union and to allow independent local telephone companies to connect to its long-distance lines. AT&T and the independents exchanged local territories creating effective monopolies in each local area. Many small towns and rural areas in the United States were and are still served by local companies that have never been part of the Bell System (e.g., in 1982, Bell system companies served 81% of the telephone lines but only 41% of the territory in the United States). AT&T, however, maintained its monopoly on long-distance service and retained control of Western Electric. The result of this regulatory structure was to transform the industry from one in which there was competition in the local areas with a monopoly in long-distance service to a fully monopolized regulated industry.

The Bell system was subject to both state and federal regulation. Regulators approved pricing structures, called tariffs, partially based on the cost of providing the service. In a telephone system, parts of the system will be used for both local and long-distance service. State regulatory commissions typically wanted to shift a portion of the cost of local service to long-distance service. This cost allocation process, in which the cost of commonly used plant is divided between local and interstate jurisdictions, is termed *separations*. AT&T opposed this practice. In 1930, the U.S. Supreme Court ruled in *Smith v. Illinois Bell* that some of the cost of the local network had to be allocated to the interstate service. In 1934, the Federal Communications Commission (FCC) was established by the Communications Act. The Act established the Joint Board. This board was composed of federal and state commissioners and was responsible for analyzing the separation of costs and making recommendations to the FCC. In 1947, the Joint Board produced the first separations manual.

In 1949, the DOJ filed an antitrust suit against Western Electric alleging monopolization of the telephone equipment market. In 1956, the DOJ sought to force AT&T to divest itself of Western Electric and to break Western Electric into three companies. The suit was settled by a consent decree in 1956 generally considered to have favored AT&T. Under the decree, the DOJ recognized that the structure of AT&T was not an illegal anticompetitive arrangement and AT&T was allowed to keep Western Electric as a separate subsidiary. However, the decree limited Western Electric to manufacturing equipment for the Bell System and to contract work for the government. Western Electric sold its small nontelephone subsidiary Westrex to Litton Industries and its holdings in Northern Electric (now known as Nortel Networks) to the public. In addition, AT&T was restricted to providing common carrier communications services. This meant that the company was restricted from entering the computer and information services markets.

During the 1950s, there was an increase in applications to the FCC for permission to build private microwave transmission systems. In 1956, the FCC began considering policies to allocate frequencies above 890 MHz. AT&T, Western Union, and the private telephone companies all argued that the use of these frequencies should be limited. Specifically, they did not want private microwave systems to operate in frequencies where common carrier service was available and they argued that the use of the frequencies should be abandoned if common carrier service became available. In 1959, in the Above 890 Decision the FCC ruled that the frequencies could be used by both common carriers and licensed private users but private users were limited to building systems for their own use. In response to this decision, AT&T filed a new tariff, Telpak, that offered large discounts for groups of private lines. This tariff substantially decreased the incentive to build private microwave systems. As was usual, Telpak prohibited resale and sharing.

However, the Telpak tariff also indicated a potential profit opportunity. A company might identify users that were too small to use the Telpak discounts, set up a microwave transmission system, and share it among those users. In 1963, Microwave Communications, Inc. (MCI) decided to move into this market. MCI filed an application with the FCC to set up a limited microwave transmission system between St. Louis and Chicago. This system would be a public system because its purpose was to sell service so the application could not be handled under the 890 Decision. Again AT&T, Western Union, and the local carriers opposed the request. In 1969, the FCC decided in favor of MCI. AT&T appealed the decision. The FCC then moved beyond the 1969 decision to a fundamental policy change and in 1971 the FCC announced the Specialized Common Carrier (SCC) decision. Under this decision individual licenses for microwave routes were still required but the decision allowed entry by firms that wanted to begin public microwave transmission service. In addition, the FCC ordered that the established common carriers be required to interconnect with any new carriers.

MCI began service between St. Louis and Chicago in January, 1972. The company expanded its service and by 1973 had a basic network to many major cities. In 1973, MCI filed a complaint with the FCC alleging that AT&T was, in general, refusing to allow interconnection and that when AT&T did allow interconnection the terms were not as favorable as those given to Western Union. Between 1973 and 1978, MCI aggressively attempted to expand its service offering beyond simple public line to foreign exchange (FX) and long-distance calling, i.e., Message Telecommunications Service (MTS). AT&T responded with tariffs and appeals of MCI's interconnection requests. In 1978, a settlement was reached that set up a structure for payments between AT&T and its competitors. The FCC also issued a decision invalidating AT&T's restrictions on resale of its services. Switched service competition was established and other long-distance rivals such as United Telecommunications (United Telecom) began to enter the market.

In 1974, both MCI and the DOJ filed antitrust suits against AT&T. The DOJ antitrust went forward and after considerable delay went to trial in 1981. In January, 1982 the suit was settled by an agreement known as the Modified Final Judgment (MFJ). Under the MFJ, AT&T was required to divest itself of its local exchange telephone service. AT&T was allowed to retain Western Electric and Bell Laboratories.

The process of breaking up AT&T took two years; formal divestiture took place on January 1, 1984, when the old Bell System was replaced by 22 Bell Operating Companies, (BOCs), 7 Regional Bell Operating Companies (RBOCs,) and AT&T. The RBOCs, commonly called Baby Bells, were Bell Atlantic, BellSouth, NYNEX, Ameritech, U.S. West, Southwestern Bell, and Pacific Telesis. AT&T's Central Services Organization became Bell Communications Research Inc. (Bellcore). Bellcore's original purpose was to provide technological expertise and research to the Bell Operating Companies. Under the MFJ the seven regional companies were made responsible for all local calling, some intrastate long-distance business, customer access to long-distance networks, and directory advertising. The RBOCs could not compete with each other in the local calling markets. The RBOCs were also restrained from entering the interstate long-distance business. They could, with the permission of the court, enter other businesses.

AT&T's post divestiture business consisted of long-distance services, services for all customer-terminal equipment then in place, research and development, and the Western Electric manufacturing company. Western Electric's charter was assumed by a new unit, AT&T Technologies. After divestiture, AT&T long-distance no longer subsidized local service and AT&T cut prices. MCI was forced to do the same and lost $448 million in 1986. It is important to note that while the MFJ forced the divestiture of AT&T it did not fully deregulate the telecommunications industry. Federal and state regulation remained pervasive and was applied to the RBOCs and to AT&T's competitive telecommunications services but not to AT&T's long-distance rivals.

While divestiture did not fully deregulate the United States telecommunications market, it did change the regulatory climate and open up opportunities. In addition, technological changes were quickly blurring the distinction between telephone networks, computer networks, and broadcast networks. These factors caused new entry as well as merger and acquisition activity to increase between 1983 and 1995. Then in 1996 the U.S. Congress passed the Telecommunications Act which effectively deregulated the United States telecommunications industry. The Telecommunications Act allowed the RBOCs to enter the long-distance market and it allowed the various long-distance providers to provide local service. This resulted in further changes in firm structure. The remainder of this section describes the effect of these changes on some of the major firms in the United States Telecommunications industry.

A. Sprint

In the early days of the railroads, many railroads had installed telegraph wires along the tracks and had developed long-distance private communications networks using these wires. Southern Pacific's switched private network telecommunications system was one of these wire networks. In 1983, GTE bought Southern Pacific's network and began replacing the copper wires with fiber-optic cables. In 1985 United Telecom, which had its own fiber-optic network, bought 30% of GTE's fiber network. In 1986, the two fiber-optic networks were combined, and a company to manage this network was formed and named U.S. Sprint. In 1989, United Telecom bought the controlling interest in U.S. Sprint from GTE and then in 1992, bought out the remaining portion of GTE's ownership. United Telecom then changed its name to Sprint Corporation.

In 1993, Sprint merged with Centel Corporation, a company specializing in local and cellular telephone services. Sprint then formed an alliance with Call-Net to market telecommunications services in Canada under the name Sprint Canada. In 1994, Sprint signed an agreement with Telefonos de Mexico to cooperate in providing communications services between the United States and Mexico. Also in 1994, the FCC began a series of auctions (still ongoing in 2001) to allocate frequencies. The next year, Sprint formed a joint venture with three major cable TV companies, Tele-Communications, Inc. (TCI), Comcast Corp., and Cox Cable, to buy the rights to Personal Communications Services (PCS) wireless licenses in 29 major United States markets. Also in 1995, Sprint entered an alliance with Deutsche Telecom of Germany and France Telecom. The three firms launched Global One to compete in the international corporate telecommunications market. In 1999, MCI WorldCom proposed a friendly acquisition of Sprint but the acquisition was blocked by the United States DOJ. In February of 2001, Deutsche Telekom announced its intention to sell its 10% stake in Sprint.

B. MCI

In 1993, MCI and British Telecommunications (BT) entered into strategic alliance and BT invested $4.3 billion to acquire a 20% stake in MCI. The next year, MCI and Grupo Financiero Banamex-Accival formed a joint venture known as Avantel to offer long-distance service in Mexico. Competition in the telecommunications industry intensified and MCI lost more than a million customers to AT&T in 1994. In 1995, MCI acquired Nationwide Cellular, an American cellular telephone company, and SHL Systemhouse, a Canadian firm specializing in corporate computer networking systems. In 1996, MCI and News Corporation formed a joint venture to offer consumers information and entertainment through a satellite system, American Sky Broadcasting (ASkyB). Similar satellite services were offered to businesses through the SkyMCI program. MCI also formed strategic alliances with Microsoft Corporation, Digital Equipment Corporation, and Intel Corporation. Later in 1996, BT purchased the remaining 80% of MCI for $20.8 billion. At the time this was the biggest foreign takeover of a United States company in history. In 1998, WorldCom Inc. purchased MCI Communications Inc. from BT for $34.7 billion.

C. WorldCom

Long Distance Discount Service (LDDS) was created in 1983 to resell AT&T long-distance service to small- and medium-size businesses in southern Mississippi. LItDS began an aggressive acquisitions strategy in 1986 purchasing small carriers in MS and TN. In 1993, LDDS acquired Metromedia Communications and Resurgens Communications Group. These acquistions extended its coverage to the entire country. The next year LDDS acquired IDB WorldCom, the fourth-ranked carrier of international long-distance calls. In early 1995, the company acquired WilTel, which owned a nationwide fiber-optic network. After the WilTel purchase, LDDS changed its name to WorldCom, Inc. In 1996, WorldCom bought MFS Communications for $12 billion. Then in 1998, WorldCom Inc. acquired MCI Communications Inc. for $34.7 billion. The combined company was renamed MCI WorldCom. In 1999, in the largest proposed takeover in United States history, MCI WorldCom announced a $115 billion acquisition of Sprint. This acquisition was blocked by the United States DOJ.

D. AT&T

In 1991, AT&T acquired NCR Corp. In 1994, the company acquired McCaw Cellular Communications of Kirkland, WA, then the largest provider of cellular phone service in the United States. McCaw Cellular was renamed AT&T Wireless Services.

In September, 1995, AT&T announced a restructuring plan. The plan divided AT&T into three separate companies: a systems and equipment company, a computer company, and a communications services company. The systems and equipment company was formed by combining AT&T's systems and technology unit and Bell Laboratories. This company was renamed Lucent Technologies. Bell Labs became the research and development arm of Lucent. As a division of AT&T, its scientists had invented the transistor, laser, solar cell, light-emitting diode (LED), cellular radio, C programming language, UNIX operating system, and many other technologies. Lucent Technologies was spun off in September, 1996. The second company consisted of the unit formed when AT&T acquired NCR in 1991. This company was spun off as NCR Corp. in January, 1997. The third company retained the AT&T name. The "new" AT&T's focus was on becoming an integrated voice and data communications company. The Bell Labs researchers who supported AT&T's pre-reorganization communications services business stayed with AT&T as the staff of the new AT&T Labs.

In 1998, AT&T acquired Teleport Communications Group (TCG). In early 1999, AT&T acquired IBM Global Networks and later that year AT&T merged with the second largest cable company in the United States, Tele-Communications Inc. (TCI). TCI was renamed AT&T Broadband and Internet Services.

In January of 2000, AT&T and British Telecom formed a joint venture, Concert. In late 2000, AT&T announced a new restructuring plan. Under this plan, which is expected to be fully implemented during 2002, AT&T will create four separate, publicly traded securities for its major business units, Broadband and Internet Service, Wireless Services, Business Services, and Consumer Services. After the restructuring is complete, AT&T Business, AT&T Wireless, and AT&T Broadband will be spun off as separate businesses while the consumer services unit will remain a part of AT&T.

E. RBOCs and Bellcore

After passage of the Telecommunications Act of 1996, two Baby Bells, Southwestern Bell and Pacific Telesis, merged in a $16.7 million deal and NYNEX bought Bell Atlantic for $22.1 billion, leaving only 5 RBOCS. In 2000, Bell Atlantic merged with GTE. As part of the merger GTE named its Internet unit Genuity and that unit was spun off.

In 1997, Bellcore was purchased by Science Application International Corporation (SAIC), and in 1999 Bellcore's name was changed to Telcordia Technologies.

IX. THE TWENTY-FIRST CENTURY

With the increased availability of IP networking services to the home, not to mention businesses, the question "Why not IP Telephony?" naturally arises. Voice over IP, or VoIP, does exist today, and many believe that VoIP will provide most voice services before the year 2010. There are many problems to solve, primarily in the area of guaranteeing quality of service, but progress is steadily being made to define specific problems that must be solved.

We have mentioned briefly that one of the main advantages of packet-switching transmission infrastructure is statistical multiplexing gain, which allows bandwidth sharing among users of the same services and across services. Many researchers view this as a prime motivation for implementing all services on packet-switched networks. However, using packet-switching as an integration strategy requires the invention and deployment of strategies that guarantee adequate quality of service in a heterogeneous environment. History tells us that people simply will not be happy with marginal voice quality; this has been demonstrated time and time again through experimental deployment of systems that provide lower voice quality for lower prices. This situation is exacerbated in multiparty teleconferencing.

Some argue that with the ability to set up circuit-switched connections blindingly fast and the advent of virtually infinite bandwidth through the worldwide deployment of fiber-optic based transmission systems, the natural evolution will be toward an all circuit-switched based infrastructure.

On the surface, an all circuit-switched approach seems to make some sense, but it has at least one major pitfall, which is that it inherently assumes allocation of specific pieces of equipment to a single service. Such an approach has an inherent limitation in that it limits the number of simultaneous services that can be provided to the same place at the same time to a fixed number. Even at this time, the limitations of such an approach as a long-term solution are obvious.

A possible solution to this dilemma is to provide circuit-switched services over a packet-based infrastructure. Such a solution is easily developed using the concept of virtual circuits and its variants. The concept of virtual circuits has been around since at least the mid-1970s, and one of its variants, *label switching*, is beginning to gain momentum today. In addition, much insight into the problems that must be solved and how to solve them was developed during the mid-1990s in the context of ATM.

The first ten years of the twenty-first century promise to be an exciting time for telecommunications from both technology and service deployment points of view.

X. TELECOMMUNICATIONS TIMELINE: TELEGRAPH, TELEPHONE, TELEVISION, COMPUTER, INTERNET: 1837–2001

1837 Morse invents telegraph.
1938 Morse files patent on telegraph.
1839 Becquerel discovers the electrochemical effects of light.
1844 Morse sends first telegraphic message.

1847 Blakewell patents the chemical telegraph.

1851 New York and Mississippi Valley Printing Telegraph Company is formed.

1854 Field forms two companies to lay transoceanic telegraph cable.

1856 New York and Mississippi Valley is reorganized and renamed Western Union Telegraph Company.

1861 Western Union builds first transcontinental telegraph line.
Caselli uses tin foil on facsimile to transmit handwriting and pictures.
2250 telegraph offices in the United States.

1866 Field's companies complete the first transatlantic telegraph cable.

1870 Edison invents multiplex telegraphy.

1876 Bell transmits first telephone message.

1877 Bell, Watson, Sanders, and Hubbard form the Bell Telephone Company as a voluntary association.

1878 Bell Telephone Company is incorporated.
First commercial telephone exchange in operation in New Haven, CT, with 21 subscribers.

1879 Western Union sells its 55 exchange system to Bell and Bell agrees to leave the telegraph business.

1882 Bell Company acquires Western Electric.

1883 Edison discovers the Edison effect.

1885 AT&T is created to finance, build, and operate Bell Telephone's long-distance service.

1890 Hollerith's method of using punched cards to tabulate data is used in United States census.

1893 Fundamental Bell patents expire in 1893 and 1894.

1896 Hollerith starts the Tabulating Machine Co.

1897 Brun invents the Cathode Ray Tube (CTR).

1899 AT&T becomes parent company for the Bell System.
Marconi's Wireless Telegraph Company, Ltd. establishes communications across English Channel.

1900 Approximately 20,000 telephone operating companies (Telcos) and 856,000 telephones in the United States.

1906 deForest invents the vacuum tube.

1907 J. P. Morgan group takes control of AT&T. Vail appointed president.

1909 AT&T purchases Western Union.
Marconi and Braun are awarded Nobel Prize for work in wireless telegraphy.

1910 Interstate Commerce Commission (ICC) given authority to regulate interstate phone service.

1911 The Tabulating Machine Co. merges with three other firms to form the Computing-Tabulating Recording Co.

1913 Bell System and the DOJ reach agreement with regard to interconnection; the Kingsbury Commitment.

1919 Westinghouse, AT&T, and United Fruit form RCA which operates the American Broadcasting Company (ABC).

1920 Valensi develops the concept of time domain multiplexing.

1923 Zworykin patents an electronic camera tube, the "ionoscope."
Westinghouse, General Electric, and AT&T begin television research.

1924 The Computing-Tabulating Recording Co. changes name to International Business Machines (IBM).

1925 Bell Telephone Laboratories is founded.
Bush at MIT builds a large-scale analog calculator; the differential analyzer.
12 million phones in service in the United States.

1926 Pulse Code Modulation (PCM) is developed by Rainey at Western Electric.
RCA creates the National Broadcasting Company (NBC).

1927 AT&T performs first long-distance TV broadcast from Washington, D.C. to New York.
RCA creates Columbia Broadcasting System (CBS).

1928 First television station starts broadcasting in Schenectady, NY.

1930 In *Smith v. Illinois*, a portion of the cost of local telephone service is allocated to long-distance.
Farnsworth patents electronic television.

1932 RCA's control of ABC and CBS is declared a violation of antitrust laws.

1934 Federal Commerce Commission (FCC) is established.

1935 First telephone call around the world.
Approximately 6700 Telcos in the United States.

1936 The first public broadcasting of television programs takes place in London.
Turing publishes "On Computable Numbers."
Zuse applies for patent on mechanical memory.

1938 AT&T introduces crossbar central office switches.
Zuse creates the Z1.

1939 Atanasoff and Berry at Iowa State University invent the first electronic digital computer, the Antanasoff-Berry Computer.
Stibtz at Bell Labs completes the Complex Number Calculator.
Hewlett and Packard found Hewlett-Packard (HP).

1941 Zuse develops the first programmable calculator, the Z3, using binary numbers and Boolean logic.
NBC and CBS are granted commercial television broadcast licenses.

The United States adopts a 525-line black and white system as the standard for TV broadcasting.

1943 Western Union and Postal Telegraph merge.
RCA sells ABC to Noble.

1944 Harvard-IBM MARK I, a large-scale calculating machine, is used in WWII by the U.S. Navy.
The Colossus, a programmable digital machine, is used to break German codes.

1945 Electronic Discrete Variable Automatic Computer (EDVAC) is created.

1946 Raytheon transmits audio via microwave link from WQXR in New York City to Boston.
Turing designs the Automatic Computing Engine.
Williams applies for patent on CRT storage device.

1947 Bardeen, Brattain, and Shockley invent the germanium point contact transistor.
Harvard MARK II goes into operation.

1948 Bell Telephone Laboratories introduces the alloy junction germanium transistor.
CBS announces development of color television.
IBM builds the Selective Sequence Electronic Calculator.

1949 Shannon of Bell Lab's publishes his seminal theory of relay logic.
Joint Board produces first separations manual.
DOJ files antitrust suit against Western Electric.

1950 Remington Rand buys Eckert-Mauchly Computer Corporation. Name is changed to UNIVAC Division of Remington Rand.

1951 The UNIVAC is the first computer to be sold commercially in the United States.

1952 The first database is implemented on RCA's Bizmac computer.
Remington Rand buys Electronic Research Associates and combines it with the UNIVAC Division.
More than 22 million TV sets in the United States.

1953 ABC merges with Paramount Theaters.

1954 Amdahl develops the first computer operating system for the IBM 704.
IBM introduces Remote Job Entry (RJE).
Sony introduces the first transistor radio.
Commercial color broadcasting begins in the United States using NTSC standards.
IBM produces and markets the IBM 650.
IBM publishes the first version of FORTRAN.
Teal at Texas Instruments (TI) discovers how to make a transistor out of silicon.

1956 AT&T enters into a consent decree with the DOJ to settle the 1949 antitrust suit.
Bell Labs scientists Bardeen, Brattain, and Shockley share the Nobel Prize in physics.
The IBM 305 RAMAC is shipped.

1957 FCC Hushaphone Decision allows foreign equipment to be attached to Bell equipment.
Russia launches the first satellite, Sputnik.
Digital Equipment Corp. (DEC) is founded.
Philco produces the TRANSAC S-2000.

1958 Scientists at Fairchild and TI developing the first integrated circuits.
TI introduces the silicon-based transistor.
Cray at Control Data Corporation (CDC) develops the first transistorized computer, Model 1604.

1959 Above 890 Decision: FCC approves private microwave communication systems.
AT&T introduces the Telpak tariff.
AT&T introduces the TH-1 1860-channel microwave system.

1960 AT&T installs the first electronic switching system in Morris, IL.
COBOL is invented.
IBM producing the IBM 1400.
3299 Telecos in the United States.

1962 AT&T introduces T-1 multiplex service in Skokie, IL.
Baron of RAND introduces the idea of distributed packet-switching networks.
Comsat is formed.
ABC requests FCC to allow domestic satellites to distribute TV programs.
Telestar 1 satellite launched.
Fairchild Semiconductor and TI begin mass-producing the integrated circuit.
Approximately 10,000 computers in service in the United States.

1963 Microwave Communications Inc. (MCI) files an application with the FCC to set up a public microwave system between St. Louis and Chicago.
Englebart at SRI patents the idea of the mouse.
The American Standard Code for Information Interchange (ASCII) is developed.

1964 IBM releases the Model 360 computer.
American Airlines and IBM introduce the Semi-Automated Business Research Environment (SABRE) system to computerize airline reservation.
Heilmeier at RCA invents the liquid crystal display.
Kennedy and Kurtz at Dartmouth University develop Beginner's Allpurpose Symbolic Instruction Language (BASIC).

1965 AT&T introduces stored program controlled switching.
Commercial satellite Early bird launched in fixed orbit.

Donald Davies coins the term "packet switching." Marill and Roberts set up first Wide Area Network (WAN) between MIT's Lincoln Lab and System Development Corporation.

2421 Telcos in the United States.

1968 Carterfone Decision; FCC rules that AT&T has to allow devices that do not cause harm to the network.

AT&T starts 56 kilobits/second service.

Noble at IBM develops the 8-inch floppy disk.

Kramer at Phillips invents the compact disk.

Noyce and Moore co-found Intel Corp.

1969 Bell Labs develops UNIX.

Department of Defense (DOD) sets up Advanced Research Projects Agency (ARPAnet) and the first four ARPAnet nodes are installed; IMP1 at UCLA, IMP2 at Stanford Research Institute, IMP3 at UC Santa Barbara, and IMP4 at the University of Utah.

Starkweather at Xerox invents the laser printer.

Control Data Corp releases the CDC 7600, the first supercomputer.

CompuServe, the first commercial online service is established.

1970 AT&T introduces the 2ESS electronic switch.

Intel introduces the 4004 4-bit microprocessor and the 1103 memory chip.

1841 Telcos in the United States.

1971 Specialized Common Carriers (SCC) Decision; FCC rules that MCI and others may provide public microwave service.

Intel designs the first microprocessor, the Intel 4004.

Wirth invents Pascal.

ARPAnet has 15 sites and averages 700,000 packets a day.

1972 MCI begins public microwave service between St. Louis and Chicago.

Ritchie at Bell Labs invents C.

1973 MCI files interconnection complaint with FCC.

The File Transfer Protocol (FTP) is introduced.

Metcalf creates Ethernet, a local area network (LAN).

1974 MCI files private antitrust suit against AT&T.

DOJ files antitrust suit against AT&T.

IBM announces System Network Architecture (SNA).

First domestic satellites are in operation.

AT&T introduces the digital subscriber loop (DSL).

Cerf and Kahn create Transmission Control Protocol (TCP).

Kilty, Merryman, and Tassel of TI patent the electronic hand-held calculator.

Intel 8080 processor becomes the industry standard.

1975 Micro Instruments Telemetry Systems ships the Altair 8800.

Bill Gates and Paul Allen start Microsoft.

1618 Telcos and 140 million phones in the United States.

1976 X.25 protocol defined for public packet-switched networks.

Jobs and Wozniak start Apple Computer.

1977 Apple introduces the Apple II.

Commodore introduces the PET.

Radio Shack introduces the TRS-80.

FCC authorizes experimental wireless telephone licenses for AT&T and Motorola.

1978 Epson introduces the TX-80 dot matrix printer.

Intel invents the 8086 processor.

The 5.25 mini-floppy becomes the industry standard.

1979 Motorola 68000 is available.

Apple II Plus is introduced.

TI 99/4 is introduced.

Hayes markets its first modem.

1980 First full deployment of Signaling System 7.

1981 DOJ suit filed in 1974 against AT&T goes to trial.

The IBM PC is introduced.

Microsoft formally incorporates.

HP introduces the first 32-bit chip.

Commodore ships the VIC20.

1982 1974 AT&T suit is settled by an agreement known as the Modified Final Judgment (MFJ) and AT&T agrees to divestiture.

AT&T introduces the 5ESS switch.

Lotus Development is founded.

Compaq Computer Corp. is founded.

Commodore 64 is introduced.

Epson introduces the first notebook computer the HX-20.

TCP and Internet Protocol (IP) are established as the protocol suite for ARPAnet.

1983 GTE buys Southern Pacific's switched private network system and replaces the copper wire with fiber-optic cables.

FCC grants commercial wireless license to AT&T.

Long Distance Discount Service (LDDS) is founded.

ARPAnet is split into Military and Civilian sections, this is the beginning of the Internet.

ARPAnet switches from NCP to TCP/IP.

Apple IIe is introduced.

1984 AT&T implements divestiture plan and the Bell system ceases to exist. In its place are 22 Bell Operating Companies (BOCs), 7 Regional Bell

Operating Companies (RBOCs), and AT&T.
AT&T retains Western Electric and Bell Labs.
British Telecommunications (BT) is privatized.
3.5-inch diskette is introduced.
Stallman starts the GNU Project.
Dell Computer is founded.
Apple Macintosh is introduced.
The Tandy 1000 is introduced.
There are approximately 1000 Internet hosts.

1985 United Telecom buys 30% of GTE's fiber-optic network.
General Electric (GE) purchases RCA.
Gateway is founded.

1986 United Telecom and GTE combine their fiber networks to form US Sprint.
GE acquires NBC.
Capital Cities acquires ABC.
Apple introduces the Mac Plus.

1987 Hayes demonstrates its Integrated Services Digital Network (ISDN) adapter, a modem for ISDN lines.
The Mac SE is introduced.
Apple introduces HyperCard.
IBM introduces the PS/2.
The number of Internet hosts passes 10,000.

1988 First transatlantic fiber-optic cable linking North America and Europe is completed.
Apple files copyright infringement lawsuit against Microsoft and HP.
Western Union Telegraph is reorganized and changes name to Western Union Corp.
Approximately 45 million PCs in use in the United States.

1989 United Telecom buys the controlling interest in U.S. Sprint.
More than 100 million computers in use worldwide and the number of Internet hosts passes 100,000.

1990 The first commercial dial-up provider of Internet access, The World, comes online.
ARPAnet ceases to exist.
First World Wide Web software is created by Berners-Lee.
Gosling creates Java.
There are more than 5 million wireless subscribers.

1991 AT&T acquires NCR Corp.
Torvalds announces Linux version 0.02.
Gopher is developed at the University of Minnesota.
The World Wide Web is launched.
Western Union Corp. sells off business units and changes name to New Valley Corp.

Over 600,000 Internet hosts.

1992 United Telecom buys remaining interest in U.S. Sprint and changes name to Sprint Corporation.
MicroSoft and IBM sever the connection between the two companies.
More than 10 million wireless subscribers.
Over 1 million Internet hosts.

1993 Sprint merges with Centel Corp.
Sprint forms an alliance with Call-Net and markets services in Canada as Canada Sprint.
BT acquires a 20% share of MCI.
LDDS acquires IDB WorldCom.
Intel releases the Pentium processor.
Court rules against Apple in 1988 suit. Apple appeals.
New Valley Corp. enters bankruptcy and sells off last of its major holdings.
FCC receives authorization to auction personal communications (PCS) licenses.
Number of Internet hosts passes 2 million and there are at least 500 HTPP servers.

1994 Sprint enters an agreement with Telefonos de Mexico.
MCI forms a joint venture, Avantel, with Grupo Financiero Banamex-Accival.
AT&T acquires McCaw Cellular Communications and renames the business AT&T Wireless.
Netscape Communications founded.
Commodore Computers files bankruptcy.
Symantec and Central Point Software merge.
Appeals court rules against Apple.
FCC begins spectrum auctions with Personal Communications Services (PCS) Narrowband auction.
IBM and Motorola announce PowerPC 601 processor.
Number of Internet hosts greater than 3 million.
2847 Local service providers in the United States.

1995 Sprint forms joint venture with Tele-Communications, Comcast, and Cox Cable to buy rights to PCS licenses. Sprint enters alliances with Deutsche Telekom and France Telecom.
MCI acquires Nationwide Cellular and SHL Systemhouse.
LDDS acquires WilTel. LDDS changes name to WorldCom Inc.
AT&T reorganizes.
Sun Microsystems introduces Java.
Westinghouse buys CBS, sells manufacturing assets and takes CBS name.
Scientific Applications International Corp. (SAIC) acquires Network Solutions.

ESCOM buys Commodore's assets.

IBM buys Lotus Development.

U.S. Patent office overturns decision to grant Hyatt patent on microprocessor. Boone is recognized as prior inventor of microprocessor.

Internet Protocol, Version 6 (IPv6) is released.

Spectrum auctions continue.

3058 Local service providers in the United States.

1996 Congress passes the Telecommunications Act which effectively deregulates the United States telecommunications industry.

Southwestern Bell and Pacific Telesis merge.

NYNEX buys Bell Atlantic.

AT&T spins off Lucent Technologies.

MCI and News Corporation form a joint venture, American Sky Broadcasting. MCI forms strategic alliances with Microsoft Corp., Digital Equipment Corp., and Intel Corp.

BT purchases remaining 80% of MCI.

WorldCom buys MFS Communications.

Gannett buys Multimedia Entertainment.

Time Warner merges with Turner Broadcasting (CNN).

U.S. West acquires control of Continental Cablevision.

News Corp. buys New World Communications Group Inc.

CBS buys Infinity Broadcasting.

Disney merges with Capital Cities/ABC.

Seagate and Conner Peripherals merge.

Advanced Micro Devices and NexGen merge.

Apple buys NeXT.

Silicon Graphics and Cray Research merge.

Intel releases Pentium Pro.

IBM and Motorola ship PowerPC 603e processor.

Spectrum auctions continue.

Number of Internet hosts passes 9 million.

3832 Local service providers in the United States.

1997 BellCore is purchased by SAIC.

AT&T spins off NCR Corp.

NEC Corp and Packard Bell merge.

Spectrum auctions continue.

Number of Internet hosts passes 16 million.

3604 Local service providers in the United States.

1998 WorldCom aquires MCI. Combined company is renamed MCI WorldCom.

Compaq acquires DEC.

AT&T acquires Teleport Communications Group.

DOJ and twenty states file antitrust case against Microsoft; in October. *U.S. v. Microsoft* begins.

IEEE ratifies 802.3z as the Gigabit Ethernet standard.

Spectrum auctions continue.

There are 4144 local service providers in the United States.

94% of United States households have telephone service, 98% have a TV, and 67% cable service, 42% have a personal computer, 74% of these have a modem, and 65% have Internet access.

1999 SAIC changes Bellcore name to Telcordia Technologies.

WorldCom attempts to acquire Sprint. Acquisition is blocked by the DOJ.

AT&T merges with Tele-Communications Inc.(TCI) and TCI is renamed AT&T Broadband and Internet Services.

AT&T acquires IBM Global Network.

Aliant merges with ALLTEL.

Global Crossing Ltd. merges with Frontier Corporation.

Disney acquires Infoseek Corp. and spins off Go.com.

AOL acquires Netscape.

Testimony in Microsoft trial ends.

Spectrum auctions continue.

More than 86 million wireless subscribers.

2000 U.S. District Court orders Microsoft split in two. Microsoft appeals.

FTC approves merger of AOL and Time Warner, FCC approval required before the merger is final.

AT&T and BT form a joint venture, Concert.

AT&T announces restructuring plan to be fully implemented by 2002. After the plan is complete AT&T Business, AT&T Wireless, and AT&T Broadband will be separate businesses.

Bell Atlantic merges with GTE. GTE's Internet unit is renamed Genuity and spun off.

CBS and Viacom merge.

Judge in Microsoft case orders the breakup of Microsoft. Microsoft appeals.

Spectrum auctions continue.

Over 100 million wireless subscribers.

2001 Up to mid-February:

FCC approves merger of AOL and Time Warner.

Deutsche Telekom AG announces intent to sell its 10% stake in Sprint.

Lucent Technologies announces plans to sell its fiber-optic unit.

SWIFT announces it has entered into an agreement to subcontract its network operations to Global Crossing.

Disposable cellular phone expected to be introduced.

Microsoft case: oral arguments begin before the
U.S. Court of Appeals.
Spectrum auctions continue.

SEE ALSO THE FOLLOWING ARTICLES

COMMUNICATION SATELLITE SYSTEMS • COMMUNICA-
TION SYSTEMS, CIVILIAN • COMPUTER NETWORKS •
DATA TRANSMISSION MEDIA • DIGITAL SPEECH PRO-
CESSING • OPTICAL FIBER COMMUNICATIONS • RADIO
SPECTRUM UTILIZATION • TELEPHONE SIGNALING SYS-
TEMS, TOUCH-TONE • VOICEBAND DATA COMMUNICA-
TIONS • WIRELESS COMMUNICATIONS • WWW (WORLD-
WIDE WEB)

BIBLIOGRAPHY

Brock, G. W. (1998). "Telecommunication Policy for the Information Age," Harvard University Press, Cambridge, Massachusetts.
Casson, H. N. (1910). "The History of the Telephone," A. C. McClurg, Chicago.
Keagy, S. (2000). "Integrating Voice and Data Networks," Cisco Press, Indianapolis.
Kurose, J. F., and Ross, K. W. (2001). "Computer Networking: A Top-Down Approach Featuring the Internet," Addison-Wesley, Reading, Massachusetts.
Lin, Y.-B., and Chlamtac, I. (2001). "Wireless and Mobile Network Architectures," John Wiley & Sons, New York.
Pildush, G. D. (2000). "Cisco ATM Solutions: Master ATM Implementation of Cisco Networks," Cisco Press, Indianapolis.
Schulzrinne, H., guest ed. (1999). "IEEE Network," Special Issue on Internet Telephony, Vol. 13, No. 3.
Valovic, T. (1992). "Corporate Networks: The Strategic Use of Telecommunications," Artech House, Boston.

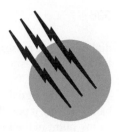

Telephone Signaling Systems, Touch-Tone

A. Michael Noll
University of Southern California

Leo Schenker
AT&T Bell Laboratories

GLOSSARY

Common control Central-office switch type of system indicating that switching connections are set up by common equipment once the complete dial information has been received.

Guard action Means for reducing the probability of talk-off.

Loop That part of the telephone plant that connects subscribers' premises to the local central office.

Multifrequency key pulsing (MFKP) Method used for signaling between telephone switches.

Progressive control *See* Step-by-step.

Station set Telephone set.

Step-by-step Central-office switching equipment of the type where the incoming dial pulses directly affect the position of switches and thus set up a talking path.

Talk-off Imitation of control signals by speech.

TOUCH-TONE dialing is a voice frequency means of signaling from telephone sets to central-office switches to indicate the destination of a call. The user interface is a set of pushbuttons. Compared to the older dialing method with the rotary dial and dial pulses, Touch-Tone signals have frequency and level characteristics such that they can be transmitted beyond central offices to any point in a telephone network. Hence, Touch-Tone signals can be used for the transmission of data after a connection has been set up. Touch-Tone dialing is widely used in North America and many other countries.

I. INTRODUCTION

Touch-Tone dialing is a method of sending signals from telephone customers' premises to central offices and beyond. It was first introduced in 1964. Compared to rotary dialing of dial pulses, its principal advantages are as follows.

1. All the signaling energy is in the voice frequency band, making it possible to transmit signaling information to any point in the telephone network to which voice can be transmitted, that is, "end to end."

2. Touch-Tone dialing is faster, reducing the dialing time for users and, equally important, reducing the holding time for central-office common equipment.

3. It provides a means for transmitting more than 10 distinct signals: 12 in all standard implementations (and the scheme is capable of 16).

4. It is a more convenient signaling method, as attested by users.

II. HISTORY OF DEVELOPMENT AND INTRODUCTION

The developmental work that led to the introduction of Touch-Tone dialing began at Bell Laboratories in the mid-1950s. This was not the first occasion on which the merits of a "push-button" dial (as it was then called) were studied. Following the introduction of operator multifrequency key pulsing (MFKP), a system of ac signaling for station equipment was designed, built, and tested in the laboratory in 1941. World War II interrupted further work, and 7 years elapsed before equipment was installed and used on an experimental basis in a small trial at Media, Pennsylvania, in 1948. This system was based on the transmission of damped oscillatory signals at two of an available six frequencies and required special receivers at the central office to detect and convert the signals. The dialing unit had mechanical linkages that plucked two of six metal reeds, each of which was resonant at a specified frequency. When a customer pushed any 1 of the 10 buttons, two reeds were plucked to form a signal coded to the corresponding digit. The energy so generated was transferred inductively to coils in the station-set network and so transmitted to the receiver at the central office. Although this mechanism was cumbersome, the performance of the equipment and the reaction of the customers pointed the way to an ultimately feasible system and indicated a favorable public response to push-button signaling. The technical and economic aspects of the system tested at Media were not attractive, and further work was deferred.

The year of the Media trial, 1948, is also noteworthy because in June of that year the discovery of the transistor was announced. Concepts that formerly were considered exotic became practical. The activity in electronic device development resulted in an entirely new array of circuit components. Miniature capacitors, ferrite inductors, precision resistors, and printed wiring became available, and electric power consumption ceased to be a formidable obstacle.

Using this new technology, a compact multifrequency oscillator, equipped with push buttons for selecting and controlling voice frequency signals, was developed in the second half of the 1950s. The oscillator was particularly adapted to the low and variable power available from the central-office battery over the range of existing loops to the station set. The oscillator design provided for control of the signal at a level high enough to be reliably detectable at the central office in the presence of the noise from numerous sources that is always present on telephone circuits, but not so high as to exceed crosstalk requirements. The concept of a four-by-four frequency code resulted in a relatively simple mechanical system at the station set.

Concurrently with these electrical and mechanical developments, human factors engineers studied pertinent psychophysical factors and performance ratings of button arrangement were made. The optimum size, spacing, travel, and operating force of the buttons were determined. It was also established that feedback of the signal tones through the telephone receiver was desirable. These studies indicated that customer dialing could be speeded up and generally facilitated by the use of a push-button operation, probably without increasing dialing irregularities seriously.

On the basis of satisfactory results in the areas of electromechanical design and psychophysical studies, a moderate-sized field trial of push-button dialing was designed and carried out in 1959. Two more trials were to follow before the system was introduced for general use. The chief objective of the first trial was to determine the effectiveness and ease of use of a modern button-operated mechanism with then-current types of switching systems, when placed in the hands of a typical segment of the public. Two locations were selected. One of these, at Elgin, Illinois, was equipped with fast-operating common control equipment; the other, at Hamden, Connecticut, used the slower step-by-step switching equipment. The necessary central-office receiver and converter equipment was provided on a "black-box" basis. At Hamden, this equipment included a digit storage facility and a dc outpulser geared to the speed of the step-by-step system. Approximately 120 customers in each central-office area were provided with Touch-Tone sets of a preliminary design.

As in the Media trial, customers in both areas were enthusiastic about this experimental service, basing their reaction on the increased speed and general ease of use. In a relatively short period, the customers achieved a dialing speed with the Touch-Tone sets that was almost twice their established rotary-dial speed. It was anticipated from laboratory tests that customer signaling irregularities would tend to occur at a higher rate with push buttons than with the slower and more familiar rotary dial, and this was confirmed in the field trial. Within a relatively short period, the Touch-Tone error rate of the trial customers showed a trend that led to the projection that, with longer experience, accuracy equal to that of rotary dialing would be achieved.

Concurrent development work on the electromechanical design of Touch-Tone dials and receiver-converter equipment resulted in designs proposed for manufacture, and a second set of trials, technical trials of this equipment, was planned and carried out in 1960. As in the Hamden and Elgin trials, two locations were selected. One of these was at Hagerstown, Maryland, served by common control switching equipment; the other was the Cave Spring office at Roanoke, Virginia, served by step-by-step equipment.

The results of this second set of trials led to the conclusion that a practical new dialing scheme could be developed based on the technical concepts of the system and components used in the technical trials. The technical performance of station sets and central-office equipment brought no surprises. Signal imitations from speech (a problem known as talk-off) and other noises were not a problem. Customers continued to be highly supportive.

Touch-Tone calling was introduced as a premium service on a market-trial basis at Findlay, Ohio, and at Greensburg, Pennsylvania, in 1961. The equipment supplied for the station installations in these areas included some minor refinements relative to that used in the technical trials, and some modifications of the dial were made to adapt the design to the full line of station sets, all of which were equipped with Touch-Tone dials. Figure 1 is a photograph of the Touch-Tone desk set used in the market trials. The appearance design was by Henry Dreyfuss. The central-office equipment also included minor refinements, but was essentially unchanged.

Central offices beyond those in the market trial were converted for Touch-Tone dialing beginning in 1963. The new method of dialing was available at the New York Worlds Fair in 1964. Today, all central offices accommodate Touch-Tone dialing.

III. TOUCH-TONE DIALING SCHEME

In the previous section, several advantages of Touch-Tone dialing over rotary dialing were listed. When the development program that led to Touch-Tone dialing began in 1953, not all of these were considered equally important. Initially, the only objective was to find a way to reduce dialing time, probably through a scheme that would involve

FIGURE 1 Early Touch-Tone desk telephone. [Courtesy of AT&T Bell Laboratories.]

a push-button person–machine interface. It was known from the Media trial that customers would be favorably disposed toward push-button dialing.

The development environment was governed by the following factors. The rotary dial had been around for several decades and was very inexpensive. The transistor had been invented a few years earlier; no mass applications of this new device existed. Power dissipation, variation of gain among devices, reliability, and cost were all issues. Thus, it was risky to consider a scheme that would involve an oscillator in millions of telephones and in millions of diverse locations. Some signaling method not requiring an active device was likely to be more practical. For example, one scheme that was studied involved the generation of damped oscillatory waves by interrupting the direct current through the coil of an inductor-capacitor tuned circuit. By switching the capacitor to 1 of 10 taps on the inductors, 10 different digit-identifying frequencies could be generated, all in the telephone voice frequency range. But such signals were highly likely to be imitated by the user's (inadvertent) speech in the dialing interval. Hence, in this scheme, each push-button operation also caused a stepwise reduction in the direct current (dc) drawn by the telephone from the central office. The damped oscillations were ignored by the central-office receiver unless accompanied by a dc step.

Very soon it became clear that it was essential to be able to transmit customers' signals "end to end," that is, from one customer's premises, through one or more central offices, to the premises of another customer, anywhere that voice can be sent. Two requirements resulted from the end-to-end signaling objective.

1. The signals must not contain an out-of-band component such as a dc step.

2. Sustained rather than damped signals must be used to maintain adequate signal-to-noise margins for the wider range of transmission losses when two customer loops are involved.

The first of these two requirements—the need for the signals to be wholly contained within the voice frequency band—also brings with it the problem of vulnerability to talk-off. The second reintroduced the uncharted domain of active devices.

A. Choice of Code

When only voice frequencies are employed, protection against talk-off must rely heavily on statistical tools. This protection is required only during interdigital intervals; speech interference with valid signals can be avoided by transmitter disablement when a push button is operated. Since signals with a simple structure are prone to frequent imitation by speech and music, some form of multifrequency code particularly difficult to imitate is indicated. If the signal frequencies are restricted in a binary fashion to being either present or absent, the greatest economy in frequency space results from the use of all combinations of N frequencies, yielding $n = 2^N$ different signals. However, some of these are no more than single frequencies and are therefore undesirable from the standpoint of talk-off. Another drawback is that as many as N frequencies must be transmitted simultaneously; these involve an N-fold sharing of a restricted amplitude range and may also be costly to generate. If $n = 10$, N needs to be at least 4. At the expense of using up more frequency range, one is led to a P-of-N code, yielding $n = N!/P!(N-P)!$ combinations. There is a statistical advantage in knowing that there are always P components in all valid signals, no more and no less.

To minimize the number of circuit elements, as well as to reduce the sharing of amplitude range, P should be as small as possible, yet larger than unity for the sake of talk-off protection. Let us, then, examine codes in which $P = 2$. If one can be found that is not readily imitated by speech or music, there is no merit in choosing P higher than 2, provided that the total number of frequencies N needed for the required number of combinations can be accommodated in the available frequency spectrum. With $P = 2$, N must be at least 5 to provide 10 combinations (for the 10 digits). With $N = 6$, 15 combinations are available. The MFKP signaling scheme makes use of a two-of-six code. There were reasons to assume that MFKP would not have adequate talk-off performance. As discussed below, a code was chosen with attributes that provided protection against talk-off by several means and with which less than 1 talk-off in 5000 calls was achieved, a degree of talk-

off performance that was deemed minimal for customer acceptance.

There are advantages in imposing the further restriction that, with $P = 2$, the frequencies for each combination fall, respectively, into two mutually exclusive frequency bands. If, for example, 15 or more combinations are required, N must be at least 8. In the four-by-four code, eight signal frequencies are divided into two groups: group A, the lower four frequencies; and group B, the upper four. Each signal is composed of one frequency from group A and one from group B, resulting in 16 signal combinations.

With a two-group arrangement, it is possible at the receiver to separate the two frequencies of a valid signal by band filtering before attempting to determine the two specific components. This separation of the two components of a signal renders reliable discrimination between valid signals and speech or noise.

B. Choice of Frequencies

Attenuation and delay distortion characteristics of typical combinations of transmission circuits were such that it is desirable to keep the frequencies of a telephone signaling system within the 700- to 1700-Hz range.

The choice of frequency spacing depends in part on the accuracies of the signal frequencies. It was expected that signals generated at the station set could be held within 1.5% of their nominal frequency values and that the pass bands of the receiver selective circuits could be maintained within 0.5% of their nominal ranges. On the basis of these numbers, the selective circuits of central-office receivers need to have recognition bands of at least ±2% about the nominal frequencies.

The standardization of amplitude at the output of the limiters used in the receiver at the central office (Fig. 2)

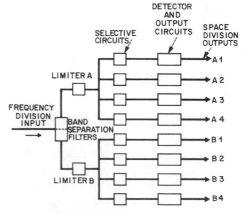

FIGURE 2 Simplified diagram of the book Touch-Tone receiver used in central-office switching systems.

permits an accurate definition of recognition bands in the receiver, independently of levels at the receiver input. As a result, frequencies may be spaced closely, approaching the recognition bandwidth of 4%. If the lowest frequency is chosen as 700 Hz, the next frequency must then be more than 728 Hz. Wider spacing makes the precise maintenance of the bandwidth less critical.

Another factor can profitably be taken into account in the selection of a frequency spacing. To reduce the probability of talk-off, the combinations of frequencies representing bona fide signals should be such that they are not readily imitated by the output from the speech transmitter. In a receiver with the guard action described, no sound composed of a multiplicity of frequencies at comparable levels is likely to produce talk-off. Thus, consonants present no problem. Vowels, however, do present a problem, as do single-frequency sounds such as whistles that are loud enough to encounter some harmonic distortion in a carbon transmitter. It was shown by H. Fletcher that an electrical analogue of the mechanism involved in the articulation of vowels is a buzzer (the vocal chords producing a fundamental and a long series of harmonics) followed by a selective network that shapes the harmonics into formants of the vowel sound. As a result, the spectrum of any sustained vowel contains a number of frequencies bearing harmonic relationships to each other. Hence, it is desirable that the pairs of frequencies representing valid signals avoid as many of these harmonic relationships as possible.

A family of frequencies that avoids a large proportion of troublesome combinations and also meets all the other requirements discussed so far is as follows, the adjacent frequencies in each group being at a fixed ratio of 21:19, with 2.5 times this interval between the groups.

Group A (Hz)	Group B (Hz)
697	1209
770	1336
852	1477
941	1633

All frequencies are essentially within the 700- to 1700-Hz range, and the spacing is adequate to accommodate the recognition bands.

The 16 pairs of frequencies representing valid signals avoid low-order ratios. This is illustrated in Fig. 3, where frequencies are plotted on two logarithmic scales: those below 1000 Hz as ordinates and those above 1000 Hz as abscissae. Any valid signal is represented by a pair of coordinates, namely, its two component frequencies. The 16 square windows represent the ±2% recognition bands required.

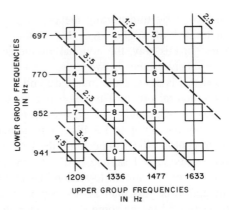

FIGURE 3 Window diagram for signal frequencies.

The diagonal lines in Fig. 3 are the loci of pairs of frequencies having simple harmonic relationships, that is, 1:2, 2:3, etc. The avoidance of these particular diagonals is beneficial because they represent the effects of harmonics not only of the corresponding order but also of higher order, for example, the third and sixth, the ninth and fifteenth, and so on. Not all applicable diagonals are shown in Fig. 3, and 18 of the 65 potentially troublesome combinations are not avoided. However, if two frequencies in speech representing two fairly high-order harmonics (e.g., 4:7 and 7:12) happen to fall at proper signal frequencies, their fundamental will be low in frequency. The odds are then good that other harmonics within the telephone speech band will together have sufficient relative intensity to bring guard action into play.

Comparable benefits with respect to talk-off can be derived from other sets of frequencies with the same geometric spacing but displaced up or down on the frequency scale. The effect of such displacement is to shift all the windows in a direction parallel to the diagonals.

C. Choice of Amplitudes

Since signaling information does not bear the redundancy of spoken words and sentences, yet must be transmitted with a high degree of reliability, it is advantageous for the signal power to be as high as permitted by the environment. A nominal combined signal power for the two frequencies of 1 dB above 1 mW at the telephone-set terminals was adopted as a realistic value, a ±3-dB tolerance about this nominal value being acceptable.

For subscriber loops, the maximum slope between 697 and 1633 Hz is about 4 dB, the attenuation increasing with the frequency. In two subscriber loops (as in end-to-end signaling), the slope may be 8 dB, although, statistically, this is unlikely. A reduction in the maximum level difference at the receiver in the two signal components can be achieved by transmitting the group B frequencies at a level 3 dB higher than that of the group A frequencies. In this way, the nominal amplitude difference at the receiver input between the two components of a valid signal is never more than 5 dB end to end and 1 dB at the central office. The nominal output powers were chosen as −3.5 and −0.5 dBm for groups A and B, respectively, adding up to 1.3 dBm. In more than 99% of station-to-central office connections, the 1000-Hz loop loss was expected to be less than 10 dB (during the early 1960s). Similarly, the station-to-station loss at 1000 Hz was estimated to be less than 27 dB in more than 99% of all connections. Making some allowance for slope and variations in the generated power, the minimum signal power at a central-office receiver was estimated to be 15.5 dBm. At a receiver involved in end-to-end signaling, the minimum power was estimated to be 32.5 dBm. Higher minimum levels can probably be expected today.

D. Signal Duration

In the development of the Touch-Tone dialing system, two important parameters had to be fixed before adequate data were available: the duration of a valid signal and the signaling rate. The probability of talk-off can be reduced by increasing the duration of the test applied to a signal by the receiver before accepting the signal as valid. It was clearly undesirable, however, to require even the most adept users to extend a push-button operation beyond an interval determined by their innate or acquired mental and physical acuity. A frame of reference for setting requirements for these intervals can be formed in two ways. First, competent typists achieve a speed of 60 words per minute, and, at five letters per word, the corresponding time interval between letters is 200 msec. Second, musical compositions occasionally include 64th notes; although the speed of performance is at the discretion of the performer or conductor, the duration of a 64th note on a piano is estimated to be rarely as short as 45 msec. A minimum of 40 msec was set as the objective for both signal and intersignal intervals, allowing for a dialing rate of more than 10 signals per second.

The allocation of a minimum of 40 msec to these intervals proved to be reasonably sound. The median tone time found in the field trials was approximately 160 msec; durations of 30 to 40 msec were observed, but only approximately 0.15% of dialing starts failed because of short tones. The median interdigital time was approximately 350 msec; no case of failure because of short interdigital intervals was observed. Fortunately, consideration never had to be given to increasing the minimum signal time to improve talk-off performance.

IV. TOUCH-TONE DIAL TELEPHONE

The scheme described in the previous section would have remained just another scheme if it had not been feasible to develop a dial that could be manufactured and maintained at a reasonable cost. The Touch-Tone dial represented a bold departure from past practices in a product line that today would be called consumer products. A proven mechanical device, honed to near-perfection as a result of extensive use and several redesigns, was to be replaced by a new mechanism and circuitry, including a relatively immature electronic active component, the germanium transistor, to be installed in millions of homes and offices.

A. Human Factors Design

Extensive tests were performed to determine customer preference, dialing speed, and error rate as various physical design parameters were changed. These parameters included button size and spacing, stroke, force displacement characteristics, and many others. One important parameter, although included in the test programs, ended up being determined by mechanical design considerations: the button arrangement. A rectangular array of three rows of three buttons, plus one button in the center of the fourth row $(3 \times 3 + 1)$, was selected because it resulted in an arrangement that required fewer electrical contacts. Of course, later, with the addition of the $*$ and $\#$ buttons, this became a 3×4 array. Fortunately, this button arrangement turned out to be close to optimal, in both performance and preference.

Consideration was given to numbering the button in the same sequence as an adding machine, that is, 7, 8, and 9 in the top row, the "Sundstrand" arrangement. User preference was the reverse order of row numbers, that is, 1, 2, and 3 in the top row. Not enough people were users of adding machines in the late fifties; handheld calculators did not exist.

User preference and performance studies, coupled with design considerations, eventually resulted in the now-familiar attributes of desk-type Touch-Tone telephones: $\frac{3}{8}$-in. square buttons, separated by $\frac{1}{4}$ in.; a $\frac{1}{8}$-in. stroke without snap action; and about $100g$ force at the bottom of the stroke.

B. Initial Mechanical Design

The push-button mechanism was the means by which the customer controlled the operation of the dual-frequency oscillator. The initial design involved a complex, yet ingenious, combination of levers and switches. In the early part of the downstroke of a push button, each tuning capacitor

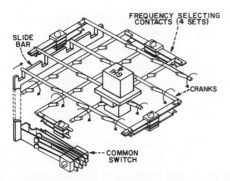

FIGURE 4 Mechanism schematic for the initial Touch-Tone caller.

FIGURE 5 Initial Touch-Tone dial assembly. [Courtesy of AT&T Bell Laboratories.]

was connected to a tap on its associated tuning coil to select the two frequencies of the signal for that button. Later in the downstroke, a switch common to all buttons was operated. The common switch had the following four functions.

1. Attenuation was inserted into the receiver circuit to reduce the level of side tone during signaling to a comfortable value. This had to occur first.

2. The transmitter path was opened to prevent speech or background noise from interfering with the signal.

3. Power was applied to the transistor oscillator.

4. The dc through the tuning coils was interrupted to initiate the signal at full amplitude by shock excitation. This had to occur last so that none of the energy stored in the tuning coils was wasted prior to enabling the transistor.

In the initial design a mechanism was conceived that kept the number of frequency selection contacts to a minimum (7), rather than to have each button operate two frequency selection contacts, which could have required 20 and, later, 24 contacts. This type of mechanism was used in the majority of Touch-Tone sets.

A schematic of the mechanism is shown in Fig. 4. Two sets of cranks operated the coil-tap switches. The cranks in the column direction operated switches connected to the high-band coil, and the cranks in the row direction operated the low-band switches. When a button was pushed, the pair of cranks crossing below that button was rotated, and one coil tap contact was made for each tuned circuit, thus selecting the two appropriate frequencies. An arm on each of the "row" cranks moved a slide bar when any one of them was rotated. Movement of this slide bar operated the common switch. Figure 5 shows the front and back of an early Touch-Tone dial.

C. Evolution of the Touch-Tone Dial

The circuitry and mechanism of the Touch-Tone dial have evolved significantly from the early versions. The availability of silicon and tantalum thin-film integrated circuits

led to a second generation of designs. Physically, this consisted of two glass substrates glued together, with about 20 leads to the crank mechanism and the speech network. Most transistors, most diodes, and some resistors were implemented in a silicon integrated circuit chip, with another smaller chip for amplitude-regulating diodes. The tuning elements were capacitors and resistors implemented in tantalum thin-film form. As technology evolved, it became possible to eliminate the separate diode chip and to combine the two glass substrates into a single ceramic substrate.

The third and current generation is based on very large-scale integration silicon technology—hundreds of devices on a chip—and is produced by a variety of manufacturers. In some cases, the very large-scale integration chips provide functions beyond the generation of the Touch-Tone signals, for example, storage of telephone numbers that can then be transmitted by pushing one or two buttons. Some of the chip designs can also generate the rotary dial-type dc pulses.

The mechanical design of Touch-Tone dials has also changed. There are no more crank-type actuating mechanisms. Rubber dome keypads or membrane keypads are used in the latest designs. The frequency selection and common switch functions are effected by the closing of a single contact under a push button and the use of logic circuitry.

V. TOUCH-TONE RECEIVER

Prior to the introduction of automatic switching, the means of "signaling" the destination of a call involved the spoken language or, more accurately, languages. In areas where more than one language was in widespread use, multilingual operators were required at switchboards. The advent of dialing simplified this situation because dialing became a universal people machine "signaling language." With the introduction of a second customer signaling scheme, the earlier multilingual requirement for operators was now imposed on the central-office signal reception equipment.

Historically, automatic switching systems can be divided into progressive control and common control systems. In progressive (step-by-step) systems, the dc pulses resulting from the operation of a rotary dial directly control a series of switches, one for each digit of the dialing sequence, to establish the talking path between calling and called customers. In common control systems the information transmitted by the dialing signals is stored in a register and action to establish a talking path is taken only after the information for the complete dialing sequence has been received. Progressive switching systems are obsolete and have been replaced by modern electronic systems, all of which use common control and registers.

A. Receiver Design

Figure 6 is a schematic diagram of the Touch-Tone receiver. A Touch-Tone signal first passes through a high-input impedance amplifier. This amplifier also includes a high-pass filter to attenuate the 60- and 180-Hz components of induced power line noise and to attenuate the dial tone, standardized as a combination of 350 and 440 Hz.

After amplification, the two signaling frequencies are separated into their respective groups by the band elimination filters.

The response of each limiter to the single Touch-Tone frequency at the band elimination filter output is a symmetrical square wave of fixed amplitude containing the fundamental and odd harmonics of the signal frequency. Each limiter drives a group of circuits used for signal frequency recognition. In each group, the tuned circuit corresponding to the incoming signaling frequency responds to the fundamental of the square wave with an amplitude sufficient to operate a detector. However, the

receiver does not deliver output signals the instant the detectors operate. Instead, it makes a check to determine the presence of one and only one valid tone in each of the two-frequency groups. This check must continue uninterrupted for a prescribed time interval before an output signal duration test is performed by the checking circuit and signal timer that are shown in the block diagram in Fig. 6.

The check for one and only one frequency in each group is achieved by a combination of simple OR and AND logic circuitry and a characteristic of the limiter permitting the operation of only one detector in a group at a time. When two frequencies of comparable amplitude are applied to a limiter input, neither frequency has sufficient energy at the limiter output to operate a detector.

Once the signal duration test is completed, the receiver is ready to deliver output signals. On minimum-length input signals, there is very little time left after tuned circuit buildup and the signal duration test for delivery of output signals. Clearly, some sort of memory is required to prolong the output signals. In the Touch-Tone receiver, this memory is achieved by providing an output timer and output gates with positive feedback to the associated detector circuits.

In addition to the channel outputs (one per signal frequency), there is a steering output that controls the advance of steering logic in the attached digit registering equipment from one digit storage position to the next. The steering lead remains energized as long as output signals are present or signal tones are present, whichever is longer. On short signals, the locked-in detectors hold the checking circuit and signal timer on so that the steering lead remains energized until after the output timer goes off. On long input signals, when the output timer goes off, the detectors remain operated from the tuned circuit outputs and hold the steering output in the operated condition until the end of the Touch-Tone pulse. This avoids double registration of long signals.

The recognition bandwidth of each channel circuit is dependent on the detector threshold level, the amplitude of the fundamental component of the limiter output, and the Q of the tuned circuit. A recognition bandwidth of approximately 5% of the nominal channel frequency is required to allow for variations in received frequencies and receiver frequency response. Knowing the required recognition bandwidth, the most important of those variables to be determined is the Q of the tuned circuit.

Receivers designed according to these principles were used in the various trials, and performance was good, including digit simulation performance, which was 1 talk-off in 10,000 calls. (The rate of digit simulation was poorer when the simulation of the six unused signals is also taken into account.)

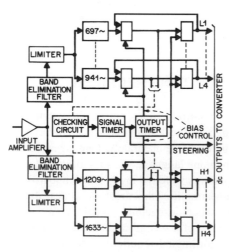

FIGURE 6 Schematic diagram of the Touch-Tone receiver.

B. Evolution of the Touch-Tone Receiver

The architecture of the receiver has essentially remained unchanged over the years. However, the physical design has changed radically, as would be expected in light of the technological advances since the 1960s. The receiver used in the various trials and in production through about 1975 used all discrete components, 50–60 transistors, coils, and so on. The very early receivers took up about 6 in. of height on 23-in.-wide racks. Soon a redesign resulted in two central-office receivers being accommodated in the same space.

During the initial phase, a receiver intended for use in PBXs was developed with a simplified architecture to save space and cost. This was justified on the basis that amplitude sensitivity and talk-off requirements could be less stringent. However, in many PBX applications this turned out to be unsatisfactory and the receivers were replaced by central-office receivers with special mounting arrangements.

Receivers introduced in 1975 had the passive LC filters replaced by active RC filters. Integrated circuits on thin-film substrates incorporating resistors were used, as well as operational amplifier chips and chip capacitors. Typical physical designs involved three 5×7-in. printed circuit boards.

Receivers in the most current applications are essentially on one silicon chip. For example, for a digital central office, where advantage is taken of the fact that the analog signals offered to the receiver have been encoded into pulse code modulation format by common input circuitry, a Touch-Tone receiver is accommodated on a 40-pin digital signal processor chip that has 45,000 transistors. The program contained in the chip emulates all the signal processing functions described earlier.

VI. CONCLUSIONS

Touch-Tone dialing, conceived in the 1950s and introduced in the 1960s in the United States, has been widely accepted by the public and is the major way most telephones signal with local switching machines.

In some countries, push-button dial phones are used that emulate the dc signals produced by rotary dials. Such instruments are usually capable of storing the information input by the actuation of the push buttons and transmitting the dc pulses at the rate at which the central office can accept them. In this way, non-tone-generating push-button schemes satisfy the needs of users for more rapid and convenient dialing, but they do not significantly reduce the interval between the start of dialing and the setting-up of a connection. In fact, they may increase the interval between the end of dialing and the setting-up of a connection. The availability of inexpensive non-tone-generating push-button dials may slow down the use of Touch-Tone when Touch-Tone-capable central offices are eventually introduced in these countries.

A major advantage of Touch-Tone is the potential for data transmission and remote control. Other services requiring interaction between the telephone user and service providers, such as banks and investment companies, are also important applications of Touch-Tone.

SEE ALSO THE FOLLOWING ARTICLES

DATA TRANSMISSION MEDIA • DIGITAL SPEECH PROCESSING • NETWORKS FOR DATA COMMUNICATION • TELECOMMUNICATIONS • VOICEBAND DATA COMMUNICATIONS

BIBLIOGRAPHY

Battista, N., Morrison, C. G., and Nash, D. H. (1963). "Signaling system and receiver for Touch-Tone calling," *IEEE Trans. Commun. Electron.* **82**.

Benson, L., Crutchfield, F. L., and Hopkins, H. F. (1963). "Application of Touch-Tone calling in the Bell system," *IEEE Trans. Commun. Electron.* **82**.

Ham, J. H., and West, F. (1963). "A Touch-Tone caller for station sets," *IEEE Trans. Commun. Electron.* **82**.

Noll, A. M. (1998). "Introduction to Telephones and Telephone Systems," 3rd ed., Artech House, Boston.

Schenker, L. (1960). "Pushbutton calling with a two-group voice frequency code," *Bell Syst. Tech. J.* **39**.

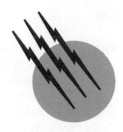

Telescopes, Optical

L. D. Barr

National Optical Astronomy Observatories (retired)

GLOSSARY

Airy disc Central portion of the diffracted image formed by a circular aperture. Contains 84% of the total energy in the diffracted image formed by an unobstructed aperture. Angular diameter $= 2.44\, \lambda/D$, where λ is wavelength and D is the unobstructed aperture diameter. First determined by G. B. Airy in 1835.

Aperture stop Physical element, usually circular, that limits the light bundle or cone of radiation that an optical system will accept on-axis from the object.

Coherency Condition existing between two beams of light when their fluctuations are closely correlated.

Diameter-to-thickness ratio Diameter of the mirror divided by its thickness. Term is generally used to denote the relative stiffness of a mirror blank: 6:1 is considered stiff, and greater than 15:1 is regarded as flexible.

Diamond turning Precision-machining process used to shape surfaces in a manner similar to lathe turning.

Material is removed from the surface with a shaped diamond tool, hence the name. Accuracies to one microinch ($\frac{1}{40}$th μm) are achievable. Size is limited by the machine, currently about 2 m.

Diffraction-limited Term applied to a telescope when the size of the Airy disc formed by the telescope exceeds the limit of seeing imposed by the atmosphere or the apparent size of the object itself.

Effective focal length (EFL) Product of the aperture diameter and the focal ratio of the converging light beam at the focal position. For a single optic, the effective focal length and the focal length are the same.

Electromagnetic radiation Energy emitted by matter with wavelike characteristics over a range of frequencies known as the electromagnetic spectrum. The shortest waves are gamma rays (<1 Å) and the longest are radio waves ($>40\,\mu$m). Visible light is in the intermediate range.

Field of view Widest angular span measured on the sky that can be imaged distinctly by the optics.

Focal ratio (f/ratio) In a converging light beam, the reciprocal of the convergence angle expressed in radians. The focal length of the focusing optic divided by its aperture size, usually its diameter.

Image quality Apparent central core size of the observed image, often expressed as an angular image diameter that contains a given percentage of the available energy. Sometimes taken to be the full width at half maximum (FWHM) value of the intensity versus angular radius function. A complete definition of image quality would include measures of all image distortions present, not just its size, but this is frequently difficult to do, hence the approximations.

Infrared For purposes of this article, wavelength region from about 0.8 to 40 μm.

Optical path distance Distance traveled by light passing through an optical system between two points along the optical path.

Seeing Measure of disturbance in the image seen through the atmosphere. Ordinarily expressed as the angular size, in arc-seconds, of a point source (a distant star) seen through the atmosphere, that is, the angular size of the blurred source. Seeing disturbances arise from air density fluctuations due to temperature variations along the line of sight.

Ultraviolet For purposes of ground-based telescopes, the wavelength region from about 3000 to 4000 Å. The remaining UV region down to about 100 Å is blocked by the earth's atmosphere.

OPTICAL TELESCOPES were devised by European spectacle makers around the year 1608. Within two years, Galileo's prominent usage of the telescope marked the beginning of a new era for astronomy and a proliferation of increasingly powerful telescopes that continues unabated today. Because it extends what the human eye can see, the optical telescope in its most restricted sense is an artificial eye. However, telescopes are not subject to the size limitation, wavelength sensitivities, or storage capabilities of the human eye and have been extended vastly beyond what even the most sensitive eye can accomplish. In this article, the word light is used to mean all electromagnetic radiation collected by telescopes but the properties of astronomical telescopes operating on the ground in the optical/IR spectral wavelength range from 3000 to about 40,000 Å (0.3–40 μm) are the principal subject. The atmosphere transmits radiation throughout much of this range. Telescopes designed for shorter wavelengths are either UV or X-ray telescopes and must operate in space. The Hubble Space Telescope is the best-known ex-

ample of a UV (and optical) telescope and is discussed in Section VII. Telescopes operating at wavelengths longer than 40 μm are in the radio-telescope category and are not considered in this article. Emphasis is placed on technical aspects of present-day telescopes rather than history.

I. TELESCOPE SIZE CONSIDERATIONS AND LIGHT-GATHERING POWER

An astronomical telescope works by capturing a sample of light emitted or reflected from a distant source and then converging that light by means of optical elements into an image resembling the original source, but appropriately sized to fit onto a light-sensitive detector (e.g., the human eye, a photographic plate, or a phototube). Figure 1 illustrates the basic telescope elements. It is customary to assume that light from a distant object on the optical axis arrives as a beam of parallel rays sufficiently large to fill the telescope entrance, as shown.

The primary light collector can be a lens, as in Fig. 1, or a curved mirror, in which case the light would be shown arriving from the opposite direction and converging after reflection. The auxiliary optics may take the form of eyepieces or additional lenses and mirrors designed to correct the image or modify the light beam. The nature and arrangement of the optical elements set limits on how efficiently the light is preserved and how faithfully the image resembles the source, both being issues of prime concern for telescope designers.

The sampled light may have traveled at light speed for a short time or for billions of years after leaving the source, which makes the telescope a unique tool for studying how the universe was in both the recent and the distant past. Images may be studied to reveal what the light source looked like, its chemistry, location, relative motion, temperature, mass, and other properties. Collecting light and forming images is usually regarded as a telescope function. Analyzing the images is done by various instruments designed for that purpose and attached to the telescope. Detectors are normally part of the instrumentation. The following discussion deals with telescopes.

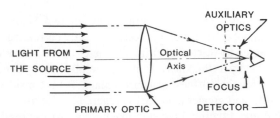

FIGURE 1 Basic telescope elements. Refractive lens could be replaced with a curved reflective mirror.

A. Telescope Size and its Effect on Images

The size of a telescope ordinarily refers to the diameter, or its approximate equivalent, for the area of the first (primary) image-forming optical element surface illuminated by the source. Thus, a 4M telescope usually signifies one with a 4-m diameter primary optic. This diameter sets a maximum limit on the instantaneous photon flux passing through the image-forming optical train. Some telescopes use flat mirrors to direct light into the telescope (e.g., solar heliostats); however, it is the size of the illuminated portion of the primary imaging optic that sets the size.

The size of a telescope determines its ability to resolve small objects. The Airy disc diameter, generally taken to be the resolution limit for images produced by a telescope, varies inversely with size. The Airy disc also increases linearly with wavelength, which means that one must use larger sized telescopes to obtain equivalent imaging resolution at longer wavelengths. This is a concern for astronomers wishing to observe objects at infrared (IR) wavelengths and also explains in part why radio telescopes, operating at even longer wavelengths, are so much larger than optical telescopes. (Radio telescopes are more easily built larger because radio wavelengths are much longer and tolerances on the "optics" are easier to meet.)

Telescopes may be used in an interferometric mode to form interference fringes from different portions of the incoming light beam. Considerable information about the source can be derived from these fringes. The separation between portions of the primary optic forming the image (fringes) is referred to as the baseline and sets a limit on fringe resolution. For a telescope with a single, round primary optic, size and maximum baseline are the same. For two telescopes directing their beams together to form a coherent image, the maximum baseline is equal to the maximum distance between light-collecting areas on the two primaries. More commonly, the center-to-center distance would be defined as the baseline, but the distance between any two image-forming areas is also a baseline. Thus, multiple-aperture systems have many baselines.

As telescope size D increases so does the physical size of the image, unless the final focal ratio F_f in the converging beam can be reduced proportionately, that is,

$$\text{final image size} = F_f \cdot D \cdot \theta,$$

where θ is the angular size of the source measured on the sky in radians. With large telescopes this can be a matter of importance when trying to match the image to a particular detector or instrument. Even for small optics, achieving focal ratios below about f/1.0 is difficult, which sets a practical limit on image size reduction for a given situation.

Another size-related effect is that larger telescopes look through wider patches of the atmosphere which usu-ally contain light-perturbing thermally turbulent (varying air density) regions that effectively set limits on seeing. Scintillation (twinkling) and image motion are caused by the turbulence. However, within a turbulent region, slowly varying isotropic subregions (also called isoplanatic patches) exist that affect the light more or less uniformly. When a telescope is sized about the same as, or smaller than, a subregion and looks through such a subregion, the instantaneous image improves because it is not affected by turbulence outside the subregion. As the subregions sweep through the telescope's field of view, the image changes in shape and position. Larger telescopes looking through many subregions integrate or combine the effects, which enlarges the combined image and effectively worsens the seeing. However, these effects diminish with increasing wavelength, which means that larger telescopes observing at IR wavelengths may have better seeing than smaller ones observing in the visible region.

Studies of atmospheric turbulence effects have given rise to the development of special devices to make optical corrections. These are sometimes called rubber mirrors or adaptive optics. An image formed from incoming light is sensed and analyzed for its apparent distortion. That information is used to control an optical element (usually a mirror) that produces an offsetting image distortion in the image-forming optical train. By controlling on a star in the isoplanatic patch with the object to be observed (so that both experience similar turbulence effects), one can, in principle, form corrected images with a large ground-based telescope that are limited only by the telescope, not the atmosphere. In practice, low light levels from stars and the relatively small isoplanatic patch sizes (typically a few arc-seconds across) have hampered usage of adaptive optics on stellar telescopes. One technique, successfully used in large telescopes, is to split off the visible wavelength portion of the light beam for image distrotion analysis while allowing the longer wavelength (IR) portion of the beam to pass through to the adaptive optic and to form the final corrected image.

B. Telescope Characteristics Related to Size

At least three general, overlapping categories related to telescope size may be defined:

1. Telescopes small enough to be portable. Sizes usually less than 1 m.

2. Mounted telescopes with monolithic primary optics. Sizes presently range up to 6 m for existing telescopes, with many 8 m telescopes either planned or under construction. Virtually all ground-based telescopes used by professional astronomers are in this category.

3. Very large telescopes with multielement primary optics. Proposed sizes range up to 25 m for ground-based telescopes. Only a few multielement telescopes have actually been built.

A fourth category could include telescopes small enough to be launched into Earth's orbit, but the possibility of an in-space assembly of components makes this distinction unimportant.

One cannot, in a short space, describe all of the telescope styles and features. Nevertheless, as one considers larger and larger telescopes, differences become apparent and a few generalizations can be postulated.

In the category of small telescopes, less than 1 m, one finds an almost unlimited variety of telescope configurations. There are few major size limitations on materials for optics. Polishing of optics can often be done manually or with the aid of simple machinery. Mechanical requirements for strength or stiffness are easily met. Adjustments and pointing can be manually performed or motorized. Weights are modest. Opportunities for uniqueness abound and are often highly prized. Single-focus operation is typical. Most of the telescopes used by amateur and professional astronomers are in this size range. Figure 2 illustrates a 40-cm telescope used by professional astronomers.

In the 1–2 m size range, a number of differences and limitations arise. Obtaining high-quality refractive optics is expensive in this range and not practical beyond. Simple three-point mechanical supports no longer suffice for the optics. The greater resolving-power potential demands higher quality optics and good star-tracking precision.

Telescope components are typically produced on large machine tools. Instrumentation is likely to be used at more than one focus position. Because of cost, the domain of the professional astronomer has been reached.

As size goes above 2 m, new issues arise. The need to compensate for self-weight deflections of the telescope becomes increasingly important to maintaining optical alignment. Flexure in the structure may affect the bearings and drive gears. Bearing journals become large enough to require special bearing designs, often of the hydrostatic oil variety. The observer may now be supported by the telescope instead of the other way around. Support of the primary optics is more complex and obtaining primary mirror blanks becomes a special, expensive task. Automated operation is typical at several focal positions. Star-tracking automatic guiders may be used to control the telescope drives, augmented by computer-based pointing correction tables. Figure 3 illustrates a 4-m telescope with all of these features.

At 5 m, the Hale Telescope on Mount Palomar is regarded as near the practical limit for equatorial-style mountings (see Section V). Altitude–azimuth (alt–az) mountings are better suited for bearing heavy rotating loads and are more compact. With computers, the variable drive speeds required with an alt–az telescope can be managed. Mounting size and the length of the telescope are basic factors in setting the size of the enclosing building. For technical reasons and lower cost, the present trend in large telescopes is toward shorter primary

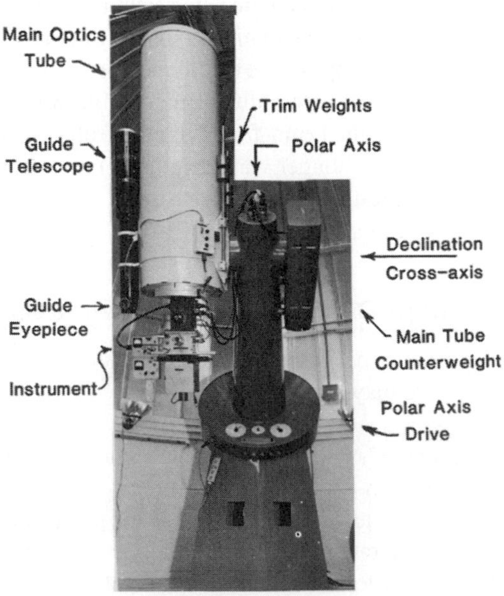

FIGURE 2 40-cm telescope on an off-axis equatorial mount. [Courtesy National Optical Astronomy Observatories, Kitt Peak.]

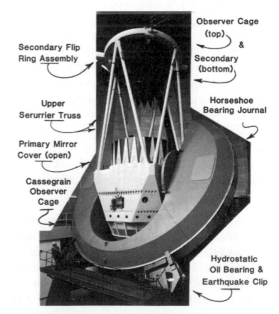

FIGURE 3 Mayall 4-m telescope, with equatorial horse-shoe yoke mountings. [Courtesy National Optical Astronomy Observatories, Kitt Peak.]

TABLE I Telescopes 3 Meters or Larger Built Since 1950

Date completed (projected)	Telescope and/or institution	Primary mirror size (m)	Primary focal ratio	Mounting style
1950	Hale Telescope, Palomar Observatory, California	5	3.3	Equatorial horseshoe yoke
1959	Lick Observatory, California	3	5.0	Equatorial fork
1973	Mayall Telescope, Kitt Peak National Observatory (KPNO), Arizona	4.0	2.7	Equatorial horseshoe yoke
1974	Cerro Tololo International Observatory (CTIO), Chile	4.0	2.7	Equatorial horseshoe yoke
1975	Anglo-Australian Telescope (AAT), Australia	3.9	3.3	Equatorial horseshoe yoke
1976	European Southern Observatory (ESO), Chile	3.6	3.0	Equatorial horseshoe yoke
1976	Large Altazimuth Telescope (BTA), Special Astrophysical Observatory, Caucusus Mtns, Russia	6.0	4.0	Alt–az
1979	Infrared Telescope Facility (IRTF), Hawaii	3.0	2.5	Equatorial English yoke
1979	Canada-France-Hawaii Telescope (CFHT), Hawaii	3.6	3.8	Equatorial horseshoe yoke
1979	United Kingdom Infrared Telescope (UKIRT), Hawaii	3.8	2.5	Equatorial English yoke
1979	Multiple Mirror Telescope (MMT) Observatory, Mt. Hopkins, Arizona	4.5[a]	Six 1.8 m[b]	Alt–az at f/2.7
1983	German-Spanish Astronomical Center, Calar Alto, Spain	3.5	3.5	Equatorial horseshoe fork
1986	Wm. Herschel Telescope, La Palma, Canary Islands	4.2	2.5	Alt–az
1989	New Technology Telescope (NTT), ESO, Chile	3.5	2.2	Alt–az
(1993)	Wisconsin-Indiana-Yale-NOAO (WIYN) Telescope, Kitt Peak, Arizona	3.5	1.75	Alt–az
(1992)	Apache Point Observatory, New Mexico	3.5	1.75	Alt–az
(1992)	Keck Ten Meter Telescope (TMT), Hawaii, Segmented Mirror Telescope (SMT). 36-element (hexagons) paraboloidal primary	10[a]	1.75	Alt–az
Proposed for Hawaii (1987) & CTIO (1999)	National Optical Astronomy Observatories (NOAO), Arizona	8	1.8	Alt–az Disc
Proposed for Chile (1997)	Magellan Project Telescope, Las Campanas, Chile	8	1.2	Alt–az Disc
(1998)	Japanese National Large Telescope (JNLT), Hawaii	8.3	1.9	Alt–az
(1998)	Columbus Project Telescope, Mt. Graham, Arizona, MMT style	11.3[a]	Two 8 m at f/1.2[b]	Alt–az "C-Ring"
(1998)	The Very Large Telescope (VLT), ESO, Chile. An array of telescopes	16[a]	Four 8.2 m at f/2[b]	Alt–az
Proposed	Russia, SMT style with a 400-element (hexagons) spherical primary.	25[a]	2.7	Alt–az

[a] Equivalent circular mirror diameter with equal light gathering area.
[b] Number, size, and f/ratio of individual primary mirror.

focal lengths and alt–az mounts. This trend is evident from Table I, which lists the telescopes 3 m in size or larger that have been built since about 1950. Also listed are the major large telescopes proposed for construction in the late 1980s and the 1990s, which will be discussed in the next section. The largest optical telescope in operation today is the Russian 6 m, which incorporates a solid, relatively thick (650 mm) primary mirror that had to be made three times in borosilicate glass and finally in a low-expansion material before it was successful. Such difficulty indicates that 6 m may be a practical limit

for that style of mirror. New approaches are needed to go beyond.

C. The New Giant Telescopes

The desire for greater light-gathering power and image resolution, especially at IR wave-lengths, continues to press astronomers to build telescopes with larger effective apertures. Costs for a given telescope style and imaging performance have historically risen nearly as the primary aperture diameter to the 2.5 power. These factors have given impetus to a number of new technology telescope designs (see Proposed Projects in Table I) that are based on one or more of the approaches discussed in the following. Computer technology plays a strong part in all of these approaches.

1. Extending the Techniques for Making Lightweight Monolithic Mirror Blanks

Sizes up to about 8 m are considered feasible, although the Russian 6 m is the largest telescope mirror produced before 1985. Further discussion on blank fabrication methods is provided in Section III. Supporting such large mirrors to form good images will be difficult without some active control of the surface figure and thermal conditions in the mirror blank.

Several American universities and the national observatories in the United States and Japan are planning telescopes of this variety.

2. Making a Large Mirror from Smaller Segments

This is also known as segmented mirror telescope, or SMT. In principle, no limit exists for the size of a mosaic of mirror segments that functions optically as a close approximation to a monolithic mirror. For coherency each segment must be precisely and continuously positioned with respect to its neighbors by means of position sensors and actuators built into its support. The segments may be hexagonal, wedge shaped, or other to avoid large gaps between segments. Practical limits arise from support structure resonances and cumulative errors of the segment positioning system. Manufacturing and testing the segments require special methods since each is likely to be a different off-axis optic that lacks a local axis of symmetry but must have a common focus with all the other segments.

The University of California and the California Institute of Technology have adopted this approach for their Keck Observatory 10-m telescope (see Fig. 4), completed in 1992. Russia also has announced plans for a 25-m SMT utilizing a spherical primary to avoid the problems of making aspheric segments.

3. Combining the Light from an Array of Telescopes

Several methods may be considered:

1. Electronic combination after the light has been received by detectors at separate telescopes. Image properties will be those due to the separate telescopes, and coherent combining is not presently possible. Strictly speaking,

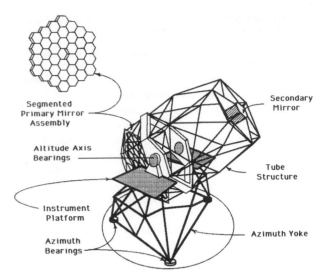

FIGURE 4 Conceptual sketch of a segmented mirror telescope (SMT) using the arrangement adopted for the Keck Ten Meter Telescope, which uses 36 hexagonal mirrors in a primary mirror mosaic to achieve light gathering power equivalent to a single 10-m diameter circular mirror. Each mirror segment must be positioned accurately with respect to its neighbors to achieve the overall optical effect of an unsegmented parabola. The gaps between the segments cause unwanted diffraction effects in the image in proportion to the area they occupy in the primary aperture; hence, they are minimized. Other segment shapes are possible, and the total telescope size is limited only by its structural stiffness and optics alignment provisions. Thus, very large SMTs are theoretically possible.

this is an instrumental technique and will not be considered further.

2. Optical combination at a single, final focus of light received at separately mounted telescopes. To maintain coherency between separate light beams, one must equalize the optical path distance (OPD) between the source and the final focus for all telescopes, a difficult condition to meet if telescopes are widely separated.

3. Placing the array of telescopes on a common mounting with a means for optically combining the separate light beams. All OPDs can be equal (theoretically), thus requiring only modest error correction to obtain coherency between telescopes. This approach is known as the multiple mirror telescope (MMT).

The simplest array of separately mounted individual telescopes is an arrangement of two on a north–south (NS) baseline with an adjustable, combined focus between them (the OPD changes occur slowly with this arrangement when observing at or near the meridian). Labeyrie pioneered this design in the 1970s at Centre d'Etudes et de Recherches Géodynamiques et Astronomiques (CERGA) in France, where he used two 25-cm telescopes on a NS variable baseline of up to 35 m to measure successfully numerous stellar diameters and binary star separations, thereby showing that coherent beam combination and the angular resolution corresponding to a long telescope baseline could be obtained. Other schemes for using arrays of separate telescopes on different baselines all require movable optics in the optical path between the telescopes to satisfy the coherency conditions, and so far none has been successfully built. However, the European Southern Observatory is building an array of four 8-m telescopes in Chile which will be known as the "Very Large Telescope" (VLT) when completed in the late 1990s. One form of the VLT is an in-line array with an "optical trombone" arrangement for equalizing OPDs. Other arrangements are also under consideration.

The MMT configuration was first used by the Smithsonian Astrophysical Observatory (SAO) and the University of Arizona (UA). The SAO/UA MMT on Mount Hopkins uses six 1.8-m image-forming telescopes arrayed in a circle around a central axis. Six images are brought to a central combined focus on the central axis, where they may be incoherently stacked, coherently combined, or used separately. The effective baseline for angular resolution (i.e., the maximum separation between reflecting areas) is 6.9 m, and the combined light-gathering power is equivalent to a single 4.5-m diameter mirror. A two-element MMT using 8-m mirrors and a 22-m baseline, illustrated in Fig. 5 and called the "Columbus Project," is under construction by the UA in collaboration with the Observatorio Astrofisico de Arcetri, Italy. Appropriately, this telescope has been dubbed the "Big Binocular."

II. OPTICAL CONFIGURATIONS

Light entering the telescope is redirected at each optical element surface until it reaches the focal region where the images are most distinct. The light-sensitive detector is customarily located in an instrument mounted at the focal region. By interchanging optics, one can create more than one focus condition; this is commonly done in large telescopes to provide places to mount additional instruments or to produce different image scales. The arrangement of optics and focal positions largely determines the required mechanical support configuration and how the telescope will be used.

The early telescopes depended solely upon the refractive power of curved transparent glass lenses to redirect the light. In general, these telescopes were plagued by chromatic aberration (rainbow images) until the invention in 1752 of achromatic lenses, which are still used today in improved forms. Curved reflective surfaces (i.e., mirrors) were developed after refractors but were not as useful until highly reflective metal coatings could be applied onto

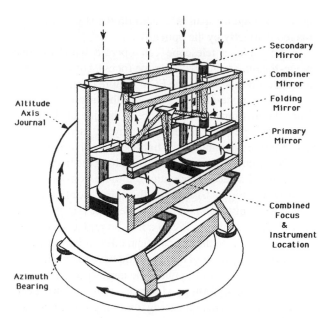

FIGURE 5 Conceptual sketch of a multiple mirror telescope (MMT) using the arrangement planned for the Columbus "Big Binocular" Telescope, which uses two 8-m diameter telescopes mounted on a common structure with provisions to combine their separate output light beams. With careful control of the optical path distances in each telescope, the resolving power of the telescope is equivalent to that of a single-mirror telescope of a size equal to the maximum separation between the primary mirrors of the two telescopes. Total light gathering power is equal to the sum of the individual telescope powers. There is no theoretical limit to the number of individual telescopes in an MMT, but the beam combining and structural provisions become increasingly complex.

glass substrates. Today, mirrors are more widely used than lenses and can generally be used to produce the same optical effects; they can be made in larger sizes, and they are without chromatic aberration. These are still the only two means used to form images in optical telescopes. Accordingly, telescopes may be refractive, reflective, or catadioptric, which is the combination of both.

A. Basic Optical Configurations: Single and Multielement

The telescope designer must specify the type, number, and location of the optical elements needed to form the desired image. The basic choices involve material selections and the shapes of the optical element surfaces. Commonly used surfaces are flats, spheres, paraboloids, ellipsoids, hyperboloids, and toroidal figures of revolution.

The optical axis is the imaginary axis around which the optical figures of revolution are rotated. Light entering the telescope parallel to this axis forms the on-axis (or zero-field) image directly on the optical axis at the focal region. The field of view (FOV) for the telescope is the widest angular span measured on the sky that can be imaged distinctly by the optics.

In principle, a telescope can operate with just one image-forming optic (i.e., at prime focus), but without additional corrector optics, the FOV is quite restricted. If the telescope is a one-element reflector, the prime focus and hence the instrument/observer are in the line of sight. For large telescopes (i.e., >3 m) this may be used to advantage, but more commonly the light beam is diverted to one side (Newtonian) or is reflected back along the line of sight by means of a secondary optic to a more convenient focus position. Figure 6 illustrates the optical configurations most commonly used in reflector telescopes. In principle, the reflectors shown in Fig. 6 could be replaced with refractors to produce the same optical effects. However, the physical arrangement of optics would have to be changed.

In practice, one tries to make the large optics as simple as possible and to form good images with the fewest elements. Other factors influencing the configuration include the following:

1. Simplifying optical fabrication. Spherical surfaces are generally the easiest to make and test. Nonsymmetric aspherics are the opposite extreme.
2. Element-to-element position control, which the telescope structure must provide. Tolerances become tighter as the focal ratio goes down.
3. Access to the focal region for viewing or mounting instrumentation. Trapped foci (e.g., the Schmidt) are more difficult to reach.
4. Compactness, which generally aids mechanical stiffness.
5. Reducing the number of surfaces to minimize light absorption and scattering losses.

Analyzing telescope optical systems requires a choice of method. The geometric optics method treats the incoming light as a bundle of rays that pass through the system while being governed by the laws of refraction and reflection. Ray-tracing methods based on geometric optics are commonly used to generate spot diagrams of the ray positions in the final image, as illustrated in Fig. 7a. More rigorous analysis based on diffraction theory is done by treating the incoming light as a continuous wave and examining its interaction with the optical system. Figure 7b is a computer-generated plot of light intensity in a diffracted image formed by a circular aperture. The central peak represents the Airy disc. For further detail on the use of these methods the reader is referred to texts on optics design.

B. Wide Field Considerations

Modern ground-based telescopes are usually designed to resolve images in the 0.25–1.0-arcsec range and to have FOV from a few arc minutes to about one degree. In general, distortions or aberrations due to the telescope optics exist in the images and are worse for larger field angles. Table II lists the basic types. Space-based telescopes (e.g., the Hubble telescope) can be built to resolve images in the 0.1 arcsec region because atmospheric seeing effects are absent, but compensation must still be made for optical aberrations. Aberrations can also arise from nonideal placement of the optical surfaces (i.e., position errors) and from nonuniform conditions in the line of sight.

To correct distortions produced by optics, the designer frequently tries to cancel aberrations produced at one surface by those produced at another. Extra optical surfaces may be introduced for just this purpose. Ingenious optical corrector designs involving both refractors and reflectors have resulted from this practice. Many of these designs are described in texts on optics under the originators' names (e.g., Ross, Baker, Wynne, Shulte, and Meinel). It is possible, however, to cancel some field aberrations by modifying the principal optical surfaces or by using basic shapes in special combination. For example, the Ritchey–Crétian telescope design for a wide FOV reduces coma by modifying the primary and secondary surfaces of a Cassegrain telescope. The Mersenne telescope cancels aberrations from the primary (a parabola) with the secondary (another parabola).

A spherical mirror with the aperture stop set at its center of curvature has no specific optical axis and forms equally good images everywhere in the field. Images of distant objects have spherical aberrations, however, and the focal region is curved. The Schmidt telescope compensates for most of the spherical aberration by means of an aspheric refractor at the center of curvature. The Maksutov telescope introduces an offsetting spherical aberration by means of a spherical meniscus refractor. Numerous variations of this approach have been devised yielding well-corrected images in field sizes of 10 deg and more. However, large fields may have other problems.

During a long observing period, images formed by a telescope with a very large FOV (i.e., ~1 deg or more) are affected differently across the field by differential refraction effects caused by the atmosphere. Color dispersion effects (i.e., chromatic blurring) effectively enlarge the images as the telescope looks through an increasing amount of atmosphere. Differential image motion also occurs, varying as a function of position in the FOV, length of observation, and telescope pointing angle. Partial chromatic correction can be made by inserting a pair of separately rotatable prisms, called Risley prisms, ahead of the focal position. Even with chromatic correction, however, images are noticeably elongated (>0.5 arcsec) at the edge of a 5-deg field compared with the on-axis image after a continuous observing period of a few hours.

TYPE	PRIMARY OPTIC	SECONDARY OPTIC	CONFIGURATION 1–PRIMARY 2–SECONDARY 3–EYEPIECES/CORRECTORS 4–FOCUS
KEPLERIAN GALILEAN (if refractive)	SPHERE or PARABOLA	NONE	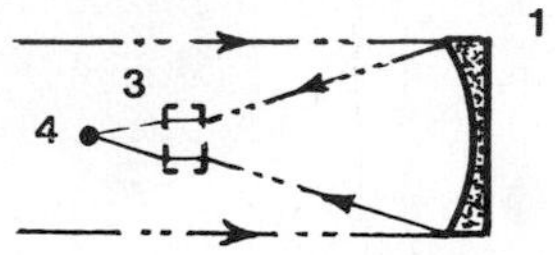
HERSCHELIAN	OFF-AXIS PARABOLA	NONE	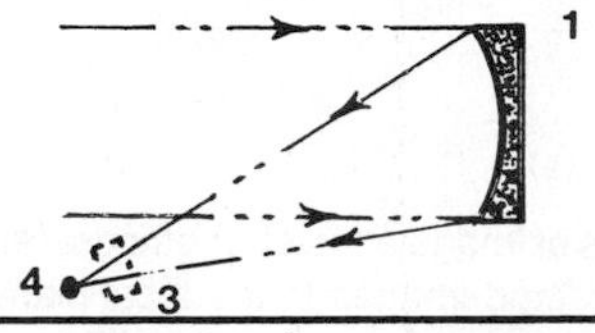
NEWTONIAN	PARABOLA	DIAGONAL FLAT	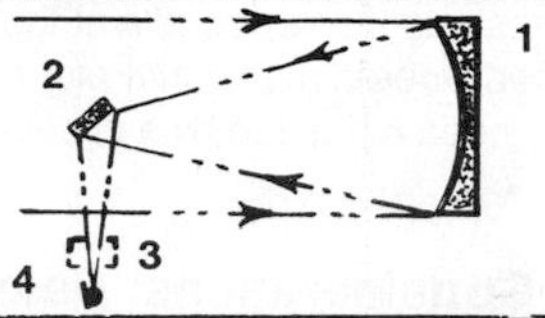
GREGORIAN	PARABOLA	ELLIPSE	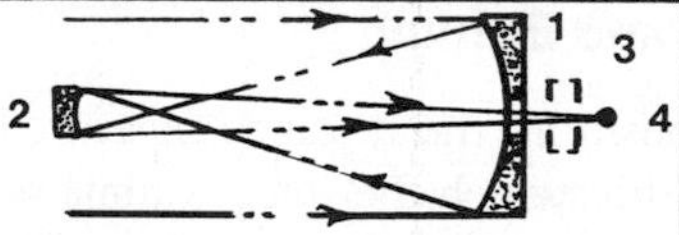
MERSENNE	PARABOLA	PARABOLA	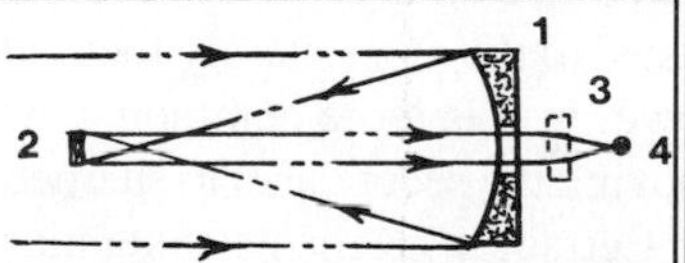
CASSEGRAIN	PARABOLA	HYPERBOLA	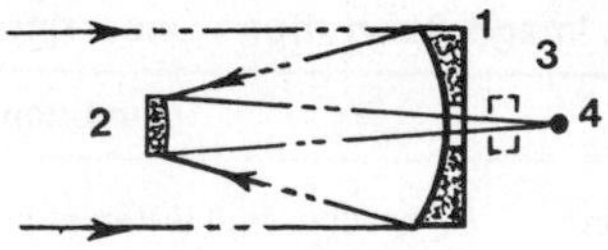
RITCHEY-CHRÉTIEN	MODIFIED PARABOLA	MODIFIED HYPERBOLA	
DALL-KIRKHAM	ELLIPSE	SPHERE	
SCHMIDT	ASPHERIC REFRACTOR	SPHERE	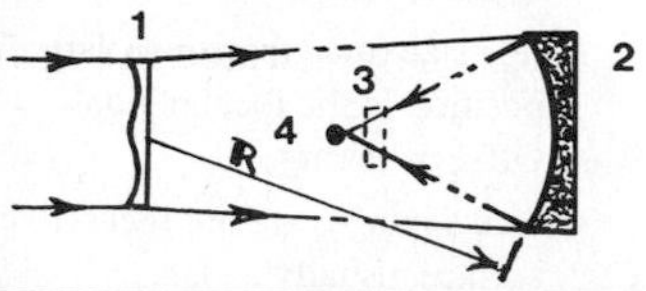
BOUWERS-MAKSUTOV	REFRACTIVE MENISCUS	SPHERE	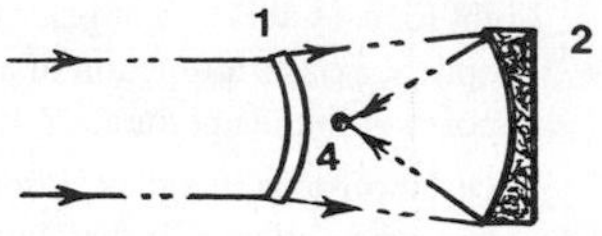

FIGURE 6 Basic optical configurations for telescopes.

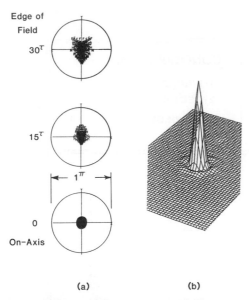

FIGURE 7 Examples of image analysis methods. (a) Typical spot diagrams of well-corrected images in the focal plane of a 1-deg field. Each image represents the ray bundle at that field location. (b) Computer-generated three-dimensional plot of intensity in a perfect image that has been diffracted by a circular aperture.

C. Instrumental Considerations: Detector Matching and Baffling

At any focal position, the distance measured along the optical axis within which the images remain acceptably defined is referred to as the depth of focus. The detector should be adjusted to the most sharply focused position in this region; however, it is more common to adjust the focus (e.g., by moving the secondary) to sharpen the images at the detector. Certain instruments containing reimaging

TABLE II Basic Image Aberrations Occurring in Telescopes

Type	Condition
Spherical aberration[a]	Light focuses at different places along the optical axis as a function of radial position in the aperture.
Coma[a]	Image size (magnification) varies with radial position in the focal region. Off-axis flaring.
Field curvature[a]	Off-axis images are not focused on the ideal surface, usually a plane.
Astigmatism[a]	Light focuses at different places along the optical axis as a function of angular position in the aperture.
Distortion[a]	Focused off-axis image is closer to or further from the optical axis than intended.
Chromatic aberration	Shift in the focused image position as a function of wavelength.

[a] Also known as Seidel aberrations.

optics (e.g., spectrographs) require only that the image be formed at the entrance to the instrument (e.g., at a slit or aperture plate). There is a practical limit to the amount of allowable focal position movement obtained by moving the secondary because optical aberrations, especially spherical aberration, are introduced when the mirror is displaced from its theoretically perfect position.

To view all or part of the focused FOV, it is common to insert a field mirror into the beam at 45 deg to divert the desired portion of the field out to an eyepiece or a TV monitoring camera. The undiverted light to be observed passes on to the instrument. The diverted light can be used for guiding purposes or to make different observations.

The angular size of the focused image (i.e., the image scale) and its sensitivity to defocusing changes is governed by the effective focal length (EFL) of the optical system that formed the image:

$$\text{image scale} = \frac{1}{\text{EFL}}$$
$$= \frac{\text{radians on the sky}}{\text{length in the focal plane}}.$$

Large EFL values produce comfortable focal depths, but the image sizes are relatively large. The physical size of the detector may place a limit on the FOV that can be accommodated for a given scale, hence a potential need for a different focal position or reimaging optics. This issue has become quite important with the advent of solid-state image detectors, known as CCDs (Charge Coupled Devices), that have much higher quantum efficiency than photographic plates but are difficult to make in large sizes (i.e., approximately 50×50 mm is considered large in 1990).

Light baffles and aperture stops are important but frequently neglected aspects of telescope design. It is common to place a light baffle just ahead of the primary to block out unwanted edge effects, in which case the baffle may become the aperture stop. In some cases, the primary surface may be reimaged further along the optical path where a light baffle can be located. This may be done either to reduce the size of the required baffle or to locate it advantageously in a controlled environment (e.g., at cryogenic temperature to reduce thermal radiation effects). In other cases, the aperture may be set by undersizing one of the optical elements that is further along the optical path. An example is an IR-optimized telescope, where the secondary is made undersized to ensure that the detector cannot see past the edge of the primary mirror. The effective light-gathering power of the telescope can be significantly reduced under these conditions.

Obstructions in the light path further reduce light-gathering power. The most common obstructions are the secondary and auxiliary optics along with the mechanical struts that support them. It is common for 10–20% of the aperture to be obstructed in this manner. Depending upon the telescope style and the instrument location, it may be necessary to use a light baffle to prevent the detector from seeing unwanted radiation. For example, a detector at the Cassegrain focus of a two-mirror telescope can see unfocused light from stars directly past the perimeter of the secondary mirror unless obstructing baffles are provided. The size and location of these baffles are generally determined empirically by ray-tracing, and it is a fact that larger FOV requires larger baffles that mean more obstruction. Thus, one cannot truly determine telescope light-gathering power until the usage is considered.

III. TELESCOPE OPTICS

Telescope performance depends fundamentally upon the quality of optics, especially the surface figures. That quality in most telescopes is a compromise between what the optical designer has specified, what the glassmakers and opticians can make, and how well the mechanical supporting structures perform. Especially for larger sizes, one should specify the optics and their supports at the same time. How the optics are to be tested should also be considered since the final figure corrections are almost always guided by test results. Optical figuring methods today range from simple manual lapping processes to sophisticated computer-controlled polishers (CCP) and direct machining using diamond tools. The remarks to follow are only a summary of a complex technical field.

A. Refractive Optics

Light passing through the surface of a refractive optic is changed in direction according to Snell's law of refraction:

$$\eta_1 \sin \theta_1 = \eta_2 \sin \theta_2,$$

where η_1 and η_2 are the refractive indices of the materials on either side of the surface and θ_1 and θ_2 the angles with respect to the surface normal of incidence and refraction. Producing a satisfactory refractor is, therefore, done by obtaining transparent glass or another transmitting material with acceptable physical uniformity and accurately shaping the surfaces through which light passes. Sometimes the optician can alter the surface to compensate for nonuniform refractive properties. Losses ordinarily occur at the surface due to scattering and also to the change in refractive index. Antireflection coatings can be applied to reduce surface transmission losses, but at the cost of restricting transmission to a specific wavelength range. Further losses occur internally by additional scattering and absorption.

Manufacturing methods for refractive optics are similar to those for reflectors.

B. Reflective Optics

Light reflecting off a surface is governed by the law of reflection:

$$\theta_1 = -\theta_2,$$

where θ_1 and θ_2 are the angles of incidence and reflection, respectively. Controlling the slope at all points on the reflecting surface is, therefore, the means for controlling where the light is directed. Furthermore, any part of the surface that is out of its proper position, measured along the light path, introduces a change at that part of the reflected wavefront (i.e., a phase change) equal to twice the magnitude of the position error. Surface accuracy is thus the prime consideration in making reflective optics. Achieving high reflectivity after that is usually done by applying a reflective coating.

The most common manufacturing method is to rough-machine or grind the mirror blank surface and progressively refine the surface with abrasive laps. Certain kinds of mirrors, especially metal ones, may be diamond turned. In this case, the accuracy of the optical surface may be governed by the turning machine, whereas ordinarily the accuracy limits are imposed principally by the optical test methods and the skill of the optician. High-quality telescope mirrors are typically polished so that most of the reflected light ($>80\%$) is concentrated in an image that is equivalent to the Airy disc or the seeing limit, whichever is larger.

C. Materials for Optics

Essential material requirements for all optical elements are related to their surfaces. One must be able to polish or machine the surfaces accurately, and afterward the blank should not distort uncontrollably. Residual stresses and unstable alloys are sources of dimensional instability to be avoided. Stresses also cause birefringence in refractors.

1. Refractors

Refractors should transmit light efficiently and uniformly throughout the operating wavelength region. However, there is no single material that transmits efficiently from 0.3 to 40 μm. One typically chooses different materials for the UV, near-IR (to $\simeq 2\ \mu$m), and far-IR regions. Glassmakers can control the index of refraction to about one part in 10^6, but only in relatively small blanks (<50-cm diameter). In larger sizes, index variations and inclusions limit the availability of good-quality refractor blanks to sizes less than 2-m diameter.

2. Reflectors

The working part of a reflector optic is usually the thin metallic layer, 1000–2000 Å thick, that reflects the light. An evaporated layer of aluminum, silver, or gold is most commonly used for this purpose, and it obviously must be uniform and adhere well to the substrate. Most of the work in making a reflector, however, is in producing the uncoated substrate or mirror blank. The choice of material and the substrate configuration are critically important to the ultimate reflector performance.

Reflector blanks, especially large ones, require special measures to maintain dimensional (surface) stability. The blank must be adequately supported to retain its shape under varying gravitational loads. It must also be stable during normal temperature cycles, and it should not heat the air in front of the mirror because that causes thermal turbulence and worsens the seeing. A mirror-to-air temperature difference less than about 0.5°C is usually acceptable. The support problem is a mechanical design consideration. Thermal stability may be approached in any of several ways:

1. Use materials with low coefficients of thermal expansion (CTE).
2. Lightweight the blank to reduce the mass and enhance its ability to reach thermal equilibrium quickly (i.e., by using thin sections, pocketed blanks, etc.). Machineability or formability of the material is important for this purpose.
3. Use materials with high thermal conductivity (e.g., aluminum, copper, or steel).
4. Use active controls for temperature or to correct for thermally induced distortion. Elastic materials with repeatable flexure characteristics are desirable. Provisions in the blank for good ventilation may be necessary.

Materials with low CTE values include borosilicate glass, fused silica, quartz, and ceramic composites. Multiple-phase (also called binary) materials have been developed that exhibit near-zero CTE values, obtained by offsetting the positive CTE contribution of one phase with the negative CTE contribution from another. Zerodur made by Schott Glaswerke, ULE by Corning Glass Works, and epoxy–carbon fiber composites are examples of multiple-phase materials having near-zero CTE over some range of operating temperature. Fiber composites usually require a fiber-free overlayer that can be polished satisfactorily.

One usually considers a lightweight mirror blank to reduce costs or to improve thermal control. This is particularly true for large telescopes. Reducing mirror weight often produces a net savings in overall telescope cost even if the lightweighted mirror blank is more expensive than a corresponding solid blank. The important initial step is to make the reflecting substrate or faceplate as thin as possible, allowing for polishing tool pressures and other external forces. One has three hypothetical design options:

1. Devise a way to support just the thin monolithic faceplate. This approach works best for small blanks. Large, thin blanks require complex supports and, possibly, a means to monitor the surface shape for active control purposes. The European Southern Observatory's 3.5 m New Technology Telescope (NTT), commissioned in 1990, is a good example of this approach. The additional complexity of the mirror support is made more worthwhile by using it to correct for image aberrations due to other causes. The NTT has resolved images of 0.33 arcsec (containing 80% of the focused light and all of the aberrations due to the telescope and atmosphere), perhaps, the best performance to date for a large ground-based telescope.
2. Divide the faceplate into small, relatively rigid segments and devise a support for each segment. Segment position sensing and control is required: a sophisticated technical task. The Keck Ten Meter Telescope uses this approach.
3. Reinforce the faceplate with a gridwork of ribs or struts, possibly connected to a backplate, to create a sandwichlike structure. One may create this kind of structure by fusing or bonding smaller pieces, by casting into a mold, or by machining away material from a solid block.

All of these approaches have been used to make lightweight reflector blanks. Making thin glass faceplates up to about 8 m is considered feasible by fusing together smaller pieces or direct casting. Titanium silicate, fused silica, quartz, and borosilicate glass are candidate materials. Castings of borosilicate glass up to 6 m (e.g., the Palomar 5-m mirror) have been produced; 8 m is considered feasible. Structured (i.e., ribbed) fused silica and titanium silicate mirrors up to 2.3 m have been produced (e.g., the Hubble Space Telescope mirror); up to 4 m is considered feasible.

Metals may also be used for lightweight mirrors but have not been widely used for large telescopes because of long-term dimensional (surface) changes. The advent of active surface control technology may alter this situation in the future. Most metals polish poorly, but this can be overcome by depositing a nickel layer on the surface to be polished. Most nonferrous metals can also be figured on diamond-turning machines; however, size is limited presently to about 2 m.

IV. SPECTRAL REGION OPTIMIZATION: GROUND-BASED TELESCOPES

The optical/IR window of the atmosphere from 0.3 to 35 μm is sufficiently broad that special telescope features are needed for good performance in certain wavelength regions. Notably, these are needed for the UV region less than about 0.4 μm and the thermal IR region centered around 10 μm. In the UV, it is difficult to maintain high efficiency because of absorption losses in the optics. When observing in the IR region, one must cope with blackbody radiation emitted at IR wavelengths by parts of the telescope in the light path, as well as by the atmosphere. Distinguishing a faint distant IR source from this nearby unwanted background radiation requires special techniques. Using such techniques, it is common for astronomers to use ground-based telescopes to observe IR sources that are more than a million times fainter than the IR emission of the atmosphere through which the source must be discerned.

A. UV Region Optimization

The obvious optimizing step for the UV region is to put the telescope into space. If the telescope is ground-based, however, one good defense against UV light losses is to use freshly coated aluminum mirror surfaces. Reflectivity values in excess of 90% can be obtained from freshly coated aluminum, but this value rapidly diminishes as the surface oxide layer develops. Protective coatings such as sapphire (Al_2O_3) or magnesium fluoride (MgF_2) can be used to inhibit oxidations. Also, multilayer coatings can be applied to the surfaces of all optics to maximize UV throughput. However, these coatings greatly diminish throughput at longer wavelengths, which leads to the tactic of mounting two or more sets of optics on turrets, each set being coated for a particular wavelength region. The desired set is rotated into place when needed. This tactic obviously works best for small optics, not the primary.

Many refractive optics materials absorb strongly in the UV. Fused silica is good low-absorbing material. If optics are cemented together, the spectral transmission of the cement should be tested. Balsam cements are to be avoided.

B. IR Region Optimizations

One cardinal rule for IR optimization is to minimize the number and sizes of emitting sources that can be seen by the detector. This includes mechanical hardware such as secondary support structures and baffles, as well as seemingly empty spaces such as the central hole in the primary mirror. All of these emit black body radiation corresponding to their temperatures.

Since the detector obviously must see the optical surfaces, another cardinal rule is to reduce the emissivity of these surfaces with a highly reflective coating. If a coating is 98% reflective, it emits only 2% of the blackbody radiation that would otherwise occur if the surface were totally nonreflective. If possible, the detector should see only reflective surfaces, and these should be receiving radiation only from the sky or other reflective surfaces in the optical train. Objects that must remain in the line of sight can also be advantageously reflective provided that they are not looking at other IR-emitting objects that could send the reflected radiation into the main beam.

Achieving these goals may require one or more of the following special telescope features:

1. Using an exchangeable secondary support structure. This enables elimination of oversize secondaries and baffles that might be needed for other kinds of observation.

2. Using the secondary mirror as the aperture stop and making it sufficiently undersized that it cannot see past the rim of the primary mirror.

3. Putting all of the secondary support (except the struts) behind the mirror so that none of the hardware is visible to the detector.

4. Placing a specially shaped (e.g., conical) reflective plug at the center of the secondary to disperse radiation emitted from the central hole region of the primary mirror.

A basic technique for ground-based IR observing is that of background subtraction. This involves alternating the pointing direction of the telescope between the object (thus generating an object-plus-background signal from the detector) and a nearby patch of sky that has no apparent object (thus generating a background-only signal). Signal subtraction then eliminates the background signal common to both sky regions. Methods for alternating between positions include (1) driving the telescope between the two positions, usually at rates below 0.1 Hz, (2) wobbling the secondary mirror at rates between 10 and 50 Hz (called chopping), and (3) using focal plane modulators such as rotating aperture plates (called focal plane chopping). With solid state CCD detectors, one may simultaneously observe the object and a nearby background patch and electronically remove the measured background through appropriate processing of the data from individual pixels on the detector.

V. MECHANICAL CONFIGURATIONS

The mechanical portion of a mounted, ground-based telescope must support the optics to the required precision, point to and track the object being observed, and support

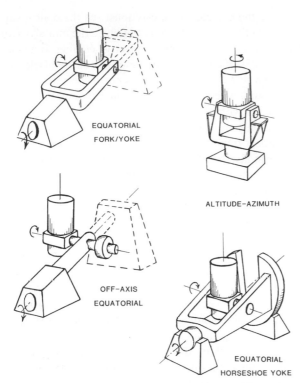

EQUATORIAL
FORK/YOKE

ALTITUDE–AZIMUTH

OFF-AXIS
EQUATORIAL

EQUATORIAL
HORSESHOE YOKE

FIGURE 8 Basic telescope mounting styles in popular use. Numerous variants on each style are in existence.

the instrumentation in accessible positions. It is customary to distinguish between the telescope (or tube), which usually supports the imaging optics, and the mounting, which points the telescope and includes the drives and bearings. Tracking motion is usually accomplished by the mounting.

Most telescope mountings incorporate two, and occasionally three, axes of rotation to enable pointing the telescope at the object to be observed and tracking it to keep it centered steadily in the FOV. In general, the rotating mass is carefully balanced around each axis to minimize driving forces and the location of each rotation axis is chosen to minimize the need for extra counterweights.

Figure 8 illustrates the basic mounting styles that are discussed in the next section.

A. Mounting Designs

A hand-held telescope is supported and pointed by the user. The user is the mounting in this case. The mechanical mounting for a telescope performs essentially the same function, except that a mechanical mounting can support heavier loads and track the object more smoothly. As a rule of thumb, short-term tracking errors in high-quality telescopes are less than 10% of the smallest resolved object that can be observed with the telescope. Smoothness

of rotation is important for long-term observations (i.e., no sudden movements) which mandates the use of high-quality bearings. Pressurized oil-film bearings (hydrostatics) are used in large telescopes for this reason.

Telescope drives range widely in style. The chief requirements are smoothness, accuracy, and the ability to move the telescope rapidly for pointing purposes (i.e., slewing) or slowly for tracking (i.e., at one revolution per day or less). Electric motor driven traction rollers, worm gears, or variants on spur gears are most commonly used. Position measuring devices (encoders) are often used to sense telescope pointing and to provide input data for automatic drive controls. Adjustments in tracking rates or pointing are accomplished either by manual control from the observer or, possibly, by star-tracking automatic guiding devices. Pointing corrections may also be based on data stored in a computer from mounting flexure and driving-error calibrations done at an earlier time. Telescope pointing accuracies to about 1 arcsec are currently possible with such corrections. Once the object is located in the FOV, the ability to track accurately is the most important consideration.

1. Equatorial Mounts

Astronomical telescopes ordinarily are used to observe stars and other objects at such great distances that they would appear stationary during an observation period if the earth did not rotate. Accordingly, the simplest telescope tracking motion is one that offsets the earth's rotation with respect to "fixed" stars (i.e., sidereal rate) and is done about a single axis parallel to the earth's north–south (N–S) polar axis. Equatorial mountings are those that have one axis of rotation (i.e., the polar or right ascension axis) set parallel to the earth's N–S axis. This axis is tilted toward the local horizontal plane (i.e., the ground) at an angle equal to local latitude. A rotatable cross-axis (also called declination axis) is needed for initial pointing and guiding corrections, but the telescope does not rotate continuously around this axis while tracking.

The varieties of equatorial mountings are limited only by the designer's imagination. The basic varieties, however, are the following:

1. Those that mount the tube to one side of the polar axis and use a counterweight on the opposite side to maintain balance. For an example, see Fig. 2. These are sometimes called off-axis or asymmetric mounts. It is also possible to mount a second telescope in place of the counterweight.

2. Those that support the tube on two sides in a balanced way to eliminate the need for a heavy counterweight. Yokes and forks are most commonly used, especially for larger telescopes. Theses are sometimes called symmetric

mounts. For an example, see Fig. 3 which shows a horse-shoe yoke mount.

2. Other Mounting Styles

The alt–az mounting is configured around a vertical (azimuth) axis of rotation and a horizontal cross-axis (the altitude or elevation axis). The altitude–altitude (alt–alt) mounting, not widely used, operates around a horizontal axis and a cross-axis that is horizontal when the telescope points at the meridian and is tilted otherwise (similar to an English yoke with its polar axis made horizontal). With either of these styles, because neither axis is parallel to the earth's rotation, it is necessary to drive both axes at variable rates to track a distant object. Furthermore, the FOV appears to rotate at the focal region, which often necessitates a derotating instrument mounting mechanism, also moving at a variable rate. These factors inhibited the use of these mountings, except for manually guided telescopes, until the advent of computers on telescopes. Computers enable second-to-second calculation of the drive rates, which is required for accurate tracking. The alt–az configuration cannot track an object passing through the local zenith because the azimuth drive rate theoretically becomes infinite at that point. In practice, alt–az telescopes are operated to within about 1 deg of zenith.

The famous Herschel 20-ft telescope, built in England in 1783, was the first large alt–az telescope. Very few were built after that, but the trend today is toward alt–az mounts (see Section I.B and Table I). The ability to support the main azimuth bearing with a solid horizontal foundation is advantageous, as is the fact that the altitude axis bearings do not change in gravity orientation. These are important considerations when bearing loads of hundreds of tons must be accommodated. The alt–alt mounting is not as suitable for carrying heavy loads because the cross-axis is usually tilted with respect to gravity.

B. Telescope Tubes and Instrument Considerations

Design of the tube begins with the optical configuration. Tube structures are designed to maintain the optics in alignment, either by being stiff enough to prevent excessive deflections or by deflecting in ways that maintain the optics in the correct relative position. The well-known Serrurier truss first used on the Palomar 5-m telescope is a much-copied example of the latter (see Fig. 9). The tube structure is normally used to support the instruments at the focal positions, sometimes along with automatic guiders, field-viewing TV monitors, calibration devices, and field derotators. Large telescopes often do not use the Serrurier truss design specifically, but the flexure of the structure is

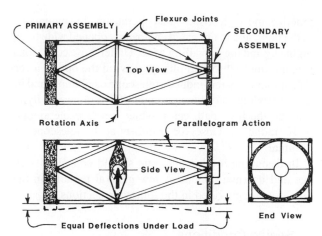

FIGURE 9 Serrurier truss used to maintain primary-to-secondary alignment as the tube rotates. Equal deflections and parallelogram action at both ends keep the optics parallel and equidistant from the original optical axis. Similar flexure is designed into most large telescope tube structures.

usually designed to perform the same function. Modern computer-based structural analysis programs, developed since about 1960 and thus not available to Serrurier, provide the tools to create lighter, stiffer structures without ambiguity about the positions of the optics.

Focal positions (i.e., instrument mount locations) on the tube obviously move as the telescope points and tracks, which can be a problem for instruments at those locations that work poorly in a varying gravity environment. In those cases, one can divert the optical beam out of the telescope tube along the cross-axis to a position on the mounting or even outside the mounting. Flat mirrors are normally used for this purpose. To reach a constant-gravity position with an equatorial mounting, one must use several mirrors to bring the converging beam out: first along the declination axis, then the polar axis, and finally to a focus off the mounting. This is known as the coudé focus and is commonly used to bring light to spectrographs that are too large to mount on the telescope tube.

One can reach a constant-gravity focus (instrument location) on an alt–az telescope by simply diverting the beam along the altitude cross-axis to the mounting structure that supports the tube. This is called the Nasmyth focus after its Scottish inventor. The instrument rides the mounting as it rotates in azimuth but does not experience a change in gravity direction.

VI. CONSIDERATIONS OF USAGE AND LOCATION

Considering the precision built into most optical telescopes, one would expect them to be sheltered carefully. In practice, most telescopes must operate on high mountains, in the dark, and in unheated enclosures opened wide to the night sky and the prevailing wind. Under these conditions, it is not unusual to find dust or dew on the optics, a certain amount of wind-induced telescope oscillation, and insects crawling into the equipment. Certain insects flying through the light path can produce a noticeable amount of IR radiation. Observer comforts at the telescope are minimal.

In designing a telescope, one should consider its usage and its environment. A few general remarks in this direction are provided in the following sections.

A. Seeing Conditions

The seeing allowed by the atmosphere above the telescope is beyond ordinary control. Compensation may be possible as discussed in Section I.A. but the choice of site largely determines how good the imaging is. Seeing conditions in the region of 0.25 arcsec or less have been measured at certain locations, but more typically, good seeing is in the 0.5–1.0 arcsec range. Beyond 2–3 arcsec, seeing is considered poor. To the extent possible, one should build the telescope to produce images equal to or better than the best anticipated seeing conditions.

Locating the telescope at high altitudes usually reduces the amount of atmosphere and water vapor that is in the line of sight (important for IR astronomy), however, the number of clear nights and the locally produced thermal turbulence should also be considered. In many locations, a cool air layer forms at night near the ground which can be disturbed by the wind and blown through the line of sight. In other cases, warm air from nearby sources can be blown through the line of sight. In either case, telescope seeing is worsened.

Other seeing disturbances can originate inside the telescope enclosure. Any source of heat (including observers) is a potential seeing disturbance. Also, any surface that looks at the night sky, and hence is cooled by radiative exchange, may be a source of cooled air that can disturb seeing if it falls through the line of sight. If possible, it is desirable to allow the telescope enclosure to be flushed out by the wind to eliminate layers and pockets of air of different temperatures. Some telescope buildings have been equipped with air blowers to aid in the process, but dumping the air well away from the building has not always been possible even though it should be done.

The study of atmospheric seeing has become a relatively advanced science, and the telescope builder is well advised to consult the experts in choosing a site or designing an enclosure. Having chosen a site, one may be guided by the truism that seeing seldom improves by disturbing Mother Nature.

B. Nighttime versus Daytime Usage

With the advent of IR astronomy, optical telescopes began to be used both day and night because the sky radiation background is only slightly worse at IR wavelengths during the day compared with night. During the day it is much harder to find guide stars, and the telescope must often point blindly (and hence, more accurately) at the objects to be observed; but much useful data can be obtained. Some problems arise from this practice, however.

A major purpose of the telescope enclosure, other than windscreening, is to keep the telescope as close as possible to the nighttime temperature during the day so that it can equalize more rapidly to the nighttime temperature at the outset of the next night's observing. Obviously, this cannot be done if the telescope enclosure has been open during the day for observation. The condition is worsened if sunlight has been allowed to fall on the telescope during the day. Accordingly, optical/IR telescopes should be designed for rapid thermal adjustment.

Some of the design options in thermal control are (1) to insulate heavy masses that cannot equalize quickly, (2) to reduce weights and masses, (3) to provide good ventilation (i.e., avoiding dead air spaces that act as insulators), (4) to make surfaces reflective so that radiative coupling to the cold night sky is minimized, and (5) to isolate or eliminate heat sources. One should also consider using parts made from materials with low thermal expansion, but these have limited value if their heating effects are allowed to spoil the telescope seeing.

C. Remote Observing on the Ground

The traditional stereotype of an astronomer is a person perched on a high stool or platform, peering through the eyepiece and carefully guiding the telescope. The modern reality is likely to be quite different. Sophisticated electronic detectors replace the eye. Automatic star-tracking guiders take over the guidance chore. The astronomer sits in a control room sometimes far away from the telescope. A TV monitor shows the FOV or, at least, that part of the field not falling on the detector. A computer logs the data and telescope conditions. The telescope is not even seen by the astronomer: It can be in the next room or even a continent away if the communication link is properly established.

The advent of space-based astronomy clearly marked the time when the astronomer and the telescope were separated. The same separation is taking place in ground-based astronomy, albeit less dramatically. Numerous demonstrations have occurred during the 1970s and 1980s in which astronomers conducted observing runs on telescopes located at distant sites. In one case, the astronomer was in

Edinburgh, Scotland, and the telescope was in Hawaii. The "first light" pictures taken with the European Southern Observatory's 3.5-m NTT in Chile were transmitted to astronomers in Munich, Germany. In both cases the connection was through a communications satellite. This trend is likely to accelerate as the cost for such connections reduces and the data transmission rates increase.

VII. OBSERVING IN SPACE: THE HUBBLE SPACE TELESCOPE

The Hubble Space Telescope, or HST, named in honor of Dr. Edwin P. Hubble, was launched by the National Aeronautics and Space Administration (NASA) into orbit 614 km above the earth on April 24, 1990, from Space Shuttle "Discovery." Launch had been delayed more than 3 years following the loss of the Space Shuttle "Challenger" in 1986. Figure 10 illustrates the HST and its major components. Although many other smaller telescopes have

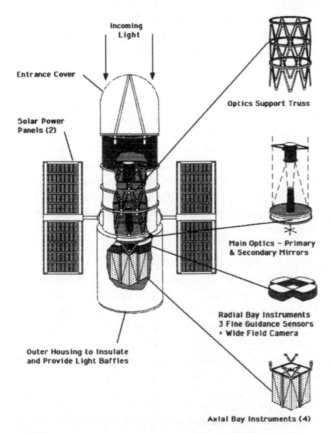

FIGURE 10 Schematic illustration of the Hubble Space Telescope (HST), showing its principal components. Light is focused behind the primary mirror and directed to the radial and axial bay instruments. Data are telemetered to the ground from antennas not shown. HST is optimized for UV region performance down to 0.12 μm wavelength and for observation of faint objects in both the UV and visible wavelength ranges.

previously been used in space, the HST represents such a large step in observing power and expected lifetime (15 years) that, for many astronomers, it marks the maturation of space-based astronomy. The discussion to follow is about the HST, but many of its operating characteristics are common to all telescopes operating in space.

The HST differs significantly from ground-based telescopes. It can observe throughout the UV region, its principal scientific justification, and it avoids the image distortions and "sky background" light emission due to the earth's atmosphere. Thus, very faint, small (presumably very distant) objects can be discerned that would be hopelessly lost in sky background seen from the ground. To produce reasonably good images at short UV wavelengths, the optics had to be substantially better than those in even the best telescopes on the ground, which results in excellent optical region images.

The HST can also observe in all regions of the IR, but its optics and structures are not cooled to minimize radiation emission, and its images are diffraction limited at IR wavelengths. In the late 1990s, NASA launched the "Space Infrared Telescope Facility," a 1-m class telescope that operates inside an enclosure cooled by liquid helium to avoid the excess radiation problem.

X-ray telescopes must operate in space like the HST, but their optics consist of near-conical shapes, appearing almost cylindrical in form, with the central axis pointed at the object. Light reflects at such steep incidence angles from these surfaces that it is not absorbed despite the high energy of the rays; hence it can be directed to a focus by properly shaping the optics.

HST operates under remote control in a hostile, high radiation, vacuum environment while producing its own power from solar panels. To avoid artificial light scattering, the nearby region of space must not be polluted with gases or debris from the telescope because no helpful breeze will blow it away. Since instrument changeovers are not possible except by visiting astronauts, HST must carry all of its instruments, essentially ready to work at all times. To maintain its orbital altitude, HST completes one "day-night" orbit of the earth every 96 min. Optical alignment must be maintained despite the rapid hot-to-cold-to-hot temperature cycling. The short observing periods, the multiplicity of instrumentation, the remote location, and the enormous number of objects to study produced a need to schedule astronomers differently than on the ground. For this purpose, a new organization, the Space Telescope Science Institute in Baltimore, Maryland, was formed in 1981 and is responsible for the scientific operation of the telescope.

The optical configuration of HST is "Ritchey-Chrétien" similar to many ground-based telescopes, with an 18 arcmin field of view. The mirrors are held in alignment by a special graphite-epoxy truss structure that is nearly dimensionally invariant under changing temperature conditions. Light from the focal plane can be diverted into four "radial bay" positions which contain three "fine guidance sensors" (FGSs) and the "wide-field camera" (WFC), an instrument for recording at all times a portion of any scene viewed by the telescope. By adjusting the pointing of the telescope, light from a chosen object can be directed into the entrance of one of four "axial bay" instruments directly behind the focal plane. When an object is being observed by an axial bay instrument, the WFC observes an adjacent region which astronomers hope will result in unexpected "serendipity discoveries." The axial bay contains the "faint object camera," two spectrographs (high and low resolution), and a photometer. All of these instruments are equipped with electronic detectors for converting light into transmittable telemetry data received on the ground by NASA's global network of radio receiving antennas.

The primary mirror blank for the HST is a sandwich structure made of Corning's "zero-expansion" ULE with an "egg-crate" style central core. It weighs about 20% as much as an equivalent solid disc. The low expansion properties of ULE enabled thin pieces to be assembled by flame welding and fusion bonding in a furnace without fear of thermal stress fracture and also avoids dimensional changes in the changing temperature environment of space. The primary mirror was later polished by the Perkin–Elmer Corporation to an accuracy enabling 80% of the focused test light (at 0.63 μm wavelength) to be concentrated into an image smaller than 0.1 arcsec across, very near the diffraction limit and an extraordinary achievement for such a large mirror.

When launched, the 2.4 m HST was the largest optical telescope designed for astronomy ever orbited and also the costliest ($2 billion). Built in the period 1977–1985, HST cost 200 times as much as typical ground-based telescopes of equivalent size built during the same period. Annual operating expenses are expected to be nearly 10% of construction costs, which makes it the world's most expensive observatory. Thus, the financial commitment by the U.S. government to HST represents major support of astronomy and was obtained only after a prolonged effort by hundreds of astronomers during the 1960s and 1970s.

One may imagine the dismay of all concerned when the HST was found soon after launch to have flawed imaging ability due to an incorrect curvature of its primary mirror. The images were blurred by spherical aberration 15 times greater than the specified 0.1 arcsec image resolution and no ground-controlled adjustment could eliminate the problem. Despite this enormous initial setback, HST has become operational. Corrective measures were taken in 1993 when a "second generation" instrument package containing a compensating lens was installed by astronauts. Until

then, astronomers used a computer to subtract the aberration from the images, a tedious but workable process. Considering the level of quality of the optics when tested individually and the imaging problems discovered later, a few words about the mirror testing are appropriate.

Testing of the HST primary required compensation of the test results for the effects of gravity which the mirror, tested on the ground, would not experience in space. This was done successfully. Additionally, the mirror was optically tested using a "null lens," which is an optic used to make the mirror seem to be a sphere that is focusing light originating at or near its center of curvature, a position that can be located with precision. In the telescope, the mirror functions as a parabola focusing light originating at infinity, much like an automobile headlamp in reverse. Simulating these conditions in an optical shop is very difficult for large mirrors, so a null lens is used. However, any errors existing in the null lens remain in the testing data and cannot easily be separated from errors in the mirror. Accordingly, opticians generally perform several different tests for comparison, searching for commonality in the results. One element of the "most accurate" HST null lens was incorrectly positioned, which caused the opticians to polish the mirror to a different curvature than was needed to function properly with the secondary mirror. The disagreement with other tests (made with other, less accurate null lenses) was discounted, and no shop tests were made with both the primary and secondary mirrors working together (as is typical for most telescopes). Unfortunately, the images formed in space by the two-mirror system for HST did contain considerable spherical aberration (the less accurate tests were right), and correction will require a new secondary mirror or an additional correcting lens. Repolishing the primary mirror would be much more expensive.

Pointing the HST, or any orbiting telescope, to a new object for observation and tracking on it steadily and accurately is challenging. Changing position is accomplished by a "reaction motor" on each of three coordinate axes. Powered acceleration of the motor rotor produces a reactive torque on the telescope, causing it to rotate in the opposite direction. On reaching the desired rotational velocity, the telescope and rotor "coast." Motion is stopped by applying power to the rotor to slow its rotation. This "slewing" action coarsely positions the telescope, enabling a pair of "fixed head" star trackers (small guide telescopes with relatively large fields of view) to identify a pair of preselected, bright, guide stars. Next, the three "fine guidance sensors" (FGS), receiving light from the main focal plane as mentioned earlier, begin searching until two have "locked on" to suitable guide stars, presumably faint, but near the object to be observed and known in position to high accuracy. After that, signals from the FGSs are used to control the reaction motors for tracking. The system is designed to control the 12-ton HST to a pointing accuracy of about 0.01 arcsec, roughly 10 times more accurate than typical ground-based telescope performance and all done without a solid foundation to react against.

Only two FGSs are needed to guide the HST. The third FGS can be used to measure star positions (i.e., to do astrometry) after the other two have locked onto stars with known position coordinates. Thus, HST is equipped to upgrade its own guide-star catalog to a precision greater than is possible from the ground. The process of pointing to a new object and locking onto guide stars requires upward of 30 min, a large penalty in observing time but a toll that must be paid to operate in space. Ground-based telescopes typically perform the same operations in less than 5 min.

The HST is an extraordinary telescope and may indeed prove to be the beginning of the era in which most professional astronomy is done in space. Dark skies are becoming hard to find on the ground, and ingenious corrections for the atmosphere will never be as good as avoiding the problem altogether, not to mention the loss of wavelength coverage on the ground. Many other orbiting telescopes are in progress, and plans already exist at NASA for a 10-m class optical space telescope, plus much larger telescopes as to operate in the millimeter and longer wavelength range.

Nevertheless, most professional astronomy is still done on the ground, as the formidable costs of working in space, exemplified by HST, must be justified, presumably by new discoveries. The location and special abilities of the HST (when its optics are finally corrected) assures this possibility. The worth of new discoveries is impossible to measure in advance, but it seems that a bright future for humanity depends upon them. Galileo would agree.

SEE ALSO THE FOLLOWING ARTICLES

ELECTROMAGNETICS • GAMMA-RAY ASTRONOMY • INFRARED ASTRONOMY • OPTICAL DIFFRACTION • ULTRAVIOLET SPACE ASTRONOMY

BIBLIOGRAPHY

Barlow, B. V. (1975). "The Astronomical Telescope," Wykeham, London.

Bell, L. (1981). "The Telescope," Dover, New York.

Burbidge, G., and Hewitt, A. (eds.) (1981) "Telescopes for the 1980s," Annual Reviews, Palo Alto, CA.

Driscoll, W. G., and Vaughan, W. (eds.) (1978). "Handbook of Optics," McGraw-Hill, New York.

King, H. C. (1979). "The History of the Telescope," Dover, New York.

Kingslake, R. (1983). "Optical System Design," Academic Press, Orlando, Florida.

Kuiper, G., and Middlehurst, B. (eds.) (1960). "Telescopes," Stars and Stellar Systems, Vol. 1. University of Chicago Press, Chicago.

Learner, R. (1981). "Astronomy through the Telescope," Van Nostrand Reinhold, New York.

Marx, S., and Pfau, W. (1982). "Observatories of the World," Van Nostrand Reinhold, New York.

Schroeder, D. J. (1987). "Astronomical Optics," Academic Press, San Diego.

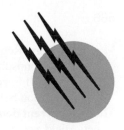

Terrestrial Atmospheric Electricity

Leslie C. Hale
Pennsylvania State University

I. Historical Note
II. Classical Global Electrical Circuit
III. Atmospheric Electrical Conductivity
IV. Thunderstorm Generator
V. Fair-Weather Field
VI. Coupling with Ionospheric, Magnetospheric, and Solar Effects
VII. Recent Developments

GLOSSARY

Air–earth current Current to the surface of the earth from the atmosphere, usually given as a current density. The fair-weather air–earth current is several picoamperes/(meter)^2 (pico $= 10^{-12}$).

Columnar resistance Total resistance of a 1-square-meter column of air between the earth and the "ionosphere." This is largely determined by the first few kilometers above the surface and is typically 1–2×10^{17} Ω.

Field changes Discrete temporal jumps in the electric field of thunderstorms coincident with lightning strokes or flashes, caused by the neutralization of charges due to lightning currents.

Galactic cosmic rays Very high-energy particles from outside the solar system, many of which penetrate to the surface of the earth and are the principal source of atmospheric ionization in the troposphere and stratosphere.

Ionospheric potential Electric potential with respect to earth of the highly conducting regions of the upper atmosphere, assumed to be equipotential (which is not always true).

Lightning flash Most spectacular visible manifestation of thunderstorms, consisting of one or more discrete lightning strokes and frequently resulting in the neutralization of tens of coulombs of previously separated charge, with the release of the order of 10^8 to 10^9 joules (J) of energy.

Magnetic storm Variations in the earth's magnetic field associated with enhanced auroral activity, usually following solar flares and solar wind enhancement (solar activity).

Encyclopedia of Physical Science and Technology, Third Edition, Volume 16

Magnetosphere Region of the earth's magnetic field above the ionosphere, which is also the site of the high-energy trapped electrons constituting the Van Allen radiation belts.

Maxwell current density Mathematical curl of the magnetic field vector **H**, equal to the vector sum of all current densities, which in the atmosphere is usually limited to conduction, convection, diffusion, lightning, and displacement current terms.

Relaxation time Characteristic time (electrical) for the atmosphere below the ionosphere, equal to the free space permittivity ε_0 divided by the electrical conductivity σ. Also known as the screening time.

Schumann resonances Characteristic frequencies of the earth–ionosphere cavity, which are approximately multiples of 7 Hz.

Solar wind Outflow of charged particles from the sun's corona. The solar wind tends to exclude galactic cosmic rays from the earth's environment.

Solenoidal field Field whose flux lines close and whose flux is divergenceless. The divergence of the curl of any vector field is zero. Any field that is the curl of a vector field is solenoidal.

Transient luminous event (TLE) Visible optical events in the upper atmosphere associated with terrestrial lightning; includes "red sprites," "blue jets," "elves," "halos," and others.

TERRESTRIAL ATMOSPHERIC ELECTRICITY is concerned with the sources of atmospheric electrification and the resulting electrical charges, fields, and currents in the vicinity of the earth. The earth and ionosphere provide highly conducting boundaries to the global electrical circuit, which consists of thunderstorm generator and the fair-weather load. Although originally conceived as a dc circuit confined principally to the lower atmosphere, it is now known to have ac components over a broad frequency range and to penetrate deeply into space, coupling with extraterrestrial phenomena.

I. HISTORICAL NOTE

In 1752 T. D'Alibard and B. Franklin independently confirmed the electrical nature of thunderclouds; in the same year L. Lemonnier observed that electrical effects occurred in fine weather. Not until the late nineteenth century was the electrical conductivity of air established by several workers, and the discovery of cosmic ray ionization by V. Hess and the postulation of the global electrical circuit by C. T. R. Wilson came in the early twentieth century. Now known as the classical global cir-

cuit, this concept has stood the test of time and is still generally agreed to be correct. However, some observations cannot be fully explained by classical theory. In any case the classical theory provides a convenient framework in which to discuss topics in atmospheric electricity and solar–terrestrial relationships of an electrical nature.

II. CLASSICAL GLOBAL ELECTRICAL CIRCUIT

The classical concept of the global electrical circuit of the earth is of an electrical system powered by upward currents from thunderstorms all over the earth (totaling $\sim$1500 A) reaching and flowing through the "ionosphere" (an equipotential conducting layer in the upper atmosphere) and returning to earth in "fairweather" regions. This viewpoint is generally attributed to C. T. R. Wilson and was used by the late R. P. Feynman as a premier example of the triumph of scientific methodology. The key evidence is that the fair-weather electric field, measured in polar, oceanic, and high-mountain regions not subject to certain local effects, generally shows a diurnal variation (frequently called the Carnegie curve, from an early expedition on which it was observed) that is relatively stationary in universal (UT) rather than local time (LT) and that shows a peak at $\sim$1900 UT. Since the atmospheric convective activity that produces thunderstorms is greater over land than over water and since thunderstorm activity peaks in the afternoon, the maximum current to the global circuit would be expected to occur when it is afternoon at the longitude of the centroid of the largest concentration of continental thunderstorm activity, which is observed. More detailed analysis shows bumps corresponding to particular continents (Fig. 1).

Considerable time averaging is implicit in these curves. They are much less stationary on a day-to-day basis and are clearly affected by various geophysical events, as described later.

The theory outlined above, frequently called the classical theory of atmospheric electricity, has stood for many decades and is still held by consensus to be generally true.

We shall examine the following topics: atmospheric electrical conductivity; the thunderstorm as a generator of both dc and ac currents to the ionosphere; the fair-weather field, including its variability; and possible coupling between global circuit elements and solar–magnetospheric effects. The author pleads guilty to over-simplification in order to discuss topics that have run to book-length references, which are recommended to the reader. Some recent observations and theory are discussed in VII.

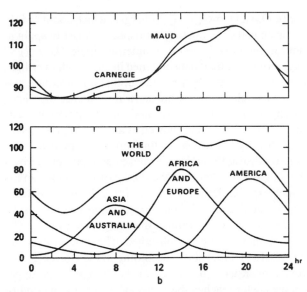

FIGURE 1 (a) Relative diurnal variation in fair-weather field in the Arctic Ocean in northern winter (Maud) and in all oceans (Carnegie). [From Parkinson, W. C., and Torrenson, O. W. (1931). "The diurnal variation of the electrical potential of the atmosphere over the oceans," *Compt. Rend. de l'Assemblée de Stockholm, 1930*, IUGG (*Sect. Terrest. Magn. Electr. Bull.* **8**, 340–345).] (b) Diurnal variation in expected thunderstorm areas in units of 10^4 km^2. [After Whipple, F. J. W., and Scrase, S. J. (1936). "Point discharge in the electric field of the earth," *Geophys. Memoirs (London)* **8**, 20.]

III. ATMOSPHERIC ELECTRICAL CONDUCTIVITY

An early theory postulated that the earth is electrically charged because it was created that way. We now know that the air has sufficient electrical conductivity that such a charge would leak off in less than an hour without a recharging mechanism.

This conductivity is due to electrically charged particles initially created by ionizing radiation, primarily galactic cosmic rays, which create ion–electron pairs. In the lower atmosphere the electrons attach rapidly to form negative ions. Subsequent photochemical reactions, attachment to and charge exchange with aerosol particles, and clustering with water molecules tend to create ions whose size varies over several orders of magnitude. Their eventual loss is due primarily to ion–ion recombination. The electrical mobility of even multiply charged aerosol particles is generally much smaller than that of small molecular ions, hence charge immobilized on aerosol particles contributes less to conductivity, causing aerosol-laden air to possess much lower conductivity. Other sources of atmospheric ionization include natural and anthropogenically induced radioactive material (e.g., soil radioactivity and radioactive gases from the earth and ^{85}Kr routinely

released from nuclear reactors) and "point discharge" or corona ions emitted from surface objects in regions of high electric fields near thunderstorms (e.g., "St. Elmo's fire").

Galactic cosmic rays in the gigaelectronvolt energy range are believed to be the principal source of atmospheric ionization in the undisturbed atmosphere below $\sim$60 km altitude. They are subject to some variability in flux with the 11-yr sunspot cycle, and from high latitudes to equatorial regions their ionization rate in the lower atmosphere decreases by a factor of $\sim$10, due to screening effects of the earth's magnetic field. Solar activity (an enhanced "solar wind") tends to decrease the galactic cosmic ray flux, and when this occurs over a few days the phenomenon is known as a Forbush decrease. Although Forbush decreases at the surface can exceed 10%, a much more common response to the frequently occurring solar activity that causes "magnetic storms" is a variation in ionization rate in the lower atmosphere of the order of 1%. Major solar proton events send large fluxes of tens of megaelectronvolt-range protons toward the earth several times per year. These events produce enhanced ionization as low as 20 km at high latitudes above $\sim$60° magnetic (referring to a coordinate system based on the earth's magnetic field).

At altitudes above $\sim$50 km the variability in electrical conductivity is much greater. Ionizing radiation (principally hydrogen Lyman-α) from the sun penetrates to about this altitude, as do high-energy electrons associated with aurora. X-rays from solar flares also penetrate the "mesosphere" (about 50 to 85 km), and *bremsstrahlung* from high-energy electrons sometimes even deeper. Conversely, aerosol clouds that form in the mesospheric polar night, in "noctilucent clouds," and to a lesser extent in the undisturbed nighttime mesosphere are expected to decrease the electrical conductivity greatly, at least at night. (In the daytime, aerosol particles exposed to the strong ionizing radiation of the upper mesosphere can be sources as well as sinks for free electrons, hence their effect on conductivity is complex.)

Figure 2 shows typical profiles of electrical conductivity with altitude under a number of conditions. The enhancement due to solar proton events is principally at high latitudes.

IV. THUNDERSTORM GENERATOR

Thunderstorms (and other electrified clouds) have generally been regarded as generators of slowly varying or "dc" current to the global circuit; lightning is of more interest as a radiator of electromagnetic energy in the kilohertz to megahertz range. It also appears that there is a substantial variability in the thunderstorm source currents on time

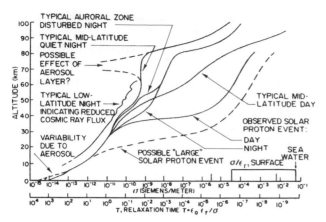

FIGURE 2 Approximate atmospheric electrical conductivity profiles under various conditions.

scales of the order of 10^{-3} to 10^3 sec, constituting a large "ac" component of current to the global circuit.

Meteorological processes in thunderstorms are believed to separate positive from negative charge. Current theories involve the separation of charge by collisional interactions between various forms of ice and water particles, with the aggregation of negative charge on larger hydrometeors and positive charge on smaller particles. The larger negative particles are affected to a greater extent by gravity, tending to fall faster, and the smaller positive particles may move more rapidly upward in regions of convective updraft. The net effect is to produce more positive charge near the top of the cloud and more negative near the bottom (Fig. 3a). Frequently a net negative charge is produced on the cloud, as positive charge may leak off more rapidly to the ionosphere due to the conductivity gradient of the atmosphere. A smaller positive charge is sometimes found near the bottom of the cloud, but the overall effect can be approximated by a "dipole" with the positive charge several kilometers above the negative. The schema of Fig. 3b has been used by numerous workers to predict dc fields and currents. The meteorological charge separation is represented by a current generator, maintaining the up-

per ($+Q_u$) and lower ($-Q_1$) charges, which continuously dissipate in the conducting atmosphere. An early application of this model used a symmetric dipole $Q_u = Q_1$ to explain electric field data obtained in an aircraft overflight (Fig. 4a). It should be pointed out that this overpass was one of the relatively few not showing evidence of at least one lightning flash (Fig. 4b) and that the passes indicating the largest currents showed evidence of many flashes (Fig. 4c). By the use of a "lightning-free" model, equations (known as the Holzer–Saxon equations) were derived that describe the fields of the model of Fig. 3b for an assumed exponential conductivity profile. In this model, using the equilibrium relation $Q = I\tau$, where $\tau = \varepsilon_0/\sigma$ is the relaxation time or lifetime of the charge, yields $Q_u < Q_1$, which is more commonly observed. It has been pointed out by J. S. Nisbet that since, according to the Holzer–Saxon equations, usually nearly all of the current to the upper charge center reaches the ionosphere, the net fraction of I_m that reaches the ionosphere in the lightning-free case is highly dependent on the height of the lower charge center. The Holzer–Saxon equations usually show that over half of I_m reaches the ionosphere and global circuit (I + GC). It should be pointed out that cloud-to-earth lightning, corona or point-discharge, and precipitation currents simply provide alternative paths for I_m to reach the I + GC; they are not independent sources. Intracloud lightning, on the other hand, reduces the current that reaches the I + GC and thus has an important effect on the dc global circuit.

It will be noted that the single lightning flash of Fig. 4b produced a major perturbation of the electric field observed above a thunderstorm, with a pulse of ~10-sec duration, a result confirmed later by other aircraft overflights. Data obtained from a rocket-launched parachute-borne probe over a thunderstorm in 1981 (Fig. 4d) revealed that pulses of similar width occurred as high as 47 km altitude. This led to the conclusion that, rather than the expected (by some) evanescent transient, these waveforms represented pulses that flow through the entire global circuit. Their source can be explained with the crude equivalent circuit of Fig. 3c. The meteorological generator I_m charges the cloud capacitance C_c directly and, in parallel, charges the cloud-to-earth capacitance C_e through the entire ionosphere and global circuit. It is then a race to see which capacitor discharges first, but since both cloud-to-earth and intracloud flashes do occur they must be charged to comparable potentials. Thus, when one capacitor discharges (creating a lightning flash), the other one dumps a comparable amount of energy into the entire ionosphere and global circuit, including the earth. It will be noted that charging and discharging C_c contributes nothing to the net dc global circuit current I_{gc}, but both the charging current and lightning discharge of C_e (causing C_c to discharge partially through the global circuit) contribute to I_{gc}.

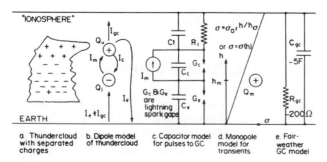

FIGURE 3 Approximate models for describing thundercloud and global circuit (GC) as electrical entities.

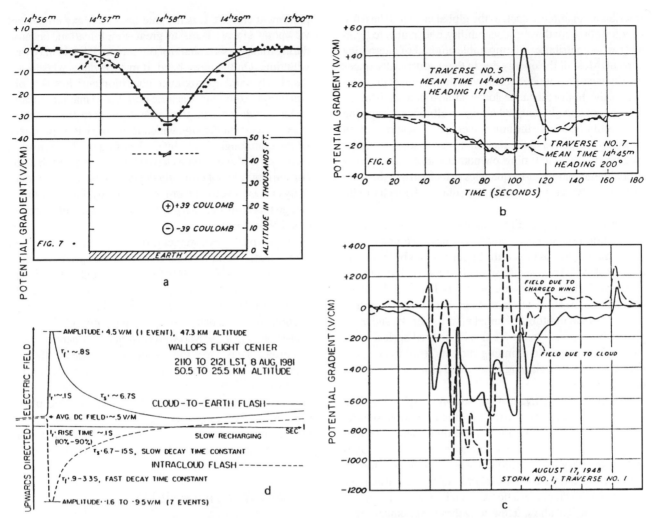

FIGURE 4 Measurements of electric fields above thunderstorms. (a) Potential gradient (negative of electric field) for aircraft traverse above thunderstorm. A, Observed; B, calculated for model shown. (b) Potential gradients over thunderstorms showing effect of single lightning flash. (c) Potential gradient over storm of large current output to global circuit. Note enhanced magnitudes and evidence of multiple lightning flashes. (d) Schema of electric fields observed at very high altitudes showing persistence of wide pulses. [Graphs (a), (b), and (c) are aircraft measurements of Gish, O. H., and Wait, G. R. (1950). *J. Geophys. Res.* **55**, 473. (d) Rocket–parachute measurements reported by Hale, L. C. (1983). *In* "Weather and Climate Responses to Solar Variations" (B. M. McCormac, ed.). Colorado Associated University Press, Boulder.]

It is noted that storms with frequent lightning supply more dc current to the I+GC (see Fig. 4). This is probably because such storms have more intense "meteorological" generators (I_m), and not because the lightning acts as a separate source.

Although the crude model of Fig. 3c predicts the general characteristics of the large pulses to the ionosphere, it does not accurately predict their shape, which is generally nonexponential, with rise times slower than what might be expected from lightning. It should be possible to predict accurately the shape of these pulses with numerical and possibly analytical modeling. A model for calculating transients is shown in Fig. 3d. It is related to the fact that

the "field changes" observed near thunderstorms can be explained by adding charge to the preflash charge distribution and calculating the fields due to the added charge, a procedure first suggested by C. T. R. Wilson. For a cloud-to-earth flash one "monopole" is added to the lower charge center, equal and opposite to the charge lowered to earth by the flash. A number of published papers have used a similar monopole model to study the transient problem, and an update on this work is provided in Section VII. Figure 3e is an approximate model for the global circuit for direct and alternating currents at frequencies below a few hertz.

It is generally agreed that several thousand simultaneously active thunderclouds contribute an average of a fraction of an ampere each to the global circuit. It has also become apparent that rapid variability, due mainly to lightning, also contributes a comparable amount of alternating current. Most of the energy due to these sources dissipates locally near the thunderstorm, with the dc power dissipating in the high-resistance lower atmosphere and the ac power capacitively coupled to higher altitudes dissipating at the altitudes where the atmospheric electrical relaxation time τ corresponds to the reciprocal of the source angular frequency $1/\omega$. Lightning phenomena also radiate electromagnetic energy at frequencies principally in the kilo- to megahertz range, forming the source of much "radio noise."

The vast majority of lightning flashes consist of several short duration "strokes," and lower negative charge (up to several coulombs) to earth. Lightning can be "normal" or "negative;" however, a small fraction of lightning is "positive." These events are generally larger and consist of a single stroke. They tend to possess longer duration currents in the lightning channel after the stroke and to be primarily responsible for TLE such as "red sprites." Because of the longer "continuing currents" (milliseconds or greater), positive lightning is regarded as more dangerous for starting fires.

The upward dc currents flow with little spreading in a rapidly increasing conductivity profile to the ionosphere, where they combine to establish the ionospheric potential and return to earth in fair-weather regions. In the ionosphere, the very lowest frequency ac currents will follow similar trajectories. The ionosphere and earth form a cavity resonator with characteristic frequencies that are multiples of $\sim$7 Hz, so that above a few hertz these "Schumann" resonances must be considered. At higher frequencies, complex modes dominate the ac propagation. However, much of the ac energy is at lower frequencies, where a simpler viewpoint prevails.

A useful viewpoint is that a relatively invariant quantity associated with thunderstorms is the Maxwell current density defined as curl H, where H is the magnetic field due to the thunderstorm currents. This quantity is divergenceless (div curl $H = 0$) and solenoidal, and hence can be followed through any circuit. In the atmosphere outside thunderclouds this generally consists of the sum of conduction current σE and displacement current $\varepsilon_0 \, \partial E/\partial t$ (E, electric field; σ, electrical conductivity; ε_0, permittivity). The divergenceless nature of the Maxwell current means that, following it around a flux tube, the magnitude of the individual components of flux may vary with their sum remaining constant. For example, it has been shown that the Maxwell current near a thunderstorm is invariant through a layer of high-conductivity air caused by point discharge (corona) ions near the earth's surface, with increased conduction current balanced by decreased dis-

placement current. It is not true that the Maxwell current viewpoint always leads to great simplification, because during transients the shape of the solenoidal field may vary with time. On the other hand, it may be dangerous to use the clearly valid approximate condition curl $E \approx 0$ to obtain unique solutions, because curl $E \equiv 0$ mathematically yields a transient solution permitting zero time-varying Maxwell current density, which is contrary to experimental data, although this solution has frequently been used incorrectly. Model studies have shown that the global circuit currents are not crucially dependent on cloud conductivity and are not, to first order, affected by the "screening" charge that forms at sharp conductivity gradients, such as cloud boundaries.

From typical measurements of the vertical electric field (Figs. 4c,d), it can be seen that, for appropriate conductivity values ($\sim$$10^{-12}$ S/m for Fig. 4c and $\sim$$10^{-10}$ S/m for Fig. 4d) the maximum displacement current density ($8.85 \times 10^{-12} \, \partial E/\partial t$) is greater than the maximum conduction current density. For the large storm of Fig. 4c the root-mean-square ac Maxwell current is comparable to the dc current. Thus, large thunderstorms may be as effective in generating ac currents as dc currents, and much more ac energy is coupled to the upper atmosphere. This ac generator mechanism may be the principal source of atmospheric noise in the ELF range between the "micropulsation" region below $\sim$1 Hz and electromagnetic radiation from lightning currents in the kilohertz range and above.

V. FAIR-WEATHER FIELD

Thunderstorm currents totaling $\sim$1500 A flow to the ionosphere, which is an approximately equipotential layer in the upper atmosphere, and return to earth in fair-weather regions. The resistance between the ionosphere and earth is $\sim$200 Ω, giving rise to an ionospheric potential of $\sim$300 kV. The vertical electric field near the earth's surface is of the order of 100 V/m, and the associated current density is typically several picoamperes/m^2. The ac component of the fair-weather field is relatively much smaller than the ac component of individual thunderstorm currents.

The upward currents from individual thunderstorms that do not return to earth or the bottom of the cloud locally flow to the ionosphere, where they combine to establish the ionospheric potential. The height of the ionosphere used to be taken, for global circuit purposes, to be $\sim$60 km, but it is now generally agreed that the horizontal currents that close the global circuit generally flow in the much higher ionosphere of the radiophysicist, above 100 km. The principle is that they tend to spread very little, flowing upward in a monotonically increasing conductivity profile, which frequently persists to the ionosphere. Above $\sim$70 km the

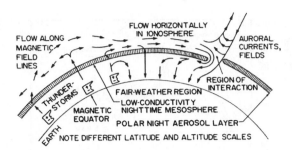

FIGURE 5 Longitudinal cross section showing global circuit current paths at night with one possibility for coupling with auroral (Birkeland) currents (morning sector).

conductivity ceases to be a simple scalar quantity but possesses directional properties, with the highest conductivity along the direction of the magnetic field, whereby the currents can flow directly into the magnetosphere along magnetic flux lines (Fig. 5). Returning along the conjugate flux line in the opposite hemisphere in a decreasing conductivity profile, the currents tend to spread very rapidly in the ionosphere. Ultimately, the largest horizontal currents are thought to flow in the region of highest conductivity transverse to magnetic field lines, which occurs between about 100 and 120 km. Here, they must merge with the very much higher magnitude ionospheric current systems (typically 10^5 A) which also flow at these altitudes. The thunderstorm currents return to earth in fair-weather regions, which include the entire surface of the earth not directly involved in thunderstorms. The ionospheric potential is approximately given by the current ($\sim$1500 A) multiplied by the total resistance between ionosphere and ground, which has been determined to be $\sim$200 Ω, and hence is $\sim$300 kV.

There are numerous factors that modify this general picture of the fair-weather field. Except in the polar regions, local convective activity in the earth's boundary layer produces a current that frequently obscures the Carnegie curve in electric fields and currents measured in the lower atmosphere over land. This is generally attributed to the effects of local atmospheric convective activity on space charge (due to an excess of ions of one sign of charge, usually positive, which is a necessary consequence of the global circuit currents in a conductivity gradient). Frequently, there is a strong correlation with water vapor, which is controlled by temperature and convective "exchange" processes. Water vapor reduces conductivity by forming ionic clusters of lower mobility. Aerosol particles from many sources (volcanoes, industry, fires, clouds, blowing dust and snow, etc.) tend to reduce the conductivity and increase the magnitude of the electric field, often with relatively constant current density. There is a clearly identified "sunrise" increase in electric field, which has also been considered to be caused by convective exchange processes.

In addition to locally induced variations in the fair-weather field, there are variations of much larger extent. There is a well-established seasonal dependence in the ionosphere potential, with a maximum in Northern Hemisphere winter. The reasons for this are not clear but are probably related to thunderstorm activity in tropical regions, which are believed to be the source of most of the global circuit current. Long-term observations of the ionospheric potential have shown variations in the range of 200 to 400 kV, probably due to variations in source current and total resistance. An expected 11-yr variation due to the effects of the sunspot cycle on galactic cosmic rays has been obscured by other factors, including aerosol particles from volcanoes. It has been predicted by W. L. Boeck that, even without nuclear disasters, the routine release of radioactive ^{85}Kr from nuclear power plants may eventually substantially affect the global circuit by increasing atmospheric conductivity, if the nuclear power industry expands. If, as some have suggested, the fair-weather conductivity (atmospheric ionization) and/or electric field play a role in initial thunderstorm development, this could potentially alter weather and climate. Some evidence exists for the effects of the release of radioactivity from nuclear reactors. In the weeks following the Chernobyl incident in May 1986, an order of magnitude enhancement in lightning frequency was observed in Central Sweden, along the path of radioactive "fallout," as compared with stationary patterns observed over many years.

The dc currents from thunderstorms add in an arithmetical manner to produce the total fair-weather current, with the majority of storms producing upward currents. The ac currents, however, tend to have random phase and add vectorially to produce a much smaller resultant, inasmuch as they are independent. This, combined with filtering due to the global circuit capacitance, means that the ac currents are not a major factor in the fair-weather electric field, although they may contribute to the noise background in the hertz to kilohertz range. (Other factors here are turbulent local space charge, fields of nearby thunderstorms, and radiated electromagnetic fields from distant lightning.) Large, independent individual events, such as "superbolts" or correlated bursts of lightning flashes, may appear as distinct perturbations of the fair-weather field.

VI. COUPLING WITH IONOSPHERIC, MAGNETOSPHERIC, AND SOLAR EFFECTS

It has been known for decades that there is coupling between elements of the global circuit and magnetospheric and solar phenomena. The modulation of galactic cosmic rays by solar activity affecting atmospheric electrical conductivity is possibly the best understood relationship, but there are many others that are accepted to various degrees. Some that cause "downward" coupling have come to the fore in the search for sun-weather relationships, but there are also cases in which thunderstorms affect the ionosphere and magnetosphere.

For many decades it has been known that lightning produces audio-frequency "whistlers." These are very low-frequency radio waves whose frequency decreases over a period of the order of 1 sec, produced by the interaction of very low-frequency radio waves from lightning with magnetospheric plasma (ionized particles) to produce amplified waves. It has been established by several groups that lightning-related events can produce precipitation or "dumping" of high-energy trapped electrons from the Van Allen radiation belts (e.g., Fig. 6). These electrons may create enough ionization enhancement in the ionosphere to affect ionospheric radio propagation measurably for periods of the order of 1 min. Such events are called Trimpi events after their discoverer, M. L. Trimpi. The data show that similar "dumping" events occur in the daytime when very low-frequency waves cannot easily propagate to the magnetosphere. It is an open question whether ac thunderstorm currents, which easily reach the magnetosphere, can trigger such events.

In the magnetosphere, the tensor (directional) electrical conductivity perpendicular to magnetic field lines is very small. Thus, "transverse" fields can "map" downward into the ionosphere. Horizontal fields, of magnetospheric or ionospheric origin, penetrate farther down into the atmosphere with an efficiency dependent on the "scale size" of the field. Fields extending over several hundred kilometers can map nearly to the surface with little attenuation. Such fields can clearly perturb the local ionospheric potential, because it can no longer be equipotential. Effects of large magnitude, however, were usually thought to be confined to auroral zone and polar cap regions. However, there is some evidence for relatively large effects penetrating to lower latitudes.

A number of both earth-based and balloon or aircraft measurements at mid-latitudes have indicated 10–30% variations in local ionospheric potential, electric field, and air–earth current, persisting for up to several days following moderate solar or magnetic events that occur more than once per month. These typically involve an approximate doubling of solar wind speed, moderate magnetic and enhanced auroral activity, and a small Forbush decrease in galactic cosmic ray flux. (The electrical events have also been correlated with movements of the earth across solar magnetic field sector boundaries. Since these also tend to be correlated with solar magnetic events, there has been some difficulty in establish-

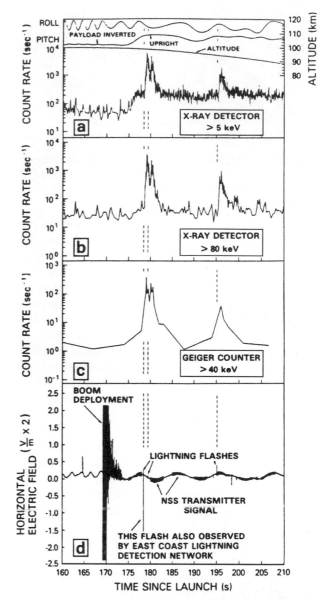

FIGURE 6 Rocket measurements showing effects of lightning on high-energy detectors, indicating lightning-induced precipitation of particles from Van Allen belts. [After Goldberg *et al.* (1986). *J. Atmos. Terr. Phys.* **48**, 293.]

ing cause and effect.) It has not been conclusively established whether the electrical effects are due to an overall change in ionospheric potential or to "external" electric fields in the atmosphere creating horizontal gradients in the local ionospheric potential. Possibly both sorts of effects are involved. Stratospheric balloon measurements have indicated horizontal fields of tens of millivolts per meter, which are much greater than readily permitted by classical global circuit theory. Carried over thousands of kilometers, these fields could produce the tens of kilovolts necessary to explain the data. An alternative explanation for the local variations in the fair-weather field is modulation of the overall ionospheric potential due to variations in the thunderstorm source current. If this could be proved true, it would represent a very important sun–weather relationship. Perhaps transient events in the magnetosphere couple to individual thunderclouds and modulate their lightning rates, thus affecting the total global generator current.

VII. RECENT DEVELOPMENTS

Much of the recent work in this field has involved the consolidation of existing knowledge, which has led in some cases to the reaffirmation of earlier theories. For example, the basic dc global circuit has undergone some rehabilitation, along with clarification.

In recent years, partisans of cloud-to-earth lightning, corona or "point discharge," and currents carried by precipitation have all argued for the relative importance of these various "sources." A consensus is developing that these are not independent sources, but different processes by which the currents originating from the basic "meteorological" generator are completed through the ionosphere and global circuit (I+GC). An exception to this "consolidation" viewpoint is the role of intra-cloud (IC) lightning, which definitely tends to weaken the global meteorological generator by discharging thunderclouds locally. Thus, IC lightning plays a critical role in determining the currents to the I+GC. (This viewpoint was first expressed to this author by Lothar Ruhnke.)

To recapitulate the basic dc global circuit, air is partially electrified by a number of different processes, and charge separation by meteorological processes produces a number of local electrical generators, primarily in thunderstorms, which act in concert to establish an "ionospheric potential" (IP). This IP then drives a "fair-weather" current back to earth in storm-free regions, with the current density determined by the IP divided by the "columnar resistance." This latter factor depends largely on aerosol loading of the atmosphere and on orography, with much larger relative currents to mountains and elevated regions such as the Antarctic Plateau. These factors have been embodied in numerous computer models (see B. K. Sapkota and N. C. Varshneya, *J. Atmos. Terr. Phys.* **52,** 1, 1990).

In the last decade or so the most (literally) spectacular observations have been of a variety of visible effects in the upper atmosphere above thunderstorms, the most colorful of which are "red sprites" observed in the nighttime mesosphere. Such phenomena were originally suggested by C. T. R. Wilson (*ca.* 1925), were first observed by J. Winckler and colleagues at the University of Minnesota

in 1989, and were later confirmed by triangulated measurements from two aircraft over the U.S. Great Plains by D. Sentman and a group from the University of Alaska. Both optical and electrical measurements have confirmed the "electrical breakdown" nature of these emissions, which generally initiate at about 75 km and spread both upward and downward, probably by "streamer" mechanisms suggested by V. Pasko and a Stanford group. Sprites are nearly always associated with "positive" lightning, possibly because such events are generally larger and tend to occur in single strokes.

The electrodynamics of such sprites depend on a number of things. First, the extremely low conductivity of the nighttime mesosphere (see Fig. 2) allows the penetration of electromagnetic energy to about 80 km. A quasi-static field due to the large Wilson monopole injected by positive lightning establishes a field between the earth and ionosphere (at about 80 km) sufficient to produce electrical breakdown at about 70 km or above. In conjunction with establishing this field, a "millisecond slow tail" is launched in the earth–ionosphere waveguide (Fig. 7). This waveform can be observed in the electric or magnetic field at thousands of kilometers horizontal distance, and

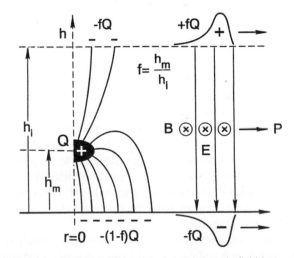

FIGURE 7 (Left) Establishment of a quasi-static field between the earth and the ionosphere at effective height h_i (typically about 80 km at night) by a Wilson monopole Q injected at height h_m by a lightning stroke. The fraction of electric field lines reaching the ionosphere is fQ, where $f = h_m/h_i$. For typical low nighttime mesospheric conductivities, such fields can persist for up to tens of milliseconds, allowing time for most sprites to develop. (Right) In order to satisfy the post-stroke boundary conditions of the field, a roughly one-millisecond "slow tail" is launched in the earth–ionosphere waveguide in the radial TEM mode (flat earth approximation). The initial charge associated with these unipolar wavelets is fQ but actually increases with distance from the source. (The polarities shown are for normal "negative" lightning; for positive lightning, the polarities of all charges are reversed.)

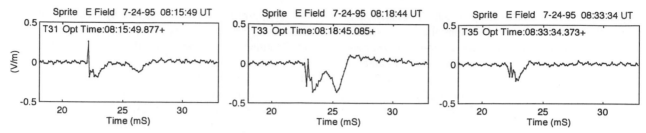

FIGURE 8 Vertical electric field waveforms observed at several thousand kilometers from related optically observed red sprites. The sprites occurred in West Texas and the fields were observed from central Pennsylvania. Similar waveforms were observed in the magnetic field. The initial slow tails in each case were due to the parent lightning, as indicated by the higher frequency components on the leading edge. The subsequent smooth tails are due to the charge separated in the sprite. Note that the middle event has less time delay and is larger in amplitude, and that the one on the right shows no detectable waveform due to the sprite itself, which is frequently the case. [Adapted from Marshall, L. H. *et al.* (1998). *J. Atmos. Solar–Terr. Phys.* **60,** 771.]

is in a mode (TEM) that can be observed at any altitude, thus not requiring balloons or other spacecraft. (This is not true for mesospheric electrical conductivity profiles; their observation during potential sprite-producing conditions would be extremely useful, but does require rocket launches.)

The development of the sprite usually takes several milliseconds after the establishment of the quasi-static field. This is probably because the breakdown process must start from a few "seed" electrons and requires many "e-folding" times to reach observable amplitude. The sprite discharge separates charge and produces a Wilson monopole at sprite altitudes, thus launching a millisecond slow tail similar to the one launched by the parent lightning. These are frequently seen on the same record (Fig. 8). Sprites are predominantly red and take a variety of shapes. They can appear over tens of kilometers in the mesosphere ($\sim$50 to 80 km altitude) and although appearing to be primarily vertical in orientation can occur over similar horizontal distances. Their shapes are so varied as to defy description, but it has been suggested that some appear to be generated by "fractal" processes.

Other such phenomena include "blue jets," which are discharges proceeding upward from a thundercloud into the stratosphere, first observed by E. Westcott and the Alaska group from aircraft and explained as propagating upward discharges by U.S. Inan and Stanford colleagues. "Elves" are much shorter discharges at higher altitudes, generally 100 km or greater, that were first observed by Fukunishi and colleagues from Tohoku University of Japan in Colorado. These events, which last only a few hundred microseconds, are generally attributed to direct breakdown caused by electromagnetic radiation from lightning.

Observations of red sprites, blue jets, elves, and a number of other phenomena that are currently under study

have largely been made from the Yucca Ridge facility of W. A. Lyons near Fort Collins, CO. This superb facility has been used over several summers by a number of groups to do coordinated studies of these interesting new phenomena. The work has been largely reported at American Geophysical Union meetings (particularly Fall) and published in numerous articles, largely in *Geophysical Research Letters*, the *Journal of Geophysical Research*, and the *Journal of Atmospheric and Solar–Terrestrial Physics*. Such phenomena are becoming known collectively as transient luminous events (TLEs).

The millisecond slow tails described in Fig. 7 may enter into other phenomena. Similar electric field waveforms have been observed to penetrate into the highly conducting ionosphere, parallel to the earth's magnetic field, and much stronger than had initially been expected according to "shielding" considerations. L. Hale has suggested that this is due to polarization of the ionospheric and magnetospheric plasma as the slow tails pass below. Furthermore, it was suggested that this polarization, which occurs on a global basis as the "slow tails" propagate with relatively little attenuation, deposits opposite polarity charge in the "conjugate" mesosphere (at the other end of a magnetic field line), where it returns toward earth. It was further suggested that such polarization is responsible for the large (up to several volts/meter) mesospheric electric fields observed frequently by Russian and U.S. groups. However, little is known about the interaction of such polarization waves with magnetospheric plasma, so the theory cannot yet be recognized as established. The situation does suggest a definitive experiment, however, observing the electric field transients coupled to the surface at a location magnetically conjugate from located lightning. Suitable venues for doing this include between Southern Africa and Central Europe and also between Australia and various locations in Asia.

SEE ALSO THE FOLLOWING ARTICLES

CLOUD PHYSICS • IONOSPHERE • RADIATION, ATMOSPHERE • THUNDERSTORMS, SEVERE

BIBLIOGRAPHY

The history of this subject and much of the earlier work is covered in:
Isräel, H. (1973). "Atmospheric Electricity," rev. ed., Keter Press, Jerusalem.

Much of the more recent work has been covered in work presented at symposia of the International Commission on Atmospheric Electricity in 1948, 1952, 1956, 1963, 1968, 1972, 1978, 1982, 1986, 1992, 1996, and 1999. Proceedings of all but the last three symposia were published, and subsequent papers have been published in the open literature, largely in the *Journal of Geophysical Research*. Current work, particularly in the field of transient luminous events, also tends to appear in *Geophysical Research Letters*.

The most comprehensive book on the subject of lightning is
Uman, M. A. (1987). "The Lightning Discharge," Academic Press, San Diego, CA.

Papers mainly on the electromagnetics involved have been presented at meetings of the International Union of Radio Science (URSI) and are frequently published in *Radio Science*.

Two papers that give an extended bibliography for many of the subjects discussed herein are
Hale, L. C. (1994). "The coupling of ELF/ULF energy from lightning and MeV particles to the middle atmosphere, ionosphere, and global circuit," *J. Geophys. Res.* **99,** 21089.

Marshall, L. H., *et al.* (1998). "Electromagnetics of sprite and elve Associated Sferics, *J. Atmos. Solar–Terr. Phys.* **60,** 771.

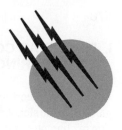

Textile Engineering

Sundaresan Jayaraman
W. W. Carr
L. Howard Olson
Georgia Institute of Technology

I. Fibers as Components of Engineering Structures
II. Textile Structures: Production and Applications
III. Trends in Textile Manufacturing

GLOSSARY

Filling (weft) Yarn that runs widthwise (at right angles to the warp) in a woven fabric; also known as pick.

Jacquard Mechanism invented at the beginning of the nineteenth century by Joseph-Marie Jacquard of Lyons, France; used for weaving elaborate fancy patterns with the help of punched cards carrying binary information. Precursor of the computer punched card.

Nonwoven Textile structure made directly from fibers that is held together by fiber entanglements or by bonding agents (adhesives or fused thermoplastic fibers).

Silver Strand of loose, untwisted fibers held together by interfiber friction. Card sliver when delivered by the carding machine, and drawn sliver when produced by the drawing machine. The fibers in the latter are closer to being parallel (oriented) than those in the former.

Slub Tiny ball of fibers on the yarn that degrades the appearance of the yarn. Sometimes specifically introduced to produce novelty yarns.

Spinning Process of producing yarns from staple fibers by inserting twist in the fine strand of fibers. In the production of man-made fibers, the extrusion of a spinning solution through a spinneret to form filaments.

Warp Yarn that runs lengthwise in a woven fabric; also referred to as end.

TEXTILE ENGINEERING is defined as the art and science of designing and creating structures from *fibers*, the fundamental textile element, and encompasses the design of processes for the production of these structures. The assembled structures have a unique combination of properties: high strength and low bending rigidity. The designing of textile processes and the production of textile structures call for an understanding of two basic scientific disciplines—physics and chemistry—and several engineering disciplines—mechanical, civil, electrical, and chemical. Thus, textile engineering can be regarded as a multidisciplinary applied field of science and engineering. The scope of textile structures has been broadened to include structures that combine high strength with high bending rigidity. These new structures, known as *composites*, find increasing applications in the fabrication of aircraft, helicopters, and space structures. Besides other textile fibers, metallic and ceramic fibers are used as the basic building blocks in the production of these structures.

I. FIBERS AS COMPONENTS OF ENGINEERING STRUCTURES

From an engineering standpoint, textile fibers can be regarded as fine slender rods. They have the unique combination of characteristics of flexibility, because of their fine diameters, and high intrinsic strength per unit weight. Their length-to-thickness ratio, known as the *aspect ratio*, is very high. Fibers of long length are known as *filaments*, whereas short fibers (inches or fractions of an inch long) are called *staple* fibers. For example, a typical cotton fiber is 1-in. long and has an aspect ratio of 1500, whereas a wool fiber, 3 in. in length, has an aspect ratio of 3000. For this reason, fibers are regarded as one-dimensional structures.

A. Classification of Fibers

Traditionally, fibers found in nature (e.g., cotton, wool, silk) have been used to meet one human basic need, namely, clothing. However, with advances in science and technology, newer fibers are being produced, and their share in the production of textile structures is growing. Thus, textile fibers can be classified according to their origin—natural or man-made. These major groups can be subdivided as shown in Table I.

Commercially, cotton is the most important textile fiber. It grows from the seed of the cotton plant, which belongs to the genus *Gossypium*. Silk is extruded by the silkworm in the form of a fine continuous filament, making it the only naturally occurring filament. Wool is obtained from the fibrous covering of the sheep. Because of the high cost, the use of silk and wool is fairly limited.

Man-made fibers are classified as either regenerated or synthetic. Regenerated fibers are produced from polymers occurring in nature (e.g., viscose rayon from the cellulose in wood pulp), whereas synthetic fibers are produced by chemical synthesis (e.g., polyester from terephthalic acid and ethylene glycol). The polymer fluid is extruded through a spinneret to form a single continuous filament. Although most man-made fibers are produced as continuous filaments, they are also used as staple fibers. Sometimes several hundred filaments (500–2000) are gathered in a ropelike structure known as a *tow*. A similar structure formed from staple fibers is referred to as *top* (e.g., a wool top).

B. Properties of Textile Fibers

The performance and end-use applications of textile structures are greatly influenced by the properties of the fundamental unit of the structure, the fiber. Table II provides a summary of the important characteristics of some typical textile fibers.

The fineness or linear density (mass/unit length) of a fiber is expressed in terms of *tex*, which has units of grams per kilometer. Typical fineness values are in the 0.1- to 1-tex range. Fiber strength and extensibility are essential for textile applications. Tenacity, expressed in grams per tex, is a measure of fiber strength, whereas breaking extension characterizes the extensibility of fibers. The extensibility of man-made fibers can be altered during the manufacturing process. For polyester it typically varies between 15 and 55%, whereas for spandex, used in swimwear, it is between 400 and 500%.

Moisture regain is defined as the ratio of the mass of absorbed water in a specimen to the mass of the dry specimen. Natural and regenerated man-made fibers are generally hydrophilic and have fairly high regains (e.g., cotton has a regain of 7%). On the other hand, most synthetic fibers are hydrophobic and have low regains (e.g., polyester has a regain of 0.4%). The moisture regain of fibers influences the wearing comfort of garments and the processing of fibers into yarns. So, cotton garments are more comfortable than those made of polyester. Polyester is highly prone to static, which causes problems in spinning. To minimize static problems, fiber producers apply antistatic finishes to man-made fibers.

Fibers are affected by light, microorganisms, and chemicals, and the degree of degradation depends on the fiber type. Cotton, for example, is attacked by mildew and fungi. Fibers are generally not affected by mild alkalis and acids. However, strong acids attack cotton, and strong alkalis can damage polyester.

TABLE I Classification of Textile Fibers

Origin	Type	Source or component	Product
Natural	Vegetable	Seed	Cotton
		Bast	Flax
		Leaf	Sisal
		Fruit	Coir
	Animal	Excreted	Silk
		Hair	Wool
	Mineral		Asbestos
Man-made	Regenerated	Protein	Casein
		Cellulose	Viscose
	Synthetic	Polyamide	Nylon
		Polyester	Dacron
		Polyacrylonitrile	Orlon
		Polyurethane	Lycra
		Polyolefin	Polypropylene
	Others	Carbon	Thornel
		Glass	Fiberglas
		Ceramic	Fiber FP

TABLE II Properties of Textile Fibers

Fiber	Diameter range (μm)	Density (g/cm³)	Tenacity[a] (g/tex)	Breaking extension[a] (%)	Moisture regain[a] (%)	Attacked by strong
Cotton	11–20	1.52	40	7	7	Acids
Wool	20–40	1.31	11.5	42	14	Alkalis
Silk	10–15	1.34	40	23.4	10	Alkalis
Viscose rayon	$\geq$12	1.46–1.54	20	20	13	Acids
Nylon 6.6	$\geq$14	1.14	35–60	15–60	4.5	Oxidizing agents
Polyester	$\geq$12	1.34	25–50	15–50	0.4	Alkalis
Orlon	$\geq$12	1.16	20–30	20–30	1.5	Alkalis
Glass	$\geq$5	2.54	76	3	0	Unaffected

[a] At 65% relative humidity and 70°F.

II. TEXTILE STRUCTURES: PRODUCTION AND APPLICATIONS

Conventional textile structures combine strength with a high degree of flexibility, and this combination sets them apart from other engineering structures. Flexibility in other engineering structures can be achieved only at the cost of a loss of strength. Discontinuities in textile structures are both necessary and inevitable. For example, a yarn spun from staple fibers has voids or air space in it ($\sim$40% voids), and the entrapped air contributes to the warmth and comfort of textile materials. In contrast, discontinuities or voids in other engineering structures act as stress raisers, which can lead to catastrophic failures. Textile structures are soft and have a fine texture, whereas other engineering structures are hard and smooth.

A. Order and Textile Structures

The fundamental goal of the different operations in the production of textile structures is the imposition of order on the mass of fibers. In the simplest of structures, *yarn*, the slender one-dimensional fibers are assembled in such a way that not only is the fiber flexibility preserved, but the resultant interlocking of the fibers provides the necessary strength to resist large-scale deformations. In essence, yarns are a linearly ordered one-dimensional assembly of fibers. Yarns are interlaced or interlooped to form two-dimensional planar structures known as *fabrics*. A typical fabric will have $\sim$10¹⁰ fibers per square yard, and so the task of the textile engineer is to ensure the correct selection and ordering of the fibers and yarns to produce a useful structure. A variation of this traditional process is the direct conversion of fiber to fabric, in which the fibers are "laid" and bonded to produce two-dimensional structures known as nonwovens. The terms *one-* and *two-*

dimensional refer to the macroscopic perception of the structure and its dimensions.

B. One-Dimensional Textile Structures

A classification of one-dimensional textile structures is shown in Table III. Yarns can be directly extruded as continuous filaments, or they can be spun from staple fibers. The different yarn types are also listed in the table. A yarn is regarded as a one-dimensional structure because its transverse dimensions are very small compared with its length.

1. Yarn

In the traditional yarn manufacturing process, a stock of fibers is opened, cleaned, and attenuated to form a fairly parallel strand of fibers. This linear assembly of fibers has no coherence, and the fibers can easily slide past one another. Coherence, and hence strength, are imparted to the fiber strand either by increasing the interfiber frictional forces or by "bonding" them together using an adhesive.

TABLE III One-Dimensional Textile Structures

Textile structure	Type	Product
Yarns	Continuous	Monofilament
		Multifilament
	Spun (staple fiber)	Homogeneous
		Blended
	Others	Textured filament
		Plied (multifold)
		Combination
		Novelty
Braids		
Ropes		
Cords		

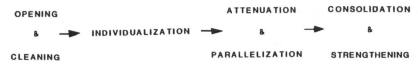

FIGURE 1 Fiber to yarn: sequence of operations.

The sequence of operations on the fibers is shown in Fig. 1. The opening and cleaning machines loosen up the fibers and, in the case of cotton, remove trash, dust, and other impurities in the fibers. Fiber individualization is brought about by the carding machine, in which the fibers are subjected to a brushing action between two metallic wire surfaces moving at vastly different speeds. The fine web of fibers delivered by the carding machine is condensed into a ropelike form known as a *sliver*. The linear density of a typical silver is in the 3- to 4-ktex range, and there are thousands of fibers in the cross section.

The yarn, in contrast, has ~100 fibers in its cross section, and its linear density is in the 5- to 50-tex range. To reduce the linear density of the sliver, the next step in the sequence is the attenuation process, also known as *drafting* or *drawing*. Drafting is carried out by passing the sliver through two pairs of rollers rotating at different speeds (Fig. 2). The first pair, through which the sliver is fed, rotates slower than the second pair, and this results in the reduction of fibers in the cross section of the material delivered by the second pair. The ratio of surface speeds is called the *draft*, and the reduction in linear density of the sliver is proportional to the draft. Drawing, while involving draft, is in fact an intermediate process used for evening out linear density variations and parallelizing fibers.

There are three principal techniques for imparting coherence and strength to the strand of fibers attenuated to the desired linear density. They are *twisting*, *wrapping*, and *bonding*.

a. Twisting. Twisting is the most commonly used technique for fiber consolidation, and *ring spinning* incor-

porates this principle. Here, the drafted strand emerging from the nip of the front rollers is led down through the *traveler*, which is rotating at high speeds (~40 m/s) on the inside flange of a ring and inserting twist (see Fig. 3). The takeup package, or *bobbin*, drags the traveler, and the difference in speed between the two causes the strand, the yarn, to wind on the package. The typical production rate is 20 m/min per spinning position. Twist can be inserted in one of two directions: clockwise (properly known as S twist) or anticlockwise (Z twist), as shown in Fig. 4. The production rates in ring spinning are limited by the fact that the yarn package has to be rotated at very high speeds (~16,000 rpm), causing traveler burnouts and excessive aerodynamic and centrifugal stresses on the yarn.

Open-end spinning utilizes the principle of twisting for yarn formation but eliminates the need to rotate the package. Here, the attenuated strand of fibers is individualized, deposited on a rotating surface, and then reassembled into a yarn at a free growing end. Hence, the name *open-end spinning* is used. In rotor spinning, the widely used open-end method, the fibers are fed into the circumferential groove in a drum rotating at high speeds (up to 80,000 rpm). The rotating free end of the yarn peels the fibers from the rotor surface, and the yarn is withdrawn simultaneously and wound onto the takeup package. The production rate is 100–120 m/min per spinning position.

b. Wrapping. Coherence and strength can be imparted to the fiber strand if it is wrapped with a continuous filament yarn (Fig. 5). The resulting wrapped yarn has no twist in the core and is bulky. In a variation of the wrapping method, the staple fibers themselves wrap around the core of fibers. The drafted strand of fibers passes through

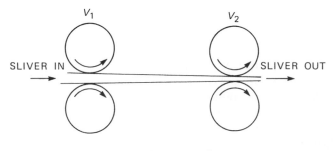

$$V_2 > V_1; \text{DRAFT} = V_2/V_1$$

FIGURE 2 Pair of drafting rollers (V_i = velocity). Note the reduction in strand thickness.

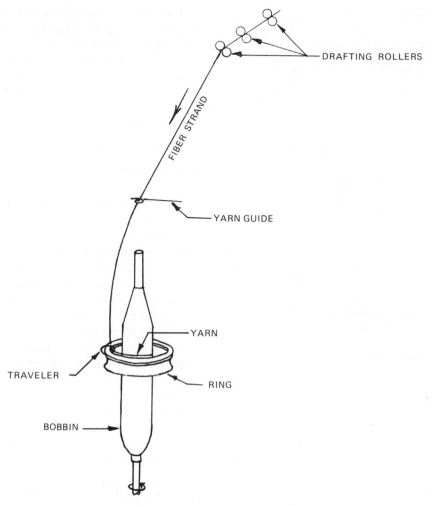

FIGURE 3 Material flow in ring spinning.

two successive twisting units that insert twist in opposite directions. As a result, the twist in the strand emerging from the second unit is "zero"; however, some of the fibers (those at the edges) that are not twisted in the first unit wrap around the core in the second and consolidate it. Thus, the yarn consists of a fairly parallel core of fibers held together by the wrapper fibers. This principle is also known as the "false-twist" method because there is no real twist in the yarn. In the commercial form, air-jet spinning, jets of air are used at the twisting zones. The production rate is ~180 m/min per spinning position.

c. Bonding. The fibers in the drafted strand can be bound together by an adhesive to form a yarn, and this is the principle used in the Twilo twistless spinning method. The yarn, although behaving like a solid rod, can withstand the rigors of fabric production, after which the adhesive is washed off to yield a fabric with a high degree of bulk.

2. Blended and Composite Yarns

Blended yarns are produced by combining two or more staple fibers, usually in the drawing process. The purpose of blending is to produce a yarn that advantageously

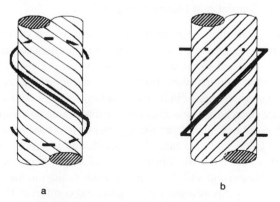

FIGURE 4 Twist in yarns. (a) S Twist; (b) Z twist.

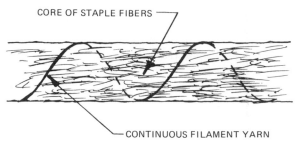

CORE OF STAPLE FIBERS

CONTINUOUS FILAMENT YARN

FIGURE 5 Yarn formation by wrapping.

combines the properties of the component fibers. For example, in a common blend like polyester–cotton, cotton provides the comfort properties, whereas polyester gives strength to the blend.

A variation of blended yarns is *core-spun yarn*, which has a continuous filament core and a staple fiber sheath. In a composite yarn, called the Bobtex ICS yarn, a filament yarn is fed through a molten polymer onto which a layer of staple fibers is subsequently applied. Very high production speeds of 1000 m/min are possible.

3. Plied and Novelty Yarns

Plied yarns are produced by twisting together two or more singles yarns. Generally, during plying, the yarns are twisted in the direction opposite to the singles-yarn twist to give a torque-balanced yarn. The balancing of twist such that the finished structure will exhibit little tendency to coil or kink on itself is done by monitoring the *twist multiplier* used in the singles yarn and in the plies. Twist multiplier is defined as the ratio of turns per inch to the square root of the cotton count. For maximum strength, the twist multiplier used in the ply structure should equal the twist multiplier of the singles yarn. For balance, the twist multiplier of the plied structure should be ~80% of the singles-yarn twist multiplier.

Yarns with special effects on their surfaces are known as *novelty* yarns. They are produced by varying the linear density across the yarn and also by introducing *slubs*.

4. Continuous Filament Yarns

The production of monofilament yarns is the same as man-made fiber production. Several monofilament yarns are combined to yield a multifilament yarn. Twist is inserted to hold the filaments together and to prevent abrasion and breakage of individual filaments. Continuous filament yarns have a smoother surface than staple fiber yarns. Thus, fabrics made from these yarns tend to be shiny and less bulky and also lack the warmth and wearing comfort associated with fabrics made of staple fiber yarns. The fiber crimp and the discontinuities (voids) in the staple fiber yarns account for the bulk and comfort of these fabrics.

Texturing is a process that modifies the structure and geometric characteristics of continuous filament yarns to impart properties normally associated with staple fiber yarns. In texturing, the continuous filament yarn is deformed (by twisting), and this deformation is set (by heating and then cooling). As a result, a bulky structure is obtained because the set deformations prevent the filaments from lying in a parallel order.

5. Yarn Structure and Properties

The linear density of staple fiber yarns is often expressed by an "indirect" system of measurement in which the length of yarn per unit weight is specified. Denoted as the *cotton count*, it is expressed as the number of hanks per pound, were a *hank* is equal to 840 yd. Thus, a 30s Yarn means that 25,200 (= 30 × 840) yd of the yarn weigh 1 lb. An individual yarn is referred to as a singles yarn; note that the *s* is used here and with indirect yarn numbers.

The geometry of the twisted yarn is shown in Fig. 6. During the yarn formation process, the fibers on the surface follow longer paths (in comparison with the fibers in the core), develop higher tensions, and displace the core fibers that are under low tension. This cyclic interchange, known as migration of fibers, produces a self-locking, yet flexible structure with a fairly regular geometry. The amount of twist inserted is determined by the fiber length and diameter, the linear density of the yarn, and the intended end use. A highly twisted yarn is hard and compact and tends to snarl, whereas a low-twist yarn is bulky and soft.

The important yarn properties are strength, elongation, appearance, and uniformity. Strength and elongation are influenced by the amount of twist inserted; appearance depends on the fineness of the fibers used and the carding process. Neps (small groups of fibers) can be introduced due to improper carding. Uniformity or variation in linear density is greatly influenced by the drafting process. A yarn with excessive linear density variation is unsuitable for most applications. An exception is novelty yarns, which are produced by varying the linear density along the length of the yarn.

6. Ropes

A rope is a collection of twisted yarns built in stages. Initially, several singles or multi-ply yarns are twisted into strands, and several of these strands are twisted in the reverse direction to yield a rope. To achieve a compact structure, proper tension should be maintained in the final twisting operation. Nylon, because of its high tensile strength, good abrasion, and flexural properties, is the most commonly used fiber for making ropes and cords. Polyester is used in applications where a stiffer or less elastic rope is needed. Large ropes are formed by plying

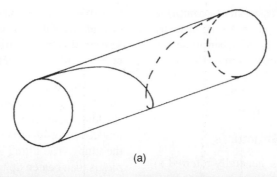

(a)

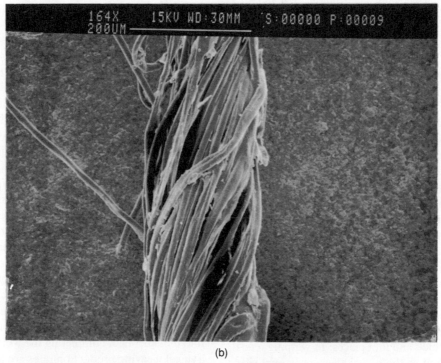

(b)

FIGURE 6 Yarn structure. (a) Twist geometry; (b) scanning electron micrograph of yarn.

additional layers of strands or smaller plied structures on top of a core.

The surface character of a rope is determined by the twist organization of the individual strands and plies of these strands in constructing the rope. A *cord* structure uses alternating twist organization. For example, a ZSZ cord is formed from Z-twisted strands (yarns) that are plied with S twist to form an intermediate structure. These are then plied with Z twist to form the final structure. Cord structures are softer than *hauser* (or *hawser*) structures, which use twist on twist in their construction. A hauser may, for example, have a ZZS twist organization.

7. Braids

A braid is essentially another one-dimensional textile structure. However, when braiding is done over a form or shaped object, as in the formation of a structural el-ement for composite material structures (e.g., a rocket nose cone), it ceases to be a one-dimensional structure. Rather than being simply wrapped in layers as may occur with rope formation, a braid involves interlacing simultaneously two opposing sets of strands to form a tube or flat tape. One or more threads or yarns are wound as a strand onto a braiding tube. A set of 16, 22, or more tubes is placed in carriers, which are driven around a circular track with each half of the set rotating in opposite directions. As carriers meet, they alternatingly deviate from the center of the circular path, inward or outward, to avoid collision.

As strands are pulled off during this process, they are interlaced. The common interlacing patterns are (1) *diamond* braid (1/1), which interlaces at each strand; (2) *regular* braid (2/2), which interlaces at every other strand and is the most common braided structure; and (3) *hercules* braid (3/3), which interlaces at every third strand.

Braids are characterized by very easy extension until the strands touch or jam against one another. A rope made by braiding will have less tendency to rotate when subjected to a tensile force than a rope made by twisting. Braiding is a slower process than twisting. Braids are used alone (e.g., shoe laces and sutures) as a protective or binder covering for a core strand and in decorative applications.

C. Two-Dimensional Textile Structures

Two-dimensional textile structures, generally referred to as fabrics, are characterized by a substantial area and relatively small thickness. Weaving and knitting are the two conventional ways of producing fabrics, whereas the production of nonwoven fabrics represents a nonconventional process.

1. Weaving and Woven Structures

The oldest method of fabric production is *weaving*, in which one set of yarns (*warp*) that runs along the length of the fabric is interlaced with another set (*filling* or *weft*) that runs across the width of the fabric. The ends of the fabric are called *selvedges*. The production of woven structures is carried out on a *loom*. While the sheet of warp yarns runs lengthwise through the loom, the filling yarn is laid singly—one length at a time.

Weaving Preparatory Processes. Analogous to the intermediate steps in the conversion of fiber to yarn, the processes in the conversion of yarn to woven fabric are winding, warping, slashing, and weaving. The purpose of winding is to convert small ring bobbins to larger packages called *cones* or *cheeses* and to remove defects in the yarn such as slubs and thick and thin places. In warping, ends from several hundred of these packages are wound side by side (under uniform tension) on a beam. The length of yarn and the width of the sheet are determined by the length and width of cloth to be woven.

During the weaving process, the warp yarns are under tension and are also subjected to the abrasive action of some of the loom parts (drop wires, *heddles*, *reed*). At the same time, there is interyarn rubbing, and the surface fibers from one yarn become entangled with the fibers on the neighboring yarns. Both situations lead to yarn breakages. The purpose of *slashing* or *sizing* is to improve the weavability of the yarns to minimize warp breaks on the loom. Yarn strength is enhanced by the use of adhesives, and their covering action entraps the protruding surface fibers, resulting in a smooth yarn. Lubricants are used to improve the pliability of the yarns. Several warp beams (giving the required number of ends in the fabric) are combined into a single sheet and led into a size bath containing adhesives, lubricants, and a solvent. Commonly used adhesives are polyvinyl alcohol and all types of starches. Wa-

ter is the general solvent, whereas mineral and vegetable oils, animal fats, and mineral waxes serve as lubricants. The sized yarns are dried on cylinders and wound onto a beam known as the weaver's beam.

a. The weaving cycle. The three primary motions on a loom are *shedding*, *picking*, and *beating up*. During shedding, some of the warp ends are raised and others are lowered according to the weave pattern to create a shed for the filling. The filling is inserted during picking, and the *pick* is then beaten up to form the cloth. In a conventional or *shuttle* loom, the filling yarn is released from a shuttle that traverses from one side of the loom to the other. Weaving speeds are measured in terms of the number of picks inserted per minute and weft insertion rate, which is equal to the product of the picks per minute and the loom width. The typical speed for a shuttle loom 1.1 m wide is 210 picks/min, giving a weft insertion rate of 231 m/min.

The large mass of the shuttle (~1 lb) and the need to accelerate and decelerate it over short distances have been factors that limit production rates. Other methods of weft insertion, in which the shuttle has been replaced, are gaining in popularity. In a projectile or gripper loom, a small gripper carries one pick length of yarn from one end to the other. Fluid jets (air and water) are also used for propelling the weft across the shed. Water-jet looms are generally used for continuous filament yarns; they cannot be used for yarns from hydrophilic fibers (like cotton) because they absorb water. In a rapier loom, metallic arms at either end of the loom are used for picking. The filling is brought from one end of the loom to the center by one arm, where it is transferred to the second rapier, which carries it to the other end, completing the weft insertion process. Typical weaving production rates are listed in Table IV.

b. Woven fabric structures. There are a wide variety of ways in which the warp and filling yarns can be interlaced. The resulting structure influences the functional and aesthetic properties of the fabric. There are three weaves, *plain*, *twill*, and *satin*, from which all other woven structures are derived. The simplest of weave is the plain weave. As shown in Fig. 7a, the first warp thread is over the first pick, under the second, over the third, and so on, whereas

TABLE IV Weaving Production Rates

Loom type	Width (m)	Speed (picks/min)	Weft insertion rate (m/min)
Shuttle	0.9–1.1	180–220	230
Projectile	2.2–5.4	380–420	1100
Air jet	1.25–3.3	900–1200	1900
Water jet	1.25–2.1	1000–1500	2100
Rapier	1.65–3.6	475–500	1200

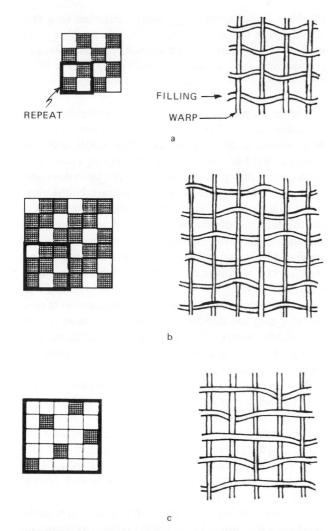

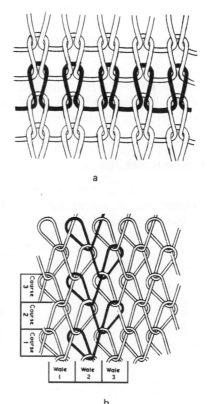

tion is used. For more elaborate patterns that repeat over several hundred ends and picks, individual control of ends is obtained using the Jacquard.

Other woven structures include pile fabrics (terry towels) produced using two sets of warp ends, triaxial fabrics woven from three sets of threads, and double cloths in which two layers of fabric are "stitched" at specific intervals.

2. Knitting and Knitted Structures

Knits are two-dimensional structures produced from yarns by forming loops in the yarn and passing successive lines of loops through one another. Whereas two sets of threads (warp and weft) are used in weaving, knitted fabrics are produced from either a warp set of yarns or a weft yarn. Hence, knitting is divided into two major areas: *weft knitting* and *warp knitting*. The term *interlooping*, in contrast to *interlacing*, describes the knitting process. Figure 8 shows the structures of weft and warp knit fabrics.

a. Weft knitting. Weft knit fabrics are produced predominantly on circular knitting machines. The simplest of the two major weft knitting machines is a jersey machine. Generally, the terms *circular knit* and *plain knit* refer to jersey goods. The loops are formed by knitting needles

FIGURE 7 Three primary weaves and corresponding interlacements. (a) Plain weave; (b) twill weave; (c) satin weave (weft faced).

the second warp thread is under the first pick, over the second, under the third, and so on. The plain weave repeats on two ends and two picks. The plain weave can be extended along the warp or filling or both directions to yield different structures.

The twill weave is characterized by diagonal lines, known as twill lines, that run at angles to the fabric. A simple 2/1 (read 2 up, 1 down) twill is shown in Fig. 7b. The steps due to the staggering of the interlacements produce the twill line. A five-end weft satin (known as sateen) is shown in Fig. 7c. The unique characteristic of the satin weave is that there are no adjacent interlacements in a repeat. In addition, they have long "floats" in either the warp or filling direction. When a low-twist bulky yarn is used for the floating thread, a soft, lustrous fabric with a luxurious feel is produced. For producing simple patterns that repeat over 36 ends and picks, a *dobby* shedding mo-

FIGURE 8 Knitted structures. (a) Weft knit; (b) warp knit.

and the jersey machine has one set of needles. Typical fabrics are hosiery, T-shirts, and sweaters.

Rib knitting machines have a second set of needles at approximately right angles to the set found in a jersey machine. They are used for the production of double-knit fabrics. In weft knits, design effects can be produced by altering needle movements to form *tuck* and *miss* stitches for texture and color patterns, respectively. Instead of a single yarn, several yarns can be used in the production of these structures. This increases the design possibilities.

b. Warp knitting.
Warp knitting involves a flat needle bed of from 6 to 15 ft in width. There are two major types of warp knitting machines: *tricot* and *raschel*. The structures of the two machines are much more functionally similar than is found in comparing jersey with rib. The warp knitting machine is somewhat more demanding mechanically to operate in terms of precision and input yarn quality demands. The tricot machine is designed to produce fabrics of generally simple design, but at a very high production rate. In contrast, the raschel machine is very versatile, producing fabrics ranging from fine laces to heavy upholstery.

Warp knits use yarn guides for each independent warp yarn beam set to deliver the yarns to the needles in a predetermined pattern. Passing yarn over the needle in warp knitting is referred to as *lapping*. The overlap across the hook of the needle can form an open or closed loop. The underlap passes to adjacent needles. Both the overlap and underlap affect the dimensionality stability of warp knit fabrics.

The two commonly used types of knitting needles are shown in Fig. 9. Latch needles are used on weft and raschel machines, whereas spring beard needles are used on tricot machines.

3. Nonwoven Structures

In a nonconventional method of producing two-dimensional structures, the intermediate (and expensive) step of yarn formation and their geometric ordering are

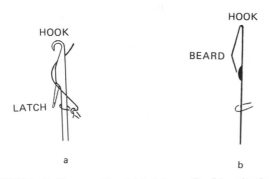

FIGURE 9 Knitting needles. (a) Latch needle; (b) spring beard needle.

eliminated, and fibers are directly converted to a fabric. Unlike a woven or knitted fabric, in which a well-controlled and definable arrangement of yarns prevails, a nonwoven structure usually consists of a web of fibers (in a stochastic arrangement) bonded together by one of several methods.

The first step in the production of nonwovens is the formation of the web. There are three types of webs: parallel-laid, cross-laid, and random-laid. When staple fibers are used, parallel-laid webs are formed by laying a number of carded webs on top of one another in the same direction, giving a highly anisotropic web. Consequently, the properties of the resulting fabric are significantly different in the two principal directions (known as *machine* and *cross* directions). To reduce this difference, a cross-laid web is produced by cross-lapping the webs. For a more isotropic web, special machines are used to produce a random arrangement of fibers, giving random-laid webs. When continuous filament yarns are used for the production of nonwovens, they are directly extruded onto a moving belt. Historically, web formation began as a wet laying process used by papermakers. The dry lay process (described earlier) predominates now.

The formed web can be bonded in one of two ways, either *adhesively* or *mechanically*. In adhesive bonding, a *thermoplastic* fiber or powder that has been blended with the web melts due to the application of heat and binds the fibers together. Alternatively, when the web is produced from continuous filament yarns, they are self-bonded and such fabrics are known as spun-bonded fabrics. Instead of a thermoplastic fiber or powder, adhesive binders such as acrylic resins are used. The web is saturated with the binder and then dried. Print bonding is the selective application of binders to the web in a desired pattern. The fibers have a greater flexibility of movement (since the proportion of binders is small), and the fabric is softer than the one produced by saturation bonding.

Mechanical bonding has its origins in the production of *felt* from wool fibers. Felts are formed by mechanical agitation of wool fibers in a wet state, which causes fiber entanglement and yields a coherent and strong structure. Finishing felt hats by hand rubbing with mercury led to the expression "as mad as a hatter." In *needle punching*, a form of mechanical bonding, barbed needles penetrate a web, causing the fibers from the surface to move *into* the structure. The embedded fibers are trapped by the neighboring fibers, giving a mechanically interlocking structure. Needle punching is usually carried out on both sides of the fabric. Since the fabrics generally tend to be weak, support yarns are used. In another variation of mechanical bonding, known as stitch bonding, yarns are knitted (usually warp knit) through the fibrous web. Arachne and Malimo are two commercially available stitch-bonded nonwoven fabrics.

D. Fabric Finishing

The fabric as it comes out of the loom or knitting machine is known as a *gray* or *greige* fabric. It is rough and stiff (due to the size) and does not absorb water, a necessary precondition for dyeing. So the fabric is subjected to a finishing process that enhances its aesthetic and functional properties. Termed *wet processing*, the typical sequence includes *desizing* for removal of the size added during slashing (not done for knitted fabrics since the yarns are not sized), *scouring* to remove surface impurities and improve the fiber's water absorption characteristics, and *bleaching*. *Heat setting* is carried out for fabrics from man-made fibers.

The fabric is then dyed to the desired shade using the class of dye appropriate to the fiber. For example, *reactive*, *vat*, and *direct* dyes are commonly used for cotton, whereas *disperse* dyes are used for polyester. *Printing*, or localized dyeing, is the selective application of dyestuff to specific areas of the fabric to produce a desired pattern. Specific chemical finishes are applied to the fabric to impart some properties such as water and soil resistance, flame retarduance, and crease recovery. Dyeing, normally a wet process, can also be carried out as a dry process.

E. Fabric Properties

The aesthetic and functional characteristics of fabrics are influenced by the finishing process, the fabric structure, and the type of fiber. The major functional properties of fabrics are strength, elongation, and resistance to tear and wrinkling. Among the surface properties of interest are resistance to abrasion, *pilling* (tendency to form fiber beads due to abrasion), and fabric handle. Air and water permeability are the major transfer properties of textile structures.

The capacity of a textile fabric to "fall" when it is hung is called *drape*, and this characteristic sets it apart from other two-dimensional structures (e.g., a sheet of paper or a metal plate). Drape is a direct consequence of the fabric's capacity to undergo shear deformation and is related to the stiffness of the fabric. Chemical finishes are used to modify the drapability of fabrics. The assessment of fabric handle (softness, drape, feel) is a purely subjective process, and attempts are being made to link the mechanical properties of fabrics (bending, tensile, shear, etc.) to the handle of fabrics. The Kawabata hand measuring system is one such commercial method.

III. TRENDS IN TEXTILE MANUFACTURING

The Jacquard card used to carry binary information into the weaving process was the precursor of the punched card for the computer and, in fact, was the inspiration for the birth of the computer card. the field of textile engineering, which triggered the computer revolution, is quite appropriately taking advantage of computers in its various activities.

A. Computers in Manufacturing

One of the primary objectives of a textile manufactuer is to produce a defect-free end product, or at least minimize the occurrence of defects. However, due to the stochastic nature of some textile processes (e.g., yarn production), precise control is not possible and consequently defects are introduced in the material. Drafting or drawing is one such operation in yarn manufacturing whereby linear density variations are introduced in the sliver. To minimize these variations, *autolevelers* operating on either the *feedback* or the *feedforward* principle are made an integral part of drawing machines. The speed of either the input or the output rollers (see Fig. 2) is changed, causing a change in the draft ration and eventually resulting in a change (correction) in the linear density of the silver.

1. Integrated Production and Quality Monitoring

Modern spinning machines, like the Murata air-jet spinning system, incorporate on-line measurement of yarn evenness; several spinning machines are linked to a central computer, enabling the operator to monitor the operation from the console. Downtime due to stoppages is computed, and machine efficiency is displayed. When the yarn package is built to the desired size, the doffing mechanism removes the package and transports it on a belt to one end of the machine, where it is picked up by another arm and placed inside a container. When the desired number of packages has been placed in the container, it is automatically boxed and rendered ready for shipping. Thus, the role of the human operator, and hence the chance of error, are minimized in the production process.

2. Sizing and Weaving

Microprocessors are also used on the slashing machine to produce a high-quality sized yarn. Parameters such as the viscosity, temperature, and pickup are monitored and regulated by these controllers. In the weaving plant, looms are monitored continuously by individual microprocessors, and the data on the type and duration of stoppages are fed into a central computer. These data are analyzed to spot trends in loom functioning and aid in the early detection of problem looms. The use of this information in dynamically assigning operators to stopped looms is presently being researched and may find its way to the weaving floor in the future. Material handling is being automated with the use of self-guided vehicles.

3. Design and Dyeing

Design workstations with high-resolution graphics terminals and color palettes enable the designer to create and view different structures before actually producing the fabric. The computer can transfer the pattern instructions to the knitting machine for production. In the area of dyeing and printing, the Millitron process from Milliken & Co. represents the state of the art in ink-jet printing. Here, programmed jets spray dyestuff in different colors and patterns onto the material. Since the patterns and colors can be varied easily, small lots can be produced with a minimum of setup time and costs.

4. Fabric Inspection

Inspection of the finished fabric for defects (e.g., broken or missing ends and picks) has long been a slow and tedious process requiring the services of an experienced operator. Defects are classified as major or minor depending on the severity. If the allowable number of defects is exceeded, the fabric is classified as *seconds* and sold at a lower price. Electronic inspection systems are replacing the human eye, making high inspection speeds possible. The fabric "ticket" (a paper containing the fabric particulars that accompanies the fabric) now incorporates not only the number of defects, but also the place of occurrence of each defect in the fabric so that the garment manufacturer can take this into account while cutting the fabric.

B. Computers in Engineering Design of Textile Structures

Apart from the traditional area of accounting and costing, computers, and especially personal computers with the huge volume of useful software, are being increasingly used in the areas of engineering design of textile structures, production calculations, inventory management, and production planning and scheduling. Figure 10 illustrates the system of constraints in the design of a woven fabric. To arrive at the necessary fabric specifications, the textile engineer must have the flexibility to analyze the effect of the different parameters (counts of warp and filling, number of warp ends and filling picks, the crimp in the warp and filling) on fabric areal density (weight per unit area). Software packages with the ability to manipulate equations are used for this purpose.

C. Textile Operations in the Future

The proliferation of inexpensive computer hardware and software will lead to the wide-spread use of

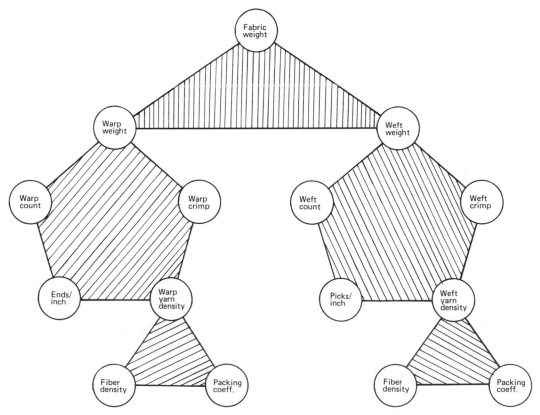

FIGURE 10 Engineering design of woven fabrics: system of constraints.

microprocessors for better control of processes. Research on knowledge-based expert systems in specific areas of textile engineering is in progress. The use of computer networks for communication within a plant and between plants situated in various places will increase. The centralization of operations such as purchasing and research and development is gaining ground. Computers and automation should become an integral part of textile operations for the continued growth of the industry.

SEE ALSO THE FOLLOWING ARTICLES

BIOPOLYMERS • COMPOSITE MATERIALS

BIBLIOGRAPHY

Acar, M. (1995). "Mechatronic Design in Textile Engineering," Kluwer Academic, Dordrecht/Norwell, Massachusetts.

Ash, M., and Ash, I. (2000). "Handbook of Textile Processing Chemicals," Synapse Information Resources, Inc., Endicott, New York.

Goswami, B. C., Martindale, J. G., and Scardino, F. L. (1977). "Textile Yarns: Technology, Structure, and Applications," Wiley (Interscience), New York.

Joseph, M. (1981). "Introductory Textile Science," 4th ed. Holt, New York.

Lord, P. R., and Mohamed, M. H. (1982). "Weaving: Conversion of Yarn to Fabric," 2nd ed. Merrow, Durham, United Kingdom.

Luneschloss, J., and Albrecht, W. (1985). "Nonwoven Bonded Fabrics," Halsted Press, Chichester, United Kingdom.

Marks, R., and Robinson, A. T. C. (1976). "Principles of Weaving," The Textile Institute, Manchester, United Kingdom.

Raheel, M. (1996). "Modern Textile Characterization Method," Dekker, New York.

Spencer, D. J. (1989). "Knitting Technology," 2nd Ed. Technomic, Lancaster, Pennsylvania.

Vigo, T. L. (1994). "Textile Processing and Properties: Preparation, Dyeing, Finishing and Performance," Elsevier, Amsterdam/New York.

Vincenti, R. (1993). "Elsevier's Textile Dictionary: In English, German, French, Italian and Spanish—CD-ROM," Elsevier, Amsterdam/New York.

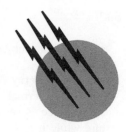

Thermal Analysis

David Dollimore, deceased

University of Toledo

GLOSSARY

Derivative thermogravimetry (DTG) Involves plotting the first derivative of the thermogravimetry with respect to either time or temperature. The DTG curve is plotted with the rate of mass loss on the ordinate, decreasing downward, and the temperature or time on the abscissa, increasing from left to right.

Differential scanning calorimetry (DSC) Two types of DSC are found in commercial instrumentation, namely, power-compensation DSC and heat-flux DSC.

Differential thermal analysis (DTA) Technique in which the temperature difference between a substance and a reference material is measured as a function of temperature while the substance and reference material are subjected to a controlled program.

Emanation thermal analysis Technique in which the release of radioactive emanation from a system is measured as a function of temperature while the system is subjected to a controlled temperature program.

Evolved gas detection (EGD) Technique in which the evolution of gas from a substance is noted when the system is subjected to a controlled temperature program. If the gas can be analyzed, then the term EGA is applied, and here the amount of product and its identity are measured as a function of temperature.

Isobaric mass-change determinations The equilibrium mass of a substance at a constant partial pressure of the volatile product(s) measured as a function of temperature while the substance is subjected to a controlled temperature program. The isobaric mass-change curve is plotted with mass on the ordinate, decreasing downward, and temperature on the abscissa, increasing from left to right.

Thermodilatometry (TDA) Technique in which the dimension of a substance under negligible load is measured as a function of temperature while the substance is subjected to a controlled temperature program.

Thermogravimetry (TG) The most widely used technique of thermal analysis, in which the mass of a substance is measured as a function of temperature while it is subjected to a controlled temperature program.

Thermomechanical analysis (TMA) Technique in which the deformation of a substance under a nonoscillatory load is measured as a function of temperature while the substance is subjected to a controlled temperature program.

Thermoparticulate analysis (TPA) Technique in which the evolved particulate material in the evolved gases is measured as a function of temperature.

THE NAME *thermal analysis* is applicd to a varicty of techniques in which the measurement of any property of a system is recorded as the system is programmed through a predetermined range of temperatures.

I. INTRODUCTION

A. Scope of Thermal Analysis

In most laboratory experiments dealing with the properties of a material or a system the properties are measured under isothermal conditions. Separate experiments are required to measure these same property at different temperatures. In thermal analysis the specified property is measured under a controlled temperature regime. The simplest temperature regime would be that of an isothermal experiment, but in most cases the temperature is raised at a predetermined rate, for example, 10°C per minute. The interpretation then involves the variation of a particular property with both temperature and time. There is, however, a decrease in labor and time which makes such studies especially interesting for industrial applications. With more complicated temperature regimes there is an ability inherent in the method to mimic industrial processes. Industries utilizing thermoanalytical methods are listed in Table 1. The plot of the physical property of the sample recorded as a function of the temperature is said to be a *thermal analysis curve*. There is still some confusion in the literature about this name, as it was initially applied to the specific technique in which the temperature of a sample was recorded against time as it was cooled down from a particular value. The use of the name in this way persists in physical chemistry textbooks where the name *thermal analysis* is used for this specific purpose. Other conditions that have to be satisfied in the practice of thermal analysis are as follows.

TABLE I Industries Utilizing Thermoanalytical Methods

Abestos industry	Polymers, plastics, and rubbers
Industrial biochemistry	
Building industry	Pharmaceuticals
Preparation of catalysts	Medicinal
Ceramics	Organic chemicals and organic metallics
Clay processing	
Explosives and pyrotechnics	Textiles
Fats, oils, and waxes	Semiconductors
Food processing	Electronics
Fuel technology	Carbon adsorbents, charcoals, and graphites
Glass industry	
Metallurgy	Expoxy laminates
Inorganic chemical industry	Composite materials
Mineral processing	Processing and quality control
Liquid crystals	

1. The physical property and the sample temperature should be measured continuously. It should be noted that the measurement of certain properties is not easily made to comply with this particular condition.
2. In practice both the property and the temperature should be recorded automatically.
3. The temperature of the sample should be altered at a predetermined rate. In many early textbooks it is stated that the sample should be cooled or heated at a uniform rate. However, the real basis of the use of these techniques is that they should be operated on a predetermined temperature regime, as this allows various parameters to be followed. For example, one may compare directly an industrial process in which the temperature is raised, then held at a particular temperature, and then raised again.

The purpose of making the measurements is to study the physical and chemical changes which occur in a system on heating. One therefore has to interpret a thermal analysis curve by relating the property measured against temperature and interpreting the changes by noting the chemical and physical events which have taken place in the system under observation. The most obvious change in a system studied in this manner is that of mass, but calorimetry experiments predate this technique and give information concerning the enthalpy changes which take place. Evolved gas analysis and detection are also the subject of many studies where systems are heated. Another group of techniques comes under the heading of *thermomechanical analysis*. These deal with dimensional changes and with properties connected with the strength of materials when subjected to temperature changes. It should be noted that this group must by definition include the measurement of the density of samples subjected to a programmed temperature variation. A recently introduced related technique is *dynamic mechanical analysis*, used to study the viscoelastic response of a sample under an oscillatory load. Other techniques are also used where the sample is under an oscillatory load.

A recent sophisticated approach is to modulate the temperature about a predetermined overall heating rate. Those who use these techniques must be subjected to a discipline which needs definitions and conventions so that the reader may understand the technique being utilized. In commercial equipment these techniques are often combined so that, in these simultaneous techniques, one material is subjected to two or even three measurement probes. Generally, thermal analysis techniques may be classified into three groups depending upon the way in which the physical property needs to be recorded.

1. The absolute value of the property itself can be measured, for example, the sample mass.
2. The differential method measures the difference between some property of the sample and that of a standard material, for example, their temperature difference.
3. The rate at which the property changes with temperature can be measured. These form the basis of derivative measurements and very often may be interpreted on a kinetic basis.

There exist national and international organizations that recommend nomenclature abbreviations and standards, and these organizations have set up committees that have formulated, in particular, a system of nomenclature which is adhered to in this chapter but is not always adhered to in journals and in certain other fields of science. These recommended nomenclatures and abbreviations are constantly under review, so the reader is advised to seek the most up-to-date recommendations in the literature.

B. Nomenclature

The recommended nomenclature has been put forward by the International Nomenclature Committee of the International Confederation for Thermal Analysis and Calorimetry (ICTAC). These recommendations are widely circulated in publications of the confederation and the International Union of Pure and Applied Chemistry (IUPAC). The most widely publicized report is the booklet "For Better Thermal Analysis," which is continually updated as required.

The recommended abbreviations of various techniques in thermal analysis are listed in Table II. The property

TABLE II Recommended Terminology for Some Thermal Analysis Techniques

Property measured	Technique name	Abbreviation
Mass	Thermogravimetry	TG
	Isobaric mass-change determination	
	Isothermal mass-change determination	
	Evolved gas detection	EGD
	Evolved gas analysis	EGA
	Derivative thermogravimetry	DTG
	Emanation thermal analysis	
	Thermoparticulate analysis	TPA
Temperature	Cooling curve[a]	
	Heating curve	
	Differential thermal analysis	DTA
Enthalpy	Differential scanning calorimetry	DSC
Dimensions	Thermodilatometry	TDA
Mechanical	Thermomechanical analysis	TMA
	Dynamic mechanical analysis	DMA

[a] Cooling curve was initially called thermal analysis. It is still called thermal analysis in many physical chemistry textbooks. Other techniques of thermal analysis measure acoustic, optical, electrical, and magnetic characteristics versus temperature. The list is not complete, as new techniques are constantly being added.

measured against temperature is indicated in Table III. This list is not complete, as new techniques are continually appearing.

The definitions follow from Tables II and III.

Thermogravimetry (TG) is the most widely used technique of thermal analysis, in which the mass of a substance is measured as a function of the temperature while it is subjected to a controlled temperature program. The record on the thermogravimetric, or TG, curve is the mass plotted against the temperature (T) or time (t) if the variation of temperature with time can be indicated as well on the same graph. In solid-state decomposition reactions, the reactant material degrades, often to be replaced by the solid product. An example of this is the decomposition of limestone to quicklime.

In the record of the mass of the solid residue against the temperature, the decomposition of the material can be followed. This may be plotted in alternative ways, as the percentage mass loss or the fractional mass loss versus the temperature or as the fractional decomposition versus the temperature.

Derivative thermogravimetry (DTG) is not really a separate technique but involves plotting the first derivative of the TG with respect to either time or temperature. The DTG curve is plotted with the rate of mass loss on

TABLE III Property Measured versus Temperature in Some Thermal Analysis Techniques

Technique	Property measured vs temperature	Instrument
Thermogravimetry (TG) and derivative thermogravimetry (DTG)	Mass and derivative of mass with respect to temperature	Thermobalance
Differential thermal analysis (DTA)	Difference in temperature between reference and sample cell	DTA apparatus
Differential scanning calorimetry (DSC)	Enthalpy (heat) flow	DSC apparatus
Evolved gas detection and analysis (EGD and EGA)	Various properties of gas	Method varies but should always be indicated
Thermodilatometry (TDA)	Length of volume	Dilatometer
Thermomechanical analysis (TMA)	Expanion under applied stress (load)	Adapted dilatometer
Dynamic mechanical analysis (DMA)	Frequency response under oscillatory stress	DMA apparatus

the ordinate, plotted downward, and the temperature or time on the abscissa, increasing from left to right. This can be achieved by an analysis of the TG curve as a separate operation in the dedicated computer part of the equipment.

Isobaric mass-change determinations refer to the equilibrium mass of a substance at a constant partial pressure of the volatile product(s) measured as a function of temperature while the substance is subjected to a controlled temperature program.

Evolved gas analysis (EGA) is a technique in which the gas evolved from a substance subjected to a controlled temperature program is analyzed. The method of analysis should always be noted.

Emanation thermal analysis is a technique in which the release of radioactive emanation from a system is measured as a function of temperature while the system is subjected to a controlled temperature program.

Thermoparticulate analysis (TPA) is a technique in which the evolved particulate material in the evolved gases is measured as a function of temperature.

Another group of techniques involves the measurement of enthalpy changes. It has already been mentioned that, historically and in physical chemistry textbooks, the term *thermal analysis* applies to the determination of cooling or heating curves: these are techniques in which the temperature of a substance is measured as the substance is either cooled down or allowed to heat up. The cooling curve is the normal technique, and again, it may be reported as the first derivative of the heating curve with respect to time from the raw experimental data while the substance is subjected to a controlled temperature regime against time. The function dT/dt should be plotted on the ordinate, and T or t on the abscissa, increasing from left to right. It is important to state whether the technique involves a cooling process or a heating process.

The two techniques *differential thermal analysis* (DTA) and *differential scanning calorimetry* (DSC) should be considered together. DTA is a technique in which the tem-

perature difference between a substance and a reference material is measured as a function of temperature while the substance and reference material are subjected to a controlled program. The plot is called a DTA curve; the temperature difference ΔT should be plotted on the ordinate, with the endotherm processes shown downward and the exotherm processes in the opposite direction, and the temperature or time on the abscissa, increasing from left to right. This technique is often applied quantitatively when the area of the peaks can be made proportional to the quantity of the material decomposing or to the enthalpy of the process. In this respect the equipment then serves as a calorimeter. The term DSC is applied to such experiments. There are two types of equipment in which the background temperature of the calorimeter is raised through a programmed temperature regime being imposed on the system. These are power-compensation DSC and heat-flux DSC. The method identified as power-compensation DSC was originally a copyright term employed by one of the instrument manufacturers, and often the term DSC is found in the literature applying just to power compensation equipment.

The members of the Nomenclature Committee have considered the distinction between quantitative DTA and heat-flux DSC: in their opinion, a system with a multiple sensor (e.g., a Calvert-type arrangement) or with a controlled heat leak (a Boersma-type arrangement) should be called heat-flux DSC. In practice, if the instrument manufacturers can show that the system operates as a calorimeter over a programmed temperature range, then they describe the equipment as DSC.

The usual method of plotting DTA results is, as already noted, with endothermic peaks shown downward on the plot and exothermic peaks shown in an upward direction. However, because a DSC is considered to measure thermodynamic quantities directly, the DSC plots are often found with the endothermic plots in an upward direction and the exothermic plots in a downward direction. This conforms with the IUPAC requirements for the presentation of thermodynamic parameters. In reporting data, therefore, the directions of the endothermic peaks and the enothermic peaks should be clearly shown.

Thermodilatometry (TDA) is a technique in which the dimension of a substance under a negligible load is measured as a function of temperature while the substance is subjected to a controlled temperature program.

The record in the thermodilatometric curve is then the dimension plotted on the ordinate, increasing upward, with temperature T or time t on the abscissa, increasing from left to right. Mention should also be made of such related techniques as *linear TDA* and *volume TDA*, which differ on the basis of the dimensions measured as indicated in their names.

Thermomechanical analysis (TMA) is a technique in which the deformation of a substance under a nonoscillatory load is measured as a function of temperature while the substance is subjected to a controlled temperature program. It is used extensively in polymer studies. The mode, as determined by the type of stress applied (compression, tension, flexture, or torsion), should always be noted. As already stated, *dynamic mechanical analysis* is a technique in which the viscoleastic response of a sample under an oscillatory stress is studied while the substance is subjected to a temperature regime. *Torsional braid analysis* is a particular case of dynamic thermomechanometry where the material is supported.

Other techniques do not always find commercial instrumentation support and are usually constructed for particular, often limited, applications. An example is *thermosonimetry*, in which the sound emitted by a substance is measured as a function of the temperature while the substance is subjected to a controlled temperature program. In an associated technique, *thermoacoustimetry*, the characteristics of imposed acoustic waves are measured as a function of the temperature after passing through a substance while the substance is subjected to a controlled temperature program.

Thermoptometry is another thermal analysis technique in which an optical characteristic of a substance is measured as a function of temperature while the substance is subjected to a controlled temperature program.

Measurements of total light, light of a specific wavelength(s), refractive index, and luminescence lead to *thermophotometry*, *thermospectrometry*, *thermorefractometry*, and *thermoluminescence*, respectively. Observations using a microscope are called *thermomicroscopy*. The microscopy observations, however, are often referred to under the general term of hot-wire microscopy. *Thermoelectrometry* is a technique in which an electrical characteristic of the substance is measured as a function of temperature. The most common measurement here is resistance, conductance, or capacitance.

Thermomagnetrometry is yet another technique of thermal analysis in which the magnetic susceptibility of a substance is measured as a function of temperature while the substance is subjected to a controlled temperature program.

Sometimes more than one technique is used in an investigation. Such multiple techniques can be classified as follows. The term *simultaneous techniques* covers the application of two or more techniques to the same sample at the same time, for example, simultaneous TG and DTA. This is becoming a widely used practice and is discussed in more detail in Section VII.

C. Symbols

The abbreviations for each technique have already been noted (see Tables II and III). In polymer studies, however, the distinction between T_g and TG may cause confusion. Here the abbreviation TG refers to thermogravimetry, while T_g represents the glass transition temperature. This has caused a number of investigators and instrument manufacturers to use TGA for TG to avoid confusion. Other aspects of the use of symbols are mentioned in the following list.

1. The international system of units (SI) should be used wherever possible.
2. The use of symbols with superscripts should be avoided.
3. The use of double subscripts should also be avoided.
4. The symbol T should be used for temperature whether expressed as degrees Celsius (°C) or as kelvins (K). For temperature intervals the symbol K or °C can be used.
5. The symbol t should be used for time, whether expressed as seconds (s), minutes (min), or hours (h).
6. The heating rate can be expressed either as dT/dt when a true derivative is intended or as β in K min^{-1} or °C min^{-1}. The heating rate so expressed need not be constant over the whole temperature range and can be positive or negative, so this should be stated.
7. The symbols m for mass and W for weight are recommended.
8. The symbol α is recommended for the fraction reacted or changed.
9. The following rules are recommended for subscripts:
   a. Where the subscript relates to an object, it should be a capital letter, e.g., M_s represents the mass of the sample, and T_R represents the temperature of the reference material.
   b. Where the subscript relates to a phenomenon occurring, it should be lowercase, e.g., T_g

represents the glass transition temperature, T_c represents the temperature of crystallization, T_m represents the melting temperature, and T_t represents the temperature of a solid-state transition.

c. Where the subscript relates to a specific point in time or a point on the curve, it should be a lowercase letter or a number, e.g., T_i represents the initial temperature, $t_{0.5}$ represents the time at which the fraction reacted is 0.5, $T_{0.3}$ represents the temperature at which the fraction reacted is 0.3, T_p represents the temperature of the peak in DTA or DSC, and T_c represents the temperature of the extrapolated onset. This can also be applied to DTG techniques.

D. Standardization

No single instrument design or set of experimental conditions is optimum for all studies. The techniques are dynamic in nature and flexible in use but produce data which may be highly dependent upon the procedure. The prime requirements for standardization can be listed as follows.

1. The provision of a common basis for relating independently acquired data.
2. The provision of the means for comparing and calibrating all available instrumentation, regardless of design.
3. The provision of the means for relating thermoanalytical data to physical and chemical properties determined by conventional isothermal procedures.

This means in effect that the geometry of the measuring system and its effects on experiments must be noted. One of the most obvious points regarding standardization is the use of proper materials to establish the validity of the thermocouple readings. Such standards would then reflect the effect of the experimental design (i.e., the geometry of the instrumentation layout), and in this way a correction can be applied to the recorded temperature. In some units, for example, the temperature measuring device is used to control the furnace temperature and is located away from the sample. This means that the sample or system under observation has a temperature record in the thermal analysis curve which is not actually the temperature of the sample; again, this must be corrected. By including curves of standard materials obtained under particular conditions, it is possible to relate and estimate such errors. It is necessary to impress on instrument manufacturers the need to locate the temperature measuring

TABLE IV Materials that Can Be Used for Temperature Calibration in Thermal Analysis[a]

Material	Transition type	Peak temperature (°C)
Polystyrene	Glass transition	~101
1-2 Dichloroethane	m.p.	−32
Cyclohexane	Transition point	−83
	m.p.	+7
Phenyl ether	m.p.	30
o-Terphenyl	m.p.	58
Potassium nitrate	Transition point	128
Indium	m.p	157
Tin	m.p.	232
Potassium perchlorate	Transition point	300
Silver sulfate	Transition point	430
Quartz	Transition point	573
Potassium sulfate	Transition point	583
Potassium chromate	Transition point	665
Barium carbonate	Transition point	810
Strontium carbonate	Transition point	925

[a] These temperature calibration materials are supplied by instrument manufacturers on request. The essential condition is that the material should be pure. The above data are generally available for determination of temperature on DTA or DSC equipment.

and control device as close to the sample as conveniently possible.

Most modern equipment has the temperature measuring device located very close to the samples. The instrument manufacturers will supply suitable reference materials on request with appropriate certification. Table IV sets out *some* suitable materials that can and have been used as temperature standards for calibration purposes. Directly these are most useful for calibrating DTA or DSC units. In DSC units knowledge of the enthalphy changes may also be required. Table V sets out the enthalpy of fusion for selected materials. Again, most DSC instrument manufacturers provide materials with their equipment for this purpose. Indium is used as a calibration material in many DSC units, and other systems undergoing phase changes can be used in a similar way.

There is a need to establish the "proper" temperature in TG experiments. Here the use of materials with magnetic transitions that can be displayed on a mass-loss curve and be referred to the temperature prove to be most useful. The reference sample is placed in the sample container and suspended within a magnetic field gradient. The magnet applying the field can be either a permanent magnet or an electromagnet which can be placed, for the purpose of calibration, near the same location. At the reference material's Curie temperature the magnetic effect diminishes to zero and the TG unit indicates an apparent mass change.

TABLE V Enthalpies of Fusion for Selected Materials

Material	Melting point (°C)	Enthalpy of fusion (J g^{-1})
Naphthalene	80.3	149
Benzoic acid	122.4	147.4
Phenacetin	134.7	173.0
Indium	156.6	28.5
Tin	231.9	59.5
Bismuth	271.3	53.1
Lead	327.5	23.1
Zinc	419.6	112.0
Aluminum	660.4	399.4
Sodium chloride	800.0	495.0
Silver	961.9	107.0
Gold	1064.4	64.0

TABLE VII Fusible Link Method for Calibrating a TG Unit

Material	Temperature (°C)		
	Observed	Corrected	Literature
Indium	159.90 ± 0.97	154.20	156.63
Lead	333.02 ± 0.91	331.05	327.50
Zinc	418.78 ± 1.080	419.68	419.58
Aluminum	652.23 ± 1.32	659.09	660.37
Silver	945.90 ± 0.52	960.25	961.93
Gold	1048.70 ± 0.87	1065.67	1064.43

Table VI indicates suitable materials recommended for use as magnetic standards. McGhie and co-workers have introduced another method for calibration of TG balance termed *fusible link temperature calibration*. In this method loops of metal are attached to the sample part of the TG balance. At their melting point these links condense to a liquid and fall off the balance (to be collected in a suitable container). The melting point is thus recorded as a sudden loss in weight. Table VII lists suitable materials for this method.

E. Reporting Thermal Analysis Data

The committee on standardization has also reported the manner in which information obtained on thermal analysis equipment should be published. Its recommendations are as follows.

To accompany each DTA, TG, EGA, and EGD or thermochemical record, the following information should be provided.

1. Substances should be identified (sample, reference, diluent) by a definite name and empirical formula or with equivalent composition data.

TABLE VI Magnetic Transition Temperatures Using ICTAC-Certified Magnetic Reference Materials GM 761

Material	Transition temperature (°C)
Permanorm 3	259.6 ± 3.7
Nickel	361.2 ± 1.3
Mumtal	403.0 ± 2.5
Permanorm 5	431.3 ± 1.6
Trafoperm	756.2 ± 1.9

2. A statement of the sources of all substances should be given, with details of their history, pretreatment, and chemical purity as far as these are known.

3. Measurement of the average rate of linear temperature change over the temperature range involved should be reported. Nonlinear temperature programs should be described in detail.

4. Identification of the sample atmosphere by pressure, composition, and purity is important and should be recorded: in particular, it should be stated whether the atmosphere was static, self-generated, or dynamic and whether it passed through or over the sample. Where applicable the ambient atmospheric pressure and humidity should also be specified. If the pressure is other than atmospheric, full details of the method of control should be given.

5. A statement of the dimension, geometry, and materials of the sample holder should be provided.

6. A statement of the method of loading should also be provided where this is applicable.

7. The abscissa scale should be identified in terms of time or temperature, and the location at which the temperature was measured should be provided. Time or temperature should be plotted so that it increases from left to right.

8. A statement of the method used to identify intermediates or final products should be given.

9. A faithful reproduction of all the original records should be available. With some instruments this is quite difficult, for the record is of mass versus time and temperature versus time, provided as two separate plots. In most modern units, however, this kind of problem can be avoided because the data are logged into a dedicated computer and automatically printed out as a graph, and the computer program then takes care of this kind of problem, giving the direct relationship between the measurement of the property selected and the temperature.

10. Identification of the apparatus used by type and/or commercial name is essential, together with details of the location of the temperature measuring thermocouples as already indicated. Again, this is important, as people using

equipment very often modify the commercial units to suit their particular needs.

There are, of course, other data that should be reported for specific techniques, and these are mentioned when dealing with these techniques.

F. Components of Thermal Analysis Instrumentation

The thermal analysis equipment enables the sample to be heated at a predetermined rate so that its temperature and one or more of its physical properties can be continuously measured and recorded. There are, therefore, three basic units involved: first, the measuring unit; second, the temperature control unit; and third, the recording unit. The simple arrangement of these components is given in Fig. 1. There is a measuring unit in all these systems and this must be in a particular position in the furnace. There must also be a system for passing a controlled atmosphere around the sample and a thermocouple proximate to the sample so that the temperature is accurately measured. The sample atmosphere is very important in liquid gas systems or in chemical reactions, and therefore, the measuring unit should be capable of operating under inert reactive gas or vacuum conditions. One point that is missed by most instrument manufacturers is the fact that the so-called vacuum systems require continuous pumping to eliminate the leak of gas into the system through various parts of the equipment. Units which hold their vacuum or which operate at controlled pressures other than atmospheric are built only by specialist instrumentation companies.

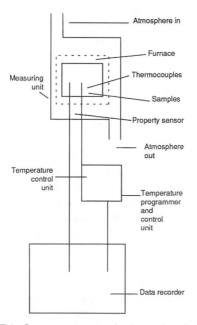

FIGURE 1 Component parts of a thermal analysis unit.

The temperature control unit can be a simple furnace and programmer. The fact that in some cases the instrumentation operates a control of the furnace temperature, rather than the sample temperature, is an important design feature which has already been noted. The recording unit receives a signal from the measuring unit and the temperature thermocouple. In most cases this is fed into a dedicated computer, and this allows the signals to be displayed in a variety of different ways, depending on the requirements of the operator.

In general, thermal analysis instruments may be divided into two groups: differential instruments, which contain the sample and a reference cell in similar environments and provide a difference signal of the properties; and derivative instruments, which note the change in the property signal versus the temperature as a proper derivative signal. However, it must be noted that the latter type of device is no longer required, as the property signal can be fed into a dedicated computer and the dedicated computer can now do the derivative calculations and provide a curve of both the property versus the temperature and the rate of change of that property versus the temperature.

1. Furnace Design

Furnaces have to be designed for particular applications since they have to be compatible with the measuring system and also with problems associated with each technique, for example, convection currents in TG. However, there are some general observations that can be made, first, regarding the thermal capacity of the furnace. If a large furnace is used, then one is going to have a range within the furnace at which a uniform temperature can be recorded. Small furnaces, however, will not have this uniform range of temperature, and the positioning of the sample in smaller furnaces becomes quite important. The second point about the size of the furnace is that large furnaces will take considerable time to reach a particular temperature and also take some time to cool down. The smaller the furnace, the easier it is to cool the temperature of the furnace back to ambient. The sample and reference material in differential measurements must also be subjected to the same temperature change. This generally involves a design feature involving both the unit in which the measurements are made and the furnace itself. One further point about furnace size and shape is that a long narrow furnace will generally give a larger uniform hot zone than a short wide furnace of similar volume. All instrumentation furnaces are electrically powered (although high-frequency inductive and infrared heating furnaces have been put on the market). The resistance wire is generally coiled around an insulating packing. The outside of the furnace is generally well insulated, although this has a bearing on the rate at

TABLE VIII   Upper Temperature Limits for Common Furnace Resistance Elements

Furnace winding(s)	Temperature limit (°C)	Atmosphere[a]
Nichrome	1000	A
Chromel A	1100	A
Tantalum	1330	B
Kanthal	1350	A
Platinum	1400	C
Platinum–10% rhodium	1500	C
Platinum–20% rhodium	1500	C
Kanthal super	1700	A
Rhodium	1800	C
Molybdenum	2200	D
Tungsten	2800	D

[a] A—an oxidizing atmosphere can be used (oxygen or air); B—a nonoxidizing atmosphere can be used (inert or vacuum); C—in these cases oxygen or air can be used at lower temperatures, but at higher temperatures an inert atmosphere is recommended; D—hydrogen should be used.

which the furnace will cool down. Because of the danger of a magnetic field interfering with the measurements of certain physical properties of the sample or system, the furnace should be noninductively wound: that is, it should have two similar windings carrying current in opposite directions so that their magnetic fields cancel, and also, the spacing of the windings should be decreased toward the end of the furnace to compensate heat losses. These points are made because, for special purposes, the operator might well require the construction of furnaces to a specific design. The resistance elements of the furnace control the temperature that can be obtained. Table VIII gives a range of resistance elements which, in theory, allows a temperature of 2800°C to be reached. Furnaces operating above a temperature of 1350°C should do so in an inert or reducing atmosphere. Graphite or silicon carbide resistance bars are used instead of metal bars in furnaces in which temperatures can go over 2000°C, and these are simple to operate, provided the restrictions on atmosphere control are noted.

The furnace should generally be mounted so that it can be moved easily and also put back in exactly the same position each time to minimize any errors due to the geometry of the system being altered by the movement which is necessary to load the sample in and to generally inspect the equipment. Where large furnaces are used, it is essential to have some kind of cooling system to allow the furnace to be programmed rapidly back to ambient temperature.

2. Temperature Programmers

Most thermal analysis results require a simple linear heating program, so that a typical instrument will allow pro-

gramming from ambient to a preset maximum temperature followed by the equipment being either switched off or programmed to cool. In some units natural cooling is used; in others there is the additional choice of maintaining the maximum temperature isothermally. The programmed linear cooling process can often be allowed to go to a preset minimum temperature, followed by the equipment being switched off, by programmed heating, or by maintenance this minimum temperature isothermally. In most modern equipment, cycling at a given rate between two preset temperatures can be achieved. It has also been found to be convenient to operate the equipment isothermally at a rapidly preset temperature or to temperature jump so that a number of isothermal intervals for a predetermined period are imposed upon the system or the sample. In other units very rapid quench cooling from a preset temperature can be achieved. It should be noted that the isothermal treatment of materials does not fall within the obvious definition of thermal analysis unless this is taken to mean that the material is heated at a zero rate, but most operators find it important because they wish to refer their heat treatment at a determined rate against more classical studies involving isothermal operation. In other applications of the work, for example, simulation of industrial processes, more complicated program temperatures are required, and with the aid of computer programming, this is generally possible. Since most thermocouples need a reference cold junction, this is now generally provided in the form of an electronic ice point reference unit.

3. Recording Systems

In the current design of thermal analysis equipment the recording system is such that the signals are received via the computer system. These signals can be obtained in the form a of a digital readout or in the form of an $X–Y$ plot. These data can be provided using software provided by the instrument manufacturer, by the dedicated computer manufacturer, or by the laboratory using the equipment. This enables the material to be subjected on the spot to various analytical processes which indicate at once the observations and the changes taking place.

G. Publications and Books on Thermal Analysis

The literature on thermal analytical methods is widely scattered because it is generally abstracted under the application heading rather than under the method heading. The American Chemical Society publishes its Chemical Abstracts, which embrace most published research papers, and all the articles which contain thermal analysis data then appear in *C.A. Select Abstracts on Thermal Analysis*. There are two international journals that publish papers on thermal analysis, *Journal of Thermal Analysis* and *Thermochimica Acta*. Other journals, however, have appeared which contain an increasing number of papers on thermal analysis; two of them are *Journal of Materials Research* and the *Journal of Analytical and Applied Pyrolysis*. The Japanese *Journal of Calorimetry and Thermal Analysis* contains material which is pertinent, and the various newsletters issued by national thermal analysis societies also contain material of interest (e.g., *ICTA Newsletter, Bulletin de l'Association Francaise de Calorimetrie et d'Analyse Thermique, NATAS Notes, Aicat Notizie*). These socieity publications serve a useful purpose by carrying information on recent developments, meetings, books, and so on that may be of interest. Every few years the proceedings of the International Conference on Thermal Analysis and Calorimetry provides an up-to-date picture of the stages of development of thermal analysis in different fields of science. These proceedings have been appearing since 1965. The journal *Analytical Chemistry* provides a 2-yearly review of highlights in the field of thermal analysis.

II. DIFFERENTIAL SCANNING CALORIMETRY AND DIFFERENTIAL THERMAL ANALYSIS

A. Introduction

In a group of techniques the changes in the heat content (enthalpy) or the specific heat of a sample are noted with respect to the temperature. The terms have been defined previously, but it is best to recall that differential thermal analysis (DTA) is a technique in which the temperature difference between a substance and a reference material is measured as a function of temperature while the substance and the reference material are subjected to a controlled temperature program. The "classical" DTA instrument is shown in Fig. 2. The record is the DTA curve, with the temperature difference (ΔT) plotted on the ordinate, with en-

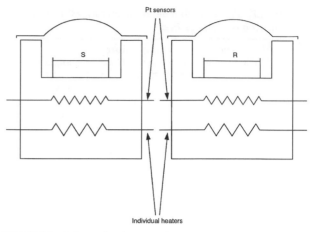

FIGURE 3 Schematic diagram of a power-compensated DSC unit.

dothermic reactions downward, and T or t on the abscissa, increasing from left ot right. The technique can be made quantitative by calibration and with the correct design of cells. Such units are labeled by instrument manufacturers as differential scanning calorimetry (DSC). The name, however, was reserved for a long time for a special method of obtaining calorimetric results, namely, by neutralizing the ΔT signal with auxiliary heaters and recording the energy required to neutralize the signal. Such equipment is shown schematically in Fig. 3 and would now be called a "power-compensated DSC." The quantity measured is the rate of change of enthalpy plotted against temperature. As already noted, the IUPAC thermodynamic convention is then to plot exothermic quantities (where the system loses energy) in a downward direction and endothermic quantities (where the system gains energy) in an upward direction. Although this is opposite to the recommended plotting of the ΔT signal, in DTA there is no real conflict, for ΔT is logically a drop in temperature for an endothermic process and a gain in temperature for an exothermic process. It is, of course, necessary to indicate in which direction endothermic and exothermic processes are shown on DTA or DSC plots. The alternative to power compensation DSC is heat-flux DSC, where the cells are designed so that the heat flux can be accurately used in the calibration process. Such a cell is shown schematically in Fig. 4. In a calibrated unit the enthalpy change in any process is proportional to the area under the curve.

B. Nomenclature for DTA

It should be noted that in DTA two cells are used, one containing the sample under investigation and the other a reference material of inert behavior. The two cells are

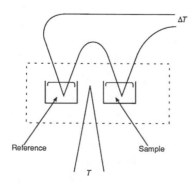

FIGURE 2 Schematic diagram of a DTA unit.

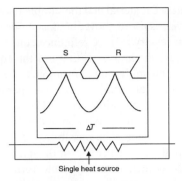

FIGURE 4 Schematic diagram of a Boersma-type DSC unit.

heated in a furnace subjected to a programmed increase or decrease in temperature. Two thermocouples, one in each cell, are "backed off" one against the other, and the difference in temperature (ΔT) between the two cells is plotted against time or temperature.

Nomenclature advice specific to DTA is as follows. The *sample* is the actual material investigated, whether diluted or undiluted. The *reference material* is a known substance, usually inactive thermally over the temperature range of interest. The term *inert material* is often used but is not recommended by the ICTA. The *specimens* are the sample and reference material. The *sample holder* is the container or support for the sample. The *reference holder* is the container or support for the reference material.

The *specimen-holder assembly* is the complete assembly in which the specimens are housed. Where the heating or cooling source is incorporated in one unit with the containers or supports for the sample and reference material, this would be regarded as part of the specimen-holder assembly. A *block* is a type of specimen-holder assembly in which a relatively large mass of material is in intimate contact with the specimens or specimen holders.

The *differential thermocouple*, or ΔT thermocouple, is the thermocouple system used to measure the temperature difference. In both DTA and TG the *temperature thermocouple* is the system used to measure temperature; its position with respect to the sample should be stated.

The *heating rate* is the rate of temperature increase (degrees per minute); likewise, the cooling rate is the rate of temperature decrease.

In DTA it must be remembered that although the ordinate is conventionally labeled ΔT, the output from the ΔT thermocouple will in most instances vary with temperature, and the measurement initially recorded is the e.m.f. output, E—that is, the conversion factor, b, in the equation $\Delta T = bE$ is not constant since $b = F(T)$, and a similar situation occurs with other sensor systems. The data fed into the computer work station, however, are corrected so that the plot presented is that of ΔT versus T.

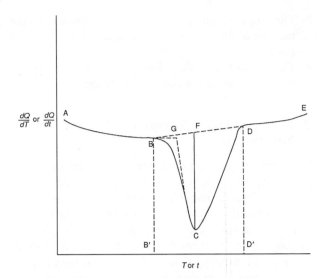

FIGURE 5 Formalized DSC signal. Note that in DTA the signal is ΔT instead of dQ/dT.

All definitions refer to a single peak such as that shown in Fig. 5. Multiple-peak systems, showing shoulders or more than one maximum or minimum, can be considered to result from the superposition of single peaks.

The baseline (AB and DE) corresponds to the portion or portions of the DTA curve for which ΔT is changing only slightly.

A peak (BCD) is that portion of the DTA curve which departs from and subsequently returns to the baseline.

An endothermic peak or endotherm is a peak where the temperature of the sample falls below that of the reference material; that is, ΔT is negative.

An exothermic peak, or exotherm, is a peak where the temperature of the sample rises above that of the reference material; that is, ΔT is positive.

The peak width (B′D′) is the time or temperature interval between the point of departure from and the point of return to the baseline. There are several ways of interpolating the baseline, and that given in Fig. 5 is only an example. The location of points B and D depends on the method of interpolation of the baseline.

The peak height is the distance, vertical to the time or temperature axis, between the interpolated baseline and the peak tip (C, in Fig. 5).

The peak area (BCDB) is the area enclosed between the peak and the interpolated baseline.

The general code of practice for recommending thermal analysis should be supplemented as follows for DTA.

1. Wherever possible, each thermal effect should be identified and supplementary supporting evidence shown.
2. The sample weight and dilution of the sample should be stated.

3. The geometry and materials of thermocouples and the location of the various thermocouples should be given.
4. The ordinate scale should indicate the deflection per degree Celsius at a specified temperature.

C. Position of the "Temperature" Thermocouple

It is normal to program the temperature to produce a linear heating rate. This, however, is not a sufficiently precise statement. One can use either the thermocouple in the reference material or the thermocouple in the sample as the indicator of temperature. In many systems the sample thermocouple is advocated as the indicator of the temperature. In the model set up by Cunningham and Wilburn there is a separate thermocouple to generate the T signal and two further thermocouples to give the ΔT signal. Using this model the same authors came to the conclusion that the temperature-measuring the thermocouple junction is to be placed in the sample. This point is also made by Mackenzie an Mitchell (1970) with reference to the above theoretical work of Wilburn and that of Grimshaw et al. (1945).

Two points can now be made. First, although it is generally recommended that the sample cell is the proper location for the temperature thermocouple, this means that a linear response is possible only while no reaction or phase change takes place. This is demonstrated in Fig. 6. Second, the peak height of the ΔT signal varies with the temperature at which T is measured.

In practice it is often found that the factors relating to the sample which are important are the particle size, the particle size distribution, the packing of the sample, the dilution of the sample with reference material (although most modern units dispense with this), and control of the atmosphere. The dilution of the sample represents an attempt to make the thermal characteristics of the two cells identical. There is, however, the possibility of reaction between the diluent reference material and the sample under investigation. However, as noted above, in most modern units dilution is not necessary and the reference cell is used empty. This is because the high sensitivity of the equipment requires only a few milligrams of sample to be studied.

Most commercial firms issue concise operating instructions and many application briefs. These should be referred to as often as needed before a deeper search of the literature is attempted. The possibility of using a single-cell model in the future is real, for the temperature of the "inert" could be calculated by some form of computer program and subtracted from the single-cell temperature reading to given the ΔT signal.

D. Theory of DTA

The main applications of DTA are (i) to describe the thermal decomposition and transitions occurring on heating a material through a programmed temperature range, (ii) to measure the heat of reaction, and (iii) to determine the kinetic parameters. The first is reasonably clear, the second requires an explanation, and the third, once the energy terms have been related to the weight changes, is similar to the treatment of kinetic data on the thermobalance.

In dealing with the measurement of the heat of reaction, the theories of DTA can be placed in two categories: (i) those which deal with heat transfer alone and (ii) those which deal with the reaction equation, that is, take into account the chemical nature of the reaction.

A simplified heat transfer theory is based on the method developed by Vold (1949).

The equation of heat balance for the cell containing the reaction is

$$CdT_1 = dH - K(T_3 - T_1)\,dt, \qquad (1)$$

and that for the cell containing the inert material is

$$CdT_2 = -K(T_3 - T_2)\,dt, \qquad (2)$$

where C is the heat capacity of each cell; T_1, T_2, and T_3 are the temperatures of the reactant, reference material, and block, respectively; K is the heat transfer coefficient between the block and the cell; and dH is the heat evolved by the reaction in time dt.

In writing these equations two assumptions are made: (i) the heat capacities of the two cells are the same and do not change during the reaction, and (ii) there is a uniform temperature throughout the sample at any instant. Otherwise a single value (T_1 and T_2) could not be written for the temperature of the material in the two cells.

Subtracting Eq. (2) from Eq. (1) gives

$$C(dT_1 - dT_2) = dH + K(T_1 - T_2)\,dt.$$

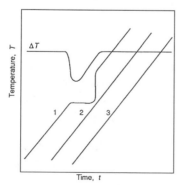

FIGURE 6 DTA plots: (1) T, sample cell; (2) T, reference cell; (3) T, third thermocouple. Scale displaced in 2 and 3.

The ΔT signal is, in fact, $T_1 - T_2$, so

$$C d\Delta T = dH - K \Delta T \, dt. \tag{3}$$

To determine the total heat of reaction it is necessary to integrate from $t = 0$ to $t = x$:

$$\Delta H = C(\Delta T - \Delta T_0) + K \int_0^x \Delta T \, dt,$$

when $t = 0$, $T = 0$, and when $t = x$, $T = 0$:

$$\Delta H = K \int_0^x \Delta T \, dt \tag{4}$$

or

$$\Delta H = KS,$$

where S is the area of the DTA peak.

The influence of physical properties on the baseline can be considered more realistically by assigning different values of C and K to each cell and considering the simple case where there is no reaction, that is, $dH = 0$; then

$$C_1 \, dT_1 = K_1(T_3 - T_1) \, dt$$

and

$$C_2 \, dT_2 = K_2(T_3 - T_2) \, dt.$$

C_1 and K_1 refer to the reactant cell, and C_2 and K_2 to the reference cell. Rearrangement gives

$$T_1 = T_3 - \frac{C_1 \, dT_1}{K_1 \, dt}$$

and

$$T_2 = T_3 - \frac{C_2 \, dT_2}{K_2 \, dt}.$$

dT_1/dt and dT_2/dt represent heating rates and should be identical, that is.

$$\frac{dT_1}{dt} = \frac{dT_2}{dt} = \frac{dT}{dt}.$$

Then

$$\Delta T = T_1 - T_2 = \frac{dT \, (K_1 C_2 - K_2 C_1)}{dt \, (K_2 K_1)} \tag{5}$$

We now have three cases.

$$\frac{C_2}{K_2} = \frac{C_1}{K_2} \quad \text{and} \quad Ck \neq f(T).$$

this is demonstrated by the zero value of ΔT (Fig. 7).

$$\frac{C_2}{K_2} > \frac{C_1}{K_1} \quad \text{and} \quad Ck \neq f(T),$$

which gives a positive constant value of ΔT (Fig. 7).

$$\frac{C_2}{K_2} < \frac{C_1}{K_1} \quad \text{and} \quad Ck \neq f(T),$$

which gives a negative constant value of ΔT (Fig. 7).

If

$$\frac{C_2}{K_2} < \frac{C_1}{K_1} \quad \text{but} \quad Ck = \text{linear } f(t)$$

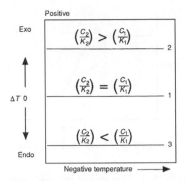

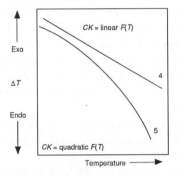

FIGURE 7 DTA baseline behavior according to Vold. Note that in the top plot CK is not a function of temperature; and in the bottom plot (C_2/K_1).

or if

$$\frac{C_2}{K_2} < \frac{C_1}{K_1} \quad \text{but} \quad Ck = \text{quadratic } f(t),$$

then the curved plot in Fig. 7 results. If, however, $C_2/K_2 > C_1/K_1$, the slopes are in the opposite direction.

For a more comprehensive treatment the publications of Wilburn and his coauthors should be consulted.

The practical tests of the use of DTA equipment in this way are to check whether the peak area is proportional to the quantity of material under examination and also to check the area under the peak for materials of a known heat of reaction. If the equipment responds properly to calibration tests of this kind, then it would seem, within the limits of the calibration range, to be proper to use it as a scanning calorimeter.

III. THERMOGRAVIMETRY

A. Introduction

As already noted, a *thermobalance* is an apparatus for weighing a sample continuously while it is being heated or cooled. The *sample* is the actual material investigated, whether diluted or undiluted. The *sample holder* is the container or support for the sample. Again, as already mentioned, in thermal analysis certain experimental items

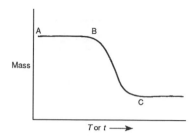

FIGURE 8 Formalized TG curve.

should be reported. The following additional details are also necessary in the reporting of TG data.

A statement must be made concerning the mass and the mass scale for the ordinate. The actual weight of the sample is reported, so a downward trend represents a mass loss. Additional scales, for example, mass loss, fractional decomposition, and molecular composition, can also be used according to the calculations which are subsequently to be made.

If derivative TG is employed, the method of obtaining the derivative should be indicated and the units of the ordinate specified.

The definitions used in TG can be illustrated by reference to the single-stage process illustrated in Fig. 8. Reference to multistage processes can be considered by thinking of the multistage process as being a series of single-stage processes.

A *plateau* (AB in Fig. 8) is that part of the TG curve where the weight is essentially constant.

The initial temperature, T_i (B in Fig. 8) is the temperature (on the Celsius or Kelvin scales) at which the cumulative weight change reaches a magnitude that the thermobalance can detect.

The final temperature, T_t (C in Fig. 8) is that temperature at which the cumulative weight change reaches a maximum. The reaction interval is that temperature difference between T_f and T_i as defined above.

B. Design Factors

The basic instrumental requirements for TG are a precision balance, a furnace capable of being programmed through a required regime of temperature change, and a computer workstation capable of programming the furnace, recording the weight change, and processing the data (see Fig. 9).

The essential requirements of an automatic and continuously recording balance are similar to those of an analytical balance and include accuracy, sensitivity, reproducibility, and capacity. In addition, a recording balance should have an adequate range of automatic weight adjustment, a high degree of mechanical and electronic stability, and a rapid response to weight changes and be unaffected by vibration.

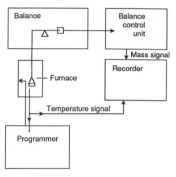

FIGURE 9 Schematic diagram of a thermobalance and control unit.

Two weighing systems need to be noted, namely, deflection and the null-point balances. There is a variety of deflection balances that can be constructed—beam type, helical spring, cantilevered beam, torsion wire, etc.—but they suffer in that the sample under observation will not remain in a fixed position in the furnace. For this reason most units employ a null-point balance.

In the use of null-point balances, a sensor must be used to detect the deviation of the balance beam from its null position, and a variety of methods may be used to detect deviations from the horizontal or vertical norm. One common system would is make use of the varying intensities of a light source impinging upon a photoelectric cell. The usual arrangement incorporates a light source, a shutter or mirror, and either single or double phototubes. The displacement of the shutter attached to the balance beam (or spring) intercepts the light beam, thus either increasing or decreasing the light intensity acting on the phototube. The resulting change in current magnitude from the phototube is then used to restore the balance to its null point.

Furnace design features and temperature measurement have already been discussed.

IV. EVOLVED GAS ANALYSIS

A. Introduction

In EGA the gases evolved from the decomposition of materials in a thermal analysis unit are analyzed. There are many methods of analyzing gases. In the past, specific chemical methods have been favored, but now instrumental methods based largely on mass spectrometric methods, chromatography, or infrared spectroscopy are generally practiced. In certain applications, however, specific chemical analysis is still used. In thermal analysis there must be an interface between the heat-treated sample and the gas detection unit. Gas analysis is rarely used in such cases by itself but more commonly combined with TG or DTA.

In an earlier section details of reporting thermal analysis were noted. In EGA, the following additional details are necessary.

1. The temperature environment of the sample during reaction should be clearly stated.
2. The ordinate scale should be identified in specific terms where possible. In general, increasing concentration of evolved gas should be plotted upward. For gas density detectors, increasing gas density should also be plotted upward.
3. The flow rate, total volume, construction, and temperature of the system between the sample and the detector should be given, together with an estimate of the time delay within the system.
4. Location of the interface between the systems for heating the sample and detecting or measuring evolved gases should be noted.
5. In the case of EGA, when the extact units are not used, the relationship between the signal magnitude and the concentration of species measured should be stated. For example, the dependence of the flame ionization signal on the number of carbon atoms and their bonding as well as on the concentration should be given.

B. Methods of Analysis

The methods of gas analysis are numerous, and any attempt at classification cannot be complete. With regard to present-day use, however, the following methods may be cited.

1. Chemical analysis, usually based on absorption of the gas or a component of the gas mixture in solution.
2. Mass spectrometric methods.
3. Infrared spectroscopy.
4. Gas chromatography.

Applied to thermal analysis there is usually the problem to consider as to how best to interface the gas detector or analysis unit with the rest of the equipment. There are also problems which may arise when the data are used for some specialized purpose, for example, to obtain kinetic data. The method of chemical analysis is rarely practiced in commercial units.

The use of an appropriate interface to sample the gas stream coming from the material under heat treatment must usually conform to the layout shown schematically in Fig. 10.

The following particular points about interfacing should be noted.

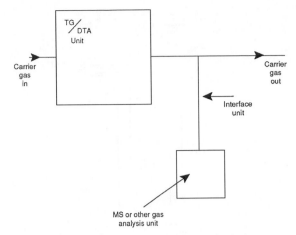

FIGURE 10 Schematic diagram showing the use of an interface in evolved gas analysis.

1. The thermal analysis unit may operate at 1 atm, but some gas analysis units, particularly mass spectrometers, operate at a very high vacuum.
2. Diffusion in the interface device may be a complicating factor.
3. Some problems of interfacing may be applicable to one particular method of gas analysis, and rather than attempt to solve a very difficult problem, it may be more convenient to choose another form of gas detector.
4. To eliminate side reactions it is necessary to put the interface as close as possible to the decomposing sample.
5. Carbon monoxide presents problems in mass spectrometry. This is because nitrogen has the same mass number as carbon monoxide, and in most mass spectrometers "ghost" peaks of nitrogen may be present. The solution is probably to use infrared detector devices for analysis of carbon monoxide.
6. Water presents problems of analysis. This is mainly because of its persistent adsorption and the difficulty in degassing it. This problem is apparent in the use of mass spectrometers and also in gas chromatography.

Prout and Tompkins (1946) devised a simple method for measuring the pressure of gas produced from a decomposing solid. They kept the total pressure very low, which allowed the kinetics of decomposition to be determined. The equipment is shown in Fig. 11, and it has been successfully applied to decomposition of potassium permanganate.

C. Mass Spectrometric Methods

In Table IX a distinction is made between *mass spectrometric thermal analysis*, in which the sample is actually located in the mass spectrometer, and *mass spectrometry coupled to either DTA, TG, or both.* The latter type is most often used by commercial instrument manufacturers. The

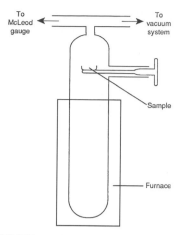

FIGURE 11 A Prout and Tompkins unit.

main problem is that of interfacing, as the mass spectrometer operates at a high vacuum, while decompositions are most often carried out at 1 atm.

The methods for the use of mass spectrometers in conjunction with thermal analysis units are as follows.

1. Collect a sample and take it to the mass spectrometer.
2. Use an interface.
3. Weigh the product gas.
4. Put the sample in the mass spectrometer.

1. Collection of Gas Samples Separately

The simplest scheme is simply to collect gas samples in glass vials and then present each vial to the mass spectrometer. If the mass spectrometer is being used for a wide variety of studies, this is the most economical approach. It has the following advantages.

1. The equipment is simple.
2. Its use is economical.
3. It is suitable for determination of kinetics. By this method the kinetics of individually evolved gas species can, in principle, be related to available surface reaction sites (not widely exploited).
4. The overall reaction can be deduced if all gas species involved are detected and evaluated.

2. Use of the Mass Spectrometer to Weight the Product Gas

In this method the gas from the reaction vessel at vacuum or at a known pressure is expanded into a large evacuated known volume. The sampling device then takes off the measured aliquot at this low acceptable pressure into the mass spectrometer. The calculation is based on the use of the Gas Laws. It has been used extensively by John Dollimore and co-workers.

A typical use of this method is a study of carbon oxidation kinetics by a temperature jump method. It should be used as a TG unit when the sample is at extremely high temperatures. This avoids complications in conventional TG equipment when used at very high temperatures.

3. Samples Placed in the Mass Spectrometer

This method has been used by Gallagher (1978) and by Price *et al.* (1980). In these units the temperature control of the sample and temperature programming is provided. The method decomposes the sample under a very high vacuum. The product gases have little time to decompose, so side reactions are eliminated. It should be noted that because of these unconventional features the results should not necessarily coincide with conventional data. The product gases are analyzed by mass number, but the solid residue cannot be determined *in situ.*

In the time-of-flight mass spectrometer equipment described by Price *et al.* (1980) the sample is decomposed in an open tube with close "line-of-sight" access to the ion source of the mass spectrometer. The sample is subjected to a linear temperature rise, with all gases rapidly removed using a powerful diffusion pump. The pressure in the ion source is always lower than 10^{-4} nm^{-2}. The ion current at the appropriate mass number is then proportional to the rate of gas evolution. If this is the only gas, it is also proportional to the mass loss on decomposition.

D. Infrared Spectroscopic Methods

Infrared spectroscopy methods are used for water vapor, CO, and CO_2. These are especially difficult gases to analyze accurately on a mass spectrometer. The infrared spectroscopy methods is also especially suited for on-stream

TABLE IX Mass Spectrometric Thermal Analysis

	Mass spectrometric analysis	Mass spectrometric thermal analysis	Mass spectrometer coupled to DTA or TG
Pressure control	That of mass spectrometer—No	That of mass spectrometer—No	Variable—Yes
Temperature control	Sample dependent—Some	Variable—Yes	Variable—Yes
Rate control	Sample dependent—No	Variable—Yes	Variable—Yes

analysis. It is not restricted to the above gases, of course, but as noted there is special merit in applying the method to them.

In an infrared radiometer, the output from an infrared source is split to pass through the sample and reference effluent streams, and the radiation intensities are then compared. The differential output is a measure of the extra radiation absorbed by the evolved gas in the sample stream. Such a system is not specific to one gas, and the response factor varies with the absorption coefficients of the component gases.

Infrared spectroscopy may be applied to EGA in two ways.

1. *Continuous monitoring of specific products.* This is a development of the infrared radiometer, in which the detector is sensitized to only one component, either by including filters (a nondispersion analyzer) or by operating at a single wavelength (a dispersion analyzer). A series of such analyzers can be used, each detecting one gas. Commercial units are available to detect CO_2, CO, and H_2O. Such methods can be used to establish kinetic parameters associated with the evolution of these gases.

2. *Special analysis of evolved gases.* When the composition or identity of the components in a gas stream needs to be established, then a spectral analysis of evolved gases is required. Obtaining such data is relatively easy but the method is noncontinuous with respect to time.

E. Gas Chromatographic Methods

There are numerous descriptions of the coupling of gas chromatography with thermal analysis units. Chiu (1968, 1970) describes a TG unit coupled with a gas chromatograph. In the use of gas chromatography the gas detector device used has to be suitable to match the gases evolved; otherwise they will escape detection. The three most commonly employed are thermal conductivity detectors, gas density detectors, and ionization detectors. The restriction on the employment of gas chromatography coupled with thermal analysis units is that the analysis is intermittent and not continuous.

V. THERMOMECHANICAL METHODS

A. Introduction

Thermodilatometry (TDA) has been defined as a technique in which a dimension of a substance under a negligible load is measured as a function of temperature while the substance is subjected to a controlled temperature pro-

gram. The record is termed the thermodilatometric curve. The dimension should be plotted on the ordinate, increasing upward, and T or t on the abscissa, increasing from left to right. Linear TDA is then the measurement of one dimension of a solid "form" against altering temperature. Volume TDA measures the change in volume of the solid or other phase versus the temperature program to which it is subjected. It should be noted that this means that any measure of change of density versus temperature of treatment falls within the definition of TDA.

Thermomechanical analysis (TMA) is a technique in which the deformation of a substance under a nonoscillatory load is measured as a function of temperature while the substance is subjected to a controlled temperature program. The mode as determined by the type of stress applied (compression, tension, flexure, or torsion), which should always be stated.

As in other cases, when reporting TMA the temperature environment of the substance should be clearly stated. The type of deformation (tensile, torsional, bending, etc.) and the dimensions, geometry, and materials of the loading elements should be noted. The ordinate scale should also be identified in specific terms where possible. For static procedures, increasing expansion, elongation, or extension and torsional displacement should be plotted upward. Increased penetration or deformation in flexure should be plotted downward. For dynamic mechanical procedures, the relative modulus and/or mechanical loss should be plotted upward.

In dealing with change in the dimensions of a sample with temperature the measurements relate to the deformation and strength of the sample. Expansion in a solid or a liquid is indicative of a decrease in surface energy. Thus, the expansion of a solid in an adsorption process may be interpreted as a decrease in surface energy.

A practical distinction between TMA and TDA is simply that in TMA some kind of stress or load is applied to the test material, while in TDA no load or stress is required. The same basic equipment may be used. In commercial equipment the temperature may go as low as $-170°C$ or as high as $1000°C$.

B. Apparatus

A wide variety of equipment is available, which includes the measurement of various properties. The list given here, with brief descriptions, is illustrative rather than comprehensive. Logically volume changes and measurement of density should be discussed under this heading. However, most commercial TMA units note only a single dimensional change on fabricated units (it is difficult to make TMA measurements on powders). The measurement of density involves techniques which would be difficult to

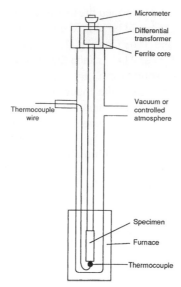

FIGURE 12 Schematic diagram of a dilatometer unit operating in a thermal analysis mode.

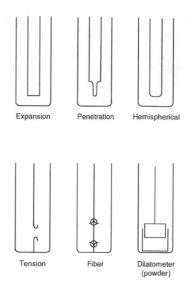

FIGURE 13 TMA probe configurations.

perform as a function of temperature. Again, logically the term *TMA* covers mechanical effects other than change in a linear dimension. Two techniques are often described: (1) TDA and (2) TMA. A third technique, dynamic mechanical analysis (DMA), is discussed in Section VI.

1. The Thermodilatometry Apparatus

A simple TDA unit is shown in Fig. 12. This simply measures expansion or contraction of a sample under test with temperature or time. The rod used in this equipment is made of a material suitable for use in the temperature range required: up to 1000°C, fused silica; up to 1200°C, porcelain; up to 1800°C, sapphire; and up to 2000°C, graphite, molybdenum, or tungsten. In the latter case a reducing or an inert gas (argon) must be used.

The coefficient of expansion against temperature should be plotted. In the simple equipment noted above, an extension rod can be used. Alternative methods of measuring the extension (or contraction) are as follows.

1. A precursor micrometer.
2. A cathetometer.
3. An interferometer.
4. From X-ray diffraction measurements.

2. The Thermomechanical Apparatus

Thermomechanical equipment represents generally an extension of the simple dilatometer principle, in which expansion or contraction under a load or a deforming stress is applied to a sample. In practice, TDA finds extensive application to ceramics and formed inorganic materials,

while TMA finds its most useful role in application to polymeric systems. Various TMA probe configurations available commercially are shown in Fig. 13.

VI. MODULATION TECHNIQUES

It has become increasingly popular to study modulation techniques. In such techniques there is a modulation of certain experimental parameters. Two methods are discussed here: dynamic mechanical analysis (DMA) and modulated DSC.

A. Dynamic Mechanical Analysis

In dynamic thermomechanometry the dynamic modulus and/or damping of a substance under an oscillatory load is measured as a function of temperature while the substance is subjected to a controlled temperature. The frequency response is then studied at various temperatures. Torsional braid analysis is a particular case of dynamic thermomechanometry in which the substance is supported on a braid. These are all sophisticated versions of thermomechanical methods. The word *dynamic* here, as noted above, means oscillatory and this term can be used as an alternative to modulation. In DMA the sample is oscillated at its resonant frequency, and an amount of energy equal to that lost by the sample is added in each cycle to keep the sample oscillation at a constant amplitude.

This technique measures the ability of materials to store dissipate mechanical energy on deformation. If a material, for example, is deformed and then released, a portion of the stored deformation energy will be returned at a rate which is a fundamental property of the material. That is, the sample goes into damped oscillation. For an ideal elastic material with a high Q (quality factor), the energy incorporated into oscillation will be equal to that introduced by deformation, with the frequency of the resultant oscillation being a function of the modulus (stiffness) of the material. Most real materials, however, do not exhibit ideal elastic behavior but, rather, exhibit viscoelastic behavior in which a portion of the deformation energy is dissipated in other forms such as heat. The greater this tendency for energy dissipation, the larger the damping of this deformation-induced oscillation. On the other hand, if this dissipated energy is continually made up (by an in-phase drive signal applied to the system), the sample will stay in continuous natural frequency (compound resonance) oscillation.

The two properties measured by DMA are the resonant frequency and energy dissipation, and these can be measured over a wide range of temperatures and moduli.

The resonant frequency obtained is related to the Young's modulus of the sample via the following equation:

$$E = \left[\frac{4\pi^2 f^2 (J - K)}{2w((L/2) + D)^2} \right] \left(\frac{L}{T} \right)^3,$$

where

E = Young's modulus,
f = DMA frequency,
J = moment of inertia of the arm,
K = spring constant of the pivot,
D = clamping distance,
w = sample width,
T = sample thickness,
L = sample length.

The energy dissipation obtained is related to properties such as impact resistance, brittleness, and noise abatement.

B. Temperature Modulated Differential Scanning Calorimetry (MDSC)

In this technique the linear temperature regime impassed in DSC is replaced by a sinusordal temperature modulation superimposed on a linear (constant) heating profile.

The program sample temperature ($T(t)$) in normal DSC is given by

$$T(t) = T_0 + \beta t,$$

where T_0 (K), β (K/min), and t (min) denote the starting temperature, linear constant heating or cooling rate, and time, respectively.

A sinusoidal modulation would then be represented by

$$T(t) = T_0 + \beta t + A_T \sin wt,$$

where A_T ($\pm$K) denotes the amplitude of the temperature modulation, w (s^{-1}) is the modulation frequency, and

$$w = \frac{2\pi}{p},$$

where p (s) is the modulation period. By using those points which lie on the linear temperature profile, a "normal" signal can be obtained together with the modulated signal. In both sets of experimental data the total heat flow at any point is given by

$$\frac{dQ}{dt} = Cp\beta + f(T_1 t),$$

where Q (J) denotes the heat, t (s) the time, Cp (J/K) the sample heat capacity, and $f(T_1 t)$ the heat flow from kinetic processes which are dependent on both temperature and time.

The linear temperature programmed DSC measures only the total heat flow. However, the sinusoidal heating profile gives the heat capacity data corresponding to the rate of temperature change. The heat capacity component of the total heat flow, $Cp\beta$, is called the reversing heat flow and the kinetic component, $f(T_1 t)$, is called the nonreversing heat flow. In one experiment the heat capacity can then be calculated using a discrete Fourier transformation by the relationship

$$Cp = K \frac{(Q_{amp})}{(T_{amp})} \frac{(P)}{(2\pi)},$$

where k denotes the heat capacity constant, Q_{amp} the heat flow amplitude, and T_{amp} the temperature amplitude.

The reversing component of the total heat flow signal ($Cp\beta$) allows the nonreversing component to be calculated using the relationship

nonreversing heat flow = total heat flow
                                            − reversing heat flow.

The technique finds advantages over normal DSC in calculation of the heat capacity and in determination of the glass transition point.

VII. SIMULTANEOUS TECHNIQUES

A. Introduction

It is becoming a common practice to apply two (or more) techniques of thermal analysis to the same sample at the same time. The most common application is that of TG and DTA. However, it should be noted that EGA is rarely used as a separate technique and is most often used in a simultaneous combination. In writing, the names of simultaneous techniques should be separated by the use of the word *and* when used in full and by a hyphen when abbreviated acceptably, e.g., simultaneous TG-DTA. Unless contrary to established practice, all abbreviations should be written in capital letters.

The other practice of thermal analysis in which more than one technique is used is *coupled simultaneous techniques*. This term covers the application of two or more techniques to the same sample when the two instruments involved are connected by an interface, for example, simultaneous DTA and mass spectrometry. The term *interface* refers to a specific piece of equipment that enables the two instruments to be joined together. In coupled simultaneous techniques as in discountinuous simultaneous techniques, the first technique mentioned refers to the first in time measurement. Thus, when a DTA instrument and a mass spectrometer are used together, then DTA-MS is correct.

The name *discountinuous simultaneous techniques* is also found, and this term covers the application of coupled techniques to the same sample when sampling for the second technique is discontinuous, for example, discontinuous simultaneous DTA and gas chromatography when selected portions of evolved volatiles are "sampled" on the gas chromatography unit. Another case of a simultaneous technique arises when two samples are subjected in the same furnace to two different thermal analysis techniques, for example, DTA and TG.

B. Advantages and Disadvantages of Simultaneous Techniques

The advantage of using simultaneous techniques lies in the nature of the samples being investigated. They are a reflection of the fact that in certain materials there is a distinct probability of different samples showing different signals. One can cite impure samples where the signal may vary with the nature and amount of the impurity. In other samples studied by simultaneous techniques such as pharmaceutical stearic acid or magnesium starate (known to contain other carboxylic acids), coal samples, and other natural products, the sample may vary from one region to another. Simultaneous tests on one sample may then represent a real advantage.

It must not be imagined that the combination of techniques should always produce the same results shown by the separate techniques, and this is particularly true for DTA and TG. Consider first the mass of material studied. Obviously the mass of the substance used in a TG run should be sufficient to note the change in mass accurately. However, the ideal mass of a sample used in DTA to achieve the best results may be much smaller than in TG. The sample holders in many DTA experiments are sealed. The use of such a technique in DTA-TG would

be to inhibit or prevent mass loss. Crucibles ideal for TG may not be ideal for DTA. Finally, the heating rate may be cited: to obtain good DTA experiments the heating rate should be reasonably fast; otherwise the peaks are broad and shallow. A slow heating rate, however, often gives the best results in TG experiments.

C. New Studies Made Possible by Simultaneous Techniques

There are cases where simultaneous techniques can lead to interpretations not possible using convention techniques separately. One such study is evaporation. In conventional DTA, with the sample in a closed crucible, such a study is impracticable. However, with the open crucible, evaporation in a TG-DTA unit shows a distinctively shaped endothermic peak; the TG for one-component systems demonstrates a zero-order process; and, using the Langmuir equation for evaporation, Price and Hawkins show that a vapor-pressure curve (partial pressure plotted against temperature) can be obtained. Calibration for temperature also becomes easy and more reliable. Combination of techniques in simultaneous units also allows interpretations of complex degradations to be studied. Thus it can be established that $Mg(NO_3)_2 \cdot 6H_2O$ melts at 90°C and then water is lost from the system, leaving an anhydrous solid salt at 350°C. This melts at 390°C and decomposes to the oxide at 600°C. DTA shows that all the processes are endothermic; gas analysis shows the loss of water and identifies the gases evolved in the final dissociation to the oxide. Hot-stage microscopy has been used to establish the melting processes.

VIII. THE ROLE OF THE COMPUTER

In some of the early designs for dedicated computers the computer was completely dedicated to the equipment and operated only with programs provided by the instrumentation company. To appreciate the role of the computer, however, it must be noted that it should play three roles.

1. To control the programs that alter the temperature regime.
2. To record the property being measured and demonstrate its variation with the temperature regime.
3. To interface this dedicated computer with computer programs or transfer the data to other computers and allow the operator in charge of the study to manipulate the data and derive appropriate parameters.

In the case of determination of kinetic data from raw experimental data, this usually means feeding such data (e.g., mass and temperature) into a spreadsheet and calculating the preexponential term and the activation energy in the Arrhenius equation.

Most modern dedicated computers fulfill these three roles and allow the operator to manipulate the basic experimental data by using his own spreadsheets or computer programs. It must be noted that instrument manufacturers supply their own programs to make sure that the use of the equipment in industry is made as easy as possible.

IX. APPLICATION OF THERMAL ANALYSIS TECHNIQUES

A. Introduction

The application of thermal analysis techniques is based on either thermodynamic considerations or the kinetics of change. It is not just the fact that the temperature is being changed that makes a choice of this kind necessary, but the kind of systems investigated. Some macromolecules are so big that phase changes which, in simple systems, would show only thermodynamic (equilibrium) features exhibit a kinetic factor. In the preceding survey certain techniques have not been reported or have not been described in detail. Only the main techniques available from more than one commercial manufacturer have been reported in some detail.

B. Thermodynamic Considerations

In a condensed system, if a phase change occurs in the heating mode where one phase is stable over a definite temperature range and the other phase is stable over, say, a higher temperature range, then the process of change at the transition temperature will be endothermic and reversible on the cooling mode with an exothermic character. This can be represented as

$$A(1) \xrightarrow[\text{heating}]{} A(2) \text{ endothermic}$$

$$A(2) \xrightarrow[\text{cooling}]{} A(1) \text{ exothermic}$$

This is true where both phase (1) and phase (2) are solids or phase (1) is solid and phase (2) is liquid. In the phase change from a liquid to a gas, the process is reversible, but because the gas is usually lost from the system, it is probably best to follow the process at different partial pressures of the vapor (p) when the relationship

$$\ln p = \frac{-\Delta H}{RT} - \text{constant} \tag{6}$$

holds, where R is the gas constant, T is the temperature of the phase change (K), and ΔH is the enthalpy of fusion. The above phase transitions can all be investigated by DTA or DSC.

If an amorphous (energy-rich) solid phase undergoes transition, however, then a stable crystal may result. The amorphorus or glassy phase is metastable, and the process is exothermic and irreversible. This can be represented as

Metastable   →   Stable form
(energy rich)   (crystalline)
Exothermic   and irreversible

Such changes are shown in both inorganic and organic systems.

The above are all examples of transitions which are of first order. However, second-order transitions can also be followed and determine the position of the baseline in DSC. This leads to the calculation of heat capacity from DSC measurements and is the basis for the establishment of the glass transition temperature (T_g) in polymers. A further extension based on thermodynamic factors is purity determination based on the lowering of the freezing point caused by the presence of impurities.

In chemical reactions certain features may be noted. Thus, carbonate decompositions are endothermic, and loss of water is also endothermic. Obviously, such processes involving weight loss can be studied using both TG and DTA/DSC. It may also be advantageous to study such processes under partial pressure (P) of the gaseous products when one may expect the relationship

$$\ln p = \frac{-\Delta H}{RT} + \text{constant}, \tag{7}$$

similar to the liquid–gas phase change, but here ΔH is the enthalpy of the reaction.

C. Kinetic Considerations

Where kinetic factors are important in studying systems using thermal analysis techniques, then the Arrhenius parameters in theory should be able to be estimated. The Arrhenius equation can be written

$$k(T) = Ae^{-E/RT}, \tag{8}$$

where $k(T)$ is the specific reaction rate constant and is a function of temperature, A is the preexponential term, E is the energy of activation, R is the gas constant, and T is the temperature (K). In the rising temperature mode we have

$$d\alpha/dt = k(T)f(\alpha), \tag{9}$$

where α is the fraction decomposed at time t, $d\alpha/dt$ is the rate of the reaction, $k(T)$ is the specific rate constant at temperature T, and $f(\alpha)$ is a score function of α describing the progress of the reaction. The temperature regime imposed on the system can be represented by

$$T = T_0 + bt, \qquad (10)$$

where T is the temperature at time t, T_0 is the starting temperature, and b is the heating rate. Combination of all these equations, noting that in linear temperature programming $b = dT/dt$, gives

$$k(T) = \frac{(d\alpha/dT)b}{f(\alpha)}, \qquad (11)$$

thus allowing k to be plotted against $1/T$ if $f(\alpha)$ can be identified, whence from the relationship

$$\ln k = \ln A - (E/RT),$$

both A and E can be calculated. The difficulty in obtaining $d\alpha/dT$ with sufficient precision has led to extensive studies into the use of an integral method. This entails the evaluation of

$$\int e^{-E/RT}\, dT,$$

and this is impossible to do analytically. Numerous methods of overcoming this difficulty have been suggested, usually based on the numerical methods of integrations or alternative similar expressions that can be analytically evaluated. The solid state is the only phase for which one has to state a prehistory, and an article by Flynn (1981) deals with this aspect. Related to these studies are the prediction of shelf life, especially of pharmaceuticals, and the proximate analysis of coal.

SEE ALSO THE FOLLOWING ARTICLES

CRITICAL DATA IN PHYSICS AND CHEMISTRY • HEAT TRANSFER • KINETICS (CHEMISTRY) • PHYSICAL CHEMISTRY • THERMODYNAMICS • THERMOMETRY

BIBLIOGRAPHY

Adamson, A. A. (1986). "A Textbook of Physical Chemistry," 3rd. ed., p. 401, Academic Press, Orlando, FL.

Atkins, P. W. (1982). "Physical Chemistry," 2nd ed., p. 298, Freeman, San Francisco.

Blaine, R. L., and Fair, P. G. (1983). *Thermochim. Acta* **67**, 233.

Boersma, L. (1955). *J. Am. Ceram. Soc.* **38**, 281.

Brown, M. E., Dollimore, D., and Galwey, A. K. (1980). Reactions in solid state. *In* "Comprehensive Chemical Kinetics" (C. H. Bamford and C. F. H. Tipper, eds.), Vol. 22, p. 99, Elsevier Science.

Charsley, E. L., Rumsey, J. A., and Warrington, S. B. (1984). *Anal. Proc.* XX, 5.

Chiu, J. (1968). *Anal. Chem.* **40**, 1516.

Chiu, J. (1970). *Thermochim. Acta* **1**, 231.

Cunningham, A. D., and Wilburn, F. W. (1970). "Differential Thermal Analysis" (R. C. Mackenzie, ed.), Vol. 1, p. 31, Academic Press, London.

David, D. J. (1964). *Anal. Chem.* **36**, 2162.

Dollimore, D., and Reading, M. (1993). Application of thermal analysis to kinetic evaluation of thermal decomposition. *In* "Treatise on Analytical Chemistry, Part I. Thermal Methods," 2nd ed., (J. D. Winefordner, D. Dollimore, and J. Dunn, eds.), Vol. 13, pp. 1–61, John Wiley & Sons, New York.

Dollimore, J., Freedman, B. H., and Quinn, D. F. (1970). *Carbon* **8**, 587.

Dunn, J. G., and Sharp, J. H. (1993). Thermogravimetry. *In* "Treatise on Analytical Chemistry, Part 1. Thermal Methods," 2nd ed. (J. D. Winefordner, D. Dollimore, and J. Dunn, eds.), Vol. 13, pp. 127–266.

Flynn, J. H. (1981). *In* "Thermal Analysis in Polymer Characterization" (E. A. Turi, ed.), p. 43, Heyden, Philadelphia.

Gallagher, P. K. (1978). *Thermochim. Acta* **26**, 175.

Galwey, A. K., and Brown, M. E. (1999). "Thermal Decomposition of Ionic Solids," p. 597, Elsevier, Amsterdam.

Grimshaw, R. W., Heaton, E., and Roberts, A. L. (1945). *Trans. Br. Ceram. Soc.* **44**, 76.

Hatakeyama, T., and Quinn, F. X. (1999). "Thermal Analysis," 2nd ed., p. 180, Wiley, Chichester.

Hill, J. O. (1991). "For Better Thermal Analysis," 3rd ed., ICTA.

Mackenzie, R. C., and Mitchell, B. D. (1970). "Differential Thermal Analysis" (R. C. Mackenzie, ed.), Vol. 1, p. 63, Academic Press, London.

Marti, E. (1972). *Thermochim. Acta* **5**, 173.

McGhie, A. R. (1983). *Anal. Chem.* **55**, 987.

McGhie, A. R., Chiv, J., Fair, P. G., and Blaine, R. L. (1983). *Thermochim. Acta* **67**, 241.

Norem, S. D., O'Neill, M. J., and Gray, A. A. (1970). *Thermochim. Acta* **1**, 29.

O'Neill, M. J. (1964). *Anal. Chem.* **36**, 1233.

Price, D., Dollimore, D., Fatemi, N. J., and Whitehead, R. (1980). *Thermochim. Acta* **42**, 517.

Prout, E. G., and Tompkins, F. C. (1946). *Trans. Faraday Soc.* **43**, 482.

Radecki, A., and Wesolowski, M. (1979). *J. Therm. Anal.* **17**, 73.

Rosenvold, R. J., Dubow, J. B., and Rajeshwar, K. (1982). *Thermochim. Acta* **53**, 321.

Shoemaker, D. P., and Garland, C. W. (1967). "Experiments in Physical Chemistry," 2nd ed., p. 177, McGraw–Hill, New York.

Vold, M. J. (1949). *Anal. Chem.* **21**, 683.

Watson, E. S. M., O'Neill, M. J., and Brenner, N. (1964). *Anal. Chem.* **36**, 1233.

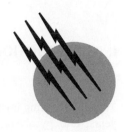

Thermal Cracking

B. L. Crynes
University of Oklahoma

Loo-Fung Tan
University of Oklahoma

Lyle F. Albright
Purdue University

GLOSSARY

Acetylenic Term describing hydrocarbons containing triple bonds, usually acetylene.

Acid gases Carbon dioxide and hydrogen sulfide, which are present in small quantities from the pyrolysis reactions.

Adiabatic Term describing an operation that occurs without the addition or removal of heat.

Endothermic reaction Reaction consuming heat as it proceeds.

Filamentous carbon Type of carbon that grows in long filaments or tubular structures on the inner walls of metal surfaces.

Hydrotreat To contact a hydrocarbon with hydrogen at moderate to high temperatures and pressures in order to perform hydrogenation reactions.

Pyrolysis gasoline Hydrocarbons formed during the pyrolysis reactions that are within the gasoline range of boiling points.

Transfer-line exchanger (TLX or TLE) Primary heat exchanger adjacent to the pyrolysis furnace.

THERMAL CRACKING, or pyrolysis, is defined as the decomposition plus rearrangement reactions of hydrocarbon molecules at high temperatures. Hydrocarbons ranging from ethane, propane, n-butane, naphthas, and gas oils are used as feedstocks in pyrolysis processes to produce ethylene plus a wide variety of by-products, including propylene, butadiene, aromatic compounds, and hydrogen. Steam is, as a rule, mixed with the hydrocarbon feedstock. Thermal cracking is sometimes referred to as steam cracking, or just cracking. The emphasis in this article is on the production of ethylene and the above-mentioned by-products.

I. INTRODUCTION

A. Historical

Thermal cracking investigations date back more than 100 years, and pyrolysis has been practiced commercially with coal (for coke production) even longer. Ethylene and propylene are obtained primarily by pyrolysis of ethane and heavier hydrocarbons. Significant amounts of butadiene and BTXs (benzene, toluene, and xylenes) are also produced in this manner. In addition, the following are produced and can be recovered if economic conditions permit: acetylene, isoprene, styrene, and hydrogen.

Ethylene and propylene are used industrially in large quantities for the production of plastics and high molecular weight polymers and as feedstocks in numerous other petrochemical processes. Before the manufacture of ethylene from light paraffins (separated from natural gas) or petroleum fractions, ethylene was produced in the laboratory, and for commercial use, from fermentation-derived ethanol. It was also produced commercially from coke oven gas as early as 1920 and for several years thereafter. The technology developed in the processing of coal and the resulting coal-derived hydrocarbons was the foundation, to a considerable extent, of thermal cracking processes that have evolved for feedstocks obtained from petroleum and natural gas. With the development of ever-larger refining operations, numerous petrochemical developments followed. The discovery of plastics, such as polyethylene, polypropylene, and polystyrene, seeded the demand for ethylene, propylene, and aromatic compounds. Considerable research was conducted in the 1980s and the 1990s to develop improved methods of producing ethylene and other olefins. Methane (main constituent of natural gas), coal, methanol, garbage, wood, and shale liquids have, for example, been used as feedstocks. Such feedstocks have found no commercial applications. The current pyrolysis processes and feedstocks will almost certainly not be replaced in the foreseeable future.

Ethylene production has increased many fold in the last 40 to 50 years. In the United States, from 1960 to 2000, ethylene production increased from about 2.6 to 30 million metric tons/year while propylene production increased from 1.2 to 14 million tons/year. The growth rates on a yearly basis have, of course, depended in this time period on economic conditions in both the United States and worldwide. In 2000, worldwide production of ethylene was about 88 million tons/year; the production capacity was 104 million tons/year. In 1960, about 70% of both the ethylene and the propylene produced was in the United States. Relative growth rates in the last few years of both ethylene and propylene production have been larger in Europe, Asia, and, more recently, the Near East. Currently, the United States produces only about 35% of the total ethylene and propylene. It should be emphasized that significant amounts of propylene are produced as a by-product in the catalytic cracking units of refineries. This propylene is sometimes separated and recovered as feedstocks to various petrochemical units. In the 1960s, studies were started relative to the interactions of reactor walls during pyrolysis reactions. More information on surface mechanisms follows later in this article.

II. MAJOR FEEDSTOCKS AND PRODUCTS

Feedstocks for various industrial pyrolysis units are natural gas liquids (ethane, propane, and n-butane) and heavier petroleum materials such as naphthas, gas oils, or even whole crude oils. In the United States, ethane and propane are the favored feedstocks due, in large part, to the availability of relatively cheap natural gas in Canada and the Arctic regions of North America; this natural gas contains significant amounts of ethane and propane. Europe has lesser amounts of ethane and propane; naphthas obtained from petroleum crude oil are favored in much of Europe. The prices of natural gas and crude oil influence the choice of the feedstock, operating conditions, and selection of a specific pyrolysis system.

Table I illustrates typical products obtained on pyrolyzing the relatively light feedstocks from ethane through butane, but significant variations occur because of the design and operating conditions employed with each light paraffin. The compositions of products obtained from naphthas, gas oils, and even heavier feedstocks differ to an even greater extent; the compositions of these heavier feeds vary over wide ranges. Tables II and III report typical

TABLE I Typical Primary Products from Light Feedstocks

Light feedstock	Product (wt%)			
	Ethane	Propane	n-Butane	i-Butane
H_2	3.7	1.6	1.5	1.1
CH_4	3.5	23.7	19.3	16.6
C_2H_2	0.4	0.8	1.1	0.7
C_2H_4	48.8	41.4	40.6	5.6
C_2H_6	40.0	3.5	3.8	0.9
C_3H_6	1.0	12.9	13.6	26.4
C_3H_8	0.03	7.0	0.5	0.4
$i\text{-}C_4H_8$				19.6
$i\text{-}C_4H_{10}$	{<0.2	{<0.8	{<1.9	20.0

TABLE II Typical Products from Heavy Feedstocks

	Product (wt%)		
Heavy feedstock	Naphtha	Gas oil	Vacuum distillate
CH_4	10.3	8.0	6.6
C_2H_4	25.8	19.5	19.4
C_2H_6	3.3	3.3	2.8
C_3H_6	16.0	14.0	13.9
C_4H_6	4.5	4.5	5.0
C_4H_8	7.9	6.4	7.0
BTX	10.0	10.7	18.9
C_5 to 200°C (not BTX)	17.0	10.0	—
Fuel Oil	3.0	21.8	25.0
$H_2 + C_2H_2 + C_3H_4 + C_3H_8$	2.2	1.8	1.4

product mixtures for these heavy feedstocks. These three tables, along with other information, suggest the following general guidelines:

1. When ethane is the feedstock, the highest yields of ethylene are achieved, often as great as 80%. When propane and heavier feedstocks are used, yields are, however, less than 50%.
2. Propylene is the major olefin obtained during isobutane pyrolysis; however, there is no known industrial unit that uses it as the feedstock. Propylene yields are often in the 12–16% range when propane, heavier normal paraffins, napthas, gas oils, and heavier petroleum feeds are pyrolyzed.
3. The heavier feedstocks produce appreciable amounts of butadiene; aromatic mixtures, commonly referred to as BTXs; and heavy nonaromatic compounds.
4. Coal oil and shale oils, containing appreciable amounts of aromatic compounds, result in correspondingly large amounts of BTXs.

TABLE III Typical Products from Nonconventional Feedstocks

	Product		
Feedstock	Coal Naphtha	Coal middistillate	Shale oil
H_2	0.8	0.7	—
CH_4	16	12	—
C_2H_4	23	14	20–22
C_3H_6	9	6	—
BTX	24	18	42–66
Fuel oil	24	47	—

III. FUNDAMENTAL AND THEORETICAL CONSIDERATIONS

A. Chemistry of Pryolysis (Gas-Phase Reactions)

Understanding the mechanisms and kinetics of pyrolysis reactions has steadily advanced along with advances in technology. In the mid-1940s, free-radical reactions, as opposed to molecular schemes, were proposed to be the primary reaction steps. Pyrolysis reactions are usually divided into initiation, propagation and isomerization, and termination steps. Until relatively recently, these reactions were often thought to occur mainly, if not exclusively, in the gas phase. Table IV explains the main gas-phase reactions in a highly simplified manner for the pyrolysis of propane. Initiation steps, which are generally rate-controlling steps, are Reactions (1) and (2) of Table IV. In Reaction (1), the ethane molecule decomposes at the C–C bond to form two methyl radicals. These radicals react via Reaction (2) to form methane (an undesired by-product) and ethyl radicals. The propagation reactions, Reactions (3) and (4), occur relatively rapidly to produce ethylene and hydrogen (the two desired products). In theory, these two reactions continue until all of the propane has reacted to form the desired products. The number of times that they repeat themselves is referred to as the chain length. Eventually, there are termination steps resulting in a net destruction of free radicals. Reaction (5), which occurs in the gas phase, is just one termination step that can occur.

Termination reactions include other reactions in addition to Reaction (5) of Table IV. There are numerous free radicals present in addition to ethyl and methyl radicals. These other free radicals can also combine or couple. The coupling reactions in the gas phase, including Reaction (5), are highly exothermic. To promote such coupling in the gas phase, a relatively heavy molecule (or third body) is likely needed to help dissipate the exothermic heat of reaction.

Termination reactions also occur when a free radical in the gas phase reacts or couples with a free radical on a solid surface. The solid coke formed as a by-product during pyrolysis is essentially pure carbon, which has numerous free radicals on its surface. The exothermic heats of such

TABLE IV Simplified Ethane Pyrolysis Model

(1) $CH_3CH_3 \rightarrow 2\ CH_3^{\bullet}$
(2) $CH_3^{\bullet} + CH_3CH_3 \rightarrow CH_4 + CH_3CH_2^{\bullet}$
(3) $CH_3CH_2^{\bullet} \rightarrow H_2C{=}CH_2 + H^{\bullet}$
(4) $H^{\bullet} + CH_3CH_3 \rightarrow CH_3CH_2^{\bullet} + H_2$
(5) $CH_3^{\bullet} + CH_3CH_2^{\bullet} \rightarrow CH_3CH_2CH_3$

termination steps are dissipated into the coke. As will be discussed later, such surface termination steps are the start of a sequence of reactions that produce more coke.

B. Kinetics of Pyrolysis

Operating conditions, and particularly temperature, have a major effect on the kinetics of pyrolysis. Typical operating conditions are as follows:

Temperature of gaseous reactants, 750–1000°C
Pressure, 2 to 6×10^5 Pa
Residence time of reaction mixture in reactor coil, 0.04–1.0 sec
Steam/hydrocarbon weight ratio, 0.25–1.0

A plot of a laboratory reactor yields versus conversion of propane is shown in Fig. 1. These conversions were obtained over a range of 700–850°C. Conversions increase rapidly with temperature for all feeds at a given residence time. Conversions depend on temperature exponentially, and so even more rapid rates of reaction occur at higher temperatures.

Especially in the past, the kinetics of pyrolyses have often been reported as being proportional to the concentration of the feed hydrocarbon raised to some power. In such cases, the reaction order tends to shift to higher values with increased conversions and temperatures. These oversimplified models fail to account for the multiple reactions and products, but are reasonably successful for predicting the reactions, especially those at low conversions.

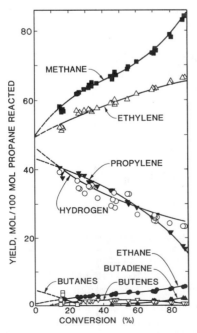

FIGURE 1 Propane conversion and yields.

Even though free-radical reaction schemes best represent the pyrolysis mechanisms, simple nth-order models for the pyrolysis of the lighter feedstocks have been proposed. The range of orders is from 0.5 to 2.0, and significant differences are often reported in the literature. The overall activation energies vary from about 50 to 95 kcal/g mol.

Pressure effects are, in general, not significant in the range of commercial pyrolysis interests. For hydrocarbon partial pressures of 0.2 to 2.0×10^5 Pa, few differences are seen. When pressures are increased to the range of 50 to 100×10^5 Pa, however, the global rate constants sometimes double.

In the past, reactor severity models were used for design purposes. These are models empirically relating some sort of reaction "severity factory" to temperature and time of reaction and/or concentrations. These can be thought of as a form of extent of conversion (severity) empirically related to reactor conditions. For example,

$$S = T \tau^a,$$

where S is the severity factor, T is the temperature, τ is the reactor residence time, and a is the empirical constant.

Steam is generally added for at least two reasons: first, it helps obtain quickly and then maintain the desired starting temperatures; second, it reduces the rate of coke collected on the inner surface of the reactor coil. Steam reacts slowly with deposited coke; nickel and iron on the coil surfaces, or present in the coke, catalyze this reaction.

The industrial furnaces are operated so that the temperature of the gaseous mixture increases steadily as it passes through the reactor coil. To obtain these increased temperatures, heat is transferred from the hot combustion gases surrounding the coil. In the inlet section of the coil, much of the heat transferred is employed to bring the gaseous mixture to the desired temperatures, and then pyrolysis starts at appreciable rates. In the latter portions of the coil, the heat transferred farther increases the temperature and, hence, increases the rates of reaction. Heat is, however, also needed to provide the large endothermic heats of reaction (for the initiation and propagation reactions, as shown in Table IV).

When lower temperatures are employed, longer residence times are required to obtain the desired conversions (or severity of pyrolysis). For the short residence time runs, in the 0.04- to 0.10-sec range, the rates of heat transfer are very large, requiring careful design of the equipment. Large temperature differences occur between the hot combustion gas surrounding the reactor coil and the reacting gases in the coil. There are substantial temperature differences across the walls of the coil and also between the inner surfaces of the coil and the reaction mixture. In those sections of the coil in which coke deposits on the inner wall, there are also substantial temperature differences between the coke and the reacting gases. Temperatures of the coil surfaces and of the coke are unfortunately not well measured or known. Yet, reactions occur on these surfaces, as will be discussed in more detail later.

Much improved understanding of both the kinetics and the mechanisms in the pyrolysis of light paraffins has been realized using mathematical models containing terms for the numerous gas-phase reaction steps. Models containing hundreds or even reportedly over a thousand steps have been developed. Values of the activation energy and the frequency factor are used for each reaction step. Such parameters can generally be found in the literature or can be approximately based on literature results. Such models with the aid of powerful computers can be solved to predict conversions, yields, and product composition when temperature profiles are employed. Some models often approximate with considerable accuracy plant and pilot plant data. A few models have even incorporated terms for surface reactions. These surface reactions produce at least some of the coke, carbon monoxide, methane, and probably other compounds. A major problem in incorporating surface reactions into a model is that surface temperatures are higher, and not well known, as compared to the gas temperatures.

Considering the initiation reactions, such as Reactions (1) and (2) of Table IV, they are obviously of most relative importance when the concentrations of the hydrocarbon feedstocks are highest, i.e., during the initial phases of pyrolysis. The propagation reactions experience their most rapid kinetics when the concentrations of the free radicals and of the hydrocarbon feedstock are both relatively high, i.e., during the intermediate stages of pyrolysis. The termination reactions become increasingly important during the final stages of pyrolysis when the concentration of the hydrocarbon feed approaches low values. In commercial reactors, the highest temperatures occur during the final stages.

Mathematical models for the pyrolysis of naphthas, gas oils, etc. are relatively empirical. The detailed analysis of such a feedstock is essentially impossible, and all heavier feedstocks have a wide range of compositions. Such heavy hydrocarbons also contain a variety of atoms often including sulfur, nitrogen, oxygen, and even various metal atoms. Nevertheless, certain models predict the kinetics of pyrolysis, conversions, yields, etc. with reasonable accuracy and help interpret mechanistic features.

C. Surfaces Reactions in Pyrolysis Coils

Laboratory results in the last several years have demonstrated the importance of reactions occurring on the inner surfaces of the pyrolysis coils and on the coke deposited on the coil surfaces.

1. Coke Formation

Coke, essentially pure carbon, is deposited on significant portions of the internal surfaces of all industrial coils. Surface reactions occur during the formation of this coke. The resulting coke is a highly undesired by-product for the following reasons:

1. The coke on the inner surface of the coil decreases heat transfer coefficients for the large amounts of heat that need to be transferred from the hot combustion gases in the furnace to the hydrocarbon-steam mixtures flowing through the coils. As a result of coke, smaller fractions of the heat of combustion are transferred to the reaction mixtures, and the already large costs of the fuel to the furnace are increased.
2. The coke deposited on the inner walls of the coil, and especially on the exit portion of the coil, increases the pressure drop of reaction gases as they flow through the coil. Hence, increased costs result, since the product gas mixture needs to be compressed as it enters the recovery and separation portion of the pyrolysis process.
3. At intervals often varying from 1 week to several months, a pyrolysis unit must be shut down in order to clean (or decoke) the coils. Such decokings sometimes require 1–2 days to complete. Obviously, as the rates of coking increase, the number of decokings per year also increases, causing the annual production rate of ethylene to decrease. Furthermore, utility and labor costs for each decoking are relatively expensive.
4. As will be discussed in more detail later, current methods of decoking contribute to decreased longevity of the metal coils. Further coking rates immediately after decoking are high for perhaps 1 day.
5. Ethylene yields are reduced. Laboratory tests at Purdue University when ethane was pyrolyzed indicate the coke formation is directly related to lower ethylene yields.

The following by-products are excellent coke precursors: acetylene, butadiene and other conjugated dienes, and aromatics. These compounds are produced mainly in the latter stages of pyrolysis. The following chemical sequences produce coke:

Mechanism 1 results in the formation of filamentous coke which has the following characteristics: is strongly attached to the metal surface of the coil; initially, has small diameters but high length; and on occasion, the filaments are hollow or tubular. In the first reaction step of formation, nickel or iron carbides are formed from a precursor such as acetylene. The carbide particles are separated from the surface as coke formation occurs at the base of the particle. Generally, the particle is incorporated at the top of the filament. Mechanism 1 obviously corrodes the coil.

Mechanism 2 results in the production initially of tar droplets suspended in the gas phase. The exact sequence of reactions differs depending on the hydrocarbon feedstock. For ethane and propane feeds, the sequence is as follows: acetylene and/or dienes combine to form simple aromatics; these aromatics react to produce polyaromatics and eventually low-viscosity tars; next, agglomeration steps form droplets suspended in the high-velocity gas stream; the low-viscosity droplets dehydrogenate because of the high temperatures to produce high-viscosity droplets and hydrogen. These droplets collide with and often collect on the solid surfaces (either the inner surface of the coil or on the coke that has already been formed or collected). Rough surfaces and particularly filamentous coke on the surface are excellent collection sites. If a low-viscosity tar droplet hits the surface, the droplets spread out on the surface. With higher viscosities, the droplets often collect on the surface and retain their spherical shapes; electron microscope photographs often show collections of droplets looking somewhat like clusters of grapes. In other cases, the high-viscosity droplets sometimes rebound from the surface like a ping-pong ball. Laboratory studies indicate that gravity affects the rate of collection of the droplets; more droplets collect on top surfaces as compared to bottom surfaces. After the droplets collect on a hot surface, all carbon–hydrogen bonds break within a few minutes to form carbon and hydrogen. As C–H bonds break, free radicals form on the surface of the coke.

Mechanism 3 has, as a first step, reactions between surface radicals on the coke and the acetylene, butadiene, and gaseous free radicals; reactions probably also occur with ethylene and propylene. Reactions with gaseous free radicals were discussed earlier as a termination step in the gas-phase reactions. When acetylene reacts with the surface radicals, aromatic structures are formed on the surface. When the C–H bonds on the surface later break, graphitic coke is formed. The cokes produced by both Mechanisms 1 and 3 tends to be highly graphitic. Microscopic photographs have shown that Mechanism 3 thickens filamentous coke and causes spherical coke particles formed by Mechanism 2 to grow in diameter.

When coils produced of high-alloy steels are used, the initial layer of coke contains an appreciable fraction of filamentous coke, as indicated by microscopic photographs of industrial coke samples. This coke is often rather porous with gas spaces around the filaments; porous coke obviously has relatively high resistances to heat transfer. Appreciable amounts of iron, nickel, and especially chromium are present in the initial layers of coke. The total amount of metals is probably in the 1–2% range. Coking rates are much higher while this coke is being formed as compared to coke produced later.

During latter stages of coking, the coke formed is generally solid and contains mainly cokes formed from Mechanisms 2 and 3. It contains less metal, perhaps in the 0.3–0.5% range; nickel, chromium, and iron are still present. On occasion, a metal fragment can be detected by the microscope with filamentous coke connected to it, i.e., a porous section of coke surrounded by solid coke.

Numerous laboratory and plant tests have been made in coils constructed of materials with nickel- or iron-free surfaces; e.g., quartz glass, silicon-coated and aluminum-coated steels, or ceramics. Filamentous coke was often completely absent, and the overall levels of coking were much reduced. Furthermore, the coking rates were essentially identical during the entire pyrolysis run.

2. Oxide Layer in Inner Surface of Coils in Furnaces

Extensive laboratory and plant data indicate that oxide layers form on the inner surfaces of high-alloy steel coils within several hours. Metal oxides and hydrogen form when metal atoms react with steam. These oxide layers may, with time, become as thick as 10–15 μ; layers of 1–5 μ often form in 1–8 h. These oxide layers, however, have a much different metal composition as compared to that of the starting steel. For a stainless steel with a composition by weight of about 45–48% iron, 30% nickel, 20% chromium, and 0.3–0.7% manganeses plus trace amounts of aluminum, titanium, niobium, carbon, etc., the surface composition, as measured by EDAX, may become approximately 65–75% chromium, 15–20% manganese, 5–8% iron, and 1–3% nickel. Occasionally, high levels of aluminum and titanium also form on the surfaces. Just below this oxide layer, a layer enriched with iron and nickel is formed. Clearly, there is a net diffusion of chromium, manganese, and sometimes aluminum and titanium toward the surface. Here, they are oxidized and form stable oxides. Because of the size and weight of the oxide molecules, further diffusion is essentially stopped. Iron and nickel form less stable oxides. The iron and nickel atoms tend to diffuse inward, forming an iron- and nickel-enriched layer below the oxide layer. Extensive experimental work performed at Purdue University clarifies the formation of these oxide layers on the surface of high-alloy steels. Gases with limited oxidative capabilities were tested at 500–1000°C. 50:1 mixtures of hydrogen:steam and of CO:steam were found to be effective gases, along with pure CO, for pretreating the steels and forming oxide layers. Incoloy 800, HK-40, and HP steels were all investigated. As measured primarily by EDAX analysis and to a limited extent by ESCA analysis, significant changes in the surface composition of Incoloy 800 occurred with as little as 0.25 h pretreatment with one of the above three gases at 800–1000°C. After about 8 h of pretreatment, the chromium content at the surface had increased in some cases by at least 3 times, and the manganese content had increased by as much as 50 times. Material balances indicate that some manganese diffused by as much as 50–100 μ to reach the surface oxide layer. The rates of diffusion in HK-40 and HP steels were, however, much slower than those in Incoloy 800.

Purdue University investigations also found that hydrogen-steam mixtures slowly gasify (or remove) coke deposited on high-alloy steel coupons at 800–1000°C, i.e., the coupons are decoked. Following such a decoking, the coupon was exposed to pyrolysis conditions and decoked. The resulting coke adhered poorly to the coupon. A similar surface was also obtained when Incoloy 800 was pretreated with CO. Assuming such surfaces were formed in industrial coils, coke formation or collection might be completely eliminated. Because of the high velocities of the gases in the coil, any coke deposited would probably be quickly eroded and removed from the surface. For high-alloy steels, the oxide layers, whether on coils or coupons, have physical properties very different than those of the base metal (or starting steel). First, the surface layer minimizes the formation of filamentous coke because of the depletion of nickel and iron at the surface. Second, and unfortunately, the surface layer is relatively brittle and rather easily spalled off (or lost). The surface layer tends to crack due to the creep in the coils plus thermal expansion and contraction. Relatively large temperature changes occur for the surface layers when decoking is started and when it is completed. Vibrations also often occur while a pyrolysis plant operates. When spalling occurs, relatively large portions of the oxide layer are lost from the surface, hence exposing the iron- and nickel-rich layer that promotes filamentous coke production. Typical decoking procedures contribute substantially to both spalling and corrosion problems and to the high rates of coke formation once pyrolysis operations resume. During decoking, oxygen is normally employed, which raises the level of metal oxides on the surface. A net diffusion of iron and nickel atoms toward or to the surface likely occurs during decoking as a result. The metal-containing particles that spall off the surface during decoking may accumulate in the bottom sections of the coil and are gradually fluidized during subsequent pyrolysis operations. Within several days of operation, some of the particles are trapped in the coke deposits.

There is no known experimental data on the fate of metal in the coke when the coke is gasified (forming carbon oxides) during decoking. Some of this metal may remain on the coil until pyrolysis operations are resumed.

In many pyrolysis units using ethane and/or propane as feeds, sulfur-containing additives are continuously added to the feed in small amounts. Typical additives include alkyl sulfides or disulfides, mercaptans, and hydrogen sulfide (H_2S). In the reactor coils, these compounds decompose, and at least some elemental sulfur is formed. Pretreatment of coils following decoking with desired amounts of these additives also reduces the level of coking when the pyrolysis operation is resumed. Such a pretreatment of the coil converts many metal oxides on the coil surface to metal sulfides; as a result, less coke is formed. It is highly important to use only optimum amounts of the additive. Too much additive results in increased amounts of coke and in excessive corrosion of the coils. Some nickel sulfides are liquids at coil temperatures and would be entrained in the gas stream. Optimum concentrations of the additive depend on the hydrocarbon used, operating conditions, and furnace used. Such additives are generally not used with naphtha or gas oil feedstocks, which generally contain sulfur compounds (often too many). Although ethylene producers hope their coils have service lives of 5–6 years, coils frequently need to be replaced sooner. That is not surprising since metal temperatures have tended to increase in the last 10 years in order to realize higher yields of ethylene. Metal temperatures in portions of the coil are currently often in the 1000–1075°C range. All of the following contribute to shorter coil life:

1. Creep in which vertical coils lengthen often by as much as several centimeters per year
2. Decreased wall thickness due to filamentous coke formation and spalling of metal oxide surfaces
3. Decreased chromium content of the high-alloy steels due to spalling and corrosion

High rates of coke formation plus frequent decokings accentuate all of the above problems. The following surface reactions occur during pyrolysis or decoking operations: oxidations to form metal oxides, reduction or partial reduction of oxides when they are contacted by the hydrocarbons in the gas phase, conversion of metal oxides to metal sulfides when sulfur-containing gases are present, conversion of metal sulfides to metal oxides due to reactions with steam or oxygen (during decoking). Such a complicated set of surface reactions promotes loss of metal atoms or compounds from the surface of the coil.

FIGURE 2 Dow Chemical ethylene plant. (Courtesy of Dow Chemical Company.)

IV. COMMERCIAL THERMAL CRACKING

A. General Process

Figure 2 is a photograph of a large, complex plant. Plants such as this contain reactor furnaces, distillation towers, heat exchangers, separators, dryers, compressors, and various other units as required by the specific feedstock and product distribution achieved. Figure 3 shows a simplified diagram of a plant.

1. Cracking Furnace and Reactor

For the primary step in the overall pyrolysis plant, the feedstock must be vaporized, if in liquid form, then mixed with steam, and finally preheated to the reactor temperature. When the feedstock (e.g., ethane or propane) is in gaseous form, vaporization is usually achieved by simple heat exchange with other product components such as condensing propylene. To vaporize liquid stocks such as naphthas, higher temperatures must be used. These feeds may be partially preheated before entering the furnace itself and then fully vaporized as they flow through the convective zone of the furnace. Typical furnace and tube geometries are shown in Fig. 4.

Typically, the tubing in the convection zone is placed in a "hairpin" fashion or series of connected U bends and is suspended horizontally. The steam that will be used as a diluent is also heated in the convection zone in other tubes. This superheated steam and the gaseous hydrocarbons are combined and further heated to ~700°C by the end of the convection section. This mixture then passes to the radiant zone, where the main pyrolysis reactions occur and where the temperature is increased to around 1000°C.

Temperature is the most important operating variable in pyrolysis reactions; the large amounts of heat must be

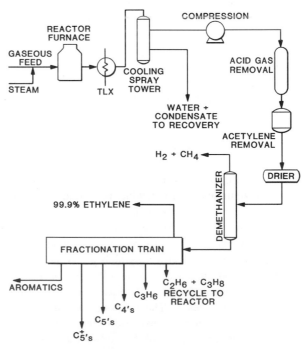

FIGURE 3 Simplified flow sheet of a general pyrolysis plant.

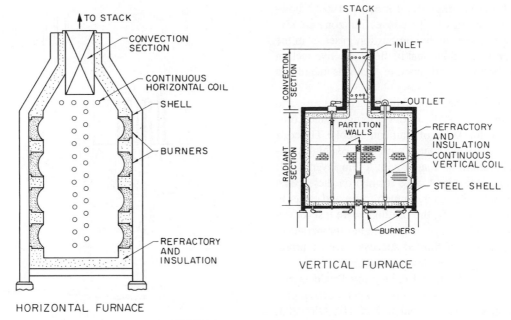

FIGURE 4 Furnace tube geometries.

transferred because most reactions are highly endothermic. As a general rule, higher temperatures (and the resulting lower residence times needed) are preferred for improved yields of ethylene, which is almost always the desired primary product. Hence, the reactor is normally fired as closely as possible to the limiting temperature of the reaction tubes.

Other variables of importance in designing these tubular pyrolysis reactors include the mass velocity (or flow velocity) of the gaseous reaction mixture in the tubes, pressure, steam-to-hydrocarbon-feedstock ratio, heat flux through the tube wall, and tube configuration and spacing. Pressure drop in the reactor is of major importance, especially because of the extremely high flow velocities normally employed.

The addition of steam to the entering feed provides several advantages, including heat sink which helps maintain higher temperatures and results in lower partial pressure of the hydrocarbons. The decreased partial pressure helps to minimize undesirable reactions. Increased amounts of steam result in increased steam-coke reactions, which result in the production of carbon oxides and in slower rates of coke formation on the metal surfaces. Typical steam requirements are shown in Table V.

High gas velocities (or high mass velocities) in the tubular reactor affect heat transfer through the boundary layer. There can be a rather large difference between the nominal process temperatures and tube skin temperatures. As already stated, there is a desire to take advantage of the tube material and to operate near its limit in order to re-

duce the actual residence time. As mass velocities increase within the reactor, higher heat transfer coefficients result. A countering effect of higher velocities, however, is increased pressure drop in the gas stream as it moves through the tube with a resultant increase in erosion at tube bends. There is a limit to the total pressure drop that can be tolerated with any given pyrolysis system from reactor tube inlet to outlet.

The reaction tubes (or coils, as they are often called) are positioned vertically in modern furnaces, but they are positioned horizontally in some furnaces of older design.

The furnaces are designed so that combustion of the fuel gas or liquid occurs around the reaction tubes. Temperatures of the combustion gases are often as high as 1200°C in the immediate vicinity of the tubes. Most heat is transferred by radiation, and the portion of the furnace in which the reaction tubes are located is often referred to as the radiant zone. The maximum permissible temperature for the metal in the tubes depends on the type of

TABLE V Typical Amounts of Steam for Commercial Pyrolysis

Feedstock	Weight ratio, steam/feed
Ethane	0.25–0.40
Propane	0.25–0.5
Naphthas	0.5–0.7
Gas oils	0.7–1.0

stainless steel, but is generally at most ~1100°C; however, metal temperatures in the convection zone are significantly lower. The effluent combustion gases from the radiant zone exchange their heat in the convective section to preheat the feedstock and generate steam, as mentioned earlier.

Major progress has been made in the last 20 years, and especially in the late 1990s, in developing coated coils, which prevent the formation of filamentous coke. The coils of 10–20 years ago were coated with surfaces highly enriched with aluminum or coated with a thin layer of silica. Much reduced levels of coke formed on these coils, and decokings were, in general, much faster and easier to accomplish. Unfortunately, the coatings were lost (or destroyed) rather quickly, especially if higher temperatures were used. Examples of limited success have occurred, but there was clearly a need for improvement.

In the late 1990s, Surface Engineering Products and Alon Surface Technologies each developed coating procedures for high-alloy steels, such as HK, HP, Alloy 803, etc., that are used in ethylene coils. As of 2001, such coated coils have been installed in about 30 furnaces worldwide. These coils have demonstrated in most, if not all, cases much improved operation, including much reduced levels of coke formation, fewer decokings, easier and faster decokings, reduced CO formation, and increased annual levels of ethylene production. Such improvements are directly related to the fact that filamentous coke formation is much reduced, if not eliminated. Furthermore, the adhesion between the coke and the coil surface appears to be rather low. Frequent tests on the coils have indicated that most coatings are still in good condition after extended times of operation. Most coils will likely have lives of 5 or more years, except for several coils operated at relatively high temperatures.

Considerable fuel is required to provide the high temperatures necessary for pyrolysis of large plants. Attention to burner design and furnace details, as well as the choice of an economic fuel of reliable supply and composition, are, of course, mandated. The recovery and use of energy are dictated by economics. The flue gases from combustion are vented through stacks with appropriate attention to air pollution caused by incomplete combustion, stack visibility, and nitrogen oxide emissions. In general, gaseous fuels are preferred over liquid forms because of the ease of flow control, more complete combustion, and higher flame temperatures.

Tubular reactors normally take one of three geometries. The most common consists of tubing of the same diameter connected by U bends. The internal diameters are often 4 in. The total length of one reactor coil may be 300 ft. Four to eight coils are commonly placed within a single furnace.

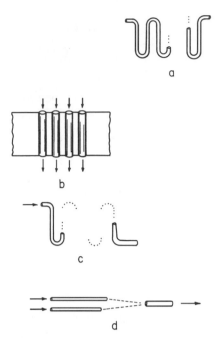

FIGURE 5 Tube designs: (a) same diameter, hairpins, which are the most common; (b) small diameter, single pass; (c) increasing diameter; (d) split coils; two small-diameter coils to one large.

Another configuration consists of the expanded and split coils that incorporate coil sections having two to three different diameters. Smaller diameter sections are connected (welded) by U bends to larger diameter sections. Often, two or three small-diameter sections arranged in parallel are connected to a single larger diameter tube (Fig. 5). Usually, the sections are positioned vertically in the furnace. The advantages of flexible residence times, reduction of pressure drops, and control of temperature in this more complex tube design are obvious. In a more recent design system, about 30–40 parallel and vertical tubes are fitted into a furnace; in this arrangement there are no U bends or any change of direction. The hydrocarbon-steam mixture simply passes through the radiant zone of the furnace and then exits into the transfer-line exchanger. These tube lengths are relatively short, up to 40 ft, and are of small diameter, up to 2 in. These furnace designs provide for extremely short residence times, less than 0.05 sec.

As indicated earlier, these configurations, almost without exception, are positioned vertically as they are suspended in the radiant zone of the furnace. Horizontally suspended tubes are more subject to buckling and sagging, especially at higher temperatures.

Multiple pyrolysis furnaces are employed in an industrial pyrolysis plant in order to maintain reasonably constant production levels, even when one furnace is shut down for decoking or maintenance repairs. The coils or tubes in a furnace or the transfer-line exchanger must

TABLE VI   General Pyrolysis Tube Conditions

Condition	Value
Tube temperature	Up to 1050°C (1922°F)
Reaction gas temperature	Up to 950°C (1742°F)
Tube inlet total pressure	3.7–6.4×10^5 Pa
Tube outlet pressure	1.7–2.4×10^5 Pa
Heat flux	20,000–30,000 Btu/h $\times$ ft^2
Residence times	0.15–0.5 sec
Mass velocities	Up to 30 lb/ft$^2 \times$ sec
Linear velocities	Up to 1000 ft/sec

normally be decoked every few months. Hence, decokings are arranged as much as possible to be staggered between the various furnaces to maintain reasonably constant production rates and to utilize the operational and maintenance personnel effectively. Typical operating conditions for a pyrolysis furnace are summarized in Table VI.

2. Transfer-Line Exchangers

Transfer-line exchangers (frequently referred to as TLXs or TLEs) recover most of the sensible heat in the product gas stream exiting from the pyrolysis furnaces. These exchangers are located as close as possible to the furnaces in order to cool the product stream quickly and to maximize heat recovery. Sometimes as much as 5–10% of the reaction can occur in the transfer line itself, which is the unheated section of tubing outside the furnace and just ahead of the TLXs. The temperature of the product stream leaving the furnace may often be 800–850°C, depending to some extent on the feedstock being used. Rapid cooling to 400–500°C is necessary to quench the reactions that destroy the more desirable products such as ethylene, propylene, and butadiene. The high-pressure steam generated in these exchangers is used to operate some of the gas compressors in the plant.

Coke formation is a problem in all TLXs, both in the cone of the TLXs and in the tubes. Coke increases the resistance to heat transfer so that less heat can be transferred (and less steam can be generated). Eventually, the thickness increases to a degree where the TLXs must be cleaned, usually manually. The reactor tubes and the TLXs must be designed and then operated such that the needs for decoking coincide for the two. As a rule, one TLX is provided for two reactor tubes.

The TLXs must be carefully designed to prevent excessive stresses in the shell or tubes when the TLX is heated to high temperatures. In addition, the design of the unit affects the amount of coke formed in the unit. Figure 6 illustrates a popular design, but others are also widely used.

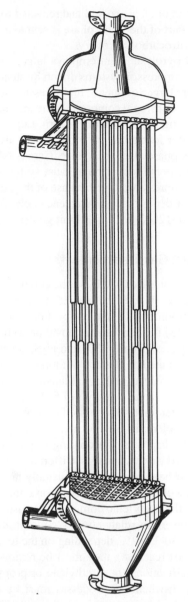

FIGURE 6   Transfer-line exchanger.

3. Product Gas Cooling and Compression

When the product gases exit from the TLX, they are still too hot for compression and must be cooled to essentially ambient temperatures. Part of the additional cooling is accomplished in some plants with a direct oil quench; the hot oil can be used to generate steam. Typically, in the final stage of cooling, a tower equipped with water sprays and a trap provides intimate contact between the water and the gaseous streams. The resultant water and liquid hydrocarbon phase (of heavier hydrocarbons) from this tower are separated in settling drums. Valuable products in both the water and oil phases are recovered and processed. The water phase is often air cooled and recycled to the scrubbing tower. Part of the water phase is sent to a distillation tower for hydrocarbon recovery.

A typical pyrolysis plant employs large, complicated, centrifugal compressor systems driven by steam turbines, which require large volumes of high-pressure steam. Four- or five-stage centrifugal compressors are the norm with interstage cooling of the product stream. Liquid hydrocarbons and water formed in the cooling steps are separated from the gas phase after each stage. Care is taken to prevent high temperatures in the exhaust systems from any particular compression stage because of the propensity to form polymer deposits from butadiene or olefins. Temperatures of 110–120°C are upper limit guides.

4. Product Gas Treatment

The hydrocarbon gases that leave the compressor are usually subjected to three additional steps: (1) the removal of acid gas components, (2) the removal of acetylenic compounds, and (3) further water removal or drying.

The acid gases, usually CO_2 and H_2S, are removed by scrubbing with diethanolamine or monoethanolamine solutions and possibly an additional caustic treatment. Older processes used caustic solutions of 5–15 wt% NaOH followed by a water wash. Of course, the spent caustic creates a disposal problem; it must be neutralized with acid and then properly disposed of according to prevailing pollution and hazard waste standards. Different column configurations have been proposed, but usually large scrubber towers with well over 30 valve-type trays are used.

Although the acetylenic compounds, mostly acetylene itself, are present in relatively low concentrations (from 2000 ppm up to 33 wt%, depending on the feedstock and temperature of reaction), they must be removed to meet product specifications; most ethylene or propylene used in various petrochemical processes must be essentially free of acetylenic compounds (1–5 ppm). These compounds are removed when present in small amounts by hydrotreating in an adiabatic reactor utilizing catalysts of nickel-molybdenum or colbalt-molybdenum supported on a high-surface-area porous substrate such as alumina. A second unit of cleanup stage treatment using a much more sensitive and expensive palladium-supported catalyst follows. There are certain liabilities with this hydrogen treatment step. Some desirable compounds, including ethylene and butadiene, are partially hydrogenated. Care must be taken in the design of this hydrotreater and in its operation to minimize the loss of desirable products and to maintain the activity of the catalyst for extended periods. Operating variables of importance in this hydrotreater include temperature, partial pressure of hydrogen (which is always present in the product stream), and residence time. An alternative to hydrogenation reactors is solvent scrubbing with dimethylformamide or acetone; the acetylene can be recovered in high yields.

The gaseous product stream is then dried to prevent ice formation during the refrigeration process. Usually, packed towers of alumina, silica gel, or molecular sieves are used. In spite of a higher initial cost, molecular sieves provide a number of advantages, including greater capacity for water and lower retention of heavy hydrocarbons. Multiple columns are utilized because a column will require regeneration approximately every 24 h. Regeneration is achieved by passing hot residue gases over the beds.

5. Product Separation

Low-temperature fractionation is the preferred method of hydrocarbon separation. Cascade refrigeration is used with propylene and ethylene, the commonly encountered refrigerants. Refrigeration compressors are large and have energy requirements essentially similar to those of the initial product gas compressors.

A demethanizer fractionation tower is frequently positioned first. This tower is often operated at $\sim 34 \times 10^5$ Pa, with temperatures low enough to obtain liquid methane. This tower is usually a tray-type column, although more recently packed towers have been introduced. The noncondensible gases (hydrogen, nitrogen, and carbon monoxide) and relatively pure methane can thus be separated from the C_2 and higher hydrocarbons.

The next unit in the separation stage is the deethanizer, which removes the C_2 hydrocarbons (ethylene and ethane) from the C_3 and higher hydrocarbons. The details of the separation train from the deethanizer through other units often vary significantly from plant to plant depending on which feedstock is utilized. For the general process described here, a depropanizer is the next unit in the train followed by a debutanizer. The former removes propane and propylene from the higher hydrocarbons and the latter removes essentially all of the C_4 hydrocarbons from the remaining C_5–C_8 hydrocarbons.

Separation of the C_2 stream to produce high-purity ethylene and ethane requires a large tower, sometimes the largest one in the plant. Separation of the C_3 stream to produce high-purity propylene and propane also requires a large tower, and in some plants it is the largest one. Separation of butadiene from the C_4 stream, if performed, is usually accomplished by extractive distillation. Aromatics are frequently recovered and separated to obtain benzene, toluene, and xylenes, especially when heavy feedstocks are used.

The distillation columns in these trains may be up to 13–18 m in diameter and 250 m high. Table VII presents typical variables for the distillation steps. The columns are

TABLE VII Information on Distillation Steps

Equipment	Overhead	Bottoms	Pressures ($\times 10^5$ Pa)	Reboiler (°C)	Reflux (°C)	Plates
Deethanizer	C_2 hydrocarbons	C_3–C_8 hydrocarbons	27	70–75	−10	40
Ethane-ethylene splitter	99.5–99.8% ethylene	Ethane	20	−5	−30	110
Depropanizer	C_3 hydrocarbons	C_4–C_8 hydrocarbons	16	105	50	40
Propane-propylene splitter	99.5% propylene	Propane	19	60	45	200

most often constructed with trays, although packed towers with demonstrated economies are now being introduced. The complex separation trains required include numerous heat exchangers to minimize energy demands. Selection of the best design and best operating conditions requires careful planning.

V. ECONOMICS

The designer of pyrolysis plants must consider numerous variables in order to achieve economic performance. The choice of a feedstock represents the most significant single variable. Plant location; plant size; and specific design details such as the separation train arrangement, compression and refrigeration configuration, and heat recovery contribute to the choices available to the designer. Table VIII lists typical contributions to the overall cost of operating an ethylene plant.

A. Feedstocks

There is often no single feedstock choice, since feedstock costs frequently vary erratically over a period of several years. A general guide to the influence of feedstock on capital investment of the entire pyrolysis for ethylene production is shown in Fig. 7. Net raw material costs for an ethylene plant often account for about 50–60% of the production costs, depending on whether the feedstock is a light material such as ethane or a heavier material such as naphtha.

B. Separation Train

The separations or fractionation equipment with its related exchangers may account for as much as 34% of the

TABLE VIII Production Costs

	Ethane feed (%)	Naphtha feed (%)
Raw materials	50–52	55–59
Utilities	24–26	21–23
Operational	5	4
Overhead	15–17	14–17

capital investment. The total cost investment of this section is generally the single greatest figure. As discussed in the preceding sections, the orientation of the various separation towers, for instance, placement of the demethanizer before or after the deethanizer, choice of refrigerants, and economy of heat exchange, influences not only the capital investment, but also the operational costs. There has been a recent trend toward the use of packed columns rather than tray columns. The former often resulted in lower pressure drops, smaller towers, and/or lower reflux ratios.

C. Cracking Reactor and Heat Recovery Equipment

About 30% of the total capital investment is the actual pyrolysis furnaces, provided that conventional reactors are used. Until nonconventional reactors fully demonstrate their economies, one cannot be certain how they will affect initial capital costs. Significant design considerations have been given to TLXs to recover the large quantities of energy available in the pyrolysis furnace exit gases other than the flue gases. Total utility costs, a significant part of which is for the fuel, frequently ranged from 22 to 27% of the production costs depending on whether the feedstock is ethane or naphtha.

In the United States, the preferred feedstocks for the production of ethylene and propylene continue to be lighter hydrocarbons such as ethane, propane, and their

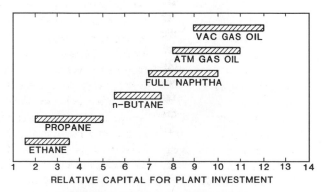

FIGURE 7 Relative capital investments.

mixtures; they are still relatively plentiful and cheap. However, there is a continuing development of reactors that will accommodate the heavier feedstocks above naphtha. This includes gas oils and whole crudes. Although experiments have been conducted to assess pyrolysis kinetics and yield structures of heavy feedstocks such as coal liquids, shale oils, tar sand bitumen, and even waste plastics and bio-materials, commercialization of these complex mixtures will not take place until far in the future when the price of conventional feedstocks becomes prohibitive.

One highly evident development of feedstock supply is the movement of olefin production to sites of inexpensive feeds such as the Mid-East. There is a gradual shifting of the world's olefin production center of gravity toward these countries.

Since 1980 there has been a marked increase in temperatures, above 1000°C, with correspondingly shorter contact times in the high-temperature zone. Contact times in the millisecond region and lower are clearly targeted. These high-severity, low-contact-time reactors offer significant advantages in processing heavier feedstocks. Until there is a major breakthrough in new tube wall material for the conventional tubular furnace reactor, temperatures and contact times for the lighter hydrocarbon feedstocks will probably not change much.

SEE ALSO THE FOLLOWING ARTICLES

ACETYLENE • CATALYSIS, INDUSTRIAL • CHEMICAL KINETICS, EXPERIMENTATION • CHEMICAL THERMODY-NAMICS • COMBUSTION • CRYOGENIC PROCESS ENGINEERING • PETROLEUM REFINING • RUBBER, SYNTHETIC • SYNTHETIC FUELS

BIBLIOGRAPHY

Albright, L. F. (1985). "Processes for Major Addition-Type Plastics and Their Monomers," Chap. 2, Krieger, Melbourne, FL.

Albright, L. F. (1988). *Oil Gas J.* August 15, 69–75; August 29, 44–48; September 19, 90–96; (1999). August 1, 35–40.

Albright, L. F., and Baker, R. T. K. (1982). "Coke Formation on Metal Surfaces," ACS Symp. Ser. 202, Am. Chem. Soc., Washington, DC.

Albright, L. F., Crynes, B. L., and Corcoran, W. H., eds. (1982). "Pyrolysis Theory and Industrial Practice," Academic Press, New York.

Albright, L. F., and Marek, J. C. (1988). *Ind. Eng. Chem. Res.* **27,** 743–759.

Baker, R. T. K., and Chludzenski, J. J. (1980). *J. Catal.* **64,** 464.

Bergeron, M. P., Maharaj, E., and McCall, T. F. (March 1999). Spring National Meeting of the American Institute of Chemical Engineers, Houston, TX.

(1999). *Chem. Eng. News.* July 15, 20–22.

(2001). *Hydrocarbon Process.* March, 21–27.

Kwilekar, A., and Bayer, G. T. (2001). *Hydrocarbon Process.* January, 80–84.

Luan, T. C. (1993). "Reduction of Coke Deposition in Ethylene Furnaces," Ph.D. thesis, Purdue University, West Lafayette, IN.

Schmidt, L. D. (1999). Personal communications, University of Minnesota.

Szechy, G., Luan, T. C., and Albright, L. F. (1992). "Novel Production Methods for Ethylene, Light Hydrocarbons, and Aromatics," Chap. 18, pp. 341–360, Dekker, New York.

Wysiekierski, A. G., Fisher, G., and Schillmoller, C. M. (1999). *Hydrocarbon Process.* January, 97–100.

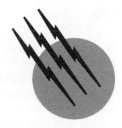

Thermionic Energy Conversion

Fred Huffman
Thermo Electron Corporation

I. Introduction
II. Fundamental Processes
III. Basic Principles of Thermionic Conversion
IV. Electrodes
V. Experimental Results
VI. Applications

GLOSSARY

Arc drop Minimum electrostatic potential difference that is required to maintain a plasma in an ignited discharge.

Barrier index Figure-of-merit parameter for characterizing the performance of a thermionic converter that is given by the sum of the arc drop and the collector work function, thus accounting for the plasma plus the collection losses. The lower this parameter, the higher is the converter performance.

Collector Cool electrode in a thermionic converter that collects the electron current.

Emitter Hot electrode in a thermionic converter from which the electron current is emitted.

Fermi level Characteristic energy of electrons in a material under thermal equilibrium at which the probability of an allowed quantum state being occupied by an electron is $\frac{1}{2}$.

Plasma Ionized gas, containing approximately equal concentrations of positive ions and electrons, that is electrically conductive and relatively field free.

Space charge Negative electrostatic barrier formed as a consequence of the finite transit time of the electrons crossing from the emitter to the collector.

Work function Heat of electron vaporization (and condensation) from a material given by the difference in electrostatic potential energy between its Fermi level and a point just outside its surface.

A THERMIONIC energy converter transforms heat into electricity by evaporating electrons from a hot emitter and condensing them on a cooler collector. This device is characterized by extremely high operating temperatures, no moving parts, modularity, and the capacity to operate inside the core of a nuclear reactor. Thermionic power generators have been built utilizing combustion, solar, radioisotope, and reactor heat sources. Although commercial applications have yet to be realized, thermionic energy conversion has the potential to (1) produce power in space reliably using nuclear reactors, (2) cogenerate electric power for a large fraction of industry unsuited to other energy converters, and (3) increase substantially the efficiency of utility power plants. Steady technical progress is bringing these possibilities closer to reality.

I. INTRODUCTION

A. Elementary Description

A thermionic converter is illustrated in Fig. 1. It consists of a hot electrode, or emitter, separated from a cooler electrode or collector by an electrical insulator. The spacing between the emitter and collector is usually a fraction of a millimeter. Electrons vaporized from the emitter cross the gap, condense on the collector, and are returned to the emitter by way of the electrical load.

The converter enclosure is hermetically sealed so that the atmosphere between the electrodes can be controlled. In the conventional thermionic converter, the interelectrode space is filled with cesium vapor from a liquid reservoir at a pressure of ~1 torr. The cesium performs two functions. First, it adsorbs on the electrodes to provide the desired electron emission properties. Second, it provides positive ions to neutralize the electron space charge so that practical current densities can be obtained from the converter.

The thermionic converter is a heat engine utilizing an electron gas as the working fluid. Thus, its efficiency cannot exceed that of the Carnot cycle. In effect, the temperature difference between the emitter and collector drives the electrons through the load.

For a given set of electrodes, the output power of the thermionic converter is a function of the emitter temperature, collector temperature, interelectrode spacing, and cesium pressure. Characteristically, thermionic converters operate at high electrode temperatures. Typical emitter temperatures range between 1600 and 2400 K and collector temperatures vary from 800 to 1100 K to provide a high current density (say, 5–10 A/cm^2) at an output potential of ~0.5 V per converter. The efficiency of converting heat to electricity is usually between 10 and 15%.

B. Historical Background

The principle of thermionic conversion derives from Edison's discovery in 1885 that current could be made to flow between two electrodes at different temperatures in a vacuum. The analysis and experimental investigation of thermionic emission from a hot electrode were performed by O. W. Richardson in 1912. W. Schlichter in 1915 recognized this means of converting heat to electricity. A patent was submitted on this topic in 1923. I. Langmuir and his associates characterized the electron and ion emission from cesium-adsorbed films on tungsten in the 1920s.

There was some interest in thermionic conversion in Russia in the 1940s. In particular, A. F. Ioffe discussed the thermionic converter as a "vacuum thermoelement" in 1949. Analytical studies of thermionic conversion that neglected the effects of space charge and collector temperature were published in the early 1950s. The first thermodynamic analysis of a thermionic converter was given by G. N. Hatsopoulos in 1956.

In the late 1950s a remarkable metamorphosis took place when several groups in the United States and Russia (working, with few exceptions, independently and without knowledge of previous or concurrent efforts) achieved efficiencies of 5 to 10% and electrical power densities of 3 to 10 W/cm^2. V. C. Wilson pioneered investigations of ignited-mode converters.

During the 1960s and early 1970s, thermionic technology development in the United States and Russia was concentrated on space reactor power systems. Remarkable progress was made in both countries during this period. In 1973, the United States terminated its space reactor program. However, Russia continued its efforts and by 1977 had tested four TOPAZ thermionic reactors.

The demise of the U.S. space thermionic program coincided with the OPEC oil action. These events provided the motivation for transferring the thermionic technology that had been developed for space to potential terrestrial applications that could conserve premium fossil fuels.

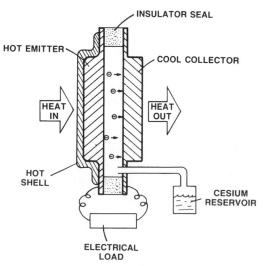

FIGURE 1 Thermionic energy converter.

By the early 1980s, the wheel had made another cycle, and the U.S. combustion thermionic conversion program was redirected to nuclear space activities. By 1985, only a token terrestrial thermionic effort existed.

II. FUNDAMENTAL PROCESSES

A. Surface Phenomena

1. Fermi Level

To a first approximation, the long-range forces between electrons in a metal are neutralized by the positive ion cores of the atoms. This implies that the metal lattice can be thought of as a box in which electrons are free to move. The number of free electrons is taken as the number of atoms times the valence of each atom. These valence electrons are responsible for electrical conduction by the metal; hence, they are termed conduction electrons, as distinguished from the electrons of the filled shells of the ion cores.

In the condensed state, the valence electrons are associated with the quantum states of the entire crystal rather than states of individual atoms. By applying the rules of statistical thermodynamics for spin $\frac{1}{2}$ particles (i.e., Fermi–Dirac statistics) the electron energy distribution $f(E)$ in a metal at a temperature T is given by the expression for the Fermi factor,

$$f(E) = \frac{1}{1 + \exp[(E - E_F)/kT]}, \tag{1}$$

where k is Boltzmann's constant ($k = 8.62 \times 10^{-5}$ eV/K) and E_F is an energy characteristic of the metal called the Fermi energy. It is numerically equal to the energy for which the Fermi factor is 0.5. At $T = 0$, E_F represents the highest energy of the electrons in the metal.

2. Work Function

A diagram of the energy levels in a conductor and the potential energy of an electron near the surface is shown in Fig. 2. To remove an electron from the metal, it must have an energy above the Fermi level by the amount of ϕ eV; this parameter is defined as the work function. The work function can be thought of as the electron heat of vaporization. If an electron is introduced into the metal from the surface, a quantity of heat, ϕ eV, will be given up. Thus, the work function is also the electron heat of condensation. It is convenient to think of the work function as a potential barrier between the inside of the metal and a point just outside (say, 500 Å) the metal surface that must be surmounted by an electron in order to escape from the metal.

Work function values vary between 4 and 5 eV for most materials. However, they can range as widely as 1.8 eV for cesium to 5.5 eV for platinum. Considerable variation of work function values for a given material can be found in the literature. These variations are due to impurities, surface films, nonuniform surfaces, and measurement techniques.

3. Thermionic Emission

Thermionic emission from a metal can be explained in terms of Fig. 2 and Eq. (1). At room temperature, few of the quantum states above the Fermi level will be filled. However, as can be seen from the Fermi factor expression, there will always be some electrons occupying the high-energy states as long as $T > 0$. Due to their random motion inside the metal, many electrons will impinge on the surface. Those electrons with kinetic energies greater than the work function barrier may escape the metal. Those that cross the boundary at the surface will transform part of their kinetic energy (i.e., ϕ eV) into potential energy.

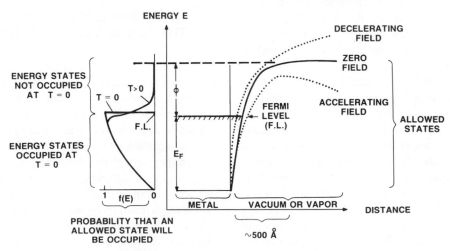

FIGURE 2 Energy levels near a metal surface.

It can be shown, either from thermodynamic considerations or from the application of statistical mechanics in connection with the quantum mechanics of electrons in metals, that the current density J of electrons emitted from a uniform surface of a pure metal of absolute temperature T can be expressed by the Richardson equation,

$$J = (1 - r)AT^2 \exp(-\phi/kT), \qquad (2)$$

where A is a constant and r is the electron reflection coefficient of the surface (of the order of 0.05). The coefficient A is composed of a combination of fundamental physical constants,

$$A = 4\pi mk^2 e/h^3 = 120 \, \text{A/cm}^2 \, \text{K}^2, \qquad (3)$$

where e is the electronic charge, m is the electronic mass, and h is Planck's constant. If the reflection coefficient is assumed to be zero, it is convenient to express the Richardson equation as

$$J = 120 \, T^2 \exp(-11606 \, \phi/T), \qquad (4)$$

where J is the current density in amperes per (centimeter)^2, ϕ is the work function in electronvolts, and T is the electrode temperature in kelvins. The value J is called the saturation current density and corresponds to a zero electric field (horizontal potential distribution in Fig. 2). A strong applied electric field changes the electron emission, because its effects are superimposed on those of the image force, and alters the shape of the potential distribution outside the electrode. The dotted-line distributions in Fig. 2 correspond to electron-accelerating and -decelerating fields. Equation (4) shows that the saturation current density of a surface increases rapidly with increasing temperature and decreasing work function.

4. Cesiated Surfaces

Thus far, all practical thermionic converters have used cesium vapor because of the remarkably fortunate combination of three major phenomena. First, Cesium atoms are adsorbed on the emitter and collector surfaces to reduce the work functions of these electrodes to values favorable for energy conversion. No other electrodes have been found that provide long life at converter temperatures with work functions suitable for energy conversion. Cesiated electrodes are quite stable since the adsorbed coverages are due to the equilibrium between the evaporating and impinging atoms. Second, some of the adsorbed cesium atoms are evaporated as positive ions, which contribute to reducing the electron space charge between the emitter and collector. Third, electron collisions with cesium atoms in the interelectrode space provide cesium ions to neutralize the space charge. Indeed, in most thermionic converters, this is the primary ion source for this purpose.

B. Plasma Considerations

A plasma is a relatively field-free region of an ionized gas that electrostatically shields itself in a distance that is small compared with other lengths of physical interest. The space charge or strong field regions on the boundary of the plasma are called sheaths. The distance over which the potential of a charge immersed in a plasma is reduced to a negligible amount can be shown, using Poisson's equation, to be

$$\lambda = \sqrt{\varepsilon_0 kT/2N_e e^2}, \qquad (5)$$

where N_e is the electron concentration and ε_0 is the permittivity of free space. This distance is called the Debye shielding distance. It gives the approximate thickness of the sheath that develops when the plasma contacts an electrode. Without such a sheath, the plasma would lose electrons much more rapidly than positive ions because of the greater electron velocity.

A plasma is in thermal equilibrium when all species (electrons, ions, and atoms) have the same average energy. At high particle densities, the collision frequency is high enough for this condition to exist. However, at the low pressures (order of 1 torr) at which a conventional thermionic converter operates, the collision frequency for electrons is low enough that they can gain significant energy between collisions. Under converter conditions, the electron temperature T_e may exceed the atom temperature T_a by a large factor. Typically, T_e is of the order of 3000 K, whereas $T_a \simeq 0.5 \, (T_E + T_C)$. A plasma cannot greatly exceed the density at which as many ions recombine locally as are produced locally. In this condition, known as local thermodynamic equilibrium, the plasma properties are equivalent to those that would exist in equilibrium with a hypothetical surface at T_e and emitting a neutral plasma.

III. BASIC PRINCIPLES OF THERMIONIC CONVERSION

An idealized potential diagram of a thermionic energy converter is shown in Fig. 3a. This diagram shows the spatial variation of the electrostatic potential perpendicular to the electrodes. Since the potential energy of the electrons in the collector is greater than that in the emitter, the collected electrons can perform work as they flow back to the emitter through the electrical load. The load voltage V is given by the difference in the Fermi levels between the emitter and collector. The emitter and collector work functions are denoted by ϕ_E and ϕ_C, respectively. The current density of the thermionically emitted electrons is given by the Richardson equation. The output power density of the converter is the product of V times the load current

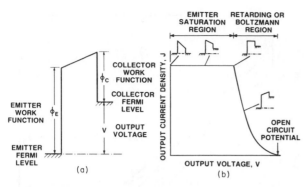

FIGURE 3 Ideal-mode converter. (a) Motive diagram; (b) current–density voltage characteristic.

density J. The $J–V$ characteristic corresponding to the "ideal-mode" potential diagram given in Fig. 3a is shown in Fig. 3b.

An analogy can be drawn between thermionic conversion and the conversion of solar heat to hydropower. In this analogy, water plays the role of electrons, the sea surface corresponds to the emitter Fermi level, the lake surface corresponds to the collector Fermi level, and the gravitational potential corresponds to the electrical potential. As water is vaporized from the sea by heat, the water vapor migrates over the mountains to a cooler region, where it condenses and falls as rain into a lake at a high altitude. The water returns through a turbine to the sea to complete the cycle by which solar energy is converted to hydropower.

A. Ideal Diode Model

The ideal diode thermionic converter model corresponds to a thermionic converter in which the emitter and collector are spaced so closely that no collisional or space charge effects take place. To reduce the complexity of the equations, ion emission effects will also be neglected. Although these assumptions do not strictly correspond to any thermionic converter, they do approach those of a very closely spaced diode operating in the vacuum mode. The ideal diode model defines the performance limit imposed by essential electron emission and heat transfer and provides a basis for comparison with practical converters.

In Fig. 3a, the energy $V + \phi_C$ must be supplied to an electron to remove it from the emitter and transport it across the gap to the collector. The electron gives up heat in the amount of ϕ_C when it is collected. The collected electron has an electrostatic potential V relative to the emitter. The electron can then be run through an electrical load to provide useful work output on its route back to the emitter.

The current density J through the load consists of J_{EC}, flowing from emitter to collector, minus J_{CE}, flowing from collector to emitter. Thus,

$$J = AT_E^2 \exp\left(-\frac{V + \phi_C}{kT_E}\right) - AT_C^2 \exp\left(-\frac{\phi_C}{kT_C}\right) \quad (6)$$

for $V + \phi_C > \phi_E$. As the output voltage is reduced such that $V + \phi_C < \phi_E$,

$$J = AT_E^2 \exp\left(-\frac{\phi_E}{kT_E}\right) - AT_C^2 \exp\left(-\frac{\phi_E - V}{kT_C}\right) \quad (7)$$

In Fig. 3b, the saturation region of the $J–V$ curve corresponds to negligible backemission. The open-circuit voltage is given by setting $J = 0$ and the short-circuit current density for the case of $V = 0$. Equations (6) and (7) can be used to construct $J–V$ curves for a range of ϕ_E, ϕ_C, T_E, and T_C values. Usually, $1600 < T_E < 2000$ K, $600 < T_C < 1200$ K, $2.4 < \phi_E < 2.8$ eV and $\phi_C \simeq 1.6$ eV.

The electrode power density is $P = JV$, and the thermal input Q_{IN} to the emitter is given by

$$Q_{IN} = Q_E + Q_R + Q_L + Q_X, \quad (8)$$

where Q_E is the heat to evaporate electrons, $J(\phi_E + 2kT_E)$ (this expression assumes negligible backemission from the collector), Q_R is the thermal radiation between electrodes, Q_L is the heat conducted down the emitter lead, and Q_X is the extraneous heat losses through structure and interelectrode vapor. Thus, the thermionic conversion efficiency is

$$\eta = JA(V - V_L)/Q_{IN}, \quad (9)$$

where A is the electrode area, V_L is the voltage loss in emitter lead ($= JAR_L$), and R_L is the resistance of emitter lead. For a given cross-sectional area of the lead, lengthening will increase V_L and decrease Q_L; shortening will decrease V_L and increase Q_L. To obtain the maximum efficiency in Eq. (9), the lead dimensions must be optimized. This optimization can be performed by setting $d\eta/dR_L = 0$. Typically, $V_L \simeq 0.1$ V.

B. Operating Modes

In practice, the ideal motive and output characteristic illustrated in Fig. 3 is never achieved because of the problems associated with the space charge barrier built up as the electrons transit from the emitter to the collector. Three of the basic approaches to circumventing the space charge problem are (1) the vacuum-mode diode, in which the interelectrode spacing is quite small; (2) the unignited-mode converter, in which the positive ions for space charge neutralization are provided by surface-contact ionization; and (3) the ignited mode, in which the positive ions are provided by thermal ionization in the interelectrode space. The potential diagrams and $J–V$ characteristics of these three operating modes are given in Figs. 4a, b, and c, respectively.

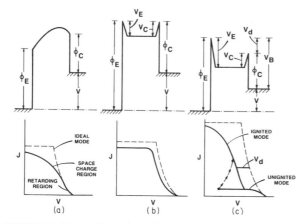

FIGURE 4 Motive diagrams and J–V characteristics for (a) vacuum, (b) unignited, and (c) ignited modes of operating a thermionic converter.

In the vacuum mode (Fig. 4a) the space charge is suppressed by making the spacing quite small—of the order of 0.01 mm. Although converters of this type have been built and operated with small-area electrodes, the practical difficulty of maintaining the necessary close spacing over large areas at high temperature differences has made the vacuum-mode diode mostly of academic interest.

Another approach to the space charge problem is to introduce low-pressure cesium vapor between the emitter and collector. This vapor is ionized when it contacts the hot emitter, provided that the emitter work function is comparable to the 3.89-eV ionization potential of cesium. A thermionic converter of this type (Fig. 4b), in which the ions for neutralizing the space charge are provided by surface-contact ionization, is said to be operating in the unignited mode. Unfortunately, surface-contact ionization is effective only at emitter temperatures that are so high (say, 2000 K) as to preclude most heat sources. To operate at more moderate temperatures, the cesium pressure must be increased to the order of 1 torr. In this case (Fig. 4c) cesium adsorption on the emitter reduces its work function so that high current densities can be achieved at significantly lower temperatures than in low-pressure unignited-mode diodes. Most of the ions for space charge neutralization are provided by electron-impact ionization in the interelectrode spacing. This ignited mode is initiated when the output voltage is lowered sufficiently. This mode of operation is that of the "conventional" thermionic converters. This ionization energy is provided at the expense of the output voltage via an "arc drop" V_d across the interelectrode plasma. Because of the high cesium pressure, the electrons are scattered many times as they cross to the collector.

For a given electrode pair, emitter temperature, collector temperature, and spacing, the J–V characteristic is a function of cesium reservoir temperature. A typical

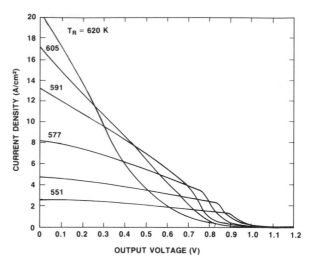

FIGURE 5 Typical output characteristics, parametric in cesium reservoir temperature. Emitter, CVD W(110); collector, noibium; $T_E = 1800$ K; $T_C = 973$ K; $d = 0.5$ mm.

J–V family, parametric in cesium reservoir temperature, is given in Fig. 5.

C. Barrier Index

The barrier index, or V_B, is a convenient parameter for characterizing thermionic converter development, comparing experimental data, and evaluating converter concepts. It is an inverse figure of merit because the lower the V_B, the higher is the converter performance. The barrier index is defined as

$$V_B = V_d + \phi_C. \qquad (10)$$

Inspection of Fig. 4c shows that, the lower the ϕ_C and V_d, the higher is the output voltage V.

The barrier index can be defined operationally. For any given emitter temperature and output current, it is possible to adjust cesium pressure, spacing, and collector temperature to maximize the power output. The spacing envelope of the optimized performance curves is shown in Fig. 6 for a converter with an emitter temperature of 1800 K. This envelope is shifted by a constant potential difference from the Boltzmann line, which represents the ideal current–voltage characteristic. This potential difference is defined as the barrier index. In Fig. 6, the barrier index of 2.1 eV represents good performance for a thermionic converter with bare metal electrodes, corresponding to $V_d \simeq 0.5$ eV and $\phi_C \simeq 1.6$ eV. The equation for the Boltzmann line is

$$J_B = A T_E^2 \exp(-V/kT_E). \qquad (11)$$

The Boltzmann line represents the idealized converter output (up to the emitter saturation current) for zero

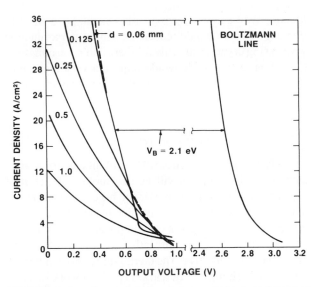

FIGURE 6 Spacing envelope of the optimized performance curves for a converter with a barrier index of 2.1 eV. Emitter, (110)W; collector, niobium; $T_E = 1800$ K; T_C and T_R, optimum.

collector work function and zero collector temperature. Thus, the barrier index incorporates the sum of the collector work function and the electron transport losses (due to ionization, scattering, electrode reflection, and electrode patchiness) into a single factor. For a real thermionic converter, the current density is

$$J = AT_E^2 \exp[-(V + V_B)/kT_E]. \qquad (12)$$

Improvements in the barrier index can be translated into either higher efficiency at a given emitter temperature or the same efficiency at a lower temperature.

The lead efficiency (i.e., the thermionic converter efficiency, which takes into account the Joule and thermal heat losses down the electrical lead) is shown in Fig. 7 as

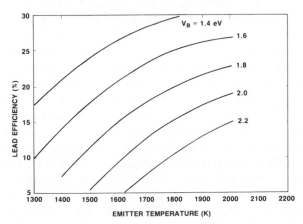

FIGURE 7 Lead efficiency as a function of emitter temperature, parametric in barrier index ($T_C = 700$–900 K; $J = 10$ A/cm^2; $\phi_C = 1.3$–1.7 eV).

a calculated function of emitter temperature for a range of barrier index values. These data emphasize that the thermionic converter is an extremely high temperature device.

IV. ELECTRODES

To operate at a practical power density and efficiency, an electrode should provide of the order of 10 A/cm^2 over a long period at emitter temperatures of 1600 to 1800 K without excessive evaporation. This requirement implies an emitter work function of 2.4 to 2.7 eV.

A. Bare Electrodes

A bare electrode is defined as a surface that does not require a vapor supply from the inter-electrode space to maintain its emission properties. No bare electrode has been found that meets the foregoing criterion. The conventional (Ba, Sr, Ca)O coating on a nickel substrate, which is widely used (e.g., radio, oscilloscope, television, microwave tubes), is not stable at the desired emitter temperatures. Lanthanum hexaboride showed promise, but its high evaporation rate and the extreme difficulty of bonding it to a substrate over a large area has eliminated it from thermionic application. For short-term experimental devices, dispenser cathodes (barium compounds impregnated in a tungsten matrix) have given useful results. However, their current density and stability do not meet the needs of a practical thermionic converter.

1. Cesium Pressure Relationship

In almost all vapor converters, the cesium pressure is adjusted by controlling the temperature of a liquid cesium reservoir. To a good approximation, the cesium vapor pressure p_{cs} in torr can be calculated from

$$p_{cs} = 2.45 \times 10^8 \, T_R^{-0.5} \exp(-8910/T_R), \qquad (13)$$

where the cesium reservoir temperature T_R is given in kelvins. It is also possible to control the pressure using cesium-intercalated graphite compounds that shift the operating reservoir temperature to higher values, which may be advantageous in some applications.

2. Rasor Plot

Experimental and theoretical results indicate that, for less than a monolayer coverage, the work function can be expressed to a good approximation as

$$\phi = f(T/T_R) \qquad (14)$$

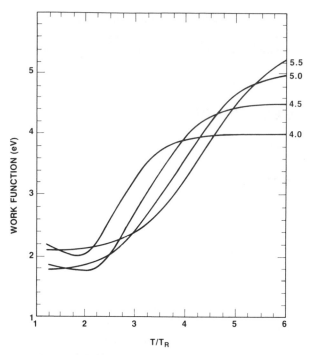

FIGURE 8 Cesiated work function of a surface versus the ratio of electrode temperature to cesium reservoir temperature, parametric in the bare work function of the surface.

for a given substrate material. This relationship is shown in Fig. 8, parametric in the bare work function ϕ_B of the substrate. Such a representation, which condenses a morass of electron emission data into a tractable form, is usually called a Rasor plot.

3. Saha–Langmuir Equation

Cesiated surfaces emit positive ions as well as electrons. The fraction of the cesium impingement rate, g, that evaporates as ions is given by the Saha–Langmuir equation,

$$g = \frac{1}{1 + 2\exp(I - \phi/kT)}, \tag{15}$$

where I is the ionization potential of the impinging vapor atoms (3.89 eV for cesium) and ϕ and T refer to the surface.

B. Additives

It can be advantageous to operate thermionic converters with additives in combination with cesium. For some applications, the improved performance can be substantial.

1. Electropositive Additives

The work function of a metal surface is reduced when the surface area is covered with a fraction of adsorbed monolayer of a more electropositive metal. Such an additive (e.g., barium) could reduce the emitter work function while using cesium primarily for thermionic ion emission. Operating in this mode would require a very high emitter temperature compared with the conventional ignited mode.

2. Electronegative Additives

The adsorption of a fractional layer of an electronegative element, such as oxygen, will increase the work function of the surface. However, Langmuir and his co-workers found that, when such a surface is operated in cesium vapor, the work function (for a given cesium pressure and substrate temperature) may be significantly lower than without the electronegative additive. The thermionic converter performance can be improved by a lower collector work function or a lower cesium pressure in the interelectrode space (or both).

This behavior is a result of the interaction of adsorbed layers with opposing dipole polarities. It should not be too surprising in view of the Rasor plot (see Fig. 8), which shows that, the higher the bare work function of a substrate, the lower is the cesiated work function for a given ratio T/T_R.

V. EXPERIMENTAL RESULTS

This section summarizes the experimental results from several classes of thermionic converters. It should provide a background for choosing the most promising paths to improved performance.

A. Vacuum-Mode Diodes

The $J–V$ characteristic of a close-spaced diode, indicating the space charge and retardingmode regions, is illustrated in Fig. 4a. Practically, it has not been possible to operate thermionic diodes at spacings appreciably closer than 0.01 mm. At such spacings, the output power density is usually less than 1.0 W/cm². Using dispenser cathodes for the emitter (1538 K) and collector (810 K), G. N. Hatsopoulos and J. Kaye obtained a current density of 1.0 A/cm² at an output potential of 0.7 V. They encountered output degradation due to evaporation of barium from the emitter to the collector, which increased the collector work function. Although the electrode stability problem can be avoided by using cesium vapor to adjust the electrode work functions, the practical problem of maintaining spacings close enough for vacuum-mode operation over large areas at high temperatures has eliminated close-spaced converters from practical applications.

To circumvent the problem of extremely small interelectrode spacings in a vacuum converter, both magnetic and electrostatic triodes have been investigated. However, neither device appears to be practical.

B. Bare Electrode Converters

1. Unignited-Mode Converters

Thermionic converters with bare electrodes have been operated in both the unignited and the ignited modes. As discussed previously, unignited-mode operation requires emitter temperatures above 2000 K. To approximate the ideal diode $J–V$ characteristics, electron–cesium scattering must be minimized by interelectrode spacings of the order of 0.1 mm. Under these conditions, the unignited-mode operation may be substantially more efficient than the ignited-mode operation, since the ions are supplied by thermal energy rather than by the arc drop, which subtracts from the output voltage.

Unfortunately, the high emitter temperatures required for unignited operation eliminate most heat sources because of materials limitations. For example, most nuclear heat sources cannot be used on a long-term basis at these temperatures because of either fuel swelling or fuel–emitter incompatibility.

2. Ignited-Mode Converters

Thus far, essentially all practical thermionic converters have operated in the ignited mode. This conventional converter has demonstrated power densities of 5 to 10 W/cm^2 and efficiencies of 10 to 15% for emitter temperatures between 1600 and 1800 K.

For a wide range of converter parameters, there exists an optimum pressure–spacing product (or Pd) of ~0.5-mm torr. For higher values of this product, the plasma loss is increased because of unnecessary electron–cesium scattering losses. From a practical viewpoint, it is difficult to fabricate large-area thermionic converters with spacings of less than 0.25 mm. However, for long-lived space reactors, system studies indicate that interelectrode spacings of up to 0.5 mm may be required to accommodate emitter distortion due to fuel swelling.

The higher the bare work function, the lower is the cesium pressure required to maintain a given electron emission. For practical spacings, the reduced cesium pressure tends toward the optimum pressure–spacing product to reduce plasma losses and improve performance. The output power density as a function of emitter temperature, parametric in the bare work function of the emitter, is given in Fig. 9. It is evident that large increases in output power density can be achieved by the use of oriented surfaces with bare work functions of greater than 4.5 eV.

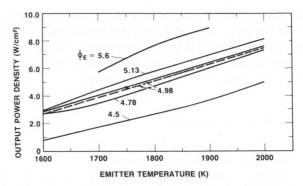

FIGURE 9 Output power density as a function of emitter temperature, parametric in the bare work function of the emitter ($J = 10$ A/cm^2; $d = 0.25$ mm).

One avenue for improving thermionic converter performance is to fabricate emitters with preferred crystalline orientations with high bare work functions.

C. Additive Diodes

Controlled additions of oxygen into thermionic converters can yield substantial improvement in power density, especially at spacings ≥ 0.5 mm. A comparison of the $J–V$ characteristics of two converters, identical except for oxide collector coating, is shown in Fig. 10. The dramatic increase in output is clear for the converter with the tungsten oxide coating on the collector. In terms of the barrier index defined earlier, $V_B \simeq 1.9$ eV for the bare tungsten collector diode. The improved performance is due to a combination of reduced collector work function and lower potential losses in the interelectrode plasma. The oxide collector supplies oxygen to the emitter so that

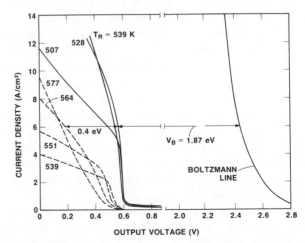

FIGURE 10 Comparison of the current density–voltage characteristics of two thermionic converters, identical except for oxide coating on collector (——, WO$_x$, $T_C = 800$ K; ---, W, $T_C = 975$ K, $T_E = 1600$ K; $d = 1$ mm).

a given current density can be obtained at a significantly lower cesium pressure. In effect, the added oxygen makes the emitter act as if it has an extremely high bare work function. As a result of the lower cesium pressure, the optimum interelectrode spacing increases.

VI. APPLICATIONS

Historically, the primary motivation for the development of thermionic converters has been their potential application to nuclear power systems in space, both reactor and radioisotope. In addition, thermionic converters are attractive for terrestrial, solar, and combustion applications.

A. Nuclear Reactors

The capacity of a thermionic converter to operate efficiently at high emitter and collector temperatures inside the core of a nuclear reactor makes it very suitable for space power applications. This combination of characteristics results in a number of fundamental and developmental advantages. For example, the necessity of radiating away all the reject heat from space conversion systems puts a premium on the thermionic converter's demonstrated capability of operating reliably and efficiently (typically, 10–15%), even at high collector temperatures (up to 1100 K). The capacity of the thermionic converter to operate inside the core of a reactor essentially eliminates the high-temperature heat transport system inherent in all out-of-core conversion systems. Thus, the thermionic reactor coolant system is at radiator temperature so that, except for the emitters, the balance of the reactor system (pumps, reflector, controls, moderator, shield, etc.) operates near the radiator temperature. Even the high-temperature emitters are distributed inside the core in "bite-sized" pellets.

A cutaway drawing of a thermionic diode that has operated inside a reactor is shown in Fig. 11. The enriched uranium oxide fuel pellets are inside the tungsten emitter, which is made by chemical vapor deposition (CVD). The outside diameter of the emitter is ∼28 mm. The collector is niobium, and the insulator seal is niobium–alumina. Thermionic diodes of this general design have operated stably inside a reactor for more than a year, and an electrically heated converter has given constant output for more than 5 years. Typically, emitter temperatures have ranged between 1700 and 2000 K.

A thermionic reactor is composed of an array of thermionic fuel elements (TFEs) in which multiple thermionic converters are connected in series inside an electrically insulating tube, much like cells in a flashlight. In principle, it is possible to design reactor systems with the thermionic converters out of core. This

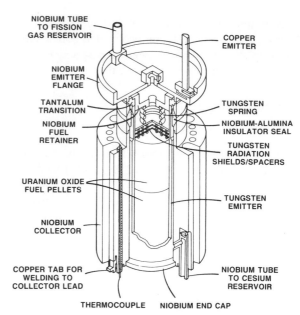

FIGURE 11 Uranium oxide-fueled thermionic diode.

design concept presents a difficult problem of electrical insulation at emitter temperature and vitiates many of the basic advantages of the TFE reactor. It is also possible to fuel the converter externally using a fuel element with a central coolant tube, the outer surface of which serves as the collector; such a design concept appears to be more limited in system power than the flashlight TFE approach.

The feasibility of the thermionic reactor system has been demonstrated by in-core converter and TFE tests in the United States, France, and Germany. As of the mid-1970s, four TOPAZ thermionic reactors had been built in Russia and tested at outputs of up to 10 kWe (kilowatts electrical).

B. Radioisotope Generators

A radioisotope thermionic generator uses the decay heat of a radioactive element. The design is usually modular. A module includes the radioisotope fuel, its encapsulation, a thermionic converter, heat transfer paths (for coupling heat from the fuel to the converter and from the converter to the radiator), hermetically sealed housing, and thermal insulation. A variety of design concepts have been considered (planar and cylindrical converters, heat pipe and conduction coupling, and several radioisotopes), and a few fueled systems have been built. However, no report of an operational use of a radioisotope thermionic generator has yet been made.

The high operating temperature of the radioisotope thermionic converter poses difficult radiological safety problems of fuel encapsulation integrity relative to launch pad accidents, launch aborts, and reentry. Since these problems are more tractable at the lower operating temperatures of radioisotope thermoelectric generators, these units have been employed for remote and space missions requiring less than a kilowatt of electrical output.

A thermionic space reactor system does not suffer a radiological safety disadvantage compared with a thermoelectric space reactor system since both systems are launched cold and neither reactor is started until a nuclear safe orbit has been achieved. The radiological hazards associated with the launch of a cold reactor system without a fission product inventroy are significantly less than those of a launch of a radioisotope thermoelectric generator.

C. Combustion Converters

The demise of the U.S. space reactor effort in 1973, along with the accompanying OPEC oil action, provided the motivation for utilizing on the ground the thermionic technology that had been developed for space. There are a number of potential terrestrial applications of thermionic conversion. These include power plant topping, cogeneration with industrial processes, and solar systems.

Although refractory metals such as tungsten and molybdenum operate stably in the vacuum of outer space at thermionic temperatures, converters operating in air or combustion atmosphere would oxidize to destruction within minutes without protection. Therefore, it is essential to develop a protective coating, or "hot shell," to isolate the refractory metals from the terrestrial environment (see Fig. 1). The thermionic converter illustrated in Fig. 12 is representative of the combustion-heated diodes that have been constructed and tested. The dome is exposed to high-temperature combustion gases. Electrons evaporated from the tungsten emitter are condensed on the mating nickel collector, which is air-cooled. A ceramic seal provides electrical insulation between the electrodes. The electrical power output is obtained between the flange and the air tube, which also functions as the collector lead.

The key component is the hot-shell–emitter (HS–EM) structure, which operates between the emitter and collector temperatures. This trilayer composite structure (tungsten–graphite–silicon carbide) is fabricated by CVD. First, the graphite is machined to the desired configuration. Next, the inside of the graphite is coated with CVD tungsten to form the emitter and its electrical lead. Finally, the outside is coated with CVD silicon carbide to protect the emitter from the combustion atmosphere.

The properties of the materials used in the HS–EM structure complement one another. Tungsten is used because of its low vapor pressure at emitter temperatures, high bare work function, low electrical resistivity, and compatibility with cesium. Silicon carbide is chosen because of its excellent high-temperature corrosion resistance, strength, thermal shock characteristics, and close match to the thermal expansion of tungsten. The graphite is selected, to match the thermal expansions of the tungsten and silicon carbide. The graphite is essential for providing good thermal shock and thermal cycle characteristics to the composite structure.

Arrays of such converters will be connected in series to provide practical output voltage. Converters of this general design have operated stably for periods of as long as 12,500 h at emitter temperatures of up to 1730 K in a natural-gas burner. The HS–EM structure has survived severe thermal shock, thermal cycle, and pressure tests.

Although combustion thermionics has made substantial progress, it is not yet ready for commercialization. The first application would probably be a thermionic burner that could be retrofit onto industrial furnaces. A

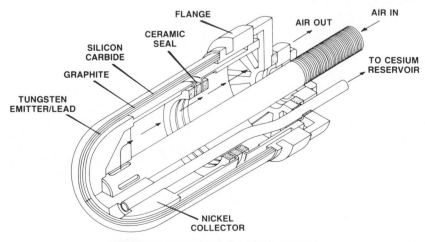

FIGURE 12 Combustion thermionic converter.

thermionic burner is a combustor whose walls are lined with thermionic converters. The emitters of the converters receive heat from the combustion gases and convert part of this heat into electricity while rejecting the balance of the heat from the collectors into the air for combustion. Operational experience with such units should provide the database for subsequent power plant topping use.

The most recent cost estimate for an installed thermionic cogeneration system was approximately \$1600/kW. Achievement of the corresponding converter cost of \$540/kW would require additional investment in converter and manufacturing development.

Relative to other advanced combustion conversion systems, such as magnetohydrodynamics (MHD), ceramic blade turbine gas turbine, and potassium Rankine cycle, thermionic development costs should be substantially lower. The cost effectiveness is a result of the modularity of thermionics, which makes it possible to perform meaningful experiments with small equipment. Thus, large investments should not be required until there is a high probability of success.

D. Solar Converters

Although the CVD silicon carbide converter was developed for combustion applications, it is very suitable for solar systems. Previous solar tests of terrestrial thermionic converters had to utilize a window so that the high-temperature refractory components could operate in an inert atmosphere. In addition to transmission losses, such windows are subject to problems of overheating and leakage.

A CVD converter has been solar-tested in a central receiver heliostat array at the Georgia Institute of Technology. The test examined heat flux cycling, control of the operating point, and mounting arrangements. The converter was mounted directly in the solar image with no cavity. The converter performance was comparable with combustion measurements made on the same diode. Thermal cycling caused no problems, and the converter showed no degradation after testing.

SEE ALSO THE FOLLOWING ARTICLES

ELECTRIC PROPULSION • NUCLEAR POWER REACTORS • PLASMA SCIENCE AND ENGINEERING

BIBLIOGRAPHY

Angrist, S. W. (1965). "Direct Energy Conversion," Allyn and Bacon, Rockleigh, New Jersey.

Baksht, F. G., et al. (1978). "Thermionic Converters and Low Temperature Plasma" (Engl. transl.). Technical Information Center/U.S. Department of Energy, Washington, DC.

Hatsopoulos, G. N., and Gyftopoulos, E. P. (1974, 1979)."Thermionic Energy Conversion," Vols. 1 and 2. MIT Press, Cambridge, Massachusetts.

Nottingham, W. B. (1956). Thermionic emission. In "Handbuch der Physik," Vol. 21. Springer-Verlag, New York.

Rasor, N. S. (1982). Thermionic energy conversion. In "Applied Atomic Collision Physics," Ch. 5, Vol. 5. Academic Press, New York.

Ure, R. W., and Huffman, F. N. (1987). Thermoelectric and thermionic conversion. In "Standard Handbook for Electrical Engineers," 12th ed. McGraw-Hill, New York.

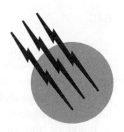

Thermodynamics

Stanley I. Sandler

University of Delaware

GLOSSARY

Activity coefficient A measure of the extent to which the fugacity of a species in a mixture departs from ideal mixture or ideal Henry's law behavior.

Equilibrium state A state in which there is no measurable change of properties and no flows.

Excess property The difference between the property in a mixture and that for an ideal mixture at the same temperature, pressure, and composition.

Homogeneous system A system of uniform properties.

Ideal mixture A mixture in which there is no change in volume, internal energy, or enthalpy of forming a mixture from its pure components at constant pressure at all temperatures and compositions.

Intensive property (or state variable) A property of a system that is independent of the mass of the system.

Multiphase system A heterogeneous system consisting of several phases, each of which is homogeneous.

Partial molar property The amount by which an extensive property of the system increases on the addition of an infinitesimal amount of a substance at constant temperature and pressure, expressed on a molar basis.

CHEMICAL THERMODYNAMICS is a science that is both simple and elegant and can be used to describe a large variety of physical and chemical phenomena at or near equilibrium. The basis of thermodynamics is a small set of laws based on experimental observation. These general laws combined with constitutive relations—that is, relations that describe how properties (for example, the density) of a substance depend on the state of the system such as its temperature and pressure—allow scientists and engineers to calculate the work and heat flows accompanying a change of state and to identify the equilibrium state.

I. THERMODYNAMIC SYSTEMS AND PROPERTIES

Thermodynamics is the study of changes that occur in some part of the universe we designate as the system; everything else is the surroundings. A real or imagined boundary may separate the system from its surroundings. A collection of properties such as temperature, pressure, composition, density, refractive index, and other properties to be discussed later characterize the thermodynamic state of a system. The state of aggregation of the system (that is, whether it is a gas, liquid or solid) is referred to as its *phase*. A system may be composed of more than one phase, in which case it is a heterogeneous system; a homogeneous system consists of only a single phase. Of most interest in thermodynamics are the changes that occur with a change in temperature, state of aggregation, composition (due to chemical reaction), and/or energy of the system.

Any element of matter contains three types of energy. First is its kinetic energy, which depends on its velocity and is given by $\frac{1}{2}mv^2$, where m is the mass and v is its center-of-mass velocity (though there may be an additional contribution due to rotational motion that we will not consider). A second contribution is the potential energy, denoted by $m\phi$ and due to gravity or electric and magnetic fields. The third, and generally the most important in thermodynamics, is the internal energy U (or internal energy per unit mass $\hat{U}$), which depends on the temperature, state of aggregation, and chemical composition of the substance. In thermodynamics, one is interested in changes in internal energy between two states of the system. For changes of state that do not involve chemical reaction, a reference state of zero internal energy can be chosen arbitrarily. However, if chemical reactions do occur, the reference state for the calculation of internal energies and other properties of each substance in the reaction must be chosen in such a way that the calculated changes on reaction equal the measured values.

There are many mechanisms by which the properties of a system can change. The mass of a system can change if mass flows into or out of the system across the system boundaries. Concentrations can change as a result of mass flows, volume changes, or chemical reaction. The energy of a system can change as a result of a number of different processes. As mass flows across the system boundary, each element of mass carries its properties, such as its internal and kinetic energy. Heat (thermal energy) can cross the system boundary by direct contact (conduction and convection) or by radiation. Work or mechanical energy can be done on a system by compressing the system boundaries, by a drive shaft that crosses the system boundaries (as in a turbine or motor), or can be added as electrical energy (in a battery or electrochemical cell). Or a system can do work on its surroundings by any of these mechanisms.

A system that does not exchange mass with its surroundings is said to be closed. A system that does not exchange thermal energy with its surroundings is referred to as an *adiabatic system*. A system that is of constant volume, adiabatic, and closed is called an *isolated system*. A system whose properties are the same throughout is referred to as a *uniform system*.

It is useful to distinguish between two types of system properties. Temperature, pressure, refractive index, and density are examples of intensive properties—properties that do not depend on the size or extent of the system. Mass, volume, and total internal energy are examples of extensive properties—properties that depend on the total size of the system. Extensive properties can be converted to intensive properties by dividing by the total mass or number of moles in the system. Volume per unit mass (reciprocal of density) and internal energy per mole are examples of intensive properties. Intensive properties are also known as *state variables*. Intensive variables per unit mass will be denoted with a $^\wedge$ (as in $\hat{V}$, to denote volume per unit mass), while those on a per mole basis are given an underbar (as in $\underline{U}$, to denote internal energy per mole). Also, $\underline{X}$, $\underline{Y}$, and $\underline{Z}$ will be used to indicate state properties such as $\underline{U}$ and $\underline{V}$, and T and P. A characteristic of a state property that is central to thermodynamic analyses is that its numerical value depends only on the state, not on the path used to get to that state. Consequently, in computing the change in value of a state property between two states, any convenient path between those states may be used, instead of the actual path.

An important experimental observation is that the specification of two independent state properties of a closed, uniform, one-component system completely fixes the values of the other state properties. For example, if two systems of the same substance in the same state of aggregation are at the same temperature and at the same pressure, all other state properties of the two systems, such as density, volume per unit mass, refractive index, internal energy per unit mass, and other properties that will be introduced shortly, will also be identical. To fix the size of the system, one must also specify the value of one extensive variable (i.e., total mass, total volume, etc.).

II. MASS AND ENERGY FLOWS AND THE EQUILIBRIUM STATE

Flows into or out of a system can be of two types. One is a forced flow, as when a pump or other device creates a continual mechanical, thermal, or chemical driving force that results in a flow of mass or energy across the boundary of a system. The other type of flow, which we refer to as a *natural flow*, occurs into or out of a system as a result of an initial difference of some property between the system and its surroundings that in time will dissipate as a result of the flow. For example, if two metal blocks of different temperatures are put in contact, a flow of heat will occur from the block of higher temperature to the one of lower temperature until an equilibrium state is reached in which both blocks have the same temperature.

An important observation is that a closed isolated system, if initially nonuniform, will eventually reach a time-invariant state that is uniform (homogeneous system) or composed of several phases, each of which is of uniform properties. Such a state of time-invariant uniformity is the equilibrium state. Systems open to natural flows will also, in time, come to equilibrium. However, a system subjected to a continuous forced flow may in time come to a time-invariant, nonuniform steady state.

The methods of thermodynamics are used to identify, describe, and sometimes predict equilibrium states. These same methods can also be used to describe nonequilibrium and steady states provided that at each point in space and time the same relations between the state properties exist as they do in equilibrium. This implies that the internal relaxation times in the fluid must be fast compared to the time scales for changes imposed upon the system.

III. LAWS OF THERMODYNAMICS

There are four laws or experimental observations on which thermodynamics is based, though they are not always referred to as such. The first observation is that in all transformations, or changes of state, total mass is conserved (note that this need not be true in nuclear reactions, but these will not be considered here.) The second observation, the first law of thermodynamics, is that in all transformations (again, except nuclear reactions) total energy is conserved. This has been known since the experiments of J. M. Joule over the period from 1837 to 1847.

The next observation, which leads to the second law, is that all systems not subject to forced flows or imposed gra-

dients (of temperature, pressure, concentration, velocity, etc.) will eventually evolve to a state of thermodynamic equilibrium. Also, systems in stable equilibrium states will not spontaneously change into a nonequilibrium state. For example, an isolated block of metal with a temperature gradient will evolve to a state of uniform temperature, but not vice versa. The third law of thermodynamics is of a different character than the first two and is mentioned later.

A. Mass Balance

After choosing a system, one can write balance equations to encompass the experimental observations above. Chemists and physicists are generally interested in the application of the laws of thermodynamics to a change of state in closed systems, while engineers are frequently interested in open systems. For generality, the equations for an open, time-varying system will be written here. The mass balance for the one-component system schematically shown in Fig. 1 is

$$\frac{dM}{dt} = \sum_{j=1}^{N} (\dot{M})_j \qquad (1)$$

where M is the total mass of the system at time t, and $(\dot{M})_j$ is the mass flow rate at the jth entry port into the system. For a mixture of C components, the total mass is the sum of the masses of each species i, $M = \sum_{i=1}^{C} M_i$ and $(\dot{M})_j = \sum_{i=1}^{C} (\dot{M}_i)_j$, where $(\dot{M}_i)_j$ is the flow rate of species i at the jth entry point. (Note that the mass balance could also be written on a molar basis; however, since the total number of moles and the number of moles of each species are not conserved on a chemical reaction, that form of the equation is a more complicated.)

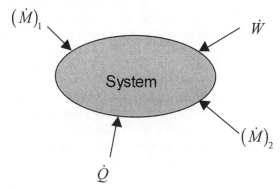

FIGURE 1 A schematic diagram of a system open to the flows of mass, heat, and work.

B. First Law

Using the sign convention that any flow that increases the energy of the system is positive, the energy balance for an open system is

$$\frac{d\left[M\left(\hat{U} + \frac{v^2}{2} + \phi\right)\right]}{dt} = \sum_{j=1}^{N}\left[\dot{M}\left(\hat{H} + \frac{v^2}{2} + \phi\right)\right]_j$$
$$+ \dot{W} + \dot{Q} - P\frac{dV}{dt} \qquad (2)$$

The term on the left is the rate of change of the total energy of the system written as a product of the mass of the system and the energy per unit mass. This includes the internal energy $\hat{U}$, the kinetic energy $v^2/2$, and the potential energy ϕ. The first term on the right accounts for the fact that each element of mass entering or leaving the system carries with it its specific enthalpy, $\hat{H} = \hat{U} + P\hat{V}$, the sum of the specific internal energy and energy due to the product of the specific volume and the pressure at the entry port. This term is summed over all entry ports. The remaining terms are the rate at which work is done on the system, $\dot{W}$, by mechanisms that do not involve a change of the system boundaries, referred to as *shaft work*; the rate at which heat or thermal energy enters the system, $\dot{Q}$; and the rate at which work is done on the system by compression or expansion of the system boundaries. A version of this equation that explicitly includes different species in multicomponent mixtures will be considered later. Also, the equation above assumes a constant pressure at the system boundary. If this is not the case, the last term is replaced by an integral over the surface of the system.

C. Second Law

To complete the formulation of thermodynamics, a balance equation is needed for another state property of the system that accounts for such experimental observations as: (1) isolated systems evolve to a state of equilibrium and not in the opposite direction, and (2) while mechanical (kinetic and potential) energy can be completely converted into heat, thermal energy can only partially be converted into mechanical energy, the rest remaining as thermal energy of a lower temperature.

Because mass, energy, and momentum are the only conserved quantities and the momentum balance is of little use in thermodynamics, the additional balance equation will be for a nonconserved property—that is, a property that can be created or destroyed in a change of state.

There are many formulations of the second law of thermodynamics to describe these observations. The one that will be used here states, by postulate, that there is a state function called the *entropy*, denoted by the symbol S (and $\hat{S}$ for entropy per unit mass), with a rate of change given by:

$$\frac{d(M\hat{S})}{dt} = \sum_{j=1}^{N}(\dot{M})_j\hat{S}_j + \frac{\dot{Q}}{T} + \dot{S}_{\text{gen}} \qquad (3)$$

where $\dot{S}_{\text{gen}}$, which is greater than or equal to zero, is the rate of entropy generation in a process due to nonuniformities, gradients, and irreversibilities in the system. It is found that $\dot{S}_{\text{gen}} = 0$ in a system at equilibrium without any internal flows, and that $\dot{S}_{\text{gen}}$ is greater than zero when such flows occur. The fact that $\dot{S}_{\text{gen}} \geq 0$ and cannot be less than zero encompasses the experimental observations above, as well as many others; indeed, $\dot{S}_{\text{gen}} \geq 0$ is the essence of the second law of thermodynamics. The third law of thermodynamics states that the entropy of all substances in the perfect crystalline state is zero at the absolute zero of temperature. This law is the basis for calculating absolute values of the entropy.

IV. CRITERIA FOR EQUILIBRIUM AND STABILITY

Consider a system that is closed (all $\dot{M} = 0$), adiabatic ($\dot{Q} = 0$), of constant volume ($dV/dt = 0$), without work flows ($\dot{W}$), and stationary (so that there are no changes in kinetic or internal energy). The mass balance, first and second law equations for this system are

$$\frac{dM}{dt} = 0; \quad M\frac{d\hat{U}}{dt} = 0; \quad M\frac{d\hat{S}}{dt} = \dot{S}_{\text{gen}} \geq 0 \qquad (4)$$

The first equation (mass balance) shows that the total mass of this system is constant, and the energy balance (first law) shows that the internal energy per unit mass is constant. The second law (entropy balance) states that the entropy of the system will increase until the system reaches the equilibrium state in which there are no internal flows so that $\dot{S}_{\text{gen}} = 0$, and $\frac{d\hat{S}}{dt} = 0$; that is, the entropy per unit mass is constant. Now, since $\hat{S}$ is increasing on the approach to equilibrium, and constant equilibrium it follows that the criterion for equilibrium is

$$\hat{S} = \text{maximum} \quad \text{for a system of constant } M, U, \text{ and } V \qquad (5a)$$

Mathematically, the equilibrium state is found by observing that for any differential change,

$$d\hat{S} = 0 \quad \text{for a system of constant } M, U, \text{ and } V$$

and

$$d^2\hat{S} < 0 \qquad (5b)$$

The first of these equations is used to identify a stationary state of the system, and the second ones ensure that the stationary state is a stable, equilibrium state (that is, a state in which the entropy is a maximum subject to the constraints, and not a minimum).

Similar arguments can be used to identify the mathematical criteria for equilibrium and stability in systems subject to other constraints. Some results are

$$\hat{A} = \hat{U} - T\hat{S} = \text{minimum} \qquad (6)$$

$$d\hat{A} = 0 \quad \text{and} \quad d^2\hat{A} > 0$$

for a system of constant M, T, and V

and

$$\hat{G} = \hat{H} - T\hat{S} = \hat{U} + P\hat{V} - T\hat{S} = \text{minimum} \qquad (7)$$

$$d\hat{G} = 0 \quad \text{and} \quad d^2\hat{G} > 0$$

for a system of constant M, T, and P

The equations above define the Gibbs free energy G and the Helmholtz free energy A.

From the first of the stability criteria above ($d^2\hat{S} \leq 0$) one can derive that, for a stable equilibrium state to exist for a pure substance, the following criteria must be met:

$$C_V = \left(\frac{\partial U}{\partial T}\right)_V > 0 \quad \text{and} \quad \left(\frac{\partial P}{\partial \underline{V}}\right)_T < 0 \qquad (8)$$

(In these equations, we have used an underbar to designate a molar property, and C_V is the constant volume heat capacity.) If these criteria are not met, the state is not a stable one, and either another state of aggregation or a two-phase system (i.e., vapor + liquid) is the equilibrium state. The stability criteria for a multicomponent mixture are much more complicated, involving derivatives of the free energy function with respect to composition.

V. PURE COMPONENT PROPERTIES

A. Interrelationships Between State Variables

The first and second laws of thermodynamics are in terms of internal energy and entropy, though the properties that are easiest to measure are temperature and pressure. In order to determine how the properties of a pure substance change with changes in temperature and pressure, consider a stationary, closed system of constant mass without any shaft work. The first and second laws for such a system (on a molar basis) are

$$\frac{dU}{dt} = \dot{Q} - P\frac{d\underline{V}}{dt} \quad \text{and} \quad \frac{d\underline{S}}{dt} = \frac{\dot{Q}}{T} + \dot{S}_{\text{gen}} \qquad (9)$$

Our interest is in the change of properties between two equilibrium states and, since any convenient path can be used for the calculation, a reversible path is used so that $\dot{S}_{\text{gen}} = 0$. Using this, and combining the two equations above, we obtain:

$$\frac{dU}{dt} = T\frac{d\underline{S}}{dt} - P\frac{d\underline{V}}{dt} \qquad (10)$$

usually written simply as $d\underline{U} = T\,d\underline{S} - P\,d\underline{V}$. By the chain rule of partial differentiation, one has

$$d\underline{X} = \left(\frac{\partial X}{\partial \underline{Y}}\right)_Z d\underline{Y} + \left(\frac{\partial X}{\partial \underline{Z}}\right)_Y d\underline{Z} \qquad (11)$$

From this equation, we find that:

$$\left(\frac{\partial U}{\partial \underline{S}}\right)_V = T; \quad \left(\frac{\partial U}{\partial \underline{V}}\right)_S = -P; \quad \left(\frac{\partial \underline{S}}{\partial \underline{V}}\right)_U = \frac{P}{T} \qquad (12)$$

Two mathematical properties for the partial derivatives of interest here are

$$\left(\frac{\partial X}{\partial \underline{Y}}\right)_Z = \frac{1}{(\partial \underline{Y}/\partial X)_Z}$$

$$\left(\frac{\partial X}{\partial \underline{Y}}\right)_Z = \left(\frac{\partial X}{\partial \underline{K}}\right)_Z \left(\frac{\partial K}{\partial \underline{Y}}\right)_Z \qquad (13)$$

Using these equations together with Eq. (12) one obtains:

$$\left(\frac{\partial U}{\partial \underline{S}}\right)_V = T = \left(\frac{\partial U}{\partial T}\right)_V \left(\frac{\partial T}{\partial \underline{S}}\right)_V = C_V \left(\frac{\partial T}{\partial \underline{S}}\right)_V$$

$$\left(\frac{\partial T}{\partial \underline{S}}\right)_V = \frac{T}{C_V} \quad \text{or} \quad \left(\frac{\partial \underline{S}}{\partial T}\right)_V = \frac{C_V}{T} \qquad (14)$$

B. Maxwell's Relations

A property of continuous mathematical functions, such as the thermodynamic properties here, is that mixed second derivatives are equal; that is,

$$\frac{\partial}{\partial \underline{X}}\bigg|_Y \left(\frac{\partial Z}{\partial \underline{Y}}\right)_X = \frac{\partial}{\partial \underline{Y}}\bigg|_X \left(\frac{\partial Z}{\partial \underline{X}}\right)_Y \qquad (15)$$

Using this property with Eq. (10), one obtains the following Maxwell relations:

$$\left(\frac{\partial \underline{S}}{\partial \underline{V}}\right)_T = \left(\frac{\partial P}{\partial T}\right)_V; \quad \left(\frac{\partial \underline{S}}{\partial P}\right)_T = -\left(\frac{\partial \underline{V}}{\partial T}\right)_P;$$

$$\left(\frac{\partial T}{\partial P}\right)_S = \left(\frac{\partial \underline{V}}{\partial \underline{S}}\right)_P; \quad \left(\frac{\partial T}{\partial \underline{V}}\right)_S = -\left(\frac{\partial P}{\partial \underline{S}}\right)_V \qquad (16)$$

Now, using the chain rule of partial differentiation and the Maxwell relations, we can write

$$dS = \left(\frac{\partial S}{\partial T}\right)_V dT + \left(\frac{\partial S}{\partial V}\right)_T dV$$

$$= \frac{C_V}{T} dT + \left(\frac{\partial P}{\partial T}\right)_V dV \qquad (17)$$

In a similar fashion, the following equations are obtained:

$$dS = \frac{C_P}{T} dT - \left(\frac{\partial V}{\partial T}\right)_P dP$$

$$dU = C_V dT + \left[T\left(\frac{\partial P}{\partial T}\right)_V - P\right] dV \qquad (18)$$

$$dH = C_P dT + \left[V - \left(\frac{\partial V}{\partial T}\right)_P\right] dP$$

where $C_P = (\frac{\partial H}{\partial T})_P$ is the constant pressure heat capacity.

C. Equations of State

Two types of information are needed to use the equations above for calculating the changes in thermodynamic properties with a change of state. First is heat capacity data. This information is usually available for each component as a function of temperature for liquids and solids or for the ideal gas state (a gas at such low pressure that interactions between the molecules are of negligible importance). The second type of information needed is an interrelation between pressure, temperature, and specific volume, that is, a volumetric equation of state (EOS). Several examples are given below:

$$PV = RT \quad \text{or} \quad Z(T, P) = \frac{PV}{RT} = 1 \quad \text{ideal gas EOS}$$

$$P = \frac{RT}{V - b} - \frac{a}{V^2}$$

or

$$Z(T, P) = \frac{PV}{RT} = \frac{V}{V - b} - \frac{a}{RTV}$$

$$\text{van der Waals EOS}$$

$$P = \frac{RT}{V - b} - \frac{a(T)}{V \cdot (V + b) + b \cdot (V - b)}$$

$$\text{Peng–Robinson EOS}$$

$$Z(T, P) = \frac{PV}{RT} = 1 + \frac{B(T)}{V} + \frac{C(T)}{V^2} + \cdots \quad \text{virial EOS}$$

$$\frac{PV}{RT} = Z(T, P) = 1 + \left(B - \frac{A}{RT} - \frac{C}{RT^3}\right)\frac{1}{V}$$

$$+ \left(b - \frac{a}{RT}\right)\frac{1}{V^2} + \frac{a\alpha}{RTV^5} + \frac{\beta}{RT^3V}\left(1 + \frac{\gamma}{V^2}\right)$$

$$\times \exp(-\gamma/V^2) \quad \text{Benedict–Webb–Rubin EOS}$$

Many other volumetric equations of state have been proposed, including more complicated ones when high accuracy is needed. In these equations, $a(T)$, $B(T)$, and $C(T)$ are functions of temperature; all other parameters are constants specific to each fluid.

The combination of heat capacity data, a volumetric equation of state, and Eqs. (17) and (18) allows the change in thermodynamic properties between any two states to be computed. However, again, a convenient path rather than the actual path is used for the calculation. For example, to compute the change in molar enthalpy between the states (P_1, T_1) and (P_2, T_2), the path followed is $(P_1, T_1) \rightarrow (P = 0, T_1) \rightarrow (P = 0, T_2) \rightarrow (P_2, T_2)$. In this way, the equation of state is used for steps 1 and 3, and the available ideal gas heat capacity is used in step 2:

$$H(T_2, P_2) - H(T_1, P_1) = \int_{P_1, T_1}^{P=0, T_1} \left[V - \left(\frac{\partial V}{\partial T}\right)_P\right] dP$$

$$+ \int_{P=0, T_1}^{P=0, T_2} C_P dT + \int_{P_0, T_2}^{P_1, T_2} \left[V - \left(\frac{\partial V}{\partial T}\right)_P\right] dP \qquad (19)$$

Similar equations are used to compute the change in other thermodynamic properties.

VI. PHASE EQUILIBRIUM IN ONE-COMPONENT SYSTEMS

A. Criterion for Phase Equilibrium

For a one-component open system with no shaft work, the first and second law equations (on a molar basis) are

$$\frac{dU}{dt} = \dot{N}H + \dot{Q} - P\frac{dV}{dt} \quad \text{and} \quad \frac{dS}{dt} = \dot{N}S + \frac{\dot{Q}}{T} + \dot{S}_{gen} \qquad (20)$$

Again, to compute property changes consider a path on which $\dot{S}_{gen} = 0$, to obtain:

$$\frac{dU}{dt} = T\frac{dS}{dt} - P\frac{dV}{dt} + G\frac{dN}{dt}$$

or simply

$$dU = T\,dS - P\,dV + G\,dN \qquad (21)$$

Analogous relations are obtained for other thermodynamic properties. For example,

$$dG = \left(\frac{\partial G}{\partial T}\right)_{P,N} dT + \left(\frac{\partial G}{\partial P}\right)_{T,N} dP + \left(\frac{\partial G}{\partial N}\right)_{T,P} dN$$

$$= -S\,dT + V\,dP + G\,dN \qquad (22)$$

To obtain the criterion for phase equilibrium in a pure fluid, consider a closed system at constant temperature and pressure, consisting of two subsystems, I and II, with mass freely transferable between them. As the composite system is closed to external mass flows,

$$dN = dN^{\mathrm{I}} + dN^{\mathrm{II}} = 0 \quad \text{or} \quad dN^{\mathrm{II}} = -dN^{\mathrm{I}} \quad (23)$$

Because the temperature and pressure are fixed and are the same in both subsystems, the change in the Gibbs free energy of the combined system accompanying an interchange of mass is

$$dG = \underline{G}^{\mathrm{I}}\, dN^{\mathrm{I}} + \underline{G}^{\mathrm{II}}\, dN^{\mathrm{II}} \quad (24)$$

At equilibrium, G is a maximum so that $dG = 0$ for all exchanges of mass. Therefore,

$$dG = 0 = \underline{G}^{\mathrm{I}}\, dN^{\mathrm{I}} + \underline{G}^{\mathrm{II}}\, dN^{\mathrm{II}} = (\underline{G}^{\mathrm{I}} - \underline{G}^{\mathrm{II}})\, dN^{\mathrm{I}}$$

This must be true for any value of dN^{I}, so that the condition for phase equilibrium is

$$\underline{G}^{\mathrm{I}}(T, P) = \underline{G}^{\mathrm{II}}(T, P) \quad \text{or equivalently}$$

$$f^{\mathrm{I}}(T, P) = f^{\mathrm{II}}(T, P) \quad (25)$$

where the fugacity, denoted by the symbol f, which is a function of temperature and pressure, is

$$\begin{aligned}
\frac{f(T, P)}{P} &= \exp\left[\frac{G(T, P) - \underline{G}^{IG}(T, P)}{RT}\right] \\
&= \exp\left[\frac{1}{RT}\int_{P=0}^{P}\left(\underline{V} - \frac{RT}{P}\right)dP\right] \\
&= \exp\left[\frac{1}{RT}\int_{\underline{V}=\infty}^{\underline{V}=ZRT/P}\left[\frac{RT}{\underline{V}} - P\right]d\underline{V}\right. \\
&\quad \left. - \ln Z(T, P) + Z(T, P) - 1\right] \quad (26)
\end{aligned}$$

It is easily shown that Eq. (25) is the condition for equilibrium for composite systems subject to other constraints (i.e., closed systems at constant U and V or constant T and V, among others).

B. Calculation of Phase Equilibrium

Figure 2 shows isotherms (lines of constant temperature) on a pressure–volume plot computed using a typical equation of state of the van der Waals form. In this diagram, $T_1 < T_2 < T_3 < T_4 < T_5$. Note that at temperatures T_1 and T_2 there are regions where $(\partial P / \partial \underline{V})_T > 0$, which violates the stability criterion of Eq. (8). Consequently, two phases (a vapor and a liquid) will form in these regions. The thermodynamic properties of the coexisting states are found by requiring that each of the temperature, pressure, and fugacity of both phases be the same. Algorithms and computer codes for such calculations appear in the applied

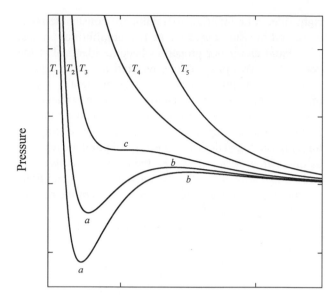

FIGURE 2 *P–V–T* plot for a typical cubic equation of state showing thermodynamically unstable regions (between points *a* and *b*). Point *c* is the critical point.

thermodynamics literature. Figure 3 is a redrawn version of the previous figure replacing the unstable region with the dome-shaped two-phase coexistence region. The left side of the dome gives the liquid properties as a function of the state variables; the vapor properties are given by the

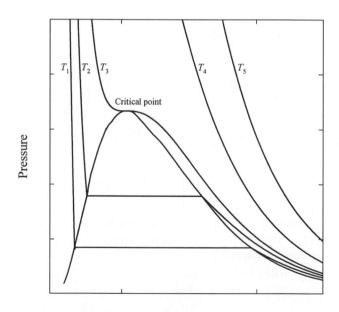

FIGURE 3 *P–V–T* plot for a cubic equation of state with the unstable region replaced with the vapor–liquid equilibrium coexistence region.

right side. A tie line (horizontal line) of constant temperature and pressure connects the two equilibrium phases. The liquid and vapor properties become identical at the peak of the two-phase dome, referred to as the *critical point*, which is point *c* in Fig. 2. Mathematically, this is the point, at which the equation of state has an inflection point, $(\partial P/\partial \underline{V})_T = (\partial^2 P/\partial \underline{V}^2)_T = 0$, and is a unique point on a pure component phase diagram. The temperature, pressure, and density at the critical point are referred as the *critical temperature*, T_c, the *critical pressure*, P_c, and the *critical volume* V_c, respectively. These conditions are frequently used to determine the values of the parameters in an equation of state.

When an equation of state is not available for a liquid, the fugacity is calculated from:

$$\frac{f(T, P)}{P} = \exp\left[\frac{1}{RT}\int_{P=0}^{P}\left(\underline{V} - \frac{RT}{P}\right)dP\right]$$

$$= \exp\left[\frac{1}{RT}\int_{P=0}^{P^{vap}(T)}\left(\underline{V}^{vap} - \frac{RT}{P}\right)dP\right.$$

$$\left. + \frac{1}{RT}\int_{P^{vap}(T)}^{P}\left(\underline{V}^{liq} - \frac{RT}{P}\right)dP\right]$$

$$= \frac{f(T, P^{vap})}{P^{vap}(T)}$$

$$\times \exp\left[\frac{1}{RT}\int_{P^{vap}(T)}^{P}\left(\underline{V}^{liq} - \frac{RT}{P}\right)dP\right]$$

$$= \frac{f(T, P^{vap})}{P^{vap}(T)}\frac{P^{vap}(T)}{P}$$

$$\times \exp\left[\frac{1}{RT}\int_{P^{vap}(T)}^{P}\underline{V}^{liq}\,dP\right]$$

or

$$f(T, P) = P^{vap}(T)\frac{f(T, P^{vap})}{P^{vap}(T)}$$

$$\times \exp\left[\frac{1}{RT}\int_{P^{vap}(T)}^{P}\underline{V}^{liq}\,dP\right] \quad (27)$$

At low vapor and total pressures, this equation reduces to $f(T, P) = P^{vap}(T)$. At higher pressures, the value of the first correction term:

$$\frac{f(T, P^{vap})}{P^{vap}(T)}$$

$$= \exp\left[\frac{1}{RT}\int_{P=0}^{P^{vap}(T)}\left(\underline{V}^{vap}(T, P) - \frac{RT}{P}\right)dP\right]$$

$$(28)$$

must be computed; note that this involves the equation of state only for the vapor. Finally, at very high pressures, the exponential term in Eq. (27), known as the *Poynting correction*, is computed using the liquid specific volume.

C. Clapeyron and Clausius–Clapeyron Equations

At equilibrium between phases, the molar Gibbs free energy is the same in both phases, that is $\underline{G}^{I}(T, P) = \underline{G}^{II}(T, P)$. For small changes in temperature, the corresponding change in the equilibrium pressure can be computed from:

$$d\underline{G}^{I}(T, P) = d\underline{G}^{II}(T, P)$$

$$\underline{V}^{I}dP - \underline{S}^{I}dT = \underline{V}^{II}dP - \underline{S}^{II}dT$$

or

$$\left(\frac{dP}{dT}\right)_{\underline{G}^{I}=\underline{G}^{II}} = \left(\frac{\underline{S}^{II} - \underline{S}^{I}}{\underline{V}^{II} - \underline{V}^{I}}\right) = \frac{1}{T}\left(\frac{\underline{H}^{II} - \underline{H}^{I}}{\underline{V}^{II} - \underline{V}^{I}}\right)$$

$$= \frac{\Delta \underline{H}}{T\Delta \underline{V}} \quad (29)$$

which is the Clapeyron equation. This equation is applicable to vapor–liquid, solid–liquid, solid–vapor, and solid–solid phase transitions. In the case of low-pressure vapor–liquid equilibrium,

$$\Delta \underline{V} = \underline{V}^{vap} - \underline{V}^{liq} \approx \underline{V}^{vap} = \frac{RT}{P}$$

so that

$$\frac{d\ln P^{vap}}{dT} = \frac{\Delta \underline{H}^{vap}}{RT^2}$$

and

$$\ln\frac{P^{vap}(T_2)}{P^{vap}(T_1)} = \int_{T_1}^{T_2}\frac{\Delta \underline{H}^{vap}}{RT^2}\,dT \quad (30)$$

which is the Clausius–Clapeyron equation. For moderate ranges of temperature, where the heat of vaporization can be considered to be approximately constant, this becomes:

$$\ln\frac{P^{vap}(T_2)}{P^{vap}(T_1)} = -\frac{\Delta \underline{H}^{vap}}{R}\left(\frac{1}{T_2} - \frac{1}{T_1}\right) \quad (31a)$$

The simpler form of this equation,

$$\ln P^{vap}(T) = A - \frac{B}{T} \quad (31b)$$

is used as the basis for correlating vapor pressure data.

VII. THERMODYNAMICS OF MIXTURES AND PHASE EQUILIBRIUM

A. Partial Molar Properties

The thermodynamic properties of a mixture are fixed once the values of two state variables (such as temperature and pressure) and the composition of the mixture are fixed. Composition can be specified by either the numbers of moles of all species or the mole fractions of all but one species (as the mole fractions must sum to one). Thus, for example, the change in the Gibbs free energy of a single-phase system of i components is

$$
dG = \left(\frac{\partial G}{\partial T}\right)_{P,N} dT + \left(\frac{\partial G}{\partial P}\right)_{T,N} dP
$$
$$
+ \sum_{i=1}^{C} \left(\frac{\partial G}{\partial N_i}\right)_{T,P,N_{j \neq i}} dN_i
$$
$$
= -S\, dT + V\, dP + \sum_{i=1}^{C} \bar{G}_i\, dN_i \qquad (32)
$$

In this equation, the notation of a partial molar property,

$$
\bar{X}_i = \left(\frac{\partial X}{\partial N_i}\right)_{T,P,N_{j \neq i}} = \left(\frac{\partial (N\underline{X})}{\partial N_i}\right)_{T,P,N_{j \neq i}} \qquad (33)
$$

has been introduced. The partial molar property $\bar{X}_i$ is the amount by which the total system property, X, changes due to the addition of an infinitesimal amount of species i at constant temperature, constant pressure, and constant number of moles of all species except i (designated by $N_{j \neq i}$). A partial molar property is a function not only of species i, but of all species in the mixture and their compositions. Indeed, a major problem in applied thermodynamics is the determination of the partial molar properties.

From Eq. (32) and the first and second laws of thermodynamics, a number of other equations can be derived. Several are listed below:

$$
dU = T\, dS - P\, dV + \sum_{i=1}^{C} \bar{G}_i\, dN_i
$$
$$
dH = T\, dS + V\, dP + \sum_{i=1}^{C} \bar{G}_i\, dN_i \qquad (34)
$$
$$
dA = -S\, dT - P\, dV + \sum_{i=1}^{C} \bar{G}_i\, dN_i
$$

Note that it is the partial molar Gibbs free energy that appears in each of these equations, which is an indication of its importance in thermodynamics. The partial molar Gibbs free energy of a species, $\bar{G}_i$ is also referred to as the chemical potential μ_i. For simplicity of notation, $\bar{G}_i$ will be used here instead of the more commonly used μ_i.

B. Criteria for Phase and Chemical Equilibrium in Mixtures

Extending the analysis of phase equilibrium used above for a pure fluid to a multicomponent, multiphase system, one obtains as the criterion for equilibrium that,

$$
\bar{G}_i^{I}(T, P, x^{I}) = \bar{G}_i^{II}(T, P, x^{II}) = \bar{G}_i^{III}(T, P, x^{III}) = \cdots \qquad (35a)
$$

or, equivalently,

$$
\bar{f}_i^{I}(T, P, x^{I}) = \bar{f}_i^{II}(T, P, x^{II}) = \bar{f}_i^{III}(T, P, x^{III}) = \cdots \qquad (35b)
$$

where x is being used to indicate the vector of mole fractions of all species present. The fugacity of species i in a mixture $\bar{f}_i$ will be discussed shortly.

Equilibrium in chemical reactions is another important area of chemical thermodynamics. The chemical reaction,

$$
\alpha A + \beta B + \cdots \Leftrightarrow \rho R + \sigma S + \cdots
$$

where α, β, etc. are the stoichiometric coefficients will be written as:

$$
\rho R + \sigma S + \cdots - \alpha A - \beta B - \cdots = 0 \qquad (36)
$$

or simply as

$$
\sum_{i=1}^{C} \nu_i I = 0
$$

The mole balance for each species in a chemical reaction can be written using the stoichiometric coefficients in the compact form,

$$
N_i = N_{i,0} + \nu_i X \qquad (37)
$$

where $N_{i,0}$ is the number of moles of species i before any reaction has occurred, and X is the molar extent of reaction, which will have the same value for all species in the reaction. The Gibbs free energy for a closed system at constant temperature and pressure is

$$
G(T, P, N) = \sum_{i=1}^{C} N_i \bar{G}_i(T, P, N)
$$
$$
= \sum_{i=1}^{C} (N_i + \nu_i X) \bar{G}_i(T, P, N) \qquad (38)
$$

where N is used to indicate the vector of mole numbers of all species present. At equilibrium in a closed system at constant temperature and pressure, G is a maximum, and $dG = 0$. Since the only variation possible is in the molar extent of reaction X, it then follows that for chemical reaction equilibrium,

$$
\sum_{i=1}^{C} \nu_i \bar{G}_i(T, P, N) = 0 \quad \text{single chemical reaction,}
$$
$$
(39a)
$$

In a multiple reaction system, defining ν_{ij} to be the stoichiometric coefficient for species i in the jth reaction, the equilibrium condition becomes:

$$\sum_{i=1}^{C} \nu_{ij} \bar{G}_i(T, P, N) = 0 \quad \text{for each reaction } j = 1, 2, \ldots$$

(39b)

In all multiple reaction systems, it is only necessary to consider a set of independent reactions—that is, a reaction set in which no reaction is a linear combination of the others.

Finally, for a system with multiple reactions and multiple phases, the criterion for equilibrium is that Eqs. (35) and (39) must be satisfied simultaneously. That is, for a state of equilibrium to exist in a multiphase, reacting system, each possible process (i.e., transfer of mass between phases or chemical reaction) must be in equilibrium for the system to be in equilibrium. This does not mean that the composition in each phase will be the same.

C. Gibbs Phase Rule

To fix the thermodynamic state of a pure-component, single-phase system, the specification of two state properties is required. Thus, the system is said to have two degrees of freedom, F. To fix the thermodynamic state of a nonreacting, C-component, single-phase system, the values of two state properties and $C-1$ mole fractions are required (the remaining mole fraction is not an independent variable as all the mole fractions must sum to one) for a total of $C+1$ variables. That is, $F = C+1$. Consider a system consisting of C components, P phases, and M independent chemical reactions. Since $C+1$ state properties are needed to fix each phase, it would appear that the system has $P(C+1)$ degrees of freedom. However, since the temperature is the same in all phases, specifying the temperature in one phase fixes its values in the $P-1$ other phases. Similarly, fixing the pressure in one phase sets its values in the $P-1$ remaining phases. That the fugacity of each species must be the same in each phase removes another $C(P-1)$ degrees of freedom. Finally, that the criterion for chemical equilibrium for each of the M independent reactions must be satisfied places another additional M constraints on the system. Therefore, the actual number of degrees of freedom is

$$F = P \cdot (C+1) - (P-1) - (P-1) - C \cdot (P-1) - M$$
$$= C - P - M + 2$$

(40)

This result is the Gibbs phase rule. It is important to note that this gives the number of state properties needed to completely specify the thermodynamic state of each

of the phases in the multicomponent, multiphase, multireaction system. However, such a specification does not give information on the relative amounts of the coexisting phases, or the total system size. Such additional information comes from the specification of the initial state and the species mass balances.

VIII. MIXTURE PHASE EQUILIBRIUM CALCULATIONS

Central to the calculation of equilibria in mixtures is the fugacity of species i in the mixture $\bar{f}_i$ which is given by:

$$\frac{\bar{f}_i(T, P)}{x_i P} = \exp\left[\frac{\bar{G}_i(T, P, x) - \bar{G}_i^{IGM}(T, P, x)}{RT} \right]$$

$$= \exp\left[\frac{1}{RT} \int_{P=0}^{P} \left(\bar{V}_i - \frac{RT}{P} \right) dP \right]$$

$$= \exp\left[\frac{1}{RT} \int_{\underline{V}=\infty}^{\underline{V}=ZRT/P} \left[\frac{RT}{\underline{V}} \right. \right.$$

$$\left. \left. - N \left(\frac{\partial P}{\partial N_i} \right)_{T, V, N_{j \neq i}} \right] d\underline{V} - \ln Z(T, P, x) \right]$$

(41)

In this equation, the superscript IGM indicates an ideal gas mixture—that is, a mixture that has the following properties:

$$PV^{IGM} = \left(\sum_{i=1}^{C} N_i \right) RT \quad \text{or} \quad P\underline{V}^{IGM} = \left(\sum_{i=1}^{C} x_i \right) RT$$

so that $\quad \bar{V}_i^{IGM}(T, P, x) = \underline{V}_i^{IG}(T, P) = RT/P$

$$\underline{U}^{IGM}(T, P, x) = \sum_{i=1}^{C} x_i \underline{U}_i^{IG}(T, P)$$

so that $\quad \bar{U}_i^{IGM}(T, P, x) = \underline{U}_i^{IG}(T, P)$

$$\underline{H}^{IGM}(P, T, x) = \sum_{i=1}^{C} x_i \underline{H}_i^{IG}(T, P)$$

so that $\quad \bar{H}_i^{IGM}(T, P, x) = \underline{H}_i^{IG}(T, P)$

(42)

$$\underline{S}^{IGM}(T, P, x) = \sum_{i=1}^{C} x_i \underline{S}_i^{IG}(T, P) - R \sum_{i=1}^{C} x_i \ln x_i$$

so that $\quad \bar{S}_i^{IGM}(T, P, x) = \underline{S}_i^{IG}(T, P) - R \ln x_i$

$$\underline{A}^{IGM}(T, P, x) = \sum_{i=1}^{C} x_i \underline{A}_i^{IG}(T, P) + RT \sum_{i=1}^{C} x_i \ln x_i$$

so that $\quad \bar{A}_i^{IGM}(T, P, x) = \underline{A}_i^{IG}(T, P) + RT \ln x_i$

$$G^{IGM}(T, P, x) = \sum_{i=1}^{C} x_i \underline{G}_i^{IG}(T, P) + RT \sum_{i=1}^{C} x_i \ln x_i$$

so that $\quad \bar{G}_i^{IGM}(T, P, x) = \underline{G}_i^{IG}(T, P) + RT \ln x_i$

Also of interest is the ideal mixture whose properties are given by:

$$V^{IM}(T, P, x) = \sum_{i=1}^{C} N_i \underline{V}_i(T, P)$$

so that $\quad \bar{V}_i^{IM}(T, P, x) = \underline{V}_i(T, P)$

$$U^{IM}(T, P, x) = \sum_{i=1}^{C} x_i \underline{U}_i(T, P)$$

so that $\quad \bar{U}_i^{IM}(T, P, x) = \underline{U}_i(T, P)$

$$H^{IM}(P, T, x) = \sum_{i=1}^{C} x_i \underline{H}_i(T, P)$$

so that $\quad \bar{H}_i^{IM}(T, P, x) = \underline{H}_i(T, P)$

$$S^{IM}(P, T, x) = \sum_{i=1}^{C} x_i \underline{S}_i(T, P) - R \sum_{i=1}^{C} x_i \ln x_i$$

so that $\quad \bar{S}_i(T, P, x) = \underline{S}_i(T, P) - R \ln x_i \quad$ (43)

$$A^{IM}(P, T, x) = \sum_{i=1}^{C} x_i \underline{A}_i(T, P) + RT \sum_{i=1}^{C} x_i \ln x_i$$

so that $\quad \bar{A}_i^{IM}(T, P, x) = \underline{A}_i(T, P) + RT \ln x_i$

$$G^{IM}(P, T, x) = \sum_{i=1}^{C} x_i \underline{G}_i(T, P) + RT \sum_{i=1}^{C} x_i \ln x_i$$

so that $\quad \bar{G}_i^{IM}(T, P, x) = \underline{G}_i(T, P) + RT \ln x_i$

While the equations for the ideal mixture appear very similar to those for the ideal gas mixture, there are two important distinctions between them. First, the *IGM* only relates to gaseous mixtures, while the *IM* is applicable to gases, liquids, and solids. Second, in the *IGM* the pure component property is that of the ideal gas at the conditions of the mixture, while in the *IM* the pure component properties are at the same temperature, pressure, and *state of aggregation* of the mixture. Note that in an ideal gas mixture,

$$\bar{V}_i^{IGM}(T, P, \underline{x}) = \frac{RT}{P} \quad \text{so that} \quad \bar{f}_i^{IGM}(T, P, \underline{x}) = x_i P$$

(44a)

while in an ideal mixture,

$$\bar{V}_i^{IM}(T, P, \underline{x}) = \underline{V}_i(T, P)$$

so that $\quad \bar{f}_i^{IM}(T, P, \underline{x}) = x_i f_i(T, P) \quad$ (44b)

That is, in the ideal mixture the fugacity of a component is the product of the mole fraction and the pure component fugacity at the same temperature, pressure, and state of aggregation as the mixture.

A. Equations of State for Mixtures

Few mixtures are ideal gas mixtures, or even ideal mixtures; consequently, there are two ways to proceed. The first method is to use an equation of state; this is the description used for all gaseous mixtures and also for some liquid mixtures, though the latter may be difficult if the chemical functionalities of the species in the mixture are very different. Generally, the same forms of equations of state described earlier are used, though the parameters in the equations are now functions of composition. For the virial equation, this composition dependence is known exactly from statistical mechanics:

$$B(T, x) = \sum_{i=1}^{C} \sum_{j=1}^{C} x_i x_j B_{ij}(T),$$

$$C(T, x) = \sum_{i=1}^{C} \sum_{j=1}^{C} \sum_{k=1}^{C} x_j x_j x_k C_{ijk}(T), \dots \quad (45)$$

where the only composition dependence is that shown explicitly. For cubic equations of state, the following mixing rules:

$$a(T, x) = \sum_{i=1}^{C} \sum_{j=1}^{C} x_i x_j a_{ij}(T), \quad b(x) = \sum_{i=1}^{C} \sum_{j=1}^{C} x_i x_j b_{ij}$$

(46)

and combining rules:

$$a_{ij}(T) = \sqrt{a_{ii}(T) a_{jj}(T)}(1 - k_{ij}), \quad b_{ij} = \frac{1}{2}(b_{ii} + b_{jj})$$

(47)

are used, where the binary interaction parameter k_{ij} is adjusted to give the best fit of experimental data. Other, more complicated mixing rules have been introduced in the last decade to better describe mixtures containing very polar compounds and species of very different functionality. There are additional mixing and combining rules for the multiparameter equations of state, and each is specific to the equation used.

B. Phase Equilibrium Calculations Using an Equation of State

If an equation of state can be used to describe both the vapor and liquid phases of a mixture, it can then be used directly for phase equilibrium calculations based on equating the fugacity of each component in each phase:

$$\bar{f}_i^L(T, P, x) = \bar{f}_i^V(T, P, y) \quad (48)$$

where the superscripts L and V indicate the vapor and liquid phases, respectively, and x and y are the vectors of their compositions. Algorithms for the computer calculation of this type of phase equilibrium calculation are available elsewhere. Because the vapor and liquid phases of hydrocarbons (together with inorganic gases such as CO_2) are well described by simple equations of state, the oil and gas industry typically does phase equilibrium calculations in this manner. Because of the limited applicability of EOS to the liquid phase of polar mixtures, the method below is commonly used for phase equilibrium calculations in the chemical industry.

C. Excess Properties and Activity Coefficients

A description that can be used for liquid and solid mixtures is based on considering any thermodynamic property to be the sum of the ideal mixture property and a second term, the excess property, that accounts for the mixture being nonideal; that is,

$$\underline{H}(T, P, x) = \underline{H}^{IM}(T, P, x) + \underline{H}^{ex}(T, P, x)$$

$$= \sum_{i=1}^{C} x_i \underline{H}(T, P) + \sum_{i=1}^{C} x_i \bar{H}_i^{ex}(T, P, x)$$

$$\underline{V}(T, P, x) = \underline{V}^{IM}(T, P, x) + \underline{V}^{ex}(T, P, x)$$

$$= \sum_{i=1}^{C} x_i \underline{V}(T, P) + \sum_{i=1}^{C} x_i \bar{V}_i^{ex}(T, P, x)$$

$$\tag{49}$$

$$\underline{G}(T, P, x) = \underline{G}^{IM}(T, P, x) + \underline{G}^{ex}(T, P, x)$$

$$= \sum_{i=1}^{C} x_i \underline{G}(T, P) + RT \sum_{i=1}^{C} x_i \ln x_i$$

$$+ \sum_{i=1}^{C} x_i \bar{G}_i^{ex}(T, P, x)$$

where

$$\bar{H}_i^{ex} = \left(\frac{\partial N\underline{H}^{ex}}{\partial N_i}\right)_{T,P,N_{j\neq i}}; \quad \bar{V}_i^{ex} = \left(\frac{\partial N\underline{V}^{ex}}{\partial N_i}\right)_{T,P,N_{j\neq i}};$$

$$\bar{G}_i^{ex} = \left(\frac{\partial N\underline{G}^{ex}}{\partial N_i}\right)_{T,P,N_{j\neq i}}; \quad \text{etc.} \tag{50}$$

Of special interest is the commonly used activity coefficient, γ, which is related to the excess partial molar Gibbs free energy as follows:

$$\bar{G}_i^{ex}(T, P, x) = RT \ln \gamma_i(T, P, x)$$

For changes in any mixture property $\theta(T, P, N)$ we can write:

$$d\theta(T, P, N) = d(N\underline{\theta}) = \sum_{i=1}^{C} N_i \, d\bar{\theta}_i + \sum_{i=1}^{C} \bar{\theta}_i \, dN_i$$

$$= N\left(\frac{\partial \underline{\theta}}{\partial T}\right)_{P,N} dT + N\left(\frac{\partial \underline{\theta}}{\partial P}\right)_{T,N} dP$$

$$+ \sum_{i=1}^{C} \bar{\theta}_i \, dN_i$$

Subtracting the two forms of the equation, and considering only changes at constant temperature and pressure, this reduces to:

$$\sum_{i=1}^{C} N_i \, d\bar{\theta} = \sum_{i=1}^{C} x_i \, d\bar{\theta}_i = 0 \tag{51a}$$

which for a binary mixture can be written as

$$x_1\left(\frac{\partial \bar{\theta}_1}{\partial x_1}\right)_{T,P} + x_2\left(\frac{\partial \bar{\theta}_2}{\partial x_1}\right)_{T,P} = 0$$

and

$$x_1\left(\frac{\partial \bar{\theta}_1^{ex}}{\partial x_1}\right)_{T,P} + x_2\left(\frac{\partial \bar{\theta}_2^{ex}}{\partial x_1}\right)_{T,P} = 0 \tag{51b}$$

since this equation is satisfied identically for the ideal mixture. Special cases of this equation are

$$x_1\left(\frac{\partial \bar{H}_1^{ex}}{\partial x_1}\right)_{T,P} + x_2\left(\frac{\partial \bar{H}_2^{ex}}{\partial x_1}\right)_{T,P} = 0;$$

$$x_1\left(\frac{\partial \bar{V}_1^{ex}}{\partial x_1}\right)_{T,P} + x_2\left(\frac{\partial \bar{V}_2^{ex}}{\partial x_1}\right)_{T,P} = 0$$

$$x_1\left(\frac{\partial \bar{G}_1^{ex}}{\partial x_1}\right)_{T,P} + x_2\left(\frac{\partial \bar{G}_2^{ex}}{\partial x_1}\right)_{T,P}$$

$$= x_1\left(\frac{\partial \ln \gamma_1}{\partial x_1}\right)_{T,P} + x_2\left(\frac{\partial \ln \gamma_2}{\partial x_1}\right)_{T,P} = 0 \tag{51c}$$

These equations, forms of the Gibbs–Duhem equation, are useful in obtaining partial molar property information from experimental data and for testing the accuracy of such data. For example, by isothermal heat-of-mixing measurements over a range of concentrations, excess enthalpy data can be obtained as follows. For a binary mixture,

$$\Delta \underline{H}^{mix} = (x_1 \bar{H}_1 + x_2 \bar{H}_2) - (x_1 \underline{H}_1 + x_2 \underline{H}_2)$$

and

$$\left(\frac{\partial \Delta \underline{H}^{mix}}{\partial x_1}\right)_{T,P} = (\bar{H}_1 - \underline{H}_1) + x_1\left(\frac{\partial \bar{H}_1}{\partial x_1}\right)_{T,P}$$

$$- (\bar{H}_2 - \underline{H}_2) + x_1\left(\frac{\partial \bar{H}_2}{\partial x_1}\right)_{T,P}$$

Using the Gibbs–Duhem equation and combining the two equations above give:

$$\Delta \underline{H}^{mix} - x_1\left(\frac{\partial \Delta \underline{H}^{mix}}{\partial x_1}\right)_{T,P} = \bar{H}_2 - \underline{H}_2$$

and

$$\Delta \underline{H}^{mix} + x_2\left(\frac{\partial \Delta \underline{H}^{mix}}{\partial x_1}\right)_{T,P} = \bar{H}_1 - \underline{H}_1 \quad (53)$$

Consequently, by having $\Delta \underline{H}^{mix}$ data as a function of composition so that the compositional derivatives can be evaluated, the partial molar enthalpies of each of the species at each composition can be obtained. If the $\Delta \underline{H}^{mix}$ data have been fitted to an equation, usually a polynomial in mole fraction, this can be done analytically. The graphical procedure shown in Fig. 4 can also be used, where the intercepts A and B then give the difference between the partial molar and pure component enthalpies at the indicated concentration. Similar procedures can be used to obtain partial molar volume data from volume change on mixing data. From vapor–liquid equilibrium data, as will be described later, activity coefficient (excess Gibbs free energy) data can be obtained. Also, if partial molar property data have been obtained experimentally, they can be tested for thermodynamic consistency by using the Gibbs–Duhem equation either differentially on a point-by-point basis or by integration over the whole dataset.

Algebraic expressions are generally used to fit excess property data as a function of composition. For example, when the two-parameter expression,

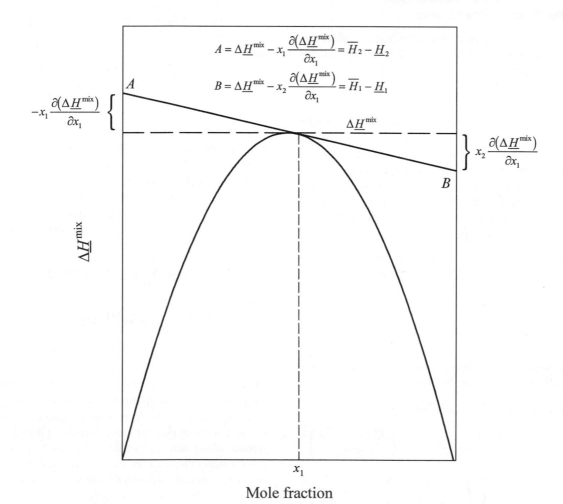

FIGURE 4 Construction illustrating how the difference between the partial molar and pure-component enthalpies can be obtained graphically at a fixed composition from a plot of $\Delta \underline{H}^{mix}$ versus composition in a binary mixture.

$$\underline{\theta}^{ex}(T, P, x) = \frac{ax_1x_2}{(x_1 + x_2b)} \tag{54a}$$

is used, one obtains, in general,

$$\bar{\theta}_1^{ex} = \frac{abx_2^2}{(x_1 + x_2b)^2} \quad \text{and} \quad \bar{\theta}_2^{ex} = \frac{ax_1^2}{(x_1 + x_2b)^2} \tag{54b}$$

and, in particular,

$$\bar{G}_1^{ex} = RT \ln \gamma_1 = \frac{abx_2^2}{(x_1 + x_2b)^2}$$

and

$$\bar{G}_2^{ex} = RT \ln \gamma_2 = \frac{ax_1^2}{(x_1 + x_2b)^2} \tag{54c}$$

which is the Van Laar model. There are many other, and more accurate, activity coefficient models in the thermodynamic literature that are used by chemists and engineers.

D. Phase Equilibrium Calculations Using Activity Coefficients

With this definition of the partial molar excess Gibbs free energy and the activity coefficient, the fugacity of a species in a liquid mixture can be computed from:

$$\bar{f}_i^L(T, P, x) = x_i \gamma_i(T, P, x) f_i^L(T, P) \tag{55}$$

where the fugacity of the pure component is equal to the vapor pressure of the pure component, $P^{vap}(T)$, if the vapor pressure and total pressure are low. If the vapor pressure is above ambient, then the fugacity at this pressure contains a correction that can be computed from the equation of state for the vapor. Also, if the total pressure is much above the pure component vapor pressure, a Poynting correction is added:

$$f_i^L(T, P) = P_i^{vap}(T) \left(\frac{f_i^L\left(T, P_i^{vap}\right)}{P_i^{vap}} \right)$$

$$\times \exp\left(\int_{P_i^{vap}(T)}^{P} \frac{V^L}{RT} dP \right) \tag{56}$$

The calculation of vapor–liquid equilibrium using activity coefficient models is then based on:

$$\bar{f}_i^L(T, P, x) = x_i \gamma_i(T, P, x) f_i^L(T, P)$$

$$= x_i \gamma_i(T, P, x) P_i^{vap}(T) \left(\frac{f_i^L\left(T, P_i^{vap}\right)}{P_i^{vap}} \right)$$

$$\times \exp\left(\int_{P_i^{vap}(T)}^{P} \frac{V^L}{RT} dP \right) = \bar{f}_i^V(T, P, y) \tag{57}$$

A common application of this equation is to vapor–liquid equilibrium at low pressures, where the vapor can be considered to be an ideal gas mixture and all pressure corrections can be neglected. This leads to the simple equation,

$$x_i \gamma_i(T, x) P_i^{vap}(T) = y_i P \tag{58}$$

relating the compositions of the vapor and liquid phases. If vapor–liquid phase equilibrium data are available, this equation can be used to obtain values of $\gamma_i(T, x)$ and, therefore, $\bar{G}_i^{ex}(T, x)$ and $\underline{G}^{ex}(T, x) = \sum x_i \bar{G}_i^{ex}(T, x) = RT \sum x_i \ln \gamma_i(T, x)$. Alternatively, if activity coefficient or $\underline{G}^{ex}$ data are available or can be predicted, the compositions of the equilibrium phases can be computed. Note that for the case of an ideal solution ($\gamma_i = 1$ for all compositions), the low-pressure vapor–liquid equilibrium relation becomes:

$$x_i P_i^{vap}(T) = y_i P \tag{59a}$$

Also, summing over all species, one then obtains for the ideal solution at low pressure:

$$P(T, \underline{x}) = \sum_{i=1}^{C} x_i P_i^{vap}(T)$$

and

$$y_i = \frac{x_i P_i^{vap}(T)}{P(T, \underline{x})} = \frac{x_i P_i^{vap}(T)}{\sum_{j=1}^{C} x_j P_j^{vap}(T)} \tag{59b}$$

(since $\sum_{i=1}^{c} y_i = 1$). The first of these equations indicates that the total pressure is a linear function of liquid-phase mole fraction. This is known as Raoult's law. The second equation establishes that the vapor and liquid compositions in an ideal solution will be different (except if, fortuitously, the vapor pressures of the components are equal).

The comparable equations for a nonideal mixture at low pressure are

$$P = \sum_{i=1}^{C} x_i \gamma_i(T, \underline{x}) P_i^{vap}(T)$$

and

$$y_i = \frac{x_i \gamma_i(T, \underline{x}) P_i^{vap}(T)}{\sum_{j=1}^{C} x_j \gamma_j(T, \underline{x}) P_j^{vap}(T)} \tag{60a}$$

Figure 5 shows the pressure versus mole fraction behavior for various mixtures. In this figure, curve 1 is for an ideal solution (i.e., Raoult's law). Curves 2 and 3 correspond to solutions with positive deviations from Raoult's law as a result of the activity coefficients of both species being greater than unity. Curves 4 and 5 are similar for the case of negative deviations from Raoult's law ($\gamma < 1$). Figure 6 is a plot of the vapor-phase mole fraction, y, versus the liquid phase mole fraction, x, for these cases. The dashed line in the figure is $x = y$.

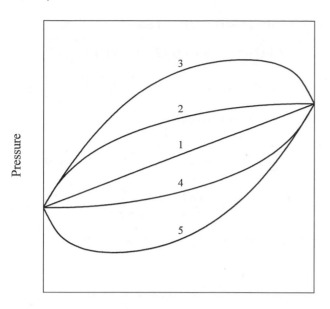

Liquid mole fraction

FIGURE 5 Pressure versus liquid composition curves for vapor–liquid equilibrium in a binary mixture. Curve 1 is for an ideal mixture (Raoult's Law). Curves 2 and 3 are for nonideal solutions in which the activity coefficients are greater than unity, and curves 4 and 5 are for nonideal solutions in which the activity coefficients are less than unity. Curves 3 and 5 are for mixtures in which the solution nonideality is sufficiently great as to result in an azeotrope.

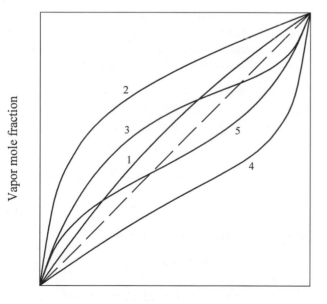

Liquid mole fraction

FIGURE 6 Liquid composition versus vapor composition (x vs. y) curves for the mixtures in Fig. 5. The dashed line is the line of $x = y$, and the point of crossing of this line is the azeotropic point.

Curve 3 in Fig. 5 is a case in which the nonideality is sufficiently great that there is a maximum in the pressure versus liquid composition curve. Mathematically, it can be shown that at this maximum the vapor and liquid compositions are identical. This is seen as a crossing of the $x = y$ line in Fig. 6. Such a point is referred to as an *azeotrope*. Curve 5 is another example of a mixture having an azeotrope, although as a result of large negative deviations from Raoult's law. Azeotropes occur as a result of solution nonidealities and are most likely to occur in mixtures of chemically dissimilar species with vapor pressures that are reasonably close. An azeotrope in a binary mixture occurs if:

$$\gamma_1(T, x_1) = \frac{P}{P_1^{vap}(T)} = \frac{x_1 P_1^{vap}(T) + x_2 P_2^{vap}(T)}{P_1^{vap}(T)}$$

and

$$\gamma_2(T, x_1) = \frac{P}{P_2^{vap}(T)} \tag{61}$$

If the azeotropic point of a mixture and the pure component vapor pressures have been measured, the two concentration-dependent activity coefficients can be calculated at this composition. This information can then be used to obtain values of the parameters in a two-parameter activity coefficient model, such as the Van Laar model discussed earlier, and then to predict values of the activity coefficients and the vapor–liquid equilibria over the whole concentration range. The occurrence of azeotropes in multicomponent mixtures is not very common. Calculations for nonideal mixtures at high pressures are considerably more complicated and are discussed in books on applied thermodynamics.

E. Henry's Law

There is an important complication that arises in the calculation of phase equilibrium with activity coefficients: To use Eq. (55) one must be able to calculate the fugacity of the pure component as a liquid at the temperature and pressure of the mixture. This is not possible, for example, if the dissolved component exists only as a gas (i.e., O_2, CO_2, etc.) or as a solid (i.e., sugar, a long-chain hydrocarbon, etc.) as a pure component at the mixture conditions. If the temperature and pressure are not very far from the melting point of the solid or boiling point of the gaseous species, Eq. (27) can still be used by extrapolation of the liquid fugacity (or vapor pressure) into the solid or gaseous states as appropriate. (Such a problem does not arise when using an equation of state, as the species fugacity in a mixture is calculated directly, not with respect to a pure component state.)

If extrapolation over a very large temperature range would be required, a different procedure is used. In this case, Eq. (53) is be replaced by:

$$\bar{f}_i^L(T, P, \underline{x}) = x_i \gamma_i^*(T, P, \underline{x}) \mathcal{H}_i(T, P) \quad (62a)$$

or

$$\bar{f}_i^L(T, P, \underline{x}) = M_i \gamma_i^\oplus(T, P, M) \mathcal{H}_i(T, P) \quad (62b)$$

depending on the concentration units used. In these two equations, forms of Henry's law, the fugacity of a gaseous or solid component dissolved in a liquid is calculated based on extrapolation of its behavior when it is highly diluted. In the first equation, the initially linear dependence of the species fugacity at high dilution is used to find the Henry's law constant $\mathcal{H}_i$. Then, the nonlinear behavior at higher concentrations is accounted for by the composition-dependent activity coefficient γ_i^*. In this description, the Henry's law constant depends on temperature and the solvent–solute pair. Also, normalization of the activity coefficient γ_i^* is different from the activity coefficient used heretofore in that its value is unity when the species is infinitely dilute, while $\gamma_i = 1$ in the pure component limit. The relation between the two is

$$\gamma_i^*(x_i) = \frac{\gamma_i(x_i)}{\gamma_i(x_i = 0)} \quad (63)$$

The second form of Henry's law, Eq. (62b), is similar but based on using molality as the concentration variable.

Both types of Henry's law coefficients are generally determined from experiment. Once values are known as a function of temperature, solvent, and solute, the phase behavior involving a solute described by Henry's law can be calculated. For example, at low total pressure, we have for the vapor–liquid equilibrium of such a component:

$$x_i \gamma_i^*(T, P, \underline{x}) \mathcal{H}_i(T, P) = y_i P = P_i$$

or

$$M_i \gamma_i^\oplus(T, P, M) \mathcal{H}_i(T, P) = y_i P = P_i \quad (64)$$

depending on the concentration variable used. At higher pressures, a Poynting correction would have to be added to the left side of both equations, and the partial pressure of the species in the vapor phase, P_i, would be replaced by its fugacity, normally calculated from an equation of state.

IX. CHEMICAL EQUILIBRIUM

The calculation of chemical equilibrium is based on Eq. (39). While the partial molar Gibbs free energy or chemical potential of each species in the mixture is needed for the calculation, what is typically available is the Gibbs free energy of formation $\Delta \underline{G}_f$ and the heat (enthalpy) of formation $\Delta \underline{H}_f$ of the pure components from their elements, generally at 25°C and 1 bar. To proceed, one writes:

$$\bar{G}_i(T, P, \underline{x}) = \underline{G}_i(T, P = 1\,\text{bar})$$

$$+ [\,\bar{G}_i(T, P, \underline{x}) - \underline{G}_i(T, P = 1\,\text{bar})]$$

$$= \underline{G}_i(T, P = 1\,\text{bar}) + RT \ln \frac{\bar{f}_i(T, P, \underline{x})}{f_i(T, P = 1\,\text{bar})} \quad (65)$$

Then, Eq. (39) can be written as:

$$\sum_{i=1}^{C} \nu_i \bar{G}_i(T, P, \underline{x}) = \sum_{i=1}^{C} \nu_i \left[\underline{G}_i(T, P = 1\,\text{bar}) \right.$$

$$\left. + RT \ln \frac{\bar{f}_i(T, P, \underline{x})}{f_i(T, P = 1\,\text{bar})} \right] = 0 \quad (66)$$

Common notation is to define the activity of each species as:

$$a_i(T, P, \underline{x}) \frac{\bar{f}_i(T, P, \underline{x})}{f_i(T, P = 1\,\text{bar})} \quad (67)$$

and to define a chemical equilibrium constant $K(T)$ from:

$$RT \ln K(T) = -\sum_{i=1}^{C} \nu_i \underline{G}_i(T, P = 1\,\text{bar})$$

$$= -\sum_{i=1}^{C} \nu_i \Delta \underline{G}_{f,i}(T, P = 1\,\text{bar}) = -\Delta G_{rxn}^o(T) \quad (68)$$

leading to:

$$K(T) = \prod_{i=1}^{C} \left(\frac{\bar{f}_i(T, P, \underline{x})}{f_i(T, P = 1\,\text{bar})} \right)^{\nu_i} = \prod_{i=1}^{C} [a_i(T, P, \underline{x})]^{\nu_i} \quad (69)$$

where $\Delta G_{rxn}^o(T)$ is the standard free energy of reaction— that is, the Gibbs free energy change that would occur between reactants in the pure component state to produce products, also as pure components. At 25°C,

$$RT \ln K(T = 25°C)$$

$$= -\sum_{i=1}^{C} \nu_i \underline{G}_i(T = 25°C, P = 1\,\text{bar})$$

$$= -\sum_{i=1}^{C} \nu_i \Delta \underline{G}_{f,i}(T = 25°C, P = 1\,\text{bar})$$

$$= -\Delta G_{rxn}^o(T = 25°C) \quad (70)$$

Also, the standard heat of reaction is

$$\Delta H_{rxn}^o(T = 25°C)$$

$$= \sum_{i=1}^{C} \nu_i \underline{H}_i(T = 25°C, P = 1\,\text{bar})$$

$$= \sum_{i=1}^{C} \nu_i \Delta \underline{H}_{f,i}(T = 25°C, P = 1\,\text{bar})$$

and

$$\Delta H^o_{rxn}(T) = \Delta H^o_{rxn}(T = 25°C)$$
$$+ \int_{T=25°C}^{T} \sum_{i=1}^{C} \nu_i C_{P,i}(T)\, dT \quad (71)$$

Then, using:

$$\frac{\partial}{\partial T}\left(\frac{G}{T}\right)_P = -\frac{H}{T^2} \quad \text{leads to}$$

$$\left(\frac{\partial \ln K(T)}{\partial T}\right)_P = \frac{\Delta H^o_{rxn}(T)}{T^2} \quad (72a)$$

and

$$\ln \frac{K(T)}{K(T = 25°C)}$$

$$= \int_{T=25°C}^{T} \frac{\Delta H^o_{rxn}(T)}{RT^2}\, dT$$

$$= \frac{\Delta H^o_{rxn}(T = 25°C)}{R}\left(\frac{1}{T} - \frac{1}{298.15}\right)$$

$$+ \int_{T=25°C}^{T} \frac{\left[\int_{T^1=25°C}^{T} \sum_{i=1}^{C} \nu_i C_{P,i}(T^1)\, dT^1\right]}{RT^2}\, dT \quad (72b)$$

For a liquid species at low and moderate pressure, and with the pure-component standard state, the activity is

$$a_i(T, P, \mathbf{x}) = \frac{\bar{f}_i^L(T, P, \mathbf{x})}{f_i^L(T, P = 1\,\text{bar})}$$

$$= \frac{x_i \gamma_i(T, P, \mathbf{x}) f_i^L(T, P)}{f_i^L(T, P = 1\,\text{bar})}$$

$$= x_i \gamma_i(T, P, \mathbf{x}) \quad (73a)$$

The activity of species in the vapor is

$$a_i(T, P, \mathbf{y}) = \frac{\bar{f}_i^V(T, P, \mathbf{y})}{f_i^V(T, P = 1\,\text{bar})} = \frac{y_i P}{1\,\text{bar}} \quad (73b)$$

where the term on the right of the expression is correct only for an ideal gas mixture. Thus, for example, the chemical equilibrium relation for the low-pressure gas-phase reaction, $H_2 + \frac{1}{2}O_2 \leftrightarrow H_2O$ is

$$K(T) = \frac{a_{H_2O}}{a_{H_2} a_{O_2}^{1/2}} = \frac{\dfrac{y_{H_2O} P}{1\,\text{bar}}}{\dfrac{y_{H_2} P}{1\,\text{bar}} \left(\dfrac{y_{O_2} P}{1\,\text{bar}}\right)^{1/2}}$$

$$= \frac{y_{H_2O}}{y_{H_2}\left(y_{O_2}\right)^{1/2}}\left(\frac{1\,\text{bar}}{P}\right)^{1/2} \quad (74)$$

which indicates that as the pressure increases, the conversion of hydrogen and oxygen to water is favored. The equilibrium relation for the low-pressure hydrogenation

of benzene to cyclohexane involving hydrogen gas and liquid benzene and cyclohexane $C_6H_6 + 3H_2 \leftrightarrow C_6H_{12}$ is

$$K(T) = \frac{a_{C_6H_{12}}}{a_{C_6H_6} a_{H_2}^3} = \frac{\dfrac{x_{C_6H_{12}} \gamma_{C_6H_{12}} f_{C_6H_{12}}^L}{f_{C_6H_{12}}^L}}{\dfrac{x_{C_6H_6} \gamma_{C_6H_{12}} f_{C_6H_6}^L}{f_{C_6H_6}^L}\left(\dfrac{y_{H_2} P}{1\,\text{bar}}\right)^3}$$

$$= \frac{x_{C_6H_{12}} \gamma_{C_6H_{12}}}{x_{C_6H_6} \gamma_{C_6H_6}}\left(\frac{1\,\text{bar}}{y_{H_2} P}\right)^3 = \frac{x_{C_6H_{12}}}{x_{C_6H_6}}\left(\frac{1\,\text{bar}}{P_{H_2}}\right)^3$$

$$(75)$$

where in the last term in this equation the activity coefficients have been omitted, as benzene and cyclohexane are so chemically similar that they are expected to form an ideal solution, and $P_{H_2} = y_{H_2} P$ is the partial pressure of hydrogen in the gas phase.

If the reaction system is closed, then the equilibrium relations have to be solved together with the mass balances. For example, suppose three moles of hydrogen and one mole of oxygen are being reacted to form water. The mass balances for this reaction give:

Species	Initial moles	Moles at equilibrium	Equilibrium mole fraction
H_2	3	$3 - X$	$\dfrac{3 - X}{4 - 0.5X}$
O_2	1	$1 - 0.5X$	$\dfrac{1 - 0.5X}{4 - 0.5X}$
H_2O	0	X	$\dfrac{X}{4 - 0.5X}$
Total moles		$4 - 0.5X$	

The chemical equilibrium relation to be solved for the molar extent of reaction X is, then,

$$K(T) = \frac{y_{H_2O}}{y_{H_2}\left(y_{O_2}\right)^{1/2}}\left(\frac{1\,\text{bar}}{P}\right)^{1/2}$$

$$= \frac{\dfrac{X}{4 - 0.5X}}{\dfrac{3 - X}{4 - 0.5X}\left(\dfrac{1 - 0.5X}{4 - 0.5X}\right)^{1/2}}\left(\frac{1\,\text{bar}}{P}\right)^{1/2}$$

$$= \frac{X(4 - 0.5X)^{1/2}}{(3 - X)(1 - 0.5X)^{1/2}}\left(\frac{1\,\text{bar}}{P}\right)^{1/2}$$

Therefore, once the temperature is specified so that value of $K(T)$ can be computed, and the pressure is fixed, the equilibrium molar extent of reaction X can be computed, and from that each of the equilibrium mole fractions.

When several reactions occur simultaneously, a similar procedure is followed in that a chemical equilibrium relation is written for each of the independent reactions, and mass balances are used for each component. The solution can be complicated since all the reactions are coupled through the mass balances; that is, the molar extent for each reaction will appear in some or all of the equilibrium relations.

When there are many reactions possible, or when there is combined chemical and phase equilibrium, calculation by direct Gibbs free energy minimization may be a better way to proceed. In this method, expressions are written for the partial molar Gibbs free energy of every component in every possible phase (which will involve the mole fractions of all species in that phase), and then a search method is used to find the state of minimum Gibbs free energy (if temperature and pressure are fixed) subject to the mass balance constraints. That is, one identifies the state in which the total Gibbs free energy is a minimum directly, rather than using chemical equilibrium constants.

X. ELECTROLYTE SOLUTIONS

Electrolyte solutions are fundamentally different from the other mixtures so far considered. One reason is that the species, such as salts, ionize in solution so that the nature of the pure component and the substance in solution is very different. Another reason is that, because the ions are charged, the interactions are much stronger and longer range than among molecules. Consequently, the solutions are much more nonideal, and the activity coefficient models used for molecules, such as the simple Van Laar model, are not applicable. Also, the anions and cations originating from a single ionizable substance are present in a fixed ratio.

Consider the ionization reaction $A_{\nu_+} B_{\nu_-} = \nu_+ A^{z+} + \nu_- B^{z-}$. Since the initial molecule has no net charge, we have

$$\nu_+ z_+ + \nu_- z_- = 0 \qquad (76a)$$

or, on a molar basis,

$$\nu_+ N_A + \nu_- N_B = 0 \qquad (76b)$$

where ν_+ and ν_- are the stoichiometric coefficients of the ions A and B in the molecule, and z_+ and z_- are their charges. By Eq. (76b) the number of moles of each ion cannot be changed independently, so the partial molar Gibbs free energy of each ion cannot be separately measured. As the total molar concentration of salt can be varied, the customary procedure is to define a mean ionic activity coefficient $\gamma_\pm$ based on Henry's law, applicable to both ions, and referenced to a hypothetical ideal one-molal solution as follows:

$$\bar{G}_{AB}(T, P, M) = \bar{G}_{AB}^{Ideal}(T, P, M = 1)$$

$$+ \nu RT \ln \left[\frac{M_\pm \gamma_\pm}{M = 1} \right] \qquad (77)$$

where $\nu = \nu_+ + \nu_-$ and $M_\pm^\nu = M_A^{\nu_+} M_B^{\nu_-}$ is the mean ionic molality. At very low ionic concentrations, the mean ionic activity coefficient $\gamma_\pm$ can be computed from the Debye–Hückel limiting law:

$$\ln \gamma_\pm = -\alpha |z_+ z_-| \sqrt{I} \qquad (78)$$

where

$$I = \frac{1}{2} \sum_{i=\text{ions}} z_i^2 M_i$$

In this equation, I is the ionic strength, the sum is over all ions in solution, and α is a temperature-dependent parameter whose value is $1.178 \, (\text{mol/L})^{-0.5}$ for water at $25°C$. At higher ionic strengths, the following empirical extensions to the limiting law have been used:

$$\ln \gamma_+ = -\frac{\alpha |z_+ z_-| \sqrt{I}}{1 + \beta \sqrt{I}}$$

and

$$\ln \gamma_\pm = -\frac{\alpha |z_+ z_-| \sqrt{I}}{1 + \beta \sqrt{I}} + \delta I \qquad (79)$$

where $\beta = 1.316 \, (\text{mol/L})^{-0.5}$ for water at $25°C$, and δ is an adjustable parameter fit to experimental data. Note that Eq. (78) and the first of Eq. (79) predict a steep and continuing decrease of $\gamma_\pm$ with increasing ionic strength, while the last of Eq. (79) correctly predicts first a decrease in $\gamma_\pm$ and then an increase with increasing ionic strength.

Since a solvent of high dielectric constant is needed for a salt to ionize, ions are not found in the vapor phase at normal conditions. However, the strong nonideality of an electrolyte solution containing ions affects vapor–liquid and reaction equilibria. For example, silver chloride is only very slightly soluble in water. The equilibrium constant for the reaction $AgCl \rightarrow Ag^+ + Cl^-$ is

$$K = \frac{a_{Ag^+} a_{Cl^-}}{a_{AgCl}} = \frac{\dfrac{M_{Ag^+}}{M = 1} \dfrac{M_{Cl^-}}{M = 1} (\gamma_\pm)^2}{1}$$

$$= M_{Ag^+} M_{Cl^-} (\gamma_\pm)^2$$

so that

$$M_{Ag^+} = \frac{K}{(\gamma_\pm)^2 M_{Cl^-}} \qquad (80)$$

The molality of the silver ion that will dissolve is affected by the addition of other ions. If a salt containing neither silver or chloride ions (e.g., KNO_3) is added to a silver chloride solution, the ionic strength of the solution will increase; this will result in a decrease in the mean ionic activity coefficient at low total ionic strength and an increase in the solubility of Ag^+ ions. Conversely, at higher ionic strength, the mean ionic activity coefficient will increase, producing a decrease in the solubility of Ag^+ ions. However, if a salt containing a Cl^- ion is added, there will be a small ionic strength effect, but a large common-ion effect resulting in a decrease in the concentration of Ag^+ ions and the solubility of AgCl. That is, because the value of the equilibrium constant is fixed, increasing the Cl^- ion concentration by addition of a Cl-containing salt will depress the Ag^+ ion concentration.

XI. COUPLED REACTIONS

For a state of equilibrium at constant temperature and pressure, the Gibbs free energy should be a minimum. If several chemical reactions occur in a system that are only linked through mass balances, then those reactions that reduce the Gibbs free energy of the system will occur, and those that increase G will not occur. There are, however, other reactions that are more closely coupled. One example is an electrolytic battery in which two electrochemical reactions occur, one of which increases the Gibbs free energy of the system while the other decreases it. When the two half cells are connected, if the sum of the two Gibbs free energy changes is negative, both reactions will occur, including the half-cell reaction that increases the Gibbs free energy system. That is, the reaction with a negative Gibbs free energy change is driving the one with a positive change.

Another example is the production of adenosine triphosphate (ATP), a molecule used to store energy in biological systems, by the phosphorylation of adenosine diphosphate (ADP), $ADP + phosphate \rightarrow ATP$. The standard-state Gibbs free energy change for this process is 29.3 kJ, so this reaction, by itself will have a very small equilibrium constant. However, by enzymatic reactions, it is coupled to the oxidation of glucose, $C_6H_{12}O_6 + 6O_2 \rightarrow 6CO_2 + 6H_2O$, with a standard state Gibbs free energy change of -2807.2 kJ, which is so large that it can drive the phosphorylation of many ADP molecules. In fact, the net overall reaction is

$$C_6H_{12}O_6 + 6O_2 + 38\,ADP + 38\,phosphate$$

$$\rightarrow 6CO_2 + 6H_2O + 38\,ATP$$

for which $\Delta G^\circ = -1756.8$ kJ. There are many other examples in biological systems of complex enzymatic reaction networks resulting in one reaction driving another.

SEE ALSO THE FOLLOWING ARTICLES

BIOENERGETICS • HEAT TRANSFER • INTERNAL COMBUSTION ENGINES • PHYSICAL CHEMISTRY • STEAM TABLES

BIBLIOGRAPHY

Pitzer, K. S. (1995). "Thermodynamics," 3rd ed., McGraw-Hill, New York.

Prausnitz, J. M., Lichtenthaler, R. N., and de Azevedo, E. G. (1999). "Molecular Thermodynamics of Fluid-Phase Equilibria," 3rd ed., Prentice-Hall, Englewood Cliffs, NJ.

Rowlinson, J. S., and Swinton, F. L. (1982). "Liquids and Liquid Mixtures," 3rd ed., Butterworths, London.

Sandler, S. I. (1999). "Chemical and Engineering Thermodynamics," 3rd ed., Wiley, New York.

Smith, J. M., Van Ness, H. C., and Abbott, M. M. (1996). "Introduction to Chemical Engineering Thermodynamics," 5th ed., McGraw-Hill, New York.

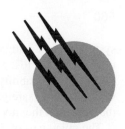

Thermoeconomics

George Tsatsaronis
Frank Cziesla

Technical University of Berlin

GLOSSARY

Exergy Useful energy or a measure for the quality of energy.
Exergy destruction Thermodynamic loss (inefficiency) due to irreversibilities within a system.
Exergetic efficiency Characterizes the performance of a component or a thermal system from the thermodynamic viewpoint.
Exergy loss Thermodynamic loss due to exergy transfer to the environment.
Thermoeconomics Exergy-aided cost minimization.

THERMOECONOMICS is the branch of thermal sciences that combines a thermodynamic (exergy) analysis with economic principles to provide the designer or operator of an energy-conversion system with information which is not available through conventional thermodynamic analysis and economic evaluation but is crucial to the design and operation of a cost-effective system. Thermoeconomics rests on the notion that exergy (available energy) is the only rational basis for assigning monetary costs to the interactions of an energy conversion system with its surroundings and to the sources of thermodynamic inefficiencies within it. Since the thermodynamic considerations of thermoeconomics are based on the exergy concept, the terms *exergoeconomics* and *thermoeconomics* can be used interchangeably.

A complete thermoeconomic analysis consists of (1) an exergy analysis, (2) an economic analysis, (3) exergy costing, and (4) a thermoeconomic evaluation. A thermoeconomic analysis is usually conducted at the system component level and calculates the costs associated with all material and energy streams in the system and the thermodynamic inefficiencies (exergy destruction) within each component. A comparison of the cost of exergy destruction with the investment cost for the same component provides useful information for improving the cost effectiveness of the component and the overall system by pinpointing the required changes in structure and parameter values. An iterative thermoeconomic evaluation and optimization is, among the currently available methods, the most effective approach for optimizing the design of a system when a mathematical optimization procedure cannot be used due to the lack of information (e.g., cost functions) and our inability to appropriately consider important factors such as safety, availability, and maintainability of the system in the modeling process.

The objectives of thermoeconomics include

- Calculating the cost of each product stream generated by a system having more than one product
- Optimizing the overall system or a specific component
- Understanding the cost-formation process and the flow of costs in a system

Thermoeconomics uses results from the synthesis, cost analysis, and simulation of thermal systems and provides useful information for the evaluation and optimization of these systems as well as for the application of artificial intelligence techniques to improve the design and operation of such systems.

I. EXERGY ANALYSIS

An exergy analysis identifies the location, the magnitude, and the sources of thermodynamic inefficiencies in a thermal system. This information, which cannot be provided by other means (e.g., an energy analysis), is very useful for improving the overall efficiency and cost effectiveness of a system or for comparing the performance of various systems (Bejan, Tsatsaronis, and Moran, 1996; Moran and Shapiro, 1998).

An exergy analysis provides, among others, the exergy of each stream in a system as well as the real "energy waste," i.e., the thermodynamic inefficiencies (exergy destruction and exergy loss), and the exergetic efficiency for each system component.

In the sign convention applied here, work done on the system and heat transferred to the system are positive. Accordingly, work done by the thermal system and heat transferred from the system are negative.

A. Exergy Components

Exergy is the maximum theoretical useful work (shaft work or electrical work) obtainable from a thermal system as it is brought into thermodynamic equilibrium with the environment while interacting with the environment only. Alternatively, exergy is the minimum theoretical work (shaft work or electrical work) required to form a quantity of matter from substances present in the environment and to bring the matter to a specified state. Hence, exergy is a measure of the departure of the state of the system from the state of the environment.

The environment is a large equilibrium system in which the state variables (T_0, p_0) and the chemical potential of the chemical components contained in it remain constant, when, in a thermodynamic process, heat and materials are exchanged between another system and the environment. It is important that no chemical reactions can take place between the environmental chemical components. The environment is free of irreversibilities, and the exergy of the environment is equal to zero. The environment is part of the surroundings of any thermal system.

In the absence of nuclear, magnetic, electrical, and surface tension effects, the *total exergy* of a system E_{sys} can be divided into four components: *physical exergy* E_{sys}^{PH}, *kinetic exergy* E^{KN}, *potential exergy* E^{PT}, and *chemical exergy* E^{CH}:

$$E_{sys} = E_{sys}^{PH} + E^{KN} + E^{PT} + E^{CH}. \quad (1)$$

The subscript *sys* distinguishes the total exergy and physical exergy of *a system* from other exergy quantities, including transfers associated with streams of matter. The total specific exergy on a mass basis e_{sys} is

$$e_{sys} = e_{sys}^{PH} + e^{KN} + e^{PT} + e^{CH}. \quad (2)$$

The *physical exergy* associated with a thermodynamic system is given by

$$E_{sys}^{PH} = (U - U_0) + p_0(V - V_0) - T_0(S - S_0), \quad (3)$$

where U, V, and S represent the internal energy, volume, and entropy of the system, respectively. The subscript 0 denotes the state of the same system at temperature T_0 and pressure p_0 of the environment.

The rate of physical exergy $\dot{E}_{ms}^{PH}$ associated with a material stream (subscript *ms*) is

$$\dot{E}_{ms}^{PH} = (\dot{H} - \dot{H}_0) - T_0(\dot{S} - \dot{S}_0), \quad (4)$$

where $\dot{H}$ and $\dot{S}$ denote the rates of enthalpy and entropy, respectively. The subscript 0 denotes property values at temperature T_0 and pressure p_0 of the environment.

Kinetic and *potential exergy* are equal to kinetic and potential energy, respectively.

$$E^{KN} = \frac{1}{2}m\vec{v}^2 \quad (5)$$

$$E^{PT} = mgz \quad (6)$$

Here, $\vec{v}$ and z denote velocity and elevation relative to coordinates in the environment ($\vec{v}_0 = 0$, $z_0 = 0$). Equations (5) and (6) can be used in conjunction with both systems and material streams.

The *chemical exergy* is the maximum useful work obtainable as the system at temperature T_0 and pressure p_0 is brought into chemical equilibrium with the environment. Thus, for calculating the chemical exergy, not only do the temperature and pressure have to be specified, but the chemical composition of the environment also has to be specified. Since our natural environment is not in equilibrium, there is a need to model an exergy-reference environment (Ahrendts, 1980; Bejan, Tsatsaronis, and Moran, 1996; Szargut, Morris, and Steward, 1988). The use of tabulated *standard chemical exergies* for substances contained in the environment at standard conditions ($T_{ref} = 298.15$ K, $p_{ref} = 1.013$ bar) facilitates the calculation of exergy values. Table I shows values of the standard chemical exergies of selected substances in two alternative exergy-reference environments. The effect of small variations in the values of T_0 and p_0

TABLE I Standard Molar Chemical Exergy $\bar{e}^{CH}$ of Various Substances at 298.15 K and p_{ref}

Substance	Formula[a]	$\bar{e}^{CH}$ (kJ/kmol)	
		Model I[b]	Model II[c]
Ammonia	$NH_3(g)$	336,684	337,900
n-Butane	$C_4H_{10}(g)$	—	2,805,800
Calcium oxide	$CaO(s)$	120,997	110,200
Calcium hydroxide	$Ca(OH)_2(s)$	63,710	53,700
Calcium carbonate	$CaCO_3(s)$	4,708	1,000
Calcium sulfate (gypsum)	$CaSO_4 \cdot 2H_2O(s)$	6,149	8,600
Carbon (graphite)	$C(s)$	404,589	410,260
Carbon dioxide	$CO_2(g)$	14,176	19,870
Carbon monoxide	$CO(g)$	269,412	275,100
Ethane	$C_2H_6(g)$	1,482,033	1,495,840
Hydrogen	$H_2(g)$	235,249	236,100
Hydrogen peroxide	$H_2O_2(g)$	133,587	—
Hydrogen sulfide	$H_2S(g)$	799,890	812,000
Methane	$CH_4(g)$	824,348	831,650
Methanol (g)	$CH_3OH(g)$	715,069	722,300
Methanol (l)	$CH_3OH(l)$	710,747	718,000
Nitrogen	$N_2(g)$	639	720
Nitrogen monoxide	$NO(g)$	88,851	88,900
Nitrogen dioxide	$NO_2(g)$	55,565	55,600
Octane	$C_8H_{18}(l)$	—	5,413,100
Oxygen	$O_2(g)$	3,951	3,970
n-Pentane	$C_5H_{12}(g)$	—	3,463,300
Propane	$C_3H_8(g)$	—	2,154,000
Sulfur	$S(s)$	598,158	609,600
Sulfur dioxide	$SO_2(g)$	301,939	313,400
Sulfur trioxide (g)	$SO_3(g)$	233,041	249,100
Sulfur trioxide (l)	$SO_3(l)$	235,743	—
Water (g)	$H_2O(g)$	8,636	9,500
Water (l)	$H_2O(l)$	45	900

[a] (g): gaseous, (l): liquid, (s): solid.
[b] Ahrendts, 1980. In this model, $p_{ref} = 1.019$ atm.
[c] Szargut, Morris, and Stewart, 1988. In this model, $p_{ref} = 1.0$ atm.

on the chemical exergy of reference substances might be neglected in practical applications.

The chemical exergy of an ideal mixture of N ideal gases $\bar{e}^{CH}_{M,ig}$ is given by

$$\bar{e}^{CH}_{M,ig} = \sum_{k=1}^{N} x_k \bar{e}^{CH}_k + \bar{R} T_0 \sum_{k=1}^{N} x_k \ln x_k, \quad (7)$$

where T_0 is the ambient temperature, $\bar{e}^{CH}_k$ is the standard molar chemical exergy of the kth substance, and x_k is the mole fraction of the kth substance in the system at T_0. For solutions of liquids, the chemical exergy $\bar{e}^{CH}_{M,l}$ can be obtained if the activity coefficients γ_k are known (Kotas, 1985):

$$\bar{e}^{CH}_{M,l} = \sum_{k=1}^{N} x_k \bar{e}^{CH}_k + \bar{R} T_0 \sum_{k=1}^{N} x_k \ln(\gamma_k x_k). \quad (8)$$

The standard chemical exergy of a substance not present in the environment can be calculated by considering a reversible reaction of the substance with other substances for which the standard chemical exergies are known. For energy-conversion processes, calculation of the exergy of fossil fuels is particularly important.

Figure 1 shows a hypothetical reversible isotherm-isobaric reactor where a fuel reacts completely at steady state with oxygen to form CO_2, SO_2, H_2O, and N_2. All substances are assumed to enter and exit unmixed at T_0, p_0. The chemical exergy of a fossil fuel $\bar{e}^{CH}_f$ on a molar basis can be derived from exergy, energy, and entropy balances for the reversible reaction (Bejan, Tsatsaronis, and Moran, 1996):

$$\bar{e}^{CH}_f = -(\Delta \bar{h}_R - T_0 \Delta \bar{s}_R) + \Delta \bar{e}^{CH} = -\Delta \bar{g}_R + \Delta \bar{e}^{CH} \quad (9)$$

with

$$\Delta \bar{h}_R = \sum_i v_i \bar{h}_i = -\bar{h}_f + \sum_k v_k \bar{h}_k = -\overline{HHV}$$

$$\Delta \bar{s}_R = \sum_i v_i \bar{s}_i = -\bar{s}_f + \sum_k v_k \bar{s}_k$$

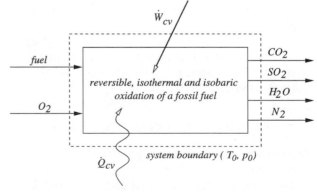

FIGURE 1 Device for evaluating the chemical exergy of a fuel.

$$\Delta \bar{g}_R = \Delta \bar{h}_R - T_0 \Delta \bar{s}_R$$

$$\Delta \bar{e}^{CH} = \sum_k \nu_k \bar{e}_k^{CH}$$

$$i = f, O_2, CO_2, H_2O, SO_2, N_2$$

$$k = O_2, CO_2, H_2O, SO_2, N_2$$

$$\nu_i \text{ and } \nu_k \geq 0 : CO_2, H_2O, SO_2, N_2$$

$$\nu_i \text{ and } \nu_k < 0 : f, O_2$$

Here, $\Delta \bar{h}_R$, $\Delta \bar{s}_R$, and $\Delta \bar{g}_R$ denote the molar enthalpy, entropy, and Gibbs function, respectively, of the reversible combustion reaction of the fuel with oxygen. $\overline{HHV}$ is the molar higher heating value of the fuel, and $\nu_k(\nu_i)$ is the stoichiometric coefficient of the kth (ith) substance in this reaction.

The higher heating value is the primary contributor to the chemical exergy of a fossil fuel. For back-of-the-envelope calculations, the molar chemical exergy of a fossil fuel $\bar{e}_f^{CH}$ may be estimated with the aid of its molar higher heating value $\overline{HHV}$:

$$\frac{\bar{e}_f^{CH}}{\overline{HHV}} \approx \begin{cases} 0.95 - 0.985 & \text{for gaseous fuels (except} \\ & \text{H}_2 \text{ and CH}_4) \\ 0.98 - 1.00 & \text{for liquid fuels} \\ 1.00 - 1.04 & \text{for solid fuels} \end{cases} \quad (10)$$

For hydrogen and methane, this ratio is 0.83 and 0.94, respectively.

The standard molar chemical exergy $\bar{e}_s^{CH}$ of any substance not present in the environment can be determined using the change in the specific Gibbs function $\Delta \bar{g}$ for the reaction of this substance with substances present in the environment (Bejan, Tsatsaronis, and Moran, 1996; Moran and Shapiro 1998):

$$\bar{e}_s^{CH} = -\Delta \bar{g} + \Delta \bar{e}^{CH} = -\sum_i \nu_i \bar{g}_i + \sum_{i \neq s} \nu_i \bar{e}_i^{CH}, \quad (11)$$

where $\bar{g}_i$, ν_i, and $\bar{e}_i^{CH}$ denote, for the ith substance, the Gibbs function at T_0 and p_0, the stoichiometric coefficient in the reaction, and the standard chemical exergy, respectively.

In solar energy applications, the ratio of the exergy of solar radiation to the total energy flux at the highest layer of the earth's atmosphere is 0.933 (Szargut, Morris, and Steward, 1988).

B. Exergy Balance and Exergy Destruction

Thermodynamic processes are governed by the laws of conservation of mass and energy. These conservation laws state that the total mass and total energy can neither be created nor destroyed in a process. However, exergy is

not generally conserved but is destroyed by irreversibilities within a system. Furthermore, exergy is lost, in general, when the energy associated with a material or energy stream is rejected to the environment.

1. Closed System Exergy Balance

The change in total exergy ($E_{sys,2} - E_{sys,1}$) of a closed system caused through transfers of energy by work and heat between the system and its surroundings is given by

$$E_{sys,2} - E_{sys,1} = E_q + E_w - E_D. \quad (12)$$

The exergy transfer E_q associated with heat transfer Q is

$$E_q = \int_1^2 \left(1 - \frac{T_0}{T_b}\right) \delta Q, \quad (13)$$

where T_b is the temperature at the system boundary at which the heat transfer occurs.

The exergy transfer E_w associated with the transfer of energy by work W is given by

$$E_w = W + p_0(V_2 - V_1). \quad (14)$$

In a process in which the system volume increases ($V_2 > V_1$), the work $p_0(V_2 - V_1)$ being done on the surroundings is unavailable for use, but it can be recovered when the system returns to its original volume V_1.

A part of the exergy supplied to a real thermal system is destroyed due to irreversibilities within the system. The exergy destruction E_D is equal to the product of entropy generation S_{gen} within the system and the temperature of the environment T_0.

$$E_D = T_0 S_{gen} \geq 0. \quad (15)$$

Hence, exergy destruction can be calculated either from the entropy generation using an entropy balance or directly from an exergy balance. E_D is equal to zero only in ideal processes.

2. Control Volume Exergy Balance

Exergy transfer across the boundary of a control volume system can be associated with material streams and with energy transfers by work or heat. The general form of the exergy balance for a control volume involving multiple inlet and outlet streams is

$$\frac{dE_{cv,sys}}{dt} = \sum_j \underbrace{\left(1 - \frac{T_0}{T_j}\right) \dot{Q}_j}_{\dot{E}_{q,j}} + \underbrace{\left(\dot{W}_{cv} + p_0 \frac{dV_{cv}}{dt}\right)}_{\dot{E}_w}$$

$$+ \sum_i \dot{E}_i - \sum_e \dot{E}_e - \dot{E}_D, \quad (16)$$

where $\dot{E}_i$ and $\dot{E}_e$ are total exergy transfer rates at inlets and outlets [see Eq. (4) for the physical exergy associated with these transfers]. The term $\dot{Q}_j$ represents the time rate of heat transfer at the location on the boundary where the temperature is T_j. The associated rate of exergy transfer $\dot{E}_{q,j}$ is given by

$$\dot{E}_{q,j} = \left(1 - \frac{T_0}{T_j}\right)\dot{Q}_j. \tag{17}$$

The exergy transfer $\dot{E}_w$ associated with the time rate of energy transfer by work $\dot{W}_{cv}$ other than flow work is

$$\dot{E}_w = \dot{W}_{cv} + p_0 \frac{dV_{cv}}{dt}. \tag{18}$$

Finally, $\dot{E}_D$ accounts for the time rate of exergy destruction due to irreversibilities within the control volume. Either the exergy balance [Eq. (16)] or $\dot{E}_D = T_0 \dot{S}_{gen}$ can be used to calculate the exergy destruction in a control volume system.

Under steady state conditions, Eq. (16) becomes

$$0 = \sum_j \dot{E}_{q,j} + \dot{W}_{cv} + \sum_i \dot{E}_i - \sum_e \dot{E}_e - \dot{E}_D. \tag{19}$$

3. Exergy Destruction

The real thermodynamic inefficiencies in a thermal system are related to exergy destruction and exergy loss. All real processes are irreversible due to effects such as chemical reaction, heat transfer through a finite temperature difference, mixing of matter at different compositions or states, unrestrained expansion, and friction. An exergy analysis identifies the system components with the highest thermodynamic inefficiencies and the processes that cause them.

In general, inefficiencies in a component should be eliminated or reduced if they do not contribute to a reduction in capital investment for the overall system or a reduction of fuel costs in another system component. Efforts for reducing the use of energy resources should be centered on components with the greatest potential for improvement. Owing to the present state of technological development, some exergy destructions and losses in a system component are unavoidable (Tsatsaronis and Park, 1999). For example, the major part of exergy destruction in a combustion processes cannot be eliminated. Only a small part can be reduced by preheating the reactants and reducing the amount of excess air.

The objective of a thermodynamic optimization is to minimize the inefficiencies, whereas the objective of a thermoeconomic optimization is to estimate the cost-optimal values of the thermodynamic inefficiencies.

Heat transfer through a finite temperature difference is irreversible. Figure 2 shows the temperature profiles for two streams passing through an adiabatic heat exchanger. The following expression for the exergy destruction $\dot{E}_{D,q}$ due to heat transfer from the hot stream 3 to the cold

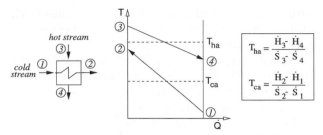

FIGURE 2 Temperature profiles and thermodynamic average temperatures for two streams passing through an adiabatic heat exchanger at constant pressure.

stream 1 can be derived (Bejan, Tsatsaronis, and Moran, 1996):

$$\dot{E}_{D,q} = T_0 \dot{Q}\frac{T_{ha} - T_{ca}}{T_{ha}T_{ca}}, \tag{20}$$

where the thermodynamic average temperatures T_{ha} and T_{ca} of the hot stream and the cold stream are given by

$$T_a = \frac{\int_i^e T\,ds}{s_e - s_i} = \frac{h_e - h_i - \int_i^e v\,dp}{s_e - s_i}. \tag{21}$$

Here, the subscripts i and e denote inlet and exit, respectively. For applications where the heat transfer occurs at constant pressure, Eq. (21) reduces to

$$T_a = \frac{h_e - h_i}{s_e - s_i} \tag{22}$$

Equation (20) shows that the difference in the average thermodynamic temperatures $(T_{ha} - T_{ca})$ is a measure for the exergy destruction. Mismatched heat capacity rates of the two streams, i.e., $(\dot{m}c_p)_h/(\dot{m}c_p)_c \neq 1$, and a finite minimum temperature difference $(\Delta T_{min} = (T_3 - T_2))$ in Fig. 2) are the causes of the thermodynamic inefficiencies. Matching streams of significantly different heat capacity rates $\dot{m}c_p$ in a heat exchanger should be avoided. Furthermore, the lower the temperature levels T_{ha} and T_{ca}, the greater the exergy destruction at the same temperature difference $(T_{ha} - T_{ca})$.

The rate of exergy destruction associated with friction $\dot{E}_{D,fr}$ can be expressed as

$$\dot{E}_{D,fr} = -\frac{T_0}{T_a} \cdot \dot{m} \cdot \int_i^e v\,dp, \tag{23}$$

where T_a is the thermodynamic average temperature of the working fluid and $\int_i^e v\,dp$ is the head loss (Bejan, Tsatsaronis, and Moran, 1996). The effect of friction is more significant at higher mass flow rates and lower temperature levels. Although exergy destruction related to friction is usually of secondary importance in thermal systems, the costs associated with the exergy destruction might be significant. Since the unit cost of electrical or mechanical power required to feed a pump, compressor, or fan is significantly higher than the unit cost of a fossil fuel, each unit of exergy destruction by frictional dissipation is relatively expensive.

Examples for the exergy loss include heat transfer to the environment ("heat losses"), dissipation of the kinetic energy of an exhaust gas, and irreversibilities due to mixing of an exhaust gas with the atmospheric air.

Guidelines for improving the use of energy resources in thermal systems by reducing the sources of thermodynamic inefficiencies are discussed by Bejan, Tsatsaronis, and Moran (1996) and by Sama (1995). However, the major contribution of an exergy analysis to the evaluation of a system comes through a thermoeconomic evaluation that considers not only the inefficiencies, but also the costs associated with these inefficiencies and the investment expenditures required to reduce the inefficiencies.

C. Exergetic Variables

The performance evaluation and the design optimization of thermal systems require a proper definition of the exergetic efficiency and a proper costing approach for each component of the system (Lazzaretto and Tsatsaronis, November 1999). The exergetic efficiency of a component is defined as the ratio between product and fuel. The product and fuel are defined by considering the desired result produced by the component and the resources expended to generate the result.

$$\varepsilon_k \equiv \frac{\dot{E}_{P,k}}{\dot{E}_{F,k}} = 1 - \frac{\dot{E}_{D,k} + \dot{E}_{L,k}}{\dot{E}_{F,k}} \tag{24}$$

An appropriately defined exergetic efficiency is the only variable that unambiguously characterizes the performance of a component from the thermodynamic viewpoint (Tsatsaronis, 1999). Table II shows the definition of fuel and product for selected system components at steady state operation.

The rate of exergy destruction in the kth component is given by

$$\dot{E}_{D,k} = \dot{E}_{F,k} - \dot{E}_{P,k} - \dot{E}_{L,k}. \tag{25}$$

Here, $\dot{E}_{L,k}$ represents the exergy loss in the kth component, which is usually zero when the component boundaries are at T_0. For the overall system, $\dot{E}_L$ includes the exergy flow rates of all streams leaving the system (see, for example, the different definitions of $\dot{E}_F$ for a fuel cell in Table II).

In addition to ε_k and $\dot{E}_{D,K}$, the thermodynamic evaluation of a system component is based on the exergy destruction ratio $y_{D,k}$, which compares the exergy destruction in the kth component with the fuel supplied to the overall system $\dot{E}_{F,tot}$:

$$y_{D,k} \equiv \frac{\dot{E}_{D,k}}{\dot{E}_{F,tot}}. \tag{26}$$

This ratio expresses the percentage of the decrease in the overall system exergetic efficiency due to the exergy destruction in the kth system component:

$$\varepsilon_{tot} = \frac{\dot{E}_{P,tot}}{\dot{E}_{F,tot}} = 1 - \sum_k y_{D,k} - \frac{\dot{E}_{L,tot}}{\dot{E}_{F,tot}}. \tag{27}$$

$\dot{E}_{D,k}$ is an absolute measure of the inefficiencies in the kth component, whereas ε_k and $y_{D,k}$ are relative measures of the same inefficiencies. In ε_k the exergy destruction within a component is related to the fuel for the same component, whereas in $y_{D,k}$ the exergy destruction in a component is related to the fuel for the overall system.

II. REVIEW OF AN ECONOMIC ANALYSIS

In the evaluation and cost optimization of an energy-conversion system, we need to compare the annual values of capital-related charges (carrying charges), fuel costs, and operating and maintenance expenses. These cost components may vary significantly within the plant economic life. Therefore, levelized annual values for all cost components should be used in the evaluation and cost optimization.

The following sections illustrate the total revenue requirement method (TRR method) which is based on procedures adopted by the Electric Power Research Institute (EPRI; 1993). This method calculates all the costs associated with a project, including a minimum required return on investment. Based on the estimated total capital investment and assumptions for economic, financial, operating, and market input parameters, the total revenue requirement is calculated on a year-by-year basis. Finally, the nonuniform annual monetary values associated with the investment, operating (excluding fuel), maintenance, and fuel costs of the system being analyzed are levelized; that is, they are converted to an equivalent series of constant payments (annuities).

An economic analysis can be conducted in current dollars by including the effect of inflation or in constant dollars by not including inflation. In general, studies involving the near term (the next 5–10 years) values are best presented in current dollars. The results of longer term studies (20–40 years) may be best presented in constant dollars.

A. Estimation of Total Capital Investment

The best cost estimates for purchased equipment can be obtained directly through vendors' quotations. The next

TABLE II Exergy Rates Associated with Fuel and Product for Defining Exergetic Efficiencies of Selected Components at Steady State Operation

Component	Schematic	Exergy rate of product $\dot{E}_P$	Exergy rate of fuel $\dot{E}_F$
Compressor, pump, quad or fan		$\dot{E}_2 - \dot{E}_1$	$\dot{E}_3$
Turbine or expander		$\dot{E}_4$	$\dot{E}_1 - \dot{E}_2 - \dot{E}_3$
Heat exchanger[a]		$\dot{E}_2 - \dot{E}_1$	$\dot{E}_3 - \dot{E}_4$
Mixing unit		$\dot{m}_1(e_3 - e_1)$	$\dot{m}_2(e_2 - e_3)$
Combustion chamber		$\dot{E}_3 - \dot{E}_2$	$\dot{E}_1 - \dot{E}_4$
Steam generator		$(\dot{E}_6 - \dot{E}_5) + (\dot{E}_8 - \dot{E}_7)$	$(\dot{E}_1 + \dot{E}_2) - (\dot{E}_3 + \dot{E}_4)$
Deaerator		$\dot{m}_2(e_3 - e_2)$	$\dot{m}_1 e_1 - (\dot{m}_3 - \dot{m}_2)e_1 - \dot{m}_4 e_4$
Compressor with cooling air extractions		$\dot{E}_3 + \dot{E}_4 + \dot{E}_5 - \dot{E}_2$	$\dot{E}_1$

continues

TABLE II (*Continued*)

Component	Schematic	Exergy rate of product $\dot{E}_P$	Exergy rate of fuel $\dot{E}_F$
Expander with cooling air supply		$\dot{E}_3$	$\dot{E}_1 + \dot{E}_4 + \dot{E}_5 - \dot{E}_2$
Fuel cell		$\dot{W} + (\dot{E}_6 - \dot{E}_5)$	Fuel cell as part of a system: $(\dot{E}_1 + \dot{E}_2) - (\dot{E}_3 + \dot{E}_4)$ Stand-alone fuel cell: $\dot{E}_1 + \dot{E}_2$
Ejector		$\dot{m}_1(e_3 - e_1)$	$\dot{m}_2(e_2 - e_3)$
Distillation column[b]		$\dot{E}_2^{CH} + \dot{E}_3^{CH} + \dot{E}_6^{CH} + \dot{E}_7^{CH}$ $- \dot{E}_1^{CH} + \dot{m}_6(e_6^{PH} - e_1^{PH})$ $+ \dot{m}_2(e_2^{PH} - e_1^{PH})$	$(\dot{E}_4 - \dot{E}_5) + \dot{m}_7(e_1^{PH} - e_7^{PH})$ $+ \dot{m}_3(e_1^{PH} - e_3^{PH})$
Gasifier		$\dot{E}_3^{CH} + (\dot{E}_3^{PH} + \dot{E}_4^{PH}$ $- \dot{E}_2^{PH} - \dot{E}_1^{PH})$	$\dot{E}_1^{CH} + \dot{E}_2^{CH} - \dot{E}_4^{CH}$
Steam reformer		$\dot{E}_3 - \dot{E}_1 - \dot{E}_2$	$\dot{E}_4 + \dot{E}_5 - \dot{E}_6$
Evaporator including steam drum		$\dot{E}_2 + \dot{E}_5 - \dot{E}_1$	$\dot{E}_3 - \dot{E}_4$

[a] These definitions assume that the purpose of the heat exchanger is to heat the cold steam ($T_1 > T_0$). If the purpose of the heat exchanger is to provide cooling ($T_3 < T_0$), then the following relations should be used: $\dot{E}_P = \dot{E}_4 - \dot{E}_3$ and $\dot{E}_F = \dot{E}_1 - \dot{E}_2$.

[b] Here, it is assumed that $e_j^{CH} > e_1^{CH}$ ($j = 2, 3, 6, 7$), $e_2^{PH} > e_1^{PH}$, $e_6^{PH} > e_1^{PH}$, $e_3^{PH} < e_1^{PH}$, $e_7^{PH} < e_1^{PH}$.

best source of cost estimates are cost values from past purchase orders, quotations from experienced professional cost estimators, or calculations using the extensive cost databases often maintained by engineering companies or company engineering departments. In addition, some commercially available software packages can assist with cost estimation.

When vendor quotations are lacking, or the cost or time requirements to prepare cost estimates are unacceptably high, the purchase costs of various equipment items can be obtained from the literature where they are usually given in the form of estimating charts. These charts have been obtained through a correlation of a large number of cost and design data. In a typical cost-estimating chart, when all available cost data are plotted versus the equipment size on a double logarithmic plot, the data correlation results in a straight line within a given capacity range. The slope of this line α represents an important cost-estimating parameter (scaling exponent) as shown by the relation

$$C_{P,Y} = C_{P,W} \left(\frac{X_Y}{X_W} \right)^\alpha, \tag{28}$$

where $C_{P,Y}$ is the purchase cost of the equipment in question, which has a size or capacity X_Y; and $C_{P,W}$ is the purchase cost of the same type of equipment in the same year but of capacity or size X_W.

In practice, the scaling exponent α does not remain constant over a large size range. For very small equipment items, size has almost no effect on cost, and the scaling exponent α is close to zero. For very large capacities, the difficulties of building and transporting the equipment may increase the cost significantly, and the scaling exponent approaches unity. Examples of scaling exponents for different plant components are provided by Bejan, Tsatsaronis, and Moran (1996). In the absence of other cost information, an exponent value of 0.6 may be used (six-tenths rule).

Predesign capital cost estimates are very often assembled from old cost data. These reference data C_{ref} can be updated to represent today's cost C_{new} by using an appropriate cost index.

$$C_{new} = C_{ref} \left(\frac{I_{new}}{I_{ref}} \right) \tag{29}$$

Here, I_{new} and I_{ref} are the values of the cost index today and in the reference year, respectively. Cost indices for special types of equipment and processes are published frequently in several journals (e.g., *Chemical Engineering*, *The Oil and Gas Journal*, and *Engineering News Record*).

B. Calculation of Revenue Requirements

The annual total revenue requirement (*TRR*, total product cost) for a system is the revenue that must be collected in a given year through the sale of all products to compensate the system operating company for all expenditures incurred in the same year and to ensure sound economic plant operation. It consists of two parts: carrying charges and expenses. Carrying charges are a general designation for charges that are related to capital investment, whereas expenses are used to define costs associated with the operation of a plant. Carrying charges (*CC*) include the following: total capital recovery; return on investment for debt, preferred stock, and common equity; income taxes; and other taxes and insurance. Examples for expenses are fuel cost (*FC*) and operating and maintenance costs (*OMC*). All annual carrying charges and expenses have to be estimated for each year over the entire economic life of a plant. Detailed information about the TRR method can be found in Bejan, Tsatsaronis, and Moran (1996) and in EPRI (1993).

C. Levelized Costs

The series of annual costs associated with carrying charges CC_j and expenses (FC_j and OMC_j) for the jth year of plant operation is not uniform. In general, carrying charges decrease while fuel costs increase with increasing years of operation (Bejan, Tsatsaronis, and Moran, 1996). A levelized value TRR_L for the total annual revenue requirement can be computed by applying a discounting factor and the capital-recovery factor CRF:

$$TRR_L = CRF \sum_{1}^{n} \frac{TRR_j}{(1 + i_{eff})^j}, \tag{30}$$

where TRR_j is the total revenue requirement in the jth year of plant operation, i_{eff} is the average annual effective discount rate (cost of money), and n denotes the plant economic life expressed in years. For Eq. (30), it is assumed that each money transaction occurs at the end of each year. The capital-recovery factor CRF is given by

$$CRF = \frac{i_{eff}(1 + i_{eff})^n}{(1 + i_{eff})^n - 1}. \tag{31}$$

If the series of payments for the annual fuel cost FC_j is uniform over time except for a constant escalation r_{FC} (i.e., $FC_j = FC_0(1 + r_{FC})^j$), then the levelized value FC_L of the series can be calculated by multiplying the fuel expenditure FC_0 at the beginning of the first year by the constant-escalation levelization factor *CELF*.

$$FC_L = FC_0 CELF = FC_0 \frac{k_{FC}\left(1 - k_{FC}^n\right)}{(1 - k_{FC})} CRF \tag{32}$$

with

$$k_{FC} = \frac{1 + r_{FC}}{1 + i_{eff}} \quad \text{and} \quad r_{FC} = \text{constant}.$$

The terms r_{FC} and CRF denote the annual escalation rate for the fuel cost and the capital-recovery factor [Eq. (31)], respectively.

Accordingly, the levelized annual operating and maintenance costs OMC_L are given by

$$OMC_L = OMC_0 CELF$$

$$= OMC_0 \frac{k_{OMC}(1 - k_{OMC}^n)}{(1 - k_{OMC})} CRF \quad (33)$$

with

$$k_{OMC} = \frac{1 + r_{OMC}}{1 + i_{eff}} \quad \text{and} \quad r_{OMC} = \text{constant}.$$

The term r_{OMC} is the nominal escalation rate for the operating and maintenance costs.

Finally, the levelized carrying charges CC_L are obtained from

$$CC_L = TRR_L - FC_L - OMC_L. \quad (34)$$

The major difference between a conventional economic analysis and an economic analysis conducted as part of a thermoeconomic analysis is that the latter is done at the plant component level. The annual carrying charges (superscript CI = capital investment) and operating and maintenance costs (superscript OM) of the total plant can be apportioned among the system components according to the contribution of the kth component to the purchased-equipment cost $PEC_{tot} = \sum_k PEC_k$ for the overall system:

$$\dot{Z}_k^{CI} = \frac{CC_L}{\tau} \frac{PEC_k}{\sum_k PEC_k}; \quad (35)$$

$$\dot{Z}_k^{OM} = \frac{OMC_L}{\tau} \frac{PEC_k}{\sum_k PEC_k}. \quad (36)$$

Here, PEC_k and τ denote the purchased-equipment cost of the kth plant component and the total annual time (in hours) of system operation at full load, respectively. The term $\dot{Z}_k$ represents the cost rate associated with capital investment and operating and maintenance expenses:

$$\dot{Z}_k = \dot{Z}_k^{CI} + \dot{Z}_k^{OM}. \quad (37)$$

The levelized cost rate of the expenditures for fuel $\dot{C}_F$ supplied to the overall system is given by

$$\dot{C}_F = \frac{FC_L}{\tau}. \quad (38)$$

Levelized costs, such as $\dot{Z}_k^{CI}$, $\dot{Z}_k^{OM}$, and $\dot{C}_F$, are used as input data for the thermoeconomic analysis.

D. Sensitivity Analysis

An economic analysis generally involves more uncertainties than a thermodynamic analysis. In the above discussion, it has been assumed that each variable in the economic analysis is known with certainty. However, many values used in the calculation are uncertain. A sensitivity analysis determines by how much a reasonable range of uncertainty assumed for each uncertain variable affects the final decision. Sensitivity studies are recommended to investigate the effect of major assumptions about values referring to future years (e.g., cost of money, inflation rate, and escalation rate of fuels) on the results of an economic analysis.

III. THERMOECONOMIC ANALYSIS

The exergy analysis yields the desired information for a complete evaluation of the design and performance of an energy system from the thermodynamic viewpoint. However, we still need to know how much the exergy destruction in a system component costs the system operator. Knowledge of this cost is very useful in improving the cost effectiveness of the system.

In a thermoeconomic analysis, evaluation and optimization, the cost rates associated with each material and energy stream are used to calculate component-related thermoeconomic variables. Thermoeconomic variables deal with investment costs and the corresponding costs associated with thermodynamic inefficiencies. Based on rational decision criteria, the required changes in structure and parameter values are identified. Thus, the cost minimization of a thermal system involves finding the optimum trade-offs between the cost rates associated with capital investment and exergy destruction.

A. Exergy Costing

In exergy costing a cost is assigned to each exergy stream. The cost rate associated with the jth material stream is expressed as the product of the stream exergy rate $\dot{E}_j$ and the average cost per exergy unit c_j:

$$\dot{C}_j = \dot{E}_j \cdot c_j = \dot{m}_j \cdot e_j \cdot c_j, \quad (39)$$

where e_j is the specific exergy on a mass basis. A cost is also assigned to the exergy transfers associated with heat or work:

$$\dot{C}_q = c_q \cdot \dot{Q}; \quad (40)$$

$$\dot{C}_w = c_w \cdot \dot{W}. \quad (41)$$

Here, c_j, c_q, and c_w denote average costs per unit of exergy in dollars per gigajoule exergy. The costs associated with each material and energy stream in a system are calculated with the aid of cost balances and auxiliary cost equations.

Sometimes it is appropriate to consider the costs of physical and chemical exergy associated with a material stream separately. By denoting the average costs per unit of physical and chemical exergy by c_j^{PH} and c_j^{CH}, respectively, the cost rate associated with stream j becomes

$$\dot{C}_j = c_j E_j = c_j^{PH} \dot{E}_j^{PH} + c_j^{CH} \dot{E}_j^{CH}$$
$$= \dot{m}_j \left(c_j^{PH} e_j^{PH} + c_j^{CH} e_j^{CH} \right). \qquad (42)$$

B. Cost Balance

Exergy costing involves cost balances usually formulated for each system component separately. A cost balance applied to the kth system component expresses that the total cost of the exiting streams equals the total cost of the entering streams plus the appropriate charges due to capital investment and operating and maintenance expenses $\dot{Z}$ (Fig. 3).

$$\sum_{j=1}^{n} \dot{C}_{j,k,in} + \underbrace{\dot{Z}_k^{CI} + \dot{Z}_k^{OM}}_{\dot{Z}_k} = \sum_{j=1}^{m} \dot{C}_{j,k,out} \qquad (43)$$

Cost balances are generally written so that all terms are positive. The rates $\dot{Z}_k^{CI}$ and $\dot{Z}_k^{OM}$ are calculated from Eqs. (35) and (36).

Introducing the cost rate expressions from Eqs. (39)–(41), we obtain

$$\sum_{j=1}^{n} (c_j \dot{E}_j)_{k,in} + \dot{Z}_k^{CI} + \dot{Z}_k^{OM} = \sum_{j=1}^{m} (c_j \dot{E}_j)_{k,out}. \qquad (44)$$

The exergy rates $\dot{E}_j$ entering and exiting the kth system component are calculated in an exergy analysis. In a thermoeconomic analysis of a component, we may assume that the costs per exergy unit c_j are known for all entering streams. These costs are known either from the components they exit or, if they are streams entering the overall system, from their purchase costs. Consequently, the unknown variables that need to be calculated with the

aid of the cost balance are the costs per exergy unit of the exiting streams. Usually, some auxiliary relations are required to calculate these costs.

In the exergetic evaluation, a fuel and a product were defined for each component of a system (see Table II). The cost flow rates associated with the fuel $\dot{C}_F$ and product $\dot{C}_P$ of a component are calculated in a similar way to the exergy flow rates $\dot{E}_F$ and $\dot{E}_P$. Then, the cost balance becomes

$$\dot{C}_P = \dot{C}_F + \dot{Z} - \dot{C}_L. \qquad (45)$$

The term $\dot{C}_L$ represents the monetary loss associated with the rejection of exergy to the environment. Table III shows the definitions of $\dot{C}_P$ and $\dot{C}_F$ for selected components at steady state.

C. Auxiliary Costing Equations

When the costs of entering streams are known, a cost balance is generally not sufficient to determine the costs of the streams exiting a component because the number of exiting streams is usually larger than one. Thus, in general, it is necessary to formulate some auxiliary equations for a component, with the number of these equations being equal to the number of exiting streams minus one. The auxiliary equations are called either F equations or P equations depending on whether the exergy stream being considered is used in the calculation of fuel or product for the component (Lazzaretto and Tsatsaronis, November 1999).

F Equations. The total cost associated with the removal of exergy from an exergy stream in a component must be equal to the cost at which the removed exergy was supplied to the same stream in upstream components. The exergy difference of this stream between inlet and outlet is considered in the definition of fuel for the component.

P Equations. Each exergy unit is supplied to any stream associated with the product of a component at the same average cost $c_{P,k}$. This cost is calculated from the cost balance and the F equations.

The following two examples illustrate the development of the F and P equations.

Example 1. Calculate the costs of the streams exiting the adiabatic turbine (subscript T) or expander in Table III. The cost balance for this component is

$$\dot{C}_1 + \dot{Z}_T = \dot{C}_2 + \dot{C}_3 + \dot{C}_4. \qquad (46)$$

By grouping together the terms associated with fuel and product, we obtain

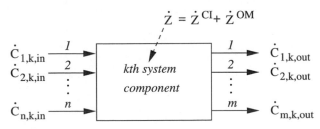

$$\dot{Z} = \dot{Z}^{CI} + \dot{Z}^{OM}$$

$\dot{C}_{1,k,in} \xrightarrow{1}$
$\dot{C}_{2,k,in} \xrightarrow{2}$ | *k*th system component | $\xrightarrow{1} \dot{C}_{1,k,out}$
$\xrightarrow{2} \dot{C}_{2,k,out}$
$\vdots$
$\dot{C}_{n,k,in} \xrightarrow{n}$ | | $\xrightarrow{m} \dot{C}_{m,k,out}$

FIGURE 3 Schematic of a system component to illustrate a cost balance.

TABLE III Cost Rates Associated with Fuel and Product for Selected Components at Steady State Operation

Component	Schematic	Cost rate of product $\dot{C}_P$	Cost rate of fuel $\dot{C}_F$
Compressor pump, or fan		$\dot{C}_2 - \dot{C}_1$	$\dot{C}_3$
Turbine or expander		$\dot{C}_4$	$\dot{C}_1 - \dot{C}_2 - \dot{C}_3$
Heat exchanger[a]		$\dot{C}_2 - \dot{C}_1$	$\dot{C}_3 - \dot{C}_4$
Mixing unit		$\dot{m}_1(c_{3,1}e_3 - c_1e_1)$ with $c_{3,1} = c_3 + \dfrac{\dot{m}_2}{\dot{m}_1}(c_3 - c_2)$	$\dot{m}_2c_2(e_2 - e_3)$
Combustion chamber		$\dot{C}_3 - \dot{C}_2$	$\dot{C}_1 - \dot{C}_4$
Steam generator		$(\dot{C}_6 - \dot{C}_5) + (\dot{C}_8 - \dot{C}_7)$	$(\dot{C}_1 + \dot{C}_2) - (\dot{C}_3 + \dot{C}_4)$
Deaerator		$\dot{m}_2(c_{3,2}e_3 - c_2e_2)$ with $c_{3,2} = c_3 + \dfrac{\dot{m}_3 - \dot{m}_2}{\dot{m}_2}(c_3 - c_1)$	$(\dot{m}_3 - \dot{m}_2)c_1(e_1 - e_3)$ $+ \dot{m}_4c_1(e_1 - e_4)$
Compressor with cooling air extractions		$\dot{C}_3 + \dot{C}_4 + \dot{C}_5 - \dot{C}_2$	$\dot{C}_1$
Expander with cooling air supply		$\dot{C}_3$	$\dot{C}_1 + \dot{C}_4 + \dot{C}_5 - \dot{C}_2$

continues

TABLE III   *(Continued)*

Component	Schematic	Cost rate of product $\dot{C}_P$	Cost rate of fuel $\dot{C}_F$
Fuel cell		Fuel cell as part of a system: $\dot{C}_7 + (\dot{C}_6 - \dot{C}_5) + (\dot{C}_3^{PH} - \dot{C}_1^{PH})$ $+ (\dot{C}_4^{PH} - \dot{C}_2^{PH})$ Stand-alone fuel cell: $\dot{C}_7 + (\dot{C}_6 - \dot{C}_5)$	Fuel cell as part of a system: $(\dot{C}_1^{CH} + \dot{C}_2^{CH}) - (\dot{C}_3^{CH} + \dot{C}_4^{CH})$ Stand-alone fuel cell: $\dot{C}_1 + \dot{C}_2$
Ejector		$\dot{m}_1(c_{3,1}e_3 - c_1 e_1)$ where $c_{3,1} = c_3 + \frac{\dot{m}_2}{\dot{m}_1}(c_3 - c_2)$	$\dot{m}_2 c_2(e_2 - e_3)$
Distillation column[b]		$\dot{C}_2^{CH} + \dot{C}_3^{CH} + \dot{C}_6^{CH} + \dot{C}_7^{CH} - \dot{C}_1^{CH}$ $+ \dot{m}_6(c_6^{PH}e_6^{PH} - c_1^{CH}e_1^{PH})$ $+ \dot{m}_2(c_2^{PH}e_2^{PH} - c_1^{CH}e_1^{PH})$	$(\dot{C}_4 - \dot{C}_5) + \dot{m}_7(c_1^{PH}e_1^{PH}$ $- c_7^{PH}e_7^{PH}) + \dot{m}_3(c_1^{PH}e_1^{PH}$ $- c_3^{PH}e_3^{PH})$
Gassifier		$\dot{C}_3^{CH} + (\dot{C}_3^{PH} + \dot{C}_4^{PH}$ $- \dot{C}_2^{PH} - \dot{C}_1^{PH})$	$\dot{C}_1^{CH} + \dot{C}_2^{CH} - \dot{C}_4^{CH}$
Steam reformer		$\dot{C}_3 - \dot{C}_1 - \dot{C}_2$	$\dot{C}_4 + \dot{C}_5 - \dot{C}_6$
Evaporator including steam drum		$\dot{C}_2 + \dot{C}_5 - \dot{C}_1$	$\dot{C}_3 - \dot{C}_4$

[a] These definitions assume that the purpose of the heat exchanger is to heat the cold steam ($T_1 > T_0$). If the purpose of the heat exchanger is to provide cooling ($T_3 < T_0$), then the following relations should be used: $\dot{C}_P = \dot{C}_4 - \dot{C}_3$ and $\dot{C}_F = \dot{C}_1 - \dot{C}_2$.

[b] Here, it is assumed that $e_j^{CH} > e_1^{CH}$ ($j = 2, 3, 6, 7$), $e_2^{PH} > e_1^{PH}$, $e_6^{PH} > e_1^{PH}$, $e_3^{PH} < e_1^{PH}$, $e_7^{PH} < e_1^{PH}$.

$$c_p \cdot \dot{E}_P = \underbrace{\dot{C}_4}_{\dot{C}_P} = \underbrace{(\dot{C}_1 - \dot{C}_2 - \dot{C}_3)}_{\dot{C}_F} + \dot{Z}_T. \qquad (47)$$

The F equation for this component states that the cost associated with each removal of exergy from the entering material stream (stream 1) must be equal to the average cost at which the removed exergy ($\dot{E}_1 - \dot{E}_2 - \dot{E}_3$) was supplied to the same stream in upstream components.

$$\dot{C}_1 - \dot{C}_2 - \dot{C}_3 = c_1 \cdot (\dot{E}_1 - \dot{E}_2 - \dot{E}_3) \qquad (48)$$

Equation (48) is equivalent to

$$c_2 = c_3 = c_1. \qquad (49)$$

The unknown cost rate $\dot{C}_4$ can be calculated with the aid of Eq. (48) or (49) and the cost balance [Eq. (47)].

Example 2. Calculate the costs of the streams exiting the adiabatic steam generator (subscript *SG*) shown in Table III, if streams 3 and 4 are (case a) rejected to the environment without additional expenses and (case b) discharged to the environment after using ash handling and gas cleaning equipment.

The cost balance for this component is

$$\dot{C}_1 + \dot{C}_2 + \dot{C}_5 + \dot{C}_7 + \dot{Z}_{SG} = \dot{C}_3 + \dot{C}_4 + \dot{C}_6 + \dot{C}_8. \quad (50)$$

By grouping together the terms associated with fuel and product, we obtain

$$c_p \cdot \dot{E}_P = \underbrace{(\dot{C}_6 - \dot{C}_5) + (\dot{C}_8 - \dot{C}_7)}_{\dot{C}_P}$$

$$= \underbrace{(\dot{C}_1 + \dot{C}_2) - (\dot{C}_3 + \dot{C}_4)}_{\dot{C}_F} + \dot{Z}_{SG}. \quad (51)$$

- Case a: The costs associated with streams 3 and 4 are calculated from *F* equations: Each exergy unit in the ash (stream 3) and in the flue gas (stream 4) is supplied by coal and oxygen at the same average cost.

$$\frac{\dot{C}_3}{\dot{E}_3} = \frac{\dot{C}_4}{\dot{E}_4} = \frac{\dot{C}_1 + \dot{C}_2}{\dot{E}_1 + \dot{E}_2} \quad (52)$$

or

$$c_3 = c_4 = \frac{\dot{C}_1 + \dot{C}_2}{\dot{E}_1 + \dot{E}_2} \quad (53)$$

The *F* equations [Eq. (52)] and the cost balance [Eq. (51)] may be used to calculate the c_P value in Eq. (51). Then, the *P* equations supply the costs associated with streams 6 and 8: each exergy unit is supplied to the main steam stream and to the reheat stream at the same average cost.

$$c_P = \frac{\dot{C}_6 - \dot{C}_5}{\dot{E}_6 - \dot{E}_5} = \frac{\dot{C}_8 - \dot{C}_7}{\dot{E}_8 - \dot{E}_7} \quad (54)$$

- Case b: All costs associated with the final disposal of streams 3 and 4 must be charged to the useful streams exiting the boiler, that is, to the main steam and the hot reheat.

The following equations are obtained from the overall system (Fig. 4) and the cost balances for ash handling (subscript *ah*) and gas cleaning (subscript *gc*):

$$\frac{\dot{C}_9}{\dot{E}_9} = \frac{\dot{C}_{10}}{\dot{E}_{10}} = \frac{\dot{C}_1 + \dot{C}_2}{\dot{E}_1 + \dot{E}_2}, \quad (55)$$

$$\dot{C}_3 = \dot{C}_9 - \dot{Z}_{ah}, \quad (56)$$

$$\dot{C}_4 = \dot{C}_{10} - \dot{Z}_{gc}. \quad (57)$$

The cost rates $\dot{C}_3$ and $\dot{C}_4$ are usually negative in case b. The unknown cost rates $\dot{C}_3$, $\dot{C}_4$, $\dot{C}_6$, $\dot{C}_8$, $\dot{C}_9$, and $\dot{C}_{10}$ are calculated from Eqs. (51) and (54)–(57).

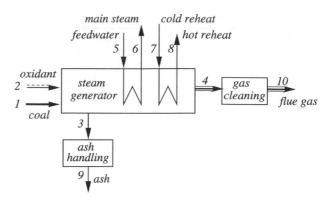

FIGURE 4 Schematic of a steam generator to illustrate case b in Example 2.

For the system components having only one exiting stream, such as compressors without extractions, pumps, fans, mixing units, and ejectors, the cost associated with this exiting stream can be calculated from the cost balance Eq. (43). When the costs of physical and chemical exergy are considered separately (e.g., for a distillation column and a gasifier in Table III), and when more than one exergy stream exits a system component, *F* or *P* equations are required, depending on whether the exiting stream being considered belongs to the fuel or to the product.

The costing equations for the turbine and the steam generator are discussed above in Examples 1 and 2. In the following, the cost balances and the required *F* and *P* equations are given for the remaining devices shown in Table III.

- Heat exchanger (*HX*):

$$(\dot{C}_2 - \dot{C}_1) = \dot{Z}_{HX} + (\dot{C}_3 - \dot{C}_4) \quad (58)$$

$$\frac{\dot{C}_4}{\dot{E}_4} = \frac{\dot{C}_3}{\dot{E}_3} \quad (59)$$

- Combustion chamber (*CC*):

$$(\dot{C}_3 - \dot{C}_2) = \dot{Z}_{CC} + (\dot{C}_1 - \dot{C}_4) \quad (60)$$

$$\frac{\dot{C}_4}{\dot{E}_4} = \frac{\dot{C}_1}{\dot{E}_1} \quad (61)$$

- Deaerator (*DA*):

$$\dot{C}_4 + \dot{C}_3 = \dot{C}_1 + \dot{C}_2 + \dot{Z}_{DA} \quad (62)$$

$$\frac{\dot{C}_4}{\dot{E}_4} = \frac{\dot{C}_1}{\dot{E}_1} \quad (63)$$

- Compressor with cooling air extractions (*C*):

$$c_3\dot{E}_3 + c_4\dot{E}_4 + c_5\dot{E}_5 - c_2\dot{E}_2 = c_1\dot{E}_1 + \dot{Z}_C \quad (64)$$

$$\frac{c_3e_3 - c_2e_2}{e_3 - e_2} = \frac{c_4e_4 - c_2e_2}{e_4 - e_2} = \frac{c_5e_5 - c_2e_2}{e_5 - e_2} \quad (65)$$

- Expander with cooling air supply (T):

$$\dot{C}_3 = (\dot{C}_1 + \dot{C}_4 + \dot{C}_5 - \dot{C}_2) + \dot{Z}_T \qquad (66)$$

$$\frac{\dot{C}_2}{\dot{E}_2} = \frac{\dot{C}_1 + \dot{C}_4 + \dot{C}_5}{\dot{E}_1 + \dot{E}_4 + \dot{E}_5} \qquad (67)$$

- Fuel cell (FC): The purpose of the fuel cell shown in Table III is to generate electricity and to supply useful thermal energy.

$$\dot{C}_7 + (\dot{C}_6 - \dot{C}_5) = (\dot{C}_1 + \dot{C}_2) - (\dot{C}_3 + \dot{C}_4) + \dot{Z}_{FC} \qquad (68)$$

The F and P equations for a fuel cell as part (component) of a system are

$$\frac{\dot{C}_3^{CH}}{\dot{E}_3^{CH}} = \frac{\dot{C}_1^{CH}}{\dot{E}_1^{CH}}; \qquad \frac{\dot{C}_4^{CH}}{\dot{E}_4^{CH}} = \frac{\dot{C}_2^{CH}}{\dot{E}_2^{CH}} \qquad (69)$$

$$\frac{\dot{C}_3^{PH} - \dot{C}_1^{PH}}{\dot{E}_3^{PH} - \dot{E}_1^{PH}} = \frac{\dot{C}_4^{PH} - \dot{C}_2^{PH}}{\dot{E}_4^{PH} - \dot{E}_2^{PH}} = \frac{\dot{C}_6 - \dot{C}_5}{\dot{E}_6 - \dot{E}_5} = \frac{\dot{C}_7}{\dot{E}_7}. \qquad (70)$$

For a stand-alone fuel cell (system), we obtain

$$\frac{\dot{C}_3}{\dot{E}_3} = \frac{\dot{C}_1}{\dot{E}_1}; \qquad \frac{\dot{C}_4}{\dot{E}_4} = \frac{\dot{C}_2}{\dot{E}_2} \qquad (71)$$

and

$$\frac{\dot{C}_6 - \dot{C}_5}{\dot{E}_6 - \dot{E}_5} = \frac{\dot{C}_7}{\dot{E}_7}. \qquad (72)$$

For a stand-alone fuel cell, the exergies of streams 3 and 4 are exergy losses .

- Distillation column (DC): The purpose of a distillation column is to separate the chemical components in the feed, i.e., to increase the chemical exergy of the substances in the feed at the expense of the physical exergy supplied to the system with the feed and in the reboiler. For this component, it is appropriate to consider the costs of physical and chemical exergy for streams 1, 2, 3, 6, and 7 separately.

The cost balance for the distillation column is given by

$$c_2^{CH} \dot{E}_2^{CH} + c_3^{CH} \dot{E}_3^{CH} + c_6^{CH} \dot{E}_6^{CH} + c_7^{CH} \dot{E}_7^{CH}$$
$$- c_1^{CH} \dot{E}_1^{CH} + \dot{m}_6 (c_6^{PH} e_6^{PH} - c_1^{PH} e_1^{PH})$$
$$+ \dot{m}_2 (c_2^{PH} e_2^{PH} - c_1^{PH} e_1^{PH}) + (c_9 \dot{E}_9 - c_8 \dot{E}_8)$$
$$= (c_4 \dot{E}_4 - c_5 \dot{E}_5) + \dot{m}_7 (c_1^{PH} e_1^{PH} - c_7^{PH} e_7^{PH})$$
$$+ \dot{m}_3 (c_1^{PH} e_1^{PH} - c_3^{PH} e_3^{PH}) + \dot{Z}_{DC}. \qquad (73)$$

The F equations are

$$c_5 = c_4, \qquad (74)$$

$$c_7^{PH} = c_3^{PH} = c_1^{PH}. \qquad (75)$$

Each exergy unit is supplied to the product at the same average cost (P equations):

$$\frac{\dot{m}_6 (c_6^{PH} e_6^{PH} - c_1^{PH} e_1^{PH})}{\dot{m}_6 (e_6^{PH} - e_1^{PH})} = \frac{\dot{m}_2 (c_2^{PH} e_2^{PH} - c_1^{PH} e_1^{PH})}{\dot{m}_2 (e_2^{PH} - e_1^{PH})}$$
$$= \frac{\dot{m}_2 (c_2^{CH} e_2^{CH} - c_1^{CH} e_1^{CH})}{\dot{m}_2 (e_2^{CH} - e_1^{CH})} = \frac{\dot{m}_3 (c_3^{CH} e_3^{CH} - c_1^{CH} e_1^{CH})}{\dot{m}_3 (e_3^{CH} - e_1^{CH})}$$
$$= \frac{\dot{m}_6 (c_6^{CH} e_6^{CH} - c_1^{CH} e_1^{CH})}{\dot{m}_6 (e_6^{CH} - e_1^{CH})} = \frac{\dot{m}_7 (c_7^{CH} e_7^{CH} - c_1^{CH} e_1^{CH})}{\dot{m}_7 (e_7^{CH} - e_1^{CH})}. \qquad (76)$$

The exergy increase of the cooling water (stream 8) is an exergy loss.

$$c_9 = c_8 \qquad (77)$$

- Gasifier (GS):

The purpose of a gasifier is to convert the chemical exergy of a solid fuel into the chemical exergy of a gaseous fuel. Due to the chemical reactions involved, the physical exergy of the material streams increase between the inlet and the outlet.

$$\dot{C}_3^{CH} + (\dot{C}_3^{PH} + \dot{C}_4^{PH} - \dot{C}_1^{PH} - \dot{C}_2^{PH})$$
$$= \dot{C}_1^{CH} + \dot{C}_2^{CH} - \dot{C}_4^{CH} + \dot{Z}_{GS} \qquad (78)$$

When the chemical and physical exergies are considered separately, the following P equation is obtained:

$$\frac{\dot{C}_3^{CH}}{\dot{E}_3^{CH}} = \frac{\dot{C}_3^{PH} + \dot{C}_4^{PH} - \dot{C}_1^{PH} - \dot{C}_2^{PH}}{\dot{E}_3^{PH} + \dot{E}_4^{PH} - \dot{E}_1^{PH} - \dot{E}_2^{PH}}, \qquad (79)$$

$$\frac{\dot{C}_4^{CH}}{\dot{E}_4^{CH}} = \frac{\dot{C}_1^{CH}}{\dot{E}_1^{CH}}. \qquad (80)$$

- Steam reformer (SR):

$$\dot{C}_3 - \dot{C}_1 - \dot{C}_2 = \dot{C}_4 + \dot{C}_5 - \dot{C}_6 + \dot{Z}_{SR}, \qquad (81)$$

$$\frac{\dot{C}_6}{\dot{E}_6} = \frac{\dot{C}_4 + \dot{C}_5}{\dot{E}_4 + \dot{E}_5}. \qquad (82)$$

- Evaporator including steam drum (EV):

$$\dot{C}_2 + \dot{C}_5 - \dot{C}_1 = \dot{C}_3 - \dot{C}_4 + \dot{Z}_{EV}, \qquad (83)$$

$$\frac{c_5 e_5 - c_1 e_1}{e_5 - e_1} = \frac{c_2 e_2 - c_1 e_1}{e_2 - e_1}. \qquad (84)$$

IV. THERMOECONOMIC EVALUATION

In a thermoeconomic evaluation, the cost per unit of exergy and the cost rates associated with each material and exergy stream are used to calculate thermoeconomic variables for each system component. From the exergy analysis we already know the exergetic variables: rate of exergy destruction $\dot{E}_{D,k}$; exergetic efficiency ε_k; and exergy destruction ratio $y_{D,k}$.

A. Thermoeconomic Variables

The definition of fuel and product for the exergetic efficiency calculation in a component leads to the cost flow rates associated with the fuel $\dot{C}_{F,k}$ and product $\dot{C}_{P,k}$ of the the kth component. $\dot{C}_{F,k}$ represents the cost flow rate at which the fuel exergy $\dot{E}_{F,k}$ is provided to the kth component. $\dot{C}_{P,k}$ is the cost flow rate associated with the product exergy $\dot{E}_{P,k}$ for the same component.

The average cost of fuel for the kth system component $c_{F,k}$ expresses the average cost at which each exergy unit of fuel (as defined in the exergetic efficiency) is supplied to the kth system component.

$$c_{F,k} = \frac{\dot{C}_{F,k}}{\dot{E}_{F,k}}. \qquad (85)$$

The value of $c_{F,k}$ depends on the relative position of the kth component in the system and on the interconnections between the kth component and the remaining components. As a general rule, the closer the kth component is to the product (fuel) stream of the overall system, usually the larger (smaller) the value of $c_{F,k}$.

Similarly, the unit cost of product $c_{P,k}$ is the average cost at which each exergy unit was supplied to the product for the kth component:

$$c_{P,k} = \frac{\dot{C}_{P,k}}{\dot{E}_{P,k}}. \qquad (86)$$

Using Eqs. (85) and (86), the cost balance for the kth system component can be written as

$$\dot{c}_{P,k}\dot{E}_{P,k} = c_{F,k}\dot{E}_{F,k} + \dot{Z}_k - \dot{C}_{L,k}. \qquad (87)$$

One of the most important aspects of exergy costing is calculating the cost of exergy destruction in each component of the energy system being considered. The cost rate $\dot{C}_{D,k}$ associated with exergy destruction $\dot{E}_{D,k}$ in the kth system component is a "hidden" cost that can be revealed only through a thermoeconomic analysis. It can be approximated by the cost of the additional fuel that needs to be supplied to this component to cover the exergy destruction and to generate the same exergy flow rate of product $\dot{E}_{P,k}$:

$$\dot{C}_{D,k} = c_{F,k}\dot{E}_{D,k}, \quad \text{when } \dot{E}_{P,k} = \text{const.} \qquad (88)$$

For most well-designed components, as the exergy destruction decreases or efficiency increases, the cost of exergy destruction $\dot{C}_{D,k}$ decreases while the capital investment $\dot{Z}_k^{CI}$ increases. The higher the $\dot{C}_{D,k}$ value is,

the lower the $\dot{Z}_k$ value. It is one of the most interesting features of thermoeconomics that the exergy destruction costs associated with a component are estimated and compared with the investment costs of the same component. This comparison facilitates the decision about the design changes that might improve the cost effectiveness of the overall system. In a thermoeconomic optimization we try to find the appropriate trade-offs between $\dot{C}_{D,k}$ and $\dot{Z}_k$. The $\dot{C}_{D,k}$ values cannot be added to calculate the $\dot{C}_D$ value for a group of components because each $\dot{C}_{D,k}$ already contains information related to the interactions among the cost of exergy destruction in upstream components. The $\dot{C}_D$ value for the group should be calculated using the average cost per exergy unit of fuel c_F for the group and the exergy destruction rate within the group.

Figure 5 shows the relation between investment cost per unit of product exergy $\dot{Z}_k^{CI}/\dot{E}_{P,k}$ and exergy destruction per unit of product exergy $\dot{E}_{D,k}/\dot{E}_{P,k}$ for the kth component. The shaded area illustrates the range of variation of the investment cost due to uncertainty and to multiple technical design solutions that might be available. The investment cost per unit of product exergy $c_{P,k}^Z$ increases with decreasing exergy destruction per unit of product exergy or with increasing exergetic efficiency. This cost behavior is exhibited by most components. The components that exhibit a decrease of $c_{P,k}^Z$ with increasing efficiency do not need to be considered in a thermoeconomic evaluation since for these components no optimization dilemma exists: Among all available solutions we would use the most efficient component that has both the lowest specific fuel expenses and the lowest specific investment cost $c_{P,k}^Z$ and, thus, the minimum $c_{P,k}$ value (Tsatsaronis and Park, 1999).

The relative cost difference r_k between the average cost per exergy unit of product and average cost per exergy unit of fuel is given by

$$r_k \equiv \frac{c_{P,k} - c_{F,k}}{c_{F,k}} = \frac{c_{F,k}(\dot{E}_{D,k} + \dot{E}_{L,k}) + (\dot{Z}_k^{CI} + \dot{Z}_k^{OM})}{c_{F,k}\dot{E}_{P,k}}$$
$$= \frac{1 - \varepsilon_k}{\varepsilon_k} + \frac{\dot{Z}_k^{CI} + \dot{Z}_k^{OM}}{c_{F,k}\dot{E}_{P,k}}. \qquad (89)$$

This equation reveals the real cost sources in the kth component, which are the capital cost $\dot{Z}_k^{CI}$, the exergy destruction within the component as expressed by $\dot{C}_{D,k}$, the operating and maintenance costs, and the exergy loss. Among these sources the first two are the most significant ones and are used for calculating the exergoeconomic factor.

The exergoeconomic factor expresses the contribution of the capital cost to the sum of capital cost and cost of exergy destruction:

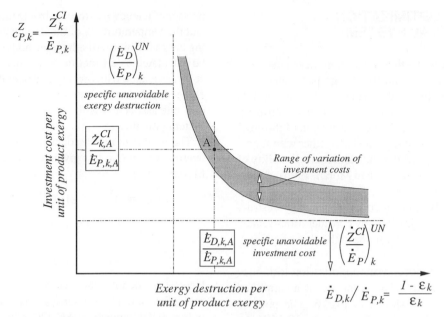

FIGURE 5 Expected relationship between investment cost and exergy destruction (or exergetic efficiency) for the kth component of a thermal system [Based on Fig. 1 in Tsatsaronis, G., and Park, M. H. (1999). "ECOS'99, Efficiency Costs, Optimization, Simulation and Environmental Aspects of Energy Systems," pp. 161–121, Tokyo, Japan, June 8–10.]

$$f_k \equiv \frac{\dot{Z}_k^{CI}}{\dot{Z}_k^{CI} + \dot{C}_{D,k}} = \frac{\dot{Z}_k^{CI}}{\dot{Z}_k^{CI} + c_{F,k}\dot{E}_{D,k}}. \qquad (90)$$

The thermoeconomic variables $\dot{Z}_k^{CI}$ and $\dot{C}_{D,k}$ provide absolut measures of the importance of the kth component, whereas the variables r_k and f_k provide relative measures of the component cost effectiveness.

B. Design Evaluation

A detailed thermoeconomic evaluation of the design of a thermal system is based on the following exergetic and thermoeconomic variables, each calculated for the kth system component:

$\dot{E}_{D,k}$	rate of exergy destruction	Eq. (25)
ε_k	exergetic efficiency	Eq. (24)
$y_{D,k}$	exergy destruction ratio	Eq. (26)
$\dot{Z}_k^{CI}$	cost rate associated with capital investment	Eqs. (35) and (37)
$\dot{C}_{D,k}$	cost rate associated with exergy destruction	Eq. (88)
r_k	relative cost difference	Eq. (89)
f_k	exergoeconomic factor	Eq. (90)

The values of all exergetic and thermoeconomic variables depend on the component type (e.g., turbine, heat exchanger, chemical reactor). For instance, the value of the exergoeconomic factor is typically between 25 and 65% for compressors and turbines.

The following guidelines may be applied to the evaluation of the kth system component to improve the cost effectiveness of the entire system:

1. Rank the components in descending order of cost importance using the sum $\dot{Z}_k^{CI} + \dot{C}_{D,k}$.
2. Consider initially design changes for the components for which the value of this sum is high.
3. Pay particular attention to components with a high relative cost difference r_k, especially when the cost rates $\dot{Z}_k^{CI}$ and $\dot{C}_{D,k}$ are high.
4. Use the exergoeconomic factor f_k to identify the major cost source (capital investment or cost of exergy destruction).

   a. If the f_k value is high, investigate whether it is cost effective to reduce the capital investment for the kth component at the expense of the component efficiency.

   b. If the f_k value is low, try to improve the component efficiency by increasing the capital investment.

5. Eliminate any subprocesses that increase the exergy destruction or exergy loss without contributing to the reduction of capital investment or of fuel costs for other components.
6. Consider improving the exergetic efficiency of a component if it has a relatively low exergetic efficiency or a relatively large value for the rate of exergy destruction, or the exergy destruction ratio.

V. ITERATIVE OPTIMIZATION OF A THERMAL SYSTEM

Design optimization of a thermal system means the modification of the structure and the design parameters of a system to minimize the total levelized cost of the system products under boundary conditions associated with available materials, financial resources, protection of the environment, and government regulation, together with safety, reliability, operability, availability, and maintainability of the system (Bejan, Tsatsaronis, and Moran, 1996). A truly optimized system is one for which the magnitude of every significant thermodynamic inefficiency (exergy destruction and exergy loss) is justified by considerations related to costs or is imposed by at least one of the above boundary conditions.

The iterative thermoeconomic optimization technique discussed here is illustrated with the aid of a simplified cogeneration system as shown in Fig. 6. The purpose of the cogeneration system is to generate 30 MW net electric power and to provide 14 kg/s saturated steam at 20 bar (stream 7). A similar cogeneration system is used in Penner and Tsatsaronis (1994) for comparing different thermoeconomic optimization techniques. It should be emphasized that the cost minimization of a real system according to Fig. 6 is significantly simpler compared with the case considered here, since in a real system engineers have to select one of the few gas turbine systems available in the market for a given capacity and then have to optimize the design of only the heat-recovery steam generator. In this example, however, to demonstrate application of the iterative thermoeconomic optimization technique with the aid of a simple system containing different components, we assume that we can freely decide about the design of each component included in the gas turbine system.

The decision variables selected for the optimization are the compressor pressure ratio p_2/p_1, the isentropic compressor efficiency η_{sc}, the isentropic turbine efficiency η_{st}, and the temperature T_3 of the combustion products entering the gas turbine. All other thermodynamic variables can be calculated as a function of the decision variables. All costs associated with owning and operating this system are charged to the product streams 7 and 10. This includes the cost rate associated with the exergy loss from the overall system (stream 5).

The total cost rate associated with the product for the overall system $\dot{C}_{P,tot}$ is given by (Bejan, Tsatsaronis, and Moran, 1996)

$$\dot{C}_{P,tot} = (\dot{C}_7 - \dot{C}_6) + \dot{C}_{10} + \dot{C}_5 + \dot{Z}_{other}$$

$$= \dot{C}_1 + \dot{C}_8 + \sum_k \dot{Z}_k + \dot{Z}_{other} \overset{!}{=} min \quad (91)$$

$$k = AC, CC, GT, HRSG$$

Here, $\dot{Z}_{other}$ denotes the levelized cost associated with the capital investment for other system components not shown in Fig. 6. The subscripts AC, CC, GT, and HRSG refer to the components air compressor, combustion chamber, gas turbine, and heat-recovery steam generator, respectively. Equation (91) represents the objective function for the cost minimization.

The iterative thermoeconomic optimization of the cogeneration system is subject to the following constraints: $6.0 \leq p_2/p_1 \leq 14.0$; $0.75 \leq \eta_{sc} \leq 0.89$; $0.8 \leq \eta_{st} \leq 0.91$; and $900 \text{ K} \leq T_3 \leq 1600 \text{ K}$. The minimum temperature difference ΔT_{min} in the heat-recovery steam generator must be at least 15 K. A first workable design (base-case design) was developed using the following values of the decision variables:

$$p_2/p_1 = 14, \quad T_3 = 1260 \text{ K}, \quad \eta_{sc} = 0.86, \quad \eta_{st} = 0.87.$$

Relevant thermodynamic and economic data as well as the values of the thermoeconomic variables for the base-case design are shown in Tables IV and V. The thermoeconomic variables are used to determine the changes in the design of each component to be implemented in the next step of the iterative optimization procedure. Knowledge about how the decision variables qualitatively affect the exergetic efficiency and the costs associated with each system component is required to determine the new values of the decision variables that improve the cost effectiveness of the overall system. Engineering judgment and critical evaluations must be used when deciding on the changes to be made from one iteration step to the next.

A. Evaluation of the First Design Case (Base-Case Design)

Table V summarizes the thermoeconomic variables and the purchased equipment costs PEC calculated for each

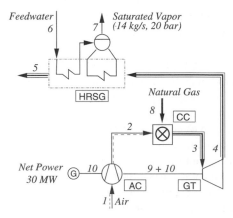

FIGURE 6 A simple cogeneration system.

TABLE IV Mass Flow Rate, Temperature, Pressure, Exergy Flow Rates, Cost Per Exergy Unit, and Cost Flow Rate for Each Stream in the Base-Case Design of the Cogeneration System

Nr.	$\dot{m}_j$ (kg/s)	T (K)	p (bar)	$\dot{E}_j^{PH}$ (MW)	$\dot{E}_j^{CH}$ (MW)	$\dot{E}_j$ (MW)	c_j ($/GJ)	$\dot{C}_j$ ($/h)
1	132.960	298.15	1.013	0.000	0.000	0.000	0.00	0
2	132.960	674.12	14.182	49.042	0.000	49.042	32.94	5815
3	134.943	1260.00	13.473	115.147	0.351	115.498	18.11	7532
4	134.943	754.35	1.066	27.518	0.351	27.869	18.11	1817
5	134.943	503.62	1.013	7.305	0.351	7.656	18.11	499
6	14.000	298.15	20.000	0.027	0.035	0.062	0.00	0
7	14.000	485.52	20.000	12.775	0.035	12.810	34.25	1579
8	1.983	298.15	20.000	0.904	101.896	102.800	4.56	1689
9						52.479	21.47	4057
10						30.000	21.47	2319

component of the cogeneration system. According to the methodology described in Section IV.B, the components are listed in order of descending value of the sum $\dot{Z}^{CI} + \dot{C}_D$. The compressor and the turbine have the highest values of the sum $\dot{Z}^{CI} + \dot{C}_D$ and are the most important components from the thermoeconomic viewpoint.

The value of the exergoeconomic factor f for the air compressor (AC) is high compared with a target value of below 65%. The f factor indicates that most of the costs associated with the air compressor are due to capital investment. Although the cost per exergy unit of fuel c_F supplied to the air compressor is the highest among all system components, the cost rate $\dot{C}_D$ associated with exergy destruction $\dot{E}_D$ in this component is relatively low. Only 6.46% of the total exergy destruction occurs in the air compressor. A decrease of the capital investment costs for the air compressor might be cost effective for the entire system even if this would decrease the exergetic efficiency ε_{AC}. These observations suggest a significant decrease in the values of p_2/p_1 and η_{sc}. Compared to the isentropic efficiency, the influence of the pressure ratio on the compressor exergetic efficiency is low. Here, a reduction of p_2/p_1 aims at a decrease of $\dot{Z}_{AC}$ only.

The value of the exergoeconomic factor f for the gas turbine expander (GT) is within the range of the target values (between 25 and 65%, Section IV.B). Therefore, no recommendations can be derived from the thermoeconomic evaluation of the gas turbine expander.

The heat-recovery steam generator (HRSG) has the highest value of the relative cost difference r. The exergoeconomic factor f indicates that most of the costs for this component are related to the cost rate associated with exergy destruction $\dot{C}_D$ [Eq. (89)]. The difference in the average thermodynamic temperature between the hot and the cold streams is a measure for the exergy destruction in the heat-recovery steam generator [Eq. (20)]. A reduction in the value of the temperature T_3 and/or an increase in the isentropic turbine efficiency η_{st} or the pressure ratio p_2/p_1 lead to a decrease in the temperature T_4. Since the temperature profile of the cold stream and the heating duty in the heat-recovery steam generator are fixed, both the average temperature difference and the temperature T_5 of the exhaust gas would decrease. These changes should result in an increase in both the exergetic efficiency and the capital investment in the heat-recovery steam generator, as well as in a decrease in the exergy loss from the overall system.

TABLE V First Design Case (Base-Case Design)[a]

Name	PEC ($10^3$$)	ϵ (%)	$\dot{E}_F$ (MW)	$\dot{E}_P$ (MW)	$\dot{E}_D$ (MW)	y_D (%)	c_F ($/GJ)	c_P ($/GJ)	$\dot{C}_D$ ($/h)	$\dot{Z}^{CI}$ ($/h)	r (%)	f (%)	$\dot{Z}^{CI}+\dot{C}_D$ ($/h)
AC	8732	93.56	52.48	49.10	3.38	3.29	21.47	32.90	261	1112	53.2	81.0	1374
GT	3282	94.12	87.63	82.48	5.15	5.01	18.11	21.47	335	418	18.5	55.4	754
HRSG	1297	63.07	20.21	12.75	7.46	7.26	18.11	34.41	486	165	90.0	25.3	652
CC	139	64.65	102.80	66.46	36.34	35.35	4.56	7.18	597	17	57.3	2.9	615

[a] $T_3 = 1260$ K, $p_2/p_1 = 14$, $\eta_{sc} = 0.86$, $\eta_{st} = 0.87$; overall system: $\varepsilon_{tot} = 41.58\%$; $\dot{C}_{P,tot} = \$4587/h$; $\dot{C}_{L,tot} = \dot{C}_5 = \$499/h$; HRSG: $\Delta T_{min} = 88$ K.

TABLE VI Sample of a Qualitative Decision Matrix for the First Design Case (Base-Case Design)[a]

System component	Objective Z_k^{CI} or ε_k	Decision variables					
		T_3	p_2/p_1	η_{sc}	η_{st}	$\dot{Z}_k^{CI}+\dot{C}_{D,k}$	
		1260	14	0.86	0.87		Initial values
AC	↓	—	↓	↓	—	1374	
GT	—	—	—	—	—	753	
HRSG	↑	↓	↑	—	↑	652	
CC	↑	↑	↑	↓	—	615	
		—	↓	↓	↑		Suggestions
		1260	12	0.84	0.88		New values

[a] The symbols ↓, ↑, and — mean a decrease, an increase, or no change, respectively, in the value of a decision variable.

The combustion chamber (CC) has the largest exergy destruction rate. However, according to the sum $\dot{Z}^{CI}+\dot{C}_D$, the economic importance of this component is rather small. Each exergy unit of the fuel is supplied to the combustion chamber at a relatively low cost. Hence, each unit of exergy destruction can be covered at the same low cost. The low value of the exergoeconomic factor f shows that $\dot{C}_{D,CC}$ is the dominant cost rate. However, only a part of the exergy destruction can be avoided by preheating the reactants or by reducing the excess air. A higher value of p_2/p_1 and/or a lower value of η_{sc} lead to an increase in the temperature T_2 of the air entering the combustion chamber, whereas a lower temperature of the combustion products T_3 reduces the amount of excess air.

Table VI shows a sample of a qualitative decision matrix for the base-case design which summarizes the suggestions from the thermoeconomic evaluation of each component. Decreasing the values of the pressure ratio p_2/p_1 and the isentropic compressor efficiency η_{sc} as well as increasing the isentropic turbine efficiency η_{st} are expected to improve the cost effectiveness of the cogeneration system. Note, that the decrease in the p_2/p_1 value contradicts the corresponding suggestions from the heat-recovery steam generator and the combustion chamber. However, changes suggested by the evaluation of a component should only be considered if they do not contradict

changes suggested by components with a significantly higher value of the sum $\dot{Z}^{CI}+\dot{C}_D$. The temperature T_3 remains unchanged, since contradictory indications are obtained from the evaluations of the heat-recovery steam generator and the combustion chamber. The values of the sum $\dot{Z}^{CI}+\dot{C}_D$ for both components are in the same range. After summarizing the results from the evaluation of the base-case design, the following new values are selected for the decision variables in the second design case: $T_3 = 12604$ K (unchanged), $p_2/p_1 = 12$, $\eta_{sc} = 0.84$, and $\eta_{st} = 0.88$.

B. Evaluation of the Second Design Case (1. Iteration)

Through the changes in the decision variables, the value of the objective function $\dot{C}_{P,tot}$ is reduced from \$4587/h to \$3913/h, and the cost rate associated with the exergy loss $\dot{C}_5$ decreased from \$499/h to \$446/h. The new values of the thermoeconomic variables are summarized in Table VII. The sum $\dot{Z}^{CI}+\dot{C}_D$ shows that the air compressor and the gas turbine expander are still the most important components from the thermoeconomic viewpoint. The importance of both components is due to the relatively high investment cost rate $\dot{Z}^{CI}$ and, to a lesser extent, to the high fuel cost c_F for these components. The

TABLE VII Second Design Case (1. Iteration)[a]

Name	PEC (10^3\$)	ϵ (%)	$\dot{E}_F$ (MW)	$\dot{E}_P$ (MW)	$\dot{E}_D$ (MW)	y_D (%)	c_F (\$/GJ)	c_P (\$/GJ)	$\dot{C}_D$ (\$/h)	$\dot{Z}^{CI}$ (\$/h)	r (%)	f (%)	$\dot{Z}^{CI}+\dot{C}_D$ (\$/h)
AC	4633	92.37	48.70	44.98	3.72	3.55	18.00	25.31	241	601	40.63	71.4	841
GT	3802	94.77	83.04	78.70	4.34	4.14	14.47	18.00	226	493	24.40	68.5	719
CC	138	64.26	104.85	67.37	37.48	35.74	4.56	7.22	616	18	58.15	2.8	634
HRSG	1245	61.62	20.69	12.75	7.94	7.57	14.47	29.00	414	161	100.45	28.1	575

[a] $T_3 = 1260$ K, $p_2/p_1 = 12$, $\eta_{sc} = 0.84$, $\eta_{st} = 0.88$; overall system: $\varepsilon_{tot} = 40.77\%$; $\dot{C}_{P,tot} = \$3913/h$; $\dot{C}_{L,tot} = \dot{C}_5 = \$446/h$; HRSG: $\Delta T_{min} = 105$ K.

cost effectiveness of these components may be improved by lowering the investment cost at the expense of the exergetic efficiency. This can be achieved by a decrease in the values of the isentropic efficiencies η_{sc} and η_{st}. For the third design, η_{sc} and η_{st} are reduced to 0.82 and 0.87, respectively.

Compared to the base-case design, the combustion chamber has now a higher relative cost importance than the heat-recovery steam generator. To reduce the cost of exergy destruction in the combustion chamber, the value of T_3 is increased to 1320 K.

Although the value of the relative cost difference r increased for the heat-recovery steam generator from the base-case design to the second design, the sum $\dot{Z}^{CI} + \dot{C}_D$ for the heat-recovery steam generator decreased. The modifications in the upstream components (AC and CC) and the interactions among the system components lead to a decrease in the cost per exergy unit of the fuel $c_{F,HRSG}$. Therefore, the cost associated with exergy destruction is reduced even if the exergy destruction rate for the heat-recovery steam generator increased in the second design. The reduced pressure ratio p_2/p_1 outweighed the effect of the increased isentropic turbine efficiency η_{st} on the temperature T_4. Hence, T_4 increased in the second design instead of an anticipated decrease. These observations show that the suggested changes based on the thermoeconomic evaluation of the other components may affect the cost effectiveness of the heat-recovery steam generator negatively. The pressure ratio p_2/p_1 remains unchanged in the third design, since the suggested changes in η_{st} and T_3 already lead to a higher temperature T_4, which results in a larger exergy destruction rate in the heat-recovery steam generator.

C. Third Design Case (2. Iteration)

As a result of the changes in the decision variables, the value of the objective function $\dot{C}_{P,tot}$ decreased to \$3484/h. Table VIII shows the new values of the thermoeconomic variables for each component. The costs associated with the compressor and the turbine are reduced significantly, whereas the sum $\dot{Z}^{CI} + \dot{C}_D$ is almost un-

changed for the combustion chamber which now has the second highest value. The values of the exergoeconomic factor f_{AC} and f_{GT} are close to their target values. Further improvements in the cost effectiveness may be achieved by slightly decreasing the isentropic efficiencies and significantly increasing the temperature T_3, even if the exergy destruction rate in the heat-recovery steam generator and the cost rate associated with the exergy loss of the overall system $\dot{C}_5$ increase.

VI. RECENT DEVELOPMENTS IN THERMOECONOMICS

Complex thermal systems cannot usually be optimized using mathematical optimization techniques. The reasons include system complexity; opportunities for structural changes not identified during model development; incomplete cost models; and inability to consider in the model additional important factors such as plant availability, maintenability, and operability. Even if mathematical techniques are applied, the process designer gains no insight into the real thermodynamic losses, the cost formation process within the thermal system, or on how the solution was obtained.

As an alternative, thermoeconomic techniques provide effective assistance in identifying, evaluating, and reducing the thermodynamic inefficiencies and the costs in a thermal system. They improve the engineer's understanding of the interactions among the system components and variables and generally reveal opportunities for design improvements that might not be detected by other methods. Therefore, the interest in applying thermoeconomics has significantly increased in the last few years. In the following, some recent developments in thermoeconomics are briefly mentioned.

To evaluate the thermodynamic performance and cost effectiveness of thermal systems and to estimate the potential for improvements it is always useful to know for the most important system components the avoidable part of exergy destruction, the cost associated with this avoidable part, and the avoidable investment cost

TABLE VIII Third Design Case (2. Iteration)[a]

Name	PEC (10^3\$)	ϵ (%)	$\dot{E}_F$ (MW)	$\dot{E}_P$ (MW)	$\dot{E}_D$ (MW)	y_D (%)	c_F (\$/GJ)	c_P (\$/GJ)	$\dot{C}_D$ (\$/h)	$\dot{Z}^{CI}$ (\$/h)	r (%)	f (%)	$\dot{Z}^{CI} + \dot{C}_D$ (\$/h)
AC	3251	91.47	46.67	42.69	3.98	3.70	15.18	20.95	218	430	38.1	66.4	648
CC	135	65.30	107.68	70.32	37.36	34.70	4.56	7.10	614	18	55.5	2.8	632
GT	2852	94.59	81.06	76.67	4.38	4.07	12.34	15.18	195	377	23.0	66.0	572
HRSG	1134	58.78	21.69	12.75	8.94	8.30	12.34	26.09	397	150	111.4	27.4	547

[a] $T_3 = 1320$ K, $p_2/p_1 = 12$, $\eta_{sc} = 0.82$, $\eta_{st} = 0.87$; overall system: $\varepsilon_{tot} = 39.70\%$; $\dot{C}_{P,tot} = $ \$3484/h; $\dot{C}_{L,tot} = \dot{C}_5 = $ \$453/h; HRSG: $\Delta T_{min} = 142$ K.

associated with each system component. Improvement efforts should then focus only on these avoidable parts of inefficiencies and costs. Figure 5 illustrates the assessment of specific unavoidable exergy destruction $\dot{E}_{D,k}^{UN}/\dot{E}_{P,k}$ and specific unavoidable investment cost $\dot{Z}_k^{UN}/\dot{E}_{P,k}$ for the kth component. The avoidable parts are obtained as the difference between total value and avoidable part. More information about this topic is provided by Tsatsaronis and Park (1999).

The design and improvement of a thermal system often involve application of heuristic rules. Due to the complexity of energy conversion systems, as well as to the uncertainty involved in some design decisions, computer programs using principles from the field of artificial intelligence and soft computing are useful tools for the process designer in improving a given design and in developing a new cost-effective thermal system. The benefits of combining knowledge-based and fuzzy approaches with an iterative thermoeconomic optimization technique are discussed together with some applications by Cziesla (2000).

Calculating the cost of each product stream generated by a thermal plant having more than one product is an important subject for which several approaches have been developed in the past. Some of these approaches use exergy-based or thermoeconomic methods; however, the results obtained by different methods may vary within a wide range. Recently, a new exergy-based approach was developed (Erlach, Tsatsaronis, and Cziesla, 2001) for (a) assigning the fuel(s) used in the overall plant to the product streams and (b) calculating the costs associated with each product stream with the aid of a thermoeconomic evaluation. This new approach is general, more objective than previous approaches, and flexible, i.e., it allows designers and operators of the plant to actively participate in the fuel and cost allocation process.

SEE ALSO THE FOLLOWING ARTICLES

Energy Efficiency Comparisons Among Countries • Energy Flows in Ecology and in the Economy • Energy Resources and Reserves • Geothermal Power Stations • Solar Thermal Power Stations • Thermodynamics

BIBLIOGRAPHY

Ahrendts, J. (1980). "Reference states," *Energy—Int. J.* **5,** 667–677.

Bejan, A., Tsatsaronis, G., and Moran, M. (1996). "Thermal Design and Optimization," Wiley, New York.

Cziesla, F. (2000). "Produktkostenminimierung beim Entwurf komplexer Energieumwandlungsanlagen mit Hilfe von wissensbasierten Methoden," No. 438, Fortschr.-Ber. VDI, Reihe 6, VDI Verlag, Düsseldorf.

Erlach, B., Tsatsaronis, G., and Cziesla, F., A new approach for assigning costs and fuels to cogeneration products. *In* "ECOS'o1, Efficiency, Costs, Optimization, Simulation and Environmental Aspects of Energy Systems," pp. 759–770, Istanbul, Turkey, July 4–6.

Kotas, T. J. (1985). "The Exergy Method of Thermal Plant Analysis," Butterworths, London.

Lazzaretto, A., and Tsatsaronis, G. (November 1999). On the calculation of efficiencies and costs in thermal systems. *In* "Proceedings of the ASME Advanced Energy Systems Division" (S. M. Aceves, S. Garimella, and R. Peterson, eds.), AES-Vol. 39, pp. 421–430, ASME, New York.

Moran, M. J., and Shapiro, H. N. (1998). "Fundamentals of Engineering Thermodynamics," Wiley, New York.

Penner, S. S., and Tsatsaronis, G., eds. (1994). "Invited papers on exergoeconomics," *Energy—Int. J.* **19**(3), 279–318.

Sama, D. A. (September 1995). "The use of the second law of thermodynamics in process design," *J. Eng. Gas Turbines Power* **117,** 179–185.

Szargut, J., Morris, D. R., and Steward, F. R. (1988). "Exergy Analysis of Thermal, Chemical, and Metallurgical Processes," Hemisphere, New York.

Electric Power Research Institute (1993). "Technical assessment guide (TAG™)," Vol. 1, TR-102276-V1R7, Electric Power Research Institute, Palo Alto, CA.

Tsatsaronis, G. (1999). Strengths and limitations of exergy analysis. *In* "Thermodynamic Optimization of Complex Energy Systems" (A. Bejan, and E. Mamut, eds.), Vol. 69, pp. 93–100, Nato Science Series, Kluwer Academic, Dordrecht/Normell, MA.

Tsatsaronis, G., and Park, M.H. (1999). On avoidable and unavoidable exergy destructions and investment costs in thermal systems. *In* "ECOS'99, Efficiency, Costs, Optimization, Simulation and Environmental Aspects of Energy Systems," pp. 116–121, Tokyo, Japan, June 8–10.

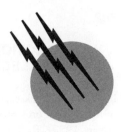

Thermoelectricity

Timothy P. Hogan
Michigan State University

I. Introduction and Basic Thermoelectric Effects
II. Thermodynamic Relationships
III. Thermodynamics of an Irreversible Process
IV. Statistical Relationships
V. Applications
VI. Summary

GLOSSARY

Boltzmann equation An equation based on the Fermi distribution equation under nonequilibrium conditions. The Boltzmann equation describes the rate of change of the distribution function due to forces, concentration gradients, and carrier scattering.

Fermi distribution A function describing the probability of occupancy of a given energy state for a system of particles based on the Pauli exclusion principle.

Fermi level The energy level which exhibits a 50% probability of being occupied.

Joule heating Heating due to I^2R losses.

Onsager relations A set of simultaneous equations that describe the macroscopic interactions between "forces" and "flows" within a thermoelectric system.

Peltier effect Absorption or evolution of thermal energy at a junction between dissimilar materials through which current flows.

Seebeck effect Open-circuit voltage generated by a circuit consisting of at least two dissimilar conductors when a temperature gradient exists within the circuit between the measuring and the reference junctions.

Thermocouple A pair of dissimilar conductors joined at one set of ends to form a measuring junction.

Thermoelectric cooler A heat pump designed from thermoelectric materials typically configured in an array as a series of thermocouples with the junction exposed.

Thermopower This is defined here as the absolute Seebeck coefficient and corresponds to the rate of change of the thermoelectric voltage with respect to the temperature of a single conductor with a temperature gradient between the ends.

Thompson effect The absorption or evolution of thermal energy from a single homogeneous conductor through which electric current flows in the presence of a temperature gradient along the conductor.

THE FIELD of thermoelectricity involves the study of characteristics resulting from electrical phenomena occurring in conjunction with a flow of heat. It includes flows of electrical current and thermal current and the interactions between them.

Encyclopedia of Physical Science and Technology, Third Edition, Volume 16

I. INTRODUCTION AND BASIC THERMOELECTRIC EFFECTS

In 1822, Seebeck reported on "the magnetic polarization of metals and ores produced by a temperature difference" (Joffe, 1957). By placing two conductors in the configuration shown in Fig. 1, Seebeck observed a deflection of the magnetic needle in his measurement apparatus (Gray, 1960). The deflection was dependent on the temperature difference between junctions and the materials used for the conductors. Shortly after this, Oersted discovered the interaction between an electric current and a magnetic needle. Many scientists subsequently researched the relationship between electric currents and magnetic fields including Ampère, Biot, Savart, Laplace, and others. It was then suggested that the observation by Seebeck was not caused by a magnetic polarization, but due to a thermoelectric current flowing in the closed-loop circuit.

Seebeck did not accept this explanation, and in an attempt to refute it, he reported measurements on a number of solid and liquid metals, alloys, minerals, and semiconductors. The magnetic polarization hypothesis was incorrect as can be seen in the open-circuit configuration of his experiment.

Experimentally, a voltage (ΔV) at the open-circuit terminals is measured when a temperature gradient exists between junctions such that

$$\Delta V = \int_{T_1}^{T_2} S_{AB} \, dT, \tag{1}$$

where S_{AB} is the Seebeck coefficient for the two conductors, which is defined as being positive when a positive voltage is measured for $T_1 < T_2$. The voltage is measured across terminals maintained at a constant temperature T_0. For this voltage to appear in the open-circuit configuration (Fig. 2), there must exist a current which flows in the closed-circuit configuration. Furthermore, in the open-circuit configuration, Seebeck would no longer observe a deflection of the magnetic needle, which is not expected if a magnetic polarization is taking effect.

The diligence of his measurements was vertified by the confirmation of his values years later by Justi and Meisner

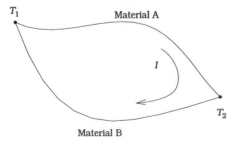

FIGURE 1 Closed-circuit Seebeck effect.

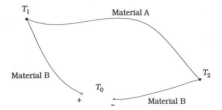

FIGURE 2 The open-circuit Seebeck effect.

as well as by Telkes, who showed, 125 years after Seebeck's measurements, that the best couple for energy conversion was formed using ZnSb and PbS, which were two materials examined by Seebeck.

Twelve years after Seebeck's discovery, a scientist and watchmaker named Jean Peltier reported a temperature anomaly at the junction of two dissimilar materials as a current was passed through the junction. It was unclear what caused this anomaly, and while Peltier attempted to explain it on the basis of the conductivities and/or hardness of the two materials, Lenz removed all doubt in 1838 with one simple experiment. By placing a droplet of water in a dimple at the junction between rods of bismuth and antimony, Lenz was able to freeze the water and subsequently melt the ice by changing the direction of current through the junction. In a way, Lenz had made the first thermoelectric cooler. The rate of heat ($\dot{Q}$) absorbed or liberated from the junction was later found to be proportional to the current, or

$$\dot{Q} = \Pi \cdot I, \tag{2}$$

where the proportionality constant (Π) was named the Peltier coefficient.

Near this time, the field of electromagnetics was being formed and captured much attention in the scientific community. Therefore, another 16 years passed before Thomson (later called Lord Kelvin) reasoned that if the current through the two junctions in Fig. 1 produced only Peltier heating, then the Peltier voltage must equal the Seebeck voltage and both must be linearly proportional to the temperature. Since this was not observed experimentally, he reasoned that there must be a third reversible process occurring. This third process is the evolution or absorption of heat whenever current is passed through a single homogeneous conductor along which a temperature gradient exists, or in equation form,

$$\dot{Q} = \Im I \frac{dT}{dx}, \tag{3}$$

where $\dot{Q}$ is the rate of heat absorbed or liberated along the conductor, $\Im$ is the Thomson coefficient, I is the current through the conductor and dT/dx is the temperature gradient maintained along the length of the conductor.

Thomson then applied the first and second laws of thermodynamics to the Seebeck, Peltier, and Thomson effects to find the Kelvin relationships

$$\Pi = S_{AB}T, \qquad (4)$$

$$\frac{dS_{AB}}{dT} = \frac{\Im_A - \Im_B}{T}, \qquad (5)$$

where the subscripts A and B correspond to the two materials in Fig. 2. The second Kelvin relation suggests that the Seebeck coefficient for two materials forming a junction can be represented as the difference between quantities based on the properties of the individual materials making up the junction. Integration of the second Kelvin relation gives

$$\int dS_{AB} = \int \frac{\Im_A - \Im_B}{T} dT = \int \frac{\Im_A}{T} dT - \int \frac{\Im_B}{T} dT \qquad (6)$$

or

$$S_{AB} = \int \frac{\Im_A}{T} dT - \int \frac{\Im_B}{T} dT. \qquad (7)$$

Defining the first term on the right-hand side as the "absolute" Seebeck coefficient of material A and the second term as the "absolute" Seebeck coefficient of material B, we find that the Seebeck coefficient for a junction is equal to the difference in "absolute" Seebeck coefficients of the individual materials making the junction. This is a very significant result, as measurements of the individual materials can be used to predict how junctions formed from various combinations of materials will behave, thus removing the need to measure every possible combination of materials. The "absolute" Seebeck coefficient or thermoelectric power of a material, hereafter referred to simply as the thermopower of the material, can be found for material A if the thermopower of material B is known or if the thermopower of material B is zero. A material in the superconducting state has a thermopower of zero, and once a material is calibrated against a superconductor, it can then be used as a reference material to measure more materials. This has been done for several pure materials such as lead, gold, and silver (Roberts, 1977; Wendling et al., 1993).

Further understanding of the basic thermoelectric properties and the relationships between them can be found through comparisons of macroscopic and microscopic derivations. The Onsager relations formulate various flows (consisting of matter or energy) as functions of the forces that drive them, thus describing macroscopic observations of materials. Another useful technique for understanding these basic thermoelectric properties utilizes semiclassical statistical mechanics to describe the microscopic processes. Comparisons between the macroscopic and the microscopic analyses can be used in deriving many useful formulas for calculating thermoelectric properties of various materials. The following sections are dedicated to developing the macroscopic and microscopic analyses.

II. THERMODYNAMIC RELATIONSHIPS

As shown by the Seebeck effect, when a temperature gradient is placed over the length of a sample, carrier flow will be predominantly from the hot side to the cold side. This indicates that a temperature gradient, ΔT, is a *force* that can cause a *flow* of carriers. It is well known that applying the *force* of an electric potential gradient, ΔV, can also induce carrier *flow*. In 1931, Onsager developed a method of relating the flows of matter or energy within a system to the forces present. In this method the forces are assumed to be sufficiently small so that a linear relationship between the forces, $\mathbf{X}_i$, and the corresponding flows, $\mathbf{J}_i$, can be written.

$$\mathbf{J}_1 = L_{11}\mathbf{X}_1 + L_{12}\mathbf{X}_2 + \cdots L_{1n}\mathbf{X}_n,$$
$$\mathbf{J}_2 = L_{21}\mathbf{X}_1 + L_{22}\mathbf{X}_2 + \cdots L_{2n}\mathbf{X}_n, \qquad (8)$$
$$\mathbf{J}_3 = L_{31}\mathbf{X}_1 + L_{32}\mathbf{X}_2 + \cdots L_{3n}\mathbf{X}_n,$$

or

$$\mathbf{J}_i = \sum_{m=1}^{n} L_{im}\mathbf{X}_m \qquad (i = 1, 2, 3, \ldots, n). \qquad (9)$$

For carrier and heat flow as described above, the Onsager relationships can be written

$$\mathbf{J} = L_{11}\nabla V + L_{12}\nabla T,$$
$$\mathbf{J}_Q = L_{21}\nabla V + L_{22}\nabla T, \qquad (10)$$

where $\mathbf{J}$ is the current density (electric charge flow), and $\mathbf{J}_Q$ is the heat flux density (heat flow). Without a temperature gradient ($\Delta T = 0$), a heat flux of zero would be expected, contrary to what Eq. (10) would indicate. It is, therefore, important to understand further the primary coefficients L_{ii} and interaction coefficients L_{ij} ($i \neq j$) linking these equations. To do so requires a consideration of the thermodynamics of an irreversible process (one in which the change in entropy is greater than zero $\Delta \bar{S} > 0$).

III. THERMODYNAMICS OF AN IRREVERSIBLE PROCESS

The general application of the Onsager relationship was derived by Harman and Honig (1967) and is summarized here. For a constant electric potential, V, throughout the sample,

$$dQ = T \, d\bar{S} = dU + P \, d\mathcal{V} - \sum_i \mu_i \, dn_i, \qquad (11)$$

where Q is the heat energy density, $\bar{S}$ is the entropy density, U is the internal energy density, P is the pressure, μ_i is the chemical potential of the particle species, and n_i is the particle density. The magnitude of the differential volume, $d\mathcal{V}$, is zero since each quantity has been specified per unit volume. This can be combined with the total energy density, E, given by

$$E = U + V \sum_i Z_i q n_i, \qquad (12)$$

where q is the magnitude of the electronic charge $[1.602 \times 10^{-19}(\text{C})]$, V is an externally applied bias, and Z_i is the number and sign of the charges on the ith particle species. For example, an electron would have charge $Z_e q$, where $Z_e = -1$. The time derivative of Eq. (12) gives

$$\frac{\partial E}{\partial t} = \frac{\partial U}{\partial t} + \frac{\partial V}{\partial t} \sum_i Z_i q n_i + V \sum_i Z_i q \frac{\partial n_i}{\partial t}. \qquad (13)$$

From Eq. (11) with $d\mathcal{V} = 0$,

$$dU = T \, d\bar{S} + \sum_i \mu_i \, dn_i. \qquad (14)$$

Taking the time derivative of (14) gives

$$\frac{\partial U}{\partial t} = T \frac{\partial \bar{S}}{\partial t} + \bar{S} \frac{\partial T}{\partial t} + \sum_i \mu_i \frac{\partial n_i}{\partial t} + \sum_i n_i \frac{\partial \mu_i}{\partial t}. \qquad (15)$$

Using (15) in (13) yields

$$\frac{\partial E}{\partial t} = T \frac{\partial \bar{S}}{\partial t} + \bar{S} \frac{\partial T}{\partial t} + \sum_i \mu_i \frac{\partial n_i}{\partial t} + \sum_i n_i \frac{\partial \mu_i}{\partial t}$$
$$+ \frac{\partial V}{\partial t} \sum_i Z_i q n_i + V \sum_i Z_i q \frac{\partial n_i}{\partial t}. \qquad (16)$$

This can be simplified by considering the Gibbs–Duhem relation (Guggenheim, 1957),

$$\bar{S} \frac{\partial T}{\partial t} + \sum_i n_i \frac{\partial \mu_i}{\partial t} = 0 \qquad (17)$$

and

$$\bar{\mu}_i = \mu_i + Z_i q V, \qquad (18)$$

where $\bar{\mu}_i$ is the electrochemical potential, μ_i is the chemical potential, and $Z_i q V$ is the electrostatic potential energy. The relationship among the chemical potential, μ, the electrochemical potential, $\bar{\mu}$, and the temperature for electrons is shown in Fig. 3, where the right side of the sample is at a potential of $-V_1$ relative to the left.

Equation (16) then reduces to

$$\frac{\partial E}{\partial t} = T \frac{\partial \bar{S}}{\partial t} + \sum_i \bar{\mu}_i \frac{\partial n_i}{\partial t} + \frac{\partial V}{\partial t} \sum_i Z_i q n_i. \qquad (19)$$

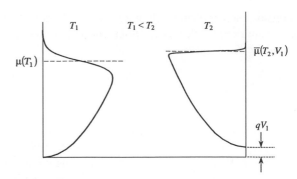

FIGURE 3 The density of states for a metal at a temperature $T_1 > 0$ K on the left and at a lower temperature, $T_2 < T_1$, on the right.

The rate of change in the particle density n_i, is governed by the *equation of continuity*,

$$\frac{\partial n_i}{\partial t} = \left(\frac{\partial n_i}{\partial t} \right)_s - \nabla \cdot \mathbf{J}_i, \qquad (20)$$

which states that the total rate of change in n_i is equal to the local particle generation rate, or source rate, minus the transport of the ith species across the boundary of the differential volume (or local system) of interest. The first term on the right-hand side of the equation is the source term and represents the particle generation (or capture) rate through chemical reactions, for example. The last term is found using Gauss's theorem,

$$\iint \mathbf{J}_i \cdot \hat{\mathbf{n}} \, dA = \iiint \nabla \cdot \mathbf{J}_i \, d\mathcal{V}, \qquad (21)$$

where $\mathbf{J}_i$ is the flux vector equal to the number of particles of type i moving past a unit cross section per unit time in the direction of $\mathbf{J}_i$, and $\hat{\mathbf{n}}$ represents a unit vector outward normal from an element of area dA on the boundary surface. This represents the total outward flux of the ith particle species from the differential volume of interest. The particle species over which the summation in Eq. (19) is evaluated includes core species, L, which form the host lattice; neutral donors, D; ionized donors, D^+; neutral acceptors, A; ionized acceptors, A^-; electrons in the conduction band, n; and holes in the valence band, p. Therefore,

$$\frac{1}{T} \sum_i \bar{\mu}_i \frac{\partial n_i}{\partial t} = \frac{1}{T} \left(\bar{\mu}_L \frac{\partial n_L}{\partial t} + \bar{\mu}_D \frac{\partial n_D}{\partial t} \right.$$
$$+ \bar{\mu}_{D^+} \frac{\partial n_{D^+}}{\partial t} + \bar{\mu}_A \frac{\partial n_A}{\partial t} - \bar{\mu}_{A^-} \frac{\partial n_{A^-}}{\partial t}$$
$$\left. + \bar{\mu}_n \frac{\partial n_n}{\partial t} + \bar{\mu}_p \frac{\partial n_p}{\partial t} \right). \qquad (22)$$

Equation (20) can now be used for each term on the right-hand side of (22). Some simplification can be readily made, however, when the species L, D, D^+, A, and A^- are assumed to be immobile such that

$$\mathbf{J}_L = \mathbf{J}_D = \mathbf{J}_{D^+} = \mathbf{J}_A = \mathbf{J}_{A^-} = 0. \tag{23}$$

Furthermore, the lattice will not be affected by local transformations, and

$$\left(\frac{\partial n_L}{\partial t} \right)_s = \frac{\partial n_L}{\partial t} = 0. \tag{24}$$

Additional relationships can be found to simplify (22) further by identifying the different mechanisms for generation of electrons, n, or holes, p, as follows:

$$\left. \begin{aligned} D &\Rightarrow D^+ + n, \\ A &\Rightarrow A^- + p, \\ &\Rightarrow n + p. \end{aligned} \right\} \quad \begin{aligned} (25) \\ (26) \\ (27) \end{aligned}$$

These reactions are reversible such that the time rate of change of ionized and unionized donors and acceptors must be considered in (22). Identifying the reactions in (25), (26), and (27) as I, II, and III, respectively, the following reaction velocities can be written

$$\left. \begin{aligned} -\left(\frac{\partial n_D}{\partial t} \right)_s &= \left(\frac{\partial n_{D^+}}{\partial t} \right)_s = \left(\frac{\partial n_n}{\partial t} \right)_{\mathrm{I}} = \nu_{\mathrm{I}}, \\ -\left(\frac{\partial n_A}{\partial t} \right)_s &= \left(\frac{\partial n_{A^-}}{\partial t} \right)_s = \left(\frac{\partial n_p}{\partial t} \right)_{\mathrm{II}} = \nu_{\mathrm{II}}, \\ \left(\frac{\partial n_n}{\partial t} \right)_s &= \left(\frac{\partial n_p}{\partial t} \right)_{\mathrm{III}} = \nu_{\mathrm{III}}. \end{aligned} \right\} \tag{28}$$

Therefore, (22) becomes

$$\begin{aligned} \frac{1}{T} \sum_i \bar{\mu}_i \frac{\partial n_i}{\partial t} = \frac{1}{T} &\left\{ \bar{\mu}_D \left(\frac{\partial n_D}{\partial t} \right)_s + \bar{\mu}_{D^+} \left(\frac{\partial n_{D^+}}{\partial t} \right)_s \right. \\ &+ \bar{\mu}_A \left(\frac{\partial n_A}{\partial t} \right)_s + \bar{\mu}_{A^-} \left(\frac{\partial n_{A^-}}{\partial t} \right)_s \\ &+ \bar{\mu}_n \left[\left(\frac{\partial n_n}{\partial t} \right)_{\mathrm{I}} + \left(\frac{\partial n_n}{\partial t} \right)_{\mathrm{III}} - \nabla \cdot \mathbf{J}_n \right] \\ &\left. + \bar{\mu}_p \left[\left(\frac{\partial n_p}{\partial t} \right)_{\mathrm{II}} + \left(\frac{\partial n_p}{\partial t} \right)_{\mathrm{III}} - \nabla \cdot \mathbf{J}_p \right] \right\}. \end{aligned} \tag{29}$$

From the relations in (28), Eq. (29) can be written in terms of the reaction velocities, ν_{I}, ν_{II}, and ν_{III} as follows:

$$\begin{aligned} \frac{1}{T} \sum_i \bar{\mu}_i \frac{\partial n_i}{\partial t} = \frac{1}{T} \{ &(-\bar{\mu}_D + \bar{\mu}_{D^+} + \bar{\mu}_n)\nu_{\mathrm{I}} + (-\bar{\mu}_A \\ &+ \bar{\mu}_{A^-} + \bar{\mu}_p)\nu_{\mathrm{II}} + (\bar{\mu}_n + \bar{\mu}_p)\nu_{\mathrm{III}} \\ &- \bar{\mu}_n \nabla \cdot \mathbf{J}_n - \bar{\mu}_p \nabla \cdot \mathbf{J}_n \} \end{aligned} \tag{30}$$

This can be substituted into (19) to give

$$\begin{aligned} \frac{\partial E}{\partial t} = T \frac{\partial \bar{S}}{\partial t} &+ \frac{1}{T} \{ A_{\mathrm{I}} \nu_{\mathrm{I}} + A_{\mathrm{II}} \nu_{\mathrm{II}} + A_{\mathrm{III}} \nu_{\mathrm{III}} - \bar{\mu}_n \nabla \cdot \mathbf{J}_n \\ &- \bar{\mu}_p \nabla \cdot \mathbf{J}_n \} + \frac{\partial V}{\partial t} \sum_i Z_i q n_i, \end{aligned} \tag{31}$$

where A_{I}, A_{II}, and A_{III} *affinities* are defined as

$$A_{\mathrm{I}} \equiv -\bar{\mu}_D + \bar{\mu}_{D^+} + \bar{\mu}_n, \qquad A_{\mathrm{II}} \equiv -\bar{\mu}_A + \bar{\mu}_{A^-} + \bar{\mu}_p,$$
$$A_{\mathrm{III}} \equiv \bar{\mu}_n + \bar{\mu}_p. \tag{32}$$

These general derivations can now be applied to more specific cases by solving for the energy flux term on the left-hand side of the equation using the appropriate approximations for the material under consideration.

A. Metals

In metals, the energy density term $\partial E / \partial t$ can be viewed as composed of four contributions.

- The rate at which an externally applied field delivers energy to the local system.
- Two terms arise from the rate of change in the electrostatic energy either due to a change in the charge concentration or due to a change in the potential, V.
- Electrons in the higher-energy states [the energies above $\mu(T_1)$ in Fig. 3] can transition to the available lower-energy states by giving up this excess energy to the lattice, resulting in a heat flux, $\mathbf{J}_Q$.

As an externally applied electric field accelerates charged carriers, they do not continue to increase in velocity as they would in free space, but attain some average drift velocity. Therefore, an internal force must exist to counterbalance the external force. This internal force is caused mainly by collisions of the carriers with the lattice, thus providing a mechanism of energy transfer from the applied electric field to the lattice. The first contribution is given by

$$\begin{aligned} \left(\frac{\partial E}{\partial t} \right)_{\mathrm{I}} &= \mathcal{E} \cdot (-n_n q \mathbf{v}_n + n_p q \mathbf{v}_p) = \mathbf{J} \cdot \mathcal{E} = -\mathbf{J} \cdot \nabla V \\ &= q(\mathbf{J}_n - \mathbf{J}_p) \cdot \nabla V, \end{aligned} \tag{33}$$

where $\mathbf{J}_n$ and $\mathbf{J}_p$ represent the particle flux densities, while $\mathbf{J}$ represents the current density such that

$$\mathbf{J} = q(\mathbf{J}_p - \mathbf{J}_n). \tag{34}$$

The electrostatic energy density is given by $V \sum_i Z_i q n_i$. The time rate of change of the electrostatic energy density is

$$\frac{\partial}{\partial t} \left(V \sum_i Z_i q n_i \right) = \frac{\partial V}{\partial t} \sum_i Z_i q n_i + V \sum_i Z_i q \frac{\partial n_i}{\partial t}. \tag{35}$$

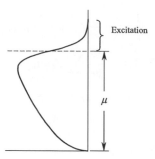

FIGURE 4 The density of states at a finite temperature. Only the excitation energy can be transferred to the lattice.

This gives the second and third contributions to the total rate of change in energy density,

$$\left(\frac{\partial E}{\partial t}\right)_{\text{II}} = \frac{\partial V}{\partial t}\sum_i Z_i q n_i, \tag{36}$$

$$\left(\frac{\partial E}{\partial t}\right)_{\text{III}} = V\sum_i Z_i q \frac{\partial n_i}{\partial t} = -V(\nabla \cdot \mathbf{J})$$
$$= qV[\nabla \cdot (\mathbf{J}_n - \mathbf{J}_p)], \tag{37}$$

where Eq. (20) was used, with the assumption of no generative sources. The fourth contribution comes from the excitation energy depicted in Fig. 4, which gives rise to a heat flux, $\mathbf{J}_Q$. Relative to the bottom of the conduction band, the total heat flux density, $\mathbf{J}_u$, is

$$\mathbf{J}_u = \mathbf{J}_Q - \frac{\mu}{q}\mathbf{J}, \tag{38}$$

thus giving the fourth contribution to energy flow through Fourier's law of heat conduction,

$$\left(\frac{\partial E}{\partial t}\right)_{\text{IV}} = -\nabla \cdot \mathbf{J}_u = -\nabla \cdot \left(\mathbf{J}_Q - \frac{\mu}{q}\mathbf{J}\right). \tag{39}$$

Summing contributions I through IV gives the total energy density rate of change as

$$\frac{\partial E}{\partial t} = -\mathbf{J} \cdot \nabla V - V(\nabla \cdot \mathbf{J}) + \frac{\partial V}{\partial t}\sum_i Z_i q n_i - \nabla \cdot \mathbf{J}_u$$

$$= -\nabla \cdot V\mathbf{J} + \frac{\partial V}{\partial t}\sum_i Z_i q n_i - \nabla \cdot \mathbf{J}_u$$

$$= \frac{\partial V}{\partial t}\sum_i Z_i q n_i - \nabla \cdot \mathbf{J}_E, \tag{40}$$

where

$$\mathbf{J}_E = \mathbf{J}_u + V\mathbf{J} \tag{41}$$

is the total energy flux density. Substituting (40) into (19) gives

$$\frac{\partial V}{\partial t}\sum_i Z_i q n_i - \nabla \cdot \mathbf{J}_E = T\frac{\partial \bar{S}}{\partial t} + \sum_i \bar{\mu}_i \frac{\partial n_i}{\partial t}$$

$$+ \frac{\partial V}{\partial t}\sum_i Z_i q n_i, \tag{42}$$

or, after cancellation and using (20), assuming no generative sources,

$$-\nabla \cdot \mathbf{J}_E = T\frac{\partial \bar{S}}{\partial t} + \frac{\bar{\mu}}{q}\nabla \cdot \mathbf{J} = T\frac{\partial \bar{S}}{\partial t} + \bar{\mu}\nabla \cdot \mathbf{J}_q. \tag{43}$$

The rate of change of entropy can, therefore, be written

$$\frac{\partial \bar{S}}{\partial t} = \frac{-\nabla \cdot \mathbf{J}_E}{T} - \frac{\bar{\mu}}{qT}\nabla \cdot \mathbf{J} = -\nabla \cdot \left(\frac{\mathbf{J}_E}{T}\right) - \nabla \cdot \left(\frac{\mathbf{J}\bar{\mu}}{qT}\right)$$

$$+ \mathbf{J}_E \cdot \nabla\left(\frac{1}{T}\right) + \mathbf{J} \cdot \nabla\left(\frac{\bar{\mu}}{qT}\right), \tag{44}$$

or using an entropy flux, $\mathbf{J}_{\bar{s}}$, defined as

$$T\mathbf{J}_{\bar{s}} = \mathbf{J}_E + \frac{\bar{\mu}}{q}\mathbf{J}, \tag{45}$$

gives

$$\frac{\partial \bar{S}}{\partial t} = \frac{\partial \bar{S}_0}{\partial t} + \frac{\partial \bar{S}_s}{\partial t}$$

$$= -\nabla \cdot \mathbf{J}_{\bar{s}} + \mathbf{J}_E \cdot \nabla\left(\frac{1}{T}\right) + \mathbf{J} \cdot \nabla\left(\frac{\bar{\mu}}{qT}\right), \tag{46}$$

where the total entropy is given by the sum of the equilibrium entropy plus additional entropy sources, or $\bar{S} = \bar{S}_0 + \bar{S}_s$. The irreversible process for which $\Delta \bar{S} = (\bar{S} - \bar{S}_0) = \bar{S}_s > 0$ then consists of the last two terms in the above equation such that

$$\frac{\partial \bar{S}_s}{\partial t} = \mathbf{J}_E \cdot \nabla\left(\frac{1}{T}\right) + \mathbf{J} \cdot \nabla\left(\frac{\bar{\mu}}{qT}\right). \tag{47}$$

Using Eq. (45) to substitute for $\mathbf{J}_E$, (47) becomes.

$$\frac{\partial \bar{S}_s}{\partial t} = \frac{-\mathbf{J}_s}{T} \cdot \nabla T + \frac{\mathbf{J}}{qT} \cdot \nabla\bar{\mu}, \tag{48}$$

or using $\bar{\mu} = \mu + qV$ along with (41) and (45) to give

$$T\mathbf{J}_{\bar{s}} = \mathbf{J}_Q, \tag{49}$$

then Eq. (48) becomes

$$\frac{\partial \bar{S}_s}{\partial t} = \frac{-\mathbf{J}_Q}{T^2} \cdot \nabla T + \frac{\mathbf{J}}{qT} \cdot \nabla\bar{\mu}. \tag{50}$$

These three equations, (47), (48), and (50), could each be written in the general form of

$$\frac{\partial \bar{S}_s}{\partial t} = \sum_i \mathbf{J}_i \cdot \mathbf{X}_i. \tag{51}$$

This is a necessary condition for using the Onsager reciprocity relation that $L_{12} = L_{21}$ in Eq. (10). Three sets of

Onsager relations can then be written, by extracting the forces, $\mathbf{X}_i$, from Eqs. (47), (48), and (50).

$$\left.\begin{aligned} \mathbf{J} &= \frac{\mathbf{Z}_{11}}{q}\nabla\left(\frac{\bar{\mu}}{T}\right) + \mathbf{Z}_{12}\nabla\left(\frac{1}{T}\right), \\ \mathbf{J}_E &= \frac{\mathbf{Z}_{21}}{q}\nabla\left(\frac{\bar{\mu}}{T}\right) + \mathbf{Z}_{22}\nabla\left(\frac{1}{T}\right), \end{aligned}\right\} \tag{52}$$

$$\left.\begin{aligned} \mathbf{J} &= \frac{\mathbf{B}_{11}}{qT}\nabla\bar{\mu} - \frac{\mathbf{B}_{12}}{T}\nabla T, \\ \mathbf{J}_{\bar{s}} &= \frac{\mathbf{B}_{21}}{qT}\nabla\bar{\mu} - \frac{\mathbf{B}_{22}}{T}\nabla T, \end{aligned}\right\} \tag{53}$$

$$\left.\begin{aligned} \mathbf{J} &= \frac{\mathbf{L}_{11}}{qT}\nabla\bar{\mu} - \frac{\mathbf{L}_{12}}{T^2}\nabla T, \\ \mathbf{J}_Q &= \frac{\mathbf{L}_{21}}{qT}\nabla\bar{\mu} - \frac{\mathbf{L}_{22}}{T^2}\nabla T, \end{aligned}\right\} \tag{54}$$

thus relating the electrical current density, $\mathbf{J}$, to the energy flux density, $\mathbf{J}_E$, the entropy flux density, $\mathbf{J}_{\bar{s}}$, and the heat flux density, $\mathbf{J}_Q$. Equations (52), (53), and (54) can now be used to identify various thermoelectric properties.

IV. STATISTICAL RELATIONSHIPS

Within crystalline materials, electron behavior can be described by the wave nature of electrons and Schrödinger's equation,

$$\nabla^2\Psi + \frac{2m}{\hbar^2}(E - V)\Psi = \frac{-\hbar}{j}\frac{\partial E}{\partial t}, \tag{55}$$

where Ψ is the electron wave function, E is the total energy, and V is the potential energy of the electrons. The solution to this equation is

$$\Psi(\mathbf{r}, t) = \psi(\mathbf{r})e^{-j\omega t}, \tag{56}$$

where Ψ is the time-independent solution to Schrödinger's equation. This solution forms a wave packet with a group velocity, $\mathbf{v}$, equal to the average velocity of the particle it describes, such that

$$\mathbf{v} = \nabla_k\omega = \frac{\partial\omega}{\partial\mathbf{k}} = \frac{1}{\hbar}\nabla_k E = \frac{1}{\hbar}\frac{\partial E}{\partial\mathbf{k}}, \tag{57}$$

where the use of Planck's relationship, $E = h\nu = \hbar\omega$, was made. Force times distance is equal to energy, or with a time derivative,

$$\mathbf{v}\cdot\mathbf{F} = \frac{\partial E}{\partial t} = \frac{1}{\hbar}\frac{\partial E}{\partial\mathbf{k}}\cdot\hbar\frac{\partial\mathbf{k}}{\partial t}, \tag{58}$$

giving

$$\mathbf{F} = \hbar\frac{\partial\mathbf{k}}{\partial t}. \tag{59}$$

The electron wave function, $\Psi(\mathbf{r}, t)$, itself does not have physical meaning, however, the product of $\Psi^*(\mathbf{r}, t)\Psi(\mathbf{r}, t)$ represents the probability of finding an electron at position $\mathbf{r}$ and time t. As a probability implies, there is a factor of uncertainty, which was quantified in 1927 by Heisenberg.

$$\Delta p_x\Delta x \geq h, \qquad \Delta p_y\Delta y \geq h, \qquad \Delta p_z\Delta z \geq h, \tag{60}$$

where Δp_x, Δp_y, and Δp_z are the momentum uncertainties in the x, y, and z directions, respectively. The positional uncertainties in the three directions are given by Δx, Δy, and Δz. It is possible to utilize these uncertainties to define the smallest volume (in real space, or momentum space) that represents a discrete electronic state. Within a cube of material with dimensions $L \times L \times L$, the maximum positional uncertainty for a given electron would be $\Delta x = \Delta y = \Delta z = L$, since the electron must be located somewhere within the cube. This would correspond to the minimum Δp_x, Δp_y, and Δp_z given by

$$\Delta p_{x_{\min}} = \frac{h}{\Delta x} = \frac{h}{L}, \qquad \Delta p_{y_{\min}} = \frac{h}{\Delta y} = \frac{h}{L},$$

$$\Delta p_{z_{\min}} = \frac{h}{\Delta z} = \frac{h}{L}. \tag{61}$$

Thus the product

$$\Delta p_{x_{\min}}\Delta p_{y_{\min}}\Delta p_{z_{\min}} = \frac{h^3}{L^3} \tag{62}$$

gives the minimum elemental volume in momentum space to represent two discrete electronic states (one for spin-up and one for spin-down). The number of states, dg, per unit volume in an element $dp_x\,dp_y\,dp_z$ of momentum space can be written

$$dg = \frac{1}{h^3}dp_x\,dp_y\,dp_z. \tag{63}$$

Schrödinger's time-independent equation for a free electron ($V = 0$) is

$$\nabla^2\psi + \frac{2m}{\hbar^2}E\psi = 0, \tag{64}$$

which has the solution

$$\psi = Ae^{j\mathbf{k}\cdot\mathbf{r}}. \tag{65}$$

Substituting back into (64) gives

$$E = \frac{\hbar^2}{2m}k^2 = \frac{p^2}{2m}, \tag{66}$$

where $p^2 = \hbar k$ was used. Within a crystal, a similar formula can be found when the concept of effective mass, m^*, is utilized to account for internal forces on the electrons due to the ion cores at each lattice point. Electrons with energies below some value E are then defined by a sphere in momentum space with radius $p = \sqrt{2m^*E}$. The number of electronic states within the material cube ($L \times L \times L$) is found by dividing the total momentum space volume by the volume per state, or

$$N = 2\left[\frac{4/3\pi r^3}{h^3/L^3}\right] = 2\left[\frac{4/3\pi(\sqrt{2m^*E})^3}{h^3/L^3}\right]$$

$$= \frac{8\pi(2m^*E)^{3/2}}{3h^3}L^3. \tag{67}$$

The factor of 2 is included to account for electrons of both spin-up and spin-down.

The density of states is defined as the number of states per unit energy per unit volume, or

$$g(E) = \frac{(dN/dE)}{L^3} = \frac{4\pi(2m^*)^{3/2}}{h^3}E^{1/2}$$

$$= \frac{1}{2\pi^2}\left(\frac{2m^*}{\hbar^2}\right)^{3/2}E^{1/2}, \tag{68}$$

where $\hbar = h/2\pi$ is Plank's reduced constant. This equation describes the number of available states for electrons to go into, but it does not describe the way the electrons fill those available states.

A. The Fermi Distribution

To determine the number of electrons in a given band, it is necessary to find the probability of a given state being occupied by an electron and then integrate over all available states. A more realistic result for metals, which does not assume spherical constant energy surfaces in **k**-space, thus allowing for the electron energy to deviate from $E = \hbar^2 k^2/2m^*$ would be found using the density of states from (63).

Within a crystalline material, charge carriers are known to follow the Pauli exclusion principle, which states that only one electrons can occupy a given energy state. The probability that an electron occupies an energy state can be found by considering a simple statistical exercise. If a system is defined to have three allowed energy levels (E_1, E_2, and E_3), two electrons, and a total energy of 4 eV as shown in Fig. 5, with the three energy levels defined as $E_1 = 1$ eV, $E_2 = 2$ eV, and $E_3 = 3$ eV, it would be expected that 80% of the time, a distribution of one electron in energy level E_1, zero electrons in E_2, and one electron in E_3, or a distribution of (1, 0, 1), would occur. The entropy of a system is related to the most probable arrangement, W_m, of the particles through Boltzmann's definition,

$$\bar{S} = k\ln W_m. \tag{69}$$

When only electrons are considered, the entropy is related to the internal energy of the system, U, the total number of electrons, N, and the volume, $\mathcal{V}$, of the system through Euler's equation

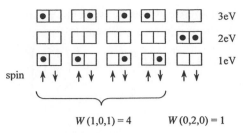

FIGURE 5 The number of ways, W, two electrons can be distributed in three energy levels to obtain a total energy of 4 eV.

$$U = T\bar{S} - P\mathcal{V} + \mu N - q\mathcal{V}N, \tag{70}$$

where V represents the internal electrostatic potential. For a simple system with just two available energy states (energy $= 0$ or energy $= E$), the probability of finding the system with energy E to that of finding it with energy 0 is

$$\frac{\mathcal{P}(E)}{\mathcal{P}(0)} = \frac{W(U_0 - E)}{W(U_0)} = \frac{e^{\bar{S}(U_0-E)/k}}{e^{\bar{S}(U_0)/k}}. \tag{71}$$

Using the approximation $\bar{S}(U_0 - E) \approx \bar{S}(U_0) - E(\frac{\partial\bar{S}}{\partial U_0})$ and

$$\frac{\partial\bar{S}}{\partial U} = \frac{1}{T} = \frac{\partial}{\partial U}\left(\frac{U + P\mathcal{V} - \mu N + q\mathcal{V}N}{T}\right) \tag{72}$$

simplifies (71) to

$$\frac{\mathcal{P}(E)}{\mathcal{P}(0)} = \frac{e^{(\bar{S}(U_0)/k)-(E/kT)}}{e^{\bar{S}(U_0)/k}} = e^{-E/kT}. \tag{73}$$

To determine the probability of a system in energy state E, and that the state is occupied by an electron, then the influence of the total number of electrons, N, must also be taken into consideration. Then the ratio of the probability that the system is occupied by one electron at energy E to the probability that the system is unoccupied with energy 0 is

$$\frac{\mathcal{P}(1, E)}{\mathcal{P}(0, 0)} = \frac{W[(U_0 - E), (N_0 - 1)]}{W[U_0, N_0]}$$

$$= \frac{e^{\bar{S}[(U_0-E),(N_0-1)]/k}}{e^{\bar{S}[U_0,N_0]/k}}. \tag{74}$$

Using $\bar{S}[(U_0 - E), (N_0 - 1)] \approx \bar{S}[U_0, N_0] - E(\partial\bar{S}/\partial U_0) - (\partial\bar{S}/\partial N_0)$ yields

$$\frac{\mathcal{P}(1, E)}{\mathcal{P}(0, 0)} = \frac{e^{(\bar{S}[U_0,N_0]/k)-(E/kT)+((\mu-q\zeta)/kT)}}{e^{\bar{S}[U_0,N_0]/k}} = e^{(E_F-E)/kT}, \tag{75}$$

where the Fermi level is defined as $E_F = \mu - q\mathcal{V}$. Since $\mathcal{P}(1, E) + \mathcal{P}(0, 0) = 1$,

$$\mathcal{P}(1, E) = f(E) = \frac{1}{1 + e^{((E_F-E)/kT)}}. \tag{76}$$

This is the Fermi–Dirac distribution and represents the probability of occupancy of an energy state in equilibrium.

B. Carrier Concentrations

In a free-electron approximation, the total number of electrons in a given energy band can then be found by integrating the product of the density of states and the probability of occupancy of the state for spherical energy surfaces:

$$n = \int_{E_{\text{bottom}}}^{E_{\text{top}}} g(E)f(E)\,dE = \frac{(2mkT)^{3/2}}{2\pi^2\hbar^3}F_{1/2}(\eta), \quad (77)$$

where $\eta = E_F/kT$, and $F_\nu(\eta)$ is the Fermi–Dirac function,

$$F_\nu(\eta) = \int_0^\infty \frac{\eta^\nu}{1+e^{x-\eta}}\,dx. \quad (78)$$

Here, the bottom of the energy band was taken to be zero energy corresponding to $\mathbf{k}=0$, and the integration was allowed to extend to ∞ since the Fermi–Dirac distribution falls to zero at high energy levels. In the degenerate limit, when $E_F \gg kT$, a series expansion of (78) leads to the following approximations for metals:

$$E_F \approx E_{F_0}\left[1 - \frac{\pi^2}{12}\left(\frac{kT}{E_{F_0}}\right)^2 + \cdots\right],$$
$$(79)$$

$$n \approx \frac{(2m)^{2/3}}{3\pi^2\hbar^3}\left(\frac{E_F}{kT}\right)^{3/2}\left[1 + \frac{\pi^2}{8}\left(\frac{kT}{E_F}\right)^2 + \cdots\right].$$

At $T=0$ K,

$$E_F = E_{F_0} = \frac{\pi^2\hbar^2}{2m}\left(\frac{3n}{\pi}\right)^{2/3}. \quad (80)$$

For nonspherical energy surfaces, the number of electrons per unit volume within an element of momentum space, $dp_x\,dp_y\,dp_z$, is found using (63) and the relation $\mathbf{p}=\hbar\mathbf{k}=\frac{h}{2\pi}\mathbf{k}$,

$$dn = \frac{2}{h^3}f(\mathbf{p},\mathbf{r})\,dp_x\,dp_y\,dp_z = \frac{1}{4\pi^3}f(\mathbf{k},\mathbf{r})\,dk_x\,dk_y\,dk_z,$$
$$(81)$$

where the factor of 2 accounts for two electrons of opposite spin. The total electron density is then found by integration.

C. The Boltzmann Function

If the material is disturbed from equilibrium, then the distribution will vary, in general, as a function of wavevector, $\mathbf{k}$, position, $\mathbf{r}$, and time, t, or $f(\mathbf{k},\mathbf{r},t)$. At a time $t+dt$, the probability that a state with wavevector $\mathbf{k}+d\mathbf{k}$ is occupied by an electron at position $\mathbf{r}+d\mathbf{r}$ can be found, using Eq. (59), to be

$$f(\mathbf{k}+d\mathbf{k},\mathbf{r}+d\mathbf{r},t+dt)$$

$$= f\left(\mathbf{k}+\frac{1}{\hbar}\mathbf{F}_t\cdot\nabla_k\,dt,\mathbf{r}+\mathbf{v}\,dt,t+dt\right). \quad (82)$$

The total rate of change of the distribution function near $\mathbf{r}$ is then

$$\frac{df}{dt} = \frac{1}{\hbar}\mathbf{F}_t\cdot\nabla_k f + \mathbf{v}\cdot\nabla_r f + \frac{\partial f}{\partial t}, \quad (83)$$

which is Boltzmann's transport equation. The first term on the right-hand side of this equation accounts for contributions from forces, $\mathbf{F}_t$, including externally applied forces, $\mathbf{F}$, and collision forces, $\mathbf{F}_c$. The middle term adds the contributions from concentration gradients, and the last term is the local changes in the distribution function about the point $\mathbf{r}$. Equation (83) is equal to zero since the total number of states in the crystal is constant, thus

$$\frac{\partial f}{\partial t} = \frac{-1}{\hbar}\mathbf{F}_t\cdot\nabla_k f - \nu\cdot\nabla_r f$$

$$= \frac{-1}{\hbar}\mathbf{F}_c\cdot\nabla_k f - \frac{1}{\hbar}\mathbf{F}\cdot\nabla_k f - \nu\cdot\nabla_r f$$

$$= \left(\frac{\partial f}{\partial t}\right)_c - \frac{1}{\hbar}\mathbf{F}\cdot\nabla_k f - \nu\cdot\nabla_r f. \quad (84)$$

With external forces applied, the distribution function, f, will be disturbed from the equilibrium value, f_0. Upon the removal of those external forces, equilibrium will be reestablished through collisions, $(\partial f/\partial t)_c$. Calculation of this collision term is a formidable task dependent largely on the scattering mechanisms for the material investigated. For small disturbances, however, a relaxation-time approximation is often used which assumes that

$$\left(\frac{\partial f}{\partial t}\right)_c = \frac{-(f-f_0)}{\tau_k} = \frac{-f_1}{\tau_k}, \quad (85)$$

where τ_k is the momentum relaxation time. In steady state, $\partial f/\partial t = 0$ and Eq. (84) becomes

$$\left.\begin{aligned} 0 &= \frac{-f_1}{\tau_k} - \frac{1}{\hbar}\mathbf{F}\cdot\nabla_k f - \mathbf{v}\cdot\nabla_r f \\ f_1 &= -\frac{\tau_k}{\hbar}\mathbf{F}\cdot\nabla_k f - \tau_k\mathbf{v}\cdot\nabla_r f. \end{aligned}\right\} \quad (86)$$

or

The electric and heat current densities are given by

$$\mathbf{J} = -q\mathbf{v}n = -q\int \mathbf{v}\,dn = -\frac{q}{4\pi^3}\int f_1(\mathbf{k})\,d\mathbf{k}, \quad (87)$$

$$\mathbf{J}_Q = \frac{1}{4\pi^3}\int \mathbf{v}(E-E_F)f_1(\mathbf{k})\,d\mathbf{k}.$$

Substituting Eq. (86) into Eqs. (87) starting with the electric current density, $\mathbf{J}$, gives

$$\mathbf{J} = -\frac{q}{4\pi^3}\int\left(\frac{-\tau_k}{\hbar}\mathbf{v}\mathbf{F}\cdot\nabla_k f - \tau_k\mathbf{v}\mathbf{v}\cdot\nabla_r f\right)d\mathbf{k}. \quad (88)$$

Assuming parabolic bands, the gradient of the distribution function in $\mathbf{k}$ space can be written

$$\nabla_k f = \frac{\partial f}{\partial E}\nabla_k E = \frac{\partial f}{\partial E}\hbar\mathbf{v}. \quad (89)$$

Also, the following can be shown by direct substitution of the Fermi–Dirac distribution (76):

$$\frac{\partial f}{\partial x} = \frac{\partial f}{\partial E} T\left[E\frac{\partial}{\partial x}\left(\frac{1}{T}\right) - \frac{\partial}{\partial x}\left(\frac{E_F}{T}\right)\right], \quad (90)$$

with similar results in the y and z directions. Substituting these results into Eq. (88) gives

$$\mathbf{J} = -\frac{q}{4\pi^3}\int\left(\frac{-\tau_k}{\hbar}\mathbf{v}\mathbf{F}\frac{\partial f}{\partial E}\hbar\mathbf{v} - \mathbf{v}\mathbf{v}\tau_k\frac{\partial f}{\partial E}T\right.$$
$$\left.\times\left[E\nabla_r\left(\frac{1}{T}\right) - \nabla_r\left(\frac{E_F}{T}\right)\right]\right)d\mathbf{k} \quad (91)$$

or

$$\mathbf{J} = \frac{q}{4\pi^3}\int\tau_k\mathbf{v}\mathbf{v}\mathbf{F}\frac{\partial f}{\partial E}\,d\mathbf{k}$$
$$+ q\int\tau_k\mathbf{v}\mathbf{v}\frac{\partial f}{\partial E}T(E - E_F)\nabla_r\left(\frac{1}{T}\right)d\mathbf{k}. \quad (92)$$

For example, the applied force might include a contribution from an external electric field $(-q\mathcal{E})$, plus a contribution caused by a temperature gradient (see Fig. 3), or in general as $\nabla\bar{\mu} = \nabla(\mu + qV) = \nabla\mu - q\mathcal{E}$. Then (92) would be

$$\mathbf{J} = \frac{q}{4\pi^3}\int\tau_k\mathbf{v}\mathbf{v}\frac{\partial f}{\partial E}(\nabla\mu - q\mathcal{E})\,d\mathbf{k}$$
$$- q\int\tau_k\mathbf{v}\mathbf{v}(E - E_F)\frac{\partial f}{\partial E}\frac{1}{T}\nabla_r T\,d\mathbf{k}, \quad (93)$$

where $\nabla_r(1/T) = -(1/T^2)\nabla_r T$ was used. The electrical current density can be simplified and put into a format similar to the Onsager relations as shown in (54) by using transport integrals defined as

$$\mathbf{K}_n = -\frac{1}{4\pi^3}\int\tau_k\mathbf{v}\mathbf{v}(E - E_F)^n\frac{\partial f_0}{\partial E}\,d\mathbf{k}, \quad (94)$$

where it is assumed that the deviations from equilibrium are small, such that $\partial f/\partial E$ in Eq. (93) may be replaced with $\partial f_0/\partial E$. This leads to an electrical current density of

$$\mathbf{J} = -q\mathbf{K}_0\nabla\bar{\mu} + \frac{q}{T}\mathbf{K}_1\nabla T. \quad (95)$$

Similarly, the heat current density, $\mathbf{J}_Q$, follows the same derivation to arrive at

$$\mathbf{J}_Q = \frac{1}{4\pi^3}\int\tau_k\mathbf{v}\mathbf{v}\frac{\partial f}{\partial E}(\nabla\mu - q\mathcal{E})(E - E_F)\,d\mathbf{k}$$
$$- \frac{1}{4\pi^3}\int\tau_k\mathbf{v}\mathbf{v}(E - E_F)^2\frac{\partial f}{\partial E}\frac{1}{T}\nabla_r T\,d\mathbf{k} \quad (96)$$

or

$$\mathbf{J}_Q = -\mathbf{K}_1\nabla\bar{\mu} + \frac{1}{T}\mathbf{K}_2\nabla T. \quad (97)$$

These derivatives form the link between the macroscopic Onsager equations and the atomistic derivations from Boltzmann's equation. Comparison of Eqs. (95) and (97) with Eq. (54) shows the following relations:

$$\left.\begin{aligned}\frac{\mathbf{L}_{11}}{qT} &= -q\mathbf{K}_0,\\[4pt]\frac{-\mathbf{L}_{12}}{T^2} &= \frac{q}{T}\mathbf{K}_1,\\[4pt]\frac{\mathbf{L}_{21}}{qT} &= -\mathbf{K}_1,\\[4pt]\frac{-\mathbf{L}_{22}}{T^2} &= \frac{1}{T}\mathbf{K}_2,\end{aligned}\right\} \quad\text{or}\quad \begin{cases}\mathbf{L}_{11} = -q^2 T\mathbf{K}_0,\\ \mathbf{L}_{12} = -qT\mathbf{K}_1,\\ \mathbf{L}_{21} = -qT\mathbf{K}_1,\\ \mathbf{L}_{22} = -T\mathbf{K}_2.\end{cases} \quad (98)$$

This shows the Onsager reciprocity relation, in that $\mathbf{L}_{12} = \mathbf{L}_{21}$. The thermoelectric properties can now be determined through an evaluation of the transport integrals and the appropriate boundary conditions of isothermal ($\nabla T = 0$), isoelectric ($\nabla V = -\mathcal{E} = 0$), static ($\mathbf{J} = 0$), or adiabatic ($\mathbf{J}_Q = 0$). For example, under isothermal conditions, where $\nabla T = 0$ and thus $\nabla\mu = 0$ (for a homogeneous metal),

$$\mathbf{J} = q^2\mathbf{K}_0\mathcal{E} = \sigma\mathcal{E}, \quad (99)$$

and the electrical conductivity is

$$\sigma = q^2\mathbf{K}_0. \quad (100)$$

The electronic contribution to the thermal conductivity is defined for static conditions as $\mathbf{J}_E|_{J=0} = -\kappa_e\nabla T$, or when $\mathbf{J}_E = \mathbf{J}_Q$ as in a one-band material, $\mathbf{J}_Q = \kappa_e\nabla T$, where (95) becomes

$$0 = -q\mathbf{K}_0\nabla\bar{\mu} + \frac{q}{T}\mathbf{K}_1\nabla T. \quad (101)$$

Solving for $\nabla\bar{\mu}$ and substituting into (97) gives

$$\mathbf{J}_Q = -\mathbf{K}_1\left(\frac{1}{T}\frac{\mathbf{K}_1}{\mathbf{K}_0}\nabla T\right) + \frac{1}{T}\mathbf{K}_2\nabla T$$
$$= \frac{1}{T}\left(\mathbf{K}_2 - \frac{\mathbf{K}_1\mathbf{K}_1}{\mathbf{K}_0}\right)\nabla T \quad (102)$$

or

$$\kappa = \frac{1}{T}\left(\mathbf{K}_2 - \frac{\mathbf{K}_1\mathbf{K}_1}{\mathbf{K}_0}\right). \quad (103)$$

The absolute Seebeck coefficient, or thermopower, $\mathbf{S}$, can also be found from the static condition, where the use of Eq. (101) gives

$$\mathbf{S} = \frac{1}{q}\frac{\nabla\bar{\mu}}{\nabla T} = \frac{1}{qT}\frac{\mathbf{K}_1}{\mathbf{K}_0}. \quad (104)$$

The Peltier coefficient, Π, can be found by evaluating the heat current density, Eq. (97), for isothermal conditions:

$$\mathbf{J}_Q = q\mathbf{K}_1\mathcal{E}. \quad (105)$$

TABLE I   Combined Results from Macroscopic and Atomistic Analysis

Thermoelectric property	Transport integral	Onsager coefficient
$\sigma = q^2 \mathbf{K}_0$	$\mathbf{K}_0 = \dfrac{\sigma}{q^2}$	$\mathbf{L}_{11} = -\sigma T$
$\mathbf{S} = \dfrac{1}{qT}\dfrac{\mathbf{K}_1}{\mathbf{K}_0}$	$\mathbf{K}_1 = T\dfrac{\sigma}{q}\mathbf{S}$	$\mathbf{L}_{12} = \mathbf{L}_{21} = -T^2 \sigma \mathbf{S}$
$\kappa_e = \dfrac{1}{T}\left(\mathbf{K}_2 - \dfrac{\mathbf{K}_1\mathbf{K}_1}{\mathbf{K}_0}\right)$	$\mathbf{K}_2 = \kappa_e T + T^2 \sigma \mathbf{S}^2$	$\mathbf{L}_{22} = -T^2 \kappa_e - T^3 \sigma \mathbf{S}^2$

Substituting for the electric field, $\mathcal{E}$, from Eq. (99) gives the direct relationship between heat current density and electric current density,

$$\mathbf{J}_Q = \frac{\mathbf{K}_1}{q\mathbf{K}_0}\mathbf{J} = \Pi\mathbf{J}, \qquad (106)$$

where the proportionality constant is simply the Peltier coefficient, Π. Comparing the Peltier coefficient (106) with the thermopower (104) leads to Kelvin's second relation:

$$\Pi = T\mathbf{S}. \qquad (107)$$

Results of the transport integrals are summarized in Table I. Thus Eqs. (54) can be rewritten in terms of the thermoelectric properties as

$$\left.\begin{aligned}
\mathbf{J} &= \frac{-\sigma}{q}\nabla\bar{\mu} + \sigma\mathbf{S}\nabla T, \\
\mathbf{J}_Q &= \frac{-T}{q}\sigma\mathbf{S}\nabla\bar{\mu} + \left(\kappa_e + T\sigma\mathbf{S}^2\right)\nabla T.
\end{aligned}\right\} \qquad (108)$$

Substitution of the transport integrals can be used to evaluate further the thermoelectric properties. Estimations can be made through a series expansion of the transport integrals using a Sommerfeld expansion,

$$\begin{aligned}
\mathbf{K}_n &= -\int_0^\infty \phi_n(E)\frac{\partial f_0}{\partial E}\,dE \\
&= \phi_n(E_F) + \frac{\pi^2}{6}(kT)^2\frac{d^2}{dE_F^2}\phi_n(E_F) + \cdots. \quad (109)
\end{aligned}$$

For the electrical conductivity,

$$\sigma = q^2\mathbf{K}_0 = -\frac{q^2}{4\pi^3}\int \tau_k\mathbf{v}\mathbf{v}\frac{\partial f_0}{\partial E}\,d\mathbf{k}. \qquad (110)$$

In its simplest form for cubic symmetry, this reduces to

$$\sigma = \frac{nq^2\tau_k}{m^*}, \qquad (111)$$

where n is the electron density with energies near E_F, and m^* is the effective mass of the electrons. Both the electron density near the Fermi level and the relaxation time are functions of energy, such that the electrical conductivity can be approximated as $\sigma = \text{const} \cdot E^\xi$, where ξ is some number.

A relationship between the electrical conductivity and the thermopower can be found by series expansion $\mathbf{K}_1$, which gives

$$\mathbf{K}_1 = -\frac{\pi^2}{3q^2}(kT)^2\frac{\partial\sigma}{\partial E}\bigg|_{E=E_F}, \qquad (112)$$

along with $\sigma = q^2\mathbf{K}_0$ and substituting into (104). This leads to the Mott–Jones equation (Barnard, 1972):

$$S_d = \frac{-\pi^2}{3}\frac{k^2 T}{q}\left(\frac{\partial\ln\sigma}{\partial E}\right)_{E_F}. \qquad (113)$$

A distinction of the diffusion thermopower, S_d, has been made here to separate it from a low-temperature effect that has not been considered above. The low-temperature effect typically appears as a peak in the measured thermopower (near 60 K for monovalent noble metals) and is the result of an increased electron–phonon interaction. When a temperature gradient exists across a crystal, heat will flow from the hot side to the cold side through lattice vibrations (phonons) and through electron flow. Various interactions among phonons, lattice defects, and electrons can be described by scattering times for each type of interaction. At high temperatures, phonon–phonon interactions are more frequent than electron–phonon interactions ($\tau_{p,p} < \tau_{p,e}$). At these high temperatures (above the Debye temperature, $T > \theta_D$), $\tau_{p,e}$ is approximately temperature independent, while $\tau_{p,p} \propto 1/T$. Under these conditions, the total thermopower is dominated by the diffusion thermopower as given in Eq. (113). At low temperatures ($T < \theta_D$), $\tau_{p,e} \propto 1/T$ and $\tau_{p,p} \propto e^{\theta_D/2T}$, therefore, as the temperature drops, $\tau_{p,p}$ increases more rapidly than $\tau_{p,e}$. When this occurs, $\tau_{p,p} > \tau_{p,e}$ and electron–phonon interactions will occur more frequently, causing electrons to "dragged" along with the phonons. This gives rise to a larger gradient of carrier concentration across the sample and is additive to the diffusion thermopower such that $S = S_d + S_g$, where S_g is the phonon-drag component of the thermopower described above. At still lower temperatures, phonon-impurity interactions can dominate, causing the magnitude of the thermopower to decrease toward zero. For the remainder of this chapter, the temperature is assumed to be much higher than the Debye temperature, such that $S \approx S_d$ and the diffusion subscript is dropped.

When the electrical conductivity can be written $\sigma = \text{const} \cdot E^\xi$, this can be used in the Mott–Jones equation to give

$$S = \frac{-\pi^2}{3}\frac{k^2 T}{qE_F}\xi = -0.0245\frac{T}{E_F}\xi\left(\frac{\mu V}{K}\right). \quad (114)$$

1. Normal Metals

In monovalent noble metals (Cu, Ag, and Au), $\xi \approx -\frac{3}{2}$ has been measured, giving the positive quantity

$$S = 0.03675\frac{T}{E_F}\left(\frac{\mu V}{K}\right). \quad (115)$$

It is instructive to compare the thermopower of noble metals to the electronic heat capacity, C_{el}, which is dependent on the density of states, $g(E_F)$, evaluated at the Fermi level. Substituting $\xi \approx -\frac{3}{2}$ into Eq. (114) gives

$$S = \frac{\pi^2}{2}\frac{k^2 T}{qE_F} \quad (116)$$

and

$$C_{el} = \frac{\pi^2}{3}g(E_F)k^2 T = \frac{\pi^2 N}{2}\frac{k^2 T}{E_F}, \quad (117)$$

where N is the total number of carriers. Then it can be seen that the electronic heat capacity per carrier is simply the electronic charge times the thermopower,

$$\frac{C_{el}}{N} = qS. \quad (118)$$

2. Transition Elements

The electronic properties of transition metals are usually considered to have contributions from two bands that overlap at the Fermi level: the s-band, from the s levels of the individual atoms, and the d-band, consisting of five individual overlapping bands. The s-band is broad and typically approximated as free electron-like, while the d-band is narrow, with a high density of states and high effective mass, thus the s electrons carry most of the current. The relaxation time is, however, greatly affected by the high density of states of the d-band. This comes about through the inverse proportionality of the relaxation time to the probability of scattering from one wavevector, $\mathbf{k}$, to another, $\mathbf{k}'$. The occupancy and availability of each of these wavevectors are, in turn, proportional to the density of states at the Fermi level. This leads to the relationship of the inverse proportionality of the relaxation time to the density of states:

$$\frac{1}{\tau} \propto g(E)|_{E=E_F}. \quad (119)$$

Due to the relatively high density of states in the d-band, the relaxation time of the highly responsive s-band electrons is dominated by s–d transitions, or

$$\frac{1}{\tau_s} \approx \frac{1}{\tau_{s-d}} \propto g_d(E)|_{E=E_F}. \quad (120)$$

Neglecting the d-band contribution to the electrical conductivity and rewriting Eq. (110) in terms of the density of states gives

$$\sigma = \frac{2}{3}q^2 v_s^2 \tau_s g_s(E)|_{E=E_F}$$
$$= \text{const} \cdot v_s^2 \frac{g_s(E)}{g_d(E)}\bigg|_{E=E_F}. \quad (121)$$

Defining the bottom of the s-band as zero energy, and the partially filled d-band in terms of the holes in the band so it can be referenced to the top of the d-band, such that E_0 is the energy at the top of the d-band, and $g_d(E) = \text{const} \cdot (E_0 - E_F)^{1/2}$, then approximating the s-band electrons as free electrons gives

$$\frac{\partial \ln \sigma}{\partial E}\bigg|_{E=E_F} = \frac{3}{2E_F} + \frac{1}{2(E_0 - E_F)}. \quad (122)$$

Typically $E_F \gg (E_0 - E_F)$, such that approximating the above equation as the second term on the right-hand side and using this in the Mott–Jones equation (113) gives

$$S = \frac{-\pi^2}{6}\frac{k^2 T}{q(E_0 - E_F)}. \quad (123)$$

Again, the electronic heat capacity can be compared to find

$$C_{el} = \frac{\pi^2}{6}\frac{Nk^2 T}{(E_0 - E_F)}, \quad (124)$$

and the relationship between the magnitude of the electronic heat capacity and the thermopower remains

$$\frac{C_{el}}{N} = qS. \quad (125)$$

3. Semimetals

The petavalent elements of As, Sb, and Bi are semimetals with rhombohedral crystal structures. This leads to nonspherical Fermi surfaces and anisotropic scattering such that $\tau \propto \mathbf{k}_x^s$ for a given crystallographic direction, where s accounts for the anisotropy. Likewise, the density of states $g(E) \propto \mathbf{k}_x^3$, and $\mathbf{k}_x \propto (E_0 - E_F)^{1/2}$. Using the density of states and the relaxation time for the electrical conductivity in an equation similar to (121) gives

$$\sigma = \text{const} \cdot (E_0 - E_F)^{(3+s)/2} \qquad (126)$$

or, in (113),

$$S \cong \frac{-\pi^2}{6} \frac{k^2 T}{q(E_0 - E_F)}(3 + s), \qquad (127)$$

where $(3 + s) < 0$. There is an exception to (127) in bismuth, which shows the expected anisotropic thermopower, but an unexpected negative thermopower ($S_\perp \approx -50 \, \mu$V/K, and $S_\| \approx -100 \, \mu$V/K at 273 K). For bismuth, a value of $\xi = (3 + s)/2$ should be used in (114) describing electron conduction, instead of using Eq. (127), which is for conduction by holes.

4. Alloys

Matthiessen's rule states that the total resistivity of an alloy formed by two metals can be found by

$$\rho = \frac{1}{\sigma} = \rho_i + \rho_j, \qquad (128)$$

where ρ_i is the resistivity of the pure solvent metal due to scattering of carriers by thermal vibrations, and ρ_j represents scattering of carriers from impurities. This rule is often used for approximations but is not widely applicable since many cases exhibit anisotropic scattering of carriers, causing a large deviation from (128). Assuming the validity of Matthiessen's rule, (113) can be written

$$S = \frac{\pi^2}{3} \frac{k^2 T}{q} \left(\frac{\partial \ln(\rho_i + \rho_j)}{\partial E} \right)_{E_F}, \qquad (129)$$

which can be written in terms of the difference between S for the alloy and the thermopower of the pure solvent metal, S_i, or $\Delta S = S - S_i$ leads to

$$\frac{\Delta S}{S} = -\frac{1 - (x_j/x_i)}{1 + (\rho_i/\rho_j)}, \qquad (130)$$

where

$$x_i = -\left(\frac{\partial \ln \rho_i}{\partial E} \right)_{E_F} \quad \text{and} \quad x_j = -\left(\frac{\partial \ln \rho_j}{\partial E} \right)_{E_F}. \qquad (131)$$

Using the Gorter–Nordheim relation for the impurity component of Matthiessen's rule, $\rho_i = CX(1 - X)$, where C is the Nordheim coefficient and X is the atomic fraction of the solute atoms in a solid solution, yields a more useful relationship:

$$S = S_j + \frac{\rho_i}{\rho}(S_i - S_j), \qquad (132)$$

where S_j is the thermopower for the impurity.

The third thermoelectric parameter listed in Table I is thermal conductivity. This can likewise be determined using (109) and substituting the transport integrals into (103). Series expansion of $\mathbf{K}_2$ gives

$$\mathbf{K}_2 = -\frac{\pi^2}{3} \frac{k^2 T^2}{q^2} \sigma. \qquad (133)$$

The thermal conductivity is given by (103), repeated here for convenience:

$$\kappa_e = \frac{1}{T} \left(\mathbf{K}_2 - \frac{\mathbf{K}_1 \mathbf{K}_1}{\mathbf{K}_0} \right). \qquad (103)$$

In metals, $(\partial/\partial E)\sigma(E)|_{E=E_F} \approx (\sigma/E_F)$, thus $\mathbf{K}_1 \approx -(\pi^2/3q^2)(kT)^2(\sigma/E_F)$, or

$$\frac{\mathbf{K}_1 \mathbf{K}_1}{\mathbf{K}_0} \approx \frac{\left[(\pi^2/3q^2)(kT)^2(\sigma/E_F) \right]^2}{\sigma/q^2}$$

$$= \left[\frac{\pi^2}{3q^2}(kT)^2\sigma \right] \left[\frac{\pi^2}{3} \frac{(kT)^2}{E_F^2} \right], \qquad (134)$$

giving

$$\kappa_e \approx \frac{\pi^2 k^2 T}{3q^2} \sigma \left(1 + \frac{\pi^2}{3} \frac{(kT)^2}{E_F^2} \right) \approx \frac{\pi^2 k^2 T}{3q^2} \sigma, \qquad (135)$$

where the approximation of $(\pi^2/3)((kT)^2/E_F^2) \ll 1$ was used, thus arriving at the Wiedemann–Franz law, or $\kappa_e/\sigma T = 2.443 \times 10^{-8}$ ((W · Ω)/K^2).

The total thermal conductivity, κ, must also include a lattice contribution, κ_L, such that

$$\kappa = \kappa_L + \kappa_e. \qquad (136)$$

The lattice thermal conductivity for metals is generally much lower than the electronic contribution.

D. Semiconductors

The above analysis is applicable to normal metals, where it is assumed that the carriers are electrons and $\nabla \mu$ is a function of temperature only. Furthermore, the Onsager relations were developed using four contributions, (33), (36), (37), and (39), to the energy density rate of change, however, two additional contributions exist for semiconductors. These contributions account for transitions of electrons across the bandgap, or the rate of change in carrier concentrations in each band, and for positional gradients of the band edges (valence band and conduction band). The last contribution could arise from temperature gradients and/or compositional variations, for example. These additional contributions have the form

$$\left(\frac{\partial E}{\partial t} \right)_V = -q\Phi_C(-\nabla \cdot \mathbf{J}_n) + q\Phi_V(-\nabla \cdot \mathbf{J}_p) \qquad (137)$$

and

$$\left(\frac{\partial E}{\partial t} \right)_{VI} = \mathbf{J}_n \cdot \nabla q\Phi_C - \mathbf{J}_p \cdot \nabla q\Phi_V, \qquad (138)$$

where $-q\Phi_C$ and $q\Phi_V$ represent the internal potential energies of the electrons and holes at the bottom of the conduction band and the top of the valence band, respectively. This leads to the Onsager relations for a two-band model, where, in a steady-state condition (Harman and Honig, 1967),

$$
\begin{aligned}
\mathbf{J}_Q &= L_{11}\mathbf{X}_Q + L_{12}\mathbf{X}_n + L_{13}\mathbf{X}_p, \\
\mathbf{J}_- &= L_{21}\mathbf{X}_Q + L_{22}\mathbf{X}_n + L_{23}\mathbf{X}_p, \\
\mathbf{J}_+ &= L_{31}\mathbf{X}_Q + L_{32}\mathbf{X}_n + L_{33}\mathbf{X}_p,
\end{aligned}
\tag{139}
$$

where

$$
\mathbf{X}_Q = -\frac{1}{T^2}\nabla T,
$$

$$
\mathbf{X}_n = -\frac{1}{T}\nabla\varphi_C + \nabla\frac{\bar{\mu}_C}{qT},
\tag{140}
$$

$$
\mathbf{X}_p = -\frac{1}{T}\nabla\varphi_V - \nabla\frac{\bar{\mu}_V}{qT},
$$

Also, $\bar{\mu}_C$ and $\bar{\mu}_V$ represent the difference between the chemical potential energy and the internal potential energy of the carriers in the two bands. The total potential energy of the carriers in an applied field for a semiconductor must include the potential energy from the field as well as the internal potential energies $-q\Phi_C$ and $q\Phi_V$, from the band edges. Contributions to the electrical current density come from electrons, $\mathbf{J}_- = -q\mathbf{J}_n$, and from holes, $\mathbf{J}_+ = q\mathbf{J}_p$, for the total current density given by $\mathbf{J} = \mathbf{J}_- + \mathbf{J}_+$. Applying the same procedure for this case as followed for metals above, with the additional consideration of the relative potential energies of the band edges using $\bar{\mu}_V = -(E_F + E_V)$ and $\bar{\mu}_C = E_F - E_C$, gives the following formula for a two-band semiconductor:

$$
\sigma = \sigma_n + \sigma_p,
$$

$$
S = \frac{S_n\sigma_n + S_p\sigma_p}{\sigma_n + \sigma_p},
\tag{141}
$$

$$
\kappa = \kappa_L + \kappa_n + \kappa_p + \frac{\sigma_n\sigma_p}{\sigma T q^2}\left[\frac{K_1^n}{K_0^n} + \frac{K_1^p}{K_0^p} + (E_C - E_V)\right]^2.
$$

Of course, as a semiconductor is doped n-type or p-type, the corresponding contributions, subscripted n or p, respectively, above will dominate. The last term in the thermal conductivity formula, when multiplied by $-\nabla T$, would relate to the transport of bandgap energy along the negative temperature gradient and is defined as an ambipolar transport mechanism.

V. APPLICATIONS

A. Thermocouples

Thermocouples are the most common application of thermoelectric materials. Application of the Seebeck coefficient (1), along with the Thompson relation (7), allows one to determine the open-circuit potential for a circuit containing temperature gradients by integrating over temperature as one traverses through the circuit from one terminal of the open circuit to the other. For example, in the circuit shown in Fig. 6 the open-circuit voltage can be written

$$
\begin{aligned}
\Delta V &= \int_{T_0}^{T_1} S_A\,dT + \int_{T_1}^{T_2} S_B\,dT + \int_{T_2}^{T_3} S_C\,dT \\
&\quad + \int_{T_3}^{T_4} S_C\,dT + \int_{T_4}^{T_5} S_D\,dT + \int_{T_5}^{T_0} S_A\,dT \\
&= \int_{T_5}^{T_1} S_A\,dT + \int_{T_1}^{T_2} S_B\,dT + \int_{T_2}^{T_4} S_C\,dT \\
&\quad + \int_{T_4}^{T_5} S_D\,dT.
\end{aligned}
\tag{142}
$$

When measuring this potential difference, care must be taken to include the contribution from the leads of the meter. This can be minimized by assuring that the thermocouple-circuit open terminals (in Fig. 6) are at a constant temperature T_0 and the terminals on the voltage meter are also at a constant temperature (not necessarily T_0).

B. Generators and Coolers

Lenz first demonstrated a thermoelectric cooler by freezing water at the junction between two conductors formed by rods of bismuth and antimony; however, a more common configuration for a thermoelectric cooler is shown in

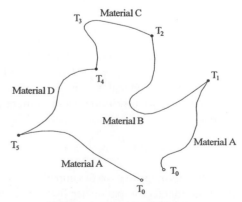

FIGURE 6 Thermocouple circuit.

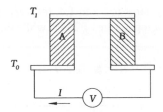

FIGURE 7 Thermoelectric cooler.

Fig. 7. Here the cooling (or warming) junction is made more accessible for device cooling (or heating).

Since current is defined as positive in the direction of positive carrier flow (hole flow), and likewise for the direction opposite to negative carrier flow (electron flow), by using one p-type leg and one n-type leg to the cooler, the highest efficiency can be achieved (Fig. 8). In this situation, all carriers flow in the same physical direction (either top to bottom or bottom to top) in both legs. Since charge carriers also carry heat as shown through the Onsager relations, heat will flow through the device in the direction of the carriers. Although the configuration shows a pn junction, these devices do not behave as diodes and electrical current is reversible. This is due to the fact that each of the legs is doped to degeneracy, or near-degeneracy, such that ohmic contacts with the metals are exhibited.

The goal in making a thermoelectric cooler is to maximize the coefficient of performance, φ, of the device, defined as

$$\varphi = \frac{\dot{Q}_0}{W}, \tag{143}$$

where $\dot{Q}_0$ is the rate of heat absorbed from the object being cooled over the amount of power, W, it takes to drive the cooler. Assuming that the thermopower of materials A and B in Fig. 7 do not vary significantly over the temperature range T_0 to T_1, then the Thompson heat may be neglected, and

$$\dot{Q}_0 = \dot{Q}_\Pi - \dot{Q}_T, \tag{144}$$

where $\dot{Q}_\Pi$ is the Peltier heat absorbed at the cold junction and $\dot{Q}_T$ is the thermal losses down the arms of the cooler. The Peltier heat absorbed is $\dot{Q}_\Pi = \Pi \cdot I$, and the thermal losses down the arms consist of thermal conduction losses, $K(T_0 - T_1)$, where K is the thermal conductance of the arms, and Joule heating losses, $\frac{1}{2}I^2R$. A factor of $\frac{1}{2}$ on the Joule heating losses is due to half of this heat flowing to the cold end and half flowing to the warm end of the cooler. Substituting gives

$$\dot{Q}_0 = \Pi \cdot I - \frac{1}{2}I^2R - K(T_0 - T_1). \tag{145}$$

Maximizing $\dot{Q}_0$ with respect to current yields $\Pi = I \cdot R$, or $I_{max} = \Pi/R$. Using the Kelvin relations,

$$I_{max} = \frac{(S_A - S_B)T_1}{R}. \tag{146}$$

In steady state, $\dot{Q}_0 = 0$, and the maximum temperature gradient $\Delta T_{max} = (T_0 - T_1)$ is

$$\Delta T_{max} = \frac{1}{2}\frac{(S_A - S_B)^2}{RK}T_1^2 = \frac{1}{2}ZT_1^2, \tag{147}$$

where Z is defined as the figure of merit for the cooler. Equation (147) clearly shows that the maximum temperature gradient is increased by choosing materials with the largest difference in thermopower values. Therefore, the logical choice is to use one n-type and one p-type material as mentioned previously.

Continuing with the evaluation of the coefficient of performance for the cooler, the power absorbed by the device is simply the product of the current and voltage supplied to the cooler, or

$$W = IV = I\{IR + (S_A - S_B)(T_0 - T_1)\}, \tag{148}$$

where the voltage across the device includes the resistive and thermoelectric voltage drops. Dividing this into $\dot{Q}_0$ yields the coefficient of performance,

$$\varphi = \frac{\Pi I - \frac{1}{2}I^2R - K(T_0 - T_1)}{I^2R + (S_A - S_B)(T_0 - T_1)I}. \tag{149}$$

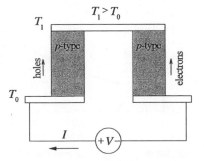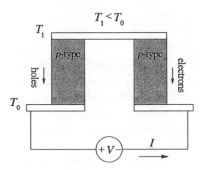

FIGURE 8 Thermoelectric cooler current flow.

Taking the derivative with respect to current and setting it equal to zero gives

$$\frac{d\varphi}{dI} = 0 = \frac{(\Pi - IR)[S\Delta T \cdot I + I^2 R] - [S\Delta T + 2IR]\left[\Pi I - \frac{1}{2}I^2 R - K(T_0 - T_1)\right]}{[S\Delta T \cdot I + I^2 R]^2}, \quad (150)$$

where the substitutions $S = (S_A - S_B)$ and $\Delta T = (T_0 - T_1)$ were used. After expansion and cancellation in the numerator,

$$\frac{d\varphi}{dI} = 0$$

$$= \frac{I^2\left[-R\Pi - \frac{1}{2}R \cdot S\Delta T\right] + I[2KR(T_0 - T_1)] + K \cdot S\Delta T^2}{[S\Delta T \cdot I + I^2 R]^2}. \quad (151)$$

Substituting $\Pi = (S_A - S_B)T_1$ for the Peltier heat removed at the cold junction gives

$$0 = I^2\left\{-R(S_A - S_B)\left[T_1 + \frac{1}{2}(T_0 - T_1)\right]\right\}$$
$$+ I[2KR(T_0 - T_1)] + K(S_A - S_B)(T_0 - T_1)^2. \quad (152)$$

Solving this quadratic equation yields the maximum coefficient of performance at the optimum current,

$$I_{opt} = \frac{(S_A - S_B)(T_0 - T_1)}{R(\sqrt{1 + Z\bar{T}} - 1)}, \quad (153)$$

where $\bar{T}$ is the average temperature $\frac{1}{2}(T_0 + T_1)$. Using this in Eq. (143) yields

$$\varphi_{opt} = \frac{T_1}{(T_0 - T_1)} \frac{\sqrt{1 + Z\bar{T}} - (T_0/T_1)}{\sqrt{1 + Z\bar{T}} + 1}, \quad (154)$$

where the first term represents the coefficient of performance for an ideal heat pump. This shows that both φ and ΔT are directly dependent on the figure of merit, Z. Thus maximizing the figure of merit for the individual materials,

$$Z = \frac{S^2}{\rho\kappa} = \frac{S^2\sigma}{\kappa}, \quad (155)$$

maximizes the efficiency of the cooler. Desirable materials have large-magnitude thermopowers, S (one n-type and one p-type), and low electrical resistivities, ρ, or, equivalently, high electrical conductivities, σ, and low thermal conductivities, κ. Since the figure of merit has units of K^{-1}, the unitless quantity of ZT is often reported.

It should also be noted that the Peltier heat, $\dot{Q}_\Pi = \Pi \cdot I$, is either absorbed or liberated based on the current direction. Therefore, the same configuration can be used as either a thermoelectric cooler or a heater.

For comparison, and to illuminate the present challenge, the coefficient of performance for standard Freon-based

refrigeration systems is 1.2 to 1.4, for a refrigerator operating at a cold temperature of 263 K while the outside (hot temperature) is at 323 K. Freon-based cooling systems have coefficients of performance that would correspond to a thermoelectric device with ZT between 3 and 4. Also shown is the COP for the present value of $ZT \sim 1$. The advantages of thermoelectric devices includes size scalability without loss of efficiency, robustness, low maintenance, a relatively small electromagnetic signature, and the ability both to heat and to cool from a single device, and they are environmentally cleaner than conventional CFC-based coolers. Many thermoelectric companies presently exist, indicating an existing market such that any increase in ZT through a new material and/or configuration could have a direct impact; however, a significant increase in the market is anticipated for an increase in ZT to 2. This, therefore, represents the current goal in Fig. 9.

These devices are heat pumps, in that it is also possible to remove the electrical power source, and force a temperature gradient across the thermoelectric device, by contacting one end of it to an external heat source. With a load connected to the device instead of the electrical power source, it then functions as a thermoelectric generator. Thus, the application of an electrical potential gradient causes the generation of a temperature gradient (thermoelectric cooler) and the application of a temperature gradient causes the generation of electrical power (thermoelectric generator).

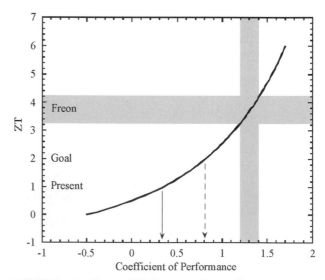

FIGURE 9 The figure of merit versus the coefficient of performance.

In the case of a generator, the efficiency, η, of the device is defined as the ratio of the power supplied to the load to the heat absorbed at the hot junction:

$$\eta = \frac{T_H - T_C}{T_H} \frac{\sqrt{1 + Z\bar{T}} - 1}{\sqrt{1 + Z\bar{T}} + (T_C/T_H)}. \qquad (156)$$

This is again dependent on the figure of merit of the device. Through Thompson's relations we can split the figure of merit for the device into a figure of merit for each of the two legs. When each of these has been maximized individually, then the total device figure of merit will also be maximized assuming that one leg is n-type and one p-type.

C. New Directions

Traditional materials used in thermoelectric devices are listed in Table II. Near-room-temperature devices have been designed largely for cooling applications, while higher-temperature materials have been generally used in electrical power generation. Research on thermoelectrics was highly active during the decade following 1954, with the United States showing a great interest in high-temperature power generation applications, such as the Si–Ge-based generators used on the satellites Voyager I and II. Recently there has been a resurgence of interest in thermoelectrics, spurred on partly by predictions of the high ZTs possible in quantum confined structures (Hicks and Dresselhaus, 1993). It was predicted that in such structures, both the electrical conductivity and the thermopower could be simultaneously increased due to the sharpening of the density of states as confinement increases from $3D \rightarrow 2D \rightarrow 1D \rightarrow 0D$ (Broido and Reinecke, 1995). The influence of such sharpening can be seen clearly within the Mott–Jones equation for thermopower (113). An indication of the effect from a rapidly varying density of states comes from mixed-valent compounds such as $CePd_3$ and $YbAl_3$, which have shown the largest power factor, σS^2, among all known materials. Unfortunately, the high thermal conductivity in these materials prevents them from having a correspondingly high figure of merit. An additional increase in ZT for quantum confined materials comes from a decrease in the thermal conductivity due to confinement barrier scattering.

Another avenue for investigating thermoelectric materials has been coined the "phonon glass electron crystal"

TABLE II The Most Widely Used TE Materials

	Z_{max} (K^{-1})	Useful range (K)	T_{max} (K)
Bi_2Te_3	3×10^{-3}	<500	300
PbTe	1.7×10^{-3}	<900	650
Si–Ge	1×10^{-3}	<1300	1100

TABLE III Desirable Material Properties for Thermoelectric Applications (Kanatzidis, 2001)

1. Many valley bands near the Fermi level, but located away from the Brillouin zone boundaries.
2. Large atomic number elements with large spin–orbit coupling.
3. Compositions with two or more elements such as ternaries and quaternaries.
4. Low average electronegativity differences between elements.
5. Large unit cells.
6. Energy gaps near 10 kT.

(PGEC) method (Slack, 1995), in which short phonon mean free paths and long electron mean free paths are simultaneously sought in a material. A suggested way for a material to exhibit PGEC behavior is making a material that incorporates cages and/or tunnels in its crystal structure large enough to accommodate an atom. The caged atom provides strong phonon scattering by rattling within the cage. Electrons would not be significantly scattered by such "rattlers" since the main crystal structure would remain intact.

Additional guidance has been provided by identifying a B parameter defined as

$$B = \gamma \frac{1}{3\pi^2} \left(\frac{2kT}{h^2} \right)^{3/2} \sqrt{m_x m_y m_z} \frac{k^2}{q\kappa_L} \mu_x, \qquad (157)$$

where γ is the degeneracy parameter (Hicks and Dresselhaus, 1993). This function should be maximized for optimal ZT. With a large number of valleys within a band, fewer carriers can exist in each valley, thus increasing the contribution to the thermopower from that valley. At the same time, the total number of carriers can be maintained for a high electrical conductivity. High degeneracy parameters are generally found in highly symmetric crystal systems. Large effective masses, or large effective mass components in the axes perpendicular to the current flow, allow for a high electrical conductivity in the direction of interest while maintaining a high B parameter. Equation (157) also indicates that high mobilities, μ_x, in the transport direction, and a low lattice thermal conductivity are also desirable. It has recently been shown that semiconductors with bandgaps of approximately 10 kT best satisfy these criteria (Mahan, 1998). Six properties of thermoelectric materials that give the best results are listed in Table III.

VI. SUMMARY

Macroscopic and atomistic derivations of the thermoelectric properties of electrical conductivity, thermoelectric power, and thermal conductivity have been presented, with approximations for various material systems. The derivations outlined have considered external forces of electric fields and temperature gradients. Additional effects are realized when more forces such as magnetic fields are included. These include Hall, magnetoresistance, Nernst, Ettingshausen, and Righi–Leduc effects. These magnetic fields also affect the operation of thermoelectric coolers, with significant enhancements of the efficiency possible under strong fields.

ACKNOWLEDGMENT

I wish to thank Sangeeta Lal, of Bihar University, for her helpful review of the manuscript.

SEE ALSO THE FOLLOWING ARTICLES

ELECTROMAGNETICS • ELECTRONS IN SOLIDS • SEMICON-DUCTOR ALLOYS • SUPERCONDUCTIVITY • THERMODY-NAMICS • THERMOMETRY

BIBLIOGRAPHY

Barnard, R. D. (1972). "Thermoelectricity in Metals and Alloys," Halsted Press (Division of John Wiley & Sons), New York.

Broido, D. A., and Reinecke, T. L. (1995). "Thermoelectric figure of merit of quantum wire superlattices," *Appl. Phys. Lett.* **67**(1), 100–102.

Gray, P. E. (1960). "The Dynamic Behavior of Thermoelectric Devices," Technology Press of the Massachusetts Institute of Technology/John Wiley & Sons, New York.

Guggenheim, E. A. (1957). "Thermodynamics," 3rd ed., North-Holland, Amsterdam.

Harman, T. C., and Honig, J. M. (1967). "Thermoelectric and Thermo-magnetic Effects and Applications," McGraw–Hill, New York.

Hicks, L. D., and Dresselhaus, M. S. (1993). "Effect of quantum-well structures on the thermoelectric figure of merit," *Phys. Rev. B* **47**(19), 12 727–12 731.

Ioffe, A. F. (1957). "Semiconductor Thermoelements and Thermoelectric Cooling," Infosearch, London.

Kanatzidis, M. G. (2001). The role of solid-state chemistry in the discovery of new thermoelectric materials. *In* "Solid State Physics" (H. Ehrenreich and F. Spaepen, eds.), Vol. 69, pp. 51–100, Academic Press, New York.

Mahan, G. D. (1998). Good thermoelectrics. *In* "Solid State Physics" (H. Ehrenreich and F. Spaepen, eds.), Vol. 51, pp. 82–157, Academic Press, New York.

Roberts, R. B. (1977). "The absolute scale of thermoelectricity," *Philos. Mag.* **36**(1), 91–107.

Slack, G. A. (1995). New materials and performance limits for thermoelectric cooling. *In* "CRC Handbook of Thermoelectrics" (D. M. Rowe, ed.), CRC Press, New York.

Wendling, N., Chaussy, J., and Mazuer, J. (1993). "Thin gold wires as reference for thermoelectric power measurements of small samples from 1.3 K to 350 K," *J. Appl. Phys.* **73**(6), 2878–2881.

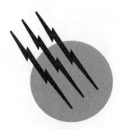

Thermoluminescence Dating

Geoff Duller
University of Wales, Aberystwyth

I. Physical Mechanism
II. Age Evaluation
III. Sample Types and Age Ranges

GLOSSARY

Annual dose The dose of ionizing radiation received by a sample per year during burial.

a-value The ratio between the luminescence signal induced by a given dose of alpha particles and that induced by the same dose of beta or gamma radiation.

Equivalent dose (ED) Dose of ionizing radiation received by a sample subsequent to the event being dated, as evaluated from measurements of natural and artificial luminescence. ED can also be denoted by the term D_E. An alternative term is palaeodose (P).

Fading Loss of luminescence during storage. Anomalous fading refers to loss in excess of the thermal fading predicted from measurement of the relevant electron trap depth and frequency factor.

Glow curve Plot of light intensity versus temperature as a sample is heated up during measurement of thermoluminescence (TL) (see Fig. 1a).

Natural luminescence The TL or OSL signal measured from a sample that is due to its exposure to ionizing radiation during burial. This is as opposed to "artificial luminescence" which is generated by exposure to radiation in the laboratory.

Optical decay curve Plot of light intensity versus time as a sample is stimulated by exposure to light during measurement of optically stimulated luminescence (OSL) (see Fig. 1b).

Palaeodose (P) See *Equivalent dose.*

Plateau test A plot of the ED calculated for a sample using different temperatures from the TL glow curve.

Preheating Annealing a sample at a fixed temperature for a fixed period of time prior to TL or OSL measurement. This has the effect of removing trapped charge from thermally unstable traps.

Residual signal The luminescence signal remaining after a zeroing event. A major advantage of OSL compared with TL is that for a given light exposure the residual signal is smaller for OSL measurements.

Saturation While exposure to radiation causes charge to accumulate in traps, the luminescence signal subsequently generated increases. However, beyond a certain point, additional radiation exposure does not cause more charge to be trapped since the trapping sites are full. At this point the sample is said to be saturated.

Zeroing Complete or partial elimination of previously acquired luminescence. It is the zeroing event that is

(a)

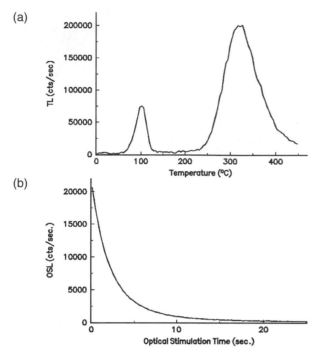

(b)

FIGURE 1 (a) A thermoluminescence (TL) glow curve showing two distinct peaks. (b) An optical decay curve generated during an OSL measurement.

normally dated by luminescence methods (see *Residual signal*).

FOLLOWING EXPOSURE to ionizing radiation (for example, alpha or beta particles, gamma radiation, or X-rays), some minerals (for example, quartz and feldspar), emit light when stimulated at some later time. The intensity of this *luminescence* is proportional to the dosage (amount) of ionizing radiation that has been absorbed since the last event that eliminated any previously acquired luminescence. For minerals in baked clay this event is firing (for example, of pottery), while for geological sediments the event is exposure to sunlight during transport and deposition. The ionizing radiation is provided by radioactive elements (e.g., ^{40}K, Th, and U) in the sample and its surroundings during burial, and cosmic rays. The age is obtained as:

$$\text{Age} = \text{ED/D} \tag{1}$$

where ED is the laboratory estimate of the radiation dose to which the sample has been exposed since the event being dated. This is known as the equivalent dose. D is the dose per year to the sample during burial (the annual dose). Luminescence measurements are used to determine ED, the equivalent dose, while emission counting or chemical measurements are used to calculate the annual dose.

I. PHYSICAL MECHANISM

A. The Basic Model

The details of the mechanism by which luminescence is produced in any given material are not well understood, and in general it is only for crystals grown in the laboratory with strict control of impurities that these details can be elucidated. However, the main features of luminescence dating can be discussed in terms of a simple model. In this, the free electrons produced by ionizing radiation are stored in traps where they remain in a metastable condition until the sample is stimulated, release being caused by increased lattice vibrations. Some of the released electrons find their way to luminescence centers, and incidental to the process of combining into such centers, photons are emitted. There are electron traps in the majority of natural ionic crystals and covalent solids, formed by defects in the lattice structure. The luminescence centers are associated with impurity atoms, and these determine the color of the luminescence emitted.

Electron traps are essentially localized regions where there is a deficit of negative charge or an excess of positive charge. The luminescence process can also result from holes being charge carriers, but in discussion it is convenient to talk only of electrons.

B. Stability

The length of time for which an electron remains trapped is one control on the age range over which luminescence dating can be applied. This period of time is determined by the energy, E, that is necessary to free the electron from the type of trap concerned and the frequency factor, s. The probability of escape per unit time is equal to the reciprocal of the lifetime, τ, where

$$\tau = s^{-1} \exp(E/kT) \tag{2}$$

where k is the Boltzmann's constant and T is the absolute temperature (degrees Kelvin). For the 325°C peak in quartz, $E = 1.69$ eV and $s = 10^{14}$ sec^{-1}, giving $\tau = 10^8$ yr at an ambient temperature of 15°C; lifetimes for other peaks occurring at temperatures above 300°C for this and other minerals are comparable.

Some minerals (for example feldspar) from some sources exhibit anomalous fading; namely, the observed lifetime is less than predicted by Eq. (2). The effect is because of an escape route not via the conduction band but directly to nearby centers. Two mechanisms have been proposed: (1) a localized transition via an excited state common to both the trap and center, and (2) wave-mechanical tunneling from the trap to the center. Reliable dating is not possible if (2) is dominant.

C. Thermal and Optical Stimulation

Measurement of a luminescence signal in the laboratory requires stimulation of the sample in order to evict charge from the metastable traps into the conduction band. This stimulation can be achieved in a number of ways, but the two used for dating are either by heating the sample to generate thermoluminescence (TL) or by exposure to light to generate optically stimulated luminescence (OSL). It is not always clear whether the charge evicted during OSL measurements can be directly linked to a specific part of that evicted during TL measurements.

For many dating applications the measurement of OSL instead of TL has a significant advantage. In dating geological materials, the event that is dated is the last exposure of the mineral grains to daylight during transport and deposition; this exposure will have reduced the luminescence signal to a low level. The effect of light upon the TL signal is variable, with some TL peaks being reduced more rapidly than others, but in almost all cases the TL signal is not reduced to zero and a considerable residual signal remains. Estimation of the residual signal at deposition then introduces an additional source of uncertainty. Through exposure to light, OSL signals tend to be reduced at least an order of magnitude more rapidly than those measured with TL, and the residual level after prolonged exposure to light is much smaller. The combined effect is that using measurements of OSL instead of TL permits dating of younger events, where the magnitude of the residual level becomes more significant, and also the dating of geological materials such as fluvial sands where the exposure to daylight at deposition may have been limited in duration.

II. AGE EVALUATION

A. Measurement of Luminescence

For dating, a luminescence signal can be obtained from both quartz and feldspars. Where grains between 100 and 300 μm are used, the quartz can be separated from the feldspars using chemical and physical processing. Alternatively, much finer grains (4 to 11 μm) may be analyzed, but in this case no attempt is made to separate different mineral types. These grain size ranges are selected because of the alpha dosimetry (see Section II.B).

Although the luminescence emitted by some bright geological minerals can be seen with the naked eye, the luminescence intensity encountered in samples to be dated is very faint. Not only is it necessary to use a highly sensitive low-noise photomultiplier, but it is also vital to discriminate against unwanted light. When measuring TL the primary concern is suppressing the incandescence generated by the sample and hotplate at high temperatures. This can

be efficiently done by the use of broadband glass filters that exclude wavelengths in the red and infrared parts of the spectrum.

For OSL measurements the situation is rather different. Optical stimulation is commonly carried out either using infrared ($\sim$880 nm) or visible wavelengths ($\sim$450–520 nm). An intense OSL signal is observed from feldspars when stimulated in either waveband, while quartz only gives a bright signal when stimulated in the visible. The optical stimulation source used is commonly emitting 10^{10} times more photons than the sample being measured. Thus, it is vital to place glass filters in front of the photomultiplier that will reject the wavelengths used for stimulation, but not those emitted by the sample.

For measurement, the mineral grains of a sample are carried on a metal disk, usually 10 mm in diameter. Automated equipment incorporating a sample changer, radioactive source, hotplate, optical stimulation sources, and a photomultiplier is frequently used.

Since many of the luminescence signals that are observed are light sensitive, all sample preparation and measurement are carried out in subdued red light. This is a wavelength that has a negligible effect upon the signals normally observed.

B. Equivalent Dose Determination

Luminescence measurements are used to estimate the radiation dose to which the mineral grains have been exposed since the event being dated—the equivalent dose (ED), or palaeodose (P) in Eq. (1). Since there is no intrinsic relationship between radiation dose and the luminescence signal subsequently measured, each sample requires individual calibration. Two basic approaches have been used to determine the ED: additive dose and regenerative dose (Fig. 2).

The additive dose procedure involves separate measurements of the luminescence signal resulting from the radiation received by the sample during burial (its natural luminescence), and of the sum of that radiation dose and known doses administered in the laboratory. A mathematical function is then fitted to these data to characterize the growth of the luminescence signal with radiation exposure. This curve is known as a growth curve (Fig. 2a). The radiation dose during burial can then be calculated by extrapolating the growth curve to the residual luminescence signal remaining after exposure to daylight.

The regenerative dose procedure also starts with measurement of the natural luminescence signal, resulting from the radiation received by the sample during burial. The sample is then exposed to daylight in order to remove the luminescence signal, and known laboratory doses are administered that regenerate the luminescence signal. A

(a)

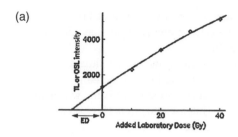

(b)

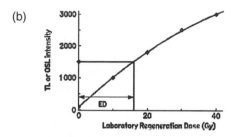

FIGURE 2 Additive dose (a) and regenerative dose (b) procedures for determination of the equivalent dose (ED).

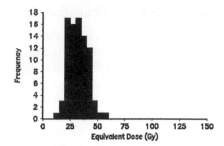

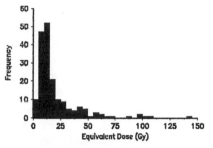

FIGURE 3 Histograms of equivalent dose (ED) values measured from single aliquots of two samples. The first sample has a normal distribution while the second sample has a broad range of ED values indicating that not all the grains in the sample were zeroed at deposition. In this case, the most likely age of deposition is that corresponding to the peak in the histogram at low values of ED.

growth curve is fitted to this regenerated dataset, and the ED is calculated by the intersection of the natural luminescence signal with the growth curve.

For TL measurements, at glow-curve temperatures for which the associated trapped electron lifetimes at ambient temperatures are long compared with the age of the sample, the ED should be independent of temperature. The existence of such an ED plateau is indicative of reliability.

For OSL measurements it is not possible to plot such a plateau since OSL measurements do not provide any information about the thermal stability of the traps from which charge is evicted. It is therefore essential to ensure that only charge from thermally stable traps will be measured during OSL procedures. This is normally achieved by annealing the sample at a fixed temperature for a fixed period of time—a procedure known as preheating—prior to OSL measurement.

C. Single Aliquot Methods

The methods of ED determination described above involve the measurement of many subsamples. Several separate subsamples, or aliquots, are used for measurement of each point on the growth curve, so that between 20 and 60 aliquots are used for each ED determination. An alternative to such multiple aliquot methods is to make all the measurements on a single aliquot. The methods used are very similar to those for multiple aliquots, with both additive dose and regenerative measurements possible.

The advantages of single aliquot methods are that there is no implicit assumption that all aliquots are equivalent, that it becomes practical to make replicate ED determina-

tions on a sample, and that one can alter the size of the sample being analyzed. In the ultimate, one can reduce the sample size to a single mineral grain.

Replicate measurements of the ED using single aliquot methods provide insight into the heterogeneity of the ED from a sample (Fig. 3). For a "simple" sample, a normal distribution is expected, with all aliquots yielding similar ED values. For some materials (for instance, fluvial sands), some grains are exposed to daylight for a limited period of time at deposition, yielding a distribution of grains with different residual signals, only some of which will yield an accurate age. Single aliquot methods can be used to measure the distribution of ED within a sample and, coupled with models of the depositional process, can be used to provide limiting ages.

An additional cause of a broad distribution in ED values can be variation in the annual radiation dose from one grain to another. Fortunately, for most samples this appears to be a relatively small source of variation.

D. Annual Radiation Dose

The annual radiation dose (D in Eq. (1)) to a sample originates from alpha, beta, and gamma radiation. By selecting certain minerals and specific grain size ranges for analysis we can simplify the calculation of the dose rate.

Because of their short range, alpha particles do not penetrate more than about 20 μm into a grain of quartz or feldspar. If we select grains in the range from 4 to 11 μm, they will have received the full alpha dose as well as the beta and gamma contributions.

Alternatively, for large (>100 μm) quartz or feldspar grains it is possible to chemically etch away the outer skin of the grains using hydrofluoric acid and thus remove the alpha contribution. For quartz grains the assumption is normally made that they are free from radioactivity and receive all their beta and gamma dose from the matrix in which they are found. For feldspars the situation is more complex since certain feldspars contain up to 14% potassium by weight. For these grains it is necessary to assess the beta dose originating within the grains. As the grain size increases this internal beta dose becomes increasingly important.

The unit of measurement for the annual radiation dose is the gray (joules deposited per kilogram). In terms of this, alpha particles—because of the high ionization density produced—are not as effective as beta or gamma radiation; the effective full alpha contribution $D'_\alpha = aD_\alpha$, where D_α is the actual alpha dose in grays, and the a-value (a) measured for each sample is typically in the range 0.1 to 0.3. The annual dose for insertion into Eq. (1) is then:

$$D = D'_\alpha + D_\beta + D_\gamma + D_c \qquad (3)$$

where the suffix indicates the type of radiation, with "c" denoting the dose due to cosmic rays. For sediments containing typical concentrations of 1% potassium, 6 ppm thorium, and 2 ppm uranium, D is of the order of 2 grays per 1000 yr for a 200-μm quartz grain.

Allowance needs to be made for the wetness of the sample during burial because the presence of water attenuates the radiation flux received. Uncertainty about average water content is a serious barrier to reducing the error limits on the age to below ±5%.

A variety of methods are used for assessment of radioactivity: neutron activation, chemical analysis, thick-source alpha and beta counting, gamma spectrometry, and beta and gamma dosimetry using special high-sensitivity phosphors (such as calcium sulfate doped with dysprosium, or aluminium oxide doped with carbon). For gamma dosimetry, a capsule of phosphor is buried onsite for about a year. An alternative method of making *in situ* measurements is with a portable gamma spectrometer based around a NaI crystal. Such devices can be used to assess both the total gamma dose rate, and the concentrations of the major radionuclides.

A potential complication in the calculation of the annual dose arises from the possibility that the uranium or thorium decay chains may not be in equilibrium; the most commonly observed cause of disequilibrium is escape of the gas radon-222 which occurs midway in the uranium-238 chain; there may also be preferential leaching of some radioelements in the chain, and the extent to which this occurs may have varied during burial. Thus, disequilibrium can be a problem for two reasons: first, different methods for assessing the total alpha, beta, and gamma dose rates may only look at a specific part of the decay chain and then assume that the remainder of the chain is in equilibrium; and, second, because the extent of disequilibrium may have varied through the period for which the sediment has been buried, causing the annual radiation dose to vary through time. Fortunately, severe disequilibrium is rare.

III. SAMPLE TYPES AND AGE RANGES

A. Baked Clay and Burned Stones

Luminescence dating can reach back to the earliest pottery, at about 10,000 yr ago, and beyond. How much beyond depends on the minerals concerned and the annual radiation dose rate at a specific site. For quartz, the onset of saturation is liable to occur around 50,000 to 100,000 yr ago, and sooner for clay of high radioactivity. TL measurements are particularly applicable in this case since the event being dated is the last heating of the sample in a kiln, oven, or fireplace.

Each sample should be at least 10 mm thick and 30 mm across; six samples per context are desirable. The error limits are usually ±5 to ±10% of the age. An important requirement is that the sample has been buried to a depth of at least 0.3 m and that samples of burial soil are available; if possible, *in situ* radioactivity measurements are also made.

The same age range limitations apply for burned stones as for baked clay; there may be additional difficulties because of heterogeneity of radioactivity and TL sensitivity. Burned flint (and chert) is not plagued in this way and is excellent material for dating; ages of several hundred thousand years have been obtained with it. The main limitation is the sparsity, on palaeolithic sites, of large enough flints that have been sufficiently heated.

B. Authenticity Testing

Luminescence dating has had a very powerful impact on testing authenticity of art ceramics. Error limits of ±25% are typical in this application, but acceptable. The wide error limits arise because of uncertainty as to D_γ and because the allowable sample size is smaller—about 100 mg of powder are obtained by drilling a small hole, 4 mm across by 4 mm deep, in an unobtrusive location. For porcelain, a 3-mm core is extracted and cut into 0.5-mm slices for measurement. It is also possible to test the authenticity of bronze heads, etc., which have a clay core baked in the casting process.

C. Aeolian Sediments: Loess and Dunes

Aeolian sediments are ideally suited to luminescence dating since one can safely assume that they were exposed to sufficient daylight at deposition to reduce the residual luminescence signal to a low level. Loess, a fine-grained, wind-blown sediment, has been dated extensively in China, the U.S., and Europe. Coastal and desert dunes are also well suited to luminescence dating for the same reason as loess. For loess the grain-size distribution is restricted, and polymineral grains from 4 to 11 μm are routinely used. By contrast, coarse grains between 100 and 300 μm can be obtained from dunes, and grains of quartz or potassium-rich feldspar can be isolated.

The oldest ages are limited by saturation of the luminescence signals. In exceptional circumstances, ages up to 800 ka have been produced, but this is only possible in environments of unusually low radioactivity. Conventionally, the limits are 100,000 yr for quartz and approximately 200,000 yr for feldspars. The youngest ages can be obtained using OSL measurements and are limited by the intrinsic brightness of the minerals concerned and the degree of exposure to daylight at deposition. For well-bleached dunes, ages of a few decades are possible.

Samples should be collected without any exposure to daylight. Sample mass is typically 250 to 500 g but is dependent on the mineralogy and grain-size distribution. Separate samples should also be taken for assessment of water content and radioactivity.

D. Fluvial, Glacial, and Colluvial Sediments

A large variety of sediments are exposed to a limited amount of daylight at deposition, so the exposure varies significantly from one grain to another. For instance, fluvially deposited sands may be well exposed to daylight in shallow rivers in Australia with intense sunlight, but poorly exposed in a deep, turbid, northern European river. Such sediments pose a challenge for luminescence dating methods since they will contain different residual signals at deposition. Included in this list are glacially derived and colluvial sediments.

Single aliquot measurements provide an independent method of assessing whether the sample was sufficiently exposed to daylight at deposition to zero the grains.

SEE ALSO THE FOLLOWING ARTICLES

DOSIMETRY • GEOLOGIC TIME • RADIATION PHYSICS • RADIOCARBON DATING • RADIOMETRIC DATING • STABLE ISOTOPES AS TRACERS OF GLOBAL CYCLES

BIBLIOGRAPHY

Aitken, M. J. (1990). "Science-Based Dating in Archaeology," Longman, London.
Aitken, M. J. (1998). "An Introduction to Optical Dating," Oxford University Press, London.
McKeever, S. W. S. (1985). "Thermoluminescence of Solids," Cambridge University Press, Cambridge, U.K.
Wagner, G. A. (1998). "Age Determination of Young Rocks and Artifacts," Springer-Verlag, Berlin.

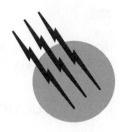

Thermometry

C. A. Swenson
Iowa State University

T. J. Quinn
Bureau International de Poids et Mesures

GLOSSARY

Fixed point Unique temperature that is associated with a well-defined thermodynamic state of a pure substance, and that generally involves two or three phases in equilibrium.

Ideal gas Assembly of noninteracting particles. Helium gas at a low pressure is a good approximation for an ideal gas.

International Temperature Scale of 1990 Internationally adopted temperature scale (abbreviated ITS-90 or T_{90}) that provides a reference for all current thermometry.

Primary thermometer Device that directly determines thermodynamic temperatures.

Secondary thermometer Instrument that is used for practical thermometry and that must be calibrated in terms of a primary thermometer.

Standard platinum resistance thermometer Carefully specified secondary thermometer that is used in the definition of the IPTS-68 over much of its range.

Thermodynamic temperature Parameter (actually, an energy) that appears in theoretical calculations of thermal effects.

MODERN THERMOMETRY extends over at least 10 decades in temperature, from the temperatures reached in nuclear cooling experiments to those achieved in nuclear explosions. At both the lowest and the highest extremes, temperatures are measured using methods that are related directly to theory and, hence, correspond to thermodynamic temperatures. At intermediate temperatures, where high accuracy is most necessary, temperatures are defined in terms of secondary thermometers (such as the standard platinum resistance thermometer) that have proved to be stable and sensitive and to have calibrations that vary smoothly with thermodynamic temperature. These instruments serve as interpolation devices between a sequence of accurately defined fixed points to which temperatures have been assigned which correspond closely to thermodynamic values. The thermometers that are used in practical situations may be more convenient to use than either thermodynamic thermometers or scale-defining secondary thermometers, may be smaller in size, and/or may be more sensitive, while lacking the smoothness and/or stability criteria.

I. INTRODUCTION

The qualitative aspects of temperature and temperature differences are synonymous with the physiological sensations of "hot" and "cold." These descriptions are ambiguous, since often it is the heat conductance or even the thermal mass of the material that is sensed, rather than its actual temperature. Hence, the temperature of a glass object always will seem to be less extreme than that of a metal object, even though the two objects are at the same temperature.

The measurement of temperature, or the science of thermometry, is made quantitative through the observation that the physical properties of materials (density, electrical resistance, and color, for instance) change reproducibly as they become "hotter" or "colder." These changes, which can be relatively large and extremely reproducible for certain well-characterized materials, allow the design and construction of practical thermometers. An important requirement in any science is that measurements made in different localities and in different ways can be related quantitatively, so an agreement on the use of standards must exist. Thermometry standards are based on the observation that certain phenomena always occur at the same, highly reproducible, temperature. The temperatures at which water freezes and then boils under a pressure of 1 atm were recognized very early as being useful thermometric "fixed points," and the Celsius (formerly called centigrade) temperature scale, t, was based on the assignment of 0 and 100°C, respectively, to these two phenomena. As described below, a number of fixed points are used today to define the currently accepted temperature scale.

Once fixed-point temperatures have been assigned, values are associated with intermediate temperatures by interpolation using a "thermometric parameter" that has been evaluated at both lower- and higher-temperature fixed points. This parameter could be, for instance, the expansion of a liquid in a glass bulb (the liquid-in-glass thermometer) or the electrical resistance of a platinum wire (the platinum resistance thermometer; PRT). Since these interpolations may give answers that depend on the material and/or the physical property involved, the standard temperature scale also must designate the type of interpolation device that is to be used. A carefully specified standard platinum resistance thermometer (SPRT) is the designated interpolation instrument over much of the intermediate temperature range, with other instruments important at the extremes of very high and very low temperatures.

The above discussion places no restrictions on what could be an arbitrary assignment of values to the various fixed points, although a "smooth" relationship between these and, for instance, the resistance of an SPRT would appear to be desirable. The concept of a characteristic thermal energy, or of a theoretical temperature, appears both in the science of thermodynamics and in theoretical calculations of thermal properties of materials. Hence, a natural additional requirement is that fixed-point temperatures (and interpolated values) coincide as closely as possible with theoretical (or thermodynamic, or absolute) temperatures, T, which will be measured in kelvins (K). This requirement can be satisfied using a "primary" thermometer, which is a practical device that can be understood completely in a theoretical sense (a gas thermometer, for instance) and that can be used experimentally to study fixed points and interpolation devices. In addition, for purely practical reasons, temperature intervals measured in kelvins and degrees centigrade should have identical numerical values. This was accomplished historically by making measurements with the primary thermometer at the two defining fixed points for the Celsius scale and by requiring that the corresponding temperature difference be exactly 100 K.

Temperatures on the Celsius scale may have either positive or negative values, since 0°C has been chosen arbitrarily, while T must always be positive, except for unusual situations, and $T = 0$ (absolute zero) has a definite meaning (see below). Once the above interval equivalence has been established, t and T will differ by an additive constant, which is the absolute temperature (in K) of the ice point. The triple point of water is much more reproducible than the ice point (see below), and the temperatures of this fixed point are defined to be 273.16 K and 0.01°C. This definition, which establishes the size of the kelvin, was based on the best data available in 1960 for the freezing and boiling points of water on the ideal gas scale. Modern measurements (see below) show that a discrepancy exists between this definition and the definition of the Celsius scale, since the temperature interval between the water freezing and the water boiling points is 99.974 K.

Standards decisions are made by the 48-nation Geneva Conference on Weights and Measures (CGPM), which meets every 4 years (1991, 1995, 1999, etc.). The CGPM acts on the advice of 18 national technical experts who form the International Committee on Weights and Measures (CIPM). The CIPM, in turn, relies heavily on the bench scientists who make up the various Consultative Committees where the actual expertise is located. Thus, it is the Consultative Committee on Thermometry (CCT) that has primary responsibility for establishing and monitoring thermometry standards through recommendations that eventually are acted upon by the CGPM. The work of the consultative committees is coordinated by the International Bureau of Weights and Measures in Sèvres, just outside Paris, France. The CCT conducts its quality-control role through exchanges of personnel and devices among laboratories and carries out carefully organized international comparisons of thermometer and fixed points. It publishes the results of these exchanges as well as the results of critical evaluations of data. The CCT was responsible for the establishment, in January 1990, of the International Temperature Scale of 1990 (ITS-90), which replaced the International Practical Temperature Scale of 1968 (IPTS-68). Standards decisions are made with great care and after much deliberation, since mistakes have a long lifetime, with, historically, changes being made only every 20 years or so.

II. STANDARDS AND CALIBRATIONS

A. Fixed Points

A useful thermometric fixed point must be reproducible from sample to sample and must exhibit a sharp, well-defined "signal" to which other measurements can be referred easily. In practice, most fixed points are associated with the properties of high-purity, single-component materials. The practical realization of a fixed point with a high accuracy requires considerable care and experience in both the setting-up and the use of the devise, and this is primarily a task for a standards laboratory. Fixed points of all kinds play such an important role in thermometry, however, that they must be a part of a discussion of temperature.

1. Triple Points

The triple point is the unique combination of temperature and pressure at which the liquid, solid, and vapor phases of a pure, single-component system coexist. The triple point of water provides an excellent; illustration of this phenomenon; Fig. 1 is a photograph of a water triple-point cell that is used to realize 273.16 K with an accuracy of 10 μK (10^{-5} K). The glass container contains only pure water, with all traces of air removed. The thermometer is inserted into the central well, around which ice is carefully frozen in a mantle, after which a narrow annulus of water is formed around this well by melting ice from the inside out. Thus, the temperature is uniquely defined since all three phases of pure water are present in equilibrium. The cell in Fig. 1 was removed from its refrigeration chamber for the photograph, but the ring of ice is present, and the thin sheath of water around the well is clearly visible.

Triple points also are important at low temperatures. These are obtained by liquefying a gas (oxygen, argon,

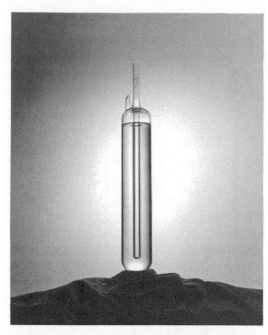

FIGURE 1 A water triple-point cell for use with PRTs. [Courtesy of Jarrett Instrument Company.]

neon, and hydrogen are examples) in a sealed system and then carefully cooling it until the solid begins to form at the triple point. Impurities in the starting material can cause changes in the triple-point temperature as the sample is frozen (or melted), and the inherent accuracy of the system (a unique definition of the temperature) is lost. Problems of contamination during gas handling are minimized with a system (Fig. 2) in which a high-purity gas at room temperature and 100 atm is sealed permanently into a carefully cleaned stainless-steel container. As this cell is cooled to the triple point, solid and liquid collect around the copper thermometer well, and the temperature can remain extremely constant as the solid is frozen and then melted. Although these cells have been in use only since 1975, they appear to be remarkably stable with time. The development of sealed triple-point cells (some of which contain several different gases in different parts of the cell) has revolutionized the ease with which low-temperature fixed points can be realized. Similar systems also have been used to obtain high-quality triple points at higher temperatures for other pure materials, with mercury, gallium, and indium metals providing examples.

2. Freezing Points

The freezing point is the temperature at which the solid begins to form from the liquid in the presence of atmospheric pressure. The freezing point of water (which defines 0°C), for instance, is approximately 0.01°C lower than the triple point, primarily because the melting temperature of water

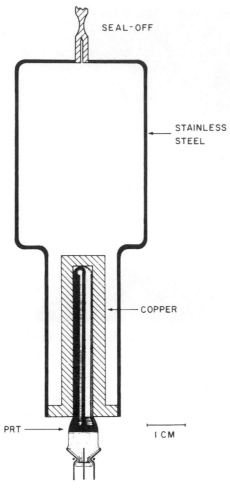

FIGURE 2 An example of the design for a sealed triple-point cell.

is depressed by the application of pressure, although it also is affected by dissolved gases and other impurities. The uncontrollable impurity effects make the freezing point of water less satisfactory as a fixed point than the triple point. To prevent ambiguities, standards thermometry is referred exclusively to the triple point of water, which is defined to be exactly −0.01°C. Melting temperatures generally increase with applied pressure, so the freezing points for most materials are higher than the triple points. Since metals tend to oxidize at high temperatures when exposed to air, atmospheric pressure may be transmitted by an inert gas, but the effect is the same. Again, as for triple points, impurities can destroy the sharpness with which the freezing point can be defined.

3. Boiling Points: Vapor Pressures

The vapor pressure of a pure substance is a unique function of the temperature, so pressure control is equivalent to temperature control. The normal boiling points of pure

substances (where the vapor pressure is 1 standard atm, or 101,325 Pa) have been used as fixed points, primarily those of water, oxygen, and hydrogen. Where possible, boiling points have been replaced as fixed points by triple points of other substances to eliminate problems due to pressure measurement and the existence of temperature gradients in the liquid. The vapor pressure–temperature relations for the liquefied helium isotopes, however, often are used directly for the calibration of other thermometers at temperatures from below 1 to 4.2 K. Reliable experimental results for the vapor pressure–temperature relation are available both for the common isotope of mass 4 (^{4}He) and for the much rarer isotope of mass 3 (^{3}He), and equations describing these form the lower temperature portion of the ITS-90. Other vapor pressure–temperature relations (hydrogen, neon, oxygen, nitrogen, oxygen) are useful as secondary standards. In this type of measurement, care must be taken to avoid temperature gradients in the liquid (a sensing bulb is preferred) and cold spots along the pressure measuring tube.

4. Superconducting Transitions

The low-temperature electrical resistance of a number of pure metals disappears abruptly at a well-defined temperature that is characteristic of the metal. These superconducting transition temperatures (T_c) have been developed by the National Institute of Standards and Technology as thermometric fixed points for temperatures from 15 mK (tungsten) to 7.2 K (lead). Early data for polycrystalline materials showed appreciable widths for the transitions, and a corresponding lack of accuracy. Later work on single crystals gives much sharper transitions. The magnitude of T_c depends on the presence of a magnetic field, so care must be taken with magnetic shielding and, also, with the magnitude of the measuring field for the noncontact mutual inductance detection method used to determine T_c.

B. Interpolation Devices

A practical interpolation device must be sensitive, capable of a high accuracy and reproducibility, and convenient to use in different environments. The temperature dependence of its thermometric parameter must be "reasonable," and understood at least qualitatively in a theoretical sense. A very carefully specified form of the platinum resistance thermometer (the SPRT) traditionally has been the interpolation instrument for international scales, and this instrument is used in the definition of the ITS-90 for temperatures from the triple point of hydrogen, 13.8033 K, to the freezing point of silver, 961.78°C. Platinum has the advantages that it can be obtained with a high purity, can be formed easily into wire, has a very high melting point,

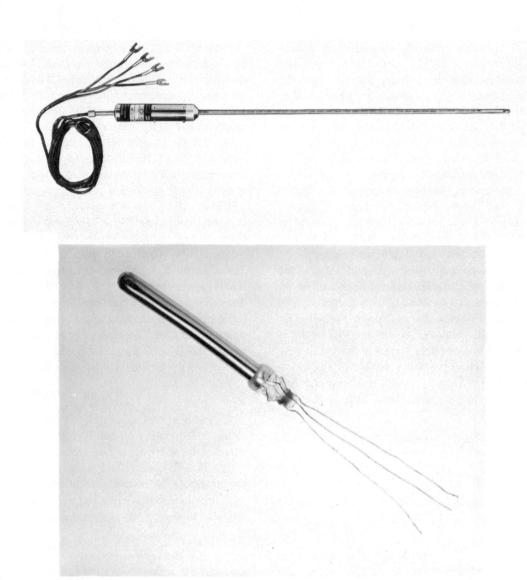

FIGURE 3 Typical standard platinum resistance thermometers. [Courtesy of Yellow Springs Instrument Company.]

and suffers little from oxidation. Many years of use have made the PRT a well-understood instrument both empirically and scientifically.

Figure 3 shows two forms of a commercially available SPRT. In each case, the fine-wire sensing element (typically 25 Ω at the triple point of water) is mounted inside a thin, roughly 6-mm-diameter, 40-mm-long platinum sheath, with a glass or fused quartz seal for introducing the electrical leads. A small amount of "air" provides thermal conductance. A four-lead design allows an unambiguous definition of the resistance of the element. The "capsule" version is intended for low-temperature use, where it can be placed in a vacuum-insulated thermometer well, as for the sealed triple-point cell of Fig. 2. The disadvantage of the capsule form is that the four leads from the resistance element are at the same temperature as the capsule, so leakage resistances between the leads can become important at temperatures greater than 200 or 300°C. The "long-stem" SPRT (Fig. 3, top) reduces this problem since the four leads leave the sealed enclosure at room temperature. Its length, however, makes this instrument impractical for use at temperatures below about 50 K. Internal electrical leakage, which even here becomes a problem for the highest temperatures (above 500°C), can be minimized through the use of long-stem thermometers with ice-point resistances as low as 0.25 Ω. The stability of an SPRT can be determined through periodic checks of its resistance when it is immersed in a triple-point cell (Fig. 1). A good SPRT will give results that are reproducible to better than 0.1 mK even when different triple-point cells are used. The resistance-temperature characteristics of PRTs are discussed specifically in Section IV.

The SPRT becomes relatively insensitive at temperatures below roughly 13.8 K, and the low-temperature calibration is very sensitive to strains that are caused by shock. Other resistance thermometers are more satisfactory for use below 13.8 K (or even 20 K), most importantly those using a rhodium–iron alloy. At the lowest temperatures, the susceptibilities of elementary magnetic systems (electronic to a few millikelvins, then nuclear) show a particularly simple temperature dependence (the Curie–Weiss law; see below) and are used both for interpolation and extrapolation. The melting curve of the helium isotope of mass 3 (^{3}He) also has a strong pressure–temperature relationship below 0.5 K and is being adopted for use as a thermometer for use down to 0.9 mK (see below).

At very high temperatures, above roughly 1000°C, the radiation emitted by a black body can be measured accurately and is used as a measure of temperature (optical pyrometry). Only a single calibration point is required for these measurements, and overlap with the PRT scales is achieved, at least in laboratory measurements. The relative intensities of lines in optical emission or absorption spectra can change with temperature as higher energy levels are excited thermally. These relative intensities can be interpreted directly in terms of T.

C. THE ITS-90

1. The Scale Definition

The currently accepted International Temperature Scale of 1990 differs appreciably from its immediate predecessor (the IPTS-68), with the magnitudes of the differences between the two scales shown in Fig. 4. The lower end of the scale now is 0.65 K rather than 13.8 K, differences from thermodynamic temperatures (especially at low temperature) are reduced to give increased smoothness, and the development of high-temperature SPRTs allows their use to the freezing point of silver (961.78°C). The discontinuity in slope at 630°C in Fig. 4 is related to the change at this temperature in the interpolation instrument which is used to define the IPTS-68. The relatively accurate and precise SPRT was used at lower temperatures, while the much less precise and stable ($\pm$0.2 K) platinum–10% rhodium/platinum thermocouple was used to the gold point.

The ITS-90 is defined in terms of the 17 fixed points in Table I, with vapor pressure–temperature relations for the helium isotopes extending the scale definition to 0.65 K. These fixed points are characterized as vapor pressure (v), triple point (tp), or freezing point (fp), with no boiling points being used. The triple point of water is assigned the exact value 273.16 K, with the relationship between the Kelvin and the Celsius temperatures defined as

$$t_{90}/°C = T_{90}/K - 273.15; \qquad (1)$$

273.15 appears here instead of 273.16 since, as discussed in the Introduction (Section I), Celsius temperatures are based on the freezing, not the triple, point of water.

The ITS-90 is described most readily in terms of the four interpolation methods (instruments) which are used to define it in four distinct but overlapping temperature ranges. These overlaps represent a change in philosophy

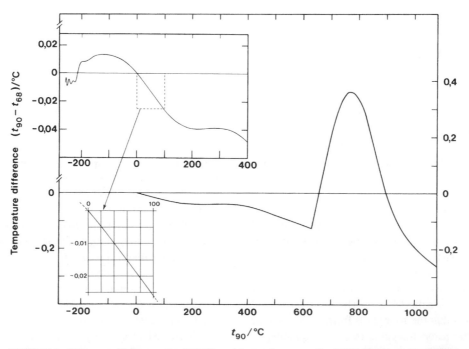

FIGURE 4 Differences between the ITS-90 and its predecessor, the IPTS-68. [From the BIPM.]

TABLE I Fixed-Point Temperatures for the ITS-90

	T_{90} (K)	t_{90} (°C)
1. Helium (v)	3 to 5	−270.15 to −268.15
2. e-Hydrogen (tp)	13.8033	−259.3467
3. e-Hydrogen (v or g)	≈17	≈−256.15
4. e-Hydrogen (v or g)	≈20.3	≈−252.85
5. Neon (tp)	24.5561	−248.5939
6. Oxygen (tp)	54.3584	−218.7916
7. Argon (tp)	83.8058	−189.3442
8. Mercury (tp)	234.3156	−38.8344
9. Water (tp)	273.16	0.01
10. Gallium (fp)	302.9146	29.7646
11. Indium (fp)	429.7485	156.5985
12. Tin (fp)	505.078	231.928
13. Zinc (fp)	692.677	419.527
14. Aluminum (fp)	933.473	660.323
15. Silver (fp)	1234.93	961.78
16. Gold (fp)	1337.33	1064.18
17. Copper (fp)	1357.77	1084.62

from the IPTS-68, since no overlap was allowed between the four ranges which defined that scale.

The low-temperature portion of the ITS-90 is divided into two regions. For the lowest temperatures (0.65 to 5 K), explicit equations are given for the vapor pressure–temperature relations for the two helium isotopes. Temperatures between 3 K and the triple point of neon (24.5561 K) are defined by an interpolating constant volume gas thermometer (see Section III.B.1), which uses either ^{4}He or ^{3}He as the working substance. A procedure is given for correcting the gas thermometer pressures (slightly) for the nonideal behavior of these gases, after which the parameters for a parabolic pressure–temperature relation are determined from the corrected pressures at fixed points 1, 2, and 5 in Table I.

The platinum resistance thermometer (an SPRT) is used to define the ITS-90 from 13.8 K (2 in Table I) to 961.78°C (the freezing point of silver; 15), with the acknowledgment that no single instrument is likely to be usable over this whole range. The characteristics of a real thermometer were used to generate an SPRT interpolation relation which, to obtain the required accuracy, is quite complex. To eliminate differences between thermometers due to different resistances, the primary variable which is used for interpolation is the dimensionless ratio of the thermometer resistance at a given temperature to its value at the triple point of water, 273.16 K,

$$W(T_{90}) = R(T_{90})/R(273.16 \text{ K}). \qquad (2)$$

The triple-point value of R typically is approximately 25 Ω for an SPRT, which will be used from the lowest temperatures to, possibly, 400°C, with smaller val-

ues (as low as 0.25 Ω) used for the highest-temperature applications.

A PRT that is acceptable for representing the ITS-90 (an SPRT) must have a high-purity, strain-free platinum element; the ITS-90 defines such an element as one for which either $W(29.7646°C) \geq 1.11807$ (the gallium triple point) or $W(−38.8344°C) \leq 0.844235$ (the mercury triple point). An SPRT that is to be used to the freezing point of silver in addition must have $W(961.78°C) \geq 4.2844$. These requirements eliminate many relatively inexpensive commercial thermometers. A practical requirement which is not stated in the scale is that an SPRT must have a reproducibility at the triple point of water after temperature cycling of better than 1 mK (preferably 0.1 mK). Thermometers which are used above the zinc point (431°C) require careful treatment because of effects due to annealing of the platinum element.

The mathematical functions that are required to describe mathematically the ITS reference interpolation relation for an SPRT are quite complex. For temperatures from 13.8 to 273.16 K, a 13-term power series is required to give $\ln[W_r(T_{90})]$ as a function of $\ln[T_{90}/273.16 \text{ K}]$, while the inverse relation, which gives T_{90} as a function of $W_r(T_{90})$, requires a 16-term power series. The corresponding power series for temperatures from 0 to 961.78°C each contain "only" 10 terms.

Only rarely will the temperature dependence of the resistance for a real thermometer, $W(T_{90})$, agree with that given by the reference function, $W_r(T_{90})$. The values of W and W_r are compared at the various fixed points, and the differences are used to determine the parameters in a deviation function which then is used together with the reference relation to obtain T_{90}. The details again are complex; an SPRT which is to be used from 13.8 to 273.16 K must be calibrated at points 2 through 9 (Table I) to determine the eight parameters in the deviation function. For a calibration which is to be used only within ±30°C of the ice point, the thermometer need only be calibrated at the mercury point, the water triple point, and the gallium point to determine two parameters for the deviation function. All in all, 11 possible subranges are defined; 4 depend on the lowest temperature below 273.16 K at which the thermometer will be used, 1 is for temperatures near 0°C, and 6 depend on the maximum temperature above 0°C at which the thermometer will be used.

A question immediately arises as to the agreement that can be expected between temperatures obtained at, for instance, −15°C, for a given thermometer which has been calibrated using five different procedures and five different sets of fixed points. This is the "uniqueness" problem. The belief is that the differences at a given temperature between calibrations using different ranges will be comparable with differences between different thermometers which are calibrated in a given range. This "nonuniqueness" will be a few tenths of a millikelvin near room temperature, less than 1 mK for the more extreme parts of the scale between 13.8 K and 420°C, and should be less than 5 mK at the highest temperatures.

The highest range of the ITS-90, above the silver point, is defined by optical pyrometry, using Planck's law to obtain the radiant emission from a black-body cavity for a given wavelength, λ, and bandwidth. The ratio of the spectral radiances at the temperature T_{90} and at the reference temperature, X, is related to the absolute temperature by

$$\frac{L_\lambda(T_{90})}{L_\lambda(T_X)} = \frac{\exp[c_2/\lambda T_{90}(X)] - 1}{\exp[c_2/\lambda T_{90}] - 1}, \qquad (3)$$

where $T_{90}(X)$ refers to any one of the silver [$T_{90}(\text{Ag}) = 1{,}234.93$ K], the gold [$T_{90}(\text{Au}) = 1{,}337.33$ K], or the copper [$T_{90}(\text{Cu}) = 1357.77$ K] freezing points. Here the optical pyrometer both defines the scale and serves as the interpolation device. The ITS-90 specifies the use of the theoretical value for the constant c_2, so there are no adjustable parameters in this relation. Proper realization of temperatures by pyrometry requires care in the design of the cavities in which the gold and the sample are located, and as with most thermometry, care must be taken to avoid systematic errors.

2. Calibration Procedures

Working thermometers (either transfer standards or working instruments) should be calibrated by following the procedures outlined in the basic ITS-90 document to reproduce the scale. In practice, this can be a cumbersome procedure, especially at low temperatures, where gas thermometry requires long-term experiments. In this temperature region, gas thermometry results will be transferred to highly stable rhodium–iron resistance thermometers, and most subsequent calibrations will be carried out in terms of "point-by-point" comparisons at thermal equilibrium between a set of standard thermometers and the unknown thermometer(s). This also may be true for higher, PRT, temperatures when calibrations are not carried out at a national standards laboratory. In this instance, "standards" which have been calibrated directly on the ITS-90 may be used as substitutes for true fixed point devices. Three standard thermometers are the useful minimum, since not more than one would be expected to show drift (instability) in any given period of time. The result is a table of temperatures and corresponding W's, with the W's converted to $R(T_{90})$ using the measured $R(273.16 \text{ K}) = R_0$ to eliminate dependence on a standard resistance value. To a first approximation, small changes in R_0 will have little effect on the $W(T_{90})$ relationship for a thermometer.

For moderate and low temperatures, the sheaths of the thermometers can be inserted in individual mounting holes in an isothermal metal block. Thermal shielding of the block, anchoring of the leads to the block, vacuum insulation, and temperature control all are important factors in such a thermometer comparator. Variable-temperature baths (oil or possibly molten salt) are used at higher temperatures where long-stem thermometers must be used. Calibrations carried out by each of the national standards laboratories can be expected to be equivalent, and to represent the ITS-90 within stated uncertainties. Other calibration sources, which generally are traceable to a national standards laboratory, generally have less rigorous controls, and care must be taken in assessing the accuracy of calibrations that are supplied. If accuracy is important, the performance of a thermometer can be spot-checked with commercially available sealed fixed-point devices, with gallium (see Table I) being most useful near room temperature. This may be particularly important when highly accurate thermometry is required for the maintenance of standards or for biological studies.

D. Electrical Measurements

High-quality electrical measurements traditionally have used very accurate dc techniques. Voltages were measured potentiometrically in terms of standard cells, while resistances were measured using Wheatstone or other types of bridges. For accurate work, a standard resistor or a resistance thermometer is designed with four terminals, two of which are for the measuring current, while the second pair, mounted just inside the current leads at each end, measures the potential drop across the resistor. If a conventional Wheatstone-type bridge technique is used, the bridge determines the sum of the resistances of the resistor and of the leads, so a separate measurement of the resistance of a pair of leads at one end of the resistor (or thermometer) must be made. Care must be taken that the lead resistances are symmetrical. These measurements can be simplified if a potentiometer is used to compare directly the potential drops across a standard resistor and the unknown for a common current. In this case, negligible current flows through the potential leads, and no lead correction is required.

In both bridge and potentiometric measurements, parasitic emfs (voltages) can exist in the lead wires and the measuring instrument, with current reversal required to eliminate their effects. In addition, since the bridge contains standard resistances of various magnitudes, these must be intercompared and recalibrated regularly to detect aging effects. The linearity of a dc potentiometer also must be calibrated at regular intervals for the same reason.

Modern semiconductor technology has caused major changes in the above procedures. First, voltmeters now routinely have extremely high input impedances (greater than 1000 MΩ) and linearities at the 10^{-6} level. Hence, most accurate electrical measurements now are made using these instruments rather than potentiometers or bridges. Modern multimeters often can be used in a four-terminal mode for resistance measurement, and most can be interfaced directly with a computer for experimental control and data acquisition.

When the highest accuracy in resistance measurement is required, variations of the potentiometer technique are used in which the accurate division of voltage levels is carried out using ratio transformers rather than resistive windings. These components are very similar to ideal transformers or inductors, with windings on a high-permeability mumetal toroid system for which the stability is determined by winding geometry rather than a physical property. The current comparator is a dc instrument in which the condition for zero magnetic flux in a core is used to determine the ratio of currents through two resistances (a standard and an unknown) when the potential drops across them are equal. The effects of parasitic voltages are eliminated by using current reversal. These instruments are in common use in standards laboratories and are capable of determining resistance ratios potentiometrically at the 10^{-8} level. This corresponds to better than 10 μK for an SPRT with a 25-Ω ice-point resistance and is better than the long-term stability of many standard resistances. It is for this reason that SPRT measurements are always expressed in terms of Eq. (2), using a direct determination of R(273.16 K).

Various alternating current bridges and potentiometers have been constructed using ratio transformer techniques. Figure 5 shows a very simple version of an ac ratio-transformer bridge. The ac voltage drop across an unknown resistor is compared with a fraction of the voltage drop across a standard resistor. This fraction, which is determined by the turns ratio, is adjusted until a null is indicated at the detector. Typically, this is a phase-sensitive detector with transformer input and a sensitivity to extremely low (nV; 10^{-9} V) voltages. This bridge is useful primarily

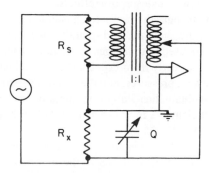

FIGURE 5 An elementary ac ratio-transformer bridge for resistance measurements.

for temperature control, since the finite input impedance of the transformer (typically 10^5 Ω at 400 Hz) causes unacceptable shunting of the reference resistor. The input impedance of the transformer can be increased greatly by sophisticated designs that use multiple cores and windings and operational amplifier feedback. As a result, accuracies of 10^{-8} are also reported for the ac measurement of a standard 25-Ω SPRT.

Although the effects of parasitic dc voltages are eliminated with ac methods, frequency-dependent lead admittance effects (due to shunt capacitances between thermometer leads) are important, and both in-phase and quadrature balance conditions must be met. This is accomplished in Fig. 5 with the variable shunt capacitor. It is for this reason that ac bridges are restricted to relatively low resistance values for the most accurate work.

III. THERMODYNAMIC TEMPERATURES

A. General Concepts

The concept of thermodynamic temperature arises from the second law of thermodynamics and the existence of reversible heat effects, such as for the isothermal compression of an ideal gas. The maximum (Carnot) efficiency for a heat engine, for example, is expressed in terms of a ratio of thermodynamic temperatures.

Developments of statistical mechanics contain a characteristic energy that is the same for all systems that are in thermal equilibrium and that increases as the internal energy of a system is increased. This characteristic energy has properties that are identical to those of temperature as it is defined in both the thermodynamic and the practical senses. This characteristic energy appears in an elementary manner in the Boltzmann factor, which determines the relative populations of two states that are separated by an energy difference ΔE,

$$N_1/N_2 = \exp(-\Delta E/k_B T). \tag{4}$$

In this expression, $k_B T$ is the characteristic energy, and k_B (as yet undetermined) is the Boltzmann constant. Equation (4) suggests that the concept of a level of temperature is purely relative. A collection of systems can be said to be at a low temperature (close to $T = 0$) if most (all) of them are in their lowest energy (ground) state, that is, if $\Delta E \gg k_B T$. Alternatively, a high temperature corresponds to an equal population of the levels. Whether or not a temperature is "high" or "low" thus depends on the characteristic energies of the system and is a purely relative concept. Absolute zero corresponds to a state at which every conceivable system is in its ground state. Negative temperatures occur when (as in some laser systems) an upper metastable level has been forced to have a larger population than a lower level.

The relationship between theoretical and practical temperatures (see Section I) has been determined most often using measurements made with an ideal gas. The experimental equation of state for such a system is written

$$PV_m = RT, \tag{5}$$

with V_m the volume per gram molecular weight of the gas, R the gas constant per mole (8.317 J/F mol-K), and T related to the Celsius scale by Eq. (1). Since a Carnot heat engine with an ideal gas as the working medium has an efficiency identical to that of a Carnot cycle, T as it appears in Eq. (5) can be chosen to be equal to thermodynamic temperatures.

Statistical mechanics as applied to an ideal gas (a collection of noninteracting particles) also gives Eq. (5), if RT is assumed to be proportional to the characteristic thermal energy of the system and to the total number of particles. The association with Eq. (4) exists through the introduction of the gas constant per molecule, the Boltzmann constant, $k_B = R/N_A$, where N_A, the Avagadro constant, is the number of molecules in a gram molecular weight of a substance. The characteristic thermal energy that appears in the Boltzmann relation is the same as that which appears in the ideal-gas law.

B. Absolute or Primary Thermometers

The use of fixed points and designated interpolation instruments would not be necessary if an absolute or primary thermometer could be used directly as a practical thermometer. A single calibration of such a thermometer at the triple point of water (273.16 K) would serve to standardize the thermometer once and for all. Unfortunately, most primary thermometers are relatively clumsy devices and may require elaborate instrumentation and possibly long equilibrium and/or measurement times.

Two exceptions are the optical pyrometer at high temperatures and the magnetic thermometer at low temperatures. In each of these cases, data are taken using the primary thermometric parameter, with this parameter related directly by theory to the absolute temperature. At intermediate temperatures, fixed points and easily used secondary thermometers must be used for the routine measurement of temperature. Primary thermometers, then, are used to establish the temperatures that are assigned to the fixed points and to test the smoothness and appropriateness of the calibration relations that are used with the secondary thermometers.

The following sections discuss briefly the various types of primary thermometers that have been used to obtain accurate thermodynamic temperatures. Gas thermometry

in various forms traditionally has been of primary importance in this area, but modern optical pyrometry has comparable importance at high temperatures, and noise and magnetic thermometry also have had important complementary roles. The existence of several approaches for a given temperature range is important to provide confidence in the relationship between theory and experiment, and to provide information about the possible existence of systematic errors.

1. Gas Thermometry

The ideal-gas law [Eq. (5)] is valid experimentally for a real gas only in the low-pressure limit, with higher-order terms (the virial coefficients, not defined here) effectively causing R to be both pressure and temperature dependent for most experimental conditions. While these terms can be calculated theoretically, most gas thermometry data are taken for a variety of pressures, and the ideal-gas limit, and, hence, the ideal-gas temperature, is achieved through an extrapolation to $P = 0$. The slope of this extrapolation gives the virial coefficients, which are useful not only for experimental design, but also for comparison with theory. The following discussion of ideal-gas thermometry is concerned, first, with conventional gas thermometry, then with the measurement of sound velocities, and, finally, with the use of capacitance or interferometric techniques. Each of these instruments should give comparable results, although the "virial coefficients" will have different forms.

Gas thermometry in the past 20 years or so has benefited from a number of innovations that have improved the accuracy of the results. Pressures are measured using free piston (dead weight) gauges that are more flexible and easier to use than mercury manometers. The thermometric gas (usually helium) is separated from the pressure-measuring system by a capacitance diaphragm gauge, which gives an accurately defined room-temperature volume and a separation of the pressure-measurement system from the working gas. In addition, residual-gas analyzers can determine when the thermometric volume has been sufficiently degassed to minimize desorption effects.

In isothermal gas thermometry, absolute measurements of the pressure, volume, and quantity of a gas (number of moles) are used with the gas constant to determine the temperature directly from Eq. (5). Data are taken isothermally at several pressures, and the results are extrapolated to $P = 0$ to obtain the ideal-gas temperature as well as the virial coefficients. A measurement at 273.16 K gives the gas constant.

A major problem in isothermal gas thermometry is determining the quantity of gas in the thermometer, since this ultimately requires the accurate measurement of a small difference between two large masses. Most often, this problem is bypassed by "filling" the thermometer to a known pressure at a standard temperature, with relative quantities of gas for subsequent fillings determined by division at this temperature between volumes that have a known ratio. The standard temperature may involve a fixed point or, for temperatures near the ice point, an SPRT that has been calibrated at the triple point of water. Since the volume of the gas for a given filling is constant for data taken on several subsequent isotherms, and the mass ratios are known very accurately, the absolute quantity of gas needs to be known only approximately. Excellent secondary thermometry is very important to reproduce the isotherm temperatures for subsequent gas thermometer fillings. The results for the isotherms (virial coefficients and temperatures) then are referenced to this standard "filling temperature."

The procedure for constant-volume gas thermometry is very much the same as that for isotherm thermometry, but detailed bulb pressure data are taken as a function of temperature for one (and possibly more) "filling" of the bulb at the standard temperature. To first order, pressure ratios are equal to temperature ratios, with thermodynamic temperatures calculated using known virial coefficients. In practice, the virial coefficients vary slowly with temperature, so a relatively few isotherm determinations can be sufficient to allow the detailed investigation of a secondary thermometer to be carried out using many data points in a constant-volume gas thermometry experiment. If the constant-volume gas thermometer is to be used in an interpolating gas thermometer mode (as for the ITS-90), the major corrections are due to the nonideality of the gas. When a nonideality correction is made using known values for the viral coefficients, the gas thermometer can be calibrated at three fixed points (near 4 and at 13.8 and 24.6 K) to give a quadratic pressure–temperature relation that corresponds to T within roughly 0.1 mK.

The velocity of sound in an ideal gas is given by

$$c^2 = (C_P/C_V)RT/M, \qquad (6)$$

where the heat capacity ratio (C_P/C_V) is 5/3 for a monatomic gas such as helium. Since times and lengths can be measured very accurately, the measurement of acoustic velocities by the detection of successive resonances in a cylindrical cavity (varying the length at constant frequency) appears to offer an ideal way to measure temperature. This is not completely correct, however, since boundary (wall and edge) effects that affect the velocity of sound are important even for the simplest case in which only one mode is present in the cavity (frequencies of a few kilohertz). These effects unfortunately become larger as the pressure is reduced. An excellent theory relates the attenuation in the gas to these velocity changes, but the situation is very complex and satisfactory

results are possible only with complete attention to detail. An alternative configuration uses a spherical resonator in which the acoustic motion of the gas is perpendicular to the wall, thus eliminating viscosity boundary layer effects. The most reliable recent determination of the gas constant, R, is based on very careful sound velocity measurements in argon as a function of pressure at 273.16 K, using a spherical resonator.

The dielectric constant and index of refraction of an ideal gas also are density dependent through the Clausius–Mossotti equation,

$$(\varepsilon_r - 1)/(\varepsilon_r + 2) = \alpha/V_m = \alpha RT/P, \qquad (7)$$

in which $\varepsilon_r (= \varepsilon/\varepsilon_0)$ is the dielectric constant and α is the molar polarizability. Equation (7) suggests that an isothermal measurement of the dielectric constant as a function of pressure should be equivalent to an isothermal gas thermometry experiment, while an experiment at constant pressure is equivalent to a constant-volume gas thermometry experiment. The dielectric constant, which is very close to unity, is most easily determined in terms of the ratio of the capacitance of a stable capacitor that contains gas at the pressure P to its capacitance when evacuated. The results that are obtained when this ratio is measured using a three-terminal ratio transformer bridge are comparable in accuracy with those from conventional gas thermometry. An advantage is that the quantity of gas in the experiment need never be known, although care must be taken in cell design to ensure that the nonnegligible changes in cell dimensions with pressure can be understood in terms of the bulk modulus of the (copper) cell construction material.

At high frequencies (those of visible light), the dielectric constant is equal to the square of the index of refraction of the gas ($\varepsilon_r = n^2$), so an interferometric experiment should also be useful as a primary thermometer. No results for this type of experiment have been reported, however.

2. Black-Body Radiation

The energy radiated from a black body is a function of both temperature and wavelength [Eq. (3)]. An ideal black body has an emissivity (and hence an absorptivity) of unity, or a zero reflectivity. The design of high-temperature black bodies to satisfy this condition requires considerable care. In practice, a usable design would consist of a long cylindrical graphite cavity with a roughened interior that is, for instance, surrounded by freezing gold to maintain isothermal conditions. The practical aspects of optical pyrometry are discussed briefly in Section IV. For the present purposes, optical pyrometry using well-defined wavelengths and sensitive detectors (so-called photon-counting techniques) can be used with Eq. (3) to measure relative temperatures with a high accuracy (better than 10 mK) at temperatures as low as the zinc point, 419.527°C. This gives a valuable relationship between the high temperature end of current gas thermometry experiments and the temperatures that are assigned to the gold and silver points.

The total energy that is radiated by a black body over all wave lengths [the integrated form of Eq. (3)] is the well-known Stefan–Boltzmann law,

$$dW/dT = \sigma T^4. \tag{8}$$

Here, $\sigma = (2\pi^5 k_B^4/15c^2h^3) = 5.67 \times 10^{-8}$ W/m^{2}K^4 is the Stefan–Boltzmann constant. Measurements of the power radiated from a black body at 273.16 K give σ directly, and, since both Planck's constant, h, and the velocity of light, c, are well known, also give the Boltzmann constant, k_B. Relative emitted powers also give temperature ratios. Total radiation measurements [Eq. (8)] have been carried out for black bodies in the range from $-130°C$ to $+100°C$ using an absorber at a low temperature (roughly 2 K) to measure the total radiant power that is emitted.

3. Noise Thermometry

Noise thermometry is another, quite different, system that can be understood completely from a theoretical standpoint and that can be realized in practice. The magnitude of the mean-square thermal noise voltage (Johnson or Nyquist noise) that is generated by thermal fluctuations of electrons across a pure electrical resistance, R, is given by

$$(V^2)_{avg} = 4k_B TR\Delta f. \tag{9}$$

This simple exact expression assumes that R is frequency independent, with the mean-square noise voltage depending on R and the bandwidth in hertz, Δf, over which the measurement is made. These measurements are difficult, since, to achieve the needed accuracy, consistent measurements must be made of the long-time average of the square of a voltage. In most instances, the results are obtained as the ratio of the mean square voltage at T to that at a standard temperature (possibly 273.16 K), so the absolute values of the voltages need not be determined. Instrumental stability is very important, however. Noise temperatures have been determined from as low as 17 mK [17×10^{-3} K, using SQUID (Superconducting Quantum Interference Device) technology] to over 1000°C. While noise thermometry is difficult to carry out in a routine fashion, the measurements involved are so different from those for gas thermometry and optical pyrometry that the results are extremely useful.

4. Magnetic Thermometry

The magnetic susceptibility of an ideal paramagnetic salt (a dilute assembly of magnetic moments) obeys Curie's law,

$$x = C/T, \tag{10}$$

where C, the Curie constant, is proportional to the number of ionic magnetic moments and their magnitudes. The magnetic moments may be due either to electronic or to nuclear effects, with a difference in magnitude of roughly 1000. Interactions between the moments eventually cause the breakdown of Eq. (10) at temperatures of the order of millikelvins (or higher) for electronic paramagnetism, and at temperatures 1000 times smaller for nuclear systems.

Magnetic thermometry involving electron spins is not strictly primary thermometry, since the number of moments in the sample cannot be determined with any precision, and Curie's law is obeyed only approximately for any real system. Magnetic interactions between the moments and complications due to the existence of excited states for the ions cause difficulties in almost every case. An ion can be chosen for which the excited states are not populated for a given experiment, with deviations due to magnetic interactions expected on theoretical grounds to give first-order corrections to Curie's law which are of the form

$$x = A + B/(T + \Delta + \delta/T). \tag{11}$$

The parameter A is due to temperature-independent diamagnetism and paramagnetism, while Δ represents effects due to surrounding moments, and δ arises because of complex spin systems. In practice, each of these parameters must be determined empirically.

While a paramagnetic salt such as cerium magnesium nitrate [CMN, $Ce_2Mg_3(NO_3)_{12} \cdot 24H_2O$] shows almost-pure Curie law behavior ($\Delta = 0.3$ mK, $\delta = 0$), the dilution of its moments and consequent small susceptibility make measurements difficult above 2 K, with a breakdown of Eq. (11) arising near 4 K due to the beginning occupation of a higher-energy state of the cerium ion. Even at low temperatures, controversy exists for CMN as to the meaning of the "nonideality" parameters, and the significance of different values of Δ for single-crystal and powdered samples. The use of SQUID technology rather than conventional ratio-transformer mutual inductance bridges allows measurements to be made with extremely small samples. Paramagnetic salts with larger susceptibilities, which are useful at higher temperatures, will have larger values for the nonideality parameters and will show deviations from even Eq. (11) at temperatures not far below 1 K.

5. Helium Melting-Pressure Thermometry

At temperatures below the lower limit of the ITS-90, 0.65 K, a new low-temperature scale is being proposed by the CCT based on the relation between the pressure and the temperature of melting ^{3}He. Although the helium melting temperature–pressure relation used in the new scale is closely related to the Clausius–Clapyron equation its temperature cannot be calculated directly from this equation with sufficient accuracy. Instead, the relation is based on experimental measurements using magnetic thermometry, noise thermometry, and nuclear-orientation thermometry. It is thus not strictly a primary thermometer. The new scale is referred to as to the "Provisional Low-Temperature Scale, 0.9 mK to 1 K: PLTS-2000." The scale is defined by the relation between the temperature of melting ^{3}He and fixed points, i.e., the minimum in the melting pressure of ^{3}He at a temperature of about 315 mK and a pressure of 2.93 MPa and at the A, A–B, and Néel transitions in ^{3}He at temperatures of about 2.44, 1.9, and 3.44 mK respectively.

6. Nuclear Orientation Thermometry

At temperatures below 100 mK or so, the splitting of nuclear energy levels in a single crystal may become comparable with the characteristic thermal energy, $k_B T$. The γ-ray emissions from the oriented nuclei then may be anisotropic, and the anisotropies can be used to determine the relative populations of these levels. In the simplest possible two-level case, Eq. (4) can be applied to obtain the temperature directly from these nuclear orientation experiments. Such measurements have been made from 10 to roughly 50 mK for radioactive cobalt of mass 60 in a single-crystal nonradioactive cobalt lattice. These have confirmed SQUID noise measurements in the assignment of absolute temperatures to the superconducting transitions of the National Bureau of Standards SRM 768 device. The energy levels of the nuclei involved must be understood in detail from other measurements before these methods can be used, but, again, it is useful that two independent measurements can be used to assign thermodynamic temperatures in an extreme region of the temperature spectrum.

7. Spectroscopic Methods

Optical spectroscopy can give information about the relative populations of excited states in a very high-temperature system, such as a plasma. This information then can be combined with the Boltzmann relation or direct theoretical calculations to obtain the temperature directly, as for nuclear orientation experiments. Again, the system must be understood theoretically, and possible complications due to interactions must be recognized. This use of spectroscopic data for primary thermometry represents the only possible means for determining extremely high temperatures.

C. The ITS-90 and Thermodynamic Temperatures

Each of the above primary thermometers has been used for at least a limited temperature region in the establishment of the ITS-90. At the lowest temperatures, the scale is based on a combination of results from magnetic, noise, and gas thermometry, with several gas thermometry experiments of most importance from liquid helium and/or liquid hydrogen temperatures to 0°C. These agree well with total-radiation experiments at temperatures above 240 K. Gas thermometry results overlap pyrometry data for temperatures from 457 to 661°C, and the comparison of an SPRT with pyrometry data provided the SPRT reference function for temperatures from 660°C to the silver point. The correspondence between the ITS-90 and thermodynamic temperatures is believed to vary from ±0.5 mK at the lowest temperatures to a maximum of ±2 mK for any temperature below 0°C. At higher temperatures, the possible difference rises from ±3 mK at the steam point to ±25 mK at 660°C. The three highest temperature reference points (based on freezing points for silver, gold, and copper) are expected to be internally consistent to within the accuracy of standards pyrometry and to have potential differences from thermodynamic temperatures of ±0.04, 0.05, and 0.06 K, respectively, which reflect the uncertainties at the primary reference temperature of 660°C. The most important characteristic of the ITS-90, however, is that it is believed to be smoothly related to T at all temperatures, with no abrupt differences in slope such as appear in Fig. 4, where, on the scale of this figure, T_{90} is identical to T.

IV. PRACTICAL THERMOMETRY

Many types of thermometers are in general use, and many more have been proposed. The following is a brief summary of the characteristics of the more common types of secondary thermometers, with no attempt made to be complete or comprehensive. The choice of a type of thermometer for a given application is somewhat arbitrary, with the deciding factors sometimes dictated by rigorous constraints but more often by personal preferences and/or prejudices. The accuracy or longevity of a thermometer calibration (a certificate or a table) should not be taken for granted when a temperature must be known within specified limits. Checks should be made, either in terms of a close-by fixed point (the freezing point of water and the triple point of gallium are particularly useful near room temperature) or by comparison with one, but preferably two or more, carefully handled, "standard" thermometer. An electrical instrument should never be relied upon to give answers that are correct to all of the significant figures that are generated in the display or in the printout, especially if important conclusions depend on these numbers.

A. Liquid-in-Glass Thermometers

These represent the oldest, and still very common, practical thermometers, although they are increasingly being replaced by low-cost electronic devices using semiconductor elements (see below) as the temperature sensor. They come in many forms and qualities with a variety of liquids, although mercury is the choice for accurate applications. A very good thermometer for use up to 100°C can be calibrated to 0.01°C or better and will remain stable at this level for a considerable period of time. Care must be taken in the use of such a thermometer, since the readings depend on the depth of immersion of the thermometer. Thus, they are most useful for measurements on liquids where a surface is defined. The disadvantage of liquid-in-glass thermometers is that they must be calibrated manually, a tedious process, and must be read by eye, with no opportunities for automated data acquisition.

B. Resistance Thermometers

Resistance thermometers, or, more strictly, thermometers for which a voltage reading depends on an applied current, quite naturally fall into two categories. The first includes pure metals and metallic alloys that exhibit a positive temperature coefficient of resistance. Alloys with very small coefficients are useful for constructing the standard resistances that must play an important role in the practical use of resistance thermometers. The second category includes primarily semiconducting materials, for which the temperature coefficient of resistance is negative. It also includes devices, such as diodes, for which the forward voltage is a function of temperature.

General considerations for the measurement of electrical resistance, discussed in Section II.D, are not repeated here. The reproducibility of a practical resistance thermometer is an important characteristic that is not always directly related to the cost. Its calibration also may depend critically on the magnitude of the measuring current, so care should be taken to follow the manufacturer's (or calibrator's) recommendations. Resistance thermometers often are used both for the control of temperature (as in a thermostat) and for the measurement of the temperature. In general, this is not a recommended procedure, since a temperature-control sensor generally is located in the

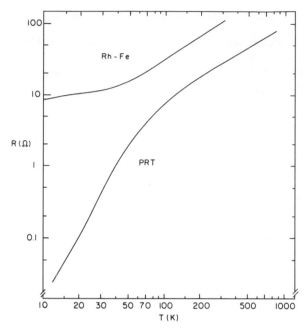

FIGURE 6 The temperature dependences of the resistances for two metallic resistance thermometers.

vicinity of the source of heat of refrigeration and will not give a true average reading for the volume that is being controlled.

1. Metallic Thermometers

The platinum resistance thermometer (PRT) is a typical metallic thermometer; the temperature dependence of the resistance that is shown in the double-logarithmic plot in Fig. 6 is characteristic of most metals. Near room temperature and above, the electrical resistance of a pure metal is associated primarily with lattice vibrations and is proportional to T, with the temperature coefficient of resistance approximately independent of temperature. Impurity effects end to dominate at low temperatures, where the resistance approaches a constant value as T approaches zero. The ratio of the room-temperature resistance to its low-temperature value (the resistance ratio) is a measure of the purity of a metal, and the ratio of 1000 for the SPRT in Fig. 6 (the nominal ice point resistance is 25 Ω) is characteristic of a very pure metal.

Industrial PRTs are constructed from a "potted" wire or a thin film bonded to a ceramic substrate. These have a characteristic resistance very similar to that of an SPRT near room temperature but have a relatively high value for the low-temperature resistance due to the quality of the platinum and also to the strains induced in fabrication. Standard calibration tables exist for these commercial PRTs for temperatures from 77 K upward, with the objective of allowing routine substitution and re-placement of thermometers as needed. One of the difficulties in using pure metallic thermometers at temperatures below 20 K is that the resistance is very sensitive to strains that are induced by shocks, so great care must be taken in handling a calibrated SPRT. Hence, a PRT that was not wound in a strain-free configuration could be expected to be relatively more unstable than the much more expensive SPRT. An additional characteristic of inexpensive PRTs is that they are primarily two-lead devices. For most applications, it is useful to attach a second pair of leads so that the resistance of the thermometer is well defined.

The temperature dependence of an alloy thermometer is also shown in Fig. 6. The primary component of this thermometer is rhodium metal, with a slight amount (0.5%) of iron added as an alloying agent. The localized magnetic moment of the iron scatters electrons very well at low temperatures and is responsible for the relatively high 10 K resistance for this thermometer, which has a nominal 100-Ω room-temperature resistance. The interaction of these iron moments with the electrons also results in an approximately linear temperature dependence for the low-temperature resistivity, in contrast with the SPRT, as shown in Fig. 7 for temperatures to 0.25 K. This thermometer is much more satisfactory than the PRT at low temperatures because of both its sensitivity and its stability. The wire is extremely stiff and difficult to fabricate

into a thermometer element. As a result, the thermometers are very insensitive to shock, and aging and annealing effects are virtually nonexistent. Rhodium thermometers, which are packaged similarly to SPRTs, now form the basis for most practical low-temperature standards thermometry. They are available also in other packages for use in practical measurements, possibly (as Fig. 6 indicates) for temperatures up to room temperature. A single thermometer that can be used with a reasonable sensitivity from 0.5 to 300 K is a very useful device.

2. Semiconductors

Figure 7 gives, along with low-temperature results for a rhodium–iron thermometer, a double-logarithmic plot of the resistance–temperature relationships for a number of low-temperature thermometers which are constructed from semiconducting materials. This presentation does not include an R-vs-T relationship for another often-used semiconducting thermometer, the thermistor (see below), which would be similar to that for the carbon–glass (CG) thermometer, but for higher temperatures.

Commercial radio resistors were used as the first semiconducting low-temperature thermometers, with the most popular being, first, those manufactured by Allen–Bradley (A-B), and, later, those manufactured by Speer. The bonding of the electrical leads to the composite material in these resistors proved to be quite rugged, and although small (occasionally large) resistance shifts occurred on subsequent coolings to liquid helium temperatures, the calibrations remained stable as long as the thermometers were kept cold. The thermometric characteristics of these two brands of resistors have the common feature that the temperature coefficient of the resistance is a smooth and monotonic function of the temperature. The details of their temperature variation are seen to be quite different, however, with the A-B resistors being very sensitive, while the Speer resistors have a reasonable resistance even at the lowest temperatures. These resistors are still used for low-temperature measurements, although improvements in their composition have changed (and downgraded) their thermometry characteristics. The carbon–glass thermometer, which uses fine carbon filaments deposited in a spongy-glass matrix, also has a well-behaved resistance–temperature characteristic, as well as a high sensitivity. This thermometer suffers from lead-attachment problems and has instabilities (minor for many purposes) that make it unsuitable for standards-type measurements. All three of these thermometers have resistances with moderate magneto-resistance characteristics so are useful for measurements in a magnetic field.

Germanium resistance thermometers consist of a small crystal of doped germanium onto which four leads (two

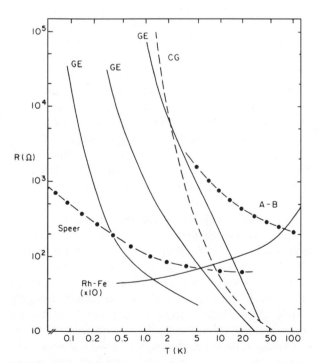

FIGURE 7 The resistance–temperature relations for several low-temperature thermometers. [The GE and CG results are through the courtesy of Lake Shore Cryotronics, Inc.]

current, two potential) are attached. These lead resistances are comparable with the sensor resistance and are similarly temperature dependent. This thermometer element is in a sealed jacket with a low pressure of exchange gas. Figure 7 shows the resistance–temperature characteristics for three of these resistors (labled GE), which are intended for different temperature ranges. The minimum usable temperature in each case is defined as that at which the resistance approaches $10^5 \, \Omega$. The shapes of the calibration curves are quite similar, with, as a crude approximation, $d \ln R / d \ln T \simeq -2$. A detailed inspection of these relations reveals a complex behavior, with a nonmonotonic temperature dependence for dR/dT, so the generation of analytical expressions for the resistance–temperature characteristic is difficult.

Germanium resistance thermometers served as the basis for low-temperature standards thermometry for many years, until rhodium–iron thermometers were introduced. The major advantages of germanium resistance thermometers for experimental work are their relatively small size, high sensitivity, and good stability. While the higher-resistance thermometers can be used up to 77 K, they cannot be used at much higher temperatures because the temperature coefficient changes sign and is positive near room temperature. Their magnetoresistance is rather high and complex, and they are seldom used for measurements in large magnetic fields. For accurate work above roughly 30 K, dc and ac calibrations of these thermometers may differ significantly, dependent on the frequency, so the measurement method corresponding to the calibration must be used.

Thermistors are two-lead sintered metal–oxide devices of a generally small mass, much smaller than any of the above thermometers'. This, combined with the high sensitivity, is their major attraction. The extreme sensitivity requires that a thermistor be chosen to work in a specific temperature range, since otherwise the resistance will be either too small or too large. They have been used at temperatures from 4.2 K (seldom) to 700°C (special design). Their stability can be quite good, especially for the bead designs, when they are handled with care.

The forward voltage of semiconducting diodes also has a well-defined dependence on temperature, which has been used to produce thermometers that are small in size and dependable. Figure 8 gives the voltage–temperature relationships for silicon and gallium arsenide diode thermometers as obtained with a 10-μA measuring current. The gallium arsenide calibration is smoother than that for the silicon diode, with the knee in the silicon curve being rather sharp. At low temperatures, the sensitivity of these thermometers can be quite good (better than 1 mK), with an accuracy and reproducibility of 0.1 K or better. At higher temperatures, these limits should

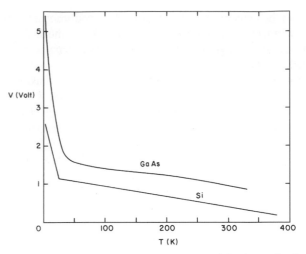

FIGURE 8 The temperature dependences of the forward voltages for two commercial diode thermometers. [Courtesy of Lake Shore Cryotronics, Inc.]

be increased by about an order of magnitude. Standard voltage–temperature relations for selected classes of these diodes allow interchange of off-the-shelf devices with anticipated low-temperature and high-temperature accuracies of 0.1 and 1 K, respectively.

C. Thermocouples

The existence of a temperature gradient in a conductor will cause a corresponding emf to be generated in this conductor which depends on the gradient (the thermoelectric effect). While this emf (or voltage) cannot be measured directly for a single conductor, the difference between the thermal emfs for two materials can be measured and can be used to measure temperatures, as in a thermocouple. When two wires of dissimilar materials are joined at each end and the ends are kept at different temperatures, a (thermoelectric) voltage will appear across a break in the circuit. This voltage will depend on the temperature difference and, also, on the difference between the thermoelectric powers of the two materials. The temperature dependence of this voltage is called the "Seebeck coefficient."

The thermocouple which was used to define the high-temperature IPTS-68 interpolation relation (platinum–10% rhodium/platinum) gives the emf (E)-vs-temperature relation, labeled S in Fig. 9. Noble-metal thermocouples typically have a relatively low sensitivity (roughly 10 μV/K) and calibrations which may change with strain and annealing. These drawbacks are compensated by the usefulness of these thermocouples for work at very high-temperatures. In time, these traditional high-temperature thermocouples may be replaced by gold–platinum and/or platinum–palladium thermocouples, which have similar

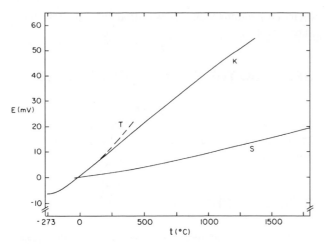

FIGURE 9 The voltage–temperature characteristics for typical noble-metal (S) and base-metal (K) thermocouples.

sensitivities but are more reproducible. More sensitive (basemetal) thermocouples are available for lower-temperature use, and two of these also are shown in Fig. 9. The type K (K) thermocouple uses nickel–chromium-vs-nickel–aluminum alloys, and the type T (T) uses copper vs a copper–nickel alloy. While Seebeck coefficients generally are very small below roughly 20 K, relatively large values (10 μV/K or so) are observed for dilute alloys (less than 0.1%) of iron in gold; these thermocouples are useful even below 1 K.

Thermocouples are convenient, especially when emfs are measured with modern semiconductor instrumentation. The reference junction generally is chosen to be at the ice point (0°C), where precautions must be taken if an ice bath is used. The junction must be electrically isolated from the bath to prevent leakage to ground, which could give false readings, and it must extend sufficiently far into the bath so that heat conduction along the wires to the junction is not important. Finally, the junction must be surrounded by melting ice (a mixture of ice and water), not cold water, since the density of water is minimum at 4°C and temperature gradients exist in water on which ice is floating. The ice bath can be replaced by an electronic device for which the output voltage simulates an ice bath and is independent of ambient temperature.

Thermocouples are relatively sensitive to their environment, and their calibration can be affected in many, sometimes subtle, ways. Annealing, oxidation, and alloying effects can change the Seebeck coefficient, while extraneous, emfs are introduced when strains and a temperature gradient coexist along a wire. Care clearly must be taken in experimental arrangements involving thermocouples, and the standard tables that exist for the various commonly used types of thermocouples must be applied

judiciously. It is important to remember that the thermal produced by a thermocouple is developed along that part of the wire passing through a temperature gradient; it has nothing to do with the junction. Consequently, strains and inhomogeneities present in that part of the wire in the temperature gradient will lead to errors in the temperature measurement.

D. Optical Pyrometry

Some of the problems involved in optical pyrometry were addressed in an earlier section, with the emissivity of the source a major concern. Commercial pyrometers have been in use for many years and have been a part of the International Temperature Scales since 1927. Early optical pyrometers matched the brightness of the radiation source with that of a filament as the filament current was varied. The temperature of the source was then calibrated directly in terms of the current through the filament. Neutral density filters are used to extend the range of these pyrometers to higher temperatures. Considerable skill is required to use these "disappearing filament" pyrometers (the filament disappears in an image of the source) reproducibly, but they are used widely in industry.

The visual instruments have been replaced in standards and, also, in most practical applications by photoelectric pyrometers, in which a silicon diode detector or a photomultiplier tube replaces the eye as the detector. These instruments have a high sensitivity and can be used with interference filters to increase their accuracy [Eq. (3)]. A major concern in optical pyrometry is that real objects do not show ideal black-body radiation characteristics but have an emittance that differs from that of a black body in a manner that can be a function of the temperature, wavelength, and surface condition. Pyrometers that operate at two or more distinct wavelengths provide at least partial compensation for these effects.

A recent development in high-temperature optical pyrometry uses a fine sapphire fiber light pipe and photoelectric detection to obtain the temperature of a system that cannot be viewed directly. The end of the fiber may be encapsulated to form a black body (producing a self-contained thermometer) or the fiber may be used to view directly the object whose temperature is to be determined. Very sensitive semiconducting infrared detectors have made possible the use of total-radiation thermometers at and above room temperature for noncontact detection of temperature changes in processing operations and even, for instance, to determine the location of "heat leaks" in the insulation of a house. The slight excess temperature associated with certain tumors in medical applications has also been detected in this way.

E. Miscellaneous Thermometry

Many other thermometric systems are useful, some for specific applications. The variation with temperature of certain quartz piezoelectric coefficients gives a thermometer with a frequency readout. Very sensitive gas thermometers can be made with pressure changes sensed by changes in the resonant frequency of tunnel-diode circuits. Glass–ceramic capacitance thermometers are unique in that they have no magnetic field dependence, so are useful for low-temperature measurements in large magnetic fields.

Superconducting technology using SQUIDs allows the detection of very small changes in magnetic flux and, hence, in the current flowing through a loop of wire. Major advantages are the high sensitivity and the capability of using small samples in, for instance, magnetic thermometry and the measurement of low voltages. They have, for example, been used with gold–iron thermocouples for high-precision temperature measurements below 1 K. SQUIDs are primarily low-temperature devices but have been applied to routine measurements at room temperature and above.

Vapor pressure thermometry, with judicious choice of working substance, allows a very high sensitivity, but only, except at liquid helium temperatures, in a narrow temperature region. Here, capacitive diaphragm gauges and other modern pressure-sensing devices replace the conventional mercury manometer and allow remote readout of the pressures involved.

SEE ALSO THE FOLLOWING ARTICLES

CRITICAL DATA IN PHYSICS AND CHEMISTRY • CRYOGENICS • HEAT TRANSFER • THERMAL ANALYSIS • THERMODYNAMICS • THERMOELECTRICITY • TIME AND FREQUENCY

BIBLIOGRAPHY

American Institute of Physics (1992). "Temperature: Its Measurement and Control in Science and Industry," Vol. 6, Proceedings of the Symposium on Temperature, AIP, New York. (See also Vol. 5 in the same series.)

Bureau International des Poids et Mesures (BIPM) (1991). "Supplementary Information for the ITS-90," BIPM, Sevres, France. (A bibliography of recent articles on thermometry from national metrology institutes can be found at the BIPM web site: www.bipm.org.)

Bureau International des Poids et Mesures (BIPM) (1996). *Metrologia* 33, No. 4, 289–425 (a special issue devoted wholly to thermometry).

Hudson, R. P. (1980). "Measurement of temperature." *Rev. Sci. Instrum.* **51,** 871.

Quinn, T. J. (1990). "Temperature," 2nd ed., Academic Press, New York.

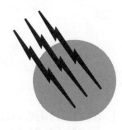

Thin-Film Transistors

Yue Kuo

Texas A&M University

GLOSSARY

Active matrix liquid crystal display Liquid crystal display with each pixel driven by an individual solid-state device, such as a thin-film transistor or diode.

Amorphous silicon Silicon with no long-range order structure, which has broad tail states and large deep states.

Dangling bond Valence electron that can form a bond but actually is not part of a bond, thus can be in a more stable state by forming a bond with elements such as hydrogen.

Deep states Gap states near the center of the energy gap, which are caused by defects.

Microcrystalline crystal silicon Silicon with very small, nanometer-sized grains imbibed in an amorphous matrix.

Off current Minimum current flowing from source to drain at a specified drain voltage.

On current Maximum current flowing from source to drain at specified gate and drain voltages.

Polycrystalline silicon Silicon-containing crystals of different orientations joined at grain boundaries.

Tail states Gap states near conduction and valence band edges.

Threshold voltage Minimum gate voltage required before an appreciable amount of drain current starts flowing through a thin-film transistor.

A THIN-FILM TRANSISTOR (TFT) is a solid-state field-effect transistor (FET), which contains three electrodes, i.e., source, drain, and gate. The operation principle of a TFT is similar to that of a metal–oxide semiconductor (MOS) transistor in a very large-scale integrated circuit (VLSIC), i.e., the magnitude of the source-to-drain current is manipulated by the voltage applied to the gate. However, due to the difference in structures, materials, and fabrication processes, the device characteristics of a TFT are very different from those of a MOS transistor. These factors widely separate their applications. For example, MOS transistors are used mainly in high-speed,

Encyclopedia of Physical Science and Technology, Third Edition, Volume 16

very small-geometry integrated circuits, while TFTs are commonly used in large-area, nonwafer displays.

I. INTRODUCTION—STATUS OF THIN-FILM TRANSISTORS

A. TFT Applications

Currently, the main application of TFTs is in the liquid crystal display (LCD). The TFT LCD is a kind of active-matrix LCD (AMLCD) of which each pixel is addressed by an individual device. Each pixel can be turned on and off separately without the influence of electrical signals of adjacent lines or pixels. In addition, since the TFT is a triode device, it has gray-scale capability, which is difficult to achieve with a diode device. Therefore, the TFT LCD has superior display qualities, such as an absence of ghost-image effects, a high switching speed, and large viewing angles, compared to other types of LCDs.

Since the beginning of their mass production in the early 1980s, TFT LCD products have been widely applied in commercial, industry, transportation, entertainment, education, and military markets. In the early stages, the TFT LCD was used mainly in high-price products such as laptop computers, projectors, instruments, and airplanes. As the production technology has matured and the yield has improved, TFT LCD products have quickly spread to other markets such as desktop computers, televisions, and games. So far, TFTs have been used in direct-view or projection displays whose size varies from less than 1 to more than 30 in. diagonally.

TFT applications are not restricted to displays. Table I lists some of the known TFT products that either are commercially available or have been fabricated into prototypes. In these applications, the TFT functions as a switch that drives a specific device, reads information from a

TABLE I   Applications of TFTs

TFT function	Examples of applications
Driving devices	Liquid crystal displays: direct view and projection
	Printers and facsimiles
	Inorganic and organic light-emitting devices
Readout devices	Two-dimensional medical imagers
	Sensors
	Radiation detectors
Integrated circuits	SRAM loading cells
	EEPROM
	High-voltage circuits
Others	Artificial retina
	Photo imagers

designated local structure, or engages in simple circuit operations. In one case, i.e., SRAMs, the TFT replaces a high-resistance resistor to control the resistivity better. Since knowledge on TFTs is available nowadays, many new applications are emerging continually.

Compared with other semiconductor transistors, such as the VLSI MOS, the major advantage of TFT technology is that the fabrication process is independent of the substrate size and materials. Currently, TFT arrays can be fabricated on a low-temperature glass substrate with a size of 1×1 m.

B. Requirements for LCD Applications

The basic requirements for a TFT are the same as those for other semiconductor transistors, i.e., a high field-effect mobility (μ_{eff}), a high on current (I_{on}), a low leakage or off current (I_{off}), and a low threshold voltage (V_{th}). Depending on the application, some of the above characteristics may be more important than others. For example, for LCD pixel driving, a low I_{off} is more desirable than a high μ_{eff}. For the circuit operation, a high μ_{eff} is most important. For low power consumption, the V_{th} has to be kept very low. There are certain unique issues regarding TFTs, such as photosensitivity and reliability, which need to be addressed. They are discussed in subsequent sections of this chapter.

II. THIN-FILM TRANSISTOR STRUCTURES

A TFT is composed of several layers of thin-film materials deposited by various methods. The substrate serves only as a supporting material for the TFT. It indirectly influences the transistor's structure, for example, through the stress mismatch or impurity distribution. Due to the flexibility of the deposition processes, TFTs can be fabricated in many different structures. In principle, as long as the semiconductor, dielectric, and conductor films can be arranged into a structure similar to that of a field-effect transistor, a TFT can be manufactured. However, even with the same structure, the TFT performance can vary drastically according to other factors such as the thin-film material properties and process conditions. These issues are discussed in subsequent sections.

Figures 1a–d show four of the most common TFT structures. The first three structures are used in the amorphous silicon (a-Si:H) TFT. The last structure is used exclusively in the polycrystalline silicon (poly-Si) TFT. Recently, the structure in Fig. 1a has also been used to prepare poly-Si TFTs.

For the a-Si:H TFT, the inverted, staggered structures, i.e., Figs. 1a and b, are more popular than the staggered structure, i.e., Fig. 1c, because the former give better

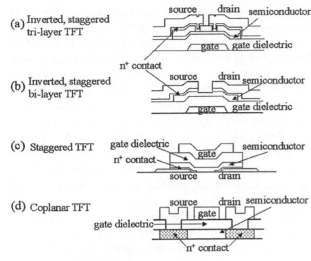

(a) Inverted, staggered tri-layer TFT

(b) Inverted, staggered bi-layer TFT

(c) Staggered TFT

(d) Coplanar TFT

FIGURE 1 Common TFT structures.

device characteristics, such as a higher μ_{eff} and lower V_{th}, than the latter. The trilayer structure, i.e., Fig. 1a, has characteristics superior to those of the bilayer structure, i.e., Fig. 1b, because the former allows the use of a very thin a-Si:H layer and has a back-channel protection film. However, the structure in Fig. 1a needs one more mask to define the back-channel passivation region than does the structure in Fig. 1b. For the mass production environment, the additional masking step can affect the throughput. Therefore, despite its inferior device performance, the bilayer structure is used in some production processes. The physical and chemical properties at the semiconductor/dielectric interfaces are critical factors that can be used to explain why certain structures give better TFT characteristics than other structures. These interface properties are also closely related to the process conditions. Detailed discussions of these factors are included in Sections III and IV.

Based on the structures in Fig. 1, many advanced TFTs have been fabricated. For example, a light-blocking layer is required to protect the a-Si:H channel area from light irradiation. If this blocking layer is made of a metal and is connected to the gate, a light-insensitive, dual-gate TFT is formed. The I_{off} of the dual-gate TFT is high because of the lack of a back-channel interface. By inserting a dielectric film into the a-Si:H layer, a vertically redundant TFT can be formed. This new TFT has a low I_{off} and an enhanced I_{on}. In another case, by splitting the gate into a U-shape, a horizontally redundant TFT can be formed. This new transistor occupies about the same space as the non-split-gate TFT but has a higher I_{on}. In addition, the channel length of a TFT can be drastically reduced to close to the a-Si:H layer thickness by forming a vertical structure where the source and drain are formed beneath and above the a-Si:H film, separately. In another design, the a-Si:H TFT can be used as a high-voltage switching device by purposely offsetting the drain region. In summary, the purpose of developing a new TFT structure is to achieve one or more of the following goals: to improve the device performance, to solve some intrinsic device problems, to obtain special functions, to reduce the area occupancy of the transistor, to form a redundant architecture, and to simplify the manufacturing process.

III. MATERIAL ISSUES IN THIN–FILM TRANSISTORS

A complete TFT involves three types of thin-film materials: semiconductors, dielectrics, and conductors. Table II lists common materials used in different device regions and some comments. Many of these materials are similar to those used in VLSI MOS transistors but with very different material properties.

TABLE II Thin-Film Materials for TFTs

Device region	Material(s)	Comments
Semiconductor layer	Hydrogenated amorphous silicon (a-Si:H)	Dangling bonds, traps, photosensitivity, etc.
	Polycrystalline silicon (poly-Si)	Grain boundaries, size, quality, etc.
	Cadmium selenide (CdSe)	Stoichiometry, donors, etc.
	Other semiconductors: Ge, GeSi, Te, Sb, organic, etc.	In the experimental stage
Gate dielectric	SiN_x (for a-Si:H TFTs); Ta_2O_5, Al_2O_3, SiN_xO_y, SiO_2 (for poly-Si TFTs)	One layer or multilayers with a critical interface layer
Passivation layer	SiN_x, SiO_2	Band bending, etc.
Interconnection layer	Al, Al alloys, Cu, refractory metals (Mo, Ti, Ta, W, MoW, etc.)	Conductivity, hillocks formation, etc.
Ohmic contact	Heavily doped n$^+$ region (a-Si:H, μc-Si, poly-Si)	Dopant concentration, efficiency, surface cleanness
Pixel electrode	Indium tin oxide (ITO), tin oxide (SnO_2), etc.	Electric conductivity, light transmittance, etc.
Substrate	Glass, plastics, etc.	Low mobile ion content, temperature and chemical resistant, transparency, etc.

Since there is no method for depositing a single crystalline film on the surface of an amorphous material, it is natural that the deposited semiconductor film has an amorphous or polycrystalline morphology. Currently amorphous silicon (a-Si:H) is the most commonly used semiconductor material because the deposition process is simple [e.g., by plasma-enhanced chemical vapor deposition (PECVD)], low-temperature (e.g., less than 350°C), and easy to scale up (e.g., to a large-area substrate). A-Si:H has the following characteristics: (1) an absence of long-range order, (2) a large number of defects within the band gap, (3) an n-type intrinsic film, and (4) photoconductivity. The high density of states (DOS) comes from defects near conduction and valence band edges (i.e., tail states) and deep within the band gap (i.e., traps). The tail states come from the lack of long-range order of the film. The trap states come from the unsaturated dangling bonds that can be partially passivated by hydrogen atoms. The band gap of a-Si:H is higher that of single-crystal silicon, i.e., 1.7 vs 1.1 eV. All a-Si:H TFTs are n-channel transistors because the as-deposited intrinsic a-Si:H film is n-type. The heavily phosphorus-doped a-Si:H (n^+) film can be deposited by introducing a large amount of phosphine (PH_3) into the deposition chamber. The n^+ film is used as the ohmic contact layer in the source and drain areas. Although a-Si:H can be doped into a p-type semiconductor, e.g., by introducing diborane (B_2H_6) gas into the feed stream, the p-type TFT is difficult to fabricate because of the high defect density near the valence band. The photoconductivity of an a-Si:H film comes from the splitting of electron-hole pairs under the exposure of light, which is also known as the Staebler–Wronski effect. The film's conductivity is increased several orders of magnitude by this effect. The a-Si:H TFT becomes highly leaky when it is illuminated by light. Several methods have been suggested to reduce the photosensitivity of the a-Si:H TFT, such as reducing the a-Si:H thickness, reducing the back-channel interface states, introducing defect centers into the film, and applying a light-blockage layer on the transistor. All of them, except for the addition of defect centers, have been implemented in commercial products.

A microcrystalline silicon (μc-Si) film is composed of small silicon crystals, e.g., smaller than 10 nm, imbibed in a matrix of amorphous silicon. The defect density of μc-Si is not much lower than that of a-Si:H even though small grains exist in the film. In addition, due to the harsh deposition conditions, there is a high interface DOS at the μc-Si/dielectric interface. Therefore, μc-Si is not a good candidate to replace a-Si:H as the semiconductor layer of a TFT. However, the n^+ μc-Si film can have a conductivity more than two orders of magnitude higher than that of the n^+ a-Si:H film, which makes it a desirable ohmic contact layer in the source and drain regions.

Poly-Si contains crystalline silicon grains in various orientations and has an energy gap similar to that of single-crystal silicon, but a large number of defects. These defects are located in two areas: at grain boundaries and within grains. Grain boundaries trap charges and form potential barriers between adjacent grains. They also serve as short paths for current. Most grain boundary defects are due to the existence of unpassivated dangling bonds. The quality of the grain is dependent on the deposition or crystallization condition of the film. The grain's bulk quality is more important than its size. For example, low-quality grains usually contain many dislocations and microtwins, which serve as defect centers. A high-performance poly-Si TFT usually contains large-size, low-defect density silicon grains. In addition to high defect densities, grain boundaries have a high impurity diffusion rate and a high oxidation rate. Therefore, grain boundary quality control is an important issue in poly-Si TFTs.

The electrical requirements of a TFT's gate dielectric are the same as those of a MOS FET's gate dielectric, i.e., low interface states and bulk film charge trapping capability. However, the material properties of these two dielectrics are very different. For example, thermally grown silicon oxide (SiO_2) is used exclusively in the MOS FET. The most popular gate dielectric for an a-Si:H TFT is the slightly nitrogen-rich silicon nitride (SiN_x), which contains a large amount of hydrogen, e.g., more than 25%. These hydrogen atoms passivate dangling silicon and nitrogen bonds in the film as well as at the interface with the a-Si:H film. For large-area a-Si:H TFT array fabrication, a dual-gate dielectric structure is often used. Although various types of dielectrics, such as SiO_x, SiO_xN_y, TaO_x, and AlO_x, have been used as the bulk film, the nonstoichiometric SiN_x always serves as the interface layer because it gives low interface states. As for the poly-Si TFT, the deposited SiO_2 is commonly used as the gate dielectric material. In certain cases, SiN_x has also been used. The material properties of these dielectric films, such as the stoichiometry, density, and hydrogen content, are changed during the subsequent thermal processing steps. For the inverted, staggered, trilayer TFT, material properties of the channel passivation layer can influence the TFT properties, such as the leakage current, through varying charge densities and interface states. The high hydrogen-content SiN_x film can be used as a hydrogen source to passivate grain boundaries of a poly-Si film.

The main material requirement for the interconnect line is the conductivity. Although copper has a very low resistivity, it is rarely used in TFT products because of material and process difficulties. For example, copper has a low adhesion force on dielectric surface, it is easy to corrode by chemicals, and the etch process is difficult to control. Recently, aluminum is getting popular as the gate interconnect material for large-area TFT arrays. However, aluminum has the hillocks formation problem, which can be solved by adding an extra element, such as zirconium, to the film or by using a capping layer. Currently, refractory metals are the most popular interconnection material. A redundant interconnect structure is commonly used for the large-area TFT array for the yield improvement. As the array size is increased and the line width is shrunk, aluminum or copper has to be used. More effective methods of solving the above material problems have to be discovered.

Indium tin oxide (ITO) is exclusively used as the pixel electrode in TFT LCD because it has a high light transmittance in the visible wavelength and a high electrical conductivity. The highly conductive ITO has an In/Sn ratio of 90/10 and is in the polycrystalline form.

Alkaline earth boroaluminosilicate glass is widely used as the substrate for large-area TFT arrays. It has a strain point around 550 to 650°C, which greatly limits the TFT process temperature. Since the a-Si:H TFT is fabricated at a much lower temperature, the substrate has little influence on transistor characteristics. However, for poly-Si TFT fabrication, glass is an important factor to consider. Even for low-temperature processes, the glass can be exposed to a high temperature for a short period of time. Glass properties, such as compaction, stress, impurity release, and surface smoothness, can be changed by the short heat pulses. Recently, there are many efforts on fabricating TFTs on plastic substrates that typically can stand a temperature lower than 150°C. The main material properties to consider for a plastic substrate are light transmittance, thermal expansion coefficient, stress, and process compatibility.

IV. DEVICE PHYSICS AND REQUIREMENTS IN THIN-FILM TRANSISTORS

The energy gap structure of a-Si:H is composed of a continuum of states near the conduction and valence edges, i.e., tail states, as well as near the midgap, i.e., deep states. Figure 2 shows diagrams of the density of states (DOS) of a-Si:H and poly-Si band gaps. For a-Si:H, the tail states are contributed by the restructured silicon bonds. The deep states come from the dangling unsaturated silicon bonds. For poly-Si, the situation is more complicated since it has a heterogeneous structure, i.e., grains and grain boundaries. The grain boundary contains high concentrations of defects including restructured and dangling bonds. Depending on the film formation process, the defect states within a grain may be low or high. Therefore, its DOS varies with

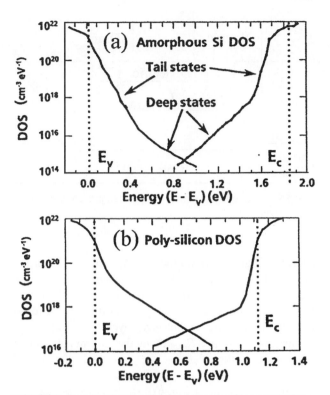

FIGURE 2 Density of states (DOS) within (a) a-Si:H and (b) Poly-Si band gaps. [From Hack, M. (1992). "Proceedings of the First Symposium on Thin Film Transistor Technologies," p. 53, Electrochemical Society, Pennington, NJ. Reproduced by permission of The Electrochemical Society, Inc.]

the location. In general, an effective DOS, which spatially averages all defects at grain boundaries and within grains, is used for poly-Si film. The poly-Si band tail profile is sharper than that of a-Si:H.

Figure 3 shows the band-bending diagram of a TFT when a positive gate voltage is applied and the drain voltage is zero. After traps in the semiconductor layer are filled, an accumulation layer is induced adjacent to the

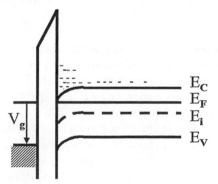

FIGURE 3 Band bending and accumulation layer formation of an n-type TFT at $V_g > 0$ and $V_{ds} = 0$ V.

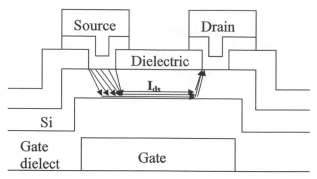

FIGURE 4 Current path in an inverted, staggered trilayer TFT in the on state.

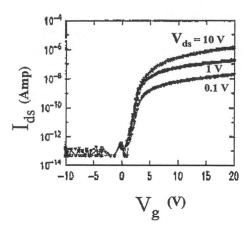

FIGURE 5 Transfer characteristic curves of an inverted, staggered trilayer a-Si:H TFT.

interface. Figure 4 shows the current path in an inverted, staggered TFT in the on state. The current flows from the source contact area to the drain contact area through a very thin region, e.g., less than 200 Å thick, adjacent to the interface of the gate dielectric layer.

The total resistance R_{total} along the current path can be expressed as follows:

$$R_{\text{total}} = R_{\text{source contact}} + R_{\text{Si layer}} + R_{\text{channel interface}}$$
$$+ R_{\text{Si layer}} + R_{\text{drain contact}}, \qquad (1)$$

where $R_{\text{source contact}}$ and R_{drain} contact are the contact resistances at the source and drain areas, R_{si}'s are the resistances of the silicon layer, and $R_{\text{channel interface}}$ is the resistance of the thin accumulation layer near the interface.

The total resistance R_{total} can be dominated by any one of the above resistances. Each resistance is a function of the thin-film material properties (e.g., Si/N ratio in SiN_x, dangling bonds in a-Si:H, interface DOS, and roughness), manufacturing processes (e.g., temperature, feed gas composition, PECVD power density), and transistor structure (e.g., bottom or top gate, bi- or trilayer). Therefore, when device properties of two TFTs are compared, these factors cannot be neglected.

For practical applications, there are five parameters that are commonly used to judge the TFT's performance: on current (I_{on}), off current (I_{off}), field-effect mobility (μ_{eff}), threshold voltage (V_{th}), and subthreshold slope (S). Figure 5 shows the transfer characteristic curves of an inverted, staggered, inverted trilayer a-Si:H TFT. Three of the above characteristics, i.e., I_{on}, I_{off}, and S, can be directly read from this curve.

I_{on} is the maximum drain current (I_{ds}) that the transistor can carry at a specified gate voltage (V_{g}) and drain voltage (V_{ds}). It is a function of the density of deep states, i.e., traps within the band gap. The a-Si:H TFT has an I_{on} several orders of magnitude lower than that of a poly-Si TFT because of the large number of trap states in the

a-Si:H layer. The I_{on} value can be improved by changing the design of the TFT, such as decreasing the channel length or increasing the channel width. The I_{on} of an a-Si:H TFT increases with the increase in temperature, while the I_{on} of a poly-Si TFT is almost independent of temperature.

I_{off} is the transistor's leakage current, which is commonly taken as the lowest I_{ds} in the transfer characteristics curve. For an a-Si:H TFT, I_{off} is a function of the back-channel interface characteristics, such as the fixed positive charge density and the interface. Its value is affected by the deposition condition and film composition of the SiN_x passivation film. A trilayer TFT usually has a lower I_{off} than the bilayer TFT because of the lower back-channel interface states. For a poly-Si TFT, the leakage current is from the Poole–Frenkel mechanism, i.e., holes tunneling out of traps near the high-voltage drain region. The grain boundary quality and grain size distribution are important factors affecting the leakage current. Conventionally, the leakage current can be reduced by passivating traps with hydrogen. Several methods, e.g., offsetting the drain region, lightly doping the drain (LDD) contact structure, and using a multigate structure, have been proved effective in further reducing the I_{off} of a poly-Si TFT. However, these methods complicate the TFT manufacturing process and may even deteriorate certain transistor characteristics.

S is a subthreshold characteristic corresponding to the amount of V_{g} increase required for a one-order of magnitude increase in I_{ds} under the condition $V_{\text{g}} < V_{\text{th}}$. The value of S is dependent on the deep states density. A small S value is desirable for the fast buildup of the accumulation layer.

Figure 6 shows the I_{ds}-vs-V_{ds} curves of an a-Si:H TFT. Both μ_{eff} and V_{th} can be derived from data in this figure using the following equations:

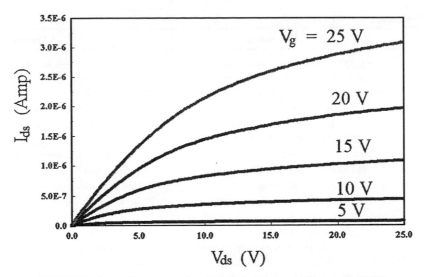

FIGURE 6 I_{ds}-vs-V_{ds} curves of an inverted, staggered trilayer a-Si:H TFT.

for linear-region parameters

$$I_{ds} = \mu_{\text{eff}} \frac{W}{L} C_{\text{ox}} \left(V_g - V_{\text{th}} - \frac{V_{ds}}{2} \right) V_{ds}; \quad (2)$$

for saturation-region parameters,

$$I_{ds} = \frac{1}{2} \mu_{\text{eff}} \left(\frac{W}{L} \right) C_{\text{ox}} (V_g - V_{\text{th}})^2. \quad (3)$$

μ_{eff} is controlled by the tail states and the gate voltage. Depending on the measurement region, μ_{eff} can be the saturation mobility, i.e., measured in the saturation region, or the linear mobility, i.e., measured in the linear region. For the LCD application, the linear mobility is critical because the pixel liquid crystal is operated in this region. However, most mobility values reported in the literature are measured in the saturation region because they are easy to measure (e.g., using a high V_{ds} and a high V_g) and the results are repeatable. V_{th} is the minimum V_g required before I_{ds} starts responding to the change in V_g apparently. It is dependent on the deep states of the film. The V_{th} value should be as low as possible to reduce the power consumption. A poly-Si TFT has a mobility one or two orders of magnitude higher than that of an a-Si:H TFT, e.g., >100 Vs vs. < 1.0 cm^2/Vs, because of its lower tail states. The former also has a lower V_{th} than the latter.

The poly-Si TFT often shows the kink effect in its I_{ds} vs. V_{ds} curves, which is a sudden increase in I_{ds} under the high-V_{ds} condition. The same phenomenon has been observed in the silicon-on-insulator (SOI) MOS transistor. It is caused by the impact ionization in the drain region under a high electric field.

Another important characteristic that can be detected from the I_{ds} vs. V_{ds} curves is the ohmic contact forma-

tion. If the source or drain contact is nonohmic, e.g., a native oxide layer exists between n$^+$ and a-Si:H, these curves will be crowded in the low-V_{ds} range. The contact resistance reduces the effectiveness of the gate voltage. A simple method to estimate the contact resistance is to draw a linear (V_{ds}/I_{ds}) vs. L (channel length) curve. The interception of the line to the y-axis $(L=0)$ is the contact resistance contributed by both source and drain.

Device degradation is another important issue in TFT technology. When a gate voltage is applied to the TFT for a period of time, the transistor characteristics deteriorate. For the a-Si:H TFT, the transfer characteristics curve shifts parallel to the positive direction, with a positive V_g bias. The change in V_{th} (ΔV_{th}) follows the power law of the bias time with a power parameter smaller than 0.5. The magnitude of ΔV_{th} is independent of the value of V_{ds}. The degradation phenomenon is composed of two mechanisms. Under the low-bias voltage condition, it is due to the state creation, e.g., bond breaking, in the a-Si:H layer. Under the high-bias voltage condition, it is due to the charge trapping in the gate dielectric layer. For the poly-Si TFT, the degradation mechanism is different. The magnitude of ΔV_{th} is dependent on the values of both V_g and V_{ds}. The change in the subthreshold slope S is dependent on the drain field strength, i.e., under the low-field condition, S is constant; under the high-field condition, S deteriorates appreciably. This is due to the defect creation in different locations within the band gap. Defect creation in TFT is directly related to the life expectancy of the device. Therefore, for any product applications, the degradation of the TFT, such as the maximum tolerable ΔV_{th}, has to be specified.

TABLE III Common Processes for TFT Fabrication

Process(es)	Function(s)	Comments
PECVD	a-Si:H TFT: gate dielectric, a-Si:H, passivation, n$^+$ ohmic contacts, etc. Poly-Si TFT: a-Si:H and SiO$_x$	Low temperature, large area, high throughput, one pump-down
Sputtering	Deposition of metals and ITO	High throughput
Wet etching	Various thin-film layers including ITO	Simple setup, low equipment cost, possible residue formation
Plasma etching	Etching of critical layers (n$^+$ layer, etc.), slope control (a-Si:H TFT gate, contact vias, etc.)	Flexible and versatile chemistry
Lithography: stepper or scanner	Alignment and exposure	Critical large-area throughput issue
Doping	a-Si:H TFT: PECVD n$^+$ a-Si:H or μc-Si	High throughput
	Poly-Si TFT: ion implantation	Critical large-area throughput issue
Thermal annealing	a-Si:H TFT: repairing of plasma etch damages	Reduction of density of states in films and at interfaces
	Poly-Si TFT: activation of dopant, removal of hydrogen, Si crystallization	
Laser annealing	Poly-Si TFT: Si crystallization, dopant activation	Low temperature, high quality
Plasma hydrogenation	a-Si:H TFT: modification of a-Si:H/gate dielectric interface, precleaning before n$^+$ deposition, etc.	Low temperature, *in situ* process
	Poly-Si TFT: passivaiton of grain boundaries	Low temperature

V. PROCESS ISSUES IN THIN-FILM TRANSISTOR FABRICATION

Basically, TFT manufacturing processes are similar to some VLSI manufacturing processes. Table III lists common processes that are used to fabricate TFTs.

Plasma-enhanced chemical vapor deposition (PECVD) is commonly used to deposit semiconductor and dielectric films. The advantages of the PECVD process are (1) the low deposition temperature (e.g., less than 350°C), (2) the variable film properties (e.g., nonstoichiometric, nitrogen-rich SiN$_x$, hydrogenated a-Si:H), (3) the low interface contamination (e.g., deposition of several layers with one pump-down), (4) the large-area capability (e.g., up to a 1×1-m glass substrate), and (5) the high throughput (e.g., high film deposition rate). All the above characteristics can be prepared with the conventional parallel-plate electrode reactor design. Since the material requirements for TFTs are different from those of MOS transistors, their process windows, which involve feed gases, temperature, pressure, power, etc., are very different.

A μc-Si can easily be prepared by PECVD under conditions similar to those for a-Si:H except for a high hydrogen concentration and a high plasma power density. Both conditions damage the surface of the underneath layer, i.e., creating a high interface DOS for the inverted, staggered TFT structure. Therefore, the μc-Si TFT usually has inferior device characteristics, such as a high V_{th} and low μ_{eff}, compared to the a-Si:H TFT. However, the harsh deposition conditions of μc-Si are effective in removing the thin native oxide on a-Si:H surface. Therefore, the ohmic contact is often formed using an n$^+$ μc-Si film. A μc-

Si film has a dual-layer structure. The bottom thin layer, e.g., less than 200 Å thick, is amorphous and the top layer is composed of microcrystals imbedded in an amorphous matrix. The crystal size and volume fraction increase with the film thickness. Although there are reports that poly-Si film can be prepared by PECVD, the deposition rate is very slow and the film quality is not good enough for TFT application.

There are other applications of the PECVD chamber. When the feed gas is hydrogen, the plasma can remove the native oxide on the source/drain vias before the deposition of an n$^+$ layer. The SiN$_x$/a-Si:H interface of an a-Si:H TFT can be modified with hydrogen, ammonia, or oxygen plasma to alter the transistor characteristics. Hydrogen plasma is also commonly used to passivate the grain boundaries of a poly-Si film, which lowers the TFT's leakage current. The PECVD reactor can be modified, e.g., by heavily biasing the substrate electrode independent of the cathode power source, to form a doping apparatus, i.e., non-mass-separation ion implanter. Although the doping efficiency is low, the process is complicated, and the power consumption is large, this equipment is effective for large area applications.

The magnetron sputtering technique is widely used to prepare all metal and ITO films for the large-area TFT array. Due to the layered structure of a TFT, the step coverage directly affects the yield. Conformal step coverage is necessary for many deposition processes. ITO is commonly deposited by reactive sputtering, i.e., the feed gas includes argon and oxygen. A high-quality ITO film is in the polycrystalline form, which can be prepared by one of the following two methods: deposited in the amorphous form and subsequently annealed to the polycrystalline form or directly deposited into the polycrystalline form. The former method is tedious but the process is easy to control. The latter method is straightforward but the process is complicated and often involves substrate heating. A polycrystalline ITO film sometimes has a two-layer structure that forms a reverse-sloped profile after a wet etching process.

Both wet and dry etching methods are used in large-area TFT array fabrication. Although the wet etching process is simple and the equipment is cheap, dry etching is necessary in certain steps. For example, a wet etching process for the n^+ layer often leaves a conductive residue that shortens the source and drain areas. A dry etching process can avoid this problem. In certain designs, several layers of different materials need to be etched as one material with a desired wall profile. Dry etching can match this requirement. However, there are some unique issues in the dry etching of a TFT. For example, since films used in TFTs have properties, e.g., stoichiometry, composition, and bond structures, different from those used in MOS transistors, their dry etching process results, such as the etch rate and selectivity, can be very different. The plasma-etched TFT can have a high V_{th} and a high leakage current due to the radiation damage mechanism. This is because traps are easily generated in PECVD a-Si:H and SiN_x films by the short-wavelength light source emitted from the plasma. This effect is not obvious in VLSI transistors, in which the films are deposited at higher temperatures by the LPCVD or thermal growth method. Fortunately, these traps can be removed with a thermal annealing step.

In some cases, a film has to be etched with a wet process because of the lack of a dry etching capability. For example, ITO is commonly etched with an aqua regia-type solution. This method creates a large undercut and the solution attacks all kinds of metals, which affects the TFT design rules and the production yield. A dry etching method can avoid these problems. However, it is very difficult to etch ITO with a plasma process because the indium and tin compounds are not volatile. There is a great deal ongoing research on this subject around the world.

A stepper is used as the lithography tool for TFT fabrication. Since the substrate size is large, the aligner has to have a large exposure field for the high throughput and a high resolution to reduce the TFT's occupancy of the pixel territory. A typical stepper has an exposure field of several square inches and a resolution higher than 3 μm. Just as in the VLSI case, the aligner's throughput is a bottleneck for the whole production process. For small-size TFT arrays, such as those used in projection displays or viewfinders, a VLSI stepper can be used since the substrate is small and the resolution needs to be high.

In poly-Si TFT fabrication, the deposition or formation of polycrystalline silicon is a critical step. Due to the restriction of the substrate material, the process temperature has to be low, e.g., below 600°C. It is difficult to deposit poly-Si film directly at such a low temperature. For example, the conventional LPCVD poly-Si deposition process requires a temperature higher than 650°C. The grain size and quality increase with the temperature. Recrystallization, in which the film is deposited in an amorphous form at a low temperature and is subsequently transformed into a polycrystalline form, is the most commonly used method. The amorphous film can be deposited by PECVD or LPCVD. A dehydrogenation step is required before the recrystallization step.

Recrystallization can be carried out with thermal or photon energy. Thermal crystallization is a solid-to-solid transformation process. The low-temperature furnace annealing method may take several days to complete the crystallization process. The crystal quality is generally poor because of the inclusion of large numbers of defects within the grain. Recently, the metal-induced crystallization method has been proved effective in reducing the crystallization time. However, it still requires more than 10 hr to complete the process, which is too slow for mass production. The rapid thermal annealing method has been proved effective in completing the crystallization process within a few minutes. However, the high-temperature duration is too long for a low-temperature glass substrate.

The excimer laser annealing (ELA) method, e.g., using XeCl, XeF, or KrF gas, produces the best-quality poly-Si for TFTs. For example, a high μ_{eff}, e.g., >200 cm^2/Vs, and a low V_{th}, e.g., <2 V, can routinely be achieved with this method. The principle of ELA is to transform the photon energy into heat that is high enough to melt silicon. Poly-Si silicon is formed after the cooling step. Since the wavelength of the laser beam is short, e.g., in the deep-UV range, all the incident energy is absorbed within the top 100 Å of the film. The pulse time is so short that the substrate is not exposed to the high heat during the process. Therefore, a low-temperature glass can be used. The ELA crystallized silicon has large, low-defect grains that are responsible for the good TFT characteristics. However, the laser crystallization process is complicated, e.g., multipath exposure, beam-to-beam overlap, substrate heating, and energy density and duration controls. It is difficult to apply ELA to a large-area substrate with a high throughput. It has been proposed that a-Si:H and poly-Si TFTs can be integrated into the same display, for example, where only the drivers are fabricated with ELA crystallized poly-Si and the pixel area is driven by a-Si:H technology. This would reduce the impact of the low-poly-Si throughput issue. However, other thin-film material properties can become an important issue in this approach because the best gate dielectric for the a-Si:H TFT is often not the best gate dielectric for the poly-Si

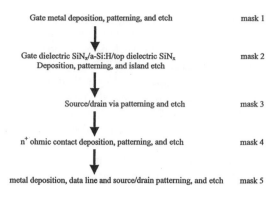

Gate metal deposition, patterning, and etch   mask 1

Gate dielectric SiN$_x$/a-Si:H/top dielectric SiN$_x$ mask 2
Deposition, patterning, and island etch

Source/drain via patterning and etch   mask 3

n$^+$ ohmic contact deposition, patterning, and etch   mask 4

metal deposition, data line and source/drain patterning, and etch   mask 5

For TFT LCD fabrication: additional steps are required for
   ITO deposition, patterning, and etch   additional mask
   Contact vias for attachment of driver IC's   additional mask
   Light block layer   additional mask

Optional simplified processes, such as
     Combination of n$^+$ and source/drain patterns
     Backlight lithography

FIGURE 7 Fabrication process of an inverted, staggered trilayer a-Si:H TFT.

TFT. In general, despite many research efforts spent on poly-Si TFT, it is still an open area requiring further investigation.

Figure 7 shows the flowchart of a simplified fabrication process for an inverted, staggered trilayer a-Si:H TFT. Five masks are required to complete the TFT. There are several key steps that affect the TFT performance, reliability, and yield: gate line slope control, deposition of the trilayer, source/drain alignment, contact resistance, and n$^+$ dry etching.

The step coverage of films deposited on the gate line is directly influenced by the profile of the gate line. A dry etch process can supply a gate line with a proper slope angle. Properties of the two interfaces of the trilayer film influence some major TFT characteristics, such as μ_{eff} and I_{off}. These interface characteristics can be manipulated by the plasma conditions of the PECVD chamber at various stages of the process. The accuracy of the source/drain-to-gate alignment is critical to the function of the TFT. A backlight lithography that uses the gate pattern as the mask guarantees a self-aligned source/drain structure. The overlap of the source/drain with the gate is small, e.g., 0.5 μm. Therefore, it is a simple and reliable process for source/drain definition and has been adapted for many production processes. The contact resistance is dependent on the cleanness of the source/drain via. Ohmic contacts, which requires the removal of the native oxide in the via region, can be obtained with a wet dip or a dry hydrogen method before n$^+$ deposition. n$^+$ residing between the source and the drain, which is often left after the wet etch

step, can cause transistor failure. A dry etching process can avoid this problem. However, a damage-repair step has to be carried out.

For the mass production of TFT arrays, a high throughput is most desirable. The simplest way to increase the throughput is to decrease the number of masks. Conventionally, the inverted, staggered, trilayer TFT requires one more mask to fabricate than the bilayer TFT. However, there are many new designs that can reduce the number of masking steps of the former to equal that of the latter. A complete TFT array for LCD application should include ITO pixels and contact pads, which increases the mask number. So far, the minimum number required to prepare a complete inverted, staggered, trilayer a-Si:H TFT array including ITO and contact patterns is four.

Figure 8 shows a simplified poly-Si TFT fabrication flowchart. The process is similar to that of a MOS transistor except for the silicon crystallization and the dopant activation steps. For small-size TFT array fabrication, quartz is used as the substrate. A high-temperature method can be used. For large-size TFT array fabrication, a low-temperature process, such as laser annealing, has to be used. The hydrogenation step can be done with a conventional or a high-density plasma reactor. It can also be carried out by annealing a high-hydrogen content PECVD SiN$_x$ film deposited on top of the TFT. Currently, there are very few large-area poly-Si TFT manufacturing processes.

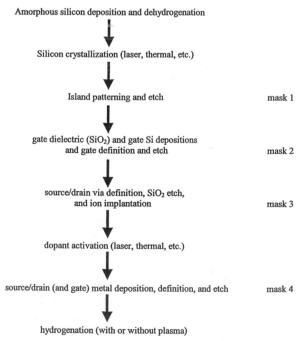

Amorphous silicon deposition and dehydrogenation

Silicon crystallization (laser, thermal, etc.)

Island patterning and etch   mask 1

gate dielectric (SiO$_2$) and gate Si depositions   mask 2
and gate definition and etch

source/drain via definition, SiO$_2$ etch,   mask 3
and ion implantation

dopant activation (laser, thermal, etc.)

source/drain (and gate) metal deposition, definition, and etch   mask 4

hydrogenation (with or without plasma)

FIGURE 8 Fabrication process of a coplanar poly-Si TFT.

The main limitation is the low-temperature crystallization step.

VI. SUMMARY AND TECHNOLOGY TRENDS

Since its invention more than 40 years, TFT has played a minor or negligible role in industries until the successful fabrication of the a-Si:H TFT. Today, TFT LCD alone is a major industry with a predicted high growth rate for the next few decades. TFT applications have been spread out from laptop computers to all major consumer and industry products.

Due to the tremendous amount of research and development activities on a-Si:H and poly-Si TFTs, the basic physics and chemistry of the thin-film materials and the device operation principles are well understood. The advantages and disadvantages of TFTs over other semiconductor devices, such as VLSI MOS transistors, have been thoroughly studied. In the long run, TFT will be a technology that complements rather than competes with VLSIC in applications requiring large-area and low-temperature substrates. Most of these applications cannot be fulfilled with non-TFT technologies.

Although the a-Si:H TFT has the disadvantage of a low field-effect mobility, it is suitable for applications that require a local switching capability but no logic functions. The poly-Si TFT has a major advantage over the a-Si:H TFT in mobility. However, a bottleneck in poly-Si TFT technology is the mass production process, especially in the low-temperature poly-Si formation step. A method for producing a high-quality poly-Si film with a high-throughput capability using a low-temperature substrate is most desirable.

SEE ALSO THE FOLLOWING ARTICLES

BIPOLAR TRANSISTORS • DIAMOND FILMS, ELECTRICAL PROPERTIES • FIELD-EFFECT TRANSISTORS • LIQUID CRYSTAL DEVICES • SUPERLATTICES

BIBLIOGRAPHY

Hack, M., Shaw, J. G., LeComber, P. G., and Willums, M. (1990). "Numerical simulations of amorphous and polycrystalline silicon thin-film transistors," *Jpn. J. Appl. Phys.* **29**, L2360–L2362.

Kuo, Y. (1995). "Plasma etching and deposition for a-Si:H thin film transistors," *J. Electrochem. Soc.* **142**, 2486–2506.

Kuo, Y. (ed.) (1992, 1994, 1996, 1998). "Proceedings of Thin Film Transistor Technologies I, II, III, and IV," Vols. 92-24, 94-35, 96-23, and 98-22, Electrochemical Society, Pennington, NJ.

Migliorato, P. (1992). Thin-film transistors. *In* "Encyclopedia of Physical Science and Technology," 3rd ed., Vol. 16, pp. 749–761, Academic Press, San Diego, CA.

Migliorato, P., and Meakin, D. B. (1987). "Materials properties and characteristics of thin-film silicon transistors for large area electronics," *Appl. Surf. Sci.* **30**, 353–371.

Powell, M. J., van Berkel C., and French, I. D. (1987). "The resolution of a-Si:H thin film transistor instability mechanism," *J. Non-Crystal. Solids* **97/98**, 321–324.

Shur, M. S., Jacunski, M. D., Slade, H. C., Owusu, A. A., Ytterdal, T., and Hack, M. (1996). SPICE models for amorphous silicon and polysilicon thin film transistors. *In* "Proceedings of Thin Film Transistor Technologies III," Vol. 96-23, pp. 242–259, Electrochemical Society, Pennington, NJ.

Street, R. A. (1991). "Hydrogenated Amorphous Silicon," Cambridge University Press, Cambridge.

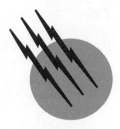

Thunderstorms, Severe

R. Jeffrey Trapp
National Oceanic and Atmospheric Administration

GLOSSARY

Adiabatic process Thermodynamic process in which there is no exchange of heat or mass between a metaphorical parcel of air and its surroundings; thus, responding to the decrease in atmospheric density with height, rising air cools adiabatically due to expansion and sinking air warms due to compression.

Dew-point temperature Temperature at which air becomes saturated (100% relative humidity) when cooled at a constant pressure and constant water-vapor content.

Gust front Leading edge of relatively cool, horizontal airflow (or outflow) that originates from the downdraft region(s) of a thunderstorm.

Hodograph Curve connecting the tips of horizontal wind vectors (in order of increasing height aboveground) that are plotted from a common origin.

Hydrometeor Liquid (rain, drizzle) or solid (hail, graupel/snow pellet, ice pellet/sleet, snow) precipitation particle formed from water vapor in the atmosphere.

Mesocyclone Region of cyclonic (counterclockwise) rotation, about a quasi-vertical axis in a supercell thunderstorm, that provides the background rotation required for tornado formation.

Multicell storm Single thunderstorm comprised of several updrafts of moderate intensity (vertical wind speeds of 20–30 m sec^{-1}), at sequential stages of evolutionary development.

Propagation (thunderstorm) Component of storm movement that deviates from individual updraft movement owing to the discrete formation of subsequent updrafts; each new updraft is on the right or left side (or "flank") of the previous updraft.

Radar, conventional Electronic ranging device that transmits radio signals and then detects returned (scattered and reflected) signals; the return signal strength (normalized for radar range) is proportional primarily to the size, type, and concentration of hydrometeors within the radar sampling volume.

Radar, Doppler Coherent radar that not only measures the strength of the received signal from hydrometeors (as does a conventional radar) but also measures the Doppler shift in transmitted frequency, due to the component of hydrometeor motion toward or away from the radar.

Radar reflectivity factor (Z_e) Product of the number of hydrometeors per cubic meter and the average sixth power of their diameters expressed as millimeters; conventionally presented in units of dBZ, defined as $10 \log_{10} Z_e$. For long-wavelength weather radar, Z_e is a measure of the reflectivity of hydrometeors.

Supercell storm Long-lived thunderstorm with an intense (vertical windspeeds of 30–50 m sec^{-1}), rotating updraft. Most prolific producer of large hail, damaging winds, and tornadoes.

Vorticity Vector measure of local rotation within fluid flow; the vertical component of vorticity (in this chapter, vertical vorticity), which represents rotation about a vertical axis, is an important quantity in the discussion of supercell storms.

Wind shear Local variations of the wind vector over some distance; in this chapter, it generally refers to variation of the horizontal wind with increasing height above the ground.

A THUNDERSTORM—a lightning-producing cumulonimbus or, in some instances, a cumulus congestus cloud—is classified as severe when it has the capability of causing significant damage to life and property on the earth's surface. The U.S. National Weather Service defines a severe thunderstorm as one that produces winds of 25.8 m sec^{-1} (50 knots) or more, hail of 1.9-cm (0.75-in.) diameter or larger, and/or a tornado. Severe thunderstorm types include the multicell storm and the supercell storm. The prominent feature of a supercell is its strong, rotating updraft, which provides the vorticity-rich environment within which tornadoes may form.

I. INTRODUCTORY REMARKS

The basic building blocks of a severe thunderstorm are vertically extensive regions of upward and downward airflow, known as updrafts and downdrafts, respectively. Typically treated as quasi-cylindrical, updrafts and downdrafts in cumulonimbus clouds have diameters that range from a few kilometers to more than 10 km, within which speeds are of the order of 10 m sec^{-1} but may exceed 50 m sec^{-1} in more intense updrafts. In the extreme example of a tornado-bearing storm, an updraft may span the ∼10-km-deep layer from cloud base (∼1 km above the ground) up to and exceeding the level of the tropopause (∼10 to 15 km above the ground); downdrafts generally are shallower.

Airflow within updrafts and downdrafts is forced to a large extent by buoyancy. Strictly, the buoyant motions (or *convection*, in the lexicon of meteorologists) arise from displacements, from a state of equilibrium, of small amounts of air (often termed "air parcels") within a stratified atmosphere. The stratified atmosphere in the neighborhood of the displacements constitutes the *environment* and is assumed to be undisturbed by the displacements. As illustrated in Section III, growth of the displacements into updrafts and downdrafts depends on the characteristics of the environment.

When viewed by weather radar, an individual thunderstorm can be likened to an elementary microscopic organism—a cell—that grows, replicates by division, etc. Thus, adapting terminology from the biological sciences, we have "multicell" storms with multiple updrafts and downdrafts, "splitting" cells in which an initial single updraft metamorphoses into two updraft/downdraft pairs, and even "supercells" (see Section IV). These thunderstorms fall into the class of "severe" when they have the potential to inflict significant damage to life and property. Although severe storms also can be associated with frequent lightning, heavy rain, and flash floods, long-standing concerns—historically by the aviation industry—of hail and high winds motivate the following emphasis on these phenomena, in addition to that on tornadoes.

II. SEVERE-THUNDERSTORM CLIMATOLOGY

A. Severe-Thunderstorm Reports

Large-scale weather systems (spanning hundreds of kilometers and lasting days) are well documented because they affect and are viewed by a large number of official observing stations at the ground. However, severe thunderstorm events are so localized (spanning tens of kilometers) and short-lived (lasting an hour or so) that they frequently do not pass over official observing stations. They may even occur in an area or at a time of day that escapes human detection. To be counted, a severe thunderstorm must be observed, perceived as a severe event, and, most importantly, reported to local weather officials. Nonmeteorological factors such as population density, public interest, and ease of reporting affect the reporting process. There also is a tendency to report the more unusual event when more than one is observed. For example, an observer may not bother to mention that a tornado was accompanied by damaging hail and winds or that strong winds accompanied large hail.

With these caveats in mind, we present the temporal and spatial distributions of (reported) severe thunderstorm events in the contiguous United States (summarized in Table I). One expects a tornado-producing supercell storm also to have damaging hail and winds. However, only 4% of the tornadoes from 1955 through 1983 were accompanied by damaging wind reports, and another

TABLE I Severe Thunderstorm Events Reported in the Contiguous United States[a]

Type	Data period	Approximate annual average	Approximate percentage
Damaging wind[b]	1955–1983	1600	50
Damaging hail	1955–1983	1000	30
Tornado	1950–1978	650	20
Severe thunderstorm	29 years	3250	100

[a] Data sources: Schaefer, J. T., National Severe Storms Forecast Center; Kelly, D. L., Schaefer, J. T., and Doswell, C. A., III (1985). *Monthly Weather Rev.* **113,** 1997–2014.

[b] Includes wind speeds (25.8 m sec^{-1} or higher) deduced from the extent of structural damage.

4% were accompanied by damaging hail reports. By assuming that hail and damaging wind occurred but are not parts of a tornado report and that damaging wind occurred but is not a part of a hail report, we can treat the reports in Table I as representing independent events. Thus, an average of about 3250 severe thunderstorm events per year was reported from the early 1950s through the early 1980s in the contiguous United States. Half of these events are due to damaging winds only. Damaging hail accounts for 30% and tornadoes account for 20% of the reports. Owing to improved reporting procedures (and other nonmeteorological factors identified above) since the early 1970s, the numbers of wind, hail, and tornado reports have been steadily increasing.

B. Seasonal Variations

Severe thunderstorms are most frequent during the spring and summer months in the contiguous United States but may occur any time during the year (Fig. 1). Nationwide, most tornadoes and hailstorms are reported in May and June; most damaging winds reports are received in June and July. Severe thunderstorm occurrence is at a minimum during the winter months.

C. Diurnal Variations

Solar heating of the earth's surface and subsequent heating of the air near the ground play a major role in the formation of convective storms. Although the maximum solar radiation is received at the ground by approximately noon (all times "local"), the air near the ground continues to warm throughout the afternoon. Consequently, the frequency of severe thunderstorm development increases markedly by early afternoon, reaching a peak in the late afternoon and early evening (2–3 hr before sunset; Fig. 2). Each of the three severe-weather phenomena of damaging wind, large hail, and tornado exhibits such a diurnal variation in frequency of occurrence. Some storms that form during the afternoon continue to be severe a few hours

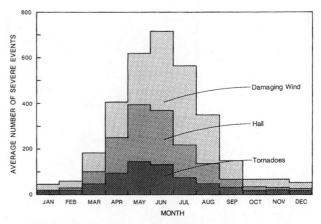

FIGURE 1 Average monthly severe thunderstorm events reported in the contiguous United States during a recent 29-year period. Damaging wind events include assumed 100% occurrences coincident with hail and tornadoes. Likewise, hail events include assumed 100% occurrences coincident with tornadoes. [Data courtesy of J. T. Schaefer, Storm Prediction Center; and from Kelly, D. L., Schaefer, J. T., and Doswell, C. A., III (1985). *Monthly Weather Rev.* **113,** 1997–2014.]

after sunset. Still fewer storms exist throughout the night; obviously, solar heating may not be implicated directly in their overnight sustenance.

D. Geographical Distribution

On average, most U.S. severe thunderstorms occur annually in a north–south-elongated region in the central portion of the country, between the Rocky Mountains and the Mississippi River; a secondary maximum extends east–west across the upper Mississippi Valley (Fig. 3).

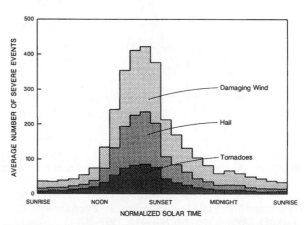

FIGURE 2 Average annual diurnal variation of severe thunderstorm events reported in the contiguous United States during a recent 29-year period. Same occurrence assumptions used as in Fig. 1. Time periods from sunrise to sunset and sunset to sunrise were separately divided into 12 time intervals. [Data courtesy of J. T. Schaefer, Storm Prediction Center; and from Kelly, D. L., Schaefer, J. T., and Doswell, C. A., III (1985). *Monthly Weather Rev.* **113,** 1997–2014.]

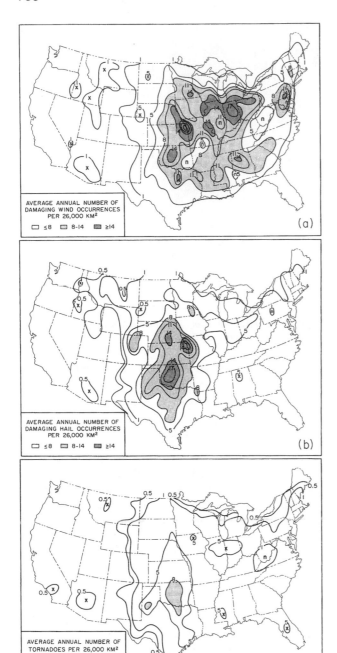

FIGURE 3 Average annual frequency of reported occurrence per 26,000 km² during a recent 29-year period for (a) damaging wind, (b) hail, and (c) tornadoes. Local maxima are indicated by x's; local minima, by n's. [Data courtesy of J. T. Schaefer, Storm Prediction Center; and from Kelly, D. L., Schaefer, J. T., and Doswell, C. A., III (1985). *Monthly Weather Rev.* **113,** 1997–2014.]

A seasonal variation in this distribution is linked essentially to the poleward (equatorward) migration in the mean position of the upper-tropospheric jet stream during Northern Hemisphere summer (winter). Accordingly, the relatively large number of damaging wind reports

(Fig. 3a) in the southeastern United States, for example, are more likely from storms during autumn–early spring, while those in the northern Great Plains and upper Mississippi Valley are more likely from storms during late spring–summer.

Two-thirds to three-quarters of all severe storms move from the southwest toward the northeast. Generally, these are the spring and early–summer severe thunderstorm events in the southern and central Great Plains, steered by the predominately southwesterly flow in the midtroposphere. A significant percentage of the summertime events in the northern Great Plains and upper Mississippi Valley, however, is associated with northwesterly flow and, hence, moves from the northwest toward the southeast.

III. METEOROLOGICAL SETTING FOR SEVERE THUNDERSTORMS

The formation of severe thunderstorms is not random but rather is regulated by the coexistence of the key ingredients of atmospheric water vapor, positive buoyancy of metaphorical "parcels" of air with respect to an ambient state (known as *buoyant instability*), and a mechanism(s) by which such vertical air motions are initiated in the lower troposphere; a "lifting mechanism" describes atmospheric phenomena such as synoptic-scale cold fronts in which horizontal convergence is concentrated at low altitudes. Given these ingredients in appropriate quantities, altitude-dependent increases and directional changes in the environmental horizontal wind (which define the *vertical wind shear*) govern, to a large degree, the physical characteristics and, hence, the "type" and severity of the thunderstorm (see Section IV).

A. Buoyant Instability of the Atmosphere

Consider a parcel of air that by definition rises in the atmosphere without mixing with the surrounding air. Decreasing air density with altitude above the ground dictates that the parcel will expand and cool adiabatically (with no loss/gain of heat to/from surroundings). If the air parcel is not saturated (relative humidity, <100%), it will cool at a rate of 9.8°C per kilometer of ascent; this process is called dry or adiabatic ascent, and this rate, the *dry adiabatic lapse rate*. If the air is saturated (100% relative humidity), it cools during ascent at a slower rate known as the *moist adiabatic lapse rate*: once the relative humidity reaches 100%, any excess water vapor condenses into cloud droplets and the concomitant release of latent heat decreases the rate of adiabatic cooling. Comparing these rates of parcel temperature change with altitude with those of the parcel's immediate environment defines the following concept of buoyant instability.

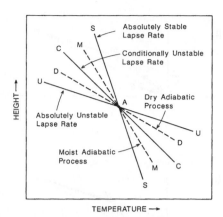

FIGURE 4 Examples of absolutely stable, conditionally unstable, and absolutely unstable environmental lapse rates (solid lines) relative to the dry and moist adiabatic process lapse rates (dashed lines) experienced by parcels of air displaced vertically from point A. Parcels follow the dry (DD) and moist (MM) adiabatic process curves.

When the parcel of air originally located at point A in Fig. 4 is displaced upward along either curves DD or MM (representing dry and moist adiabatic processes, respectively), the air becomes increasingly colder than the surrounding temperatures indicated by the SS curve. By virtue of its negative buoyancy, the parcel descends back to equilibrium point A. When the parcel is displaced downward, it becomes increasingly warmer than profile SS and positive buoyancy returns it to point A. Environmental lapse rate SS is said to be *absolutely stable* because a parcel of air displaced either dry or moist adiabatically returns to its equilibrium level.

Now consider temperature profile UU. When an air parcel at point A is displaced adiabatically upward (downward) along DD or MM, it becomes increasingly warmer (colder) than the environment and continues to rise (sink). Thus environmental lapse rate UU is said to be *absolutely unstable* because a parcel of air continues to ascend or descend once it has been displaced either moist or dry adiabatically from its equilibrium level.

It is possible for an environmental lapse rate to be both stable and unstable, depending on whether or not the air parcel is saturated. Profile CC is said to be *conditionally unstable* because it is stable for dry adiabatic processes but unstable for moist adiabatic processes.

Indices have been developed to express the instability of the environment in quantitative terms. An example is the *lifted index*, the computation of which is illustrated in Fig. 5 for an environment typical of that in which tornadic thunderstorms form. Surface air is lifted dry adiabatically (dashed curve) until it becomes saturated (at the lifted condensation level). Then it is lifted moist adiabatically to the 500-millibar (mbar) pressure level, passing, in this example, the level of free convection, above which the

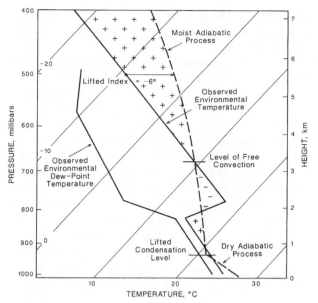

FIGURE 5 A typical tornadic thunderstorm sounding of the atmosphere plotted on a log pressure–skew temperature adiabatic chart. The thick solid lines represent environmental temperature and dew-point temperature measurements. The dashed line is the temperature of a parcel lifted from the earth's surface. Plus and minus signs indicated regions of positive and negative buoyancy experienced by the parcel. Since the parcel is 6°C warmer than the environment at 500 mbar, the lifted index (stability indicator) is defined to be −6°C. [Adapted from Fawbush, E. J., and Miller, R. C. (1953). *Bull. Am. Meteorol. Soc.* **34**, 235–244.]

parcel remains warmer than its environment. The relative warmth at the 500-mbar level is expressed in terms of the difference between the observed 500-mbar temperature and the lifted parcel's temperature: this difference defines the lifted index. In Fig. 5, the lifted index is −6°C, which indicates that the formation of severe thunderstorms is likely.

Another commonly used quantitative measure of environmental buoyant instability is the total positive buoyancy experienced by the parcel from the level of free convection (z_0) upward to the altitude where the parcel is no longer positively buoyant ($z_1 > z_0$). Such *convective available potential energy* (CAPE) may be expressed in a rudimentary form as

$$\text{CAPE} = g \int_{z_0}^{z_1} \frac{T - T'}{T} \, dz, \tag{1}$$

where g is the acceleration due to gravity (9.8 m sec^{-2}), T is the temperature (°C) of the parcel at some altitude z, and T' is the temperature (°C) of the environment at altitude z. CAPE is proportional to the gain in kinetic energy of a buoyant parcel between altitude z_0 and altitude z_1. A nominal value associated with a severe-storm environment is ~1500 m^2 sec^{-2}, but this may be larger or smaller and still be associated with a severe storm, depending on certain characteristics of the environmental wind profile that are discussed below.

The temperature profile in Fig. 5 indicates that the additional presence of a lifting mechanism is required to force an air parcel to rise through regions of negative buoyancy (between ~2 and 3 km in Fig. 5) near the temperature inversion (the layer where the temperature *increases* rather than decreases with height) up to the level of free convection. Indicated in the dewpoint temperature profile in Fig. 5 is the presence of dry air at altitudes above 2 km. This air originates in the elevated arid regions of the southwestern United States and northern Mexico. Evaporation of hydrometeors into the dry air favors the formation of vigorous, cool downdrafts that descend from middle altitudes to the ground. At the leading edge of the cool outflowing air, a gust front becomes the primary lifting mechanism for moist low-altitude air feeding the storm's updraft.

B. Environmental Wind Profile

The salient characteristics of the environmental wind profile are revealed by a hodograph, a curve connecting the tips of the horizontal wind vectors[1], plotted from a common origin, in order of ascending height. Qualitatively, a straight-line hodograph (as in Figs. 6a and b) is characteristic of the environment in which some severe thunderstorms including supercells form; a hodograph with curvature between the ground and the middle heights (as in Fig. 6c) characterizes an environment supportive primarily of supercell storms.

In the idealized, straight-line hodograph LMH in Fig. 6a, southeasterly low-altitude (L) winds turn clockwise and increase in speed with altitude, becoming southwesterly[2] at middle altitudes (M) and west–southwesterly at high altitudes (H) near the tropopause. The arrow **U** in Fig. 6a represents the motion vector for a storm's primary updraft; **U** generally is in the direction of, but at a slower speed than, the mean tropospheric wind. The arrow **S** represents the storm-motion vector, which is the vector sum of **U** and a propagation component **P** (the dashed vector in Figs. 6b and c); as illustrated in Section IV, propagation is due to the formation of a series of severe-thunderstorm updrafts, with each new one forming on the right flank of the previous one. Hence, **S** is the vector essentially of the envelope of the motions of successive updrafts. Plotted relative to the tip of **S**, storm-*relative* winds (Fig. 6b) are often invoked in discussions of the dynamics of thunderstorms, as are, at times, updraft-

---

[1]The vertical component of the environmental wind vector is negligible compared to the horizontal components.

[2]Note that meteorologists define wind direction as the direction from which the wind blows.

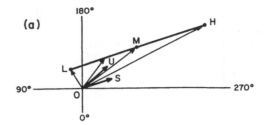

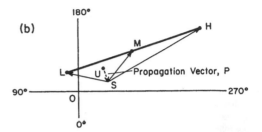

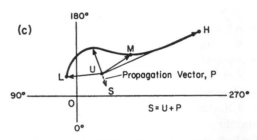

FIGURE 6 Hodographs (thick solid curves) of vertical profiles of the environmental wind constructed by connecting the tips of wind vectors (thin arrows) radiating from a common origin, O. Points L, M, and H on the hodograph indicate low-, middle-, and high-altitude wind vector locations. Thick arrows in (a) indicate updraft (**U**) and storm (**S**) motion vectors. The dashed vector in (b) and (c) is the propagation vector (**P**), representing the relative position of successive updrafts.

relative winds (plotted relative to the tip of **U**; Fig. 6c); ground-relative winds (O) are plotted in Fig. 6a.

Of importance to studies of severe thunderstorms is the vector difference between the wind at different altitudes. *Shear vector* **LM** is the difference between the middle-altitude (**OM**) and the low-altitude (**OL**) wind vectors (Fig. 6a); by definition, the shear vector at a particular height is tangent to the hodograph at that point. The magnitude of **LM** tends to correlate with the storm severity, in an environment with a sufficiently high CAPE. It is preferable, though, to consider the combined contributions of shear and CAPE, the correlation with storm severity of which is quantified through a "bulk" Richardson number,

$$\text{Ri} = \frac{\text{CAPE}}{\frac{1}{2}\bar{V}^2}, \tag{2}$$

where $\bar{V}$ is the difference between the mean, density-weighted, horizontal wind speed over the low- to middle-altitude layer (the lowest ~6 km above the ground) and the mean, horizontal wind speed within the low altitudes (the lowest ~500 m above the ground). As demonstrated in Section IV, supercell thunderstorms tend, for example, to be associated with only a narrow range of values of environmental Ri.

C. Large-Scale Environmental Features

The appropriate environmental wind, temperature, and humidity profiles required for severe-thunderstorm development is frequently found in a favored zone within large-scale (synoptic-scale) extratropical cyclones. These cyclones form in the band of strong, middle-latitude westerly winds and typically move toward the northeast during their mature stage.

An idealization of a middle-latitude cyclone is shown in Fig. 7. At the ground or surface of the earth, the cyclone is characterized by a warm front and cold front emanating from the central low-pressure area (L). The fronts

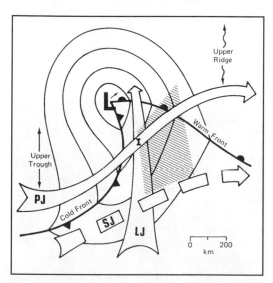

FIGURE 7 A large-scale environmental situation that favors severe-thunderstorm development. Thin curves indicate sea-level isobars (contour lines of constant pressure) surrounding the center of a low-pressure (L) of a midlatitude cyclone that is moving to the northeast (thin arrow). Broad arrows represent the low-altitude jet stream (LJ), upper-altitude polar jet stream (PJ), and still higher subtropical jet stream (SJ). The intersection (I) of LJ and PJ indicates a favored area for severe thunderstorm development owing to vertical shear of the horizontal wind and owing to the abundance of low-altitude moisture provided by the low-altitude jet stream. The hatched region represents the area of anticipated severe thunderstorms during the ensuing 6–12 hr as the large-scale low-pressure system moves to the northeast. [From Barnes, S. L., and Newton, C. W. (1986). In "Thunderstorm Morphology and Dynamics, Vol. 2. Thunderstorms: A Social, Scientific, and Technological Documentary" (E. Kessler, ed.), University of Oklahoma Press, Norman.]

represent the leading edges of warm surface air moving toward the north and northeast and of cold surface air moving toward the south and southeast. The air is warm and moist in the sector between the two fronts, a result of the northward transport of subtropical air by a low-altitude jet (LJ). In the upper troposphere, a polar jet (PJ) marks the southward extent of cold air aloft; a subtropical jet (SJ) sometimes is located south of the polar jet and at a higher altitude.

Severe thunderstorms most likely form initially near region I, where the polar jet overlies the low-altitude jet. Here, vertical shear of the horizontal wind—from south–southeasterly at the surface to very strong southwesterly in the upper troposphere—is extreme, as is the buoyant instability. As the cyclone, with its attendant jets, moves toward the northeast during the afternoon and evening, severe thunderstorms accordingly become favored in the area shaded in Fig. 7.

IV. CHARACTERISTICS OF SEVERE THUNDERSTORMS

A. Single-Cell Thunderstorms

Long-lived severe thunderstorms do not form in an environment in which the wind is essentially uniform with height (hence no vertical wind shear), even though the CAPE may be large. Such an environment, characterized by $Ri \rightarrow \infty$ does, however, allow for the development of single-cell storms. A discussion of the simpler processes associated with this storm type provides a worthwhile introduction to the more complex processes found in most severe thunderstorms.

In the initial stages in the life cycle of a single-cell storm, a cumulus cloud develops vertically, and cloud droplets in the upper regions of its updraft grow into radar-detectable hydrometeors; radar signal or *echo* returned from such a storm is contoured in Fig. 8. After ~15 min of cloud growth, the development of an indentation (termed a weak radar-echo region; WER) in the bottom of the radar reflectivity pattern indicates that the updraft has become fairly strong (at least 15–20 m sec^{-1}).

Within ~20 min, the quantity and size of hydrometeors have increased to the point where the radar reflectivity factor near the top of the updraft exceeds 50 dBZ. The rising cloud top spreads laterally and the updraft weakens upon encountering the buoyantly stable region above the tropopause, which acts much like a rigid lid to the troposphere. Lack of vertical wind shear has a noteworthy effect during this stage of the storm's evolution: larger hydrometeors aloft are not evacuated or transported away from the updraft by environmental winds stronger than

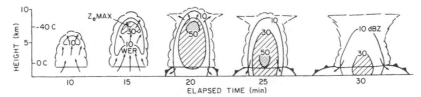

FIGURE 8 Schematic representation of the life cycle of a single-cell thunderstorm. Contours are of the equivalent radar reflectivity factor (Z_e), which is a function of the hydrometeor size, type, and concentration (proportional to rainfall rate) and which is expressed here and in subsequent figures in units of $10 \log_{10} Z_e$ (dBZ). The presence of an updraft is indicated by the weak radar-echo region (WER) indentation in the bottom of the radar reflectivity pattern. The leading edge of the surface cold-air outflow (gust front) is indicated by the bold, barbed line. [Adapted from Chisholm, A. J., and Renick, J. H. (1972). In "Hail Studies Report 72-2," pp. 24–31, Research Council of Alberta, Edmonton.]

those below, allowing the hydrometeors (in the form of rain and perhaps small hail) to fall through the updraft, slowing the updraft, and then effectively replacing it with a downrush of air. Arriving at the surface with the precipitation, such air is colder than its environment because of the evaporation of some raindrops. The associated sudden outrush of cool air produces gusty winds (which may reach severe limits), the leading edge of which forms a gust front beneath the weakening updraft and rain area. As it expands radially outward, the gust front prohibits the inflow, into the base of the updraft, of updraft-sustaining warm, unstable air. The gust front does act as the lifting mechanism for subsequent—but distinctly separate—thunderstorm formations, however. This process is evident in the photograph in Fig. 9, taken by Gemini XII astronauts over the Gulf of Mexico.

The single-cell thunderstorm is no longer active 30–45 min after its initial formation as a small cumulus cloud. Cloud droplets in the lower and middle portions of the storm begin to evaporate in response to unsaturated air that is drawn into the storm behind (above) the mass of descending precipitation. However, as may be noted in Fig. 9, portions of the upper region of the cloud that are comprised of ice crystals may continue to exist for an hour or more until all of the ice crystals sublimate (or undergo a phase change from the solid state to the gaseous state).

B. Multicell Thunderstorms

Experiments with computer-simulated thunderstorms and also observations demonstrate that multicell storms most likely occur when Ri > 30 (Fig. 10), which may be given in an environment, for example, with a 0- to 6-km shear vector magnitude of ~10 m sec^{-1} and a CAPE of ~2000 m^2 sec^{-2}.

At any particular time, the multicell storm is comprised of three or four distinct cells at various stages of development. The newest cell is consistently found on the right side of the "complex" and the oldest cell is on the left.

Resulting storm motion is significantly to the right of individual updraft–downdraft motion (Figs. 6 and 11). While individual cells within the multicell storm are short-lived, the storm as a whole can be long-lived.

The evolution and general characteristics of multicell storms are revealed by horizontal sections of an idealized multicell, plotted in Fig. 11 as a function of time and height. The following discussion focuses on "cell 3" in this multicell storm. Radar reflectivity values greater than 40 dBZ in cell 3 are shaded for emphasis; vertical cross sections in the direction of cell 3's motion are shown at the bottom of the figure.

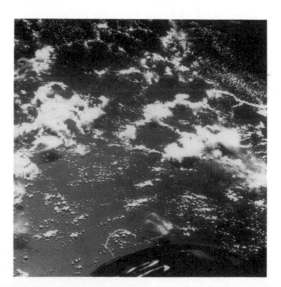

FIGURE 9 Thunderstorm activity over the Gulf of Mexico in November 1966 as photographed from the Gemini XII spacecraft (partially visible in the bottom of the picture) at an altitude of about 225 km. Each cloud-free region surrounded by a (partial) ring of clouds represents the surface pool of colder air left behind by a former thunderstorm that has totally dissipated. Careful inspection of the photograph indicates that, in a few instances, the ice-crystal cirrus clouds from the top of the former thunderstorm remain above the cloud-free region. Note that new thunderstorms are forming where two or more outflow boundaries intersect. [NASA photograph courtesy of NOAA, National Environmental Satellite, Data, and Information Service.]

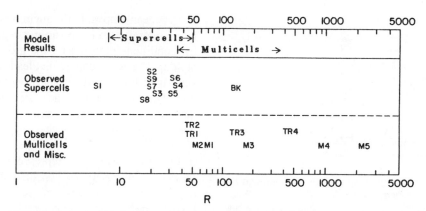

FIGURE 10 Richardson number (defined as Ri in text) of the environment of observed and numerically simulated supercell and multicell thunderstorms. S1–S9 denote observed supercell storms; M1–M9 denote observed multicell storms; TR1–TR9 denote convective storms in tropical environments. [From Weisman, M. L., and Klemp, J. B. (1986). In "Mesoscale Meteorology and Forecasting" (P. Ray, ed.), American Meteorological Society, Boston.]

At time 0 (Fig. 11), radar signals returning from hydrometeors growing in the upper portion of a new updraft (cell 3) start to appear at a height of about 9 km. This new cell is born out of the following processes: Hydrometeors forming in the middle to upper portions of the currently mature updraft (in cell 2) are carried downwind by updraft-relative environmental winds (Fig. 6c) and initially descend ahead of the updraft. As the precipitation falls, it moves toward the updraft's left (relative to its motion vector) side or "flank" in response to the updraft-relative flow. A dome of low-altitude cold air develops in the precipitation area and spreads out behind a gust front that expands beneath cell 2's updraft. As storm-relative environmental air approaches this storm's right flank, it is lifted by the gust front to form the updraft in cell 3, as illustrated in Fig. 12.

During the next 5–7 min, the increase in the radar return with height and the presence of a WER reflect the existence of a moderately strong updraft (20–30 m sec^{-1}) in the developing cell 3 (see also Fig. 12). As the cell attains its maximum height of about 12 km, light precipitation reaches the ground and a gust front begins to form as increasing amounts of downdraft air reach the surface. The gust front expands with increasing rainfall rate, causing (i) the weakening of cell 3's updraft as its lower regions are replaced by cold air and (ii) the formation of a new updraft (cell 4) to the right of cell 3. During this time period, earlier cell 2 is in its final stage of decline.

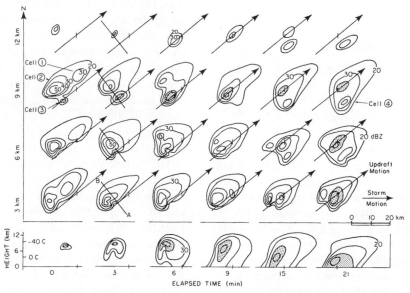

FIGURE 11 Cross sections through an idealized, multicell severe thunderstorm. Arrows through contoured Z_e indicate the direction of cell 3 (shaded) updraft motion and orientation of the vertical cross section at the bottom. Vertical cross section AB is shown in Fig. 12. [Adapted from Chisholm, A. J., and Renick, J. H. (1972). In "Hail Studies Report 72-2," pp. 24–31, Research Council of Alberta, Edmonton.]

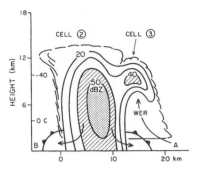

FIGURE 12 Vertical cross section AB (see Fig. 11) through an idealized, multicell severe thunderstorm. Contours are of Z_e. Bold, barbed lines show the gust front from the precipitation area of cell 2. Low-altitude environmental air approaching from the right is forced upward by the gust front to form the new cell 3; the weak echo region (WER) indicates the presence of an updraft. [Adapted from Chisholm, A. J., and Renick, J. H. (1972). In "Hail Studies Report 72-2," pp. 24–31, Research Council of Alberta, Edmonton.]

In terms of storm severity, the existence of weak to moderate shear in this case leads generally to a longer-lived storm, which may enhance the duration of damaging storm-generated surface winds. Moreover, large hailstones may grow from ice-crystal or frozen-raindrop embryos if their trajectories keep them in regions of the storm with large amounts of supercooled (unfrozen at temperatures less than $0°C$) water drops. This occurs, for example, as embryos that develop in a new cell are swept into the updraft of a mature cell. Hail then grows in the mature updraft through accretion of the supercooled water drops until the updraft no longer can suspend the hailstone, the supply of supercooled water is depleted, or the hailstone is ejected from the updraft. Tornadoes in multicell storms are rare.

C. Supercell Thunderstorms

Supercell thunderstorms typically develop in a strongly sheared (e.g., 0- to 6-km shear vector magnitude of ~ 30 m sec^{-1}) environment like those illustrated in Figs. 6 and 7. The thermodynamic environment of supercells is typified by the temperature and humidity profiles in Fig. 5. CAPE values of ~ 2500 m^2 sec^{-2} or larger are common in such an environment and suggest the potential for strong updrafts (30–50 m sec^{-1} or more). The range of combined vertical shear and buoyancy values that support most supercells is relatively small: $10 < Ri < 40$ (see Fig. 10).

The interaction between a strong updraft and strongly sheared environmental winds yields a type of storm that outwardly does not resemble a multicell storm, though the basic physical processes are the same. Unlike the multicell storm, a supercell is distinguished by a strong, long-lived, rotating updraft. Significant rotation (about a vertical axis) at middle altitudes is responsible for the storm's long life (typically several hours) and deviant motion to the right

of the mean tropospheric environmental wind. The rotating updraft also is, in varying degrees, responsible for the conditions required for the growth of large hail, the buildup of electric charge that leads to lightning, and the genesis of long-lived, strong to violent tornadoes. Additionally, strong updrafts lead to strong precipitation-driven downdrafts, which can in turn lead to damaging surface winds.

Visual characteristics of a supercell storm are depicted in Fig. 13, a schematic representation of what may be viewed by an observer a considerable distance southeast of the storm. The presence of a strong updraft is indicated both by the "overshooting" cloud top, which may extend several kilometers above the tropopause and which is visible above the expanding anvil (see Fig. 14), and by the lowering of the cloud base in the form of a "wall cloud" that marks where moist air is converging into the base of the rotating updraft. If a tornado occurs, it forms within the updraft and first becomes visible to the observer when it descends from the rotating wall cloud or perhaps as a near-ground cloud of debris prior to the development of the characteristic funnel of condensed water vapor. The main precipitation area consisting of rain and hail is found ahead (typically northeast) and to the left (typically northwest) of the updraft. Most of the "cloud-to-ground" lightning occurs within the precipitation region, but occasionally a lightning channel will exit the cloud at middle altitudes and descend to the ground in the clear air (producing an especially dangerous situation for unsuspecting humans).

Clouds that form along the storm's gust front grow in height and size as they move toward and merge with the body of the storm; this feature is called a flanking line (Fig. 14). At times, the nonrotating updraft within one of the developing clouds concentrates the vertical component of vorticity (hereafter referred to as vertical vorticity) produced by *horizontal* wind shear across the gust front, and a localized, short-lived vortex (known as a "gustnado") is generated. The presence of a vortex is indicated by a dust whirl on the ground beneath the flanking line. Although capable of inflicting some light damage, this vortex is not to be confused with the destructive tornado that forms within the storm's main rotating updraft.

The mature supercell's updraft (near the intersection of lines AB and CD in Fig. 15) has a dominant influence on the radar reflectivity structure depicted in Figs. 15 and 16. The updraft is so strong that cloud droplets do not have time to grow to radar-detectable hydrometeors until they are halfway to storm top. The resulting bounded weak echo region (BWER) extends to middle altitudes in the storm. The BWER is capped by an area of high reflectivity because the largest hydrometeors form within the upper regions of the updraft. Small hydrometeors are carried downwind by strong environmental flow, forming an anvil-shaped plume. Larger hydrometeors descend

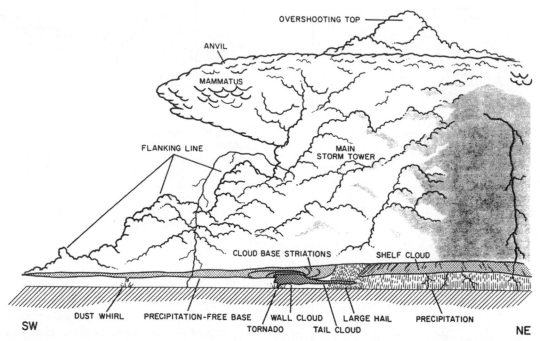

FIGURE 13 Schematic of a supercell thunderstorm as viewed by an observer on the ground, looking toward the northwest. Characteristic features are identified. [Adapted from the original; courtesy of C. A. Doswell III, National Severe Storms Laboratory.]

immediately downwind of the updraft. Responding to the environmental winds as they fall, their trajectories curve cyclonically (counterclockwise) around the updraft, forming a downdraft area on the left and left–forward flanks of the storm. Large hail, whose growth by accretion of supercooled water drops is confined to the supercell's primary updraft, is found adjacent to the updraft along the left and rear storm perimeters. The resultant curvature of the reflectivity pattern—which often takes the shape of a "hook"—

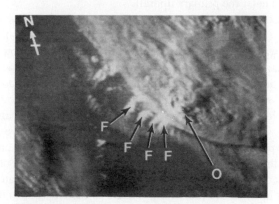

FIGURE 14 GOES-West satellite visual image of the supercell thunderstorm that produced the Wichita Falls, Texas, tornado during the afternoon of April 10, 1979. The satellite viewed the storm from the southwest from a position over the equator at 135°W longitude. The overshooting top (O) and multiple flanking lines (F's) at the rear of the storm are indicated. [Courtesy of NOAA, National Environmental Satellite, Data, and Information Service.]

around the rear of the updraft at low altitudes suggests the presence of cyclonic rotation, readily confirmed with time-lapse photography and Doppler-radar observations.

The region of cyclonically rotating air within the supercell's updraft is called a mesocyclone. As illustrated in Fig. 17, mesocyclones tend to form first at middle altitudes through what may be viewed as a two-step process in the environment characterized by the idealized straight-line hodograph. The first involves the environmental vorticity vector: the largely horizontal vorticity vector (which points to the left of the shear vector in Fig. 6a), due to the vertical shear of the environmental wind, is vertically tilted in horizontal gradients of the vertical air speed comprising the updraft. The result is counterrotating vortices on the left and right flanks of the updraft. The second step involves certain dynamics of the storm and this environment that helps induce a split of the initial updraft (Fig. 17). The new updraft on the right (left) flank of the initial updraft ultimately becomes spatially well correlated with the cyclonically (anticyclonically) rotating air as it tends to move to the right (left) of the mean tropospheric wind vector. The "right-moving" storm with its mesocyclone is the long-lived supercell that may then bear large hail and perhaps a tornado; the "left-moving" storm with its mesoanticyclone may, in some instances, continue to exist for 30 min or more but seldom produces a tornado. "Storm splitting" is unlikely in an environment with a strongly curved hodograph, in which typically only a right-moving storm forms and subsequently acquires net cyclonic rotation

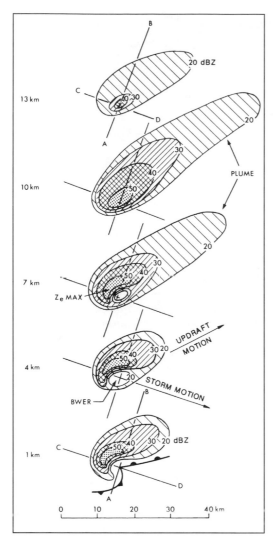

FIGURE 15 Cross sections through an idealized, supercell thunderstorm. Contours are of Z_e. A very strong updraft is indicated by the bounded weak echo region (BWER) in the reflectivity pattern. Surface gust fronts associated with the forward-flank downdraft and the rear-flank downdraft are indicated by the bold line with filled semicircular symbols and triangular symbols, respectively. Vertical cross section CD is shown in Fig. 16. [Adapted from Chisholm, A. J., and Renick, J. H. (1972). In "Hail Studies Report 72-2," pp. 24–31, Research Council of Alberta, Edmonton.]

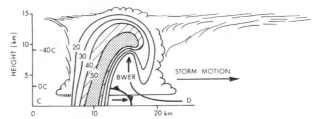

FIGURE 16 Vertical cross section CD (see Fig. 15) through an idealized, supercell thunderstorm. Contours are of Z_e. The presence of a very strong updraft is indicated by the bounded weak echo region (BWER) indentation in the reflectivity profile. The bold, barbed line shows the gust front. Low-altitude environmental air approaching from the right is forced upward by the gust front. [Adapted from Chisholm, A. J., and Renick, J. H. (1972). In "Hail Studies Report 72-2," pp. 24–31, Research Council of Alberta, Edmonton.]

in its updraft; the mesocyclone-development process with a curved hodograph also involves vertical tilting of the environmental vorticity vector.

While the overall circulation of the mesocyclone extends outward 5–10 km from its axis of rotation, peak rotational wind speeds of 20–25 m sec^{-1} typically are found at a radius of 2.5–3 km. During a mesocyclone's organizing stage, such rotational wind speeds, or alternatively vertical vorticity ($\geq 1 \times 10^{-2}$ sec^{-1}), are found primarily at middle altitudes. Some mesocyclones do not develop beyond this stage. Those that do reach "maturity"

are associated with significant vertical vorticity from the ground upward through middle altitudes. The actual process by which such mesocyclone-scale vertical vorticity is generated at lower altitudes is not clear, although observations, computer model simulations, and theory are in general agreement that the process is different than that described above. One explanation requires an elongated region of low-altitude buoyancy contrast between the air cooled by rainfall associated with the "forward-flank" (with respect to the updraft) downdraft and warm, moist, inflow air (Fig. 18); a similar region of buoyancy contrast is found to the rear and left of the updraft. Parcels of air that flow within and along the narrow zone(s) of buoyancy contrast acquire horizontal vorticity generated by virtue of a solenoidal effect (that is, through a circulation that results from the tendency of warm air to rise and cool air to sink). This vorticity is tipped into the vertical, at low altitudes, as the air parcels exit a downdraft and then enter the primary updraft.

Low-altitude, converging airstreams associated with the (i) storm-relative inflow, (ii) outflow driven by the forward-flank downdraft, and (iii) outflow driven by a "rear-flank" downdraft intensify or "stretch" the vertical vorticity of the mature mesocyclone into that of a tornado, at the location marked "T" in Fig. 18. Not all mature mesocyclones bear tornadoes, however, as alluded to in the discussion in Section V. Indeed, a delicate balance between the processes that govern the converging airstreams is one of the necessary conditions for "tornadogenesis." Consider, for example, a rear-flank downdraft and associated outflow that are too strong relative to the environmental inflow. The rear-flank gust front can then advance well into the inflow air and, ultimately, choke off the supply of moist, buoyant air to the updraft. As a consequence, the mesocyclone and the updraft to which it is coupled decay.

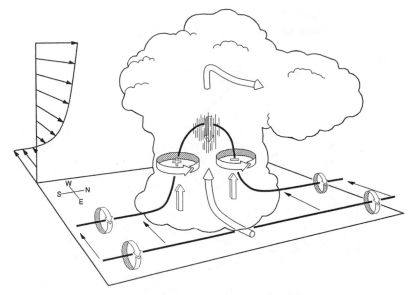

FIGURE 17 Schematic of a thunderstorm growing in an environment characterized by a straight-line hodograph. Cylindrical arrows reveal the flow within the storm. Circular-ribbon arrows depict the sense of rotation of the vortex lines (drawn as thick lines). Note the counterrotating vortices on the north and south sides of the updraft. The remaining arrows show the developing downdraft and updrafts that lead to the eventual storm split. [From Klemp, J. B. (1987). *Annu. Rev. Fluid Mech.* **19,** 369–402. Reprinted, with permission, from *Annual Review of Fluid Mechanics,* Volume 19 ©1987 by Annual Reviews.]

Note that a storm initially may allow for tornado formation, then subsequently evolve in some critical way (or move into a slightly different environment) such that the rear-flank gust front can advance and cause updraft, mesocyclone, and tornado demise. In some instances, lifting of the moist unstable air by the gust front at a location nominally southeast of the initial updraft can lead to a new updraft and storm regeneration (see Fig. 18). Since the new updraft develops generally within the region of "residual" cyclonic vertical vorticity, it rapidly concentrates the vorticity (through convergence) to produce a new mesocyclone and perhaps a new tornado. The sequence of gust-front advance, updraft/mesocyclone decay, and storm regeneration/new mesocyclone development may repeat itself several times such that a single supercell storm may in effect spawn a series of tornadoes.

D. Other Configurations

Severe thunderstorms occur most frequently in the form of multicell and supercell storms and hybrids incorporating features of both. Thus, at times, storm classification is quite arbitrary. The archetypal examples presented above provide a basis for interpreting the more complicated real-life configurations.

Consider the thunderstorms that become organized into long "squall lines." These lines can propagate as an apparent entity for a number of hours or for a few days in extreme cases. Most of the individual elements of a squall line are of the multicell type. However, some cells in the line, particularly at the south end or at breaks in the line (where there is not competition from nearby cells for the moist inflow air required for storm sustenance), may be supercellular. A broad area of light to moderate precipitation frequently is found behind the strong updrafts and downdrafts at the leading edge of the squall line.

The most common type of severe weather associated with storms in this configuration is hail and very strong "straight-line" winds that occur just to the rear of the updrafts. The latter is especially prevalent when a portion of the line accelerates forward, resulting in a convex structure that, when viewed by weather radar, is associated with a bow-shaped echo (or "bow echo"). The occasional genesis of a tornado is favored just north of the apex of such a bow echo (Fig. 19); prior to the bow-echo stage, a squall line may engender tornado formation in other locations, particularly at the southern end of the line.

Another noteworthy configuration is that of a cumulus congestus growing over a narrow zone of low-level vertical vorticity. Likened to a "vortex sheet," the vertical-vorticity zone is generated on a larger scale, for example, by terrain effects and the influence of background or planetary rotation. Stretching of the vertical vorticity in the growing cumulus congestus updraft leads to tornado development in the absence of a bona fide mesocyclone. Such a "non-mesocyclone" tornado tends to be less intense than its supercellular or mesocyclone counterpart. Nevertheless, its parent cloud—which in many regards is a single-cell

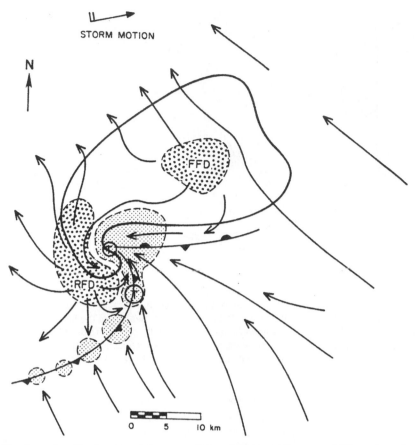

STORM MOTION

N

0   5   10 km

FIGURE 18 Structure of an idealized tornadic supercell thunderstorm at the ground. The radar echo, including the characteristic "hook" appendage, is drawn as a thick line. Arrows are streamlines of horizontal storm-relative airflow. FFD, forward-flank downdraft; RFD, rear-flank downdraft. Positions of surface gust fronts associated with the FFD and the RFD are given by the bold line with filled semicircular symbols and triangular symbols, respectively. Lightly stippled regions represent updrafts. The T within the hook echo indicates the current location of the tornado; the other T indicates the possible location of a subsequent tornado. [From Davies-Jones, R. (1986). In "Thunderstorm Morphology and Dynamics, Vol. 2. Thunderstorms: A Social, Scientific, and Technological Documentary" (E. Kessler, ed.), University of Oklahoma Press, Norman.]

thunderstorm—is certainly a member of the severe-storm family.

V. SEVERE-THUNDERSTORM FORECASTS AND WARNINGS

The temperature, humidity, and wind profiles of the atmosphere, in addition to the large-scale environmental features discussed in Section IV.C, provide clues concerning the likelihood of the occurrence of severe thunderstorms, with attendant lightning, hail, damaging winds, and/or tornadoes. In the United States, highly specialized severe-thunderstorm forecasters at the Storm Prediction Center of the National Weather Service use this and other information to alert the public to the *potential* for damaging and life-threatening storm development. The initial alert comes in the form of an increasingly accurate "watch": the percentage of verifiable severe-weather watches has increased from 63% in 1972 to 90% in 1996.

More localized "warnings" that existing storms are severe or will soon become severe are issued in the United States by individual field offices operated by the National Weather Service and depend on the visual observations of trained (and typically volunteer) storm "spotters" and on the interpretation of weather radar data. The existence of a hook-shaped appendage on the right–rear side of the radar echo at low altitudes (Fig. 15) has been used since the mid-1950s to help identify the tornado's parent mesocyclone. Unfortunately, trajectories of radar-detectable precipitation particles do not always result in hook-shaped echoes, and, many times there are similarly shaped appendages that do not represent rotation. Doppler weather radars, on the other hand, provide a positive and unambiguous identification of a mesocyclone. When a mesocyclone signature is detected from the ground to the middle or upper portions of a storm, there is virtually a 100% probability that the storm is producing (or will soon produce) damaging wind and/or hail. There is, at best, a 50%

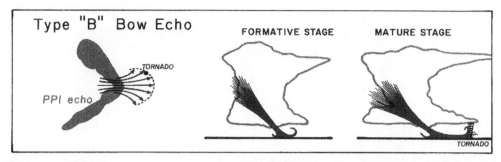

FIGURE 19 Conceptual model of tornado formation within a bow echo. Left panel: Radar echo. Note the tornado position to the north of the apex of the bow. Right panels: A vertical slice through the bow echo, in a location north of the apex. The tornado develops at the leading edge of the outflow. [From Fujita, T. T. (1985). "The Downburst," Satellite and Mesometeorology Research Project (SMRP), Department of Geophysical Sciences, University of Chicago, Chicago.]

probability that the storm will produce a tornado, but by using Doppler-radar measurements one is able to eliminate a number of other storms from the category of possible tornado producers.

As mentioned at the beginning of this article, the severe thunderstorm is such a localized phenomenon that it continues to be a challenge to forecast the expected time and locale of storm formation and, also, the anticipated storm type and severity. The key to improved forecasting is an improved knowledge of basic severe-thunderstorm processes and of the state of the atmosphere. New technological advances hold promise for such improved knowledge: During the coming decade(s), remote sensors on the ground and in space likely will provide the human forecaster as well as the numerical forecast models with more frequent and densely spaced observations of the atmospheric variables associated with storm initiation and subsequent character and severity. Once a storm forms, data from a recently implemented (and eventually upgraded) nationwide network of Doppler radars, which feed and subsequently are augmented by automated storm- and storm-attribute-detection algorithms, will continue to improve the timeliness and accuracy of severe-thunderstorm and tornado warnings.

SEE ALSO THE FOLLOWING ARTICLES

ATMOSPHERIC DIFFUSION MODELING ● ATMOSPHERIC TURBULENCE ● CLIMATOLOGY ● CLOUD PHYSICS ● METEOROLOGY, DYNAMIC (STRATOSPHERE) ● METEOROLOGY, DYNAMIC (TROPOSPHERE) ● RADAR

BIBLIOGRAPHY

Atlas, D. (ed.) (1990). "Radar in Meteorology," American Meteorological Society, Boston.

Bluestein, H. B. (1993). "Synoptic-Dynamic Meteorology in Midlatitudes. Volume II: Observations and Theory of Weather Systems," Oxford University Press, New York.

Cotton, W. R., and Anthes, R. A. (1989). "Storm and Cloud Dynamics," Academic Press, San Diego, CA.

Doviak, R. J., and Zrnic', D. S. (1993). "Doppler Radar and Weather Observations," Academic Press, Orlando, FL.

Emanuel, K. A. (1994). "Atmospheric Convection," Oxford University Press, New York.

Foote, G. B., and Knight, C. A. (eds.) (1977). "Hail: A Review of Hail Science and Hail Suppression," Meteorological Monograph, Vol. 16, No. 38, American Meteorological Society, Boston.

Houze, R. A., Jr. (1993). "Cloud Dynamics," Academic Press, San Diego, CA.

Kessler, E. (ed.) (1983a). "The Thunderstorm in Human Affairs, Vol. 1. Thunderstorms: A Social, Scientific, and Technological Documentary," 2nd ed., University of Oklahoma Press, Norman.

Kessler, E. (ed.) (1983b). "Thunderstorm Morphology and Dynamics, Vol. 2. Thunderstorms: A Social, Scientific, and Technological Documentary," 2nd ed., University of Oklahoma Press, Norman.

Kessler, E. (ed.) (1983c). "Instruments and Techniques for Thunderstorm Observation and Analysis, Vol. 3. Thunderstorms: A Social, Scientific, and Technological Documentary," 2nd ed., University of Oklahoma Press, Norman.

Ray, P. S. (ed.) (1986). "Mesoscale Meteorology and Forecasting," American Meteorological Society, Boston.

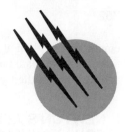

Tidal Power Systems

Ian G Bryden

Robert Gordon University

GLOSSARY

Coriolis force The apparent, or pseudo, force, which acts upon objects moving with respect to the Earth's surface as a result of the Earth's rotation. This tends to deviate motion paths to the right in the Northern Hemisphere and to the left in the Southern Hemisphere.

Ebb tide The state of the tide when the water level is falling.

Flood tide The state of the tide when the water level is rising.

Gigawatt (GW) One thousand megawatts.

Joule (J) Unit of work or energy defined as "the work done by the application of a one Newton force over a distance of one meter."

Megawatt (MW) One million watts.

Megawatt hour (MWhr) Large unit of energy, which is equivalent to the work done in one hour by a power source of one megawatt. Similarly a gigawatt hour is the work done in one hour by a power source of one gigawatt.

Turbulence Random fluctuations in pressure and flow velocity within a liquid.

Watt (W) Unit of power equivalent to one Joule (J) of work being performed every second.

THE TIDES are cyclic variations in the levels of the seas and oceans. Water currents accompany these variations in sea level, which, in some locations such as the Pentland Firth to the North of the Scottish mainland, can be extreme. Small tidal "mills" were used in Southern England, France, and Orkney, which lies to the North of the Scottish mainland, in the Middle Ages. Tidal flows in bays and estuaries offered the potential to drive cereal grinding apparatus in areas that were too low lying to allow the use of conventional water wheels. In the 20th century, the tides were seriously re-examined as potential sources of energy to power industry and commerce. The explanation of the existence of tides represented one of the greatest challenges to early oceanographers, mathematicians, and physicists. It was not until Newton that a satisfying theory emerged to explain at least some of the properties of the tides. He formulated a theory that has become known as the equilibrium theory of tides.

I. NEWTON'S APPROACH: THE EQUILIBRIUM THEORY OF TIDES

The equilibrium theory of tides gives a partial description of tidal behavior for an abstract planet Earth, which is entirely and uniformly covered by water. Consider the Earth–Moon system as shown in Fig. 1.

The Earth–Moon system rotates around a common center of mass (CoM$_s$). The radius of this circulation is given by r. The separation of the center of mass of the Earth (CoM$_e$) from the center of mass of the Moon (CoM$_m$) is given by R. If the Earth were not itself rotating, each point on, or in, the Earth would rotate about its own center of rotation. The radius of the rotation would be r. The period of rotation would be equal to the rotational period of the Earth–Moon system. This will result in acceleration towards the local center of rotation with a value of $\omega^2 r$ (ms^{-2}), where $\omega = 2\pi/T (s^{-1})$ and T is the period of rotation(s).

At the center of the Earth, the centrifugal acceleration exactly matches the gravitational acceleration. At all other points, there is an imbalance between gravitational and centrifugal effects. At point B in Fig. 1, the centrifugal effects exceed the lunar gravitational attraction. In effect, at the surface of the Earth, there will be a net flow of water from C and D to A and B. This effect results in the lunar tidal cycle. The equilibrium theory suggests the establishment of tidal bulges in the fluid surrounding the Earth as shown in Fig. 2.

The Earth rotates, and the two tidal bulges must maintain their positions with respect to the Moon. They, there-

FIGURE 2 The tidal bulge.

fore, have to travel around the Earth at the same rate as the Earth's rotation. The Moon rotates around the CoM$_s$ every 27.3 days in the same direction that the Earth rotates every 24 hours. Because the rotations are in the same direction, the net effect is that the period of the Earth's rotation, with respect to the Earth–Moon system, is 24 hours and 50 minutes. This explains why the tides are approximately an hour later each day.

During the lunar month, which is the rotational period of the Moon around the Earth, there will be variations in the lunar tide influence. The lunar orbit is not circular but is elliptical in form, and the tide-producing forces vary by approximately 40% over the month. Similarly, the Moon does not orbit around the Earth's equator. Instead there are 28° between the equator and the plane of the lunar orbit. This also results in monthly variations.

II. INFLUENCE OF THE SUN ON THE TIDES

The Earth–Sun system is also elliptical but with only a 4% difference between the maximum and minimum distance from the Earth to the Sun. The relative positions of the Earth, Moon, and Sun produce the most noticeable variations in the sizes of the tides. In the configuration shown in Fig. 3, the influences of the Moon and Sun reinforce each other to produce the large tides known as *spring tides* or *long tides*. A similar superposition also exists at the time of a full Moon. When the Sun and Moon are at 90° with respect to each other, the effect is one of cancellation, as

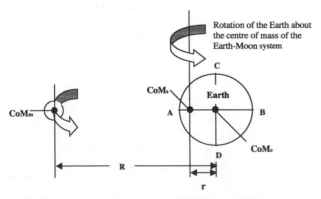

FIGURE 1 Schematic of the Earth–Moon system. CoM$_m$ = center of mass of the Moon; CoM$_e$ = center of mass of the Earth; CoM$_s$ = center of mass of the Earth–Moon system; R = distance between CoM$_m$ and CoM$_e$; and r = distance between CoM$_e$ and CoM$_s$.

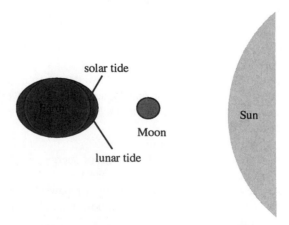

FIGURE 3 Earth, Sun, and Moon during spring tides.

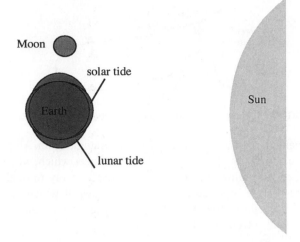

FIGURE 4 Earth, Sun, and Moon during neap tides.

shown in Fig. 4. This configuration results in *neap tides*, also know as *short tides*.

III. THE PRESENCE OF LAND AND THE RESULTING TIDAL DYNAMICS

If the Earth were covered entirely by water of a constant depth, the equilibrium theory of tides would give a perfectly reasonable description of water behavior. Fortunately, the oceans are not all of a constant depth and the presence of continents and islands severely influences the behavior of the oceans under tidal influences. The Coriolis force is a particularly important effect which, in the Northern hemisphere, diverts moving objects to the right and, in the Southern Hemisphere, diverts moving objects to the left. The influence of this force in the presence of land can be considered (as shown in Fig. 5) by visualizing water flowing into and out of a semi-enclosed basin in the Northern hemisphere under the influence of tidal effects. On the way into the channel, the water is diverted to the right towards the lower boundary. When the tidal forcing is reversed, the water is diverted towards

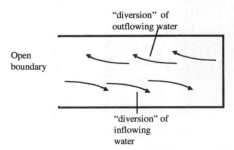

FIGURE 5 Flow of water into and out of a semi-enclosed basin the Northern hemisphere.

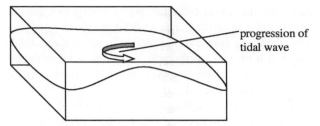

FIGURE 6 Progression of a tidal wave in a basin.

the upper boundary. This results in a substantially higher tidal range at the basin boundaries than at the center. The net result of this effect is to generate a "tidal wave" which processes anticlockwise around a point in the center of the "basin," as shown in Fig. 6.

In effect, the tides represent the terrestial manifestation of the potential and kinetic energy fluxes present in the Earth–Moon–Sun system. These fluxes are complicated by the presence of continents and other landmasses, which modify the form and phase of the tidal wave. As a result, substantially higher local fluxes occur in some regions of the world than in others. The Bay of Fundy in Canada and the Bristol Channel between England and Wales are two particularly noteworth examples of high flux regions.

IV. ENERGY AVAILABLE FROM THE TIDES

It has been estimated that the total energy from the tides which is currently dissipated through friction and drag is equivalent to 3000 GW of thermal energy worldwide. Much of this power is in inaccessible places, but up to 1000 GW are available in relatively shallow coastal regions. Estimates of the achievable worldwide electrical power capability range from about 120 GW of rated capacity to one approaching 400 GW. This is obviously a substantial energy resource, the significance of which has yet to be fully appreciated.

V. HARNESSING THE ENERGY IN THE TIDES

There are two fundamentally different approaches to exploiting tidal energy. The first is to exploit the cyclic rise and fall of the sea level using barrages and the second is to harness local tidal currents in a manner somewhat analogous to wind power.

A. Tidal Barrage Methods

There are many places in the world in which local geography results in particularly large tidal ranges. Sites of particular interest include the Bay of Fundy in Canada, which has a mean tidal range of 10 m; the Severn Estuary between England and Wales, with a mean tidal range of 8 m; and Northern France, with a mean range of 7 m. A tidal barrage power plant has been operating at La Rance (see Banal and Bichon, 1981) in Brittany since 1966. This plant, which is capable of generating 240 MW, incorporates a road crossing of the estuary. Other operational barrage sites are at Annappolis Royal in Nova Scotia (18 MW), The Bay of Kislaya near Murmansk (400 kW), and Jangxia Creek in the East China Sea (500 kW). Schemes for energy recovery have been proposed for the Bay of Fundy and for the Severn Estuary but have never been built.

1. Principles of Operation

The approach is essentially always the same. An estuary or bay with a large natural tidal range is identified and then artificially enclosed with a barrier (Fig. 7). This would typically also provide a road or rail crossing of the gap in order to maximize the economic benefit. The electrical energy is produced by allowing water to flow from one side of the barrage through low head turbines to generate electricity. Suggested modes of operation can be broken down initially into single-basin schemes and multiple-basin schemes. The simplest of these are the single-basin schemes.

2. Single-Basin Tidal Barrage Schemes

These schemes, as the name implies, require a single barrage across the estuary. There are, however, three different

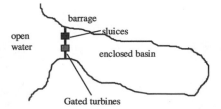

FIGURE 8 Schematic diagram single basin generation scheme.

methods of generating electricity with a single basin. All of the options involve a combination of sluices which, when open, can allow water to flow relatively freely through the barrage and gated turbines, the gates of which can be opened to allow water to flow through the turbines to generate electricity (Fig. 8).

a. Ebb Generation. During the flood tide, incoming water is allowed to flow freely through sluices into the barrage. At high tide, the sluices are closed and water is retained behind the barrage. When the water outside the barrage has fallen sufficiently to establish a substantial head between the basin and the open water, the basin water is allowed to flow out though low-head turbines and to generate electricity. The system may be considered to operate in a series of steps as shown in Fig. 9.

These phases can be represented (Fig. 10) to show the periods of generation associated with stages in the tidal cycle. Typically the water will only be allowed to flow through the turbines once the head is approximately half the tidal range. It is likely that an ebb generation system will be able to generate electricity for, at most, 40% of the tidal cycle.

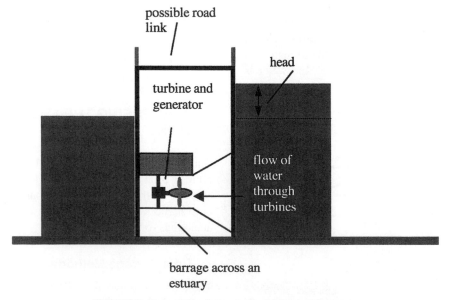

FIGURE 7 Schematic diagram of a tidal power barrage.

Flood Tide- sea water flows through sluices into the basin

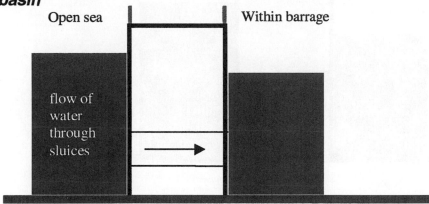

High Tide- sluices are closed to retain water in the basin

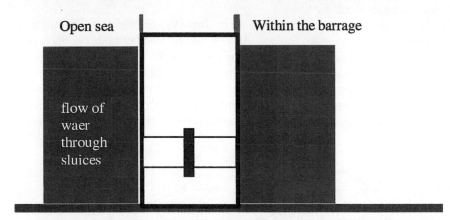

Ebb Tide(a)- water is retained in the basin to allow a useful head to develop

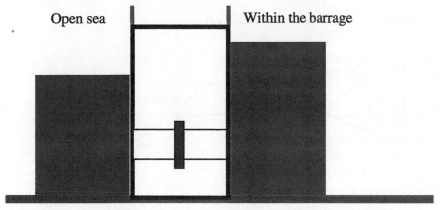

FIGURE 9 Operational steps in an ebb-generation barrage scheme.

Ebb Tide(b)- sea water flowing through the generators

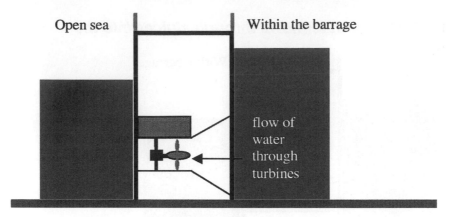

Open sea   Within the barrage

flow of water through turbines

FIGURE 9 *(Continued)*

b. Flood Generation. The sluices and turbine gates are kept closed during the flood tide to allow the water level to build up outside of the barrage. As with ebb generation, once a sufficient head has been established, the turbine gates are opened and water can, in this case, flow into the basin and generate electricity (Fig. 11).

This approach is generally viewed as less favorable than the ebb method because keeping a tidal basin at low tide for extended periods could have detrimental effects on the environment and shipping. In addition, the energy produced would be reduced, as the surface area of a basin would be larger at high tide than at low tide, which would result in rapid reductions in the head during the early stages of the generating cycle.

c. Two-Way Generation. In this mode of operation, use would be made of both the flood and ebb phases of the tide. Near the end of the flood-generation period, the sluices would be opened to allow water to get behind the barrage and would then be closed. When the level on the open-water side of the barrage had dropped sufficiently, water would be released through the turbines in

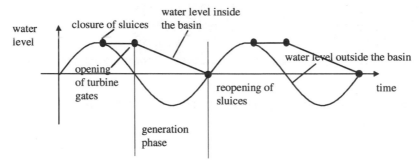

FIGURE 10 Water level inside and outside of an ebb-generation barrage.

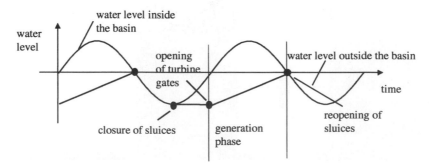

FIGURE 11 Water level inside and outside of a flood-generation barrage.

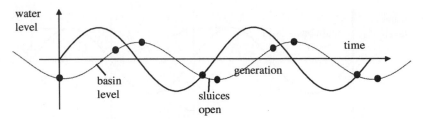

FIGURE 12 Water level inside and outside of a two-way generation barrage.

the ebb generation mode (Fig. 12). Unfortunately, computer models do not indicate that there would be a major increase in energy production. In addition, there would be additional expenses associated in a requirement for either two-way turbines or a double set to handle the two-way flow. Advantages include, however, a reduced period with no generation and a lower peak power, which would allow a reduction in the cost of the generators.

3. Optimal Design of Single-Basin Systems

This is not a simple procedure, as the construction of the barrage will inevitably change the nature of the tidal environment. It is theoretically possible, for example, for the tidal range to be very much less than it was prior to construction of a barrage. If an expensive mistake is to be avoided, it is necessary to use reliable computer models to predict future behavior as much as possible. Such models are available and can be used with a great degree of confidence.

4. Double-Basin Systems

All single-basin systems suffer from the disadvantage that they only deliver energy during part of the tidal cycle and cannot adjust their delivery period to match the requirements of consumers. Double-basin systems have been proposed to allow an element of storage and to give time control over power output levels (Fig. 13). The main basin would behave essentially like an ebb-

generation, single-basin system. A proportion of the electricity generated during the ebb phase would be used to pump water to and from the second basin to ensure that there would always be a generation capability (Fig. 14).

It is anticipated that multiple-basin systems are unlikely to become popular, as the efficiency of low-head turbines is likely to be too low to enable effective economic storage of energy. The overall efficiency of such low head storage, in terms of energy out and energy in, is unlikely to exceed 30%. It is more likely that conventional pump–storage systems will be utilized. The overall efficiencies of these systems can exceed 70%, which, considering that this is a proven technology, is likely to prove more financially attractive.

5. Possible Sites for Tidal Barrage Developments

A considerable number of sites worldwide are technically suitable for development, although whether these resources can be developed economically is yet to be conclusively determined. Some of these are listed in Table I.

6. The Environmental Impact of Basin Systems

The construction of a barrage across a tidal basin will change the environment considerably. An ebb system will reduce the time tidal sands are uncovered. This change will have considerable influences on the lives of wading birds and other creatures. The presence of a barrage

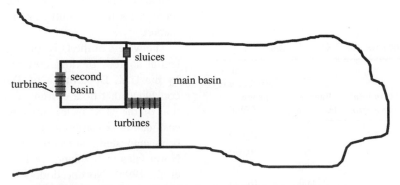

FIGURE 13 Schematic diagram of a double-basin generation system.

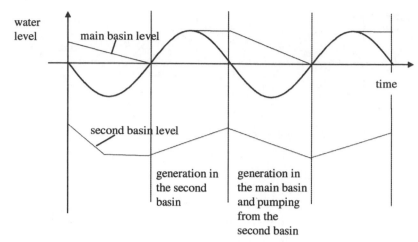

FIGURE 14 Water levels associated with a two-basin generation system.

will also influence maritime traffic, and it will always be necessary to include locks to allow vessels to pass through the barrage. This will be much easier for an ebb system, where the basin is potentially kept at a higher level, than it would be with a flood-generation system, in which the basin would be kept at a lower than natural level. The environmental impact from increasing the area of exposed sediment in an estuarial environment is difficult to quantify but likely to be detrimental (Fig. 15).

7. The Financial Implications of Tidal Barrage Development

Barrage schemes can produce a vast amount of energy. It has been suggested that the Severn Estuary could provide in excess of 8% of the U.K.'s requirement for electrical energy (DOE, 1989); however, all barrage schemes are constrained by the massive engineering operations and associated financial investment required. Construction would take several years. La Rance, for example, took 6 years, and no electricity was generated before the total project was completed. This is a major disincentive for commercial investment.

TABLE I   Resource Estimates for Selected Sites

Site	Mean tidal range (m)	Barrage length (m)	Estimated annual energy production (GWh)
Severn Estuary (U.K.)	7.0	17,000	12,900
Solway Firth (U.K.)	5.5	30,000	10,050
Bay of Fundy (Canada)	11.7	8000	11,700
Gulf of Cambay (India)	6.1	25,000	16,400

B. Tidal Current Systems

The public perception of tidal power is most definitely that of large barrage schemes such as that at La Rance or proposed for the Severn Estuary. In addition to changes in sea-surface level, however, the tides also generate water currents. In the open ocean these currents are typically very small and are measured in centimeters per second, at most. Local geographical effects can, however, result in quite massive local current speeds. In the Pentland Firth to the North of the Scottish mainland, for example, there is evidence of tidal currents exceeding 7 m/s. The kinetic energy in such a flow is considerable. Other sites, in Europe alone, with large currents include the Channel Islands and The Straits of Messina. In addition to major sites such as the Pentland Firth, numerous local sites experience very rapid currents capable of generating electricity with suitable technology.

Tidal currents represent a large and untapped energy resource. It has been estimated in a recent report for the European Commission Directorate General for Energy (CENEX, 1995) that the European Resource could represent a potential for 12.5 GW installed capacity. If even a small fraction of this potential were exploited it could represent a major contribution to the European energy market.

The waters of the U.K. offer particularly attractive potential sources of tidal current power. A study funded by the U.K. Department of Trade and Industry (1993) concluded that tidal currents could supply a major portion of the U.K.'s electricity requirements but that the cost could be prohibitive. At a regional scale it has been suggested that electricity could be generated in the North Isles of Scotland for less than 5 p/kWhr (Bryden et al., 1995). No one doubts the size of the worldwide potential of the tidal current resource but, unlike tidal

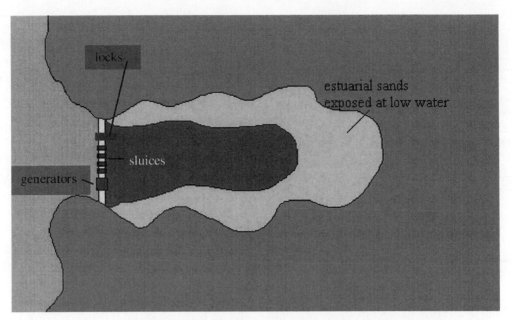

FIGURE 15 Exposure of estuarial sands at low water.

barrage power, there has not yet been a large-scale proto-type study.

1. The Technology for Tidal Current Generation

At the time of writing, there is no commercial generation of electricity from tidal currents anywhere in the world. It is anticipated, however, that the European Union will fund the first such turbine in 2000/2001 under the SeaFlow project. The device is likely to be located in South West England. There are two fundamental concepts for a current converter: vertical axis and horizontal axis turbines that mirror the development of wind turbines. In the case of the vertical axis turbine, the rotational axis of the system is perpendicular to the direction of water flow (Fig. 16). A horizontal axis turbine has the traditional form of a "fan"-type system familiar as windmills and wind-energy systems (Fig. 17). In this case, the rotational axis is parallel to the direction of the water flow.

In addition to the turbine type, there is also a question of how to fix the device in position. Is it to be suspended from a floating structure or fixed to the seabed? The floating concept has advantages of mobility and accessibility (Fig. 18). There are, however, possible problems concernig the stability of the surface pontoon and the generator/turbine. The alternative concept of fixing the turbine to the seabed could provide a stable platform, but the construction and installation costs could be very much larger (Fig. 19).

It is likely that, if tidal currents are to be commercially exploited, the generators will have to be mounted in clusters (tide farms?). If this is done, then, as with wind turbines, the devices will have to be sufficiently spread to ensure that turbulence from individual devices does not interfere with others in the cluster (Fig. 20).

Vertical Axis Turbine Concept

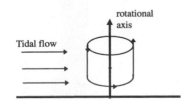

FIGURE 16 Schematic diagram of a vertical axis tidal current turbine.

Horizontal Axis Turbine Concept

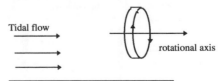

FIGURE 17 Schematic diagram of a horizontal axis tidal current turbine.

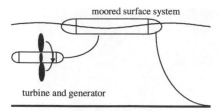

FIGURE 18 Schematic diagram of a horizontal axis tidal current turbine suspended below a floating buoy.

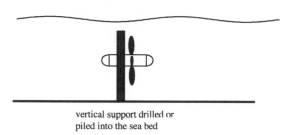

FIGURE 19 Schematic diagram seabed-mounted tidal current turbine.

2. Possible Problems

At present, there is little worldwide experience in harnessing tidal current energy. Experimental turbines have been used in Japan and Scotland but, as yet, they have never been used to generate electricity for consumption. Scottish trials in Loch Linnhe, using a floating horizontal axis system, demonstrated that the technology is feasible, while experience in Japan has been largely with vertical axis systems. Until a large-scale device is commissioned, there will be uncertainties about the long-term reliability and maintenance. Access to machines located in high-energy tidal streams will always be problematic and could prove to require expensive maintenance procedures. Similarly, the rapid tidal currents themselves could provide difficulties in transferring energy via seabed cables to the coastline. Indeed, some studies (ETSU, 1993) have suggested that the cable and its installation could be responsible for more than 20% of the total cost of a tidal current system.

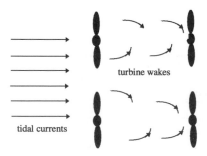

FIGURE 20 Diagram showing interactions between the wakes of tidal current turbines.

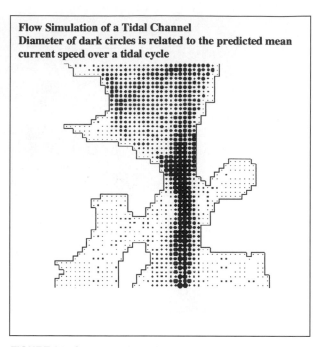

FIGURE 21 Computer simulation of tidal currents in a channel.

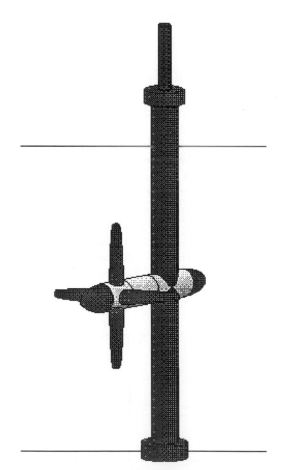

FIGURE 22 Artist's impression of a commercial tidal-current turbine.

Environmental impacts from tidal current systems should be minimal, however. Very energetic tidal channels do not tend to be home to many aquatic species and, although some species of marine mammals use tidal channels in their migration, the slow motion of the turbines and careful design should ensure minimal environmental impact.

The economics of tidal current power will depend upon careful matching of turbine performance to ensure a maximum performance in terms of investment costs. This will include careful consideration of the precise location of a turbine in an energetic tidally active location. As with tidal barrage schemes, it will be necessary to utilize computer models of flow conditions to facilitate effective location of turbines (Fig. 21).

VI. THE FUTURE OF TIDAL POWER

The high capital costs associated with tidal barrage systems are likely to restrict development of this resource in the near future. What developments do proceed in the early 21st century will most likely be associated with road and rail crossings to maximize the economic benefit. In a future in which energy costs are likely to rise, assuming that low-cost nuclear fusion or other long-term alternatives do not make unexpectedly early arrivals, then tidal barrage schemes could prove to be a major provider of strategic energy in the late 21st century and beyond. Under some local conditions, small-scale barrages might also prove attractive. The technology for tidal barrage systems is already available and there is no doubt, given the experience at La Rance, that the resource is substantial and available.

In the near future, it is likely that tidal current systems will appear in experimental form in many places around the world. Initially, these will be rated at around 300 kW

such as that planned under the Sea Flow project for the South of England. If this scheme proves a success, then the first truly commercial development should appear in the first decade of the 21st century and will probably take the form of a multiple array of fixed horizontal axis devices (Fig. 22). Current systems may not have the strategic potential of barrage systems but, in the short term at least, they do offer opportunities for supplying energy in rural coastal and island communities.

SEE ALSO THE FOLLOWING ARTICLES

COASTAL GEOLOGY • GEOTHERMAL POWER STATIONS • MOON (ASTRONOMY) • OCEAN SURFACE PROCESSES • OCEAN THERMAL ENERGY CONVERSION • PHYSICAL OCEANOGRAPHY, THERMAL STRUCTURE AND GENERAL CIRCULATION • SOLAR PHYSICS • SOLAR THERMAL POWER STATIONS • WIND POWER SYSTEMS

BIBLIOGRAPHY

Baker, C. (1991). "Tidal Power," Peter Peregrinus, London.

Banal, M., and Bichon, A. (1981). "Tidal Energy in France—The Rance Tidal Power Station: Some Results After 15 Years in Operation," Proc. of the Second Int. Symp. on Wave and Tidal Energy, Cambridge, U.K.

Bryden, I. G., et al. (1995). "Generating electricity from tidal currents in Orkney and Shetland," *Underwater Technol.* **21**(2), 1995.

CENEX (1995). "Tidal and Marine Currents Energy Exploitation," European Commission, DGXII, JOU2-CT-93-0355.

Charlier, R. H. (1983). "Alternative Energy Sources for the Centralised Generation of Electricity," Adam Hilger, Bristol.

DOE. (1989). "The Severn Barrage Project Summary: General Report," Energy Paper 57, Department of Energy, H. M. Stationery Office.

ETSU. (1993). "Tidal Stream Energy Review," ETSU T/05/00155/REP, U.K. Energy Technology Support Unit, London.

Watson, W. (1992). "Tidal power politics," *Int. J. Ambient Energy* **13**(3), 1–10.

Wilson, E. (1998). "A tide in the affairs of men," *Int. J. Ambient Energy* **9**(3), 115–117.

Tilings

Egon Schulte
Northeastern University

I. Preliminary Concepts
II. Fundamental Concepts of Tilings
III. General Considerations
IV. Plane Tilings
V. Monohedral Tilings
VI. Nonperiodic Tilings

GLOSSARY

Anisohedral tile Shape that admits a monohedral tiling but no isohedral (tile-transitive) tiling.

Aperiodic prototile set Set of prototiles that admits a tiling, but each tiling is nonperiodic.

Archimedean tiling Edge-to-edge tiling of the plane by convex regular polygons whose symmetry group is vertex transitive.

Lattice tiling Tiling by translates of a single tile such that the corresponding translation vectors form a lattice.

Monohedral tiling Tiling in which the tiles are congruent to a single prototile.

Normal tiling Tiling in which the tiles are uniformly bounded in size.

Parallelotope Convex polytope that is the prototile of a lattice tiling.

Penrose tilings Certain nonperiodic tilings of the plane with an aperiodic prototile set (discovered by Penrose).

Periodic tiling Tiling for which the symmetry group is a crystallographic group (contains translations in d independent directions, where d is the dimension of the ambient space).

Prototiles A minimal collection of shapes such that each tile in the tiling is congruent to a shape in the collection.

Space-filler Convex polytope that is the prototile of a monohedral tiling.

Voronoi region For a point x in a discrete set L, the Voronoi region of x consists of all points in space that are at least as close to x as to any other point in L.

TILINGS (or tessellations) have been investigated since antiquity. Almost all variants of the question *"How can a given space be tiled by copies of one or more shapes?"* have been studied in some form or another. The types of spaces permitted have included Euclidean spaces, hyperbolic spaces, spheres, and surfaces, and the types of shapes have varied from simple polygonal or polyhedral shapes to sets with strange topological properties. Today, tiling theory is a rapidly growing field that continues to pose challenging mathematical problems to scientists and mathematicians. The discovery of quasicrystals in 1984 has sparked a surge in interest in tiling theory because of its relevance for mathematical modeling of crystals and quasicrystals.

Encyclopedia of Physical Science and Technology, Third Edition, Volume 16

For more than 200 years, crystallographers such as Fedorov, Voronoi, Schoenflies, and Delone have greatly influenced tiling theory and the study of crystallographic groups. Another important impetus came from the geometry of numbers, notably from Minkowski's work. Hilbert's famous list of fundamental open problems in mathematics, posed in 1900, contained an unsolved problem on discrete groups and their fundamental regions that has had a strong impact on tiling theory. In the 1970s, Grünbaum and Shephard began their comprehensive work on tilings that resulted in a beautiful book on plane tilings. To date, no complete account on tilings in higher dimensional spaces is available in the literature. The recent developments on nonperiodicity of tilings are among the most exciting in tiling theory. The mathematics of aperiodic order is still in its infancy but, when developed, should be useful for understanding quasicrystals and other disordered solid materials.

The purpose of this chapter is to give a short survey on tilings in Euclidean spaces. The discussion is essentially limited to relatively well-behaved tilings and tiles.

I. PRELIMINARY CONCEPTS

It is necessary to establish some preliminary concepts before beginning the subject of tilings itself. Unless stated otherwise, the underlying space of a tiling will always be d-dimensional Euclidean space, or simply d-space, E^d. This is real d-space R^d equipped with the standard inner (scalar) product given by:

$$x \cdot y = \sum_{i=1}^{d} x_i y_i,$$

$$x = (x_1, \ldots, x_d), \quad y = (y_1, \ldots, y_d) \in E^d$$

The length of a vector x in E^d is denoted by $|x|$.

An *orthogonal transformation* of E^d is a linear self-mapping of E^d which preserves the inner product and hence lengths and distances. A *similarity transformation* σ of E^d is a self-mapping of E^d of the form:

$$\sigma(x) = c\lambda(x) + t \qquad (x \in E^d)$$

where λ is an orthogonal transformation of E^d, c a positive scalar (giving the expansion factor), and t a vector in E^d (determining the *translation part* of σ). A (Euclidean) *isometry* of E^d is a similarity transformation with $c = 1$.

Two subsets P and Q of E^d are *congruent* (respectively, *similar*), if there exists an isometry (similarity transformation) of E^d which maps P onto Q.

Let x be a point in E^d, and let $r > 0$. Then the d-dimensional (Euclidean) *ball* $B(x, r)$ of radius r centered at x consists of all points y in E^d whose distance from

x is at most r. The *unit ball* in E^d is the ball of radius 1 centered at the origin.

A subset P of E^d is *open* if every point of P admits a neighborhood (a d-dimensional ball of small positive radius) which is entirely contained in P. The union of any number of open sets is again open. A subset P of E^d is *closed* if its complement $E^d \setminus P$ in E^d is open. The intersection of any number of closed sets is again closed. A subset P of E^d is *compact* if it is closed and bounded (contained in a ball of large radius). For example, every ball $B(x, r)$ is compact.

Let P be a subset of E^d. The *interior* int(P) of P consists of those points, called the *interior points*, of P which admit a neighborhood entirely contained in P; this is the largest open set contained in P. The interior of a ball is also called an *open ball*, in contrast to the ball itself, which is a *closed ball*. The *closure* cl(P) of a subset P is the intersection of all closed subsets that contain P; this is the smallest closed subset which contains P. The *boundary* bd(P) of P is the set int(P)\cl(P), consisting of the *boundary points* of P. For example, the *unit sphere* (centered at the origin) in E^d is the boundary of the unit ball in E^d.

Two open (respectively, closed) subsets P and Q of E^d are *homeomorphic* if there exists a bijective mapping f from P onto Q such that both f and its inverse f^{-1} are continuous; such a mapping is called a *homeomorphism* from P onto Q. A subset P of E^d is called a *topological k-ball* (respectively, *topological k-sphere*) if it is homeomorphic to a Euclidean k-ball (k-sphere). A *topological disc* is a topological 2-ball.

A subset P of E^d is *convex* if, for every two points x and y in P, the line segment joining x and y is contained in P. The intersection of any number of convex sets is again convex.

Let P be a subset of E^d. The *convex hull*, conv(P), of P is the intersection of all convex sets which contain P; it is the smallest convex set which contains P.

A convex *k-polyhedron* P in E^d is the intersection of finitely many closed half-spaces in E^d which is k-dimensional. If $k = d$, then P has nonempty interior in E^d, and vice versa. A convex *k-polytope* is a bounded convex k-polyhedron. A subset of E^d is a convex polytope if and only if it is the convex hull of finitely many points in E^d. Every convex d-polytope in E^d is a topological d-ball. In applications, the tiles of a tiling will often be convex polytopes.

The boundary of a convex d-polytope P in E^d splits into finitely many lower dimensional polytopes called the *faces* of P. These include the empty set (*empty face*) and P itself as *improper* faces, of dimensions -1 and d, respectively. A *proper* face F of P is the (nonempty) intersection of P with a *supporting hyperplane* of P; this is a hyperplane H of E^d that intersects P in F such that P lies entirely in one of the two closed half-spaces bounded by H. A face of dimension 0, 1, i, or $d-1$ is also called a *vertex*, an *edge*, an *i-face*, or a *facet*, respectively. The family of all proper and improper faces of P is called the *face lattice* of P. (When ordered by set-theoretic inclusion, this partially ordered set is a lattice.) The following are two important properties of the face lattice of P: first, a face of a face is again a face (that is, if F is a face of P and G is a face of the polytope F, then G is also a face of P); and, second, the intersection of any two faces of P is again a face of P, possibly the empty face. The *boundary complex* of P, consisting of all proper faces of P (and the empty face), yields a decomposition of the boundary of P, which is a topological sphere.

Two convex d-polytopes P and Q are *combinatorially equivalent* if there exists a mapping f from the face lattice of P onto the face lattice of Q which is one-to-one and inclusion preserving; such a mapping is called a *(combinatorial) isomorphism*. An *automorphism* of P is a self-mapping of the face lattice of P which is an isomorphism. For example, a rectangle is combinatorially equivalent to a square and has the same number of automorphisms as the square.

A *parallelopiped* P is a convex d-polytope that is spanned by d linearly independent vectors $b_1, \ldots, b_d$ of E^d; that is, P is the set of linear combinations $\sum_{i=1}^{d} \lambda_i b_i$ with $0 \le \lambda_i \le 1$ for all $i = 1, \ldots, d$. The simplest example is the standard d-dimensional (unit) *cube* (or *hypercube*), obtained when $b_1, \ldots, b_d$ are the canonical base vectors of E^d. Any parallelopiped is combinatorially equivalent to the cube.

Groups are certain sets Γ with an algebraic structure. We shall not define the term in full generality but restrict ourselves to groups where the elements of the underlying set Γ are one-to-one self-mappings of another set X, such as nonsingular affine transformations of $X = E^d$, isometries of $X = E^d$, or simply bijections of X. A set Γ of this kind is called a *group* if the following two properties are satisfied: first, the composition of any two mappings in Γ is again a mapping in Γ, and, second, for every mapping in Γ, the inverse mapping also belongs to Γ. A group Γ is *finite* if the set Γ has finitely many elements; otherwise, it is *infinite*. The *order* of a finite group Γ is the number of elements of Γ.

A *subgroup* Γ' of a group Γ is a subset of Γ which itself is a group. A *(left) coset* of a subgroup Γ' is a subset of Γ of the form $\sigma\Gamma' := \{\sigma\gamma \mid \gamma \in \Gamma'\}$. The number of cosets of a subgroup Γ' is called the *index* of Γ' in Γ.

Two subgroups Γ' and Γ'' of a group Γ are said to be *conjugate subgroups* of Γ if there exists an element $\sigma \in \Gamma$ such that $\Gamma'' = \sigma\Gamma'\sigma^{-1} (= \{\sigma\gamma\sigma^{-1} \mid \gamma \in \Gamma'\})$.

Two groups Γ_1 and Γ_2 are said to be *isomorphic* if there exists a bijective mapping f from Γ_1 onto Γ_2 such that $f(\gamma\gamma') = f(\gamma)f(\gamma')$ for all $\gamma, \gamma' \in \Gamma_1$. Then f is called an *isomorphism* between Γ_1 and Γ_2.

Let P be a nonempty subset of E^d. A *symmetry* of P is an isometry which maps P onto itself. The set of all symmetries of P forms a group, the *symmetry group* $S(P)$ of P.

The set of all automorphisms of a convex polytope P also forms a group, the *automorphism group* $\Gamma(P)$ of P. Since each symmetry of a polytope maps faces to faces, the symmetry group $S(P)$ is a subgroup of $\Gamma(P)$; in general, $S(P)$ is a proper subgroup of $\Gamma(P)$.

As before, let Γ be a group of self-mappings of a set X. Then Γ is said to *act transitively*, or to be *transitive*, on X if for any two elements x and y in X there exists a mapping σ in Γ such that $\sigma(x) = y$. The *orbit* of an element x in X is the set of all images of x under mappings in Γ. Then Γ acts transitively on X if and only if X is the orbit of any (indeed, every) element x in X. For a given x in X, the subgroup of Γ consisting of the mappings that fix x is called the *stabilizer* of x in Γ.

For example, the *dihedral group* D_n, of order $2n$, is the symmetry group of the convex regular n-gon. The group consists of n rotations and n reflections, and it acts transitively on both the set X_1 of vertices and the set X_2 of edges. The subgroup of D_n consisting of the n rotations is denoted C_n and is a *cyclic* group of order n; that is, all elements are powers of one element, the rotation by $2\pi/n$ in this case. In these examples, as in many other instances described later, the original group Γ is a group of isometries of Euclidean space (in this case, the plane) X, but Γ can also be viewed as a group of self-mappings of the vertex-set X_1 or edge-set X_2 of the regular n-gon.

A subset P of E^d is called *discrete* if each point x in P has an open neighborhood that does not contain any other point of P.

A group Γ of isometries of E^d is called *discrete* (or, more exactly, is said to *act discretely*) if the orbit of every point x in E^d is a discrete subset of E^d. If Γ is a discrete group of isometries of E^d, then a *fundamental region* for Γ is an open set D of E^d satisfying the following two properties: first, the images of D under elements of Γ are mutually disjoint, and, second, the union of the closures of the images of D under Γ is the entire space E^d. Every discrete group of isometries has a fundamental region in E^d (whose boundary is a set of measure zero).

A *lattice* L in E^d is the group of all integral linear combinations of a set of d linearly independent vectors in E^d; these vectors form a *basis* of L. A lattice has many different bases. The lattices in E^d are precisely the discrete subgroups of (the additive group) E^d which contain d linearly independent vectors. By Z^d we denote the standard integer lattice in E^d consisting of all vectors with only integral coordinates.

If z is a point in E^d, then the *point reflection* in z is the affine mapping that takes each x in E^d to $2z - x$. A subset P of E^d is called *centrally symmetric* if it is invariant under the point reflection in a point z, called a *center of symmetry* of P. A set can have more than one center of symmetry.

II. FUNDAMENTAL CONCEPTS OF TILINGS

A. What Is a Tiling?

A *tiling* (or *tessellation* or *honeycomb*) $\mathcal{T}$ of Euclidean d-space E^d is a countable family of closed subsets T of E^d, the *tiles* of $\mathcal{T}$, which cover E^d without gaps and overlaps. This means that the union of all tiles of $\mathcal{T}$ is the entire space, and that any two distinct tiles do not have interior points in common. To rule out pathological situations we shall assume that the tiles of $\mathcal{T}$ are (closed) topological d-balls. In applications, the tiles will often be convex d-polytopes.

A tiling $\mathcal{T}$ of E^d is called *locally finite* if each point of E^d has a neighborhood that meets only finitely many tiles. Then, in fact, every bounded region of space meets only finitely many tiles. We will always assume that a tiling is locally finite.

A tiling $\mathcal{T}$ of E^d by topological d-balls is called *normal* if its tiles are uniformly bounded (that is, there exist positive real numbers r and R such that each tile contains a Euclidean ball of radius r and is contained in a Euclidean ball of radius R) and the intersection of every pair of tiles is a connected set. The latter condition is trivially satisfied if the tiles are convex d-polytopes. On the other hand, in a tiling by arbitrary topological d-balls, the intersection pattern of any two or more tiles can be rather complicated, so connectedness is a reasonable requirement for normality. Note that a normal tiling is necessarily locally finite. Figure 1 shows an example of a locally finite plane tiling by convex heptagons which is not normal.

We often do not distinguish two tilings $\mathcal{T}_1$ and $\mathcal{T}_2$ of E^d which are obtained from each other by a similarity transformation of E^d; such tilings are said to be *equal* or *the same* (this, of course, is an abuse of the standard notion of equality). In particular, we use the following notation. Two tilings $\mathcal{T}_1$ and $\mathcal{T}_2$ of E^d are called *congruent* (respectively, *similar*) if there is an isometry (similarity transformation) of E^d which maps the tiles of $\mathcal{T}_1$ onto the tiles of $\mathcal{T}_2$.

A central notion is that of a symmetry of a tiling. A Euclidean isometry of E^d is a *symmetry* of a tiling $\mathcal{T}$ if it maps each tile of $\mathcal{T}$ onto a tile of $\mathcal{T}$. The set of all symmetries of $\mathcal{T}$ forms a group, the *symmetry group* $S(\mathcal{T})$ of $\mathcal{T}$.

A *protoset*, or *prototile set*, of a tiling $\mathcal{T}$ of E^d is a minimal subset of tiles of $\mathcal{T}$ such that each tile of $\mathcal{T}$ is

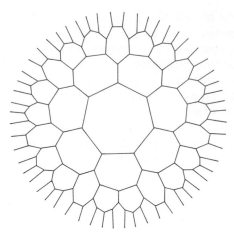

FIGURE 1 A non-normal tiling of the plane by convex heptagons, three meeting at each vertex. The tiles become longer and thinner the farther away they are from the center. [From Grünbaum, B., and Shephard, G. C. (1986). *Tilings and Patterns*, Freeman & Co., San Francisco.]

congruent to one of those in the subset. The tiles in the set are the *prototiles* of $\mathcal{T}$, and the protoset is said to *admit* the tiling $\mathcal{T}$. By abuse of notation, we also use this terminology for shapes and family of shapes that are under consideration for being prototiles or protosets of a tiling, respectively.

Tilings are closely related to packings and coverings of space. A tiling is a family of sets which cover E^d without gaps and overlaps. A *packing* is a family of sets in E^d which do not overlap; that is, no two distinct sets have interior points in common. On the other hand, a *covering* is a family of sets in E^d which leave no gaps; that is, the union of all sets is the entire space E^d. The efficiency of a packing or covering is measured by its *density*, which is a well-defined quantity provided the sets are distributed in a sufficiently regular way. The density of a packing measures the fraction of space that is covered by the sets; the density of a covering measures the average frequency with which a point of space is covered by the sets. Thus, tilings are at the same time packings and coverings of space, each with density 1.

B. Tilings by Polytopes

The best-behaved tilings of E^d are the face-to-face tilings by convex polytopes. More precisely, a tiling $\mathcal{T}$ by convex d-polytopes is said to be *face-to-face* if the intersection of any two tiles is a face of each tile, possibly the empty face. In a face-to-face tiling, the intersection of any number of tiles is a face of each of the tiles.

In a face-to-face tiling $\mathcal{T}$ of E^d by convex d-polytopes, the *i*-faces of the tiles are also called the *i-faces* of $\mathcal{T}$ ($i = 0, \ldots, d$). The *d*-faces of $\mathcal{T}$ are then the tiles of $\mathcal{T}$.

The faces of dimensions 0 and 1 are the *vertices* and *edges* of $\mathcal{T}$, and the empty set and E^d itself are considered to be *improper faces* of $\mathcal{T}$, of dimensions -1 and $d+1$, respectively. The set of all faces of $\mathcal{T}$ is called the *face lattice* of $\mathcal{T}$. (When ordered by set-theoretic inclusion, this is a lattice.) This terminology carries over to more general tilings in which the tiles are topological d-polytopes (homeomorphic images of convex d-polytopes).

Two face-to-face tilings $\mathcal{T}_1$ and $\mathcal{T}_2$ by convex d-polytopes are *combinatorially equivalent* if there exists a mapping from the face lattice of $\mathcal{T}_1$ onto the face lattice of $\mathcal{T}_2$ which is one-to-one and inclusion preserving; such a mapping is called a *(combinatorial) isomorphism*. An *automorphism* of a tiling $\mathcal{T}$ is a self-mapping of the face lattice of $\mathcal{T}$ which is an isomorphism. The set of all automorphisms of $\mathcal{T}$ forms a group, the *automorphism group* $\Gamma(\mathcal{T})$ of $\mathcal{T}$. This group is generally larger than the symmetry group of $\mathcal{T}$.

A mapping between the face lattices of $\mathcal{T}_1$ and $\mathcal{T}_2$ is called a *duality* if it is one-to-one and inclusion reserving. If such a mapping exists, then $\mathcal{T}_2$ is said to be a *dual* of $\mathcal{T}_1$. A tiling can have many metrically distinct duals but any two duals are isomorphic. Except for highly symmetrical tilings, not much is known about the existence and properties of dual tilings.

C. Tilings of the Plane

A lot more is known about tilings in the plane than about tilings in spaces of three or more dimensions. Let $\mathcal{T}$ be a (locally finite) plane tiling in E^2 by topological discs. We now introduce the notion of a ($\mathcal{T}$-induced) vertex or a ($\mathcal{T}$-induced) edge of $\mathcal{T}$. A point x in E^2 is called a *vertex* of $\mathcal{T}$ if it is contained in at least three tiles. Since the tiles are topological discs, each simple closed curve which bounds a tile T is divided into a finite number of closed arcs by the vertices of $\mathcal{T}$, where any two arcs are disjoint except for possibly vertices of $\mathcal{T}$. These arcs are said to be the *edges of the tile* T and are also referred to as *edges of* $\mathcal{T}$. Each vertex x of $\mathcal{T}$ is contained in finitely many edges of $\mathcal{T}$. The number of edges emanating from a vertex x is called the *valence* $v(x)$ of x; clearly, $v(x) \geq 3$ for each vertex x. Note that two tiles can intersect in several edges, where pairs of edges may or may not have vertices in common. Two tiles with a common edge are called *adjacents* of each other.

If a tile of $\mathcal{T}$ is a planar polygon, then the $\mathcal{T}$-induced vertices and edges of the tile will generally not coincide with the original vertices and edges of the tile. To avoid confusion we shall refer to the latter as the *corners* and *sides* of the polygonal tile. In a tiling $\mathcal{T}$ by polygons, the vertices and edges of the tiles may or may not coincide with the corners and sides of the polygons, respectively; if they do, then we call $\mathcal{T}$ an *edge-to-edge* tiling.

In higher dimensions, no general terminology has been introduced that deals with the distinction of an *a priori* facial structure and the $\mathcal{T}$-induced facial structure on the tiles of $\mathcal{T}$.

The notions of isomorphism, automorphism and duality introduced in the previous subsection carry over to general plane tilings, with the understanding that we now have inclusion preserving or inclusion reversing one-to-one mappings between the sets of all ($\mathcal{T}$-induced) vertices, edges, and tiles of the tilings. For every normal plane tiling $\mathcal{T}$ there exists a normal tiling which is dual to $\mathcal{T}$. In the dual, every vertex is an interior point of a tile of $\mathcal{T}$, every tile contains one vertex of $\mathcal{T}$, and any two vertices that correspond to a pair of adjacent tiles in $\mathcal{T}$ are joined by an edge (arc) that crosses the edge common to both tiles.

For normal plane tilings, there is a close relationship between combinatorial and topological equivalence of tilings. Two tilings of E^d are said to be of the *same topological type* or to be *topologically equivalent* if there is a homeomorphism of E^d which maps one onto the other. Two normal plane tilings are topologically equivalent if and only if they are combinatorially equivalent.

D. Monohedral Tilings

A tiling $\mathcal{T}$ of E^d is *monohedral* if all its tiles are congruent to a single set T, the *prototile* of $\mathcal{T}$. The simplest examples of monohedral tilings $\mathcal{T}$ are those in which the tiles are translates of T. If T admits such a tiling, then we say that T *tiles by translation*. In such a tiling $\mathcal{T}$, if the corresponding translation vectors form a d-dimensional lattice L in E^d, then $\mathcal{T}$ is called a *lattice tiling* (*with lattice* L). If $\mathcal{T}$ is a lattice tiling of E^d with convex d-polytopes as tiles, then the prototile T is called a *parallelotope*, or *parallelohedron* if $d = 3$. Every parallelopiped in E^d is a parallelotope, and the vectors that span it also generate the corresponding lattice.

With every lattice in E^d, and indeed with every discrete set L in E^d, is associated a tiling with tiles called *Voronoi regions* or *Dirichlet regions*. Given a point x in L, the Voronoi region $V(L, x)$ of x is the set of all points in E^d that are at least as close to x as to any other point in L; that is,

$$V(L, x) = \{y \in E^d \mid |y - x| \leq |y - z| \text{ for all } z \in L\}$$

The family of all Voronoi regions $V(L, x)$ with $x \in L$ gives a face-to-face tiling $\mathcal{T}$ of E^d by convex polyhedra (polytopes if L is a lattice), called the *Voronoi tiling* for L. Figure 2 shows the Voronoi regions for a finite set of points in the plane, including those regions that are unbounded.

If L is a lattice in E^d, then the Voronoi regions are translates of the Voronoi region $V(L) := V(L, 0)$ obtained

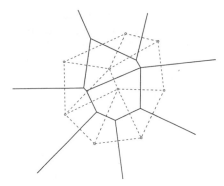

FIGURE 2 The Voronoi regions of a finite set of points in the plane. [From Schattschneider, D., and Senechal, M. (1997). *Handbook of Discrete and Computational Geometry*, (Goodman, J.E. and O'Rourke, J., eds.). CRC Press, Boca Raton.]

for $x = 0$, and $\mathcal{T}$ is a lattice tiling with $V(L)$ as prototile. In particular, the Voronoi regions are convex d-polytopes and $V(L)$ is a parallelotope. The translations by vectors in L all belong to the symmetry group of $\mathcal{T}$. The structure of the Voronoi regions of a lattice depends a great deal on the structure of the lattice itself.

The Voronoi tiling for a lattice L has a dual, called the *Delone tiling* for L, whose vertex set is L and whose tiles are called *Delone cells* (not to be confused with Delone sets defined later). The Delone cell corresponding to a vertex v of a Voronoi tiling is the convex hull of the lattice points x whose Voronoi regions $V(L, x)$ have v as a vertex.

If L is the standard integer lattice Z^d in E^d, then the Voronoi regions and Delone cells are unit cubes, and the Voronoi tiling and Delone tiling are cubical tessellations of E^d that are translates of each other.

A subset L of E^d is called a *Delone set* if there exist two positive real numbers r and R such that every ball of radius r contains at most one point of L, and every ball of radius R contains at least one point of L. The Delone sets are an important class of discrete point sets. Their Voronoi regions are again convex polytopes, but generally there is more than one shape.

There are many other important classes of monohedral tilings that we meet later on. From an artistic perspective, "spiral" tilings of the plane are among the most beautiful; see Fig. 3 for an example.

E. The Symmetry Group of a Tiling

The symmetry properties of a tiling $\mathcal{T}$ of E^d are captured by its symmetry group $S(\mathcal{T})$, which is a discrete group of Euclidean isometries of E^d. The structure of $S(\mathcal{T})$ depends essentially on its *rank* k, the number of independent translations in $S(\mathcal{T})$. Since the tiles in $\mathcal{T}$ are assumed to be topological d-balls, the group cannot contain arbitrarily small translations. In fact, the subgroup of all translations

FIGURE 3 A spiral tiling of the plane. Many other spiral tilings can be constructed with the same prototile. [From Grünbaum, B., and Shephard, G. C. (1986). *Tilings and Patterns*, Freeman & Co., San Francisco.]

in $S(\mathcal{T})$, called the *translation subgroup* of $S(\mathcal{T})$, is in one-to-one correspondence with a k-dimensional lattice in E^d consisting of the corresponding translation vectors; here, $k = 0, 1, \ldots$ or d. A tiling $\mathcal{T}$ in E^d is called *periodic* if $S(\mathcal{T})$ contains translations in d linearly independent directions (that is, $k = d$). For example, all the tilings in Fig. 4 are periodic plane tilings; the translation subgroup is generated by two independent translations. Trivially, each lattice tiling of E^d is periodic. A tiling $\mathcal{T}$ is called *nonperiodic* if $S(\mathcal{T})$ contains no translation other than the identity (that is, $k = 0$). Note that a tiling that is not periodic need not be nonperiodic.

A group G of Euclidean isometries in E^d is called a *crystallographic group* or a *space group* if it is discrete and contains translations in d independent directions. The crystallographic groups are precisely the discrete groups of isometries in E^d whose fundamental region is compact. Two discrete groups are said to be *geometrically isomorphic* if they are conjugate subgroups in the group of all nonsingular affine transformations of E^d. Clearly, geometric isomorphism of discrete groups implies isomorphism as abstract groups. But for crystallographic groups, the converse is also true; that is, two crystallographic groups are geometrically isomorphic if and only if they are isomorphic as abstract groups. If $d = 2$, 3, or 4, the number of (geometric isomorphism) types of crystallographic groups in E^d is 17, 219, or 4783, respectively. If $d \geq 5$, the number of types in E^d is finite but is not explicitly known.

Every infinite orbit of a crystallographic group is a Delone set. The Delone sets that arise from crystallographic groups are also called *regular systems of points*.

For plane tilings $\mathcal{T}$, the three possible choices for the rank k of $S(\mathcal{T})$ lead to well-known classes of plane isometry groups:

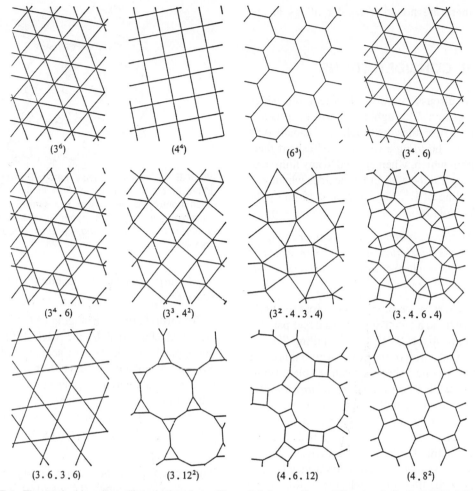

FIGURE 4 The 11 Archimedean tilings in the plane. [From Grünbaum, B., and Shephard, G. C. (1986). *Tilings and Patterns*, Freeman & Co., San Francisco.]

1. If $S(\mathcal{T})$ contains no translation other than the identity (that is, $k = 0$), then $S(\mathcal{T})$ is (isomorphic to) a cyclic group C_n or a dihedral group D_n for some $n \geq 1$. In this case, $\mathcal{T}$ has a *center*; that is, there is at least one point (exactly one if $n \geq 3$) of the plane that is left fixed by every symmetry of $\mathcal{T}$.

2. If $S(\mathcal{T})$ contains only translations in one independent direction, then $S(\mathcal{T})$ is (isomorphic to) one of seven groups called *strip groups* or *frieze groups*.

3. If $S(\mathcal{T})$ contains translations in two independent directions, then $\mathcal{T}$ is periodic and $S(\mathcal{T})$ is (isomorphic to) one of the 17 (*plane*) *crystallographic groups*, also known as *periodic groups* or *wallpaper groups*. (There are different sets of notations for these groups; the most common is that of the International Union of Crystallography, another is the orbifold notation proposed by Conway.)

Two tilings $\mathcal{T}_1$ and $\mathcal{T}_2$ of E^d are said to be of the *same symmetry type* if $S(\mathcal{T}_1)$ and $S(\mathcal{T}_2)$ are *geometrically isomorphic* as discrete groups.

An important problem in tiling theory is the classification of tilings $\mathcal{T}$ with respect to certain transitivity properties of $S(\mathcal{T})$. A tiling $\mathcal{T}$ of E^d is called *isohedral* if $S(\mathcal{T})$ acts transitively on the tiles of $\mathcal{T}$. Clearly, an isohedral tiling is monohedral but the converse is not true. A plane tiling $\mathcal{T}$ is said to be *isogonal* (*isotoxal*) if $S(\mathcal{T})$ acts transitively on the vertices (edges, respectively) of $\mathcal{T}$. Isogonality and isotoxality (and indeed, transitivity on the faces of any dimension) can also be studied more generally for face-to-face tilings $\mathcal{T}$ of E^d by convex d-polytopes.

A face-to-face tiling $\mathcal{T}$ of E^d is called *regular* if $S(\mathcal{T})$ acts transitively on the flags (maximal chains of mutually incident faces) of $\mathcal{T}$. These are the tilings with maximum possible symmetry. In the plane, there are only three regular tilings—namely, the well-known tilings by regular triangles, squares, and hexagons; see Fig. 4, where these tilings occur as (3^6), (4^4), and (6^3), respectively. For each dimension d, we also have the regular tessellation of E^d by d-cubes. Except for two additional exceptional tilings in E^4, there are no other regular tilings in any dimension.

III. GENERAL CONSIDERATIONS

In this section we discuss a number of basic results which hold in any dimension. Throughout, $\mathcal{T}$ will be a locally finite tiling of E^d whose tiles are topological d-balls (that is, discs if $d = 2$). The *tiling problem* in E^d asks if there exists an algorithm which, when applied to any finite protoset S (of topological d-balls) in E^d, decides whether or not S admits a tiling of E^d. The solution is given by the following *Undecidability Theorem*:

Theorem 1. *The tiling problem in the plane or in higher dimensional space is undecidable; that is, there exists no algorithm that decides whether or not an arbitrary finite set of prototiles admits a tiling.*

In constructing tilings we often come across the problem of having to extend a patch of tiles to a larger patch or a tiling of the entire space. By a *patch* in a tiling $\mathcal{T}$ of E^d we mean a finite collection of tiles of $\mathcal{T}$ whose union is a topological d-ball. Figure 5 shows a patch of a plane tiling.

A finite protoset S is said to *tile over arbitrarily large d-balls* in E^d if, for every d-ball B in E^d, the protoset admits a patch such that the union of the tiles in the patch contains B. A patch of tiles which covers a ball will generally not be extendable to a patch which covers a larger ball; that is, in constructing a global tiling it may be necessary to rearrange the tiles after each step. However, the following *Extension Theorem* holds:

Theorem 2. *Let S be any finite set of prototiles in E^d, each of which is a topological d-ball. If S tiles over arbi-*

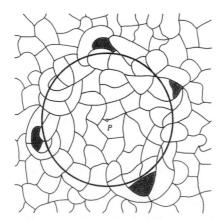

FIGURE 5 A patch of tiles in a plane tiling. The patch consists of the tiles that meet the circular disc, along with the dark-shaded tiles needed to make the arrangement a patch. [From Grünbaum, B., and Shephard, G. C. (1986). *Tilings and Patterns*, Freeman & Co., San Francisco.]

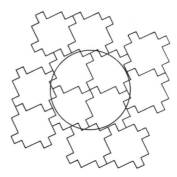

FIGURE 6 A nonextendable patch. Suitable tiles have been added to the 2×2 block of tiles in the center to make this patch nonextendable. There are similar examples based on $n \times n$ blocks for any $n \geq 2$. [From Grünbaum, B., and Shephard, G. C. (1986). *Tilings and Patterns*, Freeman & Co., San Francisco.]

trarily large d-balls, then S admits a tiling of the entire space E^d.

To illustrate that rearranging of tiles may be necessary after each step, consider the patch in Fig. 6, which is the initial patch in an increasing sequence of similar patches. No patch is obtained by extending a preceding patch, yet the extension theorem implies that the given prototile admits a tiling of the entire plane.

An important class of tilings is given by the periodic tilings. For tiles that are convex polytopes, we have the following periodicity result:

Theorem 3. *Let S be a finite set of prototiles in E^d that are convex d-polytopes. If S admits a face-to-face tiling of E^d which possesses a (nontrivial) translational symmetry, then S also admits a tiling that is periodic (and may or may not be the same as the first tiling).*

We now discuss normality of tilings in E^d. Normality has strong implications on the frequency with which tiles are distributed along the boundary of large spherical regions in space. A *spherical patch* $P(x, r)$ in a tiling is the collection of tiles whose intersection with the ball $B(x, r)$ of radius r centered at the point x is nonempty, together with any additional tiles needed to complete the patch (that is, to make the union of its tiles homeomorphic to a topological d-ball). By $t(x, r)$ we denote the number of tiles in $P(x, r)$.

The following *Normality Lemma* implies that a normal tiling cannot have "singularities" at finite points or at infinity:

Theorem 4. *In a normal tiling $\mathcal{T}$ of E^d, the ratio of the number of tiles that meet the boundary of a spherical patch to the number of tiles in the patch itself tends to zero as the radius of the patch tends to infinity. More precisely, if x is a point in E^d and $s > 0$, then:*

$$\lim_{r \to \infty} \frac{t(x, r+s) - t(x, r)}{t(x, r)} = 0$$

In a normal plane tiling, the tiles cannot have too many edges. In fact, as a consequence of the Normality Lemma we have

Theorem 5. *If all the tiles in a normal plane tiling have the same number k of edges, then $k = 3, 4, 5,$ or 6.*

Finally we note the following implication of convexity of the tiles in a tiling:

Theorem 6. *If T is a tiling of E^d with (compact) convex tiles, then necessarily each tile in T is a convex d-polyhedron (d-polytope, respectively).*

IV. PLANE TILINGS

Much of the attraction of plane tilings comes from their appearance in nature and art. They have been widely studied, and much more is known about tilings in the plane than about tilings in spaces of higher dimensions. We begin with some well-known classes of plane tilings; see Section II.C for basic notation.

A. Archimedean Tilings and Laves Tilings

Historically, tilings by (convex) regular polygons were the first kind to be investigated. If the tiles in an edge-to-edge tiling are congruent regular polygons, then the tiling must be one of the three regular tilings in the plane, by triangles, squares, or hexagons.

The Archimedean tilings are edge-to-edge tilings by regular polygons, not necessarily all of the same kind, in which all the vertices are surrounded in the same way by the tiles that contain it. More precisely, an edge-to-edge tiling T by regular polygons is said to be of *type* $(n_1.n_2. \ \ldots \ .n_r)$ if each vertex x of T is of *type* $n_1.n_2. \ \ldots \ .n_r$, meaning that, in a cyclic order, x is surrounded by an n_1-gon, an n_2-gon, and so on. Then, the type $(n_1.n_2. \ \ldots \ .n_r)$ of T is unique up to a cyclic permutation of the numbers n_i, and we will denote T itself by this symbol. The *Archimedian tilings* then are the eleven tilings described by the next theorem and depicted in Fig. 4; they were already enumerated by Kepler.

Theorem 7. *There exist precisely 11 distinct edge-to-edge tilings of the plane by convex regular polygons such that all vertices are of the same type:* (3^6), $(3^4.6)$, $(3^3.4^2)$, $(3^2.4.3.4)$, $(3.4.6.4)$, $(3.6.3.6)$, (3.12^2), (4^4), $(4.6.12)$, (4.8^2), *and* (6^3).

An edge-to-edge tiling T of the plane is called *uniform* if it is isogonal and its tiles are convex regular polygons.

Theorem 8. *The uniform plane tilings are precisely the 11 Archimedean tilings.*

A tiling T by regular polygons is called *equitransitive* if each set of mutually congruent tiles forms one transitivity class with respect to the symmetry group of T. All but one Archimedean tilings are equitransitive; the exception is $(3^4.6)$, which has two congruence classes but three transitivity classes of tiles. There are many further equitransitive tilings of both kinds, edge-to-edge or not edge-to-edge.

The Laves tilings, named after the crystallographer Laves, are monohedral tilings related to the Archimedean tilings by duality. To explain this, let T be a monohedral edge-to-edge tiling in which the tiles are convex r-gons. Call a vertex x of T *regular* if the edges emanating from x dissect a small neighbourhood of x into equiangular parts, the angle being $2\pi/v$ where v is the valence of x. If all vertices of T are regular and the vertices in each tile have valences $v_1, \ldots, v_r$ in T, then we denote T by the symbol $[v_1.v_2. \ \ldots \ .v_r]$. Again, the symbol is unique up to a cyclic permutation of its entries. The *Laves tilings* then are the eleven tilings described by the next theorem and depicted in Fig. 7:

Theorem 9. *(a) If T is a monohedral edge-to-edge tiling of the plane by convex polygons which has only regular vertices, then its symbol is one of the 11 symbols mentioned in part b. (b) To each of the eight symbols* $[3^4.6]$, $[3^2.4.3.4]$, $[3.6.3.6]$, $[3.4.6.4]$, $[3.12^2]$, $[4.6.12]$, $[4.8^2]$, *and* $[6^3]$ *corresponds a unique such tiling T; to each of* $[3^3.4^2]$ *and* $[4^4]$ *corresponds a family of such tilings depending on a single real-valued parameter; and to* $[3^6]$ *corresponds a family of tilings with two such parameters.*

The Laves tiling $[v_1.v_2. \ \ldots \ .v_r]$ and the Archimedean tiling $(v_1.v_2. \ \ldots \ .v_r)$ correspond to each other by duality. The Laves tilings are isohedral; this is analogous to the Archimedean tilings being uniform.

B. Euler's Theorem

The well-known Euler Theorem for convex polytopes in ordinary space E^3 says that, if a convex polytope has v vertices, e edges, and f two-dimensional faces, then:

$$v - e + f = 2$$

In this section we discuss its analogue and relatives for normal tilings T in the plane.

The Euler-type theorem in the plane now involves the two limits:

$$v(T) = \lim_{r \to \infty} \frac{v(x, r)}{t(x, r)} \qquad \text{and} \qquad e(T) = \lim_{r \to \infty} \frac{e(x, r)}{t(x, r)}$$

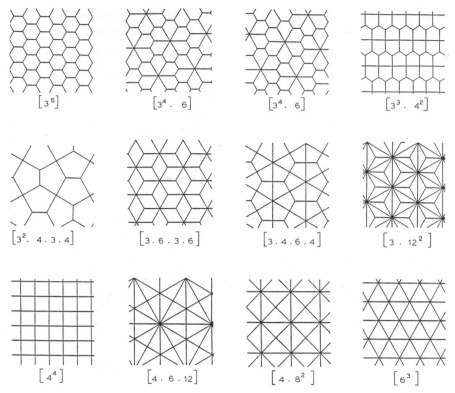

FIGURE 7 The 11 Laves tilings in the plane. [From Grünbaum, B., and Shephard, G. C. (1986). *Tilings and Patterns*, Freeman & Co., San Francisco.]

where x is a point in E^2, and $t(x, r)$, $e(x, r)$, and $v(x, r)$ denote the numbers of tiles, edges, and vertices, respectively, in the spherical patch $P(x, r)$. If these limits exist and are finite for some reference point x, then so for every point x in E^2, and the limits are independent of the reference point. Tilings for which these limits exist and are finite are called *balanced tilings*.

The following is known as *Euler's Theorem for Tilings*:

Theorem 10. *For a normal plane tiling T, if one of the limits $v(T)$ or $e(T)$ exists and is finite, then so does the other. In particular, T is balanced and*

$$v(T) - e(T) + 1 = 0$$

For periodic tilings T, this theorem can be restated as follows. Let L be the two-dimensional lattice of vectors that correspond to translations in $S(T)$. Choose any fundamental parallelogram P for L such that no vertex of T lies on a side of P, and no corner of P lies on an edge of T. Let V, E, and T denote the numbers of vertices, edges, and tiles in P, respectively, where fractions of edges and tiles are counted appropriately. Then these numbers do not depend on the choice of the fundamental parallelogram.

The following result is known as *Euler's Theorem for Periodic Plane tilings*, and corresponds to an Euler-type theorem for tessellations of the two-dimensional torus:

Theorem 11. *If T is a normal plane tiling that is periodic, then $V - E - T = 0$.*

Euler's Theorem can be extended in various ways. One such extension involves the limits,

$$v_j(T) = \lim_{r \to \infty} \frac{v_j(x, r)}{t(x, r)} \quad \text{and} \quad t_k(T) = \lim_{r \to \infty} \frac{t_k(x, r)}{t(x, r)}$$

where $v_j(x, r)$ is the number of vertices of valence j in the patch $P(x, r)$, and $t_k(x, r)$ is the number of tiles with k adjacents in $P(x, r)$. A normal tiling T is called *strongly balanced* if, for some point x, all the limits $v_j(T)$ $(j \geq 3)$ and $t_k(T)$ $(k \geq 3)$ exist; then they will be finite by normality. As before, this definition does not depend on the reference point x. In any strongly balanced tiling, $v(T) = \sum_{j \geq 3} v_j(T) \leq \infty$ and $\sum_{k \geq 3} t_k(T) = 1$; in particular, such a tiling is balanced. The following are examples of strongly balanced plane tilings: periodic tilings; tilings in which each tile has k vertices, with valences $j_1, \ldots, j_k$ (in some order); and tilings with j-valent vertices each incident with tiles that have $k_1, \ldots, k_j$ adjacents (in some order).

The equations in the next two theorems involve the relative frequences of vertices of various valences and tiles with various number of adjacents.

Theorem 12. *If T is a strongly balanced plane tiling, then:*

$$2\sum_{j\geq 3}(j-3)v_j(\mathcal{T})+\sum_{k\geq 3}(k-6)t_k(\mathcal{T})=0$$

$$\sum_{j\geq 3}(j-4)v_j(\mathcal{T})+\sum_{k\geq 3}(k-4)t_k(\mathcal{T})=0$$

$$\sum_{j\geq 3}(j-6)v_j(\mathcal{T})+2\sum_{k\geq 3}(k-3)t_k(\mathcal{T})=0$$

Theorem 13. *If $\mathcal{T}$ is a strongly balanced plane tiling, then:*

$$\frac{1}{\sum_{j\geq 3}jw_j(\mathcal{T})}+\frac{1}{\sum_{k\geq 3}kt_k(\mathcal{T})}=\frac{1}{2}$$

where $w_j(\mathcal{T}):=v_j(\mathcal{T})/v(\mathcal{T})$.

Note that the denominators on the left in Theorem 13 are the average valence of the vertices of $\mathcal{T}$ and the average number of edges of the tiles of $\mathcal{T}$. For example, for the regular tessellation of the plane by regular p-gons, q meeting at each vertex, the latter equation takes the simple form $\frac{1}{p}+\frac{1}{q}=\frac{1}{2}$. From this equation we obtain the well-known solutions $\{p, q\}=\{3, 6\}$, $\{4, 4\}$, and $\{6, 3\}$, which correspond to the tessellations by triangles, squares, or hexagons, respectively.

C. Classification by Symmetry

In this section we discuss the classification (enumeration) of plane tilings $\mathcal{T}$ by combinatorial or metrical (euclidean) symmetry. The tiles are as usual topological discs, but convexity of the tiles is not assumed. We begin with the coarser classification by combinatorial symmetry which employs the combinatorial automorphism group $\Gamma(\mathcal{T})$ and its transitivity properties on $\mathcal{T}$. As we remarked earlier, the concepts of combinatorial equivalence and topological equivalence are the same for normal plane tilings. Any automorphism in $\Gamma(\mathcal{T})$ can be realized by a homeomorphism of the plane that preserves $\mathcal{T}$, and vice versa. The classification by combinatorial symmetry is thus the same as the classification by topological symmetry.

1. Combinatorial Symmetry

Combinatorial symmetry can occur locally or globally. In a locally symmetric tiling, the combinatorial data for the neighborhoods of tiles or vertices are all the same. In a globally symmetric tiling $\mathcal{T}$, this is also true, but now the symmetry is furnished by a combinatorial automorphism group $\Gamma(\mathcal{T})$, which acts transitively on the tiles or vertices of the tiling. We first discuss symmetry properties with respect to tiles.

Let $\mathcal{T}$ be a plane tiling, and let T be a tile of $\mathcal{T}$. With T we can associate its *valence-type* $j_1.j_2. \ \ldots \ .j_k$ provided T has k vertices which, in cyclic order, have valences $j_1, j_2, \ldots, j_k$. Then $\mathcal{T}$ is said to be *homogeneous of type* $[j_1.j_2. \ \ldots \ .j_k]$ if $\mathcal{T}$ is normal and each tile of $\mathcal{T}$ has

valence-type $j_1.j_2. \ \ldots \ .j_k$. As before, these symbols for $\mathcal{T}$ and its tiles are unique up to a cyclic permutation of the entries.

A plane tiling $\mathcal{T}$ is called *homeohedral* or *combinatorially tile-transitive* if $\mathcal{T}$ is normal and $\Gamma(\mathcal{T})$ acts transitively on the tiles of $\mathcal{T}$. Each homeohedral tiling is also homogeneous.

Theorem 14. (a) *If $\mathcal{T}$ is a homogeneous plane tiling, then it has one of the 11 types* $[3^6]$, $[3^4.6]$, $[3^3.4^2]$, $[3^2.4.3.4]$, $[3.4.6.4]$, $[3.6.3.6]$, $[3.12^2]$, $[4^4]$, $[4.6.12]$, $[4.8^2]$, $[6^3]$. (b) *Each homogeneous plane tiling is homeohedral. Any two homogeneous plane tilings of the same type are combinatorially equivalent, and each type can be represented by a Laves tilings (see Theorem 9 and Fig. 7).*

We now discuss the dual concept. A tiling $\mathcal{T}$ is called *homeogonal* or *combinatorially vertex-transitive* if $\mathcal{T}$ is normal and $\Gamma(\mathcal{T})$ acts transitively on the vertices of $\mathcal{T}$. As with uniform tilings, we can associate with a homeogonal tiling $\mathcal{T}$ its *type* $(k_1.k_2. \ \ldots \ .k_j)$; here, j is the valence of any vertex x, and the j tiles that contain x have, in a suitable cyclic order, k_1 vertices, k_2 vertices, and so on. If $\mathcal{T}$ and $\mathcal{T}'$ are normal dual tilings, then $\mathcal{T}$ is homeogonal if and only if $\mathcal{T}'$ is homeohedral.

Theorem 15. (a) *If $\mathcal{T}$ is a homeogonal plane tiling, then it has one of the 11 types* (3^6), $(3^4.6)$, $(3^3.4^2)$, $(3^2.4.3.4)$, $(3.4.6.4)$, $(3.6.3.6)$, (3.12^2), (4^4), $(4.6.12)$, (4.8^2), (6^3). (b) *Any two homeogonal plane tilings of the same type are combinatorially equivalent, and each type can be represented by an Archimedean (uniform) tiling (see Theorems 7 and 8, and Fig. 4).*

A normal plane tilings $\mathcal{T}$ is said to be *homeotoxal* if its automorphism group $\Gamma(\mathcal{T})$ is edge transitive. It can be shown that there are just five "types" of homeotoxal tilings.

Normality of the tilings is essential in all these classifications. If the requirement of normality is dropped, then many further possibilities arise.

2. Metrical Symmetry

Before we investigate tilings, we introduce some terminology that applies in the wider context of classifying certain geometric objects with respect to metrical symmetries or, as we will say, *by homeomerism*. In particular, this will explain when two geometric objects are considered to be the same (of the same homeomeric type) with respect to classification purposes.

Let $\mathcal{R}$ and $\mathcal{R}'$ be two geometric objects in Euclidean space E^d, and let $S(\mathcal{R})$ and $S(\mathcal{R}')$ be their symmetry groups, respectively. Consider a homeomorphism φ of E^d which maps $\mathcal{R}$ onto $\mathcal{R}'$. Then φ is said to be *compatible with a symmetry* σ *in* $S(\mathcal{R})$ if there exists a symmetry σ' in $S(\mathcal{R}')$ such that $\sigma'\varphi = \varphi\sigma$; that is, up to the one-to-one correspondence determined by the homeomorphism φ, a symmetry σ moves elements of the object $\mathcal{R}$ in the same way as σ' moves elements of $\mathcal{R}'$. We call φ *compatible with* $S(\mathcal{R})$ if φ is compatible with each symmetry σ in $S(\mathcal{R})$.

Two geometric objects $\mathcal{R}$ and $\mathcal{R}'$ in E^d are said to be *homeomeric*, or *of the same homeomeric type*, if there exists a homeomorphism φ which maps $\mathcal{R}$ onto $\mathcal{R}'$ such that φ is compatible with $S(\mathcal{R})$ and its inverse φ^{-1} is compatible with $S(\mathcal{R}')$.

These concepts apply to the classification of normal plane tilings $\mathcal{T}$ with certain transitivity properties of their symmetry group $S(\mathcal{T})$. The following three results summarize the enumeration of all homeomeric types of tilings with a tile-transitive, vertex-transitive, or edge-transitive symmetry group, respectively, yielding the homeomeric types of isohedral, isogonal, or isotoxal tilings:

Theorem 16. *There exist precisely 81 homeomeric types of normal isohedral plane tilings. Precisely 47 types can be realized by a normal isohedral edge-to-edge tiling with convex polygonal tiles.*

Theorem 17. *There exist precisely 91 homeomeric types of normal isogonal plane tilings. Precisely 63 types can be realized by normal isogonal edge-to-edge tilings with convex polygonal tiles.*

Theorem 18. *There exist precisely 26 homeomeric types of normal isotoxal plane tilings. Precisely six types can be realized by a normal isotoxal edge-to-edge tiling with convex polygonal tiles.*

The enumeration of the tilings can be accomplished in various ways. One approach is through *incidence symbols* which encode data about the local structure of the plane tilings. For example, isohedral tilings fall into 11 combinatorial classes, typified by the Laves tilings. In an isohedral tiling, every tile is surrounded in the same way, and its vertex degree sequence is given by the symbol for the corresponding Laves tiling. The incidence symbol records (in terms of edge labels and edge orientations) how each tile meets its (congruent) neighbors. The situation is similar for isogonal and isotoxal tilings. The enumeration proceeds by identifying all possible incidence symbols.

Another algorithmic approach is through *Delaney–Dress symbols*. With every tiling is associated a "barycentric subdivision," and the symbol now stores information about the way in which the symmetry group acts on this subdivision. Then the enumeration of the tilings amounts to the enumeration of all Delaney–Dress symbols of a certain type. This approach is more flexible and allows generalizations to tilings in higher dimensions. For exam-

ple, the method has been applied to show that there are 88 combinatorial classes of periodic tilings in ordinary three-space for which the symmetry group acts transitively on the two-dimensional faces of the tiling.

V. MONOHEDRAL TILINGS

This section deals with monohedral tilings $\mathcal{T}$ in Euclidean spaces E^d of any dimension $d \geq 2$. Monohedral tilings have a single prototile. To determine which shapes occur as prototiles is one of the main open problems in tiling theory. The answer is not even known for polygonal prototiles in the plane, which necessarily cannot have more than six vertices. In its full generality, the prototile enumeration problem seems to be intractable in higher dimensions, so suitable restrictions must be imposed on the tiles or on the kind of tilings.

A. Lattice Tilings

Lattice tilings in E^d by convex d-polytopes have been widely studied. The interest in such tilings originated from crystallography and the geometry of numbers. Early contributions on the subject can be found in the works of Dirichlet, Fedorov, Minkowski, Voronoi, and Delone.

A tiling $\mathcal{T}$ by translates of a single prototile T is the simplest kind of monohedral tiling. Prototiles which admit such a tiling are very restricted. For example, if the prototile T is a convex d-polytope, then T and all its facets must necessarily be centrally symmetric. The translation vectors for the tiles in a tiling by translates need not necessarily come from a lattice. However, if a prototile tiles by translation, then it is natural to ask if it also admits a lattice tiling.

Theorem 19. *If a convex d-polytope T tiles E^d by translation, then T also admits (uniquely) a face-to-face lattice tiling of E^d and thus is a parallelotope.*

For nonconvex prototiles T, the analogue of Theorem 19 is generally not true. There are nonconvex star-shaped polyhedral sets T which tile E^d by translation but do not admit a lattice tiling. Examples have been found in almost all dimensions d, including $d = 3$.

Parallelotopes are the most basic space fillers. Simple examples are the Voronoi regions for lattices. The previous theorem is based on the following characterization of parallelohedra. A *belt* of a convex d-polytope T is a sequence of facets $F_0, F_1, \ldots, F_{k-1}, F_k = F_0$ of T, such that $F_{i-1} \cap F_i$ is a $(d-2)$-face of T for each $i = 1, \ldots, k$, and all these $(d-2)$-faces are parallel. If T is a centrally symmetric convex d-polytope with centrally symmetric facets, then every $(d-2)$-face G of T determines a belt of facets whose $(d-2)$-faces are parallel to G.

Theorem 20. *A convex d-polytope T is a parallelotope if and only if T and its facets are centrally symmetric and each belt of T contains four or six facets.*

In each dimension d, there is only a finite number of distinct combinatorial types of parallelotope. This follows from Minkowski's observation that a parallelotope in E^d can have at most $2(2^d - 1)$ facets. A complete list of parallelotopes is only known for $d \le 4$. For $d = 2, 3,$ or 4, the number of distinct combinatorial types of parallelotopes is 2, 5, or 52, respectively. Figure 8 shows representatives for the parallelotopes in the plane and ordinary space. In the plane, the parallelogram and the centrally symmetric hexagon are the only possible shapes. For $d = 3$, Minkowski's bound predicts that a parallelotope cannot have more than 14 facets; the maximum of 14 is attained by the truncated octahedron, which is the last parallelotope in Fig. 8.

Call a parallelotope T *primitive* if, in its unique face-to-face tiling $\mathcal{T}$, each vertex of $\mathcal{T}$ is contained in exactly $d + 1$ tiles. In E^2 and E^3, the centrally symmetric hexagon and truncated octahedron are the only types of primitive parallelotopes; in E^4, there are three distinct combinatorial types. The total number of combinatorial types of parallelotopes grows very fast in dimensions $d \ge 5$, and the number of primitive types is known to be at least 223 if $d = 5$. It has been conjectured (already by Voronoi) that each parallelotope is in fact an affine image of the Voronoi region for some lattice. This is known to be true in dimensions $d \le 4$, as well as for primitive parallelohedra in all dimensions.

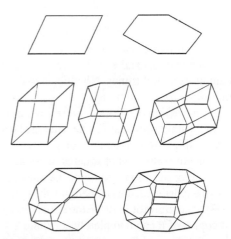

FIGURE 8 The parallelotopes in two and three dimensions. The top row shows the two possible shapes in the plane, a parallelogram, and a centrally symmetric hexagon. There are five shapes of parallelotopes in ordinary space—the parallelopiped, hexagonal prism, and rhombic dodecahedron shown in the middle row and the elongated dodecahedron and truncated octahedron shown in the bottom row. [From Schulte, E. (1993). *Handbook of Convex Geometry*, (Gruber, P. M., and Wills, J. M., eds.). Elsevier, Amsterdam.]

The d-dimensional cube obviously is a parallelotope in each dimension d. In every lattice tiling of E^d by unit d-cubes, there always is a "stack" of cubes in which each two adjacent cubes meet in a whole facet. However, for $d \ge 10$, there are nonlattice tilings of E^d by translates of unit d-cubes in which no two cubes share a whole facet. On the other hand, for $d \le 6$, any tiling of E^d by unit d-cubes must contain at least one pair of cubes that share a whole facet.

B. Space Fillers in Dimension *d*

Convex d-polytopes which are prototiles of monohedral tilings in E^d are called *space fillers* of E^d, or *plane fillers* if $d = 2$. The enumeration of all space fillers is far from being complete, even for the planar case.

Isohedral tilings are monohedral tilings with a tile-transitive symmetry group. Convex polytopes which are prototiles of isohedral tilings are called *stereohedra*. Each stereohedron is a space filler, but the converse is not true. Indeed, a famous problem by Hilbert (posed in 1900) asked whether there exists a three-dimensional polyhedral shape that admits a monohedral tiling but no isohedral tiling of E^3. Such a prototile (in any dimension) is said to be *anisohedral*. Rephrased in terms of crystallographic groups, Hilbert's problem asked whether there exists a polyhedral space-filling tile in E^3 which is not a fundamental region for a discrete group of Euclidean isometries. Hilbert's question can be reduced to the planar case by observing that three-dimensional (and higher dimensional) anisohedral prototiles can be constructed as prisms over anisohedral planar prototiles. Several authors have discovered examples of anisohedral prototiles, beginning with Reinhardt in 1928 and Heesch in 1935. In the plane, there are anisohedral convex pentagons which admit periodic monohedral tilings. The tile in Fig. 9 is

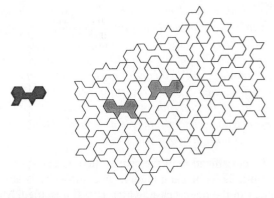

FIGURE 9 A planar anisohedral tile and its unique plane tiling. The two shaded tiles are surrounded in different ways. [From Schattschneider, D., and Senechal, M. (1997). *Handbook of Discrete and Computational Geometry*, (Goodman, J. E., and O'Rourke, J., eds.). CRC Press, Boca Raton, FL.]

anisohedral and admits a unique plane tiling; in fact, the tiling must be nonisohedral because the two shaded tiles are surrounded in different ways.

Theorem 21. *For every $d \geq 2$, there exist anisohedral space fillers that are convex d-polytopes.*

Isohedrality is a global property that a monohedral tiling may or may not have. Our next theorem is related to Hilbert's problem and describes a way to detect isohedrality of a tiling locally. We first introduce some terminology.

Let $\mathcal{T}$ be a monohedral tiling in E^d. Given a tile T of $\mathcal{T}$, the *kth corona* $C^k(T)$ of T in $\mathcal{T}$ is the set of all tiles T' in $\mathcal{T}$ for which there exists a sequence of tiles $T = T_0, T_1, \ldots, T_{m-1}, T_m = T'$ with $m \leq k$ such that $T_i \cap T_{i+1} \neq \emptyset$ for $i = 0, \ldots, m-1$. So $C^0(T)$ consists of T itself, $C^1(T)$ of the tiles that meet T, and $C^k(T)$ of the tiles that meet a tile in $C^{k-1}(T)$. It is possible that two distinct tiles T and T' of $\mathcal{T}$ have the same *kth* corona for some k; that is, $C^k(T) = C^k(T')$. A *centered kth corona* in $\mathcal{T}$ is a pair $(T, C^k(T))$, consisting of a tile T and its *kth* corona $C^k(T)$. A centered corona determines the original tile uniquely. Two centered *kth* coronas $(T, C^k(T))$ and $(T', C^k(T'))$ are said to be *congruent* (as centered coronas) if there exists an isometry γ of E^d such that $\gamma(T) = T'$ and $\gamma(C^k(T)) = C^k(T')$; here, it is not required that γ maps $\mathcal{T}$ onto itself. By $S_k(T)$ we denote the group of isometries of E^d which map the centered corona $(T, C^k(T))$ onto itself; this is the subgroup of the symmetry group $S(T)$ of the tile T which maps $C^k(T)$ onto itself. Since $S(T)$ is a finite group, the chain of subgroups $S(T) = S_0(T) \supseteq S_1(T) \supseteq \cdots \supseteq S_k(T) \supseteq \cdots$ can only contain a finite number of distinct groups.

We now have the following *Local Theorem for Tilings*, which characterizes isohedrality locally:

Theorem 22. *Let $\mathcal{T}$ be a monohedral tiling of E^d. Then $\mathcal{T}$ is isohedral if and only if there exists a positive integer k for which the following two conditions hold: first, any two centered kth coronas in $\mathcal{T}$ are congruent, and, second, $S_{k-1}(T) = S_k(T)$ for some tile (and hence all tiles) T of $\mathcal{T}$. Moreover, if the two conditions hold for k, and if T is a tile of $\mathcal{T}$, then $S_k(T)$ is the stabilizer of T in $S(\mathcal{T})$. In particular, if the prototile of $\mathcal{T}$ is asymmetric (that is, has no nontrivial symmetry), then $\mathcal{T}$ is isohedral if and only if the first condition holds with $k = 1$.*

The classification of the space fillers is one of the big open problems in tiling theory. Its complexity is already evident in the planar case, which is still unsettled. Many authors who attempted the classification for the plane believed, or even stated explicitly, that their enumeration is complete. Often this was generally accepted until a new prototile was discovered by someone else.

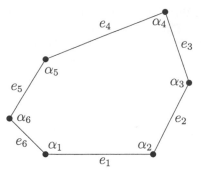

FIGURE 10 A convex hexagon with sides $e_1, \ldots, e_6$ and angles $\alpha_1, \ldots, \alpha_6$. [From Senechal, M. (1995). *Quasicrystals and Geometry*, Cambridge University Press, Cambridge.]

A plane filler must necessarily be a triangle, quadrangle, pentagon or hexagon. At present, the list of convex plane fillers comprises all triangles and all quadrangles, as well as 14 kinds of pentagons and three kinds of hexagons. To see how each triangle can tile, join two copies to form a parallelogram, and then tile the plane with congruent parallelograms. Similarly, two copies of a convex quadrangle can be joined along an edge to form a hexagon of which congruent copies tile the plane. Any convex pentagon with a pair of parallel sides is a plane filler. To describe the three kinds of hexagons that tile, label the sides of the hexagon by $e_1, \ldots, e_6$ and the angles by $\alpha_1, \ldots, \alpha_6$, as in Fig. 10. Then a convex hexagon tiles if and only if it satisfies one of the following conditions:

- $\alpha_1 + \alpha_2 + \alpha_3 = \alpha_4 + \alpha_5 + \alpha_6 = 2\pi$, $e_6 = e_3$
- $\alpha_1 + \alpha_2 + \alpha_4 = \alpha_3 + \alpha_5 + \alpha_6 = 2\pi$, $e_6 = e_3$, $e_4 = e_2$
- $\alpha_1 = \alpha_3 = \alpha_5 = 2\pi/3$, $e_6 = e_1$, $e_3 = e_2$, $e_5 = e_4$

Among these convex plane fillers, all the triangles, quadrangles, and hexagons and exactly five kinds of the pentagons admit isohedral tilings. No other convex polygons can admit isohedral plane tilings.

There is a wealth of further interesting monohedral plane tilings whose prototiles are nonconvex shapes. Examples are tilings by polyominos or polyhexes, respectively, which are composed of squares or hexagons of the regular square or hexagonal tiling of the plane. Noteworthy is also the existence of monohedral *spiral* plane tilings (see Fig. 3), whose prototiles have the remarkable property that two copies of it can completely surround a third.

The space-filler problem is most appealing in three dimensions. Many examples of space fillers have been discovered by crystallographers as Voronoi regions for suitable discrete point sets (dot patterns) in E^3. In the past, there have been several contradictory claims in the literature as to how many facets a three-dimensional space filler can have. The current world record is held by a spectacular space filler with 38 facets, which is a Voronoi region for some discrete point set (discovered using computers). Among the five Platonic solids (the regular tetrahedron, cube, octahedron, dodecahedron, and icosahedron), only the cube tiles ordinary space, in an obvious manner.

The classification of space fillers becomes even more difficult in higher dimensions. Not even the parallelotopes, which are the convex polytopes that tile by translation, have been completely enumerated for $d \geq 5$. Tilings by certain kinds of simplices have been studied, but it is still not known, even in three dimensions, which simplices tile space. There are several kinds of tetrahedra which do admit monohedral tilings of E^3, among them tetrahedra that are fundamental regions for discrete groups generated by reflections.

A more modest problem is the classification of all combinatorial types of space-filling convex polytopes in E^d. This problem is trivial for the plane, where triangles, quadrangles, pentagons, and hexagons are the only possible solutions. For dimensions $d \geq 3$ it is not even known if there are only finitely many combinatorial types of space-filling polytopes. The number of types would be finite if there was an upper bound on the number of facets of space-filling polytopes, but the existence of a general bound has not been established. The only general result available in the literature is the following theorem which provides an upper bound for the number of facets of stereohedra that admit isohedral face-to-face tilings:

Theorem 23. *If a stereohedron admits an isohedral face-to-face tiling of E^d, then its number of facets is bounded by $2^d(h - \frac{1}{2}) - 2$, where h is the index of the translation subgroup in the symmetry group of the tiling.*

The theorem gives a finite bound for the number of facets of d-dimensional stereohedra that admit an isohedral face-to-face tiling. In fact, the symmetry group of the tiling must be among the finitely many crystallographic groups in E^d, and so we know that the index h of the translation subgroup is uniformly bounded. In three dimensions, the maximum index is 48 and the bound for the number of facets is 378; it is likely that the true bound is considerably lower, possibly as low as 38.

The strong requirement of congruence of the tiles considerably restricts the various possibilities for designing monohedral tilings. If this requirement is relaxed to combinatorial equivalence, there is generally much freedom for choosing the metrical shape of the tiles to arrive at a tiling of the whole space. Call a tiling T of E^d by convex polytopes *monotypic* if each tile of T is combinatorially equivalent to a convex d-polytope T, the *combinatorial prototile* of T. In a monotypic tiling, there generally are infinitely many different metrical shapes of tiles, but the tiles are all convex and combinatorially equivalent to a single combinatorial prototile. It is not difficult to see that

the plane admits a monotypic face-to-face tiling T by convex n-gons for each $n \geq 3$. Figure 1 illustrates how such a tiling can be constructed for $n = 7$. It is a consequence of Euler's Theorem that these tilings cannot be normal if $n \geq 7$, and so infinitely many different shapes of n-gons are needed in the process.

Much less obvious are the following rather surprising results about monotypic tilings in E^3:

Theorem 24. *Every convex 3-polytope T is the combinatorial prototile of a monotypic tiling T of E^3. If all facets of T are triangles, then T can be made face-to-face.*

In other words, in three dimensions, every polytope tiles, as long as combinatorially equivalent (rather than congruent) copies are allowed. In general, these tilings will not be face-to-face. In fact, for each dimension $d \geq 3$, there are convex d-polytopes which cannot occur as combinatorial prototiles of monotypic face-to-face tilings of E^d.

VI. NONPERIODIC TILINGS

Tiling models are an important tool in the geometrical study of crystals. Crystal growth is a modular process: from a relatively tiny cluster (seed) of atoms, a crystal grows by the accretion of modules (atoms, molecules). In the geometric modeling of crystal structure, these modules are sometimes represented by space-filling polyhedra yielding a tiling of ordinary space. For nearly 200 years, until 1984, it had been an *axiom* of crystallography that the internal structure of a crystal was periodic. The presence of lattice structure in a crystal had been strongly supported by the geometry of the X-ray diffraction images of the crystal, but this "fundamental law" of crystal structure was overturned in 1984 when an alloy of aluminum and manganese was discovered that had diffraction images exhibiting icosahedral symmetry; pentagonal, and thus icosahedral, symmetry was known not to be compatible with lattice structure. Soon other crystal-like structures were found, and eventually the name "quasicrystals" emerged.

Since 1984, crystallography has gone through a process of redefining itself, as well as the objects it studies. In the world of crystals and quasicrystals, order is no longer synonymous with periodicity, but exactly how far one has to go beyond periodicity is still a big open problem. Tilings once again have become an important modeling tool, but now the interest is in nonperiodic tilings and in shapes that do not admit periodic tilings.

A. Aperiodicity

Aperiodicity is a fascinating phenomenon in tiling theory. Recall that a tiling T in Euclidean d-space E^d is called *nonperiodic* if it does not admit a nontrivial translational symmetry. For example, the spiral tiling shown in Fig. 3 is nonperiodic. There are many shapes in E^d that admit both periodic tilings and nonperiodic tilings. For example, already the d-dimensional cube admits tilings whose symmetry group contains translational symmetries in any preassigned number k of independent directions, with $k = 1, \ldots, d$. In fact, if S is any finite set of prototiles in E^d which are convex polytopes, and if S admits a face-to-face tiling of E^d which possesses a nontrivial translational symmetry, then S also admits a tiling of E^d which is periodic (see Theorem 3).

A set S of prototiles in E^d is said to be *aperiodic* if S admits a tiling of E^d, but all such tilings are nonperiodic. In discussing aperiodic prototile sets it is convenient to generalize the notion of a tiling and allow decorations or markings on the tiles. This leads to *decorated tiles* and *decorated tilings*, respectively. Here we shall not investigate these concepts in full generality but restrict ourselves to applications in the theory of nonperiodic tilings. In a typical application in two dimensions, the tiles in a prototile set S are polygons with some corners colored or with orientations on some sides (that is, the colors and orientations are the decorations). The condition then is that, in constructing a tiling with the given set of decorated prototiles, we must place tile against tile in an edge-to-edge manner, such that colored corners are placed against colored corners with the same color, and oriented sides are placed against oriented sides with the same orientation; in short, the colors at corners and the orientations on sides must match. Figure 11 shows the Penrose aperiodic prototile set consisting of two decorated tiles known as *kite* and *dart*; here the corners are colored with two colors, black and white (say), as shown, but there are no orientations given to the sides. We elaborate on this example in the next subsection.

Decorated tiles and matching rules are the price we have to pay if we want tiles of simple geometric shape. By altering the shape of the decorated tiles in a prototile set to reflect their decorations and the matching rules, we can often construct a prototile set with nondecorated tiles whose tiling properties are equivalent to those of the set of decorated tiles. However, the geometric shape of the nondecorated tiles is generally more complicated and less appealing than the underlying shape of the corresponding decorated tiles.

There is a close connection between aperiodicity and the Undecidability Theorem for tilings (see Theorem 1). In fact, the Undecidability Theorem was originally proved by Berger in 1966 by establishing the existence of an aperiodic set consisting of 20,426 so-called *Wang tiles*; these are square tiles with colored edges, which must be tiled in an edge-to-edge manner by translation only such that colors of adjacent tiles match. The existence of such a set is enough to ensure that no algorithm can exist which, for any given finite set S of prototiles, decides in a finite number of steps whether S admits a tiling or not. The number of tiles in an aperiodic set of Wang tiles has since then been reduced considerably.

It is still an open question whether there exists an aperiodic set in the plane consisting of a single decorated or nondecorated tile. Over the years, many examples of small aperiodic sets have been discovered, both in the plane and in higher dimensions. The most famous examples are the three Penrose aperiodic sets in the plane described in the next subsection; they have greatly influenced the study of aperiodicity in higher dimensional spaces. There is a pair of decorated aperiodic tiles in any dimension $d \geq 2$. There are also planar examples with only three nondecorated tiles (one hexagon and two pentagons) that force aperiodicity. In E^3, the aperiodic pair of *Penrose rhombs* depicted in Fig. 12 has been generalized to yield families of decorated (Ammann) rhombohedra which again only admit non-periodic tilings. There also exists an aperiodic set of only four decorated tetrahedral tiles in E^3 that admit tilings with global icosahedral symmetry.

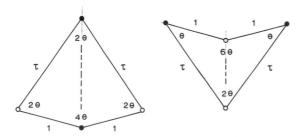

FIGURE 11 The Penrose kite and dart. The corners are colored black or white (solid or open circles, respectively), and the angles and side lengths are as indicated, with $\tau = (1 + \sqrt{5})/2$ and $\theta = \pi/5$. [From Grünbaum, B., and Shephard, G. C. (1986). *Tilings and Patterns*, Freeman & Co., San Francisco.]

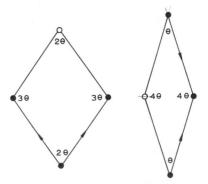

FIGURE 12 The Penrose rhombs. The rhombs have black and white corners (solid or open circles, respectively) and have orientations given to some of their sides; the angles are as indicated, with $\theta = \pi/5$. [From Grünbaum, B., and Shephard, G. C. (1986). *Tilings and Patterns*, Freeman & Co., San Francisco.]

A remarkable three-dimensional tile, known as the *Schmitt–Conway–Danzer tile*, has been discovered that is aperiodic when only direct congruent copies (that is, no mirror images) are allowed in tilings of E^3. The tile is a certain biprism with a cell-complex on its boundary, whose structure forces the tilings to have screw rotational symmetry through an irrational angle in one direction. The form of aperiodicity exhibited by this example is weaker than in the previous examples; in fact, although the tilings do not have translational symmetry, they still have a symmetry of infinite order.

B. The Penrose Tilings

A striking advance in tiling theory was Penrose's discovery in 1973 and 1974 of three aperiodic sets of decorated prototiles and matching rules in the plane, which are closely related to each other in that tilings with tiles from one set can be converted into tilings with tiles from another set. Two of these sets are depicted in Figs. 11 and 12; they are the Penrose kite and dart and the Penrose rhombs, both mentioned earlier. A *Penrose tiling* is a tiling of the entire plane that is constructed from any of the three sets by obeying the matching rules. Figure 13 shows a patch of a Penrose tiling by Penrose rhombs.

It is not difficult to make your own Penrose tiles from cardboard and build finite patches of tiles. However, even if you obey the matching rules, you will probably run into untileable regions at some stage, and you will have to remove or rearrange some tiles and try again. The decorations and matching rules are crucial in this process; in fact, the underlying geometric shapes do admit *periodic* tilings of the plane if the decorations and matching rules are ignored.

The kite and dart of Fig. 11 have sides of two lengths in the ratio $\tau{:}1$ (with $\tau = (1 + \sqrt{5})/2$ the golden ratio), and the angle θ is $\pi/5$. The corners are colored with two colors, black and white. The matching condition is that equal sides must be put together, so that the colors at the corners match. The two Penrose rhombs of Fig. 12 also have black and white corners and have orientations given to some of their sides; the angles are as indicated, where again $\theta = \pi/5$. The condition is that in constructing a tiling the colors of corners as well as lengths and orientations of sides must match.

Plane tilings by kites and darts can be transformed into plane tilings by rhombs, and vice versa. The transition from a kite and dart tiling to a rhomb tiling is illustrated in Fig. 14; the kites and darts are bisected, and the resulting triangles are merged into rhombs. The two tilings depicted in Fig. 14 are *mutually locally derivable*; that is, the tiles in either tiling can, through a process of decomposition into smaller tiles or regrouping with adjacent tiles, or a combination of both processes, form the tiles of the other. The concept of mutual local derivability links the tilings with one prototile set to those with another. Thus, characteristic properties of one prototile set translate into similar properties of the other set.

As for many other aperiodic prototile sets, the aperiodicity of the Penrose sets is based on two important transformations of tilings, namely *composition* and the corresponding inverse process of *decomposition*, which exhibit the existence of hierarchical structure. Figure 15 illustrates the composition process for a tiling by Penrose rhombs.

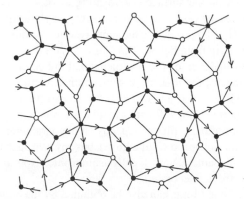

FIGURE 13 A Penrose tiling by Penrose rhombs. The tiling has been constructed according to the matching rule for Penrose tilings by Penrose rhombs (that is, the colors at corners and the orientations of sides of adjacent tiles must match). [From Grünbaum, B., and Shephard, G. C. (1986). *Tilings and Patterns*, Freeman & Co., San Francisco.]

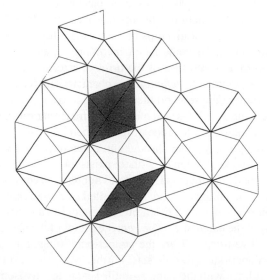

FIGURE 14 The transition from a Penrose kite and dart tiling to a Penrose rhomb tiling. The kites and darts of the original tiling are bisected, and the resulting triangles are merged into rhombs. [From Senechal, M. (1995). *Quasicrystals and Geometry*, Cambridge University Press. Cambridge.]

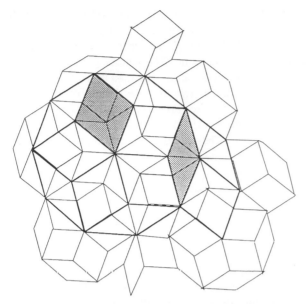

FIGURE 15 Composition for the Penrose rhombs. First, certain tiles of the original Penrose tiling are bisected to yield triangular tiles. Then, the triangular tiles and remaining rhombic tiles are regrouped with adjacent tiles to form the rhombic tiles of a new Penrose tiling of larger scale. [From Senechal, M. (1995). *Quasicrystals and Geometry*, Cambridge University Press. Cambridge.]

By composition we mean the process of taking unions of tiles (or pieces of tiles) to form larger tiles of basically the same shape as those of the original tiles, such that the decorations of the original tiles specify a matching condition equivalent to the original one. Decomposition is the basis for the process of *inflation* which, when iterated, can be used to generate arbitrarily large patches or even global tilings. Here inflation is the operation of, first, homothetically enlarging a given patch of tiles (in the case of the Penrose tiles, by some factor involving τ), and then decomposing it into tiles of the original size. Inflation and its inverse process, *deflation*, are characteristic features of many aperiodic sets and their tilings.

A tiling $\mathcal{T}$ in E^d is said to be *repetitive* if every bounded configuration of tiles appearing anywhere in $\mathcal{T}$ is repeated infinitely many times throughout $\mathcal{T}$; more precisely, for every bounded patch of $\mathcal{T}$ there exists an $r > 0$ such that every ball of radius r in E^d contains a congruent copy of the patch. Two tilings $\mathcal{T}_1$ and $\mathcal{T}_2$ are called *locally isomorphic* if for every bounded patch of $\mathcal{T}_1$ there exists a congruent copy of the patch in $\mathcal{T}_2$. The Penrose tilings of the plane are repetitive tilings, as are the nonperiodic tilings for many other aperiodic prototile sets. Furthermore, any two Penrose tilings with the same prototile set are locally isomorphic. So, for example, since there exist Penrose tilings of the plane with global (pentagonal) D_5-symmetry, there must be arbitrarily large finite patches with D_5-symmetry in any Penrose tiling with the same prototile set.

The remaining (but historically first) Penrose aperiodic set consists of six decorated prototiles and suitable matching rules.

C. The Projection Method

The projection method and its variants are important tools for the construction of nonperiodic tilings. These tilings are nondecorated.

In the *strip projection method*, the d-dimensional tilings are obtained as projection images of polyhedral surfaces that are embedded in a Euclidean "superspace" E^n. Let E be a d-dimensional linear subspace of E^n, and let $E^\perp$ be its orthogonal complement in E^n. Then, $E^n = E \oplus E^\perp$. Let π and $\pi^\perp$ denote the orthogonal projections of E^n onto E or $E^\perp$, respectively. Recall that Z^n is the vertex set of the standard tessellation $\mathcal{C}$ of E^n by n-dimensional cubes, whose lower dimensional faces are also cubical. Let C denote the n-dimensional cube whose vertex set consists of all 0–1 vectors. A vector $t \in E^n$ is said to be *regular* if the boundary of the *tube*,

$$E + C + t := \{x + c + t \mid x \in E, \ c \in C\}$$

does not have a point in common with Z^n; otherwise, z is *singular*. Note that the shape of the tube is determined by its cross-section with $E^\perp$, which is the projection image under $\pi^\perp$ of the cube $C + t$. If $d = n - 1$, the tube is a strip in E^n bounded by a pair of hyperplanes parallel to E.

Now, for every regular vector t, the d-dimensional faces of $\mathcal{C}$ that are fully contained in $E + C + t$ make up a d-dimensional polyhedral surface in E^n. This surface is tessellated by d-dimensional cubes and contains all i-dimensional faces of $\mathcal{C}$ contained in $E + C + t$, for $i = 0, \dots, d$. When restricted to the surface, the projection π onto E is one-to-one and maps the cubical polyhedral complex on the surface onto a tiling $\mathcal{T}_t$ of E. Figure 16 illustrates the case $n = 2$ and $d = 1$, which yields tilings of the real line by projecting the staircase onto the line bounding the strip from below. The tiles of $\mathcal{T}_t$ are the projection images under π of the d-dimensional cubes on the surface; that is, the tiles are d-dimensional parallelopipeds. If $E \cap Z^n = \{0\}$ (that is, if E is "totally irrational"), then the resulting tilings $\mathcal{T}_t$ with t regular are nonperiodic. The structure of these tilings depends on the parameter vector t (and the subspace E, of course), but any two such tilings (with the same E) are locally isomorphic. Note that the tilings are nondecorated.

The same tilings can also be obtained by the *canonical cut method* as follows. Let $e_1, \dots, e_n$ denote the canonical basis of E^n, and let π and $\pi^\perp$ be as above. In applications, any d of the vectors $\pi(e_i)$ in E (or equivalently, any $n - d$ of the vectors $\pi^\perp(e_i)$ in $E^\perp$), with $i = 1, \dots, n$, are linearly independent, and so we assume this from now on.

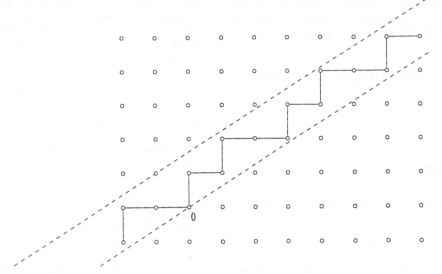

FIGURE 16 Illustration of the strip projection method for $n=2$ and $d=1$. The projection of the staircase onto the (totally irrational) line bounding the strip from below yields a nonperiodic tiling on the line by intervals of two sizes. These intervals are the projections of the horizontal or vertical line segments that are contained in the strip and connect two adjacent grid points. [From Senechal, M. (1995). *Quasicrystals and Geometry*, Cambridge University Press. Cambridge.]

We first construct a periodic tiling of E^n by prismatic tiles. For a d-element subset I of $\{1, \ldots, n\}$, let B_I and $B_I^\perp$, respectively, denote the parallelopipeds in E or $E^\perp$ that are generated by the sets of vectors $\{\pi(e_i) \mid i \in I\}$ and $\{-\pi^\perp(e_i) \mid i \notin I\}$ (the minus sign is significant). The convex n-polytope $P_I := B_I \oplus B_I^\perp$ is a "prism" with bases B_I in E and $B_I^\perp$ in $E^\perp$. Then the prisms $P_I + z$, with $z \in Z^n$ and I a d-element subset of $\{1, \ldots, n\}$, are the tiles in a periodic tiling $\mathcal{O}$ of E^n called the *oblique tiling*.

The nonperiodic tilings now are obtained as sections of $\mathcal{O}$ with certain subspaces $E + t$. More precisely, if t is regular as before, then the intersection of the affine d-dimensional subspace $E + t$ with a tile $P_I + z$ of $\mathcal{O}$ either is empty, or of the form $B_I + y + z$ with y a relative interior point of $B_I^\perp$. Thus, the intersections of $E + t$ with the tiles of $\mathcal{O}$ yield the tiles in a tiling of $E + t$ by d-dimensional parallelopipeds. Then, via projection by π onto E, this gives a tiling of E by parallelopipeds. This is the same tiling $\mathcal{T}_t$ as before. Figure 17 illustrates the case $n = 2$ and $d = 1$; now $\mathcal{O}$ is a plane tiling with two kinds of square tiles (small and large), and nonperiodic tilings on the line are obtained by cutting $\mathcal{O}$ by a (regular) translate of E.

Further properties of these tilings generally depend on the particular choice of subspace E. For example, to obtain the *generalized Penrose tilings* of the plane, take $n = 5$ and consider the group G of isometries of E^5 that permutes the coordinates like a cyclic group of order 5. Then G has three invariant subspaces, namely the line spanned by $e_1 + \cdots + e_5$, and two planes, including the plane E spanned by the vectors:

$$x_1 = (4, \sqrt{5} - 1, -\sqrt{5} - 1, -\sqrt{5} - 1, \sqrt{5} - 1)$$
$$x_2 = (\sqrt{5} - 1, 4, \sqrt{5} - 1, -\sqrt{5} - 1, -\sqrt{5} - 1)$$

The resulting nondecorated plane tilings $\mathcal{T}_t$ in E have only two prototiles, namely the nondecorated Penrose rhombs. If the (regular) parameter vector t is contained in the invariant plane E' of G distinct from E (E' is a subspace of $E^\perp$), then we obtain an ordinary nondecorated Penrose tiling by rhombs (that is, a tiling obtained from a decorated Penrose tiling by rhombs by removing the decorations). The tilings $\mathcal{T}_t$ with $t \in E'$ have the property that every finite patch in any ordinary nondecorated Penrose tiling by rhombs already occurs in one of them. It is known that

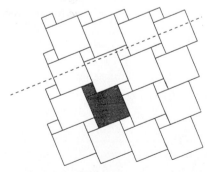

FIGURE 17 The oblique tiling $\mathcal{O}$ for $n=2$ and $d=1$. There are two kinds of square tiles in $\mathcal{O}$. A nonperiodic tiling on the line by two kinds of intervals is obtained by cutting $\mathcal{O}$ with the dotted line. [From Senechal, M. (1995). *Quasicrystals and Geometry*, Cambridge University Press. Cambridge.]

every ordinary nondecorated Penrose tiling by rhombs can be decorated by arrows in exactly one way to become an ordinary decorated Penrose tiling by rhombs.

SEE ALSO THE FOLLOWING ARTICLES

INCOMMENSURATE CRYSTALS AND QUASICRYSTALS • MATHEMATICAL MODELS

BIBLIOGRAPHY

Bezdek, K. (2000). "Space filling," *In* "Handbook of Discrete and Combinatorial Mathematics" (K. H. Rosen, ed.), pp. 824–830, CRC Press, Boca Raton, FL.

Conway, J. H., and Sloane, N. J. A. (1988). "Sphere Packings, Lattices and Groups," Springer-Verlag, New York.

Grünbaum, B., and Shephard, G. C. (1986). "Tilings and Patterns," Freeman, San Francisco, CA.

Gruber, P. M., and Lekkerkerker, C. G. (1987). "Geometry of Numbers," 2nd ed., North-Holland, Amsterdam.

Le, T. T. Q. (1995). "Local rules for quasiperiodic tilings," *In* "The Mathematics of Long-Range Aperiodic Order" (R. V. Moody, ed.), pp. 331–366, Kluwer Academic, Dordrecht/Norwell, MA.

Moody, R. V., ed. (1995). "The Mathematics of Long-Range Aperiodic Order," NATO ASI Series, Series C: Mathematical and Physical Sciences, Vol. 489, Kluwer Academic, Dordrecht/Norwell, MA.

Patera, J., ed. (1998). "Quasicrystals and Discrete Geometry," Fields Institute Monographs, Vol. 10, American Mathematical Society, Providence, RI.

Radin, C. (1999). "Miles of Tiles," Student Mathematical Library, Vol. 1, American Mathematical Society, Providence, RI.

Schattschneider, D., and Senechal, M. (1997). "Tilings," *In* "Handbook of Discrete and Computational Geometry" (J. E. Goodman and J. O'Rourke, eds.), pp. 43–62, CRC Press, Boca Raton, FL.

Schulte, E. (1993). "Tilings," *In* "Handbook of Convex Geometry" (P. M. Gruber and J. M. Wills, eds.), pp. 899–932, Elsevier, Amsterdam.

Senechal, M. (1995). "Quasicrystals and Geometry," Cambridge University Press, Cambridge, U.K.

Stein, S., and Szabo, S. (1994). "Algebra and Tiling," The Carus Mathematical Monographs, Vol. 25, Mathematical Association of Amererica, Providence, RI.

Tarasov, A. S. (1997). "Complexity of convex stereohedra," *Mathematical Notes* **61**(5), pp. 668–671.

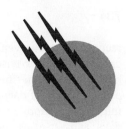

Time and Frequency

Michael A. Lombardi

National Institute of Standards and Technology

I. Concepts and History
II. Time and Frequency Measurement
III. Time and Frequency Standards
IV. Time and Frequency Transfer
V. Closing

GLOSSARY

Accuracy Degree of conformity of a measured or calculated value to its definition.

Allan deviation $\sigma_\gamma, (\tau)$ Statistic used to estimate frequency stability.

Coordinated Universal Time (UTC) International atomic time scale used by all major countries.

Nominal frequency Ideal frequency, with zero uncertainty relative to its definition.

Q Quality factor of an oscillator, estimated by dividing the resonance frequency by the resonance width.

Resonance frequency Natural frequency of an oscillator, based on the repetition of a periodic event.

Second Duration of 9,192,631,770 periods of the radiation corresponding to the transition between two hyperfine levels of the ground state of the cesium-133 atom.

Stability Statistical estimate of the frequency or time fluctuations of a signal over a given time interval.

Synchronization Process of setting two or more clocks to the same time.

Syntonization Process of setting two or more oscillators to the same frequency.

Time scale Agreed-upon system for keeping time.

THIS article is an overview of time and frequency technology. It introduces basic time and frequency concepts and describes the devices that produce time and frequency signals and information. It explains how these devices work and how they are measured. Section I introduces the basic concepts of time and frequency and provides some historical background. Section II discusses time and frequency measurements and the specifications used to state the measurement results. Section III discusses time and frequency standards. These devices are grouped into two categories: quartz and atomic oscillators. Section IV discusses time and frequency transfer, or the process of using a clock or frequency standard to measure or set a device at another location.

I. CONCEPTS AND HISTORY

Few topics in science and technology are as relevant as time and frequency. Time and frequency standards and measurements are involved in nearly every aspect of daily life and are a fundamental part of many technologies.

Time and frequency standards supply three basic types of information. The first type, *date* and *time-of-day*,

Encyclopedia of Physical Science and Technology, Third Edition, Volume 16

records when an event happened. Date and time-of-day can also be used to ensure that events are *synchronized*, or happen at the same time. Everyday life is filled with examples of the use of date and time-of-day information. Date information supplied by calendars records when birthdays, anniversaries, and other holidays are scheduled to occur. The time-of-day information supplied by watches and clocks helps keep our lives on schedule. Meeting a friend for dinner at 6 P.M. is a simple example of synchronization. If our watches agree, we should both arrive at about the same time.

Date and time-of-day information has other, more sophisticated uses. Airplanes flying in a formation require synchronized clocks. If one airplane banks or turns at the wrong time, it could result in a collision and loss of life. When a television station broadcasts a network program, it must start broadcasting the network feed at the instant it arrives. If the station and network clocks are not synchronized, part of the program is skipped. Stock market transactions require synchronized clocks so that the buyer and seller can agree on the same price at the same time. A time error of a few seconds could cost the buyer or seller many thousands of dollars. Electric power companies also use synchronization. They synchronize the clocks in their power grids, so they can instantly transfer power to the parts of the grid where it is needed most, and to avoid electrical overload.

The second type of information, *time interval*, is the duration or elapsed time between two events. Our age is simply the time interval since our birth. Most workers are paid for the time interval during which they work, usually measured in hours, weeks, or months. We pay for time interval as well—30 min on a parking meter, a 20-min cab ride, a 5-min long-distance phone call, or a 30-sec radio advertising spot.

The standard unit of time interval is the *second*. However, many applications in science and technology require the measurement of time intervals much shorter than 1 sec, such as *milliseconds* (10^{-3} sec), *microseconds* (10^{-6} sec), *nanoseconds* (10^{-9} sec), and *picoseconds* (10^{-12} sec).

The third type of information, *frequency*, is the rate of a repetitive event. If T is the period of a repetitive event, then the frequency f is its reciprocal, $1/T$. The International System of Units (SI) states that the period should be expressed as seconds (sec), and the frequency should be expressed as hertz (Hz). The frequency of electrical signals is measured in units of kilohertz (kHz), megahertz (MHz), or gigahertz (GHz), where 1 kHz equals 1000 (10^3) events per second, 1 MHz equals 1 million (10^6) events per second, and 1 GHz equals 1 billion (10^9) events per second. Many frequencies are encountered in everyday life. For example, a quartz wristwatch works by counting the oscillations of a crystal whose frequency is 32,768 Hz. When the crystal has oscillated 32,768 times, the watch records that 1 sec has elapsed. A television tuned to channel 7 receives a video signal at a frequency of 175.25 MHz. The station transmits this frequency as closely as possible, to avoid interference with signals from other stations. A computer that processes instructions at a frequency of 1 GHz might connect to the Internet using a T1 line that sends data at a frequency of 1.544 MHz.

Accurate frequency is critical to communications networks. The highest-capacity networks run at the highest frequencies. Networks use groups of oscillators that produce nearly the same frequency, so they can send data at the fastest possible rates. The process of setting multiple oscillators to the same frequency is called *syntonization*.

Of course, the three types of time and frequency information are closely related. As mentioned, the standard unit of time interval is the second. By counting seconds, we can determine the date and the time-of-day. And by counting the events per second, we can measure the frequency.

A. The Evolution of Time and Frequency Standards

All time and frequency standards are based on a *periodic event* that repeats at a constant rate. The device that produces this event is called a *resonator*. In the simple case of a pendulum clock, the pendulum is the resonator. Of course, a resonator needs an energy source before it can move back and forth. Taken together, the energy source and resonator form an *oscillator*. The oscillator runs at a rate called the *resonance frequency*. For example, a clock's pendulum can be set to swing back and forth at a rate of once per second. Counting one complete swing of the pendulum produces a time interval of 1 sec. Counting the total number of swings creates a *time scale* that establishes longer time intervals, such as minutes, hours, and days. The device that does the counting and displays or records the results is called a *clock*. The frequency uncertainty of a clock's resonator relates directly to the timing uncertainty of the clock as shown in Table I.

Throughout history, clock designers have searched for stable resonators. As early as 3500 B.C., time was kept by observing the movement of an object's shadow between sunrise and sunset. This simple clock is called a *sundial*, and the resonance frequency is based on the apparent motion of the sun. Later, water clocks, hourglasses, and calibrated candles allowed dividing the day into smaller units of time. Mechanical clocks first appeared in the early 14th century. Early models used a verge and foliet mechanism for a resonator and had an uncertainty of about 15 min/day ($\cong 1 \times 10^{-2}$).

A timekeeping breakthrough occurred with the invention of the *pendulum clock*, a technology that dominated

TABLE I Relationship of Frequency Uncertainty to Time Uncertainty

Frequency uncertainty	Measurement period	Time uncertainty
$\pm 1.00 \times 10^{-3}$	1 sec	± 1 msec
$\pm 1.00 \times 10^{-6}$	1 sec	± 1 μsec
$\pm 1.00 \times 10^{-9}$	1 sec	± 1 nsec
$\pm 2.78 \times 10^{-7}$	1 hr	± 1 msec
$\pm 2.78 \times 10^{-10}$	1 hr	± 1 μsec
$\pm 2.78 \times 10^{-13}$	1 hr	± 1 nsec
$\pm 1.16 \times 10^{-8}$	1 day	± 1 msec
$\pm 1.16 \times 10^{-11}$	1 day	± 1 μsec
$\pm 1.16 \times 10^{-14}$	1 day	± 1 nsec

timekeeping for several hundred years. Prior to the invention of the pendulum, clocks could not count minutes reliably, but pendulum clocks could count seconds. In the early 1580s, Galileo Galilei observed that a given pendulum took the same amount of time to swing completely through a wide arc as it did a small arc. Galileo wanted to apply this natural periodicity to time measurement and began work on a mechanism to keep the pendulum in motion in 1641, the year before he died. In 1656, the Dutch scientist Christiaan Huygens invented an escapement that kept the pendulum swinging. The uncertainty of Huygens's clock was less than 1 min/day ($\cong 7 \times 10^{-4}$) and later was reduced to about 10 sec/day ($\cong 1 \times 10^{-4}$). The first pendulum clocks were weight-driven, but later versions were powered by springs. In fact, Huygens is often credited with inventing the spring-and-balance wheel assembly still found in some of today's mechanical wristwatches.

Huge advances in accuracy were made by John Harrison, who built and designed a series of clocks in the 1720s that kept time to within fractions of a second per day (parts in 10^6). This performance was not improved upon until the 20th century. Harrison dedicated most of his life to solving the British navy's problem of determining longitude, by attempting to duplicate the accuracy of his land clocks at sea. He built a series of clocks (now known as H1 through H5) in the period from 1730 to about 1770. He achieved his goal with the construction of H4, a clock much smaller than its predecessors, about the size of a large pocket watch. H4 used a spring and balance wheel escapement and kept time within fractions of a second per day during several sea voyages in the 1760s.

The practical performance limit of pendulum clocks was reached in 1921, when W. H. Shortt demonstrated a clock with two pendulums, one a slave and the other a master. The slave pendulum moved the clock's hands and freed the master pendulum of tasks that would disturb its regularity. The pendulums used a battery as their power supply. The Shortt clock kept time to within a few seconds per year ($\cong 1 \times 10^{-7}$) and was used as a primary standard in the United States.

Joseph W. Horton and Warren A. Marrison of Bell Laboratories built the first clock based on a quartz crystal oscillator in 1927. By the 1940s, quartz clocks had replaced pendulums as primary laboratory standards. Quartz crystals resonate at a nearly constant frequency when an electric current is applied. Uncertainties of <100 μsec/day ($\cong 1 \times 10^{-9}$) are possible, and low-cost quartz oscillators are found in electronic circuits and inside nearly every wristwatch and wall clock.

Quartz oscillators still have shortcomings since their resonance frequency depends on the size and shape of the crystal. No two crystals can be precisely alike or produce exactly the same frequency. Quartz oscillators are also sensitive to temperature, humidity, pressure, and vibration. These limitations made them unsuitable for some high-level applications and led to the development of atomic oscillators.

In the 1930s, I. I. Rabi and his colleagues at Columbia University introduced the idea of using an atomic resonance as a frequency. The first atomic oscillator, based on the ammonia molecule, was developed at the National Bureau of Standards (now the National Institute of Standards and Technology) in 1949. A Nobel Prize was awarded in 1989 to Norman Ramsey, Hans Dehmelt, and Wolfgang Paul for their work in atomic oscillator development, and many other scientists have made significant contributions to the technology. Atomic oscillators use the quantized energy levels in atoms and molecules as the source of their resonance frequency. The laws of quantum mechanics dictate that the energies of a bound system, such as an atom, have certain discrete values. An electromagnetic field at a particular frequency can boost an atom from one energy level to a higher one. Or, an atom at a high energy level can drop to a lower level by emitting energy. The resonance frequency (f) of an atomic oscillator is the difference between the two energy levels divided by Planck's constant (h):

$$f = \frac{E_2 - E_1}{h}.$$

The principle underlying the atomic oscillator is that since all atoms of a specific element are identical, they should produce the exact same frequency when they absorb or release energy. In theory, the atom is a perfect pendulum whose oscillations are counted to measure the time interval. Quartz and the three main types of atomic oscillators (rubidium, hydrogen, and cesium) are described in detail in Section III.

Table II summarizes the evolution of time and frequency standards. The uncertainties listed for modern standards

TABLE II   The Evolution of Time and Frequency Standards

Standard	Resonator	Date of origin	Timing uncertainty (24 hr)	Frequency uncertainty (24 hr)
Sundial	Apparent motion of sun	3500 B.C.	NA	NA
Verge escapement	Verge and foliet mechanism	14th century	15 min	1×10^{-2}
Pendulum	Pendulum	1656	10 sec	7×10^{-4}
Harrison chronometer (H4)	Pendulum	1759	300 msec	3×10^{-6}
Shortt pendulum	Two pendulums, slave and master	1921	10 msec	1×10^{-7}
Quartz crystal	Quartz crystal	1927	10 μsec	1×10^{-10}
Rubidium gas cell	^{87}Rb resonance (6,834,682,608 Hz)	1958	100 nsec	1×10^{-12}
Cesium beam	^{133}Cs resonance (9,192,631,770 Hz)	1952	1 nsec	1×10^{-14}
Hydrogen maser	Hydrogen resonance (1,420,405,752 Hz)	1960	1 nsec	1×10^{-14}
Cesium fountain	^{133}Cs resonance (9,192,631,770 Hz)	1991	100 psec	1×10^{-15}

represent current (year 2000) devices, and not the original prototypes. Note that the performance of time and frequency standards has improved by 13 orders of magnitude in the past 700 years and by about 9 orders of magnitude in the past 100 years.

B. Time Scales and the International Definition of the Second

The second is one of seven base units in the International System of Units (SI). The base units are used to derive other units of physical quantities. Use of the SI means that physical quantities such as the second and hertz are defined and measured in the same way throughout the world.

There have been several definitions of the SI second. Until 1956, the definition was based on the *mean solar day*, or one revolution of the earth on its axis. The *mean solar second* was defined as 1/86,400 of the mean solar day and provided the basis for several astronomical time scales known as Universal Time (UT).

UT0: The original mean solar time scale, based on the rotation of the earth on its axis. UT0 was first kept with pendulum clocks. When quartz clocks became available, astronomers noticed errors in UT0 due to polar motion and developed the UT1 time scale.
UT1: The most widely used astronomical time scale, UT1 improves upon UT0 by correcting for longitudinal shifts of the observing station due to polar motion. Since the earth's rotational rate is not uniform the uncertainty of UT1 is about 2 to 3 msec per day.
UT2: Mostly of historical interest, UT2 is a smoothed version of UT1 that corrects for deviations in the period of the earth's rotation caused by angular momenta of the earth's core, mantle, oceans and atmosphere.

The *ephemeris second* served as the SI second from 1956 to 1967. The ephemeris second was a fraction of the tropical year, or the interval between the annual vernal equinoxes, which occur on or about March 21. The tropical year was defined as 31,556,925.9747 ephemeris sec. Determining the precise instant of the equinox is difficult, and this limited the uncertainty of Ephemeris Time (ET) to ±50 msec over a 9-year interval. ET was used mainly by astronomers and was replaced by *Terrestial Time* (TT) in 1984, equal to International Atomic Time (TAI) + 32.184 sec. The uncertainty of TT is ±10 μsec.

The era of atomic time keeping formally began in 1967, when the SI second was redefined based on the resonance frequency of the cesium atom:

The duration of 9,192,631,770 periods of the radiation corresponding to the transition between two hyperfine levels of the ground state of the cesium-133 atom.

Due to the atomic second, time interval and frequency can now be measured with less uncertainty and more resolution than any other physical quantity. Today, the best time and frequency standards can realize the SI second with uncertainties of $\cong 1 \times 10^{-15}$. Physical realizations of the other base SI units have much larger uncertainties (Table III).

International Atomic Time (TAI) is an atomic time scale that attempts to realize the SI second as closely as possible. TAI is maintained by the Bureau International des Poids et Measures (BIPM) in Sevres, France. The BIPM averages data collected from more than 200 atomic time and frequency standards located at more than 40 laboratories, including the National Institute of Standards and Technology (NIST).

Coordinated Universal Time (UTC) runs at the same rate as TAI. However, it differs from TAI by an integral number of seconds. This difference increases when *leap*

TABLE III Uncertainties of Physical Realizations of the Base SI Units

SI base unit	Physical quantity	Uncertainty
Candela	Luminous intensity	10^{-4}
Mole	Amount of substance	10^{-7}
Kelvin	Thermodynamic temperature	10^{-7}
Ampere	Electric current	10^{-8}
Kilogram	Mass	10^{-8}
Meter	Length	10^{-12}
Second	Time interval	10^{-15}

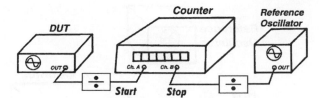

FIGURE 1 Measurement using a time interval counter.

seconds occur. When necessary, leap seconds are added to UTC on either June 30 or December 31. The purpose of adding leap seconds is to keep atomic time (UTC) within ±0.9 sec of astronomical time (UT1). Some time codes contain a UT1 correction that can be applied to UTC to obtain UT1.

Leap seconds have been added to UTC at a rate of slightly less than once per year, beginning in 1972. UT1 is currently losing about 700 to 800 msec per year with respect to UTC. This means that atomic seconds are shorter than astronomical seconds and that UTC runs faster than UT1. There are two reasons for this. The first involves the definition of the atomic second, which made it slightly shorter than the astronomical second to begin with. The second reason is that the earth's rotational rate is gradually slowing down and the astronomical second is gradually getting longer. When a positive leap second is added to UTC, the sequence of events is as follows.

23 hr 59 min 59 sec
23 hr 59 min 60 sec
0 hr 0 min 0 sec

The insertion of the leap second creates a minute that is 61 sec long. This "stops" UTC for 1 sec, so that UTI can catch up.

II. TIME AND FREQUENCY MEASUREMENT

Time and frequency measurements follow the conventions used in other areas of metrology. The frequency standard or clock being measured is called the *device under test* (DUT). The measurement compares the DUT to a *standard* or *reference*. The standard should outperform the DUT by a specified ratio, ideally by 10:1. The higher the ratio, the less averaging is required to get valid measurement results.

The test signal for time measurements is usually a pulse that occurs once per second (1 pps). The pulse width and polarity vary from device to device, but TTL levels are commonly used. The test signal for frequency measurements is usually a frequency of 1 MHz or higher, with 5 or 10 MHz being common. Frequency signals are usually sine waves but can be pulses or square waves.

This section examines the two main specifications of time and frequency measurements—*accuracy* and *stability*. It also discusses some instruments used to measure time and frequency.

A. Accuracy

Accuracy is the degree of conformity of a measured or calculated value to its definition. Accuracy is related to the offset from an ideal value. For example, *time offset* is the difference between a measured on-time pulse and an ideal on-time pulse that coincides exactly with UTC. *Frequency offset* is the difference between a measured frequency and an ideal frequency with zero uncertainty. This ideal frequency is called the *nominal frequency*.

Time offset is usually measured with a *time interval counter* (TIC) as shown in Fig. 1. A TIC has inputs for two signals. One signal starts the counter and the other signal stops it. The time interval between the start and the stop signals is measured by counting cycles from the time base oscillator. The resolution of low-cost TICs is limited to the period of their time base. For example, a TIC with a 10-MHz time base oscillator would have a resolution of 100 nsec. More elaborate TICs use interpolation schemes to detect parts of a time base cycle and have a much higher resolution—1-nsec resolution is commonplace, and even 10-psec resolution is available.

Frequency offset can be measured in either the *frequency domain* or the *time domain*. A simple frequency domain measurement involves directly counting and displaying the frequency output of the DUT with a *frequency counter*. The reference for this measurement is either the counter's internal time base oscillator, or an external time base (Fig. 2). The counter's resolution, or the number of digits it can display, limits its ability to measure frequency offset. The frequency offset is determined as

$$f(\text{offset}) = \frac{f_{\text{measured}} - f_{\text{nominal}}}{f_{\text{nominal}}},$$

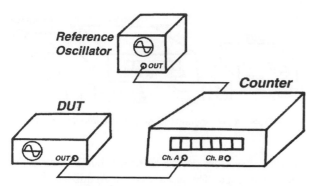

FIGURE 2 Measurement using a frequency counter.

where $f_{measured}$ is the reading from the frequency counter, and $f_{nominal}$ is the frequency labeled on the oscillator's nameplate.

Frequency offset measurements in the time domain involve a *phase comparison* between the DUT and the reference. A simple phase comparison can be made with an oscilloscope (Fig. 3). The oscilloscope will display two sine waves (Fig. 4). The top sine wave represents a signal from the DUT, and the bottom sine wave represents a signal from the reference. If the two frequencies were exactly the same, their phase relationship would not change and both would appear to be stationary on the oscilloscope display. Since the two frequencies are not exactly the same, the reference appears to be stationary and the DUT signal moves. By determining the rate of motion of the DUT signal, we can determine its frequency offset. Vertical lines have been drawn through the points where each sine wave passes through zero. The bottom of the figure shows bars whose width represents the phase difference between the signals. This difference increases or decreases to indicate whether the DUT frequency is high or low with respect to the reference.

Measuring high-accuracy signals with an oscilloscope is impractical, since the phase relationship between signals changes very slowly. More precise phase comparisons can be made with a time interval counter, using a setup similar to Fig. 1. Since frequencies like 5 or 10 MHz are usually involved, *frequency dividers* (shown in Fig. 1) or *frequency mixers* are used to convert the test frequency

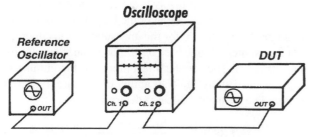

FIGURE 3 Phase comparison using an oscilloscope.

to a lower frequency. Measurements are made from the TIC, but instead of using these measurements directly, we determine the rate of change from reading to reading. This rate of change is called the phase deviation. We can estimate frequency offset as follows, where Δt is the amount of phase deviation, and T is the measurement period:

$$f(\text{offset}) = \frac{-\Delta t}{T}.$$

To illustrate, consider a measurement of $+1$ μsec of phase deviation over a measurement period of 24 hr. The unit used for measurement period (hr) must be converted to the unit used for phase deviation (μsec). The equation becomes

$$f(\text{offset}) = \frac{-\Delta t}{T} = \frac{1 \ \mu\text{sec}}{86,400,000,000 \ \mu\text{sec}}$$

$$= -1.16 \times 10^{-11}.$$

As shown, a device that accumulates 1 μsec of phase deviation/day has a frequency offset of about -1.16×10^{-11} with respect to the reference.

Dimensionless frequency offset values can be converted to units of frequency (Hz) if the nominal frequency is known. To illustrate this, consider an oscillator with a nominal frequency of 5 MHz and a frequency offset of $+1.16 \times 10^{-11}$. To find the frequency offset in hertz, multiply the nominal frequency by the offset:

$$(5 \times 10^6)(+1.16 \times 10^{-11}) = 5.80 \times 10^{-5}$$

$$= +0.0000580 \text{ Hz}.$$

Then add the offset to the nominal frequency to get the actual frequency:

$$5,000,000 \text{ Hz} + 0.0000580 \text{ Hz}$$

$$= 5,000,000.0000580 \text{ Hz}.$$

B. Stability

Stability indicates how well an oscillator can produce the same time or frequency offset over a given period of time. It does not indicate whether the time or frequency is "right" or "wrong" but only whether it stays the same. In contrast, accuracy indicates how well an oscillator has been set on time or set on frequency. To understand this difference, consider that a stable oscillator that needs adjustment might produce a frequency with a large offset. Or an unstable oscillator that was just adjusted might temporarily produce a frequency near its nominal value. Figure 5 shows the relationship between accuracy and stability.

Stability is defined as the statistical estimate of the frequency or time fluctuations of a signal over a given time

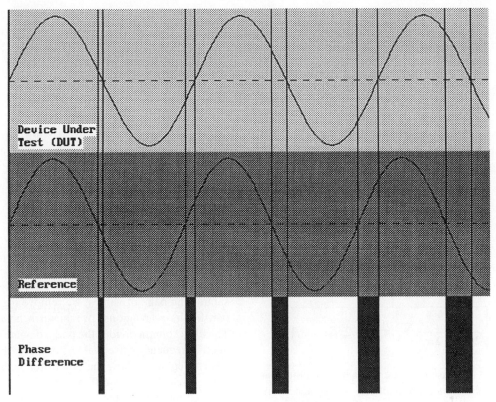

FIGURE 4 Two sine waves with a changing phase relationship.

interval. These fluctuations are measured with respect to a mean frequency or time offset. *Short-term* stability usually refers to fluctuations over intervals less than 100 sec. *Long-term* stability can refer to measurement intervals greater than 100 sec but usually refers to periods longer than 1 day.

Stability estimates can be made in either the frequency domain or the time domain, but time domain estimates are more common and are discussed in this section. To estimate frequency stability in the time domain, we can start with a series of phase measurements. The phase measurements are nonstationary, since they contain a trend contributed by the frequency offset. With nonstationary data, the mean and variance never converge to any particular

values. Instead, there is a moving mean that changes each time we add a measurement.

For these reasons, a nonclassical statistic is often used to estimate stability in the time domain. This statistic is sometimes called the *Allan variance*, but since it is the square root of the variance, its proper name is the *Allan deviation*. The equation for the Allan deviation is

$$\sigma_y(\tau) = \sqrt{\frac{1}{2(M-1)} \sum_{i=1}^{M-1} (y_{i+1} - y_i)^2},$$

where y_i is a set of frequency offset measurements that consists of individual measurements, y_1, y_2, y_3, and so on,

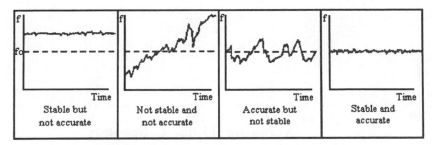

FIGURE 5 The relationship between accuracy and stability.

TABLE IV   **Using Phase Measurements to Estimate Stability**

Phase measurement (nsec), x_i	Phase deviation (nsec), Δt	Frequency offset $\Delta t / \tau (y_i)$	First difference $(y_{i+1} - y_i)$	First difference squared $(y_{i+1} - y_i)^2$
3321.44	(—)	(—)	(—)	(—)
3325.51	4.07	4.07×10^{-9}	(—)	(—)
3329.55	4.04	4.04×10^{-9}	-3×10^{-11}	9×10^{-22}
3333.60	4.05	4.05×10^{-9}	$+1 \times 10^{-11}$	1×10^{-22}
3337.65	4.05	4.06×10^{-9}	$+2 \times 10^{-11}$	4×10^{-22}
3341.69	4.04	4.04×10^{-9}	-2×10^{-11}	4×10^{-22}
3345.74	4.05	4.05×10^{-9}	$+1 \times 10^{-11}$	1×10^{-22}
3349.80	4.06	4.06×10^{-9}	$+1 \times 10^{-11}$	1×10^{-22}
3353.85	4.05	4.05×10^{-9}	-1×10^{-11}	1×10^{-22}
3357.89	4.04	4.04×10^{-9}	-1×10^{-11}	1×10^{-22}

M is the number of values in the y_i series, and the data are equally spaced in segments τ seconds long. Or

$$\sigma_y(\tau) = \sqrt{\frac{1}{2(N-2)\tau^2} \sum_{i=1}^{N-2} [x_{i+2} - 2x_{i+1} + x_i]^2},$$

where x_i is a set of phase measurements in time units that consists of individual measurements, x_1, x_2, x_3, and so on, N is the number of values in the x_i series, and the data are equally spaced in segments τ seconds long.

Table IV shows how the Allan deviation is calculated. The left column contains a series of phase measurements recorded once per second ($\tau = 1$ sec) in units of nanosec-

onds. These measurements have a trend; note that each value in the series is larger than the previous value. By subtracting pairs of values, we remove the trend and obtain the phase deviations (Δt) shown in the second column. The third column divides the phase deviation (Δt) by τ to get the frequency offset values, or the y_i data series. The last two columns show the first differences of the y_i and the squares of the first differences.

Since the sum of the squares equals 2.2×10^{-21}, the frequency stability using the first equation (at $\tau = 1$ sec) is

$$\sigma_y(\tau) = \sqrt{\frac{2.2 \times 10^{-21}}{2(9-1)}} = 1.17 \times 10^{-11}.$$

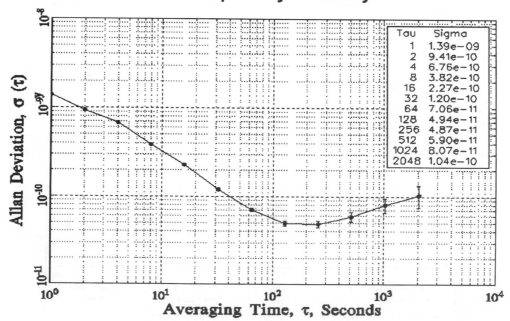

Frequency Stability

Tau	Sigma
1	1.39e−09
2	9.41e−10
4	6.76e−10
8	3.82e−10
16	2.27e−10
32	1.20e−10
64	7.06e−11
128	4.94e−11
256	4.87e−11
512	5.90e−11
1024	8.07e−11
2048	1.04e−10

FIGURE 6 A graph of frequency stability.

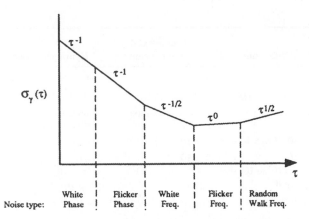

FIGURE 7 Using a frequency stability graph to identify noise types.

A graph of the Allan deviation is shown in Fig. 6. It shows the stability of the device improving as the averaging period (τ) gets longer, since some noise types can be removed by averaging. At some point, however, more averaging no longer improves the results. This point is called the *noise floor*, or the point where the remaining noise consists of nonstationary processes like aging or random walk. The device measured in Fig. 6 is has a noise floor of $\cong 5 \times 10^{-11}$ at $\tau = 100$ sec.

Five noise types are commonly discussed in the time and frequency literature: *white phase*, *flicker phase*, *white frequency*, *flicker frequency*, and *random walk frequency*. The slope of the Allan deviation line can identify the amount of averaging needed to remove these noise types (Fig. 7). Note that the Allan deviation does not distinguish between white phase noise and flicker phase noise. Several other statistics are used to estimate stability and identify noise types for various applications (Table V).

C. Uncertainty Analysis

Time and frequency metrologists must often perform an uncertainty analysis when they calibrate or measure a device. The uncertainty analysis states the measurement error with respect to a national or international standard, such as UTC(NIST) or UTC. Two simple ways to estimate measurement uncertainty are discussed here. Both use the concepts of accuracy and stability discussed above.

One common type of uncertainty analysis involves making multiple measurements and showing that a single measurement will probably fall within a stated range of values. The standard deviation (or an equivalent statistic) is usually both added and subtracted from the mean to form the upper and lower bounds of the range. The stated probability that a given measurement will fall within this range is usually 1σ (68.3%) or 2σ (95.4%).

In time and frequency metrology, the mean value is usually the accuracy (mean time or mean frequency offset), and the deviation in the mean is usually calculated using one of the statistics listed in Table IV. For example, if a device has a frequency offset of 2×10^{-9} and a 2σ stability of 2×10^{-10}, there is a 95.4% probability that the frequency offset will be between 1.8 and 2.2 parts in 10^9.

The second type of uncertainty analysis involves adding the *systematic* and *statistical* uncertainties to find the combined uncertainty. For example, consider a time signal received from a radio station where the mean path delay is measured as 9 msec (time offset), and the deviation in the path delay is measured as 0.5 msec (stability). In this example, 9 msec is the systematic uncertainty and 0.5 msec is the statistical uncertainty. For some applications, it is convenient to simply add the two numbers together and state the combined uncertainty as <10 msec.

III. TIME AND FREQUENCY STANDARDS

The stability of time and frequency standards is closely related to their quality factor, or Q. The Q of an oscillator is its resonance frequency divided by its resonance width. The resonance frequency is the natural frequency of the oscillator. The resonance width is the range of possible values where the oscillator will run. A high-Q resonator will not oscillate at all unless it is near its resonance frequency. Obviously a high resonance frequency and a narrow resonance width are both advantages when seeking a high Q.

TABLE V Statistics Used to Estimate Time and Frequency Stability and Noise Types

Name	Mathematical notation	Description
Allan deviation	$\sigma_y(\tau)$	Estimates frequency stability. Particularly suited for intermediate to long-term measurements.
Modified Allan deviation	MOD $\sigma_y(\tau)$	Estimates frequency stability. Unlike the normal Allan deviation, it can distinguish between white and flicker phase noise, which makes it more suitable for short-term stability estimates.
Time deviation	$\sigma_x(\tau)$	Used to measure time stability. Clearly identifies both white and flicker phase noise, the noise types of most interest when measuring time or phase.
Total deviation	$\sigma_{y,\,\text{TOTAL}}(\tau)$	Estimates frequency stability. Particularly suited for long-term estimates where τ exceeds 10% of the total data sample.

TABLE VI Summary of Oscillator Types

Oscillator type	Quartz		Rubidium	Commercial cesium beam	Hydrogen maser
	TCXO	OCXO			
Q	10^4 to 10^6	3.2×10^6 (5 MHz)	10^7	10^8	10^9
Resonance frequency	Various	Various	6.834682608 GHz	9.192631770 GHz	1.420405752 GHz
Leading cause of failure	None	None	Rubidium lamp (15 years +)	Cesium beam tube (3 to 25 years)	Hydrogen depletion (7 years +)
Stability, $\sigma_y(\tau)$, $\tau = 1$ sec	1×10^{-8} to 1×10^{-9}	1×10^{-12}	5×10^{-11} to 5×10^{-12}	5×10^{-11} to 5×10^{-12}	1×10^{-12}
Noise floor, $\sigma_y(\tau)$	1×10^{-9} ($\tau = 1$ to 10^2 sec)	1×10^{-12} ($\tau = 1$ to 10^2 sec)	1×10^{-12} ($\tau = 10^3$ to 10^5 sec)	1×10^{-14} ($\tau = 10^5$ to 10^7 sec)	1×10^{-15} ($\tau = 10^3$ to 10^5 sec)
Aging/year	5×10^{-7}	5×10^{-9}	1×10^{-10}	None	$\cong 1 \times 10^{-13}$
Frequency offset after warm-up	1×10^{-6}	1×10^{-8} to 1×10^{-10}	5×10^{-10} to 5×10^{-12}	5×10^{-12} to 1×10^{-14}	1×10^{-12} to 1×10^{-13}
Warm-up period	<10 sec to 1×10^{-6}	<5 min to 1×10^{-8}	<5 min to 5×10^{-10}	30 min to 5×10^{-12}	24 hr to 1×10^{-12}

Generally speaking, the higher the Q, the more stable the oscillator, since a high Q means that an oscillator will stay close to its natural resonance frequency.

This section discusses quartz oscillators, which achieve the highest Q of any mechanical-type device. It then discusses oscillators with much higher Q factors, based on the atomic resonance of rubidium, hydrogen, and cesium. The performance of each type of oscillator is summarized in Table VI.

A. Quartz Oscillators

Quartz crystal oscillators are by far the most common time and frequency standard. An estimated 2 billion (2×10^9) quartz oscillators are manufactured annually. Most are small devices built for wristwatches, clocks, and electronic circuits. However, they are also found inside test and measurement equipment, such as counters, signal generators, and oscilloscopes, and interestingly enough, inside every atomic oscillator.

A quartz crystal inside the oscillator is the resonator. It can be made of natural or synthetic quartz, but all modern devices use synthetic quartz. The crystal strains (expands or contracts) when a voltage is applied. When the voltage is reversed, the strain is reversed. This is known as the *piezoelectric effect*. Oscillation is sustained by taking a voltage signal from the resonator, amplifying it, and feeding it back to the resonator. The rate of expansion and contraction is the resonance frequency and is determined by the cut and size of the crystal. The output frequency of a quartz oscillator is either the fundamental resonance or a multiple of the resonance, called an *overtone frequency*. Most high-stability units use either the third or the fifth overtone to achieve a high Q. Overtones higher than fifth are rarely used because they make it harder to tune the device to the desired frequency. A typical Q for a quartz oscillator ranges from 10^4 to 10^6. The maximum

Q for a high-stability quartz oscillator can be estimated as $Q = 16$ million$/f$, where f is the resonance frequency in megahertz.

Environmental changes such as temperature, humidity, pressure, and vibration can change the resonance frequency of a quartz crystal, and there are several designs that reduce the environmental problems. The *oven-controlled crystal oscillator* (OCXO) encloses the crystal in a temperature-controlled chamber called an oven. When an OCXO is turned on, it goes through a "warm-up" period while the temperatures of the crystal resonator and its oven stabilize. During this time, the performance of the oscillator continuously changes until it reaches its normal operating temperature. The temperature within the oven, then remains constant, even when the outside temperature varies. An alternate solution to the temperature problem is the *temperature-compensated crystal oscillator* (TCXO). In a TCXO, the signal from a temperature sensor generates a correction voltage that is applied to a voltage-variable reactance, or varactor. The varactor then produces a frequency change equal and opposite to the frequency change produced by temperature. This technique does not work as well as oven control but is less expensive. Therefore, TCXOs are used when high stability over a wide temperature range is not required.

Quartz oscillators have excellent short-term stability. An OCXO might be stable ($\sigma_y\tau$, at $\tau = 1$ sec) to 1×10^{-12}. The limitations in short-term stability are due mainly to noise from electronic components in the oscillator circuits. Long-term stability is limited by *aging*, or a change in frequency with time due to internal changes in the oscillator. Aging is usually a nearly linear change in the resonance frequency that can be either positive or negative, and occasionally, a reversal in aging direction occurs. Aging has many possible causes including a buildup of foreign material on the crystal, changes in the oscillator circuitry, or changes in the quartz material or crystal structure. A high-quality OCXO might age at a rate of $<5 \times 10^{-9}$ per year, while a TCXO might age 100 times faster.

Due to aging and environmental factors such as temperature and vibration, it is hard to keep even the best quartz oscillators within 1×10^{-10} of their nominal frequency without constant adjustment. For this reason, atomic oscillators are used for applications that require higher long-term accuracy and stability.

B. Rubidium Oscillators

Rubidium oscillators are the lowest priced members of the atomic oscillator family. They operate at 6,834,682,608 Hz, the resonance frequency of the rubidium atom (^{87}Rb), and use the rubidium frequency to control the frequency of a quartz oscillator. A microwave signal derived from the crystal oscillator is applied to the ^{87}Rb vapor within a cell, forcing the atoms into a particular energy state. An optical beam is then pumped into the cell and is absorbed by the atoms as it forces them into a separate energy state. A photo cell detector measures how much of the beam is absorbed and tunes a quartz oscillator to a frequency that maximizes the amount of light absorption. The quartz oscillator is then locked to the resonance frequency of rubidium, and standard frequencies are derived and provided as outputs (Fig. 8).

Rubidium oscillators continue to get smaller and less expensive, and offer perhaps the best price/performance ratio of any oscillator. Their long-term stability is much better than that of a quartz oscillator and they are also smaller, more reliable, and less expensive than cesium oscillators.

The Q of a rubidium oscillator is about 10^7. The shifts in the resonance frequency are caused mainly by collisions of the rubidium atoms with other gas molecules. These shifts limit the long-term stability. Stability ($\sigma_y \tau$, at $\tau = 1$ sec) is typically 1×10^{-11}, and about 1×10^{-12} at 1 day. The frequency offset of a rubidium oscillator ranges

from 5×10^{-10} to 5×10^{-12} after a warm-up period of a few minutes, so they meet the accuracy requirements of most applications without adjustment.

C. Cesium Oscillators

Cesium oscillators are *primary frequency standards* since the SI second is defined using the resonance frequency of the cesium atom (^{133}Cs), which is 9,192,631,770 Hz. A properly working cesium oscillator should be close to its nominal frequency without adjustment, and there should be no change in frequency due to aging.

Commercially available oscillators use *cesium beam* technology. Inside a cesium oscillator, ^{133}Cs atoms are heated to a gas in an oven. Atoms from the gas leave the oven in a high-velocity beam that travels through a vacuum tube toward a pair of magnets. The magnets serve as a gate that allows only atoms of a particular magnetic energy state to pass into a microwave cavity, where they are exposed to a microwave frequency derived from a quartz oscillator. If the microwave frequency matches the resonance frequency of cesium, the cesium atoms will change their magnetic energy state.

The atomic beam then passes through another magnetic gate near the end of the tube. Those atoms that changed their energy state while passing through the microwave cavity are allowed to proceed to a detector at the end of the tube. Atoms that did not change state are deflected away from the detector. The detector produces a feedback signal that continually tunes the quartz oscillator in a way that maximizes the number of state changes so that the greatest number of atoms reaches the detector. Standard output frequencies are derived from the locked quartz oscillator (Fig. 9).

The Q of a commercial cesium standard is a few parts in 10^8. The beam tube is typically <0.5 m in length, and the atoms travel at velocities of >100 m per second inside the tube. This limits the observation time to a few

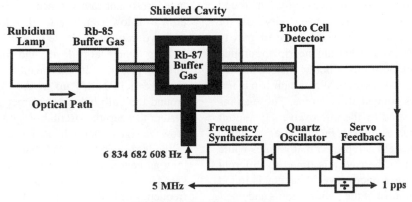

FIGURE 8 Rubidium oscillator.

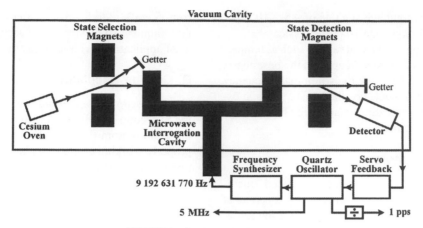

FIGURE 9 Cesium beam oscillator.

milliseconds, and the resonance width to a few hundred hertz. Stability ($\sigma_y\tau$, at $\tau = 1$ sec) is typically 5×10^{-12} and reaches a noise floor near 1×10^{-14} at about 1 day, extending out to weeks or months. The frequency offset is typically near 1×10^{-12} after a warm-up period of 30 min.

The current state-of-the-art in cesium technology is the *cesium fountain* oscillator, named after its fountain-like movement of cesium atoms. A cesium fountain named NIST-F1 serves as the primary standard of time and frequency for the United States.

A cesium fountain works by releasing a gas of cesium atoms into a vacuum chamber. Six infrared laser beams are directed at right angles to each other at the center of the chamber. The lasers gently push the cesium atoms together into a ball. In the process of creating this ball, the lasers slow down the movement of the atoms and cool them to temperatures a few thousandths of a degree above absolute zero. This reduces their thermal velocity to a few centimeters per second.

Two vertical lasers gently toss the ball upward and then all of the lasers are turned off. This little push is just enough to loft the ball about a meter high through a microwave-filled cavity. Under the influence of gravity, the ball then falls back down through the microwave cavity. The round trip up and down through the microwave cavity lasts for about 1 sec and is limited only by the force of gravity pulling the atoms to the ground. During the trip, the atomic states of the atoms might or might not be altered as they interact with the microwave signal. When their trip is finished, another laser is pointed at the atoms. Those atoms whose states were altered by the microwave signal emit photons (a state known as *fluorescence*) that are counted by a detector. This process is repeated many times while the microwave signal in the cavity is tuned to different frequencies. Eventually, a microwave frequency is found that alters the states of most of the cesium atoms and max-

imizes their fluorescence. This frequency is the cesium resonance (Fig. 10).

The Q of a cesium fountain is about 10^{10}, or about 100 times higher than a traditional cesium beam. Although the resonance frequency is the same, the resonance width is much narrower (<1 Hz), due to the longer observation times made possible by the combination of laser cooling and the fountain design. The combined frequency uncertainty of NIST-F1 is estimated at $<2 \times 10^{-15}$.

D. Hydrogen Masers

The *hydrogen maser* is the most elaborate and expensive commercially available frequency standard. The word *maser* is an acronym that stands for microwave amplification by stimulated emission of radiation. Masers operate at the resonance frequency of the hydrogen atom, which is 1,420,405,752 Hz.

A hydrogen maser works by sending hydrogen gas through a magnetic gate that allows only atoms in certain energy states to pass through. The atoms that make it through the gate enter a storage bulb surrounded by a tuned, resonant cavity. Once inside the bulb, some atoms drop to a lower energy level, releasing photons of microwave frequency. These photons stimulate other atoms to drop their energy level, and they in turn release additional photons. In this manner, a self-sustaining microwave field builds up in the bulb. The tuned cavity around the bulb helps to redirect photons back into the system to keep the oscillation going. The result is a microwave signal that is locked to the resonance frequency of the hydrogen atom and that is continually emitted as long as new atoms are fed into the system. This signal keeps a quartz crystal oscillator in step with the resonance frequency of hydrogen (Fig. 11).

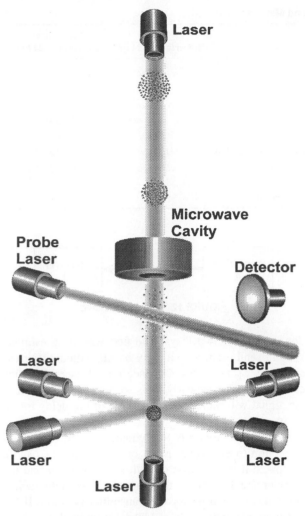

FIGURE 10 Cesium fountain oscillator.

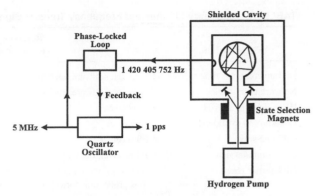

FIGURE 11 Hydrogen maser oscillator.

The resonance frequency of hydrogen is much lower than that of cesium, but the resonance width of a hydrogen maser is usually just a few hertz. Therefore, the Q is about 10^9, or at least one order of magnitude better than a commercial cesium standard. As a result, the short-term stability is better than a cesium standard for periods out to a few days—typically $<1 \times 10^{-12}$ ($\sigma_y \tau$, at $\tau = 1$ sec) and reaching a noise floor of $\cong 1 \times 10^{-15}$ after about 1 hr. However, when measured for more than a few days or weeks, a hydrogen maser might fall below a cesium oscillator's performance. The stability decreases because of changes in the cavity's resonance frequency over time.

E. Future Standards

Research conducted at NIST and other laboratories should eventually lead to frequency standards that are far more stable than current devices. Future standards might use the resonance frequency of trapped, electrically charged ions. Trapping ions and suspending them in a vacuum allows them to be isolated from disturbing influences and observed for periods of 100 sec or longer. Much of this work has been based on the mercury ion (^{199}Hg$^+$), since its resonance frequency in the microwave realm is about 40.5 GHz, or higher than that of other atoms appropriate for this trapping technique. With a resonance width of 10 mHz or less, the Q of a mercury ion standard can reach 10^{12}.

The most promising application of trapped ions is their use in optical frequency standards. These devices use ion traps that resonate at optical, rather than microwave frequencies. The resonance frequency of these devices is about 10^{15} Hz; for example, the ^{199}Hg$^+$ ion has an optical wavelength of just 282 nm. Although long observation times are difficult with this approach, experiments have shown that a resonance width of 1 Hz might eventually be possible. This means that the Q of an optical frequency standard could reach 10^{15}, several orders of magnitude higher than the best microwave experiments.

IV. TIME AND FREQUENCY TRANSFER

Many applications require clocks or oscillators at different locations to be set to the same time (*synchronization*) or the same frequency (*syntonization*). *Time and frequency transfer* techniques are used to compare and adjust clocks and oscillators at different locations. Time and frequency transfer can be as simple as setting your wristwatch to an audio time signal or as complex as controlling the frequency of oscillators in a network to parts in 10^{13}.

Time and frequency transfer can use signals broadcast through many different media, including coaxial cables, optical fiber, radio signals (at numerous places in the spectrum), telephone lines, and the Internet. Synchronization requires both an on-time pulse and a time code.

TABLE VII Summary of Time and Frequency Transfer Signals and Methods

Signal or link	Receiving equipment	Time uncertainty (24 hr)	Frequency uncertainty (24 hr)
Dial-up computer time service	Computer, software, modem, and phone line	<15 msec	NA
Network time service	Computer, software, and Internet connection	<1 sec	NA
HF radio (3 to 30 MHz)	HF receiver	1 to 20 msec	10^{-6} to 10^{-9}
LF radio (30 to 300 kHz)	LF receiver	1 to 100 μsec	10^{-10} to 10^{-12}
GPS one-way	GPS receiver	<50 nsec	$\cong 10^{-13}$
GPS common-view	GPS receiver, tracking schedule (single channel only), data link	<10 nsec	$<1 \times 10^{-13}$
GPS carrier phase	GPS carrier phase tracking receiver, orbital data for postprocessing corrections, data link	<50 nsec	$<1 \times 10^{-14}$
Two-way satellite	Receiving equipment, transmitting equipment, data link	<1 nsec	$<1 \times 10^{-14}$

Syntonization requires extracting a stable frequency from the broadcast, usually from the carrier frequency or time code.

This section discusses both the fundamentals of time and frequency transfer and the radio and network signals used. Table VII provides a summary.

A. Fundamentals of Time and Frequency Transfer

The largest contributor to time transfer uncertainty is *path delay*, or the signal delay between the transmitter and the receiver. For example, consider a radio signal broadcast over a 1000-km path. Since radio signals travel at the speed of light ($\cong 3.3$ μsec/km), we can calibrate the path by estimating the path delay as 3.3 msec and applying a 3.3-msec correction to our measurement. The more sophisticated time transfer systems are self-calibrating and automatically correct for path delay.

Path delay is not important to frequency transfer systems, since on-time pulses are not required. Instead, frequency transfer requires only a stable path where the delays remain relatively constant. The three basic types of time and frequency transfer methods are described below.

1. One-Way Method

This is the simplest and most common way to transfer time and frequency information. Information is sent from a transmitter to a receiver and is delayed by the path through the medium (Fig. 12). To get the best results, the user must estimate τ_{ab} and calibrate the path to compensate for the delay. Of course, for many applications the path delay is simply ignored. For example, if our goal is simply to synchronize a computer clock within 1 sec of UTC, there is no need to worry about a 100-msec delay through a network.

More sophisticated one-way transfer systems estimate and remove all or part of the τ_{ab} delay. This is usually

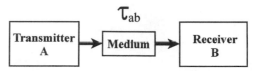

τ_{ab}

FIGURE 12 One-way transfer.

done in one of two ways. The first way is to estimate τ_{ab} and send the time out early by this amount. For example, if τ_{ab} is at least 20 msec for all users, the time can be sent 20 msec early. This advancement of the timing signal will remove at least some of the delay for all users.

A better technique is to compute τ_{ab} and to apply a correction to the broadcast. A correction for τ_{ab} can be computed if the position of both the transmitter and the receiver are known. If the transmitter is stationary, a constant can be used for the transmitter position. If the transmitter is moving (a satellite, for example), it must broadcast its position in addition to broadcasting time. The Global Positioning System provides the best of both worlds—each satellite broadcasts its position and the receiver can use coordinates from multiple satellites to compute its own position.

One-way time transfer systems often include a *time code* so that a clock can be set to the correct time-of-day. Most time codes contain the UTC hour, minute, and second. Some contain date information, a UT1 correction, and advance warning of daylight savings time and leap seconds.

2. Common-View Method

The common-view method involves a single reference transmitter (R) and two receivers (A and B). The transmitter is in common view of both receivers. Both receivers compare the simultaneously received signal to their local clock and record the data. Receiver A receives the signal over the path τ_{ra} and compares the reference to its local clock (R − Clock A). Receiver B receives the signal over

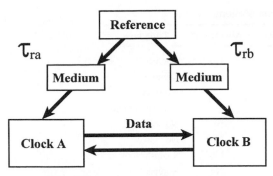

FIGURE 13 Common-view transfer.

the path τ_{rb} and records (R − Clock B). The two receivers then exchange and difference the data (Fig. 13).

Common-view directly compares two time and frequency standards. Errors from the two paths (τ_{ra} and τ_{rb}) that are common to the reference cancel out, and the uncertainty caused by path delay is nearly eliminated. The result of the measurement is (Clock A − Clock B) − ($\tau_{ra} - \tau_{rb}$).

3. Two-Way Method

The two-way method requires two users to both transmit and receive through the same medium at the same time. Sites A and B simultaneously exchange time signals through the same medium and compare the received signals with their own clocks. Site A records A − (B + τ_{ba}) and site B records B − (A + τ_{ab}), where τ_{ba} is the path delay from A to B, and τ_{ab} is the path delay from A to B. The difference between these two sets of readings produces $2(A - B) - (\tau_{ba} - \tau_{ab})$. Since the path is reciprocal ($\tau_{ab} = \tau_{ba}$), the path delay cancels out of the equation (Fig. 14).

The two-way method is used for international comparisons of time standards using spread spectrum radio signals at C- or Ku-band frequencies, and a geostationary satellite as a transponder. The stability of these comparisons is usually <500 psec ($\sigma_x\tau$, at $\tau = 1$ sec), or $<1 \times 10^{-14}$ for frequency, even when the clocks are separated by thousands of kilometers.

The two-way method is also used in telecommunications networks where transmission of a signal can be done in software. Some network and telephone time signals use a variation of two-way, called the *loop-back* method. Like

the two-way method, the loop-back method requires both users to transmit and receive, but not at the same time. For example, a signal is sent from the transmitter (A) to the receiver (B) over the path τ_{ab}. The receiver (B) then echoes or reflects the signal back to the transmitter (A) over the path τ_{ba}. The transmitter then adds the two path delays ($\tau_{ab} + \tau_{ba}$) to obtain the round-trip delay and divides this number by 2 to estimate the one-way path delay. The transmitter then advances the next time signal by the estimated one-way delay. Since users do not transmit and receive at the same time, the loop-back method has larger uncertainties than the two-way method. A reciprocal path cannot be assumed, since we do not know if the signal from A to B traveled the same path as the signal from B to A.

B. Radio Time and Frequency Transfer Signals

There are many types of radio receivers designed to receive time and frequency information. Radio clocks come in several different forms. Some are tabletop or rack-mount devices with a digital time display and a computer interface. Others are available as cards that plug directly into a computer.

The uncertainty of a radio time transfer system consists of the uncertainty of the received signal, plus delays in the receiving equipment. For example, there is cable delay between the antenna and the receiver. There are equipment delays introduced by hardware, and processing delays introduced by software. These delays must be calibrated to get the best results. When doing frequency transfer, equipment delays can be ignored if they remain relatively constant.

The following sections look at the three types of radio signals most commonly used for time and frequency transfer—high frequency (HF), low frequency (LF), and Global Positioning System (GPS) satellite signals.

1. HF Radio Signals (Including WWV and WWVH)

High-frequency (HF) radio broadcasts occupy the radio spectrum from 3 to 30 MHz. These signals are commonly used for time and frequency transfer at moderate performance levels. Some HF broadcasts provide audio time announcements and digital time codes. Other broadcasts simply provide a carrier frequency for use as a reference.

HF time and frequency stations (Table VIII) include NIST radio stations WWV and WWVH. WWV is located near Fort Collins, Colorado, and WWVH is on the island of Kauai, Hawaii. Both stations broadcast continuous time and frequency signals on 2.5, 5, 10, and 15 MHz, and WWV also broadcasts on 20 MHz. All frequencies carry the same program, and at least one frequency should be usable at all times. The stations can also be heard by

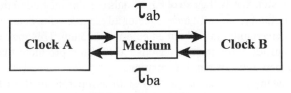

FIGURE 14 Two-way transfer.

TABLE VIII   HF Time and Frequency Broadcast Stations

Call sign	Country	Frequency(ies) (MHz)	Always on?	Language
ATA	India	10	No	English
BPM	China	2.5, 5, 10, 15	No	Chinese
BSF	Taiwan	5, 15	Yes	No voice
CHU	Canada	3.33, 7.335, 14.670	Yes	English/French
DUW21	Philippines	3.65	No	No voice
EBC	Spain	4.998, 15.006	No	No voice
HD2IOA	Ecuador	1.51, 3.81, 5, 7.6	No	Spanish
HLA	Korea	5	No	Korean
LOL1	Argentina	5, 10, 15	No	Spanish
LQB9	Argentina	8.167	No	No voice
LQC28	Argentina	17.551	No	No voice
PLC	Indonesia	11.440	No	No voice
PPEI	Brazil	8.721	No	No voice
PPR	Brazil	4.244, 8.634, 13.105, 17.194	No	No voice
RID	Russia	5.004, 10.004, 15.004	Yes	No voice
RTA	Russia	10, 15	No	No voice
RWM	Russia	4.996, 9.996, 14.996	Yes	No voice
ULW4	Uzbekistan	2.5, 5, 10	No	No voice
VNG	Australia	2.5, 5, 8.638, 12.984, 16	Yes	English
WWV	United States	2.5, 5, 10, 15, 20	Yes	English
WWVH	United States	2.5, 5, 10, 15	Yes	English
XBA	Mexico	6.976, 13.953	No	No voice
XDD	Mexico	13.043	No	No voice
XDP	Mexico	4.8	No	No voice
YVTO	Venezuela	5	Yes	Spanish

telephone; dial (303) 499-7111 for WWV and (808) 335-4363 for WWVH.

WWV and WWVH can be used in one of three modes.

- The audio portion of the broadcast includes seconds pulses or ticks, standard audio frequencies, and voice announcements of the UTC hour and minute. WWV uses a male voice, and WWVH uses a female voice.
- A binary time code is sent on a 100-Hz subcarrier at a rate of 1 bit per second. The time code contains the hour, minute, second, year, day of year, leap second, and Daylight Saving Time (DST) indicators and a UT1 correction. This code can be read and displayed by radio clocks.
- The carrier frequency can be used as a reference for the calibration of oscillators. This is done most often with the 5- and 10-MHz carrier signals, since they match the output frequencies of standard oscillators.

The time broadcast by WWV and WWVH will be late when it arrives at the user's location. The time offset depends upon the receiver's distance from the transmitter but should be <15 msec in the continental United States. A good estimate of the time offset requires knowledge of HF radio propagation. Most users receive a signal that has traveled up to the ionosphere and was reflected back to earth. Since the height of the ionosphere changes, the path delay also changes. Path delay variations limit the received frequency uncertainty to parts in 10^9 when averaged for 1 day.

HF radio stations such as WWV and WWVH are useful for low-level applications, such as the synchronization of analog and digital clocks, simple frequency calibrations, and calibrations of stopwatches and timers. However, LF and satellite signals are better choices for more demanding applications.

2. LF Radio Signals (Including WWVB)

Before the advent of satellites, low-frequency (LF) signals were the method of choice for time and frequency transfer. While the use of LF signals has diminished in the laboratory, they still have a major advantage—they can be received indoors without an external antenna. This makes them ideal for many consumer electronic products that display time-of-day information.

Many time and frequency stations operate in the LF band from 30 to 300 kHz (Table IX). These stations lack

TABLE IX   LF Time and Frequency Broadcast Stations

Call sign	Country	Frequency (kHz)	Always on?
DCF77	Germany	77.5	Yes
DGI	Germany	177	Yes
HBG	Switzerland	75	Yes
JG2AS	Japan	40	Yes
MSF	United Kingdom	60	Yes
RBU	Russia	66.666	No
RTZ	Russia	50	Yes
TDF	France	162	Yes
WWVB	United States	60	Yes

the bandwidth needed to provide voice announcements, but they often provide both an on-time pulse and a time code. The performance of the received signal is influenced by the path length and signal strength. Path length is important because the signal is divided into ground wave and sky wave. The ground wave signal is more stable. Since it travels the shortest path between the transmitter and the receiver, it arrives first and its path delay is much easier to estimate. The sky wave is reflected from the ionosphere and produces results similar to HF reception. Short paths make it possible to track the ground wave continuously. Longer paths produce a mixture of sky wave and ground wave. And over very long paths, only sky wave reception is possible.

Signal strength is also important. If the signal is weak, the receiver might search for a new cycle of the carrier to track. Each time the receiver adjusts its tracking point by one cycle, it introduces a phase step equal to the period of a carrier. For example, a cycle slip on a 60-kHz carrier introduces a 16.67-μsec phase step. However, a strong ground wave signal can produce very good results—a LF receiver that continuously tracks the same cycle of a ground wave signal can transfer frequency with an uncertainty of about 1×10^{-12} when averaged for 1 day.

NIST operates LF radio station WWVB from Fort Collins, Colorado, at a transmission frequency of 60 kHz. The station broadcasts 24 hr per day, with an effective radiated output power of 50 kW. The WWVB time code is synchronized with the 60-kHz carrier and contains the year, day of year, hour, minute, second, and flags that indicate the status of DST, leap years, and leap seconds. The time code is received and displayed by wristwatches, alarm clocks, wall clocks, and other consumer electronic products.

3. Global Positioning System (GPS)

The Global Positioning System (GPS) is a navigation system developed and operated by the U.S. Department of Defense (DoD) that is usable nearly anywhere on earth.

The system consists of a constellation of at least 24 satellites that orbit the earth at a height of 20,200 km in six fixed planes inclined 55° from the equator. The orbital period is 11 hr 58 min, which means that a satellite will pass over the same place on earth twice per day. By processing signals received from the satellites, a GPS receiver can determine its position with an uncertainty of <10 m.

The satellites broadcast on two carrier frequencies: L1 at 1575.42 MHz and L2 at 1227.6 MHz. Each satellite broadcasts a spread spectrum waveform, called a *pseudorandom noise* (PRN) code, on L1 and L2, and each satellite is identified by the PRN code it transmits. There are two types of PRN codes. The first type is a *coarse acquisition* (C/A) code, with a chip rate of 1023 chips per millisecond. The second is a *precision* (P) code, with a chip rate of 10,230 chips per millisecond. The C/A code is broadcast on L1, and the P code is broadcast on both L1 and L2. GPS reception is line-of-sight, which means that the antenna must have a clear view of the sky.

Each satellite carries either rubidium or cesium oscillators, or a combination of both. These oscillators are steered from DoD ground stations and are referenced to the United States Naval Observatory time scale, UTC(USNO), which by agreement is always within 100 nsec of UTC(NIST). The oscillators provide the reference for both the carrier and the code broadcasts.

a. GPS one-way measurements. GPS one-way measurements provide exceptional results with only a small amount of effort. A GPS receiver can automatically compute its latitude, longitude, and altitude using position data received from the satellites. The receiver can then calibrate the radio path and synchronize its on-time pulse. In addition to the on-time pulse, many receivers provide standard frequencies such as 5 or 10 MHz by steering an OCXO or rubidium oscillator using the satellite signals. GPS receivers also produce time-of-day and date information.

A quality GPS receiver calibrated for equipment delays has a timing uncertainty of about 10 nsec relative to UTC(NIST) and a frequency uncertainty of about 1×10^{-13} when averaged for 1 day.

b. GPS common-view measurements. The *common-view* method synchronizes or compares time standards or time scales at two or more locations. Common-view GPS is the primary method used by the BIPM to collect data from laboratories that contribute to TAI.

There are two types of GPS common-view measurements. *Single-channel common-view* requires a specially designed GPS receiver that can read a tracking schedule. This schedule tells the receiver when to start making measurements and which satellite to track. Another user at another location uses the same schedule and makes simultaneous measurements from the same satellite. The tracking schedule must be designed so that it chooses satellites visible to both users at reasonable elevation angles. *Multichannel common-view* does not use a schedule. The receiver simply records timing measurements from all satellites in view. In both cases, the individual measurements at each site are estimates of (Clock A − GPS) and (Clock B − GPS). If the data are exchanged, and the results are subtracted, the GPS clock drops out and an estimate of Clock A − Clock B remains. This technique allows time and frequency standards to be compared directly even when separated by thousands of kilometers. When averaged for 1 day, the timing uncertainty of GPS common-view is <5 nsec, and the frequency uncertainty is $<1 \times 10^{-13}$.

c. GPS carrier phase measurements. Used primarily for frequency transfer, this technique uses the GPS carrier frequency (1575.42 MHz) instead of the codes transmitted by the satellites. Carrier phase measurements can be one-way or common-view. Since the carrier frequency is more than 1000 times higher than the C/A code frequency, the potential resolution is much higher. However, taking advantage of the increased resolution requires making corrections to the measurements using orbital data and models of the ionosphere and troposphere. It also requires correcting for cycle slips that introduce phase shifts equal to multiples of the carrier period ($\cong 635$ psec for <1). Once the measurements are properly processed, the frequency uncertainty of common-view carrier phase measurements is $<1 \times 10^{-14}$ when averaged for 1 day.

C. Internet and Telephone Time Signals

One common use of time transfer is to synchronize computer clocks to the correct date and time-of-day. This is usually done with a time code received through an Internet or telephone connection.

1. Internet Time Signals

Internet time servers use standard timing protocols defined in a series of RFC (Request for Comments) documents. The three most common protocols are the Time Protocol, the Daytime Protocol, and the Network Time Protocol (NTP). An Internet time server waits for timing requests sent using any of these protocols and sends a time code in the correct format when a request is received.

Client software is available for all major operating systems, and most client software is compatible with either the Daytime Protocol or the NTP. Client software that uses the Simple Network Time Protocol (SNTP) makes the same timing request as an NTP client but does less processing and provides less accuracy. Table X summarizes the various protocols and their port assignments, or the port where the server "listens" for a client request.

NIST operates an Internet time service using multiple servers distributed around the United States. A list of IP addresses for the NIST servers and sample client software can be obtained from the NIST Time and Frequency Division web site: http://www.boulder.nist.gov/timefreq. The uncertainty of Internet time signals is usually <100 msec, but results vary with different computers, operating systems, and client software.

2. Telephone Time Signals

Telephone time services allow computers with analog modems to synchronize their clocks using ordinary telephone lines. These services are useful for synchronizing computers that are not on the Internet or that reside behind an Internet firewall. One example of a telephone service is NISTs Automated Computer Time Service (ACTS), (303) 494-4774.

TABLE X Internet Time Protocols

Protocol name	Document	Format	Port assignment(s)
Time protocol	RFC-868	Unformatted 32-bit binary number contains time in UTC seconds since January 1, 1900	Port 37, tcp/ip, udp/ip
Daytime protocol	RFC-867	Exact format not specified in standard. Only requirement is that the time code is sent as ASCII characters	Port 13, tcp/ip, udp/ip
Network time protocol (NTP)	RFC-1305	The server provides a data packet with a 64-bit time stamp containing the time in UTC seconds since January 1, 1900, with a resolution of 200 psec. NTP provides an accuracy of 1 to 50 msec. The client software runs continuously and gets periodic updates from the server.	Port 123, udp/ip
Simple network time protocol (SNTP)	RFC-1769	The data packet sent by the server is the same as NTP, but the client software does less processing and provides less accuracy.	Port 123, udp/ip

ACTS requires a computer, a modem, and client software. When a computer connects to ACTS it receives a time code containing the month, day, year, hour, minute, second, leap second, and DST indicators and a UT1 correction. The last character in the ACTS time code is the on-time marker (OTM). To compensate for the path delay between NIST and the user, the server sends the OTM 45 msec early. If the client returns the OTM, the server can calibrate the path using the *loop-back* method. Each time the OTM is returned, the server measures the round-trip path delay and divides this quantity by 2 to estimate the one-way path delay. This path calibration reduces the uncertainty to <15 msec.

V. CLOSING

As noted earlier, time and frequency standards and measurements have improved by about nine orders of magnitude in the past 100 years. This rapid advancement has made many new products and technologies possible. While it is impossible to predict what the future holds, we can be certain that oscillator Q's will continue to get higher, measurement uncertainties will continue to get lower, and new technologies will continue to emerge.

SEE ALSO THE FOLLOWING ARTICLES

MICROWAVE COMMUNICATIONS • QUANTUM MECHANICS • RADIO SPECTRUM UTILIZATION • REAL-TIME SYSTEMS • SIGNAL PROCESSING • TELECOMMUNICATIONS

BIBLIOGRAPHY

Allan, D. W., Ashby, N., and Hodge, C. C. (1997). "The Science of Timekeeping," Hewlett–Packard Application Note 1289, United States.

Hackman, C., and Sullivan, D. B. (eds.) (1996). "Time and Frequency Measurement," American Association of Physics Teachers, College Park, MD.

IEEE Standards Coordinating Committee 27 (1999). "IEEE Standard Definitions of Physical Quantities for Fundamental Frequency and Time Metrology—Random Instabilities," Institute of Electrical and Electronics Engineers, New York.

ITU Radiocommunication Study Group 7 (1997). "Selection and Use of Precise Frequency and Time Systems," International Telecommunications Union, Geneva, Switzerland.

Jespersen, J., and Fitz-Randolph, J. (1999). "From Sundials to Atomic Clocks: Understanding Time and Frequency," NIST Monograph 155, U.S. Government Printing Office, Washington, DC.

Kamas, G., and Lombardi, M. A. (1990). "Time and Frequency Users Manual," NIST Special Publication 559, U.S. Government Printing Office, Washington, DC.

Levine, J. (1999). "Introduction to time and frequency metrology," *Rev. Sci. Instrum.* **70**, 2567–2596.

Seidelmann, P. K. (ed.) (1992). "Explanatory Supplement to the Astronomical Almanac," University Science Books, Mill Valley, CA.

Sullivan, D. B., Allan, D. W., Howe, D. A., and Walls, F. L. (eds.) (1990). "Characterization of Clocks and Oscillators," NIST Technical Note 1337, U.S. Government Printing Office, Washington, DC.

Vig, J. R. (1992). "Introduction to Quartz Frequency Standards," United States Army Research and Development Technical Report SLCET-TR-92-1.

Walls, F. L., and Ferre-Pikal, E. S. (1999). Frequency standards, characterization. Measurement of frequency, phase noise, and amplitude noise. *In* "Wiley Encyclopedia of Electrical and Electronics Engineering," Vol. 7, pp. 767–774, Vol. 12, pp. 459–473, John Wiley and Sons, New York.

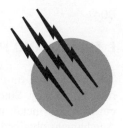

Tin and Tin Alloys

William B. Hampshire

Tin Research Institute, Inc.

GLOSSARY

Alluvial Describing a tin ore deposit where the primary (lode) deposit of tin ore has been eroded by water.

Base box Measure of tin plate surface area corresponding to 112 sheets, each 20 in. by 14 in. Tin plate gauges and tin coating thicknesses are still referred to in pounds per base box.

Bronze Alloy of copper with tin developed in prehistoric times; used for its corrosion and wear resistance.

Eutectic Two or more metallic elements at a specific composition that melts at a constant temperature, as does a pure metal. For example, an alloy of ~62% tin and 38% lead melts at 183°C, below the melting point of either tin or lead.

Fire refining Production of refined metal by a series of furnace treatments, as opposed to electrolytic refining, which is a selective dissolution-electroplating sequence.

Fusible alloy Metal alloy, usually containing at least some tin, that melts at a low temperature (183°C or below) and is used primarily because of this low melting temperature.

Pewter In the modern sense, a tin-based alloy with anti-

mony and copper additions. It is prized for its decorative appeal.

Placer Another term for an alluvial deposit.

Terneplate Mild steel base with a coating of lead–tin alloy. The tin content of the coating typically is 8–20%.

Tin plate Mild steel base with a coating of pure tin, the material from which tin cans are made.

White metals Family of tin-based (or lead-based) bearing metals, so named for their characteristic color. In the broader sense of the term, pewter is a white metal as well.

TIN is a soft, ductile, metallic element used in prehistoric times to create bronze by alloying the tin into copper. Elemental tin possesses atomic number 50, atomic weight 118.69, and chemical symbol Sn (derived from the Latin name *stannum*). Tin is in the periodic table subgroup with carbon, silicon, germanium, and lead and has some properties similar to the properties of these other elements. The applications of tin typically take advantage of the metal's low melting point, excellent corrosion resistance, and nontoxicity to living organisms. The largest of these applications are tin plate, solder, and tin chemicals.

I. MINING AND REFINING OF TIN

The practice of tin-extractive metallurgy began before recorded history. Since tin has never been known to occur as native metal, it is generally believed that Bronze Age humans obtained bronze by smelting together native copper with a tin ore, undoubtedly the tin oxide mineral called cassiterite.

Even today cassiterite is the dominant ore constituent for commercial exploitation. In some areas of the world, it still exists as primary lode deposits, but most of the production involves secondary or placer deposits, where the original lode body has been eroded by natural forces.

A. Alluvial Mining

In tropical areas of the world the secondary tin deposits are often mined by alluvial techniques. Where possible, dredges, which are large, floating processing plants, are constructed on natural or manmade lakes or, more recently, at offshore locations. Here the dredges use a bucket line or suction cutter to lift the ore from up to 50 m below the water level into the dredge, where a series of jigs, shaking tables, and flotation cells effect beneficiation of the ore, up to a level as high as 70% tin from perhaps 0.02% initially. The overburden from which the tin mineral has been removed is discharged from the rear of the dredge to settle back to the lake bottom.

Obviously the mining of such low-grade ores requires the handling of large amounts of material, and the larger dredges often are capable of processing more than 1000 tons/hr. The separation of cassiterite is energy intensive aboard a dredge, so a high throughput is necessary to justify the high capital cost of the dredge and the high operating costs.

B. Above-Water Placer Mining

Other placer deposits are found where flooding and dredging may not be economical or even possible, and these are typically worked by gravel pump mining. Typically, a pit is created and enlarged as high-pressure streams of water are directed against the sides. This water washes the tin-bearing material to a sump, where the gravel pump lifts the slurry to a processing area.

Normally, the ore is processed by pumping into a long sluice, where the cassiterite settles to the bottom and the lighter non-tin-bearing material is carried away. The partially concentrated ore can then be removed and taken to a concentrator, where jigs, shaking tables, and flotation cells are used to produce a product suitable for smelting.

It has been estimated that ~40% of the world's tin is produced in this manner. Often the operation is a modest one, perhaps run by a single family, where the miner sells

the ore to a custom concentrator, there being relatively little vertical integration in the tin mining industry.

C. Hard-Rock Mining

Primary lode deposits are found in Bolivia and Australia, for example, and these are worked by underground mining techniques. Shafts are sunk and tunnels driven from them into the lodes so that the ores can be removed. These ores are often complex in composition, requiring more expensive processes to extract the metal values.

The mining in Cornwall, England, is actually hard-rock mining, but in fact some tunnels run out under the sea. These Cornish tin mines date back centuries to Roman times. In recent years there has been renewed interest in mining there, although hard-rock mining is seldom cost competitive with other methods.

D. Refining Techniques

Often concentrates are treated before smelting to remove some of the metallic impurities. The exact nature of the treatment depends on the impurities present, but often roasting or leaching processes figure prominently. Roasting drives off certain impurities that are volatile and may convert other impurities, especially if roasted under a controlled atmosphere, to more leachable compounds. Water or acid leaching can then remove additional impurities, but of course the concentrate must then be dried before smelting.

Tin smelting can be carried out in blast furnaces, reverberatory furnaces, or electric furnaces. Regardless of the furnace used, the principles of tin smelting remain the same. The concentrate is mixed with carbon, usually as powdered anthracite, which serves as the reducing agent to change the tin oxide to metallic tin. Also added is limestone, which acts as a fluxing material.

This mixture is charged into the furnace, where it is heated to ~1400°C for ~12 hr with agitation of the bath to promote separation of the impurities from the metal. The furnace is tapped into settling pots, where the metal again sinks to the bottom and slag overflows the top. The metal can then be cast into solid form for additional refining, but even the slag must be reprocessed.

Unfortunately for tin smelters, tin oxides easily combine with silica, which is readily available in tin ores. The resulting tin silicates enter the slag in such quantity that the first slag is too valuable to be discarded. Often a separate furnace is employed, with considerable smelting experience, to remove enough tin from the first slag to leave a second slag for discarding and more impure tin metal for further refining.

Fire refining is preferred whenever the impurity content of the metal allows it. Typically, liquating is a first step and utilizes a sloped-floor furnace, heated to just above the melting point of tin. The tin-rich metallic phase melts out and runs down the slope, leaving iron, copper, and other higher melting-point impurities behind. Often liquating is followed by a poling process similar to that used in the copper industry. Poles of green wood are immersed into the metal and bring impurities to the surface, where they can be skimmed off. The resulting metal is typically ~99.8% tin.

Electrolytic refining of tin is practiced when certain impurities are present and cannot be removed by fire refining. Again as is the case for copper, impure metal is cast into anodes, which are hung in cells of electrolyte. An electrical current causes the anodes to dissolve and the tin plates out on cathodes hung nearby. The cathodes are made from electrolytically refined tin initially and, when fully plated, are removed and melted and cast into ingot form for shipping. They typically reach a purity of 99.99% or better.

II. MARKETING OF TIN

Throughout the history of its use, tin has been a strategic material. There is evidence to suggest that long-range contacts between peoples during the Bronze Age were actually fostered by a search for tin reserves. Even today, tin is a geographically significant metal, usually produced as a primary metal a long distance from its consumption. Table I lists the major tin-producing countries. The major tin consumers are shown in Table II.

TABLE I Western World Tin Production by Country (tonnes)

| Country[a] | 1995 | | 1996 | |
	Refined tin	Tin in concentrates	Refined tin	Tin in concentrates
Malaysia	41,000	6402	38,100	5175
Indonesia	44,200	46,058	48,900	51,024
Thailand	8,200	1784	11,000	1299
Brazil	17,000	19,500	17,600	19,200
Bolivia	18,000	16,300	16,700	14,802
Peru	—	22,020	500	27,002
Australia	—	7750	—	8496
Portugal	—	4616	—	4625
U.K.	—	2000	—	2000
Africa	—	3200	—	3200
Rest of West	14,000	2000	13,000	2000
Total	142,400	131,630	145,800	138,823

[a] China and Russia are major tin producers, but production data are not currently verifiable.

Note: Data are from *Tin Monitor*, published by CRU International, Ltd., London.

TABLE II Western World Consumption of Refined Tin by Country (tonnes)

Country	1995	1996
USA	35,000	36,300
Japan	28,100	26,900
Germany	19,000	19,600
U.K.	10,400	10,600
France	8200	8100
Rest of Europe	23,700	21,300
Other industrialized countries	8000	6000
Developing countries	56,500	58,000
Total	188,900	186,800

Note: Data are from *Tin Monitor*, published by CRU International, Ltd., London.

A. International Tin Agreements

The International Tin Agreement was established primarily to prevent the price fluctuations that plagued the market for many years. These price variations hurt consumers and producers alike, and both groups sought a better balance of supply and demand. The operating body for the tin agreements, currently for the Sixth International Tin Agreement, is the International Tin Research Institute (ITRI) Ltd., established in 1956. Representatives of the member countries meet to discuss the tin industry and tin markets and, when necessary, to vote to take action that will bring about a shift in the supply–demand relationship.

The nature of the ITRI is often misunderstood. It is not a "producer cartel," because both producers and consumers are represented. The producer countries are apportioned 1000 votes based on their production, and the consumer countries are apportioned 1000 votes based on their consumption, so neither group of countries by itself can force action on the other.

The primary tool available to the ITRI in influencing supply and demand is a buffer stock of tin metal or cash. The buffer stock manager buys or sells tin depending on whether the price is too rapidly declining or escalating, respectively. The exact circumstances under which the buffer stock manager must act, may act, or must not act are defined by the ITRI on the basis of the information it has on tin production costs, mining capacities, future trends, and other factors.

At the time of this writing, the ITRI has invoked its other, more serious provision for influencing the supply side, namely, export controls. By requiring the producing countries to limit their exports of tin, it is hoped that balance will be restored. This effort is extremely difficult for the producing countries, which depend greatly on tin exports for their foreign trade value.

B. Tin Markets

World prices for tin are often based on the quotations on the London Metal Exchange (LME), where tin has been traded for decades. Contracts on the LME are made for immediate delivery or for delivery in 3 months. The contract prices are used to determine the "closing prices," and there are two of these: "cash" (for immediate delivery) and "forward" (for delivery in 3 months).

Recently the Kuala Lumpur Tin Market has begun a trade in tin. This market is, of course, based on a different currency and therefore introduces the factor of currency exchange rates into the trading of tin metal. There was already a Penang price established by physical trading in concentrates at the Penang smelters. This price has been used as a guide for trading elsewhere, as have the "offerings," the number of tons of concentrates traded, an indication of supply on a daily basis.

Tin is also traded on most of the major commodity markets—for example, the COMEX in New York. Tin prices are regularly reviewed in several publications, such as the *American Metal Market* and *Iron Age*.

III. PROPERTIES OF TIN

A. Physical Properties

Table III lists some of the properties of tin. The wide range of applications of tin make use of various properties, especially the low melting point, the capacity of tin to form alloys with many other metals, its excellent corrosion resistance, and its nontoxicity. The importance of each of the properties is described in Section IV.

B. Mechanical Properties

At room temperature tin is already at more than one-half of its absolute melting point (505 K). Therefore, the metal is in the high-temperature regime of mechanical behavior. Creep, recrystallization, and grain growth may all occur readily, so the mechanical strength of tin (Table IV) is too low to permit its use as a structural material. For this reason tin is nearly always alloyed with another metal or coated onto a stronger metal to provide support. It is quite fortunate that tin coats other metals so well.

IV. APPLICATIONS OF TIN

A. Tin Plate and Canning

1. Tin Plate Manufacture

Historically the largest (~35% worldwide) and most important use of tin has been for coating mild steel sheet

TABLE III Some Physical Properties of Tin[a]

Property	Value
Atomic number	50
Atomic weight	118.69
Valencies	2, 4
Density (kg/m^3)	
β-Tin at 15°C	7.29
α-Tin at 13°C	5.77
Liquid tin at mp	6.97
Melting point (°C)	232
Vapor pressure (mm)	
At 1000 K	7.4×10^{-6}
At 1500 K	0.17
At 2000 K	30.6
At 2500 K	638
Boiling point (°C)	~2270
Latent heat of fusion (kJ/g atom)	7.08
Thermal conductivity at 20°C (W/mK)	65
Linear coefficient of thermal expansion at 0°C (°C^{-1})	19.9×10^{-6}
Surface tension at mp (mN/m)	544
Expansion on melting (%)	2.3
Electrical resistivity, β-tin at 20°C, ($\mu\Omega$ cm)	12.6
Critical temperature (°C) and pressure (atm)	3730 and 650

[a] Corrosion behavior summary: Tin stays bright in dry air at room temperatures. It is generally oxidized rather slowly in harsher atmospheric conditions. Tin is very resistant to high-purity natural fresh waters and to milk products. Tin is resistant to most dilute acid solutions (not resistant to nitric or sulfuric acids) and is resistant to some stronger acid solutions if they are oxygen free. Tin is resistant to most solvents, oils, and other chemicals, the notable exceptions being chlorine, potassium hydroxide, and sodium hydroxide. Tin is less corrosion resistant at elevated temperatures.

to make the product called tin plate. This material is then fabricated into tin cans to preserve food or for the containment of a wide variety of other products. This process is due to an ingenious materials concept that was developed

TABLE IV Some Mechanical Properties of Tin[a]

Property	Value
Tensile strength (N/mm^2)	
At 20°C, 0.4 mm/mm min	14.5
At 100°C, 0.4 mm/mm min	11.0
Young's modulus at 20°C (kN/mm^2)	49.9
Poisson's ratio	0.357
Creep strength, approx. life at 2.3 N/mm^2 and 15°C (days)	170
Brinell hardness (HB)	
At 20°C, 10 kg/5 mm × 180 sec	3.9
At 100°C, 10 kg/5 mm × 180 sec	2.3

[a] Note: The tensile properties of pure tin and many tin-based alloys are very dependent on the rate at which a load is applied.

in Bavaria in about the fourteenth century. The strength, durability, light weight, and fabricability of iron (later steel) were combined with the corrosion resistance and compatibility of a tin coating to yield an important engineering material for pots and pans, lanterns, boxes, and so on. Then when Appert invented the method of preserving foods by sterilization inside sealed containers, it seemed only natural to use tin plate as the container material.

Modern tin plate is rather different from the original material and, in fact, has evolved greatly since the 1930s. Today, cold-rolled steel in huge coils is fed into a processing line that electrolytically coats the surface with tin. At the entry end coils are welded together end to end to give a continuous feed of steel strip. The strip is passed through a series of tanks that clean, pickle, and rinse the surface.

The tanks that follow are electrolytic cells, where tin is plated on the surface. The strip may be moving at up to 600 m/min. Any one of three electrolytes may be employed for the plating process. Pure tin anodes dissolve under the influence of the applied current and plate out on the strip as it passes. Typically, both sides of the strip are plated, but the cells may be engineered to give different coating weights on the two sides, so-called differential coatings.

The plated tin shows the matte finish at this point; it is dull and somewhat whitish. To brighten the surface appearance, the tin plate is "reflowed" by passing through a unit that heats the tin coating to just above its melting point. At the same time a thin, continuous layer of iron–tin intermetallic compound is formed between the steel and the tin layer, and corrosion properties of this intermetallic improve the subsequent performance of the tin plate.

Before being recoiled at the exit end of the process line, the tin plate undergoes two more treatments. A chromate-based passivation film is developed chemically on the surface to provide stability against excessive oxidation. The strip then receives a very thin coating of oil, which helps prevent damage to the surface by rubbing during shipment. Next the strip passes an inspection station, where optical or automatic scanning equipment notes any defects in the surface. Usually, the strip is recoiled, although sometimes it is cut into sheets at this point. More often, if cut sheets are needed, they are made as a separate operation so that the tin plate line can be operated more efficiently.

The steel strip that is the basis for tin plate is made to exacting standards. The steel impurities must be kept low to optimize the corrosion resistance and fabricability of the tin plate. The strip must also have a carefully controlled thickness, often in the range of 0.15 to 0.35 mm, and must have a consistent thickness across and along the strip. The temper of the strip must be of such value as to give the required degree of stiffness and drawability required in further fabrication.

The tin coating on the surface of the steel strip must similarly meet exacting specifications. It is typically 0.0003–0.0008 mm thick on each side, with a strong tendency in recent years toward the low end of this range. The new low-tin-coating tin plates range down to $\sim$0.00007 mm for the tin thickness.

2. Two-Piece Can Manufacture

Can manufacture has also undergone many changes. With increased competition from alternative materials, tin plate cans must be made less expensively and, for beverage containers especially, two-piece can-making methods can be more efficient. The two-piece method uses circular blanks stamped from tin plate. These blanks are drawn with a punch and die arrangement to form a cup. The cup can then be redrawn to the final shape, or the cup walls can be ironed, which thins the tin plate sides and extends them up to form the final shape. The redrawing process produces a more uniform thickness overall and therefore is generally limited to smaller height-to-diameter ratios. The wall-ironed can has a relatively thicker bottom and relatively thinner sides and can be made into the taller, narrower profiles.

Huge presses can draw these cans from stock at high speeds, approaching 200 strokes per minute. This productivity puts great demands on the tin plate, since any defects in the stock may cause a tearing of the can side and the possible jamming of the press. Temper, drawability, and sheet thickness must all be held to close tolerances. The tin is a useful metallic lubricant in each process and also imparts more traditional benefits described below.

No matter which drawing process is used, the remaining steps are similar. The can top edge is trimmed to final dimension, a step required not only because of processing variables, but also owing to sheet anisotropy. The mechanical property difference between the direction parallel to the rolling direction and the direction transverse results in "ears," regular height variations along the top edge.

The next step is exterior decoration and interior coating of the can body (plus bottom) "in the round." This complex subject is discussed in Section I.A.3. As a final step, the top edge of the can may be "necked-in," a reduction in diameter that reduces the required can end diameter, which results in materials savings and facilitates stacking of the cans in shipment, storage, and display. The cans are shipped to the filling location, where the ends (tops only) are seamed on after filling.

3. Three-Piece Can Manufacture

The three-piece method of tin can manufacture is the traditional method and still accounts for the majority of cans.

The process begins with cut tin plate sheets. Food cans often have plain exteriors, to receive paper labels after filling. When the exterior is to receive decoration, however, it is usually done on flat sheets at this time, either in-house or by custom metal decorators.

Metal-decorating processes vary greatly depending on the performance requirements for the finished can. The exterior coatings are termed inks, varnishes, lacquers, or enamels, with some overlap in definitions. There may be size coats, base coats, and so on, and the processing can be roller coating or photolithography (or both). Distortion printing can be used to allow in advance for changes in shape during container forming. All these possibilities are used to good advantage by can designers to present a package that catches the attention of the consumer and conveys the sense of the quality of the product.

The interior coating, however, is undoubtedly more crucial to product preservation. In many instances, a plain tin interior allows a controlled dissolution of tin, which protects the product from deterioration in taste, color, or other organoleptic qualities. When a tin surface is not essential, any of a variety of enamels can be applied to prevent undesired product–container interactions. Even in these cases, a tin coating underneath the enamel provides a further protective barrier in case of inadvertent damage to the enamel coating.

The coated sheets of tin plate are now slit into body blanks, which are then formed into cylinders around a mandrel. The seam is typically overlapped slightly and welded together under compression using a moving consumable intermediate copper wire electrode. Rapid progress in the technology of can welding has made it possible for can manufacturers to satisfy the demand for high production rates. The tin coating can be quite thin when it is to be welded, but it is necessary for the improved electrical contact properties it provides. Welding eliminates any concerns over solder contamination of the contents and is still gaining favor for this reason.

Soldering the side seam is the older method of manufacture and is still used on nonfood packaging and some "dry" packs. For soldering, the side seam is folded into a lock seam, fluxed and pressed flat, and wiped with a 2% tin–98% lead solder. Once again this process must be capable of high speeds. In some cases prescored blanks of two or three cans are soldered as a unit, then divided into the appropriate parts. The tin coating provides an excellent solderable coating for high-speed operation, in addition to its long-established capacity to preserve the contents.

Whether welded or soldered, the can bodies may next have the side seam sprayed with a strip of special interior enamel. This strip is cured and followed by flanging—that is, bending both top edge and bottom edge out at a right angle to allow the can ends to be seamed on as

appropriate. An overall protective interior enamel coating may be applied next and cured.

Next the can bottoms, separately produced, are put into place. These have been stamped from sheet, pressed into the needed profile, coated as required, and finally edged with a sealing compound and seamed onto the bodies. Once again the tops are seamed into place after product filling.

4. Other Products

Although the preceding description concentrates on the very important application of tin plate to food and beverage containers, a wide range of other products are packed in tin plate containers—for example, paints, aerosols, specialty chemicals, and batteries. The specific details of fabrication vary, of course, but the soldering, welding, and deep drawing capacities of tin plate are all used to full advantage by the industries involved.

Tin plate also has noncontainer uses for various corrosion-resistance applications. These include gaskets, seals, automotive air and oil filters, baking pans, graters, badges, toys, and so on. Tin plate has also been given a laminated plastic coating, which allows interesting texture effects for more decorative purposes. The deep drawing capacity of tin plate has spawned widespread use as electronic shielding against electromagnetic interference in computers, televisions, video games, and similar devices.

B. Solder

Though the second most common use of tin worldwide, solder is overtaking tin plate in the developed countries, not surprisingly in proportion to the level of electronics industry development. Although the transportation industry consumes large tonnages of solder, the higher average tin content of solder used in electronics makes the latter the more important consumer of tin.

In simplest terms, solder is a mixture of tin and lead with up to 63% tin in nearly all cases. Smaller amounts of antimony, silver, copper, cadmium, bizmuth, indium, and other elements may be added for special purposes. The eutectic composition at 62% tin–38% lead melts at the lowest temperature of the tin–lead combinations, that is, at 183°C. This low temperature for metal joining has proved valuable for joining heat-sensitive electronics components with a minimal chance of heat damage.

Besides allowing joining at low temperatures, soldering also benefits from the capacity of tin to wet and alloy with a variety of useful metals of construction. Often solder is relatively inexpensive compared with other joining materials, and reasonable degrees of automation are possible. The good corrosion resistance of tin and lead in specific environments can be used advantageously by solder alloy selection.

1. Soldering Essentials

Solder alloy selection is a complex topic. To provide a few guidelines, a nearly eutectic composition of 60 to 63% tin is standard for electronics soldering. Alloys of 50% tin are traditional for plumbing and sheet metal applications. In the 20 to 40% range, the alloys are used for general engineering purposes, and the low-tin solders, up to perhaps 10% tin, are standard for can soldering, radiator soldering, and some electronics soldering in which a two-step soldering operation is required. Pure tin has been used for can side-seam soldering of milk products and baby foods and for step soldering. Antimony or silver is generally added for higher temperature soldering or for additional strength. Adding cadmium or bismuth lowers the melting point for fusible alloys and special applications.

An important design limitation of the high-tin solders involves service temperatures in the cryogenic range. Alloys of more than $\sim$20% tin undergo a dramatic ductile-to-brittle transition at about $-100°C$, so for such service temperatures low-tin solders are required.

The essential constituents of a soldered joint are the basis metals, a flux, a solder, and a source of heat. The basis metals must be clean and solderable, and if solderability of the metal is difficult (aluminum, stainless steel, case iron), it may be advisable or necessary to apply a plating of a solderable metal. Even if the metal has good solderability (copper, brass, low alloy steel), if some storage is required, a solderable coating will help preserve solderability for some time. Tin or tin–lead platings or hot-dip coatings are excellent for this purpose.

A flux is a chemical agent that removes light tarnish films on the basis metal, protects the surfaces from reoxidation during heating, and generally assists the molten solder to wet and spread over the surfaces to be soldered. Fluxes are often of proprietary compositions. For electronics they usually consist of a natural wood resin base (called "rosin base") with small amounts of halide-containing activators to improve the performance. For more difficult soldering situations and when postcleaning to remove flux residues is possible, mixtures of organic acids or of inorganic halides are used as fluxes. These more corrosive fluxes may leave corrosive residues on the soldered surfaces, and this is the reason cleaning after soldering is a necessity.

The solder composition is selected as already outlined, and for many soldering processes the solder application step is combined with the source of heating. For example, in the wave soldering process commonly used for electronics, a printed circuit board loaded with electronic components is moved by conveyor through a machine that first sprays flux on the bottom of the board or moves the board through a wide wave of flux pumped up to meet it. Next the board passes over a wave of molten solder that supplies both heat and solder to the surface. Finally, after cooling the board passes through a cleaning section. By such a process several hundred solder joints can be made in a few seconds. Dip soldering is a rather similar process.

A large number of joints are still made by hand soldering, where the solder and heat are applied separately. The solder might be in the form of wire, stick, foil, stamped preforms (small shapes customized to the specific application), or solder paste. These forms may have the flux incorporated as a core inside or may be flux-coated or, in the case of pastes, may be mixtures of flux and solder powder. The heat can then be applied with a soldering iron, with a torch, by electrical resistance heating, with a hotplate, with an oven, or by condensation of a heated fluorocarbon (called vapor-phase soldering). There are many variations in the specific details of soldering.

A growing concern among solder users is the mechanical behavior of solder joints. Traditionally, joints to be soldered were mechanically fixed to provide support, with the solder simply a filler metal to maintain electrical continuity and corrosion resistance. In recent years, however, mechanical fixing has proved time-consuming to accomplish and, in the case of delicate components, perhaps impossible. This means that the solder joint is expected to have adequate mechanical strength. In addition, there are new joining techniques, such as surface mounting, which involves direct mounting of electronic components to a circuit board with no leads on the components. In this method the solder joint must withstand the fatigue forces caused by the differences in thermal expansion when the component materials are heated and cooled in service. In fact, in nearly all soldering applications the service conditions are becoming more demanding as designers aim for more efficiency and performance.

In transportation applications, wider use of aluminum radiators will reduce the use of the low-tin solders that have become standard. The high-lead solders used for body filling are being replaced, in part, by a high-tin (nonleaded) solder. The wider use of electronics in automobiles should increase the usage of tin-rich solders.

In construction applications high-tin (nonleaded) solders are being used more for plumbing to reduce even more the very slight chance of lead contamination of water. Some of these alloys were already in use for special high-strength requirements. Overall, the use of tin in soldering shows promise of growth in the future.

C. Other Metallurgical Applications

1. Bearings

A good bearing material is one of the best examples of an engineering compromise in modern technology. The bearing must be strong enough not to deform under fluctuating loads, yet must deform enough to conform to shaft alignment variations. The bearing must be hard enough to resist wear, but soft enough to allow dirt and contamination to become embedded in the surface rather than cause additional wear. Finally, the bearing must allow a lubricating film to be maintained, yet have high corrosion resistance against many hostile environments. Not surprisingly, bearings therefore are often two-phase structures, one a hard phase for strength and hardness and the other a soft phase for conforming and embedding. The capacity of tin to hold an oil film plus its good corrosion resistance, softness, and good alloying behavior make it a natural candidate for one phase.

Tin-based white metals (often called Babbitt alloys after an inventor of 150 years ago) generally, use 7–15% antimony plus 1–5% copper, with perhaps a few percent lead or other alloying elements. The antimony and copper combine with tin to form intermetallic compounds, which provide the hard phase for the bearing material, the remaining tin-rich phase being the soft one. These alloys are still preferred for large-equipment applications (marine diesels, steam turbine bearings, earth-moving machinery, railroad applications) in which the conformability, embeddability, and corrosion resistance of tin make it indispensable. The castability and good general fabricating properties of these alloys facilitate repair and maintenance in remote locations, another important advantage.

Tin is also combined with aluminum, at contents at 6 to 40% tin, to form a family of bearing metals. The higher tin alloys among these are used for their higher strengths (load-bearing capacities) compared with the tin-based white metals, for example, in turbines and some aircraft and combustion engine applications. The lower tin alloys are used for their improved fatigue strength, as well, in high-speed engines and pumps. A disadvantage of the latter is that they must be used against a hardened shaft to prevent undue wear.

Tin bronzes make use of the copper properties in high loadings and low-speed machinery, such as rolling mills. They can be fabricated from metal powders and can therefore be impregnated with graphite or polytetrafluoroethylene, either of which can serve as a continuous-supply lubricant, allowing (otherwise) unlubricated operation, although at relatively low loads. The bronze powder parts can also be impregnated with oil as a lubricant for sealed, low-maintenance operation.

2. Bronze

Bronze is, of course, a use for tin that predates recorded history. Even today, a significant amount of tin is used to make bronzes. Often bronze castings are used as bearings in the manner just mentioned. Other bronze castings are often used for their good corrosion resistance and wear resistance, as valves, for example, in chemical processing equipment. All these castings contain 5–12% tin, with higher tin contents wherever corrosion resistance is more important. In some applications zinc additions form alloys called gunmetals, which can be more easily cast. Still higher tin contents (12–24%) are used for statuary castings, bells, and cymbals.

Wrought bronzes with 2 to 8% tin and additions of nickel, aluminum, or magnesium are used for seawater corrosion resistance, tarnish resistance, and good electrical contact performance. Some of these alloys show age hardenability, so that they can be formed readily before aging, then hardened to achieve electrical contact springiness and wear resistance with a good electrical conductivity.

3. Tin-Based Alloys

a. Pewter. The purest tin alloy with which the typical consumer may be familiar is the modern pewter alloy. Typically, the alloy is about ~92% tin, 6–7% antimony, and 1–2% copper, sometimes with bismuth or silver additions also. Pewter is eminently castable, with gravity casting in permanent molds being the traditional method for long production runs. More recently, centrifugal casting in rubber molds has become very popular for casting figurines, jewelry, belt buckles, and other objects. Pressure die casting is another useful technique for mass production and has been used for larger forms, such as drinking cups.

Also very useful for large-scale production of pewterware are spinning and deep drawing techniques. Both make use of the high ductility of pewter, which allows stretching and bending of the metal with little work hardening. In spinning, sheets or circles of pewter are turned down on a lathe, usually onto a wooden form, the spinning metal being pressed into place manually or automatically on such lathes. Goblets, teapots, and other tableware are often spun.

Intricate shapes of cast or spun items can be created by soldering individual parts together. With practice an experienced solderer can make a joint that will be invisible once it is buffed and polished. Since the pewter alloy has no lead added, the surface will not darken but will retain its luster with only minimal care.

b. Die-casting alloys. Die casting began with tin-based alloys because of their low melting points and high fluidity. Besides the pewter alloys just mentioned, a range of tin–antimony–lead alloys, with some copper added to certain alloys, are used for die casting. Various mechanical parts, such as gears and wheels and weights, are a few examples of tin die castings. Postage meter print wheels, in particular, make use of the capacity of tin alloys to form sharp details.

c. Fusible alloys. Combinations of several metals, especially tin, bismuth, lead, cadmium, and indium in various proportions, make a wide range of alloys, often eutectics, with melting points ranging from 47 up to 183°C. These are of obvious value as fusible links in fire alarms and sprinklers, temperature indicators, and devices to protect against overheating of equipment. Fusible alloys are also useful as seals for temperature-sensitive components and as castable tool holders and molds in metal working and plastics fabrication.

4. Tin in Ferrous Materials

Although tin is usually avoided as a steel impurity due to its embrittling effect, there are two uses for tin in ferrous materials. First, as an addition to cast iron, at about a 0.1% level, tin stabilizes pearlite. In so doing it reduces the tendency toward softening and dimensional change that may occur in high-temperature service, and, unlike some other pearlite stabilizers, it does not impair the machinability of the iron castings. These properties are made use of in automotive parts, such as engine blocks, crankshafts, and transmission components, and similar industrial equipment.

Tin powder has also proved useful as an addition with copper to iron powder metallurgy parts. The advantage of this application can be the use of a lower sintering temperature, but more often it is the closer dimensional control that tin allows. Pistons and connecting rods for refrigerator compressors are well-known applications for this process.

5. Pure Tin

A few applications of pure tin exist. Tin foil laminated on either side of lead foil is used for wine bottle capsules, the sheaths that fit over the corked end. Tin foil is also used for some electrical capacitors and for wrapping high-quality chocolates.

Some pure tin is still used for collapsible tubes for medicines and artists' paints, but the newest and most interesting application for pure tin is as a molten bath on which molten glass is cast, the "float glass" process. In this case the tin surface yields the optical flatness that eliminates the polishing formerly employed to make plate glass. The float glass process has quickly taken over the production of such products as windows, mirrors, and automobile windshields.

D. Tin and Tin Alloy Coatings

Since tin plate, as reviewed earlier, is such a large consumer of tin, the metal tin is undoubtedly the most commonly electroplated metal. Even without tin plate, however, the uses of electrodeposited tin and tin alloys, along with other methods for applying the coatings, are important industrially. Though the tin is often introduced into the system in the form of a chemical, it ends up in metallic form in the coating so that the reasons for specifying the coating are metallurgical.

1. Tin Electrodeposits

Nearly all the applications of tin-electroplated coatings are related to the excellent corrosion resistance of the metal. Although some are purely decorative, most of these applications have a functional nature. Often the coatings are used on food equipment, where the long association of tin with food preservation gives a sense of safety.

Tin is cathodic to steel and iron and to copper, so it is usual to plate a coating that is effectively pore free. Otherwise, rusting or corrosion may occur at the plating pores. Generally, at least about an 8-μm thickness is considered necessary, but thicker coatings are often used to allow for possible abuse in service or for more severe outdoor service. Tin coatings of ~8 μm on clean copper or copper alloys or on nickel alloys are used to keep these surfaces solderable for a year or more in normal storage. A fairly recent application is electrical contacts, where in certain designs tin has been substituted for the gold coatings previously used. The lubricating properties of tin are also useful, and tin has therefore become a standard coating for pistons and piston rings used in compressors and for the thread sections in oil-well piping.

Tin is usually plated from one of three electrolytes: alkaline plating solutions based on sodium or potassium stannate, acid baths based on stannous sulfate, or stannous fluoborate baths. Each bath has certain advantages and disadvantages compared with the others. Certain acid baths, for example, can deposit a bright tin plating by codepositing brightening additives with the tin. This technique bypasses the reflowing step sometimes incorporated to brighten the matte-finish deposits, but the brightening additives sometimes prove deleterious in later processing steps.

2. Tin–Lead Alloy Electrodeposits

Tin–lead electrodeposits, traditionally from fluoborate baths but more recently from some proprietary nonfluoborate baths as well, produce a coating that is easily soldered. This coating is, after all, essentially solder already. Tin–lead is widely used in the electronics industry for coating printed circuit boards and component leads to be soldered into those boards. It is also used by the automotive industry on radiator parts and various electrical connectors.

3. Tin–Nickel Alloy Electrodeposits

Although many tin alloy electrodeposits perform as if they were simple mixtures of the metals, the usual tin–nickel deposit is an intermetallic compound of 65% tin, deposited from a chloride–fluoride electrolyte. The coating is hard and tarnish resistant with good oil retention properties. Therefore, it has found use as an electrical contact material, for watch parts, and for precision instruments. Since it is also reasonably solderable, the plating is used in some electrical and electronic applications.

4. Tin–Zinc Alloy Electrodeposits

The tin–zinc alloy, from a stannate–cyanide bath, was originally developed as a substitute for cadmium electrodeposits. It behaves much like a mixture of tin and zinc, better than pure zinc in marine service conditions but inferior in industrial conditions. The preferred coatings are 70–85% tin.

Once again, electrical and electronic applications are common. Automotive applications, such as brake systems, make use of the resistance to hydraulic fluids.

5. Other Tin Alloy Electrodeposits

Bronze platings are important for their wear resistance and corrosion resistance in the manner outlined earlier for solid bronze. Also in the copper–tin family of alloy platings is "speculum," at ~40% tin, which resembles silver in color and is known for its tarnish resistance. The plating conditions for speculum require close, careful control, and therefore its use is not as widespread as it might be.

Tin–cadmium electrodeposits behave rather like tin–zinc and also offer good protection in marine environments. A relatively recent development is tin–cobalt alloys, which show an excellent color match to chromium coatings.

6. Other Coating Methods

a. Immersion tin. Tin and some tin alloy coatings can be deposited by simple immersion in a chemical bath. This process utilizes a replacement of surface atoms by atoms of tin as the former pass into the bath solution. The reaction ceases when the surface has been coated with tin.

Immersion tin has been used on some electronic devices as an overlayer to whiten the appearance of tin–lead electrodeposits. It has also been used to coat the interior of narrow-diameter copper pipe where electroplating would not produce coverage. Immersion tin on steel wire functions as a bonding agent for rubber or as a lubricant for drawing processes. Immersion tin on aluminum alloys provides a base for further electroplating or on aluminum pistons helps to prevent scuffing of the surface during the early stages of use.

b. Autocatalytic tin. Considerable research has been directed toward the development of an autocatalytic method for depositing tin. Such a process would coat a surface continuing past the end point of immersion coatings to provide a thick solderable layer of tin. Improvements in the speed and stability of the present formulations could make this process a commercial reality.

c. Hot-dip coatings. The first tin coatings were probably applied by hot dipping. Some Bronze Age artifacts appear to have been tin-coated by unknown processes, but surely hot dipping is the most likely candidate. Before World War II, all tin plate was made by this process, and until recent years farmers stored their milk in hot-tin-dipped milk cans awaiting shipment to the processing plant. Institutional mixing bowls and food grinders are just two examples of items that are still given a hot-dip coating of pure tin.

Tin–lead (or solder) is applied by hot dipping also. Electronics manufacturers who have seen the quality and environmental control advantages of this coating method are expressing renewed interest in both hot tin and hot solder coatings.

The method of hot dipping is not unlike soldering in that a flux is applied to remove light tarnish films on the base metal, and the part is dipped in a pot, which heats it and acts as a supply of tin (or solder) as well. The part must be drained to give a smooth coating. Of course, the operator must possess considerable skill for the handling of large or intricate shapes, but effectively any metal that can be soldered can be hot-dipped. Obviously, the metal must be capable of withstanding molten tin temperatures. Even small parts can be handled in bulk using baskets and employing a mechanical separation step to keep the parts from sticking together.

d. Terneplate. Terneplate is a mild steel strip or sheet coated, usually by automated hot dipping, with a lead–tin alloy of 8 to 20% tin. The terneplate of lower tin content is the standard material for automobile gasoline tanks and associated parts. Terneplate of higher tin contents is used in roofing materials, signs, and other outdoor-exposure

TABLE V Properties and Preparation of Some Important Inorganic Tin Compounds

Compound	Chemical formula	mp (°C)[a]	bp (°C)[a]	Specific gravity	Typical preparation
Stannic oxide	SnO_2	~1637	—	6.95	Thermal oxidation of tin metal
Stannic chloride	$SnCl_4$	−30	114	2.23	Direct chlorination of tin metal
Stannous octoate	$Sn(C_8H_{15}O_2)_2$	—	(d)	1.26	Reaction of stannous oxide and 2-ethylhexoic acid
Sodium stannate	$Na_2Sn(OH)_6$	—	—	3.03	Fusion of stannic oxide and sodium hydroxide plus leaching
Potassium stannate	$K_2Sn(OH)_6$	—	—	3.30	Fusion of stannic oxide and potassium carbonate plus leaching
Stannous chloride	$SnCl_2$	247	652	3.95	Direct chlorination of tin metal
Stannous sulfate	$SnSO_4$	360 (d)	—	4.18	Reaction of stannous oxide and sulfuric acid
Stannous fluoride	SnF_2	220	853	4.9–5.3	Reaction of metallic tin and hydrofluoric acid
Stannous oxide	SnO	1080 (d)	—	6.45	Reaction of stannous chloride and alkali followed by heating

[a] (d), Decomposes.

items. A frequently cited example of a roofing application is the Andrew Jackson home in Tennessee, which still has the terneplate roof installed in 1835.

Even a material as old as terneplate continues to evolve. For automotive applications, terneplate producers are introducing a nickel preplate on the steel before terne coating. This preplate is said to improve the corrosion resistance and general performance of the product. For roofing applications a terne-coated stainless steel strip has become available and has been termed "nearly indestructible" in outdoor exposure tests.

E. Chemical Applications: Stabilizers and Catalysts

In recent years tin chemicals have been the most rapidly growing areas of tin usage. The consumption of inorganic tin chemicals remains larger than that of organotins, but the use of the latter is expanding rapidly. The distinction

between the two is that organotin compounds have at least one tin–carbon bond (see Tables V and VI).

1. Inorganic Tin Stabilizers

Without a stabilizer, plastics, such as polyvinyl chloride (PVC), when heated or even just exposed to light for a time, will discolor or become brittle (or both). Incorporating stannous stearate or stannous oleate into the plastic prevents heat from removing HCl from the PVC, thereby stabilizing the product. These compounds may be used in food service grades.

2. Organotin Stabilizers

A range of di- and monoalkyltin compounds are also used for stabilizing PVC, and in fact this is the largest use of any organotin compounds at present. In general, organotins containing tin–sulfur bonds have proved the most effective

TABLE VI Properties of Some Important Organotin Compounds

Compound	Chemical formula	MW	mp (°C)	bp (°C)[a]	Specific gravity	Typical applications[b]
Dimethyltin dichloride	$(CH_3)_2SnCl_2$	220	107	187	—	Glass coating precursor
Dibutyltin dilaurate	$(C_4H_9)_2Sn(OOCC_{11}H_{23})_2$	632	23	400 (10 mmHg)	1.05	Catalyst
Dimethyltin bis(isooctyl-mercapto-acetate)	$(CH_3)_2Sn(SCH_2CO_2C_8H_{17})_2$	491	—	—	1.18	PVC stabilizer
Dibutyltin bis(isooctyl-mercapto-acetate)	$(C_4H_9)_2Sn(SCH_2CO_2C_8H_{17})_2$	575	—	(d)	1.11	PVC stabilizer
Bis(tributyltin) oxide	$(C_4H_9)_3SnOSn(C_4H_9)_3$	596	Less than −45	212 (10 mmHg)	1.17	Biocide, wood preservative
Triphenyltin hydroxide	$(C_6H_5)_3SnOH$	367	120	—	—	Fungicide

[a] (d), Decomposes.
[b] PVC, Polyvinyl chloride.

heat stabilizers, and dialkyltin carboxylates are used for good light stability. These compounds, too, are considered safe enough for use in food-grade packaging materials, rigid PVC piping for potable water supply, and plastic beverage bottles.

3. Catalysts

Stannic oxide is used in the petroleum industry as a heterogeneous oxidation catalyst. Stannous octoate is widely used in the production of polyurethane foams and in room-temperature-vulcanizing silicones. Stannous oxalate and some other carboxylates catalyze other chemical reactions of industrial significance.

Among the organotins, dibutyltin dilaurate and related compounds have also been used for polyurethane foams and for silicones. Some monobutyltin compounds have proved effective as catalysts for esterification reactions.

F. Chemical Applications: Biocides

Whereas mono- and diorganotin compounds possess such low toxicities that they are perfectly acceptable for food-contact plastics stabilizers, another class of organotins, the triorganotin compounds, exhibit an interesting and useful range of toxicities to living organisms. The compounds tend to be very selective in their actions depending on the organic constituent on the tin atom. For example, trimethyltin and triethyltin compounds are most toxic to mammals, so they are not used commercially. In contrast, tributyltin and triphenyltin compounds are effective against fungi and mollusks and therefore are of use as fungicides and antifouling agents.

Triphenyltin compounds have been used as fungicides since the 1960s. Not only do they combat fungi, but they also act as antifeedants, discouraging insects from feeding on the protected plants. Other triorganotins have found use in agriculture as miticides.

Bis(tributyltin) oxide has been used since 1960 for fungal control in wood, cotton, and other cellulosic materials. Water dispersions of this compound have proved effective in preventing moss and algae from growing on stonework. New organotin compounds have been developed that are more water soluble and show promise in expanding the applications of organotins in these areas.

Marine fouling is a persistent problem for commercial and naval ships, where the drag on a ship caused by the attachment of marine organisms rapidly decreases the fuel economy. Less effective antifouling paints may increase costs by requiring frequent dry-docking and consequent loss of service. Organotin-containing paints, however, have been shown to be very effective and long-lived and well worth any increased cost in initial application. Newer paint formulations incorporate the organotin into the paint polymer itself so that fresh antifouling agent is exposed as the paint is eroded—that is, as necessary to be effective.

A dialkyltin compound that has biocidal application is dibutyltin dilaurate, which is used as an antiwormer in poultry. It is typical of the selectivity of organotins that this treatment kills the worm parasites without harming the infected chickens or turkeys.

G. Other Chemical Applications

An important and promising application of tin is as a coating on glass. A thin layer of stannic chloride or monobutyltin trichloride is sprayed onto newly formed, still hot glass beverage bottles, jars, and glasses. The chemical is then converted to stannic oxide ($\sim$0.1 μm thick) by the heat, and this coating strengthens the glass, paying for itself in reduced breakage as the glass items are processed.

Thicker coatings (1 μm thick or more) of stannic oxide, sometimes combined with indium oxide, combine electrical conductivity with transparency to produce deicing glass for aircraft windows or illuminated signs or, for stannic oxide alone, precision electrical resistors. The intermediate thicknesses are the most promising, however, for in this range coatings on window glass have shown excellent insulating and heat-saving capabilities to compete with sputtered coatings but at a fraction of the price. This usage is envisioned for commercial construction and also for home use, where a significant savings in home heating costs may be effected.

Tin chemicals are also used in ceramic glazes as opacifiers, in ceramic pigments, as glass-melting electrodes, as pharmaceuticals (stannous fluoride in toothpastes), and as reducing agents. There are a variety of still more minor uses in many different industries.

Interesting new applications for organotins involve the use of mono-organotins as water-repellent agents for textiles and masonry applications and as possible flame retardants and smoke suppressants in a variety of plastics and cellulosic materials. These markets offer a huge potential for an effective product.

SEE ALSO THE FOLLOWING ARTICLES

CORROSION • ELECTROCHEMISTRY • MINERAL PROCESSING • MINING ENGINEERING

BIBLIOGRAPHY

Barry, B. T. K., and Thwaites, C. J. (1983). "Tin and Its Alloys and Compounds," Ellis Horwood, Chichester.

Evans, C. J., and Karpel, S. (1985). "Organotin Compounds in Modern Technology," Elsevier, New York.

Gielen, M., ed. (1990). "Tin-Based Antitumour Drugs," NATO ASI Series, Vol. 0, Springer–Verlag, Berlin.

Harrison, P. G., ed. (1989). "Chemistry of Tin," Blackie and Son, Glasgow.

ITRI. (1983). "Guide to Tinplate," International Tin Research Institute, London.

Lehmann, B. (1990). "Metallurgy of Tin," Lecture Notes in Earth Sciences Series, Vol. 32, Springer–Verlag, Berlin.

Leidheiser, H. (1979). "The Corrosion of Copper, Tin and Their Alloys," reprint, Krieger, Huntington, New York.

Price, J. W. (1983). "Tin and Tin-Alloy Plating," Electrochemical Publications, Ayr, Scotland.

Robertson, W. (1982). "Tin: Its Production and Marketing," Croom Helm, London.

Schumann, H., and Schumann, I. (1988). "Gmelin Handbook of Inorganic and Organometallic Chemistry," 8th ed., Gmelin Handbook Series, Vol. 16, Springer–Verlag, Berlin.

Thwaites, C. J. (1977). "Soft-Soldering Handbook," International Tin Research Institute, London.

Wright, P. A. (1983). "Extractive Metallurgy of Tin," 2nd ed., Elsevier, New York.

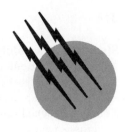

Tissue Engineering

François Berthiaume
Martin L. Yarmush

Massachusetts General Hospital, Harvard Medical School, and Shriners Hospital for Children

GLOSSARY

Allogeneic Qualifies tissues used for transplantation among different individuals of the same species.

Autologous Qualifies tissues used for transplantation to the same individual, or an identical twin, and thus not at risk of immune rejection.

Biomaterial A biocompatible material onto which cells can be cultured.

Bioreactor A device with special fittings which allows the large-scale culture of cells.

Connective tissue Tissue which primarily provides structural support in the body and is typically made of cells embedded in an extracellular matrix.

Convection A mode of transport driven by fluid flow carrying the solute particles to the cells or region where they are needed.

Differentiation The process whereby a cultured cell exhibits a greater number of characteristics reminiscent of the function and behavior of the parent tissue *in vivo*.

Diffusion A mode of transport driven by random molecular motion and which depends on the presence of a gradient of concentration of solute particles.

Endocrine cell A cell whose primary function in the body is to secrete factors which travel in the blood stream and regulate the function and metabolism of other cells.

Endothelial cell A cell which forms a selective barrier on the inner surface of blood vessels.

Epithelial cell A cell type which forms selective barriers that isolate different compartments from the rest of the body, such as the gut, stomach, or bladder.

Extracellular matrix Insoluble macromolecular network which surrounds cells in tissues.

Ligand Hormone or growth factor molecule which binds a specific receptor on the cell surface.

Morphogenesis The process whereby aggregates of cells undergo progressive reorganization into a tissue-like structure.

Receptor A specialized protein on the cell surface which binds specific hormones or growth factors and transmits this information inside the cell.

Signal transduction The process whereby the ligand–receptor binding event is intracellularly amplified and converted to a cellular response (i.e., such as cell division).

Stem cell An undifferentiated cell which has a high replicative potential and has the ability to convert into a wide variety of cell types expressing differentiated functions.

Xenogeneic Qualifies tissues used for transplantation across species.

TISSUE ENGINEERING can be defined as the application of scientific principles to the design, construction, modification, growth, and maintenance of living tissues. The main goals of tissue engineering are to help with the repair and regeneration of tissues *in vivo* and to grow tissues *in vitro* for use as models for physiological and pathophysiological studies, as well as to provide replacement parts for the body. Sometimes, tissues will repair and form scar tissue or tissue which does not exhibit a normal function and/or appearance. For example, tissue engineers have implanted polymeric tubes to promote the growth and reconnection of damaged nerves and used cultured skin grafts to cover deep burn wounds. Sometimes, the normal tissue regeneration process is too slow and temporary palliative care must be used to supply the vital missing functions to the patient. For example, tissue engineers are currently developing bioartificial liver assist devices for acute liver failure patients. Such devices may be used to buy time until a transplantable organ is available or may allow the patient's own liver function to return to recover, thereby obviating the need for liver transplantation altogether.

I. A BRIEF HISTORY OF TISSUE ENGINEERING

The use of materials derived from animal sources for making tissue replacement parts has been common practice for over 25 years with the use of bovine and porcine heart valves in cardiovascular surgery procedures. The source materials require special chemical processing and trimming before they are ready for use, all of which require extensive research and development. These devices may therefore arguably be considered the first tissue engineered devices used clinically. The function of these devices, however, is primarily a mechanical one and such implants do not significantly become repopulated with the host's cells (or, if they do, they do not significantly contribute to their function). More recently, biologically derived matrices, such as acellular human dermis (AlloDerm®, Life-Cell, Inc.), which is made by treating human cadaver skin in such a way that no cells but the extracellular matrix remain, have been used in order to promote the regeneration of tissue in deep burn wounds.

The first man-made material designed to promote cell ingrowth and permanent incorporation into the body was developed by Ioannis V. Yannas (Massachusetts Institute of Technology) and John F. Burke (Massachusetts General Hospital and Shriners Burns Hospital, Boston) in 1980. It consists of a bovine collagen–glycosaminoglycan matrix made from chemical extracts of bovine skin and shark cartilage overlaid with a thin silicone membrane. This construct is applied onto deep burn wounds to facilitate the regeneration of the dermal layer of skin, after which the silicone sheeting is removed and replaced by a skin graft. This product was approved by the U.S. Food and Drug Administration (FDA) for clinical use in 1996 and is commercialized under the name of Integra® (manufactured by Integra LifeSciences, Inc.).

Advances in cell culture techniques have also played a pivotal role in the development of tissue engineered products. A landmark discovery by Howard Green and James Rheinwald (Harvard Medical School) in 1975 is the demonstration that keratinocytes from the skin epidermis could be cultured *in vitro* using a "feeder layer" of mouse fibroblasts. Keratinocytes were harvested from patients with extensive burns and propagated *in vitro* until the available surface area of cultured skin was sufficient to use as an autologous grafting material. In 1988, Genzyme Corp. began the Epicel® service and more recently extended this concept to the propagation of chondrocytes for the treatment of cartilage defects in knee joints (Carticel®).

Other pioneering work in tissue engineering includes studies published by Eugene Bell (Massachusetts Institute of Technology) between 1979 and 1981 describing the first matrix–cell composite grown *in vitro* prior to *in vivo* implantation. Human dermal fibroblasts were seeded into collagen gels to produce a bioartificial dermis, which was then overlayed with a monolayer of epidermal cells (keratinocytes) to generate a full-thickness skin equivalent. This product, available under the name of Apligraf® (Organogenesis, Inc.), was approved by the FDA in 1998 for treating nonhealing venous ulcers and, more recently, diabetic foot ulcers. Unlike autologous skin grafts, which require several weeks to become available after harvesting the source cells from the patient, Apligraf® is made of allogeneic cells obtained from donated human foreskins and is a ready-to-use product available within a short notice. On the other hand, since Apligraf® contains allogeneic cells, it eventually becomes rejected by the recipient's immune system, and repeated treatments may be necessary until the ulcer heals on its own.

Beyond the few tissue engineered products which are currently used clinically, there are many others in the pipeline of biotechnology companies as well as research laboratories around the world. One important area of tissue

TABLE I Companies Selling Tissue Engineered Products and Products for Regenerative Medicine

Company	Product name(s)	Application(s)	Core technology
Edwards Lifesciences; Irvine, CA	Carpentier-Edwards PERIMOUNT pericardial valve, Edward Prima Plus Stentless Bioprosthesis	Heart valve replacements	Chemically treated xenogeneic heart valve tissue
Medtronics; Minneapolis, MN	Hancock® II aortic and mitral bioprostheses	Heart valve replacements	Chemically treated xenogeneic heart valve tissue
St. Jude Medical; St. Paul, MN	Toronto SPV® valve	Heart valve replacements	Chemically treated xenogeneic heart valve tissue
LifeCell; Branchburg, NJ	AlloDerm® acellular human dermis	Burn wounds	Acellular dermis from human cadavers
Regeneration Technologies; Alachua, FL	Regenapack™ regeneration template, CorlS™ and AlloAnchor™ bone-healing screws and pins	Bone wound healing	Precision-tooled natural bone matrices
Integra LifeSciences; Plainsboro, NJ	Integra® dermal regeneration template, BioMend® absorbable collagen membrane	Burn wounds, periodontal disease	Collagen-based matrices
Sulzer Medica; Winterthur, Switzerland	Ne-Osteo™ osteogenic bone-filling material	Spinal fusion, periodontal disease	Collagen-based matrices with growth factors
Curis; Cambridge, MA	OP-1 Implant™ osteogenic bone-filling material	Nonunion fractures	Collagen-based matrices with growth factors
Genzyme; Cambridge, MA	Carticel® autologous chondrocytes, Epicel® autologous keratinocytes	Cartilage defects of the knee, burn wounds	Culture of autologous keratinocytes and chondrocytes
Advanced Tissue Sciences; La Jolla, CA	TransCyte™, Dermagraft®	Burn wounds and skin ulcers	Bioreactors for three-dimensional stromal cell culture technologies
Organogenesis; Canton, MA	Apligraf®	Skin ulcers	Dermal-epidermal composites
Ortec International; New York, NY	Composite cultured skin (CSS)	Epidermolysis bullosa, skin ulcers	Dermal-epidermal composites
Circe Biomedical; Lexington, MA	HepatAssist® liver support device	Extracorporeal liver-failure treatment	Hepatocyte bioreactors
Vitagen; La Jolla, CA	ELAD™ liver support device	Extracorporeal liver-failure treatment	Cultured human hepatoma cell line
University of Berlin; Hybrid Organ GmbH, Berlin, Germany	Modular Extracorporeal Liver Support System (MELS)	Extracorporeal liver-failure treatment	Hepatocyte bioreactors

engineering still in the early stages is the development of bioartificial organs such as pancreatic islets and livers, some of which consist of complex devices that incorporate large cell numbers into novel bioreactor systems. Table I provides a more exhaustive listing of companies currently involved in the production of tissue engineered products approved for clinical use or which are in the more advanced stages of clinical trials.

The early discoveries and major advances in tissue engineering have been in large part the result of empirical studies because of our lack of understanding of the basic phenomena that control tissue formation and repair. Tissue

engineering is now also evolving as a science which uses the theoretical framework of core engineering disciplines, including thermodynamics, transport, reaction kinetics, and control theory. The basic understanding of the rules that govern tissue repair, regeneration, and development will enable one to predict the behavior and performance of more complex tissue engineered constructs, a necessary step towards the optimization of tissue engineered products. The first such studies were published by Malcom S. Steinberg (University of Princeton) in the early 1960s and describe rules governing cell-sorting phenomena in multicellular systems containing different cell types. Although the initial motivation for these studies was to undertand the mechanisms of embryonic development, the derivations are also relevant to the engineering of tissues for clinical applications. In the 1970s, several studies by Douglas A. Lauffenburger (then at the University of Illinois, now at M.I.T.) and Robert T. Tranquillo (University of Minnesota) set the stage for the modeling of intracellular signaling processes as well as cell-migration phenomena.

II. FUNDAMENTALS OF TISSUE ENGINEERING

A. Biomaterial Design

1. Materials Used in Tissue Engineering

The vast majority of mammalian cells are anchorage dependent and therefore must attach and spread onto a substrate to proliferate and function normally. While in traditional tissue culture systems two-dimensional surfaces are used to grow cells, tissue engineering often requires the use of three-dimensional matrices which allow cell ingrowth and organization reminiscent of actual tissues found *in vivo*. The choice of extracellular matrix material is highly dependent on the intended use of the tissue (whether its function is structural or biochemical or both) and on the respective roles of materials and cells in the reconstructed tissue. A list of three-dimensional materials used in tissue engineering is given in Table II.

Matrices derived of naturally occuring tissues, such as animal-derived heart valves, acellular dermis, and bone-derived matrices, are typically of allogeneic and xenogeneic origin. They are prepared via physical and chemical treatments, such as freeze-drying, cross-linking by glutaraldehyde, and detergent-mediated removal of cells, in order to enhance their physical properties and remove any antigen-bearing cells which could trigger undesirable immune responses. These materials retain the chemical composition and microarchitecture proper to the tissue that they are derived from, which can enhance their function. For example, blood vessel growth into an acellular dermis applied onto a burn wound will preferentially occur in the spaces formerly occupied by the blood vessels in the original intact tissue. Soluble factors are often retained within the matrix and can have pro-angiogenic (small intestinal mucosa) or anti-angiogenic (amniotic membrane) properties. A disadvantage of these materials is that their chemical composition is often only partially known, availability may be limited, and issues such as batch-to-batch variation and potential contamination with pathogens must be addressed on a continuous basis.

Most of the natural extracellular matrix materials, except bone, can be at least partially solubilized by chemical processing and reconstituted into three-dimensional gels of any shape or form. Although the microarchitecture is lost, these reconstituted matrices retain many chemical features of the extracellular matrix proteins including bound growth factors found in the original material. Commonly used reconstituted matrices include type I collagen

TABLE II Materials Commonly Used in Tissue Engineering

Name	Composition	Applications
Intact extracellular matrices		
Amniotic membrane	Collagen, fibronectin, laminin, GAG, growth factors	Corneal epithelium
Acellular dermis	Collagen, laminin, elastin	Skin epithelium
Small intestinal mucosa	Collagen, fibronectin, GAG, growth factors	Smooth muscle (vascular, urogenital)
Carbonate apatite (dahllite)	Calcium/magnesium carbonate/phosphate	Bone
Reconstituted extracellular matrices		
Type I collagen gel	Collagen	Skin dermis, tendon, hepatocyte
Collagen–GAG[a] complexes	Collagen, GAG	Skin dermis, tendon, nerve guidance
Engelbreth–Holm–Swarm tumor matrix gel (Matrigel)	Collagen, laminin, GAG, growth factors	Hepatocyte
Synthetic matrices		
Carbonate apatite	Calcium/magnesium carbonate/phosphate	Bone
pLA/pLGA co-polymer	Poly(lactic-*co*-glycolic) acid	Cartilage, bone, epithelium (gut, urogenital), hepatocyte
Dacron®	Polyethylene teraphtalate	Vascular endothelium
Gore-Tex®	Expanded polytetrafluoroethylene	Vascular endothelium
pHEMA/MMA co-polymer	Poly(hydroxyethyl methacrylate)	Vascular endothelium

[a] GAG = glycosaminoglycan.

which is isolated from rat tail or bovine skin by mild acid treatment. The acid solution of collagen can be induced to form a gel upon restoring a physiological pH of 7.4, which causes the polymerization of collagen molecules into a large network of fibrils. The extent of cross-linking in this collagen is very low in comparison with that of the native tissue, and as a result reconstituted collagen gels undergo rapid proteolytic degradation *in vivo*. To remediate this problem, chemical cross-linking is induced by either glutaraldehyde or dehydrothermal (vacuum and $\sim 100°C$) treatment. For example, the skin substitute Integra® is made of a mixture of solubilized collagen and glycosaminoglycans whose extent of cross-linking has been optimized to withstand the specific environment of dermal wounds such as nonhealing ulcers and deep burns.

Materials used in tissue engineering must be able to withstand physical forces to which they are subjected. These forces naturally occur in load-bearing tissues such as bone and cartilage, as well as in other applications such as blood vessels, which must have burst pressures exceeding arterial levels. Physical forces can also be generated by the cells making up the bioartificial tissue, as cells have been shown to exert tractional forces on their points of attachment. Known examples of the effect of cell tractional forces include the contraction of collagen gels by fibroblasts and the formation of "ripples" by cells placed on thin flexible silicone sheets. Specific mechanical properties are required in certain applications, especially in the case of artificial vascular grafts, which must exhibit the same compliance as that of normal blood vessels. The mismatch in compliance that often occurs between the host's vessel and the graft is believed to be be an important factor leading to artificial vascular graft failure *in vivo*.

Systems using cells that do not secrete a structurally dense extracellular matrix must rely on the synthetic matrix provided to retain their structural integrity. The matrix must be able to withstand both the weight of the cultured cells as well as tensile forces generated by cells growing on the substrate. The use of relatively fluid substrates induces different cellular morphologies than does the use of rigid surfaces. Because fluid substrates cannot oppose cell-generated forces, cell–cell adhesive forces predominate over cell–substrate adhesive forces, which leads to cell aggregation as seen with hepatocytes plated on heat-denatured collagen as opposed to type I collagen. In high-density, three-dimensional cultures, cell-generated forces may become significant as seen with fibroblasts that can dramatically reduce the volume of collagen lattices.

Naturally derived matrices provide good substrates for cell adhesion because cells express the adhesion receptors which specifically recognize and bind to extracellular matrix molecules which make up these matrices. Nevertheless, there have been considerable advances in the development of synthetic biocompatible polymers, which theoretically have an unparalleled range of physical and chemical properties. In practice, however, most tissue engineering development has been limited to using a relatively small number of man-made materials, in part due to a reluctance to expend time and money to secure regulatory approval for clinical use of untested biomaterials. The most extensively used materials in medicine are titanium and inert plastics such as Teflon® for orthopedic applications and artificial vascular graft prostheses, respectively. Typical problems encountered with artificial orthopedic materials include failure of the graft–host tissue interface in the case of bone substitutes, which may be due to an adverse reaction to the artificial material, and progressive wear-and-tear in artificial joints, which do not have the ability to regenerate and repair, unlike natural joint surfaces. Artificial vascular grafts tend to activate the blood clotting cascade and may also cause thickening of the vascular tissue near the point of attachments to the host's vascular tree. These responses do not pose a major problem for the function of large-diameter grafts (e.g., thoracic aorta), but have prevented their use as smaller vessels such as coronary bypass segments, for which the demand is very high. Finally, all artificial materials implanted *in vivo* are highly susceptible to colonization by bacteria which can form biofilms highly resistant to antibiotics. Furthermore, recent studies suggest that the function of immune cells may also be compromised on certain artificial surfaces, which reduces the ability of the host to clear infections.

To overcome the problems due to foreign-body reactions caused by artificial materials, there is currently heightened interest in the use of biocompatible polymers which naturally degrade *in vivo*. One of the best known and most commonly used synthetic biodegradable polymers in tissue engineering are the poly(lactic-*co*-glycolic) acid copolymers, which have been used in the form of biodegradable sutures for several decades. In 1988, Robert S. Langer (Massachusetts Institute of Technology) and Joseph P. Vacanti (Children's Hospital, Boston) pioneered their use in tissue engineering. Currently used as part of skin substitutes commercialized by Advanced Tissue Sciences, they are now the most widely investigated artificial biodegradable polymers in tissue engineering, with applications including cartilage, bone, and various epithelia (intestine, bladder, liver). This material hydrolyzes completely within weeks, months, or years, depending on the exact composition (in general, increased hydrophobicity correlates with a decreased degradation rate) and thickness. Cells migrating in from surrounding tissues after implantation *in vivo*, or cells directly seeded into the polymer, secrete their own extracellular matrix which gradually replaces the polymer scaffold as the latter slowly dissolves away. It is important that the degradation rate of biodegradable extracellular matrix materials, when used deliberately, must be such that the cell-generated matrix has sufficient time to form and the mechanical integrity of the tissue is maintained at all times. In the end, there are no foreign materials left in the patient to cause adverse long-term immune reactions or harbor bacterial infections.

2. Optimization of Surface Chemistry

Cells do not usually directly attach to artificial substrates, but rather to extracellular matrix proteins which are physically adsorbed (i.e., by virtue of hydrophobic and electrostatic interactions) or chemically attached (i.e., via covalent bonds) to the surface. Many polymers are highly hydrophobic and do not favor protein adsorption. Increasing substrate wettability to a certain point (water contact angle of 60° to 80°), such as by using ionized gas, increases protein adsorption and is commonly used for preparing tissue-culture-grade polystyrene Petri dishes. This process only modifies the surface of the material, and thus minimally affects its bulk mechanical properties. Highly hydrophilic surfaces are also not favorable to protein adsorption. In addition, if negatively charged, they may cause repulsive electrostatic interactions with the cells, the latter of which usually display a negative surface charge due to the presence of negative sialic acid residues on their surface glycocalyx. Conversely, coating surfaces with positively charged materials, such as poly-L-lysine, has been used to promote cell adhesion to the surface. Physisorption of proteins for which cells do not express any adhesion receptor, such as albumin, is also commonly used to prevent cell adhesion.

Physisorbed proteins are not stably bound and can be displaced by other proteins. This especially occurs in complex media such as plasma, where fibrinogen physisorption may occur within seconds, following by displacement of more slowly diffusible but "stickier" proteins (this is sometimes called the *Vroman effect*). When the surface is transferred to a different medium after protein coating, the type and amount of physisorbed proteins will change until reaching equilibrium with the proteins in solution above the surface. The time scale for desorption can extend over several hours, thus physisorption can be useful to control the initial attachment of cells at the time of seeding. Over a period of several days of culture, however, virtually all cells will have secreted significant quantities of their own extracellular matrix onto the substrate, and the initial surface properties of the material often become irrelevant.

Because physisorption is notoriously nonselective, covalent modification of substrates or chemisorption is used if it is necessary to provide more control over the type, density, and distribution of adhesive protein on the surface of the material. For this purpose, several chemical

TABLE III Ligands Used for Chemisorption of Protein to Surfaces

Ligand	Type of primary surface
Silanating reagents	Glass, silicon
Alkane thiols	Gold
Carboxylic acids	Alumina
Sulfonyl halides, carbonyldiimidazole, succinimidyl chloroformate, succimidyl esters	Synthetic polymers[a]

[a] Dacron and PTFE require chemical treatment in order to create free alcohol and carboxylic groups prior to derivatization.

processes are available depending on the type of surface to be modified (Table III). The first step involves using a reactive chemical that bonds to the surface and has a free functional group that easily reacts with free thiol, hydroxyl, carboxyl, or amine groups on proteins. This step often requires harsh chemical conditions, while the second step, which involves conjugation of the protein, can be done under physiological conditions. This approach is also suitable to graft small adhesive peptides (e.g., RGD) which otherwise would not stably bind to surfaces by physisorption (Table IV). Furthermore, physisorption sometimes leads to unexpected changes in protein activity, probably due to denaturation on the surface. For example, adsorbed fibrinogen activates and binds to platelets, unlike solution-phase fibrinogen in normal plasma or blood.

TABLE IV Compounds Used to Promote or Prevent Cell Adhesion to Surfaces

Pro-adhesive	Anti-adhesive
Extracellular matrix proteins	Polyethylene glycol
Collagen	Albumin
Fibronectin	Polyvinyl alcohol
Vitronectin	Cellulose acetate
Laminin	Agarose
	Sulfonate residues
Adhesive peptide sequences[a]	
RGD (from collagen)	
YIGSR, IKVAV (from laminin)	
REDV (endothelial-specific)	
Adhesion molecules	
Intercellular adhesion molecule-1 (ICAM-1)	
Vascular cell adhesion molecule-1 (VCAM–1)	
Platelet cell adhesion molecule-1 (PCAM-1)	
Sialyl Lewis X	

[a] Single-letter amino acid abbreviations.

The design of more sophisticated cultured tissues using more than one cell type can be enhanced by spatially controlling the seeding process. For this purpose, various methods for patterning the deposition of extracellular matrix or other cell attachment factors onto surfaces have been developed. Photolithography involves spin-coating a surface (typically silicon or glass) with an ~ 1-μm thick layer of photoresist material, exposing the coated material to ultraviolet light through a mask containing the pattern of interest, and treating the surface with a developer solution which dissolves the exposed regions of photoresist only (Fig. 1). This process leaves photoresist in previously unexposed areas of the substrate. The exposed areas of substrate can be chemically modified for attaching proteins, etc., or can be treated with hydrofluoric acid to etch the material. The etching time controls the depth of the channels created. Subsequently, the leftover photoresist is removed using an appropriate solvent, which leaves a surface patterned with different molecules and/or grooves. A disadvantage of this method is that it uses chemicals toxic to cells and generally harsh conditions which could denature proteins are used.

The etched surfaces produced by photolitography can be used to micromold various shapes in a polymer called poly(dimethylsiloxane) (PDMS). The PDMS cast faithfully reproduces the shape of the silicon or glass mold to the micron scale and can be used in various "soft lithog-

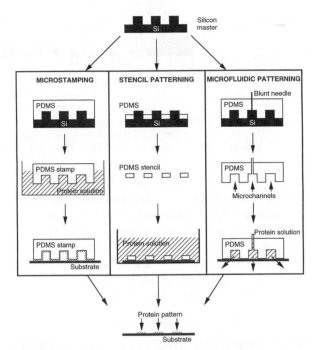

FIGURE 2 Patterning using soft lithography. The silicon master is used as a mold to create a flexible replica made of poly(dimethylsulfoxane) (PDMS). The replica can be used as a stamp to deposit protein on a substrate, as a stencil to cover up selected regions of the substrate during protein coating, or as a series of flow channels to deliver a protein-coating solution onto the substrate.

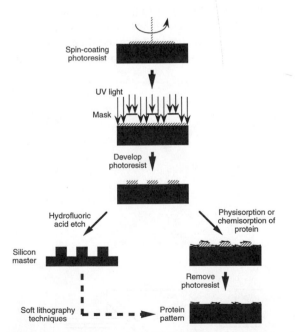

FIGURE 1 Patterning using photolithography. A silicon (or glass) wafer coated with photoresist is exposed to ultraviolet light in areas determined by a mask overlay. A developer chemical selectively removes the photoresist and the exposed areas of silicon can be either etched for use in soft lithography techniques (see Fig. 2) or coated with proteins.

raphy" techniques, including microstamping, microfluidic patterning, and stencil patterning (Fig. 2). An infinite number of identical PDMS casts can be generated from a single master mold, which makes the technique very inexpensive. Soft lithography methods can be used on virtually any type of surface, including curved surfaces, owing to the flexibility of PDMS. Another patterning method which works well at larger size scales is microprinting using laserjet technology, which can also be used to create three-dimensional structures (Fig. 3).

In using these approaches, it is important that the base material be resistant to physisorption, or the selectivity of the adhesive groups may be significantly reduced *in vivo*. A successful approach to prevent adhesion to the base material is via covalent attachment of anti-adhesive factors on the remaining functional groups.

Micropatterning is especially desirable to maximize heterotypic cell–cell interactions between a parenchymal cell such as hepatocyte and supporting or "feeder" cells such as fibroblasts. Keeping in mind that cells cultured on surfaces do not usually layer onto each other (except for malignant cancer cell lines), random seeding using a low ratio of parenchymal cells to feeder cells will achieve this goal, but at the expense of using a lot of the available

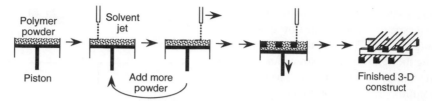

FIGURE 3 Microprinting three-dimensional scaffolds. A small jet of solvent is sprayed onto a packed bed of polymer powder to induce bonding of the powder into a solid in selected regions. Alternating solvent spraying and new additions of polymer powder eventually creates a three-dimensional shape suitable for cell culture.

surface for fibroblasts, which do not provide the desired metabolic activity. On the other hand, micropatterning techniques enable optimization of the seeding pattern of both cell types so as to ensure that each hepatocyte is near a feeder cell while minimizing the number of feeder cells. As a result, metabolic function per area of culture is increased and the ultimate size of bioreactor with the required functional capacity is reduced.

3. Fabrication of Porous Matrices

Porous matrices are often used to reconstruct connective tissues because they allow the formation of complex extracellular matrix networks responsible for the tissue's mechanical properties and the fusion of the implant with the host's tissue. Pore sizes in the range of 30 to 300 μm are the most common. Smaller pore sizes provide more surface area per volume of matrix; however, pores less than 30 μm will not allow seeding or ingrowth of the host's tissue into the matrix.

Porous materials are usually prepared by salt-leaching or freeze-drying techniques. The first method involves adding water-soluble crystals (e.g., NaCl) of size range similar to the desired pores to the melted base polymer material. After solidification of the polymer, the salt crystals in the resulting solid are dissolved by exposure to aqueous solutions, leaving a pore in the place of every crystal. An alternative approach is the use of supercritical carbon dioxide to create pores by induction of microbubble formation within the polymer.

The freeze-drying technique is based on the general principle that when freezing a solution, the solvent forms pure solid crystals while all solute materials are concentrated in the remaining unfrozen fraction. During the subsequent drying process, the solid crystals evaporate and leave pores. The morphology of the solid crystals is dependent on the physico-chemical properties of the solution, the temperature gradient at the liquid–solid interface, and the velocity of that interface. The directional solidification system shown in Fig. 4 allows one to independently control each one of these three parameters. During directional solidification, the size and shape of the crystals forming can be predicted from basic physics principles.

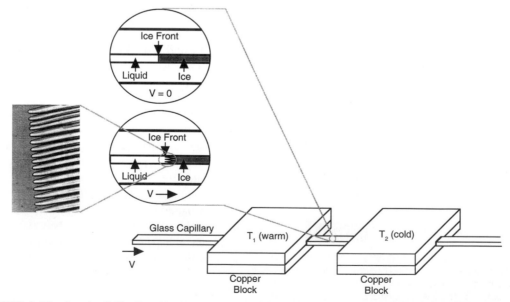

FIGURE 4 Directional solidification stage to pattern crystal formation during freezing of polymer solutions. Inset on left shows the morphology of water crystals during freezing of a collagen solution in isotonic saline with 1 mM HCl.

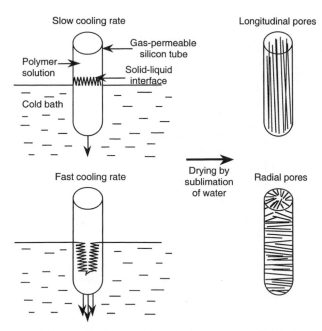

FIGURE 5 Method for creating oriented pores in cylindrical matrices. Slow cooling (top panel) promotes solidification from the solid–liquid interface, thus generating vertically oriented crystals. Fast cooling (bottom panel) promotes solidification from the walls of the tube, leading to horizontally oriented crystals. After drying, the orientation of the pores reflects that of the crystals.

TABLE V Growth Factors Commonly Used in Tissue Engineering

Growth factor	Target cell
Epidermal growth factor	Keratinocytes, hepatocytes
Hepatocyte growth factor, scatter factor	Epithelial cells
Interleukin-2	White blood cells
Platelet-derived growth factor (PDGF)	Fibroblasts, smooth muscle cells
Fibroblast growth factor	Fibroblasts, smooth muscle cells, endothelial cells
Vascular endothelial growth factor (VEGF)	Endothelial cells
Nerve growth factor	Neurons
Insulin-like growth factor	Muscle, keratinocytes
Osteogenic protein 1	Osteoblasts

Furthermore, the solid crystals tend to orient in the direction of the temperature gradient, so that the direction of the pores can be controlled as well. For practical applications, however, it is more typical to freeze solutions containing biomaterials in a bulk fashion. For example, nerve guidance tubes have been produced by immersion of suspensions of collagen–GAG complexes contained in gas-permeable silicone tubes in a cold bath. As depicted in Fig. 5, a slow rate of immersion causes the formation of crystals (and pores, eventually) predominantly oriented along the length of the tube, which is the geometry desired for this application. More rapid immersion, on the other hand, would lead to crystal growth primarily in the radial direction. It is noteworthy that the rate of freezing and the temperature gradient are difficult to control and maintain constant throughout the freezing process. Thus, typically, porous materials made by this technique exhibit nonuniform pore sizes as one moves from the surface to the center.

B. Cell Engineering

1. Growth Factors, Hormones, and Signal Transduction

Cultured cells are often included in polymer scaffolds used in tissue engineering to make up for the limited potential of the host's surrounding cells to regenerate the damaged or missing tissue. Control of the function and growth of the cells is critically important and may require the use of exogenous growth factors, which are small proteins that act as ligands binding to specific cognate receptors on target cells. Table V provides a sample list of growth factors currently used in tissue engineering. Hormones are smaller compounds which have similar effects and include small peptides as well as a range of lipid-soluble compounds derived from cholesterol and fatty acids. Hormones and growth factors may be incorporated into the scaffold itself and released over time, or the cells in the construct can be modified genetically (see section on genetic engineering) or otherwise to produce the growth factors themselves. In addition, the recipient's own tissue surrounding a tissue engineered implant may undergo an inflammatory response due to the surgical trauma or the presence of impurities (e.g., bacterial-derived lipids such as endotoxin), as well as immunogenic factors, including proteins of animal origin. There are several soluble mediators released during the course of an inflammatory response, some of which can either stimulate or suppress cell growth as well as other cellular functions in the implant.

In order to analyze, predict, and optimize the cellular response to growth factors, mathematical models can provide useful insights. The system which has been the most extensively studied with respect to the quantitative aspects of receptor-mediated signaling is that involving epidermal growth factor (EGF) binding to its receptor (Fig. 6). A feature of the early signaling events is the internalization of the EGF–EGF receptor complexes into the cell, which can then be recycled back to the cell surface or degraded within the cell. First-order kinetic mass balance equations can be derived for the species shown in Fig. 6. The binding of growth factors to receptors and ensuing

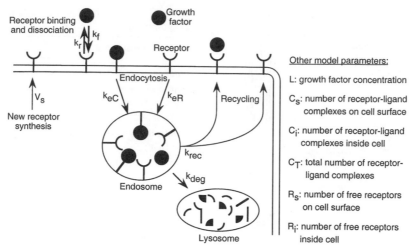

FIGURE 6 Simplified model for the binding and fate of a ligand (growth factor) binding to its receptor on the cell surface. After binding, the ligand–receptor complexes are internalized into an endosomal compartment. Its contents can then be recycled to the cell surface or fuse with a lysosome and undergo degradation. The *k* parameters shown represent first-order rate constants for each process shown.

intracellular changes occur in the time scale of seconds to minutes, and a steady state is reached within one hour. In many cases, it can be assumed that the cellular response is proportional to the total number of receptor-bound growth factor molecules. Since in most cases growth factor levels also do not change very quickly in the environment, one can assume a pseudo steady state and obtain the following results for the number of EGF–EGF receptor complexes:

$$C_s = \left(\frac{K_{ss}L}{1 + K_{ss}L} \right) \frac{V_s}{k_{eC}}, \quad K_{ss} = \frac{k_{eC}k_f}{k_{eR}(k_{rec} + k_{eC})} \tag{1}$$

$$C_i = \left(\frac{k_{eC}}{k_{deg}} \right) C_s \tag{2}$$

$$C_T = C_s + C_i \tag{3}$$

$$R_s = (k_r + k_{eC}) \frac{C_s}{k_f L} \tag{4}$$

Using accepted parameter values for this system, one can notice that the proportion of complexes that are intracellular increases with EGF concentration (Fig. 7). Furthermore, the total number of bound receptors in the cell as well as on the cell surface reaches a maximum corresponding to about 25% of the total number of receptors. The process of receptor-mediated endocytosis thus limits the number of receptors available for binding at any time, which is a well-known mechanism of downregulation of growth factor responses. Several studies have shown that a more sensitive response to a particular growth factor may be obtained using modified growth factors or cells with altered receptors which reduce the rate of internalization of the complexes. Another strategy to reduce the rate of inter-

nalization is to immobilize the growth factor to the surface of the substrate to which the cells are attached. This has been shown to work in the case of EGF and fibroblasts; however, in other systems, the activity of the growth factor may be partially or completely lost. The EGF–EGF receptor model can be further refined by taking into account the removal of growth factor from the extracellular medium, the effect of receptor clustering, several pathways operating at different rates, etc. Cells that produce their own growth factors as part of an autocrine loop can also be modeled in a similar fashion by adding appropriate terms for growth factor release and diffusion around the cells. Such a model may be useful to predict the effect of cell density on growth rate as well as other density-dependent functions.

A difficult problem which remains in this area is to relate, on a theoretical basis, the receptor–ligand binding phenomena on the cell surface to the observed cellular response. The intracellular signaling pathways typically function as a cascade of events leading to the sequential activation of intracellular signaling molecules, often by phosphorylation of specific amino acid residues on the signaling proteins. One of the final targets of this cascade may be one or several transcription factors, specialized proteins that have the ability to migrate into the cell nucleus and trigger the synthesis of new proteins or alterations in cell behavior, such as cell division. A large number of growth factors trigger the mitogen-activated protein (MAP) kinase cascade (Fig. 8), and there is evidence that similar cascades exist for other mediators. Kinetic modeling of each step as a reaction, using parameters determined through analyses in cellular extracts of the intracellular concentrations and kinetic properties of

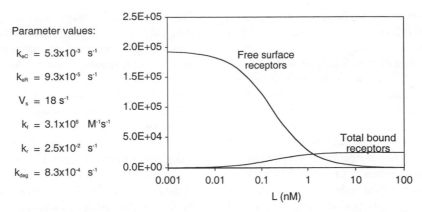

Parameter values:

$k_{eC} = 5.3 \times 10^{-3}$ s^{-1}

$k_{eR} = 9.3 \times 10^{-5}$ s^{-1}

$V_s = 18$ s^{-1}

$k_f = 3.1 \times 10^{6}$ M^{-1}s^{-1}

$k_r = 2.5 \times 10^{-2}$ s^{-1}

$k_{deg} = 8.3 \times 10^{-4}$ s^{-1}

FIGURE 7 Effect of ligand concentration on the number of remaining free receptors on the cell surface and the total number of bound receptors. Parameter values used (shown on the left of graph) are literature values for epidermal growth factor and fibroblasts.

individual components of the cascade, reveals that the cascade operates as a signal amplification system that tends to produce a switch-like behavior in the cellular response to growth factors. Thus, the response of a cell to growth factors can often be represented by a threshold model where no effect occurs below that threshold and a maximal response occurs above that threshold. When studying a cell population, however, a dose-dependent response may still be observed because cells tend to exhibit a distribution in the concentration of each effector molecule involved in the signaling cascade. It is important to note that modeling of cellular responses is still in its infancy, as there are no models yet published addressing the effect of multiple growth factors and multiple signaling cascades operating at the same time, which would be more typical of conditions used to culture cells in tissue engineering applications.

Besides soluble growth factors and hormones in the medium, there are many other parameters in the environment that influence cell growth rate and the expression of specific functions, which may be desirable or even necessary for tissue engineering applications. For example, controlling the density of adhesion sites is a potentially powerful means of controlling cell growth and function. The spreading of a cell on a surface increases together

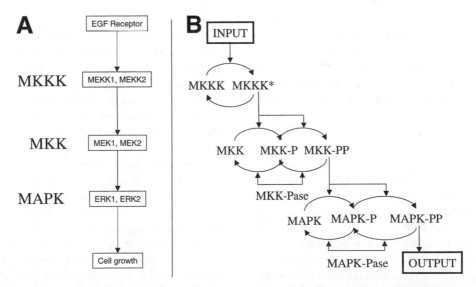

FIGURE 8 The signaling cascade triggered by a typical growth factor. MAPK = mitogen-activated protein kinase; MKK = MAPK kinase; MKKK = MKK kinase. MKKK is activated by receptor–ligand binding. Activated MKKK then phosphorylates MKK twice sequentially to activate MKK. Activated MKK then phosphorylates MAPK. MAPK activates factors (not shown) which migrate into the cell nucleus, bind the cell's DNA, and initiate cell replication. The multistep nature of the cascade causes amplification of the signal generated by the receptor–ligand binding event to trigger a switch-like cellular response. MEKK, MEK, and ERK are the names of kinases specific to the epidermal growth factor pathway.

with the surface density of extracellular matrix molecules, which is typically accompanied by an elevation in DNA synthesis and proliferation rates.

A general rule of thumb in cell culture techniques is that proliferation and differentiation are mutually exclusive. In other words, conditions promoting the expression of differentiated functions are often not optimal for replicating cells. For example, fibroblast growth factor-stimulated capillary endothelial cells plated on nonadhesive surfaces coated with decreasing concentrations of fibronectin switch from a spreading to a tubular capillary-like shape, with a concomitant reduction in cell growth. In some cases, cell differentiation can also be induced by altering the culture environment so as to mimic a subset of *in vivo* conditions. For example, keratinocytes, a type of epithelial cell which forms the epidermal component of the skin, can be propagated *in vitro* using a serum-free culture medium; a single human neonatal foreskin can provide enough cells to yield over 100 m^2 of graftable tissue. Cells in cultured epidermal sheets are not well differentiated but exposure to air while in culture or after grafting onto the host induces the formation of a stratified differentiated epidermis.

One of the challenges of tissue engineering is to produce large cell masses that are well differentiated. Although differentiated cells do not always proliferate easily *in vitro*, it may be possible to optimize culture conditions to stimulate cell propagation and then to change these conditions so that a stable and functional phenotype is exhibited by the cells. For example, chondrocytes seeded on plastic in the presence of serum proliferate but secrete a significant amount of type I collagen and small proteoglycans, which are not normally found in cartilage. Embedding these cells in an agarose gel induces the re-expression of the normal phenotype found *in vivo*, which is characterized by the production of type II collagen and deposition of large aggregating proteoglycans.

2. Genetic Engineering

While control of the extracellular environment remains the primary means of modulating cell function and proliferation in tissue engineering, it is sometimes advantageous to alter the genetic make-up of cells to extend their basic capacity to perform specific functions. Describing the techniques used for genetically altering cells is beyond the scope of this chapter, and the reader is referred to the numerous textbooks and reviews on the subject. Genetic modification of cells in tissue engineering has included the following applications: (1) expressing functions not normally present in a particular cell type or overexpressing existing functions, and (2) expressing "immortalizing" genes or genes that protect cells against death caused by apoptosis.

The first application may be part of a gene therapy protocol aimed at providing a patient who has a single enzyme deficiency (e.g., adenosine deaminase) with implantable cells to perform the missing function. Another important application is the (over)expression of angiogenic factors that promote the rapid invasion by blood vessels and vascularization of implantable tissue constructs. Immortalizing genes, such as the viral SV40 T antigen and telomerase are primarily used to promote the replication of cells typically very difficult to grow *in vitro*, such as hepatocytes, pancreatic beta cells, etc. The use of anti-apoptotic genes in tissue engineering is a relatively new trend stimulated by the difficulties of maintaining cell viability in large tissue constructs made of cells sensitive to the depletion of nutrients—for example, in the case of hepatocytes in bioartificial livers.

One of the issues raised by the use of genetic engineering in tissue engineered products is the unknown effects of persistent expression of the transgene in the implanted cells. For example, overexpressing growth factors may be beneficial to the process of growth and integration of a engineered tissue implanted in a host; however, the long-term effects of high levels of growth factors are unknown and could perhaps be detrimental. This problem may be resolved soon, however, with the advent of new molecular biology techniques that allow for the "excision" at will of the transgenes in order to restore the native state of the cells.

3. Metabolic Engineering

Metabolic engineering has been defined as the introduction of specific modifications to metabolic networks for the purpose of improving cellular properties. In recent years, metabolic engineering has gained importance in biotechnology, being used largely to improve existing processes involving the production of chemicals using microorganisms. Although less widely appreciated, metabolic engineering techniques can be applied to study physiological systems and isolated whole organs *in vivo* to elucidate the metabolic patterns that occur in different physiological states, such as fed, fasted, or in disease. Metabolic engineering techniques are also finding important uses in tissue engineering, where they can be used to monitor the metabolic response of cells and tissues to perturbations in the environment and rationally design culture media that enhance cell function and proliferation.

In metabolic engineering, the notion of cellular metabolism as a network is of central importance. Also, fundamental to metabolic engineering is the idea that metabolic processes, systemic or cellular, are coupled and as such cannot be considered separately. The major metabolic pathways (e.g., glycolysis, gluconeogenesis, pentose phosphate, urea cycle, tricarboxylic acid cycle, fatty acid synthesis and oxidation) are interrelated through common precursors and metabolic intermediates. Thus, an enhanced perspective of metabolism and cellular function can be obtained by considering a framework that incorporates all the major participating reactions, rather than a few isolated ones. Two methodologies for the characterization and analysis of cell metabolism that are especially useful for the analyses of metabolic abnormalities in human disease are metabolic flux analysis and metabolic control analysis.

Metabolic flux can be defined as the net rate of conversion of one metabolic precursor to a product. Metabolic flux analysis refers to the calculation of fluxes through metabolic pathways. Two techniques are primarily used for flux determination: (1) mass isotopomer analysis, and (2) extracellular metabolite balance models. Mass isotopomer analysis has been used extensively to quantitate fluxes in mammalian cells and tissues including brain, heart, and liver. In this approach, the body is fed, or the isolated tissue is perfused with, substrates labeled with stable isotopes (e.g., ^{13}C). A different labeling pattern of metabolites in the blood, perfusate, and/or tissue extract arises depending on the pathways utilizing these substrates. The labeling patterns are experimentally determined by nuclear magnetic resonance or mass spectroscopy. These labeling patterns are analyzed in conjunction with a mathematical model to calculate the fluxes through the various pathways which best account for the observed labeling patterns. Although isotopomer analysis is a powerful and generally noninvasive method, stably labeled compounds and the instruments required to determine the isotopomer distributions of key metabolites are relatively expensive.

Material balances of whole-body macronutrients have been used since the 19th century for evaluating bulk material processing with the body (e.g., to study the conversion of carbohydrates to fat). Material balances have

since evolved into metabolic flux balance models that can predict intracellular fluxes in complex metabolic networks. This methodology utilizes a stoichiometric model that describes the major intracellular reactions at steady state. Extracellular fluxes, which correspond to rates of consumption/production of extracellular metabolites, are experimentally determined, and intracellular fluxes are calculated based on the stoichiometric constraints of the intracellular reaction network. This approach has been used extensively to study and improve strains of microorganisms (bacteria and yeasts) of significance in biotechnology. As of now, applications of metabolic flux balance models to mammalian cell systems have been more limited but are gaining in popularity.

The starting point in this analysis is the construction of a list of steady-state material balance equations to describe the conversion of substrates to metabolic products for the biochemical system of interest. For example, if one considers a simplified scheme of amino acid metabolism in liver, one can write a set of steady-state material balance equations that represent the flow of metabolites through the network (Fig. 9). The equations contain measurable quantities (these are marked with an asterisk) which are the rates of consumption/production of extracellular metabolites. The concentrations of strictly intracellular metabolites (e.g., argininosuccinate) are assumed to be constant. In this particular case, we have eight fluxes to be determined, five of which are measurable ($F_1^*, F_2^*, F_4^*, F_7^*, F_8^*$). The five equations listed here, which relate these fluxes to each other, can be reduced to four independent equations. Thus, the system can be solved to yield the three unknown intracellular fluxes (F_1, F_2, F_4). Because the system is overdetermined, it provides an internal check for consistency of the data with each other and the assumed biochemistry.

While this method is very useful, there is a limit to the extent to which complex metabolic networks can be

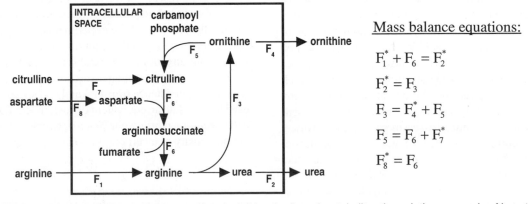

Mass balance equations:

$$F_1^* + F_6 = F_2^*$$
$$F_2^* = F_3$$
$$F_3 = F_4^* + F_5$$
$$F_5 = F_6 + F_7^*$$
$$F_8^* = F_6$$

FIGURE 9 System of mass balance equations describing the flow of metabolites through the urea cycle of hepatocytes. Measurable fluxes are labeled with a star.

elucidated by using measurements of extracellular products alone. For example, flux distribution at split points that converge at another point of the network cannot be resolved. Another limitation is that only net fluxes are determinable, while isotopic methods can sometimes resolve the rates of the forward and the backward reactions. This methodology may be particularly useful when used in combination with stable isotopes to provide fluxes that cannot be directly determined by the isotopomer analysis. Metabolic flux analysis, once validated for the particular case under study, is potentially very useful as it is noninvasive and cost effective.

Another important aspect of the metabolic network that can be investigated by metabolic engineering techniques is the "rate-controlling" enzymes of the pathway (i.e., the enzymes governing flux in the metabolic network). Over the past 30 years, several theoretical frameworks for this type of analysis have been developed. Of these, one of the most widely used is metabolic control analysis. Metabolic control analysis aims at quantifying the control that individual or groups of enzymes exert on the flux through a particular pathway by studying the response of the system to changes in nutrient levels and other factors that alter the activity of specific enzymes in the network. This analysis is generally quite difficult to perform experimentally and is often based on many assumptions; however, it provides valuable insight into the mechanisms governing metabolic adaptation to changes in the environment and a rational basis for genetically engineering cells to perform specific functions.

4. Effects of Mechanical Forces on Cells and Tissues

The environment in which cells are cultured has traditionally been defined by the presence of soluble factors such as hormones and growth factors, and the chemical properties of the surface on which they adhere and grow. In addition, certain physical forces may influence cellular function and may be used as tools to induce specific phenotypes in cells. For example, it is believed that mechanical loading plays an important role in the synthesis and deposition of extracellular matrix by cells in load-bearing tissues such as cartilage and bone *in vivo*. Thus, incorporating mechanical loading schemes in the culture environment such as cyclical compression may be beneficial. Furthermore, the mechanical loading apparatus can be coupled to sensors to provide a continuous assessment of the mechanical parameters (compressive strength and module of elasticity) of the developing tissue. Another example is the effect of fluid shear stress and uniaxial stretch, both of which induce vascular endothelial cell elongation and alignment, as well as the secretion of vasoactive compounds. Cyclic stretch has also been used to promote the fusion of

muscle myoblasts (muscle progenitor cells) into contractile myotubes containing parallel myofibers aligned in the direction of the applied force.

C. Transport Phenomena in Tissue Engineering

1. Cell Migration

Cell migration is often a critically important step in many applications of tissue engineering. For example, biomaterial implants used for nerve regeneration, cartilage, and skin wound healing require that the host's cells migrate into the matrix implant. Cell migration speed depends on a complex balance between cell tractional forces and the stickiness of the matrix to the cell. The highest cell speeds are obtained at intermediate attachment strengths, which allow the leading edge of the cell to anchor itself to the surface while the receding edge comes off the surface. A substrate with low adhesiveness does not allow the cell to form any anchors to the surface that can resist cell tractional forces, which results in poor migration. Similarly, cells "glued" to a surface that is too sticky are not able to move forward because cell–substrate bonds at the receding edge of the cell cannot be broken.

The determination of cell-migration parameters is important in order to predict the speed at which cells can invade a tissue construct. The prediction of cell migration behavior based on the knowledge of cell mechanics, interaction of cell receptors with appropriate ligands on the extracellular matrix, and function of the cytoskeleton is possible, but complicated, and involves difficult measurements. On the other hand, phenomenological cell-migration parameters can be determined via analysis of single cell trajectories or cell concentration profiles of cell populations in specific devices which allow for the visualization of cells during the migration process. The most simple model to analyze single cell trajectories is the persistent random-walk model (Fig. 10). This model assumes that cells are not restricted in their range of movement (within the duration of the experiment), they can move in any direction with equal probability, and, once they move in a certain direction, they exhibit a characteristic persistence time before they change direction. The parameters of the model, persistence time (P) and cell speed (S), can be used to calculate an equivalent diffusivity coefficient (also called *random motility coefficient*), which is a measure of the propensity of a cell population to spread:

$$D = \frac{S^2 P}{n} \qquad (5)$$

where n is the number of dimensions ($n = 2$ for a surface, $n = 3$ for a gel) where the migration occurs.

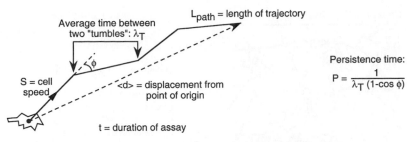

FIGURE 10 Definition of parameters that characterize single cell migration.

It is important to note that these parameters are not dependent on the geometry of the system used to measure them and thus can be used to predict cell migration in other geometries. This model chiefly applies to two-dimensional surfaces; however, it can be extended to three-dimensional matrices, in which case the effective pore size of the matrix, which may create a hindrance to the migration process, needs to be taken into account. Cell migration in a specific direction can be promoted by micropatterning tracks on a surface, which prevents cells from wandering away from the desired direction, and in three dimensions by using materials exhibiting oriented pores and/or fibers.

The above discussion relates to the process of random migration where cells do not move in a preferential direction. It is often the case, however, that soluble and insoluble factors causing cells to move in a preferential direction are present. Soluble agents that "attract" cells are called *chemotactic*, while those immobilized in the extracellular matrix are called *haptotactic*. In chemotactic or haptotactic migration, a third parameter must be determined to capture the directional preference of the migration process. This parameter is the chemotactic index (*CI*), which can be determined experimentally from single cell trajectories by the equation:

$$CI = \frac{\langle d \rangle}{L_{\text{path}}} \qquad (6)$$

where $\langle d \rangle$ is the distance of the cell from the point of origin at the beginning of the experiment, and L_{path} is the length of the path used by the cell to achieve the displacement $\langle d \rangle$. The population-relevant parameter that describes chemotaxis is the chemotaxis coefficient χ, which is calculated by the following expression:

$$\chi = \frac{S \cdot CI}{\nabla L} - \frac{1}{n}\left[\frac{d\ln P}{dL} - \frac{d\ln S}{dL}\right] \qquad (7)$$

where L and ∇L are the concentration and spatial gradient, respectively, of chemoattractant or haptotactic factor. Because chemotactic and haptotactic factors may increase cell speed, and thus increase migration via a "chemoki-

netic" or "haptokinetic" effect, the last term in Eq. (7) includes a correction for this effect.

The values of D and χ are the constitutive parameters describing cell migration in a variety of both *in vitro* and *in vivo* systems. The expression analogous to Fick's first law of diffusion for cell flux is

$$J = -D\frac{\partial C}{\partial x} + C\left(-\frac{dD}{2dL} + \chi\right)\frac{\partial L}{\partial x} \qquad (8)$$

and a cell concentration profile in any system can be derived via the continuity equation:

$$\frac{\partial C}{\partial t} = -\frac{\partial J}{\partial x} \qquad (9)$$

In real cases, Eq. (9) may need to be solved in conjunction with appropariate transport equations for the chemoattractant or haptotactic factor, which may be time varying.

As briefly discussed earlier, cells migrating on substrates exert forces that allow them to move. As a result, cells on surfaces or inside gels that are compliant can significantly alter the shape of the material. Quantitative analyses and mathematical descriptions of these phenomena allow prediction of how the cell-material construct changes shape over time. A well-known example of cell-mediated contraction is the fibroblast-populated collagen lattice, which forms the basis for some of the currently used tissue engineered skin grafts. The fibroblast-populated collagen lattice is generated by mixing fibroblasts with a chilled solution of collagen in physiological buffer, followed by exposure to 37°C to induce the gellation of the collagen. If the gel is not anchored to any surface, fibroblasts embedded in a collagen gel cause the contraction of the gel in an isotropic fashion. The contraction process can be controlled to a certain extent by mechanically restricting the motion along certain directions, which also induces a preferential alignment of the collagen fibers as well as the cells within it, which results in a nonisotropic connective tissue equivalent. Preferential alignment of cells may be important in specific applications, such as in tissue

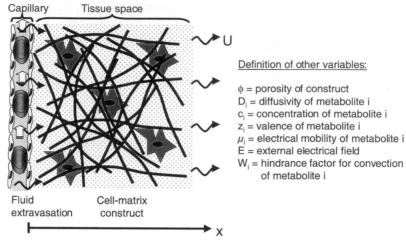

FIGURE 11 Transport of metabolites into a tissue construct implanted next to a blood vessel. U is the velocity of fluid extravasated into the tissue; see text for additional explanations.

engineered blood vessels. A potential design of such vascular grafts would have endothelial cells on the inside surface of the graft aligned with the direction of blood flow, as is observed *in vivo* in arteries. On the other hand, the appropriate direction of smooth muscle cells within the blood vessel wall would be along the circumference of the vessel in order to be able to perform their natural function of modulating the diameter of the vessel.

2. Metabolite Transport

In normal tissues, the circulatory system brings in nutrients and removes waste products. Typically, no cell *in vivo* is farther than about 100 μm, or even sometimes less, from a blood vessel. Thus, transport by diffusion does not have to occur over distances beyond 100 μm. Engineered tissue constructs are mostly devoid of any vascular system, and although there may be culturing methods allowing transport by convection throughout the cell mass, as described later in the bioreactor section, after implantation the engineered tissue is no longer perfused and the situation remains so until vascularization by angiogenesis occurs from the surrounding host's tissues. Vascularization *in situ* can be accelerated by implanting tissues that release angiogenic factors, such as fibroblast-derived growth factor and vascular endothelial growth factor. Although tissue constructs may include vascular endothelial cells, it is not yet possible to create three-dimensional vascular networks *in vitro* that are patent.

Transport through tissues can be modeled using the basic transport equations used in nonliving systems and can be useful to predict the concentration profiles of metabolites throughout engineered tissues. In the following presentation, we apply these equations with the goal of providing design criteria for tissue constructs. We consider a

tissue construct of thickness X implanted *in vivo*, assuming that this construct is avascular but surrounded with blood vessels from the hosts's tissue (Fig. 11). The fundamental equation describing the flux N of a particular species i is given by Fick's law of diffusion, to which terms to account for the electrical migration of species and convection are added:

$$N_i = \phi \left\{ -D_i \frac{\partial c_i}{\partial x} + \frac{z_i}{|z_i|} \mu_i c_i E \right\} + W_i c_i U \qquad (10)$$

To facilitate the comparison between systems of different geometries and scales, we next rewrite the previous equation using dimensionless quantities. The Péclet number is a dimensionless number defined as the ratio of nondiffusive transport (convection and electrical migration) to transport by diffusion in a particular system:

$$Pe = X \left(\frac{W_i U + \phi \frac{z_i}{|z_i|} \mu_i E}{\phi D_i} \right) \qquad (11)$$

Substituting into the flux equation yields:

$$N_i = \phi D_i \left(-\frac{\partial c_i}{\partial x} + \frac{Pe}{X} c_i \right) \qquad (12)$$

We can further simplify this equation by defining the following dimensionless variables:

$$N_i^* = \frac{N_i}{\phi D_i}, \qquad c_i^* = \frac{c_i}{X}, \qquad x^* = \frac{x}{X} \qquad (13)$$

which yields a new and simplified flux equation containing only the Péclet number as the unknown or "adjustable" parameter, the value of which depends on the system under consideration:

$$N_i^* = \left(-\frac{\partial c_i^*}{\partial x^*} + Pe \, c_i^* \right) \qquad (14)$$

The use of the dimensionless quantities facilitates comparison between systems with different geometries and scales. For example, although two systems may be different in several respects, they will behave similarly as far as metabolite transport if they have similar Péclet numbers. Furthermore, the Péclet number gives an instant idea as to which major transport mechanisms operate in the system under study. If $Pe < 1$, diffusional transport dominates, while if $Pe > 1$, convective transport and electrical migration are more important.

We now consider the typical case of a tissue engineered construct implanted *in vivo*. Although the precise values of the transport parameters are not necessarily known, it is often a useful exercise to perform an order of magnitude analysis to determine the approximate contribution of each term in the transport equation. For this purpose, we start with Eq. (12) and propose that a reasonable estimate for the concentration gradient $\partial c_i / \partial x$ is c_i / X. The flux equation becomes:

$$N_i = \phi D_i \left(-\frac{\partial c_i}{\partial x} + \frac{Pe}{X} c_i \right) \approx \phi D_i \left(\frac{c_i}{X} + \frac{Pe}{X} c_i \right)$$

$$= \frac{\phi D_i c_i}{X}(1 + Pe) \tag{15}$$

Most nutrient transport to the implant initially comes from the surrounding capillaries. These capillaries continuously leak plasma into the tissue space because the pressure inside capillaries is greater than in the tissue. Typical measured values for the capillary filtration coefficient and pressure gradient are 0.035 cm^3/min/mm Hg/100 g tissue and 27 mm Hg, respectively. Assuming capillaries are distributed evenly 100 μm (or 0.01 cm) apart, on average each capillary occupies $(0.01 \text{ cm})^3 = 10^{-6} \text{ cm}^3$. Since 100 g tissue occupy a volume of about 100 cm^3, we can estimate the flow rate and velocity of fluid exiting capillaries and flowing through the implant:

$$Q = 0.035 \frac{\text{cm}^3}{\text{min} \cdot \text{mmHg} \cdot 100 \, \text{cm}^3 \, \text{tissue}} \times 27 \, \text{mmHg}$$

$$\times \frac{1}{60 \text{s}} \times \frac{10^{-6} \text{cm}^3}{\text{capillary}} = 1.2 \times 10^{-8} \frac{\text{cm}^3}{\text{capillary} \cdot \text{s}}$$

The average surface area perfused by each capillary is 0.01 cm×0.01 cm, thus the velocity is

$$U = \frac{1.2 \times 10^{-8} \text{cm}^3/\text{s}}{(100 \times 10^{-4} \text{cm})^2} = 1.6 \times 10^{-4} \text{cm/s}$$

We next assume that the implant is 0.1 cm thick and highly porous (as would be the case for a collagen gel, for example), so that $\phi = 1$, with a mean pore size of >10 μm, which is much larger than the molecular size of transported molecules, so that $W_i = 1$. Using these values, we calculate the Péclet number and corresponding flux rate

TABLE VI Transport Parameters for a Few Metabolites Important for the Function of Tissue Engineered Constructs Implanted *in vivo*

Metabolite	D_i (cm^2/s)^a	C_i (mM)^b	Pe	N_i (μmol/cm^2/s)
Oxygen	2×10^{-5}	0.1	0.8	4×10^{-5}
Glucose	9×10^{-6}	5	1.8	120×10^{-5}
Insulin	1.5×10^{-6}	3×10^{-8}	11	5×10^{-12}

$^a D_i$ are typical values measured at 37°C (the body temperature).
$^b C_i$ are typical values found in biological fluids *in vivo* (for oxygen, arterial blood levels were used).

for oxygen, glucose, and a peptide hormone, insulin, in the tissue construct (Table VI). For the small metabolites oxygen and glucose, Péclet numbers are close to 1, which indicates that diffusive and convective transports have equal contributions. Insulin, by virtue of its molecular size, has a lower diffusivity, thus its transport is more dependent on convection. One can notice that although oxygen diffusivity is at least one order of magnitude greater that that of glucose, the transport of oxygen is almost two orders of magnitude slower. This is due to the fact that oxygen has a very low solubility in water and physiological fluids. Thus, although oxygen diffuses rapidly, it is not possible to create large gradients to provide the necessary driving force for its transport; for this reason, oxygen transport is almost always the main factor limiting the size and cell density of tissue engineered constructs.

To obtain an estimate of the maximum thickness of a tissue engineered implant based on transport considerations, we must balance metabolite delivery with consumption by the cells in the construct. For this purpose, we use the mass balance or continuity equation, which is generally written as:

$$\frac{\partial c_i}{\partial t} = -\frac{\partial N_i}{\partial x} + \sum_i (G_i - R_i) \tag{16}$$

where G_i and R_i represent the generation and consumption rates of metabolite i in the construct, respectively. Substituting the expression for the flux N_i yields:

$$\frac{\partial c_i}{\partial t} = D_i \left\{ \frac{\partial^2 c_i}{\partial x^2} + \frac{Pe}{X}\left(\frac{\partial c_i}{\partial x}\right) \right\}$$

$$+ \sum_i (G_i - R_i) \tag{17}$$

Let us now consider the specific case of oxygen transport. Furthermore, to simplify the calculations, we assume that convective transport is negligible so that $Pe = 0$. This would be a "worst-case scenario" where normal tissue perfusion is disrupted due to the surgical trauma caused by the implantation procedure itself, or the implant is encapsulated into a membrane (i.e., to protect implanted cells from the host's immune system) which does not allow convective transport into the implant. Furthermore, there is no generation of oxygen by the tissue, so $G_i = 0$. The consumption of metabolites by cells often follows first-order kinetics at low concentrations (i.e., the consumption rate is proportional to the concentration of metabolite) and progressively becomes zero order as the concentration increases. At one point, the cells are "saturated" and cannot take up more. This behavior is often described by Michaelis–Menten kinetics:

$$R_i = \frac{V_{\max}c_i}{K_M + c_i} \tag{18}$$

where $V_{\max}$ is the maximal uptake rate and K_M is the metabolite concentration at which the uptake rate is half-maximal. Substituting this expression into the continuity equation and taking into account other assumptions described above yield:

$$\frac{\partial c_i}{\partial t} = D_i \frac{\partial^2 c_i}{\partial x^2} - \frac{V_{\max}c_i}{K_M + c_i} \tag{19}$$

If we next assume steady state, we obtain:

$$0 = D_i \frac{\partial^2 c_i}{\partial x^2} - \frac{V_{\max}c_i}{K_M + c_i} \tag{20}$$

One can integrate this expression with the following boundary conditions:

$$x = 0, \qquad c_i = C_0 \tag{21}$$
$$x = X, \qquad \partial c_i / \partial x = 0 \tag{22}$$

to yield the concentration profile throughout the system. However, rather than looking at the whole concentration profile, we really only want to know if at any point within the system there will be a significant depletion of metabolite. As a rule of thumb to estimate when a cell is starving, the condition $c_i = K_M$ is often used. Figure 12 shows the results for the maximum thickness X of a construct without c_i going below K_M anywhere in the construct. Using this chart to estimate the thickness of an implantable construct containing liver parenchymal cells (hepatocytes), we find that a construct containing even a relatively low density of cells (10^7 cells/cm^3) cannot have a thickness exceeding about 500 μm (Table VII). At tissue-level cell densities of 10^8 cells/cm^3, that thickness can be as low as 100 μm, which is consistent with the *in vivo* density of capillary vessels.

3. Bioreactor Technologies

For certain applications it is necessary to maintain a large number of cells that transform an input of reactants into an output of products. This is the case for the bioartificial liver or pancreas and more recently for the production of blood cells from hematopoietic tissue. These systems require maintenance of the function of a large number of cells in a small volume. For example, a hypothetical bioar-

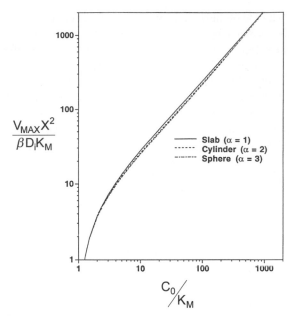

FIGURE 12 Correlation for the maximum thickness of a tissue construct to avoid nutrient depletion.

tificial liver device possessing 10% of the detoxification and protein synthesis capacity of the normal human liver (a rough estimate of the minimum processing and secretory capacities that can meet a human body's demands) would contain a total of 10^{10} adult hepatocytes. Thus, to keep the total bioreactor volume within reasonable limits (1 L or less), 10^7 cells/mL or more are required. For comparison, the normal human liver contains approximately 10^8 hepatocytes/mL. Three main types of bioreactor design have been considered in tissue engineering: (1) suspension culture methods using microcarriers, (2) cells immobilized in hollow-fiber systems, and (3) suspension culture in rotating wall vessel bioreactors.

a. Microcarrier-based systems. Microcarriers are one of the first methods used for supporting large-scale mammalian cell culture. Microcarriers are small beads

TABLE VII Maximum Thickness (X) of Tissue Constructs Estimated by Order of Magnitude Transport Analysis[a]

Geometry	10^7 cells/cm^3 X (μm)	10^8 cells/cm^3 X (μm)
Slab	300	95
Cylinder	430	135
Sphere	520	165

[a] These estimates are based on the assumptions that that surface of construct is exposed to arterial oxygen levels ($C_0 = 100$ nmol/cm^3), hypoxic damage occurs when $c_i = K_M$, and for hepatocytes $V_{\text{MAX}} = 0.4$ nmol/10^6 cells/s, $K_M = 0.6$ nmol/cm^3.

(usually less than 500 μm in diameter) with surfaces treated to support cell attachment. These beads are then maintained in suspension in medium using very low stirring speeds in order to avoid mechanical cell damage, either due to shearing forces in the liquid or due to bead–bead collisions. The surface area available per microcarrier can be increased by using porous microcarriers, where cells can migrate and proliferate within the porous matrix as well as on the microcarrier surface; furthermore, cells within the microcarrier are protected from mechanical damage. To attach cells to microcarriers, isolated cells are mixed with microcarriers in suspension. The protocol requires careful optimization of the number of cells per bead, mixing velocity (intermittent mixing may be necessary until cells are firmly attached), and supply of oxygen, which is necessary for cell attachment as the cells require energy in order to spread onto a substrate.

Bioreactor configurations using microcarriers include packed and fluidized beds. A packed bed of microcarriers consists of a column filled with microcarriers with porous plates at the inlet and outlet of the column to allow perfusion while preventing microcarrier entrainment by the flow. Reactor volume is proportional to the microcarrier diameter, thus it is advantageous to reduce the microcarrier size as much as possible. However, packed beds with small beads may clog and the cells may have a tendency to accumulate in the channels between the microcarrier surfaces. Total flow rate is mainly dependent on cell number and the nutrient uptake rate of the cells. Because oxygen is usually the limiting nutrient, the medium flow rate through the reactor is found using the following equation:

$$\text{Flow rate} = \frac{O_2 \text{ Consumption per cell} \times \text{Cell number}}{O_2 \text{ Concentration in medium}}$$
(23)

The aspect ratio of the bed (height/diameter) determines the fluid velocity through the packed bed according to the equation:

$$\text{Fluid velocity} = \frac{\text{Flow rate}}{\text{Cross-sectional area}}$$
(24)

and is adjusted so that the magnitude of fluid mechanical forces (proportional to the aspect ratio) within the bed is below damaging levels (Fig. 13). Fluidized beds differ from packed beds in that the perfusing fluid motion maintains the microcarriers in suspension. Packed-bed systems have been shown to support cell densities exceeding 10^8 cells/mL when using microporous microcarriers (500 to 850 μm in diameter). In addition, packed beads (1.5-mm diameter) have been used to entrap aggregates of hepatocytes. The latter application was shown to maintain a relatively stable level of albumin secretion (a liver-specific product) for up to 3 weeks.

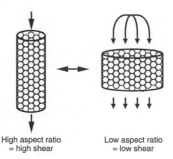

High aspect ratio = high shear   Low aspect ratio = low shear

FIGURE 13 Two possible configurations for a packed bed bioreactor of equal volumes.

b. Hollow-fiber systems. The hollow-fiber system is the most widely used type of bioreactor used in tissue engineering and in artificial organ development. It consists of a shell traversed by a large number of small-diameter tubes (Fig. 14). The cells may be placed within the fibers in the intracapillary space or on the shell side in the extracapillary space. The compartment that does not contain the cells is generally perfused with culture medium or the patient's plasma or blood. The fiber walls may provide the attaching surface for the cells and/or act as a barrier against the immune system of the host. Microcarriers have also been used as a way to provide an attachment surface for anchorage-dependent cells introduced in the shell side of hollow-fiber devices. Hollow-fiber systems can be designed to be implanted as vascular shunts, but may also be perfused with the patient's blood or plasma extracorporeally.

There are many studies on how to determine fiber dimensions, spacing, and reactor length; however, commercially available units come in a relatively limited number of sizes, usually with inner fiber diameters of 500 μm or more. Several reports in the literature describe the use of hollow-fiber systems in the development of a bioartificial pancreas, which place the islets on the shell side, while perfusing the fibers with the animal's plasma or blood. The fibers can be made relatively non-thrombogenic and of porosity sufficiently small as to avoid immune attack of the cells inside the shell. One difficulty with this configuration is that interfiber distances in the hollow-fiber device are not well controlled, so that regions within the shell space receive too little nutrients.

It may be advantageous to place cells in the lumen of small fibers because the diffusional distance between the shell (where the nutrient supply would be) and the cells is essentially equal to the fiber diameter, which is easier to control than the interfiber distance. In one configuration, cells have been suspended in a collagen solution and injected into the lumen of fibers where the collagen is allowed to gel. Contraction of the collagen lattice by the cells even creates a void in the intraluminal space, which can be perfused with hormonal supplements, etc. to enhance

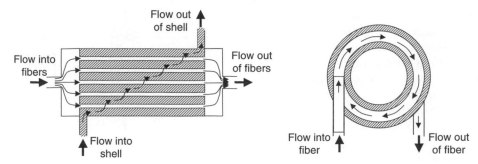

FIGURE 14 Two common types of hollow-fiber bioreactors. In the reactor shown on the left, cells may be placed either in the shell (gray color) or intrafiber space. In the reactor shown on the right, cells are typically placed in the shell space.

the viability and function of the cells, while the patient's plasma would flow on the shell side. Such a configuration has been described for the construction of a bioartificial liver using adult hepatocytes.

c. Rotating vessel wall bioreactor.

The rotating wall vessel bioreactor, a relatively new type of bioreactor system used in biotechnology, was originally developed and patented by the U.S. National Aeronautic Space Agency (NASA) to study the behavior of cells and tissues under conditions simulating low gravity on Earth. This bioreactor consists of a chamber entirely filled with culture medium containing the cells, tissue constructs, or even actual tissue explants in suspension. The chamber is rotated on a horizontal axis at a speed that approximately matches the terminal settling velocity of the cells or tissues in suspension such that they establish a fluid orbit (Fig. 15). The cells or tissues therefore never hit the bottom of the reactor or touch any of its inner surfaces. An important feature of this system is that oxygen is delivered via gas-permeable silicone membranes; no sparging of gas is necessary. The design ensures uniform hydrodynamic conditions within the bioreactor without the use of

impellers, which have been shown to cause significant cell damage in other types of bioreactors. Interestingly, data gathered on recent space missions do suggest that cells and tissues cultured in this bioreactor develop and grow as they would in low-gravity environments.

Over 50 types of cells, tissue constructs, and even tissue explants have been cultured in these bioreactors, which appear to be ideally suited to promote the expression of tissue-specific functions in the cultured cells and preserve the three-dimensional morphological characteristics of the native tissue. Thus, this system should be useful to create and maintain bioartificial tissues to be subsequently implanted *in vivo*. On the other hand, these devices would not be appropriate for tissue engineering applications requiring a combination of very high cell densities and very low liquid hold-up volumes, such as in the case of extracorporeal bioartificial livers.

D. Morphogenesis of Engineered Tissues

The quantitative difference between cell–substrate and cell–cell adhesion strength on a rigid surface dramatically affects the organization of cells on the substrate. A thermodynamic view of the problem suggests that the overall system (consisting of the cells and the extracellular support) ultimately reaches an equilibrium state when the surface free energy is minimized. According to this concept, the existence of large cell–substrate adhesion forces relative to cell–cell adhesion forces prevents cell–cell overlapping (Fig. 16). In contrast, the opposite situation would lead to cell clumping or multilayered growth on the substrate. This prediction is in agreement with the observation of cellular aggregate formation when hepatocytes are plated on a nonadherent surface as opposed to a highly adherent surface such as type I collagen.

Heterotypic cell systems or "co-culture" systems have been used for the production of skin grafts, in long-term cultures of hepatocytes, and in long-term cultures of mixed bone marrow cells. These systems take advantage of the

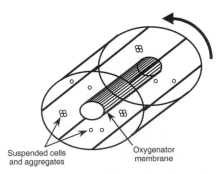

FIGURE 15 Design of a rotating wall vessel bioreactor. Cells are suspended in medium, which fills the vessel until no air bubble is left. Oxygen is delivered via a silicone membrane in the center of the vessel. The vessel rotates at a relatively low speed (∼30 rotations/min) to prevent settling of cells.

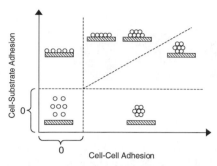

FIGURE 16 Possible configurations of cells on a flat substrate. In the absence of any adhesion, cells remain in single cell suspension (bottom left quadrant). Increasing substrate-cell adhesion causes cells to stick to the surface. Increasing cell–cell adhesion from that point causes the cells to attach to each other and form monolayers. Increasing cell–cell adhesion further promotes aggregation of the cells on the substrate.

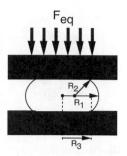

FIGURE 17 Deformation of cell aggregate during compression.

trophic factors (for the most part unknown) secreted by "feeder" cells. Greater use of different cell types used in co-culture will enable engineered cell systems to closely mimic *in vivo* organization, with potential benefits including increased cell function and viability and greater range of functions expressed by the bioartificial tissue. The organization of multicellular three-dimensional structures may not be obvious. Provided that the adherence of homotypic and heterotypic interactions is known, a thermodynamic analysis similar to that used to describe the morphology of a pure cell culture on a surface can be used to predict how cells will organize in these systems. The process of cell–cell sorting in multicellular systems may be altered by changing the composition of the medium or altering the expression of proteins mediating cell–cell adhesion via genetic engineering.

Predicting the organization of simple multicellular systems containing more than one cell type is possible if the relative cohesiveness of each cell type with respect to each other is known. The cohesiveness of a tissue is exactly the same as a tissue surface tension analogous to a liquid surface tension. The tissue surface tension σ has been measured in homogenous three-dimensional cellular aggregates by measuring the compression force required to deform an aggregate into a flattened droplet shape (Fig. 17).

The force exerted by the cell aggregate on the compression plate is given by the Laplace equation:

$$\frac{F_{eq}}{\pi R_3^2} = \sigma \left(\frac{1}{R_1} + \frac{1}{R_2} \right) \tag{25}$$

where F_{eq} is the force measured after sufficient time has been allowed for the aggregate, which behaves as a viscoelastic liquid, to relax.

When tissues A and B are combined, they will reorganize depending on the relative values of σ. At equilibrium,

they will reach a predictable configuration that minimizes the surface free energy. In the case where each cell has uniform stickiness on its entire surface, the possible configurations are illustrated in Fig. 18.

This theory has been tested and verified extensively using aggregates of embryonic cells. Furthermore, by changing the level of expression of surface adhesion molecules via hormonal induction or genetic engineering, it was possible to alter the organization of such cell aggregates in a predictable fashion. It is important to note, however, that although the theory was derived for a closed system at equilibrium, all of the cultured cell systems examined were in fact open systems because the cells were dissipating energy provided by the culture medium. Thus, care should be taken when using this approach to analyze more complex cases. In addition, further refinement to the application of the equations will be required in cases where cell adhesiveness is not uniformly distributed. This situation is common to most epithelial cells, which exhibit poor to negligible cell–cell adhesiveness on the apical surface, and as a result tend to form tubular cell structures.

E. Summary

Tissue engineering is the construction of bioartificial tissues *in vitro* as well as the *in vivo* alteration of cell growth and function via implantation of suitable cells isolated from donor tissue and biocompatible scaffold materials. Biomaterials for tissue engineering must have controlled surface chemistry, porosity, and biodegradability in order to promote optimal cell adhesion, migration, and deposition of endogenous extracellular matrix materials by the cells. Strategies to switch cells between growth and differentiation, which tend to be mutually exclusive, are used in order to provide a large cell mass that can perform specific differentiated functions required for the tissue construct. Combinations of cells and materials have the ability to reorganize themselves based on the strength of adhesion between cells and substrate and among the various cell types present in the tissue construct. Finally, tissue constructs must be intimately integrated into the

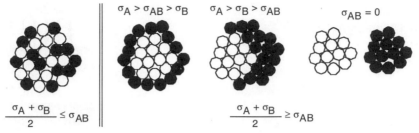

$$\sigma_A > \sigma_{AB} > \sigma_B \qquad \sigma_A > \sigma_B > \sigma_{AB} \qquad \sigma_{AB} = 0$$

$$\frac{\sigma_A + \sigma_B}{2} \leq \sigma_{AB} \qquad\qquad \frac{\sigma_A + \sigma_B}{2} \geq \sigma_{AB}$$

FIGURE 18 Possible equilibrium configurations of two cell types A and B mixed together depending on the relative tissue surface tension of each tissue. Cells will not remain mixed unless the adhesion force of A to B exceeds the average cohesion of tissues A and B (left-most case). Otherwise, cells segregate and the most cohesive tissue will tend to remain in the center while the other tissue spreads around it.

host's vascular system in order to provide efficient nutrient supply and waste removal.

III. APPLICATIONS OF TISSUE ENGINEERING

A. Connective Tissues

1. *In Vitro* Construction of Connective Tissues

Connective tissues can be reconstructed *in vitro* by incorporation of connective tissue cells within a porous biomaterial or a loose network of extracellular matrix components. The resulting geometry and organization of the tissue equivalent are similar to those of the parent tissue *in vivo*. Several different types of connective tissues have been built using this approach (Table VIII). The mechanical properties of the original biomaterial are dramatically altered by the embedded cells, due either to cell-generated forces causing contraction of the matrix material or deposition of extracellular matrix generated by the cells themselves. This remodeling is often important for the eventual function of the tissue construct.

The only engineered connective tissue equivalent currently used clinically is bioartificial skin. Bioengineered

TABLE VIII Examples of Connective Tissues Made *in vitro*

Tissue	Biomaterial	Cell type
Dermis	Collagen	Fibroblast
Vascular media	Collagen, endogenously produced matrix	Smooth muscle cell and fibroblast
Cartilage	Collagen–glycosaminoglycan complex, polylactic–glycolic copolymer	Chondrocyte
Meniscus	Collagen–glycosaminoglycan complex	Fibrochondrocyte
Bone	Calcium phosphate, polylactic–glycolic copolymer	Mesenchymal stem cell from periosteum
Tendon	Collagen	Tenocyte (tendon fibroblast)

skin has several applications, including the treatment of burn wounds and nonhealing diabetic and venous leg ulcers, as well as pressure sores. Bioartificial skin consists of a dermal equivalent made by seeding collagen gels or meshes made of biodegradable polymers with dermal fibroblasts. The dermal equivalent may have an overlay of silicone sheeting to prevent evaporative loss of water (a function normally performed by the missing epidermis). After take, the grafted material is suitable for grafting an epidermis, either in the form of a split-thickness skin graft or cultured layer of epidermal cells (keratinocytes). Bioartificial skin is also available as a complete dermal–epidermal composite comprising the dermal equivalent on top of which keratinocytes have been cultured to confluence. Differentiation of the epidermal layer into a functional epidermis with a cornified layer that exhibits a high resistance to chemical damage is induced by exposure to the air–liquid interface during the culturing process. The composites are then grafted onto the wound site in a single operation.

Metabolic engineering of connective tissue cells such as fibroblasts and smooth muscle cells via addition of ascorbic acid to the culture medium has been shown to promote the production of large quantities of extracellular matrix, such that the cells produce their own scaffolding material. This is an efficient method to generate sheets of cells that can then be layered on top of each other or rolled around a mandrill to form thicker tissue constructs. Although still in early experimental stages, this approach has been used to form tubes that can withstand significant mechanical stress and is currently being evaluated for the generation of media for bioengineered blood vessels.

Connective tissues that must bear significant loads must exhibit specific mechanical properties. This is the case of bioartificial cartilage, which would be best implanted once its mechanical properties are similar to that of the authentic tissue. Chondrocytes seeded at high density (10^7 cells/cm^3) in agarose gels retain their phenotype and remain viable for up to 6 months. In this system, the deposition of type II collagen and highly charged proteo-glycans, which are primarily responsible for the mechan-ical properties of native cartilage, is enhanced by subject-ing the tissue to cyclical mechanical compression. More recently, chondrocytes seeded in polylactic–glycolic scaf-folds and then implanted in ectopic sites *in vivo* were found to generate hyaline cartilage tissue with an overall shape similar to that of the original synthetic matrix. Thus, it may be possible to first generate the tissue of desired properties at ectopic sites and, when ready, implant it at the site requiring intervention.

2. *In Vivo* Regeneration Using Guidance Templates

It is sometimes more convenient to promote tissue regener-ation *in situ*, in which case the task of the tissue engineer is to favor wound healing and help the body overcome some of its own limitations with respect to tissue regener-ation. The first regeneration templates that became widely available are biodegradable meshes for the treatment of burn wounds. These templates are made of cross-linked collagen–glycosaminoglycan complexes and are applied to the wound site to favor regeneration of the skin dermis. The regenerated surface then provides a suitable substrate for the attachment of skin epidermal cells (keratinocytes), which can be from an autologous skin graft obtained from a donor site elsewhere on the patient or from cultured skin. In animals models, it appears that one of the main benefits of such templates is to slow down wound contraction and favor the production of new tissue resembling skin. The beneficial effect of the template depends on pore size and degradation rate. In humans, the templates appear to favor the production of normal dermis as opposed to disfiguring scar tissue.

Another area of great promise for regeneration tem-plates is to promote the reconnection of severed nerves. The natural regeneration ability of peripheral nerves is lim-ited to about 1 cm. This limitation appears to be chiefly related to the formation of scar tissue, which impedes the axonal regeneration process. To reconnect nerves over longer distances, tubes containing suitable biomaterials that promote growth of axons and inhibit scar-tissue for-mation are sutured to the ends of the nerve stumps. The material consists of a collagen–glycosaminoglycan com-posite similar to that used for skin regeneration, except for a faster degradation rate and a smaller pore size (5 μm). In addition, animal studies indicate that regeneration is better when pores are oriented along the longitudinal axis of the tube. Functional results obtained with such nerve regeneration templates in animal models approach those obtained for nerve autografts, the conventional treatment for nerve reconnection.

B. Epithelia and Endothelia

1. Secretory and Transport Functions of Epithelial and Endothelial Cells

Epithelial and endothelial cells separate different com-partments in the body; for example, endothelial cells sep-arate the intravascular from tissue space, and intestinal epithelium separates the gut lumen from the inside of the body. They control transport across these compartments, forming a selective barrier that prevents the transloca-tion of certain metabolites while favoring the transport of others, sometimes through energy-dependent processes (especially when the direction of transport is against the concentration gradient). In some cases, epithelial and en-dothelial cells also perform important secretory functions, such as the release of antithrombogenic factors by en-dothelial cells and an array of secretory and biochemi-cal functions by liver hepatocytes. Although hepatocytes *in vivo* also perform transport functions and form a sep-arate bile canalicular network, all current approaches to bioartificial liver development essentially ignore this prop-erty due to the complexity of reproducing the *in vivo* ar-rangement of hepatocytes in liver. Furthermore, it has been hypothesized that the most important hepatic functions required for survival involve secretory and biochemical functions that do not require a functional bile canalicular network. A partial list of tissue engineered endothelial and epithelial tissues is given in Table IX.

2. Tissue Constructs Using Epithelial Cells

When the barrier function of the epithelium or endothe-lium is an important component of the design of the tis-sue, cells can be cultured on a smooth surface, which al-lows cells to form a monolayer. The cells then often form tight junctions between themselves which are similar to that found *in vivo*. The barrier function can be assessed via measurement of the electrical conductivity across the monolayer or rate of leakage of proteins as well as other

TABLE IX Examples of Epithelia and Endothelia Made *in vitro*

Tissue	Cell type	Major function (s)
Vascular endothelium	Endothelial cell	Transport and secretion
Cornea	Corneal epithelial cell	Transport
Intestine	Enterocyte	Transport
Liver	Hepatocyte	Secretion
Bladder	Uroepithelial cell	Transport
Skin	Keratinocyte	Transport
Kidney	Kidney epithelial progenitor cell	Transport

relevant compounds. The surface onto which cells are grown is often shaped according to the final application. For example, bioartificial vascular grafts have been produced by seeding endothelial cells onto the luminal surface of small-diameter (6 mm or less) synthetic vascular grafts.

In cases where the ability to control transport is not an important aspect of the function of the bioengineered tissues, a simple way to maintain certain epithelial cells is via seeding in a three-dimensional matrix in a manner similar to that used for connective tissue construction. For example, hepatocytes can be maintained in porous or mesh-type matrices made of a variety of materials including polylactic–glycolic copolymer, alginate, etc., wherein they have a tendency to aggregate. Such aggregates (sometimes called *organoids*) are known to contain cells that have maintained their phenotypic stability. However, for this process to be beneficial, the aggregate size must be controlled to prevent the formation of large aggregates with anoxic cores.

3. Epithelial and Connective Tissue Composites

The long-term survival of endothelial and epithelial cells seeded onto artificial polymers is often problematic due to inflammatory responses and the poor retention of the seeded cells under *in vivo* conditions. Relevant to the case of vascular grafts, it is noteworthy that cultured endothelial cells rapidly become senescent and die *in vitro* when growth factors are removed from the culture medium. Thus, long-term retention of the cultured endothelium often requires the continual presence of such factors in the implant. A solution to this problem, which also provides a more intrinsically biocompatible approach to vascular graft production, is to seed cells on an *in vivo*-like stromal layer releasing the necessary trophic factors. Similarly, keratinocytes require either growth factors or a mesenchymal "feeder layer" of cells in order to survive and grow.

Some epithelial tissues have by the nature of the functions performed an extremely complex topology and organization, which makes it difficult to reproduce faithfully *in vitro*. For example, the intestine has on one side a layer of epithelial cells which selectively transport metabolites into the interstitial tissue space, where a large number of microscopic lymphatic and blood vessels absorb the transported metabolites. The epithelial layer is organized in the form of villi or invaginations to increase the surface area of exchange. Although it may be at some point possible to reproduce these structures "from scratch," it turns out that some of these features can be generated by the cells themselves if placed in the correct environment, in which case they display an amazing ability to reorganize themselves into functional tissues. Recent efforts in the development of bioengineered neoin-

testine use cellular aggregates isolated from intestinal tissue which are seeded onto polymer meshes shaped in a tubular form and anastomosed to the small bowel in animal models. The aggregates develop and form villus structures lined with a columnar epithelium reminiscent of the *in vivo* counterpart, and in some areas a subjacent connective tissue containing smooth muscle cells develops. Anastomosis to the existing bowel and bowel resection in the host receiving the implants have significantly improved morphogenesis and differentiation of the implant.

C. Endocrine Tissues

Major efforts in this area focus on the development of a bioartificial pancreas. Unlike other engineered tissues, the bioartificial pancreas does not need to physically integrate with the host's tissues, as its primary function is to release insulin in a controlled manner as a function of the patient's glucose levels. Furthermore, the typical patients needing insulin therapy do not have any functional islet tissue available; therefore, allogeneic or xenogeneic islet sources must be used. For these reasons, the bulk of studies on bioartificial pancreases use islets encapsulated in membrane-based devices protecting the islets from the recipient's immune system. The first devices (dating back to the 1970s) mostly consisted of hollow-fiber bioreactors in which islets were placed on the shell side of the bioreactor, and the patient's blood flowed through the fibers. Although these systems worked well over short periods of time, chronically implanted devices tended to activate blood clotting and eventually become clogged. More recently, more simple avenues have been explored, such as encapsulating islets in spherical or flat membranes placed in the host's tissues.

In theory, allogeneic implants will not trigger immune responses as long as the pore size of the capsule is small enough to prevent immune cells from the host from accessing the antigens expressed on the implanted cells. However, antigens shed from the cell surface may trigger a humoral immune response leading to the formation of antibodies. The proteins involved in this type of immune rejection include immunoglobins such as IgG and IgM, and complement molecules, the largest of which is C1q (mol wt = 410 kD, diameter = 30 nm) which is required for classical pathway complement-mediated cell damage. Binding of immunoglobins to cells in the implant without the presence of the complement system would not necessarily lead to cell damage; therefore, some investigators have claimed that a membrane with an effective pore size of less than 30 nm may suffice to protect the encapsulated cells. Smaller membrane pore sizes (<50 kD) have been typically used, however. This strategy has shown promise in animal models, with function remaining after several months to a year, although the reliability of the procedure is far from perfect. A common recurrent problem with immuno-isolated cells is the presence of a foreign-body reaction against the capsule material itself, leading to the generation, over a period of days to weeks, of a fibrotic layer around it, compromising nutrient transport and the release of insulin from the implanted cells. Besides improvements in the biocompatibility of the material, one avenue that may improve function of these devices is the use of materials and/or factors promoting the growth of blood vessels near the surface of the capsule.

It has also been suggested that the longevity of islet cells may be limited in encapsulated systems, and that integration into the host tissue may be necessary for a permanent cure. Thus, as an alternative to immuno-isolation, other approaches are currently being sought to either eliminate the antigenic proteins and polysaccharide moieties on implanted cells or interfere with the signaling pathways governing these immune responses. These studies are in fact not limited to tissue-engineered constructs, but are also under investigation for the transplantation of whole organs. For example, transgenic strains of pigs, which have a body size similar to a human and which express human surface antigens, are currently being developed.

IV. FUTURE PROSPECTS FOR TISSUE ENGINEERING

Tissue engineering is a relatively new and rapidly evolving field still in its infancy. Exciting new discoveries in biology will soon open new avenues for tissue engineers. One of these discoveries is the recent identification of stem cells. Stem cells have a high replication potential and can differentiate into a large number of different cell types. The best characterized stem cells are those of hematopoietic origin that populate the bone marrow. These cells are also found in very small numbers in the peripheral circulation. They have been cultured successfully *in vitro* on a stromal layer of connective tissue cells to produce all common blood cell lineages, including red blood cells, monocytes, lymphocytes, and platelets. More recent discoveries suggest that wound healing in specialized tissues such as muscle sometimes involves the homing of stem cells present in the circulation. While the exact nature of these cells remains to be elucidated, they open up exciting avenues for tissue engineering. For example, such stem cells could be harvested from a patient's blood, (requiring a minimally invasive procedure), grown, differentiated *in vitro* into the tissue type needed, and then implanted back into the patient. Since patients would receive their own cells, no immune suppression would be needed.

Another source of stem cells is embryos, which can be obtained at the blastocyst stage. Embryonic stem cells are totipotent, meaning that they have the potential to differentiate into any cell type found in the body. In the presence of leukemia inhibitory factor, embryonic stem cells self-renew without any loss of development potential. Otherwise, they differentiate into a wide variety of cell types, the nature of which depends on the specific factors added to the culture medium. Cloning techniques enable replacing the original DNA from the embryonic stem cell with that of a patient (extracted from one of the patient's cells such as skin). The availability of such cells could have important implications for engineering tissues made of cells that have typically lost their ability to replicate, such as neurons, lung epithelium, etc. However, serious ethical considerations will have to be resolved prior to using human embryonic stem cells in such applications. Furthermore, more progress is needed in order to increase the yield of specific cell types used in tissue engineering from stem cells.

Clinical applications for engineered tissues often require a readily available supply of a large number of cells when the need arises. Maintaining a continuous supply by culture techniques or obtaining fresh cells in large numbers from animal or human sources is clearly impractical. Thus, long-term preservation methods will be critical for the future clinical applications of tissue engineering. Cryopreservation is the most efficient method of preservation, and careful studies of the effects of freezing-associated osmotic, chemical, thermal, and mechnical stresses will be required. Although many such studies have been carried out on dissociated cells in suspension, there have been few studies on tissue constructs, which pose special challenges because the optimal freezing conditions for different cell types may not be the same, and the freezing conditions may be difficult to control uniformly in a three-dimensional system.

In summary, tissue engineering encompasses a wide spectrum of disciplines, including biological and chemical sciences, engineering sciences, and medicine. Although tissue engineering is a relatively new field, exciting applications, varying from artificial skin to treat severe burns patients to a bioartificial pancreas to treat diabetics, have in some cases reached standard clinical practice, and in others shown major advances and promising preliminary clinical results. Thus, it is not unreasonable to expect that a number of new tissue engineering approaches will enter the realm of clinical applications within the next decade. However, it should be borne in mind that clinical success relies heavily on our fundamental understanding of the many complex issues associated with reconstruction and modification of tissues as well as the development of reliable technologies for large-scale handling of tissues.

SEE ALSO THE FOLLOWING ARTICLES

BIOMATERIALS, SYNTHESIS, FABRICATION, AND AP-
PLICATIONS • BIOREACTORS • HYBRIDOMAS, GENETIC
ENGINEERING OF • MAMMALIAN CELL CULTURE •
METABOLIC ENGINEERING • POLYMERS, SYNTHESIS •
SURFACE CHEMISTRY

BIBLIOGRAPHY

Freshney, R. I. (2000). "Culture of Animal Cells. A Manual of Basic Technique," 4th ed., Wiley-Liss, New York.

Galletti, P. M., and Nerem, R. M., eds. (2000). "Prostheses and artificial organs," In The Biomedical Engineering Handbook, 2nd ed., Vol. 2, pp. 126.1–138.15, CRC Press, Boca Raton, FL.

Greco, R. S., ed. (1994). "Implantation Biology: The Host Response and Biomedical Devices," CRC Press, Boca Raton, FL.

Kreis, T., and Vale, R., eds. (1999). "Guidebook to the Extracellular Matrix, Anchor, and Adhesion Proteins," 2nd ed., Oxford University Press, London.

Lanza, R. P., Langer, R. S., and Vacanti, J. P., eds. (2000). "Principles of Tissue Engineering," 2nd ed., Academic Press, San Diego, CA.

Lauffenburger, D. A., and Lindermann, J. J. (1993). "Receptors: Models for Binding, Trafficking, and Signaling," Oxford University Press, London.

Lee, K., Berthiaume, F., Stephanopoulos, G. N., and Yarmush, M. L. (1999). "Metabolic flux analysis: a powerful tool for monitoring tissue function," Tissue Eng. 5, 347–368.

Morgan, J. R., and Yarmush, M. L., eds. (1999). "Tissue Engineering Methods and Protocols," Humana Press, Totowa, NJ.

Palsson, B. Ø., and Hubbell, J. A., eds. (2000). "Tissue engineering," In "The Biomedical Engineering Handbook," 2nd ed., Vol. 2, pp. 109. 1–125.17, CRC Press, Boca Raton, FL.

Patrick, C. W., Jr., Mikos, A. G., and McIntire, L. V., eds. (1998). "Frontiers in Tissue Engineering," Elsevier Science, New York.

Ratner, B. D., Hoffman, A. S., and Schoen, F., eds. (1997). "Biomaterials Science: An Introduction to Materials in Medicine," Academic Press, San Diego, CA.

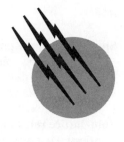

Tomography

Z. H. Cho
Korea Advanced Institute of Science

GLOSSARY

Algorithm Set of well-defined rules for solving a problem in a finite number of steps.

Coincidence detection Detection method in which an event is registered only if two photons are detected within a specified, sufficiently short time interval.

Fast Fourier transform (FFT) Highly optimized Fourier transform algorithm for digital computation.

Free induction decay (FID) Nuclear magnetic resonance signal emitted by precession of transverse magnetizations after excitation.

Gradient coils Electromagnetic coils generating magnetic fields, which are superimposed on the main magnetic field to create spatial variation in the field strength.

Gray level Discrete steps between light and dark in the image.

Linear attenuation coefficient Probabilities per unit path length that the X-ray photon will be removed from the beam. This includes the effects of photoelectric absorption, Compton scatter, and pair production.

Monochromatic Refers to an electromagnetic wave with a negligibly small region of spectrum.

Nyquist sampling criterion Criterion of the allowable maximum sampling interval that can be given to regularly spaced sampled data of a signal with bandwidth B for complete determination of the signal in its original form. It corresponds numerically to $1/(2B)$.

Pixel Abbreviation for "picture element"—a basic element in digital image.

Point-spread function (PSF) Transfer function that represents the output of a system to an infinitely high amplitude point input.

Polychromatic Antonym of monochromatic.

Radionuclide Radioactive nuclei undergoing nuclear transitions that are usually accompanied by the emission of particles or electromagnetic radiation.

RF Pulse Pulse of radio-frequency (RF) energy transmitted by an RF coil in nuclear magnetic resonance computerized tomography. Its frequency ω_0 is represented as $\omega_0 = \gamma H_0$, where γ is the gyromagnetic ratio and H_0 is the main magnetic field strength. Its shape and amplitude determine the selectivity in the frequency band and the amount of spin rotation.

Scintigraphy Imaging technique that uses an Auger (or γ) camera to visualize the distribution of radioisotopes within the human body.

Spin Property of nuclei that have an odd number of neutrons and protons. Nuclei with spin have a magnetic moment and can possess the NMR property.

Encyclopedia of Physical Science and Technology, Third Edition, Volume 16

Spin echo Signal produced by the 90°–180° RF pulse sequence. The spin-echo signal is actually conjugate symmetric to its center if its duration is short compared with T_2. Its amplitude is determined by the T_2 of the substance, excluding the effects of field inhomogeneity.

Spin–lattice relaxation (thermal or longitudinal relaxation) Phenomenon of spins going to the thermal equilibrium state with other molecules in lattice. It is characterized by the exponential time constant T_1. Also, it determines the recovery time of the longitudinal magnetization M_z.

Spin–spin relaxation (transverse relaxation) Exchange of energy of excited nuclei with other precessing nuclei. It is characterized by the exponential time constant T_2. Also, it determines the envelope of the free induction decay in a perfectly uniform magnetic field and the amplitude of the spin echo.

Superconductive magnet Magnet that requires no electrical power once the field has been established. Superconductivity is a property of some materials that have no electrical resistance when the temperature is near absolute zero. Liquid helium is generally used to maintain a low temperature.

True coincidence Event detected in coincidence without scatter of either photon, generated from an annihilated positron, in the object.

COMPUTERIZED or computed tomography (CT) is a technique of producing cross-sectional three-dimensional images from multiple views or projection data obtained with penetrating probe radiations or by other means such as magnetic field gradients, by processing those data using a computer and mathematical image reconstruction algorithms. Major applications of the basic CT concepts are medical diagnosis, industrial nondestructive testing, and other areas of the physical sciences, such as geophysical exploration.

I. INTRODUCTION

Computerized tomography is a technique by which three-dimensional (3-D) imaging of an object is made possible. The basic data are most often obtained in the form of projection data. Since the first tomographic system, known as X-ray CT, was designed by G. N. Hounsfield in 1972, many applications of the technology, based on 3-D image reconstruction from projection, have been developed. These include X-ray CT, radionuclide (isotope) emission CT (ECT), and nuclear magnetic resonance (NMR) CT. The basic forms of projection data are obtained from focused and collimated X rays, γ rays, and annihilation photons from decaying radionuclides, transmitted and reflected ultrasound beams, appropriately excited nuclear spins under a strong magnetic field, and so on. The data obtained, usually referred to as projection data, are processed by a mathematical image reconstruction algorithm using a digital computer to form an image or a set of images, each representing a slice or several slices of an object. Although direct Fourier transform image reconstruction is employed in some cases to obtain an image or a set of images, as in NMR CT, most types of image reconstruction employ some forms of projection reconstruction. Those mathematical techniques known as 3-D image reconstruction form the basis of all the CTs discussed in this article. Finally, mathematically formed or reconstructed images are displayed on a television screen or photographed by a camera attached to the system.

From the physics and engineering points of view, both 3-D image reconstruction and CT are new concepts and tools. For the first time in history, human beings are capable of visualizing the inner structures of an object noninvasively. At present, the most widely used CT is X-ray CT; it is estimated that more than 5000 X-ray CT units, each costing as much as $1 million, are in operation throughout the world. In the area of ECT, two types are under active development: single photon emission CT (SPECT) and positron emission tomography (PET). Although ECT is similar to X-ray CT, it differs in functional form; that is, X-ray CT is usually capable of visualizing anatomical details, while ECT is capable of visualizing the functional or metabolic behaviors of an object. Ultrasound CT is similar to X-ray CT and is capable of visualizing cross-sectional images. The development of ultrasound CT has been relatively slow, due mainly to the difficulties inherent in the basic properties of ultrasound, such as scattering and diffraction. The most recent and probably most exciting development in the field of CT is NMR CT. It is now capable of visualizing in three dimensions the distribution of several nuclei, such as the proton, the sodium nucleus, and the phosphorus nucleus, and thereby performing chemical imaging. As the resolution and sensitivity of NMR CT improve rapidly, it is becoming one of the most promising medical diagnostic imaging modalities.

Computerized tomography is believed to be the most important development in diagnostic imaging history since the discovery of X rays in 1895 by W. K. Roentgen. Applications of CT concepts are rapidly expanding from medical imaging to many branches of the physical sciences. The principles of X-ray CT are now being applied to many diverse areas, ranging from the examination of defects in nuclear reactor cores to the rapid inspection of automobile tires on the production line. The principles of NMR CT are being applied in the area of fluid dynamics to investigate structural changes in flow. This was not feasible in the past. Future applications of CT are unlimited; they range from the physical exploration of oil to the study of mummies excavated from ancient pyramids. Through CT, human vision has expanded beyond two dimensions to three. Until recently, it was believed that 3-D imaging was impossible, considering the fundamental nature of the physical world.

This chapter discusses in detail the mathematical foundation of CT technology, specifically image reconstruction algorithms, computers and peripherals used in CT, display systems for CT, and the physics of the data collection mechanism of various CT systems (X-ray CT, ECT, and NMR CT).

II. PRINCIPLES OF COMPUTERIZED TOMOGRAPHY

A. Mathematics and Algorithms for Image Reconstruction

Several mathematical image reconstruction algorithms used for CT have been developed since the basic concept of CT was established. These algorithms are based primarily on mathematical theory as described in detail in this section, which presents a mathematical basis for the reconstruction algorithms currently used in many CT modalities, including X-ray CT, ECT, and NMR CT. According to data processing methods, algorithms can be classified into projection reconstruction, iterative method, and direct Fourier imaging. A classification of these three general methods is given in Table I.

TABLE I Image Reconstruction Algorithms

Class	Specific method
Projection reconstruction	2-D Projection reconstruction
	Filtered backprojection (FB)
	Parallel beam mode
	Fan beam mode
	Backprojection filtering (BF)
	3-D Projection reconstruction
	True 3-D reconstruction (TTR)
	Generalized TTR
	Planar-integral projection reconstruction (PPR)
Iterative method	Algebraic reconstruction technique (ART)
	Maximum likelihood reconstruction (MLR)
Fourier reconstruction	Direct Fourier reconstruction (DFR)
	Direct Fourier imaging in NMR (DFI)

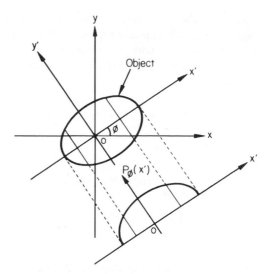

FIGURE 1 Geometry for 2-D parallel beam projection.

1. Projection Reconstruction

Since 1970, active research on image reconstruction from projection has been initiated mainly through the development of CT. This projection reconstruction is now applied to many areas of science. The most important area of its application has probably been CT; projection reconstruction has remained the basic algorithm for CT since the inception of X-ray CT in 1972.

The simplest form of projection data is illustrated in Fig. 1. The line integrals of a physical object are estimated along straight lines. Each line integral, in practice, represents a physical property in a strip with a finite width. Therefore, a set of line-integral data is obtained at each view. By repeatedly assessing the sets of data at different views, that is, around 180° or 360° with a specified angular interval $\Delta\theta$, a complete projection data set can be obtained. Each set of estimated line integrals is often called a projection or a line-integral projection. The collection of all these estimated line-integral sets around 180° or 360° is referred to as line-integral projection data or simply projections. Similarly, planar-integral projection data can also be obtained; they represent the collection of all integral values of a physical object along planes.

From a set of measured projection data or projections, an image can be formed through the use of appropriate processing algorithms. Image reconstruction from projections is the process of producing an image of a 2-D or a 3-D distribution of some physical property from the estimates of its line (or plane) integrals along a finite number of lines (or planes) of known locations. For the reconstruction of the images from projections, several algorithms can be used depending on the imaging modalities. Theoretical aspects and characteristics of these algorithms are discussed in the following subsections. Fourier transforms, which constitute the main body of image reconstruction in general, are also briefly explained in the following. Fourier transforms are basically part of a conversion process that converts spatial domain data to spatial frequency domain data. They are defined as

$$F(\omega) = \mathscr{F}_1[f(x)] = \int_{-\infty}^{\infty} f(x) \exp(-i\omega x) \, dx \quad (1)$$

$$f(x) = \mathscr{F}_1^{-1}[F(\omega)]$$

$$= \frac{1}{2\pi} \int_{-\infty}^{\infty} F(\omega) \exp(i\omega x) \, d\omega, \quad (2)$$

where $i = \sqrt{-1}$ and $\mathscr{F}_1[\cdot]$ and $\mathscr{F}_1^{-1}[\cdot]$ are the 1-D forward and inverse Fourier transform operators, respectively.

a. Two-dimensional projection reconstruction.

i. Filtered backprojection algorithm. The filtered backprojection (FB) or convolution backprojection algorithm is the most popular and most frequently used reconstruction method so far employed in CT, with the exception of NMR CT. For the mathematical formulation of this FB algorithm, there are two basic forms in existence: the parallel beam and fan beam modes.

Parallel Beam Mode. Let us assume an object distribution function $f(x, y)$ represented by the Cartesian coordinates (x, y), with rotated coordinates expressed as (x', y'). The basic data to be used in the reconstruction are projections that represent sets of line integrals of an object in various directions. The projection data $p_\phi(x')$ shown in Fig. 1 is a set of line integrals taken along the y' direction, that is,

$$p_\phi(x') = \int_{-\infty}^{\infty} f(x', y') \, dy'$$

$$= \iint\limits_{-\infty}^{\infty} f(x, y)\delta(x \cos\phi + y \sin\phi - x') \, dx \, dy,$$

$$\qquad\qquad\qquad\qquad (3)$$

where

$$\begin{bmatrix} x' \\ y' \end{bmatrix} = \begin{bmatrix} \cos\phi & \sin\phi \\ -\sin\phi & \cos\phi \end{bmatrix} \begin{bmatrix} x \\ y \end{bmatrix}. \quad (4)$$

The Fourier transform of projection data $P_\phi(\omega)$ can be related to the projection data $p_\phi(x')$ as

$$P_\phi(\omega) = \int_{-\infty}^{\infty} p_\phi(x') \exp(-i\omega x') \, dx'$$

$$= F(\omega_x, \omega_y)|_\phi = F(\omega, \phi), \quad (5)$$

where $F(\omega_x, \omega_y)$ is the 2-D Fourier transform of $f(x, y)$, $\omega_x = \omega \cos\phi$, and $\omega_y = \omega \sin\phi$ when (ω_x, ω_y) and (ω, ϕ)

represent the Cartesian and polar coordinates of (x, y) in the spatial frequency domain, respectively.

Equation (5) states that a 1-D Fourier transform of projection data at a given angle ϕ represents the 2-D Fourier transform values of the object function $f(x, y)$ in the spatial frequency domain along the radial frequency with a given angle ϕ. This is the projection theorem that plays a central role in 2-D image reconstruction. From this theorem, it can easily be shown that the object function $f(x, y)$ can be recovered as

$$f(x, y) = \frac{1}{2\pi} \int_0^\pi d\phi \left[\int_{-\infty}^{\infty} dx' \, p_\phi(x')h(x \cos\phi \right.$$

$$\left. + y \sin\phi - x') \right], \quad (6)$$

where

$$h(x') = \mathscr{F}_1^{-1}[|\omega|].$$

The convolution kernel $h(x')$ in Eq. (6) is an inverse Fourier transform of $|\omega|$, but its exact form is not realizable in practice. Therefore, several modified filter functions have been suggested. The selection of a particular filter function will affect the characteristics of the reconstructed image, that is, the desired image resolution and contrast.

Fan Beam Mode. Although the parallel beam reconstruction algorithm has been a basic tool for image reconstruction, the fan beam reconstruction algorithm is nevertheless widely used. For example, it is used in X-ray CT, due mainly to the fact that the basic data collection procedure involved is the fan beam mode. In addition, image reconstruction that utilizes the fan beam algorithm often provides better resolution with the same amount of sampled data than the parallel beam algorithm because of improved sampling at the central region.

If the fan beam projection data set is represented as $p_{\alpha_1}(\beta_1)$, where β_1 and α_1 represent the detector position and rotation angle of the beam, respectively (Fig. 2), then the relation between parallel and fan beam projection data with coordinates (x', ϕ) and (α_1, β_1) can be represented as

$$x' = R_d \sin\beta_1$$

$$\phi = \alpha_1 + \beta_1, \quad (7)$$

where R_d is the distance between the center point and the apex of the fan. Through the use of Eqs. (6) and (7) and the rotation angle $\phi = 0 \sim 2\pi$, the fan beam analogy can be derived as

$$f(x, y) = \frac{1}{4\pi} \int_0^{2\pi} d\alpha_1 \int_{-\beta_{1m}}^{\beta_{1m}} d\beta_1 \, p_{\alpha_1}(\beta_1) J(\beta_1) g$$

$$\times (\beta_1' - \beta_1) \frac{1}{V_1^2}, \quad (8)$$

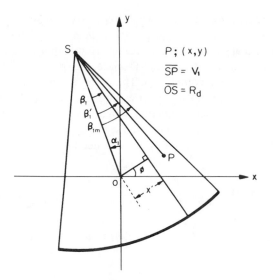

FIGURE 2 Geometry for 2-D fan beam projection.

where

$$J(\beta_1) = \left| \frac{\partial(x', \phi)}{\partial(\beta_1, \alpha_1)} \right| = R_d \cos \beta_1$$

and

$$g(\beta_1) = \left(\frac{\beta_1}{\sin \beta_1} \right)^2 h(\beta_1). \tag{9}$$

Among the variables in the above equations, V_1 is illustrated in Fig. 2 and $h(\beta_1)$ is the same one given in Eq. (6).

The fan beam reconstruction algorithm given in Eq. (8) consists of three parts: (1) weighting the projection data with $J(\beta_1)$, (2) convolution with $g(\beta_1)$, and (3) weighted backprojection with weight $1/V_1^2$.

ii. Backprojection filtering algorithm. As mentioned previously, the FB algorithm provides a high-quality image and computational efficiency. It is by far the most popular algorithm in CT. The use of this algorithm is limited, however, to the equisampled parallel or the fan beam mode of straight line-integral projection data. Therefore, the processes of rebinning and interpolation are usually required for arbitrarily oriented rays and for more general cases. It would be difficult to apply this algorithm, for example, to line-integral data along curved lines.

An alternative, backprojection filtering (BF) algorithm, has been proposed to overcome these types of problems, and this algorithm is more general as long as the point-spread function (PSF) of a backprojected image, which can be obtained from Eq. (6) without convolution kernel, follows $1/r$ characteristics, where r is the distance from the point source. Although the method appears to be attractive and general, it is rarely used because the resultant images are usually of poorer quality than those obtained through the FB algorithm. The main reasons for poor image quality are as follows:

1. Although the projection data from the finite-size object are limited in the spatial extension, the backprojected image is unlimited. Truncation of the 2-D backprojected image data for the digital processing will therefore result in image degradation.

2. The 2-D filter function to be used in the rectangular Cartesian coordinates has slope discontinuities at the boundary (near cutoff frequency), resulting in a ring artifact. By the incorporation of proper digital signal processing techniques, however, these problems can easily be resolved.

In this section, we present a BF algorithm with which high-quality images equal to the FB images can be obtained. If the PSF of the backprojected image is $1/r$, then the relation between the object density function $f(x, y)$ and the backprojected or blurred image $b(x, y)$ is

$$b(x, y) = f(x, y) ** (1/r), \tag{10}$$

where r is the distance from the source point at origin and $**$ represents the 2-D convolution. From Eq. (10), the object function $f(x, y)$ can be obtained through the inverse Fourier transform operation,

$$f(x, y) = \mathscr{F}_2^{-1}[\omega B(\omega_x, \omega_y)], \tag{11}$$

where $B = \mathscr{F}_2[b]$ and ω is the radial spatial frequency.

The computation time required in the convolution operation in the space domain is generally longer than the processing time required in the spatial frequency domain. Let us now consider Eq. (11) to be the basic form of the BF algorithm. In the implementation of this algorithm, two previously mentioned aspects should be considered: the size and form of the backprojected image and the shape of the filter function.

In conventional 2-D image processing, an image of matrix size $N \times N$ is expanded to a $2N \times 2N$ data format, in which the outsides of the $N \times N$ center array are filled with zeros to avoid the aliasing effect arising in circular convolution. A similar procedure, however, cannot be applied to the backprojected data array, because the backprojection image data are not confined to $N \times N$. Therefore, the truncation of the backprojected image data will result in a severe truncation artifact. To avoid this artifact, the image field is expanded twice; that is full $2N \times 2N$ backprojected image data are taken instead of only the $N \times N$ array from the backprojected image. The use of these full data reduces the artifact significantly.

The selection and formulation of the 2-D filter function are also important factors in determining the image quality in 2-D BF image reconstruction because the slope discontinuity at the cutoff frequency causes ring artifact. The overall reconstruction procedures of the BF algorithm are as follows:

1. Backprojection to get $2N \times 2N$ blurred image
2. Two-dimensional fast Fourier transform of the full $2N \times 2N$ data and filtering
3. Two-dimensional inverse fast Fourier transform and adoption of $N \times N$ image in the central region
4. Normalization of the image with precalculated reference data, such as the uniform disk image

b. Three-dimensional projection reconstruction.

i. True three-dimensional reconstruction with line-integral data. In the conventional reconstruction algorithm, projections of an object are taken transaxially and are used to reconstruct the object slice by slice. The stacked slices constitute the 3-D volume image of the object.

For example, in ECT, in which radionuclides are injected into the patient, the photons are emitted isotropically in 4π directions. This slice-stacking method for a volume object does not make full use of the photons because it captures photons emitted only in the direction perpendicular to the long axis of the body, constituting only a small fraction of the total emitted photons. All the emitted photons, especially in ECT, should be collected through the use of complete spherical geometry. For image reconstruction with the data collected in 4π geometry, the direct true three-dimensional reconstruction (TTR) algorithm will be required for the maximum utilization of all the available photons.

Parallel Beam Mode. Let us consider a spherical geometry that completely surrounds the object, and let us assume that all the emitted photons are captured and rearranged into 2-D parallel data sets in 4π directions. The PSF of this system is $1/r^2$, and the backprojected blurred function is given by

$$b(x, y, z) = f(x, y, z) *** (1/r^2), \qquad (12)$$

where $***$ represents a 3-D convolution operator. The object function can be obtained as

$$f(x, y, z) = \mathcal{F}_3^{-1}[\rho] *** b(x, y, z), \qquad (13)$$

where $\rho = (\omega_x^2 + \omega_y^2 + \omega_z^2)^{1/2}$ and $\mathcal{F}_3^{-1}[\cdot]$ is a 3-D inverse Fourier transform operator. Equation (13) is essentially a 3-D convolution (deconvolution) of a simple backprojected (blurred) image. This 3-D convolution process, as well as the backprojection operation, however, would require an unusually long computation time.

As an alternative approach, the FB method has been proposed. The 3-D version of the projection theorem states that the Fourier transform of the 2-D projection represents the plane data passing through the center in the frequency domain in the same direction as the projected plane of the object. This statement can be written

$$\mathcal{F}_2[p_{\theta,\phi}(x', z')] = F(\omega_{x'}, \omega_{z'}; \theta, \phi), \qquad (14)$$

where $(\omega_{x'}, \omega_{z'})$ are the spatial frequency domain coordinates of (x', z') in the direction of (θ, ϕ). If the uniform planes that pass through the center are superimposed from all possible directions, in 3-D space, the overlapped density function becomes $1/\rho$. Therefore, to obtain the 3-D object function in the Fourier domain, the 2-D Fourier transform of each projection data set should be compensated with the factor ω before the superposition.

Image function can now be obtained by

$$f(x, y, z) = \frac{1}{4\pi^2} \int_0^\pi d\theta \sin\theta$$

$$\times \int_0^{2\pi} d\phi \{ \mathcal{F}_2^{-1}[\omega] ** p_{\theta,\phi}(x', z') \}. \qquad (15)$$

Equation (15) shows a 3-D FB algorithm, which is in fact the backprojection of the filtered 2-D projection data in 3-D space. In this case, the 2-D filter function or kernel is simply $\mathcal{F}_2^{-1}[\omega]$.

ii. True three-dimensional reconstruction algorithm for generalized geometry. The ideal detector configuration for the ECT is a complete sphere. The reconstruction of such a spherically configured emission or transmission data consisting of sets of line-integral data was discussed in the preceding section. Three-dimensional images can be reconstructed by FB of the 2-D line-integral data sets in 3-D space. Practical system design, however, prohibits the construction of such a configuration when one considers the elongated shape of the human body as an example.

As a practical alternative, a truncated spherical configuration (Fig. 3) can be considered, and an algorithm suitable for such geometry has been developed. This 3-D image reconstruction algorithm is known as the TTR algorithm. The algorithm eventually will lead to a generalized algorithm for both 2-D slice reconstruction and 3-D volume image reconstruction of a complete spherical volume. Consider a sphere in which parts of the surface have been removed along the body axis to adapt to the shape of the human body. Although the sphere is truncated, this geometry retains spherical symmetry in the remaining spherical surface. Therefore, it also retains the possibility of reconstructing a true 3-D volume image, as explained in the following paragraphs.

The 3-D image within the reconstruction sphere, which has a radius of R_0, is obtained by summing all the images reconstructed at each slice orientation in which a complete set of 2-D projection data is provided. This sum image should be divided by the number of slice orientations reconstructed for normalization. Essentially, this is the basis of the reconstruction algorithm developed for truncated spherical geometry.

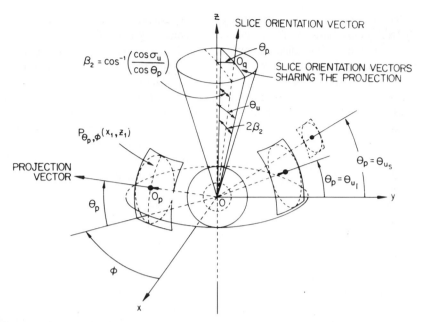

FIGURE 3 Basic geometry of the truncated spherical configuration for the development of the TTR algorithm.

For the implementation of the algorithm, all slice orientations involved in each set of projection data must be identified and the corresponding composite filter function generated. In fact, each 2-D parallel projection data set corresponding to a certain object size has a fixed number of slice orientations. This allows us to treat a certain projection data set in a unified way; that is, a 2-D projection data set can be processed with one filter function. Let us consider a 2-D projection data set projected on a direction parallel to the line OO_p in Fig. 3. Here, O_p is the center of the projection plane. As shown in the figure, these projection data are shared by slice orientations rotated around the line OO_p from $-\beta$ to β, where

$$\beta = \cos^{-1}(\cos \theta_u / \cos \theta_p). \qquad (16)$$

By use of the composite filter concept, object function $f(r)$ can be obtained with the following equation:

$$f(\mathbf{r}) = \frac{\int_0^\pi d\phi \int_{-\theta_u}^{\theta_u} d\theta_p \cos \theta_p \int_{-\beta}^{\beta} d\alpha [p_{\theta_p,\phi}(x_1, z_1) ** h(x_1, z_1; \alpha)]}{\int_0^{2\pi} d\phi \int_{-\theta_u}^{\theta_u} d\theta_p \cos \theta_p \int_{-\beta}^{\beta} d\alpha},$$

$$(17)$$

where $p_{\theta_p,\phi}(x_1, z_1)$ is the parallel projection data set at polar angle θ_p and the azimuthal angle of ϕ, and $h(x_1, z_1; \alpha)$ is the rotated form of the conventional filter function with angle α, to be applied to projection data $p_{\theta_p,\phi}$ at a slice orientation corresponding to the rotation angle variable α. Here the z axis lies in the direction of the line $\overrightarrow{OO}_q$, and

the x_1 axis is normal to both $\overrightarrow{OO}_p$ and $\overrightarrow{OO}_q$, as shown in Fig. 3.

The filter kernel $h(x_1, z_1; \alpha)$ in Eq. (17) is then given as

$$h(x_1, z_1; \alpha) = \mathscr{F}_2^{-1}[H(\omega_{x_1}, \omega_{z_1}; \alpha)]$$

and

$$H(\omega_{x_1}, \omega_{z_1}; \alpha) = |\omega_{x_2}| = \omega |\cos(\xi - \alpha)|, \qquad (18)$$

where ω_{x_2} is the rotated axis from ω_{x_1} with angle α and (ω, ξ) represents the polar coordinates of $(\omega_{x_1}, \omega_{z_1})$.

The denominator of Eq. (17) is the normalizing factor that represents the sum of all the weighting coefficients of projection data sets. The projection data set $p_{\theta_p,\phi}(x_1, z_1)$ is indepenent of α, and the convolution is a linear operation. Therefore, Eq. (17) can be further simplified as follows:

$$f(\mathbf{r}) =$$
$$\frac{\int_0^\pi d\phi \int_{-\theta_u}^{\theta_u} d\theta_p \cos \theta_p \mathscr{F}_2^{-1}[P_{\theta_p,\phi}(\omega_{x_1}, \omega_{z_1}) H_{\theta_p}(\omega_{x_1}, \omega_{z_1})]}{4\pi^2(1 - \cos \theta_u)},$$

$$(19)$$

where

$$H_\theta(\omega, \xi) = \begin{bmatrix} 2\omega \cos \xi \sin \beta, \\ 0 \le \xi \le \dfrac{\pi}{2} - \beta \\ 2\omega(1 - \sin \xi \cos \beta), \\ \dfrac{\pi}{2} - \beta < \xi \le \dfrac{\pi}{2} \end{bmatrix} \qquad (20)$$

iii. Planar-integral projection reconstruction. Fourier NMR imaging techniques suggested the possibility of exciting the entire volume of an object, thereby obtaining planar-integral data sets with which an efficient volume image reconstruction could be formed.

One-dimensional planar-integral projection data along the direction T at an angle (θ, ϕ) of a 3-D object function $f(x', y', z')$ can be given as

$$p_{\theta,\phi}(z') = \iiint_{-\infty}^{\infty} f(x', y', z')\delta(T - z')\,dx'\,dy'\,dz', \quad (21)$$

where

$$T = x \sin\theta \cos\phi + y \sin\theta \sin\phi + z \cos\theta \quad (22)$$

and (x', y', z') are the rotated Cartesian coordinates. By taking the 1-D Fourier transform of $p_{\theta,\phi}(z')$ and using Eqs. (21) and (22), one obtains

$$P_{\theta,\phi}(\rho) = \int_{-\infty}^{\infty} p_{\theta,\phi}(z') \exp(-i\rho z')\,dz'$$
$$= F(\omega_x, \omega_y, \omega_z)|_{\theta,\phi}, \quad (23)$$

where $F(\omega_x, \omega_y, \omega_z)$ is the image function in the spatial frequency domain and ρ is the radial spatial frequency. Equation (23) is another 3-D projection theorem, which states that the 1-D Fourier transform of the planar-integral data at angles θ and ϕ gives the 1-D radial frequency data in 3-D Fourier space.

To derive projection reconstruction using planar-integral data, let us first direct our attention to the FB method. Based on the FB method, recovery of the original volume image function $f(x, y, z)$ can be achieved through the inverse Fourier transform of Eq. (23),

$$f(x, y, z) = -\frac{1}{8\pi^2} \int_0^{2\pi} d\phi \int_0^{\pi} d\theta\, p_{\theta,\phi}''(T) \sin\theta, \quad (24)$$

where $p_{\theta,\phi}''(T)$ is the second derivative of $p_{\theta,\phi}(T)$ with respect to T.

2. Algebraic Reconstruction Technique and Maximum Likelihood Reconstruction

An alternative method sometimes used in image reconstruction is the iterative technique. In this category of image reconstruction, two techniques will be discussed: an algebraic method without statistical estimation and another with statistical estimation (the maximum likelihood estimation technique).

a. Algebraic reconstruction technique. The algebraic reconstruction technique (ART) was the first image reconstruction algorithm based on iterative procedures and

was the algorithm first used in the EMI brain scanner developed by Hounsfield.

The ART is not the major algorithm used in commercial scanners today, however, because for practical purposes it is inefficient. It is generally agreed that the ART is slow in computation and also that iteration cannot begin before the completion of projection data collection. The use of ART based on iterative techniques, in general, is limited, with the exception of a few special cases, such as that of limited view or limited scan angle. The basic algorithm is simple and will not be discussed in detail here.

b. Maximum likelihood reconstruction. In view of the fact that many ECTs are based on statistical phenomena, a modified ART technique that incorporates the maximum likelihood estimation has been proposed. This maximum likelihood reconstruction (MLR) technique is based on the determination of an estimate $\hat{f}$ of f, maximizing the probability or likelihood $p(n^* \mid f)$ observing the measured count n^* for the unknown distribution function f.

If each emitted photon in box b, the partition of the object, is detected in detector unit d with the probability $pr(b, d)$, $d = 1, \ldots, D$, then the estimate of the unknown $f(b)$, $b = 1, \ldots, B$ using the measured counting data $n^*(d)$ is

$$\hat{f}^{(i+1)}(b) = \hat{f}^{(i)}(b) \sum_{d=1}^{D} \frac{n^*(d)pr(b, d)}{n^{(i)}(d)}$$
$$b = 1, \ldots, B \quad (25)$$
$$n^{(i)}(d) = \sum_{b'=1}^{B} \hat{f}^{(i)}(b')pr(b', d),$$

where $\hat{f}^{(i)}(b)$ and $\hat{f}^{(i+1)}(b)$ are the (i)th and $(i + 1)$th estimates, respectively. This equation confirms that the likelihood gradually increases at each step and converges to estimate $\hat{f}(\infty)$, which has maximum likelihood. The algorithm reduces statistical noise artifacts over the Fourier transform method in ECT, such as PET, and appears to be applicable to other imaging modes.

3. Fourier Reconstruction

a. Direct Fourier reconstruction. Even though the FB algorithm is widely used because of its high image quality, it requires considerable computation, especially for the backprojection operation. One alternative algorithm is the direct Fourier domain mapping of the Fourier transform for each projection data in polar coordinates and subsequent 2-D inverse Fourier transform of the resultant Fourier domain data. This follows from the projection theorem presented in Eq. (5). Since the Fourier transform of the projection data set obtained by transaxial scanning represents the radial set of the 2-D Fourier domain object function, it is possible to form a set of polar rasters that represent the Fourier transform of the object function through 1-D Fourier transform of the projection data. In the direct Fourier reconstruction (DFR) technique, therefore, the reconstruction procedures involve the interpolation problem as well as 2-D Fourier transform, conversion of data from polar form to Cartesian form, and 2-D inverse Fourier transform of the interpolated data with the fast Fourier transform algorithm. This algorithm would be faster than the FB algorithm, but it requires an accurate and rapid polar-to-Cartesian coordinate conversion process technique, and often the accuracy of interpolation during this conversion is of critical importance to the quality of the image.

Another important point to consider in the DFR technique is the computation time involved in interpolation; 2-D interpolation usually requires a large amount of computation. An efficient interpolation scheme, based on the concentric square raster, has been developed. This new scheme is based on the Cartesian raster suitable for the fast Fourier transform. Although the DFR technique with concentric square raster sampling seems attractive, it is not easily applicable to many practical systems, such as X-ray CT or PET, due mainly to the fixed detector spacing and sampling distance. The method is useful, however, in NMR CT, where data appear in the Fourier domain, and the sampling interval is adjustable through simple variations on the magnitude of the x and y gradient fields.

b. Direct Fourier imaging in nuclear magnetic resonance. Another interesting method in the direct Fourier transform approach applicable to NMR imaging was proposed by Kumar, Welti, and Ernst and later modified by Hutchison. This method has several distinct advantages over other methods, including the simplicity of data collection, data handling, and fast image reconstruction capability. The last is due mainly to the elimination of the time-consuming backprojection operation. Although the method itself is mathematically simple, it has to be understood in conjunction with the NMR imaging method.

In 2-D NMR imaging, by virtue of NMR, the measured signal $s(t_x, t_y)$, which is known as free induction decay (FID) or echo signal, is given by

$$s(t_x, t_y) = M_0 \int\int_{-\infty}^{\infty} f(x, y) \exp(i\gamma x G_x t_x$$
$$+ i\gamma y G_y t_y) \, dx \, dy, \quad (26)$$

where M_0 is the equilibrium magnetization, $f(x, y)$ is the 2-D spin density (e.g., proton density) distribution, γ is the

gyromagnetic ratio, and G_x and G_y are the x and y directional gradient fields, respectively. As shown in Eq. (26), the measured signal $s(t_x, t_y)$ is the 2-D inverse Fourier transform of spin density $f(x, y)$. Therefore, the spin density image $\tilde{f}(\omega_x, \omega_y)$ can be reconstructed by taking the 2-D Fourier transform of the measured signal $s(t_x, t_y)$ as

$$\tilde{f}(\omega_x, \omega_y) = \int\int_{-\infty}^{\infty} s(t_x, t_y) \exp(-it_x\omega_x - it_y\omega_y) \, dt_x \, dt_y.$$
$$(27)$$

The reconstructed image $\tilde{f}(\omega_x, \omega_y)$ is then related with $f(x, y)$ as

$$\tilde{f}(\omega_x, \omega_y) = kf(\gamma G_x x, \gamma G_y y), \quad (28)$$

where k is a constant, $\omega_x = \gamma x G_x$, and $\omega_y = \gamma y G_y$. This algorithm can also be formulated through the DFR technique discussed previously. In 3-D direct Fourier imaging, the measured signal is in the 3-D inverse Fourier transform domain. Three-dimensional images, therefore, can be obtained simply through 3-D Fourier transform operation.

These algorithm, that is, DFR or direct Fourier imaging, are based on the basic properties of Fourier transform. Direct Fourier imaging is therefore valid only for NMR imaging, because the intrinsic nature of NMR imaging lies in the Fourier transform.

B. Computerized Tomography System Configuration

1. Overview

The generic term *computerized tomography* has expanded to include several different areas, including X-ray CT, ECT, and NMR CT. CT has been applied primarily in the field of diagnostic imaging. It can be divided into three major areas, depending on the source and detector mechanism: transmission CT, emission CT, and NMR CT.

Although these systems differ slightly in terms of the energy band, the principles of physical phenomena involved, and the reconstruction algorithms employed, CT systems in general, regardless of the specific modes, comprise three basic parts: data acquisition, which includes source, sensor, and controller; processing; and display.

The data acquisition part is perhaps the most important element of a CT system, it characterizes the system, that is, defines whether it is an X-ray CT or an NMR CT, and so on. The data acquisition part includes sources, detectors, controllers, data acquisition electronics, and scanner gantry. After it receives analog signals containing image information, it converts those signals to digital form and transfers them to the computing stage.

The second part is the computer system, in which the measured and quantized data are manipulated and

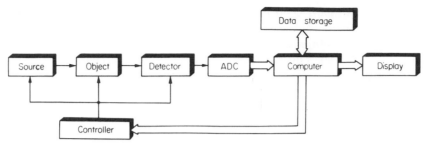

FIGURE 4 Block diagram of the general CT system.

reconstructed to create the desired images, which in turn provide information on particular structures or metabolic functions of the object under examination. The manipulation and processing of data are usually performed by general computers with array processors or by similar types of image processors, such as backprojectors, depending on the data acquisition method and reconstruction algorithm being used.

The third component is the display part, on which the observer examines and analyzes the object. A multiformat camera is commonly attached to the display console to record the pictures or images on film.

In some systems these parts are independent modular components, and in others they are integrated as a single package. A general system block diagram is shown in Fig. 4.

2. Data Acquisition

The source and sensor system with controller compose the main body of a CT system. System configuration, the related physics, and the principles of particular CTs are detailed in later sections.

It is assumed that signals containing diagnostic information can be measured in the form of electrical voltage, current, or number of photons counted. If the detected signal is not in digital form, it is converted to digital form for computer use. This process is most often performed rapidly by an analog-to-digital converter (ADC). Since all the data manipulations that follow are based on this digital representation, the performance of the ADC is vital to the overall system performance and to the final results.

An ADC is characterized by a number of parameters, including resolution, integral and differential linearities, internal noise, and operating speed. The resolution of an ADC is determined by the number of binary bits to which analog data can be digitized. For example, a 10-bit ADC has 1024 possible digitizations or output levels. Similarly, a 12-bit ADC has 4096 possible levels, while a 16-bit converter has 65,536 possible levels, and so on. The finite number of levels or steps involved in the conversion pro-

cess inevitably leads to quantization errors (noise), which can be expressed as

$$\sigma_q = \frac{\Delta}{2\sqrt{3}}, \qquad (29)$$

where σ_q is the standard deviation of the quantization error and Δ is the step size of the ADC. It is important that the quantization noise of an ADC be kept well below the level of other sources of noise within the data acquisition system.

Integral and differential linearities are other measures of imperfection arising from the component limitations in the circuitry and circuit design of the ADC.

At present, high-accuracy and high-speed ADCs of 16-bit resolution with conversion times of less than 20 to 30 μs are widely available. Ultra-high-speed conversion with high resolution is still limited, however, by the lack of high-speed and wide-range linear ADCs.

To make full use of the dynamic range of the ADC, the input signal level should be scaled and shifted to the range of the ADC. This is usually achieved through the use of a preamplifier, attenuator, and/or level shifter. It is corrected later or during data processing through the use of appropriate software or fixed hardware devices. In some cases, special converters such as log amplifiers or integrators are used depending on the detected signal and the related processing algorithms. The output of the ADC can be stored by two methods depending on the interface technique employed: (1) It can be put into the memory of the data acquisition system until a set of data has been collected, then transmitted to the computing system. (2) It can be continuously transmitted to the computing system as the conversion operation occurs. In high-speed data transmission, the direct memory access technique is often used.

An auxiliary function of the data acquisition system is buffering, in which data are temporarily stored before being sent to the computing system. Buffering also protects the computer system from the detector electronics, which often involve high voltages.

The source and detectors are usually housed in what is referred to as the scanning gantry. The gantry is a mechanical frame mounted in the system in a way that houses the patient during the scanning. The gantry aperture (an opening through which the patient moves in and out during the scanning procedure) varies from system to system, but is typically 45–66 cm in X-ray CT systems. In NMR CT, the aperture size is determined by the bore diameter of the main magnet, radio frequency (RF) coil sensitivity, and uniformity. A 60- to 120-cm air-bore magnet is common depending on whether the system is used for head only or whole body scans. In some CT systems the gantry can be tilted from the vertical position toward the front or back to allow for examination of specific cross sections of the patient.

3. Computer

The need for multidimensional image processing with more rapid reconstruction and higher image quality expedited the evolution of computer technology designed for imaging in general and for CT in particular. The amount of data memory and computations necessary for the reconstruction is large, and the trend is toward even larger amounts with higher resolution X-ray CT systems and with NMR CT. For example, to obtain a single slice image of 512×512 pixel size in X-ray CT, the required number of projections from different view angles is ~800 each, with more than 512 sample points. The use of a large number of sampling points is designed primarily to reduce interpolation errors. For 12-bit resolution, a memory of more than 5 megabits is needed to hold the measurement data. Reconstruction, in which convolution operations are often performed with fast Fourier transform, requires more than 40 million operations of multiplication and 250 million operations of addition. As an example, the number of required computations is summarized in Table II for both FB reconstruction and direct Fourier imaging.

The general structure of a CT computing system is illustrated in Fig. 5. In this figure, the measured data

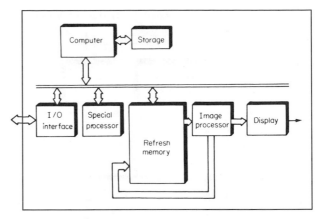

FIGURE 5 Block diagram of the general CT computational system.

are transmitted directly to a computer processing system or to an archival storage with magnetic disk, magnetic tape, or refresh memory. The measured data are transferred to a processor, either on-line or after completion of scanning, depending on the measurement speed and computer processing capability. In some CT systems, simple operations are carried out during data acquisition, and partially processed data are stored in memory or on disk. The processor can be either a general-purpose computer or special processor. The computational speed can sometimes be increased by as much as 10 to 100 times by combining special processors, such as array processors or backprojectors. They can be used most efficiently for structured data formats such as arrays or vectors.

The internal structure of the array processor comprises four functional units interconnected by internal buses (Fig. 6). The functional units are a host interface (which is system specific and provides communication with the host bus), a control processor (which controls the overall subsystem), a data memory (which acts as a data and table storage area), and a pipelined arithmetic unit (which

TABLE II **Number of Computations Required for Reconstructing a Cross Section of 512×512 Image by Convolution Backprojection and Direct Fourier Imaging**

Convolution backprojection	1024-Point FFT[a]	20,000	Multiplications and additions
	Kernel multiply	2,000	Multiplications and additions
	1024-Point IFFT[b]	20,000	Multiplications and additions
	Interpolation/view	10,000	Multiplications and additions
	Backprojection/view	262, 000	Additions
		42×10^6	Multiplications
		251×10^6	Additions
Direct Fourier imaging	Phase correction/line	2,000	Multiplications and additions
	512-Point FFT[a]	10, 000	Multiplications and additions
		11×10^6	Multiplications and additions

[a] FFT, Fast Fourier transform.

[b] IFFT, Inverse fast Fourier transform.

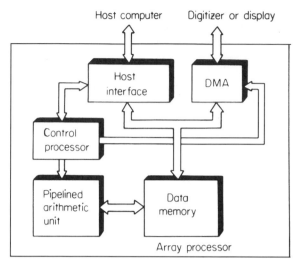

FIGURE 6 Block diagram of the internal structure of an array processor.

performs high-speed computation). In addition, an input–output interface can be used to store measured data and to transfer the reconstructed image directly to the display system without using the host computer. The speed advantage gained by the array processor is made through the parallel processing of a large number of data read from the memory of the processor, which uses its own bus, and through the use of a pipeline structure in the arithmetic unit. Distributive processing techniques can also be used to divide the processing load between the host and array processor, maximizing the efficiency of both systems.

The backprojector performs high-speed backprojection. For example, many CT systems, especially X-ray CT, are currently hindered by the backprojection operation, and therefore additional hardware computing devices like backprojectors are usually added to allow the whole image reconstruction process to be carried out almost instantaneously.

The computational speed of the special computing processor is often measured in units of million (mega) floating point operations per second (MFLOP). Through the use of array processors, more than 100 MFLOP can be easily attained. The entire processing of the data for the reconstruction depicted in Table II, for example, can be completed in only a few seconds.

Finally, reconstructed images are stored in the refresh memory, magnetic disk, or magnetic tape, depending on the amount of storage, transfer rate, access time, cost, and other factors.

Besides reconstruction, the computer provides machine control, pulse sequence control, display, and data handling. The dedicated microcomputer controller elements are therefore becoming increasingly important. They allow the implementation of highly efficient input–output

operations as well as computational structures. The host computer therefore becomes an interface with the operator and data acquisition elements in the CT scanner and related data bank.

4. Display

The display of the CT system provides a means of visual examination of the scan information obtained through the data acquisition system and computer. The CT display system is the fundamental link between the CT image and the human visual system, which provides the radiologist with maximal diagnostic information. Display monitors are usually attached to an operator console through which all the commands are made.

To facilitate visualization of small differences in tissue density, the value from each address of the display memory is first modified by some transforms before being changed to an analog signal by a digital-to-analog converter. Most present-day displays utilize 256 gray levels in accordance with human visual perception characteristics. In conventional CT display systems, a selected range of CT values is displayed uniformly over the range, which is preselected by setting window level and width. The window level control determines the midpoint of the density range, while the window width determines the width of the range. By changing the window width and level, the observer can enhance the visualization of the subject structures. In essence, the window width controls the contrast of the image, and the window level changes the density of the tissue to be displayed. Decreasing the window width usually results in a high-contrast image, but it also enhances noise.

In display systems, image processors are usually utilized to facilitate the manipulation of image enhancement and restoration (including filtering, selection of regions of interest, cut-view display, zooming, and distance measurement) without interrupting the host computer.

III. AREAS OF APPLICATION

A. X-Ray Computerized Tomography

X-Ray CT is a product of X-ray technology with advanced computer signal processing and capable of generating a cross-sectional display of the body. It is, in fact, the origin of the entire CT evolution, which began in the early 1970s.

The basic principle of X-ray CT involves X-ray generation, detection, digitization, processing, and computer image reconstruction. As the X rays pass through the body, they are attenuated at different rates by different tissues. The attenuated X rays are then collected by detectors and converted to digital impulses or data by the ADCs. These digital data are fed into a computing device for image reconstruction.

For simplicity, the pencil beam type of X ray is scanned along a line at a given direction or view. To achieve several different angles or perspectives, this scanning is repeated at each given angular view simply by rotation of both the X-ray tube and detectors.

1. Basic Physical Principles

a. Contrast mechanism and projection data.

The photon density that emerges when a narrow beam of monoenergetic photons with energy E and intensity I_0 passes through a homogeneous absorber of thickness x can be expressed as

$$I = I_0 \exp[-\mu(\rho, Z, E)x], \qquad (30)$$

where μ, ρ, and Z are the linear attenuation coefficient, density of the absorber, and atomic composition or number, respectively.

In the energy region where most commercial X-ray CT systems are being engaged for medical tomography (~ 70 keV), two types of interactions are dominant: photoelectric absorption and coherent and incoherent scattering.

In photoelectric absorption, the X-ray photon is completely absorbed by transferring all of its energy to an electron. Scattering is subdivided into two components: Rayleigh (coherent) scattering and Compton (incoherent) scattering. In Rayleigh scattering, the direction of the X-ray photon is changed but the energy is conserved. In Compton scattering, on the other hand, both direction and energy are changed. Figure 7 depicts the interactions of the X rays with water. It shows the contribution of each photon interaction to linear attenuation coefficients of different types as a function of energy. If the absorber is not homogeneous, $\mu(\rho, Z)$ is simply a space-variant function dependent on the material. By directing a monochromatic X-ray beam in the y direction, the output X-ray intensity $I(x)$ is

$$I(x) = I_0(x) \exp\left[-\int \mu(x, y)\, dy\right], \qquad (31)$$

where $I_0(x)$ and $\mu(x, y)$ are the incident X-ray intensity and X-ray attenuation coefficient, respectively. By taking the logarithm and rearranging Eq. (31), one can obtain projection data $p(x)$ as

$$p(x) = -\ln[I(x)/I_0(x)], \qquad (32)$$

where $p(x)$ is a simple integration or summation of the total attenuation coefficients along the X-ray path. In X-ray CT, the contrast is the difference in attenuation coefficients of the material involved. Since each set of projection data

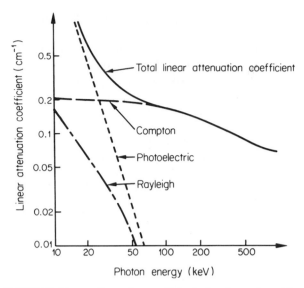

FIGURE 7 Linear attenuation coefficient of water and contribution of each interaction to the attenuation of X rays as a function of the energy.

represents the integral value of the attenuation coefficients along the path, the projection data taken at different views are the basis for tomographic image reconstruction.

b. Beam hardening.

X-Ray beams generally used in X-ray CT are not monoenergetic and have a finite spectrum. When this polychromatic X-ray beam passes through a material, X rays of different energy in the spectrum undergo different attenuation, and as a result the output energy spectrum differs from the input spectrum; that is, lower energy X rays are attenuated more heavily than higher energy X rays. This trend is accentuated if the path length is large or if material possesses components of high atomic number. The consequence of this nonuniform attentuation of the polychromatic X-ray beam is called the *beam-hardening effect*. It produces a visible artifact in the final reconstructed image. The original EMI scanner utilized compensatory measures to deal with the beam-hardening effect. A water bag was used as a compensator by surrounding the head so that the total path lengths of the X rays were always the same for all projections. Another method is to preharden the X-ray beam by passing it through an aluminum or copper filter so that the output X ray is close to the monoenergetic beam before it passes through the body.

c. X-Ray source.

Two types of X-ray sources are currently used in X-ray CT. The most simple type is the fixed-anode X-ray tube, in which the anode is cooled by oil and is continuously energized. A typical focal spot size of this type is about 2 mm × 16 mm on a 20° angle tungsten target. The relatively small heat dissipation capability associated with this type of tube limits the amount of photon flux generation, resulting in a statistically noisier image than others, for example, the rotating-anode X-ray tube, when operated continuously. The rotating-anode X-ray tube, on the other hand, allows for a large photon flux because it has much greater heat capacity.

d. Detectors.

X-Ray photons are collected by radiation detectors of various kinds, such as a scintillation crystal coupled with a photomultiplier tube (PMT). In general, the output of the detectors consists of electrical signals that are proportional to the incident X-ray energy or fluence. The most important parameters to be considered in the selection of detectors for X-ray CT are efficiency, response time (or afterglow), and linearity. Efficiency refers to the absorption and conversion efficiency of the incident X rays to electrical signals. Linearity refers to the dynamic range of the detector response. Response time refers to the speed with which the detectors can detect X-ray photons and recover to detect the next photon. This is determined by the afterglow—one of the important characteristics of detector materials for X-ray CT application.

Detector types currently in use for X-ray CT can be divided into two classes: scintillation detectors and gas ionization detectors.

i. Scintillation detector. Scintillation crystals [such as NaI(T1) and CsI(T1)] produce flashes of light as they absorb X-ray photons. The light is then converted to electrical signals by subsequent electronics. The following two types of scintillation detector systems are commonly used in X-ray CT:

Scintillation Crystal–PMT Coupled Detector. Light produced in the crystal is coupled to the photocathode of a PMT. In the PMT, photoelectrons are generated from the photocathode as the light strikes it. These electrons are multiplied through a series of cascaded dynodes in which electron multiplication processes take place. Each dynode produces more electrons than incident electrons. The multiplied or amplified electrons constitute output signals in the form of a charge or a current, which are the indicators of the energy and fluence of the incident X-ray photons on the scintillation crystal. An overall gain of a few millions is common in most PMTs. Other detector crystals, such as BGO (bismuth germanate) and CaF_2, are also used.

Scintillation Crystal–Photodiode Coupled Detector. The performance of this detector, which was developed more recently than its counterparts, has been found to be satisfactory. A typical scintillation crystal–photodiode detector comprises a CsI(T1) scintillation crystal, a *P–N* junction photodiode coupled to the crystal, and a preamplifier for the low-level signal amplification.

The incident X-ray photon is converted to visible light in the scintillation crystal, which then falls into the PN junction photodiode. Generated electron-hole pairs are collected at the junctions. Since the generated current is usually weak, a low noise preamplifier is required. The voltage output is proportional to the energy and fluence of the X-ray incidence on the detector.

ii. Gas ionization detector. Some commercial X-ray CT systems use gas ionization detectors. To improve the detector efficiency, xenon gas is pressurized by as much as 20 atm and filled in a long chamber to maximize detection efficiency. It consists of tungsten plates, which serve as anodes for electrons. When X-ray photons are incident on the detector cell, gas is ionized. This ionized gas provides current that is directly proportional to the X-ray photon energy. The advantages of the gas ionization detector are high spatial resolution capability and simplicity. In addition, the compact detector assembly can be made on a large scale by packing a few hundred equivalent detector elements that have detector widths as narrow as 1 mm. The disadvantage of the ionization detector, even with highly pressurized gas, is the low detection efficiency. Characteristics of each detector are briefly described in Table III.

e. Data acquisition and reconstruction algorithms.

At each view, detector signals from the detector system are converted to digital pulses by the ADC. These signals are collected by the computer via signal processing

TABLE III Characteristics of Detectors Used in X-ray CT

Type	Advantages	Disadvantages
NaI(T1)-PMT	High detection efficiency, 100% at 70 keV(1-in. crystal)	Afterglow; restricted dynamic range; low packing density; hygroscopic
CaF_2– PMT	No afterglow	Low detection efficiency (62% with 1-in. thick)
BGO (bismuth germanate)-PMT	No afterglow; high detection efficiency; nonhygroscopic	Low light output
CsI(T1)-photodiode	Good spectral match with available PMTs; high detection efficiency (94.5% at 120 keV with 5-mm thick)	—
Xenon gas ionization detector	Simple and compact; no afterglow; high resolution capability	Low efficiency; possible instability; slow response time

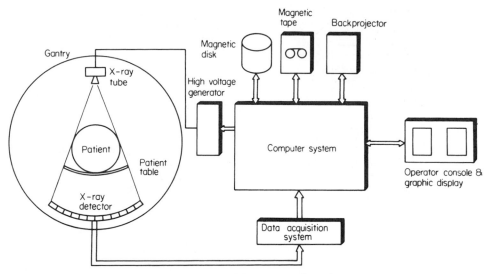

FIGURE 8 X-Ray CT system configuration.

electronics for image reconstruction. The steps of the view angle between successive views are normally ~1° or less, and a few hundred views are usually taken from each tomographic slice. After the projection data sets (i.e., line-integral projection data) are obtained, image reconstruction takes place. The time required for image reconstruction can be relatively long in comparison to data acquisition time. Therefore, the procedure usually requires special-purpose computer peripherals, such as array processors or backprojectors.

f. System configuration.   Figure 8 is a block diagram of a typical X-ray CT system currently in use. A 16- to 32-bit minicomputer system equipped with a dedicated backprojector is normally used for data acquisition, signal processing, and system control.

2. System Evolution

Although remarkable progress has been made in all aspects of X-ray CT since it was introduced in 1972 by Hounsfield at EMI, the most significant changes have been made in the area of data acquisition. These stages of progress are classified into "generations." Since 1972, the X-ray CT has evolved from the first generation to the fourth and possibly to the fifth generation, which will incorporate the latest developments in dynamic scanners.

a. First generation.   The first generation naturally entailed the first EMI scanner developed by Hounsfield. This scanner used a single pencil beam and a single detector, which translated and rotated synchronously. There was translational motion across the object being scanned, and at the end of each translational motion an

incremental 1° rotation followed in preparation for the upcoming scanning. This procedure is depicted in Fig. 9a. The collection of projection data needed for tomographic image reconstruction of a slice took several minutes.

b. Second generation.   The design of the second-generation CT incorporated a narrow-angle fan beam X ray and an array of multiple detectors. Since the diverging fan beams passing through the patient increase data collection channels, the number of angular rotations required could be reduced. Therefore, the scan time in this second generation was shortened substantially; the nominal scan time was ~20 s. The second-generation scanner still entailed translational motion as well as rotational motion to cover the object fully, however. Figure 9b illustrates the configuration of the second-generation CT scanner.

c. Third generation.   In the third generation, the fan beam angle is widened, and the fan covers the entire object to be scanned. Each projection path is defined by a matching detector, which can be either a small and narrow scintillation detector slab or a segment of a gas ionization chamber. Because the entire object is covered or encompassed, no translational movement is required. Therefore, both the X-ray tube and the detector array need only simple rotational motion around a fixed axial center. The entire 360° is usually scanned for whole data collection. Scan time is as short as 3 s. This high speed scan capability allows for synchronized rotational motion with physiological signals such as that obtainable from electrocardiograph (ECG), when the imaging of moving organs is required. The major drawback of this configuration is that the effects of the drift of the detector are cumulative so that artifacts

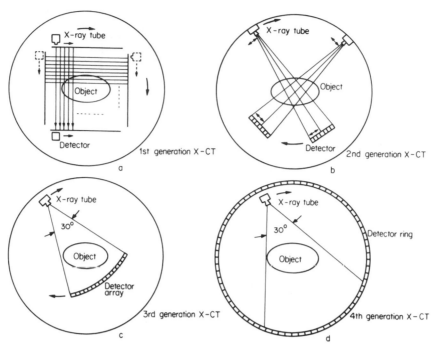

FIGURE 9 X-Ray CT system evolution from first- to fourth-generation scanners. (a) First generation; (b) second generation; (c) third generation; (d) fourth generation.

appear on the reconstructed image. Figure 9c shows the configuration of the third-generation CT scanner. Almost all third-generation scanners use pulsed X-ray sources to take advantage of significant dead time between successive views.

d. Fourth generation.

The construction of a stationary circular ring detector array is probably the ideal choice of the detector for the rotating X-ray source. A striking analogy is the circular ring PET scanners of various types developed during the 1970s. This stationary ring is the most distinct feature of the fourth-generation X-ray CT. The X-ray source rotates, but the detector array does not. A wide-angle fan beam X ray encompasses the entire patient, and 600 or more stationary detectors form a circular ring array. With this kind of configuration, detector drift is not cumulative and therefore can be corrected. The advantages of this system are similar to those of the third-generation systems, but the main drawback of the third-generation systems—drift effects—has been eliminated since detector drifts no longer accumulate over successive views. The fourth-generation systems are generally more expensive, however, due to the large number of scintillation detectors and PMTs employed. Figure 9d shows a schematic diagram of the fourth-generation X-ray CT scanner.

e. Dynamic scanner.

For the imaging of moving organs, such as the heart, ultrafast scanners are required. Typical systems of this kind are the dynamic spatial reconstructor (DSR) and cardiovascular CT (CV CT). These can be categorized as the fifth generation X-ray CT scanners.

i. Dynamic spatial reconstructor. The development of the DSR at the Mayo Clinic was completed in 1982. It can produce real-time images of body organs in motion. The DSR comprises 28 X-ray sources with 28 opposing X-ray imaging chains, which are image intensifiers coupled with X-ray detection phosphors mounted in the same gantry. The physical size of components and the required radiation flux determine the number of X-ray sources and imaging chains. As many as 240 images of adjacent slices with a thickness of 1 mm can be made from a cylindrical volume that is 38 cm in diameter and 24-cm long. A complete volume scan is achieved in 0.01 s after each of the 28 X-ray sources is pulsed in succession (for 0.34 ms). These scans can be repeated 60 times per second. A high temporal resolution image can be obtained with 28 angles of view recorded in 0.01 s if the gantry is kept stationary. For stationary objects, however, high spatial and density resolution images can be produced using all 240 views for reconstruction. Trade-offs between temporal, spatial, and density resolution can be made by selecting the appropriate subsets from the total projection data. Figure 10 depicts a schematic diagram of the DSR scanner.

ii. Cardiovascular computerized tomography. This system was proposed and developed by Boyd *et al.* at the University of California, San Francisco, with the same goals in mind, but with a more compact and physically

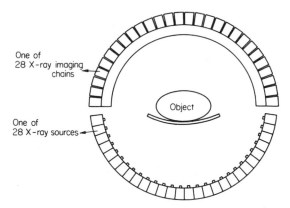

One of
28 X-ray imaging
chains

One of
28 X-ray sources

Object

FIGURE 10 First dynamic scanner DSR developed by the Mayo Clinic.

integrated design than that of the first DSR. The heart of CV CT is the electron beam scan tube and stationary scintillation crystal–photodiode coupled detector array. An accelerated and focused electron beam is deflected by a computer-controlled bending magnet to be swept along a 210° curved tungsten target ring. Four target rings are swept serially to obtain a multiple section examination. Approximately 30° of the fan-shaped sector of the X-ray beam generated at the tungsten target is detected by the detector array for image reconstruction. The detector array comprises two detector layers of a half-ring. A simple scan produces two side-by-side tomographic slices; a total of eight slices can be produced to cover a region ~9-cm deep by sweeping four targets in succession. Two adjacent tomographic slices can be obtained in 50 ms and eight slices can be obtained in 200 ms if four targets are swept serially.

Although the basic principles of X-ray CT have not changed since they were introduced, the scanning scheme, speed, and performance of X-ray CT have improved substantially. At the same time each stage of X-ray CT development has brought about new applications and widened the scope of X-ray CT. X-Ray CTs of different forms are now utilized in many diverse fields, including the inspection of tires on the production line and the detection of defects in nuclear reactor cores.

Scanning speed has evolved from several minutes to a few milliseconds, making possible dynamic scans of moving organs, such as the heart. The search for better and higher contrast and spatial resolution with shorter and shorter imaging time is expected to continue, and the growth of applications of X-ray CT to fields other than medical imaging is anticipated.

B. Emission Computerized Tomography

Emission computerized tomography is an imaging technique capable of visualizing the 3-D distribution of ra-

dionuclides in the human body. It is also capable of measuring quantitatively *in vivo* biochemical and metabolic functions. This is in contrast to transmission X-ray CT, which emphasizes visualization of the anatomical structure of the human body; that is, spatial resolution and tissue density contrast are more important. The field of ECT is one of the fastest growing areas in nuclear medicine and is rapidly replacing the conventional 2-D projection images known as scintigraphy.

Unlike transmission X-ray CT, ECT generally suffers from poor resolution or statistics. The number of photons that can be detected is usually limited, especially for the tomographic mode operation, in which the quantity of photons available is limited to a small columnlike volume for each detector or detector pair. In addition to a lack of sufficient photon statistics, emission CT is hampered by the fact that in the detection process photons emitted by the body undergo attenuation in the tissue. To provide a reliable reconstructed image, attenuation due to body tissue must be corrected.

ECT can be conveniently divided into two modes—positron emission tomography (PET) and single photon emission computerized tomography (SPECT)—according to whether the radionuclides employed undergo decay with the emission of positrons, which provide annihilation photons, or with the emission of γ rays. The former is generally considered to be more suitable for the tomographic mode, while the latter has been more readily available and has begun to show potential in tomographic imaging. In this section, these two modes, as potential tomographic imaging modalities, are detailed from a physical point of view.

1. Positron Emission Tomography

Although the potential of positron imaging was recognized as early as the 1950s, its actual tomographic mode imaging began only after X-ray CT was developed in 1972. PET is a brilliant example of a joint effort of many disciplines, including physics, electronic instrumentation, and computers. PET requires short-lived cyclotron-produced radionuclides and the appropriate labeling of these radionuclides to suitable radiopharmaceuticals that can be administered to the human body, as well as instrumentation for the detection of annihilation photons and appropriate signal processing to provide high-resolution images. Small cyclotrons are now available for use in medical facilities. Advances in the rapid chemical synthesis of radiopharmaceuticals now permit a large number of labeled compounds to be used in positron imaging. Above all, there has been phenomenal growth in PET instrumentation research and development in recent years; a few system designs are potentially capable of imaging with a resolution as high as 2 to 3 mm full width at half-maximum (FWHM).

This section is devoted to the instrumentation aspects of PET and is divided into three parts. The first part introduces some basic principles, the second deals with various physical factors affecting system performance, and the third discusses the evolution of PET system designs.

a. Basic principles.

i. Positron emitters and physics.   Positron-emitting radionuclides possess several important physical properties that make PET a unique imaging technique. Its most important property is the directionality and simultaneity of photons generated by annihilation. The emitted positron combines with a nearby electron, and two photons are generated by a phenomenon known as annihilation. These two annihilation photons, each with energy of $\sim$511 keV, are generated simultaneously (simultaneity), and they travel at $\sim$180° from one another (directionality). The nearly collinear direction of the two annihilation photons makes possible the identification of the annihilation event or the existence of positron emitters through the detection of two photons. This is usually achieved by a coincidence detection circuit that records an event only if both detectors sense annihilation photons simultaneously (solid lines, Fig. 11). Because the two detectors record coincidence events only from a volume of space defined by a column

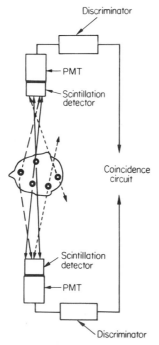

FIGURE 11 Principle of coincidence detection. True coincidence (solid line), random coincidence (dashed line), and scattered coincidence (broken line) are indicated.

or strip joining the two detectors, the physical collimation used in SPECT to confine the direction of incoming photons can be eliminated. For this reason, coincidence detection is also called electronic collimation. The total number of coincidence events detected by a given pair of detectors constitutes a measure of the integrated radioactivity (i.e., line-integral projection data) along the strip joining two detectors. From a complete set of line-integral projection data obtained from several views with detectors surrounding an object, the activity distribution within the slice can be reconstructed by using the algorithms discussed in Section II.A.

The accuracy of the spatial localization of a positron-emitting radionuclide by the coincidence detection procedure is limited, however, by two physical properties of positron annihilation: (1) The two annihilation photons are not exactly collinear, thereby creating angular uncertainty, and (2) there is uncertainty about the annihilation position of the emitted positrons, that is, uncertainty about positron ranges. The latter is strongly dependent on the kinetic energy of the emitted positron. The combined effect of these two factors introduces a fundamental uncertainty in locating the sources of the positrons. This uncertainty, depending on what kind of radionuclides are used and on detector separation, is typically about 2 to 3 mm FWHM. This value is accepted as a lower limit on the resolution that can be achieved by a positron camera. Despite these inherent spatial resolution limits, PET could potentially be the highest resolution nuclear imaging technique.

ii. Advantages of PET.   Perhaps the greatest advantage of PET over other nuclear imaging systems is the electronic collimation with which a solid angle is extended to virtually the entire object. The sensitivity (photon collection capability) gain due to this large detection solid angle that results from electronic collimation is greater than that of SPECT. This is considered to be an important advantage of PET over SPECT.

Other advantages of PET, which uses electronic collimation, are the uniformity of resolution and sensitivity over the entire range. The latter, uniform sensitivity, stems from the fact that the combined attenuation affecting a photon pair is the same, regardless of the position of the annihilation, as long as the annihilation points are inside the column defined by a detector pair. This allows us to compensate accurately for the attenuation suffered within the object. This is again a distinct advantage of PET over SPECT.

Another advantage of PET consists of the physiological aspects of most of the available positron-emitting radionuclides, which are usually of low atomic number. Among these radionuclides, ^{11}C, ^{13}N, ^{15}O, and ^{18}F are used most often in PET because of their physiological affinity in the human body and their short physical half-lives

(^{11}C, 20.34 min, ^{13}N, 9.96 min; ^{15}O, 2.05 min; ^{18}F, 110 min), which facilitate effective imaging with minimal dosages to the patient. The radionuclides ^{11}C, ^{13}N, and ^{15}O are the major components of molecules in living matter. Therefore, they are closely related to the metabolic processes in human physiology.

b. Physical factors affecting system performance.

i. Detector and related materials. The spatial resolution of the positron camera depends on the width of the detection channel and the sampling interval. The former sets the resolution limit obtainable with the system, provided that the sampling requirement or the Nyquist sampling criterion is met. The detection channel can be characterized by the detector aperture function and can be determined by simple ray tracing. At the center of a detector pair, it can be represented by a triangle with an FWHM equal to half the detector width. The aperture function toward the detector becomes trapezoidal and ultimately rectangular at the position immediately adjacent to the detector.

In most PET instruments, a standard γ-ray detector design is chosen for detector assembly. It consists of a PMT and a scintillation crystal coupled together.

Several scintillation crystals used in the past are being recommended for PET design. In the early 1970s, NaI(T1) was the most commonly used scintillation detector for PET, even though it was comparatively inefficient in high-energy 511 keV photon detection. A more recent development in scintillation detectors was the introduction of BGO, which led to substantial improvement in detection efficiency. The nonhygroscopic nature of BGO crystals further facilitates compact detector assembly or packing, which also leads to an overall increase in detection efficiency. The design of a high-resolution PET system was made possible by the introduction of BGO, since an extremely narrow slab of each detector crystal (4–5 mm in width) is considered to be the essential requirement of the high-resolution PET.

Two other crystals have been considered for PET application, CsF and BaF$_2$, both of which have very rapid detection capabilities. Therefore, these crystals have made excellent time resolution feasible. This advantage has been exploited in time-of-flight (TOF) PET systems.

Some of the physical properties of the detector materials mentioned here are listed in Table IV.

ii. Sampling. Because detector width is limited by detection efficiency, the widths of the detectors are usually kept as large as the resolution allows. With a given detector size, maximum resolution can be achieved by satisfying the Nyquist sampling criterion; that is, the sampling distance must be less than half the distance of the highest spatial resolution obtainable by that particular ring PET

TABLE IV Properties of Detector Materials Used in Positron Cameras

Detector material	NaI(T1)	BGO	CsF	BaF$_2$
Density (g/cm^3)	3.67	7.13	4.64	4.89
Linear attenuation coefficient at 511 keV (cm^{-1})	0.34	0.92	0.44	0.47
Scintillation decay time (ns)	250	300	5	0.8
Emission wavelength (Å)	4100	4800	3900	2250
Energy resolution in FWHM at 511 keV (%)	>7	>10	23	13

design, (i.e., half the detector width). In projection data obtained with a detector width of w, a maximum spatial frequency is about $2/w$, because the intrinsic resolution expected is about half the detector width. The desired sampling distance, therefore, should be less than $w/4$.

In the hexagonal- or octagonal-geometry PET system, desired sampling can easily be achieved by introducing translational and/or rotational motions (Fig. 12a). In the circular ring system, however, improving the sampling arbitrarily has been difficult, and several sampling schemes for overcoming this inherent difficulty have been proposed and implemented in experimental systems. For example, wobbling motion has been widely used (Fig. 12b). However, two sampling schemes suitable to the circular ring

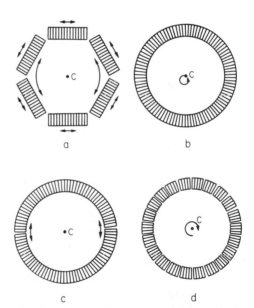

FIGURE 12 Sampling motions. (a) Hexagonal geometry with translational and rotational motions. (b) Wobbling motion applied to circular ring systems. In this scheme, the entire ring moves in a circular path (small track of circle shown at the ring center); this is 2-D motion. (c) Dichotomic motion applied to circular ring systems. In this scheme, two half-arcs, which form a ring, move back and forth in opposite directions along the circular path. Note that the motion is 1-D. (d) Positology. In this case, continuous rotation of a ring with unevenly spaced detectors provides finer sampling.

system have been developed, namely, dichotomic sampling and clam shell sampling.

Although the wobbling scheme is one of the most common methods, it is generally believed that the samplings obtained by this method are usually neither uniform nor equally spaced. To obtain more uniform and equally spaced samplings with the fewest possible number of motions, a new sampling scheme, known as dichotomic sampling, has been proposed and incorporated into an experimental system (Fig. 12c). This scheme employs two half-rings (from which the term *dichotomic* is derived), which rotate in such a way that finely sampled parallel or fan data sets can be obtained with a minimal number of scan stops. Another scheme, developed at the University of California, Berkeley, is similar to the dichotomic sampling scheme but employs a slightly different motion.

Another method for improving the sampling scheme is the positology (Fig. 12d) developed by Tanaka *et al.* in Japan. In this scheme, rapid rotational motion is employed with a nonuniformly spaced circular detector array. With this scheme, through the rotation of the entire ring, the desired sampling can be achieved.

iii. Sensitivity. Sensitivity in PET is defined as the capability of detecting the true coincidences (solid line, Fig. 11) with a given amount of radioactivity. Sensitivity has traditionally been measured with a phantom of diameter $d = 20$ cm, filled with a uniform activity concentration ρ (μCi/cm^3) (1μCi $= 37,000$ disintegrations per second). By considering several factors, including the activity in the field of view, the self-attenuation of γ rays within the phantom, the solid angle subtended by the detector array, and the detection efficiency of the array, an empirical formula for sensitivity measure is derived and given by

$$C_t = 14,500 \rho \alpha \varepsilon^2 h^2 d^2 / D \quad \text{(counts s}^{-1}\ \mu\text{Ci}^{-1}\ \text{cm}^{-3}\text{)}.$$

$$(33)$$

In this equation, α is the probability of no scatter of both annihilation photons, ε is the detector efficiency including the detector packing ratio, h is the thickness of slice to be imaged, d is the diameter of the phantom, and D is the ring diameter.

If the system contains two or more detector rings, coincidence can be measured between the detectors in the different rings (coincidences not in the planes that are perpendicular to the axis) to increase the sensitivity of a given imaging plane. Sensitivity is higher in the cross-slice planes because they involve twice as many detectors.

iv. Random coincidences. As the source of background noise in PET images, random or accidental coincidences occur when two photons emitted from two independent positions are detected within the coincidence resolving time τ (dashed line, Fig. 11). Random coinci-

dences produce a haze of background over the field of view in the reconstructed image.

For a uniform distribution of activity, a formula for the random count rate has been established and given by

$$C_r = \tau f_d C_s^2 \quad \text{(counts s}^{-1}\ \mu\text{Ci}^{-1}\ \text{cm}^{-3}\text{)},\quad (34)$$

where C_s is the single count rate for the entire ring, τ is the coincidence resolving time, and f_d is the fraction of detectors covering the object in the whole detector ring in coincidence with any one given detector, that is,

$$f_d = (2/\pi) \sin^{-1}(d/D).\quad (35)$$

Because both the single rate and true coincidence rate are proportional to the amount of activity, it is apparent from Eq. (34) that the random coincidence rate is proportional to the square of the true coincidence rate.

To reduce the random coincidences, it is imperative to minimize single counts. Because single counts can arise from both in-slice and out-of-slice annihilation events, they can be suppressed by limiting the detection channels by slice collimation or by the increase in the energy threshold so that maximum rejection of any scattered single events can be achieved. It is common practice to reduce as many annihilation events as possible from out of slice by using annular interdetector ring collimators.

Random coincidences can be partially corrected by software, provided that the single counts of each channel and the coincidence resolving time are recorded.

v. Scattered coincidences. Scattered coincidences occur when one or both γ rays resulting from an annihilation event are scattered in the medium and detected with the remaining energy of the γ rays above the energy threshold (broken line, Fig. 11). They give incorrect positional information and produce a line-spread function with long tails. The number of scattered coincidences can be reduced by setting the energy threshold level high and by using tighter interslice collimation.

vi. Correction of random and scattered coincidences. Random coincidences can be corrected by two simple procedures. One is by using the delayed coincidence measurement with the same time window and the other, mentioned earlier, is by estimating random coincidences using Eq. (34). As stated above, random coincidences can be reduced either by minimizing the coincidence time window or by reducing the activity.

Since the scattered coincidences are an inherent physical property, they cannot be easily reduced or eliminated as random coincidences. Because it is prompt in nature, differentiation between true and scatter simply by minimization of the time window is therefore difficult. In addition, the energy loss in small-angle scatter is so small that it is difficult to differentiate through the energy window. In the case of multilayer ring geometry in imaging multiple slices, scatter elimination is achieved by the use of tighter interslice collimation. However, one should admit some loss of observable volume when large-volume imaging is of importance.

vii. Attenuation correction. Attenuation correction is one of the most important parts of PET imaging. A variety of correction schemes have been developed in the past. One of the simplest and crudest ways of correcting attenuation is through the use of geometric shape, that is, by finding edge contours and using contour information to determine attenuation lengths for the subsequent correction. In this case, attenuation coefficients are customarily assumed constant. A more accurate method is to use the transmission scan information obtained by an external positron source surrounding the patient. This technique often suffers from statistical noise and thereby makes attenuation correction difficult, unless a sufficient amount of data is taken. The correction can also be made through X-ray CT by a procedure similar to that described earlier, but one should be aware that attenuation coefficients are different from those obtained with 511-keV photons.

c. Evolution of positron camera development.
There have been continuous efforts to develop PET imaging systems since the early 1950s. Some notable examples include two NaI detector systems developed by Brownell and Sweet in the early 1950s, 32 discrete NaI(Tl) detector systems developed by Rankowitz *et al.*in 1962, and the PC-I developed by the Massachusetts General Hospital (MGH) with two banks of detectors (127 detectors per bank) in 1972. Although these systems ultimately were intended for tomographic imaging, they remained quasi-tomographic machines until the introduction of the X-ray CT scanner by Hounsfield in 1972.

The first tomographic systems developed after the introduction of Hounsfield's model were the PETT (positron emission transaxial tomograph) I–III series designed by Ter-Pogossian and Phelps of St. Louis and CRTAPC (circular ring transaxial positron camera) designed by Cho *et al.* at UCLA. As a hexagonal detector array, the former employed both translational and rotational motions, while the latter remained virtually stationary. Since these developments, PET instruments have rapidly improved through the formulation of a variety of new concepts. For example, system geometry has evolved from planar to hexagonal type, from hexagonal to circular type, from single ring to multiring, and so on, Detector material has been changed from NaI(Tl) to BGO, CsF, or BaF_2. The time of flight (TOF) technique has also been introduced. Also, several commercial companies have begun to design PET systems.

i. System geometry. It is interesting to observe how system geometry has evolved since the mid-1970s. System geometry (the arrangement of detector arrays) is the most basic design choice, because it determines fundamental system performance. It can be categorized into three basic types: planar, polygonal, and circular ring. To cover the imaging volume in the axial direction, multiring systems have appeared in which several rings are stacked on one another. They provide high total sensitivity and offer $2N - 1$ (N is the number of rings) image slices simultaneously.

The trend appears to be toward more generalized circular ring types of various forms. Among the advantages of circular ring geometry are uniformity, high sensitivity due to high packing fraction, and high angular sampling capability. Because the requirement for angular sampling can be met even by the stationary ring, systems do not require further rotation to improve angular sampling. Although linear sampling has been a limiting factor in the circular ring system, various solutions to this problem have also been suggested, and their efficacy has been proved experimentally. It is also worth noting that trends are in the direction of volume imaging, either through multilayer rings or through a spherical-PET (S-PET) system to be described later. Yet another avenue, one that requires further technological development, is the TOF approach. This is considered to be an adjunct to improve system resolution.

ii. Planar system. The planar type consists of two detector planes (made of either discrete crystal arrays or position-sensitive devices, such as the Auger camera or multiwire proportional chamber) facing one another. A set of projection data can be obtained by rotating the dual planes around the patient, from which a series of image slices covering sufficient axial volume is reconstructed. A disadvantage of this type is the large number of angular rotations it requires.

MGH systems (PC-I, PC-II) and the commercial version of those systems (TCC 4200) fall into the category of discrete crystal arrays.

iii. Polygonal system. Such systems as PETT III and PETT IV by Ter-Pogossian and Phelps and ECAT and NeuroECAT by Ortec fall into this category. With the exception of the octagonal-shaped NeuroECAT, these systems are hexagonal. In this type of system, coincidence detection channels are formed between those banks opposing one another so that ring efficiency is usually limited, particularly toward the periphery of the image. An advantageous feature of the hexagonal system is the relatively simple translational and rotational sampling motions, which fulfill the requirements of uniform linear and angular samplings.

iv. Circular ring systems. A natural extension of the polygonal PET system is circular ring geometry, which provides uniformity as well as natural symmetry. The first circular ring PET system was conceived and developed by Cho *et al.* at UCLA in 1975. Various other circular ring systems were developed subsequently by Budinger *et al.* at Berkeley, by Bohm *et al.* at Stockholm, by Carroll *et al.* at Cyclotron Corporation (also at Berkeley), and by various commercial firms. Some of the undersampling problems associated with ring systems have been resolved by incorporating new sampling schemes, such as wobbling and dichotomic sampling. The ring diameter is one of the key parameters in ring system construction. If constant crystal size and spacing are maintained, the advantages of an increased diameter are as follows: improvement of resolution uniformity, decrease in the number of random coincidences, more scatter rejection, and increase in the number of view angles. On the other hand, the expected disadvantages of an increased diameter include decreased sensitivity, resolution degradation due to angular uncertainty, and increased overall system costs. Ring diameter ranging from 45 to 65 cm appears to be suitable for brain scanners, while 70–90 cm is more suitable for body scanners.

v. Time-of-flight system. Ideally, if one can detect the exact difference in the flight times of two annihilation photons, the exact position of annihilation can be located, making possible a direct mapping of activity. The TOF concept in PET appears to be useful in low-resolution PET systems, where additional TOF information can help to enhance resolution and improve the signal-to-noise ratio of the image. To obtain any significant improvement, however, the time resolution should be substantially less than 500 ps. This is usually difficult to achieve by means of existing detector and electronic technology. The development of a few fast scintillation crystals, such as CsF and BaF_2, is of interest however.

vi. Spherical-pet system. For increasing overall sensitivity and direct volume imaging capability, the spherical shape of the PET system appears to be the most effective choice. The first extensive S-PET design and preliminary study were carried out by Cho *et al.* at Columbia University. The S-PET design was initially intended for high-resolution imaging, that is, resolution of ~3 mm FWHM. To support this high resolution, it is imperative to maximize system sensitivity without impairing the true-to-random as well as scatter coincidence ratios. Therefore, spherical geometry and slice collimators focused on the system center have been incorporated. High stopping power detectors, such as BGO (with a detector width as narrow as 4 to 5 mm), have been proposed for this purpose. Special PMTs such as the rectangular PMT developed by Hamamatsu (the R2404) and new dichotomic sampling scheme have also been incorporated. In the field of image reconstruction, a TTR algorithm (discussed in Section II.A) has also been utilized. The design concept is currently under extensive evaluation from both the theoretical and experimental points of view.

2. Single Photon Emission Computerized Tomography

Although SPECT was first envisioned before the PET scanner, it is generally considered to be inferior to PET in a few critical aspects of imaging, including the capability of attenuation correction and ultimate resolution attainable with conventional collimators. Interest in SPECT has been renewed, however, due to its simplicity and availability.

In SPECT, any radioisotope that emits γ rays can be used. In contrast to annihilation photons, these γ rays are emitted as single individual photons. Isotopes common in SPECT imaging include ^{99m}Tc, ^{125}I, and ^{131}I. These are the radionuclides most often used in nuclear medicine.

Because of the nature of the isotopes used in SPECT [i.e., decay by the emission of (single) γ photons], a device for defining ray direction—the collimator—is required. This collimation process eliminates most of the γ rays emitted into 4π space. Unlike PET, this physical collimation inherently limits sensitivity.

The first rotating single-photon tomographic imaging system was developed by Kuhl and Edwards in 1963, without the benefit of modern tomographic image reconstruction techniques. Since the development of X-ray CT and PET, the instrumentation for SPECT has been extensively developed. Although most of the developments are based on the rotating Auger camera arrangements (camera-based approaches), a few systems utilize discrete detector arrays (scanner-based approaches).

The resolution of a SPECT image is directly related to the response of collimators employed and the sampling intervals in linear and angular directions. The response of the collimators determines resolution as well as system sensitivity and other factors, such as uniformity. The resolution of SPECT is poorer than that of PET, mainly due to collimation and partly due to the limited number of photons that can be collected. Axial resolution is especially dependent on the properties of collimation and depth variant.

As research into SPECT continues as expected, resolution and sensitivity will be improved through the application of new collimators and the use of new geometry, such as the triangular shape approach proposed by Lim *et al.*

C. Nuclear Magnetic Resonance Computerized Tomography

NMR CT is a 3-D imaging system that uses the NMR phenomenon as an imaging tool. Magnetic resonance is a phenomenon found in magnetic systems that possess both magnetic moment and angular momentum. The term *resonance* implies that we are in tune with a natural frequency of the magnetic system. In this case, it corresponds to the frequency of gyroscopic precession of the magnetic moment of nuclei in an external static magnetic field. Because the magnetic resonance frequencies fall typically into the radio frequency range for nuclear spins, we often use the term *radio frequency* in referring to NMR. In NMR CT, we select a region from samples and obtain spatial distributions of nuclear spins through the application of deliberately added spatial field gradients and RF signals. Thus, the cross-sectional images of an object are obtained.

The advantages of NMR CT are its nonhazardous nature, its high-resolution capability, its potential for chemically specific imaging, its capability of obtaining anatomical cross-sectional images in any direction, and its high tissue discrimination capability (high-contrast resolution among different tissues). Although it has some minor disadvantages, such as its inherently long data acquisition time due to spin–lattice relaxation time and low signal-to-noise ratio, due to its many advantages the NMR CT is rapidly becoming a major diagnostic tool. NMR CT is overcoming the problem of time-consuming data acquisition through the development of new high-speed imaging methods and is surmounting the problem of low signal-to-noise ratio through the use of high-field magnets (superconducting magnets) with a magnetic field as high as 2.0 tesla (T) (1 T is equal to 10 kG) or more.

In the early 1970s, both Lauterbur and Damadian showed that NMR spectroscopic techniques could be applied to imaging the human body and demonstrated that these techniques might eventually be applicable to diagnostic imaging techniques similar to those of X-ray CT. In 1978, Andrew demonstrated the very high resolution capability of NMR by obtaining a fine and detailed image of the submillimeter septum of a small lemon. Moore and Holland soon followed with images of the human head, demonstrating definitively the potential of NMR tomography in diagnostic imaging.

NMR tomographic images can be formed by direct mapping, projection reconstruction, or Fourier imaging. To date, two kinds of potentially useful imaging methods—direct Fourier imaging and projection reconstruction—are most widely used.

An interesting aspect of NMR imaging is its diversity in image formation, data collection, and reconstruction. Many different imaging and data processing methods are presently known and will be discussed from various points of view, such as imaging time, available field gradients, pulse strength, speed, signal-to-noise ratio, and artifacts associated with the restored object image.

Great advances have also been made in the area of instrumentation for whole-body NMR tomography. The formation of an NMR CT system requires a magnet, gradient coils, RF coils, computer and peripherals, and associated electronics. A main-field strength of 0.5 to 20 kG is used with a field gradient of 0.01 to 1 G/cm, formed by x, y, and z gradient coils. In the case of proton imaging with these strengths of magnetic field and gradients, the required RF range is approximately 2 to 85 MHz. The expected resolution in a conventional NMR imaging system depends on the field homogeneity and the available strength of field gradients.

Two main types of imaging methods—direct Fourier imaging (Kumar–Welti–Ernst method) and projection reconstruction—and the possibility of parameter imaging will be discussed in detail here. Hardware configurations and the related statistical aspects of image quality and imaging times will also be discussed briefly.

A typical NMR tomographic imaging system designed for human imaging is depicted in Fig. 13a. In this figure, a split-solenoidal type of magnet NMR CT system is shown. The sample is surrounded by an RF coil (Fig. 13b) and a gradient coil set (Fig. 13c, d, and e). The configurations of the magnet and the gradient coils may differ depending on the design scheme, but the basic concept will be similar for the majority of NMR imaging systems in the future.

1. Principles of Nuclear Magnetic Resonance Tomography

a. Nuclear magnetic resonance physics. Since NMR was discovered in 1946, it has become an indispensable analytical method and tool in chemistry and physics. Although the basic physical properties of NMR are well known and information on NMR can readily be found elsewhere, we shall discuss a few topics necessary for understanding NMR as an imaging tool.

All materials, whether organic or inorganic, contain nuclei, which are protons, neutrons, or a combination of both. Nuclei that contain an odd number of protons, neutrons, or both in combination possess a nuclear "spin" and a "magnetic moment." This situation is equivalent to the aggregation of many small magnets. In the real world many materials are composed of several nuclei and the most common nuclei with magnetic moment are ^{1}H, ^{2}H, ^{7}Li, ^{13}C, ^{23}Na, ^{31}P, and ^{127}I. Although some materials are composed of nuclei with an even number of protons and neutrons that possess no spin or magnetic moment, they often contain some nuclei with an odd number of protons or neutrons. Therefore, they are subjects of NMR imaging. For this reason, NMR is practically applicable to most solid- and liquid-phase materials. Among the many hundreds of known stable nuclei, more than 100 possess spin and magnetic moment.

When a given material is placed in a magnetic field, some of its randomly oriented nuclei experience external

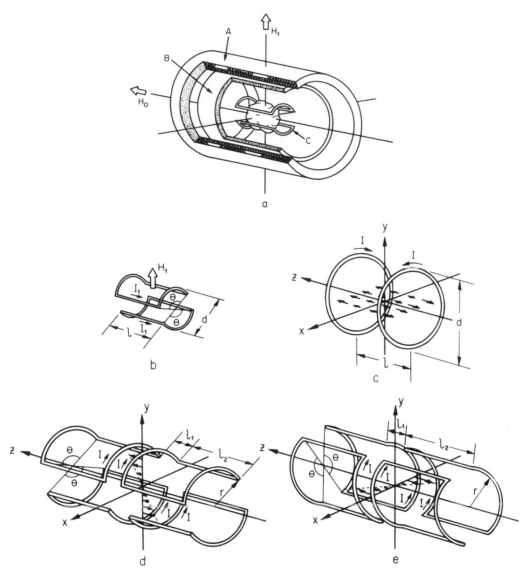

FIGURE 13 Sketch of an NMR tomograph. (a) Physical arrangement of the central part of an NMR CT system. A, Main magnet of the split-solenoidal type used in many superconducting magnets; B, gradient coil set; C, RF coil in the saddle type. (b) Saddle-shaped RF coil. Although the optimized shape for the homogeneous RF field is $\theta = 120°$ and $l/d = 2$, the shape can be changed slightly to accommodate the different sample shape or number of coil turns. (c) z-directional gradient coil. The change of the magnetic field in the z-direction is indicated by the length of the arrows. This coil is called the Maxwell pair. (d) y-directional gradient coil; often known as the Golay coil. (e) x-directional gradient coil; also of the Golay type. In (c), (d), and (e), the optimal coil shapes are $\theta = 120°$, $l/d = \sqrt{3}/2$, $l_1 = 0.78r$, and $l_2 = 2.13r$.

magnetic torque, which tends to align the nuclei in both parallel and antiparallel directions to the applied magnetic field. The fraction of magnetized nuclei in the direction parallel to the applied magnetic field is limited by thermal agitation. Therefore, it is also limited by the temperature and main magnetic field strength. Because this fraction is relatively small at room temperature, it has been a limiting factor in the sensitivity of NMR imaging. The spinning nucleus responds to the external magnetic field like a gyro-

scope precessing around the direction of the gravitational field. The rotating or precessional frequency of the spins, called the Larmor precession frequency, is proportional to the magnetic field strength.

Another important phenomenon of NMR is the creation of an energy "absorption state" (from a statistical point of view) by the applied external magnetic field. The proton has an intrinsic angular momentum or spin of $\hbar/2$, where $\hbar$ is Planck's constant divided by 2π. When proton nuclei are placed in a magnetic field, the nuclei are in two energy states $+\mu H_0$ (antiparallel) and $-\mu H_0$ (parallel to the static magnetic field H_0), where μ and H_0 are the nuclear magnetic moment and applied magnetic field, respectively (Zeeman splitting). Because at thermal equilibrium the distribution of spins in energy states follows the Boltzmann law, the lower energy state has a larger population of spins than the higher energy state. For those nuclei or protons at the $-\mu H_0$ energy state or the lower energy state whose magnetic moments are parallel to H_0, the irradiation of external electromagnetic radiation of energy E equivalent to $2\mu H_0$ tends to excite protons at the $-\mu H_0$ energy states up to the higher energy $+\mu H_0$ state. This energy is given in the form of RF magnetic field H_1. The excited protons then tend to return to their low-energy state, producing the FID signal.

Two relaxation mechanisms are associated with these excited nuclear spins: transverse, or spin–spin, relaxation and longitudinal, or spin–lattice, relaxation. It is interesting that both these relaxation times (T_1 and T_2) are sensitive to the molecular structures and environments surrounding the nuclei. For example, the mean T_1 values of normal tissues and of many malignant tissues differ substantially from one another, allowing us to differentiate malignant tissues from normal tissues in many cases (Fig. 14c). A similar tendency is observed for T_2 values. The imaging capabilities of these two important parameters, T_1 and T_2, together with the spin densities of the objects, make NMR imaging a unique, versatile, and powerful technique in diagnostic imaging. Let us now review a few of the fundamental processes involved in NMR tomographic imaging.

Although many features of NMR phenomena can be understood only by quantum mechanical considerations, a number of properties are more easily visualized by means of a classical treatment. Let us consider a magnetic moment μ in the presence of a magnetic field H_0. Figure 15a depicts the precession of proton spins in two energy states. All moments precess about H_0 at the same frequency, but without phase coherence in the x, y plane. Since the Boltzmann distribution favors the lower energy state, at equilibrium there are more nuclei aligned in the direction of H_0. The net magnetization vector M_0, which is the vector sum of μ's, is oriented along the z axis (Fig. 15b). When the net magnetization vector M_0 is at an angle θ to H_0, the net energy of the system is

$$E = -\mathbf{M}_0 \cdot \mathbf{H}_0 = -M_0 H_0 \cos\theta. \qquad (36)$$

Note that the spin system is in its lowest energy state when $\mathbf{M}_0$ is parallel to $\mathbf{H}_0$. The magnitude of the net magnetization at equilibrium is given by

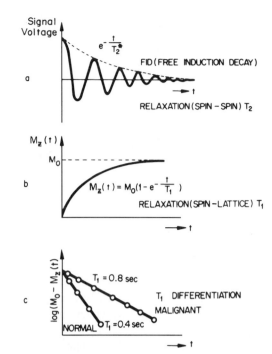

FIGURE 14 Spin–relaxation mechanisms. (a) FID signals obtained indicate a modulated decaying signal. The decay-time constant is T_2^*, which is also the effective spin–spin relaxation time. (b) Spins also decay by dissipating energy to the surroundings. This energy-dissipation mechanism is known as spin–lattice relaxation and is usually slow and decays with time constant T_1. (c) These relaxation mechanisms are expected to be used in discriminating malignant and normal tissues in NMR tomography.

$$M_0 = N(-\gamma h)^2 H_0 I(I+1)\big/3kT_0, \qquad (37)$$

where N is the number of spins, γ is the gyromagnetic ratio, I is the spin quantum number, k is the Boltzmann constant, and T_0 is the object temperature. The signal strength, which is proportional to M_0, can be increased

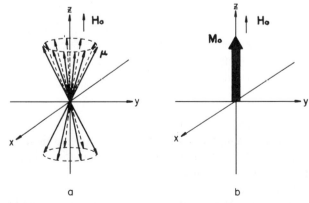

FIGURE 15 Spins in a magnetic field H_0. (a) Spins precess about H_0 in two energy states. More spins are aligned in the direction of H_0. (b) Net spin magnetization vector M_0 is given by $\mathbf{M}_0 = \Sigma\mu$. Note that, at thermal equilibrium, M_0 is along H_0.

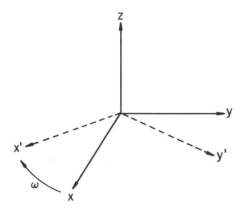

FIGURE 16 Rotating frame of reference. The coordinates (x', y', z') rotate about the z axis with the angular frequency ω. The rotating coordinates are related to the fixed coordinates (x, y, z) as $x' = x \cos \omega t + y \sin \omega t$; $y' = -x \sin \omega t + y \cos \omega t$; $z' = z$.

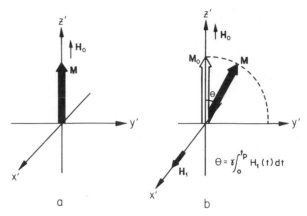

FIGURE 17 Spin magnetization in the rotating frame with and without RF pulse. (a) Spin in the absence of RF pulse. (b) Spin flip with an application of the RF field $\mathbf{H_0}$. The flipping angle θ of the magnetization is given by $\theta = \gamma \int_0^{t_p} H_1(t)dt$, where $I h_1(t)$ is the time-varying RF field intensity and t_p the length of the RF pulse. The angle θ is usually set to 90° or 180°.

by increasing the field strength H_0. Lowering T_0 would also improve the equilibrium magnetization.

Spin precession can be observed by solving the differential equation of motion (the Bloch equation), given by

$$dM_0/dt = \gamma \mathbf{M_0} \times \mathbf{H_0}. \tag{38}$$

The resulting precession of spin follows the Larmor precession frequency,

$$\omega_0 = -\gamma \mathbf{H_0}, \tag{39}$$

which is unique to each nucleus. The minus sign indicates the clockwise precession for positive γ.

In visualizing the motion of the magnetization, it is convenient to use a rotating frame of reference. Let us introduce a set of Cartesian coordinates (x', y', z') rotating about $\mathbf{H_0}$ at an angular frequency ω (Fig. 16). The magnetic field associated with this frame is called the effective magnetic field, which is given by

$$\mathbf{H_{\text{eff}}} = \mathbf{H} + \omega/\gamma. \tag{40}$$

In the absence of an RF field, $\mathbf{H} = \mathbf{H_0}$. At resonance, therefore, the fictitious field ω/γ exactly cancels $\mathbf{H}$, and $\mathbf{H_{\text{eff}}}$ becomes zero. When the static magnetic field is in the z direction and the RF field $\mathbf{H_1}$ is applied along the x' direction (in other words, $\mathbf{H_1}$ is rotating clockwise in the $x–y$ plane), the total magnetic field $\mathbf{H}$ is

$$\mathbf{H} = H_0\hat{z} + H_1(\hat{x} \cos \omega t + \hat{y} \sin \omega t), \tag{41}$$

where $\hat{x}$, $\hat{y}$, and $\hat{z}$ represent the unit vectors in the x, y, and z directions, respectively. If we insert Eq. (41) into Eq. (40), $\mathbf{H_{\text{eff}}}$ becomes

$$\mathbf{H_{\text{eff}}} = \left(H_0 - \frac{\omega}{\gamma}\right)\hat{z}' + H_1\hat{x}', \tag{42}$$

where $\hat{x}'$ and $\hat{z}'$ are the unit vectors in the x' and z' directions, respectively. At resonance ($\omega = \omega_0$), Eq. (42) can be

expressed as $\mathbf{H_{\text{eff}}} = H_1\hat{x}'$. In this case, in a rotating frame the only magnetic field is in the x' direction, and $\mathbf{M}$ precesses around the x' axis or $\mathbf{H_1}$ with frequency γH_1. For a general time-varying RF field $H_1(t)$, the flipping angle is given by

$$\theta = \gamma \int_0^{t_p} H_1(t)\,dt, \tag{43}$$

where t_p is the RF pulse duration. The application of an RF pulse, which tips the magnetization $\mathbf{M}$ into the $x–y$ plane, causes the excitation of the spin system (Fig. 17). When H_1 is applied along the x' axis for a pulse period t_p, the spin rotates or flips through an angle θ from the z' axis toward the y' axis. In general, θ is set at $\pi/2$ or π, depending on the mode of excitation and the type of NMR experiments. In the simplest case, $\theta = \pi/2$ is used to observe the maximum transverse component of magnetization.

After H_1 is turned off, the rotating magnetization induces a current into the pickup coil surrounding the object. The magnetization then relaxes, through neighboring spins and environment, to its thermal equilibrium, so that the spins realign with the original H_0 field direction. On the other hand, the transverse component of magnetization, which is related to the entropy of the system, decays through the spin–spin interaction and dephases.

In addition to the inherent spin–spin relaxation, there are other dephasing effects, such as the magnetic field inhomogeneity and field gradients. In NMR imaging, magnetic field gradients are deliberately added to resolve the spatial distribution of spin density. In fact, they produce shifts in the Larmor frequencies throughout the sample, resulting in a phase incoherency that eventually makes the composite sinusoidal signal decay more rapidly than the inherent transverse relaxation time T_2. This effective transverse relaxation time resulting from field inhomogeneity alone is expressed as

$$1/T_2^* = 1/T_2 + \gamma \Delta H/2, \qquad (44)$$

where ΔH is the field inhomogeneity, that is, the maximum deviation of magnetic field over the object region. When a field gradient is added to resolve the spatial distribution of spin density, T_2^* is further reduced to T_2^{**}, as given by

$$1/T_2^{**} = 1/T_2^* + \gamma G R, \qquad (45)$$

where G (in gauss per centimeter) is the gradient field strength and R (in centimeters) is the object diameter. The composite sinusoidal signal decaying with an effective transverse relaxation time T_2^{**} is then detected with a phase-sensitive detector. The results are similar to a decaying demodulated AM signal, as shown in Fig. 14a.

Concurrently, longitudinal or spin–lattice relaxation forces the spins to realign in the H_0 (or z) direction because it is the lowest energy state or thermal equilibrium state. Because it involves energy dissipation through the lattice, the longitudinal relaxation time T_1 is usually longer than T_2 and is related to the z component of magnetization, as stated in the following (see Fig. 14b),

$$M_z = M_0\left[1 - \left(1 - \frac{M_z'}{M_0}\right)\exp(-t/T_1)\right], \qquad (46)$$

where M_z' is the z component of magnetization at the starting time of relaxation.

The two relaxation processes work simultaneously and vary greatly depending on the characteristics of the material. In the case of tissue in field strength of 1 to 20 kG, for instance, T_1 and T_2 are of the order of 0.5 s and 50 ms, respectively; T_1 is usually larger than T_2. In Fig. 18, sequential pictures of the relaxation processes are shown.

In general, T_1, T_2, T_2^*, and T_2^{**} have the following relationship:

$$T_2^{**} \le T_2^* \le T_2 \le T_1. \qquad (47)$$

When the two relaxation mechanisms are considered, the Bloch equation can be written

$$\frac{dM_z}{dt} = \gamma(\mathbf{M}_0 \times \mathbf{H}_0)_z - \frac{M_z - M_0}{T_1}$$

$$\frac{dM_{xy}}{dt} = \gamma(\mathbf{M}_0 \times \mathbf{H}_0)_{xy} - \frac{M_{xy}}{T_2}, \qquad (48)$$

where $(\cdot)_z$ and $(\cdot)_{xy}$ represent z and x–y components, respectively. Equation (48) indicates that the magnetization components M_z and M_{xy} are independently related to the relaxation times T_1 and T_2.

Several forms of spin-echo techniques play a central and essential role in data acquisition for NMR imaging. The two basic forms of the spin-echo technique applicable to NMR imaging are the Hahn spin-echo technique and the Carr–Purcell Meiboom–Gill (CPMG) technique. In the Hahn spin-echo technique, a 90° RF pulse is applied to the direction of the x' axis, and then the magnetization vector $\mathbf{M}$ rotates to the y' axis. The spin magnetizations then dephase over time, due to field inhomogeneity or added field gradients. A subsequent 180° pulse applied along the x' axis rotates the spins around the x' axis (Fig. 19b). The spin magnetizations now continue to precess but begin to rephase. This process is equivalent to a focusing or regrowing process of the FID signal at twice the dephasing time. At this point, all the spins are completely rephased along the $-y'$ axis, but the magnetizations have now decayed by T_2 relaxation.

In the CPMG method, a 180° pulse is applied along the y' axis instead of the x' axis, so that the spin flips around

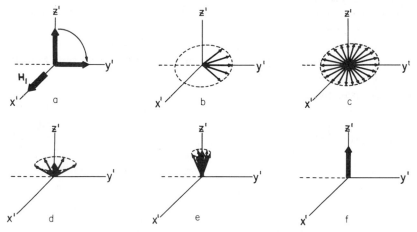

FIGURE 18 Sequential illustrations of the spin relaxation processes. (a) Spins are flipped by the RF pulse $\mathbf{H}_1$; (b) spins are dephased due to the spin–spin relaxation and field inhomogeneity; (c) FID signal decays to zero as the spins lose phase coherence; (d), (e), (f) spins relax to the original equilibrium state by the spin–lattice relaxation process.

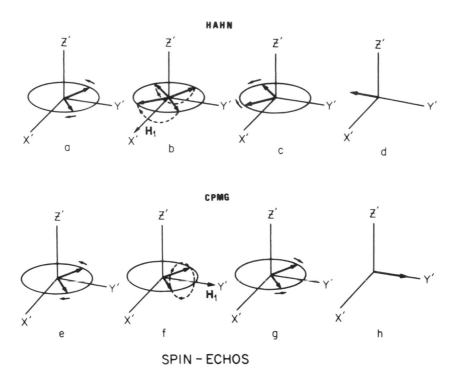

SPIN – ECHOS

FIGURE 19 Hahn spin echo: (a) Spin magnetizations dephase after 90° rotating by RF pulse; (b) 180° pulse is applied along x' axis; (c) the spins are being refocused; (d) spin echo is generated along −y' axis. Carr–Purcell and Meiboom–Gill spin echo: (e) Spins are dephased after 90° rotating by RF pulse; (f) 180° pulse is applied along y' axis; (g) the spins are being refocused; (h) spin echo is generated along y' axis.

the y' axis (Fig. 19f). Both techniques are actively used in all phases of NMR imaging to reduce several adverse effects that arise in actual data collection, such as field inhomogeneity and the effects of the gradient pulse rise time.

b. Basic theory of nuclear magnetic resonance tomography.

Conventional NMR chemistry requires a magnetic field of extreme homogeneity, in other words as uniform as possible, to reduce the frequency shift effect caused by the spatially dependent field variations. In Fourier NMR imaging, however, a field gradient or set of gradients is deliberately added to resolve the spatial distribution of spins into Fourier domain components. The basic form of signal obtained from 3-D Fourier transform NMR, which is known as FID, is expressed as

$$s(t) = M_0 \iiint\limits_{-\infty}^{\infty} f(x, y, z) \exp\left\{ i\gamma \int_0^t [xG_x(t') + yG_y(t') + zG_z(t')]\, dt' \right\} dx\, dy\, dz \quad (49)$$

where $f(x, y, z)$ is the 3-D spin density distribution and $G_x(t)$, $G_y(t)$, and $G_z(t)$ are the time-dependent field gradients along the x, y, and z axes, respectively. In Eq. (49)

the effects of T_1 and T_2 relaxation times are not included; they will be discussed in a later section. The generated FID is, in effect, a Fourier transform-domain representation of the spin density distribution. From this fundamental 3-D equation, many equations for the imaging algorithms described below can be derived.

2. Image Formation Algorithms

One of the interesting aspects of NMR imaging is that the NMR images can be formed by many different procedures, some of which are described in the following text. The discussion is limited to the techniques used most often, however.

In NMR imaging, data acquisition pulse sequences play an important role and are intimately related to the image reconstruction algorithms employed. This description of the mathematical formulations is based on the basic pulse sequence, which uses only 90° RF pulses, although in a real imaging situation the spin-echo techniques that use additional 180° RF pulses are more common.

a. Direct Fourier imaging.

i. Algorithm for direct Fourier imaging technique. This direct Fourier imaging (DFI) method was first proposed by Kumar, Welti, and Ernst (KWE). In this case, imaging can proceed through the total 3-D excitation of an object in series of time sequences. The result of the 3-D Fourier transform of those data is considered to be the 3-D spin density function or image.

In this DFI or 3-D KWE procedure, three orthogonal field gradients, G_x, G_y, and G_z, are applied in sequence after 90° RF excitation pulse at $t = 0$. The z components of the local magnetic fields are given as

$$H_z(x, y, t)$$
$$= \begin{bmatrix} H_0 + G_z z, & \text{for } 0 < t < t_z \\ H_0 + G_y y, & \text{for } t_z < t < t_z + t_y \\ H_0 + G_x x, & \text{for } t_z + t_y < t < t_z + t_y + t_x, \end{bmatrix} \quad (50)$$

where each timescale t_z and t_y is varied according to preassigned sequences, that is, $t_z \simeq t_a$, $t_y \simeq 0 \sim t_a$, where t_a is the optimal observation time of FID. The FID signal is sampled when the x gradient is applied, namely, during the t_x period. The sampled FID signal reflects the previous application of the z and y gradients by retaining the phase change caused by those gradients. For this reason, this imaging scheme is often called the phase-encoding method. The series of FID signals obtained with the various t_z's and t_y's then form a full 3-D FID signal sufficient for reconstruction of the spin density image of the entire volume.

The observed FID signal $s(t_x, t_y, t_z)$ is (neglecting the relaxation processes) expressed as

$$s(t_x, t_y, t_z) = M_0 \int\!\!\!\int\!\!\!\int_{-\infty}^{\infty} f(x, y, z) \exp[i\gamma(G_x x t_x + G_y y t_y + G_z z t_z)] \, dz \, dy \, dx. \quad (51)$$

Fourier transform of Eq. (51) results in spatial spin density function as

$$\tilde{f}(\omega_x, \omega_y, \omega_z) = \int\!\!\!\int\!\!\!\int_{-\infty}^{\infty} s(t) \exp[-i(\omega_x t_x + \omega_y t_y + \omega_z t_z)] \, dt_z \, dt_y \, dt_x. \quad (52)$$

The reconstructed image $\tilde{f}(\omega_x, \omega_y, \omega_z)$ is related to $f(x, y, z)$ as

$$\tilde{f}(\omega_x, \omega_y, \omega_z) = k f(\gamma x G_x, \gamma y G_y, \gamma z G_z), \quad (53)$$

where k is a constant.

Figure 20 shows RF and gradient pulse sequences of conventional DFI using spin echo for one-slice 2-D imaging. In this scheme, although the whole volume of an object is excited by the 90° RF pulse, only the spins in a designated slice are rephased to form an echo through the application of a narrowband 180° RF pulse and

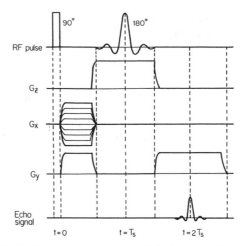

FIGURE 20 Imaging sequences of 2-D direct Fourier imaging. The slice in the z direction is selected and spin echo is used. While G_y remains constant, the intensity of G_x is varied for phase encoding. The purpose of the first part of the G_y gradient pulse is to dephase spins in the object after the nonselective 90° RF pulse. Only the spins in a designated slice are rephased by the selective 180° pulse and form the spin echo at $t = 2T_s$ on the second constant G_y gradient. The spin-echo signal is collected for image reconstruction.

z-directional selection gradient. Note that the x-directional phase encoding is achieved by varying the amplitude of the x gradient instead of varying the time interval.

ii. Time-multiplexed multislice imaging method. By using consecutive excitations of the pulse sequence shown in Fig. 20 within a suitable repetition time period, it is possible to obtain several images simply through the use of the remaining time, provided that the sum of data acquisition time of each slice is smaller than the repetition time. For example, a data acquisition time of less than 100 ms with a repetition time of 300 ms would allow three consecutive slice imagings without T_1 relaxation constraints. This multislice imaging method is a frequently used NMR imaging technique. It can also be applied to other 2-D imaging schemes, such as the line-integral projection reconstruction (LPR) technique, which is described in the following section.

b. Line-integral projection reconstruction.

i. Basic principles. Projection reconstruction using 2-D and 3-D image reconstruction algorithms is well known, especially in the areas of X-ray CT and radionuclide emission tomography, as previously discussed. Although the image can be reconstructed in several different ways, the basic forms of data collection are similar. Line-integral projection data are obtained in angular steps, by rotating the object a total of either 180° or 360°. The most familiar and convenient way to reconstruct 2-D or 3-D images is through the Fourier convolution method, which can be summarized as follows. The reconstructed 2-D image $f(x, y)$ is given by

$$f(x, y) = \int_0^\pi [p_\phi(x') * h(x')] \, d\phi, \qquad (54)$$

where $p_\phi(x')$ is projection data, $h(x')$ is the filter kernel that corrects $1/r$ blurring caused by circular symmetric linear superposition, and (x', y') is the coordinate system rotated by an angle ϕ from the original coordinates (x, y).

In Fourier transform NMR, the nuclear signal can be considered to be the inverse Fourier transform of the spatial domain spin density function. If a plane at $z = z_0$ is selected, then the FID at an angular view ϕ can be expressed as

$$s_\psi(t) - M_0 \iint\limits_{-\infty}^{\infty} f(x', y'; z_0) \exp[i\gamma x' G_{x'} t] \, dy' \, dx'.$$

$$(55)$$

Although the FID signal $s_\phi(t)$ appears in the time domain, it represents the Fourier domain projection data. Therefore, the projection data $p_\phi(x')$ are obtained through the Fourier transform of the FID signal as

$$p_\phi(x') = \mathcal{F}[s_\phi(t); t \to x']. \qquad (56)$$

The basic form of projection data obtainable in Fourier transform NMR is similar to the data obtained in X-ray CT. In Fig. 21, spin-echo signals or FIDs are obtained at different angular views through the application of the field gradient and RF excitation sequences. As a first step, all

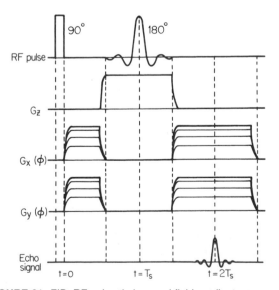

the spins in the sample are excited with a 90° RF pulse, and a subsequent 180° RF selects the slice. After spins are refocused they generate the spin-echo or FID signal (Fig. 21). After 180° or 360° rotation of projection with an appropriate step through the adjustment of the field gradients G_x and G_y, a complete projection data set sufficient for reconstruction of a slice at a given plane z_0 is obtained. At this point, 2-D image reconstruction can proceed according to Eq. (54); that is, each echo or FID signal $s_\phi(t)$ is Fourier-transformed, convolved with a filter kernel, and backprojected.

ii. Slice (plane)-encoded multislice LPR. The single-slice line-integral projection technique explained earlier can be extended to achieve multislice imaging through several encoding techniques, for example, the plane-encoding technique explained in the following paragraphs.

Let us assume that the number of planes is n. For the data set at a view ϕ_i, the same G_{xy} and G_z are applied n times, each with a different frequency composition of RF pulses. The RF pulses are specially tailored to assign desired phases to the designated slices. To obtain a complete set of view data corresponding to the n planes, the acquisition of data is repeated n times with differently composed RF pulses.

The key to this method lies in the encoding of signals according to the RF pulse sequence. A simple illustration of the encoding procedure using a coding matrix is as follows. Let the FIDs obtained at each 180° composite RF pulse sequence be $S_{\phi_0}^1(t), S_{\phi_0}^2(t), \ldots, S_{\phi_0}^n(t)$. Each FID is a composite of the line-integral projection sets, which include data from several planes at an angular view ϕ_0, that is, $s_{\phi_0 z_0}(t), s_{\phi_0 z_1}(t)$, and so on. Therefore, composite FIDs, $S_{\phi_0}^1(t), S_{\phi_0}^2(t), \ldots, S_{\phi_0}^n(t)$ can be given as

$$\mathbf{S}_{\phi_0}(t) = \begin{bmatrix} S_{\phi_0}^1(t) \\ S_{\phi_0}^2(t) \\ \vdots \\ S_{\phi_0}^n(t) \end{bmatrix} = [H_n] \begin{bmatrix} s_{\phi_0 z_0}(t) \\ s_{\phi_0 z_1}(t) \\ \vdots \\ s_{\phi_0 z_{n-1}}(t) \end{bmatrix}. \qquad (57)$$

From Eq. (57), the desired FID signal $s_{\phi_i z_i}$, which corresponds to the FID of slice z_i, can be obtained through matrix inversion.

Examples of coding matrices include the Hadamard matrix and the Fourier matrix. The advantage of this method is the statistical improvement gained as a result of the increase in total scanning time.

c. Planar-integral projection reconstruction. In the planar-integral projection reconstruction (PPR) method, both nonselective broadband 90° and spin-echo 180° pulses are applied, thus providing FID data that

FIGURE 21 FID, RF pulse timing, and field gradient sequences for the basic single-slice line-integral projection reconstruction. Gradient pulse sequence: $G_x(\phi) = G\cos\phi$; $G_y(\phi) = G\sin\phi$; $0° \leq \phi < 180°$ or $0° \leq \phi < 360°$, where G is the maximum value of the reading gradient.

originate from the entire volume. The FID signal of the total volume planar-integral projection data in the z' direction can be expressed as

$$s(t) = M_0 \int_{-\infty}^{\infty} \left[\iint_{-\infty}^{\infty} f(x', y', z') \, dx' \, dy' \right]$$
$$\times \exp[i\gamma(H_0 + z'G_{z'})t] \, dz', \qquad (58)$$

where (x', y', z') are the rotated coordinates of (x, y, z) and the z' direction coincides with the projection direction. Through phase-sensitive detection, the FID signal obtained can be written

$$s_{\theta,\phi}(t) = M_0 \int_{-\infty}^{\infty} p_{\theta,\phi}(z') \exp(i\gamma z'G_{z'}t) \, dz', \qquad (59)$$

where $p_{\theta,\phi}(z')$ is the planar-integral projection data with the angular view (θ, ϕ).

In Eq. (59), $s_{\theta,\phi}(t)$ represents the projection data in the Fourier domain. Therefore, the Fourier transform of $s_{\theta,\phi}(t)$ is spatial domain planar-integral projection data with which reconstruction can be performed (see the PPR algorithm in Section II.A).

d. Echo-planar imaging method. One of the inherent disadvantages of NMR CT is the long data acquisition time caused by constraints on spin–lattice relaxation time. The echo-planar imaging method is one of the faster imaging techniques; imaging time can be reduced to as short as 50 ms.

In this method, the FID data in the spatial frequency domain is acquired following one simple excitation through the application of the oscillating gradient during the signal reception. The image can be reconstructed through a simple 1-D Fourier transform of the obtained FID data. This method is limited by gradient driving power and speed. Nevertheless, the echo-planar technique itself or variations of the technique seem to be potential candidates for future fast imaging techniques in NMR CT.

3. Imaging Modes and Extraction of Nuclear Magnetic Resonance Parameters

The spin density $f(x, y, z)$ obtained by the various imaging methods described earlier is not a real spin density; it is weighted by T_1 or T_2 or both. Because T_1 or T_2 varies between normal and abnormal tissues, the image of spin density weighted by T_1 or T_2 has been found to be clinically useful. With this in mind, several attempts have been made to extract T_1 information, as well as spin density and T_2. The typical imaging modes and corresponding terminologies currently in use are described in the following subsections.

a. Imaging modes.

i. Saturation recovery imaging. The saturation recovery method involves simply repeating the pulse sequence at regular intervals T. The equations discussed previously are unchanged except for the replacement of $f(x, y, z)$ with $f'(x, y, z)$, which is expressed as

$$f'(x, y, z) = f(x, y, z)\{1 - \exp[-T/T_1(x, y, z)]\}. \qquad (60)$$

Note that f' is now a function of both T_1 and f.

ii. Inversion recovery imaging. Inversion recovery is similar to saturation recovery, except that the 180° RF pulse precedes the 90° RF pulse with a time interval of T_1; $f'(x, y, z)$ is related to $f(x, y, z)$ as

$$f'(x, y, z) = f(x, y, z)\{1 - 2\exp[-T_1/T_1(x, y, z)]\}. \qquad (61)$$

It is easy to see the increased dependency of the image on T_1 over that obtained in saturation recovery. This technique is often used for measuring T_1 values in tissues. Figure 22 shows the pulse sequences for 2-D inversion recovery Fourier imaging.

iii. Spin-echo imaging. Through the application of the 180° pulse following the first 90° pulse at $t = T_s$, spins are refocused at $t = 2T_s$ by the spin echo (Fig. 20). Although the spins are now refocused and coherent, the amplitude of FID decays exponentially with time constant T_2. The decayed spin density $f'(x, y, z)$ can be written

$$f'(x, y, z) = f(x, y, z) \exp[-2T_s/T_2(x, y, z)]. \qquad (62)$$

As explained earlier, the image is now weighted by T_2 as well as by T_1. By setting the appropriate T_s values, images

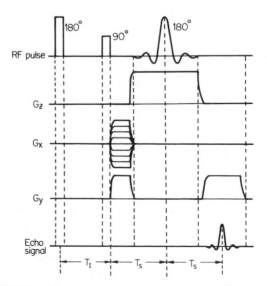

FIGURE 22 RF and gradient pulse sequences of inversion recovery direct Fourier imaging.

weighted mainly by T_2 can be obtained, provided that the repetition time is sufficiently large.

b. Parameter imaging methods.
The capability of extracting many functional parameters is one of the most important advantages of NMR CT. Flow velocity, T_1, T_2, and chemical shift are some of the interesting parameters in NMR imaging that are discussed in this section.

i. T_1 (Spin–lattice relaxation time) and T_2 (Spin–spin relaxation time). The effects of T_1 and T_2 are closely related to the NMR imaging modes. In T_1 imaging, both the saturation recovery and inversion recovery modes can be used. By varying the recovery time and observing the resulting image intensity variation, one can deduce T_1 values. Similarly, by changing the echo time, that is, varying $2T_s$ in Eq. (62) for the spin-echo method, one can obtain several images differently weighted by T_2. From the images obtained with different echo times, T_2 values of each pixel can be calculated.

ii. Flow imaging. In NMR CT, one can also measure the flow or moving velocity of nuclear spins through observation of the FID signal. In the first attempt at flow velocity measurement two RF coils were used—one for the excitation of spins and the other for reception. In this experiment, surface RF coils were used to excite and receive the signal at known locations. If the maximum signal is received at Δt seconds after the excitation with the distance Δl between two RF coils, the velocity can be estimated by $\Delta l / \Delta t$.

Several flow imaging methods have been developed. Among these, two techniques relevant to general flow measurement will be discussed: one using density information and another using phase information.

The RF and gradient pulse scheme of flow imaging using the selective saturation method uses intensity information. In this scheme, the first 90° RF pulse and the homogeneity-spoiling gradients are used to saturate the spins in the selected slice for flow imaging. The 2-D Fourier imaging sequence for the same slice follows after Δt seconds to measure the signals originating from spins that flowed in from outside the slice, where spins were not saturated. From the density change observed for several different Δt's, the flow velocity in the selection gradient direction can be determined as $\Delta z / \Delta T$, where ΔT is the minimum Δt with the maximum spin density and Δz is the slice thickness.

In another variation of flow imaging, phase information is used to measure flow velocity. Since the pixel values of an image are usually extracted by taking the real part or absolute values of the image data in complex form, it is possible to use the phase information associated with each pixel data. Let us assume that a time-varying gradient $G_x(t)$ is applied to moving spins after RF excitation. The phase

coding resulting from the time-varying gradient can be divided into two terms: the spatially coded term ϕ_s and the velocity-coded term ϕ_v, respectively. The sum appears as

$$\phi = \phi_s + \phi_v, \tag{63}$$

where

$$\phi_s = \gamma \int G_x(t)x \, dt$$

and

$$\phi_v = \gamma \int G_x(t)vt \, dt.$$

In Eq. (63), $G_x(t)$, x, and v represent the time-dependent x gradient, the x coordinate of spins at $t = 0$, and the x-directional flow velocity of the moving spins, respectively. In the flow measurement, the flow coding gradient is applied in addition to the conventional RF and gradient pulse sequences, so that the phases on the final image are changed only as a result of flow velocity. Flow velocity can be determined from the calculated phase, which is coded according to the velocity of the spins. The unique advantage of this kind of flow velocity imaging method is the capability of multidirection flow imaging by simply applying the additional flow coding gradient in the desired direction. Figure 23 shows a typical gradient waveform for flow phase coding in the x direction, in which the spatially coded phase term is canceled so that $\phi_s = 0$, while the velocity-coded phase term remains $\phi_v \neq 0$. This technique, therefore, allows us to measure velocity by simply measuring the phase, which is now purely dependent on velocity.

iii. Chemical-shift imaging. Another important aspect of NMR CT is its spectroscopic imaging capability. Before NMR CT was proposed, NMR had been

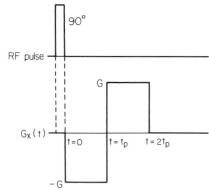

FIGURE 23 Gradient waveform for phase coding of flow velocity measurement. Note that the gradient pulsing shown effectively cancels out the spatial coding. The remaining velocity-coded phase ϕ_v can be written $\phi_v = \gamma t_p^2 Gv$, where v is the flow speed.

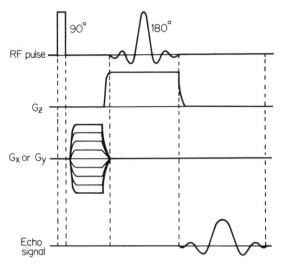

FIGURE 24 RF and gradient pulse sequences for 4-D chemical shift imaging using 2-D spatial codings. Note that the total number of coding steps for a slice is N^2.

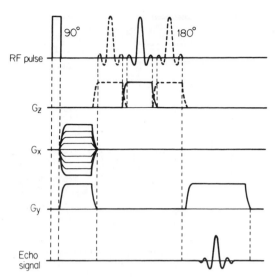

FIGURE 25 RF and gradient pulse sequences for the echo-time-encoded chemical-shift imaging. Note here that the total number of coding steps for a slice is N.

used primarily for chemical spectroscopy, in which the frequency spectrum of a specific kind of nuclei distributed over a few tens of parts per million of its Larmor precession frequency was obtained. The chemical shift was usually measured for homogeneous samples under the condition of uniform field.

In spectroscopic NMR imaging, however, the chemical spectrum for each pixel (chemical spectroscopic imaging) is to be measured neither for the homogeneous samples nor under the uniform field condition but with spatially varying gradient pulses.

A few chemical-shift imaging techniques have been proposed. An original spectroscopic imaging pulse sequence is shown in Fig. 24. The essence of this pulse scheme is the absence of the reading gradient during data acquisition. In this scheme, N^2 steps are required for a 2-D spectroscopic imaging of $N \times N$ matrix size image. In Fig. 25, a more generalized imaging sequence using echo-time encoding is shown. In this scheme, the 180° spin-echo RF pulses are applied several times, and corresponding FIDs are observed at each time. The notable difference between this scheme and the former is that here the spatial coding is identical to conventional 2-D imaging (i.e., gradient steps required are only N for $N \times N$ matrix size image), while in the former, the number of steps required is N^2. Also in the former, N determines the spectroscopic resolution. In the latter scheme, on the other hand, the number of RF time positions determines the spectroscopic resolution and, therefore, by varying the number of RF time positions, one can achieve the desired resolution. Often this step turns out to be much less than N in most *in vivo* spectroscopic imaging.

c. Other imaging methods. In addition to the imaging methods previously mentioned, there are several other imaging schemes of special form. One of these is gated or synchronized imaging, for an object that moves periodically. An example is the gated cardiac imaging of the human heart. In this case, the RF and the gradient pulse sequences are gated in synchronization with the ECG signals, and data in the different parts of the heart cycle are collected.

In the area of imaging methodology, rotating-frame zeugmatography should be noted. In this method, spatial-phase coding is achieved through the RF field gradient rather than spatial field gradients generated by the x-, y-, and z-directional gradient coils as in conventional NMR imaging. Although this method has some advantages, it is rarely used in imaging because of inherent difficulties, such as those found in the realization of the RF field gradient and associated RF coils.

4. System Configuration

The whole NMR CT system can be divided into three parts: the NMR section, the electronics, and the computer. The NMR section includes the main magnet, which provides the static main magnetic field; the gradient coil for generating magnetic field gradients; and the RF coil, which transmits and receives the RF signals. The electronic part includes a waveform synthesizer, a data acquisition component, and transmitter and receiver amplifiers. The computer performs image data processing and system control and finally displays the reconstructed images.

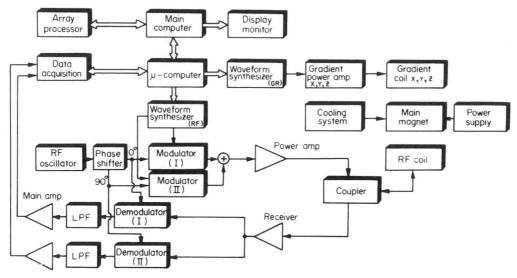

FIGURE 26 Block diagram of a typical NMR CT system.

A block diagram of a typical NMR CT system is illustrated in Fig. 26, which depicts the operation of each part. The main computer generates RF and gradient waveforms and reconstructs images after data acquisition. RF and gradient pulse waveforms generated in the main computer are transferred to a microcomputer and then to a waveform synthesizer, where data in digital form are converted to analog form. The gradient is applied to x-, y-, and z-gradient coils after being amplified in the gradient power amplifier. The RF waveform from the waveform synthesizer is modulated with the RF (reference) signal in the modulator, amplified through the power amplifier, and transferred to the RF coil via the coupler. The coupler circuit effectively switches on and off between the transmitting and receiving operations. The transmitted RF pulse excites nuclear spins in the sample. The nuclear signal induced on the RF coil by precessing spins is transferred to the receiver amplifier through the coupler. The amplified nuclear signal is demodulated with the RF reference signal and sent to the data acquisition part. Acquired nuclear signals (FIDs or echo signals) are transferred to the main computer via the microcomputer and are used for the reconstruction of the image. NMR CT systems often employ array processors for rapid image reconstruction. After reconstruction, the images are displayed on a cathode ray tube.

IV. RECENT DEVELOPMENTS

Among the many important recent developments in NMR imaging, the most notable one is the high-speed imaging using small flip angle gradient echo technique such as the steady-state free precession (SSFP) technique. This technique enables us to obtain high-quality images within a few seconds, compared to the conventional spin-echo technique, which usually requires an average of minutes or so for the imaging of a slice.

A typical pulse sequence for the fast gradient echo technique known as SSFP is illustrated in Fig. 27. The unique feature of the method, in addition to the speed advantage, is the potential of obtaining two characteristically different image data simultaneously, namely FID and echo. Here the FID image refers to the T_1 weighted image while echo image refers to the T_2 weighted image. Although the gradient echos and SSFP techniques generally suffer from

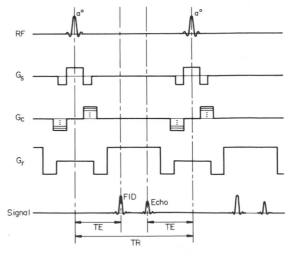

FIGURE 27 Imaging pulse sequence of the fast gradient echo imaging or its variation known as SSFP (steady-state free precession) imaging technique. With this pulse sequence, both the FID and the echo images, each of which has a characteristic contrast (namely, T_1 contrast in the FID image and strong T_2 contrast in the echo image) can be obtained simultaneously.

the susceptibility artifact, the methods nevertheless are becoming more widely used because of the advantages of speed and good image quality.

SEE ALSO THE FOLLOWING ARTICLES

IMAGE-GUIDED SURGERY • IMAGE PROCESSING • MAGNETIC RESONANCE IN MEDICINE • NONDESTRUCTIVE TESTING • NUCLEAR MAGNETIC RESONANCE (NMR) • RADIONUCLIDE IMAGING TECHNIQUES, CLINICAL • X-RAY ANALYSIS

BIBLIOGRAPHY

Barrett, H. H., and Swindell, W. (1981). "Radiological Imaging: The Theory of Image Formation, Detection, and Processing," Academic Press, New York.

Bushong, S. (2000). "Essentials of Medical Imaging: Computed Tomography," McGraw-Hill Professional, New York.

Carson, R. E. (1998). "Quantitative Functional Brain Imaging with Positron Emission Tomography," Academic Press, San Diego.

Cho, Z. H. (ed.) (1974). "Special Issue on Physical and Mathematical Aspects of 3-D Image Reconstruction," IEEE Trans. Nucl. Sci. NS-21, No. 2. Inst. Electr. Electron. Eng., New York.

Cho, Z. H. (ed.) (1976). "Special Issue on Advances in Picture Reconstruction Theory and Applications," Comput. Biol. Med. Vol. 6, No. 4. Pergamon Press, Oxford.

Cho, Z. H., and Nalcioglu, O. (eds.) (1984). "Special Issue on Physics and Engineering in Nuclear Magnetic Resonance Imaging," IEEE Trans. Nucl. Sci. NS-31, No. 4. Inst. Electr. Electron. Eng., New York.

Gardner, R. J. (1995). "Geometric Tomography," Cambridge Univ. Press, Cambridge, UK.

Herman, G. T. (1980). "Image Reconstruction from Projection," Academic Press, New York.

Herman, G., and Kuba, A. (1999). "Discrete Tomography: Foundations, Algorithms, and Applications," Birkhauser Boston, Cambridge, Massachusetts.

Kimmich, R. (1997). "NMR: Tomography, Diffusometry, Relaxometry," Springer-Verlag, Berlin/New York.

Macovski, A. (1983). "Medical Imaging Systems," Prentice-Hall, Englewood Cliffs, New Jersey.

Mansfield, P., and Morris, P. G. (1982). "NMR Imaging in Biomedicine," Academic Press, New York.

Nalcioglu, O., and Cho, Z. H. (eds.) (1984). "Selected Topics in Image Science," Springer-Verlag, Berlin/New York.

Newton, T. H., and Potts, D. G. (eds.) (1981). "Radiology of the Skull and Brain: Technical Aspects of Computed Tomography," Vol. 5. Mosby, St. Louis, Missouri.

Partain, C. L., James, A. E., Rollo, F. D., and Price, R. R. (1983). "Nuclear Magnetic Resonance (NMR) Imaging," Saunders, Philadelphia.

Yeon, K. M., Li, G.-Z., and Wol, C. (1995). "Computed Tomography: State of the Art and Future Applications," Springer-Verlag, Berlin/New York.

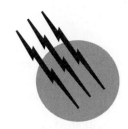

Topology, General

Taqdir Husain

McMaster University

GLOSSARY

Closure $\bar{A} = A \cup A'$ is called the closure of A, where A' is the set of all limit points of A.

Compact A topological space X is called compact if in each open covering of X there is a finite subcollection that covers X.

Connected A topological space is said to be connected if it cannot be written as the union of two disjoint nonempty open sets.

Continuous map A map f of a topological space X into a topological space Y is called continuous if for each open set Q of Y, $f^{-1}(Q) = \{x \in X : f(x) \in Q\}$ is open in X.

Covering A family $\{P_\alpha\}$ of sets is said to be a covering of $A \subset X$ if $A \subset \bigcup_\alpha A_\alpha$. If all P_α's are open, it is called an open covering.

Homeomorphism A continuous, open, and bijective map of a topological space X into a topological space Y is called a homeomorphism.

Limit point An element a is said to be a limit or accumulation point of a set A if each open set containing a contains at least one point of A other than a.

Metric d is said to define a metric for a set X if for all pairs (x, y), $x, y \in X$, $d(x, y)$ is a real number satisfying the following:

   (i) $d(x, y) \geq 0$.
   (ii) $d(x, y) = 0$, if $x = y$.
   (iii) $d(x, y) = 0$ implies $x = y$.
   (iv) $d(x, y) = d(y, x)$.
   (v) $d(x, y) \leq d(x, z) + d(z, y)$ (triangle inequality).

Neighborhood A set U is a neighborhood of a point

a if there exists an open set P such that $a \in P \subset U$.

Open set Each member of a topology $\mathcal{T}$ on a set X is called an open set.

Set A collection of "distinguished" objects is called a set.

Topological group A group G endowed with a topology is called a topological group if the map $(x, y) \to xy^{-1}$ of $G \times G$ into G is continuous.

GENERAL TOPOLOGY is a branch of mathematics that deals with the study of topological spaces and maps, especially continuous maps. This subject is regarded as a natural extension of part of classical analysis (a major field of mathematics) insofar as the study of continuous maps is concerned. Classical analysis deals primarily with functions on the real line $\mathbb{R}$ or, more generally, *n*-dimensional Euclidean spaces $\mathbb{R}^n$, in which the notions of limit and continuity are very basic. With the help of topology, these concepts can be studied more generally in topological spaces. In short, the genesis of the subject general topology is rooted in the fundamental properties of $\mathbb{R}$ or $\mathbb{R}^n$ and functions thereon.

The reader is cautioned not to confuse "topology" with "topography" and "map" with a "map" of a country. For an explanation of basic set-theoretic concepts used herein, see Table I.

TABLE I   **Symbols and Set-Theoretic Notations**

Set	A collection of distinguished objects: capital letters denote sets in the text
$\varnothing$	Null or empty set
$a \in A$	a is an element of the set A
$A \cup B$	Union of A and B
$A \cap B$	Intersection of A and B
$A \cap B = \varnothing$	A, B are disjoint
A^c	Complement of A
$A \backslash B$	The set of elements in A that are not in B
$\bigcup_{i=1}^{n} A_i$	(Finite) union of sets $A_1, A_2, \ldots, A_n$
$\bigcap_{i=1}^{n} A_i$	(Finite) intersection of sets $A_1, \ldots, A_n$
$\bigcup_{i=1}^{\infty} A_i$	Countable union of sets $\{A_1, A_2, \ldots\}$
$\bigcap_{i=1}^{\infty} A_i$	Countable intersection
$\bigcup_{\alpha \in \Gamma} A_\alpha$ or $\bigcup_\alpha A_\alpha$	(Arbitrary) union of a family $\{A_\alpha\}_{\alpha \in \Gamma}$ of sets
$\bigcap_{\alpha \in \Gamma} A_\alpha$ or $\bigcap_\alpha A_\alpha$	Arbitrary intersection
$f : X \to Y$	f is a map of X into Y
iff (or $\Leftrightarrow$)	If and only if (or implies and implied by)
$\mathbb{R}$ (or $\mathbb{C}$)	The set of all real (or complex) numbers
$\sum_{i=1}^{n} a_i$	$a_1 + a_2 + \cdots + a_n$
De Morgan's laws	$(\bigcup_\alpha A_\alpha)^c = \bigcap_\alpha A_\alpha^c$ and $(\bigcap_\alpha A_\alpha)^c = \bigcup_\alpha A_\alpha^c$

I. TOPOLOGY AND TOPOLOGICAL SPACES

A. Basic Notions of Topology and Examples

Let X be a set. A collection $\mathcal{T} = \{U_\alpha\}$ of subsets of X is said to define or is itself called a *topology* if the following axioms hold:

(i) $X, \varnothing$ belong to $\mathcal{T}$.

(ii) Finite intersections of elements in $\mathcal{T}$ are also in $\mathcal{T}$.

(iii) Arbitrary unions of elements in $\mathcal{T}$ are also in $\mathcal{T}$.

A set X endowed with a topology $\mathcal{T} = \{U_\alpha\}$ is called a *topological space* (TS) and is often denoted $(X, \mathcal{T})$. Each member of $\mathcal{T}$ is called an *open* set or $\mathcal{T}$-*open* (if the topology is to be emphasized). Thus, by definition the whole set X and the null set $\varnothing$ are open.

If the topology $\mathcal{T}$ consists of exactly two open sets $X, \varnothing$, then the topology is termed *indiscrete*. If, however, $\mathcal{T}$ consists of all subsets of X, it is called the *discrete* topology.

Let $\mathcal{T}, \mathcal{T}'$ be topologies on a set X. If each $\mathcal{T}'$-open set is also $\mathcal{T}$-open, then $\mathcal{T}$ is said to be *finer* than $\mathcal{T}'$ or, equivalently, $\mathcal{T}'$ is *coarser* than $\mathcal{T}$. Thus, the discrete topology is finer than the indiscrete one. Actually, the discrete topology is the *finest*, whereas the indiscrete one is the weakest or *coarsest* of all topologies on X.

A set C of a topological space $(X, \mathcal{T})$ is called *closed* or $\mathcal{T}$-*closed* (when the topology is to be emphasized) if the complement $C^c = X \backslash C$ of C is open. By De Morgan's laws (see Table I), arbitrary intersections and finite unions of closed sets are closed, since by definition finite intersections and arbitrary unions of open sets are open.

A subset P of a topological space $(X, \mathcal{T})$ is called a *neighborhood of a point* $x \in X$ if there is an open set U such that $x \in U \subset P$. If P itself is an open or a closed set, it is called an *open* or *closed neighborhood* of x. The collection $\mathcal{N}_x$ of all neighborhoods of x is called the *neighborhood system* at x. The following properties hold: $x \in U$ for all U in $\mathcal{N}_x$; for all U, V in $\mathcal{N}_x$, $U \cap V \in \mathcal{N}_x$; if W is any subset of X such that for some $U \in \mathcal{N}_x$, $U \subset W$, then $W \in \mathcal{N}_x$; and each U in $\mathcal{N}_x$ contains an open neighborhood of x.

Let A be a subset of a topological space $(X, \mathcal{T})$. The union of all open sets contained in A is called the *interior* of A and is denoted A^0. Clearly, $A^0 \subset A \cdot A = A^0$ iff (if and only if) A is open.

A point $x \in X$ is called a *limit* or *accumulation point* of a set $A \subset X$ if each neighborhood of x contains points of A other than x. The set of all limit points of a set A is usually denoted A' and is called the *derived set* of A. The set $\bar{A} = A \cup A'$ is called the *closure* of A. Closed sets can be characterized by closures as follows. A set A is closed iff

$A = \bar{A}$. In particular, each closed set contains all its limit points. In general, however, $A \subset \bar{A}$.

Properties of interior and closure operations are complementary and comparable: $A^0 \subset A$ (respectively, $A \subset \bar{A}$); $A^{00} = A^0$ (respectively, $\bar{\bar{A}} = \bar{A}$); and $A^0 \cap B^0 = (A \cap B)^0$ (respectively, $\bar{A} \cup \bar{B} = \overline{A \cup B}$).

The set $\partial A = \bar{A} \cap \overline{A^c}$ is called the *boundary* of A. It immediately follows that A is open (respectively, closed) iff $A \cap \partial A = \varnothing$ (respectively, $\partial A \subset A$).

A subset A of a topological space $(X, \mathcal{T})$ is said to be *dense* in X if $\bar{A} = X$. Since X is always closed, $\bar{X} = X$ implies that the whole space X is always a dense subset of itself. However, if the subset A is genuinely smaller than X, the topological space assumes some special features. A noteworthy case is the following. If A is a countable dense subset of X, then $(X, \mathcal{T})$ is called a *separable* topological space.

In practice, the collection of all open sets in a topological space is very huge. To cut it down to a smaller size without changing the topology, the notion of a base of a topology is useful. A subcollection $\mathcal{B}$ of open sets in a topological space $(X, \mathcal{T})$ is said to form a *base* of the topology if each open set of X is the union of sets from $\mathcal{B}$. To cut the size of $\mathcal{B}$ down even farther without changing the topology, one introduces the following concept. A subcollection $\mathcal{B}'$ of open sets in a topological space $(X, \mathcal{T})$ is said to form a *subbase* of the topology if the collection of all finite intersections of elements form $\mathcal{B}'$ forms a base of the topology. One notes the following facts. A collection $\mathcal{B}$ of open sets forms a base of the topology $\mathcal{T}$ iff for each $x \in X$ and each neighborhood U of x there is an element B in $\mathcal{B}$ such that $x \in B \subset U$. Furthermore, any nonempty family $\mathcal{B}'$ of subsets of a given set X such that $\mathcal{B}'$ covers X defines a unique topology on X or is a subbase of a unique topology.

If a topology $\mathcal{T}$ on X has a countable base, then $(X, \mathcal{T})$ is said to be a *second countable* (SC) space, but if each $x \in X$ has a *countable base* $\{U_n(x)\}$ of *neighborhoods*, that is, each neighborhood of x contains some $U_n(x)$, then the topological space is said to be a *first countable* (FC) space. It is clear that each second countable space is first countable, but the converse is not true. Moreover, each second countable space is separable. Again, the converse is not true. However, for a special class of topological spaces called metric space, which is introduced in Section V.A, the converse does hold. To ascertain the existence of limit points of certain subsets in a topological space, one has the following. Each uncountable subset of a second countable space has a limit point.

To make the above-mentioned abstract notions more concrete, we consider some examples. In the set $\mathbb{R}$ of all real numbers, for $a, b \in \mathbb{R}$, the subsets $(a, b) = \{x \in \mathbb{R} : a < x < b\}$ and $[a, b] = \{x \in \mathbb{R} : a \leq x \leq b\}$ are called open and closed intervals, respectively. A subset U of $\mathbb{R}$ is called open if for each $x \in U$ there is an open interval (a, b) such that $x \in (a, b) \subset U$. The collection of all open sets so designated defines a topology on $\mathbb{R}$, sometimes called the *natural* or *Euclidean topology* of $\mathbb{R}$. Clearly, each open (respectively, closed) interval of $\mathbb{R}$ is an open (respectively, closed) set. Thus, the collection of all open intervals forms a base of the natural topology of $\mathbb{R}$. Furthermore, it is easy to see that the collection of all half-open lines, $(a, \infty) = \{x \in \mathbb{R} : a < x\}$ and $(-\infty, b) = \{x \in \mathbb{R} : < b\}$, forms a subbase of the natural topology. If a, b run over rational numbers of $\mathbb{R}$, the collection $\{(a, b)\}$ of all these open intervals being a countable base of the natural topology of $\mathbb{R}$ makes $\mathbb{R}$ a second countable space. Since the set $\mathcal{Q}$ of rational numbers is a countable dense subset of $\mathbb{R}$, $\mathbb{R}$ is separable.

Apart from the natural topology of $\mathbb{R}$, as mentioned above, there are many other topologies, among which are the discrete and indiscrete. In the discrete case, each subset of $\mathbb{R}$ is open, in particular, each singleton (consisting of a single point) is open. Certainly, however, no singleton is open in the natural topology of $\mathbb{R}$. Thus, it follows that the discrete topology of $\mathbb{R}$ is strictly finer than the natural topology. $\mathbb{R}$ with the discrete topology is not separable and hence not second countable. Since for each x, $\{x\}$ forms a countable base of the neighborhood system at x, $\mathbb{R}$ with the discrete topology is first countable, but $\mathbb{R}$ with the indiscrete topology is trivial, because each singleton $\{x\}$ $(x \in \mathbb{R})$ is dense in $\mathbb{R}$.

Another important example of a useful topological space is the *n-dimensional* Euclidean real (or complex) space $\mathbb{R}^n$ (or $\mathbb{C}^n$) which is the set of all *n*-tuples $(x_1, x_2, \ldots, x_n)$ in which all $x_1, \ldots, x_n$ are real (respectively, complex) numbers. If $n = 1$, then $\mathbb{R}^1$ coincides with $\mathbb{R}$; for $n = 2$, $\mathbb{R}^2$ is called the *Euclidean plane*; for $n = 3$, $\mathbb{R}^3$ is called the *3-space*; and so on. The subsets defined by

$$B_\varepsilon(a_1, \ldots, a_n) = \{(x_1, \ldots, x_n) \in \mathbb{R}^n : (x_1 - a_1)^2 + \cdots + (x_n - a_n)^2 < \varepsilon^2\}$$

and

$$B'_\varepsilon(a_1, \ldots, a_n) = \{(x_1, \ldots, x_n) \in \mathbb{R}^n : (x_1 - a_1)^2 + \cdots + (x_n - a_n)^2 \leq \varepsilon^2\}$$

are called *open* and *closed disks* of $\mathbb{R}^n$, respectively. A subset P of $\mathbb{R}^n$ is said to be open if for each $(x_1, \ldots, x_n) \in P$ there is a real number $\varepsilon > 0$ such that $(x_1, \ldots, x_n) \in B_\varepsilon(x_1, \ldots, x_n) \subset P$. Thus, the collection of all open disks forms the base of a topology called the *Euclidean topology* of $\mathbb{R}^n$. In this topology, each open (respectively, closed) disk is an open (respectively, closed) set. Like $\mathbb{R}$, $\mathbb{R}^n$ is second countable and hence separable because the set of n-tuples $(x_1, \ldots, x_n)$ in which all $x_1, \ldots, x_n$ are rational numbers is a countable dense subset of $\mathbb{R}^n$. Here, too, the discrete topology is strictly finer than the Euclidean topology.

Another important property of $\mathbb{R}^n$ $(n \geq 1)$ that lends itself to abstraction is the following. A topological space $(X, \mathcal{T})$ is called *connected* if X cannot be written as the union of two disjoint, nonempty open sets of X. In other words, if $X = P \cup Q$ for some open sets P, Q such that $P \cap Q = \varnothing$, then one of P, Q must be a null set. $\mathbb{R}$ and $\mathbb{R}^n$ with their natural topology are connected, but not so in the discrete topology. $(X, \mathcal{T})$ is called *locally connected* if each neighborhood of each point of X contains a connected open neighborhood. Indeed, $\mathbb{R}^n$ $(n \geq 1)$ is locally connected.

A subset A of a topological space $(X, \mathcal{T})$ is called *connected* if it is so in the induced topology (scc Section III.A). The closure of a connected set is connected. Moreover, an arbitrary union of connected subsets is connected provided that no two members of the collection are disjoint. A connected subset that is not properly contained in any other connected subset is called a *component*. Clearly, if a topological space $(X, \mathcal{T})$ is connected, it is its only component. Each component of the real numbers with the discrete topology consists of a singleton, whereas $\mathbb{R}$ with the natural topology is its only component.

B. Methods of Defining Topology

The method of defining topology by means of open sets is not the only one. There are several methods used in the mathematical literature. Four of them are given below.

1. Kuratowski Axioms

One can define topology by closure operations. This method was introduced by K. Kuratowski. Let X be a set and let $\mathcal{P}(X)$ denote the collection of all subsets of X (often called the *power set* of X). Let ϕ be an operation that to each subset A of X assigns a subset of X. In other words ϕ maps $\mathcal{P}(X)$ into itself. Suppose that ϕ satisfies the so-called Kuratowski axioms: $\phi(\varnothing) = \varnothing$; $A \subset \phi(A)$; $\phi(\phi(A)) = \phi(A)$; $\phi(A \cup B) = \phi(A) \cup \phi(B)$ for all A, B in $\mathcal{P}(X)$. Then there exists a unique topology $\mathcal{T}$ on X such that $\phi(A)$ becomes the closure $\bar{A}$ of A with respect to the topology $\mathcal{T}$. Specifically, a set A is called closed with respect to $\mathcal{T}$ if $\phi(A) = A$. Hence, the complement A^c of a closed set A will, as usual, be open, and the topology $\mathcal{T}$ is thus determined.

The reader will have noticed that Kuratowski axioms are satisfied by the closure operation defined in a topological space as shown in the preceding section.

2. Neighborhood Systems

The second method of defining topology is via neighborhood systems. Let X be a set. Suppose that for each $x \in X$ there is a collection $\mathcal{U}_x$ of subsets of X satisfying the following axioms. For each $x \in X$ and each U in $\mathcal{U}_x$, $x \in U$; if a subset W of X contains some $U \in \mathcal{U}_x$, then $W \in \mathcal{U}_x$; finite intersections of elements from $\mathcal{U}_x$ are also in $\mathcal{U}_x$; and for each U in $\mathcal{U}_x$ there is a V in $\mathcal{U}_x$ such that $V \subset U$ and $U \in \mathcal{U}_y$ for all $y \in V$. Then there exists a unique topology $\mathcal{T}$ on X such that for each $x \in X$, $\mathcal{U}_x$ becomes a neighborhood system at x with respect to the topology $\mathcal{T}$.

3. Nets

Another method of defining topology is by means of nets. The notion of net generalizes the concept of sequence, which is so useful and popular in the case of real numbers. To define *net* we first need some other notions.

Let Γ denote a nonempty set. Γ is called a partially ordered set if there is a binary relation ">" (or "<") defined for elements of Γ such that whenever $\alpha > \beta$ and $\beta > \gamma$ for $\alpha, \beta, \gamma \in \Gamma$, we have $\alpha > \gamma$. (Clearly, the set of all positive integers is a partially ordered set.) A partially ordered set Γ with the binary relation $>$ is called *directed* if for $\alpha, \beta \in \Gamma$ there is a $\gamma \in \Gamma$ such that $\gamma > \alpha$, $\gamma > \beta$. Γ is called linearly ordered if for all $\alpha, \beta \in \Gamma$ either $\alpha > \beta$ or $\beta > \alpha$ and $\alpha > \beta$, $\beta > \alpha \Rightarrow \alpha = \beta$. An element $\gamma \in \Gamma$ is called an *upper bound* of a subset $\Gamma' \subset \Gamma$ if $\gamma \geq \alpha$ for all $\alpha \in \Gamma'$. γ is called the *least upper bound* if γ is the smallest upper bound of Γ'.

A map ϕ of a directed partially ordered set Γ into a set X is called a *net*. If we write $\phi(\alpha) = x_\alpha \in X$ for all $\alpha \in \Gamma$, then often a net is denoted by its range $\{x_\alpha : \alpha \in \Gamma\}$ in X. In short, one denotes a net $\{x_\alpha\}$ instead of $\{x_\alpha : \alpha \in \Gamma\}$. Clearly, if we replace Γ by $\mathbb{N}$ (positive integers), then the net $\{x_n : n \geq 1\} = \{x_n\}$ is a sequence in X. Thus, a sequence is a particular case of a net.

A net $\{x_\alpha\}$ is called a *constant net* if $x_\alpha = x$ for all $\alpha \in \Gamma$. If $\{x_n\}$ is a sequence and $n_1 < n_2 < \cdots$ are infinitely many positive integers, then $\{x_{n_k}\} = \{x_{n_1}, x_{n_2}, \ldots\}$ is called a *subsequence* of $\{x_n\}$.

A net $\{x_\alpha\}$ in a topological space $(X, \mathcal{T})$ is said to *converge* to a point $x \in X$ if for each neighborhood U of x there exists $\alpha_0 \in \Gamma$ such that for $\alpha \geq \alpha_0$ $(\alpha \in \Gamma)$, $x_\alpha \in U$. Now the closure of a set A in X can be characterized by means of nets as follows: $x \in \bar{A}$ iff there is a net $\{x_\alpha\}$ in A converging to x. Clearly, a constant net is always convergent. Moreover, a subsequence of a convergent sequence is also convergent. Without going into finer details, let us say that nets can be used to define limit points and then closures of subsets leading to the definition of topology. For details the interested reader is referred to any standard book on topology listed in the Bibliography.

4. Filters

Another method of defining a topology on a set is via filters. The notion of filters was introduced and popularized by French mathematicians, especially those belonging to the French group Bourbaki, named after a fictitious mathematician, Nicholas Bourbaki.

Let X be a nonempty set. A collection $\mathscr{F} = \{F_\alpha\}$ of subsets of X is called a *filter* if the following axioms hold. Each F_α is nonempty; for F_α, F_β in $\mathscr{F}$, $F_\alpha \cap F_\beta \in \mathscr{F}$; and if H is any subset of X such that $F_\alpha \subset H$ for some $F_\alpha \in \mathscr{F}$, then $H \in \mathscr{F}$. A collection $\mathscr{F} = \{F_\alpha\}$ of subsets F_α in X is called a *filter base* if each F_α is nonempty and if, for F_α, F_β in $\mathscr{F}$, there is F_γ in $\mathscr{F}$ such that $F_\gamma \subset F_\alpha \cap F_\beta$. Clearly, a filter base generates a filter consisting of all subsets of X, each containing a member of the filter base.

A neighborhood system $\mathscr{N}_x$ of x in a topological space $(X, \mathscr{T})$ is a filter called the *neighborhood filter*.

A filter $\mathscr{F} = \{F_\alpha\}$ in a topological space $(X, \mathscr{T})$ is said to *converge to* $x \in X$ if for each neighborhood U of x there exists $F_\alpha \in \mathscr{F}$ such that $F_\alpha \subset U$. An element $y \in X$ is called a *limit point* of a filter $\mathscr{F} = \{F_\alpha\}$ if $y \in \bar{F}_\alpha$ for all F_α in $\mathscr{F}$.

Theories of filters and nets are equivalent. Specifically, if $\{x_\alpha : \alpha \in \Gamma\}$ is a net, the collection of all $F_\alpha = \{x_\beta : \beta > \alpha, \beta \in \Gamma\}$ is a filter base. Conversely, given a filter $\mathscr{F} = \{F_\alpha\}$, it is possible to construct a net $\{x_\alpha\}$ by considering the set Γ of all pairs (x_α, F_α) in which $x_\alpha \in F_\alpha$, and by defining $(x_\alpha, F_\alpha) \leq (x_\beta, F_\beta) \Leftrightarrow F_\beta \subset F_\alpha$, we see that $\leq$ defines a partial ordering and Γ becomes a directed set because $\mathscr{F}$ is a filter. Now if we put $\phi(x_\alpha, F_\alpha) = x_\alpha$, we obtain a net $\{x_\alpha\}$. Details can be found in standard texts on topology. Thus, filters can be used to define a topology as nets do.

A filter can be contained in a *maximal* filter (one that is not a proper subcollection of a filter). A maximal filter is called an *ultrafilter*. (The existence of maximal filters is guaranteed by Zorn's lemma: Each partially ordered set in which each linearly ordered subset has a least upper bound has a maximal element.) A filter $\mathscr{F}$ in X is an ultrafilter iff for each subset A of X, one of the two sets A, A^c is in $\mathscr{F}$. Ultrafilters are useful for the study of extensions of topological spaces as well as compactness among others. We shall see one application of ultrafilters in Section V.B.

II. MAPS AND FUNCTIONS

Given two sets X and Y, if to each element x of X there corresponds a unique element y of Y, then this correspondence

is called a *map* and often written as $f : X \to Y$, meaning that $y = f(x)$ for $x \in X$. X is called the *domain* of the map f, and $f(X) = \{y \in Y : y = f(x), x \in X\}$ the *range* of f. If $f(x) = f(x')$ implies that $x = x'$ for all $x, x' \in X$, then f is called *one to one* (or *injective*). If for each $y \in Y$ there is $x \in X$ such that $y = f(x)$, then f is called *onto* (or *surjective*). A map that is both injective and surjective is called *bijective*.

In particular cases when either $Y = \mathbb{R}$ (real numbers) or $X = Y = \mathbb{R}$, a map $f : X \to Y$ is often called a *real-valued function* or simply *function*.

The notion of map or function is very basic to understanding the physical phenomena of the universe, since each physical or other kind of event for a mathematical study must be expressed as a function. Indeed, the properties of real-valued functions studied in classical analysis are used in almost all scientific studies. Here we look at some special properties of maps between topological spaces.

A. Continuous, Open Maps and Homeomorphisms

Let $(X, \mathscr{T})$ and $(Y, \mathscr{T}')$ be two topological spaces. A map $f : X \to Y$ is said to be *continuous at a point* $x_0 \in X$ if for each neighborhood V of $f(x_0)$ in Y there exists a neighborhood U of x_0 in X such that $f(x) \in V$ for all $x \in U$. If f is continuous at each point of X, it is called *continuous*.

Indeed, each continuous map is continuous at each point of X by definition, but there are functions that are continuous at one point but not at another. For example, $f(x) = 1$ if $x \geq 0$ and $= 0$ if $x < 0$ is a real-valued function on $\mathbb{R}$ that is continuous at all $x \neq 0$ but not continuous at $x = 0$.

Since the continuity of a map depends on the topologies of the topological spaces concerned and since there are several methods of defining topologies, it is natural that the continuity of a map can be expressed in more than one way. Specifically, we have the following equivalent statements:

(i) f is continuous (see the above definition).
(ii) For each open set Q of Y, $f^{-1}(Q) = \{x \in X : f(x) \in Q\}$ is open in X.
(iii) For each net $\{x_\alpha\}$ in X converging to $x \in X$, $\{f(x_\alpha)\}$ converges to $f(x)$ in Y.
(iv) For each filter $\mathscr{F} = \{F_\alpha\}$ converging to $x \in X$, $\{f(F_\alpha)\}$ converges to $f(x) \in Y$.
(v) For each subset A of X, $f(\bar{A}) \subset \overline{f(A)}$.
(vi) For each subset B of Y, $f^{-1}(\bar{B}) \subset f^{-1}(\bar{B})$.

The continuity property is transitive in the following sense. If $f : X \to Y$ and $g : Y \to Z$ are continuous maps, respectively, of topological spaces from $(X, \mathcal{T})$ into $(Y, \mathcal{T}')$ and from $(Y, \mathcal{T}')$ into $(Z, \mathcal{T}'')$, then the *composition map* $g \circ f : X \rightarrow Z$ defined by $g \circ f(x) = g(f(x))$, $x \in X$, is also continuous.

Although the continuity property of maps between topological spaces is the most popular and useful, there are other notions that complement or augment continuity.

A map f of a topological space $(X, \mathcal{T})$ into another topological space $(Y, \mathcal{T}')$ is called *open* if for each open set P of X, $f(P) = \{y = f(x) : x \in P\}$ is open in Y. It follows that f is open iff $[f(A)]^0 \subset f(A^0)$ for all subsets A of X. (Note that A^0 denotes the interior of A defined earlier.) A continuous map need not be open (e.g., $i : (\mathbb{R}, \mathcal{D}) \rightarrow (\mathbb{R}, \mathcal{T})$ in which $\mathcal{D}$ is the discrete topology and $\mathcal{T}$ the natural topology), or vice versa.

Similarly a map $f : (X, \mathcal{T}) \rightarrow (Y, \mathcal{T}')$ is called *closed* if for each closed set C of X, $f(C)$ is a closed set in Y. Here, too, we have a characterization: f is closed iff for each subset A of X, $\overline{f(A)} \subset f(\bar{A})$. Again, a continuous map need not be closed, nor is a closed map necessarily continuous.

A continuous, open, and bijective map of a topological space $(X, \mathcal{T})$ into another topological space $(Y, \mathcal{T}')$ is called a *homeomorphism*. If $f : (X, \mathcal{T}) \rightarrow (Y, \mathcal{T}')$ is a homeomorphism, then $(X, \mathcal{T})$ is said to be *homeomorphic* to $(Y, \mathcal{T}')$. In this case, there is no topological difference between the spaces $(X, \mathcal{T})$ and $(Y, \mathcal{T}')$, and hence they are sometimes called *topologically equivalent*.

It is one of the most important aspects of topological studies to discover conditions for homeomorphisms. For example, a simple result is that a bijective map between two discrete spaces is a homeomorphism. We present further examples of this kind in Section V.B.

Continuity can be used to define topology, thus adding one more method to those described earlier. Let X be a set, $(Y, \mathcal{T}')$ a topological space, and $f : X \rightarrow Y$ a map. We can endow X with a topology with respect to which f becomes continuous. Consider the collection $\mathcal{T} = \{f^{-1}(P) : P \in \mathcal{T}'\}$ of subsets of X in which $f^{-1}(P) = \{x \in X : f(x) \in P\}$ for an open set P in Y. It is routine to verify that this collection defines a topology $\mathcal{T}$ on X and f becomes continuous.

If $\{f_\alpha\}_{\alpha \in \Gamma}$ is a family of maps from a set X into a topological space $(X, \mathcal{T}')$, then the collection $\mathcal{T} = \{f_\alpha^{-1}(P) : P \in \mathcal{T}', \alpha \in \Gamma\}$ defines a topology $\mathcal{T}$ on X with respect to which each f_α becomes continuous.

B. Generalizations of Continuous Maps

If one cannot have the most desirable basic property of continuity of a map between topological spaces, one studies weaker notions to acquire more information about the maps. For this reason, a number of concepts weaker than continuity have been introduced and studied in the mathematical literature. Some of these are given here.

1. Almost Continuity

A map $f : (X, \mathcal{T}) \rightarrow (Y, \mathcal{T}')$ is said to be almost continuous at $x_0 \in X$ if for each neighborhood V of $f(x_0)$, in Y, $\overline{f^{-1}(V)}$ is a neighborhood of x_0 in X, f is called almost continuous if it is so at each point of X. (Elsewhere in the mathematical literature, what I have called an almost-continuous map is also called nearly continuous.)

Comparing the notion of almost continuity with that of continuity, we discover easily that each continuous map is almost continuous. The converse is not true, however, [e.g., $f : \mathbb{R} \rightarrow \mathbb{R}$ defined by $f(x) = 1$ or 0 according to whether x is a rational or irrational number.]

2. Closed Graphs

Let $(X, \mathcal{T}), (Y, \mathcal{T}')$ be two topological spaces. We put $X \times Y = \{(x, y) : x \in X, y \in Y\}$, which is the collection of all pairs (x, y) with first coordinate x and second coordinate y. $X \times Y$ is called the *Cartesian product* of X and Y. We endow $X \times Y$ with a topology $\mathcal{P}$, which has for a base the collection $\{U \times V\}$, in which U comes from $\mathcal{T}$ and V from $\mathcal{T}'$. The topology so defined is called the *product topology* on $X \times Y$ (see Section III.B).

If $f : X \rightarrow Y$ is a map, the subset $\{(x, y) : y = f(x), x \in X\}$ of $X \times Y$ is called the *graph of f* and is sometimes denoted G_f. If G_f is a closed subset of $X \times Y$ when the Cartesian product is endowed with the product topology, then f is said to have a *closed graph*.

Each continuous function $f : \mathbb{R} \rightarrow \mathbb{R}$ has a closed graph. However, the converse is not true [e.g., $f(x) = 1/x$ $(x \neq 0)$ and $= 0$ $(x = 0)$ has a closed graph but it is not continuous]. On the other hand, the identity map of any indiscrete space onto itself is continuous but does not have a closed graph. Thus, in general, the closed-graph property is not an exact generalization of continuity. It is true, however, if the continuous map is from a topological space into a Hausdorff space, which is introduced in Section IV.

Any result in which the closed-graph property of a map implies its continuity is called a closed-graph theorem.

3. Upper (or Lower) Semicontinuous Functions

A real-valued function f on a topological space $(X, \mathcal{T})$ is called upper (or lower) semicontinuous if for each real number α, the set $\{x \in X : f(x) < \alpha\}$ [respectively, $\{x \in X : f(x) > \alpha\}$] is an open subset of X. If we put $(-f)(x) = -f(x)$ for all $x \in X$, then f is lower semicontinuous iff $-f$ is upper semicontinuous. Indeed, each continuous real-valued function on a topological space is upper and lower semicontinuous.

4. Right- and Left-Continuous Functions

Let X be an interval of $\mathbb{R}$ and $f : X \to \mathbb{R}$ a function. f is said to be right (or left) continuous at $x_0 \in X$ if for each arbitrary real number $\varepsilon > 0$, there is a real number $\delta > 0$ such that $|f(x) - f(x_0)| < \varepsilon$ whenever $x_0 < x < x_0 + \delta$ (respectively, $x_0 - \delta < x < x_0$). Clearly, f is continuous at x_0 iff it is right and left continuous at x_0. However, a right-continuous or left-continuous function may fail to be continuous [e.g., $f(x) = [x]$, the largest integer less than or equal to $x \in \mathbb{R}$; $[x]$ is called the *step function*, which is right continuous but not continuous at integers].

III. METHODS OF CONSTRUCTING NEW TOPOLOGICAL SPACES

There are a number of methods of constructing new topological spaces from old. In this section we present some of them.

A. Subspaces and Induced Topology

Let $(X, \mathcal{T})$ be a topological space. Any subset Y of X can be made into a topological space as follows. Consider the collection $\mathcal{T}' = \{P \cap Y : P \in \mathcal{T}\}$ of subsets of Y. Then it is easy to see that this collection defines a topology of Y called the *induced* or *relative topology*. In this topology, a subset Q of Y is *relatively open* iff there is an open set P in X such that $Q = P \cap Y$. If a subset E of Y is open in X, clearly E is relatively open. However, a relatively open subset of Y need not be open in X unless Y is itself an open subset of X. A subset Y of X endowed with the induced topology $\mathcal{T}'$ is called a *subspace* of X.

A topological space $(X, \mathcal{T})$ is said to have the *hereditary property* if each subspace of X satisfies the same property as does $(X, \mathcal{T})$ (see Table III for more information).

B. Products and Product Topology

There are two types of products: finite products and infinite products. Infinite products are of two kinds: countable products and arbitrary infinite products. By forming products of topological spaces we produce new topological spaces.

1. Finite Products

Let $(X_k, \mathcal{T}_k), k = 1, 2, \ldots, n$, be n topological spaces with their respective topologies $\mathcal{T}_k$. By $\prod_{k=1}^{n} X_k = X_1 \times$ $X_2 \times \cdots \times X_n$, we mean the set of all ordered n-tuples $(x_1, x_2, \ldots, x_n)$ in which $x_k \in X_k, k = 1, 2, \ldots, n$. Here, $\prod_{k=1}^{n} X_k$ is called the *finite Cartesian product* of $X_1, X_2, \ldots, X_n$. We topologize $\prod_{k=1}^{n} X_k$ as follows. The family $\mathcal{B} = \{U_1 \times U_2 \times \cdots \times U_n : U_k$ is open in $(X_k, \mathcal{T}_k), k = 1, \ldots, n\}$ forms a base of a topology, called the *product topology* on $\prod_{k=1}^{n} X_k$. In other words, a subset P of $\prod_{k=1}^{n} X_k$ is open in the product topology iff for each point $(x_1, \ldots, x_n) \in P$ there are open sets U_k in X_k for $k = 1, 2, \ldots, n$ with $x_k \in U_k$ such that $U_1 \times U_2 \times \cdots \times U_n \subset P$. We have already considered the special case when $n = 2$.

If $X = \prod_{k=1}^{n} X_k$ is a finite product with the product topology, the map $p_k(x_1, \ldots, x_n) = x_k$ is called the *kth projection* of X into X_k. Each p_k is continuous, open, and surjective. A finite product $\prod_{k=1}^{n} X_k$ is said to have the *productive property* if X satisfies the same property as does each X_k (see Table III).

2. Infinite Products

Let A be a nonempty indexing set. For each $\alpha \in A$, let $(X_\alpha, \mathcal{T}_\alpha)$ be a topological space with the topology $\mathcal{T}_\alpha$. By $\prod_{\alpha \in A} X_\alpha$, we mean the set of all maps $x : A \to \bigcup_{\alpha \in A} X_\alpha$ such that $x(\alpha) = x_\alpha \in X_\alpha$ for all $\alpha \in A$. The $\prod_{\alpha \in A} X_\alpha$ is called a *product*. If A is a finite set, then indeed $\prod_{\alpha \in A} X_\alpha$ is a finite product, as seen above. If A is countable (i.e., there is a one-to-one map between the elements of A and those of positive integers), then $\prod_{\alpha \in A} X_\alpha$ is called the *countable product* and is often written $\prod_{k=1}^{\infty} X_k$. If A is arbitrarily infinite and not countable, the product $\prod_{\alpha \in A} X_\alpha$ is called the *arbitrary infinite product* or simply *arbitrary product*. Here, again, the map $p_\alpha : \prod_{\alpha \in A} X_\alpha \to X_\alpha$ defined by $p_\alpha(x) = x_\alpha \in X_\alpha$ is called the αth projection.

To define a topology on $\prod_{\alpha \in A} X_\alpha$, we consider the family $\mathcal{B}$ of all subsets $P = \prod_{\alpha \in A} P_\alpha$ of $\prod_{\alpha \in A} X_\alpha$ in which P_α is an open subset of X_α for a finite number of α's, say, $\alpha_1, \alpha_2, \ldots, \alpha_n$ and $P_\alpha = X_\alpha$ for all $\alpha \neq \alpha_1, \alpha_2, \ldots, \alpha_n$. Then $\mathcal{B}$ forms a base of a topology of $\prod_{\alpha \in A} X_\alpha$, called the *product topology*. With the product topology on $\prod_{\alpha \in A} X_\alpha$, each αth projection becomes continuous and open.

The product topology on $\prod_{\alpha \in A} X_\alpha$ is very useful. For example, a map f of any topological space $(X, \mathcal{T})$ into $\prod_{\alpha \in A} X_\alpha$ is continuous iff each composition map $p_\alpha \circ f : X \to X_\alpha$ is continuous.

As for finite products, one can define productive property for infinite products (see Table III).

C. Quotient Topology

Let $(X, \mathcal{T})$ be a topological space, Y an arbitrary set, and $f : X \to Y$ a surjective map. We endow Y with a topology that will make f continuous. For this, we designate a set Q of Y open whenever $f^{-1}(Q) = \{x \in X : f(x) \in Q\}$ is open in X. The collection of all such Q's defines a topology on Y, called the *quotient topology*. Clearly, if Y is endowed with the quotient topology, then $f : X \to Y$ becomes continuous.

Note that the quotient topology on Y depends on the topology of X and the map f. Indeed, one can always endow Y with the indiscrete topology to make f continuous. Among the topologies on Y that make f continuous, the quotient topology is the finest topology.

Like the product topology, the quotient topology is also very useful. For instance, suppose that Y is endowed with the quotient topology defined by $(X, \mathcal{T})$ and f. Let $(Z, \mathcal{T}')$ be a topological space and $g : Y \to Z$ a map. Then g is continuous iff $g \circ f : X \to Z$ is continuous.

IV. SEPARATION AXIOMS

The properties that appear to be obvious for real numbers are not easily available in general topological spaces. For instance, if a, b are two distinct real numbers, then for $\varepsilon = |a - b| > 0$, open intervals $A = \{x \in \mathbb{R} : |x - a| < \frac{1}{2}\varepsilon\}$ and $B = \{x \in \mathbb{R} : |x - b| < \frac{1}{2}\varepsilon\}$ are disjoint with $a \in A$ and $b \in B$. It is this kind of property that may not hold in general topological spaces. Such properties are termed *separation axioms*, also known as Alexandroff–Hopf Trennungaxioms.

One of the reasons separation axioms are necessary in the study of topological spaces is that they enable us to decide about the uniqueness of limits of convergent sequences or nets. This is not a trivial reason, for in an indiscrete space, no sequence or net has a unique limit. As a matter of fact, every sequence or net converges to every point of the indiscrete space. This abnormality makes the indiscrete spaces useless.

Now we consider the separation axioms.

A. T₀-Axiom

A topological space is said to satisfy the T_0-axiom or is called a T_0-*space* if at least one member of any pair of distinct points has an open neighborhood that does not contain the other. This axiom is not strong enough to yield any worthwhile results. Note that an indiscrete space containing more than one point is not a T_0-space.

B. T₁-Axiom

A topological space is said to satisfy the T_1-axiom or is called a T_1-space if each member of any pair of distinct points has an open neighborhood that does not contain the other. Although obviously better than the T_0-axiom, this axiom is not strong enough to guarantee the uniqueness of the limit of convergent sequences or nets. However, it does give us that each singleton (the set consisting of a single point) is a closed set and conversely (i.e., if each singleton is a closed set, the topological space must be a T_1-space). The class of T_1-spaces is hereditary and productive (see Table III).

C. T₂-Axiom

This is the most desirable separation axiom for the above-mentioned purpose and other reasons. A topological space is said to satisfy the T_2-axiom or is called a T_2-*space* or *Hausdorff space* (after F. Hausdorff) if both members of any pair of distinct points have disjoint open neighborhoods. This is indeed the weakest separation axiom that guarantees the uniqueness of the limit of a convergent sequence or net. As a matter of fact, a topological space is a Hausdorff space iff each convergent net or filter has the unique limit. The T_2-axiom is also equivalent to having the graph of the identity map closed. This class of spaces is also hereditary and productive.

There is another group of separation axioms that separate points from closed sets or closed sets from closed sets, as follows.

D. Regularity

A topological space is called regular (R) if for any closed set C and any point x not in C, there are disjoint open sets U and V such that $C \subset U$ and $x \in V$. Regularity is equivalent to the criterion that each point of the space have a base of closed neighborhoods. In general, regularity and the T_2-axiom have nothing in common. Therefore, to make them comparable, one defines T_3-*spaces* as those that are regular and satisfy the T_1-axiom. Now T_3-spaces are properly contained in the class of T_2-spaces and therefore satisfy the above-mentioned desirable property of uniqueness of limits of convergent nets in topological spaces as well. Moreover, they are hereditary and productive.

E. Complete Regularity

A topological space $(X, \mathcal{T})$ is said to be completely regular (CR) if for each closed set C of X and any point x not in C there is a continuous function $f : X \to [0, 1]$ such that $f(C) = 0$ and $f(x) = 1$. Complete regularity implies regularity but may fail to satisfy the T_2-axiom. However, if one joins the T_1-axiom with complete regularity, the resulting space does satisfy the T_2-axiom. A completely regular T_1-space is generally known as a $T_{3 1/2}$-space or *Tychonoff space* (after the Russian mathematician A. Tychonoff, who contributed a great deal to topology). Thus, each Tychonoff space is a Hausdorff space. The significance of Tychonoff spaces among others lies in the fact that on such spaces exist nonconstant continuous functions. Even on regular spaces there may not exist any nonconstant continuous function. Both classes of completely regular and Tychonoff spaces satisfy hereditary and productive properties.

F. Normality

A topological space is said to be normal (N) if for any pair of disjoint closed subsets C_1, C_2 there are disjoint open sets U_1, U_2 such that $C_1 \subset U_1$ and $C_2 \subset U_2$. A useful characterization of normality is as follows. For each closed set C and open set U with $C \subset U$, there is an open set V such that $C \subset V \subset \bar{V} \subset U$. Even this apparently stronger separation axiom does not imply a Hausdorff space or the T_2-axiom, since singletons in a normal space need not be closed sets. Thus, to make it comparable with regularity and the T_2-axiom, one attaches the T_1-axiom to normality. A normal T_1-space is generally called a T_4-*space*. Now it is clear that each T_4-space is a T_3-space and hence Hausdorff, as shown earlier.

The axiom of complete regularity mentioned between those of regularity and normality apparently looks like an odd person out, but this is not the case, thanks to Urysohn's lemma: Let C_1, C_2 be any two disjoint closed subsets of a normal space $(X, \mathcal{T})$. Then there exists a continuous function $f : X \to [0, 1]$ such that $f(C_1) = 0$ and $f(C_2) = 1$. From this it follows that a T_4-space is a Tychonoff space, which in turn implies that it is a T_3-space and hence Hausdorff.

Extensions of functions from a subspace to the whole space play an important role in almost every aspect of mathematics. Thus, the question is which of the above axioms, if any, ensures extensions. We are indebted to H. Tietze for the so-called Tietze's extension theorem: If Y is a closed subspace of a T_4-space $(X, \mathcal{T})$ and $f : Y \to \mathbb{R}$ a continuous map, there exists a continuous map $\bar{f} : X \to \mathbb{R}$ such that $\bar{f}(x) = f(x)$ for all $x \in Y$.

There are other separation axioms finer or stronger than normality for which the reader is referred to texts on topology. However, normality is sufficient enough to carry out major parts of mathematical analysis. As noted in the preceding paragraphs, of the six separation axioms, T_0 is the weakest and T_4 the strongest. Consult Table II for a complete picture of these implications. For information about whether these six classes of spaces have hereditary or productive properties (or both), as mentioned in the preceding section, see Table III.

TABLE II Interconnection between Topological Spaces[a]

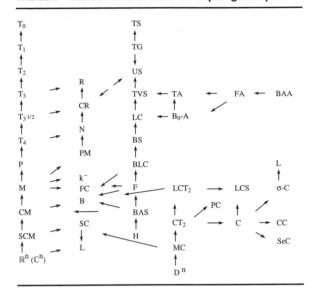

[a] A $\to$ B means that A is contained in B.

V. SPECIAL TOPOLOGICAL SPACES

In this section we consider some of the most useful topological spaces that are frequently mentioned and studied in topology and functional analysis (a modern branch of mathematics that deals simultaneously with abstract and concrete analysis).

A. Metric Spaces and Metrization

An important class of topological spaces that subsumes the classical spaces $\mathbb{R}^n$ or $\mathbb{C}^n$ is that of metric spaces.

Let X be a set. A function $d : X \times X \to \mathbb{R}$ is called a *metric* of X if d satisfies the following axioms:

(i) $d(x, y) \geq 0$ for all $x, y \in X$.
(ii) $d(x, y) = 0$ if $x = y$.
(iii) $d(x, y) = 0$ implies that $x = y$.
(iv) $d(x, y) = d(y, x)$.
(v) $d(x, y) \leq d(x, z) + d(z, y)$ (called the *triangle inequality*).

If d satisfies only (i), (ii), (iv), and (v), then d is called a *pseudometric* on X. A set X with a metric (or pseudometric) d is called a *metric* (or *pseudometric*) (M or PM, respectively) *space* and is sometimes denoted (X, d).

Each pseudometric (in particular, metric) space is indeed a topological space, for a set P in X is open if for each $x \in P$ there is a real number $r > 0$ such that the *open ball* $B_r(x) = \{y \in X : d(y, x) < r\}$ of radius r, centered at x,

TABLE III Hereditary and Productive Properties of Some Important Spaces[a]

Name	Hereditary		Productive properties		
	Arbitrary	Closed	Finite	Countable	Arbitrary
Topological spaces					
C (compact)	N	Y	Y	Y	Y
CM (compact metric)	N	Y	Y	Y	N
FC (first countable)	Y	Y	Y	Y	N
LCS (locally compact space)	N	Y	Y	N	N
M (metric)	Y	Y	Y	Y	N
P (paracompact)	N	Y	N	N	N
T_0	Y	Y	Y	Y	Y
T_1	Y	Y	Y	Y	Y
T_2	Y	Y	Y	Y	Y
T_3	Y	Y	Y	Y	Y
$T_{3 1/2}$	Y	Y	Y	Y	Y
T_4	N	Y	N	N	N
Topological vector spaces (vector subspaces)					
TVS (topological vector space)	Y	Y	Y	Y	Y
BAS (Banach space)	N	Y	Y	N	N
F (Fréchet)	N	Y	Y	Y	N
H (Hilbert)	N	Y	Y	N	N
Topological algebras (subalgebras)					
TA (topological algebra)	Y	Y	Y	Y	Y
BAA (Banach algebra)	N	Y	Y	N	N
B_0-A (B_0-algebra)	N	Y	Y	Y	N
FA (Fréchet algebra)	N	Y	Y	Y	N

[a] N (no) indicates that the property does not hold; Y (yes), that it does hold.

is completely contained in P; in symbols $x \in B_r(x) \subset P$. With this definition of open sets in a metric space (X, d), it is easily verified that the collection of all such open sets in X defines a topology called the *metric topology* of X and the collection of all open balls forms a base of the metric topology.

To see pseudometric or metric spaces from the separation-axiom point of view, note that each pseudometric (metric) space is normal (T_4-space), thus forming a telescopic chain of spaces from metric spaces up to T_0-spaces (see Table II).

If we consider the countable collection $\{B_{1/n}(x)\}$ of open balls of radius $1/n$ ($n \geq 1$, integer), centered at x, it follows that a metric space is first countable. A metric space is second countable iff it is separable.

Metric spaces enjoy an important place among topological spaces because, unlike general topological spaces, in metric spaces one can assign distances $d(x, y)$ between two points—a feature very noticeable in Euclidean spaces $\mathbb{R}^n$. Thus, one can study the notions of Cauchy sequences and uniformly continuous functions in metric spaces, which cannot be defined in general topological spaces.

A sequence $\{x_n\}$ in a metric space (X, d) is said to *converge* to $x \in X$ if for each arbitrary real number

$\varepsilon > 0$ there is a positive integer n_0 such that $d(x_n, x) < \varepsilon$ whenever $n \geq n_0$. $\{x_n\}$ is called a *Cauchy sequence* if for each real number $\varepsilon > 0$ there is a positive integer n_0 such that $d(x_n, x_m) < \varepsilon$ whenever $n, m \geq n_0$.

Every convergent sequence is a Cauchy sequence, but the converse is not true. [For example, the sequence of rational numbers ≥ 1 and $< \sqrt{2}$, approximating $\sqrt{2}$, converges to $\sqrt{2}$ (which is not a rational number) and is a Cauchy sequence in the metric space of rational numbers.] A metric space (X, d) is called *complete* if every Cauchy sequence in it is convergent. (The set $\mathbb{R}$ of all real numbers is complete by the so-called Cauchy criterion for convergence of a real sequence.) We denote complete metric spaces CM and separable complete metric spaces SCM.

One can obtain topological information about closed sets by convergent sequences in a metric space (X, d) as follows. A subset C of X is closed iff for each sequence $\{x_n\}$ in C converging to x, it follows that $x \in C$. Since metric spaces, being T_4-spaces, are Hausdorff, the limits of convergent sequences are unique. Furthermore, metric spaces clearly enjoy the hereditary property but not the general productive property (see Table III). However, if (X_k, d_k), $k = 1, 2, \ldots$, is a countable family of metric spaces, then the countable product $X = \prod_{k=1}^{\infty} X_k$ is a metric space with the metric

$$d(x, y) = \sum_{k=1}^{\infty} \frac{1}{2^k} \frac{d_k(x_k, y_k)}{1 + d_k(x_k, y_k)},$$

in which both $x = \{x_k\}$ and $y = \{y_k\}$ belong to X. In particular, the space ω of all real sequences is a metric space.

As pointed out above by the example of rational numbers, not every metric space is complete. However, there is a process, called the completion, by which an incomplete metric space can be completed. More precisely, if (X, d) is an incomplete metric space, there exists a unique complete metric space $(\hat{X}, \hat{d})$ containing a dense copy of $(X, d) \cdot (\hat{X}, \hat{d})$, called the *completion* of (X, d). For example, the set of all real numbers is the completion of the set of all rational numbers.

For metric spaces, the continuity of a map can be described as follows. Let (X, d) and (Y, d') be two metric spaces. A map $f : X \to Y$ is continuous at $x_0 \in X$ if and only if for each $\varepsilon > 0$ there is $\delta > 0$ such that for all $x \in X, d(x, x_0) < \delta$ implies that $d'(f(x), f(x_0)) < \varepsilon$. This is also equivalent to the so-called *sequential continuity* at x_0 : for all sequences $\{x_n\}$ in X converging to x_0, the sequence $\{f(x_n)\}$ converges to $f(x_0)$. A map $f : X \to Y$ is called *uniformly continuous* if for each $\varepsilon > 0$ there is $\delta > 0$ such that for all $x, y \in X, d(x, y) < \delta$ implies that $d'(f(x), f(y)) < \varepsilon$. If for all $x, y \in X$, $d'(f(x), f(y)) = d(x, y)$, then $f : X \to Y$ is called an *isometry* and thus f embeds X into Y in such a way that the metric induced from Y on $f(X)$ coincides with the metric of X. That is why it is sometimes called an *isometric embedding* of X into Y. An embedding need not be surjective. However, if $f(X) = Y$, then f is called an *isometric homeomorphism*. In this case, X and Y are set-theoretically and topologically the same.

A subset of a topological space is called an F_σ-set (respectively, G_δ-set) if it can be written as the countable union (respectively, countable intersection) of closed (respectively, open) sets. A subset A of a topological space is an F_σ-set (G_δ-set) iff its complement A^c is a G_δ-set (respectively, F_σ-set). Moreover, a countable union (countable intersection) of F_σ-sets (respectively, G_δ-sets) is an F_σ-set (respectively, G_δ-set). Also, a finite intersection (finite union) of F_σ-sets (respectively, G_δ-sets) is an F_σ-set (respectively, G_δ-set). Every closed (open) set of a metric space is a G_δ-set (respectively, F_σ-set). In particular, each closed (open) subset of $\mathbb{R}$ is a G_δ-set (respectively, an F_σ-set).

Since it is easier to grasp the topology of a metric space than that of general topological spaces, it is important to know when a topological space can be endowed with a metric topology. Indeed, indiscrete topology can never be defined by a metric, since a metric space is always Hausdorff, whereas the indiscrete space does not even satisfy the T_0-axiom. On the other hand, the discrete topology can be given by the trivial metric $d(x, y) = 1$ or 0 according to whether $x \neq y$ or $x = y$.

Whenever the topology of a topological space can be defined by a metric, the space is called *metrizable*. A nontrivial notable result in this regard is due to P. Urysohn, which goes by the name Urysohn's metrization theorem: Each second countable T_3-space is metrizable.

A separable complete metric space is called a *Polish space* and a metric space which is a continuous image of a Polish space is called a *Souslin space*. A closed subspace of a Polish (respectively, Souslin) space is also a Polish (respectively, Souslin) space. The same is true for countable products. Moreover, an open subset of a Polish space is also a Polish space. For example, $\mathbb{R}$ and the set of all irrational numbers endowed with the metric topology induced from $\mathbb{R}$ are Polish spaces. We give further properties of these spaces in the sequel.

B. Compactness and Compactifications

Another class of very important and useful topological spaces in topology as well as in analysis is that of compact spaces. Let $(X, \mathcal{T})$ be a topological space. A collection $\{P_\alpha\}$ of subsets of X is said to be a *covering* of a set $E \subset X$ if $E \subset \bigcup_\alpha P_\alpha$; in other words, each element of E belongs to some P_α. If $E = X$, then $\{P_\alpha\}$ is a covering of X if $X = \bigcup_\alpha P_\alpha$. If each member of the covering $\{P_\alpha\}$ of E is open, then $\{P_\alpha\}$ is called an *open covering* of E. If the number of sets in a covering is finite, it is called a *finite covering*.

1. Compact Spaces

A topological space $(X, \mathcal{T})$ is said to be compact (C) if from each open covering of X it is possible to extract a finite open covering of X. $(X, \mathcal{T})$ is called a *Lindelöf space* (L) if from each open covering $\{P_\alpha\}$ of X it is possible to extract a countable subcollection $\{P_{\alpha_i}, i \geq 1\}$ such that $X = \bigcup_{i=1}^{\infty} P_{\alpha_i}$. It is clear that each compact space is a Lindelöf space. However, the converse is not true (e.g., $\mathbb{R}$, with the natural topology). One of the celebrated theorems in analysis, called the Heine–Borel theorem, tells us that each closed bounded interval $[a, b]$, $-\infty < a \leq b < \infty$, of $\mathbb{R}$ is compact, but $\mathbb{R}$ itself is not. Similarly, each closed disk D^n in $\mathbb{R}^n$ is compact.

Compactness has several characterizations. Some of them are as follows. $(X, \mathcal{T})$ is compact $\Leftrightarrow$ for each family $\{C_\alpha\}$ of closed sets in X with the finite-intersection property (i.e., for any finite subcollection $\{C_{\alpha_i}\}_{i=1}^n$ of $\{C_\alpha\}$, $\bigcap_{i=1}^n C_{\alpha_i} \neq \varnothing$), we have $\bigcap_a C_\alpha \neq \varnothing \Leftrightarrow$ each net or filter of X has a cluster point $\Leftrightarrow$ each ultrafilter in X is convergent.

Since the indiscrete space has only two open sets (actually only one nonempty open set), it is always compact, although not Hausdorff. On the other hand, a discrete space is compact iff it is a finite set. A Hausdorff compact space, however, is a T_4-space and so is normal.

With regard to the hereditary property, compact spaces do not have it (see Table III), but a closed subset of a compact space is compact. A far-reaching result for application purposes is concerned with the product of compact spaces and goes by the name of the *Tychonoff theorem*: If $(X_\alpha, \mathcal{T}_\alpha) (\alpha \in A)$ is a family of compact spaces, the arbitrary product $\prod_{\alpha \in A} X_\alpha$ is also compact under the product topology. In addition, compact spaces are rich in other properties. For instance, the continuous image of a compact space is compact. Moreover, a continuous bijective map of a compact space onto a Hausdorff space is a homeomorphism. Every continuous real-valued function on a compact space is bounded; that is, there is $M > 0$ such that $|f(x)| \leq M$ for all $x \in X$.

Although a subspace of a compact space need not be compact, every subspace of a compact Hausdorff space is a Tychonoff space. This, among other properties, highlights the significance of a compact Hausdorff space. That is why it is sometimes called a *compactum*. Two esotoric examples of compacta are as follows.

a. Hilbert cube. Let $H^\omega = \{\{x_n\}, x_n \in \mathbb{R} : 0 \leq x_n \leq (1/n)\}$ with the metric $d(\{x_n\}, \{y_n\}) = \sqrt{\sum_{n=1}^\infty (x_n - y_n)^2}$. Then H^ω is a separable compact metric space, hence a second countable compactum and a Polish space. H^ω is called the Hilbert cube. It is a compact subset of the Hilbert space ℓ_2 (see Section VIII.B).

b. Cantor set. Let $A_0 = [0, 1]$, the closed unit interval. By removing the middle third of A_0 we obtain $A_1 = [0, 1/3] \cup [2/3, 1]$. Continuing this process, we obtain $A_n = [0, 1/3^n] \cup \ldots \cup [((3^n - 1)/3^n) - 1], n = 0, 1, 2, \ldots$. Each A_n being a closed subset of the compact space A_0, we obtain a nonempty compact set $\mathfrak{C} = \bigcap_{n=0}^\infty A_n$, called the *Cantor set*. Alternatively, one can describe $\mathfrak{C}$ as follows: Let $E = \{0, 2\}$ be the two-point set and let E^ω be the countable product of E by itself, which is the set of all sequences with entries 0 or 2. For each $x = \{x_i\} \in E^\omega$, put $\varphi(x) = \sum_{i=1}^\infty (x_i/3^i)$. φ is an injective map of E^ω into $[0, 1]$. It can be shown that $\mathfrak{C} = \varphi(E^\omega)$, the Cantor set. $\mathfrak{C}$ is an uncountable compactum, has no isolated points, contains no interval of $\mathbb{R}$, and is perfect (i.e., closed and dense in itself).

2. Countably and Sequentially Compact Spaces

A topological space $(X, \mathcal{T})$ is called countably compact (CC) if from each countable open covering of X it is possible to extract a finite open covering. $(X, \mathcal{T})$ is called sequentially compact (SeC) if every sequence in it contains a convergent subsequence.

It is easily seen that each compact space is countably compact as well as sequentially compact, but a countably compact or a sequentially compact space may fail to be compact. In general, there is no connection between countable compactness and sequential compactness, although the two notions coincide on T_1-spaces. A metric space is compact iff countably compact iff sequentially compact. For hereditary and productive properties, see Table III.

3. σ-Compact Spaces

A topological space is called σ-compact if it is the countable union of compact spaces. Indeed, each compact space is σ-compact, but the converse is not true (e.g., $\mathbb{R}$).

Another frequently used notion of compactness in topology is as follows.

4. Locally Compact Spaces

A topological space $(X, \mathcal{T})$ is said to be locally compact (LCS) if each point of X has a neighborhood whose closure is compact.

Since a compact space is always a closed and compact neighborhood of each of its points, it follows that each compact space is locally compact, but the converse is not true (e.g., $\mathbb{R}$). In general, a locally compact space does not satisfy any separation axiom, but a Hausdorff locally compact space is a Tychonoff space and hence a T_3-space. Note that a Hausdorff locally compact space need not be normal.

Although locally compact spaces need not have the hereditary property, a closed subspace of a locally compact space is locally compact (see Table II). Moreover, a continuous, open, and surjective image of a locally compact space is locally, compact. Locally compact spaces are good for one-point compactifications (see below).

5. Hemicompact Spaces

A topological space $(X, \mathcal{T})$ is called hemicompact if there exists a countable family $\{K_n\}$ of compact subsets K_n of X such that $X = \bigcup_{n=1}^\infty K_n$ and each compact subset of X is contained in some K_n. (K_n's are said to form a fundamental system of compact subsets of X.) Each locally compact hemicompact space is σ-compact. Each locally compact hemicompact (hence σ-compact) metrizable space is a Polish space.

6. MB Spaces

A metric space (X, d) is said to be an MB space if each closed bounded subset of X is compact. [Note that a subset B of X is bounded if the diameter $\delta(B) = \sup\{d(x, y) : x, y \in B\}$ is finite.] Clearly, each compact metric space is an MB space and each MB space is complete, separable, locally compact and σ-compact (hence hemicompact), and also a Lindelöf, Polish space. As examples, $\mathbb{R}^n$ (n-dimensional euclidean space) is an MB space, but the set of rational numbers with the induced metric from $\mathbb{R}$ is not.

Each continuous map of an MB space into a metric space carries bounded sets into bounded sets. A metric space (F, d') which is the continuous image of an MB space (E, d) such that the inverse image of a bounded set of F is bounded in E is an MB space. Indeed, a closed subset of an MB space is an MB space and, so, is the countable product of compact metric spaces.

7. Pseudocompact Spaces

A Hausdorff topological space $(X, \mathcal{T})$ is called pseudocompact (PC) if every continuous real-valued function on X is bounded (see Section V.B.1). Every Hausdorff compact or countably compact space is pseudocompact. The converse is not true, however. In particular, each closed bounded interval $[a, b]$ is pseudocompact. See Table III for other implications.

8. Paracompact Spaces

A Hausdorff topological space $(X, \mathcal{T})$ is called paracompact (P) if each open covering $\{P_\alpha\}$ of X has a *refinement* $\{Q_\beta\}$ (i.e., there exists another covering $\{Q_\beta\}$ of X such that each P_α contains some Q_β) such that $\{Q_\beta\}$ is *neighborhood finite* (i.e., for each $x \in X$ there is an open neighborhood U of x such that U intersects with only a finite number of Q_β's).

Since a finite open subcovering of an open covering is a refinement, each Hausdorff compact space is paracompact, but the converse is not true. Each paracompact space is a T_4-space (hence normal) and each metric space is paracompact. Now the significance of paracompact spaces becomes very notable. These spaces are useful for the extension of continuous functions from subspaces to the whole spaces, as well as for metrization. Paracompact spaces do not have the hereditary or productive properties, but a closed subspace of a paracompact space is paracompact.

9. Compactifications

A topological space $(\hat{X}, \hat{\mathcal{T}})$ is said to be a compactification of a topological space $(X, \mathcal{T})$ if $(\hat{X}, \hat{\mathcal{T}})$ is compact and contains a homeomorphic image of $(\hat{X}, \mathcal{T})$ that is dense in $(X, \hat{\mathcal{T}})$.

Among compactifications, two are significant for applications: one-point compactification and Stone-Čech compactification. The former is the smallest, whereas the latter is the largest in a certain sense. The former requires local compactness, whereas the latter needs complete regularity.

Let $(X, \mathcal{T})$ be a Hausdorff locally compact space. Let ω be any point outside X. Put $\hat{X} = X \cup \{\omega\}$ and endow $\hat{X}$ with the topology $\hat{\mathcal{T}}$ defined as follows. Each open subset of X is open in $\hat{X}$ (i.e., each $U \in \mathcal{T}$ is also in $\hat{\mathcal{T}}$) and any subset $\hat{U}$ of $\hat{X}$ that contains ω is in $\hat{\mathcal{T}}$ provided that $\hat{X} \backslash \hat{U}$ is compact in X. With this prescription of open sets, it is easy to verify that $\hat{\mathcal{T}}$ defines a topology such that $(\hat{X}, \hat{\mathcal{T}})$ is a Hausdorff compact space in which $(X, \mathcal{T})$ is dense. Moreover, $\hat{X} \backslash X = \{\omega\}$ consists of a single point. $(\hat{X}, \hat{\mathcal{T}})$ is called the *one-point compactification of* $(X, \mathcal{T})$.

As an example, $\mathbb{R}$ is a Hausdorff locally compact space. If we put $\hat{\mathbb{R}} = \mathbb{R} \cup \{\infty\}$, then $\hat{\mathbb{R}}$ is the one-point compactification of $\mathbb{R}$. It is easy to see that $\hat{\mathbb{R}}$ is homeomorphic with the circle: $\{(x, y) \in \mathbb{R}^2 : x^2 + y^2 = 1\}$.

For the Stone-Čech compactification, let $(X, \mathcal{T})$ be a Tychonoff space. Then the set $C(X)$ of all continuous functions from X into $[0, 1]$ contains nonconstant continuous functions, provided that X is not trivial (i.e., it consists of more than one point). If we put $I = [0, 1]$, then $I^{C(X)} = \prod_{f \in C(X)} I_f, (I_f = I)$, the infinite product is compact by the Tychonoff theorem (see Section V.B.1). If to each $x \in X$, we assign $(f(x))_{f \in C(X)} \in I^{C(X)}$, we get an injective map. The closure of this injective image, denoted βX, in the compact space $I^{C(X)}$ is also compact. βX is called the *Stone-Čech compactification* of X.

10. Perfect and Proper Maps

Let $(X, \mathcal{T})$ and $(Y, \mathcal{T}')$ be two topological spaces. A map $f : X \to Y$ is called *inversely compact* (or simply *compact*) if the inverse image $f^{-1}(\{y\})$ of each singleton $\{y\}$ in Y is compact in X. Note that a continuous map need not be compact, nor need a compact map be continuous, or closed, or open. A continuous, closed, compact, surjective map is called *perfect*.

Perfect maps transport a number of properties of their domains to their ranges. For instance, if $f : (X, \mathcal{T}) \to (Y, \mathcal{T}')$ is perfect and if X is Hausdorff or regular or completely regular or metrizable or second countable or compact or countably compact or Lindelöf or locally compact or paracompact, so is Y, respectively. Perfect maps are often used in category theory.

A map $f : (X, \mathcal{T}) \to (Y, \mathcal{T}')$ is called *proper* (à la Bourbaki) if for each compact subset B of X, $f(B)$ is compact in Y, and for each compact subset C of Y, $f^{-1}(C)$ is compact in X. Each continuous map between compacta is proper and each continuous closed compact map of a T_2-space to a T_2-space is proper.

C. Baire Spaces and *k*-Spaces

Apart from those spaces dealt with in the preceding section, there are two more classes that are frequently used in topology and functional analysis. They are Baire spaces and *k*-spaces.

1. Baire Spaces

A subset A of a topological space $(X, \mathcal{T})$ is called *nondense* or *nowhere dense* if $(\bar{A})^0 = \varnothing$ (i.e., the interior of the closure of A is empty). A countable union of nondense sets is called a *set of the first category* or *a meager* set. A set that is not of the first category is called a *set of the second category*. The complement of a set of the first category is called a *residual set*.

A topological space $(X, \mathcal{T})$ is called a *Baire space* (B) if for any countable collection of closed nondense sets $\{A_n\}$ such that $X = \bigcup_{n=1}^{\infty} A_n$, there is at least one n for which $A_n^0 \neq \varnothing$. The class of all Baire spaces includes two of the important classes of topological spaces studied above, namely, the class of Hausdorff locally compact (hence, compact) spaces and the class of all complete metric spaces. The fact that each complete metric space is a Baire space is known as the *Baire category theorem*. Moreover, each Baire space is of the second category, and each open subset of a Baire space is a Baire space.

Let τ denote a cardinal number and let $D(\tau)$ be a set of real numbers of cardinality τ. Let $B_\tau = \{\{x_i\} : x_i \in D(\tau)\}$. For $x = \{x_i\}$, $y = \{y_i\}$ in B_τ, put $d(x, y) = 1/k$, where k is the least integer i such that $x_i \neq y_i$. Then d is a metric on B_τ and the metric space (B_τ, d) is called a *Baire space* of *weight* τ. The metric topology of B_τ has a base of closed–open sets.

2. *k*-Spaces

A topological space $(X, \mathcal{T})$ is called a *k*-space if each subset C of X is closed whenever for each compact subset K of X, $C \cap K$ is closed.

All Hausdorff locally compact (hence, compact) spaces as well as metric spaces are included in the class of all *k*-spaces. However, a *k*-space need not be a Baire space. Although the class of *k*-spaces does not have the hereditary property, it is true that a closed subspace of a Hausdorff

k-space is a *k*-space. Nor does the productive property hold in the class of *k*-spaces. A notable property for *k*-spaces is as follows. Each real-valued function on a *k*-space is continuous iff its restriction on each compact subset is continuous. Clearly, this generalizes the known fact about real-valued functions defined on the set of real numbers.

A completely regular T_2-space (i.e., a Tychonoff space) $(X, \mathcal{T})$ is called a k_r-space if each real-valued function on X whose restriction to each compact subset of X is continuous is indeed continuous on X. Each *k*-space is a k_r-space, but the converse is not true. For instance, the uncountable product $\mathbb{R}^\nu$ of $\mathbb{R}$, endowed with the product topology, is a k_r-space but not a *k*-space. One of the very useful properties of *k*-spaces is that each proper map between two *k*-spaces is continuous and closed.

D. Uniform Spaces and Uniformization

There are two important reasons for studying uniform spaces. One is to define uniform continuity, and the other is to define completeness for spaces more general than metric spaces. After a little thought, it becomes obvious to the reader that these two important concepts from classical analysis cannot be expressed in general topological spaces.

Let X be any set and $X^2 = X \times X$, the Cartesian product of X by itself. Let $\mathcal{U} = \{U\}$ be a filter in X^2 satisfying the following:

(i) Each U in $\mathcal{U}$ contains the diagonal set $\{(x, y) \in X^2 : x = y\}$.
(ii) For each U in $\mathcal{U}$, $U^{-1} = \{(y, x) : (x, y) \in U\}$ is also in $\mathcal{U}$.
(iii) For each U and $\mathcal{U}$ there is a V in $\mathcal{U}$ such that $V \circ V \subset U$, in which $V \circ V = \{(x, y) \in X^2 :$ for some z both (x, z) and (z, y) are in $V\}$.

Then the filter $\mathcal{U} = \{U\}$ is called a *uniformity* and the pair $(X, \mathcal{U})$ is called a *uniform space* (US).

Each metric space (X, d) becomes a uniform space with the collection $\{U_\varepsilon\}$ of sets in X^2 as its uniformity, where $U_\varepsilon = \{(x, y) \in X^2 : d(x, y) < \varepsilon\}$ for all real numbers $\varepsilon > 0$.

Each uniform space $(X, \mathcal{U})$ can be endowed with a topology called the *uniform topology* defined as follows. A subset P of X is called open in the uniform topology if for each $x \in P$ there is U in $\mathcal{U}$ such that $\{y \in X : (x, y) \in U\} \subset P$. Each uniform space with the uniform topology is completely regular. Thus, if the uniform topology is Hausdorff, the uniform space becomes a Tychonoff space.

A map f of a uniform space $(X, \mathcal{U})$ into another uniform space $(Y, \mathcal{V})$ is called *uniformly continuous* if for each

V in $\mathcal{V}$ there exists U in $\mathcal{U}$ such that $(f(x), f(y)) \in V$ whenever $(x, y) \in U$. In particular, for metric spaces $(X, d), (Y, d')$, the uniform continuity of f has the following formulation. For each $\varepsilon > 0$ there exists $\delta > 0$ such that $d'(f(x), f(y)) < \varepsilon$ whenever $d(x, y) < \delta$.

Indeed, each uniformly continuous map is continuous. However, the converse is not true [e.g., $f : \mathbb{R} \to \mathbb{R}$ such that $f(x) = x^2$]. However, if $(X, \mathcal{T})$ is a compact uniform space, each real-valued continuous function on X is uniformly continuous. In particular, each continuous map $f : [a, b] \to \mathbb{R}$ is uniformly continuous.

A filter $\{F_\alpha\}$ (or net $\{x_\alpha\}$) in a uniform space $(X, \mathcal{U})$ is called a *Cauchy filter* (or *Cauchy net*) if for each U in $\mathcal{U}$ there exists F_α such that the set $\{(x, y) : x, y \in F_\alpha\} \subset U$ [or there exists α_0 such that for all $\alpha, \beta \geq \alpha_0, (x_0, x_\beta) \in U]$. Indeed, each convergent filter (or net) is a Cauchy filter (or net), but not conversely. Whenever the converse holds, the uniform space is said to be *complete*. As for metric spaces, each incomplete Hausdorff uniform space $(X, \mathcal{T})$ can be completed to a complete Hausdorff uniform space $(\tilde{X}, \tilde{\mathcal{T}})$. $(\tilde{X}, \tilde{\mathcal{T}})$ is called the *completion of* $(X, \mathcal{T})$.

The class of all complete Hausdorff uniform spaces includes all compact Hausdorff uniform spaces and is productive but not hereditary. However, a closed subset of a complete Hausdorff uniform space is complete.

A topological space $(X, \mathcal{T})$ is said to be *uniformizable* if there exists a uniformity $\mathcal{U}$ with respect to which $\mathcal{T}$ becomes the uniform topology. Indeed, each metric space is uniformizable, as shown above. An important characterization of uniformization is the following. A topological space $(X, \mathcal{T})$ is uniformizable iff it is completely regular iff $\mathcal{T}$ is defined by a family $\{d_\alpha\}$ of pseudometrics (see Sections II.A and V.A). A topological space whose topology is defined by a family of pseudometrics is called a *gauge space*.

E. Sequential Spaces

First, we note that a sequence $\{x_n\}$ in a nonmetric topological space $(X, \mathcal{T})$ is said to converge to a point $x \in X$ if for each neighborhood U of x there is a positive integer n_0, depending upon U, such that for all $n \geq n_0, x_n \in U$. Indeed, this definition is equivalent to the convergence of a sequence in a metric space given in Section V.A.

A topological space $(X, \mathcal{T})$ is called a *sequential space* if for any subset A of $x, A \neq \bar{A}$, there is a sequence $\{x_n\}$ in A converging to a point of $\bar{A} \setminus A$. Each metrizable space (in particular, $\mathbb{R}^n$) is a sequential space, but $\mathbb{R}^\mathbb{R}$ (see Section VII), the set of all real functions on $\mathbb{R}$ endowed with the product topology, is not.

For a subset A of a sequential space, the set of all points $x \in X$ for which there exists a sequence in A converging to x, is called the *sequential closure* of A and is de-

noted $\bar{A}^s$. It is easily seen that $\bar{\varnothing}^s = \varnothing, A \subset \bar{A}^s \subset \bar{A}$, and $\overline{A \cup B}^s = \bar{A}^s \cup \bar{B}^s$ for all subsets A, B of X. In general, however, $\overline{\overline{A}^s}^s \neq \bar{A}^s$, i.e., the sequential closure operation is not idempotent, unlike the topological closure, where we have $\bar{\bar{A}} = \bar{A}$.

A topological space $(X, \mathcal{T})$ is called a Fréchet–Urysohn space if $\overline{\bar{A}^s} = \bar{A}$ for all subsets A of X. Clearly, each Fréchet–Urysohn space is a sequential space but the converse is not true. A useful characterization of a Fréchet–Urysohn space is: A topological space $(X, \mathcal{T})$ is a Fréchet–Urysohn space iff each subspace of $(X, \mathcal{T})$ is a sequential space. A first countable (hence metrizable) space is a Fréchet–Urysohn space. Each continuous map $f : (X, T) \to (Y, T')$ is sequentially continuous, but the converse is not true. However, if $(X, \mathcal{T})$ is a sequential space, then each sequential map of $(X, \mathcal{T})$ into $(Y, \mathcal{T}')$ is continuous. Indeed, on metric spaces, sequential continuity coincides with the usual concept of continuity as noted above (see Section V.A).

F. Proximity Spaces

Let $\mathcal{P}(X)$ denote the class of all subsets of a set X. Let δ define a binary relation on $\mathcal{P}(X)$. If $A\delta B$, then A is said to be *near* B for A, B in $\mathcal{P}(X)$ and $A \cancel{\delta} B (\cancel{\delta},$ the negation of δ) means that A is *distant* from B. The pair (X, δ) is called a *proximity space* if the following axioms hold:

(i) $A\delta B \Leftrightarrow B\delta A$.
(ii) $\{x\}\delta\{x\}$.
(iii) $A\delta(B \cup C) \Leftrightarrow A\delta B$ or $A\delta C$.
(iv) $A \cancel{\delta} \varnothing$ for all $A \subseteq X$.
(v) $A \cancel{\delta} B \Rightarrow$ there is $C \in \mathcal{P}(X)$ such that $A \cancel{\delta} C$ and $B \cancel{\delta}(X \setminus C)$, where A, B, C are subsets of X and $x \in X$. δ is called a *proximity* for X.

A proximity δ for (X, δ) induces a topology, called the *proximity topology*, on X as follows. Let $\mathcal{T}_\delta$ denote the collection of all subsets P of X such that for each $x \in P$, $\{x\} \cancel{\delta} P^c$. The collection $\mathcal{T}_\delta$ satisfies all the properties of a topology and $(X, \mathcal{T}_\delta)$ becomes a completely regular topological space.

For a proximity space $(X, \delta), d(A, B) = 0 \Leftrightarrow A\delta B$ gives a metric, where $d(A, B) = \inf\{d(a, b) : a \in A, b \in B\}$. On the other hand, if (X, d) is a metric space, then the relation defined by $A\delta_d B \Leftrightarrow d(A, B) = 0$ gives a proximity δ_d and (X, δ_d) becomes a proximity space. Moreover, the proximity topology induced by δ_d coincides with the original metric topology of (X, d). Thus each metric space is a proximity space.

Similarly, if $(X, \mathcal{U})$ is a Hausdorff uniform space, then the relation δ_u, defined by $A\delta_u B \Leftrightarrow$ for each $U \in \mathcal{U}$ there exists an element $a \in A$ and an element $b \in B$ such that $(a, b) \in U$ gives a proximity relation on X. On the other hand, for each proximity space (X, δ), let $U_{A,B} = [(A \times B) \cup (B \times A)]^c$. Then the collection $\mathcal{U}_\delta = \{U_{A,B} : A\not\delta B\}$ forms a subbase of a uniformity for X. Moreover, the proximity derived from this uniformity coincides with the given proximity δ. Thus each Hausdorff uniform space admits a proximity called the *uniform proximity*.

The collection of all proximities on a set X can be ordered by inclusion. A proximity δ_1 is larger than a proximity δ_2 if $A\delta_1 B \Rightarrow A\delta_2 B$ for all $A, B \in \mathcal{P}(X)$. Under this ordering, there is a largest proximity which can be described as follows: $A\not\delta B \Leftrightarrow A \cap B = \varnothing$. The largest proximity induces a discrete topology.

For any proximity space (X, δ), $\mathrm{Cl}_\delta A = \{x \in X : \{x\}\delta A\}$ defines a closure operation on X which gives the same proximity topology. Moreover, $A\delta B \Leftrightarrow \mathrm{Cl}_\delta A\delta \mathrm{Cl}_\delta B$. If (X, δ) is a compact proximity space, then $A\delta B \Leftrightarrow \mathrm{Cl}_\delta A \cap \mathrm{Cl}_\delta B \neq \varnothing$. Every Hausdorff proximity space (X, δ) is a dense subspace of a unique compact Hausdorff space αX such that $A\delta B$ in X iff $\mathrm{Cl}_\delta A \cap \mathrm{Cl}_\delta B \neq \varnothing$ where $\mathrm{Cl}_\delta A$ is taken in αX. αX is called the *Smirnov compactification* of X.

G. Connected and Locally Connected Spaces

For the definition of these spaces, see Section I.A. In addition to the properties of these spaces given in that section, we have the following.

A continuous map of a topological space into another carries connected sets into connected sets. A product of connected spaces is connected. If A is a connected subset of a topological space and B is any set such that $A \subset B \subset \bar{A}$, then B is connected. Any interval of $\mathbb{R}$, including $\mathbb{R}$ itself, is connected. Each convex subset of a real topological vector space (see Section VIII.B) is connected. An interesting characterization of connected spaces is: A topological space $(X, \mathcal{T})$ is connected if and only if every continuous map f of X into a discrete space Y is constant, i.e., $f(X) = \{y\}, y \in Y$.

A characterization of locally connected spaces is: A topological space is locally connected if and only if the components of its open sets are open. Local connectedness is preserved under closed maps.

A connected, metrizable compactum $(X, \mathcal{T})$ is sometimes called a *continuum*. Let $(X, \mathcal{T})$ be a continuum and $(Y, \mathcal{T}')$ a Hausdorff space. If $f : (X, \mathcal{T}) \to (Y, \mathcal{T}')$ is a continuous surjective map, then $(Y, \mathcal{T}')$ is also a continuum. Further, if $(Y, \mathcal{T}')$ is locally connected, so is $(Y, \mathcal{T}')$.

A topological space each of whose components consists of a singleton is called *totally disconnected*. For example, the set of integers is totally disconnected.

H. Extensions and Embeddings

Recall Tietz's extension theorem (Section IV), which states that each continuous function from a closed subset Y of a normal space $(X, \mathcal{T})$ into $[0, 1]$ can be extended to a continuous function of X into $[0, 1]$. It can be shown that $[0, 1]$ can be replaced by any interval $[a, b]$, $-\infty < a < b < \infty$. Actually, $[a, b]$ can be replaced by unit ball $B_n = \{(x_1, \ldots, x_n), x_i \in \mathbb{R} : \sum_{i=1}^n |x_i^2| \leq 1\}$ of $\mathbb{R}^n$. Furthermore, similarly, a continuous map $f : Y \to S^n = \{(x_1, \ldots, x_n), x_i \in \mathbb{R} : \sum_{i=1}^n |x_i^2| = 1\}$ (unit sphere of $\mathbb{R}^n$) extends to a continuous map on a neighborhood of Y.

We note that the given function $f : Y \to [0, 1]$ in Tietz's extension theorem is bounded. It is possible to replace $[0, 1]$ by $\mathbb{R}$, i.e., the theorem is true even for unbounded functions. Furthermore, normality of $(X, \mathcal{T})$ can be relaxed under certain conditions.

Let Y be a compact subset of a Tychonoff space $(X, \mathcal{T})$ (note that each normal T_1-space is a Tychonoff space) and $f : Y \to [0, 1]$ a continuous function, then there exists a continuous function $\tilde{f} : (X, \mathcal{T}) \to [0, 1]$ such that $\tilde{f}(x) = f(x)$ for all $x \in Y$. Further, each continuous bounded function on a Tychonoff space $(X, \mathcal{T})$ extends to its Stone-Čech compactification βX. In this result, the implicit assumption of the range of f being in $[a, b]$ can be weakened under suitable conditions. For instance, a continuous map f of a Tychonoff space $(X, \mathcal{T})$ to a compactum Y can be extended to a continuous map $\tilde{f} : \beta X \to Y$ such that $\tilde{f}(x) = f(x)$ for $x \in X$.

Let $(X, \mathcal{U})$ be a uniform space with its uniform topology induced by $\mathcal{U}$, and Y a dense subspace of X. Then each uniformly continuous map f of Y into a complete Hausdorff uniform space Z extends to a uniformly continuous map $\tilde{f} : X \to Z$ such that $\tilde{f}(x) = f(x)$ for all $x \in Y$.

Whenever there is a bijective map between two sets, they are deemed to be set-theoretically the same. If there is a surjective homeomorphism between two topological spaces, they are regarded to be the same set-theoretically as well as topologically. However, if a homeomorphism is into, i.e., if there is a continuous open injective map f of a topological space $(X, \mathcal{T})$ into $(Y, \mathcal{T}')$, then f is called an *embedding* of X into Y. In this case, $f(X)$ need not be equal to Y but X is homeomorphic to $f(X)$ (a subset of Y) with the topology induced from Y. Embeddings of abstract topological spaces into given or known topological spaces are of paramount importance and interests.

Simple examples of embeddings are as follows: Each metric space can be embedded into a complete metric space (its completion). In particular, the set of rational numbers is embedded into reals. Each Hausdorff uniform space can be embedded into a complete Hausdorff uniform space. Somewhat more sophisticated embeddings are as follows. Each second countable T_3-space can be embedded into the Hilbert cube. Furthermore, a Hausdorff space embeds into the Cantor perfect set if

define the ordering in I^2 as follows: $(a_1, b_1) \leq (a_2, b_2)$ if and only if $a_1 < a_2$ or $a_1 = a_2$ and $b_1 < b_2$. This is called *dexicographic ordering* on I^2. Thus one obtains an order topology on I^2. It is possible to show from this example that a subspace of an ordered space need not be an ordered space.

We end this section with two diagrams indicating the interconnection of some spaces studied above. (As usual, $A \to B$ means that A is contained in B.)

$$
\begin{array}{ccc}
& \text{locally compact } T_2 & \\
\nearrow & & \searrow \\
\text{Compact } T_2 & & \text{Baire} \to k\text{-space} \to k_r\text{-space.} \\
\searrow & & \nearrow \\
& \text{complete metric} &
\end{array}
\tag{1}
$$

$$
\begin{array}{ccc}
\text{Metric compact} \to \text{compact} \to \left\{ \begin{array}{c} \text{locally compact} \\ \sigma\text{-compact} \end{array} \right\} \to & \text{paracompact} \\
\downarrow & \downarrow \\
\begin{array}{c} \text{locally compact} \\ \text{2nd countable} \end{array} \to \text{Polish} \to \text{Souslin} & \to \text{metric} \to \text{normal.}
\end{array}
\tag{2}
$$

and only if it has a countable base of closed–open sets. A metric space (X, d) can be embedded into a Baire space $B(\tau)$ of weight τ if and only if X has a base $\mathscr{B}$ of closed–open sets such that the cardinality of $\mathscr{B}$ is less than or equal to τ. Each uniform space embeds in a product of pseudometric spaces and each Hausdorff uniform space embeds into a product of metric spaces.

Topological spaces having a base of closed–open sets are called *0-dimensional*. Every 0-dimensional T_0-space is completely regular. In a compact 0-dimensional space, two disjoint closed subsets can be separated by two disjoint open sets whose union is the whole space. Every locally compact totally disconnected space is 0-dimensional.

I. Topologies Defined by an Order

Let $(X, \leq)$ be a linearly ordered set, i.e., for $x, y \in X$, "$x \leq y$" means that "x is less than or equal to y," and for all $x, y \in X, x \neq y$, either $x < y$ or $y < x$. The notation $(x, y) = \{z \in X : x < z < y\}$ is called an "interval" in X. The collection of all such intervals forms a base of a topology on X, called the *order topology*, and $(X, \leq)$ endowed with the order topology is called an *ordered space*. $\mathbb{R}$ is an ordered space. An ordered space which is sequential is first countable. A discrete space is an ordered space. Unlike metric spaces, a separable ordered space need not be second countable.

$\mathbb{R}^2$ with the usual metric topology is not an ordered space. Let $I^2 = [0, 1] \times [0, 1]$ be the unit square in $\mathbb{R}^2$. We

As pointed out above, a continuous function may not have a fixed point in general. However, if the space is restricted, one has a different situation. Specifically, if $D^n = \{(x_1, \ldots, x_n) \in \mathbb{R}^n : x_1^2 + \cdots + x_n^2 \leq 1\}$ is the closed n-dimensional disk, each continuous map $f : D^n \leftarrow D^n$ has a fixed point but not necessarily unique. This is known as the Brouwer fixed point theorem in the mathematical literature. This result has many extensions in functional analysis.

VI. FIXED POINTS

The information regarding those points that remain fixed under certain transformations is very useful. It can be put to work in many fields (e.g., differential equations and algebraic topology).

Let X be a set and $f : X \to X$ a self-map. $x \in X$ is called a *fixed point* of f if $f(x) = x$. Indeed, the identity map $i : X \to X, i(x) = x$, keeps every point of X fixed. On the other hand, $f : \mathbb{R} \to \mathbb{R}$ when $f(x) = e^x + 1$ has no fixed point. Neither of these extreme cases has useful applications. The case where there exists a unique fixed point is of utmost significance. The result that guarantees the existence and uniqueness of a fixed point is known as the contraction principle.

Let (X, d) be a metric space. A map $f : X \to X$ is called a *contraction* if there exists $\alpha, 0 \leq \alpha < 1$, such that for all $x, y \in X$ we have

$$d(f(x), f(y)) \leq \alpha d(x, y).$$

It is obvious that a contraction map is uniformly continuous (hence, continuous), but not conversely.

A. Contraction Principle

Each contraction map $f:(x, d) \to (X, d)$ of a complete metric space (X, d) has a unique fixed point.

Some important extensions are as follows: If C is a convex compact subset of a normed space (see Section VIII.B for the definitions), then a continuous map $f: C \to C$ has a fixed point (Banach–Schauder theorem). This is generalized by Tychonoff: Let C be a convex compact subset of a locally convex space (see Section VIII.B); then a continuous map $f: C \to C$ has a fixed point. An important contribution to fixed point theorems is due to Markov and Kakutani: Let C be a convex compact subset of a locally convex space. Let $\{f_\alpha: \alpha \subset \Gamma\}$ be a family of continuous maps from C to C such that each f_α is *affine* [i.e., for all $x, y \in C$ and $0 < t < 1$, $f_\alpha(tx + (1 - t)y) = tf_\alpha(x) + (1 - t)f_\alpha(y)$] and the family is commuting [i.e., $f_\alpha \circ f_\beta(x) = f_\beta \circ f_\alpha(x)$ for all $x \in C$]. Then there is $x_0 \in C$ such that $f_\alpha(x_0) = x_0$ for all $\alpha \in \Gamma$, i.e., the family $\{f_\alpha\}$ has a common fixed point.

VII. FUNCTION SPACES

So far we have been concerned with the properties of topological spaces and maps. In this section we want to topologize certain sets of functions. One might say that so far we have been working on the ground floor and now we want to move up to the first floor.

Let X, Y be two nonempty sets. Y^X denotes the set of *all* maps $f: X \to Y$. If X, Y are topological spaces, then $C(X, Y)$ denotes the set of all continuous maps of X into Y. Clearly, $C(X, Y)$ is a subset of Y^X. Both Y^X and $C(X, Y)$ with some appropriate topologies are called function spaces. Several topologies are very useful. Two of them are point-open topology and compact-open topology.

For each $x \in X$ and each open set V in Y, let $T(x, V) = \{f \in Y^X: f(x) \in V\}$. As x runs over X and V over all open subsets of Y, the collection $\{T(x, V)\}$ forms a subbase of a topology called the *point-open topology* and is denoted $\mathcal{T}_p$. Similarly, if K is a compact subset of X and V an open subset of Y, then

$$T(K, V) = \{f \in Y^X: f(x) \in V \text{ for all } x \in K\}.$$

The topology (denoted $\mathcal{T}_c$) having the collection $\{T(K, V)\}$ as a subbase is called the *compact-open topology*. Since each singleton is compact, it follows that $\mathcal{T}_c$ is finer than $\mathcal{T}_p$. Hence, the induced topology $\mathcal{T}_c$ on $C(X, Y)$ is finer than the induced topology $\mathcal{T}_p$. If Y is a Hausdorff

space, both $(Y^X; \mathcal{T}_p/\mathcal{T}_c)$ and $(C(X, Y), \mathcal{T}_p/\mathcal{T}_c)$ are also Hausdorff spaces.

The following three cases are noteworthy. If (X, d) is a metric locally compact space and (Y, d') a complete metric space, then $(C(X, Y), \mathcal{T}_c)$ is a complete uniform space. If (X, d) is a compact metric space and (Y, d') a complete metric, then $(C(X, Y), \mathcal{T}_c)$ is a complete metric space with metric $d^+(f, g) =$ the least upper bound of $d'(f(x), g(x))$, $x \in X$ for $f, g \in C(X, Y)$. Thus, for a compact metric space (X, d), $C(X) = C(X, \mathbb{R})$ is a complete metric space. The same is true for $C[a, b]$.

Hereafter, both $C(X, \mathbb{R})$ and $C(X, \mathbb{C})$ are denoted $C(X)$, the space of all real- or complex-valued continuous functions on the topological space $(X, \mathcal{T})$. Indeed, if X is compact, then each $f \in C(X)$ is bounded and $\|f\| =$ the least upper bound of $\{|f(x)|, x \in X\}$ is finite. $\|f\|$ is called the *norm* of f and $d(f, g) = \|f - g\|$ gives a metric on $C(X)$ which is complete. For each fixed $x \in X$, $\varphi_x(f) = f(x)$, $f \in C(X)$, defines a map of $C(X)$ into $\mathbb{R}$ or $\mathbb{C}$. φ_x is called an *evaluation map*. The map φ_x is useful. We come back to it later.

In some particular cases, the topological space X can be embedded in $C(X)$. For instance, let (X, d) be a metric space and $C_b(X)$, the space of all continuous bounded functions on X. $C_b(X)$ is a metric space with the metric $d^+(f, g) =$ the least upper bound of $\{|f(x) - g(x)| : x \in X\}$, $f, g \in C_b(X)$. Now let $x_0 \in X$ be a fixed element, and for each $y \in X$, set $f_y(x) = d(x, y) - d(x, x_0)$, $x \in X$. Then for each $y \in X$, $f_y \in C_b(X)$ and the map $y \to f_y$ of X into $C_b(X)$ is an isometry. Thus X can be isometrically embedded in $C_b(X)$. If X is compact, then $C_b(X) = C(X)$ and so each compact metric space can be isometrically embedded in $C(X)$.

If X, Y are topological spaces, then a continuous map $f: X \to Y$ induces a map $f^*: C(Y) \to C(X)$ by $f^*(g) = g \circ f$ (composition map). It is interesting to know what properties of f induce similar properties for f^*. An important scenario occurs when both X and Y are compact, viz., X, Y are homeomorphic if and only if $C(X), C(Y)$ are isometrically homeomorphic.

VIII. MARRIAGE OF TOPOLOGY AND ALGEBRA

Algebraic operations such as addition, subtraction, multiplication, and division play an important role in algebra. Topology, however, is not primarily concerned with them. When algebra and topology are married in the sense that a set with algebraic operations is endowed with a topology, new and interesting theories emerge. For instance, in the field of functional analysis (a major branch of mathematics), the topics of topological groups, topological vector spaces, and topological algebras, among others, are the outcome of this marriage.

Among many topics in algebra, one specifically comes across the following algebraic structures: semigroups, groups, vector spaces, and algebras. By endowing them with topologies so that the underlying algebraic operations are continuous, one obtains interesting areas of mathematics.

A. Topological Semigroups and Groups

A set S with an associative binary operation [i.e., for all $x, y \in S, x \circ y \in S$ and $x \circ y \circ z = (x \circ y) \circ z = x \circ (y \circ z)$, in which "$\circ$" denotes the binary operation] is called a *semigroup*. A semigroup S with a topology $\mathcal{T}$ is called a *topological semigroup* (TS) if the map $(x, y) \to x \circ y$ of $S \times S$ into S is continuous.

A semigroup G is called a *group* if G has an *identity* $e (e \circ x = x \circ e = x$ for $x \in G$) and each $x \in G$ has an *inverse* x^{-1} ($x^{-1} \circ x = x \circ x^{-1} = e$). According to whether "$\circ$" is "$+$" or "$\times$," the group is called *additive* or *multiplicative*. If $x \circ y = y \circ x$, for all $x, y \in G$, the group G is called *Abelian* or *commutative*. A map f of a group G into a group H is called a *homomorphism* if $f(xy) = f(x) f(y)$. A bijective homomorphism of a group G into a group H is called an *isomorphism*.

A group G with a topology is called a *topological group* (TG) if the map $(x, y) \to xy^{-1}$ of $G \times G$ into G is continuous.

Each topological group is a uniform space, hence completely regular. If the topology satisfies the T_0-axiom, the topological group becomes a Tychonoff space, hence Hausdorff. A Hausdorff topological group is metrizable iff the neighborhood system at its identity has a countable base.

The class of all topological groups satisfies arbitrary productive properties. Moreover, the closure of each subgroup (i.e., a subset H of G such that H by its own right is a group under induced algebraic operations) is a topological group and each open subgroup is closed.

Let H be an invariant ($xH = Hx$ for all $x \in G$) subgroup of a group G. To each $x \in G$ we associate xH, called the *coset*. Let G/H denote the set of all cosets. With multiplication $(xH)(yH) = xyH$ and identity $H, G/H$ becomes a group called the *quotient* or *factor* group. The map $\phi : x \to xH$ of G onto G/H is called the *quotient* or *canonical* map; ϕ is a homomorphism because $\phi(xy) = \phi(x) \phi(y)$. If G is a topological group, we can endow G/H with the quotient topology (see Section III), making ϕ continuous. It turns out that ϕ is a continuous, open, and surjective homomorphism of the topological group G into the topological group G/H. The quotient topology on G/H is Hausdorff iff H is closed.

1. Examples

$\mathbb{R}^n$ ($n \geq 1$) with coordinatewise addition $(x_1, \ldots, x_n) + (y_1, \ldots, y_n) = (x_1 + y_1, \ldots, x_n + y_n)$ is an additive Abelian group and has identity $(0, \ldots, 0)$. With the euclidean topology defined in Section 1.A, $\mathbb{R}^n$ is an Abelian additive topological group. In particular, the set $\mathbb{R}$ of all real numbers is an Abelian additive topological group in which $\mathbb{Z}$, the subgroup of all integers, is a closed invariant subgroup. Thus, the quotient group $\mathbb{R}/\mathbb{Z}$ is a topological group. We denote $\mathbb{R}/\mathbb{Z}$ by $\mathbb{T}$. $\mathbb{T}$ is actually isomorphic with the circle group $\{e^{it} : 0 \leq t \leq 2\pi, i = \sqrt{-1}\}$. Hence, $\mathbb{T}$ is a compact Abelian topological group.

If G is an Abelian topological group, the set G' of all continuous homomorphisms $\phi : G \to \mathbb{T}$ is called the *dual group* of G. We can endow G' with the compact-open topology $\mathcal{T}_c$ (see Section VII). If G is a Hausdorff locally compact Abelian topological group, then $(G', \mathcal{T}_c)$ is also a Hausdorff locally compact Abelian topological group. Repeating the process, we see that G'' (the dual group of G') with the compact-open topology is also a locally compact Abelian topological group. A celebrated result called the Pontrjagin duality theorem tells us that G and G'' are isomorphic and homeomorphic. Furthermore, if G is compact (discrete), G' is discrete (compact).

If algebraic operations in a topological group with some additional structure are differentiable instead of being only continuous, it leads to the theory of Lie groups, which form an important branch of mathematics.

B. Topological Vector Spaces, Banach Spaces, and Hilbert Spaces

An algebraic system that is richer than groups is vector space. All scalars involved in this section are either real or complex numbers.

An Abelian additive group E is called a *real* or *complex* (depending on which scalars are used throughout) *linear* or *vector space* if for all $x \in E$ and scalar $\lambda, \lambda x \in E$ satisfies the following: $\lambda(x + y) = \lambda x + \lambda y$; $(\lambda + \mu)x = \lambda x + \mu x$; $\lambda(\mu x) = (\lambda \mu)x$ for all scalars λ, μ and all $x, y \in E$; $1x = x$ and $0x = 0$, the identity of the additive group E. If $F \subset E$ and F is a vector space in its own right over the same scalars as those of E, then F is called a *linear subspace* of E. A subset C of E is called *circled* if for scalars $\lambda, |\lambda| \leq 1$ and $x \in C, \lambda x \in C$. A subset C of a vector space E is called *convex* if for all $x, y \in C$ and $0 \leq \lambda \leq 1, \lambda x + (1 - \lambda)y \in C$. C is *absorbing* if for all $x \in E$ there is $\alpha_0 > 0$ such that for $\lambda, |\lambda| \geq \alpha_0, x \in \lambda C$.

A linear or vector space E endowed with a topology $\mathcal{T}$ is called a *topological linear* or *vector space* (TVS) if the maps $(x, y) \to x + y$ and $(\lambda, x) \to \lambda x$ of $E \times E$ into E and of $K \times E$ into E, respectively, are continuous, in which the field K of real numbers or complex numbers is endowed with its natural topology. If there exists a base of convex neighborhoods of identity 0 in E, then E is called a *locally convex* (LC) space.

Clearly, each topological vector space is an Abelian additive topological group, hence it is a uniform space and, therefore, completely regular. Hence, a Hausdorff topological vector space is a Tychonoff space. As for topological groups, a Hausdorff topological vector space is metrizable iff the neighborhood system at 0 has a countable base. Among the metrizable topological vector spaces, there is a very distinguished and useful subclass of spaces called the normable spaces defined below.

Let E be a real (or complex) vector space. A map $p : E \to K$ is called a *functional*. If $p(x) \geq 0$ for all $x \in E$ such that $p(\lambda x) = |\lambda| p(x)$ and $p(x + y) \leq p(x) + p(y)$, then p is called a *seminorm*. If $p(x) = 0 \Leftrightarrow x = 0$, then p is called a *norm*, denoted $p(x) = \|x\|$. A seminorm p on a vector space E defines a pseudometric $d(x, y) = p(x - y)$. If p is a norm, d becomes a metric. If the topology of a topological vector space is induced by a *norm*, it is called a *normable* topological vector space. If a normable topological vector space is complete in its induced metric topology, it is called a *Banach space* (BAS). Each Banach space is a complete metric locally convex space called a *Fréchet space* (F). A Hausdorff locally convex space is called a *barreled space* (BS) if, in it, each *barrel* (convex, circled, absorbing, and closed set) is a neighborhood of 0. Each Baire locally convex (BLC) space is a barreled space, and each Fréchet space is a Baire space.

A map f of a vector space E into another vector space F is called *linear* if $f(\alpha x + \beta y) = \alpha f(x) + \beta f(y)$ for all $x, y \in E$ and scalars α, β. A functional f on a normed space E is called *bounded* if there is a real number $M > 0$ such that $|f(x)| \leq M \|x\|$ for all $x \in E$. An interesting but simple fact is that a linear functional on a normed space is bounded iff it is continuous.

To exhibit the richness that results from the marriage of topology and vector spaces, we cite a few prototypical results.

The first is the so-called Hahn–Banach extension theorem: Let F be a linear subspace of a vector space E. Let p be a seminorm on E, and f a linear functional on F such that $|f(x)| \leq p(x)$ for all $x \in F$. Then there exists a linear functional $\tilde{f}$ on E such that $\tilde{f}(x) = f(x)$ for all $x \in F$ and $|\tilde{f}(x)| \leq p(x)$ for all $x \in E$. It is worth comparing this result with Tietze's extension theorem given in Section IV.

The heart of functional analysis, especially that of topological vector spaces, lies in the so-called *twins* of functional analysis, popularly known as the open-mapping and closed-graph theorems.

1. Open-Mapping Theorem

Each continuous linear surjective map of a Fréchet (in particular, Banach) space onto a barreled (in particular, Fréchet or Banach) space is open.

2. Closed-Graph Theorem

Each linear map of a barreled (in particular, Fréchet or Banach) space into a Fréchet (or Banach) space with closed graph is continuous.

Another important result in topological vector spaces is as follows. If $\{f_n\}$ is a sequence of continuous linear functionals on a barreled (or Fréchet or Banach) space E such that $f(x) = \lim_n f_n(x), x \in E$, then f is also linear and continuous. (This is called the Banach–Steinhaus theorem.)

If E is a real or complex topological vector space, the set E' of all continuous linear functionals on E is called the *dual* of E. If E is a normed space, so is E' with the norm $\|f\| = $ the least upper bound of $\{|f(x)| : \|x\| \leq 1\}$. Actually E' is a Banach space regardless of E being a complete or incomplete normed space. In special cases E' can be determined by the points of E, as shown below.

A normed space E is called a *pre-Hilbert* or *an inner product space* if the norm $\| \cdot \|$ satisfies the so-called *parallelogram law*: $\|x + y\|^2 + \|x - y\|^2 = 2(\|x\|^2 + \|y\|^2)$. The parallelogram law implies the existence of a bilinear functional $\langle , \rangle : E \times E \to K$ satisfying the following properties: $\langle x, x \rangle \geq 0; \langle x, x \rangle = 0 \Leftrightarrow x = 0; \langle x + y, z \rangle = \langle x, z \rangle + \langle y, z \rangle$; and $\langle x, y \rangle = \langle y, x \rangle$ (or $\overline{\langle y, x \rangle}$ for the complex scalars). We see that $\|x\| = +\sqrt{\langle x, x \rangle}$ gives a norm. If a pre-Hilbert space is complete, it is called a *Hilbert space* (H). For a Hilbert space H, its dual H' can be identified with the points of H as demonstrated by the so-called Riesz representation theorem: If H is a Hilbert space, then for each $f \in H'$ there exists a unique element $y_f \in H$ such that $f(x) = \langle x, y_f \rangle, x \in H$ with $\|y_f\| = \|f\|$. (See Table II for the interrelation of various topological spaces.)

Hereafter, we assume all topological vector spaces to be Hausdorff. For any vector space E over $\mathbb{R}$ or $\mathbb{C}$, E^* denotes the set of all real or complex linear functionals on E, called the *algebraic dual* of E. If E is a TVS, E' denotes the set of all continuous linear functionals on E, called the *topological dual* or simply the *dual* of E as mentioned above. If E is LC and $E \neq \{0\}$, then $E' \neq \{0\}$ and clearly $E' \subset E^* \subset \mathbb{R}^E$ (or $\mathbb{C}^E$), where $\mathbb{R}^E$ carries the pointwise convergence topology. The topology induced from $\mathbb{R}^E$ to E' is called the weak-star or w^*-*topology*. Under this topology all maps $f \to f(x)$ (for each fixed x, i.e., evaluation maps) are continuous.

Similarly, the coarsest topology on E which makes all the maps $x \to f(x)$ (for any fixed $f \in E'$) continuous is called the *weak topology* $\sigma(E, E')$ on E, which is coarser than the initial topology of E. The space E with the weak topology $\sigma(E, E')$ becomes an LC space. The finest locally convex topology on E that gives the same dual E' of E is called the *Mackey topology* $\tau(E, E')$, which is finer than the initial topology $\mathcal{T}$ of E. Thus we have $\sigma(E, E') \subset \mathcal{T} \subset \tau(E, E')$. An LC space with the Mackey topology [i.e., $\mathcal{T} = \tau(E, E')$] is called a *Mackey space*. Every barreled (hence Fréchet, Banach, or Hilbert) space is a Mackey space. However, there exist Mackey spaces which are not Fréchet spaces. The weak, weak*, and Mackey topologies are extensively used in functional analysis. Here we list only a few samples of their usage.

Let E be a Banach space; then for each weakly compact subset A of E, the convex closure $\bar{c}_o(A)$ of A is also weakly compact, where $\bar{c}_o(A)$ is the intersection of all closed convex subsets of E containing A. Further, a weakly closed subset A of E is weakly sequentially compact if and only if A is weakly compact. This is known as the Eberlein theorem. Note that in general topological spaces sequential compactness is not equivalent to compactness. Furthermore, if A is a closed convex subset of E such that for each $f \in E'$ there is $x_f \in A$ with $|f(x_f)| =$ the least upper bound of $\{|f(x)| : x \in A\}$, then A is weakly compact. This is known as the James theorem.

As pointed out earlier, the dual E' of a normed space E is a Banach space with the norm $\|f\|$. The w^*-topology on E' is coarser than this norm topology. The unit ball $\{f \in E' : \|f\| \leq 1\}$ of E' is w^*-compact (Alaoglu's theorem) but not norm compact in general. *Actually, the unit ball of a normed space E is norm compact iff E is finite-dimensional*, i.e., E is homeomorphic to $\mathbb{R}^n$ (or $\mathbb{C}^n$) for some finite positive integer n.

Since the dual E' of a normed space E is a normed (actually Banach) space, we can consider the dual E'' of E'. E'' is called the *bidual* of E. Clearly E'' is also a Banach space. There is a natural embedding of E in E''. Put $x''(f) = f(x), f \in E'$, for each $x \in E$. Then it can be verified that $x'' \in E''$ for each $x \in E$ and so $x \to x''$ gives a mapping of E into E'' such that $\|x''\| = \|x\|$. Thus E is embedded into E'' isometrically. In general this embedding is not surjective, i.e., $E \subset E''$, $E \neq E''$. Whenever $E = E''$, E is called *reflexive*. Since E'' is a Banach space, it follows that a reflexive normed space must be a Banach space. By virtue of the Riesz representation theorem, each Hilbert space is reflexive. Here are some examples of reflexive and nonreflexive spaces. Let $1 \leq p < \infty$ be a given real number. Let $\ell_p = \{\{a_i\} : \sum_{i=1}^{\infty} |a_i|^p < \infty\}$. Then ℓ_p is a Banach space with the norm $\|\{a_i\}\|_p = \{\sum_{i=1}^{\infty} |a_i|^p\}^{\frac{1}{p}}$. For all $p's$, $1 < p < \infty$, ℓ_p is a reflexive Banach space; note that ℓ_2, being a Hilbert space, is reflexive, but ℓ_1 is not reflexive. Note that for $1 < p < \infty$, $\ell'_p = \ell_q$, where

$(1/p) + (1/q) = 1$ and $\ell'_1 = \ell_\infty$, the space of all bounded sequences.

For a Tychonoff space $(X, \mathcal{T})$, $C(X) = (C(X), \mathcal{T}_c)$ denotes the set of all real or complex continuous functions on X, endowed with the compact-open topology $\mathcal{T}_c$ (see Section VII). One sees that $C(X)$ is an LC space. It is not, in general, a metrizable, or a complete, or a barreled or a Fréchet or a Banach space. As shown above, if X is compact, then $C(X)$ is a Banach space. Conversely, if $C(X)$ is a Banach space, then indeed X is compact. This immediately suggests an interplay of topological properties of X and $C(X)$.

To display this duet between X and $C(X)$ briefly, we need the following concept. Let $\mathcal{L} \in C'(X)$, the dual of $C(X)$. The smallest compact subset A of X such that for all $f \in C(X)$ with $f(A) = 0$ implies $\mathcal{L}(f) = 0$ is called the *support* of $\mathcal{L}$, denoted supp$\mathcal{L}$. If B' is a subset of $C'(X)$, then supp$B' = \text{Cl}\{\cup (\text{supp}\mathcal{L} : \mathcal{L} \in B')\}$ is the support of B'. Now we have the following:

- $C(X)$ is a Banach space iff X is compact.
- $C(X)$ is metrizable iff X is hemicompact.
- $C(X)$ is complete iff X is a k_r-space.
- $C(X)$ is a Fréchet space iff X is a hemicompact, k_r-space.
- $C(X)$ is a Mackey space iff suppB' is compact for each convex circled w^*-compact subset B' of $C'(X)$.
- $C(X)$ is a barreled space iff X is a μ-space [i.e., for each w^*-bounded subset B' of $C'(X)$, suppB' is compact].

C. Topological Algebras

An algebraic structure richer than vector spaces is what is known as algebra. A real or complex vector space A is called an *algebra* if there is some multiplication defined on A, that is, for all $x, y \in A$, $xy \in A$ satisfying the following axioms: $x(y + z) = xy + xz$; $(x + y)z = xz + yz$; $\lambda(xy) = (\lambda x)y = x(\lambda y)$; $x(yz) = (xy)z = xyz$ for all $x, y, z \in A$ and scalar λ. An algebra A is called *commutative* if $xy = yx$ for all $x, y \in A$. A has an identity e if $ex = xe = x$ for all $x \in A$. An element $x \in A$ is called *invertible* if there is $y \in A$ such that $xy = yx = e$. y is called the *inverse* of x and written $y = x^{-1}$. An algebra in which each nonzero element has an inverse is called a *division* algebra. An operation $* : x \to x^*$ of A onto A is called an *involution* if $(x + y)^* = x^* + y^*$, $(\lambda x)^* = \bar{\lambda} x^*$, $(xy)^* = y^* x^*$, and $x^{**} = x$.

An algebra A with a topology $\mathcal{T}$ is called a *topological algebra* (TA) if the maps $(x, y) \to x + y$, $(\lambda, x) \to \lambda x$, and $(x, y) \to xy$ are continuous. It is clear that each topological algebra is a topological vector space. Hence, the results pertaining to topological vector spaces can be used for topological algebras. If the topology is given by a norm on an algebra, it is called a *normed algebra*, provided that $\|xy\| \leq \|x\|\|y\|$. A complete normed algebra is called a *Banach algebra* (BAA). A Banach algebra with an involution * is called a B*-algebra if $\|xx^*\| = \|x\|^2$ for all x. We deal with these algebras in the next section.

If the topology of a topological algebra is locally convex, it is called a *locally convex algebra*. A complete metric locally convex algebra is called a B_0-*algebra* (B_0-A). Clearly, each Banach algebra is a B_0-algebra.

Examples of nonnormed algebras also abound in the literature. For example, for any Tychonoff space $(X, \mathcal{T})$, the set $C(X)$ of all continuous real or complex functions forms an algebra with pointwise operations, i.e.,

$$(f + g)(x) = f(x) + g(x), \qquad (\lambda f)(x) = \lambda f(x),$$

$$(fg)(x) = f(x)g(x),$$

where λ is a real or complex scalar. With the compact-open topology T_c, $C(X)$ is actually a locally convex algebra. Among many applications associated with $C(X)$, there is a celebrated result called the Stone–Weierstrass theorem: Let A be a subalgebra of $C(X)$ such that

(i) for all $x, y \in X$, $x \neq y$, there is $f \in A$ with $f(x) \neq f(y)$, (i.e., A separates points of X),
(ii) for each $x \in X$, there is $f \in A$ with $f(x) \neq 0$, and
(iii) for each $f \in A$, the complex conjugate $\bar{f}$ [i.e., $\bar{f}(x) = \overline{f(x)}$] is in A.

Then A is dense in $C(X)$ [i.e., $\bar{A} = C(X)$]. If, in addition, A is a closed subalgebra of $C(X)$, then $A = C(X)$. Note that condition (iii) is redundant for real algebras. From this theorem, one derives the classical Weierstrass theorem: Each continuous real function on $[a, b]$, $-\infty < a < b < \infty$, can be approximated by polynomials.

IX. BANACH ALGEBRAS

Banach algebras form an important subclass of topological algebras because they have a richer structure owing to the norm. Hence, they have more useful applications. The theory of Banach algebras can be used to advantage in mathematical analysis, Fourier series, representation theory, and other significant areas of mathematics.

An important result revealing a basic fact about Banach division algebra is the so-called Gelfand–Mazur theorem: A complex Banach division algebra is isomorphically homeomorphic with the algebra of complex numbers.

A functional f on an algebra A is called *multiplicative* if $f(xy) = f(x)f(y)$. A simple but beautiful result regarding

multiplicative functionals is as follows. Each multiplicative linear functional on a Banach algebra is continuous. It is beautiful because an algebraic hypothesis (namely, linearity and multiplicativity) gives a topological conclusion (namely, the continuity). It is not known if this beautiful result is true for Fréchet algebras [i.e., those B_0-algebras whose topology is defined by a sequence of seminorms $\{p_n\}$ such that $p_n(xy) \leq p_n(x)p_n(y)$ for all $x, y \in A$].

If $\Delta(A)$ denotes the set of all nonzero multiplicative linear functionals on a commutative Banach algebra A, the above result says that $\Delta(A) \subset A'$, the dual of A. The set $\Delta(A)$ endowed with the $\mathcal{T}_p$ topology is called the *maximal ideal space* of A. $\Delta(A)$ is locally compact, but if the Banach algebra has identity, $\Delta(A)$ is a compact Hausdorff space.

Banach algebras are often grouped into three classes: function algebras, group algebras, and operator algebras. Function algebras are the targets of study in topology, whereas group algebras are studied in harmonic analysis and operator algebras in operator theory. The latter two constitute large fields of mathematics.

Here we consider only an instance of function algebras. Let $(X, \mathcal{T})$ be a compact Hausdorff space and $C(X)$ the set of all continuous complex-valued functions on X. With algebraic operations $(f + g)(x) = f(x) + g(x)$, $(\lambda f)(x) = \lambda f(x)$, $(fg)(x) = f(x)g(x)$, $C(X)$ becomes a complex algebra. If we put $\|f\| = $ the least upper bound of $\{|f(x)|, x \in X\}$, we have a norm on $C(X)$ with $\|fg\| \leq \|f\|\|g\|$. Using the last result in Section VII, $C(X)$ is complete in the norm topology. In other words, $C(X)$ is a Banach algebra. It is commutative and has identity $1 (1(x) = 1)$. Moreover, there is an involution $f^*(x) = \overline{f(x)}$, complex conjugate. One can verify that $\|ff^*\| = \|f\|^2$. Thus, $C(X)$ is a B*-algebra. It was a crowning achievement of I. Gelfand and M. Naimark (Russian mathematicians) to show that each commutative B*-algebra with identity is isometric (i.e., norm goes into norm) to $C(X)$ for some compact Hausdorff space X. Actually, it turns out that $X = \Delta(A)$, the maximal ideal space of A.

X. ALGEBRAIC TOPOLOGY

Certain properties of topological spaces, such as "invariance" and "congruence," can, under suitable conditions, be expressed in terms of groups—specifically homotopy and homology groups—associated with spaces. The study of this phenomenon constitutes part of an important field of mathematics called algebraic topology. Among many fascinating topics in this area, there are two noteworthy theories: homotopy and homology.

A. Homotopy

Let X, Y be two topological spaces and $I = [0, 1]$. Two maps $f, g: X \to Y$ are said to be *homotopic* (written $f \sim g$) if there exists a continuous map $\phi: X \times I \to Y$ such that $\phi(x, 0) = f(x)$ and $\phi(x, 1) = g(x)$ for $x \in X$. If g is a constant map [i.e., $g(x) = y_0 \in Y$ for all $x \in X$] and $f \sim g$, then f is called *null homotopic* and written $f \sim 0$.

In some cases each map is null homotopic. For instance, if $f: X \to \mathbb{R}^n$ or $g: \mathbb{R}^n \to Y$, then $f \sim 0$ and $g \sim 0$. However, it fails if we replace $\mathbb{R}^n$ by $S^{n-1} = \{(x_1, \ldots, x_n) \in \mathbb{R}^n : x_1^2 + \cdots + x_n^2 = 1\}$, the n-dimensional sphere.

The relation $f \sim g$ is an equivalence relation ($f \sim f; f \sim g \Rightarrow g \sim f; f \sim g$ and $g \sim h \Rightarrow f - h$). Thus, the homotopic relation decomposes $C(X, Y)$ into disjoint equivalence classes $\dot{f} (g \in \dot{f} \Leftrightarrow f \sim g)$ called the *homotopic classes*, which are denoted $[X, Y]$. If $\Psi: X \to Y$ is continuous, then for any topological space Z, there exists a map $\Psi^*: [Y, Z] \to [X, Z]$ defined by $\Psi^*(f) = f \circ \Psi$, $f \in [Y, Z]$. Thus, the topological properties of $\Psi: X \to Y$ can be studied by means of the properties of Ψ^*, and vice versa.

Homotopy is actually the study of the extension of maps, to which the reader has been introduced in previous sections. For $f \sim g$ iff the map $\phi: (X \times \{0\}) \cup (X \times \{1\}) \to Y$ defined by $\phi(x, 0) = f(x), \phi(x, 1) = g(x), x \in X$ can be extended to a continuous map $\tilde{\phi}: X \times I \to Y$. Such an extension in a particular case has a special name.

Two topological spaces X, Y are said to be *homotopic* if there exist continuous maps $f: X \to Y$ and $g: Y \to X$ such that the composition maps $f \circ g$ and $g \circ f$ are homotopic to the identity maps of Y and X, respectively. The symbol "$X \approx Y$" means that X is homotopic to Y.

It is easy to see that the relation of being homotopic is an equivalence relation on the class of all topological spaces. It follows easily that if X is homeomorphic with Y, then $X \approx Y$. But the converse need not be true. For example, $\mathbb{R}^n$ is homotopic to $\{0\}$ but is not homeomorphic.

The equivalence relation of being "homotopic" decomposes the class of all topological spaces into disjoint equivalence classes of homotopic spaces.

A property of a topological space is called a *topological* (respectively, *homotopic*) *invariant* if it remains *unaltered* under homeomorphisms (respectively, homotopic maps). Most topological invariants are not *homotopic invariants*. But some of them are.

A topological space (X, T) is called *pathwise connected* if for each pair $a, b \in X, a \neq b$, there is a path $f: I \to X$ such that $f(0) = a$, $f(1) = b$. $\mathbb{R}^n$, S^n (for $n \geq 1$) are pathwise connected. Pathwise connectedness is a homotopic invariant.

1. Retracts

A subset A of a topological space $(X, \mathcal{T})$ is said to be a retract of X if the identity map $i: A \to A(i(x) = x)$ can be extended to a continuous map $f: X \to A$ so that $f(x) = x$ for all $x \in A$. In this case f is called a *retract*.

Each retract of a Hausdorff topological space is closed. Each closed unit disk $D^n = \{(x_1, \ldots, x_n) \in \mathbb{R}^n : x_1^2 + \cdots + x_n^2 \leq 1\}$ is a retract of $\mathbb{R}^n$ and each sphere S^{n-1} is a retract of $\mathbb{R}^n \setminus \{0\}$.

A topological space Y is called an AR for *absolute retract* [or *AR* (*normal*)] if for any topological (respectively, normal) space X and a closed subjset $A \subset X$, each continuous map $A \to Y$ has a continuous extension to a map $X \to Y$. As shown in Tietze's extension theorem (Section IV), $\mathbb{R}$ is an AR (normal).

A subset E of a topological space $(X, \mathcal{T})$ is said to be *deformable* into a subset $F \subset X$ if the identity map $i: E \to E$ is homotopic in X to a map of E into F. Deformability is *topologically invariant*; that is, if $\phi: X \to Y$ is a homeomorphism, then $\phi(E)$ is deformable to $\phi(F)$ whenever E is deformable to F. For example, $\mathbb{R}^n \setminus \{0\}$ is deformable to S^{n-1}.

2. Degree

Another property of a map that is invariant under homotopy is the degree of a self-map of S^n. If $f: S^n \to S^n$ is a continuous map, there exists an integer $D(f)$ (positive, negative, or zero) called the *degree* of f. In the simple case of $n = 1$, the degree of $f: S^1 \to S^1$ (note that S^1 is the circle) represents the number of time and sense that the image point $f(x)$ rotates around S^1 when x rotates in one oriented direction of S^1. For example, the degree of $f(z) = z^n$ when f maps S^1 into S^1 is n. It follows that if $f, g: S^n \to S^n$, then $f \sim g \Leftrightarrow D(f) = D(g)$. From this or otherwise. Brouwer derived the fact that the identity map of S^n is not null homotopic. However, if $m < n$ and $f: S^m \to S^n$, then $f \sim 0$. If $m > n$, the situation is in general uncertain. However, if $n > 1$, then $f: S^n \to S^1$ is null homotopic. If $f: S^n \to S^n$ is *antipodal* [$f(-x) = -f(x)$], $n \geq 0$, then $D(f)$ is an odd number and hence f cannot be null homotopic.

3. Jordan Curves

To study further properties of $\mathbb{R}^n$, we say that a subset A of $\mathbb{R}^n$ *separates* $\mathbb{R}^n$ if $\mathbb{R}^n \setminus A$ is not connected (see Section I). For example, the circle S^1 separates the plane $\mathbb{R}^2$. Actually the *Jordan separation theorem* states that every homeomorphic image of S^{n-1} into $\mathbb{R}^n$ separates $\mathbb{R}^n$. Furthermore, it can be derived that if $m \neq n$, then $\mathbb{R}^n$ is not homeomorphic with $\mathbb{R}^m$. (This is called the *invariance of dimension theorem*.) In the particular case of $\mathbb{R}^2$, every homeomorphic image of S^1 separates $\mathbb{R}^2$ in exactly two components. (This is the *Jordan curve theorem*.)

4. Paths and Loops

A continuous map p of $I = [0, 1]$ into a topological space $(X, \mathcal{T})$ is called a *path* in X. $p(0) = a$ is called the *starting point* of p, and $p(1) = b$ the end point of p. If $p(0) = a = p(1) = b$, then p is called a *loop*. The set of all loops having their starting and end points at a is denoted $\mathscr{P}(X, a)$ and is called the *loop space*. Clearly, $\mathscr{P}(X, a)$ is a subset of $C(I, X)$ and so it can be given the compact-open topology $\mathcal{T}_c$, (see Section VIII). The study of loop spaces can be subsumed under a general structure called the H-structure.

5. H-Structures

A topological space X with a continuous map $\eta : X \times X \to X$ is said to have an H-structure if there exists a fixed point $a \in X$ making the maps $x \to \eta(x, a)$ and $x \to \eta(a, x)$ homotopic to the identity map of X. A topological space that carries an H-structure is called an *H-space*. For example, each topological group (see Section VIII.A) carries an H-structure, and so does a loop space $\mathscr{P}(X, a)$.

In general the point a in the H-structure is not unique. Let $P = \{a: \text{the maps } x \to \eta(x, a) \text{ and } x \to \eta(a, x) \text{ are homotopic to the identity map of } X\}$. P is called a *path component* of X and is said to be the *principal component* of the H-space. If Comp X denotes the discrete space of all path components, it forms a group, called the *fundamental group*, of X at a. In general, the fundamental group is not Abelian. However, if X is *path-connected* [i.e., for all pairs $a, b, a \neq b$, in X, there is a path $p: I \to X$ with $p(0) = a$, $p(1) = b$], then the fundamental group is Abelian.

In modern algebraic topology, the study of H-structures can be subsumed under *fiber structure*.

6. Fiber Spaces

A triple (X, p, B), consisting of two topological spaces, X and B, and a continuous surjective map $p: X \to B$, is called a *fiber structure*. X is called a *total* (or *fibered*) space, B a *base* space, and p a *projection*. We say that (X, p, B) is a fiber structure over B, and for each $b \in B$, the set $p^{-1}(b)$ is called the *fiber* over b.

Let Y be a topological space and $f: Y \to B$ a continuous function. If there exists a continuous function $\tilde{f}: X \to B$ such that $p \circ \tilde{f} = f$, then $\tilde{f}$ is called the *lifting* or *covering* of f. In particular, a lifting of the identity $i: B \to B$ is called a *cross section*.

Let $\mathscr{A}$ be a class of topological spaces. A fiber structure (X, p, b) is called a *fiber space* (or *fibration*) for the class $\mathscr{A}$ if for each Y in $\mathscr{A}$, each continuous map $f: Y \times \{0\} \to X$, and each homotopy $\varphi: Y \times I \to B$ of $p \circ f$, there exists a homotopy $\tilde{\varphi}$ of f covering φ. A fiber space for the class of all topological spaces is called a Hurewicz *fibration*.

It can be shown that if both X and B are metric spaces and if (X, p, B) is a fibration for the class of all metric spaces, then (X, p, B) is a Hurewicz fibration.

For further details and application of these ideas, the interested reader can consult any appropriate book on algebraic topology; some relevant texts are listed in the Bibliography.

B. Homology

Homology can be studied in a more general way than we do here. We are concerned with the homology of polyhedra in $\mathbb{R}^n$. Usually such a study is part of the so-called combinatorial topology, which is a part of algebraic topology.

Let $A = \{a_0, a_1, \ldots, a_m\}$ be a set of $m + 1$ $(m \leq n)$ points in $\mathbb{R}^n$ such that vectors $\{a_i - a_0 : 1 \leq i \leq m\}$ are linearly independent. ($\{x_0, \ldots, x_m\}$ are linearly independent if $\sum_{i=0}^m \lambda_i x_i = 0 \Rightarrow \lambda_i = 0$, for $i = 0, \ldots, m$.) The convex hull $\Delta_m = \{\sum_{i=0}^m \lambda_i a_i : \lambda_i \geq 0, \sum_{i=0}^m \lambda_i = 1\}$ of A is called an *m-dimensional simplex*. The points in A are called the *vertices* of Δ_m. If $m = 1$, then Δ_1 is a *straight line segment*. If $m = 2$, Δ_2 is a *triangle*, Δ_3 a *tetrahedron*, and so on.

If $r < m$, the simplex Δ_r with vertices $\{a_0, \ldots, a_{r-1}, a_{r+1}, \ldots, a_m\}$ is called the *r-face* of Δ_m.

A finite collection K of simplexes in $\mathbb{R}^n$ is called a *geometric complex* or just a *complex* if every face of each simplex is also in K and every two simplexes, if they intersect, intersect in a common face. If a complex K contains at least one m-dimensional simplex but not an $(m + 1)$-dimensional one, then K is said to be an *m-complex*. The point set of a complex is called a *polyhedron*.

If one gives an ordering to the vertices of a simplex, the simplex is said to have an *orientation* and is called an *oriented simplex*.

Let $\Delta_m^1, \Delta_m^2, \ldots, \Delta_m^{k(m)}$ denote a set of arbitrarily oriented m-simplexes in a complex K. Then $C_m = \sum_{i=1}^{k(m)} g_i \Delta_m^i$ (where g_i are integers) is called an *m-chain*. The set of all m-chains forms an Abelian additive group. If Δ_{m+1} is an $(m + 1)$-dimensional simplex, $\partial \Delta_{m+1}$ denotes the sum of all m-dimensional faces. The ∂ is called the boundary of Δ_{m+1}. One can verify that $\partial^2 = 0$. An m-chain C_m is called an *m-cycle* if $\partial C_m = 0$. Let Z_m denote the set of all m-cycles. Z_m is a subgroup of the group of m-chains. An m-cycle is called *homologous to zero* if it is the boundary of an $(m + 1)$-chain. The set of all

m-cycles that are homologous to zero forms a subgroup of Z_m. The quotient or factor group $Z_m/H_m = B_m$ is called the *m-dimensional homology group or Betti group* of the complex K.

Betti groups are topologically invariant that is, if the two polyhedra K and K' are homeomorphic, their respective Betti groups are isomorphic.

Using group theory from modern algebra, it can be claimed that each Betti group has a finite number of generators. These generators are of two kinds: One kind gives an infinite cyclic group, and the other kind gives groups of finite order. The number of generators of the first kind is called the *rank* of the Betti group, whereas the order of each generator of the second kind is called the *torsion coefficient* of the Betti group. These numbers are related by the so-called *Euler–Poincaré formula*:

Let K be an n-complex. Let $k(m)$ denote the number of m-simplexes of K, and $p(m)$ the m-dimensional Betti number for $m = 0, \ldots, n$. Then $\sum_{m=0}^{n}(-1)^m k(m) = \sum_{m=0}^{n}(-1)^m p(m)$. The number $\chi(K) = \sum_{m=0}^{n}(-1)^m p(m)$ is called the *Euler characteristic* of the complex K.

1. Example

For Δ_2, the triangle joining the vertices $\{a_0, a_1, a_2\}$ in $\mathbb{R}^2$, $k(0) = 3$, $k(1) = 3$, $k(2) = 1$, and so $\chi(\Delta_2) = 3 - 3 + 1 = 1$.

There are a number of fascinating topics in algebraic topology that have not been touched here, for example, dimension theory, fiber bundles, and manifolds. The interested reader can consult the Bibliography.

SEE ALSO THE FOLLOWING ARTICLES

ALGEBRA, ABSTRACT • COMPLEX ANALYSIS • CONVEX SETS • DISTRIBUTED PARAMETER SYSTEMS • KNOTS • MANIFOLD GEOMETRY • SET THEORY

BIBLIOGRAPHY

Armstrong, M. A. (1983). "Basic Topology," Springer-Verlag, New York.

Gamelin, T. W., and Greene, R. E. (1983). "Introduction to Topology," Saunders, Philadelphia.

Hochschild, G. P. (1981). "Basic Theory of Algebraic Groups and Lie Algebras," Springer-Verlag, New York.

Husain, T. (1977). "Topology and Maps," Plenum, New York.

Husain, T. (1981). "Introduction to Topological Groups" (reprint), Kreiger, Huntington, NY.

Husain, T. (1983). "Multiplicative Functionals on Topological Algebras," Pitman, Boston.

Husemoeller, D. (1982). "Fibre Bundles," Springer-Verlag, New York.

Rourke, C. P., and Sanderson, B. J. (1982). "Introduction to Piecewise Linear Topology," Springer-Verlag, New York.

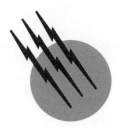

Toxicology in Forensic Science

Olaf H. Drummer
Monash University

GLOSSARY

Amphetamines A class of drugs that act as powerful stimulants with actions similar to norepinephrine.

Benzodiazepines A class of over 50 drugs that act as minor tranquilizers and hypnotics and include diazepam, temazepam, flunitrazepam, oxazepam, and alprazolam etc.

Central nervous system (CNS) The part of the body including the brain and spinal cord.

Confirmatory testing The use of a second test to confirm the intial test such that there is no doubt over the presence of the substance in the sample.

Ethical drugs Drugs that are available by prescription or legally over the counter.

Exhibits Items brought to a laboratory that form part of the evidence in a case; can be tissue specimens or physical items such as drug powders, tablets, syringes, etc.

Forensic toxicology The application of toxicology to the needs of the law.

Initial testing The use of screening tests to establish the likely presence of drugs or drug classes in a sample.

Opioids A class of drugs related to morphine, often also known as opiates.

TOXICOLOGY in the context of forensic science deals with the detection of drugs and other chemicals in situations involving legal proceedings. This discipline is best called *forensic toxicology*; however, the term *analytical toxicology* is also often applied to this subspecialty of toxicology. A forensic toxicologist is concerned with the detection of drugs or poisons in samples and is capable of

TABLE I   Types of Specimens Collected and Their Principal Applications

Specimen	Application
Blood (or plasma or serum)	Most commonly used for drug detection and is the preferred single specimen for perpetrators and victims of crimes, death investigation cases, drivers of motor vehicles suspected of drug use
Urine	Used in all cases as a screening specimen for drugs of abuse, and is the preferred specimen for workplace, rehabilitation, and corrections drug testing programs, as well as in sports drug testing programs
Liver and other tissues from deceased persons	Used to supplement blood toxicology in some death investigations, particularly when body is decomposed and when the interpretation of blood results are equivocal
Hair	Used when a longer period of drug exposure is required (1–6 months) to complement other information on drug use
Sweat and saliva	Occasionally used as alternatives to other forms of testing when onsite[a] results are required in workplace and correctional or rehabilitation settings
Breath	Used extensively to establish presence of alcohol in drivers of motor vehicles and in workplace settings

[a] Onsite refers to an initial detection of drugs in specimens at the point of collection, rather than waiting for a result from a laboratory.

defending the results in a court of law. This distinction from an ordinary analytical toxicologist is important, as a conventional toxicologist is mainly concerned with the detection of substances and may not understand the specific medico-legal requirements in forensic cases. A forensic toxicologist also is able to assist legal proceedings in the interpretation of the significance of the results obtained.

I. APPLICATIONS OF FORENSIC TOXICOLOGY

Forensic toxicology has a number of applications. It provides clinicians with information of a possible drug taken in overdose or authorities investigating a sudden death or poisoning with the possible substances(s) used.

Toxicology testing is also important in victims of crime, or in persons apprehended for a crime. Drugs may have been given by the assailant to reduce consciousness of the victim, such as in rape cases. These drugs include the benzodiazepines (e.g., Rohypnol, Valium, Ativan, etc.) and gamma-hydroxybutyrate (GHB). Toxicology also establishes if any drug was used by the victim that may have affected consciousness or behavior. Defendants arrested shortly after allegedly committing a violent crime may be under the influence of drugs. Alcohol and drugs are also commonly targeted in drivers suspected of driving under the influence.

Ultimately, toxicology testing results will assist the investigator, pathologist, coroner, or medical examiner in establishing evidence of drug use or refuting the use of relevant drugs. This latter application is essential, as few drugs leave any visible trace of their presence in a person.

Forensic toxicology is also used in employment drug testing, rehabilitation settings, and in human performance testing. The detection of drugs of abuse in potential employees prior to being hired is becoming an important application of toxicology. Attempts to exclude drugs from prisons and to aid in rehabilitation of drug-dependent persons are other applications of toxicology. Human perfor-

mance testing relates to the detection of drugs that might have improved performance in athletic events. This testing may even apply to racing animals such as horses, camels, dogs, etc. Specimens used in these cases are usually urine, although hair is being increasingly used to provide a greater window of opportunity in workplace settings.

II. SPECIMENS

A wide variety of specimens can be collected to assess possible drug use. These include blood (or sometimes plasma or serum*), urine, breath, saliva, sweat in living persons, or a range of other tissues from bodies at autopsy during death investigations. In cadavers the most common specimens, after blood and urine, are liver, bile, and vitreous humour. Muscle, brain, bone, and fat do have uses in certain types of cases when blood is unavailable due to decomposition or following suspicion of unusual poisons. The principal applications of some specimens are summarized in Table I.

III. CHAIN OF CUSTODY

Courts and other legal processes usually require proof that the laboratory has taken all reasonable precautions against unwanted tampering or alteration of the evidence. This applies to specimens and to physical exhibits used by the laboratory for toxicology investigations. Consequently, it is essential that the correct identifying details are recorded on the exhibit or specimen container and an adequate record is kept of persons in possession of the exhibit(s). Alternatively, when couriers are used to transport exhibits, the exhibit must be adequately sealed to prevent unauthorized tampering.

*Obtained from blood by either centrifugation to remove red blood cells (plasma) or other types of separation processes to obtain plasma or serum (fluid obtained after blood has clotted).

TABLE II Drugs Commonly Targeted in Forensic Toxicology Investigations

Most common drugs	Alcohol
	Amphetamines, benzodiazepines, cannabis, cocaine and opiates
Common ethical drugs	Antidepressants including tricyclics, serotonin reuptake inhibitors, and monoamine oxidase inhibitors
	Antipsychotics drugs such as phenothiazines, haloperidol, olanzapine, or clozapine
	Other analgesics and anti-inflammatory agents
	Digoxin, anti-arrhythmic drugs, anti-hypertensive drugs, and many other cardiovascular drugs
Less common drugs	GHB, LSD, and other hallucinogens
	Anabolic steroids and other performance-enhancing drugs
	Barbiturates and older sedatives and hypnotics such as methaqualone
	Volatile substances such as butane, or gasoline
	Various domestic, industrial, and agricultural poisons such as organophosphates, or solvents

IV. WHAT CHEMICALS SHOULD BE TARGETED?

It is common to find a variety of ethical and illicit drugs or unusual poisons. Worldwide experience also shows that forensic cases often involve more than one drug substance. High rates of multiple drug use are found in deaths from misuse of drugs and also in perpetrators of violent crimes.

It is also well known by forensic toxicologists that the information provided to the laboratory concerning possible drug use may not agree with what is actually detected. It is therefore strongly recommended that laboratories provide a systematic approach to their analyses and include as wide a range of common ethical and illicit drugs as feasible. A laboratory using this approach would normally include a range of screening methods often incorporating both chromatographic and immunological techniques. Drug classes such as alcohol, analgesics, opioid and non-opioid narcotics, amphetamines, antidepressants, benzodiazepines, barbiturates, cannabis, cocaine, major tranquilizers (antipsychotic drugs), and other CNS-depressant drugs should be included (Table II).

The incorporation of a reasonably complete range of drugs in any testing protocol is important, as many of these drugs are mood altering and can therefore affect behavior as well as the health of an individual. Persons using benzodiazepines, for example, will be further affected by alcohol and other CNS-active drugs. The toxic concentrations of drugs are also influenced by the presence of other potentially toxic drugs. For example, the toxicity of heroin is affected by the concomitant use of alcohol and other CNS-depressant drugs.

V. TECHNIQUES USED

The range of techniques available to detect drugs in specimens or physical exhibits varies from commercial kit-based immunoassays and traditional thin-layer chromatography (TLC) to instrumental separation techniques such as high-performance liquid chromatography (HPLC), gas chromatography (GC), and capillary electrophoresis (CE). Mass spectrometry (MS) is the definitive technique used to establish proof of structure of an unknown substance and can be linked to GC, HPLC, and more recently to CE.

The use of appropriate extraction techniques is critical to all analytical methods. Three main types of extractions are used: liquid–liquid, solid-phase, and direct injection. Traditionally, liquid techniques have been favored in which a blood or urine specimen is treated with a buffer of an appropriate pH followed by a solvent capable of partitioning the drug out of the matrix. Solvents used include chloroform, diethyl ether, ethyl acetate, toluene, hexane, various alcohols, and butyl chloride and mixtures thereof. The solvent is then isolated from the mixture and either cleaned up by another extraction process or evaporated to dryness.

Solid-phase techniques are becoming increasingly favored, as they offer the ability to extract substances of widely differing polarity more readily than with liquid techniques.

Direct-injection techniques into either GC or HPLC instruments bypass the extraction step and can offer a very rapid analytical process. In GC, solid-phase microextraction (SPE) can be used, while HPLC tends to require use of pre-columns that are backflushed with the use of column-switching valves.

VI. INITIAL TESTS AND CONFIRMATION

The process of conducting toxicology in forensic science is similar to other analytical disciplines, in that sufficiently suitable analytical techniques need to be employed that are appropriately validated. The foremost goal is the need to provide a substantial proof of the presence of a substance(s). The use of conventional GC, TLC, or HPLC by themselves would not normally be sufficient to accept unequivocal proof of the presence of a chemical substance.

Two or more independent tests are normally required, or the use of a more powerful analytical test, such as mass spectrometry (MS) may often be preferred. Because of the need to perform a rigorous analysis, the analytical schema is often broken up into two steps. The identification stage is termed the screening or initial test, while the second analytical test is the confirmation process. The confirmation process often also provides a quantitative measure of how much substance was present in the sample; otherwise, a separate test is required to quantify the amount of substance present in the specimen (see later).

In all processes, it is important that no analytical inconsistency appears, else a result may be invalidated. For example, in the identification of amphetamine in a blood specimen, an immunoassay positive to the amphetamine class is expected to be positive for one or more amphetamines in the confirmation assay. The apparent detection of a drug in one analytical assay but not in another means that the drug was not confirmed, providing both assays are capable of detecting this drug or one or more members of a drug class.

While MS is the preferred technique for confirmation of drugs and poisons, some substances display poor mass spectral definition. Compounds with base ions at mass/charge ratios of less than 100 or with common ions such as m/z 105 and with little or no ions in the higher mass range are not recommended for confirmation by MS alone. Derivitization of a functional group to produce improved mass spectral properties can often be successful. Common derivatives include acyl esters, silyl ethers, etc. Alternatively, reliance on other chromatographic procedures can provide adequate confirmation. It is important when using any chromatographic procedure (HPLC, GS, CE, etc.) that the retention time of the substance being identified matches that of an authentic standard.

Some apparent analytical inconsistencies may provide important forensic information. For example, if a result for opiates is negative in urine, but positive in blood, it is quite likely that heroin* was administered shortly before death and the metabolites had not yet been excreted. This situation is often found in acute sudden death among heroin users in whom substantial urinary excretion has not yet occurred.

VII. QUALITY ASSURANCE AND VALIDATION

Essential components of any form of toxicological testing are validation and quality assurance. It is important that the testing method used is appropriately validated; that is, it has been shown to accurately and precisely detect particular substance(s), there is little or interference from other drugs or from the matrix, and a useful detection limit has been established. Moreover, it is essential that the method is rugged and will allow any suitably trained analyst to conduct the procedure and achieve the same results as another analyst. To achieve these aims, it will be necessary to trial the method in the laboratory with specimens of varying quality before full validation can be achieved.

It is recommended to include internal quality controls with each batch of samples to enable an internal check of the reliability of each assay. These controls contain known drugs at known concentrations. Suitable acceptance criteria are required for these controls before results of unknown cases can be accepted and released to a client. Acceptance criteria vary depending on the analyte and application. For example, blood alcohol estimations have acceptance criteria less than 5%,* while postmortem blood procedures may be 10 to 20%.

An important feature of analytical assays in forensic toxicology is the use of internal standards. These are drugs with chemical and physical characteristics similar to the drug(s) being analyzed and, when added at the start of the extraction procedure, provide an ability to negate the effects of variable or low recoveries from the matrix. Hence, even when recoveries are low, the ratios of analyte and drug are essentially the same as for situations of higher recovery. An ideal recovery marker is when the internal standard is a deuterated analog of the analyte. When deuterated internal standards are used, it may not be necessary to match the calibration standards with the same matrix as the unknown samples. It is important, however, that absolute recoveries are reasonable (i.e., at least over 30%). This ensures less variability between samples and optimizes the detection limit.

From time to time, it will be important to run unknown samples prepared by another laboratory or by a person not directly involved in laboratory work to establish proficiency. These are known as proficiency programs or quality assurance programs. These trials are often conducted with many other laboratories conducting similar work and provide an independent assessment of the proficiency of the laboratory to detect (and quantify) specific drugs. The performance of the laboratory should be regularly assessed from these results and any corrective action implemented, if appropriate. This process provides a measure of continuous improvement, an essential characteristic of any laboratory.

---

*Heroin is rapidly metabolized to morphine.

*Normally, the coefficient of variation (CV) of the mean is calculated as a standard deviation divided by the mean of the result.

VIII. REPORTS

Once an analysis is complete a report must be issued to the client(s) which accurately details the analytical findings. These results should indicate the type of tests conducted, the analytical method used (HPLC, GC–MS, etc.), on which specimens the analyses were conducted, and of course the result(s). The result(s) should be unambiguous, using such terms as "detected" or "not detected." The use of the term "not present" should be avoided, as it implies no possibility of the substance being present. A toxicologist can rarely be so definitive and can only indicate that a substance was not detected at a certain threshold concentration. For this reason, a detection limit alongside tests for specific substances should be provided for "not detected" results.

For quantitative results consistency in units is advised and should not be given with more significant digits than the accuracy will allow. For example, there is no point in reporting a result for blood morphine as 0.162 mg/L when the accuracy and precision of the method is ±20%. A result of 0.16 mg/L would suffice.

For drug-screening results, it is advisable to provide clients with an indication of the range of substances a method is capable of detecting and some indication of the detection limit, such as "at least therapeutic concentrations" or "only supra-therapeutic concentrations."

In postmortem cases, all reports should indicate the site of blood sampling and provide (where relevant) some comment on the possibility of postmortem artefacts such as redistribution. By incorporating these comments, those reading the report are less likely to unwittingly misinterpret the results.

IX. INTERPRETATION OF TOXICOLOGICAL RESULTS

Interpretation of any toxicological result is complex. Consideration must be given to the circumstances of the case, and in particular what significance may be drawn from the toxicology. For example, the finding of a drug in potentially toxic concentrations in a person killed by a gunshot wound to the head cannot reasonably lead to the conclusion that the drug caused the death. On the other hand, the absence of an obvious anatomical cause of death will lead investigators to consider the role of any drug use. Considerations must include the chronicity of drug use, the likely time of ingestion, the route of ingestion, the age and health of the person (e.g., presence of heart, liver, or kidney disease), the use of other active substances, and even genetic factors that may lead to an altered metabolism.

X. ARTEFACTS IN ANALYSIS

A. Stability of Drugs

Chemical instability occurs for a number of drugs and metabolites that will alter the concentration and even cause the drug to disappear if storage conditions are not optimal. This will occur at room temperature and even sometimes when specimens are stored frozen at $-20°C$.

Alcohol will be lost to evaporation unless sealed tubes are used or specimens are stored at $-80°C$; however, alcohol can also be produced by bacterial action on glucose and other sugars found in blood. The use of potassium fluoride as a preservative (minimum 1% w/v) is required to prevent bacterial activity for up to one month after collection when the sample is stored at $4°C$.

B. Bioconversion

A number of drugs can undergo chemical changes in a body after death. These chemical changes can be either metabolically mediated or caused by spontaneous degradative processes. For example, the metabolism of heroin to morphine occurs in life, in blood and other tissues following collection, or even *in situ* when a person has died. For this reason, heroin or the immediate 6-acetyl morphine are rarely detected in blood. Morphine is therefore the target drug. Aspirin is also converted rapidly to salicylate by hydrolytic mechanisms. Most drugs activated by de-esterification or hydrolysis will be subject to similar processes.

Nitro-containing drugs, such as the benzodiazepines, nitrazepam, or flunitrazepam, are also rapidly biotransformed after death to their respective amino metabolites by the action of certain types of bacteria. Toxicologists must therefore target their analyses to these transformation products rather than the parent drug.

Sulfur-containing drugs, such as dothiepin, thiopental, or thioridazine, are also subject to bacterial attack during the postmortem interval, leading to progressive losses over time. Of course, the parallel process of tissue loss will also affect the tissue concentration during putrefaction.

C. Redistribution

The process of death imparts a number of other special processes that affect the collection and analysis of specimens obtained at autopsy. These include postmortem redistribution in which the concentration of a drug in blood has been affected by diffusion of drug from neighboring tissue sites and organs such as stomach contents. This is minimized, but not arrested, by using peripheral blood from the femoral region. Even liver concentrations are affected

TABLE III Likely Extent of Postmortem Redistribution for Selected Drugs[a]

Drug/drug class	Likely extent of postmortem redistribution
Acetaminophen (paracetamol)	Low
Alcohol (ethanol)	Low
Barbiturates	Low to moderate
Benzodiazepines	Low to moderate
Cocaine	Low
Digoxin	Very high
Methadone	Low to moderate
Morphine	Low
Phenothiazines	Moderate to high
Propoxyphene	Very high
Salicylate	Low
Serotonin reuptake inhibitors	Low to moderate
Tricyclic antidepressants	High

[a] These changes should only be used as a guide, as the environmental conditions, length of time from death to specimen collection, and quality of specimen can affect the extent of redistribution.

Note: Low = up to 20% elevation; moderate = 21–50%; high = 50–200%; very high > 200%.

by diffusion from intestinal contents or from incomplete circulation and distribution within the liver.

This process is particularly significant for drugs with high lipid solubility, as these drugs tend to show concentration differences in tissues and blood. Table III shows the extent of these changes for selected drugs when comparisons are made between blood collected from the heart and that collected from the femoral region.

The femoral blood is least subject to redistribution after death; however, drugs with much higher concentrations in muscular tissue will still diffuse through the vessel walls and elevate the neighboring blood concentrations. If the femoral vessels are not tied off from the vena cava and aorta, then the process of drawing blood can also extract blood from the abdominal cavity that has been contaminated from diffusion of gastric and intestinal contents. It is therefore advisable to reduce these processes by collecting blood specimens as soon as possible after death from the femoral region with blood vessels tied off to reduce contamination.

XI. COURT TESTIMONY AND EXPERTISE

Forensic toxicologists and other professionals called to give evidence in court should consider that much of their technical evidence is beyond the ready comprehension of lay people in juries, legal counsel, and judges. Restricting one's testimony to understandable language and simple concepts is highly recommended.

A further problem relates to an assumption often made by legal counsel (and indeed other parties) that a toxicological investigation was exhaustive and all drugs and poisons were excluded in the testing processes. Most toxicology performed is restricted to a few analytical tests for a range of "common drugs and poisons," unless the client has made a request to examine for (additional) specific chemicals. Analysts should make courts aware of the actual testing conducted and provide a list of substances incorporated in the investigation. Importantly, advice on any limitations applied to the interpretation of the analytical results should be provided (e.g., poor-quality specimens or postmortem artifacts). Above all, toxicologists must restrict their evidence to those areas for which they claim expertise. Stretching their expertise to apparently assist the court can lead to incorrect or misleading evidence and damage the reputation of the expert.

SEE ALSO THE FOLLOWING ARTICLES

ANALYTICAL CHEMISTRY • DNA TESTING IN FORENSIC SCIENCE • ENVIRONMENTAL TOXICOLOGY • MASS SPECTROMETRY IN FORENSIC SCIENCE • ORGANIC CHEMISTRY, COMPOUND DETECTION • SPECTROSCOPY IN FORENSIC SCIENCE

BIBLIOGRAPHY

Baselt, R. H., and Cravey, R. H. (1996). "Disposition of Toxic Drugs and Chemicals in Man," 4th ed., Year Book Medical Publishers, Chicago.

de Zeeuw, R. A. (1997). "Drug screening in biological fluids: the need for a systematic approach," *J. Chromatogr.* **689,** 71–79.

Drummer, O. H. (1998). "Adverse drug reactions." *In* "The Inquest Handbook" (H. Selby, ed.), The Federation Press, Leichhardt, NSW Australia.

Drummer, O. H. (1999). "Review: chromatographic screening techniques in systematic toxicological analysis," *J. Chromatogr.* **733,** 27–45.

Freckleton, I., and Selby, H. (1993). "Expert Evidence," LBS Information Services, Sydney, Australia.

International Association of Forensic Toxicologists (TIAFT). (2001). http://www.tiaft.org.

Karch, S. (1998). "Drug Abuse Handbook," CRC Press, Boca Raton, FL.

Levine, B. (1999). "Principles of Forensic Toxicology," AACC Press, Washington, D.C.

Maurer, H. H. (1992). "Systematic toxicological analysis of drugs and their metabolites by gas chromatography-mass spectrometry," *J. Chromatogr.* **118,** 3–42.

Moffatt, A. C., ed. (1986). "Clarke's Isolation and Identification of Drugs," The Pharmaceutical Press, London.

Siegel, J., ed. (2000). "Encyclopedia of Forensic Science," Academic Press, London.

Society of Forensic Toxicologists (SOFT). (2001). http://www.soft-tox.org.

United Nations. (1995). "Recommended Methods for the Detection

and Assay of Heroin, Cannabinoids, Cocaine, Amp
Methamphetamine and Ring-Substituted Derivatives in Biological
Specimens," U.N. Publ. No. ST/NAR/27, United Nations International
Drug Control Programme, Vienna, Austria.

United Nations. (1997). "Recommended Methods for the Detection
and Assay of Barbiturates and Benzodiazepines in Biological Specimens," U.N. Publ. No. ST/NAR/28, United Nations International
Drug Control Programme, Vienna, Austria.

ISBN 0-12-227426-1

90038